FIFTH EDITION

Diagnostic
Medical
Parasitology

FIFTH EDITION
Diagnostic Medical Parasitology

Lynne Shore Garcia, M.S., MT, CLS, F(AAM)
LSG & Associates, Santa Monica, California

ASM
PRESS

WASHINGTON, D.C.

Address editorial correspondence to ASM Press, 1752 N St. NW, Washington, DC
20036-2904, USA

Send orders to ASM Press, P.O. Box 605, Herndon, VA 20172, USA
Phone: 800-546-2416; 703-661-1593
Fax: 703-661-1501
E-mail: books @asmusa.org
Online: estore.asm.org

Library of Congress Cataloging-in-Publication Data

Garcia, Lynne Shore.
    Diagnostic medical parasitology / Lynne Shore Garcia. — 5th ed.
        p. ; cm.
    Includes bibliographical references and index.
    ISBN 1-55581-380-1
    1. Diagnostic parasitology. I. Title.
    [DNLM: 1. Parasitic Diseases—diagnosis. WC 695 G216da 2007]

    QR255.G37   2007
    616.9´6075—dc22   2006047737

10 9 8 7 6 5 4 3 2 1

Cover figure: *Enterocytozoon bieneusi* intestinal infection (cover image copyright
Dennis Kunkel Microscopy, Inc.).

Dedication

As with the first four editions, I dedicate this book to Marietta Voge, a truly rare individual who was widely recognized as one of the world's leading parasitologists. During her years as a diagnostic and research parasitologist at the University of California, Los Angeles, she touched the lives of many students and staff in a very special way. She was always more than willing to share her expertise with all who asked and volunteered this help over the years whenever contacted. She was always willing to donate a considerable amount of her personal time as a volunteer for various medical projects throughout the world.

She was a very special individual to work with, always interested in the person as well as the problem at hand. Her areas of teaching extended far beyond science. Whatever subject she was interested in received her total enthusiasm and dedication, and she had an exceptional ability to deal with detailed work. Her sense of fairness and professional integrity were remarkable; these ideals were shared with all who came in contact with her.

Her contributions to the field of diagnostic parasitology were numerous and included many classes, seminars, papers, and textbooks. The importance of working with Dr. Voge is hard to put into words. She was unique in her ability to allow a student to grow, both scientifically and personally. She could guide without constraints, teach without formal lectures, counsel without being judgmental, challenge without being unrealistic, tease without being cruel, and always be supportive regardless of the situation. She expected much from her students and employees and yet always gave considerably more than she received.

Scientific information gained from our association with her was invaluable; however, her impact on our lives was considerably more than scientific. She was always available for consultations and just to talk. She left all of us with a sense of having personally matured as a result of knowing and working with her over the years. She is missed by all of us, and yet her contributions in terms of teaching, consultations, volunteer work, professionalism, and friendship will remain with us forever.

I also dedicate this book to John Lawrence. He was an extraordinary individual, and without his original encouragement and assistance, the first edition of the book would never have been written.

Contents

Preface

During the past few years, the field of diagnostic medical parasitology has seen dramatic changes, including newly recognized parasites, emerging pathogens in new geographic areas, bioterrorism considerations and requirements, alternative techniques required by new regulatory requirements, reevaluation of diagnostic test options and ordering algorithms, continuing changes in the laboratory test menus, implementation of testing based on molecular techniques, reporting formats, coding and billing requirements, managed care relevancy, increased need for consultation and educational initiatives for clients, and an overall increased awareness of parasitic infections from a worldwide perspective. We have seen organisms like the microsporidia change from the status of "unusual parasitic infection" to being widely recognized as causing some of the most important infections in both immunocompetent and compromised patients. More sensitive diagnostic methods for organism detection in stool specimens are now commercially available for *Entamoeba histolytica*, *Entamoeba histolytica/E. dispar*, *Giardia lamblia*, *Cryptosporidium parvum*, and *Trichomonas vaginalis*. Reagents are actively being developed for other organisms such as *Dientamoeba fragilis* and the microsporidia. We have seen *Cyclospora cayetanensis* coccidia become well recognized as the cause of diarrhea in immunocompetent and immunocompromised humans. We continue to see new disease presentations in compromised patients; a good example is granulomatous amebic encephalitis caused by *Acanthamoeba* spp., *Sappinia diploidea*, and *Balamuthia mandrillaris*. With the expansion of transplantation options, many parasites are potential threats to patients who are undergoing immunosuppression, and they must be considered within the context of this patient group. Transfusion-associated transmission of potential parasitic pathogens continues to be problematic. Transfusion in general is becoming more widely recognized as a source of infection, and donors are also more likely to come from many areas of the world where parasitic infections are endemic.

With expanding regulatory requirements related to the disposal of chemicals, laboratories are continuing to review the use of mercury compounds as specimen fixatives and learning to become familiar with organism morphology when using substitute compounds. Permanent staining of fecal smears confirms that none of the substitute fixatives provide results of the same quality found with the use of mercuric chloride-based fixatives. However, the key issue is whether the intestinal parasites can be identified using these alternative fixatives, not how "perfect" they look. Many fixative options are now available, including single-vial collection systems, some of which are coupled with their own stains. Requirements also mandate that any laboratory using formalin must have formalin vapor monitored as both an 8 h time weighted average and 15 min readings. Most laboratories are now familiar with the regulations on protection of health care workers from blood and other body fluids and have implemented specific changes that are no longer optional. Although laboratories were already using many of the safety recommendations, these regulations delineate in detail what must be done and documented. Regulatory information based on new shipping requirements is also included.

On the basis of excellent suggestions and comments regarding features of the fourth edition of this book, the following changes have been made in this edition: a new chapter has been added that contains a large number of parasite medical case histories (case history, study questions, correct answer and discussion, and illustrative material); some of the life cycles have been redrawn, and new life cycles have been added; algorithms have been expanded; new tables and figures have been added throughout the book; additional drawings and photographs have been added; extensive updated text information is included, all of which was taken from a comprehensive literature review of all aspects of diagnostic medical parasitology; additional examples of unusual parasitic infections are included; the chapter on arthropods has been expanded and includes additional photographs and drawings and expanded text; the chapter on the immunology of parasitic infections has been enlarged, and updated information on both antigen and antibody detection methods continues to be included in this edition; the chapter on histologic identification of parasites has been greatly expanded with diagrams of various parasites and their visual presentations in tissue sections along with greatly enhanced legends for all images; diagnostic methods using newer immunoassay and "dipstick" technology are described; and the chapter on quality control has been expanded to include information on instrumentation and equipment, safety regulations, quality control and quality systems information, continuous quality improvement, and managed care considerations. The appendixes have been expanded to contain more information on artifacts; expanded lists and photographs of products and commercial suppliers; algorithms for ordering specific tests that complement the ova and parasite examination; flowcharts for processing stool specimens; quality control recording sheets for use in the laboratory; and general references and relevant websites. One of the most important expanded areas of the fifth edition is found in appendix 8. This section contains information that has been published within months prior to the final printing of this edition. This "late breaking" synopsis of very recent publications can assist the reader in having access to the latest information available. I encourage you to review this section as you read various chapters throughout the book.

The approach to the fifth edition of the book is similar to that for the first four editions. My objective is to provide the user with clear, concise, well organized, clinically relevant, cost effective, and practical quality procedures for use in the clinical laboratory setting. To use and fully understand these methods for the parasites discussed, it is imperative that the user also understand information related to life cycle, morphology, clinical disease, pathogenesis, diagnosis, treatment, epidemiology, and prevention. My intent is to provide a comprehensive discussion of both aspects of the field of diagnostic medical parasitology: first, a comprehensive discussion of the individual parasites, and second, relevant diagnostic methods designed to detect and identify the organisms present. I believe that the book fulfills these objectives and provides readers, whether they are laboratorians, physicians, or other health care professionals, with not only comprehensive but also very practical information.

It is also important for readers to understand that there are many diagnostic test options available to the clinical laboratory; not every laboratory will approach the diagnosis of parasitic infections in the same way. The key to quality and clinically relevant diagnostic work is a thorough understanding of the pros and cons of each option and how various options may or may not be relevant for one's particular geographic area, laboratory size and range of expertise, client base, number and type of patients seen, personnel expertise and availability, equipment availability, educational initiatives, and communication options, just to name a few variables. However, it is also important to understand the regulations and technical recommendations that govern and guide this type of laboratory work; many of these guidelines are related to coding and reimbursement, proficiency testing, and overall clinical relevance.

The use of product names is not intended to endorse specific products or to exclude substitute products. Also, because of possible advances and changes in the therapy of parasitic infections, independent verification of drugs and drug dosages is always recommended. The diagnostic procedures are intended for laboratory use only by qualified and experienced individuals or by the personnel under their direct supervision. Every effort has been made to ensure accuracy; however, ASM Press and I encourage you to submit to us any suggestions, comments, and information on errors found.

Acknowledgments

I would like to express my thanks to the many colleagues and students who have helped shape my perspective regarding the field of medical parasitology over the years, especially Yost Amrein, Bruce Anderson, Michael Arrowood, Lawrence Ash, Gordon Ball, Ralph Barr, Marilyn Bartlett, Kenneth Borchardt, Peter Boreham, Emilio Bouza, Thomas Brewer, Sandra Bullock-Iacullo, Ann Call, David Casemore, Francis Chan, Ray Chan, John Christie, Frank Cox, William Current, Peter Deplazes, J. P. Dubey, Mark Eberhard, Ronald Fayer, Sydney Finegold, Ana Flisser, Jacob Frenkel, Thomas Fritsche, Hector Garcia, Raj Gill, Robert Goldsmith, Thomas Hanscheid, George Healy, Barbara Herwaldt, Donald Heyneman, George Hillyer, Peter Horen, Peter Hotez, David John, Stephanie Johnston, Irving Kagan, Ray Kaplan, John Kessel, Jay Keystone, Mary Klassen-Fischer, Elmer Koneman, Jaime LaBarca, William Lewis, Andrea Linscott, Earl Long, Alex Macias, Edward Markell, Marilyn Marshall, Mae Melvin, Michael Miller, Anthony Moody, William Murray, Ronald Neafie, Ron Neimeister, Ann Nelson, Susan Novak, Thomas Nutman, Thomas Orihel, Ynes Ortega, Robert Owen, Josephine Palmer, Graeme Paltridge, William Petri, Kathy Powers, Paul Prociv, Gary Procop, Fred Rachford, Sharon Reed, William Rogers, Jon Rosenblatt, Norbert Ryan, Peter Schantz, James Seidel, Nicholas Serafy, Irwin Sherman, Robyn Shimizu, James Smith, Rosemary Soave, Deborah Stenzel, Charles Sterling, Alex Sulzer, Sam Telford, R. C. A. Thompson, Peter Traynor, Jerrold Turner, Govinda Visvesvara, Marietta Voge, Susanne Wahlquist, Kenneth Walls, Rainer Weber, Wilfred Weinstein, John Williams, John Wilson, Marianna Wilson, Martin Wolfe, Lihua Xiao, Charles and Wiladene Zierdt, and many others I may have forgotten to mention specifically. If the information contained in this edition provides help to those in the field of microbiology, I will have succeeded in passing on this composite knowledge to the next generation of students and teachers.

Special thanks go to Graeme Paltridge, who reviewed the third edition and offered suggestions for improvements for subsequent editions, and to Sharon Belkin for her additional illustrations for this edition. I also thank Ronald Neafie from the Armed Forces Institute of Pathology for providing many photographs to illustrate several areas of the book, particularly the information on histological identification of parasites, and Herman Zaiman for providing slides that he has prepared and/or edited from many contributors worldwide.

I would like to acknowledge the excellent contributions of David A. Bruckner, my coauthor for the first three editions. I sincerely hope that the fifth edition will match the high standards that were set for the previous editions.

I also thank members of the editorial staff of ASM Press, including Susan Birch, Jeff Holtmeier, and our copy editor, Yvonne Strong; they are outstanding professionals and made my job not only challenging but fun.

Above all, my very special thanks go to my husband, John, for his love and support for the many projects that I have been involved in over the years. I could never have undertaken these challenges without his help and understanding.

PART I

Clinically Important Human Parasites

Philosophy and Approach to Diagnostic Parasitology

The basic approach to diagnostic parasitology should be no different from that used in other areas of microbiology. There are guidelines published by the American Society for Microbiology (10, 11), the American Society of Parasitologists (8), the American Society for Medical Technology (16), the College of American Pathologists (7), and the Clinical and Laboratory Standards Institute (formerly NCCLS) (1, 2, 13–15) that contain recommended procedures for this field. If these general guidelines and recommendations are not followed, there is some question as to the qualifications of the laboratory performing the diagnostic work. At the very least, the clinician should be informed about the limitations of the procedures that are being used. These guidelines are also accompanied by specific regulations for a number of laboratory issues and include the Clinical Laboratory Improvement Act of 1988 and requirements related to safety and protection of employees from blood or bloodborne pathogens (standard precautions) (1, 3–6, 12).

Because it is difficult for the medical staff to maintain expertise in every available diagnostic procedure within microbiology, it is mandatory that close communication exist between the laboratory and clinicians. This type of frequent and complete communication, particularly concerning appropriate test orders and the clinical relevance of any diagnostic procedure within the context of total patient care and quality assurance, is very important. Therapeutic intervention often depends on results obtained from these procedures; therefore, the clinician must be aware of the limitations of each test method and the results obtained. This information becomes particularly important when one is discussing the patient's history and the recommended number and types of specimens to be submitted for examination.

During the past few years, there has been an increased awareness of the importance of having trained and qualified personnel perform these diagnostic procedures. There has been a concerted effort among many individuals and institutions in this country to upgrade the level of teaching and to bring to the medical community's attention the need for individuals who are familiar with diagnostic parasitology. With many laboratories decreasing staff size as a cost containment measure, we are also seeing more "generalists" who are rotating throughout many sections of the laboratory, not just microbiology. Although necessary because of managed-care constraints and continued growth of capitated contracts, this approach contributes to the difficulties in maintaining well-trained staff in some of the specialty areas of microbiology. It becomes even more important to provide well-written laboratory protocols and to standardize test methods for consistency. There has also been increased awareness within the medical community of the need for additional training in the area of infectious diseases for the clinician and laboratory technician alike. This need has been reflected in the number of workshops, seminars, and publications that are available. The integration of information among all members of the health care team has certainly improved in terms of overall patient care.

The field of microbiology has taken on additional relevance and importance for a number of other reasons. Improved means of travel has made the world a smaller place. An individual's chances of exposure to parasites not endemic to his or her homeland and the possibility of acquiring or transmitting certain infections have been increasing. These facts emphasize the need for taking a correct and complete history from a patient. It is important to be aware of the organisms commonly found within certain areas of the world and the makeup of the patient population being serviced at any particular health facility.

The most important step in the diagnosis of parasitic infections is the selection and submission of the appropriate clinical specimen within specified time lines and according to set protocols.

It is also important for the physician to know the efficacy of any diagnostic technique for parasite recovery and eventual diagnosis. Our approach to testing is undergoing continuous review, particularly within the current health care environment and cost containment initiatives. The issue of patient care becomes particularly important when we begin to examine the number and types of compromised patients now being seen in all facilities. The increased publicity concerning patients with AIDS has led to a greater awareness of parasitic infections in the compromised patient, regardless of the original cause of the immune deficiencies. Many of these patients with immune system defects are particularly at risk, whether because of previously acquired infections that have remained latent for many years or because of susceptibility to new infections. Many of these infections may present with unusual symptoms, and some are relatively new disease entities (microsporidiosis, granulomatous amebic encephalitis, and infection with *Cyclospora cayetanensis*).

Often in other areas of microbiology, therapy is begun on the basis of patient history and symptoms. This approach is generally not recommended or used in cases of parasitic infection. Thus, understanding of the characteristics of any parasitic infection (general geographic range, life cycle, clinical disease, diagnostic methods, therapy, epidemiology,

and control) and the use of appropriate diagnostic procedures accompanied by a complete understanding of the limitations of each procedure become very important. Because of this approach to patient care, general consensus among individuals within the field of diagnostic medical parasitology is that the use of certain incomplete procedures may result in incorrect information for the physician and may ultimately compromise patient care.

The main emphasis should be on the importance of understanding and recognizing potential parasitic infections, submitting the appropriate number and type of clinical specimens, knowing what procedures may provide confirmation of the diagnosis, and recognizing the implications and limitations of information provided to the physician. With the current emphasis on the development and use of immunoassay methods, understanding the benefits and limitations of these procedures will be critical to patient care outcomes. If there is an incomplete understanding of the requirements for high-quality diagnostic testing, incomplete information will be transmitted to the clinician (9). It is the responsibility of both the laboratory and the clinician to develop a greater awareness of the importance of these requirements.

References

1. **Clinical and Laboratory Standards Institute.** 2005. *Protection of Laboratory Workers from Occupationally Acquired Infection*, 3rd ed. Approved guideline M29-A3. Clinical and Laboratory Standards Institute, Wayne, Pa.
2. **Clinical and Laboratory Standards Institute.** 2005. *Procedures for the Recovery and Identification of Parasites from the Intestinal Tract*, 2nd ed. Approved guideline M28-A2. Clinical and Laboratory Standards Institute, Wayne, Pa.
3. **Code of Federal Regulations.** 1987. Update May 27, 1992. Title 29, parts 1910.1200 and 1910.1296. U.S. Government Printing Office, Washington, D.C.
4. **Code of Federal Regulations.** 1989. Title 29, part 1910.106. U.S. Government Printing Office, Washington, D.C.
5. **Code of Federal Regulations.** 1989. Title 29, part 1910.1200. U.S. Government Printing Office, Washington, D.C.
6. **Code of Federal Regulations.** 1989. Title 29, part 1910.1450. U.S. Government Printing Office, Washington, D.C.
7. **College of American Pathologists.** 2006. Commission on Laboratory Accreditation Inspection Checklist. College of American Pathologists, Chicago, Ill.
8. **Committee on Education, American Society of Parasitologists.** 1977. Procedures suggested for use in examination of clinical specimens for parasitic infection. *J. Parasitol.* **63:**959–960.
9. **Garcia, L. S.** 1999. *Practical Guide to Diagnostic Parasitology.* ASM Press, Washington, D.C.
10. **Isenberg, H. D.** (ed.). 2004. *Clinical Microbiology Procedures Handbook*, 2nd ed., p. 9.0.1–9.10.8.3. ASM Press, Washington, D.C.
11. **Isenberg, H. D.** (ed.). 1995. *Essential Procedures for Clinical Microbiology.* ASM Press, Washington, D.C.

12. **Joint Commission for the Accreditation of Healthcare Organizations.** 1987. *Monitoring and Evaluation of Pathology and Medical Laboratory Services.* Joint Commission for the Accreditation of Healthcare Organizations, Chicago, Ill.

13. **National Committee for Clinical Laboratory Standards.** 2002. *Clinical Laboratory Waste Management*, 2nd ed. Approved guideline GP5-A2. NCCLS, Wayne, Pa.

14. **National Committee for Clinical Laboratory Standards.** 2002. *Clinical Laboratory Procedure Manuals*, 4th ed. Approved guideline GP2-A4. NCCLS, Wayne, Pa.

15. **National Committee for Clinical Laboratory Standards.** 2000. *Laboratory Diagnosis of Blood-Borne Parasitic Diseases.* Approved guideline M15-A. NCCLS, Wayne, Pa.

16. **Parasitology Subcommittee, Microbiology Section of Scientific Assembly, American Society for Medical Technology.** 1978. Recommended procedures for the examination of clinical specimens submitted for the diagnosis of parasitic infections. *Am. J. Med. Technol.* **44:**1101–1106.

Intestinal Protozoa: Amebae

Entamoeba histolytica

Entamoeba histolytica was first described by Losch after being isolated in Russia from a patient with dysenteric stools. Although Losch found the organisms in human colonic ulcers at autopsy and was able to induce dysentery in a dog that was inoculated rectally with the patient's stools, he failed to recognize the causal relationship. This organism was eventually more fully investigated and differentiated from *Entamoeba coli* and *Entamoeba hartmanni* with respect to both morphology and pathogenesis (15). Based on the genome, this organism has a number of metabolic adaptations shared with two other amitochondrial protest pathogens: *Giardia lamblia* and *Trichomonas vaginalis*. These adaptations include the use of oxidative stress enzymes usually associated with anaerobic prokaryotes, as well as reduction or elimination of most mitochondrial metabolic pathways. There is evidence for lateral gene transfer of bacterial genes into the *E. histolytica* genome, a large number of novel receptor kinases, and expansions of a variety of gene families, including those associated with virulence (25).

E. histolytica has been recovered worldwide and is more prevalent in the tropics and subtropics than in cooler climates. However, infection rates in unsanitary conditions in temperate and colder climates have been found to equal those seen in the tropics. The World Health Organization reported that *E. histolytica* causes approximately 50 million cases and 110,000 deaths annually (51). Only malaria and schistosomiasis surpass amebiasis as leading parasitic causes of death (41).

At least 10 amebae are found in the mouth or intestinal lumen (*E. histolytica*, *E. dispar*, *E. moshkovskii*, *E. hartmanni*, *E. coli*, *E. polecki*, *E. gingivalis*, *E. nana*, *Iodamoeba bütschlii*, and *Blastocystis hominis*). However, of these, only *E. histolytica* and *B. hominis* have been considered to be pathogenic.

E. histolytica versus E. dispar

Although a large number of people throughout the world are infected, only a small percentage will develop clinical symptoms. Morbidity and mortality due to *E. histolytica* vary, depending on the geographic area, the organism strain, and the patient's immune status. For many years, the issue of pathogenicity has been very controversial, with, essentially, two points of view. Some thought that what was called *E. histolytica* was really two separate species of *Entamoeba*, one being pathogenic and causing invasive disease and the other being nonpathogenic and causing mild or asymptomatic infections. Others thought that all organisms designated *E. histolytica* were potentially pathogenic, with symptoms depending on the result of host or environmental factors, including the intestinal flora.

In 1961, with the development of successful axenic culture methods requiring no bacterial coculture, sufficient organisms could be obtained for additional studies. Approximately 15 years later, reports indicated that *E. histolytica* clinical isolates could be classified into groups by using starch gel electrophoresis and review of banding patterns related to specific isoenzymes (15). The four isoenzymes are glucophosphate isomerase, phosphoglucomutase, malate dehydrogenase, and hexokinase. Sargeaunt concluded from this work that there are pathogenic and nonpathogenic strains (zymodemes) of *E. histolytica* that can be differentiated by isoenzyme analysis. On the basis of analysis of thousands of clinical isolates, he also concluded that the zymodeme patterns were probably genetic rather than phenotypic.

Others argued that the bacterial flora present in the gut played a role in the potential pathogenicity of *E. histolytica*. Some of the in vitro culture studies indicated that bacteria enhance the virulence of *E. histolytica*. During the 1980s and early 1990s, these issues were discussed with a number of questions still remaining unanswered (30).

According to Diamond and Clark (8), the most logical explanation of their failure to obtain conversion of "nonpathogenic" *E. histolytica* isolates to the "pathogenic" form during axenization of the amebae was culture contamination. In reviewing this possibility, they used a method based on analysis of stable DNA polymorphisms that allows the identification of individual pathogenic isolates. The DNA patterns obtained with the "converted" amebae were identical to those of reference isolates present in the laboratory at the time of conversion (8). Diamond and Clark also found that transfer of very few pathogenic organisms was necessary for the pathogens to become established in a nonpathogenic culture. Within the context of these studies, cross-contamination fully explained the conversion and the designation of specific pathogenic *E. histolytica* and nonpathogenic *E. dispar* was confirmed (8).

Studies confirming the differences between the two organisms include direct sequencing of the PCR-amplified small-subunit rRNA gene of *E. dispar* and the design of primers for rapid differentiation from *E. histolytica*, differences in the *E. histolytica* and *E. dispar* phosphoglucomutases, molecular biology of the differences in the hexokinase isoenzyme, differences in the secretion of acid

phosphatase, and the detection of DNA sequences unique to *E. histolytica* (35, 49, 57).

Genetic differences may not totally explain the overall differences. Among 79 *E. histolytica* isolates obtained from different geographic regions, a total of 53 different genotypes were observed among 63 isolates. The most extensive variations among the four loci were found in the serine-rich *E. histolytica* protein (SREHP) locus. However, no association between clinical presentation and the SREHP genotype was demonstrated (17).

Research has centered on explaining the molecular differences between the pathogenic *E. histolytica* and the nonpathogenic *E. dispar*. However, the molecules considered the most important for host tissue destruction (amebapore, galactose/N-acetylgalactosamine inhibitable lectin, and cysteine proteases) can be found in both organisms. Pathogenicity differences may be related to the composition and properties of the surface coat components (or pathogen-associated molecular patterns) and the ability of the innate immune response to recognize these components, thus eliminating the organisms (6). Targets of the host immune system modulation appear to be both neutrophils and macrophages, which are unable to abort the infection even when present at the site of the lesion (10). The total body of evidence supports differentiation of the pathogenic *E. histolytica* from the nonpathogenic *E. dispar* and their classification as two distinct species (15).

Life Cycle and Morphology

The life cycle is shown in Figure 2.1. The cyst form is the infective form for humans.

Morphology of Trophozoites (Table 2.1). Living trophozoites vary in size from about 12 to 60 µm in diameter. Organisms recovered from diarrheic or dysenteric stools are generally larger than those in a formed stool from

Figure 2.1 Life cycle of *Entamoeba histolytica*.

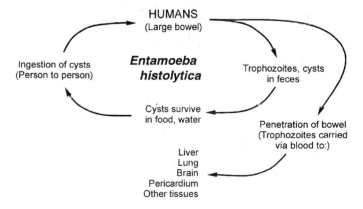

Table 2.1 Intestinal protozoa: trophozoites of common amebae

Characteristic	Entamoeba histolytica	Entamoeba dispar/ Entamoeba moshkovskii	Entamoeba hartmanni	Entamoeba coli	Endolimax nana	Iodamoeba bütschlii
Size[a] (diam or length)	12–60 µm (usual range, 15–20 µm); invasive forms may be over 20 µm	Same as E. histolytica	5–12 µm (usual range, 8–10 µm)	15–50 µm (usual range, 20–25 µm)	6–12 µm (usual range, 8–10 µm)	8–20 µm (usual range, 12–15 µm)
Motility	Progressive, with hyaline, fingerlike pseudopodia; motility may be rapid	Same as E. histolytica	Usually nonprogressive	Sluggish nondirectional, with blunt, granular pseudopodia	Sluggish, usually nonprogressive	Sluggish, usually nonprogressive
Nucleus: no. and visibility	Difficult to see in unstained preparations; 1 nucleus	Same as E. histolytica	Usually not seen in unstained preparations; 1 nucleus	Often visible in unstained preparations; 1 nucleus	Occasionally visible in unstained preparations; 1 nucleus	Usually not visible in unstained preparations; 1 nucleus
Peripheral chromatin (stained)	Fine granules, uniform in size and usually evenly distributed; may have beaded appearance	Same as E. histolytica	Nucleus may stain more darkly than in E. histolytica/E. dispar, although morphology is similar; chromatin may appear as solid ring rather than beaded	May be clumped and unevenly arranged on the membrane; may also appear as solid, dark ring with no beads or clumps	Usually no peripheral chromatin; nuclear chromatin may be quite variable	Usually no peripheral chromatin
Karyosome (stained)	Small, usually compact; centrally located but may also be eccentric	Same as E. histolytica	Usually small and compact; may be centrally located or eccentric	Large, not compact; may or may not be eccentric; may be diffuse and darkly stained	Large, irregularly shaped; may appear "blotlike"; many nuclear variations are common; may mimic E. hartmanni or D. fragilis	Large, may be surrounded by refractile granules that are difficult to see ("basket nucleus")
Cytoplasm appearance (stained)	Finely granular, "ground-glass" appearance; clear differentiation of ectoplasm and endoplasm; if present, vacuoles are usually small	Same as E. histolytica	Finely granular	Granular, with little differentiation into ectoplasm and endoplasm; usually vacuolated	Granular, vacuolated	Granular, may be heavily vacuolated
Inclusions (stained)	Noninvasive organism may contain bacteria; presence of RBCs diagnostic	Organisms usually contain bacteria; RBCs not present in cytoplasm	May contain bacteria; no RBCs	Bacteria, yeasts, other debris	Bacteria	Bacteria

[a] Wet preparation measurements (on permanent stains, organisms usually measure 1 to 2 µm less).

an asymptomatic individual (Figures 2.2 and 2.3). The motility has been described as rapid and unidirectional, with pseudopods forming quickly in response to the conditions around the organism. The motility may appear to be sporadic. Although this characteristic motility is often described, it is rare to diagnose amebiasis on the basis of motility seen in a direct mount. The cytoplasm is differentiated into a clear outer ectoplasm and a more granular inner endoplasm.

When the organism is examined on a permanent stained smear (trichrome or iron hematoxylin), the morphological characteristics of *E. histolytica*/*E. dispar* are readily seen. The nucleus is characterized by having evenly arranged chromatin on the nuclear membrane and having a small, compact, centrally located karyosome. The cytoplasm is usually described as finely granular with few ingested bacteria or debris in vacuoles. In organisms isolated from a patient with dysentery, red blood cells (RBCs) may be visible in the cytoplasm, and this feature is diagnostic for *E. histolytica*. Most often, infection with *E. histolytica*/*E. dispar* is diagnosed on the basis of organism morphology without the presence of RBCs.

Figure 2.2 (1) Trophozoite of *Entamoeba histolytica*; (2) trophozoite of *E. histolytica*/*E. dispar*/*E. moshkovskii*; (3 and 4) early cysts of *E. histolytica*/*E. dispar*/*E. moshkovskii*; (5 to 7) cysts of *E. histolytica*/*E. dispar*/*E. moshkovskii*; (8 and 9) trophozoites of *Entamoeba coli*; (10 and 11) early cysts of *E. coli*; (12 to 14) cysts of *E. coli*; (15 and 16) trophozoites of *Entamoeba hartmanni*; (17 and 18) cysts of *E. hartmanni*.

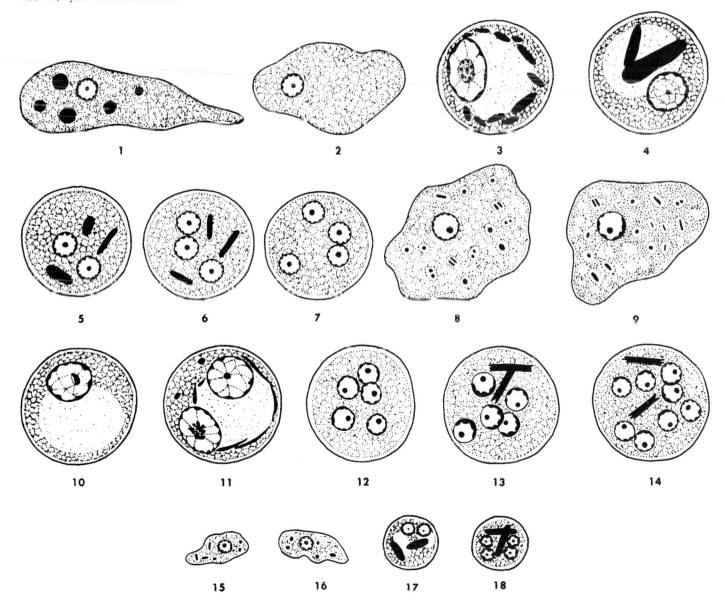

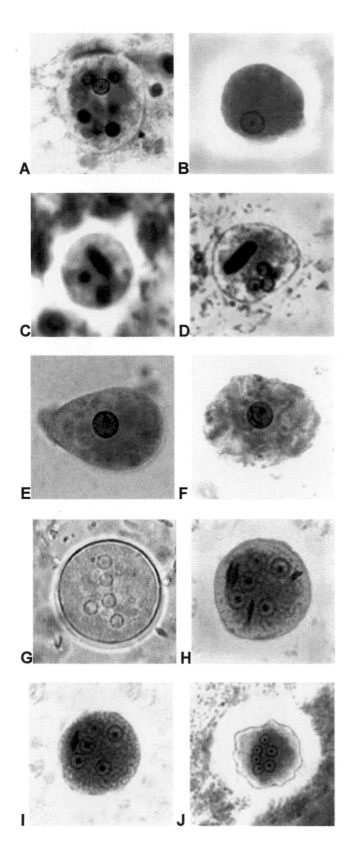

Morphology of Cysts (Table 2.2). For reasons that are not specifically known, the trophozoites may condense into a round mass (precyst), and a thin wall is secreted around the immature cyst. There may be two types of inclusions within this immature cyst, a glycogen mass and highly refractile chromatoidal bars with smooth, rounded edges. As the cysts mature (metacyst), there is nuclear division with the production of four nuclei; on rare occasions, eight nuclei are produced and the cysts range in size from 10 to 20 μm. Often, as the cyst matures, the glycogen completely disappears; the chromatoidals may also be absent in the mature cyst. Cyst morphology does not differentiate between *E. histolytica* and *E. dispar.* Cyst formation occurs only within the intestinal tract; once the stool has left the body, cyst formation does not occur. The one-, two-, and four-nucleated cysts are infective and represent the mode of transmission from one host to another (Figures 2.1 to 2.3).

Infection. After cyst ingestion, no changes occur in an acidic environment; however, once the pH becomes neutral or slightly alkaline, the encysted organism becomes active, with the outcome being four separate trophozoites (small, metacystic trophozoites). These organisms develop into the normal trophozoites when they become established in the large intestine.

Clinical Disease

The presentations of disease are seen with invasion of the intestinal mucosa and/or dissemination to other organs, the most common being the liver. However, it is estimated that only a small proportion (2 to 8%) of infected individuals will have invasive disease beyond the lumen of the bowel. Also, organisms may be spontaneously eliminated with no disease symptoms.

Asymptomatic Infection. Individuals harboring *E. histolytica* may have either a negative or weak antibody titer and negative stools for occult blood and may be passing cysts that can be detected if the routine ova and parasite examination is performed. Although trophozoites may also be found, they do not contain any phagocytized RBCs and cannot be differentiated from *E. dispar.* Isoenzyme analyses of organisms isolated from asymptomatic individuals generally indicate that the isolates belong to nonpathogens (*E. dispar*). Generally, asymptomatic patients never become symptomatic and may excrete cysts for a short period.

Figure 2.3 (A) Trophozoite of *Entamoeba histolytica*; (B) trophozoite of *Entamoeba histolytica/E. dispar/E. moshkovskii*; (C and D) cysts of *E. histolytica/E. dispar/E. moshkovskii*; (E and F) trophozoites of *Entamoeba coli*; (G to J) cysts of *E. coli.* Note the chromatoidal bars in panels H and I and the shrunken cyst (which is often seen in permanent stained smears) in panel J.

Table 2.2 Intestinal protozoa: cysts of common amebae

Characteristic	Entamoeba histolytica/ Entamoeba dispar/ Entamoeba moshkovskii	Entamoeba hartmanni	Entamoeba coli	Endolimax nana	Iodamoeba bütschlii
Size[a] (diam or length)	10–20 μm (usual range, 12–15 μm)	5–10 μm (usual range, 6–8 μm)	10–35 μm (usual range, 15–25 μm)	5–10 μm (usual range, 6–8 μm)	5–20 μm (usual range, 10–12 μm)
Shape	Usually spherical	Usually spherical	Usually spherical; may be oval, triangular, or other shapes; may be distorted on permanent stained slide because of inadequate fixative penetration	Usually oval, may be round	May vary from oval to round; cyst may collapse because of large glycogen vacuole space
Nucleus: no. and visibility	Mature cyst, 4; immature, 1 or 2 nuclei; nuclear characteristics difficult to see on wet preparation	Mature cyst, 4; immature, 1 or 2 nuclei; two nucleated cysts very common	Mature cyst, 8; occasionally 16 or more nuclei are seen; immature cysts with 2 or more nuclei are occasionally seen	Mature cyst, 4; immature cysts, 2; very rarely seen and may resemble cysts of Enteromonas hominis	Mature cyst, 1
Peripheral chromatin (stained)	Peripheral chromatin present; fine, uniform granules, evenly distributed; nuclear characteristics may not be as clearly visible as in trophozoite	Fine granules evenly distributed on the membrane; nuclear characteristics may be difficult to see	Coarsely granular; may be clumped and unevenly arranged on membrane; nuclear characteristics not as clearly defined as in trophozoite; may resemble E. histolytica/E. dispar	No peripheral chromatin	No peripheral chromatin
Karyosome (stained)	Small, compact, usually centrally located but occasionally eccentric	Small, compact, usually centrally located	Large, may or may not be compact and/or eccentric; occasionally centrally located	Smaller than karyosome seen in trophozoites, but generally larger than those of the genus Entamoeba	Large, usually eccentric refractile granules may be on one side of karyosome ("basket nucleus")
Cytoplasm, chromatoidal bodies (stained)	May be present; bodies usually elongate, with blunt, rounded, smooth edges; may be round or oval	Usually present; bodies usually elongate, with blunt, rounded, smooth edges; may be round or oval	May be present (less frequently than in E. histolytica/E. dispar); splinter shaped with rough, pointed ends	Rare chromatoidal bodies present; occasionally small granules or inclusions seen; fine linear chromatoidals may be faintly visible on well-stained smears	No chromatoidal bodies present; small granules occasionally present
Glycogen (stained with iodine)	May be diffuse or absent in mature cyst; clumped chromatin mass may be present in early cysts (stains reddish brown with iodine)	May or may not be present as in E. histolytica/ E. dispar	May be diffuse or absent in mature cyst; clumped mass occasionally seen in mature cysts (stains reddish brown with iodine)	Usually diffuse if present (stains reddish brown with iodine)	Large, compact, well-defined mass (stains reddish brown with iodine)

[a] Wet preparation measurements (on permanent stains, organisms usually measure 1 to 2 μm less).

This pattern is found for patients infected with either nonpathogenic or pathogenic species.

Intestinal Disease. Although the exact mode of mucosal penetration is not known, microscopy studies suggest that amebae have enzymes which lyse host tissue, possibly from lysosomes on the surface of the amebae or from ruptured organisms. Amebic ulcers most often develop in the cecum, appendix, or adjacent portion of the ascending colon; however, they can also be found in the sigmoidorectal area. From these primary sites, other lesions may occur. Ulcers are usually raised, with a small opening on the mucosal surface and a larger area of destruction below the surface, i.e., "flask shaped." The mucosal lining may appear normal between ulcers.

Although the incubation period varies from a few days to a much longer time, in an area where *E. histolytica* is endemic it is impossible to determine exactly when the exposure took place. Normally, the time frame ranges from 1 to 4 weeks. Certainly not every patient infected with *E. histolytica* will develop invasive amebiasis; the outcome will depend on the interaction between parasite virulence factors and the host response.

There are four clinical forms of invasive intestinal amebiasis, all of which tend to be acute. They include dysentery or bloody diarrhea, fulminating colitis, amebic appendicitis, and ameboma of the colon. Dysentery and diarrhea account for 90% of cases of invasive intestinal amebiasis (10). Symptoms may range from none to those mimicking ulcerative colitis. Patients with colicky abdominal pain, frequent bowel movements, and tenesmus may present with a gradual onset of disease. With the onset of dysentery, bowel movements characterized by blood-tinged mucus are frequent (up to 10 per day). Although dysentery may last for months, it usually varies from severe to mild over that time and may lead to weight loss and prostration. In patients with severe cases, symptoms may begin very suddenly and include profuse diarrhea (over 10 stools per day), fever, dehydration, and electrolyte imbalances. If the clinical picture is acute, the illness may mimic appendicitis, cholecystitis, intestinal obstruction, or diverticulitis.

Hepatic Disease. Blood flow draining the intestine tends to return to the liver, most commonly the upper right lobe (Figure 2.4). The organisms present in the submucosa can therefore be carried via the bloodstream to the liver. Onset of symptoms may be gradual or sudden; upper right abdominal pain with fever from 38 to 39°C is the most consistent finding. Weakness, weight loss, cough, and sweating are less commonly seen. There tends to be hepatomegaly with tenderness; however, liver function tests may be normal or slightly abnormal, with jaundice being very rare. There may be changes at the base of the right lung owing to the elevated

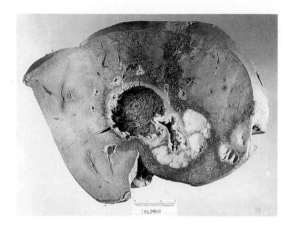

Figure 2.4 Liver containing multiple amebic abscesses. The necrotic tissue has been removed from one portion, leaving a cavity. ×0.4. (Armed Forces Institute of Pathology photograph.)

diaphragm. The abscess can be visualized radiologically, sonically, or by radionuclear scan, and the majority of patients have a single abscess in the right lobe of the liver. The most common complication is rupture of the abscess into the pleural space. An abscess can also extend into the peritoneum and through the skin. Hematogenous spread to the brain, as well as to the lungs, pericardium, and other sites, is possible. For patients living in areas of endemic infection and patients with a relevant travel history, amebic abscess should be suspected in those with spiking fever, weight loss, and abdominal pain in the upper right quadrant or epigastrium and those with tenderness in the liver area. The presence of leukocytosis, a high alkaline phosphatase level, and an elevated right diaphragm suggests a hepatic abscess. Pyogenic and amebic liver abscesses are the two most common hepatic abscesses. The severity of a pyogenic abscess depends on the bacterial source and the patient's underlying condition. Amebic abscess tends to be more prevalent in individuals with suppressed cell-mediated immunity, men, and younger individuals (20, 24, 32, 42, 56).

E. histolytica cysts and trophozoites are found in the stools of only a few patients with liver abscess. About 60% of these patients have no intestinal symptoms or any history of dysentery.

Although hepatic abscess is secondary to earlier colonization in the wall of the large intestine and is the most common site for extraintestinal amebiasis, *E. histolytica* has been seen in almost every soft organ and tissue of the body, including the lungs, heart, skin, and brain. Amebiasis of the lungs usually develops as an extension of a hepatic abscess that has ruptured through the diaphragm; however, it can also originate as a secondary site from the original infection in the intestine. Typically, the patient experiences expectoration of liver-colored sputum. In a very low percentage of cases, the pericardium can be involved; this is more likely to occur in patients with an abscess in the left lobe of the liver.

Amebiasis of the skin usually develops as a perianal extension of acute amebic colitis or on the abdominal wall due to rupture or open drainage of an internal intestinal or hepatic lesion (36). Amebiasis cutis may also occur as a venereal infection of the penis after anal intercourse.

Amebic abscess in the brain is quite rare and usually arises from either hepatic or pulmonary involvement. Other rare ectopic sites include the spleen, adrenals, kidneys, ureters, urinary bladder, urethra, clitoris, and pericardium.

Pathogenesis

It has been estimated that E. histolytica infection kills more than 100,000 people each year. Key to the pathogenesis of this organism is its ability to directly lyse host cells and cause tissue destruction. Amebic lesions show evidence of cell lysis, tissue necrosis, and damage to the extracellular matrix. There is evidence that E. histolytica trophozoites interact with the host through a series of steps: adhesion to the target cell, phagocytosis, and cytopathic effect (6, 10, 45, 47). Numerous other parasite factors also play a role. From the perspective of the host, E. histolytica induces both humoral and cellular immune responses, with cell-mediated immunity representing the major human host defense against this complement-resistant cytolytic protozoan.

Our understanding of pathogenesis has expanded dramatically within the past few years. We now know that invasive strains of E. histolytica are resistant to complement-mediated lysis. These organisms activate complement via cleavage of C3 by an extracellular cysteine proteinase. Cysteine proteinases are important virulence factors encoded by at least three genes, one of which is unique to invasive strains (41). The amebic cysteine proteinases are homologous to proteinases released by transformed cells and probably represent a common mechanism of tissue invasion. The proteinase activity of amebae shows excellent correlation with their virulence. Cysteine proteinase can cleave the extracellular structural matrix and can degrade fibronectin and laminin, as well as type I collagen; this suggests that degradation of basement membrane is an important step in tissue invasion (41). Hemoglobin-degrading cysteine proteases also play an important role in iron acquisition by E. histolytica. Differences in the pathogenesis of E. histolytica and E. dispar also help explain the epidemiology, clinical syndromes, and pathology of amebiasis. Antibodies to the neutral cysteine proteinase have been detected in 83% of patients with invasive E. histolytica infections but not in patients with noninvasive E. dispar infections (46). Apparently, inhibition of the expression of cysteine proteinases does not affect E. histolytica cytopathic or hemolytic activity but inhibits phagocytosis. There is

recent evidence that a cysteine protease (EhCP112) and a protein with an adherence domain (EhADH112) form the E. histolytica 112-kDa adhesin.

The first evidence of amebic pathology is depletion of the intestinal mucus and disruption of the epithelial barrier, which results from the action of cysteine proteinases as well as other factors (6, 41). Clinical isolates of E. histolytica in culture release 10- to 1,000-fold more cysteine proteinases than do E. dispar isolates. Trophozoites use cysteine proteinases to cleave and penetrate the extracellular matrix (collagen, elastin, fibrinogen, and laminin) in order to cause invasive disease (41). These enzymes also interfere with the host immune system; hence, the data supporting a key role of cysteine proteinases in virulence are quite compelling, especially since specific genes for cysteine proteinases appear to be present and overexpressed in E. histolytica and absent or underexpressed in E. dispar.

Several molecules that may be involved in the parasite-host interaction processes have been identified. A lectin inhibitable by galactose and N-acetyl D galactosamine (Gal/GalNAc) is present on the trophozoite surface. Adherence of E. histolytica trophozoites to colonic mucin, epithelium, and other target cells is mediated by the amebic Gal/GalNAc lectin (39). Lectin activity in the parasite is regulated by inside-out signaling; this inside-out signaling via the lectin cytoplasmic domain may control the extracellular adhesive activity of the amebic lectin and provides evidence of the role of lectin in virulence (54). Studies also confirm that dominant-negative mutations in the Gal/GalNAc lectin affect adhesion and cytolysis while mutations in meromyosin affect cytoskeletal function (14). Myosin IB also is involved in phagocytosis of human erythrocytes by E. histolytica (55). Recently, a distinct β_2 integrin-like molecule with a potential role in cellular adherence was identified. This amebic β_2 integrin appeared to be distinct from the amebic Gal/GalNAc lectin based on recombinant expression, amebic colocalization, and enzyme-linked immunosorbent assay (ELISA) studies. The presence of integrin-dependent binding may allow trophozoites to opportunistically adhere to activated intestinal epithelium or vascular endothelium during amebic colitis or hepatic abscess (38).

There are also correlations between differences in the cell surface phosphorylated glycolipids and virulence; lipophosphoglycan and lipophosphopeptidoglycan molecules are abundant in virulent strains of E. histolytica and are absent or present at low levels in avirulent E. histolytica strains and E. dispar (31).

Various amebic pore-forming proteins are known, of which the 5-kDa protein (amebapore) has been extensively studied. These proteins can insert into the lipid bilayers of target cells, forming ion channels (10).

Data obtained by using human myeloid cells as target cells in vitro have revealed that the target cells die due to necrosis with cell swelling, rupture of plasma membranes, and release of cell contents including nucleic acids. These findings confirm the notion that these particular target cells undergo necrosis when killed by viable trophozoites, as well as by isolated amebapores (10).

The role of zinc in the pathogenicity of *E. histolytica* may also be relevant. Apparently, zinc alters the functionality of the amebae in vitro by causing a decrease in replication and adhesion and in vivo by inhibiting amebic pathogenicity.

Diagnosis

Routine Diagnostic Procedures. The laboratory diagnosis of amebiasis, particularly intestinal amebiasis, depends on a number of procedures, any one of which can prevent organism recovery if not performed properly (Table 2.3). Organism detection depends on collection of the correct specimens, the number of specimens submitted, processing methods and diagnostic tests used, and examination by personnel who are well trained in identification of protozoa. This diagnosis can be one of the more difficult to achieve. Lack of appropriate training and diagnostic testing may lead to missed infections. In some cases, false positives may be due to identification of human white blood cells as amebae. A complete discussion of specimen collection and submission to the laboratory is presented in chapter 26.

The standard ova and parasite examination is the recommended procedure for recovery and identification of *E. histolytica* in stool specimens. Microscopic examination of a direct saline wet mount may reveal motile trophozoites, which may contain RBCs. However, the number of times these trophozoites with RBCs are present is limited. In many patients who do not present with acute dysentery, trophozoites may be present but do not contain RBCs and the organisms may be *E. histolytica* or *E. dispar*. An asymptomatic individual may have few trophozoites and possibly only cysts in the stool. Although the concentration technique is helpful in demonstrating cysts, *the most important technique for the recovery and identification of protozoan organisms is the permanent stained smear* (normally stained with trichrome or iron hematoxylin). A minimum of three specimens collected over a time frame of not more than 10 days is recommended (21).

Note that if the permanent stained smear is quite dark, a delicate *E. histolytica/E. dispar* organism may appear more like *Entamoeba coli*; on a very thin, pale smear, *E. coli* can appear more like *E. histolytica/E. dispar*. Also, when the organisms are dying or have been poorly fixed, they appear more highly vacuolated, thus looking like *E. coli* trophozoites rather than *E. histolytica/E. dispar*.

If slides are prepared properly and examined carefully, sigmoidoscopy specimens may be very helpful. At least six areas of the mucosa should be sampled. Smears from these areas should be examined after permanent staining. This does not take the place of the recommended minimum of three stool specimens submitted for ova and parasite examinations (direct, concentration, and permanent stained smear). Refer to Algorithm 2.1 for detailed information.

Definitive diagnosis of liver abscess can be achieved by identification of organisms from liver aspirate material. However, this procedure is rarely performed, and often the specimen obtained is not collected properly. Aspirated material must be aliquoted into several different containers as it is removed from the abscess; amebae may be found only in the last portion of the aspirated material, theoretically material from the abscess wall, not necrotic debris from the abscess center. Refer to Algorithm 2.2 for detailed information.

Table 2.3 Intestinal amebae: recommended diagnostic procedures

Organism	Recommended specimen	Diagnostic procedure
Entamoeba histolytica	Stool	Complete ova and parasite exam
		Immunoassays for antigen detection
	Sigmoidoscopy material	Permanent stain
	Hepatic abscess aspirate	Digestion/wet preparations
Entamoeba dispar	Stool	Complete ova and parasite exam
Entamoeba moshkovskii	Stool	Complete ova and parasite exam
Entamoeba hartmanni	Stool	Complete ova and parasite exam
Entamoeba coli	Stool	Complete ova and parasite exam
Entamoeba polecki	Stool	Complete ova and parasite exam
Entamoeba gingivalis	Gingival scrapings	Wet preparations and permanent stains
	Pyorrheal material	Wet preparations and permanent stains
Endolimax nana	Stool	Complete ova and parasite exam
Iodamoeba bütschlii	Stool	Complete ova and parasite exam
Blastocystis hominis	Stool	Complete ova and parasite exam

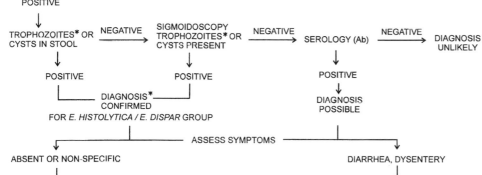

Algorithm 2.1 Intestinal amebiasis. Fecal immunoassay kits can be used on human stool specimens to identify organisms in the group *E. histolytica/E. dispar* or to differentiate *E. histolytica* from *E. dispar*. Currently available reagents require fresh or frozen specimens only. *, trophozoites containing RBCs confirmatory for *E. histolytica*; trophozoites with no ingested RBCs and/or cysts confirm only the *E. histolytica/E. dispar* group.

Algorithm 2.2 Amebic liver abscess. *, trophozoites containing RBCs confirmatory for *E. histolytica*; trophozoites with no ingested RBCs and/or cysts confirm only the *E. histolytica/E. dispar* group.

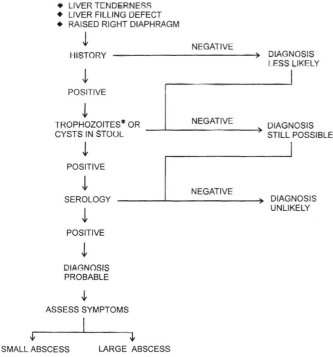

Immunodetection. A number of enzyme immunoassay reagents are commercially available, and their specificity and sensitivity provide excellent options for the clinical laboratory (Table 2.4) (2, 3, 5, 12, 15, 18, 53). These tests can determine the presence of the *E. histolytica/E. dispar* group as being distinct from the rest of the *Entamoeba* species such as nonpathogenic *E. coli* or *E. hartmanni*. Other test reagents can actually separate *E. histolytica* from *E. dispar*. The majority of these procedures use the enzyme-linked immunosorbent assay (ELISA) or enzyme immunoassay (EIA) formats (11). However, in some cases they have been found to be less sensitive than PCR (15) and in other instances they have been found equal to PCR (11). Another product is available as a cartridge format that uses an immunochromatographic strip-based detection system for the *E. histolytica/E. dispar* group, *Giardia lamblia*, and *Cryptosporidium* spp. (51). These commercially available reagent kits require fresh or frozen stools; formalin has been found to interfere with the *Entamoeba* reagents. Any available antigen detection system should always be reviewed for compatibility with stools submitted in preservatives rather than fresh specimens. There are limitations when using kits that include organisms in the genus *Entamoeba* (51). However, almost all reagent kits for the detection of *G. lamblia* and *Cryptosporidium* spp. can be used with formalin based stool preservatives, as well as fresh or frozen specimens.

To provide clinically relevant information to physicians for treatment of patients infected with pathogenic *E. histolytica*, methods involving monoclonal antibodies, purified antigens, or DNA probes are now available. Reagents have been developed to differentiate pathogenic

Table 2.4   Commercial assays used to identify *E. histolytica*

Assay	Manufacturer	Sensitivity (%)	Specificity (%)	Comments[a]
Antigen detection TechLab *E. histolytica* II	TechLab, Blacksburg, Va.	100	~95	EIA; requires fresh or frozen stool; reagents differentiate between *E. histolytica* and *E. dispar*
ProSpecT *Entamoeba histolytica* microplate assay	Remel, Lenexa, Kans.	90.3	97.7	EIA; identification of the *E. histolytica/E. dispar* group; fresh or preserved specimens acceptable
Entamoeba CELISA-*PATH*	Cellabs Pty Ltd., Brookvale, Australia	94	100	EIA; no specific information per fresh or preserved specimens
Wampole *E. histolytica* test	Wampole Laboratories, Cranbury, N.J.	94.7		EIA; requires fresh or frozen stool; reagents differentiate between *E. histolytica* and *E. dispar*
Merlin Optimun S ELISA	Merlin Diagnostika, Bernheim-Hersel, Germany	100		EIA; no specific information per fresh or preserved specimens
Triage parasite panel	BIOSITE Diagnostics, San Diego, Calif.	~83–96	~99–100	Lateral-flow cartridge; identification of the *E. histolytica/E. dispar* group; fresh or frozen specimens required
Entamoeba IHA	Biotrin Int., Heidelberg, Germany			EIA; no specific information per fresh or preserved specimens
Antibody detection IHA Cellognost Amoebiasis	Behring Diagnostics, Marburg, Germany	~72–100	~91	IHA
Light Diagnostics Amebiasis (*Entamoeba histolytica*) EIA	Chemicon, Int., Temecula, Calif.			EIA; USA
Amebiasis Serology Microwell EIA	IVD Research, Carlsbad, Calif.	92.5	91.3	EIA; USA
BLA-Bichrolatex-Amibe	Fumouze Diagnostics, Levallois-Perret, France	98.3	96.1	LA
Ambiase HAI FUMOUZE	Fumouze Diagnostics, Levallois-Perret, France	93.4	97.5	IHA
The Melotest Amoebiasis EIA	Melotec, S.A., Barcelona, Spain			EIA

[a] EIA, enzyme immunoassay; IHA, indirect hemagglutination; LA, latex agglutination.

E. histolytica from nonpathogenic *E. dispar* (5, 11, 15, 19). Although some of these reagents are now available, routine use in clinical laboratories may not be accepted because of cost, limited numbers of cases, and infrequent identification of organisms in the genus *Entamoeba*. Reliable probes based on the hypothesis of zymodeme stability and two separate species (one pathogenic and one nonpathogenic) offer significant improvements over current serologic or direct detection methods; these probes may also help eliminate false-positive results due to misidentification of human cells or other nonpathogenic protozoa in patient specimens.

The laboratory now has the ability to routinely indicate to the physician whether the *E. histolytica/E. dispar* organisms seen in the stool specimen are pathogenic *E. histolytica* or nonpathogenic *E. dispar*. Without the use of these types of reagents, the only way to morphologically identify true pathogenic *E. histolytica* would be to observe the rare presence of trophozoites containing ingested RBCs. In the event the laboratory does not use these reagents, the presence of *E. histolytica/E. dispar* should be reported to the physician, along with some additional commentary related to the newer information on pathogenicity. Ultimately, the physician will have to decide on the issue of therapy on the basis of the patient's clinical condition. However, without more accurate information on the pathogenicity of organisms seen in clinical specimens, from a public health point of view it may be appropriate to treat all infected patients. Also, depending on each state's requirements, pathogenic *E. histolytica*, as well as *E. histolytica*/*E. dispar*, is generally reported to the Public Health Facility (County).

Antibody Detection. Serologic testing for intestinal disease is normally not recommended unless the patient has true dysentery; even in these cases, the titer (indirect hemagglutination as an example) may be low and thus difficult to interpret. The definitive diagnosis of intestinal amebiasis should not be made without demonstrating the organisms. For patients suspected of having extraintestinal disease, serologic tests are much more relevant (5, 19, 58). Indirect hemagglutination and indirect fluorescent-antibody tests have been reported positive with titers of ≥1:256 and ≥1:200, respectively, in almost 100% of cases of amebic liver abscess (19, 51). Positive serologic results, in addition to clinical findings, make the diagnosis highly probable. In the absence of STAT serologic tests for amebiasis (tests with very short turnaround times for results), the decision about the identity of the causative agent often must be made on clinical grounds and on the basis of results of other diagnostic tests such as scans.

Serologic tests based on recombinant *E. histolytica* antigens, using SREHP and the 170-kDa subunit of the galactose-specific adhesin, have been developed. More than 50% of the patients examined had become seronegative by one or both recombinant tests within 180 days of their diagnosis of amebic liver abscess. In contrast, all patients remained seropositive by a standard conventional indirect-hemagglutination test at more than 6 months after presentation (3, 5). These findings may make the recombinant-antigen tests more useful for the serologic diagnosis of amebiasis in areas of endemic infection.

Recombinant human monoclonal antibody Fab fragments specific for *E. histolytica* have also been developed. These antibodies may be applicable for distinguishing *E. histolytica* from *E. dispar* and for use in the serodiagnosis of amebiasis (48).

In a study in Bangladesh, 199 (69%) of children tested were positive at least once using the *E. histolytica* antigen detection test. Although the incidence rate was not related to baseline nutritional status, age, sex, blood group, or duration of breast feeding, it was significantly related to the baseline *E. histolytica* anti-lectin immunoglobulin IgG status of the children. Children who had recovered from a diarrheal episode with *E. histolytica*,

but not *E. dispar*, had half the chance of developing subsequent *E. histolytica*-associated diarrhea, consistent with the development of species-specific acquired immunity (18).

In another study in Brazil, 30% of a sample population was colonized with *E. histolytica*. However, no correlation between seropositivity for anti-GalNAc lectin antibody and colonization was found. These results suggest that colonization does not necessarily produce immunity to reinfection (5).

Histology. A histologic diagnosis of amebiasis can be made when the trophozoites within the tissue are identified. Organisms must be differentiated from host cells, particularly histiocytes and ganglion cells. Periodic acid-Schiff staining is often used to help locate the organisms. The organisms appear bright pink with a green-blue background (depending on the counterstain used). Hematoxylin and eosin staining also allows the typical morphology to be seen, thus allowing accurate identification. As a result of sectioning, some organisms exhibit the evenly arranged nuclear chromatin with the central karyosome and some no longer contain the nucleus.

KEY POINTS—LABORATORY DIAGNOSIS

Entamoeba histolytica

1. When routine ova and parasite examinations are ordered, three specimens (stool) should be submitted for the diagnosis of intestinal amebiasis.
2. Any examination for parasites in stool specimens must include the use of a permanent stained smear (even on formed stool).
3. Presumptive identification on a wet preparation must be confirmed by using the permanent stained smear. Without using *E. histolytica*-specific antigen detection tests, the presence of the organism(s) should be reported as follows: Trophozoites containing ingested RBCs within the cytoplasm should be reported as the true pathogen ("*Entamoeba histolytica* trophozoites present"). However, if cysts or trophozoites containing no ingested RBCs are seen, the report should indicate that the *Entamoeba* group is present ("*Entamoeba histolytica*/*E. dispar* group trophozoites and cysts present").
4. The six smears should be prepared at sigmoidoscopy but should not take the place of the ova and parasite examination.
5. Immunoassay procedures can be performed to confirm the presence of the *E. histolytica*/*E. dispar* group or can be used to differentiate pathogenic *E. histolytica* from nonpathogenic *E. dispar*. Remember

that some of these reagent kits may require fresh or frozen stools only; formalin-preserved stool or stool containing polyvinyl alcohol are not acceptable.

6. The serologic test result for antibody may *or may not* be positive in intestinal disease and is much more likely to be positive in extraintestinal disease.

Treatment

Over the years, there has been much controversy concerning the possibility of pathogenic and nonpathogenic strains of *E. histolytica* and whether every patient found to harbor this parasite should be treated. Brumpt is credited with the first suggestion that *E. histolytica* consists of two morphologically identical organisms, one pathogenic and invasive (*E. histolytica*) and the other a harmless commensal (*E. dispar*). While carriers usually harbor nonpathogenic *E. dispar*, pathogenic *E. histolytica* also may be found in these individuals. At present, tests which routinely differentiate between *E. histolytica* and *E. dispar* are not used routinely by all diagnostic laboratories. The diagnosis of *E. histolytica/E. dispar* infection is most often based on organism morphology. For this reason, in general, patients in the United States who harbor these organisms may be treated, regardless of the presence or absence of symptoms.

There are two classes of drugs used in the treatment of amebic infections: luminal amebicides such as iodoquinol or diloxanide furoate and tissue amebicides such as metronidazole, chloroquine, or dehydroemetine. Because differences in drug efficacy exist, it is important that the laboratory report for the physician indicate whether cysts, trophozoites, or both are present in the stool specimen.

Asymptomatic Infection.
Patients found to have true *E. histolytica* in the intestinal tract, even if they are asymptomatic, should be treated to eliminate the organisms. Both diloxanide furoate and iodoquinol or paromomycin are available for treatment of patients who have cysts in the lumen of the gut. A study encompassing 14 years of experience of using diloxanide furoate for treating asymptomatic cyst passers in the United States indicates that the drug is safe and effective and may be particularly well tolerated in children (29). In general, these treatments are ineffective against extraintestinal disease. If the patient is passing both trophozoites and cysts, the recommended treatment is metronidazole plus iodoquinol (29).

Mild to Moderate Disease.
For patients with mild to moderate disease, metronidazole (Flagyl) should be used when tissue invasion occurs, regardless of the tissue involved. Drugs directed against the lumen organisms should also be used in these instances.

Severe Intestinal Disease.
Metronidazole plus one of the luminal drugs should be used for therapy.

Hepatic Disease.
Metronidazole plus one of the luminal drugs should be used to treat hepatic disease. Some other combinations can also be used; some contain emetine, in which case the patient must be monitored very carefully for possible cardiotoxicity. The importance of using both luminal and tissue amebicides to treat patients with amebic liver abscess is emphasized. Asymptomatic colonization may be present with the true pathogen, *E. histolytica*. In patients treated with metronidazole (tissue amebicide), there is generally 100% clinical response to the hepatic lesions; however, failure to eliminate the organism from the bowel can lead to second bouts with invasive disease and intestinal colonization. Also, these carriers constitute a public health hazard since they are shedding infective cysts. Specific drugs and dosages are discussed in chapter 25.

Epidemiology and Prevention

Infections with *E. histolytica* are worldwide in distribution and are generally most prevalent in the tropics. In 1984, 500 million people were estimated to be infected with *E. histolytica*, and 40 to 50 million of these had extensive symptoms including colitis or extraintestinal abscesses. Prevalence figures for the United States are generally thought to be <5%. It has also been estimated that for every case of invasive disease diagnosed, there are at least 10 to 20 asymptomatic individuals excreting infective cysts.

Population groups with a higher incidence of amebiasis include recent immigrants and refugees from South and Central America and from Southeast Asia. Residents in southeastern and southwestern parts of the United States also tend to have more infections with intestinal parasites, as do groups such as patients in mental institutions.

Following broad social changes seen in the late 1960s, the open expression of homosexuality, increased numbers of sexual contacts, increased frequency of sexual activities, and anonymity of sexual partners contributed to dramatic increases in the prevalence of sexually transmitted organisms, including *E. histolytica*. Although infections with this organism are usually associated with poor sanitation and underdeveloped areas of the world, epidemiologic studies in both Europe and the United States in the 1970s documented that sexual transmission of amebiasis occurred mainly among urban homosexual men. With the increase in sexually transmitted pathogens, a number of clinical syndromes have been recognized; these clinical presentations have been grouped together and referred to as "the gay bowel syndrome." Many published reports confirm that *E. histolytica* is one of the major pathogens in the gay bowel syndrome. It is also well known that the clinical presentations within the homosexual community often differ from those seen in the heterosexual population.

Epidemiologic evidence of sexual transmission of *E. histolytica* has grown significantly since the early 1970s, particularly in areas such as New York City and San Francisco. In San Francisco, the incidence of reported symptomatic intestinal amebiasis among homosexual men between 20 and 39 years of age has increased over 1,000% during the last 10 years. Although percentages vary, it appears that approximately 30% of urban homosexual men may be infected with *E. histolytica*, a sharp increase over the estimated rate of less than 5% seen in the general population within the United States.

Direct oral-anal contact (anilingus) is a common practice among homosexual men, leading to both fecal exposure and oral contact with a variety of intestinal pathogens. Although anilingus has often been listed as a key risk factor in potential exposure, transmission can also occur during oral-genital sex after anal intercourse has occurred. It has also been shown that active heterosexuals can acquire infection with *E. histolytica* through sexual activities that provide an opportunity for fecal-oral contamination. The key factor is not necessarily homosexuality but the frequency of sexual activity and the potential for fecal-oral contact.

It is important to remember than in earlier studies in which "*E. histolytica*" was specifically mentioned, reference is actually being made to the *E. histolytica/E. dispar* group rather than the actual pathogen, *E. histolytica*. Genotyping of *E. histolytica* isolates could serve as a tool to fingerprint individual isolates, thus helping to determine geographic origins of isolates and routes of transmission (16).

With the advent of AIDS and the subsequent modifications in sexual practices within homosexual communities, the incidence of *E. histolytica* infection is thought to be much lower than before. In recent years, the coccidian parasites *Isospora belli* and *Cryptosporidium* spp. and the microsporidia *Enterocytozoon bieneusi* and *Encephalitozoon intestinalis* have become much more of a problem in patients with AIDS.

There are certain urban areas (Mexico City, Mexico; Medellin, Colombia; and Durban, South Africa) where the incidence of invasive disease is considerably higher than in the rest of the world. Contributing factors (not well delineated) may include poor nutrition, tropical climate, decreased immunologic competence of the host, stress, altered bacterial flora in the colon, traumatic injuries to the colonic mucosa, alcoholism, and genetic factors.

Humans are the reservoir host for *E. histolytica* and can transmit the infection to other humans, primates, dogs, cats, and possibly pigs. However, contact with sewage-contaminated water provides another route of infection; therefore, amebiasis can be considered a zoonotic waterborne infection as well as being transmitted through human-to-human transmission (43). The cyst stages are very resistant to environmental conditions and can remain viable in the soil for 8 days at 28 to 34°C, 40 days at 2 to 6°C, and 60 days at 0°C. Cysts are normally removed by sand filtration or destroyed by 200 ppm of iodine, 5 to 10% acetic acid, or boiling. The asymptomatic cyst passer who is a food handler is generally thought to play the most important role in transmission.

A colonization-blocking vaccine could eliminate *E. histolytica* as a cause of human disease, particularly since humans serve as the only significant reservoir host (46). The application of molecular biological techniques has led to the identification and structural characterization of three amebic antigens, SREHP, the 170-kDa subunit of the Gal/GalNAc binding lectin, and the 29-kDa cysteine-rich protein, which all show promise as recombinant antigen-based vaccines to prevent amebiasis (46). An immunogenic dodecapeptide derived from the SREHP molecule has been genetically fused to the B subunit of cholera toxin to create a recombinant protein capable of inducing both antiamebic and anti-cholera toxin antibodies when administered by the oral route. Continued research should lead to the development of a cost-effective oral combination "enteric-pathogen" vaccine capable of inducing protective mucosal immune responses to several important enteric diseases including amebiasis (46).

Entamoeba dispar

Until relatively recently, an unexplained aspect of the epidemiology of amebiasis was the number of asymptomatic individuals who were passing cysts but had no clinical evidence of invasive disease. Extensive work involving isoenzyme analysis and molecular techniques has demonstrated conclusively that there are two genetically distinct but morphologically identical species of what was formerly known as *E. histolytica*. After many years of debate, the species that is associated with amebiasis in humans is now classified as *E. histolytica*. The other species, which is more common but is not capable of causing invasive disease, is called *E. dispar*. Studies confirming the differences between the two organisms include direct sequencing of the PCR-amplified small-subunit rRNA gene of *E. dispar* and the design of primers for rapid differentiation from *E. histolytica*, detection of differences in the phosphoglucomutases and cysteine proteinases from *E. histolytica* and *E. dispar*, molecular biology of the differences in the hexokinase isoenzyme pattern, detection of differences in the secretion of acid phosphatase, and detection of DNA sequences unique to *E. histolytica* (2, 8, 10, 14, 30, 31, 39, 41, 42, 46, 48, 53, 57).

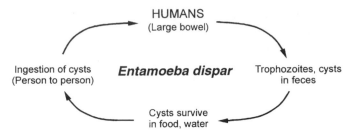

HUMANS
(Large bowel)

Ingestion of cysts
(Person to person)

Entamoeba dispar

Trophozoites, cysts
in feces

Cysts survive
in food, water

Figure 2.5 Life cycle of *Entamoeba dispar* (and *E. moshkovskii*).

Life Cycle and Morphology

The life cycle is essentially identical to that of *Entamoeba coli* or any of the other nonpathogenic intestinal protozoa, and the cyst form is the infective form for humans (Figure 2.5).

Morphology of Trophozoites (Table 2.1).

Living trophozoites vary in diameter from about 12 to 60 μm. Motility has been described as rapid and unidirectional, with pseudopods forming quickly in response to the conditions around the organism; it may appear to be sporadic. Although this characteristic is often described, it is rare to diagnose these organisms on the basis of motility seen in a direct wet mount. The cytoplasm is differentiated into a clear outer ectoplasm and a more granular inner endoplasm. Based on the recent ability to culture these organisms in axenic culture systems and on light and electron microscopy studies, there may be some morphologic differences between *E. histolytica* and *E. dispar*. However, these differences would not be recognized by routine diagnostic methods such as the permanent stained smear. If organisms were seen that were consistent with *E. histolytica/E. dispar* on the permanent stained smear, the laboratory report would indicate that fact and would be written as "*Entamoeba histolytica/E. dispar*."

When the organism is examined on a permanent stained smear (trichrome or iron hematoxylin), the morphological characteristics are easily seen. The nucleus generally has evenly arranged chromatin on the nuclear membrane and has a small, compact, centrally located karyosome. The cytoplasm is described as finely granular with few ingested bacteria or debris in vacuoles. Ingested RBCs are never seen in the trophozoites; if ingested RBCs are seen, this finding identifies the organism as *E. histolytica*, not *E. dispar*.

Morphology of Cysts (Table 2.2).

Within the immature cyst, there may be two types of inclusions, a glycogen mass and highly refractile chromatoidal bars with smooth, rounded edges. As the cysts mature, nuclear division occurs with the production of four nuclei; the cysts range in diameter from 10 to 20 μm. Usually, as the cyst matures, the glycogen disappears and chromatoidal bars may or may not be present. Cyst morphology does not differentiate between *E. histolytica* and *E. dispar*. Cyst formation occurs only within the intestinal tract; once the stool has been passed from the body, cyst formation does not occur. The one-, two-, and four-nucleated cysts are infective and represent the mode of transmission from one host to another (Figures 2.1 to 2.3).

Clinical Disease

Prior to the separation of *E. histolytica* into two distinct species (*E. histolytica* and *E. dispar*), it was suggested that some *E. histolytica* strains were pathogenic and some were not. Now that *E. dispar* has been classified as a nonpathogenic organism, totally separated from the pathogenic *E. histolytica*, treatment is usually not recommended. However, this recommendation is based on the separation of the two species by molecular means, not by morphology alone. Since few laboratories will be routinely identifying the two *Entamoeba* species to the level of *E. histolytica* or *E. dispar*, the clinician will have to decide on the basis of clinical findings whether the patient has true pathogenic *E. histolytica* or nonpathogenic *E. dispar* infection. The laboratory report will merely indicate the presence of the "*Entamoeba histolytica/E. dispar*" group.

Diagnosis

Routine Diagnostic Procedures.

Infection with *E. dispar* cannot be diagnosed on the basis of morphologic criteria alone, even from the permanent stained smear. When the organism is seen in the routine ova and parasite examination, it cannot be differentiated from *E. histolytica* unless trophozoites are seen to contain RBCs in the cytoplasm, a finding that is diagnostic for *E. histolytica*. Using routine diagnostic methods and not immunoassay procedures specific for *E. histolytica*, morphologic criteria would provide characteristics of both; the laboratory report would indicate "*Entamoeba histolytica/E. dispar*."

Immunodetection.

Immunoassay procedures are available that can determine the presence of the *E. histolytica/E. dispar* group as being distinct from the rest of the *Entamoeba* species such as nonpathogenic *E. coli* or *E. hartmanni*. Other test reagents can actually separate *E. histolytica* from *E. dispar*. Most of these procedures use the ELISA or EIA formats (8). In some cases they have been found to be less sensitive than PCR (15), and in other instances they have been found equal to PCR (11). Another product is available as a cartridge format that uses an immunochromatographic strip-based detection system for the *E. histolytica*/*E. dispar* group, *G. lamblia*, and *Cryptosporidium* spp. (51). These commercially available reagent kits require fresh or frozen stools; formalin has been found to interfere with the *Entamoeba* reagents. Any available antigen detection system should always be reviewed for compatibility with stools submitted in preservatives rather than fresh specimens.

Epidemiology and Prevention

The mode of transmission is similar to that found with other protozoa and is related to the ingestion of infective cysts in contaminated food or water. Based on differences in numbers between symptomatic patients with amebiasis (*E. histolytica*) and those who remain asymptomatic (*E. dispar*), it is obvious that many more patients are infected with nonpathogenic *E. dispar* than with pathogenic *E. histolytica*. Since travel from temperate zones to tropical and subtropical areas continues to become more common, it is necessary to consider imported amebiasis in many areas of the world where the infection is normally seen less frequently than infections with the nonpathogenic *E. dispar*.

Entamoeba moshkovskii

Entamoeba moshkovskii is found worldwide and is generally considered to be a free-living ameba. Based on microscopic morphology, this organism is indistinguishable from *E. histolytica* and *E. dispar*, except in cases of invasive disease when *E. histolytica* contains ingested RBCs. Although first isolated from sewage, *E. moshkovskii* can also be found in clean riverine sediments to brackish coastal pools. Apparently, there are some differences that separate this organism from *E. histolytic* and *E. dispar*. However, these differences pertain to physiology rather than morphology; *E. moshkovskii* is osmotolerant, can be cultured at room temperature, and is resistant to emetine (1).

Life Cycle and Morphology

The life cycle is essentially identical to that of *E. dispar*, and morphological differences are minimal to none (Figure 2.5). In wet preparations, trophozoites usually range in size from 15 to 20 µm and cysts normally range in size from 12 to 15 µm. It is important to remember that on the permanent stained smear there is a certain amount of artificial shrinkage due to dehydration; therefore, all of the organisms, including pathogenic *E. histolytica*, may be somewhat smaller (from 1 to 1.5 µm) than the sizes quoted for the wet-preparation measurements.

Morphology of Trophozoites (Table 2.1).

Trophozoites do not ingest RBCs, and the motility is similar to that of both *E. histolytica* and *E. dispar* (Figure 2.3). Nuclear and cytoplasmic characteristics are very similar to those seen in *E. histolytica*; however, trophozoites of *E. moshkovskii* do not contain ingested RBCs.

Morphology of Cysts (Table 2.2).

Nuclear characteristics and chromatoidal bars are similar to those in *E. histolytica* and *E. dispar* (Figure 2.3).

Clinical Disease

Although human isolates have been found in a number of geographic areas (North America, Italy, South Africa, and Bangladesh), they have not been associated with disease. Because *E. moshkovskii* is generally accepted as being nonpathogenic, treatment is usually not recommended. However, few studies have been undertaken to actually identify infections with this organism. Unfortunately, it is quite likely that infections with this ameba have been diagnosed as being caused by *E. histolytica* and that the patients have been treated unnecessarily with antiamebic chemotherapy.

Diagnosis

Because *E. moshkovskii* can be easily confused with other amebae, particularly in a wet preparation, it is almost mandatory that the final identification be obtained from the permanent stained smear. However, without the use of molecular tools to specifically identify this organism, the results are usually reported as "*Entamoeba histolytica*/*E. dispar* group" with no mention of *E. moshkovskii*. The true prevalence of this organism may be much higher than suspected; this may also explain some of the microscopy-positive/antigen-negative results obtained when using the *Entamoeba* test kit.

Epidemiology and Prevention

The mode of transmission is similar to that for other protozoa and is related to ingestion of cysts in contaminated food or water. In areas where accurate identifications have been made, the prevalence of *E. moshkovskii* may be similar to or higher than that of *E. histolytica* (1). Using PCR techniques, in one study from Bangladesh *E. dispar*-infected children were almost twice as likely to have a mixed infection with *E. moshkovskii* (35%) than with *E. dispar* and *E. histolytica* (18%) or with *E. dispar* alone (18%) (1). Future epidemiologic studies of *E. histolytica* infection should include the capability of diagnosing all three species individually (*E. histolytica*, *E. dispar*, and *E. moshkovskii*).

Entamoeba hartmanni

Entamoeba hartmanni is found worldwide and until 1957 was generally considered to be a small race of *E. histolytica*. Consequently, the prevalence figures prior to that time are thought to be inaccurate. It is now well accepted that this ameba is a separate species from *E. histolytica*.

Life Cycle and Morphology

The life cycle is essentially identical to that of *E. dispar*, and morphological differences involve size, although there is even an overlap in size between the two species (Figure 2.6). In wet preparations, trophozoites range in size from 4 to 12 μm and cysts range in size from 5 to 10 μm. It is important to remember that on the permanent stained smear there is a certain amount of artificial shrinkage due to dehydration; thus, all of the organisms, including pathogenic *E. histolytica*, may be somewhat smaller (from 1 to 1.5 μm) than the sizes quoted for the wet-preparation measurements.

Morphology of Trophozoites (Table 2.1).

Trophozoites do not ingest RBCs, and the motility is usually less rapid (Figures 2.2 and 2.7). The nuclear and cytoplasmic characteristics are very similar to those seen in *E. histolytica*, with two exceptions. Frequently, cysts contain only one or two nuclei, even though the mature cyst contains four nuclei. Mature cysts of *E. hartmanni* also tend to retain their chromatoidal bars, a characteristic which is often not seen in *E. histolytica/E. dispar*.

Morphology of Cysts (Table 2.2).

Chromatoidal bars are similar to those in *E. histolytica* and *E. dispar* but are smaller and more numerous (Figures 2.2 and 2.7). Because differentiation between *E. hartmanni* and *E. histolytica/E. dispar* at the species level depends primarily on size, it is mandatory that laboratories use calibrated microscopes which are rechecked periodically for accuracy.

Clinical Disease

Although early studies incriminated *E. hartmanni* as a possible animal pathogen and suggested that it could

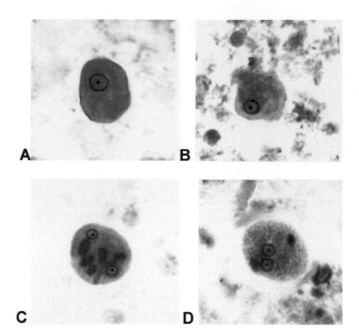

Figure 2.7 (A and B) Trophozoites of *Entamoeba hartmanni*; (C and D) cysts of *E. hartmanni*.

transform from small "nonpathogenic" strains to large "pathogenic" ones, none of this information has been adequately confirmed by subsequent studies. Because *E. hartmanni* is generally accepted as being nonpathogenic, treatment is usually not recommended.

Diagnosis

Because *E. hartmanni* can be easily confused with other small amebae, particularly in a wet preparation, it is almost mandatory that the final identification be obtained from the permanent stained smear. Accurate measurement of organisms also confirms the tentative visual diagnosis (Table 2.3).

Epidemiology and Prevention

The mode of transmission is similar to that found with other protozoa and is related to ingestion of cysts in contaminated food or water. In areas where accurate identifications have been made, the prevalence of *E. hartmanni* is similar to that of *E. histolytica*.

Entamoeba coli

Although *Entamoeba coli* was probably recognized and described in the late 1800s, it was not until 1913 that the organism was proven to be nonpathogenic in human volunteers. *E. coli* is worldwide in distribution. Like *E. histolytica*, it tends to be more common in the warmer climates, but it has also been found in colder areas such as Alaska.

Figure 2.6 Life cycle of *Entamoeba hartmanni*, *Entamoeba coli*, *Entamoeba polecki*, *Endolimax nana*, and *Iodamoeba bütschlii*.

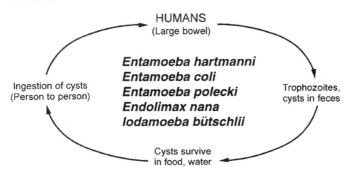

Life Cycle and Morphology

The life cycle is identical to that of *E. dispar* (Figure 2.6).

Morphology of Trophozoites (Table 2.1). Living trophozoites tend to be somewhat larger than those of *E. histolytica* or *E. dispar* and range from 15 to 50 µm (Figures 2.2 and 2.3). Motility has been described as sluggish with broad, short pseudopods. It may be extremely difficult to differentiate nonpathogenic *E. coli* from pathogenic *E. histolytica* in wet preparations. On the permanent stained smear, the cytoplasm appears granular with few to numerous vacuoles containing bacteria, yeasts, and other food materials. The nucleus has a moderately large karyosome that is frequently eccentric in position. The chromatin on the nuclear membrane is usually clumped and irregular in placement. If there are RBCs in the intestinal tract, *E. coli* may ingest these rather than bacteria; occasionally, the cytoplasm also contains ingested spores of *Sphaerita* sp. and possibly a *Giardia lamblia* cyst.

Morphology of Cysts (Table 2.2). Again, for reasons that are not specifically known, trophozoites discharge their undigested food and begin to round up prior to precyst and cyst formation. Early cysts usually contain a dense glycogen mass and may also contain chromatoidal bars, which tend to be splinter shaped and irregular. Eventually, the nuclei divide until the mature cyst, containing eight nuclei, is formed (Figures 2.2 and 2.3). Occasionally, the number of nuclei reaches 16. The cysts measure 10 to 35 µm and almost always lose their chromatoidal bars as they mature. For some reason, as the cyst of *E. coli* matures, it becomes more resistant to fixation with various preservatives. Therefore, the cyst may be seen on the wet preparation but not on the permanent stained smear. Occasionally, on trichrome smears, the cysts appear distorted and somewhat pink. This is not an indication of poor reagents or techniques but, rather, an indication that these cysts do not fix well with the routinely used preservatives. Better fixation and more detailed morphology can be obtained by heating some of the fixatives prior to specimen preservation, although for most laboratories, this is not a practical approach.

Studies suggest that after cyst ingestion, the metacyst undergoes division of the cytoplasm, thus becoming metacystic trophozoites that will grow and divide within the lumen of the intestine. It is interesting that the total number of trophozoites formed from the mature cyst usually is fewer than eight.

Diagnosis

With the exception of the mature cyst, the morphologies of *E. histolytica*, *E. dispar*, *E. moshkovskii*, and *E. coli* are similar in most of the stages. Therefore, it is very important to examine permanent stained smears, even if a tentative identification has been made from a wet preparation examination. Specific treatment is not recommended for this nonpathogen (Table 2.3). Consequently, correct differentiation between the four species is critical to good patient care. Since the four species are acquired in the same way, it is also important to remember that all four can be found in the same patient. If few *E. histolytica*/*E. dispar* organisms are present among many *E. coli* organisms, additional searching and/or the use of species-specific immunoassay testing may be necessary to correctly identify the different species.

Epidemiology and Prevention

The mode of transmission is ingestion of cysts from contaminated food or water. Apparently, the infection is readily acquired, and in some warmer climates or areas with primitive hygiene conditions, the infection rate can be quite high. Prevention depends on adequate disposal of human excreta and improved personal hygiene (preventive measures which apply to the majority of the intestinal protozoa).

Entamoeba polecki

Entamoeba polecki was originally found in the intestines of pigs and monkeys and has also been found as a human parasite (Figure 2.6). In certain areas of the world, such as Papua New Guinea, it is the most common intestinal ameba of humans. Few cases are reported, possibly in part because the organism resembles *E. histolytica*/*E. dispar*, *E. moshkovskii*, and *E. coli*. The trophozoites resemble *E. coli* in that the cytoplasm is granular, containing ingested bacteria, and the motility tends to be sluggish (Table 2.5). The nuclear morphology is almost a composite of those of *E. histolytica*/*E. dispar*, *E. moshkovskii*, and *E. coli*. Without some of the cyst stages for comparison, it would be very difficult to identify this organism to the species level on the basis of the trophozoite alone. The cyst normally has only a single nucleus, chromatoidal material like that seen in *E. histolytica*, and frequently an inclusion body (Figure 2.8; Table 2.6). This mass tends to be round or oval and is not sharply defined on the edges. The material, which is not glycogen, remains on the permanent stained smear and stains less intensely than nuclear material or chromatoidal bars. Differentiation of this organism from either *E. histolytica* or *E. coli* is rarely accomplished on the basis of a wet preparation examination. The size on the permanent stained smear would range from 10 to 12 µm for the trophozoite and from 5 to 11 µm for the cyst. To date, there is no evidence of pathogenicity in humans; therefore, no therapy is recommended.

Table 2.5   Intestinal protozoa: trophozoites of less common amebae

Characteristic	Entamoeba polecki	Entamoeba gingivalis
Size (diam or length)	10–12 µm	5–20 µm (avg, 10–15 µm)
Motility	Usually nonprogressive, sluggish (like *E. coli*)	Multiple pseudopodia, vary from long and lobose to short and blunt
Nucleus: no. and visibility	Can occasionally be seen on a wet preparation; intermediate between *E. histolytica* and *E. coli*; 1 nucleus	Similar to *E. histolytica*, 1 nucleus
Peripheral chromatin (stained)	Fine granules (may be interspersed with large granules) evenly arranged on membrane; chromatin may also be clumped at one or both edges of membrane	Fine granules, closely packed
Karyosome (stained)	Small, usually centrally located	Small, well defined, usually centrally located
Cytoplasm appearance (stained)	Finely granular	Finely granular
Inclusions (stained)	May contain ingested bacteria	Ingested epithelial cells and host leukocytes

Figure 2.8   (Upper) Cyst of *Entamoeba polecki* (illustration); (lower) cyst of *E. polecki* (note the large inclusion in the cytoplasm as seen in the drawing).

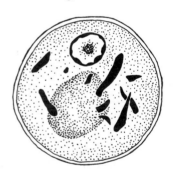

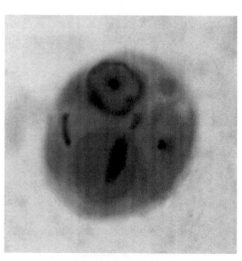

Entamoeba gingivalis

Entamoeba gingivalis was the first parasitic ameba of humans to be described. It was recovered from the soft tartar between the teeth. It has also been recovered from the tonsillar crypts and can multiply in bronchial mucus, thus appearing in sputum (Figure 2.9). Since morphologically it is very similar to *E. histolytica*, it is important to make the correct identification from a sputum specimen, that is, *E. gingivalis*, which is considered to be a nonpathogen, rather than *E. histolytica* from a possible pulmonary abscess. The trophozoite measures approximately 5 to 15 µm, and the cytoplasm most often contains ingested

Table 2.6   Intestinal protozoa: cysts of less common amebae

Characteristic	Entamoeba polecki	Entamoeba gingivalis
Size (diam or length)	5–11 µm	No known cyst stage
Shape	Usually spherical	
Nucleus: no. and visibility	Mature cyst, 1; may be visible in wet preparations (rarely 2 or 4 nuclei)	
Peripheral chromatin (stained)	Similar to that seen in trophozoite	
Karyosome	Similar to that seen in trophozoite	
Cytoplasm (chromatoidal bodies, glycogen, inclusions [stained])	Abundant, angular pointed ends; threadlike chromatoidals may be present; half of cysts contain spherical or ovoid "inclusion mass"	

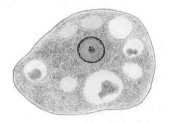

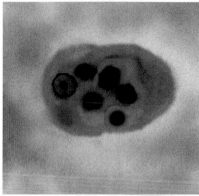

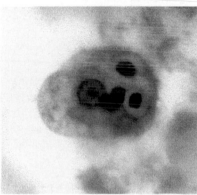

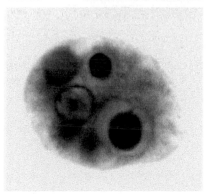

Figure 2.9 (Upper) Trophozoite of *Entamoeba gingivalis* (illustration by Sharon Belkin); (lower three images) trophozoites of *E. gingivalis* (note the ingested polymorphonuclear leukocytes).

leukocytes (Table 2.5). On the permanent stained smear, nuclear fragments of the white cells can be seen within the food vacuoles, which are usually larger than the vacuoles seen in *E. histolytica* (Figure 2.9). This fact will help to differentiate the two, since *E. gingivalis* is the only species that ingests white cells. No cysts are formed by this species (Table 2.6). Although these amebae are most often recovered from patients with pyorrhea alveolaris, *E. gingivalis* is still considered to be nonpathogenic.

Organisms identified as *E. gingivalis* have been recovered in vaginal and cervical smears from women using intrauterine devices; the organisms spontaneously disappeared after removal of the devices. In an unusual case, *E. gingivalis* was identified in a left upper neck nodule by fine-needle aspiration. Apparently, the patient had an increased number of amebae within the oral cavity secondary to radiation therapy. Radiation therapy may have contributed to a fistula tract between the oral cavity and the surgical incision site, thus resulting in the formation of a small inflammatory nodule in the upper portion of the neck (37). Generally, no treatment is indicated, regardless of the body site from which the organisms are recovered. The infection does suggest a need for better oral hygiene and can be prevented by proper care of teeth and gums.

Although *E. gingivalis* is considered a nonpathogen, it has been found in the oral cavities of human immunodeficiency virus type 1 (HIV-1)-infected patients with periodontal disease. Its presence was not related to the degree of immunodeficiency but to the HIV diagnosis (26). In another study, the risk for severe gingival inflammation and/or increased probing depth of the gums was not associated with the presence of *E. gingivalis* (40).

Endolimax nana

Endolimax nana is one of the smaller nonpathogenic amebae and was distinguished as a separate ameba around 1908. It is also worldwide in distribution and is seen in most populations at least as frequently as *E. coli*.

Life Cycle and Morphology

E. nana has the same stages in the life cycle as does *E. dispar*: trophozoite, precyst, cyst, and metacystic forms (Figure 2.6). The trophozoite usually measures from 6 to 12 µm, with the normal range being 8 to 10 µm (Table 2.1).

Motility has been described as sluggish and nonprogressive with blunt, hyaline pseudopods. In the permanent stained smear, the nucleus is more easily seen. There is normally no peripheral chromatin on the nuclear membrane, and the karyosome tends to be large, with either a central or eccentric location within the nucleus (Figures 2.10 and 2.11). There is tremendous nuclear variation in this organism, and occasionally it

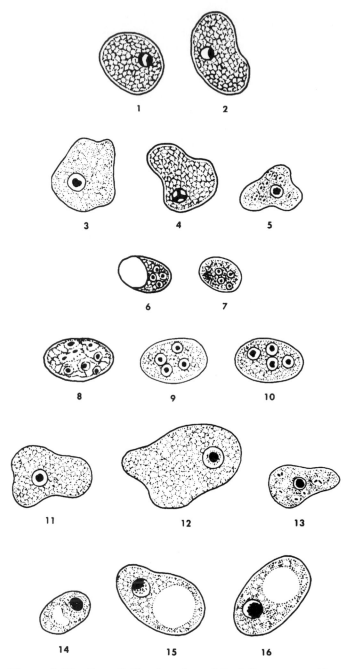

Figure 2.10 (1 to 5) Trophozoites of *Endolimax nana*; (6 to 10) cysts of *E. nana*; (11 to 13) trophozoites of *Iodamoeba bütschlii*; (14 to 16) cysts of *I. bütschlii.*

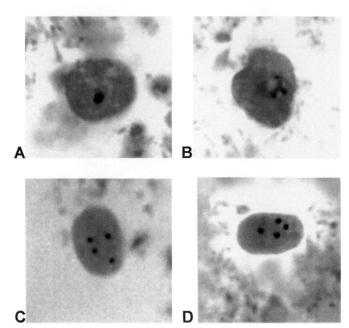

Figure 2.11 (A and B) Trophozoites of *Endolimax nana* (note the nuclear variation in panel B); (C and D) cysts of *E. nana.*

mimics *Dientamoeba fragilis* and *E. hartmanni*. The more organisms there are on the smear, the more likely it is that some of them will mimic other species of the amebae. The cytoplasm may have small vacuoles containing ingested debris or bacteria.

Cysts usually measure 5 to 10 µm, with a normal range of 6 to 8 µm (Table 2.2). In some instances, cysts

as large as 14 µm have been seen. The cyst is usually oval to round, with the mature cyst containing four nuclei. Occasionally, very small, slightly curved chromatoidal bars are present (Figures 2.10 and 2.11). It is unusual to see the two-nucleated stage, and there are frequently both trophozoites and cysts in clinical specimens.

Diagnosis

Although *E. nana* is a nonpathogen and no therapy is recommended, it is still important to differentiate it from other amebae, some of which are pathogenic. Because these organisms are small, the definitive diagnosis of *E. nana* may have to be made on the basis of the permanent stained smear (Table 2.3). If the original fixation is good, the cysts may be identified on the basis of the wet preparation such as the concentration sediment or flotation surface film. The four nuclear karyosomes appear very refractile in the wet preparation.

Epidemiology and Prevention

As with the other organisms that have been discussed, the mode of transmission is ingestion of the mature cysts in contaminated food or water. The cysts of *E. nana* tend to be less resistant to desiccation than do those of *E. coli*. This organism is also found in the same areas of the world as are the other amebae, that is, in warm, moist climates and in other areas where there is a low standard of personal hygiene and poor sanitary conditions. Prevention would depend on improved personal hygiene and an overall upgrading of sanitary conditions and waste disposal.

Iodamoeba bütschlii

Iodamoeba bütschlii, one of the nonpathogenic amebae, is also worldwide in distribution and was identified in the early 1900s. The original species name of "williamsi" probably referred to atypical cysts of *E. coli*. Although it is found in areas where the other amebae have been recovered, its incidence is not as high as that of *E. coli* or *E. nana*. One of the most striking morphological features of this organism is the large glycogen vacuole which appears in the cyst and readily stains with iodine on a wet preparation smear.

Life Cycle and Morphology

The stages within the life cycle are exactly the same as those seen with *E. nana*. The trophozoite normally varies from 8 to 20 μm and is described as having fairly active motility in a fresh stool preparation (Table 2.1). The cytoplasm is rather granular, containing numerous vacuoles with ingested debris and bacteria. In general, the vacuolated cytoplasm is more obvious than that seen in *E. nana* trophozoites. The nucleus has a large karyosome, which can be either centrally located or eccentric (Figures 2.10 and 2.12). On the permanent stained smear, the nucleus may appear to have a halo. Chromatin granules fan out around the karyosome. If the granules are on one side, the nucleus may appear to have the "basket nucleus" arrangement of chromatin, which is often found in the cyst stage. The trophozoites of *I. bütschlii* and *E. nana* may be very similar and difficult to differentiate at the species level, even on the permanent stained smear. Both organisms are considered to be nonpathogens, and *E. nana* is recovered in clinical specimens much more frequently than *I. bütschlii*.

The cysts rarely have a classic round appearance but tend to be somewhat oval (Table 2.2). The glycogen

Figure 2.12 (A) Trophozoite of *Iodamoeba bütschlii*; (B) cyst of *I. bütschlii* (note the large glycogen vacuole).

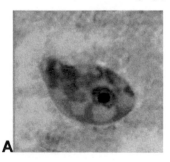
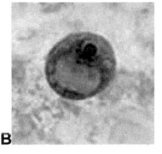

vacuole is so large that occasionally the cyst collapses on itself. Since there is no nuclear multiplication in the cyst form, the mature cyst contains only a single nucleus. The cysts measure approximately 5 to 20 μm and are rarely confused with those of other amebae (Figures 2.10 and 2.12).

Diagnosis

Although the cyst can often be identified from the wet smear preparation, particularly if the vacuole is stained with iodine, the trophozoites may be difficult to detect and identify without the permanent stained smear (Table 2.3).

Epidemiology and Prevention

The mode of transmission is ingestion of infective cysts in contaminated food or water. Prevention would depend on improved personal hygiene and sanitary conditions.

Blastocystis hominis

Blastocystis hominis, an inhabitant of the human intestinal tract, was first described in 1912. Over the years, this organism been classified as an organism related to *Blastomyces* spp., the cyst form of a flagellate, a yeast of the genus *Schizosaccharomyces*, and a member of the stramenopiles (examples are brown algae such as kelp, diatoms, slime nets, and water molds). However, currently this organism is classified as a protozoan (52). It has been suggested that it be placed in a new class, Blastocystea, and a new order, Blastocystida. Analysis of 10 stocks of *B. hominis* isolated from human stools revealed two distinct groups of organisms. Proteins of the two groups were immunologically distinct, and hybridization studies showed that the DNA contents of the two groups were different. Other studies of 61 isolates suggest that there may be four separate serologic groups (33). Further studies are under way to determine whether these groups should be classified as separate species and whether there is any epidemiologic significance related to the two groups (4, 5, 7, 22, 27, 34, 50, 59). Based on PCR-based genotype classification data, there may be as many as 12 or more different species within the genus (34). Confirmation of the existence of these species and determination of their pathogenic status may also explain why some patients are asymptomatic and some have clinical symptoms.

Life Cycle and Morphology

B. hominis is capable of pseudopod extension and retraction, reproduces by binary fission or sporulation, and has a membrane-bound central body (previously called a vacuole) that takes up 90% of the cell and functions in

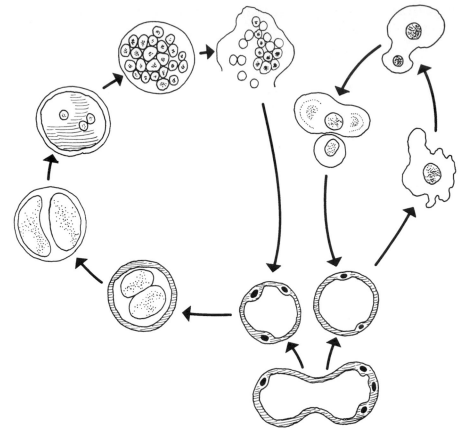

Figure 2.13 Life cycle of *Blastocystis hominis*. The central-body forms (or vacuolar forms) can be seen at the bottom of the life cycle (generally seen in clinical specimens); the development of the granular forms can be seen in the left portion of the life cycle; the ameboid forms can be seen in the right portion of the life cycle. Both the granular and ameboid forms can arise from the vacuolated forms (central-body forms); multiplication can occur through binary fission (vacuolar and granular forms), budding (ameboid forms), schizogony (central-body form), and sporulation (less common in the vacuolated form). Modified from reference 60.

sexual and asexual reproduction (Figures 2.13 to 2.15; Table 2.7). Other structures with unknown function are not yet defined. The organism is strictly anaerobic, normally requires bacteria for growth, and is capable of ingesting bacteria and other debris.

The classic form that is usually seen in the human stool specimen varies tremendously in size, from 6 to 40 μm, and is characterized by a large central body, which may be involved with carbohydrate and lipid storage (visually like a large vacuole) (Figures 2.14 and 2.15; Table 2.7). The more amebic form is occasionally seen in diarrheal fluid but may be extremely difficult to recognize. Generally, *B. hominis* is identified on the basis of the more typical round form with the central body.

Figure 2.14 (A and B) Central-body forms of *Blastocystis hominis* (note the multiple nuclei around the edges, outside the central-body area).

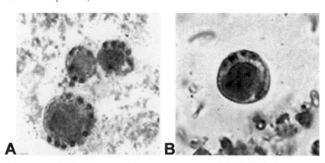

Figure 2.15 *Blastocystis hominis* classical shape showing binary fission and the large central "vacuole."

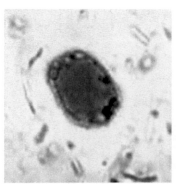

Table 2.7   Morphological criteria used to identify *Blastocystis hominis*

Shape and size	Other features
Organisms are generally round, measure approximately 6–40 µm, and are usually characterized by a large central body (looks like a large vacuole) surrounded by small, multiple nuclei; central-body area can stain various colors (trichrome) or remain clear	The more amebic form can be seen in diarrheal fluid but is difficult to identify; because of variation in size, may be confused with various yeast cells

Both thin- and thick-walled cysts have been confirmed. It is suggested that the thin-walled cysts are autoinfectious, leading to multiplication of the organism in the intestinal tract. The thick-walled cysts are probably responsible for external transmission via the fecal-oral route. This type of life cycle might explain the high percentage of positive carriers in many studies, showing that the percentage of patients infected with *B. hominis* is often much higher than the percentage of those infected with other intestinal protozoa (34).

Clinical Disease

When *B. hominis* is present in the absence of other pathogenic parasites, bacteria, or viruses, it may be the cause of diarrhea, cramps, nausea, fever, vomiting, and abdominal pain and may require therapy. In one recent study of patients with irritable bowel syndrome, there was a set of patients in whom the presence of *B. hominis* did not appear to be incidental (13). Possible relationships between *B. hominis* and intestinal obstruction and perhaps even infective arthritis have been suggested. In patients with other underlying conditions, the symptoms may be more pronounced (22). The incidence of this organism appears to be higher than suspected in stools submitted for parasite examination. For symptomatic patients in whom no other etiologic agent has been identified, *B. hominis* should certainly be considered a possible candidate. Other recent studies suggest that when a symptomatic *B. hominis* infection responds to therapy, the improvement probably represents elimination of some other undetected pathogenic organism (*E. histolytica*, *G. lamblia*, or *D. fragilis*) (28). Data from other geographic areas indicate that although it is commonly seen in stool samples, *B. hominis* is thought to be nonpathogenic (44). Although for a number of years the true role of this organism in terms of colonization or disease was still somewhat controversial, it is now generally considered a causative agent of intestinal disease (9, 34). It has also been suggested that ribodemes I, III, and VI may be responsible for gastrointestinal symptoms (23).

In a study of 1,216 adults, including immunocompromised patients, institutionalized psychiatric or elderly subjects, immigrants from developing countries, travelers to developing tropical countries, and controls, the results showed a high prevalence of parasites in all the risk groups studied, and *B. hominis* was the parasite most frequently detected in each studied group. *B. hominis* showed a significant correlation with gastrointestinal symptoms only when detected in the group including subjects with a severe immunodepression (7).

Diagnosis

Routine stool examinations are very effective in recovering and identifying *B. hominis* (Table 2.7), although the permanent stained smear is the procedure of choice since the examination of wet preparations may not easily reveal the organism. If the fresh stool is rinsed in water before fixation (for the concentration method), *B. hominis* organisms, other than the cysts, are destroyed, thus possibly yielding a false-negative report (21). The organisms should be quantitated on the report form, i.e., as rare, few, moderate, or many. It is also important to remember that other possible pathogens should be adequately ruled out before a patient is treated for *B. hominis*.

Both ELISA and fluorescent-antibody tests have been developed for detection of serum antibody to *B. hominis* infections (23, 61). By using ELISA and a threshold dilution of 1/50, 27 of 30 serum samples (28 patients with no protozoan parasites other than *B. hominis*) were positive at 1/50 or higher; 42 control blood bank serum samples were all negative at 1/50 (61). A strong antibody response is consistent with the ability of this organism to cause symptoms. Also, demonstration of serum antibody production both during and after *B. hominis* symptomatic disease provides immunological evidence for the pathogenic role for this protozoan, although it may take 2 years or more with chronic infections before a serologic response is detected (23).

KEY POINTS—LABORATORY DIAGNOSIS

Blastocystis hominis

1. Although the intestinal protozoa are not quantitated routinely on the report, *B. hominis* should be quantitated (rare, few, moderate, many).

2. A minimum of three ova and parasite examinations (including the permanent stained smear) and/or some of the immunoassays should be performed to ensure that no other protozoa are found before attributing symptoms to *B. hominis*.

3. Other possible microbial agents should also be considered.

4. Remember that there is still controversy regarding quantitation of these organisms on the laboratory report form. Although there was thought to be correlation between numbers of organisms present and patient symptoms, some physicians have seen symptomatic patients in whom even small numbers of *B. hominis* are present, apparently with no other causative agents present in the stool specimen(s). For this reason, quantitation of the organisms is still recommended. However, it is important to communicate with your physician clients so that they know how you are reporting and/or quantitating this organism. The approach must be determined by each laboratory after discussions with physician clients.

Treatment

Although there is not a great deal of clinical evidence, there have been studies on the in vitro susceptibility of *B. hominis* to numerous drugs (60). At present, metronidazole (Flagyl) appears to be the most appropriate drug. Diiodohydroxyquin (Yodoxin) has also been effective, and dosage schedules for these two drugs are as recommended for other intestinal protozoa. The development of a new drug sensitivity assay may improve our ability to scientifically evaluate the activities of various drugs against this organism.

Epidemiology and Prevention

From present information, it appears that *B. hominis* is transmitted via the fecal-oral route through contaminated food or water. Although other possible modes of transmission are not defined, the incidence and apparent worldwide distribution indicate the traditional route of infection. Recent studies suggest the existence of numerous zoonotic isolates, with frequent animal-to-human and human-to-animal transmissions, and of a large potential reservoir in animals for infections in humans. Prevention would probably involve improved personal hygiene and sanitary conditions (34, 52).

Very specific prevention measures for all of the intestinal protozoa can be seen in Table 2.8. These recommendations are relevant for all organisms that can be acquired through the ingestion of food or water contaminated with fecal material.

Table 2.8   Measures for the prevention of infection with intestinal protozoa

1. Wash hands with soap and water after using the toilet and before handling food.

2. Avoid water or raw food that may be contaminated.

3. Wash and peel all raw vegetables and fruits before eating.

4. When traveling in countries where the water supply may be unsafe, avoid drinking unboiled tap water and avoid uncooked foods washed with unboiled tap water. Bottled or canned carbonated beverages, seltzers, pasteurized fruit drinks, and steaming hot coffee and tea are safe to drink.

5. If you work in a child care center where you change diapers, be sure to wash your hands thoroughly with plenty of soap and warm water after every diaper change, even if you wear gloves.

References

1. Ali, I. K. M., M. B. Hossain, S. Roy, P. F. Ayeh-Kuni, W. A. Petri, Jr., R. Haque, and C. G. Clark. 2003. *Entamoeba moshkovskii* infections in children in Bangladesh. *Emerg. Infect. Dis.* **9:**580–584.

2. Blessmann, J., H. Buss, P. A. T. Nu, B. T. Dinh, Q. T. V. Ngo, A. L. Van, M. D. Alla, T. F. H. G. Jackson, J. I. Ravdin, and E. Tannich. 2002. Real-time PCR for detection and differentiation of *Entamoeba histolytica* and *Entamoeba dispar* in fecal samples. *J. Clin. Microbiol.* **40:**4413–4417.

3. Blessmann, J., L. P. Van, P. A. T. Nu, H. D. Thi, B. Muller-Myhsok, H. Buss, and E. Tannish. 2002. Epidemiology of amebiasis in a region of high incidence of amebic liver abscess in central Vietnam. *Am. J. Trop. Med. Hyg.* **66:**578–583.

4. Boreham, P. F. L., J. A. Upcroft, and L. A. Dunn. 1992. Protein and DNA evidence for 2 demes of *Blastocystis hominis* from humans. *Int. J. Parasitol.* **22:**49–53.

5. Braga, L. L. B. C., M. L. Gomes, M. W. Da Silva, F. E. Facanha, Jr., L. Fiuza, and B. J. Mann. 2001. Household epidemiology of *Entamoeba histolytica* infection in an urban community in Northeastern Brazil. *Am. J. Trop. Med. Hyg.* **65:**268–271.

6. Campos-Rodriguez, R., and A. Jarillo-Luna. 2005. The pathogenicity of *Entamoeba histolytica* is related to the capacity of evading innate immunity. *Parasite Immunol.* **27:**1–8.

7. Cirioni, O., A. Giacometti, D. Drenaggi, F. Ancarani, and G. Scalise. 1999. Prevalence and clinical relevance of *Blastocystis hominis* in diverse patient cohorts. *Eur. J. Epidemiol.* **15:**389–393.

8. Diamond, L. S., and C. G. Clark. 1993. A redescription of *Entamoeba histolytica* Schaudinn, 1903 (emended Walker, 1911), separating it from *Entamoeba dispar* Brumpt, 1925. *J. Eukaryot. Microbiol.* **40:**340–344.

9. Doyle, P. W., M. M. Helgason, R. G. Mathias, and E. M. Proctor. 1990. Epidemiology and pathogenicity of *Blastocystis hominis. J. Clin. Microbiol.* **28:**116–121.

10. Espinosa-Cantellano, M., and A. Martinez-Palomo. 2000. Pathogenesis of intestinal amebiasis: from molecules to disease. *Clin. Microbiol. Rev.* **13:**318–331.

11. Furrows, S. J., A. H. Moody, and P. L. Chiodini. 2004. Comparison of PCR and antigen detection methods for diagnosis of *Entamoeba histolytica* infection. *J. Clin. Pathol.* 57:1264–1266.

12. Gatti, S., G. Swierczynski, F. Robinson, M. Anselmi, J. Corrales, J. Moreira, G. Montalvo, A. Bruno, R. Maserati, Z. Bisoffi, and M. Scaglia. 2002. Amebic infections due to the *Entamoeba histolytica-Entamoeba dispar* complex: a study of the incidence in a remote rural area of Ecuador. *Am. J. Trop. Med. Hyg.* 67:123–127.

13. Giacometti, A., O. Cirioni, A. Fiorentini, M. Fortuna, and G. Scalise. 1999. Irritable bowel syndrome in patients with *Blastocystis hominis* infection. *Eur. J. Clin. Microbiol. Infect. Dis.* 18:436–439.

14. Gilchrist, C. A., and W. A. Petri. 1999. Virulence factors of *Entamoeba histolytica*. *Curr. Opin. Microbiol.* 2:433–437.

15. Gonin, P., and L. Trudel. 2003. Detection and differentiation of *Entamoeba histolytica* and *Entamoeba dispar* isolates in clinical samples by PCR and enzyme-linked immunosorbent assay. *J. Clin. Microbiol.* 41:237–241.

16. Haghighi, A., S. Kobayashi, T. Takeuchi, G. Masuda, and T. Nozaki. 2002. Remarkable genetic polymorphism among *Entamoeba histolytica* isolates from a limited geographic area. *J. Clin. Microbiol.* 40:4081–4090.

17. Haghighi, A., S. Kobayashi, T. Takeuchi, N. Thammapalerd, and T. Nozaki. 2003. Geographic diversity among genotypes of *Entamoeba histolytica* field isolates. *J. Clin. Microbiol.* 41:3748–3756.

18. Haque, R., D. Mondal, B. D. Kirkpatrick, S. Akther, B. M. Farr, R. B. Sack, and W. A. Petri, Jr. 2003. Epidemiologic and clinical characteristics of acute diarrhea with emphasis on *Entamoeba histolytica* infections in preschool children in an urban slum of Dhaka, Bangladesh. *Am. J. Trop. Med. Hyg.* 69:398–405.

19. Hira, P. R., J. Iqbal, F. A. R. Philip, S. Grover, E. D'Almeida, and A. A. Al-Eneizi. 2001. Invasive amebiasis: challenges in diagnosis in a non-endemic country (Kuwait). *Am. J. Trop. Med. Hyg.* 65:341–345.

20. Hung, C. C., H. Y. Deng, W. H. Hsiao, S. M. Hsieh, C. F. Hsiao, M. Y. Chen, S. C. Chang, and K. E. Su. 2005. Invasive amebiasis as an emerging parasitic disease in patients with human immunodeficiency virus type 1 infection in Taiwan. *Arch. Intern. Med.* 165:409–415.

21. Isenberg, H. D. (ed.). 2004. *Clinical Microbiology Procedures Handbook*, 2nd ed., p. 9.0.1–9.10.8.3. ASM Press, Washington, D.C.

22. Kaneda, Y., N. Horiki, X. Cheng, Y. Fujita, M. Maruyama, and H. Tachibana. 2001. Ribodemes of *Blastocystis hominis* isolated in Japan. *Am. J. Trop. Med. Hyg.* 65:393–396.

23. Kaneda, Y., N. Horiki, X. Cheng, H. Tachibana, and Y. Tsutsumi. 2000. Serologic response to *Blastocystis hominis* infection in asymptomatic individuals. *Tokai J. Exp. Clin. Med.* 25:51–56.

24. Kurland, J. E., and O. S. Brann. 2004. Pyogenic and amebic liver abscesses. *Curr. Gastroenterol. Rep.* 6:273–279.

25. Loftus, B., I. Anderson, R. Davies, U. C. Alsmaark, J. Samuelson, P. Amedeo, P. Roncaglia, M. Berriman, R. P. Hirt, B. J. Mann, T. Nozaki, B. Suh, M. Pop, M. Duchene, J. Ackers, E. Tannich, M. Leippe, M. Hofer, I. Bruchhaus, U. Willhoeft, A. Bhattacharya, T. Chillingworth, C. Churcher, Z. Hance, B. Harris, D. Harris, K. Jagels, S. Moule, K. Mungall, D. Ormond, R. Squares, S. Whitehead, M. A. Quail, E. Rabbinowitsch, H. Norbertczak, C. Price, Z. Wang, N. Guillen, C. Gilchrist, S. E. Stroup, S. Bhattacharya, A. Lohia, P. G. Foster, T. Sicheritz-Ponten, C. Weber, U. Singh, C. Mukherjee, N. M. El-Sayed, W. A. Petri, Jr., C. G. Clark, T. M. Embley, B. Barrell, C. M. Fraser, and N. Hall. 2005. The genome of the protist parasite *Entamoeba histolytica*. *Nature* 433:865–868.

26. Lucht, E., B. Evengard, J. Skott, P. Pehrson, and C. E. Nord. 1998. *Entamoeba gingivalis* in human immunodeficiency virus type 1-infected patients with periodontal disease. *Clin. Infect. Dis.* 27:471–473.

27. Mansour, N. S., E. M. Mikhail, N. A. Elmasry, A. G. Sabry, and E. W. Mohareb. 1995. Biochemical characterization of human isolates of *Blastocystis hominis*. *J. Med. Microbiol.* 42:304–307.

28. Markell, E. K., and M. P. Udkow. 1986. *Blastocystis hominis*: pathogen or fellow traveler? *Am. J. Trop. Med. Hyg.* 35:1023–1026.

29. McAuley, J. B., B. L. Herwaldt, S. L. Stokes, J. A. Becher, J. M. Roberts, M. K. Michelson, and D. D. Juranek. 1992. Diloxanide furoate for treating asymptomatic *Entamoeba histolytica* cyst passers: 14 years' experience in the United States. *Clin. Infect. Dis.* 15:464–468.

30. Mirelman, D. 1992. Pathogenic versus nonpathogenic *Entamoeba histolytica*. *Infect. Agents Dis.* 1:15–18.

31. Moody, S., S. Becker, Y. Nuchamovitz, and D. Mirelman. 1997. Virulent and avirulent *Entamoeba histolytica* and *E. dispar* differ in their cell surface phosphorylated glycolipids. *Parasitology* 114:95–104.

32. Moran, P., G. Rico, M. Ramiro, H. Olvera, F. Ramos, E. Gonzalez, A. Valadez, O. Curiel, E. I. Melendro, and C. Ximenez. 2002. Defective production of reactive oxygen intermediates (ROI) in a patient with recurrent amebic liver abscess. *Am. J. Trop. Med. Hyg.* 67:632–635.

33. Muller, H. E. 1994. Four serologically different groups within the species *Blastocystis hominis*. *Zentbl. Bakteriol.—Int. J. Med. Microbiol. Virol. Parasitol. Infect. Dis.* 280:403–408.

34. Noel, C., F. Dufernez, D. Gerbod, V. P. Edgcomb, P. Delgado-Viscongliosi, L. Ho, M. Singh, R. Wintjens, M. L. Sogin, M. Capron, R. Pierce, L. Zenner, and E. Viscogliosi. 2005. Molecular phylogenies of *Blastocystis* isolates from different hosts: implications for genetic diversity, identification of species, and zoonosis. *J. Clin. Microbiol.* 43:348–355.

35. Novati, S., M. Sironi, S. Granata, A. Bruno, S. Gatti, M. Scaglia, and C. Bandi. 1996. Direct sequencing of the PCR amplified SSU rRNA gene of *Entamoeba dispar* and the design of primers for rapid differentiation from *Entamoeba histolytica*. *Parasitology* 112:363–369.

36. Parshad, S., P. S. Grover, A. Sharma, D. K. Verma, and A. Sharma. 2002. Primary cutaneous amoebiasis: case report with review of the literature. *Int. J. Dermatol.* 41:676–680.

37. Perez-Jaffe, L., R. Katz, and P. K. Gupta. 1998. *Entamoeba gingivalis* identified in a left upper neck nodule by fine-needle aspiration: a case report. *Diagn. Cytopathol.* 18:458–461.

38. **Pillai, D. R., and K. C. Kain.** 2005. *Entamoeba histolytica:* identification of a distinct β₂-integrin-like molecule with a potential role in cellular adherence. *Exp. Parasitol.* **109:**135–142.

39. **Pillai, D. R., P. S. K. Wan, W. C. W. Yau, J. I. Ravdin, and K. C. Kain.** 1999. The cysteine-rich region of the *Entamoeba histolytica* adherence lectin (170-kilodalton subunit) is sufficient for high-affinity Gal/GalNAc-specific finding in vitro. *Infect. Immun.* **67:**3836–3841.

40. **Pomes, C. E., W. A. Bretz, A. de Leon, R. Aguirre, E. Milian, and E. S. Chaves.** 2000. Risk indicators for periodontal disease in Guatemalan adolescents. *Braz. Dent. J.* **11:**49–57.

41. **Que, X., and S. L. Reed.** 2000. Cysteine proteinases and the pathogenesis of amebiasis. *Clin. Microbiol. Rev.* **13:**196–206.

42. **Sanchez-Guillen, M. D. C., R. Perez-Fuentes, H. Salgado-Rosas, A. Ruiz-Arguelles, J. Ackers, A. Shire, and P. Talamas-Rohana.** 2002. Differentiation of *Entamoeba histolytica/Entamoeba dispar* by PCR and their correlation with humoral and cellular immunity in individuals with clinical variants of amoebiasis. *Am. J. Trop. Med. Hyg.* **66:**731–737.

43. **Schuster, F. L., and G. S. Visvesvara.** 2004. Amebae and ciliated protozoa as causal agents of waterborne zoonotic disease. *Vet. Parasitol.* **126:**91–120.

44. **Shlim, D. R., C. W. Hoge, R. Rajah, J. G. Rabold, and P. Echeverria.** 1995. Is *Blastocystis hominis* a cause of diarrhea in travellers? A prospective controlled study in Nepal. *Clin. Infect. Dis.* **21:**97–101.

45. **Stanley, S. L., Jr.** 2003. Amoebiasis. *Lancet* **361:**1025–1034.

46. **Stanley, S. L., Jr.** 1997. Progress towards development of a vaccine for amebiasis. *Clin. Microbiol. Rev.* **10:**637–649.

47. **Stauffer, W., and J. I. Ravdin.** 2003. *Entamoeba histolytica:* an update. *Curr. Opin. Infect. Dis.* **16:**479–485.

48. **Tachibana, H., X. J. Cheng, K. Watanabe, M. Takehoshi, F. Maeda, S. Aotsuka, Y. Kaneda, T. Takeuchi, and S. Ihara.** 1999. Preparation of recombinant human monoclonal antibody Fab fragments specific for *Entamoeba histolytica*. *Clin. Diagn. Lab. Immunol.* **6:**383–387.

49. **Talamus-Rohana, P., M. M. Aguirre-Garcia, M. Anaya-Ruiz, and J. L. Rosales-Encina.** 1999. *Entamoeba dispar* contains but does not secrete acid phosphatase as does *Entamoeba histolytica*. *Exp. Parasitol.* **92:**219–222.

50. **Tan, K. S.** 2004. *Blastocystis* in humans and animals: new insights using modern methodologies. *Vet. Parasitol.* **126:**121–144.

51. **Tanyuksel, M., and W. A. Petri, Jr.** 2003. Laboratory diagnosis of amebiasis. *Clin. Microbiol. Rev.* **16:**713–729.

52. **Thathaisong, U., J. Worapong, M. Mungthin, P. Tan-Ariya, K. Viputtigul, A. Sudatis, A. Noonai, and S. Leelayoova.** 2003. *Blastocystis* isolates from a pig and a horse are closely related to *Blastocystis hominis*. *J. Clin. Microbiol.* **41:**967–975.

53. **Verweij, J. J., D. Laeijendecker, E. A. T. Brienen, L. van Lieshout, and A. M. Polderman.** 2003. Detection and identification of *Entamoeba* species in stool samples by a reverse line hybridization assay. *J. Clin. Microbiol.* **41:**5041–5045.

54. **Vines, R. R., G. Ramakrishnan, J. B. Rogers, I. A. Lockhart, B. J. Mann, and W. A. Petri.** 1998. Regulation of adherence and virulence by the *Entamoeba histolytica* lectin cytoplasmic domain, which contains a β₂-integrin motif. *Mol. Biol. Cell* **9:**2069–2079.

55. **Voigt, H., J. C. Olivo, P. Sansonetti, and N. Guillen.** 1999. Myosin IB from *Entamoeba histolytica* is involved in phagocytosis of human erythrocytes. *J. Cell Sci.* **112:**1191–1201.

56. **Wells, C. D., and M. Arguedas.** 2004. Amebic liver abscess. *South. Med. J.* **97:**673–682.

57. **Willhoeft, U., H. Buss, and E. Tannich.** 1999. DNA sequences corresponding to the ariel gene family of *Entamoeba histolytica* are not present in *E. dispar*. *Parasitol. Res.* **85:**787–789.

58. **Wilson, M., P. Schantz, and N. Pieniazek.** 1995. Diagnosis of parasitic infections: immunologic and molecular methods, p. 1159–1170. *In* P. R. Murray, E. J. Baron, M. A. Pfaller, F. C. Tenover, and R. H. Yolken (ed.), *Manual of Clinical Microbiology*, 6th ed. American Society for Microbiology, Washington, D.C.

59. **Yoshikawa, H., N. Abe, M. Iwasawa, S. Kitano, I. Nagano, Z. Wu, and Y. Takahashi.** 2000. Genomic analysis of *Blastocystis hominis* strains isolated from two long-term health care facilities. *J. Clin. Microbiol.* **38:**1324–1330.

60. **Zierdt, C. H.** 1991. *Blastocystis hominis*—past and future. *Clin. Microbiol. Rev.* **4:**61–79.

61. **Zierdt, C. H., W. S. Zierdt, and B. Nagy.** 1995. Enzyme-linked immunosorbent assay for detection of serum antibody to *Blastocystis hominis* in symptomatic infections. *J. Parasitol.* **81:**127–129.

3

Intestinal Protozoa: Flagellates and Ciliates

Giardia lamblia

The flagellate *Giardia lamblia* (syn. *Giardia intestinalis, Giardia duodenalis*) was first discovered by Leeuwenhoek in 1681 in his own stool specimens but was not described until 1859 by Lambl. The organism was named after Professor A. Giard of Paris and Dr. F. Lambl of Prague (6). *G. lamblia* is worldwide in distribution and apparently is more prevalent in children than in adults and more common in warm climates than in cool ones. It is the most commonly diagnosed flagellate in the intestinal tract, and it may be the most commonly diagnosed intestinal protozoan in some areas of the world.

Although various criteria, including host specificity, various body dimensions, and variations in structure, have been used to differentiate species of *Giardia*, there is still considerable debate over the appropriate classification and nomenclature regarding this group of organisms. On the basis of work by Filice related to structural variations, three groups have been proposed: amphibian *Giardia* spp. (represented by *G. agilis*), the muris group from rodents and birds (represented by *G. muris*), and the intestinalis group from a variety of mammals (including humans), birds, and reptiles (represented by *G. duodenalis*) (1). During the 1980s, the name *G. duodenalis* was supported and in the 1990s *G. intestinalis* was supported by various investigators. Despite disagreement concerning the various species names, all three continue to be used to describe this organism. Meyer prefers to use *Giardia duodenalis*, which is often followed by the name of the animal from which the organism was obtained (47). Within the United States, the term *Giardia lamblia* has been commonly used for many years and will continue to be used throughout this text to refer to those organisms found in humans and other mammals; this designation will, we hope, eliminate any confusion, since the majority of health workers are used to this name and continue to report the presence of the organism by using the term "*Giardia lamblia*" (Table 3.1).

Molecular classification methods have been helpful in understanding the pathogenesis and epidemiology of human and animal *Giardia* infections. These studies have confirmed the division of *G. lamblia* human isolates into two major genotypes (Table 3.2). Adam has proposed that genotype A be accepted as the designation for Nash group 1 (A-1), Nash group 2 (A-2), assemblage A, and the Polish isolates. Genotype B would designate Nash group 3, assemblage B,

Table 3.1 *Giardia* species[a]

Species	Hosts	Morphology — Light microscopy	Morphology — Electron microscopy	Molecular data
G. agilis	Amphibians	Long, slender; long teardrop-shaped median body	Focal contacts by the lateral crest of the ventral disk, the ventrolateral flange, the lateral shield, and by numerous microvillus-like appendages found along the lateral border of the trophozoite	NA[b]
G. ardeae	Herons	Pear shaped; 1 or 2 transverse, claw hammer-shaped median bodies	Ventral disk and caudal flagellum similar to *G. muris*	Closer to *G. lamblia* than to *G. muris*
G. lamblia	Humans and many other mammals	Pear shaped; 1 or 2 transverse, claw hammer-shaped median bodies; sucking disk shorter than half the body length	Nuclei have a defined position, and fibrils perform an anchoring system; median bodies vary in number, shape, and position, are found in mitotic and interphasic trophozoites, are present in about 80% of the cells, and are not completely free in the cells	Clade with multiple genotypes
G. microti	Voles and muskrats	Pear shaped; 1 or 2 transverse, claw hammer-shaped median bodies	Cysts contain 2 trophozoites with mature ventral disks	Similar to *G. lamblia* genotypes
G. muris	Rodents	Short and rounded; small round median body		Distant from *G. lamblia*
G. psittaci	Psittacine birds	Pear shaped; 1 or 2 transverse, claw hammer-shaped median bodies	Incomplete ventrolateral flange, no marginal groove bordering adhesive disk	NA

[a] Adapted from references 1, 5, 32, and 65.
[b] NA, not available.

and the isolates from Belgium. He feels it may also be appropriate to use a similar designation for assemblages C through G, since these also reflect both genetic and host differences (1).

Life Cycle and Morphology

The life cycle is seen in Figure 3.1. Both the trophozoite and the cyst are included in the life cycle of *G. lamblia*. Trophozoites divide by means of longitudinal binary fission, producing two daughter trophozoites. The most common location of the organisms is in the crypts within the duodenum. The trophozoites are the intestinal dwelling stage and attach to the epithelium of the host villi by means of the ventral disk. The attachment is substantial and results in disk "impression prints" when the organism detaches from the surface of the epithelium. Trophozoites may remain attached or detach from the mucosal surface. Since the epithelial surface sloughs off the tip of the villous every 72 h, apparently the trophozoites detach at that time.

There are several hypotheses on possible mechanisms for attachment by the ventral disk; these include microtubule mediation, hydrodynamic action, contractile protein activity, and the interaction of lectins with surface-bound sugars. Because there is conflicting information, the attachment process probably depends on multiple mechanisms. A study by Katelaris et al. indicates that attachment to enterocyte-like differentiated Caco-2 cells is primarily by cytoskeletal mechanisms, which can be inhibited by interfering with contractile filaments and microtubules; attachment by mannose-binding lectin also seems to mediate binding (38).

For reasons which are not fully known, cyst formation takes place as the organisms move down through the jejunum after exposure to biliary secretions. The trophozoites retract the flagella into the axonemes, the cytoplasm becomes condensed, and the cyst wall is secreted. As the cyst matures, the internal structures are doubled, so that when excystation occurs the cytoplasm divides, thus producing two trophozoites. Excystation would normally occur in the duodenum or appropriate culture medium (Figure 3.1) (1).

Figure 3.1   Life cycle of *Giardia lamblia*.

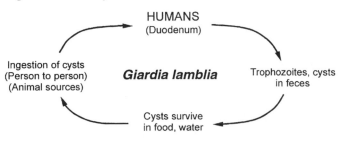

Table 3.2 *Giardia* genotypes[a]

Adam designation	Nash group	Mayrhofer assemblage	Origin	Hosts
Genotype A-1	1	A (group 1)	Poland	Human, beaver, cat, lemur, sheep, calf, dog, chinchilla, alpaca, horse, pig, cow
Genotype A-2	2	A (group 2)		Human, beaver
Genotype B	3	B (groups 3 and 4)	Belgium	Human, beaver, guinea pig, dog, monkey
		C		Dog
		D		Dog
		E (or A-livestock)		Cow, sheep, alpaca, goat, pig
		F		Cat
		G		Rat

[a] Adapted from references 1 and 65.

The process of encystment consists of two morphologically different stages (intracellular and extracellular) and requires approximately 16 h to complete. The extracellular phase is initiated with the appearance of cyst wall antigen on small protrusions of the trophozoite membrane, which becomes enlarged over time to form caplike structures ranging up to 100 nm in diameter. Caplike structures involved with filament growth have been detected over the entire surface of the trophozoite including the adhesive disk and flagella. Encysting cells round up, lose attachment to the substratum, and become enclosed in a layer of filaments. The internal portion includes two trophozoites with four nuclei. If methods could be found to interfere with the development of the filamentous layer of the cyst wall, regulation of viable *Giardia* cyst release from the host could be modified or eliminated (2).

Changes observed in *G. lamblia* lipids (increased fatty acid unsaturation and the accumulation of storage lipids) are consistent with parasite differentiation into a cyst stage that can survive outside the host at reduced temperatures and with reduced available nutrient resources. *G. lamblia* not only has the capacity to synthesize isoprenoid lipids de novo but also can metabolize fatty acids by the addition of double bonds (1). Studies have also shown that bile acids found in high concentrations in the small intestine enhance trophozoite survival, multiplication, and differentiation into the cyst stage (16). Results also indicate that cholesterol starvation is necessary and sufficient for the stimulation of *Giardia* encystations in vitro and probably in the intestines of mammalian hosts (44).

It has been determined that oxygen uptake rate and metronidazole sensitivity in *Giardia* cysts are greatly reduced compared with those in trophozoites. The resistance to metronidazole may be due to a change in metabolic flux away from the pyruvate ferredoxin oxireductase pathway.

However, after the induction of excystation, oxygen uptake increases exponentially during the following 30 min and metronidazole sensitivity returns within 15 min and is fully restored within 30 min (1). Excystation occurs with exposure to the contents of the proximal small intestine; the process is quite rapid. Externally, the flagella emerge through the cyst wall, followed by the entire trophozoite. Recent findings on several aspects of *Giardia* differentiation can be found in the review by Adam (1). The process is divided into three distinct parts: (i) the stimulus for encystations and the regulation of encystations-specific gene expression, (ii) the synthesis and intracellular transport of cyst wall components, and (iii) the assembly of the extracellular cell wall.

Although *Giardia* has many features of the prokaryotes (anaerobic metabolism, an enzyme that reduces oxygen directly to water, cysteine as the "keeper" of the redox balance, a plasmid, and toxin-like genes), unlike prokaryotes it also has a sophisticated, highly developed cytoskeleton, bounded nuclei, linear chromosomes capped with telomeric repeats, and telomere positional regulation of gene expression (5, 64). Studies have also clarified the pathway of formation of ethanol, one of the major end products of the fermentative metabolism of the amitochondriate protist, *G. lamblia*. In contrast to most eukaryotes, in which ethanol formation proceeds from pyruvate via acetaldehyde, the *G. lamblia* pathway starts from acetyl coenzyme A, a more distal product of extended glycolysis (53).

Morphology of Trophozoites (Table 3.3). The trophozoite is usually described as being teardrop shaped from the front, with the posterior end being pointed (Figures 3.2 to 3.4). If one examines the trophozoite from the side, it resembles the curved portion of a spoon. The concave portion is the area of the sucking disk (Figure 3.2). There are four pairs of flagella, two nuclei, two axonemes, and two slightly curved bodies called the median bodies. The

Table 3.3 Intestinal protozoa: trophozoites of flagellates

Characteristic	Shape and size	Motility	No. of nuclei and visibility	No. of flagella (usually difficult to see)	Other features
Dientamoeba fragilis	Shaped like amebae; 5–15 μm; usual range, 9–12 μm	Usually nonprogressive; pseudopodia are angular, serrated, or broad lobed and almost transparent	Percentage may vary, but 40% of organisms have 1 nucleus and 60% have 2 nuclei; not visible in unstained preparations; no peripheral chromatin; karyosome is composed of a cluster of 4–8 granules	No visible flagella	Cytoplasm finely granular and may be vacuolated with ingested bacteria, yeasts, and other debris; may be great variation in size and shape on a single smear
Giardia lamblia	Pear shaped; length, 10–20 μm; width, 5–15 μm	"Falling-leaf" motility may be difficult to see if organism is in mucus	2; not visible in unstained mounts	4 lateral, 2 ventral, 2 caudal	Sucking disc occupying 1/2–3/4 of ventral surface; pear-shaped front view; spoon-shaped side view
Chilomastix mesnili	Pear shaped; length, 6–24 μm (usual range, 10–15 μm); width, 4–8 μm	Stiff, rotary	1; not visible in unstained mounts	3 anterior, 1 in cytostome	Prominent cytostome extending 1/3–1/2 length of body; spiral groove across ventral surface
Pentatrichomonas hominis (*Trichomonas hominis*)	Pear shaped; length, 5–15 μm (usual range, 7–9 μm); width, 7–10 μm	Jerky, rapid	1; not visible in unstained mounts	3–5 anterior, 1 posterior	Undulating membrane extends length of the body; posterior flagellum extends free beyond end of body
Trichomonas tenax	Pear shaped; length, 5–12 μm (avg, 6.5–7.5 μm); width, 7–9 μm	Jerky, rapid	1; not visible in unstained mounts	4 anterior, 1 posterior	Seen only in preparations from mouth; axostyle (slender rod) protrudes beyond the posterior end and may be visible; posterior flagellum extends only halfway down body and there is no free end
Enteromonas hominis	Oval; length, 4–10 μm (usual range, 8–9 μm); width, 5–6 μm	Jerky	1; not visible in unstained mounts	3 anterior, 1 posterior	One side of body flattened; posterior flagellum extends free posteriorly or laterally
Retortamonas intestinalis	Pear shaped or oval; length, 4–9 μm (usual range, 6–7 μm); width, 3–4 μm	Jerky	1; not visible in unstained mounts	1 anterior, 1 posterior	Prominent cytostome extending approximately 1/2 length of body

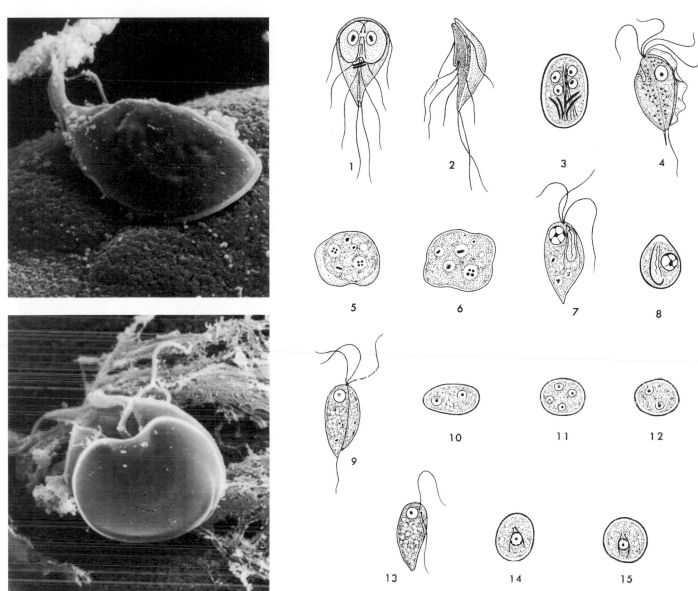

Figure 3.2 *Giardia lamblia* trophozoites on the mucosal surface. (Scanning electron micrographs courtesy of Marietta Voge.)

Figure 3.3 (1) Trophozoite of *Giardia lamblia* (front view); (2) trophozoite of *G. lamblia* (side view); (3) cyst of *G. lamblia*; (4) trophozoite of *Pentatrichomonas hominis*; (5) trophozoite of *Dientamoeba fragilis* (two fragmented nuclei); (6) trophozoite of *D. fragilis* (one fragmented nucleus); (7) trophozoite of *Chilomastix mesnili*; (8) cyst of *C. mesnili*; (9) trophozoite of *Enteromonas hominis*; (10–12) cysts of *E. hominis*; (13) trophozoite of *Retortamonas intestinalis*; (14 and 15) cysts of *R. intestinalis*. (Illustrations by Sharon Belkin.)

trophozoites usually measure 10 to 20 μm in length and 5 to 15 μm in width (Figures 3.2 to 3.4).

Morphology of Cysts (Table 3.4). The cysts may be either round or oval, and they contain four nuclei, axonemes, and median bodies (Figures 3.3 and 3.4). Often, some cysts appear to be shrunk or distorted, and one may see two halos, one around the cyst wall itself and one inside the cyst wall around the shrunken organism. The halo effect around the outside of the cyst is particularly visible on the permanent stained smear. Cysts normally measure 11 to 14 μm in length and 7 to 10 μm in width.

Clinical Disease

From available data, it appears that the incubation time for giardiasis ranges from approximately 12 to 20 days.

Table 3.4 Intestinal protozoa: cysts of flagellates

Species	Size	Shape	No. of nuclei	Other features
Dientamoeba fragilis, Pentatrichomonas hominis (Trichomonas hominis), Trichomonas tenax	No cyst stage			
Giardia lamblia	Length, 8–19 µm (usual range, 11–14 µm); width, 7–10 µm	Oval, ellipsoidal, or may appear round	4; not distinct in unstained preparations; usually located at one end	Longitudinal fibers in cysts may be visible in unstained preparations; deep-staining median bodies usually lie across the longitudinal fibers; there is often shrinkage, and the cytoplasm pulls away from the cyst wall; there may also be a halo effect around the outside of the cyst wall because of shrinkage caused by dehydrating reagents
Chilomastix mesnili	Length, 6–10 µm (usual range, 7–9 µm); width, 4–6 µm	Lemon shaped with anterior hyaline knob	1; not distinct in unstained preparations	Cytostome with supporting fibrils, usually visible in stained preparation; curved fibril along side of cytostome usually referred to as "shepherd's crook"
Enteromonas hominis	Length, 4–10 µm (usual range, 6–8 µm); width, 4–6 µm	Elongate or oval	1–4; usually 2 lying at opposite ends of cyst; not visible in unstained mounts	Resembles *E. nana* cyst; fibrils or flagella usually not seen
Retortamonas intestinalis	Length, 4–9 µm (usual range, 4–7 µm); width, 5 µm	Pear shaped or slightly lemon shaped	1; not visible in unstained mounts	Resembles *Chilomastix* cyst; shadow outline of cytostome with supporting fibrils extends above nucleus; bird beak fibril arrangement

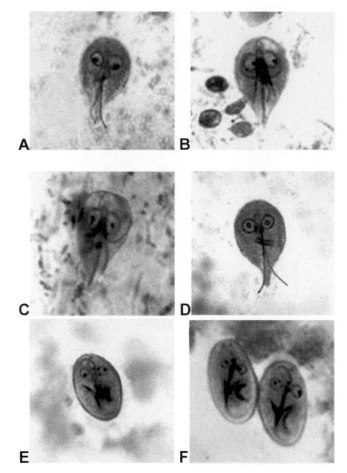

Figure 3.4 (A to D) Trophozoites of *Giardia lamblia*; (E and F) cysts of *G. lamblia.*

Because the acute stage usually lasts only a few days, giardiasis may not be recognized as the cause but may mimic acute viral enteritis, bacillary dysentery, bacterial or other food poisoning, acute intestinal amebiasis, or "traveler's diarrhea" (toxigenic *Escherichia coli*). However, the type of diarrhea plus the lack of blood, mucus, and cellular exudate is consistent with giardiasis.

Although the organisms in the crypts of the duodenal mucosa may reach very high densities, they may not cause any pathology. The organisms feed on the mucous secretions and do not penetrate the mucosa as *G. muris* has been shown to do in mice, from which trophozoites have been recovered in the deeper tissues. Although organisms have been seen in biopsy material obtained from inside the intestinal mucosa, others have been seen attached only to the epithelium. For some reason, in symptomatic patients there may be irritation of the mucosal lining, increased mucus secretion, and dehydration (68).

A study by Chen et al. has demonstrated a *G. duodenalis* cysteine-rich surface protein (CRP136) that has 57%

homology to the product of the gene encoding the precursor of the sarafotoxins, a group of snake toxins from the burrowing adder that are known to cause symptoms similar to those of humans acutely infected with *Giardia* species (10). Thus, CRP136 represents the first evidence for a potential *Giardia* toxin.

Onset may be accompanied by nausea, anorexia, malaise, low-grade fever, and chills, and there may be a sudden onset of explosive, watery, foul-smelling diarrhea. Other symptoms include epigastric pain, flatulence, and diarrhea with increased amounts of fat and mucus in the stool but no blood (68). Weight loss often accompanies these symptoms. Although there is speculation that the organisms coating the mucosal lining may act to prevent fat absorption, this does not completely explain why the uptake of other substances normally absorbed at other intestinal levels is prevented. During the acute phase of giardiasis, jejunal active-antigen uptake is increased, leading to a delayed recruitment of mucosal and connective tissue mast cells. These changes may play a role in the increased incidence of hypersensitivity reactions and allergic disease associated with *Giardia* infection (31). Severe malabsorption has also been linked with isolated levothyroxine malabsorption, leading to severe hypothyroidism (54) and secondary impairment of pancreatic function (8). In both cases, treatment with metronidazole led to complete remission of symptoms. Occasionally, the gallbladder is also involved, causing gallbladder colic and jaundice. *G. lamblia* has also been identified in bronchoalveolar lavage fluid (59).

The acute phase is often followed by a subacute or chronic phase. Symptoms in these patients include recurrent, brief episodes of loose, foul-smelling stools; there may be increased distention and foul flatus. Between passing the mushy stools, the patient may have normal stools or may be constipated. Abdominal discomfort continues to include marked distention and belching with a rotten-egg taste. Chronic disease must be differentiated from amebiasis; disease caused by other intestinal parasites such as *Dientamoeba fragilis*, *Cryptosporidium* spp., *Cyclospora cayetanensis*, *Isospora belli*, and *Strongyloides stercoralis*; and inflammatory bowel disease and irritable colon. On the basis of symptoms such as upper intestinal discomfort, heartburn, and belching, giardiasis must also be differentiated from duodenal ulcer, hiatal hernia, and gallbladder and pancreatic disease.

Ophthalmic examinations were performed on 141 children with active or past giardiasis. Salt-and-pepper retinal changes (with normal electroretinographic findings) were diagnosed in 28 (19.9%) of the children, all of whom were consistently younger than those with normal retinas. Although these findings indicate that asymptomatic, nonprogressive retinal lesions were particularly common in younger children with giardiasis, the risk does not seem to be related to the severity of the infection, its duration, or the use of metronidazole but may reflect a genetic disposition (13).

Recent studies document antigenic variation with surface antigen changes during human infections with *G. lamblia*; although the biological importance of this work is not clear, it suggests that this variation may provide a mechanism for the organism to escape the host immune response (48). The variant-specific surface proteins are a family of related, highly unusual proteins that cover the surface of the organism; the variant-specific surface proteins are resistant to the effects of intestinal proteases, thus allowing the parasites to survive in the protease-rich small intestine. Antigenic variation at the surface membrane of trophozoites occurs frequently; one would suspect that the higher the rate of change, the more likely that a chronic infection would be seen. However, certain surface antigens that allow the organisms to survive better in the intestinal tract may not be immunologically selected.

Although patients with symptomatic giardiasis usually have no underlying abnormality of serum immunoglobulins, a high incidence of giardiasis occurs in patients with immunodeficiency syndromes, particularly in those with common variable hypogammaglobulinemia (19). Giardiasis was the most common cause of diarrhea in these patients and was associated with mild to severe villous atrophy. Successful treatment of giardiasis led to symptomatic cure and improvement in mucosal abnormalities, with the exception of nodular lymphoid hyperplasia.

With the advent of AIDS, there was speculation that *G. lamblia* might be an important pathogen in this group of individuals. However, clinical findings to date do not seem to confirm this possibility (56). Despite a suppressed immune system in AIDS patients, the immune response to *Giardia* does not seem to be very different from that seen in non-AIDS patients (19). It does not appear that drugs such as corticosteroids, cyclosporin A, and other immunosuppressive agents of cell-mediated immunity affect the outcome of *Giardia* infections in humans.

Diagnosis

Routine Diagnostic Procedures. Routine stool examinations are normally recommended for the recovery and identification of intestinal protozoa (11). However, in the case of *G. lamblia*, because the organisms are attached so securely to the mucosa by means of the sucking disk, a series of even five or six stool samples may be examined without recovering the organisms. The organisms also tend to be passed in the stool on a cyclical basis. The Entero-Test

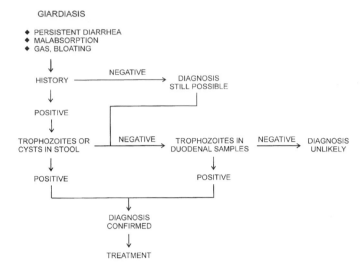

GIARDIASIS

- PERSISTENT DIARRHEA
- MALABSORPTION
- GAS, BLOATING

Algorithm 3.1   Giardiasis. Fecal immunoassay kits can be used on human stool specimens to confirm the presence of *Giardia* spp. Currently available reagents can be used with fresh, frozen, or preserved specimens (not those containing polyvinyl alcohol). If fluorescent testing is performed, the specimen should not be frozen (the actual organisms are disrupted during the freeze-thaw cycle); however, fresh or preserved specimens are acceptable. Check with each manufacturer for compatibility of the fixative with the proposed test; not all fixatives are compatible with all types of fecal immunoassays.

capsule can be helpful in recovering the organisms, as can the duodenal aspirate. Although cysts can often be identified on the wet stool preparation, many infections may be missed without the examination of a permanent stained smear (11, 12). If material from the string test (Entero-Test) or mucus from a duodenal aspirate is submitted, it should be examined as a wet preparation for motility; however, motility may be represented by nothing more than a slight flutter of the flagella because the organism is caught up in the mucus. After diagnosis, the rest of the positive material can be preserved as a permanent stain.

Fluoroscopy may reveal hypermotility at the duodenal and jejunal levels, and X rays may reveal mucosal defects. If the cause is *G. lamblia*, the organisms may be recovered in a series of stool examinations. However, because giardiasis may not produce any symptoms at all, demonstration of the organism in symptomatic patients may not rule out other possibilities such as peptic ulcer, celiac disease of some other etiology, strongyloidiasis, and possibly carcinoma. Refer to Algorithm 3.1 for detailed information.

Antigen Detection.   The development of fecal immunoassays for the detection of *Giardia* and/or *Cryptosporidium* antigen in stool has dramatically improved the sensitivity over that seen with the routine ova and parasite examination and/or modified acid-fast stains (46) (Table 3.5).

Procedures involving the enzyme-linked immunosorbent assay (ELISA) have been developed to detect *Giardia* antigen in feces. The ELISA is at least as sensitive as microscopic wet examinations (3, 4, 20, 21). Fluorescent methods with monoclonal antibodies have also proven to be extremely sensitive and specific in detecting *G. lamblia* in fecal specimens (17, 20, 21, 24). A number of studies have compared the newer immunoassay products, with excellent and comparable results (3, 4, 17, 20–24, 37, 58). Other products are available as a cartridge format that uses an immunochromatographic strip-based detection system for *G. lamblia* and/or *Cryptosporidium* spp. (23, 37). Another cartridge format is used to test for the *Entamoeba histolytica*/*Entamoeba dispar* group, *Giardia lamblia*, and *Cryptosporidium* spp. This commercially available reagent kit requires fresh or frozen stools, since formalin interferes with the *Entamoeba* reagents (22). Any available antigen detection system should always be reviewed for compatibility with stools submitted in preservatives rather than fresh specimens. There are limitations when using kits that include organisms in the genus *Entamoeba*. However, commercial reagent kits for the detection of *Giardia* and *Cryptosporidium* can be used with formalin-based stool preservatives, as well as fresh or frozen specimens. Many of these cartridge format tests are now available that detect both *G. lamblia* and *C. parvum*, provide an answer within 10 min, and are equal to or better than other immunoassays in their sensitivity and specificity. Many of these newer methods are being used to test patients suspected of having giardiasis or those who may be involved in an outbreak situation. The detection of antigen in stool or visual identification of organisms by using monoclonal antibody reagents indicates current infection. With the increase in *Giardia* infections and awareness of particular situations such as nursery school settings, these detection assays are rapid and reliable immunodiagnostic procedures. However, it has been well documented that the use of a fecal immunoassay does not eliminate the need to analyze multiple stool specimens for sensitive detection of *G. lamblia* (30).

Antibody Detection.   Unfortunately, serodiagnostic procedures for giardiasis do not yet fulfill the criteria necessary for wide clinical use, particularly since they may indicate either past or present infection (19). Studies comparing serum antibody responses to giardiasis in symptomatic and asymptomatic patients revealed that antigen recognition by anti-parasite immunoglobulin M (IgM), IgA, IgG1, and IgG3 of these patients, determined by immunoblotting, was heterogeneous and revealed only minor differences in the responses of the two groups (19, 57). In a study investigating the secretory immune response to membrane antigens during giardiasis in

Table 3.5 Representative commercially available kits for immunodetection of *Giardia lamblia*

Diagnostic kit	Manufacturer and/or distributor[a]	Type of test[b]	Comments
ProSpecT			
Microplate assay	Remel	EIA	Different EIA formats (contact company for
Rapid assay, combination with *Cryptosporidium*	Remel	EIA	additional information)
GiardEIA	Antibodies Inc.	EIA	
PARA-TECT *Giardia*	Medical Chemical	EIA	Both DFA and EIA formats available; DFA is
PARA-TECT combination with *Cryptosporidium* spp.	Medical Chemical	DFA	combination reagent (*Giardia/Cryptosporidium*)
MeriFluor, combination with *Cryptosporidium* spp.	Meridian Bioscience	DFA	DFA is combination reagent (*Giardia/Cryptosporidium*)
Giardia Cel	Cellabs	DFA	
Giardia CELISA	Cellabs	EIA	
Crypto/Giardia Cel	Cellabs	DFA	DFA is combination reagent (*Giardia/Cryptosporidium*)
Giardia II	TechLab	EIA	
Giardia Test	Wampole	EIA	
Triage Parasite Panel, combination test with *Cryptosporidium* and *Entamoeba histolytica/E. dispar* group	Biosite Diagnostics, Inc.	Rapid, cartridge (lateral flow)	Requires fresh or frozen stool; combination test with *Cryptosporidium* and *Entamoeba histolytica/E. dispar* group
ColorPAC *Giardia/Cryptosporidium*	Becton Dickinson	Rapid, cartridge (lateral flow)	Can be used with fresh, frozen, or formalin-preserved stool; combination test with *Cryptosporidium* (some other fixatives also acceptable—check with company)
SIMPLE-READ *Giardia*	Medical Chemical	Rapid, cartridge (lateral flow)	Can be used with fresh, frozen, or formalin-preserved stool (some other fixatives also acceptable—check with company)
SIMPLE-READ *Cryptosporidium*	Medical Chemical	Rapid, cartridge (lateral flow)	Can be used with fresh, frozen, or formalin-preserved stool (some other fixatives also acceptable—check with company)
ImmunoCard STAT! *Cryptosporidium/Giardia*	Meridian Bioscience	Rapid, cartridge (lateral flow)	Can be used with fresh, frozen, or formalin-preserved stool; combination test with *Cryptosporidium* (some other fixatives also acceptable—check with company)
XPECT *Giardia*	Remel	Rapid, cartridge (lateral flow)	Can be used with fresh, frozen, or formalin-preserved stool (some other fixatives also acceptable—check with company)
XPECT *Giardia/Cryptosporidium*	Remel	Rapid, cartridge (lateral flow)	Can be used with fresh, frozen, or formalin-preserved stool; combination test with *Cryptosporidium* (some other fixatives also acceptable—check with company)

[a] Antibodies Incorporated, P.O. Box 1560, Davis, CA 95617-1560; Cellabs, P.O. Box 421, Brookvale, NSW 2100, Australia; Becton Dickinson Diagnostic Systems, 7 Loveton Circle, Sparks, MD 21152; Biosite Diagnostics, 11030 Roselle St., San Diego, CA 92121; Medical Chemical Corporation, 19430 Van Ness Ave., Torrance, CA 90501; Meridian Bioscience, 3471 River Hills Drive, Cincinnati, OH 45244; Remel, 12076 Santa Fe Drive, Lenexa, KS 66215; TechLab, 2001 Kraft Drive, Blacksburg, VA 24060-6358; Wampole Laboratories, 2 Research Way, Princeton, NJ 08540.
[b] EIA, enzyme immunoassay; DFA, direct fluorescent antibody; Rapid, cartridge, immunochromatographic lateral flow.

humans, the saliva samples showed a heterogeneous response to the membrane fraction when they were assayed by immunoblotting. Among the antigens recognized by patient saliva samples, those of 170, 105, 92, 66, 32, 29, and 14 kDa stood out since they were not recognized by saliva samples from healthy individuals. These findings may become more relevant in future studies of protection from or diagnosis of *G. lamblia* infections (52). Since *Giardia* trophozoites rarely invade the tissues or stimulate the systemic immune response, serodiagnostic assays fail to show differences in serum antibody responses between symptomatic and asymptomatic patients and are not recommended.

Histology. Trophozoites are detectable in the duodenum and proximal jejunum of infected patients. Although there has been some suggestion of mucosal invasion by this parasite, mucosal invasion has generally been found in areas where necrosis or mechanical trauma was present. In reviewing descriptions of general mucosal appearances, there can be a spectrum of change from normal to almost complete villous atrophy, with a greater density of inflammatory infiltrate in the lamina propria when villous atrophy is present. Apparently, patients with giardiasis also have reduced mucosal surface areas compared with controls, and the degree of abnormality correlates with indices of malabsorption. Enumeration of lamina propria plasma cells of different immunoglobulin classes reveals no clear patterns, however, suggesting that *Giardia* infection initially provokes IgM and IgE synthesis, with relative suppression of local IgA production.

Histologic changes in the mucosal architecture in immunodeficient patients with giardiasis also range from mild to severe villous atrophy. Nodular lymphoid hyperplasia has been reported, as well as mixed lesions. Again, the amount of villous damage seems to correlate with the degree of malabsorption. It appears that giardiasis produces a more severe degree of villous damage in patients with hypogammaglobulinemia. Giardiasis does not appear to be an important pathogen in patients with AIDS, although the infection has certainly been found in this group and in homosexual men.

In a study of 567 patients with giardiasis, a retrospective analysis of cases reported between 1988 and 1994 was performed. On histologic slides, trophozoites were found in duodenal (82.5%) and jejunal (2.1%) mucosa and also in gastric antral (8.7%) and ileal (12.1%) mucosa but rarely in the colon (0.4%). An entirely normal light-microscopic appearance of the duodenal mucosa was found in 462 patients (93.6%). Mild villous shortening and mild inflammation of the lamina propria were observed in the duodenal mucosa in 3.7% of the patients. Since the histology of the small-bowel mucosa is inconspicuous in

the majority of patients with giardiasis, careful review of biopsy specimens must be performed (49).

KEY POINTS—LABORATORY DIAGNOSIS

Giardia lamblia

1. Even if a series of three stool specimens is submitted and examined correctly using the ova and parasite examination, the organisms may not be recovered and identified.
2. Motility on wet preparations may be difficult to see because the organisms may be caught up in mucus. Do not use the ethyl acetate step in the concentration procedure if the specimen contains a lot of mucus.
3. Any examination for parasites in stool specimens must include the use of a permanent stained smear (even on formed stool).
4. Duodenal drainage (centrifuge and examine sediment; do not use the ethyl acetate step) and/or the use of the Entero-Test capsule may be very helpful in organism recovery. However, this technique does not take the place of the ova and parasite examination.
5. It has been well documented that the use of a fecal immunoassay (enzyme immunoassay, fluorecent-antibody assay, or rapid cartridge) does not eliminate the need to analyze multiple stool specimens for sensitive detection of *G. lamblia*; two different stool samples are recommended unless the first specimen is positive.
6. When performing the enzyme immunoassay, make sure that the wells are washed thoroughly according to directions; in order to avoid false-positive results, do not skip any wash steps. Make sure that the stream of buffer goes directly into the wells. Also, make sure that you use a wash bottle with a small opening, so that you have to squeeze the bottle to get the fluid to squirt directly into the wells.
7. When the directions tell you to "slap" the tray down onto some paper towels to remove all the fluid, make sure you slap it several times. Don't be too gentle; the cups will not fall out of the holder.
8. It is recommended that the fluorescent-antibody assay procedures be performed using centrifuged stool (500 × *g* for 10 min); prepare thin (rather than thick) smears and perform the rinse step *gently*. Remember that not all clinical specimens will provide the 3+ to 4+ fluorescence that we often see in the positive control. Also, from time to time, you may see fluorescing bacteria and/or some yeast in certain patient specimens. This is not particularly common, but the shapes can be distinguished from *Giardia* cysts and/or *Cryptosporidium* oocysts.

9. When using the cartridge (lateral-flow) test, if the stool is too thick, addition of reagents will not thin it out enough. If the specimen poured into the well remains too thick, the fluid will not flow up the membrane. If your specimens arrive in fixative and there is no fluid at the top of the vial overlaying the stool, this means that the vial may have been overfilled with stool. These specimens will have to be diluted with the appropriate diluent before being tested. It is always important to see the control line indicated as positive all the way across the membrane, not just at the edges. *A positive test result may be much lighter than the control line*; *this is normal.* At the cutoff time to read the result, any color visible in the test area should be interpreted as a positive. Do not read/interpret the results after the time indicated in the directions; you may get a false-positive result.

Treatment

The incidence of giardiasis worldwide may be as high as 1 billion cases. If giardiasis is diagnosed, the patient should be treated (Table 3.6); more complete dosage information can be found in chapter 25. Current recommended treatments include the nitroheterocyclic drugs tinidazole, metronidazole, and furazolidone, the substituted acridine quinacrine, and the benzimidazole albendazole. Paromomycin is also used, as is nitazoxanide. Treatment failures have been reported with all the common antigiardial agents, and resistance to these drugs has been demonstrated in the laboratory. Clinical resistance has been reported, and includes situations in which patients have failed to respond to both metronidazole and albendazole treatments (25, 69). Metronidazole is not recommended for pregnant women; although not well absorbed and not highly effective, paromomycin may be used to treat giardiasis in pregnancy (25). Tinidazole has also been used and has proven more effective than metronidazole as a single dose (25). Furazolidone is another option, but it has been reported to be mutagenic and carcinogenic. Within the United States, metronidazole is the only member of the nitroimidazole class available to treat giardiasis. Additional information can be found in chapter 25. Metabolic pathways related to anaerobic metabolism (25) and the presence of phosphatidylglycerol and other glycerol-based phospholipids (26) may also serve as potential targets for chemotherapy of giardiasis.

A new and simple colorimetric method has been determined for determining in vitro activity against *G. lamblia*. The microtiter plate assay is based on the nucleoside hydrolase activity released from the organism by lysis. Action of the nucleoside hydrolase on the substrate analog, 4-nitrophenyl β-D-ribofuranoside, gives rise to a colored product which can be determined by the change in absorbance. Values obtained using this method for metronidazole, tinidazole, and furazolidone were consistent with published values. This simple method, which does not involve radioisotopes or complex instrumentation, provided a convenient method for screening potential antigiardial agents (36).

Wheat germ (WG) agglutinin inhibits excystation and trophozoite growth in vitro, as well as reducing cyst passage in mice. In a double-masked, placebo-controlled study of dietary supplementation with WG, there appeared to be no clinically important differences between the two groups. However, symptoms appear to have resolved more rapidly in those taking WG in addition to metronidazole, and the WG was well tolerated in both the symptomatic and asymptomatic groups. Thus, in combination with antiprotozoal drugs, WG may be able to modify symptomatic giardiasis (28).

In the absence of a parasitologic diagnosis, the treatment of suspected giardiasis is a common question with no clear-cut answer. The approach depends on the alternatives and the degree of suspicion of giardiasis, both of which vary among patients and physicians. However, it is not recommended that treatment be given without a good parasitologic workup, particularly since relief of symptoms does not allow a retrospective diagnosis of giardiasis; the most commonly used drug, metronidazole, targets other organisms besides *G. lamblia*.

Another question involves treatment of asymptomatic patients. Generally, it is recommended that all cases of proven giardiasis be treated because the infection may cause subclinical malabsorption, symptoms are often periodic and may appear later, and a carrier is a potential source of infection for others. Certainly, in areas of the world where infection rates, as well as the prospect of reinfection, are extremely high, the benefit-per-cost ratio would also have to be examined.

Epidemiology and Prevention

Transmission is by ingestion of viable cysts. Although contaminated food or drink may be the source, intimate contact with an infected individual may also provide the infection mechanism (33, 41, 55, 60, 70). This organism tends to be found more frequently in children or in groups that live in close quarters (14, 40). Often, there are outbreaks due to poor sanitation facilities or breakdowns as evidenced by infections of travelers and campers (42, 43). There is also an increase in the prevalence of giardiasis in the male homosexual population, probably because of anal and/or oral sexual practices (25).

Although seasonal patterns have been identified for some infectious diseases, limited information is available for giardiasis. Some data suggest an association with the cooler, wetter months of the year; this is not surprising if

Table 3.6 Recommendations for therapy for giardiasis[a]

Clinical picture	Drug and duration of treatment	Side effects
Symptomatic infection, USA		
Adult and pediatric	Metronidazole, 5–7 days	Headache, vertigo, nausea, and metallic taste, urticaria; disulfiram-like reaction with alcohol ingestion; rare side effects include pancreatitis, central nervous system toxicity, reversible neutropenia, peripheral neuropathy, T-wave flattening with prolonged use; mutagenic, carcinogenic?
Alternatives	Furazolidone, 7–10 days	Nausea, vomiting, diarrhea; brown discoloration of urine; disulfiram-like reaction with alcohol ingestion; reacts unfavorably with monoamine oxidase inhibitors; mild hemolysis in glucose 6-phosphate dehydrogenase deficiency; carcinogenic?
	or quinacrine, 5–7 days	Nausea and vomiting, dizziness, headache; yellow/orange discoloration of skin, mucous membranes; rare side effects include toxic psychosis
	or albendazole, 5–7 days	Anorexia, constipation; rare side effects include reversible neutropenia and elevated liver function tests; teratogenic?
Symptomatic infection, non-USA		
Adult and pediatric	Tinidazole (single dose)	As for metronidazole
	or ornidazole (single dose)	As for metronidazole
Pregnancy		
First trimester	Paromomycin, 5–10 days	Ototoxicity and nephrotoxicity with systemic administration
Second and third trimesters	Paromomycin, 5–10 days	
	or metronidazole, 5–7 days	
Resistant infection or relapse	Drug of different class or combination nitroimidazole plus quinacrine, 2 wk or more	See above

[a] Adapted from references 1, 14, 25, 68, and 69.

one considers the issue of environmental conditions advantageous to cyst survival.

It has been documented that certain occupations may place an individual at risk for infection; such at-risk individuals include sewage and irrigation workers, who may become exposed to infective cysts. In situations where young children are grouped together, such as nursery schools, there may be an increased incidence of exposure and subsequent infection of both children and staff.

Most experts agree that the single most effective practice that prevents the spread of infection in the child care setting is thorough handwashing by the children, staff, and visitors. Specific guidelines can be found in Table 3.7. Rubbing the hands together under running water is the most important part of washing away infectious organisms. Premoistened towelettes or wipes and waterless hand cleaners should not be used as a substitute for washing hands with soap and running water. These guidelines are not limited to giardiasis but include all potentially infectious organisms.

The possible association between gastric acidity and giardiasis has been debated for some time. Evidence indi-

Table 3.7 Guidelines for hand washing

Children

    When they arrive at the child care setting

    Immediately before and after eating

    After using the toilet or having their diapers changed

    Before using water tables

    After playing on the playground

    After handling pets, pet cages, or other pet objects such as toys

    Whenever hands are visibly dirty

    Before going home

Staff, visitors

    When they arrive at the child care setting

    Immediately before handling food, preparing bottles, or feeding children

    After using the toilet, assisting a child in using the toilet, or changing diapers

    After contact with a child's body fluids, including wet or soiled diapers, runny noses, spit, and vomit

    After handling pets, pet cages, or other pet objects such as toys

    Whenever hands are visibly dirty or after cleaning up a child, the room, bathroom items, or toys

    After removing gloves for any purpose (if gloves are worn, hands should be washed immediately after the gloves are removed, even if the hands appear to be clean; use of gloves alone will not prevent contamination of hands or spread of organisms and should not be considered a substitute for handwashing)

    Before giving or applying medication or ointment to a child or self

    Before going home

cates that decreased gastric acid production may predispose people to *Giardia* infection. It is thought that normal gastric acidity acts as a barrier to the establishment of an infection; patients who have had a gastrectomy are prone to *Giardia* infection. Although achlorhydria is associated with blood group A and some evidence suggests that members of this group are more susceptible to giardiasis, subsequent evidence has not confirmed this information. Since reduction in gastric acid also occurs as a result of malnutrition, these factors may be linked and, as a group, increase the susceptibility to infection with this organism. This link between malnutrition and giardiasis may also be explained by the impairment of the host immune system.

The issue of whether breast milk modifies *Giardia* infection has also been discussed in terms of different interpretations of the results. A lower incidence of giardiasis in children younger than 6 months may be related to an association with breast-feeding and some protection against infection via secretory IgA; however, the lower incidence may also be related to decreased exposure to *Giardia* organisms in breast-fed infants.

Giardiasis is one of the more common causes of traveler's diarrhea and has been recorded in all parts of the world. It has also been speculated that visitors in areas where *Giardia* infections are endemic are more likely to present with symptoms than are individuals who live in the area; this difference is most probably due to the development of immunity of residents from prior and possibly continued exposure to the organism.

There have been a number of outbreaks attributed to either resort or municipal water supplies in Oregon, Colorado, Utah, Washington, New Hampshire, and New York (25, 41, 55, 60, 70). High rates of infection were also reported for hikers and campers who drank stream water. Because some of these areas were remote from human habitation, infected wild animals, especially beaver, are suspected (18).

During the past few years, this infection has received much publicity. With increased travel, there has been a definite increase in symptomatic giardiasis within the United States (2). Various surveys show infection rates of 2 to 15% in various parts of the world.

Because of the potential for wild-animal and possibly domestic-animal reservoir hosts, measures in addition to personal hygiene and improved sanitary measures have to be considered (17). Iodine has been recommended as an effective disinfectant for drinking water, but it must be used as directed (71, 72). Filtration systems have also been recommended, although they have certain drawbacks, such as clogging.

Water Testing. Over the past few years, concern over safe drinking water has increased, and these issues have

Table 3.8 Questions for consideration when testing well water

---

1. Are other members of your family or users of your well water ill?
   (If yes, the well water may be a source of infection.)

2. Is your well located at the bottom of a hill or is it considered shallow?
   (If either of these is true, runoff from rainwater or floodwater may be draining into the well, causing contamination.)

3. Is your well in a rural area where animals graze?
   (Well water can be contaminated with animal waste seepage contaminating the groundwater. This can occur if the well has cracked casings, is poorly constructed, or is too shallow.)

---

led to the development of a new industry related to water treatment and testing. Untreated surface water is more likely than groundwater to contain organisms, because of the possibility of direct contamination with animal feces, treated and untreated human sewage, or fecal runoff from adjacent land after heavy rain or snow melt. Wells that have been poorly located, constructed, or damaged have also been implicated in outbreaks. Specific points to consider when testing well water are listed in Table 3.8; because specific testing for *Giardia* is difficult and expensive, the well water can also be tested for fecal coliforms as an indicator of fecal contamination. In some areas of the world, the percent positivity for water supplies can be very high for both *Cryptosporidium* and *Giardia* in both raw water and treated water samples (40).

The conventional filtration process used by most surface water treatment plants usually includes several steps: coagulation-flocculation, sedimentation, and filtration. During the first step, a chemical coagulant is added to the water to cause small suspended particles to stick together to form larger particles (flocculation), which either settle to the bottom (sedimentation) or are removed by filtration, primarily through fine sand. The coagulation-flocculation step is critical for successful removal of chlorine-resistant protozoa (1 to 20 μm in size), which could slip through the 50- to 70-μm spaces between grains of sand in a water treatment filter. Unfortunately, there are a number of variables that influence adequate filtration (amount of coagulant added, degree of mixing of coagulant with water, and changes in raw water pH, temperature, or turbidity). Conventional water filtration should trap nearly all protozoan parasites, including *Giardia* and *Cryptosporidium*, if appropriate procedures are followed (7).

Environmental sampling methods were originally developed to assist in the investigation of suspected waterborne outbreaks of infection. The sample volumes generally recommended are 100 liters for source water and 1,000 liters or more for finished water. The usual procedure involves filtration through 1-μm polypropylene filter cartridges or fiberglass-resin cartridges. The filters are then eluted in a laboratory with detergent solutions, and recovered particulates are concentrated by centrifugation. A flotation purification step using density gradients is then used to separate the organisms from inorganic materials and debris. Currently, the most commonly used method for examining purified material for protozoa is an antibody-based immunofluorescence assay. This approach has several limitations, including the lack of information on the infectivity or viability of cysts or oocysts; the lack of indication of the host species or origin of the organisms; the chance of false identification of algal or other protozoal species; poor recovery efficiency; the labor-intensive procedures; and the need for skilled microscopists. It also appears that different organism numbers from the same water source can be attributed to method artifact rather than true differences in counts (50). A newer method for the determination of *Giardia* cyst viability in environmental and fecal samples involves fluorogenic dye staining and differential interference contrast microscopy (62). This method has proved significantly better than the routine counting methods being used.

Newer methods that have been tried include a portable differential continuous-flow centrifuge for concentration of *Cryptosporidium* oocysts and *Giardia* cysts from large volumes of water (72). The effect of electroporation (very short pulses of high-voltage electricity) alone and in combination with various chemicals has been examined. While electroporation had a minor effect on oocyst and cyst survival, the combination of electrical and chemical treatment produced superior inactivation, particularly when chlorine, hydrogen peroxide, or potassium permanganate was used (29). Additional kinetic and optimum treatment combinations continue to be developed. The use of immunomagnetic beads to separate *Giardia* cysts from complex matrixes of environmental surface waters followed by DNA release and PCR amplification of the target giardin gene has improved the reliability of detection of *Giardia* with increased sensitivity (45).

Studies of the epidemic and endemic seroprevalence of antibodies to *Cryptosporidium* and *Giardia* in residents of three communities with different drinking water supplies were undertaken to review possible differences or similarities (34). The findings were consistent with a lower risk of exposure from drinking water obtained from deep well sources and increased seropositivity associated with an outbreak situation. Although intriguing associations were found between seroprevalence, outbreak-related serologic data, and patterns of parasite contamination of drinking water, further studies are required to validate this approach to risk assessment of waterborne parasitic infections at the community level.

Phagocytosis of *Giardia* cysts by hemocytes of the Asian freshwater clam, *Corbicula fluminea*, has been documented, and the mean number of cysts ingested per hemocyte increased significantly over time. These findings may result in this biological system being used to serve as an indication of contamination of wastewaters and agricultural drainage with *Giardia* cysts (27). Also, it is well known that rotifers thrive in environments such as wastewater and domestic sewage; they have been found to ingest *Giardia* cysts in an artificial setting. If this finding is consistent with results from natural environmental studies, perhaps the information will further explain the environmental dispersion of the cysts and suggest how this dispersion could be curtailed (63).

Characterization of *Giardia* Isolates.   Isoenzyme studies, designed primarily to assist in organism identification and classification, have also provided additional information regarding pathogenicity, implication in waterborne outbreaks, and human disease. In one study in which isoenzyme patterns of 32 *Giardia* isolates obtained from both humans and animals were examined, there was no obvious correlation between clinical symptoms and isoenzyme patterns. Isolates from asymptomatic individuals were found in the same zymodemes (isoenzyme groups) as were isolates from symptomatic hosts. This study also confirmed previous observations regarding genetic heterogeneity and demonstrated significant differences between isolates from within a single region and other widely separated geographic locations (1). Australian data on two *Giardia* demes indicated that both appeared to be pathogenic and were derived from children with similar chronic symptoms (64).

Various studies have published data suggesting that zoonotic transmission between humans and animals is, at times, more likely or less likely to occur (65). One study of canine and human isolates indicated that zoonotic transmission of *Giardia* between humans and dogs does not occur frequently in these communities (32). Data also suggest that *Giardia* isolates of human and rabbit origins do not have strict host specificity and that cross-transmission may occur (1).

Dientamoeba fragilis

Dientamoeba fragilis was first seen in 1909 but was not described until 1918. In 1940, Dobell recognized the close morphologic similarities between *D. fragilis* and *Histomonas meleagridis*, the ameboflagellate parasite of turkeys. A key scientific advance was made in 1934 when Tyzzer reported that *H. meleagridis* is transmitted in the eggs of *Heterakis*, the cecal worm of chickens and turkeys, a fact that has relevance to the life cycle of *D. fragilis* (6). On the basis of electron microscopy studies, *D. fragilis* has been reclassified as an ameboflagellate rather than an ameba and is closely related to *Histomonas* and *Trichomonas* spp. (35). It has a cosmopolitan distribution, and past surveys demonstrate incidence rates of 1.4 to 19%. Much higher incidence figures have been reported for mental institution inmates, missionaries, and Native Americans in Arizona. *D. fragilis* tends to be common in some pediatric populations, and incidence figures are higher for patients younger than 20 years in some studies. Currently, it is in the phylum Parabasala, class Trichomonadea, family Trichomonadidae, and possibly the subfamily Trichrichomonidinae. This classification will probably undergo change in the near future.

Life Cycle and Morphology

The life cycle and mode of transmission of *D. fragilis* are not known, although transmission via helminth eggs such as those of *Ascaris* and *Enterobius* spp. has been postulated (35) (Figure 3.5). The cyst stage has not been confirmed to date (Tables 3.3 and 3.4).

Figure 3.5   Life cycle of *Dientamoeba fragilis*.

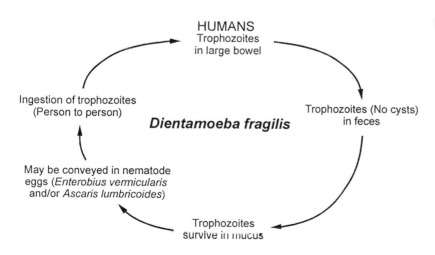

HUMANS
Trophozoites
in large bowel

Ingestion of trophozoites
(Person to person)

Dientamoeba fragilis

Trophozoites (No cysts)
in feces

May be conveyed in nematode
eggs (*Enterobius vermicularis*
and/or *Ascaris lumbricoides*)

Trophozoites
survive in mucus

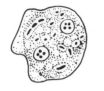

Figure 3.6 Trophozoites of *Dientamoeba fragilis*. (Illustration by Sharon Belkin.)

The trophozoite is characterized as having one (20 to 40%) or two (60 to 80%) nuclei. The nuclear chromatin is usually fragmented into three to five granules, and there is normally no peripheral chromatin on the nuclear membrane (Figures 3.6 and 3.7). In some organisms, the nuclear chromatin tends to mimic that of *Endolimax nana*, *Entamoeba hartmanni*, or even *Chilomastix mesnili*, particularly if the organisms are overstained. The cytoplasm is usually vacuolated and may contain ingested debris as well as some large, uniform granules. The cytoplasm can also appear uniform and clean with few inclusions. There can also be considerable size and shape variation among organisms, even on a single smear.

Clinical Disease

Although its pathogenic status is still not well defined, *D. fragilis* has been associated with a wide range of symptoms (15, 35, 39, 51, 61, 66, 67). The significance of two genetically distinct forms of *D. fragilis* may ultimately serve to clarify the issues of virulence and clinical perceptions regarding pathogenicity (35). Case reports of children infected with *D. fragilis* reveal a number of

Figure 3.7 (A, B) Trophozoites of *Dientamoeba fragilis*. (A) The trophozoite contains a single nucleus fragmented into four parts; (B) the trophozoite contains two nuclei, both of which are also fragmented.

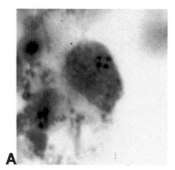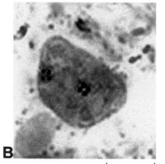

10 μm

symptoms, including intermittent diarrhea, abdominal pain, nausea, anorexia, malaise, fatigue, poor weight gain, and unexplained eosinophilia. The most common symptoms in patients infected with this parasite appear to be intermittent diarrhea and fatigue (39). In some patients, both the organism and the symptoms persist or reappear until appropriate treatment is initiated.

Eleven pediatric patients, seven of whom had peripheral eosinophilia and a history of recent travel, were diagnosed with *D. fragilis* infection and reported symptoms of anorexia, intermittent vomiting, abdominal pain, and diarrhea. Based on findings in these patients, including bovine protein allergy and eosinophilic colitis, the authors recommended that *D. fragilis* should be included in the differential diagnosis of chronic diarrhea and eosinophilic colitis. The identification of this pathogen requires clinical awareness of epidemiologic risk factors and presenting complaints, as well as proper laboratory permanent staining procedures essential to proper identification (14).

Diagnosis

Routine Diagnostic Procedures. Clinicians should include infection with *D. fragilis* in their differential diagnosis of patients presenting with abdominal pain, diarrhea, unexplained flatulence, nausea, and vomiting. Diagnosis of *D. fragilis* infections depends on proper collection and processing techniques (a minimum of three fecal specimens) (11). Although the survival time for this parasite has been reported as 24 to 48 h, the survival time in terms of morphology is limited, and stool specimens must be examined immediately or preserved in a suitable fixative soon after defecation (35). It is particularly important that permanently stained smears of stool material be examined with an oil immersion lens (100×). These organisms have been recovered in formed stool; therefore, a permanent stained smear must be prepared for every stool sample submitted for a parasite examination. Organisms seen in direct wet mounts may appear as refractile, round forms; the nuclear structure cannot be seen without examination of the permanent stained smear.

Antigen Detection. Although fecal immunoassays for antigen detection are not yet available commercially, they have been developed using the immunofluorescence format (35). Studies using the enzyme immunoassay method are also under way. It is anticipated that at publication time these assays may be available commercially, since preliminary results look very promising. The potential for detection of DNA from feces is also being investigated; certainly the development of rapid, specific, and sensitive tests would be extremely helpful within the diagnostic laboratory setting (35).

Antibody Detection. Using an indirect immunofluorescence assay, Chan et al. (9, 35) found that serum samples from patients with confirmed *D. fragilis* infections had positive titers of 80, and all 12 matched controls had positive titers ranging from 20 to 160. The specificity of this assay was reinforced by immunoblotting 20 representative serum samples against *D. fragilis*; in all 17 indirect immunofluorescence-positive serum samples, a 39-kDa protein band of *D. fragilis* was identified.

KEY POINTS—LABORATORY DIAGNOSIS

Dientamoeba fragilis

1. A minimum of three specimens (stool) should be submitted for the diagnosis of *Dientamoeba* infections.

2. Since there is no known cyst stage, these organisms will not be seen on a wet preparation. Consequently, it is mandatory that a permanent stained smear be included in the ova and parasite examination. Trophozoites with either one or two nuclei can be found in the same specimen; there may also be tremendous size variation among the organisms seen in a single smear.

3. The trophozoite forms have been recovered from formed stool, hence the need to perform the permanent stained smear on specimens other than liquid or soft stools.

4. Organisms with a single nucleus can easily be confused with *Endolimax nana* or *Entamoeba hartmanni*, both of which are considered nonpathogens.

Treatment

Clinical improvement has been observed in adults receiving tetracycline; symptomatic relief has been observed in children receiving either diiodohydroxyquin, metronidazole, or tetracycline. Current recommendations include iodoquinol, paromomycin, or tetracycline. Since symptomatic relief has been observed to follow appropriate therapy, *D. fragilis* is probably pathogenic in infected individuals who are symptomatic. Although limited studies have been undertaken regarding the efficacy of various therapies, information continues to support the fact that the elimination of this organism from symptomatic patients leads to clinical improvement. Additional information can be found in chapter 25.

Epidemiology and Prevention

As reported for many of the intestinal protozoa, *D. fragilis* is worldwide in distribution. It is suspected that the true incidence of this infection is considerably higher than reported, particularly since many laboratories do not yet emphasize diagnostic methods such as the permanent stained smear that would confirm the diagnosis (67). Infection with *D. fragilis* is particularly common in Canada. This parasite has been associated with diarrhea and abdominal pain in some patients. In looking at seroprevalence data for a large group of randomly selected patients aged from 6 months to 19 years, 172 (91%) of 189 gave positive results at titers of 1:10 or higher. The specificity of the indirect-immunofluorescence assay was reinforced by immunoblotting of 20 representative serum samples against *D. fragilis*. Findings over a 5-year period indicate that *D. fragilis* was the most common protozoan, followed closely by *G. lamblia* and more distantly by *C. parvum*. Previous data reported 85.6% for *G. lamblia* and 86% for *C. parvum* (9).

Since fecal-oral transmission has not been documented, it is difficult to speculate on preventive measures. However, if transmission does occur from the ingestion of certain helminth eggs, the use of hygiene and sanitary measures to prevent contamination with fecal material would be appropriate. There is speculation that *D. fragilis* may be infrequently recovered and identified; low incidence or absence from survey studies may be due to poor laboratory techniques and a general lack of knowledge about the organism (11, 35).

Pentatrichomonas hominis (*Trichomonas hominis*)

It is generally thought that *Trichomonas hominis* was first observed by Davaine in 1854 and described by him with the species name *hominis*. Although other protozoologists doubt that he actually saw this organism, the species name has been accepted. The genus name has been changed to *Pentatrichomonas* based on the five anterior flagella and a granular parabasal body. Although most readers are familiar with the generic term *Trichomonas*, this name will be changed to *Pentatrichomonas* for this edition of the book.

P. hominis is probably the most commonly identified flagellate other than *G. lamblia* and *D. fragilis*. It has been recovered from all parts of the world, in both warm and temperate climates. It is considered to be nonpathogenic, although it is often recovered from diarrheic stools (Tables 3.3 and 3.4). Statistics generated over the years include college students (0.16%); rural Georgia (0.3%); New Orleans clinic population in 1935 (1.2%); Peking, China, clinic patients (3.5%); and a Cali, Colombia, survey (1956 to 1961) (2.0 to 13.5%) (6).

Life Cycle and Morphology

There is no known cyst stage (Figure 3.8). The trophozoites live in the cecal area of the large intestine and feed on bacteria. The organism is not considered to be invasive. The trophozoite measures 5 to 15 µm in length and 7 to 10 µm in width. It has a pyriform shape and has both an axostyle and undulating membrane, which help in the

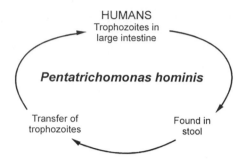

Figure 3.8 Life cycle of *Pentatrichomonas hominis.*

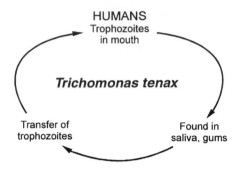

Figure 3.9 Life cycle of *Trichomonas tenax.*

identification (Figure 3.3). The undulating membrane extends the entire length of the body, in contrast to that seen in *Trichomonas vaginalis.*

Diagnosis

In freshly passed stool specimens, particularly in unformed stool, the motility may be visible. In a wet preparation, one specifically looks for the flagellar movement from the undulating membrane and the presence of the axostyle. These small flagellates are difficult to stain and may even be missed on a permanent stained smear, especially if the stain is pale. In more formed stools, the trophozoites may round up (not encysted) and may not exhibit the characteristic motility seen in more active forms. When preserved fecal specimens are received in the laboratory, the permanent stained smear is the most sensitive method for finding and identifying *P. hominis.*

Epidemiology and Prevention

Since there is no known cyst stage, transmission probably occurs in the trophic form. If ingested in a protecting substance such as milk, these organisms can apparently survive passage through the stomach and small intestine in patients with achlorhydria. *P. hominis* cannot be transplanted into the vagina, the natural habitat of *T. vaginalis.* The incidence of this organism is relatively low, but it tends to be recovered more often than *Enteromonas hominis* or *Retortamonas intestinalis.* The infection is diagnosed more often in warm climates and in children rather than adults. Because of the fecal-oral transmission route, preventive measures should emphasize improved hygienic and sanitary conditions.

Trichomonas tenax

Trichomonas tenax was first recovered from the mouth, specifically in tartar from the teeth. It is worldwide in distribution. There is no known cyst stage (Tables 3.3 and 3.4; Figure 3.9). The trophozoite has a pyriform shape and is

smaller and more slender than that of *P. hominis.* The undulating membrane extends for the length of the body, like that seen in *P. hominis.* There is also the typical axostyle and a single nucleus. *T. tenax* cannot survive passage through the stomach and cannot be established in the vagina. Although the exact mode of transmission is not known, it is assumed to be by direct contact or use of contaminated dishes and glasses. Prevalence rates vary from 0 to 25%, depending on opportunities for exposure and on oral hygiene. Although *T. tenax* is considered a harmless commensal in the mouth, there are reports of respiratory infections and thoracic abscesses, particularly in patients with underlying cancers or other lung diseases. The majority of these cases were reported from western Europe. Diagnosis is based on recovery of the organisms from the teeth, gums, or tonsillar crypts, and no therapy is indicated. Better oral hygiene will rapidly eliminate the infection.

Chilomastix mesnili

Chilomastix mesnili tends to have a cosmopolitan distribution, although it is found more frequently in warm climates. It has both trophozoite and cyst stages and is somewhat more easily identified than are some of the smaller flagellates such as *E. hominis* and *R. intestinalis* (Tables 3.3 and 3.4; Figure 3.10). The trophozoite is described as being

Figure 3.10 Life cycle of *Chilomastix mesnili, Enteromonas hominis,* and *Retortamonas intestinalis.*

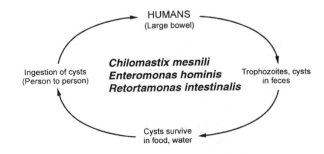

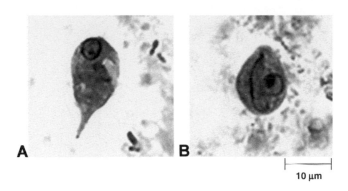

Figure 3.11   (A) Trophozoite of *Chilomastix mesnili* (note the feeding groove; flagella are not visible); (B) cyst of *C. mesnili* (note the "lemon or pear" shape and the curved fibril, the shepherd's crook).

pear shaped, measuring 6 to 24 µm in length and 4 to 8 µm in width. There is a single nucleus and a distinct oral groove or cytostome close to the nucleus (Figures 3.3 and 3.11). Flagella are difficult to see unless there is obvious motility in a wet preparation. The morphology can be confirmed by using a permanent stained smear, particularly when the cytostome is visible. The cysts are pear or lemon shaped, and they range from 6 to 10 µm in length and 4 to 6 µm in width (Figure 3.11). There is also a single nucleus in the cyst and the typical curved cytostomal fibril, which is called the shepherd's crook. Again, more definitive morphology can be seen on a permanent stain. *C. mesnili* normally lives in the cecal region of the large intestine, where the organisms feed on bacteria and debris. It is considered to be a nonpathogen, and no treatment is recommended. Since transmission is through ingestion of infective cysts, prevention depends on improved personal hygiene and upgraded sanitary conditions.

Enteromonas hominis

The flagellate *Enteromonas hominis* is considered to be nonpathogenic and has been reported from both warm and temperate climates (Tables 3.3 and 3.4; Figure 3.10). The trophozoites are somewhat pear shaped, measuring approximately 4 to 10 µm by 3 to 6 µm (Figure 3.3). There is no cytostome, and the flagella are rarely visible unless motile organisms are seen. The cyst measures approximately 6 to 10 µm by 4 to 6 µm and tends to be oval. There are two nuclei, and the cyst may mimic a two-nucleated *Endolimax nana* cyst. However, *E. nana* cysts containing two nuclei are quite rare in most clinical specimens. Also, *E. hominis* is not often reported, perhaps in part because of the small size of the organism and the difficulties in identification. Even on a permanent stained smear, this organism is often difficult to identify accurately.

Infection is considered to be caused by ingestion of the infective cysts. No symptoms have been attributed to this infection, and therapy is not indicated.

Retortamonas intestinalis

Retortamonas intestinalis has been recovered from both warm and temperate areas of the world (Figure 3.10). It has also been found in certain groups such as mental hospital patients. Like *E. hominis*, it is not commonly found in patient stools (Tables 3.3 and 3.4). *R. intestinalis* is also a small flagellate with trophozoites that are elongate pyriform or ovoidal (Figure 3.3). They measure 4 to 9 µm in length by 3 to 4 µm in width. This flagellate does have a cytostome, which may be difficult to see, even in a permanent stained smear. The cysts are somewhat pear shaped, measuring 4 to 9 µm in length and 4 to 6 µm in width (Figure 3.3). Both the trophozoite and the cyst have a single nucleus. This organism is rarely found in clinical specimens, possibly in part because of the difficulties in accurate identification of this and other smaller flagellates, even when the permanent stain technique is used. Infection is through ingestion of the cysts, and improved hygienic conditions would certainly prevent spread of the infection. *R. intestinalis* is considered a nonpathogen; thus, no therapy is indicated.

Balantidium coli

Balantidium coli was first isolated by Malmsten in 1857 from the dysenteric stools of two patients and was soon also observed by Leuckart in 1861 and Stein in 1862, who transferred the species to the genus *Balantidium* (prior genera were *Paramoecium*, *Leukophyra*, and *Holophyra*) (6).

B. coli is widely distributed in hogs, particularly in warm and temperate climates, and in monkeys in the tropics. Human infection is found in warmer climates, sporadically in cooler areas, and in institutionalized groups with low levels of personal hygiene. It is rarely recovered in clinical specimens within the United States.

Life Cycle and Morphology

B. coli is the only pathogenic ciliate and is the largest of the protozoa that parasitize humans (Tables 3.9 and 3.10; Figure 3.12). Both the trophozoite and cyst forms are found. The trophozoite is quite large, oval, and covered with short cilia; it measures approximately 50 to 100 µm in length and 40 to 70 µm in width (Figures 3.13 and 3.14). The organism can easily be seen in a wet preparation on lower power. The anterior end is somewhat pointed and has a cytostome present; in contrast, the posterior end is broadly rounded. The cytoplasm contains many vacuoles with ingested bacteria and debris. There are two nuclei within the trophozoite, a very

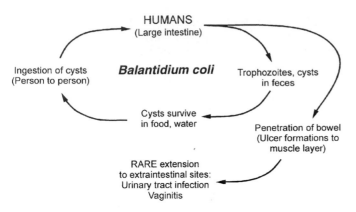

Figure 3.12   Life cycle of *Balantidium coli*.

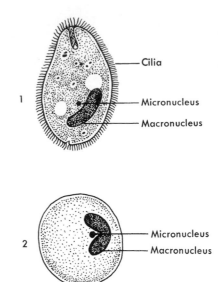

Figure 3.13   (1) Trophozoite of *Balantidium coli*; (2) cyst of *B. coli*. (Illustration by Sharon Belkin.)

large bean-shaped macronucleus and a smaller round micronucleus. The organisms normally live in the large intestine.

The cyst is formed as the trophozoite moves down the intestine. Nuclear division does not occur in the cyst; therefore, only two nuclei are present, the macronucleus and the micronucleus (Figures 3.13 and 3.14). The cysts measure from 50 to 70 μm in diameter.

Clinical Disease

Some individuals with *B. coli* infections are totally asymptomatic, whereas others have symptoms of severe dysentery similar to those seen in patients with amebiasis. Symptoms usually include diarrhea or dysentery, tenesmus, nausea, vomiting, anorexia, and headache. Insomnia, muscular weakness, and weight loss have also been reported. The diarrhea may persist for weeks to months prior to the development of dysentery. There may be tremendous fluid loss, with a type of diarrhea similar to that seen in cholera or in some coccidial or microsporidial infections.

B. coli has the potential to invade tissue. On contact with the mucosa, *B. coli* may penetrate the mucosa with cellular infiltration in the area of the developing ulcer

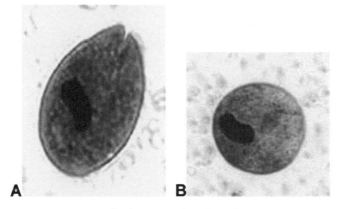

Figure 3.14   (A) Trophozoite of *Balantidium coli* (note the large bean-shaped macronucleus and evidence of cilia around the organism); (B) cyst of *B. coli* (note the large bean-shaped macronucleus).

Table 3.9 Intestinal protozoa: ciliate

Balantidium coli	Shape and size	Motility	No. of nuclei	Other features
Trophozoite	Ovoid with tapering anterior end; 50–100 μm in length; 40–70 μm in width (usual range, 40–50 μm)	Ciliates: rotary, boring; may be rapid	1 large kidney-shaped macronucleus; 1 small, round micronucleus, which is difficult to see even in the stained smear; macronucleus may be visible in unstained preparation	Body covered with cilia, which tend to be longer near cytostome; cytoplasm may be vacuolated
Cyst	Spherical or oval; 50–70 μm (usual range, 50–55 μm)		1 large macronucleus visible in unstained preparation; micronucleus difficult to see	Macronucleus and contractile vacuole are visible in young cysts; in older cysts, internal structure appears granular; cilia difficult to see within the cyst wall

Table 3.10 Intestinal flagellates and ciliates: recommended diagnostic procedures

Organism	Specimen	Diagnostic procedure
Giardia lamblia	Stool	Complete ova and parasite exam, fluorescent antibody, enzyme and immunochromatographic immunoassays
	Duodenal contents	Wet preparation and permanent stains
	Entero-Test	Wet preparation and permanent stains
	Aspirate	Wet preparation and permanent stains
	Biopsy	Routine histology
Dientamoeba fragilis	Stool	Complete ova and parasite exam
Enteromonas hominis	Stool	Complete ova and parasite exam
Retortamonas intestinalis	Stool	Complete ova and parasite exam
Chilomastix mesnili	Stool	Complete ova and parasite exam
Pentatrichomonas hominis (*Trichomonas hominis*)	Stool	Complete ova and parasite exam
Trichomonas tenax	Gingival scrapings	Wet preparation and permanent stains
	Pyorrheal material	Wet preparation and permanent stains
Balantidium coli	Stool	Complete ova and parasite exam
	Biopsy	Routine histology

(Figure 3.15). Some of the abscess formations may extend to the muscular layer. The ulcers may vary in shape, and the ulcer bed may be full of pus and necrotic debris. Although the number of cases is small, extraintestinal disease has been reported (peritonitis, urinary tract infection, and inflammatory vaginitis).

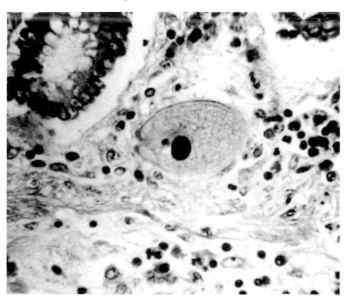

Figure 3.15 Trophozoite of *Balantidium coli* in tissue. Note the cytostome (oral groove) and a portion of the macronucleus. The cilia are not clearly visible.

Diagnosis

Routine stool examinations, particularly wet-preparation examinations of fresh and concentrated material, will demonstrate the organisms. Organism recognition and identification on a permanent stained smear may be very difficult. These protozoa are so large that they tend to stain very darkly, thus obscuring any internal morphology. *B. coli* organisms may even be confused with helminth eggs because of their size, particularly when the cilia are not visible. The recovery of *B. coli* from specimens within the United States is rare. Nonetheless, laboratories are expected to be able to identify these organisms and are called upon to do so with proficiency-testing specimens.

KEY POINTS—LABORATORY DIAGNOSIS

Balantidium coli

1. Since the organisms are so large, they can frequently be seen under low power (100×), particularly in a concentration sediment wet mount.
2. If wet mounts are examined using high dry power (400×), the organisms may be confused with vegetable cells (both cells and parasites are larger than other intestinal protozoa). The cilia tend to be short and can be missed on microscopic examination.
3. Although this infection is very rare within the United States, all laboratories may receive positive specimens for proficiency testing.
4. These organisms do not stain well (too large and thick) on the permanent stained smear and

can be confused with fecal debris (including helminth eggs), hence the need to make the diagnosis by using wet smears (from direct mounts or concentrate sediment).

Treatment

For treatment of *B. coli*, tetracycline is the drug of choice, although it is considered investigational for this infection. Iodoquinol or metronidazole may be used as alternatives. Additional information can be found in chapter 25.

Epidemiology and Prevention

Although *B. coli* has been found in many different simian hosts, domestic hogs probably serve as the most important reservoir host for human infection. In areas where pigs are the main domestic animal, the incidence of human infection can be quite high. Particularly susceptible to infection are persons working as pig farmers or in slaughterhouses (28% infection in New Guinea). Human infection is fairly rare in temperate areas, although once the infection is established, it can develop into an epidemic, particularly in areas with poor environmental sanitation and personal hygiene. This situation has been seen in mental hospitals in the United States. Preventive measures involve increased attention to personal hygiene and sanitation measures, since the mode of transmission is ingestion of infective cysts through contaminated food or water.

References

1. Adam, R. D. 2001. Biology of *Giardia lamblia*. *Clin. Microbiol. Rev.* **14:**447–475.
2. Addis, D. G., J. P. Davis, J. M. Roberts, and E. E. Mast. 1992. Epidemiology of giardiasis in Wisconsin: increasing incidence of reported cases and unexplained seasonal trends. *Am. J. Trop. Med. Hyg.* **47:**13–19.
3. Addis, D. G., H. M. Mathews, J. M. Stewart, S. P. Walquist, R. M. Williams, R. J. Finton, H. C. Spencer, and D. D. Juranek. 1991. Evaluation of a commercially available enzyme-linked immunosorbent assay for *Giardia lamblia* antigen in stool. *J. Clin. Microbiol.* **29:**1137–1142.
4. Aldeen, W. E., K. Carroll, A. Robison, M. Morrison, and D. Hale. 1998. Comparison of nine commercially available enzyme-linked immunosorbent assays for detection of *Giardia lamblia* in fecal specimens. *J. Clin. Microbiol.* **36:**1338–1340.
5. Ali, S. A., and D. R. Hill. 2003. *Giardia intestinalis*. *Curr. Opin. Infect. Dis.* **16:**453–460.
6. Beaver, P. C., R. C. Jung, and E. W. Cupp. 1984. *Clinical Parasitology*, 9th ed. Lea & Febiger, Philadelphia, Pa.
7. Betancourt, W. Q., and J. B. Rose. 2004. Drinking water treatment processes for removal of *Cryptosporidium* and *Giardia*. *Vet. Parasitol.* **126:**219–234.
8. Carroccio, A., G. Montalto, G. Iacono, S. Ippolito, M. Soresi, and A. Notarbartolo. 1997. Secondary impairment

of pancreatic function as a cause of severe malabsorption in intestinal giardiasis: a case report. *Am. J. Trop. Med. Hyg.* **56:**599–602.
9. Chan, F., N. Stewart, M. Guan, I. Robb, L. Fuite, I. Chan, F. Diazmitoma, J. King, N. MacDonald, and A. MacKenzie. 1996. Prevalence of *Dientamoeba fragilis* antibodies in children and recognition of a 39 kDa immunodominant protein antigen of the organism. *Eur. J. Clin. Microbiol. Infect. Dis.* **15:**950–954.
10. Chen, N., J. A. Upcroft, and P. Upcroft. 1995. A *Giardia duodenalis* gene encoding a protein with multiple repeats of a toxin homologue. *Parasitology* **111:**423–431.
11. Clinical Laboratory Standards Institute. 2005. *Procedures for the Recovery and Identification of Parasites from the Intestinal Tract.* Approved guideline M28-A2. Clinical Laboratory Standards Institute, Wayne, Pa.
12. Collins, J. P., K. F. Keller, and L. Brown. 1978. "Ghost" forms of *Giardia lamblia* cysts initially misdiagnosed as *Isospora*. *Am. J. Trop. Med. Hyg.* **27:**334–335.
13. Corsi, A., C. Nucci, D. Knafelz, D. Bulgarini, L. Dilorio, A. Polito, F. Derisi, F. A. Morini, and F. M. Paone. 1998. Ocular changes associated with *Giardia lamblia* infection in children. *Br. J. Ophthalmol.* **82:**59–62.
14. Craft, J. C. 1982. *Giardia* and giardiasis in children. *Pediatr. Infect. Dis. J.* **1:**196–211.
15. Cuffari, C., L. Oligny, and E. G. Seldman. 1998. *Dientamoeba fragilis* masquerading as allergic colitis. *J. Pediatr. Gastroenterol. Nutr.* **26:**16–20.
16. Das, S., C. D. Schteingart, A. F. Hofmann, D. S. Reiner, S. B. Aley, and F. D. Gillin. 1997. *Giardia lamblia*: evidence for carrier-mediated uptake and release of conjugated bile acids. *Exp. Parasitol.* **87:**133–141.
17. Deng, M. Q., and D. O. Cliver. 1999. Improved immunofluorescence assay for detection of *Giardia* and *Cryptosporidium* from asymptomatic adult corvine animals. *Parasitol. Res.* **85:**733–736.
18. Dykes, A. C., D. D. Juranek, R. A. Lorenz, S. Sinclair, W. Jakubowski, and R. Davies. 1980. Municipal waterborne giardiasis. An epidemiologic investigation. *Ann. Intern. Med.* **93:**165–170.
19. Faubert, G. 2000. Immune response to *Giardia duodenalis*. *Clin. Microbiol. Rev.* **13:**35–54.
20. Fedorko, D. P., E. C. Williams, N. A. Nelson, L. B. Calhoun, and S. S. Yan. 2000. Performance of three enzyme immunoassays and two direct fluorescence assays for detection of *Giardia lamblia* in stool specimens preserved in ECOFIX. *J. Clin. Microbiol.* **38:**2781–2783.
21. Garcia, L. S., and R. Y. Shimizu. 1997. Evaluation of nine immunoassay kits (enzyme immunoassay and direct fluorescence) for detection of *Giardia lamblia* and *Cryptosporidium parvum* in human fecal specimens. *J. Clin. Microbiol.* **35:**1526–1529.
22. Garcia, L. S., R. Y. Shimizu, and C. N. Bernard. 2000. Detection of *Giardia lamblia*, *Entamoeba histolytica/Entamoeba dispar*, and *Cryptosporidium parvum* antigens in human fecal specimens using the Triage Parasite Panel enzyme immunoassay. *J. Clin. Microbiol.* **38:**3337–3340.
23. Garcia, L. S., R. Y. Shimizu, S. Novak, M. Carroll, and F. Chan. 2003. Commercial assay for detection of *Giardia*

lamblia and *Cryptosporidium parvum* antigens in human fecal specimens by rapid solid-phase qualitative immuno-chromatography. *J. Clin. Microbiol.* 41:209–212.

24. Garcia, L. S., A. C. Shum, and D. A. Bruckner. 1992. Evaluation of a new monoclonal antibody combination reagent for direct fluorescence detection of *Giardia* cysts and *Cryptosporidium* oocysts in human fecal specimens. *J. Clin. Microbiol.* 30:3255–3257.

25. Gardner, T. B., and D. R. Hill. 2001. Treatment of giardiasis. *Clin. Microbiol. Rev.* 14:114–128.

26. Gibson, G. R., D. Ramirez, J. Maier, C. Castillo, and S. Das. 1999. *Giardia lamblia*: incorporation of free and conjugated fatty acids into glycerol-based phospholipids. *Exp. Parasitol.* 92:1–11.

27. Graczyk, T. K., M. R. Cranfield, and D. B. Conn. 1997. In vitro phagocytosis of *Giardia duodenalis* cysts by he-mocytes of the Asian freshwater clam *Corbicula fluminea*. *Parasitol. Res.* 83:743–745.

28. Grant, J., S. Mahanty, A. Khadir, J. D. MacLean, E. Kokoskin, B. Yeager, L. Joseph, J. Diaz, E. Gotuzzo, N. Mainville, and B. J. Ward. 2001. Wheat germ supple-ment reduces cyst and trophozoite passage in people with giardiasis. *Am. J. Trop. Med. Hyg.* 65:705–710.

29. Haas, C. N., and D. Aturaliye. 1999. Semi-quantitative characterization of electroporation-assisted disinfection processes for inactivation of *Giardia* and *Cryptosporidium*. *J. Appl. Microbiol.* 86:899–905.

30. Hanson, K. L., and C. P. Cartwright. 2001. Use of an en-zyme immunoassay does not eliminate the need to analyze multiple stool specimens for sensitive detection of *Giardia lamblia*. *J. Clin. Microbiol.* 39:474-477.

31. Hardin, J. A., A. G. Buret, M. E. Olson, M. H. Kimm, and D. G. Gall. 1997. Mast cell hyperplasia and increased macromolecular uptake in an animal model of giardiasis. *J. Parasitol.* 83:908–912.

32. Hopkins, R. M., B. P. Meloni, D. M. Groth, J. D. Wether-all, J. A. Reynoldson, and R. C. A. Thompson. 1997. Ri-bosomal RNA sequencing reveals differences between the genotypes of *Giardia* isolates recovered from humans and dogs living in the same locality. *J. Parasitol.* 83:44–51.

33. Hsu, B. M., C. P. Huang, G. Y. Jiang, and C. L. L. Hsu. 1999. The prevalence of *Giardia* and *Cryptosporidium* in Taiwan water supplies. *J. Toxicol. Environ. Health Part A* 57:149–160.

34. Isaac-Renton, J., J. Blatherwick, W. R. Bowie, M. Fyfe, M. Khan, A. Li, A. King, M. McLean, L. Medd, W. Moorehead, C. S. Ong, and W. Robertson. 1999. Epidemic and endemic seroprevalence of antibodies to *Cryptosporidium* and *Giardia* in residents of three communities with different drinking water supplies. *Am. J. Trop. Med. Hyg.* 60:578–583.

35. Johnson, E. H., J. J. Windsor, and C. G. Clark. 2004. Emerging from obscurity: biological, clinical, and diagnos-tic aspects of *Dientamoeba fragilis*. *Clin. Microbiol. Rev.* 17:553–570.

36. Kang, E. Q., K. Clinch, R. H. Furneaux, J. E. Harvey, P. F. Schofield, and A. M. Gero. 1998. A novel and simple colorimetric method for screening *Giardia intestinalis* and anti-giardial activity in vitro. *Parasitology* 117:229–234.

37. Katanik, M. T., S. K. Schneider, J. E. Rosenblatt, G. S. Hall, and G. W. Procop. 2001. Evaluation of ColorPAC

Giardia/Cryptosporidium Rapid Assay and ProSpecT *Giardia/Cryptosporidium* Microplate Assay for detection of *Giardia* and *Cryptosporidium* in fecal specimens. *J. Clin. Microbiol.* 39:4523–4525.

38. Katelaris, P. H., A. Naeem, and M. J. G. Farthing. 1995. Attachment of *Giardia lamblia* trophozoites to a cultured human intestinal cell line. *Gut* 37:512–518.

39. Kean, B. H., and C. L. Malloch. 1966. The neglected ameba: *Dientamoeba fragilis*: a report of 100 "pure" infections. *Am. J. Dig. Dis.* 11:735–745.

40. Keystone, J. S., S. Karjden, and M. R. Warren. 1978. Person-to person transmission of *Giardia lamblia* in day-care nurseries. *Can. Med. Assoc. J.* 119:242–244.

41. Kirner, J. C. 1978. Waterborne outbreak of giardiasis in Camas, Washington. *J. Am. Water Works Assoc.* 70:35–40.

42. Knaus, W. A. 1974. Reassurance about Russian giardiasis. *N. Engl. J. Med.* 291:156.

43. Lopez, C. E., D. D. Juranek, S. P. Sinclair, and M. A. Schultz. 1978. Giardiasis in American travelers to Madeira Island, Portugal. *Am. J. Trop. Med. Hyg.* 27:1128–1132.

44. Lujan, H. D., M. R. Mowatt, L. G. Byrd, and T. E. Nash. 1996. Cholesterol starvation induces differentiation of the intestinal parasite *Giardia lamblia*. *Proc. Natl. Acad. Sci. USA* 93:7628–7633.

45. Mahbubani, M. H., F. W. Schaefer, D. D. Jones, and A. K. Bej. 1998. Detection of *Giardia* in environmen-tal waters by immuno-PCR amplification methods. *Curr. Microbiol.* 36:107–113.

46. Mank, T. G., J. O. M. Zaat, A. M. Deelder, J. T. M. Vaneijk, and A. M. Polderman. 1997. Sensitivity of microscopy ver-sus enzyme immunoassay in the laboratory diagnosis of giardiasis. *Eur. J. Clin. Microbiol. Infect. Dis.* 16:615–619.

47. Meyer, E. A. 1990. Pathology and pathogenesis of the intestinal mucosal damage in giardiasis, p. 155–173. *In* A. Ferguson, J. Gillon, and G. Munro (ed.), *Human Parasitic Diseases*. Elsevier Biomedical Press, New York, N.Y.

48. Nash, T. E. 2002. Surface antigenic variation in *Giardia lamblia*. 2002. *Mol. Microbiol.* 45:585–590.

49. Oberhuber, G., N. Kastner, and M. Stolte. 1997. Giardiasis: a historic analysis of 657 cases. *Scand. J. Gastroenterol.* 32:48–51.

50. Payment, P., A. Berte, and C. Fleury. 1997. Sources of variation in isolation rate of *Giardia lamblia* cysts and their homogeneous distribution in river water entering a water treatment plant. *Can. J. Microbiol.* 43:687–689.

51. Peek, R., F. R. Reedeker, and T. van Gool. 2004. Direct amplification and genotyping of *Dientamoeba fragilis* from human stool specimens. *J. Clin. Microbiol.* 42:631–635.

52. Rosales-Borjas, D. M., J. Diaz-Rivadeneyra, A. Donaleyva, S. A. Zambrano-Villa, C. Mascaro, A. Osuna, and L. Ortiz-Ortiz. 1998. Secretory immune response to membrane antigens during *Giardia lamblia* infection in humans. *Infect. Immun.* 66:756–759.

53. Sanchez, L. B. 1998. Aldehyde dehydrogenase (CoA-acetylating) and the mechanism of ethanol formation in the amitochondriate protist, *Giardia lamblia*. *Arch. Biochem. Biophys.* 354:57–64.

54. Seppel, T., F. Rose, and R. Schlaghecke. 1996. Chronic intestinal giardiasis with isolated levothyroxine malabsorp-tion as reason for severe hypothyroidism—implications for

localization of thyroid hormone absorption in the gut. *Exp. Clin. Endocrinol. Diabetes* **104:**180–182.

55. Shaw, P. K., R. E. Brodsky, D. D. Lyman, B. T. Wood, C. P. Hibler, G. R. Healy, K. I. E. MacLeod, W. Stahl, and M. G. Schultz. 1977. A community wide outbreak of giardiasis with evidence of transmission by a municipal water supply. *Ann. Intern. Med.* **87:**426–432.

56. Smith, P. D., H. C. Lane, V. J. Gill, J. F. Manischewitz, G. V. Quinnan, A. S. Fauci, and H. Masur. 1988. Intestinal infections in patients with the acquired immunodeficiency syndrome (AIDS). *Ann. Intern. Med.* **108:**328–333.

57. Soliman, M. M., R. Taghi-Kilani, A. F. A. Abou-Shady, S. A. A. El-Mageid, A. A. Handousa, M. M. Hegazi, and M. Belosevic. 1998. Comparison of serum antibody responses to *Giardia lamblia* of symptomatic and asymptomatic patients. *Am. J. Trop. Med. Hyg.* **58:**232–239.

58. Sterling, C. R., R. M. Kutob, M. J. Gizinski, M. Verastequi, and L. Stetzenbach. 1988. *Giardia* detection using monoclonal antibodies recognizing determinants of in vitro derived cysts, p. 219–222. *In* P. Wallis and B. Hammond (ed.), *Advances in* Giardia *Research*. University of Calgary Press, Calgary, Alberta, Canada.

59. Stevens, W. J., and P. A. Vermeire. 1981. *Giardia lamblia* in bronchoalveolar lavage fluid. *Thorax* **36:**875.

60. Stuart, J. M., H. J. Orr, F. G. Warburton, S. Jeyakanth, C. Pugh, I. Morris, J. Sarangi, and G. Nichols. 2003. Risk factors for sporadic giardiasis: a case-control study in southwestern England. *Emerg. Infect. Dis.* **9:**229–233.

61. Swerdlow, M. A., and R. B. Burrows. 1955. *Dientamoeba fragilis*, an intestinal pathogen. *JAMA* **157:**176–178.

62. Thiriat, L., F. Sidaner, and J Schwartzbrod. 1998. Determination of *Giardia* cyst viability in environmental and faecal samples by immunofluorescence, fluorogenic dye staining and differential interference contrast microscopy. *Lett. Appl. Microbiol.* **26:**237–242.

63. Trout, J. M., E. J. Walsh, and R. Fayer. 2002. Rotifers ingest *Giardia* cysts. *J. Parasitol.* **88:**1038–1040.

64. Upcroft, J., and P. Upcroft. 1998. My favorite cell: *Giardia. Bioessays* **20:**256–263.

65. Van Keulen, H., P. Macechko, S. Wade, S. Schaaf, P. Wallis, and S. Erlandsen. 2002. Presence of human *Giardia* in domestic, farm and wild animals, and environmental samples suggests a zoonotic potential for giardiasis. *Vet. Parasitol.* **108:**97–107.

66. Wenrich, D. H. 1937. Studies on *Dientamoeba fragilis* (protozoa). 11. Report of unusual morphology in one case with suggestions as to pathogenicity. *J. Parasitol.* **23:**183–196.

67. Windsor, J. J., and E. H. Johnson. 1999. *Dientamoeba fragilis*: the unflagellated human flagellate. *Br. J. Biomed. Sci.* **56:**293–306.

68. Wolfe, M. S. 1992. Giardiasis. *Clin. Microbiol. Rev.* **5:**93–100.

69. Wright, J. M., L. A. Dunn, P. Upcroft, and J. A. Upcroft. 2003. Efficacy of antigiardial drugs. *Expert Opin. Drug Saf.* **2:**529–541.

70. Wright, R. A., H. C. Spencer, R. E. Brodsy, and T. M. Vernon. 1977. Giardiasis in Colorado: an epidemiologic study. *Am. J. Epidemiol.* **105:**330–336.

71. Zemlyn, S., W. W. Wilson, and P. A. Hillweg. 1981. A caution on iodine water purification. *West. J. Med.* **135:**166–167.

72. Zuckerman, U., R. Armon, S. Tzipori, and D. Gold. 1999. Evaluation of a portable differential continuous flow centrifuge for concentration of *Cryptosporidium* oocysts and *Giardia* cysts from water. *J. Appl. Microbiol.* **86:**955–961.

Intestinal Protozoa (Coccidia and Microsporidia) and Algae

Coccidia

Cryptosporidium spp.

Cyclospora cayetanensis

Isospora (Cystoisospora) belli

Sarcocystis spp.

Microsporidia

Algae (*Prototheca*)

Coccidia

Although most readers understand the main categories related to parasitology classification, the coccidia and microsporidia can be somewhat confusing; a few comments as review may be helpful. The major phyla of protozoa include Sarcomastigophora (includes the amebae and flagellates), the Apicomplexa (malaria parasites, coccidia, and *Babesia*), the Microspora (the microsporidia), and the Ciliophora (*Balantidium coli*). Several protozoan genera are included in the phylum Apicomplexa and are referred to as coccidia. All organisms included in the Apicomplexa are unicellular with an apical complex typically composed of polar rings, rhoptries, micronemes, and usually a conoid; subpellicular microtubules and micropores are common. These structures can be seen in electron microscopy studies and are used to help classify the various coccidia. Genera that develop in the gastrointestinal tract of vertebrates throughout their entire life cycle include *Eimeria*, *Isospora*, *Cyclospora*, and *Cryptosporidium*. Those that are capable of or require extraintestinal development are referred to as cyst-forming coccidia and include *Besnoitia*, *Caryospora*, *Frenkelia*, *Hammondia*, *Neospora*, *Sarcocystis*, and *Toxoplasma*. The coccidian genera that cause disease in humans include *Cryptosporidium*, *Cyclospora*, *Isospora*, *Sarcocystis*, and *Toxoplasma* (see chapter 6 for a discussion of *Toxoplasma*).

Cryptosporidium spp.

The first reported description of *Cryptosporidium parvum* was in 1907 in the gastric crypts of a laboratory mouse (Tyzzer). Subsequently, it has been found in chickens, turkeys, mice, rats, guinea pigs, horses, pigs, calves, sheep, rhesus monkeys, dogs, cats, and humans (Table 4.1) (6, 12, 40). Recent information also supports previous suggestions that cryptosporidiosis is a zoonosis, is not host specific, and is transmitted via the fecal-oral route (40). What was previously called *C. parvum* and was thought to be the primary *Cryptosporidium* species infecting humans is now classified as two separate species, *C. parvum* (mammals, including humans) and *C. hominis* (primarily humans) (6, 32, 41) (Table 4.1). Differentiation of these two species based on oocyst morphology is not possible.

This infection is now well recognized as causing disease in humans, particularly in those who are in some way immunosuppressed or immunodeficient.

Table 4.1 Current species names within the genus *Cryptosporidium*[a]

Species	Main host(s)	Less common host(s)
C. andersoni	Cattle, Bactrian camels	Sheep
C. baileyi	Chicken, turkeys	Cockatiels, quails, ostriches, ducks
C. canis	Dogs	Humans
C. felis	Cats	Humans, cattle
C. galli	Finches, chicken, capercalles, grosbeaks	
C. hominis	Humans, monkeys	Dugongs, sheep
C. meleagridis	Turkeys, humans	Parrots
C. molnari	Fish	
C. muris	Rodents, Bactrian camels	Humans, rock hyrax, mountain goats
C. parvum	Cattle, sheep, goats, humans	Deer, mice, pigs
C. saurophilum	Lizards	Snakes
C. serpentis	Snakes, lizards	
C. wrairi	Guinea pigs	

[a] Adapted from reference 40.

Cryptosporidium has been implicated as one of the more important opportunistic agents seen in patients with AIDS (6, 12, 22, 40). Unfortunately, at the time of this writing, no totally effective therapy for cryptosporidiosis has been identified, despite the testing of over 100 drugs. Consequently, detection of this parasite in immunocompromised hosts, especially those with AIDS, usually carries a poor prognosis. Also, reports of respiratory tract and biliary tree infections confirm that the developmental stages of this organism are not always confined to the gastrointestinal tract.

Currently, the literature contains over 3,000 published papers related to cryptosporidiosis, including information on nosocomial transmission, day care center outbreaks, and a number of waterborne outbreaks. This is remarkable, since fewer than 30 publications were available prior to 1980. In response to this rapidly expanding area of parasitology, there have been definite improvements in diagnostic procedures, particularly those involving the newer immunoassay detection kits (15, 22, 24, 40).

The pathogenesis of cryptosporidial diarrhea, even in patients with AIDS, is not fully understood. However, patients with no intestinal symptoms who are passing formed stool containing few oocysts contrast sharply with those who are symptomatic with watery diarrhea and are passing large numbers of oocysts. These findings imply a relationship between intestinal dysfunction and injury and the number of organisms infecting the intestinal mucosa.

Age and immune status at the time of primary exposure to *Cryptosporidium* do not appear to be primary factors influencing the susceptibility of humans to infection. Symptomatic intestinal and respiratory cryptosporidiosis has been seen in both immunocompetent and immuno-deficient patients of all ages. However, once the primary infection has been established, the immune status of the host plays a very important role in determining the length and severity of the illness. People who are immunocompetent usually develop a short-term, self-limited diarrhea lasting approximately 2 weeks.

In contrast, those who are immunocompromised initially develop the same type of illness; however, it becomes more severe with time and results in a prolonged, life-threatening, cholera-like illness. These severe infections have been seen in patients undergoing immunosuppressive chemotherapy with drugs that affect both T- and B-lymphocyte function, in patients with hypogammaglobulinemia, and in those with AIDS. From many observations of these types of patients, the difference in outcome can probably be explained by the development, in the immunocompetent host, of an immune response sufficient to eradicate the parasite from the infected individual. This explanation is also supported by the fact that individuals whose immunosuppressive therapy has been discontinued will rapidly clear the body of *Cryptosporidium* once their immune system function is restored. Studies also suggest that the overall immune response to cryptosporidiosis is probably an antibody-dependent, cell-mediated, cytotoxic effect of unknown mechanism.

Life Cycle and Morphology

Cryptosporidium differs from other coccidia infecting warm-blooded vertebrates in that its developmental stages do not occur deep within host cells but are confined to an intracellular, extracytoplasmic location. Each stage is within a parasitophorous vacuole of host cell origin;

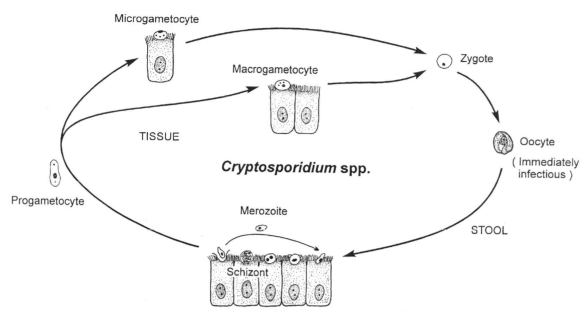

Figure 4.1   Life cycle of *Cryptosporidium* spp. illustrating various stages. (Illustration by Sharon Belkin.)

however, the vacuole containing the organism is located at the microvillous surface of the host cell. Details of the life cycle can be seen in Figure 4.1.

The presence of a thin-walled, autoinfective oocyst may explain why a small inoculum can lead to an overwhelming infection in a susceptible host and why immunosuppressed patients may have persistent, life-threatening infections in the absence of documentation of repeated exposure to oocysts (Figure 4.2). The stages found on the microvillous surface measure 1 μm, and the oocysts recovered in stool specimens measure 4 to 6 μm (Table 4.2).

Each intracellular stage of *Cryptosporidium* resides within a parasitophorous vacuole within the microvillous region of the host cell. Oocysts undergo sporogony while they are in the host cells and are immediately infective when passed in the stool. This is in contrast to the oocysts of *Isospora belli*, which do not sporulate until they are passed from the host and are exposed to oxygen and temperatures below 37°C. Approximately 20% of the oocysts of *Cryptosporidium* do not form the thick, two-layered, environmentally resistant oocyst wall. The four sporozoites within this autoinfective stage are surrounded by a single-unit membrane. After being released from a host cell, this membrane ruptures and the invasive sporozoites penetrate the microvillous region of other cells within the intestine and reinitiate the life cycle. As mentioned above, these thin-walled oocysts that can recycle are thought to be responsible for the development of severe, life-threatening disease in immunocompromised patients, even those who are no longer exposed to the environmentally resistant oocysts.

The prepatent period from the time of ingestion of infective oocysts to completion of the life cycle with excretion of newly developed oocysts in the human is approximately 4 to 22 days. The patent period during excretion of oocysts can range from 1 to 20 days in humans. Previous studies have suggested that persons infected with *Cryptosporidium* develop antibody responses to several antigens. Results now suggest that characteristic antibody responses develop following infection and that persons with preexisting antibodies may be less likely to develop illness, particularly when infected with low oocyst doses (22, 40).

Oocysts recovered in clinical specimens are difficult to see without special staining techniques, such as the modified acid-fast, Kinyoun, and Giemsa methods, or the newer immunoassay methods. The four sporozoites may be seen within the oocyst wall in some of the organisms, although they are not always visible in freshly passed specimens.

Information Based on Electron Microscopy. The oocyst wall has distinct inner and outer layers and is unique in having a suture at one end; the suture dissolves during excystation, opening the wall, through which the sporozoites leave the oocyst. Excystation of sporozoites is enhanced by exposure to reducing conditions followed by exposure to pancreatic enzymes and/or bile salts. Recent data suggest that a novel parasite protein, CP47, may play an important role in attachment of sporozoites to the host cell; this protein exhibits significant binding affinity for the surfaces of both human and animal ileal

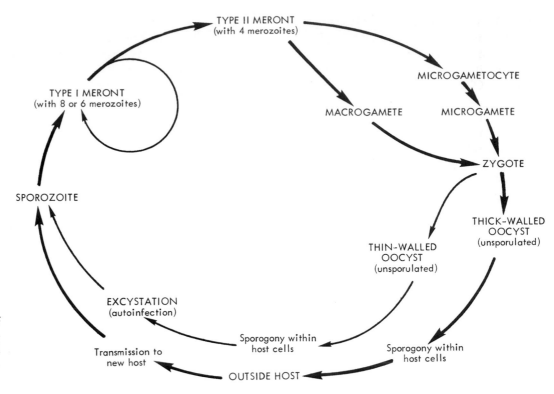

Figure 4.2 Life cycle of *Cryptosporidium* spp. (Adapted from W. L. Current and L. S. Garcia, *Clin. Microbiol. Rev.* 4:325–358, 1991.)

Table 4.2 Coccidia

Species	Shape and size	Other features
Cryptosporidium spp.	Oocyst generally round, 4–6 μm, each mature oocyst containing sporozoites	Oocyst usual diagnostic stage in stool; various other stages in life cycle can be seen in biopsy specimens taken from gastrointestinal tract (brush border of epithelial cells, intestinal tract) and possibly other tissues (respiratory tract, biliary tract)
Cyclospora cayetanensis	Organisms generally round, 8–10 μm; they mimic *Cryptosporidium* spp. (acid fast) but are larger	In wet smears they look like nonrefractile spheres; they will also autofluoresce with epifluorescence; are acid-fast variable from no color to light pink to deep red; those that do not stain may appear wrinkled; in a trichrome-stained stool smear, they will appear as clear, round, somewhat wrinkled objects; they cause diarrhea in both immunocompetent and immunosuppressed patients
Isospora belli	Ellipsoidal oocyst; usual range, 20–30 μm long and 10–19 μm wide; sporocysts rarely seen broken out of oocysts but measure 9–11 μm	Mature oocyst contains 2 sporocysts with 4 sporozoites each; usual diagnostic stage in feces is immature oocyst containing spherical mass of protoplasm (diarrhea, intestinal tract)
Sarcocystis hominis, *S. suihominis*, *S. bovihominis*	Oocyst thin-walled and contains 2 mature sporocysts, each containing 4 sporozoites; thin oocyst wall frequently ruptures; ovoid sporocysts each measure 9–16 μm long and 7.5–12 μm wide	Thin-walled oocyst or ovoid sporocysts occur in stool (intestinal tract)
S. "*lindemanni*"	Shapes and sizes of skeletal and cardiac muscle sarcocysts vary considerably	Sarcocysts contain several hundred to several thousand trophozoites, each of which measures 4–9 μm in width and 12–16 μm in length; the sarcocysts may also be divided into compartments by septa, not seen in *Toxoplasma* cysts (tissue/muscle)

cells. Both sporozoites and merozoites appear similar to those of other coccidia, with organelles such as the pellicle, rhoptries, micronemes, electron-dense granules, nucleus, ribosomes, subpellicular microtubules, and apical rings. Once they are inside the parasitophorous vacuole, changes in the apex of the host cell and the parasite result in an attachment or feeder organelle. The fertilized macrogamont, or zygote, develops into an oocyst with either the thick wall (80%, which are passed in the stool) or the thin wall (20%, which remain internal, causing internal autoinfection). The zygotes that develop into the thin-walled oocysts do not contain the characteristic wall-forming bodies. The thick-walled oocysts are characterized by having a thin, moderately coarse outer layer and a finely granular inner layer. Between these two layers of the oocyst wall is an electron-lucent zone that consists of the two oocyst membranes sandwiched between the outer and inner layers of the oocyst wall. The outer layer of the wall is continuous and of a uniform thickness, while the inner layer contains a suture at one pole.

Clinical Disease

Immunocompetent Individuals. Clinical symptoms include nausea, low-grade fever, abdominal cramps, anorexia, and 5 to 10 watery, frothy bowel movements per day, which may be followed by constipation. Some patients may present with diarrhea as described above, and others may have relatively few symptoms, particularly later in the course of the infection. In patients with the typical watery diarrhea, the stool specimen contains very little fecal material but consists mainly of water and mucus flecks. Often the organisms are entrapped in the mucus, and diagnostic procedures are performed accordingly. Generally, a patient with a normal immune system has a self-limited infection; however, patients who are compromised may have a chronic infection with a wide range of symptoms (asymptomatic to severe).

Occasionally, these patients require fluid replacement and the diarrhea persists for more than 2 weeks. This is particularly true of infants, in whom excessive fluid loss may last for over 3 weeks. In general, when CD4$^+$ cells are present at levels greater than 200/µl, infections are acute and resolve in approximately 2 weeks; however, when the CD4$^+$ cell count drops below 200/µl, the infection may be chronic and may not resolve.

Based on questions that arose from the 1993 Milwaukee outbreak, regarding appropriate management and prognosis of patients with inflammatory bowel disease who are infected with *Cryptosporidium*, it is clear that cryptosporidiosis may present as an acute relapse of inflammatory bowel disease and responds to standard therapy. Antibiotics confer no benefit. It appears that immunosuppressive therapy does not predispose to chronic or severe illness in these patients. Cryptosporidiosis may present with acute findings initially mimicking Crohn's disease. The total cost of outbreak-associated illness during the Milwaukee outbreak was $96.2 million: $31.7 million in medical costs and $64.6 million in productivity losses. The average total costs for persons with mild, moderate, and severe illness were $116, $475, and $7,808, respectively. The authors also comment that these cost-of-illness estimates are conservative (8).

Failure to thrive has also been attributed to chronic cryptosporidiosis in infants. Since diarrheal illness is a major cause of morbidity and mortality in young children living in developing countries, it is likely that cryptosporidiosis plays a major role in the overall health status of these children. Apparently, cryptosporidiosis has a lasting adverse effect on linear growth (height), especially when acquired during infancy and when children are stunted before they become infected. It has also been suggested that *Cryptosporidium* may be implicated in the respiratory disease that often accompanies diarrheal illness in malnourished children. Those with respiratory infections tend to have coughing, wheezing, croup, hoarseness, and shortness of breath.

Histologically, enteric infections are characterized by villous atrophy, enlarged crypts, and infiltration of the lamina propria by inflammatory cells. Respiratory infections have been characterized by infiltration of the lamina propria by inflammatory cells, loss of cilia, hyperplasia, and hypertrophy of epithelial cells. Similar inflammatory cell infiltration has also been reported for infections involving other organs.

Immunocompromised Individuals. *AIDS (intestinal disease).* The duration and severity of diarrheal illness depend on the immune status of the patient. Most severely immunocompromised patients cannot overcome the infection, the illness becomes progressively worse with time, and the sequelae may be a major factor leading to death. The length and severity of illness may also depend on the ability to reverse the immunosuppression. In these patients, *Cryptosporidium* infections are not always confined to the gastrointestinal tract; additional symptoms (respiratory problems, cholecystitis, hepatitis, and pancreatitis) have been associated with extraintestinal infections. While the clinical features of sclerosing cholangitis secondary to opportunistic infections of the biliary tree in patients with AIDS are well known, the mechanisms by which pathogens such as *Cryptosporidium* actually cause disease are unclear.

In reviewing data from patients with gastric cryptosporidiosis, no clear correlation was found between endoscopy and histologic findings; however, a close correlation between the intensity of the infection and the degree of histologic alterations was observed (22). These findings suggest that in patients with AIDS, cryptosporidiosis, and severe immunodepression, upper endoscopy with random gastric biopsies should be performed, even in the absence of endoscopically identified lesions. The diagnosis of gastric cryptosporidiosis relies on histologic findings; pathologists need to be aware that *Cryptosporidium* organisms are found mainly in areas showing reactive hyperplasia.

When reviewing data from AIDS patients with four clinical syndromes, i.e., chronic diarrhea (36%), cholera-like disease (33%), transient diarrhea (15%), and relapsing illness (15%), there is no statistically significant correlation between the histologic intensity of infection and the clinical severity of illness. However, infected patients had a much shorter survival time than did *Cryptosporidium*-negative AIDS patients, and survival was independent of sex, race, or injection drug use (22).

Patients with CD4 counts of 180/µl or greater can usually clear the infection with *Cryptosporidium* over a period of 7 days to 1 month. These data are important in predicting the natural progression of the infection and in designing therapeutic trials. Frequent diagnosis of *Cryptosporidium* infection in patients with neoplasia and diarrhea who are undergoing chemotherapy suggests that for this patient group it is appropriate to routinely submit fecal specimens for examination for oocysts and that the cost of laboratory testing is justified.

AIDS (biliary tract disease). Sclerosing cholangitis is well known as a complication of AIDS. Direct cytopathic effects are noted in infected monolayers (human biliary epithelial cell line) and are associated with widespread programmed cell death (apoptosis) beginning within hours after exposure to the organism. This finding of specific cytopathic invasion of biliary epithelia by *Cryptosporidium* may be relevant to the pathogenesis and possible therapy of secondary sclerosing cholangitis seen in AIDS patients with biliary cryptosporidiosis. It appears that more severely immunocompromised individuals are more likely to exhibit biliary tract disease.

AIDS (pancreatitis). In AIDS patients with pancreatitis, cellular changes are generally not severe, and these patients exhibit hyperplastic squamous metaplasia. The patients usually present with abdominal pain resistant to analgesics, elevated serum amylase levels, and abnormalities on sonography and computed tomography. It is difficult to clarify the impact of cryptosporidiosis-related pancreatic disease, and this disease complication is apparently not linked to significant morbidity.

AIDS (respiratory tract disease). Because many patients with respiratory cryptosporidiosis also have other pathogens present, it is difficult to determine the significance of this complication in AIDS patients. Certainly for patients with respiratory symptoms, it is important to examine sputum, as well as stool specimens to confirm the diagnosis.

Primary immunodeficiency diseases. Many different primary immunodeficiency diseases have been described; in general, they can be categorized as follows: combined immunodeficiencies (which impact both T and B lymphocytes); antibody deficiencies; complement deficiencies, and defects in phagocytes (decreased number and function). In general, the number of these cases reported in relation to cryptosporidiosis is few; however, the most serious immunodeficiency in terms of risk is severe combined immunodeficiency syndrome. These patients are at risk for disseminated disease; often the prognosis is poor.

Malignant disease. Although cryptosporidiosis can be seen in patients with malignant disease, infection with *Cryptosporidium* spp. does not appear to pose a special risk. Exceptions to this general statement seem to involve leukemia and other hematologic malignancies (22). Certainly there is more interest in patients who are candidates for bone marrow transplantation, particularly when assessing possible risks related to cryptosporidiosis.

Solid-organ transplant patients. Cryptosporidiosis has been found in both liver and kidney transplant recipients, as well as following small bowel transplantation (22, 35). However, in most cases the cryptosporidiosis was not unusually severe and did not involve dissemination to extraintestinal sites.

Diagnosis

Previously, most human cases were diagnosed after examination of small or large bowel biopsy material, often under both light and electron microscopy (Figure 4.3). However, because biopsies may miss the infected area of the mucosa, cases have recently been diagnosed by recovering oocysts from fecal material by a flotation method; several other previously reported cases were also diagnosed by using fecal concentrates (Table 4.3). Fresh fecal specimens can be concentrated by using Sheather's sugar solution, and coverslip preparations can be examined by both phase-contrast and bright-field microscopy.

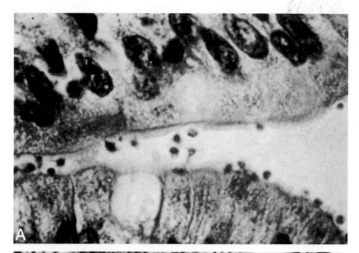

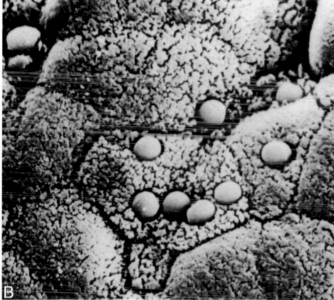

Figure 4.3 *Cryptosporidium* spp. (A) Organisms on the brush border of the mucosal surface. (Courtesy of Wilfred M. Weinstein.) (B) Organisms on the mucosal surface, human tissue (scanning electron micrograph, ×4,000). (Courtesy of Marietta Voge.)

Table 4.3 Intestinal coccidia: recommended diagnostic procedures[a]

Organism	Specimen	Diagnostic procedure[a]
Cryptosporidium spp.	Stool	Modified acid-fast stains, fluorescent stains, immunoassays
	Sputum	Immunoassays Modified acid-fast stains
	Scraping	Modified acid-fast stains or routine histology
	Biopsy specimen	Routine histology
Cyclospora cayetanensis	Stool	Modified acid-fast stains, autofluorescence
	Biopsy specimen	Routine histology
Isospora belli	Stool	Concentration sedimentation and modified acid-fast stains
	Biopsy specimen	Routine histology
Sarcocystis spp.	Stool	Concentration sedimentation and special stains
	Biopsy	Routine histology

[a] When using immunoassay reagents (FA) or modified acid-fast stains for organism detection, these procedures should be performed with centrifuged stool (500 × g for 10 min). When using immunoassay reagents for antigen detection, centrifugation of the fecal specimen is not required. Procedure directions must be checked, and compatibility issues between the kit reagents and stool preservatives must be kept in mind.

The oocysts will float, and one must focus directly under the glass coverslip. Most laboratories no longer use this approach and have switched to modified acid-fast stains performed on concentrated stool sediment or the newer immunoassays (15, 23, 24).

Although examination of flotation material by phase-contrast microscopy has proven to be an excellent procedure for the recovery and identification of *Cryptosporidium* oocysts, many laboratories have neither access to such equipment nor experience with phase-contrast microscopy. Also, since the organism is considered to be infectious to laboratory personnel, fixed specimens can be processed by using several acid-fast stains, many of which are very satisfactory in demonstrating the organisms in stool material (Figures 4.4 and 4.5). The more formed and normal the specimen, the better the chance for artifact material that can be confused with *Cryptosporidium* spp. It is important to remember that the number of oocysts is directly correlated with the consistency of the stool; the more diarrheic the stool, the more oocysts are present.

For the diagnosis of cryptosporidiosis, stool and other body fluid specimens should be submitted as fresh material or in 5 or 10% formalin, sodium acetate-acetic acid-formalin, or one of the newer single-vial stool collection systems (polyvinyl alcohol prepared with zinc sulfate-based Schaudinn's fixative). Fixed specimens are recommended because of potential biohazard considerations. Some research laboratories routinely use potassium dichromate solution (2 to 3% [wt/vol] in water) as a storage medium to preserve oocyst viability; it is not a fixative and is not recommended for the routine clinical laboratory. Either fresh or preserved specimens can be

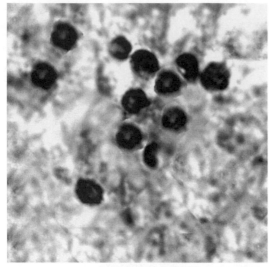

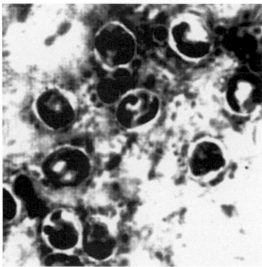

Figure 4.4 *Cryptosporidium* oocysts from stool, stained with modified acid-fast stain. Note that in the bottom image some of the sporozoites are visible within the oocyst wall.

examined using the routine stool formalin-ethyl acetate concentration and one of the modified acid-fast stains or the newer immunoassay kit reagents (Table 4.4). It is recommended that the stool specimen be centrifuged at $500 \times g$ for 10 min prior to use of any of the stains or fluorescent-antibody (FA) immunoassay reagents; centrifugation under these conditions ensures maximum recovery of the oocysts. Centrifugation at lower speeds and for shorter times (often mentioned in some procedures) may not guarantee recovery of the oocysts. Uncentrifuged fresh, frozen, or fixed fecal material can be used for the antigen detection immunoassays; these procedures do not rely on visual identification of the oocysts (FA) but

instead rely on antigen detection. The selection of fresh or frozen stool rather than fixed specimens depends on testing parameters; currently, if the test reagents or kit format includes *Entamoeba histolytica/E. dispar* or *E. histolytica* alone, the test format requires fresh or frozen stool. If the test format includes *Cryptosporidium* and/or *Giardia*, fresh, frozen, OR preserved stools can be used. *However, it is always important to check with the manufacturer to confirm that the particular stool fixative you are currently using will be compatible with the immunoassay test reagents.* It is also important to review sensitivity and specificity data prior to test selection; false-positive results have been reported with some reagent test kits. Multiple stool specimens may have to be examined to diagnose the infection; this is particularly true when dealing with formed stool specimens.

Various screening approaches have been used, and the approach varies from testing all specimens submitted for a routine stool examination to testing limited risk groups or testing only on request. There are numerous pros and cons to each approach; each laboratory will have to select its own approach on the basis of the patient base, client preferences, single versus batch testing, opportunity for in-service education for clients, availability of personnel, level of personnel training and expertise, clinical relevance, and cost. Prevalence data tend to be low unless specific risk groups are screened. A comprehensive approach would be to screen all symptomatic patients; however, even this approach often results in the identification of very few patients and is certainly not particularly cost-effective. A selective approach involving all compromised patients or compromised patients with diarrhea might be more reasonable. However, this assumes that the laboratory will always have access to this type of patient information, an assumption that is not realistic. Many laboratories have elected to provide in-service education to their physician clients and will perform these procedures only on request. Options for test ordering can be seen in Table 4.5.

Although there is evidence for regional and seasonal variability of cryptosporidiosis, these possibilities will probably have little impact on the laboratory's overall approach to diagnostic testing.

Respiratory cryptosporidiosis has been found in AIDS patients. Sputum specimens from immunodeficient patients with undiagnosed respiratory illness should be submitted in 10% formalin and examined for *Cryptosporidium* oocysts by the same techniques as used for stool samples. Infection of the gallbladder and biliary tree should result in oocysts being passed in the stool.

Preliminary results from serologic methods for the diagnosis of cryptosporidiosis appear promising; however,

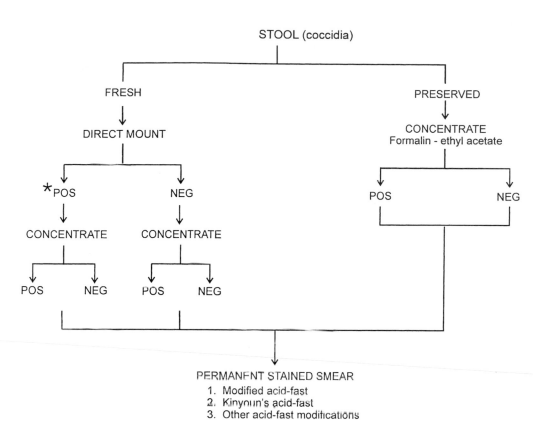

Figure 4.5 Diagnostic work flow diagram leading from submission of stool specimen to final permanent-stained smear for *Cryptosporidium* spp. and *Cyclospora cayetanensis*. *Identification of coccidia may be difficult in a direct wet mount unless there are numerous oocysts present. In many laboratories, the use of fecal immunoassays has replaced acid-fast staining techniques for *Cryptosporidium* spp. The immunoassays can be performed on unconcentrated material (EIA, lateral-flow rapid cartridge) or concentrated material (FA assay).

additional patients must be studied before the actual diagnostic potential of these methods can be evaluated. Serologic tests have been very useful in estimating the prevalence of the infection in different parts of the world. Some of the data reveal very interesting associations between seroprevalence, outbreak-related laboratory serologic data, and patterns of parasite contamination of drinking water (12).

The use of immunoassays has proven to be very helpful in providing a more sensitive method of detecting organisms in stool specimens. A direct FA procedure with excellent specificity and sensitivity has been developed and results in a significantly increased detection rate over conventional staining methods. Enzyme immunoassay (EIA) also provides excellent specificity and sensitivity for laboratories using this approach (15, 24). Some of these reagents, particularly the combination direct-FA product used to identify both *Giardia* cysts and *Cryptosporidium* oocysts, are being widely used in water testing and outbreak situations (3, 12, 18–20, 30, 42).

Flow cytometry methods for the quantitation of *Cryptosporidium* oocysts in stool specimens have been developed as an alternative approach. Studies indicate that the results are approximately 10 times more sensitive than those of conventional immunofluorescence assays.

However, this approach is somewhat impractical for most clinical laboratories.

PCR technology also offers alternatives to conventional diagnosis of *Cryptosporidium* in both clinical and environmental samples. Compared with microscopic examination by conventional acid-fast staining procedures, PCR is more sensitive and easier to interpret but requires more "hands-on" time and expertise, as well as being more expensive. An important advantage of PCR is the ability to directly differentiate between different *Cryptosporidium* genotypes, which is important in outbreak situations. Sensitivity, specificity, ability to genotype, ease of use, and adaptability to batch testing make PCR a useful tool for future diagnostic and molecular epidemiologic studies.

KEY POINTS—LABORATORY DIAGNOSIS

Cryptosporidium **spp.**

1. If the specimen represents the typically watery diarrhea, numerous organisms should be caught up in mucus.

2. The more normal the stool (semiformed, formed), the fewer the organisms and the more artifact material that will be present. It is unlikely that the oocysts will be seen in a wet preparation unless the parasite load is

Table 4.4 Commercially available kits for immunodetection of *Cryptosporidium* spp.

Organism and kit name	Manufacturer and/or distributor	Type of test	Comments (URLs are given only the first time the company appears in the table)
Cryptosporidium spp.			The tests detect *C. hominis*, different *C. parvum* genotypes, and other species depending on the intensity of the infection
ProSpecT Cryptosporidium microplate assay	Remel (www.remel.com)	EIA	Can be used with fresh, frozen, or formalin-preserved stool
XPECT *Cryptosporidium* kit	Remel	Cartridge	
Cryptosporidium antigen ELISA	Medical Chemical (www.med-chem.com)	EIA	Contact manufacturer
Crypto-CELISA	Cellabs (www.cellabs.com)	EIA	Can be used with fresh, frozen, or formalin-preserved stool
Crypto Cel	Cellabs	DFA	
Cryptosporidium test	TechLab (www.techlabinc.com)	EIA	
Cryptosporidium test	Wampole (www.wampolelabs.com)	EIA	
Combination tests: *Cryptosporidium* and *Giardia*			
ProSpecT Giardia/Cryptosporidium microplate assay	Remel	EIA	Can be used with fresh, frozen, or formalin-preserved stool
MERIFLUOR Cryptosporidium/Giardia	Meridian Bioscience (www.meridianbioscience.com)	DFA	
Cryptosporidium/Giardia DFA	Medical Chemical	DFA	Contact manufacturer
Crypto/Giardia-Cel	Cellabs	DFA	Contact manufacturer
ColorPAC *Giardia/Cryptosporidium* rapid assay	Becton-Dickinson (www.bd.com)	Cartridge device, lateral flow	Can be used with fresh, frozen, or formalin-preserved stool
ImmunoCard STAT! *Cryptosporidium/Giardia*	Meridian Bioscience	Cartridge device, lateral flow	Can be used with fresh, frozen, or formalin-preserved stool
SIMPLE-READ *Cryptosporidium*	Medical Chemical	Cartridge device, lateral flow	Can be used with fresh, frozen, or formalin-preserved stool
Xpect Giardia/Cryptosporidium	Remel	Cartridge device, lateral flow	Can be used with fresh, frozen, or formalin-preserved stool
Combination tests: *Cryptosporidium*, *Giardia*, and *Entamoeba*			
Triage Parasite Panel	BIOSITE Diagnostics, Inc. (www.biosite.com)	Cartridge device, lateral flow	Requires fresh or frozen stool; combination test with *Giardia* and *E. histolytica/E. dispar* group; does not differentiate between *E. histolytica* and *E. dispar*

quite heavy; in this case, suspicious objects seen in the wet mount may be suggestive of coccidian oocysts.

3. Several concentrates and at least five or six modified acid-fast smears should be examined before a patient is considered negative, especially if the patient has been receiving experimental medications.

4. The normal permanent stains (trichrome, iron hematoxylin) do not adequately stain *Cryptosporidium* spp., although oocysts are occasionally seen in these preparations, especially those from patients with heavy infections.

5. Although the organisms can be stained with auramine-rhodamine stains, they should be confirmed by using acid-fa st stains or the immunoassay reagents (most sensitive). This is particularly true if the stool contains other cells or a lot of artifact material (more normal stool consistency).

6. When using immunoassay reagents, it is important to remember that the stool specimen fixative and immunoassay reagents must be compatible. Concentrated stool sediment ($500 \times g$ for 10 min) should always be used for staining tests (modified acid-fast, etc.)

Table 4.5 Testing options for the diagnosis of cryptosporidiosis[a]

History	Pros	Cons
No travel outside United States		
Test every stool		
Modified acid-fast stain	Identifies infected patients	May not detect carriers or patients with small number of organisms
Immunoassay	Is more sensitive than the modified acid-fast stains and may identify patients who are carriers with low parasite numbers	High cost; may not be clinically relevant to screen all patients, particularly if there are small numbers in the patient base
Test on order only (after appropriate education of the physician client base)		
Geographic area with little to no reported cases	Is clinically relevant and more cost-effective	Still necessary to set up linkages with public health, water companies, etc., so that appropriate communication can serve as a link and notice of possible early outbreak situation
Geographic area with reported cases and history of outbreak situations	Education of the physician client base targets potential patient risk groups and potential problem situations	Still necessary to continue linkages with public health, water companies, etc., so that appropriate communication can serve as a link and notice of possible early outbreak situation
Travel outside United States		
Order routine ova and parasite examinations (minimum of two; must include the concentration and permanent stained smear)	Facilitates detection and identification of possible other relevant pathogens. If patient has become asymptomatic by the time the ova and parasite examination results are available (negative results), no additional testing is required	Routine ova and parasite exam will not reveal the presence of *Cryptosporidium* (with rare exceptions based on organism numbers and method selection) If the patient remains symptomatic after the ova and parasite examination results (negative results) are available, more sensitive immunoassay procedures can be performed
Order routine ova and parasite examinations and immunoassay for *Cryptosporidium*	This approach may be more clinically relevant in terms of turnaround time issues	May result in excess testing

[a] It is important to remember that laboratories in different geographic regions may approach this testing differently. The important point is to ensure that physicians and laboratory personnel communicate regarding the pros and cons of each approach. Preliminary education, selection of the most clinically relevant approach, client education and communication, and periodic review of testing options and approaches will result in more cost-effective and better patient care.

and FA immunoassay because the tests are detecting the oocysts. However, when immunoassay reagents are used to detect *Cryptosporidium* antigen, the specimen does not have to be concentrated before being tested. Check specific kit directions.

7. Sputum specimens should be submitted in 5 or 10% formalin and processed in the same way as a stool sample.

8. It is important to remember that microsporidia are also present in approximately 30% of severely immunocompromised patients ($\leq$100 CD4$^+$ cells) who have cryptosporidiosis. The diagnostic procedures for the identification of *Cryptosporidium* spp. are not appropriate for the identification of microsporidial spores. Modified trichrome stains and optical brightening agents (calcofluor white) can be used for that purpose (23).

When examining histology preparations, developmental stages in the life cycle of *Cryptosporidium* spp. can be found at all levels of the intestinal tract, with the jejunum being the most heavily infected site. Routine hematoxylin and eosin staining is sufficient to demonstrate these parasites. Under regular light microscopy, the organisms are visible as small (~1 to 3 μm), round structures aligned along the brush border. They are intracellular but extracytoplasmic and are found in parasitophorous vacuoles. Developmental stages are more difficult to identify without using transmission electron microscopy. It is also important to remember that *Cryptosporidium* spp. have been found in other body sites, primarily the lungs, as a disseminated infection in severely immunocompromised patients. Within tissue, confusion with *Cyclospora cayetanensis* is unlikely, because the oocysts of this coccidian are approximately 8 to 10 μm and the developmental stages occur within a vacuole at the luminal end of the enterocyte, rather than at the brush border. Developmental stages of *Isospora belli* also occur within the enterocyte and so should not be confused with *Cryptosporidium* spp.

Newer methods involving PCR have also been used to detect *Cryptosporidium* DNA in fixed, paraffin-embedded tissue (30). This approach could provide a sensitive and specific method for detection of parasite material in paraffin-embedded tissues and could be very valuable in retrospective studies of archival material. PCR has also been used to detect viable *Cryptosporidium* oocysts. Using PCR, confirmation of the presence of *Cryptosporidium* spp. in environmental studies has become more widely used (6, 14, 16, 31, 42).

During the past few years, the issues of water quality and water testing have become more important and controversial. Since the late 1980s, most of the water utilities have either set up in-house water quality testing laboratories or contracted with commercial water laboratories for the recovery and identification of *Cryptosporidium* and *Giardia* species. In a recent study, two methods were evaluated for the recovery of *Cryptosporidium* and *Giardia* species. The American Society for Testing and Materials method involves sampling a minimum of 100 liters of water through a polypropylene yard cartridge filter, extracting the particulates, flotation concentrating the extracted particulates on a Percoll-sucrose gradient with the commercially available *Cryptosporidium/ Giardia* combination FA reagent, and reviewing the slides for the presence of *Cryptosporidium* oocysts and/or *Giardia* cysts. This method tends to be labor-intensive, complex, and lengthy and depends on the skill and experience of the personnel and the quality control measures in use within the laboratory. Recovery rates tend to be low with this approach. The second method uses sampling by membrane filtration, Percoll-Percoll step gradient, and immunofluorescent staining; this method was characterized by higher recovery rates in all three types of waters tested: raw surface water, partially treated water from a flocculation basin, and filtered water. When seeded specimens were used, much higher recovery rates were obtained when the flotation step was eliminated from the protocol. Oocyst and cyst recovery rates decreased in both methods as the turbidity of the water increased. When smaller tubes were used for flotation, the membrane filter method was less time-consuming and cheaper. However, the lack of a confirmatory step might be a problem, particularly if cross-reacting algae are present in water samples. A number of other test options, including PCR, are currently being used (3, 12, 19, 20, 30, 40, 42). Unfortunately, these procedures emphasize organism detection; the issues of organism viability and infectivity have not yet been totally resolved.

Treatment

Cryptosporidiosis tends to be self-limiting in patients who have an intact immune system. For patients who are receiving immunosuppressive agents, one method of therapy would be to discontinue such a regimen. Other approaches with specific therapeutic drugs have been tried, but to date the results are still somewhat controversial (Table 4.6). Tissue culture systems with cell lines such as human ileocecal adenocarcinoma (HCT-8) cells and animal models that support the entire life cycle have been developed, and these may provide excellent opportunities for in vitro drug studies.

Unfortunately, even the studies reported for oral treatment with spiramycin, a macrolide antibiotic related to erythromycin, have been inconclusive. Studies indicate a statistically significant reduction in median oocyst excretion after 2 weeks of therapy with paromomycin. Several studies have reported success for a combination of paromomycin (1 g twice daily) and azithromycin (600 mg daily) for 4 weeks followed by paromomycin monotherapy for an additional 8 weeks. Data indicate that severely immunocompromised AIDS patients with refractory cryptosporidiosis may show a variable response to letrazuril, with a high rate of relapse and rash as a major side effect. Nitazoxanide has also been used and is currently being tested for its efficacy against chronic cryptosporidiosis in human clinical trials. Ongoing studies with immunologic intervention hold promise, but this approach is still being evaluated in clinical trials.

One approach that has had a dramatic impact on cryptosporidiosis in AIDS patients is highly active antiretroviral therapy leading to an increased CD4 count. Resolution of the diarrhea resulting from cryptosporidiosis is apparently related to the enhanced CD4 count rather than to any change in the viral load or any therapeutic impact of the drugs themselves. Thus, it appears that cellular immunity is critical in clearing *Cryptosporidium* infection in these patients.

Epidemiology and Prevention

Information on the epidemiology of *Cryptosporidium* continues to expand, since its potential pathogenicity in humans is now well documented (Tables 4.7 to 4.9; Figures 4.6 to 4.8). The organism is transmitted by oocysts that are usually fully sporulated and infective at the time they are passed in stool. The oocysts are resistant to some of the disinfectants that are routinely used in medical care facilities. Exactly which of the oocyst layers is primarily responsible for the mechanical rigidity and resistance to disruption exhibited by the intact oocyst is not known. However, it has been suggested that the central glycolipid/lipoprotein layer and the inner filamentous layer may provide rigidity and elasticity to the wall.

Table 4.6 Chemotherapeutic agents and supportive therapy for the treatment of human cryptosporidiosis[a]

Chemotherapeutic agent[b]	Comments
Macrolides	
Spiramycin (oral)	Although a number of studies have been reported, divergent results are seen. Placebo results were often comparable to those obtained from patients receiving the drug. Absorption seems to be decreased in the presence of food. Decrease in diarrhea and oocyst excretion was seen in immunocompetent infants randomized to receive spiramycin (100 mg/kg/day) or placebo. A positive clinical response, as well as organism eradication, has also been reported in a group of AIDS patients.
Spiramycin (intravenous)	This approach was taken to circumvent absorption problems. Data revealed a statistically significant drop in oocyst count compared to placebo; however, administration of intravenous spiramycin was associated with paresthesias, taste perversion, nausea, and vomiting. At doses of >75 mg/kg/day there were cases of severe colitis due to intestinal injury.
Azithromycin	This drug has a long half-life (6 to 8 h) and achieves high concentrations in tissue, particularly in the biliary tree and gallbladder. Although decreased oocyst counts have been obtained from patients with high drug levels in serum, marginal improvement has been seen in other patients. A lactose-free form of azithromycin in 300-mg tablets is available through Pfizer's compassionate-use program. There may be a therapeutic effect on biliary tree infection, particularly in patients receiving intravenous azithromycin. Intravenous azithromycin at doses up to 2 g daily for 2 weeks is well tolerated.
Clarithromycin	Little information is available on other macrolides, including clarithromycin, roxithromycin, and dirithromycin; however, studies are in progress.
Benzeneacetonitrile derivatives	
Diclazuril	Interest stems from its efficacy for *Eimeria*; however, reports indicate that doses ranging from 50 to 800 mg were no more efficacious than placebo. This may have been due to poor drug bioavailability.
Letrazuril	Poor absorption of diclazuril led to the development of letrazuril. Although bioavailability was improved, patient clinical improvement was minimal. Data also suggest that this drug may interfere with modified acid-fast staining methods.
Miscellaneous agents	
Paromomycin (Humatin)	Paromomycin is synonymous with aminosidine (marketed outside the United States). It is an oligosaccharide aminoglycoside related to kanamycin; it achieves high concentrations in the colon but is poorly absorbed. Studies indicate that the drug effect is probably static rather than cidal for *Cryptosporidium*. Although decreased bowel movements and oocyst shedding have been reported, a small number of patients developed biliary tract disease and some of these required cholecystectomy. Despite conflicting results, paromomycin currently is used as the first-line drug in treating cryptosporidiosis. Since patients enter treatment trials after paromomycin failure, the question of delay and possible spread to the biliary tree has been raised.
Nitazoxanide	This is a nitrothiazole benzamide with a wide spectrum of activity against coccidia, amebae, nematodes, cestodes, and trematodes. Preliminary data appear promising, with suggested activity against cryptosporidial infection in AIDS patients. The agent appears to be well tolerated. Additional trials are under way.
Difluoromethylornithine (DFMO) (Eflornithine)	Although some success with this agent against *Cryptosporidium* in AIDS patients has been reported, serious toxicity has been documented (bone marrow suppression, gastrointestinal intolerance, and hearing impairment).
Atovaquone (Mepron)	This drug is an antimalarial that has been evaluated against *Pneumocystis jiroveci*, *Toxoplasma gondii*, and *Cryptosporidium* in AIDS patients. Preliminary results do not look promising.
Supportive therapeutic agents	
Nonspecific antidiarrheal agents (Kaopectate, Imodium, Lomotil, Pepto-Bismol, opiates)	These agents are often helpful, but individual regimens need to be developed for each patient; safety in patients with cryptosporidiosis is not known.
Octreotide (Sandostatin)	This is a synthetic cyclic octapeptide analog of somatostatin; anecdotal reports indicate some success in treating AIDS-related diarrhea. However, most patients had neither cryptosporidial diarrhea nor other identifiable pathogens.

[a] Adapted from B. L. Blagburn and R. Soave, Prophylaxis and chemotherapy: human and animal, p. 112–115, *in* R. Fayer (ed.), Cryptosporidium *and Cryptosporidiosis*, ASM Press, Washington, D.C., 1997, and reference 22.

[b] Although other agents have been tried (zidovudine, recombinant interleukin-2), any improvement is most probably due to increased patient immune function rather than to the antiparasitic activity of the drug.

Oocysts are susceptible to ammonia, 10% formalin in saline, freeze-drying, and exposure to temperatures above 65°C for 30 min. Commercial bleach in a 50% solution is also effective (Table 4.10). For most of the chemicals that have been tried for disinfection, effective concentrations are not practical outside the laboratory and high concentrations that significantly reduce oocyst infectivity are either very expensive or quite toxic. Unfortunately, oocysts of *Cryptosporidium* have shown high levels of resistance to most commercial disinfectants, including iodine water purification tablets (4). Although chlorine and related compounds can dramatically reduce the ability of oocysts to excyst or infect, high concentrations and long exposure times are required, making this approach impractical. Although rare, some stool specimen collection kits contain a vial of potassium dichromate specifically for *Cryptosporidium* spp., and the oocysts remain viable in this solution for at least 12 months.

Almost all oocysts exposed to soils that are frozen at −10°C become inactivated within 50 days; inactivation occurs regardless of the presence or absence of freeze-thaw cycles (25). Although oocysts that are purified from fecal material prior to cryopreservation tend to lose viability and infectivity for tissue culture and mice, when the oocysts are cryopreserved in feces a certain number remain viable and maintain infectivity. Based on these studies, fecal material appears to protect the oocysts during cryopreservation stored at −20°C for at least 30 days (26).

Studies have shown that calves and perhaps other animals serve as potential sources of human infections. Contact with these animals may be an unrecognized cause of gastroenteritis in humans in both rural and urban settings. Direct person-to-person transmission is also likely and may occur through direct or indirect contact with stool material. Direct transmission may occur during sexual practices involving oral-anal contact. Indirect transmission may occur through exposure to positive specimens

Table 4.7 Biological factors which impact the epidemiology of *Cryptosporidium*[a]

Small (4–6 μm), environmentally resistant, and fully sporulated/infectious oocysts when passed
Large livestock animal and human reservoir host populations
Ubiquitous, able to cross-infect multiple host species (not species specific)
Low infective dose (≤10–100 oocysts)
Large multiplication capability (>10^{10}) in a single host
Resistance to disinfectants
Lack of effective therapy; resistance to available drugs

[a] Adapted from D. P. Casemore, S. E. Wright, and R. L. Coop, Cryptosporidiosis—human and animal epidemiology, p. 66–73, *in* R. Fayer (ed.), Cryptosporidium *and* Cryptosporidiosis, ASM Press, Washington, D.C., 1997.

Table 4.8 Risk factors for acquisition of cryptosporidiosis

Risk factor	Examples
Deficient immunity	AIDS, other acquired or congenital immunodepression syndormes (e.g., boys with mutation of the CD154 gene, congenital X-linked immunodeficiency with hyper-IgM), immunosuppression, malnutrition
Zoonotic contact	Leisure activities such as camping, backpacking, farm visits
Occupational exposure	Veterinary, agricultural, nursing or medical, laboratory, day care centers
Poor sanitary conditions	Drinking water and food, inadequate sewage or waste disposal, inadequate fly control
Exposure to untreated water (surface or recreational)	Leisure activities
Exposure to inadequately treated water supply	Inadequate treatment or breakdown in treatment process
Consumption of raw foods	Unpasteurized milk, raw meat
Travel	Travel from developed to underdeveloped countries and from urban to rural areas
Age (infants, young children)	Weaning, teething, pica, finger sucking, wearing diapers; poor infection control
Contact with case of diarrhea	Day care, household, parents

Table 4.9 Factors related to potential outbreaks of cryptosporidiosis

Factor	Examples
Factors unrelated to water	Sufficient number of parasites (viability, infectivity, virulence)
	Sufficient number of susceptible individuals for ingestion of oocysts leading to primary infection
	Sufficient number of susceptible individuals for whom contact with primary case leads to secondary transmission
	Adequate epidemiologic and laboratory surveillance and reporting
Water-related factors	Abnormal weather conditions (heavy rainfall); increased load for water treatment facilities
	High turbidity of surface water; increased challenge for water treatment facilities
	Breakdown in integrity of groundwater aquifers (prolonged rain after drought)
	Suboptimal water treatment functions
	Problems with finished-water distribution

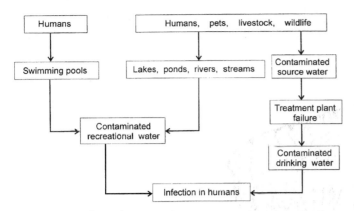

Figure 4.6 Flow diagram demonstrating sources of water-borne transmission of *Cryptosporidium* oocysts.

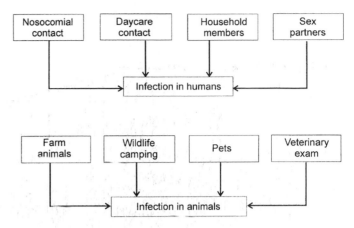

Figure 4.8 Flow diagram demonstrating sources of close-contact transmission of *Cryptosporidium* oocysts.

in a laboratory setting or from contaminated surfaces or food or water (3, 6, 8, 12, 14–22, 27, 31, 40, 42).

In January and February 1987, cryptosporidiosis was associated with an estimated 13,000 cases of gastroenteritis in Carroll County, Ga. *Cryptosporidium* oocysts were identified in the stools of 39% of the persons examined during the outbreak; a randomized telephone survey indicated probable attack rates of 54% within the city of Carrollton and 40% overall for the county. The only significant risk factor associated with illness was exposure to the public water supply, which was filtered and chlorinated. It was interesting that according to records kept during the outbreak, the treatment facility was operating within Environmental Protection Agency guidelines. Additional reports of waterborne outbreaks have appeared in the past few years (Table 4.11). Another massive waterborne outbreak was reported in Milwaukee,

Wis. Contamination of the public water supply during March and April 1993 resulted in approximately 300,000 infections. Infection during this outbreak appeared to be more severe than in previous reports of large case series. The risk of secondary infection in a household was low

Figure 4.7 Flow diagram demonstrating sources of food-borne transmission of *Cryptosporidium* oocysts.

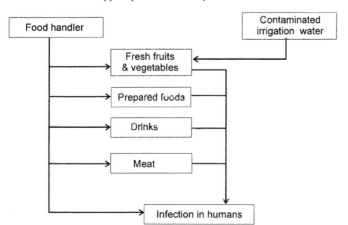

Table 4.10 Physical disinfection of *Cryptosporidium* oocysts[a]

Agent	Conditions	Results
Heat	50–55°C, 5 min	Noninfectious (in vivo, mice)
	45°C, 20 min	Noninfectious (in vivo, mice)
	60°C, 6 min	Noninfectious (in vivo, mice)
	59.7°C, 5 min	Infectious (in vivo, mice)
	64.2°C, 5 min	Noninfectious (in vivo, mice)
	67.5°C, 1 min	Infectious (in vivo, mice)
	72.4°C, 1 min	Noninfectious (in vivo, mice)
Freezing	−196°C, 10 min	Noninfectious (in vivo, mice)
	−70°C, 1 h	Noninfectious (in vivo, mice)
	−20°C, 1 day, 3 days	Noninfectious (in vivo, mice)
	−20°C, 8 h	Infectious (in vivo, mice)
	−15°C, 1 wk	Noninfectious (in vivo, mice)
	−15°C, 24 h	Infectious (in vivo, mice)
	−10°C, 1 wk	Infectious (in vivo, mice)
	Liquid nitrogen	100% reduced (excystation, dyes)
	−22°C, <32 days	98% reduced (excystation, dyes)
UV light	15,000 mW/s, 2 h	Infectious (in vivo, mice)
	15,000 mW/s, 2.5 h	Noninfectious (in vivo, mice)
	80 mW/s cm^{-2}	90% reduced (excystation)
	120 mW/s cm^{-2}	99% reduced (excystation)
	8,748 mW/s cm^{-2}	100% reduced (excystation, dyes)
Drying	Air dried, 2 h	97% reduced (excystation, dyes)
	Air dried, 4 h	100% reduced (excystation, dyes)
	Air dried in feces, 1–4 days	Noninfectious (in vivo, mice)

[a]Adapted from R. Fayer, C. A. Speer, and J. P. Dubey, The general biology of *Cryptosporidium*, p. 30–33, *in* R. Fayer (ed.), Cryptosporidium *and Cryptosporidiosis*, ASM Press, Washington, D.C., 1997.

Table 4.11 Waterborne outbreaks within the United States associated with water intended for drinking[a]

Parasite	No. of outbreaks (no. of cases) in:										
	1984	1985	1986–1988	1989–1990	1991–1992	1993–1994	1995–1996	1997–1998	1999–2000	2001–2002	Total[b]
Cryptosporidium spp.	1 (117)		1 (13,000)		3 (3,551)	5 (403,246)		2 (1,400+)	1 (5)	1 (10)	14 (421,329)
C. cayetanensis				1 (21)		2 (23)					3 (44)
G. lamblia	6 (879)	3 (741)	9 (1,169)	7 (697)	4 (123)	5 (385)	2 (1,460)	4 (159)	6 (52)	3 (18)	49 (5,683)
E. histolytica	1 (4)										1 (4)

[a] Data from the CDC *Surveillance for Waterborne-Disease Outbreaks—United States* (usually provided on a 2-year cycle with a 2- to 3-year delay).
[b] In June/July, 2004, approximately 50 people were infected with *C. cayetanensis* from the ingestion of snow peas (7).

(5%) when the index case involved an adult. Also, the recurrence of watery diarrhea after apparent recovery was a frequent occurrence among visitors with laboratory-confirmed cryptosporidiosis (39%) and among visitors and residents with clinical infection (21%).

This increase in the number of reported waterborne disease outbreaks associated with *Cryptosporidium* spp. can be attributed to improved techniques for oocyst recovery and identification. The use of high-volume filters, immunofluorescence, and molecular detection methods has resulted in the demonstration of oocysts in surface and drinking waters and in sewage effluents from many geographic regions (3, 12, 19, 20, 30, 42).

The immunofluorescence method can be used to examine large volumes of water, is relatively specific for *Giardia* and *Cryptosporidium* spp., and permits examination of the specimen by epifluorescence and phase-contrast or differential interference contrast microscopy (Figure 4.9). However, disadvantages of the method include low recovery percentages (5 to 25%), long processing times (1–2 days to 1–2 weeks), the need for highly trained per-

Figure 4.9 Procedure for water testing for the recovery of *Giardia* cysts and *Cryptosporidium* oocysts in water. (Adapted from J.B. Rose, J.T. Lisle, and M. LeChavalier, Waterborne cryptosporidiosis: incidence, outbreaks, and treatment strategies, p. 95–100, *in* R. Fayer (ed.), Cryptosporidium *and* Cryptosporidiosis, ASM Press, Washington, D.C., 1997.

- Water sample (example: 400 liters of surface water; 1,000 liters of groundwater)
- Filter through large polypropylene cartridge filter (pore size, 1 μm)
- Cut filter in half lengthwise
- Suspend fibers in 0.1% SDS–0.1% Tween 80 buffer
- Hand wash or homogenize with large stomacher (to release organisms from polypropylene filter fibers)
- Centrifuge and combine sedimented material into one pellet
- Select appropriate volume, sonicate for 10 min
- Underlay or overlay with Percoll-sucrose flotation medium (specific gravity, 1.10)
- Centrifuge at 1,050 × *g* for 10 min
- Draw off top layer, dilute to 50 ml, centrifuge at 1,050 × *g* for 10 min
- Layer flotation surface sample on 25-mm filter (pore size, 0.2 μm)
- Add monoclonal antiserum mixture (*Giardia* and *Cryptosporidium*)
- Wash
- Add labeling reagent
- Wash
- Dehydrate with alcohol series
- Clear filter with 2% DABCO–glycerol solution
- Examine using epifluorescence, phase-contrast, or differential interference contrast microscopy

sonnel, the high cost per sample (approximately $300), the inability to identify viable or virulent organisms, and cross-reactivity with nonhuman *Cryptosporidium* organisms. Unfortunately, the more the specimen or water sample is manipulated during processing, the more likely it is that organisms will be lost in the process, thus accounting for the low recovery percentages.

Incorporation of the use of fluorogenic vital dyes to determine oocyst viability appears to offer some improvements; however, the positive relationship between vital-dye reactions and excystation may not always correlate directly with viability or infectivity. A number of other methods have been used in an effort to improve organism recovery and to determine the viability and infectivity of the oocysts recovered. These include the use of a fluorescence in situ hybridization technique for labeling of *Cryptosporidium* oocysts in water samples, species- and strain-specific typing of parasites in clinical and environmental samples, use of PCR for the detection of *Cryptosporidium* oocysts in municipal water samples, use of cell culture to determine the infectivity of *Cryptosporidium* oocysts, immunomagnetic separation coupled with immunofluorescence microscopy for oocyst detection, a combination of immunomagnetic separation and integrated cell culture-PCR, and use of β-tubulin mRNA as a marker of *Cryptosporidium* oocyst viability.

Data indicate that viable *Cryptosporidium* oocysts exposed to chlorine or other oxidizing conditions may lack identifying epitopes. This finding may be important for the water industry, where naturally occurring oxidizing conditions or sanitizing treatments could produce viable oocysts that are undetectable by standard protocols. Detailed knowledge of the structure and stability of the *Cryptosporidium* oocyst wall may ultimately contribute to the development of water treatment options that will reduce the viability of the intact oocyst and thus the waterborne dissemination of the parasite.

The importance of agricultural wastewater and runoff, particularly from lambs and calves, is also now recognized as a potential source of infective *Cryptosporidium* oocysts (42). It has been shown that in healthy adults with no serologic evidence of past infection with *Cryptosporidium*, a low dose of oocysts is sufficient to cause infection. With healthy human volunteers, the median infective dose was 132 oocysts and one-fifth of the volunteers became infected with a dose as low as 30 oocysts. Specific guidelines are available for the prevention of cryptosporidiosis, particularly where drinking water is involved. Information can be seen in Tables 4.12 to 4.14.

Various surveys indicate prevalence rates of 1 to 2% in Europe, 0.6 to 4.3% in North America, and 3 to 20% in other areas of the world (Asia, Australia, Africa, and Central and South America). It has been suggested that the infection rate for individuals with diarrheal illness is 2.2% for persons living in industrialized countries and 8.5% for those residing in developing countries. Available data indicate that there are 250 million to 500 million *Cryptosporidium* infections annually among persons living in Asia, Africa, and Latin America. Generally, young children tend to have higher infection rates and all patients appear to have symptoms. Available serologic data support the prevalence rates, with figures from Europe and North America normally between 25 and 35% and those from South America indicating rates of up to 64%.

The epidemiologic considerations for cryptosporidiosis emphasize transmission by environmentally resistant oocysts, numerous potential reservoir hosts for zoonotic transmission, documentation of person-to-person transmission in day care centers, nosocomial transmission within the health care setting, occurrence of asymptomatic infections (infective, carrier state), widespread environmental distribution resulting in the probability of waterborne transmission, and the link between cryptosporidiosis and severe, life-threatening disease in individuals with impaired immune function. Prevalence data from the Centers for Disease Control and Prevention for patients with AIDS are thought to be an underestimate of the actual prevalence (2 to 5%), particularly when other figures indicate percentages of 15% (National Institutes of Health, Bethesda, Md.) and 11% (Great Britain). Surveillance systems currently under discussion can be seen in Table 4.15.

As our understanding of the epidemiology of cryptosporidiosis expands, it is obvious that there are numerous ways for the resistant oocysts to become disseminated. Some of these recent examples have generated enhanced interest in and concern about additional sources of infectious oocysts and the potential for accidental human consumption through contaminated food and/or water. Specific examples can be seen in Table 4.16 and Figures 4.6 and 4.10.

Cyclospora cayetanensis

In recent years, human cyclosporiasis has emerged as an important infection, with a number of outbreaks being reported in the United States and Canada. Prior to 1995, only one outbreak of cyclosporiasis had been reported in the United States; however, from May through August 1996, more than 1,400 cases were reported from 20 states, Washington, D.C., and two Canadian provinces (7, 22, 36). Understanding of the biology and epidemiology of *C. cayetanensis* is limited and has been complicated by a lack of information about the parasite's

Table 4.12 Prevention of cryptosporidiosis: recommendations (particularly important for patients who are immunocompromised)[a]

Preventive measure	Comments
Handwashing	Prior to eating and food preparation; after touching children in diapers; after touching clothing, bedding, toilets, bed pans from anyone with diarrhea; after gardening; after touching pets or other animals; after touching or coming in contact with anything contaminated with human or animal stool (includes dirt); after removing gloves to perform any of the above activities.
Practicing safe sex	Always wash hands after touching partner's anal or rectal area.
Avoiding touching farm animals	Particularly important for contact with young animals; wash hands thoroughly; avoid touching stool of any animal; after potential exposure, clean shoes (HIV-infected persons should not perform this task).
Avoiding pets' stools	If HIV infected, avoid cleaning litter boxes or cages or stool disposal (very important if animals are younger than 6 months); have puppies or kittens (if younger than 6 months) tested for presence of *Cryptosporidium* before bringing them home; have any pet tested that has diarrhea.
Swimming	Avoid swallowing water when in lakes, rivers, pools, or hot tubs; organisms are not killed by routine chlorination; avoid swimming in polluted ocean water, since oocysts can survive for several days in salt water.
Washing and cooking food	Thoroughly wash all fruits and vegetables if eating uncooked; use safe water (see below) for washing food; peel fruit; avoid unpasteurized milk or dairy products. Cooking kills *Cryptosporidium*, so cooked or packaged food is safe unless handled by someone infected with the organism.
Drinking water	Do not drink directly from lakes, rivers, streams, or springs; may want to avoid tap water (including refrigerator ice-maker ice and chilled water); boil water (rolling boil for 1 min is sufficient to kill organisms); filter with appropriate filters (look for "reverse osmosis," "absolute 1 micron," "Standard 53," and the words "cyst reduction" or "cyst removal"); a home distiller can be used. Store filtered water as for boiled water.
Bottled water	Bottled water labels reading "well water," "artesian well water," "spring water," or "mineral water" do not guarantee that the water does not contain *Cryptosporidium*. However, water that comes from protected well or spring water sources is less likely to be contaminated than bottled water from rivers and lakes.
Drinks	Canned or bottled soda, seltzer, fruit drinks, and steaming hot tea and coffee are safe; fountain drinks, fruit drinks mixed with tap water from concentrate, and iced tea or coffee may not be safe.

[a] Adapted from information prepared by the interagency Working Group on Waterborne Cryptosporidiosis.

Table 4.13 Filter options for removal of *Cryptosporidium* oocysts: label information[a]

Label information for filters designed to remove *Cryptosporidium*[b]	Label information for filters that may not be designed for removal of *Cryptosporidium*
Reverse osmosis (with or without NSF testing)	*Nominal* pore size of 1 micron or smaller
Absolute pore size of 1 micron or smaller (with or without NSF testing)	One-micron filter
	Effective against *Giardia*
	Effective against parasites
Tested and certified by NSF Standard 53 for cyst removal	Carbon filter
	Water purifier
Tested and certified by NSF Standard 53 for cyst reduction	EPA approved (Caution: EPA does not approve or test filters)
	EPA registered (Caution: EPA does not register filters for *Cryptosporidium* removal)
	Activated carbon
	Removes chlorine
	Ultraviolet light
	Pentiodide resins
	Water softener

[a] Adapted from information prepared by the interagency Working Group on Waterborne Cryptosporidiosis.

[b] Any of the four comments in this column on a package label indicate that the filter should be able to remove *Cryptosporidium*.

Table 4.14 Bottled water: label information[a]

Label information for bottled water that has been processed by a method effective for removal of *Cryptosporidium*	Label information for bottled water that may not have been processed by a method effective for removal of *Cryptosporidium*
Reverse osmosis treated	Filtered
Distilled	Microfiltered
Filtered through an *absolute* 1-micron or smaller filter	Carbon filtered
	Particle filtered
One micron *absolute*	Multimedia filtered
	Ozonated
	Ozone treated
	Ultraviolet light treated
	Activated carbon treated
	Carbon dioxide treated
	Ion exchange treated
	Deionized
	Purified
	Chlorinated

[a] Adapted from information prepared by the interagency Working Group on Waterborne Cryptosporidiosis.

Table 4.15 Possible surveillance approaches for the establishment of baseline data on the occurrence of cryptosporidiosis

Surveillance approach	Comments
Make cryptosporidiosis reportable to CDC	Applies to each state or city; provides legal authority for collecting information
Monitor sales of antidiarrheal medications	Local pharmacy computerized databases (provide historical as well as current diarrheal illness information)
Monitor logs maintained by HMOs and hospitals (complaints of diarrheal illness)	May be particularly useful if keyed to zip codes, particularly in case of waterborne transmission
Monitor incidence of diarrheal illness in nursing homes	Can monitor differences between use of municipal and bottled water; could also serve to establish baseline data
Monitor laboratory data for *Cryptosporidium* spp.	More likely to provide information on AIDS (or other immunocompromised) patients; may not be best way to obtain information on general public (excessive cost)
Monitor tap water in selected cities or locations	Intensive surveillance might be recommended in approximately 10 cities known to have *Cryptosporidium* spp. in their finished water, again to establish some baseline data
Establish methods of rapid reporting and communication among relevant agencies	Education of physicians, public, etc., vital to the understanding of epidemiology of cryptosporidiosis

Table 4.16 Potential sources for *Cryptosporidium* oocysts related to accidental human ingestion of contaminated food and/or water

Transmission agent	Comments
Eastern oysters (*Crassostrea virginica*)	Oysters collected from Maryland tributaries of the Chesapeake Bay contain infectious oocysts; 7 sites in Chesapeake Bay area are positive for infected oysters
Atlantic coast commercial shellfish	*Cryptosporidium* spp. found in commercial shellfish from 64.9% of sites examined along the Atlantic coast, using either microscopy or molecular testing
Clams (Italy, seawater)	Oocysts detected in 23 of 32 pools of clams
Clams (California, freshwater)	*C. parvum* DNA in clams from 3 different riverine ecosystems confirmed with PCR and DNA sequence analysis
Clams (*Tapes decussates*, Spain)	Oocysts found in siphons, gills, stomach, digestive diverticula, intestine; oocysts found in branchial mucus and in interfilamentary spaces suggest repeated filtrations and the possibility of infective oocysts
Bent mussels (*Ischadium recurvum*)	Mussels may be useful as biological indicator of water contamination with *Cryptosporidium* oocysts; found in same areas indicated above for oysters
Mussels (*Mytilus galloprovincialis*)	Oocysts survive in seawater for at least 1 year; filtered out by benthic mussels, retaining infectivity up to 14 days
Canada geese (migrating birds)	Infectious oocysts disseminated by migrating waterfowl
House flies (*Musca domestica*)	Adult and larval stages of house flies can mechanically carry oocysts in their digestive tracts and deposit them on external surfaces
Filth flies (Calliphoridae, Sarcophagidae, Muscidae)	Wild-caught flies (>90%) harbored *Cryptosporidium* oocysts; oocysts carried internally and externally—potential for mechanical transmission of infectious oocysts
Filth flies (Muscidae, Calliphoridae, Lauxaniidae, Anthomyiidae)	80% of all *Cryptosporidium* oocysts viable; >69% of *Giardia lamblia* cysts viable; potential for mechanical transmission of infectious oocysts high
Dung beetles (*Anoplotrupes stercorosus*, *Aphodius rufus*, *Onthophagus fracticornis*)	Although many oocysts pass safely through mouthparts and gastrointestinal tracts of the beetles, most are destroyed; however, coprophagous insects are considered important as agents of both control and dissemination
Free-living soil nematode (*Caenorhabditis elegans*)	*C. elegans* can ingest and excrete oocysts; the role of free-living nematodes in contamination requires further study

origins, possible animal reservoir hosts, and relationship to other coccidia (28, 36).

During the past few years, there have been several outbreaks of diarrhea associated with *C. cayetanensis*; the distribution is worldwide (United States, Caribbean, Central and South America, Southeast Asia, Eastern Europe, Australia, Nepal). These organisms are acid-fast variable and have been found in the feces of immunocompetent travelers to developing countries, immunocompetent subjects with no travel history, and patients with AIDS. Patients reported symptoms of a "flu-like" illness with nausea, vomiting, anorexia, weight loss, and explosive diarrhea lasting 1 to 3 weeks. Cumulative evidence suggests that outbreaks in the United States and Canada during the spring months of 1996 and 1997 were related to the importation and ingestion of Guatemalan raspberries. It is quite likely that the outbreak reported from Florida in 1995 was also attributed to contaminated food. The cases reported in all three outbreaks probably represented only a small fraction of those that occurred.

These organisms have now been identified as coccidia in the genus *Cyclospora* and were first discovered in the late 1880s but not connected with human infections.

Life Cycle and Morphology

The oocysts that had previously been recovered from human stool specimens were immature, so the structure of the mature oocyst was not seen in those specimens. It takes approximately 5 days or more for oocyst maturation, so the mature stage may not have been seen in human specimens. Evidence obtained from excystation experiments indicates that the oocyst contains two sporocysts, each containing two sporozoites, a pattern which places these organisms in the coccidian genus *Cyclospora*. Electron microscopy (EM) confirmed the presence of characteristic organelles for coccidian organisms of the phylum Apicomplexa. The name *Cyclospora cayetanensis* sp. nov. has been proposed for this newly described disease-producing coccidian organism from humans. The species name comes from the university where it was initially studied (Universidad Peruana Cayetano Heredia).

Two types of meronts and sexual stages have been seen in jejunal enterocytes in biopsy specimens from infected patients. These findings confirm that the entire life cycle can be completed within a single host (Figure 4.11). Unsporulated oocysts are passed in the stool, and sporulation occurs within approximately 7 to 13 days. Complete sporulation produces two sporocysts that rupture to reveal two crescent-shaped sporozoites measuring 1.2 by 9.0 µm.

The transmission of *Cyclospora* is thought to be fecal-oral, although direct person-to-person transmission has not been well documented and may not be a factor, since sporulation takes a number of days. Outbreaks linked to contaminated water and various types of fresh produce (raspberries, basil, baby lettuce leaves, and snow peas) have been reported (Table 4.17). Also, information on potential reservoir hosts has yet to be defined; however, it appears that in some areas the human is the only host.

Clinical Disease

Developmental stages of *C. cayetanensis* usually occur within epithelial cells of the jejunum and lower portion of the duodenum. *Cyclospora* infection reveals characteristics of a small bowel pathogen, including upper gastrointestinal symptoms, malabsorption of D-xylose, weight loss, and moderate to marked erythema of the distal duodenum. Histopathology in small bowel biopsy specimens reveals acute and chronic inflammation, partial villous atrophy, and crypt hyperplasia.

The incubation period is approximately 2 to 11 days after exposure. There is generally 1 day of malaise and

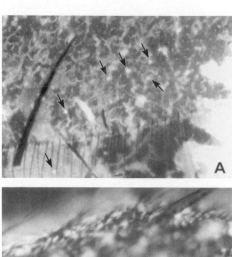

Figure 4.10 *Cryptosporidium* oocysts (arrows) detected by immunofluorescent antibodies in fecal spots and on the exoskeleton of house flies (*Musca domestica*) exposed to bovine diarrheal feces containing *Cryptosporidium* oocysts. (A) Fly fecal spot. (B) Leg (tibia) with *Cryptosporidium* oocyst captured by the hairs. (C) Posterior margin of the wing; note the *Cryptosporidium* oocyst captured by the wing bristles. (Photograph courtesy of T. K. Graczyk, with permission.)

low-grade fever, with rapid onset of diarrhea of up to seven stools per day. There may also be fatigue, anorexia, vomiting, myalgia, and weight loss with remission of self-limiting diarrhea in 3 to 4 days, followed by relapses lasting from 4 to 7 weeks. In patients with AIDS, symptoms may persist for as long as 12 weeks; biliary disease has also been detected in this group. Diarrhea alternating with constipation has also been reported; this is not uncommon in a number of protozoal gastrointestinal infections. Clinical clues include unexplained prolonged diarrheal

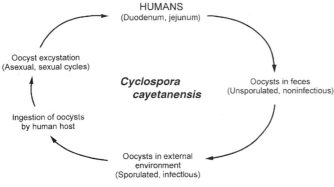

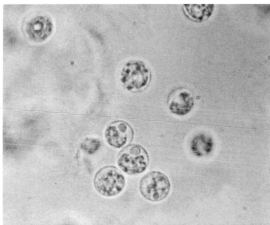

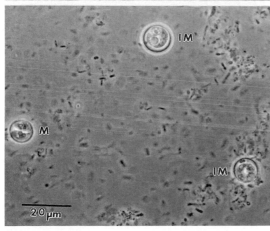

Figure 4.11   (Top) *Cyclospora cayetanensis* life cycle. (Middle and bottom) *C. cayetanensis* in wet preparation, phase contrast. M, mature oocyst; IM, immature oocyst. (Courtesy of Susan Novak [middle] and Charles R. Sterling [bottom].)

illness during the summer in any patient and in persons returning from tropical areas. The majority of infected individuals had intermittent diarrhea for 2 to 3 weeks, and many complained of intense fatigue, as well as anorexia and myalgia, during the illness. The clinical presentation of patients infected with this organism is similar to that of patients infected with *Cryptosporidium*.

Diagnosis

When examined in formalin-preserved human stool specimens and by EM, these organisms exhibited characteristics of a cyanobacterium. The organisms stained orange with safranin and were modified acid-fast variable, with some organisms staining deep red with a mottled appearance but no internal organization (38). Preliminary EM revealed internal structures that resembled those of the cyanobacteria. In clean wet mounts, the organisms are seen as nonrefractile spheres and are acid-fast variable with the modified acid-fast stain; those that are unstained appear as glassy, wrinkled spheres (Figures 4.11 to 4.14). Modified acid-fast stains stain the oocysts light pink to deep red, and some will contain granules or have a bubbly appearance. It is very important to be aware of these organisms when the modified acid-fast stain is used, because *Cryptosporidium* spp. and other similar but larger structures (approximately twice the size of *Cryptosporidium* oocysts) are seen in the stained smear. It is important for laboratories to measure all acid-fast oocysts, particularly if they appear to be somewhat larger than those of *Cryptosporidium*. The oocysts autofluoresce strong green (450 to 490 DM excitation filter) or intense blue (365 DM excitation filter) under UV epifluorescence. It is strongly recommended that during concentration (formalin-ethyl acetate) of stool specimens, the centrifugation be carried out for 10 min at 500 × *g*.

Although modified acid-fast stains are the most commonly used stains to identify the oocysts in stool, a safranin-based stain that tends to uniformly stain *Cyclospora* oocysts is also recommended (38).

Another diagnostic option involves the use of flow cytometry. This approach appears to be a useful alternative to microscopy, particularly for screening large numbers of clinical specimens for *Cyclospora* oocysts in an outbreak situation (11).

KEY POINTS—LABORATORY DIAGNOSIS

Cyclospora cayetanensis

1. Specimens may appear to be larger than normal *Cryptosporidium* oocysts; be sure to measure these organisms to confirm their size.

2. Fecal specimens should be concentrated prior to staining; centrifugation for 10 min at 500 × *g* is recommended.

3. On wet smears, the oocysts appear as nonrefractile spheres and, depending on the filters used, autofluoresce green to blue with epifluorescence. Autofluorescence may be 2+ to 4+.

4. On modified acid-stained smears, the oocysts stain from light pink to deep red, and some contain

Table 4.17 Outbreaks of *Cyclospora cayetanensis* infection

Outbreak	Implicated cause	Comments
July 1990 (11 cases), Chicago hospital	Tap water from the physicians' dormitory was probably the source of the outbreak	This is the first reported outbreak of diarrhea associated with *Cyclospora* in the United States
1995 (14 cases), Florida	Fresh raspberries and strawberries	Given the cumulative evidence and the occurrence in 1996 and 1997 of outbreaks in North America associated with consumption of Guatemalan raspberries, food-borne transmission of *Cyclospora* was likely in 1995 in Florida as well
23 May 1996 (38 cases), South Carolina	Fresh raspberries	Imported from Guatemala
26 June 1996 (48–54 possible cases), Virginia	Basil-pesto pasta salad	No raspberries or mesclun lettuce was served at the luncheon
June and July 1996 (185 possible cases), 25 clusters in northern Virginia, Washington D.C., and Baltimore, Md.	Food containing fresh basil	All basil was associated with the same company
1996 (180 cases), Florida	Fresh raspberries	Imported from Guatemala
1996 (1,465 cases), 20 states	Fresh berries, berries with or without fresh raspberries	As few as five Guatemalan farms could have accounted for the 25 events; the raspberries could be traced to a single exporter per event
April and May 1997 (80 cases), California, Florida, Nevada, New York, and Texas	Fresh raspberries, mesclun (baby lettuce leaves, known as spring mix, field greens, or baby greens)	These 80 cases a part of the next listing for 1997 of a total of 1,450 cases
1997 (1,450 cases), United States and Canada	Fresh raspberries, mesclun	Most cases were linked to ingestion of raspberries from Guatemala; exportation of fresh raspberries from Guatemala was voluntarily suspended at the end of May 1997
1998 (94 cases), Boston, Mass.	Dessert containing raspberries, strawberries, blackberries, and blueberries	No specific commentary on source of fresh berries, but probably imported
July 1999 (62 cases), Missouri	Fresh basil	Chicken pasta salad and tomato basil salad; grown either in Mexico or United States
June 2000 (54 cases)	Fresh raspberries	Wedding cake, which had a cream filling that included raspberries imported from Guatemala
2001 (17 cases), British Columbia	Fresh Thai basil	Importation was traced from the United States
June and July 2004 (96 cases), Pennsylvania	Pasta salad containing raw snow peas	Snow peas imported from Guatemala

granules or have a bubbly appearance. Those that do not stain may appear wrinkled. A strong decolorizer should not be used; a 1% sulfuric acid solution is recommended and also works well for modified acid-fast stains for *Cryptosporidium* spp. and/or *Isospora belli*. The original 3 to 5% sulfuric acid is usually too strong for *Cyclospora* and removes too much color. Even with the 1% acid decolorizer, some oocysts may appear clear or very pale.

5. If the oocysts are seen in a regular trichrome-stained smear of stool, they may appear as clear, round, somewhat wrinkled objects that are larger than *Cryptosporidium* spp.
6. It is very important to measure these oocysts; *Cyclospora* oocysts measure 8 to 10 µm, and *Cryptosporidium* oocysts measure from 4 to 6 µm.
7. In the modified safranin technique, the oocysts uniformly stain a brilliant reddish orange if fecal smears

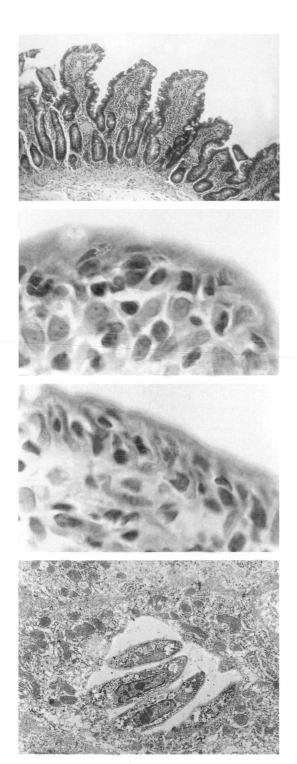

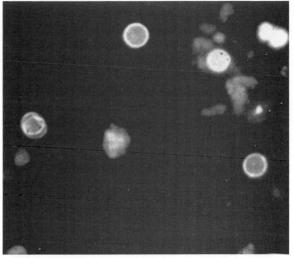

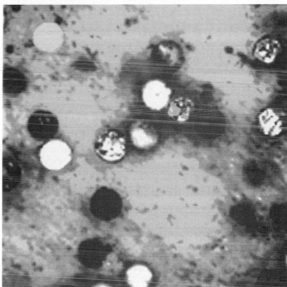

Figure 4.13 (Upper) *Cyclospora cayetanensis* autofluorescence. (Lower) Modified acid-fast stain of *Cyclospora cayetanensis* (note that some oocysts stain quite dark, while some have not retained the stain; this is typical, and the organisms are described as "modified acid-fast variable").

Figure 4.12 Human jejunal biopsy sections containing *Cyclospora cayetanensis* (courtesy of Ynes Ortega). Hematoxylin-and-eosin-stained sections. (Top) Low magnification (×10). Infected tissue showing villus atrophy and widening of villi. (Second from top) Magnification, ×100. Multiple intracellular forms. (Third from top) Magnification, ×100. Meront. (Bottom) Transmission electron microscopy. Meront with four merozoites.

are heated in a microwave oven during staining. The stained slide can also be examined by epifluorescence microscopy first, and suspect oocysts can be confirmed by bright-field microscopy.

Treatment

Apparently, patients do not respond to conventional antimicrobial therapies. Some patients have been treated symptomatically with antidiarrheal preparations and have obtained some relief; however, the disease appears to be

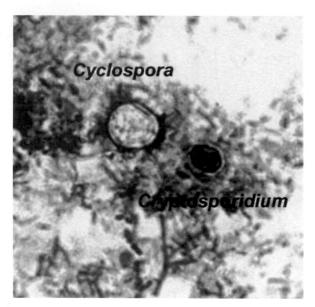

Figure 4.14 Modified acid-fast stain (note the *Cryptosporidium* [4 to 6 µm] and *Cyclospora cayetanensis* [8 to 10 µm] oocysts); the *Cyclospora* oocyst has not retained the stain.

self-limiting within a few weeks. Trimethroprim-sulfamethoxazole (TMP-SMX) is currently the drug of choice (adults: 760/800 mg given orally twice daily for 7 days; children: 5 mg of TMP/kg plus 25 mg of SMX/kg given orally twice daily for 7 days); relief of symptoms has been seen by 1 to 3 days posttreatment. Patients with AIDS may need higher doses and long-term maintenance treatment. However, symptoms recur within 1 to 3 months posttreatment in over 40% of patients. In a study with TMP-SMX in Nepal, shedding of oocysts and symptoms disappeared after 7 days of treatment (28).

Epidemiology and Prevention

Published reports indicate that individuals of all ages, including both those who are immunocompetent and those who are immunosuppressed, can become infected. Travelers to areas with endemic *Cyclospora* infection should be aware of the increased risk of infection during the late spring and summer. Since this parasite, like *Cryptosporidium*, is not killed by routine chlorination, drinking water treated by halogenation may not be safe. Boiling is recommended. Hot drinks such as coffee or tea are recommended, rather than placing any reliance on iodine treatment of the water. Fresh fruits and vegetables should be thoroughly washed and/or peeled prior to eating. Although prophylaxis has not been adequately investigated and is not recommended, specific circumstances might warrant this approach (e.g., for human immunodeficiency virus [HIV]-positive individuals visiting these areas). In Peru, infections with this organism have shown

some seasonal variation, with peaks during April to June. This pattern is similar to that seen in *Cryptosporidium* infections in Peru. Preliminary data and extrapolation from what we currently know about cryptosporidiosis suggest that modes of transmission may be similar, particularly considering waterborne transmission (Table 4.11).

As our understanding of the epidemiology of cyclosporiasis expands, it is obvious that there may be numerous ways for the resistant oocysts to become disseminated. Some of these recent examples identified for *Cryptosporidium* spp. have generated enhanced interest in and concern about additional sources of infectious oocysts and the potential for accidental human consumption through contaminated food and/or water. These examples may be just as relevant for *Cyclospora*. Specific examples for *Cryptosporidium* can be seen in Table 4.16. Recovery of waterborne oocysts of *C. cayetanensis* in Asian freshwater clams (*Corbicula fluminea*) has been reported (28, 36).

Isospora (*Cystoisospora*) belli

Isospora belli was first described by Virchow in 1860 but not named until 1923. Molecular biology studies have demonstrated that *Isospora* spp., particularly those from primates and carnivores, are more closely related to the Sarcocystiidae than to the Eimeriidae. This has required the transfer of the species into the family Sarcocystiidae and the genus *Cystoisospora*. However, throughout the remainder of the chapter, the organism is referred to as *Isospora belli* and the disease as isosporiasis. Transition to the new genus name will probably occur during the next couple of years. Although isosporiasis has been found in various parts of the world, certain tropical areas in the Western hemisphere appear to contain some well-defined locations of endemic infections. These organisms can infect both adults and children, and intestinal involvement and symptoms are generally transient unless the patient is immunocompromised. *I. belli* has also been implicated in traveler's diarrhea (2). However, unlike *C. cayetanensis*, large outbreaks of isosporiasis have not been reported.

Life Cycle and Morphology

Schizogonic and sporogonic stages in the life cycle of *I. belli* have been found in human intestinal mucosal biopsy specimens. Development in the intestine usually occurs within the epithelial cells of the distal duodenum and proximal jejunum. Eventually, oocysts are passed in the stool; they are long and oval and measure 20 to 33 µm by 10 to 19 µm (Table 4.2). Usually the oocyst contains only one immature sporont, but two may be present (Figures 4.15 and 4.16). Continued development occurs outside the body with the development of two mature sporocysts,

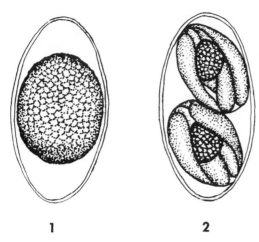

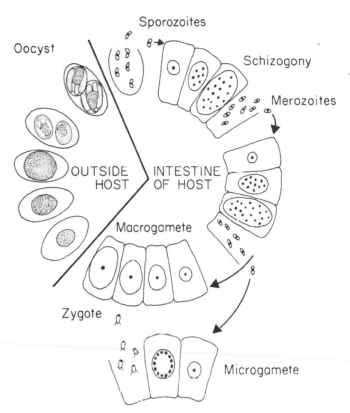

Figure 4.15 (1) Immature oocyst of *Isospora belli*; (2) mature oocyst of *I. belli*. (Illustration by Nobuko Kitamura.)

each containing four sporozoites, which can be recovered from the fecal specimen (Figures 4.17 and 4.18). The sporulated oocyst is the infective stage that will excyst in the small intestine, releasing the sporozoites, which penetrate the mucosal cells and initiate the life cycle. The life cycle stages, i.e., schizonts, merozoites, gametocytes,

Figure 4.17 Life cycle of *Isospora belli*. (Illustration by Gwen Gloege.)

Figure 4.16 (Upper) Immature oocyst of *Isospora belli*, wet preparation. (Lower) Immature oocyst of *I. belli*, modified acid-fast stain (note that the entire oocyst has stained).

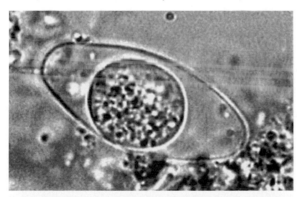

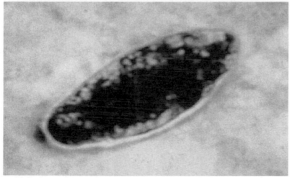

gametes, and oocysts, are structurally similar to those seen in the other coccidia.

The patent period is not known but may be as short as 15 days. Chronic infections develop in some patients, and oocysts can be shed for several months to years. In one particular case, an immunocompetent individual had symptoms for 26 years and *I. belli* was recovered in stool a number of times over a 10-year period.

Clinical Disease

Clinical symptoms include diarrhea, which may last for long periods (months to years), weight loss, abdominal colic, and fever; diarrhea is the main symptom. Bowel movements (usually 6 to 10 per day) are watery to soft, foamy, and offensive smelling, suggesting a malabsorption process. Eosinophilia is found in many patients, recurrences are quite common, and the disease is more severe in infants and young children.

Patients who are immunosuppressed, particularly those with AIDS, often present with profuse diarrhea associated with weakness, anorexia, and weight loss. In one patient with a well-documented infection of long standing, a series of biopsies showed a markedly abnormal mucosa with short villi, hypertrophied crypts, and

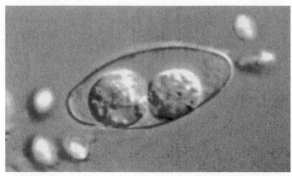

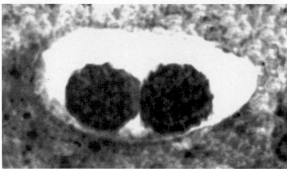

Figure 4.18 (Upper) Mature oocyst of *Isospora belli*, wet preparation. (Lower) Mature oocyst of *I. belli*, modified acid-fast stain.

infiltration of the lamina propria with eosinophils, neutrophils, and round cells. It has been recommended that physicians consider *I. belli* in AIDS patients with diarrhea who have immigrated from or traveled to Latin America, are Hispanics born in the United States, are young adults, or have not received prophylaxis with TMP-SMX for *Pneumocystis jiroveci*. It has also been recommended that AIDS patients traveling to Latin America and other developing countries be advised of the waterborne and food-borne transmission of *I. belli*, and they may want to consider chemoprophylaxis.

Extraintestinal infections in AIDS patients have been reported. One patient was a 38-year-old white man who presented with progressive dyspnea and fever; he also complained of dysphagia, nausea, vomiting, and brown watery diarrhea (eight or nine episodes daily) and had lost 20 lb in 2 months. He was diagnosed with *P. jiroveci* pneumonia and oropharyngeal candidiasis and was treated with TMP-SMX and pentamidine. His condition improved, and he was discharged; however, he was readmitted complaining of nausea, vomiting, and diarrhea; he was diagnosed with giardiasis and treated. Five months later he was diagnosed as having *I. belli* and *Entamoeba histolytica* infection and was treated. Three months later he presented with dyspnea, fever, diarrhea, and generalized wasting; he was diagnosed with cytomegalovirus pneumonia and died 2 weeks later. At autopsy,

microscopic findings associated with *I. belli* infection were seen in the lymph nodes and walls of both the small and large intestines. Intracellular zoites were seen in the cytoplasm of histiocytes. Each organism was surrounded by a thick eosinophilic cyst wall in routine histologic preparations with hematoxylin and eosin, and the cyst wall was periodic acid-Schiff (PAS) positive. Examination of the intestinal tissues revealed intraepithelial asexual and sexual stages of *I. belli*, as well as some merozoites that appeared to be in cells of the lamina propria.

The second case involved a 30-year-old black woman who was living in France but was originally from Burkina Faso. Initially she was symptomatic with fever, diarrhea, and weight loss. She was diagnosed with esophageal candidiasis and *I. belli* infection. She was treated with TMP-SMX and given maintenance therapy but suffered episodes of recurrent infection over the next 3 years. Examination of samples collected at autopsy revealed stages of *I. belli* in the intestine, mesenteric and mediastinal lymph nodes, liver, and spleen. The extraintestinal stages were always observed as single organisms that did not stain well with acid-fast stains. Massive infection with plasmacytosis and some eosinophils, but no granulomatous reaction, was observed.

In a third AIDS patient who presented with watery, nonbloody diarrhea and fever, examination of small intestine biopsy specimens revealed merozoites in the intestinal lumen, lamina propria, and lymphatic channels. Finding the merozoites within the lymphatic channels documents a means of dissemination to lymph nodes and other tissues. Additional studies of extraintestinal tissue cysts have identified early tissue cysts that lack a developed cyst wall, demonstrating that more than one tissue cyst can occupy a host cell, describing the distribution of micronemes and the shedding of zoite membranes, and identifying tubular structures in the inner tissue cyst wall and inner compartment.

Charcot-Leyden crystals derived from eosinophils have also been found in the stools of patients with *I. belli* infection. The diarrhea and other symptoms may continue in compromised patients, even those on immunosuppressive therapy, when the regimen of therapy is discontinued. This infection has been found in homosexual men, all of whom were immunosuppressed and had several months of diarrhea.

Diagnosis

Examination of a fecal specimen for the oocysts is recommended. However, wet-preparation examination of fresh material either as the direct smear or as concentrated material is recommended rather than the permanent stained smear (Figure 4.5; Table 4.3). The oocysts are very pale and transparent and can easily be overlooked.

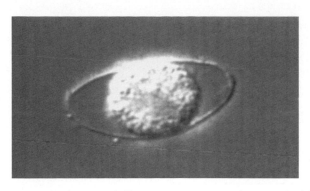

Figure 4.19 *Isospora belli* immature oocyst photographed by phase-contrast microscopy.

They can also be very difficult to see if the concentration sediment is from polyvinyl alcohol-preserved stool. The light level should be reduced, and additional contrast should be obtained with the microscope for optimal examination conditions (Figures 4.16, 4.18, and 4.19). It is also quite possible to have a positive biopsy specimen but not recover the oocysts in the stool because of the small numbers of organisms present. These organisms are acid fast and can also be demonstrated by using auramine-rhodamine stains. Organisms tentatively identified by using auramine-rhodamine stains should be confirmed by wet smear examination or acid-fast stains, particularly if the stool contains other cells or excess artifact material (more normal stool consistency).

KEY POINTS—LABORATORY DIAGNOSIS

Isospora belli

1. The oocysts are more easily recovered and identified when wet preparations are examined (direct smear, concentration smear).
2. Oocysts recovered in a concentrated sediment from polyvinyl alcohol-preserved stool may be very difficult to detect (the oocyst wall is very difficult to see).
3. A biopsy can be positive while no organisms are seen in the stool. This is not necessarily due to poor-quality laboratory work but may reflect normal findings.
4. Although these organisms can be stained with auramine-rhodamine stains, they should be confirmed by wet smears or modified acid-fast stains.

Treatment

Effective eradication of the parasites has been achieved with co-trimoxazole, TMP-SMX, pyrimethamine-sulfadiazine, primaquine phosphate-nitrofurantoin, and primaquine phosphate-chloroquine phosphate (1, 2). Other drugs proven to be ineffective include dithiazanine,

tetracycline, metronidazole, phanquone, and quinacrine hydrochloride. The drug of choice is TMP-SMX, which is classified as an investigational drug for treatment of this infection. TMP (160 mg)-SMX (800 mg) is given every 6 h for 10 days and then twice a day for 3 weeks. Pyrimethamine alone (50 to 75 mg daily) has cured infections in patients allergic to sulfonamides. In immunosuppressed patients with recurrent or persistent infection, therapy must be continued indefinitely.

Epidemiology and Prevention

I. belli is thought to be the only species of *Isospora* that infects humans, and no other reservoir hosts are recognized for this infection. Transmission is through ingestion of water or food contaminated with mature, sporulated oocysts. Sexual transmission by direct oral contact with the anus or perineum has also been postulated, although this mode of transmission is probably much less common. The oocysts are very resistant to environmental conditions and may remain viable for months if kept cool and moist; oocysts usually mature within 48 h following stool evacuation and are then infectious. It has been speculated that diagnostic methods for laboratory examinations may tend to miss the organisms when they are present. Since transmission is via the infective oocysts, prevention centers on improved personal hygiene measures and sanitary conditions to eliminate possible fecal-oral transmission from contaminated food, water, and possibly environmental surfaces.

Sarcocystis spp.

The organism once known as *Isospora hominis* is now recognized as being a part of the life cycle of *Sarcocystis* spp. (Figure 4.20). Two well-described species are *Sarcocystis bovihominis* (cattle) and *S. suihominis* (pigs). Some publications refer to *S. bovihominis* as *S. hominis*. When uncooked meat from these infected animals is ingested by humans, gamogony can occur within the intestinal cells, with the eventual production of the sporocysts in stool. In this case, humans who have ingested meat containing the mature sarcocysts serve as the definitive hosts (2, 13, 34). There have been reports of fever, severe diarrhea, abdominal pain, and weight loss in immunocompromised hosts, although the number of patients with these symptoms has been quite small.

Life Cycle and Morphology

Species of *Sarcocystis* have an obligatory two-host life cycle. Intermediate hosts such as herbivores and omnivores become infected through ingestion of sporocysts excreted in the feces of the definitive hosts such as carnivores and omnivores. The definitive hosts become

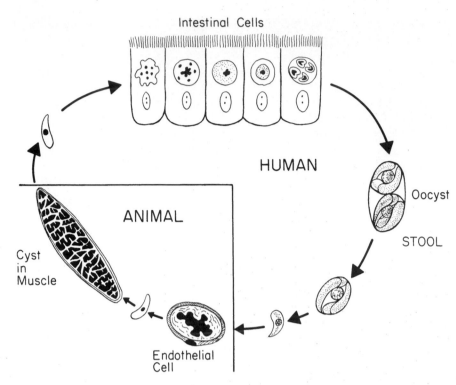

Figure 4.20 Life cycle of *Sarcocystis* spp. (Illustration by Gwen Gloege.)

Table 4.18 *Sarcocystis* development and disease symptoms in humans[a]

Topic	Muscle	Intestine
Infection source	Water or food contaminated with infected feces containing oocysts from unknown carnivore or omnivore	Raw or poorly cooked meat
Infective stage	Oocyst or free sporocysts	Sarcocyst containing bradyzoites
Developmental stages	Intravascular schizonts (not seen); intramuscular sarcocysts	Sexual stages found in lamina propria; oocysts excreted in feces
Incubation time from ingestion of infective stage to disease symptoms	Weeks to months, lasting months to years	3–6 h, lasting approximately 36 h
Disease symptoms	Musculoskeletal pain, fever, rash, cardiomyopathy, bronchospasm, subcutaneous swelling	Nausea, anorexia, vomiting, stomach ache, bloat, diarrhea, dyspnea, and tachycardia; may also include abdominal pain and eosinophilia; severe symptoms may include segmental necrotizing enteritis (rare and not well confirmed)
Diagnosis	Biopsy specimen(s) contains sarcocysts; antibodies to bradyzoites; routine histology on biopsy specimens; wall may be PAS positive, but stain variability is common	Oocysts or sporocysts in feces, beginning about 5–12 days after ingestion of raw or undercooked meat; *S. hominis* (beef) is excreted 14–18 days after meat ingestion, while *S. suihominis* (pork) is excreted 11–13 days after ingesting meat
Therapy	Co-trimoxazole, furazolidone, albendazole, anticoccidials, pyrimethamine, anti-inflammatories	None

[a] Adapted from references 2, 13, and 34.

infected through ingestion of mature cysts found in muscles of the intermediate hosts. In some intermediate hosts, such as cattle and sheep, all adult animals may be infected.

The sporocysts that are recovered in the stool are broadly oval and slightly tapered at the ends, measure 9 to 16 µm, and contain four mature sporozoites and the residual body (Table 4.2). Normally, two sporocysts are contained within the oocyst (similar to that seen with *I. belli*); however, in *Sarcocystis* infections, the sporocysts are released from the oocyst and normally are seen singly (Figure 4.20). They tend to be larger than *Cryptosporidium* oocysts, which contain four sporozoites, so there should be no confusion between the two. The oocysts are fully sporulated when passed in the stool.

Clinical Disease

When humans accidentally ingest oocysts from other animal stool sources, the sarcocysts that develop in human muscle apparently do little if any harm (schizogony) (2, 34). The prepatent period in humans is from 9 to 10 days. There is essentially no inflammatory response to these stages in the muscle and no conclusive evidence of pathogenicity of the mature sarcocyst. However, symptoms have been seen in some patients, probably associated with disintegration of the sarcocysts (Table 4.18). Painful muscle swellings, measuring 1 to 3 cm in diameter, initially associated with erythema of the overlying skin in various parts of the body, occur periodically and last for 2 days to 2 weeks. Some patients also have fever, diffuse myalgia, muscle tenderness, weakness, eosinophilia, and bronchospasm. No apparent clinical manifestations have been found in patients in whom myocardial *Sarcocystis* infection was detected accidentally by autopsy. A number of different morphological types of skeletal and cardiac muscle sarcocysts have been recovered in humans. No specific therapy is known for this type of infection; however, corticosteroids should reduce the allergic inflammatory reactions occurring after cyst rupture.

Intact sarcocysts in skeletal or cardiac muscle of humans measure up to 100 by 325 µm and are usually not accompanied by an inflammatory reaction (Figure 4.21). Each sarcocyst contains many bradyzoites, approximately 7 to 16 µm long. Inflammation follows disintegration of the cysts and death of the intracystic bradyzoites. Vasculitis is seen in the muscle and subcutaneous tissues. Histologic findings include myositis with vasculitis and sometimes myonecrosis.

When humans serve as definitive hosts, prevention involves adequate cooking of beef and pork (34); when humans are intermediate hosts, preventive measures involve careful disposal of animal feces that may contain

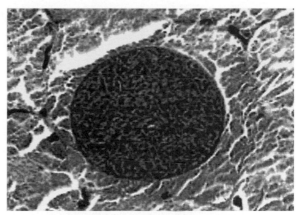

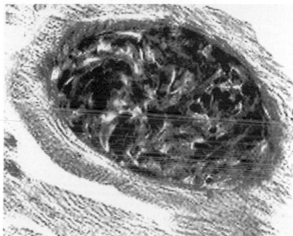

Figure 4.21   (Upper) *Sarcocystis* spp. in human muscle. (Lower) Higher magnification showing individual elongate bradyzoites.

the infective sporocysts. This may be impossible in the wilderness areas, where wild animals may serve as reservoir hosts for many of the different types of organisms that have been grouped under the term *Sarcocystis* "*lindemanni*." However, this name is no longer used. No specific therapy is known for the muscle stages of *Sarcocystis* spp. Various sizes of protozoan cysts, oocysts, and spores that occur in human stool specimens are given in Table 4.19.

Diagnosis

Presumptive diagnosis is often based on the patient symptoms, including a history of having ingested raw or poorly cooked meat. Confirmation of the diagnosis may depend on several stool examinations, some of which may reveal sporocysts. They are generally passed beginning 14 to 18 days after beef ingestion and about 11 to 13 days after pork ingestion. Some researchers think that the zinc sulfate fecal flotation method is preferred to the formalin-ethyl acetate sedimentation method for sporocyst and/or oocyst recovery (13). Sporocysts of *S. hominis* tend to be

Table 4.19 Protozoan cysts, oocysts, sporocysts, and spores seen in human feces[a]

Organism	Shape	Size (µm)	Features	Comments
Balantidium coli	Spherical to oval	50–70	Large macronucleus, thick wall	Although cilia are within the cyst wall, they are more difficult to see than the cilia surrounding the trophozoite form
Isospora belli	Oval, more tapered ends	20–33 by 10–19	Generally immature oocysts passed, but some containing two sporocysts may be visible	Not infectious when passed; best diagnostic approach is wet mount of concentrated sediment; some recommend flotation rather than sedimentation concentration
Sarcocystis hominis				
Oocysts	Spherical	15–19 by 15–20	Contain 2 sporocysts, but oocyst wall appears to be missing	Generally infectious when passed
Sporocysts	Oval	15–19 by 8–10	Contain 4 sporozoites	Infectious
Sarcocystis suihominis				
Oocysts	Spherical	15–19 by 15–20	Contain 2 sporocysts, but oocyst wall appears to be missing	Generally infectious when passed
Sporocysts	Oval	15–19 by 8–10	Contain 4 sporozoites	Infectious
Entamoeba histolytica/E. dispar	Spherical	12–15		Precyst contains single nucleus (generally), while mature cyst contain 4 nuclei and possible chromatoidal bars with rounded ends
Giardia lamblia	Spherical to oval	11–14	Contains four nuclei, curved median bodies and linear axonemes	Some cysts appear to be found; important to focus up and down to see morphologic characteristics; infective when passed
Cyclospora cayetanensis	Spherical	8–10	May appear as "wrinkled cellophane"	Not infectious when passed
Cryptosporidium hominis and *C. parvum*	Nearly spherical	4–6	Sporozoites may be difficult to see; modified acid-fast stains recommended	Infectious when passed
Encephalitozoon spp. and *Enterocytozoon bieneusi*	Oval, mimic small yeast cells and/ or bacteria	1.5–3.5	Horizontal or diagonal stripe (polar tubule) may be difficult to see in every spore	Infectious when passed

[a] Other pathogenic protozoa, *Dientamoeba fragilis* (no cyst form) and *Blastocystis hominis*, may overlap *Giardia* and *Entamoeba* spp. in size and may appear spherical to somewhat oval.

more elongated than those of *S. suihominis*, but the two may be difficult to differentiate.

Intramuscular infection would be suspected based on symptoms, including persistent myalgia, episodic weakness, subcutaneous nodules, dermatomyositis, eosinophilia, and elevated muscle creatinine kinase levels. Often these symptoms, coupled with a history of travel to or living in tropical locations, suggest the next step, which would be a muscle biopsy. Sarcocysts in biopsy specimens can be identified by microscopy of routine histologic sections stained with hematoxylin and eosin. Stain variability is common, and PAS positivity of the sarcocyst wall is not always seen.

Sarcocystis can be detected in meat by using routine histologic methods. Most sarcocysts in humans have been identified in skeletal and cardiac muscle; however, muscles in the larynx, pharynx, and upper esophagus have also been involved. Artificial digestion fluids can also be used on the meat source; after centrifugation, the sediment can be examined microscopically for the presence of bradyzoites.

KEY POINTS—LABORATORY DIAGNOSIS

Sarcocystis spp.

1. The oocysts are more easily recovered and identified in wet preparations (direct smear, concentration smear; flotation is recommended over sedimentation concentration methods).
2. The sporocysts are released from the oocyst and are seen singly (they are larger than *Cryptosporidium* oocysts).

3. Sarcocysts in muscle can be identified by routine histologic methods; however, there is a lot of staining variability (hematoxylin and eosin stains; PAS-positive sarcocyst walls may not be easily visible).

Treatment

Intestinal sarcocystosis is self-limiting and often asymptomatic and lasts for a short time. Although co-trimoxazole and furazolidone have been used, their effectiveness is still questionable. Currently, no therapy has been approved for treatment for myositis, vasculitis, or related lesions in humans. The efficacy of albendazole has not been demonstrated in controlled studies. Although drugs such as pyrimethamine and others used to treat toxoplasmosis have been considered, the efficacy of such drugs has not been documented. It is also unknown if immunosuppressives might be effective in reducing the inflammatory reactions seen in vasculitis or myositis. Without more definitive data, no recommended course of therapy is currently available.

Epidemiology and Prevention

Apparently, multiple ruminant species serve as intermediate hosts of *Sarcocystis* spp. and as sources of infection for humans, and humans serve as intermediate hosts for several species of *Sarcocystis*, perhaps by ingesting sporocysts excreted by predators of nonhuman primates. Humans, baboons, and rhesus monkeys serve as definitive hosts for *S. hominis*, and humans, chimpanzees, and rhesus and cynomolgus monkeys serve as definitive hosts for *S. suihominis*. *S. hominis* occurs worldwide in areas where cattle or buffalo have access to human feces and humans ingest raw or undercooked beef. *S. suihominis* probably occurs worldwide in areas where swine have access to human feces and humans ingest raw or undercooked pork.

Thoroughly cooking or freezing meat to kill bradyzoites within the sarcocysts will prevent intestinal sarcocystosis. Freezing at –4 and –20°C for 48 and 24 h, respectively, is sufficient to kill bradyzoites. It has been documented that rare roast beef and hamburger purchased from a supermarket contained bradyzoites infectious for dogs whereas cooked products were not infectious for dogs.

Prevention of infection in food animals depends on eliminating the sporocyst stage from human feces in contaminated water, feed, and bedding. Also, to prevent humans from serving as intermediate hosts, ingestion of sporocysts must be prevented; the most likely source is contaminated water for drinking or washing foods. Boiling is the best method to guarantee disinfection. Thorough washing or cooking of foods is also recommended.

Microsporidia

Although the microsporidia are true eukaryotes (i.e., they possess a typical eukaryotic nucleus, endomembrane system, and cytoskeleton), they also display molecular and cytological characteristics of prokaryotes, including features of the translational apparatus and genome size (range of bacteria), and lack recognizable mitochondria, peroxisomes, and typical Golgi membranes. Phylogenetic studies have confirmed that the microsporidia evolved from the fungi, being most closely related to the zygomycetes (28). Other features that are shared with fungi include the presence of chitin and trehalose, similarities in cell cycles, and certain gene organizations. Microsporidia are now considered to be highly derived fungi that underwent genetic and functional losses, thus resulting in one of the smallest eukaryotic genomes known. However, at this point, clinical and diagnostic issues and responsibilities may remain with the parasitologists, and these organisms will be maintained as a part of this parasitology text. We may be in a transition stage, similar to that seen with *Pneumocystis jiroveci* as it was moved from the parasites to fungi in terms of classification status.

During the early part of the 19th century, an epidemic disease was causing tremendous problems in the silkworm industry, particularly in Italy and France; the disease was called pébrine, which means "pepper disease." In 1863 Louis Pasteur was asked to study this problem, and by 1870 he had published his findings, in which he described a method for controlling and preventing the disease. This work represents the first scientific study of a disease caused by a protozoan, a microsporidian parasite, which resulted in practical control measures. Nägeli named the agent *Nosema bombycis*; however, he placed it in the Schizomycetes (at that time consisting of yeast and bacteria). A similar infection is seen in honeybee hives and is caused by *Nosema apis*. In 1882, Balbiani suggested the order Microsporida to handle *N. bombycis*, the only microsporidian known at the time. It was not until 1976 that the phylum Microspora was created, and it was updated in 1992 to propose that Microsporidia Balbiani 1882 should be the acknowledged phylum name, as well as the correct author and date (2).

The microsporidia are obligate intracellular organisms that have been recognized in a variety of animals, particularly invertebrates. Typical sizes range from 1.5 to 5 µm wide and 2 to 7 µm long; unfortunately, the organisms found in humans tend to be quite small, ranging from 1.5 to 2 µm long. Until recently, awareness and understanding of human infections have been marginal; only with increased understanding of AIDS within the immunosuppressed population has attention been focused on these organisms. Limited availability of EM capability has also played a role in the inability to recognize and diagnose these infections. However, with the introduction of newer diagnostic methods, the ability to identify these parasites has definitely improved. The organisms are characterized by having spores containing a polar tubule, which is an extrusion mechanism for injecting the infective spore contents into host cells. To date, seven genera have been recognized in humans: the more common, *Encephalitozoon* and *Enterocytozoon*, and the less common, *Brachiola*, *Pleistophora*, *Trachipleistophora*, *Vittaforma*, and "*Microsporidium*," a catch-all genus for organisms that have not yet been classified (or may never be classified due to a lack of specimen). Classification criteria include spore size, configuration of the nuclei within the spores and developing forms, the number of polar tubule coils within the spore, and the relationship between the organism and host cell.

Microsporidiosis is an important emerging opportunistic infection in HIV-infected patients, and several hundred patients with chronic diarrhea have been seen worldwide. Patients infected with *Cryptosporidium* spp. may also have concurrent infections with microsporidia, and this number may approach 30%. These findings emphasize the importance of considering both organisms as potential causative agents of diarrhea in compromised patients, particularly those with HIV. However, with the current use of antiretroviral combination therapy, the same decrease seen with cryptosporidiosis in this patient group has also been seen with microsporidiosis. Microsporidia also cause disease in organ transplant recipients, children, travelers, contact lens wearers, and the elderly (35).

Life Cycle and Morphology

The spore is the only life cycle stage able to survive outside of the host cell and is the infective stage. The spore normally reaches the new host through ingestion, although other routes of transmission have been identified.

As seen in Figure 4.22, infection occurs with the introduction of infective sporoplasm through the polar tubule into the host cell. The microsporidia multiply extensively within the host cell cytoplasm; the life cycle includes repeated divisions by binary fission (merogony) or multiple fission (schizogony) and spore production (sporogony). Both merogony and sporogony can occur in the same cell at the same time. During sporogony, a thick spore wall is formed, providing environmental protection for this infectious stage of the parasite. An example of infection potential is illustrated by *Enterocytozoon bieneusi*, an intestinal pathogen. The spores are released into the intestinal lumen and are passed in the stool. These spores are environmentally resistant and can then be ingested by other hosts. There is also evidence for inhalation of spores and evidence in animals that suggests that

human microsporidiosis may also be transmitted via the rectal route (28).

Clinical Disease

Microsporidia have been recognized as causing disease in animals as early as the 1920s but were not recognized as agents of human disease until the AIDS pandemic began in the mid-1980s. Several earlier human cases had been reported but were thought to be very unusual. Currently, there are 7 genera and 13 species of microsporidia that have been implicated in human infection; with more than 1,000 species of microsporidia, one would expect confirmation of additional human parasites in the future (Table 4.20) (9, 28).

Enterocytozoon bieneusi. Although currently there are seven genera, one of which is the catch-all genus "*Microsporidium*," human infections are more common with some than with others. A number of cases of infection with *E. bieneusi* have been found in AIDS patients. Chronic intractable diarrhea, fever, malaise, and weight loss are symptoms of *E. bieneusi* infections and are similar to those seen in patients with cryptosporidiosis or isosporiasis. These patients have already been diagnosed with AIDS, and they tend to have four to eight watery, nonbloody stools each day, which can be accompanied

Figure 4.22 Life cycle of microsporidia. (From C. H. Gardiner, R. Fayer, and J. P. Dubey, *An Atlas of Protozoan Parasites in Animal Tissues*, U.S. Department of Agriculture handbook no. 651, U.S. Department of Agriculture, Washington, D.C., 1988.)

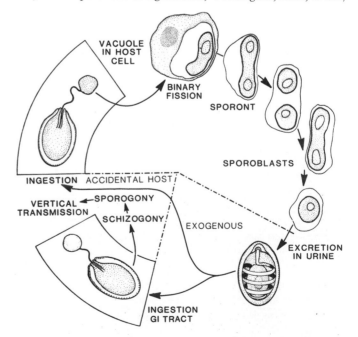

Table 4.20 Microsporidia causing human infection[a]

Species	Nonhuman hosts	Sources of human infection	Sites of infection	Comments
Common				
Enterocytozoon bieneusi	Pigs, nonhuman primates, nonmammals (chickens, pigeons)	Fecal-oral; oral-oral, inhalation, ingestion, direct human-to-human; organ transplants	Small intestine, gallbladder, liver, disseminated (respiratory system)	Short-term culture only; 3 strains identified but not named; AIDS patients with chronic diarrhea, immunocompetent individuals; more than 50 genotypes described (17 infect humans); zoonotic and nonzoonotic genotypes
Encephalitozoon cuniculi				
Strain I (rabbit)	Rabbits, rodents, carnivores, monkeys	Probably ingestion	Cerebrospinal fluid, intestine, respiratory, kidneys, nasal epithelium, eyes	Definitely zoonotic; spores highly resistant in environment (months under humid conditions); waterborne infections feasible
Strain II (mouse)	Mice, wild rats, blue fox	Not relevant	Not relevant	Not identified in humans
Strain III (dog)	Dog, tamarins (may have been imported from United States)	Probably ingestion	Kidneys, adrenal glands, ovaries, brain, heart, spleen, lungs, lymph nodes	Case occurred in Italy; AIDS (may be first report of infection involving ovaries); identified in humans in United States (not confirmed in dogs related to cases)
Encephalitozoon hellem	Psittacine birds	No epidemiologic studies; surface waters not confirmed; probably ingestion, aerosols, or oral or ocular direct inoculation; for drug addicts, contaminated syringes suggested but not proven	Disseminated, ocular HIV-positive patients; patients with eye infections owned birds	Asymptomatic respiratory infections reported; bronchoalveolar lavage in HIV-seronegative patients; stools from travelers (coinfected with *E. bieneusi*); *E. hellem* isolates from the United States and Europe are different populations
Encephalitozoon intestinalis	Dogs, donkeys, pigs, cows, goats, free-ranging gorillas	Probably ingestion, inhalation, direct inoculation; waterborne transmission suspected	Small intestine, disseminated; immunocompromised and immunocompetent	Second most common species infecting humans; double infections with *E. bieneusi* occur
Uncommon				
Brachiola algerae	Mosquitoes	Possible arthropod-borne	Eyes, muscle; possible dissemination; immunocompromised and immunocompetent	Formerly *Nosema algerae*
Brachiola connori	None identified	Unknown	Disseminated	Formerly *Nosema connori*; immunocompromised child (SCID)
Brachiola vesicularum	None identified	Unknown	Corneal stroma, skeletal muscle	Probably a *Nosema*-like species; non-HIV patient
Microsporidium africanum	None identified	Unknown	Corneal ulcer of African woman	Catch-all genus for organisms that could not be identified; HIV-seronegative individual at autopsy
Microsporidium ceylonensis	None identified	Unknown	Corneum	Boy gored by goat (injured eye), but may not be relevant
Pleistophora ronneafiei	Fish, reptiles; possible animal reservoir not defined	Unknown	Muscle	Non-HIV and HIV positive; *Pleistophora* spp. (2 cases) may actually be *Trachipleistophora* spp.
Trachipleistophora anthropophthera	None identified, but may be insects	Unknown	Brain, heart, kidneys, eyes	AIDS patients
Trachipleistophora hominis	None identified, but may be insects	Unknown, but probably direct inoculation	Muscle, nasal epithelium, conjunctiva, keratitis	AIDS patient (severe sinusitis, myositis); immunocompetent patient
Vittaforma corneae	Appear to be ubiquitous	Waterborne very likely; some intestinal infections may reflect intestinal passage of spores rather than infection	Eyes, corneal stroma; possible kidneys	First human isolate established in an in vitro culture; non-HIV patient; formerly *Nosema corneum*

[a] Adapted from E. S. Didier and G. T. Bessinger, Host-parasite relationships in microsporidiosis: animal models and immunology, p. 225–229, *in* M. Wittner and L. M. Weiss (ed.), *The Microsporidia and Microsporidiosis*, ASM Press, Washington, D.C., 1999; additional data from references 5, 9, and 28.

by nausea and anorexia. There may be dehydration with mild hypokalemia and hypomagnesia, as well as D-xylose and fat malabsorption. The patients tend to be severely immunodeficient, with a CD4 count always below 200 and often below 100. A dual infection with *E. bieneusi* and another microsporidian that infects the small intestine has also been reported. *E. bieneusi* infection has been implicated in AIDS-related sclerosing cholangitis. However, demonstration of *E. bieneusi* spores in extraepithelial tissues does not always appear to be associated with development of subsequent systemic infection.

E. bieneusi spores have been identified in the sputum and bronchoalveolar lavage fluid, as well as stool samples, from a patient with a 2-year history of intestinal microsporidiosis. Although no pulmonary pathology could be established in this or one other reported case, it is well established that *E. bieneusi* is capable of colonizing the respiratory tract, and these clinical specimens may reveal the presence of spores. Using transmission EM, multiorgan microsporidiosis due to *E. bieneusi* has been diagnosed in an HIV-infected patient; organisms were recovered in stools, duodenal biopsy specimens, nasal discharge, and sputum.

Infection with this organism has also been found in individuals with intact immune function; symptoms were self-limited, and the diarrheal disease resolved within 2 weeks. These cases suggest that *E. bieneusi* may be more commonly associated with sporadic diarrheal disease than was previously suspected and that the immune system may play a role in the control of this organism within the intestine. It is also quite possible that *E. bieneusi* may persist as an asymptomatic infection in immunocompetent individuals.

Encephalitozoon spp. Both *Encephalitozoon cuniculi* and *E. hellem* have been isolated from human infections, the first species from the central nervous system and the second from the eye. The first case occurred in a 9-year-old Japanese boy who presented with neurologic symptoms such as convulsions, vomiting, headaches, fever, and periods of unconsciousness. The spores were recovered in the cerebrospinal fluid and urine. This particular patient was treated with sulfisoxazole and recovered. Another case of *E. cuniculi* was in a 2-year-old child with a similar neurologic illness; the spores were recovered from the urine. A case of keratoconjunctivitis and chronic sinusitis due to infection with *E. cuniculi* in an AIDS patient has been reported; diagnosis was confirmed by EM. Currently, there are at least three *E. cuniculi* strains which may become more important in the epidemiology of human infections. As additional isolates are characterized, further classification changes will probably occur. Several eye infections with *E. hellem* have been found in

AIDS patients, including the first reported case of *E. hellem* infecting not only sinuses and conjunctivae but also the nasal epithelium. In the first case of disseminated *E. hellem* infection in an AIDS patient, a complete autopsy revealed organisms in the eyes, urinary tract, and respiratory tract. The presence of numerous organisms within the lining epithelium of almost the entire length of the tracheobronchial tree was very suggestive of respiratory acquisition. It is also interesting that although *E. cuniculi* and *E. hellem* differ biochemically and immunologically, their fine structure and development are indistinguishable. Some of the eye infections have also suggested a topical route rather than dissemination. Dual infection with *E. hellem* and *Vittaforma corneae* has also been found in an AIDS patient; this is the second isolation of *V. corneae* and the first description of urinary tract infection due to *V. corneae* in a patient with AIDS.

Encephalitozoon (Septata) intestinalis. *Encephalitozoon (Septata) intestinalis* infects primarily small intestinal enterocytes, but infection does not remain confined to epithelial cells. *E. intestinalis* is also found in lamina propria macrophages, fibroblasts, and endothelial cells. Dissemination to the kidneys, lower airways, and biliary tract appears to occur via infected macrophages (29). Fortunately, these infections tend to respond to therapy with albendazole, unlike the infections with *E. bieneusi*.

Brachiola (Nosema) spp. *Brachiola connori* (formerly referred to as *Nosema connori*) has been identified in human tissues, with the spores being oval and measuring approximately 2 by 4 μm. The polar tubule has about 11 coils. The single human case occurred in a 4-month-old infant with combined immunodeficiency disease with a disseminated, fatal infection. Parasites were found in the myocardium, diaphragm, arterial walls, kidney tubules, adrenal cortex, liver, and lungs. Concurrently, this patient also had *P. jiroveci* pneumonia.

B. algerae (formerly *N. algerae*) is apparently transmitted by mosquitoes. Infections have been found in the eye and muscle, possibly from a disseminated case. Both immunocompromised (HIV-negative, but receiving immunosuppressive therapy for rheumatoid arthritis) and immunocompetent patients have become infected. This organism has been known as an insect pathogen for some time. It appears that an initial infection of the eye allows the parasite to adapt to conditions within the mammalian host, thus spreading via macrophages to internal organs.

B. vesicularum is probably a *Nosema*-like species found in non-HIV-infected patients. Corneal stroma and skeletal muscle were involved in a case in an AIDS patient. Although it was thought that *B. algerae* and *B. vesicularum* might be identical, there are ultrastructural differences between the two organisms. Thus, the designation of two species is currently accepted.

Microsporidium spp. *Microsporidium africanum* were identified in the histiocytes of a corneal ulcer. Although it was first thought to be a *Nosema* sp., electron micrographs confirmed the differences and it was designated *M. africanum*. No information on other hosts is available.

M. *ceylonensis* was found in an eye infection in a child who had been gored by a goat and suffered an eye injury. *Nosema helminthorum* was considered a possibility, since the organism is a parasite of goat tapeworms; however, since a generic classification could not be completed, it was assigned to the genus *Microsporidium*. Since no additional infections have been seen and there is no tissue left for further study, this genus name, *Microsporidium*, will probably remain. Since it is now accepted, you will see it written as *Microsporidium* rather than "*Microsporidium*."

Pleistophora spp. Microsporidia of the genus *Pleistophora* have rarely been identified in humans; however, three cases in immunocompetent patients have been reported. In these cases, atrophic and degenerating muscle fibers were full of spores, which were seen in clusters of about 12 organisms, each cluster enclosed by an enveloping membrane, the pansporoblastic membrane. The spores were oval, measuring approximately 2.8 by 3.4 μm. The disease was characterized by a 7-month history of progressive generalized muscle weakness and contractions in addition to fever, generalized lymphadenopathy, and an 18-kg weight loss. Although the patient was diagnosed as having AIDS, the presence of HIV was never demonstrated. Early biopsies demonstrated atrophic and degenerating muscle fibers; later biopsies showed fibrosis and scarring. *Pleistophora* microsporidia have been identified in a confirmed AIDS patient. Ultrastructural studies of specimens from one of the cases led to the classification of *P. ronneafiei*; no information is available about possible animal reservoirs.

Trachipleistophora spp. *Trachipleistophora hominis* causes severe myositis and sinusitis and was first detected in an AIDS patient. To date, no natural hosts other than humans have been identified, although when inoculated intraperitoneally into athymic mice, the organisms infect the skeletal muscle, as seen in human infections. *T. anthropophthera* is the most recently described microsporidian in humans and has been found in brain tissue of two AIDS patients at autopsy; the organisms were also found in the kidneys and heart.

Vittaforma corneae. It has been recently proposed that the human microsporidian *Nosema corneum* be reclassified in a new genus, *Vittaforma corneae*. This organism has been documented in both immunocompromised (urinary tract) and immunocompetent (ocular) patients. These organisms are found everywhere, but no specific natural common hosts have been identified. PCR and sequencing data from surface water studies suggest that these organisms may be present and that they can also be present in human feces; it has been suggested that this might represent intestinal passage of microsporidial spores, although these patients were symptomatic with diarrhea (29).

Infections in Immunocompetent Hosts. In studies of rabbits and mice, the majority of immunologically competent hosts that become infected with microsporidia develop chronic and persistent infections with few clinical signs or symptoms. Spore shedding has been described as sporadic, and parasite replication seems to be regulated by the host immune system. However, if these animals are intentionally immunosuppressed, more severe symptoms are seen. On the basis of serologic studies, it is very likely that immunocompetent humans have persistent or chronic infections with microsporidia. This is also supported by the fact that patients with microsporidiosis who are HIV-seronegative show clinical resolution of their infections after a few weeks. As improved diagnostic methods, such as PCR, are developed and used, verification of human subclinical microsporidial infections should become possible.

Infections in Immunocompromised Hosts. In the immunocompromised host, microsporidial infection may lead to overwhelming disease and death. The first cases of human microsporidiosis were identified in children with impaired immune systems, and infections have been widely recognized and studied in individuals with AIDS, primarily those with fewer than 100 CD4+ T lymphocytes. Infection has also been recognized in organ transplant recipients who are intentionally immunosuppressed before and after transplantation.

Diagnosis

Routine Histology. A number of techniques are available for recovery and identification of microsporidia in clinical specimens (Tables 4.21 and 4.22). Although the organisms have been identified in routine histologic preparations, they do not tend to stain predictably if at all (Figure 4.23). Occasionally, the spores take on a refractile gold appearance in formalin-fixed, paraffin-embedded, routine hematoxylin-and-eosin-stained sections. Some of the difficulty may be attributed to the use of formalin; alternative fixatives are currently being tried. Spores are occasionally seen very well by using PAS stain, silver stains,

Table 4.21   Microsporidia: general information

Genus	Body site	Diagnosis	Comments
Brachiola, Encephalitozoon, Enterocytozoon, Microsporidium, Pleistophora, Trachipleistophora, Vittaforma (*Enterocytozoon bieneusi* and *Encephalitozoon intestinalis* are the most common pathogenic species recovered in humans)	All body organs, including eyes	Routine histology with hematoxylin and eosin (fair); tissue Gram, acid fast, PAS, silver recommended (spores); other specimens (stool, urine, etc.), modified trichrome stains, optical brightening agents, poly- or monoclonal antibody-based kits (FA, EIA); animal inoculation not recommended—laboratory animals may carry occult infections; EM may be required; some can be cultured, but results are not consistent among the various genera and species	These organisms have been found as insect or other animal parasites; route of infection may be ingestion, inhalation, or direct inoculation (eye); well documented as emerging opportunistic infection in AIDS and other immunocompromised patients; name changes and reclassifications continue to occur; also refer to Table 4.20

Table 4.22   Microsporidia: recommended diagnostic techniques

Technique[a]	Use[b]	Comments[a]
Light microscopy		
Stool specimens (more artifacts)		
Modified trichrome	++	Reliable, available; light infections difficult to identify
Giemsa	−	Not recommended for routine use, hard to read
Optical brightening agents	++	Calcofluor, Fungi-Fluor, Uvitex 2B; sensitive but nonspecific
IF technique	++	Commercial availability limits use; products in development
Other body fluids (fewer artifacts)		
Modified trichrome	++	Reliable, available; light infections difficult to identify
Giemsa	+	Urine, conjunctival swab, BAL, CSF, duodenal aspirate
Optical brightening agents	++	Calcofluor, Fungi-Fluor, Uvitex 2B; sensitive but nonspecific
IF technique	++	Commercial availability limits use; products in development
Routine histology		
Hematoxylin and eosin	+	Sensitivity uncertain with small parasite numbers
PAS	+	Controversy over effectiveness
Modified Gram stains (Brown Brenn, Brown-Hopps)	++	Sensitive, generally recommended
Giemsa	+	Sensitivity uncertain with small parasite numbers
Warthin-Starry	+	Not standardized, may not be necessary
Modified trichrome	++	Reliable, sensitive
IF technique	++	Commercial availability limits use; products in development
Electron microscopy		
Body fluids	+	Specific, sensitivity unknown; used for identification to species level
Tissue sections	++	"Gold standard" for confirmation, but sensitivity lower than for detection of spores in stool or urine; used for identification to species level
Serologic antibody detection	−	Reagents not commercially available, preliminary results controversial
Cultures	−	Generally used in the research setting; continued advances in culture options and organism survival/growth
PCR	−	Availability limited to research laboratories; studies ongoing

[a] IF, immunofluorescence detection procedure; BAL, bronchoalveolar lavage; CSF, cerebrospinal fluid.

[b] −, not available or recommended for routine use; +, reported; ++, techniques in general use (probably most widely used).

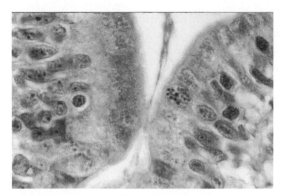

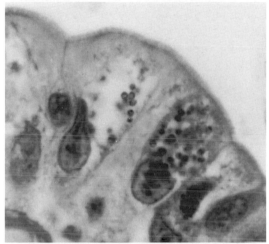

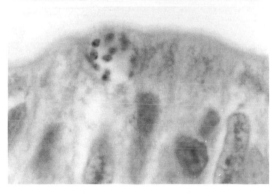

Figure 4.23 Spores in enterocyte. Note the position of the spores between the cell nucleus and the lumen of the intestine. Top photograph taken with a 40× objective; middle and bottom photographs taken with a 100× oil immersion lens. Note the dark-staining dot within the spores in the bottom photograph. The tissue was stained with Giemsa stain.

or acid-fast stains. Modified Gram stains have also been found to be sensitive although not always standardized. The spore has a small, PAS-positive posterior body, the spore coat stains with silver, and the spores are acid-fast variable (Figures 4.23 to 4.26). There is also evidence to indicate that specimens from plastic-embedded tissues are seen more easily, regardless of the fixative used. Tissue

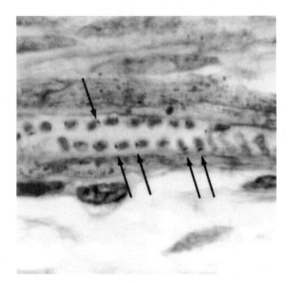

Figure 4.24 Section of appendix showing *Nosema* spores in muscularis. The anterior end of the spore has a PAS-positive granule (arrows) (PAS, ×1,260). (Armed Forces Institute of Pathology photograph.)

examination by EM techniques is still considered the best approach (Figure 4.27); however, this option is not available to all laboratories, and the sensitivity of EM may not be equal to that of other methods when examining stool or urine.

Techniques that do not require tissue embedding are now becoming more popular. Touch preparations of fresh biopsy material that are air dried, methanol fixed, and Giemsa stained have been used; however, screening must be performed at ×1,000 magnification. Another

Figure 4.25 *Nosema* spores in muscularis of jejunum (GMS, ×1,260). (Armed Forces Institute of Pathology photograph.)

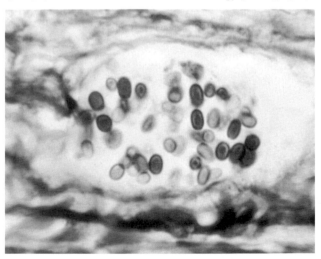

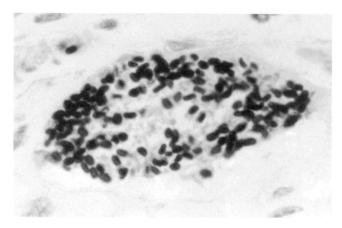

Figure 4.26 Heart. Some *Nosema* spores are acid fast (photographed with green filter; Ziehl-Neelsen, ×1,080). (Armed Forces Institute of Pathology photograph.)

study, using cytocentrifugation followed by Giemsa staining, found that 27% of 55 AIDS patients with chronic diarrhea were positive for *E. bieneusi*; all results were confirmed by EM.

Routine Clinical Specimens. *Modified trichrome stains.* Other, more recent options recommend using modified trichrome stains in which the chromotrope 2R component added to the stain is 10 times the concentration normally used in the routine trichrome stain for stool (9, 23, 29). It is important to remember that the stool preparations must be very thin, the staining time must be 90 min, and the slide must be examined at ×1,000 (or higher) magnification. Unfortunately, there are many objects within stool material that are oval, stain pinkish with trichrome, and measure approximately 1.5 to 3 μm.

Figure 4.27 Spores in cytoplasm of intact jejunal enterocyte of a man with intractable diarrhea and malabsorption; transmission electron micrograph of a jejunal suction biopsy specimen. Note the dark oval developing spores; at higher magnification, some of the elements of the polar tubules would be visible. (Courtesy of R. L. Owen; from M. J. G. Farthing and G. I. Keusch (ed.), *Enteric Infection: Mechanisms, Manifestations and Management*, Chapman & Hall, Ltd., London, 1987.)

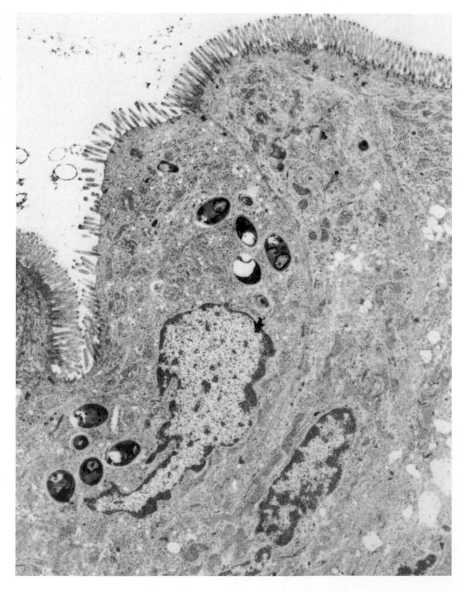

If this stain is used for the identification of microsporidia in stool, positive control material should be available for comparison. Additional modifications of this method include the use of heat and a shorter staining time (23). There is also some evidence to indicate that pretreatment of fecal specimens (1:1) with 10% KOH may provide a better-quality smear to examine when using the modified trichrome stains (Figure 4.28). Unfortunately, there are always situations in which artifact material can be confused with actual spores (Figure 4.29).

Optical brightening agents. Another approach involves the use of chemofluorescent agents (optical brightening agents) such as calcofluor, Fungi-Fluor, or Uvitex 2B (Figure 4.28). These reagents are sensitive but nonspecific; objects other than microsporidial spores will also fluoresce. This is a particular problem when examining stool specimens; both false-positive and false-negative results have been seen. When these reagents are used

Figure 4.28 Microsporidian spores. (A) Spores stained with modified trichrome stain (Ryan blue) (note the diagonal or horizontal "stripes" that are evidence of the polar tubule); (B) calcofluor white staining of urine sediment (note the small oval intracellular and extracellular spores) (this is a nonspecific stain but is more likely to represent a true positive than when seen in a stool specimen containing many fluorescing artifacts).

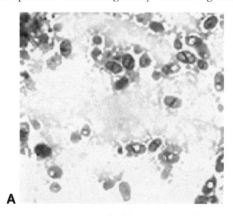

A

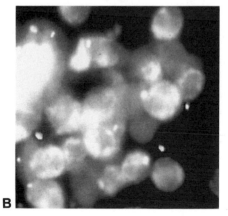

B

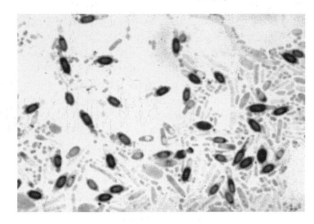

Figure 4.29 Artifacts stained with a modified trichrome stain (Ryan blue). Note the shape and staining characteristics are not consistent with microsporidial spores; also note the lack of horizontal or diagonal "stripes" representing the polar tubule.

with other body fluids, particularly urine, the interpretation of results is much easier than when they are used with stool. There is now general agreement that this method is recommended as a simple, sensitive screening method for the detection of microsporidial spores in stool specimens.

Antigen detection. The newest approach for the identification of spores in clinical specimens uses antisera in an indirect FA procedure (Figure 4.30). Fluorescing microsporidial spores were distinguished by a darker cell wall and by internal visualization of the polar tubule as diagonal lines or cross lines within the cell. In another study with the same antiserum, 9 (30%) of 27 patients who

Figure 4.30 Microsporidian spores. This image shows a positive FA test using *Encephalitozoon*-specific reagent. Note the fluorescing spores, indicating the organisms are within the genus *Encephalitozoon*.

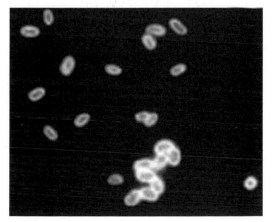

had already been diagnosed as having cryptosporidiosis (8 AIDS patients and 1 non-AIDS but immunodeficient patient) were found to have *Encephalitozoon* spp. in the stool. Although there is some cross-reactivity with bacteria, this technique offers a more sensitive approach than routine staining methods currently available for the examination of stool specimens. Very specific reagents are in various stages of development and clinical testing and should provide some additional, more sensitive methods than are currently available at this time. Some of these reagents are also being used in conjunction with flow cytometry. As clinicians begin to suspect these infections and laboratorians become more familiar with the diagnostic methods, the number of positive patients, particularly those who are immunocompromised, may increase dramatically. Current recommendations also suggest that multiple diagnostic methods may be necessary to diagnose microsporidiosis, particularly when examining fecal specimens (23).

Molecular Methods. Molecular studies of the microsporidia have been limited to date (33, 39). Few genes have been located, and DNA sequence studies have been reported for just a few of the identified human pathogens. Microsporidia have extremely small genomes, very similar to those seen in bacteria. This may be due to the early divergence of microsporidia or may be related to the highly adapted parasitic life-style of these organisms. Although PCR methods have been tried in the research setting, commercial products are not yet available. For HIV-infected patients suspected of having microsporidiosis but with negative stool and biopsy results, PCR testing of intestinal biopsy specimens may be helpful in diagnosing the infection. Application of molecular techniques to diagnosis, identification of isolates to the species level, and classification will certainly continue to expand during the next few years. These advances should include newer diagnostic reagent kits that will simplify molecular testing in the routine clinical laboratory setting, particularly tests applicable to the stool specimen.

Cell Culture. The use of in vitro culture methods continues to provide confirmatory as well as diagnostic information. This approach has been instrumental in the development of immunologic reagents for diagnosis and species differentiation. In vitro culture has also been used to assess the efficacy of antimicrobial agents on several microsporidian species including *Encephalitozoon cuniculi*, *E. hellem*, and *E. intestinalis* (29). However, the use of cell culture in routine clinical diagnosis is not practical and is generally reserved for research facilities. A number of cell lines have been used, including monkey and rabbit kidney cells (Vero and RK13), human fetal lung fibroblasts (MRC-5), and MDCK cells. Although a number of the microsporidia have been cultured using these techniques, one of the most common human microsporidial pathogens, *Enterocytozoon bieneusi*, has been propagated only in short-term cultures (6 months). The inability to grow this organism in continuous culture may reflect additional nutritional requirements that have not yet been identified.

Serologic Testing. A variety of serologic tests (carbon immunoassay, indirect immunofluorescent-antibody test, enzyme-linked immunosorbent assay, counterimmunoelectrophoresis, and Western blotting) have been used to detect immunoglobulin G (IgG) and IgM antibodies to microsporidia, particularly to *E. cuniculi*. The indirect immunofluorescent-antibody test and the enzyme-linked immunosorbent assay have been the most useful because of the simplicity of the test methods; however, the actual sensitivity and specificity of all the available tests have not yet been defined. Antibodies to *E. cuniculi* and *E. intestinalis* have been found in humans both with and without HIV infection; whether this represents true infection, cross-reactivity with other species, or nonspecific reactions is not clear. Serologic surveys for antibodies to *E. cuniculi* appear to show a link between exposure to a tropical environment and microsporidiosis. Unfortunately, there are no serologic assays available for *E. bieneusi* due to the lack of success in long-term culture and development of available antigens. Whether antibodies to this organism will cross-react with those to *E. cuniculi* is not known. At present, the available serologic data are interesting but the tests are not useful as diagnostic tools for the diagnosis of microsporidiosis.

Although *E. bieneusi* appears to be the most common human microsporidian, serologic evidence is based primarily on results for *E. cuniculi* as the parasite antigen. Whether these surveys reveal actual infections with this organism or whether there is cross-reactivity with other microsporidia is unknown. However, the data are highly suggestive that latent infections with microsporidia occur in a number of groups, ranging from patients with other infections such as tuberculosis, typhoid fever, leprosy, malaria, schistosomiasis, toxoplasmosis, Chagas' disease, and toxocariasis to normal individuals and a group of homosexual men. Microsporidian infections in immunocompetent patients may result in a self-cure following mild symptoms over a short time frame, a situation similar to that seen with both cryptosporidiosis and isosporiasis. As awareness of this infection increases and more sensitive diagnostic techniques are developed, we may find that these infections are not uncommon in immunocompetent hosts.

Microsporidia

1. Approximately 30% of AIDS patients and those with cryptosporidiosis may also have infections with microsporidia. Microsporidian infections must be considered for this group of patients with chronic diarrhea.

2. The modified trichrome staining procedure for stool may be difficult to interpret without positive controls to review. Make sure that the material on the slides is very thin, the smear is stained for the recommended time, and the smear is examined under oil immersion (total magnification of at least ×1,000). When using various stains, optical brightening agents, or experimental immunoassay reagents where diagnosis is based on seeing the actual spores, it is recommended that the fecal specimen be concentrated and centrifuged at 500 × g for 10 min. Personal experience in our laboratory has demonstrated a larger number of spores in the concentrated sediment than in the unconcentrated specimen *or in specimens that have been centrifuged at a lower speed or for a shorter time.*

3. The optical brightening agents (calcofluor, Fungi-Fluor, Uvitex 2B) provide a sensitive screening method, but the results are nonspecific. False-positive results have been reported because of fluorescent artifact material.

4. As immunoassay reagents become commercially available, we may see diagnostic procedures with greater specificity and sensitivity. Reagents are currently being used in a few institutions; these preparations are generally easier to read than are the routine stains.

5. Touch preparations can be methanol fixed and stained with Giemsa.

6. Plastic-embedded tissues stained with PAS, silver, acid-fast, and routine hematoxylin-eosin stains generally stain better than paraffin-embedded tissues. This finding may be related to the use of formalin as a tissue fixative. Tissue Gram stains (Brown Brenn, Brown-Hopps) are also highly recommended

7. Be sure to gain some experience in examining these preparations before sending out patient specimen results. This work almost mandates the use of positive control material, regardless of which technique(s) is used. Once a laboratory has gained experience with these methods, it is appropriate to begin accepting clinical specimens.

8. EM is the "gold standard" (very specific) for confirming infection and for attempting to classify the organisms seen in tissues. However, EM procedures may not be as sensitive as some of the other available methods.

9. It is recommended that more than one procedure be used when working with stool specimens (modified trichrome, optical brightening agent); this is primarily because of the large number of artifacts present in stool. Urine and other body fluid specimens are much easier to examine for the presence of microsporidial spores, since less artifact material is seen in these specimens than in stool.

Treatment

Although a number of drugs have been tried, results have been variable. Agents such as metronidazole, itraconazole, octreotide, primaquine, Lomotil, sulfasalazine, loperamide, and albendazole have been used in various patients. In some cases, the diarrhea subsided; however, biopsy specimens showed the continued presence of organisms, which were probably *Enterocytozoon bieneusi.* Over the past few years, confirmatory evidence indicates that a complete parasitological cure is possible with albendazole. However, these patients had disseminated infections with *Encephalitozoon* (*Septata*) *intestinalis*; albendazole, a benzimidazole that binds to β-tubulin, appears to be very effective in treating this particular organism (1). Currently, the recommended dose for ocular (*Encephalitozoon cuniculi, Vittaforma corneae*), intestinal (*Enterocytozoon bieneusi, Encephalitozoon* [*Septata*] *intestinalis*), and disseminated (*Encephalitozoon hellem, E. cuniculi, E. intestinalis, Pleistophora* spp., *Trachipleistophora* spp.) infections with albendazole is 400 mg orally twice a day. In vitro, albendazole has activity at concentrations of <0.1 μg/ml against all of the Encephalitozoonidae. Although apparently static rather than cidal effects are seen with *E. bieneusi* infections, treatment with albendazole results in reduction of symptoms in as many as 50% of patients infected with this organism. Albendazole as a systemic agent is recommended when the organisms have been confirmed in urine or nasal smears.

Fumagillin (soluble salt Fumidil B) has activity against microsporidia, and solutions applied topically have been used in corneal infections. The effects of this drug are static rather than cidal, and relapses of infection occur when the treatment has been discontinued. In one study, the efficacy of fumagillin was measured by clearance of *E. bieneusi* from stools and intestinal biopsy specimens; four patients who received fumagillin remained free of *E. bieneusi* after a mean follow-up of 10 months.

Itraconazole can also be recommended to treat ocular, nasal, and paranasal sinus infection caused by *E. cuniculi* parasites when albendazole treatment fails. In a patient with sinusitis and keratoconjunctivitis, the symptoms completely resolved with itraconazole treatment (200 mg/day for 8 weeks) after albendazole therapy (400 mg/day for 6 weeks) was unsuccessful.

Epidemiology and Prevention

The sources of human infections are not yet totally defined. From what we currently know, possibilities include human-to-human and animal-to-human transmission (29). Although there is some speculation that insect microsporidia can infect humans, this is still unknown. Many questions relating to reservoir hosts and possible congenital infections are still unanswered. Primary infection can occur by inhalation or ingestion of spores from environmental sources or by zoonotic transmission. There is serologic evidence to indicate widespread occurrence of antibodies to *E. cuniculi* in humans. Recent data indicate that the presence of *Encephalitozoon intestinalis* was confirmed in tertiary sewage effluent, surface water, and groundwater; *Enterocytozoon bieneusi* was confirmed in surface water; and *Vittaforma corneae* was confirmed in tertiary effluent. This study represents the first confirmation, to the species level, of human-pathogenic microsporidia in water, indicating that these parasites may be waterborne pathogens.

The presence of infective spores in human clinical specimens suggests that precautions when handling body fluids and personal hygiene measures such as hand washing may be important in preventing primary infections in the health care setting. However, comprehensive guidelines for disease prevention will require more definitive information regarding sources of infection and modes of transmission.

Algae (*Prototheca*)

Although algae are not parasites, it is important to know what they are and the relationship of opportunistic infections with these organisms to the compromised patient. The genus *Prototheca* contains several species, the most prevalent of which is *Prototheca wickerhamii*. These organisms are achlorophyllic algae found in the slime flux of trees and freshwater environments. Cutaneous infections were reported more than 30 years ago, and subsequent reports describe other manifestations including systemic disease. The other species implicated in human infections is *P. zopfii* (10, 37). Susceptibility to infection is not well defined but may involve the inability of neutrophils to effectively kill phagocytosed organisms. Approximately 100 cases have been reported throughout the temperate and tropical areas of all continents. Protothecosis is being reported with increasing frequency, and within the United States the majority of cases have been reported from the southeastern states (10).

Life Cycle and Morphology

Prototheca spp. are unicellular organisms that reproduce asexually by internal septation and cleavage; they then rupture, releasing endospores. The sporangium contains 2 to 16 sporangiospores, which develop into mature endosporulating cells. No sexual reproductive stages have been identified. These organisms are thought to be achlorophyllous mutants of green algae in the genus *Chlorella*.

Clinical Disease

The overall incubation time is not known but is generally considered to be about 2 weeks from the time of trauma and possible implantation of infectious organisms. The majority of infections involve the skin and underlying structures, such as the olecranon bursa (10, 37); however, disseminated infection may also occur. Cutaneous or bursal infection may result from accidental inoculation or from trauma, including surgical. One patient developed nasopharyngeal ulceration due to *P. wickerhamii* after prolonged endotracheal intubation. Bursal infection presents with indolent swelling and tenderness. Cutaneous lesions have been described as painless papules, papulonodules, diffuse erythema, plaques (may be verrucous), and eczematoid or ulcerated areas. Immunocompromised patients may have more severe lesions. In addition to disseminated infection, peritonitis has been found in patients undergoing chronic ambulatory peritoneal dialysis and in the blood of patients with central venous access infections. Infections have also been seen in the gallbladder, peritoneum, and liver in a patient presenting with symptoms consistent with sclerosing cholangitis.

One 80-year-old woman receiving 2 mg of dexamethasone daily presented with tenderness in an area of diffuse erythema of the right arm, forearm, and dorsum of the hand together with several 4- to 6-mm flesh-colored papules. Biopsy revealed granulomatous inflammation with many organisms present in the dermis (10, 37). These organisms stained with both PAS and Gomori's methenamine-silver (GMS) stain and were approximately 5 to 15 μm in diameter. They contained multiple internal septa. Another patient presented with recalcitrant ulcerated papules and plaques on both legs.

This infection has also been found in an HIV-positive patient (37). The patient presented with a nodule in the extensor face of the thumb. A routine excision was performed, and histology revealed numerous granulomas, some with central fibrinoid necrosis. Large numbers of prototheca were found in PAS and GMS stains, showing endosporulation with diagnostic morula- or daisy-like sporangia. On the basis of the relative number of polymorphonuclear leukocytes, lymphocytes, and plasma cells present, the cellular response was considered minimal. The organisms were found extracellularly or within macrophages and giant cells. Protothecal meningitis has also been found in an AIDS patient. In general, these cases are seen in patients who are immunocompromised, either from other underlying diseases or from immunosuppressive therapy (10).

Diagnosis

These algae are easily recovered on routine culture media; however, they are inhibited by cycloheximide, which is present in many selective fungal media. The optimal temperature for growth is 30°C, and these algae grow as white to creamy yeast-like colonies on Sabouraud's medium. Microscopic examination of culture material reveals the same structures as seen in tissue, including spherical sporangia containing multiple sporangiospores (endospores) (Figures 4.31 to 4.33). The organisms have been described as looking like a spoked wheel. Macroscopically, the cells are variable in size and shape and do not bud or form hyphae. Asexual reproduction occurs through release of sporangiospores from sporangia. Sporangia of *P. zopfii* range from 14 to 16 μm in diameter,

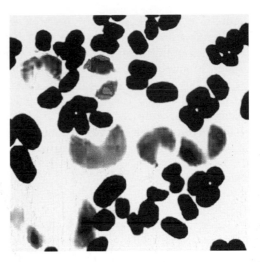

Figure 4.33 *Prototheca zopfii*. Ruptured sporangia showing the released sporangiospores (oval to cylindrical).

and those of *P. wickerhamii* are 7 to 13 μm. Cells of *P. wickerhamii* tend to be round, while those of *P. zopfii* are oval or cylindrical. Both species assimilate glucose and galactose; *P. wickerhamii* also assimilates trehalose. These organisms are nonfermentative.

In histologic preparations, the typical tissue response includes a granulomatous inflammation with multinuclear giant cell formation. The organisms may be numerous or few and can be seen as single cells or clusters. The GMS stain is recommended. In tissue the sporangia are usually larger (30 μm) than those seen in culture (*P. wickerhamii*, 7 to 13 μm; *P. zopfii*, 14 to 16 μm).

Figure 4.31 Sporangia containing sporangiospores (endospores) of *Prototheca wickerhamii*. (Illustration by Sharon Belkin.)

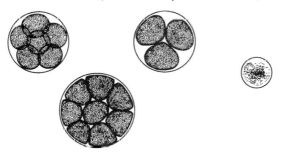

Figure 4.32 Sporangia containing developing sporangiospores (endospores) of *Prototheca* spp.

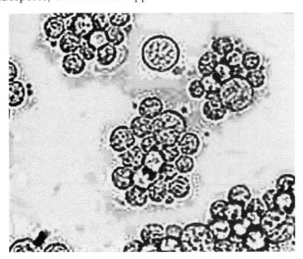

Treatment

Most of the protothecal infections are indolent and very slow to heal. Although surgical debridement or excision may be necessary and effective, most infections are treated with antifungal agents. Combined therapy with tetracycline and amphotericin B appears to be effective. Infections that are multifocal or visceral or occur in an immunocompromised host require amphotericin B or combined therapy. Recent reports indicate that itraconazole is effective; however, healing may be gradual. Infections caused by *Prototheca* spp. tend to be resistant to 5-fluorocytosine.

Epidemiology and Prevention

Prototheca spp. have been isolated from a number of different environmental sites, including slime flux of trees, freshwater and seawater, soil, and sewage. Organisms have also been isolated from dogs, cats, cattle, sheep, deer, and some wild animals. *Prototheca* spp. have been recovered from human feces, sputum, skin scrapings, and fingernails, even in cases with no true infectious process. Although infections caused by *Prototheca* spp. are relatively uncommon, the true incidence worldwide may be underestimated. These infections should always be suspected in cases where the skin and subcutaneous tissues are involved and/or olecranon bursitis is present with indolent swelling and tenderness. An awareness of the possibility of infection with these organisms is critical to accurate diagnosis and prompt treatment.

References

1. Abramowicz, M. (ed). 2004. Drugs for parasitic infections. *Med. Lett. Drugs Ther.* **46**:1–12.

2. Beaver, P. C., R. C. Jung, and E. W. Cupp. 1984. *Clinical Parasitology*, 9th ed. Lea & Febiger, Philadelphia, Pa.

3. Betancourt, W. Q., and J. B. Rose. 2004. Drinking water treatment processes for removal of *Cryptosporidium* and *Giardia*. *Vet. Parasitol.* **126**:219–234.

4. Butkus, M. A., J. A. Starke, D. D. Bowman, M. Labare, E. A. Fogarty, A. Lucio-Forster, J. Barbi, M. B. Jenkins, and M. Pavlo. 2005. Do iodine water purification tablets provide an effective barrier against *Cryptosporidium parvum*? *Mil. Med.* **170**:83–86.

5. Cali, A., and P. M. Takvorian. 2003. Ultrastructure and development of *Pleistophora ronneafiei* n. sp., a Microsporidium (Protista) in the skeletal muscle of an immune-compromised individual. *J. Eukaryot. Microbiol.* **50**:77–85.

6. Cama, V. A., C. Bern, I. M. Sulaiman, R. H. Gilman, E. Ticona, A. Vivar, V. Kawai, D. L. Vargas, and L. Xiao. 2003. *Cryptosporidium* species and genotypes in HIV-positive patients in Lima, Peru. *J. Eukaryot. Microbiol.* **50** (Suppl):531–533.

7. Centers for Disease Control and Prevention. 2004. Outbreak of cyclosporiasis associated with snow peas—Pennsylvania, 2004. *Morb. Mortal. Wkly. Rep.* **53**:1–3.

8. Corso, P. S., M. H. Kramer, K. A. Blair, D. G. Addiss, J. P. Davis, and A. C. Haddix. 2003. Cost of illness in the 1993 waterborne *Cryptosporidium* outbreak, Milwaukee, Wisconsin. *Emerg. Infect. Dis.* **9**:426–431.

9. Didier, E. S. 2005. Microsporidiosis: an emerging and opportunistic infection in humans and animals. *Acta Trop.* **94**:61–76.

10. DiPersio, J. R. 2001. *Prototheca* and prothothecosis. *Clin. Microbiol. Newsl.* **23**:115–119.

11. Dixon, B. R., J. M. Bussey, L. J. Parrington, and M. Parenteau. 2005. Detection of *Cyclospora cayetanensis* oocysts in human fecal specimens by flow cytometry. *J. Clin. Microbiol.* **43**:2375–2379.

12. Fayer, R. 2004. *Cryptosporidium*: a waterborne zoonotic parasite. *Vet. Parasitol.* **126**:37–56.

13. Fayer, R. 2004. *Sarcocystis* spp. in human infections. *Clin. Microbiol. Rev.* **17**:894–902.

14. Fayer, R., J. M. Trout, E. J. Lewis, M. Santin, L. Zhou, A. A. Lal, and L. Xiao. 2003. Contamination of Atlantic coast commercial shellfish with *Cryptosporidium*. *Parasitol. Res.* **89**:141–145.

15. Garcia, L. S., R. Y. Shimizu, S. Novak, M. Carroll, and F. Chan. 2003. Commercial assay for detection of *Giardia lamblia* and *Cryptosporidium parvum* antigens in human fecal specimens by rapid solid-phase qualitative immunochromatography. *J. Clin. Microbiol.* **41**:209–212.

16. Giangaspero, A., U. Molini, R. Iorio, D. Traversa, B. Paoletti, and C. Giansante. 2005. *Cryptosporidium parvum* oocysts in seawater clams (*Chameleagallina*) in Italy. *Prev. Vet. Med.* **69**:203–212.

17. Gomez-Couso, H., F. Freire-Santos, G. A. Hernandez-Cordova, and M. E. Ares-Mazas. 2005. A histological study of the transit of *Cryptosporidium parvum* oocysts through clams (*Tapes decussates*). *Int. J. Food Microbiol.* **102**:57–62.

18. Graczyk, T. K., B. H. Grimes, R. Knight, A. J. Da Silva, N. J. Peiniazek, and D. A. Veal. 2003. Detection of *Cryptosporidium parvum* and *Giardia lamblia* carried by synanthropic flies by combined fluorescent in situ hybridization and a monoclonal antibody. *Am. J. Trop. Med. Hyg.* **68**:228–232.

19. Harwood, V. J., A. D. Levine, T. M. Scott, V. Chivukula, J. Lukasik, S. R. Farrah, and J. B. Rose. 2005. Validity of the indicator organism paradigm for pathogen reduction in reclaimed water and public health protection. *Appl. Environ. Microbiol.* **71**:3163–3170.

20. Hsu, B. M. 2003. Evaluation of analyzing methods for *Giardia* and *Cryptosporidium* in a Taiwan water treatment plant. *J. Parasitol.* **89**:369–371.

21. Huamanchay, O., L. Genzlinger, M. Iglesias, and Y. R. Ortega. 2004. Ingestion of *Cryptosporidium* oocysts by *Caenorhabditis elegans*. *J. Parasitol.* **90**:1176–1178.

22. Hunter, P. R., and G. Nichols. 2002. Epidemiology and clinical features of *Cryptosporidium* infection in immunocompromised patients. *Clin. Microbiol. Rev.* **15**:145–154.

23. Isenberg, H. D. (ed.). 2004. *Clinical Microbiology Procedures Handbook*, 2nd ed. ASM Press, Washington, D.C.

24. Katanik, M. T., S. K. Schneider, J. E. Rosenblatt, G. S. Hall, and G. W. Procop. 2001. Evaluation of ColorPAC *Giardia/Cryptosporidium* rapid assay and ProSpecT *Giardia/Cryptosporidium* microplate assay for detection of *Giardia* and *Cryptosporidium* in fecal specimens. *J. Clin. Microbiol.* **39**:4523–4525.

25. Kato, S., M. B. Jenkins, E. A. Fogarty, and D. D. Bowman. 2002. Effects of freeze-thaw events on the viability of *Cryptosporidium parvum* oocysts in soil. *J. Parasitol.* **88**:718–722.

26. Kim, H. C., and M. C. Healey. 2001. Infectivity of *Cryptosporidium parvum* oocysts following cryopreservation. *J. Parasitol.* **87**:1191–1194.

27. Kniel, K. E., and M. C. Jenkins. 2005. Detection of *Cryptosporidium parvum* oocysts on fresh vegetables and herbs using antibodies specific for a *Cryptosporidium parvum* viral antigen. *J. Food Prot.* **68**:1093–1096.

28. Mansfield, L. S., and A. A. Gajadhar. 2004. *Cyclospora cayetanensis*, a food- and waterborne coccidian parasite. *Vet. Parasitol.* **126**:73–90.

29. Mathis, A., R. Weber, and P. Deplazes. 2005. Zoonotic potential of the microsporidia. *Clin. Microbiol. Rev.* **18**:423–445.

30. **McCuin, R. M., and J. L. Clancy.** 2005. Methods for the recovery, isolation and detection of *Cryptosporidium* oocysts in wastewaters. *J. Microbiol. Methods* **63:**73–88.

31. **Miller, W. A., E. R. Atwill, I. A. Gardner, M. A. Miller, H. M. Fritz, R. P. Hedrick, A. C. Melli, N. M. Barnes, and P. A. Conrad.** 2005. Clams (*Corbicula fluminea*) as bioindicators of fecal contamination with *Cryptosporidium* and *Giardia* spp. in freshwater ecosystems in California. *Int. J. Parasitol.* **35:**673–684.

32. **Morgan-Ryan, U. M., A. Fall, L. A. Ward, N. Hijjawi, I. Sulaiman, R. Fayer, R. C. Thompson, M. Olson, A. Lal, and L. Xiao.** 2002. *Cryptosporidium hominis* n. sp. (Apicomplexan: Cryptosporidiidae) from *Homo sapiens*. *J. Eukaryot. Microbiol.* **49:**433–440.

33. **Notermans, D. W., R. Peek, M. D. de Jong, E. M. Wentink-Bonnema, R. Boom, and T. van Gool.** 2005. Detection and identification of *Enterocytozoon bieneusi* and *Encephalitozoon* species in stool and urine specimens by PCR and differential hybridization. *J. Clin. Microbiol.* **43:**610–614.

34. **Pena, H. F., S. Ogassawara, and I. L. Sinhorini.** 2001. Occurrence of cattle *Sarcocystis* species in raw kibbe from Arabian food establishments in the city of Sao Paulo, Brazil, and experimental transmission to humans. *J. Parasitol.* **87:**1459–1465.

35. **Pozio, E., F. Rivasi, and S. M. Caccio.** 2004. Infection with *Cryptosporidium hominis* and reinfection with *Cryptosporidium parvum* in a transplanted ileum. *APMIS* **112:**309–313.

36. **Shields, J. M., and B. H. Olson.** 2003. *Cyclospora cayetanensis*: a review of an emerging parasitic coccidian. *Int. J. Parasitol.* **33:**371–391.

37. **Thiele, D., and A. Bergmann.** 2002. Protothecosis in human medicine. *Int. J. Hyg. Environ. Health* **204:**297–302.

38. **Visvesvara, G. S., H. Moura, E. Kovacs-Nace, S. Wallace, and M. L. Eberhard.** 1997. Uniform staining of *Cyclospora* oocysts in fecal smears by a modified safranin technique with microwave heating. *J. Clin. Microbiol.* **35:**730–733.

39. **Wolk, D. M., S. K. Schneider, N. L. Wengenack, L. M. Sloan, and J. E. Rosenblatt.** 2002. Real-time PCR method for detection of *Encephalitozoon intestinalis* from stool specimens. *J. Clin. Microbiol.* **40:**3922–3928.

40. **Xiao, L., R. Fayer, U. Ryan, and S. J. Upton.** 2004. *Cryptosporidium* taxonomy: recent advances and implications for public health. *Clin. Microbiol. Rev.* **17:**72–97.

41. **Xu, P., G. Widmer, Y. Wang, L. S. Ozaki, J. M. Alves, M. G. Serrano, D. Puiu, P. Manque, D. Akiyoshi, A. J. Mackey, W. R. Pearson, P. H. Dear, A. T. Bankier, D. L. Peterson, M. S. Abrahamsen, V. Kapur, S. Tzipori, and G. A. Buck.** 2004. The genome of *Cryptosporidium hominis*. *Nature* **431:**1107–1112.

42. **Zhou, L., A. Singh, J. Jiang, and L. Xiao.** 2003. Molecular surveillance of *Cryptosporidium* spp. in raw wastewater in Milwaukee: implications for understanding outbreak occurrence and transmission dynamics. *J. Clin. Microbiol.* **41:**5254–5257.

Free-Living Amebae

Naegleria fowleri

Acanthamoeba spp.

Balamuthia mandrillaris

Sappinia diploidea

Infections caused by small, free-living amebae belonging to the genera *Naegleria*, *Acanthamoeba*, *Balamuthia*, and *Sappinia* are generally not very well known or recognized clinically. Also, methods for laboratory diagnosis are unfamiliar and not routinely offered by most laboratories. However, infections caused by these organisms include primary amebic meningoencephalitis (PAM) caused by *Naegleria fowleri*; granulomatous amebic encephalitis (GAE) caused by *Acanthamoeba* spp. and *Balamuthia mandrillaris*; and amebic keratitis, cutaneous lesions, and sinusitis caused by *Acanthamoeba* spp. (Table 5.1). Another organism, recently identified, has been linked to encephalitis; *Sappinia diploidea* has been confirmed as a newly recognized human pathogen. Effective therapy remains a problem with these infections. When free-living, these organisms feed on bacteria and nutrients in moist soil and in freshwater and marine water. The development of diagnostic molecular methods represents a specific and potentially rapid method for the identification of some of these organisms soon after their primary isolation from the environment (6, 7, 18).

Although it is well known that both free-living and pathogenic protozoa harbor a variety of endosymbiotic bacteria, it is unclear what role these organisms play in terms of host survival, infectivity, and invasiveness. The study of parasites and symbionts of free-living amebae is relatively new, and many of the terms used to describe these relationships are listed in Table 5.2 (10, 15). Serologic evidence that some of the endosymbionts within the free-living amebae might be human pathogens has led to additional studies of these relationships. The ameba-resistant bacteria are varied; some are natural hosts of free-living amebae, some have been recovered by amebic coculture, and some are resistant to a particular amebic host in vitro. Some bacteria are obligate intracellular organisms, others are facultative intracellular bacteria, and some are extracellular. While some of these bacteria are established human pathogens, some are emerging potential pathogens which were formerly considered environmental species. The presence of intracellular bacteria has been detected in axenically grown *Acanthamoeba* isolates. *Legionella pneumophila* can infect, multiply within, and kill both *Naegleria* and *Acanthamoeba* amebae. These amebae may be natural hosts for *Legionella* organisms as well as for other bacteria such as *Campylobacter jejuni*, *Listeria monocytogenes*, *Vibrio cholerae*, *Mycobacterium leprae*, *Burkholderia cepacia*, *Burkholderia pickettii*, and *Pseudomonas aeruginosa* (3, 10, 15). Because of their well-known resistance to chlorine, the amebic cysts are considered to be vectors for these intracellular

Table 5.1   Free-living amebae causing disease in humans[a]

Characteristic	Acanthamoeba			Balamuthia	Naegleria
Disease parameter	GAE	*Acanthamoeba* keratitis	Cutaneous lesions, sinusitis	GAE	PAM
General disease description	Chronic, protracted, slowly progressive CNS infection (may involve lungs); generally occurs in individuals with underlying diseases	Painful, progressive, sight-threatening corneal disease; patients generally immunocompetent	Most common in patients with AIDS, with or without CNS involvement, and in those receiving immunosuppressive therapy for organ transplantation	Chronic, protracted, slowly progressive CNS infection (may involve lungs); generally occurs in individuals with underlying diseases	Rare but nearly always fatal infection; migration of amebae to brain through olfactory nerve; symptoms can mimic bacterial meningitis; death usually occurs 3–7 days after onset of symptoms; clinical suspicion based on history critical
Entry into the body	Olfactory epithelium, respiratory tract, skin, sinuses	Corneal abrasion	Skin, sinuses, respiratory tract	Olfactory epithelium, skin, respiratory tract	Olfactory epithelium
Incubation	Weeks to months	Days	Weeks to months	Weeks to months	Days
Clinical symptoms	Confusion, headache, stiff neck, irritability	Blurred vision, photophobia, inflammation, corneal ring, pain	Skin lesions, nodules, sinus lesions, sinusitis	Slurred speech, muscle weakness, headache, nausea, seizures	Headache, nausea, vomiting, confusion, fever, stiff neck, seizures, coma
Disease pathology	Focal necrosis, granulomas	Corneal ulceration	Granulomatous reaction in skin, inflammation	Multiple necrotic foci, inflammation, cerebral edema	Hemorrhagic necrosis
Diagnostic methods[b]	Brain biopsy, CSF smear/wet prep, culture, indirect immunofluorescence of tissue, PCR	Corneal scrapings or biopsy, stain with calcofluor white, culture, confocal microscopy	Skin lesion biopsy, culture, indirect immunofluorescence of tissue	Brain biopsy, culture on mammalian cells, indirect immunofluorescence of tissue	Brain biopsy, CSF wet prep, culture, indirect immunofluorescence of tissue, PCR

[a] Table information adapted from references 5, 12, 15, 17, 20, and 22.
[b] Indirect immunofluorescence on tissue and PCR methods available from the Centers for Disease Control and Prevention.

Table 5.2   Definitions pertaining to the relationships among free-living amebae and other organisms[a]

Description	Definition
Adaptation	Selection pressure = change in an organism
Ameba-resistant microorganisms	Microorganisms that evolved to resist destruction by amebae, but being neither internalized nor killed while within the amebae
Ameba-resistant bacteria	Bacteria that evolved to resist killing by amebae
Character	Phenotypic characteristics of an organism
Crib	"Baby bed"—free-living amebae that serve as a reservoir and incubator of new antibiotic-resistant bacteria = adaptation to human macrophages
Commensalism	Symbiosis—one organism benefits and the other neither is harmed nor benefits
Endosymbiont	Symbiont that lives within another organism
Endosymbiotic	During portion of life cycle or throughout; no lysis activity
Lytic	Lyses the host cell
Mutualism	Symbiosis in which both organisms benefit
Parasite	Benefits from an association but is harmful to the host
Symbiont	Lives in close contact with another organism during most of life cycle
Symbiosis	Two organisms living in close contact for most of their life spans
Symbiosis island	Gene cluster that confers symbiotic traits, which may be transferred horizontally
Trojan horse	Protozoan "horse" that may bring ameba-resistant organisms into the human host
Virulence	Pathogenicity

[a] Adapted from references 10 and 15.

bacteria. This can have tremendous significance for any hospital where the water source is contaminated with free-living amebae. There is also evidence to suggest that the presence of amebae in domestic water supplies may provide growth conditions that enhance the pathogenicity of the organisms for the human host (23). These amebae are also acceptable hosts for *Chlamydia pneumoniae*, echoviruses, and polioviruses (10).

Naegleria fowleri

Culbertson and his colleagues in 1958 were the first to develop the concept that free-living soil and water amebae could cause disease in humans. A number of fatal cases of acute meningoencephalitis have been reported since that time (20, 21). Almost all patients whose cases were documented in the years immediately following this awareness of amebic infections had a history of swimming in freshwater lakes or ponds a few days prior to the onset of symptoms.

Infections of the central nervous system (CNS) caused by free-living amebae have been recognized only since the mid-1960s, and our understanding of this disease process is still incomplete. One type of meningoencephalitis (PAM) is a fulminant and rapidly fatal disease that affects mainly children and young adults. The disease closely resembles bacterial meningitis but is caused by *N. fowleri*, an organism found in moist soil and freshwater habitats. Close to 200 cases of PAM have occurred worldwide, and approximately 90 of those cases have been reported from the United States. Until recently, it was thought that this infection was limited to humans; however, infections have also been reported in other animals.

Life Cycle and Morphology

There are both trophozoite and cyst stages in the life cycle, with the stage primarily depending on environmental conditions (Figure 5.1). Trophozoites can be found in water or moist soil and can be maintained in tissue culture or other artificial media.

Morphology of Trophozoites.

The trophozoites can occur in two forms, ameboid and flagellate (Figures 5.2 through 5.4). Motility can be observed in hanging-drop preparations from cultures of cerebrospinal fluid (CSF); the ameboid form (the only form recognized in humans) is elongate with a broad anterior end and tapered posterior end (Table 5.3). The size ranges from 7 to 35 µm. The diameter of the rounded forms is usually 15 µm. There is a large, central karyosome and no peripheral nuclear chromatin. The cytoplasm is somewhat granular and contains vacuoles (Figure 5.2). The ameboid-form

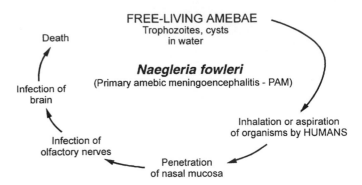

Figure 5.1 Life cycle of *Naegleria fowleri*.

organisms change to the transient, pear-shaped flagellate form when they are transferred from culture or teased from tissue into water and maintained at a temperature of 27 to 37°C. The change may occur very quickly (within a few hours) or may take as long as 20 h. The flagellate form has two flagella at the broad end. Motility is typical, with either spinning or jerky movements. These flagellate forms do not divide, but when the flagella are lost, the ameboid forms resume reproduction.

Morphology of Cysts.

Cysts from nature and from agar cultures look the same and have a single nucleus almost identical to that seen in the trophozoite. They are generally round, measuring from 7 to 15 µm, and there is a thick double wall (Figure 5.2; Table 5.3).

Figure 5.2 Diagram of trophozoite. (Upper row) Flagellate and cyst forms of *Naegleria fowleri*; (lower row) trophozoite and cyst of *Acanthamoeba* spp. (Illustration by Sharon Belkin.)

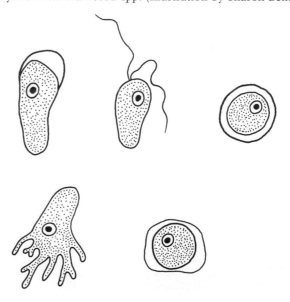

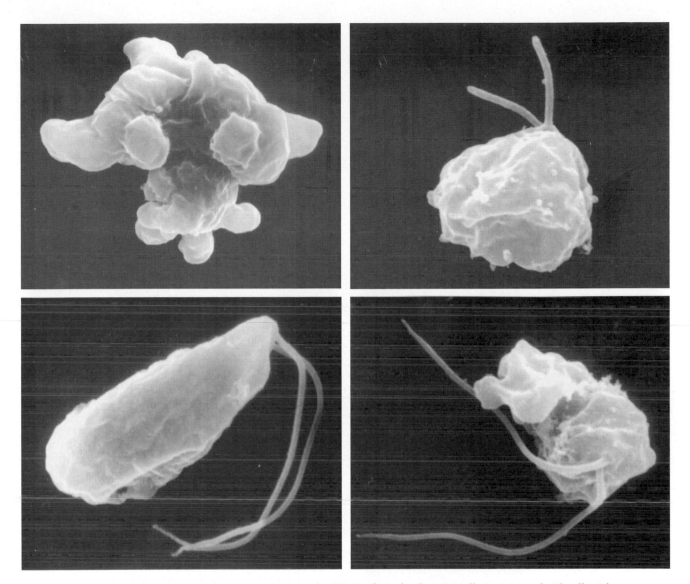

Figure 5.3   (Upper left) Scanning electronic micrograph (SEM) of *N. fowleri* (Lovell strain) ameboid cell with numerous blunt pseudopods. (Upper right) SEM of *N. fowleri* (Lovell strain) preflagellate, rounded and producing two short, blunt flagella. (Lower left) SEM of *N. fowleri* (Lovell strain) flagellate with two long, pointed flagella emerging from the anterior rostrum. (Lower right) SEM of *N. fowleri* (Lovell strain) reverting flagellate with two long, pointed flagella. (Photographs courtesy of D. T. John. Reprinted from B. L. Cable and D. T. John, Conditions for maximum enflagellation in *Naegleria fowleri*, *J. Protozool.* 33:467–472, 1986, with permission.)

Table 5.3   Comparison of *Naegleria fowleri* and *Acanthamoeba* spp.

Characteristic	*Naegleria fowleri*	*Acanthamoeba* spp.
Trophozoite	Biphasic (amebic) and flagellate forms; 8–15 (7–35) μm; lobate pseudopodia (amebic form)	Large (15–25 μm): no flagella; filiform pseudopodia
Cysts	Not present in tissue; small, smooth, rounded; 7–15 μm	Present in tissue; large with wrinkled double wall
Growth on media	Require living cells (bacteria or cell culture); do not grow with >0.4% NaCl	May grow without bacteria; not affected by 0.85% NaCl
Appearance in tissue[a]	Smaller than *Acanthamoeba* spp.; dense endoplasm; less distinct nuclear staining	Large; rounded; less endoplasm; nucleus more distinct

[a] *Entamoeba histolytica* has a delicate nuclear membrane and a small, pale-staining nucleolus. Freshwater amebae have a distinct nuclear membrane and a large, deep-staining nucleolus.

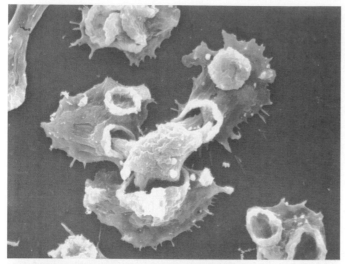

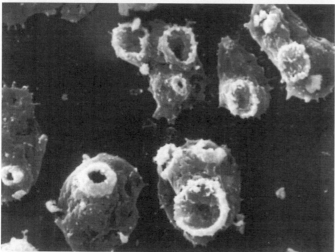

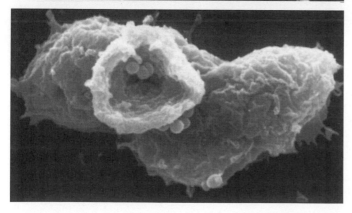

Figure 5.4 (Top) Three amebae of the pathogenic free-living ameba *N. fowleri* attacking and engulfing a fourth, presumably dead, ameba. The amebae are using sucker-like structures (amebostomes) in this novel form of phagocytosis. (Photograph courtesy of D. T. John. *In* D. T. John, T. B. Cole, Jr., and F. M. Marciano-Cabral, Sucker-like structures on the pathogenic amoeba *Naegleria fowleri*, *Appl. Environ. Microbiol.* **47:**12–14, 1984). (Middle) *N. fowleri* trophozoites; (bottom) *N. fowleri* trophozoite ingesting starch granules. (Middle and bottom images courtesy of F. M. Marciano-Cabral.)

Clinical Disease

Amebic meningoencephalitis caused by *N. fowleri* is an acute, suppurative infection of the brain and meninges (Figure 5.5). With extremely rare exceptions, the disease is rapidly fatal in humans. The period between contact with the organism and onset of clinical symptoms such as fever, headache, and rhinitis may vary from 2 to 3 days to as long as 7 to 15 days.

The amebae may enter the nasal cavity by inhalation or aspiration of water, dust, or aerosols containing the trophozoites or cysts. The organisms then penetrate the nasal mucosa, probably through phagocytosis of the olfactory epithelium cells, and migrate via the olfactory nerves to the brain (5, 20). Data suggest that *N. fowleri* directly ingests brain tissue by producing food cups or amebostomes, in addition to producing a contact-dependent cytolysis which is mediated by a heat-stable hemolytic protein, heat-labile cytolysis, and/or phospholipase enzymes (20). Cysts of *N. fowleri* are generally not seen in brain tissue.

Early symptoms include vague upper respiratory distress, headache, lethargy, and occasionally olfactory problems. The acute phase includes sore throat, stuffy blocked or discharging nose, and severe headache. Progressive symptoms include pyrexia, vomiting, and stiffness of the neck. Mental confusion and coma usually occur approximately 3 to 5 days prior to death. The cause of death is usually cardiorespiratory arrest and pulmonary edema.

PAM can resemble acute purulent bacterial meningitis, and these conditions may be difficult to differentiate, particularly in the early stages. The CSF may have a predominantly polymorphonuclear leukocytosis, increased protein concentrations, and decreased glucose concentration like those seen with bacterial meningitis. Unfortunately, if the CSF Gram stain is interpreted incorrectly (identification of bacteria as a false positive), the resulting antibacterial therapy has no impact on the amebae and the patient will usually die within several days.

Extensive tissue damage occurs along the path of amebic invasion; the nasopharyngeal mucosa shows ulceration, and the olfactory nerves are inflamed and necrotic. Hemorrhagic necrosis is concentrated in the region of the olfactory bulbs and the base of the brain. Organisms can be found in the meninges, perivascular spaces, and sanguinopurulent exudates.

The first case of organ transplantation from a donor who had died of undiagnosed *N. fowleri* infection was reported in 1997 (20). While no subsequent amebic infections occurred in the three organ recipients, this report emphasizes the need for adequate evaluation of the benefits and risks of transplanting tissues from persons whose illness might have been caused by an infectious

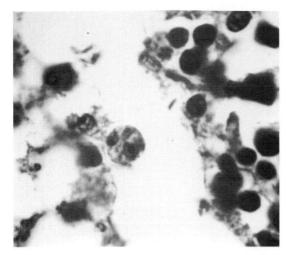

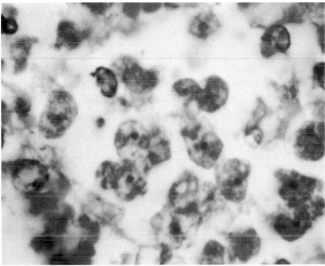

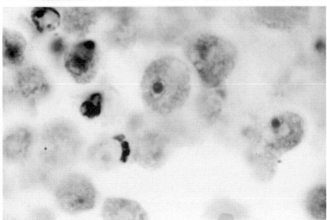

Figure 5.5 *Naegleria fowleri* organisms in brain tissue from a human with PAM.

agent. Apparently, the Centers for Disease Control and Prevention are also aware of another instance in which kidneys were transplanted prior to determination that the donor was infected with *N. fowleri* (20). At present, the risk of transmission of *N. fowleri* by donor organs has not been clarified, and no practical test is available to ensure that donor organs are organism free. Also, no prophylactic drug regimen to treat transplant recipients has been established.

Diagnosis

Clinical and laboratory data usually cannot be used to differentiate pyogenic meningitis from PAM, and so the diagnosis may have to be reached by a process of elimination. A high index of suspicion is often mandatory for early diagnosis (20). Although most cases are associated with exposure to contaminated water through swimming or bathing, this is not always the case. The rapidly fatal course of 3 to 6 days after the beginning of symptoms (with an incubation period of 1 day to 2 weeks) requires early diagnosis and immediate chemotherapy if the patient is to survive.

Analysis of the CSF shows decreased glucose and increased protein concentrations. Leukocytes may range from several hundred to >20,000 cells/mm^3. Gram stains and bacterial cultures of CSF are negative; however, the Gram stain background can incorrectly be identified as bacteria, thus leading to incorrect therapy for the patient.

A definite diagnosis could be made by demonstration of the amebae in the CSF or in biopsy specimens. Either CSF or sedimented CSF should be placed on a slide, under a coverslip, and observed for motile trophozoites; smears can also be stained with Wright's or Giemsa stain. CSF, exudate, or tissue fragments can be examined by light microscopy or phase-contrast microscopy. Care must be taken not to mistake leukocytes for actual organisms or vice versa. It is very easy to confuse leukocytes and amebae, particularly when one is examining CSF by using a counting chamber, hence the recommendation to use just a regular slide and coverslip. Motility may vary, and so the main differential characteristic is the spherical nucleus with a large karyosome.

Specimens should never be refrigerated prior to examination. When centrifuging the CSF, low speeds (250 × *g*) should be used so that the trophozoites are not damaged. Although bright-field microscopy with reduced light is acceptable, phase microscopy, if available, is recommended. Use of smears stained with Giemsa or Wright's stain or a Giemsa-Wright's stain combination can also be helpful. If *N. fowleri* is the causative agent, trophozoites only are normally seen. If the infecting organism is *Acanthamoeba*, cysts may also be seen in specimens from patients with CNS infection. Unfortunately, most cases are diagnosed at autopsy; confirmation of these tissue findings must include culture and/or special staining with monoclonal reagents in indirect fluorescent-antibody procedures. Organisms can also be cultured on nonnutrient agar plated with *Escherichia coli.*

In cases of presumptive pyogenic meningitis in which no bacteria are identified in the CSF, the computed tomography appearance of basal arachnoiditis (obliteration of basal cisterns in the precontrast scan with marked enhancement after the administration of intravenous contrast medium) should alert the staff to the possibility of acute PAM.

The amebae can be identified in histologic preparations by indirect immunofluorescence and immunoperoxidase techniques. The organism in tissue sections looks very much like an *Iodamoeba bütschlii* trophozoite, with a very large karyosome and no peripheral nuclear chromatin; the organisms can also be seen with routine histologic stains.

In infections possibly caused by *Acanthamoeba* spp., periodic acid-Schiff stains the cyst wall red and methenamine silver stains the cyst black; these stains may be helpful. Normally, *Naegleria* and *Acanthamoeba* isolates are identified to the species level by a reference laboratory, such as the Centers for Disease Control and Prevention, using indirect fluorescent-antibody procedures with a monoclonal or polyclonal antibody. The formalin stability of at least one of the epitopes may be useful in detecting *N. fowleri* in fixed biopsy specimens and in investigating the pathologic process (20).

In general, serologic tests have not been helpful in the diagnosis of this infection. The disease progresses so rapidly that the patient is unable to mount an immune response.

KEY POINTS—LABORATORY DIAGNOSIS
Naegleria fowleri

1. Never refrigerate the specimen(s) prior to examination.
2. Beware of the false-positive Gram stain, especially since PAM usually mimics bacterial meningitis (CSF contains increased numbers of leukocytes, increased protein concentrations, and decreased sugar concentrations).
3. The *Naegleria* trophozoites mimic leukocytes in a counting chamber. Motility is more likely to be seen if a drop of CSF is placed directly on a slide and a coverslip is added. Phase-contrast optics are recommended; however, if regular bright-field microscopy is used, the light level should be low. The slide can also be warmed to 35°C to stimulate trophozoite motility.

4. Organisms can be cultured on nonnutrient agar plated with *Escherichia coli.*
5. Low-speed centrifugation is recommended for CSF (150 × *g* for 5 min).
6. Incubate the plate in room air at 35 to 37°C for tissues from the CNS and at 30°C for tissues from other sites.
7. Using the low-power objective (10×), observe the plates daily for 7 days.
8. Remember that *B. mandrillaris* does not grow on agar plates seeded with bacteria.
9. Standard precautions should be used when handling these specimens and cultures. Procedures should be performed in a biological safety cabinet.

Treatment

Many antimicrobial and antiparasitic drugs have been screened for in vitro and in vivo activity against *N. fowleri.* Although *N. fowleri* is very sensitive to amphotericin B in vitro, only a few patients have recovered after receiving intrathecal and intravenous injections of this drug alone or in combination with miconazole (20). One case within California in which the patient was successfully treated with amphotericin B, miconazole, and rifampin has been documented. There was a synergistic effect with miconazole and amphotericin B in vitro; however, rifampin was ineffective. Dexamethasone and phenytoin were also given to this patient to decrease intracranial pressure and seizure, respectively. *Naegleria* infections have also been treated successfully with amphotericin B, rifampin, and chloramphenicol; amphotericin B, oral rifampin, and oral ketoconazole; and amphotericin B alone (1, 20, 21). Delay in diagnosis and the fulminant nature of PAM result in few survivors; this is unfortunate since *N. fowleri* is quite sensitive to the antifungal agent amphotericin B.

Epidemiology and Prevention

Nearly 200 presumptive or proven cases of PAM have been reported in the literature, including cases from the United States, Ireland, England, Belgium, Czechoslovakia, Australia, New Zealand, Brazil, and Zambia. Clinical patient histories indicate exposure to the organism via freshwater lakes or swimming pools shortly before onset; patients had been previously healthy with no specific underlying problems (20). Pathogenic *Naegleria* organisms have also been isolated from nasal passages of individuals with no history of water exposure, suggesting the possibility of airborne exposure.

The first isolations of the environmental strains of pathogenic *N. fowleri* were reported from water and soil in Australia and from sewage sludge samples in India. Detection of *N. fowleri* in heated discharge water has been reported in Belgium and Poland. Since then, there have been additional reports describing isolations of virulent or avirulent strains of *N. fowleri* from the environment. Studies in Belgium have clearly indicated that *N. fowleri* strains are present in artificially heated waters (power plant warm discharge), and studies in the United States have indicated that virulent strains are also found in lakes in geographic latitudes with water temperatures of 14 to 35°C that are totally isolated from any source of thermal discharge (20). These data suggest that the presence of pathogenic or potentially pathogenic amebic strains may depend on both climate and modification of the natural environment. The ability of the cysts to survive under various environmental conditions has been investigated by several workers. These findings suggest that *N. fowleri* cysts produced in the warm summer months may survive the winter and are capable of growth during the following summer.

General preventive measures include public awareness of potential hazards of contaminated water. It has been recommended that warm discharge water not be used for sports and recreational purposes (20), particularly since DNA restriction fragment profiles of environmental strains and human isolates were homogeneous.

In more recent studies, 30 strains of *N. fowleri* were investigated by the randomly amplified polymorphic DNA (RAPD) method. The data confirmed earlier work indicating that RAPD variation is not correlated with geographic origin. In particular, Mexican strains belong to the variant previously detected in Asia, Europe, and the United States. In France, strains gave RAPD patterns identical to those of the Japanese strains. All of these strains, in addition to another strain from France, exhibited similarities to South Pacific strains. The results confirmed the presence of many variants in Europe, while only two variants were detected in the United States; these two variants were different from the South Pacific variants (20).

Acanthamoeba spp.

Formerly, the genus *Acanthamoeba* was included in the genus *Hartmanella*. In 1975, Sawyer and Griffin proposed a new family, Acanthamoebidae, for the spiny amebae, with *Acanthamoeba* as the type genus and *Acanthamoeba castellanii* as the type species (5). The most characteristic feature of this genus is the presence of spine-like pseudopods called acanthapodia. Several species of *Acanthamoeba* (*A. culbertsoni*, *A. castellanii*, *A. polyphaga*, *A. astronyxis*, *A. healyi*, and *A. divionensis*) cause GAE, primarily in immunosuppressed, chronically ill, or otherwise debilitated persons. These patients tend to have no relevant history involving

exposure to recreational freshwater. *Acanthamoeba* spp. also cause amebic keratitis, and it is estimated that to date approximately 1,000 cases of *Acanthamoeba* keratitis have been seen in the United States. Species of *Acanthamoeba* linked to keratitis include *A. castellanii*, *A. polyphaga*, *A. hatchetti*, *A. culbertsoni*, *A. rhysodes*, *A. griffini*, *A. quina*, and *A. lugdunensis* (15). The number of cases worldwide has also been increasing; apparently, the incidence of *Acanthamoeba* keratitis in the United Kingdom is 15 times that in the United States and 7 times that in Holland (13).

The pathogenesis of *Acanthamoeba* infection involves the ability of the organism to invade tissues. This capability depends on adherence to mucosal surfaces, migration through tissues, and the release of oxygen radicals and proteases that can destroy connective tissue. Studies have found that alkaline cysteine proteinases were more active in pathogenic strains whereas serine proteases were found in both pathogenic and nonpathogenic strains (15).

Humoral immunity and complement activation appear to be the primary defense mechanisms against *Acanthamoeba* infections; serum antibodies are mainly the immunoglobulin M (IgM) and IgG classes. Both antibody and complement promote recognition of the organisms by phagocytic cells. The amebae are destroyed by neutrophils, which are activated by lymphokines to accelerate the respiratory burst and to release lysosomal enzymes. Macrophages and neutrophils are the major inflammatory cell types found in tissues surrounding amebae. Although the cornea is infiltrated by neutrophils and there is a measurable humoral response, the disease will progress to corneal opacity or perforation without therapy in the immunocompetent patient. It has been suggested that the organisms mask their antigens from cellular immune responses, thus suppressing macrophage functions and/or preventing neutrophil activity. Total serum immunoglobulins do not differ significantly between healthy subjects and patients with *Acanthamoeba* keratitis, but levels of ameba-specific tear IgA were significantly lower in patients with keratitis (15). Although *Acanthamoeba* infections stimulate a granulomatous response, the response in AIDS patients is minimal or absent, findings which are consistent with the poor immune response in these patients.

Another type of meningoencephalitis, GAE, caused by freshwater amebae is less well defined and may occur as a subacute or chronic disease with focal granulomatous lesions in the brain (Table 5.4). The route of CNS invasion is thought to be hematogenous, with the primary site being the skin or lungs. In this infection, both trophozoites and cysts can be found in the CNS lesions. An acute-onset case of fever, headache, and pain in the neck preceded by 2 days of lethargy has also been documented (15). The causative organisms are probably *Acanthamoeba* spp. in

Table 5.4 *Acanthamoeba* infection: clinical manifestations of patients with AIDS[a]

General	Neurologic	Nasal	Cutaneous	Musculoskeletal
Fever	Headache	Congestion	Papule	Abscess
Chills	Mental changes	Discharge	Lump	
Fatigue	Seizures	Tenderness	Disseminated nodules	
Weight loss	Motor deficit	Ulcer	Foot ulcer	
	Cranial nerve involvement	Hemorrhage	Abscesses	
	Meningeal signs	Chronic sinusitis		
	Sensory deficits			

[a] Adapted from J. P. Sison, C. A. Kemper, M. Loveless, D. McShane, G. S. Visvesvara, and S. C. Deresinski, Disseminated *Acanthamoeba* infection in patients with AIDS: case reports and review, *Clin. Infect. Dis.* 20:1207–1216, 1995.

most cases, but it is possible that other genera are involved (15, 20). In the acute case mentioned above, *A. culbertsoni* was repeatedly found in and cultured from the CSF. The patient responded dramatically to a combination therapy of penicillin and chloramphenicol. Cases of GAE have been reported in chronically ill or immunologically impaired hosts; however, some patients apparently have no definite predisposing factor or immunodeficiency. Conditions associated with GAE include malignancies, systemic lupus erythematosus, human immunodeficiency virus infection, Hodgkin's disease, skin ulcers, liver disease, pneumonitis, diabetes mellitus, renal failure, rhinitis, pharyngitis, and tuberculosis. Predisposing factors include alcoholism, drug abuse, steroid treatment, pregnancy, systemic lupus erythematosus, hematologic disorders, AIDS, cancer chemotherapy, radiation therapy, and organ transplantation. This infection has become more widely recognized in AIDS patients, particularly those with a low CD4[+] cell count.

The *Acanthamoeba* group also causes keratitis and corneal ulceration. There are a number of published cases that emphasize the need for clinicians to consider as-canthamoebic infection in the differential diagnosis of eye infections that fail to respond to bacterial, fungal, or viral therapy. These infections are often due to direct exposure of the eyes to contaminated materials or solutions. It is now recognized that the wearing of contact lenses is the leading risk factor for keratitis. Conditions which are linked with disease include the use of homemade saline solutions, poor contact lens hygiene, and corneal abrasions.

Life Cycle and Morphology

Unlike *N. fowleri*, *Acanthamoeba* spp. do not have a flagellate stage in the life cycle, only the trophozoite and cyst (Figures 5.6 and 5.7) (Table 5.3).

Morphology of Trophozoites. Motile organisms have spine-like pseudopods; however, progressive movement is usually not very evident (Figures 5.2 and 5.8). There is a wide range (25 to 40 μm), with the average diameter of the trophozoites being 30 μm. The nucleus has the typical large karyosome, like that seen in *N. fowleri* (Figures 5.2 and 5.9 to 5.11). This morphology can be seen on a wet preparation. In the environment, these organisms feed on bacteria, algae, and yeast; however, they can also exist on liquid nutrients taken up through pinocytosis

Figure 5.6 Life cycle of *Acanthamoeba*, the etiologic agent of amebic keratitis.

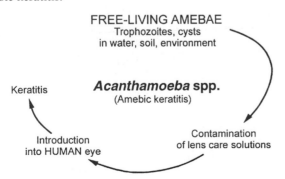

Figure 5.7 Life cycle of *Acanthamoeba* and *Balamuthia mandrillaris*, the etiologic agents of GAE.

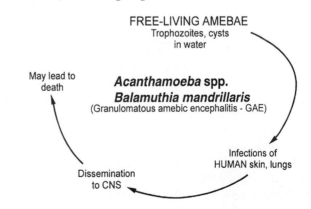

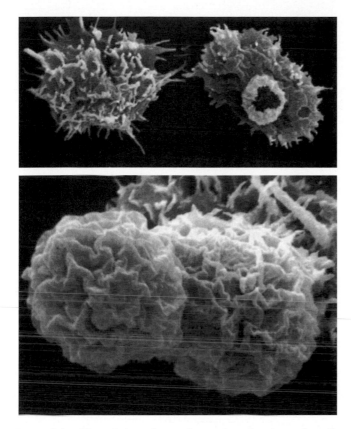

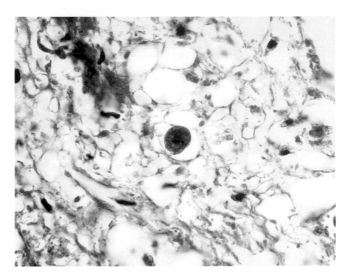

Figure 5.10 *Acanthamoeba* trophozoite in the subcutis of an AIDS patient (hematoxylin and eosin; magnification, ×562). (Armed Forces Institute of Pathology photograph.)

Figure 5.8 (Upper) *Acanthamoeba* trophozoites (note the spiky acanthapodia). (Lower) *Acanthamoeba* cysts. (Images courtesy of F. M. Marciano-Cabral.)

(15). Organelles typically found in higher eukaryotic cells have been identified and include a Golgi complex, smooth and rough endoplasmic reticula, free ribosomes, digestive vacuoles, mitochondria, and microtubules.

Morphology of Cysts.   The cysts are usually round with a single nucleus, also having the large karyosome seen in the trophozoite nucleus. The double wall is usually visible, with the slightly wrinkled outer cyst wall and what has been described as a polyhedral inner cyst wall (15) (Figures 5.2, 5.9, 5.12, and 5.13). This cyst morphology

Figure 5.9 *Acanthamoeba* cysts and trophozoites in the subcutis of an AIDS patient (hematoxylin and eosin; magnification, ×112). (Armed Forces Institute of Pathology photograph.)

Figure 5.11 *Acanthamoeba* trophozoite in the subcutis of an AIDS patient (hematoxylin and eosin; magnification, ×2,298). (Armed Forces Institute of Pathology photograph.)

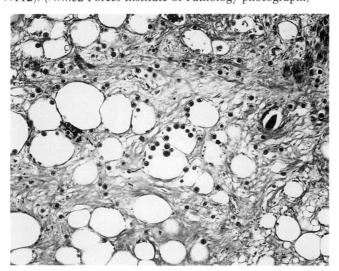

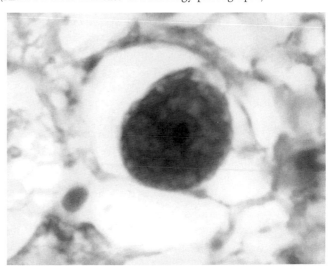

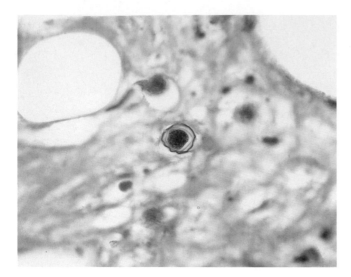

Figure 5.12 *Acanthamoeba* cyst in the subcutis of an AIDS patient (hematoxylin and eosin; magnification, ×562). (Armed Forces Institute of Pathology photograph.)

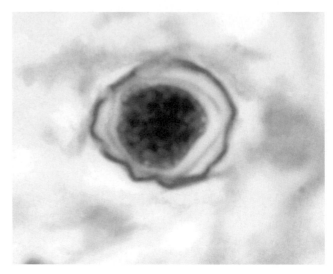

Figure 5.13 *Acanthamoeba* cyst in the subcutis of an AIDS patient (hematoxylin and eosin; magnification, ×2,456). (Armed Forces Institute of Pathology photograph.)

can be seen in organisms cultured on agar plates. Cyst formation occurs under adverse environmental conditions; cysts are resistant to biocides, chlorination, and antibiotics and can also survive low temperatures (0 to 2°C).

Clinical Diseases

GAE. Meningoencephalitis caused by *Acanthamoeba* spp. may present as an acute suppurative inflammation of the brain and meninges like that seen with *N. fowleri* infection. However, *Acanthamoeba* spp. are generally reported to cause a more chronic protracted slowly progressive form of meningoencephalitis (GAE). Due to the confusion in the earlier literature regarding the nomenclature of *Acanthamoeba* and *Hartmanella* spp., these organisms have been referred to in the past as belonging to the *Hartmanella-Acanthamoeba* group. However, since no true *Hartmanella* species has been found to be pathogenic to humans, references to *Hartmanella* in human tissues should be revised to read *Acanthamoeba* or *Balamuthia* spp. (14, 15).

The incubation period of GAE is unknown; several weeks or months are probably necessary to establish the disease. The clinical course tends to be subacute or chronic and is usually associated with trauma or underlying disease, not swimming. GAE is characterized by confusion, dizziness, drowsiness, nausea, vomiting, headache, lethargy, stiff neck, seizures, and sometimes hemiparesis. Within the CNS, the cerebral hemispheres are the most likely tissue to be involved. There may be edema and hemorrhagic necrosis within the temporal, parietal, and occipital lobes. A chronic inflammatory exudate can be seen over the cortex, mainly of polymor-

phonuclear leukocytes and mononuclear cells. Unlike in PAM caused by *N. fowleri*, both trophozoites and cysts are found throughout the tissue. Granulomatous inflammation necrosis and thrombosed vessels are seen in the brain. Multinucleated giant cells forming granulomas can be seen in immunocompetent patients, while granuloma formation is generally poor or lacking in immunocompromised individuals. It is unknown whether brain necrosis is due to direct tissue destruction caused by the organisms, to inflammatory cytokines such as interleukin-1 or tumor necrosis factor, or both. Also, dissemination to other tissues such as the liver, kidneys, trachea, and adrenals can occur in immunocompromised individuals; more unusual sites also include the ears and necrotic bone from a bone graft of the mandible (15).

In the differential diagnosis, other space-occupying lesions of the CNS (tumor, abscess, fungal infection, etc.) must also be considered. Some of the predisposing factors have included Hodgkin's disease, diabetes, alcoholism, pregnancy, and steroid therapy. Organisms have also been found in the adrenal gland, brain, eyes, kidneys, liver, pancreas, skin, spleen, thyroid gland, and uterus.

Amebic meningoencephalitis can also be an unusual complication of bone marrow transplantation. A case has been reported in a patient with non-Hodgkin's lymphoma after autologous stem cell transplantation. Leg weakness, fever, and urinary retention developed 69 days after transplantation. The patient then developed fever, seizures, rapid deterioration of mental functions, and hypercapneic respiratory failure. Magnetic resonance imaging demonstrated a ring-enhancing lesion at the level of thoracic spines 11 and 12. Pleocytosis was seen in the CSF. On

Table 5.5 *Acanthamoeba* infection: examples of clinical features in specific patients with AIDS[a]

Other illnesses[b]	Initial diagnosis	Clinical syndrome
Chronic sinusitis, PCP, disseminated MAC, CMV retinitis	Disseminated fungal infection	Suspected GAE, disseminated cutaneous infection
PCP, esophageal candidiasis, recurrent sinusitis	Cat scratch disease or vasculitis	Disseminated cutaneous infection
Toxoplasma encephalitis, CMV retinitis, giardiasis	Kaposi's sarcoma	Nasal, disseminated cutaneous infection, possible GAE
Cryptococcal meningitis, oral candidiasis, disseminated herpes zoster	*Toxoplasma* encephalitis or lymphoma	GAE
PCP, MAC infection, Kaposi's sarcoma	*Acanthamoeba* infection	Nasal ulcer (single nodule), intramuscular abscess
Inflammatory demyelinating polyneuropathy, esophageal candidiasis, varicella zoster, dendritic keratitis	*Acanthamoeba* infection	Disseminated cutaneous infection, possible GAE

[a] Adapted from J. P. Sison, C. A. Kemper, M. Loveless, D. McShane, G. S. Visvesvara, and S. C. Deresinski, Disseminated *Acanthamoeba* infection in patients with AIDS: case reports and review, *Clin. Infect. Dis.* 20:1207–1216, 1995.

[b] CMV, cytomegalovirus; MAC, *Mycobacterium avium* complex; PCP, *Pneumocystis* pneumonia.

autopsy, a subacute meningoencephalitis secondary to *A. culbertsoni* infection was seen (15).

AIDS Patients. Although the first case of an *Acanthamoeba* infection in a patient with AIDS was in 1986, many additional cases have been reported since that time. *Acanthamoeba* spp. are now well accepted as opportunistic pathogens in immunocompromised patients with AIDS. Unfortunately, most of the cases have resulted in death and have been confirmed postmortem. These individuals usually have advanced HIV disease with a very low CD4 T-cell count, generally less than 200/mm³. Often the patients expire within a month after onset of CNS symptoms. Unfortunately, the diagnosis of this infection requires a high index of suspicion, since both clinical and histologic findings may mimic those of disseminated fungal or algal disease (Tables 5.4 and 5.5). The primary site of infection is thought to be the sinuses and lungs. The skin is also thought to be a possible portal of entry. Clinical manifestations of patients with AIDS and *Acanthamoeba* infection can be reviewed in Table 5.4. They may include general complaints such as fever and chills, nasal congestion or lesions, neurologic symptoms, and musculoskeletal and cutaneous lesions. Some patients, especially those with AIDS, can develop erythematous nodules, chronic ulcerative skin lesions, or abscesses. Although CSF lymphocytosis is seen in non-AIDS patients, the CSF may reveal no cells in HIV-positive patients. These patients also tend to exhibit chronic sinusitis, otitis, and cutaneous lesions (Figure 5.14). Patients with CNS symptoms and AIDS may be misdiagnosed with toxoplasmosis, although serologic tests for *Toxoplasma* may be negative. Other patients have been misdiagnosed with CNS vasculitis, squamous cell carcinoma, or bacterial meningitis.

Both PAM and GAE carry a grave prognosis and, although considered rare, should be considered in the differential diagnosis of patients who present with appropriate histories and imaging findings including nonspecific brain edema on computed tomography in PAM and focal

Figure 5.14 *Acanthamoeba* infection in a 28-year-old Mexican woman who had a skin lesion for 6 months. She developed CNS symptoms including headache and nausea. (From A Pictorial Presentation of Parasites: a cooperative collection prepared and/or edited by H. Zaiman. Photograph courtesy of G. H. Healy.)

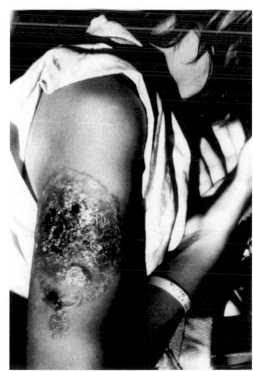

punctate enhancing lesions in the posterior cranial fossa on T1-weighted magnetic resonance imaging in GAE. The initial symptoms are commonly fever and headache. It is important to remember that in patients thought to have toxoplasmosis, review of the cranial computed tomogram could not distinguish between the two etiologic agents *Toxoplasma* and/or *Acanthamoeba*.

In patients with disseminated disease, the duration of infection from onset to death can range from 7 days to 5 months. Once neurologic symptoms became obvious, the survival time is generally about a month. Since early therapy can alter the clinical outcome, recognition of this disease as a potential cause of the patient's presentation is critical. It is important to initiate appropriate diagnostic testing for confirmation of the causative agent as soon as possible.

Keratitis.   Keratitis, uveitis, and corneal ulceration have been associated with *Acanthamoeba* spp. Infections have been seen in both hard- and soft-lens wearers, and particular attention has been paid to soft-lens disinfection systems, including homemade saline solutions. Heat disinfection overall appears to be more effective than cold disinfection methods in killing *Acanthamoeba* trophozoites and cysts. Recent studies indicate that not all commercial contact lens disinfectant solutions are equal in their ability to kill *A. castellanii* (4). Solutions containing hydrogen peroxide or chlorhexidine-thimerosal were active against both trophozoites and cysts. A benzalkonium chloride-based solution was effective only against trophozoites, and solutions containing polyaminopropyl biguanide or polyquaternium 1 were completely ineffective. Unfortunately, a lack of standard methods for testing disinfection solutions may also complicate the issue, although the use of a more reproducible approach may be helpful (4). Obviously, the cyst form of the parasite is much more resistant, and the need to test any solution for both trophozoites and cysts is critical. In addition to the active ingredient(s), it is important to consider adequate exposure times. In general, products containing hydrogen peroxide were the most active, demonstrating the best minimum trophozoite amebicidal concentration for both trophozoites (3 min) and cysts (9 h). It is always important to remember that different products containing hydrogen peroxide at the same concentration may have different killing properties; one report indicated that when hydrogen peroxide was used according to the manufacturer's directions (30 min of contact time), it was not effective against *A. castellanii* trophozoites and cysts. An excellent discussion of the manufacturer's minimum recommended disinfection time for contact lens solutions can be found in reference 4. In order for a disinfectant to be cysticidal, it must be able to gain access to the trophozoite contained within the cyst wall. The most likely route would be the pores in the double cell wall; however, it is thought that these pores (ostioles)

are plugged with mucopolysaccharide. Thus, the disinfectant would need to be able to disrupt these "plugs" in order to reach the trophozoite.

The onset of corneal infection with *Acanthamoeba* spp. can vary tremendously; however, two factors often appear to be involved: trauma and contaminated water. When corneal abrasions occur, the disease process is usually more rapid, with ulceration, corneal infiltration, iritis, scleritis, severe pain, hypopyon (pus in the anterior chamber), and loss of vision. When this process occurs in an individual who wears contact lenses, the onset is more gradual but the results are often the same. A contact lens can act as a mechanical vector for transport of amebae present in the storage case onto the cornea. Subsequent multiplication and invasion of the tissue may occur. Another consideration involves the potential infection of the nasal cavity via lacrimal drainage, a condition that probably causes no problems in a healthy individual. However, it remains unknown whether the presence of these organisms in the nasopharynx of an immunocompromised individual increases the risk of GAE.

Decreased corneal sensation has contributed to the misdiagnosis of *Acanthamoeba* keratitis as herpes simplex keratitis; this mistake can also be attributed to the presence of irregular epithelial lesions, stromal infiltrative keratitis, and edema, which are commonly seen in herpes simplex keratitis. *Acanthamoeba* keratitis may be present as a secondary or opportunistic infection in patients with herpes simplex keratitis. Unfortunately, as a result, treatment can be delayed for 2 weeks to 3 months. The presence of nonhealing corneal ulcers and the presence of ring infiltrates are also clinical signs that alert the ophthalmologist to the possibility of amebic infection.

It is important to consider *Acanthamoeba* infection in the differential diagnosis of uveitis in patients with AIDS. It is also important to remember that incisional keratotomy may predispose the cornea to delayed-onset infectious keratitis; *Acanthamoeba* should always be considered as a possible cause of infection, and clinical specimens should be cultured in refractory cases.

One of the mechanisms of *Acanthamoeba* adhesion to the corneal surface may involve interactions between the mannose binding protein of *Acanthamoeba* and mannose-containing glycoproteins on the surface of the corneal epithelium (15, 20). It appears that enhanced growth is seen in surface-attached amebae and that cocontamination of lens systems with bacteria may be a prime factor in the development of amebic keratitis. Data also suggest that obligate bacterial endosymbionts are able to enhance the amebic pathogenic potential in vitro, although the exact mechanisms have not yet been identified (10). There are also data to show that monoclonal anti-*Acanthamoeba* IgA antibodies can protect against *Acanthamoeba* keratitis, suggesting that this occurs by inhibiting adhesion of the amebae to the corneal epithelium (15).

Acanthamoeba spp. elaborate a variety of proteases, which may facilitate cytolysis of the corneal epithelium, invasion of the extracellular matrix, and dissolution of the corneal stromal matrix (20). *Acanthamoeba* spp. are also resistant to complement lysis. The ability of *Acanthamoeba* spp. to activate the alternative complement pathway but to resist complement-mediated cellular lysis can be attributed to both the release of a transport-dependent extracellular matrix and the presence of complement-inhibitory surface proteins.

Although previous studies have shown that early diagnosis and therapy are definitely related to outcome, differences among patients with and without contact lenses should also be considered. The mean time to diagnosis could be significantly longer in patients who are not contact lens users than in those who are, primarily due to the differences in the index of suspicion regarding the link between contact lens use and *Acanthamoeba* keratitis. These differences could be directly related to poor outcome, with patients who did not wear contact lenses having a very poor visual acuity or penetrating keratoplasty compared with patients who wore contact lenses. It is important to remember that all patients with unresponsive microbial keratitis, even those who do not wear contact lenses, should be evaluated for possible *Acanthamoeba* infection.

Cutaneous Lesions. Cutaneous infections are more common in patients with AIDS, regardless of the presence or absence of CNS involvement. The disease includes the presence of hard erythematous nodules or skin ulcers (Figures 5.14 and 5.15). The early lesions appear as firm papulonodules that drain purulent material; these lesions then develop into nonhealing indurated ulcers. Although disseminated skin lesions may be the first sign of *Acanthamoeba* infection, it is unclear whether the skin lesions represent a primary focus or may result from hematogenous spread from other body sites. Although the mortality rate for these individuals without CNS involvement is around 75%, it increases to 100% if cutaneous infection is accompanied by CNS disease (15).

Histology reveals necrosis surrounded by inflammatory cells, vasculitis, and the presence of trophozoites and cysts of the amebae. However, these findings can also mimic skin lesions caused by fungi, viruses, mycobacteria, or foreign-body reactions. The amebae can also resemble yeast forms of *Blastomyces dermatitidis*, sporangia of *Rhinosporidium seeberi*, *Cryptococcus neoformans*, or *Prototheca wickerhamii*. Because cutaneous acanthamoebiasis has been misdiagnosed as a number of different conditions, it is important to repeat the examination of biopsy material with a different specimen if the first results are negative and no amebae are seen.

Diagnosis

Laboratory examinations similar to those for *N. fowleri* can be used to recover and identify these organisms; the one exception is recovery by culture, which has not proven to be as effective with GAE patients infected with *B. mandrillaris*. The most effective approach uses nonnutrient agar plates with Page's saline and an overlay growth of *Escherichia coli* on which the amebae feed (15, 20) (Figure 5.16). There is also evidence that phosphate-buffered saline can be used. Specimens transported in ameba-saline (5.0 ml) and filtered through 13-mm, 0.22-µm-pore-size cellulose acetate and nitrate filters (Millipore Corp., Bedford, Mass.) have also been acceptable for organism recovery. The filter is then placed in the center of the nonnutrient agar plate seeded with *E. coli*. Tissue stains are also effective, and cysts can be stained with Gomori's silver methenamine, periodic acid-Schiff, and calcofluor white (15, 20).

Various improved medium formulations have been developed and provide some additional options. *Acanthamoeba* isolates have been recovered from clinical specimens inoculated onto buffered charcoal-yeast extract agar (BCYE), nonnutrient agar with live or dead *E. coli*, and tryptic soy agar (TSA) with 5% horse or sheep blood. Good recovery of trophozoites has been obtained on BCYE, TSA with rabbit blood, TSA with horse blood, and Remel TSA with sheep blood. BBL TSA with horse blood or rabbit blood provided good recovery of cysts. All species of live or dead bacteria (*Enterobacter aerogenes*, *E. coli*, *Klebsiella pneumoniae*, *Pseudomonas aeruginosa*, *Serratia marcescens*, *Staphylococcus aureus*,

Figure 5.15 Microscopic view of *Acanthamoeba* amebae in skin. Note the typical large karyosome. (From A Pictorial Presentation of Parasites: a cooperative collection prepared and/or edited by H. Zaiman. Photograph courtesy of G. H. Healy.)

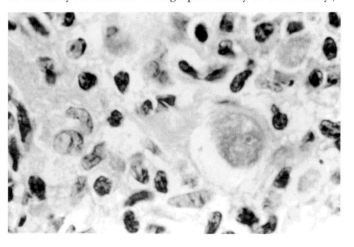

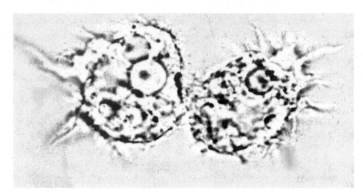
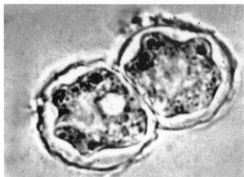

Figure 5.16 (Left) *Acanthamoeba* trophozoites in wet preparation (note the "spiky" pseudopods, called acanthapodia); (right) *Acanthamoeba* cyst in wet preparation (note the double-wall appearance).

and *Stenotrophomonas maltophilia*) yielded good recovery of trophozoites; however, only nonnutrient agar with live *P. aeruginosa*, live *E. aerogenes*, or live *S. maltophilia* gave good recovery of cysts. TSA with rabbit blood or horse blood, BCYE, and nonnutrient agar prepared with live *P. aeruginosa*, *E. aerogenes*, or *S. maltophilia* offered optimal recovery of *Acanthamoeba*. Filtration of specimens through cellulose membranes and cultivation on nonnutrient agar plates covered with *E. coli* was compared with culture of centrifuged specimens (15). Significantly higher parasite detection could be obtained with samples processed by filtration (100% of specimens containing 10 parasites and 65% of specimens containing 2 parasites compared with 8 and 0%, respectively, with centrifugation).

Cryopreservation studies have also found that the highest viability of trophozoites of *A. culbertsoni* occurs when dimethyl sulfoxide is used at a final concentration of 10% and an equilibrium temperature of 4°C. Gross cultural or morphological changes were not noted in trophozoites thawed from frozen suspension; the organisms were grown in serum-casein-glucose-yeast extract medium (2).

Identification of acanthamoebae in ocular and other tissues can be difficult, even for trained microscopists; in histologic preparations the organisms look much like keratoplasts, as well as neutrophils and monocytes. It has been estimated that up to 70% of clinical *Acanthamoeba* keratitis cases are misdiagnosed as viral keratitis. Also, the mean time to diagnosis of keratitis due to *Acanthamoeba* infection can average 2.5 weeks longer for non-contact lens wearers than for contact lens users. Certainly the availability of rapid, accurate, and relatively simple diagnostic tests would increase the timely application of appropriate chemotherapy. Genetic markers have also been developed and used to identify pathogenic *Acanthamoeba* strains associated with nonkeratitis infections (8).

Cutaneous lesion material can be inoculated onto various growth media, including nonnutrient agar plates

with a bacterial overlay. Inoculation of mammalian cell culture monolayers is also possible, as for GAE. Histologic stains such as hematoxylin and eosin, periodic acid-Schiff, or calcofluor white can be used. Also, DNA-based molecular methods such as restriction fragment length polymorphism and randomly amplified DNA analysis can be used.

CSF or bronchoalveolar lavage fluid cytospin preparations can be used to look for amebae in patients with GAE or respiratory symptoms. The characteristic morphology of the *Acanthamoeba* trophozoites, such as the prominent nucleolus, contractile vacuole, and cytoplasmic vacuoles, can be seen more easily using trichrome or hematoxylin and eosin stains on fixed preparations after cytocentrifugation.

Although detection of antibodies to *Acanthamoeba* spp. has been confirmed, the usefulness of serologic tests for the diagnosis of these infections has not been proven. Also, these procedures are not generally available. Further work related to the issue of false-positive and false-negative results and interpretive guidelines regarding past exposure or present infection must be performed before these diagnostic procedures can be routinely used.

KEY POINTS—LABORATORY DIAGNOSIS

Acanthamoeba **spp.**

1. As with *N. fowleri*, beware of a false-positive Gram stain (the leukocyte count in CSF may be elevated).
2. Examine clinical material on a slide with a coverslip; do not use a counting chamber (organisms in CSF look like leukocytes).
3. Nonnutrient agar with Page's saline and an *E. coli* overlay can be tried for culture recovery.
4. Various other media containing different agar bases, some of which contain horse, sheep, or rabbit blood, are also available.

5. Tissue stains are also effective, and the use of cal-cofluor white to visualize the double-walled cyst is recommended.

6. **Refer to chapter 32 for specific methods and recommendations.**

Treatment

Disseminated Infections. Trophozoites and cysts of *Acanthamoeba* isolates vary in their sensitivity to antimicrobial agents. They are sensitive in vitro to ketoconazole, pentamidine, hydroxystilbamidine, paromomycin, 5-fluorocytosine, polymyxin, sulfadiazine, trimethoprim-sulfamethoxazole, azithromycin, and extracts of medicinal plants and, especially, to combinations of these drugs (15, 21). Increased awareness of these infections and early therapy play a large role in patient outcomes. Although steroids have been used in the past to treat cerebral edema and inflammation in CNS infections, this approach should not be used since the steroids tend to exacerbate the *Acanthamoeba* infection. Encephalitis has been treated, somewhat successfully, with antimicrobial combinations including sterol-targeting azoles (clotrimazole, miconazole, ketoconazole, fluconazole, and itraconazole), pentamidine isethionate, 5-fluorocy-tosine, and sulfadiazine. The use of drug combinations helps address resistance that may exist or occur during treatment. In vitro testing confirms strain and species differences in sensitivity, so that no single drug is effective against all amebae (16).

Amebic Keratitis. Because the infection may be misdiagnosed and/or therapy may be delayed, total blindness may result in the infected eye(s). However, if the infection is correctly diagnosed early and the epithelium alone is involved, debridement may be sufficient for cure. This approach not only removes organisms but also enhances the delivery of topical medications. However, if treatment is delayed and the organisms invade the cornea or tissues below the cornea, therapy may need to be continued for many months to a year or longer. These patients must be monitored continually because the cysts tend to be quite resistant to therapy; treatment failures tend to be fairly common. In vitro susceptibility testing may be very helpful in these cases (16).

Cutaneous Disease. Cutaneous infections with subsequent dissemination to other body sites have been found in patients who are immunocompromised due to HIV infection or preparation for organ transplantation. As with other disease manifestations, a number of drug combinations have been tried, some of which were successful.

Table 5.6 *Acanthamoeba* infections, treatment, and positive patient outcomes[a]

Immunocompetent or underlying disease	Amebic disease involvement	Treatment[b]	Outcome
Immunocompetent	CNS (2 patients)	TMP/SMZ + KC + rifampin	Survived
AIDS	Cutaneous/sinusitis	Ketoconazole + i.v. FC	Resolution of lesions; no CNS involvement
AIDS	Cutaneous/sinusitis	Debridement of sinuses + pentam + topical chlorhexidine + 2% KC; pentam toxicity, maintenance on FC	Survived
AIDS	CNS lesion	Sulfadiazine + pyrimethamine, flucon + sulfadiazine; one localized brain lesion excised	Survived
AIDS	Rhinosinusitis	Removal of all diseased nasal mucosa, pentam, levofloxacin, FC, AmpB	Survived
HIV positive	Sinusitis	Debridement of sinuses, IT + gentamicin nasal wash	Resolution of lesions
HIV positive	Sinusitis/cutaneous, lobular panniculitis	Surgical debridement + IT, azithromycin, 5FC, rifampin	Survived
Renal transplant	Cutaneous lesions, IgA deficiency	Pentamidine + chlorhexidine + 2% KC cream; maintenance on oral IT	Resolution of lesions; no CNS involvement
Lung transplant	Cutaneous	AmpB + i.v. pentam + IT + chlorhex/KC cream + azithro + FC + pentam, maintenance FX + clarithromycin	Survived, but died 6 months after *Acanthamoeba* diagnosis

[a] Based on references 1, 15, 16, 17, 20, and 21.

[b] FC, fluorocytosine; IT, itraconazole; azithro, azithromycin; AmpB, amphotericin B; KC, ketoconazole; TMP/SMZ, trimethoprim-sulfamethoxazole; pentam, pentamidine; flucon, fluconazole; chlorhex, chlorhexidine; i.v., intravenous.

However, not all combinations are effective; toxicity with certain drugs also becomes a consideration (Table 5.6).

In a case of skin ulcers due to *Acanthamoeba* in an immunodeficient individual who had undergone kidney transplantation, the patient was successfully treated after extended therapy over a period of 8 months. He received topical as well as systemic administration of combination therapy. The skin ulcers were cleaned twice daily with chlorhexidine gluconate solution, and 2% ketoconazole cream was applied topically. He also received pentamidine isethionate intravenously for 1 month and was then given oral itraconazole therapy for 8 months. This resulted in complete healing of the ulcers (15).

Successful treatment of *Acanthamoeba* infections has been limited; specific examples of patients, disease manifestations, treatment, and outcomes are seen in Table 5.6.

Epidemiology and Prevention

General preventive measures are similar to those for *N. fowleri*. Several disinfectants have been tested against both *Naegleria* and *Acanthamoeba* spp. The conclusion is that *Naegleria* spp. are generally susceptible to swimming-pool levels of chlorine but *Acanthamoeba* spp. are more resistant (15). The recovery of *Acanthamoeba* spp. in nasal isolates and pharyngeal swabs may indicate human introduction of the organisms into swimming pools. With the increasing reports of infection in wearers of soft contact lenses, it will be important to carefully consider the effectiveness of various contact lens disinfection systems. A number of factors probably play a role in the increased incidence of infection: a large number of human immunodeficiency virus-infected individuals and more patients undergoing cancer chemotherapy or immunosuppressive therapy for organ transplantation.

Mitochondrial DNA fingerprinting of *Acanthamoeba* spp. from the American Type Culture Collection and environmental sources has confirmed that approximately half of the 35 isolates displayed fingerprints similar to those of clinical isolates. Comparisons with other published mitochondrial DNA fingerprints indicated that two groups have counterparts in Europe and Japan and in Europe and Australia. These data provide strong evidence that the most common clinical isolates do have counterparts that are geographically widespread and can be isolated from the environment.

Acanthamoeba cysts can remain viable in water at 4°C for 24 years. In animal studies, it has been shown that some of the isolates completely lost their virulence only after 8 years of in vitro cultivation. On the basis of these results, one can assume that cyst viability may be as long as 25 years and that these cysts can maintain their invasive properties (15, 20).

Both failure to disinfect daily-wear soft contact lenses and the use of chlorine-release lens disinfection systems, which have little effect against the organism, are major risk factors for *Acanthamoeba* keratitis. These risks have been very common in disposable-lens use. Some investigators think that over 80% of cases of *Acanthamoeba* keratitis could be avoided by the use of lens disinfection systems that are effective against the organism.

Current recommendations for contact lens wearers include (i) regular scrubbing of contact lens case interiors to disrupt biofilms, (ii) exposure of the contact lens case to very hot water (at least 70°C) to kill *Acanthamoeba* contaminants, (iii) allowing the contact lens case to air dry between uses, (iv) use of a two-step system if hydrogen peroxide disinfection is preferred, and (v) regular replacement of the contact lens case. It has also been documented that some bottled waters contain large numbers of potential ocular pathogens and that, as a general recommendation, bottled water is not safe for routine use with contact lenses.

There have been a number of studies reviewing the potential for carriage of pathogenic bacteria within free-living amebae. In one study in Germany, *Legionella*-contaminated hot water systems and moist sanitary areas in six hospitals were sampled for amebae. Amebae were detected in 29 of 56 hot-water samples and 23 of 49 swabs; six *Acanthamoeba* species isolated from the moist areas were considered potential pathogens (10). It has also been shown that *Acanthamoeba* species can produce vesicles containing live *Legionella pneumophila* cells. Such vesicles may be agents for the transmission of legionellosis associated with cooling towers, and the risk may be underestimated by plate count methods since each vesicle could contain several hundred bacteria. Data also demonstrate that obligate anaerobic bacteria can survive and multiply in *Acanthamoeba* spp. under aerobic conditions, thus suggesting a previously undescribed mechanism for the spread, replication, and persistence of obligate anaerobes in the environment (10). *E. coli* O157 also survives and replicates within *A. polyphaga*. These amebae may serve as an important environmental reservoir for transmission of *E. coli* O157 and other pathogens (10).

Balamuthia mandrillaris

The free-living ameba *Balamuthia mandrillaris* is relatively uncommon and was originally thought to be another harmless soil organism, unlikely or unable to infect mammals. However, since *B. mandrillaris* was first discovered in a pregnant mandrill at the San Diego Wild Animal Park that died of meningoencephalitis, a number of primates including gorillas, gibbons, baboons, orangutans, and monkeys, as well as a sheep and a horse, have died of CNS infection caused by this organism. Over 90 cases of human amebic encephalitis worldwide have been

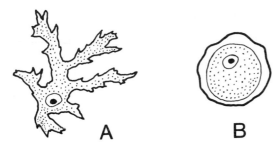

Figure 5.17 (A) Trophozoite and (B) cyst of *Balamuthia mandrillaris*. (Illustration by Sharon Belkin.)

identified, with about half of the cases diagnosed within the United States (19). Death can occur from 1 week to several months after the onset of stroke-like symptoms, which can mimic other conditions including brain stem glioma. Patients eventually die of a massive CNS infection. One individual who survived the infection was left with severe neurologic deficits.

Life Cycle and Morphology

The life cycle is similar to that seen with *Acanthamoeba* spp.; like *Acanthamoeba* spp., *Balamuthia* does not have a flagellated stage in its life cycle as do organisms classified as *N. fowleri* (Figure 5.7).

Morphology of Trophozoites. Trophozoites of *B. mandrillaris* are usually irregular in shape, and actively feeding amebae may measure from 12 to 60 µm in length (normal, 30 µm) (Figure 5.17). In tissue culture, broad pseudopodia are usually seen; however, as the monolayer cells are destroyed, the trophozoites develop fingerlike pseudopodia.

Morphology of Cysts. Cysts are usually spherical and measure from 6 to 30 µm in diameter (Figure 5.17). Under electron microscopy, the cysts are characterized by having three layers in the cyst wall: an outer wrinkled ectocyst, a middle structureless mesocyst, and an inner thin endocyst. Under light microscopy, they appear to have two walls, an outer irregular wall and an inner round wall.

Clinical Disease

The disease is very similar to GAE caused by *Acanthamoeba* spp. and has an unknown incubation period (Figures 5.18 to 5.21). The clinical course tends to be subacute or chronic and is usually not associated with swimming in freshwater. No characteristic clinical symptoms, laboratory findings, or radiologic indicators have been found to be diagnostic for GAE. The neuroimaging findings show heterogeneous, hyperdense, nonenhancing, space-occupying lesions. Whether single or multiple, they involve mainly the cerebral cortex and subcortical white matter. These findings suggest a CNS

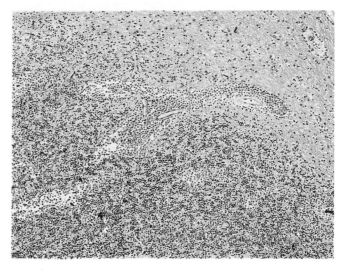

Figure 5.18 *Balamuthia mandrillaris* trophozoites in brain (hematoxylin and eosin; magnification, ×74). (Armed Forces Institute of Pathology photograph.)

neoplasm, tuberculoma, or septic infarcts. Patients complain of headaches, nausea, vomiting, fever, visual disturbances, dysphagias, seizures, and hemiparesis. There may also be a wide range in terms of the clinical course, from a few days to several months.

In immunocompetent hosts, an inflammatory response is mounted and amebae are surrounded by macrophages, lymphocytes, and neutrophils. However, with rare exceptions, these patients also tend to die of severe CNS disease.

Figure 5.19 *Balamuthia mandrillaris* trophozoites in neutrophilic exudate of brain (hematoxylin and eosin; magnification, ×112). (Armed Forces Institute of Pathology photograph.)

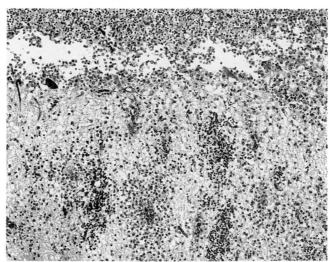

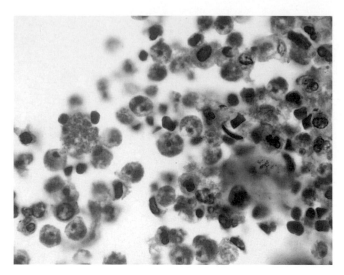

Figure 5.20 *Balamuthia mandrillaris* trophozoites in brain (hematoxylin and eosin; magnification, ×742). (Armed Forces Institute of Pathology photograph.)

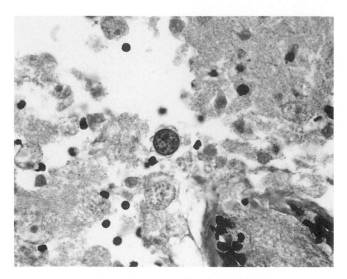

Figure 5.21 *Balamuthia mandrillaris* cyst with outer, wrinkled cyst wall in brain (hematoxylin and eosin; magnification, ×562). (Armed Forces Institute of Pathology photograph.)

Both trophozoites and cysts of *B. mandrillaris* are found in many of the same CNS tissues as are *Acanthamoeba* spp. Although differentiation of these two organisms in tissue by light microscopy is difficult, *B. mandrillaris* appears to have more than one nucleolus in the nucleus in some tissue sections (6). Generally, electron microscopy and histochemical methods are required for definitive identification of *B. mandrillaris*. An immunofluorescence test using species-specific sera is the most reliable means of distinguishing between *Acanthamoeba* and *Balamuthia* spp. (5).

Diagnosis

It is interesting that attempts to isolate the organisms from some humans with GAE have not been successful, and *B. mandrillaris* has also been shown not to grow well on *E. coli*-seeded nonnutrient agar plates. In the diagnostic laboratory, these organisms can be cultured in mammalian cell cultures; some success has been obtained with monkey kidney cells and with MRC, HEp-2, and diploid macrophage cell lines. Using human brain microvascular endothelial cells, investigators have cultured *B. mandrillaris* postmortem from brain and CSF from a case of granulomatous amebic meningoencephalitis (11). A cell-free growth medium is also now available (5). The trophozoites are characterized by extensive branching and a single nucleus (occasionally binucleate forms are seen) with a central karyosome. Occasionally, a few elongated forms with several contractile vacuoles are seen. The cysts have a single nucleus (occasionally, binucleate forms are seen) and have the typical double wall, with the outer wall being thick and irregular.

Serum antibodies to *B. mandrillaris* have been found in both adults and children; however, testing is not currently available for routine use. Although the mouse model is rarely used and is somewhat impractical, mice infected with *Acanthamoeba* spp. or *B. mandrillaris* may die of acute disease within 5 to 7 days or chronic disease after several weeks. In either case, amebae can be seen in the mouse brain after culture or histologic examination.

Based on sequence analysis of mitochondrial small-subunit rRNA genes, primers have been developed that amplify a *Balamuthia*-specific PCR product. These primers will be useful not only for retrospective analyses of preserved tissues but also for possible organism identification in vivo (7). Genotyping studies indicate that lethal infections caused by *B. mandrillaris* are due to a single species with a global distribution. Apparently, there is no correlation between a particular mitochondrial sequence and the genus of infected vertebrate. The mitochondrial sequence from the mandrill isolate is identical to that obtained from a human (6).

Treatment

In vitro studies indicate that *B. mandrillaris* is susceptible to pentamidine isethiocyanate and that patients with this infection may benefit from this treatment. Other studies indicate that ketoconazole, propamidine isethionate, clotrimazole, and certain biguanides have amebicidal activity (12, 21). In one immunocompetent patient who survived, treatment included combination antibiotics (pentamidine, 300 mg intravenously once a day; sulfadiazine, 1.5 g four times a day; fluconazole, 400 mg once a day; and clarithromycin, 500 mg three times a day) (14). As of April 2004, there have been no neurologic sequelae in this patient.

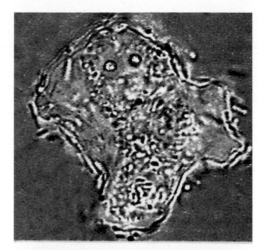

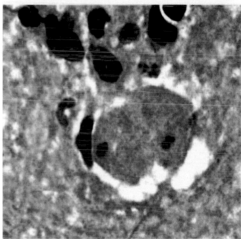

Figure 5.22 *Sappinia diploidea.* (Upper) Trophozoite in a wet preparation (note the two nuclei in the upper portion of the organism. (Lower) Two trophozoites in brain tissue (note the trophozoite on the right has the two nuclei clearly visible). (Photographs courtesy of Govinda Visvesvara, Centers for Disease Control and Prevention.)

Table 5.7 Summary of information on the newly recognized pathogenic free-living ameba, *Sappinia diploidea*[a]

Characteristic	Comments
Patient	38-year-old man
Symptoms	Visual disturbances, headache, seizure
Brain scan	Solitary tumor-like mass in posterior left temporal lobe
Surgical excision	Necrotizing hemorrhagic inflammation
Histologic findings	Central necrotic and hemorrhagic inflammations; acute and chronic inflammatory cells without granulomas or eosinophils; trophozoites present in viable brain parenchyma on mass periphery without inflammatory response
Organisms present	Amebic trophozoites
Trophozoite characteristics	Large (40–70 µm), 2 nuclei per trophozoite
Stained trophozoites	Bright; Giemsa, periodic acid-Schiff stains
Cysts	Not found, but possibility not excluded
Therapy after surgery	Azithromycin, pentamidine, itraconazole, flucytosine

[a] Information from reference 9.

Epidemiology and Prevention

Current recommendations are similar to those for *Acanthamoeba* spp. Information on the environment of *B. mandrillaris* is scarce; however, it is assumed that its habitat is similar to that of other free-living amebae. Although attempts to isolate *B. mandrillaris* from soil samples related to previous encephalitis cases were unsuccessful, there was always some question regarding the designation "free-living" amebae. However, following the death of a 3-year-old Northern California resident from amebic encephalitis caused by *Balamuthia*, environmental samples were collected around the child's home and outdoor play areas. An ameba, consistent with *Balamuthia*, was isolated from the soil found in a potted plant in the home (19). Thus, the isolation of *Balamuthia* amebas from soil affirms its status as a free-living ameba; like *Acanthamoeba* and *Naegleria*, *Balamuthia* is also an opportunistic pathogen.

From the information presented above, it is clear that *B. mandrillaris*, formerly regarded as having no pathogenic potential, will continue to be identified as the etiologic agent of fatal meningoencephalitis in humans as well as animals (Figures 5.18 to 5.21). This opportunistic infection may also continue to cause disease in individuals with AIDS.

Sappinia diploidea

Sappinia diploidea is a newly recognized human pathogen, causing amebic encephalitis. The infection has been well documented in a single patient (9). Trophozoites measure from 40 to 70 µm and have a distinctive double nucleus (Figure 5.22). Transmission electron microscopy confirms that the trophozoites contain two nuclei attached to each other by connecting perpendicular filaments. The trophozoites ingested host blood cells and stained brightly with Giemsa and periodic acid-Schiff stains. Cysts were not seen, but their presence in the human host was not excluded. These and other morphologic characteristics led to the diagnosis of *S. diploidea*, a demonstrated human pathogen for the first time.

The patient was a 38-year-old man who complained of visual disturbances and headache; he also had a seizure.

Brain scans revealed a solitary tumor-like mass in the posterior left temporal lobe. On biopsy, the mass demonstrated necrotizing hemorrhagic inflammation, which contained free-living amebae. Postoperatively, the patient was treated with a sequential regimen of antiamebic drugs, including azithromycin, pentamidine, itraconazole, and flucytosine. Characteristics of this case are summarized in Table 5.7.

References

1. Abramowicz, M. (ed). 2004. Drugs for parasitic infections. *Med. Lett. Drugs Ther.* **46:**1–12.

2. Alejandre-Aguilar, R., M. L. Calvo-Méndez, B. Nogueda-Torres, and F. de la Jara-Alcocer. 1998. Maintenance of *Acanthamoeba culbertsoni* by cryopreservation. *J. Parasitol.* **84:**1261–1264.

3. Axelsson-Olsson, D., J. Waldenstrom, T. Broman, B. Olsen, and M. Holmberg. 2005. Protozoan *Acanthamoeba polyphaga* as a potential reservoir for *Campylobacter jejuni*. *Appl. Environ. Microbiol.* **71:**987-992.

4. Beattie, T. K., D. V. Seal, A. Tomlinson, A. K. McFadyen, and A. M. Grimason. 2003. Determination of amoebicidal activities of multipurpose contact lens solutions by using a most probable number enumeration technique. *J. Clin. Microbiol.* **41:**2992–3000.

5. Beaver, P. C., R. C. Jung, and E. W. Cupp. 1984. *Clinical Parasitology*, 9th ed. Lea & Febiger, Philadelphia, Pa.

6. Booton, G. C., J. R. Carmichael, G. S. Visvesvara, T. J. Byers, and P. A. Fuerst. 2003. Genotyping of *Balamuthia mandrillaris* based on nuclear 18S and mitochondrial 16S rRNA genes. *Am. J. Trop. Med. Hyg.* **68:**65–69.

7. Booton, G. C., J. R. Carmichael, G. S. Visvesvara, T. J. Byers, and P. A. Fuerst. 2003. Identification of *Balamuthia mandrillaris* by PCR assay using the mitochondrial 16S rRNA gene as a target. *J. Clin. Microbiol.* **41:**453–455.

8. Booton, G. C., G. S. Visvesvara, T. J. Byers, D. J. Kelly, and P. A. Fuerst. 2005. Identification and distribution of *Acanthamoeba* species genotypes associated with nonkeratitis infections. *J. Clin. Microbiol.* **43:**1689–1693.

9. Gelman, B. B., V. Popov, G. Chaljub, R. Nader, S. J. Rauf, H. W. Nauta, and G. S. Visvesvara. 2003. Neuropathological and ultrastructural features of amebic encephalitis caused by *Sappinia diploidea. J. Neuropathol. Exp. Neurol.* **62:**990–998.

10. Greub, G., and D. Raoult. 2004. Microorganisms resistant to free-living amoebae. *Clin. Microbiol. Rev.* **17:**413–433.

11. Jayasekera, S., J. Sissons, J. Tucker, C. Rogers, D. Nolder, D. Warhurst, S. Alsan, J. M. White, E. M. Higgins, and N. A. Khan. 2004. Post-mortem culture of *Balamuthia mandrillaris* from the brain and cerebrospinal fluid of a case of granulomatous amoebic meningoencephalitis using human brain microvascular endothelial cells. *J. Med. Microbiol.* **53:**1007–1012.

12. Jung, S., R. L. Schelper, G. S. Visvesvara, and H. T. Chang. 2004. *Balamuthia mandrillaris* meningoencephalitis in an immunocompetent patient: an unusual clinical course and a favorable outcome. *Arch. Pathol. Lab. Med.* **128:**466–468.

13. Kilvington, S., T. Gray, J. Dart, N. Morlet, J. R. Beeching, D. G. Frazer, and M. Matheson. 2004. *Acanthamoeba* keratitis: the role of domestic tap water contamination in the United Kingdom. *Investig. Ophthalmol. Vis. Sci.* **45:**165–169.

14. Kinnear, F. B. 2003. Cytopathogenicity of *Acanthamoeba, Vahlkampfia* and *Hartmannella*: quantative and qualitative in vitro studies on keratocytes. *J. Infect.* **46:**228–237.

15. Marciano-Cabral, F., and G. Cabral. 2003. *Acanthamoeba* spp. as agents of disease in humans. *Clin. Microbiol. Rev.* **16:**273–307.

16. McBride, J., P. R. Ingram, F. L. Henriquez, and C. W. Roberts. 2005. Development of colorimetric microtiter plate assay for assessment of antimicrobials against *Acanthamoeba. J. Clin. Microbiol.* **43:**629–634.

17. Paltiel, M., E. Powell, J. Lynch, B. Baranowski, and C. Martins. 2004. Disseminated cutaneous acanthamebiasis: a case report and review of the literature. *Cutis* **73:**241–248.

18. Pasricha, G., S. Sharma, P. Garg, and R. K. Aggarwal. 2003. Use of 18S rRNA gene-based PCR assay for diagnosis of *Acanthamoeba* keratitis in non-contact lens wearers in India. *J. Clin. Microbiol.* **41:**3206–3211.

19. Schuster, F. L., T. H. Dunnebacke, G. C. Booton, S. Yagi, C. K. Kohnmeier, C. Glaser, D. Vugia, A. Bakardjiev, P. Azimi, M. Maddux-Gonzalez, A. J. Martinez, and G. S. Visvesvara. 2003. Environmental isolation of *Balamuthia mandrillaris* associated with a case of amebic encephalitis. *J. Clin. Microbiol.* **41:**3175–3180.

20. Schuster, F. L., and G. S. Visvesvara. 2004. Free-living amoebae as opportunistic and non-opportunistic pathogens of humans and animals. *Int. J. Parasitol.* **34:**1001–1027.

21. Schuster, F. L., and G. S. Visvesvara. 2004. Opportunistic amoebae: challenges in prophylaxis and treatment. *Drug Resist. Updat.* **7:**41–51.

22. Shenoy, S., G. Wilson, H. V. Prashanth, K. Vidyalakshmi, B. Dhanashree, and R. Bharath. 2002. Primary meningoencephalitis by *Naegleria fowleri*: first reported case from Mangalore, South India. *J. Clin. Microbiol.* **40:**309–310.

23. Thomas, V., T. Bouchez, V. Nicolas, S. Robert, J. F. Loret, and Y. Levi. 2004. Amoebae in domestic water systems: resistance to disinfection treatments and implication in Legionella persistence. *J. Appl. Microbiol.* **97:**950–963.

6

Protozoa from Other Body Sites

Trichomonas vaginalis

Toxoplasma gondii

Trichomonas vaginalis

Trichomonas vaginalis was first isolated in 1836 from the purulent discharge from the female genital organs. The infection occurs worldwide and has been found in the population whenever the appropriate specimens have been examined for the presence of the organism. It is estimated that 5 million women and 1 million men in the United States have trichomoniasis, with the annual incidence of new cases estimated at 7.4 million (35). The annual incidence of trichomoniasis worldwide is estimated to be more than 170 million cases, which does not include the number of asymptomatic cases that are not treated. In North America, more than eight million new cases are reported yearly, with an estimated rate of asymptomatic cases being as high as 50% (35). Trichomoniasis is now the primary nonviral sexually transmitted disease worldwide. Infection with *T. vaginalis* has major health consequences for women, including complications in pregnancy, association with cervical cancer, and predisposition to human immunodeficiency virus (HIV) infection.

Life Cycle and Morphology

T. vaginalis has only the trophozoite stage in its life cycle, and it is very similar in morphology to other trichomonads (Figure 6.1; Table 6.1). The trophozoite is 7 to 23 μm long and 5 to 15 μm wide. The axostyle is usually extremely obvious, and the undulating membrane stops halfway down the side of the trophozoite (Figure 6.2). The nuclear chromatin is uniformly distributed, and there are many siderophil granules that are particularly evident along the axostyle. The axostyle is thought to anchor the parasite to vaginal epithelial cells. The cytoskeleton is composed of tubulin and actin fibers.

There are five flagella, with the fifth being incorporated within the undulating membrane. Different conditions alter the shape of the trophozoites; in axenic culture the trophozoite tends to be more pear shaped and oval, but it may appear more ameboid when attached to vaginal epithelial cells. When environmental conditions are unfavorable, the organisms may tend to round up and internalize the flagella. Although some workers have suggested that this may be a preliminary or pseudocyst form, it is probably just a degenerating trophozoite form.

The life cycle has been described but remains poorly understood. In growth phase culture, large, round forms of the organism are seen, some without

123

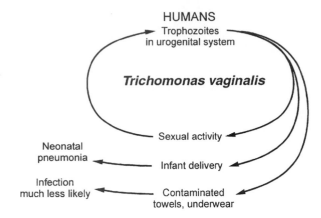

Figure 6.1 Life cycle of *Trichomonas vaginalis*.

Table 6.1 Characteristics of *Trichomonas vaginalis*

Shape and size	Pear shaped; 7–23 µm long (avg, 13 µm), 5–15 µm wide
Motility	Jerky, rapid
No. of nuclei and visibility	1; not visible in unstained mounts
No. of flagella (usually difficult to see)	3–5 anterior, 1 posterior
Other features	Seen in urine, urethral discharges, and vaginal smears; undulating membrane extends half the length of the body; no free posterior flagellum; axostyle easily seen
Infective stage	Trophozoite
Usual location	Vagina (male, urethra)
Striking clinical findings	Leukorrhea, pruritus vulva (thin, white urethral discharge in male)
Other sites of infection	Urethra (prostate in male)
Stage usually recovered during clinical phase	Trophozoite only (no cyst)

flagella, some with flagella and a dividing nucleus, and some with flagella and multiple nuclei. Although there are different opinions about these forms, they are probably developmental stages that appear prior to the typical mononuclear flagellates that generally reproduce by longitudinal binary fission.

Glucose Metabolism. *T. vaginalis* is a primitive eukaryotic organism; however, its metabolism is similar to that of primitive anaerobic bacteria (33). Carbohydrate metabolism is described as being fermentative under both anaerobic and aerobic conditions because glucose is incompletely oxidized. Carbohydrate metabolism occurs in the cytoplasm and an organelle called the hydrogenosome, which is similar to the mitochondria of higher eukaryotes. Within the cytoplasm, glucose is converted to phosphoenolpyruvate and subsequently to pyruvate via the typical Embden-Meyerhoff-Parnas pathway. Pyruvate, which is generated

through glycolysis, undergoes further metabolism in the hydrogenosome. The pyruvate:ferredoxin oxidoreductase has retained diversity in the usage of alternative keto acids for energy production by using a wide variety of substrates. Also, *T. vaginalis* has alternative enzymes that are active in metronidazole-resistant parasites; these alternative pathways are clearly important for the survival of the organism (33). Although *T. vaginalis* is not very energy efficient, it can adapt its metabolism to variable carbon sources. The organism has a high maintenance energy, expending up to half its carbon flow on maintaining internal homeostasis. This ability is critical since the vaginal environment is constantly changing in terms of pH, hormones, menses, and the nutrient supply (33).

Lipid Metabolism. *T. vaginalis* contains cholesterol, phosphatidylethanolamine, phosphatidylcholine, and sphingomyelin as its major phospholipids; apparently the organism is unable to synthesize fatty acids and sterols and must rely on exogenous sources of lipid moieties to survive. It has been postulated that de novo glycolipid and glycophosphosphingolipid synthesis occurs in trichomonads.

Amino Acid Metabolism. Although carbohydrates are the preferred source of nutrients, amino acids can sustain survival and growth. Under normal culture conditions, *T. vaginalis* consumes large amounts of arginine and smaller amounts of methionine for energy production. There appears to be some form of equilibrium that is maintained between the free amino acids within the cell and its external environment.

Figure 6.2 *Trichomonas vaginalis* trophozoite. (Illustration by Sharon Belkin.)

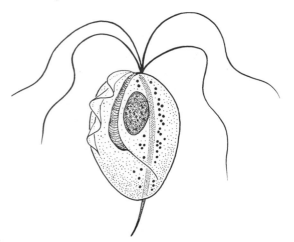

Nucleotide Metabolism. The organism is unable to synthesize purines and pyrimidines; adenine and guanine or their nucleosides are required for growth, as are thymidine, cytidine, uracil, and/or uridine (33). It also appears that there may be two nucleoside carriers, a characteristic that has also been found in *Leishmania donovani.*

Humans are the only natural host for *T. vaginalis*, and normal body sites for these organisms include the vagina and prostate. Apparently, the organisms feed on the mucosal surface of the vagina, where bacteria and leukocytes are found (35). The preferred pH for good growth in females is slightly alkaline or acidic (pH 6.0 to 6.3 optimal), not the normal pH of the healthy vagina. Although the organisms can be recovered in urine, in urethral discharge, or after prostatic massage, the pH preference of the organisms in the male has not been determined. Often, the organisms are recovered in the centrifuged urine sediment from both male and female patients.

Clinical Disease

T. vaginalis is site specific and usually cannot survive outside the urogenital system. Although the incubation period is not well defined, in vitro studies provide a range of 4 to 28 days. After introduction, proliferation begins, with resulting inflammation and large numbers of trophozoites in the tissues and the secretions. As the infection becomes more chronic, the purulent discharge diminishes, with a decrease in the number of organisms. The onset of symptoms such as vaginal or vulval pruritus and discharge is often sudden and occurs during or after menstruation as a result of the increased vaginal acidity. In acute infections, diffuse vulvitis is seen and is due to copious leukorrhea. The discharge is frothy, yellow or green, and mucopurulent; however, only about 10 to 12% of women exhibit this frothy discharge. Small punctate hemorrhagic spots can be seen on the vaginal and cervical mucosa; this has been called a "strawberry appearance" and is seen in about 2% of patients. In general, symptoms include vaginal discharge (42%), odor (50%), and edema or erythema (22 to 37%) (35). Other complaints include dysuria and lower abdominal pain.

In chronic infections, symptoms may be very mild with pruritus and some pain during sexual intercourse, while the vaginal secretion may be scanty and mixed with mucus. Individuals with these symptoms are the major source of transmission. From 25 to 50% of infected women may be asymptomatic and have a normal vaginal pH of 3.8 to 4.2 and normal vaginal flora. Although there is a carrier form, about 50% of these women will develop clinical symptoms during the following 6 months.

Although vaginitis is the most common finding in women with trichomoniasis, other complications include pyosalpinx (distention of a fallopian tube with pus), endometritis, infertility, low birth weight, and cervical erosion. There is also an association with increased HIV transmission and cervical dysplasia (35, 36). In a study of Malawian men, there was no difference in the rate of HIV seropositivity in men with and without *T. vaginalis* infection. However, for HIV-positive men with symptomatic urethritis, the median HIV RNA concentration in seminal plasma was significantly higher in patients with trichomoniasis than in those without trichomoniasis (35).

About 20% of women with vaginal trichomoniasis have dysuria, a symptom that may occur before any additional symptoms appear. Infection in the male may be latent, with essentially no symptoms, or may be present as self-limited, persistent, or recurring urethritis. *T. vaginalis* has been detected in 10 to 20% of subjects with nonspecific urethritis and in 20 to 30% of those whose sexual partners had vaginitis. Once established, the infection persists for a long time in females but only for a short time (usually about 10 days or less) in males. *T. vaginalis* is the cause of 11% of all cases of nongonococcal urethritis in males (33).

Respiratory distress was found in a full-term, normal male infant with *T. vaginalis* who had severe respiratory difficulties after delivery. A wet preparation of thick, white sputum showed few leukocytes and motile flagellates, which were identified as *T. vaginalis*. This study supports previous data indicating that this organism may cause neonatal pneumonia (8, 33, 35).

Factors that impact the inflammatory response are not clearly defined but may include hormonal levels, coexisting vaginal flora, and the strain and concentration of the organisms present. Lymphocytes from patients with active trichomoniasis show a proliferative response when incubated in the presence of secretory and cellular products of *T. vaginalis*. This finding indicates that delayed hypersensitivity reactions may act to modify inflammatory responses.

Several mechanisms are used by *T. vaginalis* to maintain the infection in the female urogenital tract. Recognition and binding to vaginal epithelial cells by adhesins that resemble metabolic enzymes is the first step (Table 6.2). Immune responses are evaded by masking the organisms with host proteins, shedding of trichomonad proteins into the secretions, and degradation of all immunoglobulin subclasses and complement (33). Parasites undergo phenotypic variation between surface and cytoplasmic expression of a number of immunogenic proteins against which host antibody is directed. Phenotypic variation occurs in response to iron-regulated proteins, virus-induced and -repressed proteins, erythrocyte binding proteins and hemolysins, proteinases, and adhesins and contributes to the organism's overall antigenic diversity and complexity (40). Subpopulations of trichomonads express high- or low-iron-regulated proteins depending on

Table 6.2 Virulence factors associated with *Trichomonas vaginalis* pathogenesis[a]

Capability	Virulence factor(s)
Nutrient acquisition and adherence	Receptors for nutrient acquisition Adhesin that is time, temperature, pH dependent Adhesin proteins, which are receptors to host extracellular matrix proteins Carbohydrates, which provide basis for ligand-receptor binding Presence of cysteine proteinases required Microfilaments concentrated on side of parasite that attaches Laminin may be target for adhesion
Cysteine proteinases	Enzymes that are lytic factors in hemolysis of erythrocytes Required for adherence to epithelial cells
Contact-independent mechanisms	Cell detaching factor (CDF) causes cytopathic effects in cell culture (not seen in nonpathogen *Pentatrichomonas hominis*) CDF levels correlate with severity of clinical disease
Damage of target cell plasma membrane	Pores created in erythrocyte membranes
Interaction with vaginal flora	Normal vaginal pH 4.5; *T. vaginalis* thrives in less acidic pH (pH > 5) Rise in pH produces concomitant reduction or loss of protective lactobacilli
Immune system evasion	Surface coating with host proteins Shedding of immunogens Cysteine proteinase degradation of IgG and IgA and complement Epitope phenotypic variation Continuous secretion of large amounts of highly immunogenic soluble antigens short-circuits antibody response
Regulation of virulence genes	General nutrient limitations Environmental factors (iron) Parasite factors (double-stranded RNA virus)

[a] Adapted from references 34, 35, and 40.

the iron status of microenvironments within the vagina. Thus, the expression of virulence genes is controlled by both parasite and host environmental factors. Populations of *T. vaginalis* from fresh clinical isolates probably comprise a heterogeneous population, thus guaranteeing that some organisms will be able to infect the host. Based on current knowledge, it appears that interference with trichomonad mucin receptors and proteinases may be a strategy to prevent colonization with this parasite.

Diagnosis

Unfortunately, clinical manifestations are not reliable criteria for the diagnosis of trichomoniasis; demonstration of the parasite is required for accurate diagnosis of infection. It was demonstrated in the early 1980s that if the classic features such as "strawberry cervix" and frothy discharge were used alone in the diagnosis of trichomoniasis, 88% of infected women would not be diagnosed and 29% of uninfected women would be incorrectly diagnosed as having trichomoniasis (33).

Wet Mounts. The identification of *T. vaginalis* is usually based on the examination of wet preparations of vaginal and urethral discharges and prostatic secretions. This examination must be performed within 10 to 20 min of sample collection, or the organisms lose motility and

may be very difficult to see. Several specimens may have to be examined to detect the organisms, and the sensitivity of this technique may vary between 40 and >80%. The specimen should be diluted with a drop of saline and examined under low power with reduced illumination for the presence of actively moving organisms; urine sediment can be examined in the same way. As the jerky motility of the trophozoite diminishes, it may be possible to see the movement of the undulating membrane, particularly under high dry power. Specimens should never be refrigerated.

Since the morphology of nonpathogenic *Pentatrichomonas hominis* from stool samples is very similar to that of pathogenic *T. vaginalis*, it is important to ensure that the specimen is not contaminated with fecal material.

Stained Smears. Diagnostic tests other than wet preparations, such as permanent stains and fluorescent stains, can also be used. Organisms may be difficult to recognize in permanent stains; however, if a dry smear is submitted to the laboratory, Giemsa or Papanicolaou stain can be used. Chronic *Trichomonas* infections may cause atypical cellular changes that can be misinterpreted, particularly on the Papanicolaou smear. The organisms are routinely missed on Gram stains. Unfortunately, *T. vaginalis* does not always appear in its typical pear-shaped forms but instead may resemble polymorphonuclear leukocytes.

The number of false-positive and false-negative results reported on the basis of stained smears strongly suggests that the results should be confirmed by observation of motile organisms either from the direct wet mount or from appropriate culture media (35).

Culture. If culture techniques are used, it is mandatory that the specimen be collected correctly, immediately inoculated into the proper medium, and properly incubated. If these requirements are not met, the findings may represent a false-negative result. Although culture is more sensitive than wet mounts, this approach is not routinely used because of cost and inconvenience. To improve these parameters, a more convenient plastic envelope method has been developed (4) (Figure 6.3). This envelope approach allows both immediate examination and culture in one self-contained system. The system is commercially avail-

Figure 6.3 InPouch TV diagnostic system for culturing *Trichomonas vaginalis*. (A) The swab containing a specimen from the patient is being inserted into the liquid medium within the plastic pouch; (B) the top of the plastic pouch is folded over and sealed with the tabs; (C) the pouch is inserted into the plastic holder; (D) the pouch (in the holder) is placed onto the microscope stage for examination (organism motility). (Photographs courtesy of BIOMED Diagnostics.) See also chapter 32.

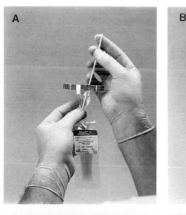
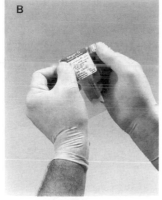

able as the InPouch TV (BIOMED Diagnostics, San Jose, Calif.), which serves as the specimen transport container, the growth chamber during incubation, and the "slide" during microscopy. Once it is inoculated, it requires no opening for examination, and positive growth occurs within 5 days. The sensitivity of this system is reported to be superior to that of other available culture methods (4, 33, 35). Also, no detectable differences were seen with the timing of the inoculation of the top or bottom pouch (4). This delayed-inoculation technique allows the initial reading of the wet mount and then inoculation of the culture pouch if the wet mount is negative.

Various studies have been published regarding newer medium formulations and comparisons, including the use of modified Columbia agar (37). Although the results of this study were excellent, the specimens were processed at bedside, an approach that is often not practical for specimen collection. Cell culture methods use a variety of cell lines to recover *T. vaginalis* from clinical specimens, and they include McCoy cells. This has been reported to detect the organism at a concentration as low as 3 organisms/ml (33). However, this approach is not as practical as other available culture options and is rarely used in the routine clinical laboratory setting. Similarities in the results of these various diagnostic methods, particularly culture methods, have to be reviewed within the context of delivery systems and compliance with the proper use of transport options.

Antigen Detection. Screening patients for the presence of *T. vaginalis* has been somewhat limited by the available tests, many of which tend to be insensitive or expensive or require delays before the result is reported. A rapid diagnostic test for *T. vaginalis* infection has been developed and has proven to be much more sensitive than the wet-mount examination (Table 6.3). When compared with culture (InPouch), the XenoStrip-Tv (Xenotope Diagnostics, Inc., San Antonio, Tex.) was 76.7% sensitive and 99.8% specific and 79.4% sensitive and 97.1% specific in two different studies (24). Rapid test performance did vary with the day of the culture-positive result, with a 71% decline in XenoStrip sensitivity for every additional day of delay until cultures were positive. This product is easy to use, is more sensitive than the wet mount, and should be considered as an alternative to the wet mount, particularly for point-of-care diagnosis when microscopy may be impractical (34).

Another rapid detection system for *T. vaginalis* uses an immunochromatographic capillary-flow (dipstick) assay, which provides results in 10 min. Compared with wet mount and culture, this system, OSOM Trichomonas Rapid Test (Genzyme Diagnostics, Cambridge, Mass.), showed 83.3% sensitivity and 98.8% specificity, while the wet mount has a sensitivity of 71.4% and a specificity of 100% (Figure 6.4). This test also detected *T. vaginalis* in

Table 6.3 Summary of commercially available kits for immunodetection of *Trichomonas vaginalis*

Diagnostic kit	Manufacturer and/or distributor[a]	Type of test	Sensitivity (%), specificity (%)	Comments
Affirm VPIII	Becton-Dickinson	DNA probe	93, 99	*Trichomonas vaginalis, Gardnerella vaginalis, Candida* spp.
Quik-Trich	PanBio Integrated Diagnostics	Latex agglutination	95, 99	*Trichomonas vaginalis*
OSOM Trichomonas Rapid Test	Genzyme Diagnostics	Dipstick	83.3, 98.8	*Trichomonas vaginalis*; *T. vaginalis* detected in samples that required 48 to 72 h of incubation prior to becoming positive
Trichomonas vaginalis DFA kit	Light Diagnostics— Chemicon	DFA[b]	97, 99	96.1% agreement with wet mount, but more positives found with DFA
XenoStrip-Tv	Xenotope Diagnostics	Dipstick	76.7, 99.8	*Trichomonas vaginalis*

[a] Becton Dickinson Diagnostic Systems, 7 Loveton Circle, Sparks, MD 21152; Chemicon International, Inc., 28835 Single Oak Drive, Temecula, CA 92590; Genzyme Diagnostics, One Kendall Square, Cambridge, MA 02139; PanBio Integrated Diagnostics, 1756 Sulfur Springs Road, Baltimore, MD 21227; Xenotope Diagnostics, Inc., 3463 Magic Drive #350, San Antonio, TX 78229.
[b] DFA, direct fluorescent antibody.

samples that required 48 to 72 h of incubation before becoming culture positive (17). Compared to culture alone, OSOM was 83.5% sensitive and 99.0% specific.

Antibody Detection. Various techniques have been used to demonstrate antibodies against *T. vaginalis*; they include agglutination, complement fixation, indirect hemagglutination, gel diffusion, fluorescent-antibody assay, and enzyme-linked immunosorbent assay (ELISA). However, there are a number of reasons why this approach has not been used routinely. In some cases, an antibody response is not seen, due to test sensitivity or the timing of the test procedure. Also, antitrichomonal

Figure 6.4 *Trichomonas vaginalis*, direct antigen detection strip used for rapid testing (OSOM Trichomonas Rapid Test). (Courtesy of Genzyme Diagnostics, Cambridge, Mass.)

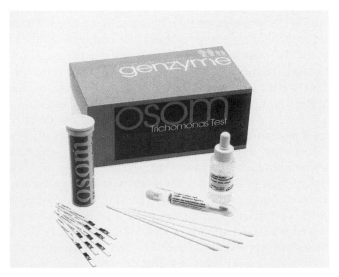

antibodies persist for long periods after therapy, and so current and past infections cannot be differentiated.

Recombinant-DNA Techniques. The use of PCR methods has led to improvements in *T. vaginalis* detection; nonviable organisms and cells and target sequences can also be detected. In addition to various culture and transport options, there are currently several other products available, including the Affirm VPIII probe (Becton Dickinson, Cockeysville, Md.), the Quik-Tri/Can latex agglutination test (PanBio Integrated Diagnostics, Baltimore, Md.), and the T.VAG DFA (Light Diagnostics, Temecula, Calif.). The Affirm VP system uses synthetic oligonucleotide probes for the detection of both *Gardnerella vaginalis* and *T. vaginalis* from a single vaginal swab (33). A PCR-based ELISA with urine as the clinical specimen gave sensitivity and specificity data (compared with wet mount or culture) as 90.8 and 93.4%, respectively, in women and 92.7 and 88.6%, respectively, in men (21, 22). For clinical settings in which vaginal specimens are not available and culture is not an option, urine-based PCR-ELISA may be another option. In some settings, self-obtained vaginal swabs may be useful for PCR, and it appears that PCR may be more sensitive than culture for the diagnosis of *T. vaginalis* in males (35). Based on your patient population, client base, number of requests, and cost, these tests may be relevant for your laboratory test menu.

KEY POINTS—LABORATORY DIAGNOSIS

Trichomonas vaginalis

1. Wet-preparation (motile organisms) examination is recommended; the specimen should be examined immediately. If this is not possible, other methods should be considered.

2. Fluorescent stains and culture have also been recommended.
3. The newer rapid methods may not be as sensitive as culture or PCR, but organisms can be detected, often 48 to 72 h prior to cultures becoming positive for *T. vaginalis*.
4. If dry smears are submitted to the laboratory, they can be fixed with absolute methanol, allowed to dry, and stained like a thin blood smear (Giemsa stain).
5. Serologic tests are not routinely available.

Treatment

Metronidazole is recommended for the treatment of urogenital trichomoniasis, although resistance to both metronidazole and other 5-nitroimidazoles has been reported (1). It is also recommended that all sexual partners of the patient be treated simultaneously to avoid immediate reinfection. Metronidazole-resistant *T. vaginalis* has been implicated in an increased number of cases; unfortunately, metronidazole is currently the only drug approved for the treatment of trichomoniasis in the United States (8). Although vaginal metronidazole creams and pessaries are available, their cure rate compared with oral drug administration is poor. Metronidazole can be taken as a single or multiple doses, as well as being administered intravenously. Additional information on therapy can be found in chapter 25.

A recent review reported that metronidazole reduced the risk of persistent infection in women with *T. vaginalis* infection but increased the incidence of preterm birth (31). Neonatal infection depends on maternal estrogen levels, which wane after the third to sixth week of life; after this time, the infection may disappear. However, if required, treatment of neonates should be postponed until the sixth to eighth week of age and then given if the infant is symptomatic (33).

Although the cure rate is excellent, treatment failure can be due to several factors including noncompliance and reinfection. Published case reports of *T. vaginalis* strains that are clinically resistant to metronidazole are increasing; in the late 1990s, it was estimated that approximately 5 to 10% of all patient isolates demonstrated some level of resistance to metronidazole (13). Resistance can be linked to progressive decrease and eventual loss of the pyruvate:ferredoxin oxidoreductase, so that the drug-activating process is averted. The development of resistance in *T. vaginalis* is also accompanied by decreased expression of ferredoxin (8). Refractory cases are defined as cases in which two standard courses of therapy fail to cure the patient, and these cases can be very difficult to treat. Refractory cases are treated with higher (often double) doses of metronidazole for an extended period; this is effective approximately 80% of the time. Treatment protocols are now available for marginal, low, moderate, and high levels of metronidazole resistance. Therefore, the need for metronidazole susceptibility testing has become more important for clinically relevant patient care. Different methods are used as pharmacological screening procedures for new drugs, including simple cell enumeration. However, this approach is highly subjective. Another approach, which involves determination of acid phosphatase activity as the screening procedure, has been recommended. Compared with microscopy procedures, there are some clear advantages to these newer colorimetric methods, including greater speed and efficiency and less subjectivity (28). Other metabolic routes of *T. vaginalis* have also been used to develop susceptibility testing methods for determining antitrichomonal activity (28).

Tinidazole, a 5-nitroimidazole, is currently being used outside of the United States; this drug is delivered more effectively to vaginal secretions at levels found in serum. However, potential cross-resistance among nitroimidazoles is a concern (8, 33). Other potential drugs include niridazole, nitazoxanide, sulfimidazole, nifuratel, berberine sulfate, furazolidone, mebendazole, butoconazole, benzoisothiazolinon, and gynalgin, most of which have shown some promise, but research is in the preliminary stages (8, 33).

Epidemiology and Prevention

Trichomoniasis has been found in every continent and climate, with no season variability. It has a cosmopolitan distribution within all racial groups and throughout all socioeconomic levels. The annual incidence of trichomoniasis is estimated to be more than 170 million cases worldwide. The incidence can exceed 50% in patients attending sexually transmitted disease clinics and depends on age, sexual activity, number of sexual partners, other sexually transmitted diseases, sexual customs, phase of the menstrual cycle, and sensitivity of patient examination and specimen collection and testing.

Humans are the only natural host for *T. vaginalis*. Risk factors for trichomoniasis have also been investigated. The associations between HIV and trichomoniasis may relate to increased shedding of HIV due to local inflammation produced by *T. vaginalis*, increased susceptibility to HIV as a result of breaks in the mucosal barriers, a higher prevalence of sexually transmitted diseases in HIV-infected individuals, and/or an increased susceptibility to HIV due to immunosuppression. Transmission of HIV is enhanced by coinfection with *T. vaginalis*, and treatment of trichomoniasis reduces the levels of HIV RNA (35).

A symbiotic relationship between *T. vaginalis* and *Mycoplasma hominis* has been documented; such a symbiosis may indicate a potential role of *T. vaginalis* in the transmission of bacterial infection (9). These findings suggest that this relationship between the two organisms may provide the bacteria with the ability to resist the host defense mechanisms and therapy.

Infection is acquired primarily through sexual intercourse, hence the need to diagnose and treat asymptomatic males. The organism can survive for some time in a moist environment such as damp towels and underclothes; however, this mode of transmission is thought to be very rare. Evidence implicating *T. vaginalis* as a potential contributor to poor outcomes in both women and men continues to accumulate. In women, this infection may play a role in the development of cervical neoplasia, postoperative infections, and potential problems with pregnancy. It is also seen as a factor in atypical pelvic inflammatory disease and infertility. In men, trichomoniasis causes nongonoccocal urethritis and contributes to male infertility. The importance of this infection needs to be recognized beyond the limits of the sexually transmitted disease clinics (36).

Vaccines.   As with many sexually transmitted diseases, infection with *T. vaginalis* does not produce long-term immunity. Although the antibody response to this organism is well documented, the antibodies provide only limited protection, and antibody titers disappear after therapy. Vaccine development has shown some promise; potential targets include a 100-kDa protein that is immunogenic across a number of *T. vaginalis* isolates, as well as essential adherence molecules such as adhesins, mucinases, and cysteine proteinases (8). Although several vaccine trials were held during the 1960s and the late 1970s, subsequent studies led to doubts about the efficacy of the vaccines. Later studies using a murine model have shown promise, and a bovine *Tritrichomonas foetus* model in cattle has also yielded promising results. A whole-cell *T. foetus* vaccine is now commercially available for cattle and provides protection, as well as accelerating the eradication of infection.

Toxoplasma gondii

Toxoplasma gondii is worldwide in distribution, is closely related to other coccidia, and also has certain similarities to malarial parasites. The parasites were first discovered in 1908, by Charles Nicolle and Louis Manceaux at the Pasteur Institute, in the North African rodent called the gundi, hence the species name *gondii*. At the same time, Alfonso Splendore described the organism from laboratory rabbits at the Portuguese Hospital in São Paulo, Brazil. However, the complete life cycle was not determined until 1970 (5). Although serologic evidence indicates a high rate of human exposure to the organism, the disease itself is relatively rare. *T. gondii* can infect many vertebrates as

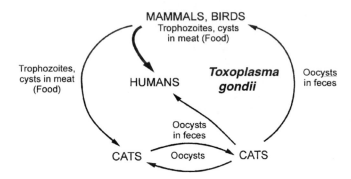

Figure 6.5   Life cycle of *Toxoplasma gondii*. The active forms (trophozoites) are also referred to as tachyzoites, while the resting stages within the cysts are called bradyzoites (very slow-growing trophozoites).

well as humans, but the definitive host is the house cat and other members of the Felidae family.

Life Cycle and Morphology

The life cycle is shown in Figure 6.5. Humans can acquire the infection in several ways, either by the accidental ingestion of oocysts shed in cat feces or by the ingestion of rare or raw meats, particularly pork, lamb, or venison. Transmission can also occur in utero, by transfusion, and during organ transplantation. Toxoplasmosis should be considered as a possible cause of multiorgan failure after liver transplantation. The inclusion of serologic testing should be considered in routine donor testing. The most common means of infection is probably through ingestion of rare or raw meats. The oocysts shed in cat feces are 9 to 11 µm wide by 11 to 14 µm long and contain two sporocysts, each containing four sporozoites (Figure 6.6). Cats shed oocysts after ingestion of any of the three infectious stages of *T. gondii*: tachyzoites in groups, bradyzoites in tissue cysts, and sporozoites in oocysts (11). Cats become infected from eating birds, small rodents, or other sources of raw meat. In the tachyzoite-induced life cycle of *T. gondii* in cats, the animals can shed up to 360 million oocysts per day and will shed for 4 to 6 days (11).

There are three infectious stages of *T. gondii*: the tachyzoites (in groups or clones), the bradyzoites (in tissue cysts), and the sporozoites (in oocysts). Tachyzoites rapidly multiply in any cell of the intermediate host (many other animals and humans) and in nonintestinal epithelial cells of the definitive host (cats). Bradyzoites are found within the tissue cysts and multiply very slowly; the cyst may contain few to hundreds of organisms, and intramuscular cysts may reach 100 µm in size. Although the tissue cysts may develop in visceral organs such as the lungs, liver, and kidneys, they are more prevalent in neural and muscular tissues, including the brain, eyes, and skeletal and cardiac muscle. Intact tissue cysts can persist for the life of the host and do not cause an inflammatory response (5).

The organisms are obligate intracellular parasites and are found in two forms in humans. The actively proliferating trophozoites or tachyzoites are usually seen in the early, more acute phases of the infection. The resting forms or cysts are found primarily in muscle and brain, probably as a result of the host immune response. The cysts contain the more slowly growing trophozoites or bradyzoites. Bradyzoites encyst approximately 8 to 10 days after entry into the host and differ from tachyzoites in being more resistant to pepsin, having a longer generation time, containing cytoplasmic vacuoles which may serve as carbohydrate stores, and being the only stage to initiate the enteroepithelial cycle and transform into oocysts in the feline intestine.

The trophozoites (tachyzoites) are crescent shaped and are 2 to 3 µm wide by 4 to 8 µm long (Figures 6.7 and 6.8). One end tends to be more rounded than the other.

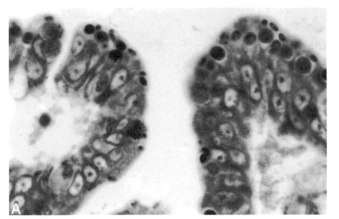

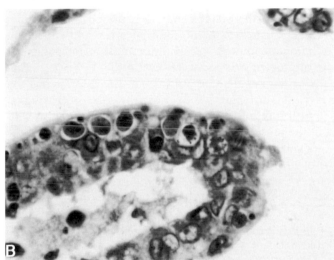

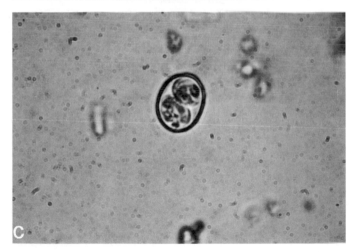

Figure 6.6 *Toxoplasma gondii* in the intestine of a cat. (A) Bradyzoites (×1,000); (B) gametocytes (×1,000); (C) oocyst containing two sporocysts with sporozoites (×1,600). (Courtesy of J. K. Frenkel, University of Kansas School of Medicine, Kansas City.)

Figure 6.7 Forms of *Toxoplasma gondii* found in humans. (A) Trophozoites (tachyzoites) seen within a cell; (B) cyst containing the bradyzoites (illustration by Sharon Belkin; based on illustration from H. H. Najarian, *Textbook of Medical Parasitology*, The Williams & Wilkins Co., Baltimore, Md., 1967); (C) tachyzoites recovered from mouse peritoneal fluid.

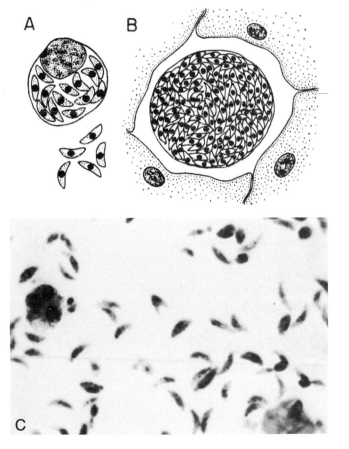

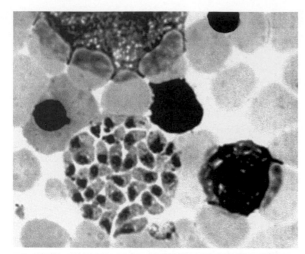

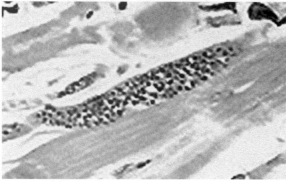

Figure 6.8 Forms of *Toxoplasma gondii* found in humans. (Upper) Tachyzoites seen within bone marrow. (Lower) Bradyzoites seen in human tissue.

Giemsa is the stain of choice; the cytoplasm stains pale blue and the nucleus stains red and is situated toward the broad end of the organism.

The cysts are formed in chronic infections, and the organisms within the cyst wall are strongly periodic acid-Schiff positive (Figure 6.7). During the acute phase, there may be groups of tachyzoites that appear to be cysts; however, they are not strongly periodic acid-Schiff positive and have been termed pseudocysts.

The cycle occurring in cats includes asexual multiplication and sexual reproduction in the small intestinal epithelial cells (5, 11). The final stage of the reproductive cycle is the oocyst, which is passed in the feces. After defecation, development of infective sporozoites inside the oocysts requires 1 to 5 days and depends on environmental conditions. The developmental cycle in the cat takes approximately 19 to 48 days after infection with the oocysts but only 3 to 10 days after the ingestion of meat infected with cysts, e.g., a mouse.

The number of cats in the United States is estimated at around 70 million, with the number of feral cats probably being greater than that of domestic cats. Of the cats examined within the United States, <1% had *T. gondii* oocysts in their feces. However, this may be because the period of shedding is usually only 1 to 2 weeks during the life of the cat and the oocysts are quite small, measuring approximately 10 μm. Consequently, the use of routine fecal examinations to determine whether cats are positive for *T. gondii* is somewhat limited and is not particularly helpful in determining whether the animal is shedding oocysts.

Clinical Disease

When the tachyzoites are actively proliferating, they invade cells adjacent to the original infected cell as it ruptures. This process creates continually expanding focal lesions. Once the cysts are formed, the process becomes quiescent, with little or no multiplication and spread. In the immunocompromised or immunodeficient patient, a cyst rupture or primary exposure to the organisms often leads to lesions. The organisms can be disseminated via the lymphatics and the bloodstream to other tissues. Disintegration of cysts may give rise to clinical encephalitis in the presence of apparently adequate immunity.

The large number of people who are serologically positive for *T. gondii* suggests that the majority of infections are benign, with most people exhibiting few (e.g., cold or light case of the flu) or no symptoms. The most severe symptoms are seen with congenital, transplacental infections or infections in compromised patients (Table 6.4). Toxoplasmosis can be categorized into four groups: (i) disease acquired in immunocompetent patients, (ii) disease acquired or reactivated in immunosuppressed or immunodeficient patients, (iii) congenital disease, and (iv) ocular disease.

Acquired Infections in Immunocompetent Individuals.   In approximately 90% of cases, no clinical symptoms are seen during the acute infection. However, 10 to 20% of patients with acute infection may develop painless cervical lymphadenopathy. This presentation is benign and self-limited, with symptoms resolving within weeks to months. Acute visceral manifestations are seen in rare cases. In one study from French Guiana over a period of 6.5 years, 13

Table 6.4 People at risk for severe toxoplasmosis

Category	Comments
Infants born to mothers who are first exposed to *Toxoplasma* infection several months before or during pregnancy	Mothers who are first exposed to *Toxoplasma* more than 6 months before becoming pregnant are not likely to pass the infection to their children
Persons with severely weakened immune systems	Infection that occurred at any time during life can reactivate

of 16 patients had consumed game in the 2 weeks prior to the onset of symptoms and in 8 cases the game was undercooked (7). Although other cases have been reported during the last decade, very few were reported in detail. In reviewing the possible link between *T. gondii* and schizophrenia, it appears that some cases of acute toxoplasmosis in adults are associated with psychiatric symptoms such as delusions and hallucinations (39).

Infections in Immunocompromised Patients. Infections in immunocompromised patients can lead to severe complications; underlying conditions that may influence the course of the disease include various malignancies (such as Hodgkin's disease, non-Hodgkin's lymphomas, leukemias, and solid tumors), collagen vascular disease, organ transplantation, and AIDS. In immunocompromised patients, the central nervous system (CNS) is primarily involved, with diffuse encephalopathy, meningoencephalitis, or cerebral mass lesions. More than 50% of these patients show an altered mental state, motor impairment, seizures, abnormal reflexes, and other neurologic sequelae. Studies show that even in these groups, most patients receiving chemotherapy for toxoplasmosis will improve significantly or have complete remission. However, in those with AIDS, therapy must be continued for long periods to maintain a clinical response.

In genotyping studies, it appears that the type of *T. gondii* strain does not predominantly influence the pathogenesis of toxoplasmosis in immunocompromised patients. Thus, regardless of the strain genotype, there remains the need for specific prophylaxis in these patients infected by *T. gondii* (16).

Transplant recipients. In transplant recipients, disease severity depends on previous exposure to *T. gondii* by the donor and recipient, the type of organ transplanted, and the level of immunosuppression of the patient. Disease can be due to reactivation of a latent infection or an acute primary infection acquired directly from the transplanted organ (43). Stem cell transplant recipients are particularly susceptible to severe toxoplasmosis, primarily due to reactivation of a previously acquired latent infection. If stem cell transplant recipients have a positive serologic test prior to transplantation, they are known to be at risk for severe disseminated toxoplasmosis (29). All potential transplant recipients should be tested for *Toxoplasma*-specific immunoglobulin G (IgG) antibodies to determine their antibody status. An individual with acute acquired infection often produces detectable IgG and IgM antibodies, while those with reactivation may or may not have an increase in the levels of IgG antibodies and normally do not demonstrate an IgM response. Seronegative cardiac transplant recipients who receive an organ from a sero-positive donor may develop toxoplasmic myocarditis; this disease presentation may also mimic organ rejection.

AIDS patients. Patients who become infected with *T. gondii* risk developing disease when their CD4$^+$ T-lymphocyte count falls below 100,000/ml. Fever and malaise usually precede the first neurologic symptoms; headache, confusion, seizures, or other focal signs strongly suggest the diagnosis of toxoplasmosis.

Toxoplasma encephalitis (TE) was reported as a life-threatening opportunistic infection among patients with AIDS prior to the use of highly active antiretroviral therapies. This condition is fatal if untreated. Most AIDS patients with TE have detectable IgG antibodies to *T. gondii*; however, <5% may not have antibody in their serum. Psychiatric manifestations of *T. gondii* are also seen in immunocompromised individuals with AIDS in whom latent infections have become reactivated. Altered mental status occurs in approximately 60% of patients, with symptoms including delusions, auditory hallucinations, and thought disorders (39).

Disseminated toxoplasmosis should be considered in the differential diagnosis of immunocompromised patients with culture-negative sepsis, particularly if combined with neurologic, respiratory, or unexplained skin lesions. Examination of Wright's-stained peripheral blood smears or antitoxoplasma immunoperoxidase studies of skin biopsy specimens may be diagnostic (5).

In histologically confirmed cases of TE when the majority of the patients received some therapy, the overall prognosis has been poor, with the median survival following initiation of therapy being 4 months. No beneficial or harmful effects could be attributed to the use of corticosteroids, and toxicity attributed to pyrimethamine-sulfonamide therapy was reported for many patients. A clinical relapse of TE has been seen in approximately half of the patients discharged from the hospital.

Although an inflammatory process is observed during TE in HIV-positive patients, the major cause of CNS lesions is probably uncontrolled parasite multiplication rather than immunopathologic changes. Also, *T. gondii* is capable of enhancing HIV-1 replication within reservoir host cells and, simultaneously, HIV-1 itself undermines acquired immunity to the parasite, promoting reactivation of chronic toxoplasmosis.

Respiratory disease due to *T. gondii* has also been recognized in immunocompromised patients, and the few cases that have been found in HIV-positive patients have been in association with CNS disease. The finding of pulmonary toxoplasmosis in the absence of neurologic findings in HIV-positive patients emphasizes the importance of considering this organism as the cause of pulmonary disease in such patients with respiratory symptoms.

Congenital Infections. Congenital infections result from the transfer of parasites from the mother to the fetus when she acquires a primary infection during pregnancy. In 90% of patients, no clinical symptoms are apparent during acute infection. Congenital infections may be particularly severe if the mother acquires the infection during the first or second trimester of pregnancy. At birth or soon thereafter, symptoms in these infants may include retinochoroiditis, cerebral calcification, and occasionally hydrocephalus or microcephaly. Symptoms of congenital CNS involvement may not appear until several years later.

The risk of congenital toxoplasmosis depends on the timing of the mother's acute infection. Transmission to the fetus increases with gestation age: 15 to 25% in the first trimester, 30 to 45% in the second, and 60 to 65% in the third. However, the severity of congenital disease decreases with gestation age. Signs of infection at delivery are present in 21 to 28% of those infected in the second trimester and in ≤11% of those infected in the third trimester. Overall, 10% are born with severe disease. Congenital transmission can occur even if the mother is immune, although this is rare; reinfection of the mother during pregnancy is possible, particularly if she comes in contact with large numbers of infective cysts and/or oocysts.

The characteristic symptoms of hydrocephalus, cerebral calcifications, and chorioretinitis resulting in mental retardation, epilepsy, and impaired vision represent the most severe form of the disease. Cerebral lesions may calcify, providing retrospective signs of congenital infection.

In the case of seroconversion by the mother during pregnancy, prenatal diagnosis is usually performed by analysis of amniotic fluid. However, it is important to remember that despite the use of advanced diagnostic methods, some cases of congenital toxoplasmosis cannot be detected early. This demonstrates the importance of careful follow-up of newborns who are at risk. Treatment of the mother may reduce the incidence of congenital infection; therefore, rapid and accurate diagnosis should be emphasized.

Ocular Infections. Chorioretinitis in immunocompetent patients is generally due to an earlier congenital infection. Patients may be asymptomatic until the second or third decade; at that point, cysts may rupture, with lesions then developing in the eye. The number of people who develop chorioretinitis later in life is unknown but may represent over two-thirds of the total. Also, up to 30% of patients relapse after treatment. Chorioretinitis is usually bilateral in patients with congenitally acquired infection and is generally unilateral in patients with recently acquired infection.

Diagnosis

Diagnosis is usually made by various serologic procedures. Other procedures include performing PCR, examining biopsy specimens, buffy coat cells, or cerebrospinal fluid, and isolating the organism in tissue culture or in laboratory animals (Tables 6.5 and 6.6). It is difficult to demonstrate the parasite in lymph node biopsy specimens, and while some authorities think the histologic appearance is very characteristic, others consider the histologic changes to be nonspecific. However, because many individuals have been exposed to *T. gondii* and may have cysts within the tissues, recovery of organisms from tissue culture or animal inoculation may be misleading, since the organisms may be isolated but may not be the etiologic agent of disease. For this reason, serologic tests are often recommended as the diagnostic approach of choice. However, two representative situations in which the detection of organisms may be very significant are (i) tachyzoite-positive smears and/or tissue cultures inoculated from cerebrospinal fluid and (ii) in patients with acute pulmonary disease, the demonstration of tachyzoites in Giemsa-stained smears of bronchoalveolar lavage fluid, with some tachyzoites being extracellular and some being intracellular.

Emphasis on pediatric patients, particularly neonates, is seen in the extensive literature on serologic testing. An algorithm for the diagnosis and treatment of neonatal toxoplasmosis is seen in Figure 6.9; a summary of risk management issues in pregnancy can be seen in Figure 6.10. An understanding of the difficulties in test interpretation is helpful. First, all newborns from mothers who are antibody positive will have passively transferred maternal IgG. These titers can be quite high but may not indicate infection in the baby. The detection of IgM antibodies, which do not cross the placental barrier, provides a much more accurate indication of infection in the newborn. There can still be situations in which a false-negative (IgG saturation of antigen receptors) or false-positive (presence of rheumatoid factor and antinuclear antibody) result is possible. There is evidence to indicate that in certain instances, the presence of IgM antibody titers in single serum specimens cannot be used to indicate recent exposure (14, 43). Interpretation of serologic tests for IgM can be seen in Figure 6.11.

The Sabin-Feldman dye test, one of the first methods used to diagnose toxoplasmosis, is based on the fact that in the presence of immune serum, *T. gondii* loses its affinity for methylene blue stain. Because this test uses live *Toxoplasma* trophozoites, however, most laboratories do not provide it routinely. A specific guide to the interpretation of serologic titers for toxoplasmosis is provided in Table 6.7.

The serologic diagnosis of toxoplasmosis is very complex and has been discussed extensively in the literature; a number of additional procedures, some of which are

Table 6.5 Clinical use of immunodiagnostic tests[a]

Questions to answer	Tests to use	Interpretation	Comments
Is the patient currently infected?	IgG assays	If IgG positive = infected If IgG negative, retest in 3 weeks if acute infection suspected	A single specimen is satisfactory to determine antibody status Situations where baseline information is helpful: (i) women of childbearing age tested prior to conception (ii) before initiation of immunosuppressive therapy (iii) after determination of HIV-1-positive status (iv) organ donors prior to transplantation
Has the patient recently acquired the infection?	IgM and IgA assays	If IgG positive and IgM negative = infected for more than 6 months If IgG positive and IgM positive = infection within last 2 years, or false-positive IgM result	Specimens drawn 2 wk apart may or may not be helpful; often an increase in IgG or IgM titer is not possible; titers may already have reached a plateau when first sample is drawn
How recently was the infection acquired by the patient?	IgG avidity, differential agglutination	If IgG avidity high = infected at least 12 wk earlier If IgG avidity low = recent infection possible; some people have persistent low IgG avidity for months after infection	If IgG avidity low, draw second sample 3 weeks after first sample; send both specimens to *Toxoplasma* reference laboratory for confirmation of IgG, IgM, and IgG avidity results and possibly differential agglutination, IgA, and IgE testing
Is evidence of active infection present?	PCR, culture, antigen detection	*Toxoplasma* isolated from 95% of placentas of congenitally infected newborns when mother has not been treated and from approximately 81% when the mother has received therapy	Attempts should be made to isolate *T. gondii* from the placenta, amniotic fluid, and cord blood to confirm the diagnosis

[a] Adapted from references 10 and 43.

automated, have been developed and reported in the last few years (32). These methods include enzyme immunoassays, some of which are automated (20), ELISAs (14, 30, 42, 43), direct agglutination (43), an immunosorbent agglutination assay (43), an indirect-immunofluorescence assay, immunocapture (43), and immunoblotting (15). Evaluations of various commercial assays have also been discussed (Table 6.7) (43).

A great deal of molecular biology-based work involving PCR for the diagnosis of toxoplasmosis and strain typing has also been done (2, 15, 19, 29, 38, 41, 43). This technology may be useful in situations where knowledge of recent infection is necessary. PCR has also been used to detect *T. gondii* in deparaffinized ocular tissue sections, cerebrospinal fluid, blood, and bronchoalveolar fluid; however, the absence of *Toxoplasma* DNA does not exclude toxoplasmosis. PCR can be very helpful in evaluating fetal infection; certainly the time needed to detect parasites is much shorter than when using mouse or tissue culture approaches. Sensitivity depends on the clinical presentation; it is higher when confirming congenital toxoplasmosis from amniotic fluid but lower in patients with infection reactivation such as HIV-positive patients.

Antigen detection has improved during the last few years; however, this approach still appears to lack sensitivity when used with human samples. Antigen detection may be helpful in diagnosing the compromised patient in whom antibody titers are low or absent. This serologic approach can also clarify low titers, i.e., those found in early acute infection or chronic infection. This test capability can also be very helpful in patients with monoclonal gammopathies, whose titers to *T. gondii* may be extremely high but may not be causing the clinical condition. See Algorithm 6.1.

KEY POINTS—LABORATORY DIAGNOSIS

Toxoplasma gondii

1. Serologic results are normally used to diagnose toxoplasmosis. Results of all other procedures must be interpreted in light of the serologic findings.
2. *T. gondii* identified in histologic preparations or isolated in animals or in tissue culture systems may or may not be the causative agent of the symptoms (many individuals harbor the organisms in their tissues but have no symptoms).

Table 6.6 General comments on the performance and interpretation of serologic results for the diagnosis of toxoplasmosis[a]

1. Specific IgM normally develops early, within 1–2 wk after primary infection.

2. Specific IgG normally develops within 4 wk after primary infection.

3. Production and increase in titers usually peak within 4–8 wk, but in individual cases the increase in IgM may continue for some weeks and IgG for several months.

4. The antibody result from a single serum sample gives no clear indication of when the infection occurred.

5. A rise in titer (IgG) should be demonstrated between 2 or more samples tested in the same test run; a 4-tube antibody increase is required to support the diagnosis of acute febrile toxoplasmosis. If serum is taken during late infection or for the presence of lymphadenopathy, titers in excess of 1:1,000 should be present.

6. The presence of IgM does *not* confirm a very recent infection; a positive IgM result has low predictive value for identifying primary *T. gondii* infection.

7. The amount of specific IgM may decrease to below the detection level in less than 3 mo after infection; however, IgM may be detectable for many months after the primary infection (can be detected as long as a year or more after primary infection). IgM antibody can persist for 6 mo in the conventional test but as long as 6 yr in the capture test.

8. IgG avidity gradually increases after primary infection. Low or equivocal avidity results can persist for months to >1 yr after primary infection. High-avidity test results seen in individuals infected for at least 3 to 5 mo.

9. For congenital infections, an IgG titer of >1:1,000 indicates a tentative diagnosis, pending confirmation of a positive IgM test, the exclusion of rheumatoid factor and antinuclear antibody, and the isolation of *Toxoplasma*, if possible. Passively transferred antibody shows a 10-fold decay every 3 mo; during infection, high antibody titers are stable or will increase.

10. Specific IgA results are not particularly helpful. A negative result does not exclude, nor does a positive result confirm, a recent primary infection.

11. Termination of pregnancy may be offered to women who seroconvert in the first 8 wk of pregnancy and to those infected in the first 22 wk of pregnancy when fetal infection is confirmed. A more conservative approach is based on abortion only if there is ultrasonographic evidence of hydrocephalus.

12. In the context of recurrent pregnancy loss, intravenous immunoglobulin therapy can cause confusion in the interpretation of serologic results for the diagnosis of toxoplasmosis; the appearance of anti-*Toxoplasma* antibodies may lead to the conclusion that seroconversion has occurred. Also, if positive serologic results are seen, the woman may be incorrectly considered immune.

[a] Adapted from references 7, 14, 20, 30, and 43.

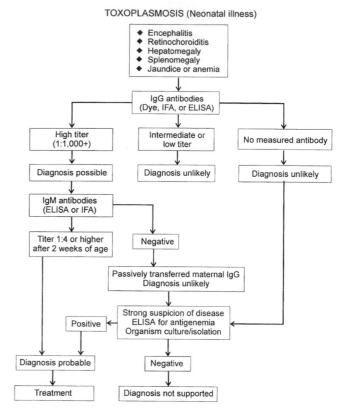

Figure 6.9 Chart for the diagnosis and treatment of neonatal toxoplasmosis. (Adapted from reference 14.)

Treatment

It is generally recommended that the following conditions be treated: clinically active disease, diagnosed congenital toxoplasmosis, and disease in symptomatic compromised patients. Therapy for pregnant patients who acquire the infection and for newborns with *Toxoplasma* antibody is somewhat controversial. However, prophylactic therapy is often recommended for the newborn until it can be demonstrated that IgM antibody is not present. Spiramycin has been used to treat women infected in pregnancy, since it is potentially less toxic for the fetus. Apparently this drug is concentrated in the placenta but does not cross the barrier freely, and drug levels in the neonate are low. Spiramycin has also been used to treat congenitally infected infants every other month, alternating with sulfadiazine-pyrimethamine treatment.

Drugs that have been found to be effective, either alone or in combination, include sulfonamides and pyrimethamine (actively multiplying tachyzoites). Clindamycin and spiramycin have also been used in certain circumstances.

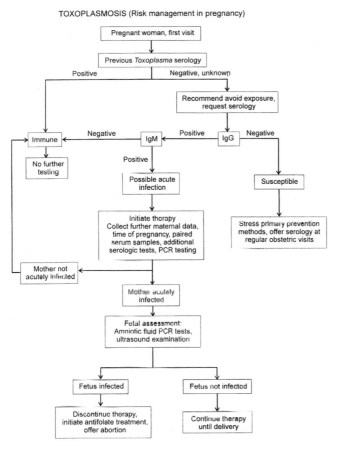

Figure 6.10 Toxoplasmosis: risk management in pregnancy.

Approximately 10 to 50% of patients who are seropositive for antibodies to *T. gondii* have CD4$^+$ T-lymphocyte counts of <100/mm^3. Detection of IgG antibodies indicates prior infection and possible presence of tissue cysts. Since TE may result from reactivation of tissue cysts, prophylaxis with TMP-SMX or dapsone plus pyrimethamine with leucovorin is recommended for these patients. HIV-infected persons who are seronegative for *Toxoplasma* IgG should be counseled to protect themselves from primary infection by eating well-cooked meats and washing their hands after possible soil contact. Cats kept as pets should be fed commercial or well-cooked food, should be kept indoors, and should have their litter box changed each day. An algorithm for prevention of TE can be seen in Figure 6.12. TE continues to occur among patients not receiving prophylaxis. It is estimated that at least half of the cases could be prevented by better motivation of physicians and increased compliance by patients.

The management of ocular toxoplasmosis depends on the severity of the clinical presentation. Most specialists treat patients whose visual acuity has progressed to worse than 20/200, who have lesions in the peripapillary, perifoveal, or maculopapillary bundle regions, or who have severe vitreous inflammation. However, most do not treat patients whose visual acuity is 20/20, who have lesions in the far peripheral retina, or who have trace to mild vitreous inflammation. Various therapeutic regimens are used; however, almost all specialists, when asked, indicated that they used corticosteroids in the initial treatment regimen. In general, most indicated that they would modify treatment during pregnancy, in newborns, and in AIDS patients. Additional information can be found in chapter 25.

Although a combination of pyrimethamine and a sulfonamide is the standard regimen for treating TE in AIDS patients, the use of these drugs can be contraindicated in patients with a history of bone marrow suppression and severe allergic reactions to sulfonamides. It is also important to realize the potential danger of using high doses of sulfadiazine, which may lead to the development of sulfadiazine-induced crystalluria; to prevent crystalluria, patients should be adequately hydrated and their urinary pH should be maintained above 7.5. The use of oral clindamycin and pyrimethamine is effective (1). Another combination regimen includes pyrimethamine-clarithromycin, which is comparable to the conventional therapeutic approach to acute TE in AIDS patients. Recent data also suggest that a combination of trimethoprim and sulfamethoxazole (TMP-SMX, co-trimoxazole) may be as effective as pyrimethamine-sulfadiazine in AIDS patients with TE. It is interesting that treatment of cerebral toxoplasmosis protects against *Pneumocystis* pneumonia in AIDS patients; the decreased risk is probably due to chronic suppressive treatment with pyrimethamine and sulfonamides.

Epidemiology and Prevention

Human infection can be acquired through ingestion or handling of infected meat or through ingestion of infective oocysts, which can remain viable within cool, moist soil for a year or longer (Table 6.8). Certainly, hand washing is highly recommended when there has been potential exposure to the oocysts. It is recommended that meat be cooked so that the internal temperature reaches 150°F (66°C). The question of the house cat always arises; indoor cats fed on dry, canned, or boiled food are unlikely to be infected. However, cats that go outside and have access to other infected animals (birds or mice) may shed oocysts in their feces; therefore, other preventive measures, including changing the litter box daily and disinfecting the pan with boiling water, have been recommended. Also, the feces should not be placed onto the soil but, rather, should be disposed of either in the toilet or within bags.

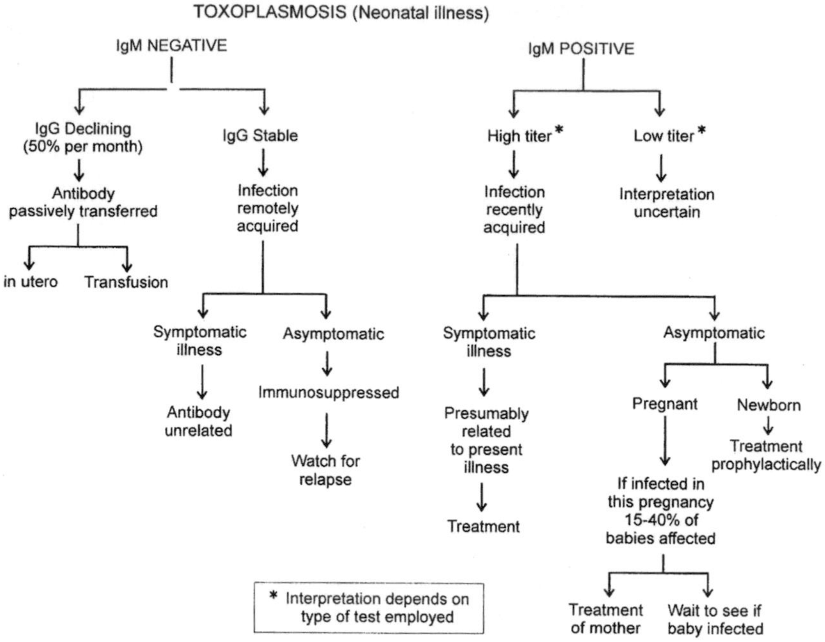

TOXOPLASMOSIS (Neonatal illness)

IgM NEGATIVE

- IgG Declining (50% per month)
  - Antibody passively transferred
    - in utero
    - Transfusion
- IgG Stable
  - Infection remotely acquired
    - Symptomatic illness
      - Antibody unrelated
    - Asymptomatic
      - Immunosuppressed
        - Watch for relapse

IgM POSITIVE

- High titer*
  - Infection recently acquired
    - Symptomatic illness
      - Presumably related to present illness
        - Treatment
    - Asymptomatic
      - Pregnant
        - If infected in this pregnancy 15-40% of babies affected
          - Treatment of mother
          - Wait to see if baby infected
      - Newborn
        - Treatment prophylactically
- Low titer*
  - Interpretation uncertain

* Interpretation depends on type of test employed

Figure 6.11   Chart for the analysis of conventional antibody tests for *Toxoplasma gondii* by the IgM technique. Interpretation depends on the type of test used. In the IgM-indirect immunofluorescence assay test, a titer of 1:64 would support a diagnosis of infection acquired 1 to 3 months earlier. In the double-sandwich IgM-ELISA (IgM capture), titers of 1:256 and higher should be expected in the first 4 months of infection. (Adapted from references 10, 14, and 43.)

Table 6.7 Guide to interpretation of serologic tests for the diagnosis of toxoplasmosis[a]

Clinical problem[b]	Infection status[c] of patients with titer[d] of:				
	0 (negative)	1:2–1:16	1:32–1:128	1:256–1:512	≥1:1,024
Asymptomatic	Susceptible to infection	Infection in the past, probably immune	Infection in the past, probably immune	Infection in the past, probably immune	Recent infection or reinfection
Pregnancy	Susceptible to infection	Infection in the past, probably immune	Infection in the past, probably immune	Infection in the past, probably immune	Need to monitor neonate for infection
Newborn, asymptomatic or with jaundice	No	No	Diagnosis unlikely	Diagnosis unlikely	Possible, perform testing for IgM
Newborn with encephalitis	No	No	No	Diagnosis unlikely	Diagnosis may be correct if titer is stable or rises; perform testing for IgM
Lymphadenopathy	No	No	No	Diagnosis unlikely	Diagnosis possible
Fever with pneumonia, myocarditis, or hepatitis	No	No	Diagnosis unlikely	Diagnosis unlikely	Diagnosis possible; perform testing for IgM
Retinochoroiditis	No	Diagnosis possible	Diagnosis possible	Diagnosis possible	Diagnosis possible
Encephalitis	No	No	Diagnosis unlikely	Diagnosis unlikely	Possible; perform testing for IgM
Encephalitis (immunosuppressed patient)	Nondiagnostic	Diagnosis unlikely (may be tranfusion related)	Diagnosis possible	Diagnosis possible	Diagnosis likely; perform testing for IgM

[a] Adapted from reference 14.
[b] Patients are immunocompetent except where stated otherwise.
[c] Susceptible to infection: the patient is not immune and is susceptible to a primary infection. No: with this antibody titer, toxoplasmosis should not be diagnosed. Diagnosis unlikely: these titers would be the lowest found in typical forms of the syndrome. Diagnosis possible: these titers would be characteristic but not necessarily diagnostic. Diagnosis likely: this titer would probably be diagnostic for toxoplasmosis. Perform testing for IgM: the presence of antibody may be informative.
[d] Dye test or indirect fluorescent-antibody test for IgG.

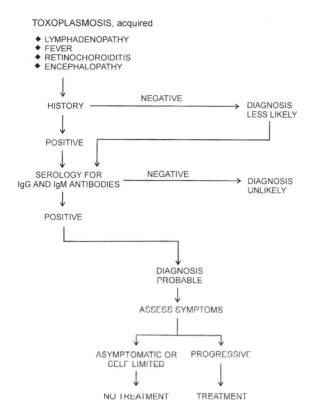

Algorithm 6.1   Toxoplasmosis.

Figure 6.12   Chart for the prevention of TE in HIV-positive patients; the approach is based on results of the *Toxoplasma* IgG serology. (Adapted from references 10, 14, and 43.)

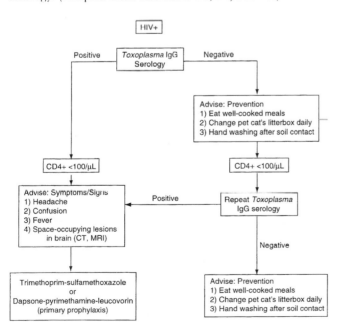

Table 6.8 Prevention of infection with *Toxoplasma gondii*

Prevention	Comments
Do not eat raw or under-cooked meat or eggs	An example would be hollandaise sauce, prepared from raw eggs; pork, lamb, and venison particularly in-volved (6, 25, 27)
Cured and smoked meat and sausages are considered safe	As an additional precaution, freeze at −20°C for a minimum of 1 day; how-ever, some sausages are eaten almost raw after adding salt and spices (inci-dence of *T. gondii* in certain popula-tions can be very high)
Consume only pasteurized or ultra-heat-treated milk and dairy products	Remember to consider ingredients of dairy products
When handling raw meat, avoid contaminating face, other food, utensils, cutting boards, or countertops	Hand washing and cleaning utensils, sinks, cutting boards, and countertops sufficient
Wash or peel fruits prior to eating them uncooked	Washing of smooth fruits is much easier than of berries; irradiation at 0.5 kGy is effective in "killing" coccidian oocysts on fruits and vegetables (23)
Control insect pests and their access to food	Keep foods covered and/or refrigerated
Wear gloves when garden-ing or handling sand where cats usually defecate	Wash hands after removing gloves
If living with cats, change litter boxes every day and rinse them with boiling water	In one day, oocysts in cat feces will not have time to mature—extended periods in litter boxes will lead to fully infec-tious oocysts
Monitor and limit cat food sources	Infection in cats can be reduced by not allowing them to eat raw meat, birds, and rodents; use dry or canned cat food; in some areas of the world, ingestion of infected chicken viscera can lead to oocyst shedding and spread of *T. gondii* in the environment (12)
Avoid handling stray cats and kittens	May be outdoor cats or may have been fed raw meat
Administer feline *T. gondii* vaccine to farm cats on swine operations	Reduction of exposure of finishing pigs to *T. gondii*; promising intervention to reduce risk of contaminating pork

Until recently, waterborne transmission of *T. gondii* was considered uncommon, but several large outbreaks linked to contaminated municipal water sources have led to a reassessment of this potential mode of transmission (3, 10). Considering the importance of using autofluores-cence in environmental detection of oocysts, it has now been established that *T. gondii* oocysts fluoresce with the same pale blue color as can be seen with *Cyclospora cayetanensis* and other organisms. Apparently, morphologic differences among coccidian oocysts provide fluorescent details sufficient for identification when testing environmental water samples (26).

Since primary infection with *T. gondii* leads to specific and definitive protection against reinfection, the feasibility of developing a vaccine is now being investigated. To protect humans, indirect procedures such as vaccination of the definitive host, i.e., the cat, or direct procedures by immunization with nonliving vaccines may be required.

References

1. Abramowicz, M. (ed). 2004. Drugs for parasitic infections. *Med. Lett. Drugs Ther.* **46:**1–12.

2. Ajzenberg, D., A. Dumètre, and M. Dardé. 2005. Multiplex PCR for typing strains of *Toxoplasma gondii*. *J. Clin. Microbiol.* **43:**1940–1943.

3. Bahia-Oliveira, L. M., J. L. Jones, J. Azevedo-Silva, C. C. Alves, F. Orefice, and D. G. Addiss. 2003. Highly endemic, waterborne toxoplasmosis in north Rio de Janeiro state, Brazil. *Emerg. Infect. Dis.* **9:**55–62.

4. Barenfanger, J., C. Drake, and C. Hanson. 2002. Timing of inoculation of the pouch makes no difference in increased detection of *Trichomonas vaginalis* by the InPouch TV method. *J. Clin. Microbiol.* **40:**1387–1389.

5. Beaver, P. C., R. C. Jung, and E. W. Cupp. 1984. *Clinical Parasitology*, 9th ed. Lea & Febiger, Philadelphia, Pa.

6. Cañon-Franco, W. A., L. E. O. Yai, A. M. Joppert, C. E. Souza, S. R. N. D'Auria, J. P. Dubey, and S. M. Gennari. 2003. Seroprevalence of *Toxoplasma gondii* antibodies in the rodent capybara (*Hidrochoeris hidrochoeris*) from Brazil. *J. Parasitol.* **89:**850.

7. Carme, B., F. Bissuel, D. Ajzenberg, R. Bouyne, C. Aznar, M. Demar, S. Bichat, D. Louvel, A. M. Bourbigot, C. Peneau, P. Neron, and M. L. Dardé. 2002. Severe acquired toxoplasmosis in immunocompetent adult patients in French Guiana. *J. Clin. Microbiol.* **40:**4037–4044.

8. Cudmore, S. L., K. L. Delgaty, S. F. Hayward-McClelland, D. P. Petrin, and G. E. Garber. 2004. Treatment of infections caused by metronidazole-resistant *Trichomonas vaginalis*. *Clin. Microbiol. Rev.* **17:**783–793.

9. Dessi, D., G. Delogu, E. Emonte, M. R. Catania, P. L. Fiori, and P. Rappelli. 2005. Long-term survival and intracellular replication of *Mycoplasma hominis* in *Trichomonas vaginalis* cells: potential role of the protozoan in transmitting bacterial infection. *Infect. Immun.* **73:**1180–1186.

10. Dubey, J. P. 2004. Toxoplasmosis—a waterborne zoonosis. *Vet. Parasitol.* **126:**57–72.

11. Dubey, J. P. 2002. Tachyzoite-induced life cycle of *Toxoplasma gondii* in cats. *J. Parasitol.* **88:**713–717.

12. Dubey, J. P., D. H. Graham, E. Dahl, C. Sreekumar, T. Lehmann, M. F. Davis, and T. Y. Morishita. 2003. *Toxoplasma gondii* isolates from free-ranging chickens from the United States. *J. Parasitol.* **89:**1060–1062.

13. Dunne, R. L., L. A. Dunn, P. Upcroft, P. J. O'Donoghue, and J. A. Upcroft. 2003. Drug resistance in the sexually transmitted protozoan *Trichomonas vaginalis*. *Cell Res.* **13:**239–249.

14. Frenkel, J. K., and J. L. Fishback. 2000. Toxoplasmosis, p. 691–701. *In* G. T. Strickland (ed.), *Hunters Tropical Medicine and Emerging Infectious Diseases*, 8th ed. The W. B. Saunders Co., Philadelphia, Pa.

15. Garweg, J. G., S. L. Garweg, F. Flueckiger, P. Jacquier, and M. Boehnke. 2004. Aqueous humor and serum immunoblotting for immunoglobulin types G, A, M, and E in cases of human ocular toxoplasmosis. *J. Clin. Microbiol.* **42:**4593–4598.

16. Honore, S., A. Couvelard, Y. J. Garin, C. Bedel, D. Henin, M. L. Darde, and F. Derouin. 2000. Genotyping of *Toxoplasma gondii* strains from immunocompromised patients. *Pathol. Biol.* **48:**541–547.

17. Huppert, J. S., B. E. Batteiger, P. Braslins, J. A. Feldman, M. M. Hobbs, H. Z. Sankey, A. C. Sena, and K. A. Wendel. 2005. Use of an immunochromatographic assay for rapid detection of *Trichomonas vaginalis* in vaginal specimens. *J. Clin. Microbiol.* **43:**684–687.

18. Jones, J. L., D. Kruszon-Moran, and M. Wilson. 2003. *Toxoplasma gondii* infection in the United States, 1999–2000. *Emerg. Infect. Dis.* **9:**1371–1374.

19. Joseph, P., M. M. Calderón, R. H. Gilman, M. L. Quispe, J. Cok, E. Ticona, V. Chavez, J. A. Jiminez, M. C. Chang, M. J. Lopez, and C. A. Evans. 2002. Optimization and evaluation of a PCR assay for detecting toxoplasmic encephalitis in patients with AIDS. *J. Clin. Microbiol.* **40:**4499–4503.

20. Kaul, R., P. Chen, and S. R. Binder. 2004. Detection of immunoglobulin M antibodies specific for *Toxoplasma gondii* with increased selectivity for recently acquired infections. *J. Clin. Microbiol.* **42:**5705–5709.

21. Kaydos, S. C., H. Swygard, S. L. Wise, A. C. Sena, P. A. Leone, W. C. Miller, M. S. Cohen, and M. M. Hobbs. 2002. Development and validation of a PCR-based enzyme-linked immunosorbent assay with urine for use in clinical research settings to detect *Trichomonas vaginalis* in women. *J. Clin. Microbiol.* **40:**89–95.

22. Kaydos-Daniels, S. C., W. C. Miller, I. Hoffman, T. Banda, W. Dzinyemba, F. Martinson, M. S. Cohen, and M. M. Hobbs. 2003. Validation of a urine-based PCR-enzyme-linked immunosorbent assay for use in clinical research settings to detect *Trichomonas vaginalis* in men. *J. Clin. Microbiol.* **41:**318–323.

23. Kniel, K. E., D. S. Lindsay, S. S. Sumner, C. R. Hackney, M. D. Pierson, and J. P. Dubey. 2002. Examination of attachment and survival of *Toxoplasma gondii* oocysts on raspberries and blueberries. *J. Parasitol.* **88:**790–793.

24. Kurth, A., W. L. H. Whittington, M. R. Golden, K. K. Thomas, K. K. Holmes, and J. R. Schwebke. 2004. Performance of a new, rapid assay for detection of *Trichomonas vaginalis*. *J. Clin. Microbiol.* **42:**2940–2943.

25. Kutz, S. J., B. T. Elkin, D. Panayi, and J. P. Dubey. 2001. Prevalence of *Toxoplasma gondii* antibodies in barrenground caribou (*Rangifer tarandus groenlandicus*) from the Canadian Arctic. *J. Parasitol.* **87:**439–442.

26. Lindquist, H. D. A., J. W. Bennett, J. D. Hester, M. W. Ware, J. P. Dubey, and W. V. Everson. 2003. Autofluorescence

of *Toxoplasma gondii* and related coccidian oocysts. *J. Parasitol.* **89**:865–867.

27. **Lindsay, D. S., M. V. Collins, S. M. Mitchell, C. N. Wetch, A. C. Rosypal, G. J. Flick, A. M. Zajac, A. Lindquist, and J. P. Dubey.** 2004. Survival of *Toxoplasma gondii* oocysts in Eastern oysters (*Crassostrea virginica*). *J. Parasitol.* **90**:1054–1057.

28. **Martinez-Grueiro, M. M., D. Montero-Pereira, C. Giménez-Pardo, J. J. Nogal-Ruiz, J. A. Escario, and A. Gómez-Barrio.** 2003. *Trichomonas vaginalis*: determination of acid phosphatase activity as a pharmacological screening procedure. *J. Parasitol.* **89**:1076–1077.

29. **Menotti, J., G. Vilela, S. Romand, Y. J. Garin, L. Ades, E. Gluckman, F. Derouin, and P. Ribaud.** 2003. Comparison of PCR-enzyme-linked immunosorbent assay and real-time PCR assay for diagnosis of an unusual case of cerebral toxoplasmosis in a stem cell transplant recipient. *J. Clin. Microbiol.* **41**:5313–5316.

30. **Montoya, J. G., H. B. Huffman, and J. S. Remington.** 2004. Evaluation of the immunoglobulin G avidity test for diagnosis of toxoplasmic lymphadenopathy. *J. Clin. Microbiol.* **42**:4627–4631.

31. **Okun, N., K. A. Gronau, and M. E. Hannah.** 2005. Antibiotics for bacterial vaginosis or *Trichomonas vaginalis* in pregnancy: a systematic review. *Obstet. Gynecol.* **105**:857–868.

32. **Petersen, E., M. V. Borobio, E. Guy, O. Liesenfeld, V. Meroni, A. Naessens, E. Spranzi, and P. Thulliez.** 2005. European multicenter study of the LIAISON automated diagnostic system for determination of *Toxoplasma gondii*-specific immunoglobulin G (IgG) and IgM and the IgG avidity index. *J. Clin. Microbiol.* **43**:1570–1574.

33. **Petrin, D., K. Delgary, R. Bhatt, and G. Garber.** 1998. Clinical and microbiological aspects of *Trichomonas vaginalis*. *Clin. Microbiol. Rev.* **11**:300–317.

34. **Pillay, A., J. Lewis, and R. C. Ballard.** 2004. Evaluation of Xenostrip-Tv, a rapid diagnostic test for *Trichomonas vaginalis* infection. *J. Clin. Microbiol.* **42**:3853–3856.

35. **Schwebke, J. R., and D. Burgess.** 2004. Trichomoniasis. *Clin. Microbiol. Rev.* **17**:794–803.

36. **Soper, D.** 2004. Trichomoniasis: under control or undercontrolled? *Am. J. Obstet. Gynecol.* **190**:281–290.

37. **Stary, A, A Kuchkinka-Koch, and L. Teodorowicz.** 2002. Detection of *Trichomonas vaginalis* on modified Columbia agar in the routine laboratory. *J. Clin. Microbiol.* **40**:3277–3280.

38. **Switaj, K., A. Master, M. Skrzypczak, and P. Zaborowski.** 2005. Recent trends in molecular diagnostics for *Toxoplasma gondii* infections. *Clin. Microbiol. Infect.* **11**:170–176.

39. **Torrey, E. F., and R. H. Yolken.** 2003. *Toxoplasma gondii* and schizophrenia. *Emerg. Infect. Dis.* **9**:1375–1380.

40. **Vargas-Villarreal, J., B. D. Mata-Cárdenas, R. Palacios-Corona, F. González-Salazar, E. I. Cortes Gutierrez, H. G. Martinez-Rodríguez, and S. Said-Fernández.** 2005. *Trichomonas vaginalis*: identification of soluble and membrane-associated phospholipase A_1 and A_2 activities with direct and indirect hemolytic effects. *J. Parasitol.* **91**:5–11.

41. **Vidal, J. E., F. A. Colombo, A. C. Penalva de Oliveira, R. Rocaccia, and V. L. Pereira-Chioccola.** 2004. PCR assay using cerebrospinal fluid for diagnosis of cerebral toxoplasmosis in Brazilian AIDS patients. *J. Clin. Microbiol.* **42**:4765–4768.

42. **Villard, O., D. Filisetti, F. Roch-Deries, J. Garweg, J. Flament, and E. Candolfi.** 2003. Comparison of enzyme-linked immunosorbent assay, immunoblotting, and PCR for diagnosis of toxoplasmic chorioretinitis. *J. Clin. Microbiol.* **41**:3537–3541.

43. **Wilson, M., and J. B. McAuley.** 2003. *Clinical Use and Interpretation of Serologic Tests for* Toxoplasma gondii. Standard M36-A. NCCLS, Wayne, Pa.

7

Malaria and Babesiosis

Malaria

Babesiosis

Malaria

Malaria represents one of the oldest documented diseases of humans, and even today, organisms in the genus *Plasmodium* kill more people than do any other infectious microorganisms. Egyptian mummies more than 3,000 years old have been found with enlarged spleens, most probably due to malaria. Deadly periodic fevers and splenomegaly have been mentioned, dating from as early as 2700 B.C., in both Egyptian and Chinese writings.

Based on the lack of malaria records in the New World prior to the arrival of European explorers, conquistadors, and colonists, it is assumed that they imported *Plasmodium malariae* and *P. vivax* to the Americas. Apparently, the arrival of *P. falciparum* coincided with the importation of African slaves, and by the early 1800s malaria was found worldwide.

Malaria has probably had a greater impact on world history than any other infectious disease. It has been responsible for the outcome of wars, population movements, and the growth and development of various nations throughout the world. Before the American Civil War, malaria was found as far north as southern Canada; however, there has been a gradual decline, and by the early 1950s it was no longer an endemic disease within the United States. It is still a very common disease in many parts of the world, particularly in tropical and subtropical areas (Table 7.1).

Of the four most common species that infect humans, *P. vivax* and *P. falciparum* account for 95% of infections. Some estimates indicate that *P. vivax* may account for 80% of the infections. This species also has the widest distribution, extending throughout the tropics, subtropics, and temperate zones. *P. falciparum* is generally confined to the tropics, *P. malariae* is sporadically distributed, and *P. ovale* is confined mainly to central West Africa and some South Pacific islands.

We usually associate malaria with patients having a history of travel within an area where malaria is endemic. However, other situations that may result in infection involve the receipt of blood transfusions, use of hypodermic needles contaminated by prior use (as with, for example, drug addicts), possibly congenital infection, and transmission within the United States by indigenous mosquitoes that acquired the parasites from imported infections (27).

With the exception of tuberculosis, malaria kills more people than any other communicable disease in the world. More than 300 million to 500 million

Table 7.1   Countries reported by the Centers for Disease Control and Prevention to have areas of endemic malaria

Afghanistan	Guatemala	Panama
Algeria	Guinea	Papua New Guinea
Angola	Guinea-Bissau	Paraguay
Argentina	Guyana	Peru
Bangladesh	Haiti	Philippines
Belize	Honduras	Rwanda
Benin	India	São Tomé and Principe
Bhutan	Indonesia	Saudi Arabia
Bolivia	Iran	Senegal
Botswana	Iraq	Sierra Leone
Brazil	Jordan	Solomon Islands
Burkina Faso	Kenya	Somalia
Burundi	Lao People's Democratic Republic	South Africa
Cambodia	Liberia	Sri Lanka
Cameroon	Libyan Arab Jamahiriya (formerly Libyan	Sudan
Central African Republic	Arab Republic)	Surinam
Chad	Madagascar	Swaziland
China, People's Republic	Malawi	Syrian Arab Republic
Colombia	Malaysia	Tanzania
Comoros	Maldives	Thailand
Congo	Mali	Togo
Costa Rica	Mauritania	Tunisia
Cote d'Ivoire (formerly Ivory Coast)	Mauritius	Turkey
Djibouti	Mayotte	Uganda
Dominican Republic	Mexico	USSR (former)[a]
Ecuador	Morocco	United Arab Emirates (formerly Trucial
Egypt	Mozambique	Sheikdoms)
El Salvador	Myanmar (formerly Burma)	Vanuatu
Equatorial Guinea	Namibia	Venezuela
Eritrea	Nepal	Vietnam
Ethiopia	Nicaragua	Yemen
French Guiana	Niger	Yemen, Democratic
Gabon	Nigeria	Zaire
Gambia	Oman	Zambia
Ghana	Pakistan	Zimbabwe (formerly Rhodesia)

[a] Armenia, Azerbaijan, Belarus, Estonia, Georgia, Kazakstan, Kyrgyzstan, Latvia, Lithuania, Moldova, Russia, Tajikistan, Turkmenistan, Ukraine, Uzbekistan, and in a few areas adjacent to Afghanistan and Iran.

individuals throughout the world are infected with malaria, and 1.5 million to 2.7 million people a year, most of whom are children, are being killed by the disease. More people are dying from malaria today than 30 years ago. Malaria is endemic in over 90 countries in which 2,400 million people live; this is 40% of the world's population (66). The African countries, where 90% of the malaria deaths occur, are ravaged by warfare, and the richer nations that fund most malaria research are cutting budgets. Malaria prevention is difficult, and no drug is universally effective. Vaccine development studies are under way, but malarial vaccines are not yet in general use. Although over 1,000 cases of malaria are reported to the Centers

for Disease Control and Prevention each year, the actual number of cases may be much larger. There has also been a definite increase in the number of cases of *P. falciparum* malaria reported, which may be related to increased resistance to chloroquine.

It is important for both physicians and laboratorians within areas where malaria is not endemic to be aware of the difficulties related to malaria diagnosis and the fact that symptoms are often nonspecific and may mimic other problems. It is important for physicians to recognize that travelers are susceptible to malarial infection when they return to a country where malaria is endemic, and they should receive prophylactic medication. It is also important to remember that 80% of the *P. falciparum* cases acquired by American civilians are contracted in sub-Saharan Africa; currently there are 40 cities in Africa with over 1 million inhabitants, and by 2025 over 800 million people will live in urban areas. Malaria has always been a rural disease in Africa; however, it appears that the urban poor are at a much higher risk of contracting malaria than was previously recognized. These changes also have the potential to affect travelers to Africa.

With the tremendous increase in the number of people traveling from the tropics to malaria-free areas, the number of imported malaria cases is also on the rise. There have been reports of imported infected mosquitoes transmitting the infection among people who live or work near international airports. It is also possible that mosquitoes can reach areas far removed from the airports (84). This situation has been termed "airport malaria," i.e., malaria that is acquired through the bite of an infected anopheline mosquito by persons with apparently no risk factors for the disease. Unfortunately, unless a careful history is obtained, the diagnosis of malaria can be missed or delayed. Tests to exclude malaria should be considered for patients who work or live near an international airport and who present with an acute febrile illness. The potential danger of disseminating the mosquito vectors of malaria via aircraft is well recognized; however, modern disinfection procedures have not yet eliminated the risk of vector transportation. Not only can insects survive nonpressurized air travel, but also they may be transported further by car or other means after arriving at the airport. Contraction of airport malaria is certainly an unusual event, particularly considering that the number of air travelers is well over 1 billion annually. However, in addition to asking a patient, "Where have you been, and when were you there?" one should ask, "Where do you live?"

Potential contributing factors to malaria-related fatalities in U.S. travelers include (i) failure to seek pretravel advice, (ii) failure to prescribe correct chemoprophylaxis, (iii) failure to obtain prescribed chemoprophylaxis, (iv) failure to adhere to the chemoprophylaxis regimen, (v) failure to seek medical care promptly for their illness, (vi) failure to obtain an adequate patient history, (vii) delay in diagnosis of malaria, and (viii) delay in initiating treatment of malaria (78). One or more of these potential problems can lead to the death of the patient.

The most recent group of cases of imported malaria within the United States military has been reported to be in U.S. Army Rangers returning from Afghanistan. There were 38 cases represented in this group, with all diagnosed infections being due to *P. vivax*. It is interesting that the self-reported compliance rates were 52% for weekly chemoprophylaxis, 41% for postdeployment chemoprophylaxis, 31% for both weekly and postde-ployment chemoprophylaxis, 82% for treating uniforms with permethrin, and 29% for application of insect repellent (58).

Life Cycle and Morphology

The vector for malaria is the female anopheline mosquito. When the vector takes a blood meal, sporozoites contained in the salivary glands of the mosquito are discharged into the puncture wound (Figures 7.1 and 7.2) (10). Within an hour, these infective stages are carried via the blood to the liver, where they penetrate hepatocytes and begin to grow, thus initiating the pre-erythrocytic or primary exoerythrocytic cycle. Detailed study of sporozoite entry into the hepatocytes indicates that the process involves parasite-encoded surface proteins and host molecules. The sporozoites become round or oval and begin dividing repeatedly. This schizogony results in large numbers of exoerythrocytic merozoites. Once these merozoites leave the liver, they invade the red blood cells (RBCs), thus initiating the erythrocytic cycle. It has been reported that a secondary or dormant schizogony may occur in *P. vivax* and *P. ovale* organisms, which remain quiescent in the liver until a later time. These resting stages have been termed hypnozoites (60). Delayed schizogony does not occur in *P. falciparum* and probably does not occur in *P. malariae*.

The situation in which the RBC infection is not eliminated by the immune system or by therapy and the numbers in the RBCs begin to increase again with subsequent clinical symptoms is called a recrudescence. All species may cause a recrudescence. The situation in which the erythrocytic infection is eliminated and a relapse occurs later because of a new invasion of the RBCs from liver merozoites, called a recurrence or true relapse, theoretically occurs only in *P. vivax* and *P. ovale* infection.

Once the RBCs and reticulocytes have been invaded, the parasites grow and feed on hemoglobin. Within the RBC, the merozoite (or young trophozoite) is vacuolated, ring shaped, more or less ameboid, and uninucleate. The excess protein, an iron porphyrin, and hematin left over from the metabolism of hemoglobin combine to form malarial pigment (some workers use the term "hemozoin"; however, the term "malarial pigment" is used throughout this chapter).

Once the nucleus begins to divide, the trophozoite is called a developing schizont. The mature schizont contains merozoites (whose number depends on the species), which are released into the bloodstream. Many of the merozoites are destroyed by the immune system, but others invade RBCs, in which a new cycle of erythrocytic schizogony begins.

After several erythrocytic generations, some of the merozoites do not become schizonts but, rather, begin

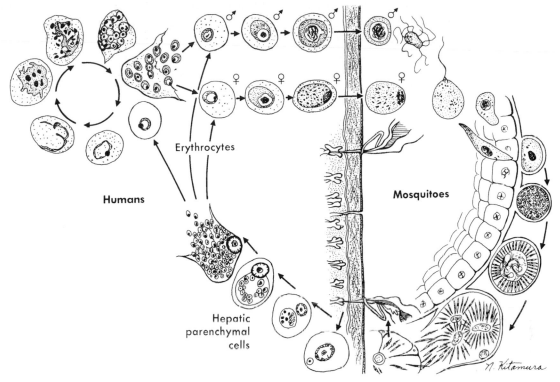

Figure 7.1   Life cycle of *Plasmodium* spp. (Illustration by Nobuko Kitamura; adapted from Wilcox, A., *Manual for the Microscopical Diagnosis of Malaria in Man*, U.S. Public Health Service bulletin 180, National Institutes of Health, Bethesda, Md., 1951.)

to undergo development into the male and female gametocytes. Whether this development is predetermined genetically or as a response to some specific stimulus is unknown (8). It is likely that cells are committed to sexual development in the preceding round of asexual schizogony within the preerythrocytic schizont rather than differentiating following invasion of an RBC by uncommitted merozoites. Data also suggest that genetic and environmental factors may predispose *Plasmodium* spp. to switch from asexual to sexual development.

Apparently, sexual commitment is highest in merozoites derived directly from the liver schizont and there is a progressive loss of commitment with increasing numbers of asexual cycles. Environmental stimuli may relate directly to an increase in parasite density; there may also be some direct links between sublethal doses of chloroquine and an increase in gametocytogenesis, particularly in chloroquine-resistant parasites.

The asexual and sexual forms just described circulate in the bloodstream during infections by three

Figure 7.2   Life cycle of malaria.

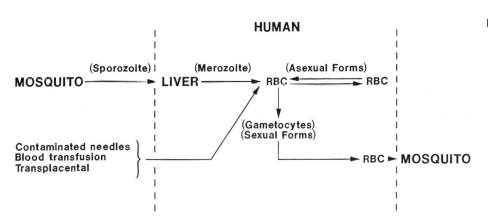

of the *Plasmodium* species. However, in *P. falciparum* infections, as the parasite continues to grow, the RBC membrane becomes sticky and the cells tend to adhere to the endothelial lining of the capillaries of the internal organs. Thus, only the ring forms and the gametocytes (occasionally mature schizonts) normally appear in the peripheral blood.

If gametocytes are ingested when the mosquito takes a blood meal, they mature into gametes while in the mosquito gut. The male microgametes undergo nuclear division by a process called exflagellation, in which the microgametes break out of the RBC, become motile, and penetrate the female macrogamete, with the fertilized stage being called the zygote (Figure 7.3). This zygote then becomes elongate and motile and is called the ookinete. This stage migrates to the mosquito midgut, secretes a thin wall, and grows into the oocyst, which extends into the insect's hemocele. Within a few days to 2 weeks, the oocyst matures, with the formation of hundreds of sporozoites. When the oocyst ruptures, the sporozoites are released into the hemocele and dispersed throughout the body, and some make their way into the salivary glands. When the mosquito next takes a blood meal, the sporozoites are injected with saliva into the host. Unfortunately, when blood tubes are held at room temperature, especially with the cap removed, the process of exflagellation can occur within the blood. Subsequently, when the blood films are prepared, these microgametocytes can be confused with *Borrelia* spp. (11, 94).

Figure 7.3 Scanning electron micrograph of exflagellating male gametocyte. (Courtesy of Robert E. Sinden.)

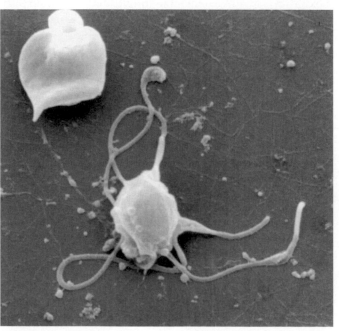

Clinical Disease

From the time of the original mosquito bite until a week or more later, the patient remains asymptomatic. During this time, the organisms are undergoing multiplication (preerythrocytic cycle) in the liver. When the liver merozoites invade the RBCs, several broods begin to develop; however, one will eventually dominate and suppress the others, thus beginning the process of periodicity. Once the cycle is synchronized, the simultaneous rupture of a large number of RBCs and liberation of metabolic waste by-products into the bloodstream precipitate the paroxysms of malaria.

Clinical symptoms of malaria include anemia, splenomegaly, and the classic paroxysm, with its cold stage, fever, and sweats. Although the febrile paroxysms strongly suggest infection, many patients who are seen in medical facilities, particularly in the early stages of the infection, do not exhibit the typical fever pattern. They may have fever or several small, random peaks each day. Since the symptoms associated with malaria are so nonspecific, the diagnosis should be considered in any symptomatic patient with a history of travel to an area where malaria is endemic. During the primary infection in a nonimmune host, the early fever episodes can affect density-dependent regulation of the parasite population, maintaining cycles of parasitemia and promoting synchronous parasite growth. The typical paroxysm begins with the cold stage and rigors lasting 1 to 2 h. During the next few hours, the patient spikes a high fever and feels very hot, and the skin is warm and dry. The last several hours are characterized by marked sweating and a subsequent drop in body temperature to normal or subnormal.

Anemia seen in malarial infections can be caused by a number of mechanisms, including (i) direct RBC lysis as a function of the life cycle of the parasite, (ii) splenic removal of both infected and uninfected RBCs (coated with immune complexes), (iii) autoimmune lysis of coated infected and uninfected RBCs, (iv) decreased incorporation of iron into heme, (v) increased fragility of RBCs, and (vi) decreased RBC production from bone marrow suppression.

Malaria can mimic many other diseases, such as gastroenteritis, pneumonia, meningitis, encephalitis, or hepatitis. Other possible symptoms include lethargy, anorexia, nausea, vomiting, diarrhea, and headache. Leukopenia can also be seen in malaria, as can an occasional elevated white blood cell count with a left shift. Eosinophilia and thrombocytopenia may also be seen but are much less frequent. A clinical comparison of some of the features of the four different malarias is presented in Table 7.2. Information on the pathogenesis of malaria is provided in Figure 7.4 and Tables 7.3 to 7.5.

Plasmodium vivax. The primary clinical attack usually occurs 7 to 10 days after infection, although there are strain differences, with a much longer incubation period being possible. In some patients, symptoms such as headache, photophobia, muscle aches, anorexia, nausea, and sometimes vomiting occur before organisms can be detected in the bloodstream. In other patients, the parasites can be found in the bloodstream several days before symptoms appear.

During the first few days, the patient may not exhibit a typical paroxysm pattern but, rather, may have a steady low-grade fever or an irregular remittent fever pattern. Once the typical paroxysms begin, after an irregular periodicity, a regular 48-h cycle is established. An untreated primary attack may last from 3 weeks to 2 months or longer. The paroxysms become less severe and more irregular in frequency and then stop altogether. In 50% of patients, relapses occur after weeks, months, or up to 5 years (or more).

Severe complications are rare in P. vivax infections, although coma and sudden death or other symptoms of cerebral involvement have been reported. These patients can exhibit cerebral malaria, renal failure, circulatory collapse, severe anemia, hemoglobinuria, abnormal bleeding, acute respiratory distress syndrome, and jaundice (57, 94). Studies have confirmed that these were not mixed infections with P. falciparum but single-species infections with P. vivax.

Since P. vivax infects only the reticulocytes, the parasitemia is usually limited to around 2 to 5% of the available RBCs. Splenomegaly occurs during the first few weeks of infection, and the spleen progresses from being soft and palpable to hard, with continued enlargement during a chronic infection. If the infection is treated during the early phases, the spleen returns to its normal size.

Leukopenia is usually present; however, leukocytosis may be present during the febrile episodes. Concentrations of total plasma proteins are unchanged, although the albumin level may be low and the globulin fraction may be elevated. The increase in the concentration of gamma globulins is caused by the development of antibodies. The level of potassium in serum may also be increased as a result of RBC lysis.

Table 7.2 *Plasmodium* spp.: clinical characteristics of the four infections[a]

Infection parameter	P. vivax	P. ovale	P. malariae	P. falciparum	Comments
Incubation period	8–17 days	10–17 days	18–40 days	8–11 days	May be extended for months to years
Prodromal symptoms					In all 4 species, may mimic symptoms seen with influenza
Severity	Mild to moderate	Mild	Mild to moderate	Mild	
Initial fever pattern	Irregular (48 h)	Irregular (48 h)	Regular (72 h)	Continuous-remittent (48 h)	Lack of regular periodicity may not suggest malaria as a potential diagnosis, particularly in patients who present to the emergency room early in the course of the infection
Symptom periodicity	48 h	48 h	72 h	36–48 h	
Initial paroxysm					
Severity	Moderate to severe	Mild	Moderate to severe	Severe	
Mean duration	10 h	10 h	11 h	16–36 h	
Duration of untreated primary attack	3–8+ wk	2–3 wk	3–24 wk	2–3 wk	
Duration of untreated infection	5–7 yr	12 mo	20+ yr	6–17 mo	
Parasitemia limitations	Young RBCs	Young RBCs	Old RBCs	All RBCs	
Anemia	Mild to moderate	Mild	Mild to moderate	Severe	
CNS involvement	Rare	Possible	Rare	Very common	
Nephrotic syndrome	Possible	Rare	Very common	Rare	

[a]Adapted from E. K. Markell and M. Voge, *Medical Parasitology*, 5th ed., The W. B. Saunders Co., Philadelphia, Pa., 1981.

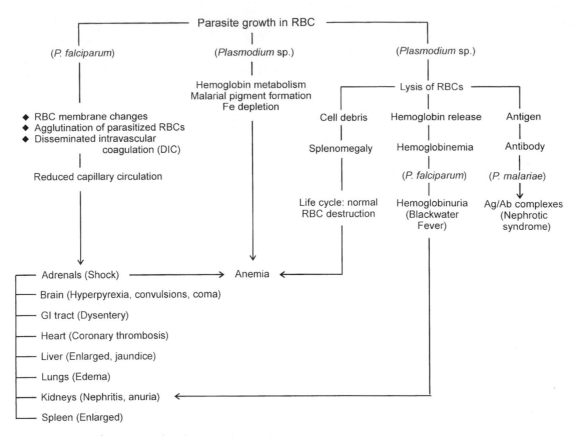

Figure 7.4   Pathogenesis of malaria. (Adapted from E. K. Markell and M. Voge, *Medical Parasitology*, 5th ed., The W. B. Saunders Co., Philadelphia, Pa., 1981.)

Table 7.3   Malaria pathophysiology[a]

Factor	Comments[b]
Parasite virulence factors	Multiplication capacity: rapidly expanding biomass outstrips host defenses; immunity reduces multiplication
	RBC selectivity: age of RBC determines level of parasitemia; less so for *P. falciparum*
	Cytoadherence, rosetting: all four species can induce rosetting; *P. falciparum* causes cytoadherence
	Cytokine release: *P. falciparum* varies in ability to induce TNF release, even those derived from single genotype
	Antigenicity: in areas of endemic infection, parasites not recognized by the host will have relative growth advantage
	Drug resistance: antimalarial drugs widely available, self-medication common, selective pressure against drug-sensitive parasites
Parasite multiplication	At the onset of fever, approximaely 10^8 parasites in the body; corresponds to 20–20,000 parasites/µl of blood; in areas of endemic infection, patients may tolerate parasitemias up to 10,000 / µl of blood without feeling ill; only *P. falciparum* has ability to infect all RBCs and may exceed 50%
Parasite biomass	Loose positive correlation between parasitemia and prognosis in *P. falciparum* malaria; positive correlation between number of sequestered paraasitized RBCs and severity of *P. falciparum* malaria
Toxicity and cytokines	Question of true toxin vs paroxysmal cytokine release due to nonspecific reaction to amount of particulate debris released into the circulation at schizont rupture; high concentrations of cytokines appear to be harmful, while lower levels probably benefit the patient; they are responsible for the typical symptoms of fever and malaise
RBC deformability	As the parasite grows, the normally flexible biconcave disk becomes more spherical and rigid, and the surface becomes irregular and covered with small protrusions; reduction in deformability results from reduced membrane fluidity, increasing sphericity, and enlarging, viscous intraerythrocytic parasite; also occurs in uninfected RBCs in acute malaria

(continued)

Table 7.3 *(continued)*

Factor	Comments[b]
Sequestration and pathology	RBCs adhere to microvascular endothelium (cytoadherence) and disappear from peripheral circulation (i.e., sequestration); greatest involvement is in brain, heart, kidney, intestines, and adipose tissue, least prominent is in skin; consequences of microcirculatory obstruction are reduced oxygen/substrate supply (anaerobic glycolysis and lactic acidosis)
Coma	Coma associated with 15–20% mortality; however, both adults and children recover without sequelae; thought to be metabolic or anesthetic encephalopathy (may involve neurotransmitter abnormalities and NO)
Intracranial pressure	Mild generalized increase in systemic vascular permeability in severe malaria; majority of patients with cerebral malaria do not have cerebral edema; all children with cerebral malaria (at some time) have elevated ICP; background rise in ICP probably results from increase in cerebral blood volume, independent of permeability
Immunological factors	Unlikely that cerebral malaria is due to immune-mediated damage; fatal falciparum malaria has lack of extravascular pathology; in cases of hypergammaglobulinemia, most antibody is not directed against malaria antigens; description suggests a "smoke screen" with broad-spectrum and nonspecific activation that interferes with the development of a specific cellular immune response; exception may be nephrotic syndrome caused by *P. malariae* and antigen-antibody immune complex formation in the kidneys
Renal failure	Basic pathology is acute tubular necrosis; urine sediment unremarkable, sodium level in urine low, blood pressure normal; no proof, but renal microvascular obstruction due to sequestration in kidneys may be responsible; glomerulonephritis is very rare in acute malaria, and renal necrosis does not occur
Pulmonary edema	Pulmonary function usually normal; pregnant women with severe falciparum malaria at risk for sudden, acute pulmonary edema (resulting from a sudden increase in pulmonary capillary permeability); cause not known
Cardiovascular function	Pump function not a problem in severe malaria; may be autonomic dysfunction in acute malaria (postural hypotension associated with impaired cardioacceleration), but may be common to all febrile infections
Fluid space; electrolyte changes	Patients may be dehydrated; mild hyponatremia and hypochloremia are common, but potassium levels are usually normal (in spite of hemolysis); after rehydration, total body water and extracellular volume are normal
Anemia	In general, degree of anemia corresponds to duration and severity of infection; obligatory destruction of RBCs due to life cycle, accelerated destruction of nonparasitized RBCs correlates with disease severity; condition is compounded by bone marrow dysfunction; factors that trigger removal of nonparasitized RBCs not fully explained; RBC survival reduced in malaria (not impacted by corticosteroids); reduced oxygen carriage; past explanation of death due to congestive cardiac failure may be due to tissue hypoxia and severe metabolic acidosis
Coagulopathy and thrombocytopenia	Accelerated coagulation cascade activity with accelerated fibrinogen turnover; moderate thrombocytopenia common in *P. falciparum* and *P. vivax* malaria, caused by splenic clearance; DIC no longer considered important in pathogenesis of malaria
Blackwater fever	Massive intravascular hemolysis and passage of dark red, brown, or usually black urine; occurs when patients with G6PD deficiency take oxidant drugs (whether or not they have malaria), when G6PD-deficient patients have malaria and receive treatment with quinine or artemisinin derivatives, and when some patients with normal G6PD levels are treated with quinine or artemisinin derivatives; blackwater fever is not usually accompanied by renal failure, although renal impairment is common
Spleen	Rapid splenic enlargement for removal of parasitized RBCs; may also modulate cytoadherence, but mechanism is not known
Gastrointestinal tract	Minor ulceration of stomach and duodenum common in severe malaria; may be reduced splanchnic perfusion (malabsorption of sugars, fats, and amino acids); may be increased gut permeability
Liver	Hepatic dysfunction usually seen in severe malaria; jaundice common (hemolytic, hepatic, cholestatic components); reduced clotting factor synthesis, metabolic drug clearance, and biliary excretion; failure of glyconeogenesis (lactic acidosis and hypoglycemia); liver blood flow may be reduced or may be elevated
Acidosis	Consistent feature; arterial, capillary, venous, and CSF lactate concentrations rise in direct proportion to disease severity; majority of lactic acid produced in malaria derives from host rather than parasite
Hypoglycemia	Closely associated with hyperlactatemia; hypoglycemia leads to nervous system dysfunction, and is associated with residual neurological deficit in survivors of cerebral malaria
Placenta	Pregnancy increases susceptibility to malaria (suppression of systemic and placental cell-mediated immune responses); intense sequestration of *P. falciparum*-infected RBCs in placenta and maternal anemia; leads to placental insufficiency and fetal retarded growth; maternal mortality approaches 50%, and fetal death usual; high incidence of cerebral malaria, hypoglycemia, and pulmonary edema; *P. vivax* also linked to reductions in birth weight, suggesting that systemic effects of malaria may be more important than previously recognized
Coincident infections	Patients vulnerable to other infections, including other species of malaria; relationship to HIV also questionable; chronic hepatitis B virus carriage also associated with increased risk of severe malaria

[a] Adapted from references 18, 66, and 107.
[b] TNF, tumor necrosis factor; ICP, intracranial pressure; G6PD, glucose-6-phosphate dehydrogenase; NO, nitric oxide; DIC, disseminated intravasular coagulopathy; CSF, cerebrospinal fluid.

Table 7.4   *Plasmodium* pathogenesis and sequelae

Clinical situation and condition	Comments[a]	Reference(s)
Chronic malaria in West Africa	West Africa; SLVL and hyperreactive malarial splenomegaly have clinical and immunological features in common. Dysregulated immune response to repeated malaria infections results in a stimulated, proliferating pool of B cells; environmental and other factors promote the development of SLVL. Several infectious agents have been associated with B-cell non-Hodgkin's lymphomas.	7
Clinically severe malaria; 489 patients in Zimbabwe	Malaria and blood group A patients had lower hemoglobin levels; also at more risk of coma than patients in other blood groups.	31
64 *P. falciparum* patients in eastern Sudan; plasma samples and seasonal agglutination of infected RBCs	First demonstration of marked seasonal fluctuations in capacity of sera to agglutinate parasitized RBCs; may reflect drop in levels of agglutinating antibodies between seasons or shifts in antigens being recognized from one transmitting season to the next.	34
Cerebral malaria (*P. falciparum*); 250,000 Vietnam veterans	Cerebral malaria results in multiple, major, substantially underappreciated neuropsychiatric symptoms iin Vietnam veterans (poor dichotic listening, "personality changes," depression, and, in some cases, partial seizure-like symptoms). History of malaria needs to be assessed in workup of any Vietnam veteran; positive response could dramatically change diagnosis and therapy.	105
Bone marrow transplantation in recipient of multiple blood transfusions	*P. falciparum* trophozoites and gametocytes found on peripheral blood smears.	99
Problem with donors living in high-risk areas	11/420 patients (2.6%) positive for *P. falciparum* (mean time 22 + 44 days).	103
Both donor and recipient checked prior to BMT	Recommendation for treatment prior to harvest, conditioning of cells, regardless of smear results.	64
Case of unrelated donor	Case of *P. vivax* infection, donor asymptomatic 11 months prior to donation.	76
Relapse of primary unidentified infection	Case of *P. vivax* infection after BMT.	87
Case of pancytopenia after 70 days	Case of *P. vivax* infection after BMT; always consider malaria in the differential.	82
Severe *P. falciparum* malaria; 24 adults (NO)	No support for the hypothesis that excessive NO production contributes to the pathogenesis of severe falciparum malaria. However, local changes in NO production (CNS) might not be reflected in total production or levels in the CSF.	24, 26
NO production varies throughout childhood (African children)	NO levels high in infancy, decrease after first year of life, then increase again after 5 yr of age. Elevated NO in both infants and older children may be related to age and malaria infection, respectively, and may be one of the mediators of anti-disease immunity in these two age groups.	3
NO production related to anemia (Tanzania)	NO production was inversely associated with hemoglobin concentration; thus, it is not a predictor of anemia after correction of age/gender data.	2
Cerebral malaria (*P. falciparum*)	Identification of rare *P. falciparum* parasites expressing a panadhesive phenotype linked to RBC rosetting (correlate of cerebral malaria).	30
	Parasitized RBCs may bind integrin $\alpha_V\beta_3$ on microvascular endothelial cells; sequestration leads to cerebral malaria.	92
	Binding of parasitized RBCs to cerebral endothelial cells lead to functional changes in the blood-brain barrier in cerebral malaria.	15
Adhesion phenomena (*P. falciparum*)	Cytoadherence begins at 12 h and peaks at 28 h; rosette formation (16–20 h); rosette formation and immunoglobulin binding rate of about 50% (24–36 h).	100
Molecular mechanisms of cytoadherence (*P. falciparum*)	Cytoadherence may activate intracellular signaling pathways in both endothelial cells and infected RBCs, leading to gene expression of mediators such as cytokines, which could modify the outcome of the infection.	47, 85
RBC deformability (*P. falciparum*)	Reduced deformability of uninfected RBCs at high shear stresses and subsequent splenic removal of these cells may be an important contributor to anemia of severe malaria.	24
Eosinophil activity (*P. falciparum*)	Appears that low eosinophil counts may be due to tissue sequestration and destruction rather than decreased production; levels of eosinophilia granule proteins in plasma correlated with levels of TNF and soluble IL-2 receptor, implicating inflammatory responses and T-cell activation as causes of eosinophil activation.	62

(continued)

Table 7.4 *(continued)*

Clinical situation and condition	Comments[a]	Reference(s)
Fatal malaria (*P. falciparum*) in African children	Three clinical syndromes: severe anemia, cerebral malaria, malaria-associated hyperpnea (increased rate and depth of breathing); "respiratory distress" may be misleading, malaria hyperpneic syndrome (severe systemic acidosis) may not be recognized in conscious children.	75
Frequent parasitemias with *P. falciparum* (asymptomatic); children in Tanzania	Asymptomatic (especially polyclonal) *P. falciparum* infection protects against clinical disease due to new parasites.	28
Maternal and infant mortality/ (*P. falciparum*)	Accumulation of large numbers of infected RBCs in maternal blood spaces of placenta may be mediated by adhesion of infected RBCs to molecules on the syncytiotrophoblast surface. Parasites infecting pregnant women may be antigenically distinct from those common in childhood disease.	9
P. vivax malaria	Although *P. vivax* is seldom fatal, it is a major cause of morbidity. *P. vivax*-infected RBCs form rosettes, like those seen in *P. falciparum* infections.	19

[a] SLVL, splenic lymphoma with villous lymphocytes; BMT, bone marrow transplantation; TNF, tumor necrosis factor; IL-2, interleukin-2; NO, nitric oxide; CFS, cerebrospinal fluid.

Table 7.5 Immunopathology of malaria[a]

Pathology	Identified or possible mechanisms[b]
Autoimmunity	Autoantibodies
	Antinuclear antibodies
Immunosuppression	Antigenic competition
	Decreased lymphocyte production
	Disruption of spleen function
	Macrophage dysfunction
	Polyclonal activation and immune system "exhaustion"
Hypergammaglobulinemia	Ag-induced cytokine production (IL-6)
Anemia	Anti-RBC antibodies
	Decrease in RBC production (effect of TNF)
	Increased RBC destruction
	Iron deficiency
Thrombocytopenia	Increased platelet removal
	Coating of platelets with malarial antigen; subsequent platelet destruction
Hyperactive malarial splenomegaly	Genetic predisposition
	Hypogammaglobulinemia
	Chronic increase of lymphocyte proliferation
Nephrotic syndrome	Immune complex (Ag-Ab) deposition on kidney glomeruli
	Autoimmunity

[a] Adapted from references 49 and 107.
[b] Ag, antigen; Ab, antibody; IL, interleukin; TNF, tumor necrosis factor.

Plasmodium ovale. Although *P. ovale* and *P. vivax* infections are clinically similar, *P. ovale* malaria is usually less severe, tends to relapse less frequently, and usually ends with spontaneous recovery, often after no more than 6 to 10 paroxysms. The incubation period is similar to that seen in *P. vivax* malaria, but the frequency and severity of the symptoms are much lower, with a lower fever and a lack of typical rigors. *P. ovale* infects only the reticulocytes (as does *P. vivax*), so that the parasitemia is generally limited to around 2 to 5% of the available RBCs (21). The geographic range has been thought to be limited to tropical Africa, the Middle East, Papua New Guinea, and Irian Jaya in Indonesia. However, infections with *P. ovale* in Southeast Asia may be the cause of benign and relapsing malaria in this region. In both Southeast Asia and Africa, two different types of *P. ovale* circulate in humans. Human infections with variant-type *P. ovale* are associated with higher parasitemias and thus have possible clinical relevance (109).

Plasmodium malariae. *P. malariae* invades primarily the older RBCs, so that the number of infected cells is somewhat limited. The incubation period between infection and symptoms may be much longer than that seen with *P. vivax* or *P. ovale* malaria, ranging from about 27 to 40 days. Parasites can be found in the bloodstream several days before the initial attack, and the prodromal symptoms may resemble those of *P. vivax* malaria. A regular periodicity is seen from the beginning, with a more severe paroxysm, including a longer cold stage and more severe symptoms during the hot stage. Collapse during the sweating phase is not uncommon.

Proteinuria is common in patients with *P. malariae* infection; in children it may be associated with clinical signs of the nephrotic syndrome. It has been suggested that kidney problems may result from deposition within the glomeruli of circulating antigen-antibody complexes in an antigen excess situation seen with a chronic infection. Apparently, the nephrotic syndrome associated with *P. malariae* infections is unaffected by the administration of steroids. A membranoproliferative type of glomerulonephritis with relatively sparse proliferation of endothelial and mesangial cells is the most common type of lesion seen in *P. malariae* infection. Granular deposits of immunoglobulin M (IgM), IgG, and C3 are seen in immunofluorescence assays (104). Since chronic glomerular disease associated with *P. malariae* infections is usually not reversible with therapy, genetic and environmental factors may play a role in the nephrotic syndrome.

The infection may end with spontaneous recovery, or there may be a recrudescence or series of recrudescences over many years. These patients are left with a latent infection and persisting low-grade parasitemia for many, many years.

Plasmodium falciparum. *P. falciparum* tends to invade all ages of RBCs, and the proportion of infected cells may exceed 50%. Schizogony occurs in the internal organs (spleen, liver, bone marrow, etc.) rather than in the circulating blood. Ischemia caused by the plugging of vessels within these organs by masses of parasitized RBCs produces various symptoms, depending on the organ involved (Figure 7.5). It has been suggested that a decrease in the ability of the RBCs to change shape when passing through capillaries or the splenic filter may lead to plugging of the vessels.

Figure 7.5 *Plasmodium falciparum*-parasitized blood cells plugging the capillaries in brain tissue. (From A Pictorial Presentation of Parasites: A cooperative collection prepared and/or edited by H. Zaiman.)

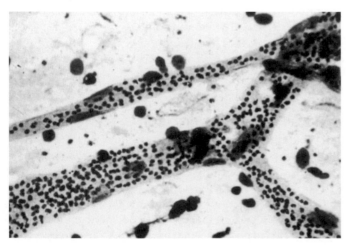

The onset of a *P. falciparum* malaria attack occurs 8 to 12 days after infection and is preceded by 3 to 4 days of vague symptoms such as aches, pains, headache, fatigue, anorexia, or nausea. The onset is characterized by fever, a more severe headache, and nausea and vomiting, with occasional severe epigastric pain. There may be only a feeling of chilliness at the onset of fever. Periodicity of the cycle is not established during the early stages, and the presumptive diagnosis may be totally unrelated to a possible malaria infection. If the fever does develop a synchronous cycle, it is usually a cycle of somewhat less than 48 h.

An untreated primary attack of *P. falciparum* malaria usually ends within 2 to 3 weeks. True relapses from the liver do not occur, and after a year, recrudescences are rare.

Severe or fatal complications of *P. falciparum* malaria can occur at any time during the infection and are related to the plugging of vessels in the internal organs, with the symptoms depending on the organ(s) involved (Table 7.3). The severity of the complications in a malaria infection may not correlate with the parasitemia seen in the peripheral blood, particularly in *P. falciparum* infections. Acute lung injury is more likely to occur in patients with very severe, multisystemic *P. falciparum* malaria than in other malaria patients. When patients present with acute lung injury and septic shock, bacterial coinfection should be suspected and treated empirically.

Disseminated intravascular coagulation is a rare complication of malaria; it is associated with a high parasite burden, pulmonary edema, rapidly developing anemia, and cerebral and renal complications. Vascular endothelial damage from endotoxins and bound parasitized blood cells may lead to clot formation in small vessels.

Cerebral malaria is most often seen in *P. falciparum* malaria, although it can occur in the other types as well. If the onset is gradual, the patient may become disoriented or violent or may develop severe headaches and pass into coma. Some patients, even those who exhibit no prior symptoms, may suddenly become comatose. Physical signs of central nervous system (CNS) involvement are quite variable, and there is no real correlation between the severity of the symptoms and the peripheral blood parasitemia. It has been shown that patients with cerebral malaria were infected with RBC rosette-forming *P. falciparum* and that plasma from these patients generally had no antirosetting activity. A rosette usually consists of a parasitized RBC surrounded by three or more uninfected RBCs (Figure 7.6A). Interaction with adjacent uninfected RBCs in rosettes appears to be mediated by knobs seen on the parasitized RBC (Figure 7.6B) (56). In contrast, *P. falciparum* parasites from patients with mild malaria lacked the rosetting phenotype or had a much lower rosetting

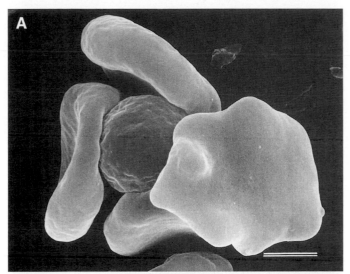

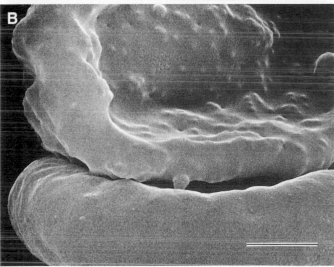

Figure 7.6 (A) Scanning electron micrograph of a rosette consisting of a central parasitized RBC surrounded by several attached uninfected RBCs (original magnification, ×19,000; bar, 1 μm). (B) Scanning electron micrograph of the interaction of a parasitized RBC with an uninfected RBC that appears to be mediated by protrusion of knobs (original magnification, ×53,000; bar, 0.5 μm). (From reference 86 with permission.)

rate. Also, antirosetting activity has been detected in the plasma of these patients. These findings strongly support the idea that RBC rosetting contributes to the pathogenesis of cerebral malaria while antirosetting antibodies offer protection against these clinical sequelae.

Interplay among many factors, including free oxygen radicals and other mediators of inflammation, is likely to be involved in the pathogenesis of cerebral malaria (20, 66). Sequestration often predisposes to cerebral symptoms but in ways not necessarily restricted to obstruction of blood flow. Tumor necrosis factor alpha

and other pyrogenic cytokines play a role in the regulation of parasite density in the host; cerebral malaria is also thought to be related to excessive tumor necrosis factor alpha.

Magnetic resonance imaging of the brains of patients with cerebral malaria shows that the volume of the brain is increased. This increased volume probably results from sequestration of parasitized RBCs and compensatory vasodilatation rather than from edema. Brain stem herniation may occur, but its temporal relation to brain death in patients with cerebral malaria is still unclear. Ultrastructural study of the brain in patients with fatal *P. falciparum* infections reveals that sequestration of parasitized RBCs in cerebral microvessels is much higher in the brains of patients with cerebral malaria (cerebrum, cerebellum, and medulla oblongata) than in the brains of patients with noncerebral malaria. There were 26.6 times more parasitized RBCs in the brain microvasculature than in the peripheral blood (66).

Extreme fevers, 107°F (ca. 41.7°C) or higher, may develop in a relatively uncomplicated attack of malaria or may develop as another manifestation of cerebral malaria. Without vigorous therapy, the patient usually dies. Cerebral malaria is considered to be the most serious complication and the major cause of death from infection with *P. falciparum*; this complication accounts for up to 10% of all *P. falciparum*-infected patients admitted to the hospital and for 80% of fatal cases.

Bilious remittent fever involves the liver, with symptoms including abdominal pain, nausea, and severe and persistent vomiting, with the vomitus containing evidence of bile or fresh blood. There may also be severe diarrhea or dysentery with resulting dehydration. The liver is large and tender, the skin becomes icteric, and the urine contains bile.

Diarrhea or dysentery without liver involvement or jaundice is also seen, with malabsorption characterized by decreased absorption of D-xylose and vitamin B₁₂ and a low carotene level. Jejunal biopsy reveals edema, round cell infiltration of the lamina propria, and blunting of villi, which have also been found on postmortem examinations.

Algid malaria symptoms include circulatory collapse; low blood pressure; hypothermia; rapid, thready pulse; and pale, cold, clammy skin. There may be severe abdominal pain, vomiting and diarrhea, and muscular cramps. At postmortem examination, the adrenal glands are congested, necrotic, and hemorrhagic. In the subacute stage and after recovery, some patients develop signs resembling Addison's disease.

Although infrequent, blackwater fever is most often associated with *P. falciparum* malaria. Usually, there is a history of previous malarial attacks. Sudden, intravascular hemolysis results in a dramatic color change in the urine (acidic urine with a high methemoglobin content). Since the hemoglobinuria does not necessarily give rise to symptoms, the onset may be missed. Quinine sensitivity has been postulated as one possible factor leading to blackwater fever. Another has been the presence of antibodies acting as hemolysins against RBC antigens. The parasitemia may be relatively low in patients with blackwater fever. With the increased use of quinine to treat chloroquine-resistant patients who live in areas with malaria, more cases of blackwater fever may be seen in the future.

Acute renal failure may also occur unassociated with blackwater fever and hemolytic anemia. Renal anoxia is thought to lead to tubular necrosis. The nephrotic syndrome, caused by acute glomerulonephritis, has also been found in some patients with *P. falciparum* malaria.

Independent risk factors contributing to anemia in falciparum malaria in patients presenting with anemia on admission include age younger than 5 years, a palpable spleen, a palpable liver, recrudescent infections, being female, a history of prolonged history of illness (>2 days) before admission, and pure *P. falciparum* infections rather than mixed infections with *P. falciparum* and *P. vivax*. It appears that coinfection with *P. vivax* can modify the severity of the disease.

In African children with malaria, the presence of impaired consciousness or respiratory distress can identify those at high risk of death. Biochemical evidence of hepatic and renal dysfunction also suggests a poor prognosis; however, in contrast to adults with malaria, children with severe malaria do not always have acute renal failure.

Using several different approaches, it is estimated that approximately 100,000 infant deaths in Africa could be attributed to malaria during pregnancy, resulting in low-birth-weight babies (38). Antimalarial intermittent preventive treatment during pregnancy has been estimated to prevent the death of 22,000 children younger than 5 years through reduced numbers of preterm deliveries. However, if one assumes an efficacy rate of 80% in preventing placental malaria, the number of deaths prevented may be closer to 80,000.

Mixed Infections. Coincident infection with more than one species of malaria is more common than previously suspected. As an example, in Thailand, up to 30% of patients who present with severe *P. falciparum* malaria also are infected with *P. vivax*. Based on overlapping geographic areas where both *P. falciparum* and *P. vivax* are found, it is not surprising that dual infections occur. In Africa, dual infections with *P. falciparum* and *P. malariae* have also been found. Although the

figures may be underestimates, in Gambian children the prevalence of mixed-species infection varies from <1% to >60%. There has even been a report of a rare quadruple malaria infection in a remote area in the western half of New Guinea Island (Irian Jaya Province, Indonesia), and this infection was confirmed using nested PCR and species-specific primer pairs (81). In Africa, *P. ovale* occurs commonly in areas with Duffy-negative populations who are refractory to *P. vivax*, thus appearing to prevent the natural coexistence of the two species. However, the Duffy blood group-negative trait that protects many African populations from *P. vivax* infection is considered very rare or nonexistent in New Guinea. Since *P. ovale* is considered a very unusual finding outside of Africa, the natural occurrence of *P. ovale* and *P. vivax* together in the same patient is considered a very rare finding.

Although antagonism and predominance of one *Plasmodium* species over the other in the human host has been described, differences in the methods and sensitivity of detection may have complicated the interpretation of results. It is generally accepted that *P. falciparum* predominates over *P. vivax* in mixed infections. *P. ovale* has been found in mixed infections and has been established in persons carrying *P. falciparum*, *P. vivax*, and *P. malariae* at the same time. Simultaneous patent infections with three or four *Plasmodium* species are not entirely rare findings in the blood of healthy carriers in regions of hyperendemic infection.

Interaction between *P. falciparum* and *P. vivax* is often reported to be suppressive. There are reciprocal seasonal changes in the prevalence of the two species that cannot be explained by changes in vector behavior, antimalarial drug use, or age. Changes may be due to competition for host cells, since *P. falciparum* causes dyserythropoiesis, leading to a decreased number of reticulocytes available for invasion by *P. vivax*. It is possible that some cross-species immunity may be involved. Also, patients with a mixed *P. falciparum*-*P. vivax* infection were four times less likely to exhibit patent gametocytemia than were those with a *P. falciparum* infection alone. Thus, coinfection with *P. vivax* may attenuate infections with *P. falciparum*. Also, there appear to be no limitations to vector infections with multiple species of malaria, and transmission of more than one species to humans has been documented.

Detection of mixed infections can be difficult due to different levels of parasitemia, low organism densities, and confusion among various morphologic criteria for identification to the species level; these problems have been well documented. Using PCR methods, it is likely that a higher detection rate of chronic and mixed malarial species will be possible (81).

Diagnosis

Although malaria is no longer endemic within the United States, it is considered to be life-threatening, and laboratory requests for blood smear examination and organism identification should be treated as "STAT" requests. Malaria is usually associated with patients having a history of travel within an area where malaria is endemic, although other routes of infection are well documented (Algorithm 7.1).

Frequently, for a number of different reasons, organism recovery and subsequent identification are more difficult than the textbooks imply. It is very important that this fact be recognized, particularly when one is dealing with a possibly fatal infection with *P. falciparum*.

Patient Information. When requests for malarial smears are received in the laboratory, some patient history information should be made available to the laboratorian. This information should include the following.

1. Where has the patient been, and what was the date of return to the United States? ("Where do you live?")
2. Has malaria ever been diagnosed in the patient before? If so, what species was identified?
3. What medication (prophylaxis or otherwise) has the patient received, and how often? When was the last dose taken?
4. Has the patient ever received a blood transfusion? Is there a possibility of other needle transmission (drug user)?

5. When was the blood specimen drawn, and was the patient symptomatic at the time? Is there any evidence of a fever periodicity?

Answers to such questions may help eliminate the possibility of infection with *P. falciparum*, usually the only species that can rapidly lead to death.

Conventional Microscopy. Often, when the diagnosis of malaria is considered, only a single blood specimen is submitted to the laboratory for examination; however, single films or specimens cannot be relied on to exclude the diagnosis, especially when partial prophylactic medication or therapy is used. Partial use of antimalarial agents may be responsible for reducing the numbers of organisms in the peripheral blood, thus leading to a blood smear that contains few organisms, which then reflects a low parasitemia when in fact serious disease is present. Patients with a relapse case or an early primary case may also have few organisms in the blood smear (Table 7.6).

It is recommended that both thick and thin blood films be prepared on admission of the patient, and at least 200 to 300 oil immersion fields should be examined on both films before a negative report is issued (74). Since one set of negative films will not rule out malaria, additional blood specimens should be examined over a 36-h time frame. Although Giemsa stain is recommended for all parasitic blood work, the organisms can also be seen if other blood stains, such as Wright's stain, are used (Tables

Algorithm 7.1 Malaria.

MALARIA

- FEVER AND CHILLS
- SPLENOMEGALY
- ANEMIA

Table 7.6 Parasitemia determined from conventional light microscopy: clinical correlation[a]

Parasitemia (%)	No. of parasites/µl	Clinical correlation[b]
0.0001–0.0004	5–20	Number of organisms that are required for a positive TBF (sensitivity) Note (TBF): Examination of 100 TBF fields (0.25 µl) may miss up to 20% of infections (sensitivity, 80–90%); at least 300 TBF fields should be examined before reporting a negative result Note (THBF): Examination of 100 THBF fields (0.005 µl); at least 300 THBF fields should be examined before reporting a negative result; *both* TBF and THBF should be examined for every specimen submitted for a suspect malaria case. **One set (TBF + THBF) of negative blood films does not rule out a malaria infection.**
0.002	100	Patients may be symptomatic below this level
0.2	10,000	Level above which immune patients will exhibit symptoms
2	100,000	Maximum parasitemia of *P. vivax* and *P. ovale* (infect young RBCs only) (rarely may exceed 2%)
2–5	100,000–250,000	Hyperparasitemia, severe malaria,[c] increased mortality
10	500,000	Exchange transfusion may be considered, high mortality

[a] Adapted from references 39 and 108.
[b] TBF, thick blood film; THBF, thin blood film.
[c] WHO criteria for severe malaria are parasitemia of >10,000/ µl and severe anemia (hemoglobin, <5 g/liter). Prognosis is poor if >20% of parasites are pigment-containing trophozoites and schizonts and/or if >5% of neutrophils contain visible pigment.

7.7 and 7.8; Figures 7.7 to 7.11). Blood collected with the use of EDTA anticoagulant is acceptable; however, if the blood remains in the tube for any length of time, true stippling may not be visible within the infected RBCs (*P. vivax*, as an example). Also, when using anticoagulants, it is important to remember that the proper ratio between blood and anticoagulant is necessary for good organism morphology. Heparin can also be used, but EDTA is preferred. Finger stick blood is recommended, particularly when the volume of blood required is minimal (i.e., when no other hematologic procedures have been ordered). The blood should be free flowing when taken for smear preparation and should not be contaminated with alcohol used to clean the finger prior to the stick.

Accurate species diagnosis is essential for good patient management, since identification to the species level may determine which drug or combination of drugs will be indicated. Some patients with *P. falciparum* infections may not yet have the crescent-shaped gametocytes in the blood. Low parasitemias with the delicate ring forms may be missed; consequently, oil immersion examination at ×1,000 is mandatory.

Parasite density generally correlates with disease severity, but peripheral parasitemia does not always reflect the number of sequestered organisms. Malaria pigment may serve as a peripheral indicator of parasite biomass, since the pigment can be seen within monocytes and polymorphonuclear leukocytes during light microscopy examination. The presence of pigment has been strongly associated with more severe disease than occurs with uncomplicated cases of malaria (Figure 7.12). Pigmented neutrophils (polymorphonuclear leukocytes, monocytes) have been associated with cerebral malaria and with death in children with severe malaria (67).

Table 7.7 Malaria characteristics

Organism	Characteristics
Plasmodium vivax (benign tertian malaria)	1. 48-h cycle 2. Tends to infect young cells 3. Enlarged RBCs 4. Schüffner's dots after 8–10 h 5. Delicate ring 6. Very ameboid trophozoite 7. Mature schizont contains 12–24 merozoites
Plasmodium malariae (quartan malaria)	1. 72-h cycle (long incubation period) 2. Tends to infect old cells 3. Normal-size RBCs 4. No stippling 5. Thick ring, large nucleus 6. Trophozoite tends to form "bands" across the cell 7. Mature schizont contains 6–12 merozoites
Plasmodium ovale	1. 48-h cycle 2. Tends to infect young cells 3. Enlarged RBCs with fimbriated edges (oval) 4. Schüffner's dots appear in the beginning (in RBCs with very young ring forms, in contrast to *P. vivax*) 5. Smaller ring than *P. vivax*) 6. Trophozoite less ameboid than that of *P. vivax* 7. Mature schizont contains avg of 8 merozoites
Plasmodium falciparum (malignant tertian malaria)	1. 36–48-h cycle 2. Tends to infect any cell regardless of age; thus, very heavy infection may result 3. All sizes of RBCs 4. No Schüffner's dots (Maurer's dots: may be larger, bluish, single dots) 5. Multiple rings/cell (only young rings, gametocytes, and occasional mature schizonts seen in peripheral blood) 6. Delicate rings; may have two dots of chromatin/ring, appliqué or accolé forms 7. Crescent-shaped gametocytes

Table 7.8 Plasmodia in Giemsa-stained thin blood films

Characteristic	Plasmodium vivax	Plasmodium malariae	Plasmodium falciparum	Plasmodium ovale
Persistence of exoerythrocytic cycle	Yes	No	No	Yes
Relapses	Yes	No, but long-term recrudescences are recognized	No long-term relapses	Possible, but usually spontaneous recovery
Time of cycle (h)	44–48	72	36–48	48
Appearance of parasitized RBC (size and shape)	1.5 to 2 times larger than normal; oval to normal; may be normal size until ring fills half of cell	Normal shape; size may be normal or slightly smaller	Both normal	60% of cells larger than normal and oval; 20% have irregular, frayed edges
Schüffner's dots (eosinophilic stippling)	Usually present in all cells except early ring forms	None	None; comma-like red dots (Maurer's dots) occasionally present	Present in all stages including early ring forms; dots may be larger and darker than in P. vivax
Color of cytoplasm	Decolorized, pale	Normal	Normal, blush tinge at times	Decolorized, pale
Multiple rings/cell	Occasional	Rare	Common	Occasional
All developmental stages present in peripheral blood	All stages present	Ring forms few, as ring stage brief; mostly growing and mature trophozoites and schizonts	Young ring forms and no older stages; few gametocytes	All stages present
Appearance of young trophozoite (early ring form)	Ring 1/3 diameter of cell; cytoplasmic circle around vacuole; heavy chromatin dot	Ring often smaller than in P. vivax, occupying 1/8 of cell; heavy chromatin dot; vacuole at times "filled in"; pigment forms early	Delicate, small ring with small chromatin dot (frequently 2); scanty cytoplasm around small vacuoles; sometimes at edge of red cell (appliqué form) or filamentous slender form; may have multiple rings per cell	Ring larger and more ameboid than in P. vivax; otherwise similar to P. vivax
Appearance of growing trophozoite	Multishaped irregular ameboid parasite; streamers of cytoplasm close to large chromatin dot; vacuole retained until close to maturity; increasing amounts of brown pigment	Nonameboid rounded or band-shaped solid forms; chromatin may be hidden by coarse dark brown pigment	Heavy ring forms; fine pigment grains	Ring shape maintained until late in development; nonameboid compared with P. vivax

(continued)

Table 7.8 Plasmodia in Giemsa-stained thin blood films (*continued*)

Characteristic	*Plasmodium vivax*	*Plasmodium malariae*	*Plasmodium falciparum*	*Plasmodium ovale*
Appearance of mature trophozoite	Irregular ameboid mass; 1 or more small vacuoles retained until schizont stage; fills almost entire cell; fine brown pigment	Vacuoles disappear early; cytoplasm compact, oval, band shaped, or nearly round, almost filling cell; chromatin may be hidden by peripheral coarse dark brown pigment	Not seen in peripheral blood (except in severe infections); development of all phases following ring form occurs in capillaries of viscera	Compact; vacuoles disappear; pigment dark brown, less abundant than in *P. malariae*
Appearance of schizont (presegmenter)	Progressive chromatin division; cytoplastic bands containing clumps of brown pigment	Similar to *P. vivax* except smaller; darker, larger pigment granules peripheral or central	Not seen in peripheral blood (see above)	Smaller and more compact than *P. vivax*
Appearance of mature schizont	Merozoites, 16 (12–24), each with chromatin and cytoplasm, filling entire RBC, which can hardly be seen	8 (6–12) merozoites in rosettes or irregular clusters filling normal-size cells, which can hardly be seen; central arrangement of brown-green pigment	Not seen in peripheral blood (see above)	3/4 of cells occupied by 8 (8–12) merozoites in rosettes or irregular clusters
Appearance of macrogametocyte	Rounded or oval homogeneous cytoplasm; diffuse delicate light brown pigment throughout parasite; eccentric compact chromatin	Similar to *P. vivax*, but fewer; pigment darker and more coarse	Sex differentiation difficult; "crescent" or "sausage" shapes characteristic; may appear in "showers," black pigment near chromatin dot, which is often central	Smaller than *P. vivax*
Appearance of microgametocyte	Large pink to purple chromatin mass surrounded by pale or colorless halo; evenly distributed pigment	Similar to *P. vivax*, but fewer; pigment darker and more coarse	Same as macrogametocyte (described above)	Smaller than *P. vivax*
Main criteria	Large RBC; trophozoite irregular; pigment usually present; Schüffner's dots not always present; several phases of growth seen in one smear; gametocytes appear as early as day 3	RBCs normal in size and color; trophozoites compact, stain usually intense, band forms not always seen; coarse pigment; no stippling of RBCs; gametocytes appear after a few weeks	Development following ring stage takes place in blood vessels of internal organs; delicate ring forms and crescent-shaped gametocytes are the only forms normally seen in peripheral blood; gametocytes appear after 7–10 days	RBC enlarged, oval, with fimbriated edges; Schüffner's dots seen in all stages; gametocytes appear after 4 days or as late as 18 days

Malaria is one of the few parasitic infections considered to be immediately life-threatening, and a patient with the diagnosis of P. falciparum *malaria should be considered a medical emergency because the disease can be rapidly fatal. Any laboratory providing the expertise to identify malarial parasites should do so on a 24-h basis, 7 days per week.*

Patients with malaria may appear for diagnostic blood work when least expected. Laboratory personnel should be aware of the "STAT" nature of such requests and the importance of obtaining some specific patient history information. The typical textbook presentation of the blood smears may not be seen by the technologist. It is very important that the smears be examined at length and under oil immersion. The most important thing to remember is that even though a low parasitemia may be present on the blood smears, the patient may still be faced with a serious, life-threatening disease. See Algorithm 7.1 for additional information.

Alternative Methods. Giemsa-stained thick and thin blood films have been used to diagnose malaria for many years and have always been considered the "gold standard." However, a number of alternative methods have been developed, including different approaches to microscopy, flow cytometry, biochemical methods, immunoassay, and molecular methods (Table 7.9). The aim of these procedures has been to reduce cost, reduce requirements for expensive equipment, increase sensitivity, and provide simple, rapid methods that do not require conventional microscopy (22, 32, 37, 39, 51–53, 72, 77, 79).

Microhematocrit centrifugation, using the QBC tube (glass capillary tube and closely fitting plastic insert [QBC malaria blood tubes; Becton Dickinson, Sparks, Md.]), has been used for the detection of blood parasites. At the end of centrifugation of 50 to 110 µl of capillary or venous blood (5 min in a QBC centrifuge [$14,387 \times g$]), parasites or RBCs containing parasites are concentrated into a small, 1- to 2-mm region near the top of the RBC column and are held close to the wall of the tube by the plastic float, thereby making them readily visible by microscopy. Tubes precoated with acridine orange provide a stain that induces fluorescence in the parasites. This method automatically prepares a concentrated smear that represents the distance between the float and the walls of the tube. The tube is placed into a plastic holder, a drop of oil is applied to the top of the hematocrit tube, a coverslip is added, and the tube is examined with a 40× to 60× oil immersion objective (the objective must have a working distance of 0.3 mm or greater).

Note. Although a malaria infection can be detected by the microhematocrit centrifugation method (which is more sensitive than use of the thick or thin blood smear), *appropriate thick and thin blood films must be examined to accurately identify the species causing the infection.*

Another approach uses the fluorescent dye benzothiocarboxypurine (Becton Dickinson, Mountain View, Calif.), which intensely stains nucleic acid. In fresh blood, the dye easily penetrates the RBC membrane and stains malarial parasites, with the asexual forms staining more intensely than the gametocytes. Under these circumstances, the white blood cell nuclei do not stain. However, when fixed, the dye penetrates and stains both malarial parasites and white blood cells. The advantages of this alternative to Giemsa-stained blood films, particularly in field laboratories, are the speed of staining and evaluation and the relatively low level of training necessary to provide consistent results. Another fluorescent dye that is used is 4',6-diamidine-2-phenylindolo-propidium iodide. However, the one disadvantage for field work is the requirement for a fluorescence microscope. Additional work is continuing on the use of benzothiocarboxypurine, and a stable prestained slide system that eliminates the need for liquid-stain application is now available.

With the increase in drug-resistant *P. falciparum* cases in the tropics, it is no longer feasible to treat febrile patients empirically with inexpensive yet effective antimalarial agents. The substitution of newer, more expensive drugs mandates rapid, accurate, and inexpensive diagnostic procedures so that directed therapy can be provided. A *P. falciparum* antigen detection system, the ParaSight-F test (Becton Dickinson Europe), has been very effective in field trials and detected the histidine-rich protein 2 (HRP-2). This procedure is based on an antigen capture approach and has been incorporated in a dipstick format; the entire test takes approximately 10 min (Figure 7.13). Another test that detects HRP-2, also in the dipstick format, is the ICT Malaria *Pf* (AMRAD ICT). This test appears to perform well and is comparable to the ParaSight-F test; however, the false-positive rate appears to be lower in this procedure. The third dipstick format procedure, OptiMAL (Flow Inc.), uses monoclonal antibodies against species-specific parasite lactate dehydrogenase (pLDH). Its sensitivity and specificity appear to be similar to those of the HRP-2 assays, with similar limitations at lower parasitemias. An advantage of this method is its use in follow-up for posttherapy patients to confirm a cure; the test detects only viable parasites.

Other methods that have been developed include a dot blot assay that provides significant improvements over previously reported DNA-based procedures for the diagnosis of malaria. Another method permits direct detection of *P. falciparum* by using a specific DNA probe after PCR amplification of target DNA sequences. For PCR amplification, blood samples were lysed and filtered onto filter paper, and after being dried, a portion of the filter paper was added directly to the PCR mixture. PCR products were detected by using a nonisotopically labeled probe. Monoclonal antibodies have also been used with good specificity and relatively good sensitivity.

P. vivax   P. malariae   P. ovale   P. falciparum

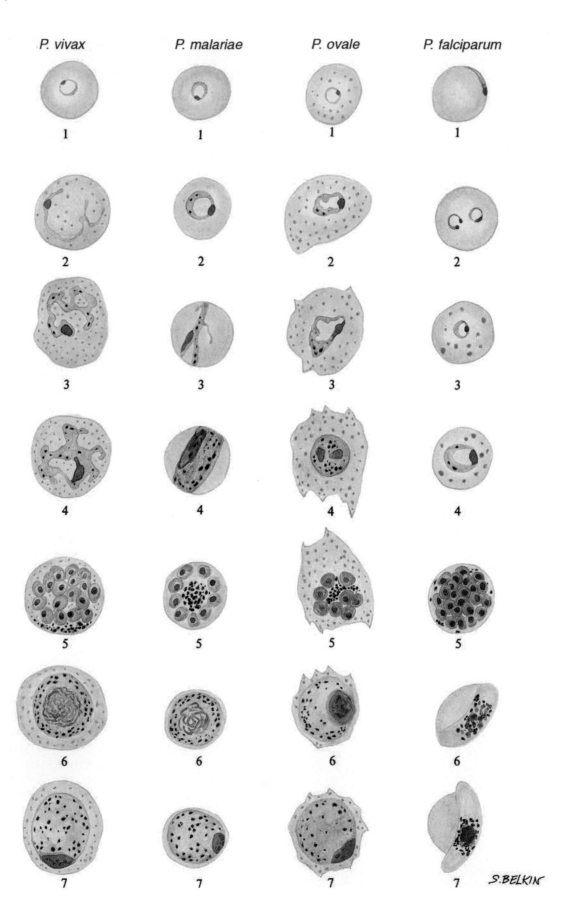

S.BELKIN

A number of recent studies have reported on the benefits of using PCR for detection of malaria; the high sensitivity, rapidity, and simplicity of some of the methods should be applicable to large-scale epidemiology studies, follow-up of drug treatment, and immunization trials. Detection is possible with as few as 5 or 10 parasites per µl of blood, confirming that PCR detects many more cases of low-level parasitemia than do thick blood films (12, 89).

Note. Potential diagnostic problems with the use of automated differential instruments have been reported (33, 106). Some cases of malaria, as well as *Babesia* infection, have been completely missed by these methods. The number of fields scanned by a technologist on instrument-read smears is quite small; thus, failure to detect a light parasitemia is almost guaranteed. In both cases, after diagnosis had been made on the basis of smears submitted to the parasitology division of the laboratory, all previous smears examined by the automated system were reviewed and found to be positive for parasites. Failure to make the diagnosis resulted in delayed therapy. Although these instruments are not designed to detect intracellular blood parasites, the inability of the automated systems to discriminate between uninfected RBCs and those infected with parasites may pose serious diagnostic problems in situations where the parasitemia is ≥0.5%.

Microscopic examination of a peripheral blood smear occasionally reveals ookinetes of *P. vivax* (or another of the *Plasmodium* species); this represents a potential "artifact" that could lead to the incorrect diagnosis of *P. falciparum* infection (Figure 7.11). The ookinetes tend to resemble the crescent-shaped gametocytes seen in infections with *P. falciparum*. The appearance of these stages in the peripheral blood, rather than the mosquito gut, was probably due to the delay between blood collection and smear preparation. It has been documented that some gametocytes (*P. falciparum*) exflagellate at room temperature (pH 7.4). In this case, the microgametes could resemble spirochetes (94).

KEY POINTS—LABORATORY DIAGNOSIS

Malaria

1. Blood films should be prepared on admission of the patient. A fever pattern may not be apparent early in the course of the infection.
2. Both thick and thin blood films should be prepared. At least 200 to 300 oil immersion fields on both thick and thin films should be examined before the specimen is considered negative.
3. Wright's, Wright-Giemsa, Giemsa, or a rapid stain can be used, although Giemsa is the stain of choice for blood parasites; the majority of the original organism descriptions were based on Giemsa stain. However, if the white blood cells appear to be well stained, any blood parasites present will also be well stained.
4. Malarial parasites may be missed with the use of automated differential instruments. Even with technologist review of the smears, a light parasitemia is very likely to be missed.
5. The number of oil immersion fields examined may have to be increased if the patient has had any prophylactic medication during the past 48 h (the number of infected cells may be decreased).
6. *One negative set of blood smears does not rule out malaria. Quantitate organisms from every positive blood specimen.*
7. In spite of new technology, serial thick-film parasite counts are a simple, cheap, rapid, and reliable method for identifying patients at high risk of recrudescence due to drug resistance and treatment failure.
8. If you are using any of the alternative methods, make sure you thoroughly understand the pros and cons of each compared with the thick and thin blood film methods; see chapter 31 for additional information on diagnostic methods for blood parasites.

Figure 7.7 Morphology of malaria parasites. *Plasmodium vivax*: (1) early trophozoite (ring form); (2) late trophozoite with Schüffner's dots (note enlarged RBC); (3) late trophozoite with ameboid cytoplasm (very typical of *P. vivax*); (4) late trophozoite with ameboid cytoplasm; (5) mature schizont with 18 merozoites and clumped pigment; (6) microgametocyte with dispersed chromatin; (7) macrogametocyte with compact chromatin. *Plasmodium malariae*: (1) early trophozoite (ring form); (2) early trophozoite with thick cytoplasm; (3) early trophozoite (band form); (4) late trophozoite (band form) with heavy pigment; (5) mature schizont with nine merozoites arranged in a rosette; (6) microgametocyte with dispersed chromatin; (7) macrogametocyte with compact chromatin. *Plasmodium ovale*: (1) early trophozoite (ring form) with Schüffner's dots; (2) early trophozoite (note enlarged RBC); (3) late trophozoite in RBC with fimbriated edges; (4) developing schizont with irregularly shaped RBC; (5) mature schizont with eight merozoites arranged irregularly; (6) microgametocyte with dispersed chromatin; (7) macrogametocyte with compact chromatin. *Plasmodium falciparum*: (1) early trophozoite (accolé or appliqué form); (2) early trophozoite (one ring is in the headphone configuration with double chromatin dots); (3) early trophozoite with Maurer's dots; (4) late trophozoite with larger ring and Maurer's dots; (5) mature schizont with 24 merozoites; (6) microgametocyte with dispersed chromatin; (7) macrogametocyte with compact chromatin. *Note*: Without the appliqué form, Schüffner's dots, multiple rings per cell, and other developing stages, differentiation among the species can be very difficult. It is obvious that the early rings of all four species can mimic one another very easily. *Remember: One set of negative blood films cannot rule out a malaria infection.*

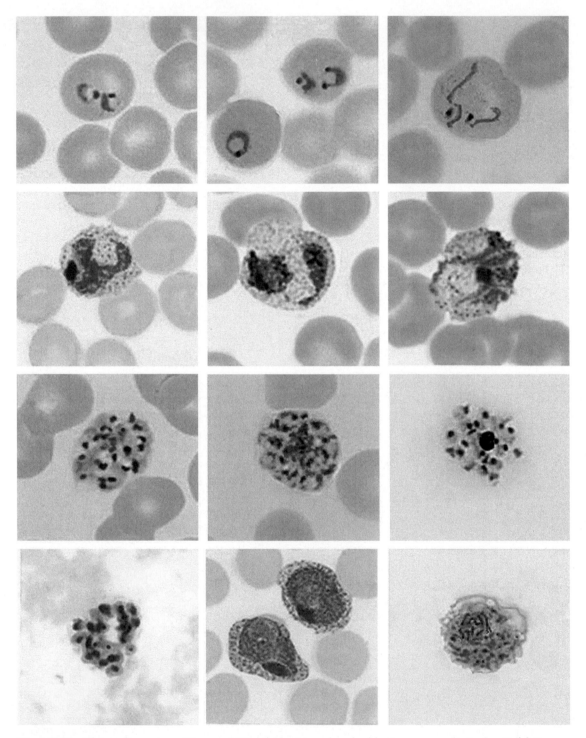

Figure 7.8   *Plasmodium vivax.* (Top row) Ring forms (note that double rings are not limited to *P. falciparum* and that the RBC is beginning to enlarge as the rings grow). (Second row) Growing trophozoites (note the ameboid nature of the trophozoites and the presence of Schüffner's dots). (Third row) Developing schizonts (note the number of merozoites). (Bottom row) Mature schizonts, female gametocytes, male gametocyte.

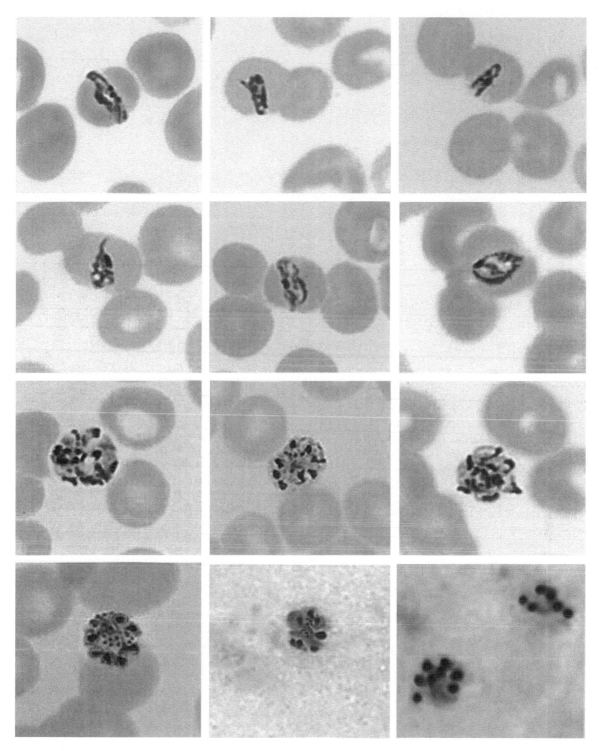

Figure 7.9 *Plasmodium malariae*. (Top row and second row) Band form developing trophozoites (note that the RBC is normal or slightly smaller than normal). (Third row) Developing schizonts (note the small number of merozoites). (Bottom row) Mature schizonts in a thick blood film.

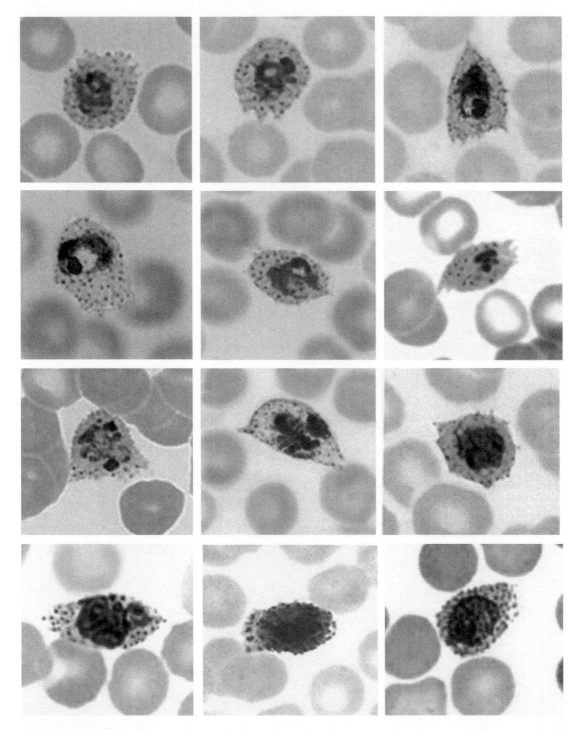

Figure 7.10 *Plasmodium ovale*. (Top three rows) Developing trophozoites (note the nonameboid trophozoites, presence of Schüffner's dots, oval RBCs, and fimbriated edges of infected red cells). (Bottom row) Developing schizonts and gametocytes.

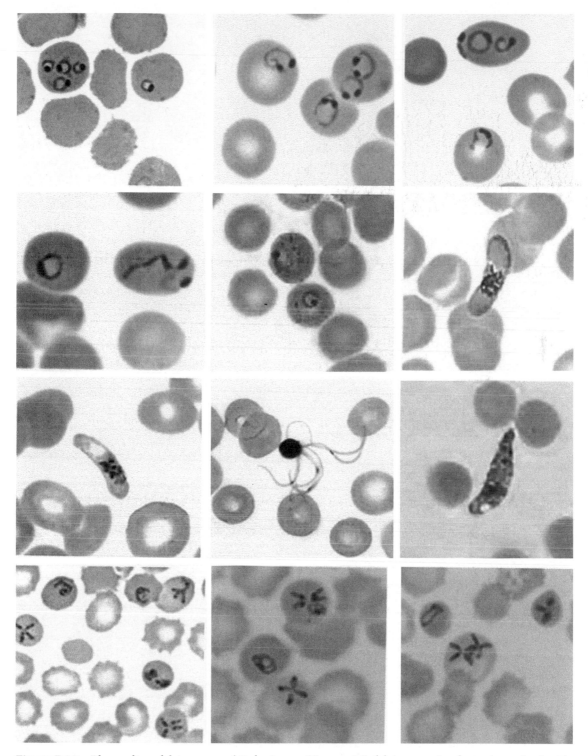

Figure 7.11 *Plasmodium falciparum* and *Babesia* spp. (Top row) *P. falciparum* ring forms (note the double rings per cell, appliqué or accolé form, and headphone appearance of some rings). (Second row) Ring forms and gametocyte (note Maurer's clefts and crescent-shaped gametocyte). (Third row) Gametocyte, exflagellation of male gametocyte, and developing ookinete. (Bottom row) *Babesia* spp. (note the Maltese cross configuration of rings; these are not seen in every species of *Babesia*).

Treatment

Malaria has become a more common health problem during the last few years, both in residents of areas where the disease is endemic and in travelers returning to areas where it is not endemic. Therapy has become more complex because of the increase in resistance of *P. falciparum* to a variety of drugs and because of advances in treatment of severe disease complications (Tables 7.10 to 7.12). Antimalarial drugs are classified by the stage of malaria against which they are effective. These drugs are often referred to as tissue schizonticides (which kill tissue schizonts), blood schizonticides (which kill blood schizonts), gametocytocides (which kill gametocytes), and sporonticides (which prevent formation of sporozoites within the mosquito). To ensure that proper therapy is given, it is important for the clinician to know what species of *Plasmodium* is involved, the estimated parasitemia, and the geographic and travel history of the patient to determine the area where infection was probably acquired and the possibility of drug resistance related to that geographic area. The use of oral or parenteral therapy will be determined by the clinical status of the patient. A number of studies related to the molecular genetics of drug resistance, particularly in *P. falciparum* malaria and modeling of the development of resistance of *P. falciparum* to antimalarial drugs, are ongoing (Tables 7.13 to 7.15).

Development of resistance to the inexpensive antimalarials, such as chloroquine and Fansidar (pyrimethamine in combination with sulfadoxine), has had a tremendous financial and public health impact on developing countries. In the late 1980s, chloroquine-resistant *P. falciparum* completely covered Africa. Resistance to pyrimethamine has also spread rapidly since the introduction of this drug. Fansidar is no longer effective in many areas of Indochina, Brazil, and Africa; also, severe allergic reactions have limited its use for chemoprophylaxis.

Although chloroquine-resistant *P. falciparum* is well recognized, chloroquine has generally been effective for both chemoprophylaxis and treatment of *P. vivax, P. ovale*, and *P. malariae* infections. Cases of chloroquine-resistant *P. vivax* (CRPV) malaria have now been identified in Papua New Guinea, Indonesia, Malaysia, Vietnam, and Myanmar. CRPV is now endemic in Malaysia, Myanmar, and Vietnam. However, there is no evidence of CRPV from Thailand, Cambodia, Laos, China, the Korean Peninsula, or the Philippine archipelago (Baird). Although CRPV has been documented in a few areas within South America (Guyana, Colombia, Brazil, Peru), it is present at a very low frequency, with a risk of infection at <5%. Population-based epidemiologic studies are necessary to define the geographic range and prevalence of CRPV. It is not clear whether this development represents a newly emerging public health problem; however, clinicians need

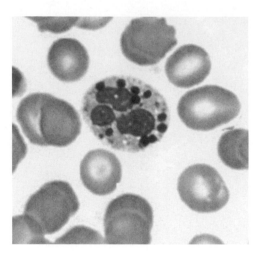

Figure 7.12 Intracytoplasmic malaria pigment seen within a polymorphonuclear leukocyte in a specimen from a patient with malaria. (Courtesy of Loyda Oduber, Jaime LaRoche, Pitágoras Ureña, and Cinthia Batista, Clinical Hospital San Fernando, Panama.)

to be aware that possible therapeutic or prophylactic failures may result when chloroquine is used for *P. vivax* infections in people who have been in Indonesia or Oceania.

Suppressive Therapy. Chemoprophylactic agents to prevent clinical symptoms are given to individuals who are going into areas where malaria is endemic (78). These

Figure 7.13 ParaSight-F test kit showing a negative test strip with a reagent control mark (A) and a positive test strip with a reagent control mark above the positive test result (B).

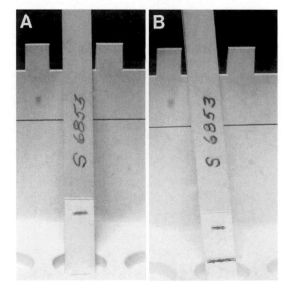

Table 7.9   Alternative approaches to malaria diagnosis[a]

Principle of the method	Method	Comments	Pros	Cons
Giemsa-stained blood films, light microscopy	TBF and THBF, after cytocentrifugation (100 µl of lysed blood centrifuged onto slide, stained with Giemsa)	Traditional method; commercially available (used in histopathology labs)	TBF used to screen larger amount of blood; THBF allows visualization of parasite within RBC; generally thought to be more sensitive than rapid antigen tests, particularly for ID to the species level	TBF unable to see parasite within RBC; THBF provides less blood per exam
Fluorescent DNA-RNA stains	TBF and THBF films (e.g., acridine orange)		Sensitivity good	Requires expertise in reading
	After centrifugation (QBC)	Commercially available	Saves time, fairly sensitive	Expensive, species ID may be difficult
	Flow cytometry	Commercially available; low sensitivity	Automated counts	Low sensitivity due to "background noise"
Molecular methods	DNA-RNA hybridization	Poor sensitivity	"New approach"	Low sensitivity, cumbersome procedures
	PCR	Sensitive	Detect small numbers; species ID with mixed infections; used with blood spot collection	Lengthy procedure; automation may solve this problem
	Real-time PCR (QT-PCR)	Sensitive	Sensitivity of 20 parasites/ml of blood	Requires 500 µl of blood; takes 12 h longer than QT-NASBA
	Real-time nucleic acid sequence-based amplification (QT-NASBA)	Sensitive	Sensitivity of 20 parasites/ml of blood	Requires 50 µl of blood; takes 12 h less than QT-PCR; lends itself to testing large numbers of samples
Malaria pigment detection	Dark-field microscopy	Inferior to TBF	Pigment in WBCs sensitive indicator of prognosis	Not as sensitive as TBF
	Automated blood cell analyzers	Commercially available; detection of cases in patients not suspected as having malaria	Detects malaria pigment (WBCs, schizonts, gametocytes)	Nonspecific findings per species and parasitemia
	Depolarizing monocytes containing malarial pigment	Related to duration of symptoms before testing	Malarial pigment in monocytes	More obvious if parasitemia is $\geq 0.5\%$
	Pseudoreticulocytosis	Related to *P. falciparum*	Nuclear material from intraerythrocytic parasites	More obvious if parasitemia is $\geq 0.5\%$
Antigen detection	HRP-2 assays: ParaSight-F	Commercially available	Simple, rapid dipstick format; good correlation with conventional microscopy; may detect *P. falciparum* (chloroquine resistant) after patient treated for *P. vivax* (mixed infections)	*P. falciparum* only; drop in sensitivity at low levels of parasitemia ($<100/\mu l$); misses cases of only mature gametocytes (no HRP-2); false positives seen
	ICT Malaria *Pf*	Commercially available	Simple, rapid dipstick format; lower false-positive rate; good sensitivity, especially when coupled with microscopy	Some reagents must be stored at 2–8°C; cost may be a factor; more sensitive for placental malaria than microscopy
	MAKROmed	Commercially available	Antigen capture immuno-chromatographic strip format; more sensitive than conventional microscopy	Highly sensitive in detecting placental malaria, lower than expected specificity
	pLDH OptiMAL	Commercially available; test positive only with viable organisms	Simple, rapid dipstick format; also picks up *P. vivax*; few false positives; good test of cure (picks up viable parasites only)	Similar limitations at low parasitemias ($<0.01\%$); mixed-infection ID difficult; cannot replace microscopy

[a] Adapted from references 22, 32, 37, 39, 51–53, 69, 71, 72, 77, 79, 89, and 106.
[b] HRP-2, histidine-rich protein 2; ID, identification; pLDH, parasite lactate dehydrogenase; RBC, red blood cell; TBF, thick blood film; THBF, thin blood film; WBCs, white blood cells.

Table 7.10 Antimalarial drugs and their actions[a,b]

Stage affected	Antimalarial action by interference with:					
	Folic acid synthesis	Folic acid metabolism	Nucleic acid metabolism	Iron-dependent free-radical generation	Parasite components	Unknown
Erythrocytic cycle (schizonts)	Sulfonamides Sulfones	Proguanil Pyrimethamine	Quinine[c] Amodiaquine[e]	Artemisinins	Chloroquine[d] (food vacuole function) Atovaquone (acts on mictochondria)	
Erythrocytic cycle (gametocytes)		Proguanil	Proguanil			Primaquine[f]
Exoerythrocytic cycle		Pyrimethamine	Pyrimethamine			Primaquine

[a] Format adapted from E. K. Markell and M. Voge, *Medical Parasitology*, 5th ed., The W. B. Saunders Co., Philadelphia, Pa., 1981; information updated from references 97 and 110.

[b] Not every drug is included in this table; however, additional information can be seen in the references mentioned above.

[c] Affects the formation of malarial pigment; may also block ion transport to the parasite; inhibits production of tumor necrosis factor.

[d] Interferes with the normal function of the food vacuole and hence with maturation through the trophozoite stage; specific mechanisms include (i) binding to free heme to prevent its incorporation into malaria pigment, (ii) inhibiting aspartic and cysteine protease activity, and (iii) raising vacuolar pH above the pH optima of the aspartic and cysteine proteases in the food vacuole. Inhibition of DNA synthesis is no longer considered a likely mechanism for chloroqine action against the parasite.

[e] Amodiaquine binds heme and inhibits heme polymerization; this is an action similar to that seen with chloroquine.

[f] Primaquine is effective against gametocytes of all four species, unlike some of the other drugs which are partially effective; following administration of primaquine, documented swelling of mitochondria in malarial parasites.

Table 7.11 *Plasmodium falciparum* drugs and therapeutic use[a]

Drug	Mechanisms of resistance	Clinical pharmacokinetics	Therapeutic use	Adverse effects and drug interactions
4-Aminoquinolones				
Chloroquine	Reduced cellular accumulation of drug; drug efflux mechanism; *pfcrt* major chloroquine resistance gene	Rapidly absorbed after oral, intramuscular, or subcutaneous administration; half cleared by kidneys; terminal elimination half-life, 1–2 mo	*P. vivax, P. ovale, P. malariae*; uncomplicated *P. falciparum* in semi-immune patients; nonimmune patients require other drugs due to risks of drug failure	Well tolerated; >250 µg/ml causes dizziness, headache, diplopia, nausea, and exacerbation of epilepsy; cumulative dose of >100 g over time leads to irreversible retinopathy
Amodiaquine	Cross-resistance between this and chloroquine; may also be *pfcrt* gene	In contrast to chloroquine, extensively converted to metabolite	Higher efficacy in much of Africa; used alone or in combination with sulfadoxine-pyrimethamine	Hepatitis and agranulocytosis during prophylaxis; no longer recommended; metabolic antigen-generated immune response may occur
Antifolate drugs and combinations				Severe allergic reactions to sulfa drugs well known; dapsone has concentration-related problems (allergy, problems with G6PD deficiency); pyrimethamine, proguanil, chlorproguanil have little serious toxicity other than folate deficiency
Sulfadoxine-pyrimethamine	Both long elimination half-lives, causing selective pressure for resistance; mutations in *dhfr* and *dhps* genes	Pyrimethamine well absorbed after oral or intramuscular administration; sulfa drugs given for synergistic effects with pyrimethamine or chlorproguanil	Used to treat chloroquine-resistant infections in Africa; severe skin reactions mean that it has no prophylaxis use	
Proguanil	Prophylaxis, but deteriorating clinical response; short half-life		Alone for prophylaxis; fixed-ratio combination with atovaquone (Malarone) (see below)	
Chlorproguanil and dapsone	Used for treatment; short half-life			

(continued)

Table 7.11   (*continued*)

Drug	Mechanisms of resistance	Clinical pharmacokinetics	Therapeutic use	Adverse effects and drug interactions
Quinine	Less potent than chloroquine, small therapeutic range; resistance rare in Africa; not major clinical problem	Oral bioavailability high; intramuscular half-life varies at 10 to 40 min; extensive hepatic biotransformation	Used in combination with antibiotics (tetracycline) in Southeast Asia; parenteral first choice for severe malaria	Tinnitus, deafness, headache, nausea, visual disturbances in the majority of patients receiving therapeutic levels; however, dose reduction not recommended; more serious sequelae can be seen
Artemisinin group	Artemether, artesunate, dihydroartemisinin semisynthetic derivatives of plant extract; high-cost drugs	Rapidly hydrolyzed in vivo to dihydroartemisinin; short elimination half-lives; artemether absorption slower, more variable if administered intramuscularly or intravenously	Used to treat severe malaria and as combination therapy	Generally safe and well tolerated; main concern is reproductive safety (long-bone shortening); cannot be used in first trimester (WHO)

[a] Adapted from references 97, 110, and 112.

Table 7.12   Antimalarial drugs and associated problems[a]

Drug	Associated problems[b]
Artemisinin (qinghaosu)	Recrudescence, neurotoxicity; short half-life
Atovaquone (Mepron)	Limited experience; rapid development of resistance when used alone; recrudescence common
Azithromycin	Limited use; efficacy to be determined
Chloroquine (Aralen)	Resistance worldwide except in Central America; appears to increase *P. falciparum* gametocytogenesis
Doxycycline	Phototoxicity; not used in pregnant women and children; gastrointestinal intolerance
Fansidar	Severe allergic reactions; resistance worldwide
Halofantrine (Halfan)	Cardiotoxicity, poor absorption, sporadic resistance
Mefloquine (Lariam)	Psychoses; resistance in Indochina and Africa
Primaquine	Narrow therapeutic index; not used in G6PD-deficient patients; resistance may be a problem
Proguanil (Paludrine)	Mouth ulcers; resistance worldwide
Quinine sulfate	Tinnitus; resistance in Brazil and Indochina
Quinidine gluconate	Limited availability; requires cardiac monitoring

[a] Adapted from references 16, 70, and 83.
[b] G6PD, glucose-6-phosphate dehydrogenase.

Table 7.13   Malaria resistance[a]

Resistance definition	Comments
Sensitive	From initiation of therapy, asexual parasites are cleared by day 6; no evidence of recrudescence up to day 28
	Peripheral blood films appear to go from positive to negative very quickly (can be a change from one draw to the second draw 6 h later)
Resistance type I (RI)	From initiation of therapy, asexual parasites have cleared for at least 2 consecutive days (the latest day being day 6); recrudescence follows
	Parasite count initially drops and blood films appear to be negative; patient should be monitored for a period of days, particularly if drug-resistant *P. falciparum* suspected
Resistance type II (RII)	Within 48 h of initiation of therapy, marked reduction of asexual parasitemia to <25% of pretreatment count; however, no subsequent disappearance of parasitemia (smear positive on day 6)
	Patient appears to be improving; parasite count drops, but blood films always appear positive
Resistance type III (RIII)	Modest reduction in parasitemia may be seen; no change or increase in parasitemia seen during first 48 h after treatment; no clearing of asexual parasites
	In some cases, parasite count continues to increase with no visible decrease at any time; blood films show overall parasite increase

[a] Adapted from references 49 and 70.

Table 7.14   Areas with reported chloroquine-resistant *Plasmodium falciparum*[a]

Africa	Asia
Angola	Afghanistan (May–Nov)
Benin	Bhutan (southern areas)
Botswana (Nov–June, northern areas)	Cambodia[b]
Burkina Faso	China (Hainan Island and southern provinces)
Burundi	Indonesia[c]
Cameroon	Iran (Mar–Nov)
Central African Republic	Lao People's Democratic Republic[d]
Chad	Malaysia
Comoros	Myanmar (formerly Burma)
Congo	Nepal (areas other than Kathmandu, northern high-altitude areas)
Cote d'Ivoire	
Djibouti	Oman
Equatorial Guinea	Philippines (Luzon, Basilan, Mindoro Palawan, and Mindanao islands; Sulu Archipelago)
Ethiopia (some areas)	
Gabon	Saudi Arabia (some areas)
Gambia	Sri Lanka (some areas)
Ghana	Thailand
Guinea	Vietnam
Guinea-Bissau	Yemen (Sept–Feb, some areas)
Kenya	
Liberia	
Madagascar	**Oceania**
Malawi	Papua New Guinea
Mali	Solomon Islands
Mauritania (some areas)	Vanuatu
Mozambique	
Namibia (Nov–June, northern rural areas)	**Indian subcontinent**[b]
Niger	Bangladesh (north and east)
Nigeria	India
Rwanda	Pakistan (Rawalpindi)
São Tomé and Principe	
Senegal	
Sierra Leone	**South America**
Somalia	Bolivia
South Africa (northern and eastern Transvaal)	Brazil[e]
Sudan	Colombia
Swaziland (northern and eastern plains)	Ecuador[f]
Tanzania (some areas)	French Guiana
Togo	Guyana
Uganda	Panama (east of the Canal Zone including the San Blas Islands)
Zaire	
Zambia (all year in Zambezi valley, Nov–June rest of country below 1,200 m)	Peru (northern provinces)
	Suriname (some areas)
Zimbabwe	Venezuela (some areas)

[a] There is no malaria risk in urban areas unless otherwise indicated. This table should be used in conjunction with the text in determining appropriate prophylaxis. (From *Morb. Mortal. Wkly. Rep.* **34**[14], 1985.) Also see reference 14.

[b] Malaria risk exists in most urban areas.

[c] Malaria risk exists in urban areas of Timor and Kalimantan provinces. Irian Jaya should be considered as Oceania.

[d] Malaria risk exists in all urban areas except Vientiane.

[e] Malaria risk exists in urban areas of interior Amazon River region.

[f] Malaria risk exists in urban areas of Esmeralda, Manabi, El Oro, and Guayas provinces (including the city of Guayaquil). From J. S. Seidel (ed.), *The Pediatric Clinics of North America: Pediatric Infections* **32**:4, 1985.

Table 7.15 Drugs available in the United States for treatment of malaria*a*

Tablets for oral use
  Atovaquone-proguanil (Malarone), 250 mg of atovaquone/100 mg of proguanil
  Chloroquine phosphate (Aralen and generics), 500 mg (300-mg base), 1,000 mg (600-mg base)
  Hydroxychloroquine sulfate (Plaquenil), 400 mg (310-mg base)
  Mefloquine (Lariam and generics), 500 mg (456-mg base), 750 mg (684-mg base)
  Primaquine phosphate, 52.6 mg (30-mg base)
  Pyrimethamine (Daraprim), 25 mg
  Quinine, 180 mg, 325 mg
  Quinine sulfate, 650 mg (542-mg base)

Ampoules for parenteral use
  Chloroquine hydrochloride (200-mg chloroquine base, 5-ml vial) and quinine dihydrochloride
  Quinidine gluconate, 800 mg (500 mg of quinidine) in 10 ml of sterile water

*a*Adapted from CDC *Guidelines for Treatment of Malaria in the United States* (23 March 2005) (www.cdc.gov/malaria/pdf/treatmenttable.pdf). Also see Table 7.17 and chapter 25.

drugs are effective only against the erythrocytic forms and do not prevent the person from getting malaria; i.e., they do not prevent sporozoites from entering the liver and beginning the preerythrocytic cycle of development. The most commonly used drugs for this purpose are quinine, chloroquine, hydroxychloroquine, and amodiaquine, all of which are effective when the organisms enter the RBCs from the liver and begin the erythrocytic cycle. Fansidar is recommended for prophylaxis only for travelers who are staying for a long time in high-transmission areas where choroquine-resistant malaria is endemic. This recommendation was based on revisions in response to increased numbers of adverse reactions to Fansidar prophylaxis, including severe mucocutaneous reactions, some of which were fatal (Table 7.12). In general, chloroquine is still the choice for prophylaxis against *P. falciparum*, but it is not totally effective against *P. vivax*. Resistance to Fansidar was reported in Thailand as early as 1981. Quinine is normally used in these situations; however, quinine-resistant *P. falciparum* has been reported in Africa, with little to no information on previous chloroquine intake. A combination of quinine and tetracycline is recommended for chloroquine- and Fansidar-resistant *P. falciparum*. Areas with chloroquine-resistant *P. falciparum* are listed in Table 7.14, and drug information is presented in Tables 7.15 and 7.16; specific dosage information is given in chapter 25.

Summary of Prophylaxis. With the introduction of chloroquine, with its high potency, long half-life, and low toxicity, continued research into the chemoprophylactic potential of other antimalarial agents such as primaquine was drastically reduced. Resistance to the newer chemoprophylactic drugs such as chloroquine, sulfadoxine/pyrimethamine (Fansidar), proguanil, mefloquine, and doxycycline continues to increase, with many of these agents now becoming clinically useless. Other antimalarial agents such as quinine, halofantrine, or the artimisinine deriva-

tives are too toxic or have too short a half-life to be used for prophylaxis. A summary of antimalarial drugs used for prophylaxis and their activity on different life cycle stages of the malaria parasite can be seen in Table 7.10.

Perhaps the most common approach to chemoprophylaxis in travelers to areas with multidrug-resistant *P. falciparum* is the use of doxycycline. However, this tetracycline derivative must be taken daily, and its use must be continued for 4 weeks after the patient has left the area. Potential side effects include diarrhea, upper gastrointestinal upset, and phototoxicity. Patients may also be predisposed to oral or vaginal candidiasis, and the drug cannot be given during pregnancy or to young children. Also, breakthrough can occur in individuals who have missed only a single dose.

Radical Cure. The radical-cure approach to therapy eradicates all malarial organisms, both the liver and the RBC stages, from the body. Therapy is usually given to individuals who have returned from areas where malaria is endemic; it prevents relapses with *P. vivax* or *P. ovale* infection, although relapses with both *P. vivax* and *P. ovale* infections occasionally occur after treatment with primaquine (8). The gametocytes are also eliminated, thus stopping the chain of transmission to the mosquito vector. The drugs used are primaquine and other 8-aminoquinolones. Treatment with primaquine is usually not necessary for malarial cases acquired by transfusion or contaminated needles or passed from mother to child as a congenital infection. More specific information can be found in chapter 25.

An increase in the incidence of malaria caused by *P. falciparum* is well recognized, and delays in diagnosis and subsequent therapy may result in complicated or fatal cases, particularly in nonimmune hosts. Some of the complications of *P. falciparum* malaria that have been discussed include cerebral malaria, renal failure,

TABLE 7.16 Drugs used for the prevention of malaria[a]

Drug	Usage	Adult dosage	Pediatric dosage	Comments
Atovaquone/proguanil[b] (Malarone)	Primary prophylaxis in areas with chloroquine-resistant or mefloquine-resistant *Plasmodium falciparum*	250 mg of atovaquone/100 mg of proguanil hydrochloride; 1 adult tablet orally daily	62.5 mg of atovaquone/25 mg of proguanil hydrochloride 11–20 kg: 1 tablet 21–30 kg: 2 tablets 31–40 kg: 3 tablets ≥40 kg: 1 adult tablet daily	Contraindicated in persons with severe renal impairment; should be taken with food or a milky drink; not recommended for children <11 kg, pregnant women, or women breast-feeding infants weighing <11 kg.
Chloroquine phosphate (Aralen and generic)[c]	Primary prophylaxis in areas where chloroquine-sensitive *P. falciparum* is endemic	300 mg of base (500 mg of salt) orally once/week	5 mg/kg (base), 8.3 mg/kg (salt) orally once/week, up to maximum adult dose of 300 mg of base	Safe to take during pregnancy, including first trimester; may exacerbate psoriasis.
Doxycycline (many brand names and generic)[d]	Primary prophylaxis in areas with chloroquine-resistant or mefloquine-resistant *P. falciparum*	100 mg orally once/day	>8 years of age: 2 mg/kg orally/day up to maximum adult dose of 100 mg/day	Contraindicated in children <8 years of age; teeth may become permanently stained; may cause photosensitivity; contraindicated during pregnancy.
Hydroxychloroquine sulfate (Plaquenil)[e]	Alternative to chloroquine for primary prophylaxis only in areas with chloroquine-sensitive *P. falciparum*	310 mg of base (400 mg of salt) orally once/week	5 mg/kg (base), 6.5 mg/kg (salt) orally once/week, up to maximum adult dose of 310 mg of base	May be better tolerated than chloroquine; see chloroquine comment.
Mefloquine (Lariam and generic)[a]	Primary prophylaxis in areas with chloroquine-resistant *P. falciparum*	228 mg of base (250 mg of salt) orally once/week (tablet = 250 mg of salt)	5–10 kg: 1/8 tablet orally once/week 10–20 kg: ¼ tablet once/week 20–30 kg: ½ tablet once/week 30–45 kg: ¾ tablet once/week >45 kg: 1 tablet once/week The recommended dose of mefloquine is 5 mg/kg of body weight once weekly. Approximate tablet fraction is based on this dosage. Exact doses for children weighing less than 10 kg should be prepared by a pharmacist.	Contraindicated in persons allergic to mefloquine and in those with active depression or a previous history of depression, generalized anxiety disorder, psychosis, schizophrenia, other major psychiatric disorders, or seizures. Not recommended for persons with cardiac conduction abnormalities.
Primaquine (for primary prophylaxis)	An option for primary prophylaxis in special circumstances. Call Malaria Hotline (770-488-7788) for additional information.	30 mg base (52.6 mg salt) orally daily	0.6 mg/kg base (1.0 mg/kg salt) up to adult dose orally daily	Contraindicated in persons with G6PD[f] deficiency. Also contraindicated during pregnancy and lactation unless the infant being breast-fed has a documented normal G6PD level. Use in consultation with malaria experts.
Primaquine (for terminal prophylaxis)	Used for terminal prophylaxis to decrease the risk of relapses of *P. vivax* and *P. ovale* infection	30 mg of base (52.6 mg of salt) orally once/day for 14 days after departure from the malarious area Note: The recommended dose of primaquine for terminal prophylaxis has been increased from 15 mg to 30 mg for adults	0.6 mg/kg base (1.0 mg/kg salt) up to adult dose orally once/day for 14 days after departure from the malarious area Note: The recommended dose of primaquine for terminal prophylaxis has been increased from 0.3 mg/kg to 0.6 mg/kg for children	Indicated for persons who have had prolonged exposure to *P. vivax* and *P. ovale* or both. Contraindicated in persons with G6PD deficiency. Also contraindicated during pregnancy and lactation unless the infant being breast-fed has a documented normal G6PD level.

Atovaquone-proguanil (Malarone); self-treatment drug	Self-treatment drug to be used if professional medical care is not available within 24 h; medical care should be sought *immediately* after treatment	4 tablets (each dose contains 1,000 mg of atovaquone and 400 mg of proguanil) orally as a single daily dose for 3 consecutive days. Seek medical care immediately after treatment	Daily dose to be taken for 3 consecutive days using **adult-strength tablets:** 11–20 kg: 1 tablet 21–30 kg: 2 tablets 31–40 kg: 3 tablets ≥41 kg: 4 tablets **Seek medical care immediately after treatment**	Contraindicated in persons with severe renal impairment (creatinine clearance, <30 ml/min). Not recommended for self-treatment in persons on atovaquone-proguanil prophylaxis. Not currently recommended for children weighing <11 kg, pregnant women, or women breast-feeding infants <11 kg. Malaria can be fatal; if a traveler develops a fever or other flu-like symptoms and professional medical care is not available within 24 h, a self-treatment dose of atovaquone-proguanil is recommended

[a] From Centers for Disease Control and Prevention, Travel Information (http://www.cdc.gov/travel/malariadrugs2.htm).

[b] Take the first dose 1 to 2 days before travel to malarious areas; take doses daily, at the same time each day, while in the malarious area; take doses daily for 7 days after leaving such areas. Mefloquine should be taken on a full stomach, such as after dinner; most travelers have few, if any, side effects. However, side effects can include abdominal pain, nausea, vomiting, and headache.

[c] Take the first dose 1 to 2 weeks prior to arrival in the malarious area; continue once a week on the same day of the week during travel to malarious areas and for 4 weeks after leaving such areas. Side effects could include gastrointestinal disturbance, headache, dizziness, blurred vision, insomnia, and pruritus, but these symptoms usually do not require discontinuation of the drug.

[d] Doxycycline should be taken on a full stomach, such as after dinner; do not lie down for 1 h to minimize reflux; can predispose women to vaginal yeast infections; should *not* be used in pregnancy, by children younger than 8 years, or by persons allergic to tetracyclines. Doxycycline can also cause sun sensitivity, usually manifested as an exaggerated sunburn reaction. Persons taking this drug should use sunscreen that absorbs long-wave UVA radiation.

[e] Used for travelers to Africa who cannot take the more effective drugs mefloquine or doxycycline; less-effective combination than taking mefloquine or doxycycline and may put them at higher risk for malaria (see also comments under footnote c).

[f] G6FD, glucose-6-phosphate dehydrogenase.

respiratory problems, disseminated intravascular coagulation, blackwater fever, and hepatic dysfunction. Death is directly related to the level of parasitemia and onset of complications, with significant mortality related to >5% parasitemia in spite of appropriate parenteral therapy and supportive care. In these situations, the role of exchange blood transfusion may be appropriate for the following reasons: (i) rapid reduction of the parasitemia, (ii) decreased risk of severe intravascular hemolysis, (iii) improved blood flow, and (iv) improved oxygen-carrying capacity. Powell and Grima have published an excellent review of the use of exchange transfusion for treatment of malaria (80).

Exposure to both placental malaria and maternal human immunodeficiency virus (HIV) infection increases postnatal mortality to a greater degree than does the independent risk associated with exposure to either maternal HIV or placental malaria infection alone. Therefore, malaria chemoprophylaxis during pregnancy could decrease the impact of HIV transmission from mother to infant (Table 7.17).

New Drug Development. Past approaches used to identify potential new antimalarials involved empirical screening of many natural plant extracts, as well as synthesis of new analogs of compounds known to have activity against microbes, tumors, and/or malarial parasites. Primary studies involve in vitro cultures and animal models; secondary and tertiary screening is then undertaken. Very few compounds are selected for clinical evaluation due to financial, compound formulation, stability, toxicity, or efficacy problems encountered in preliminary testing. New approaches involving computerized structure-based drug design and molecular modeling represent significant improvements. Molecular biology methods have become very important and widely used in this development process.

Drug candidates are currently being identified and developed for further testing. With the continued increase in the number of multiple-drug-resistant strains of *P. falciparum,* these drug development efforts have resulted from renewed commitment and have been moved up the priority list. However, the process of antimalarial drug discovery, design, and testing is very long and complex, not to mention expensive. In the United States, a national average of 12 to 15 years and an investment of 200 million to 500 million dollars per licensed drug are expended.

There seem to be inconsistencies between malaria control initiatives and actual policy development and implementation. Current consensus seems to be that a policy change is urgent when high-level resistance occurs in 40% or more of treated patients, when parasitological

Table 7.17   Current issues related to maternal malaria, transmission, pathogenesis, and therapy options

Issue	Comments	Reference(s)
Naturally acquired *Plasmodium knowlesi* in human (Thailand)	The occurrence of simian malaria in humans emphasizes the possible importance of wild primate populations in malaria-endemic and disease transmission areas.	55
Perinatal mother-to-child transmission (MTCT) of HIV	Placental malaria controlled at low density may cause increase in immune responses that protect against MTCT. Uncontrolled, high-density malaria may disrupt placenta structure and generate antigen stimulus to HIV replication, thus increasing the risk for MTCT.	4
Chloroquine resistance in *P. vivax*	Chloroquine has been the drug of choice for *P. vivax* since 1946. No genetic mutations have been linked to chloroquine resistance in *P. vivax* infections. Some recommend mefloquine as alternative; daily doxycycline is also an option when combined with chloroquine. Malarone plus primaquine is the only available therapy with proven efficacy (>90%).	5
Chloroquine resistance in *P. falciparum*	By 1973 in Thailand, chloroquine replaced by a combination of sulfadoxine and pyrimethamine (SP); >10 African countries also switched their first-line drug to SP. In 1985 SP was replaced by mefloquine; resistance to mefloquine led to introduction of artemisinin as combination drug in the mid-1990s. New drugs cost more, have more side effects, take longer time for cure, and have more compliance problems than chloroquine. Another option would be to combine SP (better parasitologic response than chloroquine) with chloroquine (more rapid clinical response).	29, 50
Atovaquone-proguanil resistance in nonimmune North American traveler to West Africa (*P. falciparum*)	Atovaquone-proguanil is a useful agent for treating uncomplicated *P. falciparum* malaria (oral, short course, low side effects). PCR and sequence analysis confirmed the acquisition of a point mutation in the cytochrome *b* gene of the recrudescent isolate (known to confer high-level resistance to atovaquone by reducing the binding affinity of atovaquone). Also, in this case, suboptimal therapy may have played a role in the emergence of resistance.	61
Rapid elimination of asexual forms of *P. falciparum* (most effective way to prevent development of gametocytes)	Primaquine (mefloquine-primaquine) is not effective in the eradication of gametocytes in the primary infection and in the prevention of subsequent gametocytemia compared with artesunate therapy (mefloquine-artesunate). Recrudescent infections lead to subsequent gametocytemia, which drives drug resistance.	95
Distinguishing *P. falciparum* treatment failures from reinfections: PCR genotyping	Study in Tanzania: all samples contained multiple *P. falciparum* infections; primary and recrudescent parasites analyzed by PCR; some of the recrudescing parasites originated from new infections.	68
Fake antimalarials in Southeast Asia (major impediment to malaria control)	Myanmar, Lao People's Democratic Republic, Vietnam, Cambodia, Thailand: 53% of tablet packs labeled as "artesunate" contained no artesunate; 9% of mefloquine samples contained <10% of the expected amount of active ingredient.	25
Recrudescence in artesunate-treated *P. falciparum* cases (depends on parasite burden, not on parasite factors)	Artemisinin derivatives are first-line drugs in Thailand; no documented resistance to artemisinin. Patients with admission parasitemias of >10,000/µl had a 9-fold higher likelihood of recrudescence than did patients with lower parasitemias; recrudescence not due to resistance but to admission parasitemia.	54

response is poor, and when the costs of treatment failures are higher than those of treatments with a newer drug. Improved cooperation is needed between the private and public sectors regarding overall delivery of health care relevant to malaria control issues (73).

Overall Patient Care.   It is very important to remember that patients who present to a medical clinic, office, etc., that does not have expertise in tropical medicine tend to receive suboptimal treatment. Malaria may not be recognized, and the diagnosis may be delayed until it is no longer possible to save the patient. Improvements in obtaining the patient history, symptom recognition, presumptive diagnosis development, and treatment of malaria are mandatory to prevent severe illness and death among travelers.

Epidemiology and Prevention
Malaria is primarily a rural disease and is transmitted by the female anopheline mosquito. There are great

Table 7.18 Malaria epidemics in recent years: climatic change, population displacement, control policy failures, drug resistance[a]

Geographic area	Comments
Central America	Agricultural development, irrigation changes, colonization, insecticide resistance have led to overall increase in malaria
South America	
Amazon rain forest	New settlement and mining has led to >500,000 cases/year in Brazil (>50% of malaria cases in the Americas) and 6,000–10,000 deaths
Peru	Prolonged rains, abnormally high temperatures, and long, humid summers have led to an increase in malaria
Africa	
Savanna and forest	Increasing chloroquine resistance; >50% of population is infected; and malaria is main cause of death in children, killing 1 in 20 younger than 5 yr.
Dry savanna and desert fringe	Increased rains and population movements have led to epidemics
Cities	Severe increased drug resistance, inadequate sanitation, overextended services, and increasing deaths in young adults are all factors
Burundi, Rwanda, neighboring countries	Mass refugee problem has led to worsening overall malaria situation
Botswana, Namibia, Swaziland, Zimbabwe	Exceptional rains and population movements have led to an overall increase in malaria
East African highlands and Madagascar	Cyclical epidemics related to meteorological changes, complicated by agriculture changes, deforestation, and population movements
Ethiopia	Repeated epidemics in highlands due to degradation of environment, drought and famine, and large-scale resettlement approaches
Azerbaijan, Iraq, Tajikistan, Turkey	Civil disturbances, population displacement, and intensified agricultural activities have led to an epidemic resurgence of malaria
Afghanistan	Interruption of control and displacement of populations due to war are responsible for 2 million to 3 million cases annually
South Asia: middle	Increased epidemics in tribal, forest, and hill areas; periodic epidemics in dry lands of northwestern India and Sri Lanka (pooling of river sources, environment conducive to vector breeding); over 2.5 million cases/yr
Cambodia, Lao People's Democratic Republic, Myanmar, Thailand, Vietnam	500,000 cases/yr; rapidly increasing risk in frontier areas of economic activity, often including illicit mining and civil unrest; most severe drug resistance in the world
Papua New Guinea, Philippines; Solomon Islands, Vanuatu	300,000 cases/yr, released to colonization of new areas

[a] Adapted from the WHO/CTD Health Map, 1997, and reference 101.

variations in vector susceptibility to infection with the parasite, with many variations being related to differences in parasite strain. Even when the vector is present in an area, an average number of bites per person per day must be sustained or the infection gradually dies out. This critical level can be influenced by a number of factors, including the vector preference for human blood and habitation and the duration of infection in a specific area. Once an area is clear of the infection, there may also be a drop in population immunity, a situation that may lead to a severe epidemic if the infection is reintroduced into the population (101) (Table 7.18).

Immunity. There are some genetic alterations in the RBCs that confer natural immunity to malaria. Changes in the RBC surface interfere with attachment and invasion of merozoites. Changes in hemoglobin or intracellular enzymes interfere with parasite growth and multiplication (Table 7.19).

Duffy antigen-negative RBCs lack surface receptors for *P. vivax* invasion. Many West Africans and some American blacks are Duffy antigen negative, which may explain the low incidence of *P. vivax* in West Africa. In other areas of Africa, *P. vivax* is much more prevalent.

Partial resistance is seen in individuals with the sickle cell trait and in those with sickle cell anemia. Resistance is

Table 7.19 Host genetic factors in resistance to malaria[a]

Factor[b]	Comments
Hemoglobin disorders	
Sickling disorders	Includes sickle cell disease (SS), heterozygous states for sickle cell gene and those for hemoglobin C (SC) or β-thalassemia (S-thal); carrier state for sickle cell gene is AS; widely distributed throughout tropical Africa, parts of the Mediterranean, the Middle East, and central India; hemoglobin S confers protection against *P. falciparum* malaria (the parasite cannot complete life cycle due to cell sickling and destruction; reduced oxygen levels result in diminished parasite growth; reduced rosetting of RBCs in sickle carriers; protection also related to hemoglobin structure and parasite inability to metabolize)
Hemoglobins C and E	Data relating to selective advantage against malaria less convincing than for hemoglobin S; reduced parasite invasion and impaired growth have been documented; data for hemoglobin E have shown inconsistent results
Hemoglobin F	Growth of *P. falciparum* reduced in presence of hemoglobin F, thought to be due to hemoglobin itself rather than RBC properties; higher levels of hemoglobin F during first year of life might offer protection (newborn infants and adults with persistent fetal hemoglobin production)
Thalassemia (α and β)	Distribution of β-thalassemia coincides with that of malaria; 70% reduction in clinical malaria and 50% reduction in risk; although not seen in all geographic areas, in Papua New Guinea high protective effect of homozygous state for α-thalassemia against complications of *P. falciparum* malaria; babies under 2 yr (homozygous for α^+-thalassemia) had higher frequency of both *P. vivax* and *P. falciparum* but were resistant from 2 yr of age
Erythrocyte polymorphisms	
Glucose-6-phosphate dehydrogenase deficiency	Both female heterozygotes and male hemizygotes have reduced risk (around 50%) of developing severe malaria
Duffy-negative RBCs	Resistant to invasion by *P. vivax*
Ovalocytosis	Patients subject to severe malarial infection with high parasitemias; however, there is strong protection against cerebral malaria; structural change in RBC membrane interferes with binding of infected RBCs to vascular endothelium
Immunogenetic variants	
HLA genes	Each protective HLA type associated with 40–50% decrease in risk; these HLA types are common
HLA class I	HLA-A, -B, -C determine specificities of CD8[+] T cells (cytotoxic, major role in defense against intracellular pathogens); HLA-B35 frequency reduced in children with cerebral malaria and those with severe malarial anemia
HLA class II	HLA-DR, -DQ, -DP determine specificities of CD4[+] T cells that secrete cytokines and provide T-cell help for antibody production and action of other T cells; HLA-DRB1*1302–HLA-DQB1*1501 frequently reduced in children with severe malarial anemia
Cytokine, other immune response genes	
TNF	Major mediator of malaria fever; TNF is increased in children with cerebral malaria and markedly so in children with fatal cerebral malaria
MBL	Deficiency may be associated with increased susceptibility to infectious diseases; effect on malaria may be small to none
CD35	Also called complement receptor 1 (CR1); plays a role in rosetting; African variant of CR1 may protect against severe malaria

[a] Adapted from reference 45.
[b] HLA, human leukocyte antigen; TNF, tumor necrosis factor; MBL, mannose-binding lectin.

related to the sickling of hemoglobin S (HbS)-containing RBCs. Other factors include the formation of deoxyhemoglobin aggregates within the cells, the loss of potassium from sickled cells, and the fact that the parasites actually cause HbS cells to sickle.

Resistance to *P. falciparum* is also seen in glucose-6-phosphate dehydrogenase-deficient cells. Partial immunity to malarial infection is seen in areas where malaria is endemic when HbC, HbE, β-thalassemia, and pyruvate kinase deficiencies exist.

Figure 7.14 Bed nets used in malaria prevention. (A) Nets being dipped in a suspension of insecticide following purchase. For people to appreciate the value of the insecticide treatment, it is important for them to see the process. The nets are wrung out, placed in a plastic bag, and returned to the owner. Later, in the house, the net is laid directly on the bed and allowed to dry out. This must be done indoors to avoid direct exposure to the sun, which degrades the insecticide. (B) Net in place in the home. The occupant of the house in the Bagamoyo BedNet Project, Tanzania, is preparing for rest. The net is a rectangular design, which provides a better ventilating effect than the conical variety. (C) Rectangular bed net being set up outdoors to enable the collection of mosquitoes by the Centers for Disease Control and Prevention Light Trap suspended on the left. Nets for this use are not treated with insecticide. Mosquitoes attracted to the person sleeping under the net are attracted to the light and enter the trap. In the morning, the trap is removed and the collection will indicate which mosquito species are hunting outdoors for blood meals. Similar collections can be made indoors. (Courtesy of Clive Shiff, Johns Hopkins University, with permission.)

Infants are also relatively immune to malarial infections during the first year of life as a result of the presence of a large percentage of HbF, passive immunity from maternal antibodies, and diets deficient in *p*-aminobenzoic acid.

Both cellular and humoral immunities play a part in protection. However, immunity is species specific and in some cases even strain specific. It has also been noted that acute malarial infections can cause immunosuppression. Actual impairment of immune responses to vaccination after acute malaria has been documented.

Insecticide-Impregnated Bed Nets and Clothing. Insecticide-impregnated bed nets have been widely used with various degrees of success. Important factors for the success of any bed net program include insect susceptibility to the particular insecticide used (often pyrethroids), high coverage with impregnated bed nets, high malaria incidence, good community participation at an acceptable sustained level, high mosquito densities when people go to bed, and a high proportion of *P. falciparum*. Bed net impregnation appeared to be somewhat more effective than DDT spraying (Figure 7.14) (65, 90, 111).

The use of permethrin-impregnated clothing has also been evaluated and has been found to reduce the incidence of both malaria and leishmaniasis by a large percentage in the absence of any other interventions. In addition to use by the military (93), permethrin-impregnated clothing for travelers has a number of advantages. First, this insecticide is not a repellent but actually kills mosquitoes, flies, and ticks on contact. Also, impregnated clothing fibers remain effective even after 3 months of repeated machine washing. Apparently, this approach, in which the clothing is soaked with the appropriate solutions, works better than use of an aerosol spray. It is recommended that travelers treat two sets of clothing to wear in high-risk situations. Permethrin liquid is available in the United States as Permanone.

Vaccines. The development of techniques for in vitro culture of *P. falciparum* has enhanced the ability to study parasite metabolism and mechanisms of attachment and invasion. Other studies involving genetic engineering techniques and serologic immunofluorescence tests have led to the development of genetic manipulation of *Escherichia coli* to produce proteins that are identical to the receptor protein on the parasite necessary for RBC attachment. With the use of these and other, newer techniques, progress toward vaccine production may lead to effective protection against malarial infections (6, 35, 36, 40, 63, 98).

Recently, increased funding through private and public partnerships has led the industrial sector to become more involved in vaccine research. There are now at least three vaccine candidates being studied in pediatric populations in field settings.

The Perfect Vaccine. A perfect malaria vaccine will induce immune responses to every stage of the complicated life cycle of *Plasmodium*. Different and distinct immune mechanisms operate against these different stages in the life cycle. Therefore, a multistage vaccine must be a multi-immune response vaccine. Development of these types of vaccines is required to reduce significantly the increasing numbers of malaria infections and deaths seen each year (35).

Pre-erythrocytic stage as vaccine target. A pre-erythrocytic-stage vaccine would prevent sporozoites from invading the liver or prevent liver-stage parasites from developing to maturity and releasing infective merozoites, or both. This type of vaccine would prevent disease (absence of RBC cycle and clinical symptoms) and the transmission of malaria (no gametocyte formation).

Erythrocytic stage as vaccine target. The asexual RBC stage vaccine would prevent or reduce morbidity or mortality by parasite load reduction. Neutralizing antibodies to parasite products (released during schizont rupture) or

inhibiting cytoadherence of infected RBCs in severe forms of malaria would thus reduce morbidity and mortality.

Sexual stage as vaccine target. A vaccine designed to induce immune responses to the sexual-stage antigens would provide no protection to the infected patient but would dramatically limit or eliminate transmission of the parasite within the community (it would destroy gametocytes, prevent fertilization, and prevent development of the parasite within the mosquito).

Multistage, multivalent, multi-immune response vaccine. One of the main advantages of the multistage vaccine is its ability to combat parasite variation. Organism variants that evade the immune mechanisms operating at one of the early stages in the life cycle would most probably be susceptible to one of the later-stage immune mechanisms. Antigenic characteristics of the parasite vary with the life cycle stages, and most antigens are not expressed during all stages but may be present only during a single stage. This mandates either total removal of the parasite in one form within the life cycle or partial removal of various stages, with the overall result being parasite elimination (40).

For a malaria vaccine to induce optimal and sustainable protection, both cellular and humoral responses will have to be elicited against multiple targets from the different life cycle stages. It appears that global malaria vaccine development will move forward on several fronts, hopefully leading to effective products available for clinical use (35, 40).

Overall Control of Malaria. The major determinants related to malaria transmission include (i) the parasite, with its many biological options and genetic diversity; (ii) the vector, with tremendous differences in behavior and transmission capability; (iii) the human host, with biological, behavioral, political, and social ramifications; and (iv) the environment, with tremendous changes such as altitude, temperature, rainfall, and global warming (Table 7.20). Relevant concepts for the control of mosquito transmission of human malaria parasites are listed in Table 7.21; examination of this aspect of malarial transmission control provides an excellent example of the complexities that exist.

Another critical aspect of control involves the safety of the blood supply. A number of factors impact the blood supply worldwide, including government policies and regulations, professional society standards, the donor pool (voluntary and/or paid), the ability of health care personnel to order blood when necessary, infectious-disease screening, posttransfusion adverse effects monitoring, and product-related side effects (88). A number of constraints will continue to impact the development of a global strategy for the control of malaria (Table 7.22).

Table 7.20 Epidemiology of malaria[a]

Characteristic	Stable malaria	Unstable malaria
Transmission pattern	Occurs throughout the year; uniform intensity; pattern repeated yearly	Seasonal transmission intensity variable; dramatic epidemics likely
Community immunity	Strong resistance due to intense transmission	Reduced resistance due to low-level and variable transmission intensity
Age group affected	Mainly young children	All age groups
Control	Difficult	Easier
Major determinants relevant to malarial transmission	Parasite—different biology, genetic diversity Vector—air temperature, relative humidity, breeding sites, density, feeding times, resting places, life cycle Human host—genetic factors, acquired immunity, general health status, behavior patterns, political/social factors Environment—altitude, temperature, rainfall, flow rate of rivers and streams, global warming, colonization, group migrations, agricultural and building projects, deforestation	

[a] Adapted from reference 49.

Table 7.21 Relevant concepts for the control of mosquito transmission of human malaria parasites[a]

Concept	Discussion	Control relevance
Only mosquitoes in the genus *Anopheles* support the sporogonic development of human malaria parasites	A high degree of specificity exists between mosquito species and malaria parasites. Approximately 50 of the 500 *Anopheles* species mosquitoes are competent vectors of human malaria.	It is important to correctly identify vector species in nature and to target vector control to coincide with blood-feeding and resting behaviors of the vector species.
Infected humans carrying gametocyte-stage parasites are the "reservoir of infection"	The probability that mosquitoes will become infected depends on the number and quality of gametocytes present and ingested during blood-feeding.	Most antimalarials have no direct effects on gametocytes; thus, treated individuals may remain infectious for the mosquito vectors.
The rate of *Plasmodium* development in the mosquito depends primarily on temperature	The rate of development from ingested gametocytes to sporozoites in the salivary glands depends on temperature. Temperature also determines mosquito survival and the probability of mosquito infection.	These relationships play a large role in parasite survival and parasite growth, impacting both the probability and intensity of transmission.
A female mosquito can remain infective for life	The female anopheline mosquito is capable of transmitting sporozoites during each consecutive blood-feeding instance, occasionally to multiple individuals during each feeding cycle.	Mosquito control has multiple parameters for consideration.
Epidemiology of human infection associated with vector populations	Worldwide patterns of epidemics are associated with environmental and climatic factors that impact vector distribution, abundance, and blood-feeding behavior.	Personnel knowledgeable about environmental issues and the relationship between the environment and vector control are vital.
Parasite transmission intensity depends on both the numbers of biting vectors and the proportion of vectors carrying sporozoites in the salivary glands	Intensity of transmission depends on the entomological inoculation rate (EIR), which is the product of the biting rate and the proportion of mosquitoes with sporozoites. EIRs are expressed in terms of average numbers of infective bites per person per unit of time.	EIRs in Africa range from 1 to >1,000 infective bites/yr. EIRs of >5 bites/year are associated with malaria prevalence rates of >70% in human populations.
Both the frequency of sporozoite inoculation and dose per infectious bite are important determinants of human infection	Sporozoites inoculated by only a single mosquito can cause human malaria and serious illness, even death.	In areas of Africa where residents are exposed to <5 bites/year, some of the highest incidences of severe disease can be seen.
Marked reduction in transmission intensity is required to lower the prevalence of human malarial infection	An infection acquired by a single individual can be transmitted via mosquitoes to many uninfected individuals in a short time.	In some areas, EIR reductions of >95% are required before the malaria prevalence drops.
From a public health point of view, there are no acceptable levels of malaria transmission	Multiple factors are required for control efforts to be successful, including an integrated approach: drug therapy, vector control, and public health education at the community level.	Future efforts will include vaccines, new antimalarial drugs, and better vector control means. Public health will emphasize "control" rather than "eradication."

[a] Adapted from reference 10.

Table 7.22   Global strategy for malaria control: constraints[a]

Problem	Comments
Misleading assumptions Malaria control is a vertical program Existing tools will not allow impact on morbidity and mortality Malarial control is sole responsibility of health sector Malaria in Africa is uniform and can be controlled by "magic bullet"	Misconceptions have led to reduced political and financial commitment at local, national, and international levels. Accelerated control programs have not been achieved due to lack of funding.
Shortage of people with appropriate knowledge	Little knowledge and understanding of malaria epidemiology and planning/management of control efforts. Peripheral health services not equipped for correct and adequate diagnosis and treatment. Very few countries have operational training programs in place.
Drug resistance; shortage of antimalarial drugs	Complicates the development of national antimalarial drug policies and provision of adequate disease management. Poor-quality antimalarial drugs; no mechanisms to ensure quality, availability, and rational use of affordable and effective drugs.
Poor collaboration	Inter- and intracollaborative efforts within and among countries poor. Surveillance systems and information exchange mechanisms limited; improvement is required for successful implementation.
Limited commitment from private sector	New links between research and development and malaria control needs are minimal; limited commitment to development of new tools for control. However, various initiatives aiming to develop malaria clinical research have recently been launched. Donators are public or international (Global Fund, Roll Back Malaria Initiative, National Institutes of Health [NIH], European Developing Countries Clinical Trial Partnership [EDCTP] program), as well as private (Bill & Melinda Gates Foundation). These substantial funds should enhance the research of new antimalarial drugs and large-scale, adequately designed trials.

[a] Adapted from reference 101.

Babesiosis

When blood parasites are discussed, malaria is usually considered the most important and well-known infection. However, there is another organism that also causes human infection. Recognition of the spectrum of disease caused by organisms of the genus *Babesia* has recently resulted in an increase in the number of cases reported. Increased interest has also been generated because the infection can be transmitted via transfusion, can cause serious illness, particularly in compromised patients, and is difficult to treat (48).

The *Babesia* blood parasite was first discovered in cattle in 1888, and in 1893 it was recognized as the cause of Texas cattle fever transmitted by ticks (8). Cases of babesiosis have been documented worldwide, and several outbreaks in humans have occurred in the northeastern United States, particularly in Long Island, Cape Cod, and the islands off the East Coast (48). Although there are many species of *Babesia*, *Babesia microti* is the cause of most human infections in the United States.

A case reported from California occurred in a 24-year-old man who had been splenectomized in 1985 for an immune system-mediated thrombocytopenia and who was a U.S. Army officer stationed at Fort Ord in Monterey County. This is the third reported case of presumed indigenous transmission of babesiosis in patients in California, all of whom had been previously splenectomized.

Over the past few years, one patient from Washington and four from California were identified as being infected with a similar protozoal parasite, very similar in morphology to *Babesia* species; the Washington isolate was designated WA1. All four patients had undergone splenectomy, two as a result of trauma and two for medical reasons. Two of the four patients had complications, and one died. Piroplasm-specific nuclear small-subunit ribosomal DNA was recovered from the blood of the four patients by PCR amplification. Genetic sequence analysis showed that the organisms were almost identical. Phylogenic analysis showed that this strain is more closely related to a known canine pathogen (*B. gibsoni*) and to *Theileria* species than to most members of the genus *Babesia*. Serum from three of the patients was reactive to WA1 but not to *B. microti* antigen. It now appears that there is a newly identified *Babesia*-like organism which causes clinical infections in the western United States. The spectrum of illness ranges from asymptomatic infection to fulminant, fatal disease (48). The first case of transfusion-transmitted babesiosis with the WA1 type has been reported from Washington state.

Studies in New York state indicate evidence of exposure to five tick-borne pathogens, including *B. microti*, in a high-risk population in the northeastern United States. There was also evidence of possible dual infection in some patients (46). In the same area, a cluster of transfusion-associated babesiosis cases was traced to a single asymptomatic donor; six patients were exposed, and three of these developed parasitemia (23). Environmental conditions necessary for the transmission of *B. microti* now appear to be present in areas of New Jersey, and physicians in central New Jersey should be aware of this potential threat.

A total of 10 cases of babesiosis have been reviewed, most of which were probably acquired in northwestern Wisconsin. Of 10 cases, 3 were fatal and occurred in elderly patients, who died after experiencing complications (44). These cases also serve to emphasize the importance of this potentially fatal zoonosis, which is not limited to the northeastern United States. In another case from western Wisconsin, a woman who had not traveled outside of Wisconsin was hospitalized and found to be infected with both *B. microti* and *Borrelia burgdorferi* (96).

Babesia spp. are grouped into the small *Babesia* spp. (1.0 to 2.5 μm), which include *B. gibsoni*, *B. microti*, and *B. rodhaini*, and the large *Babesia* spp. (2.5 to 5.0 μm) which include *B. bovis*, *B. caballi*, and *B. canis*. These phenotypic classifications are, for the most part, consistent with genetic characterization based on nuclear small-subunit ribosomal DNA sequences. The small *Babesia* spp. are more closely related to *Theileria* spp. than are the larger organisms. The one exception is *B. divergens*, which appears small on blood smears but is genetically related to the large *Babesia* spp. The two primary pathogens of humans are *B. microti* and *B. divergens*, along with the unnamed species WA1 (Washington), CA1 (California), and MO1 (Missouri) (42).

Life Cycle and Morphology

These organisms have at least three stages of reproduction within the life cycle: (i) gamogony (formation and fusion of gametes inside the tick gut), (ii) sporogony (asexual reproduction in salivary glands), and (iii) merogony (asexual reproduction in the vertebrate host). Infection is transmitted by several species of ticks in which the sexual multiplication cycle occurs. Several thousand sporozoites can be found in the dermis surrounding the tick's mouth during the final hours of attachment and subsequent feeding. When a tick takes a blood meal, the infective forms are introduced into the human host (Figure 7.15). The organisms infect the RBCs, in which they appear as pleomorphic, ringlike structures (Figure 7.16). They resemble the early trophozoite (ring) forms of malarial parasites, particularly *P. falciparum*. The organisms measure 1.0 to 5.0 μm, the RBCs are not enlarged or pale, and the cells do not contain stippling. Malarial pigment is never seen. The early form contains very little cytoplasm and has a very small nucleus. In mature forms, two or more chromatin dots may be seen. Occasionally, a tetrad formation, referred to as a Maltese cross, is seen (Figure 7.16).

Clinical Disease

It is suspected that the frequency of *B. microti* and WA1 infection in the United States is greater than the number of cases reported. Between 1970 and 1991, 136 cases were reported in New York, while between 1969 and 1998, 160 cases were reported in Nantucket. Babesiosis

Figure 7.15 Life cycle of *Babesia* spp. (Illustration by Gwen Gloege.)

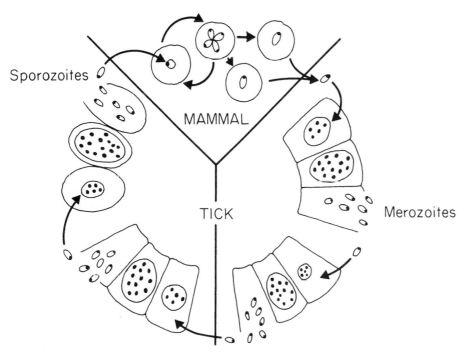

Sporozoites

MAMMAL

TICK

Merozoites

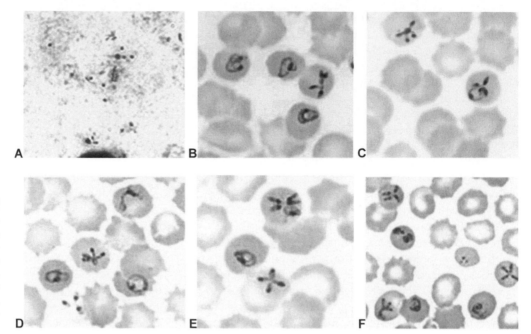

Figure 7.16 Six examples of blood films containing the ringlike forms of *Babesia* spp. (A) Thick blood film containing numerous *Babesia* rings; (B to F) various ring forms, multiple rings per cell, some rings present outside of the red blood cells, and the typical image of the "Maltese cross" configuration of the four rings.

is usually a self-limiting disease; there are probably also asymptomatic and undiagnosed carriers. The mortality rate has been reported at about 5% (48).

However, infections caused by *B. divergens* in Europe are far more serious, with a mortality rate of 42%. Fatal cases of *B. divergens* have been reported in France, Britain, Ireland, Spain, Sweden, Switzerland, the former Yugoslavia, and the former USSR (113) (Table 7.23). In two patients from Austria and Italy, the DNA sequences were identical; however, in phylogenetic analysis, the organism clusters with *B. odocoilei*, a parasite of white-tailed deer that is related to *B. divergens* (41). Although human babesiosis is relatively new to Japan, it appears that a new type of *B. microti*-like parasite, the Hobetsu type, is the major type among Japanese wild rodents. Seroepidemiologic surveys of humans are under way to estimate the prevalence of *Babesia* infection among Japanese (102).

In the first four human cases (before 1969) within the United States, the patients had been splenectomized and three died as a result of the infection. Since then, other cases have been found in patients with intact spleens. In the patients from Nantucket Island, Mass., who were not splenectomized, symptoms began 10 to 20 days after a tick bite and continued for several weeks. The symptoms started with general malaise, which was followed by fever, shaking chills, profuse sweating, arthralgias, myalgias, fatigue, and weakness. Hepatosplenomegaly was present, and five patients had slightly elevated serum bilirubin and transaminase levels as a result of hemolytic anemia.

During the summer of 1980, six patients from Shelter Island and eastern Long Island were diagnosed as having babesiosis caused by *B. microti*. Symptoms lasted for 19 to 24 days in five patients and included fever, shaking chills, dark urine, and headache, as well as anorexia, malaise, and lethargy. The one patient in the group who had been splenectomized also had the most severe illness. The parasitemia could not be detected in one patient, was 5% in four of the six, and was 80 to 90% in the splenectomized patient. Hamster inoculation produced a patent parasitemia in all cases.

By the end of 1991, 13 cases of babesiosis had been reported in Connecticut, this being the largest number of human cases reported on the mainland United States (1). Information from the patients suggests that 12 infections were acquired via tick bites and 1 was acquired from a blood transfusion. The ages of the patients ranged from 61 to 95 years for those with tick-acquired infections. Two patients died of active infections, and one died of chronic obstructive pulmonary disease. Indirect fluorescent-antibody titers ranged from 11,024 to 14,096, and five of eight patients had significant IgG or IgM titers (1,640 to 15,120) to *Borrelia burgdorferi*. *Babesia microti* from the blood of 7 of 12 patients tested was isolated in Syrian hamsters. It is now clear that babesiosis is endemic in New Jersey, as well as other areas of the northeast (43).

Babesia infections described in California and in other parts of the world are quite different from those seen in the northeastern United States, where the infection is most often subclinical. Infections in California and Europe tend to present as a fulminant, febrile, hemolytic disease affecting splenectomized or immunosuppressed individuals.

Table 7.23 Similarities and differences in *Babesia* spp.[a]

Species[b]	Geographic location	Vector	Incubation period	Disease	Comments
Babesia divergens	Europe; relatively rare, but 42% mortality	*Ixodes ricinus*	1–3 wk	Sudden appearance, hemoglobinuria followed by jaundice; shock-like picture	Most patients splenectomized (frequent contact with cattle); bovine pathogen; same vector for Lyme disease
EU1 (*B. divergens*-like)	Italy, Austria	Not known, but probably *I. ricinus*	One patient had possible tick exposure 2 wk prior to symptoms	Fever, chills, confusion, jaundice	Both patients splenectomized; organism most closely related to *B. odocoilei* (infects white-tailed deer)
Babesia microti	United States: northeastern and Great Lakes regions; range probably expanding to include as far south as New Jersey, parts of Europe, and Japan; mortality rate for clinically apparent infections 5%	*Ixodes scapularis* (*I. dammini*)	1–6 wk; up to 3 mo	Most cases are mild and resolve on their own; most severe cases have fulminating malaria-like infection; malaise, chills, myalgia, anemia, fatigue, fever; hepatomegaly and splenomegaly may be present; hemolytic anemia several days to few months (more common in elderly or asplenic individuals)	Age range 3 wk to 86 yr; most within 50–60 yr; most are normosplenic; Japanese strains appear to be *B. microti*-like
WA1	Washington	Not known, but *Dermacentor variabilis, Ornithodoros coriaceus,* and *Ixodes pacificus* found in areas where WA1 found	Not known	Splenectomized patients	Related to canine pathogen, *B. gibsoni*, but WA1 does not seem to infect dogs; highly virulent in hamsters
Babesia divergens-like	Washington	Not known	Not known	Asplenic 82-year-old male with high parasitemia (41.4%) and multiorgan dysfunction including renal failure; survived	Patient's serum reacted to *B. divergens* but not to *B. microti* or WA1
CA1	California	Not known	Not known	Splenectomized patient; 2 patients had complicated courses, and 1 died	This strain is more closely related to *B. gibsoni* and to *Theileria* species
MO1	Missouri	Not known	Not known	Parasitemia in 3–4%; systemic lupus erythematosus, diagnosed in 1979, taking prednisone; splenectomy in 1979	*B. divergens*-like; 72 year old asplenic male
PB-1	Baboon colonies	Not known	Not known	Potential problems with organ xenotransplantation to human recipients	Most closely related to *B. microti*; may be organism previously described as *Entopolypoides*; issues include prevention of potential zoonotic transmission

[a] Adopted from references 41, 42, 96, 102, 113.
[b] CA, California; EU, European Union; MO, Missouri; WA, Washington.

Work suggests that species other than *B. microti* or the bovine parasites were involved (17).

The serum of an 82-year-old splenectomized man from Washington reacted to *B. divergens* but not to *B. microti* or WA1 antigens. This case emphasizes the point that unusual parasites may be detectable by blood film examination but not by serologic or molecular testing for *B. microti* or WA1-type parasites (42). This is important to remember, since new organisms similar but not identical to a number of the well-known *Babesia* spp. may be identified in patients in different parts of the world. Without the manual examination of blood films, these organisms may be missed.

A case of transfusion-induced babesiosis accompanied by disseminated intravascular coagulopathy in an elderly patient was reported in 1982. Symptoms included fever, chills, nausea, arthralgias, and lethargy, which began after the patient received 2 units of packed RBCs during surgery. One of the donors had a high indirect fluorescent-antibody titer against *B. microti*, but the infection was confirmed only by hamster inoculation.

Diagnosis

Most documented human patients have had a low parasitemia; thus, both thick and thin blood films stained with Giemsa stain must be examined. When the organisms cannot be demonstrated in a suspected patient, blood can be inoculated intraperitoneally into a hamster or gerbil. The organisms can usually be demonstrated within a few days, with the same morphology as that seen in humans.

Serologic studies have been reported; however, cross-reactivity has been seen among different species of *Babesia* and several species of *Plasmodium*. False-positive results have been seen in patients with connective tissue disorders such as systemic lupus erythematosus and rheumatoid arthritis. Titers of 1:128 to 1:256 rarely represent false positives; however, blood donor screening at a 1:64 titer may result in occasional false positives. False-negative results can be seen in immunosuppressed patients or in patients from whom blood specimens are drawn early in the course of the infection. Use of the indirect fluorescent-antibody test is helpful in the diagnosis of *B. microti* infections, particularly chronic infections. This test is specific and sensitive and is the currently recommended serologic method. Antibody can usually be detected when patients are first diagnosed; titers can remain elevated for 13 months to 6 years.

Note. Potential diagnostic problems with the use of automated differential instruments have been reported. Cases of babesiosis and malaria have been missed when blood smears were examined by these methods. The number of fields scanned by a technologist on instrument-read smears is quite small; therefore, a light parasitemia will almost surely be missed. Failure to make the diagnosis results in a delay of a number of days before therapy is instituted. Although the majority of these instruments are not designed to detect intracellular blood parasites, routine use of the automated systems may pose serious diagnostic problems, particularly if the suspected diagnosis is not conveyed to the laboratory.

KEY POINTS—LABORATORY DIAGNOSIS
Babesiosis

1. Blood films should be prepared on admission of the patient. Travel history and/or possible exposure to tick-infested areas, recent blood transfusion, or splenectomy information is very important.
2. Both thick and thin blood films should be prepared. At least 200 to 300 oil immersion fields should be examined before the smears are considered negative.
3. Wright's, Wright-Giemsa, Giemsa, or rapid stain can be used, although Giemsa is the stain of choice for blood parasites.
4. *Babesia* parasites may be missed by automated differential instruments. Even with technologist review of the smears, a light parasitemia is very likely to be missed.
5. There does not have to be any significant history of travel outside the United States. If organisms are seen, they mimic *P. falciparum* rings.
6. No morphological stages other than ring forms will be seen. The classic arrangement of the four rings (Maltese cross) is not always seen.
7. *One negative set of blood smears does not rule out a* Babesia *infection.*
8. The quantitative buffy coat system can also be used and has shown 100% correlation with Giemsa-stained peripheral blood smears for the detection of parasitemia in patients with babesiosis. However, confirmation of morphologic differences between *Babesia* and malarial rings might require examination of Giemsa-stained blood films.
9. The indirect fluorescent-antibody test and possibly PCR are recommended. Hamster inoculation is helpful but is neither realistic nor practical for most diagnostic laboratories.

Treatment

Often, babesiosis can be effectively managed with supportive care. Currently, the combination of clindamycin plus quinine has been recommended as the standard regimen for human babesial infection. However, failure to eliminate the organisms in some immunocompromised and HIV-infected persons has been reported (91). Based on recent findings, the use of azithromycin in combination with quinine may be considered as an alternative therapy for human babesiosis, especially in a situation of failure to respond to standard regimens (91). It also appears that azithromycin combined with atovaquone is another possibility (86).

Any person with babesiosis acquired in Europe should be treated on an emergency basis. Such patients should receive prompt specific therapy to reduce parasitemia and to prevent the extensive hemolysis and renal failure that may follow. Massive exchange transfusion may be required due to the rapidly increasing parasitemia characteristic of the disease with *B. divergens*. The element of time becomes critical; with the onset of hemoglobinuria, rapidly increasing intravascular hemolysis leads to renal failure. Treatment for these seriously ill patients can include exchange transfusion, intravenous clindamycin, and oral quinine. Atovaquone has been suggested as an option, as have chloroquine, tetracycline, primaquine, sulfadiazine, and pyrimethamine, with variable results.

Epidemiology and Prevention

The public health notice that was distributed to the Nantucket Board of Health indicated that *Babesia* parasites are transmitted from mice to humans primarily by the bite of the deer tick *Ixodes scapularis* and that individuals with the infection may be asymptomatic. The notice indicated that prevention still relies on the avoidance of ticks or their prompt removal once detected, since there are apparently no fully effective tick repellents. If symptoms appear 1 to 2 weeks after a tick bite, a physician should be consulted. In a study of Block Island, R.I., from 1991 through 2000, the incidence of babesial infection appeared to be at least 10-fold greater than previously recognized elsewhere (59).

Another problem is related to blood donors. The duration of parasitemia following asymptomatic or symptomatic infection is not known. In areas where the infection is endemic, exclusion of blood donors who are febrile, who are known to have had babesiosis, or who have an anti-*Babesia* titer of 116 or higher may reduce the risk. However, until a practical, effective, and sensitive screening method is developed, cases of transfusion-induced babesiosis will probably continue to occur.

It is now well established that under blood-banking conditions (4°C for 30 days), *B. microti* can remain infective and that transfusion-acquired infection with this parasite could occur during the normal storage time for blood.

The most common and accepted approach used to protect humans from infection involves methods to reduce the tick density. A number of methods have been used, including spraying acaricidal emulsions on vegetation. Repellents are useful for personal protection; however, daily examination of the body surface of a person who may have come in contact with ticks is critical. If ticks are found, they should be removed immediately to take advantage of the time lag between attachment to the human and transmission of infection (usually 50 to 60 h).

References

1. **Anderson, J. F., E. D. Mintz, J. J. Gadbaw, and L. A. Magnarelli.** 1991. *Babesia microti*, human babesiosis, and *Borrelia burgdorferi* in Connecticut. *J. Clin. Microbiol.* **29:**2779–2783.

2. **Anstey, N. M., D. L. Granger, M. Y. Hassanali, E. D. Mwaikambo, P. E. Duffy, and J. B. Weinberg.** 1999. Nitric oxide, malaria, and anemia: inverse relationship between nitric oxide production and hemoglobin concentration in asymptomatic, malaria-exposed children. *Am. J. Trop. Med. Hyg.* **61:**249–252.

3. **Anstey, N. M., J. B. Weinberg, Z. Wang, E. D. Mwaikambo, P. E. Duffy, and D. L. Granger.** 1999. Effects of age and parasitemia on nitric oxide production/leukocyte nitric oxide synthase type 2 expression in asymptomatic, malaria-exposed children. *Am. J. Trop. Med. Hyg.* **61:**253–258.

4. **Ayisi, J. G., A. M. van Eijk, R. D. Newman, F. O. ter Kuile, Y. P. Shi, C. Yang, M. S. Kolszak, J. A. Otieno, A. O. Misore, P. A. Kager, R. B. Lal, R. W. Steketee, and B. L. Nahlen.** 2004. Maternal malaria and perinatal HIV transmission, Western Kenya. *Emerg. Infect. Dis.* **10:**643–652.

5. **Baird, J. K.** 2004. Chloroquine resistance in *Plasmodium vivax*. *Antimicrob. Agents Chemother.* **48:**4075–4083.

6. **Ballou, W. R., M. Arevalo-Herrera, D. Carucci, T. L. Richie, G. Corradin, C. Diggs, P. Druilhe, B. K. Giersing, A. Saul, D. G. Heppner, K. E. Kester, D. E. Lanar, J. Lyon, V. S. Hill, W. Pan, and J. D. Cohen.** 2004. Update on the clinical development of candidate malaria vaccines. *Am. J. Trop. Med. Hyg.* **71S:**239–247.

7. **Bates, I., and G. Beduaddo.** 1998. Chronic malaria and splenic lymphoma: clues to understanding lymphoma evolution. *Leukemia* **11:**2162–2167.

8. **Beaver, P. C., R. C. Jung, and E. W. Cupp.** 1984. *Clinical Parasitology*, 9th ed. Lea & Febiger, Philadelphia, Pa.

9. **Beeson, J. G., G. V. Brown, M. E. Molyneux, C. Mhango, F. Dzinjalamala, and S. J. Rogerson.** 1999. *Plasmodium falciparum* isolates from infected pregnant women and children are associated with distinct adhesive and antigenic properties. *J. Infect. Dis.* **180:**464–472.

10. **Beier, J. C., and J. P. Vanderberg.** 1998. Sporogonic development in the mosquito, p. 49–61. *In* I. W. Sherman (ed.), *Malaria: Parasite Biology, Pathogenesis, and Protection.* ASM Press, Washington, D.C.

11. **Berger, S. A., and L. David.** 2005. Pseudo-borreliosis in patients with malaria. *Am. J. Trop. Med. Hyg.* **73:**207–209.

12. **Berry, A., R. Fabre, F. Benoit-Vical, S. Cassaing, and J. F. Magnaval.** 2005. Contribution of PCR-based methods to diagnosis and management of imported malaria. *Med. Trop.* **65:**176–183.

13. **Bertagnolio, S., E. Tacconelli, G. Camilli, and M. Tumbarello.** 2001. Case report: retinopathy after malaria prophylaxis with chloroquine. *Am. J. Trop. Med. Hyg.* **65:**637–638.

14. **Bloland, P. B.** 2001. *Drug Resistance in Malaria*. WHO/CDS/CSR/DRS/2001.4. World Health Organization. Geneva, Switzerland.

15. **Brown, H., T. T. Hien, N. Day, N. T. H. Mai, L. V. Chuong, T. T. H. Chau, P. P. Loc, N. H. Phu, D. Bethell, J. Farrar, K. Gatter, N. White, and G. Turner.** 1999. Evidence of blood-brain barrier dysfunction in human cerebral malaria. *Neuropathol. Appl. Neurobiol.* **25:**331–341.

16. **Buckling, A., L. C. Ranford-Cartwright, A. Miles, and A. F. Read.** 1999. Chloroquine increases *Plasmodium*

falciparum gametocytogenesis in vitro. *Parasitology* **118**:339–346.

17. **California Department of Health Services.** 1992. Babesiosis in California. *Calif. Morbid.*, 24 January.

18. **Chen, Q., M. Schlichtherle, and M. Wahlgren.** 2001. Molecular aspects of severe malaria. *Clin. Microbiol. Rev.* **13**:439–450.

19. **Chotivanich, K. T., S. Pukrittayakamee, J. A. Simpson, N. J. White, and R. Udomsangpetch.** 1998. Characteristics of *Plasmodium vivax*-infected erythrocyte rosettes. *Am. J. Trop. Med. Hyg.* **59**:73–76.

20. **Clark, I. A., L. M. Alleva, A. C. Mills, and W. B. Cowden.** 2004. Pathogenesis of malaria and clinically similar conditions. *Clin. Microbiol. Rev.* **17**:509–539.

21. **Collins, W. E., and G. M. Jeffery.** 2005. *Plasmodium ovale*: parasite and disease. *Clin. Microbiol. Rev.* **18**:570–581.

22. **De Monbrison F., P. Gerome, J. F. Chaulet, M. Wallon, S. Picot, and F. Peyron.** 2004. Comparative diagnostic performance of two commercial rapid tests for malaria in a non-endemic area. *Eur. J. Clin. Microbiol. Infect. Dis.* **10**:784–786.

23. **Dobroszycki, J., B. L. Herwaldt, F. Boctor, J. R. Miller, J. Linden, M. L. Eberhard, J. J. Yoon, N. M. Ali, H. B. Tanowitz, F. Graham, L. M. Weiss, and M. Wittner.** 1999. A cluster of transfusion-associated babesiosis cases traced to a single asymptomatic donor. *JAMA* **281**:927–930.

24. **Dondorp, A. M., B. J. Angus, K. Chotivanich, K. Silamut, R. Ruangveerayuth, M. R. Hardeman, P. A. Kager, J. Vreeken, and N. J. White.** 1999. Red blood cell deformability as a predictor of anemia in severe falciparum malaria. *Am. J. Trop. Med. Hyg.* **60**:733–737.

25. **Dondorp, A. M., P. N. Newton, M. Mayxay, W. Van Damme, F. M. Smithuis, S. Yeung, A. Petit, A. J. Lynam, A Johnson, T. T. Hien, R. McGready, J. J. Farrar, S. Looareesuwan, N. P. Day, M. D. Green, and N. J. White.** 2004. Fake antimalarials in Southeast Asia are a major impediment to malaria control: multinational cross-sectional survey on the prevalence of fake antimalarials. *Trop. Med. Int. Health* **9**:1241–1246.

26. **Dondorp, A. M., T. Planche, E. E. Debel, B. J. Angus, K. T. Chotivanich, K. Silamut, J. A. Romjn, R. Ruangveerayuth, F. J. Hoek, P. A. Kager, J. Vreeken, and N. J. White.** 1998. Nitric oxides in plasma, urine, and cerebrospinal fluid in patients with severe falciparum malaria. *Am. J. Trop. Med. Hyg.* **59**:497–502.

27. **Eliades, M. J., S. Shah, P. Nguyen-Dinh, R. D. Newman, A. M. Barber, P. Nguyen-Dinh, J. M. Roberts, S. Mali, M. E. Parise, A. M. Barber, and R. Steketee.** 2005. Malaria surveillance—United States, 2003. *Morb. Mortal. Wkly. Rep. Surveill. Summ.* **54**:25–40.

28. **Farnert, A., I. Rooth, A. Svensson, G. Snounou, and A. Bjorkman.** 1999. Complexity of *Plasmodium falciparum* infections is consistent over time and protects against clinical disease in Tanzanian children. *J. Infect. Dis.* **179**:989–995.

29. **Farooq, U., and R. C. Mahajan.** 2004. Drug resistance in malaria. *J. Vector Borne Dis.* **41**:45–53.

30. **Fernandez, V., C. J. Treutiger, G. B. Nash, and M. Wahlgren.** 1998. Multiple adhesive phenotypes linked to rosetting binding of erythrocytes in *Plasmodium falciparum* malaria. *Infect. Immun.* **66**:2969–2975.

31. **Fischer, P. R., and P. Boone.** 1998. Short report: severe malaria associated with blood group. *Am. J. Trop. Med. Hyg.* **58**:122–123.

32. **Forney, J. R., C. Wongsrichanalai, A. J. Magill, L. G. Craig, J. Sirichaisinthop, C. T. Bautista, R. S. Miller, C. F. Ockenhouse, K. E. Kester, N. E. Aronson, E. M. Andersen, H. A. Quino-Ascurra, C. Vidal, K. A. Moran, C. K. Murray, C. C. DeWitt, D. G. Heppner, K. C. Kain, W. R. Ballou, and R. A. Gasser, Jr.** 2003. Devices for rapid diagnosis of malaria: evaluation of prototype assays that detect *Plasmodium falciparum* histidine-rich protein 2 and a *Plasmodium vivax*-specific antigen. *J. Clin. Microbiol.* **41**:2358–2366.

33. **Garcia, L. S., R. Y. Shimizu, and D. A. Bruckner.** 1986. Blood parasites: problems in diagnosis using automated differential instrumentation. *Diagn. Microbiol. Infect. Dis.* **4**:173–176.

34. **Giha, H. A., T. G. Theander, T. Staalso, C. Roper, I. M. Elhassan, H. Babiker, G. M. H. Satti, D. E. Arnot, and L. Hviid.** 1998. Seasonal variation in agglutination of *Plasmodium falciparum*-infected erythrocytes. *Am. J. Trop. Med. Hyg.* **58**:399–405.

35. **Good, M. F.** 2005. Genetically modified *Plasmodium* highlights the potential of whole parasite vaccine strategies. *Trends Immunol.* **26**:295–297.

36. **Good, M. F.** 2005. Vaccine-induced immunity to malaria parasites and the need for novel strategies. *Trends Parasitol.* **21**:29–34.

37. **Grobusch, M. P., T. Hanscheid, K. Gobels, H. Slevogt, T. Zoller, G. Rogler, and D. Teichmann.** 2003. Comparison of three antigen detection tests for diagnosis and follow-up of falciparum malaria in travelers returning to Berlin, Germany. *Parasitol. Res.* **89**:354–357.

38. **Guyatt, H. L., and R. W. Snow.** 2004. Impact of malaria during pregnancy on low birth weight in sub-Saharan Africa. *Clin. Microbiol. Rev.* **17**:760–769.

39. **Hanscheid, T.** 1999. Diagnosis of malaria: a review of alternatives to conventional microscopy. *Clin. Lab. Haematol.* **21**:235–245.

40. **Heppner, D. G., Jr., K. E. Kester, C. F. Ockenhouse, N. Tornieporth, O. Ofori, J. A. Lyon, V. A. Stewart, P. Dubois, D. E. Lanar, U. Krzych, P. Moris, E. Angov, J. F. Cummings, A. Leach, B. T. Hall, S. Dutta, R. Schwenk, C. Hillier, A. Barbosa, L. A. Ware, L. Nair, C. A. Darko, M. R. Withers, B. Ogutu, M. E. Polnemus, M. Fukuda, S. Pichyangkul, M. Gettyacamin, C. Diggs, L. Soisson, J. Milman, M. C. Dubois, N. Garcon, K. Tucker, J. Wittes, C. V. Plowe, M. A. Thera, O. K. Duombo, M. G. Pau, J. Goudsmit, W. R. Ballou, and J. Cohen.** 2005. Towards an RTS,S-based, multi-stage, multi-antigen vaccine against falciparum malaria: progress at the Walter Reed Army Institute for Research. *Vaccine* **23**:2243–2250.

41. **Herwaldt, B. L., S. Cacciò, F. Gherlinzoni, H. Aspöck, S. B. Slemenda, P. Piccaluga, G. Martinelli, R. Edelhofer, U. Hollenstein, G. Poletti, S. Pampiglione, K. Löschenberger, S. Tura, and N. J. Pieniazek.** 2003. Molecular characterization of a non-*Babesia divergens* organism causing zoonotic babesiosis in Europe. *Emerg. Infect. Dis.* **9**:942–948.

42. **Herwaldt, B. L., G. deBruyn, N. J. Pieniazek, M. Homer, K. H. Lofy, S. B. Slemenda, T. R. Fritsche, D. H. Persing, and A. P. Limaye.** *Babesia divergens*-like infection, Washington state. *Emerg Infect. Dis.* **10**:622–629.

43. Herwaldt, B. L., P. C. McGovern, M. P. Gerwel, R. M. Easton, and R. R. MacGregor. 2003. Endemic babesiosis in another eastern state: New Jersey. *Emerg. Infect. Dis.* 9:184–188.

44. Herwaldt, B. L., F. E. Springs, P. P. Roberts, M. L. Eberhard, K. Case, D. H. Persing, and W. A. Agger. 1995. Babesiosis in Wisconsin: a potentially fatal disease. *Am. J. Trop. Med. Hyg.* 53:146–151.

45. Hill, A. V. S., and D. J. Weatherall. 1998. Host genetic factors in resistance to malaria, p. 445–455. *In* I. W. Sherman (ed.), *Malaria: Parasite Biology, Pathogenesis, and Protection.* ASM Press, Washington, D.C.

46. Hilton, E., J. DeVoti, J. L. Benach, M. L. Halluska, D. J. White, H. Paxton, and J. S. Dumler. 1999. Seroprevalence and seroconversion for tickborne diseases in a high-risk population in the northeast United States. *Am. J. Med.* 106:404–409.

47. Ho, M., and N. J. White. 1999. Molecular mechanisms of cytoadherence in malaria. *Am. J. Physiol.* 45C:1231–1242.

48. Homer, M. J., I. Aguilar-Delfin, S. R. Telford III, P. J. Krause, and D. H. Persing. 2000. Babesiosis. *Clin. Microbiol. Rev.* 13:451–469.

49. Hommel, M., and H. M. Gilles. 1998. Malaria, p. 361–409. *In* F. E. G. Cox, J. P. Krier, and D. Wakelin (ed.), *Topley & Wilson's Microbiology and Microbial Infections,* 9th ed., Arnold, London, United Kingdom.

50. Hugosson, D. Tarimo, M. Troye-Blomberg, S. M. Montgomery, Z. Premji, and A. Björkman. 2003. Antipyretic, parasitologic, and immunologic effects of combining sulfadoxine/pyrimethamine with chloroquine or paracetamol for treating uncomplicated *Plasmodium falciparum* malaria. *Am. J. Trop. Med. Hyg.* 69:366–371.

51. Iqbal, J., N. Khalid, and P. R. Hira. 2002. Comparison of two commercial assays with expert microscopy for confirmation of symptomatically diagnosed malaria. *J. Clin. Microbiol.* 40:4675–4678.

52. Iqbal, J., A. Muneer, N. Khalid, and M. A. Ahmed. 2003. Performance of the Optimal test for malaria diagnosis among suspected malaria patients at the rural health centers. *Am. J. Trop. Med. Hyg.* 68:624–628.

53. Iqbal, J., A. Siddique, M. Jameel, and P. R. Hira. 2004. Persistent histidine-rich protein 2, parasite lactate dehydrogenase, and panmalarial antigen reactivity after clearance of *Plasmodium falciparum* monoinfection. *J. Clin. Microbiol.* 42:4237–4241.

54. Ittarat, W., A. L. Pickard, P. Rattanasinganchan, P. Wilairatana, S. Looareesuwan, K. Emery, J. Low, R. Udonsangpetch, and S. R. Meshnick. 2003. Recrudescence in artesunate-treated patients with falciparum malaria is dependent on parasite burden, not on parasite factors. *Am. J. Trop. Med. Hyg.* 68:147–152.

55. Jongwutiwes, S., C. Putaporntip, T. Iwasaki, T. Sata, and H. Kanbara. 2004. Naturally acquired *Plasmodium knowlesi* malaria in human, Thailand. *Emerg. Infect. Dis.* 10:2211–2213.

56. Kawai, S., S. Kano, and M. Suzuki. 1995. Rosette formation by *Plasmodium coatneyi*-infected erythrocytes of the Japanese macaque (*Macaca fuscata*). *Am. J. Trop. Med. Hyg.* 53:295–299.

57. Kochar, D. K., V. Saxena, N. Singh, S. K. Kochar, S. V. Kumar, and A. Das. 2005. *Plasmodium vivax* malaria. *Emerg. Infect. Dis.* 11:132–134.

58. Kotwal, R. S., R. B. Wenzel, R. A. Sterling, W. D. Porter, N. N. Jordan, and B. P. Petrucelli. 2005. An outbreak of malaria in U.S. Army Rangers returning from Afghanistan. *JAMA* 293:212–216. (Erratum, 293:678.)

59. Krause, P. J., K. McKay, J. Gadbaw, D. Christianson, L. Closter, T. Lepore, S. R. Telford III, V. Sikand, R. Ryan, D. Persing, J. D. Radolf, A. Spielman, and the Tick-Borne Infection Study Group. 2003. Increasing health burden of human babesiosis in endemic sites. *Am. J. Trop. Med. Hyg.* 68:431–436.

60. Krotoski, W. A., P. C. C. Garnham, R. S. Bray, D. M. Krotoski, R. Killick-Kendrick, C. C. Draper, G. A. T. Targett, and M. W. Guy. 1982. Observations on early and late post sporozoite tissue stages in primate malaria. 1. Discovery of a new latent form of *Plasmodium cynomolgi* (the hypnozoite), and failure to detect hepatic forms within the first 24 hours after infection. *Am. J. Trop. Med. Hyg.* 31:24–35.

61. Kuhn, S., M. J. Gill, and K. C. Kain. 2005. Emergence of atovaquone-proguanil resistance during treatment of *Plasmodium falciparum* malaria acquired by a non-immune North American traveler to West Africa. *Am. J. Trop. Med. Hyg.* 72:407–409.

62. Kurthals, J. A. L., C. M. Reimert, E. Tette, S. K. Dunyo, K. A. Koran, B. D. Akanmori, F. K. Nkrumah, and L. Hviid. 1998. Increased eosinophil activity in acute *Plasmodium falciparum* infection—association with cerebral malaria. *Clin. Exp. Immunol.* 112:303–307.

63. Kwiatkowski, D., and K. Marsh. 1998. Development of a malaria vaccine. *Lancet* 350:1696–1701.

64. Lefere, F., C. Besson, A. Datry, P. Chaibi, V. Leblond, J. L. Binet, and L. Sutton. 1996. Transmission of *Plasmodium falciparum* by allogeneic bone marrow transplantation. *Bone Marrow Transplant.* 18:473–474.

65. Lindblade, K. A., T. P. Eisele, J. E. Gimnig, J. A. Alaii, F. Odhiambo, F. O. ter Kuile, W. A. Hawley, K. A. Wannemuehler, P. A. Phillips-Howard, D. H. Rosen, B. L. Nahlen, D. J. Terlouw, K. Adazu, J. M. Vulule, and L. Slutsker. 2004. Sustainability of reductions in malaria transmission and infant mortality in western Kenya with use of insecticide-treated bednets: 4 to 6 years of follow-up. *JAMA* 291:2639–2641.

66. Lou, J., R. Lucas, and G. E. Grau. 2001. Pathogenesis of cerebral malaria: recent experimental data and possible applications for humans. *Clin. Microbiol. Rev.* 14:810–820.

67. Lyke, K. E., D. A. Diallo, A. Dicko, A. Kone, D. Coulibaly, A. Guindo, Y. Cissoko, L. Sangare, S. Coulibaly, B. Dakouo, T. E. Taylor, O. K. Doungo, and C. V. Plowe. 2003. Association of intraleukocytic *Plasmodium falciparum* malaria pigment with disease severity, clinical manifestations, and prognosis in severe malaria. *Am. J. Trop. Med. Hyg.* 69:253–259.

68. Magesa, S. M., K. Y. Mdira, A. Färnert, P. E. Simonsen, I. C. Bygbjerg, and P. H. Jakobsen. 2001. Distinguishing *Plasmodium falciparum* treatment failures from re-infections by using polymerase chain reaction genotyping in a holoendemic area in northeastern Tanzania. *Am. J. Trop. Med. Hyg.* 65:477–483.

69. Mayxay, M., S. Pukrittayakamee, K. Chotivanich, M. Imwong, S. Looareesuwan, and N. J. White. 2001. Identification of cryptic coinfection with *Plasmodium*

falciparum in patients presenting with vivax malaria. *Am. J. Trop. Med. Hyg.* **65**:588–592.

70. **Milhous, W. K., and D. E. Kyle.** 1998. Introduction to the modes of action of and mechanisms of resistance to antimalarials, p. 303–316. *In* I. W. Sherman (ed.), *Malaria: Parasite Biology, Pathogenesis, and Protection.* ASM Press, Washington, D.C.

71. **Mockenhaupt, F. P., U. Ulmen, C. von Gaertner, G. Bedu-Addo, and U. Bienzle.** 2002. Diagnosis of placental malaria. *J. Clin. Microbiol.* **40**:306–308.

72. **Moody, A.** 2002. Rapid diagnostic tests for malaria parasites. *Clin. Microbiol. Rev.* **15**:66-78.

73. **Mwenesi, H. A.** 2005. Social science research in malaria prevention, management and control in the last two decades: an overview. *Acta Trop.* **95**:292–297.

74. **National Committee for Clinical Laboratory Standards.** 2000. *Laboratory Diagnosis of Blood-Borne Parasitic Diseases.* Approved guideline M15-A. NCCLS, Wayne, Pa.

75. **Newton, C. R. J. C., T. E. Taylor, and R. O. Whitten.** 1998. Pathophysiology of fatal falciparum malaria in African children. *Am. J. Trop. Med. Hyg.* **58**:673–683.

76. **O'Donnell, J., J. M. Goldman, K. Wagner, G. Ehinger, N. Martin, M. Leahy, N. Kariuki, and I. Roberts.** 1998. Donor-derived *Plasmodium vivax* infection following volunteer unrelated bone marrow transplantation. *Bone Marrow Transplant.* **21**:313–314.

77. **Palmer, C. J., J. A. Bonilla, D. A. Bruckner, E. D. Barnett, N. S. Miller, M. A. Haseeb, J. R. Masci, and W. M. Stauffer.** 2003. Multicenter study to evaluate the OptiMAL test for rapid diagnosis of malaria in U.S. hospitals. *J. Clin. Microbiol.* **41**:5178–5182.

78. **Petersen, E.** 2004. Malaria chemoprophylaxis: when should we use it and what are the options? *Expert Rev. Anti-Infect. Ther.* **2**:119–132.

79. **Playford, E. G., and J. Walker.** 2002. Evaluation of the ICT malaria P.f/P.v and the OptiMAL rapid diagnostic tests for malaria in febrile returned travelers. *J. Clin. Microbiol.* **40**:4166–4171.

80. **Powell, V. I., and K. Grima.** 2002. Exchange transfusion for malaria and *Babesia* infection. *Transfus. Med. Rev.* **16**:239–250.

81. **Purnomo, A. Solihin, E. Gomez-Saladin, and M. J. Bangs.** 1999. Rare quadruple malaria infection in Irian Jaya Indonesia. *J. Parasitol.* **85**:574–579.

82. **Raina, V., A. Sharma, S. Gujral, and R. Kumar.** 1998. *Plasmodium vivax* causing pancytopenia after allogeneic blood stem cell transplantation in CML. *Bone Marrow Transplant.* **22**:205–206.

83. **Ringwald, P., F. S. Meche, and L. K. Basco.** 1999. Short report: effects of pyronaridine on gametocytes in patients with acute uncomplicated falciparum malaria. *Am. J. Trop. Med. Hyg.* **61**:446–448.

84. **Robert, L. L., P. D. Santos-Ciminera, R. G. Andre, G. W. Schultz, P. G. Lawyer, J. Nigro, P. Masuoka, R. A. Wirtz, J. Neely, D. Gaines, C. E. Cannon, D. Pettit, C. W. Garvey, D. Goodfriend, and D. R. Roberts.** 2005. *Plasmodium*-infected anopheles mosquitoes collected in Virginia and Maryland following local transmission of *Plasmodium vivax* malaria in Loudoun County, Virginia. *J. Am. Mosq. Control Assoc.* **21**:187–193.

85. **Rogerson, S. J., R. Tembenu, C. Dobano, S. Plitt, T. E. Taylor, and M. E. Molyneux.** 1999. Cytoadherence characteristics of *Plasmodium falciparum*-infected erythrocytes from Malawian children with severe and uncomplicated malaria. *Am. J. Trop. Med. Hyg.* **61**:467–472.

86. **Rosenblatt, J. E.** 1999. Antiparasitic agents. *Mayo Clin. Proc.* **74**:1161–1175.

87. **Salutari, P., S. Sica, P. Chiusolo, G. Micciulli, P. Plaisant, A. Nacci, A. Antinori, and G. Leone.** 1996. *Plasmodium vivax* malaria after autologous bone marrow transplantation: an unusual complication. *Bone Marrow Transplant.* **18**:805–806.

88. **Schmunis, G. A., and J. R. Cruz.** 2005. Safety of the blood supply in Latin America. *Clin. Microbiol. Rev.* **18**:12–29.

89. **Schneider, P., L. Wolters, G. Schoone, H. Schallig, P. Sillekens, R. Hermsen, and R. Sauerwein.** 2005. Real-time nucleic acid sequence-based amplification is more convenient than real-time PCR for quantification of *Plasmodium falciparum. J. Clin. Microbiol.* **43**:402–405.

90. **Shiff, C.** 2002. Integrated approach to malaria control. *Clin. Microbiol. Rev.* **15**:278–293.

91. **Shih, C. M., and C. C. Wang.** 1998. Ability of azithromycin in combination with quinine for the elimination of babesial infection in humans. *Am. J. Trop. Med. Hyg.* **59**:509–512.

92. **Siano, J. P., K. K. Grady, P. Mellet, and T. M. Wick.** 1998. Short report. *Plasmodium falciparum*: cytoadherence to $\alpha_v\beta_3$ on human microvascular endothelial cells. *Am. J. Trop. Med. Hyg.* **59**:77–79.

93. **Soto, J., F. Medina, N. Dember, and J. Berman.** 1995. Efficacy of permethrin-impregnated uniforms in the prevention of malaria and leishmaniasis in Colombian soldiers. *Clin. Infect. Dis.* **21**:599–602.

94. **Spudick, J. M., L. S. Garcia, D. M. Graham, and D. A. Haake.** 2005. Diagnostic and therapeutic pitfalls associated with primaquine-tolerant *Plasmodium vivax. J. Clin. Microbiol.* **43**:978–981.

95. **Supattamongkol, Y., S. Chindarat, S. Silpasakorn, S. Chaikachonpatd, K. Lim, K. Chanthapakajee, N. Kaewkaukul, and V. Thamlikitkul.** 2003. The efficacy of combined mefloquine-artesunate versus mefloquine-primaquine on subsequent development of *Plasmodium falciparum* gametocytemia. *Am. J. Trop. Med. Hyg.* **68**:620–623.

96. **Sweeney, C. J., M. Ghassemi, W. A. Agger, and D. H. Persing.** 1998. Coinfection with *Babesia microti* and *Borrelia burgdorferi* in a western Wisconsin resident. *Mayo Clin. Proc.* **73**:338–341.

97. **Talisuna, A. O., P. Bloland, and U. D'Alessandro.** 2004. History, dynamics, and public health importance of malaria parasite resistance. *Clin. Microbiol. Rev.* **17**:235–254.

98. **Todryk, S. M., and M. Walther.** 2005. Building better T-cell-inducing malaria vaccines. *Immunology* **115**:163–169.

99. **Tran, V. B., V. B. Tran, and K. H. Lin.** 1997. Malaria infection after allogeneic bone marrow transplantation in a child with thalassemia. *Bone Marrow Transplant.* **19**:1259–1260.

100. **Treutiger, C. J., J. Carlson, C. Scholander, and M. Wahlgren.** 1998. The time course of cytoadhesion, immunoglobulin

binding, rosette formation, and serum-induced agglutination of *Plasmodium falciparum*-infected erythrocytes. *Am. J. Trop. Med. Hyg.* **59**:202–207.

101. **Trigg, P. I., and A. V. Kondrachine,** 1998. The current global malaria situation, p. 11–22. *In* I. W. Sherman (ed.), *Malaria: Parasite Biology, Pathogenesis, and Protection.* ASM Press, Washington, D.C.

102. **Tsuji, M., Q. Wei, A. Zamoto, C. Morita, S. Arai, T. Shiota, M. Fujimagari, A. Itagaki, H. Fujita, and C. Ishihara.** 2001. Human babesiosis in Japan: epizooiologic survey of rodent reservoir and isolation of new type of *Babesia microti*-like parasite. *J. Clin. Microbiol.* **39**:4316–4322.

103. **Turkmen, A., M. S. Sever, T. Ecder, A. Yildiz, A. E. Aydin, R. Erkoc, H. Eraksoy, U. Eldegez, and E. Ark.** 1997. Posttransplant malaria. *Transplantation* **62**:1521–1523.

104. **van Velthuysen, M. L. F., and S. Florquin.** 2000. Glomerulopathy associated with parasitic infections. *Clin. Microbiol. Rev.* **13**:55–66.

105. **Varney, N. R., R. J. Roberts, J. A. Springer, S. K. Connell, and P. S. Wood.** 1997. Neuropsychiatric sequelae of cerebral malaria in Vietnam veterans. *J. Nerv. Ment. Dis.* **185**:695–703.

106. **Wever, P. C., Y. M. C. Henskens, P. A. Kager, J. Dankert, and T. van Gool.** 2002. Detection of imported malaria with the Cell-Dyn 4000 hematology analyzer. *J. Clin. Microbiol.* **40**:4729–4731.

107. **White, N. J.** 1998. Malaria pathophysiology, p. 371–385. *In* I. W. Sherman (ed.), *Malaria: Parasite Biology, Pathogenesis, and Protection.* ASM Press, Washington, D.C.

108. **Wilkinson, R. J., J. L. Brown, J. L., G. Pasvol, P. L. Chiodini, and R. N. Davidson.** 1994. Severe falciparum malaria: predicting the effect of exchange transfusion. *Q. J. Med.* **87**:553–557.

109. **Win, T. T., A. Jalloh, I. S. Tantular, T. Tsuboi, M. U. Ferreira, M. Kimura, and F. Kawamoto.** 2004. Molecular analysis of *Plasmodium ovale* variants. *Emerg. Infect. Dis.* **10**:1235–1240.

110. **Winstanley, P., S. Ward, R. Snow, and A. Breckenridge.** 2004. Therapy of falciparum malaria in sub-Saharan Africa: from molecule to policy. *Clin. Microbiol. Rev.* **17**:612–637.

111. **World Health Organization.** 2005. *World Malaria Report, 2005.* World Health Organization, Geneva, Switzerland. (http://rbm.who.int/wmr2005/html/exsummary_en.htm)

112. **Yeung, S., W. Pingtavornpinyo, I. M. Hastings, A. J. Mills, and N. J. White.** 2004. Antimalarial drug resistance, artemisinin-based combination therapy, and the contribution of modeling to elucidating policy choices. *Am. J. Trop. Med. Hyg.* **71**:179–186.

113. **Zintl, A., G. Mulcahy, H. E. Skerrett, S. M. Taylor, and J. S. Gray.** 2003. *Babesia divergens,* a bovine blood parasite of veterinary and zoonotic importance. *Clin. Microbiol. Rev.* **16**:622–636.

Leishmaniasis

Old World leishmaniasis

Cutaneous

Visceral

New World leishmaniasis

Cutaneous

Visceral

Leishmania spp. are protozoa belonging to the family Trypanosomatidae (Table 8.1). They are obligate intracellular parasites that are transmitted to the mammalian host by the bites of infected sand flies. On the basis of development in sand flies, the genus *Leishmania* has been divided into two subgenera. In the subgenus *Leishmania*, the development or organisms is restricted to the anterior portion of the alimentary tract of the sand fly, while in the subgenus *Viannia*, the parasites develop in the midgut and hindgut of the sand fly.

Leishmaniasis is mainly a zoonosis, although in certain areas of the world there is primarily human-vector-human transmission. The World Health Organization estimates that 1.5 million cases of cutaneous leishmaniasis (CL) and 500,000 cases of visceral leishmaniasis (VL) occur every year in 88 countries. Estimates indicate that there are approximately 350 million people at risk for acquiring leishmaniasis, with 12 million currently infected (25).

Leishmaniasis refers to a diverse group of diseases, with the spectrum of disease depending on the infecting species. Disease syndromes range from self-healing cutaneous lesions to debilitating mucocutaneous infections, subclinical viscerotropic dissemination, and fatal visceral involvement. Since disease presentations vary considerably, a well-defined classification system based on clinical findings is difficult and sometimes confusing (6, 14, 16, 20, 30, 39, 59–61, 82). With recent outbreaks in many areas of the world, including Brazil, India, Italy, Spain, Sudan, and Kenya, leishmaniasis has become more widely recognized as an important emerging infectious disease in many developed as well as underdeveloped countries.

Cases of leishmaniasis are diagnosed each year in the United States and can be attributed to immigrants from countries with endemic infection, military personnel, and American travelers. Another concern is the potential for more infections occurring in areas of endemic infection in Texas and Arizona (37, 50). Organisms can remain latent for years, and even when the potential exposure history is in the past, leishmaniasis should be considered, particularly in immunocompromised patients such as those infected with human immunodeficiency virus (HIV) (19, 21, 53, 58, 61, 72, 76).

The taxonomy of the *Leishmania* spp. is controversial and reflects the development of multiple parameters on which classification is based. Previously, organisms were classified based on the clinical disease picture and geographic

Table 8.1   Taxonomic classification of *Leishmania* parasites pathogenic for humans[a]

Taxon	Old World[b]				New World[c]			
Family				Trypanosomatidae				
Genus				*Leishmania*				
Subgenus		*Leishmania*			*Leishmania*		*Viannia*	
Complex	*L. major*	*L. tropica*	*L. aethiopica*	*L. donovani*	*L. donovani*	*L. mexicana*	*L. braziliensis*	*L. guyanensis*
Species	*L. major*	*L. tropica*	*L. aethiopica*	*L. donovani*	*L. chagasi*	*L. mexicana*	*L. braziliensis*	*L. guyanensis*
				L. infantum		*L. venezuelensis*	*L. peruviana*	*L. panamensis*
						L. garnhami	*L. colombiensis*	
						L. amazonensis	*L. lainsoni*	
						L. pifanoi	*L. shawi*	
							L. naiffi	
Hybrid species[d]							*L. braziliensis/L. panamensis*	
							L. braziliensis/L. guyanensis	

[a] The classification scheme is based on biochemical and molecular characteristics.
[b] Animal parasites of the Old World (*L. arabica*, *L. turanica*, and *L. gerbilli*) are not listed.
[c] Animal parasites of the New World (*L. aristedesi*, *L. enrietti*, *L. forattinii*, *L. deanei*, *L. equatorensis*, and *L. bertigi*) are not shown; although human infection with these species has not yet been identified, there is often a time lag between the discovery of new animal genera and confirmation of human infection. In many cases, the relevant vectors have been found to bite humans as well as the normal animal hosts.
[d] Strains with phenotypic and genotypic characters of two species have been recorded in different geographic areas of Latin America; either they represent strains that originated from a common ancestor or they are the result of genetic exchange.

locations; with the development of criteria based on biochemical and genetic data, the taxonomic groupings no longer coincide with recognized disease entities.

Old World Leishmaniasis: Cutaneous Leishmaniasis

Leishmania major, *L. tropica*, *L. aethiopica*, and, rarely, *L. infantum* cause cutaneous disease in the Old World; disease manifestations include nodular and ulcerative skin lesions (Table 8.2; Figure 8.1). Local geographic names for CL include Oriental sore, Baghdad boil, Delhi boil, Biskra button, and Aleppo evil. Although Wright was credited with the first detailed description of the cutaneous form of the disease, it had already been described by Borovsky in 1898 in an obscure military medical journal.

Life Cycle and Morphology

The life cycle can be seen in Figure 8.2. The parasite is found in two morphologic forms in the life cycle, the amastigotes and the promastigotes. The amastigotes are small (3 to 5 μm in diameter), ovoid, nonmotile intracellular forms, while the promastigotes are elongated, motile, extracellular stages (1.5 to 3.5 by 15 to 20 μm with a single free flagellum 15 to 28 μm long) (Figure 8.3). The form introduced into the skin of the mammalian host by the sand fly is the promastigote. The organism is engulfed by reticuloendothelial cells (RE cells) of the mammalian host, where the parasite transforms into the intracellular amastigote form (Leishman Donovan body). Apparently, this process takes approximately 12 to 24 h. The large nucleus and small kinetoplast in the amastigotes can be seen in tissue after staining with Giemsa or Wright's stain. The short intracytoplasmic portion of the flagellum can also be seen within some of the amastigotes. The amastigotes multiply by binary fission within the macrophage until the cell is destroyed, and the released parasites are phagocytized by other RE cells or ingested by sand flies. This phase is chronic and can last from months to years, even a lifetime, depending on the species involved. In CL, the amastigotes remain confined to the skin, usually as skin lesions such as raised papules or ulcers. Lesions usually occur on exposed parts of the body such as the face, hands, feet, arms, and legs; uncommon sites include the ears, tongue, and eyelids.

During a blood meal taken by the sand fly (genus *Phlebotomus*), amastigotes are ingested and transform into the promastigote stage. Promastigotes multiply by longitudinal fission in the insect gut. Stages found in the sand fly vary from rounded or stumpy forms to elongated, highly motile metacyclic promastigotes. The metacyclic promastigotes migrate to the hypostome of the sand fly, where they are inoculated into humans when the sand fly

Table 8.2   Old World cutaneous leishmaniasis, subgenus *Leishmania*

Species and sand fly vector	Human disease and lesion	Geographic distribution	Important hosts	Comments
L. major: *Phlebotomus papatasi, P. duboscqi, P. salehi*	Cutaneous leishmaniasis (Oriental sore; rural, zoonotic); lesions may number >100; wet lesion with crusting; necrotic and exudative	North Africa, sub-Saharan Africa, Central and West Asia	Great gerbil (*Rhombomys opimus*), fat sand rat (*Psammomys obesus*)	Primarily an infection of desert rodents; gerbil infection rate may reach 30%; village inhabitants near gerbil burrows may have infection prevalence rates of 100%; major problem for rural settlers in Israel, military personnel in the Sinai Desert, Israel, Iraq, and Iran
L. tropica: *Phlebotomus sergenti, P. papatasi, P. chaudaudi*	Cutaneous leishmaniasis (Oriental sore; urban, anthroponotic); lesions usually ~2; lesions more swollen, less necrotic; less exudate than with *L. major*; leishmaniasis recidivans also seen (chronic lesions on face or other exposed areas, persist for years)	Middle East, Mediterranean littoral, India, Pakistan, Central Asia, Iran, Iraq	Dogs and humans	Major epidemic of anthroponotic CL has occurred in war-ravaged Kabul, Afghanistan, with thousands of cases identified; responsible for viscerotropic syndrome in a small group of American military personnel infected in the Persian Gulf War; extensive scarring can lead to disfigurement; can last for up to 40 years
L. aethiopica: *Phlebotomus longipes, P. pedifer*	Cutaneous leishmaniasis, diffuse cutaneous leishmaniasis; lesions more swollen, less necrotic; barely exudative	Ethiopia, Kenya, bordering countries	Rock hyraxes (*Heterohyrax brucei, Procavia* spp.)	Human disease usually self-healing, small number develop nonhealing disseminated CL

takes its next blood meal. Various surface molecules on the promastigote, such as gp63 and lipophosphoglycans, help bind the parasite to the host macrophage receptors, thus allowing it to be phagocytized. Depending on the species, the duration of the life cycle in the sand fly varies from 4 to 18 days.

Sand flies are pool feeders and possess cutting mouthparts that slice into the skin, allowing a small pool of blood to form. Thus, although they are small flies (2 to 3 mm), the bite can be painful. In most endemic foci, the vast majority of sand flies are uninfected; however, those that are infected are very efficient vectors. Infected flies probably remain so for life, although this is just a matter of weeks. Under optimal conditions, transformation of the amastigotes to promastigotes generally takes 5 to 7 days; by that time, the female phlebotomine fly is ready to take her next blood meal.

Leishmaniavirus is a double-stranded RNA virus that persistently infects some strains of *Leishmania*. There is a great deal of interest in the possibility that the presence of this virus may be able to alter the parasite phenotype and may affect disease pathogenesis. The virus has been detected in cultured parasites and has also been detected in human biopsy samples prior to manipulation in culture (68).

Clinical Disease

The incubation period of CL is usually 2 to 8 weeks but can be as long as several years. The first sign of cutaneous disease is a lesion (generally a firm, painless papule) at the bite site. Although a single lesion may appear insignificant, multiple lesions of *L. major* or disfiguring facial lesions of *L. tropica* may be psychologically or physically devastating. In general, all lesions on a patient have a similar appearance and progress at the same speed.

The original lesion may remain as a flattened plaque or may progress, with the surface becoming covered with fine, papery scales. These scales are dry at first but later become moist and adherent, covering a shallow ulcer. As the ulcer enlarges, it produces exudate and may develop a crust. The edge of the lesion is usually raised. Depending on the species, satellite lesions are common and merge with the original lesion (Figure 8.4).

***L. major* (Rural Disease).**   *L. major* causes rural disease, a variety of CL. Multiple lesions are present and are

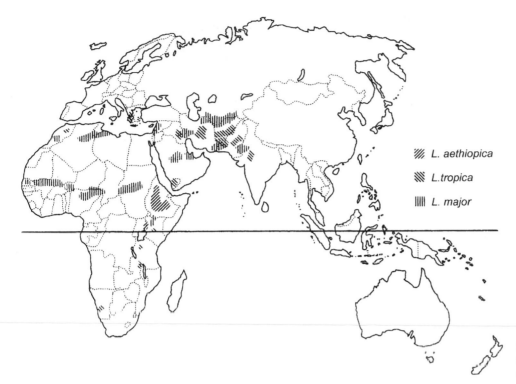

L. aethiopica

L. tropica

L. major

Figure 8.1 Geographic distribution of the agents of Old World cutaneous leishmaniasis.

accompanied by marked inflammation and crusting. The center of the lesion tends to be necrotic and exudative, forming a loose crust. The lesions tend to mature and heal relatively quickly, often lasting only a few months. Lymphatic spread may occur in *L. major* infections, with subcutaneous nodules in a linear distribution and regional lymphadenopathy; if the initial lesion is on the hand, this clinical presentation may resemble sporotrichosis. *L. major* infections may result in severe scarring.

From August 2002 to February 2004, over 500 cases of parasitologically confirmed cases of CL were found in U.S. military personnel serving in Afghanistan, Iraq, and Kuwait. The majority of these cases were probably acquired in Iraq, with *L. major* being confirmed for most of the cases through isoenzyme electrophoresis of cultured

parasites (16, 83). Based on the data from Fort Campbell, Ky., approximately 1% of troops returning from Iraq were diagnosed with CL, most by laboratory confirmation, including PCR.

L. tropica (Urban Disease). The lesions of urban disease (a type of CL) caused by *L. tropica* occur singly or two at a time, develop more slowly than those caused by *L. major*, and can take a year or more to heal. They tend to be more swollen and less necrotic, with less exudate and a thicker crust.

L. tropica (Leishmaniasis Recidivans). Leishmaniasis recidivans is generally associated with *L. tropica*. Small, nonulcerating lesions begin to appear, mainly on the

Figure 8.2 *Leishmania* life cycle.

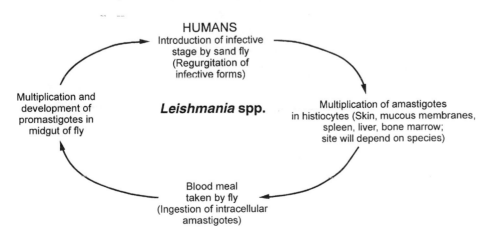

HUMANS
Introduction of infective
stage by sand fly
(Regurgitation of
infective forms)

Multiplication and
development of
promastigotes in
midgut of fly

***Leishmania* spp.**

Multiplication of amastigotes
in histiocytes (Skin, mucous membranes,
spleen, liver, bone marrow;
site will depend on species)

Blood meal
taken by fly
(Ingestion of intracellular
amastigotes)

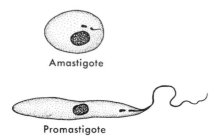

Figure 8.3 *Leishmania* stages found in human and insect hosts. (Illustration by Nobuko Kitamura.)

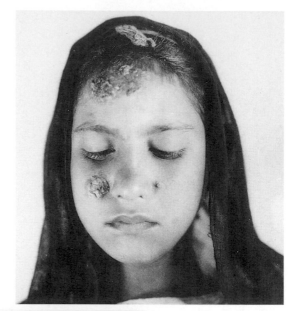

margins of healed lesions, and continue to expand the limits of the original scar. Disfigurement due to extensive scarring can develop. This condition is thought to be associated with a strong cell-mediated immune response and tends to occur primarily in Iran and Iraq. Although there is a strong immune response on the part of the host, not all amastigotes are eliminated, and the infection can last for up to 40 years. The patient's physical appearance, as well as histologic sections, resembles cutaneous tuberculosis; therefore, the name "lupoid leishmaniasis" has also been used. Parasites may be very difficult to demonstrate; however, culture or animal inoculation has been successful. The leishmanin skin test (LST) (Montenegro test) is strongly positive in these patients.

***L. aethiopica* (Cutaneous Leishmaniasis).** Lesions of CL caused by *L. aethiopica* are the least inflamed but are more swollen. There may be very little exudate, but scaling and exfoliation of the dermis occur. These lesions tend to remain for years prior to complete healing.

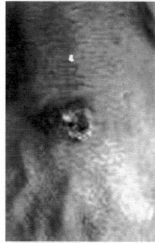

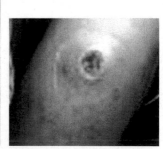

***L. aethiopica* (Diffuse Cutaneous Leishmaniasis).** Diffuse "disseminated" CL (DCL) is an anergic form of CL in which neither humoral nor cell-mediated immune responses are functional; however, delayed hypersensitivity to tuberculin appears to be normal. DCL is caused by *L. aethiopica* and is seen in Ethiopia and Kenya; approximately 100 cases have been reported. The parasites are found in the skin, and the disease is characterized by the formation of thickened skin in plaques, papules, or multiple nodules. This condition is similar to that associated with the specific anergy in lepromatous leprosy, and DCL was originally confused with and thought to be leprosy. Although lesions are usually seen on the face, nose, limbs, and buttocks, they can occur over the entire body. The nodules are often described as soft and fleshy, while those of leprosy are generally more indurated. Although organisms have rarely been recovered from blood or bone marrow, there are no visceral lesions associated with this form of leishmaniasis. Treatment is difficult,

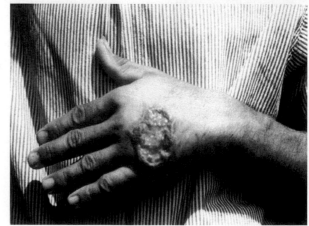

Figure 8.4 (Top) Cutaneous leishmaniasis facial lesion. (Armed Forces Institute of Pathology photograph.) (Middle) Examples of cutaneous lesions on the hand and arm. (Bottom) Extensive cutaneous lesion on the hand.

and the infection may last for the life of the patient. The LST remains negative unless treatment is effective and recovery occurs.

Diagnosis

In areas of the world where physicians are very familiar with leishmaniasis, the diagnosis may be made on clinical grounds. However, in other areas of the world, where the disease is rare, the condition may not be recognized. Definitive diagnosis depends on demonstrating the amastigotes in tissue specimens or the promastigotes in culture (Table 8.3). CL may have to be differentiated from a number of other lesions and diseases, including basal cell carcinoma, tuberculosis, various mycoses, cheloid, and lepromatous leprosy.

Specimen Collection. CL normally presents with no visceral involvement or systemic disease. The ability to detect parasites in aspirates, scrapings, or biopsy specimens depends on the number of amastigotes present, the level of the host immune response, the absence or presence of bacterial and/or fungal contamination within the lesion, and whether the specimen is collected from an active or healing lesion. If the patient has multiple lesions, specimens should be collected from the more recent or active lesions. These lesions should be thoroughly cleaned with 70% alcohol, and necrotic debris should be removed to prevent the risk of bacterial and/or fungal contamination of the specimen. This is mandatory if cultures are to be used for isolation and growth of the parasites; excess bacterial or fungal contamination prevents development of the promastigote forms. Also, the

specimen should be taken from the advancing margin of the lesion; the central portion of the ulcer contains nothing but necrotic debris.

If specimens are collected by aspiration, scraping, or biopsy, local anesthesia may be used to reduce patient discomfort. The specimen of choice would be a collection of several punch biopsy specimens taken from the most active lesion areas. Biopsy specimens can be divided and used for cultures and touch preparations; some material should always be saved and submitted for routine histology. If biopsy specimens need to be divided, material for culture should be obtained by using sterile techniques and instruments. The biopsy specimen can be used to make impression smears (touch preparations); these smears should be prepared after portions of the specimen have been placed in culture media by using a sterile technique. After any excess blood is cleaned off, a horizontal cut is made through the biopsy core and the cut surface is gently touched to glass slides; multiple slides should be prepared. Once material has been set up for culture and touch preparations have been made, the remainder of the tissue can be sent to the pathology laboratory for routine processing. It is important to remember that the amastigotes are more difficult to identify in tissue sections than in touch preparations; they appear smaller and may be cut at different angles.

Another option involving sterile techniques is to make a perpendicular slit in the margin of the lesion. After blood is removed with sterile gauze, the blade can be used to scrape the inner walls of the slit. Material can be inoculated into culture media and can be smeared onto several glass slides; the use of multiple slides is recommended to increase the chances of seeing the

Table 8.3 Features of Old World human leishmanial infections related to demonstration of the parasite[a]

Organism	Disease type[b]	Humoral antibodies	Delayed hypersensitivity to leishmanin	Parasite present in clinical specimen	Self-cure	Recommended biopsy specimen
Old World CL						
L. major	CL	Present	Present	Present	Rapid	Skin
L. tropica	CL	Variable	Present	Present	Yes	Skin
	LR	Variable	Strong	Few	Slow	Skin
L. aethiopica	CL	Variable	Weak	Present	Slow	Skin
	DCL	Variable	Absent	Abundant	No	Skin
Old World VL						
L. donovani	VL	Abundant	Absent	Present	Rare	Bone marrow, spleen
	CL (primary lesion)	Variable	Present	Present	Yes	Skin
	PKDL	Variable	Variable	Variable	Variable	Skin
L. infantum	VL	Abundant	Absent	Present	Variable	Bone marrow, spleen

[a] For culture, specimens must be collected aseptically, and in older lesions the parasites may be scant and difficult to recover. Animal isolation involves hamsters and culture media (Novy-MacNeal-Nicolle medium and Schneider's *Drosophila* medium with 30% fetal bovine serum). Serologic testing is most suitable for VL, has little value for CL, and has limited value for MCL. The Montenegro test (LST) is a delayed-hypersensitivity reaction to intradermal injection of cultured parasites.

[b] Additional information on relevant diagnostic methods can be found in chapters 31 and 32.

parasites. Fine-needle aspiration can also be performed with a sterile syringe containing sterile saline (0.1 ml) and a 26-gauge needle. The needle is inserted into the outer border of the lesion and rotated several times, after which saline is injected and tissue fluid is aspirated back into the needle. Material can be submitted for culture, and/or smears can then be prepared for staining. Additional techniques and specific methods can be found in chapters 30 to 32.

Specimen Staining. All smears to be Giemsa stained should be air dried and fixed with 100% methanol prior to staining. Other blood stains can also be used. The amastigotes are found within the macrophages or close to disrupted cells. Although this intracellular form may mimic other intracellular parasites (*Histoplasma capsulatum* and some of the microsporidia), the presence of the intracytoplasmic kinetoplast confirms the identification. After staining, the cytoplasm appears light blue and the nucleus and kinetoplast appear red to purple. In very early and older lesions, very few organisms may be present, so that prolonged examination of the slide may be required.

Culture Media. Media used for the growth of *Leishmania* spp. will also support the growth of bacteria and fungi, hence the need for aseptic techniques in collecting the clinical specimen. If cultures become overgrown with contaminating organisms, the parasites will not survive. It is also important to release the organisms from tissue prior to culture; mincing is recommended. However, tissue grinders are not recommended because of possible damage to the organisms. Culture media that have been used for the recovery and growth of leishmaniae include Novy-MacNeal-Nicolle medium and Schneider's *Drosophila* medium supplemented with 30% fetal bovine serum (see chapter 32). A number of newer medium formulations have also been developed and may prove very useful. These include a completely defined medium without serum and macromolecules (52), a semisynthetic fetal calf serum-free liquid medium (2), a new liquid medium with less expensive chemicals (40), and a sensitive new microculture method involving microcapillary tubes (3). Patient cultures should not be set unless the laboratory maintains specific organism strains for quality control checks. Both the control and patient cultures should be examined twice a week for the first 2 weeks and once a week thereafter for up to 4 weeks before they are reported as negative. Promastigote stages can be detected microscopically in wet mounts taken from centrifuged culture fluid. This material can also be stained with Giemsa stain to facilitate observation at a higher magnification. A number of molecular methods have been developed for species identification of the promastigotes, including the use of

DNA probes, isoenzyme analysis, PCR, and monoclonal antibodies (18, 46, 62, 84).

Animal Inoculation. Animal inoculation can be helpful when only a small number of organisms is present; however, most laboratories no longer consider this approach practical. Golden hamsters are inoculated intranasally, and the major argument against this approach is that it may take several months before the animal gives positive responses. A combination of tissue smears, culture, animal inoculation, and histopathology may be required for optimal diagnosis.

Skin Test. The LST (Montenegro test) is useful for epidemiologic population surveys to detect groups at risk for infection. However, the test is not species specific and must be interpreted with caution, particularly since some patients may have more than one type of leishmaniasis. Apparently, cross-reactions do not occur with African or American trypanosomiasis. The LST involves a delayed-hypersensitivity reaction to a suspension of killed promastigotes. The reaction is read at 48 and 72 h after injection. Positive reactions are generally seen in patients with cutaneous lesions; however, those with DCL give negative reactions.

Serologic Tests. Although available for epidemiology and/or research studies, serologic tests are not very useful for the diagnosis of cutaneous leishmaniasis. Reagent availability is usually limited to areas of endemic infection. Antibody detection is variable in patients with CL; if antibody is present, titers are usually low, and cross-reactions have been seen in patients with leprosy. Patients infected with the viscerotropic form of *L. tropica* seen during Operation Desert Storm have had low or undetectable antibody titers (26).

KEY POINTS—LABORATORY DIAGNOSIS

Old World Leishmaniasis: Cutaneous Leishmaniasis

1. For a complete laboratory diagnosis, microscopic examination of Giemsa-stained touch preparations and cultures is recommended.
2. Aspirates or biopsy specimens should be taken from the margin or base of the most active lesion (papule or ulcer). The lesion should be cleaned before the sample is collected, to reduce the chances of contamination with fungi or bacteria.
3. Multiple slides should be prepared for examination.
4. Tissue imprints or smears should be stained with Giemsa stain (or one of the stains commonly used for blood smears) (see chapter 31).

5. Amastigote stages should be found within macrophages or close to disrupted cells.
6. If cultures are to be performed, specimens must be taken aseptically and control organism cultures should be set up at the same time (see chapter 32).
7. Cultures should be checked weekly for 4 weeks before they are declared negative.

Treatment

Most Old World CL lesions are self-healing over a period of a few months to several years. This healing process confers immunity to individuals living within the immediate area of endemic infection; therefore, treatment is not recommended unless disfiguring scarring is a possibility. Chemotherapy should be given to patients with lesions on the face, particularly if the lesions are multiple or large. Pentavalent antimonials (sodium stibogluconate [Pentostam] and meglumine antimoniate [Glucantime]) are usually effective but must be given in adequate doses for the complete recommended time frame, often weeks or months (see chapter 25). Toxicity does not usually present a problem. Specific chemotherapy, including systemic administration, is used much less frequently for Old World leishmaniasis than for the New World disease presentations. The lesions of leishmaniasis recidivans respond to Pentostam at doses used for treatment of VL, and local heat therapy and intralesional injections are commonly used in the Middle East. In treating DCL, pentamidine, given as a single weekly injection for at least 4 months longer than it takes to eliminate parasites from slit skin smears, seems to be a good compromise between efficacy and toxicity. Cutaneous lesions, whether treated or not, should always be cleaned and covered to prevent secondary infections and vector access to the lesion.

In addition to chemotherapy, other more local treatment options have included cryotherapy, heat, and surgical excision of lesions (83). Organisms within the *L. tropica* group do not survive in temperatures above 37°C, and local heat therapy has proven to be effective. The temperature within the lesion should reach 40 to 42°C for 12 h at a time; this can be accomplished by using an infrared heat lamp. The use of hot water in baths or saunas has been helpful in cases of DCL, particularly since the response to chemotherapy is poor.

The administration of Pentostam as direct injections into the lesion has also been used. The dose is divided into four parts and injected into four different quadrants of the lesion (59). Another approach involves the application of a topical formulation of 15% paromomycin–12% methylbenzethonium chloride in soft white paraffin (P ointment or the "El-On" preparation), which decreases the healing time by half. Since this preparation causes multiple local reactions, a number of other preparations have been tried, some effective and some no better than the placebo. The exact formula and optimal dose have not yet been finalized; however, formulations containing paromomycin appear to be very promising (73).

Epidemiology and Prevention

L. major, the cause of rural Oriental sore, is a zoonotic parasite and is found primarily in desert rodents such as gerbils that live in the dry desert areas of central Asia, in southern Russia, throughout Iran and Pakistan, in parts of Iraq, and in northwestern India and China. In areas of Libya, Saudi Arabia, and Israel, the rat is an important host. The infection is maintained in the reservoir population by sand flies that frequent rodent burrows. Human disease is maintained through the vectors *Phlebotomus papatasi* in Sudan and *P. duboscqi* elsewhere.

L. tropica, the cause of urban CL, has been isolated from dogs and rats; however, the human is considered to be the primary reservoir and source of infection. Various phlebotomine sand flies (*P. sergenti* and *P. papatasi*) transmit the infection, depending on the specific geographic area; areas of endemic infection include southern France, Italy, some Mediterranean islands, the Middle East, Greece, Pakistan, and northwestern India. In the first half of the 20th century, 40 to 70% of Europeans living in Delhi became infected. A major epidemic of CL due to *L. tropica* has occurred in war-ravaged Kabul, Afghanistan, with thousands of cases identified in the displaced population.

L. aethiopica is also a zoonotic parasite, primarily involving the rock hyrax (*Procavia habessinica*) and the tree hyrax (*Heterohyrax brucei*). The organism is found in the Rift Valley in Ethiopia and Kenya. As human habitation moves into deforested mountain areas, exposure to the hyrax becomes much greater and coincides with the vector, *P. longipes*, while *P. pedifer* is implicated in Kenya.

Preventive measures involve vector control; residual spraying with DDT as a means of malaria control was very helpful. However, with the decline and total cessation of spraying in many areas, the leishmaniasis has returned. The infection can also be transmitted by direct contact with infected lesions and mechanically through bites by the stable or dog fly (*Stomoxys* spp.); lesions should be kept covered to prevent this type of transmission by primary or secondary vectors. Application of insect repellents to the skin has also been effective. Although reservoir control has been unsuccessful in most areas, disease incidence has been reduced by poisoning or deep plowing gerbil burrows.

In areas of Russia, Israel, Iran, Iraq, and Jordan where CL is endemic, vaccination has been used for many years. Exudate from a naturally acquired lesion can be inoculated onto an inconspicuous area of the body of a nonimmune person. Another approach is to use live, attenuated *L. major* promastigotes, which are inoculated intracutaneously. The infection is allowed to develop and is usually limited to a single lesion; self-cure proceeds as in a natural infection. The immunized individual will become a carrier and source of infection until the lesion heals; a very small percentage of those vaccinated will develop lesions requiring treatment. Lifelong immunity to the homologous species develops and is comparable to that following a natural infection.

Old World Leishmaniasis: Visceral Leishmaniasis

VL (kala azar or dumdum fever) is caused by *L. donovani* and *L. infantum* in the Old World (Figure 8.5). *L. tropica* also causes classical kala azar in India and causes viscerotropic leishmaniasis in the Middle East. The two main species tend to cause similar diseases, although *L. infantum* mainly affects young children and has a greater tendency to cause lymph node enlargement. The course of the disease and outcome depend on the general health status of the patient at the time of infection; apparently malnutrition exacerbates *L. infantum* infection (9). Dermotropic strains of *L. infantum* are known to cause visceral infection in immunocompromised patients with AIDS. VL is very prevalent among AIDS patients in certain geographic areas (e.g., southern

Spain), with a high proportion of cases being subclinical. Like other opportunistic infections, subclinical VL can be found at any stage of HIV-1 infection, but symptomatic cases appear mainly when severe immunosuppression is present. HIV-*Leishmania* coinfection is being seen more and more frequently in the Mediterranean basin, especially in Spain, France, and Italy.

Life Cycle and Morphology

The life cycle and morphology of the organisms are essentially the same as for those causing Old World CL (Figure 8.2). However, amastigotes can be found throughout the body and are not confined to the macrophages of the skin. Organisms inoculated into the bite site by the sand fly are engulfed by tissue macrophages, which then enter the bloodstream and are carried to RE centers such as the bone marrow, spleen, and liver (Figures 8.6 and 8.7).

Clinical Disease

Visceral Leishmaniasis. Once the organisms have been carried to the bone marrow, spleen, and liver, a granulomatous cell-mediated immune response occurs that can result in subclinical disease and self-cure or in the clinical syndrome of VL (Table 8.4). The incubation period usually ranges from 2 to 6 months; however, it can take a few days or weeks to several years. When giving a history, many patients do not remember having a primary skin lesion. In areas of endemic infection, the onset may be gradual, with vague symptoms and a general feeling of ill health. The

Figure 8.5 Geographic distribution of the agents of Old World visceral leishmaniasis.

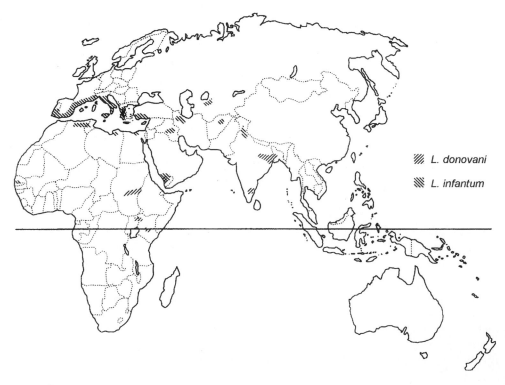

L. donovani

L. infantum

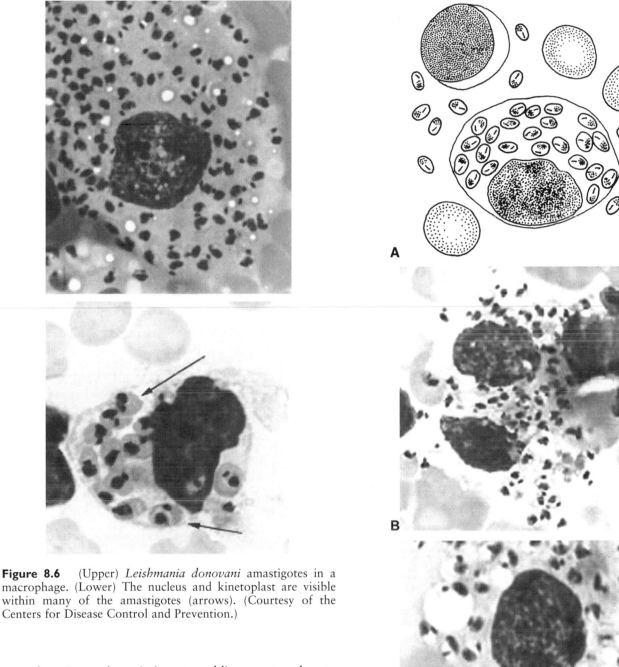

Figure 8.6 (Upper) *Leishmania donovani* amastigotes in a macrophage. (Lower) The nucleus and kinetoplast are visible within many of the amastigotes (arrows). (Courtesy of the Centers for Disease Control and Prevention.)

onset in naïve patients (migrants, soldiers, or travelers to areas of endemic infection) may be acute, with high fever, chills, anorexia, malaise, weight loss, and, frequently, diarrhea. This syndrome can easily be confused with typhoid fever, malaria, or other febrile illnesses caused by bacteria or viruses. Death may occur after a few weeks or several years. Generally, untreated VL leads to death; secondary bacterial and viral infections are also common in these patients.

In areas where VL is endemic, many people develop a positive LST with no history of actual clinical disease; these individuals appear to be resistant to infection with *L. donovani*. The difference between an asymptomatic

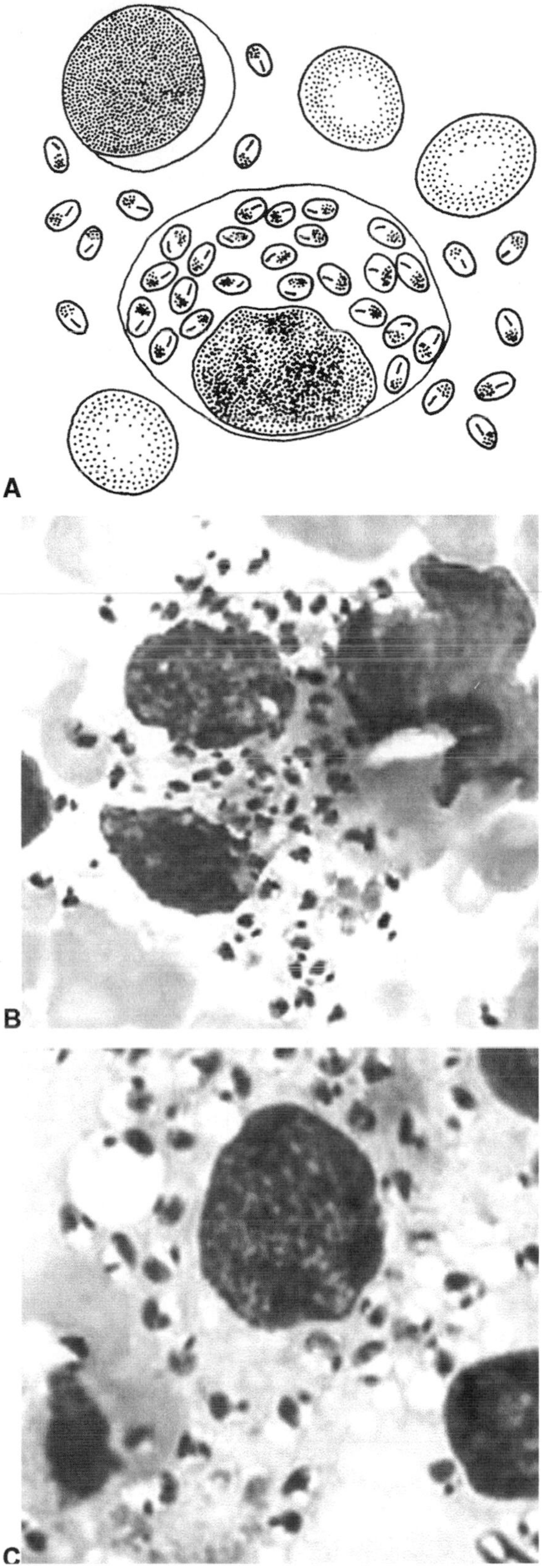

Figure 8.7 (A) *Leishmania donovani* amastigotes in bone marrow (illustration by Sharon Belkin); (B and C) *L. donovani* amastigotes in touch preparations (tissue imprints).

Table 8.4 Old World visceral leishmaniasis

Species and sand fly vector	Human disease and lesion	Geographic distribution	Important hosts	Comments
L. donovani: *P. argentipes*, *P. orientalis*, *P. martini*	Visceral leishmaniasis, kala azar, post-kala azar dermal leishmaniasis	Indian subcontinent, East Africa	Humans	Characterized by irregular fever, splenomegaly, hepatomegaly, weight loss, pancytopenia, and hypergam-maglobulinemia; statistically significant differences between number of organisms in peripheral blood taken during the evening and during the day (66 and 46%, respectively), which coincides with biting habit of sand fly vector (night biting) (69)
L. infantum: *P. ariasi*, *P. perniciosus*	Infantile visceral leishmaniasis, cutaneous leishmaniasis (rare)	Mediterranean basin, central and western Asia	Domestic dogs	Mainly affects young children and tends to cause lymph node enlargement; poor nutrition exacerbates infection

infection and subclinical disease is difficult to establish, particularly since many of these patients do not seek medical care. However, "viscerotropic" leishmaniasis may be more common than was previously thought and should be considered in individuals with unexplained mild symptoms (cough, malaise, chronic fatigue, abdominal pain, intermittent fevers, and diarrhea) in areas of endemic infection or in travelers with a history of exposure.

Immunology. With parasitization of RE cells, there is a hyperplasia of RE cells from the liver, spleen, bone marrow, lymph nodes, mucosa of the small intestine, and other lymphoid tissues. Hematopoiesis is initially normal and then becomes depressed, with a reduction in the life span of erythrocytes and leukocytes, causing anemia and granulocytopenia. RE cell proliferation may lead to massive hypertrophy of the liver and spleen, both of which may return to normal size after successful treatment. In response to infection, there are decreases in T-cell blastogenesis production of interleukin-1 and -2, gamma interferon, and tumor necrosis factor. With successful therapy, these responses return to normal (27). Infection results in suppression of cell-mediated immunity, leading to uncontrolled reproduction of the parasite and dissemination. There is increased production of globulins, with high levels of immunoglobulin G (IgG) and IgM. The major target for the humoral immune response is heat shock proteins; Hsp70 was recognized in 92% of persons with VL (59). Circulating immune complexes involving IgA, IgG, and IgM can be found in the serum. Delayed-hypersensitivity responses to skin test antigens are suppressed or absent. With resolution of infection, the delayed-hypersensitivity reaction returns, and patients recovering from infection are considered immune to reinfection.

Clinical signs. Fever may occur at irregular intervals and, once established, may be seen as a double (dromedary) or triple fever peak each day. The patient is often weak and appears emaciated (Figure 8.8). Common clinical signs include a nontender enlarged liver with Kupffer cells containing many amastigotes and little or no cellular reaction, a markedly enlarged spleen, femoral and inguinal lymphadenopathy, and occasional acute abdominal pain or vague abdominal discomfort. In progressive disease, splenic infarcts are common; in acute cases, the spleen is smooth and friable; while in more chronic cases, the spleen is firm. Hair changes (thinning, dryness, hypopigmentation, and loss of curl) may be seen. Edema of the legs, jaundice, petechiae, and purpura have also occasionally been seen.

Skin. Skin changes are often seen on the face, hands, feet, and abdomen, particularly in India, where patients acquire an earth-gray color; this darkening of the skin apparently gave rise to the name kala azar (black sickness). In Africa, skin lesions often accompany VL and may appear very different from one patient to another. Some lesions are diffuse, warty, and nonulcerated and contain variable numbers of amastigotes.

Mouth and nasopharynx. Patients with VL in the Sudan and sometimes in East Africa and India have mucosal lesions in the mouth and nasopharynx. The lesions may appear as nodules or ulcers; the gums, palate, tongue, or lips can be involved. Lesions in the nasal mucosa can result in perforation of the septum, while nasopharyngeal and laryngeal lesions are characterized by mucosal swelling and subsequent hoarseness. Parasites tend to be more numerous in oral than in nasopharyngeal lesions;

however, they can be demonstrated in nasal and pharyngeal fluids.

Laboratory findings. Patients generally have anemia (normocytic and normochromic, assuming that there is no underlying iron deficiency), neutropenia, relative lymphocytosis, an almost complete absence of eosinophils, thrombocytopenia, and hypergammaglobulinemia (polyclonal B-cell activation). Liver enzyme levels may be mildly elevated, but elevations of bilirubin levels are uncommon. Malabsorption may be a severe problem in children with VL and can lead to complications or death.

Infections in immunocompromised individuals. During the last decade, leishmaniasis has emerged as one of the first and most important opportunistic infections in HIV-infected patients. Over 1,600 coinfection cases have been recorded, the majority from Europe, where 7 to 17% of HIV-positive individuals with fever have amastigotes. This finding strongly suggests that individuals infected with *Leishmania* spp. without symptoms will express symptoms of leishmaniasis if they become immunosuppressed. Based on the patient's general health status and prior exposure to the infection, leishmaniasis may be seen as an early opportunistic infection or as a complication late in the course of AIDS. In HIV-infected patients, VL occurs in late stages of HIV disease and often has a relapsing course (58). Secondary prophylaxis reduces the risk of relapse. VL in the HIV-infected population should be included in the Centers for Disease Control and Prevention clinical category C for the definition of AIDS in the same way that other geographically specific opportunistic infections are included. The majority of AIDS patients present with the classical picture of VL, but asymptomatic CL, mucocutaneous leishmaniasis, DCL, and post-kala-azar dermal leishmaniasis can be seen (these are usually caused by *L. infantum*). Cutaneous lesions of VL are being found increasingly frequently in patients with HIV infection, and their significance is still somewhat unclear. Lesions often do not present a uniform or specific appearance and have been seen as erythematous papules and hypopigmented macules on the dorsa of the hands, feet, and elbows; small subcutaneous nodules on the thighs; and erythematoviolaceous, scaly plaques on the face. Also, the digestive and respiratory tracts are often parasitized (5), as well as the pleura and peritoneum. Lesions can also be coinfected with other organisms, as evidenced by a case in which VL and *Mycobacterium avium-intracellulare* were seen in the same specimens from bone marrow and duodenal mucosa.

CD4 T-lymphocyte counts are usually $<50/mm^3$ and are almost always $<200/mm^3$ (64). However, many physicians think that VL does not cause severe symptoms in these patients but is just one of several opportunistic infections that may coinfect AIDS patients. However, VL can manifest atypical aspects in HIV-positive patients depending on the degree of immunosuppression and should therefore be listed among AIDS-defining conditions. The most common symptoms are fever, splenomegaly, hepatomegaly, and pancytopenia.

Data from Spain provide some interesting outcome figures regarding coinfection compared with leishmanial infections in non-HIV-infected persons. In spite of a good initial response to treatment for VL, 60.6% of the HIV-coinfected patients had relapsed by the end of 1 year. Mortality from the first episode was 18.5%, and 24% died in the first month after diagnosis of any VL episode. The mean survival of the 29 patients who died was 10.27 months. Survival in patients with and without AIDS at the time of the first episode of VL was compared at 30 months and found to be 53.7 and 30.5%, respectively. A diagnosis of AIDS at the time of the first episode of VL and thrombocytopenia were the only risk factors found to be related to survival. Certainly in AIDS patients, VL is a recurrent disease that is highly prevalent and whose clinical course is modified by HIV (41). VL is very prevalent among HIV-1-infected patients in southern Spain, with a high proportion of cases being subclinical. Symptomatic cases appear mainly when severe immunosuppression is

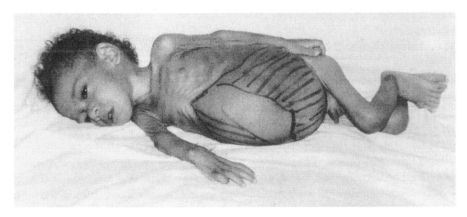

Figure 8.8 Cachexia (wasting) seen in severe visceral leishmaniasis. (Courtesy of Richard D. Pearson.)

present. There is also an association between VL, the male sex, and intravenous drug use.

In HIV-1-infected patients with active VL, the sensitivity of a peripheral blood smear is about 50%, due to a high parasitemia in these patients. However, in patients with HIV-1 and subclinical VL, the sensitivity of a routine blood smear is <10% (23). Unfortunately, in these patients serology and *Leishmania* skin tests have low sensitivities and are not very helpful (51). Cultures have proven to be effective in diagnosing VL in patients who are coinfected with HIV-1 (24).

Post-Kala Azar Dermal Leishmaniasis. Post-kala azar dermal leishmaniasis (PKDL) was first detected in patients with VL in India, occurs in up to 20% of these patients, has been called dermal leishmanoid, and is associated with *L. donovani* only. PKDL has also been seen in >50% of Sudanese VL patients treated in the current epidemic. In India, skin lesions may appear 2 to 10 years after successful therapy for VL; in East Africa, lesions appear within a few months; and in the Sudan, onset also occurs in about 2 months. Macules and papules usually appear first around the mouth and spread to the face and then to the extensor surfaces of the arms, the trunk, and sometimes the legs. In the beginning they look like small hypopigmented patches; these then enlarge and may progress to nodules that often resemble leprosy. The lesions are somewhat delicate but do not tend to ulcerate unless traumatized. In some cases, PKDL is seen in patients with no history of visceral disease. Some patients with past or concomitant PKDL also have ocular lesions including conjunctivitis, blepharitis, and anterior uveitis; these conditions all respond to antimonial therapy, including steroid and atropine eyedrops.

Treatment of PKDL with antimonials usually brings improvement; however, the skin changes persist indefinitely. The course of therapy is generally longer than that required for VL and may require 120 days. In Africa, patients with PKDL may self-cure. It is important to remember that these patients may serve as reservoirs of infection during interendemic cycles, since some patients with PKDL tend to have large numbers of parasites in the skin. Persistent lesions currently require daily injections of sodium stibogluconate for 2 to 4 months, and even then the treatment may not be successful. Treatment with liposomal amphotericin B (AmBisome) has been found effective and is considered less nephrotoxic than nonliposomal amphotericin B because it specifically targets the macrophages in which the *Leishmania* parasites develop. Apparently, there are also intrinsic differences in the antibodies generated in the sera from patients with PKDL and VL (54). Diagnostic methods would be identical to those used for the isolation and identification of amastigotes found in CL.

Diagnosis

A clinical diagnosis of Old World VL is often made based on clinical findings and local epidemiologic factors. However, confirmation of the diagnosis requires demonstration of the amastigotes in tissues or clinical specimens or the promastigotes in culture (Algorithm 8.1). Detection of parasite genetic material or antigen detection also suffices for confirmation. Leishmaniasis should be suspected in individuals who have resided in or traveled to areas where the disease is endemic. The diagnosis would be supported by findings of remittent fevers, hypergammaglobulinemia with anemia, circulating immune complexes, rheumatoid factors, weight loss, leukemia, and splenomegaly. The differential diagnosis of late-stage VL is limited to hematologic and lymphatic malignancies. The diagnosis of early disease is more difficult, and the differential would include malaria, African trypanosomiasis, brucellosis, enteric fevers, bacterial endocarditis, generalized histoplasmosis, chronic myelocytic leukemia, Hodgkin's disease and other lymphomas, sarcoidosis, hepatic cirrhosis, and tuberculosis. Also, patients with multiple myeloma and Waldenström's macroglobulinemia have monoclonal hypergammaglobulinemia. Mixed cryoglobulinemia secondary to VL has also been documented.

Blood. Examination of buffy coat smears shows that there is a significant difference in recovery between samples collected during the day (46%) and those collected

Algorithm 8.1 Visceral leishmaniasis.

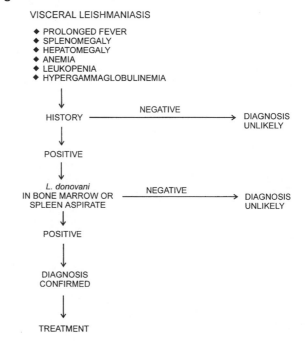

VISCERAL LEISHMANIASIS

- ◆ PROLONGED FEVER
- ◆ SPLENOMEGALY
- ◆ HEPATOMEGALY
- ◆ ANEMIA
- ◆ LEUKOPENIA
- ◆ HYPERGAMMAGLOBULINEMIA

↓

HISTORY ——— NEGATIVE ———→ DIAGNOSIS UNLIKELY

↓

POSITIVE

↓

L. donovani IN BONE MARROW OR SPLEEN ASPIRATE ——— NEGATIVE ———→ DIAGNOSIS UNLIKELY

↓

POSITIVE

↓

DIAGNOSIS CONFIRMED

↓

TREATMENT

during the night (66%) (69). Prior to an invasive procedure, blood specimens should be collected, keeping in mind the amastigote diurnal periodicity.

Tissue Aspirates. Although splenic puncture yields the highest rate of organism detection (98%), it carries a high degree of risk for the patient. Certainly deaths have occurred after this procedure, probably due to splenic laceration. Patients with coagulation disorders should not undergo splenic puncture. Other specimens include lymph node aspirates, liver biopsy specimens, sternal or iliac crest bone marrow, and buffy coat preparations of venous blood. Bone marrow aspirate is generally used in lieu of splenic puncture; however, smears from bone marrow usually contain fewer organisms, with a positivity rate of 80 to 85%. As discussed earlier in this chapter, it is important to prepare multiple smears for examination. Although amastigotes can be seen in buffy coat smears in Indian and Kenyan patients with VL, it is unusual to make the diagnosis by this method.

Culture and Animal Inoculation. Culture and animal inoculation studies are also helpful. Culture requirements, including specimen collection by an aseptic technique and the use of quality control strains, were discussed earlier in this chapter and are also addressed in chapter 32. A method for rapidly isolating promastigotes from the culture fluid has been developed using the stain acridine orange. This type of vital staining combines the advantages of direct microscopic examination with more accurate morphological imaging of the stained parasites. If available, hamsters should be inoculated intraperitoneally and killed after 2 to 3 months, and the liver and spleen should be checked for parasites (microscopic examination and/or cultures).

PCR and Other Antigen Detection Methods. PCR methods have excellent sensitivity and specificity for direct detection, for identification of causative species, and for assessment of treatment efficacy (18, 47, 48, 67, 70, 71, 84). A rapid immunochromatographic dipstick test using the recombinant RK39 antigen has become available for the qualitative detection of total anti-*Leishmania* immunoglobulins. The test uses antigen-impregnated nitrocellulose paper strips. Results from one study were very promising, with 100% sensitivity and 98% specificity (78). The RK39 strip test is ideal for rapid reliable field diagnosis of visceral leishmaniasis (80). The test has high sensitivity and specificity; however, it remains positive long after treatment (up to 3 years).

Formol-Gel Test (Hypergammaglobulinemia). In patients with kala azar, there is a characteristic hypergammaglobulinemia, including both IgG and IgM. This

fact has been used as the basis for a diagnostic procedure called the formol-gel test, which is used in areas of endemic infection. The test can be used as a screen; most patients with active disease are positive. A drop of formaldehyde is added to a test tube containing 1 ml of patient serum, which is observed for opacity and a stiff jelly consistency within 3 min to 24 h. Results also indicate that a specific IgE response is useful in the diagnosis of active VL and may be used to evaluate the response to treatment (10).

Skin Test. The Montenegro (LST) skin test is described above. In patients with active VL, this delayed-hypersensitivity test is negative. However, within several months to a year after recovery, the skin test reverts to positive. It is of limited value because it becomes positive late in the course of disease, and false-positive results are common.

Serologic Tests. A number of serologic diagnostic methods have been developed worldwide, although they are not widely available. In general, sensitivities are >90% but specificities are lower. False-positive results can be seen in patients with other infectious diseases, in patients treated for VL (in whom antibodies persist for months), and in some asymptomatic VL patients. There are currently four tests in use, the direct agglutination test (DAT), the indirect fluorescent-antibody test, counterimmunoelectrophoresis, and enzyme-linked immunosorbent assay (ELISA). DAT appears to be the best choice as a diagnostic tool since it is very specific and does not require expensive equipment or reagents (11). However, patients may still have positive sera 1.5 to 5 years after treatment (56). DAT has also been adapted to filter paper sampling, and the filter paper eluates compared well with their homologous sera and correlated strongly with respect to antibody titers (63). DAT based on axenic amastigote antigen provided 100% sensitivity and specificity, making it particularly useful for diagnosis of macular PKDL cases, which are often missed by the RK39 strip test. Thus, DAT provides a simple, reliable, and inexpensive test for PKDL diagnosis with potential applicability under field conditions (74). However, newer studies using ELISA formats indicate excellent sensitivity and specificity and may be very helpful in diagnosis of early infections (17, 66).

KEY POINTS—LABORATORY DIAGNOSIS

Old World Leishmaniasis: Visceral Leishmaniasis

1. For a complete laboratory diagnosis, microscopic examination of Giemsa-stained touch preparations and cultures are recommended, as well as PCR and animal inoculation if available.

2. Due to potential problems with splenic puncture, bone marrow specimens may be recommended. Although not that productive, examination of buffy coat preparations is recommended prior to a more invasive procedure. When collecting blood specimens, remember the diurnal periodicity.

3. Multiple slides should be prepared for examination from aspirates and/or biopsy specimens.

4. Tissue imprints or smears should be stained with Giemsa stain (or one of the stains commonly used for blood smears) (see chapter 31).

5. Amastigote stages should be found within RE cells or close to disrupted cells.

6. If cultures are to be performed, specimens must be collected aseptically and control organism cultures should be set up at the same time (see chapter 32).

7. Cultures should be checked weekly for 4 weeks before they are declared negative.

8. Patients with kala azar have hypergammaglobulinemia; the formol-gel procedure may be helpful.

9. Serologic tests are helpful, and newer procedures demonstrate increased sensitivity and specificity; this should be considered if parasites cannot be detected in clinical specimens.

Treatment

Patients should receive supportive care, especially if they are malnourished or have other infections such as pneumonia, tuberculosis, or other chronic conditions. Since the mid-1930s, pentavalent antimony compounds have been the drugs of choice for the treatment of VL. However, with the first reports of primary treatment failures in the mid-1990s, additional drugs have been used and include lipid-associated amphotericin B for Mediterranean and Indian VL (54). VL in renal transplant recipients has been successfully treated with liposomal amphotericin B (AmBisome). The disadvantages of using Pentostam and Glucantime include the need for intravenous administration and hospitalization, cardiotoxicity, pancreatitis, treatment failures, and increasing resistance (79). The dosing regimen for treating VL is also associated with a high rate of serious side effects in HIV-infected patients (22). Therapy combining either paromomycin or allupurinol with pentavalent antimonials has also been used with some success.

There is no effective orally administered medication for any *Leishmania* infection. However, studies with miltefosine appear very promising. This drug is a phosphocholine analogue that affects cell-signaling pathways and membrane synthesis. When 100 mg of miltefosine was given per day (approximately 2.5 mg/kg of body weight/day) for 4 weeks, 29 (97%) of 30 patients with at least moderate (2+) *Leishmania* in their splenic aspirates were cured. Gastrointestinal side effects were frequent but were only mild to moderate in severity, and no patient discontinued therapy. The use of orally administered miltefosine appears to be an effective treatment for Indian VL (35, 77).

Epidemiology and Prevention

The epidemiology of VL depends on a number of factors including the interaction between sand flies, reservoir hosts, and susceptible humans. In the two main areas of endemicity, the Ganges river basin and the southern Sudan, high infection rates and many deaths have been documented during the last 10 years. Factors which affect the number of infections and the transmission include population movements, often related to forced migration, poor general health, and poverty, all of which are seen in both geographic areas. The disease is a zoonosis, with the exception of India, where humans serve as the major reservoir host and the relevant vector is *P. argentipes*. Dogs serve as the main reservoir host around the Mediterranean, and the infection is transmitted through bites of infected sand flies such as *P. perniciosus*, *P. major*, *P. simici*, and *P. longicuspis*. Disease in this area is usually found in infants and children; young dogs are very susceptible to *L. infantum* and often die of the disease. Dogs are also important reservoirs in certain parts of China. In southern France and central Italy, infected foxes appear to be the reservoir, with *P. ariasi* and *P. perfiliewi* serving as vectors. Jackals also probably serve as reservoir hosts in the Middle East and central Asia. The epidemiology of VL in Africa is somewhat confusing, since the presence of epidemics suggests a human reservoir but dogs have also been found to be infected. The main vector in Kenya appears to be *P. martini*, and in the Sudan the vector is *P. orientalis*.

L. donovani occurs in Africa and Asia, and in areas where transmission is somewhat sporadic, wild canids and various rodents are natural reservoirs of the disease. *L. infantum* is seen in Africa, Europe, the Mediterranean area, and southwest Asia. Humans are accidental hosts, with dogs and other members of the Canidae acting as natural reservoirs (28).

Although the usual route of transmission is through the bite of infected sand flies, parasites can be transmitted in other ways including blood transfusion, sexual contact, congenital transmission, and occupational exposure. However, for the most part, transmission is through the normal vector cycle. Apparently, *L. infantum* can circulate intermittently and at low density in the blood of healthy seropositive individuals, who appear to be asymptomatic carriers; this certainly has implications for the safety of the blood supply in certain areas of the world, including southern France.

In areas where humans serve as the main reservoir host, treatment of active cases is likely to interrupt the cycle of transmission. Reservoir control has proven to be somewhat sporadic, and there is little indication that it has any impact on overall infection rates. Residual insecticide spraying, primarily for malaria vector control, was relatively successful; however, when spraying of DDT was discontinued, the incidence of VL began to increase again. Vaccines hold a great deal of promise, but currently no commercial products are available (33).

New World Leishmaniasis: Cutaneous Leishmaniasis

New World CL extends from southern Texas and Arizona to Latin and South America and the Caribbean islands (37). As seen in Old World leishmaniasis, there is a wide range of disease presentation, and separation based on clinical signs and symptoms is not possible (Figures 8.9 and 8.10). Organisms in this group belong to the subgenera *Leishmania* and *Viannia* (Table 8.1). Within the subgenus *Leishmania* is the *L. mexicana* complex (Table 8.5), and within the subgenus *Viannia* are the *L. braziliensis* complex (Table 8.6) and the *L. guyanensis* complex (Table 8.7). It is very likely that taxonomy will continue to change as molecular methods reveal genetic differences and similarities among the various genera and species. Common names for New World CL include uta (Peru), dicera de Baurid (Brazil), chiclero ulcer or bay sore (Mexico), and pian bois or forest yaws (Guyana). Collectively, this group has also been called American CL. This disease is quite old and has even been depicted on pottery from Peru and Ecuador that has been dated ca. AD 400 to 900.

Life Cycle and Morphology

The life cycle and organism morphology are essentially the same as for Old World CL (Figure 8.2). However, in some cases, destructive oral, nasal, or pharyngeal lesions may develop, causing mucocutaneous disease. In New World leishmaniasis, the sand fly vector is in the genus *Lutzomyia* or *Psychodopygus* rather than *Phlebotomus*. The genus *Leishmania* is divided into its two subgenera, *Leishmania* and *Viannia*, depending on the location of the development of the promastigote within the sand fly. Promastigotes from the subgenus *Viannia* develop in the midgut and hindgut of sand flies, while promastigotes from the subgenus *Leishmania* develop in the anterior portion of the alimentary tract of the sand fly (31).

Clinical Disease

Cutaneous Leishmaniasis. The lesions of New World CL are very similar to those seen with Old World cutaneous disease. Weeks to months after infection, an erythematous, often pruritic papule develops at the bite site. This papule may become scaly and enlarge, developing a central ulcer surrounded by a raised margin. Disease progression at this point will vary depending on the species involved. Lesions may be single or multiple and usually occur on exposed areas

Figure 8.9 *Leishmania.* Healed facial scar, "Seal of the Forest" (Costa Rica). (From A Pictorial Presentation of Parasites: A cooperative collection prepared and/or edited by H. Zaiman.)

of the body. Many lesions may self-heal over a period of approximately 6 months (Figure 8.9); however, lesions on the ear occur in 40% of the patients and are chronic, lasting many years. Ulcerations on the ears may be quite destructive but cause the patient few problems. *L. mexicana* lesions are usually single, self-limiting

Figure 8.10 Facial lesion on a 9-month-old girl. (Courtesy of Yoshihisa Hashiguchi.)

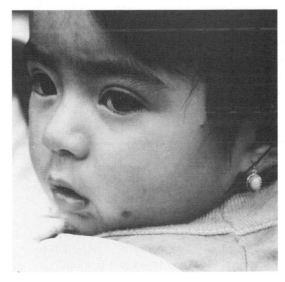

Table 8.5   New World cutaneous leishmaniasis subgenus *Leishmania*, complex *Leishmania mexicana*

Species and sand fly vector	Human disease and lesion	Geographic distribution	Important hosts	Comments
L. mexicana: Lutzomyia olmeca, L. anthophora (Texas and Arizona)	Cutaneous leishmaniasis, chiclero ulcer	Belize, Guatemala, Honduras, Costa Rica, Mexico, Texas, Arizona (Panama, South America questionable); Southern Plains of Texas (wood rat [*Neotoma micropus*]); Arizona (white-throated wood rat [*Neotoma albigula*])	Forest rodents	Prolonged exposure of "chicleros," who live in forest collecting chewing gum latex from chicle trees (30% infected during first year); timber cutters, road builders, agricultural workers
L. venezuelensis: Lutzomyia olmeca bicolor	Cutaneous leishmaniasis	Venezuela (humans only)	Humans	Poor long-term growth in blood agar medium; wild mammalian host not yet discovered; found in horses and the domestic car
L. garnhami: Lutzomyia youngi (not confirmed)	Cutaneous leishmaniasis	Venezuelan Andes	Humans, marsupial (*Didelphis marsupialis*)	Identical to *L. amazonensis* by isoenzyme profiles, but clear differences between nontranscribed rDNA intergenic spacer sequences; still difference of opinion regarding validity of species designation
L. amazonensis: Lutzomyia flaviscutellata	Cutaneous leishmaniasis, diffuse cutaneous leishmaniasis	Amazon region of Brazil, Bolivia, Colombia, French Guyana, Paraguay	Forest rodents (primary); marsupials, foxes (secondary)	Human disease uncommon; vector nocturnal and not very anthropophilic, lives in swampy areas with few inhabitants; DCL similar to that seen in Ethiopia (Old World); specific anergy to leishmanin skin test
L. pifanoi: Lutzomyia flaviscutellata	Diffuse cutaneous leishmaniasis	Venezuela (states of Yaracuy, Lara, and Miranda)	Humans	Some confusion regarding actual taxonomy; wild mammalian host not yet discovered; DCL similar to that seen in Ethiopia (Old World); specific anergy to leishmanin skin test

cutaneous papules, nodules, or ulcers that are painless and are found on the face and ears (Figures 8.10 and 8.11). Other infections that are usually associated with single cutaneous lesions include those due to *L. colombiensis*, *L. amazonensis*, *L. garnhami*, *L. lainsoni*, *L. peruviana*, and *L. venezuelensis*. Multiple lesions caused by spread along lymphatics are common with *L. guyanensis* infections, and subcutaneous lymphatic nodules can be seen with *L. panamensis*; the lesions can mimic those seen in sporotrichosis. The differential diagnosis of New World CL includes sporotrichosis, blastomycosis, yaws, syphilis, cutaneous tuberculosis, *Mycobacterium marinum* infection, and dermatologic cancers.

Mucocutaneous Leishmaniasis. Cutaneous lesions caused by *L. braziliensis* usually require a longer period for self-cure, taking up to 18 months or longer, and they can begin with lymphadenopathy. Approximately 1 to 3% of patients infected with *L. braziliensis* develop metastatic spread of the organisms to the nasal, pharyngeal, and buccal mucosa (mucocutaneous leishmaniasis [ML]). ML, also known as espundia, may occur months to years after the initial cutaneous lesion has healed partially or completely. However, in many cases, mild mucosal lesions are present concomitantly with active cutaneous lesions. Mucosal disease tends to occur more frequently in males than females, and the patients are usually 10 to 30 years of age.

ML may begin with erythema and edema of the involved mucosal tissues, along with nasal inflammation, stuffiness, difficulty in breathing through the nose, and occasional bleeding. Following the early symptoms, ulcerations covered with a mucopurulent exudate may develop (Figure 8.12). There is often severe erosion and destruction of the soft tissues and cartilage, leading to loss of the

Table 8.6   New World cutaneous leishmaniasis, subgenus *Viannia,* complex *L. braziliensis*

Species and sand fly vector	Human disease and lesion	Geographic distribution	Important hosts	Comments
L. braziliensis: Lutzomyia wellcomei, L. intermedius, L. pessoai	Cutaneous leishmaniasis, mucocutaneous leishmaniasis	Brazil, Guatemala, Belize, Honduras, Nicaragua, Peru, Ecuador, Bolivia, Costa Rica, Panama, Colombia, Venezuela, Paraguay, Argentina	Forest rodents, humans	Vector feeds primarily during daylight hours; common in farming communities in newly cleared forest areas and in road construction and mining workers; also known as nariz de anta (tapir nose) and espundia
L. peruviana: Lutzomyia peruensis, L. verrucarum	Cutaneous leishmaniasis	Peru, western slopes of the Andes, inter-Andean valleys; 900–3000 m above sea level	Humans, dogs	Frequent in school children, commonly resulting in extensive facial scars; ulcers self-healing; known as uta; in some villages 90% of people are infected or scarred
L. colombiensis: Lutzomyia hartmanni, L. gomezi, Psychodopygus panamensis	Cutaneous leishmaniasis	Colombia, Panama, Venezuela, probably into Brazil, Peruvian lowlands	Humans, single isolate from the sloth (*Choloepus hoffmanni*)	Single to multiple ulcerating skin lesions; ML has not been seen
L. lainsoni: Lutzomyia ubiquitalis	Cutaneous leishmaniasis	Amazon region of northern Brazil, probably elsewhere	Humans, rodent (*Agouti paca*)	Multiple lesions have also been seen; ML has not been seen
L. shawi: Lutzomyia whitmani	Cutaneous leishmaniasis	Amazon region of northern Brazil, south of the Amazon River	Humans, monkeys, sloths, coatimundi	Multiple lesions have also been seen; ML has not been seen
L. naiffi: Psychodopygus ayrozai	Cutaneous leishmaniasis	Brazilian states of Pará and Amazonas	Humans, nine-banded armadillo (*Dasypus novemcinctus*)	Multiple lesions have also been seen; ML has not been seen

Table 8.7   New World cutaneous leishmaniasis, subgenus *Viannia,* complex *Leishmania guyanensis*

Species and sand fly vector	Human disease and lesion	Geographic distribution	Important hosts	Comments
L. guyanensis: Lutzomyia umbratilis (high infection rate, up to 7%)	Cutaneous leishmaniasis	Brazil north of Amazon River, Guyanas; range extends into Ecuador, Venezuela, lowland forests of Peru	Humans, two-toed sloth, lesser anteater, rodents, opossums	Reservoir infections inapparent, organisms in normal skin and viscera; vector lives in forest canopy and tree trunks; timing right for deforestation (early-morning biting); Brazil nut and fruit collectors, topographers, visiting botanists, zoologists, tourists affected; also known as pian bois or forest yaws
L. panamensis: Lutzomyia trapidoi, L. ylephiletor, L. gomezi, Psychodopygus panamensis	Cutaneous leishmaniasis	Panama and the Canal Zone, western and central Colombia, Ecuador, Venezuela, Costa Rica, Honduras, Nicaragua	Humans, sloths (two-toed and three-toed), occasionally other animals	Frequent in American military personnel; rare cases of ML

lips, soft parts of the nose, and soft palate. The lesions are chronic and progressive and can lead to death due to secondary infection or aspiration pneumonia. In the non-ulcerative form, there may be hypertrophy of the tissues of the upper lip and nose, leading to the development of the "tapir nose." Unfortunately for diagnosis, parasites are few and difficult to find in tissues involved in ML. Although this disease can be extremely disfiguring, the mortality rate

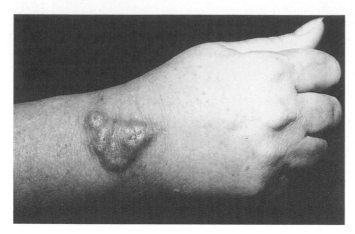

Figure 8.11   Large erythematous nodule with early ulceration on the right wrist, due to *Leishmania mexicana*. (Courtesy of Bonnie B. Furner.)

is low. New World ML mimics paracoccidioidomycosis, histoplasmosis, tuberculosis, syphilis, malignant tumors, and lethal midline granuloma.

Diffuse Cutaneous Leishmaniasis.   Disseminated DCL is an anergic form of CL in which neither humoral nor cell-mediated immune responses are functional but in which delayed hypersensitivity to tuberculin appears to be normal. DCL is seen most frequently in Venezuela, Brazil, Mexico, and the Dominican Republic. Individuals infected with *L. pifanoi* and 30% of those infected with *L. amazonensis* are likely to develop DCL (Figure 8.13).

The parasites are found in the skin, and the disease is characterized by the formation of thickened skin in plaques, papules, or multiple nodules. Although lesions are usually seen on the face, nose, limbs, and buttocks, they can occur over the entire body (Figure 8.14). The nodules are often described as soft and fleshy, while those of leprosy are generally more indurated. Although organisms have rarely been recovered from blood or bone marrow, there are no visceral lesions associated with this form of leishmaniasis. Treatment is difficult, and the infection may last for the life of the patient. The LST remains negative unless treatment is effective and recovery occurs. Nodular DCL will have to be differentiated from lepromatous leprosy.

Diagnosis

In areas of the world where physicians are very familiar with leishmaniasis, the diagnosis may be made on clinical grounds. However, in areas of the world where the disease is rare, the condition may not be recognized as leishmaniasis. Definitive diagnosis depends on demonstrating the amastigotes in tissue specimens or the promastigotes in culture (Table 8.8). More recent results suggest that PCR is a valuable tool for the diagnosis of leishmaniasis on a routine basis and can provide valuable epidemiologic informa-

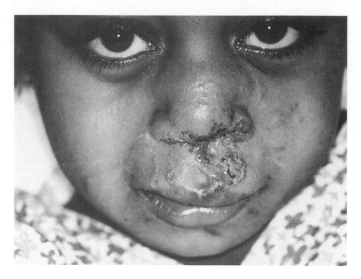

Figure 8.12   Mucocutaneous lesions on a 7-year-old girl with disseminated leishmaniasis. (Courtesy of Yoshihisa Hashiguchi.)

tion in areas of endemic infection (15, 43, 64, 65). CL may have to be differentiated from a number of other lesions and diseases, including basal cell carcinoma, tuberculosis, various mycoses, cheloid, and lepromatous leprosy.

Specimen Collection

CL normally presents with no visceral involvement or systemic disease. The ability to detect parasites in aspirates, scrapings, or biopsy specimens depends on the number of amastigotes present, the level of the host immune response, the absence or presence of bacterial and/or fungal contamination within the lesion, and whether the specimen is collected from an active or healing lesion. If the patient has multiple lesions, specimens should be collected from the more recent or active lesions. These lesions should be thoroughly cleaned with 70% alcohol, and necrotic debris should be removed to prevent the risk of bacterial and/or fungal contamination of the specimen. Also, the specimen should be taken from the advancing margin of the lesion; the central portion of the ulcer contains nothing but necrotic debris unless portions of the lesion are taken below the crater area.

The specimen of choice would be a collection of several punch biopsy specimens taken from the most active lesion areas. Biopsy specimens can be divided and used for cultures and touch preparations; some material should always be saved and submitted for routine histologic testing. The biopsy specimen can be used to make impression smears (touch preparations); these smears should be prepared after portions of the specimen have been placed in culture media by using a sterile technique. After any excess blood is cleaned off, a horizontal cut is made through the biopsy specimen core and the cut surface is gently touched

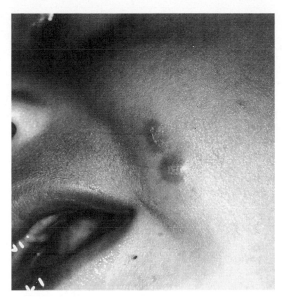

Figure 8.13 Two erythematous facial lesions on the right cheek of a 5-month-old boy infected with *Leishmania pifanoi*. (Courtesy of Yoshihisa Hashiguchi.)

to glass slides; multiple slides should be prepared. Once material has been set up for culture and touch preparations have been made, the remainder of the tissue can be sent to the pathology laboratory for routine processing.

Another option involving sterile techniques is to make a perpendicular slit in the margin of the lesion. After blood is removed using sterile gauze, the blade can be used to scrape the inner walls of the slit. Material can be inoculated into culture media and can be smeared onto several glass slides; multiple slides are recommended to increase the chances of seeing the parasites. Fine-needle aspiration can also be performed with a sterile syringe containing sterile saline (0.1 ml)

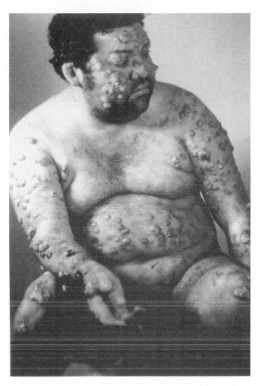

Figure 8.14 Diffuse anergic cutaneous leishmaniasis due to *Leishmania mexicana* in a patient of Dr. Convit (Venezuela). (From A Pictorial Presentation of Parasites: A cooperative collection prepared and/or edited by H. Zaiman.)

and a 26-gauge needle. The needle is inserted into the outer border of the lesion and rotated several times, after which saline is injected and tissue fluid is aspirated back into the needle. Material can be submitted for culture, and/or smears can be prepared for staining. Additional techniques and specific methods can be found in chapters 30 to 32.

Table 8.8 Features of New World human leishmanial infections related to demonstration of the parasite[a]

Organism	Disease type[b]	Humoral antibodies	Delayed hypersensitivity to leishmanin	Parasite present in clinical specimen	Self-cure	Recommended biopsy specimen
New World CL						
L. mexicana	CL	Present	Present	Present	Yes	Skin
complex	DCL	Variable	Absent	Abundant	No	Skin
L. braziliensis	CL	Present	Present	Present	Yes	Skin
complex	ML	Variable	Present	Few	No	Skin, mucous membranes
L. guyanensis complex	CL	Present	Present	Present	Slow	Skin
New World VL						
L. chagasi	VL	Abundant	Absent	Present	Possible	Bone marrow, spleen

a For culture, specimens must be collected aseptically, and in older lesions the parasites may be scant and difficult to recover. Animal isolation involves hamsters and culture media (Novy-MacNeal-Nicolle medium and Schneider's *Drosophila* medium with 30% fetal bovine serum). Serologic testing is most suitable for VL, has little value for CL, and has limited value for MCL. The Montenegro test (LST) is a delayed-hypersensitivity reaction to intradermal injection of cultured parasites.

b Additional information on relevant diagnostic methods can be found in chapters 31 and 32.

Specimen Staining. All smears to be Giemsa stained should be air dried and fixed with 100% methanol prior to staining. The amastigotes will be found within the macrophages or close to disrupted cells. In very early and older lesions, very few organisms may be present, so that prolonged examination of the slide may be required.

Culture Media. It is important to release the organisms from tissue prior to culture; mincing is recommended. However, tissue grinders are not recommended because of possible damage to the organisms. Culture media that have been used for the recovery and growth of the leishmaniae include Novy-MacNeal-Nicolle medium and Schneider's *Drosophila* medium supplemented with 30% fetal bovine serum (see chapter 32). Patient cultures should not be set unless the laboratory maintains specific organism strains for quality control checks. Both the control and patient cultures should be examined twice a week for the first 2 weeks and once a week thereafter for up to 4 weeks before they are reported as negative. Promastigote stages can be detected microscopically in wet mounts taken from centrifuged culture fluid. This material can also be stained with Giemsa stain to facilitate observation at a higher magnification.

Animal Inoculation. Animal inoculation can be helpful when only a small number of organisms are present; however, most laboratories no longer consider this approach practical. Golden hamsters are inoculated intranasally, and the major argument against this approach is that it may take several months before the animals give positive responses. A combination of tissue smears, culture, animal inoculation, and histopathology may be required for optimal diagnosis.

Skin Test. The LST (Montenegro test) is not species specific and must be interpreted with caution, particularly since some patients may have more than one type of leishmaniasis. Positive reactions are generally seen in patients with cutaneous lesions; however, those with DCL give negative reactions.

Serologic Tests. Standard serologic tests are not particularly useful for the diagnosis of New World CL or ML. Low titers are seen, tests are likely to be negative in early disease, and overall interpretation is difficult at best. However, if a patient is positive, serial measurement of antibody titers may be helpful in tracking the outcome of therapy. The development and use of recombinant antigens for specific and sensitive serodiagnosis of CL have led to improved diagnostic results (34).

Treatment

In contrast to Old World CL, systemic chemotherapy is recommended for New World CL. Differences between the two types of CL tend to justify this recommendation and include cases with multiple lesions; the long-term, chronic nature of some infections; and the tendency for disease progression and mucosal involvement in the absence of chemotherapy. Time to healing has been improved in studies in which recombinant human granulocyte-macrophage colony-stimulating factor coupled with stibogluconate has been used (4). Studies also show that the combination of allopurinol and stibogluconate is superior to stibogluconate alone and can be used as an adjunct to stibogluconate or other oral agents in the treatment of CL (45). However, a combination of allopurinol and stibogluconate used for the treatment of ML provided no clinical benefit. It also appears that low-dose CL therapy using an antimony regimen (5 mg/kg/day for 30 days) may work just as well as the larger recommended dose (20 mg/kg/day for 3 to 4 weeks), particularly for children and the elderly (55).

The most promising drug found so far is an anticancer compound, miltefosine, which has undergone experimental and clinical trials and is 94 to 97% effective. However, the drug is contraindicated during pregnancy and may cause severe gastrointestinal side effects. Also, the cost is another limiting factor. Other drugs such as paromomycin, allopurinol, and sitamaquine have been reported to have variable cure rates. Because of these limitations, a combination therapy, preferably coupled with specific parasite enzyme inhibitors, may be the most effective approach (75).

In Mexico, CL occurs in 17 of 32 states and is considered a serious public health problem. Treatment with a localized current field radiofrequency device to generate precisely controlled heat as an alternative to expensive drug treatment appears to be effective and convenient for use in primary health care facilities. Lesions were first anesthetized with 1% lidocaine hydrochloride and moistened with normal saline solution. Treatment consisted of a single application of heat that produced 50°C for 30 s. After 4 weeks, 95% of 122 patients were totally cured (including those with infected ear cartilage), and at 8 weeks, 90% of 191 patients were cured (81).

ML is the most serious clinical form of CL and, once diagnosed, requires prolonged and close supervision of therapy; incomplete treatment can lead to severe sequelae and even death. For many years, mucosal involvement of leishmaniasis in America was attributed to *L. braziliensis* only. However, dissemination via blood or lymph or extension to mucosal tissue has been documented for *L. panamensis*, *L. guyanensis*, *L. mexicana*, and *L. amazonensis*. It is now accepted that most of the New World *Leishmania* species are capable of causing a wide spectrum of disease, depending on the immune status of the host, parasite strain, and anatomic exposure. The simultaneous presentation of both mucosal and active cutaneous lesions contrasts with the classical descriptions of ML caused by *L. braziliensis*, thus emphasizing the importance of early diagnosis of mucosal disease by examination of the mucosa in all patients with CL. In some cases, treatment with both conventional and experimental regimens may not be effective against ML (44).

Epidemiology and Prevention

***L. mexicana* Complex.**   *L. mexicana* (Table 8.5) causes chiclero ulcer and is common throughout Mexico, Belize, Guatemala, and the southern United States. It is transmitted among forest rodents by vectors that are not particularly attracted to humans. However, prolonged exposure of many forest workers leads to a high infection rate in these groups. The range of endemic infection has now been expanded within the United States to include areas in Arizona that are farther north than the southern Texas site previously identified. Apparently, wood rats serve as the primary wild reservoir hosts in these two states.

L. venezuelensis is found in the Lara and Yaracuy States, Venezuela, and has been recovered from humans, horses, and domestic cats; no wild reservoir hosts have been identified. Although it develops rapidly in hamsters (animal inoculation), it does not grow well in the blood-based media routinely used for culture.

Recovery of *L. garnhami* is limited to the Venezuelan Andes. Although the human is a known host, a single infection has also been recorded in the opossum, *Didelphis marsupialis*.

L. amazonensis has been found in Bolivia, Brazil, Colombia, French Guyana, and Paraguay and is likely to be found in other South American geographic locations where the appropriate vector is located. It is primarily an infection of forest rodents; however, marsupials and foxes can also be reservoir hosts. The vector tends to be nocturnal; this is probably the reason why human disease is not as common as with other species.

L. pifanoi is limited to the states of Yaracuy, Lara, and Miranda within Venezeula, and humans are the only known mammalian host. The wild mammalian host has not yet been identified.

***L. (Viannia) braziliensis* Complex.**   Unfortunately, the geographic distribution of *L. braziliensis* (Table 8.6) is not well defined, but infections have been reported from most Latin American countries including Argentina, Belize, Bolivia, Brazil, Colombia, Costa Rica, Ecuador, French Guyana, Guatemala, Honduras, Nicaragua, Panama, Paraguay, Peru, Suriname, and Venezeula. Many researchers think that it is unlikely that this broad a geographic range is accurate. The situation is even more confusing when ecologic and epidemiologic features of CL and ML differ in different regions. Humans are the main hosts, and wild hosts include sylvatic mammals such as rodents and marsupials. Infections have also been seen in dogs, mules, horses, and, rarely, cats. Extensive deforestation has been implicated in areas with suspected domestic animal transmission. Because there are questions regarding the reports of geographic range, it is difficult to accurately define vectors with proven transmission related to *L. braziliensis*. Human infections tend to be common in farming communities that develop along with deforestation, road construction, and mining.

L. peruviana is found in Peru on the western slopes of the Andes, including the inter-Andean valleys, generally at 900 to 3,000 m above sea level. It probably has a wider geographic range than reported. Transmission occurs in mountainous areas with little vegetation, and wild reservoir hosts probably include rodents and marsupials. Suspect vectors have been identified, but confirmation is often difficult. Human infection is usually contracted in and around the dwelling, and in some villages, the combined rate of present and past infection reaches 90%.

L. colombiensis has been found in Colombia, Panama, and Venezuela but probably has a wider geographic range, extending into Brazil and Peru, where appropriate wild reservoir hosts and vectors are found. Humans are the main host; however, a single isolate has been found from a sloth in Panama.

L. lainsoni occurs in the Amazon region of northern Brazil; however, the geographic range probably extends beyond these limits, as does the wild reservoir, *Agouti paca*, and the sand fly vectors. Evidence indicates that the vector, *Lutzomyia ubiquitalis*, does not like human blood, probably accounting for the low infection rate with *L. lainsoni*.

L. shawi is also found in the Amazon region of northern Brazil, south of the Amazon river. Humans are recognized hosts, and wild reservoir hosts include monkeys, sloths, and the coatimundi, as well as probably a few others.

L. naiffi infections have been recorded from the Brazilian states of Pará and Amazonas. Like many of the other species where the wild reservoir hosts and appropriate vectors are found beyond these recorded boundaries, it is strongly suspected that the infection also has a broader range into other areas of Brazil and surrounding countries. The only recorded wild reservoir host is the nine-banded armadillo, in which there is a high infection rate in what appears to be normal skin and viscera. Like *L. lainsoni*, the vector of *L. naiffi* does not prefer human blood, hence the low human infection rate. However, other vectors may be implicated and do prefer human blood.

***L. (Viannia) guyanensis* Complex.**   *L. guyanensis* (Table 8.7) is a very common cause of CL in Brazil, north of the Amazon river, and the Guyanas. Infections have also been recorded from Ecuador, Venezuela, and the lowland forests of Peru. Humans are known hosts, as are wild mammalian reservoir hosts such as the sloth and lesser anteater; other rodents and opossums have also been implicated. Like the wild reservoir hosts of *L. naiffi*, large numbers of parasites are located in what appears to be normal skin and viscera. The vector thrives in primary forest land, rather than in secondary forests or plantings of nonindigenous trees, and there may be a high human infection rate in such areas. Disease tends to be common in forest workers.

L. panamensis is found in Panama and the Canal Zone and has been reported as causing frequent infection in American military personnel stationed in these areas, as well as in agricultural workers. It has also been found in western and central Colombia, Ecuador, Venezuela, Costa Rica, Honduras, and Nicaragua. Humans are known hosts, and the epidemiology is similar to that seen with *L. guyanensis*. The major wild reservoir host is the two-toed sloth; fewer infections have been found in two genera of the three-toed sloths that live in the forest canopy, as do the vectors.

Vaccines. A. single-dose vaccine with killed antigen, using recombinant human interleukin-12 and alum as adjuvants, appears to be safe and fully effective in a primate model of New World CL. This information extends the murine data to primates, thus providing a basis for further human trials (36). Field trials of a vaccine against New World CL have also been tried in Ecuador, with very promising results after 12 months of follow-up (8). Other studies are under way and should provide results on the potency and stability of various candidate vaccines against New World CL (49).

New World Leishmaniasis: Visceral Leishmaniasis

New World VL (sometimes called American VL [AVL]) is caused by *L. chagasi* and has many similarities to Old World VL caused by *L. infantum*, including the fact that the infection is seen primarily in children and that both infections cause visceral and cutaneous disease.

Life Cycle and Morphology

The life cycle and morphology of *L. chagasi* are essentially the same as those of the species causing Old World VL (Figure 8.2). Amastigotes can be found throughout the body and are not confined to the macrophages of the skin. Organisms inoculated into the bite site by the sand fly are engulfed by tissue macrophages, which then enter the bloodstream and are carried to RE cell centers such as the bone marrow, spleen, and liver.

Clinical Disease

L. chagasi causes subclinical infections and AVL, which is potentially fatal if not treated, and it has recently been associated with atypical cutaneous leishmaniasis (ACL) in Central America, particularly in Honduras (Table 8.9). Typically, AVL is associated primarily with malnourished children younger than 5 years, whereas the ACL presentations are seen mainly in children older than 5 years and young adults. Data indicate that this atypical disease presentation may be more common than was previously thought and may affect up to 10% of a local population (12). The fact that the same organism can cause AVL and ACL suggests that the host immune response may play a role in the outcome. The disease presentation includes nonulcerated cutaneous nodules.

Diagnosis

Confirmatory diagnosis depends on the demonstration of parasites, either amastigotes in stained smears of spleen, bone marrow, or lymph node aspirates or promastigotes in culture (Algorithm 8.1). Although tissue aspirates should contain many parasites, this is not always the case. If only a few are present, culture may not be successful. Intraperitoneal inoculation of aspirate material into hamsters is by far the most reliable means of isolating *L. chagasi*; however, there may be long delays before parasites are detected in the animals. PCR will probably become the test of choice, as with other types of leishmaniasis, but the practicality and cost issues mean that routine methods will continue to be the most likely approach, particularly since the majority of cases are seen in more remote rural areas.

Treatment

Although the treatment regimens are similar to those used for Old World VL, some patients are unresponsive to the recommended antimonial dosages; this has been seen in Bahia, Brazil. Alternatives would include increased dosages and/or the use of second-line drugs such as pentamidine. Unfortunately, serious side effects also increase when increased dosages are used. A review of available therapy for both Old World and New World VL is given in Table 8.10.

Table 8.9 New World (American) visceral leishmaniasis

Species and sand fly vector	Human disease and lesion	Geographic distribution	Important hosts	Comments
L. chagasi: *Lutzomyia longipalpis*	Visceral leishmaniasis, kala azar, post-kala azar dermal leishmaniasis, atypical cutaneous leishmaniasis	Most of Latin America (Argentina, Bolivia, Brazil, Colombia, Ecuador, Paraguay, Suriname, Venezuela, Guatemala, Guadeloupe, Honduras, Martinique, Mexico, El Salvador, Costa Rica)	Humans, dogs, foxes	Closely resemble *L. infantum* infections in Old World; fatal outcome unless treated; the term American visceral leishmaniasis (AVL) is preferred to kala azar (reserved for Indian VL)[a]

[a] Although there is some controversy about whether *L. chagasi* is indigenous to the Americas or whether AVL is due to *L. infantum* which was introduced into the New World by immigrants in post-Columbian times via infected dogs, there appears to be strong evidence for the former hypothesis (38).

Table 8.10   Specific drugs for the treatment of visceral leishmaniasis (see also chapter 25)

Drug	Dosage	Side effects[a]	Comments[a]
Pentavalent antimonials (SbV) Sodium stibogluconate (SSG) (Pentostam) Sodium antimony gluconate Meglumine antimoniate (Glucantime)	Pentostam for VL (India, Kenya): 20 mg/kg/day (28 days Kenya, 40 days India); upper limit of safe, daily dose not as relevant, with the exception of AIDS patients	i.m. injections painful; recommend slow injection with 23- to 26-gauge needle; dilution of daily dose 1:10 with 5% DW; nausea, anorexia, abdominal pain, malaise, headache, arthralgias, myalgias, lethargy (beginning 7–10 days into therapy); VL patients tolerate SbV better than do CL or ML patients; symptoms resolve after therapy; serum amylase and lipase levels rise (peak at 7–14 days); may need to interrupt therapy for 5-fold increase in amylase and 15-fold increase in lipase levels; ECG abnormalities occur in 50% of patients receiving SSG	Variation in total antimony (Sb) content; cannot assume that one drug will work like another; lot variations may contribute to safety and efficacy differences; storage conditions important (4°C in the dark optimal) (79)
Amphotericin B deoxycholate Lipid-associated amphotericin B Liposomal amphotericin B (AmBisome) Cholesterol dispersion (Amphotec) Amphoterican B lipid complex (Abelcet)	Slow i.v. over 4–6 h, starting at 0.1 mg/kg daily, increasing to 1 mg/kg every 2 days until total dose of 20 mg/kg has been given; AmBisome regimen is 3 mg/kg daily on days 1–5, sixth dose on day 14, seventh dose on day 21 (total dose is 21 mg/kg); regimen for immunosuppressed patients is 4 mg/kg daily on days 1–5, 10, 17, 24, 31, and 38 (total dose of 40 mg/kg); in this group relapse is common	Fever, chills, thrombophlebitis, renal insufficiency, anemia, hypokalemia, anorexia, nausea; caution when used on patients previously receiving SbV, who may be predisposed to cardiac events; for treatment of patients with VL within the United States, AmBisome is recommended and should be used unless contraindicated	Used in cases of VL resistant to SbV compounds; cure rates in VL clinically resistant to SbV and pentamidine approach 100%; all drugs available in United States have been approved for other conditions; AmBisome now also approved for VL; AmBisome effective and safe for VL (Brazil, India, Kenya, Mediterranean); lipid-associated amphotericin B preparations probably not more effective than amphotericin B deoxycholate but can be given at higher doses with less toxicity; also more expensive
Pentamidine (as isethionate or methane sulfonate salt)	4 mg/kg daily 3 times/wk for 24 injections (8 wk); regimens of 4–6 mo have also been used	Pain, induration, sterile abscesses at injection site; hypotensive reactions if drug given too quickly; hypoglycemia diabetes may occur; high-dose, long-duration therapy more toxic than SbV and amphotericin B, giving it a limited role in treatment of VL	Provided as sterile powder, must be dissolved in DW; administered i.m. or slow i.v.; tends to be effective in Kenya with little toxicity; resistance in India now common; addition of SbV to pentamidine does not increase the cure rate
Aminosidine (paromomycin)	Parenteral (Kenya): 15 mg/kg/day for mean of 19 days (79% cure); parenteral (India): 20 mg/kg/day for 21 days (97% cure); accepted monotherapy for Indian VL at 16 mg/kg/day for 21 days	Possible auditory nerve damage and renal damage; vertigo and pancreatitis when given i.v.	Aminoglycoside with broad antiparasitic activity; if used in combination with standard doses of SbV, shorter duration (17–20 days) possible (Kenyan, Sudanese, Indian VL)
Immunotherapy	Has been used as monotherapy and adjunct therapy in SbV-resistant and primary VL in Kenya, Brazil, and India	Fever, chills, myalgias, fatigue, headache, depression	IFN-γ should be considered when faced with SbV-resistant VL and in selected cases of relapse; expensive (1, 7, 27, 29, 32)

[a] DW, distilled water (sterile); IFN-γ, gamma interferon; i.m., intramuscular; i.v., intravenous; ECG, electrocardiogram.

Epidemiology and Prevention

The public health implications of widespread *L. chagasi* infections in Central and South America may be quite broad. Parasites that cause ACL appear to be genetically very similar to those that cause AVL. This suggests that with diminished immune capacity, cutaneous infection may cause visceral disease, particularly since it is well known that this occurs in AIDS patients. ACL cases may also serve as reservoirs for vector transmission, leading to additional cases of visceral disease. Certainly treatment of these ACL patients would be recommended as a preventive measure for reduction of the number of potential future cases. Subcutaneous nodules are often misdiagnosed and treated inappropriately. Also, ACL may serve as a marker for the possible occurrence of AVL in a particular area and could be used as an indicator to more closely monitor for the presence of AVL.

In areas where humans serve as the main reservoir host, treatment of active cases is likely to interrupt the cycle of transmission. Reservoir control has proven to be somewhat sporadic, and there is little indication that it has any impact on overall infection rates. Residual insecticide spraying, primarily for malaria vector control, was relatively successful; however, when spraying of DDT was discontinued, the incidence of AVL began to increase again. Vaccines hold a great deal of promise, but currently no commercial products are available. As the incidence of disease continues to increase, particularly in Brazil, the importance of reviewing procedures for acceptance of blood donors and subsequent transfusions is becoming more relevant and important. Currently, blood is not routinely monitored for antibodies; however, serologic studies would serve as a first approach to the study of transmission of leishmanial organisms through blood transfusion. A seroreactivity of 9% has been found, increasing to 25% in a periurban focus of AVL infection (42).

Control programs for AVL often include elimination or treatment of infected dogs. Although attempting to remove seropositive dogs is insufficient as an eradication measure, data suggest that elimination of the majority of seropositive dogs may affect the cumulative incidence of seroconversion in dogs temporarily and may also diminish the incidence of human cases of AVL (57). Since one of the current strategies is based on detection and destruction of infected dogs, the highly predictive, sensitive, and specific fucose-mannose ligand ELISA represents a useful tool for field control of New World VL (13).

References

1. Albrecht, H., H. J. Stellbrink, G. Gross, B. Berg, U. Helmchen, and H. Mensing. 1994. Treatment of atypical leishmaniasis with interferon gamma resulting in progression of Kaposi's sarcoma in an AIDS patient. *Clin. Investig.* **72:**1041–1047.

2. Ali, S. A., J. Iqbal, B. Ahmad, and M. Masoon. 1998. A semisynthetic fetal calf serum-free liquid medium for in vitro cultivation of *Leishmania* promastigotes. *Am. J. Trop. Med. Hyg.* **59:**163–165.

3. Allahverdiyev, A. M., S. Uzun, M. Bagirova, M. Durdu, and H. R. Memisoglu. 2004. A sensitive new microculture method for diagnosis of cutaneous leishmaniasis. *Am. J. Trop. Med. Hyg.* **70:**294–297.

4. Almeida, R., A. D'Oliveira, P. Machado, O. Bacellar, A. I. Ko, A. R. de Jesus, N. Mobashery, J. B. Santos, and E. M. Carvalho. 1999. Randomized, double-blind study of stibogluconate plus human granulocyte macrophage colony-stimulating factor versus stibogluconate alone in the treatment of cutaneous leishmaniasis. *J. Infect. Dis.* **180:**1735–1737.

5. Alvar, J., C. Canavate, B. Guttierrez-Solar, M. Jimenez, F. Laguna, R. Lopez-Velez, R. Molina, and J. Moreno. 1997. *Leishmania* and human immunodeficiency virus coinfection: the first 10 years. *Clin. Microbiol. Rev.* **10:**298–319.

6. Andrade-Narvaez, F. J., S. Medina-Peralta, A. Vargas-Gonzalez, S. B. Canto-Lara, and S. Estrada-Parra. 2005. The histopathology of cutaneous leishmaniasis due to *Leishmania* (*Leishmania*) *mexicana* in the Yucatan peninsula, Mexico. *Rev. Inst. Med. Trop. Sao Paulo* **47:**191–194.

7. Arana, B. A., T. R. Navin, F. E. Arana, J. D. Berman, and F. Rosenkaimer. 1994. Efficacy of a short course (10 days) of high-dose meglumine antimonate with or without interferon in treating cutaneous leishmaniasis in Guatemala. *Clin. Infect. Dis.* **18:**381–384.

8. Armijos, R. X., M. M. Weigel, H. Aviles, R. Maldonado, and J. Racines. 1998. Field trial of a vaccine against new world cutaneous leishmaniasis in an at-risk child population: safety, immunogenicity, and efficacy during the first 12 months of follow-up. *J. Infect. Dis.* **177:**1352–1357.

9. Ashford, R. W., and P. A. Bates. 1998. Leishmaniasis in the Old World, p. 214–240. *In* F. E. G. Cox, J. P. Krier, and D. Wakelin (ed.), *Topley & Wilson's Microbiology and Microbial Infections*, 9th ed. Arnold, London, United Kingdom.

10. Atta, A. M., A. Doliveira, J. Correa, M. L. B. Atta, R. P. Almeida, and E. M. Carvalho. 1998. Anti-leishmanial IgE antibodies: a marker of active disease in visceral leishmaniasis. *Am. J. Trop. Med. Hyg.* **59:**426–430.

11. Bagchi, A. K., S. Tiwari, S. Gupta, and J. C. Katiyar. 1998. The latex agglutination test: standardization and comparison with direct agglutination and dot-ELISA in the diagnosis of visceral leishmaniasis in India. *Ann. Trop. Med. Parasitol.* **92:**159–163.

12. Belli, A., D. Garcia, X. Palacios, B. Rodriguez, S. Valle, E. Videa, E. Tinoco, F. Marin, and E. Harris. 1999. Widespread atypical cutaneous leishmaniasis caused by *Leishmania* (*L.*) *chagasi* in Nicaragua. *Am. J. Trop. Med. Hyg.* **61:**380–385.

13. Cabrera, G. P. B., V. O. Da Silva, R. T. Da Costa, A. B. Reis, W. M. O. Genaro, and C. B. Palatink-de-Sousa. 1999.

The fucose-mannose ligand-ELISA in the diagnosis and prognosis of canine visceral leishmaniasis in Brazil. *Am. J. Trop. Med. Hyg.* **61:**296–301.

14. **Calvopina, M., E. A. Gomez, H. Uezato, H. Kato, S. Nonaka, and Y. Hashiguchi.** 2005. Atypical clinical variants in New World cutaneous leishmaniasis: disseminated, erysipeloid, and recidiva cutis due to *Leishmania (V.) panamensis. Am. J. Trop. Med. Hyg.* **73:**281–284.

15. **Castilho, T. M., J. J. Shaw, and L. M. Floeter-Winter.** 2003. New PCR assay using glucose-6-phosphate dehydrogenase for identification of *Leishmania* species. *J. Clin. Microbiol.* **41:**540–546.

16. **Centers for Disease Control and Prevention.** 2004. Update: cutaneous leishmaniasis in U.S. military personnel—Southwest/Central Asia, 2002–2004. *Morb. Mortal. Wkly. Rep.* **53:**264–265.

17. **Chatterjee, M., C. L. Jaffe, S. Sundar, D. Basu, S. Sen, and C. Mandal.** 1999. Diagnostic and prognostic potential of a comparative enzyme-linked immunosorbent assay for leishmaniasis in India. *Clin. Diagn. Lab. Immunol.* **6:**550–554.

18. **Chiurillo, M. A., M. Sachdeva, V. S. Dole, Y. Yepes, E. Miliani, L. Vázquez, A. Rojas, G. Crisante, P. Guevara, N. Añez, R. Madhubala, and J. L. Ramírez.** 2001. Detection of *Leishmania* causing visceral leishmaniasis in the old and new worlds by a polymerase chain reaction assay based on telomeric sequences. *Am. J. Trop. Med. Hyg.* **65:**573–582.

19. **Couppie, P., E. Clyti, M. Sobesky, F. Bissuel, P. del Giudice, D. Sainte-Marie, J. P. Dedet, B. Carme, and R. Pradinaud.** 2004. Comparative study of cutaneous leishmaniasis in human immunodeficiency virus (HIV)-infected patients and non-HIV-infected patients in French Guiana. *Br. J. Dermatol.* **151:**1165–1171.

20. **Dedet, J. P., and F. Pratlong.** 2000. *Leishmania, Trypanosoma* and monoxenous trypanosomatids as emerging opportunistic agents. *J. Eukaryot. Microbiol.* **47:**37–39.

21. **de la Rosa, R., J. A. Pineda, J. Delgado, J. Macías, F. Morillas, J. A. Mira, A. Sánchez-Quijano, M. Leal, and E. Lissen.** 2002. Incidence of and risk factors for symptomatic visceral leishmaniasis among human immunodeficiency virus type 1-infected patients from Spain in the era of highly active antiretroviral therapy. *J. Clin. Microbiol.* **40:**762–767.

22. **Delgado, J., J. Macias, J. A. Pineda, J. E. Corzo, M. P. Gonzalez-Moreno, R. De la Rosa, A. Sanchez-Quijano, M. Leal, and E. Lissen.** 1999. High frequency of serious side effects from meglumine antimoniate given without an upper limit dose for the treatment of visceral leishmaniasis in human immunodeficiency virus type-1-infected patients. *Am. J. Trop. Med. Hyg.* **61:**766–769.

23. **Delgado, J., J. A. Pineda, J. Macias, C. Regordan, J. A. Gallardo, M. Leal, A. Sanches-Quijano, and E. Lissen.** 1998. Low sensitivity of peripheral blood smear for diagnosis of subclinical visceral leishmaniasis in human immunodeficiency virus type 1-infected patients. *J. Clin. Microbiol.* **36:**315–316.

24. **Dereure, J., F. Pratlong, J. Reynes, D. Basset, P. Bastian, and J. P. Dedet.** 1998. Haemoculture as a tool for diagnosing visceral leishmaniasis in HIV-negative and HIV-positive patients: interest for parasite identification. *Bull. W. H. O.* **76:**203–206.

25. **Desjeux, P.** 2004. Leishmaniasis: current situation and new perspectives. *Comp. Immunol. Microbiol. Infect. Dis.* **27:**305–318.

26. **Dillon, D. C., C. H. Day, J. A. Whittle, A. J. Magill, and S. G. Reed.** 1995. Characterization of a *Leishmania tropica* antigen that detects immune responses in Desert Storm viscerotropic leishmaniasis patients. *Proc. Natl. Acad. Sci. USA* **92:**7981–7985.

27. **Falcoff, E., N. J. Taranto, C. E. Remondegui, J. P. Dedet, L. M. Canini, C. M. Ripoll, L. Dimier-David, F. Vargas, L. A. Gimenez, J. G. Bernabo, and O. A. Bottasso.** 1994. Clinical healing of antimony-resistant cutaneous or mucocutaneous leishmaniasis following the combined administration of interferon gamma and pentavalent antimonial compounds. *Trans. R. Soc. Trop. Med. Hyg.* **88:**95–97.

28. **Gavgani, A. S. M., H. Mohite, G. H. Edrissian, M. Mohebali, and C. R. Davies.** 2002. Domestic dog ownership in Iran is a risk factor for human infection with *Leishmania infantum. Am. J. Trop. Med. Hyg.* **67:**511–515.

29. **Ghalib, H. W., J. A. Whittle, M. Kubin, F. A. Hashim, A.M. El-Hassan, K. H. Grabstein, G. Trinchieri, and S. G. Reed.** 1995. IL-12 enhances Th1-type responses in human *Leishmania donovani* infections. *J. Immunol.* **154:**4623–4629.

30. **Gontijo, C. M., R. S. Pacheco, F. Orefice, E. Lasmar, E. S. Silva, and M. N. Melo.** 2002. Concurrent cutaneous, visceral and ocular leishmaniasis caused by *Leishmania (Viannia) braziliensis* in a kidney transplant patient. *Mem. Inst. Oswaldo Cruz* **97:**751–753.

31. **Grimaldi, G., Jr., and R. B. Tesh.** 1993. Leishmaniases of the new world: current concepts and implications for future research. *Clin. Microbiol. Rev.* **6:**230–250.

32. **Guler, M. L., J. D. Gorham, C. Hsieh, A. J. Mackey, R. G. Steen, W. F. Dietrich, and K. M. Murphy.** 1996. Genetic susceptibility to *Leishmania*: IL-12 responsiveness in T_H1 cell development. *Science* **271:**984–987.

33. **Handman, E.** 2001. Leishmaniasis: current status of vaccine development. *Clin. Microbiol. Rev.* **14:**229–243.

34. **Isaza, D. M., M. Restrepo, and W. Mosca.** 1997. Immunoblot analysis of *Leishmania panamensis* antigens in sera of patients with American cutaneous leishmaniasis. *J. Clin. Microbiol.* **35:**3043–3047.

35. **Jha, T. K., S. Sundar, C. P. Thakur, P. Bachmann, J. Karbwang, C. Fischer, A. Voss, and J. Berman.** 1999. Miltefosine, an oral agent, for the treatment of Indian visceral leishmaniasis. *N. Engl J. Med.* **341:**1795–1800.

36. **Kenney, R. T., D. L. Sacks, J. P. Sypek, L. Vilela, A. A. Gam, and K. Evans-Davis.** 1999. Protective immunity using recombinant human IL-12 and alum as adjuvants in a primate model of cutaneous leishmaniasis. *J. Immunol.* **163:**4481–4488.

37. **Kerr, S. F., C. P. McHugh, and R. Merkelz.** 1999. Short report. A focus of *Leishmania mexicana* near Tucson, Arizona. *Am. J. Trop. Med. Hyg.* **61:**378–379.

38. **Lainson, R., and J. J. Shaw.** 1998. New World leishmaniasis—the neotropical *Leishmania* species, p. 241–266. *In* F. E. G. Cox, J. P. Krier, and D. Wakelin (ed.), *Topley & Wilson's Microbiology and Microbial Infections*, 9th ed. Arnold, London, United Kingdom.

39. Lesho, E. P., G. Wortmann, R. Neafie, and N. Aronson. 2005. Nonhealing skin lesions in a sailor and a journalist returning from Iraq. *Cleve. Clin. J. Med.* **72:**93–94.

40. Limoncu, M. E., I. C. Balcioglu, K. Yereli, Y. Ozbel, and A. Ozbilgin. 1997. A new experimental in vitro culture medium for cultivation of *Leishmania* species. *J. Clin. Microbiol.* **35:**2430–2431.

41. Lopez-Velez, R., J. A. Perez-Molina, A. Guerrero, F. Baquero, J. Villar-Rubia, L. Escribano, C. Bellas, F. Perez-Corral, and J. Alvar. 1998. Clinicoepidemiologic characteristics, prognostic factors and survival analysis of patients coinfected with human immunodeficiency virus and *Leishmania* in an area of Madrid, Spain. *Am. J. Trop. Med. Hyg.* **58:**436–443.

42. Luz, K. G., V. O. Da Silva, E. M. Gomes, F. C. S. Machado, M. A. F. Araujo, H. E. M. Fonseca, T. C. Friere, J. B. D'Almeida, M. Palatnik, and C. B. Palatnik-de-Sousa. 1997. Prevalence of anti-*Leishmania donovani* antibody among Brazilian blood donors and multiply transfused hemodialysis patients. *Am. J. Trop. Med. Hyg.* **57:**168–171.

43. Marques, M. J., A. C. Volpini, O. Genaro, W. Mayrink, and A. J. Romanha. 2001. Simple form of clinical sample preservation and *Leishmania* DNA extraction from human lesions for diagnosis of American cutaneous leishmaniasis via polymerase chain reaction. *Am. J. Trop. Med. Hyg.* **65:**902–906.

44. Marsden, P. D., H. A. Lessa, M. R. F. Olveira, G. A. S. Romero, J. G. Marotti, R. N. R. Sampaio, A. Barral, E. M. Carvalho, C. C. Cuba, A. V. Magalhaes, and V. O. Macedo. 1998. Clinical observations of unresponsive mucosal leishmaniasis. *Am. J. Trop. Med. Hyg.* **59:**543–545.

45. Martinez, S., M. Gonzalez, and M. E. Vernaza. 1997. Treatment of cutaneous leishmaniasis with allopurinol and stibogluconate. *Clin. Infect. Dis.* **24:**165–169.

46. Martín-Sánchez, J., J. A. Pineda, F. Morillas-Márquez, J. A. García-García, C. Acedo, and J. Macías. 2004. Detection of *Leishmania infantum* kinetoplast DNA in peripheral blood from asymptomatic individuals at risk for parenterally transmitted infections: relationship between polymerase chain reaction results and other *Leishmania* infection markers. *Am. J. Trop. Med. Hyg.* **70:**545–548.

47. Mathis, A., and P. Deplazes. 1995. PCR and in vitro cultivation for detection of *Leishmania* spp. in diagnostic samples from humans and dogs. *J. Clin. Microbiol.* **33:**1145–1149.

48. Maurya, R., R. K. Singh, B. Kumar, P. Salotra, M. Rai, and S. Sundar. 2005. Evaluation of PCR for diagnosis of Indian kala-azar and assessment of cure. *J. Clin. Microbiol.* **43:**3038–3041.

49. Mayrink, W., J. Pinto, C. da Costa, V. Toledo, T. Guimaraes, O. Genaro, and L. Vilela. 1999. Short report. Evaluation of the potency and stability of a candidate vaccine against American cutaneous leishmaniasis. *Am. J. Trop Med. Hyg.* **61:**294–295.

50. McHugh, C. P., P. C. Melby, and S. G LaFon. 1996. .Leishmaniasis in Texas: epidemiology and clinical aspects of human cases. *Am. J. Trop. Med. Hyg.* **55:**547–555.

51. Medrano, F. J., C. Canavate, M. Leal, C. Rey, E. Lissen, and J. Alvar. 1998. The role of serology in the diagnosis and prognosis of visceral leishmaniasis in patients coinfected with human immunodeficiency virus type-1. *Am. J. Trop. Med. Hyg.* **59:**155–162.

52. Merlen, T., D. Serano, N. Brajon, F. Rostand, and J. L. Lemesre. 1999. *Leishmania* spp.: completely defined medium without serum and macromolecules (CDM/LP) for the continuous in vitro cultivation of infective promastigote forms. *Am. J. Trop. Med. Hyg.* **60:**41–50.

53. Mira, J. A., J. E. Corzo, A. Rivero, J. Macias, F. L. De Leon, J. Torre-Cisneros, J. Gomez-Mateos, R. Jurado, and J. A. Pineda. 2004. Frequency of visceral leishmaniasis relapses in human immunodeficiency virus-infected patients receiving highly active antiretroviral therapy. *Am. J. Trop. Med. Hyg.* **70:**298–301.

54. Musa, A. M., E. A. Khalil, F. A. Mahgoub, S. Hamad, A. M. Elkadaru, and A. M. El Hassan. 2005. Efficacy of liposomal amphotericin B (AmBisome) in the treatment of persistent post-kala-azar dermal leishmaniasis (PKDL). *Ann. Trop. Med. Parasitol.* **99:**563–569.

55. Oliveira-Neto, M. P., A. S. M. Mattos, S. C. Goncalves-Costa, and C. Pirmez. 1997. A low-dose antimony treatment in 159 patients with American cutaneous leishmaniasis: extensive follow-up studies (up to 10 years). *Am. J. Trop. Med. Hyg.* **57:**651–655.

56. Oskam, L., J. L. Nieuwenhuijs, and A. Hailu. 1999. Evaluation of the direct agglutination test (DAT) using freeze-dried antigen for the detection of anti-*Leishmania* antibodies in stored sera from various patient groups in Ethiopia. *Trans. R. Soc. Trop. Med. Hyg.* **93:**275–277.

57. Palatnik-de-Sousa, C. B., W. R. dos Santos, J. C. Franca-Silva, R. T. da Costa, A. B. Reis, M. Palatnik, W. Mayrink, and O. Genaro. 2001. Impact of canine control on the epidemiology of canine and human visceral leishmaniasis in Brazil. *Am. J. Trop. Med. Hyg.* **65:**510–517.

58. Pasquau, F., J. Ena, R. Sanchez, J. M. Cuadrado, C. Amador, J. Flores, C. Benito, C. Redondo, J. Lacruz, V. Abril, J. Onofre, and the Leishmania HIV Mediterranean Co-Operative Group. 2005. Leishmaniasis as an opportunistic infection in HIV-infected patients: determinants of relapse and mortality in a collaborative study of 228 episodes in a Mediterranean region. *Eur. J. Clin. Microbiol. Infect. Dis.* **24:**411–418.

59. Pearson, R. D., S. M. B. Jeronimo, and A. Q. Sousa. 2001. Leishmaniasis, p. 287–313. In S. H. Gillespie and R. D. Pearson (ed.), *Principles and Practice of Clinical Parasitology.* John Wiley & Sons, Ltd., Chichester, United Kingdom.

60. Pearson, R. D., and A. Q. Sousa. 1995. Clinical spectrum of leishmaniasis. *Clin. Infect. Dis.* **22:**1–13.

61. Postigo, C., R. Llamas, C. Zarco, R. Rubio, F. Pulido, J. R. Costa, and L. Iglesias. 1998. Cutaneous lesions in patients with visceral leishmaniasis and HIV infection. *J. Infect.* **35:**265–268.

62. Pratlong, F., J. A. Rioux, P. Marty, F. Faraut-Gambarelli, J. Dereure, G. Lanotte, and J. Dedet. 2004. Isoenzymatic analysis of 712 strains of *Leishmania infantum* in the south of France and relationship of enzymatic polymorphism to clinical and epidemiological features. *J. Clin. Microbiol.* **42:**4077–4082.

63. **Rab, M. A., and D. A. Evans.** 1998. Detection of anti-*Leishmania* antibodies in blood collected on filter paper by the direct agglutination test. *Trans. R. Soc. Trop. Med. Hyg.* **91:**713–715.

64. **Ramirez, J. R., S. Agudelo, C. Muskus, J. F. Alzate, C. Berberich, D. Barker, and I. D. Velez.** 2000. Diagnosis of cutaneous leishmaniasis in Colombia: the sampling site within lesions influences the sensitivity of parasitologic diagnosis. *J. Clin. Microbiol.* **38:**3768–3773.

65. **Rodrigues, E. H. G., M. E. F. de Brito, M. G. Mendonca, R. P. Werkhäuser, E. M. Coutinho, W. V. Souza, M. Albuquerque, M. L. Jardim, and F. G. C. Abath.** 2002. Evaluation of PCR for diagnosis of cutaneous leishmaniasis in an area of endemicity in Northeastern Brazil. *J. Clin. Microbiol.* **40:**3572–3576.

66. **Ryan, J. R., A. M. Smithyman, G. Rajasekariah, L. Hochberg, J. M. Stiteler, and S. K. Martin.** 2002. Enzyme-linked immunosorbent assay based on soluble promastigote antigen detects immunoglobulin M (IgM) and IgG antibodies in sera from cases of visceral and cutaneous leishmaniasis. *J. Clin. Microbiol.* **40:**1037–1043.

67. **Saha, S., T. Mazumdar, K. Anam, R. Ravindran, B. Bairagi, B. Saha, R. Goswami, N. Pramanik, S. K. Guha, S. Kar, D. Banerjee, and N. Ali.** 2005. *Leishmania* promastigote membrane antigen-based enzyme-linked immunosorbent assay and immunoblotting for differential diagnosis of Indian post-kala-azar dermal leishmaniasis. *J. Clin. Microbiol.* **43:**1269–1277.

68. **Saiz, M., A. Llanos-Cuentas, J. Echevarria, N. Roncal, M. Cruz, M. T. Muniz, C. Lucas, D. F. Wirth, S. Scheffter, A. J. Magill, and J. L. Patterson.** 1998. Short report. Detection of leishmaniavirus in human biopsy samples of leishmaniasis from Peru. *Am. J. Trop. Med. Hyg.* **58:** 192–194.

69. **Saran, R., M. C. Sharma, A. K. Gupta, S. P. Sinha, and S. K. Kar.** 1998. Diurnal periodicity of *Leishmania* amastigotes in peripheral blood of Indian kala-azar patients. *Acta Trop.* **68:**357–360.

70. **Schulz, A., K. Mellenthin, G. Schönian, B. Fleischer, and C. Drosten.** 2003. Detection, differentiation, and quantitation of pathogenic *Leishmania* organisms by a fluorescence resonance energy transfer-based real-time PCR assay. *J. Clin. Microbiol.* **41:**1529–1535.

71. **Selvapandiyan, A., K. Stabler, N. A. Ansari, S. Kerby, J. Riemenschneider, P. Salotra, R. Duncan, and H. L. Nakhasi.** 2005. A novel semiquantitative fluorescence-based multiplex polymerase chain reaction assay for rapid simultaneous detection of bacterial and parasitic pathogens from blood. *J. Mol. Diagn.* **7:**268–275.

72. **Sharma, S. K., T. Kadhiravan, A. Banga, T. Goyal, I. Bhatia, and P. K. Saha.** 2004. Spectrum of clinical disease in a series of 135 hospitalized HIV-infected patients from north India. *BMC Infect. Dis.* **4:**52.

73. **Shazad, B., B. Abbaszadeh, and A. Khamesipour.** 2005. Comparison of topical paromomycin sulfate (twice/day) with intralesional meglumine antimoniate for the treatment of cutaneous leishmaniasis caused by *L. major. Eur. J. Dermatol.* **15:**85–87.

74. **Singh, R. B., V. S. Raju, R. K. Jain, and P. Salotra.** 2005. Potential of direct agglutination test based on promastigote and amastigote antigens for serodiagnosis of post-kala-azar dermal leishmaniasis. *Clin. Diagn. Lab. Immunol.* **12:**1191–1194.

75. **Singh, S., and R. Sivakumar.** 2004. Challenges and new discoveries in the treatment of leishmaniasis. *J. Infect. Chemother.* **10:**307–315.

76. **Sinha, P. K., K. Pandey, and S. K. Bhattacharya.** 2005. Diagnosis and management of leishmania/HIV co-infection. *Indian J. Med. Res.* **121:**407–414.

77. **Sundar, S., L. B. Gupta, M. K. Makharia, M. K. Singh, A. Voss, F. Rosenkaimer, J. Engel, and H. W. Murray.** 1999. Oral treatment of visceral leishmaniasis with miltefosine. *Ann. Trop. Med. Parasitol.* **93:**589–597.

78. **Sundar, S., S. G. Reed, V. P. Singh, P. C. K. Kumar, and H. W. Murray.** 1998. Rapid accurate field diagnosis of Indian visceral leishmaniasis. *Lancet* **351:**563–565.

79. **Sundar, S., P. R. Sinha, N. K. Agrawal, R. Srivastava, P. M. Rainey, J. D. Berman, H. W. Murray, and V. P. Singh.** 1998. A cluster of cases of severe cardiotoxicity among kala-azar patients treated with a high-osmolarity lot of sodium antimony gluconate. *Am. J. Trop. Med. Hyg.* **59:**139–143.

80. **Toz, S. O., K. Chang, Y. Ozbel, and M. Z. Alkan.** 2004. Diagnostic value of RK39 dipstick in zoonotic visceral leishmaniasis in Turkey. *J. Parasitol.* **90:**1484–1486.

81. **Velasco-Castrejon, O., B. C. Walton, B. Rivas-Sanchez, M. F. Garcia, G. J. Lazaro, O. Hobart, S. Roldan, J. Floriani-Verdugo, A. Mingulas-Saldana, and R. Berzaluce.** 1997. Treatment of cutaneous leishmaniasis with localized current field (radio frequency) in Tabasco, Mexico. *Am. J. Trop. Med. Hyg.* **57:**309–312.

82. **Weina, P. J., R. C. Neafie, G. Wortmann, M. Polhemus, and N. E. Aronson.** 2004. Old world leishmaniasis: an emerging infection among deployed U.S. military and civilian workers. *Clin. Infect. Dis.* **11:**1674–1680.

83. **Willard, R. J., A. M. Jeffcoat, P. M. Benson, and D. S. Walsh.** 2005. Cutaneous leishmaniasis in soldiers from Fort Campbell, Kentucky, returning from Operation Iraqi Freedom highlights diagnostic and therapeutic options. *J. Am. Acad. Dermatol.* **52:**977–987.

84. **Wortmann, G., C. Sweeney, H. Houng, N. Aronson, J. Stiteler, J. Jackson, and C. Ockenhouse.** 2001. Rapid diagnosis of leishmaniasis by fluorogenic polymerase chain reaction. *Am. J. Trop. Med. Hyg.* **65:**583–587.

Trypanosomiasis

Trypanosomes are hemoflagellates of the family Trypanosomatidae that live in the blood and tissues of their human hosts (Table 9.1). The African trypanosomes belong to the subgenus *Trypanozoon* and as a group are referred to as the *Trypanosoma brucei* complex. *T. brucei brucei* strains are parasites of domestic and wild animals and are not known to be infectious for humans. Human infections are caused by *T. brucei gambiense* (West African trypanosomiasis) and *T. brucei rhodesiense* (East African trypanosomiasis). Infections caused by *T. brucei rhodesiense* are much more fulminant than those caused by *T. brucei gambiense*, the parasitemia is much higher, and the disease has a much faster progression. It is not possible to differentiate the three trypomastigotes of *T. brucei gambiense*, *T. brucei rhodesiense*, and *T. brucei brucei* on the basis of morphology, and all three organisms can be recovered in animal reservoir hosts. In the past, differentiation was based on clinical signs and geographic area of isolation. Differentiation is now based on isoenzyme characteristics and DNA and RNA methods in addition to the aforementioned factors. It is estimated that 60 million people in 36 countries are at risk, and there are approximately 50,000 new cases per year of African trypanosomiasis (49). Based on available information from western and central Africa, there appears to be no increased risk of African trypanosomiasis among those infected with human immunodeficiency virus (HIV); however, this has not been totally confirmed. There are also unanswered questions concerning disease presentations and diagnostic test results in patients who are coinfected with both organisms.

Trypanosomes infecting humans in the Americas belong to the subgenus *Tejaraia* (*T. rangeli*) and *Schizotrypanum* (*T. cruzi*). *T. rangeli* infections are asymptomatic, with no evidence of pathology; however, *T. cruzi* infections (American trypanosomiasis) can cause considerable morbidity and mortality.

African Trypanosomiasis

Trypanosoma brucei gambiense

Although the first descriptions of human sleeping sickness were made by European colonists in the late 19th and early 20th centuries, Atkins described a "sleepy distemper" in 1742 and David Livingstone described a "fly disease" in

Table 9.1   Classification of trypanosomes infecting humans

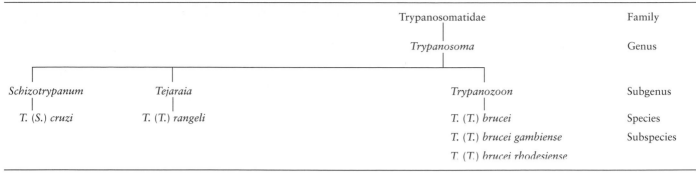

			Trypanosomatidae		Family
			Trypanosoma		Genus
Schizotrypanum	*Tejaraia*		*Trypanozoon*		Subgenus
T. (S.) cruzi	T. (T.) rangeli		T. (T.) brucei		Species
			T. (T.) brucei gambiense		Subspecies
			T. (T.) brucei rhodesiense		

1857 (49). In 1901, *T. brucei gambiense* was seen in the blood of a European steamboat captain working on the Gambia River. After Forde's discovery in 1901, Dutton in 1902 proposed the name *Trypanosoma gambiense*. Although a trypanosome was discovered in 1903 in the spinal fluid of a patient suffering from sleeping sickness and Castellani named it *Trypanosoma ugandense*, it was soon clear that the two organisms were identical. By 1909, the mode of transmission by the tsetse fly, *Glossina palpalis*, was confirmed. The primary area of endemic infection with *T. brucei gambiense* coincides with the tsetse fly belt through the heart of Africa, where 300,000 to 500,000 people may be infected in western and central Africa (Figure 9.1). The disease remains largely unchecked due to the sophisticated biological defense mechanisms adopted by the parasite, its tolerance to therapy, and the socioeconomic realities in a region where priorities may not include adequate surveillance and control (49).

Life Cycle and Morphology

Trypanosomal forms are ingested by the tsetse fly (*Glossina* spp.) when a blood meal is taken (Figures 9.2 and 9.3). Once the short, stumpy trypomastigote reaches the midgut of the tsetse fly, it transforms into a long, slender procyclic stage. The organisms multiply in the lumen of the midgut and hindgut of the fly. After approximately 2 weeks, the organisms migrate back to salivary glands through the hypopharynx and salivary ducts, where the organisms attach to the epithelial cells of the salivary ducts and then transform to their epimastigote forms. In the epimastigote forms, the nucleus is posterior to the kinetoplast, in contrast to the trypomastigote, in which the nucleus is anterior to the kinetoplast (Figure 9.2). There is continued multiplication

Figure 9.1   Distribution of trypanosomes in Africa. (Armed Forces Institute of Pathology photograph.)

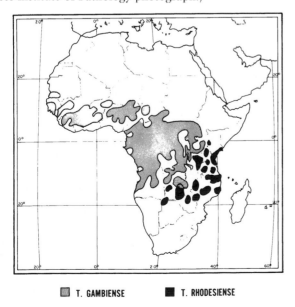

T. GAMBIENSE   T. RHODESIENSE

Figure 9.2 Life cycle stages of trypanosomes. Amastigotes have been confirmed in the life cycle of *Trypanosoma cruzi* as one of the stages in the human host, but this stage in humans in African trypanosomiasis is only suspected and not confirmed. (Illustration by Nobuko Kitamura.)

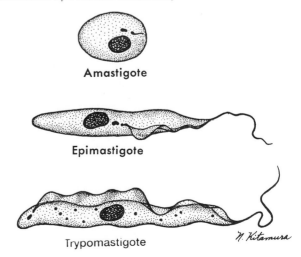

Amastigote

Epimastigote

Trypomastigote

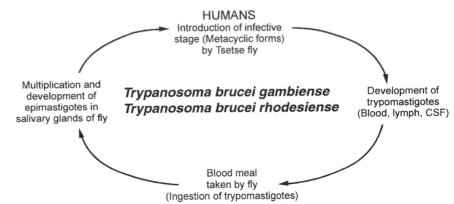

HUMANS
Introduction of infective
stage (Metacyclic forms)
by Tsetse fly

Multiplication and
development of
epimastigotes in
salivary glands of fly

Trypanosoma brucei gambiense
Trypanosoma brucei rhodesiense

Development of
trypomastigotes
(Blood, lymph, CSF)

Blood meal
taken by fly
(Ingestion of trypomastigotes)

Figure 9.3 Life cycle of *Trypanosoma brucei gambiense* and *T. brucei rhodesiense.*

within the salivary gland, and metacyclic (infective) forms develop from the epimastigotes in 2 to 5 days. With development of the metacyclic forms, the tsetse fly becomes infective. During the act of feeding, the fly introduces the metacyclic trypanosomal forms into the next victim in saliva injected into the puncture wound. The entire developmental cycle in the fly takes about 3 weeks, and once infected, the tsetse fly remains infected for life.

The trypanosomal (trypomastigote) forms can be found in the blood, cerebrospinal fluid (CSF), lymph node aspirates, and fluid aspirated from the trypanosomal chancre (if one forms at the site of the tsetse fly bite). The trypomastigote forms multiply by longitudinal binary fission. The organism is highly pleomorphic, having a variety of trypanosomal forms in the same blood smear. The forms range from long, slender-bodied organisms with a long flagellum (trypomastigote) that attain a length of 30 μm or more to short, fat, stumpy forms without a free flagellum that are approximately 15 μm long. The short, stumpy forms do not divide in the bloodstream but are the infective stage for the tsetse fly.

In blood, the trypanosomes move rapidly among the red blood cells. An undulating membrane and flagellum may be seen with slower-moving organisms. The trypomastigote forms are 14 to 33 μm long and 1.5 to 3.5 μm wide (Figure 9.4). With Giemsa or Wright's stain, the granular cytoplasm stains pale blue and contains dark blue granules and possibly vacuoles. The centrally located nucleus stains reddish. At the posterior end of the organism is the kinetoplast, which also stains reddish, and the remaining intracytoplasmic flagellum (axoneme), which may not be noticeable. The flagellum arises from the kinetoplast, as does the undulating membrane. The flagellum runs along the edge of the undulating membrane until the undulating membrane merges with the trypanosome body at the anterior end of the organism. At this point, the flagellum becomes free to extend beyond the body (Figure 9.2). Although the trypomastigote is the diagnostic stage seen in clinical specimens, amastigotes have been seen in the choroid plexus in mice.

A unique feature of African trypanosomes is their ability to change the surface coat of the outer membrane of the trypomastigote, helping to evade the host immune response. The trypomastigote surface is covered with a dense coat of approximately 10^7 molecules of the membrane-form variant surface glycoprotein (VSG). There are approximately 100 to 1,000 genes in the genome, responsible for encoding as many as 1,000 different VSGs. More than 100 serotypes have been detected in a single infection. It is postulated that the trypomastigote changes its coat about every 5 to 7 days (antigenic variation). This change is responsible for successive waves of parasitemia every 7 to 14 days and allows the parasite to evade the host humoral immune response (32). There is no evidence that the host immune system induces the VSG switches. During antigenic switching, VSG is both internalized by the trypomastigote and released or shed into the blood. This shed VSG is most probably responsible for the immune dysfunctions noticed during infections. There is a release of gamma interferon (which stimulates parasite growth), suppression of interleukin-2, and hypergammaglobulinemia. Based on studies using *T. brucei*, bloodstream forms when aggregated in the presence of antibodies can subsequently disaggregate, and this mechanism may function to aid survival of the trypomastigotes in the presence of antibody in the host prior to the occurrence of a VSG switching event (74). The disaggregation appears to be strictly energy dependent, and the organisms remain motile, metabolically active, and infective following this process.

Antigenic variation, however, is probably only one of several mechanisms enabling these parasites to thrive in spite of the host immune defenses. The ability to grow at high levels of gamma interferon, to avoid complement-mediated destruction, and to vary their susceptibility to a subclass of human high-density lipoproteins may also facilitate organism survival (29).

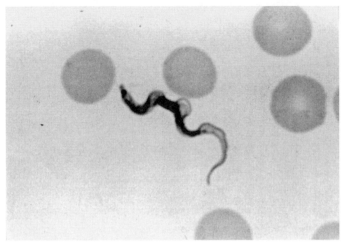

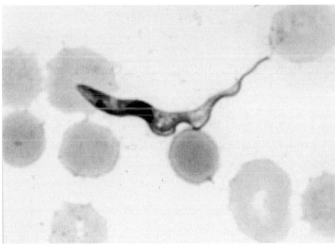

Figure 9.4 *Trypanosoma brucei gambiense* trypomastigotes in peripheral blood smears. Note the small kinetoplast.

The genome of *T. brucei* has been defined with a complete analysis of the 11 megabase-sized chromosomes. The 26-Mb genome contains 9,068 predicted genes, including approximately 900 pseudogenes and approximately 1,700 *T. brucei*-specific genes. There are 806 VSG genes. Based on this research, a number of potential drug targets have been identified (14).

Clinical Disease

In general, African trypanosomiasis caused by *T. brucei gambiense* (West African sleeping sickness) has a long, mild, chronic course that ends fatally with central nervous system (CNS) involvement after several years' duration. This is unlike the disease caused by *T. brucei rhodesiense* (East African sleeping sickness), which has a short course and ends fatally within a year.

After the host has been bitten by an infected tsetse fly, metacyclic trypomastigote stages are introduced into the skin, where they multiply and set up a local inflammatory reaction. A nodule or chancre at the site (3 to 4 cm) may develop after a few days. However, this primary lesion will resolve spontaneously within 1 to 2 weeks. The chancre is seen frequently in white Europeans but rarely in patients indigenous to an area where the disease is endemic. Trypomastigotes may be detected in fluid aspirated from the ulcer. The trypomastigotes enter the bloodstream, causing a symptom-free low-grade parasitemia that may continue for months. This is considered stage I disease, where the patient can have systemic trypanosomiasis without CNS involvement. During this time, the parasites may be difficult to detect, even by thick blood film examinations. The infection may self-cure during this period without development of symptoms or lymph node invasion.

Symptoms may occur months to years after infection. The first distinct symptoms appear on invasion of the lymph nodes and are followed by the onset of remittent, irregular fevers with night sweats. Headaches, malaise, and anorexia frequently accompany the fevers. The febrile period of up to a week is followed by an afebrile period of variable duration and then another febrile period, and so on. Many trypomastigotes may be found in the circulating blood during fevers, but few are seen during afebrile periods. Lymphadenopathy is a consistent feature of Gambian trypanosomiasis. Enlarged lymph nodes are soft, painless, and nontender. Although any lymph node may be affected, posterior cervical regions are most frequently involved (Winterbottom's sign) (Figure 9.5). Trypomastigotes can be aspirated from the enlarged lymph nodes. In addition to lymph node involvement, the spleen and liver become enlarged. With Gambian trypanosomiasis, the blood-lymphatic stage may last for years before the sleeping sickness syndrome occurs.

Transient edema is frequently seen and occurs in the face, hands, feet, and other periarticular areas. Pruritus is common, and in light-skinned individuals, an irregular erythematous rash, suggestive of erythema multiforme, can be seen. The rash occurs 6 to 8 weeks after the onset of infection and is typically located on the trunk, shoulders, buttocks, and thighs. It usually lasts only a few hours, and it appears and disappears during the febrile period. Patients may experience delayed sensation to pain (Kerandel's sign). Laboratory findings include anemia, granulocytopenia, reduction in platelet counts, increased sedimentation rate, polyclonal B-cell activation with a marked increase in the levels of immunoglobulin M (IgM), heterophile and anti-DNA antibodies, rheumatoid factor, and circulating immune complex in serum. The sustained high IgM levels are a result of the parasite producing variable antigen types which allow the organism to evade the patient's defense system. *In an immunocompetent host,*

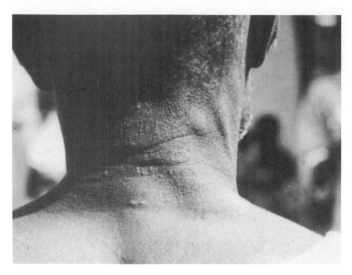

Figure 9.5 Winterbottom's sign (enlarged cervical lymph nodes in the posterior cervical triangle). (Armed Forces Institute of Pathology photograph.)

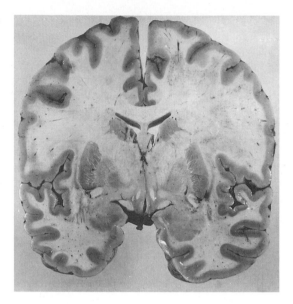

Figure 9.6 Coronal section of brain showing congested white matter and petechial hemorrhages. (Armed Forces Institute of Pathology photograph.)

the absence of elevated IgM levels in serum rules out trypanosomiasis.

Once trypomastigotes invade the CNS, the sleeping sickness stage of the infection is initiated (stage II disease). The trypomastigotes are found primarily in the frontal lobes, pons, and medulla. Behavioral and personality changes are seen during CNS invasion. Gambian trypanosomiasis is characterized by steady progressive meningoencephalitis, apathy, confusion, fatigue, coordination loss, and somnolence. In the terminal phase of the disease, the patient becomes emaciated and progresses to profound coma and death, usually from secondary infection. Immunosuppression in patients with Gambian trypanosomiasis is manifested by decreases in both cellular and humoral immunity.

With invasion of the CNS (Figure 9.6), there is a progressive leptomeningitis, and CSF findings include increased protein levels, elevated IgM levels, lymphocytosis, and morular cells of Mott (Figure 9.7). Morular (mulberry) cells are altered plasma cells whose cytoplasm is filled with proteinaceous droplets (Figure 9.8). Morular cells are not seen in all patients; however, they are characteristic of African trypanosomiasis.

In humans with African trypanosomiasis, autoantibodies have been found, including autoantibodies directed to red blood cells, smooth muscle, liver and cardiolipid nucleic acids (DNA and RNA), and intermediate filaments of the cytoskeleton in smooth muscle cells. Autoantibodies to components of CNS myelin have also been found, as have those to neurofilament proteins in sera and CSF from untreated patients with stage II disease and CNS involvement (7). The CSF antibodies to neurofilament proteins were IgM in 22 of 25 patients and IgG in 8 of 25.

Diagnosis

Physical findings and clinical history are very important in establishing the diagnosis. Diagnostic symptoms include irregular fever, enlargement of the lymph nodes (particularly those of the posterior triangle of the neck [Winterbottom's sign]), delayed sensation to pain (Kerandel's sign), and erythematous skin rashes. Definitive diagnosis depends on demonstration of trypomastigotes in blood, lymph node aspirates, sternum bone marrow, and CSF. Due to the greater frequency of high parasitemia, there is a better chance of detecting organisms in body fluids in infections caused by *T. brucei rhodesiense* than in those caused by *T. brucei gambiense*. Because of periodicity, parasite numbers in the blood may vary; therefore, multiple specimens should be collected and a number of techniques should be used to detect the trypomastigotes (Figure 9.4; Tables 9.2 and 9.3). Health care personnel must adhere to standard precautions when handling specimens from patients with suspected cases of African trypanosomiasis because the trypomastigotes are highly infectious.

Chancre Aspirate.   Organisms can be obtained from the chancre by using puncture techniques; the fluid is then examined as wet mounts and Giemsa-stained smears. This approach is seldom used; patients are generally not seen until after the chancre has already healed.

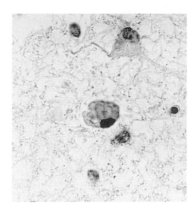

Figure 9.7 Morular cell in the hypothalamus. (Armed Forces Institute of Pathology photograph.)

Lymph Node Aspirate. When enlarged cervical lymph nodes are seen, they can be punctured and the aspirate can be examined as a wet preparation for the presence of motile trypomastigotes. The aspirated material can also be stained using Giemsa or other blood stains. Depending on a number of factors, the sensitivity ranges from about 40 to 80% (49).

Blood. Blood can be collected from either finger stick or venipuncture. Venous blood should be collected in a tube containing EDTA. Multiple slides should be made for examination, and multiple blood examinations should be done before trypanosomiasis is ruled out. Parasites are found in large numbers in the blood during the febrile period and in small numbers when the patient is afebrile. In addition to thin and thick blood films, a buffy coat concentration method is recommended to detect the parasites. Parasites can be detected on thin blood films with a detection limit at approximately 1 parasite/200 microscopic fields (high dry power magnification, ×400); they can be detected on thick blood smears when the numbers are greater than 2,000/ml and when they are greater than 100/ml with hematocrit capillary tube concentration.

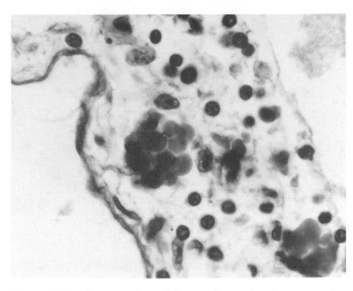

Figure 9.8 Close-up view of the meninges showing a morular cell. (Armed Forces Institute of Pathology photograph.)

Microhematocrit Centrifugation. The original microhematocrit centrifugation technique uses capillary tubes containing anticoagulant, filled with finger stick blood. The bottom of the tube is sealed with plasticine, and the tube is centrifuged in a hematocrit centrifuge for about 8 min. The trypomastigotes are concentrated at the buffy coat layer. The tubes are then placed in a special holder and examined through the glass at low magnification for motile organisms (21).

Microhematocrit centrifugation with use of the QBC malaria tube (glass capillary tube and closely fitting plastic insert; CBC malaria blood tubes [Becton Dickinson, Tropical Disease Diagnostics, Sparks, Md.]) has been used for the detection of blood parasites. At the end of centrifugation of 50 to 60 μl of capillary or venous blood (5 min in a QBC centrifuge; 14,387 × g), parasites or erythrocytes containing parasites are concentrated into a small, 1- to 2-mm region near the top of the RBC column and are held close to the wall of the tube by the plastic

Table 9.2 Characteristics of East and West African trypanosomiasis

Characteristic	East African	West African
Organism	*Trypanosoma brucei rhodesiense*	*Trypanosoma brucei gambiense*
Vector	Tsetse fly, *Glossina morsitans* group	Tsetse fly, *Glossina palpalis* group
Primary reservoirs	Animals (wild and domestic)	Humans
Illness	Acute (early CNS invasion), <9 mo	Chronic (late CNS invasions), months to years
Lymphadenopathy	Minimal	Prominent
Parasitemia	High	Low
Epidemiology	Anthropozoonosis, game parks	Anthroponosis, rural populations
Diagnostic stage	Trypomastigote	Trypomastigote
Recommended specimens	Chancre aspirate, lymph node aspirate, blood, CSF	Chancre aspirate, lymph node aspirate, blood, CSF

Table 9.3 Methods available for the diagnosis of African trypanosomiasis (*T. brucei gambiense*)

Method	Description	Comments
Chancre aspirate	Puncture; fluid obtained is examined as wet preparation and Giemsa-stained smears (other blood stains are acceptable)	Not commonly used; patients are often seen after the chancre has already healed.
Lymph node aspirate	Puncture; fluid obtained is examined as wet preparation and Giemsa-stained smears (other blood stains are acceptable)	Sensitivity varies between 40 and 80%; yield is very low with seronegative patients. Not recommended unless clinical signs are present.
Blood	Both wet films and stained thick and thin blood films can be examined	Organism motility among red blood cells enhances detection; sensitivity about 1 organism per 200 microscopic fields (~10,000 trypomastigotes/ml) (using high dry objective). Stained thick blood film improves sensitivity with detection threshold of ~5,000 trypomastigotes/ml. Organism morphology better on thin blood film. Giemsa or other blood stains are acceptable.
Microhematocrit centrifugation	Centrifugation of capillary tubes containing anticoagulant filled with finger stick blood (end of tube sealed with plastic clay); tube examined under microscope, and organisms are concentrated at buffy coat layer	
Microhematocrit centrifugation (QBC)	High-speed centrifugation of blood in capillary tubes containing acridine orange; motile organisms identified by fluorescent nuclei and kinetoplasts in the buffy coat layer	Current production of QBC kit has been stopped; however, capillary tube production has been resumed (21). The technique is used sporadically within the United States (cost is a factor).
Mini-anion-exchange centrifugation	Separation of trypomastigotes from venous blood using anion-exchange chromatography; subsequent concentration at tip of glass tube	Sensitivity of <100 organisms/ml; procedure is very time-consuming. The columns are produced in Belgium. Method validation is under way.
Cerebrospinal fluid	White blood cell count used to determine disease stage; cell-counting chambers should have a capacity of 1 μl (Fuchs-Rosenthal and Neubauer chambers); do not dilute CSF	Patients (trypomastigotes in lymph node or blood) with >20 cells/μl of CSF are treated as having second-stage disease; therapeutic approach depends on the geographic areas and drug selection. Trypomastigotes in CSF are confirmatory for second-stage disease; sensitivity is increased by centrifugation.
Antigen detection	A simple and rapid test, the card indirect agglutination trypanosomiasis test (TrypTect CIATT) is available, primarily in areas of endemic infection; an ELISA method has been used to detect antigen in serum and CSF	Good sensitivity but possible specificity issues with CIATT; ELISA antigen is rapidly cleared after therapy.
Antibody detection	CATT fast/simple agglutination assay for *T. brucei gambiense*-specific antibodies; latex agglutination is another option; immunofluorescence excellent (sensitivity and specificity), but lower with filter paper eluates; ELISA available for serum, filter paper eluates, and CSF	Sensitivity reported on whole blood (87–98%); negative predictive value is excellent for mass screening. Both false positives and false negatives have been reported. Can be performed with blood-impregnated filter paper. Latex agglutination shows higher specificity but lower sensitivity than fluorescence. Latex is also more specific than CATT on diluted specimens.
Molecular methods	A number of methods have been reported; however, cost and technical issues have prevented their use in the routine clinical laboratory	Some methods appear to be very sensitive; simplification of the method may lead to methods adaptable to routine clinical laboratory use.

float, thereby making them readily visible by microscopy. Tubes precoated with acridine orange provide a stain which induces fluorescence in the parasites. This method automatically prepares a concentrated smear, which represents the distance between the float and the walls of the tube. Once the tube is placed into the plastic holder (Paraviewer) and immersion oil is applied onto the top of the hematocrit tube (no coverslip is necessary), the tube is examined with a 40× to 60× oil immersion objective (there must be a working distance of 0.3 mm or greater) (refer to chapter 31).

The range of sensitivity for the QBC test has been as low as 55% up to >90% compared with microscopy. While most think the method is rapid, some think there are some disadvantages such as (i) the high cost of capillary tubes and equipment, (ii) problems in species identification and quantification, and (iii) technical problems, broken capillaries, and the fact that the capillaries cannot be stored for later reference. This may not be the most appropriate approach for a laboratory that receives a small number of specimens for blood parasite testing, but it could be helpful in other settings. Apparently the QBC kit is no longer in production, but capillary tube manufacture has been resumed (19).

Mini-Anion-Exchange Centrifugation. The mini-anion-exchange method allows the separation of trypomastigotes from venous blood by anion-exchange chromatography, with subsequent low-speed centrifugation into the tip of the glass tube. The tube is then examined under the microscope with a sensitivity of <100 organisms/ml. Validation of this approach is under way (21).

Cerebrospinal Fluid. If CSF is examined, a volume greater than 1 ml, preferably 5 ml or more, should be collected. CSF examination must be conducted with centrifuged sediments. The presence of Mott cells in the CSF is pathognomonic for trypanosomiasis. Mott cells are rarely observed when there are few white blood cells in the CSF. When trypomastigotes are present in undetectable numbers in the blood, they may be seen in aspirates of inflamed lymph nodes; however, attempts to demonstrate them histopathologically are not practical. Blood and CSF specimens should be examined during therapy to evaluate the response and again 1 to 2 months after therapy. Patients treated for CNS disease should be monitored clinically for 2 to 3 years after completion of therapy to detect a relapse.

Antigen Detection. An enzyme-linked immunosorbent assay (ELISA) method has been used to detect antigen in serum and CSF (73). It was noted that antigen was rapidly cleared from the circulation following successful therapy. This method could also be used for clinical staging of dis-

ease to determine whether there was CNS infection and for follow-up to therapy.

A simple and rapid test, the card indirect agglutination trypanosomiasis test (TrypTect CIATT), is available, primarily in areas of endemic infection, for the detection of circulating antigens in persons with African trypanosomiasis. The test is normally performed on a drop of freshly collected heparinized blood and is followed by a more specific confirmation test on diluted blood, plasma, or serum. The sensitivity of the test (95.8% for *T. brucei gambiense* and 97.7% for *T. brucei rhodesiense*) is significantly higher than that of lymph node puncture, microhematocrit centrifugation, and CSF examination after single and double centrifugation (72). Its specificity is excellent, and it has a high positive predictive value. A complement-mediated prozone phenomenon may occur, causing lower test sensitivity at lower sample dilutions. A simple solution is the addition of a Ca^{2+} chelating agent such as EDTA to the diluted blood, plasma, or serum (77).

Antibody Detection. The card agglutination test for trypanosomiasis (CATT/*T. b. gambiense*) is more sensitive than methods that require visualization of the trypomastigotes with clinical specimens and can be performed with blood, plasma, or serum. The test is produced by the Institute of Tropical Medicine in Antwerp, Belgium, and is widely used for mass population screening. This approach can also be used with blood-impregnated filter paper. It has been recommended that treatment with pentamidine should be considered for all serologically suspected individuals with a CATT-plasma end dilution titer of ≥1:16 in areas of moderate to high disease prevalence (20).

A semiquantitative ELISA, using VSG of *T. brucei gambiense* as the antigen, has been developed for the detection of antibodies of different immunoglobulin isotypes in serum and CSF of patients with sleeping sickness. In serum and CSF, a drastic increase in the level of IgG, basically IgG1, as well as in the IgM level, was seen; an increase in the IgG3 level was also seen, while the IgA level remained relatively normal. Measurement of immunoglobulin and trypanosome-specific antibody concentrations in serum and CSF allows calculation of intrathecal antibody synthesis and may be a possible tool for the determination of clinical stages of sleeping sickness (63). Both latex agglutination and immunofluorescence tests are also available; however, sensitivity results have not been as good with the latex approach as with fluorescence. Specificity with the latex test appears to be better than the CATT with diluted blood; use of the latex test also decreased the workload and costs (79, 93).

Serologic techniques which have been widely used for epidemiologic screening include indirect fluorescent-antibody assays, ELISA, the indirect hemagglutination test, and the card agglutination trypanosomiasis test (see above). A major serodiagnostic problem in areas where the disease is endemic is that many in the population have elevated antibody levels due to exposure to animal trypanosomes that are noninfectious to humans. Serum and CSF IgM concentrations are of diagnostic value. CSF antibody titers should be interpreted with caution because of the lack of reference values and the possibility that the CSF contains serum as the result of a traumatic tap.

Molecular Methods. Accurate identification of trypanosome species, which is necessary to clarify the epidemiology of human and animal African trypanosomiasis, has been difficult. Great progress has been made over the last 10 years through the application of DNA probe technology, although this has also revealed greater complexity than was previously supposed (39). *T. brucei* organisms have been detected in human blood by using a nonradioactive branched-DNA-based method. Buffy coat specimens are used, and the results agree with those obtained using buffy coat microscopy. The assay can be performed with crude lysates without the need for extensive sample preparation. Also, because it amplifies the signal from a target molecule rather than the target itself, this technique is not prone to the artifact problems that can be seen with PCR-based tests (48). The development of a peptide nucleic acid fluorescence in situ hybridization probe appears to offer an excellent diagnostic tool. By combining this test with a cytospin step, the limit of detection from blood can be improved to 5 parasites/ml, a level of sensitivity that equals an optimal PCR detection limit with blood specimens (84). A newer PCR technique targets the gene encoding the small ribosomal subunit in order to identify and differentiate all clinically important African trypanosome species and some subspecies. This method appears to be more economical, simple, and sensitive than some other screening methods, and it yields more detailed information (24). Other real-time PCR options appear to be suitable for use within the routine clinical laboratory (12). The loop-mediated isothermal amplification (LAMP) reaction is a method that amplifies DNA with high specificity, efficiency, and rapidity under isothermal conditions with only simple incubators. An added advantage of LAMP over PCR-based methods is that DNA amplification can be monitored spectrophotometrically and/or with the naked eye without the use of dyes (58).

Animal Inoculation and Culture. *T. brucei gambiense* isolation in small laboratory animals is usually unsuccessful; in contrast, *T. brucei rhodesiense* readily infects animals. Cultivation is not practical for most di-

agnostic laboratories but is more successful than animal inoculation. See Algorithm 9.1.

KEY POINTS—LABORATORY DIAGNOSIS

Trypanosoma brucei gambiense

1. Trypomastigotes are highly infectious, and health care workers must use blood-borne-pathogen precautions.
2. Trypomastigotes may be detected in aspirates of the chancre and enlarged lymph nodes in addition to blood and CSF.
3. Concentration techniques, such as centrifugation of CSF and blood, should be used in addition to thin and thick smears (Giemsa or Wright's stain).
4. Trypomastigotes are present in largest numbers in the blood during febrile periods.
5. Examinations of multiple daily blood samples may be necessary to detect the parasite.
6. Blood and CSF specimens should be examined during therapy and again 1 to 2 months posttherapy.

Treatment

All drugs currently used in the therapy of African trypanosomiasis are toxic and require prolonged administration. Treatment should be started as soon as possible and is based on the patient's symptoms and laboratory findings. The choice of antiparasitic drug depends on whether the CNS is infected. Suramin or pentamidine isethionate can be used when the CNS is not infected. Although pentamidine is usually listed as an alternative drug, it appears that in early CNS involvement, the results (94% cure rate in a group of 52 patients observed for 2 years) are comparable to those obtained with melarsoprol or eflornithine in terms of tolerance and availability (31). Pentamidine isethionate does not cross the blood-brain barrier and is administered intramuscularly; side effects include an immediate hypotensive reaction, nausea, vomiting, and Herxheimer-type (inflammatory) reactions. Suramin (Bayer 205), a diamidine, is effective in treating the hemolymphatic stage and CNS disease of *T. brucei gambiense*. When suramin is used, a test dose should be given to the patient to ensure that the drug can be tolerated. It should not be given to patients with renal disease, and if protein, red blood cells, or casts are detected in the urine, treatment should be stopped. Suramin is given intravenously; its side effects include nausea, vomiting, loss of consciousness, seizures, pruritus, edema, and hepatitis. Although rare, fatalities have been reported during suramin therapy. Suramin can be obtained under an Investigational New Drug (IND) protocol from the CDC Drug Service, Centers for Disease Control and Prevention, Atlanta, Ga., 30333;

AFRICAN TRYPANOSOMIASIS

```
          ◆ SKIN ULCER
          ◆ FEVER, LYMPHADENOPATHY,
            RASH
          ◆ MENTAL CHANGES

                    ↓
                          NEGATIVE        DIAGNOSIS
        HISTORY    ─────────────────→     UNLIKELY
                    ↓
        POSITIVE
                    ↓
     TRYPOMASTIGOTES       NEGATIVE       DIAGNOSIS
     IN BLOOD SMEARS  ──────────────→     UNLIKELY
                    ↓
        POSITIVE
                    ↓
        DIAGNOSIS
        CONFIRMED
                    │
          ┌─────────┴─────────┐
          ↓                   ↓
   INFECTION IN BLOOD    INFECTION IN CNS
     TREATMENT I           TREATMENT II
```

Algorithm 9.1   African trypanosomiasis.

404-639-3670 (evenings, weekends, or holidays: 404-639-2888) (1).

Melarsoprol, a toxic trivalent arsenic derivative, is effective for both hemolymphatic and neural stages but is recommended for treatment of late-stage sleeping sickness. It is administered intravenously, and its side effects include nausea, vomiting, encephalopathy, and exfoliative dermatitis. Melarsoprol has been used for CNS involvement for years; however, relapse occurs in up to 6% of patients. Arsenic-related encephalopathy occurs in 10% of patients, and prednisolone treatment may help reduce the incidence. The mortality due to this side effect depends on whether supportive treatment is given, but it is often above 50% (67). Resistance of both *T. brucei gambiense* and *T. brucei rhodesiense* to melarsoprol has been noted.

Eflornithine (DL-α-difluoromethylornithine; DFMO) has been used for more than 10 years to treat melarsoprol-resistant *T. brucei gambiense* infection with or without CNS involvement; treatment is started intravenously and is followed by oral therapy. In one study, 47 patients with a relapse following a first treatment were teated with a 7-day course of intravenous eflornithine (100 mg/kg every 6 h) and monitored for 2 years; the failure rate was 6.5% (54). This approach appears to provide adequate treatment in cases of Gambian trypanosomiasis relapsing after treatment with another drug. Disadvantages of eflornithine therapy include the prolonged course of treatment and frequent side effects of diarrhea and anemia. With adequate treatment, non-CNS disease has a cure rate of >80%. All

treated patients should be followed up twice yearly for at least 2 to 3 years. Specific drug dosage information is provided in chapter 25; however, apparently eflornithine is no longer commercially available but can be obtained under an IND protocol from the CDC Drug Service (1, 49).

Nifurtimox (Lampit), a nitrofuran, has been used to treat patients infected with *T. brucei gambiense* who do not respond to melarsoprol. The relapse rate is quite high when nifurtimox is used alone, but the drug is quite effective when used in combination with melarsoprol (97). It is given orally. Side effects include confusion, tremor, vertigo, anorexia, and weight loss. This drug is not recommended for the treatment of *T. brucei rhodesiense* infections. Nifurtimox can be obtained under an IND protocol from the CDC Drug Service (1).

Epidemiology and Prevention

T. brucei gambiense is transmitted from person to person by the bite of the tsetse fly (*G. palpalis* and *G. tachinoides*) after infectivity develops within the insect. The development cycle in the fly, depending on temperature and moisture, varies from 12 to 30 days and averages 20 days. Fewer than 10% of tsetse flies become infective after obtaining blood from infected patients, and vertical transmission from infected fly to offspring is not known to occur. *G. palpalis* and *G. tachinoides* can be found in areas of thick shrubbery and trees near the banks of rivers, streams, or water holes; therefore, transmission readily occurs when people frequent these areas. Both female and male tsetse flies can transmit the infection. In addition to biological transmission, the tsetse fly may mechanically transmit the infection with its proboscis if it bites an uninfected person within a few hours of biting an infected person. Congenital transmission in humans has also been documented.

Although *T. brucei gambiense* can be transmitted to animals, no animal reservoirs for West African sleeping sickness have been documented. Trypanosomal strains isolated from hartebeest, kob, chickens, dogs, cows, and domestic pigs in West Africa are identical to those isolated from humans in the same area. These findings suggest that animals may serve as reservoirs; however, there is no indication that these animal trypanosomal forms are directly transmitted to humans and that a patent infection will develop. Epidemiologic evidence suggests that in areas of endemic infection, transmission may be entirely from human to human. Asymptomatic individuals are thought to be the residual reservoir of the disease.

In areas where West African sleeping sickness is endemic, *T. brucei gambiense* has been controlled by eliminating the parasite through regular population-screening programs. This approach effectively reduces the prevalence of the disease to low levels; however, with interruptions in surveillance, resurgence of the disease will occur. An example of resurgence of West African trypanosomiasis was well documented in southern Sudan, where the disease prevalence found was among the highest ever documented.

Although game parks in Tanzania (Tarangire and Serengeti National Parks) were thought to be low-risk areas for African trypanosomiasis, during 2001 nine patients from these areas were diagnosed with *T. brucei gambiense* infection. All of the South African patients but one were European nationals. During their trip or shortly after their return, the patients, all febrile, were seen by general practitioners or emergency departments. Most patients were seen during the primary disease stage; however, several showed cerebral manifestations of the secondary stage. Most patients also showed a typical skin lesion, the trypanosome chancre. Diagnosis was established by examining thin and thick blood films. Although three patients had multiorgan failure, and specific medication was difficult to obtain, drug treatment proved successful in all but one patient, who died (53).

At present, there are no areas where *T. brucei gambiense* and *T. brucei rhodesiense* are endemic together. Although foci of both organisms exist in Uganda, they appear to be geographically separate. Based on evidence of the spread of cattle carrying *T. brucei rhodesiense* further north, a merger of the two foci could create tremendous problems since the diagnosis and treatment of the two diseases differ significantly (19, 26).

Vector control measures have met with limited success. Methods have included clearing streams of underbrush, eliminating breeding grounds, using insecticides, using fly traps, and releasing sterile male flies. Because tsetse flies have a low reproductive rate and have not developed insecticide resistance, they are very tempting control targets. The most effective control measures include an integrated approach to reduce the human reservoir of infection and to use insecticides and fly traps.

In regions where the disease is endemic, natives appear to be more resistant to infection than are new arrivals to the area, even though there is no evidence of acquired immunity. West African sleeping sickness affects primarily the rural populations, and tourists are rarely infected. Chemoprophylaxis is not recommended because of drug toxicity, and vaccines are unavailable. It may be possible to develop a vaccine, because immunity to reinfection occurs with *T. brucei gambiense*.

Tourists are usually not at great risk unless they are traveling and spending long periods in rural areas of western and central Africa. Persons visiting areas where the disease is endemic should avoid tsetse fly bites by wearing protective clothing (long-sleeved shirts and long pants); khaki or olive-colored clothing is optimal, since the tsetse fly is attracted to bright colors and very dark colors. Because heavy clothing is not always practical owing to heat and humidity, other measures, including the use of insect repellents, bed netting, and screens, are recommended. It has also been recommended not to ride in the back of jeeps, pickup trucks, or other open vehicles; the tsetse fly is attracted to the dust created by moving vehicles and wild animals. Bushes should be avoided, since the tsetse fly is less active during the hot period of the day and rests in bushes but will bite if disturbed.

Trypanosoma brucei rhodesiense

T. brucei rhodesiense is closely related to *T. brucei gambiense*, and the two subspecies are morphologically indistinguishable. Stephans and Fantham in 1910 discovered *T. brucei rhodesiense* in the blood of a patient in Rhodesia with symptoms consistent with "sleeping sickness." In 1912, this organism was shown to be transmitted by a different species of tsetse fly (*Glossina morsitans*) from that associated with *T. brucei gambiense* (*G. palpalis*). *T. brucei rhodesiense* was also differentiated from *T. brucei gambiense* on the basis of virulence in rats and detection of a morphologic variant not previously found in *T. brucei gambiense*. Many investigators think that *T. brucei rhodesiense* is the human strain of *T. brucei*, a common trypanosome that is found in many game animals and causes a fatal disease known as nagana in cattle. In 1961, Ormerod indicated that the original strains of *T. brucei rhodesiense* probably arose in the Zambesi basins as mutants from *T. brucei* (11). In the early 1900s, human volunteer inoculation studies with *T. brucei* resulted in no human infections. Currently, it is accepted that there are two subspecies within the *T. brucei* species, and they have been designated *T. brucei gambiense* and *T. brucei rhodesiense*. *T. brucei rhodesiense* (which causes Rhodesian trypanosomiasis or East African sleeping sickness) is more limited in distribution than *T. brucei gambiense*, being found only in central East Africa, where the disease has been responsible for some of the most serious obstacles to economic and social development in Africa. Within this area, the vast majority of tsetse flies prefer animal blood, which therefore limits the raising of livestock. The infection has a greater morbidity and mortality than does *T. brucei gambiense* infection, and game animals, such as the bushbuck, and cattle are natural reservoir hosts. Biochemical evidence shows that the human-infective *T. brucei rhodesiense* differs from the human-infective *T. brucei brucei* because of its resistance to lysis by components in human serum. This resistance can be acquired by sensitive strains and is caused by the decreased internalization of a trypanolytic factor. Thus, *T. brucei rhodesiense* appears to be a host range variant of *T. brucei brucei* (49). While human- and non-human-infective strains are genetically closely related, very little gene flow appears to be occurring between these two host range variants in natural populations. Restricted gene flow could be due to a lack of genetic exchange caused by lack of opportunity or some type of biological barrier. Evidence also suggests that genetic exchange can occur in the tsetse fly; however, large-scale studies of *T. brucei* genotypes in tsetse flies are difficult, especially since only 0.1% of the flies are usually infected (49).

Life Cycle and Morphology

The life cycle and reproduction of *T. brucei rhodesiense* are similar to those of *T. brucei gambiense* (Figure 9.3). The tsetse fly vectors for transmission of *T. brucei rhodesiense* are *G. pallidipes*, *G. morsitans*, and *G. tachinoides*. The morphology of *T. brucei rhodesiense* in tsetse flies and humans is similar to that of *T. brucei gambiense*, except that when the parasite is inoculated into laboratory animals, the posterior nucleate forms are more common

Clinical Disease

T. brucei rhodesiense produces a more rapid, fulminating disease than does *T. brucei gambiense*. Fever, severe headaches, irritability, extreme fatigue, swollen lymph nodes, and aching muscles and joints are common symptoms. Progressive confusion, personality changes, slurred speech, seizures, and difficulty in walking and talking occur as the organisms invade the CNS. The early stages of the pathologic process parallel those of *T. brucei gambiense* infections; however, the disease progresses more rapidly, such that death may occur before there is extensive CNS involvement even though CNS invasion occurs early. The incubation period is short, often within 1 to 4 weeks, with trypomastigotes being more numerous and appearing earlier in the blood. Lymph node involvement is less pronounced, and Winterbottom's sign may be absent. Febrile paroxysms are more frequent, and the patients are more anemic and more likely to develop myocarditis or jaundice. Some patients may develop persistent tachycardia, and death may result from arrhythmia and congestive heart failure due to pancarditis. Myocarditis may develop in patients with Gambian trypanosomiasis but is more common and severe with the Rhodesian form. Review of a splenectomized patient with Rhodesian trypanosomiasis and CNS involvement revealed that the onset of symptoms, laboratory studies, and disease progression did not differ from those for reported cases in healthy individuals (65). However, the role of the spleen in trypanosomiasis is not well understood.

Diagnosis

The techniques used are similar to those for detecting Gambian trypanosomiasis; however, trypomastigotes are more numerous in the blood in the Rhodesian form (Table 9.2). A differential diagnosis may be possible

on geographic grounds, but there are areas where the two forms overlap. Because infection with *T. brucei rhodesiense* is so fulminant, the immune response with humoral antibody will be weaker than that for *T. brucei gambiense* infections; therefore, serodiagnosis may be difficult. Blood and CSF specimens should be examined during and 1 to 2 months after therapy.

T. brucei rhodesiense is more adaptable to cultivation than is *T. brucei gambiense*. A number of media have been described; however, cultivation is not a practical diagnostic approach. See Algorithm 9.1, Table 9.4, and the "Key Points—Laboratory Diagnosis" presented below.

KEY POINTS—LABORATORY DIAGNOSIS

Trypanosoma brucei rhodesiense

1. Trypomastigotes are highly infectious, and health care workers must use blood-borne-pathogen precautions.
2. Trypomastigotes may be detected in aspirates of the chancre and enlarged lymph nodes in addition to blood and CSF; trypomastigotes tend to be more numerous in the blood in Rhodesian trypanosomiasis, and diagnosis is normally confirmed from blood examination.
3. Concentration techniques, such as centrifugation of CSF and blood, should be used in addition to thin and thick smears (Giemsa or Wright's stain). However, concentration methods may not be required, since organisms tend to be numerous in the blood.
4. Trypomastigotes are present in the blood in largest numbers during febrile periods.
5. Examinations of multiple daily blood samples may be necessary to detect the parasite.
6. Blood and CSF specimens should be examined during therapy and 1 to 2 months posttherapy.
7. Culture is possible using well-described media but is probably not practical for most diagnostic laboratories; again, the use of blood-borne-pathogen precautions is mandatory.

Treatment

Because East African trypanosomiasis is such a serious illness, treatment should be started as soon as possible and is based on the patient's symptoms and laboratory findings. Within the United States, medication and therapeutic advice is available through the CDC. Without treatment, nearly all patients develop neural involvement and die; however, even in treated patients, in those with CNS infection, cure rates are above 80%. The choice of antiparasitic drug depends on whether the CNS is infected; usually patients are hospitalized during treatment. Suramin or pentamidine isethionate can be used early in the course of

Table 9.4 Methods available for the diagnosis of African trypanosomiasis (*T. brucei rhodesiense*)

Method	Description	Comments
Chancre aspirate	Puncture; fluid obtained is examined as wet preparation and Giemsa-stained smears (other blood stains are acceptable)	Not commonly used; patients are often seen after the chancre has already healed.
Lymph node aspirate	Puncture; fluid obtained is examined as wet preparation and Giemsa-stained smears (other blood stains are acceptable)	Normally not required, disease presents as acute febrile illness.
Blood	Both wet films and stained thick and thin blood films can be examined	Organism motility among red blood cells enhances detection; trypomastigotes are generally numerous. Organism morphology better on thin blood film. Giemsa or other blood stains acceptable.
Microhematocrit centrifugation	Centrifugation of capillary tubes containing anticoagulant filled with finger stick blood (end of tube sealed with plastic clay); tube examined under microscope; and organisms are concentrated at buffy coat layer	Simple method, widely used for epidemiology studies; based on large numbers of organisms in *T. brucei rhodesiense* infections. Buffy coat should be highly motile with organisms.
Microhematocrit centrifugation (QBC)	High-speed centrifugation of blood in capillary tubes containing acridine orange; motile organisms identified by fluorescent nuclei and kinetoplasts in the buffy coat layer	Current production of QBC kit has been stopped; however, capillary tube production has been resumed (21). The technique is used sporadically within the United States (cost is a factor).
Mini-anion-exchange centrifugation	Separation of trypomastigotes from venous blood using anion-exchange chromatography; subsequent concentration at tip of glass tube	Sensitivity of <100 organisms/ml; procedure is very time-consuming. The columns are produced in Belgium. Method validation is under way. Not required for diagnosis of *T. brucei rhodesiense* (many organisms present).
Cerebrospinal fluid	White blood cell count used to determine disease stage; cell-counting chambers should have a capacity of 1 µl (Fuchs-Rosenthal and Neubauer chambers); do not dilute CSF	Patients (trypomastigotes in lymph node or blood) with >20 cells/µl of CSF are treated as having second-stage disease; therapeutic approach depends on the geographic areas and drug selection. Trypomastigotes in CSF are confirmatory for second-stage disease; sensitivity is increased by centrifugation.
Antigen detection	Not relevant for *T. brucei rhodesiense* due to large numbers of trypomastigotes in blood	Diagnostic method much more relevant for *T. brucei gambiense* infections.
Antibody detection	No equivalent to CATT available for screening for *T. brucei rhodesiense*; development and testing under way for agglutination card tests and particle agglutination tests; immunofluorescence tests and ELISA for antibody available, but the sensitivity varies	Immunofluorescence tests and ELISA available in reference centers (highly skilled individuals are required for testing and interpretation).
Molecular methods	Molecular markers (serum resistance-associated gene, *T. brucei rhodesiense*; receptor-like flagellar pocket glycoprotein, *T. brucei gambiense*) may be helpful in monitoring the distribution of both species	Could also be helpful for targeting therapy for cattle infected with *T. brucei rhodesiense*, as well as distinguishing species in humans exposed to both organisms.

the infection, when the CNS is not infected. Pentamidine isethionate does not cross the blood-brain barrier and is administered intramuscularly; side effects include an immediate hypotensive reaction, nausea, vomiting, and Herxheimer-type (inflammatory) reactions.

Suramin is the drug of choice for treating the hemolymphatic stage *of T. brucei rhodesiense* infections. When suramin is used, a test dose should be given to the patient to ensure that the drug can be tolerated. It should not be given to patients with renal disease, and if protein, red blood cells, or casts are detected in the urine, treatment should be stopped. Suramin is given intravenously; its side effects include nausea, vomiting, loss of consciousness, seizures, pruritus, edema, and hepatitis. Although rare, fatalities have been reported during suramin therapy. Suramin can be obtained under an IND protocol from the CDC Drug Service, Centers for Disease Control and Prevention, Atlanta, Ga., 30333; 404–639–3670 (evenings, weekends, or holidays: 404–639–2888) (1).

Melarsoprol is capable of penetrating the blood-brain barrier and is the drug of choice when *T. brucei rhodesiense* CNS disease is suspected; it is effective for both hemolymphatic and neural stages. Melarsoprol has been used for CNS involvement for years; however, relapse occurs in up to 6% of patients. Arsenic-related encephalopathy occurs in 10% of patients, and prednisolone treatment may help reduce the incidence. The mortality due to this side effect depends on whether supportive treatment is given, but it is often above 50% (67). Resistance of both *T. brucei gambiense* and *T. brucei rhodesiense* to melarsoprol has been noted. Melarsoprol can be obtained under an IND protocol from the CDC Drug Service (1). All treated patients should be followed up twice yearly for at least 2 to 3 years. Specific drug dosage information is provided in chapter 25.

Epidemiology, Epidemics, and Prevention

The incidence of *T. brucei rhodesiense* infections is considerably lower than that of *T. brucei gambiense* infections, and the extent of distribution is smaller. East African trypanosomiasis is usually found in woodland and savannah areas away from human habitation. The tsetse fly vectors of Rhodesian trypanosomiasis are game feeders that may incidentally transmit the disease from human to human or animal to human. Because the infection results in acute rather than chronic disease, asymptomatic carriers are not a source of transmission as they are in Gambian trypanosomiasis. The disease is an occupational hazard for persons working in game reserves and a threat to visitors to game parks. *T. brucei rhodesiense* has been isolated from a variety of game animals (bushbuck, hartebeest, and lion) and domestic animals (cattle and sheep), a group that may prove to be the more important group of reservoir animals.

The incidence of East African trypanosomiasis is characterized by short epidemics interspersed with long periods of low (often undetectable) endemicity (50). The reasons for such outbreak situations have been widely discussed and debated. Studies of the composition of trypanosome strains present during epidemics have been directed at answering the following questions. (i) Are human-infective trypanosomes (*T. brucei rhodesiense*) distinct from those infecting animals (*T. brucei brucei*)? (ii) Are new strains responsible for the outbreaks? (iii) Are animals an important reservoir for strains that cause human disease? (iv) Can the occurrence of new outbreaks be traced to the movement of new "virulent" strains into the area?

Distinguishing the Two Organisms. Field data using restriction fragment length polymorphism analysis of repetitive DNAs and isoenzymes confirm that the subspecific distinction of *T. brucei rhodesiense* and *T. brucei brucei* appears to be correct (50). Although isoenzyme analysis had incorrectly detected taxonomy identity caused by convergent evolution, DNA patterns in restriction fragment length polymorphism indicate that strains of *T. brucei rhodesiense* from Zambia and from Kenya and Uganda are very different and have independent origins.

Animal Reservoirs of Sleeping Sickness. In the past, it was assumed that wild animals, particularly bushbuck, were the most important reservoir host of the parasites. More recently, cattle have been implicated in several outbreaks (50). Data show that between 21 and 33% of infected domestic animals (particularly cattle) can be infected with trypanosomes that are also infective for humans during an epidemic. Probabilities of transmission indicate that a fly infected with *T. brucei rhodesiense* is five times more likely to have picked up that infection from domestic cattle than from a human.

"New Strains" and Epidemics. In looking at changes in the ecology of a region, the appearance of a new strain, and the generation of new zymodemes by genetic exchange, it appears that there is little evidence to support the idea of new strains as a mechanism for the generation of epidemics but that the spread of a single existing strain group is more likely to be responsible.

Movement of People and Trypanosomes. It was originally thought that *T. brucei rhodesiense* originated in Zambia in the early 1900s and spread north to cause epidemics in Uganda and Kenya in the 1940s. However, analysis of trypanosomes from various locales does not support this idea. It is more likely that the origins of Rhodesian sleeping sickness in the Uganda foci were there all the time and actually coexisted with the outbreaks in Zambia (50).

Model for the Origin and Maintenance of Disease Foci in East Africa. A model can be proposed as follows: (i) each sleeping sickness focus has a single strain of human-infective trypanosome associated with it; (ii) this strain is maintained within the ecosystem of the focus, even in times of endemicity; and (iii) when favorable epidemiologic conditions are present, this strain generates an epidemic (49). A number of stages can be involved: (i) a new genotype arises that can be maintained in the population; (ii) once a transmission cycle is established, selection may maintain a single human-infective strain or a small number of such strains; and (iii) epidemic selection of this strain may occur when the tsetse fly biting rate increases. When the overall density of the infected human population increases, a full-blown epidemic caused by a single trypanosome strain occurs. A period of endemicity then follows as control measures decrease the number of cases. The human-infective strains are left circulating until such time as ecologic changes trigger another epidemic.

Control Measures.   Because Rhodesian trypanosomiasis is a zoonosis, control measures are more difficult to attain than with the Gambian form. In the past, reduction or elimination of game animals has been a major control method, although more recent information suggests that cattle are more relevant as reservoir hosts. Other measures used were the reduction of human contact in areas of endemic infection, reduction of vegetation around human settlements, insecticide spraying, use of fly traps, and prophylactic treatment of domestic animals. Tourists are usually not at great risk unless they are traveling and spending long periods in rural areas of eastern and central Africa. Persons visiting areas where the disease is endemic should avoid tsetse fly bites by wearing protective clothing (long-sleeved shirts and long pants); khaki or olive-colored clothing is optimal, since the tsetse fly is attracted to bright colors and very dark colors. Because heavy clothing is not always practical owing to heat and humidity, other measures, including the use of insect repellents, bed netting, and screens, are recommended. It has also been recommended not to ride in the back of jeeps, pickup trucks, or other open vehicles; the tsetse fly is attracted to the dust created by moving vehicles and wild animals. Bushes should be avoided, since the tsetse fly is less active during the hot period of the day and rests in bushes but will bite if disturbed.

With constant political change in Africa, vector control, mass treatment, prophylaxis, and population movement tracking will remain very difficult. Unfortunately, constant monitoring of areas of endemic infection is very difficult to accomplish and trained personnel continue to be scarce as other issues take priority.

American Trypanosomiasis

Trypanosoma cruzi

American trypanosomiasis (Chagas' disease) is a zoonosis caused by *T. cruzi*, which was discovered in the intestine of a triatomid bug in Brazil in 1909 by Carlos Chagas, who described the entire life cycle in reservoir hosts. After infected bugs were allowed to feed on a monkey, the trypomastigote form was found in the blood of the animal. Chagas then found the organisms in the blood of a child who had fever, anemia, and enlargement of the lymph nodes; he proved that the parasites were the cause of this common illness endemic in areas of Brazil. It is

interesting that this represents the first example where an animal parasite causing the disease and the insect vector were discovered prior to the disease itself. Based on his assumption that the organism multiplies by schizogony, Chagas named the parasite *Schizotrypanum cruzi*, a name that is still used by some. *T. cruzi* causes an acute or chronic parasitemia and invades the cells of many organs (e.g., the heart, esophagus, and colon). Chagas' disease has been found in a 15th- or 16th-century Peruvian mummy with megacolon and megaesophagus. It is one of the major health problems in Latin American countries.

It is estimated that 100 million persons are at risk of infection; between 16 million to 18 million are actually infected. There are approximately 50,000 deaths per year due to Chagas' disease. In certain areas of endemic infection, approximately 10% of all adult deaths are due to Chagas' disease. The southern United States, particularly Texas, is also now identified as having a number of cases. The geographic distribution of *T. cruzi* infection overlaps that of *T. rangeli* infection; therefore, the trypomastigotes may be misidentified (Table 9.5).

Two evolutionary lineages have been identified in *T. cruzi*, *T. cruzi* I and II. Prior to 2005, epidemiological and immunological data suggested that chronic infections occurring in Brazil and Argentina were caused primarily by *T. cruzi* II strains. Using PCR techniques, this hypothesis has been confirmed, and future studies can establish which strains are responsible for Chagas' disease in other geographic locations (36, 98).

Life Cycle and Morphology

The disease is transmitted to humans through the bite wound caused by reduviid bugs (triatomids, kissing bugs, or conenose bugs) (Figure 9.9). Humans are infected when metacyclic trypomastigotes are released with the feces while the insect is taking a blood meal and the feces are rubbed or scratched into the bite wound or onto mucosal surfaces such as eyes or mouth, an action stimulated by the allergic reaction to the insect's saliva. The organisms can also be transmitted as congenital infections, by blood transfusion, or by organ transplantation. On entry into the body, the metacyclic forms invade local tissues, transform to the amastigote stage, and begin to multiply within the cells. The local inflammatory process continues, forming the primary lesion, the chagoma, which blocks the lymphatic capillaries and causes edema.

A number of receptors have been identified that are involved in the invasion process of the trypomastigote into the host cell. In humans, *T. cruzi* can be found in two forms, amastigotes and trypomastigotes (Figures 9.2, 9.10, and 9.11). The trypomastigote does not divide in the blood but carries the infection to all parts of the body.

Table 9.5 Characteristics of American trypanosomiais

Characteristic	*Trypanosoma cruzi*	*Trypanosoma rangeli*
Vector	Reduviid bug	Reduviid bug
Primary reservoirs	Opossums, dogs, cats, wild rodents	Wild rodents
Illness	Symptomatic (acute, chronic)	Asymptomatic
Diagnostic stage		
Blood	Trypomastigote	Trypomastigote
Tissue	Amastigote	None
Recommended specimens	Blood, lymph node aspirate, chagoma	Blood, but organisms rarely recovered

The amastigote form multiplies within virtually any cell, preferring cells of mesenchymal origin such as reticuloendothelial, myocardial, adipose, and neuroglial cells. The parasites also occur in histiocytes of cutaneous tissue and in cells of the epidermis, as well as in the intestinal mucous membrane.

The trypomastigote (Figure 9.10) is spindle shaped and approximately 20 μm long, and it characteristically assumes a C or U shape in stained blood films. Trypomastigotes occur in the blood in two forms, a long slender form and a short stubby one. The nucleus is situated in the center of the body, with a large oval kinetoplast located at the posterior extremity. The kinetoplast consists of a small blepharoplast and a large oval parabasal body. A flagellum arises from the blepharoplast and extends along the outer edge of an undulating membrane until it reaches the anterior end of the body, where it projects as a free flagellum. When the trypomastigotes are stained with Giemsa stain, the cytoplasm stains blue and the nucleus, kinetoplast, and flagellum stain red or violet.

On invasion of a cell, the trypomastigote loses its flagellum and undulating membrane and divides by binary fission to form an amastigote (Figure 9.11). The amastigote continues to divide and eventually fills and destroys the infected cell. Both amastigote and trypomastigote forms are released from the cell. The amastigote is indistinguishable from those found in leishmanial infections. It is 2 to 6 μm in diameter and contains a large nucleus and rod-shaped kinetoplast that stains red or violet with Giemsa stain. The cytoplasm stains blue. Only the trypomastigotes are found free in the peripheral blood.

Trypomastigotes are ingested by the reduviid bug as it obtains a blood meal. The trypomastigote transforms into an epimastigote that multiplies in the posterior portion of the midgut (Figure 9.12). After 8 to 10 days, metacyclic trypomastigotes develop from the epimastigotes and are passed in the feces to infect humans when rubbed into the insect's puncture wound or rubbed onto exposed mucous membranes.

Clinical Disease

In addition to contracting *T. cruzi* infections through the insect's bite wound or exposed mucous membranes, persons can be infected by blood transfusion, organ transplantation, placental transfer, and accidental ingestion of parasitized reduviid bugs. The clinical syndromes associated with Chagas' disease can be broken down into acute, indeterminate, and chronic stages. The acute stage is the result of the first encounter of the patient with the parasite, whereas the chronic phase is the result of late sequelae. In children younger than 5 years, the disease is seen in its severest form, whereas in older children and adults, the disease is milder and is commonly diagnosed in the subacute or chronic form rather than in the acute form. Overall, the incubation period in humans is about 7 to 14 days but is somewhat longer in some patients.

Early Disease. A localized inflammatory reaction of variable intensity may ensue at the infection site. In most cases, the reaction is mild and may not be apparent. An erythematous subcutaneous nodule (chagoma), seen

Figure 9.9 Life cycle of *Trypanosoma cruzi*.

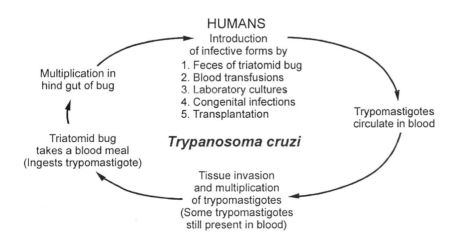

HUMANS
Introduction
of infective forms by
1. Feces of triatomid bug
2. Blood transfusions
3. Laboratory cultures
4. Congenital infections
5. Transplantation

Trypanosoma cruzi

Multiplication in hind gut of bug

Triatomid bug takes a blood meal (Ingests trypomastigote)

Trypomastigotes circulate in blood

Tissue invasion and multiplication of trypomastigotes (Some trypomastigotes still present in blood)

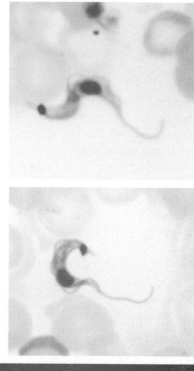

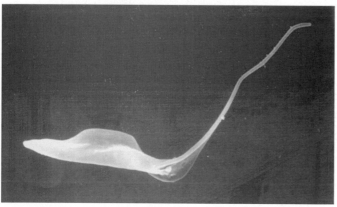

Figure 9.10 (Top and middle) *Trypanosoma cruzi* trypomastigotes in a peripheral blood smear. Note the large kinetoplast. (Bottom) Scanning electron micrograph of a trypomastigote of *T. cruzi.* (Bottom photograph courtesy of David T. John.)

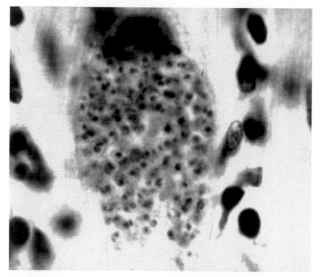

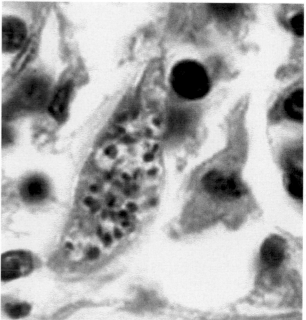

Figure 9.11 *Trypanosoma cruzi* amastigotes in cardiac muscle.

most frequently on the face, may form. The chagoma is painful and may take 2 to 3 months to subside. In early stages of infection, amastigotes or trypomastigotes may be aspirated from the chagoma; if the route of inoculation is the ocular mucosa, edema of the eyelids and conjunctivitis may occur (Romaña's sign) (Figure 9.13). The edema does not pit on pressure, and the skin is dry. Although Romaña's sign is thought to be almost pathognomonic for early Chagas' disease, in an analysis of 300 patients with early disease, unilateral periorbital edema was found in only 48.7% of patients (11). The infective stages spread to the regional lymph nodes, which become enlarged,

hard, and tender. Similar chagoma lesions may develop in other body areas as the infection spreads hematogenously. Trypomastigotes appear in the blood about 10 days after infection and persist through the acute phase. These stages are rare or absent during the chronic phase.

Acute-Stage Disease. Acute-stage symptoms only occur in about 1% of patients, are usually seen in younger children, and are less obvious in older individuals because of the nonspecific nature of the symptoms and the lack of availability of health care. In general, symptoms last for 4 to 8 weeks and then may subside, even without therapy.

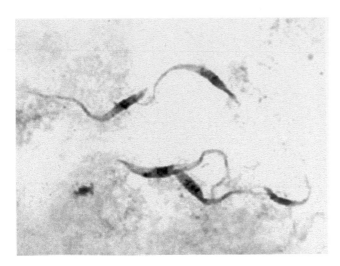

Figure 9.12 Epimastigotes of *Trypanosoma cruzi*.

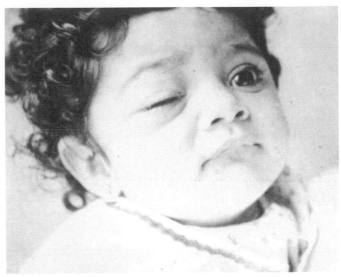

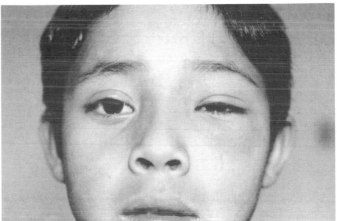

Figure 9.13 (Upper) Female child with Romaña's sign. (Armed Forces Institute of Pathology photograph.) (Lower) Male child with Romaña's sign. (From A Pictorial Presentation of Parasites: A cooperative collection prepared and/or edited by H. Zaiman.)

Acute systemic signs occur around week 2 to 3 of infection and are characterized by high fevers, which may be intermittent, remitting, or continuous; hepatosplenomegaly; myalgia; erythematous rash; acute myocarditis; lymphadenopathy; keratitis; and subcutaneous edema of the face, legs, and feet. There may be signs of CNS involvement including meningoencephalitis, which has a very poor prognosis. Myocarditis is manifested by electrocardiographic changes, tachycardia, chest pain, and weakness. Amastigotes proliferate within and destroy the cardiac muscle cells, leading to conduction defects and a loss of heart contractility (Figures 9.14 to 9.17). Death may occur due to myocardial insufficiency or cardiac arrest. In infants and very young children with acute Chagas' disease, swelling of the brain can develop, causing death. Also, in some areas of endemic infection, a high frequency of early electrocardiographic abnormalities in children who are *T. cruzi* seropositive suggests a rapid evolution from infection to disease. Under such conditions, a public health chemotherapy program targeted to this population would be recommended (27).

Probably all patients with acute Chagas' disease show some degree of myocarditis, which is apparently directly related to the presence of parasitic antigens. Apparently, both CD4 and CD8 cells participate in the pathogenesis of myocarditis, possibly leading to chronic heart failure (37). The numbers of CD4 and CD8 cells may actually correlate better with myocarditis than does the presence of parasite antigens.

Indeterminate Stage. Patients in the acute stage of Chagas' disease may die within a few weeks or months, may recover, or may enter the chronic stage of infection. The patient may have a subpatent parasitemia and will develop antibodies to a variety of *T. cruzi* antigens. Chronic Chagas' disease is diagnosed more commonly than the acute disease. The chronic stage may be initially asymptomatic (indeterminate stage), and even though trypomastigotes are seldom seen in peripheral blood, transmission by blood transfusion is a serious problem in areas where the disease is endemic. There are regional variations in disease severity, which may account for the geographic differences in morbidity. Approximately 8 to 10 weeks after infection, the indeterminate stage begins, during which the patients do not have any symptoms. Patients infected with low-virulence strains may remain infected for years with few to no problems.

Chronic Stage. Chronic Chagas' disease demonstrates close interaction between the parasite and the host, causing

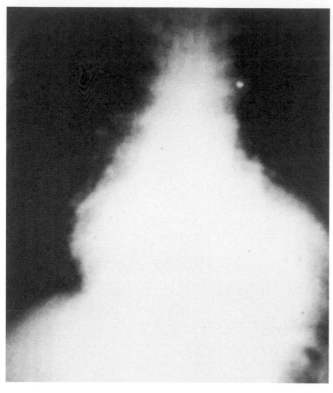

Figure 9.14 An enlarged heart in a 46-year-old man 13 days after onset of acute Chagas' disease. (Armed Forces Institute of Pathology photograph.)

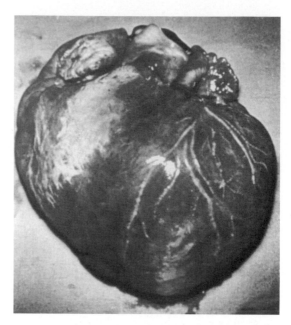

Figure 9.15 Acute chagasic myocarditis. (Original photograph by B. H. Kean; from A Pictorial Presentation of Parasites: A cooperative collection prepared and/or edited by H. Zaiman.)

a number of different clinical syndromes. Approximately 30% of patients develop chronic Chagas' disease and experience cardiomyopathy, megacolon, and megaesophagus. Symptoms of the chronic stage are related to the damage sustained during the acute stage of the disease, the state of the host's immune system, and the inflammatory response. Other contributing factors include the host genetic background, environmental and social factors, the genetic composition of the parasite, mixed infections, and reinfections (16). The inflammatory response undergoes periods of exacerbation and is probably responsible for neuronal damage, microcirculatory changes, heart matrix deformations, and possible cardiac failure (51). Selenium deficiency has also been linked to some cardiomyopathies and appears to be a biological marker for Chagas' disease and related to progressive pathology (86).

Chronic Chagas' disease may develop years or decades after undetected infection or after the diagnosis of acute disease. The most frequent clinical sign of chronic Chagas' disease is cardiomyopathy manifested by cardiomegaly and conduction changes. Myocardial damage typical of catecholamine cardiotoxicity has been described as an explanation for sudden patient death. Death has been attributed to focal denervation with regional asynergy and consequent compensatory adrenergic stimulus with myotoxicity and

malignant arrhythmia (9). A review of the potential influence of associated risk factors such as smoking, alcoholism, obesity, and hypertension on the establishment of chronic chagasic cardiomyopathy showed that these factors apparently were not associated with increased risk (13).

Autoimmunity Hypothesis (Pros and Cons). The merits of the autoimmunity hypothesis for the pathogenesis of Chagas' disease have been debated for years. Although several auto- or cross-reactive antibodies have been identified in chagasic patients, no convincing evidence exists for their ability to induce the lesions seen in infected patients (55). Because so few organisms are isolated or seen in heart tissue, much of the cardiac tissue destruction is thought to be related to autoimmune antibodies, possibly due to cross-reactivity with related *T. cruzi* antigens (molecular mimicry) (40–43, 52, 62). There is massive lymphocyte polyclonal activation, which results in the generation of autoantibodies cross-reacting with parasites and host tissues. Degenerative abnormalities occur in cardiac, esophageal, and colon tissues, leading to megasyndrome. The mechanism that triggers the megasyndrome is unknown. An antibody that reacts with endocardium, vascular structures, and interstitium of striated muscle (EVI antibody) has been detected in patients with Chagas' disease. The EVI antibody has been detected in a significant number of patients with chagasic cardiomyopathy. Antibodies that react with the Schwann sheaths of somatic and autonomic peripheral nerves have also been detected. A correlation of prevalence and serologic

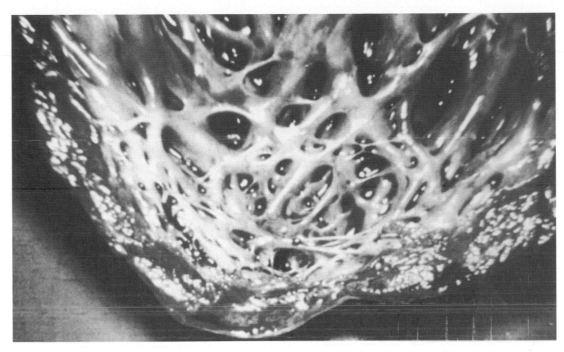

Figure 9.16 Open chagasic heart with thinning of wall. (Original photograph by B. H. Kean; from A Pictorial Presentation of Parasites: A cooperative collection prepared and/or edited by H. Zaiman.)

titers was detected for the antinerve and EVI antibodies. These antibodies could be absorbed from the sera by using *T. cruzi* epimastigotes, which suggests the presence of cross-reacting antibodies between parasite and hosts. These antibodies may play a large role in the denervation that occurs with megaesophagus, megacolon, and cardiomegaly.

Abnormalities of the coronary microcirculation have been seen in patients with acute chagasic myocarditis and may cause transient ischemia consistent with the hypothesis of a microvascular factor as a cause of myocardial changes. These changes are linked to endothelium structure and integrity. If the protective capability of the endothelial cells is reduced, vasospasm and platelet aggregation within coronary microvessels may lead to focal pathology.

Various mechanisms have been proposed for the pathogenesis of chronic chagasic cardioneuromyopathy. Chronic interaction of chagasic IgG with myocardial muscarinic acetylcholine receptors, behaving as a muscarinic agonist, might lead to cell dysfunction or tissue damage. Also, these antibodies could produce desensitization, internalization, or degradation of cardiac muscarinic acetylcholine receptors, explaining the progressive blockade of these receptors in the myocardium with parasympathetic denervation. Data also suggest that cardiac autonomic impairment and early myocardial damage involving the right ventricle are independent phenomena in Chagas' disease (66). Cross-reactive antibodies to cardiac and parasite immunogenic glycolipids have been

detected and may be involved in Chagas' disease autoimmunity (95). Increases in the levels of asialoganglioside- and monosialoganglioside-reactive antibodies have also been detected in patients with chronic Chagas' disease (6). Molecular mimicry has also been proposed on the basis of data related to cross-reactive antibodies between cardiac myosin and the *T. cruzi* B13 protein.

Human chronic Chagas' cardiomyopathy appears to fulfill many of the criteria for autoimmune diseases, similar to insulin-dependent diabetes mellitus or rheumatoid arthritis. However, validation of the target antigens must involve the induction of cardiac lesions after immunization or passive transfer of antigen-specific T cells. Aberrant T-cell activation may also play a role in cardiac injury (30). Not all researchers agree that autoantibodies play a role in pathogenesis (8, 76). Also, some propose a multifactorial or "combined" theory to explain the sequence of events leading to chronic myocarditis. The actual clinical course may vary from heart failure to a slow but continuing loss of cardiac function, with possible ventricular rupture and thromboemboli.

Tissue Involvement Other than Cardiomyopathy. Although it is less common than cardiac involvement, patients from certain areas are more likely to have dilation of the digestive tract with or without cardiomyopathy. These symptoms are most frequently seen in the esophagus and colon as a result of neuronal destruction (Figures 9.18 and 9.19). Megaesophagus characterized by dysphagia, chest

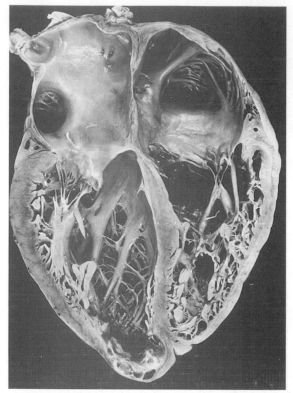

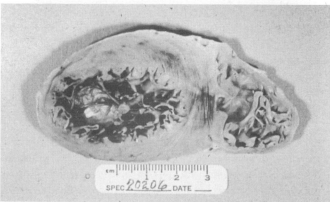

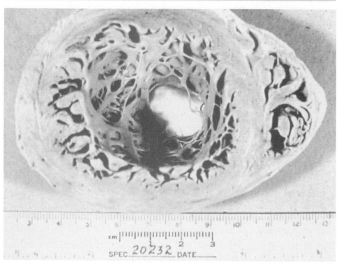

pain, regurgitation, and malnutrition is related to loss of contractility of the lower esophagus. Hypersalivation may occur, leading to aspiration with repeated bouts of aspiration pneumonia. Leiomyosarcoma of the esophagus in a patient with chagasic megaesophagus has been reported (2). Megacolon results in constipation, abdominal pain, and the inability to discharge feces. In some individuals, there may be acute obstruction leading to perforation, septicemia, and death. Cases of colon cancer have also been associated with chagasic megacolon.

Congenital Transmission.   Congenital transmission can occur in both the acute and chronic stages of the disease. Common clinical findings of congenital infection are stillbirth, low birth weight, myocarditis, neurologic alterations, and death shortly after birth. Infants of seropositive mothers should be monitored for up to 1 year after birth. Monitoring should include examination of blood for parasites and serologic tests.

Immunocompromised Patients.   Chagas' disease has been found in HIV-positive patients, and neurologic sequelae due to *T. cruzi* have been major findings in these patients (64). Individuals who were previously infected with *T. cruzi* and later become positive for HIV are at risk of reactivation of their Chagas' disease. These patients develop a severe multifocal or diffuse meningoencephalitis with numerous tissue parasites (89). Meningoencephalitis is a rare event in non-AIDS patients and is seen most frequently in immunocompetent children younger than 4 years. Because of the immunodeficiency associated with HIV infections, concomitant infection with *T. cruzi* may be difficult to recognize, particularly in individuals who have moved to areas where the disease is not endemic. *T. cruzi* infection may also have a protracted asymptomatic course in immunosuppressed HIV patients (25). Although a positive blood smear has been considered the key indicator of Chagas' disease reactivation in immunocompromised patients with chronic disease, this finding may occur late, rather than early, in the reactivation process (90). Apparently, there is no association between given

Figure 9.17   (Top) Posterior half of a heart with chronic chagasic cardiopathy. Note the left ventricle dilation and apical aneurysm. (Armed Forces Institute of Pathology photograph.) (Middle) Aneurysmal dilatation and thinning of the apical myocardium, plus marked concentric muscular hypertrophy. (Bottom) Same heart with thin apical myocardium. Note the light shining through the thinned muscle. (Middle and bottom photographs by J. H. Edgecome; from A Pictorial Presentation of Parasites: A cooperative collection prepared and/or edited by H. Zaiman.)

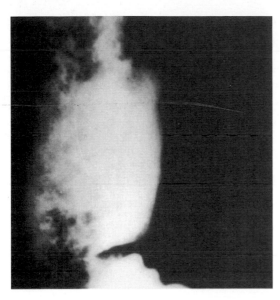

Figure 9.18 Chagasic megaesophagus. (Original photography by D. N. Reifsnyder; from A Pictorial Presentation of Parasites: A cooperative collection prepared and/or edited by H. Zaiman.)

Figure 9.19 Megacolon in an adult male with Chagas' disease. (Armed Forces Institute of Pathology photograph.)

T. cruzi genotypes and specific clinical forms of Chagas' disease-HIV associations (81).

Recrudescence of *T. cruzi* infections in immunosuppressed patients, particularly transplantation patients, is a grave concern. Transplant recipients can also become infected through receipt of infected organs (19). For patients with end-stage Chagas' cardiomyopathy, heart transplantation is an option which has had variable success (28). Reactivation of the disease with the development of cutaneous lesions has been seen. Bone marrow transplant recipients are also at risk of Chagas' disease due to reactivation or transfusion. Prophylactic treatment of these patients has led to favorable outcomes.

In studies of the Indians in western Paraguay, there appears to be a link between *T. cruzi* seropositivity and human T-cell leukemia virus type 2 (HTLV-2) infection. In this particular group, an individual infected with HTLV-2 is 2.28 times more likely to be *T. cruzi* positive than an HTLV-2-negative person. The public health significance of these findings is emphasized by the fact that approximately 18 million people are infected with *T. cruzi* in the Americas and the frequency of HTLV-2 infection continues to increase (35).

The high levels of IgM present in patients with African trypanosomiasis are not found in patients with either acute or chronic Chagas' disease. Antigenic variation, which is characteristic of infections with African trypanosomes, is less common in *T. cruzi* disease. In spite of humoral and cellular immunity against *T. cruzi*, the infection is able to persist in the host. As occurs in African trypanosomiasis,

there is suppression of the immune response, with decreases in the levels of gamma interferon and interleukin-2, which help regulate cellular and humoral responses.

Diagnosis

Health care personnel working with specimens from patients suspected of having Chagas' disease must use standard precautions. Trypomastigotes are highly infectious, and certain strains of the parasite are more virulent than others. The diagnosis of Chagas' disease should be considered if there is a history of consistent exposure to reduviid bug bites, residence in or travel to areas where the disease is endemic, laboratory accident, or recent blood transfusion in an area where the disease is endemic. Chagomas appear similar to injury caused by trauma, insect bites, or bacterial and mycotic infections. The differential diagnosis of acute Chagas' disease should include brucellosis, endocarditis, salmonellosis, schistosomiasis, toxoplasmosis, tuberculosis, connective tissue diseases, and leukemia. Chronic Chagas' disease with cardiomyopathy may be confused with endocarditis, ischemic heart disease, and changes associated with rheumatic heart disease.

Routine Laboratory Methods. Acute Chagas' disease should be suspected in any individual from an area where the disease is endemic who develops an acute febrile illness with lymphadenopathy and myocarditis. The presence of a chagoma or Romaña's sign may be diagnostic; however, allergic reactions to insect bites may produce similar lesions. The definitive diagnosis depends on demonstration of trypomastigotes in the blood, amastigote stages in tissues, or positive serologic reactions (Table 9.5). Methods used for examination of blood are similar to those used for the diagnosis of African trypanosomiasis discussed earlier in this chapter. Trypomastigote stages (Figure 9.10) may be easily detected in the blood in young children; however, in chronic disease, this stage is rare or absent except during febrile exacerbations. Trypomastigotes may be detected in blood by using thin and thick blood films or by buffy coat concentration techniques. The most sensitive methods for detection of trypomastigotes are concentration techniques such as the Strout method of buffy coat preparations (see chapter 31 for a discussion of concentration methods). With the Strout method, blood is collected without anticoagulant and allowed to clot. The serum is centrifuged at low speed to remove the remaining blood cells and then at a higher speed ($600 \times g$) to concentrate the parasites in the sediment. The stain of choice is Giemsa for both trypomastigote and amastigote stages.

Because of the overlap of geographic distribution, infections with *T. rangeli* trypomastigotes may have to be differentiated from those with *T. cruzi* trypomastigotes. *T. cruzi* trypomastigotes are usually C or U shaped on fixed blood films and have a large oval kinetoplast at the posterior end. *T. rangeli* trypomastigotes have a smaller kinetoplast a short distance from the posterior end, and there is no amastigote stage. In areas where kala azar occurs, amastigote stages (Figure 9.11) appear similar, and infections with *Leishmania donovani* and *T. cruzi* must be differentiated. This can be done by PCR, immunoblotting, culture (epimastigote in *T. cruzi* versus promastigote in *L. donovani*), serologic tests, animal inoculation, or xenodiagnosis.

Aspirates from chagomas and enlarged lymph nodes can be examined for amastigotes and trypomastigotes. Histologic examination of biopsy specimens may also be done. Aspirates, blood, and tissues can also be cultured, which is valuable in detecting low-grade parasitemias. The medium of choice is Novy-MacNeal-Nicolle medium. Cultures should be incubated at 25°C and observed for epimastigote stages for up to 30 days before they are considered negative. If available, laboratory animals (rats or mice) can be inoculated and the blood can be observed for trypomastigotes.

In the chronic stage of Chagas' disease, trypomastigotes are very rare or absent in the peripheral blood except during febrile exacerbations. Diagnosis depends primarily on culture, xenodiagnosis, or serologic tests. Some individuals with chronic Chagas' disease may have a depressed humoral immune response, being serologically negative. This response has been correlated with specific zymodemes. Chronic disease should be considered in individuals from areas of endemic infection who show signs of cardiomegaly, cardiac conduction defects, severe constipation, or dysphagia.

Follow-up blood specimens should be reexamined 1 to 2 months after therapy. If available, xenodiagnosis may be done. The patient may also be monitored through serologic tests and electrocardiograms (see Algorithm 9.2).

PCR. Although not routinely available except in specialized centers, PCR has been used to detect positive patients with as few as one trypomastigote in 20 ml of blood (4, 17, 18, 45). PCR may be very useful for the diagnosis of patients with chronic Chagas' disease because of the lack of sensitivity and specificity of serologic tests and the lack of sensitivity of xenodiagnosis. PCR testing may be very useful for monitoring patients who have received therapy in order to validate cures (38, 87). It has also been used to diagnose congenital infections by detecting *T. cruzi* in neonatal blood (96).

Immunoassays. Immunoassays have been used to detect antigens in the urine and sera of patients with congenital infections and those with chronic Chagas' disease (15, 22, 23, 75, 78). Determination of antigenuria can be valuable for early diagnosis of Chagas' disease and also for diagnosis of chronic cases in patients with conflicting serologic test results. A highly sensitive and specific chemiluminescent ELISA has been developed for blood bank screening and to monitor patients who are undergoing chemotherapy (3). Another ELISA using a recombinant antigen consisting of four different peptides has also been developed for screening blood donors and for epidemiologic studies and diagnosis (34).

Xenodiagnosis. In areas of endemic infection where reduviid bugs are readily available, xenodiagnosis can be used to detect light infections; this technique is most valuable for chronic infections when there are few trypomastigotes in the blood. Trypanosome-free bugs are allowed to feed on individuals suspected of having Chagas' disease. If organisms are present in the blood meal, the parasites multiply and can be detected in the bug's intestinal contents, which should be examined monthly for flagellated forms over a period of 3 months. To optimize this method, 20 third- or fourth-instar nymphs per day are allowed to feed on a patient for three consecutive days. A significant disadvantage of this technique is that not all triatomids are equally susceptible to infection and development of the epimastigote stage. Individuals who have been previously bitten by triatomids may be sensitized to their salivary secretions and may develop an anaphylactic reaction. Because the epimastigotes of *T. cruzi* and *T. rangeli* have heterogeneous morphologies, it may not be possible to make a definitive diagnosis microscopically.

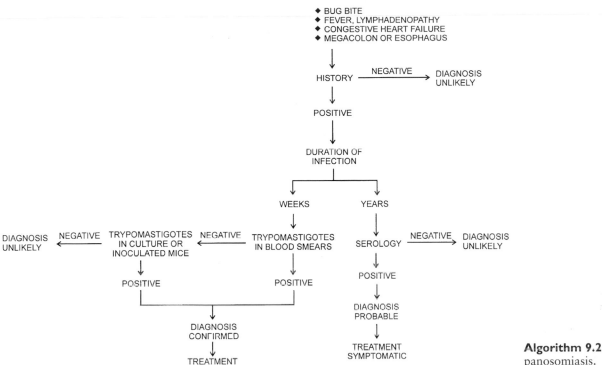

AMERICAN TRYPANOSOMIASIS

Algorithm 9.2 American trypanosomiasis.

Alternative diagnostic techniques such as PCR and use of a monoclonal antibody may be needed, pa rticularly in areas where there is geographic overlap (71).

Serologic Testing. Serologic tests used for the diagnosis of Chagas' disease include complement fixation (Guerreiro-Machado test), indirect fluorescent-antibody tests, indirect hemagglutination tests, and ELISA (57, 59, 61). The use of synthetic peptides and recombinant antigens has improved the sensitivity and specificity of these diagnostic techniques, particularly in the diagnosis of congenital disease by using IgM and IgA antibody detection and in the evaluation of cure (80, 94). Depending on the antigens used, cross-reactions have been noted to occur in patients with *T. rangeli* infection, leishmaniasis, syphilis, toxoplasmosis, hepatitis, leprosy, schistosomiasis, infectious mononucleosis, systemic lupus erythematosus, and rheumatoid arthritis. The Western blot method has been recommended for confirmatory serologic diagnosis of Chagas' disease (85). Another potential confirmatory assay using preserved protein antigens of *T. cruzi* discriminates chagasic from nonchagasic infections and may be useful for elucidating inconclusive results obtained by standard serologic tests; this is particularly relevant to blood bank testing (68). Long-lasting antibodies detected by a *trans*-sialidase inhibition assay and antibody re-

sponses to certain heat shock proteins can serve as sensitive markers for previous *T. cruzi* infection in patients who are parasite free and serologically cured (56). ELISA procedures have also been used to detect specific antibodies in naturally infected dogs (92).

Although Bolivian donor screening studies using indirect hemagglutination tests, indirect fluorescent-antibody tests, and four different ELISAs demonstrated 96.5 to 100% sensitivity and 87.0 to 98.9% specificity, use of a single test results in unacceptable numbers of false-negative samples in areas of highly endemic infection or in at-risk populations. Use of two tests would miss only 1 infected unit per 10,000 screened; however, the selection of tests depends on costs and feasibility (82).

KEY POINTS—LABORATORY DIAGNOSIS

Trypanosoma cruzi

1. Laboratory workers should use blood-borne-pathogen precautions when examining blood from Chagas' disease patients, because the trypomastigotes are infective.
2. Trypomastigotes are prevalent in the blood of patients with acute Chagas' disease; however, organism numbers are much smaller in the indeterminate and chronic stages of the infection.
3. *T. rangeli* cannot be differentiated from *T. cruzi* on

the basis of parasite morphology; patient information regarding geographic exposure is required for more appropriate interpretation of laboratory results.

4. In addition to thin and thick blood smears, concentration methods should be used to concentrate the trypomastigotes in the blood.

5. Immunoassays for antigen detection are now available and are highly sensitive and specific. Alternative methods such as culture and serologic testing can be used; however, these approaches may not be feasible without the use of a reference laboratory.

Treatment

Although numerous drugs have been tried, including those used to treat African trypanosomiasis and leishmaniasis, few have proven to be effective for therapy of Chagas' disease. In acute and congenital Chagas' disease and infections caused by laboratory accidents, treatment should be administered as soon as possible, even though in some cases symptoms are self-limited. Drug therapy has little effect on reducing the progression of chronic Chagas' disease.

Nifurtimox (Lampit), a nitrofurfurylidine derivative, is tolerated better in younger than older patients and should not be used during pregnancy. It reduces the duration and severity of illness and decreases mortality due to acute and congenital Chagas' disease. Reversible gastrointestinal, cutaneous, and neurologic adverse effects are common. Treatment success varies from one country to another, possibly indicating differences in the susceptibility of strains of *T. cruzi*. There is no indication that treatment of patients with chronic Chagas' disease is beneficial. Nifurtimox must be taken orally for prolonged periods, and there can be severe side effects including abdominal pain, nausea, vomiting, anorexia, and neurologic symptoms.

Benznidazole (RO-7-1051, Rochagan, Radanil), an imidazole derivative, is effective in reducing or suppressing parasites in the acute stages of disease but has limited capacity to produce a parasitic cure. It appears to be slightly more active and better tolerated than nifurtimox (33). Benznidazole is taken orally for prolonged periods and, similarly to nifurtimox, has little effect on changing clinical manifestations or reducing the progression of chronic Chagas' disease. Side effects include peripheral polyneuropathy, abdominal pain, nausea, vomiting, and severe skin reactions.

Allopurinol, a purine analog, was found in limited clinical trials to be as effective as nifurtimox and benznidazole in treating Chagas' disease. The drug is taken orally, and its side effects include skin rashes in patients with renal impairment, epigastric pain, transient diarrhea, and pruritus. In a study using itraconazole or allopurinol in patients with chronic Chagas' disease, parasitologic cure was evident in 44% of those treated with allopurinol and 53% of those treated with itraconazole. Electrocardiographic evaluation showed normalization in 36.5 and 48.2%, respectively, of patients with chronic or recent cardiopathy (5).

Symptoms associated with megaesophagus and megacolon may be treated with dietary measures or may require surgery. Patients with chronic chagasic heart disease may receive supportive therapy or be managed in some cases with pacemakers. With the use of a low cyclosporin dosage, 80% survival at 24 months has been reported for heart transplantation (67).

Epidemiology and Prevention

Chagas' disease is a zoonosis occurring throughout American continents and involves reduviid bugs living in close association with human reservoirs (dogs, cats, armadillos, opossums, raccoons, and rodents) (Table 9.5) (47, 83, 98). The most ubiquitous sylvatic reservoir host is the opossum, *Didelphus*, which is found throughout much of the range of *T. cruzi* in the Americas. Multiple nesting or resting sites of the opossum encompass many types of triatomine habitat. High *T. cruzi* prevalence rates are partly due to the fact that opossums eat triatomines and may also transmit infection via anal gland secretions. Sylvatic cycles of *T. cruzi* transmission extend from southern Argentina and Chile to northern California. Housing conditions are extremely important in transmission; the prevalence and incidence of infection are very high in human dwellings where the vector has adapted to living in the mud and in thatch walls and palm leaf roofs. The reduviid bug has easy access to humans to obtain blood meals and transmit the infection in this type of dwelling. A significant reduction in transmission was noted to occur in houses with walls made of plaster where cracks and crevices were covered, in contrast to houses with mud or thatch walls. Human infections occur mainly in rural areas where poor sanitary and socioeconomic conditions and poor housing provide excellent breeding places for reduviid bugs. These conditions allow maximum contact between the vector and humans. Although 12 species of reduviids occur within the United States, they have not adapted themselves to household habitation (10). Humans should avoid sleeping in thatch, mud, or adobe houses; bed nets should be used by persons sleeping in these types of houses. Travelers planning to stay in hotels, resorts, or other well-constructed housing facilities are not at high risk for contracting Chagas' disease. Insecticides can be used to kill the vectors and reduce the risk of transmission. Also, one must remember that in some countries, the blood supply may not always be screened for Chagas' disease and so blood transfusions may carry a risk of infection. The severity of Chagas' disease varies with the geographic area and may be related to strain differences in *T. cruzi*. Efforts have been made to characterize isolates based on enzyme profiles and have resulted in subpopulation classifications or schizodemes (zymodemes) (69, 70).

Transmission to humans is highly dependent on the defecation habits of the insect vector. In areas where the local species of reduviid bug does not ordinarily defecate while feeding, there are no human infections. This may explain why there are few human infections in the United States, even though sylvatic infections are known to occur in southern states. A number of autochthonous cases have been reported in the United States, in both Texas and California. The reduviid species involved in transmitting the infection to humans vary with the geographic area. Various species involved include *Panstrongylus megistus*, *Rhodnius prolixus*, *Triatoma brasiliensis*, *T. infestans*, *T. protracta*, *T. guasayana*, and *T. sordida* (Table 9.6).

Until recently, control of Chagas' disease has been mainly through the use of insecticides to eliminate the reduviid vector. In certain areas, insecticide resistance in triatomids has been noted. In addition to residual insecticide-spraying programs, construction of reduviid-proof dwellings and education are essential for effective control programs. Improvements in unsanitary living conditions, plastering of walls to obtain a smooth, crack-free surface, and replacement of palm-thatched roofs with metal roofing have been shown to considerably reduce the number of reduviids in houses. Although control of Chagas' disease is feasible, few countries have initiated control programs because of both political and economic constraints.

Transmission can occur through organ transplants and blood transfusion (91). Some of the countries in areas where the disease is endemic have laws mandating serologic testing of blood donors. Financial constraints may hinder the implementation of these laws, and there is a lack of standardization of currently available serologic tests. An alternative approach to serologic screening in the United States, because of a lack of Food and Drug Administration-approved tests, is the use of a questionnaire to identify prospective donors who may have resided in high-risk areas (60). This approach may not be practical, since even in vector-free areas in countries with endemic infection a significant portion of the donor blood units were positive for antibodies to Chagas' disease. In areas where seroprevalence is high, rather than discarding all positive blood units, laboratorians add gentian violet to the units and store them at 4°C for 24 h before use to kill the organism.

Table 9.6 *Trypanosoma cruzi*: principal vectors

Species	Geographic area
Panstrongylus megistus	Eastern seaboard of Brazil
Rhodnius prolixus	Venezuela, Colombia, French Guiana, Guyana, Suriname; most important vector in northern South America and Central America
Triatoma brasiliensis	Central and eastern Brazil
T. dimidiata	Ecuador, Colombia, western Central America
T. infestans	Brazil, Bolivia, Peru, Chile, Argentina, Paraguay, Uruguay
T. sordida	Brazil, Argentina, Bolivia, Paraguay, Uruguay; often associated with chickens
T. guasayana	Argentina

Evaluation of electrocardiographic (ECG) changes among urban workers in Sao Paulo, Brazil, showed that there were 2.2% positive sera for *T. cruzi* among 27,081 workers. A much higher percentage of workers with ECG abnormalities (42.7%) were seropositive for *T. cruzi* compared with the percentage of workers with ECG abnormalities (19.8%) who were seronegative. Based on these data, it appears that the high frequency of ECG abnormalities emphasizes the importance of providing medical assistance to this group (44).

Laboratory personnel working with *T. cruzi* must be aware of the hazard involved in handling the highly infectious trypomastigote form. Precautions would include the use of gloves and eye protection. All clinical specimens must be handled by using standard precautions.

Trypanosoma rangeli

T. rangeli was first described by Tejera during an examination of intestinal contents from the reduviid *R. prolixus*. The infection in human blood was first found by DeLeon in Guatemalan children. *T. rangeli* infections are encountered in many areas where *T. cruzi* also occurs; consequently, the two parasites may be confused and must be differentiated morphologically (Table 9.5). No pathology is associated with *T. rangeli* infections. In a study conducted in Venezuela, the relatively high frequency of *T. rangeli* infections detected in patients with acute-stage Chagas' disease (18.6%) complicated the diagnosis of *T. cruzi* infection. More than 2,600 cases of human infection have been documented in Venezuela,

Guatemala, Panama, Colombia, El Salvador, Costa Rica, Peru, and Brazil, the majority of which have been in Venezuela and Guatemala (46).

Life Cycle and Morphology

After the reduviid bug is infected during a blood meal, *T. rangeli* migrates from the intestine to the hemolymph and then to the salivary glands, where infective metacyclic forms can be found (Figure 9.20). Transmission to humans is through the bite of the reduviid (triatomid) bug *R. prolixus* and related species by inoculation of metacyclic trypomastigotes into the wound from the infected saliva. Infections with *T. rangeli* and *T. cruzi* can coexist in the reduviid vector. The trypomastigote is approximately 30 μm long and has a free flagellum (Figure 9.21). The nucleus is anterior to the middle of the body, and the small kinetoplast is subterminal.

Clinical Disease

Human infections are apparently asymptomatic, and trypomastigotes have been noted in the blood for longer than a year. No evidence of pathogenicity has been detected in human volunteers (26).

Diagnosis

Trypomastigotes can be detected in the blood of infected patients by using thin and thick blood smears and buffy coat concentration techniques (Table 9.5). The parasites can be stained with Giemsa or Wright's stain (see the discussion of *T. cruzi* for a description of the trypomastigote). Infections can also be detected by xenodiagnosis, in which the reduviid vector is allowed to obtain a blood meal from an infected patient and then the hemolymph and salivary glands are examined for epimastigotes and metacyclic trypomastigotes. In addition, blood can be cultured (Tobie's medium and Novy-MacNeal-Nicolle medium) or injected into laboratory animals (mice) and examined for epimastigotes and trypomastigotes, respectively. Although there are no serologic tests for *T. rangeli*, serologic cross-reactions have been noted to occur with tests for *T. cruzi*. A 48-kDa protein has been

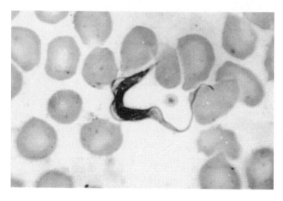

Figure 9.21 *Trypanosoma rangeli* trypomastigote in a peripheral blood smear. Note the small kinetoplast. (Armed Forces Institute of Pathology photograph.)

identified that is a specific and abundant intracellular antigen for *T. rangeli* epimastigotes. Results suggest that this antigen may be a useful marker for the identification and characterization of *T. rangeli* isolates (88).

KEY POINTS—LABORATORY DIAGNOSIS

Trypanosoma rangeli

1. *T. rangeli* trypomastigotes can be detected in the peripheral blood by using thin and thick blood smears and concentration techniques; however, in practice *T. rangeli* is rarely seen in human blood (69).
2. *T. rangeli* trypomastigotes are morphologically similar to African trypanosome trypomastigotes.
3. *T. rangeli* infections can occur together with *T. cruzi* infections; therefore, *T. rangeli* trypomastigotes must be differentiated from trypomastigotes of *T. cruzi* in the peripheral blood.

Epidemiology and Prevention

T. rangeli infections have been found in both Central and South America and often overlap with *T. cruzi* infections. In some areas, *T. rangeli* infections are five to six times more frequent than *T. cruzi* infections in the population.

Figure 9.20 Life cycle of *Trypanosoma rangeli*.

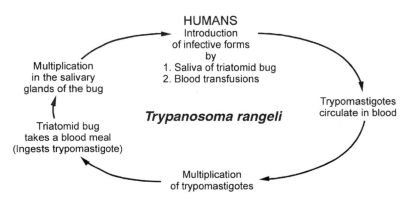

Animals naturally infected with *T. rangeli* include opossums, rodents, anteaters, raccoons, dogs, and nonhuman primates. The organisms appear to be harmless to animals; however, they do reduce the life span of the reduviid vector (26).

References

1. Abramowicz, M. (ed.) 2004. Drugs for parasitic infections. *Med. Lett. Drugs Ther.* **46:**1–12.

2. Adad, S. J., R. M. Etchebehere, E. M. Hayashi, R. K. Asai, P. Fernandes, C. F. C. Macedo, and E. Crema. 1999. Leiomyosarcoma of the esophagus in a patient with chagasic megaesophagus: case report and literature review. *Am. J. Trop. Med. Hyg.* **60:**879–881.

3. Almeida, I. C., D. T. Covas, L. M. T. Soussumi, and L. R. Travassos. 1997. A highly sensitive and specific chemiluminescent enzyme-linked immunosorbent assay for diagnosis of active *Trypanosoma cruzi* infection. *Transfusion* **37:**850–857.

4. Antas, P. R. Z., N. Medrano-Mercado, F. Torrico, R. Ugarete-Fernandez, F. Gómez, R. C. Oliveira, A. C. L. Chaves, A. J. Romanha, and T. C. Araujo-Jorge. 1999. Early, intermediate, and late acute stages in Chagas' disease: a study combining anti-galactose IgG, specific serodiagnosis, and polymerase chain reaction analysis. *Am. J. Trop. Med. Hyg.* **61:**308–314.

5. Apt, W., X. Aguilera, A. Arribada, C. Pérez, C. Miranda, G. Sanchez, I. Zulantay, P. Cortex, J. Rodriguez, and D. Juri. 1998. Treatment of chronic Chagas' disease with itraconazole and allopurinol. *Am. J. Trop. Med. Hyg.* **59:**133–138.

6. Avila, J. L., M. Rojas, and A. Avila. 1998. Increase in asialoganglioside- and monosialoganglioside-reactive antibodies in chronic Chagas' disease patients. *Am. J. Trop. Med. Hyg.* **58:**338–342.

7. Ayed, Z., I. Brindel, B. Bouteille, N. Van Meirvenne, F. Doua, D. Houinato, M. Dumas, and M. Jauberteau. 1997. Detection and characterization of autoantibodies directed against neurofilament proteins in human African trypanosomiasis. *Am. J. Trop. Med. Hyg.* **57:**1–6.

8. Baig, M. K., O. Salomoni, A. L. P. Caforio, J. H. Goldman, M. Amuchastegui, T. Caiero, and W. J. McKenna. 1997. Human chagasic disease is not associated with an antiheart humoral response. *Am. J. Cardiol.* **79:**1135.

9. Baroldi, G., S. J. M. Oliveira, and M. D. Silver. 1997. Sudden and unexpected death in clinically 'silent' Chagas' disease. A hypothesis. *Int. J. Cardiol.* **58:**263–268.

10. Beard, C. B., G. Pye, F. J. Steurer, R. Rodriguez, R. Campman, A. T. Peterson, J. Ramsey, R. A. Wirtz, and L. E. Robinson. 2003. Chagas disease in a domestic transmission in Southern Texas, USA. *Emerg. Infect. Dis.* **9:**103–105.

11. Beaver, P. C., R. C. Jung, and E. W. Cupp. 1984. *Clinical Parasitology*, 9th ed., p. 55–100. Lea & Febiger, Philadelphia, Pa.

12. Becker, S., J. R. Franco, P. P. Simarro, A. Stich, P. M. Abel, and D. Steverding. 2004. Real-time PCR for detection of *Trypanosoma brucei* in human blood samples. *Diagn. Microbiol. Infect. Dis.* **50:**193–199.

13. Berra, H., F. Carnevali, S. Revelli, H. Moreno, S. M. Pezzotto, J. C. Morini, and O. Bottasso. 1998. Electrocardiographic alterations in chronically *Trypanosoma cruzi*-infected persons exposed to cardiovascular factors. *Arch. Med. Res.* **29:**241–246.

14. Berriman, M., E. Ghedin, C. Hertz-Fowler, G. Blandin, H. Renauld, D. C. Bartholomew, N. J. Lennard, E. Caler, N. E. Hamlin, B. Hass, U. Bohme, et al. 2005. The genome of the African trypanosome *Trypanosoma brucei*. *Science* **309:**416–422.

15. Bertot, G. M., R. S. Corral, M. Fresno, C. Rodriguez, A. M. Katzin, and S. Grinstein. 1998. *Trypanosoma cruzi* tubulin eliminated in the urine of the infected host. *J. Parasitol.* **84:**608–614.

16. Campbell, D. A., S. J. Westenberger, and N. R. Sturm. 2004. The determinants of Chagas' disease: connecting parasite and host genetics. *Curr. Mol. Med.* **4:**549–562.

17. Carriazo, C. S., A. Sembaj, A. M. Aguerri, J. M. Requena, C. Alonso, J. Búa, A. Ruiz, E. Segura, and J. M. Barral. 1998. Polymerase chain reaction procedure to detect *Trypanosoma cruzi* in blood samples from chronic chagasic patients. *Diagn. Microbiol. Infect. Dis.* **30:**183–186.

18. Carriazo, C. S., A. Sembaj, A. M. Aguerri, J. M. Requena, C. Alonso, J. Búa, A. Ruiz, E. Segura, and J. M. Barral. 1998. Polymerase chain reaction procedure to detect *Trypanosoma cruzi* in blood samples from chronic chagasic patients. *Diagn. Microbiol. Infect. Dis.* **30:**183–186.

19. Centers for Disease Control and Prevention. 2002. Chagas' disease after organ transplantation—United States, 2001. *Morb. Mortal. Wkly. Rep.* **51:**210–212.

20. Chappuis, F., E. Stivanello, K. Adams, S. Kidane, A. Pittet, and P. A. Bovier. 2004. Card agglutination test for trypanosomiasis (CATT) end-dilution titer and cerebrospinal fluid cell count as predictors of human African trypanosomiasis (*Trypanosoma brucei gambiense*) among serologically suspected individuals in southern Sudan. *Am. J. Trop. Med. Hyg.* **71:**313–317.

21. Chappuis, F., L. Loutan, P. Simarro, V. Lejon, and P. Büscher. 2005. Options for field diagnosis of human African trypanosomiasis. *Clin. Microbiol. Rev.* **18:**133–146.

22. Corral, R. S., J. M. Alchch, and H. L. Freilij. 1998. Presence of IgM antibodies to *Trypanosoma cruzi* urinary antigen in sera from patients with acute Chagas' disease. *Int. J. Parasitol.* **28:**589–594.

23. Corral, R. S., A. Orn, and S. Grinstein. 1992. Detection of soluble exoantigens of *Trypanosoma cruzi* by a dot-immunobinding assay. *Am. J. Trop. Med. Hyg.* **46:**31–38.

24. Cox, A., A. Tilley, F. McOdimba, J. Fyfe, M. Eisler, G. Hide, and S. Welburn. 2005. A PCR based assay for detection and differentiation of African trypanosome species in blood. *Exp. Parasitol.* **111:**24–29.

25. Da Cruz, A. M., R. P. Igreja. W. Dantas, A. C. Junqueira, R. S. Pacheco, A. J. Silva-Goncalves, and C. Pirmez. 2004. Long-term follow up of co-infected HIV and *Trypanosoma cruzi* Brazilian patients. *Trans. R. Soc. Trop. Med. Hyg.* **98:**728–733.

26. D'Alessandro, A. 1976. Biology of *Trypanosoma* (*Herpetosoma*) *rangeli* tejera, 1920, p. 327–403. *In* W. H. R. Lumsden and D. A. Evans (ed.), *Biology of the Kinetoplastida*. Academic Press, Inc., New York, N.Y.

27. de Andrade, A. L. S. S., F. Zicker, A. Rassi, A. G. Rassi, R. M. Oliveira, S. A. Silva, S. S. de Andrade, and C. M. T. Martelli. 1998. Early electrocardiographic abnormalities in *Trypanosoma cruzi*-seropositive children. *Am. J. Trop. Med. Hyg.* **59:**530–534.

28. Decarvalho, V. B., E. F. L. Sousa, J. H. A. Vila, J. P. Dasilva, M. R. Caiado, S. R. D. Araujo, R. Macruz, and E. J. Zerbani. 1996. Heart transplantation in Chagas' disease—10 years after the initial experience. *Circulation* 94:1815–1817.

29. Donelson, J. E., K. L. Hill, and N. M. A. Elsayed. 1998. Multiple mechanisms of immune evasion by African trypanosomes. *Mol. Biochem. Parasitol.* 91:51–66.

30. DosReis, G. A., C. G. Freire-de-Lima, M. P. Nunes, and M. F. Lopes. 2005. The importance of aberrant T-cell responses in Chagas' disease. *Trends Parasitol.* 21:237–243.

31. Doua, F., T. W. Miezan, J. R. S. Singaro, F. B. Yapo, and T. Baltz. 1997. The efficacy of pentamidine in the treatment of early-late stage *Trypanosoma brucei gambiense* trypanosomiasis. *Am. J. Trop. Med. Hyg.* 55:586–588.

32. Dubois, M. E., K. P. Demick, and J. M. Mansfield. 2005. Trypanosomes expressing a mosaic variant surface glycoprotein coat escape early detection by the immune system. *Infect. Immun.* 73:2690–2697.

33. Estani, S. S., E. L. Segura, A. M. Ruiz, E. Valazquez, B. M. Porcel, and C. Yampotis. 1998. Efficacy of chemotherapy with benznidazole in children in the indeterminate phase of Chagas' disease. *Am. J. Trop. Med. Hyg.* 59:526–529.

34. Ferreira, A. W., Z. R. Belem, E. A. Lemos, S. G. Reed, and A. Campos-Neto. 2001. Enzyme-linked immunosorbent assay for serological diagnosis of Chagas' disease employing a *Trypanosoma cruzi* recombinant antigen that consists of four different peptides. *J. Clin. Microbiol.* 39:4390–4395.

35. Ferrer, J. F., E. Esteban, A. Nurua, S. Gutierrez, S. Dube, B. Poiesz, L. Feldman, M. A. Basombrio, and D. Galligan. 2003. Association and epidemiologic features of *Trypanosoma cruzi* and human T cell lymphotropic A virus type II in inhabitants of the Paraguayan Gran Chaco. *Am. J. Trop. Med. Hyg.* 68:235–241.

36. Freitas, J. M., E. Lages-Silva, E. Crema, S. D. Pena, and A. M. Macedo. 2005. Real time PCR strategy for the identification of major lineages of *Trypanosoma cruzi* directly in chronically infected human tissues. *Int. J. Parasitol.* 35:411–417.

37. Fuenmayor, C., M. L. Higuchi, H. Carrasco, H. Parada, P. Gutierrez, V. Aiello, and S. Palimino. 2005. Acute Chagas' disease: immunohistochemical characteristics of T cell infiltrate and its relationship with *T. cruzi* parasitic antigens. *Acta Cardiol.* 60:33–37.

38. Galvão, L. M. C., E. Chiari, A. M. Macedo, A. O. Luquetti, S. A. Silva, and A. L. S. S. Andrade. 2003. PCR assay for monitoring *Trypanosoma cruzi* parasitemia in childhood after specific chemotherapy. *J. Clin. Microbiol.* 41: 5066–5070.

39. Gibson, W. 2002. Epidemiology and diagnosis of African trypanosomiasis using DNA probes. *Trans. R. Soc. Trop. Med. Hyg.* 96:S141–S143.

40. Goin, J. C., E. S. Borda, S. Auger, R. Storino, and L. Sterin-Borda. 1999. Cardiac M-2 muscarinic cholinoceptor activation by human chagasic autoantibodies: association with bradycardia. *Heart* 82:273–278.

41. Goin, J. C., C. P. Leiros, E. Borda, and L. Sterin-Borda. 1997. Interaction of chagasic autoantibodies with the third extracellular domain of the human heart muscarinic receptor. Functional and pathological implications. *Med. Buenos Aires* 56:699–704.

42. Goin, J. C., C. P. Leiros, E. Borda, and L. Sterin-Borda. 1997. Interaction of human chagasic IgG with the second extracellular loop of the human heart muscarinic acetylcholine receptor: functional and pathological implications. *FASEB J.* 11:77–83.

43. Goin, J. C., L. Sterin-Borda, C. R. Bilder, L. M. Varrica, G. Iantorno, M. C. Rios, and E. Borda. 1999. Functional implications of circulating muscarinic cholinergic receptor autoantibodies in chagasic patients with achalasia. *Gastroenterology* 117:798–805.

44. Goldbaum, M., F. Y. Ajimura, J. Litvoc, S. A. Carvalho, and J. Eluf-Neto. 2004. American trypanosomiasis and electrocardiographic alterations among industrial workers in Sao Paulo, Brazil. *Rev. Inst. Med. Trop. Sao Paulo* 46:299–302.

45. Gomes, M. L., L. M. C. Galvao, A. M. Macedo, S. D. J. Pena, and E. Chiari. 1999. Chagas' disease diagnosis: comparative analysis of parasitologic, molecular, and serologic methods. *Am. J. Trop. Med. Hyg.* 60:205–210.

46. Guhl, F., and G. A. Vallejo. 2003. *Trypanosoma (Herpetosoma) rangeli* Tejera, 1920: an updated review. *Mem. Inst. Oswaldo Cruz* 98:435–442.

47. Gurtier, R. E., J. E. Cohen, M. C. Cecere, M. A. Lauricella, R. Chuit, and E. L. Segura. 1998. Influence of humans and domestic animals on the household prevalence of *Trypanosoma cruzi* in *Triatoma infestans* populations in northwest Argentina. *Am. J. Trop. Med. Hyg.* 58:748–758.

48. Harris, E., J. Detmer, J. Dungan, F. Doua, T. White, J. A. Kolberg, M. S. Urdea, and N. Agabian. 1996. Detection of *Trypanosoma brucei* spp. in human blood by a nonradioactive branched DNA-based technique. *J. Clin. Microbiol.* 34:2401–2407.

49. Hide, G. 1999. History of sleeping sickness in East Africa. *Clin. Microbiol. Rev.* 12:112–125.

50. Hide, G., A. Tait, I. Maudlin, and S. C. Welburn. 1996. The origins, dynamics and generation of *Trypanosoma brucei rhodesiense* epidemics in East Africa. *Parasitol. Today* 12:50–55.

51. Higuchi, M. L., L. A. Benvenuti, R. M. Martinas, and M. Metzger. 2003. Pathophysiology of the heart in Chagas' disease: current status and new developments. *Cardiovasc. Res.* 60:96–107.

52. Huang, H., T. M. Calderon, J. W. Berman, V. L. Braunstein, L. M. Weiss, M. Wittner, and H. B. Tanowitz. 1999. Infection of endothelial cells with *Trypanosoma cruzi* activates NF-κB and induces vascular adhesion molecule expression. *Infect. Immun.* 67:5434–5440.

53. Jelinek, T., Z. Bisoffi, L. Bonazzi, P. van Thiel, U. Bronner, A. de Frey, S. G. Gundersen, P. McWhinney, D. Ripamonti, and the European Network on Imported Infectious Disease Surveillance. 2002. Cluster of African trypanosomiasis in travelers to Tanzanian national parks. *Emerg. Infect. Dis.* 8:634–635.

54. Khonde N., J. Pepin, and B. Mpia. 1997. A seven day course of eflornithine for relapsing *Trypanosoma brucei gambiense* sleeping sickness. *Trans. R. Soc. Trop. Med. Hyg.* 91:212–213.

55. Kierszenbaum, F. 1999. Chagas' disease and the autoimmunity hypothesis. *Clin. Microbiol. Rev.* 12:210–212.

56. Krautz, G. M., J. D. Peterson, L. M. Godsel, A. U. Krettli, and D. M. Engman. 1998. Human antibody responses to *Trypanosoma cruzi* 70-kD heat-shock proteins. *Am. J. Trop. Med. Hyg.* 58:137–143.

57. Krieger, M. A., E. Almeida, W. Oelemann, J. S. Lafaille, J. B. Perreira, H. Krieger, M. R. Carvalho, and S. Goldenberg. 1992. Use of recombinant antigens for the accurate immunodiagnosis of Chagas' disease. *Am. J. Trop. Med. Hyg.* **46:**427–434.

58. Kuboki, N., N. Inoue, T. Sakurai, F. di Cello, D. J. Grab, H. Suzuki, C. Sugimoto, and I. Igarashi. 2003. Loop-mediated isothermal amplification for detection of African trypanosomiasis. *J. Clin. Microbiol.* **41:**5517–5524.

59. Leiby, D. A., R. M. Herron, Jr., E. J. Read, B. A. Lenes, and R. J. Stumpf. 2002. *Trypanosoma cruzi* in Los Angeles and Miami blood donors: impact of evolving donor demographics on seroprevalence and implications for transfusion transmission. *Transfusion* **42:**549–555.

60. Leiby, D. A., E. J. Read, B. A. Lenes, A. J. Yund, R. J. Stumpf, L. V. Kirchhoff, and R. Y. Dodd. 1997. Seroepidemiology of *Trypanosoma cruzi*, etiologic agent of Chagas' disease, in US blood donors. *J. Infect. Dis.* **176:**1047–1052.

61. Leiby, D. A., S. Wendel, D. T. Takaoka, R. M. Fachini, L. C. Oliveira, and M. A. Tibbals. 2000. Serologic testing for *Trypanosoma cruzi*: comparison of radioimmuno-precipitation assay with commercially available indirect immunofluorescence assay, indirect hemagglutination assay, and enzyme-linked immunosorbent assay kits. *J. Clin. Microbiol.* **38:**639–642.

62. Leiros, C. P., L. Sterin-Borda, E. S. Borda, J. C. Goin, and M. M. Hosey. 1997. Desensitization and sequestration of human m2 muscarinic acetylcholine receptors by autoantibodies from patients with Chagas' disease. *J. Biol. Chem.* **272:**12989–12993.

63. Lejon, V., P. Buscher, E. Magnus, A. Moons, I. Wouters, and N. Van Meirvenne. 1998. A semi-quantitative ELISA for detection of *Trypanosoma brucei gambiense* specific antibodies in serum and cerebrospinal fluid of sleeping sickness patients. *Acta Trop.* **69:**151–164.

64. Madalosso, G., A. C. Pellini, M. J. Vasconcelos, A. F. Ribeiro, L. Weissmann, G. S. O. Filho, A. C. Penalva de Oliveira, and J. F. Vidal. 2004. Chagasic meningoencephalitis: case report of a recently included AIDS-defining illness in Brazil. *Rev. Inst. Med. Trop. Sao Paulo* **46:**199–202.

65. Malesker, M. A., D. Boken, T. A. Ruma, P. J. Vuchetich, P. J. Murphy, and P. W. Smith. 1999. Rhodesian trypanosomiasis in a splenectomized patient. *Am. J. Trop. Med. Hyg.* **61:**428–430.

66. Marin-Neto, J. A., G. Bromberg-Marin, A. Pazin, M. V. Simoes, and B. C. Maciel. 1998. Cardiac autonomic impairment and early myocardial damage involving the right ventricle are independent phenomena in Chagas' disease. *Int. J. Cardiol.* **65:**261–269.

67. Martín-Rabadán, P., and E. Bouza. 1999. Blood and tissue protozoa, p. 8.34.1–8.34.14. *In* D. Armstrong and J. Cohen (ed.), *Infectious Diseases.* Mosby International, London, United Kingdom.

68. Mendes, R. P., S. Hoshinoshimizu, A. M. M. Dasilva, I. Mota, R. A. G. Heredia, A. O. Luquetti, and P. G. Leser. 1997. Serological diagnosis of Chagas' disease: a potential confirmatory assay using preserved protein antigens of *Trypanosoma cruzi. J. Clin. Microbiol.* **35:**1829–1834.

69. Miles, M. A. 1998. New World trypanosomiasis, p. 283–302. *In* L. Collier, A. Balows, and M. Sussman (ed.), *Topley and Wilson's Microbiology and Microbial Infections*, 9th ed., vol. 5. Arnold, London, United Kingdom.

70. Montamat, E. E., G. M. L. D'Oro, R. H. Gallerano, R. Sosa, and A. Blanco. 1996. Characterization of *Trypanosoma cruzi* populations by zymodemes: correlation with clinical picture. *Am. J. Trop. Med. Hyg.* **55:**625–628.

71. Murthy, V. K., K. M. Dibbern, and D. A. Campbell. 1992. PCR amplification of mini-exon genes differentiates *Trypanosoma cruzi* from *Trypanosoma rangeli. Mol. Cell. Probes* **6:**237–243.

72. Nantulya, V. M. 1997. TrypTect CIATT®—a card indirect agglutination trypanosomiasis test for diagnosis of *Trypanosoma brucei gambiense* and *T. brucei rhodesiense* infections. *Trans. R. Soc. Trop. Med. Hyg.* **91:**551–553.

73. Nantulya, V. M., F. Doua, and S. Molisho. 1992. Diagnosis of *Trypanosoma brucei gambiense* sleeping sickness using an antigen detection enzyme-linked immunosorbent assay. *Trans. R. Soc. Trop. Med. Hyg.* **86:**42–45.

74. O'Beirne, C., C. M. Lowry, and H. P. Voorheis. 1998. Both IgM and IgG anti-VSG antibodies initiate a cycle of aggregation-disaggregation of bloodstream forms of *Trypanosoma brucei* without damage to the parasite. *Mol. Biochem. Parasitol.* **91:**165–193.

75. Oelemann, W. M. R., M. D. M. Teixeira, G. C. V. Dacosta, J. Borges-Pereira, J. A. F. Decastro, J. R. Coura, and J. M. Peralta. 1998. Evaluation of three commercial enzyme-linked immunosorbent assays for diagnosis of Chagas' disease. *J. Clin. Microbiol.* **36:**2423–2427.

76. Olivares-Villagomez, D., T. L. McCurley, C. L. Vnencak-Jones, R. Correa-Oliveira, D. G. Colley, and C. E. Carter. 1998. Polymerase chain reaction amplification of three different *Trypanosoma cruzi* DNA sequences from human chagasic cardiac tissue. *Am. J. Trop. Med. Hyg.* **59:**563–570.

77. Pansaerts, R., N. Van Meirvenne, E. Magnus, and L. Verheist. 1998. Increased sensitivity of the card agglutination test CATT/*Trypanosoma brucei gambiense* by inhibition of complement. *Acta Trop.* **70:**349–354.

78. Partel, C. D., and C. L. Rossi. 1998. A rapid, quantitative enzyme-linked immunosorbent assay (ELISA) for the immunodiagnosis of Chagas' disease. *Immunol. Investig.* **27:**89–96.

79. Penchenier, L., P. Grebaut, F. Njokou, V. E. Eyenga, and P. Buscher. 2003. Evaluation of LATEX/T.b. gambiense for mass screening of *Trypanosoma brucei gambiense* sleeping sickness in Central Africa. *Acta Trop.* **85:**31–37.

80. Peralta, J. M., M. G. M. Teixeira, W. G. Shreffler, J. B. Pereira, J. M. Burns, Jr., P. R. Sleath, and S. G. Reed. 1994. Serodiagnosis of Chagas' disease by enzyme-linked immunosorbent assay using two synthetic peptides as antigens. *J. Clin. Microbiol.* **32:**971–974.

81. Perez-Ramirez, L., C. Barnabé, A. M. C. Sartori, M. S. Ferreira, J. E. Tolezano, E. V. Nunes, M. K. Burgarelli, A. C. Silva, M. A. Shikanai-Yasuda, J. N. Lima, A. M. da Cruz, O. C. Oliveira, C. Guilherme, B. Bastrenta, and M. Tibayrenc. 1999. Clinical analysis and parasite genetic diversity in human immunodeficiency virus/Chagas' disease coinfections in Brazil. *Am. J. Trop. Med. Hyg.* **61:**198–206.

82. Pirard, M., N. Iihoshi, M. Boelaert, P. Basanta, F. Lopez, and P. van der Stuyft. 2005. The validity of serologic tests

for *Trypanosoma cruzi* and the effectiveness of transfusional screening strategies in a hyperendemic region. *Transfusion* 45:554–561.

83. Pung, O. J., C. W. Banks, D. N. Jones, and M. W. Krissinger. 1995. *Trypanosoma cruzi* in wild raccoons, opossums and triatomine bugs in southeast Georgia, U.S.A. *J. Parasitol.* 81:324–326.

84. Radwanska, M., S. Magez, H. Perry-O'Keefe, H. Stender, J. Coull, J. M. Sternberg, P. Büscher, and J. J. Hyldig-Nielsen. 2002. Direct detection and identification of African trypanosomes by fluorescence in situ hybridization with peptide nucleic acid probes. *J. Clin. Microbiol.* 40:4295–4297.

85. Reiche, E. M. V., M. Cavazzana, Jr., H. Okamura, E. C. Tagata, S. I. Jankevicius, and J. V. Jankevicius. 1998. Evaluation of the Western blot in the confirmatory serologic diagnosis of Chagas' disease. *Am. J. Trop. Med. Hyg.* 59:750–756.

86. Rivera, M. T., A. P. De Souza, A. H. M. Moreno, S. S. Xavier, J. A. S. Gomes, M. O. C. Rocha, R. Correa-Oliveira, J. Neve, J. Vanderpas, and T. C. Araujo-Jorge. 2002. Progressive Chagas' cardiomyopathy is associated with low selenium levels. *Am. J. Trop. Med. Hyg.* 66:706–712.

87. Russomando, G., M. M. C. de Tomassone, I. de Guillen, N. Acosta, N. Vera, M. Almiron, N. Candia, M. J. Calcena, and A. Figueredo. 1998. Treatment of congenital Chagas' disease diagnosed and followed up by the polymerase chain reaction. *Am. J. Trop. Med. Hyg.* 59:487–491.

88. Saldana, A., R. A. Harris, A. Örn, and O. E. Sousa. 1998. *Trypanosoma rangeli*: identification and purification of a 48-kDa-specific antigen. *J. Parasitol.* 84:67–73.

89. Sartori, A. M. C., H. H Caiaffa-Filho, R. C. Bezerra, C. D. S. Guilherme, M. H. Lopes, and M. A. Shikanai-Yasuda. 2002. Exacerbation of HIV viral load simultaneous with asymptomatic reactivation of chronic Chagas' disease. *Am. J. Trop. Med. Hyg.* 67:521–523.

90. Sartori, A. M. C., M. H. Lopes, L. A. Benvenuti, B. Caramelli, A. O. di Pietro, E. V. Nunes, L. P. Ramirez, and M. A. Shikanai-Yasuda. 1998. Reactivation of Chagas' disease in a human immunodeficiency virus-infected patient leading to severe heart disease with a late positive direct microscopic examination of the blood. *Am. J. Trop. Med. Hyg.* 59:784–786.

91. Schmunis, G. A., and J. R. Cruz. 2005. Safety of the blood supply in Latin America. *Clin. Microbiol. Rev.* 18:12–29.

92. Shadomy, S. V., S. C. Waring, O. A. Marins-Filho, R. C. Oliveira, and C. L. Chappell. 2004. Combined use of enzyme-linked immunosorbent assay and flow cytometry to detect antibodies to *Trypanosoma cruzi* in domestic canines in Texas. *Clin. Diagn. Lab. Immunol.* 11:313–319.

93. Truc, P., V. Lejon, E. Magnus, V. Janonneau, A. Nangouma, D. Verloo, L. Penchenier, and P. Buscher. 2002. Evaluation of the micro-CATT, CATT/*Trypanosoma brucei gambiense*, and LATEX/*T.b. gambiense* methods for serodiagnosis and surveillance of human African trypanosomiasis in West and Central Africa. *Bull. W. H. O.* 80:882–886.

94. Umezawa, E. S., S. F. Bastos, M. E. Camargo, L. M. Yamauchi, M. R. Santos, A. Gonzalez, B. Zingales, M. J. Levin, O. Sousa, R. Rangel-Aldao, and J. F. da Silveira. 1999. Evaluation of recombinant antigens for serodiagnosis of Chagas' disease in South and Central America. *J. Clin. Microbiol.* 37:1554–1560.

95. Vermelho, A. B., M. D. L. Demeirelles, M. C. Pereira, G. Pohlentz, and E. Barretobergter. 1997. Heart muscle cells share common neutral glycosphingolipids with *Trypanosoma cruzi*. *Acta Trop.* 64:131–143.

96. Virreira, M., F. Torrico, C. Truyens, C. Alonso-Vega, M. Solano, Y. Carlier, and M. Svoboda. 2003. Comparison of polymerase chain reaction methods for reliable and easy detection of congenital *Trypanosoma cruzi* infection. *Am. J. Trop. Med. Hyg.* 68:574–582.

97. Wery, M. 1994. Drugs used in the treatment of sleeping sickness (human African trypanosomiasis: HAT). *Int. J. Antimicrob. Agents* 4:227–238.

98. Yabsley, M. J., and G. P. Noblet. 2002. Biological and molecular characterization of a raccoon isolate of *Trypanosoma cruzi* from South Carolina. *J. Parasitol.* 88:1273–1276.

10

Intestinal Nematodes

The largest number of helminth organisms that parasitize humans is found in the roundworm group. Several roundworms are important in causing disease; some of these have other mammalian hosts, and some are considered to be pathogenic only under certain circumstances. Consequently, there is great variation in life cycle stages and pathologic sequelae found in humans.

Nematodes are unsegmented helminths with bilateral symmetry, have a fully functional digestive tract, are usually long and cylindrical, and vary from a few millimeters to over a meter long. Their numbers per patient vary considerably; however, worm size and numbers do not necessarily correlate with symptoms or pathologic changes.

All nematodes that are parasitic in humans have separate sexes, with the male usually being smaller than the female. Egg production varies considerably from species to species but tends to be consistent within a specific group. The number of eggs produced per day can range from a few (*Strongyloides stercoralis*) to more than 200,000 (*Ascaris lumbricoides*).

The life cycle of nematodes has five successive stages: four larval stages and the adult. In most cases, the third-stage larva is the infective stage. The eggs and larvae of nematodes living in the intestinal tract are passed outside the body in the feces or may be deposited on the perianal skin by the female worm. Eggs of certain species are almost fully embryonated, while other species may require an extended period of egg embryonation in the soil. In some cases, the eggs are infective when swallowed and other eggs will hatch in the soil, thus initiating infection by larval penetration of the human skin rather than egg ingestion.

As Table 10.1 shows, nematodes are long-lived worms. Usually, the host response to the presence of these worms is directly related to the worm burden. Diagnosis of these infections depends on the recovery and identification of adult worms, eggs, or larvae; specific information is provided in Table 10.2.

World Health Organization estimates suggest that over 3.5 billion people carry nematode infections; filarial infections are not included in this figure. With these numbers in mind, it is not uncommon for almost any laboratory within the United States to occasionally recover helminth eggs, larvae, or adults from human specimens. Unfortunately, lowering the intensity of infection in a host population through the use of chemotherapy may produce minimal declines in transmission relative to its initial endemic level (17).

249

Table 10.1 Normal life spans of the most common intestinal nematodes

Nematode	Life span	Comments
Ascaris lumbricoides	1–2 yr	Infection may be aborted by spontaneous passage of adult worms
Enterobius vermicularis	Several months to years	Reinfection due to both self-infection and outside sources is extremely common
Trichuris trichiura	Several years	Often accompanies *Ascaris* infection (both are acquired by egg ingestion from contaminated soil)
Hookworms		
Necator americanus	4–20 yr	Symptoms are directly related to worm burden; many infections are asymptomatic
Ancylostoma duodenale	5–7 yr	
Strongyloides stercoralis	30+ yr	Autoinfection capability can lead to dissemination and the hyperinfection syndrome in the compromised host

It is well known that large numbers of people in sub-Saharan Africa are infected with both human immuno-deficiency virus type 1 (HIV-1) and intestinal helminths. There may be detrimental immunologic effects from coinfection with these two pathogens (11, 43). Based on recent studies, the high helminth prevalence and possible adverse interactions between helminths and HIV suggest that helminth diagnosis and treatment should be a part of the routine care of HIV-infected patients (50).

Ascaris lumbricoides

The number of people infected with *Ascaris lumbricoides* worldwide is probably second only to the number infected with the pinworm, *Enterobius vermicularis*. *A. lumbricoides* was well known in Roman times as *Lumbricus teres* (confused with the common earthworm) and has probably been infecting humans for thousands of years. It is prevalent in moist, warm climates but can also survive in the temperate zones. In most countries of Central and South America, the average infection rates range up to approximately 45%. Also, foci of high prevalence among young children still persist in the southeastern United States (Table 10.3).

Life Cycle and Morphology

The adult worms are cylindrical, with a tapering anterior end. They are the largest of the common nematode parasites of humans; females measure 20 to 35 cm long, and males are 15 to 31 cm long, with a curved posterior

end (Figure 10.1). Also, the three well-developed lips are characteristic of this group.

The life cycle of *A. lumbricoides* is usually referred to as an "indirect cycle" since there is extensive migration throughout the body prior to adult worm maturation and egg production. Infection in humans is acquired through ingestion of the embryonated eggs from contaminated soil (Figure 10.2). On ingestion, the eggs hatch in the stomach and duodenum, where the larvae actively penetrate the intestinal wall; they are then carried to the right heart via the hepatic portal circulation. Larvae within the eggs undergo one or possibly two molts prior to penetration of the intestinal wall. From the right heart they are carried into the pulmonary circulation, where they are filtered out by the capillaries. After approximately 10 days in the lung, the larvae break into the alveoli, migrate via the bronchi until they reach the trachea and pharynx, and then are swallowed. The worms then mature and mate in the intestine, with the eventual production of eggs, which are passed in the stool. The entire developmental process from egg ingestion to egg passage from the adult female takes from 8 to 12 weeks. During her life span, she may deposit a total of 60,000,000 or more eggs.

Both unfertilized and fertilized eggs are passed (Figures 10.3 and 10.4). Often only female worms are recovered from the intestine. Fertilized eggs become infective within 2 weeks if they are in moist, warm soil, where they may remain viable for months or even years. The fertilized egg is broadly oval, with a thick, mammillated coat, usually bile stained a golden brown. These eggs

Table 10.2 The most common intestinal nematodes

Characteristic	Ascaris lumbricoides	Enterobius vermicularis	Trichuris trichiura	Hookworms (Necator americanus, Ancylostoma duodenale)	Trichostrongylus spp.	Strongyloides stercoralis
Usual time to infective stage	2–3 wk in soil; second-stage larva in egg	4–6 h; first-stage larva in egg	2–3 wk in soil; first-stage larva in egg	5–7 days in soil; free, third-stage larva	3–5 days in soil; free, third-stage larva	5–7 days in soil; free, third-stage larva
Mode of infection	Ingestion of infective egg	Ingestion of infective egg	Ingestion of infective egg	Skin penetration by N. americanus; ingestion and skin penetration by A. duodenale	Ingestion of third-stage larva	Skin penetration
Development and location in human host	Obligatory larval migration through liver and lungs; adults in small intestine	Direct development to adult in intestinal tract; adults in cecum, appendix, colon, and rectum	Direct development to adult in intestinal tract; adults in cecum, appendix, and colon	Larval migration through lungs; adults attached to mucosa of small intestine	Direct development to adult in intestinal tract; adults in small intestine	Larval migration through lungs; adult females in mucosal epithelium of small intestine; autoinfection may occur
Prepatent period	2 mo	3–4 wk	3 mo	5–8 wk	2–3 wk	2–4 wk
Normal life span	Up to 1 yr or slightly longer	1–2 mo	Up to 15 yr or more; usually 5–10 yr	Up to 15 yr or more; usually 5–10 yr	Up to 1 yr or slightly longer	Up to many years (30+)
Diagnosis by usual means	Bile-stained, mammillated, thick-shelled eggs (45–75 by 35–50 μm) in 1-cell stage in feces; infertile eggs (85–95 by 43–47 μm) have thinner shells and distorted mammillations; mature or immature adults may be found in feces or may spontaneously migrate out of the anus, mouth, or nares	Smooth, thick-shelled eggs (50–60 by 20–32 μm) in cellulose tape preparations; rarely seen in feces; adult or immature worms may be found in feces	Unembryonated, bile-stained, thick-shelled eggs (50–54 by 20–23 μm) with mucoid plugs at each end, in feces	Thin-shelled eggs (56–75 by 36–40 μm) in 4- to 16-cell stage in feces	Large, thin-shelled eggs (73–95 by 40–50 μm), tapered at one end, in feces; inner membrane of egg frequently wrinkled; ovum already in advanced cleavage when passed	First-stage larvae (108–380 by 14–20 μm) in feces; larvae have a short buccal cavity and a prominent, conspicuous genital primordium

(continued)

Table 10.2 The most common intestinal nematodes (*continued*)

Characteristic	Ascaris lumbricoides	Enterobius vermicularis	Trichuris trichiura	Hookworms (Necator americanus, Ancylostoma duodenale)	Trichostrongylus spp.	Strongyloides stercoralis
Diagnostic problems	Fertile eggs may lose outer mammillated layer (decorticate eggs); infertile eggs may be difficult to recognize; also, will not float in usual solution of ZnSO$_4$ (sp gr, 1.18) used for concentration	Eggs not usually seen in feces; cellulose tape method should be used to demonstrate eggs from perianal region	Rarely presents a problem; routine stool examination	Eggs of the 2 species are indistinguishable; if eggs hatch in feces owing to delay in examination, these first-stage larvae must be differentiated from *Strongyloides* larvae	May be confused with hookworm eggs	Larvae may be passed sporadically and may be found only by culture, concentration procedures, or use of Entero-Test or duodenal intubation
Clinical notes	Owing to potential migration of adult worms (as a result of fever, drugs, anesthetics), all infections should be treated; pulmonary symptoms may be present during larval migration (prior to egg recovery in the stool); eosinophils present but not impressive	Generally, only symptomatic patients treated because of high reinfection rate; there may or may not be eosinophilia	Light infections usually not treated; patients may be asymptomatic, with eggs an incidental finding; *Ascaris* and *Trichuris* infections often found together; moderate eosinophilia in heavy infections (usually will not exceed 15%)	Skin penetration by larvae produces allergic reaction (ground itch, cutaneous larva migrans); pulmonary symptoms usually present only in heavy infection; iron deficiency anemia and eosinophilia up to 70% may be present	Rarely seen in the United States; common in the Orient, Europe, Middle East, and Africa; light infections usually not treated	Hyperinfections may lead to death in the compromised or immunosuppressed host; patient may become symptomatic many years after original infection (without additional exposure); eosinophilia 10–40% or higher

Table 10.3 Factors related to continuing *Ascaris lumbricoides* infections in urban communities in developing countries and DALYs lost to a range of causes, including ascariasis[a]

Factor	Value
Estimates for 1990[b]	
Population	2,500 million
Daily fecal output of population	500,000 tons
No. infected with *A. lumbricoides*	1,000 million
Daily fecal output contaminated with eggs of *A. lumbricoides*	200,000 tons
Daily discharge of eggs of *A. lumbricoides*	2×10^{14} eggs
Population of urban communities	750 million
Daily fecal output of urban population	150,000 tons
Number of urban people infected with *A. lumbricoides*	300 million
Daily fecal output contaminated with eggs of *A. lumbricoides* in urban communities	60,000 tons
Daily discharge of eggs of *A. lumbricoides*	6×10^{13} eggs
Estimates for urban communities in the year 2000[b, c]	
Population	2,200 million
Daily fecal output	440,000 tons
No. infected with *A. lumbricoides*	880 million[d]
Daily fecal output contaminated with eggs of *A. lumbricoides*	1,760,000 tons
Daily discharge of eggs of *A. lumbricoides*	1.76×10^{14} eggs
Comparison of DALYs[e] lost to a range of causes globally	
Total intestinal helminths	39 million
Ascaris lumbricoides	10.5 million
Trichuris trichiura	6.4 million
Hookworms	22.1 million
Malaria	35.7 million
Schistosomiasis	4.5 million
Tuberculosis	46.5 million
Measles	34.1 million
Vitamin A deficiency	11.8 million
Diabetes mellitus	8 million
Total maternal causes (hemorrhage, sepsis, eclampsia, hypertension, obstructed labor, and abortion)	29.7 million
Motor vehicle accidents	31.7 million

[a] Adapted from I. Coombs and D. W. T. Crompton. How much human helminthiasis is there in the world? *J. Parasitol.* **85:**397–403, 1999; D. A. P. Bundy, This wormy world—then and now, *Parasitol. Today* **13:**407–408, 1997; and M. S. Chan, The global burden of intestinal nematode infections—fifty years on, *Parasitol. Today* **13:**438–443, 1997.

[b] The world's population in 1990 was about 5,000 million, with about half living in developing countries (about 70% rural and 30% urban). About 1,000 million people in developing countries are infected with *A. lumbricoides*. About 200 g of stool is produced per person daily. About 1,000 eggs of *A. lumbricoides* are present daily in each gram of stool from an infected person.

[c] It is assumed that the prevalence of *A. lumbricoides* will not have changed and that, overall, the proportion of infected people in rural and urban communities will be the same.

[d] Probably underestimated and could be as high as 1,565 million.

[e] DALY, disability-adjusted life year; based on disabilities experienced translated into years of health lost.

Figure 10.1   Adult *Ascaris lumbricoides* (male).

measure up to 75 μm long and 50 μm wide. Unfertilized eggs are usually more oval, measure up to 90 μm long, and may have a pronounced mammillated coat or an extremely minimal mammillated layer. Often, both types of eggs are found in the same stool specimen. The total absence of fertilized eggs means that only female worms are present in the intestine.

The occurrence of *A. lumbricoides* usually peaks in childhood or early adolescence. Although the percentage of humans harboring *A. lumbricoides* is high, little is known about these infections in pregnant women and the possible impact on the fetus. In a clinical and epidemiologic study of an infant with diarrhea and failure to thrive, the child became ill at 7 days of age, symptoms became worse at 21 days of age, and *Ascaris* eggs were found in the stool at 40 days of age. The minimum

prepatent period after ingestion of the infective eggs is 60 days. After consideration of other possibilities, this case is very suggestive of congenital transmission. Further studies are necessary, but if confirmed, these findings will certainly reinforce the recommendations for antiparasitic treatment of pregnant women to prevent neonatal infection.

Clinical Disease

Infection with *A. lumbricoides* continues to be a significant health problem throughout the world. Prevalence data in one of the first studies of its kind demonstrate that roundworm infection occurs at a high level in the population examined. There is a high within-individual correlation over time, indicating that there is individual predisposition to infection with *A. lumbricoides*. There is unequivocal evidence for a genetic component, accounting for 30 to 50% of the variation in worm burden. Shared environmental effects account for 3 to 13% of the total phenotypic variance. The remarkable consistency of the results through time and across different measures of the burden phenotype strongly indicate their validity and suggest the utility of future evaluations focused on identifying the specific genes responsible for these sizable genetic effects. Many of the parasitic helminth infections lend themselves to genetic analysis, and future studies may lead to significant improvements in pharmacologic agents (77).

Pathogenesis caused by *Ascaris* infections is attributed to (i) the host immune response, (ii) effects of larval migration, (iii) mechanical effects of the adult worms, and (iv) nutritional deficiencies due to the presence of the adult worms. Although the initial passage of larvae through the liver and lungs usually elicits no symptoms, there can be signs of pneumonitis if the number of larvae is quite large. When the larvae break out of the lung tissue and into the alveoli, there may be some damage to the bronchial epithelium.

Figure 10.2   *Ascaris lumbricoides* life cycle.

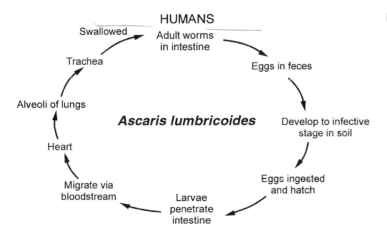

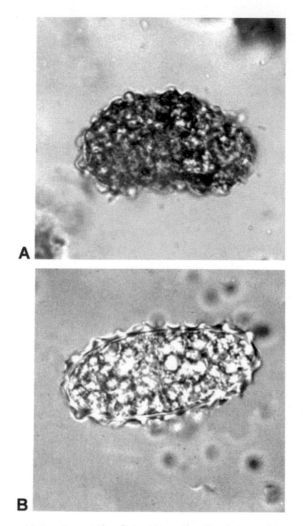

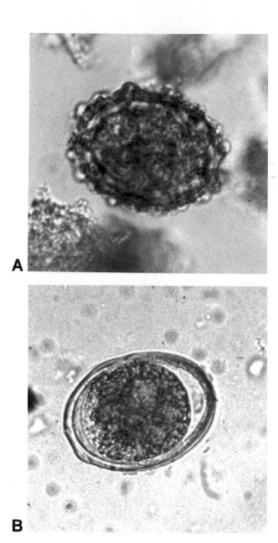

Figure 10.3 *Ascaris lumbricoides* unfertilized eggs; these eggs can also be decorticate (loss of the bumpy coat), as can the fertilized egg seen in Figure 10.4.

Figure 10.4 (A) *Ascaris lumbricoides,* fertilized egg; (B) *A. lumbricoides,* fertilized decorticate egg.

With reinfection and subsequent larval migrations, there may be intense tissue reactions, even with small numbers of larvae (Figure 10.5). There may be pronounced tissue reaction around the larvae in the liver and lungs, with infiltration of eosinophils, macrophages, and epithelioid cells. This condition has been called *Ascaris* pneumonitis and is accompanied by an allergic reaction consisting of dyspnea, a dry or productive cough, wheezing or coarse rales, fever (39.9 to 40.0°C), transient eosinophilia, and a chest X ray suggestive of viral pneumonia. This picture of transient pulmonary infiltrates that clear within a couple of weeks and are associated with peripheral eosinophilia is frequently called Loeffler's syndrome. In addition to eosinophils and Charcot-Leyden crystals, the sputum may contain larvae, although this finding is not common and examination of gastric washings may be more helpful. Asthma and urticaria may continue during the intestinal phase of ascariasis.

Eosinophilic gastroenteritis is an inflammatory disease characterized by eosinophilic infiltration of the gastrointestinal tract accompanied by varying abdominal symptoms and usually by peripheral eosinophilia. Although the precise etiology of this condition has not been determined, allergies to certain allergens such as foods, drugs, and parasites have been repeatedly proposed as causing the disease. In one rare case, a woman who had extensive eosinophilic infiltration in the descending and rectal colon, with a high titer to immunoglobulin G (IgG) antibody against *Ascaris suum*, was treated successfully with prednisolone (71).

A case of hepatic ascariasis occurring in a 3-year-old child with acute lymphoblastic leukemia suggests that *Ascaris* worms not only can enter the biliary tree and cause hepatic ascariasis but also potentially can carry pathogens on their surface and provide an alternate

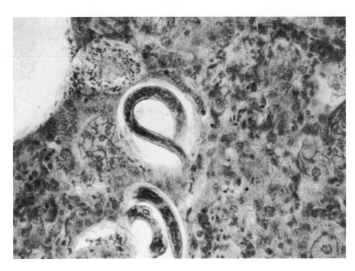

Figure 10.5 *Ascaris lumbricoides* larvae in lung tissue. (From a Pictorial Presentation of Parasites: A cooperative collection prepared and/or edited by H. Zaiman.)

portal of entry for severe infections in neutropenic patients. At autopsy, two hepatic parenchymal abscesses were seen in the right lobe, each containing *Ascaris* worms, which were also found in the duodenum, common bile duct, and right intrahepatic duct. Cultures also grew *Klebsiella* spp., *Enterobacter cloacae*, and *Candida albicans* (J. A. Lecciones, M. S. Cortet, S. Tungol, C. D. V. Macalunga, and M. A. Cruz, *Abstr. 91st Gen. Meet. Am. Soc. Microbiol. 1991*, abstr. L-43, 1991). Another case involved a 25-year-old woman with sickle cell trait and a normal delivery 3 months earlier who presented with acute obstructive cholangitis with septicemia and multiple hepatic abscesses. After removal of 60 adult *Ascaris* worms, surgical drainage of the hepatic abscesses, and wide-spectrum antibiotic therapy, the patient recovered. This case also points up the potential complications that can be seen with *Ascaris* infections, particularly when they occur in areas where this infection is not endemic. Endoscopy was successful in 19 of 25 patients resistant to medical therapy. Surgery remains important in the management of ascariasis complicated by biliary or pancreatic strictures and stones or by the presence of worms in the gallbladder (10).

The presence of adult worms in the intestine usually causes no difficulties unless the worm burden is very heavy; however, because of the tendency of the adult worms to migrate, even a single worm can cause serious sequelae. Worm migration may occur as a result of stimuli such as fever (usually over 38.9°C), the use of general anesthesia, or other abnormal conditions. This migration may result in intestinal blockage; entry into the bile duct, pancreatic duct, or other small spaces; or

entry into the liver or peritoneal cavity. The worms can also migrate out of the anus or come out the mouth or nose. Other body sites, such as the kidneys, appendix, or pleural cavity, have been involved. A Swiss patient with acute pancreatitis and no evidence of gallstones or history of alcohol abuse died after a short fulminant illness. Autopsy findings revealed an *Ascaris* worm impacted within the ampulla of Vater, a reminder that nonindigenous causes of biliary tract obstruction need to be considered (49).

Through analysis of published reports, epidemiologic aspects of *Ascaris*-induced intestinal obstruction were identified. In 9 studies of at least 100 patients admitted to the hospital with ascariasis, intestinal obstruction was the single most common complication and accounted for 38 to 88% of all complications. Both the proportion and number of cases per 1,000 population were significantly related to the local prevalence of ascariasis. In 12 studies of at least 30 patients with *Ascaris*-induced intestinal obstruction, the case fatality rates ranged from 0 to 8.6%. The mean age of patients with obstruction was ≥5 years in six of seven studies in which age was specified (25).

In children, particularly those younger than 5 years, there may be severe nutritional impairment related to the worm burden. Directly measurable effects include increased fecal nitrogen and fecal fat and impaired carbohydrate absorption, all of which would return to normal with elimination of the adult worms. Worms can also be spontaneously passed without any therapy.

An association between helminthic infection and educational achievement has long been recognized. In one study, children treated with mebendazole for ascariasis showed significant improvement in several test scores, learning ability, concentration, and eye-hand coordination after 5 months of treatment (33). In another study, low test results in language, social, gross motor, and fine motor skills were associated with low weight for age, and lower language test scores were associated with infections with intestinal parasites and *Ascaris* in particular. Developmental disabilities are a significant and frequently undetected health problem in developing countries; malnutrition associated with intestinal helminth infections may be an important contributory factor to these disabilities (53).

Immunology and Host Resistance

Gastrointestinal nematode infections cause vigorous immune responses, although the mechanisms involved are not fully understood. A number of circulating immunoglobulins have been identified in *A. lumbricoides* infections, including IgG1 to IgG4, IgA, and IgE. The levels of IgG4 and IgE are elevated in helminth infections, and antiparasite IgE responses are associated with infection resistance. However, apparently IgG4 antibody can block IgE-mediated immunity and thus can block allergic responses in humans (11, 76).

The T-cell component of the immune system plays an important role in resistance to gastrointestinal nematodes. Different nematode species may stimulate the same Th2 cascade, and functional resistance to certain species depends on different subsets of mediators and effectors within the cascade. Functional effectors at the base of the Th2 cascade probably include eosinophilia, IgA, IgE, proliferation of mast cells, and increased mucus secretion. Physiological changes in the intestine, such as increased mucosal permeability and smooth muscle contractivity, may produce an environment in which worms become trapped in mucus and are swept from the gut; these changes are driven by interleukin-4 (IL-4) and IL-13.

It is well recognized that some helminth infections, including ascariasis, can suppress host immune responses. Worms may also subvert host immunity through the stimulation of inappropriate effectors. Various excretory-secretory products from parasitic nematodes can also modulate host immune responses. These products may interfere with immune effectors or influence mediators involved in induction and control. Overall, there remain many questions related to the deworming of human populations and the effect this could have on host susceptibility to certain pathogens (e.g., HIV) and on pathogen-induced immunopathology (11).

Diagnosis

In the larval migration phase of infection, diagnosis can be made by finding the larvae in sputum or in gastric washings. The typical Loeffler's syndrome is more likely to be seen in areas where transmission is highly seasonal.

During the intestinal phase, the diagnosis can be made by finding the eggs (unfertilized or fertilized) or adult worms in the stool (Figures 10.3 and 10.4). The eggs are most easily seen on a direct wet smear or a wet preparation of the concentration sediment.

Caution. Remember that unfertilized *Ascaris* eggs do not float with use of the zinc sulfate flotation concentration method (the eggs are too heavy). Also, if too much iodine is added to the wet preparations, the eggs may look like very dark debris. Eggs may be very difficult to identify on a permanent stained smear because of stain retention and asymmetric shape.

Intestinal disease can often be diagnosed from radiographic studies of the gastrointestinal tract, in which the worm intestinal tract may be visualized (Figures 10.6 and 10.7). This may be particularly obvious when two worms are lying parallel, like "trolley car lines" (Figure 10.8) (9).

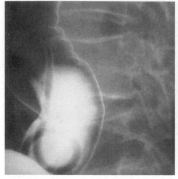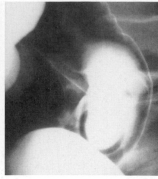

Figure 10.6 *Ascaris lumbricoides.* Two spot films from an upper gastrointestinal barium study show that white barium in the small bowel outlines an adult worm that has displaced the barium and appears dark. (From a Pictorial Presentation of Parasites: A cooperative collection prepared and/or edited by H. Zaiman.)

Other involved body sites may present specific symptoms indicative of bowel obstruction, biliary or pancreatic duct blockage, appendicitis, or peritonitis (Figure 10.9). Since acute abdomen caused by ascariasis is more commonly seen in developing countries, clinicians in developed countries may not consider this potential problem when seeing a patient with similar symptoms. Increased awareness of ascariasis presenting as an acute abdomen is required for optimal patient care; this may be possible only for an experienced radiologist (64). Therapeutic measures are related to specific symptoms and involved areas. Refer to Algorithm 10.1 for detailed information.

KEY POINTS—LABORATORY DIAGNOSIS

Ascaris lumbricoides

1. Both fertilized and unfertilized eggs can easily be recovered by the sedimentation concentration method. (Unfertilized eggs do not float with use of the zinc sulfate flotation concentration method.)

2. Because of the potential problems caused by migration of the adult worms, patients who are undergoing elective surgery and receiving general anesthetic should be checked for the presence of *A. lumbricoides* if there is any possible exposure history (some anesthetics stimulate the worms to migrate). Usually, a single stool examination will suffice to rule out the infection.

3. Larvae can be recovered from sputum (larval migration through the lungs); however, this is not a common finding.

4. Eggs may be very difficult to recognize on a permanent stained smear. They are usually very darkly stained and may be mistaken for debris.

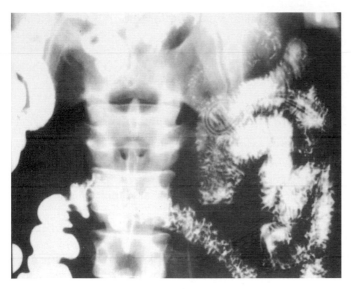

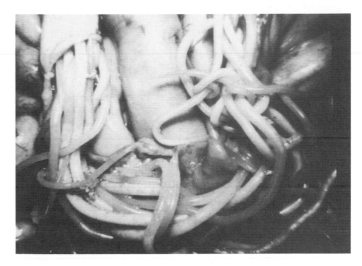

Figure 10.9 *Ascaris lumbricoides*, adult worms in the gut. (From A Pictorial Presentation of Parasites: A cooperative collection prepared and/or edited by H. Zaiman.)

Figure 10.7 *Ascaris lumbricoides*. Small bowel barium study shows adult worms that present as longitudinal dark lines against the white barium in the small bowel. The worms have ingested the barium, and their guts appear as white lines between the dark body outlines. (From a Pictorial Presentation of Parasites: A cooperative collection prepared and/or edited by H. Zaiman.)

Treatment

There are a number of modern anthelmintic agents that represent a vast improvement over some of the older remedies, and none of the recommended drugs require pre- or posttreatment purging or fasting. Mebendazole is considered the drug of choice for both children and adults; ivermectin and albendazole are alternatives but are con-

Figure 10.8 *Ascaris lumbricoides*, paired worms in the gut (X ray). (From A Pictorial Presentation of Parasites: A cooperative collection prepared and/or edited by H. Zaiman.)

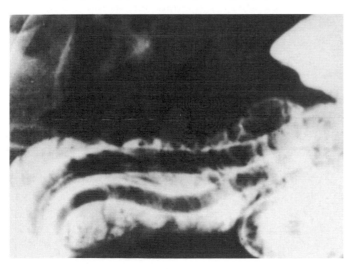

sidered investigational for this condition by the U.S. Food and Drug Administration. Specific drugs and dosages are listed in chapter 25. Although these drugs are effective in eliminating the adult worms, there is no conclusive evidence that they are effective against the larval migration phase of the infection. If other body sites are involved, surgical intervention may be necessary.

The prognosis for treated intestinal disease is excellent; however, in cases of perforation or surgical procedures, there may be complications. Also, the prognosis may be extremely poor in cases of massive larval migration through the lungs.

Some individuals seem to be susceptible to heavy infections while others are not; deworming has a greater impact on the intensity of infection than on its prevalence; and mass chemotherapy is likely to be a more effective means of controlling morbidity than is selective treatment of heavily infected individuals only. Enhancing coverage and child-targeted treatment represent a more cost-effective approach than increasing the frequency of treatment. A study to compare the effects of mass, targeted, and selective chemotherapy with levamisole as an intervention for the control of ascariasis in areas of endemic infection revealed that in terms of cost per 1,000-egg reduction in intensity and cost per person treated, the mass and targeted approaches were considerably more cost-effective than the selective approach (36).

Epidemiology and Prevention

Since the ultimate transmission of ascariasis depends on fecal contamination of the soil, the use of appropriate sanitary facilities is the primary means of prevention. There are apparently no practical means of killing the eggs while they are in the soil, especially when they are

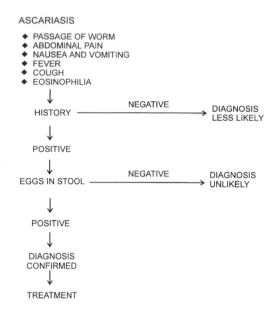

ASCARIASIS

- ◆ PASSAGE OF WORM
- ◆ ABDOMINAL PAIN
- ◆ NAUSEA AND VOMITING
- ◆ FEVER
- ◆ COUGH
- ◆ EOSINOPHILIA

↓

HISTORY ——— NEGATIVE ——→ DIAGNOSIS LESS LIKELY

↓

POSITIVE

↓

EGGS IN STOOL ——— NEGATIVE ——→ DIAGNOSIS UNLIKELY

↓

POSITIVE

↓

DIAGNOSIS CONFIRMED

↓

TREATMENT

Algorithm 10.1   Ascariasis.

in clay soil under favorable environmental conditions (warmth and moisture). In some areas of the world where infections are common, some mass population treatment plans have been used with great success, even in areas with high reinfection rates. The use of human feces, or "night soil," for fertilization of crops should be recognized as a potential hazard. Any vegetables or fruits from such fields cannot be eaten raw or unprocessed. Even with proper pretreatment of night soil, *Ascaris* eggs remain viable and infective more often than eggs of any other helminth species.

In looking at the health of schoolchildren, four significant environmental influences on helminth infections have been identified: (i) an inadequate water supply, (ii) availability of school canteens, (iii) regular water/sanitation maintenance regimes, and (iv) overcrowded classrooms. Helminth infections are strongly associated with anemia, stunting and low weight, and the environmental conditions mentioned above. Although mass anthelmintic drug campaigns have taken place, reinfection is common and drug therapy alone is insufficient for elimination of the problem, particularly in a school environment (38).

Using *A. suum* as a model for *A. lumbricoides*, the ability of sludge treatment processes to kill the eggs of parasitic roundworms was examined. Both unembryonated and embryonated eggs were used, with interesting results. After 1 week in an anaerobic sludge digester, 95% of *A. suum* eggs produced two-cell larvae in vitro, with 86% progressing to motile larvae; after 5 weeks in the digester, 51% progressed to motile larvae. Between 42 and 49% of eggs stored in a sludge lagoon for 29 weeks

were viable and able to develop motile larvae. Of embryonated eggs prior to treatment, >98% survived up to 5 weeks in an anaerobic sludge digester. More than 90% of embryonated eggs survived for 29 weeks in the sludge lagoon and were able to develop motile larvae (40).

Enterobius vermicularis

Enterobius vermicularis has been known since ancient times and has been studied extensively through the years; its original name was *Oxyuris vermicularis*. *E. vermicularis* is thought to cause the world's most common human parasitic infection. It has been said, "You had this infection as a child; you have it now; or you will get it again when you have children!" The infection is more prevalent in the cool and temperate zones, where people tend to bathe less often and change their underclothes less frequently. Prevalence in children can be high, a fact that has been recorded despite the difficulties in confirming the infection.

Although there has been some discussion of *Enterobius gregorii* being a distinct species of pinworm, it appears that this organism may actually be a young stage of *E. vermicularis*. A series of 849 male pinworms from a single individual were examined. Based on spicule morphology, the worms were classified into different groups. However, various transitional forms were observed in the spicule morphology in the worms with intermediate body size between *E. vermicularis* and *E. gregorii*, showing that the basal portion of the spicule of *E. vermicularis* develops after completion of the *E. gregorii*-type basal portion (34).

Life Cycle and Morphology

The female worm measures 8 to 13 mm long by 0.3 to 0.5 mm wide and has a pointed tail (hence the name "pinworm") (Figure 10.10). The male is much smaller, measuring 2 to 5 mm long by 0.1 to 0.2 mm wide, and has a curved caudal end. Infection in humans is initiated by the ingestion of infective eggs, which hatch in

Figure 10.10   *Enterobius vermicularis*, adult female pinworm. (Photomicrograph by Zane Price.)

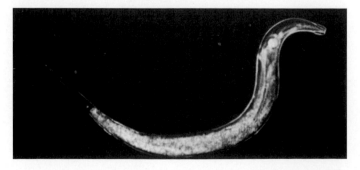

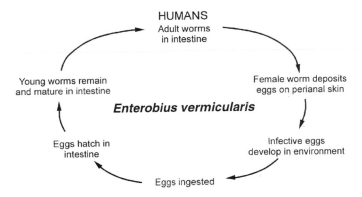

Figure 10.11 *Enterobius vermicularis* life cycle.

the intestine (cecal region), where they develop into the adult worms (Figure 10.11). It probably takes about 1 month for the female to mature and begin egg production. After fertilization of the female worms, the males usually die and may be passed out in the stool. In gravid females, almost the entire body is filled with eggs. At this point, the female migrates down the colon and out of the anus, where the eggs are deposited on the perianal and perineal skin. Occasionally, the female worm migrates into the vagina. It is speculated that after egg deposition, the female worm returns to the intestine; however, this has not been proven (9). Occasionally, when the bolus of stool passes out of the anus, adult worms become attached to the stool and can be found on the surface. Adult worms can also sometimes be picked up on the Scotch tape preparations used to diagnose this infection. Although egg deposition usually does not occur in the intestine, some eggs may be recovered in the stool. The eggs are fully embryonated and infective within a few hours. Although transmission is often attributed to the

Figure 10.12 *Enterobius vermicularis* eggs on a Scotch tape preparation.

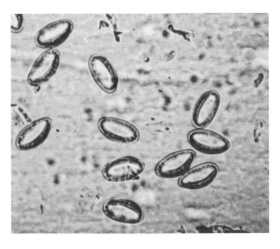

ingestion of infective eggs by nail biting and inadequate hand washing, airborne eggs can also be inhaled and ingested.

The eggs have been described as being shaped like footballs with one side flattened. They are oval, compressed laterally, and flattened on one side and measure 50 to 60 µm long by 20 to 30 µm wide (Figure 10.12).

Clinical Disease

Pinworm infection is the most common helminth infection within the United States and Western Europe; prevalence rates in some areas may be as high as 30 to 50%. The most striking symptom of this infection is pruritus, which is caused by the migration of the female worms from the anus onto the perianal skin before egg deposition. The sometimes intense itching results in scratching and occasional scarification. In most infected people, this may be the only symptom, and many individuals remain asymptomatic.

The degree of infection varies tremendously from patient to patient. As many as 5,000 worms have been removed from a single patient; however, most cases average less than 1 migrating worm per evening. Women are symptomatic three times as often as men, and young people are symptomatic more frequently than older people.

Eosinophilia may or may not be present. Although tissue invasion has been attributed to the pinworm, these cases are not numerous. Infections tend to be more common in children and occur more often in females than in males. In heavily infected females, there may be a mucoid vaginal discharge, with subsequent migration of the worms into the vagina, uterus, or fallopian tubes, where they become encapsulated. Other symptoms have also been attributed to pinworm infection, particularly in children; they include nervousness, insomnia, nightmares, and even convulsions (9).

In one report, a homosexual man presented with severe abdominal pain and hemorrhagic colitis, eosinophilic inflammation of the ileum and colon, and numerous unidentifiable larval nematodes in the stool. By using morphologic characteristics and molecular cloning of nematode rRNA genes, the parasites were identified as larvae of *E. vermicularis*; these larvae are rarely seen and are not thought to cause disease. The authors state that occult enterobiasis is widely prevalent and may be a cause of unexplained eosinophilic enterocolitis (48). Although enterobiasis is an uncommon cause of acute appendicitis in children within the United States, this infection may be associated with acute appendicitis, chronic appendicitis, or ruptured appendicitis or with no significant symptoms (3).

Other uncommon ectopic sites have included the peritoneal cavity, lungs, liver, urinary tract, and natal cleft. Pathologic examination usually reveals chronic granulomatous inflammation with or without central necrosis, which is surrounded by polymorphonuclear neutrophilic leukocytes, eosinophils, and fibroblasts. Macrophages, giant cells, epithelioid cells, and Charcot-Leyden crystals may also be present. During a case of suspected transverse colon carcinoma, histologic examination of the mass revealed eggs of *E. vermicularis* embedded in granulomatous tissue in the submucosa of the colon; no malignancy was found (46). Apparently, this is the first report of enterobiasis presenting as colon carcinoma.

Diagnosis

Although the patient history of anal itching, irritability, and insomnia may suggest a pinworm infection, diagnosis depends on demonstrating the eggs or adult worms. This is normally accomplished by sampling the perianal and perineal skin with cellulose tape (Scotch tape), which is applied sticky side down to the skin. The tape is transferred to a glass slide and examined under the microscope for the presence of eggs or adult worms (see chapter 29). Commercial paddles are also available for the collection of eggs and/or adult pinworms. Eggs are rarely found in the stool (approximately 5% of the time), and sampling of the perianal folds yields more accurate results. Since the female worms migrate on a sporadic basis, a series of four to six consecutive tapes may be necessary to demonstrate the infection. The tapes are used late in the evening, when the patient has been sleeping for several hours, or first thing in the morning before the patient takes a shower or goes to the bathroom. These samples can be taken from children at home and transported to the laboratory for examination. Refer to Algorithm 10.2.

Algorithm 10.2 Enterobiasis.

In cases of ectopic infection, diagnosis usually requires biopsy and histologic examination. Although rare, histologic examination of tissue is particularly important in differentiating malignancy from parasitic infection.

KEY POINTS—LABORATORY DIAGNOSIS

Enterobius vermicularis

1. The cellulose tape (Scotch tape) preparation is recommended as the diagnostic test of choice (a minimum of four to six consecutive negative tapes is required to rule out the infection). Commercial paddles or other collection devices are also acceptable. Although the paraffin swab is also an option, this method is seldom used.
2. Adult worms may be found on or under the surface of the stool specimen, particularly in children. The adult worms can also be found on the tapes.
3. Eggs are occasionally recovered in stool, but this is an incidental finding and not the specimen of choice.

Treatment

Although several drugs such as pyrantel pamoate or mebendazole are extremely effective in eliminating the worms, the decision to treat should be based on evidence of infection and consideration of whether the patient is symptomatic. Repeat reinfection is always possible; thus, repeated treatment may be necessary to ensure that the patient is free of infection. Repeat treatment after about 2 weeks is routinely recommended to eliminate any infection acquired from ingestion of eggs that remain in the environment following the first treatment. Asymptomatic individuals are rarely treated, particularly in a group situation in which more susceptible individuals are not present. Treatment often includes counseling for the parents, who may be very upset at learning that their children have "worms." They may not realize how prevalent the infection is, particularly in children, and that many children never have any symptoms or sequelae of the infection. In a group situation, often only one or several individuals will be symptomatic and continue to be infected. Treatment of the entire group, followed by retreatment of the more susceptible individuals, may provide an infection-free group for a longer period.

Epidemiology and Prevention

Since this infection is so common and transmission is so easy (anus-to-mouth contamination, soiled nightclothes, airborne eggs, and contaminated furniture, toys, and other objects), prevention is marginal. Improved personal and group hygiene combined with group therapy can be helpful. It is also recommended that children sleep in closed garments and maintain short, clean fingernails. Sunlight and UV lamp radiation will destroy the eggs in the environment, and dry heat can be used to sterilize metal toys. The eggs are not killed by the level of chlorination used in swimming pools. Recent studies also reveal that *Enterobius* infestation can occur in cockroaches, which may serve as potential reservoirs for these helminths and have considerable public health significance (14). *Under most circumstances, total prevention is neither realistic nor possible.*

Trichuris trichiura

Trichuris trichiura was first described by Linnaeus in 1771 and was studied extensively in the late 1880s, and early 1900s. Two early names include *Trichocephalus trichiurus* and *Trichocephalus dispar*. Well-preserved *Trichuris* eggs were found in feces in the rectum of a 9-year-old Inca girl whose frozen body was recovered in a stone building at the 17,568-ft altitude near Santiago, Chile; it was calculated that the child had died approximately 450 years earlier (9). In 1991, a well-preserved mummy of a 25- to 30-year-old man was found in a glacier in the Alps at an altitude of 3,200 m; it was estimated that the man had died about 5,200 to 5,300 years earlier. In studies of material from the colon, preservation and processing of material in sodium acetate-acetic acid-formalin fixative revealed the presence of *T. trichiura* eggs, the oldest finding of *Trichuris* in a human (3200 to 3300 BC) (6).

Infection with *T. trichiura* is more common in warm, moist areas of the world and is often seen in conjunction with *Ascaris* infections. Worm burdens vary considerably, and individuals with few worms are unaffected by the presence of these parasites. Prevalence rates of 20 to 25% have been reported from the southern United States (9).

Life Cycle and Morphology

Whipworms are much larger than pinworms, measuring 35 to 50 mm long (female) or 30 to 45 mm long

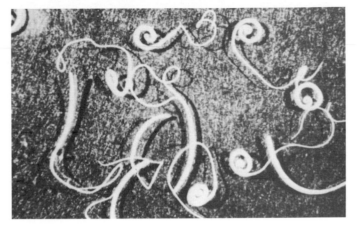

Figure 10.13 *Trichuris trichiura* adult worms. (From A Pictorial Presentation of Parasites: A cooperative collection prepared and/or edited by H. Zaiman.)

(male); the male has a 360° coil at the caudal extremity (Figure 10.13). The adult worms are rarely recovered from the stool, since they are attached to the wall of the intestine. The head portion of the worm is very thin and is embedded in the mucosa, while the posterior end is much thicker and lies free in the lumen of the large intestine. The large, posterior end has been described as the whip handle while the thin, anterior end is the whip itself, hence the name whipworm.

Human infection is acquired through ingestion of the fully embryonated eggs from the soil (Figures 10.14 and 10.15). The eggs hatch in the small intestine and eventually attach to the mucosa in the large intestine. The adults mature in about 3 months and begin egg production. The eggs are barrel shaped with clear, mucoid-appearing polar plugs (Figures 10.15 and 10.16). They measure 50 to 54 µm long and 22 to 23 µm wide. They are passed in the unsegmented stage and require 10 to 14 days in moist soil for embryonation to occur. Distorted eggs that are much larger than normal have been seen following therapy with mebendazole and with other drugs. This is not a common

Figure 10.14 *Trichuris trichiura* life cycle.

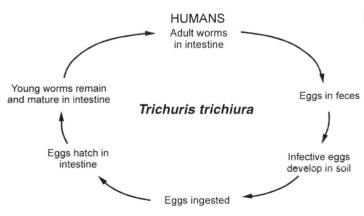

HUMANS
Adult worms
in intestine

Young worms remain
and mature in intestine

Trichuris trichiura

Eggs in feces

Eggs hatch in
intestine

Infective eggs
develop in soil

Eggs ingested

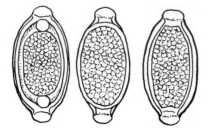

Figure 10.15 *Trichuris trichiura* eggs. (Illustration by Sharon Belkin.)

finding but is something to consider if distorted eggs are seen. There are also some reports in the literature that *T. vulpis* (dog whipworm) eggs have been recovered in human stools (67). These eggs tend to be larger (70 to 80 μm long by 30 to 42 μm wide) and have prominent but small polar plugs compared with those of *T. trichiura*. Several studies have also confirmed that larger eggs among others of normal size have been recovered in human stool in cases where the larger egg size was not associated with treatment. Studies also confirm that these larger eggs can develop to the infective stage (9).

Clinical Disease

The differential diagnosis of chronic diarrhea requires consideration of several diseases, including celiac disease, inflammatory bowel disease, and irritable bowel syndrome. Patients infected with *T. trichiura* may present with a chronic dysentery-like syndrome if they have a massive infestation leading to anemia and growth retardation. In some cases, diarrhea may last for years without blood and mucus. However, once blood is evident, medical intervention may occur; in some cases the diagnosis requires colonoscopy. Also, in these cases, prolonged therapy may be necessary to eliminate the parasites (28).

Mechanical damage to the mucosa and the allergic response by the host appear to be the main reasons for any abnormalities associated with *T. trichiura* infection and are definitely related to the worm burden, the length of the infection, and the age and overall health status of the host. Although the worm is actually threaded into the epithelium of the cecum, damage produced by this process is minimal unless there is dysentery, during which the mucosa may be edematous and friable, and there may be abdominal cramps, severe rectal tenesmus, and rectal prolapse (Figures 10.17 and 10.18). The dysentery seen in severe infections may mimic that seen in amebiasis. Most infections are light to moderate, with minimal or no symptoms.

Hypochromic anemia may be seen in patients with prolonged, massive infection; however, the anemia is due

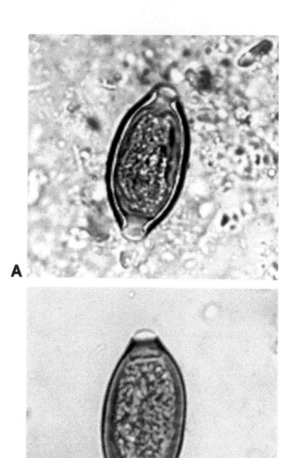

Figure 10.16 *Trichuris trichiura* eggs; note the two polar plugs, one at each end of the egg.

to malnutrition and blood loss from the friable colon and is not related to blood ingestion by the parasite. In a study of 409 Jamaican schoolchildren, iron deficiency anemia was associated with *Trichuris* infections involving over 10,000 eggs/g but not with less severe infections (56). Although eosinophils and Charcot-Leyden crystals are present in the stool in patients with dysentery, a peripheral eosinophilia on the differential smear is not always seen and the degree of eosinophilia may not correlate with the severity of infection (it rarely exceeds 15%). Heavy infections tend to be rare in developed countries, and complications requiring surgical intervention are also rarely described. A case of colonic obstruction and perforation has been described in an 84-year-old woman. She was originally admitted for a chest infection; 2 days after admission, she suffered from nausea and vomiting, followed by bowel stoppage 2 days later. A partial right-sided ileocolectomy was performed; pathology

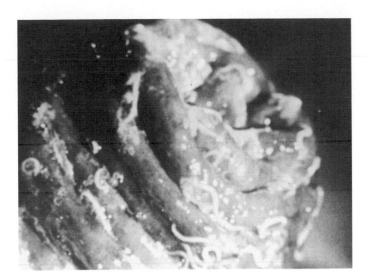

Figure 10.17 *Trichuris trichiura*, rectal prolapse.

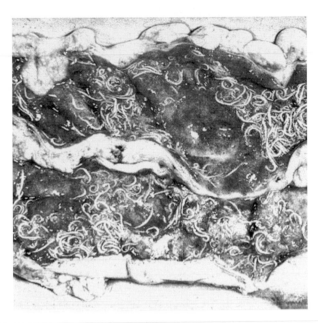

Figure 10.18 Numerous adult *Trichuris trichiura* worms attached to the colonic mucosa of a patient who died of trichuriasis. (From A Pictorial Presentation of Parasites. A cooperative collection prepared and/or edited by H. Zaiman.)

revealed a heavy infestation with *T. trichiura*, causing a pseudotumor following a proliferative inflammatory response (7).

There are few data on mucosal immune responses to intestinal helminths in humans, particularly involving the IgE system, which is thought to be important in parasite expulsion. Cooper et al. (22) did find evidence of an IgE-mediated immune mucosal response to *Trichuris* infections, but it was not sufficient to cause parasite expulsion. On examination of rectal biopsy specimens, children with *Trichuris* dysentery syndrome had significantly greater numbers of mast cells and cells with surface IgE than did the control group. Unanswered questions involve possible antibody-mediated immunity and protection against invading larvae and whether *Trichuris* dysentery syndrome develops in some children because they have a defect in antiparasite cell-mediated immunity. Probably both cellular and humoral responses are required to eliminate the worms from the colon.

A study in England was designed to determine whether *T. trichiura* dysentery is associated with evidence of a systemic inflammatory response and whether the plasma protein disturbance has special characteristics compared with the situation in uninfected children in the area of endemic infection (21). C-reactive protein, α-antitrypsin, total globulin, and fibronectin levels and plasma viscosity were significantly higher in children with *T. trichiura* dysentery. There appeared to be an acute-phase response in intense trichuriasis and a specific elevation of the fibronectin level in plasma. Also, plasma viscosity remained abnormally high 6 months after treatment, although it was lower than at diagnosis.

It appears that an increase in the rate of epithelial cell turnover in the large intestine may lead to the expulsion of *Trichuris*. This increased rate of epithelial cell movement is under immune control by the cytokine IL-13 and the chemokine CXCL 10 (19). Such a host protective mechanism may help explain the roles played by the intestinal epithelium in mucosal defense.

Trichuris suis Therapy for Active Ulcerative Colitis

Inflammatory bowel disease, including Crohn's disease, probably occurs from a failure to downregulate a chronic Th1 intestinal inflammatory process. Induction of a Th2 immune response by intestinal helminths reduces the Th1 inflammatory process. Ulcerative colitis is more common in Western industrialized countries than in underdeveloped countries, particularly those where helminth infections are common. People with helminth infections exhibit altered immunologic antigen responses. It has also been shown in animal models that helminths prevent or improve colitis through the induction of regulatory T cells and modulatory cytokines. The use of *Trichuris suis* in the therapy of ulcerative colitis has been a controversial and debated issue for several years. A recent randomized, double-blind, placebo-controlled trial conducted at the University of Iowa and selected private practices has shown improvement in 43.3% of patients treated with ova compared with 16.7% who received placebo.

Although the benefit was temporary in some patients who received a single dose of 2,500 *T. suis* eggs, improvement could be sustained with maintenance therapy every 3 weeks. Consequently, this approach appears to be safe and effective in patients with active colitis (70).

Diagnosis

Most whipworm infections can be easily diagnosed by finding the characteristic eggs in the stool. These eggs should be quantitated (rare, few, moderate, many), since light infections usually cause no problems and do not require therapy.

Note. *T. trichiura* eggs submitted in stool preserved with polyvinyl alcohol do not concentrate as well as those preserved in formalin. *However, the very small numbers of eggs that might be missed in a concentrate obtained from polyvinyl alcohol-preserved stool material are not clinically significant.* If the specimen was collected early in the infection, when egg production was minimal, eggs would be seen in later specimens (if they were submitted because of patient symptoms).

Although dysentery with *T. trichiura* and dysentery with *Entamoeba histolytica* are very similar, whipworm dysentery is usually more chronic, associated with malnutrition, and likely to cause rectal prolapse. Recovery and identification of the eggs or protozoan trophozoites would differentiate the two infections. Whipworm infections in heavily infected children may be accompanied by infection with *E. histolytica* and enteropathogenic bacteria. In severe infections, the adult worms are usually visible on the rectal mucosa. Refer to Algorithm 10.3.

Algorithm 10.3 Trichuriasis.

TRICHURIASIS

- ◆ DIARRHEA
- ◆ ABDOMINAL COMPLAINTS
- ◆ RECTAL PROLAPSE

↓

HISTORY ——— NEGATIVE ———→ DIAGNOSIS LESS LIKELY

↓

POSITIVE

↓

EGGS IN STOOL ——— NEGATIVE ———→ DIAGNOSIS UNLIKELY

↓

POSITIVE

↓

DIAGNOSIS CONFIRMED

↓

TREATMENT

KEY POINTS—LABORATORY DIAGNOSIS

Trichuris trichiura

1. Eggs are normally recovered by using the routine ova and parasite examination from the direct wet mount or the concentration procedure. Egg recovery is generally not as good when concentrations are performed using polyvinyl alcohol-preserved fecal specimens; formalin or other non-polyvinyl alcohol preservatives are recommended.
2. These eggs should be quantitated on the laboratory report (rare, few, moderate, many), since light infections may not require treatment.
3. Adult worms are very rarely seen.
4. Most infections are mild; however, in very heavy infections, dysentery may have to be differentiated from that caused by *E. histolytica*.
5. The eggs can usually be identified from the permanent stained smear, but morphology is more easily seen in the wet mount preparations (either direct wet mounts or from concentrated sediment).

Treatment

Several available drugs are very effective, including mebendazole and albendazole. Exact information and dosages are provided in chapter 25. Therapy may not be necessary, depending on the number of eggs recovered in the stool. In investigating the relationship between whipworm infection and cognition in a large group of Jamaican children, a recent study concluded that albendazole treatment of children with mild to moderate *T. trichiura* infections produced little benefit in cognition if they were adequately nourished; however, children who were undernourished were more likely to benefit (66).

Epidemiology and Prevention

The geographic range of *T. trichiura* is similar to that of *A. lumbricoides*, and the two infections are often found together in the same host. Infection rates are highest in children; the eggs contaminate the soil where the children play, and the children reinfect themselves with eggs transferred from the soil to their mouths. The eggs do not survive dry conditions or intense cold.

Adequate disposal of feces reduces both the number of infections and the worm burdens. This approach is particularly important when one is dealing with children who may be used to defecating on the ground.

Capillaria philippinensis

The first proven case of human infection with *Capillaria philippinensis* occurred in 1963 in a patient from the Philippines who died 3 days after admission to the hos-

Figure 10.19 *Capillaria philippinensis* eggs.

pital with a diagnosis of malabsorption syndrome (9). Although the significance was not recognized until 4 years later, *C. philippinensis* eggs were found in the stools and autopsy showed parasitism of the large and small intestines. In 1967, health authorities recognized infections causing severe symptoms and death in adult males, an infection that is now known to be widely distributed in the Northern Luzon area of the Philippines. Since that time, the disease has become widespread in Thailand and cases have been reported in Japan, Taiwan, Iran, and Egypt.

The adult female worms are 2.5 to 4.3 mm long, while males are 2.3 to 3.17 mm long. These worms live in the mucosa of the small bowel, most commonly the jejunum. Human infection is initiated by the ingestion of raw fish; the infective larvae are located in the mucosa of the fish intestine. Specific details of the life cycle are not clearly known; however, there does seem to be an internal autoinfection capability (23). The female worms produce eggs that resemble those of *T. trichiura*. They are thick shelled, with less prominent polar plugs, and the shell is striated (Figure 10.19). The eggs are somewhat smaller than those of *T. trichiura*, measure 36 to 45 by 20 μm, and require 10 to 14 days in the soil to embryonate and 3 weeks to develop into the infective form in fish.

Life Cycle and Morphology

Although the exact mode of transmission is unknown, experimental infection is transmitted through small fish that serve as the intermediate host; often, whole, small fish may be ingested. Development to the infective stage in the fish takes at least 3 weeks. In areas of the Philippines where this infection occurs, people also eat raw shrimp, crabs, and snails. They also tend to defecate in the fields or water where the fish, shrimp, crabs, and snails are obtained, thus completing the life cycle. The worms live burrowed into the mucosa of the small bowel, mainly the jejunum. The adult worms are small. The male has ventrolateral caudal expansions and a very long, smooth

spicular sheath. The female has two almost equal parts, with the anterior containing the esophagus and stichosome (esophageal glands) and the posterior containing the intestine and reproductive system. Females produce eggs with thin shells and free larvae, as well as the typical thick-shelled eggs that pass in the stool. All stages of development are seen in the human host, and internal autoinfection is a normal part of the life cycle. Eggs that are passed in stool have been described as "peanut shaped," with flattened bipolar plugs and striated shells, measuring 36 to 45 μm long by 20 μm wide and somewhat resembling *T. trichiura* eggs.

Clinical Disease

Symptoms are related to the worm burden; with large numbers of worms, there may be intestinal malabsorption and fluid loss along with electrolyte and plasma protein imbalance. Most of the abnormality is found in the small intestine, where the wall is thickened and indurated and contains many larval and adult worms. Watery stools are passed (up to eight per day), with fluid loss of several liters. Patients lose weight rapidly and develop muscle wasting, abdominal distention, and edema. Death from pneumonia, heart failure, hypokalemia, or cerebral edema may occur within several weeks to a few months (29). In some cases, patients reported chronic abdominal pain and diarrhea over a period of many months prior to diagnosis. On gastroduodenoscopy and subsequent histology, the jejunal mucosa revealed flattened villi, crypt proliferation, acute inflammation, and eosinophilic granulomata.

Diagnosis

Diagnosis is based on recovery and identification of the eggs in the stool, which might also contain larvae or adult worms. Knowledge of the geographic range would also provide specific clues to a possible infection. Unfortunately, without some index of suspicion, patients may go for many months with the cause of their symptoms remaining unexplained.

Treatment

Some of the more common therapeutic agents, including mebendazole and albendazole, are effective in treating this infection and result in dramatic improvement in the patient's condition within the first day. Specific drug and dosage information is provided in chapter 25.

Epidemiology and Prevention

Although the life cycle is not completely known, the geographic range is known and expanding. Raw infected fish, shrimp, crabs, and snails may serve to transmit the larvae. In areas where this infection is endemic, adequate cooking of all suspect foods should reduce the number of cases.

Infections have now been described in the Philippines, Thailand, Japan, Iran, Egypt, Taiwan, and Indonesia.

Hookworms (*Ancylostoma duodenale* and *Necator americanus*)

Although the "Old World" hookworm, *Ancylostoma duodenale*, was mentioned in ancient Egypt in 1600 BC and in the Persian literature in approximately AD 1000, the first good description was written by Dubini in 1843. Diagnosis was confirmed in the late 1800s by finding eggs in stool, particularly in patients suffering from "miner's anemia." Final delineation of the complete life cycle was completed in 1897 by Arthur Looss from both his accidental self-infection and additional studies on dogs. Although the geographic ranges of *A. duodenale* and the "New World" hookworm *Necator americanus* used to be much more distinct, nowadays there are many areas of the world where both are found. *N. americanus* was described by Stiles in 1902 as a new species and was soon found to be widely distributed throughout the Western Hemisphere (see below); it is known as the "American hookworm" or the "American murderer."

Hookworm infections are found in moist, warm areas and are responsible for much human disease, although they cause more morbidity than actual mortality. Although there are distinct morphologic differences between the two most common human hookworms (adult worms), the diagnostic stage (egg) is essentially identical for both. Globally, about 900 million people harbor hookworms. In looking at various hookworm isolates, molecular biology-based studies have been very helpful in determining correct classification relationships. Studies of the nucleotide sequences of the second internal transcribed spacer of rDNA of adult worms of *N. americanus* from Togo, Africa, and Sarawak, Malaysia, suggest that there is population variation in the sequence of *N. americanus* or that *N. americanus* from the two countries may represent genetically distinct but morphologically similar species. However, comparison of the sequence differences among other hookworm species supports genetically distinct species (58). As these types of studies continue to examine former classification schemes, we may see generic and species name changes in future publications.

Life Cycle and Morphology

Adult males measure 7 to 11 mm long by 0.4 to 0.5 mm wide. The adult *Ancylostoma* worm tends to be larger than the *Necator* worm. Adult worms are rarely seen, since they remain firmly attached to the intestinal mucosa by means of well-developed mouth parts (teeth in *A. duodenale* and cutting plates in *N. americanus*) (Figure 10.20).

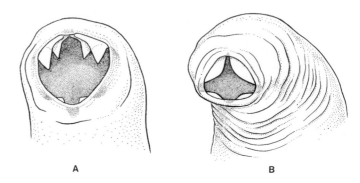

Figure 10.20   (A) *Ancylostoma duodenale* mouth parts (teeth); (B) *Necator americanus* mouth parts (cutting plates). (Illustration by Sharon Belkin; adapted as a composite from a number of photographs, including those in E. K. Markell, M. Voge, and D. T. John, *Medical Parasitology*, 7th ed., The W. B. Saunders Co., Philadelphia, Pa., 1992.)

Infection in humans is acquired through active skin penetration of filariform larvae from the soil (Figure 10.21). The larvae are normally in a motionless, resting position; stimuli for movement include touch, vibration, water currents, heat, light, human breath, and skin extracts such as fatty acids (32). During skin penetration, infective hookworm larvae encounter hyaluronic acid; hookworm hyaluronidase activity has been confirmed and can facilitate passage of the infective larvae through the epidermis and dermis during larval migration (37). The infective larvae of *N. americanus* secrete various classes of proteolytic enzymes with two overall pH optima of 6.5 and 8.5. Since skin penetration is mandatory for the infective larvae, the effect of each of these enzyme classes against macromolecules derived from human skin has been examined. Larval secretions were shown to degrade collagen types I, III, IV, and V, fibronectin, laminin, and

Figure 10.21   Hookworm life cycle.

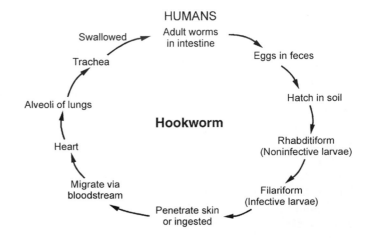

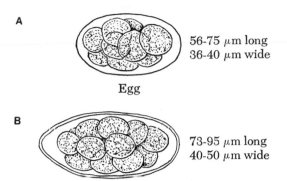

A   56-75 μm long
36-40 μm wide

Egg

B   73-95 μm long
40-50 μm wide

Figure 10.22 (A) Hookworm egg; (B) *Trichostrongylus* egg. (Illustration by Sharon Belkin.)

elastin. All the skin macromolecules tested were hydrolyzed by aspartyl proteinase activity. Studies of the effect of proteinase inhibitors revealed that larval penetration was significantly inhibited only by pepstatin A, confirming the importance of the aspartyl proteinase activity during skin penetration (12). After skin penetration, the larvae are carried by the venules to the right heart and then to the lungs. Larvae break into the alveoli, migrate via the bronchi until they reach the trachea and pharynx, and are then swallowed, bringing them to the small intestine, where they reside. They attach to the mucosa by using a temporary mouth structure, mature sexually, and finally develop the permanent characteristic mouth structure that they use to attach to the mucosa. The females begin to deposit eggs about 5 months after the initial infection, although the prepatent period may be 6 to 10 months. If mature filariform larvae of *A. duodenale* are swallowed, they can develop into mature worms in the intestine without migrating through the lungs. A *Necator* infection produced by percutaneous exposure of a volunteer to three larvae was monitored by periodic egg counts for 18 years and 4 months (8). Until this report, the longest infection recorded was 15 years.

The eggs are usually in the early cleavage stage when passed in the stool (Figures 10.22 and 10.23). They are oval with broadly rounded ends and measure approximately 60 μm long by 40 μm wide. They characteristically have a clear space between the developing embryo and the thin eggshell. Although hookworm eggs and eggs of *Trichostrongylus* spp. are similar, specific size differences can be seen in Figure 10.24. Egg survival and larval development are maximum in moist, shady, warm soil (sandy loam), where larvae hatch from the eggs within 1 to 2 days. The infective filariform larvae develop within 5 to 8 days and may remain viable in the soil for several weeks.

A review by Schad (61) of hookworm developmental biology provides some excellent information on the relationship between arrested development of hookworms

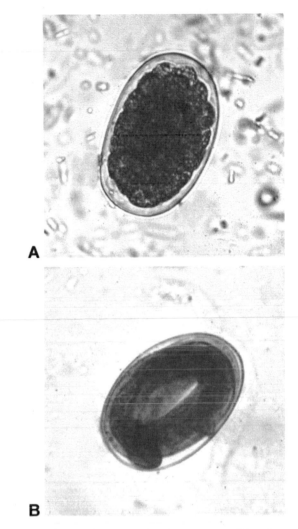

Figure 10.23 (A) Immature hookworm egg; (B) embryonated hookworm egg.

and vertical transmission. Epidemiologic evidence supports the theory that *A. duodenale* may enter a period of arrested parasitic development for approximately 8 months, as a mechanism allowing the species with the shorter adult life span to survive the dry climatic season in the host. Experimental evidence is based on two self-induced *A. duodenale* infections with very long prepatent periods of 40 weeks (hypobiotic state), about five times the normally expected prepatent period. Clinical evidence involved a group that was exposed to infection and after 11 days left the area where the infection was endemic; hypobiotic larvae were shown to have remained in 8 of 15 patients, with adult and immature worms being expelled after repeated therapy 253 days after the only known exposure. Arrested development is a prerequisite for vertical transmission, and there is mounting evidence that congenital infection can occur (61). Apparently, arrested

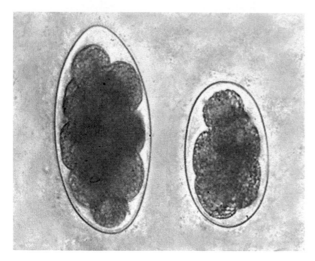

Figure 10.24 (Left) *Trichostrongylus orientalis*, immature egg; (right) hookworm, immature egg.

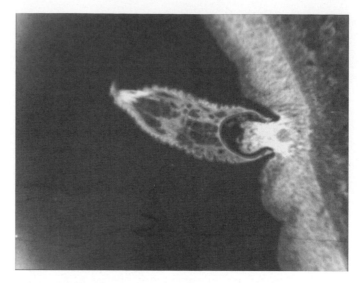

Figure 10.25 Hookworm, section of a worm attached to the mucosa. (From A Pictorial Presentation of Parasites: A cooperative collection prepared and/or edited by H. Zaiman.)

larvae tend to be highly resistant to most therapeutic agents; control programs that do not take into account this aspect of the life cycle will not achieve optimal results.

Clinical Disease

Initial symptoms after larval penetration of the skin often depend on the number of larvae involved. There may be minimal to severe pruritus and possible secondary infection if the lesions become vesicular and are opened by scratching. The development of vesicles from the erythematous papular rash is called ground itch.

Any pneumonitis due to migrating larvae depends on the number of larvae present. These larvae do not cause the same level of sensitization seen with *Ascaris* or *Strongyloides* infection.

Symptoms from the intestinal phase of the infection are caused by (i) necrosis of the intestinal tissue within the adult worm mouth and (ii) blood loss by direct ingestion of blood by the worms and continued blood loss from the original attachment site, possibly as a result of anticoagulant secreted by the worm (Figure 10.25).

Patients with acute infections involving many worms may experience fatigue, nausea, vomiting, abdominal pain, diarrhea with black to red stools (depending on the level of blood loss), weakness, and pallor. As in many other parasitic infections, heavy worm burdens in young children may have serious sequelae, including death. During this acute intestinal phase, there is an increased peripheral eosinophilia.

In chronic infections, the main symptom is iron deficiency anemia (microcytic, hypochromic) with pallor, edema of the face and feet, listlessness, and hemoglobin levels of 5 g/dl or less. There may be cardiomegaly and both mental and physical retardation.

The hookworms cause gastrointestinal blood loss, and in clinical studies significant blood losses have been reported with *A. duodenale*. However, there has been no evidence that endemic *A. duodenale* infection has a greater impact than *N. americanus* infection on the iron status of populations. In a sample of 525 schoolchildren in Tanzania, the degree of anemia and iron deficiency associated with the two hookworm species was compared at the individual and community school levels. Hemoglobin and ferritin concentrations decreased with increasing proportions of *A. duodenale*. In children infected with only *N. americanus* larvae, the prevalence of anemia was 60.5% and the prevalence of low ferritin levels was 33.1%, while in children with at least 50% *A. duodenale* larvae, the respective prevalences were 80.6 and 58.9%. When children were grouped by prevalence of *A. duodenale* at the school level, children from high-prevalence (≥20%) schools had significantly worse iron deficiency and anemia than did children from low-prevalence schools. Since the species of hookworm being transmitted within a community influences the burden of iron deficiency anemia in the community, this factor should be considered in prioritizing and planning programs for hookworm and anemia control (1). It is also well documented that there is a relationship between the hookworm worm burden and anemia during pregnancy; women with moderate or heavy hookworm infections were more likely to suffer from anemia than were women having light infections. These findings support routine therapy within prenatal care programs in areas where infection is highly endemic (45).

Egg counts of 5/mg of stool are rarely clinically significant, counts of more than 20/mg are usually associated with symptoms, and counts of 50/mg or more represent very heavy worm burdens.

Immunology and Host Resistance

There is a mixed Th1-Th2 response in hookworm infection with subsequent immunosuppression of gamma interferon responses. There is also evidence for protective immunity in hookworm infection, including antilarval IL-5- and IgE-dependent mechanisms (55). Antihookworm antibodies show cross-reactivity with *A. lumbricoides* and *Schistosoma mansoni*, while IgG4 or IgE responses tend to be more specific. The levels of polyclonal IgG in serum and intestinal IgG, IgM, and IgE in the intestines are also elevated, while intestinal IgA levels are reduced.

Eosinophilia is common, usually develops 25 to 35 days after exposure, and peaks about a month (*N. americanus*) to 2 months (*A. duodenale*) later. Proliferative lymphocyte responses tend to be variable, while cytokine results indicate that hookworm infection produces a mixed Th1-Th2 response. There is significant production of Th1 (gamma interferon and IL-12) and Th2 (IL-4, IL-5, and IL-13) cytokines. There are also high levels of IL-10 and tumor necrosis factor alpha.

Hookworm-mediated immunosuppression has relevant implications for vaccine development. Unless preexisting infections are cleared by therapy, vaccine-induced immunity may be susceptible to immunosuppression. Also, if the vaccine is only partially effective, vaccine recipients may be susceptible to reinfection, which might further suppress vaccine efficacy. Careful selection of vaccine candidates may prevent this problem. Currently, the ASP-2 protein secreted by infective larvae of *N. americanus* is under development as a recombinant vaccine (30).

There is evidence for a protective role of helminth infection against asthma and malaria; however, helminth infection may increase susceptibility to HIV/AIDS or tuberculosis. People with hookworm infection are also more likely to be infected with *A. lumbricoides* and *T. trichiura*, findings that can only partially be explained by overlapping areas of endemic infection and potential exposure.

Diagnosis

Definitive diagnosis of hookworm infection depends on demonstration of the eggs in stool, especially since the symptoms cannot be differentiated from malnutrition (Figures 10.22 and 10.23). The eggs are best seen in the direct smear or the concentration sediment; they are distorted on the permanent stained smear.

Note. If the stool specimen is stored at room temperature (no preservative) for more than 24 h, the larvae continue

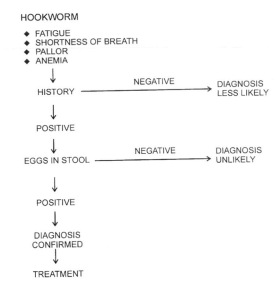

Algorithm 10.4 Hookworm.

to mature and hatch. These larvae must be differentiated from *Strongyloides* larvae, since therapy is often quite different for the two infections.

Many infections seen in the United States are not severe, and eggs in the stool may represent an incidental finding. Radiographic findings include intestinal hypermotility, proximal jejunal dilatation, and coarsening of the mucosal folds (9). Refer to Algorithm 10.4.

In one study examining stools for helminth eggs by the Kato-Katz method and by a modified Harada-Mori nematode larval culture method, there was an 81.8% agreement between the two methods for the detection of hookworm infection, an infection that was endemic in the study population and highly prevalent in pregnant women in the area (51).

KEY POINTS—LABORATORY DIAGNOSIS

Hookworms (*Ancylostoma duodenale* and *Necator americanus*)

1. Eggs are normally passed in the stool in the unembryonated state (usually about an 8- to 32-cell stage of development). There are typically a thin shell and clear space between the developing embryo and the shell.
2. If the stool remains unpreserved for over 24 h, the eggs may continue to develop and the larvae may hatch. These larvae must be differentiated from those of *S. stercoralis*, since therapies for the two infections are different.
3. The eggs are normally distorted on the permanent stained smear, and morphology is more easily seen in wet preparations.

Treatment

Several effective therapeutic agents, including mebendazole, pyrantel pamoate, and albendazole, are available in the United States. Additional studies support the data showing that despite its relatively low cost and wide availability, pyrantel pamoate at a single-dose rate of 10 mg/kg of body weight is not effective against *A. duodenale* and should not be considered a drug of choice at this dose rate. In contrast, albendazole (at a single 400-mg dose) cleared hookworm infections completely (57). Confirmatory studies also indicate that in the southern region of Mali, single-dose treatment with mebendazole is not an effective treatment for hookworm infections and that despite its relatively low cost and wide availability, mebendazole should not be considered a drug of choice in the mass treatment of hookworm infections (18). See chapter 25 for specific drug and dosage recommendations.

Adequate nutrition can prevent and overcome disease symptoms but not the infection itself. Apparently, adequate absorption of iron and nutrients is not possible in heavy infections, even with extensive dietary supplements.

Epidemiology and Prevention

Sanitary disposal of feces is the primary means of infection control. This is sometimes difficult in poor rural communities where sanitary facilities are minimal or absent. An overall prevention approach also requires educational programs. The wearing of shoes and the use of soil sterilization have not proven to be practical.

Several factors influence hookworm prevalence: infection in the human population, defecation onto the soil, acceptable environmental conditions, and human contact with the infective larvae in soil. Environmental conditions include temperature, rainfall pattern, and the presence of open, sandy soil.

A significant increase in the prevalence of hookworm infection has been seen in an area of Haiti where intestinal parasites are common but hookworm has not been common. Changing environmental conditions, specifically deforestation and subsequent silting of a local river, have caused periodic flooding with deposition of a layer of sandy loam topsoil and increased soil moisture. These conditions, all of which are conducive to hookworm transmission, have allowed hookworm to reemerge as an important human pathogen in this area. This example emphasizes the value of longitudinal surveillance data for monitoring disease prevalence. Shifts in the prevalence of infectious diseases can be caused by environmental changes, including planned human activity, or can be an indirect consequence of political strife, and these factors should always be considered when changes in infectious disease patterns are detected.

A. duodenale (Old World hookworm) is found primarily in southern Europe, the north coast of Africa, northern India, northern China, and Japan. *N. americanus* (New World hookworm) is found throughout the southern United States, the Caribbean, Central America, northern South America, central and southern Africa, southern Asia, Melanesia, and Polynesia. In some areas such as northern Ghana, both hookworms are present; mixed infections have been confirmed using PCR (24).

Trichostrongylus spp.

Trichostrongylus nematodes are commonly found in herbivores throughout the world. Various species have been found in humans, and some are more clinically important than others. Throughout the world, there are certain areas of endemic infection where the percentage of positive patients is substantial, and at least 10 species have been found to be human parasites. Human infections tend to be prevalent in Iran.

Life Cycle and Morphology

Trichostrongylus spp. are small worms, similar to hookworms, and live embedded in the mucosa of the small intestine. Unlike the adult hookworms, the adult worms have no distinct buccal capsule with special mouth parts (teeth or cutting plates).

Infection in humans is acquired through ingestion of the infective larvae contaminating plant material. After reaching the small intestine, the larvae mature in 3 to 4 weeks without any migratory pathway through the lungs. The eggs are very similar to those of hookworms, being oval and somewhat longer, with the ends being more pointed than in hookworm eggs (Figures 10.22 and 10.24). The eggs may hatch within 24 h under favorable conditions (warm, moist soil) and develop into infective larvae after about 60 h.

Clinical Disease

Symptoms are related to the worm burden and damage to the intestinal mucosa. Hemorrhage and desquamation may occur (similar to findings in hookworm infection); however, symptoms are usually not clinically significant unless several hundred worms are present. Patients present with epigastric pain, diarrhea, anorexia, nausea, dizziness, and generalized fatigue or malaise. Heavy worm loads may lead to the development of anemia and cholecystitis, as the worms enter the biliary tract. A *Trichostrongylus* infestation has been seen masquerading as conditioning toxicity of the gut in a bone marrow transplant recipient, another reason to consider all possibilities in immunocompromised patients (15).

Diagnosis

The definitive diagnosis can be made by identification of eggs in the stool. Hatched larvae can also be differentiated from those of hookworms and *S. stercoralis* (Figure 10.26).

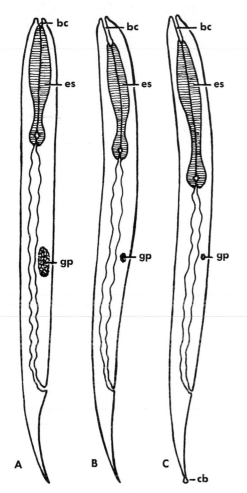

Figure 10.26   Rhabditiform larvae. (A) *Strongyloides stercoralis*; (B) hookworm; (C) *Trichostrongylus* sp. Abbreviations: bc, buccal cavity; es, esophagus; gp, genital primordium; cb, head like swelling of caudal tip. (Illustration by Nobuko Kitamura; from L. S. Garcia, *in* S. M. Finegold and E. J. Baron, *Diagnostic Microbiology*, 7th ed., The C. V. Mosby Co., St. Louis, Mo., 1986.)

Treatment

Several anthelmintic agents are available, including pyrantel pamoate, mebendazole, and albendazole, some of which are also used for the other nematode infections. Specific drug and dosage information is presented in chapter 25.

Epidemiology and Prevention

Since the infection is found in herbivores throughout the world, any grazing areas are constantly being reinfected. In some areas of the world where animal dung is formed into solid masses, dried, and burned for fuel, this practice may be important in the transmission of animal *Trichostrongylus* to humans. The main preventive measures are adequate cleaning and cooking of all vegetables prior to eating.

Strongyloides spp.

Strongyloides stercoralis was first found in 1876 in the feces of French soldiers with diarrhea who were returning from Indochina. A series of workers dealt with the classification, life cycle differentiation, and possible pathogenesis of this worm during the early 1900s. Early work led to demonstration of the internal autoinfective portion of the life cycle, which has become extremely important, particularly when one is working with immunocompromised patients. *S. stercoralis* is most commonly found in warm areas but can survive in colder climates. The infection is most often found to overlap the geographic range of hookworm infections and has been reported as the most commonly diagnosed helminth infection at the University of Kentucky Medical School. It is very important to consider this infection in military personnel and travelers who, many years earlier, were in an area where the infection is endemic. More than 30 to 40 years after acquisition of the original infection, persistent, undiagnosed disease can be found in these individuals. If for any reason they become immunocompromised, the result can be disseminated disease leading to the hyperinfection syndrome and death.

Life Cycle and Morphology

The life cycle of *S. stercoralis* is one of the more complex cycles and can be seen in Figure 10.27. Human infection is acquired by skin penetration of the filariform larvae (infective larvae) from the soil. These larvae are long and slender (up to 630 µm long by 16 µm wide) and may remain viable in soil or water for several days. After penetration of the skin, the larvae are carried via the cutaneous blood vessels to the lungs, where they break out of the pulmonary capillaries into the alveoli. They then migrate via the respiratory tree to the trachea and pharynx, are swallowed, and enter the mucosa in the duodenum and upper jejunum. Development of the female worms to adults usually takes about 2 weeks, after which the females begin egg production by parthenogenesis. The eggs are oval and thin shelled, and they measure 50 to 58 µm long by 30 to 34 µm wide (generally a bit smaller than hookworm eggs). They hatch, and the rhabditiform larvae (noninfective larvae) pass out of the intestinal tract in the feces. These larvae are passed out onto the soil in the feces and develop into free-living male and female worms, eventually producing infective filariform larvae (egg, noninfective larvae, and infective larvae). In temperate climates, the free-living male and female worms do not develop; however, the rhabditiform larvae that pass out in the stool develop into the filariform (infective) larvae, which are then ready to infect the next host through skin penetration.

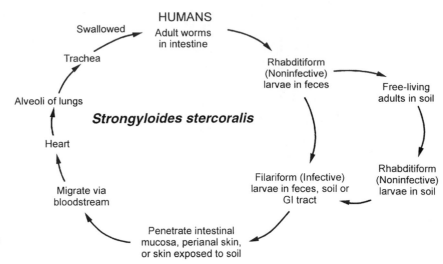

HUMANS

Strongyloides stercoralis

Swallowed — Adult worms in intestine

Trachea

Alveoli of lungs

Heart

Migrate via bloodstream

Penetrate intestinal mucosa, perianal skin, or skin exposed to soil

Rhabditiform (Noninfective) larvae in feces

Free-living adults in soil

Rhabditiform (Noninfective) larvae in soil

Filariform (Infective) larvae in feces, soil or GI tract

Figure 10.27 *Strongyloides stercoralis* life cycle.

S. fuelleborni is also transmitted primarily by skin contact. However, investigations into the high prevalence of *S. fuelleborni* in African children have led to the postpartum identification of larvae in breast milk. In Papua New Guinea, high rates of infection (up to 65%) in children younger then 3 months suggest that transmammary infection might also occur in this area, although the parasite has not been identified in breast milk. Also, *S. fuelleborni* has been identified as the causative agent of "swollen belly syndrome," a fulminant and fatal enteritis that has been well documented in infants younger than 6 months in areas of highly endemic infection in Papua New Guinea. It has been suggested that transmammary infection followed by early external autoinfection from dirty diapers may result in rapid multiplication of worms in these young children (4).

In situations in which autoinfection occurs, some of the rhabditiform larvae that are within the intestine develop into the filariform larvae while passing through the bowel. These larvae can then reinfect the host by (i) invading the intestinal mucosa, traveling via the portal system to the lungs, and returning to the intestine or (ii) being passed out in the feces and penetrating the host on reaching the perianal or perineal skin.

The rhabditiform larvae that normally pass out in the stool measure up to 380 µm long by 20 µm wide, with a muscular esophagus (club-shaped anterior, then a restriction, and a posterior bulb) (Figure 10.28). There is a genital primordium packet of cells, which is fairly obvious and can be seen about two-thirds of the way back from the anterior end (Figure 10.29). One of the key morphologic differences between these larvae and those of hookworm is the length of the mouth opening (buccal capsule). The opening in the rhabditiform larvae of *S. stercoralis* is extremely short (only a few micrometers),

while the mouth opening in hookworm rhabditiform larvae is approximately three times as long (Figures 10.26 and 10.28). These differences can be seen by examining the larvae under the microscope with the low (10×) or high dry (40×) objectives.

Strongyloides spp. have the ability to switch between alternative free-living developmental pathways in response to changing environmental conditions. Anterior chemosensory neurons (amphidial neurons) are thought to respond to environmental cues and, via signal transduction pathways, control the direction of larval development. Larvae that pass to the exterior may develop directly to infectivity by the homogonic route or may develop by the heterogonic route to establish a population of soil-dwelling dioecious adult worms.

Figure 10.28 Buccal cavity (rhabditiform larvae). (A) *Strongyloides stercoralis*; (B) hookworm.

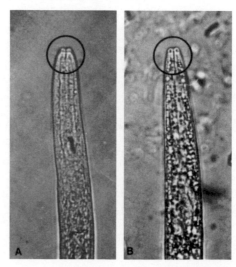

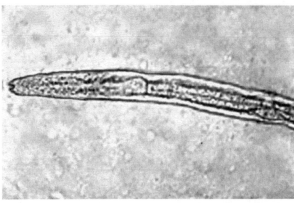

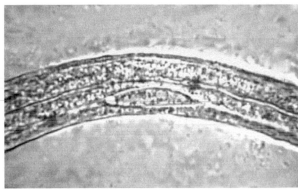

Figure 10.29 (Upper) *Strongyloides stercoralis* rhabditiform larva; note the packet of genital primordial cells toward the right side of the image. (Lower) *S. stercoralis* rhabditiform larva; note the packet of genital primordial cells in the middle of the field (enlarged).

The infective larvae produced by either route are environmentally resistant, developmentally arrested worms that can develop further only after entering the host. Studies confirm that neurons homologous to the two classes (ASF and ASI) do occur in *S. stercoralis* and that they control the direction of free-living development in this species (5). Apparently, the free-living cycle develops most frequently in tropical climates with moist soil while the parasitic life cycle may predominate in more temperate climates.

Review of the Life Cycle. The life cycle of *S. stercoralis* contains the following phases, some or all of which may be present, depending on environmental conditions:

1. Free-living forms in the soil that can produce infective larvae (capable of human skin penetration, common in the moist, warm tropics).
2. Noninfective rhabditiform larvae that pass in the stool and transform into infective filariform larvae while in the soil (capable of human skin penetration, common in temperate zones and responsible for most human infections).

3. Autoinfection, which involves development of filariform larvae in or on the body (may account for continued infection 30 to 40 years after the host has left the original area where the infection is endemic) (26).

Comments on the Life Cycle. Although it is generally accepted that the life cycle presented above is correct, the skin-penetrating larvae travel from the skin to the intestine via an obligatory route that includes the blood, lungs, trachea, and upper gastrointestinal tract (the standard pulmonary route). It is also assumed that following auto-infection, the larvae follow the same route to return to the duodenum, where they mature. However, review information and data from Schad et al. (62) provide evidence to indicate that the pulmonary route is just one of several possible pathways to the duodenum, regardless of whether the larvae penetrated the skin or the intestine. Direct sampling of the larvae within the trachea gave an estimate of the number of larvae within the duodenum, presumably transmitted by the pulmonary route, that was insufficient to account for the number actually found there. In spite of some caveats, these studies definitely challenge the current accepted understanding of the life cycle. The pulmonary migration route for all skin-penetrating nematodes may not be as universally applicable as was once thought (62).

Clinical Disease

The pathologic changes associated with strongyloidiasis can vary both in severity and in the areas of the body involved. Some individuals may remain totally asymptomatic, with the only abnormal clinical finding being a peripheral eosinophilia (75).

Considering the life cycle, the three areas of primary involvement would be the skin (cutaneous), the lungs (pulmonary), and the intestinal mucosa (intestinal).

Cutaneous. Initial skin penetration usually causes very little reaction, although there may be some pruritus and erythema if the number of penetrating larvae is large. In some patients, the reaction at the site of larval penetration may last several weeks. With repeated infections, the patient may mount an allergic response that prevents the parasite from completing the life cycle. The larvae may be limited to skin migration (larva migrans). The term "larva currens" (racing larva) was proposed in 1958 and is now generally accepted for cases of strongyloidiasis in which there is one or more rapidly progressing linear urticarial tracks starting near the anus. There is speculation that some of these cases may involve larvae of other species of *Strongyloides*. These tracks may progress as fast as 10 cm/h, with an intermittent movement, usually on the thighs. Onset is sudden, and the lesions may disappear within 12 to 18 h (9).

Pulmonary. Larval migration through the lungs may stimulate symptoms, depending on how many larvae are present and the intensity of the host immune response. Some patients may be asymptomatic, while others may present with pneumonia. With a heavy infective dose or in the hyperinfection syndrome, individuals often develop cough, shortness of breath, wheezing, fever, and transient pulmonary infiltrates (Loeffler's syndrome). There have been cases reported in which the larvae were even found in the sputum during this larval migration.

Intestinal. Although the timing may vary from person to person, gastrointestinal symptoms such as diarrhea, constipation, anorexia, and abdominal pain usually begin about 2 weeks after infection, with larvae being detectable in the stool after 3 to 4 weeks. In heavy infections, the intestinal mucosa may be severely damaged, with sloughing of tissue, although this type of damage is unusual (Figure 10.30). Parasitized individuals may show villous atrophy and crypt hyperplasia. There is infiltration of the lamina propria by IgA-positive plasma cells and of the epithelium by intraepithelial lymphocytes. Infection is also associated with increased expression of secretory component and decreased expression of HLA-DR in epithelial cells. These changes tend to correlate with the degree of clinical severity. In a group of immunocompetent patients with chronic *S. stercoralis* infection, no abnormality of mucosal structure and no increase in the levels of nonspecific inflammatory cells were found. Likewise, there was no increase in the numbers of mucosal T cells or macrophages (74). The symptoms may mimic peptic ulcer, with abdominal pain, which may be localized in the

Figure 10.30 Mucosal damage caused by larval penetration in strongyloidiasis. (From A Pictorial Presentation of Parasites: A cooperative collection prepared and/or edited by H. Zaiman.)

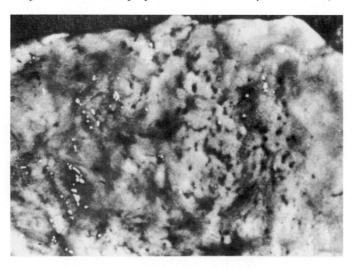

right upper quadrant. Radiographic findings may mimic those of Crohn's disease of the proximal small intestine. In immunocompetent patients, there is leukocytosis with peripheral eosinophilia of 50 to 75%, while in patients with chronic cases, the eosinophilia may be much lower. Some of these chronic infections have lasted for over 30 years as a result of the autoinfective capability of the larvae. One case of chronic strongyloidiasis persisted for approximately 65 years.

Chronic Infection. Chronic infections are usually asymptomatic; however, there may be symptoms that are unrelated to autoinfection. These symptoms include intermittent vomiting, diarrhea, constipation, and loud gurgling noises produced by movement of gas, fluid, or both in the gastrointestinal tract. Anal pruritus, urticaria and larval currens rashes, recurrent asthma, and nephrotic syndrome have also been associated with chronic disease. With development of the hyperinfection syndrome, intestinal obstruction, ileus, and gastrointestinal bleeding can also occur.

Hyperinfection Syndrome. Autoinfection is probably the mechanism responsible for long-term infections that persist years after the person has left the area where the infection is endemic. The parasite and host reach a status quo so that neither suffers any serious damage. If for any reason this equilibrium is disturbed and the individual becomes immunosuppressed, the infection proliferates, with large numbers of larvae being produced and found in every tissue of the body, including the gastric mucosa. Several conditions, including increased use of immunosuppressive therapy, predispose an individual to the hyperinfection syndrome (42). However, the distinction between autoinfection and hyperinfection tends to be quantitative rather than being specifically defined (Table 10.4).

In contrast to the usual *Strongyloides* hyperinfection syndrome seen in patients with some forms of immunodeficiency, in which small intestinal and pulmonary symptoms are seen, in one group of patients only a characteristic transmural eosinophilic granulomatous inflammation, affecting mostly the colonic wall and clinically mimicking ulcerative colitis or Crohn's disease, has been found. This eosinophilic granulomatous enterocolitis apparently results from a florid inflammatory response by eosinophils, histiocytes, and giant cells, with formation of granulomas that destroy the larvae entering the colon (31).

Strongyloides hyperinfection and dissemination are recognized complications in kidney transplant recipients; however, strongyloidiasis in renal transplant recipients receiving cyclosporin A has not been described, nor have any cases of strongyloidiasis complicating heart, liver, lung, or pancreas transplantation been published

Table 10.4 Strongyloidiasis hyperinfection syndrome[a]

Body site	Symptoms	Diagnosis	Comments
Gastrointestinal tract	Crampy or bloating abdominal pain; watery diarrhea, constipation, anorexia, weight loss, difficulty swallowing, sore throat, nausea, vomiting, gastrointestinal bleeding; ileus and small bowel obstruction; diffuse abdominal tenderness, hypoactive bowel sounds; hypo-albuminemia; hypokalemia	Direct stool exam shows numerous rhabditiform and filariform larvae; occasionally adult worms and eggs; mucosal ulceration common in small intestine	Symptoms are very common, but nonspecific; some patients are asymptomatic
Cardiopulmonary system	Cough, wheezing, choking sensation, hoarseness, chest pain (pleuritic); hemoptysis; palpitations, atrial fibrilliation, dyspnea, respiratory collapse (rare)	Direct sputum exam shows rhabditiform and filariform larvae; occasionally eggs; chest X rays show bilateral or focal interstitial infiltrates	If condition goes undiagnosed and steroids are given, symptoms may improve temporarily
Skin and connective tissue	Pruritic linear streaks on lower trunk, thighs, and buttocks; petechial and purpuric rashes; evidence of vasculitis	Larvae can be seen in skin biopsy	Associated gram-negative bacterial sepsis and/or disseminated intravascular coagulation may be present
Central nervous system	Meningeal symptoms; hyponatremia and meningitis; signs of aseptic meningitis in spinal fluid (pleocytosis, elevated protein, normal glucose, negative cultures); gram-negative bacterial infection	Larvae recovered in spinal fluid	Typical gram-negative bacteria isolated include *Escherichia coli*, *Proteus mirabilis*, *Klebsiella pneumoniae*, *Enterococcus faecalis*, and *Streptococcus bovis*; immunosuppressed patients may develop *Candida* infections
Other infected organs	Liver: elevated liver enzymes	Larvae may be seen in various tissues	Other body sites include mesenteric lymph nodes, gallbladder, diaphragm, heart, pancreas, muscle, kidneys, and ovaries

[a]Adapted from reference 42.

prior to this case. One possible explanation for this observation is the antiparasitic activity of cyclosporin A found in animal studies. In one renal transplant patient, *Strongyloides* hyperinfection occurred immediately after cyclosporin A treatment was discontinued. Approximately 44 days after transplantation, ongoing cellular rejection was confirmed via renal biopsy and cyclosporin A treatment was discontinued according to protocol. His earlier history was unremarkable, and several stool specimens examined for parasites were negative; he had never had any eosinophilia. Based on respiratory symptoms, he underwent bronchoscopy, and *S. stercoralis* filariform larvae were recovered from sputum, bronchoalveolar lavage fluid, and three stool specimens. After treatment with thiabendazole, the patient remained on a prophylactic regimen of thiabendazole at 25 mg/kg twice a day for 2 days every month. In situations where information on the donor is not readily available (possible strongyloidiasis), the recipient should be monitored for evidence of clinical strongyloidiasis and prophylaxis may be justified, since most episodes of

hyperinfection occur during the first 3 months following transplantation. Also, discontinuation of cyclosporin A treatment in a setting in which no prior screening for *Strongyloides* had taken place might place a patient at risk for dissemination (54). Specific conditions associated with immunosuppression and strongyloidiasis can be seen in Table 10.5 (20, 42, 59).

The possibility of hyperinfection also puts patients with asthma and intestinal strongyloidiasis at significant risk if their lung disease is treated with systemically administered corticosteroids. Apparently, fewer than 10 days of therapy with corticosteroids has led to disseminated strongyloidiasis (16, 42). This potential situation demonstrates the need to screen all asthmatics who come from areas where *S. stercoralis* is endemic, including the southeastern United States, for the presence of this parasite. Various immunosuppressive drugs that have been associated with hyperinfection include the glucocorticoids (both high and low doses), vinca alkaloids (vincristine and vinblastine), and others, although in many cases glucocorticoids were given at the same time (20, 42, 59).

Table 10.5 Strongyloidiasis: associated conditions related to immunosuppression[a]

Condition	Description	Comments
Immunosuppressive drugs		
Glucocorticoids	Both high- and low-dose steroids; can begin as early as 20 days after therapy started; suppression of eosinophilia and lymphocyte activation	Most directly linked with hyperinfection syndrome; used for lupus, lymphoma, rheumatoid arthritis, leprosy, polymyositis, corneal ulcer, and Bell's palsy
Vinca alkaloids	Vincristine; decrease intestinal motility, longer time for larvae to change from noninfectious to infectious	Several reports, but information anecdotal
Cyclosporin A	No proven association; some evidence that drug may be anthelmintic	Widespread use, but does not appear to play a role in hyperinfection
Other drugs	Azathioprine, cyclophosphamide, antithymocyte globulin, anti-CD_w, chlorambucil, 6-mercaptopurine, methotrexate, bleomycin, adriamycin, doxorubicin, daunorubicin, ifosfamide, melphalan, carmustine, VP16, mitoxantrone, and total-body irradiation	Because glucocorticoids were administered at the same time, hyperinfection link to these drugs may or may not be correct
Hematologic malignancies	Case reports of hyperinfection syndrome prior to therapy with immunosuppressives	Chronic parasite-induced inflammation may contribute to lymphoma
Transplant recipients		
Kidney	Majority of transplant hyperinfection cases seen with kidneys	Probably caused by glucocorticoids in response to organ rejection
Heart	Case of graft rejection	Patient treated with increased doses of corticosteroids
Bone marrow	Few cases reported	Supposition that drugs (bisulfan, cyclophosphamide) may prevent egg hatching; therefore, hyperinfection would be rare in these cases
HTLV-1	Associated with increased strongyloidiasis, failure of conventional therapy, and hyperinfection syndrome	Suggested that serum IgE confers some protection against heavy infections and response may be inhibited by HTLV-1 infection
Hypogammaglobulinemia	Seen in case of multiple myeloma and nephrotic syndrome	Immunity to infective-stage larvae not protective against autoinfective larvae

[a]Adapted from references 20, 42, 59, 63, and 65.

In addition to the actual tissue damage caused by the migrating larvae, the patient may die of sepsis and/or meningitis, primarily as a result of the intestinal flora. Cerebrospinal fluid cultures have grown *Escherichia coli*, *Proteus mirabilis*, *Klebsiella pneumoniae*, *Enterococcus faecalis*, and *Streptococcus bovis*, as well as other isolates. Migrating larvae have been found in CSF, meningeal vessels, the dura, and the epidural, subdural, and subarachnoid spaces (42). Other causes of death may include peritonitis, brain damage, or respiratory failure.

Few cases of reactive arthritis have been reported, and the number of cases may be underestimated. A case of reactive arthritis combined with uveitis associated with a long-standing, heavy *Strongyloides* infection was found in a 32-year-old patient positive for human T-cell leukemia virus type 1 (HTLV-1). Treatment with thiabendazole and ivermectin resulted in rapid improvement. A case of myelitis associated with eosinophilia in blood

and cerebrospinal fluid was identified as being caused by *S. stercoralis*; apparently, this is the first report of myelitis in the course of strongyloidiasis (35).

HTLV-1 Infection. The prevalence of *S. stercoralis* is significantly higher in HTLV-1 carriers than HTLV-1 noncarriers in Japan and elsewhere. Since cellular immunity plays a major role in the host defense against strongyloidiasis, infection with HTLV-1 appears to change the immune system capability, leading to severe clinical manifestations and disseminated disease (13). Hypereosinophilia, which is common in patients with mild strongyloidiasis, is often not seen in these patients and may be inhibited by the HTLV-1 infection. The relationship between *S. stercoralis*, infection with HTLV-1, and serum IgE levels reveals some very interesting information. The prevalence of *Strongyloides* infection is significantly higher in HTLV-1 carriers than in those without HTLV-1 infection. The level of IgE is low in HTLV-1 carriers and is significantly lower in HTLV-1 carriers than in noncarriers among people with *Strongyloides* infection. This information suggests that serum IgE may confer some degree of protection against heavy infections with *Strongyloides* and that this response may be inhibited by HTLV-1 infection.

HIV Infection. Although it is well known that many people are coinfected with HIV and *S. stercoralis*, few cases of hyperinfection have been found in this patient group, many of whom previously received steroids for *Pneumocystis* pneumonia or non-Hodgkin's lymphoma. Due to documented cases of posttreatment hyperinfection after therapy for intestinal infection, questions remain regarding screening of HIV-positive individuals for strongyloidiasis and treating infected patients with anthelmintics.

S. fuelleborni Infection. Children suffering from the swollen belly syndrome caused by *S. fuelleborni* usually present with protein-losing enteropathy, abdominal distention, and respiratory distress at about 8 to 10 weeks of age. The diarrhea is generally not severe, and there is no significant fever. Severe hypoproteinemia without proteinuria is very characteristic and leads to peripheral edema and ascites. Egg counts are often greater than 100,000/g of stool. Untreated, the condition is associated with a high mortality (4).

Diagnosis

Confirmation of the infection depends on recovery and identification of the adult worms, larvae, or eggs from the stool, duodenal material, or sputum. Fecal specimens may not be positive, even after the routine ova and parasite examination, including the concentration procedure, is performed. Numbers of larvae found in the stool also vary from day to day. It is well known that the detection of *S. stercoralis* larvae in the stools may be very difficult, particularly in patients with chronic, low-level infection (Table 10.6). This problem can be attributed to irregular and low output of the larvae, as well as the sensitivity of current diagnostic methods. Patterns of *S. stercoralis* detection vary widely among infected patients, and intermittent larval shedding can lead to inflated estimates of drug efficacy. It is important to remember that prior to entering a patient in a clinical trial of drug efficacy, it has been recommended that four consecutive stool specimens should be examined for *S. stercoralis* to determine the patient's infection status.

Some suggest that sampling of duodenal contents may recover larvae if stools are negative, while others think that duodenal sampling may not be effective. Special techniques that have been recommended for the recovery of duodenal contents include the Entero-Test capsule (see chapter 29). There are also special concentration techniques (Baermann) and methods for larva culture (Harada-Mori, petri dish), which are discussed in detail in chapter 28. These techniques provide a means of recovery of larvae from large amounts of stool material, and often the filariform larvae are recovered. To differentiate the *Strongyloides* filariform larvae from those of the hookworm, one should examine the tail under the microscope. *Strongyloides* filariform larvae have a slit in the tail, while hookworm filariform larvae have a pointed tail (Figure 10.31). Refer to Algorithm 10.5.

Algorithm 10.5 Strongyloidiasis.

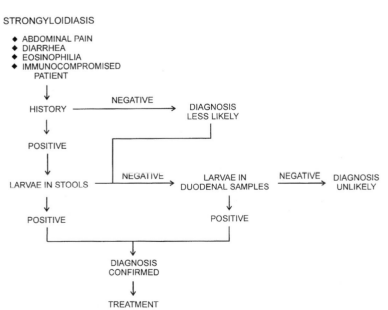

STRONGYLOIDIASIS

♦ ABDOMINAL PAIN
♦ DIARRHEA
♦ EOSINOPHILIA
♦ IMMUNOCOMPROMISED
   PATIENT

Table 10.6 Diagnostic methods for strongyloidiasis[a]

Method	Pros	Cons	Comments
Direct wet mount	Simple to perform	Sensitivity very low	Other methods much more sensitive
Concentration wet mount (formalin ethyl acetate concentration)	Simple to perform; centrifugation at $500 \times g$ for 10 min	Sensitivity low for a single stool sample	Four stools recommended to determine patient's infection status
Permanent stained smear (trichrome, iron-hematoxylin)	Performed as a routine part of the ova and parasite examination	Larval morphology may be distorted	If anything seen, reexamine concentration sediments
Baermann concentrate	Examination of much larger amount of stool	Messy to set up, possible exposure to infectious larvae	Handle with gloves; decontaminate glassware, etc., when test completed
Harada-Mori, agar plate concentration	Considered "mini" Baermann procedures; relatively simple to set up	Will require checking for several days; not as sensitive as Baermann concentrate	Not routinely performed, but can be helpful
Agar plate culture	Often considered the most sensitive method for larval recovery and identification	Must be examined for 5 days before calling culture negative	More sensitive than formalin ethyl acetate concentration if larvae are present at <50/g of stool
Duodenal contents	Some studies indicate good sensitivity for larval recovery; use of the Entero-Test capsule helpful	Sensitivity debated; duodenal intubation invasive	Multiple stool exams may be more helpful, particularly if agar plate cultures set up
Serologic tests	More widely available; sensitive but not always specific	May not distinguish present from past infection; sensitivity decreased in patients with HTLV-1 infections and hematologic malignancies	Sensitivity may be lower in travelers, but helpful in immigrants with chronic infection; gelatin particle agglutination reported more sensitive than ELISA
Skin test	Immediate hypersensitivity	Coinfection with HTLV-1 decreases sensitivity	Further developmental work required

[a]Adapted from references 39, 42, 52, 68, and 69.

Agar plate cultures are also recommended and tend to be more sensitive than some of the other available diagnostic methods (60); stool is placed onto agar plates, and the plates are sealed to prevent accidental infections and held for 2 days at room temperature. As the larvae crawl over the agar, they carry bacteria with them, thus creating visible tracks over the agar. The plates are examined under the microscope for confirmation of larvae, the surface of the agar is then washed with 10% formalin, and final confirmation of larval identification is made via wet examination of the sediment from the formalin washings (see chapter 28). The agar consists of 1.5% agar, 0.5% meat extract, 1.0% peptone, and 0.5% NaCl. This approach is thought to be more sensitive and has become more widely used (2, 39, 41, 44, 60).

Becoming more widely available, the application of enzyme-linked immunosorbent assay (ELISA), immunoblotting, and gelatin particle agglutination methods has been investigated for both the diagnosis and postchemotherapy evaluation of human strongyloidiasis (47, 68).

These methods appear to be both sensitive and specific and have demonstrated significant decreases in antibody levels after therapy. However, when treatment was not effective, the individuals did not show a significant fall in antibody titers. More detailed studies to determine appropriate intervals for serologic evaluation and criteria for successful cure are required before these tests become more widely used.

Note. Debilitated or immunocompromised patients should always be suspected of having strongyloidiasis, particularly if they experience unexplained bouts of diarrhea and abdominal pain, repeated episodes of sepsis or meningitis with intestinal bacteria, or unexplained eosinophilia. However, a recent comparative study of the occurrence of *Strongyloides* spp. in 554 AIDS and 142 non-AIDS patients demonstrated similar prevalences of infection in the two groups, thus indicating no significant statistical differences (27). Nevertheless, diagnosing intestinal parasites in HIV/AIDS patients is necessary, especially in those

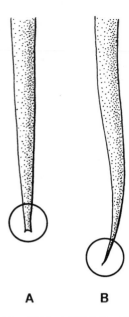

A   **B**

Figure 10.31 Tail (filariform larvae). (A) *Strongyloides stercoralis*; (B) hookworm.

reported to be chronic alcoholics and those who are not receiving antiretroviral treatment. It is also important to check pre-hematopoietic stem cell transplantation patients in areas of endemic infection in the United States (59).

In contrast to *S. stercoralis*, the diagnosis of *S. fuelleborni* infection is made by finding eggs, rather than larvae, in the stool. Children with swollen belly syndrome tend to have very heavy infections and excrete large numbers of eggs daily. The eggs are oval and approximately 50 to 70 μm long.

KEY POINTS—LABORATORY DIAGNOSIS

Strongyloides stercoralis

1. Recovery of the rhabditiform (noninfective) larvae is normally from the stool concentrate. *Caution: Filariform (infective) larvae can also be recovered in the stool.* A minimum of four stools are recommended before indicating that the patient is not infected (routine formalin, ethyl acetate sedimentation concentration); use centrifugation at $500 \times g$ for 10 min.
2. If the stool specimens are negative, examination of duodenal contents is recommended (duodenal aspirates, Entero-Test capsule); however, the overall sensitivity of the method varies.
3. Various concentrates (Baermann) and cultures (Harada-Mori, petri dish) can also be used for larval recovery.
4. Eggs are rarely seen in the stool but may be recovered from duodenal contents.
5. In very heavy infections, eggs (less common), larvae (both types), and adult worms may be recovered in the stool.
6. If agar plates are used (culture, method of choice), they must be dried sufficiently to eliminate free water on the agar prior to use. If fresh agar plates have just been prepared, the drying time may be longer than if prepackaged commercial plates, which are shipped with little to no free moisture on the surface of the agar, are used. They need to be examined at 24 h and 3 and 5 days.

Treatment

At present, the recommended drug of choice is ivermectin; however, thiabendazole has been used very successfully for many years; specific dosages are provided in chapter 25. In some cases, repeated therapy is necessary, and the cure rates vary (42). Patients with the hyperinfection syndrome should be hospitalized during therapy for proper monitoring. Thiabendazole has been used to treat children with swollen belly syndrome due to *S. fuelleborni* infection. This approach seems to be effective, but reinfection is very likely to occur in areas of endemic infection.

Side effects with thiabendazole occur in most patients and include nausea, foul-smelling urine, neuropsychiatric effects, malaise, and dizziness. Rectal administration of ivermectin has also been reported and may be helpful in patients who cannot absorb or tolerate oral therapy (72).

Mebendazole has also been used to treat strongyloidiasis; however, it is poorly absorbed. Albendazole has also been used to treat patients with strongyloidiasis, as well as those with the hyperinfection syndrome. Few side effects have been seen, and this drug is an excellent alternative to ivermectin. It is important to remember that HTLV-1 infection may be a possibility in patients with uncomplicated intestinal strongyloidiasis who do not respond to standard therapy (73).

Epidemiology and Prevention

Contact with contaminated infective soil, feces, or surface water should be avoided. Communities where close living conditions and poor sanitation facilities exist, including both rural and urban areas in the developing world, often have high prevalence rates of strongyloidiasis. Also, closed communities such as institutions for the mentally handicapped may have high rates of infection and transmission. The geographic distribution is quite broad and includes both tropical and temperate climates. It is important to remember that internal autoinfection allows for maintenance of the parasite within the host for years following the initial exposure. Since all infected individuals are at risk for hyperinfection and disseminated disease, the identification of people who may have contracted their infection many years earlier is of great clinical significance. Individuals found to have the infection should be treated. *All patients who are going to receive immunosuppressive drugs should be screened for strongyloidiasis before therapy.*

References

1. **Albonico, M., R. J. Stolzfus, L. Savioli, J. M. Tielsch, H. M. Chwaya, E. Ercole, and G. Cancrini.** 1998. Epidemiological evidence for a differential effect of hookworm species, *Ancylostoma duodenale* or *Necator americanus*, on iron status of children. *Int. J. Epidemiol.* **27:**530–537.

2. **Arakaki, T., M. Iwanaga, F. Kinjo, A. Saito, R. Asato, and T. Ikeshiro.** 1990. Efficacy of agar-plate culture in detection of *Strongyloides stercoralis* infection. *J. Parasitol.* **76:**425–428.

3. **Arca, M. J., R. L. Gates, J. I. Groner, S. Hammond, and D. A. Caniano.** 2004. Clinical manifestations of appendiceal pinworms in children: an institutional experience and a review of the literature. *Pediatr. Surg. Int.* **20:**372–375.

4. **Ashford, R. W., and G. Barnish.** 1990. *Strongyloides fuelleborni* and similar parasites in animals and man, p. 271–286. *In* I. E. Grove (ed.), *Strongyloidiasis: a Major Roundworm Infection of Man.* Taylor and Francis, London, United Kingdom.

5. **Ashton, F. T., V. M. Bhopale, D. Holt, G. Smith, and G. A. Schad.** 1998. Developmental switching in the parasitic nematode *Strongyloides stercoralis* is controlled by the ASF and ASI amphidial neurons. *J. Parasitol.* **84:**691–695.

6. **Aspöck, H., H. Auer, and O. Picher.** 1996. *Trichuris trichiura* eggs in the Neolithic glacier mummy from the Alps. *Parasitol. Today* **12:**255–256.

7. **Bahon, J., J. Poirriez, C. Creusy, A. N. Edriss, J. P. Laget, and E. Deicas.** 1997. Colonic obstruction and perforation related to heavy *Trichuris trichiura* infestation. *J. Clin. Pathol.* **50:**615–616.

8. **Beaver, P. C.** 1988. Light, long-lasting *Necator* infection in a volunteer. *Am. J. Trop. Med. Hyg.* **39:**369–373.

9. **Beaver, P. C., R. C. Jung, and E. W. Cupp.** 1984. *Clinical Parasitology*, 9th ed. Lea & Febiger, Philadelphia, Pa.

10. **Beckingham, I. J., S. N. R. Cullis, J. E. J. Krige, P. C. Bornman, and J. Terblanche.** 1998. Management of hepatobiliary and pancreatic *Ascaris* infestation in adults after failed medical treatment. *Br. J. Surg.* **85:**907–910.

11. **Bradley, J. E., and J. A. Jackson.** 2004. Immunity, immunoregulation and the ecology of trichuriasis and ascariasis. *Parasite Immunol.* **26:**429–444.

12. **Brown, A., N. Girod, E. E. Billett, and D. I. Pritchard.** 1999. *Necator americanus* (human hookworm) aspartyl proteinases and digestion of skin macromolecules during skin penetration. *Am. J. Trop. Med. Hyg.* **60:**840.

13. **Carvalho, E. M., and P. A. Da Fonseca.** 2004. Epidemiological and clinical interaction between HTLV-1 and *Strongyloides stercoralis. Parasite Immunol.* **26:**487–497.

14. **Chan, O. T., E. K. Lee, J. M. Hardman, and J. J. Navin.** 2004. The cockroach as a host for *Trichinella* and *Enterobius vermicularis*: implications for public health. *Hawaii Med. J.* **63:**74–77.

15. **Chim, C. S., W. K. Luk, and K. Y. Yuen.** 1997. *Trichostrongylus* infestation masquerading as conditioning toxicity of the gut in bone marrow transplantation. *Bone Marrow Transplant.* **19:**955–956.

16. **Chu, E., W. L. Whitlock, and R. A. Dietrich.** 1990. Pulmonary hyperinfection syndrome with *Strongyloides stercoralis. Chest* **97:**1475–1477.

17. **Churcher, T. S., N. M. Ferguson, and M. G. Basanez.** 2005. Density dependence and overdispersion in the transmission of helminth parasites. *Parasitology* **131:**121–132.

18. **Clercq, D. D., M. Sacko, J. Behnke, F. Gilbert, P. Dorny, and J. Vercruysse.** 1997. Failure of mebendazole in treatment of human hookworm infections in the southern region of Mali. *Am. J. Trop. Med. Hyg.* **57:**25–30.

19. **Cliffe, L. J., N. E. Humphreys, T. E. Lane, C. S. Potten, C. Booth, and R. K. Grencis.** 2005. Accelerated intestinal epithelial cell turnover: a new mechanism of parasite expulsion. *Science* **308:**1463–1465.

20. **Concha, R., W. Harrington, Jr., and A. I. Rogers.** 2005. Intestinal strongyloidiasis: recognition, management, and determinants of outcome. *J. Clin. Gastroenterol.* **39:**203–211.

21. **Cooper, E. S., D. D. Ramdath, C. Whytealleng, S. Howell, and B. E. Serjeant.** 1997. Plasma proteins in children with *Trichuris* dysentery syndrome. *J. Clin. Pathol.* **50:**236–240.

22. **Cooper, E. S., J. Spencer, C. A. M. Whyte-Alleng, O. Cromwell, P. Whitney, S. Venugopal, D. A. P. Bundy, B. Haynes, and T. T. MacDonald.** 1991. Immediate hypersensitivity in colon of children with chronic *Trichuris trichiura* dysentery. *Lancet* **338:**1104–1107.

23. **Cross, J. H.** 1992. Intestinal capillariasis. *Clin. Microbiol. Rev.* **5:**120–129.

24. **De Gruijter, J. M., L. van Leishout, R. B. Gasser, J. J. Verweij, E. A. Barienen, J. B. Ziem, L. Yelifari, and A. M. Polderman.** 2005. Polymerase chain reaction-based differential diagnosis of *Ancylostoma duodenale* and *Necator americanus* infections in humans in northern Ghana. *Trop. Med. Int. Health* **10:**574–580.

25. **Desilva, N. R., H. L. Guyatt, and D. A. P. Bundy.** 1997. Morbidity and mortality due to *Ascaris*-induced intestinal obstruction. *Trans. R. Soc. Trop. Med. Hyg.* **91:**31–36.

26. **de Silva, S., P. Saykao, H. Kelly, C. R. MacIntyre, N. Ryan, J. Leydon, and B. A. Biggs.** 2002. Chronic *Strongyloides stercoralis* in Laotian immigrants and refugees 7–20 years after resettlement in Australia. *Epidemiol. Infect.* **128:**439–444.

27. **Dias, R. M. D. S., A. C. S. Mangini, D. M. A. G. V. Torres, S. A. G. Vellosa, M. I. P. G. Dasilva, R. M. Dasilva, M. O. A. Correa, and C. Coletti.** 1992. *Strongyloides stercoralis* in patients with acquired immunodeficiency syndrome (AIDS). *Rev. Inst. Med. Trop. Sao Paulo* **34:**15–17. (In Portuguese.)

28. **Diniz-Santos, D. R., J. Jambeiro, R. R. Mascarenhas, and L. R. Silva.** 2005. Massive *Trichuris trichiura* infection as a cause of chronic bloody diarrhea in a child. *J. Trop. Pediatr.* 0:731 Jul. 6. (Epub ahead of print.) **52:**66–68.

29. **el-Karaksy, H., M. el-Shabrawi, N. Mohsen, M. Kotb, N. el-Koofy, and N. el-Deeb.** 2004. *Capillaria philippinensis*: a cause of fatal diarrhea in one of two infected Egyptian sisters. *J. Trop. Pediatr.* **50:**57–60.

30. **Goud, G. N., M. E. Bottazzi, B. Zhan, S. Mendez, V. Deumic, J. Plieskatt, S. Liu, Y. Wang, L. Bueno,**

R. Fujiwara, A. Samuel, S. Y. Ahn, M. Solanki, O. A. Asojo, J. Wang, J. M. Bethony, A. Loukas, M. Roy, and P. J. Hotez. 2005. Expression of the *Necator americanus* hookworm larval antigen Na-ASP-2 in *Pichia pastoris* and purification of the recombinant protein for use in human clinical trials. *Vaccine* 23:4754–4764.

31. Gutierrez, Y., P. Bhatia, S. T. Garbadawala, J. R. Dobson, T. M. Wallace, and T. E. Carey. 1996. *Strongyloides stercoralis* eosinophilic granulomatous enterocolitis. *Am. J. Surg. Pathol.* 20:603–612.

32. Haas, W., B. Haberl, Syafruddin, I. Idris, D. Kallert, S. Kersten, P. Stiegeler, and Syafruddin. 2005. Behavioural strategies used by the hookworms *Necator americanus* and *Ancylostoma duodenale* to find, recognize and invade the human host. *Parasitol. Res.* 95:30–39.

33. Hadidjaja, P., E. Bonang, M. A. Suyardi, S. A. N. Abidin, I. S. Ismid, and S. S. Margono. 1998. The effect of intervention methods on nutritional status and cognitive function of primary school children infected with *Ascaris lumbricoides*. *Am. J. Trop. Med. Hyg.* 59:791–795.

34. Hasegawa, H., Y. Takao, M. Nakao, T. Fukuma, O. Tsuruta, and K. Ide. 1998. Is *Enterobius gregorii* Hugot 1983 (Nematoda: Oxyuridae) a distinct species? *J. Parasitol.* 84:131–134.

35. Henon, H., P. Vermersch, E. Dutoit, F. Dubois, D. Camus, and H. Petit. 1995. Myelitis caused by *Strongyloides stercoralis*. *Rev. Neurol.* 151:139–141.

36. Holland, C. V., E. O'Shea, S. O. Asaolu, O. Turley, and D. W. T. Crompton. 1996. A cost-effectiveness analysis of anthelminthic intervention for community control of soil-transmitted helminth infection: levamisole and *Ascaris lumbricoides*. *J. Parasitol.* 82:527–530.

37. Hotez, P. J., S. Narasimhan, J. Haggerty, L. Milstone, V. Bhopale, G. A. Schad, and F. F. Richards. 1992. Hyaluronidase from infective *Ancylostoma* hookworm larvae and its possible function as a virulence factor in tissue invasion and in cutaneous larva migrans. *Infect. Immun.* 60:1018–1023.

38. Hughes, R. G., D. S. Sharp, M. C. Hughes, S. Akau'ola, P. Heinsbroek, R. Velayudhan, D. Schulz, K. Palmer, T. Cavalli-Sforza, and G. Galea. 2004. Environmental influences on helminthiasis and nutritional status among Pacific schoolchildren. *Int. J. Environ. Health Res.* 14:163–177.

39. Intapan, P. M., W. Maleewong, T. Wongsaroj, S. Singthong, and N. Morakote. 2005. Comparison of the quantitative formalin ethyl acetate concentration technique and agar plate culture for diagnosis of human strongyloidiasis. *J. Clin. Microbiol.* 43:1932–1933.

40. Johnson, P. W., R. Dixon, and A. D. Ross. 1998. An in-vitro test for assessing the viability of *Ascaris suum* eggs exposed to various sewage treatment processes. *Int. J. Parasitol.* 28:627–633.

41. Jongwutiwes, S., M. Charoenkorn, P. Sitthichareonchai, P. Akaraborvorn, and C. Putaporntip. 1999. Increased sensitivity of routine laboratory detection of *Strongyloides stercoralis* and hookworm by agar-plate culture. *Trans. R. Soc. Trop. Med. Hyg.* 93:398–400.

42. Keiser, P. B., and T. B. Nutman. 2004. *Strongyloides stercoralis* in the immunocompromised population. *Clin. Microbiol. Rev.* 17:208–217.

43. King, E. M., H. T. Kim, N. T. Dang, E. Michael, L. Drake, C. Needham, R. Haque, D. A. Bundy, and J. P. Webster. 2005. Immuno-epidemiology of *Ascaris lumbricoides* infection in a high transmission community: antibody responses and their impact on current and future infection intensity. *Parasite Immunol.* 27:89–96.

44. Koga, K. S., C. Kasuya, K. Khamboonruang, M. Sukhavat, M. Ieda, N. Takatsuka, K. Kita, and H. Ohtomo. 1991. A modified agar plate method for detection of *Strongyloides stercoralis*. *Am. J. Trop. Med. Hyg.* 45:518–521.

45. Larocque, R., M. Casapia, E. Gotuzzo, and T. W. Gyorkos. 2005. Relationship between intensity of soil-transmitted helminth infections and anemia during pregnancy. *Am. J. Trop. Med. Hyg.* 73:783–789.

46. Lee, S., K. Hwang, W. Tsai, C. Lin, and N. Lee. 2002. Detection of *Enterobius vermicularis* eggs in the submucosa of the transverse colon of a man presenting with colon carcinoma. *Am. J. Trop. Med. Hyg.* 67:546–548.

47. Lindo, J. F., D. J. Conway, N. S. Atkins, A. E. Bianco, R. D. Robinson, and D. A. P. Bundy. 1994. Prospective evaluation of enzyme-linked immunosorbent assay and immunoblot methods for the diagnosis of endemic *Strongyloides stercoralis* infection. *Am. J. Trop. Med. Hyg.* 51:175–179.

48. Liu, L. X., J. Chi, M. P. Upton, and L. R. Ash. 1995. Eosinophilic colitis associated with larvae of the pinworm *Enterobius vermicularis*. *Lancet* 346:410–412.

49. Maddern, G. J., A. R. Dennison, and L. H. Blumgart. 1992. Fatal *Ascaris* pancreatitis—an uncommon problem in the West. *Gut* 33:402–403.

50. Modjarrad, K., I. Zulu, D. T. Redden, L. Njobvu, D. O. Freedman, and S. H. Vermund. 2005. Prevalence and predictors of intestinal helminth infections among human immunodeficiency virus type 1-infected adults in an urban African setting. *Am. J. Trop. Med. Hyg.* 73:777–782.

51. Navitsky, R. C., M. L. Dreyfuss, J. Shrestha, S. K. Khatry, R. J. Stolzfus, and M. Albonico. 1998. *Ancylostoma duodenale* is responsible for hookworm infections among pregnant women in the rural plains of Nepal. *J. Parasitol.* 84:647–651.

52. Neva, F. A., A. A. Gam, C. Maxwell, and L. L. Pelletier. 2001. Skin test antigens for immediate hypersensitivity prepared from infective larvae of *Strongyloides stercoralis*. *Am. J. Trop. Med. Hyg.* 65:567–572.

53. Oberhelman, R. A., E. S. Guerrero, M. L. Fernandez, M. Silio, D. Mercado, N. Comiskey, G. Ihenacho, and R. Mera. 1998. Correlations between intestinal parasitosis, physical growth, and psychomotor development among infants and children from rural Nicaragua. *Am. J. Trop. Med. Hyg.* 58:470–475.

54. Palau, L. A., and G. A. Pankey. 1997. *Strongyloides* hyperinfection in a renal transplant recipient receiving cyclosporine: possible *Strongyloides stercoralis* transmission by kidney transplant. *Am. J. Trop. Med. Hyg.* 57:413–415.

55. Quinnell, R. J., J. Bethony, and D. I. Pritchard. 2004. The immunoepidemiology of human hookworm infection. *Parasite Immunol.* 26:443–454.

56. Ramdath, D. D., D. T. Simeon, M. S. Wong, and S. M. Grantham-McGregor. 1995. Iron status of school children with varying intensities of *Trichuris trichiura* infection. *Parasitology* 110:347–351.

57. Reynoldson, J. A., J. M. Behnke, L. J. Pallant, M. G. Macnish, F. Gilbert, S. Giles, R. J. Spargo, and R. C. A. Thompson. 1997. Failure of pyrantel in treatment of human hookworm infections (*Ancylostoma duodenale*) in

the Kimberley region of North West Australia. *Acta Trop.* 68:301–312.

58. Romstad, A., R. B. Gasser, P. Nansen, A. M. Polderman, and N. B. Chilton. 1998. *Necator americanus* (Nematoda: Ancylostomatidae) from Africa and Malaysia have different ITS-2 rDNA sequences. *Int. J. Parasitol.* 28:611–615.

59. Safdar, A., K. Malathum, S. J. Rodriguez, R. Husni, and K. V. Rolston. 2004. Strongyloidiasis in patients at a comprehensive cancer center in the United States. *Cancer* 100:1531–1536.

60. Sato, Y., J. Kobayashi, H. Toma, and Y. Shiroma. 1995. Efficacy of stool examination for detection of *Strongyloides stercoralis. Am. J. Trop. Med. Hyg.* 53:248–250.

61. Schad, G. A. 1991. Hooked on hookworm: 25 years of attachment. *J. Parasitol.* 77:179–186.

62. Schad, G. A., L. M. Aikens, and G. Smith. 1989. *Strongyloides stercoralis*: is there a canonical migratory route through the host? *J. Parasitol.* 75:740–749.

63. Schaeffer, M. W., J. F. Buell, M. Gupta, G. D. Conway, S. A. Ashter, and L. E. Wagoner. 2004. *Strongyloides* hyperinfection syndrome after heart transplantation: case report and review of the literature. *J. Heart Lung Transplant.* 23:905–911.

64. Schulze, S. M., R. J. Chokshi, M. Edavettal, and E. Tarasov. 2005. Acute abdomen secondary to *Ascaris lumbricoides* infestation of the small bowel. *Am. Surg.* 71:505–507.

65. Seet, R. C. S., L. G. Lau, and P. A. Tambyah. 2005. *Strongyloides* hyperinfection and hypogammaglobulinemia. *Clin. Diagn. Lab. Immunol.* 12:680–682.

66. Simeon, D. T., S. M. Grantham-McGregor, and M. S. Wong. 1995. *Trichuris trichiura* infection and cognition in children: results of a randomized clinical trial. *Parasitology* 110:457–464.

67. Singh, S., J. C. Samantaray, N. Singh, G. B. Das, and I. C. Verma. 1993. *Trichuris vulpis* infection in an Indian tribal population. *J. Parasitol.* 79:457–458.

68. Sithithaworn, J., P. Sithithaworn, T. Janrungsopa, K. Suvatanadecha, K. Ando, and M. R. Haswell-Elkins. 2005. Comparative assessment of the gelatin particle agglutination test and an enzyme-linked immunosorbent assay for diagnosis of strongyloidiasis. *J. Clin. Microbiol.* 43:3278–3282.

69. Sudarshi, S., R. Stumpfle, M. Armstrong, T. Ellman, S. Parton, P. Krishnan, P. L. Chiodini, and C. J. Whitty. 2003. Clinical presentation and diagnostic sensitivity of laboratory tests for *Strongyloides stercoralis* in travelers compared with immigrants in a non-endemic country. *Trop. Med. Int. Health* 8:728–732.

70. Summers, R. W., D. E. Elliott, J. F. Urban, Jr., R. A. Thompson, and J. V. Weinstock. 2005. *Trichuris suis* therapy for active ulcerative colitis: a randomized controlled trial. *Gastroenterology* 128:825–832.

71. Takayama, Y., S. Kamimura, J. Suzumiya, K. Oh, M. Okumura, H. Akahane, H. Maruyama, Y. Nawa, T. Ohkawara, and M. Kikuchi. 1997. Eosinophilic colitis with high antibody titre against *Ascaris suum. J. Gastroenterol. Hepatol.* 12:204–206.

72. Tarr, P. E., P. S. Miele, K. S. Peregoy, M. A. Smith, F. A. Neva, and D. R. Lucey. 2003. Case report: rectal administration of ivermectin to a patient with *Strongyloides* hyperinfection syndrome. *Am. J. Trop. Med. Hyg.* 68:453–455.

73. Terashima, A., H. Alvarez, R. Tello, R. Infante, D. O. Freedman, and E. Gotuzzo. 2002. Treatment failure in intestinal strongyloidiasis: an indicator of HTLV-I infection. *Int. J. Infect. Dis.* 6:28–30.

74. Trajman, A., T. T. MacDonald, and C. C. S. Elia. 1997. Intestinal immune cells in *Strongyloides stercoralis* infection. *J. Clin. Pathol.* 50:991–995.

75. Tsai, H. C., S. S. Lee, Y. C. Liu, W. R. Lin, C. K. Huang, Y. S. Chen, S. R. Wann, T. H. Tsai, H. H. Lin, M. Y. Yen, C. M. Yen, and E. R. Chen. 2002. Clinical manifestations of strongyloidiasis in southern Taiwan. *J. Microbiol. Immunol. Infect.* 35:29–36.

76. Turner, J. D., H. Faulkner, J. Kamgno, M. W. Kennedy, J. Behnke, M. Boussinesq, and J. E. Bradley. 2005. Allergen-specific IgE and IgG4 are markers of resistance and susceptibility in a human intestinal nematode infection. *Microbes Infect.* 7:990–996.

77. Williams-Blangero, S., J. Subedi, R. P. Upadhayay, D. B. Manral, D. R. Rai, B. Jha, E. S. Robinson, and J. Blangero. 1999. Genetic analysis of susceptibility to infection with *Ascaris lumbricoides. Am. J. Trop. Med. Hyg.* 60:921–926.

Tissue Nematodes

Trichinella spp.

Baylisascaris procyonis

Lagochilascaris minor

Toxocara canis and *T. cati* (visceral larva migrans and ocular larva migrans)

Ancylostoma braziliense and *A. caninum* (cutaneous larva migrans)

Human eosinophilic enteritis

Dracunculus medinensis

Angiostrongylus (Parastrongylus) cantonensis (cerebral angiostrongyliasis)

Angiostrongylus (Parastrongylus) costaricensis (abdominal angiostrongyliasis)

Gnathostoma spinigerum

Gnathostoma doloresi, G. nipponicum, G. hispidum, and *G. binucleatum*

Anisakis simplex, A. physetesis, Pseudoterranova decipiens, Contracaecum osculatum, Hysterothylacium aduncum, and *Porrocaecum reticulatum* (larval nematodes acquired from saltwater fish)

Capillaria hepatica

Thelazia spp.

Trichinella spp.

Although *Trichinella spiralis* was first seen in human tissue at autopsy in the early 1800s, it was not until 1860 that Freidrich von Zenker concluded that the infection resulted from eating raw sausage (10). The consumption of rare or raw pork as the cause of trichinosis was experimentally proved a few years later. By the 1900s, trichinosis was definitely recognized as a public health problem. This particular infection has a cosmopolitan distribution but is more important in the United States and Europe than in the tropics or the Orient. The prevalence in autopsies within the United States has declined from 15.9% of human diaphragms studied at autopsy from 1931 to 1944 to 4.5% from 1948 to 1963; in recent years, the prevalence has decreased to 2.2% and the mortality associated with this infection has decreased to less than 1%. In 1990, only 105 cases of human trichinosis were reported in the United States, and by 1994 the number had dropped to 35 (53). During 1997 to 2001 the incidence decreased to a median of 12 cases annually and no reported deaths (77). However, in many areas of the world, trichinosis remains a problem (55).

During the 5-year period from 1997 to 2001, 72 cases were reported to the Centers for Disease Control and Prevention. Of these, 31 cases were associated with eating wild game: 29 with bear meat, 1 with cougar meat, and 1 with wild-boar meat (77). In comparison, only 12 cases were associated with eating commercial pork products; 4 of these cases were traced to a foreign source. Nine cases were associated with eating noncommercial pork from home-raised or direct-from-farm swine where U.S. commercial pork production industry standards and regulations are not applicable.

Studies of isolates of *Trichinella* spp. from Arctic, temperate, and tropical areas have confirmed that there are major differences related to their genetic structure and overall biology. Various species are involved, depending on the geographic area. *Trichinella* forms a complex of species, all of which appear to be the same morphologically but, based on DNA studies and comparative features, are actually quite different (Tables 11.1 and 11.2) (5). There are now 11 recognized *Trichinella* species or genotypes, *T. spiralis, T. nativa, T. nelsoni, T. britovi, T. pseudospiralis, T. murrelli,* T6, T8, T9, *T. papuae,* and *T. zimbabwensis* (37, 45, 69, 71, 92). *Trichinella* is quite different from many other helminths because all stages of development (adult and larva) occur within a single host. More than

Table 11.1 Characteristics of *Trichinella* species

Characteristic	*T. spiralis*	*T. pseudospiralis*	*T. nativa*	*T. nelsoni*	*T. britovi*
Geographic distribution	Worldwide	Worldwide	Arctic	Equatorial Africa, southern Europe	Temperate zone of Palearctic region
Major animal host reservoirs	Pigs, rats, horses (see Table 11.6)	Birds, raccoons, rodents	Polar bears, walrus, whales, seals, squirrels, dogs, wolves, foxes, horses (see Table 11.6)	Bush pig, warthog, lions, leopards, cheetahs, hyenas	Foxes, wolves, wild bears, horses (see Table 11.6), free-ranging swine
Human infection	Most common	First human outbreak reported (Thailand) (37)	Common in the Arctic	Less common	Primarily sylvatic
Infectivity for:					
Humans	High	Moderate	High	High	Moderate
Swine	High	Low	Low	Low	Low
Rats	High	Moderate	Low	Low	Low
Mice (*Mus*)	High	Moderate	Low	Low	Low
Chickens	No	Yes	No	No	No
Pathogenicity for humans	High	Low to moderate	High	Low	Moderate
Resistance to freezing	Low	Low	High	Low	Low
Time for nurse cell development (days)	16–37	Absent (no capsule forms around L1)	20–30	34–60	24–42
Availability of diagnostic DNA probes	Yes	Yes	Yes	Yes	Yes
Unique alloenzyme markers	6	12	2	4	1
Studies					
Pig infectivity (31, 41) / High antibody responses regardless of infection	Yes / Yes	Not done	No / Yes	Not done	Yes / Yes
T. pseudospiralis in sedentary night birds of prey from Central Italy (70)		First documented report in animals in Western Europe			
T. papuae, new species from New Guinea (64, 72)		Lack nurse cell; total length of larva is 1/3 longer than *T. pseudospiralis*			
Highly conserved genes (*tsmyd-1, tsJS*) (44)	Yes	Yes			Yes
New genotype *Trichinella* T9 (Japanese isolates) (58)					Japanese strains

Table 11.2 *Trichinella* spp.: encapsulated and nonencapsulated species[a]

Type	Animals infected	Geographic location	Comments
Encapsulated	Mammals		Encapsulated species/genotypes infect only mammals; cluster into 4 main geographic groups
T. spiralis		Cosmopolitan	
T. britovi		Temperate regions	
T. murrelli		Temperate regions	
Trichinella T8		Temperate regions	
Trichinella T9		Temperate regions	
T. nativa		Arctic region	
Trichinella T6		Arctic region	
T. nelsoni		Equatorial region	
Nonencapsulated			
T. pseudospiralis	Mammals, birds	Cosmopolitan	Palearctic, Nearctic, and Australian regions
T. papuae	Wild pigs, saltwater crocodiles	Equatorial region	Can complete life cycle in both poikilothermic and homoiothermic animals
T. zimbabwensis	Farmed Nile crocodiles	Equatorial region	Can complete life cycle in both poikilothermic and homoiothermic animals

[a] Data from references 45, 69, and 71.

100 different mammals are susceptible to infection, and the cysts can remain viable and infectious for many years, even in decaying muscle tissue. These factors ensure successful transmission and the survival of the parasite.

Life Cycle and Morphology

Human infection is initiated by the ingestion of raw or poorly cooked pork, bear, walrus, or horse meat or meat from other mammals (carnivores and omnivores) containing viable, infective larvae (Figure 11.1; Table 11.3). The tissue is digested in the stomach, and the first-stage larvae (L1) are resistant to gastric juice. The excysted larvae then invade the intestinal mucosa, develop through four larval stages within about 36 h, mature, and mate by the second day. By the sixth day of infection, the female worms begin to deposit motile larvae, which are carried by the intestinal lymphatic system or mesenteric venules to the body tissues, primarily striated muscle (Figures 11.2 and 11.3). Deposition of larvae continues for approximately 4 to 16 weeks, with each female producing up to 1,500 larvae in the nonimmune host. Newborn larvae can penetrate almost any tissue but can continue their development only in striated muscle cells. With the exception of *T. pseudospiralis*, *T. papuae*, and *T. zimbabwensis*, invasion of striated muscle cells stimulates the development of nurse cells (Table 11.1) (55). As the larvae begin to coil, the nurse cell completes the formation of the cyst within about 2 to 3 weeks. Within the human host, the cyst measures about 400 by 260 µm, and within the cyst, the coiled larva measures 800 to 1,000 µm in length. At this point, the larvae are fully infective.

The very active muscles including the diaphragm, muscles of the larynx, tongue, jaws, neck, and ribs, the biceps, and the gastrocnemius, which have the greatest blood supply, are invaded. The encysted larvae may remain viable for many years, although calcification can occur within less than a year. As few as 5 larvae/g of body muscle can cause death, although 1,000 larvae/g have been recovered from individuals who died from causes other than trichinosis.

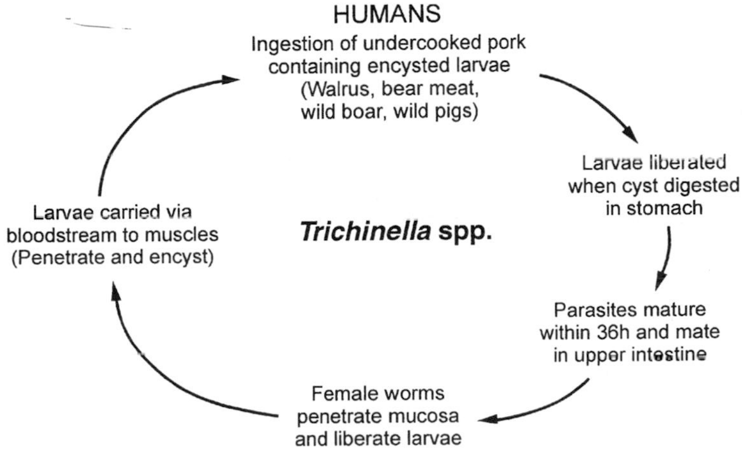

HUMANS
Ingestion of undercooked pork containing encysted larvae (Walrus, bear meat, wild boar, wild pigs)

Larvae liberated when cyst digested in stomach

Trichinella spp.

Parasites mature within 36h and mate in upper intestine

Female worms penetrate mucosa and liberate larvae

Larvae carried via bloodstream to muscles (Penetrate and encyst)

Figure 11.1 Life cycle of *Trichinella* spp. A number of infected meat sources other than pork are relevant for the various species of *Trichinella*.

Table 11.3 Tissue nematodes

Name	How acquired	Location in body	Symptoms	Diagnosis
Trichinella spiralis (*T. nativa*) (*T. nelsoni*) (*T. britovi*) (*T. pseudospiralis*) (*T. murrelli*) (*Trichinella* T6) (*Trichinella* T8) (*Trichinella* T9) (*T. papuae*) (*T. zimbabwensis*)	Ingestion of raw or rare meats (pork, bear, walrus, other carnivores and/or omnivores)	Active muscles contain encysted larvae (diaphragm, tongue, larynx, neck, ribs, biceps, gastrocnemius)	Diarrhea (larval migration through intestinal mucosa), nausea, abdominal cramps, general malaise; muscle invasion: periorbital edema, pain, swelling, weakness, difficulties in swallowing, breathing, etc.; most severe symptom is myocarditis; high eosinophilia (20–90%)	Biopsy or autopsy specimen (muscle) compression smear or routine histology; artificial digestion of muscle to release larvae (larvae are very infective and precaution should be taken); serologic testing can be very helpful
Baylisascaris procyonis	Ingestion of viable eggs in the soil (most probably from raccoon feces)	CNS and eye contain larvae	Eosinophilic meningitis, unilateral neuroretinitis	Biopsy or autopsy specimen, routine histology; eggs from raccoon feces measure 80 μm long by 65 μm wide, have a thick shell with a finely granulated surface, and resemble *Ascaris lumbricoides* eggs
Lagochilascaris minor	Life cycle and route of human infection unknown; suspect ingestion of viable eggs in the soil	Adult worms, larvae, and eggs occur in life cycle within human lesions (neck, throat, nasal sinuses, tonsillar tissue, mastoids, brain, lungs)	Pustule swelling, pus in lesions; chronic granulomatous inflammation	Identification of adult worms, larvae, or eggs from lesions, sinus tracts, or biopsy or autopsy specimens
Toxocara canis and *T. cati* (visceral and ocular larva migrans)	Ingestion of infective eggs (dog/cat ascarids) from fecal material in the soil	Usually the liver; migratory pathway may include the lungs and even back to the intestine	Migration of larvae may cause inflammation and granuloma formation; there may be fever, hepatomegaly, pulmonary infiltrates, cough, and neurologic symptoms; high eosinophilia (up to 90%; 20–50% common)	Confirmation at autopsy; serologic test (ocular fluids as well as serum if eye involved)
Ancylostoma braziliense and A. caninum (cutaneous larva migrans)	Skin penetration of filariform/infective larvae of dog/cat hookworms; infection can also occur via ingestion of infective larvae	Larval migration in the skin produces linear/raised/vesicular tracts; can be on any area of the body	Intense itching, pneumonitis (if larvae migrate to deeper tissues)	Picture of linear tracts; possible removal of larva from tunnel

Organism	Source/transmission	Location in body	Symptoms	Diagnosis
Dracunculus medinensis (fiery serpent)	Ingestion of infected copepod/water flea (*Cyclops*)	Adult worms develop in deep connective tissue; gravid female migrates to feet and ankles (can occur anywhere), where blister forms for larval deposition into the water through the ruptured blister on the skin	Before blister formation: erythema, tenderness, urticarial rash, intense itching, nausea, vomiting, diarrhea, or asthmatic attacks; if secondary infection occurs, there may be cellulitis, arthritis, myositis, etc.	Formation of cutaneous lesion with appearance of adult female worm depositing larvae into the water; calcified worms can also be found on X ray
Angiostrongylus cantonensis (eosinophilic meningitis) (cerebral)	Accidental ingestion of infective larvae in slugs, snails, or land planarians	Brain tissue, eye (rare), lung tissue (rare)	Severe headache, convulsions, limb weakness, paresthesia, vomiting, fever, eosinophilia up to 90%	Presumptive: severe headache, meningitis or meningoencephalitis, fever, ocular involvement; definitive: examination of tissues (surgical specimens)
Angiostrongylus costaricensis (abdominal)	Accidental ingestion of slugs, often on contaminated salad vegetables	Bowel wall	Pain, tenderness, palpable tumor-like mass in right lower quadrant, fever, diarrhea, vomiting, eosinophilia (60%), and leukocytosis	Worm recovery and clinical history
Gnathostoma spinigerum	Ingestion of raw, poorly cooked, pickled freshwater fish or chicken (and other birds), frogs, or snakes	Migration of larvae in deep cutaneous or subcutaneous tissues (may appear anywhere), eyes, or CSF (less common)	Migratory swellings (hard, nonpitting) with inflammation, redness, pain	Worm recovery and clinical history
Anisakis, Contracaecum, Pseudoterranova, Hysterothylacium, and *Porrocaecum* spp.	Ingestion of raw, pickled, salted, or smoked saltwater fish	Wall of gastrointestinal tract	Nausea, vomiting; may mimic gastric/duodenal ulcer, carcinoma, appendicitis; stool positive for occult blood	Worm recovery and clinical history
Capillaria hepatica	Accidental ingestion of eggs from soil	Liver	May mimic hepatitis, amebic abscess, or other infections involving the liver	Histologic identification
Thelazia spp.	Larval deposition by flies	Conjunctival sacs/migrating over cornea	Excessive lacrimation, itching, pain (feeling of foreign object in eye)	Worm recovery (from eye) and identification

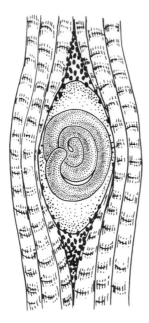

Figure 11.2 *Trichinella* spp., encysted larva in muscle. (Illustration by Sharon Belkin.)

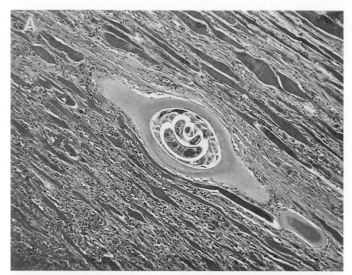

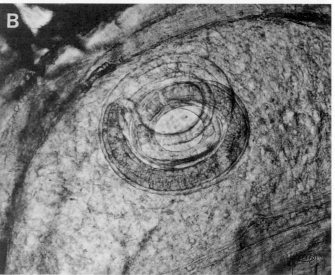

Figure 11.3 (A) *Trichinella spiralis*, encysted larva in muscle (×85). Adjacent muscle has been replaced by fibrous tissue and is infiltrated with chronic inflammatory cells. (Armed Forces Institute of Pathology photograph.) (B) Coiled larva of *T. spiralis* in a teased muscle preparation (wet mount, ×193). (Armed Forces Institute of Pathology photograph.)

There are species differences in low-temperature (freezing) survival, infectivity, and capsule formation (55). Also, studies on a pig farm indicate that even in the absence of a known source of infected meat (garbage containing meat scraps or dead animals), the rat population maintained the infection, probably through cannibalism. Consequently, to reduce transmission of *T. spiralis* between rats and swine, rat populations in an agricultural ecosystem must be controlled. It is also important to limit access to the farmyard by wild and feral animals.

Although recommendations have been made to use several species designations, some publications still use the single species designation *Trichinella spiralis* (Tables 11.2 and 11.3). Genetic relationships among many *Trichinella* isolates are currently being assessed by dot blot hybridization, restriction endonuclease, and gel electrophoresis techniques. On the basis of the presence of repetitive DNA sequences in the *Trichinella* genome, distinctive banding patterns have been seen among the isolates, and taxonomic changes will continue to occur.

Clinical Disease

Pathologic changes due to trichinosis can be classified as (i) intestinal effects and (ii) muscle penetration and larva encapsulation (Table 11.4). Any damage caused in either phase of the infection is usually based on the original number of ingested cysts; however, other factors such as the patient's general health, age, and size also play a

role in the disease outcome. Symptoms of trichinosis are generally separated into three phases, with phase 1 being related to the presence of the parasite in the host prior to muscle invasion and phase 2 being related to the inflammatory and allergic reactions due to muscle invasion. There may also be an incubation period of up to 50 days. Phase 3 is the convalescent phase or chronic period (Table 11.5).

Symptoms that may develop within the first 24 h include diarrhea, nausea, abdominal cramps, and general malaise, all of which may suggest food poisoning, particularly if several people are involved. Studies also indicate

Table 11.4 *Trichinella spiralis*: life cycle stages and clinical conditions[a]

Stage in life cycle	Time after infection when symptoms begin	Clinical condition
Excysted larvae enter intestinal mucosa	2–4 h	
	24 h	Gastrointestinal symptoms
Worms mature and mate	30 h	
Females deposit larvae/muscle invasion begins	Day 6	
	Day 7	Facial edema/fever
Heaviest muscle invasion	Day 10	Maximum fever (40–41°C)
	Day 11	Muscle inflammation/pain
Decrease in larval deposition	Day 14	Eosinophilia/antibody
Larvae differentiated	Day 17	
	Day 20	Maximum eosinophilia
Encapsulation of larvae	Day 21	Myocarditis/neurologic symptoms
Intestine free of adult worms	Day 23	
	Day 26	Respiratory symptoms
Encapsulation almost complete	Mo 1	
	Mo 2	Fever subsides
Adult worms die	Mo 3	Death from myocarditis or encephalitis
Cyst calcification begins	Mo 6	Slow convalescence
	Mo 8	Myocarditis/neurologic symptoms subside
Cyst calcification usually complete	Yr 1	
Most larvae still viable within calcified cyst	Yr 6	

[a]Adapted from reference 10.

that the diarrhea can be prolonged, lasting up to 14 weeks (average, 5.8 weeks) with few or no muscle symptoms. It is still unknown whether this clinical presentation is related to variant biological behavior of Arctic *Trichinella* organisms, to previous exposure to the parasite, or to other factors.

During muscle invasion, there may be fever, facial (particularly periorbital) edema, and muscle pain, swelling, and weakness. The extraocular muscles are usually the first to be involved, followed by the muscles of the jaw and neck, limb flexors, and back. Muscle damage may cause problems in chewing, swallowing, breathing, etc., depending on which muscles are involved. The most severe symptom is myocarditis, which usually develops after the third week; death may occur between the fourth and eighth weeks. Other severe symptoms, which can occur at the same time, may involve the central nervous system (CNS). Although *Trichinella* encephalitis is rare, it is life-threatening. Technological advances such as the computed tomogram, angiogram,

and electroencephalogram are of no diagnostic assistance and probably add nothing to traditional diagnostic information, which includes eosinophilia, sedimentation rate, and muscle biopsy.

It is estimated that 10 to 20% of the patients with trichinosis have CNS involvement and that the mortality rate may reach 50% in these patients if they are not treated. Symptoms may mimic those of polyneuritis, acute anterior poliomyelitis, myasthenia gravis, meningitis, encephalitis, dermatomyositis, and polyarteritis nodosa. There may be focal paresis or paralysis (quadriplegia to single muscle group).

Peripheral eosinophilia of at least 20%, often over 50%, and possibly up to 90% is present during the muscle invasion phase of the infection. Fever can also be present at this time and can persist for several days to weeks, depending on the intensity of the infection. However, once the larvae begin to encapsulate, patient symptoms subside; eventually the cyst wall and larvae calcify.

Table 11.5 Trichinosis: incubation period, larval numbers, and degree of illness

Characteristic	Mild disease	Moderate disease	Severe disease	Abortive disease
No. of larvae/g of muscle	10, probably subclinical	50–500	1,000 or more	Less than 10
Incubation period (days)	21	16	7	30
Intestinal phase (phase 1) (days 2–7)[a]	Nausea, abdominal aches, cramps, loss of appetite, vomiting, mild fever, mild diarrhea or constipation; frontal headaches, dizziness, weakness	As with mild disease, symptoms may mimic the flu	Same symptoms as indicated in mild to moderate disease; diarrhea may be severe	Patient may be asymptomatic
Muscle invasion phase (phase 2) (days 9–28)[b]	Penetration of larvae initiates inflammatory response (extraocular muscles) in masseters; muscles of the larynx, tongue, diaphragm, and neck; intercostals; and muscular attachments to tendons and joints; headache, fainting, urticaria, splinter hemorrhages beneath the fingernails and toenails, conjunctivitis, loss of appetite, hoarseness, dysphagia, dyspnea, and edema of the legs may also occur; range of symptoms is based on number of larvae and general health, age, size of patient		Muscular pain, facial edema (swelling of eyelids), fever, chills, eosinophilia, tachycardia, coma, respiratory difficulties; neurologic symptoms may be severe (may stimulate meningitis); myocarditis is a serious complication, may lead to congestive heart failure	Symptoms may or may not be seen
Convalescent phase (phase 3)	Decrease in muscular symptoms, beginning in the second month; fever and itching subside		Evidence of congestive heart failure may appear (if patient becomes active too soon)	Change in symptoms may or may not be obvious

[a] Symptoms reflect mucosal irritation.
[b] Encystment occurs after day 14.

In an outbreak in Spain, 44 members of eight families were examined. Various people had suggestive symptoms (10 of 44), hypereosinophilia (20 of 44), and positive serologic test results (15 of 44). Three groups could be identified according to the home-prepared product each had ingested (pork sausage, blood pudding, and loin). In these cases, the common source of all infections was the poorly cooked pork sausage, since the blood pudding is boiled for a long time at high temperature and the loin is always served thoroughly fried or roasted. Twelve months later, all had a normal eosinophil count and negative serologic test results (86). Another outbreak in Spain involved 38 people, 15 of whom were hospitalized after the ingestion of sausage made from uninspected wild boar meat and infected pork. Almost all patients had myalgias, about half reported diarrhea and/or vomiting, 75% reported periorbital edema, and 76% had fever. Sixteen patients were positive for *T. britovi* by indirect fluorescent-antibody test (IFAT) and 20 were positive by Western blotting (33). The ingestion of wild boar containing *T. pseudospiralis* in France and bear meat containing *T. nativa* from New York and Tennessee has also been implicated in recent outbreaks (16, 75).

There have also been a number of outbreaks due to consumption of horse meat; these outbreaks were caused by different species of *Trichinella* and were associated with differences in clinical symptoms. Although most human infections have been attributed to *T. spiralis*, these outbreaks clearly demonstrate that different species produce different clinical syndromes (Table 11.6).

Diagnosis

Depending on the severity of the infection, trichinosis can mimic many other conditions. Most mild cases with a small loading dose of infective larvae may present with flu-like symptoms. Unless the clinician recognizes an appropriate history, fever, myalgia, periorbital edema, and/or rising eosinophilia (50% or higher), the cause may go undetected. Often, the first clue is the patient's history of possible ingestion of raw or rare pork or other infected meat. There may also be other individuals from the same group with similar symptoms. Trichinosis should always be included in the differential diagnosis of any patient with periorbital edema, fever, myositis, and eosinophilia, regardless of whether a complete history of consumption of raw or poorly cooked pork is available. If present, subconjunctival and subungual splinter hemorrhages also add support for such a presumptive diagnosis. If the meat consumption history is incomplete, food poisoning, intestinal flu, or typhoid may be suspected. It is very rare to

Table 11.6   Outbreaks due to consumption of *Trichinella* spp. in horse meat[a]

Characteristic	Outbreak[b]					
	Italy, 1975 (*T. britovi*)	France, 1976 (*T. spiralis* probably)	France, 1985 (*T. nativa*)	France, 1985 (*T. spiralis*)	Italy, 1986 (*T. britovi*)	France, 1993 (*T. spiralis*)[c]
No. of cases	89	125	343	396	161	444
Percentage of patients with:						
Fever	+	65	90	85	70	81
Myalgia	+	59	93	88	67	82
Weakness		NE	87	77	NE	NE
Facial edema	+	57	58	84	62	75
Diarrhea	+	16	50	41	21	35
Vomiting		8	NE	NE	9	NE
Headache		60	58	51	66	NE
Rash		5	44	11	4	NE
Ocular involvement		31	28	34	26	NE
Mortality rate	0%	0%	0 %	0.4%	0%	0%

[a] From T. Ancelle, J. Dupouy-Camet, J. C. Desenclos, E. Maillot, S. Savage-Houze, F. Charlet, J. Druckner, and A. Moren, A multifocal outbreak of trichinellosis linked to horse meat imported from North America to France in 1993, *Am. J. Trop. Med. Hyg.* 59:615–619, 1998; K. D. Murrell, Trichinosis, *in* G. T. Strickland (ed.), *Hunter's Tropical Medicine and Emerging Infectious Diseases*, 8th ed., The W. B. Saunders Co., Philadelphia, Pa., 2000; K. D. Murrell and F. Bruschi, Trichinellosis, *In* T. Sun (ed.), *Progress in Clinical Parasitology*, vol. 4, CRC Press, Inc., Boca Raton, Fla., 2000.
[b] +, symptoms present, actual patient numbers not provided; NE, not evaluated.
[c] Severe neurotrichinosis, 1.4%; cardiac symptoms or electrocardiograph changes, 4.7%.

recover adult worms or larvae from stool or other body fluids (blood, cerebrospinal fluid [CSF], etc.), even if the patient has diarrhea.

Muscle biopsy (gastrocnemius, deltoid, and biceps) specimens may be examined by compressing the tissue between two slides and checking the preparation under a microscope at low power (10× objective). However, this method does not provide positive results until 2 to 3 weeks after the onset of the illness. It is also important to remember that not all species form the capsule (Figure 11.4). Muscle specimens or samples of the suspect meat can also be examined by using an artificial digestion technique to release the larvae. These techniques are described in detail in chapter 30.

Serologic tests are also very helpful, the standard two being the enzyme immunoassay (EIA) and the bentonite flocculation test, which are recommended for trichinosis. The EIA is used for routine screening, and all EIA-positive specimens are tested by bentonite flocculation for confirmation. A positive reaction in both tests indicates infection with *T. spiralis* within the last few years. Often, antibody levels are not detectable within the first month postinfection. The titers tend to peak in the second or third months postinfection and then decline over a period of a few years. The IFAT has also been used to track the course of disease after infection. Antibody detection using IFAT was reported at 70.2% of patients 1 week after onset of disease and increased to 91, 94.3,

and 100% 2, 3, and 4 weeks, respectively, after the onset. The antibody detection decreased to 25% 4 months after therapy (90). More information on serologic testing is presented in chapter 22. Refer to Algorithm 11.1 for a detailed review.

A dot enzyme-linked immunosorbent assay (ELISA) with purified antigens has been developed for detecting *T. spiralis* in swine. This test is as sensitive as an ELISA with excretory-secretory products as the antigen and Western immunoblot analysis and is nearly as specific as

Figure 11.4   Unencapsulated *Trichinella pseudospiralis*. (Image courtesy of the Centers for Disease Control and Prevention, Atlanta, Ga.)

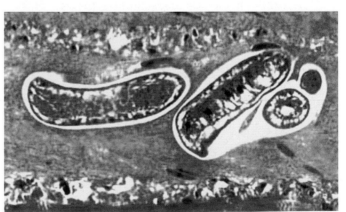

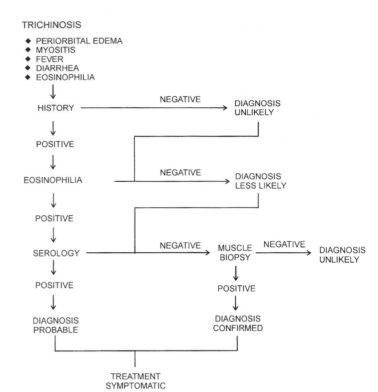

Algorithm 11.1   Diagnosis of trichinosis.

the Western blot. Also, the dot ELISA is much easier to perform than is a Western blot analysis (83).

A modified double-sandwich ELISA has been developed using polyclonal antibodies against larval somatic antigens. In experimental studies, the first detection of coproantigen occurred as early as day 1 postinfection, peaking on day 7 and then disappearing by week 3. These results were confirmed using the coagglutination test (Co-A). Based on these studies, this approach could be used to confirm early infection in humans (11).

KEY POINTS—LABORATORY DIAGNOSIS

Trichinella spiralis

1. The history and clinical findings may suggest possible trichinosis (consumption of rare or raw infected meat). Remember to check hematology results for a possible eosinophilia (can reach 50% or higher).
2. Using compression slides, examination of suspect meat may reveal larvae (artificial digestion procedure) (see chapter 30). Note that not all species of larvae form the capsule; however, the unencapsulated larvae can still be seen in a "squash" preparation of biopsy material.
3. Larvae or adult worms are rarely recovered in fecal specimens during the intestinal phase (diarrhea).

4. Examination of muscle tissue obtained at biopsy may confirm the diagnosis (tissue compression between two slides or the artificial digestion technique).
5. Serologic tests for antibody detection may be very helpful; coproantigen detection tests are being developed.

Treatment

Therapy depends on the phase of the disease, the immune status of the patient, and the intensity and length of the disease. For the early phase of infection, the objective is to reduce the number of larvae that will invade the muscles. Once larval invasion of the muscles has occurred, the objective becomes to reduce muscle damage. The current recommendation for the gastrointestinal phase is the use of mebendazole (200 to 400 mg/day for 3 to 5 days) or albendazole (400 mg/day); after the first 3 to 5 days of mebendazole therapy, the dose is changed (400 to 500 mg/day) and given for 10 days. For acute, severe infection, steroids are recommended (prednisolone, 40 to 60 mg/day) along with mebendazole (5 mg/kg/day) or albendazole (400 to 800 mg twice a day for 8 to 14 days). For moderate or mild infection, steroids can be given as required; once fever and allergic signs diminish, steroid administration can be discontinued. Unfortunately, the disease is often not diagnosed until well after muscle invasion has begun. At this point, supportive therapy may be the only option. Summary information can be found in chapter 25.

Epidemiology and Prevention

The *Trichinella* cycle that is maintained in nature occurs among cannibalistic and carrion-feeding carnivores. While domestic pigs and rats tend to be secondary hosts, the majority of human infections have been from infected pork. In reviewing data on trichinellosis in the United States from 1997 to 2001, pork was implicated in 30% of the cases and wild game was implicated in 43% (77). Sausage was the most frequently implicated pork product and was often from noncommercial sources (77). While cases acquired from pork consumption continue to decline, the proportion of cases acquired from wild game meat has increased. However, the absolute numbers remain at about 9 to 12 per year. Continued multiple-case outbreaks and the identification of nonpork sources of infection require ongoing education and control measures.

In other parts of the world, the infected-meat source statistics vary; in the former Soviet Union, >90% of the cases have been attributed to the ingestion of poorly cooked bear and wild boar meat. Factors contributing to the slow decline of trichinellosis incidence in Russia and to the increase in the number of cases originating from wild-animal meat include the distribution and consumption of veterinary-uncontrolled pork, poaching and distribution of wild-animal meat, and poor compliance with regulations (65). Twenty-seven outbreaks of human disease occurred in China between 1964 and 2004 and were associated with mutton, dog, and game meat. However, the quarantine of infected meat is not mandatory in China (89). Although outbreaks are rare in Israel, outbreaks have been detected in immigrant agricultural workers; infected wild-boar meat was implicated (49). In tropical Africa, the infected meat source tends to be bush pigs and warthogs. Most infections in Central and South America have been associated with domestic pigs. It is difficult to say with certainty exactly which animals may be infected throughout the world; very few comprehensive studies of wild animals have been attempted. In one study of *Trichinella* infection in wildlife in the southwestern United States, the range of *T. murrelli* was extended from previous reports limiting this species to the eastern United States. Thus, *Trichinella* infection is now documented in three states bordering Mexico, New Mexico, Arizona, and Texas (73). Information is also now available confirming the presence of anti-*Trichinella* antibodies in a human population living in Papua New Guinea (58, 64).

Although *T. spiralis* and *T. pseudospiralis* are found worldwide, the other species tend to have a more narrow

Table 11.7 Trichinellosis: prevention and control measures

Potential infection source (reference)	Prevention and control measures
Meat preparation (55)	1. Cook meat products until the juices run clear or to an internal temperature of 160°F (71°C).
	2. Freeze pork less than 6 in. thick for 20 days at 5°F (−17°C) to kill any encysted larvae. Other options would be −10°F (−23°C) for 10 days or −20°F (−29°C) for 6 days. However, remember that *T. nativa* in bear meat probably survives freezing for a year or longer.
	3. Cook wild game meat thoroughly. Freezing wild game meats, unlike freezing pork, even for long periods, may not effectively kill all encysted larvae.
	4. Clean meat grinders after preparing ground meats.
	5. Curing (salting), drying, smoking, or microwaving meat does not consistently kill infective larvae.
Pig farms (68)	1. Barriers must be in place to prevent entrance of rodents and other potential hosts into the pigsty and food store.
	2. New animals should not be admitted to the farm prior to serologic testing for antibodies to *Trichinella* spp.[a]
	3. Procedures for sanitary disposal of dead animals must be used at all times.
	4. No raw or improperly heated swill or waste food containing meat may be present at the farm.
	5. No rubbish dumps should be located in the immediate vicinity of the farm.

[a] The use of serologic tests has been an important technological advance, not only for the diagnosis of human infection but also for the identification of infected animals (76, 82).

geographic range (Table 11.2). *T. nativa* and *Trichinella* T6 are found primarily in the Arctic regions; *T. britovi*, *T. murrelli*, *Trichinella* T8, and *Trichinella* T9 are found in the temperate zones; and *T. nelsoni*, *T. papuae*, and *T. zimbabwensis* are found in equatorial areas.

Countries in the European Union, Eastern European countries, and the former members of the Soviet Union require direct inspection of pork, using microscopic examination of small tissue samples of pig diaphragm or examination of pooled digested tissue samples; within the United States, the U.S. Department of Agriculture requires strict standards for the freezing, cooking, and curing of pork and pork products (40, 68). A number of excellent preventive measures have been identified for implementation during the preparation of potentially infected meat sources, as well as control measures for commercial pig farms (Table 11.7). Recent information also confirms the need to review the intentional feeding of animal products and kitchen waste to horses, a high-risk practice which requires implementation of regulations to ensure that such feeds are rendered safe for horses, as is currently required for products fed to swine (56).

In 1981, the U.S. Department of Agriculture issued a news release that suggested that microwave cooking might not kill the larvae. On the basis of a number of subsequent studies, the current recommendation states that "*all parts* of pork muscle tissue must be heated to a temperature not lower than 137°F (58.3°C)" (55). It has been recommended that an internal meat thermometer be used when cooking pork; the meat can be tested after being removed from the microwave oven if the oven is not equipped with an internal thermometer. Reduction in the number of cases is due primarily to regulations requiring heat treatment of garbage and low-temperature storage of the meat. Occasional outbreaks are frequently due to problems with feeding, processing, and cooking of pigs raised for home use.

Baylisascaris procyonis

Although *Baylisascaris procyonis* was first isolated from raccoons in 1931 in New York, it was also recognized in raccoons in Europe. The genus was defined in 1968 and was named after. H. A. Baylis, who had been with the British Museum of Natural History in London.

It was first recognized as causing neural larva migrans (NLM) in rodents and then recognized as being able to produce serious NLM in >100 different species of birds and mammals. The potential for causing human disease was considered by earlier parasitologists but has been recognized only during the last twenty years. Approximately 15 confirmed human cases have been documented (32).

Life Cycle and Morphology

B. procyonis is an ascarid normally found in raccoons (*Procyon lotor*), has a normal ascarid-like life cycle, causes a very serious zoonotic disease in humans, and is

most often reported from North America (Figure 11.5) (32). Raccoons are infected by ingesting infective eggs and by eating larvae encysted in the tissue of intermediate hosts, such as rodents, rabbits, and birds. The larvae then penetrate the mucosa of the small intestine and develop there before reentering the intestinal lumen to mature. Raccoons may be infected by up to 60 worms, and young animals have a higher prevalence of infection. These adult worms produce >150,000 eggs/worm/day; infected raccoons can shed as many as 250,000 eggs/g of feces. This level of egg production can lead to significant environmental contamination.

B. procyonis eggs are somewhat oval, are dark brown, and measure from 63 to 88 μm by 50 to 70 μm. The eggs contain a single-celled embryo and a thick shell with a finely granular surface; they are not infective immediately after being passed but can survive in moist soil for years (Figure 11.6). *Toxocara* eggs tend to be somewhat larger and have a coarsely pitted thick shell.

Human infections result from ingestion of eggs that are passed in very large numbers (millions of eggs/day) in the feces of infected raccoons; the human then becomes the accidental intermediate host. Once ingested, the eggs hatch in the intestinal tract, releasing the immature larvae. Rather than developing to adult worms as occurs in the raccoon, the larvae begin to migrate extensively throughout the body tissues, causing visceral larva migrans (VLM) and/or NLM. Although this infec-

Figure 11.5 Life cycle of *Baylisascaris procyonis*. Accidental infection in the human leads to visceral larva migrans (VLM), ocular larva migrans (OLM), and/or neural larva migrans (NLM), which are very serious diseases that can cause death. Infections are seen primarily in very young children.

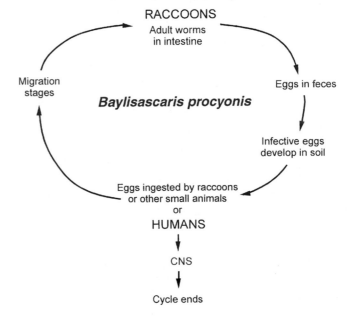

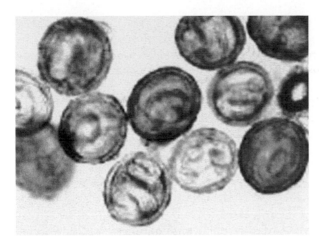

Figure 11.6 *Baylisascaris procyonis* eggs; note the larvae within the egg shells. (From A Pictorial Presentation of Parasites: A cooperative collection prepared and/or edited by H. Zaiman.)

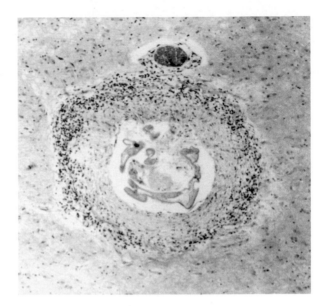

Figure 11.7 *Baylisascaris procyonis*, CNS atrophy, sclerosis, and larval granuloma in the brain in a 2-year-old boy who died following 14 months of CNS problems. (From A Pictorial Presentation of Parasites: A cooperative collection prepared and/or edited by H. Zaiman. Photograph courtesy of K. R. Kazacos.)

tion presents as acute fulminant eosinophilic meningoencephalitis, two features of the life cycle are somewhat different from those of other helminths causing larva migrans: there is targeted migration to the CNS and continued growth of the larvae to a much larger size within the CNS. Despite different courses of therapy, there are no documented neurologically intact survivors of this infection.

Clinical Disease

The first confirmed cases of human *B. procyonis* infection were described in the 1980s. Risk factors have been identified as contact with infected raccoons, their feces, or a contaminated environment. Geophagia or pica has also been identified as a potential risk, which is often seen in children younger than 2 years. Tissue damage is caused by the actual larval migration, as well as an intense inflammatory reaction (Figures 11.7 and 11.8). Unlike other helminth larvae that cause VLM, the larvae of *B. procyonis* continue to grow during the migratory phase of the life cycle and can reach lengths of 2 mm. In addition to continued growth, the larvae tend to exhibit very vigorous migratory behavior and remain viable for long periods. They tend to invade the eyes, causing ocular larva migrans (OLM), which has also been found in immunocompetent adults, and the spinal cord and brain, causing NLM, found primarily in infants and young children. Larval migration can cause permanent neurologic damage, blindness, or death. It is also important to realize that with the absence of a definitive diagnostic test, the prevalence of subclinical cases is totally unknown. However, it is probably higher than is currently recognized. Unfortunately, even with clinical cases, most patients are diagnosed only after there has already been severe CNS damage.

Patients can present with eosinophilic meningoencephalitis or unilateral neuroretinitis; the presentation probably depends on the number of eggs that were ingested. Larvae tend to enter the CNS approximately 1 to 4 weeks after infection, and the disease may progress very rapidly. OLM may present with a broad range of symptoms, including chronic endophthalmitis with retinal detachment, posterior pole granuloma, vitreous abscess, pars planitis, optic neuritis, keratitis, uveitis, iritis, hypopyon, and meandering retinal tracks containing larvae. A new clinical entity, diffuse unilateral subacute neuroretinitis (DUSN), is also caused by *B. procyonis* (51). DUSN is a form of OLM characterized by progressive unilateral visual loss, retinal pigmentation, and optic nerve anatomy changes, all of which lead to severe ocular damage.

The neural form may present with symptoms ranging from mild neuropsychologic problems to seizure, convulsions, ataxia, coma, and death. Patients may exhibit sudden lethargy, irritability, loss of muscle coordination, decreased head control, spasmodic contractions of the neck muscles, stupor, nystagmus, obtundation, coma, hypotonia, and hyperreflexia. Infants who have survived meningoencephalitis demonstrate sequelae including hemiparesis, inability to sit or stand, ocular muscle paralysis, cortical blindness, and severely delayed development. Survivors are left in a persistent vegetative state or with severe neurologic deficits, including blindness, all of which can require extensive supportive care.

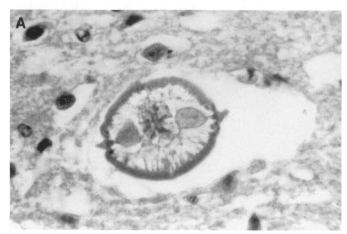

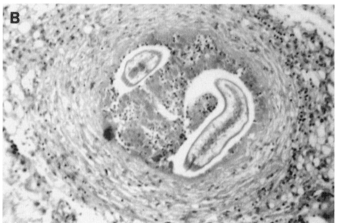

Figure 11.8 (A) *Baylisascaris procyonis*, larva in the brain of an 18-month-old boy who died after a 3-week bout of CNS disease. (B) Mesenteric granuloma from the same patient. (From A Pictorial Presentation of Parasites: A cooperative collection prepared and/or edited by H. Zaiman. Photograph courtesy of K. R. Kazacos.)

Diagnosis

Human infections with *B. procyonis* are rare and are often diagnosed by a process of elimination; when all other recognized causes of larva migrans have been explored, *B. procyonis* may be considered. There are approximately 15 documented cases, almost all of which have been seen in young children. Results obtained from routine hematologic and cerebrospinal fluid examinations are usually consistent with a parasitic infection but are nonspecific. Definitive diagnosis requires identification of the larvae in tissues; however, this can be difficult depending on the body site. Cross sections of larvae tend to measure 60 to 70 μm, and the larvae have prominent, single lateral alae and paired, conical excretory columns (smaller than central intestine) (Figure 11.8A). Ocular examinations may reveal retinal lesions, larval tracks, or migrating larvae; these findings may provide the tentative diagnosis

of a helminth infection. Using IFAT, ELISA, and Western blotting, anti-*Baylisascaris* antibodies can be detected in cerebrospinal fluid. Currently, the only source of serologic testing is the Department of Veterinary Pathobiology at Purdue University, West Lafayette, Ind. [K. Kazacos, phone (765) 494-7558] (32). Acute- and convalescent-phase titers demonstrate severalfold increases in both serum and CSF antibody levels; there is no cross-reactivity with *Toxocara*. With the availability of a reliable serologic test, there is less need to perform a brain biopsy. *In most cases, the clinical history provides the main clues.*

The liver and lungs do not tend to be involved in *Baylisascaris* infections, but cerebral lesions are often detected. Well-formed granulomas can be seen in any part of the nervous system. The damage tends to be prominent, even showing tracks with tissue disruption. Differential diagnosis findings are summarized in Table 11.8.

Treatment

There is no effective cure for *B. procyonis* infection; treatment is symptomatic and involves systemic corticosteroids and anthelmintic agents. Unfortunately, NLM is usually not responsive to anthelmintic therapy; by the time the diagnosis is made, extensive damage has already taken place. Drugs that can be tried include albendazole, mebendazole, thiabendazole, levamisole, diethylcarbamazine, and ivermectin. Experimental data from animal studies indicate that albendazole and diethylcarbamazine may have the best CSF penetration and larvicidal activity. Unfortunately, since the diagnosis does not occur until the onset of symptoms, larval invasion of the CNS has already taken place. Therefore, treatment is started late in the course of the infection.

Considering the potential outcomes associated with delaying treatment until symptoms begin, prophylactic treatment for asymptomatic children exposed to raccoons or contaminated environments is now being considered. Since the anthelmintics do not appear to be problematic, prophylaxis seems appropriate for specific individuals. Albendazole has been used in these cases; treatment has been discontinued when environmental testing has proven to be negative.

Ocular infections have been treated successfully by using laser photocoagulation therapy to destroy the intraretinal larvae (1). Systemic corticosteroids have been used to decrease or prevent resulting intraocular inflammatory responses from the killed larvae.

Epidemiology and Prevention

The relationship between *B. procyonis* infection, raccoons, and humans has now been well defined. Recent reports of patent *B. procyonis* infections in dogs have caused concern because of the potential for expanded

Table 11.8 Differential diagnosis of *Baylisascaris procyonis* infection

Infection	Geographic areas	Clinical findings	Comments
Baylisascaris procyonis	Americas from Canada to Panama, Europe, Japan, Soviet Union, Asia	Peripheral and CSF eosinophilia, meningoencephalitis	Eosinophilic meningoencephalitis; history of pica or geophagia, age of patient extremely important in suspecting infection
Toxocara spp.	Cosmopolitan	VLM	Uncommon cause of NLM; antibodies in blood and CSF
Angiostrongylus cantonensis	Southeast Asia, China, Japan, Jamaica, Western Pacific Islands, Hawaii, Madagascar, Cuba, Egypt, Puerto Rico, New Orleans, Nigeria, eastern Australia, Africa	Eosinophilic meningitis	Relatively benign course, good prognosis
Gnathostoma spinigerum	Southeast Asia, China, Japan	Severe neurologic sequelae, myeloencephalitis, focal cerebral hemorrhage (with xanthochromia), radiculopathy, migrating cutaneous swellings	Poor prognosis

Less common causes of eosinophilic meningoencephalitis in humans
  Neurocysticercosis
  Paragonimiasis
  Toxocariasis
  Neurotrichinosis
  Schistosomiasis
	Cosmopolitan in many areas; more narrow geographic areas would apply to schistosomiasis	Could mimic infection with *B. procyonis*	History critical in suspecting any of these infections

Coccidioides immitis	Southwestern United States, Central and South America	Intense basilar enhancement, hydrocephalus, acute infarction on neuroimaging, positive CSF serology	Disseminated disease, most common cause of eosinophilic meningitis in United States
Acute disseminated encephalomyelitis	Cosmopolitan	Acute encephalopathy, cerebral white matter changes on neuroimaging	Monophasic nonprogressive illness, more discrete multifocal grey and white matter abnormalities on neuroimaging; generally good prognosis

human exposure. Groups of raccoons tend to defecate in common areas called latrines, which tend to be present off the ground in fallen logs (firewood may be contaminated), rocky outcroppings, and trees. In areas that have been carefully investigated such as Pacific Grove, Calif., many latrines are present and are located directly on the ground, on roofs, in attics, and on steps and fences. These findings suggest that a very large number of raccoons are present in this location. The eggs remain viable in the soil for extended periods, often years. The eggs also have a sticky surface coating that causes them to adhere to objects, including human hands and toys. Apparently, incineration or soaking the feces with volatile solvents such as mixtures of xylene and ethanol appears to be the only means of killing the eggs (32). In some cases, removal and disposal of several inches of topsoil may also be indicated. Recognition of this new human infection and prevention of the establishment of raccoon latrine sites around areas of human habitation and recreational use are critical to successful control.

This infection is of great public health concern and has the potential to cause extensive damage in the human host, particularly young children (32). Risk is highest for young children or infants with pica or geophagia; these individuals need to be kept away from potentially contaminated areas. Raccoons should be discouraged from visiting yards by refraining from putting out food and by not leaving dog food uncovered and available. Keeping pet raccoons, particularly in homes with young children, should be discouraged.

Although *B. procyonis* was thought to be absent from many regions, it is now becoming clear that where raccoons are found, *B. procyonis* is also likely to be present. The infection has also been found in a 3-day-old domestic lamb, suggesting that mammalian fetuses, in general, should be added to the list of those at risk for NLM due to ascarid parasites (4). The increasing number and recognition of cases highlight the critical importance of controlling and preventing this potentially devastating zoonotic infection (29). Certainly education for the public will be paramount in helping to determine at-risk populations and preventing additional infections.

Lagochilascaris minor

Although human infections with *Lagochilascaris minor*, which is normally found in the small intestine of the cloudy leopard, have been documented, neither the natural life cycle nor the route of human infection is known (48). Human cases have been recorded from the West Indies, Suriname, Costa Rica, Mexico, Venezuela, Colombia, and Brazil. The cases of human infection are characterized by lesions in the oropharynx and other soft tissues in the head and neck. The first sign of infection may be a pustule, usually on the neck, which increases in size to a large swelling. After the skin breaks, living adults, larvae, and eggs are expelled at intervals in the pus. Eggs are continually developing into larvae and then into adults in the tissues at the base of the abscess; thus, the life cycle is a continuing process. Adult male worms measure up to 9 mm long, and the females are about 1.5 cm; the eggs resemble those of *Toxocara cati* and measure 45 to 65 μm by 59 to 73 μm. Irregularly shaped pits are present on the egg surface; 20 to 32 pits surround the equator of an egg (Figure 11.9). The area of induration increases, with the development of sinus tracts. Chronic granulomatous inflammation may last for months. Other tissues that have been involved include the throat, nasal sinuses, tonsillar tissue and mastoids, brain, and lungs. Although thiabendazole has been tried, efficacy was not really documented. Albendazole has also been tried, but it was possible to confirm only transitory elimination of adult worms.

A study in which wild rodents were used as experimental intermediate hosts of *L. minor* has provided some additional information about the life cycle and epidemiology of this parasite (66). After inoculation of infective eggs, larvae were found in viscera, skeletal muscle, and adipose and subcutaneous tissues from all rodents. Adult worms were recovered in the cervical region, rhinopharynx, and oropharynx of domestic cats fed the rodent tissues. Based on this study, it appears that (i) wild rodents act as intermediate hosts; (ii) under natural conditions rodents could act as either intermediate hosts or paratenic hosts of *L. minor*; (iii) despite the occurrence of an autoinfection cycle in felines (definitive hosts), the cycle is completed only when intermediate hosts are provided; and (iv) in the wild, rodents could serve as a source of infection for humans since they are frequently used as food (guinea pigs and agoutis as examples) in regions with the highest incidence of human lagochilascariasis. Thus, humans are accidental hosts, possibly becoming infected by eating raw meat from wild rodents containing L3 larvae. It is quite possible that autoinfection also occurs; this possibility is suggested by the chronicity of the disease in patients over a number of years.

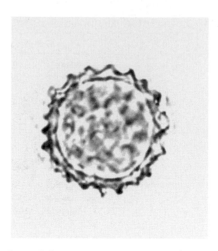

Figure 11.9 *Lagochilascaris minor* egg; note the pits in the egg shell (somewhat similar to *Toxocara* eggs). (From A Pictorial Presentation of Parasites: A cooperative collection prepared and/or edited by H. Zaiman.)

Toxocara canis and T. cati (Visceral Larva Migrans and Ocular Larva Migrans)

The VLM syndrome was described by Beaver and colleagues in New Orleans in 1952. This syndrome is caused by the migration of larvae of *Toxocara canis*, *T. cati*, and some other animal helminths. Within 10 years of the initial report, more than 2,000 cases had been reported from 48 countries and from every region of the United States. The disease, frequently seen in young children, usually does not cause severe problems, although it persists for months to more than a year. One serious possible complication is invasion of the eye (OLM), often resulting in a granulomatous reaction in the retina. Larva migrans caused by *Toxocara* spp. are widely recognized as zoonotic infections throughout the world and may be much more common than previously thought. Infection rates in dogs have been reported to be 2 to 90%, and the highest rates are seen in puppies as a result of transmission from their dams. The overall incidence of infected dogs older than 6 months is probably less than 10% (79).

Life Cycle and Morphology

Humans acquire the infection by ingesting infective eggs of the dog (primarily) or cat ascarid *T. canis* or *T. cati* (Figures 11.10 and 11.11). Puppies are often infected by vertical transfer of larvae from their dams transplacentally or lactogenically, and egg shedding by puppies can begin as early as 2 weeks of age. In cats, lactogenic but not transplacental transmission occurs. Young kittens and puppies tend to recover from the infections between 3 and 6 months of age. Infections in older animals are acquired by the ingestion of infective eggs from the soil or ingestion of larvae in infected rodents, birds, or other paratenic hosts. Eggs are shed in the feces and take about 2 to 3 weeks to mature and become infective. After the eggs are accidentally ingested by a human, the larvae hatch in the small intestine, penetrating the intestinal mucosa and migrating to the liver. Migratory routes include the lungs and/or other parts of the body, or the larvae may remain in the liver. During this migration, the larvae do not mature, even if they make their way back into the intestine. The larvae are usually <0.5 mm long and 20 μm wide. Information also implicates the ingestion of uncooked meats as a potential cause of human toxocariasis, with possibly the first North American case following ingestion of raw lamb liver being reported by Salem and Schantz (78). Although transplacental and lactogenic transmission

has not been reported to occur in humans, in a study of *Toxocara* titers in maternal and cord blood, 6 (35%) of 17 mothers in the *Toxocara* antibody-positive group had previously miscarried compared with 3 (8.6%) of 35 *Toxocara*-negative mothers (84).

Most infections are probably asymptomatic and/or go unrecognized. VLM tends to be seen in younger children, around the age of 3 years, while OLM is more likely to occur in older children, around 8 years. However, this does not tend to be the case with *B. procyonis*, where most cases of VLM, NLM, and OLM have been identified in very young children.

Although VLM is generally associated with *T. canis* or *T. cati*, other helminths have also been implicated in disease. *B. procyonis* has been implicated in VLM, NLM, and OLM in humans. An excellent review of the diagnostic morphology of four larval ascaridoid nematodes that may cause VLM includes identification keys for *Toxascaris leonina*, *B. procyonis*, *Lagochilascaris sprenti*, and *Hexametra leidyi* (12). If discovered in tissue sections, the four species of ascaridoid larvae described in the study by Bowman (12) can be differentiated from other known ascaridoids that may cause VLM.

Clinical Disease

Clinical symptoms depend on the number of migrating larvae and the tissue or tissues involved. Infections may range from asymptomatic to severe disease. Larvae often remain in the liver and/or lungs, where they become encapsulated in dense fibrous tissue. Other larvae may continue

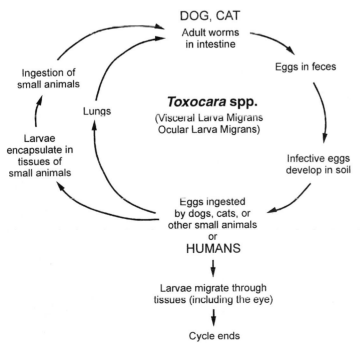

Figure 11.10   Life cycle of *Toxocara* spp., the cause of visceral larva migrans (VLM) and ocular larva migrans (OLM).

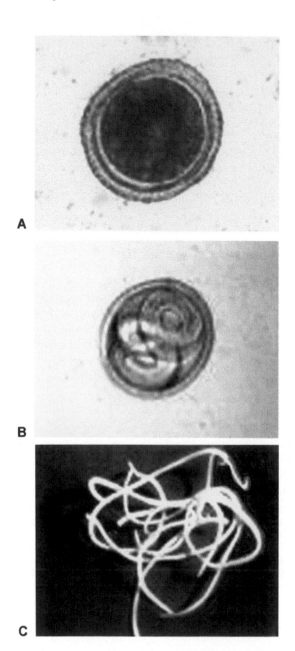

Figure 11.11 *Toxocara canis*: (A) immature egg; (B) mature egg containing larva; (C) adult worms. (From A Pictorial Presentation of Parasites: A cooperative collection prepared and/or edited by H. Zaiman.)

to migrate throughout the body, causing inflammation and granuloma formation. The most outstanding feature of the disease is a high peripheral eosinophilia, which may reach 90%. The overall severity of the clinical picture depends on the initial dose of infective eggs. As few as 200 *T. canis* larvae in small children may produce a peripheral eosinophilia of 20 to 40% for more than a year, with no other detectable symptoms. Patients with 50% eosinophilia usually have symptoms, which might include fever, hepatomegaly, hyperglobulinemia, pulmonary infiltrates, cough, neurologic

disturbances, and endophthalmitis. Although a rare complication of toxocariasis, CNS involvement can cause seizures, neuropsychiatric symptoms, or encephalopathy (8, 88).

The VLM syndrome is most commonly seen in children 1 to 4 years of age; however, rare cases have been seen in adults, in whom the illness includes mild pulmonary involvement suggestive of VLM. Severe bronchospasm resulting in respiratory failure was not reported until 1992; the case was confirmed by ELISA (30).

A report from Mexico City reviewed six adult patients with toxocariasis presenting with rheumatic symptoms, including lower-extremity nodules, edema suggestive of thrombophlebitis, and synovitis of a knee without effusion. Some of the patients reported having a prolonged, nonproductive cough, generalized pruritus, and migratory cutaneous lesions. One patient with monarthritis subsequently developed orchitis. All patients had an eosinophilia of 14 to 20%. One patient was biopsy positive, two had positive serologic test results, and the diagnosis was never confirmed in the other three (43). In another case, a 17-year-old boy with palpable purpura, oligoarthritis, acute abdominal pain, microhematuria, and cutaneous vasculitis was found to have toxocariasis with a clinical history including hypereosinophilia and domestic contact with a puppy (34). The infection was confirmed using serologic tests, and complete spontaneous resolution occurred within a few days.

The relationship between asthma and covert toxocariasis remains unclear; however, in a 1999 study, the seroprevalence of anti-*T. canis* antigen (E/S antigen) was 26.3% in asthmatic patients and 4.5% in the controls. All asthmatic patients with anti-*Toxocara* immunoglobulin E (IgE) had cutaneous reactivity to E/S antigen. Therefore, the authors concluded that asthmatic patients with anti-*Toxocara* IgE and IgG were experiencing a covert toxocariasis (52). Significantly elevated levels of IgE/anti-IgE immune complexes have been detected in sera of patients with symptomatic disease, including VLM and OLM (61). While specific IgG may act via antibody-dependent cell-mediated cytotoxicity mechanisms, IgE/anti-IgE immune complexes may participate in VLM and OLM by inducing type III hypersensitivity.

Evidence suggests that ocular disease can occur in the absence of systemic involvement and vice versa for VLM. Although these facts may be explained by possible strain differences of *Toxocara* spp., VLM may reflect the consequences of the host inflammatory response to waves of migrating larvae while OLM may occur in individuals who have not become sensitized (27).

Patients with symptoms that do not fall into the more strict categories of VLM or OLM are often described as having covert toxocariasis. These cases are characterized by symptoms including abdominal pain, anorexia, behavior disturbances, cervical adenitis, wheezing, limb pains, and fever (79).

Diagnosis

VLM symptoms caused by *Toxocara* spp. must be differentiated from those caused by other tissue-migrating helminths (ascarids, hookworm, filariae, *Strongyloides* spp., and *Trichinella* spp.), as well as other hypereosinophilic syndromes. OLM may be confused with retinoblastoma, ocular tumors, developmental anomalies, exudative retinitis, trauma, and other childhood eye problems. It is important to remember that in OLM, peripheral eosinophilia may be absent. OLM should be considered in any child with unilateral vision loss and strabismus who has raised, unilateral, whitish or gray lesions in the fundus. VLM should be suspected in any pediatric patient with an unexplained febrile illness and eosinophilia. If the patient has a history of pica and there is hepatosplenomegaly and multisystem disease, then VLM becomes even more likely.

The suspected diagnosis can be confirmed only by identification of larvae in autopsy or biopsy specimens. However, if children are found to have *Ascaris* or *Trichuris* infections, one might suspect toxocariasis, since all three infections are transmitted via ingestion of contaminated soil. Since biopsy specimens are usually not recommended, serologic testing has become widely accepted as the most appropriate approach. In patients with presumptive ocular toxocariasis, higher antibody titers have been detected in the aqueous humor than in the serum, suggesting localized antibody production. The EIA is recommended and uses third-stage larva secretory antigen. The test is highly specific, and there are no cross-reactions with sera obtained from patients infected with other commonly occurring human parasites. The diagnostic titers vary between VLM throughout the body and OLM. A titer of 1:32 is considered diagnostic for VLM, while a titer of 1:8 is considered diagnostic for OLM. This lower titer for OLM raises the possibility of a false-positive diagnosis in a patient who has an asymptomatic *Toxocara* infection and an ocular disease due to other etiology. Measurement of antibody levels in ocular fluid should increase the specificity of the ELISA, yielding a better definitive diagnosis for the patient. The serologic tests become very important when one is trying to differentiate OLM from retinoblastoma, which may have serious consequences. Serologic testing for toxocariasis is recommended in patients with Fuchs heterochromic cyclitis and retinal scars in the absence of toxoplasmosis (85). Refer to Algorithm 11.2 for a detailed review.

A commercial ELISA kit has been evaluated and found to have an overall diagnostic sensitivity of 91%

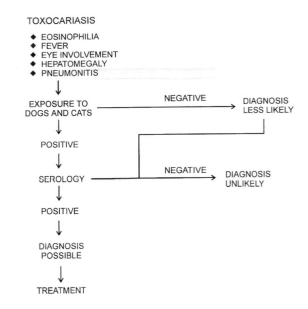

Algorithm 11.2 Diagnosis of toxocariasis.

and specificity of 86%, with cross-reactivity being seen with sera from patients with strongyloidiasis, trichinellosis, filariasis, and fascioliasis. However, because these infections tend to be infrequent, potential cross-reactivity may not be relevant unless the geographic area includes a high prevalence of any of the above-mentioned infections (36).

Although many cases of VLM are diagnosed by serologic testing, toxocariasis has generally been defined as an infection with *Toxocara* spp., with no attempt to identify the species involved. Studies involving preabsorption of patient sera with cross-reacting antigens and review of follow-up reactivity changes in the sera have confirmed the ability to specifically identify the infecting parasite as *T. canis* or *T. cati*. This ability to distinguish between the two species should be helpful in further biological, epidemiologic, and clinical studies of toxocariasis (57). In cases of encephalitis and myelitis with CSF eosinophilia, parasitic infection should be suspected and appropriate serologic tests should be performed (54). Other more unusual presentations have included thrombocytosis and eosinophilic pleural effusion (6, 39). The probability of hepatic toxocariasis can be further evaluated using imaging techniques and ultrasonography. Findings would include focal ill-defined hepatic lesions, hepatosplenomegaly, biliary dilatation, sludge, and periportal lymph node enlargement (38).

Although the currently recommended serologic test for toxocariasis is EIA, when interpreting serologic results a measurable titer does not always represent current infection. A small percentage of the U.S. population

(2.8%) exhibits a positive titer that reflects the prevalence of asymptomatic toxocariasis.

In one study, pseudocystic transformation of the peripheral vitreous appeared to be a rather specific and sensitive ultrasonographic biomicroscopic sign in patients with presumed peripheral toxocariasis, and this finding will aid diagnosis in difficult cases (87). In a group of 48 patients with diverse inflammatory conditions of the retroiridal space who were examined by ultrasonographic biomicroscopy, no characteristic *Toxocara*-associated pseudocystic images were seen.

KEY POINTS—LABORATORY DIAGNOSIS

Toxocara canis and *T. cati* (Visceral and Ocular Larva Migrans)

1. Biopsy specimens are usually not recommended.
2. Serologic tests are recommended. Serum samples can be sent to the appropriate state Department of Public Health (check the applicable state submission requirements). These specimens are often sent to the Centers for Disease Control and Prevention. History information is required, and each sample must be specified as "serum" *or* "eye fluid," so that a correct interpretation of the results can be made.
3. Tissue specimens containing larvae can be referred to a reference center (university or Armed Forces Institute of Pathology, Washington, D.C.).

Treatment

Diethylcarbamazine, thiabendazole, ivermectin, and albendazole are effective in some cases but not in others. Corticosteroids may also be given to patients with VLM or OLM. Destruction of the larva by photocoagulation is recommended when the larva is visible in the eye. (Specific drug and dosage information is provided in chapter 25.) Even when the eye is involved, the prognosis is usually favorable, particularly when a prompt diagnosis is made and treatment is effective. Albendazole is the treatment of choice. Although mebendazole is poorly absorbed outside the gastrointestinal tract, it has been used with some success (27).

Epidemiology and Prevention

Given that (i) *Toxocara* worms are commonly found in dogs and cats, (ii) puppies and kittens are infected early in life, and (iii) pets and children are often found in the same household, it is not surprising that the combination of small children playing in contaminated soil and pets passing large numbers of infective eggs leads to VLM and/or OLM. However, it is important to remember that this disease can also occur in adults (43). The eggs become infective after about 3 weeks and remain viable in the soil for months. Examination of soil from parks and playgrounds in various areas of the world has demonstrated infective *Toxocara* eggs that contribute to the high infection rate seen in dogs (20).

One preventive measure includes worming dogs and cats periodically with mebendazole to keep them free of worms. Another recommendation is preventing children from eating dirt, particularly soil that could be contaminated by neighborhood or family pets. Proper curbing of dogs in the street during defecation has also been recommended. Another approach involves protection of sandboxes in public parks from *Toxocara* egg contamination. The recommendation is to cover the sandboxes with clear vinyl sheets at night and on rainy days.

In summary, the following preventive measures should be emphasized: regular deworming of dogs and cats, beginning at 2 weeks of age; removal of cat and dog feces in places adjacent to homes and children's playgrounds; ensuring that children's sandboxes are covered when not being used; regular hand washing after handling soil and before eating; and teaching children not to put dirty objects into their mouths.

Ancylostoma braziliense and *A. caninum* (Cutaneous Larva Migrans)

Cutaneous larva migrans (CLM), also called creeping eruption, was recognized as a clinical syndrome before the 1800s. Various reports were published during the late 1800s; however, it was not until 1926 that the most common etiologic agent of CLM in the southern United States was found to be *Ancylostoma braziliense*, a very common hookworm of dogs and cats. *A. caninum*, the common hookworm of dogs, has been implicated in cases of CLM. Other species are also capable of producing CLM, although they are less common than *A. braziliense*.

Life Cycle and Morphology

Infection in humans is acquired through skin penetration by infective larvae from the soil (Figure 11.12). These larvae can also cause infection when ingested. When the larvae penetrate the skin, they produce pruritic papules, which after several days become linear tracks that are elevated and vesicular (Figure 11.13). Movement by the larvae in the tunnel may extend the track several millimeters each day.

Clinical Disease

Within a few hours after larval penetration of the skin, an itching red papule develops. As the worm begins to migrate from the area of the papule, a serpiginous track

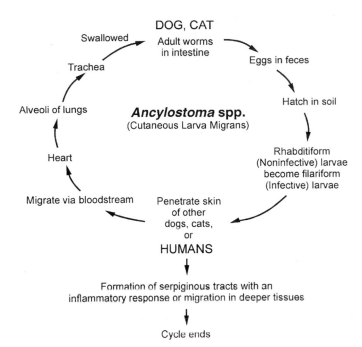

Figure 11.12 Life cycle of *Ancylostoma* spp., the cause of cutaneous larva migrans (CLM).

appears. The surrounding tissues are edematous and very inflamed. The larva continues to migrate several centimeters each day, and the older portion of the track dries and becomes scarred. This process is associated with severe pruritus, and scratching can lead to secondary infection. Larvae that first enter the skin and cause creeping eruption may later migrate to the deeper tissues (lungs). Deeper-tissue migration may lead to pneumonitis with larval recovery in the sputum (13). A peripheral eosinophilia, as well as many eosinophils and Charcot-Leyden crystals in the sputum, may also be present.

Figure 11.13 Linear tracks caused by migration of *Ancylostoma* spp. (cutaneous larva migrans [CLM]). (From a Pictorial Presentation of Parasites: A cooperative collection prepared and/or edited by H. Zaiman.)

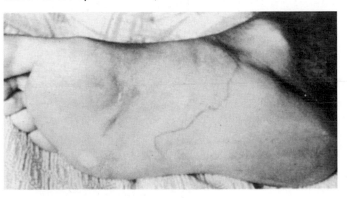

Diagnosis

Diagnosis can usually be made on the basis of the characteristic linear tunnels or tracks and a history of possible exposure; however, other organisms (less common) can also cause CLM.

KEY POINTS—LABORATORY DIAGNOSIS
***Ancylostoma braziliense* and *A. caninum* (Cutaneous Larva Migrans)**   1. Diagnosis is usually based on possible exposure history and/or the presence of the linear tracks. Biopsy is not recommended. However, newer PCR methods for the detection and identification of larvae in human tissues may provide improved test results.   2. There may be an elevated eosinophilia (peripheral or sputum).

Treatment

Treatment is generally carried out with thiabendazole, which can be administered either by mouth or topically. Specific drug dosages are provided in chapter 25. Symptoms can also be treated with antihistamines, antipruritic agents, sedatives, and/or topical anesthetics.

Epidemiology and Prevention

Most infections are acquired from contact with larvae in moist or sandy soil. Such areas include beaches and sandboxes. Dogs and cats tend to defecate in such areas, providing a perfect situation for accidental infection with the filariform larvae. Specific preventive measures include covering all sandboxes when they are not being used, keeping dogs and cats away from beaches, and periodic deworming of domestic dogs and cats.

Human Eosinophilic Enteritis

As a general definition, primary eosinophilic gastrointestinal disorders selectively affect the gastrointestinal tract, with eosinophil-rich inflammation, in the absence of known causes of eosinophilia, including parasitic infections, drug reactions, and malignancy. Becoming more common, these disorders include eosinophilic esophagitis, eosinophilic gastritis, eosinophilic gastroenteritis, eosinophilic enteritis, and eosinophilic colitis. It is well known that eosinophils are important members of the gastrointestinal mucosal immune system and that eosinophilic gastrointestinal problems involve mechanisms that include IgE-mediated and delayed Th2-type responses.

Segmental eosinophilic inflammation of the gastrointestinal tract may occur as an isolated condition or part of a multisystem problem. During the past 20 years, an increasing number of cases have been reported in northern Queensland, Australia (21, 74). All of the patients were Caucasians, and they ranged in age from 16 to 72 years, with no previous illness. They presented with severe abdominal pain, occasional diarrhea, weight loss, and dark stools; all cases were associated with eosinophilia and elevated levels of IgE in serum.

The conclusion that the etiologic agent was *A. caninum* was determined for the following reasons. A single adult *A. caninum* worm was found within a segment of inflamed ileum of one patient. Human hookworms do not occur in urban Australia, and no hookworm eggs were being passed in the stool. In contrast, all of the patients were closely associated with dogs, almost all of which were infected with hookworms. Also, all patients treated with anthelmintic agents responded with a return to normal peripheral blood eosinophil counts. The similarities between this case and the 33 previously reported cases (21) implicate *A. caninum* as the cause of eosinophilic enteritis (EE) in that group of patients. It has been speculated that *A. caninum* causes human EE by inducing allergic responses to its secretions, including cysteine proteinases, which are involved in pathogenesis in other parasites. Recent immunologic studies involving ELISA and Western blotting for IgG and IgE antibodies to excretory-secretory antigens from adult *A. caninum* also suggest that this parasite is a major cause of EE and peripheral blood eosinophilia (PE) (47).

The combination of an affluent, rapidly growing Caucasian population with large numbers of infected dogs as pets, human exposure to infective larvae in a tropical climate with appropriate temperature and humidity, and advanced medical facilities led to the recognition of this association between the parasite infection and subsequent human disease. It also appears that climate directly influences the rate of human enteric infection by canine hookworms (22). Although there are other causes of EE, this disease entity may become more commonly recognized in other areas of the world, thus confirming the causative agent as the common dog hookworm (74).

Mebendazole, pyrantel pamoate, and albendazole have been used for the treatment of EE. However, unless there is an awareness of this relatively new syndrome, the causative agent may not be considered in the differential diagnosis.

Dracunculus medinensis

Some people speculate that the fiery serpent of biblical times was, in fact, *Dracunculus medinensis* (48). The clinical syndrome was well known in ancient Egypt and during the Greek and Roman periods (10). The contemporary term, guinea worm disease, derives its name from a European explorer who named the disease for the geographic area in which it was found, along the western African coast. The staff of Aesculapius, Roman god of medicine, may have originated from the ancient, still used procedure of removing the adult worm by slowly winding it around a stick (Figure 11.14). Although the worms are very long and thin, they are not true filarial worms but, rather, are grouped in their own order.

Life Cycle and Morphology

Human infection is acquired from ingestion of infected copepods (*Cyclops* water fleas) (Figure 11.15). The released larvae penetrate the duodenal mucosa and develop in the loose connective tissue. The possibility also exists that paratenic hosts, such as tadpoles and frogs, are important means of transporting infective larvae of *Dracunculus* species up the food chain, thus facilitating transmission to the definitive hosts.

The worms are very long, with the females measuring up to 1 m in length by 2 mm in width. The male is much smaller and inconspicuous (2 cm long). The worms mature in the deep connective tissue, and the females migrate to the subcutaneous tissues when they are gravid and contain coiled uteri filled with rhabditiform larvae. Maturation takes approximately 1 year. At this stage

Figure 11.14 *Dracunculus medinensis*, blister on leg (contains female worm); note removal of the adult worm. (Illustration by Sharon Belkin.)

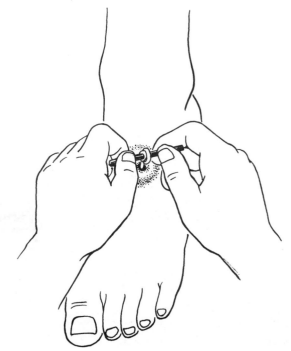

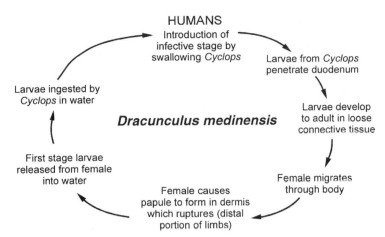

HUMANS
Introduction of
infective stage by
swallowing *Cyclops*

Larvae from *Cyclops*
penetrate duodenum

Larvae ingested by
Cyclops in water

Dracunculus medinensis

Larvae develop
to adult in loose
connective tissue

First stage larvae
released from female
into water

Female migrates
through body

Female causes
papule to form in dermis
which ruptures (distal
portion of limbs)

Figure 11.15 Life cycle of *Dracunculus medinensis*, the cause of guinea worm disease.

in the life cycle, the female migrates to the skin and a papule is formed in the dermis, usually by the ankles or feet (although papules can be anywhere on the body). The papule changes into a blister within 24 h to several days. Eventually, the blister ulcerates, and on contact with freshwater, a portion of the uterus prolapses through the worm's body wall, bursts open, and discharges thousands of larvae into the water (Figures 11.16 to 11.20). This may happen several times until all of the larvae are discharged. The larvae are then ingested by an appropriate species of *Cyclops*. Development takes about 8 days before the larvae are infective for humans.

Although the adult worms are often described as creamy white, there are reports of red worms that appear to be female *D. medinensis*. These infections occurred in an area of Pakistan where the incidence of guinea worm in 1988 was 15%. Unfortunately, examination of histologic sections was unable to determine the cause of

the red color; however, blood was excluded as a possible cause.

Clinical Disease

After ingestion of an infected copepod, no specific pathologic changes are associated with larval penetration into the deep connective tissues and maturation of the worms. Once the gravid female begins to migrate to the skin, there may be some erythema and tenderness in the area where the blister will form. Several hours before blister formation, the patient may exhibit some systemic reactions, including an urticarial rash, intense pruritus, nausea, vomiting, diarrhea, or asthmatic attacks. The lesion

Figure 11.17 *Dracunculus medinensis*, abscess caused by developing worm. (From A Pictorial Presentation of Parasites: A cooperative collection prepared and/or edited by H. Zaiman.)

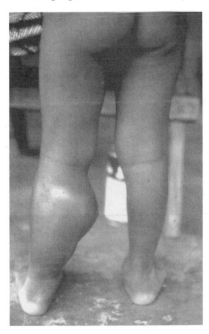

Figure 11.16 *Dracunculus medinensis*, adult worm in subcutaneous tissues below the breast. (From A Pictorial Presentation of Parasites: A cooperative collection prepared and/or edited by H. Zaiman. Photograph courtesy of J. Donges.)

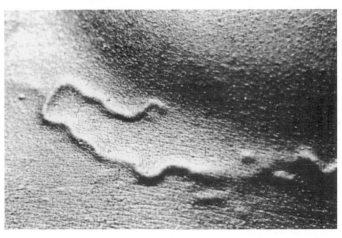

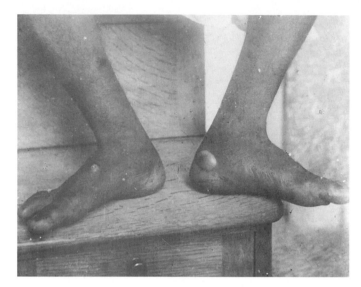

Figure 11.18 *Dracunculus medinensis* infection. Blisters develop in the skin, usually on the lower extremities, when the gravid female worm approaches the surface of the skin. (Armed Forces Institute of Pathology photograph.)

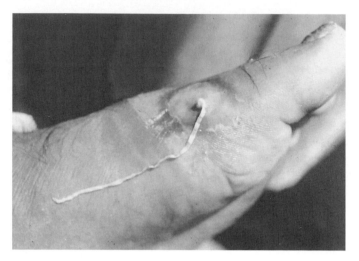

Figure 11.20 Extraction of *Dracunculus medinensis*. The stick has been removed, and larvae have been squeezed from the flattened exterior portion of the worm. (Armed Forces Institute of Pathology photograph; contributed by Everett L. Schiller, Johns Hopkins University School of Public Health, Baltimore, Md.)

develops as a reddish papule, measuring 2 to 7 cm in diameter. Symptoms usually subside when the lesion ruptures, discharging both the larvae and worm metabolites.

In a 1995 study in Nigeria, 1,200 people were surveyed for dracunculiasis. Many (982 [82%] of 1,200) were infected, and most infections involved the lower limbs (98%). Worms were also seen emerging from the umbilicus, groin, palm, wrist, and upper arm. Of the 982 infected individuals, 206 (21%) were totally incapacitated, 193 (20%) were seriously disabled, 431

Figure 11.19 Female *Dracunculus medinensis* worm being removed by winding on a stick. (Armed Forces Institute of Pathology photograph; contributed by Everett L. Schiller, Johns Hopkins University School of Public Health, Baltimore, Md.)

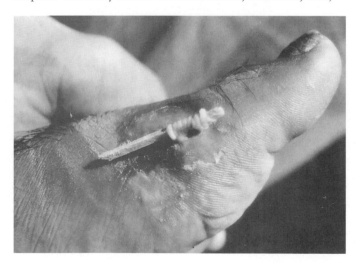

(44%) were mildly incapacitated, and 152 (16%) were unaffected (62).

If the worms are removed at this time, healing usually occurs with no problems. If the worm is damaged or broken during removal, there may be an intense inflammatory reaction with possible cellulitis along the worm's migratory track. If secondary infection occurs, there may be serious sequelae, including arthritis, synovitis, and other symptoms, depending on the site of the lesion.

Diagnosis

Diagnosis can be confirmed at the time the cutaneous lesion forms, with subsequent appearance of the adult worm. Infected lesions must be distinguished from carbuncles, deep cellulitis, focal myositis or periostitis, and even rheumatism. Calcified worms may also be found in subcutaneous tissues by radiography. They may appear as linear densities (up to 25 cm), tightly coiled structures, or sometimes nodules.

Treatment

For centuries, the worms have been removed by slowly being wound around a stick (Figures 11.14 and 11.19). This approach works well unless the worm is accidentally broken and secondary infection occurs. Allergic manifestations can be decreased by using epinephrine (10).

Four drugs have been used with various degrees of success: niridazole, thiabendazole, metronidazole, and mebendazole. The action seems to involve suppression of inflammation rather than any specific effect on the adult worms, although 400 to 500 mg/day for 6 days has been reported to kill the worms directly (1).

The prognosis is usually quite good unless there are complications. Specific drug dosage information is provided in chapter 25.

Epidemiology and Prevention

Disease transmission depends on several factors: (i) water sources where *Cyclops* spp. breed, (ii) direct contact between infected humans and the water source, and (iii) use of this water source for drinking. In various parts of the world, certain types of water sources (e.g., step wells in India, covered cisterns in Iran, and ponds in Ghana) provide all of these transmission requirements. The disease can be eliminated within 1 to 2 years by provision of safe drinking water.

In 1986, approximately 3.5 million cases of dracunculiasis occurred in 20 countries, and 120 million people were at risk for the disease. The target date for eradication of guinea worm infection of 1995 had been set by the African Regional Office of the World Health Organization and accepted by the United Nations Children's Fund and the United Nations Development Program. Although much progress has been achieved, some of the remaining obstacles to eradication include civil wars in Africa and apathy of some officials (15). By the end of 2004, Asia was free from dracunculiasis, while all remaining countries in Africa with endemic infection reported a 50% reduction in cases from 2003 and 2004. During 2005, Ghana and Sudan have reported 95% of the world's cases; however, the continued lack of security in some areas has delayed case reporting and disease interventions. With the settlement of Sudan's civil war and a new challenge grant from the Bill & Melinda Gates Foundation, completion of eradication efforts by 2009 in Sudan and earlier elsewhere may become reality. It is hoped that this infection will eventually join smallpox as one that can actually be eradicated from the world (17, 35).

Angiostrongylus (*Parastrongylus*) *cantonensis* (Cerebral Angiostrongyliasis)

Human infection with the rat lungworm *Angiostrongylus cantonensis* was first detected in 1945 in a 15-year-old Taiwanese boy with suspected meningitis. It has been recognized in the Pacific areas for many years, with Thailand, Tahiti, and Taiwan being areas of high endemicity. Sporadic cases have also been reported in other parts of the world, including Australia, Fiji, Sri Lanka, Egypt, Madagascar, Central America, and Cuba. The first case in the United States was reported in 1995 (60). Apparently, many gastropods in New Orleans are competent hosts for *A. cantonensis*. Also, the presence of infected rats and primates (at the Audubon Zoo) indicates that there is a reservoir of infection in New Orleans. The infection is associated with eosinophilic meningitis and sometimes eye involvement (3). Currently there are four species within the genus: *A. cantonensis*, *A. costaricensis*, *A. malaysiensis*, and *A. mackerrase* (these four are now placed in the new genus *Parastrongylus*). Since most parasitologists may continue to use the genus name *Angiostrongylus*, this generic designation is used throughout this chapter.

Life Cycle and Morphology

Mature adult worms of *A. cantonensis* inhabit the pulmonary arteries of a wide variety of rodents, primarily those within the genera *Rattus* and *Bandicota*. Eggs laid by the female lodge in the pulmonary arteries. On hatching, the first-stage larvae enter the alveolar space, migrate up the trachea and down the alimentary tract, and are excreted in the feces. Terrestrial snails, slugs, and aquatic snails serve as intermediate hosts, either by first-stage larval penetration of tissues or by the ingestion of contaminated rodent feces. Larval development continues in the mollusk, where third-stage, infective larvae develop within about 2 weeks. When ingested by rodents, the infective third-stage larvae migrate to the brain via the circulation and develop into fourth-stage larvae and then young adults within 4 weeks. They then go to the subarachnoid space, enter the venous system, and arrive in the pulmonary arteries, where sexual maturity occurs within another 2 weeks.

Human infection begins with the accidental ingestion of infective larvae in several species of slugs, snails, or land planarians (Figure 11.21). Ingestion of infected raw paratenic hosts, including fish, amphibians, reptiles, crustaceans, and vegetables contaminated with larvae, also leads to infection of the human host. Survival of the fifth-stage larvae is not certain; some probably die in the brain and spinal cord while some reach the eye chamber; very few probably reach the lungs. In the natural life cycle, infective larvae have also been found in land crabs, coconut crabs, and freshwater prawns, which are often consumed raw in the Pacific Islands. In summary, human infection can originate through the following: by ingestion of L3 larvae in raw or undercooked intermediate or paratenic hosts, by drinking infected water, by oral contact with hands contaminated with mollusk larvae, or possibly through the skin.

The worms are very thin and delicate, measuring 17 to 25 mm long by 0.26 to 0.36 mm wide. The young adults within the brain tissue are approximately 2 mm long.

Clinical Disease

The incubation period is normally around 20 days but may be twice that. The main symptom in all reported

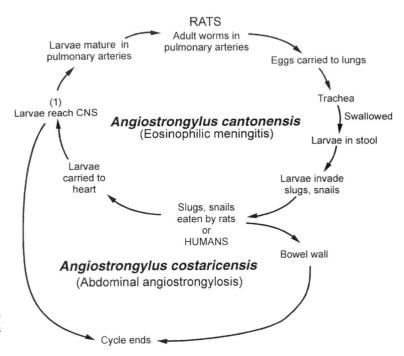

Figure 11.21 Life cycle of *Angiostrongylus* spp. [see (1) in figure]. Humans present with encephalitis when worms die in the brain tissue.

cases was severe headache. Other symptoms include convulsions, weakness of the limbs, paresthesia, vomiting, constipation, nausea, anorexia, facial paralysis, neck stiffness, and fever. Pulmonary symptoms are usually absent, although immature adult worms have been seen in lung tissue sections. Recovery of larvae in sputum or stool has not been reported. The spinal fluid usually contains white blood cells (100 to 2,000/mm³ with many eosinophils). There is often a peripheral eosinophilia with moderate leukocytosis (Figure 11.22). In most cases, the disease is self-limiting and the patient recovers within a month.

Eye involvement is characterized by visual impairment, pain, possible retinal hemorrhage, and retinal detachment. Living worms have been removed in some cases (Figure 11.23). Ocular angiostrongyliasis has been reported from Thailand, Taiwan, Vietnam, Indonesia, Japan, Papua New Guinea, and Sri Lanka. Worms have been found in the anterior chamber, retina, and other sites, causing multiple symptoms and rare cases of blindness.

The patient from New Orleans was an 11-year-old boy who presented with myalgia, headache, low-grade fever, and vomiting. He had no travel history but admitted

Figure 11.22 *Angiostrongylus cantonensis* in the central canal of the spinal cord. (From A Pictorial Presentation of Parasites: A cooperative collection prepared and/or edited by H. Zaiman.)

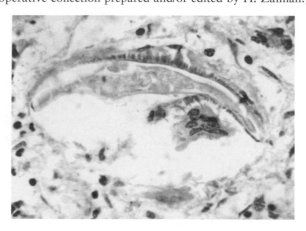

Figure 11.23 *Angiostrongylus cantonensis* adult male worm in the eye. (From A Pictorial Presentation of Parasites: A cooperative collection prepared and/or edited by H. Zaiman.)

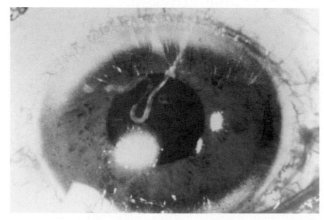

that, on a dare, he had eaten a raw snail from the street a few weeks earlier. While serologic tests for VLM and *Toxoplasma gondii* were negative, the results for *A. cantonensis* were positive (60).

Most patients, including the one from New Orleans, recover uneventfully and do not require hospitalization. Symptoms gradually disappear, with the meningeal problems resolving first, followed by the visual abnormalities and finally by the paresthesia.

Diagnosis

A presumptive diagnosis can be made in areas where infections are endemic on the basis of symptoms of severe headache, meningitis, or meningoencephalitis, with fever and ocular involvement. A peripheral eosinophilia and eosinophils in the CSF would also be highly suggestive. Lesions can also be seen in the brain by computed tomography. Larvae or young adult worms can often be recovered in the CSF, and the serologic ELISA can also provide confirmation. Both adult and young *A. cantonensis* worm antigens purified by immunoaffinity chromatography have been used to detect antibody in serum and CSF by ELISA. Infected patients had increased levels of IgG, IgA, IgM, and IgE, with higher IgM and IgE levels in serum than in CSF. Both worm antigens were highly sensitive in ELISA for serum antibodies but less so in tests for antibodies in CSF (91).

Treatment

If the worm is found in the eye, surgical removal of the worm is normally recommended. Anthelmintic agents are usually not used, although both mebendazole and thiabendazole have been tried. Mebendazole is currently the drug of choice (1). These drug trials have been inconclusive, and symptomatic therapy is normally recommended. However, since pathogenesis is frequently ascribed to dead or dying worms, anthelmintics should be used with caution. Corticosteroids may also eliminate some symptoms.

Epidemiology and Prevention

The lack of host specificity, the natural mobility of rats, and the expansion of the geographic range of the large African land snail have all contributed to the spread of this infection throughout the tropical and subtropical areas of the world. It is often difficult to identify the specific source of human infections; however, awareness of the various possible hosts may decrease the number of infections. Angiostrongyliasis is an emerging public health problem in mainland China (19). No overall control measures have been recommended. However, control of the spread of infected rats and mollusks to areas where *A. cantonensis* is not endemic will help restrict the geographic range.

Angiostrongylus (Parastrongylus) costaricensis (Abdominal Angiostrongyliasis)

At the time *Angiostrongylus costaricensis* was described in 1971, the natural host was unknown. The cotton rat, the black rat, and a number of other rodents harbor the adult worms, while various slugs harbor the larvae. Human infections are most common in Costa Rica (where there are about 600 per year) but have been reported in other areas of Mexico and Central and South America.

Life Cycle and Morphology

The life cycle is similar to that of *A. cantonensis*, with human infection being initiated by accidental ingestion of the appropriate slug, frequently on contaminated salad vegetables (Figure 11.21). This infection is called abdominal angiostrongyliasis, and the worms cause inflammatory lesions of the bowel wall.

The eggs are oval and about 90 μm long, have a thin shell, and are unembryonated (they may be embryonated in humans, with some larvae being released from the egg). The adult worms measure 42 by 350 mm (females) and 22 by 140 mm (males).

Clinical Disease

Abdominal angiostrongyliasis is found mainly in children under 13 years, and some groups have reported that two-thirds of these are male. The appendix is often involved; however, the worms can also be found in the terminal ileus, cecum, ascending colon, regional lymph nodes, and mesenteric arteries. There may be inflammation, thrombosis, and regional necrosis, with granulomas and areas of eosinophilic infiltrates around eggs and larvae in various stages of development. The most common symptoms are pain and tenderness, with a palpable mass in the lower right quadrant, along with fever and possibly vomiting and diarrhea. Occasionally, the worms are present in the liver; the symptoms may mimic those of VLM. Leukocytosis is present, with eosinophilia of up to 80%. Clinical symptoms occur about 2 weeks after infection and include abdominal pain in the right iliac fossa and right flank, fever, anorexia, vomiting, diarrhea, and constipation.

Diagnosis

Eggs or larvae may be seen in tissue sections, with most specimens being diagnosed on microscopic findings. Without histologic sections, the diagnosis is made on clinical grounds. Radiology may reveal abnormalities in the terminal ileum, cecum, and ascending colon. Contrast medium studies show spasticity, filling defects, and irritability at the cecum and ascending colon. When the liver is involved, there may be leukocytosis and eosinophilia, as well as elevated liver enzyme levels.

Treatment

Several drugs have been tried; the drug of choice is thiabendazole, with another option being mebendazole (1).

Epidemiology and Prevention

The ingestion of raw slugs in the areas of endemicity is considered strictly accidental. Most of the infections reported from Costa Rica have been in children, with a higher incidence in boys than in girls. Prevention involves rodent control to break the normal parasite cycle, as well as thorough washing of vegetables and other foods prior to consumption.

Recently, *A. costaricensis* has been reported in several primates at the Miami MetroZoo (siamang) and the Monkey Jungle in Miami (Ma's night monkeys). Also *A. costaricensis* was found in an opossum trapped in the MetroZoo, as well as in four raccoons near the MetroZoo. These are the first records from all four species of hosts. The primates were zoo born and the raccoons and opossum were native, thus indicating that this parasite is now endemic at these two sites (51a).

Gnathostoma spinigerum

Although many *Gnathostoma* species have been mentioned in the literature, *G. spinigerum* is considered to be the most medically important. This parasite is normally found in dogs and cats; the largest number

of infections, both in humans and in reservoir hosts, is found in Thailand (10). Areas of endemicity include China, the Philippines, and other areas in the Far East. However, the first record of a confirmed case of *G. doloresi* infection has been reported from Japan; the parasite was dissected from the skin and was identified as a third-stage larva. Although the entire life cycle is not fully understood, the patient reported eating raw brook trout about 2 months before the onset of the creeping eruption (59). Another case report from Japan describes colonic ileus due to nodular lesions caused by *G. doloresi* (80).

Life Cycle and Morphology

Within the definitive host (cats, dogs, and some wild carnivores), the adult worm lies coiled in a tumor-like mass in the stomach wall. Eggs are extruded from the stomach lesions and are passed out with the feces. These eggs hatch 10 to 12 days after reaching water, releasing first-stage larvae. The larvae are then ingested by copepods, where they develop into second-stage larvae within about 2 weeks. When an infected copepod is then ingested by any of the many intermediate hosts (fish, amphibians, reptiles, birds, and mammals), the third-stage larvae encyst. Once the intermediate host is ingested by the definitive host, the parasites become localized in the stomach wall, where they mature in 2 to 12 months (14).

Human infections are acquired by the ingestion of raw or poorly cooked or pickled freshwater fish, chicken and other birds, frogs, or snakes (Figure 11.24). There is also speculation that human infection can occur from the ingestion of copepods containing the advanced third-stage

Figure 11.24 Life cycle of *Gnathostoma spinigerum*.

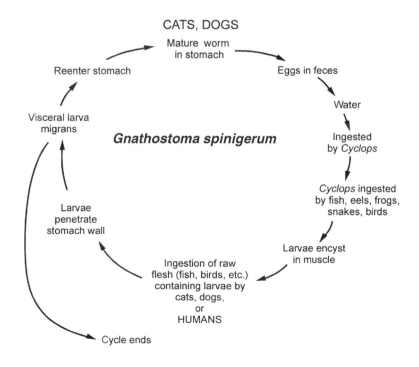

larvae or by actual skin penetration of larvae from handling infected meat. Three cases of parasitologically proven gnathostomiasis in neonates have been reported; these infections were presumably transmitted prenatally or perinatally from mother to infant (14).

The adult worms, which are found in the stomach lesions, are 25 to 54 mm (females) and 11 to 25 mm (males) long. The anterior half of the worm is covered with leaflike spines (Figures 11.25 and 11.26).

Clinical Disease

In most cases, the incubation period is difficult to determine. The disease is usually seen in two very different clinical forms, larval gnathostomiasis and eosinophilic myeloencephalitis.

Larval Gnathostomiasis. Cutaneous and visceral forms are seen, with the visceral larval migration including pulmonary, gastrointestinal, urogenital, ocular, otorhinolaryngeal, and cerebral tissues. In the human host, the larvae do not mature into adults in the wall of the stomach but, rather, migrate throughout the body. Several days after ingestion of the larvae, penetration of the intestinal wall may lead to

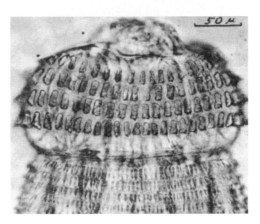

Figure 11.26 Head of adult gnathostoma. (Armed Forces Institute of Pathology photograph.)

epigastric pain, vomiting, and anorexia. These symptoms subside as the larvae begin to migrate through the tissues in deep cutaneous or subcutaneous tunnels. Evidence of this migration appears either as lesions similar to those found in patients with CLM or, more commonly, as migratory swellings with inflammation, redness, and pain. This swelling is hard and nonpitting and may last for several weeks. After it disappears, it may reappear in a location not far from the first swelling, which can be on the upper extremity, shoulder, neck, thorax, face, scalp, abdominal wall, thigh, or foot. Sometimes the lesions are painless, and sometimes there is pain and pruritus. These swellings probably result from the allergic response of the host to the presence of the worms, and an eosinophilia of 35 to 80% is reported in patients with cutaneous involvement.

If worm migration occurs relatively close to the surface of the body, creeping eruption, cutaneous abscesses, or nodules may be seen. Visceral migration can lead to serious sequelae, depending on the organ(s) affected (Table 11.3).

Eosinophilic Gnathostomiasis. Neurologic manifestations are seen as a complex group of symptoms called eosinophilic myeloencephalitis. More serious symptoms can occur if the eyes or CNS is involved. The larval migration occurs along a peripheral nerve, into the spinal cord, and then into the brain. Symptoms include pain, paralysis, seizures, coma, and death. Sudden severe headache and sensory impairment followed by coma can occur and mimics a cerebrovascular accident. In rare instances, cutaneous migratory swellings and neurologic symptoms occur in the same patient. In areas of endemic infection, cerebral hemorrhage in a younger individual should indicate possible infection with *Gnathostoma* spp. The spinal fluid may be xanthochromic or bloody. Ocular invasion probably occurs via the optic nerve, although penetration through the sclera may be possible.

Figure 11.25 *Gnathostoma spinigerum*, adult worm. [Illustration by Sharon Belkin; adapted from reference 10; originally adapted from I. Miyazaki, Studies on *Gnathostoma* occurring in Japan (Nematoda: Gnathostomidae). II. Life history of *Gnathostoma* and morphological comparison of its larval forms, *Kyushu Mem. Med. Sci.* 5:123–140, 1954.]

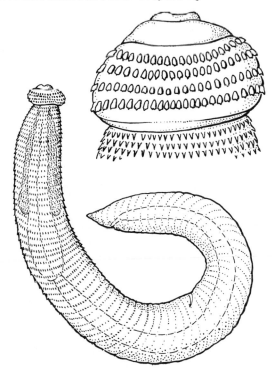

Diagnosis

Patient symptoms may also suggest sparganosis, cutaneous paragonimiasis, CLM, and myiasis. Spinal fluid that is bloody or xanthochromic and contains many eosinophils may be suggestive of infection with *A. cantonensis*. A definitive diagnosis depends on recovery and identification of the worms. Newer serologic methods may also be helpful.

Treatment

The only effective therapy is surgical incision of the lesion and removal of the worm. Albendazole has also been recommended as an adjunct to surgical removal (1). Worm removal from the eye may also prevent CNS invasion. The prognosis is usually good unless the CNS is involved. Additional therapeutic information is presented in chapter 25.

Epidemiology and Prevention

Most infections are probably caused by the ingestion of raw or poorly cooked fish, domestic ducks, and chickens. The larvae do not survive cooking, and they can be killed by immersion in strong vinegar for at least 5 h. Unfortunately, soaking in lime juice or storage at 4°C for 1 month does not kill the larvae. Untreated groundwater is also a potential problem and source of transmission because it can contain infected copepods. Once the life cycle is established in an area, potential infection can become a serious public health issue for residents. Increases in world travel and the importation of food require heightened awareness of this infection, particularly in areas like Europe and North America, where the infection is not endemic.

Gnathostoma doloresi, *G. nipponicum*, *G. hispidum*, and *G. binucleatum*

Other *Gnathostoma* species causing human infection include *G. doloresi*, *G. nipponicum*, *G. hispidum*, and *G. binucleatum*. The general life cycle is identical to that of *G. spinigerum*, with slight variations related to the second, paratenic, and definitive hosts. Patient symptoms are also similar to those seen with *G. spinigerum* infections. In Japan, freshwater fish are the most likely source of *G. doloresi*, *G. hispidum*, and *G. nipponicum* (2). In Myanmar, an outbreak among Korean emigrants was reported; 38 of 60 individuals became symptomatic after ingesting raw freshwater catfish, freshwater bream, and snake-headed fish in a local Korean restaurant (18). Although positive serologic test results were based on the use of *G. doloresi* antigen, *G. spinigerum* was suspected as the causative agent and is known to be present in the Yangon area of Myanmar.

An increasing number of human cases have also been reported in Sinaloa, Mexico, most of which occurred in persons who had eaten raw fish dishes such as "cebiche." A report confirming five cases from Mexico described the first known outbreak of acute gnathostomiasis on the American continent (28). These patients were seropositive to *G. doloresi* antigen. Five species of fish and four species of ichthyophagous birds collected from three lakes and a nearby estuary were infected with third-stage larvae of *G. binucleatum*, a species found in Ecuador and Mexico.

Anisakis simplex, *A. physeteris*, *Pseudoterranova decipiens*, *Contracaecum osculatum*, *Hysterothylacium aduncum*, and *Porrocaecum reticulatum* (Larval Nematodes Acquired from Saltwater Fish)

Anisakiasis was first recognized and reported in the Netherlands. Since this infection was reported in Japan in 1965, hundreds of Japanese cases have been documented, as have several in the United States (10). Up to 1990, more than 12,000 cases were reported from Japan and only 519 cases were reported in 19 countries outside of Japan. Less than 100 cases have been reported from the United States; however, this infection is probably misdiagnosed and underreported. With the tremendous increase in the popularity of sushi and sashimi, it is likely that the number of case reports will increase over the next few years. It is well recognized that human infection can occur from the ingestion of raw or poorly cooked marine fish or squid. Anisakid larvae known to cause human infections include *Anisakis simplex*, *A. physeteris*, *Pseudoterranova decipiens*, *Contracaecum osculatum*, *Hysterothylacium aduncum*, and *Porrocaecum reticulatum*. Within this group, *A. simplex* and *P. decipiens* are considered the most important human parasites.

Life Cycle and Morphology

The primary hosts of *Anisakis* are dolphin, porpoise, and whale; those of *Pseudoterranova* are seal, fur seal, walrus, and sea lion. These sea mammals ingest third-stage larvae, which penetrate the gastric mucosa and develop into adult male and female worms. The worms live in clusters, with their anterior ends embedded in the gastric wall. Eggs are then passed out into the sea, where the second-stage larvae are ingested by small marine crustacea (krill) and develop into third-stage larvae. These are then transmitted from krill to fish or from fish to fish, etc., via the normal food chain. Generally, these larvae reside in the viscera of the fish; however, some fish contain larvae within the muscle. More than 150 species of fish can serve as intermediate hosts. Herring, salmon, mackerel, cod, and squid tend to transmit *Anisakis* infection, while cod, halibut, flatfish, greenling, and red snapper can transmit *Pseudoterranova*.

Human infection is acquired by the ingestion of raw, pickled, salted, or smoked saltwater fish or squid (Figure 11.27). The larvae often penetrate into the walls of the digestive tract (frequently the stomach), where they become

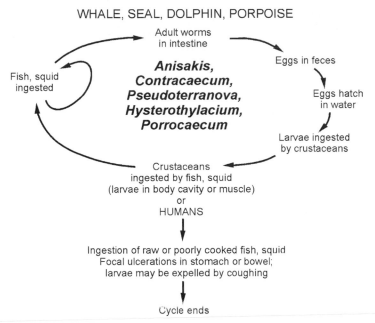

WHALE, SEAL, DOLPHIN, PORPOISE

Adult worms
in intestine

*Anisakis,
Contracaecum,
Pseudoterranova,
Hysterothylacium,
Porrocaecum*

Fish, squid
ingested

Eggs in feces

Eggs hatch
in water

Larvae ingested
by crustaceans

Crustaceans
ingested by fish, squid
(larvae in body cavity or muscle)
or
HUMANS

Ingestion of raw or poorly cooked fish, squid
Focal ulcerations in stomach or bowel;
larvae may be expelled by coughing

Cycle ends

Figure 11.27 Life cycle of *Anisakis, Contracaecum, Pseudoterranova, Hysterothylacium,* and *Porrocaecum* spp. (larval nematodes acquired from saltwater fish).

embedded in eosinophilic granulomas (Figure 11.28). Occasionally, the throat is involved. These large larvae (third stage) measure 1 to 3 cm long or more by 1 mm wide. Histologic sections are characterized by the large body size, moderately thick cuticle, and large lateral cords that extend into the body cavity.

Clinical Disease

The clinical manifestations of anisakiasis are varied, depending on the site of penetration of the larvae. There may be acute gastric presentation, which is the most commonly recognized clinical syndrome. Intestinal anisakiasis is also seen, sometimes with acute symptoms and sometimes with a mild, chronic presentation; intestinal disease develops within 2 days after infection and occurs most often in the ilial region. Occasionally ectopic disease occurs, where the larvae are found outside of their usual location, usually elsewhere in the gastrointestinal tract. Within North America, one of the most common presentations has been described as the "tingling-throat syndrome" and is often associated with infection by *Pseudoterranova* larvae. In these cases, the patient may even feel the worm in the oropharynx or proximal esophagus. The patient often coughs up the worm, which is then submitted to the laboratory for identification. Larvae in other true ectopic sites are rare, with the abdominal cavity being the most common; mild symptoms are usually the case, and the larvae may be found at surgery for totally unrelated causes. Apparently *P. decipiens* tends to be less invasive than *A. simplex* and is usually expelled by vomiting. *A. simplex* larvae tend to penetrate the gastrointestinal wall, invading the abdominal cavity.

It is well known that *A. simplex* can cause allergic reactions in sensitized patients. At present, a nonseafood diet is recommended for any patients with any kind of *A. simplex* allergy. However, it appears that patients can tolerate the ingestion of seafood when the parasites are dead and noninfective. It has also been suggested that immunologic methods to detect specific antibodies against *Anisakis* should be used routinely before eosinophilic gastroenteritis is diagnosed as the primary disorder. Data indicate that a Th2 mechanism plays an important role in the inflammatory infiltrate produced by the attachment of parasites to the gastrointestinal wall (26).

Figure 11.28 Scanning electron micrograph of an *Anisakis simplex* larva penetrating human stomach tissue in vitro. Note the tunnel created in the mucosa and the burrow in the submucosa. (Reprinted with permission from J. A. Sakanari and J. H. McKerrow, Identification of the secreted neutral proteases from *Anisakis simplex*, J. Parasitol. 76:625–630, 1990.)

There may be nausea or vomiting, often within 24 h after ingestion of raw marine fish (Figures 11.28 and 11.29). Depending on the location of the larvae, infections can mimic gastric or duodenal ulcer, carcinoma, appendicitis, or other conditions requiring surgery. There is usually a low-grade eosinophilia (10% or less) and a positive result for occult blood in the stool.

There are also two reported cases of pulmonary anisakiasis. One of the cases involved a 22-year-old man in Japan, who developed high fever, respiratory distress, and pleural effusion after consumption of raw fish. A parasitic infection was suspected, and various immunoserologic tests were performed. Since extragastrointestinal anisakiasis was strongly suspected, this diagnosis was confirmed by a microplate ELISA and Western blot analysis with a monoclonal antibody (50).

Worms have been recovered or seen after surgery for intestinal obstruction, in eosinophilic granulomas, from a portion of resected small intestine, during gastroscopic examination, in vomitus, and in histologic sections.

Figure 11.29 (A) *Anisakis* in fish flesh. (B) Larval nematodes in fish viscera. (From A Pictorial Presentation of Parasites: A cooperative collection prepared and/or edited by H. Zaiman. Photograph courtesy of L. A. Jensen.)

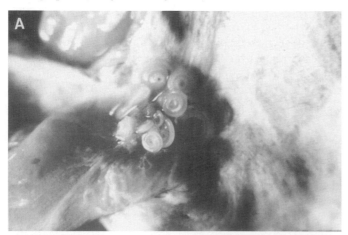

Diagnosis

A presumptive diagnosis can be made on the basis of the patient's food history. Definitive identification is based on larval recovery or histologic examination of infected tissue. Although serologic reagents have been developed, they are not commercially available (46). Molecular biology-based methods may also provide some additional diagnostic tools. A rise in the levels of total and specific IgE in the first month after an allergic reaction, consistent with the patient's history of gastroallergic anisakiasis, can provide valuable information, particularly if the parasite cannot be seen by fiber-optic gastroscopy (23).

Treatment

There is no recommended therapy other than removal of the larvae, often through surgery. Gastric endoscopy is usually effective in larval location and removal. If the diagnosis is confirmed and there is no ileus, then surgical intervention can be avoided; the larvae will die and become absorbed within several weeks.

Epidemiology and Prevention

Although the distribution of anisakid larvae in infected marine fish is worldwide, within the United States salmon and Pacific rockfish (red snapper) are implicated most often in transmission. Although *Anisakis* normally has a marine life cycle, *A. simplex* and other anisakid parasites have been found in populations of river otter in the Pacific Northwest. This has been linked to the ingestion of shad during their spawning runs and outmigration. Thus, consumption of shad that are infected with anisakid worms may be confirmed as an emerging parasitic disease of veterinary and human medical concern (81).

Raw, pickled, salted, or smoked marine fish should be avoided. All fish intended for raw, partly cooked, or marinated consumption should be blast-frozen to –35°C (–31°F) or below for 15 h or be normally frozen to –23°C (–10°F) or below for 7 days. This disease could be totally prevented by thorough cooking of all marine fish. Also, sushi served at professional sushi bars and restaurants is rarely responsible for infections. Generally in these settings, fish other than salmon, cod, mackerel, herring, whiting, and haddock are used for sushi preparation.

It is also interesting that even under severe conditions, *P. decipiens* is a well-established parasite of the Antarctic fauna. This cosmopolitan species can complete its life cycle even at subzero temperatures (67).

During the last few years, the finding of allergic hypersensitivity symptoms in anisakiasis has emphasized that this widespread etiologic agent can induce acute symptoms as well as chronic urticaria. Additional reports will certainly confirm the importance of this infection as a potentially growing public health problem (9, 24, 25).

Capillaria hepatica

Capillaria hepatica infection is commonly found in rats, other rodents, and other mammals. Human cases have been reported from various parts of the world, including the United States. Infection occurs via accidental ingestion of eggs from the soil. These eggs hatch and are carried via the portal system to the liver, where the larvae mature in approximately 4 weeks and begin to deposit eggs in the liver parenchyma (Figure 11.30). Symptoms of this infection mimic those of hepatitis, amebic liver abscess, trichinosis, VLM, Loeffler's syndrome, Hodgkin's disease, and histoplasmosis. In the first case reported from Maine, the patient presented with a subacute history of severe abdominal pain, fevers, and weight loss. After open laparotomy for resection of the hepatic mass and treatment with thiabendazole, he recovered; the source of the infection was unknown but was probably accidental ingestion of soil contaminated with mature *Capillaria* eggs. (42)

In a true human infection, no eggs are found in the stool. Diagnosis requires histologic examination. Eggs in liver biopsy specimens can be identified on the basis of their characteristic morphology. The recent development of an IFAT may lend itself to testing of human sera for the detection of early *C. hepatica* infection (7).

Note. In cases of spurious infection, in which infected animal liver has been ingested, *C. hepatica* eggs may be passed in the stool. These eggs measure 51 to 68 μm long by 30 to 35 μm wide and resemble those of *C. philippinensis* (45 by 21 μm), which can be seen in the stool in true human infection (see chapter 10).

Figure 11.30 *Capillaria hepatica* eggs in liver. (From A Pictorial Presentation of Parasites: A cooperative collection prepared and/or edited by H. Zaiman.) For information on intestinal capillariasis with *Capillaria philippinensis*, see chapter 10.

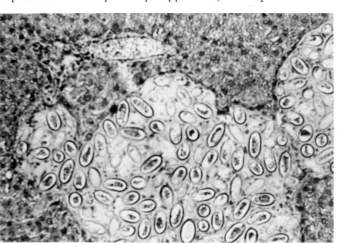

Thelazia spp.

Thelazia spp. have been recovered from the human conjunctiva and may damage the eye tissues. The worms are 1 to 1.5 cm long and are threadlike. They may be found in the conjunctival sac or lachrymal glands or migrating over the cornea. Symptoms are mild and include excessive lacrimation, itching, or pain (as with a foreign object in the eye). After use of a topical anesthetic, the worms can be safely removed with forceps (63).

Thelazia californiensis has been reported from California, and the larvae are transmitted by a fly belonging to the genus *Fannia*.

References

1. **Abramowicz, M. (ed.).** 2004. Drugs for parasitic infections. *Med. Lett. Drugs Ther.* **46**:1–12.
2. **Akahane, H., M. Sano, and M. Kobayashi.** 1998. Three cases of human gnathostomiasis caused by *Gnathostoma hispidum*, with particular reference to the identification of parasitic larvae. *Southeast Asian J. Trop. Med. Public Health* **29**:611–614.
3. **Alicata, J. E.** 1991. The discovery of *Angiostrongylus cantonensis* as a cause of human eosinophilic meningitis. *Parasitol. Today* **7**:151–153.
4. **Anderson, B. C.** 1999. Congenital *Baylisascaris* sp. larval migrans in a newborn lamb. *J. Parasitol.* **85**:128–129.
5. **Appleyard, G. D., D. Zarlenga, E. Pozio, and A. A. Gajadhar.** 1999. Differentiation of *Trichinella* antigen of larval *Trichinella pseudospiralis*. *Parasitol. Res.* **85**:685–691.
6. **Ashwath, M. L., D. R. Robinson, and H. P. Katner.** 2004. A presumptive case of toxocariasis associated with eosinophilic pleural effusion: case report and literature review. *Am. J. Trop. Med. Hyg.* **71**:764.
7. **Assis, B. C., L. M. Cunha, A. P. Baptista, and Z. A. Andrade.** 2004. A contribution to the diagnosis of *Capillaria hepatica* infection by indirect immunofluorescence test. *Mem. Inst. Oswaldo Cruz* **99**:173–177.
8. **Bachli, H., J. C. Minet, and O. Gratzl.** 2004. Cerebral toxocariasis: a possible cause of epileptic seizure in children. *Childs Nerv. Syst.* **20**:468–472.
9. **Baeza, M. L., A. Rodriguez, V. Matheu, M. Rubio, P. Tornero, M. de Barrio, T. Herrero, M. Santoalalla, and J. M. Zubeldia.** 2004. Characterization of allergens secreted by *Anisakis simplex* parasite: clinical relevance in comparison with somatic allergens. *Clin. Exp. Allergy* **34**:296–302.
10. **Beaver, P. C., R. C. Jung, and E. W. Cupp.** 1984. *Clinical Parasitology*, 9th ed. Lea & Febiger, Philadelphia, Pa.
11. **Boulos, L. M., I. R. Ibrahim, A. Y. Negm, and S. M. Aly.** 2001. Detection of coproantigen in early trichinellosis. *Parasite* **8**:S136–S139.
12. **Bowman, D. D.** 1987. Diagnostic morphology of four larval ascaridoid nematodes that may cause visceral larva migrans: *Toxascaris leonina*, *Baylisascaris procyonis*, *Lagochilascaris sprenti*, and *Hexametra leidyi*. *J. Parasitol.* **73**:1198–1215.
13. **Brenner, M. A., and M. B. Patel.** 2003. Cutaneous larva migrans: the creeping eruption. *Cutis* **72**:111–115.

14. **Bunnag, T.** 2000. Gnathostomiasis, p. 790–792. *In* G. T. Strickland (ed.), *Hunter's Tropical Medicine and Emerging Infectious Diseases*, 8th ed. The W. B. Saunders Co., Philadelphia, Pa.

15. **Cairncross, S., R. Muller, and N. Zagaria.** 2002. Dracunculiasis (Guinea worm disease) and the eradication initiative. *Clin. Microbiol. Rev.* **15:**223–246.

16. **Centers for Disease Control and Prevention.** 2003. Trichinellosis associated with bear meat—New York and Tennessee, 2003. *Morb. Mortal. Wkly. Rep.* **53:** 606–610.

17. **Centers for Disease Control and Prevention.** 2005. Progress toward global eradication of dracunculiasis, January 2004–July 2005. *Morb. Mortal. Wkly. Rep.* **54:** 1075–1077.

18. **Chai, J. Y., E. T. Han, E. H. Shin, J. H. Park. J. P. Chu, M. Hirota, F. Nakamura-Uchiyama, and Y. Nawa.** 2003. An outbreak of gnathostomiasis among Korean emigrants in Myanmar. *Am. J. Trop. Med. Hyg.* **69:**67–73.

19. **Chen, X. G., H. Li, and Z. R. Lun.** 2005. Angiostrongyliasis, mainland China. *Emerg. Infect. Dis.* **11:**1645–1647.

20. **Chorazy, M. L., and D. J. Richardson.** 2005. A survey of environmental contamination with ascarid ova, Wallingford, Connecticut. *Vector Borne Zoonotic Dis.* **5:**33–39.

21. **Croese, J.** 1988. Eosinophilic enteritis—a recent North Queensland experience. *Aust. N. Z. J. Med.* **18:**848–853.

22. **Croese, J.** 1995. Seasonal influence on human enteric infection by *Ancylostoma caninum*. *Am. J. Trop. Med. Hyg.* **53:**158–161.

23. **Daschner, A., A. Alonso-Gomez, T. Caballero, J. M. Suarez-De-Parga, and M. C. Lopez-Serrano.** 1999. Usefulness of early serial measurement of specific and total immunoglobulin E in the diagnosis of gastroallergic anisakiasis. *Clin. Exp. Allergy* **29:**1260–1264.

24. **Daschner, A., and C. Y. Pascual.** 2005. *Anisakis simplex*: sensitization and clinical allergy. *Curr. Opin. Allergy Clin. Immunol.* **5:**281–285.

25. **Daschner, A., F. Vega de la Osada, and C. Y. Pascual.** 2005. Allergy and parasites reevaluated: wide-scale induction of chronic urticaria by the ubiquitous fish-nematode *Anisakis simplex* in an endemic region. *Allergol. Immunopathol.* **33:**31–37.

26. **del Pozo, V., I. Arrieta, T. Tunon, I. Cortegano, B. Gomez, B. Cardaba, S. Gallardo, M. Rogo, G. Renedo, P. Palomino, A. I. Tabar, and C. Lahoz.** 1999. Immunopathogenesis of human gastrointestinal infection by *Anisakis simplex*. *J. Allergy Clin. Immunol.* **104:**637–643.

27. **Despommier, D.** 2003. Toxocariasis: clinical aspects, epidemiology, medical ecology, and molecular aspects. *Clin. Microbiol. Rev.* **16:**265–272.

28. **Diaz-Camacho, S. P., K. Willms, M. C. de la Cruz-Otero, M. L. Zazueta-Ramos, S. Bayliss-Gaxiola, R. Castro-Valazquez, I. Osuna-Ramirez, A. Bojorquez-Contreras, E. H. Torres-Montoya, and S. Sanchez-Gonzales.** 2003. Acute outbreak of gnathostomiasis in a fishing community in Sinaloa, Mexico. *Parasitol. Int.* **52:**133–140.

29. **Eberhard, M. L., E. K. Nace, K. Y. Won, G. A. Punkosdy, H. S. Bishop, and S. P. Johnston.** 2003. *Baylisascaris procyonis* in the metropolitan Atlanta area. *Emerg. Infect. Dis.* **9:**1636–1637.

30. **Feldman, G. J., and H. W. Parker.** 1992. Visceral larva migrans associated with the hypereosinophilic syndrome and the onset of severe asthma. *Ann. Intern. Med.* **116:** 838–840.

31. **Gamble, H. R., E. Pozio, J. R. Lichtenfels, D. S. Zarlenga, and D. E. Hill.** 2005. *Trichinella pseudospiralis* from a wild pig in Texas. *Vet. Parasitol.* **132:**147–150.

32. **Gavin, P. J., K. R. Kazacos, and S. T. Shulman.** 2005. Baylisascariasis. *Clin. Microbiol. Rev.* **18:**703–718.

33. **Gomez-Garcia, V., J. Hernandez-Quero, and M. Rodriguez-Osorio.** 2003. Short report: human infection with *Trichinella britovi* in Granada, Spain. *Am. J. Trop. Med. Hyg.* **68:**463–464.

34. **Hamidou, M. A., B. Gueglio, E. Cassagneau, D. Trewick, and J. Y. Grolleau.** 1999. Henoch-Schonlein purpura associated with *Toxocara canis* infection. *J. Rheumatol.* **26:**443–445.

35. **Hopkins, D. R., E. Ruiz-Tiben, P. Downs, P. C. Withers, Jr., and J. H. Maguire.** 2005. Dracunculiasis eradication: the final inch. *Am. J. Trop. Med. Hyg.* **73:**669–675.

36. **Jacquier, P., B. Gottstein, Y. Stingelin, and J. Eckert.** 1991. Immunodiagnosis of toxocarosis in humans: evaluation of a new enzyme-linked immunosorbent assay kit. *J. Clin. Microbiol.* **29:**1831–1835.

37. **Jongwutiwes, S., N. Chantachum, P. Kraivichian, P. Sirlyasatien, C. Putaporntip, A. Tamburrini, G. Larosa, C. Sreesunpasirikul, P. Yingyourd, and E. Pozio.** 1998. First outbreak of human trichinellosis caused by *Trichinella pseudospiralis*. *Clin. Infect. Dis.* **26:**111–115.

38. **Kabaalioglu, A., K. Ceken, E. Alimoglu, R. Saba, and A. Apaydin.** 2005. Hepatic toxocariasis: US, CT and MRI findings. *Ultraschall. Med.* **26:**329–332.

39. **Kagialis-Girard, S., V. Mialou, M. Ffrench, S. Dupuis-Girod, M. P. Pages, and Y. Bertrand.** 2005. Thrombocytosis and toxocariasis: report of two pediatric cases. *Pediatr. Blood Cancer* **44:**190–192.

40. **Kapel, C. M.** 2005. Changes in the EU legislation on *Trichinella* inspection—new challenges in the epidemiology. *Vet. Parasitol.* **132:**189–194.

41. **Kapel, C. M. O., P. Webster, P. Lind, E. Posio, S. A. Henrik-sen, K. D. Murrell, and P. Nansen.** 1998. *Trichinella spiralis*, *T. britovi*, and *T. nativa*: infectivity, larval distribution in muscle, and antibody response after experimental infection in pigs. *Parasitol. Res.* **84:**264–271.

42. **Klenzak, J., A. Mattia, A. Valenti, and J. Goldberg.** 2005. Hepatic capillariasis in Maine presenting as a hepatic mass. *Am. J. Trop. Med. Hyg.* **72:**651–653.

43. **Kraus, A., X. Valencia, A. R. Cabral, and G. de la Vega.** 1995. Visceral larva migrans mimicking rheumatic diseases. *J. Rheumatol.* **22:**497–500.

44. **Kuratli, S., J. G. Lindh, B. Gottstein, D. F. Smith, and B. Connolly.** 1999. *Trichinella* spp.: differential expression of two genes in the muscle larva of encapsulating and non-encapsulating species. *Exp. Parasitol.* **93:**153–159.

45. **La Rosa, G., G. Marucci, and E. Pozio.** 2003. Biochemical analysis of encapsulated and non-encapsulated species of *Trichinella* (Nematoda, Trichinellidae) from cold- and warm-blooded animals reveals a high genetic divergence in the genus. *Parasitol. Res.* **91:**462–466.

46. **Lorenzo, S., R. Iglesias, M. T. Audicana, R. Garcia-Vellaescusa, F. Pardo, M. L. Sanmartin, and F. M. Ubeira.**

1999. Human immunoglobulin isotype profiles produced in response to antigens recognized by monoclonal antibodies specific to *Anisakis simplex*. *Clin. Exp. Allergy* **29:** 1095–1101.

47. **Loukas, A., J. Opderbeeck, J. Croese, and P. Prociv.** 1994. Immunologic incrimination of *Ancylostoma caninum* as a human enteric pathogen. *Am. J. Trop. Med. Hyg.* 50:69–77.

48. **Markell, E. K., M. Voge, and D. T. John.** 1992. *Medical Parasitology*, 7th ed. The W. B. Saunders Co., Philadelphia, Pa.

49. **Marva, E., A. Markovics, M. Gdalevich, N. Asor, C. Sadik, and A. Leventhal.** 2005. Trichinellosis outbreak. *Emerg. Infect. Dis.* 11:1979–1981.

50. **Matsuoka, H., T. Nakama, H. Kisanuki, H. Uno, N. Tachibana, H. Tsubouchi, Y. Horii, and Y. Nawa.** 1994. A case report of serologically diagnosed pulmonary anisakiasis with pleural effusion and multiple lesions. *Am. J. Trop. Med. Hyg.* 51:819–822.

51. **Mets, M. B., A. G. Noble, S. Basti, P. Gavin, A. T. Davis, S. T. Shulman, and K. R. Kozacos.** 2003. Eye findings of diffuse unilateral subacute neuroretinitis and multiple choroidal infiltrates associated with neural larva migrans due to *Baylisascaris procyonis*. *Am. J. Ophthalmol.* **135:** 888–890.

51a. **Miller, C. L., J. M. Kinsella, M. M. Garner, S. Evans, P. A. Gullett, and R. E. Schmidt,** 2006, Endemic infections of *Parastrongylus* (=*Angiostrongylus*) *costaricensis* in two species of nonhuman primates, raccoons, and an opossum from Miami, Florida. *J. Parasitol.* 92:406–408.

52. **Minvielle, M. C., G. Niedfeld, M. L. Ciarmela, A. De Falco, H. Ghiani, and J. A. Basualdo.** 1999. Asthma and covert toxocariasis. *Med. Buenos Aires* 59:243–248.

53. **Moorhead, A., P. E. Grunenwald, V. J. Dietz, and P. M. Schantz.** 1999. Trichinellosis in the United States, 1991–1996: declining but not gone. *Am. J. Trop. Med. Hyg.* 60:66–69.

54. **Moreira-Silva, S. F., M. G. Rodrigues, J. L. Pimenta, C. P. Gomes, L. H. Freire, and F. E. L. Pereira.** 2004. Toxocariasis of the central nervous system: with report of two cases. *Rev. Soc. Bras. Med. Trop.* 37:169–174.

55. **Murrell, K. D.** 2000. Trichinosis, p. 780–787. *In* G. T. Strickland (ed.), *Hunter's Tropical Medicine and Emerging Infectious Diseases*, 8th ed. The W. B. Saunders Co., Philadelphia, Pa.

56. **Murrell, K. D., M. Djordjevic, K. Cuperlovic, L. J. Sofronic, M. Savic, M. Djordjevic, and S. Damjanovic.** 2004. Epidemiology of *Trichinella* infection in the horse: the risk from animal product feeding practices. *Vet. Parasitol.* 123:223–233.

57. **Nagakura, K., S. Kanno, H. Tachibana, Y. Kaneda, M. Ohkido, K. Kondo, and H. Inoue.** 1990. Serologic differentiation between *Toxocara canis* and *Toxocara cati*. *J. Infect. Dis.* 162:1418–1419.

58. **Nagano, I., Z. Wu, A. Matsuo, E. Pozio, and Y. Takahashi.** 1999. Identification of *Trichinella* isolates by polymerase chain reaction-restriction fragment length polymorphism of the mitochondrial cytochrome c-oxidase subunit I gene. *Int. J. Parasitol.* 29:1113–1120.

59. **Nawa, Y., J. Imai, K. Ogata, and K. Otsuka.** 1989. The first record of a confirmed human case of *Gnathostoma doloresi* infection. *J. Parasitol.* 75:166–169.

60. **New, D., M. D. Little, and J. Cross.** 1995. *Angiostrongylus cantonensis* infection from eating raw snails. *N. Engl. J. Med.* 332:1105–1106.

61. **Obwaller, A., E. Jensen-Jarolim, H. Auer, A. Huber, D. Kraft, and H. Aspock.** 1998. *Toxocara* infestations in humans: symptomatic course of toxocariasis correlates significantly with levels of IgE/anti-IgE immune complexes. *Parasite Immunol.* 20:311–317.

62. **Okoye, S. N., C. O. E. Onwuliri, and J. C. Anosike.** 1995. A survey of predilection sites and the degree of disability associated with guineaworm (*Dracunculus medinensis*). *Int. J. Parasitol.* 25:1127–1129.

63. **Otranto, D., and D. Traversa.** 2005. *Thelazia* eyeworm: an original endo-and ecto-parasitic nematode. *Trends Parasitol.* 21:1–4.

64. **Owen, I. L., E. Pozio, A. Tamburrini, R. T. Danaya, F. Bruschi, and M. A. G. Morales.** 2001. Focus of human trichinellosis in Papua New Guinea. *Am. J. Trop. Med. Hyg.* 65:553–557.

65. **Ozeretskovskaya, N. N., L. G. Mikhailova, T. P. Sabgaida, and A. S. Dovgalev.** 2005. New trends and clinical patterns of human trichinellosis in Russia at the beginning of the XXI century. *Vet. Parasitol.* 132:167–171.

66. **Paco, J. M., D. M. B. Campos, and J. A. de Oliveira.** 1999. Wild rodents as experimental intermediate hosts of *Lagochilascaris minor* Leiper, 1909. *Mem. Inst. Oswaldo Cruz* 94:441–449.

67. **Palm, H. W.** 1999. Ecology of *Pseudoterranova decipiens* (Krabbe, 1878) (Nematoda: Anisakidae) from Antarctic waters. *Parasitol. Res.* 85:638–646.

68. **Pozio, E.** 1998. Trichinellosis in the European Union: epidemiology, ecology and economic impact. *Parasitol. Today* 14:35–40.

69. **Pozio, E.** 2005. The broad spectrum of *Trichinella* hosts: from cold- to warm-blooded animals. *Vet. Parasitol.* 132:3–11.

70. **Pozio, E., M. Goffredo, R. Fico, and G. La Rosa.** 1999. *Trichinella pseudospiralis* in sedentary night-birds of prey from Central Italy. *J. Parasitol.* 85:759–761.

71. **Pozio, E., G. Marucci, A. Casulli, L. Sacchi, S. Mukaratirwa, C. M. Foggin, and G. La Rosa.** 2004. *Trichinella papuae* and *Trichinella zimbabwensis* induce infection in experimentally infected varans, caimans, pythons, and turtles. *Parasitology* 128:333–342.

72. **Pozio, E., I. L. Owen, G. La Rosa, L. Sacchi, P. Rossi, and S. Corona.** 1999. *Trichinella papuae* n.sp. (Nematoda), a new non-encapsulated species from domestic and sylvatic swine of Papua, New Guinea. *Int. J. Parasitol.* 29:1825–1839.

73. **Pozio, E., D. B. Pence, G. La Rosa, A. Casulli, and S. E. Henke.** 2001. *Trichinella* infection in wildlife of the southwestern United States. *J. Parasitol.* 87:1208–1210.

74. **Prociv, P., and J. Croese.** 1990. Human eosinophilic enteritis caused by dog hookworm *Ancylostoma caninum*. *Lancet* 335:1299–1302.

75. **Ranque, S., B. Faugère, E. Pozio, G. La Rosa, A. Tamburrini, J. F. Pellissier, and P. Brouqui.** 2000. *Trichinella pseudospiralis* outbreak in France. *Emerg. Infect. Dis.* 6:543–547.

76. **Rodriguez-Osorio, M., J. M. Abad, T. de Haro, R. Villa-Real, and V. Gomez-Garcia.** 1999. Human trichinellosis in southern Spain: serologic and epidemiologic study. *Am. J. Trop. Med. Hyg.* 61:834–837.

77. **Roy, S. L., A. S. Lopez, and P. M. Schantz.** 2003. Trichinellosis surveillance—United States, 1997–2001. *Morb. Mortal. Wkly. Rep.* **52:**1–8.

78. **Salem, G., and P. Schantz.** 1992. Toxocaral visceral larva migrans after ingestion of raw lamb liver. *Clin. Infect. Dis.* **15:**743–744.

79. **Schantz, P. M.** 2000, Toxocariasis, p. 787–787. *In* G. T. Strickland (ed.), *Hunter's Tropical Medicine and Emerging Infectious Diseases*, 8th ed. The W. B. Saunders Co., Philadelphia, Pa.

80. **Seguchi, K., M. Matsuno, H. Kataoka, T. Kobayashi, H. Maruyama, H. Itoh, M. Koono, and Y. Nawa.** 1995. A case report of colonic ileus due to eosinophilic nodular lesions caused by *Gnathostoma doloresi* infection. *Am. J. Trop. Med. Hyg.* **53:**263–266.

81. **Shields, B. A., P. Bird, W. J. Liss, K. L. Groves, R. Olsen, and P. A. Rossignol.** 2002. The nematode *Anisakis simplex* in American shad (*Alosa sapidissima*) in two Oregon rivers. *J. Parasitol.* **88:**1033–1035.

82. **Shin, S. S., F. Elvinger, A. K. Prestwood, and J. R. Cole.** 1997. Exposure of swine to *Trichinella spiralis* antigen as determined by consecutive ELISAs and Western blot. *J. Parasitol.* **83:**430–433.

83. **Su, X., and A. K. Prestwood.** 1991. A dot-ELISA mimicry western blot test for the detection of swine trichinellosis. *J. Parasitol.* **77:**76–82.

84. **Taylor, M. R. H., P. O'Connor, A. R. Hinson, and H. V. Smith.** 1996. *Toxocara* titers in maternal and cord blood. *J. Infect.* **32:**231–233.

85. **Teyssot, N., N. Cassoux, P. Lehoang, and B. Bodaghi.** 2005. Fuchs heterochromic cyclitis and ocular toxocariasis. *Am. J. Ophthalmol.* **139:**915–916.

86. **Tiberio, G., G. Lanzas, M. I. Galarza, J. Sanchez, I. Quilez, and V. Martinez Artola.** 1995. Short report: an outbreak of trichinosis in Navarra, Spain. *Am. J. Trop. Med. Hyg.* **53:**241–242.

87. **Tran, V. T., L. Lumbroso, P. LeHoang, and C. P. Herbort.** 1999. Ultrasound biomicroscopy in peripheral retinovitreal toxocariasis. *Am. J. Ophthalmol.* **127:**607–609.

88. **Vidal, J. E., J. Sztajnbok, and A. C. Seguro.** 2003. Eosinophilic meningoencephalitis due to *Toxocara canis*: case report and review of the literature. *Am. J. Trop. Med. Hyg.* **69:**341–343.

89. **Wang, Z. Q., J. Cui, and L. J. Shen.** 2005. The epidemiology of animal trichinellosis in China. *Vet. J.* 12 Sept. [Epub ahead of print]

90. **Wang, Z. Q., J. Cui, F. Wu, F. R. Mao, and X. X. Jin.** 1998. Epidemiological, clinical and serological studies on trichinellosis in Henan Province, China. *Acta Trop.* **71:**255–268.

91. **Yen, C.-M., and E.-R. Chen.** 1991. Detection of antibodies to *Angiostrongylus cantonensis* in serum and cerebrospinal fluid of patients with eosinophilic meningitis. *Int. J. Parasitol.* **21:**17–21.

92. **Yera, H., S. Andiva, C. Perret, D. Limonne, P. Boireau, and J. Dupouy-Camet.** 2003. Development and evaluation of a Western blot kit for diagnosis of human trichinellosis. *Clin. Diagn. Lab. Immunol.* **10:**793–796.

12

Filarial Nematodes

The filarial nematodes are a group of arthropod-borne worms that reside in the subcutaneous tissues, deep connective tissues, lymphatic system, or body cavities of humans. Some adult filarial worms can survive in the human host for many years, causing a number of chronic and debilitating symptoms, including inflammatory reactions. The female worms produce large numbers of larvae called microfilariae, which are highly motile, threadlike prelarvae that in some species maintain the egg membrane as a sheath; these are called sheathed forms, while those that rupture the egg membrane are called unsheathed forms. Once released by the female worm, microfilariae can be detected in the peripheral blood or cutaneous tissues, depending on the species. The microfilariae, which may survive for 1 to 2 years, are not infective for other vertebrate hosts, nor do they undergo any further development in the vertebrate host. The infections are transmitted to humans by the bites of obligate blood-sucking arthropods that had become infected through ingesting larvae (microfilariae) contained in a blood meal obtained from a mammalian host (Table 12.1). Each parasite has a complex life cycle, and human infections are not easily established unless there is intense and prolonged exposure to infective larvae. After exposure, it may take years before significant pathologic changes in the human host are evident. The asymptomatic incubation period can be from 6 months to 3 years; therefore, the chances of eliciting a relevant patient history are rare, at best. For this reason, some physicians recommend routine posttravel serologic testing for patients who have been exposed, even years before.

Depending on the species, microfilariae may exhibit periodicity in the circulation. A circadian fluctuation in which the largest number of microfilariae occurs in the blood at night is called nocturnal periodicity; these microfilariae circulate in the peripheral circulation between 9 p.m. and 2 a.m. Some species are nonperiodic or diurnal. In the latter form, the microfilariae circulate at somewhat constant levels during the day and night. Subperiodic or nocturnally subperiodic microfilariae are those that can be detected in the blood throughout the day but are detected at higher levels during the late afternoon or at night. Only a small number of the total microfilarial population is found in the circulating blood, even at peak levels for microfilariae normally found in the blood. When not in the peripheral blood, the microfilariae are found primarily in the capillaries and blood vessels of the lungs. The basis for filarial periodicity is unknown; however, it may be an adaptation to the biting habits of the relevant vector.

Table 12.1 Characteristics of species causing human filariasis

Species	Distribution	Vector	Location of adult	Features of microfilariae				
				Location	Periodicity	Sheath and appearance	Length (µm) (range)	Tail nuclei
Wuchereria bancrofti	Tropics and subtropics worldwide; mainly India, China, Indonesia (nocturnal); Eastern Pacific (subperiodic)	Mosquito (*Culex, Aedes, Anopheles*)	Lymphatic	Blood, hydrocele fluid	Nocturnal, subperiodic	Present; graceful, sweeping curves; blue/pale with Giemsa	260 (244–296)	Do not extend to tip of tail; tail tapers to delicate point; terminal nucleus elongate
Brugia malayi	Southeast Asia, Indonesia, India (nocturnal); Indonesia, Southeast Asia (subperiodic)	Mosquito (*Mansonia, Anopheles, Coquilletidia*)	Lymphatic	Blood	Nocturnal, subperiodic	Present; stiff, with secondary kinks; pink with Giemsa	220 (177–230)	Subterminal and terminal; tail often constricted between 2 terminal nuclei; terminal nucleus round
Brugia timori	Islands of Timor and Lesser Sunda in Indonesia	Mosquito (*Anopheles*)	Lymphatic	Blood	Nocturnal	Present; tapering gradually; nearly invisible	310 (290–325)	Subterminal and terminal; no tail constriction between 2 terminal nuclei; terminal nucleus elongate to oval
Loa loa	Africa	Deerfly (*Chrysops*)	Subcutaneous	Blood	Diurnal	Present	275 (250–300)	Continuous to tip of tail; pointed tail; terminal nucleus elongate
Mansonella perstans	South and Central America, Africa	Biting midge (*Culicoides*)	Body cavities, mesentery, perirenal	Blood	None	Absent	195 (190–200)	Continuous to tip of tail; bluntly rounded tail
Mansonella ozzardi	South and Central America, Caribbean	Biting midge (*Culicoides*), blackfly (*Simulium*)	Subcutaneous, possible body cavities	Blood	None	Absent	200 (173–240)	Do not extend to tip of tail; pointed tail
Mansonella streptocerca	West and Central Africa	Biting midge (*Culicoides*)	Subcutaneous	Skin	None	Absent	210 (180–240)	Single row to tip of tail, tail curved "shepherd's crook"
Onchocerca volvulus	South and Central America, Africa	Blackfly (*Simulium*)	Subcutaneous	Skin	None	Absent	254 (221–287)	Do not extend to tip of tail; finely pointed tail
Dirofilaria immitis	Japan, Australia, United States	Mosquito	Pulmonary nodules (coin lesion)	Immature worms have been found in the lungs of humans				
Other *Dirofilaria* spp.	See comment	Mosquito	Subcutaneous tissues	*D. tenuis* is found in raccoons, *D. repens* is found in dogs and cats; subcutaneous dirofilariasis has been found in the United States, Africa, Asia, Europe, and South America; rare gravid *D. tenuis* worms have been found in humans.				

Microfilarial periodicity may be altered by reversing the working and sleeping habits of the host. This reversal may take up to a week to complete. For species exhibiting nocturnal periodicity, it has been noted that the insect vector bites primarily at night, whereas in areas of nonperiodic disease, the vector bites mainly during the day.

Basic Life Cycle

All of the human filarial infections have the same basic life cycle, including five larval molts, three in the insect vector and two in the human host. Infection begins with the bite of an infected arthropod vector.

1. Infective L3 larvae are deposited into the skin or blood of the human via the bite of the vector. Larvae pass through the puncture wound and then reach the lymphatic system.
2. A 6- or 12-month period is required for sexual maturity to occur; during this time, the parasite undergoes two molts and the adult worm develops (L5). The adult worms reside in the lymphatics or the lymph nodes.
3. The adult female worm gives birth to L1 larvae called microfilariae. Up to 50,000 microfilariae per day can be produced by the female worm.
4. Microfilariae are ingested by the arthropod vector; two molts occur, and the infective L3 larvae develop and are capable of infecting the human host.

The Endosymbiont

Infection with *Wuchereria bancrofti*, *Brugia malayi*, or *Brugia timori* causes lymphatic filariasis, while *Onchocerca volvulus* causes "river blindness," which can lead to severe ocular involvement and blindness. These parasites contain an endosymbiotic alpha-proteobacterium of the genus *Wolbachia*; these bacteria are *Rickettsia*-like, matrilineally inherited, obligate intracellular bacteria. They are required for larval worm development and adult worm viability and fertility. *Wolbachia* spp. have also been implicated in the pathogenesis of filariasis, including inflammatory responses related to filarial chemotherapy and death of the parasites. These bacteria have also been linked to the onset of lymphedema and blindness (40, 55, 116). Female adult worms carry the largest load of bacteria, most of which are concentrated around the reproductive structures.

Filarial and *Wolbachia* antigens activate the release of proinflammatory and chemotactic cytokines, which induce cellular infiltration and amplification of the inflammatory process. Toll-like receptors (TLRs), especially TLR4, appear to play a role in this process. Repeated and extensive exposure to filarial and *Wolbachia* antigens can lead to scarring and fibrosis in lymphatic filariasis and skin thickening and corneal scarring in onchocerciasis.

Like other bacteria in the family Rickettsiaceae, *Wolbachia* is sensitive to the tetracyclines (tetracycline and doxycycline) and to azithromycin and rifampin. The use of anti-*Wolbachia* antibiotics is promising as a means to improve the success of other mass treatment programs (55). In one study using doxycycline treatment on the major cause of lymphatic filariasis, *Wuchereria bancrofti*, an 8-week course was found to be a safe and well-tolerated treatment for lymphatic filariasis, with significant activity against adult worms and microfilaraemia (117).

Human Pathogens

There are eight filarial species in which the human is the definitive host, six of which are thought to be pathogenic. *W. bancrofti*, *B. malayi*, and *B. timori* all cause lymphatic filariasis. *Loa loa* causes Calabar swellings and allergic manifestations. *Mansonella streptocerca* causes skin disease. *O. volvulus* causes dermatitis and eye lesions. *Mansonella ozzardi* and *Mansonella perstans* are thought to be nonpathogenic.

An estimated 300 million people, primarily in India, Southeast Asia, and sub-Saharan Africa, live in areas where lymphatic filariasis is endemic, and at least 130 million people are probably currently infected (25). Globally, approximately 2 billion people are at risk for lymphatic filariasis (31). It is estimated that genital disease is present in 25 million infected men and that lymphedema, or elephantiasis, of the leg is present in 15 million people, most of whom are women. Approximately 30 million people are exposed to onchocerciasis, with probably at least 18 million being infected, 270,000 of whom are blind. Conditions associated with human filariasis have a tremendous impact on the physical health, economic well-being, and quality of life of infected persons and the communities in which they live.

It is interesting that in some areas of endemic infection, people do not accept the mosquito theory of transmission but believe in other physical, spiritual, and hereditary causes of lymphatic filariasis. Manifestations are often treated with herbal preparations that are used orally, smeared on the affected parts after scarification, or given as an enema. Although the etiology is thought to be different, the perception of the developmental process of elephantiasis closely parallels medical understanding (4).

Wuchereria bancrofti

Ancient Egyptian, Hindu, and Persian physicians (600 BC) were the first to note elephantiasis, which was probably due to *W. bancrofti*. Demarquay discovered microfilariae

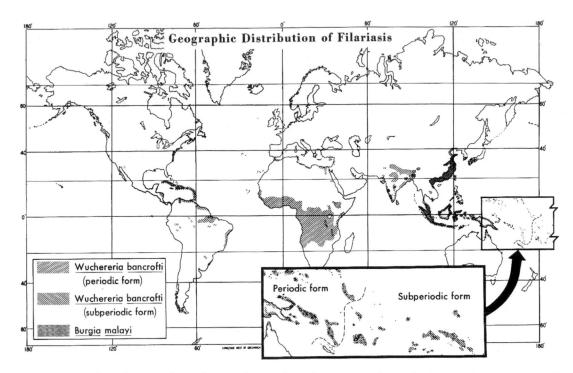

Figure 12.1   Distribution of *Wuchereria bancrofti* and *Brugia malayi* infections in humans. (Armed Forces Institute of Pathology photograph.)

in hydrocele fluid in 1863, Wucherer found organisms in chylous urine in 1868, Lewis found the microfilariae in blood in 1872, and the adult worm was found in a lymphatic abscess by Bancroft in 1872. By 1900, the entire life cycle of *W. bancrofti* had been elucidated; however, it was Patrick Manson's discoveries that showed for the first time that an arthropod was a vector for a parasitic disease. He had clarified the uptake of microfilariae by *Culex* mosquitoes and their subsequent maturation to infective forms. *W. bancrofti* infections are widely distributed throughout the tropics and subtropics (Figure 12.1). Humans are the only known reservoir hosts. At one time

there was an endemic focus in the region of Charleston, S.C., which was probably related to the slave trade.

Life Cycle and Morphology

Humans and mosquitoes are necessary to complete the life cycle of *W. bancrofti* (Figure 12.2). The intermediate host, a mosquito, acquires the infection by ingestion of microfilariae in the blood meal. The major vectors in urban and semiurban areas are culicine mosquitoes, while anophelines are involved in the rural areas of Africa and *Aedes* is found in the Pacific Islands. An individual within an area of endemic infection is exposed to approximately

Figure 12.2   Life cycles of *Wuchereria bancrofti* and *Brugia malayi*.

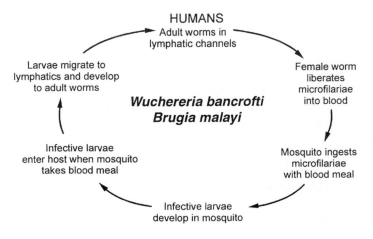

50 to 300 infectious larvae each year. Within hours after their arrival in the mosquito stomach, the microfilariae lose their sheaths. The larvae then penetrate the wall of the gut, migrate to the thoracic muscles, and develop into infective (filariform) larvae over a period of 7 to 21 days. The larvae migrate to the labella (distal end of the proboscis) of the mosquito and enter the skin of the definitive host through the puncture wound when a blood meal is taken. The infective larvae enter the peripheral lymphatic system and migrate to lymph vessels distal to the lymph nodes, where they grow to mature female and male adults and mate. Thousands of developing embryos can be found within the uteri of the female. Microfilariae are released from the gravid female and can be detected in the peripheral circulation in 8 to 12 months postinfection; however, filariasis without microfilaremia is not uncommon.

Adult worms are minute and threadlike, have a smooth cuticle, and are found in the lymph nodes and lymphatic channels. Adult males are about 40 mm long by 0.1 mm in diameter. Adult females are 80 to 100 mm long by 0.24 to 0.30 mm in diameter. To perpetuate their life cycle, sheathed microfilariae invade the blood; sometimes they can also be found in hydrocele fluid and chylous urine. The microfilariae range from 244 to 296 µm long and actively move about in the lymph or blood. The microfilaria has a sheath, and the body nuclei do not extend to the tip of the tail (Figures 12.3 to 12.6). Stains such as Giemsa, Wright's, or Delafield's hematoxylin have been used to help differentiate morphological features and thus identify the microfilariae to the species level. Stained microfilariae are about 245 to 300 µm long.

In most regions of the world where filariasis is endemic, *W. bancrofti* microfilariae exhibit periodicity, either nocturnal or subperiodic. In areas of nocturnal

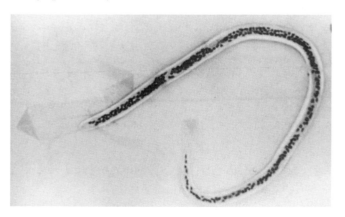

Figure 12.3 *Wuchereria bancrofti* microfilaria in a peripheral thick blood film. (From A Pictorial Presentation of Parasites: A cooperative collection prepared and/or edited by H. Zaiman. Photograph courtesy of L. Garcia.)

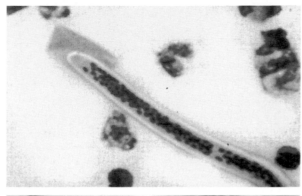

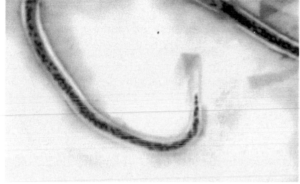

Figure 12.4 *Wuchereria bancrofti* microfilaria. (Upper) Head of microfilaria; note the sheath and very clearly delineated nuclei. (Lower) Tail of microfilaria; note that the sheath and the tail nuclei do not go all the way to the end of the tail. (From A Pictorial Presentation of Parasites: A cooperative collection prepared and/or edited by H. Zaiman. Photograph courtesy of L. Garcia.)

periodicity, the microfilariae are in their highest concentrations in the circulating blood at night, generally between 9 to 10 p.m. and 2 to 4 a.m., and are scant or absent during the daylight hours. A subperiodic form of filariasis occurs in the Pacific region, where humans exhibit a microfilaremia all the time but the largest numbers are detected between noon and 8 p.m.

Clinical Disease

Lymphatic Disease. Estimates of 82 to 142 days for the prepatent period have been obtained from experimental human infections, and manifestations of filariasis can range from none to severe, depending on host factors and parasite strains (Algorithm 12.1). Early symptoms of filariasis include high fevers (filarial or elephantoid fever), lymphangitis, and lymphadenitis. Filarial fever usually begins with a high fever and chills that last 1 to 5 days before spontaneously subsiding, and in many cases, patients with filarial fevers do not have a microfilaremia. The lymphangitis extends in a distal direction from the affected nodes where the filarial worms reside

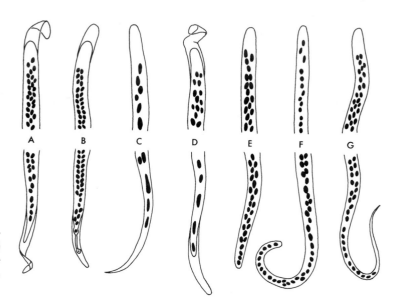

Figure 12.5 Characteristics of human microfilariae. (A) *Wuchereria bancrofti*; (B) *Brugia malayi*; (C) *Onchocerca volvulus*; (D) *Loa loa*; (E) *Mansonella perstans*; (F) *Mansonella streptocerca*; (G) *Mansonella ozzardi*. (Illustration by Nobuko Kitamura.)

(Figure 12.7). Lymphadenitis and lymphangitis develop in the lower extremities more commonly than in the upper. However, in some areas of endemic infection such as northeastern Brazil, factors other than filarial worms can be the cause of subclinical pathology of the leg lymphatics and are not specific for bancroftian filariasis (71). In addition to limbs, there can be genital (almost exclusively a feature of *W. bancrofti* infection) and breast involvement. The lymph nodes most often affected are the epitrochlears and femorals. The nodes are firm, discrete, and tender and tend to remain enlarged, while the lymph

Figure 12.6 Key to microfilariae commonly found in the blood.

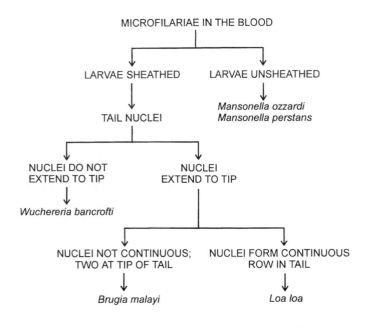

vessel is indurated and inflamed. The overlying skin is tense, erythematous, and hot, and the surrounding area is edematous. Occasionally, abscesses form at the lymph node or along the lymphatic system and may take 2 to 3 months to heal.

In lymphatic vessels that harbor the worms, an inflammatory reaction includes the infiltration of plasma cells, eosinophils, and macrophages in and around the affected vessel. Repeated inflammatory attacks lead to hyperplasia of the endothelium in addition to cellular infiltration. There is an increase in hydrostatic pressure resulting from the damage to the lymph vessel, which causes an increase in vessel wall permeability. The chronic leakage of fluid containing high levels of protein into the surrounding tissues (lymphedema) results in the hard or brawny edema, with thickening and verrucous changes in the skin, also known as elephantiasis (Figure 12.8). Histologic examination of the skin shows hyperkeratosis and acanthosis, with scarring and a loss of dermal elasticity. Regional lymph nodes proximal to the worms eventually become fibrotic, and the proximal lymph vessels become stenotic and obstructed, leading to the development of collateral lymph channels. Elephantoid tissue is a mixture of lymph and fat in a matrix of fibrotic tissue. There is the danger of secondary bacterial, particularly streptococcal, or fungal infection in patients with lymph stasis. A specific syndrome, acute dermatolymphangioadenitis, is not caused by filarial worms per se but probably results from secondary bacterial infections (27).

Obstructed Lymphatics.   Obstructed genital lymphatics may lead to hydrocele or scrotal lymphedema (112). In one area of endemicity, approximately 11% of those

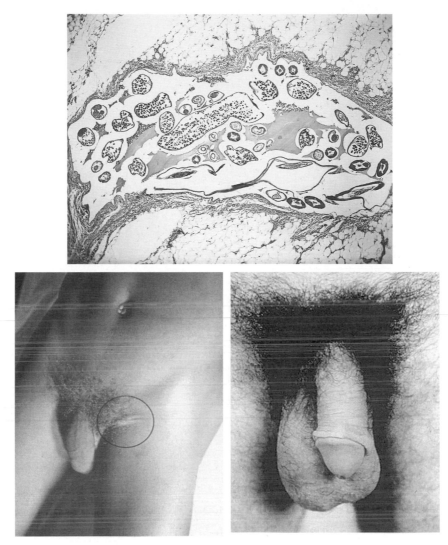

Figure 12.7 (Upper) *Wuchereria bancrofti* female in lymph node (cross section). (Armed Forces Institute of Pathology photograph.) (Lower left) Inguinal nodes enlarged due to filariasis. (Adapted from the Public Health Image Library, Centers for Disease Control and Prevention.) (Lower right) Scrotal lymphangitis due to filariasis. (Adapted from the Public Health Image Library, Centers for Disease Control and Prevention.)

at risk developed lymphedema (112). Many of these individuals progressed through sequential phases: uninfected, microfilaremic, and amicrofilaremic to irreversible obstructive lymphatic pathology. Early lymphatic damage was noted in 90% of the men at risk presenting with hydrocele, whereas 90% of the women at risk had no detectable clinical signs of filariasis on physical examination.

Lymph Varices. Lymph varices near the scrotal surface are readily seen and may rupture. Obstruction of the retroperitoneal lymphatics may cause the renal lymphatics to rupture into the urinary tract, leading to chyluria. Chyluria is often prominent in the morning and is intermittent. Almost half of the microfilaremia patients have renal abnormalities characterized by proteinuria and hematuria.

Acute Filarial Lymphangitis. Acute filarial lymphangitis occurs when the adult worms die, leading to severe localized inflammation. This syndrome appears to be uncommon in untreated individuals (28). The dead worms may be absorbed, or they may calcify or produce abscesses. The inflammatory reaction evoked by the filarial infection is a result of host reactions to the products of the developing worm or its death. Microfilariae do not appear to be responsible for the major sequelae

BANCROFTIAN AND MALAYAN FILARIASIS

- ◆ LYMPHANGITIS
- ◆ LYMPHADENOPATHY
- ◆ EPIDIDYMITIS
- ◆ HYDROCELE
- ◆ CHYLURIA
- ◆ ELEPHANTIASIS

HISTORY ———NEGATIVE———→ DIAGNOSIS LESS LIKELY

↓

POSITIVE

↓

MICROFILARIAE IN BLOOD, FLUIDS ———NEGATIVE———→ WORMS ON BIOPSY ———NEGATIVE———→ DIAGNOSIS POSSIBLE

↓ POSITIVE   ↓ POSITIVE   ↓ ASSESS SYMPTOMS

DIAGNOSIS CONFIRMED   RECURRENT INFLAMMATORY EPISODES   CHRONIC OBSTRUCTIVE LESIONS

↓ TREATMENT ←———   TREATMENT SYMPTOMATIC

Algorithm 12.1 Bancroftian and Malayan filariasis.

of lymphatic filariasis. The immune response to infection varies considerably, with the antibody response being lowest in asymptomatic microfilaremic patients and highest in amicrofilaremic patients. Chronic disease is also strongly associated with acute-disease incidence, and microfilaremia has a statistically significant association with acute disease in the leg, arm, and breast but not the scrotum (7).

Amicrofilaremic Patients. Some patients may harbor adult worms without a peripheral microfilaremia, or the microfilaremia may be so low that it cannot be detected by the usual laboratory procedures. Other patients may have a heavy microfilaremia but be clinically asymptomatic. Many individuals who are thought to be microfilaremic but clinically asymptomatic may have symptoms of elephantiasis. Microfilariae are more easily detected in individuals with hydrocele than in individuals with symptoms of elephantiasis. Even though microfilaremia appears to be rare in children younger than 5 years, children born to mothers harboring microfilaria are more likely to be microfilaremic than are children born to afilaremic mothers, suggesting in utero acquisition of tolerance.

Tropical Eosinophilia. Many patients infected with lymphatic filaria do not have microfilaremia. Some may be experiencing tropical eosinophilia (Weingarten's) syndrome. This syndrome is characterized by pulmonary infiltrates, peripheral eosinophilia, cough, asthmatic attacks (especially at night), and a history of prolonged residence in the tropics. These patients have high peripheral eosinophilia counts, high immunoglobulin E (IgE) levels, and high antifilarial antibody titers. Microfilariae are not detected in the peripheral blood but may be found in lung biopsy specimens. On treatment with diethylcarbamazine (DEC), there is usually a prompt response and the symptoms disappear.

Infection in Women of Reproductive Age. In a study reviewing pretreatment microfilarial densities, there was a reduction among women of approximately reproductive age; this reduction was statistically significant. However, a comparison of pregnant women and controls showed no evidence that the reduction was specifically related to pregnancy. It has been suggested that the observed reduction in microfilarial intensity may result from hormonal changes associated with female reproduction, possibly in combination with other factors (6).

Immunologic Responses. Asymptomatic microfilaremic individuals generally have a marked Th1 hyporesponsiveness but a strong Th2 response. In contrast, amicrofilaremic individuals with lymphatic pathology have strong Th1 and Th2 responses. There is a decrease in the gamma interferon (IFN-γ) level and a significant rise

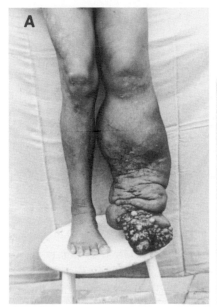
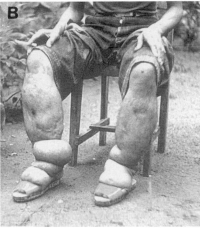

Figure 12.8 *Wuchereria bancrofti* elephantiasis of lower extremities. (Panel A from A Pictorial Presentation of Parasites: A cooperative collection prepared and/or edited by H. Zaiman; panel B adapted from the Public Health Image Library, Centers for Disease Control and Prevention.)

in interleukin-4 (IL-4) production in infected patients, suggesting a profound form of anergy. In infected individuals, the levels of IgG and IgE are elevated. The highest levels of IgE were found in microfilaremic individuals with acute manifestations of filariasis or hydrocele, while the highest concentrations of parasite-specific IgE were found in amicrofilaremic individuals. This may suggest a protective effect for IgE, because the highest ratios of specific to total IgE are found in patients with tropical pulmonary eosinophilia. The predominant serum antibody response is IgG4. Filarial antigenemia, IgG4 antibody, and household infection are significant risk factors for microfilaremia (123). The ratio of IgG4 to IgE is highest in asymptomatic microfilaremic individuals, whereas chronic disease (elephantiasis) is associated with elevated levels of IgG1 (3, 68). Chronic disease is associated with a specific IgG1 to IgG3 response rather than a Th1 or IgE response. IgG4 levels decline with successful therapy. For signs and symptoms to develop, prolonged exposure to infected mosquitoes is required, and even then the host response is variable.

In French Polynesia, individuals harboring adult worms had a higher filaria-specific IgG4 level but lower IgG3 and IgE levels than did individuals without adult worms, independent of the presence of microfilariae. Low filaria-specific IgG1 and IgG2 levels were associated with the presence of microfilariae but independent of the presence of adult worms. The filarial antibody responses were associated with the parasitological status of individuals but not with clinical symptoms such as hydrocele or limb lymphangitis or elephantiasis. The reduction of filaria-specific immunoglobulin levels was higher after treatment with DEC than with ivermectin, a finding that reflects drug efficacy for the adult worms (76). There appears to be a clear association between IgG2 indirect fluorescent antibody positivity and a negative microfilarial status and negative specific circulating-antigen status. These findings suggest that specific antisheath antibodies are associated with an immunologic resistance mechanism that is expressed with highest prevalence in young individuals prior to the development of patent microfilaremia (109).

Children and Adolescents. There is evidence that children acquire infection early in life and that antifilarial antibody responses may peak in early childhood. Apparently, helminth-specific immune responses acquired during gestation persist into childhood, and this prenatal sensitization biases T-cell immunity induced by BCG vaccination away from IFN-γ responses associated with protection against mycobacterial infection (69).

Often, little attention is paid to the signs and symptoms of bancroftian filariasis in childhood. There appears to be a predominance of lymph node involvement, where the adult worms are located in the vessels of draining lymph nodes, often the inguinal nodes. There are no signs of inflammation in the overlying skin. In the adult population, adult worms are usually found in extranodal lymphatic vessels.

Certainly, in areas of endemic infection, filarial infection should be considered in children and adolescents with adenopathy. Ultrasound and other noninvasive methods should be considered; biopsy is probably not necessary (33).

Diagnosis

Routine Blood Tests. Filarial infections should always be considered in the differential diagnosis of individuals who have resided in or migrated from areas where filariasis is endemic and have had a significant insect bite history. In areas of endemic infection around the world, the presumptive diagnosis of filarial infections is frequently based on clinical evidence; however, definitive diagnosis is based on the detection of microfilariae, primarily in the blood. Microfilariae may also be found in hydrocele fluid and urine, particularly in patients who have high microfilaremias or who have been treated recently with DEC.

When a patient is suspected of having filariasis, the clinical history helps determine the most appropriate specimen and collection time. The optimal time for drawing blood to detect nocturnal periodic *W. bancrofti* infections is between 10 p.m. and 4 a.m. Blood used to detect subperiodic *W. bancrofti* may be drawn any time, although peak microfilaremias occur in the late afternoon. Finger prick, earlobe, or venous blood (using EDTA anticoagulant [purple-top tube]) may be taken for direct wet, thin, and thick blood smears. Blood films may be stained with Giemsa or Delafield's hematoxylin stain; however, Giemsa stain does not stain the microfilarial sheath adequately, although hematoxylin stains do so (see chapter 31). Examination of a thin blood film for microfilariae should include low-power review of the entire film, not just the feathered edge. Sheathed microfilariae often lose their sheath when drying on thick films.

Because of the insensitivity of the wet, thin, and thick blood films, larger blood volumes are recommended for examination to detect low microfilaremias. Also, it may be somewhat impractical to obtain blood from a patient late at night or in the early-morning hours. Microfilariae concentrate in the peripheral capillaries; therefore, the numbers of microfilariae in finger prick blood may be similar to those in venous blood. In certain areas, a primary problem with obtaining venous blood is the unwillingness of people to submit to phlebotomy.

Concentration Methods. The Knott technique allows up to 1 ml of blood to be concentrated per 10 ml of 2% formalin so that the sediment can be examined as a wet preparation or stained for microfilariae. Another concentration technique involves the use of polycarbonate filters (Nuclepore) to trap microfilariae on the filter after the red blood cells have been lysed (30). The 3- to 5-μm-pore-size filters may be examined directly on a microscope slide because the filters are transparent when wet. Infection has been rapidly diagnosed by the detection of microfilaria using a microhematocrit tube coated with acridine orange (QBC method) (see chapter 31). The major disadvantage of the QBC method is the necessity for a fluorescence microscope.

In addition to concentration techniques to detect low levels of microfilariae, a provocative test has been used to induce microfilariae to appear in the blood through treatment with a single dose of DEC. Blood should be drawn within 15 min to 1 h after administration of DEC. A fall rather than a rise in microfilaremia may be detected with the subperiodic strain of *W. bancrofti* after administration of DEC.

Only the microfilariae of *W. bancrofti*, *B. malayi*, and *L. loa* have sheaths; all others are nonsheathed. In addition to the presence or absence of a sheath, the body nuclei should be closely examined for their morphology and location in the tail of the microfilaria. Histopathologic examination of a tissue biopsy specimen may be used for diagnosis to reveal adult worms and microfilariae. Patients who have been treated for lymphatic filariasis should have blood specimens reexamined for microfilariae 2 to 6 weeks posttherapy.

Serologic Methods. A wide variety of serologic tests have been used for diagnostic and epidemiologic purposes. Except for patients not native to the area of endemicity, immunodiagnostic tests are of limited value. Current tests lack both sensitivity and specificity, and most people from the region of endemicity have a positive serologic response. This response may be due to exposure to nonhuman filarial antigens from infected mosquitoes; also, filarial antigens may cross-react with antibodies to other parasitic diseases. The immune response of the filaria-specific immunoglobulin IgG4 is considerably increased in active filarial infections. The detection of IgG4 antibodies also reduces the cross-reactivity of nonfilarial antibodies (120). The IgG2 level appears to be increased in patients with elephantiasis (94). Studies also suggest that specific antisheath antibodies are associated with an immunologic resistance mechanism that in the endemic community is expressed with highest prevalence in young individuals before development of patent microfilaremia (111). Patients with tropical pulmonary eosinophilia have a marked IgE response in addition to elevated eosinophilia.

The recombinant antigen BmR1 has been extensively employed in both enzyme-linked immunosorbent assay and immunochromatographic rapid dipstick (*Brugia* Rapid) formats for the specific and sensitive detection of

IgG4 antibodies against the lymphatic filarial parasites *B. malayi* and *B. timori*; however, in sera of individuals infected with *W. bancrofti* the IgG4 reactivity to BmR1 is variable and thus is not recommended for detection (34, 78). A rapid-format, qualitative flowthrough immuno-filtration test has been developed for the identification of total IgG antibodies to recombinant filarial antigen WbSXP-1. This test system employs colloidal gold-protein A reagent as the antibody capture reagent. In a study of 1,230 sera, the sensitivity of the test was 90.8% for patients with proven brugian filariasis (*n* = 70) and 91.4% for patients with proven bancroftian filariasis (*n* = 140). This rapid test is performed with reagents that can be stored at room temperature, is user friendly, and is applicable for field use as an initial screening method. It is also recommended for epidemiologic monitoring of bancroftian and brugian filarial infections (11).

Antigen Detection. Because of the lack of sensitivity of direct-detection methods and problems with obtaining blood specimens within correct time frames, detection of circulating antigens has been used diagnostically (97). Antigen detection has also been used to monitor therapy. Assays for the detection of circulating antigen of *W. bancrofti* rely on two diagnostic tests, an Og4C3 enzyme-linked immunosorbent assay (Trop Bio Og4C3 antigen test; Trop Bio) and an immunochromatographic card test (NOW ICT; Binax). The sensitivity of the Trop Bio test is close to 100%, while the specificity is about 99 to 100%. The ICT test has a sensitivity of 96 to 100% and a specificity of 95 to 100% (109, 122). In looking at the possibility of false-positive test results with the NOW card test, it is clear that the test performs well if the card results are read at the 10-min time limit; false positives may occur if the cards are held and read after 10 min (110). Studies using both of these methods emphasize the benefits of testing to determine the estimation of circulating filarial antigen in a population (23, 82, 90).

PCR Amplification. PCR assays are becoming useful diagnostic tools because they can discriminate between past and present infection and can be used to monitor therapy and to detect and differentiate multiple filarial infections (35, 53, 66, 73). An evaluation of the diagnostic capability of PCR using diurnally collected sputum from patients with bancroftian filariasis has been carried out; the data were excellent, and this approach appears to have great potential (1). A seminested PCR has been used to detect *W. bancrofti* in carriers and in long-term storage blood samples, with no cross-reactivity related to *B. malayi*, *Plasmodium falciparum*, or *P. vivax* (58). A PCR-restriction fragment length polymorphism (PCR-RFLP)-

based method to detect and differentiate a broad range of filarial species in a single PCR has recently been developed. This approach provides differentiation to the species level for *W. bancrofti*, *B. malayi*, *B. pahangi*, *Dirofilaria immitis*, and *D. repens*. The PCR-RFLP of ITS1 rRNA genes cloned from *W. bancrofti*, *B. malayi*, and *B. pahangi* as universal priming sites will be useful in diagnosing and differentiating filarial parasites in humans, animal reservoir hosts, and mosquito vectors (79).

Ultrasonography. Infection with *W. bancrofti* not only affects the structure and function of lymphatic vessels but also is associated with extralymphatic pathology and disease. Since it is now possible to detect living adult worms by ultrasonography, a great deal of emphasis is placed on lymphatic pathology (32). However, the finding of renal damage in asymptomatic microfilaremic carriers has led to increased recognition of the importance of extralymphatic complications in bancroftian filariasis (28, 29). Filarial pathology of the male genitalia is apparently underreported if physical examination alone is used. Compared with ultrasonography, the sensitivity of physical examination in one study was only 44.3% (119). Ultrasonography is also very helpful in assessing lymphatic filariasis in children after treatment and in women (41, 70). In one study, the surprisingly large number of worm nests detectable in microfilaremic women suggests that ultrasonography would be appropriate for diagnosis and posttherapy monitoring of female patients infected with *W. bancrofti* (70). Often the "filaria dance sign" can be seen; this is a characteristic pattern of movement exhibited by adult worms that has been described as rapid and random.

KEY POINTS — LABORATORY DIAGNOSIS

Filariasis

1. A travel and geographic history should be obtained to maximize the best type of specimen and optimal collection time for the filarial infection suspected.
2. In addition to multiple thin and thick blood films, Knott or membrane concentration techniques should be used to detect microfilariae normally found in the peripheral blood.
3. It is important to examine every portion of the thin and thick blood films; microfilariae are often found at the outside edges or in the original drop from which the thin film was "pulled."
4. Giemsa stain does not stain the *W. bancrofti* sheath as well as a hematoxylin-based stain (Delafield's hematoxylin).
5. Serologic tests are more meaningful in patients

who have not resided in the areas of endemicity for extended periods.

6. Antigen detection tests are commercially available and may be very helpful in the detection of circulating filarial antigens.

7. PCR may prove to be valuable in the diagnosis of lymphatic filariasis; however, these procedures are often limited to research facilities.

8. Ultrasonography has proven to be very valuable in assessing lymphatic filariasis in both adults and children; this approach can be much more sensitive than a physical examination alone.

Treatment

It is recognized that some patients with lymphatic filariasis may be asymptomatic; however, if they have microfilaria in the blood, they probably have subclinical disease. Early treatment of these patients is recommended to prevent additional disease manifestations such as lymphatic damage. For patients who appear to have no circulating microfilariae but have a confirmed presence of adult worms, treatment with DEC and albendazole is recommended (56).

Diethylcarbamazine. The two important drugs in the treatment of filariasis are DEC and ivermectin, which may be used in combination with albendazole (see chapter 25) (Table 12.2) (42, 43). Treatment differs depending on the objective, interruption of transmission by microfilaria suppression or treatment of an individual patient. DEC, a piperazine derivative, rapidly kills microfilariae and can kill some but not all adult worms. The effect on adult worms is apparently related to the total dose given, with spaced doses over a period of months to years probably being superior to a more concentrated dose. DEC can be administered orally. Side effects from rapid destruction of microfilariae and possibly adult worms include fever, urticaria, and lymphangitis. These reactions may be controlled by administering antihistamines or by giving low initial doses of DEC and gradually increasing the dose. Nonspecific adverse reactions to DEC treatment include headache, nausea, vomiting, generalized malaise, and vertigo. These symptoms develop within a day or two and may last for several days. A wide variety of treatment regimens have been used to optimize the antifilarial activity of DEC (72, 96). In areas of endemicity, prolonged low-dose administration has successfully reduced transmission and pathologic sequelae (123). Drug combinations containing DEC are the most effective against microfilarial prevalence and intensity relative to single drugs or other combinations (118).

Ivermectin. Ivermectin (Mectizan; 22,23-dihydroavermectin B$_1$) is a semisynthetic macrocyclic lactone derivative of avermectin. The drug is taken orally and has been successfully used either alone or in combination with DEC to treat *W. bancrofti* infections (65, 99, 121). The mechanism of action involves chloride ion permeability. Few severe side effects, good microfilaricidal activity, and single-dose oral administration have encouraged its use. In individuals with higher microfilaremia levels, more severe side effects include fever, headache, myalgia, sore throat, and cough. Although there is a rapid and complete clearance of microfilariae in patients treated with one dose of ivermectin in contrast to those treated with DEC, the microfilaria counts begin to rise again after a few months (20). The rate of increase is variable and may be related to the variability of the parasite or host response. At higher therapeutic doses, ivermectin may have cidal activity against the adult worms. It appears that albendazole given in combination with ivermectin as a single dose is more effective in clearing microfilariae than is ivermectin alone and also has the benefit of reducing intestinal helminth infections (12, 59). A significant decrease in the prevalence of *W. bancrofti* infection in mosquitoes has been seen following the addition of albendazole to annual ivermectin-based mass treatments in Nigeria (101).

Tissue Removal and Hydrocele Drainage. Surgical removal of elephantoid tissue has been satisfactory for scrotal elephantiasis but not for effects on the extremities. The use of elastic pressure bandages may help to reduce the induration, but the underlying fibrosis is unaffected. Hydrocele drainage can be repeated or managed surgically; drug treatment should be emphasized for patients with active infection.

Posttherapy Allergic Responses. Microfilaricidal chemotherapy with either DEC or ivermectin induces an acute inflammatory reaction due to an IL-5-dependent increase of eosinophil numbers, eosinophil degranulation, and elevated levels of proinflammatory cytokines such as IL-6, tumor necrosis factor alpha, and IFN-γ. Long-term effects of drug-induced clearance of microfilariae include a reversal of the parasite-specific hyporesponsiveness with restoration of Th1 cell-mediated responses and resistance to reinfection (124). Also, high levels of nitrite and nitrate in serum are released during microfilarial clearance, and sustained elevated levels have been observed 6 months after chemotherapy, suggesting a role of nitric oxide (NO) in the elimination of microfilariae. There is also evidence to indicate that although microfilaremic individuals may regain the ability to produce IFN-γ to parasite antigens posttreatment, they subsequently become hyporesponsive to microfilaria-containing antigens; this appears to be

Table 12.2 Treatment options for bancroftian filariasis

Drugs	Findings	Comments	Reference
DEC plus iodine added to salt	Reduced microfilaremia and transmission to low levels	Well accepted by community, no reported side effects	43
DEC, mass administration: 3 options	Long-term effect on microfilarial intensities	At the end of 10 yr, community microfilaremia levels approached pretreatment levels; suppressive effect most pronounced in (iii)	72
(i) 12-day treatment	Only 11% of pretreatment		
(ii) Semiannual single-dose treatment	Only 13% of pretreatment		
(iii) Monthly low dose	Only 2% of pretreatment		
DEC, single dose, 6 mg/kg	Reduced blood microfilarial counts, clearance of 69% of subjects; reduced filarial antigen levels	Low-endemicity setting; mass treatment with DEC alone may be sufficient to interrupt transmission	96
DEC plus albendazole	Combination is highly effective against adult *W. bancrofti*; reduced filarial antigen levels	Healthy Egyptian adults with microfilaria counts of >80/ml	56
DEC plus albendazole	Albendazole and DEC effective in treating intestinal helminths, more effective than DEC alone in treating *W. bancrofti*	Combining chemotherapy of intestinal helminths and lymphatic filariasis; no increase in severity of adverse reactions	42
DEC plus albendazole	Average microfilarial intensity decrease of 4.6% of pretreatment values	Most effective against microfilarial intensity relative to single drugs	118
DEC plus ivermectin	Average microfilarial intensity decrease of 0.7% of pretreatment values	Most effective against microfilarial intensity relative to single drugs	118
Ivermectin	Ivermectin annual therapy for onchocerciasis did not interrupt transmission of *W. bancrofti*	Question of whether mass ivermectin administration for onchocerciasis would also impact lymphatic filariasis in areas where both organisms are endemic	100
Ivermectin	In villages where onchocerciasis is hyperendemic, after 14 yr of ivermectin no *W. bancrofti* in population; in adjacent villages, prevalence of 3% found	Despite long period of ivermectin, *Mansonella perstans* did not appear to respond in this setting	65
Ivermectin plus albendazole	After the third round of mass drug administrations, significant decreases in mosquito infections and infectivity seen	The combination of albendazole and ivermectin appears to be superior to ivermectin alone for reducing the frequency of *W. bancrofti* infection in mosquitoes	101
Ivermectin plus albendazole	Average microfilarial intensity decrease of 12.7% of pretreatment values	Relative efficacies of drug combinations not well documented, but provide valuable estimates of drug effect using existing data	118
Ivermectin plus albendazole	No significant differences in severity of posttreatment reactions in those coinfected with *M. perstans*	Coinfection with *M. perstans* does not alter posttreatment reaction profiles to single-dose ivermectin-albendazole in *W. bancrofti* infection in this region	59
Doxycycline	*Wolbachia* endosymbionts are vital for larval development and adult-worm fertility and viability; this dependency on the bacterium for survival of the parasites provides a new approach to treat filariasis with antibiotics; doxycycline almost completely eliminated microfilaremia at 8–14 mo follow-up; ultrasonography detected adult worms in only 6 (22%) of 27 individuals treated with doxycycline compared with 24 (88%) of 27 treated with placebo at 14 mo after the start of treatment; filarial antigenemia in the doxycycline group fell to about half of that before treatment; adverse events were few and mild	An 8-wk course of doxycycline is safe and well tolerated for lymphatic filariasis, with significant activity against adult worms and microfilaremia	117

independent of the recurrence of microfilaremia and the response to nonparasite antigens (49).

Epidemiology and Prevention

W. bancrofti infections are widely distributed throughout the tropical and subtropical regions of Africa, Asia, Central and South America, the Caribbean islands, and the Pacific Islands. This species infects an estimated 115 million individuals; however, based on diagnostic test limitations, the actual number may be twice as great. *Anopheles* and *Culex* mosquitoes are night-biting vectors for the nocturnally periodic *W. bancrofti*, while the subperiodic strain is transmitted by day-biting *Aedes* mosquitoes. In areas of endemicity, exposure begins early in childhood, with microfilaria rates increasing with age, although the infection may not be clinically apparent.

Since these helminths do not multiply within the human host, infection levels are related directly to the number of infective larvae to which humans are exposed. Thus, prevalence and infection levels in an area of endemicity depend on the intensity of transmission. Factors that influence transmission include vector breeding sites, parasite longevity, levels of microfilariae in the blood when the vector bites, microfilarial load in the population, prevalence of lymphatic pathology, and immunologic tolerance. Transmission is quite inefficient, and nonresidents who move into areas of endemicity rarely develop patent infections; however, they may develop signs of acute disease (60).

Individual protection involves the use of insect repellents and bed netting; however, long-term protection has been sought through vector control and the use of chemotherapy. To carry out vector control properly, one must identify the species responsible for transmission, including its feeding, flight, and breeding habits. Because identification of filarial species in infected vectors by microscopy is difficult, DNA hybridization has been used to help identify mosquito populations involved in transmission. Wet pit latrines are favorite breeding sites for mosquitoes, and sanitation improvement to reduce breeding places in urbanized communities where the infections are endemic is a very effective means of control. Although vector control and mass treatment have been successful in small isolated areas, these approaches have been difficult in large areas.

In areas of endemic infection, DEC has been given to the inhabitants at weekly and monthly intervals over long periods to reduce transmission. The drug has even been incorporated into table salt for continuous low-dose administration to the population. Unfortunately, substantial increases in microfilaremia in humans and higher infection rates in mosquitoes in these areas have been noted when mass DEC chemoprophylaxis has been stopped (19).

Ivermectin inhibits the exsheathment of microfilariae in mosquitoes (98). This action prevents biological development within the mosquito, thereby disrupting the chain of transmission. Vaccine development strategies have focused on unique filarial proteins that are critical for biological functions; however, there are no vaccines in general use today (22, 50).

Brugia malayi

Microfilariae of *B. malayi* were first observed by Lichtenstein in blood films from natives in Indonesia, and Brug described the microfilariae as a new species in 1927. Rao and Maplestone recovered adult worms from the forearm of a patient in India in 1940, giving the first descriptions of *B. malayi* adults. The geographic distribution of *B. malayi* overlaps that of *W. bancrofti* in certain areas (Figure 12.1). Both species are considered to be lymphatic filariae; however, *Brugia* spp. are much more widely adapted to animal hosts other than humans.

Life Cycle and Morphology

Although the two species can be differentiated morphologically, the life cycle of *B. malayi* is similar to that of *W. bancrofti* (Figure 12.2). However, *Brugia* has a shorter development time in the mosquito vector and the time from infection to appearance of microfilariae may be as short as 3 to 4 months. The adult worms inhabit the lymphatics, and the females give birth to sheathed microfilariae. The microfilariae differ from those of *W. bancrofti* by having two terminal nuclei that are distinctly separated from the other nuclei in the tail (Figures 12.5 and 12.6). The last terminal nucleus is quite small and is found at the tip of the tail (Figures 12.9 and 12.10). The microfilariae range from 177 to 230 μm in length. The intermediate host is a mosquito that may be infected with a periodic or subperiodic strain, depending on the geographic area.

Figure 12.9 *Brugia malayi.* (Left) Head of microfilaria; note that the sheath is not that clearly visible. (Right) Tail of microfilaria; note the sheath and the two terminal tail nuclei.

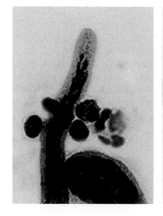 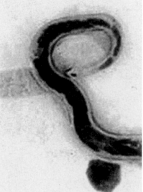

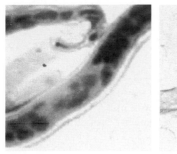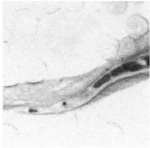

Figure 12.10 *Brugia malayi.* Both images show the microfilarial tail. Note the sheath and two terminal tail nuclei with the bulge in the tail.

Clinical Disease

The clinical pathology of *B. malayi* infections in humans is similar to that of *W. bancrofti.* Clinical manifestations usually develop months or years after infection, and many of the patients are asymptomatic even when they have microfilaremia. Lymphangitis and filarial abscesses occur with a greater degree of frequency than in *W. bancrofti* infections. If elephantiasis occurs, the swelling is normally restricted to the lower extremities below the knee. Sclerotic cordlike lymphatics and enlarged nodes in the arms and legs are common; urogenital involvement with chyluria does not occur. In disease caused by *B. malayi*, episodes of prolonged fever, adenolymphangitis, abscesses of affected lymph nodes, and local residual scarring occur quite frequently. Chronic lymphedema or elephantiasis, as seen in bancroftian filariasis, does not occur frequently.

Cytokine analysis revealed that in an area where brugian filariasis is endemic, IFN-γ was most affected by the shifts in microfilarial densities; proliferative responses remained low throughout the study period in microfilaremic individuals; however, in those with no demonstrable microfilariae, the responses were inversely related to the microfilarial densities (104). In contrast, IL-4 responses showed little correlation with changes in parasite densities.

Diagnosis

The diagnostic methods are similar to those for *W. bancrofti* and include thin and thick blood films, wet preparations and concentrations, PCR, ultrasonography, and antigen and antibody detection. There are nocturnally periodic and nocturnally subperiodic strains, so travel history can be helpful in determining optimal specimen collection times. Giemsa stains the sheath of *B. malayi* (will stain pink) but does not stain the sheath of *B. timori*, a species found in the islands near Indonesia. Apparently, the sheath is absent in 50% of periodic microfilariae but in only 5%

of subperiodic microfilariae (Figures 12.5 and 12.6; Table 12.1; Algorithm 12.1). A DNA hybridization assay has been developed to identify microfilariae trapped on nitrocellulose membranes but has not been used in field trials (93).

Treatment

The frequency of severe reactions is greater in *B. malayi* infections than in *W. bancrofti* infections. Initiation of treatment with lower doses of DEC and use of anti-inflammatory drugs reduces the adverse reactions. Ivermectin has also been used to treat *B. malayi* infections. Side effects are attributed to host responses to the death of microfilariae rather than directly to the drug. Single doses are effective in clearing or reducing the microfilaremic burden for up to 6 months (107). However, a gradual decrease in microfilaria occurs over several weeks to 15 to 20% of pretreatment levels; this response is much slower than that seen with *W. bancrofti* and microfilaria clearance. The combination of DEC and albendazole, both well-tested drugs, offers a new option for countries such as India, where there is no onchocerciasis or loiasis and where ivermectin may not be readily available; the efficacy of albendazole against intestinal helminths would provide additional benefits (106). Since acute attacks of adenolymphangitis (ADL) contribute to the overall morbidity with brugian filariasis, studies now indicate that foot care seems to play the most important role in the prevention of ADL attacks. Benefits may accrue from local or systemic antibiotic use in patients with high grades of edema, whereas antifilarials are ineffective in the prevention of ADL attacks (108).

Epidemiology and Prevention

There are two types of *B. malayi*, one nocturnally periodic and the other nocturnally subperiodic. The periodic strains, which have no animal reservoirs, are widely distributed in Asia, whereas the subperiodic form has been noted in Malaysia, Indonesia, and the Philippines. The principal mosquito vectors are *Mansonia* spp.; however, in certain areas *Anopheles* and *Aedes* spp. may be important. There are a number of reservoir hosts for *B. malayi* infections, including humans, dogs, cats, and monkeys.

Any prevention program must consider reservoir hosts other than humans in the control of *B. malayi*. Measures used for control and prevention have been similar to those used for *W. bancrofti*.

Brugia timori

Human infections in the Indonesian island chain (Timor) were first reported in 1964, and the filarial worms were recognized in 1965 by David and Edeson as being

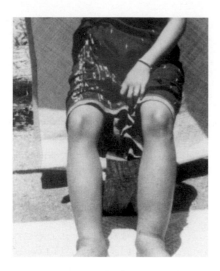

Figure 12.11 *Brugia timori* elephantiasis of the lower extremity. (From A Pictorial Presentation of Parasites: A cooperative collection prepared and/or edited by H. Zaiman. Photograph courtesy of D. Dennis, Kuala Lumpur.)

distinctly different from *B. malayi* (24, 26). *B. timori* infections produce symptoms and pathology similar to those found in *W. bancrofti* and *B. malayi* infections, and the life cycle is similar. However, there appear to be more abscesses related to *B. timori* infections. As with *B. malayi* infections, if elephantiasis occurs, the swelling is normally restricted to the lower extremities below the knee (Figure 12.11).

The microfilariae have nocturnal periodicity and are transmitted by *Anopheles* mosquitoes. *B. timori* microfilariae are morphologically similar to *B. malayi* microfilariae, except that on average they are longer (310 versus 220 μm), the cephalic space (length from the tip of the microfilaria head to the first body nucleus) has a length/width ratio of 31 versus 21, and the sheath does not stain with Giemsa stain as well as the *B. malayi* sheath does (Table 12.1). Also, there tend to be five to eight, rather than four or five, contiguous nuclei

Figure 12.12 *Brugia timori* microfilarial tail. Note the two terminal tail nuclei with no bulge in the tail.

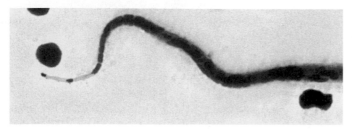

in the tail, and little to no bulge is seen in the cuticle (Figure 12.12).

The Brugia rapid test is an immunochromatographic dipstick test used to detect IgG4 antibodies that are reactive with a recombinant *B. malayi* antigen. This approach has been used successfully for the detection of *B. timori* antibodies and has been recommended as a monitoring tool for *B. timori* in the framework of the Global Program to Eliminate Lymphatic Filariasis (36, 113). Treatment is similar to that for *B. malayi* infections (36, 83).

Zoonotic *Brugia* Infections (American Brugian Filariasis)

A number of species of *Brugia* have been implicated in zoonotic transmission from animals to humans worldwide, including the United States. Some species normally infect dogs, cats, rabbits, and raccoons. Human infection with *Brugia* spp. is also referred to as brugian filariasis, zoonotic brugian filariasis, or brugian zoonosis. Mosquitoes may accidentally transmit the infective stages by feeding on an infected animal and then on a human approximately 2 weeks later (85, 86). Fewer than 50 cases have been reported, most of which have been documented since 1977. Most of the cases in the United States have been seen in the northeastern part of the country and have been due to *B. beaveri* from the raccoon (*Procyon lotor*) and possibly bobcat and mink and *B. leporis* from swamp rabbits (*Sylvilagus aquaticus*) and eastern cottontails (*Silvilagus floridanus alacer*) (71a). Cases have been reported from Malaysia, where differentiation of this infection from other endemic infections is difficult; reports of infections, most of which are probably due to *B. guyanensis* infecting the raccoon-like coatimundi (*Nasua nasua*) and the weasel-like grison (*Grison vittatus*), have also come from South America, specifically Colombia, Brazil, Ecuador, and Peru.

Patients with American brugian filariasis rarely exhibit any systemic symptoms. Usually patients present with a tender mass in the cervical, axillary, or inguinal region; the mass may have been present for a few weeks before the patient sought medical help. A number of different conditions may be suspected, including an enlarged lymph node, lymphoma, lipoma, and papilloma. Specific areas of the body that have been involved include the pulmonary artery, the groin, the area behind the ear, the chest wall, and the neck. Once the mass is removed and examined histologically, a nematode worm is often found lodged in a dilated lymphatic vessel in the capsule, cortex, or interior of the node. The worm is often living and the structure is well preserved; however, if the worm is dead, it is probably surrounded by a granulomatous reaction. Once the mass is removed, no further treatment is required. *B. beaveri* and *B. leporis* are the only described species known to occur in the United States; they are commonly found in the raccoon in Louisiana, and microfilariae have been recovered from bobcats in Louisiana and Florida.

In Malaysia, *B. pahangi* is a common parasite of dogs and cats and has been implicated in cases of lymphangitis and lymphadenitis. It may be very difficult to differentiate infections with *B. pahangi* from endemic infections with *B. malayi*, particularly since they have the same mosquito vectors and the microfilariae are very similar to those of *B. malayi*.

Tropical Pulmonary Eosinophilia

Tropical pulmonary eosinophilia (TPE) is associated with *W. bancrofti* or *B. malayi* infections causing diffuse pulmonary infiltrates with significant local and systemic eosinophilia and high levels of polyclonal and parasite-specific IgE (60). Migration of microfilariae through the pulmonary blood vessels with subsequent allergic hypersensitivity responses is thought to be the cause of this syndrome. The epidemiology of this condition is not well understood, particularly since there can be other causes. It seems to occur in long-term residents but not in travelers or residents who have been in the area of endemicity for a short time or only a few years. TPE can also occur many years after susceptible individuals have left the area. The majority of cases have been reported in India, Pakistan, Sri Lanka, Brazil, Guyana, and Southeast Asia.

Studies have shown numerous pulmonary granulomas, possibly indicating trapping of the parasites within the lungs. Because no blood-borne microfilariae are present in these patients, the syndrome may be due to an intense inflammatory response to the migration of microfilariae through the pulmonary blood vessels. No apparent reason for this outcome in a small number of patients has yet been confirmed.

TPE usually occurs in young men, and the onset is gradual, progressing over a period of months. Patients present with a nonproductive cough, wheezing, fever, generalized malaise, fatigue, and weight loss. Symptoms tend to occur at night and may be associated with the release of microfilariae that have become trapped in the lungs. The chest X ray usually shows diffuse interstitial or reticulonodular infiltrates 1 to 3 mm in diameter; increased bronchovascular markings are also seen.

Diagnosis is based on clinical, epidemiologic, and laboratory findings. Successful resolution of symptoms after treatment with DEC generally confirms the diagnosis; the same dosage schedule as for bancroftian filariasis is used. Failure to respond to DEC strongly suggests another diagnosis, including allergic granulomatosis with angitis, Wegener's granulomatosis, periarteritis nodosa, allergic bronchopulmonary aspergillosis, idiopathic hypereosinophilic syndrome, and severe asthma. TPE is an important diagnostic consideration in patients with eosinophilia, respiratory symptoms, and a history of exposure to this disease. Without therapy, TPE can lead to chronic and progressive respiratory disease and then to death. Prompt recognition and treatment with DEC is critical to patient outcome (15).

Loa loa

Loa loa, the African eye worm, was first noted in the eye of a Negro girl in the West Indies in 1770. In 1895, Argyll-Robertson described the adult worms that he extracted from the eye of a woman who had resided at Old Calabar in West Africa. The adult worms migrate through the subcutaneous tissues, producing intermittent "Calabar swellings," in addition to migrating beneath the conjunctiva. Approximately 13 million people are infected with *L. loa* in Central and West Africa. The vectors are biting flies (*Chrysops*) and are known as red flies in Africa.

Life Cycle and Morphology

The adult male and female worms live and migrate in the subcutaneous and deep connective tissues, and the microfilariae are found in the blood, where they can be ingested by mango flies or deerflies (*Chrysops* spp.) (Figure 12.13). Once ingested by the fly, the microfilariae of *L. loa* lose their sheaths, penetrate the gut wall, migrate to the fat body, undergo two molts, and become infective in approximately 10 to 12 days. Humans are infected when bitten by infected flies, and the infective larvae enter the skin through the bite wound. Development into adult worms takes about 6 to 12 months, and they can survive up to 17 years. The females measure 50 to 70 by 0.5 mm, while the males measure 30 to 35 by 0.3 to 0.4 mm. The microfilariae frequently are not detected in the blood until years after the adult worms are noted. The microfilariae have a diurnal periodicity whose peak occurs about midday; the remainder of the time they can be found in pulmonary capillaries. The microfilariae are sheathed and are 250 to 300 μm long. When stained, the body nuclei are continuous to the tip of the tail (Figures 12.5, 12.6, 12.14, and 12.15). The developmental cycle in the mango fly or deerfly is similar to the development of *W. bancrofti* in the mosquito.

Clinical Disease

The *Chrysops* bite results in erythema, swelling, and itching, symptoms which can worsen with the presence

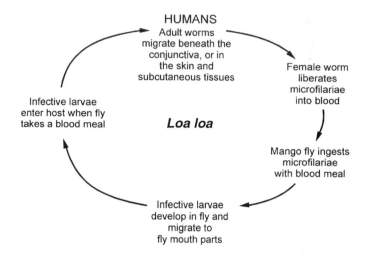

Figure 12.13 Life cycle of *Loa loa*.

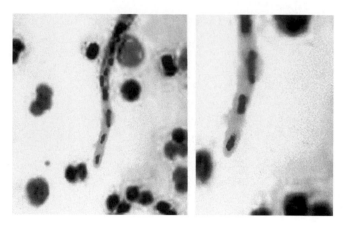

Figure 12.15 *Loa loa*. (Left) Microfilarial tail; note that the tail nuclei go to the tip of the tail. (Right) Microfilarial tail enlarged; note that the tail nuclei go to the tip of the tail.

of infective larvae. The adult worms normally migrate through the subcutaneous tissue at a rate of about 1 cm/min. This migration is not painful and is seldom noticed unless the worm is passing over the bridge of the nose or through the conjunctiva of the eye (Figures 12.16 to 12.18). Many patients with active *L. loa* infections do not have a microfilaremia. The most common pathologic sequelae associated with *L. loa* infections are Calabar or fugitive swellings (angioedema). Calabar swellings are localized subcutaneous edemas, a type of inflammatory reaction brought about by a host response to the worm

Figure 12.14 *Loa loa* microfilariae in blood. Note the presence of the sheath and the tail nuclei positioned to the end of the tail.

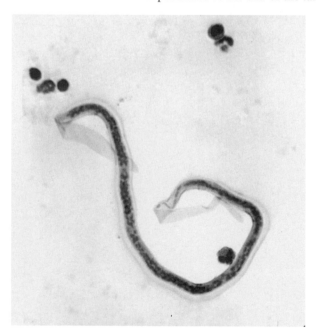

or its metabolic products. Calabar swellings may be found anywhere on the body but predominate on the extremities. The swellings develop rapidly over a few hours and may be preceded by localized pain, pruritus, and urticaria; the swellings usually last 1 to 3 days. Serious complications due to loiasis have included cardiomyopathy, encephalopathy, nephropathy, and pleural effusions. Even with low doses of DEC to initiate treatment of patients from the area of endemicity followed by gradually increasing doses combined with corticosteroids, microfilaremia encephalitis is frequently fatal (61). Host genetic factors appear to be involved in acquiring loiasis in areas of endemic infection, because individuals who are not microfilaremic by the age of 30 have a low probability of becoming positive later in life (46). A Th2-type immune response has been found in microfilaremic *L. loa*-infected patients (125), and IgE antibodies recognized more antigens in the microfilarial stage than in the adult worms (5)

Residents from areas of endemicity have been generally noted to have asymptomatic microfilaremia (90% of residents) and occasional episodic Calabar swellings (16% of residents), moderate eosinophilia, and variable antibody levels. Infections in nonresidents were more often symptomatic. Due to increased international travel, the diagnosis of loiasis must be considered in patients who have visited these areas and have a history of eosinophilia, migratory angioedema, and urticarial vasculitis (95). Approximately 10% are microfilaremic and have frequently reoccurring Calabar swellings, hypereosinophilia of over 70%, increased levels of IgE, and very high antibody titers (80). However, complications such as renal disease, encephalopathy, and cardiomyopathy have been found (80).

The most serious complications include meningoencephalitis, occurring in patients with large numbers of circulating microfilariae who have undergone treatment

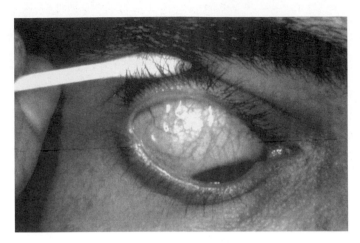

Figure 12.16 *Loa loa.* Subconjunctival nematode. (From A Pictorial Presentation of Parasites: A cooperative collection prepared and/or edited by H. Zaiman. Photograph courtesy of D. Gendelman.)

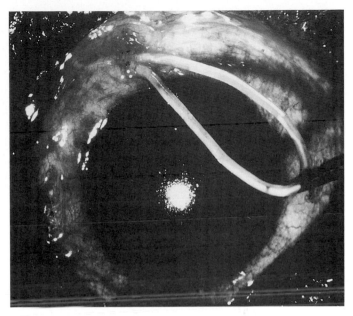

Figure 12.18 *Loa loa* adult extracted from the eye. (Armed Forces Institute of Pathology photograph.)

with DEC; immune complex glomerulonephritis; and endomyocardial fibrosis.

Diagnosis

Since the microfilariae of *L. loa* do not appear in the blood until years after the adult worms or the host

Figure 12.17 *Loa loa* beneath the conjunctiva of the eye. (Armed Forces Institute of Pathology photograph.)

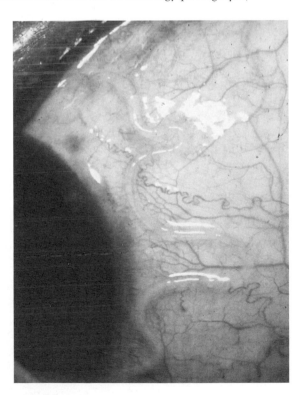

reaction to infection is noted, the diagnosis is frequently made on the basis of clinical history (Calabar swellings, worm migrations through the eye, eosinophilia, and residence in an area of endemicity). Microfilariae can be detected in the blood by using methods similar to those used to detect *W. bancrofti* or *B. malayi*. Microfilariae are shed by the females on a random schedule; therefore, multiple samples over a period of days may have to be tested. Individuals not from the area of endemicity usually are amicrofilaremic (63). The microfilariae exhibit diurnal periodicity; therefore, blood samples should be collected during daytime, preferably between 10 a.m. and 3 p.m. The sheathed microfilariae must be differentiated from those of *W. bancrofti* and *B. malayi* (Figures 12.3 to 12.5, 12.14, 12.15, and 12.19; Table 12.1). Demonstration of the sheathed microfilariae with nuclei extending to the tip of the tail is sufficient for the diagnosis (Figures 12.14 and 12.15). The infection can also be diagnosed by detection or removal of adults from the eye (Algorithm 12.2)

Occasionally, in asymptomatic individuals, microfilariae can be recovered from peripheral blood films, urine, sputum, cerebrospinal fluid, cervicovaginal smears, blood vessel biopsy specimens, or autopsy specimens (71a). Although Giemsa stain does not reveal the sheath, Delafield's hematoxylin may do so and is recommended. Biopsy specimens of Calabar swellings prior to therapy rarely demonstrate the presence of adult worms.

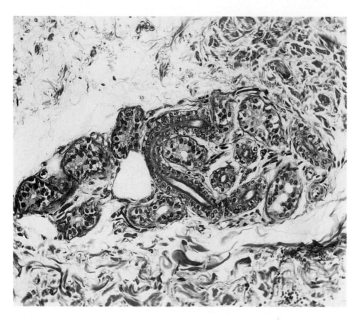

Figure 12.19 *Loa loa* microfilaria in the capillaries of the dermis. (Armed Forces Institute of Pathology photograph.)

Treatment

Surgical removal of the adult worms as they are migrating across the bridge of the nose or through the conjunctiva is relatively simple (Figure 12.18). DEC is an effective treatment; however, in patients with a heavy microfilaremia, anti-inflammatory drugs may also have to be administered to reduce severe side effects. The most serious sequelae are neurologic complications resembling the encephalitis that is commonly associated with a high micro-

Algorithm 12.2 Loiasis.

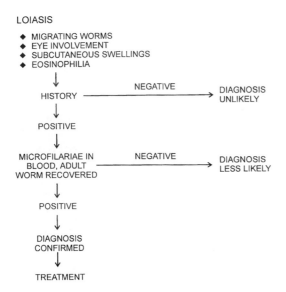

filaremia. If neurologic disorders develop, it is mandatory that therapy with DEC be stopped. Some individuals may require multiple courses of therapy to be clinically cured. Although DEC is curative in most individuals, relapses can occur within the first 12 months and up to 8 years after treatment (64). In areas of endemic infection, DEC at a prophylactic dose of 300 mg weekly is effective in preventing loiasis in long-term visitors (2). The occurrence of serious adverse effects following DEC treatment of onchocerciasis in areas where *L. loa* is endemic has been increasingly reported over the past decade. These include a severely disabling, and potentially fatal, encephalopathy, which appears to correlate with a high load of *L. loa* microfilariae (>30,000/ml).

Both mebendazole and ivermectin have been used to treat loiasis. The use of mebendazole is not promising, whereas ivermectin is effective in reducing microfilaremia (57). Albendazole, a benzimidazole derivative related to mebendazole, is effective in reducing microfilaremia, and there is a lower risk of encephalopathy; however, repeated courses of therapy may be required (2, 62). Individuals treated with albendazole have a significant reduction in the levels of eosinophils, antifilarial IgG, and IgG4. Although there were no cures, it is possible that albendazole directly affects the adult stages.

Epidemiology and Prevention

Loiasis is endemic to the Central and West African rain forests and may infect as many as 13 million people. Disease distribution depends on the vectors, which breed in wet mud at the side of streams under the rain forest canopy. Flies appear to be attracted by the movement of people or vehicles, as well as rising smoke. An excellent example of vector-host interactive sites would include rubber plantations with a dense high canopy and human workers. The *Chrysops* flies are more common in the rainy season, and exposure differences result in higher infection rates in adults rather than children. Although vector control is one means of prevention, the inaccessibility of breeding sites has proven to be a problem. Clearing of growth around houses and use of screens and protective clothing have all been helpful.

In general, control of loiasis also depends to a great degree on chemoprophylaxis with DEC. Although other therapeutic approaches have been tried, including mebendazole and ivermectin, problems remain with both options.

Mansonella ozzardi

Mansonella ozzardi infections are confined to the New World in Central and South America and the Caribbean islands. Areas of endemic infection in South America include northern Argentina, Bolivia, Brazil, Colombia,

Ecuador, and Peru. The infection was first noted by Ozzard in an Indian in South America and was described by Manson. Humans are the only known natural definitive hosts, and prevalence rates may reach 70% in certain areas.

Life Cycle and Morphology

The adult male and female worms live in the subcutaneous and connective tissues, and the microfilariae are found in the blood, where they can be ingested by *Culicoides* spp. (midges) or *Simulium* spp. (blackflies), the insect vectors. The microfilariae are nonperiodic and unsheathed, with body nuclei that do not extend to the tip of the tail, and are 173 to 240 µm long (Figures 12.5, 12.6, and 12.20). Microfilariae of *M. ozzardi* resemble those of *M. perstans*; however, the tail of *M. ozzardi* tends to be pointed and slightly flexed and has a longer caudal space (Figure 12.20) (71a). Apparently, humans are the only recognized natural definitive hosts.

Clinical Disease

There is no definitive information available regarding the incubation period in humans. The infections generally produce no symptoms (West Indies); however, they may cause adenopathy in the lower extremities, knee and ankle pain, pruritus, urticaria, eosinophilia, and fever (Amazon region). Although the worms are generally found in the thoracic and peritoneal cavities and possibly the lymphatics, there is generally no inflammation. Microfilariae were detected in pericardial and pleural fluid from a patient with adenocarcinoma (103). It might be possible to find microfilariae throughout the body of an immunocompromised patient. A recent cross-sectional

Figure 12.20 *Mansonella ozzardi* microfilaria. Note that the microfilaria is unsheathed and that the tail is pointed and slightly flexed and has a longer caudal space than *M. perstans*.

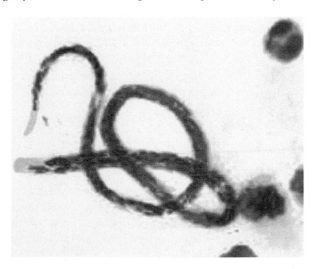

survey in the Chaco region of Bolivia showed that 26% (77 of 296) and 0.7% (2 of 298) of the rural populations of two village areas harbored *M. ozzardi* microfilariae. No sex differences were noted, and the lowest prevalence was in 0- to 14-year-old children. There was a sharp increase in 25- to 34-year-old adults, and the prevalence continued to increase in the older age groups. Microfilaremia also increased with age. No significant association between clinical symptoms (fever, skin rash, pruritus, headache, lymphedema, elephantiasis, and articular pain) and microfilaremia was seen (10).

Diagnosis

Methods similar to those used for *W. bancrofti* and *B. malayi* are used to detect microfilariae in the blood. The microfilariae are nonperiodic and must be differentiated from other blood-borne microfilariae (Figures 12.5, 12.6, and 12.20; Table 12.1). Since the microfilariae are nonperiodic, blood specimens can be taken at any time during the day.

Treatment

Asymptomatic patients need not be treated. Ivermectin has been successfully used to treat infections (57); however, DEC treatment may not be effective. The overall prognosis is excellent.

Epidemiology and Prevention

The infection is indigenous in the Caribbean islands and in Central and South America. *Culicoides* spp. (midges) are the main vectors except in the Amazon basin area of Brazil and Colombia, where blackflies (*Simulium* spp.) are the vectors. Humans are the only known reservoir hosts. The infection may be prevented by the use of insecticides to prevent fly bites. Because of the lack of public health significance of this infection, control programs have not been designed.

Mansonella perstans

Mansonella perstans was previously known as *Dipetalonema perstans* or *Acanthocheilonema perstans*. Manson discovered the microfilariae in a blood smear taken from an African patient, and Daniels described the adult worm collected from a South American Indian. In certain areas, infection rates of more than 50% have been reported.

Life Cycle and Morphology

Adult worms are found in the body cavities (peritoneal and pleural), and their microfilariae are found in the blood, where they are ingested by *Culicoides* spp. (midges), the insect vectors. The microfilariae are nonperiodic and unsheathed, with body nuclei that extend to the tip of the tail, and are 190 to 200 µm long (Figures 12.5, 12.6, and

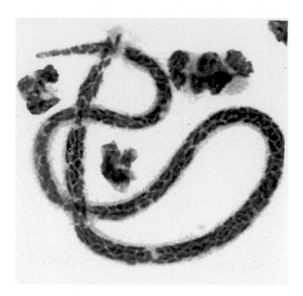

Figure 12.21   *Mansonella perstans* microfilaria. Note that the microfilaria is unsheathed and that there is a cephalic space and very little caudal space.

12.21). There is no caudal space seen in microfilariae of *M. perstans*, and the posterior end tends to be blunt.

Clinical Disease

Even though the adult worms inhabit the body cavities, there is minimal literature on the pathology caused by this infection. Live adults appear to produce little or no host reaction; however, eosinophilia is common. Pruritus, abdominal pain, urticaria, and Calabar-like swellings on the arms, shoulders, and face have been described (77). *M. perstans* filariasis has been found in the bone marrow of a human immunodeficiency virus (HIV)-infected patient (74). Ocular symptoms include periorbital edema and conjunctival irritation. In Uganda, acute periorbital inflammation is common; this condition is known as bung-eye or bulge-eye (71a). These types of changes are also reported from Nigeria and Sudan.

Diagnosis

The methods used to detect microfilariae in the blood are similar to those used to detect *W. bancrofti* and *B. malayi* microfilariae. The microfilariae are nonperiodic and must be differentiated from other blood-borne microfilariae (Figures 12.5, 12.6, and 12.21; Table 12.1). Although these microfilariae are nonperiodic, larger numbers can be found in blood collected at night.

Treatment

Reports on the effectiveness of DEC are contradictory; the microfilariae are eliminated, but repeated therapy is re-

quired. However, mebendazole has been used effectively. Ivermectin is not effective in treating *M. perstans* infections (57). Dead worms release substances that are very antigenic and cause a dramatic immune response; the worms are surrounded by inflammatory exudates, and there may be abscesses, granulomas, and scarring. In a recent study, six different therapeutic protocols were undertaken for the treatment. DEC reduced microfilarial density but seldom eliminated the infection after a single treatment. Mebendazole was more effective than DEC in eliminating the infection, while ivermectin and praziquantel showed no modification of microfilarial concentration. Thiabendazole demonstrated a small but significant activity against the infection, while combination treatments with DEC and mebendazole showed significantly higher activity than did single drugs alone (17).

Epidemiology and Prevention

The infection occurs in tropical Africa, the West Indies, and along the Atlantic coast in Central and South America. *Culicoides* spp. (midges) are the insect vectors, and humans, gorillas, and monkeys are the known reservoir hosts. The intermediate host (*Culicoides* spp.) breeds in swamps and jungles, and so eradication or control measures are not practical. Insecticides can be used to prevent fly bites.

Mansonella streptocerca

Until recently, *M. streptocerca* was named *Dipetalonema streptocerca*. It was also previously called *Acanthocheilonema* or *Tetrapetalonema*. Microfilariae were first noted in skin snips obtained from a West African native in 1922, and the adult worms were described in the 1970s.

Life Cycle and Morphology

Adult female and male worms live in the dermal tissues, and the microfilariae are found in the skin, where they can be ingested by *Culicoides* spp. (midges), the insect vectors. The microfilariae are unsheathed, with body nuclei that extend to the tip of the tail, and are 180 to 240 μm long (Figure 12.5). When they are immobile, the microfilariae curve the tail to assume a "shepherd's crook" configuration.

Clinical Disease

Although unconfirmed, the incubation period is suspected to be several months. There are limited reports on the symptoms and pathology associated with *M. streptocerca* infections, and many patients are asymptomatic. Pruritic dermatitis with hypopigmented macules on the skin and lymphadenopathy have been common complaints. Many patients have inguinal adenopathy.

In areas where leprosy is endemic, the hypopigmented dermal lesions of streptocerciasis are often misdiagnosed as leprosy; however, no sensory changes are seen, in contrast to leprosy. It is recommended that these lesions be examined histologically for confirmation of either filariasis, leprosy, or both.

Diagnosis

Any patient from the area of endemicity who presents with hypopigmented macules and pruritus should be suspected of having the infection. However, this appearance can be confused with leprosy, and some patients are misdiagnosed and inappropriately treated for leprosy for long periods. The diagnosis is made by the detection of microfilariae from skin snips taken over the scapula and examined as wet mounts. Orihel (84) noted that the microfilarial tails are split rather than blunt, as they were previously described. These microfilariae must be differentiated from O. volvulus, which also can be detected in the skin (Figure 12.5; Table 12.1).

Microfilaria can be found primarily in the upper third of the dermis in the dermal collagen. Changes in the skin include sclerosis of dermal papillae, edema, incontinence of melanin, fibrosis, and cellular infiltrates around blood vessels, and the lymphatics are frequently dilated. In untreated patients, adult worms are occasionally found in the dermis; they do not tend to cause an immune response when alive.

Treatment

The infection responds to treatment with DEC; however, a complaint from most patients is the intense pruritus that develops. The long-term effect of a single oral ivermectin dose of 150 µg/kg of body weight on M. streptocerca was studied in western Uganda. The results indicate that ivermectin is highly effective against M. streptocerca

and that a single dose leads to a sustained suppression of microfilariae in the skin. In Africa, ivermectin is used for mass treatment to control O. volvulus and W. bancrofti. Since these filarial infections are often coendemic with M. streptocerca, the treated population may receive suppression benefits for M. streptocerca microfilaria as well (37).

Epidemiology and Prevention

The infection has been found in western and central sub-Saharan Africa, as far east as western Uganda, and as far south as northern Angola. The incidence may reach 90% in certain areas of dense forests of equatorial Congo. Both monkeys and humans harbor the infection; however, there is speculation that the species infecting monkeys may differ from those infecting humans. Biting midges (*Culicoides* spp.) are the insect vectors. There are no control programs aimed specifically at M. streptocerca or *Culicoides* spp. Insecticides can be used to prevent fly bites.

Onchocerca volvulus

Onchocerca volvulus microfilariae were first found by O'Neil in 1875, and the adult worms were detected by Leuckart in 1893 in nodules removed from the scalp and chest of a West African native. Onchocerciasis, which is also known as river blindness, is a major public health problem because the infection is a leading cause of blindness in the world and may also cause significant disfigurement of the skin. Currently, 17.7 million people are infected with O. volvulus; of these, 270,000 are blind and 500,000 have severe visual impairment (52). Unfortunately, these figures are probably underestimates.

Life Cycle and Morphology

Adult female and male worms lie within fibrous tissue capsules in the dermis and subcutaneous tissues.

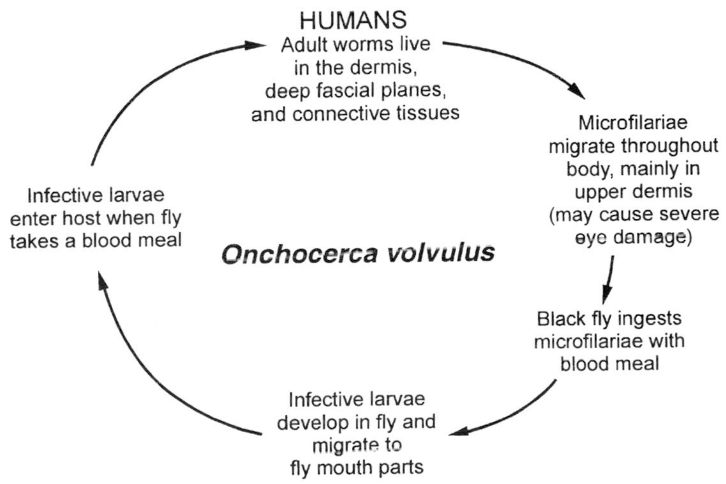

Figure 12.22 Life cycle of *Onchocerca volvulus*.

The
to $
cm
fila
and
Sim
blac
con
skir
the
infe
bos
mea
infe
(Fig
der
are
the
The
12 ₁
gra
beg
mea
15 :

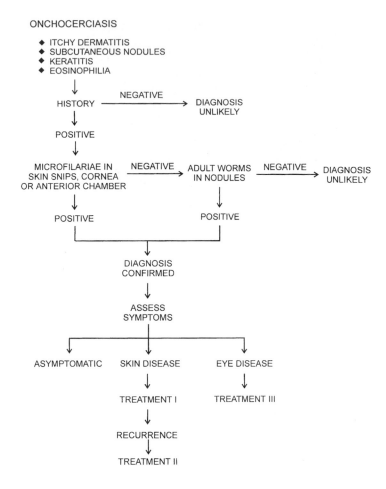

ONCHOCERCIASIS

♦ ITCHY DERMATITIS
♦ SUBCUTANEOUS NODULES
♦ KERATITIS
♦ EOSINOPHILIA

HISTORY ————— NEGATIVE ————→ DIAGNOSIS UNLIKELY

POSITIVE

MICROFILARIAE IN SKIN SNIPS, CORNEA OR ANTERIOR CHAMBER ——— NEGATIVE ——→ ADULT WORMS IN NODULES ——— NEGATIVE ——→ DIAGNOSIS UNLIKELY

POSITIVE   POSITIVE

DIAGNOSIS CONFIRMED

ASSESS SYMPTOMS

ASYMPTOMATIC   SKIN DISEASE   EYE DISEASE

TREATMENT I   TREATMENT III

RECURRENCE

TREATMENT II

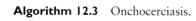

Algorithm 12.3 Onchocerciasis.

between their knees for about 10 min. Once they sit up, the microfilariae become concentrated behind the central cornea and are easier to see and count.

The Mazzotti test can be used when results with skin snips are negative; however, there may be severe reactions in patients with heavy microfilarial burdens. A single oral dose of 25 to 50 mg of DEC is given; individuals often develop intense pruritus within a few hours. The itching can be severe and may be accompanied by erythema, edema, and papules; it usually subsides within a few days. Topical application of DEC in a patch test eliminates some of these problems; the appearance of a localized inflammatory reaction at the site of the patch indicates a positive test. The Mazzotti test should be limited to patients with light infections who have no microfilariae in the eyes.

The use of recombinant antigens for serologic diagnosis can help overcome the problems associated with poor sensitivity and specificity, particularly in areas where other human filarial infections overlap with those caused by *O. volvulus* (114). A cocktail of recombinant *O. volvulus* antigens for serologic diagnosis with the potential to predict the endemicity of onchocerciasis infection has been developed (16). When a sentinel population of 5- to 15-

year-old children was used to compare communities, the serologic index was sufficiently sensitive to measure the changes in endemicity that would be required for control programs. However, serologic testing cannot distinguish between current and past infection and tends to be more important in monitoring specific groups.

Treatment

Surgical removal of detectable nodules, especially in the head region, is simple and easy to perform. Nodule removal from the head, particularly in children, reduces the risk and severity of ocular complications. Removal of adult worms decreases the exposure of the host to microfilariae, which helps decrease the risk of complications. Because many nodules are nonpalpable or the adults may be freely migrating, the removal of nodules may have little effect on the total microfilarial or worm burden. Removal of enlarged regional lymph glands has been traditional therapy for sowda in Yemen.

Ivermectin is the drug of choice for treating onchocerciasis (100). It markedly reduces microfilaria numbers within 48 h and blocks the release of microfilariae from the uterus of the adult female worm for up to 6 months.

In areas where leprosy is endemic, the hypopigmented dermal lesions of streptocerciasis are often misdiagnosed as leprosy; however, no sensory changes are seen, in contrast to leprosy. It is recommended that these lesions be examined histologically for confirmation of either filariasis, leprosy, or both.

Diagnosis

Any patient from the area of endemicity who presents with hypopigmented macules and pruritus should be suspected of having the infection. However, this appearance can be confused with leprosy, and some patients are misdiagnosed and inappropriately treated for leprosy for long periods. The diagnosis is made by the detection of microfilariae from skin snips taken over the scapula and examined as wet mounts. Orihel (84) noted that the microfilarial tails are split rather than blunt, as they were previously described. These microfilariae must be differentiated from *O. volvulus*, which also can be detected in the skin (Figure 12.5; Table 12.1).

Microfilaria can be found primarily in the upper third of the dermis in the dermal collagen. Changes in the skin include sclerosis of dermal papillae, edema, incontinence of melanin, fibrosis, and cellular infiltrates around blood vessels, and the lymphatics are frequently dilated. In untreated patients, adult worms are occasionally found in the dermis; they do not tend to cause an immune response when alive.

Treatment

The infection responds to treatment with DEC; however, a complaint from most patients is the intense pruritus that develops. The long term effect of a single oral ivermectin dose of 150 µg/kg of body weight on *M. streptocerca* was studied in western Uganda. The results indicate that ivermectin is highly effective against *M. streptocerca*

and that a single dose leads to a sustained suppression of microfilariae in the skin. In Africa, ivermectin is used for mass treatment to control *O. volvulus* and *W. bancrofti*. Since these filarial infections are often coendemic with *M. streptocerca*, the treated population may receive suppression benefits for *M. streptocerca* microfilaria as well (37).

Epidemiology and Prevention

The infection has been found in western and central sub-Saharan Africa, as far east as western Uganda, and as far south as northern Angola. The incidence may reach 90% in certain areas of dense forests of equatorial Congo. Both monkeys and humans harbor the infection; however, there is speculation that the species infecting monkeys may differ from those infecting humans. Biting midges (*Culicoides* spp.) are the insect vectors. There are no control programs aimed specifically at *M. streptocerca* or *Culicoides* spp. Insecticides can be used to prevent fly bites.

Onchocerca volvulus

Onchocerca volvulus microfilariae were first found by O'Neil in 1875, and the adult worms were detected by Leuckart in 1893 in nodules removed from the scalp and chest of a West African native. Onchocerciasis, which is also known as river blindness, is a major public health problem because the infection is a leading cause of blindness in the world and may also cause significant disfigurement of the skin. Currently, 17.7 million people are infected with *O. volvulus*; of these, 270,000 are blind and 500,000 have severe visual impairment (52). Unfortunately, these figures are probably underestimates.

Life Cycle and Morphology

Adult female and male worms lie within fibrous tissue capsules in the dermis and subcutaneous tissues.

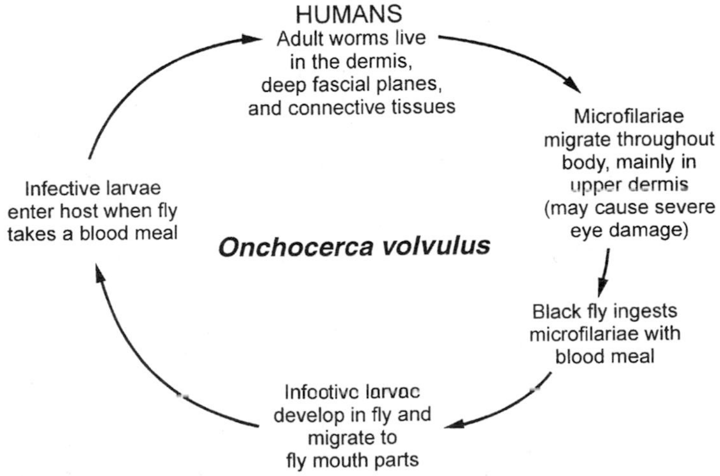

Figure 12.22 Life cycle of *Onchocerca volvulus*.

HUMANS
Adult worms live
in the dermis,
deep fascial planes,
and connective tissues

Microfilariae
migrate throughout
body, mainly in
upper dermis
(may cause severe
eye damage)

Infective larvae
enter host when fly
takes a blood meal

Onchocerca volvulus

Black fly ingests
microfilariae with
blood meal

Infective larvae
develop in fly and
migrate to
fly mouth parts

The female adult is very long and thin, measuring up to 50 cm by 0.3 mm, while the adult male is 2.5 to 5 cm by 125 to 200 μm. Gravid females produce microfilariae, which invade the subcutaneous tissues, skin, and eyes. The infection is transmitted by a species of *Simulium* (blackfly or buffalo gnat). Apparently the blackflies are pool feeders, and the saliva probably contains anticoagulants, as well as attractants for the skin-dwelling microfilariae (21). After being ingested, the microfilariae migrate to the flight muscles, become infective in about 6 to 12 days, and migrate to the proboscis, where they are transmitted when the next blood meal is taken. Humans are infected when bitten by the infected fly, and larvae are deposited into the bite site (Figure 12.22). Microfilariae are normally found in the dermis and rarely in the blood, sputum, or urine. They are unsheathed, with body nuclei that do not extend to the tail tip, and are 221 to 360 μm long (Figure 12.5). The infective larvae undergo two molts during the next 12 months prior to becoming mature adult worms. The gravid females produce thousands of microfilariae daily beginning about 12 to 15 months after infection. The mean reproductive life span of the adult worm is up to 15 years, with maximum reproduction during the first

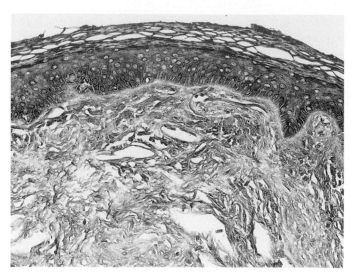

Figure 12.24 *Onchocerca volvulus* microfilariae in the dermis. (Armed Forces Institute of Pathology photograph.)

5 years of patency; the microfilariae may live for up to 15 years (71a).

Clinical Disease

The prepatent period in humans is 3 to 15 months. Light infections may produce no symptoms, and worms may be free in the tissue; however, in heavy infections, major disease manifestations include dermatitis, onchocercomas (subcutaneous nodules containing adult worms), lymphadenitis, and blindness. Even though there is an inflammatory reaction that causes the formation of a fibrotic capsule (onchocercoma) around the adult, the main pathology appears to be directed against the microfilariae (67) (Figures 12.23 and 12.24).

Onchocercomas. Light to moderate infection may produce an itchy rash, while heavy infections (up to 200 million microfilariae) may lead to severe sequelae. Heavy infections are the result of continued exposure to infective larvae over a period of many years. The onchocercoma is a firm, round, and nontender subcutaneous nodule that contains the adult worms. In Africa, the nodules are commonly found on the trunk (particularly the hip area) or limbs, while in Mexico and Guatemala, they are frequently found on the scalp (Figures 12.25 to 12.27).

Skin Changes. Some individuals with onchocerciasis have clinically normal skin, whereas others have pruritus and disfiguring skin lesions. The pruritus may be so severe as to cause sleeplessness, fatigue, and weakness, symptoms that may affect the individual's ability to work and interact socially. Scratching often leads to ulcers, bleeding,

Figure 12.23 *Onchocerca volvulus* microfilariae in the skin (cross section). (From A Pictorial Presentation of Parasites: A cooperative collection prepared and/or edited by H. Zaiman.)

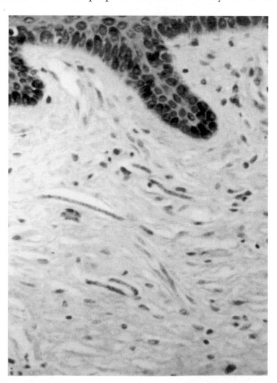

Figure 12.25 Onchocerciasis. The nodule on the patient's head contains adult worms of *Onchocerca volvulus*. (From A Pictorial Presentation of Parasites: A cooperative collection prepared and/or edited by H. Zaiman.)

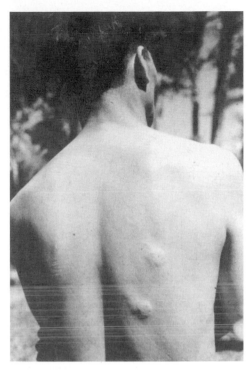

Figure 12.26 Onchocerciasis. Nodules on the back prior to surgical removal. (From A Pictorial Presentation of Parasites: A cooperative collection prepared and/or edited by H. Zaiman.)

and secondary bacterial and/or fungal infections. The skin may be painful, hot, and edematous. This inflammatory condition may recur numerous times, eventually resulting in permanent thickening of the skin. Acute attacks of onchodermatitis have resulted in a purplish skin discoloration on the trunk and upper limbs known as mal morado in Central America. Chronically infected skin loses its elasticity and becomes hypertrophic or thickened. As a result of atrophy and loss of skin elasticity, patients may develop premature exaggerated wrinkling of the skin (Figure 12.28). In Central America, these skin changes are frequently seen as noticeable thickening of the earlobes and thickening of facial skin to mimic leonine facies (seen in lepromatous leprosy). In Africa, the same skin changes occurring in the hip region, usually around the inguinal lymph nodes, produce a condition known as hanging groin, which may predispose infected individuals to inguinal and femoral hernias. Lymphadenopathy in the inguinal and femoral areas is common among Africans (Figure 12.29).

Skin may become hypo- or hyperpigmented. In Africa, patients may exhibit localized areas of spotty depigmentation surrounded by slightly hyperpigmented zones occurring mainly on the shins; this condition is known as leopard skin (Figure 12.30). These white areas have an absence of melanin, and the prevalence of skin depigmentation has been proposed as a marker for endemic onchocerciasis.

Sowda is a chronic hyperreactive form of onchocerciasis occurring in a subset of infected individuals, usually in Arabia, Ecuador, Guatemala, Sudan, and West Africa. It is characterized by a severe papular dermatitis usually localized to one limb, typically a leg, with darkening of the skin. The number of microfilariae in the skin is extremely small, and there is extensive follicular hyperplasia of the regional lymph nodes. Individuals with sowda have significantly higher antibody titers, including IgE and antigen-specific IgG, than do individuals with generalized onchocerciasis (45). These skin lesions improve after therapy. It is possible that this form of onchocerciasis is linked to the formation of autoantibodies to defensins and the hyperimmune response to the parasite.

Ocular Involvement. A major complication of onchocerciasis has been ocular lesions leading to bilateral blindness. The prevalence of blindness depends on the organism strain and the prevalence of disease within an area, while the degree of eye involvement depends on the duration and severity of infection (Figure 12.31). Visual problems can include fluffy corneal opacities, sclerosing keratitis, and iridocyclitis leading to glaucoma and cataracts; chorioretinitis and optic atrophy also occur. Eye damage is caused by living and dead microfilariae in the cornea, anterior and posterior chambers, iris, retrolental space, vitreous, choroid, retina, sclera, and optic nerve.

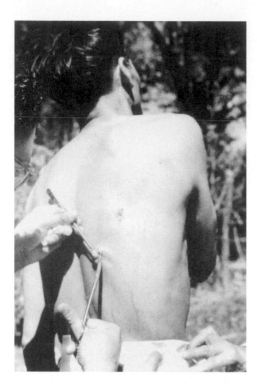

Figure 12.27 Onchocerciasis. Surgical removal of a nodule on the back. (From A Pictorial Presentation of Parasites: A cooperative collection prepared and/or edited by H. Zaiman.)

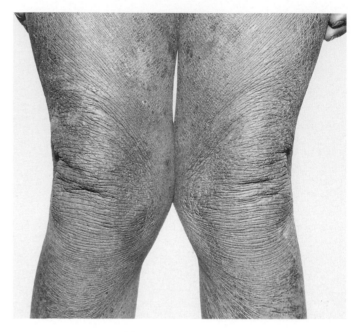

Figure 12.28 Wrinkling of the skin due to onchocerciasis. (Armed Forces Institute of Pathology photograph; published in D. H. Connor, N. E. Morrison, F. Kerdel-Vegas, H. A. Berkoff, F. Johnson, R. Tunnicliffe, F. C. Failings, L. N. Hale, and K. Lindquist, *Hum. Pathol.* **1:**553–579, 1970.)

Progressive pathology generally develops over time, with resulting blindness seen in adults.

Ocular lesions are a result of the host immune response to the microfilariae. Circulating antibodies to retinal antigens have been found in patients with onchocercal retinopathy (102, 126). Whether these autoantibodies play an appreciable role in ocular pathology is unknown. Chorioretinal changes have been noted in 75% of the population in areas of West Africa where the infection is endemic.

In West Africa, the severity of ocular pathology is based on strain differences as confirmed by DNA sequence determination. Classification of these strains based on sequence differences correlates very well with pathologic differences. PCR testing has been used to map the distribution of the blinding and nonblinding strains in Nigeria (52, 81). Preliminary studies indicated that certain species of blackfly transmitted the different strains; delineation of these transmission zones has become more difficult, and transition zones appear to contain both strains and/or hybrids. Ocular problems such as severe sclerosing keratitis, anterior uveitis, and blindness are more prevalent in the savanna regions of West Africa but much less common in the rain forest areas.

Immune System Effects. In individuals with microfiladermia, there is a lack of a cellular immune response;

however, there is a Th2 response, which is most rigorous in patients with the sowda form of onchocercal dermatitis. Individuals with chronic onchocerciasis have an abundance of T cells producing IL-4 and IL-5, and they have markedly high levels of antigen-specific IgE and IgG4 antibodies and eosinophilia (47, 88).

Figure 12.29 Hanging groin due to *Onchocerca volvulus* femoral lymphadenitis. (Armed Forces Institute of Pathology photograph.)

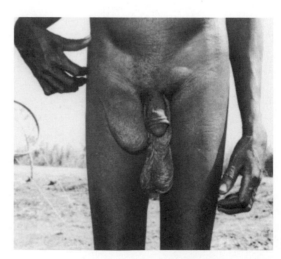

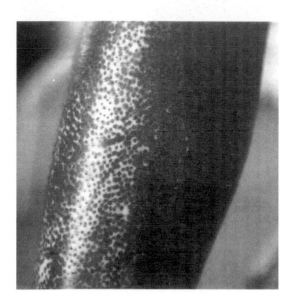

Figure 12.30 Depigmentation due to onchocerciasis. (Armed Forces Institute of Pathology photograph.)

Recent studies suggest a role for mast cells and tissue basophiles in the pathogenesis of onchocerciasis. The IgE bound to the plasma membrane of these mediator-rich cells could contribute to the inflammatory reaction; parasite-soluble components and cuticular proteins are inducers of histamine release. Possible recognition of cuticle internal matrix could occur during molting and after attack by immune responses directed against other targets. If this is true, histamine-containing cells could participate in the body's response to the parasite. The

Figure 12.31 Keratitis due to onchocerciasis. (Armed Forces Institute of Pathology photograph.)

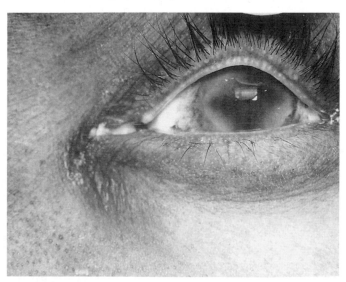

chronic inflammatory reaction seen in onchocerciasis may be mediated by mechanisms involving IgE. Product release from basophiles or mast cells acting on eosinophils could be very relevant to the outcome of the infection (48). It also appears that the eosinophil chemokine eotaxin plays a role in recruitment of eosinophils to the dermis in early-stage onchocercal skin disease (91).

In Uganda, as well as in other areas of Africa, *O. volvulus* and HIV are two immunocompromising infectious agents of major health concern. Cellular immune responses in patients coinfected with these two agents suggest an HIV-related lack of specific reactivity to *O. volvulus* antigen and impairment of IL-4 and IL-5 production in addition to the lack of an IFN-γ response to antigenic stimulation (105).

Diagnosis

In areas of endemic infection, the clinical diagnosis of onchocerciasis is not difficult when individuals present with typical features such as hanging groin, leopard skin, skin atrophy, or subcutaneous nodules. However, skin lesions must be differentiated from scabies, insect bites, streptocerciasis, contact dermatitis, hypersensitivity reactions, traumatic or inflammatory depigmentation, tuberculoid leprosy, dermatomycoses, and treponematoses.

Microfilariae are commonly detected in the skin and must be differentiated from *M. streptocerca* microfilariae, which are also found in the skin (Figure 12.5; Table 12.1). Multiple skin snips are recommended, with the sensitivity of six snips per person being estimated at 91.6% (115). Infrequently, microfilariae are found in the urine, blood, or sputum, particularly after initiation of DEC treatment. Skin snips should be examined 3 to 6 months after therapy to determine whether the treatment was successful (Algorithm 12.3). PCR has been used successfully with skin snips to detect infections. The PCR method is significantly more sensitive than routine microscopic methods in detecting microfilariae in skin snips and would be more reliable in detecting active infections than are serologic assays (128). Microscopic examination of skin snips cannot detect prepatent or low-level infections.

The diagnosis can also be confirmed by finding onchocercomas and adult worms within fibrotic tissues. Ultrasonography can distinguish onchocercomas from lymph nodes, lipomas, fibromas, and a foreign-body granuloma. A typical nodule appears as a central homogeneous area containing some dense particles with a lateral acoustic shadow.

In addition, slit-lamp examination of the eye to detect and visualize microfilariae in the anterior chamber or retroillumination to detect corneal microfilariae can be performed. Patients are instructed to place their heads

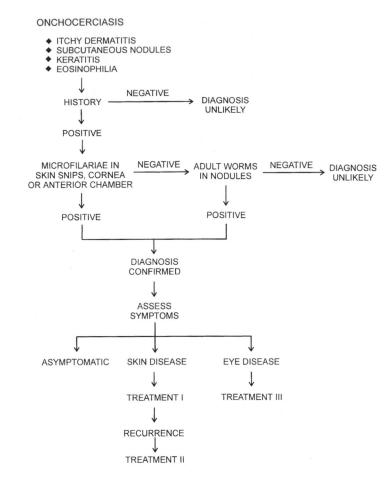

Algorithm 12.3 Onchocerciasis.

between their knees for about 10 min. Once they sit up, the microfilariae become concentrated behind the central cornea and are easier to see and count.

The Mazzotti test can be used when results with skin snips are negative; however, there may be severe reactions in patients with heavy microfilarial burdens. A single oral dose of 25 to 50 mg of DEC is given; individuals often develop intense pruritus within a few hours. The itching can be severe and may be accompanied by erythema, edema, and papules; it usually subsides within a few days. Topical application of DEC in a patch test eliminates some of these problems; the appearance of a localized inflammatory reaction at the site of the patch indicates a positive test. The Mazzotti test should be limited to patients with light infections who have no microfilariae in the eyes.

The use of recombinant antigens for serologic diagnosis can help overcome the problems associated with poor sensitivity and specificity, particularly in areas where other human filarial infections overlap with those caused by *O. volvulus* (114). A cocktail of recombinant *O. volvulus* antigens for serologic diagnosis with the potential to predict the endemicity of onchocerciasis infection has been developed (16). When a sentinel population of 5- to 15-

year-old children was used to compare communities, the serologic index was sufficiently sensitive to measure the changes in endemicity that would be required for control programs. However, serologic testing cannot distinguish between current and past infection and tends to be more important in monitoring specific groups.

Treatment

Surgical removal of detectable nodules, especially in the head region, is simple and easy to perform. Nodule removal from the head, particularly in children, reduces the risk and severity of ocular complications. Removal of adult worms decreases the exposure of the host to microfilariae, which helps decrease the risk of complications. Because many nodules are nonpalpable or the adults may be freely migrating, the removal of nodules may have little effect on the total microfilarial or worm burden. Removal of enlarged regional lymph glands has been traditional therapy for sowda in Yemen.

Ivermectin is the drug of choice for treating onchocerciasis (100). It markedly reduces microfilaria numbers within 48 h and blocks the release of microfilariae from the uterus of the adult female worm for up to 6 months.

Following treatment, microfilariae disappear from the skin and can be detected in large numbers in the regional lymph nodes, where there is a high concentration of eosinophils. Ivermectin is well tolerated, but its side effects include pruritus, skin edema, and arthralgia; however, these effects are not as severe as those noted with DEC treatment. Ivermectin can also cause renal glomerular and tubular disturbances in patients with onchocerciasis; however, these are minor and do not seem to be clinically relevant (18). Although ivermectin significantly reduces the microfilarial burden in the host, thereby reducing dermatitis, it has no effect on depigmentation and scarring. Ivermectin treatment even in single doses reduces the prevalence of ocular microfilariae, including the number found in the anterior chamber of the eye, resulting in improvement of ocular lesions and visual acuity. When used in multiple doses, ivermectin is not only microfilaricidal but also macrofilaricidal (i.e., it kills adult worms). However, total doses of ivermectin up to 1,600 µg/kg are not significantly more effective than 150 µg/kg in killing the adult worms; more effective drugs against the adult worms are still needed (8). Ivermectin is contraindicated for pregnant women, children younger than 5 years, children weighing less than 33 lbs (15 kg), and patients with concurrent illness. Even though pruritus was frequent, no ocular or systemic complications were noted.

DEC is an effective microfilaricidal drug, but it may precipitate serious dermal and systemic complications as well as ocular damage in heavily infected patients. It has little or no effect on adult worms. The side effects resulting from the death of the microfilariae can be reduced by the use of anti-inflammatory drugs. Another method commonly employed is to gradually increase the DEC doses to therapeutic levels.

Suramin is effective in eliminating adult filarial worms as well as the microfilariae; however, it must be administered systemically and is very nephrotoxic. DEC and suramin in combination have been effective and well tolerated.

Amocarzine (CGP 6140) is a methylpiperazine derivative of amoscanate. Individuals treated with amocarzine had a rapid drop in the number of skin microfilariae, and the low level was maintained for over 2 years. A large number of adult female worms (65 to 73%) either were killed or became moribund (127). Amocarzine affects both mitochondrial function, by blocking NADH-dependent respiration, and acetylcholinesterase activity. Toxicity associated with amocarzine treatment included dizziness, impaired coordination, and Mazzotti-type reactions consisting mainly of pruritus and rash.

Mebendazole and flubendazole (benzimidazole derivatives) are effective microfilaricidal drugs; they are useful alternatives to DEC, and fewer ocular side effects were noted. Mebendazole is poorly absorbed and requires multiple doses. It is teratogenic and embryotoxic in laboratory animals and is contraindicated in pregnancy. Although flubendazole is not teratogenic, a complication is ulceration of the injection site (121).

Epidemiology and Prevention

Onchocerciasis is widely distributed throughout Central Africa, occurs in Saudi Arabia and Yemen, and is found in areas of Brazil, Colombia, Venezuela, Ecuador, Guatemala, and Mexico. An estimated 85.6 million people are considered at risk for onchocerciasis worldwide. There appear to be no known important animal reservoirs, although natural infections in gorillas have been found in Africa.

Simulium spp. (blackfly or buffalo gnat) are the insect vectors, and they are usually found near rivers and streams. Many species require fast-flowing streams for larval development. River blindness is the common name for onchocerciasis as a result of the frequency of blindness in people living along rivers and streams in areas of endemic infection. In some areas, control programs may be misdirected because of the inability to distinguish animal from human onchocercal larval forms in infected flies. DNA probes to distinguish larval forms could help in targeting the areas where human infections are most frequent. Pool screen PCR for estimating the prevalence of *O. volvulus* infection in the vectors has been used in epidemiologic surveillance for infection (127).

Treatment of streams and rivers with insecticides has successfully reduced or eliminated transmission. A significant problem in vector control has been the development of resistance to larvicides. A solution to this problem is the rotational use of different larvicides during the treatment campaign. Because of the long life span of the adult worms (15 years or more), campaigns must be continued for extended periods.

In areas where application of insecticides would be difficult, surgical removal of nodules has been used as a means of controlling as well as preventing blindness. Chemotherapy as a means of control has not been previously used because of severe side effects from microfilarial death. Because ivermectin can be given orally, the number of doses is limited in comparison with DEC, and compliance is better than with DEC because ivermectin has fewer side effects. Ivermectin therapy could be used as a control measure to reduce transmission in areas of endemicity. The disadvantages of this approach are cost, the fact that ivermectin is not effective against adult worms unless used in multiple doses, and the finding that in areas where *L. loa* also exists, treatment may induce encephalitis in *L. loa*-infected individuals. In communities where ivermectin has been used on a large scale with multiple doses over time, significant reductions in the number of infected flies

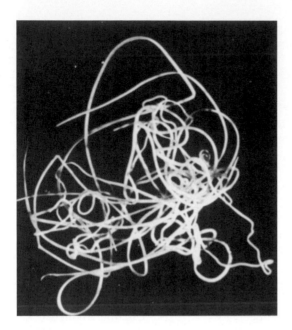

Figure 12.32 Adult *Dirofilaria immitis*. (From A Pictorial Presentation of Parasites: A cooperative collection prepared and/or edited by H. Zaiman.)

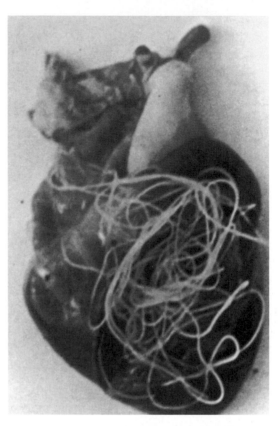

Figure 12.33 Adult *Dirofilaria immitis* within a canine heart. (Original photograph by S. Kume; from A Pictorial Presentation of Parasites: A cooperative collection prepared and/or by H. Zaiman.)

and the number of infective larvae transmitted have been noted. Significant decreases in the prevalence of onchocerciasis in children have been noted when mass treatment with ivermectin has been carried out in areas of endemic infection. The use of ivermectin in control and eradication programs holds considerable promise. A search for vaccine candidate immunogens is under way.

Dirofilaria Dirofilaria and *Dirofilaria Nochtiella* spp.

Currently, there are two subgenera, *Dirofilaria Dirofilaria* and *Dirofilaria Nochtiella*, based on the presence or absence of external longitudinal cuticular ridges (71a). *D. Dirofilaria immitis* is referred to in this volume as *D. immitis*. The others in the *Dirofilaria Nochtiella* species are designated *Dirofilaria* and the species name. A number of these species, such as *D. tenuis*, *D. repens*, *D. ursi*, *D. striata*, and *D. subdermata*, as well as *Brugia* spp., are animal filariae causing subcutaneous or conjunctival nodules in humans (9, 14, 44, 51, 54, 86, 87). Almost all of the worms detected in humans have been nongravid, with the exception of *Brugia* spp. detected in immunodeficient patients and in the right cervical lymph node lymphatic vessel of a normal human and a *Dirofilaria* sp. found in the right lung of a patient with lymphoid leukemia (9, 13). It is possible that the release of microfilariae into the vascular system is suppressed in normal human hosts.

D. immitis, the dog heartworm, causes a common zoonotic filarial infection in dogs in the tropical, subtropical, and warm temperate regions of the world. The infection was prevalent predominantly in the southern coastal and southeastern United States; however, the zone of endemicity has been increasing northward and to the west. Pulmonary lesions are the most common findings in humans infected with *D. immitis*. *D. tenuis* is a parasite of the raccoon within the United States; *D. repens* is found in dogs and cats in Europe, Africa, and Asia and is the cause of human infections in these geographic areas; *D. ursi* infects bears in Canada, the northern United States, and Japan; *D. subdermata* infects porcupines in North America; and *D. striata* infects wild cats such as bobcats, Florida panthers, ocelots, and margays in North and South America. (Table 12.3).

Life Cycle and Morphology

Adult female and male worms reside in the right heart of dogs, and microfilariae are found in the blood, where they are ingested by mosquitoes or, in the case of *D. ursi*, *Simulium* blackflies (Figures 12.32 and 12.33). After biological development of the larvae in the mosquito,

Table 12.3 *Dirofilaria* species infecting humans

Species	Length and diam		Geographic area	Clinical findings	Comments
	Females	**Males**			
D. immitis	230–310 mm, 350 μm	120–190 mm, 300 μm	Throughout the world; tropics to temperate areas	Small pulmonary infarcts usually asymptomatic and appear as "coin lesions"—usually only a single lesion	Humans most often infected in United States, Japan, Europe, Australia; no external longitudinal cuticular ridges
D. tenuis	80–130 mm, 260–360 μm	40–48 mm, 190–260 μm	Subcutaneous parasite of raccoon in United States	Gravid worm in thigh and abdominal wall (2 patients)	External longitudinal cuticular ridges; usually low and rounded
D. repens	100–170 mm, 460–650 μm	50–70 mm, 370–450 μm	Subcutaneous parasite of dogs, cats in Europe, Africa, Asia	Can cause pulmonary disease as with *D. immitis*; also recovered from subcutaneous tissue; subcutaneous (normally nodular) and the submucosa (nodular or not)	The most commonly affected areas are the head, the thoracic wall, and the upper limbs; most infections seen in Italy; many cases probably not diagnosed, and patients recover spontaneously; external ridges appear beaded
D. ursi	117–224 mm, 460–700 μm	51–86 mm, 330–480 μm	Parasitizes bears in Canada, northern United States, and Japan	Usual location of nodules containing the worms was the scalp or a covered part of the upper body where blackflies, the intermediate hosts of *D. ursi*, normally feed; 5 reports from along U.S.-Canadian border	Case reports indicate worms like either *D. ursi* or *D. subdermata*; distinct longitudinal cuticular ridges, cuticular ridges fewer and farther apart, regularly and widely spaced on the outer surface, usually evident even when the worms are necrotic
D. subdermata	117–185 mm, 440–660 μm	41–66 mm, 280–420 μm	Infects porcupines in North America	May see inflamed subcutaneous nodules (painful, erythematous, sometimes migratory)	Very difficult to differentiate female *D. ursi* from *D. subdermata*; now called *D. ursi*-like
D. striata	250–360 mm, 440–500 μm	80–120 mm, 350–380 μm	Bobcats in Florida, panthers, ocelots, margays in North and South America	Rare, but produces subcutaneous lesions; also recovered from eye in one child	Ridges may mimic lateral alae

dogs and humans may be infected with the infective larvae when the mosquito takes a blood meal. Development to a mature adult takes approximately 180 days in the dog. In humans, the worms do not reach maturity and no microfilariae can be detected.

Adult worms in the *Nochtiella* subgenus live in the subcutaneous tissues of their hosts, although in some cases the worms migrate to the pulmonary vessels. There is often only a single worm; therefore, there is no production of microfilariae.

Clinical Disease

The first reported case in humans was recorded in 1941. Many of the patients were asymptomatic, and pulmonary nodules (coin lesions) were noted on routine X-ray examinations or at autopsy (75). Symptomatic patients exhibited signs and symptoms leading to chest examination, with common findings of chest discomfort, cough, fever, chills, malaise, and occasional hemoptysis. Infections were more frequent in males than in females, and most individuals were 40 to 60 years old.

When the worms lodge in the pulmonary artery branches, they cause an infarct. These lesions are usually on the periphery of the lungs and are sharply defined (coin lesion). There is a central necrotic area surrounded by a granulomatous inflammation and a fibrous wall. Dead or dying worms may be found in the lesion (Figures 12.34 and 12.35).

In one review, 39 patients were discussed. All were Americans between 8 and 80 years old and included 23 men and 16 women (38). Twenty-two of the patients (56%) were asymptomatic, and the pulmonary nodule was discovered on chest X rays during a routine physical examination. Seventeen patients (44%) presented with general complaints or respiratory symptoms. Only 10% of the patients had a peripheral eosinophilia. Most lesions (76%) were seen in the right lung, with a predilection for the lower lobe. Five

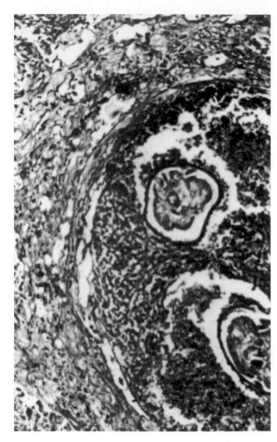

Figure 12.34 Pulmonary nodule with *Dirofilaria immitis* (cross section). (Original photograph by G. Healy; from A Pictorial Presentation of Parasites: A cooperative collection prepared and/or edited by H. Zaiman.)

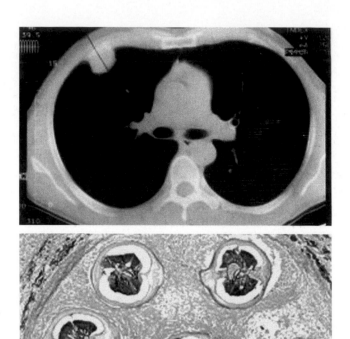

Figure 12.35 Dirofilariasis. (Upper) Computed tomogram image with fine needle in left pulmonary round focus; round focus in right lung periphery. (Lower) Histology of resected "coin lesion" with immature *Dirofilaria* in the peripheral pulmonary artery; silver methenamine stain. (From R. Fueter and J.-O. Gebbers, *Schweiz. Med. Wochenschr.* **127**:2014, 1997, with permission).

patients had multifocal nodules. All patients underwent thoracotomy with excisional lung biopsy or lobectomy. In all cases, histology revealed a histiocyte-rimmed necrotic nodule containing fragments of a partially degenerated *D. immitis* worm, while adjacent lung parenchyma showed morphological features consistent with other pulmonary processes such as extrinsic allergic alveolitis and/or pulmonary vasculitis syndromes. This group of cases highlights the morphologic variation seen in human pulmonary dirofilariasis and emphasizes the need to consider this diagnosis in all cases of necrotizing granulomas of the lung.

Diagnosis

Microfilariae cannot be found in the blood or tissues, and serologic results are of little value because of the lack of sensitivity and specificity of this method. Some patients have a moderate eosinophilia. Diagnosis can be confirmed by the identification of worms in surgical or autopsy specimens. Because the immature larvae may be detected in only a few microscopic sections, careful histologic examination is necessary. The cuticle of nematodes contains chitin, which

can be stained with nonspecific whiteners such as calcofluor white. *Dirofilaria* larvae stained with calcofluor white can be easily recognized in tissue sections, whereas the parasite may be difficult to identify using routine histologic stains.

In a very extensive study of over 90 cases of human infection with *D. repens*, the authors commented on the diagnostic difficulties involved with histologic diagnosis (89). As the worms begin to die within the nodule, the invasion of inflammatory cells begins at the worm's orifices (mouth, vulva, anus, and cloaca) and gradually spreads throughout its body. It is recommended that the nodule be sectioned at different points in order to review morphologic features that may still be recognizable. This review also contains a series of color photographs in which the nematodes are in an excellent state of preservation, while other photographs illustrate the range of disintegration of the worm within the nodule (Figure 12.36).

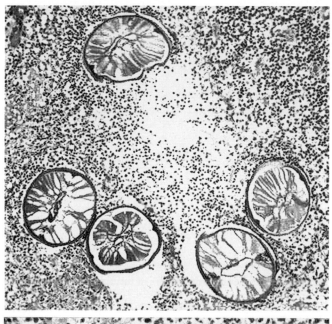

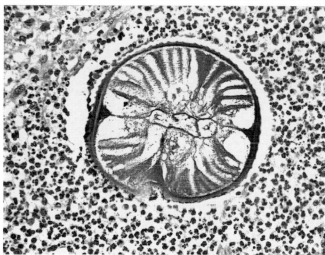

Figure 12.36 *Dirofilaria tenuis.* (Upper) Cross section from right flank muscle. (Lower) Cross section enlarged. Note the key characteristics: thick cuticle; low, rounded, wavy external longitudinal cuticular ridges; prominent lateral chords; prominent somatic musculature; and location within soft tissue. (Photographs courtesy of Marjorie R. Fowler, with permission.)

Treatment

Lesions identified in the lungs are frequently removed when cancer is suspected (39). The correct diagnosis can be confirmed by examination of histologic preparations after surgery. Identification of the worm in biopsy specimens or extraction of the worm from the lesion will confirm the diagnosis. Surgical removal of the worms is the only known treatment. No chemotherapeutic agents are used, since they appear to be ineffective.

Epidemiology and Prevention

Although at least 15 species of mosquitoes can carry the infective forms and the infection is considered an emerging zoonosis, probably fewer than 1,000 cases of human infection have been described. However, it is quite possible that many infections go undiagnosed and the number of cases is larger than documented. Canines, cats, foxes, muskrats, otters, and sea lions are natural hosts. The infection in animals has been noted in the United States, Africa, Mediterranean countries, Southeast Asia, and the Pacific Islands. The prevalence of infection in dogs in the southeastern United States may be as high as 40%.

References

1. **Abbasi, I., J. Githure, J. J. Ochola, R. Agure, D. K. Koech, R. M. Ramzy, S. A. Williams, and J. Hamburger.** 1999. Diagnosis of *Wuchereria bancrofti* infection by the polymerase chain reaction employing patients' sputum. *Parasitol. Res.* **85:**844–849.

2. **Abramowicz, M. (ed.).** 2004. Drugs for parasitic infections. *Med. Lett. Drugs Ther.* **46:**1–12.

3. **Addiss, D. G., K. A. Dimock, M. L. Eberhard, and P. J. Lammie.** 1995. Clinical parasitologic and immunologic observations of patients with hydrocele and elephantiasis in an area with endemic lymphatic filariasis. *J. Infect. Dis.* **171:**755–758.

4. **Ahorlu, C. K., S. K. Dunyo, K. A. Koram, F. K. Nkrumah, J. Aagaard-Hansen, and P. E. Simonsen.** 1999. Lymphatic filariasis related perceptions and practices on the coast of Ghana: implications for prevention and control. *Acta Trop.* **73:**251–261.

5. **Akue, J. P., M. Hommel, and E. Devaney.** 1998. IgG subclass recognition of *Loa loa* antigens and their correlation with clinical status in individuals from Gabon. *Parasite Immunol.* **20:**387–393.

6. **Alexander, N. D. E., and B. T. Grenfell.** 1999. The effect of pregnancy on *Wuchereria bancrofti* microfilarial load in humans. *Parasitology* **119:**151–156.

7. **Alexander, N. D. E., R. T. Perry, Z. B. Dimber, P. J. Hyun, M. P. Alpers, and J. W. Kazura.** 1999. Acute disease episodes in a *Wuchereria bancrofti*-endemic area of Papua, New Guinea. *Am. J. Trop. Med. Hyg.* **61:**319–324.

8. **Awadzi, K., S. K. Attah, E. T. Addy, N. O. Opoku, and B. T. Quartey.** 1999. The effects of high-dose ivermectin regimens on *Onchocerca volvulus* in onchocerciasis patients. *Trans. R. Soc. Trop. Med. Hyg.* **93:**189–194.

9. **Baird, J. K., and R. C. Neafie.** 1988. South American brugian filariasis: report of a human infection acquired in Peru. *Am. J. Trop. Med. Hyg.* **39:**185–188.

10. **Bartoloni, A., G. Cancrini, F. Bartalesi, D. Marcolin, M. Roselli, C. C. Arce, and A. J. Hall.** 1999. *Mansonella ozzardi* infection in Bolivia: prevalence and clinical associations in the Chaco region. *Am. J. Trop. Med. Hyg.* **61:**830–833.

11. **Baskar, L. K., T. R. Srikanth, S. Suba, H. C. Mody, P. K. Desai, and P. Kaliraj.** 2004. Development and evaluation of a rapid flow-through immuno filtration test using recombinant filarial antigen for diagnosis of brugian and bancroftian filariasis. *Microbiol. Immunol.* **48:**519–525.

12. Beach, M. J., T. G. Streit, D. G. Addiss, R. Prospere, J. M. Roberts, and P. J. Lammie. 1999. Assessment of combined ivermectin and albendazole for treatment of intestinal helminth and *Wuchereria bancrofti* infections in Haitian schoolchildren. *Am. J. Trop. Med. Hyg.* **60:**479–486.

13. Beaver, P. C., T. C. Orihel, and G. Leonard. 1990. Pulmonary dirofilariasis: restudy of worms reported gravid. *Am. J. Trop. Med. Hyg.* **43:**167–169.

14. Beaver, P. C., J. S. Wolfson, M. A. Waldron, M. N. Swartz, G. W. Evans, and J. Adler. 1987. *Dirofilaria ursi*-like parasites acquired by humans in the northern United States and Canada: report of two cases and a brief review. *Am. J. Trop. Med. Hyg.* **37:**357–362.

15. Boggild, A. K., J. S. Keystone, and K. C. Kain. 2004. Tropical pulmonary eosinophilia: a case series in a setting of nonendemicity. *Clin. Infect. Dis.* **39:**1123–1128.

16. Bradley, J. E., B. M. Atogho, L. Elson, G. R. Stewart, and M. Boussinesq. 1998. A cocktail of recombinant *Onchocerca volvulus* antigens for serologic diagnosis with the potential to predict the endemicity of onchocerciasis infection. *Am. J. Trop. Med. Hyg.* **59:**877–882.

17. Bregani, E. R., A. Rovellini, N. Mbaidoum, and M. G. Magnini. 2005. Comparison of different anthelminthic drug regimens against *Mansonella perstans* filariasis. *Trans. R. Soc. Trop. Med. Hyg.* 25 Oct. [Epub ahead of print.] **100:**458–463.

18. Burchard, G. D., T. Kubica, F. W. Tischendorf, T. Kruppa, and N. W. Brattig. 1999. Analysis of renal function in onchocerciasis patients before and after therapy. *Am. J. Trop. Med. Hyg.* **60:**980–986.

19. Cartel, J. L., N. L. Nguyen, A. Speigel, J. P. Moulia-Pelat, R. Plichart, P. M. V. Martin, A. B. Manuellan, and F. Lardeux. 1992. *Wuchereria bancrofti* infection in human and mosquito populations of a Polynesian village ten years after interruption of mass chemoprophylaxis with diethylcarbamazine. *Trans. R. Soc. Trop. Med. Hyg.* **86:**414–416.

20. Chodakewitz, J. 1995. Ivermectin and lymphatic filariasis: a clinical update. *Parasitol. Today* **11:**233–235.

21. Cooper, P. J., and T. B. Nutman. 2000. Onchocerciasis, p. 756–769. *In* G. T. Strickland (ed.), *Hunter's Tropical Medicine and Emerging Infectious Diseases*, 8th ed. The W. B. Saunders Co., Philadelphia, Pa.

22. Dabir, P., S. Dabir, B. V. Siva Prasad, and M. V. Reddy. 2005. Isolation and analysis of partial cDNA sequence coding for superoxide dismutase in *Wuchereria bancrofti*. *Infect. Genet. Evol.* 28 Sept. [Epub ahead of print.] **6:**287–291.

23. Das, D., S. Kumar, P. K. Sahoo, and A. P. Dash. 2005. A survey of bancroftian filariasis for microfilariae and circulating antigenaemia in two villages of Madhya Pradesh. *Indian J. Med. Res.* **121:**771–775.

24. David, H. L., and J. F. B. Edeson. 1965. Filariasis in Portuguese Timor, with observations on a new microfilaria found in man. *Ann. Trop. Med. Parasitol.* **59:**193–204.

25. de Almeida, A. B., and D. O. Freedman. 1999. Epidemiology and immunopathology of bancroftian filariasis. *Microbes Infect.* **1:**1015–1022.

26. Dennis, D. T., F. Partono, P. S. Atmosoedjono, and J. S. Saroso. 1976. Timor filariasis: epidemiologic and clinical features in a defined community. *Am. J. Trop. Med. Hyg.* **25:**797–802.

27. Dreyer, G., P. Dreyer, and W. F. Piessens. 1999. Extralymphatic disease due to bancroftian filariasis. *Braz. J. Med. Biol. Res.* **32:**1467–1472.

28. Dreyer, G., Z. Medeiros, M. J. Netto, N. C. Leal, L. G. de Castro, and W. F. Piessens. 1999. Acute attacks in the extremities of persons living in an area endemic for bancroftian filariasis: differentiation of the two syndromes. *Trans. R. Soc. Trop. Med. Hyg.* **93:**413–417.

29. Dreyer, G., A. Santos, J. Noroes, F. Amaral, and D. Addiss. 1998. Ultrasonographic detection of living adult *Wuchereria bancrofti* using a 3.5-MHz transducer. *Am. J. Trop. Med. Hyg.* **59:**399–403.

30. Eberhard, M. L., and P. J. Lammie. 1991. Laboratory diagnosis of filariasis. *Clin. Lab. Med.* **11:**977–1010.

31. Erlanger, T. E., J. Keiser, M. Caldas De Castro, R. Bos, B. H. Singer, M. Tanner, and J. Utzinger. 2005. Effect of water resource development and management on lymphatic filariasis, and estimates of populations at risk. *Am. J. Trop. Med. Hyg.* **73:**523–533.

32. Faris, R., O. Hussain, M. El Setouhy, R. M. R. Ramzy, and G. J. Weil. 1998. Bancroftian filariasis in Egypt: visualization of adult worms and subclinical lymphatic pathology by scrotal ultrasound. *Am. J. Trop. Med. Hyg.* **59:**864–867.

33. Figueredo-Silva, J., and G. Dreyer. 2005. Bancroftian filariasis in children and adolescents; clinical-pathological observations in 22 cases from an endemic area. *Ann. Trop. Med. Parasitol.* **99:**759–769.

34. Fischer, P., I. Bonow, T. Supali, P. Ruckert, and N. Rahmah. 2005. Detection of filarial-specific IgG4 antibodies and filarial DNA, for the screening of blood spots for *Brugia timori*. *Ann. Trop. Med. Parasitol.* **99:**53–60.

35. Fischer, P., X. L. Liu, M. Lizotte-Waniewski, I. H. Kamal, R. M. R. Ramzy, and S. A. Williams. 1999. Development of a quantitative, competitive polymerase chain reaction enzyme linked immunosorbent assay for the detection of *Wuchereria bancrofti* DNA. *Parasitol. Res.* **85:**176–183.

36. Fischer, P., T. Supali, and R. M. Maizels. 2004. Lymphatic filariasis and *Brugia timori*: prospects for elimination. *Trends Parasitol.* **20:**351–355.

37. Fischer, P., E. Tukesiga, and D. W. Buttner. 1999. Long-term suppression of *Mansonella streptocerca* microfilariae after treatment with ivermectin. *J. Infect. Dis.* **180:**1403–1405.

38. Flieder, D. B., and C. A. Moran. 1999. Pulmonary dirofilariasis: a clinicopathologic study of 41 lesions in 39 patients. *Hum. Pathol.* **30:**251–256.

39. Foroulis, C. N., L. Khaldi, N. Desimonas, and G. Kalafati. 2005. Pulmonary dirofilariasis mimicking lung tumor with chest wall and mediastinal invasion. *Thorac. Cardiovasc. Surg.* **53:**173–175.

40. Foster, J. M., S. Kumar, M. B. Ganatra, I. H. Kamal, J. Ware, J. Ingram, J. Pope-Chappell, D. Guiliano, C. Whitton, J. Daub, M. L. Blaster, and B. E. Slatko. 2004. Construction of bacterial artificial chromosome libraries from the parasitic nematode *Brugia malayi* and physical mapping of the genome of its *Wolbachia* endosymbionts. *Int. J. Parasitol.* **34:**733–746.

41. Fox, L. M., B. W. Furness, J. K. Haser, J. M. Brissau, J. Louis-Charles, S. F. Wilson, D. G. Addiss, P. J. Lammie, and M. J. Beach. 2005. Ultrasonographic examination of Haitian children with lymphatic filariasis: a longitudinal

assessment in the context of antifilarial drug treatment. *Am. J. Trop. Med. Hyg.* **72**:642–648.

42. Fox, L. M., B. W. Furness, J. K. Haser, D. Desire, J. M. Brissau, M. D. Milord, J. Lafontant, P. J. Lammie, and M. J. Beach. 2005. Tolerance and efficacy of combined diethylcarbamazine and albendazole for treatment of *Wuchereria bancrofti* and intestinal helminth infections in Haitian children. *Am. J. Trop. Med. Hyg.* **73**:115–121.

43. Freeman, A. R., P. J. Lammie, R. Houston, M. D. LaPointe, T. G. Streit, P. L. Jooste, J. M. Brissau, J. G. Lafontant, and D. G. Addiss. 2001. A community-based trial for the control of lymphatic filariasis and iodine deficiency using salt fortified with diethylcarbamazine and iodine. *Am. J. Trop. Med. Hyg.* **65**:865–871.

44. Fuentes, I., A. Cascales, J. M. Ros, C. Sansano, J. L. Gonzalez-Arribas, and J. Alvar. 1994. Human subcutaneous dirofilariasis caused by *Dirofilaria repens* in Ibiza, Spain. *Am. J. Trop. Med. Hyg.* **51**:401–404.

45. Gallin, M. Y., A. B. Jacobi, D. W. Buttner, O. Schonberger, T. Marti, and K. D. Erttmann. 1995. Human autoantibody to defensin: disease association with hyperreactive onchocerciasis (Sowda). *J. Exp. Med.* **182**:41–47.

46. Garcia, A., L. Abel, M. Cot, S. Ranque, P. Richard, M. Boussinesq, and J. P. Chippaux. 1995. Longitudinal survey of *Loa loa* filariasis in Southern Cameroon: long-term stability and factors influencing individual microfilarial status. *Am. J. Trop. Med. Hyg.* **52**:370–375.

47. Garraud, O., C. Nkenfou, J. E. Bradley, F. B. Perler, and T. B. Nutman. 1995. Identification of recombinant filarial proteins capable of inducing polyclonal and antigen-specific IgE and IgG4 antibodies. *J. Immunol.* **155**:1316–1325.

48. González-Muñoz, M., T. Gárate, S. Puente, M. Subirats, and I. Moneo. 1999. Induction of histamine release in parasitized individuals by somatic and cuticular antigens from *Onchocerca volvulus*. *Am. J. Trop. Med. Hyg.* **60**:974–979.

49. Gopinath, R., L. E. Hanna, V. Kumaraswami, S. V. P. Pillai, V. Kavitha, V. Vijayasekaran, A. Rajasekharan, and T. B. Nutman. 1999. Long term persistence of cellular hyporesponsiveness to filarial antigens after clearance of microfilaremia. *Am. J. Trop. Med. Hyg.* **60**:848–853.

50. Grieve, R. B., N. Wisnewski, G. R. Frank, and C. A. Tripp. 1995. Vaccine research and development for the prevention of filarial nematode infections, p. 737–768. *In* M. F. Powell and M. J. Newman (ed.), *Vaccine Design: the Subunit and Adjuvant Approach.* Plenum Press, New York, N.Y.

51. Gutierrez, Y., I. Misselevich, M. Fradis, L. Podoshin, and J. H. Boss. 1995. *Dirofilaria repens* infection in Northern Israel. *Am. J. Surg. Pathol.* **19**:1088–1091.

52. Hall, L. R., and E. Pearlman. 1999. Pathogenesis of onchocercal keratitis (river blindness). *Clin. Microbiol. Rev.* **12**:445–453.

53. Harnett, W., J. E. Bradley, and T. Garate. 1998. Molecular and immunodiagnosis of human filarial nematode infections. *Parasitology* **117**:S59–S71.

54. Herzberg, A. J., P. R. Boyd, and Y. Gutierrez. 1995. Subcutaneous dirofilariasis in Collier County, Florida, U.S.A. *Am. J. Surg. Pathol.* **19**:934–939.

55. Hise, A. G., I. Gillette-Ferguson, and E. Pearlman. 2004. The role of endosymbiotic *Wolbachia* bacteria in filarial disease. *Cell. Microbiol.* **6**:97–104.

56. Hussein, O., M. E. Setouhy, E. S. Ahmed, A. M. Kandil, R. M. Ramzy, H. Helmy, and G. J. Weil. 2004. Duplex Doppler sonographic assessment of the effects of diethylcarbamazine and albendazole therapy on adult filarial worms and adjacent host tissues in Bancroftian filariasis. *Am. J. Trop. Med. Hyg.* **71**:471–477.

57. Jenkins, D. C. 1990. Ivermectin in the treatment of filarial and other nematode diseases of man. *Trop. Dis. Bull.* **87**:R1–R9.

58. Kanjanavas, P., P. Tan-ariya, P. Khawsak, A. Pakpitcharoen, S. Phantana, and K. Chansiri. 2005. Detection of lymphatic *Wuchereria bancrofti* in carriers and long-term storage blood samples using semi-nested PCR. *Mol. Cell. Probes* **19**:169–172.

59. Keiser, P. B., Y. I. Coulibaly, F. Keita, D. Traore, A. Diallo, D. A. Diallo, R. T. Semnani, O. K. Dounbo, S. F. Traore, A. D. Klion, and T. B. Nutman. 2003. Clinical characteristics of post-treatment reactions to ivermectin/albendazole for *Wuchereria bancrofti* in a region co-endemic for *Mansonella perstans*. *Am. J. Trop. Med. Hyg.* **69**:331–335.

60. King, C. L., and D. O. Freedman. 2000. Filariasis, p. 740–775. *In* G. T. Strickland (ed.), *Hunter's Tropical Medicine and Emerging Infectious Diseases*, 8th ed. The W. B. Saunders Co., Philadelphia, Pa.

61. Klion, A. 2000. Loiasis, p. 754–756. *In* G. T. Strickland (ed.), *Hunter's Tropical Medicine and Emerging Infectious Diseases*, 8th ed. The W. B. Saunders Co., Philadelphia, Pa.

62. Klion, A. D., J. Horton, and T. B. Nutman. 1999. Albendazole therapy for loiasis refractory to diethylcarbamazine treatment. *Clin. Infect. Dis.* **29**:680–682.

63. Klion, A. D., A. Massougbodji, B. Sadeler, E. A. Ottesen, and T. B. Nutman. 1991. Loiasis in endemic and nonendemic populations: immunologically mediated differences in clinical presentation. *J. Infect. Dis.* **63**:1318–1325.

64. Klion, A. D., E. A. Ottesen, and T. B. Nutman. 1994. Effectiveness of diethylcarbamazine in treating loiasis acquired by expatriate visitors to endemic regions: long-term follow-up. *J. Infect. Dis.* **169**:604–610.

65. Kyelem, D., J. Medlock, S. Sanou, M. Bonkoungou, B. Boatin, and D. H. Molyneux. 2005. Short communication: impact of long-term (14 years) bi-annual ivermectin treatment on *Wuchereria bancrofti* microfilaraemia. *Trop. Med. Int. Health* **10**:1002–1004.

66. Lizotte, M. P., T. Supali, F. Partono, and S. S. Williams. 1994. A polymerase chain reaction assay for the detection of *Brugia malayi* in blood. *Am. J. Trop. Med. Hyg.* **51**:314–321.

67. MacKenzie, C. D., J. F. Williams, R. H. Guderian, and J. O'Day. 1987. Clinical responses in human onchocerciasis: parasitological and immunological implications, p. 46–72. *In* D. Evered and S. Clark (ed.), *Filariasis*. John Wiley & Sons, Inc., New York, N.Y.

68. Maizels, R. M., E. Sartono, A. Kurniawan, F. Partono, M. E. Selkirk, and M. Yazdanbakhsh. 1995. T-cell activation and the balance of antibody isotypes in human lymphatic filariasis. *Parasitol. Today* **11**:50–56.

69. Malhotra, I., P. Mungai, A. Wamachi, J. Kioko, J. H. Ouma, J. W. Kazura, and C. L. King. 1999. Helminth- and bacillus Calmette-Guérin-induced immunity in children sensitized in utero to filariasis and schistosomiasis. *J. Immunol.* **162**:6843–6848.

70. **Mand, S., A. Debrah, L. Batsa, O. Adjei, and A. Hoerauf.** 2004. Reliable and frequent detection of adult *Wuchereria bancrofti* in Ghanian women by ultrasonography. *Trop. Med. Int. Health* **9:**1111–1114.

71. **Marchetti, F., W. F. Piessens, Z. Medeiros, and G. Dreyer.** 1998. Abnormalities of the leg lymphatics are not specific for bancroftian filariasis. *Trans. R. Soc. Trop. Med. Hyg.* **92:**650–652.

71a.**Meyers, W. M., R. C. Neafie, A. M. Marty, and D. J. Wear (ed.).** 2000. *Pathology of Infectious Diseases*, vol. 1. *Helminthiases*, p. 245–306. Armed Forces Institute of Pathology, Washington, D.C.

72. **Meyrowitsch, D. W., P. E. Simonsen, and S. M. Magesa.** 2004. Long-term effect of three different strategies for mass diethylcarbamazine administration in bancroftian filariasis: follow-up at 10 years after treatment. *Trans. R. Soc. Trop. Med. Hyg.* **98:**627–634.

73. **Mishra, K., D. K. Raj, A. P. Dash, and R. K. Hazra.** 2005. Combined detection of *Brugia malayi* and *Wuchereria bancrofti* using single PCR. *Acta Trop.* **93:**233–237.

74. **Molina, A., T. Cabezas, and J. Gimenez.** 1999. *Mansonella perstans* filariasis in the bone marrow of an HIV patient. *Haematologica* **84:**861.

75. **Nicholson, C. P., M. S. Allen, V. F. Trastek, H. D. Tazelaar, and P. C. Pairolero.** 1992. *Dirofilaria immitis*: a rare, increasing cause of pulmonary nodules. *Mayo Clin. Proc.* **67:**646–650.

76. **Nicolas, L., S. Langy, C. Plichart, and X. Desparis.** 1999. Filarial antibody responses in *Wuchereria bancrofti* transmission area are related to parasitological but not clinical status. *Parasite Immunol.* **21:**73–80.

77. **Noireau, F., and G. Pichon.** 1992. Population dynamics of *Loa loa* and *Mansonella perstans* infection in individuals living in an endemic area of the Congo. *Am. J. Trop. Med. Hyg.* **46:**672–676.

78. **Noordin, R., R. A. Aziz, and B. Ravindran.** 2004. Homologs of the *Brugia malayi* diagnostic antigen BmR1 are present in other filarial parasites but induce different humoral immune responses. *Filaria J.* **3:**10.

79. **Nuchprayoon, S., A. Junpee, Y. Poovorawan, and A. L. Scott.** 2005. Detection and differentiation of filarial parasites by universal primers and polymerase chain reaction-restriction fragment length polymorphism analysis. *Am. J. Trop. Med. Hyg.* **73:**895–900.

80. **Nutman, T. B., K. D. Miller, M. Mulligan, and E. A. Ottesen.** 1986. *Loa loa* infection in temporary residents of endemic regions: recognition of a hyperresponsive syndrome with characteristic clinical manifestations. *J. Infect. Dis.* **154:**10–18.

81. **Ogunrinade, A., D. Boakye, A. Merriweather, and T. R. Unnasch.** 1999. Distribution of the blinding and nonblinding strains of *Onchocerca volvulus* in Nigeria. *J. Infect. Dis.* **179:**1577–1579.

82. **Onapa, A. W., P. E. Simonsen, I. Baehr, and E. M. Pedersen.** 2005. Rapid assessment of the geographical distribution of lymphatic filariasis in Uganda, by screening of schoolchildren for circulating filarial antigens. *Ann. Trop. Med. Parasitol.* **99:**141–153.

83. **Oqueka, T., T. Supali, I. S. Ismid, Purnomo, P. Ruckert, M. Bradley, and P. Fischer.** 2005. Impact of two rounds of mass drug administration using diethylcarbamazine combined with albendazole on the prevalence of *Brugia timori* and of intestinal helminths on Alor Island, Indonesia. *Filaria J.* **4:**5.

84. **Orihel, T. C.** 1984. The tail of the *Mansonella streptocerca* microfilaria. *Am. J. Trop. Med. Hyg.* **33:**1278.

85. **Orihel, T. C., and L. R. Ash.** 1995. *Parasites in Human Tissues*, p. 164–165. ASCP Press, Chicago, Ill.

86. **Orihel, T. C., and P. C. Beaver.** 1989. Zoonotic *Brugia* infections in North and South America. *Am. J. Trop. Med. Hyg.* **40:**638–647.

87. **Orihel, T. C., and E. K. Isberg.** 1990. *Dirofilaria striata* infection in a North Carolina child. *Am. J. Trop. Med. Hyg.* **42:**124–126.

88. **Ottesen, E. A.** 1994. Immune responsiveness and the pathogenesis of human onchocerciasis. *J. Infect. Dis.* **171:**659–671.

89. **Pampiglione, S., F. Rivasi, and G. Canestri-Trotti.** 1999. Pitfalls and difficulties in histological diagnosis of human dirofilariasis due to *Dirofilaria* (*Nochtiella*) *repens*. *Diagn. Microbiol. Infect. Dis.* **34:**57–64.

90. **Pani, S. P., S. L. Hoti, P. Vanamail, and L. K. Das.** 2004. Comparison of an immunochromatographic card test with night blood smear examination for detection of *Wuchereria bancrofti* microfilaria carriers. *Natl. Med. J. India.* **17:**304–306.

91. **Pearlman, E., L. Toe, B. A. Boatin, A. A. Gilles, A. W. Higgins, and T. R. Unnasch.** 1999. Eotaxin expression in *Onchocerca volvulus*-induced dermatitis after topical application of diethylcarbamazine. *J. Infect. Dis.* **180:**1394–1397.

92. **Plaisier, A. P., W. C. Cao, G. J. Van Oortmarssen, and J. D. F. Habbema.** 1999. Efficacy of ivermectin in the treatment of *Wuchereria bancrofti* infection: a model-based analysis of trial results. *Parasitology* **119:**385–394.

93. **Poole, C. B., and S. A. Williams.** 1990. A rapid DNA assay for the species-specific detection and quantification of *Brugia* in blood samples. *Mol. Biochem. Parasitol.* **40:**129–136.

94. **Rahmah, N., A. K. Anuar, R. Karim, R. Mehdi, B. Sinniah, and A. W. Omar.** 1994. Potential use of IgG2-ELISA in the diagnosis of chronic elephantiasis and IgG4-ELISA in the follow-up of microfilaraemic patients infected with *Brugia malayi*. *Biochem. Biophys. Res. Commun.* **205:**202–207.

95. **Rakita, R. M., A. C. White, Jr., and M. A. Kielhofner.** 1993. *Loa loa* infections as a cause of migratory angioedema: report of three cases from the Texas Medical Center. *Clin. Infect. Dis.* **17:**691–694.

96. **Ramzy, R. M. R., M. El Setouhy, H. Helmy, A. M. Kandil, E. S. Ahmed, H. A. Farid, R. Faris, and G. J. Weil.** 2002. The impact of single-dose diethylcarbamazine treatment of bancroftian filariasis in a low-endemicity setting in Egypt. *Am. J. Trop. Med. Hyg.* **67:**196–200.

97. **Ramzy, R. M. R., O. N. Hafez, A. M. Gad, R. Faris, M. Harb, A. A. Buck, and G. J. Weil.** 1994. Efficient assessment of filariasis endemicity by screening for filarial antigenaemia in a sentinel population. *Trans. R. Soc. Trop. Med. Hyg.* **88:**41–44.

98. **Rao, U. R., A. C. Vickery, B. H. Kwa, and J. K. Nayar.** 1992. *Brugia malayi*: ivermectin inhibits the exsheathment of microfilariae. *Am. J. Trop. Med. Hyg.* **46:**183–188.

99. Richards, F. O., Jr., M. L. Eberhard, R. T. Bryan, D. F. McNeeley, P. J. Lammie, M. B. NcNeeley, Y. Bernard, A. W. Hightower, and H. C. Spencer. 1991. Comparison of high dose ivermectin and diethylcarbamazine for activity against bancroftian filariasis in Haiti. *Am. J. Trop. Med. Hyg.* **44:**3–10.

100. Richards, F. O., Jr., A. Eigege, D. Pam, A. Kal, A. Lenhart, J. O. Oneyka, M. Y. Jinadu, and E. S. Miri. 2005. Mass ivermectin treatment for onchocerciasis: lack of evidence for collateral impact on transmission of *Wuchereria bancrofti* in areas of co-endemicity. *Filaria J.* **4:**6.

101. Richards, F. O., Jr., D. D. Pam, A. Kal, G. Y. Gerlong, J. Onyeka, Y. Sambo, J. Danboyi, B. Ibrahim, A. Terranella, D. Kumbak, A. Dakul, A. Lenhart, L. Rakers, J. Umaru, S. Amadiegwu, P. C. Withers, Jr., H. Mafuyai, M. Y. Jinadu, E. S. Miri, and A. Eigege. 2005. Significant decrease in the prevalence of *Wuchereria bancrofti* infection in anophyline mosquitoes following the addition of albendazole to annual, ivermectin-based, mass treatments in Nigeria. *Ann. Trop. Med. Parasitol.* **99:**155–164.

102. Rokeach, L. A., P. A. Zimmerman, and T. R. Unnasch. 1994. Epitopes of the *Onchocerca volvulus* RAL1 antigen, a member of the clareticulin family of proteins, recognized by sera from patients with onchocerciasis. *Infect. Immun.* **62:**3696–3704.

103. Roncoroni, A. J., J. Altieri, and S. Lopez. 1991. Tissue invasion by *Mansonella ozzardi* in a patient with adenocarcinoma. *Chest* **100:**592.

104. Sartono, E., C. Lopriore, Y. C. M. Kruize, A. Kurniawan-Atmadja, R. M. Maizels, and M. Yazdanbakhsh. 1999. Reversal in microfilarial density and T cell responses in human lymphatic filariasis. *Parasite Immunol.* **21:**565–571.

105. Sentongo, E., T. Rubaale, D. W. Buttner, and N. W. Brattig. 1998. T cell responses in coinfection with *Onchocerca volvulus* and the human immunodeficiency virus type 1. *Parasite Immunol.* **20:**431–439.

106. Shenoy, R. K., S. Dalia, A. John, T. K. Suma, and V. Kumaraswami. 1999. Treatment of the microfilaraemia of asymptomatic brugian filariasis with single doses of ivermectin, diethylcarbamazine or albendazole, in various combinations. *Ann. Trop. Med. Parasitol.* **93:**643–651.

107. Shenoy, R. K., V. Kumarswami, K. Rajan, S. Thankom, and Jalajokumari. 1992. Ivermectin for the treatment of periodic Malayan filariasis: a study of efficacy and side effects following single oral dose and retreatment at six months. *Ann. Trop. Med. Parasitol.* **86:**271–278.

108. Shenoy, R. K., V. Kumaraswami, T. K. Suma, K. Rajan, and G. Radhakuttyamma. 1999. A double-blind, placebo-controlled study of the efficacy of oral penicillin, diethylcarbamazine or local treatment of the affected limb in preventing acute adenolymphangitis in lymphoedema caused by brugian filariasis. *Ann. Trop. Med. Parasitol.* **93:**367–377.

109. Simonsen, P. E., and S. K. Dunyo. 1999. Comparative evaluation of three new tools for diagnosis of bancroftian filariasis based on detection of specific circulating antigens. *Trans. R. Soc. Trop. Med. Hyg.* **93:**278–282.

110. Simonsen, P. E., and S. M. Magesa. 2004. Observations on false positive reactions in the rapid NOW Filariasis card test. *Trop. Med. Int. Health* **9:**1200–1202.

111. Simonsen, P. E., and D. W. Meyrowitsch. 1998. Bancroftian filariasis in Tanzania: specific antibody responses in relation to long-term observations on microfilaremia. *Am. J. Trop. Med. Hyg.* **59:**667–672.

112. Srividya, A., S. P. Pani, P. K. Rajagopalan, D. A. P. Bundy, and B. T. Grenfell. 1991. The dynamics of infection and disease in bancroftian filariasis. *Trans. R. Soc. Trop. Med. Hyg.* **85:**255–259.

113. Supali, T., N. Rahmah, Y. Djuardi, E. Sartono, P. Ruckert, and P. Fischer. 2004. Detection of filarial-specific IgG4 antibodies using Brugia Rapid test in individuals from an area highly endemic for *Brugia timori*. *Acta Trop.* **90:**255–261.

114. Tawill, S. A., W. Kipp, R. Lucius, M. Gallin, K. D. Erttmann, and D. W. Buttner. 1995. Immunodiagnostic studies on *Onchocerca volvulus* and *Mansonella perstans* infections using a recombinant 33kDa *O. volvulus* protein (Ov33). *Trans. R. Soc. Trop. Med. Hyg.* **89:**51–54.

115. Taylor, H. R., B. Munoz, F. Keyvan-Larijani, and B. M. Greene. 1989. Reliability of detection of microfilariae in skin snips in the diagnosis of onchocerciasis. *Am. J. Trop. Med. Hyg.* **41:**467–471.

116. Taylor, M. J., C. Bandi, and A. Hoerauf. 2005. *Wolbachia* bacterial endosymbionts of filarial nematodes. *Adv. Parasitol.* **60:**245–284.

117. Taylor, M. J., W. H. Makunde, H. F. McGarry, J. D. Turner, S. Mand, and A. Hoerauf. 2005. Macrofilaricidal activity after doxycycline treatment of *Wuchereria bancrofti*: a double-blind, randomized placebo-controlled trial. *Lancet* **365:**2116–2121.

118. Tisch, D. J., E. Michael, and J. W. Kazura. 2005. Mass chemotherapy options to control lymphatic filariasis: a systematic review. *Lancet Infect. Dis.* **5:**514–523.

119. Tobian, A. A. R., N. Tarongka, M. Baisor, M. Bockarie, J. W. Kazura, and C. L. King. 2003. Sensitivity and specificity of ultrasound detection and risk factors for filarial-associated hydroceles. *Am. J. Trop. Med. Hyg.* **68:**638–642.

120. Turner, P., B. Copeman, D Gerisi, and R. Speare. 1993. A comparison of the Og4C3 antigen capture ELISA, the Knott test, an IgG4 assay and clinical signs, in the diagnosis of Bancroftian filariasis. *Trop. Med. Parasitol.* **44:**45–48.

121. Waa, E. A. V. 1991. Chemotherapy of filariasis. *Parasitol. Today* **7:**194–199.

122. Weil, G. J., P. J. Lammie, and N. Weiss. 1997. The ICT Filariasis test: a rapid-format antigen test for diagnosis of bancroftian filariasis. *Parasitol. Today* **13:**401–404.

123. Weil, G. J., R. M. R. Ramzy, M. El Setouhy, A. M. Kandil, E. S. Ahmed, and R. Faris. 1999. A longitudinal study of Bancroftian filariasis in the Nile Delta of Egypt: baseline data and one-year follow-up. *Am. J. Trop. Med. Hyg.* **61:**53–58.

124. Winkler, S., I. El Menyawi, K. F. Linnau, and W. Graninger. 1998. Short report: total serum levels of the nitric oxide derivatives nitrite/nitrate during microfilarial clearance in human filarial disease. *Am. J. Trop. Med. Hyg.* **59:**523–525.

125. Winkler, S., M. Willheim, K. Baier, A. Aichelburg, P. G. Kremsner, and W. Graninger. 1999. Increased frequency

of Th2-type cytokine-producing T cells in microfilaremic loiasis. *Am. J. Trop. Med. Hyg.* **60:**680–686.

126. **World Health Organization.** 1995. Onchocerciasis and its control. *WHO Tech. Rep. Ser. no. 852.*

127. **Yameogo, L., L. Toe, J. M. Hougard, B. A. Boutin, and T. R. Unnasch.** 1999. Pool screen polymerase chain reaction for estimating the prevalence of *Onchocerca volvulus* infection in *Simulium damnosum* sensu lato: results of a field trial in an area subject to successful vector control. *Am. J. Trop. Med. Hyg.* **60:**124–128.

128. **Zimmerman, P. A., R. H. Guderian, E. Aruajo, L. Elson, P. Phadke, J. Kubofcik, and T. B. Nutman.** 1994. Polymerase chain reaction-based diagnosis of *Onchocerca volvulus* infection: improved detection of patients with onchocerciasis. *J. Infect. Dis.* **169:**686–689.

13

Intestinal Cestodes

Adult cestodes, or tapeworms, live attached to the mucosa in the small intestine and absorb food from the host intestine. The attachment organ is called the scolex, to which is attached a chain of segments or proglottids called the strobila. Each proglottid contains a male and female reproductive system. The proglottids are classified as immature, mature, or gravid (the latter are found at the end of the strobila and contain the fully developed uterus full of eggs). The uterine structure in the gravid proglottids is often used as the main criterion for identification. The eggs and/or scolex can also be used to identify a cestode to the species level (Algorithm 13.1).

Cestodes have complex life cycles that usually involve both the intermediate and definitive hosts. In some infections, humans serve as only the definitive hosts, with the adult worm in the intestine (*Diphyllobothrium latum*, *Taenia saginata*, *Hymenolepis diminuta*, and *Dipylidium caninum*). In other cases, humans can serve as both the definitive and intermediate hosts (*T. solium* and *H. nana*).

Tapeworms have been recorded in writings from about 1500 BC and are among the earliest known human parasites. Galen, in the second century, recognized *Taenia*, and other tapeworms have been recognized since the late 1500s (3). Apparently, eggs of *Diphyllobothrium* spp. have been found in Alaska and date from the 13th and 17th centuries (8). It was not until 1993 that a new human species was identified and described, *Taenia asiatica*; however, there is some discussion regarding whether this is actually a separate species or a subspecies of *T. saginata*.

The exact prevalence of human intestinal tapeworms is not really known, but estimates are that as many as 100 million people are infected with *T. saginata* and *T. solium* while approximately 75 million are probably infected with *H. nana*. Certainly, the presence of adult tapeworms in the intestinal tract is not life-threatening; however, infection with larval tapeworms can be very serious and can lead to death.

Diphyllobothrium latum

D. latum belongs to the pseudophyllidean tapeworm group, which is characterized by having a scolex with two bothria (sucking organs) rather than the typical four suckers seen in the *Taenia* tapeworms. The distribution of this worm

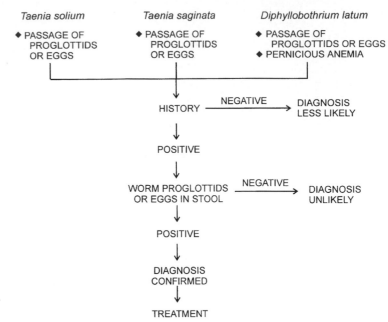

Taenia solium   *Taenia saginata*   *Diphyllobothrium latum*

◆ PASSAGE OF   ◆ PASSAGE OF   ◆ PASSAGE OF
   PROGLOTTIDS   PROGLOTTIDS   PROGLOTTIDS OR EGGS
   OR EGGS   OR EGGS   ◆ PERNICIOUS ANEMIA

HISTORY ——— NEGATIVE ——→ DIAGNOSIS
                                         LESS LIKELY

POSITIVE

WORM PROGLOTTIDS ——— NEGATIVE ——→ DIAGNOSIS
OR EGGS IN STOOL   UNLIKELY

POSITIVE

DIAGNOSIS
CONFIRMED

TREATMENT

Algorithm 13.1   *Taenia solium, T. saginata,* and *Diphyllobothrium latum.*

is worldwide, with various increased outbreaks reported from time to time. Areas of the world where a high incidence of infection with *Diphyllobothrium* spp. has been reported to occur include the Baltic countries, countries within the former USSR, and, especially, Finland; infections are also present in tropical Africa and parts of Asia, as well as North and South America. *Diphyllobothrium* spp. are not host specific. *D. dendriticum* occurs in fish-eating birds and mammals and can be found in certain human populations in Alaska, while *D. ursi* occurs in bears and humans in Alaska and Canada. Occasionally, *D. pacificum* is found in humans, although it is normally found in pinnipeds off the coast of Peru. Infections have also been found in New Guinea and Australia. Currently there are 13 species in the genus, most of which can infect humans.

In reviewing data from Europe during the last 20 years, results indicate that several dozen cases of diphyllobothriasis have been reported each year from Finland and Sweden, that there have been numerous cases in the French- or Italian-speaking areas around subalpine lakes, and that sporadic cases have been observed in Austria, Spain, Greece, Romania, Poland, and Norway. Over 30 cases have been identified on the Swiss shores of Lake Maggiore since 1990, and 70 cases have been reported on the Swiss and French shores of Lake Leman between 1993 and 2002. Between 8 and 12% of perch fillets from Lake Leman and 7.8% of perch from Lake Maggiore were found to contain larvae. Sources of infection include marinated fish fillets in northern Europe, "carpaccio di persico" in northern Italy, and perch and charr consumed raw or undercooked around Lake Leman (20).

The infection is acquired by the ingestion of raw, poorly cooked, or pickled freshwater fish and has been associated with a condition similar to pernicious anemia (Table 13.1).

Life Cycle and Morphology

The life cycle can be seen in Figure 13.1. Infection with the adult worm is acquired by the ingestion of raw, poorly cooked, or pickled freshwater fish (pike, perch, lawyer, salmon, trout, whitefish, grayling, ruff, turbot, etc.) containing the encysted plerocercoid larvae. After ingestion, the worm matures, with egg production beginning in about the fifth or sixth week. The adult worm reaches a length of 10 m or more and may contain up to 3,000 proglottids.

The scolex of *D. latum* is elongate and spoon shaped and has two long sucking grooves, one on the dorsal surface and the other on the ventral surface (Figure 13.2). The mature and gravid proglottids are wider than they are long, with the main reproductive structures (mainly the uterus) located in the center of the gravid proglottid. This configuration of the uterine structure has been called a rosette (Figures 13.3 and 13.4). Identification to the species level is usually based on this typical morphology of the gravid proglottids. Both eggs and proglottids may be found in the stool. Often, a partial chain of proglottids may be passed (a few inches to several feet).

The eggs are broadly oval and operculated and measure 58 to 75 μm by 40 to 50 μm (Figure 13.5). After developing for 2 weeks in freshwater, the eggs hatch and the ciliated, coracidium larvae are ingested by the first intermediate host, the copepod (Figure 13.6). The copepods, containing the second larval stage (procercoid), are then ingested by

Table 13.1 Cestode parasites (intestinal)

Characteristic	Diphyllobothrium latum	Taenia saginata	Taenia saginata asiatica	Taenia solium	Hymenolepis nana	Hymenolepis diminuta	Dipylidium caninum
Intermediate hosts (common)	Two: copepods and fish	One: cattle	One: pig	One: pig	One: various arthropods (beetles, fleas) or none	One: various arthropods (beetles, fleas)	One: various arthropods (fleas, dog lice)
Mode of infection	Ingestion of plerocercoid (sparganum) in flesh of infected fish	Ingestion of cysticercus in infected beef	Ingestion of cysticercus in infected tissues, primarily liver	Ingestion of cysticercus in infected pork	Ingestion of cysticercoid in infected arthropod or by direct ingestion of egg; autoinfection may also occur	Ingestion of cysticercoid in infected arthropod	Ingestion of cysticercoid in fleas, lice
Prepatent period	3–5 wk	10–12 wk	10–12 wk	5–12 wk	2–3 wk	~3 wk	3–4 wk
Normal life span	Up to 25 yr	Up to 25 yr	Up to 25 yr	Up to 25 yr	Perhaps many years owing to autoinfection	Usually <1 yr	Usually <1 yr
Length	4–10 m	4–12 m	4–3 m	1.5–8 m	2.5–4.0 cm	20–60 cm	10–70 cm
Scolex	Spatulate, 3 by 1 mm; no rostellum or hooklets; has 2 shallow grooves (bothria)	Quadrate, 1–2 mm in diam; no rostellum or hooklets; 4 suckers	Quadrate, 1–2 mm in diam; has rostellum but no hooklets; 4 suckers	Quadrate, 1 mm in diam; has rostellum and hooklets; 4 suckers	Knoblike but not usually seen; has rostellum and hooklets; 4 suckers	Knoblike but not usually seen; has rostellum but no hooklets; 4 suckers	0.2–0.5 mm in diam; has conical/retractile rostellum armed with 4–7 rows of small hooklets; 4 suckers
Usual means of diagnosis	Ovoid, operculate yellow-brown eggs (58–75 µm by 40–50 µm) in feces; egg usually has small knob at aбopercular end; proglottids may be passed, usually in chain of segments (few centimeters to 0.5 m long); proglottids wider than long (3 by 11 mm) and have rosette-shaped central uterus	Gravid proglottids in feces; they are longer than wide (19 by 17 mm) and have 15–20 lateral branches on each side of the central uterine stem; they usually appear singly; spheroidal yellow-brown, thick-shelled eggs (31–43 µm in diam) containing an oncosphere; may be found in feces	Gravid proglottids in feces; they are longer than wide (10 by 4 mm) and have ≥20 lateral branches on each side of central uterine stem; spheroidal yellow-brown, thick-shelled eggs (31–43) µm in diam containing an oncosphere; may be found in feces	Gravid proglottids in feces; they are longer than wide (11 by 5 mm) and have 7–13 lateral branches on each side of central uterine stem; usually appear in chain of 5–6 segments; spheroidal yellow-brown, thick-shelled eggs (31–43 µm) containing an oncosphere; may be found in feces	Nearly spheroidal pale, thin-shelled eggs (30–47 µm in diam) in feces; oncosphere surrounded by rigid membrane, which has 2 polar thickenings from which 4–8 filaments (polar filaments) extend into the space between the oncosphere and the thin, outer shell	Large, ovoid, yellowish, moderately thick-shelled eggs (70–85 µm by 60–80 µm) in feces; egg contains an oncosphere but no polar filaments	Gravid proglottids (8–23 mm long) containing compartmented clusters of eggs in feces; proglottids have genital pores at both lateral margins; occasionally individual oncospheres (20–33 µm in diam) visible in feces

(continued)

Table 13.1 Cestode parasites (intestinal) *(continued)*

Characteristic	Diphyllobothrium latum	Taenia saginata	Taenia saginata asiatica	Taenia solium	Hymenolepis nana	Hymenolepis diminuta	Dipylidium caninum
Diagnostic problems or notes	Eggs are sometimes confused with eggs of *Paragonimus* spp.; eggs are unembryonated when passed in feces	Eggs are identical to those of *T. solium*; ordinarily can distinguish between species only by examination of gravid proglottids; eggs can be confused with pollen grains (handle all proglottids with extreme care)	Eggs are identical to those of *T. solium* and *T. saginata*; ordinarily can distinguish between species only by examination of gravid proglottids; proglottids much smaller than those of *T. saginata* or *T. solium*; eggs can be confused with pollen grains (handle all proglottids with extreme care)	Eggs are identical to those of *T. saginata*; eggs less likely to be found in feces than with *T. saginata* (handle all proglottids with extreme care as *T. solium* eggs are infective to humans)	Sometimes confused with eggs of *H. diminuta*; rodents serve as reservoir hosts	Should not be confused with *H. nana*, as eggs lack polar filaments; rodents serve as reservoir hosts	Gravid proglottids resemble rice grains (dry) or cucumber seeds (moist); dogs and cats serve as reservoir hosts

fish, which may be ingested by larger fish. In this situation, the final fish intermediate host may contain many plerocercoid larvae, which initiate the infection with the adult worm when ingested by humans (Figure 13.7).

Clinical Disease

Symptoms depend on a number of variables: the number of worms present, the amounts and types of by-products produced by the worm, the patient's reaction to such by-products, and the absorption of various metabolites by the worms. Occasionally intestinal obstruction, diarrhea, abdominal pain, or anemia occurs. If the worm is attached at the jejunal level, there may be a vitamin B_{12} deficiency, which resembles pernicious anemia and develops in a very small percentage of persons harboring the tapeworm. Heavy or long-term infections with *D. latum* may cause megaloblastic anemia due to parasite-mediated dissociation of the vitamin B_{12}-intrinsic factor complex within the gut lumen, making B_{12} unavailable to the host (46). This clinical picture is seen most frequently in Finland, where individuals tend to have a genetic predisposition to pernicious anemia. In patients without this genetic predisposition, symptoms of a *D. latum* infection may be absent or minimal, consisting of a slight leukocytosis with eosinophilia.

Most patients are actually asymptomatic; however, the following symptoms were found more frequently in those who were infected than in the control group: fatigue, diarrhea, dizziness, weakness, numbness of extremities, and a sensation of hunger. No differences were found in terms of abdominal pain. Symptoms occur in approximately 22% of infections. No proglottids are routinely passed in the stool as is seen with *Taenia* infections, although occasionally a string of proglottids is passed or vomited.

About 40% of patients infected with *D. latum* have reduced vitamin B_{12} levels, but fewer than 2% develop anemia. In cases of severe anemia, there may be neurologic symptoms such as weakness, numbness, paresthesia, movement and coordination problems, and impairment of deep sensibilities. Both the anemia and neurologic problems are reversed by administration of vitamin B_{12} and do not reoccur after the worm has been eliminated. This type of severe anemia is rarely seen, probably as a result of overall improvements in general nutrition of the population in the area of endemic infection.

Diagnosis

Diagnosis is usually based on the recovery and identification of the characteristic eggs or proglottids. If the egg operculum is difficult to see, the coverslip of the wet preparation can be tapped and the pressure may cause the operculum to pop open, thus making it more visible. The eggs are unembryonated at the time they are passed in the stool. Proglottids are often passed in chains (a few inches to several feet), and this is a clue to *D. latum*. The overall

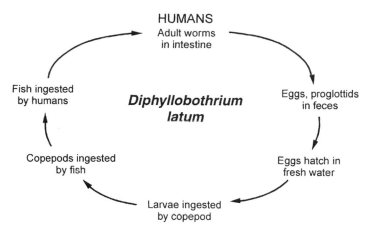

Diphyllobothrium latum

Fish ingested
by humans

Eggs, proglottids
in feces

Copepods ingested
by fish

Eggs hatch in
fresh water

Larvae ingested
by copepod

Figure 13.1   Life cycle of *Diphyllobothrium latum*.

proglottid morphology with the rosette uterine structure also facilitates identification.

Although differentiation is not generally a problem, the eggs of *D. latum* look very much like those of *Nanophyetus salmincola*, a trematode that is commonly found in the Pacific Northwest region of the United States and in eastern Siberia. *N. salmincola* eggs have been described as being broadly oval, operculate, relatively thick-shelled, and yellowish; they measure 60 to 80 µm by 34 to 50 µm, a direct overlap with the size range of *D. latum*. However, human infection with *N. salmincola* is generally incidental and uncommon.

KEY POINTS—LABORATORY DIAGNOSIS
Diphyllobothrium latum

1. Careful examination of the eggs should reveal the operculum. If it is difficult to see, it can sometimes be popped open by tapping on the coverslip of the wet preparation. The light should be somewhat reduced to allow the operculum to be seen more easily.

Figure 13.2   Scolex of *Diphyllobothrium latum*.

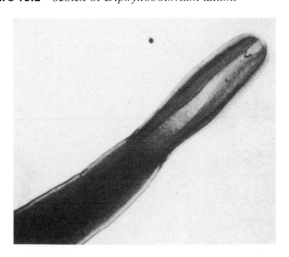

2. Often, the proglottids are passed in chains (a few inches to a few feet). Also, the gravid proglottids are much wider than long, with the uterine structure being seen in the middle of the proglottid (rosette).

3. Handling of the specimens presents no danger in terms of cysticercosis; however, all fecal specimens should be considered potentially infectious (with other organisms).

Treatment

The use of both praziquantel and niclosamide has been recommended (see chapter 25). A single dose of praziquantel (5 to 10 mg/kg of body weight) for both adults and children eliminates the infection in >90% of patients. Niclosamide is given as a single dose of 2 g (four tablets chewed thoroughly). A single dose of 50 mg/kg is recommended for children. Niclosamide may not be available within the United States. After therapy, if the worm has not been spontaneously passed within 2 h, a saline purgation can be used.

Epidemiology and Prevention

Although *D. latum* infections have been detected in other mammals, in areas where human infection has become

Figure 13.3   Gravid proglottids. (1) *Taenia saginata*; (2) *T. solium*; (3) *Diphyllobothrium latum*; (4) *Dipylidium caninum*.

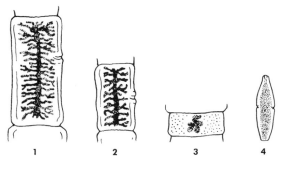

1   2   3   4

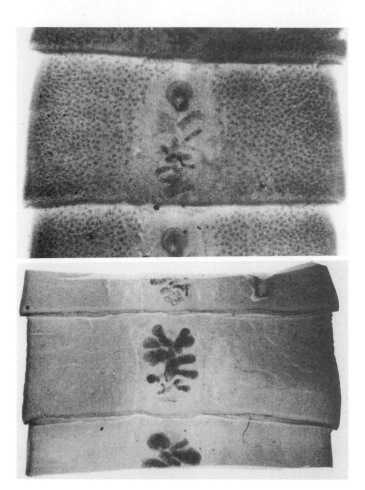

Figure 13.4 Gravid proglottids of *Diphyllobothrium latum*. (From A Pictorial Presentation of Parasites: A cooperative collection prepared and/or edited by H. Zaiman. Slides from J. F. Mueller.)

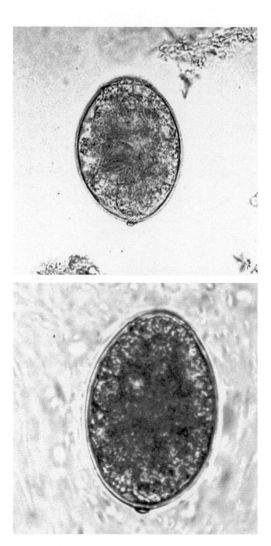

Figure 13.5 *Diphyllobothrium latum* eggs. Note the small knob at the abopercular end (end opposite the operculum); also note that the operculum does not fit into "shoulders" like the *Paragonimus* eggs but presents a smooth outline on the egg shell.

rare, the natural transmission cycle from mammals other than humans does not seem to be sustained. It is quite possible to acquire the infection from the ingestion of infected raw freshwater fish that has been shipped under refrigeration to areas where the infection is not endemic. Preventive measures would include thorough cooking of all freshwater fish or freezing for 24 to 48 h at –18°C. This infection has been called the Jewish housewives' disease, since the individual preparing the food may sample the dish (e.g., gefilte fish) prior to cooking and acquire the infection. Other groups who tend to eat raw or insufficiently cooked fish include the Russians, Finns, and Scandinavians (7). Raw fish marinated in lime juice (ceviche) is also a source of infection (*D. pacificum*) in Latin America. Since the domestic dog can serve as a reservoir host, infected dogs should be periodically treated. Other factors contributing to the parasitic cycle include the continued dumping of wastewater into lakes and possible animal reservoirs.

Taenia solium

T. solium infection may have been recognized since biblical times, and the life cycle was delineated in the mid-1850s (7). The tapeworm is found in many parts of the world and is considered an important human parasite where raw or poorly cooked pork is eaten (Table 13.1).

Cysticercosis infections with *T. solium* larvae are relatively common in certain parts of the world. However, cysticercosis is becoming more common within the United States; the infection is associated with Hispanic ethnicity, immigrant status, exposure to areas where neurocysticercosis is endemic, and the American Southwest (15, 43, 50, 58). Serum immunoblotting shows that the seroprevalence of cysticercosis and infection with the adult tapeworm in a

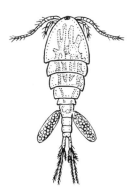

Figure 13.6 Copepod. (Illustration by Sharon Belkin.)

rural southern California population approximates the prevalence in some areas of Latin America where the infection is endemic (15). Also, an increasing number of cases has been reported in the United States over the past 50 years (58). This extraintestinal form of the infection is far more serious than the presence of the adult worm in the intestine.

Life Cycle and Morphology

The life cycle can be seen in Figure 13.8. Infection with the adult worm is initiated by the ingestion of raw or poorly cooked pork containing encysted *T. solium* larvae (Figure 13.9). The larva is digested out of the meat in the stomach, and the tapeworm head evaginates in the upper small intestine, attaches to the intestinal mucosa, and grows to the adult worm within 5 to 12 weeks. Although usually a single worm is present in the intestine, multiple worms may occur. The adult worm reaches a length of 2 to 7 m and may survive for 25 years or more (7).

Figure 13.7 *Diphyllobothrium latum*, encysted pleurocercoids in cherry salmon fillet. (From A Pictorial Presentation of Parasites: A cooperative collection prepared and/or edited by H. Zaiman. Photograph courtesy of T. Oshima, Yokohama City University.)

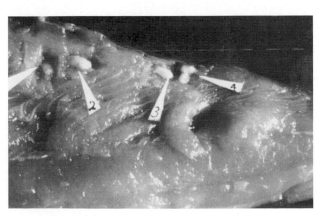

The attachment organ, or scolex, has four large suckers with a rounded rostellum containing a double row of 25 to 30 small hooks (Figure 13.10). The neck region is short and about half the width of the scolex. The total number of proglottids is less than 1,000, with the immature proglottids being wider than long, the mature proglottids being approximately square, and the gravid segments being longer than wide.

Identification to the species level is usually based on the number of lateral uterine branches in the gravid proglottid (Figures 13.3 and 13.11). The main lateral branches should be counted where they come off the main central stem. Only one side is counted, and there are between 7 and 13 branches, with an average of 9.

The terminal gravid proglottids break off from the main strobila and are passed out in the stool or actually migrate out of the bowel. In some cases, three or four attached proglottids pass out in the stool. Eggs usually pass from the uterus through the ruptured wall, where the proglottids break off from the strobila. The eggs are round or slightly oval (31 to 43 µm), have a thick, striated shell, and contain a six-hooked embryo or oncosphere (Figure 13.12). These eggs may remain viable in the soil for weeks.

On ingestion by hogs or humans, the eggs hatch in the duodenum or jejunum after exposure to gastric juice in the stomach. The released oncospheres penetrate the intestinal wall, are carried via the mesenteric venules throughout the body, and are filtered out in the subcutaneous and intramuscular tissues, the eyes, the brain, and other body sites. This cysticercus is an ovoid, milky-white bladder with the head invaginated into the bladder (Figure 13.13). The bladder worms measure 5 mm long by 8 to 10 mm wide; however, in the brain, they may reach a volume of 60 ml. The host generally produces a fibrous capsule around the bladders, unless they are located in the brain, particularly the ventricles.

Clinical Disease

Adult Worm. The presence of the adult worm usually causes no problems other than slight irritation at the site of attachment or vague abdominal symptoms (hunger pains, indigestion, diarrhea, and/or constipation). There may be a low-grade eosinophilia, usually under 15%, and increased levels of immunoglobulin E (IgE) in serum.

Cysticercosis. The larval forms of *T. solium* have been called *Cysticercus cellulosae*; however, this term is not taxonomically correct and is used infrequently. These cysticerci have been recovered from all areas of the body, and symptoms depend on the particular body site involved. The presence of cysticerci in the brain represents the most frequent parasitic infection of the human nervous system and

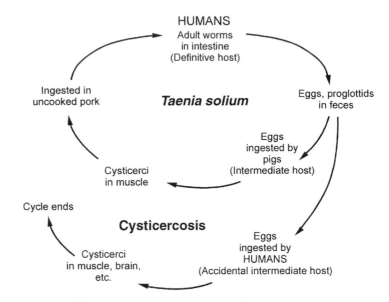

Figure 13.8 Life cycle of *Taenia solium*.

Figure 13.9 (A) Cysticerci in muscle (pork). (From A Pictorial Presentation of Parasites: A cooperative collection prepared and/or edited by H. Zaiman.) (B) Cysticerci in muscle (dog). (Courtesy of W. Jann Brown and Marietta Voge, University of California, Los Angeles.)

the most common cause of adult-onset epilepsy throughout the world (13, 26, 43). For unknown reasons, in Latin America it is unusual for both brain and muscle cysticercosis to occur in the same patient, with fewer than 6% of patients having cysticerci in both sites. However, elsewhere in the world, subcutaneous involvement by *T. solium* cysticerci has been detected in as many as 78.5% of patients with cerebral cysticercosis (12). Possible reasons for such differences may include (i) the immune status of the patient, (ii) the human leukocyte antigen type, (iii) the nutritional status of the patient, (iv) the burden of eggs infecting the patient, and (v) a difference in the strains of *T. solium*. It is also rare for a patient with cysticercosis to harbor adult *T. solium* worms in the intestine. Neurocysticercosis

Figure 13.10 Scolex of *Taenia solium*.

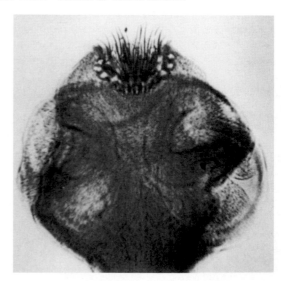

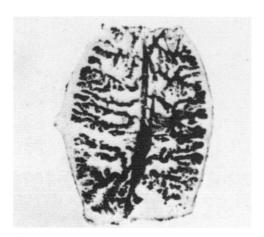

Figure 13.11   Gravid proglottid of *Taenia solium*.

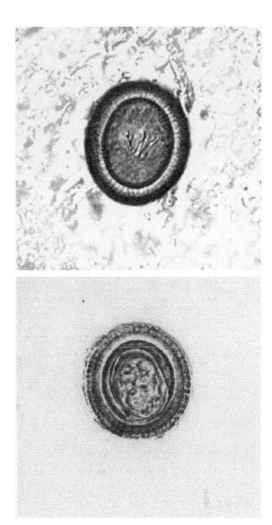

Figure 13.12 *Taenia* eggs. Note the striated shell and the presence of the six-hooked oncosphere within the thick egg shell. The hooks may not be visible in every egg, particularly in specimens that have been stored for long periods in preservatives; however, in clinical specimens submitted to the laboratory in fixatives for timely examination, the hooklets are generally visible within the egg shell. Without special staining, the eggs of *Taenia* spp. cannot be identified to the species level (*T. solium*, *T. saginata*, *T. asiatica*).

usually affects males and females of all ages, with a peak incidence between 30 and 50 years of age.

Neurocysticercosis is the most common parasitic disease of the central nervous system; in some countries the prevalence exceeds 10% and is responsible for up to 50% of late-onset epilepsy cases (43). Factors related to neurocysticercosis pathology include (i) the individual immune response to the presence of the parasite, from tolerance to a severe inflammatory response; (ii) the location of the parasite in the brain; (iii) the number of cysticerci present; and (iv) the stage or age of the lesions (live cysts, inflammatory exudates, granulomas, calcifications, and residual fibrosis).

When the larvae are found in the brain, symptoms can result from actual larval invasion of the brain tissue and/or death of the organism, which stimulates tissue reactions around the larvae (Figure 13.14). It has been recommended that in every case of epilepsy occurring in a patient with no family history of fits and no personal history of fits in childhood, the probability of cysticercosis should be entertained (39). Although studies indicate the presence of epileptiform seizures in patients with cysticercosis, other symptoms, including abnormal behavior, transient paresis, intermittent obstructive hydrocephalus, disequilibrium, meningoencephalitis, and visual problems, may also be present. Spinal fluid from patients with cerebral cysticercosis shows possible eosinophilia and pleocytosis (the most common cell type is the lymphocyte), with atypical mononuclear lymphoid cells. The most distinctive features of the infection in the spinal fluid are atypical, reactive lymphoid cells with a mixed cellular population. A study by Sotelo et al. (54) of 753 cases of neurocysticercosis reviewed inactive and active disease, the frequency of signs and symptoms, and cerebrospinal fluid (CSF) findings. More than 50% of all patients exhibit at least two of these clinical presentations (54) (Table 13.2). Although rare, the

occurrence of secondary parkinsonian tremors in previously shunted patients with cysticercosis suggests obstruction and requires immediate evaluation (33).

In a study from Mexico, the frequency of cerebral arteritis in subarachnoid cysticercosis was higher than previously reported. Involvement of middle-sized vessels is a common occurrence, even in patients without clinical evidence of ischemia (6).

Although rare, a case of acute leukemia and massive cardiopulmonary cysticercosis has been reported (38). In this case, pulmonary cysticercosis may have been caused by the immunosuppression.

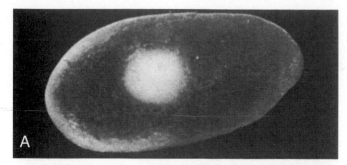

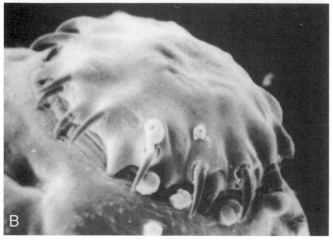

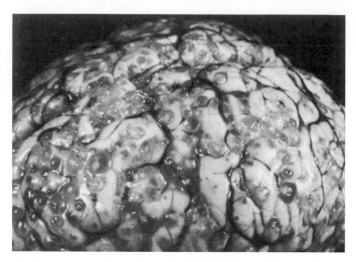

Figure 13.14 Gross specimen of brain containing many cysticerci. (From A Pictorial Presentation of Parasites: A cooperative collection prepared and/or edited by H. Zaiman.)

Figure 13.13 (A) Cysticercus with head invaginated into the bladder. (From A Pictorial Presentation of Parasites: A cooperative collection prepared and/or edited by H. Zaiman.) (B) Scanning electron micrograph of scolex from cysticercus. (Photograph courtesy of U.S. Department of Agriculture).

Cysticerci in the eye can be found in various areas, including in the eyelid and under the conjunctiva. The larvae are unencapsulated when found in the vitreous or anterior chamber and will actively change shape, casting shadows in front of the retina (Figure 13.15). Unless the organism is removed, permanent eye damage may occur, particularly if the parasite dies, thus stimulating an intense inflammatory reaction.

A single case of cysticercosis involving the scrotum has been reported; the patient's presentation was that of testicular torsion requiring surgical exploration. The patient was a 9-year-old Hispanic boy who was born and raised in the United States and had no history of foreign travel. Histologic testing was confirmatory, as were the serologic results (4).

Based on fine-needle aspiration cytology, five cases of cysticercosis of the tongue and buccal mucosa in children have been reported. Symptoms included gradually increasing nodular swelling; cysticercosis was not

Table 13.2 Classification of neurocysticercosis[a]

Form	Occurrence in 753 cases (%)
Active	
Arachnoiditis[b] (headache, vertigo, vomiting, cranial nerve dysfunction)	46.2
Hydrocephalus[c] secondary to meningeal inflammation (intracranial hypertension, ataxic gait, intellectual deterioration)	25.7
Cysts within brain itself (epilepsy, headache)	13.2
Brain infarction secondary to vasculitis (unilateral pyramidal signs, intracranial hypertension, psychiatric disturbances)	2.3
Mass effect due to a large cyst or clump of cysts (tumorlike signs, epilepsy)	1.0
Inactive	
Granulomas or calcifications within the brain (epilepsy, usually partial seizures)	57.6
Hydrocephalus[c] secondary to meningeal fibrosis (intellectual deterioration, cranial nerve dysfunction, intracranial hypertension, Parinaud's sign[d])	3.8

[a] Adapted from reference 54.
[b] Inflammation of the intermediate membrane (between pia mater and dura mater) which encloses the brain and spinal cord.
[c] Increased accumulation of cerebrospinal fluid within the ventricles of the brain, usually caused by interference with normal circulation of fluid.
[d] Paralysis of conjugate upward movement of the eyes without paralysis of convergence; associated with lesions of the midbrain.

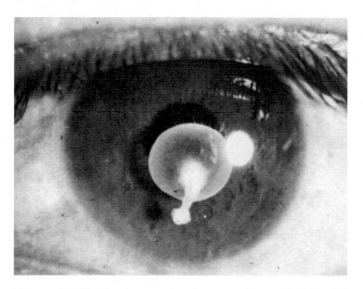

Figure 13.15 Cysticercus in the eye. (From A Pictorial Presentation of Parasites: A cooperative collection prepared and/or edited by H. Zaiman. Photograph courtesy of R. Delgado y Garnica, Hospital Infantil de Tacubaya, Arq. Carlos Laxo y Gaviota, Mexico D.F.)

considered in any of these cases. Since patients with these lesions may be seen by a dentist rather than a physician, cysticercosis should be included in the differential diagnosis of solitary nodular lesions in the oral cavity, particularly in young patients, in areas of endemic infection (48).

There appear to be cases where reactivation of neurocysticercosis occurs, often several years after the patient was considered cured. Several years earlier, one patient had undergone treatment and was presumed to be cured of active disease. Computed tomographic (CT) and magnetic resonance imaging (MRI) studies performed 3 months prior to presentation revealed multiple intracerebral calcified lesions consistent with resolved neurocysticercosis. The patient presented with frontal headaches that occurred intermittently and were associated with vomiting, slowed mentation, and episodic visual scintillations. On presentation, additional imaging studies were performed and showed the same calcified lesions; however, these lesions were now surrounded by large areas of edema (52). Although this patient's symptoms resolved spontaneously within several days, she received a 28-day course of albendazole, cimetidine, and dexamethasone, with subsequent resolution of the cerebral edema. Another earlier report describes five patients whose CT scans showed an area of brain swelling surrounding previously inactive calcifications (16). There are several possible explanations for these relapse cases. Cerebral edema may have resulted from the death of worms that were incompletely calcified and still

viable. There may have been antigen release from dead, calcified parasites. Another explanation, which might be more realistic, suggests that a spontaneous increase in the immune response could have been triggered by antigen release from live or dying parasites. Based on these cases, it is important to remember that patients with calcified lesions may present with late-onset symptoms associated with marked inflammation and cerebral edema surrounding these lesions (26). Retreatment with anthelmintics may or may not be required; corticosteroid therapy may be sufficient.

Larger numbers of cysticerci usually correlate with more serious disease. In general, patients with <20 cysticerci within the parenchyma and the absence of hydrocephalus have a more positive prognosis than do patients who have many basal or ventricular lesions accompanied by hydrocephalus. Also, it is not clear how many asymptomatic patients with intracranial cysticerci eventually develop symptoms or how many symptomatic patients spontaneously recover.

Cysticercus, Racemose Type. The racemose, or proliferating, form of the cysticercus is composed of several bladders that are connected and are often found in the brain, particularly the fourth ventricle and subarachnoid space (Figures 13.16 and 13.17). Within these bladders, no scolices are found and the growth may resemble that of a metastatic tumor. Other larval cestodes may also have a racemose growth pattern, and it may be very difficult to differentiate a *T. solium* racemose cysticercus from other types histologically. The prognosis for such infections is very poor. Racemose cysticerci are usually not found in infected children and may be either a degenerative form or formed in response to infection at a different anatomic

Figure 13.16 Racemose form of *Taenia solium* cysticercus, left temporal lobe (filmy, white material). (Courtesy of W. Jann Brown and Marietta Voge, University of California, Los Angeles.)

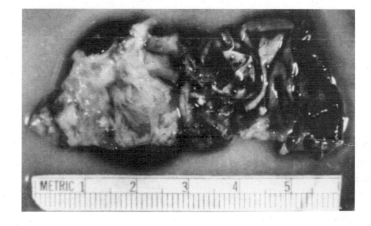

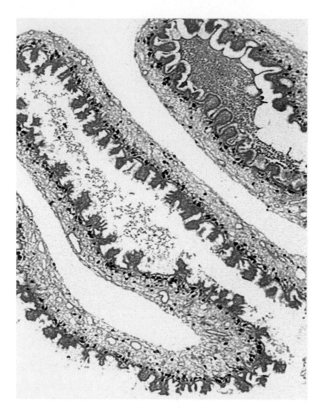

Figure 13.17 *Taenia solium.* Section of proliferating bladder wall of racemose cysticercus, demonstrating multiple layers. (From A Pictorial Presentation of Parasites: A cooperative collection prepared and/or edited by H. Zaiman.)

site. Cervical spinal cysticercosis is usually associated with the racemose type within the posterior fossa.

Although infection with both human immunodeficiency virus (HIV) and *T. solium* would be expected to occur more frequently due to the increased frequency of HIV in areas where cysticercosis is endemic, very little is known about the impact of HIV on the racemose type of cysticercosis. Very large cysts and the racemose type of neurocysticercosis appear to be more frequent in HIV-positive patients; these findings may be related to an impaired cell-mediated immune response (18).

Diagnosis

Adult Worms.   Identification of adult worms to the species level is not possible from the eggs because *T. solium* and *T. saginata* eggs look identical. Identification is usually based on the recovery and examination of the gravid proglottids, in which the main uterine lateral branches are counted after India ink injection staining or some other clearing mechanism (7 to 13 branches for *T. solium* and 15 to 20 for *T. saginata*) (Figure 13.3). If the scolex is recovered after therapy (purgation may be necessary to recover the intact attachment organ), the presence of the

four suckers and the armed rostellum with hooks will differentiate the worm from *T. saginata*. There are some other differences in the mature proglottids; however, these require staining and clearing of the proglottid and may be much more difficult to see.

It is well known that the proglottids of *T. saginata* can actively migrate from the anus. However, proglottids of *T. solium* are actively motile in the stool, so they would appear to be capable of migrating from the anus as well. Since these proglottids contain infective eggs and cannot be identified to the species level at first glance, it is very important that personnel use precautions to prevent accidental infection with *T. solium* eggs, which may lead to cysticercosis. It is also very important that all patients be instructed to use good personal hygiene if they are found to have *Taenia* tapeworm infection, since there may be a time lag between diagnosis and initiation of therapy and a danger of autoinfection with the eggs or possible infection of others.

Using DNA differential diagnosis by base excision sequence scanning thymine-base reader analysis with mitochondrial genes, four distinct *Taenia* spp. can be identified: *T. saginata*, *T. asiatica*, and two genotypes of *T. solium*. Thus, if nucleotide databases are available, this system provides a useful tool for the identification of taeniid specimens from patients and for the study of genetic variations of these helminths (61).

Coproantigen Detection.   Preliminary enzyme-linked immunosorbent assay (ELISA) studies for the diagnostic detection of *T. solium* and *T. saginata* coproantigens appear to be very promising (1, 2, 19, 36). The ability to detect human taeniasis by coproantigen ELISA in the absence of eggs in the stool would represent a significant advance in diagnosis. Also, the ability to reliably identify *Taenia* carriers in areas where infections are endemic would improve epidemiologic studies (47, 59). High concentrations of coproantigens are present until approximately 6 to 17 days after treatment. With the newer immunologic and biological methods available, test kits applicable under laboratory and field conditions may be developed and refined in the near future. By using *T. solium* and *T. saginata* DNA probes, a rapid, highly sensitive and specific dot blot assay for the detection of *T. solium* has been developed (11).

Cysticercosis.   The diagnosis of cysticercosis usually depends on surgical removal of the parasite and microscopic examination for the presence of suckers and hooks on the scolex. Multiple larvae are frequently present, and the presence of cysticerci in the subcutaneous or muscle tissues indicates that the brain is probably also involved. The calcified larvae can be readily seen on X-ray examination (Figure 13.18). CT scans or MRI

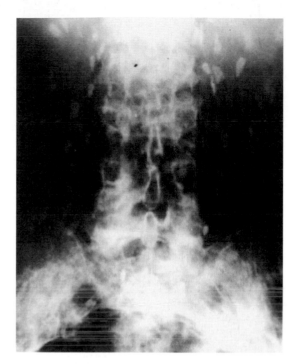

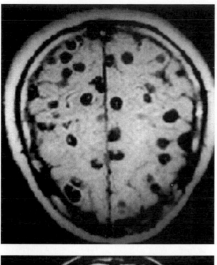

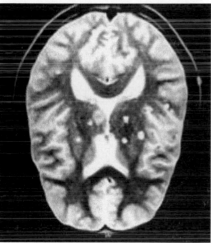

Figure 13.18 Calcified cysticerci in tissue. (Armed Forces Institute of Pathology photograph.)

may reveal the presence of lesions in the brain (Figure 13.19). Although neurocysticercosis may be suspected, particularly in a patient living in an area of endemicity, the multiple neurologic syndromes and the various clinical presentations make it difficult to confirm the diagnosis, even with CT or MRI techniques. The primary consideration may be brain tumor. If the racemose form is present in the brain, the CT scan cannot distinguish the lesion from that caused by other tumors. Ultrafast CT has been used successfully to diagnose a case of cysticercosis involving the myocardium (14). Ocular cysticercosis can usually be diagnosed by visual identification of the movements and morphology of the larval worm.

Fine-needle aspiration cytology has also been recommended as a cost-effective procedure, since it eliminates the need for open biopsy (32, 48). The characteristic cytomorphology of parasitic tegument and parenchyma has proven to be very helpful. Review of fine-needle aspiration smears from 258 patients with cysticercosis showed that Charcot-Leyden crystals were conspicuous by their absence, although cholesterol crystals were often seen (5). Absence of Charcot-Leyden crystals in cysticercosis could be attributed to the high lipid content of these lesions, as indicated by the frequent occurrence of cholesterol crystals.

Cysticercosis Serologic Testing. The immunoblot assay for neurocysticercosis has been a definite improvement in antibody detection and is available at the Centers

Figure 13.19 Neurocysticercosis. Scanning images of cysticerci in brain. (Upper, adapted from the Public Health Image Library, Centers for Disease Control and Prevention; lower, from A Pictorial Presentation of Parasites: A cooperative collection prepared and/or edited by H. Zaiman.)

for Disease Control and Prevention (29, 37). A semipurified abstract of *T. solium* cysticerci is used as the antigen, with seven major glycoprotein bands (50 to 13 kDa); a serum reaction with any one of these bands is indicative of cysticercosis. The sensitivity is 98%, and the specificity is 100% (37). Both a single parietal cyst and subcutaneous infections without neurologic involvement are usually detected by this procedure. This test is more sensitive than the ELISA procedure, especially when using serum (45). However, in another study using cerebrospinal fluid, the ELISA results were positive for 22 of 24 patients with active neurocysticercosis; however, six cases with calcified cysts and one case in transition were both negative. Thus, a significant difference was seen between active and inactive neurocysticercosis (41).

An enzyme-linked immunoelectrotransfer blot (EITB) assay using blood dried on filter paper has proven to be a convenient, economical way of handling field samples for epidemiologic surveys of cysticercosis, particularly in underdeveloped countries, where refrigeration and venipuncture represent problems with specimen collection and handling. There is no decrease in sensitivity compared with traditional serum samples used in the EITB assay (30).

Using the EITB assay on CSF samples, the sensitivity of a recent study was 86% (27). Patients with hydrocephalus had higher levels of circulating antigen, and there was no difference between antigen levels in CSF taken before and 14 days after therapy. There was a high correlation between the number of live cysts detected by CT and the number and intensity of antibody reactions. In contrast, there was a negative correlation with the number of enhancing lesions seen using CT, supporting the hypothesis that enhancing lesions correspond to the dead stage of the parasite.

KEY POINTS—LABORATORY DIAGNOSIS

Taenia solium

1. Preliminary examination of the proglottid may not allow identification without clearing or injection of the uterine branches with India ink.

   Note. Because of the possible danger of egg ingestion, specimens and proglottids should be handled with extreme care. The proglottid uterine structure will be packed with viable, infectious eggs, even if the proglottids are received in fixative.

2. The eggs look identical to those of *T. saginata*, and no identification to the species level can be performed with eggs alone without special procedures.

3. *Any patient with* Taenia *eggs recovered in the stool should be cautioned to use good hygiene. These patients should be treated as soon as possible to avoid the potential danger of accidental infection with the eggs, which may lead to cysticercosis (*T. solium*).*

4. Although serologic reagents for antibody detection and antigen detection tests have been developed, not all reagents produce sensitive and specific results. Consultation with the Centers for Disease Control and Prevention and/or your state public health laboratories may be helpful. Efforts made to use the immunodominant bands in ELISA are expected to yield a sensitive but easier and cheaper test for neurocysticercosis (55).

Treatment

Adult Worm. The use of praziquantel or niclosamide has been recommended (see chapter 25). Since there is always the possibility of cysticercosis via autoinfection, patients should be treated as soon as they are diagnosed.

Cysticercosis. Surgical removal is recommended when possible. In patients with ocular cysticercosis, cyst removal rather than enucleation is recommended. To prevent loss of the eye, it is recommended that the cysticercus be removed while it is still living; antiparasitic drugs may lead to a loss of vision secondary to intraocular inflammatory reaction. Excisional biopsy is recommended for subconjunctival cysticercosis. Idiopathic cystic myositis can present like extraocular muscle cysticercosis but can be differentiated by resolution with corticosteroid treatment. Therapy for orbital cysticercosis with oral albendazole and corticosteroids can prevent recurrent inflammation and improve ocular mobility (51).

The introduction of new drugs for cysticercosis has significantly changed the approach to patient management. Two drugs, praziquantel and albendazole, are effective against the cysticerci (53, 55). Praziquantel is rapidly absorbed and reaches its peak concentration in plasma in 1.5 to 2 h; it can cross the blood-brain barrier, and the concentration in CSF is approximately 24% of that obtained in the plasma (55). Apparently, the drug action damages the worm tegument to produce spastic paralysis of the scolex. Albendazole is a broad-spectrum anthelmintic drug; the anticysticercal impact is by inhibition of glucose uptake by parasitic membranes. In a study of 114 patients with parenchymal brain cysticercosis, albendazole was found to be more effective than praziquantel for both a full and a short course of therapy. Also, the length of albendazole therapy could be shortened without loss of efficacy; in contrast, a short course of praziquantel caused a 12% reduction in drug effectiveness. From the results of this study, an 8-day course of albendazole is recommended as treatment for parenchymal brain cysticercosis; a 15-day course of praziquantel could subsequently be used in patients demonstrating partial response to albendazole (53). The issue of medical or surgical therapy for subarachnoid or ventricular cysticercosis is controversial; however, successful therapy with albendazole has been reported for a patient with cysticerci in both subarachnoid and ventricular locations (17).

The prognosis for the patient is excellent when the adult worm is present, good when the cysticerci can be surgically removed or medically treated, and poor when the racemose form of the parasite is present, particularly in the brain. However, treatment of hydrocephalus secondary to cerebral racemose cysticercosis using endoscopic third ventriculostomy appears to be an effective and safe technique; the efficacy of endoscopic third ventriculostomy will be evaluated in additional studies (34).

Epidemiology and Prevention

Although other animals occasionally harbor the cysticercus stage, poorly cooked or raw pork is the usual source of human infection for the development of the adult worm. Three cases of cysticercosis in black bears in three northern California counties have been confirmed as due to *T. solium* (56). The number and geographic separation of the cases indicate that infection was not due to a single contaminated source but to human fecal material contaminated with *T. solium* eggs. How much of a threat this human-to-bear transmission poses to the bear population and/or to hunters (cysticercosis) is unknown, as is the reverse situation, bear-to-human transmission (adult worm). Human infection with cysticerci can involve thousands of organisms obtained from various sources: ingestion of *T. solium* eggs in contaminated food or water; self-infection from the presence of the adult worm in the intestine; and possibly internal autoinfection, in which the eggs come in contact with the stomach acid, thus possibly allowing hatching and penetration of the larvae into the tissues.

Prevention involves awareness of the infection route and the use of good sanitary and personal hygiene measures. The ingestion of vegetables fertilized with sewage should also be avoided. Adequate requirements for human sewage disposal should be followed. One of the most important means of prevention is the adequate cooking of pork and pork products; thorough cooking at 65°C kills the cysticerci; however, salting or pickling of the meat is usually not effective. In pig carcasses held at 4°C, more than 80% of the cysticerci were viable between days 1 and 25 (24). These findings confirm that eating raw or undercooked pork or viscera of pigs following refrigeration at 4°C for less than 30 days can lead to infection with *T. solium*. Effective immune protection of pigs against cysticercosis through vaccination has also been reported (42). The use of the EITB assay in sentinel piglets has been used to determine environmental contamination with *T. solium* eggs. This approach permits an indirect assessment of human risk and could be useful in monitoring the efficacy of intervention programs (28).

For a number of reasons, *T. solium* is a candidate for control: (i) neurocysticercosis has a tremendous impact on human health; (ii) since *T. solium* is the only source of cysticercosis for both humans and pigs, it may be epidemiologically controllable; (iii) effective and practical therapeutic intervention is available; (iv) international commitment to the use of control measures is present; and (v) the cost is acceptable. Through education, allocation of funds, recognition of both the human and veterinary issues, implementation of control measures, and chemotherapy, morbidity and mortality may be drastically reduced throughout the world (10, 44, 49).

Taenia saginata

T. saginata was apparently differentiated from *T. solium* in the late 1700s; however, cattle were not identified as the intermediate host until 1863. This infection has a worldwide distribution and is generally much more common than *T. solium* infection, particularly in the United States (Table 13.1). The overall impact on human health is much smaller than that seen with *T. solium*, since cysticercosis with *T. saginata* is apparently quite rare.

Life Cycle and Morphology

The life cycle is very similar to that of *T. solium* (Figure 13.20). Infection with the adult worm is initiated by the ingestion of raw or poorly cooked beef containing encysted *T. saginata* larvae. As with *T. solium*, the larva is digested out of the meat in the stomach, and the tapeworm evaginates in the upper small intestine and attaches to the intestinal mucosa, where the adult worm matures within 5 to 12 weeks. The adult worm can reach 25 m but often measures only about half this length. Although a single worm is usually found, multiple worms can be present.

The scolex is "unarmed" and has four suckers with no hooks (Figure 13.21). The proglottids usually number 1,000 to 2,000, with the mature proglottids being broader than long and the gravid proglottids being narrower and longer. Identification to the species level is usually based on the

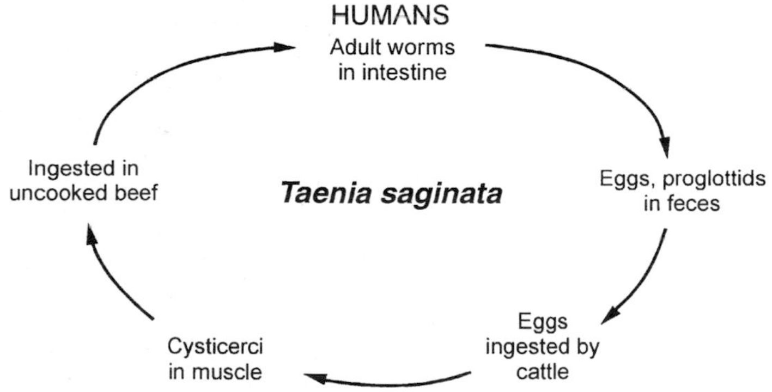

HUMANS
Adult worms
in intestine

Ingested in
uncooked beef

Taenia saginata

Eggs, proglottids
in feces

Cysticerci
in muscle

Eggs
ingested by
cattle

Figure 13.20   Life cycle of *Taenia saginata*. The same life cycle is applicable to *T. asiatica* (except that the eggs are contained primarily in the liver of the pig).

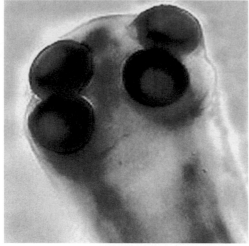

Figure 13.21 Scolex of *Taenia saginata*.

number of main uterine lateral branches, which are counted on one side of the gravid proglottid. There are between 15 and 20 branches, with an average of 18 (Figure 13.22).

Gravid proglottids often crawl from the anus during the day, when the host is most active. These proglottids may crawl under the stool when a specimen is submitted unpreserved in the stool collection carton. The eggs cannot be distinguished from those of *T. solium*. They are round to slightly oval, measure 31 to 43 μm, have a thick, striated shell, and contain the six-hooked embryo (oncosphere) (Figure 13.12). The eggs can remain viable in the soil for days to weeks.

On ingestion by cattle, the oncospheres hatch in the duodenum, penetrate the intestinal wall, and are carried via the lymphatics or bloodstream, where they are filtered out in the striated muscle. They then develop into the bladder worm or cysticercus within approximately 70 days. The mature cysticercus measures 7.5 to 10 mm wide by 4 to 6 mm long and contains the immature scolex, which has no hooks (unarmed). Other animals found to harbor cysticerci include the buffalo, giraffe, llama, and possibly reindeer (7). Actual cases of human cysticercosis with *T. saginata*

Figure 13.22 Gravid proglottid of *Taenia saginata*.

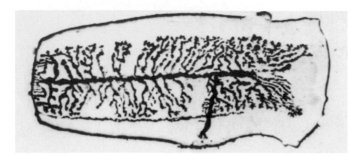

are rare in the literature, and there is speculation that some reported cases have been inaccurately diagnosed.

Clinical Disease

There are usually few symptoms associated with the presence of the adult worm in the intestine. Although rare symptoms (obstruction, diarrhea, hunger pains, weight loss, and appendicitis) have been reported, the most common complaint is the discomfort and embarrassment caused by the proglottids crawling from the anus. This occurrence may be the first clue that the patient has a tapeworm infection. Occasionally, the proglottids are also seen on the surface of the stool after it is passed.

Diagnosis

Since the eggs of *T. saginata* and *T. solium* look identical, identification to the species level is normally based on the recovery and examination of gravid proglottids, in which the main uterine lateral branches are counted (15 to 20 for *T. saginata* and 7 to 13 for *T. solium*). Often, the gravid proglottids of *T. saginata* are somewhat larger than those of *T. solium*; however, this difference may be minimal or impossible to detect when looking at fresh or preserved proglottids. If the scolex is recovered after therapy (which may require purgation), there will be four suckers and no hooks (Figure 13.21).

KEY POINTS—LABORATORY DIAGNOSIS

Taenia saginata

1. Preliminary examination of the gravid proglottid may not allow identification without clearing or injection of the uterine branches with India ink.

   Note. Since there is always the possible danger that the proglottid is *T. solium*, with the inherent problem of egg ingestion and cysticercosis, all specimens should be handled with extreme caution.

2. The eggs look identical to those of *T. solium*, so identification to the species level cannot be performed with eggs alone.

3. Although gravid proglottids of *T. saginata* are often somewhat larger than those of *T. solium*, this difference may be very minimal or impossible to detect.

4. *Any patient with* Taenia *eggs recovered in the stool should be cautioned to use good hygiene. These patients should be treated as soon as possible to avoid the potential danger of accidental infection with the eggs, which may lead to cysticercosis (*T. solium*).*

Treatment

The use of praziquantel or niclosamide has been recommended (see chapter 25). The same general approach used for *T. solium* is used to eradicate the adult worm. Therapy is usually very effective; however, if the proglottids begin to reappear in the stool or crawl from the anus, retreatment is necessary.

Epidemiology and Prevention

Since human infection is acquired from the ingestion of raw or poorly cooked beef, all beef prepared for human consumption should be inspected for the presence of cysticerci. Thorough cooking of beef provides complete protection. A reliable method to detect infected cattle is needed, particularly because of public health concerns. The method of choice for routine serologic surveillance of cysticercosis is probably ELISA; however, previous antigenic reagents have lacked the necessary sensitivity and/or specificity. To improve the sensitivity and specificity of an existing IgM monoclonal antibody (MAb)-based ELISA for the detection of circulating antigen in the sera of cattle infected with *T. saginata* metacestodes, a modified sandwich ELISA was developed. When the original IgM MAb-based ELISA was compared with the IgG MAb-based ELISA using heat-treated sera from animals harboring more than 50 living metacestodes of *T. saginata*, the sensitivity increased from 56% with the former assay to 92% with the latter. Only a few animals carrying fewer than 50 cysts were detected by both the ELISA using IgG MAbs and that using IgM MAbs. The specificity of the IgM and IgG MAb-based ELISAs was 93.4 and 98.7%, respectively (57). To minimize the chance of infection, cattle should not be allowed to graze on ground contaminated by human sewage. Egg viability can range from a couple of weeks in untreated sewage to approximately 5 months on grass. According to Beaver et al. (7), the rate of bovine cysticercosis in the United States reached 0.37% in 1930 and has remained around this level in federally inspected cattle.

Taenia saginata asiatica (Asian *Taenia* or *Taenia asiatica*)

Recent epidemiologic studies of taeniasis in Southeast Asia indicate the presence of a third form of human *Taenia* spp. which can be distinguished from *T. saginata* and *T. solium*. This newly recognized cestode was originally called the Taiwan *Taenia* sp. and was first detected in Taiwanese aboriginals. However, it is now referred to as the Asian *Taenia* sp. or *T. asiatica* since it has also been found in a number of other Asian countries. Genetic studies have been used to determine whether the Asian *Taenia* sp. should be considered a new species or a subspecies, strain, or variant of *T. saginata*, which it closely resembles. The use of PCR-restriction fragment length polymorphism determined the cestode to be different from either *T. solium* or *T. saginata*. However, mitochondrial and nuclear sequence comparisons indicate that the Asian *Taenia* sp. is more closely related to *T. saginata*. These results support earlier conclusions that the Asian *Taenia* sp. is a genetically distinct entity but is closely related to *T. saginata*. Although there is no total agreement, its taxonomic classification as a subspecies or strain of *T. saginata* may be changed to a new species designation (9, 21, 22, 25, 31). Because of this close relationship with *T. saginata*, any public health significance would be unlikely, since human cysticercosis is caused almost exclusively by *T. solium*.

Life Cycle and Morphology

The life cycle is similar to that of *T. saginata* (Figure 13.20). However, cysticerci of *T. asiatica* develop in the liver of pigs, cattle, and goats, rather than the muscles of cattle. The cysticerci are much smaller than those of *T. saginata* or *T. solium* and measure approximately 0.5 to 1.7 mm by 0.5 to 2.0 mm. The protoscolex is armed, but the hooklets are lost in the mature worm (Figure 13.23). The adult worm matures in 10 to 12 weeks, like *T. saginata*, but is smaller, measuring 4 to 8 m long, and contains 300 to 1,000 proglottids (Figure 13.24). In some indigenous populations in Taiwan, 10 to 20% of the people are infected, primarily from the ingestion of pork meat and viscera. Another difference appears to be the presence of multiple worms, whereas infections with *T. saginata* are usually limited to a single worm (23). In one area of Taiwan during the late 1990s, the rate of *T. asiatica* infection was 11% and the annual economic loss was calculated at US$18,673,495. These and other figures suggest that taeniasis not only is a

Figure 13.23 *Taenia saginata asiatica* (*Taenia asiatica*). (Left) Scolex (unarmed, no hooks) from cysticercus; (right) scolex from adult worm. (Photographs courtesy of Ping-Chin Fan, with permission.)

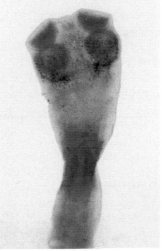

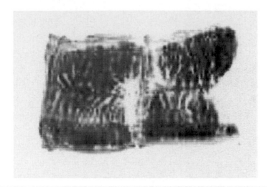

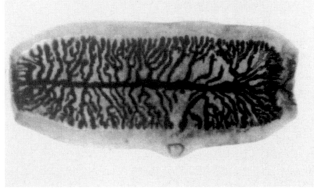

Figure 13.24 *Taenia saginata asiatica* (*Taenia asiatica*). (Upper) Gravid proglottid. (Lower) Gravid proglottid (note the large number of uterine branches and the similarity of this proglottid to the one of *Taenia saginata* in Figure 13.22). (Photographs courtesy of Ping-Chin Fan, with permission.)

significant public health problem, but also leads to severe economic losses in East Asia.

Clinical Disease

Infected individuals may not be aware of the infection and may remain asymptomatic. However, they may complain of having abdominal pain, nausea, weakness, weight loss, changes in appetite, and headache. These symptoms suggest that the tapeworm causes irritation; biopsies of the mucosa may reveal inflammatory changes. Eosinophilia is seen in some patients.

Diagnosis

Since the morphologies of *T. asiatica* and *T. saginata* are so similar, the diagnosis may rely on the patient's history. Have they ingested undercooked or raw beef (*T. saginata*) or raw pig liver (*T. asiatica*)? Although wartlike formations can be found on the surface of the larval worm, they are not always visible in sections (Figure 13.25).

Treatment

The use of praziquantel or niclosamide has been recommended (see chapter 25). The same general approach used for *T. solium* and *T. saginata* is used to eradicate the adult worm. Therapy is usually very effective; however, if the

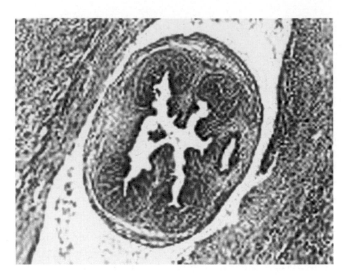

Figure 13.25 *Taenia saginata asiatica* (*Taenia asiatica*). Cross section of cysticercus. (Photograph courtesy of Ping-Chin Fan, with permission.)

proglottids begin to reappear in the stool or crawl from the anus, retreatment is necessary.

Epidemiology and Prevention

In areas where *Taenia* tapeworms are endemic in Korea, *T. saginata* is dominant over *T. solium*; however, pig consumption is considerably higher then beef consumption. By examination of data published between 1963 and 1999, the ratio of armed versus unarmed tapeworms in humans was estimated at approximately 1:5. The ratio of pig eaters versus cattle eaters, however, was approximately 5:1. This inconsistency was explained with the recently described *T. asiatica*, which infects humans through the eating of raw pig viscera, primarily liver. Considering that the scolex of *T. asiatica* is unarmed and the consumption of visceral organs of pigs is higher than beef consumption, this information provides a clearer understanding of *T. asiatica* infection together with coexisting *T. saginata* and *T. solium* infection in Korea (22). Prevention depends on a thorough understanding of the epidemiology of this infection.

Hymenolepis nana

H. nana has been called the dwarf tapeworm and has a worldwide distribution. The fact that an intermediate host is not required in the life cycle was determined in the late 1800s by Grassi and Rovelli (7). For this reason, *H. nana* has been considered to be the most common tapeworm throughout the world, and it is estimated that 50 million to 75 million people worldwide are infected. The infection is most commonly seen in children, although adults are also infected (Table 13.1). There has been some discussion regarding placing *H. nana* in the genus *Rodentolepis*.

Life Cycle and Morphology

The life cycle can be seen in Figure 13.26. Infection is usually acquired by the ingestion of infective *H. nana* eggs, primarily from human stool (direct life cycle). The eggs hatch in the stomach or small intestine, and the liberated larvae, or oncospheres, penetrate the villi in the upper small intestine. The larvae develop into the cysticercoid stage in the tissue and migrate back into the lumen of the small intestine, where they attach to the mucosa (Figure 13.27). The adult worms mature within several weeks. They are very small compared with worms of the *Taenia* species and measure up to 40 mm long. The more worms that are present, the shorter the total length of each worm. Insect intermediate hosts include grain- and flour-eating beetles such as *Tribolium* and *Tenebrio*; fleas such as *Pulex irritans*, *Xenopsylla cheopis*, and *Ctenocephalides canis*; and moths. The indirect life cycle is initiated when the insect ingests an infective *H. nana* egg. Within the insect body cavity, the oncosphere contained within the egg becomes free of the egg membranes and transforms into the cysticercoid larval form, which is infective when the insect is ingested by the human host.

The scolex has four suckers and a retractable, prominent rostellum with hooks. The adult worm is rarely seen in the stool. The eggs are released by disintegration of the gravid proglottids, pass out in the stool, and are immediately infectious. They are round to oval with a thin shell and measure 30 to 47 µm in diameter. The oncosphere has two polar thickenings from which arise polar filaments that lie between the oncosphere and the shell (Figure 13.28). Although accidental ingestion of the insect intermediate host can result in development of the adult worms, this mode of infection is probably rare. There is evidence that auto- or hyperinfection occurs when eggs spontaneously hatch in the small intestine and initiate a new life cycle. Although adult worms live for only about a year, some patients can carry a large worm burden for many years, possibly due to the autoinfection capability within the life cycle or repeated reinfection.

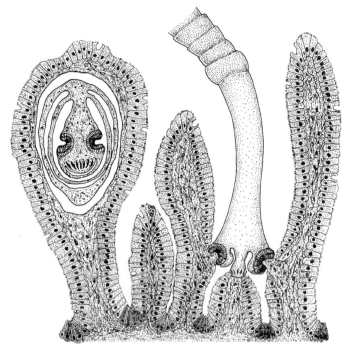

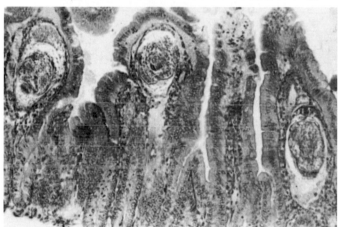

Figure 13.27   *Hymenolepis nana*. (Upper) Diagram of both a cysticercoid and adult worm in the small intestine. (Illustration by Sharon Belkin.) (Lower) Cysticercoids of *H. nana* in the villi of the small intestine. (From A Pictorial Presentation of Parasites: A cooperative collection prepared and/or edited by H. Zaiman.)

Figure 13.26   Life cycle of *Hymenolepis nana*.

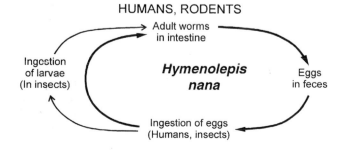

Clinical Disease

Within the human host, the adult worms tend to be located in the upper portion of the ileum. An infection with *H. nana* may cause no symptoms, even with a heavy worm burden. Some patients complain of headache, dizziness, anorexia, abdominal pain, diarrhea, or possibly irritability, particularly if the worm burden is 1,000 to 2,000 worms. Some patients have a low-grade eosinophilia of 5% or more. When young children have heavy infections, they may have loose stools or even diarrhea containing mucus. Persistent, diffuse abdominal pain seems to be the most common symptom.

Heavy human infection may be attributed to internal autoinfection, in which the eggs hatch in the intestine and follow the normal life cycle to the adult worm. This autoinfection feature of the life cycle may lead to complications in immunocompromised patients (35).

Diagnosis

Since the adult worm or proglottids are rarely seen in the stool, the diagnosis is based on recovery and identification of the characteristic eggs containing polar filaments (Figure 13.28). They are most easily seen and identified from fresh specimens or those preserved in formalin-based fixatives. Specimens preserved in polyvinyl alcohol contain eggs that do not exhibit morphologic characteristics that are as well delineated as those seen in formalin-fixed specimens. These thin-shelled eggs also tend to collapse on the permanent stained smear and may be difficult to identify in this type of preparation; the direct wet film or concentration wet mount is recommended.

KEY POINTS—LABORATORY DIAGNOSIS

Hymenolepis nana

1. Adult worms or proglottids are rarely seen in the stool.
2. Eggs with the characteristic thin shell, six-hooked oncosphere, and polar filaments are diagnostic.
3. Egg morphology is more easily seen in fresh specimens or those preserved in formalin-based fixatives.
4. The eggs are infectious, and unpreserved stools should be handled with caution.

Treatment

Niclosamide has been widely used for years. However, the use of praziquantel in a single oral dose is apparently more effective and kills the cysticercoid as well as the adult worm. The recommended dose for adults and children is 25 mg/kg in a single dose (see chapter 25).

Figure 13.28 (Left) Egg of *Hymenolepis nana* (note the thin egg shell, the six-hooked oncosphere, and the polar filaments that lie between the oncosphere and egg shell). (Right) Egg of *Hymenolepis diminuta* (note the thin egg shell, the six-hooked oncosphere, and the lack of polar filaments.)

Epidemiology and Prevention

H. nana is the only human tapeworm in which an intermediate host is not necessary and transmission is from person to person. Children are usually infected more often than adults. Since infection is from person to person via the eggs, good personal hygiene is an important preventive measure. Infection from rats and mice is always a possibility, as is the accidental ingestion of infective insect intermediate hosts.

Hymenolepis diminuta

Although *H. diminuta* is commonly found in rats and mice, it is infrequently found in humans. It has a worldwide distribution in normal hosts, and fewer than 500 human cases, primarily from India, the former Soviet Union, Japan, Italy, and certain areas of the southern United States (Tennessee, Georgia, and Texas), have been reported (7, 60) (Table 13.1).

Life Cycle and Morphology

The life cycle is very similar to the indirect life cycle of *H. nana*, and the infection is considered a true zoonosis; the arthropod intermediate host is not optional but obligatory (Figure 13.29). A number of different arthropods (lepidopterans, earwigs, myriapods, larval fleas, and various beetles) can serve as intermediate hosts. After egg ingestion by the arthropods, the cysticercoid stage forms, which is similar in morphology to the cysticercoid stage of *H. nana*. Infection by accidental ingestion of the infected arthropod containing cysticercoids results in development of the adult worm in the intestine of the human host. In contrast to *H. nana*, the scolex of *H. diminuta* has no hooks on the rostellum. Multiple worms may be present, each being shorter than if a single worm were present. Although some of the adult worms (measuring 20 to 60 cm long) are occasionally spontaneously passed in the stool, usually the eggs are recovered and identified. They are morphologically similar to the eggs of *H. nana* and measure from 60 to 79 μm. They contain the six-hooked oncosphere and a clear area between the oncosphere and the shell. This area contains a gelatinous matrix that does not have polar filaments arising from the polar thickenings, as is seen in *H. nana* (Figure 13.28).

Clinical Disease

As with most tapeworms, the infection is usually tolerated very well by the host, with few if any symptoms. Most infections have been reported from children younger than 3 years; however, infected adults have also been reported. Symptoms that have been reported include diarrhea, anorexia, nausea, headache, and dizziness; they are most likely to be seen in infected children with a heavy infection.

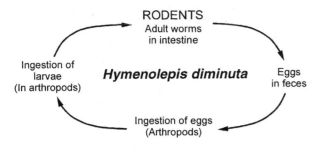

Figure 13.29   Life cycle of *Hymenolepis diminuta*.

Diagnosis

Since the proglottids usually disintegrate in the gut, diagnosis is based on recovery and identification of the characteristic eggs. The similarity to *H. nana* eggs is striking; however, there are no polar filaments in the eggs of *H. diminuta* (Figure 13.28).

KEY POINTS—LABORATORY DIAGNOSIS

Hymenolepis diminuta

1. Although human infection is rare, diagnosis can be based on recovery and identification of the characteristic eggs (thin shell, presence of six-hooked oncosphere, and no polar filaments).
2. The eggs are occasionally submitted to laboratories for identification as a part of various proficiency testing programs. The important point is to be able to differentiate these eggs from those of *H. nana*.
3. Unlike *H. nana* eggs, the eggs of *H. diminuta* are not infectious from person to person.

Treatment

Praziquantel is recommended as a single dose of 5 to 10 mg/kg of body weight (see chapter 25).

Epidemiology and Prevention

Since the infection is acquired from accidental ingestion of the infected intermediate arthropod hosts, avoidance of this type of exposure is recommended. Possible situations include swallowing ectoparasites from the rodent host or accidentally ingesting beetles in precooked cereals. Rat control programs might also decrease the possibility of human exposure.

Dipylidium caninum

D. caninum is commonly found throughout the world in dogs and cats, both domestic and wild. Human infections have also been reported from many areas of the world, including the United States. Most reported infections have been in children (Table 13.1). A case reported in 2003 appears to be the first case reported in the American pathology literature during the last 36 years (MEDLINE database, 1966 to 2002) (40).

Life Cycle and Morphology

The life cycle is very similar to that of *H. diminuta*, in which the arthropod is an obligatory intermediate host (Figure 13.30). The adult worms are found in the dog or cat intestine, and gravid proglottids separate from the strobila and may migrate singly or in short chains out of the anus. The adult worms measure 10 to 70 cm long and have a scolex with four suckers and an armed rostellum. The single proglottids have been described as looking like cucumber seeds when moist and like rice grains when dry (Figure 13.31). Groups of eggs (egg packets) may be found in the stool (Figure 13.32). Each egg measures from 25 to 40 μm and contains the six-hooked oncosphere. The eggs are ingested by the larval stages of the dog, cat, or human flea, in which they develop into cysticercoid larvae. When these fleas are then ingested by the definitive host (dogs, cats, or humans), the adult worm develops within 3 to 4 weeks.

Clinical Disease

The symptoms are related to worm burden; however, in most patients (usually children), complaints may consist of indigestion and appetite loss. The infection may first be noticed when proglottids migrate from the anus. Symptoms have been reported and can include abdominal pain, diarrhea, urticaria, and pruritus ani. A moderate eosinophilia may also be present.

Diagnosis

Diagnosis depends on the recovery and identification of the characteristic egg packets or the proglottids. The proglottids appear like white cucumber seeds when fresh and rice grains when dry.

Figure 13.30   Life cycle of *Dipylidium caninum*.

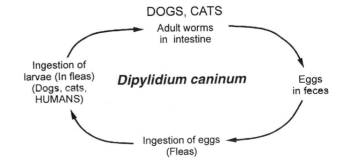

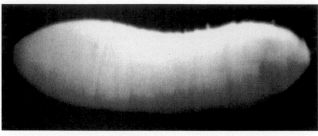

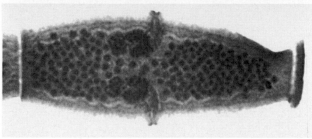

Figure 13.31 Proglottids of *Dipylidium caninum*. (Upper) Fresh proglottid that may resemble a cucumber seed (note the shape and light color); when the proglottids dry, they may resemble rice grains. (Lower) Stained proglottid. (Photographs from the Centers for Disease Control and Prevention Program: Laboratory Identification of Parasites of Public Health Concern.)

Figure 13.32 *Dipylidium caninum* egg packets. (Upper) Diagram showing the packet of *Taenia*-like eggs. (Lower) Egg packet in wet mount preparation.

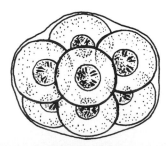

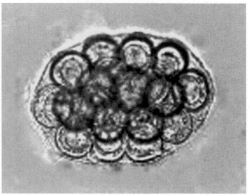

KEY POINTS—LABORATORY DIAGNOSIS

Dipylidium caninum

1. The egg packets are very characteristic and are frequently used to make the diagnosis.
2. Proglottids can be identified on the basis of shape (cucumber seeds when fresh, rice grains when dry) and the presence of egg packets that can be recovered from a ruptured proglottid.

Treatment

A single dose of praziquantel (10 mg/kg of body weight) for both adults and children is recommended (see chapter 25).

Epidemiology and Prevention

Most human cases have been in children, indicating that children may be more likely to accidentally swallow the infected fleas or may be more susceptible to infection. Periodic administration of anthelmintic agents to dogs and cats and use of flea powders will help reduce the risk of infection.

References

1. **Allan, J. C., G. Avila, J. Garcia Noval, A. Flisser, and P. S. Craig.** 1990. Immunodiagnosis of taeniasis by coproantigen detection. *Parasitology* **101:**473–477.
2. **Allan, J. C., and P. S. Craig.** 2005. Coproantigens in taeniasis and echinococcosis. *Parasitol. Int.* 5 Dec. [Epub ahead of print.]
3. **Andreassen, J.** 1998. Intestinal tapeworms, p. 520–537. *In* F. E. G. Cox, J. P. Krier, and D. Wakelin (ed.), *Topley & Wilson's Microbiology and Microbial Infections*, 9th ed. Arnold, London, United Kingdom.
4. **Andrews, R., and W. Mason.** 1987. Cysticercosis presenting as acute scrotal pain and swelling. *Pediatr. Infect. Dis. J.* **6:**942–943.
5. **Arora, V. K., N. Singh, and A. Bhatia.** 1997. Charcot-Leyden crystals in fine needle aspiration cytology. *Acta Cytol.* **41:**409–412.
6. **Barinagarrementeria, F., and C. Cantu.** 1998. Frequency of cerebral arteritis in subarachnoid cysticercosis: an angiographic study. *Stroke* **29:**123–125.
7. **Beaver, P. C., R. C. Jung, and E. W. Cupp.** 1984. *Clinical Parasitology*, 9th ed. Lea & Febiger, Philadelphia, Pa.
8. **Bouchet, F., D. Lefevre, D. West, and D. Corbett.** 1999. First paleoparasitological analysis of a midden in the Aleutian Islands (Alaska): results and limits. *J. Parasitol.* **85:**369–372.
9. **Bowles, J., and D. P. McManus.** 1994. Genetic characterization of the Asian *Taenia*, a newly described taeniid cestode of humans. *Am. J. Trop. Med. Hyg.* **50:**33–44.
10. **Cao, W., C. P. B. Vanderploeg, J. Xu, C. Gao, L. Ge, and J. D. F. Habbema.** 1997. Risk factors for human cysticercosis morbidity: a population-based case-control study. *Epidemiol. Infect.* **119:**231–235.
11. **Chapman, A., V. Vallejo, K. G. Mossie, D. Ortiz, N. Agabian, and A. Flisser.** 1995. Isolation and characterization of species-specific DNA probes from *Taenia solium*

and *Taenia saginata* and their use in an egg detection assay. *J. Clin. Microbiol.* 33:1283–1288.

12. Cruz, I., M. E. Cruz, W. Teran, P. M. Schantz, V. Tsang, and M. Barry. 1994. Human subcutaneous *Taenia solium* cysticercosis in an Andean population with neurocysticercosis. *Am. J. Trop. Med. Hyg.* 51:405–407.

13. Cruz, M. E., P. M. Schantz, I. Cruz, P. Espinosa, P. M. Preux, A. Cruz, W. Benitez, V. C. W. Tsang, J. Fermoso, and M. Dumas. 1999. Epilepsy and neurocysticercosis in an Andean community. *Int. J. Epidemiol.* 28:799–803.

14. Cutrone, J. A., D. Georgiou, K. Gilgomez, and B. H. Brundage. 1995. Myocardial cysticercosis detected by ultrafast CT. *Chest* 108:1752–1754.

15. DeGiorgio, C., S. Pietsch-Escueta, V. Tsang, G. Corral-Leyva, L. Ng, M. T. Medina, S. Astudillo, N. Padilla, P. Leyva, L. Martinez, J. Noh, M. Levine, R. del Villasenor, and F. Sorvillo. 2005. Sero-prevalence of *Taenia solium* cysticercosis and *Taenia solium* taeniasis in California, USA. *Acta Neurol. Scand.* 111:84–88.

16. del Brutto, O. H. 1994. Prognostic factors for seizure recurrence after withdrawal of antiepileptic drugs in patients with neurocysticercosis. *Neurology* 44:1706–1709.

17. del Brutto, O. H., and J. Sotelo. 1990. Albendazole therapy for subarachnoid and ventricular cysticercosis. *J. Neurosurg.* 72:816–817.

18. Delobel, P., A. Signate, M. El Guidj, P. Couppie, M. Gueye, D. Smadja, and R. Pradinaud. 2004. Unusual form of neurocysticercosis associated with HIV infection. *Eur. J. Neurol.* 11:55–58.

19. Deplazes, P., J. Eckert, Z. S. Pawlowski, L. Machowska, and B. Gottstein. 1991. An enzyme-linked immunosorbent assay for diagnostic detection of *Taenia saginata* copro-antigens in humans. *Trans. R. Soc. Trop. Med. Hyg.* 85:391–396.

20. Dupouy-Camet, J., and R. Peduzzi. 2004. Current situation of human diphyllobothriasis in Europe. *Euro Surveill.* 9:31–35.

21. Eom, K. S., H. K. Jeon, Y. Kong, U. W. Hwang, Y. Yang, X. Li, L. Xu, Z. Feng, Z. S. Pawlowski, and H. J. Rim. 2002. Identification of *Taenia asiatica* in China: molecular, morphological, and epidemiological analysis of a Huzhai isolate. *J. Parasitol.* 88:758–764.

22. Eom, K. S., and H. J. Rim. 2001. Epidemiological understanding of *Taenia* tapeworm infections with special reference to *Taenia asiatica* in Korea. *Korean J. Parasitol.* 39:267–283.

23. Fan, P. C. 1997. Annual economic loss caused by *Taenia saginata asiatica* taeniasis in East Asia. *Parasitol. Today* 13:194–196.

24. Fan, P. C., Y. X. Ma, C. H. Kuo, and W. C. Chung. 1998. Survival of *Taenia solium* cysticerci in carcasses of pigs kept at 4 C. *J. Parasitol.* 84:174–175.

25. Flisser, A., A. E. Viniegra, L. Aguilar-Vega, A. Garza-Rodriguez, P. Maravilla, and G. Avila. 2004. Portrait of human tapeworms. *J. Parasitol.* 90:914–916.

26. Garcia, H. H., O. H. Del Brutto, T. E. Nash, A. C. White, Jr., V. C. Tsang, and R. H. Gilman. 2005. New concepts in the diagnosis and management of neurocysticercosis (*Taenia solium*). *Am. J. Trop. Med. Hyg.* 72:3–9.

27. Garcia, H. H., L. J. S. Harrison, R. M. E. Parkhouse, T. Montenegro, S. M. Martinez, V. C. W. Tsang, and R. H. Gilman. 1998. A specific antigen-detection ELISA for the diagnosis of human neurocysticercosis. *Trans. R. Soc. Trop. Med. Hyg.* 92:411–414.

28. Gonzalez, A. E., R. Gilman, H. H. Garcia, J. McDonald, K. Kacena, V. C. W. Tsang, J. B. Pilcher, F. Suarez, C. Gavidia, E. Miranda, and the Cysticercosis Working Group in Peru. 1994. Use of sentinel pigs to monitor environmental *Taenia solium* contamination. *Am. J. Trop. Med. Hyg.* 51:847–850.

29. Ito, A., A. Plancarte, L. Ma, Y. Kong, A. Flisser, S. Y. Cho, Y. H. Liu, S. Kamhawi, M. W. Lightowlers, and P. M. Schantz. 1998. Novel antigens for neurocysticercosis: simple method for preparation and evaluation for serodiagnosis. *Am. J. Trop. Med. Hyg.* 59:291–294.

30. Jafri, H. S., F. Torrico, J. C. Noh, R. T. Bryan, F. Balderrama, J. B. Pilcher, and V. C. W. Tsang. 1998. Application of the enzyme-linked immunoelectrotransfer blot to filter paper blood spots to estimate seroprevalence of cysticercosis in Bolivia. *Am. J. Trop. Med. Hyg.* 58:313–315.

31. Jeon, H. K., K. H. Lee, K. H. Kim, U. W. Hwang, and K. S. Eom. 2005. Complete sequence and structure of the mitochondrial genome of the human tapeworm, *Taenia asiatica* (Platyhelminthes; Cestoda). *Parasitology* 130: 717–726.

32. Kamal, M. M., and S. V. Grover. 1995. Cytomorphology of subcutaneous cysticercosis—a report of 10 cases. *Acta Cytol.* 39:809–812.

33. Keane, J. R. 1995. Tremor as the result of shunt obstruction: four patients with cysticercosis and secondary parkinsonism: report of four cases. *Neurosurgery* 37:520–522.

34. Lapergue, B., H. Hosseini, M. Liance, C. Russo, and P. Decq. 2005. Hydrocephalus and racemose cysticercosis: surgical alternative by endoscopic third ventriculostomy. *Neurochirurgie* 51:481–488.

35. Lucas, S. B., O. Hassounah, R. Muller, and M. J. Doenhoff. 1980. Abnormal development of *Hymenolepis nana* larvae in immunosuppressed mice. *J. Helminthol.* 54:75–82.

36. Maass, M., E. Delgado, and J. Knobloch. 1991. Detection of *Taenia solium* antigens in merthiolate-formalin preserved stool samples. *Trop. Med. Parasitol.* 42:112–114.

37. Maddison, S. E. 1991. Serodiagnosis of parasitic diseases. *Clin. Microbiol. Rev.* 4:457–469.

38. Mauad, T., C. N. Battlehner, C. L. Bedrikow, V. L. Capelozzi, and P. H. N. Saldiva. 1997. Case report. Massive cardiopulmonary cysticercosis in a leukemic patient. *Pathol. Res. Pract.* 193:527–529.

39. Medina, M. T., E. Rosas, F. Rubio-Donnadieu, and J. Sotelo. 1990. Neurocysticercosis as the main cause of late-onset epilepsy in Mexico. *Arch. Intern. Med.* 450:325–327.

40. Molina, C. P., J. Ogburn, and P. Adegboyega. 2003. Infection by *Dipylidium caninum* in an infant. *Arch. Pathol. Lab. Med.* 127:e157–e159.

41. Molinari, J. L., E. Garcia-Mendoza, Y. De la Garza, J. A. Ramirez, J. Sotelo, and P. Tato. 2002. Discrimination between active and inactive neurocysticercosis by metacestode excretory/secretory antigens of *Taenia solium* in an enzyme-linked immunosorbent assay. *Am. J. Trop. Med. Hyg.* 66:777–781.

42. Nascimento, E., J. O. Costa, M. P. Guimaraes, and C. A. P. Tavares. 1995. Effective immune protection of pigs against cysticercosis. *Vet. Immunol. Immunopathol.* 45:127–137.

43. Ong, S., D. A. Talan, G. J. Moran, W. Mower, M. Newdow, V. C. W. Tsang, R. W. Pinner, and the Emergency ID Net Study Group. 2002. Laboratory diagnosis of human neurocysticercosis: double-blind comparison of enzyme-linked immunosorbent assay and electroimmunotransfer blot assay. *Emerg. Infect. Dis.* 8:608–613.

44. Pawlowski, Z. S. 1990. Perspectives on the control of *Taenia solium. Parasitol. Today* 6:371–373.

45. Proano-Narvaez, J. V., A. Meza-Lucas, O. Mata-Ruiz, R. C. Garcia-Jeronimo, and D. Correa. 2002. Laboratory diagnosis of human neurocysticercosis: double-blind comparison of enzyme-linked immunosorbent assay and electroimmunotransfer blot assay. *J. Clin. Microbiol.* 40:2115–2118.

46. Sampaio, J. L. M., V. P. de Andrade, M. D. C. Lucas, L. Fung, S. M. B. Gagliardi, S. R. P. Santos, C. M. F. Mendes, M. B. D. P. Eduardo, and T. Dick. 2005. Diphyllobothriasis, Brazil. *Emerg. Infect. Dis.* 11:1598–1600.

47. Sanchez, A. L., M. T. Medina, and I. Ljungstrom. 1998. Prevalence of taeniasis and cysticercosis in a population of urban residence in Honduras. *Acta Trop.* 69:141–149.

48. Saran, R. K., V. Rattan, A. Rajwanshi, R. Nijkawan, and S. K. Gupta. 1998. Cysticercosis of the oral cavity: report of five cases and a review of literature. *Int. J. Paediatr. Dent.* 8:273–278.

49. Sarti, E., A. Flisser, P. M. Schantz, M. Gleizer, M. Loya, A. Plancarte, G. Avila, J. Allan, P. Craig, M. Bronfman, and P. Wueyaratne. 1997. Development and evaluation of a health education intervention against *Taenia solium* in a rural community in Mexico. *Am. J. Trop. Med. Hyg.* 56:127–132.

50. Schantz, P. M., and V. C. Tsang. 2003. The U.S. Centers for Disease Control and Prevention (CDC) and research and control of cysticercosis. *Acta Trop.* 87:161–163.

51. Sekhar, G. C., and B. N. Lemke. 1999. Orbital cysticercosis. *Ophthalmology* 104:1599–1604.

52. Sheth, T. N., C. Lee, W. Kucharczyk, and J. Keystone. 1999. Reactivation of neurocysticercosis: case report. *Am. J. Trop. Med. Hyg.* 60:664–667.

53. Sotelo, J., O. H. del Brutto, P. Penagos, F. Escobedo, B. Torres, J. Rodriguea-Carbajal, and F. Rubio-Donnadieu. 1990. Comparison of therapeutic regimen of anticysticercal drugs for parenchymal brain cysticercosis. *J. Neurol.* 237:69–72.

54. Sotelo, J., V. Guerrero, and F. Rubiol. 1985. Neurocysticercosis: a new classification based on active and inactive forms. *Arch. Intern. Med.* 145:422–445.

55. Takayanagui, O. M. 2004. Therapy for neurocysticercosis. *Expert Rev. Neurother.* 4:129–139.

56. Theis, J. H., M. Cleary, M. Syvanen, A. Gilson, P. Swift, J. Banks, and E. Johnson. 1996. DNA-confirmed *Taenia solium* cysticercosis in black bears (*Ursus americanus*) from California. *Am. J. Trop. Med. Hyg.* 55:456–458.

57. Van Kerckhoven, I., W. Vansteenkiste, M. Claes, S. Geerts, and J. Brandt. 1998. Improved detection of circulating antigen in cattle infected with *Taenia saginata* metacestodes. *Vet. Parasitol.* 76:269–274.

58. Wallin, M. T., and J. F. Kurtzke. 2005. Neurocysticercosis in the United States: review of an important emerging infection. *Neurology* 63:1559–1564.

59. Wilkins, P. P., J. C. Allan, M. Verastegui, M. Acosta, A. G. Eason, H. H. Garcia, A. E. Gonzalez, R. H. Gilman, and V. C. W. Tsang. 1999. Development of a serologic assay to detect *Taenia solium* taeniasis. *Am. J. Trop. Med. Hyg.* 60:199–204.

60. Wiwanitkit, V. 2004. Overview of *Hymenolepis diminuta* infection among Thai patients. *Med. Gen. Med.* 6:7

61. Yamasaki, H., M. Nakao, Y. Sako, K. Nakaya, M. O. Sato, W. Mamuti, M. Okamoto, and A. Ito. 2002. DNA differential diagnosis of human taeniid cestodes by base excision sequence scanning thymine-base reader analysis with mitochondrial genes. *J. Clin. Microbiol.* 40:3818–3821.

14

Tissue Cestodes: Larval Forms

Echinococcus granulosus (cystic disease, hydatid disease)

Echinococcus multilocularis (alveolar disease, hydatid disease)

Echinococcus oligarthrus and *Echinococcus vogeli* (polycystic hydatid disease)

Taenia (Multiceps) spp. (*Taenia multiceps, Taenia serialis*) (coenurosis)

Spirometra mansonoides and *Diphyllobothrium* spp. (sparganosis)

NOTE: *Taenia solium* (cysticercosis) is discussed in chapter 13.

Echinococcus granulosus (Cystic Disease, Hydatid Disease)

Currently, four species are recognized within the genus *Echinococcus*: *E. granulosus* (which causes cystic disease), *E. multilocularis* (which causes alveolar disease), *E. vogeli* (which causes polycystic disease), and *E. oligarthrus* (which causes polycystic disease) (Table 14.1). There is clear evidence of strain variation in *E. granulosus*, and 10 different strains have now been identified (Table 14.2) (33, 34, 71). However, evidence for strain variation for the other species of *Echinococcus* does not exist. Based on historical information, hydatid cysts were recognized by Hippocrates, Aretaeus, and Galen. There was speculation concerning the possible relationship between human and animal hydatid cysts during the 1600s and 1700s. The adult worms and various aspects of the life cycle were studied in the late 1800s.

The areas of the world involved in sheep and cattle raising tend to be the areas where infections are endemic; they even include the Basque sheep farmers in California. The number of infections in both animals and humans has decreased over the years as a result of education and various control measures. However, figures suggest that the surgical incidence in some areas in Central Asia is now greater than 10 in 100,000 (up to 27 in 100,000 in Tadjikistan), and many of the cases are in children and the unemployed (131). The rates of infection in major livestock species such as sheep have also increased, with a doubling of prevalence, to a figure approaching 25% in some areas.

There is also evidence to indicate a reemergence of canine *E. granulosus* infection in Wales. This has been associated with the large-scale outdoor slaughter of sheep during the 2001 foot-and-mouth disease outbreak in the United Kingdom and the suspected increased risk for transmission between sheep and dogs. The data suggest that allowing dogs to roam free and infrequent dosing of dogs with praziquantel has contributed to the increase; thus, increased transmission to humans appears probable in this setting (16).

Life Cycle and Morphology

The life cycle of *E. granulosus* is presented in Figure 14.1. The adult worms are very small (3 to 6 mm long) and consist of a scolex, a neck, and only a single proglottid at each stage of development (immature, mature, and gravid) (Figure 14.2). There may be several hundred worms in the intestine of the

Table 14.1 Tissue cestodes

Characteristic	Echinococcus granulosus	Echinococcus multilocularis	Echinococcus vogeli, E. oligarthrus	Taenia (Multiceps) spp.	Spirometra and Diphyllobothrium spp.
Disease	Hydatid disease (cystic)	Hydatid disease (alveolar)	Hydatid disease (polycystic)	Coenurosis	Sparganosis
Geographic location of the parasite	Worldwide	North America, northern and central Eurasia	Central and South America (85%) in Brazil, Colombia, Ecuador, Argentina	Worldwide	Worldwide; more common in China, Japan, Southeast Asia
Definitive host(s)	Domestic dog, wild canids (coyote, dingo, red fox, etc.)	Red fox, Arctic fox, raccoon dog, coyote, domestic dog, cat	Bush dog, domestic dog	Dogs, other canids	Dogs, cats
Acquired by	Egg ingestion (dogs)	Egg ingestion (foxes, cats)	Egg ingestion (dogs [E. vogeli], felids [E. oligarthrus])	Egg ingestion (dogs, other canids)	(i) Ingestion of infected Cyclops spp. (procercoid); (ii) ingestion of raw infected flesh of amphibians, reptiles, birds, mammals (spargana); (iii) local application of raw infected flesh as a poultice (spargana)
Intermediate host(s)	Primarily ungulates (sheep, cattle, swine, horses); also marsupials	Rodents (voles, lemmings, shrews, mice), other small mammals	Paca, agouti, spiny rat	Sheep, goats, cattle, horses, lagomorphs, rodents	Frogs, mammals
Stage of organism found in tissue	Larval form (fluid-filled unilocular hydatid cyst), contains protoscolices and daughter cysts, limiting membrane; 1–>15 cm; visceral, primarily liver and lungs	Larval form (alveolar hydatid cyst), no limiting membrane, usually sterile with no protoscolices; visceral, primarily liver	Larval form (fluid-filled polycystic hydatid cyst), scolex visible in wet mounts, large hooklets (38–46 µm long) and small hooklets (30–37 µm long); visceral, liver, abdomen, lungs (E. vogeli); orbits, heart (E. oligarthrus), vesicles partitioned by septa, protoscolices present	Larval form (intermediate between cysticercus and hydatid cyst)	More elongate, wormlike structure, no suckers or hooks; resemble narrow tapeworm proglottids, motile
Type of growth in humans	Concentric expansion	Exogenous proliferation, tumorlike; similar to metastatic growth	Exogenous and endogenous proliferation (E. vogeli); expansive, no indication of exogenous proliferation (E. oligarthrus)	Multiple scolices, but no daughter cysts	Most tissues have been involved; depends on the site of the poultice application
Location in body	Liver (60%), lungs (20%), kidneys (4%), muscle (4%), spleen (3%), soft tissues (3%), brain (2%), bones (1%), other (1%)	All sites as for E. granulosus; most common site is liver, metastases in lungs, brain, bones, etc.	Liver, lungs (15%) (E. vogeli); eye and heart (E. oligarthrus)	Most often in central nervous system	Most tissues have been involved; depends on the site of the poultice application
Symptoms	Depends on cyst location, usually mechanical from enlarging cyst; may also be allergic reactions from cyst fluid leakage	Hepatic disease resembles slow growing mucoid carcinoma (no fever); hepato- and splenomegaly, jaundice, ascites	Hepatomegaly, palpable peritoneal masses, jaundice (E. vogeli)	Like space-occupying lesion in the central nervous system, similar to tumor, rarely muscles or subcutaneous tissues	Edema, pain, irritation, inflammation, toxemia, eye damage, elephantiasis if lymphatics involved, slowly growing, tender, subcutaneous nodules (may be migratory), ocular sparganosis
Treatment	Surgical removal, albendazole, praziquantel	Albendazole, praziquantel; surgery not recommended	Surgery plus albendazole; difficult to treat	Rarely diagnosed preoperatively	Surgical removal and drainage

Table 14.2 *Echinococcus granulosus* strains[a]

Strain (genotype, G)	Geographic distribution	Definitive host	Intermediate host	Infective for humans
G1: common sheep strain	Europe, Middle East, Africa, Iran, India, Nepal, China, Russia, Australian mainland, Tasmania, New Zealand, United States, South America, Sardinia	Dog, fox, dingo, jackal, hyena	Sheep, cattle, pig, camel, goat, macropods	Yes
G2: Tasmanian sheep strain	Tasmania, Argentina, Europe	Dog, fox	Sheep, cattle?	Yes
G3: buffalo strain	Asia	Dog, fox?	Buffalo, cattle?	?
G4: horse strain (*E. equinus*)	Europe, Middle East, South Africa (New Zealand?, United States?)	Dog	Horse, other equines	No?
G5: cattle strain (*E. ortleppi*)	Europe, South Africa, India, Nepal, Sri Lanka, Russia, South America?	Dog	Cattle, buffalo, sheep, goat	Yes
G6: camel strain	Middle East, Iran, Africa, China, Nepal, Argentina	Dog	Camel, goat, cattle	Yes
G7: pig strain	Poland, Slovakia, Ukraine, Russia, Argentina, Europe, Sardinia	Dog	Pig	Yes
G8: cervid strain	North America, Eurasia	Wolf, dog	Cervids	Yes
G9: ?	Poland	?	?	?
G?:	Africa	Lion	Zebra, wildebeest, warthog, bushpig, buffalo, various antelope species, giraffe?, hippopotamus?	?
G10: Fennoscandian cervid strain	Finland	Wolf, dog	Cervids; closely related to G5, G6, and G7 strains	Probably

[a] Adapted from references 7, 33, 53, 54, 71, 85, 107, 110, and 135.

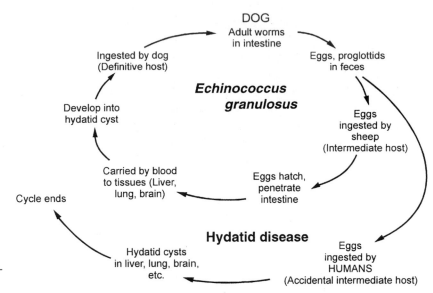

Figure 14.1   Life cycle of *Echinococcus granulosus* (cystic echinococcosis).

Figure 14.2   Adult worm of *Echinococcus granulosus*. (A) Diagram. (B) Whole mount (note that only three proglottids are present in the adult worm).

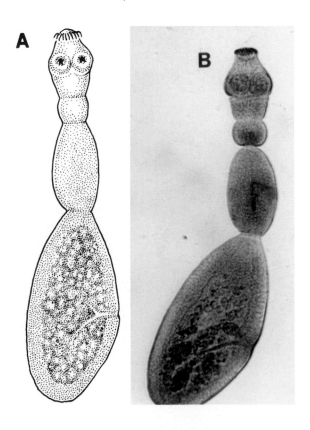

canine host (usually the dog). The worms may survive in the host for up to 20 months, and each gravid proglottid contains few eggs (100 to 1,500), which are much larger than those of some of the other tapeworms. After the gravid proglottids and eggs are passed in the feces, they may be swallowed by an intermediate host, including humans. After exposure to gastric and intestinal enzymes, the eggs hatch in the duodenum. The released oncospheres penetrate the intestine and are carried via the bloodstream, where they are filtered out in the various organs. The most common site in humans is the liver (60 to 70% of cases). Although not all oncospheres survive this passage through the body, those that do survive begin to grow and develop. Usually by the fifth month, the wall of the hydatid cyst has become differentiated into an outer friable, laminated, nonnucleated layer and an inner nucleated germinal layer. Various daughter cysts (brood capsules) bud off from the inner germinal layer and may remain attached or float free in the interior of the fluid-filled cyst. The individual scolices bud off from the inner wall of the daughter cysts; these scolices and free daughter cysts are called hydatid sand. Each scolex normally invaginates to protect the hooklets (Figures 14.3 and 14.4). Although not every cyst produces daughter cysts and/or scolices, this general tissue organization is called a unilocular cyst, in which the cyst contents are held within a single limiting cyst wall. On average, the cysts increase in size by about 1 cm per month; this growth rate was reported for a series of multiple intracranial hydatid cysts caused by *E. granulosus* (37).

When animals that serve as intermediate hosts are slaughtered, the viscera may not be disposed of properly and may be consumed by animals that serve as definitive hosts. The adult worms then develop in the intestine of the definitive host.

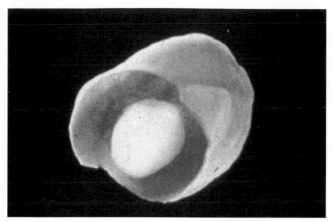

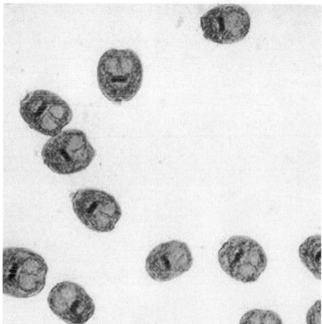

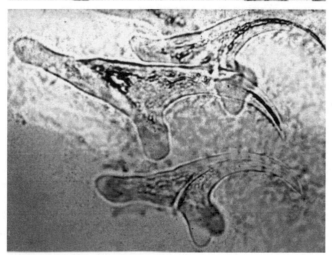

Figure 14.3 *Echinococcus granulosus.* (Top) Daughter cyst within the main cyst wall. (Middle) Hydatid sand (immature scolices). (Bottom) Individual hooklets. (Top image from A Pictorial Presentation of Parasites: A cooperative collection prepared and/or edited by H. Zaiman.)

Based on differences in the host specificity of the larval stage, at least 10 different strains of *E. granulosus* have been identified (Table 14.2). The strains differ in their intermediate host spectrum, geographic distribution, adult and larval stage morphologies, and protoscolex production. Also, at least seven of the strains are infective for humans. The northern form is found in the holoarctic zones of the tundra and boreal forest of North America and Eurasia. The larval stage occurs primarily in moose, elk, and reindeer, while the adult worm is found in wolves and sometimes in dogs and other wild canids. The European form of *E. granulosus* includes domestic animals, primarily sheep as the main intermediate host and dogs as the main definitive host. Although mixed infections with *E. granulosus* and *E. multilocularis* are rare, they have been reported from areas in China, a recognized area where echinococcosis is hyperendemic (143).

The risk of infection depends to a high degree on the association between humans and dogs, with the exception of the lion strain. Those at high risk include populations where dogs are used to herd sheep and are also intimate members of the family, often having unrestricted access to the house and family members. Cystic echinococcosis has been recorded in 21 of China's 31 provinces, autonomous regions, and municipalities (approximately 87% of the territory). This infection constitutes one of the major health problems in this part of the world. Hydatid disease caused by *E. granulosus* is a zoonosis of major public health concern throughout Latin America, particularly in the Andean and South Cone regions. Cystic echinococcosis is also widely found throughout the region comprising Arab North Africa and the Middle East. In areas of endemic infection around the world, the practice of giving raw viscera of slaughtered livestock to the dogs enhances transmission; in areas where this practice has been curtailed, prevalence figures have decreased.

Clinical Disease

Hydatid disease in humans is potentially dangerous; however, cyst size and organ location greatly influence the outcome. Clinical symptoms may appear after an incubation period of several months to years.

Liver Cysts. The majority of hydatid cysts occur in the liver, causing symptoms that may include chronic abdominal discomfort, occasionally with a palpable or visible abdominal mass. Liver cysts tend to occur more frequently in the right lobe. Symptoms can include pain, hepatomegaly, cholestasis, biliary cirrhosis, portal hypertension, and ascites. If a cyst becomes infected with bacteria, it resembles an abscess. Severe symptoms can occur if the cyst ruptures spontaneously, from trauma, or during surgery. In addition to the spread of tissue from which additional hydatid cysts can grow, there may be serious allergic reactions,

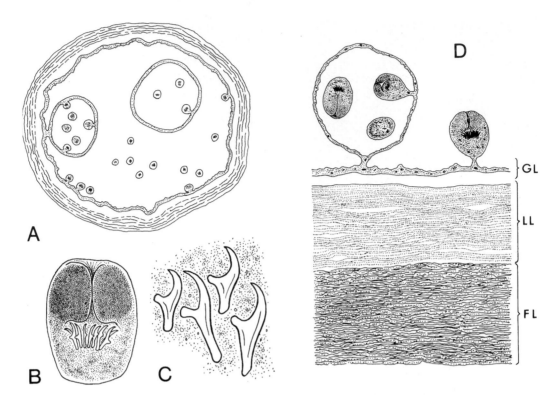

Figure 14.4 *Echinococcus granulosus* hydatid cyst. (A) Diagram of unilocular cyst (limiting membrane). (B) Diagram of single immature worm. (C) Diagram of hooklets. (D) Diagram of laminated layers, daughter cysts, and scolices. (Illustration by Sharon Belkin.) GL, germinal layer; LL, laminated layer; FL, fibrous layer.

including rash, anaphylactic shock, or death. Cyst rupture can also lead to cholangitis and cholestasis (9). Morbidity and mortality rates associated with perforated hydatid cysts are higher than those associated with nonperforated cysts. Specific disease sites and complications can be seen in Table 14.3. Some unilocular cysts may remain undetected for many years until they become large enough to crowd other organs (Figure 14.5).

Lung Cysts. Cysts in the lungs are usually asymptomatic until they become large enough to cause chronic cough, shortness of breath, or chest pain (Figure 14.6). Cyst rupture can lead to expectoration of hydatid fluid or membranes, followed by the development of infection and a lung abscess. If the rupture occurs into the lung, it may cause pneumothorax and empyema, allergic reactions, and even anaphylactic shock.

Other Sites. The spleen and kidneys are the next most common intra-abdominal organs, occurring in 3 to 5% of patients, and cardiac and cerebral cysts have been found in about 1 to 1.5% of patients (70). Even relatively small cysts can cause severe damage if they are located in a vital area or in bone (osseous cysts). Diagnostic procedures for

the detection of liver or lung cysts should always be considered when cysts are detected in more unusual or rare body sites (Figure 14.7).

Cyst Location Related to Patient Age. It is well known that hydatid cysts are most commonly found in the liver and lungs, but certain locations have been reported to be more prevalent in children and/or young adults. Lung, brain, spinal, and orbital hydatid cysts have been more commonly seen in younger patients. It is thought that age somehow alters the host-parasite relationship and thus affects the organ distribution of the cysts (148).

Hydatid Disease and Cancer. The simultaneous occurrence of hydatid disease and cancer is relatively rare. In a patient who had liver disease as well as acute leukemia, therapy for his leukemia over a 3-month time frame had no effect on the size and contents of the liver hydatid cyst. The hydatid serologic titers remained unchanged, and the cyst classification remained the same (4). Hodgkin's lymphoma and cystic hydatid disease as a dual infection has been reported; however, no residual and/or new cyst formation was seen on follow-up after surgery and subsequent therapy for the lymphoma (145).

Table 14.3 Examples of hydatid disease, complications, and/or outcomes

Body site, description	Comments	Reference(s)
Liver, asymptomatic disease	Despite a limited number of cases, most asymptomatic liver hydatid patients (75%) remain symptom free for >10 yr, regardless of cyst size or type. Such carriers are probably at low risk for developing complications.	39
Liver, surgical treatment and review of 304 cases	238 had cyst in right lobe, 41 in left lobe, 25 in both lobes; 45 had multiple hepatic cysts, 18 had coexisting cysts in other intra-abdominal sites. For management of infection, capitonnage, omentoplasty, cyst excision, segmentectomy, or cystoenterostomy are all superior to tube drainage.	6
Liver, right hepatic vein	Defect on wall of right hepatic vein, as well as 3 ruptures into the bile duct. Unusual presentation.	22
Liver, spontaneous rupture of cyst into biliary tract	Resolution of hydatid cyst by spontaneous rupture into the biliary tract; impaction of hydatid material into common bile duct was relieved endoscopically; patient was treated with 2 courses of mebendazole.	5, 69
Liver, intrabiliary rupture of cyst	Ideal treatment is traditional lavage of biliary tree followed by radical treatment of cyst and free drainage of bile ducts; hepatectomy and cystojejunostomy can be avoided.	65
Liver, cyst rupture directly into the left colon	Rupture of hydatid cyst into a hollow area such as the colon is extremely rare; partial drainage of the cyst was demonstrated by creation of an air-fluid level.	76
Intra-abdominal organs, spontaneous cyst rupture	Two case reports with severe spontaneous systemic anaphylactic reaction due to cyst rupture (without known precipitating factors such as trauma, accidental). Emphasizes importance of high index of suspicion for hydatid etiology for idiopathic anaphylactic reactions in areas of endemic infection.	88
Pancreas, rupture into main pancreatic duct	Reports a case of recurrent pancreatitis due to rupture of a solitary hydatid cyst of the pancreas into the main pancreatic duct.	121
Lung, mean age of patients with giant hydatid cysts younger than patients with normal-size cysts	Increase in cyst diameter correlated with higher lung tissue elasticity and delay in diagnosis due to delayed symptoms; review of 47 patients.	47
Lung, secondary pleural disease (usually primary, not secondary)	Presented 4 yr after treatment for a pyopneumothorax caused by rupture of pulmonary cyst near the pleural space; treatment with albendazole; surgery ruled out due to severe chronic obstructive pulmonary disease.	1
Unusual intrathoracic, extrapulmonary hydatid cysts	Sites included fissure, pleural cavity, chest wall, mediastinum, myocardium, diaphragm; total resection in 18/22 patients; no deaths after 1 yr.	94
Aorta, causing occlusion of aorta and both iliac arteries	Patient had previous surgery for a paraspinal hydatid cyst; presented with intermittent limping. CT and MRI showed multiple cysts in soft tissues of the back, retroperitoneum, and lumen of aorta and iliac arteries.	86
Heart, causing tamponade	Patient complained of severe dyspnea, thoracic pain, dizziness, short period of unconsciousness. Preliminary serology negative, low eosinophilia (5%); hydatid dismissed. Later, positive serology and eosinophilia of 59% found. Surgery and therapy with albendazole were used. With "cardiac tumor," negative serology and absence of eosinophilia do not rule out hydatid disease.	13
Heart, death caused by hydatid embolism	Rare case, limited to the heart; asymptomatic course until sudden death caused by arterial embolism. Note: no antibodies were detected by indirect hemagglutination.	57
Heart, cardiac involvement is usually by only 0.5 to 3% of hydatid cysts	Patient evaluated for intracardiac cyst; surgically excised; diagnosis confirmed by Giemsa staining of hooklets and scolices in cyst fluid. Patient was young (4-yr-old male).	77

(continued)

Table 14.3   Examples of hydatid disease, complications, and/or outcomes *(continued)*

Body site, description	Comments	Reference(s)
Heart, in the septum interventriculare	Unusual location for hydatid cyst; positive serology, surgery rejected due to severe underlying hypertensive cardiopathy (60-yr-old man).	14
Heart (primary cyst), cerebral cysts (secondary cysts)	Rare case in 7-yr-old child; symptoms of raised intracranial pressure; underwent 9 operations over 8 years because primary ruptured cyst in myocardium of left ventricle was detected; concurrent therapy with mebendazole (would not have been successful without removal of intracardiac hydatid cyst).	134
Heart, cyst located in septum and right ventricular cavity	Cyst located exclusively in the heart, first manifested as anaphylactic shock of unknown origin, required immediate surgical treatment due to severe hemodynamic compromise.	69
Orbital hydatid cyst	Emphasis on the superiority of MRI over CT for orbital cysts.	43
Central nervous system	Multiple hydatid cysts in cerebral subarachnoid space; based on clinical findings, latex test results, cerebrospinal fluid exam, and MRI findings.	132
Spine, progressive paraplegia	Rare hydatid disease manifestation. Disease misdiagnosed as tuberculous spondylitis. Patient presented with increasing weakness in lower limbs; treated for tuberculous spondylitis. Presented 1.5 yr later and underwent surgery and chemotherapy, paraparesis resolved. Early decompressive surgery with stabilization of the spine plus chemotherapy is treatment of choice.	11
Spine, uncommon but significant	Review of 28 reports; only 14 patients had pulmonary or other organ involvement; most patients misdiagnosed as Pott's disease (tuberculous spondylitis); surgical intervention followed by chemotherapy; hydatid disease should be considered when radiologic findings suggest spinal infections or tumors.	133
Spine, rare case	Intraspinal extradural hydatid disease of the thoracic region with spinal cord compression; diagnosis on MRI findings; patient recovered completely after surgery.	10
Gluteus muscle	Unusual location for hydatid cyst.	18
Ovary	Asymptomatic right adnexal mass; vaginal ultrasound revealed cyst in posterior cul-de-sac adjacent to the right ovary, with internal septa resembling a maze or an onion slide structure. Color Doppler revealed simple echogenic pattern; in contrast, ovarian carcinomas have a complex internal structure.	27
Tongue	Unusual site (5 reported cases in the literature); additional liver and cyst lesions found and treated with albendazole.	111

Immune Reactions.   During the life of the cyst, there may be small fluid leaks into the systemic circulation that sensitize the patient. Later, if the cyst should burst or if there is a large fluid leak, serious allergic sequelae, including anaphylactic reactions, may occur. Release of cyst tissue may lead to abscess formation, emboli, and/or the development of additional young cysts at secondary sites.

Although publications have described autoimmune phenomena in patients with hydatid cysts, apparently there is no association between hydatid infection and the level of autoantibodies to a broad range of self antigens such as antinuclear antibodies, tissue-specific autoantibodies, and rheumatoid factor. No significant differences have been found between autoantibody levels in the sera from patients and from controls.

Echinococcus-specific immunoglobulin E (IgE) is present in most of the patients and is associated with the severity of disease. Specific histamine release by circulating basophils stimulated with *E. granulosus* antigens is present in all patients with cystic echinococcosis and alveolar echinococcosis. *Echinococcus* allergens include (i) AgB 12-kDa subunit, a protease inhibitor and a potent Th2 inducer; (ii) Ag5, a serine protease; (iii) EA 21, a specific cyclophilin, with homology to other types of cyclophilins; and (iv) Eg EF-1 beta/delta, an elongation factor with homology to *Strongyloides stercoralis* EF, which has the same IgE epitope. Clarification of the immunology of echinococcosis in humans has led to new therapy options, such as immunomodulation using alpha interferon. Also, a Th2-driven immunological response and an interleukin-10

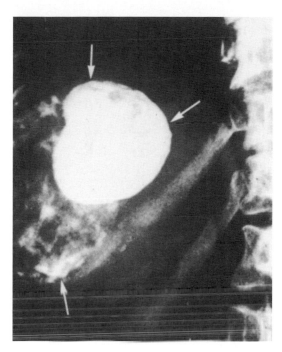

Figure 14.5 Hydatid cysts in liver. This patient had three cysts (marked by arrows) in the right lobe. Two were completely calcified, and the large one in the background was partially calcified. (From V. Zaman, *Atlas of Medical Parasitology*, Lea & Febiger, Philadelphia, Pa., 1979.)

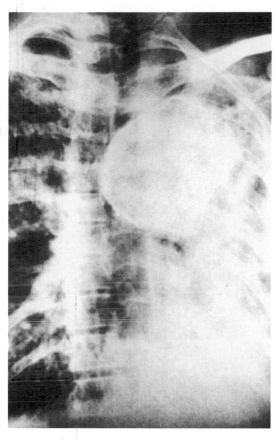

Figure 14.6 Hydatid cyst in the lung. This patient had a single large cyst in the left lung. There was collapse and consolidation in part of the left lung as a result of pressure on the bronchus. Partial pneumonectomy was required to remove the cyst. (From V. Zaman, *Atlas of Medical Parasitology*, Lea & Febiger, Philadelphia, Pa., 1979.)

(IL-10)-related tolerance state are common characteristics of atopic allergy and echinococcosis (138).

Diagnosis

In an area of endemic infection, exposure to sheepdogs, the presence of an enlarged liver, the presence of a palpable mass in the right upper quadrant of the abdomen, expectoration of a salty-tasting fluid with hemoptysis, and a round lesion on chest radiographs all support the presumptive diagnosis of hydatid disease in either the liver or the lungs. Hydatid cysts in the liver may be confused with congenital liver cysts, amebic abscess, or hepatic tumors. Pulmonary symptoms may be confused with tuberculosis.

Hydatid cysts should be considered in patients with abdominal masses with no clearly defined diagnosis. Although eosinophilia is present in 20 to 25% of patients, it is merely suggestive. Many asymptomatic cysts are first discovered after radiologic studies. The cyst usually has a well-defined margin with occasional fluid level markings. These studies can also be helpful in diagnosing osseous involvement. Scans may also demonstrate a space-occupying lesion, particularly in the liver. If the cyst is large and located in the abdomen, a thrill is sometimes detected.

Serologic tests are available, including the enzyme-linked immunotransfer blot (EITB) test, which apparently offers greater sensitivity and specificity than do the enzyme-linked immunosorbent assay (ELISA) and arc-5 double-diffusion assay (DD5); when the tests were run simultaneously, the largest number of cases was detected by a combination of the EITB and DD5 tests (136). Newer enzyme immunoassay procedures appear to provide greater than 90% sensitivity and specificity compared with EIBT (125). Specific antibodies in serum samples from patients with hydatid disease have been recognized by immunoblotting; recognition of 8- and 116-kDa hydatid antigens by a patient's serum sample seems to be a specific test confirming a clinical diagnosis (56). Evaluation of a monoclonal antibody-based competition ELISA for the diagnosis of human hydatidosis also provides a highly sensitive (92.8%) and specific screening system for human hydatid disease diagnosis (75). The use of recombinant antigens for immunodiagnosis of cystic echinococcosis has yielded an overall sensitivity of 92.2% and a specificity of 95.4%, excellent levels considering that the study was undertaken with a large panel of 896 human sera (Table 14.4) (75). Additional information is provided in chapter 22.

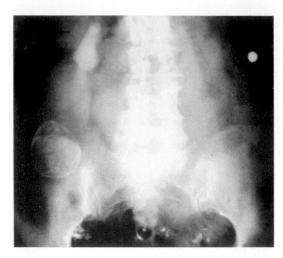

Figure 14.7 *Echinococcus granulosus* calcified liver cyst. The cyst is visible at the upper edge of the right pelvis; the dense pearlike structure is the gallbladder. (From A Pictorial Presentation of Parasites: A cooperative collection prepared and/or edited by H. Zaiman.)

A newer method, the hydatid antigen dot immunobinding assay, consists of incubation of a serum sample with a textile colloidal dye and a nitrocellulose stick to which the hydatid antigen has been bound. The presence of parasite-specific antibodies leads to dyeing of the stick reactive area, and a colored spot appears. This method showed good correlation with other available methods (87).

Serologic differentiation between cystic and alveolar echinococcosis, involving recombinant larval antigens, has also been used (50). Studies show good correlation between radiologic and serologic detection; however, serologic detection is more specific but less sensitive than imagery (5).

Once the cyst is discovered and surgical removal is selected as the approach, some of the cyst fluid can be aspirated and submitted for microscopic examination to detect the presence of hydatid sand, thus confirming the diagnosis. This procedure is definitely not without risk because of possible fluid and/or tissue leakage or dissemination. Cyst aspiration is usually performed at the time of surgery.

Hydatid sand is not always present. Also, if the cyst is old, the daughter cysts and/or scolices may have disintegrated, so that only the hooklets are left. These may be difficult to find and identify if there is debris within the cyst (Figure 14.3).

To detect hydatid sand, a drop of centrifuged fluid is placed on a slide and then another slide is placed on top of the drop. The two slides are then rubbed back and forth over the fluid. The grating of the hooks on the glass may be felt and heard (hydatid sand sounds like glass grating on sand grains). Of course, if the individual scolices are intact, routine microscopic examination of the centrifuged fluid as

Table 14.4 Antibody detection tests for human cystic and alveolar echinococcosis[a]

Type of echinococcosis test	Antigen	Sensitivity (%)	Specificity (%)	Cross-reactions
Cystic echinococcosis				
IgG ELISA	Crude *E. granulosus* cyst fluid	80->99	61.7	Cestodes (89%), trematodes (30%), nematodes (39%)
	Antigen B (native or synthetic peptide)	63-92	85-93	Alveolar echinococcosis
IgG4 ELISA	Crude cyst fluid	61-67	>99	Alveolar echinococcosis only
EITB	Crude cyst fluid	71	>98[b]	*T. solium* cysticercosis only
	Antigen B fraction	92	100	None
	Antigen B subunit	34-36	>90	
Alveolar echinococcosis				
IgG ELISA	Crude cyst fluid	97.1	61.7	See above
	E. multilocularis PLUS	97.1	98.9	Cystic echinococcosis (2.5%)
	Em2/Em2G11	89.3	100	Cystic echinococcosis (5.6%)
	EmII/3-10	86.4	98.4	Cystic echinococcosis (6.5%)
IgG4 ELISA	Crude cyst fluid	48-67	>99	Cystic echinococcosis
EITB	Em18	97	100	None
	Glycoproteins	70-90	>95	

[a] Adapted from references 44–46, 82, 106, 124, 142, and 150.
[b] Tested with 80 to 184 sera from patients with different parasitic infections (excluding *Echinococcus*-infected patients).

a wet mount will confirm the diagnosis. If the cyst is sterile (no daughter cysts or scolices), the diagnosis could be confirmed histologically from the cyst wall. Refer to Algorithm 14.1 and Tables 14.4 and 14.5. If the material is purulent, place a few milliliters of fluid into a tube and add concentrated hydrochloric acid (HCl). Either boil for 10 min or leave at room temperature for approximately 1 h until the material is quite fluid. Centrifuge and examine the material thoroughly for the refractile hooklets, which survive the HCl treatment intact. Along with the cellular and other mucoid material, the scolices are destroyed; however, the hooklets are liberated and can be seen and identified (personal communication from Graeme Paltridge). Additional information on diagnostic methods can be seen in Table 14.5.

KEY POINTS—LABORATORY DIAGNOSIS

Echinococcus granulosus (Cystic Disease)

1. Presumptive diagnosis may be based on history, radiographic studies, or scans.
2. Additional supportive data may be acquired from immunologic tests.
3. Microscopic examination of hydatid cyst fluid may reveal the hydatid sand or, under certain circumstances, just the hooklets.
4. If the cyst fluid is thick or purulent, it may have to be subjected to a digestion procedure prior to examination; the hooklets survive this treatment and can then be seen and identified.
5. Light, fluorescence, and epifluorescence microscopy can be used to visualize the hooklets; some approaches require staining, and some do not.

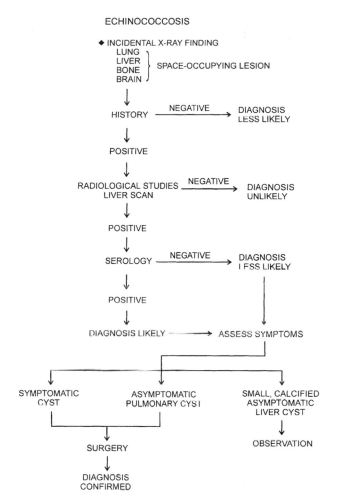

Algorithm 14.1 Echinococcosis.

Treatment

A number of options are available for treatment of cystic echinococcosis, including surgery, puncture-aspiration-injection-reaspiration (PAIR), and chemotherapy. Surgery is generally considered the treatment of choice for a complete cure, although this approach is limited to unilocular cysts in operable body sites. In cases where multiple cysts are present in several different sites or in patients with a high surgical risk, PAIR and chemotherapy are considered appropriate options, either together or separately.

Surgery. Because there is always the risk of cyst fluid spillage during surgery, pre- and/or postoperative use of chemotherapy may be appropriate. Chemotherapy with albendazole or mebendazole offers the advantage of reducing the risk of disease recurrence and intraperitoneal seeding of infection that may develop via cyst rupture and spillage occurring spontaneously or during surgery or needle drainage. Surgery is normally reserved for patients with hydatid

cysts refractory to PAIR due to advanced hydatid disease or secondary bacterial infection or for patients with cyst-biliary communication or obstruction. Histologic images of cystic echinococcosis can be seen in chapter 23.

Puncture-Aspiration-Injection-Reaspiration. PAIR is minimally invasive and includes the following steps: (i) under ultrasonographic guidance, percutaneous puncture of the cyst; (ii) aspiration of 10 to 15 ml of the cyst fluid; (iii) injection of a parasitocidal solution (95% ethanol in a volume one-third the amount of fluid aspirated); and (iv) reaspiration of the fluid after 5 min (33). Although hypertonic NaCl (15% [final concentration] in the cyst fluid) can be used instead of 95% ethanol, reaspiration of the fluid should not be performed until after 15 to 20 min. PAIR appears to have a higher incidence of cure, lower rates of complications, mortality, and disease recurrence, and fewer hospitalization days compared with surgery (126). Also, for patients with hepatic echinococcosis who fail to respond to chemotherapy alone, PAIR

Table 14.5 Procedures for the diagnosis of echinococcosis

Procedure	Comments	Reference
Cellular and humoral responses to purified antigens	Lymphoproliferative assay could be used as an additional tool for the diagnosis of hydatid disease, particularly in seronegative cases; unsuitable for effective monitoring of treatment or surgery.	62
E. granulosus hydatid fluid antigen-ELISA	Diagnostic sensitivity, 91%; specificity, 82% (with cross-reactions in tumor patients [6%] and in patients with other parasitic diseases). Immunoblotting (8, 29, or 34 kDa) provided a 99% discrimination between seropositive cystic hydatid disease cases and noncestode parasitic infections or malignancies.	100
E. granulosus (cystic hydatid disease) cyst fluid (antigen B-rich fraction), *E. multilocularis* (alveolar hydatid disease), protoscolex extract (Em18-rich fraction)	Antigen B (8 kDa) not species specific for *E. granulosus* (genus specific; Em18 antigen is a serologic marker for differentiation of *E. granulosus* and *E. multilocularis*)	52
Enzyme-linked immunotransfer blot (EITB)	Utility of discriminant analysis applied to the serologic results was more sensitive than conventional serologic diagnosis; detected 100% of patients with active hydatid cyst with a specificity of 100%. However, failed to detect 1 case of uncomplicated hyaline pulmonary hydatid cyst.	40
Pre- and posttreatment sera tested for circulating hydatid antigen using coagglutination and countercurrent immunoelectrophoresis and for antibody using indirect hemagglutination	Serial pre- and postoperative or chemotherapy estimation of antigen valuable as an index of cure or of continuing hydatid infection.	102
Countercurrent immunoelectrophoresis of urine for hydatid antigen	Antigen detected in 44% with surgically confirmed disease, 40% with ultrasonography-proven disease, and 57% with presumptive disease. No cross-reactivity, but 8% false positives. Simple, rapid, and noninvasive method of diagnosis.	96
Countercurrent immunoelectrophoresis of hydatid fluid	Test is moderately sensitive (79%) and 100% specific; although sensitivity is not very high, test is simple, inexpensive, and rapid.	102
Coagglutination for detection of hydatid antigen in fluid aspirated from hydatid cysts	Scolices and hooklets seen in <50% of aspirated samples; hydatid antigen demonstrated in all samples with 100% sensitivity; provides alternative to microscopy for confirmation of hydatid etiology.	96
FNA biopsy; detection of *E. granulosus* antigen 5 in liver cyst biopsy specimens	Sensitivity of microscopy after FNA biopsy, 64% (16/25); antigen 5 results were positive in those who were microscopy positive (confirmed) but in only 11% of suspected patients.	97
Fiber-optic bronchoscopy in diagnosis of complicated pulmonary unilocular cystic hydatidosis	Review of 1,726 cases. While clinical, radiologic, and laboratory findings are presumptive, fiber-optic bronchoscopy and cytologic-histopathologic exam of material obtained are conclusive for diagnosis.	117
Dark-field microscopy compared with cytochemical stains for diagnosis of complicated pulmonary unilocular cystic hydatidosis	Superior to cytochemical stains, providing increased sensitivity of cytologic detection of true hydatid elements such as hooklets and laminated membrane fragments.	95
Visualization of hydatid elements; several methods	Ryan and modified Baxby (steaming-hot 1% safranin for 2 min, malachite green for 30 s) stains recommended for examination under transmitted light; Ziehl-Neelsen stain under green excitation light (excitation, 546 nm; long pass, 590 nm) or violet light with no staining for fluorescence; epifluorescence microscopy convenient for samples concentrated by filtration.	20
Sandwich ELISA for coproantigen in formalin- and heat-treated fecal material from dogs	Sensitivity of 100% and specificity of 96%.	81

is a safe and effective procedure. It is recommended for treatment of hepatic cysts of ≥5 cm in diameter, for cysts with daughter cysts, for cysts with detached membranes, and for multiple cysts accessible to puncture. Contraindications are for cysts communicating with the biliary tree, hepatic cysts with high surgical risk or in inaccessible locations, cysts free in the abdominal cavity, and cysts in the lungs, heart, brain, or spine (33).

To prevent chemical cholangitis, it is critical that PAIR not be used to treat cysts with biliary communication. Cyst aspirates must be checked for traces of bilirubin; if it is found in the cyst fluid, the PAIR protocol must be stopped immediately.

Percutaneous Thermal Ablation. Percutaneous thermal ablation of the cyst germinal layer using a radiofrequency ablation device has proven to be a newer option that may find greater use in the future (33). One advantage of this approach is that it avoids injection of parasitocidal fluids into the hydatid cyst.

Chemotherapy. During the last few years, experience with the benzimidazole derivatives mebendazole and albendazole suggests that under certain circumstances, chemotherapy may replace surgery. This approach is recommended for patients with inoperable cystic disease and those with multiple cysts in several body sites; however, chemotherapy for bone disease is less effective. Although relapses after chemotherapy have been reported, patients are usually sensitive to retreatment (3, 33, 98). In general, side effects are mild.

Immunologic markers indicating the effectiveness of pharmacologic treatments with albendazole and mebendazole in human hydatid disease have been reviewed. These studies show that IgE may be a useful marker of therapeutic success in patients with pretreatment specific IgE antibodies, since the levels of these antibodies decrease rapidly in those who respond fully to therapy. IgG subclass responses and differential immunoglobulin in subclass binding pattern to hydatid antigens may also be useful (105). Specific dosage information is provided in chapter 25 (Table 14.6).

Epidemiology and Prevention

The percentage of infected hosts varies throughout the world, but human infection is still much less common than infection of any of the reservoir hosts. Areas of endemicity of growing concern are the sheep- and cattle-raising areas of Argentina, Uruguay, southern Brazil, and Chile (8) (Table 14.7). There are two life cycles of *E. granulosus*, the first of which can be domestic, involving the domestic dog as the definitive host and various species of domestic ungulates as intermediate hosts.

The second type of life cycle is sylvatic, in which wild carnivores serve as definitive hosts and wild ungulates serve as intermediate hosts. The dog-sheep strain is the most common and important throughout the world. There are often overlapping domestic and sylvatic cycles, all of which present special situations and challenges related to disease control.

Most human hydatid cysts are probably acquired in childhood and may continue to grow undetected for many years. Several preventive measures are considered mandatory if the incidence is to decrease. First, all infected viscera from slaughtered animals (intermediate hosts containing hydatid cysts) must be disposed of so that dogs do not have access to this material. Also, personal hygiene must be emphasized to prevent accidental ingestion of the infective eggs from soil contaminated with dog feces, particularly since the eggs are very resistant to disinfectants. In many areas, extensive control measures have also been taken to reduce hydatid disease (Table 14.8). Certainly, educational programs are also very important.

ELISAs for *Echinococcus* coproantigen have been developed to detect infection in animals (33). Antigen could be detected for 6 months in fecal samples stored in 5% formalin. Improvements in sensitivity in patients with low worm burdens and modifications for field application may prove to be very helpful in future epidemiologic studies.

Echinococcus multilocularis (Alveolar Disease, Hydatid Disease)

For many years, pathologists recognized the difference between the unilocular type of human hydatid cyst and the alveolar form. Some workers thought that the alveolar form was a variant species, while others thought that it might arise from an early unilocular cyst. Investigations in Alaska and Germany served to define the characteristics of alveolar hydatid disease caused by *E. multilocularis* (Table 14.1). Alveolar hydatid disease is the most lethal of helminthic diseases, with radical surgery still being the only curative therapy. However, resection has been possible in only 25 to 57% of patients. Most cases are diagnosed in rural residents, suggesting direct or indirect contact with fecal material from infected foxes or dogs. The spread of infections and the increasing numbers of documented cases in the former USSR, China, Japan, and Alaska are causing concern among public health personnel in these regions. There is some variation among isolates from North America and Eurasia; however, there appear to be no actual genetic strain differences (46).

Table 14.6 Additional information on therapy for hydatid disease

Therapy	Comments	Reference
Albendazole vs. mebendazole (3–6 month continuous cycles)	448 patients with 929 hydatid cysts; 74% of cysts showed degenerative changes at the end of therapy, more frequent in albendazole-treated than mebendazole-treated patients (82 and 56%, respectively). Relapse percentages similar. Cysts recurred most often within 2 yr after therapy.	38
Albendazole vs. placebo	20 patients with 179 pulmonary cysts; 800 mg of albendazole daily in 3 cycles of 6 wk (2 wk between cycles); 91% (10/11) of treatment group showed cure (5) or improvement (5); 25% of placebo group showed improvement but no cure; 71% of cysts in treatment group (88/124) showed improvement compared to 15% (4/26) in placebo group.	61
Albendazole	12 patients, cystic hydatid disease of liver, continuous treatment for 6 courses of 4 wk with a 2-wk drug-free interval between cycles; at 24 mo, continuous therapy better than discontinuous therapy.	78
Albendazole, cerebral cyst	Cerebral cyst treated with albendazole; follow-up showed complete disappearance of the cysts with residual focal calcification on CT and presumed gliosis on MRI.	55
Mebendazole, dosage and duration	53 patients, single-, multiple-, and/or multiorgan hydatid cysts (30–70 mg/kg/day) over various periods (6–24 mo); treatment failure in 40%, cure in 38%, intermediate result in 23%; cure rates increased both with dosage and duration; daily dosages of 60–70, 50, and 30–40 mg/kg cured 48, 33, and 25%, respectively.	139
Mebendazole, intracystic application of solution	2 patients with liver cyst; first report of paired therapy; concentration of 2.4 µg/ml injected into cyst plus 50 mg/kg/day oral treatment; at end of 6- to 9-mo period, total recovery observed.	36
Surgery vs. percutaneous drainage or prolonged chemotherapy	63 patients; if CT findings include evidence of exophytic (growth out from the original lesion or dilated ducts in the vicinity of the cyst), surgery is indicated and should not be delayed when either sign is encountered.	72
Percutaneous drainage combined with albendazole	Percutaneous drainage, combined with albendazole therapy, is an effective and safe alternative to surgery for treatment of uncomplicated hydatid cysts of the liver; requires shorter hospital stay.	63
Laparoscopic partial pericystectomy	Laparoscopic partial pericystectomy is a promising new surgical approach to hepatic hydatid disease; practical without increasing the risk of intra-abdominal spillage of scolices; some cysts may not be accessible by this procedure; approach not possible for *E. multilocularis*.	128
Pulmonary cyst; percutaneous drainage and albendazole treatment	Aspiration plus chemotherapy common for liver cysts; not as commonly reported for pulmonary cysts; other options for pulmonary cysts should also be considered.	83
One-stage surgical approach for bilateral lung and liver hydatid cysts	Simultaneous combined resection of hydatid cysts in one stage through midsternotomy along with laparotomy or transdiaphragmatic removal of liver cysts.	25
Endoscopic stenting for postoperative biliary strictures	13 cases; endoscopic stenting is safe in treatment of postoperative benign biliary strictures secondary to hepatic hydatid disease.	146
Orbital hydatid cyst; percutaneous treatment	The cyst was treated percutaneously under ultrasonographic guidance with aspiration, 15% hypertonic saline injection, and reaspiration; no complications; patient asymptomatic after 9 mo; same follow-up at 21 mo; approach may be a safe and effective alternative to surgery.	2
Cardiac and pericardiac disease	14 patients; intraoperative surface echocardiography is critical for diagnosis and planning management of successful surgery; definitive treatment is surgical removal of the cyst.	12
Pancreas	Patient with primary hydatid cyst in head of the pancreas, with obstructive jaundice due to extrinsic compression of the intrapancreatic portion of the bile duct; percutaneous drainage used with hypertonic (20%) saline.	144
Evaluation of scolicidal agents in experimental model	Mouse model; hydrogen peroxide and povidone iodine show a greater protoscolicidal effect than does simple cleansing with physiological saline, hypertonic saline, or praziquantel.	41

Table 14.7 Examples of epidemiology of cystic hydatid disease (*E. granulosus*)

Geographic location	Comments	Reference(s)
Northern Libya	20,220 people screened; use of ultrasonography combined with serology as a mass screening approach; of ultrasonography-positive cases, 69% seropositive (ELISA with hydatid cyst fluid antigen B); cystic echinococcosis increased significantly with age; 25% of ultrasonography-negative members of families with an index case were seropositive.	122
Iran	Majority of *E. granulosus*-infected livestock animals can potentially act as reservoirs of human infection; important implications for hydatid control and public health.	149
Turkana, Kenya	Ultrasonography of 260 sheep and 320 goats; 9.2% of sheep and 2.5% of goats positive.	84
Turkana, Kenya	Ultrasonography of 16 sheep and 284 goats; sensitivity and specificity of ultrasonography in sheep and goats were 54 and 98%, respectively.	112
Argentina	PCR-restriction fragment length polymorphism characterized 4 distinct genotypes of 33 isolates; first report of G2 and G6 genotypes in humans—may have consequences for human health.	109
Peru	High prevalence of human and animal echinococcosis in village in central Peruvian Andes; field studies include coproantigen enzyme immunoassay and arecoline purging used for study of canine disease; EITB used in sheep; overall prevalence was 9.3% in humans, 46% in canines, and 65% in sheep.	89, 90
La Paloma, Central Uruguay	Prevalence of human infection was 5.6%; almost 20% of dogs were infected; human cystic echinococcosis highly endemic in Uruguay, one of the highest prevalence rates in the world; mass screening by ultrasonographic scanning with confirmatory serologic testing is an effective approach to detection at community level.	21
Florida, Uruguay	Ultrasonography of 9,515 patients with liver lesions revealed 156 positive; sensitivity of ELISA and latex agglutination serology compared with ultrasonography was 48 and 28%, respectively; specificity was >85%. No correlation with dog ownership or home slaughter of sheep, but offal disposal was important.	17
Northern Israel	IgG ELISA, using both crude and purified antigens, very useful for seroepidemiologic screening for hydatid disease; the condition is an emerging disease in this area.	35
Bulgaria	Review of control measures over periods spanning 46 yr. After a period of effective control measures, economic changes led to removal of control structures; the incidence rose, providing convincing evidence that cessation of control measures can lead to intensification in the transmission of *E. granulosus* and to resurgence of disease to former levels.	130

Table 14.8 Control measures for cystic hydatid disease

Approach	Comments	Reference
Vaccination of intermediate host (sheep)	Immunization of sheep with specific purified protein results in 92% resistance; immunization using fusion protein has given 97 and 98% resistance to a challenge infection; shelf life is 12 mo; 2 injections given 1 mo apart effective for at least 12 mo.	48
Vaccination of intermediate host (sheep)	Cloned recombinant antigen from parasite oncosphere; sheep protected (mean, 96–98%); potential for human vaccination.	74
Vaccination of intermediate host (sheep)	EG95 recombinant vaccine; 86% of vaccinated sheep completely free of viable hydatid cysts approximately 1 yr after challenge infection; number of viable cysts reduced by 99% compared with unvaccinated controls.	73
Treatment of sheep with oxfendazole	215 adult sheep; in daily, weekly, and monthly groups receiving treatment, 100, 97, and 78% of sheep, respectively, were cured or improved compared with 35% in the control group; daily dosing at 30 mg of oxfendazole per kg proved highly toxic (24% death rate in daily group).	31

Life Cycle and Morphology

The life cycle is essentially identical to that of *E. granulosus*, with the following exceptions: the definitive hosts in the sylvatic cycle tend to be dogs, cats, wolves, and foxes while the intermediate hosts have been identified as voles, field mice, house mice, ground squirrels, and shrews (Figure 14.8). Laboratory infections have also been maintained in deer mice and gerbils.

The adult worms are smaller than those of *E. granulosus* (1.2 to 3.7 mm versus 3.0 to 6.0 mm long). Also, the eggs are *Taenia*-like and very resistant to cold.

The cyst itself is composed of many irregular cavities with little or no fluid, rare or no free scolices, and often central necrosis and cavitation of the lesion (Figure 14.9). The human lesion is an example of persistent parasitic growth at a low level rather than active growth or malignancy.

Clinical Disease

Although the alveolar form of hydatid disease has been found in other tissues, the liver is the most common site. The disease may resemble a slowly growing carcinoma and may cause symptoms of intrahepatic portal hypertension. In humans there are several stages of the infection; they have been described as initial, progressive, advanced, stable, and abortive course (33) (Table 14.9).

Proliferation of the cyst is most active at the advancing margin, while the interior of the cyst structure may deteriorate and appear to be a necrotic abscess filled with pus. The lesions tend to spread by direct extension into the surrounding tissues and can also be transported to other body sites via the lymphatic or hematogenous routes. In about 2% of cases, pieces of the germinal membrane can metastasize to other body sites, including the brain, lungs, and mediastinum. Since the human is not a normal host,

the typical brood capsules, scolices, and calcareous corpuscles may not develop (Figure 14.9). In the human, the laminated membrane may surround vesicles that are void of scolices and other internal structures. The lesions caused by *E. multilocularis* may appear as firm, solid, cancerlike masses, and the outer layer has a poorly defined margin that continues to grow and spread throughout the tissue.

In well-established disease, there may be hepatomegaly, splenomegaly, and finally jaundice and ascites. Patient symptoms include right upper quadrant pain and the presence of a palpable hepatic mass. There may also be shortness of breath, as well as central nervous system symptoms from brain metastases. Because the growth of the alveolar cyst is very slow, some patients may be infected for 20 or 30 years before exhibiting any symptoms. In patients who remain untreated, the mortality rate can exceed 90% (80).

Studies directed at the metacestode developmental stage of *E. multilocularis* have confirmed that the 14-3-3 protein is found in the germinal layer and is expressed about 10-fold in this stage compared with the adult worm. These results, taken together with current knowledge of the 14-3-3 protein family, suggest that this parasite molecule may contribute to the promotion of the progressive, potentially unlimited growth behavior of the organism within the host tissue (123). Studies also suggest that viable *E. multilocularis* vesicles induce significant cellular production of cytokines and chemokines. In patients with viable alveolar hydatid disease, these immune mediators may enhance antibody-dependent cellular effector mechanisms against proliferating metacestodes (30).

Both humoral and cell-mediated immune responses occur. In the mouse model, there is an initial Th1-dominant cytokine response, including IL-2 and gamma interferon

Figure 14.8 Life cycle of *Echinococcus multilocularis* (alveolar hydatid disease).

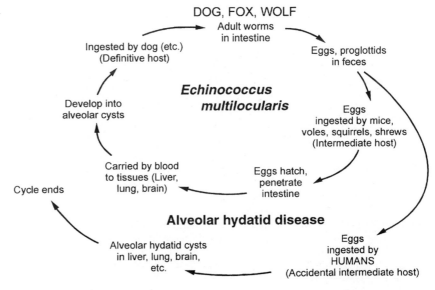

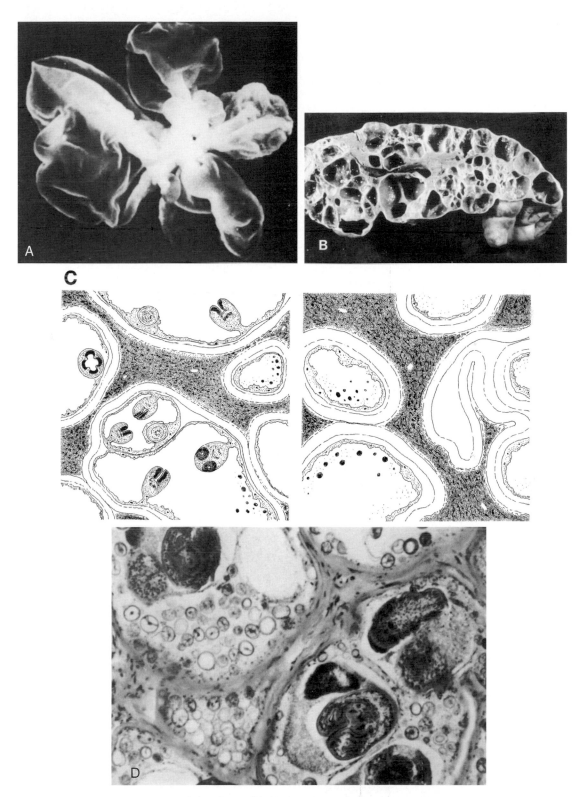

Figure 14.9 *Echinococcus multilocularis*. (A) Entire cyst. (B) Multicystic nature of the parasite in an experimental infection in a muskrat. (From A Pictorial Presentation of Parasites: A cooperative collection prepared and/or edited by H. Zaiman. Photograph courtesy of R. L. Rausche.) (C) Diagram of a histologic section (the left side represents the normal animal intermediate host in which scolices are found; the right side demonstrates the cyst structure within the human accidental intermediate host in which the scolices may not develop). (D) Histologic section. (Illustration by Sharon Belkin.)

Table 14.9 Infection phases: alveolar echinococcosis

Disease phase	Symptoms	Comments
Initial	Asymptomatic	Spontaneous cure or become progressive: incubation period 5–15 yr
Progressive	Abdominal pain, jaundice, hepatomegaly, fever, anemia, weight loss, pleural pain	Symptoms occur when metacestode infiltrates larger parts of liver or impacts organ functions
Advanced	Severe hepatic dysfunction, portal hypertension	Disease duration variable between weeks and years; mortality in both treated and untreated patients can be high
Stable	Patients on long-term chemotherapy	Parasite growth inhibited
Abortive	Seen in asymptomatic patients	Parasite dies and the cyst is calcified

(IFN-γ). This response is associated with diminished parasite growth. Rapid organism growth was associated with secretion of Th1- and Th2-type cytokines including IL-5 and IL-10. In patients with progressive disease, IL-10 was more pronounced and IL-10 levels in serum were higher than in patients with the abortive form (42).

Diagnosis

Imaging Techniques. Ultrasonography, in association with computed tomography (CT), is recommended; magnetic resonance imaging (MRI) can also be helpful. These imaging options can detect the majority of lesions, including those that have calcified. In a recent study, contrast-enhanced power Doppler ultrasonography was compared with CT for the diagnosis of alveolar echinococcosis. The three-phase helical CT was able to visualize the vascularization associated with *E. multilocularis* lesions in the liver, while neither unenhanced nor contrast-enhanced ultrasonography in power mode was suitable for this purpose (66).

Immunodiagnostic Tests. The diagnosis of alveolar hydatid disease may be difficult unless there is a presumptive diagnosis based on history. Often, the correct diagnosis is missed even at autopsy. Serologic procedures used for unilocular hydatid disease may be helpful (Table 14.4) (33, 150). The metacestodes of *E. multilocularis* contain glycolipids which are recognized by serum antibodies from patients with alveolar hydatid disease. The ELISA procedure results were confirmed by immunostaining on thin-layer chromatography (99). This approach may also provide another procedure for the diagnosis of alveolar hydatid disease. The use of monoclonal antibodies in a sandwich ELISA has also proven valuable in the diagnosis of alveolar hydatid disease, particularly when one is distinguishing this infection from those caused by other helminths. In general, one of the difficulties with any complex parasite involves the cross-reactivity seen when crude antigen extracts or antigen fractions from different isolates are used for immunodiagnosis. This approach results in a lack of test standardization and inability to interpret and compare test results. Because serodiagnosis fails in 5 to 10% of cases and radiologic techniques are difficult to interpret in some cases, PCR has been developed for the detection of *Echinococcus*-specific mRNA from fine-needle aspiration (FNA) biopsy specimens (60).

Fine-Needle Aspiration. FNA biopsy findings for hepatic *E. multilocularis* have been reviewed; there appear to be some basic diagnostic criteria for this approach (19). Cytologic smears were stained with May-Grunwald-Giemsa and periodic acid-Schiff (PAS) stains. Tissue sections were stained with hematoxylin and eosin and with PAS. In tissue sections, homogeneous, thin cystic structures of various dimensions were strongly PAS positive. Mucoid material within the cystic structures was also PAS positive. Wide coagulative necrosis was seen in all cases, and there were foreign-body-type giant cells at the periphery of the lesion in some cases. In all of the cytologic smears, there were necrotic ground, PAS-positive hyaline cuticular structures, and mucoid globules; in some smears, there were also foreign-body-type giant cells.

In another case where the patient presented with a multicystic mass in the pancreas, FNA biopsy revealed some scolices, free hooklets, and fragments of the laminated layer, findings which are pathognomonic for echinococcosis. In this case, the patient had pancreatic alveolar hydatid disease (28). Serologic testing using an ELISA with Em2(plus) antigen showed high antibody reactivity with the patient serum, a finding indicative of an *E. multilocularis* infection. Diagnosis was confirmed by molecular analysis of the cytologic material using PCR and a direct fluorescent-antibody assay.

KEY POINTS—LABORATORY DIAGNOSIS

Echinococcus multilocularis (Alveolar Disease, Hydatid Disease)

1. Serologic results may be helpful (the same procedures as used for unilocular disease).
2. Histologic identification may be unsuccessful at both the biopsy and autopsy phases of the diagnosis

as a result of nonrecognition of the etiologic agent or insufficient tissue.

3. Imaging techniques can be very helpful, particularly when coupled with immunodiagnostic tests for confirmation.

Treatment

Surgery. Total surgical removal of an alveolar hydatid cyst is almost impossible, since the cyst morphology does not have a limiting capsule. However, in cases diagnosed in the early stages, radical surgery may result in a cure. In more advanced cases of alveolar disease, the effectiveness of radical surgery drops dramatically. Liver transplantation is generally not recommended, and there remains a risk of metastasis after surgery.

Chemotherapy. In addition to surgery, the same drugs that have been used for *E. granulosus* have been tried. Mebendazole and albendazole are the drugs of choice for treating alveolar echinococcosis. However, if one considers cost, albendazole reduces costs by about 40% and is easier for patients to take (104). In some situations, the data indicate that mebendazole therapy is parasitostatic rather than parasitocidal; however, even with recurrence, the patients responded favorably to reintroduction of chemotherapy. Because IFN-γ and IFN-α levels are decreased in patients with echinococcosis, the addition of IFN-γ to standard therapy with mebendazole has been reported to halt disease progression (120). Specific drug information is provided in chapter 25. The overall prognosis for this infection is usually very grave. However, spontaneous death of *E. multilocularis* cysts in humans has been reported. Additional information can be found in Table 14.10.

Assessment of chemotherapy can be difficult and is usually based on the following criteria: (i) long-term patient survival rates, (ii) lack of disease progression, (iii) improved clinical status, (iv) slowed or stopped parasite proliferation, and (v) absence of disease recurrence (32).

Epidemiology and Prevention

A number of variables play roles in the epidemiology of *E. multilocularis* infection (Table 14.11). Also, there are serious questions regarding the emergence and geographic spread of this infection. It is found in at least 30 countries, including Russia, Kazakhstan, China, parts of Europe, and North America. Although it appears that areas of endemic infection are increasing in size, it is difficult to tell whether the data reflect actual changes in the parasite range or merely clarification of range extensions that were already present but unrecognized (59, 127).

Handling of infective definitive hosts or ingestion of contaminated food or water probably initiates the infection, while accidental ingestion of soil, berries, or other vegetables contaminated with eggs may be less important in transmission. However, it is important that the dangers of playing in the woods, picking berries, etc., in areas of endemicity that are frequented by wolves, foxes, and dogs be recognized, although routes of infection are difficult to prove due to the long incubation periods for the infection and lack of detailed information (32, 67, 108, 129, 137). Educational programs should be initiated early with children, since most infections are presumed to be acquired in childhood. There are probably additional factors that influence human infections but remain unknown (59, 68).

The use of coproantigen testing by a sandwich ELISA method to detect infection in foxes is more sensitive than routine egg detection assays and has been recommended for diagnosis in the definitive host of *E. multilocularis* (24, 92, 93). Coprodiagnosis by PCR is an excellent alternative to necropsy, since the sensitivity of necropsy is no higher than 76%. The PCR system is an alternative, sensitive method for the routine diagnosis of *E. multilocularis* in carnivores (29).

Chemotherapy with praziquantel has the potential to reduce the prevalence of *E. multilocularis* in wild foxes. However, the question of its long-term efficacy and other unresolved problems such as cost need to be addressed by additional studies before routine application of this baiting method can be recommended (49, 118, 140).

Echinococcus oligarthrus and *Echinococcus vogeli* (Polycystic Hydatid Disease)

Life Cycle and Morphology

E. oligarthrus and *E. vogeli* are the only indigenous species known in the neotropical area of the New World (101). The *E. oligarthrus* life cycle involves wild felids such as pumas, jaguars, and wild cats as definitive hosts, while the intermediate hosts harboring the larval forms (hydatid cysts) include the paca, agouti, spiny rats, and opossums. Further studies to determine which other wild animals might be involved in the life cycle are under way. The definitive hosts of *E. vogeli* include the bush dog, while the intermediate hosts are the same as for *E. oligarthrus*.

Clinical Disease

In a case of human hydatid disease caused by *E. oligarthrus* (26), the patient presented with proptosis of the left eye and persistent headache; a CT scan revealed a single orbital, retro-ocular cyst approximately the same

Table 14.10 Additional information on therapy for alveolar hydatid disease

Therapy	Comments	Reference
Gamma knife radiosurgery and albendazole	Patient had inoperable cerebral alveolar hydatid cysts; procedure performed in 2 sessions; repeated courses of albendazole given concurrently; MRI follow-up showed marked shrinkage of the irradiated cystic structures and initially increased perifocal edema; at 3-yr follow-up, the lesion, perifocal edema, and neurologic symptoms all decreased; patient now stable with minimal neurologic symptoms.	119
Albendazole plus dipeptide methyl ester (Phe-Phe-OMe)	Animal model (gerbils); histologic and ultrastructural studies showed that Phe-Phe-OMe increases the effect of albendazole; dipeptide has a low acute toxicity and is available at a low cost; studies under way to optimize the combination of Phe-Phe-OMe and albendazole.	116
Liver transplantation	21 patients received liver transplants for incurable alveolar echinococcosis; primary-disease recurrence is not rare after this treatment; immunosuppressive therapy may favor larval growth in extrahepatic sites, requiring extensive extrahepatic radiological checkup prior to transplantation; benzimidazole seems to stabilize residual foci; anti-Em2 ELISA did not appear useful in detecting recurrence, but ELISA using crude heterologous antigen (*E. granulosus*) allowed early diagnosis of residual disease.	15
Follow-up after treatment	11 patients with nonresectable lesions, treated with albendazole for 17–69 mo; various responses (2 cures, 5 stabilized, 3 with recurrences, 1 treatment failure); serum follow-up using Em2(plus)-ELISA and Em18-Western blot was useful in evaluating and predicting efficacy of chemotherapy.	79
Follow-up after treatment	Group of patients (cured, stabilized, aggravated) receiving therapy (mebendazole, albendazole, and isoprinosine); purified alkaline phosphatase antigen (pAP-Ag) is recommended for diagnosis and follow-up.	115
Amphotericin B	Used for benzimidazole intolerance or treatment failure (mebendazole, albendazole); may be used as salvage treatment, suppresses parasite growth; long-term treatment required.	103

size as the eye and thought to be a tumor. During surgery, fluid and internal tissues were taken from the cyst, which was also later removed. A few protoscolices and hooklets were found, and *E. oligarthrus* was identified (Figure 14.10). The first three known cases of *E. oligarthrus* have been documented, with the first two involving infection in the eye and the third involving infection in the heart.

E. vogeli is the etiologic agent of polycystic disease in humans in areas such as Brazil, Colombia, Venezuela, Panama, and Ecuador. Nine patients from Brazil were found to have polycystic hydatid disease, and the diagnosis was based on the shape and dimensions of the rostellar hooks. The liver was the most common site, followed by the lungs, mesentery, spleen, and pancreas. The main symptoms included abdominal pain, hepatomegaly, jaundice, weight loss, anemia, fever, hemoptysis, palpable abdominal masses, and signs of portal hypertension. The most common clinical presentation involved the abdomen, with hard, round masses in or connected to the liver. In 25% of cases, there were also signs of portal hypertension; all of these patients died, either from the disease or from surgical complications. Asymptomatic

individuals accounted for 10% of the cases. Hepatic calcifications were seen in four patients. Patients may have calcified, round structures in the liver, suggestive of calcified polycystic hydatids. This disease is chronic and can last for over 20 years; in advanced cases the mortality is high.

Fewer cases of *E. oligarthrus* infection in humans have been documented; to date, four cases have been published, three orbital and one cardiac (33, 91).

To date, approximately 80 cases of polycystic hydatid disease have been recorded, with more than 85% occurring in Brazil, Colombia, Ecuador, and Argentina. Patient ages have ranged from 6 to 78 years (median, 44 years), with one-third of the patients being younger than 22 years (23).

Diagnosis

In 80% of the patients, the lesions were in the liver alone or in both the liver and other organs. In the remainder of the cases, they were located in the lung or other single sites. Patients were diagnosed by various imaging techniques (radiography, ultrasonography, CT scan). Serologic tests did not always confirm the

Table 14.11 Variables in the epidemiology of alveolar echinococcosis[a]

Life cycle phase	Epidemiologic factors	Comments
Definitive hosts, adult form of the parasite	Animal species: Distribution, population size, interactions with prey animals, feeding habits, behavior, habitat characteristics, relation to human habits; Parasite: Prevalence, intensity and curation of infection, egg production, host immunity	Arctic region: Arctic fox is the sole definitive host (highly susceptible); voles are a major intermediate host and prey of foxes (infection prevalence in voles can reach 40%); Sub-Arctic region: Landscape, biotopes, climate more varied; more intermediate hosts, more than one definitive host
Parasite egg survival	Survival and dispersal of eggs, climate and weather conditions	Detailed knowledge of egg production, survival, dispersal not clearly defined; eggs infective for about 1 yr if kept moist and cool; −70°C for 4 days or −80°C for 3 days required for egg inactivation
Intermediate hosts, metacestode	Animal species: Longevity, distribution, population size, feeding habits, role as prey animals, habitat; Parasite: Prevalence, level of infection, protoscolex production, host immunity	Transmission highest if one or two species of rodents are dominant, populations reach high densities, rodents are prey animals for foxes; areas of permanent grasslands are important
Humans, other hosts	Egg exposure, hygienic behavior, relationship to carnivores, susceptibility, resistance, immunity	Handling infected definitive hosts, ingestion of contaminated food or water

[a] Adapted from references 32, 33, 58, and 59.

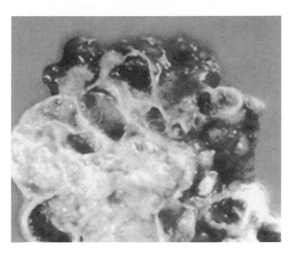

Figure 14.10 Polycystic hydatid disease.

infection; species identifications were based on morphological characteristics of the hooklets from protoscolices. Aspiration cytology has also been reported and can be very helpful in confirming the cause of the "mass" (64). Another option is DNA analysis by PCR using protoscolices. In cases where the organism was identified to the species level, the ratio of *E. vogeli* to *E. oligarthrus* was approximately 10:1.

Treatment

Albendazole therapy has resulted in clinical improvement, with resolution or reduction in the size of the cysts. Because surgical resection is difficult at best, some physicians think that albendazole therapy should be tried prior to surgical intervention. However, a combination of surgery and albendazole therapy has also been advocated.

Epidemiology and Prevention

Human polycystic echinococcosis is now recognized as a distinct disease entity. Although the cestodes that cause polycystic hydatid disease are indigenous to the humid tropical forests in Central and northern South America, *E. oligarthrus* has now been found in North America (114). This is the first time that the adult worm has been recovered from a bobcat. The isolate was identified from San Fernando, Tamaulipas State, in northeast Mexico; this location is approximately 100 km south of Brownsville, Tex. Apparently, infection risk exists in a larger geographic area than was previously suspected.

The paca is widely hunted and used for food in northern South America, and local people tend to feed the viscera of these animals to their dogs. Based on this common practice, it is assumed that the dog is probably the primary source of infection for humans in this and other areas of endemic infection.

Taenia (*Multiceps*) spp. (*Taenia multiceps, Taenia serialis*) (Coenurosis)

Although the adult tapeworms resemble *Taenia* spp., the larval form is a coenurus, a unilocular cyst with the same configuration as a cysticercus but containing multiple scolices. Human cases have been reported worldwide; however, identification to the species level is often very difficult if not impossible (Table 14.1).

Life Cycle and Morphology

The dog, wolf, and fox are the definitive hosts for *Taenia* (*Multiceps*) spp. (Figure 14.11). Infection of the intermediate host (herbivores) occurs on ingestion of infective eggs from the adult tapeworm in the carnivore intestine. The oncosphere hatches in the intestine, penetrates the intestinal wall, and is carried via the bloodstream to various parts of the body, primarily the central nervous system. In sheep, one of the more common herbivore intermediate hosts, the presence of the coenurus in the central nervous system causes a disease known as gid (unstable gait or giddiness). When these infected herbivore tissues are ingested by a carnivore, each scolex within the coenurus has the potential to become an adult worm in the carnivore's intestine.

The coenurus structure is intermediate between that of a cysticercus with a single scolex and that of a hydatid cyst with both multiple scolices and daughter cysts. The coenurus contains multiple scolices but no daughter cysts (Figures 14.12 to 14.14).

Clinical Disease

In humans, most coenuri have been located in the brain or spinal cord, but they can also be found in the subcutaneous tissues (51). Symptoms have been those associated with any space-occupying lesion, including headache, vomiting, paraplegia, hemiplegia, aphasia, seizures, and eye symptoms (if the coenurus is present in the eye).

Diagnosis

Diagnosis is based on gross and histologic examinations of the coenurus after surgical removal.

Treatment

Surgical removal, if possible, is recommended. Although praziquantel has been effective in killing these parasites, there is some concern about the marked inflammation which occurs as a result of treatment (51).

Epidemiology and Prevention

An understanding of the manner of transmission among carnivores and herbivores suggests the importance of proper disposal of carcasses of potentially infected herbivores. Also, good personal hygiene for those in close contact with dogs in areas of the world where herbivores (cattle, goats, and sheep) are raised is recommended.

Spirometra mansonoides and *Diphyllobothrium* spp. (Sparganosis)

The larval forms (spargana) of *Spirometra mansonoides* and *Diphyllobothrium* spp. were first recognized in humans from tissues removed at autopsy. In areas where infections are endemic, similar forms can be found in the subcutaneous connective tissue and between the muscles in frogs, lizards, snakes, birds, and certain mammals. When these spargana are fed to the definitive hosts (dogs and cats), adult tapeworms of several related species of *Spirometra* develop. The general name for these spargana

Figure 14.11   Life cycle of *Taenia* (*Multiceps*) spp.

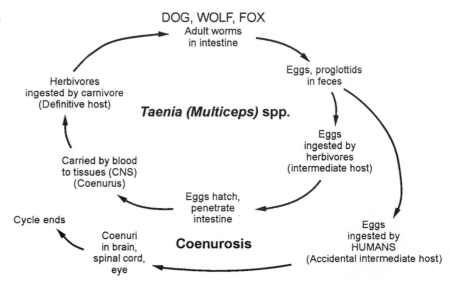

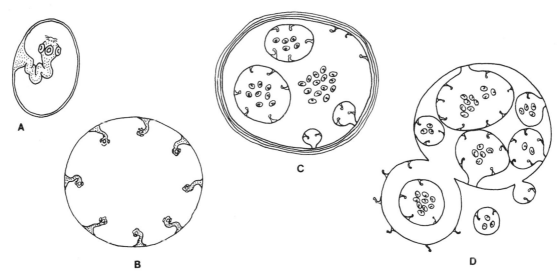

Figure 14.12 Representative cysts. (A) Cysticercus; (B) coenurus; (C) hydatid; (D) multilocular hydatid. (Illustration by Sharon Belkin.)

Figure 14.13 Coenurosis. (Upper) Diagram of cyst. (Illustration by Sharon Belkin.) (Lower left) Actual cyst. (Armed Forces Institute of Pathology photograph.) (Lower right) Actual cyst. (From A Pictorial Presentation of Parasites: A cooperative collection prepared and/or edited by H. Zaiman. Photograph courtesy of J. Harper III and L. C. Marcus.)

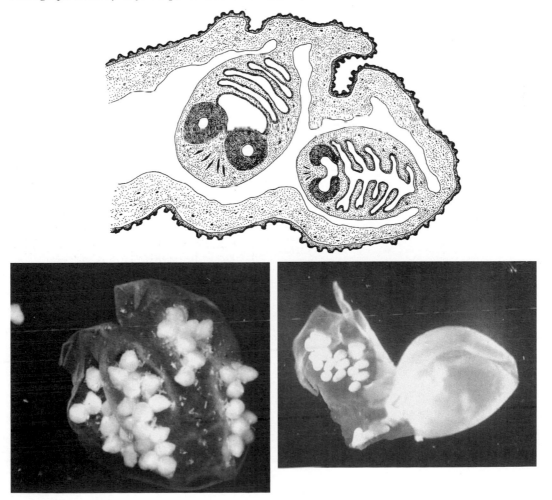

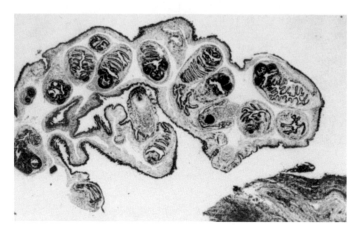

Figure 14.14 Coenurosis: cross section of cyst. (Armed Forces Institute of Pathology photograph.)

is *Spirometra mansonoides*. Although cases have been reported worldwide, sparganosis is most common in China, Japan, and Southeast Asia (Table 14.1).

Life Cycle and Morphology

Spargana. Eggs passed from adult tapeworms in the dog or cat intestine hatch in freshwater and are ingested by *Cyclops* spp., in which they develop into the procercoid larva. When the *Cyclops* spp. are eaten by a second intermediate host (fish, snakes, amphibians, or mammals), the procercoid larva penetrates the intestinal wall,

migrates to the tissues, and develops into the larva (the sparganum). On ingestion by a dog or cat, the sparganum develops into the adult tapeworm, thus completing the life cycle (Figure 14.15).

Human infection can occur from (i) ingestion of infected *Cyclops* spp. (the procercoid stage penetrates the intestinal wall, migrates into tissues, and develops into the sparganum), (ii) ingestion of raw infected flesh of one of the second intermediate hosts (fish, snakes, amphibians, and mammals), or (iii) local application of raw, infected flesh of one of the second intermediate hosts to the skin, conjunctiva, or vagina (the spargana migrate from vertebrate host tissues into human tissue). The spargana are white, ribbonlike worms measuring from a few millimeters to several centimeters long.

Aberrant Spargana. *Sparganum proliferum* is a rare tapeworm larva that grows by continuous branching and budding. The normal tissue organization and symmetry are no longer visible. The lesion gradually expands and may ulcerate to form subcutaneous or cutaneous lesions. The adult form of this organism is unknown.

Clinical Disease

Spargana. Most patients have slowly growing, tender, subcutaneous nodules. These nodules may also be migratory (147). Ocular sparganosis is accompanied by pruritus, pain, lacrimation, and edematous swelling of the eyelid (which may look much like Romaña's sign

Figure 14.15 Life cycle of *Spirometra mansonoides* and *Diphyllobothrium* spp. (sparganosis).

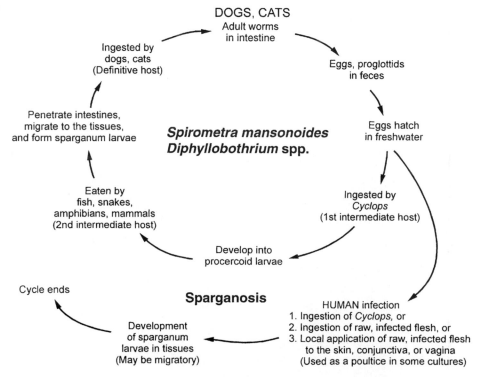

in Chagas' disease). Depending on the location of the sparganum, there may be elephantiasis (lymph channels), peritonitis (intestinal perforation), or brain abscess. Most patients have leukocytosis and eosinophilia. The sparganum has been reported to live for more than 9 years. In a recent study in Thailand, 17 cases were ocular, 10 were subcutaneous, 5 involved the central nervous system, 1 was auricular, 1 was pulmonary, 1 was intraosseous, and 1 was intraperitoneal. Risk factors included a history of drinking impure water, eating frog or snake meat, or using frog or snake meat as a poultice; some patients had multiple risk factors. Most of these patients presented with superficial ocular mass lesions (141).

Aberrant Spargana. In the few documented cases of *S. proliferum* infections, thousands of spargana have been recovered in subcutaneous tissues, intermuscular fasciae, intestinal wall, mesentery, kidneys, lungs, heart, brain, and bone.

Diagnosis

Spargana. Usually, diagnosis is not made until drainage of the lesion, surgical removal of the worms, and gross and/or microscopic examination of the tissue have been performed. The more elongate, wormlike structure of the sparganum can usually be distinguished from the bladder structure of a cysticercus or coenurus, since the last two have suckers and hooklets, both of which are lacking in a sparganum. The parasites are usually recovered alive and are approximately several centimeters in length (26). They are glistening white and opaque; they resemble narrow tapeworm proglottids; and they are motile, similar to the motility seen with a *Taenia saginata* proglottid.

Aberrant Spargana. This rare form of sparganosis can be diagnosed by histologic examination. The tissue may appear to have many separate worms; however, each cross section probably represents a continuous branching worm.

KEY POINTS—LABORATORY DIAGNOSIS

Spirometra mansonoides and *Diphyllobothrium* spp. (Sparganosis)

1. The diagnosis is usually based on the identification of the worm after surgical removal.
2. The absence of suckers and hooklets will usually differentiate spargana from a cysticercus or coenurus.

Treatment

Surgical removal is the recommended approach. Drug therapy is usually not effective; however, treatment with praziquantel is recommended. In the proliferating type of infection, surgical removal is very difficult if not impossible.

Epidemiology and Prevention

Considering the possible routes of human infection, all drinking water in areas of endemicity should be boiled or filtered to prevent accidental ingestion of *Cyclops* spp. The ingestion of raw tadpole, frog, snake, fowl, or mammalian flesh should be avoided in these areas. Educational information on the dangers of local application of raw, infected animal flesh (poulticing) to humans, particularly those with ulcerated lesions, should be disseminated.

References

1. Aguilar, X., J. Fernandez-Muixi, R. Magarolas, A. Sauri, F. Vidal, and C. Richart. 1998. An unusual presentation of secondary pleural hydatidosis. *Eur. Respir. J.* **11:**243–245.
2. Akhan, O., S. Bilgic, D. Akata, H. Kiratli, and M. N. Ozmen. 1998. Percutaneous treatment of an orbital hydatid cyst: a new therapeutic approach. *Am. J. Ophthalmol.* **125:**877–879.
3. Akhan, O., M. N. Ozmen, A. Dincer, I. Sayek, and A. Gocmen. 1996. Liver hydatid disease: long-term results of percutaneous treatment. *Radiology* **198:**259–264.
4. Ali, R., F. Ozkalemkas, V. Ozkocaman, T. Ozcelik, H. Akalin, A. Ozkan, Y. Altundal, and A. Tunali. 2005. Hydatid disease in acute leukemia: effect of anticancer treatment on echinococcosis. *Microbes Infect.* **7:**1073–1076.
5. Babba, H., A. Messedi, S. Masmoudi, M. Zribi, R. Grillot, P. Ambroise-Thomas, I. Beyrouti, and Y. Sahnoun. 1994. Diagnosis of human hydatidosis: comparison between imagery and six serologic techniques. *Am. J. Trop. Med. Hyg.* **50:**64–68.
6. Balik, A. A., M. Basoglu, F. Celebi, D. Oren, K. Y. Polat, S. S. Atamanalp, and M. N. Akcay. 1999. Surgical treatment of hydatid disease of the liver—review of 304 cases. *Arch. Surg.* **134:**166–169.
7. Bart, J. M., S. Morariu, J. Knapp, M. S. Ilie, M. Pitulescu, A. Anghel, I. Cosoroaba, and R. Parroux. 2006. Genetic typing of *Echinococcus granulosus* in Romania. *Parasitol. Res.* **98:**130–137.
8. Beaver, P. C., R. C. Jung, and E. W. Cupp. 1984. *Clinical Parasitology*, 9th ed. Lea & Febiger, Philadelphia, Pa.
9. Becker, K., T. Frieling, A. Saleh, and D. Haussinger. 1997. Resolution of hydatid liver cyst by spontaneous rupture into the biliary tract. *J. Hepatol.* **26:**1408–1412.
10. Berk, C., E. Ciftci, and A. Erdogan. 1998. MRI in primary intraspinal extradural hydatid disease: case report. *Neuroradiology* **40:**390–392.
11. Bhojraj, S. Y., and N. R. Shetty. 1999. Primary hydatid disease of the spine: an unusual cause of progressive paraplegia—case report and review of the literature. *J. Neurosurg.* **91**(Suppl.):216–218.
12. Birincioglu, C. L., H. Bardakei, S. A. Kucuker, A. T. Ulus, K. Arda, B. Yamak, and O. Tasdemir. 1999. A clinical dilemma: cardiac and pericardiac echinococcosis. *Ann. Thorac. Surg.* **68:**1290–1294.

13. Borner, H., S. Demertzis, A. Heisel, G. Berg, G. Schneider, and H. J. Schafers. 1999. Acute cardiac tamponade due to echinococcal cyst of the heart. *Z. Kardiol.* **88:**1028–1033.

14. Brechignac, X., I. Durieu, M. Perinetti, L. Geriniere, C. Richalet, and D. V. Durand. 1997. Hydatid cyst of the heart. *Presse Med.* **26:**663–665.

15. Bresson-Hadni, S., S. Koch, I. Beurton, D. A. Vuitton, B. Bartholomot, S. Hrusovsky, B. Heyd, D. Lenys, A. Minello, M. C. Becker, C. Vanlemmens, G. A. Mantion, and J. P. Miguet. 1999. Primary disease recurrence after liver transplantation for alveolar echinococcosis: long-term evaluation in 15 patients. *Hepatology* **30:**857–864.

16. Buishi, I., T. Walters, Z. Guildea, P. Craig, and S. Palmer. 2005. Reemergence of canine *Echinococcus granulosus* infection, Wales. *Emerg. Infect. Dis.* **11:**568–571.

17. Carmona, C., R. Perdomo, A. Carbo, C. Alvarez, J. Monti, R. Grauert, D. Stern, G. Perera, S. Lloyd, R. Bazini, M. A. Gemmell, and L. Yarzabal. 1998. Risk factors associated with human cystic echinococcosis in Florida, Uruguay: results of a mass screening study using ultrasound and serology. *Am. J. Trop. Med. Hyg.* **58:**599–605.

18. Casero, R. D., M. G. Costas, and E. Menso. 1996. An unusual case of hydatid disease: localization to the gluteus muscle. *Clin. Infect. Dis.* **23:**395–396.

19. Ciftcioglu, M. A., M. I. Yildirgan, M. N. Akcay, A. Reis, M. Safali, and E. Aktas. 1997. Fine needle aspiration biopsy in hepatic *Echinococcus multilocularis*. *Acta Cytol.* **41:**649–652.

20. Clavel, A., M. Varea, O. Doiz, L. Lopez, J. Quilez, F. J. Castillo, C. Rubio, and R. Gomez-Lus. 1999. Visualization of hydatid elements: comparison of several techniques. *J. Clin. Microbiol.* **37:**1561–1563.

21. Cohen, H., E. Paolillo, R. Bonifacino, B. Botta, L. Parada, P. Cabrera, K. Showden, R. Gasser, R. Tessier, L. Dibarboure, H. Wen, J. C. Allan, H. S. D. Alfaro, M. T. Rogan, and P. S. Craig. 1998. Human cystic echinococcosis in a Uruguayan community: a sonographic, serologic, and epidemiologic study. *Am. J. Trop. Med. Hyg.* **59:**620–627.

22. Coskun, I., M. Esenturk, and Y. Yoruk. 1996. The rupture of hepatic hydatid disease into the right hepatic vein and bile ducts: a case report. *Hepato-Gastroenterology* **43:**1006–1008.

23. D'Alessandro, A. 1997. Polycystic echinococcosis in tropical America: *Echinococcus vogeli* and *E. oligarthrus*. *Acta Trop.* **67:**43–65.

24. Deplazes, P., P. Alther, I. Tanner, R. C. A. Thompson, and J. Eckert. 1999. *Echinococcus multilocularis* coproantigen detection by enzyme-linked immunosorbent assay in fox, dog, and cat populations. *J. Parasitol.* **85:**115–121.

25. Dhaliwal, R. S., and M. S. Kalkat. 1997. One-stage surgical procedure for bilateral lung and liver hydatid cysts. *Ann. Thorac. Surg.* **64:**338–341.

26. Diaz, J. F., and R. H. Gilman. 2000. Sparganosis, p. 876–877. *In* G. T. Strickland (ed.), *Hunter's Tropical Medicine and Emerging Infectious Diseases*, 8th ed. The W. B. Saunders Co., Philadelphia, Pa.

27. Diaz-Recasens, J., A. Garcia-Enguidanos, I. Munoz, and R. S. de la Cuesta. 1998. Ultrasonographic appearance of an *Echinococcus* ovarian cyst. *Obstet. Gynecol.* **91:**841–842.

28. Diebold-Berger, S., H. Khan, B. Gottstein, E. Puget, J. L. Frossard, and S. Remadi. 1997. Cytologic diagnosis of isolated pancreatic alveolar hydatid disease with immunologic and PCR analyses—a case report. *Acta Cytol.* **41:**1381–1386.

29. Dinkel, A., M. von Nickisch-Rosenegk, B. Bilger, M. Merli, R. Lucius, and T. Romig. 1998. Detection of *Echinococcus multilocularis* in the definitive host: coprodiagnosis by PCR as an alternative to necropsy. *J. Clin. Microbiol.* **36:**1871–1876.

30. Dreweck, C. M., P. T. Soboslay, H. Schulz-Key, B. Gottstein, and P. Kern. 1999. Cytokine and chemokine secretion by human peripheral blood cells in response to viable *Echinococcus multilocularis* metacestode vesicles. *Parasite Immunol.* **21:**433–438.

31. Dueger, E. L., P. L. Moro, and R. H. Gilman. 1999. Oxfendazole treatment of sheep with naturally acquired hydatid disease. *Antimicrob. Agents Chemother.* **43:**2263–2267.

32. Eckert, J., and P. Deplazes. 1999. Alveolar echinococcosis in humans: the current situation in Central Europe and the need for countermeasures. *Parasitol. Today* **15:**315–319.

33. Eckert, J., and P. Deplazes. 2004. Biological, epidemiological, and clinical aspects of echinococcosis, a zoonosis of increasing concern. *Clin. Microbiol. Rev.* **17:**107–135.

34. Eckert, J., and R. C. A. Thompson. 1997. Intraspecific variation of *Echinococcus granulosus* and related species with emphasis on their infectivity to humans. *Acta Trop.* **64:**19–34.

35. Elon, J., E. Khaleel, Y. Malsha, J. Nahmias, P. Schantz, R. Sneir, R. Benismail, M. Furth, and G. Hoida. 1997. *Echinococcus granulosus*: a seroepidemiological survey in northern Israel using an enzyme-linked immunosorbent assay. *Trans. R. Soc. Trop. Med. Hyg.* **91:**529–532.

36. Erzurumlu, K., M. Sahin, M. B. Selcuk, C. Yildiz, and M. Kesim. 1996. Intracystic application of mebendazole solution in the treatment of liver hydatid disease—preliminary report of two cases. *Eur. Surg. Res.* **28:**466–470.

37. Eviliyaoglu, C., M. Yuksel, B. Gul, E. Kaptanoglu, and M. Yaman. 1998. Growth rate of multiple intracranial hydatid cysts assessed by CT from the time of embolisation. *Neuroradiology* **40:**387–389.

38. Franchi, C., B. Di Vico, and A. Teggi. 1999. Long-term evaluation of patients with hydatidosis treated with benzimidazole carbamates. *Clin. Infect. Dis.* **29:**304–309.

39. Frider, B., E. Larrieu, and M. Odriozola. 1999. Long-term outcome of asymptomatic liver hydatidosis. *J. Hepatol.* **30:**228–231.

40. Gadea, I., G. Ayala, M. T. Diego, A. Cunat, and J. G. de Lomas. 1999. Immunological diagnosis of human cystic echinococcosis: utility of discriminant analysis applied to the enzyme-linked immunoelectrotransfer blot. *Clin. Diagn. Lab. Immunol.* **6:**504–508.

41. Garcia, J. I. L., E. Alonso, J. Gonzalez-Uriarte, and D. R. Romano. 1997. Evaluation of scolicidal agents in an experimental hydatid disease model. *Eur. Surg. Res.* **29:**202–208.

42. Godot, V., S. Harraga, I. Beurton, C. Pater, M. E. Sarciron, B. Gottstein, and D. A. Vuitton. 2000. Resistance/susceptibility to *Echinococcus multilocularis* infection and cytokine profile in humans. I. Comparisons of patients with progressive and abortive lesions. *Clin. Exp. Immunol.* **121:**484–490.

43. Gokcek, C., A. Gokcek, M. A. Bayar, S. Tanrikulu, and Z. Buharali. 1997. Orbital hydatid cyst: CT and MRI. *Neuroradiology* **39**:512–515.

44. Gottstein, B. 2000. Immunodiagnosis of infections with cestodes, p. 347–373. *In* Y. H. Hui, S. A. Sattar, K. D. Murrell, W.-K. Nip, and P. S. Stanfield (ed.), *Foodborne Disease Handbook*. Marcel Dekker, Inc., New York, N.Y.

45. Grimm, F., F. E. Maly, J. A. Lü, and R. Ilano. 1998. Analysis of specific immunoglobulin G subclass antibodies for serological diagnosis of echinococcosis by a standard enzyme-linked immunosorbent assay. *Clin. Diagn. Lab. Immunol.* **5**:613–616.

46. Haag, K. L., A. Zaha, A. M. Araujo, and B. Gottstein. 1997. Reduced genetic variability within coding and noncoding regions of the *Echinococcus multilocularis* genome. *Parasitology* **115**:521–529.

47. Halezeroglu, S., M. Celik, A. Uysal, C. Senol, M. Keles, and B. Arman. 1997. Giant hydatid cysts of the lung. *J. Thorac. Cardiovasc. Surg.* **113**:712–717.

48. Heath, D. D., and B. Holcman. 1997. Vaccination against *Echinococcus* in perspective. *Acta Trop.* **67**:37–41.

49. Hegglin, D., P. I. Ward, and P. Deplazes. 2003. Anthelmintic baiting of foxes against urban contamination with *Echinococcus multilocularis*. *Emerg. Infect. Dis.* **9**:1266–1272.

50. Helbig, M., P. Frosch, P. Kern, and M. Frosch. 1993. Serological differentiation between cystic and alveolar echinococcosis by use of recombinant larval antigens. *J. Clin. Microbiol.* **31**:3211–3215.

51. Ing, M. B., P. M. Schantz, and J. A. Turner. 1998. Human coenurosis in North America: case reports and review. *Clin. Infect. Dis.* **27**:519–523.

52. Ito, A., L. Ma, P. M. Schantz, B. Gottstein, Y. H. Liu, J. J. Chai, S. K. Abdel-Hafez, N. Altintas, D. D. Joshi, M. W. Lightowlers, and Z. S. Pawlowski. 1999. Differential serodiagnosis for cystic and alveolar echinococcosis using fractions of *Echinococcus granulosus* cyst fluid (antigen B) and *E. multilocularis* protoscolex (EM18). *Am. J. Trop. Med. Hyg.* **60**:188–192.

53. Jenkins, D. J. 2005. *Echinococcus granulosus* in Australia, widespread and doing well! *Parasitol. Int.* 10 Dec. [Epub ahead of print.] **55**:S203–S206.

54. Jenkins, D. J., and C. N. Macpherson. 2003. Transmission ecology of *Echinococcus* in wild-life in Australia and Africa. *Parasitology* **127**:S63–S72.

55. Kalaitzoglou, I., A. Drevelengas, A. Petridis, and P. Palladas. 1998. Albendazole treatment of cerebral hydatid disease: evaluation of results with CT and MRI. *Neuroradiology* **40**:36–39.

56. Kanwar, J. R., S. P. Kaushik, I. M. S. Sawhney, M. S. Kamboj, S. K. Mehta, and V. K. Vinayak. 1992. Specific antibodies in serum of patients with hydatidosis recognized by immunoblotting. *J. Med. Microbiol.* **36**:46–51.

57. Keil, W., H. Pandratz, A. Szabados, and C. Baur. 1997. Sudden death caused by hydatid embolism. *Dtsch. Med. Wochenschr.* **122**:293–296.

58. Kern, P., A. Ammon, M. Kron, G. Sinn, S. Sander, L. R. Petersen, W. Gaus, and P. Kern. 2004. Risk factors for alveolar echinococcosis in humans. *Emerg. Infect. Dis.* **10**:2088–2093.

59. Kern, P., K. Bardonnet, E. Renner, H. Auer, Z. Pawlowski, R. W. Ammann, D. A. Vuitton, P. Kern, and the European Echinococcosis Registry. 2003. European echinococcosis registry: human alveolar echinococcosis, Europe, 1982–2000. *Emerg. Infect. Dis.* **9**:343–349.

60. Kern, P., P. Frosch, M. Helbig, J. G. Wechsler, S. Usadel, K. Beckh, R. Kunz, R. Lucius, and M. Frosch. 1995. Diagnosis of *Echinococcus multilocularis* infection by reverse-transcription polymerase chain reaction. *Gastroenterology* **109**:596–600.

61. Keshmiri, M., H. Baharvahdat, S. H. Fattahi, B. Davachi, R. H. Dabiri, H. Baradaran, T. Ghiasi, M. T. Rajab-imashhadi, and F. Rajabzadeh. 1999. A placebo controlled study of albendazole in the treatment of pulmonary echinococcosis. *Eur. Respir. J.* **14**:503–507.

62. Kharebov, A., J. Nahmias, and J. El-On. 1997. Cellular and humoral immune responses of hydatidosis patients to *Echinococcus granulosus* purified antigens. *Am. J. Trop. Med. Hyg.* **57**:619–625.

63. Khuroo, M. S., N. A. Wani, G. Javid, B. A. Khan, G. N. Yattoo, A. H. Shah, and S. G. Jeelani. 1997. Percutaneous drainage compared with surgery for hepatic hydatid cysts. *N. Engl. J. Med.* **337**:881–887.

64. Kini, Y., S. Shariff, and V. Nirmala. 1997. Aspiration cytology of *Echinococcus oligarthrus*—a case report. *Acta Cytol.* **41**:544–548.

65. Kornaros, S. E., and T. A. Aboulnour. 1996. Frank intrabiliary rupture of hydatid cyst: diagnosis and treatment. *J. Am. Coll. Surg.* **183**:466–470.

66. Kratzer, W., S. Reuter, K. Hirschbuehl, A. R. Ehrhardt, R. A. Mason, M. M. Haenle, P. Kern, and A. Gabelmann. 2005. Comparison of contrast-enhanced power Doppler ultrasound (Levovist) and computed tomography in alveolar echinococcosis. *Abdom. Imaging* 20 Jan. [Epub ahead of print.] **30**:286–290.

67. Kreidl, P., F. Allerberger, G. Judmaier, H. Auer, H. Aspock, and A. J. Hall. 1998. Domestic pets as risk factors for alveolar hydatid disease in Austria. *Am. J. Epidemiol.* **147**:978–981.

68. Kruse, H., A. M. Kirkemo, and K. Handeland. 2004. Wildlife as source of zoonotic infections. *Emerg. Infect. Dis.* **12**:2067–2072.

69. Laghi, A., A. Teggi, P. Pavone, C. Franchi, F. De Rosa, and R. Passariello. 1998. Intrabiliary rupture of hepatic hydatid cysts: diagnosis by use of magnetic resonance cholangiography. *Clin. Infect. Dis.* **26**:1465–1467.

70. Laglera, S., M. A. Garcia-Enguita, F. Martinez-Gutierrez, J. P. Ortega, A. Gutierrez-Rodriguez, and A. Urieta. 1997. A case of cardiac hydatidosis. *Br. J. Anaesth.* **79**:671–673.

71. Lavikainen, A., M. J. Lehtinen, T. Meri, V. Hirvela-Koski, and S. Meri. 2003. Molecular genetic characterization of the Fennoscandian cervid strain, a new genotypic group (G10) of *Echinococcus granulosus*. *Parasitology* **127**:207–215.

72. Lewall, D. B., and P. Nyak. 1998. Hydatid cysts of the liver: two cautionary signs. *Br. J. Radiol.* **71**:37–41.

73. Lightowlers, M. W., O. Jensen, E. Fernandez, J. A. Iriarte, D. J. Woollard, C. G. Cauci, D. J. Jenkins, and D. D. Heath. 1999. Vaccination trials in Australia and Argentina confirm the effectiveness of the EG95 hydatid vaccine in sheep. *Int. J. Parasitol.* **29**:531–534.

74. Lightowlers, M. W., S. B. Lawrence, C. G. Cauci, J. Young, M. J. Ralston, D. Maas, and D. D. Heath. 1996. Vaccination against hydatidosis using a defined recombinant antigen. *Parasite Immunol.* **18**:457–462.

75. Liu, D., M. W. Lightowlers, and M. D. Rickard. 1992. Evaluation of a monoclonal antibody-based competition ELISA for the diagnosis of human hydatidosis. *Parasitology* **104**:357–361.

76. Locasto, A., S. Salerno, M. Grisanti, and G. Mastrandrea. 1997. Hydatid cyst of the liver communicating with the left colon. *Br. J. Radiol.* **70**:650–651.

77. Lopez-Rios, F., A. Perez-Barrios, and P. P. de Agustin. 1997. Primary cardiac hydatid cyst in a child—cytologic diagnosis of a case. *Acta Cytol.* **41**:1387–1390.

78. Luchi, S., A. Vincenti, F. Messina, M. Parenti, A. Scasso, and A. Campatelli. 1997. Albendazole treatment of human hydatid disease. *Scand. J. Infect. Dis.* **29**:165–167.

79. Ma, L., A. Ito, Y. H. Liu, Y. Q. Yao, D. G. Yu, and Y. T. Chen. 1997. Alveolar echinococcosis: Em2(plus)-ELISA™ and Em18-Western blots for follow-up after treatment with albendazole. *Trans. R. Soc. Trop. Med. Hyg.* **91**:476–478.

80. Malczewski, A., B. Rocki, A. Ramisz, and J. Eckert. 1995. *Echinococcus multilocularis* (Cestoda), the causative agent of alveolar echinococcosis in humans: first record in Poland. *J. Parasitol.* **81**:318–321.

81. Malgor, R., N. Nonaka, I. Basmadjian, H. Sakai, B. Carambula, Y. Oku, C. Carmona, and M. Kamiya. 1998. Coproantigen detection in dogs experimentally and naturally infected with *Echinococcus granulosus* by a monoclonal antibody-based enzyme-linked immunosorbent assay. *Int. J. Parasitol.* **27**:1605–1612.

82. Mamuti, W., Y. Sako, M. Nakao, N. Xiao, K. Nakaya, Y. Ishikawa, Y. Hiroshi, M. W. Lightowlers, and A. Ito. 2005. Recent advances in characterization of *Echinococcus* antigen B. *Parasitol. Int.* 14 Dec. [Epub ahead of print.] **55**:S57–S62.

83. Mawhorter, S., B. Temeck, R. Chang, H. Pass, and T. Nash. 1997. Nonsurgical therapy for pulmonary hydatid cyst disease. *Chest* **112**:1432–1436.

84. Maxson, A. D., T. M. Wachira, E. E. Zeyhle, A. Fine, T. W. Mwangi, and G. Smith. 1997. The use of ultrasound to study the prevalence of hydatid cysts in the right lung and liver of sheep and goats in Turkana, Kenya. *Int. J. Parasitol.* **26**:1335–1338.

85. McManus, D. P., L. Zhang, L. J. Castrodale, T. H. Le, M. Pearson, and D. Blair. 2002. Short report: molecular genetic characterization of an unusually severe case of hydatid disease in Alaska caused by the cervid strain of *Echinococcus granulosus*. *Am. J. Trop. Med. Hyg.* **67**:296–298.

86. Men, S., C. Yucesoy, T. R. Edguer, and B. Hekimoglu. 1999. Intraaortic growth of hydatid cysts causing occlusion of the aorta and of both iliac arteries: case report. *Radiology* **213**:192–194.

87. Mistrello, G., M. Gentili, P. Falagiani, D. Roncarolo, G. Riva, and M. Tinelli. 1995. Dot immunobinding assay as a new diagnostic test for human hydatid disease. *Immunol. Lett.* **47**:79–85.

88. Mooraki, A., M. H. Rahbar, and B. Bastani. 1996. Spontaneous systemic anaphylaxis as an unusual presentation of hydatid cyst: report of two cases. *Am. J. Trop. Med. Hyg.* **55**:302–303.

89. Moro, P. L., N. Bonifacio, R. H. Gilman, L. Lopera, B. Silva, R. Takumoto, M. Verastegui, and L. Cabrera. 1999. Field diagnosis of *Echinococcus granulosus* infection among intermediate and definitive hosts in an endemic focus of human cystic echinococcosis. *Trans. R. Soc. Trop. Med. Hyg.* **93**:611–615.

90. Moro, P. L., J. McDonald, R. H. Gilman, B. Silva, M. Verastegui, V. Malqui, G. Lescano, N. Falcon, G. Montes, and H. Bazalar. 1998. Epidemiology of *Echinococcus granulosus* infection in the central Peruvian Andes. *Bull. W. H. O.* **75**:553–561.

91. Murthy, R., S. G. Honavar, G. K. Vemuganti, M. Naik, and S. Burman. 2005. Polycystic echinococcosis of the orbit. *Am. J. Ophthalmol.* **140**:561–563.

92. Nonaka, N., M. Iida, K. Yagi, T. Ito, H. K. Ooi, Y. Oku, and M. Kamiya. 1997. Time course of coproantigen excretion in *Echinococcus multilocularis* infections in foxes and an alternative definitive host, golden hamsters. *Int. J. Parasitol.* **26**:1271–1278.

93. Nonaka, N., H. Tsukada, N. Abe, Y. Oku, and M. Kamiya. 1998. Monitoring of *Echinococcus multilocularis* infection in red foxes in Shiretoko, Japan, by coproantigen detection. *Parasitology* **117**:193–200.

94. Oguzkaya, F., Y. Akcali, C. Kahraman, N. Emirogullari, M. Bilgin, and A. Sahin. 1997. Unusually located hydatid cysts: intrathoracic but extrapulmonary. *Ann. Thorac. Surg.* **64**:334–337.

95. Oztek, I., H. Baloglu, D. Demirel, A. Saygi, K. Balkanli, and B. Arman. 1997. Cytologic diagnosis of complicated pulmonary unilocular cystic hydatidosis—a study of 131 cases. *Acta Cytol.* **41**:1159–1166.

96. Parija, S. C., P. T. Ravinder, and M. Shariff. 1996. Detection of hydatid antigen by co-agglutination in fluid samples from hydatid cysts. *Trans. R. Soc. Trop. Med. Hyg.* **90**:255–256.

97. Paul, M., and J. Stefaniak. 1997. Detection of specific *Echinococcus granulosus* antigen 5 in liver cyst bioptate from human patients. *Acta Trop.* **64**:65–77.

98. Pawlowski, Z. S., J. Eckert, D. A. Vuitton, R. W. Ammann, P. Kern, P. S. Craig, F. K. Dar, F. De Rosa, C. Filice, B. Gottstein, F. Grimm, C. N. L. Macpherson, N. Sato, T. Todorov, J. Uchino, W. von Sinner, and H. Wen. 2001. Echinococcosis in humans: clinical aspects, diagnosis and treatment, p. 20–66. *In* J. Eckert, M. A. Gemmell, F.-X. Meslin, and Z. S. Pawlowski (ed.), *WHO/OIE Manual on Echinococcosis in Humans and Animals: a Public Health Problem of Global Concern*. World Organisation for Animal Health, Paris, France.

99. Persat, F., C. Vincent, M. Mojon, and A. F. Petavy. 1991. Detection of antibodies against glycolipids of *Echinococcus multilocularis* metacestodes in sera of patients with alveolar hydatid disease. *Parasite Immunol.* **13**:379–389.

100. Poretti, D., E. Felleisen, F. Grimm, M. Pfister, F. Teuscher, C. Zuercher, J. Reichen, and B. Gottstein. 1999. Differential immunodiagnosis between cystic hydatid disease and other cross-reactive pathologies. *Am. J. Trop. Med. Hyg.* **60**:193–198.

101. Rausch, R. L., and A. D'Alessandro. 2002. The epidemiology of echinococcosis caused by *Echinococcus oligarthrus* and *E. vogeli* in the Neotropics, p. 107–113. *In* P. Craig and Z. Pawlowski (ed.), *Cestode Zoonoses: Echinococcosis*

and Cysticercosis, an Emergent and Global Problem. IOS Press, Amsterdam, The Netherlands.

102. **Ravinder, P. T., and S. C. Parija.** 1997. Countercurrent immunoelectrophoresis test for detection of hydatid antigen in the fluid from hydatid cysts: a preliminary report. *Acta Trop.* **66:**169–173.

103. **Reuter, S., A. Buck, O. Grebe, K. Nussie-Kugele, P. Kern, and B. J. Manfras.** 2003. Salvage treatment with amphotericin B in progressive human alveolar echinococcosis. *Antimicrob. Agents Chemother.* **47:**3586–3591.

104. **Reuter, S., B. Jensen, K. Buttenschoen, W. Kratzer, and P. Kern.** 2000. Benzimidazoles in the treatment of alveolar echinococcosis: a comparative study and review of the literature. *J. Antimicrob. Chemother.* **46:**451–456.

105. **Rigano, R., E. Profumo, S. Ioppolo, S. Notargiacomo, E. Ortona, A. Teggi, and A. Siracusano.** 1995. Immunological markers indicating the effectiveness of pharmacological treatment in human hydatid disease. *Clin. Exp. Immunol.* **102:**281–285.

106. **Rogan, M. T., and P. S. Craig.** 2002. Immunological approaches for transmission and epidemiological studies in cestode zoonoses—the role of serology in human infection, p. 135–145. *In* P. Craig and Z. Pawlowski (ed.), *Cestode Zoonoses: Echinococcosis and Cysticercosis, an Emergent and Global Problem.* IOS Press, Amsterdam, The Netherlands.

107. **Romig, T., A. Dinkel, and U. Mackenstedt.** 2005. The present situation of echinococcosis in Europe. *Parasitol. Int.* 10 Dec. [Epub ahead of print.] **55:**S187–S191.

108. **Romig, T., W. Kratzer, P. Kimmig, M. Frosch, W. Gaus, W. A. Flegel, B. Gottstein, R. Lucius, and P. Kern.** 1999. An epidemiologic survey of human alveolar echinococcosis in southwestern Germany. *Am. J. Trop. Med. Hyg.* **61:**566–573.

109. **Rosenzvit, M. C., L. H. Zhang, L. Kamenetzky, S. G. Canova, E. A. Guarnera, and D. P. McManus.** 1999. Genetic variation and epidemiology of *Echinococcus granulosus* in Argentina. *Parasitology* **118:**523–530.

110. **Sadjjadi, S. M.** 2005. Present situation of echinococcosis in the Middle East and Arabic North Africa. *Parasitol. Int.* 5 Dec. [Epub ahead of print]

111. **Saez, J., P. Pinto, W. Apt, and I. Zulantay.** 2001. Cystic echinococcosis of the tongue leading to diagnosis of multiple localizations. *Am. J. Trop. Med. Hyg.* **65:**338–340.

112. **Sage, A. M., T. M. Wachira, E. E. Zeyhle, E. P. Weber, E. Njoroge, and G. Smith.** 1998. Evaluation of diagnostic ultrasound as a mass screening technique for the detection of hydatid cysts in the liver and lung of sheep and goats. *Int. J. Parasitol.* **28:**349–353.

113. **Salama, H., M. F. Abdel-Wahab, and G. T. Strickland.** 1996. Diagnosis and treatment of hepatic hydatid cysts with the aid of echo-guided percutaneous cyst puncture. *Clin. Infect. Dis.* **21:**1372–1376.

114. **Salinas-Lopez, N., F. Jimenez-Guzman, and A. Cruz-Reyes.** 1996. Presence of *Echinococcus oligarthrus* (Diesing, 1863) Luhe, 1910 in *Lynx rufus texensis* Allen, 1895 from San Fernando, Tamaulipas State in northeast Mexico. *Int. J. Parasitol.* **26:**793–796.

115. **Sarciron, M. E., S. Bresson-Hadni, M. Mercier, P. Lawton, C. Duranton, D. Lenys, A. F. Petavy, and D. A. Vuitton.** 1997. Antibodies against *Echinococcus multilocularis*

alkaline phosphatase as markers for the specific diagnosis and the serological monitoring of alveolar echinococcosis. *Parasite Immunol.* **19:**61–68.

116. **Sarciron, M. E., N. Walchshofer, S. Walbaum, C. Arsac, J. Descotes, A. F. Petavy, and J. Paris.** 1997. Increases in the effects of albendazole on *Echinococcus multilocularis* metacestodes by the dipeptide methyl ester (Phe-Phe-OMe). *Am. J. Trop. Med. Hyg.* **56:**226–230.

117. **Saygi, A., I. Oztek, M. Guder, F. Sungun, and B. Arman.** 1997. Value of fibreoptic bronchoscopy in the diagnosis of complicated pulmonary unilocular cystic hydatidosis. *Eur. Respir. J.* **10:**811–814.

118. **Schelling, U., W. Frank, R. Will, T. Romig, and R. Lucius.** 1997. Chemotherapy with praziquantel has the potential to reduce the prevalence of *Echinococcus multilocularis* in wild foxes (*Vulpes vulpes*). *Ann. Trop. Med. Parasitol.* **91:**179–186.

119. **Schmid, M., G. Pendl, H. Samonigg, G. Ranner, S. Eustacchio, and E. C. Reisinger.** 1998. Gamma knife radiosurgery and albendazole for cerebral alveolar hydatid disease. *Clin. Infect. Dis.* **26:**1379–1382.

120. **Schmid, M., H. Samonigg, H. Stoger, H. Auer, M. H. J. Sternthal, M. Wilders-Truschnig, and E. C. Reisinger.** 1995. Use of interferon gamma and mebendazole to stop the progression of alveolar hydatid disease: case report. *Clin. Infect. Dis.* **20:**1543–1546.

121. **Sebbag, H., C. Partensky, J. Roche, T. Ponchon, and A. Martins.** 1999. Recurrent pancreatitis due to rupture of a solitary hydatid cyst of the pancreas into the main pancreatic duct. *Gastroenterol. Clin. Biol.* **23:**793–794.

122. **Shambesh, M. A., P. S. Craig, C. N. L. Macpherson, M. T. Rogan, A. M. Gusbi, and E. F. Echtuish.** 1999. An extensive ultrasound and serologic study to investigate the prevalence of human cystic echinococcosis in northern Libya. *Am. J. Trop. Med. Hyg.* **60:**462–468.

123. **Siles-Lucas, M., R. S. J. Felleisen, A. Hemphill, W. Wilson, and B. Gottstein.** 1998. Stage-specific expression of the 14-3-3 gene in *Echinococcus multilocularis*. *Mol. Biochem. Parasitol.* **91:**281–293.

124. **Siles-Lucas, M. M., and B. Gottstein.** 2001. Molecular tools for the diagnosis of cystic and alveolar echinococcosis. *Trop. Med. Int. Health* **6:**463–475.

125. **Sloan, L., S. Schneider, and J. Rosenblatt.** 1995. Evaluation of enzyme-linked immunoassay for serological diagnosis of cysticercosis. *J. Clin. Microbiol.* **33:**3124–3128.

126. **Smego, R. A., and P. Sebanego.** 2005. Treatment options for hepatic cystic echinococcosis. *Int. J. Infect. Dis.* **9:**69–76.

127. **Sréter, T., Z. Széll, Z. Egyed, and I. Varga.** 2003. *Echinococcus multilocularis*: an emerging pathogen in Hungary and Central Eastern Europe? *Emerg. Infect. Dis.* **9:**384–386.

128. **Strauss, M., J. Schmidt, H. Boedeker, H. Zirngibl, and K. W. Jauch.** 1999. Laparoscopic partial pericystectomy of *Echinococcus granulosus* cysts in the liver. *Hepato-Gastroenterology* **46:**2540–2544.

129. **Tackmann, K., U. Loschner, H. Mix, C. Staubach, H. H. Thulke, and F. J. Conraths.** 1998. Spatial distribution patterns of *Echinococcus multilocularis* (Leukart 1863) (Cestoda: Cyclophyllidea: Taeniidae) among red foxes in an endemic focus in Brandenburg, Germany. *Epidemiol. Infect.* **120:**101–109.

130. Todorov, T., and V. Boeva. 1999. Human echinococcosis in Bulgaria: a comparative epidemiological analysis. *Bull. W. H. O.* **77:**110–118.

131. Torgerson, P. R., B. Oguljahan, A. E. Miminov, R. R. Karaeva, O. T. Kuttubaev, M. Aminjanov, and B. Shaikenov. 2005. Present situation of cystic echinococcosis in Central Asia. *Parasitol. Int.* 14 Dec. [Epub ahead of print.] **55:**S207–S212.

132. Tsitouridis, J., A. S. Dimitriadis, and E. Kazana. 1997. MR in cisternal hydatid cysts. *Am. J. Neuroradiol.* **18:**1586–1587.

133. Turgut, M. 1997. Hydatid disease of the spine: a survey study from Turkey. *Infection* **25:**221–226.

134. Turgut, M., K. Benli, and M. Eryilmaz. 1997. Secondary multiple intracranial hydatid cysts caused by intracerebral embolism of cardiac echinococcosis: an exceptional case of hydatidosis—case report. *J. Neurosurg.* **86:**714–718.

135. Varcasia, A., S. Canu, M. W. Lightowlers, A. Scala, and G. Garippa. 2005. Molecular characterization of *Echinococcus granulosus* strains in Sardinia. *Parasitol. Res.* 2 Dec. [Epub ahead of print]

136. Verastegui, M., P. Moro, A. Guevara, T. Rodriguez, E. Miranda, and R. H. Gilman. 1992. Enzyme-linked immunoelectrotransfer blot test for diagnosis of human hydatid disease. *J. Clin. Microbiol.* **30:**1557–1561.

137. Viel, J. F., P. Giraudoux, V. Abrial, and S. Bresson-Hadni. 1999. Water vole (*Arvicola terrestris* Scherman) density as risk factor for human alveolar echinococcosis. *Am. J. Trop. Med. Hyg.* **61:**559–565.

138. Vuitton, D. A. 2004. Echinococcosis and allergy. *Clin. Rev. Allergy Immunol.* **26:**93–104.

139. Vutova, K., G. Mechkov, P. Vachkov, R. Petkov, P. Georgiev, S. Handjiev, A. Ivanov, and T. Todorov. 1999. Effect of mebendazole on human cystic echinococcosis: the role of dosage and treatment duration. *Ann. Trop. Med. Parasitol.* **93:**357–365.

140. Wei, J., F. Cheng, Q. Qun, Nurbek, S. D. Xu, L. F. Sun, X. K. Han, Muhan, L. L. Han, Irixiati, P. Jie, K. J. Zhang, Islayin, and J. J. Chai. 2005. Epidemiological evaluations of the efficacy of slow-released praziquantel-medicated bars for dogs in the prevention and control of cystic echinococcosis in man and animals. *Parasitol. Int.* **54:**231–236.

141. Wiwanitkit, V. 2005. A review of human sparganosis in Thailand. *Int. J. Infect. Dis.* **9:**312–316.

142. Xiao, N., W. Mamuti, H. Yamasaki, Y. Sako, M. Nakao, K. Nakaya, B. Gottstein, P. M. Schantz, M. W. Lightowlers, P. S. Craig, and A. Ito. 2003. Evaluation of use of recombinant Em18 and affinity-purified Em18 for serological differentiation of alveolar echinococcosis from cystic echinococcosis and other parasitic infections. *J. Clin. Microbiol.* **41:**3351–3353.

143. Yang, Y. R., X. Z. Liu, D. A. Vuitton, B. Bartholomot, Y. H. Wang, A. Ito, P. S. Craig, and D. P. McManus. 2005. Simultaneous alveolar and cystic echinococcosis of the liver. *Trans. R. Soc. Trop. Med. Hyg.* 3 Dec. [Epub ahead of print]

144. Yattoo, G. N., M. S. Khuroo, S. A. Zargar, F. A. Bhat, and B. A. Sofi. 1999. Case report: percutaneous drainage of the pancreatic head hydatid cyst with obstructive jaundice. *J. Gastroenterol. Hepatol.* **14:**931–934.

145. Yavuz, G., S. Emir, E. Unal, N. Tacyildiz, H. Gencgonul, A. Yagmurlu, S. Fitoz, and S. Erekul. 2004. Coexistence of Hodgkin lymphoma and cyst hydatic disease of the liver. *Pediatr. Hematol. Oncol.* **21:**95–99.

146. Yimaz, U., B. Sakin, S. Boyacioglu, U. Saritas, T. Cumhar, and M. Akoglu. 1998. Management of post-operative biliary strictures secondary to hepatic hydatid disease by endoscopic stenting. *Hepato-Gastroenterology* **45:**65–69.

147. Yoon, K. C., M. S. Seo, S. W. Park, and Y. G. Park. 2004. Eyelid sparganosis. *Am. J. Ophthalmol.* **138:**873–875.

148. Zahawi, H. M., O. K. Hameed, and A. A. Abalkhail. 1999. The possible role of age of the human host in determining the localization of hydatid cysts. *Ann. Trop. Med. Parasitol.* **93:**621–627.

149. Zhang, L., A. Eslami, S. H. Hosseini, and D. P. McManus. 1998. Indication of the presence of two distinct strains of *Echinococcus granulosus* in Iran by mitochondrial DNA markers. *Am. J. Trop. Med. Hyg.* **59:**171–174.

150. Zhang, W., J. Li, and D. P. McManus. 2003. Concepts in immunology and diagnosis of hydatid disease. *Clin. Microbiol. Rev.* **16:**18–36.

15

Intestinal Trematodes

The intestinal trematodes, or flukes, are parasites of vertebrates; they are dorsoventrally flattened and hermaphroditic and require one or more intermediate hosts. The adult worms vary in size from the barely visible (*Heterophyes heterophyes*) to the very large (*Fasciolopsis buski*). To complete the life cycle, specific species of intermediate hosts must be available for trematode development. All of the intestinal trematodes require a freshwater snail to serve as an intermediate host. These infections are food borne and are emerging as a major public health problem, with more than 50 million people being infected (Algorithm 15.1) (17). The true economic impact of these infections is difficult to assess. Intestinal trematode infections can be found in Southeast Asia, the Far East, the Middle East, and North Africa.

Eggs deposited by the adult worms are passed to the outside in the feces (Figure 15.1). All intestinal trematode eggs have an operculum, or "lid," from which the miracidium larva can escape. Eggs hatch in freshwater, and these larvae must find their way into the snail (intermediate host) through penetration of the snail tissues; in some cases, the snail ingests the eggs before they hatch. A series of developmental stages occur within the snail, eventually producing cercariae, which are released from the snail. These cercariae then encyst on aquatic plant material or encyst in the tissues of freshwater mollusks or fish and become metacercariae. When the plants or mollusks are ingested raw, the metacercariae then infect the human host, where they excyst in the small intestine and develop into mature adult trematodes within the intestinal tract.

Adult worms vary in shape and size and can be distinguished from one another (Table 15.1; Figure 15.2). The adult worms are hermaphroditic, containing both male and female reproductive systems.

Fasciolopsis buski

Fasciolopsiasis was first noted by Busk in 1843, when worms were detected in the duodenum of a deceased East Indian sailor (2). *F. buski* has also been known as the giant intestinal fluke and is one of the largest parasites to infect humans, measuring 20 to 75 mm in length, 8 to 20 mm in width, and 0.5 to 3 mm in thickness.

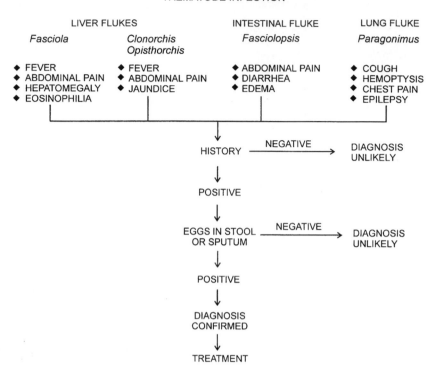

TREMATODE INFECTION

Algorithm 15.1   Trematode infection.

Life Cycle and Morphology

The adult worms live in the small intestine of pigs and humans, where the worms lay unembryonated eggs that are then passed from the intestinal lumen with the feces. The eggs are ellipsoidal, operculate, and yellow-brown. They measure 130 to 140 μm by 80 to 85 μm, with the operculum found at the more pointed end of the transparent eggshell (Figures 15.3 and 15.4; Table 15.1). Depending on the temperature, the eggs embryonate within 3 to 7 weeks. Once in the water, the mature miracidium hatches from the egg and tries to locate the appropriate snail species to infect (*Segmentina* and *Hippeutis* spp.). Once the miracidium has penetrated the soft tissues of the snail (first intermediate host), it begins to develop into a first-generation sporocyst. The sporocyst is an elongated sac, without distinct internal structures, in which germ balls proliferate. These germ balls develop into rediae that contain a mouth, pharynx, blind cecum, and birth pore. Within the rediae, the germ balls again proliferate, developing into cercariae. Unlike some of the other intestinal trematodes, *Fasciolopsis* rediae develop a second generation of rediae before forming cercariae. On reaching

Figure 15.1   Life cycle of intestinal trematodes.

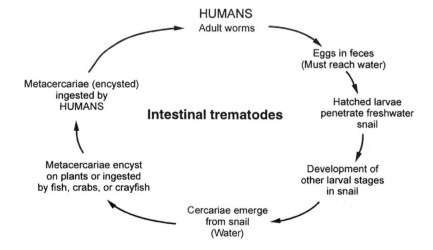

Table 15.1 Characteristics of intestinal trematodes

Species	Distribution	Agent of infection	Reservoir host	Egg		Comments
				Size (μm)	Morphology	
Fasciolopsis buski	Far East	Water chestnut, bamboo shoots, water caltrop	Dogs, pigs, rabbits	130–140 by 80–85	Unembryonated operculated	The less mature the egg, the more difficult it may be to see the actual operculum—it blends into the shell outline, and the "breaks" in the shell may be difficult to identify. This egg has no opercular shoulders, and so it is difficult to see where the operculum "breaks" in the shell actually occur.
Echinostoma ilocanum	Far East	Mollusks	Rats, dogs	86–116 by 59–69	Unembryonated operculated	Although *E. ilocanum* eggs are considerably smaller than *F. buski* eggs, some strains produce eggs that overlap in size; the operculum may be difficult to see (as indicated above for *F. buski*)
Heterophyes heterophyes	Far East, Middle East	Freshwater fish	Fish-eating mammals	27–30 by 15–17	Embryonated operculated with very subtle opercular shoulders; no "seated" operculum	These eggs have very inconspicuous opercular shoulders and, unlike *C. sinensis*, lack the "seated" operculum and knob at the abopercular end. *H. heterophyes* eggs are some of the smallest trematode eggs seen and may be missed using the low power of the microscope, and high dry power may be required for identification.
Metagonimus yokogawai	Far East, former USSR, Israel, Spain	Freshwater fish	Fish-eating mammals	26–28 by 15–17	Embryonated operculated with minimal opercular shoulders; no "seated" operculum	These eggs have inconspicuous opercular shoulders but a more obvious operculum than *H. heterophyes* eggs and are some of the smallest trematode eggs seen. They do not really have the "seated" operculum and knob at the abopercular end like those seen in *C. sinensis* eggs. They may be missed using the low power of the microscope, and high dry power may be required for identification.
Gastrodiscoides hominis	Far East, Middle East, former USSR	Freshwater fish	Pigs, deer mice, rats	60–70 by 150	Unembryonated operculated	The egg of *G. hominis* tends to be more slender than that of *F. buski*, but they are very much alike; this egg has no opercular shoulders, and so it is difficult to see where the operculum "breaks" in the shell actually occur.

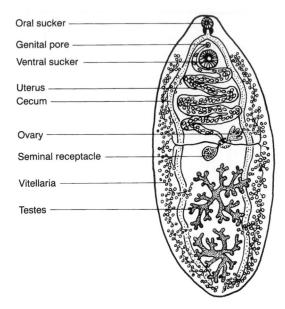

Oral sucker
Genital pore
Ventral sucker
Uterus
Cecum
Ovary
Seminal receptacle
Vitellaria
Testes

Figure 15.2 Diagram of a general intestinal/liver trematode. (Illustration by Sharon Belkin.)

maturity, the cercariae escape from the snail host into the water. The cercariae encyst on water plants such as water caltrops, water chestnuts, and water bamboo, where they develop into metacercariae in approximately 4 weeks. Humans become infected by ingesting the raw or under-cooked plants containing the metacercariae. The metacer-cariae excyst, attach to the duodenal or jejunal mucosa, and develop into adult worms within 3 months (Figure 15.5). The adult life span seldom exceeds 6 months to a year (2, 11, 12).

F. buski is the largest of the intestinal trematodes and attaches to the duodenal and jejunal walls (Figure 15.6). The adult worms are fleshy, dark red, and elongate-ovoid and have no cephalic cone structures like that seen in the liver fluke, *Fasciola hepatica*. The large egg tends to be yellowish brown, has a clear, thin shell with a small operculum but no opercular shoulders, and is nearly identical to that of *F. hepatica* and *F. gigantica*. The less

Figure 15.3 *Fasciolopsis buski* egg. (Illustration by Nobuko Kitamura.)

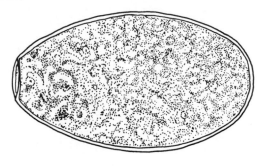

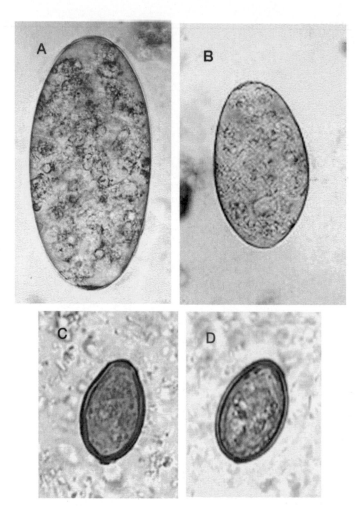

Figure 15.4 Intestinal trematode eggs. (A) *Fasciolopsis buski*; (B) *Echinostoma ilocanum*; (C) *Heterophyes heterophyes*; (D) *Metagonimus yokogawai*. Note that the eggs in panels C and D are shown at a higher magnification to demonstrate very minor differences; the operculum of *H. heterophyes* is less pronounced than that of *Clonorchis sinensis* (see chapter 16), and the operculum outline of *M. yokogawai* is the least obvious of the three small trematode eggs that are often confused (*C. sinensis*, *H. heterophyes*, and *M. yokogawai*).

mature the egg, the more difficult it is to see the operculum "breaks" in the shell outline.

Clinical Disease

In light infections, the adults inhabit the duodenum and jejunum; in heavy infections, they may be found in the stomach and most of the intestinal tract. The attachment of worms to the mucosal wall produces local inflammation with hypersecretion of mucus, hemorrhage, ulceration, and possible abscess formation. In heavy in-fections, the worms may cause bowel obstruction, acute ileus, and absorption of toxic or allergic worm metabo-lites, producing general edema and ascites. Edema of

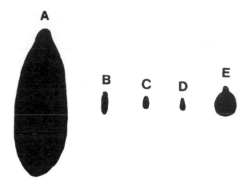

Figure 15.5 Outline of the sizes of adult intestinal trematodes. (A) *Fasciolopsis buski*; (B) *Echinostoma ilocanum*; (C) *Heterophyes heterophyes*; (D) *Metagonimus yokogawai*; (E) *Gastrodiscoides hominis*.

the face may occur as a result of hypoalbuminemia secondary to malabsorption or protein-losing enteropathy. Occasionally, vitamin B_{12} absorption is impaired, with resulting low vitamin B_{12} levels. A marked eosinophilia and leukocytosis are commonly seen. Few symptoms are associated with light infections, but in heavier infections, the patient may experience abdominal pain and

Figure 15.6 *Fasciolopsis buski* adult worms. (Left) Unpreserved adult worm (fleshy, dark red, elongate-ovoid with no cephalic cone as seen in *F. hepatica*; the adult *F. buski* worms measure 20 to 75 mm in length by 8 to 20 mm in width, and 0.5 to 3 mm in thickness); (right) stained and flattened adult worm (from anterior to posterior characterized by a large ventral sucker, coiled uterus, branched ovary, and two branched testes filling most of the posterior section of the worm; the lateral fields are filled with vitellaria).

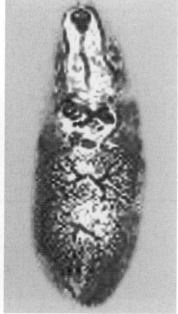

diarrhea. In heavy infections, the stools are profuse and yellow-green and contain increased amounts of undigested food, suggesting a malabsorption process. The symptoms may be confused with those of giardiasis or peptic ulcer or with other causes of bowel obstruction. Depending on the worm burden, the infection can be associated with severe cachexia and prostration and can lead to death (13).

Diagnosis

The diagnosis is suggested by the clinical features in areas where infections are endemic and is confirmed by detecting the eggs in the stool. The zinc sulfate flotation concentration is not recommended for concentrating trematode eggs because of the operculum and the fragility of the eggshell. The high specific gravity of the zinc sulfate causes the operculum to "pop" open, the eggshell fills with fluid, and the egg sinks to the bottom of the tube, where it may appear to be very distorted. Therefore, when this concentration procedure is used, both the upper flotation layer and the sediment must be examined. The formalin-ethyl acetate sedimentation concentration is recommended for routine fecal concentration. The sedimented material can be examined with or without iodine. The eggs of *Echinostoma ilocanum*, *Fasciola hepatica*, *F. buski*, and *Gastrodiscoides hominis* are similar in size and shape; therefore, an exact identification cannot be made from examining the eggs. It is possible to detect adult worms in the stool in heavy infections when they lose their ability to remain attached to the intestinal mucosa.

KEY POINTS—LABORATORY DIAGNOSIS

Fasciolopsis buski

1. The eggs are found in the stool; rarely, adult worms are found during heavy infections.
2. The eggs of *F. buski*, *E. ilocanum*, *G. hominis*, *F. hepatica*, and *F. gigantica* appear almost identical and are very difficult to differentiate from one another. When passed, the eggs do not contain mature larvae but instead contain undifferentiated embryos.
3. The formalin-ethyl acetate sedimentation concentration is recommended for egg recovery; operculated eggs do not float in the zinc sulfate flotation concentration method.
4. The less mature the egg, the more difficult it may be to see the actual operculum—it blends into the shell outline, and the "breaks" in the shell may be difficult to identify. This egg has no opercular shoulders; therefore, it is difficult to see where the operculum "breaks" in the shell actually occur.

Treatment

Although different drugs have been used for therapy, the drug of choice is praziquantel, an isoquinoline derivative that is administered orally (1). The drug is well tolerated; however, there may be some side effects, including abdominal pain, headache, dizziness, nausea, drowsiness, pruritus, and myalgia. These side effects usually disappear within 48 h but may be more pronounced in heavily infected individuals. The drug is administered as a single 15-mg/kg dose after the evening meal or before going to bed. A second regimen consists of 75 mg/kg/day in three doses for 1 day. Apparently, children tend to tolerate the drug better than adults. Contraindications for use include ocular cysticercosis, cerebral paragonimiasis, pregnancy, and physically demanding tasks where dizziness or drowsiness might put the patient at risk.

An alternative drug is niclosamide (Niclocide), a salicylamide derivative; alcohol should be avoided during treatment. The drug may be given for 1 or 2 days at 40 mg/kg/day (up to a maximum of 4 g) (3). This drug is minimally absorbed from the gastrointestinal tract, and side effects include nausea, vomiting, diarrhea, and abdominal pain (see chapter 25).

Epidemiology and Prevention

F. buski reservoir hosts include dogs, pigs, and rabbits. The infection is commonly found in Bangladesh, Cambodia, central and southern China, India, Indonesia, Laos, Malaysia, Pakistan, Taiwan, Thailand, and Vietnam; it has also been reported from Japan. Drainage of farm waste, use of manure for cultivation, and defecation in or near ponds or lakes that contain snails from the family Planorbidae (*Gyraulus*, *Polypylis*, *Segmentina*, or *Hippeutis* spp.), with water plants acting as vectors, allow the life cycle of the worm to continue. Metacercariae encyst on freshwater vegetation, such as water chestnuts, bamboo shoots, or water caltrops, and the infection is acquired when these infested plants are consumed raw or the outer coat is peeled off the nut with the teeth, resulting in accidental ingestion.

To prevent the infection, plants should be cooked or immersed in boiling water for a few seconds before they are eaten or peeled. In areas of endemicity, the use of unsterilized night soil for fertilizer should be prohibited. If these safeguards were used, the risk of infection would be considerably decreased.

Fasciolopsiasis remains a public health problem despite changes in eating habits, agricultural practices, health education, industrialization, and environmental approaches. The disease occurs focally and is most prevalent in school-age children. In areas of endemicity, the prevalence of infection in children ranges from 57% in mainland China to 25% in Taiwan and from 50% in Bangladesh and 60% in India to 10% in Thailand.

Control programs are not fully successful because of long-standing traditions of eating raw aquatic plants and using untreated water. Control measures are also impacted by social and economic factors, an expanding free-food market, a lack of sufficient food inspection and sanitation, and declining economic conditions (6).

Echinostoma ilocanum

A number of species of echinostomes have been reported to infect humans. Most of the species are found in oriental countries; *E. ilocanum* is the most important species. The infection was first noted by Garrison in 1907 (2).

Life Cycle and Morphology

Adult worms are attached to the mucosal wall of the small intestine and lay eggs, which are passed from the intestinal lumen in the feces (Figure 15.7). The eggs are immature, ellipsoidal, yellow-brown, and operculated. They measure 86 to 116 μm by 58 to 69 μm (Table 15.1; Figure 15.4). The miracidia take 1 to 2 weeks to mature in the environment before they hatch from the eggs and infect the snail intermediate host. Two or more generations

Figure 15.7 *Echinostoma ilocanum* adult worms. (Upper) Stained adult worm characterized by having a circumoral disk with a crown of spines surrounding the small oral sucker. (Lower) Enlarged image of the crown of spines (surrounding the oral sucker) and the large ventral sucker. The living worm is reddish gray and measures 2.5 to 6.5 mm in length by 1 to 1.35 mm in width.

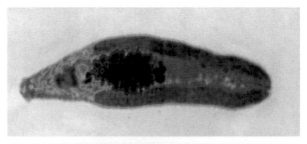

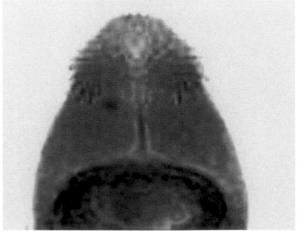

of rediae are produced before cercariae develop. These cercariae encyst in freshwater mollusks, and humans are infected by eating these mollusks raw rather than cooked. The metacercariae hatch in the intestine, attach to the mucosal wall, and develop into mature adult worms.

The adults are less than 1 cm long and 0.2 cm wide and are reddish gray (Figure 15.5). The anterior end of the adult worm has a circumoral disk with a crown of spines surrounding the oral sucker (Figure 15.7).

Clinical Disease

With light infections, the patient may be asymptomatic and the adult worms cause only minor problems other than localized inflammation. In heavy infections, the worms can produce catarrhal inflammation and mild ulceration and the patient may experience diarrhea and abdominal pain.

Diagnosis

Eggs of *E. ilocanum* can be detected in the stool. They are the same size (some strains) and shape as the eggs of *F. hepatica* and *F. buski*. Because of the overlap in size range, exact identification cannot be made from the eggs; therefore, the diagnosis may have to be made on the basis of patient history or clinical findings.

KEY POINTS—LABORATORY DIAGNOSIS

Echinostoma ilocanum

1. The eggs are found in the stool; rarely, adult worms are found during heavy infections.
2. The eggs of *E. ilocanum*, *G. hominis*, *F. buski*, *F. hepatica*, and *F. gigantica* appear almost identical and are very difficult to differentiate from one another. When passed, the eggs do not contain mature larvae but instead contain undifferentiated embryos.
3. The formalin-ethyl acetate sedimentation concentration is recommended for egg recovery; operculated eggs do not float in the zinc sulfate flotation concentration method.
4. The less mature the egg, the more difficult it may be to see the actual operculum—it blends into the shell outline, and the "breaks" in the shell may be difficult to identify. This egg has no opercular shoulders; therefore, it is difficult to see where the operculum "breaks" in the shell actually occur.

Treatment

Praziquantel is the drug of choice and can be given in a single dose of 40 mg/kg at bedtime. Another option would be albendazole at 400 mg twice a day for 3 days (see chapter 25).

Epidemiology and Prevention

Information concerning *E. ilocanum* infections and epidemiology is limited. Both rats and dogs have been found to be infected in areas where infections are endemic. The infection has been reported from China, Indonesia, Thailand, Taiwan, the Philippines, and the Celebes. In some areas in northeastern Thailand, the infection rate may be as high as 50%. In areas of the Philippines with high levels of endemic infection (Luzon, Mindanao, and Leyte), local infection rates from 1 to 44% have been reported (3). In China, the infection appears to be limited to dogs, with an infection rate of 14% being reported. The miracidia must infect a susceptible species of snail from the family Planorbidae (*Gyraulus* or *Hippeutis* spp.). The cercariae can encyst in a number of freshwater mollusks, particularly snails, which, if eaten raw, will allow the metacercariae to excyst in the intestinal tract and develop to adult worms. The infection can be prevented in areas of endemicity by restricting the use of night soil for fertilizer and eating cooked rather than raw mollusks.

Human infections with other echinostome species have been reported; however, actual infection rates are unknown or estimated. Some of these species have been identified and are listed in Table 15.2. For a number of these species, the food source is often fish rather than mollusks.

Heterophyes heterophyes

H. heterophyes is one of the smallest trematodes to infect humans and is acquired through the ingestion of pickled or uncooked fish. The heterophyids are capable of parasitizing birds and mammals, and infection was first reported by Bilharz in 1851 (14).

Life Cycle and Morphology

The adult worms measure 1.0 to 1.7 mm in length by 0.3 to 0.4 mm in width, are gray, and have a broadly rounded posterior end (Figures 15.5 and 15.8). They lay embryonated eggs in the intestinal lumen, and the eggs are discharged into the environment via the feces. The eggs are small, brownish yellow, and operculate, and they possess very subtle opercular shoulders. They measure 27 to 30 µm by 15 to 17 µm (Table 15.1; Figure 15.4). The eggs are ingested by the snail intermediate host before the miracidium hatches. A sporocyst and one or two redia generations develop in the snail before the production and release of cercariae. The cercariae encyst under the scales or in the flesh of freshwater fish. Humans become infected by eating raw, pickled, or poorly cooked fish containing the metacercariae (9). The metacercariae excyst in the intestinal tract, attach to the walls of the small intestine, and grow to maturity. The adult worm, in addition to an oral and ventral sucker, contains a third sucker, the genital sucker, surrounding the genital pore.

Table 15.2 Other echinostome infections reported from humans[a]

Trematode	Location	Host	No. and location of spines	Infection acquired
Echinostoma malayanum	Northeastern and northern Thailand	Rats, dogs	43 crown	Freshwater snails, usually cooked
Echinostoma revolutum	Taiwan (3–7% infection rate)	Ducks, geese, rats	37 circumoral	Raw mollusks
Echinostoma lindoense	Indonesia (infection may no longer be present in this area [Lindu Valley]); found in Brazil	Rats, mice	37 collar	Raw mollusks
Hypoderaeum conoideum	Northeastern Thailand	Ducks, other fowl, rats, humans	Two rows of 45–53	Salted or raw snails or clams
Echinostoma cinetorchis, Psilorchis hominis[b]	Japan, Taiwan			Tadpoles, frogs
Euparyphium melis, E. beaveri	Romania, China; may also be in Douglas Lake, Mich. (not studied)		27 crown	Tadpoles
Hypoderaeum cibiuduyn	Northeast Thailand (55% infection rate)	Ducks, geese, fowl		Raw snails
Artyfechinostomum mehrai	India (2 cases)	Pig	39–42 collar	Raw snails
Echinostoma recurvatum	Taiwan, Indonesia, Egypt	Rats	45 crown	Tadpoles, frogs
Echinoparyphium paraulum[c]	Former USSR	Ducks, geese, swans, doves		
Himasthla muehlensi	New York City		32 circumoral	Raw clams
Echinochasmus perfoliatus	Hungary, Italy, Romania, former USSR, Far East	Dogs, cats, pigs, foxes	24 circumoral	Raw freshwater fish

[a] Data from reference 2.
[b] May or may not be an echinostome; specimens may have lost crown spines.
[c] May be the same as *Echinostoma revolutum.*

Clinical Disease

Following ingestion of metacercariae, the prepatent period is approximately 9 days. The adult worms cause little damage to the intestinal tract mucosa except for a mild inflammatory reaction. In heavy infections, the worms cause abdominal pain, mucous diarrhea, and ulceration of the intestinal wall. Because of the small size of the eggs and the fact that the adult may attach itself deeply into the intestinal wall, some of the eggs may end up in the general circulation of the host. The eggs may provoke pathologic lesions, particularly in the heart and brain (5, 7). In the Philippines, 15% of fatal heart disease may be a result of heterophyid myocarditis. Neurologic manifestations due to adult worms or eggs in the brain have been reported (13).

Diagnosis

The diagnosis is suggested by detecting the eggs in the stool. Because the eggs of *H. heterophyes* are similar in shape and size to those of *Metagonimus yokogawai* and *Clonorchis sinensis*, differentiation must be made on the basis of patient history, clinical findings, or recovery of adult worms in the stool.

In Kobe, Japan, *H. katsuradai*, a closely related species that is broader and more rounded, with a very large ventral sucker and smaller eggs (25 to 26 μm by 14 to 15 μm), has been recovered after therapy in patients with diarrhea. Apparently, infection is acquired from the ingestion of raw mullet.

KEY POINTS—LABORATORY DIAGNOSIS

Heterophyes heterophyes

1. The eggs are found in the stool; rarely, adult worms are found in patients with heavy infections after therapy or at autopsy.
2. The eggs of *H. heterophyes*, *M. yokogawai*, and *C. sinensis* appear almost identical and are very difficult to differentiate from one another. When passed, each egg contains a miracidium larva, but the eggs are ingested by the snail before they hatch.

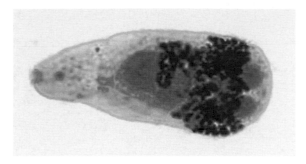

Figure 15.8 *Heterophyes heterophyes* adult worm. These worms tend to have a broadly rounded posterior end, measure approximately 1 to 1.7 mm in length by 0.3 to 0.4 mm in width, and tend to be gray.

3. The formalin-ethyl acetate sedimentation concentration is recommended for egg recovery; operculated eggs do not float in the zinc sulfate flotation concentration method.

4. The eggs have very inconspicuous opercular shoulders and, unlike *C. sinensis*, lack the "seated" operculum and knob at the abopercular end. *H. heterophyes* eggs are some of the smallest trematode eggs seen and may be missed using the low power of the microscope (10× oculars and 10× objective); high dry power may be required for identification (10× oculars and 40× objective).

Treatment
Praziquantel is the drug of choice and can be given in a single dose of 15 to 25 mg/kg at bedtime (see chapter 25).

Epidemiology and Prevention
Numerous fish-eating mammals, including dogs, cats, and birds, act as reservoirs (15). The infection is found in China, Egypt, India, Iran, Israel, Japan, Korea, Sudan, Taiwan, the Philippines, Tunisia, and Turkey. Snails serving as the first intermediate host include *Pironella* and *Cerithidea* spp., and a variety of freshwater fish can serve as the second intermediate hosts (4, 9).

Humans become infected by ingestion of parasitized fish that is raw, inadequately cooked, or improperly pickled or salted. Metacercariae can survive for up to 7 days in salted fish. Off the coast of Israel a brackish-water fish, *Mugil capito*, can be heavily infected with *H. heterophyes*, with metacercariae numbering 2,300 to 6,000 per g of fish. The life cycle could be disrupted by improved sanitary conditions and educational information about proper sewage disposal away from ponds or lakes where the intermediate hosts reside.

Figure 15.9 *Metagonimus yokogawai* adult worm. These worms are quite small like *Heterophyes* and measure 1 to 2.5 mm by 0.4 to 0.75 mm. The ventral sucker is deflected to the right of the midline and is visible in this image.

Metagonimus yokogawai

M. yokogawai was first discovered by Katsurada in 1911 in an infected human (16). The infection is found most often in the Far East and clinically resembles that of *H. heterophyes*.

Life Cycle and Morphology
The adult worm measures 1.0 to 2.5 mm by 0.4 to 0.8 mm, attaches to the mucosal wall of the small intestine, and lays embryonated eggs that are passed out with the feces into the environment (Figure 15.9). The egg is small, brownish yellow, and operculate, with minimal opercular shoulders; it measures 26 to 28 μm by 15 to 17 μm (Table 15.1). The eggs are ingested by the snail intermediate host before the miracidium is liberated. On infection of the snail, a sporocyst generation and two redia generations develop before cercariae are liberated. Cercariae encyst under the scales or in the flesh of freshwater fish, which are eaten uncooked or pickled by humans. The metacercariae excyst in the intestinal tract, attach to the mucosal wall, and develop to mature worms. A *Metagonimus* adult is slightly larger than a *Heterophyes* adult and has a ventral sucker that is deflected to the right of the midline axis. The genital pore is attached to the outer rim of the ventral sucker.

Clinical Disease
Symptoms and pathologic changes are similar to those produced by *H. heterophyes* and depend largely on the worm burden of the host. Eggs may infiltrate into the intestinal capillaries and lymphatics and be carried to the myocardium, brain, spinal cord, and other tissues, where emboli or granulomatous reactions may occur.

Diagnosis
The diagnosis is based on recovery of the eggs in the feces. Because the eggs of *M. yokogawai* are similar in shape and size to those of *H. heterophyes* and *C. sinensis*, the

definitive diagnosis must be made on the basis of clinical findings, patient history, or recovery of adult worms after therapy or at autopsy.

KEY POINTS—LABORATORY DIAGNOSIS

Metagonimus yokogawai

1. The eggs are found in the stool; rarely, adult worms are found in patients with heavy infections after therapy or at autopsy.
2. The eggs of *M. yokogawai*, *H. heterophyes*, and *C. sinensis* appear almost identical and are very difficult to differentiate from one another. When passed, each egg contains a miracidium larva, but the eggs are ingested by the snail before they hatch.
3. The formalin-ethyl acetate sedimentation concentration is recommended for egg recovery; operculated eggs do not float in the zinc sulfate flotation concentration method.
4. The eggs have inconspicuous opercular shoulders but a more obvious operculum than *H. heterophyes* eggs and are some of the smallest trematode eggs seen. *M. yokogawai* eggs do not really have the "seated" operculum and knob at the abopercular end like those seen in *C. sinensis* eggs. They may be missed using the low power of the microscope (10× oculars and 10× objective), and high dry power may be required for identification (10× oculars and 40× objective).

Treatment

Praziquantel is the drug of choice and can be given in a single dose of 15 to 25 mg/kg at bedtime (see chapter 25).

Epidemiology and Prevention

Reservoir hosts are similar to those of *H. heterophyes*. The infection is found in China, Indonesia, Israel, Korea, Japan, Russia, Spain, the Balkans, and Taiwan. *M. yokogawai* is considered the most common intestinal fluke infection in the Far East; the prevalence tends to be high in Japan, Korea, and Taiwan, with rates of 2 to 50% being reported from Japan. Freshwater snails (*Semisulcospira* spp.) act as the first intermediate hosts, and a variety of freshwater fish act as the second intermediate hosts.

Preventive measures are similar to those proposed for *H. heterophyes*. Improved education and sanitary standards would greatly reduce the risk of infections.

Other heterophyid species have been found in areas of Asia where people eat raw or poorly cooked freshwater fish. A list of these trematodes can be found in Table 15.3.

Aquaculture of fish has increased dramatically from 5.3% in 1970 to 32.2% in 2000 (8). Freshwater production has increased greatly and now accounts for 45.1% of the total aquaculture production. The production of grass carp, an important species that serves as an intermediate host for food-borne trematodes, has increased from >10,000 tons in 1950 to >3 million tons in 2002. This growth must be monitored for potential problems related to increased disease in which infection is transmitted through the ingestion of raw or poorly cooked fish.

Table 15.3 Other heterophyid infections reported in humans[a]

Organism	Comments
Centrocestus formosanus	Reservoir hosts are fish-eating birds and mammals; reported from Taiwan and mainland China
Haplorchis pumilio, *H. yokogawai*	Reservoir hosts are fish-eating birds and mammals; reported from Taiwan, the Philippines, Indonesia, and Thailand
Metagonimus minutus, *Diorchitrema formosanum*, *D. amplicaecale*	Reservoir hosts are fish-eating birds and mammals; reported from Taiwan
Stellantchasmus falcatus	Reservoir hosts are fish-eating birds and mammals; reported from Hawaii (from eating raw mullet), Japan, the Philippines, and Thailand
Pygidiopsis summa	Reservoir hosts are fish-eating birds and mammals; reported from Korea (up to 4,000 worms have been recovered from patients)

[a] Intestinal lesions are similar to those produced by *H. heterophyes* and *M. yokogawai*. Eggs of *Haplorchis yokogawai*, *H. pumilio*, and *H. taichui* have been found in cardiac lesions but rarely cause complications such as vascular occlusion. Treatment with praziquantel in a single dose of 15 to 25 mg/kg is effective.

Gastrodiscoides hominis

Gastrodiscoides hominis was first recovered in the late 1800s by Lewis and McConnell from an Indian patient suffering from diarrhea (2). The infection is commonly found in Assam, India. Although gastrodiscoidiasis was initially supposed to be restricted to Asian countries, this infection is now being reported in African countries (10).

Life Cycle and Morphology

The complete life cycle is unknown but is probably similar to that of other amphistome trematodes; aquatic plants, crustaceans, and amphibians are thought to be involved. The adult worm is bright pink, can be found attached to the cecum and ascending colon in humans, and may produce a mucous diarrhea. In India, the planorbid snail *Helicorbis coenosus* serves as an intermediate host for this infection.

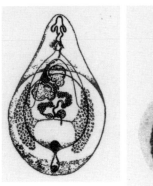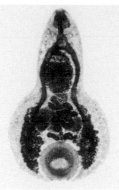

Figure 15.10 *Gastrodiscoides hominis* adult worms. Note the pyriform shape and the deep concavity on the ventral surface, which contains a large sucker. The adult worms measure 8 to 14 mm in length by 4 to 5 mm in width and are usually bright pink. (Illustration by Sharon Belkin. Adapted from references 2 and 3.)

The living worm measures approximately 4 to 5 mm by 8 to 14 mm, is usually bright pink, and is pyriform in outline with a conical anterior (Figures 15.5 and 15.10). The anterior portion measures about 2 mm. The ventral surface has a prominent acetabulum that bears a characteristic notch. The eggs are greenish brown, operculated, and immature when laid; they measure 60 to 70 µm by 150 µm.

Clinical Disease

The adult worms live in the cecum and ascending colon; symptoms are related to the overall worm burden. In light infections the patient may be asymptomatic, while in heavier infections there may be a mucous diarrhea; lymphocytes, plasma cells, and eosinophils infiltrate the mucosa and submucosa (3). At the attachment site, papular lesions with surface desquamation may lead to necrosis.

Diagnosis

The diagnosis is based on recovery of eggs in the feces. They resemble the eggs of *F. buski* but tend to be more slender. A definitive diagnosis could be made on the basis of clinical findings, patient history, geographic area, or recovery of adult worms after therapy or at autopsy.

KEY POINTS—LABORATORY DIAGNOSIS
Gastrodiscoides hominis

1. The eggs are found in the stool; rarely, adult worms are found after therapy or at autopsy in patients with heavy infections.
2. The eggs of *G. hominis*, *F. buski*, *E. ilocanum*, *F. hepatica*, and *F. gigantica* appear very much

alike and can be difficult to differentiate from one another. When passed, these eggs do not contain mature larvae but instead contain undifferentiated embryos. The egg of *G. hominis* tends to be a bit smaller and more slender than that of *F. buski*, but they are very much alike.
3. The formalin-ethyl acetate sedimentation concentration is recommended for egg recovery; operculated eggs do not float in the zinc sulfate flotation concentration method.
4. The less mature the egg, the more difficult it may be to see the actual operculum—it blends into the shell outline, and the "breaks" in the shell may be difficult to identify. This egg has no opercular shoulders, and so it is difficult to see where the operculum "breaks" in the shell actually occur.

Treatment

Praziquantel is the drug of choice and is administered as a single 15-mg/kg dose after the evening meal or at bedtime. A second regimen consists of 75 mg/kg/day in three doses for 1 day. Apparently, children tend to tolerate the drug better than adults (see chapter 25).

Epidemiology and Prevention

G. hominis is commonly found in Asia and has been reported from Assan, Bihar, and Orissa in India, as well as from Vietnam, Burma, China, the Philippines, Thailand, Kazakhstan, and Guyana. One study in India found a 41% prevalence rate in which pigs were the reservoir host. Other reservoir hosts include deer mice in Malaysia and rats in Indonesia, Japan, and Thailand.

References
1. **Abramowicz, M. (ed.).** 2004. Drugs for parasitic infections. *Med. Lett. Drugs Ther.* **46:**1–12.
2. **Beaver, P. C., R. C. Jung, and E. W. Cupp.** 1984. *Clinical Parasitology*, 9th ed. Lea & Febiger, Philadelphia, Pa.
3. **Bunnag, D., J. H. Cross, and T. Bunnag.** 2000. Intestinal fluke infections, p. 832–840. *In* G. T. Strickland (ed.), *Hunter's Tropical Medicine and Emerging Infectious Diseases*, 8th ed. The W. B. Saunders Co., Philadelphia, Pa.
4. **Cort, W. H., and S. Yokogawa.** 1921. A new human trematode from Japan. *J. Parasitol.* **8:**66–69.
5. **Deschiens, R., H. Collomb, and J. Demarchi.** 1958. Distomatose cerebrale a *Heterophyes heterophyes*, p. 265. *In Abstracts of the 6th International Congress on Tropical Medicine and Malaria.*
6. **Graczyk, T. K., R. H. Gilman, and B. Fried.** 2001. Fasciolopsiasis: is it a controllable food-borne disease? *Parasitol. Res.* **87:**80–83.
7. **Kean, B. H., and R. C. Breslau.** 1964. *Parasites of the Human Heart*, p. 95–103. Grune & Stratton, New York, N.Y.

8. **Keiser, J., and J. Utzinger.** 2005. Emerging foodborne trematodiasis. *Emerg. Infect. Dis.* **11:**1507–1514.

9. **Khalil, M.** 1933. The life history of the human trematode parasite, *Heterophyes heterophyes. Lancet* **ii:**537.

10. **Mas-Coma, S., M. D. Bargues, and M. A. Valero.** 2005. Fascioliasis and other plant-borne trematode zoonoses. *Int. J. Parasitol.* **35:**1255–1278.

11. **Nakagawa, K.** 1921. On the life cycle of *Fasciolopsis buski* (Lankester). *Kitasato Arch. Exp. Med.* **4:**159–167.

12. **Nakagawa, K.** 1922. The development of *Fasciolopsis buski* Lankester. *J. Parasitol.* **8:**161–166.

13. **Sen-Hai, Y., and K. E. Mott.** 1994. Epidemiology and morbidity of food-borne intestinal trematode infections. *Trop. Dis. Bull.* **91:**R126–R150.

14. **von Siebold, T.** 1852. Beitrage zur Helminthographia Humana. *Z. Wiss. Zool.* **4:**53–76.

15. **Wells, W. H., and B. H. Randall.** 1956. New hosts for trematodes of the genus *Heterophyes* in Egypt. *J. Parasitol.* **42:**287–292.

16. **Witenberg, G.** 1929. Studies on the trematode—family Heterophyidae. *Ann. Trop. Med.* **23:**131–239.

17. **World Health Organization.** 1995. Control of foodborne trematode infections. *WHO Tech. Rep. Ser.* 849.

16

Liver and Lung Trematodes

Clonorchis, Opisthorchis, and *Fasciola* spp. are trematodes that parasitize the biliary ducts of humans. *Clonorchis* and *Opisthorchis* spp. are narrow, elongate worms that localize in the more distal, smaller ducts of the biliary tree. *Fasciola hepatica*, because of its much larger size, resides in the larger bile ducts and gallbladder.

Paragonimus spp., also known as lung flukes, cause paragonimiasis in humans. The adults encapsulate in the lungs and are occasionally found in other body sites.

The infections caused by the liver and lung trematodes are food borne and have considerable economic and public health impact. It is estimated that more than 50 million people have acquired foodborne trematode infections. It is estimated that 601 million, 293 million, 91 million, and 80 million people are at risk of infection with *Clonorchis sinensis, Paragonimus* spp., *Fasciola* spp., and *Opisthorchis* spp., respectively (23). Of great public health concern is cholangiocarcinoma associated with *Clonorchis* and *Opisthorchis* infections, severe liver disease associated with *Fasciola* infections, and the misdiagnosis of tuberculosis in those infected with *Paragonimus* spp.

Liver Flukes

Clonorchis sinensis

McConnell was the first to describe the adult worms and pathologic changes caused by *C. sinensis*, in a Chinese patient who died in Calcutta, India. The complete life cycle was developed in a number of publications by Iijima in 1887, Saito in 1898, Kobayashi in 1914, and Muto in 1918 (9). *C. sinensis* is also known as the Chinese or oriental liver fluke (Table 16.1).

Life Cycle and Morphology

Adult worms deposit eggs in the bile ducts, and the eggs are discharged with the bile fluid into the feces and passed out into the environment (Figure 16.1). The eggs are fully embryonated when laid and measure 28 to 35 µm by 12 to 19 µm (Figure 16.2). The eggs are ovoid, with a thick, pale brownish yellow shell and an operculum. There are distinct opercular shoulders surrounding the operculum. Frequently, the eggs contain a comma-shaped appendage at

Table 16.1 Characteristics of liver and lung trematodes

Species	Distribution	Agent of infection	Reservoir host	Size (μm)	Morphology	Egg Comments
Clonorchis sinensis	Far East	Uncooked fish	Dogs, cats, other fish-eating mammals	28–35 by 12–19	Embryonated, operculated	Very prominent opercular shoulders; has a "seated" operculum and knob at the abopercular end.
Opisthorchis viverrini	Northern Thailand, Laos	Uncooked fish	Dogs, cats, other fish-eating mammals	19–29 by 12–17	Embryonated, operculated	Prominent opercular shoulders; has a "seated" operculum and may or may not have a knob at the abopercular end; eggs tend to be broader with less prominent shoulders than *C. sinensis* eggs.
Opisthorchis felineus	Poland, Germany, Russian Federation, Kazakhstan, western Siberia	Uncooked fish	Dogs, cats, other fish-eating mammals	28–30 by 11–16	Embryonated, operculated	Prominent opercular shoulders; has a "seated" operculum and may or may not have a knob at the abopercular end; eggs tend to be broader with less prominent shoulders than *C. sinensis* eggs.
Fasciola hepatica	Worldwide; mixed *F. hepatica* and *F. gigantica* infections have been reported from Pakistan	Uncooked water plants	Herbivores	130–150 by 63–90	Unembryonated, operculated	The less mature the egg, the more difficult it may be to see the actual operculum—it blends into the shell outline, and the "breaks" in the shell may be difficult to identify. This egg has no opercular shoulders, and so it is difficult to see where the operculum "breaks" in the shell actually occur. Eggs resemble those of *F. buski*, *E. ilocanum*, *F. gigantica*, and *G. hominis*; eggs may have thickening at the abopercular end of the shell (unlike eggs of *F. buski*).
Fasciola gigantica	Africa, southern Europe, southern United States, Hawaii, former Soviet Union, Middle East, Southeast Asia	Uncooked water plants	Herbivores	160–190 by 70–90; egg is larger than *F. hepatica*	Unembryonated, operculated	The egg is larger than that of *F. hepatica*; both *F. hepatica* and *F. gigantica* worms have been found in ectopic sites.
Dicrocoelium dendriticum, *D. hospes*, *Eurytrema pancreaticum*	Europe, Turkey, northern Africa, Far East, China, Japan, North and South America	Ants, grasshoppers, crickets	Cattle, sheep, deer, water buffalo	38–45 by 22–30; dark brown	Embryonated, operculated, thick shell	Eggs have a thick, dark brown shell and essentially no opercular shoulders; they cannot be differentiated from each other.
Paragonimus westermani	Far East, Africa	Crabs, crayfish	Dogs, cats, tigers, lions	80–120 by 45–65	Unembryonated, operculated	Eggs have a moderately thick, dark golden-brown shell, a prominent operculum, opercular shoulders, and a thickened abopercular end; may be confused with *D. latum* eggs (smaller, abopercular knob).
P. mexicanus	Central and South America	Crabs	Opossum, cats, dogs	60–70 by 42–56	Unembryonated, operculated	Thin, irregular undulations on outer shell; eggs have a prominent operculum, opercular shoulders, and a thickened abopercular end; may be confused with *D. latum* eggs (smaller, abopercular knob); eggs smaller than *P. westermani*; also golden-brown shell.

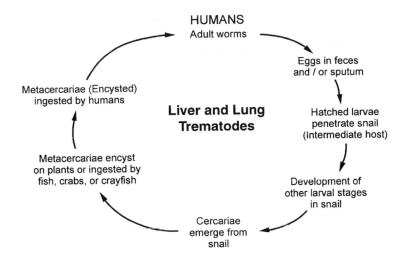

HUMANS
Adult worms

Eggs in feces
and / or sputum

Metacercariae (Encysted)
ingested by humans

**Liver and Lung
Trematodes**

Hatched larvae
penetrate snail
(Intermediate host)

Metacercariae encyst
on plants or ingested by
fish, crabs, or crayfish

Development of
other larval stages
in snail

Cercariae
emerge from
snail

Figure 16.1 Life cycle of the liver and lung trematodes.

the abopercular end. Eggs are ingested by the snail host, at which time the miracidium hatches to infect the snail. More than 100 species of snails can serve as an intermediate host for *C. sinensis*. The snails generally live in ponds used for commercial fish raising, lakes, and slow-moving water, and the overall infection rate is relatively low (12). Sporocyst and redia generations are produced before cercariae are released to encyst in the skin or flesh of freshwater fish. Humans become infected by ingesting the metacercariae in uncooked fish.

Metacercariae excyst in the duodenum, enter the common bile duct, and travel to the distal bile capillaries, where the worms mature (Figures 16.3 and 16.4). The life cycle takes approximately 3 months to complete in humans. Reservoir hosts include dogs, cats, pigs, mink, rats, and other fish-eating mammals.

Clinical Disease

In general, the complications of clonorchiasis are the result of biliary obstruction. As the worms mature in the distal bile ducts, an inflammatory response is seen in the biliary epithelium. The extent of pathologic changes is related to the intensity and duration of infection (Algorithm 16.1). The lesions are confined mainly to the biliary system and are the result of mechanical irritation and toxins produced by the worms. In light infections, there appears to be little or no change in liver parenchyma, whereas heavy infections cause thickening and localized dilations of the bile ducts with hyperplasia of the mucinous glands. As a result, the biliary tract may become obstructed, causing bile retention, infiltration of lymphocytes and eosinophils, and fibrosis (Figure 16.5). The adenomatous changes may persist for many years in patients with light infections. The infections have been associated with obstructive jaundice, which may be aggravated by biliary stones and liver abscesses. Many patients infected with *C. sinensis* have recurrent pyogenic cholangitis. There is no direct evidence that infection with *C. sinensis* causes chronic bacterial infection. Acute pancreatitis, cholecystitis, and cholelithiasis may be the result of worm

Figure 16.2 Eggs of *Clonorchis sinensis*. Note the range of sizes and shapes.

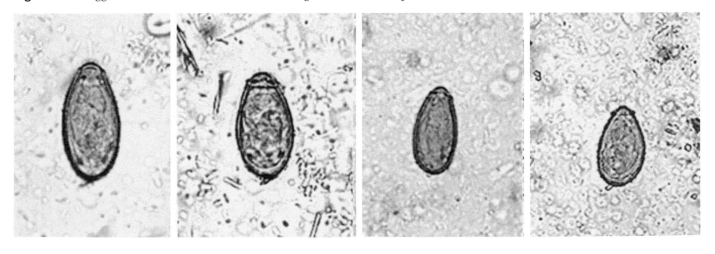

Figure 16.3 *Clonorchis sinensis* adult (10 to 25 mm long by 3 to 5 mm wide). (Armed Forces Institute of Pathology photograph.)

invasion. Cirrhosis is probably related to malnutrition rather than parasitic infections. Computed tomographic (CT) evaluation has shown that in the presence of diffuse mild intrahepatic bile duct dilation, enlargement of the body or tail (or both) of the pancreas, with a cluster of small cystic changes within the pancreatic parenchyma, provides strong evidence of *C. sinensis* pancreatitis (25). This infection may continue in the human host for more than 20 years.

C. sinensis has been linked to neoplasms of the bile duct and to cholangiocarcinoma (1, 15), which is most frequently observed in areas where clonorchiasis is endemic. Cholangiocarcinoma is a malignant tumor that arises from the bile duct epithelium and is the second most prevalent liver cancer after hepatocellular carcinoma. The tumor usually occurs in patients 60 to 80 years of age and rarely in patients younger than 40 years. There appears to be no direct link between infection and carcinoma, although one of the first steps in malignant transformation may be induced by the biliary tract hyperplasia caused by the worms. Patients with primary sclerosing cholangitis (PSC) have a substantial predisposition for bile duct carcinoma. Although the exact mechanisms are not well defined, long-standing inflammation can lead to cholangiocarcinoma in patients with chronic *C. sinensis* infection. PSC is an uncommon disease, characterized by stricturing, fibrosis, and

Figure 16.4 Adult liver and lung trematodes. (A) *Fasciola hepatica*; (B) *Paragonimus westermani*; (C) *Clonorchis sinensis*; (D) *Opisthorchis viverrini* and *O. felineus*.

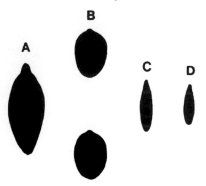

inflammation of the biliary tree, which is closely associated with chronic inflammatory bowel disease, particularly ulcerative colitis. The only effective treatment for PSC is orthotopic liver transplantation, which in the absence of cholangiocarcinoma has a 5-year survival rate of 89%. However, patients with cholangiocarcinoma who undergo liver transplantation have a high risk of recurrence and poor survival statistics. Between 10 and 20% of patients with PSC will go on to develop a cholangiocarcinoma (14). Therefore, identification of patients with deteriorating liver function prior to the development of cholangiocarcinoma remains an important goal in the management of PSC, particularly in areas where *C. sinensis* infection is endemic (18). It is also probable that cofactors play a role; liver flukes are promoters and not initiators of cholangiocarcinoma (1). It has been suggested that regular biliary cytologic sampling to detect dysplasia can predict the development of cholangiocarcinoma; however, this approach is unlikely to be widely used. In the Far East, forms of chronic inflammation associated with cholangiocarcinoma include infestation with either of the liver flukes, *C. sinensis* or *Opisthorchis viverrini*.

Immunoglobulin E (IgE) and *C. sinensis*-specific IgE levels in serum are elevated in infected individuals. In patients with acute infections, there is an increase in IgM levels in serum followed by increases in IgA and IgG levels. In patients with chronic infections, the IgA level returns to normal while the IgG and IgM levels remain elevated.

In light infections, the patient generally experiences no symptoms. In heavier infections acquired over time, the patient may experience dull pain and abdominal discomfort that may last for 1 to 2 h, often in the afternoon. As the disease progresses, the duration of pain lengthens and the pain may become so severe that the patient is unable to work. Patients who have had the disease for a long time show liver enlargement with some degree of functional impairment that is secondary to biliary obstruction. Acute infections caused by ingestion of large numbers of metacercariae cause fever, chills, diarrhea, epigastric pain, enlarged tender liver, and possibly jaundice within a month of ingestion. The acute symptoms last for about 1 month and subside at about the time when eggs are detected in the stool. In chronic infections, cholangitis, cholelithiasis, pancreatitis, and cholangiocarcinoma are common complications and can lead to death (15, 41).

Diagnosis

Individuals who have lived in areas where infections are endemic, have a history of eating raw fish, and have symptoms of upper abdominal pain, indigestion, diarrhea, and hepatomegaly may be suspected of having liver trematode infection. Eggs can be detected in the feces by direct microscopy or by examination of fecal concentrates.

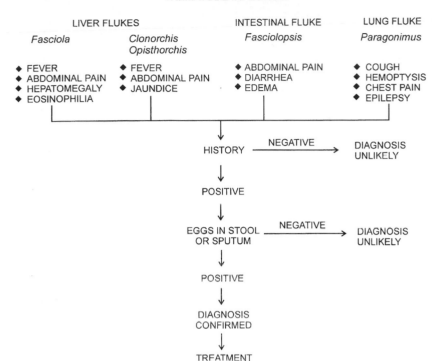

TREMATODE INFECTION

LIVER FLUKES

Fasciola   *Clonorchis*
                    Opisthorchis

INTESTINAL FLUKE

Fasciolopsis

LUNG FLUKE

Paragonimus

- ◆ FEVER
- ◆ ABDOMINAL PAIN
- ◆ HEPATOMEGALY
- ◆ EOSINOPHILIA

- ◆ FEVER
- ◆ ABDOMINAL PAIN
- ◆ JAUNDICE

- ◆ ABDOMINAL PAIN
- ◆ DIARRHEA
- ◆ EDEMA

- ◆ COUGH
- ◆ HEMOPTYSIS
- ◆ CHEST PAIN
- ◆ EPILEPSY

HISTORY — NEGATIVE → DIAGNOSIS UNLIKELY

POSITIVE

EGGS IN STOOL OR SPUTUM — NEGATIVE → DIAGNOSIS UNLIKELY

POSITIVE

DIAGNOSIS CONFIRMED

TREATMENT

Algorithm 16.1   Trematode infection.

Sedimentation concentration is recommended since these eggs have opercula and do not float in the zinc sulfate flotation concentration method. Sedimented material can be examined using saline or D'Antoni's iodine or both. It is important not to add too much iodine; if the eggs stain too darkly, they may resemble debris. In light infections, egg counts may be low and so multiple specimens may be required to confirm their presence. The eggs of *C. sinensis* (Figure 16.2) and *O. viverrini* are similar in size and shape to those of *Heterophyes heterophyes* and *Metagonimus yokogawai* and cannot be readily differentiated from them. If a patient has not resided in or recently visited areas where infections are endemic, the infection is probably due to *C. sinensis* or *O. viverrini*. The infection may be confirmed by detecting eggs in the bile fluid (duodenal aspirate), by recovering adult worms, or from the clinical history. Some strains of *C. sinensis* produce eggs that have a comma-shaped appendage at the abopercular end. Multiple egg measurements are usually required to determine size differences; however, absolute identification among the small trematode eggs can be very difficult. Eggs can also be seen in duodenal drainage material or, if the Entero-Test capsule is used, in the mucus removed from the string. Definitive identification usually requires examination of adult worms recovered after therapy or during surgery or autopsy.

Eggs are not found in stool specimens from patients with biliary obstruction; needle aspiration, surgery, or autopsy specimens may be required to confirm their presence. In these patients, biliary obstruction must be differentiated from enlarged gallbladder, cholangitis with jaundice, liver carcinoma, and cholangiocarcinoma. Cholangiography, ultrasonography, and liver scans may reveal lesions consistent with liver fluke infection.

Clonorchiasis is often diagnosed incidentally during abdominal ultrasonography, since symptoms tend to be nonspecific. However, none of these methods are any more sensitive than fecal examination because of their limited sensitivity and specificity (27).

In general, immunologic procedures are not readily available; however, they have been developed and are used in the research setting. A multiplex PCR approach has proven to be species specific, sensitive, and rapid in the accurate diagnosis of clonorchiasis and/or opisthorchiasis. This test can be used for the detection of metacercariae in infected fish or adult worms or eggs from patients in areas of endemic infection (26).

KEY POINTS—LABORATORY DIAGNOSIS

Clonorchis sinensis

1. Multiple stool specimens may be required to find the eggs.
2. The sedimentation concentration method should be used; because the eggs are operculated, they do not float in the zinc sulfate flotation concentration method.

3. It is important not to add too much iodine to the wet preparation, or the small eggs will stain very darkly and resemble debris.

4. Although the wet preparation can be examined using the 10× (low-power) objective, these eggs measure approximately 30 μm and can be easily missed. It is recommended that the 40× (high dry) objective be used for the wet preparation examination.

5. Eggs of *C. sinensis* can resemble those of many other intestinal and liver trematodes.

6. The wet preparation should not be so thick that the small eggs are obscured by normal stool debris.

7. Some strains of *C. sinensis* and *O. viverrini* produce eggs that have a comma-shaped appendage at the abopercular end; this finding generally differentiates the liver fluke eggs from those of the intestinal flukes, *Heterophyes heterophyes* and *Metagonimus yokogawai*.

8. If duodenal sampling is performed (drainage or the Entero-Test), eggs may be found in this material.

9. If available, antigen or antibody detection may be helpful in confirming the infection.

Treatment

Praziquantel is the drug of choice for treatment of *C. sinensis* infections. It has very few or limited side effects but is not recommended for use during pregnancy. The recommended dose is 75 mg/kg after meals three times a day for 1 or 2 days (2). These regimens yielded 85 and 100% cure rates, respectively. The majority of patients do well, with the exception of those who have pyogenic cholangitis, obstructive jaundice, or cholangiocarcinoma. Albendazole may be used as an alternative drug. Once cholangiocarcinoma is diagnosed, complete resection with negative histologic margins provides the greatest chance for long-term survival. Unfortunately, cholangiocarcinoma responds poorly to chemotherapy.

Epidemiology and Prevention

Humans, dogs, cats, and other fish-eating mammals are reservoir hosts and, in some areas, the sole cause of continual transmission. The infection is found in China, Japan, Korea, Malaysia, Singapore, Taiwan, and Vietnam, and it is estimated that 7 million individuals are infected, with 4.7 million in China and approximately 1 million in Korea (12). Natives of Hawaii also harbor the infection, which is contracted through the ingestion of frozen, pickled, or dried fish imported from areas of endemicity. Human infections result from the widely prevalent practice of consumption of metacercaria-infested freshwater fish eaten uncooked, pickled, smoked, salted, or dried. The traditional eating habits which are a part of a cultural

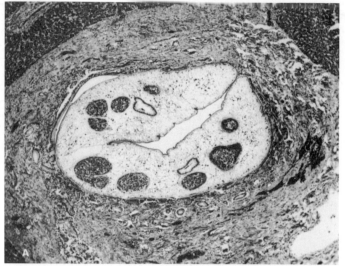

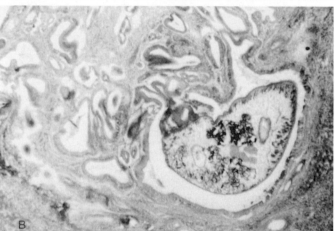

Figure 16.5 *Clonorchis sinensis* adult in a biliary duct (cross section). (Panel A, Armed Forces Institute of Pathology photograph; panel B, from A Pictorial Presentation of Parasites: A cooperative collection prepared and/or edited by H. Zaiman.)

heritage are very difficult to change. In the past few years, clonorchiasis has often been found in refugee immigrants from areas of endemic infection who are resident in the United States, Canada, France, or Australia. Participants in laboratory proficiency testing programs have improved their ability to identify these eggs in challenge stool samples, thus increasing the probability that patients who are infected with *C. sinensis* will be diagnosed.

The life cycle is perpetuated by humans or reservoir hosts defecating in or near freshwater sources containing suitable species of snails. In areas where fish cultivation is of economic importance, toilets have been built over ponds to provide a source of fish food. Operculate snails noted to be infected are *Parafossarulus*, *Bulinus*, *Semisulcospira*, *Alocinma*, and *Melanoides* spp. Metacercariae can encyst in numerous species of freshwater fish.

Aquaculture of fish has increased dramatically from 5.3% in 1970 to 32.2% in 2000 (23). Freshwater fish production has increased dramatically and now accounts for 45.1% of the total aquaculture production. The production of grass carp, an important species that serves as an intermediate host for food-borne trematodes, has increased from >10,000 tons in 1950 to >3 million tons in 2002. Growth of this industry must be monitored for potential problems related to an increased incidence of disease in which infection is transmitted through the ingestion of raw or poorly cooked fish.

The life cycle can be broken and infection can be prevented in humans by thorough cooking of all freshwater fish. Although cultural habits are difficult to change, public health education can be used to modify them. Night soil used without disinfection for fertilizer should not be applied in lakes or ponds containing susceptible snails. It has been suggested that night soil should be stored prior to use; eggs of *C. sinensis* die within 2 days when stored at 26°C.

Opisthorchis viverrini

O. viverrini has been reported to infect a significant portion of the population in northern Thailand and Laos. In some areas, the prevalence was 100% for age groups older than 10 years, and it is estimated that 10 million people are infected (19) (Table 16.1). Although the overall prevalence rate has been reported to be 35%, there are some areas where it is greater than 90%. There is an association between infection with *O. viverrini* and carcinoma of the bile duct epithelium (cholangiocarcinoma) (39).

Life Cycle and Morphology

Adult worms in distal bile ducts produce eggs that are carried by the bile to the intestinal lumen and to the outside environment in the feces (Figures 16.1, 16.4, and 16.6). The eggs are brownish yellow, oval, and operculated; they have opercular shoulders and may have a comma-shaped appendage at the abopercular end. The egg size averages 27 by 15 μm. The life cycle is similar to that of *C. sinensis*, requiring developmental periods in susceptible snails and finally in freshwater fish, in which the metacercariae encyst. Metacercariae excyst in the duodenum of the mammalian host and migrate to the intrahepatic bile ducts, where they develop to mature adults. The adult worms are transparent and leaf-shaped and measure approximately 8 to 12 mm long (Figure 16.4). The length-to-width ratio is similar to that of *C. sinensis*; therefore, the two can be confused. The life span of *O. viverrini* can exceed 20 years.

Clinical Disease

As with *C. sinensis*, the pathologic changes are confined mainly to the biliary tract system and morbidity is significantly associated with the worm burden of the host. Intensity of infection is correlated with clinical signs of abdominal pain, flatulence, weakness, hepatomegaly, cholangitis, chronic cholecystitis, cholelithiasis, and obstructive jaundice. Patients with heavy worm burdens may have severe cirrhosis, ascites, pedal edema, and acute abdominal pain. There may be severe jaundice, secondary infection of the biliary system (as seen with *C. sinensis*), cholangitis, and hepatomegaly. The bilirubin transaminase level tends to be elevated, while the serum albumin is low.

A large number of individuals with *O. viverrini* infection have cholangiocarcinomas, which were also seen at autopsy. The biliary epithelium may be highly susceptible to malignant transformation because of chronic proliferation due to *O. viverrini* infection. There is a significant relationship between the intensity of infection and cholangiocarcinoma, and there is a high prevalence of the disease among males (1, 39). There is a definite relationship between a high risk of liver cancer and infestation with *O. viverrini* in northeastern Thailand. However, in other areas of the country where the prevalence of liver cancer is high, there is little to no exposure to *O. viverrini* infection. Case-control studies suggest that exposure to exogenous and possibly endogenous nitrosamines in food or tobacco in betel nut cigarettes may play a role in the development of hepatocellular carcinoma, while infestation with *O. viverrini* and chemical interaction of nitrosamines may also be etiologic factors in the development of cholangiocarcinoma (30). Carcinomas of the biliary tract are rare cancers developing from the epithelial or blast-like cells lining the bile ducts. A variety of known predisposing factors, including infection with *O. viverrini*, have been identified. Chronic inflammatory processes, generation of active oxygen radicals, altered cellular detoxification mechanisms, activation of oncogenes, functional loss of tumor suppressor genes, and dysregulation of cell proliferation and apoptotic mechanisms have been identified as important contributors to the development of cholangiocarcinomas (21). Approximately 20,000 worms have been recovered from a single individual at autopsy.

Elevated IgE levels in serum of up to three or four times normal can be detected in infected patients (42). There is an early rise in the serum levels of IgM followed by elevations of IgA and IgG. In chronic infections, the

Figure 16.6 *Opisthorchis viverrini* adult trematode.

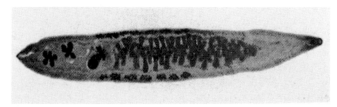

IgG and IgM levels remain elevated whereas the IgA levels return to normal (43).

Diagnosis

Individuals who have lived in areas where infections are endemic, have a history of eating raw fish, and have symptoms of right upper abdominal pain, indigestion, diarrhea, and hepatomegaly may be suspected of having liver trematode infection. Pain usually appears in the late afternoon and lasts several hours. Eggs can be detected in the feces by direct microscopy or by examination of fecal concentrates. Sedimentation concentration is recommended since these eggs have opercula and do not float in the zinc sulfate flotation concentration method. Sedimented material can be examined using saline or D'Antoni's iodine or both. It is important not to add too much iodine; if the eggs stain too darkly, they may resemble debris. In light infections, egg counts may be low and so multiple specimens may be required to confirm their presence. The eggs of *O. viverrini* and *C. sinensis* are similar in size and shape to those of *H. heterophyes* and *M. yokogawai* and cannot be readily differentiated from them. If a patient has not resided in or recently visited areas where infections are endemic, the infection is probably due to *O. viverrini* or *C. sinensis*. The diagnosis can be confirmed by detecting eggs in the bile fluid (duodenal aspirate), by recovering adult worms, or from the clinical history. Like *C. sinensis*, some strains of *O. viverrini* produce eggs that have a comma-shaped appendage at the abopercular end. Differentiation is difficult without a patient history. Multiple egg measurements are usually required to determine size differences; however, absolute identification among the small trematode eggs can be very difficult. Eggs can also be seen in duodenal drainage material or if the Entero-Test capsule is used. Definitive identification usually requires examination of adult worms recovered after therapy or during surgery or autopsy.

Eggs are not found in stool specimens from patients with biliary obstruction, so needle aspiration, surgery, or autopsy specimens may be required to confirm their presence. In these patients, biliary obstruction must be differentiated from enlarged gallbladder, cholangitis with jaundice, liver carcinoma, and cholangiocarcinoma. Cholangiography, ultrasonography, and liver scans may reveal lesions consistent with liver fluke infection. However, over 75% of patients with this infection tend to have normal ultrasonograms.

DNA hybridization and detection of antigen have been used to detect infections in stool (36). DNA hybridization and antigen detection methods have not been extensively tested in areas where infections are endemic but may be slightly better than classical methods of detection (37). Although serologic diagnostic tests for detection of antibodies have been developed, they are not widely available, are not standardized, and give false-positive results due to cross-reactivity with other parasitic diseases (5).

KEY POINTS—LABORATORY DIAGNOSIS

Opisthorchis viverrini

1. Multiple stool specimens may be required to find the eggs.
2. The sedimentation concentration method should be used; because the eggs are operculated, they do not float in the zinc sulfate flotation concentration method.
3. It is important not to add too much iodine to the wet preparation, or the small eggs will stain very darkly and resemble debris.
4. Although the wet preparation can be examined using the 10× (low-power) objective, these eggs measure approximately 30 μm and can be easily missed. It is recommended that the 40× (high dry) objective be used for the wet preparation examination.
5. Eggs of *O. viverrini* can resemble those of many other intestinal and liver trematodes.
6. The wet preparation should not be so thick that the small eggs are obscured by normal stool debris.
7. Some strains of *O. viverrini* and *C. sinensis* produce eggs that have a comma-shaped appendage at the abopercular end; this finding differentiates the liver fluke eggs from those of the intestinal flukes, *Heterophyes heterophyes* and *Metagonimus yokogawai*.
8. If duodenal sampling is performed (drainage or the Entero-Test), eggs may be found in this material.
9. If available, antigen or antibody detection may be helpful in confirming the infection.

Treatment

In patients with asymptomatic and mild to moderate cases, 75 mg of praziquantel per kg given three times after meals in a single day or a single dose of 40 mg/kg is 100 and 90% effective, respectively (2). In those with heavy infections, a single dose of 50 mg/kg produces a cure rate of 97% (12). No eggs are recovered in the stool after about a week; however, clinical symptoms and gallbladder dysfunction may take several months to resolve. Any side effects can be minimized by administration of the single-dose regimen at bedtime.

Mebendazole has also been given at 30 mg/kg/day for 3 or 4 weeks, with cure rates of 89 and 94%, respectively. No adult worms were recovered after therapy (12).

It is important to remember that only 5 to 10% of cases are relieved with praziquantel alone and that relapsing cholangitis and obstructive jaundice may require the use of antimicrobials. For complicated cases, surgery may also be required. Unfortunately, cholangiocarcinoma associated with *O. viverrini* infection carries a poor prognosis.

Epidemiology and Prevention

Dogs, cats, and other fish-eating mammals including humans are reservoir hosts. The infection has been reported to occur in Cambodia, Laos, and Thailand. In Southeast Asia, close to 70% of the population may be infected in some villages. Human infections result from the widely prevalent practice of consumption of metacercaria-infested freshwater fish eaten uncooked, pickled, smoked, salted, or dried. *Bithynia* snails are the first intermediate hosts, and freshwater fish are the second intermediate hosts. The metacercarial cysts are somewhat less resistant to pickling than those of *C. sinensis*; however, thorough cooking of fish is the only guaranteed method of prevention.

Public health education should be used to advise the at-risk population in areas of endemicity about the hazards of eating uncooked fish. This approach has not met with much success because of the difficulty in disrupting established cultural and dietary traditions. Also, defecation in or near ponds or lakes should be prevented, as should the application of night soil where the intermediate hosts are abundant. Community-based therapy of opisthorchiasis will significantly decrease the prevalence of infection; however, therapy alone will not solve the problem because of the high rates of reinfection due to cultural habits. Mass treatment can be undertaken with praziquantel as a single dose of 40 to 50 mg/kg at bedtime.

Opisthorchis felineus

Opisthorchis felineus infections in humans have been reported from Poland, Germany, the Russian Federation, and Kazakhstan, with the largest area of endemicity being found in western Siberia. Over 1.5 million people are infected, with some prevalence rates reaching 95%. The first report of a human infection was in 1892.

Life Cycle and Morphology

The life cycle can be seen in Figure 16.1. The eggs are elongate and ovoid, with an operculum that resembles that seen in *O. viverrini*. They are generally light yellowish brown and measure approximately 30 by 11 µm. Like the eggs of some strains of *C. sinensis* and *O. viverrini*, some eggs have a small tubercular thickening ("comma") at the abopercular end. Although the miracidium larva is

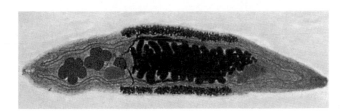

Figure 16.7 *Opisthorchis felineus* adult trematode.

fully developed when the egg is laid, hatching occurs only after ingestion by the snail intermediate host. Apparently, the only known molluscan host is the snail *Cordiella* (*Bithynia*), in which the first- and second-generation sporocysts and rediae develop. The cercariae leave the snail host and then encyst within the tissues of freshwater fish in the carp family. More than 20 species from 17 genera of freshwater fish serve as second intermediate hosts. Humans, cats, and other reservoir hosts become infected after ingestion of raw or poorly cooked fish. The metacercariae excyst in the duodenum, migrate to the distal bile passages, attach, and begin egg production within 3 to 4 weeks. The adult worm lives attached in the biliary and pancreatic ducts of the host. The adult worm is lancet-shaped and measures 7 to 12 mm in length, which is larger than the adult worms of *O. viverrini* (Figures 16.4 and 16.7).

Clinical Disease

Although this infection tends to have an incubation period of several weeks, symptoms may appear at the end of the first week. Like infections with other liver flukes, the degree of clinical involvement depends on the extent and duration of the infection. With a low worm burden (up to 50 or 60 worms), there may be localized damage in the distal bile ducts but the liver function remains normal. With a worm count up to 1,000, there may be hepatomegaly and jaundice; the pancreas may also be involved, which can result in digestive problems. Like some of the other liver trematode infections, it is quite likely that chronic infection will lead to cholangiocarcinoma. Overall, the symptoms and clinical sequelae are similar to those seen with *O. viverrini* infection. Acute symptoms can include abdominal pain, urticaria, dizziness, and fever.

Diagnosis

The infection is diagnosed by finding the eggs in the stool; however, they resemble eggs of the other liver trematodes such as *C. sinensis* and *O. viverrini*. The eggs have been described as being slightly more narrow (30 by 12 µm) than *C. sinensis* eggs and are more regularly ovoid without a clear shoulder at the operculum. Eggs can also be seen in duodenal drainage material or if the Entero-Test capsule is used.

Opisthorchis felineus

1. Multiple stool specimens may be required to find the eggs.
2. The sedimentation concentration method should be used; because the eggs are operculated, they do not float in the zinc sulfate flotation concentration method.
3. It is important not to add too much iodine to the wet preparation, or the small eggs will stain very darkly and resemble debris.
4. Although the wet preparation can be examined using the 10× (low-power) objective, these eggs measure approximately 30 µm and can be easily missed. It is recommended that the 40× (high dry) objective be used for the wet preparation examination.
5. Eggs of *O. felineus* can resemble those of many other intestinal and liver trematodes; however, they tend to be somewhat more narrow than the other eggs.
6. The wet preparation should not be so thick that the small eggs are obscured by normal stool debris.
7. Like some strains of *C. sinensis* and *O. viverrini*, *O. felineus* usually produces eggs that have a comma-shaped appendage at the abopercular end; this finding generally differentiates the liver fluke eggs from those of the intestinal flukes, *Heterophyes heterophyes* and *Metagonimus yokogawai*.
8. If duodenal sampling is performed (drainage or the Entero-Test), eggs may be found in this material.
9. If available, antigen or antibody detection may be helpful in confirming the infection.

Treatment

A single praziquantel dose of 40 mg/kg after a meal has been recommended. Although side effects are usually mild, they tend to occur in a number of patients and include abdominal pain, vomiting, diarrhea, lassitude, myalgia, headache, and rashes. Any side effects can be minimized by administration of the single-dose regimen at bedtime.

Epidemiology and Prevention

In central, eastern, and southeastern Europe, the Russian Federation, and Kazakhstan, the dog and cat, the red, silver, and polar foxes, domestic and wild swine, Norway rat, water rat, wolverine, marten, beaver, rabbit, polecat, Caspian seal, bearded seal, and lion (in captivity) are infected (5). In some areas of the Ukraine, there is an 82% infection rate, primarily from eating fish on the first day of salting. There is a 15 to 68% infection rate in the middle Urals, with an 84% infection rate in cats. There

has been as high as 90% infection with metacercariae in carp in the Tyumen Region of western Siberia (32). Cats are an important reservoir host, with a 43% infection rate being reported from the Ukraine.

Fasciola hepatica

Fascioliasis (sheep liver fluke disease) is primarily a zoonotic disease that causes liver infections with adult flukes. Over 2.4 million people are infected with this parasite worldwide, including Europe, the Americas, northern Asia, Oceania, Africa, New Zealand, Tasmania, Great Britain, Iceland, Cyprus, Corsica, Sardinia, Sicily, Japan, Papua New Guinea, the Philippines, and the Caribbean. This infection has been recognized by the World Health Organization as an important health problem, particularly in the Andean countries of Peru, Bolivia, and Chile (17). *F. hepatica* was the first trematode to be described and the first for which the entire life cycle was defined; it is interesting that approximately 500 years passed between the description of the parasite and the clarification of its life cycle. The first recorded information on *F. hepatica* infections was provided by de Brie in 1379, when he described the effects of certain types of water plants on sheep that had eaten them. The complete life cycle was established by the investigations of Leuckart and Thomas in 1883, independently of one another (9). The most common definitive hosts are sheep; however, other herbivores such as goats, cattle, horses, camels, hogs, vicuna, rabbits, and deer are often infected as well. In sheep, the migratory phase produces such extensive liver parenchyma damage that the disease is known as liver rot.

Although the disease has a cosmopolitan distribution, only a single case of autochthonous infection in a human in the United States has been reported (33) (Table 16.1).

Life Cycle and Morphology

Adult worms, which may live for many years in the bile ducts, produce eggs that are carried by the bile fluid into the intestinal lumen and passed into the environment with the feces (Figures 16.1, 16.4, and 16.8). The eggs are unembryonated, operculated, large, ovoid, and brownish yellow and measure 130 to 150 µm by 63 to 90 µm (Figure 16.8). The miracidium develops within 1 to 2 weeks in water from 22 to 26°C and escapes from the egg to infect the snail intermediate host, *Lymnaea* sp. These snails are amphibious. Within 4 to 7 weeks, cercariae are liberated from the snail after the production of a sporocyst generation and two or three redia generations. Cercariae encyst on water vegetation, e.g., watercress. Humans are infected by ingestion of uncooked aquatic vegetation on which metacercariae are encysted. Metacercariae excyst in the duodenum and migrate through the intestinal wall into the peritoneal cavity. The larvae enter the liver by penetrating the capsule (Glisson's capsule) and wander through the liver parenchyma for up to 9 weeks. The larvae finally enter the bile ducts, where they mature and produce eggs, which are passed out in the feces. The adult worms can attain a length of 30 mm and a width of about 13 mm and can live for more than 10 years (Figures 16.4 and 16.8).

Clinical Disease

The incubation period for fascioliasis can range from a few days to a few months. The patient's symptoms reflect the phase of the infection, as well as the number of parasites present in the host. In the acute phase, symptoms may be present over a period of weeks to months. Metacercarial larvae do not cause significant pathologic damage until they begin to migrate through the liver parenchyma. The amount of damage depends on the worm burden of the host. Linear lesions of 1 cm or greater can be found. Hyperplasia of the bile ducts occurs, possibly as a result of toxins produced by

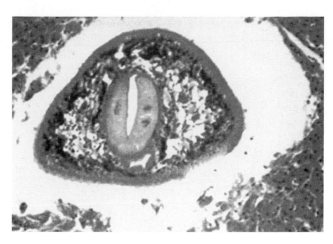

Figure 16.9 *Fasciola hepatica* adult worm in a bile duct; note the U-shaped worm within the bile duct with surrounding tissues.

Figure 16.8 *Fasciola hepatica* adult worm (upper left), egg (upper right), and living worm (lower). The operculum can be difficult to see; this egg resembles that of *F. buski* (intestinal trematodes) and *F. gigantica* (liver trematodes), although the egg of *F. gigantica* is larger. The *F. hepatica* adult worm is 20 to 30 mm long by 8 to 13 mm wide.

the larvae. Symptoms associated with this migratory phase have included fever, epigastric and right upper quadrant pain, intestinal complaints, and urticaria, although other patients remain asymptomatic. Asymptomatic infections appear to be more common in Peru (12). Other symptoms include loss of appetite, flatulence, nausea, diarrhea, and a nonproductive cough. Patients may also have hepatomegaly, splenomegaly, ascites, chest signs, and jaundice. Leukocytosis, eosinophilia, and mild to moderate anemia are found in many patients. Levels of IgG, IgM, and IgE in serum are usually elevated.

In the more chronic phases of the disease, the patient generally has few to no symptoms once the flukes have lodged in the biliary passages (Figure 16.9). However, there may be some epigastric and right upper quadrant pain, diarrhea, nausea, vomiting, hepatomegaly, and jaundice. If the flukes are found in the extrahepatic biliary ducts, symptoms may mimic those seen in cholelithiasis. In the chronic phase, there tend to be some liver function abnormalities, as well as eosinophilia (Figure 16.10). Larvae may be found in ectopic foci after penetrating the peritoneal cavity. Worms in human infections have been discovered in many areas of the body other than the liver. Other body sites include the intestinal wall, lungs, heart, brain, and skin. Symptoms mimic those seen with visceral larva migrans and include vague abdominal pain. Unfortunately, surgery may be required to determine the exact cause of the symptoms.

Once the worms have established themselves in the bile ducts and matured, they cause considerable damage from mechanical irritation and metabolic by-products as well as obstruction. The degree of pathologic change depends on the number of flukes penetrating the liver. The infection produces hyperplasia of the biliary epithelium and fibrosis of the ducts with portal or total biliary obstruction. The gallbladder undergoes

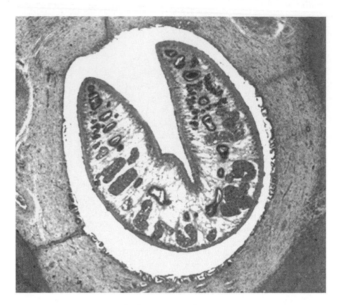

Figure 16.10 *Fasciola hepatica* adult worm in the liver; note the U-shaped worm with surrounding hepatic tissue, with a cellular response seen around the worm.

similar pathologic changes and may even harbor adult worms. Adult worms may reinvade the liver parenchyma, producing abscesses.

Symptoms and signs of infection during the late stages after egg production has begun are those associated with biliary obstruction and cholangitis. Acute epigastric pain, fever, pruritus, jaundice, hepatomegaly, and eosinophilia are common.

In areas of endemicity where uncooked goat and sheep livers may be eaten, such as Lebanon, adult worms may attach to the pharyngeal mucosa, causing suffocation (halzoun syndrome). This condition is temporary, although it may cause considerable discomfort. The adult worms may lodge on the pharyngeal mucosa, causing edema and congestion of the soft palate, pharynx, larynx, nasal fossae, and eustachian tube. The symptoms include dyspnea, dysphagia, deafness, and occasionally suffocation. It has also been suggested that a number of these cases may be caused by infection with larval linguatulids rather than adult worms of *F. hepatica*.

Diagnosis

Stool examinations and enzyme-linked immunosorbent assay (ELISA) are recommended for presumptive diagnosis, while radiographic techniques and magnetic resonance imaging are used for confirmation and follow-up. Although they may be helpful, invasive techniques such as percutaneous cholangiography, endoscopic retrograde cholangiography, and liver biopsy are not always essential (6). However, liver biopsy may be very important in diagnosing fascioliasis, particularly for asymptomatic disease (3).

The eggs of *F. hepatica*, *Fasciolopsis buski*, *Echinostoma ilocanum*, and *Gastrodiscoides hominis* are similar in size and shape. Differentiation may be difficult without a patient history. Patients may be symptomatic during the first weeks of infection, but no eggs are found in the stool until the worms mature, which takes 8 weeks (20). Multiple stool examinations may be needed to detect light infections.

Eggs may be detected in the stools of individuals who have eaten *F. hepatica*-infected liver, thereby yielding an erroneous laboratory result (not a true human infection, but what is called a "spurious" infection). True infection and spurious infection can be differentiated by maintaining the patient on a liver-free diet for at least 3 days. If the patient continues to pass eggs in the stool, the infection is probably genuine.

Serologic diagnosis with *F. hepatica* excretion-secretion antigens has been successfully used in areas of endemicity to detect patient antibodies (20). This approach includes ELISA, counterelectrophoresis, and enzyme-linked immunoelectrotransfer blots. Serologic tests can be used to monitor treated patients to ensure successful therapy (8). Serologic cross-reactions have been noted in patients infected with schistosomiasis. Serologic assays are the tests of choice for the diagnosis of acute and chronic fascioliasis when few eggs can be found in the stool or when large populations are to be screened (35).

KEY POINTS—LABORATORY DIAGNOSIS

Fasciola hepatica

1. Multiple stool specimens may be required to find the trematode eggs. In a very light infection, eggs may not be recovered; however, the patient might be asymptomatic.
2. The sedimentation concentration method should be used; because the eggs are operculated, they do not float in the zinc sulfate flotation concentration method.
3. It is important not to add too much iodine to the wet preparation, or the eggs will stain very darkly and resemble debris.
4. The wet preparation can be examined using the 10× (low-power) objective; these eggs are large enough that they can usually be seen using this magnification.
5. Eggs of *F. hepatica* can resemble those *Fasciola gigantica* (liver fluke) and *Fasciolopsis buski* (intestinal fluke).
6. The wet preparation should not be so thick that the eggs are obscured by normal stool debris.
7. If available, antigen or antibody detection may be helpful in confirming the infection.

Treatment

Although praziquantel is sometimes effective at a dose of 25 mg/kg taken after each meal for 2 days, it does not appear to be effective in treating patients in Egypt. Treatment with biothionol at 30 to 50 mg/kg on alternate days for 10 to 15 doses is recommended. Triclabendazole at 10 mg/kg as a single dose is also recommended, but this dose may not be as easily obtained. If a single dose has failed, the use of triclabendazole at 20 mg/kg is recommended (16); the drug is given orally in single or multiple doses and has few side effects (8). It acts by inhibiting protein synthesis in *F. hepatica* and will probably become the drug of choice; however, it is not currently available in the United States unless possibly through a compounding pharmacy (2).

Epidemiology and Prevention

Fascioliasis is a cosmopolitan disease that is found where there is close association of livestock, humans, and snails. The largest numbers of infections have been reported from Bolivia, Ecuador, Egypt, France, Iran, Peru, and Portugal. Although animal fascioliasis is endemic throughout the Americas, human infections are rare in most countries (29). Reservoir hosts include herbivores such as cattle, goats, and sheep. Animal fascioliasis is a major veterinary problem in Europe. The infection is contracted by the ingestion of metacercariae encysted on uncooked water plants, such as watercress. Watercress is the main source of infection; however, in many countries, wild watercress is touted as a healthy natural food. In many areas of the world, animal manure is used as the primary fertilizer for cultivation of watercress.

In surveying potential risk factors in the Northern Peruvian Altiplano, the principal exposure factor for *F. hepatica* infection was drinking alfalfa juice. Human fascioliasis should be suspected when a patient from an area in Peru where livestock is raised presents with recurrent episodes of jaundice and has a history of alfalfa juice or aquatic-plant ingestion or has eosinophilia (28).

Prevention may be accomplished by public health education in areas where infections are endemic, stressing the dangers of eating watercress grown in the wild where animals and snails are abundant. Other measures have included killing snails, treating infected animals, and draining pasture lands.

Fasciola gigantica

Fasciola gigantica is closely related to *F. hepatica*. It is also a common parasite of cattle, camels, and other herbivores in Africa and of herbivores in some Pacific islands. Human infections have been reported in a number of areas of endemicity. Generally, *F. hepatica* is found in temperate zones and is the predominant species in Europe,

the Americas, and Oceania, while *F. gigantica* is better adapted to tropical and aquatic environments and is the predominant species in Africa.

Life Cycle and Morphology

The life cycle is similar to that of *F. hepatica*; however, the snail hosts of *F. gigantica* are aquatic rather than amphibious like the first intermediate host for *F. hepatica*. Humans become infected through ingestion of water plants which carry the infective metacercariae. Apparently, developmental stages of *F. gigantica* grow more slowly, survive longer at high temperatures, and are more susceptible to drying than those of *F. hepatica*. The adult worm resembles *F. hepatica* but is somewhat more lanceolate, with a less distinct cephalic cone (Figure 16.11). The eggs are very difficult to differentiate from those of *F. hepatica* or *Fasciolopsis buski*; however, they tend to be larger (160 to 190 µm by 70 to 90 µm) (Figure 16.11).

Clinical Disease

The clinical symptoms of *F. gigantica* infection are very similar to those of *F. hepatica* infection and depend on the worm burden. The prepatent period between infection and the presence of adult worms in the bile ducts is 9 to 12 weeks. Patients may experience fever, nausea, vomiting, abdominal pain, hepatomegaly, hepatic tenderness, and eosinophilia. As in many light trematode infections,

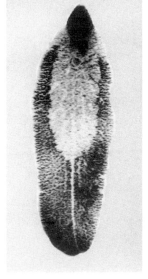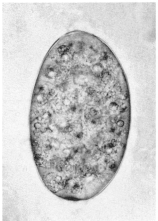

Figure 16.11 *Fasciola gigantica* adult worm (left) and egg (right). The operculum can be difficult to see; this egg resembles that of *F. buski* (intestinal trematodes) and *F. hepatica* (liver trematodes), although the egg of *F. gigantica* is larger (160 to 190 µm by 70 to 90 µm). The *F. gigantica* adult worm is 25 to 75 mm long by 12 mm wide. (From A Pictorial Presentation of Parasites: A cooperative collection prepared and/or edited by H. Zaiman.)

there may be vague symptoms or the patient may remain asymptomatic. Abscess or tumorlike reactions have also been reported to occur in subcutaneous tissues or in the liver (12).

Diagnosis

The eggs can be found in the stool; however, they may be absent more often than in infections with *F. hepatica*, so that multiple stool examinations may be required to demonstrate the eggs. Although these eggs are larger than those of *F. hepatica* or *F. buski*, they are very similar in shape. Recovery of adult flukes at surgery would confirm the diagnosis. Other diagnostic options discussed for *F. hepatica* are also recommended for this infection.

KEY POINTS—LABORATORY DIAGNOSIS

Fasciola gigantica

1. Multiple stool specimens may be required to find the trematode eggs. In a very light infection, eggs may not be recovered; however, the patient might be asymptomatic.
2. The sedimentation concentration method should be used; because the eggs are operculated, they do not float in the zinc sulfate flotation concentration method.
3. It is important not to add too much iodine to the wet preparation, or the eggs will stain very darkly and resemble debris.
4. The wet preparation can be examined using the 10× (low-power) objective; these eggs are large enough that they can usually be seen under this magnification.
5. Eggs of *F. gigantica* can resemble those *F. hepatica* (liver fluke) and *F. buski* (intestinal fluke); however, they are somewhat larger and could probably be differentiated by careful measurement. Unfortunately, many people might not consider this particular species and thus might not even measure the eggs.
6. The wet preparation should not be so thick that the eggs are obscured by normal stool debris.
7. If available, antigen or antibody detection may be helpful in confirming the infection.

Treatment

Although praziquantel is sometimes effective at a dose of 25 mg/kg taken after each meal for 2 days, it is not always effective. Treatment with biothionol at 30 to 50 mg/kg on alternate days for 10 to 15 doses is recommended. Triclabendazole at 10 mg/kg as a single dose is also recommended, but this drug may not be as easily obtained.

If a single dose has failed, the use of triclabendazole at 20 mg/kg is recommended (16). It is not currently available in the United States unless possibly through a compounding pharmacy (2).

Epidemiology and Prevention

Like *F. hepatica*, infection can occur in a wide range of herbivorous mammals when they ingest infected water plants or drink water contaminated with metacercariae. In some areas, the rate of infection in these animal hosts is quite high: in China, the rates are 50% for cattle, 45% for goats, and 33% for water buffalo; in Iraq, the rates are 71% for water buffalo and 27% for cattle; in northeastern Thailand, the rate is 60% for cattle (12). In various surveys, the occurrence of infections in cattle in Zambia was approximately 61%; in Tanzania, the incidence in cattle in traditional, large-scale dairy, and small-scale dairy herds was 63, 46.2, and 28.4% (24, 34). Studies in Australia testing a commercially available ELISA for antibody detection have been reported; the results indicate this test will be very valuable in screening cattle and sheep for infections with *F. hepatica* and probably also *F. gigantica* (31).

Human infections have also been reported in specific areas where animal infection is relatively uncommon; these include Zimbabwe, Uganda, Tashkent, Iraq, Vietnam, Hawaii, and some areas within Thailand.

Less Common Liver Flukes

Dicrocoelium dendriticum, *Dicrocoelium hospes*, and *Eurytrema pancreaticum*

Dicrocoelium dendriticum (lancet fluke, lanceolate fluke) is commonly found in the biliary passages of sheep, deer, and other herbivores and omnivores in Europe, Poland, Turkey, northern Africa, northern Asia, Turkistan, parts of the Far East, and North and South America; as an example, in southern Poland, 80% of the sheep are reported to be infected. Although many human infections have been reported, most are probably spurious infections acquired from eating infected raw sheep liver. However, true human infections have been reported from Europe, Egypt, Iran, Nigeria, Ivory Coast, and China. Approximately 30 cases of human infection with *D. hospes* have been reported from Ghana, Sierra Leone, Nigeria, and Democratic Republic of Congo. Although *D. hospes* has been thought to be a polymorph of *D. dendriticum*, ultrastructural studies can now distinguish *D. hospes* from other species in the family Dicrocoeliidae (4).

Eurytrema pancreaticum is usually found in the pancreatic ducts of hogs in southern China and of herbivores such as cattle, sheep, goats, monkeys, and camels in the

Orient and Brazil. True infection in humans has been reported in China and Japan.

Life Cycle and Morphology

The life cycle is similar to that of the other liver trematodes. However, in this case, the snail intermediate host is a land snail. The cercariae are released from the snail after rains follow a long period of dry weather. They are released from the snail's respiratory chamber as slime balls that are left behind on grass as the snail crawls along the ground or on plants. The ant (*Formica fusca*) is the required second intermediate host for *D. dendriticum* within the United States, while other ants serve this function in other areas. The second intermediate hosts for *E. pancreaticum* are either tree crickets or grasshoppers. Human infection is acquired through accidental ingestion of ants, primarily on fresh herbs or plants used for human consumption. Accidental ingestion of infected crickets or grasshoppers can also result in human infection. The metacercariae excyst and migrate to the biliary passage (for *D. dendriticum*) or the pancreatic ducts (for *E. pancreaticum*), where they then become adult flukes.

The adult worms of *D. dendriticum* are lancet-shaped, flat, and transparent and measure 5 to 15 mm long by 1.5 to 2.5 mm wide (Figure 16.12). The eggs are thick-shelled, operculate, and deep golden brown and measure 38 to 45 µm by 22 to 30 µm; the eggs of the two flukes cannot be differentiated (Figure 16.12). The eggs are embryonated when passed and are resistant to drying. The adult worm of *E. pancreaticum* is 8 to 16 mm long by 5 to 9 mm wide and tends to be more ovate and broader than that of *D. dendriticum* (Figure 16.13).

Clinical Disease

Although the life cycle is similar to that of *F. hepatica*, the pathogenic effects are less severe and patients may report mild symptoms. Symptoms include chronic constipation and flatulent dyspepsia. In patients with heavy infections, there may be jaundice with an enlarged liver. There may also be vomiting and diarrhea, as well as systemic toxemia. Eosinophilia tends to be absent in this infection.

Diagnosis

Diagnosis is based on recovery of the eggs in the stool, bile, or duodenal drainage specimens. It is important to remember that any presumptive true infection must be differentiated from spurious infections acquired from the ingestion of infected raw animal liver. True infection can be distinguished from spurious infection by maintaining the patient on a liver-free diet for at least 3 days. If the patient continues to pass eggs in the stool, the infection is

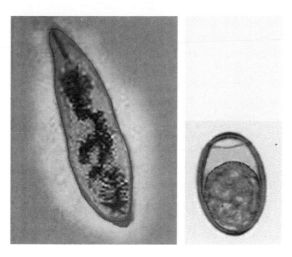

Figure 16.12 *Dicrocoelium dendriticum* adult worm (left) and egg (right). Adapted from reference 9.

probably genuine. Definitive diagnosis would require the recovery of adult worms at surgery or autopsy.

KEY POINTS—LABORATORY DIAGNOSIS

Dicrocoelium dendriticum, Dicrocoelium hospes, and Eurytrema pancreaticum

1. Multiple stool specimens may be required to find the eggs; bile and duodenal specimens can also be examined for the eggs.
2. The sedimentation concentration method should be used; because the eggs are operculated, they do not float in the zinc sulfate flotation concentration method.

Figure 16.13 *Eurytrema pancreaticum* adult worm (left) and egg (right). Adapted from reference 9.

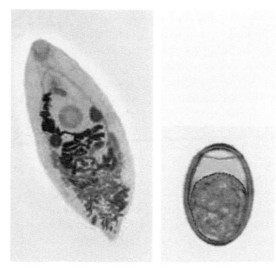

3. It is important not to add too much iodine to the wet preparation, or the small eggs will stain very darkly and resemble debris.

4. Although the wet preparation can be examined using the 10× (low-power) objective, these eggs measure approximately 40 µm and can be easily missed. It is recommended that the 40× (high dry) objective be used for the wet preparation examination.

5. Eggs of *D. dendriticum*, *D. hospes*, and *E. pancreaticum* can resemble those of many other small intestinal and liver trematodes such as *C. sinensis*, *O. viverrini*, *H. heterophyes*, and *M. yokogawai*.

6. The wet preparation should not be so thick that the small eggs are obscured by normal stool debris.

7. If duodenal sampling is performed (drainage or the Entero-Test), eggs may be found in this material.

8. If available, antigen or antibody detection may be helpful in confirming the infection.

Treatment

Praziquantel is the drug of choice, and the same dosage is recommended as that used for infection with *Opisthorchis* infection. Triclabendazole, an imidazole derivative, may also be effective.

Epidemiology and Prevention

Because infection in humans is usually accidental, few preventive measures are effective. Use of wild herbs and grasses as food should be avoided unless they are carefully washed. Crickets or grasshoppers could also be accidentally ingested if they were within plant material that had not been carefully washed.

Lung Flukes

Paragonimus spp.

Paragonimiasis is a disease of humans and carnivores in which adult flukes of the genus *Paragonimus* are found in the lungs. Naterer first detected the lung fluke in 1828; however, Kerbert detected flukes in the lungs of a Bengal tiger in 1878 and named the worms *Distoma westermani*, after the director of the Amsterdam Zoo, Westerman. Both Manson and von Baelz in 1880 reported finding eggs in the sputum of humans, and Yokogawa in 1915 and Nakagawai in 1916 completely described the life cycle. Since 1899, the name *Paragonimus westermani* has been used (Table 16.1). *P. mexicanus* is an important human pathogen in Central and South America, while *P. kellicotti* infections are found in North and South America. Most paragonimiasis infections are caused by *P. westermani*, which is the focus of this section (Table 16.2).

Life Cycle and Morphology

The life cycle is shown in Figure 16.1. The adult worm is a plump, ovoid, reddish brown fluke found encapsulated in the lung. Eggs deposited by the worms are ovoid, brownish yellow, unembryonated, and thick-shelled, with an operculum at one end and opercular shoulders. The eggs measure 80 to 120 µm by 45 to 65 µm (Figure 16.14). *P. westermani* eggs are often confused with *Diphyllobothrium* eggs because they are operculated, unembryonated, and somewhat similar in size. However, unlike *Diphyllobothrium* eggs, *P. westermani* eggs have opercular shoulders and a thickened shell at the abopercular end (Figure 16.15). Eggs escape from the encapsulated tissue through the bronchioles, are coughed up and voided in the sputum, or are swallowed and passed out in the feces. The eggs hatch in the water in 2 to 3 weeks, releasing a miracidium to infect a susceptible snail host. Cercariae are released after sporocyst and redia generations. Crabs and crayfish are infected by cercariae via the gill chamber or on ingestion of an infected snail. Cercariae encyst in the gill vessels and muscles. Humans are infected by ingesting uncooked crabs or crayfish containing metacercariae (Fig. 16.16). The metacercariae excyst in the duodenum and migrate through the intestinal wall into the abdominal cavity. The larvae migrate around or through the diaphragm into the pleural cavity and the lungs; they mature to adults in the vicinity of the bronchioles, where they discharge their eggs into the bronchial secretions. Although these worms are hermaphroditic, two worms are usually required for fertilization to occur. The worms can live as long at 20 years, but most die after about 6 years.

Although we generally think of infected raw crabs or crayfish as the most common sources of infection, a number of other foods and cultural practices can be implicated as well, including the ingestion of metacercariae in the flesh of paratenic hosts (Figure 16.16). Some of these are listed in Table 16.3.

Clinical Disease

Migration of the larval forms through the intestinal wall into the abdominal cavity is generally not associated with any significant pathologic changes or symptoms. If the larvae remain in the abdomen, some patients may have abdominal pain, intra-abdominal masses, tenderness, fever, diarrhea, nausea, vomiting, and eosinophilia. Once the larvae have reached the peritoneal cavity, they begin to migrate through organs and tissues, producing localized hemorrhage and leukocytic infiltrates.

Pulmonary Disease. When the worms finally reach the lungs and mature, a pronounced tissue reaction occurs, with infiltration of eosinophils and neutrophils. A fibrotic

Table 16.2 Human paragonimiasis

Species	Geographic distribution	Disease (mode of transmission)	Mean egg size (µm) and characteristics
P. westermani[a]	Asia	Pulmonary (crabs, crayfish, freshwater shrimp)	85–100 by 47; abopercular thickening; oval with barrel-shaped center, operculum with opercular shoulders
P. heterotremus[a] (*P. truanshanensis*)	China, Laos, Thailand	Pulmonary (crabs, raw shrimp salad in Thailand)	86 by 48, uniform thickness
P. mexicanus[a] (*P. peruvianus*, *P. ecuadoriensis*, *P. caliensis*)	Mexico, Central and South America	Pulmonary (crabs)	79 by 48; eggshell thin, irregularly undulated
P. africanus	Cameroon, Congo, Nigeria, Liberia, Guinea, The Gambia, and Côte d'Ivoire	Pulmonary (crabs)	90 by 50; tend to be more narrow, abopercular thickening
P. kellicotti	North and South America	Pulmonary (crabs, crayfish)	75–118 by 48–68; shell 2–3 µm thick; abopercular thickening; tapers more sharply than *P. westermani*
P. miyazakii[b]	Japan	Pleural; high frequency of pleural effusion, pleural exudate, marked eosinophilia (crabs, raw juice of crabs, crayfish)	75 by 43; shell very thin; thinner at abopercular end
P. westermani filipinus (*P. philippinensis*)	Philippines	Pulmonary (crabs)	79 by 50; abopercular thickening
P. skrjabini (*P. szechuanensis*)[b]	China	Pleural, subcutaneous nodules (crabs)	75 by 46; egg slightly asymmetrical, small knob at abopercular end
P. hueitungensis	China	Migratory, subcutaneous nodules (crabs); high leukocytosis, elevated eosinophilia	75 by 46; widest part of egg slightly anterior to middle; thin shell, visible opercular shoulders; small knob at abopercular end
P. uterobilateralis	Cameroon, Guinea, Liberia, Nigeria, The Gambia, Gabon, Sudan, Côte d'Ivoire, and southern Africa	Pulmonary (crabs)	70 by 45; abopercular thickening of shell

[a] Pathogenic organisms most frequently isolated from humans; synonyms are given in parentheses.

[b] It has been suggested that in some areas in China and Japan, *P. skrjabini* and *P. miyazakii* should be referred to as *P. skrjabini miyazakii*. Some workers regard the following as synonyms of *P. skrjabini*: *P. miyazakii*, *P. szechuanensis*, and *P. hueitungensis*. Examination of specimens and reclassification studies are under way (10).

capsule forms around the worm. The cysts contain purulent fluid with flecks or "iron filings" composed of brownish yellow eggs. Many of the cysts perforate into the bronchioles, releasing their contents of eggs, necrotic debris, metabolic by-products, and blood into the respiratory tract. The eggs may also enter the pulmonary tissue, or they may be carried by the circulatory system to other body sites, where they cause a granulomatous reaction (Figure 16.17).

Symptoms of paragonimiasis depend largely on the worm burden of the host and are usually insidious in onset and mild in patients with chronic infections. Light infections may be asymptomatic, although peripheral blood eosinophilia and lung lesions may be noted on X-ray examination. As the cysts rupture, a cough develops, with increased production of viscous blood-tinged sputum (rusty sputum, which may have a foul fish odor) and increasing chest pain. The patient may experience increasing dyspnea with chronic bronchitis and be misdiagnosed as having tuberculosis or bronchial asthma (40). The individual generally has a moderately high peripheral blood eosinophilia and leukocytosis, with elevated levels of IgG and IgE in serum. Although some patients exhibit symptoms continuously, others may remain asymptomatic for weeks to months between periods of hemoptysis.

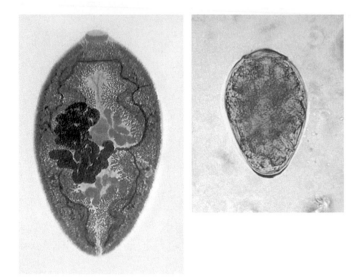

Figure 16.14 *Paragonimus westermani* adult worm (left) and egg (right). The opercular shoulders are clearly visible at the top of the egg, as is the thickened abopercular end.

Cerebral Disease. Larval forms may end up in many ectopic sites other than the lungs; ectopic infections are generally associated with *P. heterotremus*, *P. mexicanus*, and occasionally *P. westermani*. Cysts have been detected in the liver, intestinal wall, muscles, peritoneum, and brain (Figure 16.18). The most serious consequences of paragonimiasis are the cerebral complications, which are commonly found in younger age groups. Unlike adult flukes in other extrapulmonary sites, worms found in the brain usually contain eggs. The worms probably migrate from ruptured lung cysts and travel through the soft tissues surrounding veins

Figure 16.15 Photographs (top) and drawings (bottom) of *Diphyllobothrium latum* egg (A) and *Paragonimus westermani* egg (B). (Illustration by Nobuko Kitamura.)

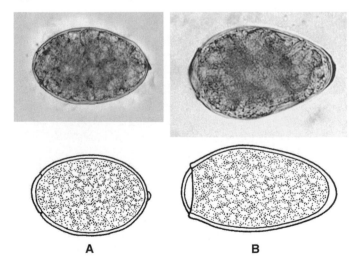

A   **B**

Figure 16.16 Crab second intermediate host for *Paragonimus* spp.

into the brain area. The worms eventually encapsulate, but before being walled off they cause necrosis within the brain tissue, as well as possible cerebral hemorrhage, edema, and meningitis.

Most patients with cerebral or other extrapulmonary lesions have an associated lung lesion or a history of lung

Table 16.3 Possible means of acquiring infection with *Paragonimus* spp.[a]

Geographic area	Food or cultural practice
Asia	Ingestion of raw or inadequately cooked crabs, crayfish, and shrimp
China	"Drunken crab" (immersion of live crabs in wine for 12 h); raw crab sauce or jam; crayfish curd
Thailand	Kung Ten (raw shrimp salad); Nam Prik Poo (crab sauce)
Philippines	Sinugba (roasted crab); Kinilao (raw crab)
Korea	Ke Jang (crab immersed in soy sauce)
Korea and Japan	Raw juice of crabs or crayfish used in traditional medicine (treatment of measles, diarrhea, and urticaria)
Japan	Ingestion of raw infected wild boar meat (paratenic host)
Africa	Ingestion of raw crustaceans to increase fertility
Any of the above	Contaminated utensils, chopping blocks, and cloths used in food preparation

[a] Data from reference 12.

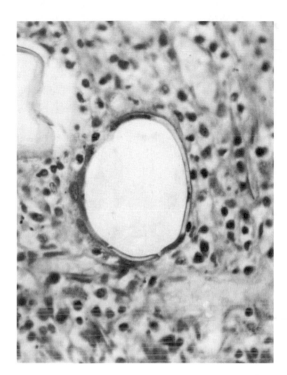

Figure 16.17 *Paragonimus westermani* egg in the lung (cross section). (Armed Forces Institute of Pathology photograph.)

disease. Symptoms include fever, headache, nausea, vomiting, visual disturbances, motor weakness, localized or generalized paralysis, and possibly death. More than half of these patients also exhibit personality changes, possible disorientation, and a general decline in cognitive function. Depending on the central nervous system location, symptoms may also include paraplegia, sensory loss, or vision problems. Specific sites can include the cerebral cortex, cerebellum, basal ganglia, medulla oblongata, and spinal cord (11, 32). When a worm dies, the lesion cavity becomes filled with necrotic material. Cerebral paragonimiasis can be difficult to differentiate from brain disease caused by other parasites such as *Schistosoma japonicum*, *Gnathostoma spinigerum*, or *Angiostrongylus cantonensis*, as well as other infectious agents such as bacteria or viruses.

Paragonimiasis in Other Body Sites. Abdominal or subcutaneous masses are frequently seen in patients with infections caused by *P. skrjabini*, *P. heterotremus*, *P. hueitungensis*, or *P. mexicanus*. Migratory subcutaneous nodules occur in 20 to 60% of patients with *P. skrjabini* infection and approximately 10% of patients with *P. westermani* infection. The nodules are firm, tender, a few millimeters to 10 cm, and somewhat irritating. They are often located in the lower abdomen, inguinal region, and thigh and are somewhat mobile. Ulcers or abscesses can

also occur in the skin or subcutaneous tissue. Involvement of the mastoid area has been seen in *P. africanus* infections. Other body sites that have been infected include the breast, lymph nodes, heart, pericardium, mediastinum, kidneys, adrenal gland, omentum, bone marrow, stomach wall, bladder, spleen, pancreas, and reproductive organs. While ectopic lesions are usually thought to be caused by worm migration, dissemination of eggs to other body sites can also be responsible for the pathology.

In China, *P. skrjabini* does not develop to the adult worm stage, but the disease is characterized by the presence of trematode larva migrans (12). These migratory subcutaneous nodules are associated with a high eosinophilia, as well as necrotic liver lesions. Brain involvement is common and is associated with subarachnoid hemorrhage.

Figure 16.18 (Upper) *Paragonimus westermani* egg in brain granuloma (cross section). (Armed Forces Institute of Pathology photograph.) (Lower) *P. westermani* eggs in brain tissue. (From A Pictorial Presentation of Parasites: A cooperative collection prepared and/or edited by H. Zaiman.)

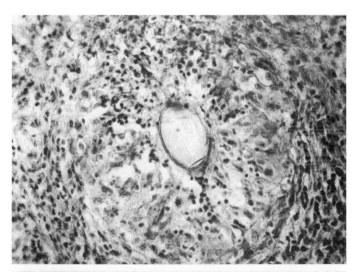

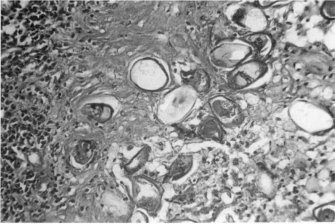

Diagnosis

Laboratory Tests. Individuals with symptoms of chronic cough, vague chest pains, and hemoptysis who have resided in an area where infections are endemic and have a history of eating raw crayfish or crabs should be suspected of having paragonimiasis. *Paragonimus* eggs can be detected in the sputum and the stool, and concomitant examinations should be performed to improve the overall detection rate. In many individuals in whom the infection is eventually confirmed, small numbers of eggs are present intermittently in the sputum and feces. For patients with light infections, up to seven sputum examinations have been recommended. Frequently, pulmonary paragonimiasis is misdiagnosed as pulmonary tuberculosis. The Ziehl-Neelsen method for detecting mycobacteria destroys *Paragonimus* eggs (40). It is important to remember that the typical findings of cough, hemoptysis, and eggs in the feces or sputum may be absent in patients with ectopic or pleural *Paragonimus* infection.

Radiography and Scans. Skull radiographs generally reveal round or oval cystic calcifications; they have been described as looking like bubbles. CT scans reveal multilocular cysts with edema, migration tracks, peripheral density, bronchial wall thickening, and centrilobular nodules; ring enhancement may be visible after the use of contrast. Magnetic resonance imaging (MRI) demonstrates areas of granulomatous inflammation surrounding the lesions; MRI of the brain may reveal multiple conglomerated iso- or low-signal-intensity round nodules with peripheral rim enhancement. In the liver, CT and MR images may reveal a cluster of small cysts with rim enhancement in the subcapsular area. This finding seems to be characteristic for hepatic paragonimiasis. Solitary nodular lesions can often mimic lung cancer, tuberculosis, or fungal diseases. When a pulmonary mass lesion or empyema is detected in patients who live in areas where paragonimiasis is endemic, paragonimiasis should always be included in the differential diagnosis of lung diseases.

Immunodiagnostic Tests. In addition to the methods described in the Diagnosis section, immunodiagnostic tests have been used to diagnose pulmonary and extrapulmonary infections. Complement fixation has been used to diagnose active infections; however, the test becomes negative soon after the death of the worms. An immunoblot assay for the detection of antibodies to *P. westermani* by using adult worm homogenates is highly sensitive and specific (38). Only one false-positive result has been detected by this method, in a patient with a *Schistosoma haematobium* infection. ELISA with adult excretory-secretory antigens is very sensitive for detecting parasite-specific IgG and IgE (22). Pleural effusion fluid is more suitable than serum for detection of infections. Dot ELISA has also been used to detect parasite-specific antigen in human sera (45). Monoclonal antibodies used to detect species-specific and stage-specific antigens are highly sensitive and specific for the detection of active infections (44). The serologic assays are available in areas of endemicity or in specialized diagnostic centers. Most of these assays involve nonstandardized reagents and have not been used in clinical trials. An intradermal test is also available for epidemiologic surveys in areas of endemicity; however, positive reactions persist for years after parasitologic cures.

KEY POINTS—LABORATORY DIAGNOSIS

Paragonimus spp.

1. *Paragonimus* eggs may be confused with *Diphyllobothrium latum* eggs because of similarities in their size and shape. However, most *Paragonimus* eggs have opercular shoulders and a marked thickening at the abopercular end, unlike *D. latum* eggs.

2. The sedimentation concentration method should be used; because the eggs are operculated, they do not float in the zinc sulfate flotation concentration method.

3. It is important not to add too much iodine to the wet preparation, or the eggs will stain very darkly and resemble debris.

4. The wet preparation can be examined using the 10× (low-power) objective; these eggs are large enough that they can usually be seen under this magnification.

5. The wet preparation should not be so thick that the eggs are obscured by normal stool debris.

6. When looking at sputum specimens for eggs, it is necessary to carefully examine any blood-tinged flecks for eggs; the egg clusters have been described as looking like iron filings.

7. In light infections, multiple stool and sputum specimens may be needed before the eggs are detected.

8. If available, antigen or antibody detection may be helpful in confirming the infection.

Treatment

The drug of choice is praziquantel at 25 mg/kg three times at 4-h intervals after meals for 2 or 3 days. Very few patients require retreatment; 100% cure is usually obtained except for certain patients with a heavy worm burden. There is dramatic improvement in symptoms, and they are usually gone within a few months. The dose required in cerebral paragonimiasis is generally higher and may have to be adjusted relative to the clinical outcome; convulsions and coma have been seen. These patients should be hospitalized for therapy and should be monitored closely; corticosteroids may also have to be given when treating patients with praziquantel. Bithionol is an alternative drug, which is available under an Investigational New Drug protocol from the Centers for Disease Control and Prevention Drug Service (2).

Triclabendazole at 10 mg/kg as a single dose or two doses is also recommended, but this dose may not be as easy to obtain. Patients should be examined at 3 months to determine the need for retreatment. If a single dose has failed, the use of triclabendazole at 20 mg/kg is recommended (13). Triclabendazole is not currently available in the United States unless possibly available through a compounding pharmacy (2).

Pulmonary paragonimiasis is rarely fatal; however, cerebral disease is characterized by chronic morbidity and symptoms including epilepsy, dementia, and other neurologic sequelae. About 5% of patients with cerebral disease die due to hemorrhage in the first 2 years of the disease (12).

Epidemiology and Prevention

There are approximately 200 million people at risk and 22 million people with paragonimiasis worldwide; 10 species are recognized as causing human disease (40) (Table 16.2). *P. westermani* is the most common human pathogen.

Reservoir hosts include dogs and cats in areas of endemic infection (the Far East and Africa). Eggs expectorated or passed in the feces in the vicinity of lakes or streams where the intermediate hosts live serve as the source of infection. Snail intermediate hosts include *Semisulcospira* and *Brotia* spp. In Ecuador, the intermediate host for *P. mexicanus* is *Aroapyrgus* sp., a freshwater snail belonging to the family Hydrobiidae (7). Humans contract the infection through ingestion of raw, undercooked, pickled, or wine-soaked crabs or crayfish; they may also become infected by the ingestion of uncooked meat from wild animals, such as wild boar. The migrating larvae in this meat may pass through the intestinal wall and continue their developmental cycle when eaten. Raw juice from crushed crayfish used as a home remedy for the treatment of measles has been a significant source of infection and may be a cause of cerebral paragonimiasis in children (43).

In most areas, the disease is primarily one of crab- or crayfish-eating mammals, with humans being incidental hosts. Measures devoted to public health education, warning people of the dangers of eating uncooked crabs or crayfish from areas of endemic infection, and sanitary care of utensils and fingers used to prepare food would reduce the risk of infection. Because the cycle depends on animals, the elimination of human paragonimiasis would have little effect on the overall prevalence of the disease.

References

1. **Abdel-Rahim, A. Y.** 2001. Parasitic infections and hepatic neoplasia. *Dig. Dis.* **19:**288–291.
2. **Abramowicz, M. (ed.).** 2004. Drugs for parasitic infections. *Med. Lett. Drugs Ther.* **46:**1–12.
3. **Adachi, S., K. Kotani, T. Shimizu, K. Tanaka, T. Shimizu, and K. Okada.** 2005. Asymptomatic fascioliasis. *Intern. Med.* **44:**1013–1015.
4. **Agostini, S., J. Miquel, P. I. Ndiaye, and B. Marchand.** 2005. *Dicrocoelium hospes* Looss, 1907 (Digenea, Dicrocoeliidae): spermiogenesis, mature spermatozoon and ultrastructural comparative study. *Parasitol. Res.* **96:**38–48.
5. **Akai, P. S., S. Pungpak, W. Chaicumpa, V. Kitikoon, Y. Ruangkunaporn, D. Bunnag, and A. D. Befus.** 1995. Serum antibody responses in opisthorchiasis. *Int. J. Parasitol.* **25:**971–973.
6. **Aksoy, D. Y., U. Kerimoglu, A. Oto, S. Erguven, S. Arslan, S. Unal, F. Batman, and Y. Bayraktar.** 2005. Infection with *Fasciola hepatica. Clin. Microbiol. Infect.* **11:**859–861.
7. **Amunarriz, M.** 1991. Intermediate hosts of *Paragonimus* in the eastern amazonic region of Ecuador. *Trop. Med. Parasitol.* **42:**164–166.
8. **Apt, W., X. Aguilera, F. Vega, C. Miranda, I. Zulantay, C. Perez, M. Gabor, and P. Apt.** 1995. Treatment of chronic fascioliasis with triclabendazole: drug efficacy and serologic response. *Am. J. Trop. Med. Hyg.* **52:**532–535.
9. **Beaver, P. C., R. C. Jung, and E. W. Cupp.** 1984. *Clinical Parasitology*, 9th ed. Lea & Febiger, Philadelphia, Pa.
10. **Blair, D., Z. Chang, M. Chen, A. Cui, B. Wu, T. Agatsuma, M. Iwagami, D. Corlis, C. Fu, and X. Zhan.** 2005. *Paragonimus skrjabini* Chen, 1959 (Digenea: Paragonimidae) and related species in eastern Asia: a combined molecular and morphological approach to identification and taxonomy. *Syst. Parasitol.* **60:**1–21.
11. **Brown, W. J., and M. Voge.** 1982. *Neuropathology of Parasitic Infections.* Oxford University Press, Oxford, United Kingdom.
12. **Bunnag, D., J. H. Cross, and T. Bunnag.** 2000. Liver fluke infections, p. 840–847. *In* G. T. Strickland (ed.), *Hunter's Tropical Medicine and Emerging Infectious Diseases*, 8th ed. The W. B. Saunders Co., Philadelphia, Pa.
13. **Calvopina, M., R. H. Guderian, W. Paredes, and P. J. Cooper.** 2003. Comparison of two single-day regimens of triclabendazole for the treatment of human pulmonary paragonimiasis. *Trans. R. Soc. Trop. Med. Hyg.* **97:**451–454.
14. **Chapman, R. W.** 1999. Risk factors for biliary tract carcinogenesis. *Ann. Oncol.* **10:**S308–S311.
15. **Choi, B. I., J. K. Han, S. T. Hong, and K. H Lee.** 2004. Clonorchiasis and cholangiocarcinoma: etiologic relationship and imaging diagnosis. *Clin. Microbiol. Rev.* **17:**540–552.
16. **Dauchy, F. A., P. Vincendeau, and F. Livermann.** 2005. Eight cases of fascioliasis: clinical and microbiological features. *Med. Mal. Infect.* 22 Nov. [Epub ahead of print.] **36:**42–46.

17. Fuentes, M. V., S. Sainz-Elipe, P. Nieto, J. B. Malone, and S. Mas-Coma. 2005. Geographical information systems risk assessment models for zoonotic fascioliasis in the South American Andes region. *Parassitologia* 47:151–156.

18. Harrison, P. M. 1999. Prevention of bile duct cancer in primary sclerosing cholangitis. *Ann. Oncol.* 10:S208–S211.

19. Haswell-Elkins, M. R., P. Sithithaworn, and D. Elkins. 1992. *Opisthorchis viverrini* and cholangiocarcinoma in northeast Thailand. *Parasitol. Today* 8:86–89.

20. Hillyer, G. V., M. S. de Galanes, J. Rodriguez-Perez, J. Bjorland, M. S. de Lagrava, S. R. Guzman, and R. T. Bryan. 1992. Use of the Falcon™ assay screening test–enzyme-linked immunosorbent assay (FAST-ELISA) and the enzyme-linked immunoelectrotransfer blot (EITB) to determine the prevalence of fascioliasis in the Bolivian altiplano. *Am. J. Trop. Med. Hyg.* 46:603–609.

21. Holzinger, F., K. Z'graggen, and M. W. Buchler. 1999. Mechanisms of biliary carcinogenesis: a pathogenetic multistage cascade towards cholangiocarcinoma. *Ann. Oncol.* 10:S122–S126.

22. Ikeda, T., Y. Oikawa, M. Owhashi, and Y. Nawa. 1992. Parasite-specific IgE and IgG levels in the serum and pleural effusion of *Paragonimiasis westermani* patients. *Am. J. Trop. Med. Hyg.* 47:104–107.

23. Keiser, J., and J. Utzinger. 2005. Emerging foodborne trematodiasis. *Emerg. Infect. Dis.* 11:1507–1514.

24. Keyyu, J. D., A. A. Kassuku, L. P. Msalilwa, J. Monrad, and N. C. Kyvsgaard. 2006. Cross-sectional prevalence of helminth infections in cattle on traditional, small-scale, and large-scale dairy farms in Iringa District, Tanzania. *Vet. Res. Commun.* 30:45–55.

25. Kim, Y. H. 1999. Pancreatitis in association with *Clonorchis sinensis* infestation: CT evaluation. *Am. J. Roentgenol.* 172:1293–1296.

26. Le, T. H., D. N. Van, D. Blair, P. Sithithaworn, and D. P. McManus. 2005. *Clonorchis sinensis* and *Opisthorchis viverrini*: development of a mitrochondrial-based multiplex PCR for their identification and discrimination. *Exp. Parasitol.* 24 Nov. [Epub ahead of print.] 112:109–114.

27. Lim, J. H. 1990. Radiologic findings of clonorchiasis. *Am. J. Roentgenol.* 155:1001–1008.

28. Marcos, L., V. Maco, F. Samalvides, A. Terashima, J. R. Espinoza, and E. Gotuzzo. 2005. Risk factors for *Fasciola hepatica* infection in children: a case-control study. *Trans. R. Soc. Trop. Med. Hyg.* 100:158–166.

29. Mas-Coma, S. 2005. Epidemiology of fascioliasis in human endemic areas. *J. Helminthol.* 79:207–216.

30. Mitacek, E. J., K. D. Brunnemann, M. Suttajit, N. Martin, T. Limsila, H. Ohshima, and L. S. Caplan. 1999. Exposure to N-nitroso compounds in a population of high liver cancer regions in Thailand: volatile nitrosamine (VNA) levels in Thai food. *Food Chem. Toxicol.* 37:297–305.

31. Molloy, J. B., G. R. Anderson, T. I. Fletcher, J. Landmann, and B. C. Knight. 2005. Evaluation of a commercially available enzyme-linked immunosorbent assay for detecting antibodies to *Fasciola hepatica* and *Fasciola gigantica* in cattle, sheep, and buffaloes in Australia. *Vet. Parasitol.* 130:207–212.

32. Muller, R. 1975. *Worms and Disease: a Manual of Medical Helminthology*, p. 25–31. William Heinemann Medical Books Ltd., London, United Kingdom.

33. Norton, R. A., and L. Monroe. 1961. Infection by *Fasciola hepatica* acquired in California. *Gastroenterology* 41:46–48.

34. Phiri, A. M., I. K. Phiri, C. S. Sikasunge, and J. Monrad. 2005. Prevalence of fasciolosis in Zambian cattle observed at selected abattoirs with emphasis on age, sex and origin. *J. Vet. Med. Ser. B* 52:414–416.

35. Shaker, Z. A., Z. A. Demerdash, W. A. Mansour, H. I. Hassanein, H. G. el Baz, and H. I. el Gindy. 1994. Evaluation of specific *Fasciola* antigen in the immunodiagnosis of human fascioliasis in Egypt. *J. Egypt. Soc. Parasitol.* 24:463–470.

36. Sirisinha, S., R. Chawengkirttikul, M. R. Haswell-Elkins, D. B. Elkins, S. Kaewkes, and P. Sithithaworn. 1995. Evaluation of a monoclonal antibody-based enzyme linked immunosorbent assay for the diagnosis of *Opisthorchis viverrini* infection in an endemic area. *Am. J. Trop. Med. Hyg.* 52:521–524.

37. Sirisinha, S., R. Chawengkirttikul, R. Sermswan, S. Amornpant, S. Mongkolsuk, and S. Panyim. 1991. Detection of *Opisthorchis viverrini* by monoclonal antibody-based ELISA and DNA hybridization. *Am. J. Trop. Med. Hyg.* 44:140–145.

38. Slemenda, S. B., S. E. Maddison, E. C. Jong, and D. D. Moore. 1988. Diagnosis of paragonimiasis by immunoblot. *Am. J. Trop. Med. Hyg.* 39:469–471.

39. Sripa, B. 2003. Pathobiology of opisthorchiasis: an update. *Acta Trop.* 88:209–220.

40. Toscano, C., Y. S. Hai, P. Nunn, and K. E. Mott. 1995. Paragonimiasis and tuberculosis, diagnostic confusion: a review of the literature. *Trop. Dis. Bull.* 92:R1–R26.

41. Watanapa, P., and W. B. Watanapa. 2002. Liver fluke-associated cholangiocarcinoma. *Br. J. Surg.* 89:962–970.

42. Woolf, A., J. Green, J. A. Levine, E. G. Estevez, N. Weatherly, E. Rosenberg, and T. Frothingham. 1984. A clinical study of Laotian refugees infected with *Clonorchis sinensis* or *Opisthorchis viverrini*. *Am. J. Trop. Med. Hyg.* 33:1279–1280.

43. World Health Organization. 1995. Control of foodborne trematode infections. *WHO Tech. Rep. Ser. 849*.

44. Zhang, Z., Y. Zhang, Z. Shi, K. Sheng, L. Lui, Z. Hu, and W. F. Piessens. 1993. Diagnosis of active *Paragonimus westermani* infections with a monoclonal antibody-based antigen detection assay. *Am. J. Trop. Med. Hyg.* 49:329–334.

45. Zihao, Z., S. Yiping, and W. F. Piessens. 1991. Characterization of stage-specific antigens of *Paragonimus westermani*. *Am. J. Trop. Med. Hyg.* 44:108–115.

Blood Trematodes: Schistosomes

Schistosoma mansoni

Schistosoma japonicum

Schistosoma mekongi

Schistosoma haematobium

Schistosoma intercalatum

Schistosomes belong to the phylum Platyhelminthes, family Schistosomatidae, and are a group of digenetic, dioecious trematodes requiring definitive and intermediate hosts to complete their life cycles. Four species are important agents of human disease: *Schistosoma mansoni*, *S. japonicum*, *S. mekongi*, and *S. haematobium*. *S. intercalatum* is of less epidemiologic importance. Schistosomiasis affects between 200 million and 300 million people in 77 countries throughout the world and is a significant cause of disease in areas of endemic infections. In Egypt, approximately 20% of the population is infected; prevalence rates in some villages have been estimated to be 85%. The number of infected individuals in China is estimated to be 1.52 million. Although only about 10% of infected people have serious disease, this represents 20 million to 30 million individuals worldwide. Approximately half of the remaining 180 million to 270 million infected individuals have symptoms.

The earliest known instance of schistosomiasis was found in Egyptian mummies of the predynastic period (3100 BC), using enzyme-linked immunosorbent assay (ELISA) to detect circulating anodic antigen (43). Previously, eggs were detected in the kidneys of mummies from the Twentieth Dynasty (1250 BC to 1000 BC). Fujii in 1947 described schistosomiasis caused by *S. japonicum*, and Bilharz in 1951 noted *S. haematobium* infections in Cairo. Sambon proposed the name for *S. mansoni*. Schistosome infections in the New World probably began with the African slave trade in the Americas during the 16th and 17th centuries.

Schistosomes are somewhat different from other human trematodes since they (i) have two sexes, (ii) live in the blood vessels, (iii) have nonoperculated eggs, and (iv) have no encysted metacercarial stage in the life cycle. The human is the definitive host for *S. mansoni*, *S. japonicum*, and *S. haematobium*; for *S. mekongi* and *S. malayi* (both similar to *S. japonicum*); and for *S. intercalatum*. *S. mattheei*, which causes infections in sheep, cattle, and horses, also infects humans and can cause disease. Other schistosomes have been found in humans but do not tend to cause any pathology. Also, cercariae from birds and mammals can penetrate human skin but cannot complete the life cycle and tend to die without migrating or maturing; however, they do cause cercarial dermatitis.

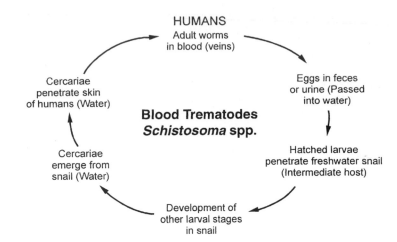

Figure 17.1   Life cycle of schistosomes.

Schistosoma mansoni

In one of the earliest reports on African schistosomiasis, Theodor Bilharz first noted the disease but mistakenly described two types of eggs, one with a lateral spine and one with a terminal spine, as occurring in the same adult worm. Manson was the first to note schistosomiasis in the New World, but it was Sambon who adequately described the differences between intestinal and urinary tract schistosomiasis and proposed the name *S. mansoni*. *S. mansoni* was probably established in the Western Hemisphere as a result of the African slave trade.

In Egypt, hepatitis C virus (HCV) infection, along with *S. mansoni*, is the main cause of chronic liver disease and liver cirrhosis. HCV genotype 4a is highly prevalent in Egypt. Schistosomiasis, HCV, HBV, and HGV are all associated with liver disease. However, coinfection with two or more of these infectious agents may potentiate the pathogenesis of the liver disease (21).

Life Cycle and Morphology

Humans become infected by penetration of cercariae through intact skin rather than by ingestion of metacercariae (Figure 17.1); penetration can occur within as little as 5 min after initial skin contact. Cercariae consist of a body with glands containing a proteolytic enzyme (elastase), which is used to penetrate skin, and a bifurcated tail that is lost when the cercariae penetrate the skin (Figure 17.2). Once the cercariae have successfully entered the host, the organism is termed a schistosomulum. After about 48 h, the schistosomulum migrates through the tissues and finally invades a blood vessel. On entry into the blood vessels, the schistosomulum is carried to the lungs and then the liver. Once within the liver sinusoids, the worms begin to mature into adults. The adults of *S. mansoni* are found in the inferior mesenteric veins. The worms form pairs (male and female), with the female lying in the

gynecophoral canal of the male (Figures 17.3 and 17.4). Sexual maturity of female schistosomes depends on the presence of mature male worms. The female worm has selective gene expression in the reproductive tract in the presence of male worms, whereas females separated from

Figure 17.2   Schistosome cercariae. (Top, illustration by Sharon Belkin.)

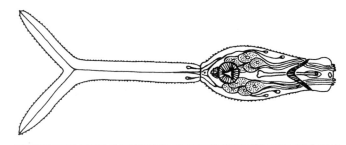

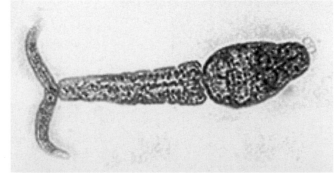

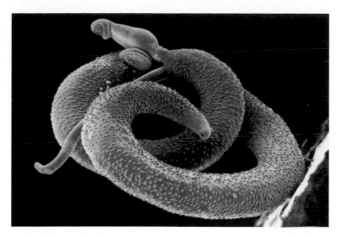

Figure 17.3 Scanning electron micrograph of male and female schistosome in copula.

male worms lose this capability. Exploitation of these genetic differences may be of value in drug research and in attempts to reduce host morbidity. Prior to egg production, migration and maturation generally require about 4 to 6 weeks.

Humans are the only important definitive hosts of *S. mansoni* (Table 17.1). Fully embryonated eggs without an operculum (114 to 180 µm by 45 to 73 µm) are passed

Figure 17.4 *Schistosoma mansoni* male and female in copula; note the female within the male worm.

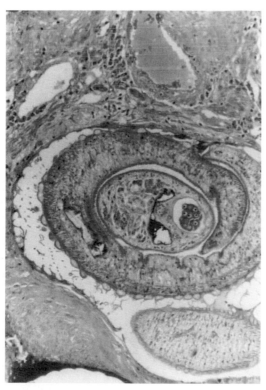

in the stool (Figures 17.5 and 17.6). The eggs are light yellowish brown and elongate, and they possess a large lateral spine projecting from the side near one end of the egg. The eggshell is acid fast when stained by a modified Ziehl-Neelsen technique. On reaching water, the eggs hatch, liberating miracidia (Figure 17.7), which must penetrate a suitable snail host (*Biomphalaria* spp.). Cercariae liberated from the infected snail infect humans when the latter come in contact with cercaria-infested water. A period of 4 weeks or more from the time of first infection with a miracidium is required before the snail sheds cercariae. On infection, the cercariae lose their tails, and the body (now called a schistosomulum) migrates through the lungs to the liver, where it develops to a mature adult in the mesenteric veins of the large bowel. Some worms are also found in the superior mesenteric veins, vesical plexus, and intrahepatic portion of the portal veins.

The adult male and female can reach lengths of 1.2 and 1.6 cm, respectively. They have oral and ventral (acetabulum) suckers at the anterior end. The worm surface is a tegument containing a syncytium of cells. The male worm's body, which is flattened behind the ventral sucker, appears cylindrical as it curves to form the gynecophoral canal to clasp the female worm. The female worm is long, slender, and cylindrical in cross section (Figure 17.4). While held in the gynecophoral canal of the male, the female ingests 10 times more red blood cells than does the male. The tegument of the male contains many prominent tuberculations, while the tegument of the female is devoid of the tuberculations (Figure 17.3). The uterus of the female is short and usually contains only one egg at a time. After mating, the female leaves the male and migrates against the flow of blood to the small venules of the intestine. Initial egg production begins 4 to 7 weeks after infection. *S. mansoni* mature females produce 100 to 300 eggs per day.

The eggs are immature when first laid and take approximately 8 to 10 days to develop a mature miracidium. They are nonoperculate and contain a lateral spine. Egg deposition takes place intravascularly. Many of the eggs laid are swept away and become lodged in the microvasculature of the liver and other organs. About half of the eggs swept away by the bloodstream become embedded in the mesenteric venule wall. The presence of eggs in the tissues stimulates granuloma formation; the eggs die, calcify, and are eventually absorbed by the host (Figures 17.8 and 17.9).

The eggs that are not trapped in the tissues will continue the normal life cycle. The process of maturation and tissue penetration takes about 8 to 10 days; the eggs work their way through the tissues into the lumen of the intestine to be released from the body in the feces. Egg migration through the tissues is facilitated by the release of enzymes

Table 17.1 Characteristics of blood trematodes

Species	Distribution	Reservoir hosts	Intermediate snail host genus	Diagnostic specimen or test	Egg Size (μm)	Egg Morphology	Comments
S. mansoni	Africa, Madagascar, West Indies, Suriname, Brazil, Venezuela	Humans, nonhuman primates	*Biomphalaria*	Stool, rectal biopsy, serology	114–180 by 45–73	Elongate, prominent lateral spine; acid-fast positive	In a wet preparation, egg may be turned so lateral spine is not visible; evidence of flame cell activity is proof of viability; can use hatching test (unpreserved specimens) to confirm; occasionally found in urine
S. japonicum	China, Indonesia, Japan, Philippines	Dogs, cats, cattle, water buffalo, pigs	*Oncomelania*	Stool, rectal biopsy, serology	70–100 by 50–65	Oval, minute lateral spine; acid-fast positive	Eggs can mimic debris in the wet preparation; small spine may be very hard to see; debris often clings to the surface of the egg; less likely to be found in urine, but possible
S. mekongi	Mekong River basin	Humans, dogs, rodents	*Lithoglyphopsis*	Stool, rectal biopsy, serology	50–65 by 30–55	Subspherical, minute lateral spine	Looks very much like *S. japonicum* egg; lateral spine may be difficult to see
S. haematobium	Africa, Middle East, India, Portugal	Humans	*Bulinus*	Urine, stool (some cases), bladder biopsy, serology	112–170 by 40–70	Elongate, terminal spine; acid-fast negative	If present, usually easy to identify; however, urine sedimentation can be time-consuming; can use membrane filter approach as well; eggs occasionally found in stool; hatching test on urine (unpreserved) sediment is relevant
S. intercalatum	Central and western Africa	Humans	*Bulinus*	Stool, rectal biopsy, serology	140–240 by 50–85	Elongate, terminal spine; acid-fast positive	Egg resembles *S. haematobium* egg but is found in stool rather than urine

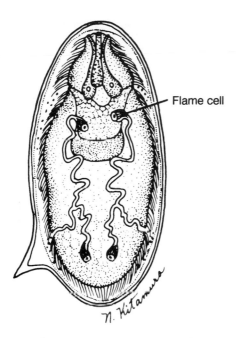

Flame cell

Figure 17.5 *Schistosoma mansoni* egg with a mature miracidium. (Illustration by Nobuko Kitamura; adapted from Figure 27.15A *in* P. C. Beaver, R. C. Jung, and E. W. Cupp, *Clinical Parasitology*, 9th ed., Lea & Febiger, Philadelphia, Pa., 1984.)

by the miracidium through the eggshell. These enzymes help digest the tissue. Eggs making their way through the tissues into the lumen contain a mature miracidium that is released when the eggs are in contact with water (Figure 17.7). The actively swimming miracidium seeks a suitable snail host, which it penetrates. Within the snail, the miracidium develops into a mother sporocyst, which in turn produces daughter sporocysts. The daughter sporocysts develop in the hepatopancreas of the snail and produce cercariae (Figure 17.2). The miracidium, which may be either female or male, gives rise to cercariae of the same sex. This portion of the cycle in the intermediate snail host takes approximately 3 to 5 weeks. The cercariae are released from the snail into freshwater, where they may infect humans by penetrating the skin. Stimuli for cercarial shedding of *S. mansoni* cercariae include bright, direct sunlight and water temperatures of 24 to 30°C, leading to higher transmission rates during the summer. *S. mansoni* cercariae move vertically, alternating between migrations toward the surface and sinking. The cercariae attach to the skin of the definitive host via their ventral or oral suckers. Approximately 40% of the cercariae that penetrate the skin will go on to become adult worms; if the cercariae do not find a suitable host, they usually die within 48 to 72 h.

Clinical Disease

Disease syndromes associated with schistosomiasis are related to the stage of infection, previous host exposure,

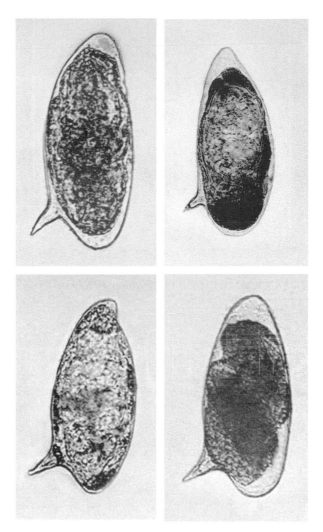

Figure 17.6 *Schistosoma mansoni* eggs. Note the large lateral spine.

worm burden, and host response. Syndromes include cercarial dermatitis, acute schistosomiasis (Katayama fever), and related tissue changes resulting from egg deposition. It has been noted that schistosomiasis exerts disruptive influences on the nutritional reserves and growth of humans from middle childhood through adolescence.

Schistosome Cercarial Dermatitis. Cercarial dermatitis follows skin penetration by cercariae, and the reaction may be due partly to previous host sensitization. Few clinical manifestations are associated with primary exposure, but both humoral and cellular immune responses are elicited on subsequent exposure. After cercarial skin penetration, petechial hemorrhages with edema and pruritus occur. The subsequent maculopapular rash, which may become vesicular, may last 36 h or more. Cercarial dermatitis is common with *S. mansoni* infections. Dermatitis is a constant feature of human infection with avian

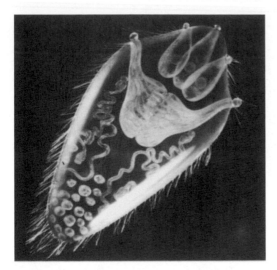

Figure 17.7 *Schistosoma* miracidium. (Armed Forces Institute of Pathology photograph.)

schistosomes, with cercarial death occurring in the subcutaneous tissues and immediate hypersensitivity reactions occurring at the invasion sites. This condition is known as swimmer's itch. Avian schistosomiasis can be found in the Great Lakes and Delaware Bay regions in the United States and in Lake Geneva. Any body of water, typical of bay regions with freshwater or saltwater, where infected snails are found can be linked to nonhuman cercarial dermatitis. Previous contact with cercariae will lead to a more immediate intense immune response.

Acute Disease: Katayama Fever. Acute schistosomiasis (Katayama fever) is associated with heavy primary infections and the initiation of egg production throughout areas of high transmission risk. The signs and symptoms resemble those of serum sickness. Characteristic symptoms include high fever, hepatosplenomegaly, lymphadenopathy, eosinophilia, and dysentery. This clinical syndrome occurs a few weeks after the primary infection, particularly with *S. mansoni*; however, it is infrequent in inhabitants of areas where infections are endemic. Although widely described, this syndrome does not appear to be as common as it once was, nor are its immunology and pathology well understood.

Immune Response. After infection and the transformation of cercariae to schistosomula, the schistosomulum expresses antigens on its surface that evoke a host immune response that provides some degree of resistance to reinfection. As the worms mature, they become less antigenic and more resistant to the immune system-mediated killing mechanisms of the host. The worms are able to incorporate host antigen onto their surface, thus preventing the host from recognizing these parasites as foreign.

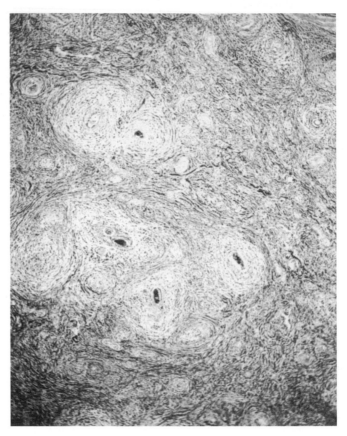

Figure 17.8 Hepatic fibrosis with multiple schistosome egg granulomas. (Armed Forces Institute of Pathology photograph.)

Because schistosomes are incapable of synthesizing fatty acids or cholesterol, they must bind low-density lipoproteins onto their surface. These bound low-density lipoproteins may be essential in allowing the parasite to evade the host immune response. The adult worms do not appear to cause any damage to the host, nor do they evoke much of an immune response. Immunoglobulin E (IgE) antibodies and eosinophilia appear to play a major role in the host immune response and resistance to reinfection (14). In infected individuals, there is a marked increase in the levels of IgE and IgG4 against various parasitic antigens (36). Individuals develop an age-acquired resistance as they are exposed to infection (16). These antibodies have also been associated with resistance to reinfection. In acute schistosomiasis, the levels of IgA and IgM correlate with the intensity of infection (54). The detection of IgA might be useful for the differentiation of acute and chronic schistosomiasis. High IgM levels can be detected in both acute and chronic disease.

Chronic Schistosomiasis. After production of eggs by the adult worms (Figure 17.6), the eggs become trapped in the fine venules and are able to pass through the tissues, escaping into the intestine or, less commonly, the bladder.

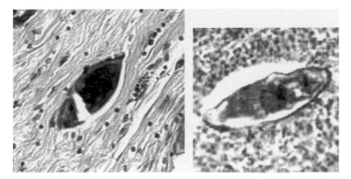

Figure 17.9 Schistosome eggs in tissue. (Left) *S. haematobium*, (right) *S. mansoni*.

The eggs liberate a number of soluble antigens, evoking minute abscesses, which facilitate their passage into the lumen. The passage of eggs through the wall of the intestine or bladder leads to symptoms that correlate with the worm burden of the host, including fever, abdominal pain, liver tenderness, urticaria, and general malaise. In *S. mansoni* infection, blood and mucus are detected in the stools and the patient may have diarrhea or dysentery.

Intestinal disease. As eggs are deposited in the tissues, the antigenic substances released by the eggs invoke a host immune response that includes the formation of granulomas around the eggs trapped in the tissues. Many eggs are retained in the intestinal tissue. Cellular infiltrates include lymphocytes, eosinophils, macrophages, and fibroblasts. Granuloma aggregation in the intestinal tract wall is the major cause of pathologic change. The intestinal wall becomes inflamed, thickened, and fibrotic, leading to mechanical obstruction. Apparently, adhesion molecules are important elements in schistosomal intestinal granuloma formation (28). Intestinal schistosomiasis is common in *S. mansoni* infections but can also occur in *S. haematobium* infections.

Hepatosplenic disease. Although many eggs remain where they are deposited, others are swept into the circulation and filtered out in the liver, leading to hepatosplenic schistosomiasis. Hepatosplenomegaly is common in patients with *S. mansoni* chronic infections. In some areas, *S. mansoni*, *S. japonicum*, and viral hepatitis are the most common causes of chronic liver disease. Patients with concomitant viral hepatitis may develop a more severe form of hepatosplenic schistosomiasis. Eggs deposited in the portal triads of the liver stimulate a granulomatous response, leading to continuous fibrosis of the periportal tissue (Figure 17.8). The fibrotic tissue is white and hard and has been referred to as Symmer's pipe-stem fibrosis (Figure 17.10). However, Symmer's pipe-stem fibrosis is now known as fibro-obstructive hepatic schistosomiasis

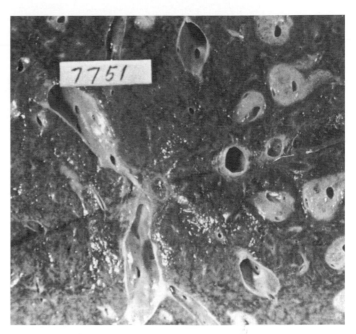

Figure 17.10 Pipe-stem fibrosis in the liver as a result of *Schistosoma mansoni*. (Original photograph by L. Millman; from A Pictorial Presentation of Parasites: A cooperative collection prepared and/or edited by H. Zaiman.)

(56). The liver increases in size as a result of the lesions induced by the eggs. The major effect of the liver changes is the blockage of portal blood flow, resulting in increased portal hypertension and the development of collateral circulation, such as esophageal varices (Figure 17.11). Ascites may be evident, depending on the degree of liver obstruction. The spleen becomes congested and increases greatly in size (Figure 17.12). The presence of large quantities of antigenic material in the spleen could suggest an important role for this organ in antigen clearance and might provide an additional explanation for the hepatosplenomegaly seen in *S. mansoni*-infected children.

Other body sites and sequelae. Collateral circulation allows the eggs to be carried to the lungs, leading to fibrosis of the pulmonary bed. In addition to being deposited in the liver and lungs, eggs may be trapped in other tissues, such as the brain, spinal cord, spleen, pancreas, and myocardium. Eggs trapped in the brain have caused focal and generalized seizures. Spinal cord schistosomiasis has been found in patients with *S. mansoni* infections; the myelopathy is probably the result of hypersensitization to the eggs. Myelopathies occur infrequently, and eggs are commonly found in the spinal cord of patients without eliciting symptoms.

Pulmonary involvement is seen with all forms of schistosomiasis in patients with heavy worm burdens. Schistosomal lung disease is usually seen only after the development of collateral circulation in infections with

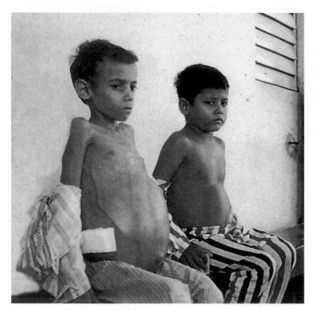

Figure 17.11 Schistosomiasis. Note the engorgement of collateral circulation of the abdomen. (From A Pictorial Presentation of Parasites: A cooperative collection prepared and/or edited by H. Zaiman.)

S. mansoni and *S. japonicum.* The subsequent development of pulmonary hypertension leads to cor pulmonale (right-sided heart failure). Signs of the disease include fatigue, cough with possible hemoptysis, palpitations, dyspnea on exertion, right ventricular hypertrophy, and pulmonary artery dilation.

Figure 17.12 *Schistosoma mansoni* infection showing liver and splenic enlargement; 800 worms were removed from this patient. (From A Pictorial Presentation of Parasites: A cooperative collection prepared and/or edited by H. Zaiman. Photograph courtesy of R. Goldsmith and B. H. Kean.)

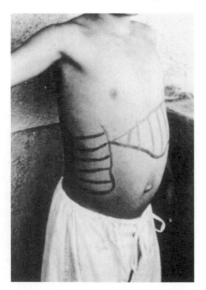

Chronic bacteremic infections with *Salmonella* spp. have been associated with *S. mansoni* infections. Common features of this association between *Salmonella* and *Schistosoma* infections include a long history of recurrent febrile illness, frequent isolation of *Salmonella* spp. from the blood, and chronic active schistosomiasis. The *Salmonella-Schistosoma* infection association is seen most frequently in males between the ages of 15 and 30 years. If the underlying schistosome infection is not treated, the *Salmonella* infection will continue to recur.

The level of circulating immune complexes appears to be correlated with the worm burden of the host. These immune complexes may be deposited on the basement membrane and glomerular capillaries, causing nephrosclerosis and kidney failure. The severity of kidney disease is associated with the worm burden and the degree and duration of hepatic fibrosis. The relationship between renal and hepatic disease may be the result of development of a collateral circulation, which would interfere with the normal clearance of circulating antigens and immune complexes by Kupffer cells.

Diagnosis

Schistosomiasis should be suspected in individuals who have had significant contact with water in areas of endemic infection. Symptoms of acute schistosomiasis include fever, diarrhea, abdominal pain, weight loss, and eosinophilia. Symptoms of chronic disease include chronic diarrhea, abdominal pain, and hepatomegaly or hepatosplenomegaly.

Diagnostic Procedures. Specific diagnosis of schistosomiasis by detection of eggs in stool or urine specimens is possible only after egg production has begun. Eggs may be found in feces as early as 5 weeks after infection. The ease of egg detection depends on the worm burden and the duration of the infection. Patients with a low worm burden or old (chronic) infections may have very few eggs in the feces or urine, and the infections may not be confirmed due to insensitive diagnostic methods. Multiple stool or urine examinations should be performed for any individual suspected of having schistosomiasis. Occasionally, *S. mansoni* eggs are detected in the urine; adult worms may be found in vessels that are not their normal habitat, and this finding is known as "crossover."

S. mansoni eggs are yellowish brown and measure 114 to 180 μm long by 45 to 73 μm wide (Figure 17.6; Table 17.1). The eggs are elongate and ovoid and have a large lateral spine projecting near one end. Direct detection or concentration techniques can be used to detect eggs in the stool or urine. Direct microscopic examination of stool smears is not very sensitive but may be useful for screening purposes. The Kato thick smear is a simple and sensitive quantitation technique that has been used successfully in the field. In one study, more than 50% of the eggs were missed by the sedimentation technique; the geometric mean egg count was 94 eggs/g when two Kato-Katz smears were used and 43 eggs/g when the sedimentation technique was used (15). However, no details of the centrifugation speed and time were given. Another study indicated that examination of fewer samples collected on different days was more effective than examination of more slides from one stool specimen for accurate estimation of the real infection status.

The zinc sulfate concentration technique is not recommended for schistosome eggs. The eggs rupture and do not float but instead are found in the sediment. The formalin-ethyl acetate technique is recommended for concentrating eggs; however, because it involves fixation, it cannot be used to detect egg viability.

In chronic infections in which the worm burden is light, hatching tests can be performed. The stool specimens are diluted with nonchlorinated water in a sedimentation flask or a beaker. The sides of the flask or beaker are covered with aluminum foil to prevent light from passing through. A light source is used to project a perpendicular light beam through the water at the top. Miracidia that hatch from the live eggs will concentrate in the light and can be detected swimming around. This motility can easily be observed with a hand lens. Aliquots of the surface water can be transferred to a small petri dish and observed under a dissecting microscope for miracidia. Observation periods should be frequent because of the limited life span of the miracidia. Ideally, observations should be made every 30 min over a period of 4 h. The hatching test is designed to mimic the conditions in nature with spring water and sunlight (see chapter 28).

Biopsy Specimens.   Rectal biopsy specimens have been particularly useful in detecting eggs in patients with light, chronic, or inactive infections (Figure 17.13). The biopsy tissue can be crushed between two glass slides. This technique is more effective than histologic examination and allows assessment of the species and viability of the eggs. The viability of the eggs can be determined by closely observing the miracidium for flame cell activity. Each miracidium contains two pairs of flame cells, and these are actively beating in live miracidia (Figure 17.7). Eggs can also be recovered from tissue by digestion in 4% KOH at 37°C for 6 to 18 h and examination of the sediment. *S. mansoni* eggshells can be stained acid fast with a modified Ziehl-Neelsen stain. This technique has been used in tissue sections to differentiate *S. mansoni* eggs from *S. haematobium* eggs, which are not acid-fast positive.

Antibody Detection.   A large number of serologic tests have been used in the diagnosis of schistosomiasis (69);

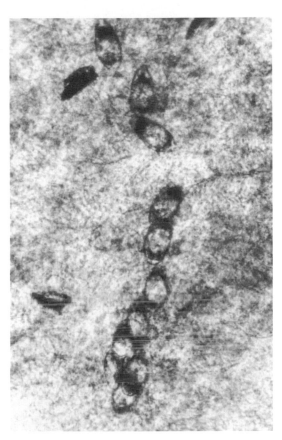

Figure 17.13 *Schistosoma mansoni* eggs in crushed rectal biopsy tissue. (Armed Forces Institute of Pathology photograph.)

however, many have not been particularly useful because of cross-reactions with other helminth infections, continuation of elevated titers long after successful treatment, and slow immunologic response of the host. Because of the complex life cycle, a large number of antigens have been used in serologic tests. Only a few have been useful. The most frequently used tests are the circumoval precipitin test, the cercaria-Hullen reaction, the indirect fluorescent-antibody test, the indirect hemagglutination test, and ELISA (57, 69). See Algorithm 17.1.

ELISA methods using soluble adult worm or egg antigens for the detection of antibodies have been used as screening tests, and infection has then been confirmed by the enzyme-linked immunoelectrotransfer blot (EITB) test (69). Individuals infected with schistosomes have elevated IgE and IgG4 responses, as do patients with other helminthic infections. Few studies have investigated the persistence of antibodies over time after therapy (69). It is possible that high titers posttherapy reflect treatment failure and reinfection.

Antigen Detection.   ELISA has also been used to detect circulating schistosome antigens in the serum and urine of

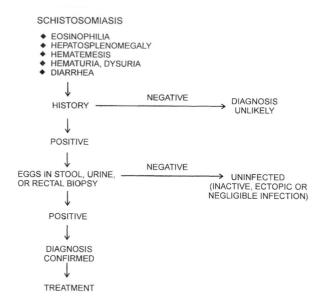

SCHISTOSOMIASIS

- ◆ EOSINOPHILIA
- ◆ HEPATOSPLENOMEGALY
- ◆ HEMATEMESIS
- ◆ HEMATURIA, DYSURIA
- ◆ DIARRHEA

HISTORY ———— NEGATIVE ————→ DIAGNOSIS UNLIKELY

POSITIVE

EGGS IN STOOL, URINE, ———— NEGATIVE ————→ UNINFECTED
OR RECTAL BIOPSY   (INACTIVE, ECTOPIC OR
                                               NEGLIGIBLE INFECTION)

POSITIVE

DIAGNOSIS CONFIRMED

TREATMENT

Algorithm 17.1   Schistosomiasis.

infected patients and may be one of the preferred methods of diagnosis (5, 51, 76). In addition to diagnosis of infection, antigen levels have been used to assess disease severity in terms of intensity of infection and to monitor the efficacy of therapy (51, 76) (Table 17.2). Individuals treated with praziquantel experienced a rapid clearance of circulating anodic (CAA) and cathodic (CCA) antigens; however, in individuals treated with other therapeutic drugs, clearance may take much longer. A dipstick assay based on the anti-CCA sandwich ELISA, on a nitrocellulose format, can detect CCA in the urine of *S. mansoni*-infected patients with a sensitivity of 92%. The results were at least as good as those obtained by ELISA or the Kato-Katz fecal smear examination. This rapid, simple approach may lend itself to more extensive use in future screening programs (64, 65, 73). The combination of urine CCA and serum CAA for the detection of circulating antigens and the combination of the *S. haematobium* adult worm microsomal antigens (HAMA) FAST-ELISA and the HAMA EITB test for detecting antibodies has significantly improved the sensitivity of antibody and antigen detection (3).

When a soluble egg antigen (SEA) monoclonal antibody (MAb) was used, circulating schistosome antigen was undetectable in serum from patients with other parasitic infections and in 94% of serum samples and 84% of urine samples from *S. mansoni*-infected patients; however, it was present in serum (90.5%) and urine (94%) samples from patients with *S. haematobium* infection. A significant reduction in circulating schistosome antigen levels was detected in serum and urine samples approximately 12 weeks after praziquantel therapy (58). No cross-reactivity with *S. mansoni* was seen.

KEY POINTS—LABORATORY DIAGNOSIS

Schistosoma mansoni

1. Eggs cannot be detected in stool until the worms mature (may take 4 to 7 weeks after initial infection).
2. In very light or chronic infections, eggs may be very difficult to detect in stool; therefore, multiple stool examinations may be required. Biopsy and/or immunologic tests for antigen or antibody may be helpful in diagnosing infection in these patients.
3. The zinc sulfate flotation concentration method should not be used; the eggs do not float. The sedimentation method is recommended.
4. It may be necessary to tap the coverslip to move the eggs; the lateral spine may not be visible if the egg is turned on its side.
5. Occasionally, eggs of *S. mansoni* are detected in the urine (crossover phenomenon).
6. Patients who have been treated should have follow-up ova and parasite examinations for up to 1 year to evaluate treatment.
7. In active infections, the eggs should contain live or mature miracidia. Examination for the confirmation of flame cell activity must be performed with fresh specimens, using the microscopic wet mount test or the hatching test; no preservatives can be used prior to these two tests.
8. During the hatching test, the light must not come too close to the surface of the water, since excess heat can kill the liberated larvae. Also, the water should be examined about every 30 min for up to 4 h.

Treatment

Praziquantel. Praziquantel, a pyrazinoquinolone, is the drug of choice for treating schistosome infections (2). The current recommendation is 40 mg/kg/day in two doses for 1 day. Praziquantel can also be taken orally in a single dose and is well tolerated by the host. Side effects, which are usually mild and transient, include abdominal discomfort, dizziness, drowsiness, headache, fever, and loose stools. These effects usually disappear within 48 h. In heavily infected individuals, the side effects, although transient, may be very severe (62). Split doses have been recommended for treating heavy infections, but this may be a concern for population-based therapy, where individuals may not present themselves for follow-up therapy. Data suggest that a single dose of praziquantel (40 mg/kg of body weight) may have a longer-lasting effect than previously thought, possibly more than 2.5 years (18). Praziquantel has little effect on immature stages of the parasite, and this affects the cure rates as the immature

Table 17.2 Improved immunologic testing for the diagnosis of *S. mansoni* infection

Test	Comments	Reference
ELISA, IgG, IgM, IgA	Crude adult worm extract as an ELISA antigen provides a serologic method with high sensitivity and specificity for the diagnosis of acute and chronic schistosomiasis and for serologic distinction between the two forms of the disease.	72
Immunomagnetic beads with 1 ml of urine	Assay method uses larger sample volumes than ELISA; enhanced detection can be achieved. Valuable in areas with low prevalence or assessment of cure after therapy.	47
ELISA for cercarial elastase	Cercarial elastase is one of the first proteins released in the host by invading cercariae. Detection of antibodies to this protein indicates exposure but not necessarily active infection.	55
ELISA using egg antigen (CEF6)	In areas of endemicity there was a positive correlation between egg intensity and ELISA value. In other areas, specificity was 90%; false-positive results may have been due to inapparent or cured infection.	19
APIA	Solid-phase alkaline phosphatase immunocapture assay for detection of human IgG antibodies to the alkaline phosphatase of *S. mansoni* adults was evaluated. The method appears to be specific for *S. mansoni* and is useful in areas of endemicity where mixed infections occur.	10
Immunoblot	Sm31/32 protein fraction is highly immunogenic; may be a useful serologic marker for diagnosing and differentiating between acute and chronic infection.	71
Fast dot-ELISA	C5C4 MAb is used for rapid and simple diagnosis; the 63-kDa circulating antigen was detected in 92% of 330 urine samples; the false positive rate in 130 noninfected individuals was 16%.	4
Indirect hemagglutination combined with ELISA	Erythrocytes coated with *S. mansoni* adult worm antigens and ELISA using *S. mansoni* egg antigens; combined use of these two tests led to high degrees of sensitivity and specificity for *S. mansoni*, *S. haematobium*, and combined infections.	75
Antigen detection in urine or serum	Test provides a quantitatively more stable diagnosis of *S. mansoni* than fecal egg counts based on a single stool exam.	50
Antigen detection in fecal specimens	Compared with the Kato-Katz parasitologic exam, PCR has a sensitivity of 96.7% and a specificity of 88%.	52
Antigen detection in urine using dipstick or reagent strip test	Lateral flow on nitrocellulose strip; detection of CCA; sensitivity equal to CCA ELISA; specificity of 96%.	73
Antigen detection in urine using dipstick	Commercially available antigen capture dipstick; detection of CCA; sensitivity of 83% and specificity of 81%; may be too expensive for wide use (US $2.60).	64

parasites develop to maturity and egg production. The mechanism of action may involve glutathione *S*-transferase, a detoxifying enzyme found in many helminths. Cure rates have been reported to be from 70 to 95%. Praziquantel damages the tegument of the worm, thereby exposing the antigens to parasite-specific antibodies. The use of immunoglobulins along with praziquantel has been proposed to increase the efficacy of treatment.

Although the exact mechanism by which praziquantel kills the parasite is not well understood, it is known that the immune response of the host plays an important role in drug efficacy. It is possible that disease states of humans that lead to immunodeficiencies, such as infection with human immunodeficiency virus type 1 (HIV-1), may render praziquantel less effective in treating schistosomiasis. However, studies indicate that persons with HIV-1 infection can be treated effectively for schistosomiasis with praziquantel (30).

Repeated praziquantel therapy can lead to improvement of liver morbidity and prevention of hepatic fibrosis, even in a well-established focus of hyperendemic infection. These results were based on a parasitologic, clinical, and ultrasonographic longitudinal study in a focus of hyper-endemic *S. mansoni* infection. Support is also emphasized for programs that encourage people to seek examination and treatment on their own. This type of chemotherapy-based control strategy is called passive chemotherapy.

Reduced efficacy of praziquantel in treating human *S. mansoni* infections and infected animal models has been reported (27). Drug resistance to praziquantel has been demonstrated in the murine model, and there have been unsubstantiated field study reports of resistance. However, resistance to praziquantel in schistosomes is now considered a fact in strains selected in the laboratory. In at least two human populations in Senegal and Egypt with endemic infections, resistance or tolerance to praziquantel has been confirmed. Although the extent of the problem is unclear, it is important to determine the prevalence and significance of the problem and to develop alternate control strategies.

Oxamniquine.   Oxamniquine, a nitroquinolone, is active only against *S. mansoni* infections and is very effective against the strains in the Americas and West Africa when given at 15 to 20 mg/kg of body weight. North and East African strains of *S. mansoni* are less sensitive to this drug, and doses of up to 60 mg/kg given over 2 to 3 days may be required to achieve the same results. The drug is more active against male than female worms and has little effect against immature worms. Side effects include abnormal electroencephalograms and dizziness. Oral treatment has reduced egg counts by 95%, with 70 to 100% cure rates. Despite being tolerant to praziquantel, a Senegal isolate of *S. mansoni* was fully susceptible to oxamniquine (17). Unfortunately, this drug is relatively expensive, making it more difficult to use in control programs, particularly if a higher dosage is required. Although mass therapy with this drug can lead to reduction of schistosomiasis morbidity, approximately 68% of patients do not show improvement in their condition (11). It is also important to remember that oxamniquine is contraindicated in pregnancy.

Summary.   Patients who have been treated for schistosomiasis should be monitored posttherapy to determine the effectiveness of the treatment. Depending on the drug, follow-up stool or urine examinations may be needed for as long as 1 year posttherapy. It is not safe to assume that one course of therapy or disappearance of symptoms is indicative of cure. Optimal posttreatment reevaluation times have not been established. Patients, particularly those who were heavily parasitized, should be monitored for a reduction in the number of excreted and viable eggs within a few months of therapy.

For patients with secondary bacterial infections, appropriate antibiotics may also be needed to treat systemic infections. Patients suffering from severe Katayama fever and transverse myelitis may require steroid treatment.

A complete review of the management and therapy of schistosomiasis can be found in Table 17.3.

Epidemiology and Prevention

S. mansoni infections occur in western and central Africa, Egypt, Madagascar, and the Arabian peninsula. As a consequence of the African slave trade, the infection was established in the Western Hemisphere in Brazil, Suriname, Venezuela, and the West Indies. Human infections are nearly all derived from human sources, even though non-human primates, insectivores, marsupials, and rodents can carry the infection. The infection is transmitted by *Biomphalaria* sp., an aquatic snail.

The geographic distribution of the disease depends on the distribution of the intermediate snail hosts and the opportunity to infect humans and snails. Schistosomiasis is transmitted from infected people defecating or urinating in or near water where the appropriate snail host resides. Infections can persist indefinitely in humans. Most infected individuals have low worm burdens, but a few have very heavy infections. It is this latter group that probably makes the greatest contribution to the dissemination of schistosomiasis. The highest rates of infection have been found in children; the greatest cercarial exposure usually occurs in boys aged 5 to 10 years. Older children may have less recreational exposure but are more likely to be exposed while performing chores such as agricultural activities, washing dishes and clothes, and bathing younger siblings. Although adults also have less recreational exposure to water, they also have partial acquired immunity.

Information on the presence of cercariae in water is important when only one of many snails is infected yet capable of shedding enough cercariae to maintain high endemicity. The use of sentinel mice for this purpose is inaccurate, time-consuming, and costly. Filtration is also limited and is associated with many problems. A PCR assay for the detection of *S. mansoni* cercariae in water has been developed. The high sensitivity of this test permits the detection of a single cercaria; a single cercaria could be detected in repeated tests of water filtrates (22).

Studies indicate that there was little transmission of *S. mansoni* in Puerto Rico during the first half of the 1990s and confirm physician anecdotal data that no new infections have been seen during the past few years (24). Recent trends indicate that human schistosomiasis is disappearing from Puerto Rico (23a). The Puerto Rico Public Health Department reports that from 1990 through 26 September 2005, only 21 cases were reported. The last single case was in 2005 in a 69-year-old male from the east coast, this case probably being an "old" case. Although low transmission is still possible in Puerto Rico, data suggest that human schistosomiasis has been eliminated from this area. However, in other areas the news is not as good. The emergence of *S. mansoni* in Upper Egypt

Table 17.3   Management and therapy of schistosomiasis[a]

Condition or disease manifestation	Approach	Comments
Schistosome cercarial dermatitis	Self-limited and may go undiagnosed; in severe cases, topical corticosteroids and parenteral antihistamines reduce symptoms	Other names include swimmer's itch, schistosome dermatitis, clam digger's itch, sawah itch, and koganbyo; symptoms will vary due to host susceptibility, schistosome type, and patient's previous exposure history
Katayama fever	Corticosteroids for life-threatening disease; therapy according to schistosome species	Tourists who returned home within a month of exposure; history of water contact, possible swimmer's itch, serum sickness-like disease are clues for possible schistosomiasis
Gastrointestinal disease	**S. mansoni:** Praziquantel (40 mg/kg) single dose or oxamniquine (15 mg/kg) single dose for adults (20 mg/kg for children); North and East Africa (60 mg/kg) over 2–3 days in 3 or 4 equally divided doses	Many complications of intestinal granulomas and/or polyp disease are the same as for inflammatory bowel disease (Crohn's and ulcerative colitis); misdiagnosis could have serious sequelae due to potential use of steroid therapy
	S. japonicum: Praziquantel (60 mg/kg) 2 or 3 doses in a single day	
	Colonic polyps: Colonoscopy to confirm extent of disease; chemotherapy; surgery may be required for fibrotic obstructive bowel lesion; intussusception (barium enema should be used with care to avoid perforation); rectal prolapse/chemotherapy; anorectal fistulas/abscess, surgical drainage and antibiotic therapy	Seen in 10–15% of some Middle Eastern populations; symptoms are those of focal granulomatous large bowel disease; may mimic chronic severe mucohemorrhagic diarrhea; residual fibrotic scarring may lead to permanent partial bowel obstruction or wall deformation
Hepatosplenic disease	**S. mansoni:** Praziquantel or oxamniquine; ultrasonography to define extent of disease	May be genetic predisposition to hepatosplenic syndrome; jaundice characteristically absent; liver cell function generally normal until terminal stages
	S. japonicum: Praziquantel; ultrasonography	
	Hypersplenism: Therapy	Moderate and generally tolerable to patient
	Ascites: Low-salt diet and diuretics	End-stage disease can involve massive ascites; usually irreversible
	Bleeding esophageal varices: Esophogoscopy to confirm diagnosis (transfusion and esophageal sclerotherapy; surgical shunt for portal hypertension in severe cases)	Liver sonography can estimate degree of portal hypertension (measure distention of portal vein)
Urinary disease	**Uncomplicated:** Praziquantel (40 mg/kg) single dose; ultrasonography or intravenous pyelogram to detect silent obstructive disease	Most infections are mild and asymptomatic
	Obstructive uropathy: Chemotherapy; may require surgery	Most lesions occur in lower third of ureters or bladder; pyelonephritis/gram-negative sepsis common
	Urinary tract infection: Antibiotics plus antischistosomal therapy; pyelonephritis may require parenteral antibiotics	Primarily gram-negative infections
	Renal failure: Hemodialysis or renal transplantation	Bilateral ureteral obstruction common; may be accompanied by *Salmonella* bacteriuria and/or bacteremia
	Bladder cancer: Radial surgery, chemotherapy usually only palliative	Usually lags 10–20 yr behind the peak *S. haematobium* infection; bladder calcifications may resolve over years if the patient remains schistosome free
Cardiopulmonary disease	**Larval pneumonitis:** Self-limited course; may require short-course therapy with steroids if respiratory distress present	Occurs days to weeks after heavy cercarial exposure; symptoms last 2–4 wk (marked eosinophilia)
	Reactionary pneumonitis: Discontinue therapy until resolution; may require short-course therapy with steroids if respiratory distress is present	Löffler-like syndrome seen during therapy of heavy infections
	Cor pulmonale: Antischistosome therapy; symptomatic therapy	Heart affected after exposure to chronic abnormal pressure gradients; antischistosomal therapy prevents disease progression but does not lead to improvement

(continued)

Table 17.3 Management and therapy of schistosomiasis*a* *(continued)*

Condition or disease manifestation	Approach	Comments
Central nervous system disease	**Praziquantel therapy:** steroids recommended to reduce inflammation and size of space-occupying lesions	
	Cerebral: Focal, generalized seizures	Most occur with *S. japonicum* (2–4% of cases); parietal lobe is the most common site
	Spinal: Transverse myelitis of various degrees	Most occur with *S. mansoni*, also *S. haematobium*; lesions may be reversible
Miscellane ous	*Salmonella*-**schistosome syndrome:** Simultaneous antischistosomal and antibacterial therapy	Occurs most commonly in males 15–30 yr old; antibiotic response is dramatic; disease remains if schistosome disease is not also treated
	Hepatitis B or C schistosome disease: Antischistosomal therapy; transmission precautions for hepatitis B or C	Prognosis more guarded with dual infections

a Adapted from reference 65.

has not followed any particular pattern; however, the high prevalence detected in certain areas could be associated with the increasing abundance of the snail vector seen in the canals. Underreporting may also be associated with changing surveillance data. Water resource development projects involving the construction of artificial bodies of water and canals for irrigation have contributed to the extension of the disease. The migration of infected persons into virgin areas where suitable snail hosts reside also contributes to the spread of schistosomiasis. These two important factors make control programs difficult.

Schistosoma japonicum

Fujii in 1847 published the first clinical description of *S. japonicum* infections in his Katayama memoir. Katsurada in 1904 described the ova and adult worms that he detected in dogs, cats, and humans and named the parasite *S. japonicum*. Also known as the Oriental blood fluke, the parasite is confined to the Far East and has been separated from *S. mekongi*, a more recently described species also found in the Far East. There is enough genetic variation in *S. japonicum* strains isolated from different areas of endemic infection that there may be separate species.

Life Cycle and Morphology

In addition to infections in humans, *S. japonicum* infections can be found in many mammals exposed to infected water; thus, there are many reservoir hosts, in contrast to other major human schistosomes (Table 17.1). Fully embryonated eggs without an operculum (70 to 100 µm by 50 to 65 µm) are passed in the feces. The eggs are smaller and somewhat more spherical than those of *S. haematobium* and *S. mansoni*. Near one end of the egg is a minute lateral spine; this may be absent in some strains (Figures 17.14 and 17.15).

The adult worms resemble those of *S. haematobium* and *S. mansoni*; however, they lack integumentary tuberculations. *S. japonicum* adults are found in the radicles of the superior mesenteric vein draining the small intestine (Figure 17.16). The female worm may contain as many as 50 eggs at one time in the uterus. Because of the higher egg production (3,000 eggs per worm pair per day) and the smaller egg size, an infection with just a few worm pairs can be very serious because the eggs, free in the general circulation, can be filtered out in the liver, the lungs, and even the central nervous system. The miracidium, released from the egg on contact with water, must find a suitable snail host (*Oncomelania* spp.) to continue the life cycle. Cercariae of *S. japonicum* tend to attach themselves to the water surface film, where they remain unless disturbed.

Clinical Disease

The pathology caused by *S. japonicum* is similar to that caused by *S. mansoni* infections and may include cercarial dermatitis as the first sequence in the disease process (Table 17.4). Although the severity is dependent on the worm burden, the disease has a poor prognosis because of the higher egg production capacity of *S. japonicum* females and the smaller eggs. *S. japonicum* infections have been associated with hepatocellular carcinoma and colorectal cancer. The mechanism of this schistosome-associated carcinogenic enhancement is unknown. The adult worms also have a greater tendency to be found in ectopic sites. Although adult worms have never been detected in the brain, cerebral schistosomiasis is a distinct syndrome that occurs with *S. japonicum* infections. The entire intestinal tract may be involved, but the large intestine is often the primary site and the disease is more severe there.

Acute Disease: Katayama Fever. Acute schistosomiasis (Katayama fever) is associated with heavy primary

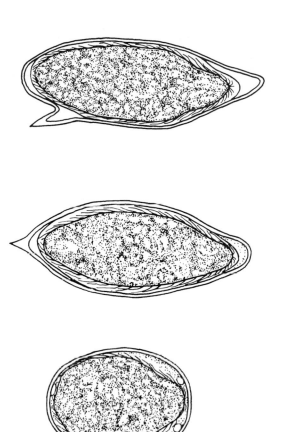

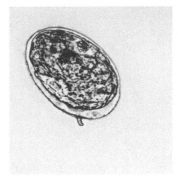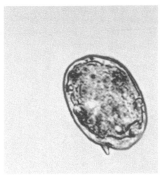

Figure 17.15 *Schistosoma japonicum* eggs. Note the small lateral spine.

Figure 17.14 *Schistosoma mansoni* (top), *S. haematobium* (middle), and *S. japonicum* (bottom) eggs. Note how much smaller the *S. japonicum* egg normally is than those of the other two species, and note the small size of the lateral spine. (Illustration by Nobuko Kitamura.)

fibrosis and stenosis may occur with thickening of the mesentery.

Hepatosplenic Disease.   Eggs that do not reach the intestinal lumen may be trapped or swept up by the portal blood flow. The eggs tend to accumulate in the venules within the portal triads, especially the left lobe; this is followed by granulomatous inflammation, fibrosis, venous obstruction, portal hypertension, and splenomegaly. Hepatosplenomegaly results from the granulomatous response surrounding the eggs. Although splenomegaly generally occurs during the early stages of chronic disease, the enlargement is due to tissue hyperplasia rather than granulomatous inflammation; thus schistosome eggs are rarely found in the spleen. In China, splenectomy is frequently used for severe hypersplenism (56).

After extended chronic disease, gross periportal fibrosis occurs with portal hypertension and secondary

infections and the initiation of egg production. The signs and symptoms include high fever, nocturnal fever peaks, coughing, generalized muscle pain, headache, hepatomegaly, lymphadenopathy, eosinophilia, and dysentery. Splenomegaly also occurs in approximately one-third of the patients, while all patients have eosinophilia. Diffuse pulmonary infiltrates are found; some patients have signs of meningoencephalitis. Generally, disease onset occurs about 40 or more days after exposure to a first infection or a large infection (56).

Intestinal Disease.   As egg clusters accumulate, mucosal inflammation, hyperplasia, ulceration, microabscess formation, blood loss, and pseudopolyposis may occur. Lower abdominal pain is common and may be colicky and in the left lower quadrant. Diarrhea alternating with constipation is also common, with either visible blood, occult blood, or both. In severe chronic disease, bowel

Figure 17.16 *Schistosoma japonicum* paired adult male and female in a mesenteric vein. (From A Pictorial Presentation of Parasites: A cooperative collection prepared and/or edited by H. Zaiman.)

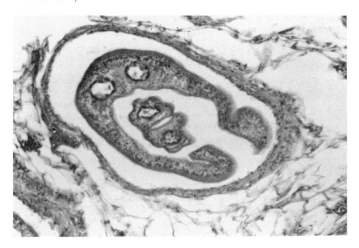

Table 17.4   Pathogenesis, pathology, and symptoms seen with *S. japonicum* infection

Stage of infection	Comments
Incubation (migration, maturation)	
Dermatitis	Associated with penetration of cercaria; urticarial rash; diarrhea at the end of this period
Lungs	Traumatic infiltrative changes, at times with hemorrhages, local accumulation of eosinophils, epithelioid cells, giant cells around pulmonary blood vessels (in wake of migrating larvae)
Liver	Acute hepatitis during larval growth in intrahepatic portal blood vessels
Small intestine	Hyperemia in wall of small intestine following arrival and during maturation of worms in superior mesenteric venules
Blood	Marked increase in number of circulating eosinophils
Egg deposition and extrusion	
Small intestine	Trauma with hemorrhage as eggs laid by female worms escape from venules through intestinal submucosa and mucosa into intestinal lumen; may be profuse dysentery; acute attack subsides in 3–10 wk; pseudotubercles form around eggs; repair occurs with scar tissue formation; giant cells develop within pseudotubercles, which become transformed to fibrous nodules
Blood	Blood picture changes from leukocytosis and hypereosinophilia to that of secondary anemia
Liver, spleen, lymph nodes	Liver increases in size as result of lesions around eggs; spleen becomes congested, fibrous reticulum increases, becomes greatly enlarged; mesenteric lymph nodes lose active lymphoid tissue
Tissue proliferation and repair	
Liver, spleen	Liver gradually decreases in size as parenchyma becomes replaced by scar tissue; spleen continues to enlarge; thoracic viscera are pushed upward; patient is weak with marked dyspnea on slight exertion; dilatation of superficial veins of abdomen and thorax is common
Intestine	Intestinal wall becomes thickened and fibrosed; papillomata develop on mucosa
Other problems	Ascites may be present; physical and mental retardation may occur in children; anemia is more pronounced
Blood vessels	Mesenteric and portal vessels become thrombosed; portal fibrosis due to egg deposition around the portal triads; severe fibrosis with white and hard tissue, frequently referred to as Symmer's fibrosis or pipe-stem fibrosis (fibro-obstructive hepatic schistosomiasis)
Cerebral disease	Patients classified as acute, chronic with granulomas or diffuse lesions, psychotic, or as having portal system encephalopathies; 2–4% (in some areas, this figure is higher) of all patients with *S. japonicum* schistosomiasis have acute or chronic cerebral involvement, including epileptic seizures

esophageal varices. Histology reveals wide bands of fibrous tissue extending along the portal tracts, hence the term "pipe-stem fibrosis," also seen with *S. mansoni* infections. These sequelae are more serious with *S. japonicum* infections, due primarily to the high daily egg production by the female worms. The parenchymal form of hepatic fibrosis occurs most commonly with *S. japonicum* infection and may be a subclinical find in groups who have been treated repeatedly and undergone overt liver disease resolution.

Central Nervous System Disease.   Although not every patient with eggs in the central nervous system is symptomatic, symptoms can include Jacksonian convulsions and grand mal seizures. The use of praziquantel has led to improvement in over 75% of the cases in which it has been used. It is also unclear why Chinese schistosomiasis is associated with ectopic cerebral lesions while African disease tends to be associated with ectopic spinal lesions.

Cancer and *S. japonicum* Infection.   The incidence of colorectal cancer and *S. japonicum* infection correlates with the prevalence, intensity, and length of inflammatory bowel symptoms. This correlation has been reported from both China and Japan. In a study published in 2005 from rural China, schistosomiasis was thought to be responsible for 24% of colon cancers and 27% of liver cancer within that study group (53).

In a study investigating the relationship between schistosomiasis due to *S. japonicum* and liver cancer and cirrhosis in an area where no cases of schistosomiasis have been reported since 1978, schistosome eggs were found in the livers of newly deceased liver cancer and cirrhosis patients. The chronic effect of *S. japonicum* could contribute to the high mortality rate due to liver cancer in this area.

However, there are certainly other issues that need to be considered, such as HBV, HCV, and alcohol intake (67).

Viral, Bacterial, and Alcohol Associations.   Although hepatitis B has been reported to be more frequent in patients with schistosomiasis, the evidence is not that strong. Chronic bacteremic infections with *Salmonella* spp. have been associated with *S. haematobium*, *S. japonicum*, and *S. mansoni* infections (65). Common features of this association between *Salmonella* and *Schistosoma* infections include a long history of recurrent febrile illness, frequent isolation of *Salmonella* spp. from the blood, and chronic active schistosomiasis. The *Salmonella-Schistosoma* infection association is seen most frequently in males between the ages of 15 and 30 years. If the underlying schistosome infection is not treated, the *Salmonella* infection will continue to recur. However, this association appears to be rare for *S. japonicum* compared with the other schistosome species. There is evidence that there may be some association between alcoholic cirrhosis and hepatic fibrosis; however, details of specific mechanisms for this association are minimal.

Diagnosis

Schistosomiasis should be suspected in individuals who have had significant contact with water in areas of endemic infection. Symptoms of acute schistosomiasis include fever, diarrhea, abdominal pain, weight loss, and eosinophilia. Symptoms of chronic disease include chronic diarrhea, abdominal pain, and hepatomegaly or hepatosplenomegaly.

Diagnostic Procedures.   Specific diagnosis of schistosomiasis by detection of eggs in stool specimens is possible only after egg production has begun. Eggs may be found in feces as early as 5 weeks after infection. The ease of egg detection depends on the worm burden and the duration of the infection. Patients with low worm burden or old (chronic) infections may have very few eggs in the feces or urine, and the infections are often not diagnosed because of the limits of sensitivity of examination techniques. Multiple stool examinations should be performed for any individual suspected of having schistosomiasis. Occasionally eggs of *S. japonicum* or *S. mansoni* are detected in the urine. The eggs of *S. japonicum* are more spherical than those of *S. mansoni* or *S. haematobium*; they measure 70 to 100 μm by 50 to 65 μm (Figures 17.14 and 17.15). A minute lateral spine may be found at one end of the egg; however, it is absent in some strains of *S. japonicum*.

Data from China suggest that a single Kato-Katz smear may be quite adequate for detecting moderately to heavily infected individuals (>100 eggs/g of stool); however,

for estimating the true prevalence of infection within a community or for obtaining an accurate estimate of egg excretion, multiple Kato-Katz smears are recommended. It should be noted that large variations in egg counts and the lack of sensitivity of the Kato-Katz method, particularly in light and moderate infections, have been reported (81). Eggs also appear to be nonrandomly distributed in stool specimens between the center and surface of the stool; stirring the specimen prior to sampling decreased the count variability. Direct detection or concentration techniques can be used to detect eggs in stool. Direct microscopic examination of stool smears is not very sensitive but may be useful for screening purposes. The Kato thick smear is a simple and sensitive quantitation technique that has been used successfully in the field.

The zinc sulfate concentration technique is not recommended for schistosome eggs. The eggs will rupture and do not float but instead are found in the sediment. The formalin-ethyl acetate technique is recommended for concentrating eggs; however, because it involves fixation, it cannot be used to detect egg viability.

In chronic infections in which the worm burden is light, hatching tests can be performed. The miracidium hatching test has been widely used in China for many years; however, only 50 to 70% of the eggs will hatch, thus presenting problems for detection of light infections. The stool specimens are diluted with nonchlorinated water in a sedimentation flask or a beaker. The sides of the flask or beaker are covered with aluminum foil to prevent light from passing through. A light source is used to project a perpendicular light beam through the water at the top. Miracidia that hatch from the live eggs become concentrated in the light and can be detected swimming around. This motility can easily be observed with a hand lens. Aliquots of the surface water can be transferred to a small petri dish and observed under a dissecting microscope for miracidia. Observation periods should be frequent because of the limited life span of the miracidia. Ideally, observations should be made every 30 min over a period of 4 h. The hatching test is designed to mimic the conditions in nature with spring water and sunlight.

Biopsy Specimens.   Although rectal biopsy specimens can be examined, this approach is neither practical nor convenient for population surveys. Also, it is difficult to differentiate dead from living eggs; these findings can lead to excessive therapy.

Sonography.   Infection with *S. japonicum* currently affects about 1 million people, mainly in the Philippines and China. Individuals with chronic infection may develop liver fibrosis with portal hypertension, potentially resulting in life-threatening upper gastrointestinal hemorrhage.

Apparently, there is no clear evidence about which sonographic features of infection with *S. japonicum* are related to portal hypertension and thus to a more likely fatal outcome. Using Doppler sonography, a recent study demonstrated that measurements of portal perfusion are a prognostic indicator of the risk of gastrointestinal hemorrhage and overall survival of these patients (31). This study also confirmed the suitability of Doppler sonography for research under tropical field conditions.

Antibody Tests. A large number of serologic tests have been used in the diagnosis of schistosomiasis (69); however, most have not been particularly useful because of cross-reactions with other helminth infections, continuation of elevated titers long after successful treatment, and slow immunologic response of the host. Because of the complex life cycle, a large number of antigens have been used in serologic tests. Only a few have been useful. The most frequently used tests are the circumoval precipitin test, the cercaria-Hullen reaction, the indirect fluorescent-antibody test, the indirect hemagglutination test, and ELISA (57, 69). See Algorithm 17.1.

ELISA methods using soluble adult worm or egg antigens for the detection of antibodies have been used as screening tests, and infection has then been confirmed by the EITB test (69). Individuals infected with schistosomes have elevated IgE and IgG4 responses, as in those with other helminthic infections. Few studies have investigated the persistence of antibodies over time after therapy (69). It is possible that high titers posttherapy reflect treatment failure and reinfection. In a more recent study, the overall diagnostic effectiveness of the keyhole limpet hemocyanin dot-ELISA and the soluble egg antigen dot-ELISA for the detection of *S. japonicum* infection in regions of China with endemic infection was 97 and 92%, respectively (38).

ELISA has also been used to detect circulating schistosome antigens in the serum and urine of infected patients and may be the preferred method of diagnosis (51, 76). In addition to diagnosis of infection, antigen levels have been used to assess disease severity in terms of intensity of infection and to monitor the efficacy of therapy (51, 76). Individuals treated with praziquantel experience a rapid clearance of CAA and CCA; however, with other therapeutic drugs, clearance may take much longer.

Antigen Tests. The sensitivity of antigen tests is usually higher than that of routine fecal examinations for eggs; however, the specificity is lower. A rapid one-step enzyme immunoassay has been reported to have a sensitivity of 95% and a specificity of 100% (56). Unfortunately, antigen detection methods are not widely used in China. Limitations include the cost per case detected, complicated

protocols, expensive reagents with a short shelf life, and a lack of complete understanding of the mechanics of circulating schistosome antigens. However, the potential advantages of the simple dipstick test(s) may lead to improved antigen testing, and the tests may become more widely available.

KEY POINTS—LABORATORY DIAGNOSIS

Schistosoma japonicum

1. Eggs cannot be detected in stool until the worms mature (may take 5 to 6 weeks after initial infection).
2. In very light or chronic infections, eggs may be very difficult to detect in stool; therefore, multiple stool examinations may be required. Biopsy and/or immunologic tests may be helpful in diagnosing infection in these patients.
3. The zinc sulfate flotation concentration method should not be used, since the eggs do not float; the sedimentation method is recommended.
4. It may be necessary to tap the coverslip to move the eggs; the small lateral spine may be difficult to see if debris is deposited on the egg (this is common).
5. Occasionally, eggs of *S. japonicum* are detected in the urine.
6. Patients who have been treated should have follow-up ova and parasite examinations for up to 1 year to evaluate treatment.
7. In active infections, the eggs should contain live or mature miracidia. Examination for the confirmation of flame cell activity must be performed on fresh specimens; no preservatives can be used prior to this test or the hatching test.
8. During the hatching test, the light must not come too close to the surface of the water, since excess heat can kill the liberated larvae. Also, the water should be examined about every 30 min for up to 4 h.

Treatment

Praziquantel is the drug of choice for treating infection with *S. japonicum*; the adult and pediatric doses are the same (60 mg/kg/day in three doses within 1 day). Side effects, which are usually mild and transient, include abdominal discomfort, dizziness, drowsiness, headache, fever, and loose stools. These effects usually disappear within 48 h. In heavily infected individuals, the side effects, although transient, may be very severe (62). Split doses have been recommended for treating heavy infections, but this may be a concern for population-based therapy, where individuals may not present themselves for follow-up therapy. Praziquantel has little effect on immature stages of the parasite, and this affects the cure rates as the immature parasites develop to maturity and egg production. The mechanism of action may involve glutathione *S*-transferase, a detoxifying enzyme found in many helminths. Cure rates have been reported to be from 70 to 95%. Praziquantel damages the tegument of the worm, thereby exposing the antigens to parasite-specific antibodies. The use of immunoglobulins along with praziquantel has been proposed to increase the efficacy of treatment.

Epidemiology and Prevention

S. japonicum infections occur in the Far East, including China, Indonesia, Japan, and the Philippines. There is a focus of enzootic infection in Taiwan; however, there is no evidence of locally acquired human infections. In addition to humans, numerous animals (cats, cattle, dogs, goats, horses, sheep, pigs, mice, rats, and water buffaloes) are naturally infected in many areas of endemic infection. The infection is transmitted by *Oncomelania* snails, which are operculated, amphibious, and able to survive prolonged periods of desiccation. On contact with water, the snails become active again and shed infective cercariae. The widespread use of night soil as fertilizer significantly increases the occupational hazards of the disease for rice farmers. There is evidence that congenital *S. japonicum* infection of pigs can occur if sows are infected during mid to late pregnancy; this may have implications not only for pigs but also for other mammalian hosts of schistosomes, including humans (77). Praziquantel tends to be highly effective against *S. japonicum* in pigs without causing pathologic side effects in the liver.

The geographic distribution of the disease depends on the distribution of the intermediate snail hosts and the opportunity to infect humans and snails. Schistosomiasis is transmitted from infected people defecating or urinating in or near water where the appropriate snail host resides. Infections can persist indefinitely in humans. Most infected individuals have low worm burdens, but a few have very heavy infections. It is the latter group that probably makes the greatest contribution to the dissemination of schistosomiasis. The highest rates of infection have been found in children; the greatest cercarial exposure usually occurs in boys aged 5 to 10 years. Older children may have less recreational exposure but are more likely to be exposed while performing chores such as agricultural activities, washing dishes and clothes, and bathing younger siblings. Although adults also have less recreational exposure to water, they also have partial acquired immunity.

Dongting Lake is located in Hunan Province in southern China. Asian schistosomiasis has been endemic in this region for centuries, with a significant impact on the public health of the local population. Even with dedicated control programs, the snail habitats are huge (1,768 km^2) and are increasing at a rate of 34.7 km^2 annually. It is anticipated that the construction of the Three Gorges Super Dam, the largest engineering project ever undertaken and due for completion in 2009, will substantially increase the range of the snail habitats and hence increase the number of new cases of schistosomiasis. In many areas, human reinfections after praziquantel therapy remain very high (up to 20%), primarily due to occupational water contact. Efforts aimed at control or eradication of the infection in the future will be very complex (39).

Domestic cattle and buffalo are the most important reservoir hosts in China, with humans and animals frequently coming in contact at water sources, which lead to increased transmission. Although praziquantel is used for periodic treatment of bovines, the animals are prone to reinfection.

Vaccine Development. The development of an effective vaccine against *S. japonicum* has been under way for some time. Selection and manufacture of the final product will be difficult, particularly regarding funding issues related to Good Manufacturing Practice scale-up for extensive veterinary coverage and to support any future human trials. The final product will probably be a mixture of several molecules and will be used in an integrated program of control that would also include health education and targeted chemotherapy (42). A number of immunogens have been identified which could be used for vaccines, and vaccine development is ongoing (23, 41, 79). A number of other initiatives are also under way and include newer drugs, improved diagnostics, disease risk prediction, and transmission control using satellite-based remote sensing (70).

Schistosoma mekongi

Schistosomiasis resembling the disease caused by *S. japonicum* has been known to exist in the lower Mekong River basin since 1957. Further investigation has shown the etiologic agent to be a new species, *S. mekongi*; this species is responsible for most of the cases reported from mainland Indochina and has a prevalence of up to 25% in some areas.

Life Cycle and Morphology

Dogs as well as humans may serve as definitive hosts (Table 17.1). The adult worms are found in the mesenteric veins and resemble those of *S. japonicum*, although they lack tuberculations. Egg production occurs within 4 to 6 weeks after infection, and adult females may have many eggs in the uterus. The eggs of *S. mekongi* are similar in shape to those of *S. japonicum* but are smaller. They are subspherical, with a size range of 30 to 55 μm by 50 to 65 μm, and they have a small lateral spine near one end (Figure 17.17). Fully embryonated eggs without an operculum are passed in the feces. When the egg comes in contact with freshwater, the miracidium is released. The miracidium must infect the aquatic snail *Lithoglyphopsis aperta* to continue the life cycle.

Clinical Disease

Few clinical studies have been conducted in areas of endemic infection; however, the pathologic effects of *S. mekongi* infections appear similar to those of other human intestinal schistosomes. *S. mekongi* appears to be no more or less pathogenic than *S. japonicum*, with hepatosplenomegaly and portal venous hypertension being the most important serious sequelae (Table 17.5). Brain involvement of the infection has also been confirmed (26).

Diagnosis

Schistosomiasis should be suspected in individuals who have had significant contact with water in areas of endemic infection, particularly along the Mekong River in Laos and Cambodia. Symptoms of acute schistosomiasis include fever, diarrhea, abdominal pain, weight loss, and eosinophilia. Symptoms of chronic disease include chronic diarrhea, abdominal pain, and hepatomegaly or hepatosplenomegaly.

Diagnostic Procedures. Specific diagnosis of schistosomiasis by detection of eggs in stool specimens is possible only after egg production has begun. Eggs may be found in feces as early as 5 weeks after infection. The ease of egg detection depends on the worm burden and the duration of the infection. Patients with low worm burden or old (chronic) infections may have very few eggs in the feces or urine, and the infections are often not diagnosed because of the limits of sensitivity of examination techniques. Multiple stool examinations should be performed for any individual suspected of having schistosomiasis. The eggs

of *S. mekongi* are more round and tend to be smaller than those of *S. japonicum*; they measure 50 to 65 μm by 30 to 55 μm (Figure 17.17). A minute lateral spine is found at one end of the egg; however, it may be very difficult to see.

Direct detection or concentration techniques can be used to detect eggs in stool. Direct microscopic examination of stool smears is not very sensitive but may be useful for screening purposes. The Kato or Kato-Katz thick smears are simple and sensitive quantitation techniques that have been used successfully; however, sensitivity is lacking in light infections.

The zinc sulfate concentration technique is not recommended for schistosome eggs. The eggs rupture and do not float but instead are found in the sediment. The formalin-ethyl acetate technique is recommended for concentrating eggs; however, because it involves fixation, it cannot be used to detect egg viability.

For diagnosis of chronic infections in which the worm burden is light, hatching tests can be performed. See chapter 28 and the discussion under *Schistosoma mansoni*.

Antibody Tests. A large number of serologic tests have been used in the diagnosis of schistosomiasis (69); however, most have not been particularly useful because of cross-reactions with other helminth infections, continuation of elevated titers long after successful treatment, and slow immunologic response of the host. Because of the complex life cycle, a large number of antigens have been used in serologic tests. Only a few have been useful. The most frequently used tests are the circumoval precipitin test, the cercaria-Hullen reaction, the indirect fluorescent-antibody test, the indirect hemagglutination test, and ELISA (57, 69). See Algorithm 17.1.

ELISA methods using soluble adult worm or egg antigens for the detection of antibodies have been used as screening tests, and infection has been confirmed by the EITB test (69). Individuals infected with schistosomes have elevated IgE and IgG4 responses, as in those with other helminthic infections. Few studies have investigated the persistence of antibodies over time after therapy (69). It is possible that high titers posttherapy reflect treatment failure and reinfection.

The dipstick dye immunoassay was developed in China for the detection of antibodies against *S. japonicum*. This assay has also been tested in the diagnosis of *S. mekongi* infections in Cambodia and Laos. When testing Cambodians, the sensitivity was 97.1%; when testing patients from Laos, the sensitivity was 98.6%. Cross-reactivity was found in patients infected with *Opisthorchis viverrini*; interpretation may therefore be more difficult in areas where both *S. mekongi* and *O. viverrini* are endemic (82).

Figure 17.17 *Schistosoma mekongi* eggs. Note the subspherical shape and smaller size than *S. japonicum* eggs.

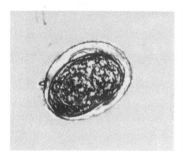
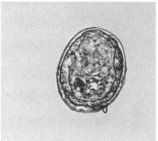

Table 17.5 Distribution of infection and morbidity due to *Schistosoma mekongi* in northern Cambodia

Information	Comments
Study population 1	Household surveys in villages on east and west banks of Mekong River (1,396 participants)
Diagnostic method	Single stool examination (Kato-Katz smear)
Overall infection prevalence	49.3%
Infection intensity	118.2 eggs/g of stool
Prevalence and intensity distribution	No differences between east and west shores of Mekong (see below)
Morbidity	Severe morbidity common; hepatomegaly (48.7%), splenomegaly (26.8%), visible diverted circulation (7.2%), ascites (0.1%); significantly more hepatomegaly, splenomegaly, diverted circulation seen on west bank
Prevalence details	West bank, 71.8%; east bank, 71.9%; age group most affected was 10–14 yr; west bank, 172.4 eggs/g; east bank, 194.2 eggs/g
Peak age group (10–14 yr)	West bank, hepatomegaly (88.1%); east bank, hepatomegaly (82.8%)
Study population 2	20 primary schools on east and west banks (2,391 participants, age 6–16 yr)
Diagnostic method	Single stool examination (Kato-Katz smear)
Overall infection prevalence	40.0% (range 7.7–72.9% per school)
Infection intensity	110.1 eggs/g of stool (range 26.7–187.5 eggs/g per school)
Prevalence and intensity distribution	Significantly higher in schools on the east bank
Morbidity	Severe morbidity common; hepatomegaly (55.2%), splenomegaly (23.6%), visible diverted circulation (4.1%), ascites (0.5%), reported blood (26.7%) and mucus (24.3%) in stool; significantly more hepatomegaly, splenomegaly, diverted circulation, and blood in stool in schools on the east bank
Prevalence details	No differences in sex for prevalence and intensity of infection or prevalence of hepatomegaly; boys more likely to have splenomegaly, ascites, and bloody stools; overall prevalence and intensity of infection highly associated in this study population

Antigen Tests. ELISA has also been used to detect circulating schistosome antigens in the serum and urine of infected patients and may be the preferred method of diagnosis (5, 51, 76). In addition to diagnosis of infection, antigen levels have been used to assess disease severity in terms of intensity of infection and to monitor the efficacy of therapy (51, 76). Individuals treated with praziquantel experienced a rapid clearance of CAA and CCA; however, with other therapeutic drugs, clearance may take much longer.

KEY POINTS—LABORATORY DIAGNOSIS
Schistosoma mekongi
1. Eggs cannot be detected in stool until the worms mature (may take 4 to 6 weeks after initial infection).
2. In very light or chronic infections, eggs may be very difficult to detect in stool; therefore, multiple stool examinations may be required. Biopsy and/or immunologic tests may be helpful in diagnosing infection in these patients.
3. The zinc sulfate flotation concentration method should not be used; the sedimentation method is recommended.
4. It may be necessary to tap the coverslip to move the eggs; the small lateral spine may be difficult to see if debris is deposited on the egg (this is common).
5. These eggs tend to be more round and are smaller than those of *S. japonicum*.
6. Patients who have been treated should have follow-up ova and parasite examinations for up to 1 year to evaluate treatment.
7. In active infections, the eggs should contain live or mature miracidia. Examination for the confirmation of flame cell activity must be performed on fresh specimens by using the wet mount or hatching test; no preservatives can be used prior to the wet mount examination or the hatching test.
8. During the hatching test, the light must not come too close to the surface of the water, since excess heat can kill the liberated larvae. Also, the water should be examined about every 30 min for up to 4 h.

Treatment
Praziquantel is the drug of choice for treating infection with *S. mekongi*; the adult and pediatric dose is the same

(60 mg/kg/day in three doses within 1 day). Side effects may include abdominal discomfort, dizziness, drowsiness, headache, fever, and loose stools; however, these effects usually disappear within 48 h. In heavily infected individuals, the side effects may be more severe. Split doses have been recommended for treating heavy infections, but this may be a concern for population-based therapy, where individuals may not present themselves for follow-up therapy. Praziquantel has little effect on immature stages of the parasite. Cure rates have been reported to be from 70 to 95%. Praziquantel damages the tegument of the worm, exposing the antigens to parasite-specific antibodies. In treating the patient with brain involvement mentioned above, the response to praziquantel was excellent but repeated courses of corticosteroid therapy were required to suppress recurring neurologic symptoms (26).

Epidemiology and Prevention

S. mekongi infections occur in the Mekong River basin in Cambodia, Laos, and Thailand. Dogs and rodents also harbor the infection, which presents a problem for prevention of the spread of infection. The snail *Lithoglyphopsis aperta*, a member of the family Hydrobiidae, is aquatic, lives on solid debris in water, has a limited distribution, and is responsible for transmission of the disease. Although the snail host is sensitive to common molluscicides, control measures have not been undertaken on a concerted basis.

A baseline epidemiologic survey of *S. mekongi* was conducted along the Mekong River in northern Cambodia; this comprehensive study documents for the first time the public health importance of schistosomiasis mekongi in this area and the high level of morbidity associated with the infection (Table 17.5) (63). Schistosomiasis control in both Laos and Cambodia was based on universal treatment campaigns and resulted in a dramatic decrease of the prevalence of the infection and in morbidity control. However, even if the disease and the infection have been satisfactorily controlled, transmission still occurs, and in very limited areas the prevalence reaches rates of more than 15%. Today, 60,000 people are estimated to still be at risk of infection in Laos and about 80,000 are at risk in Cambodia.

Schistosoma haematobium

Urinary schistosomiasis was common in the Nile River valley and has been found during examination of the kidneys of Twentieth Dynasty Egyptian mummies. Bilharz in 1851 documented the disease when he recovered adult worms from the veins of autopsy cadavers. Leiper in 1915 definitively proved the existence of two separate species,

S. haematobium and *S. mansoni*, which until that time had been confused with one another.

Life Cycle and Morphology

Humans are the only significant reservoir hosts of *S. haematobium* (Table 17.1). Fully embryonated eggs without an operculum (112 to 170 μm by 40 to 70 μm) escape from the body in the urine. The eggs are light yellowish brown and contain a conspicuous terminal spine (Figures 17.14 and 17.18).

The adult male worm contains minute integumentary tuberculations, smaller than those found on adult *S. mansoni* males. *S. haematobium* adults reside in the vesical and pelvic plexuses of the venous circulation. The adult female can contain 20 to 100 eggs in the uterus at one time. In addition to the vesical and pelvic plexuses, oviposition may occur in the rectal venules (Figure 17.19). To maintain the life cycle, the miracidium must find a suitable snail host (*Bulinus* sp.) when it is released from the egg in freshwater. The immature larval worms, after reaching the liver, may migrate via the inferior mesenteric veins to the rectal vein in order to mature. However, they most commonly migrate through the hemorrhoidal and pudendal veins to the vesical and pelvic plexuses. Within 3 months of exposure to cercariae, egg production begins.

Clinical Disease

Urogenital Disease. Light infections with urinary schistosomiasis usually produce no symptoms; however, in early disease there may be dysuria and hematuria due to cystitis from deposited eggs. Depending on the worm burden, symptoms may include dysuria, frequency, and terminal or total hematuria. Hematuria is so common that in some areas of endemic infection this phenomenon was considered to be analogous to menarche in girls (6). Symptoms are usually not seen for 3 to 6 months and

Figure 17.18 *Schistosoma haematobium* eggs. Note the terminal spine and slight size variations.

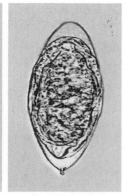

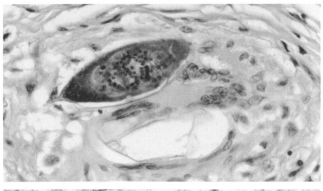
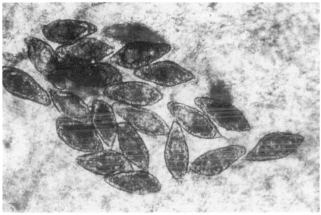

Figure 17.19 *Schistosoma haematobium.* (Upper) Egg in the appendix. Note the terminal spine. (Lower) Eggs in a rectal biopsy specimen. (From A Pictorial Presentation of Parasites: A cooperative collection prepared and/or edited by H. Zaiman.)

may take a year or more to develop. Physical examination is usually normal, but urinalysis may reveal many red blood cells and a few white blood cells on microscopic examination; reagent strip results may indicate hematuria and proteinuria. Chronic disease may lead to major diseases, including obstructive uropathy, chronic bacteriuria, bladder carcinoma, and bladder calcification (Table 17.6).

Eggs are most highly concentrated in the tissues of the bladder and lower ureter (Figures 17.20 and 17.21). As the eggs become trapped in the tissues, granulomas and pseudoabscesses form, leading to fibrosis and ulceration. With extensive fibrosis, the bladder loses its contractility. The urethra frequently is occluded because of hyperplasia, polyp formation, and discharge of purulent debris plugs from the bladder. The ureters are also frequently involved, and obstruction can cause urine reflux, hydronephrosis, retrograde infections, and renal failure. Heavy infections in males may involve the penis, resulting in elephantiasis due to blockage of the scrotal lymphatics by egg deposition. Detection of *S. haematobium* eggs in 43% of semen samples with increased levels of eosinophil

cationic protein suggests that the male genital organs are frequently affected (37).

Up to 75% of women with urinary schistosomiasis have *S. haematobium* eggs in the genitalia; urinary schistosomiasis is associated with sandy patches in the lower genital tract. Eggs have also been detected in biopsy specimens from vaginal mucosa. Female genital schistosomiasis (FGS) is often associated with eggs in the cervix, vagina, and/or vulva, as seen during examination of a wet cervical biopsy specimen crushed between two glass slides. There is significant correlation between the size of the genital lesions and the number of eggs per square millimeter of crushed tissue. Women with FGS also tend to have more tumors in the vulva than do women with schistosomiasis limited to the urinary tract (35). In one study, only 43% of the patients with FGS had hematuria. Also, since FGS frequently exists in women with scanty or no egg excretion in the urine and since this disease represents both an individual and public health hazard in areas of endemic infection, mass treatment targeted to women of childbearing age should be a consideration (49). *S. haematobium* eggs have been associated with homogeneous yellow sandy patches, mucosal bleeding, and abnormal blood vessels. However, the finding of eggs was not a predictor for ulcers, papillomata, leukoplakia, polyps, or cell atypia (34). Genital pathology due to sequestered *S. haematobium* ova is partially reversible 2 to 9 weeks after the adult worms have been killed by praziquantel. A reduction in inflammatory responses can be detected in histologic sections and vaginal lavage fluid. However, these sandy patch lesions associated with contact bleeding and vessel abnormalities may be refractive to treatment for up to 12 months and may be an important risk factor for potential acquisition and transmission of HIV (33).

Obstructive Uropathy.   Egg deposits in the ureter walls are the general causes of obstructed urinary flow; conditions leading to obstruction include polypoid patches, fibrosis, ureteritis cystica, calculi, and the total number of eggs present. Often these patients remain asymptomatic other than having symptoms related to cystitis or pain from the ureters.

Bladder Calcification.   Calcified eggs produce the typical image, which consists of confluent sandy patches as seen in Figure 17.21 (left). The tissues of the bladder are not calcified and may continue to function normally; however, the visual radiologic picture suggests a heavy infection. Calcified eggs can also be seen in the ureters and seminal vesicles and less often in the colon.

Bladder Carcinoma.   Carcinoma of the bladder has been frequently noted in patients infected with *S. haematobium*.

Table 17.6   Disease manifestations seen with *Schistosoma haematobium* infection

Disease manifestation	Comments
Obstructive uropathy	Hydronephrosis, hydroureter, pyelonephritis, renal failure often silent clinically but can be demonstrated by intravenous pyelogram, ultrasonography, or renogram function testing; most lesions occur in lower third of ureters or bladder
Pyelonephritis	Secondary to obstruction; gram-negative sepsis; may be the leading cause of mortality in urinary disease
Renal failure	Bilateral ureteral obstruction common; may be the direct result of obstruction, pyelonephritis, or ureterovesicular urinary reflux
Chronic bacteriuria	Infection rates up to 5% in schoolboys with *S. haematobium* infection; coliforms are responsible, with *Salmonella* being the most common in some studies; treatment usually unsuccessful unless schistosome infection is also treated and schistosomal mucosal lesions are eliminated
Bladder carcinoma	Bladder irritation, gross hematuria, weight loss, metastasis (inguinal, femoral, and retroperitoneal lymph nodes) tend to lag 10–20 yr behind age at which *S. haematobium* infection peaks; most cases diagnosed in patients aged 40–49 yr; biopsy is the usual method of confirmation; urinary cytology may be helpful
Bladder calcification	The "fetal head" sign is seen in abdominal X rays and is caused by calcium deposits around schistosome eggs in the bladder and uterine walls (Figure 17.20); sonography can also be used to see calcification in the bladder walls; although bladder loses contractility due to fibrotic walls, it does not become rigid; calcification resolves over years if patient remains free of schistosomiasis
Cutaneous disease	Has been previously reported but is an unusual presentation; skin lesions, appearing 3 yr after exposure, have been seen as the only manifestation of schistosomiasis; routine biopsy confirmed infection; awareness of this infection when evaluating patients with unusual skin lesions who have traveled in areas of endemicity is recommended (12)
Other body sites	Cerebral schistosomiasis has also been reported (49)

Many factors have been suggested as agents promoting schistosome-associated bladder cancer (Table 17.7). *N*-Nitroso compounds in association with secondary bacterial infections of the urinary tract may contribute to the high prevalence of bladder cancer (44). Bladder cancer is the most prevalent cancer in Egyptians (Figure 17.21). Many of the tumors involve the posterior wall of the bladder and are noted to occur more frequently in males than in females. The extent of *S. haematobium*

infection plays a significant role in the induction of different types of carcinoma, since squamous cell carcinoma is usually associated with moderate and/or high worm burdens while transitional cell carcinoma occurs more frequently in areas associated with lighter parasite loads. The predominance of squamous cell carcinoma in urinary bladder tissues in patients with schistosomiasis is probably due to continuous exposure to the larger quantities of carcinogens (*N*-nitroso compounds) in urine in patients with the disease (45).

Other Body Sites. Periportal fibrosis with hepatomegaly and splenomegaly has been noted in patients in areas where infections are endemic (1). Splenic enlargement has been correlated with the intensity of *S. haematobium* infection. Cardiopulmonary disease can develop in patients with hepatosplenic schistosomiasis; eggs are carried via the mesenteric veins to the lungs through systemic collateral veins. Schistosomal cor pulmonale tends to be rare in infections with *S. haematobium*; and eggs can be found in lung tissue at autopsy but are generally not clinically relevant. Although any *Schistosoma* species may involve the central nervous system, *S. mansoni* and *S. haematobium* more commonly involve the spinal cord than the brain. Severe disease presents as transverse myelitis.

As with *S. mansoni*, chronic bacteremic infections with *Salmonella* are seen, particularly in cases of chronic

Figure 17.20 Calcified *Schistosoma haematobium* eggs in a crushed sliver of fixed tissue from the bladder of an Egyptian patient. Unstained, ×239. (Armed Forces Institute of Pathology photograph.)

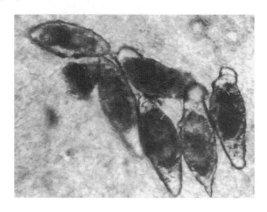

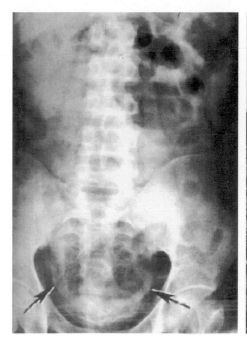
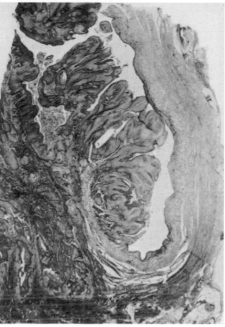

Figure 17.21 *Schistosoma haematobium*: (Left) X ray of the pelvis in a 24-year-old Egyptian with schistosomal obstructive uropathy. Linear calcification is seen in the base of the bladder and along the lower ureters, which are dilated (arrows). (Armed Forces Institute of Pathology photograph. Courtesy of Marcia Young.) (Right) Urinary bladder. An exophytic squamous cell carcinoma is present in the central portion of the photograph, and infiltrating squamous cell cancer extends to the left. Magnification, ×4. (Armed Forces Institute of Pathology photograph.)

active schistosomiasis. Also, schistosomal appendicitis appears to be specific to infection with *S. haematobium* and is an uncommon cause of appendicitis in areas where schistosomiasis is not endemic. Therapy requires antischistosomal medication in addition to surgery (13, 61).

Diagnosis

Diagnostic Procedures. *S. haematobium* eggs are usually detected in the urine, although in heavy infections they may also be found in the stools. The terminal hematuria portion of the urine specimen may contain numerous eggs trapped in the mucus and pus. Peak egg excretion occurs between noon and 3 p.m. Samples collected during this time, or during a 24-h urine collection without preservatives, may be used for examination. Urine can be examined under a microscope after sedimentation or centrifugation. It is important to use saline and not water for the concentration procedures; this will avoid hatching of the eggs. Nuclepore filtration is an excellent method for the concentration of eggs in urine. Some data indicate that egg output in urine is an accurate method of confirming the diagnosis and shows less day-to-day variation than in ELISA detection of schistosome circulating antigens in urine (74).

Use of reagent strips to detect hematuria and proteinuria in areas of endemic infection has proven to be an effective screening technique for *S. haematobium* infections (7). The assay correctly identified 87% of egg-negative controls and 98% of parasitologically confirmed cases, including those in six individuals who had been treated but

not cured. Also, antigen was detected in the urine of three individuals from whom two specimens had to be examined microscopically to confirm the infection. This suggests that the test is more sensitive than microscopy. To increase the sensitivity of the reagent strip tests, the tests should be performed during the period of peak egg excretion in the urine (noon to 3 p.m.). Although the use of reagent strips allows one to rapidly screen a population at risk, other conditions can yield a positive result. The strips should be used in conjunction with microscopic examination or serologic testing for confirmation of infection. Reagent strips coated with MAbs to CCA were developed for testing urine specimens for *S. haematobium*. The levels of CCA are much higher in the urine than are the levels of CAA. Although not as sensitive as ELISA methods, this technique is suitable for screening large populations in areas of endemic infection. Eosinophilia has also been used to diagnose infections, with a positive predictive value of >80%.

Antibody Detection.   A large number of serologic tests have been used in the diagnosis of schistosomiasis (69); however, many have not been particularly useful because of cross-reactions with other helminth infections, continuation of elevated titers long after successful treatment, and slow immunologic response of the host. Because of the complex life cycle, a large number of antigens have been used in serologic tests. Only a few have been useful. The most frequently used tests are the circumoval precipitin test, the cercaria-Hullen reaction, the indirect fluorescent-antibody test, the indirect hemagglutination test, and ELISA (69, 75). See Algorithm 17.1.

Table 17.7   Mechanisms of bladder carcinogenesis due to *Schistosoma haematobium*

Mechanism	Comments
Inflammatory cells (IC)	Macrophages and neutrophils are important sources of endogenous oxygen radicals that are implicated in the formation of carcinogenic *N*-nitrosamines; IC induce mutations, sister chromatid exchanges, and DNA strand breaks; increased numbers of IC may enhance carcinogenic potential of aromatic amines by increasing their rate of activation
Microorganisms	Hospitalized patients (39–66%) had bacteriuria; interaction between schistosomiasis and bacterial infection appears to depend on gender (more common in males than females); there may be a symbiotic relationship between worms and bacteria; bacterial infection of the urinary tract increases the risk of bladder cancer in patients with problems other than schistosomiasis; individuals with bacterial cystitis and schistosomiasis are potentially more exposed to nitrate and/or nitrite, increasing the risk of in situ formation of *N*-nitrosamines
Genetic changes	Specific genes associated with neoplastic progression in schistosomal bladder cancer include the activation of H-*ras* (the *ras* gene family consists of frequently activated oncogenes), inactivation of *p53*, and inactivation of the retinoblastoma gene; alterations in genes and/or proteins can change function, leading to uncontrolled cell growth and tumor formation
	Mutation of the coding sequence of the *p53* tumor suppressor gene is the most frequent genetic alteration in a number of human malignancies; 6 of 7 Egyptian bladder cancer patients had *p53* mutations; the frequency of *p53* mutations varies with different grades of schistosomiasis-associated bladder cancer; alterations in the *p53* gene may modulate gene expression for regulation of relevant DNA repair processes, as well as cell division and death by apoptosis
	Bcl-2 is overexpressed in some schistosomiasis-associated bladder cancers (expressed at high levels in squamous cell carcinoma but not transitional cell carcinoma); may be upregulated in later stages of tumor progression; *Bcl-2* may also be related to tumor cell lineage
	A 9p gene, possibly *CDKN2*, may contribute to the development of the majority of schistosomiasis-associated bladder tumors, but the genes on 9q play a much less important role (59)
	Detection of early genetic changes may be valuable in identifying high-risk individuals
Diet	Dietary fat is consistently high on a list of associated factors (attenuated when other diet components are included); small amounts of *N*-nitroso compounds are found in Egyptian food (cheese, fava beans, raw salted fish), all of which are consumed almost daily
Carcinogen metabolism	Increased levels of enzymes responsible for activation of carcinogenic *N*-nitroso compounds, aromatic amines, and polycyclic aromatic hydrocarbons have been found in schistosomiasis; may increase and prolong bladder exposure to DNA-damaging agents; nitrite levels are about 20 times higher in *S. haematobium*-infected patients with bladder cancer than in those without bladder cancer; these compounds may play a role not only in initiation of the carcinogenic process, but also in its progression
Carcinogens, consequences of DNA damage	Damage to bladder tissue DNA may be responsible for initiation of bladder cancer

ELISA methods using soluble adult worm or egg antigens for the detection of antibodies have been used as screening tests, and infection has then been confirmed by the EITB test (69). Individuals infected with schistosomes have elevated IgE and IgG4 responses, as in those with other helminthic infections. Few studies have investigated the persistence of antibodies over time after therapy (69). It is possible that high titers posttherapy reflect treatment failure and reinfection.

Antigen Tests. Based on 92% specificity and 90% sensitivity, an ELISA using MAb 290-2E6-A is one of the most promising assays for immunodiagnosis of *S. haematobium* infections (46). Another study of children from Kenya indicates that the SEA ELISA compared well with microhematuria within egg count classes and with egg counts within hematuria classes (29). Levels of SEA and egg output have shown similar correlations with sonographically detectable pathology; these correlations were better than the correlation between hematuria and pathology (29). Another antigen capture ELISA using MAb 128C3/3/21 has shown a correlation between the level of circulating antigen and the intensity of infection as measured by egg counts. There appears to be a direct relationship between antigen level and disease severity, as monitored by sonography. Also, a purified recombinant serine protease inhibitor that is species specific for *S. haematobium* has been used in serologic assays.

Using the SEA MAb, circulating schistosome antigen was undetectable in serum from patients with other parasitic infections and was detected in 90.5% of serum samples and 94% of urine samples from *S. haematobium*-infected patients. A significant reduction in circulating schistosome antigen levels was detected in serum and urine samples 12 weeks after praziquantel therapy (58). No cross-reactivity with *S. mansoni* was seen.

A MAb-based dipstick assay has been used with a sensitivity of 98.8% and specificity of 53.6%. Although the detection of microhematuria and proteinuria had a higher specificity, the sensitivity was much lower than that of the dipstick assay (8).

KEY POINTS—LABORATORY DIAGNOSIS

Schistosoma haematobium

1. Eggs cannot be detected in urine until the worms mature (may take up to 3 months after initial infection).

2. In very light or chronic infections, eggs may be very difficult to detect in urine; therefore, multiple urine examinations may be required. Biopsy and/or immunologic tests may be helpful in diagnosing infection in these patients.

3. Both 24-h and spot urine samples should be examined as wet mounts (after concentration using no preservatives [use saline so hatching will not occur]); the urine specimens should be collected with no preservatives. These eggs are also occasionally recovered in stool, and so both urine and stool specimens should be examined.

4. The membrane filtration technique using Nuclepore filters can be very helpful in diagnosing infection with *S. haematobium*.

5. Patients who have been treated should have follow-up ova and parasite examinations for up to 1 year to evaluate treatment.

6. In active infections, the eggs should contain live or mature miracidia. Examination for the confirmation of flame cell activity must be performed on fresh specimens using the wet mount or hatching test; no preservatives can be used prior to the wet mount or the hatching test.

7. During the hatching test, the light must not come too close to the surface of the water, since excess heat can kill the liberated larvae. Also, the water should be examined about every 30 min for up to 4 h.

8. It is important to remember that the small and less commonly seen miracidium larva of *S. haematobium* may be present in the urine; motility in unpreserved specimens or stained morphology could confirm this diagnosis.

9. It is also possible to see *S. haematobium* eggs in semen specimens, even when repeated urinary and fecal examinations and serologic tests are negative.

Treatment

Praziquantel is the drug of choice for treating infections with *S. haematobium*. It can be taken orally in a single dose and is well tolerated by the host. Side effects, which are usually mild and transient, include abdominal discomfort, dizziness, drowsiness, headache, fever, and loose stools. These effects usually disappear within 48 h. In heavily infected individuals, the side effects, although transient, may be very severe (62). Split doses have been recommended for treating heavy infections, but this may be a concern for population-based therapy, where individuals may not present themselves for follow-up therapy. Data suggest that a single dose of praziquantel (40 mg/kg of body weight) may have a longer-lasting effect than was previously thought: possibly more than 2.5 years (18). In certain population-based treatment programs, it appears that a 20-mg/kg dose of praziquantel may be sufficient to provide control of morbidity due to urinary schistosomiasis (32). Praziquantel has little effect on immature stages of the parasite, and this will affect the cure rates as the immature parasites develop to maturity and egg production. The mechanism of action may involve glutathione *S*-transferase, a detoxifying enzyme found in many helminths. Cure rates have been reported to be 70 to 95%. Praziquantel damages the tegument of the worm, thereby exposing the antigens to parasite-specific antibodies. Adults are more resistant to schistosome infection than are children, and the switch to an "adult" response suggests that praziquantel treatment may have an immunizing effect, with benefits beyond a transient reduction in infection levels (40). Treatment can significantly reduce urinary tract morbidity despite reinfection; important risk factors in adulthood are cumulative intensity and duration of infection during early adolescence (66). Data also indicate that therapy given in childhood or adolescence decreases the risk for some but not all manifestations of *S. haematobium* infection in later adult life (48).

Although the exact mechanism by which praziquantel kills the parasite is not well understood, it is known that the immune response of the host plays an important role in drug efficacy. It is possible that disease states of humans that lead to immunodeficiencies, such as infection with HIV-1, may render praziquantel less effective in treating schistosomiasis. However, recent studies indicate that persons with HIV-1 infection can be treated effectively for schistosomiasis with praziquantel (30, 33).

Therapeutic failure of praziquantel in the treatment of *S. haematobium* infection in Brazilians returning from Africa has been reported. Patients were treated with a single dose of praziquantel (40 mg/kg, single dose). Two additional treatments were given, and cystoscopy between 6 and 24 months after each treatment confirmed the presence of viable eggs. These findings confirm the possible need for repeated therapy, as well as using the urinary bladder biopsy as the criterion of cure (60).

The efficacy of praziquantel during the incubation period of schistosomiasis has recently been studied for the first time (19a). In a group of 18 tourists infected with *S. haematobium* in Mali, the authors observed the efficacy of praziquantel given at different phases of the infection. Early praziquantel treatment from days 10 to 15 after exposure was less effective than later treatment (days 28 to 40) in preventing acute schistosomiasis, while neither treatment effectively prevented chronic schistosomiasis.

Patients with *S. haematobium* infection are suitable recipients for kidney transplantation, although they are at risk for urologic complications. Long-term urologic follow-up is recommended, including urethrocystoscopy, due to the increased risk for bladder cancer.

Metrifonate, an organophosphorus cholinesterase inhibitor, is an alternative to praziquantel for the treatment of *S. haematobium* infections. The drug is given orally over a number of weeks. Side effects, which are mild and transient, include abdominal discomfort, diarrhea, dizziness, headache, vomiting, and weakness. Metrifonate should not be given to patients who have been recently exposed to insecticides or who are using therapeutic agents that can potentiate cholinesterase inhibition. It has been noted that the dorsal tegumental surface of adult male schistosomes contains acetylcholinesterase and nicotinic acetylcholinesterase receptors, which may be the target of metrifonate. Because the acetylcholine receptor expression is increased when the worms pair and become sexually mature, it appears that acetylcholine plays a vital function in nutrient transport.

Epidemiology and Prevention

S. haematobium infections occur in Africa, Asia Minor, Cyprus, the islands off the African east coast, and southern Portugal; there is a focus of endemic infection in India. Humans appear to be the only important reservoir hosts, although naturally infected monkeys, baboons, and chimpanzees have been found. The intermediate snail host, *Bulinus* sp., can survive in the mud when the water dries up. The snails retain their infectivity and resume shedding cercariae when the rainy season begins. In some areas, cercariae are found in areas where infected snails are absent, and sometimes cercariae are absent in the presence of infected snails.

Data suggest that fewer than 1 in 100 contacts result in infection and fewer than 1 in 1,000 result in egg output.

This suggests that there may be substantial attrition of invading cercariae even in naïve individuals (78).

In many areas of endemic infection, awareness of urinary schistosomiasis and its symptom of blood in the urine is high, but specific knowledge about the parasite, its vector, and the interaction between them in the life cycle is lacking. Activities that require behavior and attitude modification can be identified and encouraged as components in the control of schistosomiasis.

An *S. haematobium*-derived recombinant glutathione *S*-transferase (Sh28GST) vaccine is capable of significantly reducing worm fecundity in experimentally infected primates. Although Freund's complete adjuvant induced higher levels of protection, the efficacy of *Mycobacterium bovis* as an adjuvant appeared sufficient to justify consideration of its use in a vaccine against human urogenital schistosomiasis (9).

A summary of the control methods that can be used for all types of schistosomiasis is given in Table 17.8. All of these initiatives are coupled with public health communication and training for populations in areas of endemic infection.

Schistosoma intercalatum

S. intercalatum infections were first noted by Fisher in the early 1930s, and the species has been extensively described. The infection has been found in central and western Africa. In areas where both *S. haematobium* and *S. intercalatum* are endemic, hybridization with the two trematodes appears to be occurring; natural hybrids have been found in Loum, Cameroon (68). Also, it appears that the competitive exclusion between *S. mansoni* and *S. intercalatum* may be an important factor restricting the distribution of *S. intercalatum* in Africa. *S. intercalatum* causes an intestinal form of schistosomiasis similar to that caused by *S. mansoni* but characterized by lesions, mainly situated at the rectum and sigmoid level. Urinary schistosomiasis caused by *S. haematobium* has a tendency to supplant recto-sigmoidal schistosomiasis caused by *S. intercalatum*, especially in endemic areas where hybridization between the two species of schistosomes is occurring.

Life Cycle and Morphology

The life cycle is similar to that of the other schistosomes. Adult worms are found in the mesenteric veins, and eggs are shed in the feces. Eggs are fully embryonated without an operculum and resemble *S. haematobium* eggs in shape, having a terminal spine (Table 17.1; Figure 17.22). The eggs are 140 to 240 μm by 50 to 85 μm in size and can be differentiated from *S. haematobium* eggs since they are Ziehl-Neelsen acid-fast positive. This technique cannot be totally relied on, because if hybridization occurred between *S. intercalatum* and *S. haematobium*, the eggs identified from the urine by

Table 17.8 Schistosomiasis control measures

Approach	Comments
Chemotherapy	
Mass	Most appropriate in population with prevalence rates of >30%; all members of population treated (no attempt at individual diagnosis) (25)
Targeted population	Relies on identification of individual moderate to heavy egg excretors; follow-up for evaluation of cure
Snail control	
Molluscicide use	Niclosamide commonly used; well suited for arid areas (seasonal infections) and small habitats; unfortunately, snail populations can reestablish themselves within 3 mo of molluscicide application
Biologic control	Use of natural predators of snail hosts (ducks, fish, turtles, fungi, other snails); not realistic on a mass level
Environmental modification	Prevention and/or reduction of breeding; site modification to produce detrimental conditions; cementing or closing over irrigation ditches; increasing canal water flow
Reduction of water contact and contamination	
Domestic water supplies	Clean water availability; however, population may continue to use contaminated water sources
Sanitary excreta disposal	Mandatory; combined with health education on biological transmission and importance of control measures; improvement in living standards is important
Vaccination	Continued problem of reinfection is a problem due to slow maturation of natural resistance; several antigens have reached an advanced stage of development and are undergoing testing; vaccination will be used in conjunction with all other control methods; vaccines can be used to attenuate infection and/or to prevent pathology and/or to reduce worm burdens (80)

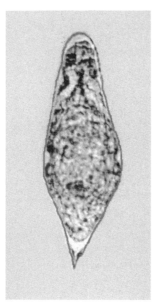

Figure 17.22 *Schistosoma intercalatum* egg. The miracidium tends to be somewhat narrower in the middle, giving it an hourglass shape. (From A Pictorial Presentation of Parasites: A cooperative collection prepared and/or edited by H. Zaiman.)

DNA analysis would be Ziehl-Neelsen stain positive. The snail intermediate hosts are *Bulinus forskalis* and *B. africanus*.

Clinical Disease

Clinical manifestations are similar to those of *S. mansoni* infections. Pathology is confined to the mesenteric-portal system; rectal bleeding appears to be the most common finding. Other symptoms can include pain in the left iliac fossa with tenesmus, anorexia, nausea, abdominal pain, and diarrhea. There may also be changes in liver function tests. The prevalence and intensity of infection were noted to increase during the first two decades of life and gradually decrease thereafter. As with other species of *Schistosoma*, there is an association between *Salmonella* bacteremia and infection with *S. intercalatum*. Without treatment for schistosomiasis, the *Salmonella* infection cannot be cured. Complications of *S. intercalatum* infection include severe rectal or genital lesions.

Diagnosis

The infection can be diagnosed by finding the characteristic eggs in feces or in squash preparations of tissues from the rectum; occasionally, eggs are also found in the urine. The miracidium often appears to be more narrow in the middle, giving it an "hourglass" shape.

KEY POINTS—LABORATORY DIAGNOSIS

Schistosoma intercalatum

1. Eggs cannot be detected in stool until the worms mature (may take from 4 to 6 weeks after initial infection).
2. In very light or chronic infections, eggs may be very difficult to detect in stool; therefore, multiple stool examinations may be required. Biopsy and/or immunology may be helpful in these patients.
3. The zinc sulfate flotation concentration method should not be used; the sedimentation method is recommended.
4. It may be necessary to tap the coverslip to move the eggs; the terminal spine may be difficult to see if debris is covering the egg (this may occur).
5. Occasionally eggs of *S. intercalatum* are detected in the urine.
6. Patients who have been treated should have follow-up ova and parasite examinations for up to 1 year to evaluate treatment.
7. In active infections, the eggs should contain live or mature miracidia. Examination for the confirmation of flame cell activity must be performed on fresh specimens, using a wet mount or the hatching test; no preservatives can be used prior to the wet mount examination or the hatching test.
8. During the hatching test, the light must not come too close to the surface of the water, since excess heat can kill the liberated larvae. Also, the water should be examined about every 30 min for up to 4 h.

Treatment

Praziquantel is the drug of choice for treating infections with *S. intercalatum*. It can be taken orally in a single dose and is well tolerated by the host. Side effects, which are usually mild and transient, include abdominal discomfort, dizziness, drowsiness, headache, fever, and loose stools. These effects usually disappear within 48 h. A single dose of praziquantel (40 mg/kg of body weight) has been recommended.

Epidemiology and Prevention

S. intercalatum infections occur in the same geographic areas as *S. mansoni* infections, i.e., western and central Africa (Cameroon, Equatorial Guinea, Gabon, Nigeria, Democratic Republic of Congo, and Sao Tome), Egypt, Madagascar, and the Arabian peninsula. Human infections are nearly all derived from human sources, even though nonhuman primates, insectivores, marsupials, and rodents can carry the infection. Disease spread is linked to seasonal migration of nomad populations, as well as migrant African laborers. In regions of endemic infection, prevalences of 5 to 25% have been found. The infection is transmitted by *Bulinas* sp., an aquatic snail.

Review of 333 reports related to imported cases of schistosomiasis in Europe showed that male patients accounted for 64% of all cases. The average age of all patients was 29.5 years. The majority of patients were of European origin (53%). Europeans traveled predominantly for tourism (52%). Travel for people from areas of endemic infection was related to immigration and refuge (51%) and visits to relatives and friends (28%). The majority of infections were acquired in Africa; 92 infections were clearly attributable to *S. haematobium*, 130 to *S. mansoni*, and 4 to *S. intercalatum*. Praziquantel was the only treatment used, and no deaths were recorded (20).

References

1. **Abdel-Wahab, M. F., G. Esmat, I. Ramzy, R. Fouad, M. Abdel-Rahman, A. Yosery, S. Narooz, and G. T. Strickland.** 1992. *Schistosoma haematobium* infection in Egyptian schoolchildren: demonstration of both hepatic and urinary tract morbidity by ultrasonography. *Trans. R. Soc. Trop. Med. Hyg.* **86:**406–409.
2. **Abramowicz, M. (ed.).** 2004. Drugs for parasitic infections. *Med. Lett. Drugs Ther.* **46:**1–12.
3. **Al-Sherbiny, M. M., A. M. Osman, K. Hancock, A. M. Deelder, and V. C. Tsang.** 1999. Application of immunodiagnostic assays: detection of antibodies and circulating antigens in human schistosomiasis and correlation with clinical findings. *Am. J. Trop. Med. Hyg.* **60:**960–966.
4. **Attallah, A. M., E. Yones, H. Ismail, S. A. El Masry, A. Tabll, A. A. Elenein, and N. A. El Ghawalby.** 1999. Immunochemical characterization and diagnostic potential of a 63-kilodalton *Schistosoma* antigen. *Am. J. Trop. Med. Hyg.* **60:**493–497.
5. **Barsoum, I. S., K. A. Kamal, S. Bassily, A. M. Deelder, and D. G. Colley.** 1991. Diagnosis of human schistosomiasis by detection of circulating cathodic antigen with a monoclonal antibody. *J. Infect. Dis.* **164:**1010–1013.
6. **Bausch, D., and B. L. Cline.** 1995. The impact of control measures on urinary schistosomiasis in primary school children in northern Cameroon: a unique opportunity for controlled observations. *Am. J. Trop. Med. Hyg.* **53:**577–580.
7. **Bosompem, K. M., I. Ayi, W. K. Anyan, T. Arishima, F. K. Nkrumah, and S. Kojima.** 1997. A monoclonal antibody-based dipstick assay for diagnosis of urinary schistosomiasis. *Trans. R. Soc. Trop. Med. Hyg.* **91:**554–556.
8. **Bosompem, K. M., O. Owusu, E. O. Okanla, and S. Kojima.** 2004. Applicability of a monoclonal antibody-based dipstick in diagnosis of urinary schistosomiasis in the Central Region of Ghana. *Trop. Med. Int. Health* **9:**991–996.
9. **Boulanger, D., A. Warter, B. Sellin, V. Lindner, R. J. Pierce, J. P. Chippaux, and A. Capron.** 1999. Vaccine potential of a recombinant glutathione *S*-transferase cloned from *Schistosoma haematobium* in primates experimentally infected with an homologous challenge. *Vaccine* **17:**319–326.
10. **Cesari, I. M., A. Ferrer, M. Kombila, E. Pichard, C. Decam, L. S. Qiu, D. Bout, and D. Richard-Lenoble.** 1998. Specificity of the solid phase alkaline phosphatase immunocapture assay for the diagnosis of human *Schistosoma mansoni* infection. *Trans. R. Soc. Trop. Med. Hyg.* **92:**38–39.

11. Cota, G. F., R. A. Pinto-Silva, C. M. Antunes, and J. R. Lambertucci. 2006. Ultrasound and clinical investigation of hepatosplenic schistosomiasis: evaluation of splenomegaly and liver fibrosis four years after mass chemotherapy with oxamniquine. *Am. J. Trop. Med. Hyg.* **74**:103–107.

12. Davis-Reed, L., and J. H. Theis. 2000. Cutaneous schistosomiasis: report of a case and review of the literature. *J. Am. Acad. Dermatol.* **42**:678–680.

13. Doudier, B., P. Parola, J. P. Dales, N. Linzberger, P. Brouqui, and J. Delmont. 2004. Schistosomiasis as an unusual cause of appendicitis. *Clin. Microbiol. Infect.* **10**:89–91.

14. Dunne, D. W., A. E. Butterworth, A. J. G. Fulford, H. C. Kariuki, J. G. Langley, J. H. Ouma, A. Capron, and R. J. Pierce. 1992. Immunity after treatment of human schistosomiasis: association between IgE antibodies to adult worm antigens and resistance to reinfection. *Eur. J. Immunol.* **22**:1483–1494.

15. Ebrahimm A., H. Elmorshedy, E. Omer, S. Eldaly, and R. Barakat. 1998. Evaluation of the Kato-Katz thick smear and formol ether sedimentation techniques for quantitative diagnosis of *Schistosoma mansoni* infection. *Am. J. Trop. Med. Hyg.* **57**:706–708.

16. Etard, J., M. Audibert, and A. Dabo. 1995. Age-acquired resistance and predisposition to reinfection with *Schistosoma haematobium* after treatment with praziquantel in Mali. *Am. J. Trop. Med. Hyg.* **52**:549–558.

17. Fallon, P. G., J. S. Mubarak, R. E. Fookes, M. Niang, A. E. Butterworth, R. F. Sturrock, and M. J. Doenhoff. 1997. *Schistosoma mansoni*: maturation rate and drug susceptibility of different geographic isolates. *Exp. Parasitol.* **86**:29–36.

18. Frenzel, K., L. Grigull, E. Odongo-Aginya, C. M. Ndugwa, T. Loroni-Lakwo, U. Schweigmann, U. Vester, N. Spann-Brucker, and E. Dochring. 1999. Evidence for a long-term effect of a single dose of praziquantel on *Schistosoma mansoni*-induced hepatosplenic lesions in northern Uganda. *Am. J. Trop. Med. Hyg.* **60**:927–931.

19. Ghandour, A. M., K. Tricker, M. J. Doenhoff, A. A. Alrobai, and A. A. Banaja. 1997. An enzyme-linked immunosorbent assay using *Schistosoma mansoni* purified egg antigen for the diagnosis of schistosomiasis in Saudi Arabia. *Trans. R. Soc. Trop. Med. Hyg.* **91**:287–289.

19a. Grandiere-Perez, L., S. Ansart, L. Paris, A. Faussart, S. Jaureguiberry, J. P. Grivois, E. Klement, F. Bricaire, M. Danis, and E. Caumes. 2006. Efficacy of praziquantel during the incubation and invasive phase of *Schistosoma haematobium* schistosomiasis in 18 travelers. *Am J. Trop. Med. Hyg.* **74**:814–818.

20. Grobusch, M. P., N. Muhlberger, T. Jelinek, Z. Bisoffi, M. Corachan, G. Harms, A. Matteeli, G. Fry, C. Hatz, I. Gjorup, M. L. Schmid, J. Knobloch, S. Puente, U. Brunner, A. Kapaun, J. Clerinx, L. N. Nielsen, K. Fleischer, J. Beran, S. da Cunha, M. Schulze, B. Myrvang, and U. Hellgren. 2003. Imported schistosomiasis in Europe: sentinel surveillance data from TropNetEurop. *J. Travel Med.* **10**:164–169.

21. Halim, A. B., R. F. Garry, S. Dash, and M. A. Gerber. 1999. Effect of schistosomiasis and hepatitis on liver disease. *Am. J. Trop. Med. Hyg.* **60**:915–920.

22. Hamburger, J., X. Yu-Xin, R. M. Ramzy, J. Jourdane, and A. Ruppel. 1998. Development and laboratory evaluation of a polymerase chain reaction for monitoring *Schistosoma mansoni* infestation of water. *Am. J. Trop. Med. Hyg.* **59**:468–473.

23. Hewitson, J. P., P. A. Hamblin, and A. P. Mountford. 2005. Immunity induced by the radiation-attenuated schistosome vaccine. *Parasite Immunol.* **27**:271–280.

23a. Hillyer, G. V. 2005. The rise and fall of bilharzia in Puerto Rico: its centennial 1904–2004. *P. R. Health Sci. J.* **24**:225–235.

24. Hillyer, G. V., V. C. W. Tsang, B. E. Vivas-Gonzalez, J. Noh, L. H. Ahn, and V. Vorndam. 1999. Age-specific decrease in seroprevalence of schistosomiasis in Puerto Rico. *Am. J. Trop. Med. Hyg.* **60**:313–318.

25. Homeida, M. M. A., I. A. Eltoum, M. M. Ali, S. M. Suliaman, E. A. Elobied, M. Mansour, A. M. Saad, and J. L. Bennett. 1996. The effectiveness of annual versus biennial mass chemotherapy in reducing morbidity to schistosomiasis: a prospective study in Gezira-Mamagil, Sudan. *Am. J. Trop. Med. Hyg.* **54**:140–145.

26. Houston, S., K. Kowalewska-Grochowska, S. Naik, J. McKean, E. S. Johnson, and K. Warren. 2004. First report of *Schistosoma mekongi* infection with brain involvement. *Clin. Infect. Dis.* **38**:e1–e6.

27. Ismail, M., S. Botros, A. Metwally, S. William, A. Farghally, L. F. Tao, T. A. Day, and J. L. Bennett. 1999. Resistance to praziquantel: direct evidence from *Schistosoma mansoni* isolated from Egyptian villagers. *Am. J. Trop. Med. Hyg.* **60**:932–935.

28. Jacobs, W., J. J. Bogers, J. P. Timmermans, A. M. Deelder, and E. A. Vanmarck. 1998. Adhesion molecules in intestinal *Schistosoma mansoni* infection. *Parasitol. Res.* **84**:276–280.

29. Kahama, A. I., A. E. Odek, R. W. Kihara, B. J. Vennervald, Y. Kombe, T. Nkulila, C. F. Hatz, J. H. Ouma, and A. M. Deelder. 1999. Urine circulating soluble egg antigen in relation to egg counts, hematuria, and urinary tract pathology before and after treatment in children infected with *Schistosoma haematobium* in Kenya. *Am. J. Trop. Med. Hyg.* **61**:215–219.

30. Karanja, D. M. S., A. E. Boyer, M. Strand, D. G. Colley, B. L. Nahlen, J. H. Ouma, and W. E. Secor. 1998. Studies on schistosomiasis in western Kenya: II. Efficacy of praziquantel for treatment of schistosomiasis in persons coinfected with human immunodeficiency virus-1. *Am. J. Trop. Med. Hyg.* **59**:307–311.

31. Kardorff, R., R. M. Olveda, L. P. Acosta, U. J. Duebbelde, G. D. Aligui, N. J. Alcorn, and E. Doehring. 1999. Hepatosplenic morbidity in schistosomiasis japonica: evaluation with Doppler sonography. *Am. J. Trop. Med. Hyg.* **60**:954–959.

32. King, C. H., E. M. Muchiri, P. Mungai, J. H. Ouma, H. Kadzo, P. Magak, and D. K. Koech. 2002. Randomized comparison of low-dose versus standard-dose praziquantel therapy in treatment of urinary tract morbidity due to *Schistosoma haematobium* infection. *Am. J. Trop. Med. Hyg.* **66**:725–730.

33. Kjetland, E. F., T. Mduluza, P. D. Ndhlovu, E. Gomo, L. Gwanzura, N. Midzi, P. R. Mason, H. Friis, and S. G. Gundersen. 2006. Genital schistosomiasis in women: a clinical 12-month in vivo study following treatment with praziquantel. *Trans. R. Soc. Trop. Med. Hyg.* 5 Jan. [Epub ahead of print.]

34. Kjetland, E. F., P. D. Ndhlovu, T. Mduluza, E. Gomo, L. Gwanzura, P. R. Mason, E. N. Kurewa, N. Midzi, H. Friis, and S. G. Gundersen. 2005. Simple clinical manifestations of genital *Schistosoma haematobium* infection in rural Zimbabwean women. *Am. J. Trop. Med. Hyg.* **72**:311–319.

35. Kjetland, E. F., G. Poggensee, G. Hellinggiese, J. Richter, A. Sjaastad, L. Chitsulo, N. Kumwenda, S. G. Gundersen, I. Drantz, and H. Feldmeier. 1997. Female genital schistosomiasis due to *Schistosoma haematobium*—clinical and parasitological findings in women in rural Malawi. *Acta Trop.* **62:**239–255.

36. Langley, J. G., H. C. Kariuki, A. P. Hammersley, J. H. Ouma, A. E. Butterworth, and D. W. Dunne. 1994. Human IgG subclass responses and subclass restriction to *Schistosoma mansoni* egg antigens. *Immunology* **83:**651–658.

37. Leutscher, P., C. E. Ramarokoto, C. Reimert, H. Feldmeier, P. Esterre, and B. J. Vennervald. 2000. Community-based study of genital schistosomiasis in men from Madagascar. *Lancet* **355:**117–118.

38. Li, Y. S., A. G. P. Ross, Y. Li, Y. K. He, X. S. Luo, and D. P. McManus. 1997. Serological diagnosis of *Schistosoma japonicum* infections in China. *Trans. R. Soc. Trop. Med. Hyg.* **91:**19–21.

39. Li, Y. S., A. C. Sleigh, A. G. Ross, G. M. Williams, M. Tanner, and D. P. McManus. 2000. Epidemiology of *Schistosoma japonicum* in China: morbidity and strategies for control in the Dongting Lake region. *Int. J. Parasitol.* **30:**273–281.

40. Matupi, F., P. D. Kdhlovu, P. Hagan, J. T. Spicer, T. Mduluza, C. M. Turner, S. K. Chandiwana, and M. E. Woolhouse. 1998. Chemotherapy accelerates the development of acquired immune responses to *Schistosoma haematobium* infection. *J. Infect. Dis.* **178:**289–293.

41. McManus, D. P. 2005. Prospects for development of a transmission blocking vaccine against *Schistosoma japonicum*. *Parasite Immunol.* **27:**297–308.

42. McManus, D. P. 2000. A vaccine against Asian schistosomiasis: the story unfolds. *Int. J. Parasitol.* **30:**265–271.

43. Miller, R. L., G. L. Armelagos, S. Ikram, N. de Jonge, F. W. Krijger, and A. M. Deelder. 1992. Palaeoepidemiology of schistosoma infections in mummies. *Br. Med. J.* **304:**555–556.

44. Mohsen, M. A. A., A. A. Hassan, S. M. El-Sewedy, T. Aboul-Azm, C. Magagnotti, R. Fanelli, and L. Airoldi. 1999. Biomonitoring of N-nitroso compounds, nitrite and nitrate in the urine of Egyptian bladder cancer patients with or without *Schistosoma haematobium* infection. *Int. J. Cancer* **82:**789–794.

45. Mostafa, M. H., S. A. Sheweita, and P. J. O'Connor. 1999. Relationship between schistosomiasis and bladder cancer. *Clin. Microbiol. Rev.* **12:**97–111.

46. Nibbeling, H. A. M., A. I. Kahama, R. J. M. Van Zeyl, and A. M. Deelder. 1998. Use of monoclonal antibodies prepared against *Schistosoma mansoni* hatching fluid antigens for demonstration of *Schistosoma haematobium* circulating egg antigens in urine. *Am. J. Trop. Med. Hyg.* **58:**543–550.

47. Nibbeling, H. A. M., L. Vanetten, Y. E. Fillie, and A. M. Deelder. 1997. Enhanced detection of *Schistosoma* circulating antigens by testing 1 ml urine samples using immunomagnetic beads. *Acta Trop.* **66:**85–92.

48. Ouma, J. H., C. H. King, E. M. Muchiri, P. Mingai, D. K. Koech, E. Ireri, P. Magak, and H. Kadzo. 2005. Late benefits 10–18 years after drug therapy for infection with *Schistosoma haematobium* in Kwale District, Coast Province, Kenya. *Am. J. Trop. Med. Hyg.* **73:**359–364.

49. Poggensee, G., I. Kiwelu, M. Saria, J. Richter, I. Krantz, and H. Feldmeier. 1998. Schistosomiasis of the lower reproductive tract without egg excretion in urine. *Am. J. Trop. Med. Hyg.* **59:**782–783.

50. Polman, J., D. Engels, L. Fathers, A. M. Deelder, and B. Gryseels. 1998. Day-to-day fluctuation of schistosome circulating antigen levels in serum and urine of humans infected with *Schistosoma mansoni* in Burundi. *Am. J. Trop. Med. Hyg.* **59:**150–154

51. Polman, K., F. F. Stelma, B. Gryseels, G. J. van Dam, I. Talla, M. Niang, L. van Lieshout, and A. M. Deelder. 1995. Epidemiologic application of circulating antigen detection in a recent *Schistosoma mansoni* focus in northern Senegal. *Am. J. Trop. Med. Hyg.* **53:**152–157.

52. Pontes, L. A., M. C. Oliveira, N. Katz, E. Dias-Neto, and A. Rabello. 2003. Comparison of a polymerase chain reaction and the Kato-Katz technique for diagnosing infection with *Schistosoma mansoni*. *Am. J. Trop. Med. Hyg.* **68:**652–656.

53. Qiu, D. C., A. E. Hubbard, B. Zhong, Y. Zhang, and R. C. Spear. 2005. A matched, case-control study of the association between *Schistosoma japonicum* and liver and colon cancers, in rural China. *Ann. Trop. Med. Parasitol.* **99:**47–52.

54. Rabello, A. L. T., M. M. A. Gracia, R. A. P. da Silva, R. S. Rocha, A. Chaves, and N. Katz. 1995. Humoral immune responses in acute schistosomiasis mansoni: relations to morbidity. *Clin. Infect. Dis.* **21:**608–615.

55. Ramzy, R. M. R., R. Faris, M. Bahgat, H. Helmy, C. Franklin, and J. H. McKerrow. 1997. Evaluation of a stage-specific proteolytic enzyme of *Schistosoma mansoni* as a marker of exposure. *Am. J. Trop. Med. Hyg.* **56:**668–673.

56. Ross, A. G. P., A. C. Sleigh, Y. Li, G. M. Davis, G. M. Williams, Z. Jiang, Z. Feng, and D. P. McManus. 2001. Schistosomiasis in the People's Republic of China: prospects and challenges for the 21st century. *Clin. Microbiol. Rev.* **14:**270–295.

57. Rossi, C. L., V. C. W. Tsang, and J. B. Pilcher. 1991. Rapid, low-technology field- and laboratory-applicable enzyme-linked immunosorbent assays for immunodiagnosis of *Schistosoma mansoni*. *J. Clin. Microbiol.* **29:**1836–1841.

58. Salah, F., Z. Demerdash, Z. Shaker, A. El Bassiouny, G. El Attar, S. Ismail, N. Badir, A. S. El Din, and M. Mansour. 2000. A monoclonal antibody against *Schistosoma haematobium* soluble egg antigen: efficacy for diagnosis and monitoring of cure of *S. haematobium* infection. *Parasitol. Res.* **86:**74–80.

59. Shaw, M. E., P. A. Elder, A. Abbas, and M. A. Knowles. 1999. Partial allelotype of schistosomiasis-associated bladder cancer. *Int. J. Cancer* **80:**656–661.

60. Silva, I. M., R. Thiengo, M. J. Conceicao, L. Rey, H. L. Lenzi, F. E. Pereira, and P. C. Ribeiro. 2005. Therapeutic failure of praziquantel in the treatment of *Schistosoma haematobium* infection in Brazilians returning from Africa. *Mem. Inst. Oswaldo Cruz* **100:**445–449.

61. Spicher, V. M., B. Genin, A. R. Jordan, L. Rubbia-Brandt, and C. Le Coultre. 2004. Peritoneal schistosomiasis: an unusual laparoscopic finding. *J. Pediatr. Surg.* **39:**631–633.

62. Stelma, F. F., I. Talla, S. Sow, A. Kongs, M. Niang, K. Polman, A. M. Deelder, and B. Gryseels. 1995. Efficacy and side effects of praziquantel in an epidemic focus of *Schistosoma mansoni*. *Am. J. Trop. Med. Hyg.* **53:**167–170.

63. Stich, A. H., S. Biays, P. Odermatt, C. Men, C. Saem, K. Sokha, C. S. Ly, P. Legros, M. Philips, J. D. Lormand, and M. Tanner. 1999. Foci of schistosomiasis mekongi,

Northern Cambodia. II. Distribution of infection and morbidity. *Trop. Med. Int. Health* **4:**674–685.

64. Stothard, J. R. N. B. Kabatereine, E. M. Tukahebwa, F. Kazibwe, D. Rollinson, W. Mathieson, J. P. Webster, and A. Fenwick. 2005. Use of circulating cathodic antigen (CCA) dipsticks for detection of intestinal and urinary schistosomiasis. *Acta Trop.* 27 Dec. [Epub ahead of print.]

65. Strickland, G. T., and B. L. Ramirez. 2000. Schistosomiasis, p. 805–832. *In* G. T. Strickland (ed.), *Hunter's Tropical Medicine and Emerging Infectious Diseases*, 8th ed. The W. B. Saunders Co., Philadelphia, Pa.

66. Subramanian, A. K., P. Mungai, J. H. Ouma, P. Magak, C. H. King, A. A. F. Mahmoud, and C. L. King. 1999. Long-term suppression of adult bladder morbidity and severe hydronephrosis following selective population chemotherapy for *Schistosoma haematobium. Am. J. Trop. Med. Hyg.* **61:**476–481.

67. Takemura, Y., S. Kikuchi, and Y. Inaba. 1998. Epidemiologic study of the relationship between schistosomiasis due to *Schistosoma japonicum* and liver cancer/cirrhosis. *Am. J. Trop. Med. Hyg.* **59:**551–556.

68. Tchuente, L. A. T., V. R. Southgate, J. Vercruysse, A. Kaukas, R. Kane, M. P. Mulumba, J. R. Pages, and J. Jourdane. 1997. Epidemiological and genetic observations on human schistosomiasis in Kinshasa, Zaire. *Trans. R. Soc. Trop. Med. Hyg.* **91:**263–269.

69. Tsang, V. C. W., and P. P. Wilkins. 1991. Immunodiagnosis of schistosomiasis. Screen with FAST-ELISA and confirm with immunoblot. *Clin. Lab. Med.* **11:**1029–1039.

70. Utzinger, J., X. N. Zhou, M. G. Chen, and R. Berquist. 2005. Conquering schistosomiasis in China: the long march. *Acta Trop.* **96:**69–96.

71. Valli, L. C. P., H. Y. Kanamura, R. M. Da Silva, R. Ribeiro-Rodriques, and R. Dietze. 1999. Schistosomiasis mansoni: immunoblot analysis to diagnose and differentiate recent and chronic infection. *Am. J. Trop. Med. Hyg.* **61:**302–307.

72. Valli, L. C. P., H. Y. Kanamura, R. M. Da Silva, M. I. P. G. Silva, S. A. G. Vellosa, and E. T. Garcia. 1997. Efficacy of an enzyme-linked immunosorbent assay in the diagnosis of and serologic distinction between acute and chronic *Schistosoma mansoni* infection. *Am. J. Trop. Med. Hyg.* **57:**358–362.

73. van Dam, G. J., J. H. Wicheers, T. M. F. Ferreira, D. Ghati, A. van Amerongen, and A. M. Deelder. 2004. Diagnosis of schistosomiasis by reagent strip test for detection of circulating cathodic antigen. *J. Clin. Microbiol.* **42:**5458–5461.

74. van Etten, L., P. G. Kremsner, F. W. Kruger, and A. M. Deelder. 1997. Day-to-day variation of egg output and schistosome circulating antigens in urine of *Schistosoma haematobium*-infected school children from Gabon and follow-up after chemotherapy. *Am. J. Trop. Med. Hyg.* **57:**337–341.

75. van Gool, T., H. Vetter, T. Vervoort, M. J. Doenhoff, J. Wetsteyn, and D. Overbosch. 2002. Serodiagnosis of imported schistosomiasis by a combination of a commercial indirect hemagglutination test with *Schistosoma mansoni* adult worm antigens and an enzyme-linked immunosorbent assay with *S. mansoni* egg antigens. *J. Clin. Microbiol.* **40:**3432–3437.

76. van Lieshout, L., A. M. Polderman, S. J. de Vlas, P. de Caluwe, F. W. Krijger, B. Gryseels, and A. M. Deelder. 1995. Analysis of worm burden variation in human *Schistosoma mansoni* infections by determination of serum levels of circulating anodic and circulating cathodic antigen. *J. Infect. Dis.* **172:**1336–1342.

77. Willingham, A. L., III, M. V. Johansen, H. O. Bogh, A. Ito, J. Andreassen, R. Lindberg, N. O. Christensen, and P. Nansen. 1999. Short report: congenital transmission of *Schistosoma japonicum* in pigs. *Am. J. Trop. Med. Hyg.* **60:**311–312.

78. Woolhouse, M. E., F. Mutapi, P. D. Ndhlovu, S. K. Chandiwana, and P. Hagan. 2000. Exposure, infection and immune responses to *Schistosoma haematobium* in young children. *Parasitology* **120:**37–44.

79. Wu, Z. D., Z. Y. Lu, and X. B. Yu. 2005. Development of a vaccine against *Schistosoma japonicum* in China: a review. *Acta Trop.* **96:**106–116.

80. Yole, D. S., G. D. F. Reid, and R. A. Wilson. 1996. Protection against *Schistosoma mansoni* and associated immune responses induced in the vervet monkey *Cercopithecus aethiops* by the irradiated cercaria vaccine. *Am. J. Trop. Med. Hyg.* **54:**265–270.

81. Yu, J., S. J. Devlas, H. C. Yuan, and B. Gryseels. 1998. Variations in fecal *Schistosoma japonicum* egg counts. *Am. J. Trop. Med. Hyg.* **59:**370–375.

82. Zhu, Y. C., D. Socheat, K. Bounlu, Y. S. Liang, M. Sinuon, S. Insisiengmay, W. He, M. Xu, W. Z. Shi, and R. Berquist. 2005. Application of dipstick dye immunoassay (DDIA) kit for the diagnosis of schistosomiasis mekongi. *Acta Trop.* **96:**137–141.

18

Unusual Parasitic Infections

Aquatic Protist

Rhinosporidium seeberi

Protozoa

Myxozoan parasites

Nematodes

Oesophagostomum spp.

Eustrongylides spp.

Mermis nigrescens

Micronema deletrix

Dioctophyma renale

Ternidens deminutus

Mammomonogamus laryngeus (*Syngamus laryngeus*)

Ascaris suum

Gongylonema pulchrum

Haycocknema perplexum

Cestodes

Diplogonoporus spp.

Bertiella studeri

Inermicapsifer madagascariensis

Raillietina celebensis

Mesocestoides spp.

Taenia crassiceps

Trematodes

Alaria americana

Plagiorchis spp.

Neodiplostomum seoulense

Spelotrema brevicaeca

Brachylaima sp.

Troglotrema salmincola

Stellantchasmus falcatus

Phaneropsolus bonnei and *Prosthodendrium molenkempi*

Phaneropsolus spinicirrus

Haplorchis taichui

Gymnophalloides seoi

Metorchis conjunctus (North American liver fluke)

Schistosoma mattheei

Philophthalmus lacrimosus

Achillurbainia spp.

Pentastomids

Armillifer spp., *Linguatula serrata*, and *Sebekia* spp.

Acanthocephalans

Macracanthorhynchus hirudinaceus and *Moniliformis moniliformis*

Most parasites infecting humans are recognized as being able to survive easily within the human host and to possibly cause disease, but there are some organisms that only rarely infect humans. Perhaps humans are only accidental hosts. In any case, because these infections, being so rare, are not reported frequently, such parasites may not be considered possible etiologic agents. It is important to at least be aware of such parasites and of the possible disease process created by these relatively benign organisms, because they can become a problem in immunocompromised patients. Examples of these unusual infections are described in Table 18.1.

Aquatic Protist

Rhinosporidium seeberi

Rhinosporidiosis is a granulomatous disease of humans and animals that is characterized by slow-growing tumorlike polyps, affecting primarily the nasal mucosa and ocular conjunctiva. The causative agent is *Rhinosporidium seeberi*, an organism that has undergone several reclassifications. Although it was thought to be a fungus or protozoan, it is now placed in a novel clade of aquatic protistan parasites, the Mezomycetozoea; this occurs at the divergence between animals and fungi (32, 71). Since this organism cannot be grown in synthetic media or human or animal cell lines, diagnosis is established by seeing the organism in tissue biopsy specimens. Two stages are seen in tissue: the larger endosporulating spherical form, or sporangia, and the smaller trophocyte (Figure 18.1). The genus *Rhinosporidium* may include multiple host-specific strains (83).

The sporangium is the mature form and measures 100 to 350 µm; contained within the sporangium are numerous endospores (sporangiospores). The zonal arrangement of sporangiospores is characteristic, with the immature ones forming a crescent-shaped mass at the periphery, the maturing ones arranged sequentially toward the center, and the mature ones in the center. This arrangement of the sporangiospores separates *R. seeberi* from other spherical endosporulating organisms seen in tissue.

Rhinosporidiosis has been found in young men between 20 and 40 years old; the lesions tend to be slow-growing polyps or tumorlike masses, usually of the nasal mucosa or conjunctiva. Lesions have been found in the paranasal

Table 18.1 Unusual parasitic infections

Organism	Body site	Diagnosis	Comments
Aquatic protist			
Rhinosporidium seeberi	Nasopharynx, ocular conjunctiva, genitalia	Identification of sporangium and/or trophocytes in tissue biopsy specimens	Cannot be grown in synthetic media or animal/human tissue cell lines; classification has been difficult over a number of years
Protozoa			
Myxozoa	Intestine	Recovery of spores in stool	The morphology can mimic normal human spermatozoa, causing confusion with possible sexual abuse situations; true human infection is still somewhat unclear
Nematodes			
Oesophagostomum spp.	Intestine or abdominal cavity (nodular mass lesions)	Examination of nodule (dissection or aspirate); definitive diagnosis depends on identification of worm (may be difficult)	Carcinoma, tuberculosis, or Crohn's disease often suspected; mass often attached to bowel wall or ileum or colon; infection via ingestion of larvae
Eustrongylides spp.	Abdominal cavity	Identification of larvae (bright red)	Rare but may increase with consumption of raw fish; infective larval form also found in amphibians and reptiles
Mermis nigrescens	Mouth	Identification of larvae or adults	Soil nematodes with eggs deposited on plants and larvae in grasshoppers; adult worm accidentally transmitted to infant, probably by dog or cat
Micronema deletrix	Brain (single case)	Routine histologic test with consultation	Free-living worms living in soil, manure, humus
Dioctophyma renale	Primarily right kidney (may be in left kidney or body cavities)	Identification of worm in autopsy specimens or living worms passed from the urethra	Route of infection probably ingestion of raw freshwater fish and frogs; widely reported in dogs
Ternidens deminutus	Large intestine	Identification of eggs in stool; ulcers and/or cystic nodules may be seen in the large intestine	Third-stage larvae will develop in soil; probably infective for humans; life cycle not known
Mammomonogamus laryngeus	Trachea	Identification of Y-shaped male and female worms (in copulo) from coughed-up sputum or from eggs in sputum or stool	Third-stage larvae thought to be infective for the vertebrate host; normal host cattle, felines; life cycle not known
Ascaris suum	Intestine or liver, lungs; intestine but worms do not mature (can be serious disease)	Identification based on size and shape of denticles and patient history of exposure to pigs raised on farms; dose required for symptoms of pneumonitis very minimal; large doses can lead to severe disease	Ingestion of infective eggs probably more common than thought, particularly in people who live and/or work on farms where pigs are raised
Gongylonema pulchrum	Mucosa or submucosa of buccal cavity; esophageal mucosa	Identification based on worm recovery and egg morphology	Accidental ingestion of insect intermediate hosts, possibly cockroaches or other infected insects
Haycocknema perplexum	Skeletal muscle fibers	Identification of adults or larvae in muscles	Symptoms of progressive muscle weakness, eosinophilia, active polymyositis

(continued)

Table 18.1 Unusual parasitic infections *(continued)*

Organism	Body site	Diagnosis	Comments
Cestodes			
Diplogonoporus spp.	Intestine	Identification of proglottids or eggs (similar to those of *Diphyllobothrium latum*)	Ingestion of infected fish; usual parasites of whales
Bertiella spp.	Intestine	Identification of chains of proglottids or very characteristic eggs	Accidental ingestion of mites containing the infective form; most infections seen in children
Inermicapsifer madagascariensis	Intestine	Identification of proglottids, eggs, or egg packets in stool	Very difficult to differentiate from *Raillietina* spp.; patient is asymptomatic to mild symptoms; usually seen in children; infection probably from ingestion of arthropod vector (unknown)
Raillietina celebensis	Intestine	Identification of proglottids, eggs, or egg packets in stool	Intermediate hosts are ants (accidental ingestion)
Mesocestoides spp.	Intestine	Identification of proglottids in stool	Infection probably due to ingestion of raw meat of some animal (unknown) carrying the intermediate larval form in the tissue; usually found in children; symptoms minimal
Taenia crassiceps	Eye, subcutaneous paravertebral infiltrate (AIDS)	Identification of small, white, spherical masses (2–3 mm) of larval worms	Common parasite of the red fox; in animals, larval forms found under skin and in body cavities; human cases very rare but very serious
Trematodes			
Alaria americana	Disseminated throughout many tissues including eye, organs, central nervous system; can be very serious disease	Recovery of mesocercariae at surgery or autopsy	Ingestion of inadequately cooked frogs or frog legs
Plagiorchis spp.	Intestine	Recovery of adult worms and/or eggs in stool	Ingestion of insect grubs containing the infective stage; raw fish also implicated in transmission
Neodiplostomum seoulense	Intestine	Recovery of adult worms and/or eggs in stool	Associated with ingestion of raw or poorly cooked snake meat, frogs, or house rats
Spelotrema brevicaeca	Intestine or disseminated to other tissues	Recovery of adult worms in stool and/or eggs in stool or tissues	Ingestion of raw crabs containing infective metacercariae; reported to cause severe illness in compromised patients
Brachylaima sp.	Intestine	Recovery of adult worms and/or eggs in stool	Ingestion of raw snails containing infective metacercariae; may be mild abdominal pain and diarrhea
Troglotrema salmincola	Intestine	Recovery of adult worms and/or eggs in stool	Ingestion of raw fish viscera and/or giant salamander; rickettsia associated with infection causes severe illness in dogs, may also be human pathogen

(continued)

Table 18.1 *(continued)*

Organism	Body site	Diagnosis	Comments
Stellantchasmus falcatus	Intestine	Recovery of adult worms and/or eggs in stool	Ingestion of raw brackish-water fish; reported from Hawaii after ingestion of raw mullet
Phaneropsolus bonnei, Prosthodendrium molenkempi	Intestine	Recovery of adult worms and/or eggs in stool; eggs resemble those of *Opisthorchis viverrini* or each other	Ingestion of raw small fish infected with naiads (water nymphs of dragonflies)
Phaneropsolus spinicirrus	Intestine	Recovery of adult worms and/or eggs in stool	Second species of genus found in humans; adults small (651 µm long by 502 µm wide); eggs 29 µm long by 14 µm wide, smooth, operculate (some with shoulders)
Haplorchis taichui	Small intestine	Identification of eggs or adult worms	Musosal ulceration, mucosal and submucosal hemorrhages, fusion and shortening of the villi, chronic inflammation, and fibrosis of the submucosa
Gymnophalloides seoi	Intestine	Recovery of adult worms and/or eggs in stool	New human intestinal trematode; southwestern island of Korea; bivalves serve as first or second intermediate hosts; eggs smaller than *Clonorchis* eggs
Metorchis conjunctus (North American liver fluke)	Biliary tree	Recovery of adult flukes and/or eggs in stool	Ingestion of raw white sucker fish (sashimi) containing infective metacercariae
Schistosoma mattheei	Venules of the bladder and/or intestine	Recovery of eggs in urine and/or stool; eggs resemble those of *S. intercalatum*	Skin penetration of infective cercaria from freshwater snail
Philophthalmus lacrimosus	Conjunctiva	Identification of adult worm from eye	Main symptom is the feeling of having a foreign body sensation in the eye
Achillurbainia spp.	Granulomata, retroauricular cysts or abscesses	Identification of eggs or adult worms from biopsy specimens	Symptoms may or may not be present
Pentastomids (tongue worms) *Armillifer* spp., *Linguatula serrata*	Infective larvae die in situ (tissues, including the eye); ingestion of mature larvae may lead to development of adults in nasopharyngeal tissues (halzoun syndrome)	Identification of the worm/larvae in biopsy or autopsy specimen	Most infections acquired from ingestion of raw, infected snakes; liver, lung, spleen involved; often not discovered until autopsy; lung involvement may cause congestion
Sebekia spp.	Dermatitis, serpiginous burrow, erythema	Larvae in burrow	Adult parasites of respiratory tract of reptiles, immature forms found in viscera, muscles, along spinal cord of freshwater fish
Acanthocephalans *Macracanthorhynchus hirudinaceus*	Intestine	Identification of adult worms; immature worms very difficult to identify	Infections acquired by ingestion of raw beetles; clinical manifestations are intestinal obstruction and perforation; eggs often not seen in feces
Moniliformis moniliformis	Intestine	Identification of adult worms; immature worms very difficult to identify; eggs can be used to identify the helminth	Infections acquired by ingestion of raw cockroaches; symptoms can be diarrhea, pallor, anorexia, weight loss, weakness

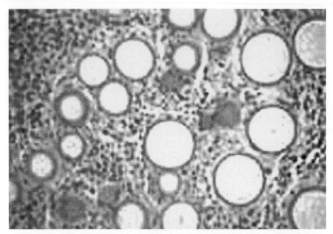

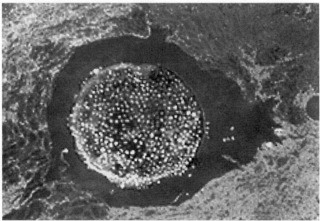

Figure 18.1 Histology of rhinosporidiosis; human nasal polyp. (Upper) Immature *Rhinosporidium seeberi* trophocytes; note the thick walls and the surrounding inflammatory cells. (Lower) Sporangium containing endospores. (Images from the Centers for Disease Control and Prevention.)

sinuses, larynx, and external genitalia. Dissemination appears to be rare. Symptoms include nasal obstruction and epistaxis. Surgical excision of the lesions is generally recommended as the treatment of choice; if total excision is not possible, recurrences are typical.

Over 90% of rhinosporidiosis cases occur primarily in India and Sri Lanka; however, cases have been reported from the Americas, Europe, Africa, and Asia. Three cases have been reported from the southeastern United States. Various studies have linked infection to swimming or bathing in freshwater ponds, lakes, or rivers (32, 71).

Protozoa

Myxozoan Parasites

Myxozoans are commonly found as parasites in aquatic vertebrates (especially fish) and invertebrates. They are recognized as a separate phylum with two classes: the Myxosporea, which infect fish, amphibians, and reptiles, and the Actinosporea, which infect aquatic invertebrates (primarily oligochaetes and sipunculids) (59, 60). Classification has been difficult, and it has been recommended that a number of factors be used in making taxonomic decisions, including determination of 18S rRNA gene sequences combined with information on tissue tropism, host species infected, and developmental cycles in the fish and alternate host (when known). There are over 1,000 species, and new ones are being described each year; however, many of the life cycles have not yet been defined. Alternation between actinosporean stages in oligochaetes and myxosporean stages in fish may occur (59, 95). These organisms parasitize a variety of fish tissues, where they produce pseudocysts that contain hundreds of thousands of small spores. As marine netpan farming of salmonid fish continues to expand in many countries, more unusual host-parasite relationships may be seen in the future (49).

Many of these myxozoan spores contain chitin in their spore coat, and the spores can be passed intact through the vertebrate gastrointestinal tract (68). There is no evidence that these organisms cause human disease; however, spores have been recovered in stool samples from patients with gastrointestinal tract symptoms, including abdominal pain and/or diarrhea (9). There is also speculation that these parasites could be potentially infectious and pathogenic in immunocompromised patients. Coinfections with other parasites such as *Cryptosporidium* spp., *Isospora belli*, *Giardia lamblia*, *Hymenolepis nana*, *Strongyloides stercoralis*, and *Ascaris lumbricoides* have been found, thus complicating the determination of the pathogenic role of *Myxobolus* spp.

Spores of a myxosporean parasite (*Henneguya salminicola*) have been detected in fecal samples from two human patients (66), in one case being mistaken for human spermatozoa and leading to suspicion of sexual abuse (Figure 18.2). The spores were passed undigested in feces after the consumption of infected salmon and were not thought to have any causal relationship to the patient's symptoms. True infections with myxosporean parasites in humans have not previously been recorded, other than this case. Organisms resembling human spermatozoa have been seen in fecal concentrates at high dry magnification (×400). They measured about 40 μm in total length with a teardrop-shaped spore body of approximately 10 μm, which tapered gradually to one or two caudal filaments. Most of the spores had two pyriform polar capsules situated in the anterior region of the spore body. Although routine iron hematoxylin and trichrome staining failed to stain the spores, the modified trichrome stain was effective.

A second report discussed the detection of a second myxosporean species in stool samples from three

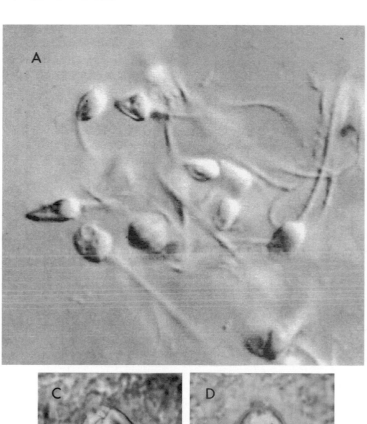

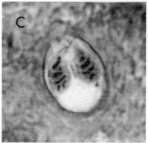

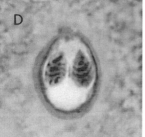

Figure 18.2 (A) Normal human spermatozoa. (B) Myxosporean spore from salmon. Note the similarities between panels A and B. (C and D) Two spores demonstrating the polar filaments. (Reprinted from reference 66 with permission.)

additional patients (9). It is very likely that additional cases will be recorded; it is important that clinicians and microbiologists understand the possible presence of myxozoan spores in stool samples, particularly in patients with a high dietary intake of fish.

In an iodine preparation, the spores appear as dark pyriform bodies, while in modified acid-fast stains, the spore wall and enclosed sporoplasmic and capsulogenic cells stain acid fast. When the fecal specimen is stained with trichrome stain, the spores appear as darkly stained pear-shaped bodies measuring 11 to 12 μm by 7 to 8 μm with two polar capsules each measuring 5 by 2 μm. It is important to use the modified acid-fast stains, since these methods allow the visualization of the convolutions of the polar filament.

Nematodes

Oesophagostomum spp.

The roundworms of *Oesophagostomum* spp. infect the intestine or abdominal cavity of humans or animals, normally ruminants, primates, or swine. The disease is most common in the tropics or subtropics, with approximately 60 human cases being reported, primarily from Africa. Several species of *Oesophagostomum* have been implicated in human disease; however, accurate species identification is very difficult if one cannot examine an intact whole worm. The life cycle is seen in Figure 18.3. Usually, the worms are unable to complete their life cycle in humans; however, there have been instances in which stunted

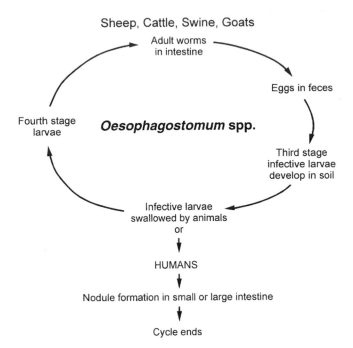

Sheep, Cattle, Swine, Goats

Oesophagostomum spp.

Adult worms
in intestine

Eggs in feces

Fourth stage
larvae

Third stage
infective larvae
develop in soil

Infective larvae
swallowed by animals
or

HUMANS

Nodule formation in small or large intestine

Cycle ends

Figure 18.3   Life cycle of *Oesophagostomum* spp.

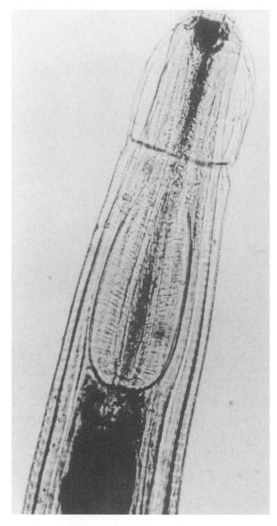

Figure 18.4   Anterior end of an adult *Oesophagostomum bifurcum* worm recovered from human feces. Note the cephalic groove, characteristic of the genus (×115). (Reprinted from reference 75 with permission.)

adult worms have been recovered. In northern Togo and Ghana, this infection is found in humans, with the recovery of both adult males and females passed in the stool after therapy (Figures 18.4 and 18.5) (97). The parasite is *Oesophagostomum bifurcum*, which apparently does not require an animal reservoir. The eggs passed in the stool closely resemble those of hookworm (75), and both infections are found in this population (Figure 18.6).

The infection can develop in humans at any age. Most of the worms recovered from humans are immature. Symptoms include abdominal pain, with localized tenderness (usually right lower quadrant). There may be gradual onset of pain, with a low-grade fever and no diarrhea, vomiting, or anorexia (37). Careful examination of the patient may reveal a mass in the abdominal cavity, adhering to the abdominal wall (Figure 18.7) (1). Although these nodular mass lesions usually do not cause intestinal obstruction, other complications, such as hernia, intussusception, or worm penetration of the bowel wall, may occur (36). The mass is often attached to the bowel wall of the ileum or colon (91). Nodules usually measure 4 to 6 cm. Often carcinoma, tuberculosis, or Crohn's disease may be suspected. In one patient, the worm produced a cutaneous lesion, the first report of a lesion outside of the gastrointestinal tract and without evidence of bowel infection (78).

If oesophagostomiasis is suspected, confirmation can be made by immediate examination of the nodule, either by nodule dissection (Figure 18.8) or by examination of aspirate material. Definitive diagnosis depends on identifying the whole worm or serial sections of it. At present,

surgical resection of the involved tissue or removal of the worm from the nodules is the only means of treatment. However, both pyrantel pamoate and albendazole have been effective in removing adult worms from the intestine (76). It is very likely that a single dose of albendazole will cure an *O. bifurcum* infection (96).

Oesophagostomum spp., as well as hookworm infections, are hyperendemic among the population in northern Togo and Ghana. In northern Togo, all the villages examined contained residents who were infected with hookworm, with a patchy distribution of *Oesophagostomum* spp. Women were infected with *O. bifurcum* more often than men were, while infections with hookworm were the opposite. The prevalence and intensity of infection with both parasites were clearly age dependent. It is estimated that more than 100,000 people

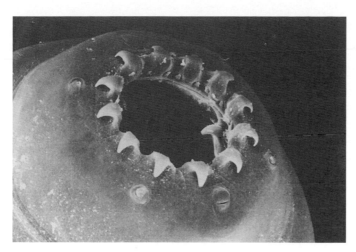

Figure 18.5 Scanning electron micrograph of the cephalic end of a female *Oesophagostomum bifurcum* worm recovered from human feces. Note the inner and outer leaf crowns with 12 and 24 (sometimes 10 and 20) lips, respectively, and the papillae (×630). (Photograph by the Department of Electron Microscopy, Medical Faculty, Leiden University; reprinted from reference 75 with permission.)

in Togo are infected with O. *bifurcum* and more than 230,000 are infected with hookworm (73).

Using a panel of 155 well-defined fecal and DNA samples, a new PCR assay achieved a sensitivity of 94.6% and a specificity of 100% in the identification of O. *bifurcum*. This PCR assay will be useful for the diagnosis of O. *bifurcum* infection, as well as being a molecular biology tool to help clarify the epidemiology of human disease with this parasite (90). When using routine diagnostic methods to estimate the intensity of infection, it is sufficient to make repeated coprocultures from only a single stool sample; collection of multiple samples on subsequent days does not give better estimates of the individual infection status (72). It is also important to remember that coproculture can also yield larvae of *Oesophagostomum* and *Ternidens* species, as well as the more common hookworm and *Strongyloides stercoralis* (45).

Figure 18.6 *Oesophagostomum bifurcum* egg. This egg cannot be differentiated from that of the hookworm.

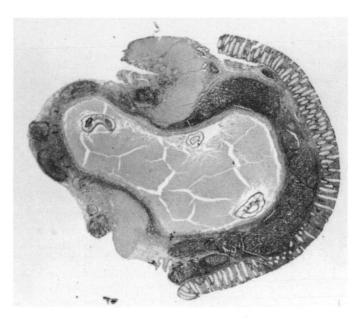

Figure 18.7 Nodular lesion of oesophagostomiasis seen as a paraumbilical mass in an otherwise healthy 5-year-old girl in Nigeria. At laparotomy, the lesion was found to have arisen in the ascending colon and was adherent to the muscles and fascia of the anterior abdominal wall (×11). (Armed Forces Institute of Pathology photograph.)

Eustrongylides spp.

The larvae of *Eustrongylides* nematodes are found in raw fish and in amphibians and reptiles. Although not all fish hosts are known, it was assumed that, in one

Figure 18.8 Surgically resected nodule of oesophagostomiasis. Note the worm protruding into the cavity. Specific identification of these worms is difficult, because intact specimens are rarely obtained. (Armed Forces Institute of Pathology photograph.)

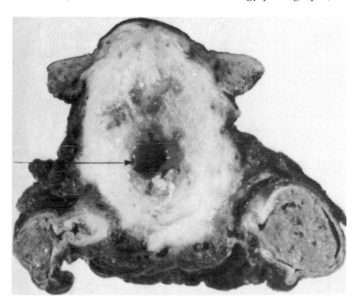

case, the patient most probably acquired the infection by eating raw fish (sashimi and sushi) approximately once a month. He usually ingested the fish in restaurants; however, the day prior to admission, he had eaten raw fish of an unknown species from a neighborhood fish market. He presented with a history of pain in the right lower quadrant. On the basis of the symptoms, the patient underwent an appendectomy, at which time the appendix and distal ileum were found to be normal. However, prior to closure of the incision, a pinkish red worm was seen crawling onto the surgical drapes from the abdominal cavity. The patient's recovery was unremarkable. On the basis of larval morphology, the worm was determined to be in the *Eustrongylides* group, which includes parasites that are commonly found in fish-eating birds (92). A second patient also presented with symptoms simulating acute appendicitis. The patient had consumed two live minnows obtained from Big Timber Creek, Belmawr, N.J. The parasite *Eustrongylides ignotus* was recovered surgically from this patient (69). Human infection with *Eustrongylides* spp. is certainly uncommon, but with the increased consumption of raw or improperly cooked fish, additional cases may be seen. The majority of cases have been related to anisakine larvae such as those of *Anisakis*, *Phocanema*, *Contracaecum*, and *Pseudoterranova* spp., which are indigenous to marine fish found off both coasts of the United States as well as in Japan and Europe.

Mermis nigrescens

Mermis nigrescens nematodes are found in the soil and have a unique life cycle. After mating and egg maturation, the females leave the soil and deposit eggs on the aboveground portions of plants. The normal hosts are grasshoppers, which become infected by ingesting eggs while feeding on vegetation. After completing their development in the grasshoppers, the worms emerge and enter the soil, where mating occurs. One specimen that was removed from the mouth of a 1-month-old infant was identified as a gravid female *M. nigrescens* (Figure 18.9). On the basis of the parasite life cycle, it was assumed that the female was ovipositing on vegetation and was carried, perhaps on the fur of a dog or cat present in the house,

Figure 18.9 (1) Head of *Mermis nigrescens* removed from the mouth of an infant. Note the mouth opening (arrow), lip papilla (L), and cephalic papillae (C). Bar, 80 μm. (2) Eggs of *M. nigrescens* inside the uterus of the specimen. Note the thick, double-walled shell surrounding the fully developed embryos. Arrows point to poles containing collapsed byssi. Bar, 55 μm. (Reprinted from reference 74 with permission.)

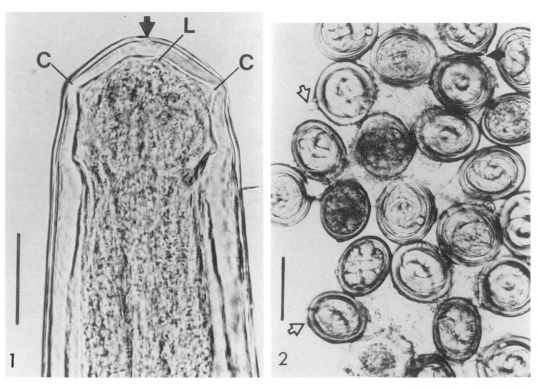

to the vicinity of the child. The parasite was possibly attracted to the moisture of the mouth (74).

Micronema deletrix

Micronema deletrix roundworms are free-living and are normally found in soil, manure, or decaying humus. *M. deletrix* is reported to have caused a fatal meningoencephalomyelitis in a 5-year-old boy who accidentally fell into a manure spreader and passed through the mechanism. He sustained multiple severe wounds and fractures, all of which had been heavily contaminated with manure. Although his lesions appeared to heal satisfactorily, 18 days after the injury he became lethargic. Lumbar puncture revealed over 300 cells per ml and equal numbers of lymphocytes and macrophages. The child died of meningoencephalitis 24 days after the accident (42). The autopsy revealed extensive inflammatory foci in the brain, each of which contained various developmental stages of the worms (Figure 18.10) (8).

Figure 18.10 Longitudinal section through the rhabditiform esophagus in the anterior end of a mature female *Micronema deletrix* worm in the brain. The procorpus (pc), metacorpus (mc), isthmus (is), bulb (bu), and intestine (in) are indicated (×1,040). (Armed Forces Institute of Pathology photograph; published in *Can. J. Neurol. Sci.* **2**:125, 1975.)

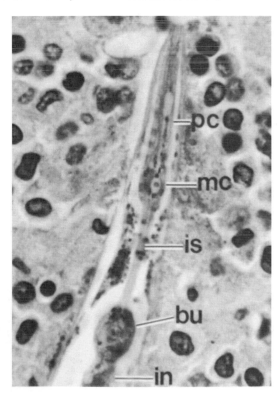

A second fatal case of human meningoencephalitis caused by *M. deletrix* has been reported. The patient had severe granulomatous meningoencephalitis with many foci of hemorrhage and encephalomalacia. Many adult female and larval worms were found in the brain. The organisms probably live in decaying organic material but occasionally parasitize the kidneys, nose, or central nervous system of the horse. In its parasitic form, the nematode reproduces parthenogenetically (82).

Dioctophyma renale

The large (14 to 20 cm long by 46 mm wide) roundworm *Dioctophyma renale* is widely reported to have been isolated from dogs. Infection in humans is acquired from the ingestion of raw freshwater fish and frogs. The adult worms usually live in the pelvis of the right kidney or in body cavities. The kidney parenchyma is gradually destroyed, leaving only the capsule. In all but 1 of the approximately 13 reported human cases, the kidney was involved (5). Infections were discovered at autopsy, by detection of worms migrating from the urethra, by detection of a worm discharged through the skin over an abscessed kidney (1), and by recovery of eggs in the urine (1) (Figure 18.11).

A 50-year-old Chinese man was found to have a retroperitoneal mass in the right upper quadrant of the abdomen by ultrasonography and computed tomography. At surgery, a hemorrhagic cyst was seen at the upper pole of the right kidney adjacent to the adrenal gland. Histologic examination revealed that the granulomatous cyst wall was filled with eggs and cross sections of parasites that were subsequently identified as *D. renale*. The eggs had a birefringent striated double wall (Figure

Figure 18.11 *Dioctophyma renale* egg. This egg measures 60 to 80 μm by 39 to 46 μm and has a thick shell that appears wrinkled; the ends of the egg are usually lighter in color.

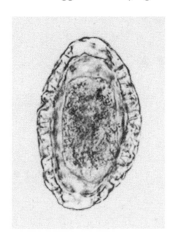

18.11). The right kidney was intact, and no eggs were seen in the urine (88).

Liesegang rings (LRs) in tissue are periodic precipitation zones from supersaturated solutions in colloidal systems. They are formed by a process involving interplay of diffusion, nucleation, flocculation or precipitation, and supersaturation. In a description of LRs from the lesions of 29 patients, the LRs formed in cysts or in fibrotic, inflamed, or necrotic tissue. Special stains and energy-dispersive radiographic analysis or scanning electron microscopy revealed that some LRs contained calcium, iron (hemosiderin), silicon, and sulfur. Some pathologists have mistaken LRs for eggs, larvae, or adults of *D. renale*. It is important to remember that these LRs can be mistaken for parasites and have been found in tissues and from renal cyst fluid (48, 89).

Ternidens deminutus

The nematode *Ternidens deminutus* is found as a parasite of the large intestine of several simian hosts and is occasionally found in humans (45). The worms inhabit the wall of the large intestine and may produce ulcers or cystic nodules (4). The geographic range includes Africa, India, and Indonesia; human infections have been reported in Southeast and South Africa. Although it is apparently fairly common in natives of Zimbabwe and has been found in large numbers in certain specific communities, it is not found in Europeans (4). Morphologically, *T. deminutus* resembles a hookworm; however, the anterior buccal capsule is guarded by a double crown of 22 stout bristles. The oval eggs closely resemble those of human hookworms but are larger (84 by 51 µm) (Figure 18.12). Third-stage larvae develop in soil or Harada-Mori cultures in 8 to 10 days; however, the complete life cycle is unknown. Newly identified genetic markers provide a foundation for an improved PCR-based diagnosis of these infections (79). Both thiabendazole and pyrantel pamoate produce cure rates over 90%.

Figure 18.12 *Ternidens deminutus* egg. This egg measures 85 by 50 µm and closely resembles the egg of hookworm and *Oesophagostomum*; however, it is larger.

Mammomonogamus laryngeus (Syngamus laryngeus)

Mammomonogamus laryngeus is commonly found in the upper respiratory tract of some mammals, including cattle and felines. There have been approximately 100 cases of human infection, with almost all cases being reported from the Caribbean islands; more than half of the cases occurred in Martinique or in France in people who had traveled to Martinique (4, 70). Other reports have mentioned Puerto Rico, Dominica, St. Lucia, Trinidad, Guyana, Brazil, and the Philippines. Although third-stage larvae are probably infective for the vertebrate host, little is known about the life cycle.

The worm is red to reddish brown because of the ingestion of blood (70). The male and female worms are permanently joined in copula and present a Y shape. The short arm of the Y is the male worm, and the longer arm and base of the Y is the female worm (Figure 18.13.) Both worms have a thick-walled buccal capsule which is armed in the inner base with eight small teeth. Male worms measure 3.0 to 6.3 mm long and 360 to 380 µm wide. Female worms measure 8.7 to 23.5 mm long and 550 to 570 µm wide. The eggs are oval, have a thick cortical shell, and measure 78 to 95 µm long by 42 to 54 µm wide (Figure 18.14).

The worms in the trachea are often associated with a severe, nonproductive cough; the cough is often accompanied by hemoptysis and occasionally asthma, possibly because of obstruction of the bronchi or bronchioles (22, 23, 70). Eosinophilia may or may not be present. Diagnosis can be performed by examination either of

Figure 18.13 *Mammomonogamus laryngeus* (*Syngamus trachea*) worms in the trachea of a turkey; note the typical Y configuration of the adult male and female worms. (From A Pictorial Presentation of Parasites: A cooperative collection prepared and/or edited by H. Zaiman.)

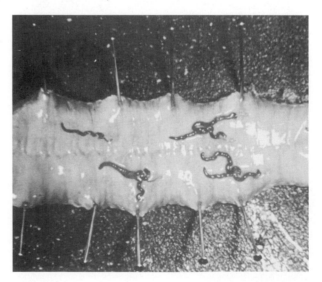

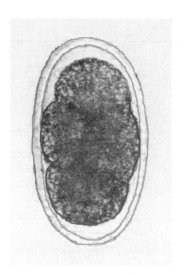

Figure 18.14 *Mammomonogamus laryngeus* egg. This egg measures 78 to 95 µm by 42 to 54 µm.

worms or eggs coughed up in the sputum or of eggs that may be swallowed and recovered in the stool. Direct visual examination of the throat and/or endoscopy may also reveal the presence of worms; worms can be removed by using forceps or surgical techniques. There have been no controlled studies of specific therapy of human infections with *M. laryngeus*.

M. laryngeus is a common veterinary parasite, and the adult worms are normally found in the laryngotracheal region of ruminants. The eggs are coughed up and swallowed and can be found in the sputum or stool. Once the eggs are outside of the host, they undergo development until the larva becomes infective; the larva can remain in the egg or be released into the environment. Ruminants become infected through the ingestion of food or water contaminated with infective larvae. Carrier hosts such as earthworms, snails, or arthropods may also play a role in infection. Apparently, the larvae are released in the digestive tract, cross the intestinal wall to the mesenteric veins, and migrate to the tracheolaryngeal area, where the adult worms mate and begin reproduction (70).

Ascaris suum

Human infection with adult *A. suum* is occasionally documented, based primarily on very specific worm morphology and relevant history with potential exposure to pigs. Generally, *Ascaris lumbricoides* and *A. suum* are morphologically so similar that individual worms usually cannot be identified as one or the other.

Of much greater concern are the human infections with *A. suum* in which the larvae cause visceral larva migrans (VLM) and invade the liver, lungs, and the intestine but do not mature to adult worms (65). In some cases,

encephalopathy can also be associated with VLM caused by *A. suum* (44). In one case, the patient suffered from drowsiness, quadriparesis, eosinophilia, and elevated immunoglobulin E levels in serum. Magnetic resonance imaging revealed multiple cerebral cortical and white matter lesions, and serologic test results indicated recent infection with *A. suum*.

Patients with VLM have respiratory symptoms; although eosinophilia is marked, no eggs or larvae have been found in stool specimens. Multiple small mass lesions can be seen on magnetic resonance imaging; they all disappear after anthelmintic therapy. Although liver tissue biopsy specimens showed marked infiltrations by eosinophils in the portal tracts and hepatic sinusoids, no larvae or eggs were seen (38). Hepatic nodules have been located mainly in periportal or subcapsular regions, which may represent periportal eosinophilic granuloma (47).

The number of infective eggs required to initiate an infection is apparently quite small; massive infections with *A. suum* can cause severe disease (5). However, the actual significance of human infection with this nematode is not known. Among people who visit or live on farms where pigs are raised, *A. suum* may be a significant cause of illness.

Gongylonema pulchrum

The thin nematode *Gongylonema pulchrum* is a common parasite of ruminants and has been diagnosed in pigs, bears, hedgehogs, monkeys, and occasionally humans. Human cases have been reported from Europe, Morocco, the former USSR, China, New Zealand, Sri Lanka, and the United States. Human infection is acquired by accidental ingestion of infected insects. Pathology in human cases is due to maturing and adult worms migrating through the mucosa and submucosa of the buccal cavity and the esophageal mucosa. Worms have been recovered from the lips, gums, hard and soft palate, tonsils, and angle of the jaw. In some cases, blood is present. In a case report from the United States, a 41-year-old woman living in New York City presented with a history of a sensation of a "moving organism" in her mouth for more than a year (27). On two occasions she removed worms from her mouth, once from the lip and once from the gum. The patient also reported a history of reflux-like symptoms that may have been associated with esophageal involvement. The worm extracted from the lip was identified as a mature female *Gongylonema* worm, probably *G. pulchrum*. The anterior was very thin and threadlike, with a diameter of 0.5 mm; it was covered with typical bosses (scutes) arranged in longitudinal rows (Figure 18.15). The posterior end of the worm was considerably wider. The eggs measured 50 by 28 µm.

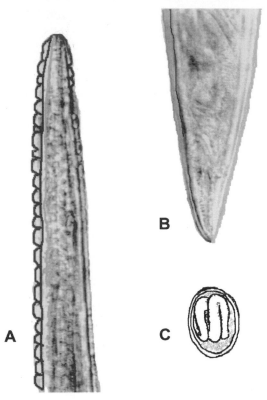

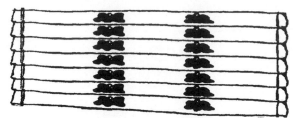

Figure 18.16   *Diplogonoporus* tapeworm proglottids. Note the width of the proglottids and their very short length; also note the two genital organs in each proglottid.

A new species, *Echinococcus shiquicus*, has been identified from China. Although no human infections have yet been identified, the possibility of infection certainly exists (93).

Diplogonoporus spp.

Diplogonoporus grandis and *D. balaenopterae* have been identified as tapeworm parasites of humans. This tapeworm is normally found in whales but has been isolated from approximately 50 Japanese patients (5). The adult worm measures 1.4 to 5.9 m long and has gravid proglottids that are very wide (15 to 25 mm) and very short (0.54 mm). There are two sets of genital organs in each proglottid (Figure 18.16). The eggs are operculated, their morphology is very similar to that of eggs of *Diphyllobothrium latum*, and they measure 63 to 68 μm by 50 μm (Figure 18.17).

A more recent case has been reported from Spain, in which a 58-year-old man who consumed raw fish passed a short chain of gravid proglottids. After therapy with 50 mg of paromomycin sulfate per kg divided into three doses, all given on the same day, the worm was finally evacuated. The man had not been out of Zaragoza Province in Spain for approximately 20 years, and the infection was probably associated with the importation

Figure 18.15   *Gongylonema pulchrum.* (A) Anterior end of the worm. (B) Posterior end of the worm. (C) Egg. (Panel A adapted from H. B. Ward, *Gongylonema* in the role of a human parasite. *J. Parasitol.*, 1916. Panel B adapted from reference 4. Panel C adapted from reference 27.)

Haycocknema perplexum

Adult male and female nematodes and larvae were recovered from myofibers after biopsy of the right vastus lateralis muscle. This infection in a man in Tasmania, Australia, was associated with a polymyositis (85). Two other potential cases have been reported, one of which was also in a patient from Tasmania (24). A muspiceoid nematode causing polymyositis was involved in these cases. These minute adult worms are very slender (<35 μm) and measure approximately 300 μm (male) and 425 μm (female) in length. The adult worms are very small and may be mistaken for larvae.

Muscle weakness, eosinophilia, and biopsy results have been consistent with an active polymyositis. While treatment with prednisone produced no improvement, subsequent treatment with albendazole led to immediate improvement (34).

Cestodes

Although *Echinococcus oligarthrus* and *Echinococcus vogeli* were covered in this chapter in previous editions of this book, they have been moved to chapter 14 (61).

Figure 18.17   *Diplogonoporus* egg. This egg measures 63 to 68 μm by 50 μm; note the resemblance to the *Diphyllobothrium latum* egg.

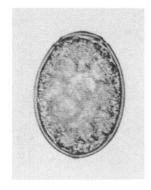

of fish into Spain. Outside of the Far East, this is the first confirmed case of diplogonoporiasis. The viability of the infective larval stage is evidence of resistance to export/import conditions (20).

Bertiella studeri

The tapeworm *Bertiella studeri* was originally recovered from an orangutan in Borneo and has been isolated from approximately 30 patients from India, Sumatra, Java, Borneo, Vietnam, Singapore, East Africa, St. Kitts, British West Indies, Philippines, Malaysia, Sri Lanka, and Yemen. The worm is relatively small, measuring up to 300 mm long and 10 mm wide. The proglottids are shed about 20 at a time; each mature proglottid is approximately 6 mm broad by 0.75 mm long. The eggs have an irregular, oval shape and measure 45 to 46 µm by 49 to 50 µm. There is a delicate middle envelope and an inner shell with a bicornuate protrusion on one side (Figure 18.18).

Infection is acquired by the accidental ingestion of mites containing the cysticercoid larval stage of the worm. After therapy with niclosamide or other anthelmintics, the worms are evacuated; they may also be passed spontaneously, as seen with other tapeworms. Patients may be asymptomatic or complain of vague intestinal disturbances. Some symptoms include intermittent epigastric pain after meals, accompanied by nausea, diarrhea, and anorexia, with no fever or weight loss (7). The eggs are very characteristic, and identification can be confirmed

Figure 18.18 *Bertiella studeri*. (A) Scolex. (B) Egg. (C) Mature proglottid. Note the description of the egg in the text related to the illustration. (Adapted from an original drawing by R. Blanchard, in E. C. Faust, *Human Helminthology*, Lea & Febiger, Philadelphia, Pa., 1929, and reference 4.)

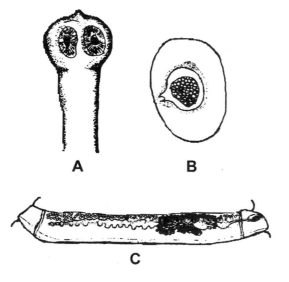

by finding the eggs or a chain of proglottids in the stool. Most of the confirmed human infections have been found in children. Although both niclosamide (Yomesan) and albendazole have been used for therapy, treatment is not always successful (94).

Human infections with *B. mucronata* have also been found; however, fewer than 10 cases have been recorded. One infection has been recorded in a child in Fargo, N.D., who may have come in contact with an imported monkey (5).

Inermicapsifer madagascariensis

The tapeworm *Inermicapsifer madagascariensis* is often seen in humans in Havana, Cuba; in fact, outside of Africa, where it causes a disease of rodents, it may be restricted to humans. Although the life cycle has not been established, an arthropod vector is thought to be involved in human infections, which are most common in children aged 1 to 3 years. Symptoms are vague, but there may be anorexia, weight loss, abdominal pain, or irritability (34). Diagnosis is performed by finding the rice grain-like proglottids; however, these proglottids, eggs, or egg capsules are extremely similar to those of *Raillietina* spp. Specific treatment with praziquantel has been reported to be effective in eliminating the tapeworm (35).

Raillietina celebensis

The tapeworm *Raillietina celebensis*, common in rats, has also been found in humans, mostly children. The intermediate hosts are various species of ants. Usually, the infection causes no symptoms. The proglottids are small (3 mm or less) and resemble those of *I. madagascariensis*.

The adult worm is approximately 40 cm long; each egg packet contains one to four large eggs measuring 90 by 46 µm. Any symptoms that might be present tend to disappear after treatment with niclosamide.

A number of other species in the genus *Raillietina* have been implicated in human disease, although the number of cases has been small. *R. demerariensis* has been found in approximately 5% of patients in the rural areas near Quito, Ecuador; cases have also been reported from Cuba. The adult worms measure up to 60 cm long by 3 mm wide and have around 5,000 proglottids (5). The individual proglottids resemble rice grains and contain about 200 to 250 egg packets, each measuring about 200 µm in diameter. The oval eggs contain very conspicuous hooklets. *R. asiatica*, *R. garrisoni*, and *R. siriraji* all cause infection in humans, and, again, children rather than adults are affected. Accidental ingestion of arthropod intermediate hosts is implicated in transmission.

Mesocestoides spp.

Approximately 27 cases of human infection have been reported, probably acquired from ingestion of the larval form (tetrathyridium) in inadequately cooked meat from some animal that serves as the intermediate host. The tapeworm in one case measured 40 cm by 2 mm. The infected child was irritable, had a poor appetite and abdominal colic, and was anemic (16). A second case was reported from Korea (28). The patient, who was a farm worker, complained of abdominal pain and passed a number of proglottids in the stool over a period of several months. The proglottids were described as looking like sesame seeds. Adult worms were recovered after therapy with niclosamide and consisted of 32 complete worms. This is probably the heaviest worm burden reported in a human case. The patient had a history of frequently eating raw viscera of chickens. The seventh case has been reported in the United States; the patient was a 19-month-old boy from Alexandria, La. The infection was probably food borne, associated with local dietary customs in the Acadian and Creole communities of Louisiana, including eating sausage containing wild-animal viscera (33).

Taenia crassiceps

Taenia crassiceps is a common tapeworm of the red fox. Larval forms are generally found under the skin and in the body cavities in animals. Human cases are quite rare. An infection was reported in a 33-year-old patient with AIDS and a history of *Pneumocystis jiroveci* pneumonia and cerebral toxoplasmosis. The patient developed a subcutaneous paravertebral infiltrate that resembled a hematoma. During a period of several weeks, the lesion spread such that almost the entire back was covered. Approximately 3 weeks after admission, the infiltrate ruptured spontaneously, releasing blood and many white spherical masses (2 to 3 mm in diameter). The masses were identified as the larval form of *T. crassiceps*. After therapy with mebendazole and praziquantel, the infiltrate regressed. Praziquantel treatment was continued for 10 weeks, at which time the patient stopped taking the medication. At the end of 4 months, the lesion recurred but was successfully treated with an additional course of mebendazole and praziquantel (50).

A second case in an AIDS patient has been reported from France; this case involved invasion of subcutaneous and muscular tissues by the larval forms of *T. crassiceps* (17). The authors emphasized the problems in differentiating this infection from cysticercosis, coenurosis, sparganosis, and hydatidosis. The proliferative growth of the larvae, recurrence of lesions after surgical removal, and possible invasion of other tissues carry a poor prognosis for this group of immunocompromised patients. The biological

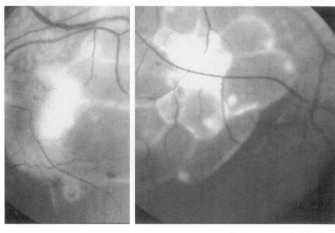

Figure 18.19   *Taenia crassiceps*. Retinal photographs showing budding cysticercus in situ. (From A Pictorial Presentation of Parasites: A cooperative collection prepared and/or edited by H. Zaiman. Photograph courtesy of Renio Freeman.)

properties of *T. crassiceps* cysticerci and the epidemiologic characteristics of pandemic AIDS could indicate the potential for new cases of this disease in humans (31).

Another case illustrates a non-AIDS-associated infection with *T. crassiceps* (2). A healthy 15-year-old girl developed an iridocyclitis in her right eye. A contractable living parasite measuring about 2 mm could be seen in the anterior chamber, while the rest of the eye was not involved (Figure 18.19). The patient reported having close contact with a young dog in the family. The larval form was removed alive but was not developed enough to have the characteristic scolex (Figure 18.20). Western blot results provided sufficient indirect evidence to confirm the identity of the organism as *T. crassiceps*. After follow-up for 2 years, no further evidence of ocular or general infestations with this parasite could be found.

The method for obtaining antigenic extracts from *T. crassiceps* cysticerci and their cross-reactivity with *T. solium* cysticerci antigens have made them an interest-

Figure 18.20   *Taenia crassiceps* cysticercus after surgical removal from the eye. (From A Pictorial Presentation of Parasites: A cooperative collection prepared and/or edited by H. Zaiman. Photograph courtesy of Renio Freeman.)

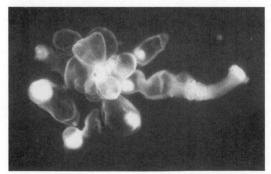

ing antigen source for diagnostic work. The use of 18- and 14-kDa *T. crassiceps* immunoaffinity-purified proteins for detection of anti-cysticercus antibodies in cerebrospinal fluid and/or serum samples by using an enzyme-linked immunosorbent assay system showed good performance and a high specificity for serum samples, dispensing with the use of confirmatory tests, such as immunoblotting, for checking specificity (29).

Trematodes

Alaria americana

Alaria americana is a trematode that is normally found in the intestines of wild mammalian carnivores such as wolves, foxes, raccoons, bobcats, and skunks. The eggs are passed in the stool, hatch in water, and invade freshwater snails of the genus *Helisoma*. Cercariae then leave the snail and infect frog tadpoles. The first case of disseminated human infection by the mesocercarial stage of this worm was found in a 24-year old man in Canada who died 8 days after the onset of his illness (30). The infection was probably acquired through ingestion of inadequately cooked frogs, which serve as intermediate hosts for this parasite. The infection was diagnosed by lung biopsy, and mesocercariae were found in the stomach wall, lymph nodes, liver, myocardium, pancreas and surrounding adipose tissue, spleen, kidney, lungs, brain, and spinal cord at autopsy (Figures 18.21 and 18.22). Apparently, there was no host reaction to the parasites. Granulomas were present in the stomach wall, lymph nodes, and liver, but no worms were recovered from these sites.

Two cases of eye infections with mesocercariae of *Alaria* were found in unrelated Asian men who had unilateral decreased vision (67). Both patients had pigmentary tracks in the retina, areas of active and/or healed retinitis, and other signs of diffuse unilateral subacute neuroretinitis (Figure 18.23). The worms were seen in the

Figure 18.21 *Alaria americana*. Living mesocercaria removed from a human liver at autopsy. (From A Pictorial Presentation of Parasites: A cooperative collection prepared and/or edited by H. Zaiman. Photograph courtesy of R. S. Freeman.)

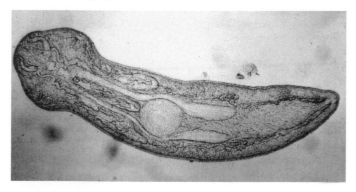

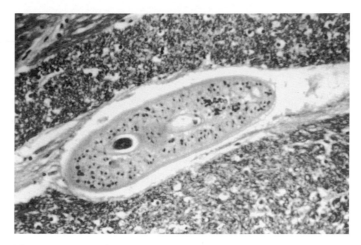

Figure 18.22 *Alaria americana*. Mesocercaria in the patient's pons. (From A Pictorial Presentation of Parasites: A cooperative collection prepared and/or edited by H. Zaiman. Photograph courtesy of R. S. Freeman.)

retinas and vitreous several years after the original infection was thought to have occurred. The probable source of infection was ingestion of undercooked frogs' legs containing the infective stage. In addition to the prolonged intraocular infection, the mesocercariae caused diffuse unilateral subacute neuroretinitis, a condition thought to be caused only by intraocular nematode larvae.

Plagiorchis spp.

Plagiorchis spp. cause human intestinal infections that have been confirmed in the Philippines, Indonesia, Korea, Thailand, and Japan. The infection is acquired from the ingestion of grubs of certain insects thought to be

Figure 18.23 *Alaria americana* in the human eye. Note the relationship to the optic disk and macula. (From A Pictorial Presentation of Parasites: A cooperative collection prepared and/or edited by H. Zaiman. Photograph courtesy of R. S. Freeman.)

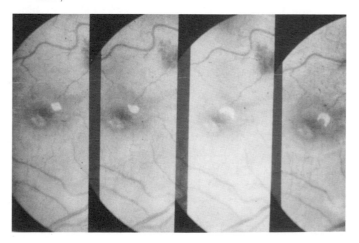

the second intermediate host for this fluke. Ingestion of raw or poorly cooked freshwater fish containing the metacercariae was implicated in the case in Korea (41). Metacercariae obtained by an artificial ingestion technique used on freshwater fish were administered to a Sprague-Dawley rat; on day 8 postinfection, one *Plagiorchis muris* worm was recovered from the small intestine of the rat. Infections with multiple helminths from northeast Thailand have included *P. harinasutai*, as well as other unusual trematodes (77). Confirmation of these infections may be difficult since the eggs resemble those of *Opisthorchis viverrini* and *Clonorchis sinensis*.

Neodiplostomum seoulense

Neodiplostomum seoulense is known to infect humans in Korea and has been linked to snake consumption. The parasites can normally be found in house rats, frogs and their tadpoles, and several terrestrial snakes. Symptoms have been noted in human infections and include epigastric pain, diarrhea, fever, and eosinophilia. The eggs can be differentiated from those of *Paragonimus*, *Echinostoma*, and *Fasciola*; they measure 81 to 102 μm long by 51 to 63 μm wide, are golden brown and bilaterally symmetrical, have an oblique opercular margin, and are immature when passed in the stool. In animal studies, this parasite appears to have the potential to kill most strains of mice by causing irreversible damage to their intestine (51). In follow-up studies of intestinal parasitic infections of the Army soldiers in Whachon-gun, Korea, additional cases of infection with *N. seoulense* were identified (43). Additional molluscan intermediate hosts for *N. seoulense* continue to be identified (19).

Spelotrema brevicaeca

Spelotrema brevicaeca has been reported to cause infections in humans in the Philippines. The eggs of these trematodes are suspected of causing lesions in the heart, brain, and spinal cord of patients dying of acute cardiac dilation. Although the complete life cycle is not known, the encysted metacercarial stage has been found in the crab *Cararius maenas*, which is the implicated food source. The adult worms are small, measuring 0.5 to 0.7 mm long by 0.3 to 0.4 mm wide, and have a spinose cuticle. The eggs are quite small (15 to 16 μm long by 9 to 10 μm wide). The infection can be confirmed by finding the eggs in stool or tissues or by finding adult worms in the stool after therapy.

Brachylaima sp.

Two 21-month-old children from the same rural area of South Australia presented 18 months apart with mild abdominal pain and diarrhea; trematode eggs of *Brachylaima* sp. were found in the stool (13). Therapy with praziquantel resulted in cessation of the symptoms and elimination of eggs in the stool. The life cycle involving the common house mouse, poultry, and introduced European helicid snails is well established in South Australia. Both children had been seen eating raw snails; examination of the snails from the environment confirmed that they were infected with metacercariae of this brachylaimid trematode.

In 1998, a 78-year-old woman presented with an 18-month history of intermittent diarrhea. Eggs of a brachylaimid trematode were seen in the stool for a week. After treatment with praziquantel, a degenerated adult fluke was recovered from the stool and subsequently identified as *Brachylaima* sp. She lived in South Australia and consumed vegetables grown in her garden, which was infested with helicid snails. These snails, commonly infected with brachylaimid intermediate larval stages, are considered to be the source of human infection (12). *B. cribbi* is the first brachylaimid known to have infected humans and is probably of European origin; the intermediate host snails were all introduced into Australia from Europe. This life cycle was established in the laboratory from eggs recovered from human feces in Australia (11).

Troglotrema salmincola

The trematode *Troglotrema salmincola* has been associated with poisoning in dogs eating raw salmon or trout on the Pacific coast of North America and Canada. Normal hosts would include the coyote, fox, raccoon, mink, and lynx in North America and aborigines of eastern Siberia. The adult worms are quite small, measuring 0.8 to 1.1 mm long by 0.3 to 0.5 mm wide. The eggs are broadly oval, operculated, thick-shelled, and yellow and measure 60 to 80 μm long by 34 to 50 μm wide. When salmon and trout, as well as other fish and the Pacific giant salamander, are eaten raw, metacercariae in their kidneys cause infection in the intestine of the definitive host (the dog). A rickettsia is associated with the fluke infection and has caused severe and fatal symptoms in dogs, coyotes, and foxes; this rickettsia is also possibly a human pathogen. Praziquantel is the recommended therapy for human infection with this trematode.

Stellantchasmus falcatus

Stellantchasmus falcatus infections in two humans have been reported from Korea. Vague abdominal discomfort and hunger pains were reported. In both patients, adult worms were recovered in the stool after treatment with praziquantel (39, 84). Both patients reported eating several kinds of brackish-water fish. Based on these case histories, the total number of human infections remains at <10 to date. This infection has also been documented in a patient eating raw mullet in Hawaii (10). Another infection linked to the ingestion of raw mullet is caused by a *Phagicola* sp. A patient in Brazil complained of mild intestinal pain, and examination of the stool specimen revealed trematode eggs in the stool. The patient also had an 8% eosinophilia. After treatment with praziquantel, the symptoms disappeared (18). Another infection linked to the ingestion of raw brackish-water or freshwater fish in Korea is caused by *Heterophyopsis continua*, a trematode whose consumption results in vague abdominal complaints (40).

Phaneropsolus bonnei and *Prosthodendrium molenkempi*

The two small trematodes *Phaneropsolus bonnei* and *Prosthodendrium molenkempi* have been recovered in large numbers from the intestine of humans; however, there are no well-defined intestinal pathologic or clinical findings. These infections are documented and endemic among rural populations in northeastern Thailand and Laos; in some areas the prevalence is as high as 10 to 40%. Infection occurs after ingestion of raw or poorly cooked small fish contaminated with infected naiads (water nymphs of dragonflies). Diagnosis can be confirmed by finding adult worms or eggs in the stool; however, the eggs may be difficult to differentiate from each other and from those of *O. viverrini*. *O. viverrini* and several species of heterophyid and lecithodendriid flukes are endemic in two riverside localities in Laos, suggesting that the intensity of infection and the relative proportion of fluke species vary by locality along the Mekong River basin (15). Staining using a 1% (wt/vol) concentration of potassium permanganate for 1 min is useful in differentiating the egg surface morphology of *O. viverrini*, *Haplorchis taichui*, and *P. bonnei* eggs (86).

Phaneropsolus spinicirrus

Members of the *Phaneropsolus* group of trematodes have been found in reptiles, birds, and mammals; however, only *Phaneropsolus bonnei* was reported to have been isolated from humans (58). Adult worms of *P. spinicirrus* sp. nov. were recovered after treatment with praziquantel followed by a magnesium sulfate purgative 5 h after medication. All stools passed over the next 48 h were preserved in 10% formalin and strained through a 120-mesh screen (200-μm openings), and the strained material was examined with a stereoscopic microscope. The adult worms were stained with Mayer's hemalum, dehydrated, cleared, and mounted for examination. Differences in morphologic character-

Figure 18.24 *Phaneropsolus spinicirrus* isolated from a patient. (1) Ventral view of the holotype. (2) Ventral view of the structure of the terminal portion of the reproductive system. c, cirrus; cs, cirrus sac; gp, genital pore; mt, metraterm; os, oral sucker; pg, prostate gland, sv, seminal vesicle. (Reprinted from reference 46 with permission.)

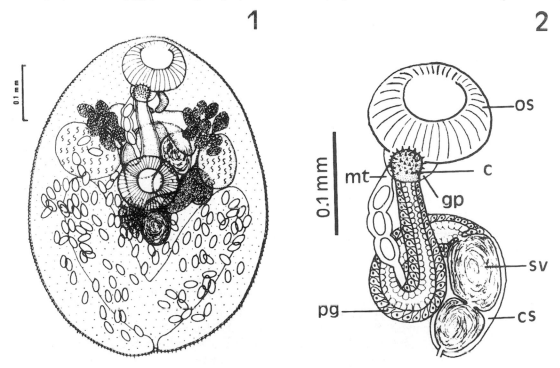

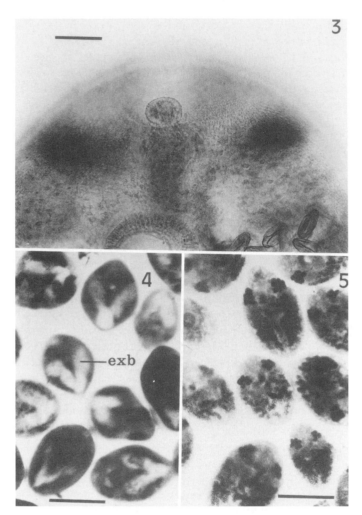

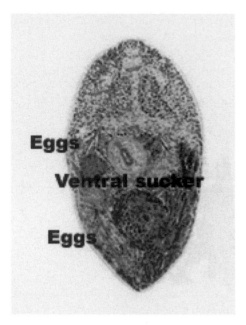

Figure 18.26 *Haplorchis taichui.* Note the pyriform adult trematode, which is covered with scale-like spines on the tegument; eggs can be seen to the right of the labels, while the ventral sucker can be seen immediately above the label. The ventral sucker has a crescentric group of 12 to 16 hollow spines (scleritis) (not visible in this image). (Armed Forces Institute of Pathology photograph.)

Figure 18.25 Photographs of two species of *Phaneropsolus* collected from the same patient. (3) The anterior part of *Phaneropsolus spinicirrus*, showing the spinose cirrus and the tegumental spines. Bar, 0.05 mm. (4) Unstained adults of *P. spinicirrus*, showing the larger excretory bladder (exb) and the more conspicuous border. Bar, 0.5 mm. (5) Unstained adults of *P. bonnei*. Bar, 0.5 mm. (Reprinted from reference 46 with permission.)

istics that provided evidence for the identification of *P. spinicirrus* rather than *P. bonnei* included more conspicuous spines on the cirrus, a larger body and internal organs, conical spines on the tegument, an ovoid to bilobed ovary, and a large, V-shaped excretory bladder (46). The structure of the tegumental spines and the characteristic excretory bladder allow the two species to be recognized, even in unstained specimens (Figures 18.24 and 18.25).

Haplorchis taichui

Haplorchis taichui is a minute human intestinal trematode that parasitizes the small intestine of humans and other mammals (Figure 18.26). Pathologic changes seen in three patients from Thailand include mucosal ulceration, mucosal and submucosal hemorrhages, fusion and shortening of the villi, chronic inflammation, and fibrosis of the submucosa (87).

An exceptionally high rate of infection (36%) has been identified in human populations on Mindanao Island, southern Philippines. The source of infection has been linked to consumption of raw or undercooked freshwater fish containing infective metacercariae. Clinical symptoms include upper abdominal discomfort or pain and loud gurgling noises produced by the movement of gas, fluid, or both in the alimentary canal, sounds which are audible at a distance. Other reported symptoms were nausea, chronic diarrhea, and weight loss. Symptoms rapidly resolved after treatment with praziquantel (6). The adult fluke is pyriform; eggs and the ventral sucker can generally be seen (6).

Gymnophalloides seoi

Gymnophalloides seoi is a new human intestinal trematode that was first found in a Korean woman who complained of epigastric discomfort. This trematode is highly prevalent among the inhabitants of a seashore village on a southwestern island of the Republic of Korea. Normally, it is a parasite of shorebirds and is thought to be the only known species in the family Gymnophallidae that can infect hu-

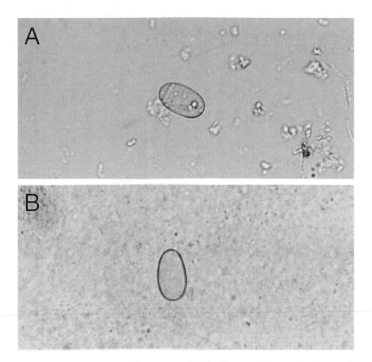

Figure 18.27 Eggs of *Gymnophalloides seoi* from the feces of a patient, detected by formalin-ether sedimentation (unstained) (A) and cellophane thick-smear technique (malachite green stained) (B). (Reprinted from reference 57 with permission.)

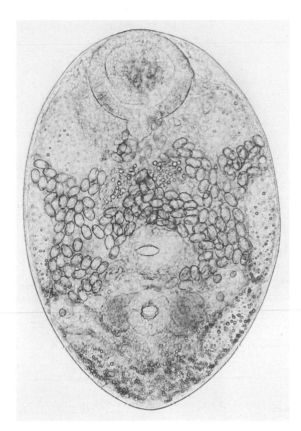

Figure 18.28 Specimen of *Gymnophalloides seoi* (ventral view, unstained) showing various structures within this fluke. (Reprinted from reference 57 with permission.)

mans. Various bivalves (clams) are known to serve as the first or second intermediate hosts for some of these trematodes. Symptoms are generally somewhat nonspecific and include abdominal pain, diarrhea, indigestion, anorexia, easy fatigability, and weakness. On the basis of fecal examinations, 70 (71.4%) of 98 were found to be infected with various intestinal parasites. *G. seoi* eggs were found in 48 (49.0%) of 98 patients. Individual worm burdens, as measured by collection of adult flukes after treatment and purgation, ranged from 106 to 26,373 specimens. The average number of worms per patient was 3,326. There were no sex- or age-related differences. This study indicates that *G. seoi* is a human intestinal trematode under natural conditions and does not represent just an accidental infection (56, 57). The eggs of *G. seoi* measure 19 to 25 μm in length (smaller than *C. sinensis*) and have a thin, transparent shell (Figure 18.27). The adult worm measures approximately 0.4 mm long and 0.25 mm wide (Figure 18.28).

Metorchis conjunctus (North American Liver Fluke)

A common-source outbreak of acute clinical illness among 19 people who ate raw fish (sashimi) prepared from the white sucker has been reported from Canada.

The sashimi contained infective metacercariae of *Metorchis conjunctus*. The adult fluke lives within the biliary tree of the definitive host. Symptoms included persistent upper abdominal pain, low-grade fever, high eosinophil counts, and increased liver enzyme levels with an obstructive pattern. After 15 days, trematode eggs were recovered in the stool specimens; morphologically they were identical to those of *O. viverrini*. The symptoms lasted several weeks, but the patients responded rapidly to therapy with praziquantel. Necropsy of golden hamsters that had been fed metacercariae from the fish revealed adult flukes identified as *M. conjunctus*. Apparently, this is a newly emerging human disease and represents the first report of a common-source outbreak of an acute illness caused by these North American liver flukes (63). Although some patients remain asymptomatic, the prevalence in some isolated communities within Canada reaches 20% (62). Because fish parasites infecting humans within the United States and Canada are becoming more important and prevalent, a composite list has been provided (Table 18.2).

Table 18.2 Human parasites acquired from the ingestion of raw or poorly cooked fish[a,b]

Parasite	Fish host(s)
Nematodes	
Anisakis simplex	Salmonids, tuna, herring, mackerel, squid
Pseudoterranova decipiens	Cod, pollack, haddock
Eustrongylides spp.	Killifish, estuarine species
Dioctophyma renale	Freshwater fish, estuarine species
Cestodes	
Diphyllobothriurn latum	Salmonids, pike, perch, burbot
Diphyllobothrium dendriticum	Salmonids
Diphyllobothrium ursi	Salmonids
Diphyllobothrium spp.[c]	Salmonids
Schistocephalus solidus	Sticklebacks
Trematodes	
Troglotrema (Nanophyetus) salmincola	Salmonids
Metorchis conjunctus	White sucker
Haplorchis taichui	Freshwater fish
Plagiorchis spp.	Freshwater fish
Cryptocotyle lingua	Flounder
Amphimerus pseudofelineus	Freshwater fish

[a] Adapted from reference 62.
[b] Not every parasite mentioned in this table has been included in this chapter on unusual parasites; often case reports are very limited. Parasites listed in this table are geographically relevant to the United States and Canada, and the table does not include all parasites discussed in the chapter.
[c] The taxonomy of the genus *Diphyllobothrium* remains unclear.

Schistosoma mattheei

Schistosoma mattheei causes infections in sheep, cattle, and horses; human infections, which caused mild disease, have also been reported from South Africa. The eggs resemble those of *Schistosoma intercalatum*, have a terminal spine, and measure 120 to 180 µm in length. Clinical specimens include urine and stool; eggs can be recovered in both. Although human infections tend to be mild, the effect of this infection on animals is more serious.

The pathology of human disease is similar to that seen in other schistosome diseases and can include egg granulomas in the large intestine and liver, mild hepatosplenomegaly, and granulomatous bowel disease. There tends to be coinfection with either *S. haematobium* or *S. mansoni*. Patients may present with mild mucoid diarrhea in which blood is present, crampy abdominal pain, and general nonspecific complaints; many patients remain asymptomatic. Praziquantel is recommended for therapy.

Enzyme electrophoresis indicates that most, if not all, *S. mattheei* eggs passed in the urine in human infections derive from *S. mattheei* females in copulo with *S. haematobium* males. It appears that *S. mattheei* males do not reach sexual maturity in humans; however, the *S. haematobium* × *S. mattheei* males possibly do (52). Studies indicate that is it highly unlikely that *S. mattheei* and *S. haematobium* will evolve into a single species; it does not seem likely that the virulence of the parental species will be influenced (53).

Philophthalmus lacrimosus

Philophthalmus lacrimosus is an avian eye trematode, found primarily in the eyes of wild and domestic birds. A number of mammals, including humans, can serve as the intermediate as well as the definitive host; however, human cases are quite rare. In cases of conjunctivitis, one of the symptoms is the feeling of having a foreign-body sensation in the eye. In one patient from Mexico, a small live parasite was found in the connective tissue of the conjunctiva and was removed surgically (Figure 18.29) (54). A recent case has been reported from Thailand (90a).

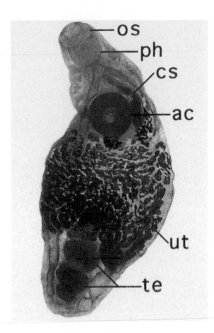

Figure 18.29 Adult *Philophthalmus* sp. trematode. Note oral sucker (os), pharynx (ph), cirrus sac (cs), acetabulum (ventral sucker) (ac), uterus (ut), and testes (te). (Armed Forces Institute of Pathology photograph.)

Achillurbainia spp.

Achillurbainia spp. are trematodes (similar to *Paragonimus* spp.) that are normally found in the maxillary sinuses of the opossum. They have been isolated from humans. In one case of surgical inguinal hernia repair, massive numbers of small granulomata on the omentum and other peritoneal surfaces were found to contain trematode eggs like those of *A. recondita* (3). *Achillurbainia* eggs and/or adult worms have been found in retroauricular cysts or abscesses in individuals in West Africa and Southeast Asia.

Pentastomids

Armillifer spp., *Linguatula serrata*, and *Sebekia* spp.

The organisms are found in a separate phylum, Pentastomida, and are called tongue worms. Human infections have been reported from Africa, Europe, Asia, and the Americas (14). A representative life cycle is seen in Figure 18.30. The pentastomids are a group of parasites classified between the arthropods and annelids (26).

When humans serve as the intermediate hosts, the infective larvae die in situ. However, when mature larvae (often encysted) are ingested, they may migrate from the stomach, attach themselves to nasopharyngeal tissues, develop into adult pentastomids, and produce symptoms of the halzoun syndrome. Symptoms include throat discomfort, paroxysmal coughing, sneezing, and occasionally dysphagia and vomiting (81).

Occasionally, pentastomid larvae infect the eye. Inflammation is minimal, and infection is probably the result of direct eye contact with water containing pentastomid eggs (25). In one case report, a 34-year-old woman in Guayaquil, Ecuador, presented with a 2-month history of ocular pain on the right side, with conjunctivitis and visual difficulties due to the presence of a shadow (55). Apparently, when she lay down on her right side, her vision became clear. Examination revealed a mobile body in the anterior chamber of the eye; the body was removed, fixed, stained, and mounted for more detailed microscopic examination. The parasite was identified as a third-stage instar larval form of *Linguatula serrata*. This case represents the first report in Ecuador and South America.

A case of dermatitis in a 31-year-old Costa Rican woman was reported to be caused by a pentastomid

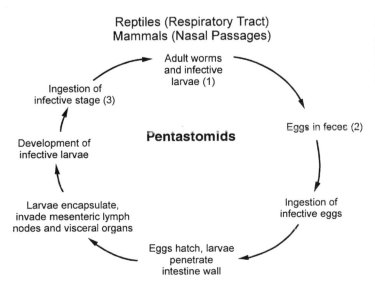

Figure 18.30 Life cycle of the tongue worms (pentastomids) *Armillifer* and *Linguatula* spp. (1) Humans ingest raw or inadequately cooked snake or lizard; (2) humans ingest eggs on vegetation or in water; (3) humans ingest the flesh of intermediate hosts containing infective larvae.

larva belonging to the genus *Sebekia*, apparently the first reported case of infection of a human by this genus. The adult worms are parasites found in the respiratory tract of reptiles, and the larval forms are found in the viscera and muscles and along the spinal cord of freshwater fish. The patient, who was pregnant, presented with a "strong itch" in the lower right quadrant of her abdomen and a "biting sensation," particularly at night. She had a serpiginous burrow surrounded by an intense erythematous zone on her lower abdomen and a 30% eosinophilia. A white, active larva approximately 1 cm long was removed from the burrow. On the basis of morphologic features, including spines on the hooks and other details, the larva was identified as being in the genus *Sebekia*. Although there is no way to determine how the patient became infected or why the larva migrated to the skin as does larva migrans, the possibility of ingesting water containing eggs of the parasites or ingesting raw or improperly cooked fish could explain the subsequent infection (64).

Diagnosis is made by identifying the pentastomid in a biopsy specimen or at autopsy (Figures 18.31 and 18.32). Treatment is usually not necessary; however, surgical removal of the parasite is recommended if the patient is symptomatic.

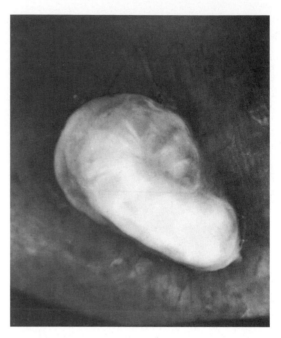

Figure 18.32 Encysted *Armillifer armillatus* larva on the surface of the liver of a 55-year-old Zairian. The cyst contains a C-shaped larva (×12). (Armed Forces Institute of Pathology photograph.)

Figure 18.31 Nonencysted mature *Armillifer armillatus* larva attached to the abdominal surface of a human diaphragm (×4). (Armed Forces Institute of Pathology photograph.)

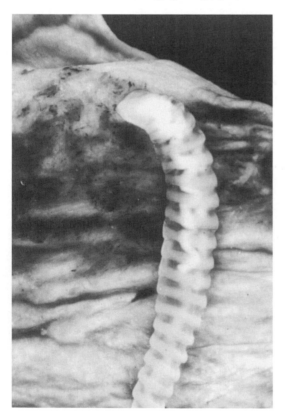

Acanthocephalans

Macracanthorhynchus hirudinaceus and *Moniliformis moniliformis*

The acanthocephalans (thorny-headed worms) were formerly included in the same phylum as the nematodes. These worms get their name from the proboscis, which is covered with spines (Figure 18.33). The life cycle is seen in Figure 18.34.

Very few cases have been reported in the literature; however, reported symptoms include abdominal pain and tenderness, anorexia, and nausea. In some cases, adult worms were passed in the stool (Figure 18.35) (76, 80).

In the last few years, a number of human infections with *Macracanthorhynchus hirudinaceus* have been reported from China. The infections have been primarily in children, probably acquired from the ingestion of raw beetles. Clinical manifestations include intestinal obstruction and perforation, with perforation occurring in over 200 cases. No eggs were found in the stool specimens of these patients. In many cases, the worms were recovered at surgery or autopsy. Signs and symptoms of infection with *Moniliformis moniliformis* include serious gastrointestinal pain, diarrhea, exhaustion, somnolence, and tinnitus; vomiting, jaundice, cough, and a protuberant abdomen have also been reported (21). In

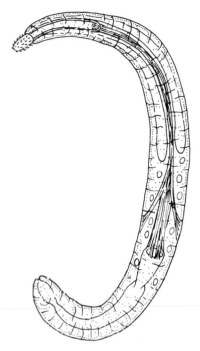

Figure 18.33 Thorny-headed worm. Note the proboscis, which is covered with spines. (Illustration by Sharon Belkin.)

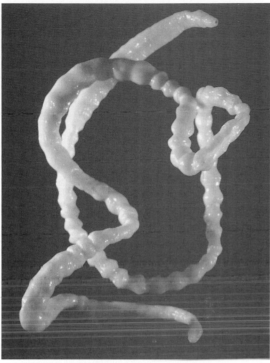

Figure 18.35 *Moniliformis moniliformis*, complete adult specimen, approximately 133 mm long (×5). (Armed Forces Institute of Pathology photograph.)

children, the infection has resulted in retarded development and irritability (5).

In some areas of China, eggs have been recovered in 16 to 60% of pigs examined (5); in pigs, this worm can be found worldwide. In Figure 18.36, the proboscis of *M. hirudinaceus* is seen attached to the pig bowel wall. The insertion of the toothed proboscis, along with the inflammation, ulceration, and hemorrhage caused, can

Figure 18.36 *Macracanthorhynchus hirudinaceus* proboscis attached to pig bowel wall. The insertion of the toothed proboscis, along with the inflammation, ulceration, and hemorrhage caused, are seen in this low-power view. (From A Pictorial Presentation of Parasites: A cooperative collection prepared and/or edited by H. Zaiman.)

Figure 18.34 Life cycle of the acanthocephalans (thorny-headed worms).

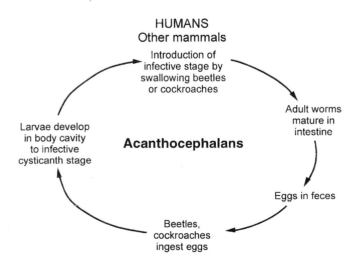

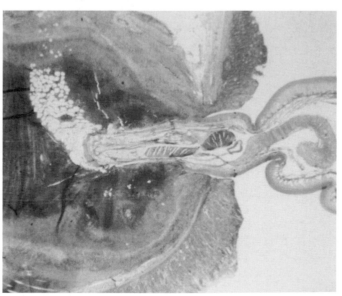

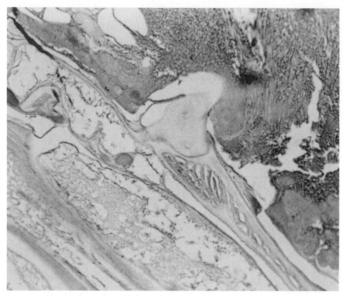

Figure 18.37 *Macracanthorhynchus hirudinaceus.* A single tooth on the proboscis is visible at this higher magnification. (From A Pictorial Presentation of Parasites: A cooperative collection prepared and/or edited by H. Zaiman.)

be seen. In Figure 18.37, a higher-magnification photograph shows a single tooth on the proboscis that is visible. The eggs of *M. hirudinaceus* and *M. moniliformis* can be seen in Figures 18.38 and 18.39. Both eggs contain a fully developed acanthor larva, and the rostellar hooks are also visible.

Figure 18.38 *Macracanthorhynchus hirudinaceus* egg. This egg measures 80 to 100 μm by 50 μm, has a thick, dark brown, textured shell, and contains a developed acanthor larva with hooks. (From A Pictorial Presentation of Parasites: A cooperative collection prepared and/or edited by H. Zaiman.)

Figure 18.39 *Moniliformis moniliformis* egg. This egg measures 90 to 125 μm by 65 μm and contains a fully developed acanthor larva with visible hooks. (From A Pictorial Presentation of Parasites: A cooperative collection prepared and/or edited by H. Zaiman.)

References

1. **Anthony, P. P., and I. W. McAdam.** 1972. Helminthic pseudotumours of the bowel: thirty-four cases of helminthoma. *Gut* **13:**8.

2. **Arocker-Mettinger, E., V. Huber-Spitzy, H. Auer, G. Grabner, and M. Stur.** 1992. *Taenia crassiceps* in the anterior chamber of the human eye. A case report. *Klin. Monatsbl. Augenheilkd.* **201:**34–37.

3. **Beaver, P. C., R. A. Duron, and M. D. Little.** 1977. Trematode eggs in the peritoneal cavity of man in Honduras. *Am. J. Trop. Med. Hyg.* **26:**684–687.

4. **Beaver, P. C., R. C. Jung, and E. W. Cupp.** 1984. *Clinical Parasitology,* 9th ed. Lea & Febiger, Philadelphia, Pa.

5. **Beaver, P. C., and J. H. Theis.** 1979. Dioctophymatid larval nematode in a subcutaneous nodule from a man in California. *Am. J. Trop. Med. Hyg.* **28:**206–212.

6. **Belizario, V. Y., Jr., W. U. de Leon, M. J. J. Bersabe, Purnomo, J. K. Baird, and M. J. Bangs.** 2004. A focus of human infection by *Haplorchis taichui* (Trematoda: Heterophyidae) in the southern Philippines. *J. Parasitol.* **90:**1165–1169.

7. **Bhagwant, S.** 2004. Human *Bertiella studeri* (family Anoplocephalidae) infection of probable southeast Asian origin in Mauritian children and an adult. *Am. J. Trop. Med. Hyg.* **70:**225–228.

8. **Binford, C. H., and D. H. Connor.** 1976. *Pathology of Tropical and Extraordinary Diseases.* Armed Forces Institute of Pathology, Washington, D.C.

9. **Boreham, R. E., S. Hendrick, P. J. O'Donoghue, and D. J. Stenzel.** 1998. Incidental finding of *Myxobolus* spores (Protozoa: Myxozoa) in stool samples from patients with gastrointestinal symptoms. *J. Clin. Microbiol.* **36:**3728–3730.

10. **Bunnag, D., J. H. Cross, and T. Bunnag.** 2000. Intestinal fluke infections, p. 832–840. *In* G. T. Strickland (ed.), *Hunter's Tropical Medicine and Emerging Infectious Diseases,* 8th ed. The W. B. Saunders Co., Philadelphia, Pa.

11. **Butcher, A. R., and D. I. Grove.** 2001. Description of the life-cycle stages of *Brachylaima cribbi* n. sp. (Digenea: Brachylaimidae) derived from eggs recovered from human faeces in Australia. *Syst. Parasitol.* **49**:211–221.

12. **Butcher, A. R., P. Parasuramar, C. S. Thompson, and D. I. Grove.** 1998. First report of the isolation of an adult worm of the genus *Brachylaima* (Digenea: Brachylaimidae), from the gastrointestinal tract of a human. *Int. J. Parasitol.* **28**:607–610.

13. **Butcher, A. R., G. A. Talbot, R. E. Norton, M. D. Kirk, T. H. Cribb, J. R. Forsyth, B. Knight, and A. S. Cameron.** 1996. Locally acquired *Brachylaima* sp. (Digenea: Brachylaimidae) intestinal fluke infection in two South Australian infants. *Med. J. Aust.* **164**:475–478.

14. **Cannon, D. A.** 1942. Linguatulid infestation of man. *Ann. Trop. Med.* **36**:160–166.

15. **Chai, J. Y., J. H. Park, E. T. Han, S. M. Guk, E. H. Shin, A. Lin, J. L. Kim, W. M. Sohn, T. S. Yong, K. S. Eom, D. Y. Min, E. H. Hwang, B. Phommasack, B. Insisiengmay, and H. J. Rim.** 2005. Mixed infections with *Opisthorchis viverrini* and intestinal flukes in residents of Vientiane municipality and Saravane province in Laos. *J. Helminthol.* **79**:283–289.

16. **Chandler, A. C.** 1942. First record of a case of human infection with tapeworms of the genus, *Mesocestoides*. *Am. J. Trop. Med.* **22**:493–497.

17. **Chermette, R., J. Bussieras, J. Marionneau, E. Boyer, C. Roubin, B. Prophette, H. Maillard, and B. Fabiani.** 1995. Invasive cysticercosis due to *Taenia crassiceps* in an AIDS patient. *Bull. Acad. Natl. Med.* **179**:777–780.

18. **Chieffi, P. P., O. H. Leite, R. M. Dias, D. M. Torres, and A. C. Mangini.** 1990. Human parasitism by *Phagicola* sp. (Trematoda, Heterophyidae) in Cananeia, Sao Paulo state, Brazil. *Rev. Inst. Med. Trop. Sao Paulo* **32**:285–288.

19. **Chung, P. R., Y. Jung, and D. S. Kim.** 1997. Planorbid snails as potential molluscan intermediate host of a human intestinal fluke, *Neodiplostomum seoulensis* (Trematoda: Diplostomatidae) in Korea. *Southeast Asian J. Trop. Med. Public Health* **28**:S201–S208.

20. **Clavel, A., M. D. Bargues, F. J. Castillo, M. D. Rubio, and S. Mas-Coma.** 1997. Diplogonoporiasis presumably introduced into Spain: first confirmed case of human infection acquired outside the Far East. *Am. J. Trop. Med. Hyg.* **57**:317–320.

21. **Counselman, K., C. Field, G. Lea, B. Nickol, and R. Neafie.** 1989. *Moniliformis moniliformis* from a child in Florida. *Am. J. Trop. Med. Hyg.* **41**:88–90.

22. **Cunnac, M., J. F. Magnaval, D. Cayarci, and P. Leophonte.** 1988. Three cases of human synganiasis in Guadeloupe. *Rev. Pneumol. Clin.* **44**:140–142.

23. **da Costa, J. C., M. L. Delgado, P. Vieira, A. Alfonso, B. Conde, and J. H. Cross.** 2005. Syngamoniasis in tourist. *Emerg. Infect. Dis.* **11**:1976–1977.

24. **Dennett, X., S. J. Siejka, J. R. H. Andrews, I. Beveridge, and D. M. Spratt.** 1998. Polymyositis caused by a new genus of nematode. *Med. J. Aust.* **168**:226–227.

25. **DeWeese, M.W., W. F. Murrah, and S. B. Caruthers.** 1962. Case report of a tongue worm (*Linguatula serrata*) in the anterior chamber. *Arch. Ophthalmol.* **68**:587.

26. **Drabick, J. J.** 1987. Pentastomiasis. *Rev. Infect. Dis.* **9**:1087–1094.

27. **Eberhard, M. L., and C. Busillo.** 1999. Human *Gongylonema* infection in a resident of New York City. *Am. J. Trop. Med. Hyg.* **61**:51–52.

28. **Eom, K. S., S. H. Kim, and H. J. Rim.** 1992. Second case of human infection with *Mesocestoides lineatus* in Korea. *Kisaengchung hak Chapchi* **30**:147–150.

29. **Espindola, N. M., A. H. Iha, I. Fernandes, O. M. Takayanagui, L. D. R. Machago, J. A. Livramento, A. A. M. Maia, J. M. Peralta, and A. J. Vaz.** 2005. Cysticercosis immunodiagnosis using 18- and 14-kilodalton proteins from *Taenia crassiceps* cysticercus antigens obtained by immunoaffinity chromatography. *J. Clin. Microbiol.* **43**:3178–3184.

30. **Fernandes, B. J., J. D. Cooper, J. B. Cullen, R. S. Freeman, A. C. Ritchie, A. A. Scott, and P. F. Stuart.** 1976. Systemic infection with *Alaria americana* (Trematoda). *Can. Med. Assoc. J.* **115**:1111–1114.

31. **Francois, A., L. Favennec, C. Cambon-Michot, I. Gueit, N. Biga, F. Tron, P. Brasseur, and J. Hemet.** 1998. *Taenia crassiceps* invasive cysticercosis: a new human pathogen in acquired immunodeficiency syndrome? *Am. J. Surg. Pathol.* **22**:488–492.

32. **Fredricks, D. N., J. A. Jolley, P. W. Lepp, J. C. Kosek, and D. A. Relman.** 2000. *Rhinosporidium seeberi*: a human pathogen from a novel group of aquatic protistan parasites. *Emerg. Infect. Dis.* **6**:273–282.

33. **Fuentes, M. V., M. T. Galan-Puchades, and J. B. Malone.** 2003. A new case report of human *Mesocestoides* infection in the United States. *Am. J. Trop. Med. Hyg.* **68**:566–567.

34. **Goldsmith, J. M., and M. Muir.** 1972. *Inermicapsifer madagascariensis* (Davaine, 1870) Baer, 1956 (Platyhelminths; Cestoda) as a parasite of man in Rhodesia. *Cent. Afr. J. Med.* **18**:205–207.

35. **Gonzalez, N. I., J. M. Diaz, and F. F. Nunez.** 1996. Infection by *Inermicapsifer madagascariensis* (Davaine, 1870); Baer, 1956. A report of 2 cases. *Rev. Cubana Med. Trop.* **48**:224–226.

36. **Gordon, J. A., C. M. Ross, and H. Affleck.** 1969. Abdominal emergency due to an oesophagostome. *Ann. Trop. Med. Parasitol.* **63**:161–164.

37. **Haaf, E., and A. H. van Soest.** 1964. Oesophagostomiasis in man in North Ghana. *Trop. Geogr. Med.* **49**:49.

38. **Hayashi, K., H. Tahara, K. Yamashita, K. Kuroki, R. Matsushita, S. Yamamoto, T. Hori, S. Hirono, Y. Nawa, and H. Tsubouchi.** 1999. Hepatic imaging studies on patients with visceral larva migrans due to probable *Ascaris suum* infection. *Abdom. Imaging* **24**:465–469.

39. **Hong, S. J.** 2000. A human case of *Stellantchasmus falcatus* infection in Korea. *Korean J. Parasitol.* **38**:25–27.

40. **Hong, S. J., C. K. Chung, D. H. Lee, and H. C. Woo.** 1996. One human case of natural infection by *Heterophyopsis continua* and three other species of intestinal trematodes. *Korean J. Parasitol.* **34**:87–89.

41. **Hong, S. J., H. C. Woo, and J. Y. Chai.** 1996. A human case of *Plagiorchis muris* (Tanabe, 1922: Digenea) infection in the Republic of Korea: freshwater fish as a possible source of infection. *J. Parasitol.* **82**:647–649.

42. **Hoogstraten, J., and W. C. Young.** 1975. Meningoencephalomyelitis due to the saprophagous nematode *Micronema deletrix*. *Can. J. Neurol. Sci.* **21**:121–126.

43. **Huh, S., S. U. Lee, and S. C. Huh.** 1994. A follow-up examination of intestinal parasitic infections of the Army

soldiers in Whachon-gun, Korea. *Korean J. Parasitol.* **32:**61–63.

44. Inatomi, Y., T. Murakami, M. Tokunaga, K. Ishiwata, Y. Nawa, and M. Uchino. 1999. Encephalopathy caused by visceral larva migrans due to *Ascaris suum. J. Neurol. Sci.* **164:**195–199.

45. Jozefzoon, L. M. E., and B. F. J. Oostburg. 1994. Detection of hookworm and hookworm-like larvae in human fecocultures in Suriname. *Am. J. Trop. Med. Hyg.* **51:**501–505.

46. Kaewkes, S., D. B. Elkins, and M. R. Haswell-Elkins. 1991. *Phaneropsolus spinicirrus* n.sp. (Digenea: Lecithodendriidae), a human parasite in Thailand. *J. Parasitol.* **77:**514–516.

47. Kakihara, D., K. Yoshimitsu, K. Ishigami, H. Irie, H. Aibe, T. Tajima, K. Shinozaki, A. Nishie, T. Nakayama, K. Hayashida, M. Nakamuta, H. Nawata, and H. Honda. 2004. Liver lesions of visceral larva migrans due to *Ascaris suum* infection: CT findings. *Abdom. Imaging* **29:**598–602.

48. Katz, L. B., and H. Ehya. 1990. Liesegang rings in renal cyst fluid. *Diagn. Cytopathol.* **6:**197–200.

49. Kent, M. L. 2000. Marine netpen farming leads to infections with some unusual parasites. *Int. J. Parasitol.* **30:**321–326.

50. Klinker, H., K. Tintelnot, R. Joeres, J. Muller, U. Gross, H. Schmidtrotte, P. Landwehr, and E. Richter. 1992. *Taenia crassiceps* infection in AIDS. *Dtsch. Med. Wochenschr.* **117:**133–138.

51. Kook, J., Y. Nawa, S. H. Lee, and J. Y. Chai. 1998. Pathogenicity and lethality of a minute intestinal fluke, *Neodiplostomum seoulense*, to various strains of mice. *J. Parasitol.* **84:**1178–1183.

52. Kruger, F. J. 1990. Frequency and possible consequences of hybridization between *Schistosoma haematobium* and *S. mattheei* in the Eastern Transvaal Lowveld. *J. Helminthol.* **64:**333–336.

53. Kruger, F. J., and A. C. Evans. 1990. Do all human urinary infections with *Schistosoma mattheei* represent hybridization between *S. haematobium* and *S. mattheei*? *J. Helminthol.* **64:**330–332.

54. Lamothe-Argumedo, R., S. P. Diaz-Camacho, and Y. Nawa. 2003. The first human case in Mexico of conjunctivitis caused by the avian parasite, *Philophthalmus lacrimosus. J. Parasitol.* **89:**183–185.

55. Lazo, R. F., E. Hidalgo, J. E. Lazo, A. Bermeo, M. Llaguno, J. Murillo, and V. P. A. Teixeira. 1999. Ocular linguatuliasis in Ecuador: case report and morphometric study of the larva of *Linguatula serrata. Am. J. Trop. Med. Hyg.* **60:**405–409.

56. Lee, S. H., and J. Y. Chai. 2001. A review of *Gymnophalloides seoi* (Digenea: Gymnophallidae) and human infections in the Republic of Korea. *Korean J. Parasitol.* **39:**85–118.

57. Lee, S. H., J. Y. Chai, H. J. Lee, S. T. Hong, J. R. Yu, W. M. Sohn, W. G. Kho, M. H. Choi, and Y. J. Lim. 1994. High prevalence of *Gymnophalloides seoi* infection in a village on a southwestern island of the Republic of Korea. *Am. J. Trop. Med. Hyg.* **51:**281–285.

58. Lie, K. J. 1951. Some human flukes from Indonesia. *Doc. Neerl. Indones. Morb. Trop.* **3:**105–116.

59. Lom, J. 1990. Phylum Myxozoa, p. 36–52. *In* L. Margulis, J. O. Corliss, M. Melkonian, and D. J. Chapman (ed.), *Handbook of Protoctista.* Jones and Bartlett Publishers, Boston, Mass.

60. Lom, J., and I. Dykova. 1994. *Protozoan Parasites of Fishes*, p. 159–235. Elsevier Science Publishers B.V., Amsterdam, The Netherlands.

61. Lopera, R. D., R. D. Melendez, I. Fernandez, J. Sirit, and M. P. Perera. 1989. Orbital hydatid cyst of *Echinococcus oligarthrus* in a human in Venezuela. *J. Parasitol.* **75:** 467–470.

62. MacLean, J. D. 1998. The North American liver fluke, *Metorchis conjunctus*, p. 243–256. *In* W. M. Scheld, W. A. Craig, and J. M. Hughes (ed.), *Emerging Infections*, 2nd ed. ASM Press, Washington, D.C.

63. MacLean, J. D., J. R. Arthur, B. J. Ward, T. W. Gyorkos, M. A. Curtis, and E. Kokoskin. 1996. Common-source outbreak of acute infection due to the North American liver fluke *Metorchis conjunctus. Lancet* **347:**154–158.

64. Mairena, H., M. Solano, and W. Venegas. 1989. Human dermatitis caused by a nymph of *Sebekia. Am. J. Trop. Med. Hyg.* **41:**352–354.

65. Maruyama, H., Y. Nawa, S. Noda, and T. Mimori. 1997. An outbreak of ascariasis with marked eosinophilia in the southern part of Kyushu District, Japan, caused by infection with swine ascaris. *Southeast Asian J. Trop. Med. Public Health* **28:**194–196.

66. McClelland, R. S., D. M. Murphy, and D. K. Cone. 1997. Report of spores of *Henneguya salminicola* (Myxozoa) in human stool specimens: possible source of confusion with human spermatozoa. *J. Clin. Microbiol.* **35:**2815–2818.

67. McDonald, H. R., K. R. Kazacos, H. Schatz, and R. N. Johnson. 1994. Two cases of intraocular infection with *Alaria* mesocercaria. *Am. J. Ophthalmol.* **117:**447–455.

68. McDonald, T. E., and L. Margolis. 1995. Synopsis of the parasites of fishes of Canada: supplement (1978–1993). *Can. Spec. Publ. Fish. Aquat. Sci.* **122:**1–265.

69. Narr, L. L., J. G. O'Donnell, B. Libster, P. Alessi, and D. Abraham. 1996. Eustrongylidiasis—a parasitic infection acquired by eating live minnows. *J. Am. Osteopath Assoc.* **96:**400–402.

70. Nosanchuk, J. S., S. E. Wade, and M. Landolf. 1995. Case report and description of parasite in *Mammomonogamus laryngeus* (human syngamosis) infection. *J. Clin. Microbiol.* **33:**998–1000.

71. Pfaller, M. A., and D. J. Diekema. 2005. Unusual fungal and pseudofungal infections of humans. *J. Clin. Microbiol.* **43:**1495–1504.

72. Pit, D. S., W. De Graaf, H. Snoek, S. J. De Vlas, S. M. Baeta, and A. M. Polderman. 1999. Diagnosis of *Oesophagostomum bifurcum* and hookworm infection in humans: day-to-day and within-specimen variation of larval counts. *Parasitology* **118:**283–288.

73. Pit, D. S., F. E. Rijcken, E. C. Raspoort, S. M. Baeta, and A. M. Polderman. 1999. Geographic distribution and epidemiology of *Oesophagostomum bifurcum* and hookworm infections in humans in Togo. *Am. J. Trop. Med. Hyg.* **61:**951–955.

74. Poinar, G. O., Jr., and E. P. Hoberg. 1988. *Mermis nigrescens* (Mermithidae: Nematoda) recovered from the mouth of a child. *Am. J. Trop. Med. Hyg.* **39:**478–479.

75. Polderman, A. M., H. P. Krepel, S. Baeta, J. Blotkamp, and P. Gigase. 1991. Oesophagostomiasis, a common infection

of man in northern Togo and Ghana. *Am. J. Trop. Med. Hyg.* **44**:336–344.

76. Prociv, P., J. Walker, L. J. Crompton, and S. C. Tristram. 1990. First record of human acanthocephalan infections in Australia. *Med. J. Aust.* **152**:215–216.

77. Radomyos, P., B. Radomyos, and A. Tungtrongchitr. 1994. Multi-infection with helminths in adults from northeast Thailand as determined by post-treatment fecal examination of adult worms. *Trop. Med. Parasitol.* **45**:133–135.

78. Ross, R. A., D. I. Gibson, and E. A. Harris. 1989. Cutaneous oesophagostomiasis in man. *J. Helminthol.* **63**:261–262.

79. Schindler, A. R., J. M. de Gruijter, A. M. Polderman, and R. B. Gasser. 2005. Definition of genetic markers in nuclear ribosomal DNA for a neglected parasite of primates, *Ternidens deminutus* (Nematoda: Strongylida)—diagnostic and epidemiological implications. *Parasitology* **131**:539–546.

80. Schmidt, G. D. 1971. Acanthocephalan infections of man with two new records. *J. Parasitol.* **57**:582–584.

81. Self, J. T. 1972. Pentastomiasis: host responses to larval and nymphal infections. *Trans. Am. Microsc. Soc.* **91**:2.

82. Shadduck, J. A., J. Ubelaker, and V. Q. Telford. 1979. *Micronema deletrix* meningoencephalitis in an adult man. *Am. J. Clin. Pathol.* **72**:640–643.

83. Silva, V., C. N. Pereira, L. Ajello, and L. Mendoza. 2005. Molecular evidence for multiple host-specific strains in the genus *Rhinosporidium*. *J. Clin. Microbiol.* **43**:1865–1868.

84. Sohn, W. M., J. Y. Chai, and S. H. Lee. 1989. A human case of *Stellantchasmus falcatus* infection. *Kisaengchung hak Chapchi* **27**:277–279.

85. Spratt, D. M., I. Beveridge, J. R. H. Andrews, and X. Dennett. 1999. *Haycocknema perplexum* n. g., n. sp. (Nematoda: Robertdollfusidae): an intramyofibre parasite in man. *Syst. Parasitol.* **43**:123–131.

86. Sukontason, K., S. Piangjai, K. Sukontason, and U. Chaithong. 1999. Potassium permanganate staining for differentiation the surface morphology of *Opisthorchis viverrini*, *Haplorchis taichui*, and *Phaneropsolus bonnei* eggs. *Southeast Asian J. Trop. Med. Public Health* **30**:371–374.

87. Sukontason, K., P. Unpunyo, K. L. Sukontason, and S. Piangjai. 2005. Evidence of *Hailorchis taichui* infection as pathogenic parasite: three case reports. *Scand. J. Infect. Dis.* **37**:388–390.

88. Sun, T., A. Turnbull, P. H. Lieberman, and S. S. Sternberg. 1986. Giant kidney worm (*Dioctophyma renale*) infection mimicking retroperitoneal neoplasm. *Am. J. Surg. Pathol.* **10**:508–512.

89. Tuur, S. M., A. M. Nelson, D. W. Gibson, R. C. Neafie, F. B. Johnson, F. K. Mostofi, and D. H. Connor. 1987. Liesegang rings in tissue. How to distinguish Liesegang rings from the giant kidney worm, *Dioctophyma renale*. *Am. J. Surg. Pathol.* **11**:598–605.

90. Vermeij, J. J., A. M. Polderman, M. C. Wimmenhove, and R. B. Gasser. 2000. PCR assay for the specific amplification of *Oesophagostomum bifurcum* DNA from human faeces. *Int. J. Parasitol.* **30**:137–142.

90a. Waikagul, J., P. Dekumyoy, T. Yoonuan, and R. Praevanit. 2006. Conjuctiva philophthalmosis: a case report in Thailand. *Am. J. Trop. Med. Hyg.* **74**:848–849.

91. Welchman, J. R. 1966. Helminthic abscess of the bowel. *Br. J. Radiol.* **39**:372.

92. Wittner, M., J. W. Turner, G. Jacquette, L. R. Ash, M. P. Salgo, and H. B. Tanowitz. 1989. Eustrongylidiasis—a parasitic infection acquired by eating sushi. *N. Engl. J. Med.* **320**:1124–1126.

93. Xiao, N., J. Qiu, M. Nakao, T. Li, W. Yang, X. Chen, P. M. Schantz, P. S. Craig, and A. Ito. 2005. *Echinococcus shiquicus*, a new species from the Qinghai-Tibet plateau region of China: discovery and epidemiological implications. *Parasitol. Int.* 4 Dec. [Epub ahead of print.]

94. Xuan, L. T., M. T. Anantaphruti, P. A. Tuan, X. L. Tu, and T. V. Hien. 2003. The first human infection with *Bertiella studeri* in Vietnam. *Southeast Asian J. Trop. Med. Public Health* **34**:298–300.

95. Yokoyama, H., K. Ogawa, and H. Wakabayashi. 1995. *Myxobolus cultus* n. sp. (Myxosporea: Myxobolidae) in the goldfish *Carassius auratus* transformed from the actinosporean stage in the oligochaete *Branchiura sowerbyi*. *J. Parasitol.* **81**:446–451.

96. Ziem, J. B., I. M. Kettenis, A. Bayita, E. A. Brienen, S. Dittoh, J. Horton, A. Olsen, P. Magnussen, and A. M. Polderman. 2004. The short-term impact of albendazole treatment on *Oesophagostomum bifurcum* and hookworm infections in northern Ghana. *Ann. Trop. Med. Parasitol.* **98**:385–390.

97. Ziem, J. B., A. Olsen, P. Magnussen, J. Horton, E. Agongo, R. B. Geskus, and A. M. Polderman. 2006. Distribution and clustering of *Oesophagostomum bifurcum* and hookworm infections in northern Ghana. *Parasitology* **132**:525–534.

Parasitic Infections in the Compromised Host

This chapter discusses some of the representative opportunistic organisms that can cause disease in immunocompromised patients. Any parasitic infection in the immunocompromised host may cause more severe symptoms; however, the organisms presented in this chapter have been identified as causing severe disease in this population.

The body has three types of host defense mechanisms: surface and mechanical factors, the humoral immune system, and the cellular immune system. Within these three groups, further distinction can be made between first-line (nonspecific) innate immunity and second-line (specific) adaptive immunity. By definition, a compromised host is one in whom normal defense mechanisms are impaired (e.g., AIDS), absent (e.g., congenital deficiencies), or bypassed (e.g., penetration of the skin barrier). These patients are becoming more common in most medical facilities and represent a growing problem in terms of diagnosis and subsequent therapy.

As seen in Table 19.1, first-line surface and mechanical factors include the skin, the cough reflex, ciliary motion (respiratory epithelium), and flushing action (urine); secondary defense mechanisms include immunoglobulins such as immunoglobulin A (IgA). In the humoral portion of the immune system, first-line defenses include complement, lysozyme, fibronectin, and interferon; secondary defense mechanisms include antibodies such as IgM and IgG. The cellular portion of the immune system includes first-line defenses such as phagocytes (polymorphonuclear leukocytes [neutrophils and eosinophils] and monocytes) and natural killer cells (T cells); the second line of defense is defined as cell-mediated immunity. Specific problems and potential causes that can arise include penetration of the skin barrier (needle penetrations, burns); splenectomy (impaired IgM antibody production); aspiration of stomach contents, impaired cough reflex, impaired microbicidal clearance by alveolar macrophages, defects in complement activity, etc. (alcoholism); neutropenia (use of immunosuppressive drugs); problems with inhibited phagocytosis (rheumatic diseases); neutropenia, cell-mediated immunity defects, and antibody response defects (renal failure); skin and mucous membrane defects due to poor protein production as well as diminished IgA levels in the gut (malnutrition); and diminished humoral and cellular immunity defenses (AIDS, other underlying infectious diseases). To determine possible host defense defects, a number of procedures can be used (Table 19.2).

506

Table 19.1 Host defense mechanisms

Immunity type	Factor		
	Surface and mechanical	**Humoral (antibody dependent)**	**Cellular (antibody independent)**
Nonspecific	Skin (intact) (intravenous lines, obstruction of ureters, bronchi) Cough reflex Ciliary motion in respiratory epithelium Flushing action (urine)	Complement Lysozyme Fibronectin Interferon	Phagocytes (neutrophils) Chemotaxis Phagocytosis (monocytes, eosinophils) Natural killer cells
Specific	Immunoglobulins (IgA) (antibody found in secretions)	Antibodies B cells (IgG, IgM) (found in serum and extravascular spaces) IgE (affinity for mast cells, basophils, and eosinophils)	Cell-mediated immunity Sensitized cells (cytotoxic T cells, activated macrophages, natural killer cells)
Additional comments		IgG and IgM fix complement; initiate complement-mediated lysis of the target cell. IgG and IgE bind to macrophages, eosinophils, and mast cells via the Fc portion and to the target cell via the Fab portion	Macrophage functions include detection of organism invasion, restriction of organism spread, recruitment of immune cells, action as accessory cells in lymphocyte activation and as effector cells in cell-mediated immunity

Humans have very effective defense mechanisms to protect themselves against foreign invaders. One system consists of surface and mechanical barriers, pH, temperature, phagocytosis, and nonspecific inflammation (innate resistance). The other system is induced and includes specific products that recognize foreign invaders (adaptive immunity). The two major components of the adaptive immune response are humoral (antibody) and cellular (sensitized cells); B lymphocytes (B cells) are responsible for the humoral response, and T lymphocytes (T cells) are responsible for the cellular response. The human immune system is very complex, with different classes of antibody and various subsets of T cells that have different functions as immune effectors. Unfortunately, the immune system is not always protective and under certain circumstances can be directed toward the human body, not the foreign invader.

Within the context of this background information on the immune system, various parasitic infections are discussed, with emphasis on infections in the immunocompromised patient. With the continued increase in the number of these patients, every institution will eventually provide medical care for someone in this category and laboratories will continue to perform more procedures for the diagnosis of opportunistic infections. Diagnostic procedures for the diagnosis of parasitic infections in the compromised host are presented in Table 19.3, and clinical findings in normal and compromised hosts are described in Table 19.4.

Parasitic infections in individuals with a normal immune system are not uncommon. Not only are they unpleasant and debilitating, but they can be fatal. However, individuals with immune system defects have an abnormally high susceptibility to infections with nonvirulent and minimally pathogenic organisms. Many of these individuals contract parasitic infections in addition to suffering numerous infectious episodes with bacterial, viral, and fungal organisms. Immune system deficiencies can be attributed to congenital absence, abnormal development, malignancy, therapy with cytotoxic drugs, irradiation, or infections, such as with human immunodeficiency virus (HIV). Of particular interest are patients with AIDS.

Although any parasitic infection may be more severe in the immunosuppressed host, certain organisms tend to produce greater pathologic sequelae in these patients, while other organisms occur with a higher frequency in individuals with certain immune system deficiencies. Among the implicated organisms discussed in this chapter are *Entamoeba histolytica*, free-living amebae, *Giardia lamblia*, *Toxoplasma gondii*, *Cryptosporidium* spp., *Cyclospora cayetanensis*, *Isospora belli*, *Sarcocystis* spp., microsporidia, *Leishmania* spp., *Strongyloides stercoralis*, and *Sarcoptes scabiei* (the agent of crusted scabies). It is important for the laboratorian and clinician to be aware of problems that these organisms can cause in compromised patients and to be aware of the proper diagnostic techniques and their clinical relevance.

Table 19.2 Selected procedures for determination of host defense defects[a]

Defect	Procedure	
	Screening	**Diagnostic**
Inflammation	White blood cell count and differential C$_{50}$ hemolytic complement	Complement levels
		Inhibitor levels
		Complement-fixing immune complexes
Phagocytosis	White blood cell count and morphology	Phagocytic index
	Immunophenotyping	Bactericidal tests
		Nitroblue tetrazolium test
		Cell surface glyproteins
		Lymph node biopsy
Humoral antibody	Immunoelectrophoresis	IgE levels in serum
	Immunoglobulin levels in serum	Secretory IgA
		B-cell levels, subsets
		Isohemagglutinins
		Purine-metabolizing enzymes
Cellular sensitivity	Skin tests	Sheep erythrocyte rosettes
	Quantitation of T cells and T-cell subsets	Mitogen responses
		Lymph node biopsy
		CT, MRI of thymus

[a] Adapted from reference 99.

Entamoeba histolytica

Entamoeba histolytica is the cause of amebiasis (Tables 19.3 and 19.4; Figure 19.1). Two types of disease, reviewed here, are particularly helpful when one is discussing serologic test results. "Intestinal disease" indicates that the organisms are confined to the gastrointestinal tract, with no gross or microscopic invasion through the mucosal lining. "Extraintestinal disease" denotes invasion of the mucosal lining of the gastrointestinal tract, the organisms having passed through the mucosal lining, entered the bloodstream, and been carried to other body tissues, particularly the liver. In patients in whom the organisms have begun to invade the mucosal lining, interpretation of serologic tests may be difficult. In this situation, the terms "intestinal" and "extraintestinal" do not strictly apply. These differences should be understood when one is discussing the clinical interpretation of serologic tests for amebiasis.

Although a large number of people throughout the world are infected with this organism, only a small percentage develop clinical symptoms. Morbidity and mortality due to *E. histolytica* vary depending on the geographic area, organism strain, and patient immune status. For many years, the issue of pathogenicity has been very controversial; some thought that what was called *E. histolytica* was really two separate species of *Entamoeba*, one being pathogenic and causing invasive disease and the other being nonpathogenic and causing mild or asymptomatic infections. Others thought that all organisms designated *E. histolytica* were potentially pathogenic, with symptoms depending on the result of host or environmental factors, including the intestinal flora.

In 1961, with the development of successful axenic culture methods requiring no bacterial coculture, sufficient organisms could be obtained for additional studies. Approximately 15 years later, reports indicated that *E. histolytica* clinical isolates could be classified into groups by starch gel electrophoresis and review of banding patterns related to specific isoenzymes. Sargeaunt concluded from this work that there are pathogenic and nonpathogenic strains (zymodemes) of *E. histolytica* that can be differentiated by isoenzyme analysis. On the basis of analysis of thousands of clinical isolates, he also concluded that the zymodeme patterns were probably genetic rather than phenotypic.

Research emphasis has been on the molecular differences between pathogenic *E. histolytica* and nonpathogenic *E. dispar*. However, the molecules considered the most important for host tissue destruction (amebapore, galactose/N-acetylgalactosamine-inhibitable lectin, and cysteine proteases) can be found in both organisms. Pathogenicity differences may be related to the composi-

Table 19.3 Diagnosing parasitic infection in the compromised host

Organism	Size	Specimen	Procedure	Method of examination[a]	Laboratory findings[a]	Comments[a,b]
Entamoeba histolytica Intestinal disease	Trophozoite: usual range, 15–20 μm; range, 10–60 μm. Cyst: usual range, 12–15 μm; range, 8.5–20 μm. *Note:* These sizes refer to organisms seen on the permanent stained smear. There may be as much as 1–1.5 μm of shrinkage of organisms as a result of reagents.	Stool	O&P examination (minimum of concentration and permanent stained smear)	Concentration (high dry, ×400)	Trophozoites or cysts	A series of 3 stools (collected every other day) is recommended for the O&P examination.
Note: Pathogenic *E. histolytica* can now be differentiated from nonpathogenic *E. dispar* using immunoassays; morphologic characteristics are identical, *E. histolytica* containing RBCs in the trophozoite could be identified as pathogenic true *E. histolytica.*			Immunoassays	Depends on format	Presence or absence of antigen	Can detect *Entamoeba histolytica* or the *E. histolytica/E. dispar* group
		Sigmoidoscopic biopsy or scrapings	Permanent stained smear and routine histologic test	Permanent stain (oil immersion, ×1,000)	Trophozoites	The permanent stained smear is mandatory for a complete O&P examination.
		Serum	Amebic serologic tests	IHA, IFA, CIE, ID, ELISA	Not recommended for intestinal disease; negative or possible low titer, indicating present or past exposure	At least 2 different serologic tests (methods) should be performed so results can be compared.
			Radiographic studies	Barium	Intestinal abnormalities (may also be normal)	
Extraintestinal disease	Trophozoite stage only	Stool	O&P examination	See above	No organisms recovered in about 50% of cases	There may be no evidence of GI disease before or during extraintestinal invasion.
		Sigmoidoscopic biopsy or scrapings	Permanent stained smear and routine histologic test	See above	No organisms recovered in about 50% of cases	
		Serum	Amebic serologic tests (antibody); IHA, EIA reagents available	See above	Positive titers may be lower in compromised patients	At least 2 serologic tests (methods) should be performed so that results can be compared.
			Scans		Evidence of space-occupying lesion	
Free-living amebae *Naegleria fowleri*	Biphasic (amebic and flagellate forms). Trophozoite: usual range, 8–15 μm, amebic form, lobate pseudopodia. Cyst: not present in tissue, small, smooth, rounded.	CSF	Centrifugation, wet or stained preparation; agar plates with bacterial overlay (culture)	Wet: low (×100) and high dry (×400); stained: oil immersion (×500, or ×1,000); culture: low (×100) and high dry (×400) on material from culture plate surface	Trophozoites in specimens; trophozoites and cysts in culture	Since all these organisms are normally free-living, they will survive shipment via mail to any laboratory for subsequent examination and/or culture; various culture temperatures are recommended in different situations (refer to chapter 32).

(continued)

Table 19.3 Diagnosing parasitic infection in the compromised host (continued)

Organism	Size	Specimen	Procedure	Method of examination[a]	Laboratory findings[a]	Comments[a,b]
Acanthamoeba spp.	Trophozoite: large, 15–25 μm, amebic form, filiform pseudopodia. Cyst: present in tissue, large with wrinkled double wall.	CSF, eye specimens, cutaneous biopsy	Centrifugation, wet or stained preparation; agar plates with bacterial overlay (culture); histology	As above, histology and routine microscopic exam	Trophozoites and/or cysts in specimens and culture	Agar plates with bacterial overlay (culture) more sensitive than wet preparations
Balamuthia mandrillaris	Trophozoite: large, 12–60 μm, amebic form, fingerlike pseudopodia. Cyst: large, 6–30 μm, with wrinkled double wall.	CSF, eye specimens, cutaneous biopsy	Centrifugation, wet or stained preparation; cell culture only, will not grow on routine agar culture with bacterial overlay; histology	All of the above, including cell culture if available	Trophozoites and/or cysts in specimens and cell culture	Becoming more widely recognized as causing same types of disease as Acanthamoeba spp.
Giardia lamblia	Trophozoite: usual range, 10–12 μm long by 5–10 μm wide. Cyst: usual range, 11–14 μm long by 7–10 μm wide	Stool	O&P examination	Concentration (high dry, ×400), permanent stain (oil immersion, ×1,000)	Trophozoites or cysts (may not always be detected)	The parasite may not be recovered from stool even after 4 or 5 examinations on different specimens. This does not mean that the diagnostic techniques are inadequate or that the organisms are missed by inexperienced personnel. Additional diagnostic techniques (including biopsy) may have to be used. Immunoassays available and now being used as orderable tests (FA, EIA, immunochromatographic formats).
			Immunoassays	Depends on format	Presence or absence of antigen or actual organisms	
		Duodenal aspirate	Wet preparations, permanent stained smears	See above	Trophozoites	
		Entero-Test capsule	Wet preparations, permanent stained smears	See above	Trophozoites	
		Biopsy	Routine histologic tests and special stains		Possible upper GI tract changes and/or trophozoites	
		Serum	Serologic tests	Experimental (not generally available)		

Organism	Description	Specimen	Test	Method/Stain	Finding	Comments
Toxoplasma gondii	Tachyzoite: 4–6 μm long by 2–3 μm wide. Bradyzoite: individual tissue cysts may contain many hundreds of organisms. Cyst: may range up to 50–75 μm.	Serum	Serologies	IHA, IFA, ELISA, Sabin-Feldman dye test	Results are often sufficient to confirm diagnosis	Patients with positive IgM serologic test results may have Epstein-Barr virus infections. There may be nonspecific stimulation of other antibody-producing cells. Although organism isolation techniques with animals and/or tissue culture may be used, these results must be evaluated within the context of the total clinical picture.
		Biopsy	Routine histologic test		Possible recovery of bradyzoites	Organism may not be etiologic agent of disease
			Tissue culture or animal isolation		Possible recovery of tachyzoites	Organism may not be etiologic agent of disease
		Spinal fluid	Organism isolation (direct, tissue culture, or animal isolation)		Presence of tachyzoites	Considered confirmatory
			Serologic tests		Confirmatory if higher in CSF than in serum	Rare finding
Cryptosporidium spp.	Developmental stages within brush border of intestine Cells: 1–2 μm; oocysts in stool: 4–6 μm.	Stool	Acid-fast stains (may need modified KOH digestions)	Permanent stain (high dry, ×400; confirm on oil immersion, ×1,000)	Oocysts visible	Regardless of staining technique used on stool, the more normal the stool specimen (formed), the greater the chances that artifacts will present problems in identifying organisms and fewer organisms will be present. Specimens representing classic diarrhea will allow easier recovery and identification of organisms. Numbers of organisms will vary from day to day.
			Immunoassays	Depends on format	Presence or absence of antigen or actual oocysts	
		Intestinal mucosal biopsy	Routine histologic test	Hematoxylin and eosin stain	Developmental stages (1 μm) along brush border	
			EM	Routine EM	May see developmental stages or possibly oocysts	Very specific but not very sensitive approach
		Serologic specimen		Experimental only (not generally available)		Used for epidemiologic studies

(continued)

Table 19.3 Diagnosing parasitic infection in the compromised host (*continued*)

Organism	Size	Specimen	Procedure	Method of examination[a]	Laboratory findings[a]	Comments[a,b]
Cyclospora cayeta-nensis	Oocyst: 8–10 μm	Stool	Acid-fast stains (modified)	Permanent stain (high dry, ×400; confirm on oil immersion, ×1,000)	Oocysts visible	Owing to morphologic similarities to *Cryptosporidium* spp., it is important to measure the oocysts. There is tremendous staining variability of the oocysts with acid-fast stains (from clear to deep purple). Oocysts appear wrinkled. Oocysts in stool are unsporulated and do not contain any internal definition or structure.
			Wet preparation examinations; oocysts will auto-fluoresce	Green with 450–490 DM excitation filter, blue with 365 DM excitation filter. Wet preparation (high dry, ×400).	Oocysts visible as green or blue circles (interior of oocyst does not fluoresce) (1+ to 3+)	Used as a screening method
		Intestinal muco-sal biopsy	Routine histologic test	Hematoxylin and eosin stain	Organisms resembling coccidia found in jejunal enterocytes	Developmental location will differentiate from *Cryptosporidium* spp.
			EM	Routine EM	May see developmental stages or possibly oocysts	
Isospora (Cystoisospora) belli	Oocyst: 30 by 20 μm, development within the intestinal mucosal cells; oocysts in stool range from undivided mass of protoplasm to those containing fully developed sporocysts and sporozoites	Stool	O&P examination (concentration is important; PVA distorts morphology)	Concentration (low, ×100, and high dry, ×400)	Oocysts (may be in various stages of development)	Although these organisms are thought to be rare, some workers speculate that oocysts are occasionally missed during a stool examination. There is also the possibility that a biopsy specimen is positive while the stool examination is negative.
			Permanent stained smear	Acid-fast stains	Oocysts (as above). High dry, ×400.	
		Intestinal muco-sal biopsy	Routine histologic test	Hematoxylin and eosin stain, high dry (×400)	Developmental stages within the mucosal cells	

Organism	Description	Specimen	Test	Look for	Comments
Sarcocystis suihominis, S. bovihominis	Oocyst thin walled, contains 2 mature sporocysts, each with 4 sporozoites; sporocysts are 9–16 μm long by 7.5–12 μm wide	Stool	O&P examination (concentration important; PVA may distort morphology)	Oocysts or sporocysts	Not commonly seen; size can overlap that of *I. belli*
Other *Sarcocystis* spp.	Shapes and sizes of skeletal and cardiac muscle sarcocysts vary	Tissue biopsy	Routine histology	Developmental stages within tissue cells	Sarcocysts contain several hundred to thousands of trophozoites, each measuring 4–9 μm wide by 12–16 μm long; sarcocysts may be divided into compartments by septa (not seen in *Toxoplasma* cysts).
Microsporidia (*Brachiola, Vittaforma, Encephalitozoon, Enterocytozoon, Pleistophora, Trachipleistophora, "Microsporidium"* spp.)[a]	Spore, 1–4 μm; dividing forms within epithelial cells	Tissue biopsy	Routine histologic test	Spores, dividing forms	It is sometimes difficult to find and identify the organisms by routine histologic techniques. Because of the difficulties in identification at the light microscopy level, EM methods have been recommended. Organisms have been identified in stool, although current procedures may be somewhat difficult to interpret.
			Metheramine-silver, acid-fast, PAS-positive granule, Giemsa, hematoxylin and eosin		
			EM (Routine EM)	Spores, dividing forms	Specific but insensitive
		Stool	Permanent stained smear (Modified trichrome)	Spores	Artifacts confusing
			EM (Routine EM)		
			Direct detection methods (IFA, not yet available commercially)	Spores	Specific but insensitive
		Serologic sample	IFA, CF, microagglutination (Experimental, not generally available)	Titers	
Leishmania spp.	Amastigotes in tissue measure 1–3 μm; tend to appear larger when released from cells, such as a touch or squash preparation; will appear as dots within the cells in routine histology preparations	Blood specimen (looking at buffy coat cells); tissue biopsy (bone marrow most common)	Buffy coat preparations, centrifugation, smear preparation; culture (Novy-MacNeal-Nicolle medium) of bone marrow aspirates; Microscopic examination (high dry, ×400 power and ×1,000 oil immersion)	Presence of amastigotes in macrophages and/or tissues; presence of promastigotes (motile, flagellated) in culture	Multiple specimens may be required; cultures must not become contaminated or results may represent a false negative; controls must accompany all patient cultures.

(continued)

Table 19.3 Diagnosing parasitic infection in the compromised host (*continued*)

Organism	Size	Specimen	Procedure	Method of examination[a]	Laboratory findings[a]	Comments[a,b]
Strongyloides stercoralis	Larval stage: 180–380 by 14–20 μm, in feces; larvae (rhabditiform) have a short buccal cavity and a prominent genital primordium; both noninfective and infective larvae, adults, and eggs may be found in duodenal specimens	Stool	O&P examination (concentration important)	Direct smears and concentration (low, ×100; confirm on high dry, ×400)	Rhabditiform larvae (infective filariform larvae may also be found)	Caution should be used when working with fresh specimens that may contain infective filariform larvae. *Strongyloides* infections may be difficult to diagnose from stool specimens. Examination of material from the duodenum may be necessary, and concentrations or cultures can also be used.
		Duodenal aspirate	Baermann concentrate, Harada-Mori and petri dish cultures	As above	As above	
			Wet preparations (direct smears)	As above	Larvae (usually rhabditiform)	
		Entero-Test	Wet preparations (direct smears)	As above	Larvae (as above)	
Crusted (Norwegian) scabies (*Sarcoptes scabiei*)	Mites are microscopic; live in cutaneous burrows where female deposits eggs; mites range from 215 to 390 μm in length; eggs are 170 μm long by 92 μm wide; fecal pellets are yellow-brown	Skin scrapings	Use skin-scraping technique (see chapter 24)	Prepare slides of skin-scraping material and examine microscopically using low power (×100) and high power (×400)	Mites, eggs, or scybala (fecal pellets) may be present	Highly infectious; most common skin sites are interdigital spaces, backs of the hands, elbows, axillae, groin, breasts, umbilicus, penis, shoulder blades, small of the back, buttocks; may see kerotic excrescences on the body; itching may be absent.

[a] BAL, bronchoalveolar lavage; CF, complement fixation; CIE, counterimmunoelectrophoresis, EIA, enzyme immunoassay; EM, electron microscopy; GI, gastrointestinal; O&P, ova and parasite; PVA, polyvinyl alcohol; RBC, red blood cells.

[b] The use of immunoassays for the direct and indirect detection of some of these parasites has become much more common. Reagents are available for *Entamoeba histolytica/E. dispar*, *Entamoeba histolytica*, *Giardia lamblia*, *Cryptosporidium* spp., and the microsporidia (if not yet available, should be approved in the near future).

[c] These genera have been reported in humans; organisms that have not yet been classified to genus have been placed in the catchall genus called *Microsporidium*.

Table 19.4 Parasitic infections: clinical findings in normal and compromised hosts

Organism	Normal host	Compromised host
Entamoeba histolytica	Asymptomatic to chronic or acute colitis; extraintestinal disease may also occur (primary site: right upper lobe of liver)	Diminished immune system capacity may lead to extraintestinal disease
Free-living amebae	Patients tend to have eye infections with *Acanthamoeba* spp.; linked to poor lens care	Primary amebic meningoencephalitis (PAM); granulomatous amebic encephalitis (GAE)
Giardia lamblia	Asymptomatic to malabsorption syndrome	Certain immunodeficiencies tend to predispose an individual to infection
Toxoplasma gondii	Approximately 50% of individuals have antibody and organisms in tissue but are asymptomatic	Disease in compromised host tends to involve the central nervous system with various neurological symptoms; *Toxoplasma* encephalitis (TE)
Cryptosporidium spp.	Self-limiting infection with diarrhea and abdominal pain	Owing to autoinfective nature of life cycle, is not self-limiting, may produce fluid loss of over 10 liters/day, and there may be multisystem involvement; no known totally effective therapy
Cyclospora cayetanensis	Self-limiting infection with diarrhea (3–4 days), with relapses common	Diarrhea may persist for 12 wk or more; biliary disease has also been reported in this group, particularly in those with AIDS
Isospora (*Cystoisospora*) *belli*	Self-limiting infection with mild diarrhea or no symptoms	May lead to severe diarrhea, abdominal pain, and possible death (rare case reports); diagnosis is occasionally missed because of nonrecognition of the oocyst stage; is not seen when concentrated from polyvinyl alcohol fixative
Sarcocystis spp.	Self-limiting infection with diarrhea, or mild symptoms	Symptoms may be more severe and last longer
Microsporidia (*Brachiola*, *Vittaforma*, *Encephalitozoon*, *Enterocytozoon*, *Pleistophora*, *Trachipleistophora*, "*Microsporidium*" spp.)	Little known about these infections in the normal host; serologic evidence suggests infections may be more common than recognized	Can infect various parts of the body; diagnosis often depends on histologic examination of tissues; routine examination of clinical specimens (stool, urine, etc.) becoming more common; can probably cause death
Leishmania spp.	Asymptomatic to mild disease	More serious manifestations of visceral leishmaniasis; some cutaneous species will manifest visceral disease; difficult to treat and manage; definite coinfection with AIDS
Strongyloides stercoralis	Asymptomatic to mild abdominal complaints; can remain latent for many years owing to low-level infection maintained by internal autoinfective life cycle	Can result in disseminated disease (hyperinfection syndrome due to autoinfective nature of life cycle); abdominal pain, pneumonitis, sepsis-meningitis with gram-negative bacilli, eosinophilia; distinct link to some leukemias and lymphomas; can be fatal
Crusted (Norwegian) scabies (*Sarcoptes scabiei*)	Infections can range from asymptomatic to moderate itching	Severe infection with reduced itching response; hundreds or thousands of mites on body; infection very easily transferred to others; secondary infection very common

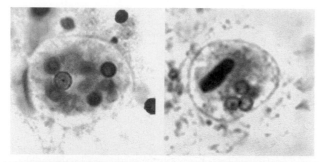

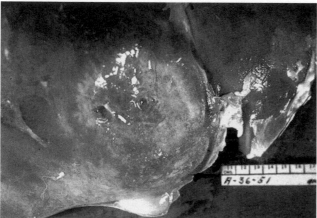

Figure 19.1 (Upper left) *Entamoeba histolytica* trophozoite; (upper right) *Entamoeba histolytica/E. dispar* cyst; (lower) gross specimen, amebic liver abscess.

tion and properties of the surface coat components (or pathogen-associated molecular patterns) and the ability of the innate immune response to recognize these components, thus eliminating the organisms. Targets of the host immune system modulation appear to be both neutrophils and macrophages, which are unable to abort the infection even when present at the site of the lesion. The total body of evidence supports the classification of the pathogenic *E. histolytica* and the nonpathogenic *E. dispar* as two distinct species. See chapter 2 for additional information.

Intestinal Disease

In patients with intestinal disease, symptoms range from none to acute or chronic amebic colitis, both of which can mimic inflammatory bowel disease. A number of factors influence the outcome of an amebic infection, including the host immune system response. At one time, it was thought that invasive strains were found only in tropical areas of the world; however, it is now apparent that there are strains within the United States that cause hepatic abscess in patients who have no history of travel outside the United States.

In examining data from AIDS patients with diarrhea in India, the authors conclude that the presence of different parasites in 62 (56.4%) of 110 stool specimens from patients with AIDS indicates that their specific diagnosis is essential. This will help initiate therapy to reduce the morbidity and mortality due to these pathogens in such patients (51).

The diagnosis of amebiasis begins with an ova and parasite examination. This procedure includes the following: the direct saline mount, which is designed to allow the mobility of the organisms to be seen; the concentration procedure, which provides a method to recover helminth eggs and larvae and protozoan cysts; and the permanent stained smear, which is the most important technique for diagnosis of the intestinal protozoa. Although motility can be seen on direct wet preparations, the material must be fresh, and a diagnosis of amebic infection should *never* be made solely from this type of examination. At the very least, the concentration procedure and permanent stained smear should be performed on every stool sample that is submitted to the laboratory for an ova and parasite examination (48).

Stools should normally be submitted to the laboratory on an every-other-day basis for a period of no more than 10 days. Although for many years the recommended minimum number of stools to be examined has been three, some laboratories are accepting two specimens. When intestinal amebiasis is suspected, six specimens may be submitted over a 15-day period. Multiple specimens are recommended because populations of intestinal protozoa tend to be cyclic. That is, the specimen may be negative on collection day 1 but positive by collection day 3. It is also possible for only a nonpathogen to be found on collection day 1 but pathogenic organisms to be recovered from the second or third specimen. If nonpathogenic protozoa are found, there is always the possibility of finding pathogenic organisms on subsequent examinations.

Presumptive identification of the organism as *Entamoeba histolytica/E. dispar* is based on the typical nuclear morphology (evenly arranged chromatin on the membrane and the presence of a small, central karyosome). The presence of red blood cells within the trophozoite cytoplasm would provide definitive identification as true *E. histolytica* (rather than nonpathogenic *E. dispar*); however, red blood cells must also be present in the background of the smear if they are reported to be inside the cytoplasm of the organism. In general, if the patient is symptomatic, one assumes that the causative agent is pathogenic *E. histolytica* (assuming that the cause is related to the presence of intestinal protozoa).

The cyst of *E. histolytica/E. dispar* may contain chromatoidal bars with smooth, rounded edges. Even though the mature cyst contains four nuclei, these are not always visible, and the cyst can often be identified from the chromatoidal bars alone.

In a wet preparation, such as the direct smear or concentration sediment, the trophozoites and cysts of *E. histolytica* measure >12 and >10 µm, respectively. Because of artificial shrinkage during the preparation of permanent stained smears, the sizes are diminished somewhat, with the trophozoites measuring at least 10.5 to 11 µm and the cysts measuring at least 8.5 to 9 µm. Organisms measuring less than these limits and containing morphologic characteristics similar to those of *E. histolytica* would be identified as *E. hartmanni*, a nonpathogenic ameba.

One of the most important things to remember is that some human cells found in the stool can mimic *E. histolytica*. Macrophages or monocytes can look like the trophozoite form of *E. histolytica*, and polymorphonuclear leukocytes, when they have been in the stool for a while, can mimic the four-nucleus *E. histolytica* mature cyst.

Other tests include sigmoidoscopy, in which both scrapings and smears can be submitted for permanent stains. Biopsy specimens should also be submitted for histopathologic studies. Radiographic examination with barium is helpful; however, the presence of barium in the stool makes the ova and parasite examination very difficult to perform. Ova and parasite examinations should be done before barium studies or at least 1 week to 10 days afterward. The sensitivity of serologic tests for antibody in patients with intestinal amebiasis varies, and a negative serologic test does not necessarily rule out the presence of amebic infection. Serologic tests for antibody are not recommended for the diagnosis of intestinal amebiasis.

Tests for antigen detection in the stool are much more relevant for intestinal disease. Reagents are currently available for antigen detection of the *E. histolytica/E. dispar* group and specifically for *E. histolytica* (107). Fresh or frozen stool is required for the procedure; PCR, isoenzyme analysis, and antigen detection are available for confirmation of *E. histolytica* infection. However, the antigen detection method is both rapid and technically simple.

If *E. histolytica* organisms are not seen in the stool and the final diagnosis is inflammatory bowel disease, the patient may receive corticosteroid therapy. Immunosuppression may predispose the patient to more severe disease if the true pathogen, *E. histolytica*, is present and not detected.

Extraintestinal Disease

When organisms invade the mucosal lining and are carried via the bloodstream to the liver (extraintestinal or hepatic disease), a somewhat different approach to diagnosis is necessary. Both fecal and sigmoidoscopic examinations for organisms are negative in approximately half of the patients with extraintestinal amebiasis. Both radioactive and computed tomographic (CT) scans are available to assist the clinician in defining the presence of an abscess.

Lesions in the liver may range from less than a few centimeters to several inches in diameter. In either case, if the patient has a normal humoral immune system, serologic tests for antibody are generally positive.

Serologic tests for extraintestinal amebiasis are generally very specific and quite sensitive. However, in some parts of the country, sensitivity may present a problem in patients previously exposed to *E. histolytica*. In these individuals, a low to moderate titer may represent past infection rather than current disease. Also, patients with diminished immunoglobulin levels may present with low or negative serologic titers (Table 19.3). Various serologic techniques include complement fixation, indirect hemagglutination assay (IHA), counterimmunoelectrophoresis, latex agglutination, indirect fluorescent-antibody (IFA) assay, and enzyme-linked immunosorbent assay (ELISA), as well as tests for the detection of antigen in the stool. It is generally recommended that two different procedures be used so that the results can be compared. One approach is to use both qualitative screen and quantitative titer procedures. Commercial suppliers of kits are limited, and few laboratories routinely provide this type of testing. Serum for serologic tests should be sent to the state and local public health laboratories if the test is not routinely available.

AIDS Patients. Invasive amebiasis appears to be an emerging parasitic disease in patients infected with HIV in areas where amebic infection is endemic. Medical, microbiological, and histopathologic records of 296 HIV-infected patients and serologic data from IHAs of samples from 126 HIV-infected patients were reviewed to identify cases of invasive amebiasis. An IHA titer of 1:128 was considered positive. Of these 296 HIV-infected patients, 18 (6.1%) were diagnosed with invasive amebiasis. Clinical manifestations included amebic colitis (13 patients), amebic liver abscess (9 patients), both colitis and abscess (4 patients), and pleural effusion (2 patients). Invasive amebiasis was the initial presentation of HIV infection in nine patients. Of the 18 patients diagnosed with invasive amebiasis, 13 had an IHA titer of ≥1:128. The sensitivity of the IHA in the diagnosis of invasive amebiasis was 72.2%, and the specificity was 99.1%. The positive predictive value of IHA for invasive amebiasis in this patient population was 92.9% and the negative predictive value was 95.5% (45). It appears that invasive amebiasis is becoming more important as one of the opportunistic parasitic infections in patients with HIV infection in areas of endemicity for amebiasis. In this type of area of endemicity, the IHA should be considered a relevant diagnostic approach.

In a study from Mexico, the prevalence estimated with PCR data showed that *E. histolytica* infection was more common in the HIV-positive/AIDS group (25.32%), than in HIV-negative contacts (18.46%). *E. histolytica* and *E. dispar* infection was more frequent in HIV-positive/AIDS patients (13.3%) than in HIV-negative contacts (0.7%). However, *E. histolytica* and/or *E. dispar* infection was highly prevalent in HIV-positive/AIDS patients (34.1%) without evidence of recent or current invasive disease. Contacts of HIV-positive/AIDS patients who were infected with *E. histolytica* were asymptomatic cyst passers. These data indicate that *E. histolytica* strains prevalent in the studied community appear to be of low pathogenic potential within the compromised patient (71).

However, in another study, 49 (5.2%) of 951 HIV-infected persons had 51 episodes of invasive amebiasis. A high IHA titer was detected in 39 (6.2%) of 634 HIV-infected persons compared with 10 (2.3%) of 429 uninfected controls with gastrointestinal symptoms and 0 of 178 uninfected healthy controls. Stool specimens from 40 (12.1%) of 332 HIV-infected persons and 2 (1.4%) of 144 uninfected healthy controls were positive for *E. histolytica* or *E. dispar* antigen. Of the 40 antigen-positive stool specimens from HIV-infected persons, 10 (25.0%) contained *E. histolytica*. Therefore, persons infected with HIV in Taiwan are at increased risk for invasive amebiasis and exhibit a relatively high frequency of elevated antibody titers and intestinal colonization with *E. histolytica* (45).

High-risk areas for acquiring amebiasis include Mexico, the western portion of South America, western Africa, South Africa, parts of the Middle East, and South and Southeast Asia. Invasive disease is much more common in these areas of the world.

Free-Living Amebae

Infections caused by small, free-living amebae are becoming recognized clinically as important parasitic pathogens, particularly in immunocompromised patients. Primary amebic meningoencephalitis (PAM) caused by *Naegleria fowleri* and granulomatous amebic encephalitis (GAE) caused by *Acanthamoeba* spp. and *Balamuthia mandrillaris* (including cases in patients with AIDS) are now well recognized (Figures 19.2 and 19.3). The most recently identified group of infections caused by these organisms is *Acanthamoeba* keratitis, which is related primarily to poor lens care in those who wear contact lenses. When free-living, these organisms feed on bacteria and nutrients in moist soil and in fresh and marine waters.

Although it is well known that both free-living and pathogenic protozoa harbor a variety of endosymbiotic bacteria, it is unclear what role these organisms play

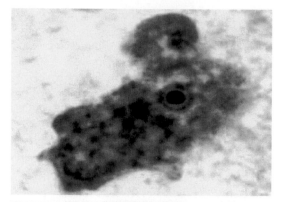
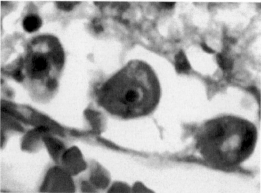
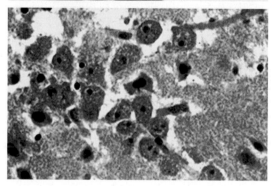

Figure 19.2 (Top) *Naegleria fowleri* trophozoite (note the large karyosome within the nucleus); (middle) *Naegleria fowleri* trophozoites within brain tissue; (bottom) *Balamuthia mandrillaris* in brain tissue. (Bottom, Armed Forces Institute of Pathology photograph.)

in terms of host survival, infectivity, and invasiveness. *Legionella pneumophila* can infect, multiply within, and kill both *Naegleria* and *Acanthamoeba* amebae. These amebae may be natural hosts for *Legionella*, as well as for other bacteria such as *Listeria monocytogenes*, *Vibrio cholerae*, *Mycobacterium leprae*, and *Pseudomonas aeruginosa*. Because of their well-known resistance to chlorine, the amebic cysts are considered to be vectors for these intracellular bacteria. This can have tremendous significance for any hospital where the water source is contaminated with free-living amebae. There

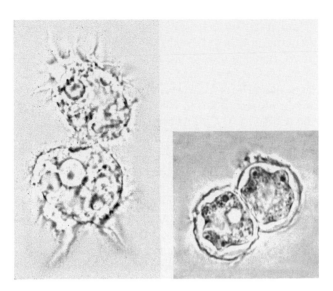

Figure 19.3 (Left) *Acanthamoeba* trophozoites (note the sharp, spiky pseudopodia); (right) *Acanthamoeba* cysts (note the hexagonal double wall).

is also evidence to suggest that the presence of amebae in domestic water supplies may provide growth conditions that enhance the pathogenicity of the organism for the human host. These amebae are also acceptable hosts for echoviruses and polioviruses. The epidemiology, immunology, protozoology, pathology, and clinical features of the infections produced by these free-living protozoa differ considerably.

Naegleria fowleri

Central nervous system (CNS) infections caused by free-living amebae have been recognized only since the mid-1960s. *N. fowleri* causes PAM, a fulminant and rapidly fatal disease that affects mainly children and young adults (Figure 19.2). The disease closely resembles bacterial meningitis but is caused by this free-living ameba, an organism found in moist soil and freshwater habitats. Although PAM is usually associated with healthy, young individuals with a history of recent water-related sport activities, it is an important infection to consider in both immunocompetent and immunocompromised individuals. However, almost all infections have been seen in otherwise healthy individuals. Close to 200 cases of PAM have occurred worldwide, and approximately 90 of those cases have been reported from the United States.

Life Cycle.   There are both trophozoite and cyst stages in the life cycle, and the stage depends primarily on environmental conditions. Trophozoites can be found in water or moist soil and can be maintained in tissue culture or other artificial media. The trophozoites can occur in two forms, ameboid and flagellate. Motility can be observed in hanging-drop preparations from cultures of cerebrospinal fluid (CSF); the ameboid form (the only form recognized in humans) is elongate with a broad anterior end and a tapered posterior end. The size ranges from 7 to 20 µm. The diameter of the rounded forms is usually 15 µm. There is a large central karyosome and no peripheral nuclear chromatin. The cytoplasm is somewhat granular and contains vacuoles. The ameboid-form organisms change to the flagellate form when they are transferred from culture or teased from tissue into water and maintained at a temperature of 27 to 37°C. The flagellate form is pear shaped, with two flagella at the broad end. Motility is typical, with either spinning or jerky movements. These flagellate forms do not divide, but when the flagella are lost, the ameboid forms resume reproduction. Cysts from nature and from agar cultures look the same and have a single nucleus almost identical to that seen in the trophozoite. The cysts are generally round, measuring 7 to 10 µm in diameter, with a thick double wall.

Amebic Meningoencephalitis.   Amebic meningoencephalitis caused by *N. fowleri* is an acute, suppurative infection of the brain and meninges. In humans, with extremely rare exceptions, the disease is rapidly fatal. The period between contact with the organism and onset of clinical symptoms such as fever, headache, and rhinitis may vary from 2–3 days to as long as 7–15 days.

The amebae may enter the nasal cavity by inhalation or aspiration of water, dust, or aerosols containing the trophozoites or cysts. The organisms then penetrate the nasal mucosa, probably through phagocytosis of the olfactory epithelial cells, and migrate via the olfactory nerves to the brain. Early symptoms include vague upper respiratory distress, headache, lethargy, and occasionally olfactory problems. Cysts of *N. fowleri* are generally not seen in brain tissue. The acute phase includes sore throat; stuffy, blocked, or discharging nose; and severe headache. Progressive symptoms include pyrexia, vomiting, and stiffness of the neck. Mental confusion and coma usually occur approximately 3 to 5 days prior to death. The cause of death is usually cardiorespiratory arrest and pulmonary edema.

Organ Transplantation.   The first case of organ transplantation from a donor who had died of undiagnosed *N. fowleri* infection was reported in 1997 (96, 97). While no subsequent amebic infections occurred in the three organ recipients, this report emphasizes the need for adequate evaluation of the benefits and risks of transplanting tissues from persons whose illness might have been caused by an infectious agent. Apparently, the Centers for

Disease Control and Prevention are also aware of another instance in which kidneys were transplanted prior to determination that the donor was infected with *N. fowleri* (96, 97). At present, the risk of transmission of *N. fowleri* by donor organs has not been clarified, and no practical test is available to ensure that donor organs are organism free. Also, no prophylactic drug regimen to treat transplant recipients has been established.

Diagnosis. PAM can resemble acute purulent bacterial meningitis and may be difficult to differentiate, particularly in the early stages. The CSF may have the predominantly polymorphonuclear leukocytosis, increased protein concentration, and decreased glucose concentration that are seen with bacterial meningitis. Unfortunately, if the CSF Gram stain is interpreted incorrectly (identification of bacteria as a false-positive result), the antibacterial therapy will have no effect on the amebae and the patient will usually die within several days. Extensive tissue damage occurs along the path of amebic invasion; the nasopharyngeal mucosa shows ulceration, and the olfactory nerves are inflamed and necrotic. Hemorrhagic necrosis is concentrated in the region of the olfactory bulbs and the base of the brain. Organisms can be found in the meninges, perivascular spaces, and sanguinopurulent exudates.

Clinical and laboratory data usually cannot be used to differentiate pyogenic meningitis from PAM, and so the diagnosis may have to be reached by a process of elimination. A high index of suspicion is often mandatory for early diagnosis. Although most cases are associated with exposure to contaminated water through swimming or bathing, this is not always the case. The rapidly fatal course of 4 to 6 days after the beginning of symptoms (with an incubation period of 1 day to 2 weeks) requires early diagnosis and immediate chemotherapy if the patient is to survive.

Analysis of the CSF shows decreased glucose and increased protein concentrations. Leukocyte counts may range from several hundred to >20,000 cells/mm^3. Gram stains and bacterial cultures of CSF are negative. A definite diagnosis could be made by demonstration of the amebae in the CSF or in biopsy specimens. Either CSF or sedimented CSF should be placed on a slide under a coverslip and observed for motile trophozoites; smears can also be stained with Wright's or Giemsa stain. CSF, exudate, or tissue fragments can be examined by light microscopy or phase-contrast microscopy. Care must be taken not to mistake leukocytes for actual organisms or vice versa. It is very easy to confuse leukocytes and amebae, particularly when one is examining CSF by using a counting chamber, hence the recommendation to use just a regular slide and coverslip. Motility may vary, so that the main differential characteristic is the spherical nucleus with a large karyosome.

Unfortunately, most cases are diagnosed at autopsy; confirmation of these tissue findings must include culture and/or special staining using monoclonal reagents in IFA procedures. Organisms can also be cultured on nonnutrient agar plated with *Escherichia coli*. In cases of presumptive pyogenic meningitis in which no bacteria are identified in the CSF, the CT appearance of basal arachnoiditis (obliteration of basal cisterns in the precontrast scan with marked enhancement after the administration of intravenous contrast medium) should indicate the possibility of acute PAM.

The amebae can be identified in histology preparations by using indirect immunofluorescence and immunoperoxidase techniques. The organism in tissue sections looks very much like an *Iodamoeba bütschlii* trophozoite, with a very large karyosome and no peripheral nuclear chromatin; the organisms can also be seen with use of routine histologic stains. In general, serologic tests have not been helpful in the diagnosis of this infection. The disease progresses so rapidly that the patient is unable to mount an immune response.

Acanthamoeba spp.

Granulomatous Amebic Encephalitis. Another type of meningoencephalitis, GAE, caused by freshwater amebae may occur as a subacute or chronic disease with focal granulomatous lesions in the brain (Figure 19.3) (13, 60). The route of CNS invasion is thought to be hematogenous, with the primary site being the skin or lungs. In this infection, both trophozoites and cysts can be found in the CNS lesions. A case involving acute onset of fever, headache, and pain in the neck preceded by 2 days of lethargy has been described. Conditions associated with GAE include skin ulcers, liver disease, pneumonitis, diabetes mellitus, renal failure, rhinitis, pharyngitis, and tuberculosis (76). Predisposing factors include alcoholism, pregnancy, systemic lupus erythematosus, hematologic disorders, AIDS, chemotherapy, radiation therapy, and steroid treatment. This infection has become more widely recognized in AIDS patients, particularly those with a low CD4$^+$ cell count. Although *Acanthamoeba* infections stimulate a granulomatous response, the response in AIDS patients is minimal or absent, consistent with the poor immune response in these patients. Examples of specific infections seen in immunocompromised patients are listed in Table 19.5.

The *Acanthamoeba* group also causes keratitis and corneal ulceration. There are a number of published cases, which emphasizes the need for clinicians to consider acanthamoebic infection in the differential diagnosis of eye infections that fail to respond to bacterial, fungal, or viral therapy. These infections are often due to direct eye exposure to contaminated materials or solutions.

Table 19.5 *Acanthamoeba* infections in immunocompromised patients

Underlying condition[a]	Disease manifestation	Comments	Reference(s)
AIDS	Cutaneous nonhealing lesion	Cutaneous disease in the absence of CNS involvement is increasingly recognized	109
AIDS	Granulomatous uveitis	*Acanthamoeba* infection should be considered in the differential diagnosis of uveitis in patients with AIDS	41
AIDS	Osteomyelitis	Disseminated disease has been described (the usual sites are skin, sinus, and brain); the patient had a 6-mo history of cutaneous infection	98
AIDS	Disseminated disease	Patients had cutaneous manifestations but no CNS involvement; lesions included pustules, subcutaneous and deep dermal nodules, and ulcers, primarily on extremities and face; high index of suspicion necessary for both dermatologist and dermatopathologist	94, 109
AIDS	GAE	Primary foci in lungs, skin; inflammatory response is chronic, with granulomatous response characterized by presence of multinucleated giant cells; though nonspecific, imaging can support the diagnosis and direct biopsy results; physician should be aware of potential diagnosis in AIDS patients with an insidious encephalitis and cerebral cognitive abnormalities, with or without focal motor signs	60
SLE therapy	GAE	Lung probably primary focus; infectious encephalitis difficult to differentiate from a flare-up of CNS lupus	57
Immunosuppressive drugs for BMT	GAE	Autologous stem cell transplantation; leg weakness, fever, and urinary retention developed 69 days after transplantation; patient then developed fever, seizures, and rapid deterioration of mental functions	60
Renal transplant	Osteomyelitis, widespread cutaneous lesions	Both trophozoites and cysts found at autopsy in skin and bone specimens; no causative agent identified prior to death	105

[a] SLE, systemic lupus erythematosus; BMT, bone marrow transplantation.

Unlike *N. fowleri*, *Acanthamoeba* spp. do not have a flagellate stage in the life cycle, only the trophozoite and cyst (Figure 19.3). Motile organisms have spinelike pseudopods; however, progressive movement is usually not very evident. There is a wide range in size, with the average diameter of the trophozoites being 30 µm. The nucleus has the typical large karyosome, like that seen in *N. fowleri*. This morphology can be seen on a wet preparation. The cysts are usually round with a single nucleus, also having the large karyosome seen in the trophozoite nucleus. The double wall is usually visible, with the slightly wrinkled outer cyst wall and what has been described as a polyhedral inner cyst wall. This cyst morphology can be seen in organisms cultured on agar plates.

Laboratory examinations similar to those for *N. fowleri* can be used to recover and identify these organisms; the one exception is recovery by culture, which has not proven to be as effective with GAE cases. The most effective approach uses nonnutrient agar plates with Page's saline and an overlay growth of *E. coli* on which the amebae feed. There is also evidence to indicate that phosphate-buffered saline can be used. Specimens transported in ameba saline (5.0 ml) and filtered through 13-mm-diameter, 0.22-µm-pore-size cellulose acetate and nitrate filters (Millipore Corp., Bedford, Mass.) have also been acceptable for organism recovery. The filter is then placed in the center of the nonnutrient agar plate seeded with *E. coli*. Tissue stains are also effective, and cysts can be stained with Gomori's methenamine-silver, periodic acid-Schiff (PAS), and calcofluor white.

Although detection of antibodies to *Acanthamoeba* spp. has been confirmed, the usefulness of serologic tests for the diagnosis of these infections has not been proven. Also, these procedures are not generally available. Further work related to the issue of false-positive and false-negative results and interpretive guidelines regarding past exposure or present infection need to be pursued before these diagnostic procedures can be routinely used.

Balamuthia mandrillaris

The leptomyxid ameba *Balamuthia mandrillaris* is relatively uncommon and was originally thought to be another harmless soil organism, unlikely or unable to infect mammals. However, in studies from the Centers for Disease Control and Prevention, antisera to the leptomyxid amebae have been shown to react in indirect-immunofluorescence assays with tissues from both animals and humans with GAE. There have been over 70 cases worldwide, some of which have occurred in AIDS patients (96, 97, 112). The disease is very similar to GAE caused by *Acanthamoeba* spp. and has an unknown incubation period. The clinical course tends to be subacute or chronic and is usually not associated with swimming in freshwater. No characteristic clinical symptoms, laboratory findings, or radiologic indicators have been found to be diagnostic for GAE. The neuroimaging findings show heterogeneous, hyperdense, nonenhancing, space-occupying lesions (Figure 19.2). Whether single or multiple, they involve mainly the cerebral cortex and subcortical white matter. These finding suggest a CNS neoplasm, tuberculoma, or septic infarcts. Patients complain of headaches, nausea, vomiting, fever, visual disturbances, dysphagias, seizures, and hemiparesis. There may also be a wide range in terms of the clinical course, from a few days to several months. Death can occur from a week to several months after the onset of stroke-like symptoms that can mimic other conditions. Both trophozoites and cysts of *B. mandrillaris* are found in many of the same CNS tissues as are *Acanthamoeba* organisms. Although differentiation of these two organisms in tissue by light microscopy is difficult, in some tissue sections *B. mandrillaris* appears to have more than one nucleolus in the nucleus. Generally, electron microscopy and histochemical methods are required for definitive identification of *B. mandrillaris*.

Attempts to isolate the organisms from some humans with GAE have not been successful, and leptomyxid amebae do not grow well on *E. coli*-seeded nonnutrient agar plates. In the diagnostic laboratory, these organisms can be cultured using mammalian cell cultures; some success has been obtained with monkey kidney cells and with MRC, HEp-2, and diploid macrophage cell lines. The trophozoites have extensive branching and a single nucleus (occasionally binucleate forms are seen) with a central karyosome, and they measure 15 to 60 μm. Occasionally a few elongated forms with several contractile vacuoles are seen. The cysts have a single nucleus (occasionally binucleate forms are seen), have the typical double wall with the outer wall being thick and irregular, and measure 15 to 30 μm. Using primary cultures of human brain microvascular endothelial cells, amebae have been isolated from the brain and CSF (50). Serum antibodies to *B. mandrillaris* have been found in both adults and children;

however, testing is generally limited to the Centers for Disease Control and Prevention, where PCR has also been used for the detection of mitochondrial 16S rRNA gene DNA from the amebae in clinical specimens (117).

From the information presented above, it is clear that *B. mandrillaris*, formerly regarded as having no pathogenic potential, will continue to be identified as the etiologic agent of fatal meningoencephalitis in humans as well as animals. This opportunistic infection may also continue to cause disease in individuals with AIDS.

Giardia lamblia

Giardia lamblia is a flagellate commonly found in many parts of the world (Tables 19.3 and 19.4; Figure 19.4). An infection with this organism can be transmitted via the cyst form in both water and food contaminated with fecal material, as can *E. histolytica* infections. Various conditions that have been associated with giardiasis in the compromised patient include hypogammaglobulinemia, protein or caloric malnutrition, previous gastrectomy, histocompatibility antigen HLA YB-12, gastric achlorhydria, blood group A, differences in mucolytic proteins in Ig deficiencies, and reduced secretory IgA levels in the gut (Table 19.2). Generally, adults and children are mentioned in relation to giardiasis; however, it is important to remember that the elderly, in both tropical and temperate zones, are also susceptible. These patients tend to be somewhat immunodepressed due to age and may be overlooked as having giardiasis.

Although patients with HIV infection have also been found to have giardiasis, the infection does not appear to be more severe among this group, regardless of the CD4+ cell count (52). In some groups, the prevalence of *G. lamblia* is greater for HIV-infected subjects (30, 113). Both humoral and cellular immune responses play a role in acquired immunity, and the mechanisms involved remain somewhat unclear. However, the immune response to intestinal parasites, including *G. lamblia*, might be a risk factor for HIV/AIDS and tuberculosis (2). It is well known that diagnosing intestinal parasites in HIV/AIDS patients is necessary, especially in those who are chronic alcoholics or are not receiving antiretroviral treatment (17, 39, 70, 102).

Symptoms of giardiasis may range from none to the malabsorption syndrome, characterized by gas, bloating, frothy green foul-smelling stools, and abdominal pain. The organism is reasonably easy to identify. The trophozoite form is teardrop shaped, and the flagella and sucking disk are easily visible. On side view, the trophozoite often looks somewhat like the curved end of a spoon. Trophozoites tend to stain pale, while the cysts usually stain darker and are easier to identify. Cysts are round

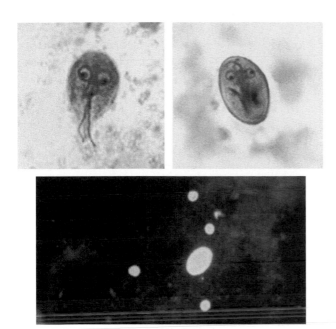

Figure 19.4 (Upper left) *Giardia lamblia* trophozoite (note two nuclei, curved median bodies, and linear axonemes); (upper right) *Giardia lamblia* cyst; (lower) *Giardia lamblia* cyst (large) and *Cryptosporidium* oocysts (small) demonstrating fluorescence in the fecal FA immunoassay (note that the background demonstrates use of the counterstain).

to oval, with long axonemes and curved median bodies being visible. Normally, cysts can be seen and identified on wet mount preparations from concentrated material and confirmed by using the permanent stained smear.

In contrast to *E. histolytica*, *G. lamblia* inhabits the duodenal area of the intestine, tends to adhere very tightly to the mucosa, and can be very difficult to recover, even after a series of five or six stool examinations. For this reason, other techniques, such as the Entero-Test string capsule, duodenal aspirate, or biopsy, may have to be used. Fecal immunoassays are also recommended for the diagnosis of giardiasis.

Toxoplasma gondii

Toxoplasma gondii is worldwide in distribution, and serologic data indicate that human infections are common, although most infections are benign or cause no symptoms (Tables 19.3 and 19.4). Serologic evidence from throughout the United States indicates that anywhere from 20 to 70% of the population are serologically positive for *T. gondii*. The percentage of serologically positive patients tends to be higher in the tropical areas of the world.

Toxoplasmosis can vary from an asymptomatic, self-limiting infection to a fatal disease, as seen in patients

with congenital infections or in debilitated patients in whom underlying conditions may influence the final outcome of the infection (Table 19.2). In immunocompromised patients, the infection most often involves the nervous system, with diffuse encephalopathy, meningoencephalitis, or cerebral mass lesions (Figure 19.5). In one study, multiplex nested PCR analysis was used to determine the genotypes of parasites in CSF; 8 of 10 human HIV-positive patients had infections with type I strains, even though this lineage is normally uncommon in humans and animals (54). Although gastric toxoplasmosis is rare, in some patients with AIDS it is the primary manifestation (33, 67).

Underlying conditions associated with toxoplasmosis in the compromised host include various types of malignancies, such as Hodgkin's disease; non-Hodgkin's lymphomas, leukemias, or solid tumors; collagen vascular disease; and organ transplantation. We must also add to this category patients with AIDS. Over 50% of such patients with toxoplasmosis have alterations of mental status, motor impairments, seizures, abnormal reflexes, and other neurologic symptoms. Studies have shown that even among compromised patients, 80% of those receiving chemotherapy for toxoplasmosis experienced significant clinical improvement or complete remission; however, in those with AIDS, therapy must be continued for long periods to maintain a clinical response.

Toxoplasma encephalitis has been reported as a life-threatening opportunistic infection among patients with AIDS. The overall prognosis can be poor, with the median survival following initiation of therapy being 4 months. A clinical relapse of *Toxoplasma* encephalitis has been reported to occur in 50% of patients discharged from hospital. In AIDS patients who may not receive prophylaxis for *Pneumocystis jiroveci* pneumonia, almost half of the *Toxoplasma*-seropositive patients can develop CNS toxoplasmosis. This figure is considerably higher than previously thought for the development of CNS involvement.

Figure 19.5 *Toxoplasma gondii* in bone marrow.

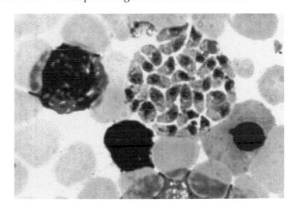

Statistics indicate that, overall, approximately 2.5% of AIDS patients have been diagnosed with cerebral toxoplasmosis. Information on toxoplasmosis and the immunocompromised host can be seen in Table 19.6.

Humans usually become infected with *T. gondii* by ingesting raw or poorly cooked meats. Another means of transmission is ingestion of infected oocysts found in cat feces. It has been recommended that litter boxes be cleaned daily, with feces disposed of in the toilet and the pans disinfected with boiling water. Since we know that transplacental transmission can occur during an acute but perhaps undiagnosed maternal infection, pregnant women, unless they have serologic evidence of previous infection, should avoid contact with cats whose source of food is not controlled and should not empty litter boxes. Uterine or neonatal infections can give rise to a number of serious sequelae, including microcephaly and hydrocephalus.

In humans, *T. gondii* is found in two different stages. The actively proliferating intracellular forms are called trophozoites or tachyzoites and are crescent shaped. Many different tissues may be parasitized by these organisms, particularly lung, heart, lymphoid organ, and CNS tissues. The resting forms or cyst stages are found in the tissues and contain the more slowly developing bradyzoites.

Toxoplasmosis can be diagnosed by using various serologic procedures, histologic findings from the examination of biopsy specimens, and isolation of the organism, either in a tissue culture system or by animal inoculation. Individuals with positive serologic tests have been exposed to *Toxoplasma* organisms and may have the resting stages within their tissues. This is why histologic identification of the organism and its recovery either in a tissue culture system or by animal inoculation may be misleading, since the organisms could be isolated yet might not be the etiologic agent of disease. Thus, serologic tests have been recommended as the diagnostic tests of choice. One of the first serologic tests was the Sabin-Feldman dye test, a reaction based on the fact that serum antibodies to *T. gondii* will affect the organisms so that they become refractory to staining with methylene blue. However, since this test requires the use of living *Toxoplasma* trophozoites, most laboratories do not provide it for clinical use.

In an immunodeficient or immunosuppressed patient, a presumptive diagnosis of toxoplasmosis can be made by observing an elevated serologic titer and the presence of the clinical syndrome, which would include neurologic symptoms. However, in certain patients with monoclonal gammopathies, titers to *T. gondii* may be extremely high but may not reflect the cause of the clinical condition. One of the most common diagnostic tests used for toxoplasmosis is the IFA procedure. Definitive diagnosis is usually made when there is a rising titer, an IgM-IFA titer

of at least 1:64, or an IgM-ELISA titer of 1:256. A finding that the Sabin-Feldman dye reaction is higher in CSF than in serum is also significant but not that common. An elevated serologic titer in conjunction with demonstration of the actual trophozoites or isolation of the organisms from cerebrospinal fluid is very significant.

Cryptosporidium spp.

Although it has been recognized as an animal parasite for many years, one of the newer parasitic organisms infecting humans, particularly compromised patients, is *Cryptosporidium*, a cause of diarrhea (Tables 19.3 and 19.4; Figure 19.6). *Cryptosporidium* differs from other coccidia that infect warm-blooded vertebrates in that the developmental stages do not occur deep within host cells but are confined to an intracellular, extracytoplasmic location. Each stage is within a parasitophorous vacuole of host cell origin. What was previously called *C. parvum* and was thought to be the primary *Cryptosporidium* species infecting humans is now classified as two separate species, *C. parvum* (mammals, including humans) and *C. hominis* (primarily humans). Differentiation of these two species based on oocyst morphology is not possible.

Studies have shown that calves and perhaps other animals may serve as potential sources of human infections. Kittens, rodents, and puppies are possible reservoir hosts, since they are easily infected with human *Cryptosporidium*. Person-to-person transmission is also likely and may occur through direct or indirect contact with stool material in the environment and the hospital setting.

This organism is worldwide in distribution, and although the first cases in the literature indicated that the patients were immunosuppressed or immunodeficient, current literature reports numerous infections and outbreaks in patients with a normal immune system capability. In patients with a normal immune system capability, cryptosporidiosis is self-limiting. However, in certain compromised patients, the diarrhea is very severe and prolonged, with death being the eventual outcome (Tables 19.4 and 19.7). Clinical symptoms include nausea, low-grade fever, abdominal cramps, anorexia, and 5 to 10 watery stools per day. Fluid loss in these patients is significant and has been reported to be as great as 17 liters/day.

Although many therapeutic regimens have been tried, there is no completely satisfactory therapy for cryptosporidiosis in humans. Paromomycin, clarithromycin, nitazoxanide, and hyperimmune bovine colostrum are among the drugs being tried in different regimens; the current recommendation is nitazoxanide (1). Specific recommendations for immunocompromised patients in

Table 19.6 *Toxoplasma gondii* infections in immunocompromised patients

Underlying condition	Disease manifestation and complication[a]	Comments[a]	Reference
AIDS and lymphoma	Cerebral toxoplasmosis	Regional CBV (rCBV) determined by perfusion MRI; rCBV reduced throughout toxoplasmosis lesions; all active lymphomas had increased rCBV results; differences were significant; perfusion MRI is a rapid, non-invasive tool that may allow differentiation between cerebral lymphoma and toxoplasmosis	28
AIDS and lymphoma	Cerebral toxoplasmosis	Knowledge of patient's serostatus does not aid discrimination between lymphoma and toxoplasmosis; single lesions on MRI with focal accumulation of thallium-201 strongly suggest lymphoma; multiple lesions on MRI with uptake ratios of ≥2.9 also suggest lymphoma; uptake ratios of <2.1 do not aid in discrimination	69
AIDS	CNS disease	Diagnosis of CNS toxoplasmosis often incorrect; another diagnosis is most likely in patients who are anti-*Toxoplasma* seronegative or who are receiving prophylactic TMP-SMX	63
AIDS	CNS disease and intracranial mass lesions	Neurologic disease is first indication of AIDS in 10–20% of HIV-seropositive individuals; over half will present with intracranial mass lesion; toxoplasmosis is the most common cause in developed countries; data indicate that it is also the most common cause in underdeveloped countries	9
AIDS	CNS disease and gastrointestinal mastocytosis	Simultaneous occurrence of both conditions and regression after treatment for toxoplasmosis; may suggest a possible relationship between mast cell proliferation and parasitic infection	56
AIDS	Chorea	Chorea not unusual in AIDS; however, its causes are variable; AIDS-related disease should be considered in young patients presenting with chorea with no family history of movement disorders	84
AIDS	Risk factors	While risk factors are still somewhat unclear, specific prophylaxis for patients with low CD4 cell counts and high *Toxoplasma* Ab titers is recommended	7
AIDS	Emerging disease in India (TE)	Management protocol in AIDS patients critical to the diagnosis of TE in patients with CNS symptoms; increasing numbers of AIDS patients require constant consideration of opportunistic infections	65
AIDS	Immunoblot profiles (TE)	In patients with *Toxoplasma* Ab (ELISA), IgG 27 band or IgG 31 band on Western blot is highly indicative of TE, independent of each other; Western blot positivity provides useful data for improving diagnosis of presumptive TE in HIV-infected patients with suspected TE and positive *Toxoplasma* serology	89
AIDS	Long-term outcome (TE)	Cerebral infections like TE may negatively influence HIV-1 activity so far latent in the brain	6
AIDS	Sulfadiazine and obstructive nephropathy	Patients receiving sulfadiazine for cerebral toxoplasmosis can develop obstructive nephropathy; will resolve with alkalic hydration and discontinuation of sulfadiazine	18
AIDS	High or low doses of TMP-SMX	High doses of TMP-SMX appear to be more effective than low doses for decreased risk for toxoplasmosis in HIV-infected patients; rifampin may reduce the efficacy of TMP-SMX	92
Immunosuppressive drugs for BMT	Cerebral toxoplasmosis	Clinical diagnosis confirmed by PCR on CSF and blood leukocytes; retrospective testing indicated PCR was positive 52 days prior to onset of clinical symptoms; results highlight value of PCR in early detection of cerebral toxoplasmosis	55

(continued)

Table 19.6 *Toxoplasma gondii* infections in immunocompromised patients *(continued)*

Underlying condition[a]	Disease manifestation and complication[a]	Comments[a]	Reference
Immunosuppressive drugs for BMT	TE	In group of 27 BMT patients (allogeneic only) who developed CNS infections, the most common (74%) was TE; one group showed edema but no enhancement, while one group showed the typical MRI pattern with the exception of frequent hemorrhagic transformation	64
Immunosuppressive drugs for BMT	Cerebral toxoplasmosis reactivation	*T. gondii* infection reactivation occurs predominantly in patients after allogeneic BMT; tends to occur within 3 mo after transplantation; late cerebral toxoplasmosis reactivation may be triggered by a course of corticosteroids (administered for graft-versus-host disease); risk factors should signal reinstitution of prophylactic treatment with TMP-SMX	121
Immunosuppressive drugs for BMT	MRI problems	In patients with low leukocyte count (poor immunologic response) and/or high doses of immunosuppressive therapy, typical contrast enhancement may be absent (poor visualization of cerebral lesions)	26
SLE therapy	Cerebral toxoplasmosis	Symptoms of fever, confused state, convulsions consistent with lupus cerebritis; patient died of cerebral toxoplasmosis, which was confused with lupus cerebritis	120
Non-Hodgkin's lymphoma	TE	Involvement of mediastinum, para-aortic lymph nodes, pleura, peritoneum, bone marrow; patient deteriorated; CSF smear revealed tachyzoites; TE should be considered a potential cause of consciousness disturbance in patients with malignant lymphoma	119

[a] Ab, antibody; BMT, bone marrow transplantation; CBV, cerebral blood volume; MRI, magnetic resonance imaging; SLE, systemic lupus erythematosus; TE, *Toxoplasma* encephalitis; TMP-SMX, trimethoprim-sulfamethoxazole.

avoiding potential risk situations for acquiring infection with *Cryptosporidium* spp. can be found in chapter 4 and Table 19.8.

Since the organisms live in the microvillous surface of the mucosal epithelial cells of the intestine, diagnostic techniques have included histologic examination of intestinal biopsy material, scanning and transmission electron microscopic examination of duodenal aspirates and biopsy specimens, and the examination of stool specimens for oocysts.

Figure 19.6 *Cryptosporidium* oocysts stained using the modified acid-fast stain (note the spherical shape; oocysts measure 4 to 6 μm).

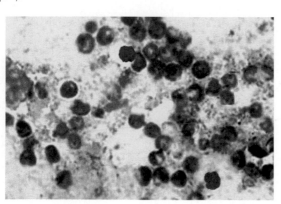

Although biopsy techniques have been used to recover and identify the organisms in tissue, one problem with a tissue specimen is that the specific area where the organisms reside may not be subjected to biopsy. The examination of stool material for oocysts by using an acid-fast technique or immunoassays allows one to screen a sample from the entire intestinal tract. It is considerably cheaper to examine a stool specimen than to perform endoscopy or histologic testing. *Cryptosporidium* oocysts in the stool range from 4 to 6 μm in diameter and can be very difficult to identify. One must also remember that even in patients with typical watery diarrhea, the numbers of organisms passed can be quite variable.

Respiratory cryptosporidiosis has also been detected in AIDS patients. Sputum specimens should be submitted in 10% formalin or as a fresh specimen and can be examined by the same techniques as used for stool samples.

Some techniques have included sugar flotation, formalin sedimentation, Giemsa stain, trichrome, PAS, silver methenamine, acridine orange, auramine-rhodamine, iodine, modified acid-fast, Kinyoun and Ziehl-Neelsen acid-fast, and immunoassay techniques. The most widely used techniques are modified acid-fast stains, Kinyoun's methods, and fecal immunoassays. Specimens preserved in polyvinyl alcohol do not stain well in the acid-fast techniques.

Table 19.7 *Cryptosporidium* infections in immunocompromised patients

Underlying condition	Disease manifestation	Comments	Reference
AIDS	Diarrhea and possible dissemination	Parasites associated with 14% of acute diarrheal episodes (≤28 days) and 35% of chronic episodes (>28 days); parasite associated with diarrhea in 14% of patients when CD4$^+$ counts are ≥200 cells/μl and in 27% when counts are <200 cells/μl; *Cryptosporidium* associated with 5% of acute episodes and 13% of chronic episodes of diarrhea; *Cryptosporidium* likely to be associated with diarrhea as immunosuppression worsens and with chronic rather than acute diarrhea	77
AIDS	Diarrhea	Attack rate 14% among HIV-negative individuals and 31% among HIV-positive individuals (varies due to CD4 cell count); chronic symptoms in 15% (<150 CD4 cells/μl at onset); 14% had antibody before outbreak, 51% developed specific antibodies during outbreak; clinical manifestations of cryptosporidiosis strongly influenced by level of HIV-induced immunosuppression	88
AIDS	Pancreatitis	Acute or chronic pancreatitis associated with *Cryptosporidium*, infection; abdominal pain resistant to analgesics, increased amylase level in serum; abnormalities on both sonography and CT; papillary stenosis in all patients; oocysts found in both bile and stool; viral cultures negative; several antibiotic protocols ineffective	15
AIDS	Gastric involvement	Of 24 individuals, 34% positive for *Cryptosporidium*, 16 (89%) of these had parasites in gastric epithelium; most patients did not show specific symptoms indicating presence in stomach; gastric involvement probably more frequent than expected; no clear correlation between gastric location and related clinical or pathologic features	93
AIDS	Diarrhea	22 *Cryptosporidium* isolates (Kenya, Switzerland, United States); 4 distinct genotypes identified (*C. parvum*, human genotype; *C. parvum*, cattle genotype; *C. felis*; and *C. meleagridis*, turkey); immunocompromised individuals susceptible to wide range of *Cryptosporidium* spp. and genotypes; full public health ramifications not known	73
AIDS	Diarrhea	*C. parvum* genotype subgroupings correlate to some extent with infectivity and suggest that additional heterogeneity is present within the subgroups	114
AIDS	Diarrhea	Treatment of HIV-1-associated microsporidiosis and cryptosporidiosis with combination antiretroviral therapy; combination therapy that includes protease inhibitors can restore immunity to both organisms and result in clinical, microbiological, and histologic responses; persistent CD8 cell and macrophage infiltrate and short time to relapse with declining CD4 lymphocyte counts suggest that neither infection was eradicated	16
AIDS	Diarrhea due to contaminated tap water	Risk assessment model applied to New York City population; calculated number of tap water-related cases per year was 6 (non-AIDS subgroups) or 34 (AIDS subgroups)	83
AIDS, pediatrics	Diarrhea	Some 81% of HIV-infected children with persistent diarrhea excreted *Cryptosporidium*, microsporidia, or both organisms, compared with only 10% of children with persistent diarrhea testing negative for HIV; 74% of isolates were *C. hominis*, 17% were *C. parvum*, and 8% were a mixture of the 2 or others	111
All causes	Diarrhea due to zoonotic transmission	Presence of *Cryptosporidium* in cats (5%) (in addition to other organisms such as *Giardia* spp., *Toxocara cati*, *Salmonella*, *Campylobacter*); enteric zoonotic organisms detected in total of 13% of cats studied; cats in homes of immunocompromised humans should be evaluated for zoonotic organisms	42
Hodgkin's lymphoma	Diarrhea	Immunocompromised prevalence of 14.3%; control group had 5.6%	91
Immunosuppressive therapy for pediatric oncology	Diarrhea	In contrast to studies of adult oncology patients, pediatric oncology patients in some areas appear to be at low risk for cryptosporidiosis	12

(continued)

Table 19.7 *Cryptosporidium* infections in immunocompromised patients *(continued)*

Underlying condition	Disease manifestation	Comments	Reference
Immunosuppressive drugs for kidney transplantation, chronic renal insufficiency	Diarrhea	*Cryptosporidium* found in 35% (renal transplant group), 25% (patients on hemodialysis), and 17% (control group with systemic arterial hypertension) of patients; when number of positive fecal samples are taken into account, there is a significantly higher frequency in renal transplant patients	110
Immunosuppressive drugs for kidney transplantation	Diarrhea	*Cryptosporidium* present in 19% of patients, *Blastocystis hominis* in 39%; symptomatic infections with Cryptosporidium significantly higher in transplant patients than controls	78
Immunosuppressive drugs for liver transplantation	Diarrhea	Primary hypogammaglobulinemia, chronic hepatitis C, hepatic failure; 14 mo posttransplantation, *Cryptosporidium* diarrhea; no response to therapy; patient died	10
Immunosuppressive drugs for solid-organ transplantation[a]	Diarrhea	Included 3 liver transplants and 1 small bowel transplant; all 4 patients resolved cryptosporidiosis; while cryptosporidiosis is a nonlethal complication, it allows physician to monitor degree of immunosuppression and relationship to infection	34

[a] Excludes kidney transplantation.

Because other organisms may also fluoresce when auramine or the auramine-rhodamine combination is used, presumptive positive specimens should be confirmed by using one of the acid-fast methods. Both *Cryptosporidium* and *Isospora* spp. fluoresce with auramine-rhodamine and are acid-fast positive. Oocysts are fully mature and infective at the time they are passed in the stool. One can occasionally see the four sporozoites with the dark-staining residual body in the fully developed oocyst.

Immunoassay procedures for the direct detection of *Cryptosporidium* antigen or oocysts in fecal specimens have proven to be much more sensitive than the routine acid-fast stains. Since acid-fast stains do not always consistently stain all oocysts, the increased sensitivity and specificity of these immunoassays provide excellent testing methods. Enzyme immunoassay, solid-phase immunochromatographic assay in cartridge format, and FA methods are currently available. The choice of method depends on a number of variables and differs among diagnostic laboratories.

Since *Cryptosporidium* has been recovered from patients with AIDS, it has been recommended that stool specimens be submitted in 5 or 10% formalin, sodium acetate-acetic acid-formalin, or the newer single-vial collection systems for processing. However, each laboratory must confirm that the fixatives selected are compatible with diagnostic procedures performed, including the ova and parasite examination and immunoassays; not all single-vial systems are compatible with diagnostic immunoassay reagents. Laboratories should use standard precautions when working with these stool specimens. When specimens are centrifuged, capped centrifuge tubes should be used to prevent the formation of aerosols. Also, if possible, specimen processing should be performed in a biological safety cabinet.

Cyclospora cayetanensis

Cyclospora cayetanensis organisms are acid-fast-variable coccidia and have been found in the feces of immunocompetent travelers to developing countries, immunocompetent subjects with no travel history, and patients with AIDS (Figure 19.7). The life cycle has been confirmed, but information on potential reservoir hosts has yet to be defined; however, it appears that in some areas the human is the only host (Table 19.9). Outbreaks linked to contaminated water and various types of fresh produce (raspberries, basil, baby lettuce leaves, and snow peas) have been reported.

Oocyst maturation takes approximately 5 days or more, so the mature stage may not have been seen in human specimens. The oocyst contains two sporocysts, each containing two sporozoites, a pattern which places these organisms in the coccidian genus *Cyclospora*. Unsporulated oocysts are passed in the stool, and sporulation occurs within approximately 1 to 2 weeks. In patients who have *Cyclospora* oocysts in their stool specimens, two types of meronts and sexual stages have been found within the jejunal enterocytes.

There is generally 1 day of malaise and low-grade fever, with rapid onset of diarrhea of up to seven stools per day. There may also be fatigue, anorexia, vomiting, myalgia, and weight loss with remission of self-limiting diarrhea in 3 to 4 days followed by relapses lasting from 4 to 7 weeks. In patients with AIDS, symptoms may persist for as long as 12 weeks; biliary disease has also been

Table 19.8 Prevention of cryptosporidiosis in immunocompromised patients[a]

Preventive measure	Comments
Hand washing	Prior to eating and food preparation; after touching children in diapers; after touching clothing, bedding, toilets, or bed pans from anyone with diarrhea; after gardening; after touching pets or other animals; after touching or coming in contact with anything contaminated with human or animal stools (includes dirt); after removing gloves to perform any of the above activities
Safe sex	Always wash hands after touching partner's anal or rectal area
Avoidance of farm animals	Particularly important for contact with young animals (especially lambs and calves); wash hands thoroughly; avoid touching stools of any animal; after potential exposure, clean shoes (avoid having someone who is HIV infected perform this task)
Avoidance of pet stools	If HIV infected, avoid cleaning litter boxes or cages and disposing of stools; very important to avoid if animals are less than 6 months old; have puppies or kittens (if less than 6 months old) tested for *Cryptosporidium* before bringing them home; adult cats have also been found to be positive; have any pet tested that develops diarrhea
Precautions while swimming	Avoid swallowing water when in lakes, rivers, pools, or hot tubs; organisms not killed by routine chlorination; oocysts can also survive several days in salt water, so avoid swimming in polluted ocean water
Washing and cooking food	Wash well all fruits and vegetables if eating uncooked; use safe water for washing food; peel fruit; avoid unpasteurized milk or dairy products; cooking kills *Cryptosporidium*; cooked or packaged food is safe unless handled by someone infected with the organism
Water for drinking	Do not drink directly from lakes, rivers, streams, or springs; may want to avoid unboiled tap water (includes water and/or refrigerator ice-maker ice); boil water (rolling boil for 1 min sufficient to kill organisms), filter with appropriate filters (look for "reverse osmosis," "absolute 1 micron," "Standard 53," and the words "cyst reduction" or "cyst removal"); home distiller can be used; store water as for boiled water
Bottled water	Bottled water labels reading "well water," "artesian well water," "spring water," or "mineral water" do not guarantee that the water does not contain *Cryptosporidium*; however, water that comes from protected well or spring water sources is less likely to be contaminated than bottled water from rivers and lakes
Drinks	Canned or bottled soda, seltzer, fruit drinks, and steaming hot tea and coffee are safe; fountain drinks, fruit drinks mixed with tap water from concentrate, and iced tea or coffee may not be safe

[a] Adapted from information prepared by the interagency Working Group on Waterborne Cryptosporidiosis.

found in this group. Diarrhea alternating with constipation has also been reported; this is not uncommon in a number of protozoal gastrointestinal infections. Clinical clues include unexplained prolonged diarrheal illness during the summer in any patient and in persons returning from tropical areas. The clinical presentation of patients infected with this organism is similar to that of patients infected with *Cryptosporidium* spp.

In clean wet mounts, the organisms are seen as nonrefractile spheres and are acid-fast variable with the modified acid-fast stain; those that are unstained appear as glassy, wrinkled spheres. Modified acid-fast stains stain the oocysts from light pink to deep red, and some contain granules or have a bubbly appearance. It is very important to be aware of these organisms when the modified acid-fast stain is used for *Cryptosporidium* spp. and other similar but larger structures (approximately twice the size of *Cryptosporidium* oocysts) are seen in the stained smear. It is important for laboratories to measure all acid-fast oocysts, particularly if they appear to be somewhat larger

than those of *Cryptosporidium* spp. The oocysts autofluoresce green (1+ to 3+) (450 to 490 DM excitation filter) or blue (1+ to 3+) (365 DM excitation filter) under UV epifluorescence. It is strongly recommended that, during concentration (formalin/ethyl acetate) of stool specimens, the centrifugation time and speed be 10 min at $500 \times g$.

Apparently, patients do not respond to conventional antimicrobial therapies. Some patients have been treated symptomatically with antidiarrheal preparations and have obtained some relief; however, the disease appears to be self limiting within a few weeks. Trimethoprim-sulfamethoxazole is currently the drug of choice; relief of symptoms has been seen in 1 to 3 days posttreatment. However, symptoms recur within 1 to 3 months posttreatment in over 40% of the patients. Although prophylaxis with trimethoprim-sulfamethoxazole is not generally recommended, it might be worth considering for an HIV-positive person visiting an area of endemic infection like Nepal during the summer, when the risk of infection tends to be the highest.

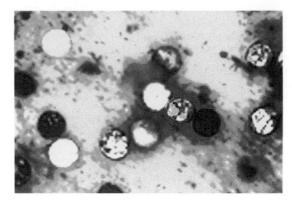

Figure 19.7 *Cyclospora cayetanensis* oocysts stained using the modified acid-fast stain (note the spherical shape; oocysts measure 8 to 10 µm; some oocysts do not stain, thus the organisms are said to be "modified acid-fast variable").

Published reports indicate that individuals of all ages, including immunocompetent and immunosuppressed individuals, can become infected. In Peru, infections with this organism have shown some seasonal variation, with peaks during April to June. This pattern is similar to that seen for *Cryptosporidium* infections in Peru. Preliminary data and extrapolation from what we currently know about cryptosporidiosis suggest that modes of transmission may be similar, particularly considering waterborne transmission.

Isospora (Cystoisospora) belli

An organism similar to *Cryptosporidium* that causes diarrhea in the compromised host is *Isospora belli* (Tables 19.3 and 19.4; Figure 19.8). Transition to the new genus name will probably occur during the next couple of years. Although *Isospora* infections have been found in many areas of the world, certain tropical areas in South America appear to contain some well-defined locations of endemic infections. These organisms can infect both adults and children. Intestinal tract involvement and symptoms are generally transient unless the patient is immunocompromised (Tables 19.4 and 19.10). In a group of AIDS patients in Brazil, 2% were found to be positive for *I. belli* infection. In reviewing prevalence data for this and other parasites in this group, there did not appear to be any association between CD4$^+$ cell counts and any particular parasite. However, the data supported the value of using standard fecal examinations with this group of infected patients, even in the absence of diarrhea, since examinations can be easily performed, have a low cost, and can disclose treatable conditions (17).

Relapse tends to be common in both immunocompetent and immunosuppressed patients and is thought to be associated with extraintestinal stages (32). These unizoite tissue cysts in extraintestinal sites have been confirmed to be present in the liver, spleen, and mesenteric lymphoid tissues in AIDS patients (32, 68).

Table 19.9 *Cyclospora cayetanensis*: updated information[a]

Information	Comments
Disease	
Haiti	Fewer infections in winter; summer temperatures may influence seasonality of infection
Nepal	Oocysts identified in sewage water, vegetable washings
Egypt	HIV-negative patient with weight loss, cough with purulent sputum, and dyspnea; suggested tuberculosis, *Cyclospora* confirmed by nested PCR; 2 cases on record; organism must be considered a new respiratory system pathogen
India	Important to test for intestinal parasites in HIV-positive patients
Reservoir hosts	
Haiti	Pigs, cattle, horses, goats, dogs, cats, guinea pigs, chickens, ducks, turkeys, pigeons negative; humans may be the only natural host
Disease risk factors	
Guatemala	U.S. infections linked to importation of Guatemalan raspberries; risk factors for workers include drinking water source, sewage drainage, proximity to chickens/other fowl; risk factors for children include contact with soil
Egypt	Sporulated and unsporulated oocysts found in tap water and lettuce heads; patients with Hodgkin's lymphoma on chemotherapy at risk
Peru	Oocysts found in wastewater; may be link to oocysts on vegetables
North America	Linked to raspberries, mesclun, basil, snow peas; over 3,600 cases since 1995

[a] Adapted from references 46, 59, 70, and 91.

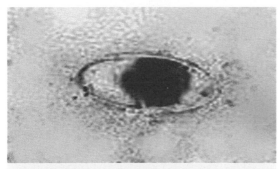

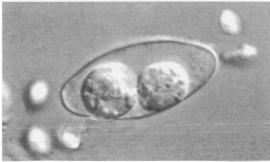

Figure 19.8 *Isospora (Cystoisospora) belli.* (Upper) Immature oocyst (contains single sporocyst) stained using modified acid-fast stain; (lower) more mature oocyst (contains two sporocysts) in a saline wet mount of stool concentration sediment.

I. belli infection is frequent in patients with AIDS in tropical areas. It has also been reported to occur in patients with other immunodepressive diseases, such as lymphoblastic leukemia, adult T-cell leukemia, and Hodgkin's disease. Non-Hodgkin's lymphoma with chronic diarrhea due to *I. belli* has also been reported (90). *I. belli* can cause severe chronic diarrhea in patients with malignancies whose country of origin is in an area of endemic infection. Isosporiasis should be suspected in HIV-infected patients from tropical countries presenting with diarrhea, weight loss, eosinophilia, and low CD4+ cell counts.

Since *I. belli* infects both immunocompetent and immunocompromised adults and children, symptoms range from none to severe diarrhea. Persistent nonbloody diarrhea, similar to that seen with microsporidiosis or cryptosporidiosis, is the major symptom seen in immunocompromised individuals. Pathology with blunting of the villi, villous atrophy, and collagen deposition in the lamina propria is commonly seen. Extraintestinal infections have been detected in AIDS patients, revealing infections in lymph nodes, liver, spleen, and biliary system.

The life cycle of *I. belli* is generally considered to be similar to the life cycles of other coccidia; however, *I. belli* is thought to infect only humans (see chapter 4). Oocysts are passed in the stool and are usually underdeveloped

and noninfectious when passed. These oocysts are most often seen in wet preparation, such as the direct smear or the concentration sediment.

There has been much speculation over the years that many cases of *Isospora* infection go undetected and that the incidence may be higher than that reported in the literature. However, it is also possible to have a positive biopsy specimen and not recover the oocysts in the stool because of the small numbers of organisms present. The oocyst wall in *I. belli* remains intact, and development can range from little differentiation to the presence of two well-defined sporocysts within the oocyst wall.

Sarcocystis spp.

Sarcocystis spp. require two hosts for completion of the life cycle. Two well-described species are *Sarcocystis bovihominis* (cattle) and *S. suihominis* (pigs). When raw or poorly cooked meat from infected animals is ingested by humans, gamogony occurs within the intestinal cells, with the production of sporocysts in the stool. In this case, humans who have ingested meat containing the mature sarcocysts serve as the definitive hosts. There have been reports of fever, severe diarrhea, abdominal pain, and weight loss in compromised hosts. Eosinophilic enteritis and ulcerative obstructive enterocolitis may be seen as occasional complications. In one group of Thai laborers, a 26.6% prevalence of *Sarcocystis* infection was seen. This high prevalence rate was indicative of the local habit of eating raw beef and pork; sarcocystosis could be a significant food-borne zoonotic infection in Thailand (115).

The sporocysts that are recovered in the stool are broadly oval, measuring 9 by 16 μm and containing four mature sporozoites and the residual body. Normally two sporocysts are contained within the oocyst (similar to *I. belli*); however, in *Sarcocystis* infections, the sporocysts are released from the oocyst and normally are seen singly. They tend to be larger than *Cryptosporidium* oocysts, which contain four sporozoites, and so there should be no confusion between the two (29). Differentiation of encysted protozoan parasites in human feces is described in Table 19.11.

When humans serve as the accidental intermediate host, they accidentally ingest oocysts from other animal stool sources. The sarcocysts that develop in human muscle (schizogony) apparently do little if any harm (Figure 19.9). There is essentially no inflammatory response to these stages in the muscle and no conclusive evidence of pathogenicity of the mature sarcocyst. A number of different morphologic types of skeletal and cardiac muscle sarcocysts have been recovered in humans; sarcocysts have also been found in muscles of the larynx, pharynx,

Table 19.10 *Isospora (Cystoisospora) belli*: parasite development and disease

Characteristic	Description	Comments
Infection source	Transmitted by infective oocysts in fecal-oral transmission	Excystation induced by body temperature with bile or bile salts and trypsin
Infective stage	Mature oocysts containing sporocysts with internal sporozoites; not infectious when passed	Oocyst wall is split, thus releasing sporozoites
Developmental stages	Zoites and meronts develop within epithelial cells; eventually merozoites form macrogametes and microgametes	Small intestine; in immunocompromised, parasite development occurs in biliary epithelial cells, as well; parasite cysts also seen in extraintestinal sites such as lymph nodes, liver, and spleen
Immunocompromised host	Recurrence of infection in AIDS patients following successful chemotherapy	May be due to activation of cysts in extraintestinal sites
Cellular and molecular biology	Parasite material difficult to obtain; culture difficult, complete development difficult	Overall little known about biology of this parasite
Clinical aspects	Infects both immunocompetent and immunocompromised patients; 54% have peripheral eosinophilia; vomiting, steatorrhea, headache, fever, malaise; symptoms in children can be severe	Most information comes from disease as seen in Chile; persistent nonbloody diarrhea like that seen with microsporidiosis and cryptosporidiosis
Immunology	CD4 T cells involved	Infection more severe in AIDS patients (low CD4 counts)
Pathology	Moderate to severe villous atrophy; collagen deposition in the lamina propria	Both sexual and asexual stages seen in epithelium, enclosed by parasitophorous vacuole
Diagnosis	Routine fecal ova and parasite examination; concentration important	Modified acid-fast stains helpful; auramine-rhodamine; unstained oocysts autofluorescent; intestinal biopsy (routine stains)
Epidemiology	Transmission most common in tropical zones (subequatorial Africa, Caribbean, parts of South America)	Prevalence in populations with high HIV infection quite variable; percentages vary from 3% to over 20%; rarely seen in HIV-positive children
Treatment	Trimethoprim-sulfamethoxazole recommended	An approved drug, but considered investigational for this condition by the Food and Drug Administration

and upper esophagus. Generally, histopathologic diagnoses are myositis with vasculitis and sometimes myonecrosis. No specific therapy is known for this type of infection. Characteristics of this infection can be seen in Table 19.12. In another four cases, the patients presented with lumps, pain in the limbs, or a discharging sinus. Histologic examination of biopsy specimens in all cases revealed characteristic cysts of *Sarcocystis* spp.

For infections in which humans serve as definitive hosts, prevention involves adequate cooking of beef and pork; for infections in which humans are intermediate hosts, preventive measures involve careful disposal of animal feces possibly containing the infective sporocysts. This may be impossible in the wilderness areas, where wild animals may serve as reservoir hosts for many of the different types of organisms that have been grouped under the term

"*Sarcocystis lindemanni.*" However, evidence of multiple morphologically different cysts indicates that there are probably several species of *Sarcocystis* that cause human infections. Therefore, the name is no longer used. No specific therapy is known for the muscle stages of *Sarcocystis* spp., and there is no report of prophylaxis in humans.

Microsporidia

The microsporidia are obligate intracellular parasites that have been detected in a variety of animals, particularly invertebrates; the organisms found in humans tend to be quite small, ranging from 1.5 to 3 µm. Until recently, awareness and understanding of human infections have been marginal; only with increased understanding of AIDS within the immunosuppressed

Table 19.11 Encysted pathogenic protozoan parasites seen in human feces

Protozoan parasite	Size (μm)	Shape	Features
Sarcocystis hominis and *S. suihominis*			Excreted sporulated (immediately infectious)
Oocysts	15–19 by 15–20	Spherical	Contain 2 sporocysts
Sporocysts	15–19 by 8–10	Oval	Contain 4 sporozoites
Isospora (*Cystoisospora*) *belli*	20–33 by 10–19	Ovoid with tapered ends	Excreted unsporulated (not immediately infectious)
Cyclospora cayetanensis	7.7–10	Spherical	Excreted unsporulated (not immediately infectious)
Cryptosporidium hominis and *C. parvum*	4–6	Nearly spherical	Excreted sporulated (immediately infectious); some internal sporozoites difficult to see
Giardia lamblia	11–14	Spherical to ovoid to ellipsoid	4 nuclei, curved median bodies, linear axonemes (immediately infectious)
Entamoeba histolytica and *E. dispar* (*E. histolytica/E. dispar* group)	10–15	Spherical	Precyst generally contains single nucleus (karyosome with peripheral nuclear chromatin); mature cyst contains 4 nuclei and possible chromatoidal bars (with round, smooth ends) (mature cyst immediately infectious)
Blastocystis hominis (currently classified with the protozoa)	6–40	Spherical to somewhat oval	Central body form with nuclei around the central body area (immediately infectious as far as known)
Balantidium coli	50–70	Spherical to oval	Large macronucleus, thick wall (infectious)

population has attention been focused on these organisms (Tables 19.3, 19.4, and 19.13; Figure 19.10). Limited availability of electron microscopy capability has also played a role in our inability to recognize and diagnose these infections.

Phylogenetic studies have confirmed that the microsporidia evolved from the fungi, being most closely related to the zygomycetes (66). Features shared with fungi include the presence of chitin and trehalose, similarities in cell cycles, and certain gene organizations. Microsporidia are now considered to be highly derived fungi that underwent genetic and functional losses, thus resulting in one of the smallest eukaryotic genomes known. However, at this point, clinical and diagnostic issues and responsibilities may remain with the parasitologists, and these organisms are maintained as a part of this parasitology text. We may be in a transition stage, similar to that seen with *Pneumocystis jiroveci* as it was moved from the parasites to the fungi in terms of classification status.

The organisms are characterized by having spores containing a polar tubule, which is an extrusion mechanism for injecting the infective spore contents into host cells. To date, seven genera have been recognized in humans: the more common, *Encephalitozoon* and *Enterocytozoon*, and the less common, *Brachiola*, *Pleistophora*, *Trachipleistophora*, *Vittaforma*, and "*Microsporidium*," a catch-all genus for organisms that have not yet been classified (or may never be classified due to a lack of specimen). Classification criteria include spore size, configuration of the nuclei within the spores and developing forms, the number of polar tubule coils within the spore, and the relationship between the organism and host cell. Infection occurs with the introduction of infective sporoplasm through the polar tubule into the host cell. The microsporidia multiply extensively within the host cell cytoplasm; the life cycle includes repeated divisions by binary fission (merogony) or multiple fission (schizogony) and spore production (sporogony). Both merogony and sporogony can occur in the same cell at the same time. During sporogony, a thick spore wall is formed, providing environmental protection for this infectious stage of the parasite. An example of infection potential is illustrated by *Enterocytozoon bieneusi*, an intestinal pathogen. The spores are released into the intestinal lumen and are passed in the stool. These spores are environmentally resistant and can then be ingested by other hosts.

Encephalitozoon

Both *Encephalitozoon cuniculi* and *E. hellem* have been isolated from human infections, the first species from the CNS and the second from the eye. The first case occurred in a 9-year-old Japanese boy who presented with neurologic symptoms such as convulsions, vomiting,

Table 19.12 *Sarcocystis*: parasite development and disease[a]

Characteristic	Infection in muscle	Infection in intestine
Infection source	Food or water contaminated with feces from unknown carnivore or omnivore	Raw or undercooked meat
Infective stage	Oocyst or free sporocysts	Sarcocyst containing bradyzoites
Developmental stages	Intravascular schizonts (not seen); intramuscular sarcocysts	Sexual stages in lamina propria; oocysts passed in feces
Incubation period from ingestion to symptoms	Weeks to months; can last months to years	3–6 h, lasting 36 h
Symptoms	Musculoskeletal pain, fever, fleeting pruritic rash, transient lymphadenopathy, subcutaneous nodules associated with eosinophilia, cardiomyopathy, and bronchospasm	Nausea, loss of appetite, vomiting, stomach ache, bloat, diarrhea, dyspnea, and tachycardia
Diagnosis	Biopsy specimen containing sarcocyst; antibodies to bradyzoites (routine histology)	Oocysts or sporocysts in feces, beginning 5–12 days after ingestion (routine ova and parasite examination, particularly concentration)
Therapy (none approved)	Co-trimoxazole, furazolidone, albendazole, anticoccidials, pyrimethamine, anti-inflammatories	None

[a] Adapted from reference 29.

headaches, fever, and periods of unconsciousness (66). The spores were recovered in the CSF and in urine. This particular patient was treated with sulfisoxazole and recovered. Another case of *E. cuniculi* infection was in a 2-year-old child with a similar neurologic illness; the spores were recovered from the urine. A recent case of keratoconjunctivitis and chronic sinusitis due to infection with *E. cuniculi* has been found in an AIDS patient; the diagnosis was confirmed by electron microscopy (66). Several eye infections with *E. hellem* have been found in AIDS patients, including the first reported case of *E. hellem* infecting not only sinuses and conjunctivae but also the nasal epithelium. In the first case of disseminated *E. hellem* infection in an AIDS patient, a complete autopsy revealed organisms in the eyes, urinary tract, and respiratory tract. The presence of numerous organisms within the lining epithelium of almost the entire length of the tracheobronchial tree was very suggestive of respiratory acquisition. It is also interesting that although *E. cuniculi* and *E. hellem* differ biochemically and immunologically, their fine structure and development are indistinguishable. Some of the eye infections have also suggested a topical route rather than dissemination.

Brachiola spp.

Brachiola connori (formerly referred to as *Nosema connori*) was identified in human tissues in a 4-month-old infant with combined immunodeficiency disease who had contracted a disseminated, fatal infection. Parasites were found in the myocardium, diaphragm, arterial walls, kidney tubules, adrenal cortex, liver, and lungs. Concurrently, this patient also had *P. jiroveci* pneumonia. *B. algerae* (formerly *Nosema algerae*) is apparently transmitted by a mosquito. Infections have been found in the eye and muscle, possibly from a disseminated case. Both immunocompromised (HIV-negative, but receiving immunosuppressive therapy for rheumatoid arthritis) and immunocompetent patients have become infected. *B. vesicularum* has also been associated with AIDS and myositis. Muscle tissue obtained at biopsy and examined by light and electron microscopy revealed organisms developing in direct contact with muscle cell cytoplasm and

Figure 19.9 *Sarcocystis* sp. in muscle tissue (note the bradyzoites contained within the sarcocyst).

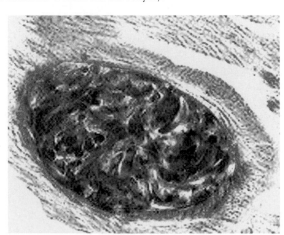

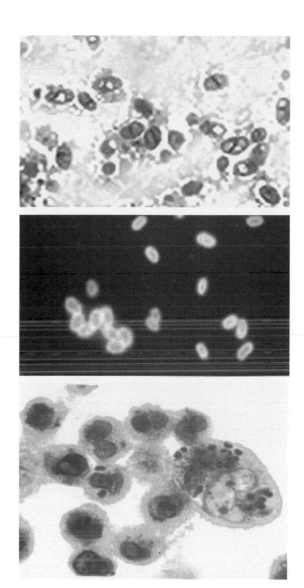

Figure 19.10 Microsporidia. (Top) Microsporidial spores seen in fecal specimen (concentration sediment) stained with Ryan modified trichrome stain (note the horizontal line through some of the spores, representing the presence of the polar tubule); (middle) spores stained with specific FA reagents for the detection of *Encephalitozoon* spp.; (bottom) spores within a urine sediment after staining with Ryan modified trichrome stain (taken at a lower magnification).

fibers. Apparently, no other tissue types were infected. Clinical and histologic clearing of the infection was achieved after treatment with albendazole and itraconazole.

Pleistophora and *Trachipleistophora* spp.

Microsporidia in the genera *Pleistophora* and *Trachipleistophora* have rarely been identified in humans. However, when *Pleistophora* was found, atrophic and degenerating muscle fibers were full of spores which

were seen in clusters of about 12 organisms, each cluster enclosed by an enveloping membrane, the pansporoblastic membrane. The disease was characterized by a 7-month history of progressive generalized muscle weakness and contractions, in addition to fever, generalized lymphadenopathy, and an 18-kg weight loss. Although the patient was diagnosed as having AIDS, the presence of HIV was never demonstrated. Specimens from early biopsies demonstrated atrophic and degenerating muscle fibers; specimens from later biopsies showed fibrosis and scarring. *Trachipleistophora hominis* causes severe myositis and sinusitis and was first found in an AIDS patient. To date, no natural hosts other than humans have been identified; however, when inoculated intraperitoneally into athymic mice, the organisms infect the skeletal muscle, as seen in human infections. *T. antropophtera* is the most recently described microsporidian in humans and has been found in brain tissue of two AIDS patients at autopsy; organisms were also found in the kidneys and heart.

Enterocytozoon bieneusi

Although currently there are seven genera, one of which is the catchall genus "*Microsporidium*," human infections with some genera are more common than with others. A number of infections with *Enterocytozoon bieneusi* have been found in AIDS patients. Chronic intractable diarrhea, fever, malaise, and weight loss are symptoms of *E. bieneusi* infections and are similar to those seen with cryptosporidiosis or isosporiasis. These patients have already been diagnosed with AIDS and each day have four to eight watery, nonbloody stools, which can be accompanied by nausea and anorexia. There may be dehydration with mild hypokalemia and hypomagnesia, as well as D-xylose and fat malabsorption. The patients tend to be severely immunodeficient, with a CD4 count always below 200/mm^3 and often below 100/mm^3. *E. bieneusi* infection has also been implicated in AIDS-related sclerosing cholangitis.

Dual infections with *E. bieneusi* and *Encephalitozoon intestinalis* have been reported. *Encephalitozoon intestinalis* infects primarily small intestinal enterocytes, but infection does not remain confined to epithelial cells. *E. intestinalis* is also found in lamina propria macrophages, fibroblasts, and endothelial cells. Dissemination to the kidneys, lower airways, and biliary tract appears to occur via infected macrophages. Fortunately, these infections tend to respond to therapy with albendazole, unlike the infections with *E. bieneusi*.

A number of techniques for recovery and identification of microsporidia in clinical specimens are available. Although the organisms can be identified in routine histologic preparations, they do not tend to stain predictably. Occasionally, the spores take on a refractile

Table 19.13 Microsporidia: updated information (AIDS patients)

Findings in AIDS patients	Comments[a]	Reference
Intestinal microsporidiosis	High proportion of patients remained symptomatic after 6 mo (54.8% with persistent diarrhea, 51.2% with weight loss); predictors include high HIV RNA viral load and no initiation of protease inhibitor therapy	23
Children in Spain	1.2% overall and 2% of those with diarrhea (*n* = 83) had microsporidiosis; different data from those previously reported for adults (13.9%) in the same area, although adult group showed more severely depressed CD4 lymphocyte counts than children did	25
Children in Uganda	Some 81% of HIV-infected children with persistent diarrhea excreted *Cryptosporidium*, *Enterocytozoon bieneusi*, or both organisms, compared with only 10% of children with persistent diarrhea testing negative for HIV; 74% of isolates were *C. hominis* (the anthroponotic species), 17% were *C. parvum* (the zoonotic species), and 8% were a mixture of the 2 or others	111
Patients with or without diarrhea in Germany	Prevalence of intestinal microsporidiosis varies from 7 to 50%; most common enteropathogen in 18/50 patients (36%) with diarrhea and 2/47 (4.3%) without diarrhea; microsporidia present in 60% of patients with chronic diarrhea and 5.9% of patients with acute diarrhea (*E. bieneusi* in 18, *E. intestinalis* in 2); prevalence in this group one of highest in the world	103
Mucosal abnormalities	Mechanisms appear to be malabsorption caused by reduction of absorptive mucosal surface and impairment of enterocyte function	95
Epidemiology		
Waterborne outbreak	Of 338 individuals, 261 had HIV, 16 were transplant recipients, and 61 had other sources of immunosuppression; disease appears to be endemic in HIV-positive persons; 200 people involved in waterborne outbreak; no evidence of fecal contamination of water; outbreak ended several months before antiprotcasc cra	21
Risk factors	Intestinal microsporidiosis in persons with HIV infection and ≤200 CD4 cells/mm³ is associated with male homosexuality and swimming in pools, suggesting fecal-oral transmission, including sexual and waterborne routes	47
New disease manifestation		
Nodular skin disease	36-yr-old woman presented with late-stage AIDS and disseminated, nodular cutaneous lesions and underlying osteomyelitis; disseminated microsporidial infection with *Encephalitozoon*-like species diagnosed by EM; long-term clindamycin therapy cured cutaneous infection; response to clindamycin merits further study	53
Current clinical syndromes	Enteropathy, keratoconjunctivitis, sinusitis, tracheobronchitis, encephalitis, interstitial nephritis, hepatitis, cholecystitis, osteomyelitis, myositis; *E. bieneusi* is the most common cause of intestinal disease (can also disseminate); *E. (Septata) intestinalis* is associated with disseminated and intestinal disease	58
Modification of clinical course	Of 37 AIDS patients with *E. bieneusi* diarrhea, 15 showed clearance (40.5%); peripheral blood CD4 cell count of ≥100/mm³, use of two or more antiretroviral medications, and use of protease inhibitor statistically associated with time to clearance of parasite; albendazole not associated with parasite eradication	19

[a] EM, electron microscopy.

gold appearance in formalin-fixed, paraffin-embedded, routine hematoxylin-and-eosin-stained sections. Some of the difficulty may be attributed to the use of formalin; alternative fixatives are currently being tried. Spores are occasionally seen very well by using the PAS stain, the methenamine-silver stain, tissue Gram stains, or acid-fast stains. The spore has a small, PAS-positive posterior body, the spore coat stains with silver, and the spores are acid-fast variable. There is also evidence to indicate that specimens from plastic-embedded tissues are seen more easily, regardless of the fixative used. Tissue examination by electron microscopy is still considered the best approach; however, this option is not available to all laboratories and is not that sensitive.

Techniques that do not require tissue embedding are also popular. Touch preparations of fresh biopsy material that are air dried, methanol fixed, and Giemsa stained have been used; however, screening must be performed using the 100× oil immersion objective. Another study, involving cytocentrifugation followed by Giemsa staining, found that 27% of 55 AIDS patients with chronic diarrhea were positive for *E. bieneusi*; all results were confirmed by electron microscopy.

Routine staining of clinical specimens (stool, urine, eye specimens) can be performed using a modified trichrome stain in which the chromotrope 2R component added to the stain is 10 times the concentration normally used in the routine trichrome stain for stool. The stool preparations must be *very* thin, the staining time is 90 min, and the slide must be examined at ×1,000 (or higher) magnification. Unfortunately, there are many objects within stool material that are oval, stain pinkish with trichrome, and measure approximately 1.5 to 3 µm. If this stain is used for the identification of microsporidia in stool, positive control material should be available for comparison. The newest approach for the identification of spores in clinical specimens uses antisera in an IFA procedure. Fluorescing microsporidial spores were distinguished by a darker cell wall and by internal visualization of the polar tubule as diagonal lines or cross lines within the cell. In another study with this same antiserum, 9 (33%) of 27 patients diagnosed as having cryptosporidiosis (8 AIDS patients and 1 non-AIDS but immunodeficient patient) were found to have *E. bieneusi* in the stool. Although there is some cross-reactivity with bacteria, this technique offers a more sensitive approach than routine staining methods currently available for the examination of stool specimens. As clinicians begin to suspect these infections and laboratorians become more familiar with the diagnostic methods, the number of positive patients, particularly those who are immunocompromised, may increase dramatically.

Although *E. bieneusi* appears to be the most common human microsporidian, serologic evidence is based primarily on results for *Encephalitozoon cuniculi* as the parasite antigen. Whether these surveys reveal actual infections with this organism or whether there is cross-reactivity with other microsporidians is unknown. However, the data strongly suggest that latent infections with microsporidia occur in a number of groups, ranging from patients with other infections such as tuberculosis, typhoid fever, leprosy, malaria, schistosomiasis, toxoplasmosis, Chagas' disease, and toxocariasis to normal individuals and a group of homosexual men (44). Serologic evidence of human infection with the microsporidia has been reviewed elsewhere (44). Microsporidian infections in immunocompetent patients may result in a self-cure following mild symptoms over a short time frame, a situation similar to that seen with both cryptosporidiosis and isosporiasis. As awareness of this infection increases and more sensitive diagnostic techniques are developed, we may find that these infections are not uncommon in immunocompetent hosts.

The vast majority of human cases of microsporidiosis have been reported during the past 20 years, in patients with HIV/AIDS, while only relatively rare cases have been described in immunocompetent individuals. However, microsporidial infections are being increasingly found in patients following solid-organ transplantation, where the main symptom has been diarrhea. The first case of pulmonary microsporidial infection in an allogeneic bone marrow transplant recipient in the United States and only the second case in the world has recently been reported (79).

Leishmania spp.

Visceral leishmaniasis (VL) continues to increase in frequency in patients who are HIV positive. The majority of cases of leishmaniasis in AIDS patients have been seen in France, Spain, Italy, and Portugal. Like other opportunistic infections, subclinical VL can be found at any stage of HIV-1 infection, but symptomatic cases appear mainly when severe immunosuppression is present. HIV *Leishmania* coinfection is being seen more and more frequently in the Mediterranean basin (Table 19.14; Figure 19.11).

Over 850 coinfection cases have been recorded where 7 to 17% of HIV-positive individuals with fever have amastigotes. This finding strongly suggests that individuals infected with *Leishmania* spp. without symptoms will express symptoms of leishmaniasis if they become immunosuppressed. Based on the patient's general health status and prior exposure to the infection, leishmaniasis may be seen as an early opportunistic infection or may be seen as a complication late in the course of AIDS. Clinical signs of VL usually occur in the later stages of AIDS. The majority of AIDS patients present with the classic picture of VL; however, asymptomatic cutaneous, mucocutaneous, diffuse cutaneous, and post-kala azar dermal leishmaniasis (usually caused by *L. infantum*) can be seen. Cutaneous lesions in VL are being detected increasingly frequently in patients with HIV infection, and their significance is still somewhat unclear. Lesions often do not present a uniform or specific appearance and have been seen as erythematous papules and hypopigmented macules on the dorsa of the hands, feet, and elbows; small subcutaneous nodules on the thighs; and erythematoviolaceous, scaly plaques on the face. Also, the digestive and respiratory tracts are often parasitized, as well as the pleura and peritoneum. Lesions can also be coinfected with other organisms, such as *Mycobacterium avium-intracellulare*.

CD4 T-lymphocyte counts are usually less than 50/mm^3 and are almost always less than 200/mm^3. Many physicians think that VL does not cause severe symptoms in these patients but is just one of several opportunistic infections that may coinfect AIDS patients. However, VL can manifest atypical aspects in HIV-positive patients depending on the degree of immunosuppression, and it

Table 19.14 *Leishmania* infections in immunocompromised patients

Underlying condition	Disease manifestation[a]	Comments[a]	Reference
AIDS	VL (southern Spain)	HIV-1 carriers (291) underwent bone marrow aspiration (regardless of symptoms); 32 (11%) showed VL and 13 of these (41%) had subclinical cases; female sex negatively associated with VL; i.v. drug use high, but not independent association; high proportion of cases subclinical; symptomatic cases evident when deep immunosuppression is present	85
AIDS	VL (Mediterranean area)	*Leishmania*-HIV coinfection common in Mediterranean area; most common species is *L. infantum*; most frequent if CD4 count is <200/mm³ and in parenteral drug addicts; chronic development with relapses is frequent; most effective diagnostic method is bone marrow culture (NNN); low sensitivity for serologies; PCR useful in asymptomatic cases, for therapeutic control, and in relapses; treatment similar to that in immunocompetent patients; generally necessary to use secondary prophylaxis; mortality rate is high (~25%)	72
AIDS	VL and possible dissemination	WHO estimates that 2–9 in 100 HIV patients in Africa and Asia will develop *Leishmania*-HIV coinfection; VL most common; unusual locations suggest multiorgan spreading in absence of host immune response; high incidence of recurrence; follow-up required; clear picture of prophylaxis not yet developed	62
AIDS	VL	Since 1990, 1,616 cases of coinfection reported, mainly from southern Europe (Spain, southern France, Italy); coinfected patients mainly young adults, i.v. drug users; 272 isolates showed 18 different zymodemes, 10 of which were found only during HIV coinfection; new foci emerging (Brazil and East Africa)	24
AIDS	VL (and mucocutaneous disease in areas where *L. donovani* is not endemic)	15 AIDS patients in Germany, all of whom were severely immunocompromised; 1 was diagnosed at autopsy; 1 mucocutaneous case was diagnosed by nasal biopsy; others were diagnosed by bone marrow (13/13), liver (3/3), and gastrointestinal mucosa (4/4) biopsies; serology positive in only 6/13; VL must be ruled out in every patient with fever and/or pancytopenia and appropriate travel history	4
AIDS	Case mimicking VL	HIV-positive patient presented with picture of VL coinfection; bone marrow biopsy performed on admission; amastigote forms seen in bone marrow, and organisms grew in culture as promastigotes; molecular analysis revealed that the organisms were not *Leishmania*, but were *Leptomonas pulexsimulantis*, a trypanosomatid found in dog fleas; thus, opportunistic infection with an insect trypanosomatid was suspected and confirmed	80
Immunosuppressive drugs for organ transplantation or cancer	VL	Immunosuppressive therapy results in reactivation of asymptomatic VL or facilitates new infections	24
AIDS	CL (Brazil)	IRIS has been found in AIDS patients since the introduction of HAART; syndrome is characterized by clinical manifestations of opportunistic infections when signs of immune reconstitution are observed during therapy; leishmaniasis, suggestive of HAART-induced IRIS, in 2 patients with AIDS was seen; after beginning HAART, 1 patient presented with disseminated CL lesions,	87

(continued)

Table 19.14 *(continued)*

Underlying condition	Disease manifestation[a]	Comments[a]	Reference
AIDS *(continued)*	CL (Brazil)	whereas the other patient's preexisting lesions worsened and became more extensive; however, their CD4+ T-cell counts were increasing and viral loads were decreasing significantly; the lesions healed with anti-*Leishmania* therapy	
AIDS	CL (French Guiana)	In French Guiana, CL in moderately immunosuppressed HIV-infected subjects (CD4+ T-cell count, >200/mm3) is characterized by a higher rate of recurrence or reinfection and is more difficult to treat than that in HIV-negative subjects	22
AIDS	CL (Italy)	Patient was HIV infected, an i.v. drug user, with an unusual disseminated CL after an initial visceral disease and after a 13-month maintenance treatment with liposomal amphotericin; severe concurrent immunosuppression probably played a role in leading to this atypical cutaneous form, characterized by diffuse, nonulcerated, nonscabby maculopapular lesions	14

[a] CL, cutaneous leishmaniasis; HAART, highly active antiretroviral therapy; IRIS, immune reconstitution inflammatory syndrome; i.v., intravenous; NNN, Novy-MacNeal-Nicolle medium; WHO, World Health Organization.

should be listed among AIDS-defining conditions. The most common symptoms are fever, splenomegaly, hepatomegaly, and pancytopenia.

Data from Spain provide some interesting outcome figures regarding *Leishmania* infections in persons with or without HIV infections. In spite of a good initial response to treatment for VL, 60.6% of the patients had relapsed by the end of 1 year. Mortality from the first episode was 18.5%, and 24% died in the first month after diagnosis of any VL episode. The mean survival of the 29 patients who died was 10.27 months. Survival in patients with and without AIDS at the time of the first episode of VL was compared at 30 months and found to be 30.5 and 53.7%, respectively. A diagnosis of AIDS

Figure 19.11 *Leishmania donovani* in bone marrow (note the individual amastigotes, each one containing the bar and nucleus); specimen stained using Giemsa stain.

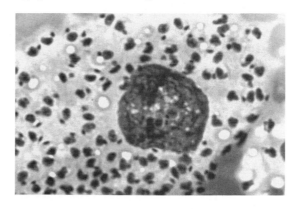

at the time of the first episode of VL and thrombocytopenia were the only risk factors found to be related to survival. Certainly in AIDS patients, VL is a recurrent disease that is highly prevalent and whose clinical course is modified by HIV. VL is a very prevalent disease among HIV-1-infected patients in southern Spain, with a high proportion of cases being subclinical. Symptomatic cases appear mainly when severe immunosuppression is present. There is also an association between VL and the male sex and intravenous drug use. In HIV-infected patients, VL occurs in late stages of HIV disease and often has a relapsing course. Secondary prophylaxis reduces the risk of relapse.

In HIV-1-infected patients with active VL, the sensitivity of a peripheral blood smear is about 50%, because of the high parasitemia in these patients. However, in HIV-1-infected patients with subclinical VL, the sensitivity of a routine blood smear is <10%. Unfortunately, in these patients serology and *Leishmania* skin tests have low sensitivities and are little help. Cultures have proven to be effective in diagnosis of VL in patients who are coinfected with HIV-1.

A clinical diagnosis of Old World VL is often made based on clinical findings and local epidemiologic factors. However, confirmation of the diagnosis requires demonstration of the amastigotes in tissues or clinical specimens (blood, buffy coat preparations) or the promastigotes in culture. Detection of parasite genetic material or antigen detection is also sufficient for confirmation. Leishmaniasis should be suspected in individuals who have resided in or traveled to areas where the disease is endemic. The diagnosis would be supported by findings of remittent fevers, hypergammaglobulinemia with anemia, circulating immune complexes, rheumatoid factors, weight loss, leukemia, and splenomegaly. The differential diagnosis of late-stage VL is limited to hematologic and lymphatic malignancies. The diagnosis of early disease is more difficult, and the differential would include malaria, brucellosis, enteric fevers, bacterial endocarditis, generalized histoplasmosis, chronic myelocytic leukemia, Hodgkin's disease and other lymphomas, sarcoidosis, hepatic cirrhosis, and tuberculosis. Also, patients with multiple myeloma and Waldenström's macroglobulinemia have monoclonal hypergammaglobulinemia. Mixed cryoglobulinemia secondary to VL has also been reported.

Although the usual route of transmission is through the bite of infected sand flies, parasites can be transmitted in other ways including blood transfusion, sexual contact, congenital transmission, occupational exposure, and shared needles. Apparently, *L. infantum* can circulate intermittently and at low density in the blood of healthy seropositive individuals, who appear to be asymptomatic carriers; this certainly has implications for the safety of the blood supply in certain areas of the world, including southern France, where coinfection with HIV is a problem.

Strongyloides stercoralis

In the life cycle of the intestinal nematode *Strongyloides stercoralis*, the larvae migrate through the heart and lungs, pass up the trachea, are swallowed, and finally grow to maturity in the gastrointestinal tract (see chapter 10) (Tables 19.3, 19.4, and 19.15; Figure 19.12). One significant feature of the *Strongyloides* life cycle is an internal autoinfection capability. Noninfective rhabditiform larvae, which normally pass out in the stool, may transform to infective filariform larvae while still within the gastrointestinal tract or perhaps on the perianal surface. These larvae can then penetrate the bowel wall or skin and reinitiate the life cycle. This cycle can maintain itself at a very low level over a period of many years in an individual without causing symptoms. The only unusual finding in the patient may be an unexplained eosinophilia. Patients who contracted the original infection many years earlier can become severely ill when they become immunosuppressed (Table 19.4). In the compromised patient, the number of larvae and adult worms may increase rapidly, leading to the hyperinfection syndrome and disseminated strongyloidiasis.

Patients in whom systemic *S. stercoralis* infections have been found include those with various leukemias and lymphomas, chronic infections such as leprosy and tuberculosis, and miscellaneous conditions such as kidney disease, organ transplantation, asthma, systemic lupus erythematosus, and various fungal infections. Other conditions include malnutrition, alcoholism, chronic renal failure, and achlorhydria. When migrating *Strongyloides* larvae begin to increase in numbers in the compromised host, the patient may present with abdominal complaints and there may be repeated episodes of unexplained bacteremia or meningitis with enteric bacteria. This occurs when the larvae penetrate the bowel and reinitiate the cycle. In doing so, they carry members of the bowel flora with them. The other consistent finding in these patients is an unexplained eosinophilia, which may range from 10 to 50%. Like *G. lamblia*, *S. stercoralis* resides in the duodenum, and recovery of the larvae in the stool is difficult in patients with low worm burdens. For this reason, additional techniques like the Entero-Test string capsule and the duodenal aspirate may be used (see chapter 29). Other techniques for concentrating and recovering *Strongyloides* larvae include the Harada-Mori and petri dish culture techniques, as well as the Baermann apparatus.

In individuals with preserved immune function, direct development of *S. stercoralis* is favored, whereas in individuals with lesser immune function, indirect development is relatively more common. These results may explain the notable absence of disseminated strongyloidiasis in patients with advanced HIV disease. Because disseminated infection requires the direct development of infective larvae in the gut, the observed tendency toward indirect development in individuals immunosuppressed by advancing HIV disease is not consistent with the promotion of disseminated infection.

The typical rhabditiform larvae of *S. stercoralis* are characterized by a short buccal capsule or mouth opening and the presence of a genital primordial packet of cells. In contrast, the rhabditiform larvae of hookworm have a much longer buccal capsule and essentially no genital primordial cells present.

Whenever one is working with material from a patient suspected of having strongyloidiasis, one should use extreme care because of possible filariform larvae in the specimen. Gloves must be worn to prevent skin penetration by these larval forms.

As we prolong life for more individuals, the medical community must become aware of some of the parasitic infections that may pose a threat to the life of the compromised patient. Awareness is the first step in preventing more serious disease.

Crusted (Norwegian) Scabies

Scabies is a highly contagious infestation of the itch mite, *Sarcoptes scabiei*. Symptoms are related to pruritic dermatitis, and apparently transmission can occur through

Table 19.15   *Strongyloides stercoralis* infections in immunocompromised patients

Underlying condition	Disease manifestation	Comments	Reference
AIDS and non-Hodgkin's lymphoma	Disseminated disease	After chemotherapy, patient complained of retrosternal pain, dysphagia, dry cough, and upper abdominal discomfort; eosinophilia was 41%; recurrence of lymphoma ruled out; treatment for disseminated atypical mycobacterial infection; biopsy exam of duodenal fluid revealed worms and larvae; albendazole therapy successful	75
Idiopathic pulmonary fibrosis therapy	Disseminated disease	64-year-old man with pulmonary fibrosis was admitted for shortness of breath and cough; immunosuppressed due to steroids; repeated sputum and stool positive for *S. stercoralis*; died suddenly from severe cardio-respiratory failure while on mebendazole therapy	61
Nephrotic syndrome therapy	Disseminated disease	Patient with nephrotic syndrome developed disseminated strongyloi-diasis after steroid therapy; renal biopsy, time course, resolution of nephrotic syndrome after thiabendazole suggested causal relationship between chronic strongyloidiasis and nephrotic syndrome; early diagno-sis important to good clinical outcome	118
Nephrotic syndrome therapy	Glomerulonephritis	Glomerulonephritis documented	116
Nephrotic syndrome therapy	Disseminated disease	Problems after administration of steroids	74
Chronic renal failure	Intestinal disease	Vomiting, diarrhea predominant symptoms; albendazole effective	104
Immunosuppressive drugs for kidney transplantation	Disseminated disease	Development of strongyloidiasis in renal transplant recipients follow-ing cyclosporin A has not been described; *Strongyloides* hyperinfection occurred immediately after cyclosporin A was discontinued; possible transmission by kidney transplant	81
Adult HTLV-1 infection	Intestinal or dissemi-nated disease	Onset usually follows long period of viral latency; strongyloidiasis is considered a cofactor of leukemogenesis; hypereosinophilia is also seen, could be associated with parasites or leukemic process; cofactors might be important in HTLV-1-associated leukemogenesis; hypereosinophilia affects prognosis	86
Adult HTLV-1 infection	Intestinal disease	Prevalence of both HTLV-1 and *Strongyloides* increased with age, espe-cially over 50 yr; prevalence of *S. stercoralis* infection significantly high-er in HTLV-1 carriers; IgE level low in HTLV-1 carriers and significantly lower in HTLV-1 carriers with *Strongyloides* infection	40
Adult HTLV-1 infection	Disseminated disease	Strongyloidiasis progresses to hyperinfection or disseminated disease in carriers of HTLV-1; larvae detected in stool; meningitis responded to therapy; larvae still present after thiabendazole; coinfection with *S. ster-coralis* is a predisposing factor to meningitis; association may warrant preventive antiparasitic treatment in patients infected with HTLV-1	31
Adult HTLV-1 infection (Peru)	Disseminated disease including meningitis	*Strongyloides* hyperinfection among Peruvian patients highly associated with HTLV-1 infection	36
HTLV-1 infection and ATLL[a]	Intestinal or dissemi-nated disease	ATLL represents ~50% of T-cell lymphomas in Martinique; *Strongyloides* infection during ATLL is linked with high response rate to chemotherapy and prolonged survival; p53 overexpression seen in ~50% of cases of aggressive ATLL from Martinique and is associated with resistance to chemotherapy and short-term survival	3
HTLV-1 infection	Intestinal disease	*S. stercoralis* hyperinfection syndrome and therapeutic failure in appar-ently healthy patients with nondisseminated strongyloidiasis may be markers of HTLV-1 infection; HTLV-1 coinfection may also impact clinical course of scabies and HIV disease	35

(continued)

Table 19.15 *Strongyloides stercoralis* infections in immunocompromised patients *(continued)*

Underlying condition	Disease manifestation	Comments	Reference
Advanced small noncleaved cell lymphoma	Disseminated disease	*S. stercoralis* infection in child mimics relapse (the child was on intensive multiagent chemotherapy for lymphoma, and the treatment was successful in this case); *S. stercoralis* infection in adults being treated for malignant lymphoma is frequently fatal	94
SLE[a] therapy (active disease)	Disseminated disease	Death due to massive pulmonary hemorrhage induced by filariform infection; diagnosed from bronchoalveolar lavage fluid 2 days before death; importance of checking patients for possible parasitic infections prior to starting therapy, even in areas without endemic infection	100
Cancer chemotherapy (Mexico)	Intestinal disease	In epidemiologic settings with high prevalence of intestinal parasitic infections, stool exams should be performed and antiparasitic therapy provided before initiating chemotherapy	38
Ribavirin plus interferon therapy for chronic HCV[a] infection	Disseminated disease	Ribavirin may have immunomodulatory and immunosuppressive action; severe strongyloidiasis reported after use of ribavirin and interferon therapy for HCV infection	82
Steroids	Intestinal disease	In patients on corticosteroids, vomiting and diarrhea are key symptoms; duodenal mucosa varied from being normal to having severe ulceration; symptoms may be attributed to underlying disease	104
Steroids for chronic idiopathic thrombocytopenia	Disseminated disease	Diagnosis difficult due to lack of diarrhea and eosinophilia; negative multiple stool specimens; upper endoscopy revealed larvae, as did sputum specimens; death in spite of thiabendazole therapy	106
Single-dose steroids prior to radiosurgery	Disseminated disease	Patient received single dose of dexamethasone before stereotactic radiosurgery; fatal disseminated case of strongyloidiasis	108
Steroids	Intestinal disease (mimicking ulcerative colitis), possibly leading to disseminated disease	*Strongyloides* infection mimics ulcerative colitis over long period; accurate diagnosis is essential	104
Immunosuppressive drugs	Intestinal disease	Patient with chronic *S. stercoralis* infection initially presented with generalized prurigo nodularis and lichen simplex chronicus (skin nodules, thickened skin)	49
Diabetes mellitus and malnutrition	Disseminated disease	Repeated vomiting and upper gastrointestinal bleeding; unexplained leukocytosis; pulmonary infiltrates, gastric aspirate leukocytosis, presence of adult worms, eggs, and filariform and rhabditiform larvae in stool specimens; patients died while undergoing treatment with mebendazole	43
Diabetes mellitus and eosinophilic pleural effusion	Disseminated disease	Fever, exudative type of eosinophilic pleural effusion; bronchial washings contained *S. stercoralis* larvae; patient history extremely important	27
Unknown	Intestinal disease	Patient admitted for progressive weight loss, abdominal bloating; gastroscopic exam revealed mucosal prepyloric elevations in gastric mucosa; gastric strongyloidiasis confirmed by presence of adult worms, as well as ova and rhabditiform larvae; histologically confirmed case of gastric disease	101
None (immunocompetent patient)	Eosinophilic pleural effusion	Helminth infections should be considered in cases of eosinophilic pleural effusion	37
None (immunocompetent patient)	Colitis	Diarrhea, weight loss, microcytic anemia; colonoscopy revealed pancolitis and intense eosinophilic infiltrates; biopsy revealed *S. stercoralis* larvae	5
None (immunocompetent patient)	Colitis	Poor sensitivity of stool exams, careful search needs to be made in all patients before diagnosis of idiopathic eosinophilic colitis; steroid therapy may have fatal outcome	20

[a] ATLL, adult T-cell leukemia/lymphoma; HCV, hepatitis C virus; SLE, systemic lupus erythematosus.

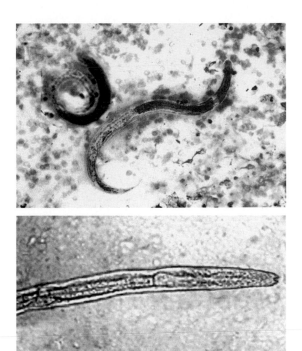

Figure 19.12 *Strongyloides stercoralis.* (Upper) Rhabditiform larvae seen in bronchoalveolar lavage fluid specimen (larvae can also be seen in sputum in heavy infections or in the hyperinfection syndrome); (lower) rhabditiform larva from fecal concentration sediment (note the short mouth opening/buccal capsule and the packet of genital primordial cells at the bottom left).

person-to-person direct skin contact or via fomites (Table 19.3; Figure 19.13). Outbreaks of scabies in hospitals or institutions, particularly for the elderly or retarded, have been documented worldwide. These outbreaks are particularly severe when associated with Norwegian or crusted scabies; such patients may be infected with thousands of mites on the skin. This infection has been linked with cellular immune deficiencies, including AIDS. Other potential problems include a decreased awareness of itching and the use of topical corticosteroids to reduce itching. Because of the number of mites present, these patients are extremely contagious, from sloughing skin, direct contact, and environmental contamination. Another potential problem is the fact that the symptoms mimic those of many other dermatologic conditions. Burrows may not be evident, and hyperkeratotic, crusted, scaling, fissured plaques are present over the scalp, face, and back, with associated gross nail thickening. The number of documented reports, particularly in AIDS patients, suggests that crusted scabies may become more common in the future.

Crusted scabies should be considered in the differential diagnosis of a generalized cutaneous eruption in

a human T-cell leukemia virus type 1 (HTLV-1)-positive patient. Patients with crusted scabies from an HTLV-1-endemic population should be tested for a possible HTLV-1 infection. Infection with HTLV-1 is an important cofactor related to Norwegian scabies in Peru. Testing for HTLV-1 in all Norwegian scabies cases is highly recommended, especially when no other risk factors are apparent. These patients may be at increased risk of progressing to adult T-cell leukemia/lymphoma (8, 11). Definitive parasitic diagnosis can be difficult; difficulties in management have led to renewed interest in both scabies and pediculosis. The diagnosis of scabies should always be considered in patients with advanced malignancies and associated pruritus.

The diagnosis of Norwegian scabies can easily be missed. Serpiginous tracks were noted on the surface of Sabouraud's dextrose agar used for fungal culture of the skin scrapings from an elderly long-term-care facility resident. This unusual laboratory manifestation alerted clinical microbiologists to the possible diagnosis of scabies. Although many microbiology laboratories are aware of these unusual findings, personnel can forget to consider scabies in such situations.

Ivermectin is increasingly being used to treat scabies, especially crusted (Norwegian) scabies. However, treatment failures, recrudescence, and reinfection can occur, even after administration of multiple doses. Clinical and in vitro evidence of ivermectin resistance has been found in several patients with multiple recurrences of crusted scabies who had previously received 30 to more then 50 doses of ivermectin over a 4-year time frame. As predicted, ivermectin resistance in scabies mites can develop after intensive ivermectin use.

Figure 19.13 *Sarcoptes scabiei* "itch mite." (Left) Mite from skin scraping preparation (note the four pairs of legs); (right) hand of an individual with severe scabies (Norwegian scabies). (Right, Armed Forces Institute of Pathology photograph.)

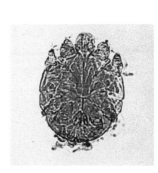
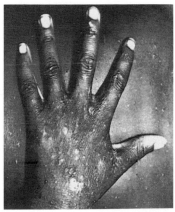

References

1. Abramowicz, M. (ed.). 2004. Drugs for parasitic infections. *Med. Lett. Drugs Ther.* **46:**1–12.

2. Adams, V. J., M. B. Markus, J. F. Adams, E. Jordaan, B. Curtis, M. A. Dhansay, C. C. Obihara, and J. E. Fincham. 2005. Paradoxical helminthiasis and giardiasis in Cape Town, South Africa: epidemiology and control. *Afr. Health Sci.* **5:**276–280.

3. Agape, P., M. C. Copin, M. Cavrois, G. Panelatti, Y. Plumelle, M. Ossondo-Landeau, D. Quist, N. Grossat, B. Gosselin, P. Fexaux, and E. Wattel. 1999. Implication of HTLV-I infection, strongyloidiasis, and P53 overexpression in the development, response to treatment, and evolution of non-Hodgkin's lymphomas in an endemic area (Martinique, French West Indies). *J. Acquir. Immune Defic. Syndr. Hum. Retrovirol.* **20:**394–402.

4. Albrecht, H., I. Sobottka, C. Emminger, H. Jablonowski, G. Just, A. Stoehr, T. Kubin, B. Salzberger, T. Lutz, and J. van Lunzen. 1996. Visceral leishmaniasis emerging as an important opportunistic infection in HIV-infected persons living in areas nonendemic for *Leishmania donovani*. *Arch. Pathol. Lab. Med.* **120:**189–198.

5. Al Samman, M., S. Haque, and J. D. Long. 1999. Strongyloidiasis colitis: a case report and review of the literature. *J. Clin. Gastroenterol.* **28:**77–80.

6. Arendt, G., H. J. von Giesen, H. Hefter, E. Neuen-Jacob, H. Roick, and H. Jablonowski. 1999. Long-term course and outcome in AIDS patients with cerebral toxoplasmosis. *Acta Neurol. Scand.* **100:**178–184.

7. Belanger, F., F. Derouin, L. Grangeot-Keros, L. Meyer, and HEMOCO and SEROCO Study Groups. 1999. Incidence and risk factors of toxoplasmosis in a cohort of human immunodeficiency virus-infected patients: 1988–1995. *Clin. Infect. Dis.* **28:**575–581.

8. Bergman, J. N., W. A. Dodd, M. J. Trotter, J. J. Oger, and J. P. Dutz. 1999. Crusted scabies in association with human T-cell lymphotropic virus. *J. Cutan. Med. Surg.* **3:**148–152.

9. Bhigjee, A. I., K. Naidoo, V. B. Patel, D. Govender, and the Neuroscience AIDS Research Group. 1999. Intracranial mass lesions in HIV-positive patients—the KwaZulu/Natal experience. *S. Afr. Med. J.* **89:**1284–1288.

10. Bjoro, K., E. Schrumpf, A. Bergan, T. Haaland, K. Skaug, and S. S. Froland. 1998. Liver transplantation for endstage hepatitis C cirrhosis in a patient with primary hypogammaglobulinaemia. *Scand. J. Infect. Dis.* **30:**520–522.

11. Blas, M., F. Bravo, W. Castillo, W. J. Castillo, R. Ballona, P. Navarro, J. Catacora, R. Cairampoma, and E. Gotuzzo. 2005. Norwegian scabies in Peru: the impact of human T cell lymphotropic virus type I infection. *Am. J. Trop. Med. Hyg.* **72:**855–857.

12. Burgner, D., N. Pikos, G. Eagles, A. McCarthy, and M. Stevens. 1999. Epidemiology of *Cryptosporidium parvum* in symptomatic paediatric oncology patients. *J. Paediatr. Child Health* **35:**300–302.

13. Cabral, G. A., and F. Marciano-Cabral. 2004. Cannabinoid-mediated exacerbation of brain infection by opportunistic amebae. *J. Neuroimmunol.* **147:**127–130.

14. Calza, L., A. D'Antuono, G. Marinacci, R. Manfredi, V. Colangeli, B. Passarini, R. Orioli, O. Varoli, and F. Chiodo. 2004. Disseminated cutaneous leishmaniasis after visceral disease in a patient with AIDS. *J. Am. Acad. Dermatol.* **50:**461–465.

15. Calzetti, C., G. Magnani, D. Confalonieri, A. Capelli, S. Moneta, P. Scognamiglio, and F. Fiaccadori. 1997. Pancreatitis caused by *Cryptosporidium parvum* in patients with severe immunodeficiency related to HIV infection. *Ann. Ital. Med. Int.* **12:**63–66.

16. Carr, A., D. Marriott, A. Field, E. Vasek, and D. A. Cooper. 1998. Treatment of HIV-1-associated microsporidiosis and cryptosporidiosis with combination antiretroviral therapy. *Lancet* **351:**256–261.

17. Cimerman, S., B. Cimerman, and D. S. Lewi. 1999. Prevalence of intestinal parasitic infections in patients with acquired immunodeficiency syndrome in Brazil. *Int. J. Infect. Dis.* **3:**203–206.

18. Colebunders, R., K. Depraetere, E. De Droogh, A. Kamper, B. Corthout, and E. Bottiau. 1999. Obstructive nephropathy due to sulfa crystals in two HIV seropositive patients treated with sulfadiazine. *JBR-BTR* **82:**153–154.

19. Conteas, C. N., O. G. Berlin, C. E. Speck, S. S. Pandhumas, M. J. Lariviere, and C. Fu. 1998. Modification of the clinical course of intestinal microsporidiosis in acquired immunodeficiency syndrome patients by immune status and anti-human immunodeficiency virus therapy. *Am. J. Trop. Med. Hyg.* **58:**555–558.

20. Corsetti, M., G. Basilisco, R. Pometta, M. Allocca, and D. Conte. 1999. Mistaken diagnosis of eosinophilic colitis. *Ital. J. Gastroenterol. Hepatol.* **31:**607–609.

21. Cotte, L., M. Rabodonirina, F. Chapuis, F. Bailly, F. Bissuel, C. Raynal, P. Gelas, F. Persat, M. A. Piens, and C. Trepo. 1999. Waterborne outbreak of intestinal microsporidiosis in persons with and without human immunodeficiency virus infection. *J. Infect. Dis.* **180:**2003–2008.

22. Couppie, P., E. Clyti, M. Sobesky, F. Bissuel, P. Del Giudice, D. Sainte-Marie, J. P. Dedet, B. Carme, and R. Pradinaud. 2004. Comparative study of cutaneous leishmaniasis in human immunodeficiency virus (HIV)-infected patients and non-HIV-infected patients in French Guiana. *Br. J. Dermatol.* **151:**1165–1171.

23. Dascomb, K., R. Clark, J. Aberg, J. Pulvirenti, R. G. Hewitt, P. Kissinger, and E. S. Didier. 1999. Natural history of intestinal microsporidiosis among patients infected with human immunodeficiency virus. *J. Clin. Microbiol.* **37:**3421–3422.

24. Dedet, J. P., and F. Pratlong. 2000. *Leishmania, Trypanosoma* and monoxenous trypanosomatids as emerging opportunistic agents. *J. Eukaryot. Microbiol.* **47:**37–39.

25. Del Aguila, C., R. Navajas, D. Gurbindo, J. T. Ramos, M. J. Mellado, S. Fenoy, M. A. Munoz-Fernandez, M. Subirats, J. Ruiz, and N. J. Pieniazek. 1997. Microsporidiosis in HIV-positive children in Madrid (Spain). *J. Eukaryot. Microbiol.* **44:**84S–85S.

26. Dietrich, U., M. Maschke, A. Dorfler, M. Prumbaum, and M. Forsting. 2000. MRI of intracranial toxoplasmosis after bone marrow transplantation. *Neuroradiology* **42:**14–18.

27. Emad, A. 1999. Exudative eosinophilic pleural effusion due to *Strongyloides stercoralis* in a diabetic man. *South. Med. J.* **92:**58–60.

28. Ernst, T. M., L. Chang, M. D. Witt, H. A. Aronow, M. E. Cornford, I. Walot, and M. A. Goldberg. 1998.

Cerebral toxoplasmosis and lymphoma in AIDS: perfusion MR imaging experience in 13 patients. *Radiology* **208**:663–669.

29. Fayer, R. 2004. *Sarcocystis* spp. in human infections. *Clin. Microbiol. Rev.* **17**:894–902.

30. Feitosa, G., A. C. Bandeira, D. P. Sampaio, R. Badaro, and C. Brites. 2001. High prevalence of giardiasis and strongyloidiasis among HIV-infected patients in Bahia, Brazil. *Braz. J. Infect. Dis.* **5**:339–344.

31. Foucan, L., I. Genevier, I. Lamaury, and M. Strobel. 1997. Aseptic purulent meningitis in two patients co-infected by HTLV-1 and *Strongyloides stercoralis*. *Med. Trop.* **57**:262–264.

32. Frenkel, J. K., M. B. Silva, J. Saldanha, M. L. de Silva, V. D. Correia Filho, C. H. Barata, E. Lages, L. E. Ramirez, and A. Prata. 2003. *Isospora belli* infection: observation of unicellular cysts in mesenteric lymphoid tissues of a Brazilian patient with AIDS and animal inoculation. *J. Eukaryot. Microbiol.* **50**(Suppl.):682–684.

33. Ganji, M., A. Tan, M. I. Maitar, C. M. Weldon-Linne, E. Weisenberg, and D. P. Rhone. 2003. Gastric toxoplasmosis in a patient with acquired immunodeficiency syndrome. A case report and review of the literature. *Arch. Pathol. Lab. Med.* **127**:732–734.

34. Gerber, D. A., M. Green, R. Jaffe, D. Greenberg, G. Mazariegos, and J. Reyes. 2000. Cryptosporidial infections after solid organ transplantation in children. *Pediatr. Transplant.* **4**:50–55.

35. Gotuzzo, E., C. Arango, A. de Queiroz-Campos, and R. E. Isturiz. 2000. Human T-cell lymphotropic virus-I in Latin America. *Infect. Dis. Clin. North Am.* **14**:211–239.

36. Gotuzzo, E., A. Terashima, H. Alvarez, R. Tello, R. Infante, D. M. Watts, and D. O. Freedman. 1999. *Strongyloides stercoralis* hyperinfection associated with human T cell lymphotropic virus type-1 infection in Peru. *Am. J. Trop. Med. Hyg.* **60**:146–149.

37. Goyal, S. B. 1998. Intestinal strongyloidiasis manifesting as eosinophilic pleural effusion. *South. Med. J.* **91**:768–769.

38. Guarner, J., T. Matilde-Nava, R. Villasenor-Flores, and G. Sanchez-Mejorada. 1997. Frequency of intestinal parasites in adult cancer patients in Mexico. *Arch. Med. Res.* **28**:219–222.

39. Guk, S. M., M. Seo, Y. K. Park, M. D. Oh, K. W. Choe, J. L. Kim, M. H. Choi, S. T. Hong, and J. Y. Chai. 2005. Parasitic infections in HIV-infected patients who visited Seoul National University Hospital during the period 1995–2003. *Korean J. Parasitol.* **43**:1–5.

40. Hayashi, J., Y. Kishihara, E. Yoshimura, N. Furusyo, K. Yamaji, Y. Kawakami, H. Murakami, and S. Sashiwagi. 1997. Correlation between human T cell lymphotropic virus type-1 and *Strongyloides stercoralis* infections and serum immunoglobulin E responses in residents of Okinawa, Japan. *Am. J. Trop. Med. Hyg.* **56**:71–75.

41. Heffler, K. F., T. J. Eckhardt, A. C. Reboli, and D. Stieritz. 1996. *Acanthamoeba* endophthalmitis in acquired immunodeficiency syndrome. *Am. J. Ophthalmol.* **122**:584–586.

42. Hill, S. L., J. M. Cheney, G. F. Taton-Allen, J. S. Reif, C. Bruns, and M. R. Lappin. 2000. Prevalence of enteric zoonotic organisms in cats. *J. Am. Vet. Med. Assoc.* **216**:687–692.

43. Ho, P. L., W. K. Luk, A. C. Chan, and K. Y. Yuen. 1997. Two cases of fatal strongyloidiasis in Hong Kong. *Pathology* **29**:324–326.

44. Hollister, W. S., E. U. Canning, and A. Willcox. 1991. Evidence for widespread occurrence of antibodies to *Encephalitozoon cuniculi* (Microspora) in man provided by ELISA and other serological tests. *Parasitology* **102**:33–43.

45. Hung, C. C., H. Y. Deng, W. H. Hsiao, S. M. Hsieh, C. F. Hsiao, M. Y. Chen, S. C. Chang, and K. E. Su. 2005. Invasive amebiasis as an emerging parasitic disease in patients with human immunodeficiency virus type 1 infection in Taiwan. *Arch. Intern. Med.* **165**:409–415.

46. Hussein, E. M., A. H. Abdul-Manaem, and S. L. el-Attary. 2005. *Cyclospora cayetanensis* oocysts in sputum of a patient with active pulmonary tuberculosis: case report in Ismailia, Egypt. *J. Egypt. Soc. Parasitol.* **35**:787–793.

47. Hutin, Y. J., M. N. Sombardier, O. Liguory, C. Sarfati, F. Derouin, J. Modai, and J. M. Molina. 1998. Risk factors for intestinal microsporidiosis in patients with human immunodeficiency virus infection: a case-control study. *J. Infect. Dis.* **178**:904–907.

48. Isenberg, H. D. (ed.). 2004. *Clinical Microbiology Procedures Handbook*, 2nd ed., p. 9.0.1–9.10.8.3. ASM Press, Washington, D.C.

49. Jacob, C. I., and S. F. Patten. 1999. *Strongyloides stercoralis* infection presenting as generalized prurigo nodularis and lichen simplex chronicus. *J. Am. Acad. Dermatol.* **41**:357–361.

50. Jayasekera, S., J. Sissons, J. Tucker, C. Rogers, D. Nolder, D. Warhurst, S. Alsam, J. W. White, E. M. Higgins, and N. A. Khan. 2004. Post-mortem culture of *Balamuthia mandrillaris* from the brain and cerebrospinal fluid of a case of granulomatous amoebic meningoencephalitis, using human brain microvascular endothelial cells. *J. Med. Microbiol.* **53**:1007–1012.

51. Joshi, M., A. S. Chowdhary, P. J. Dalal, and J. K. Maniar. 2002. Parasitic diarrhoea in patients with AIDS. *Natl. Med. J. India* **15**:72–74.

52. Kaminsky, R. G., R. J. Soto, A. Campa, and M. K. Baum. 2004. Intestinal parasitic infections and eosinophilia in a human immunedeficiency virus positive population in Honduras. *Mem. Inst. Oswaldo Cruz* **99**:773–778.

53. Kester, K. E., G. W. Turiansky, and P. L. McEvoy. 1998. Nodular cutaneous microsporidiosis in a patient with AIDS and successful treatment with long-term oral clindamycin therapy. *Ann. Intern. Med.* **128**:911–914.

54. Khan, A., C. Su, M. German, G. A. Storch, D. B. Clifford, and L. D. Sibley. 2005. Genotyping of *Toxoplasma gondii* strains from immunocompromised patients reveals high prevalence of type I strains. *J. Clin. Microbiol.* **43**:5881–5887.

55. Khoury, H., D. Adkins, R. Brown, L. Goodnough, M. Gokden, T. Roberts, G. Storch, and J. DiPersio. 1999. Successful treatment of cerebral toxoplasmosis in a marrow transplant recipient: contribution of a PCR test in diagnosis and early detection. *Bone Marrow Transplant.* **23**:409–411.

56. Koeppel, M. C., R. Abitan, C. Angeli, J. Lafon, J. Pelletier, and J. Sayag. 1998. Cutaneous and gastrointestinal mastocytosis associated with cerebral toxoplasmosis. *Br. J. Dermatol.* **139**:881–884.

57. Koide, J., E. Okusawa, T. Ito, S. Mori, T. Takeuchi, S. Itoyama, and T. Abe. 1998. Granulomatous amoebic encephalitis caused by *Acanthamoeba* in a patient with systemic lupus erythematosus. *Clin. Rheumatol.* **17:**329–332.

58. Kotler, D. P., and J. M. Orenstein. 1998. Clinical syndromes associated with microsporidiosis. *Adv. Parasitol.* **40:**321–349.

59. Mandell, L. A. 1990. Infections in the compromised host. *J. Int. Med. Res.* **18:**177–190.

60. Marciano-Cabral, F., and G. Cabral. 2003. *Acanthamoeba* spp. as agents of disease in humans. *Clin. Microbiol. Rev.* **16:**273–307.

61. Mariotta, S., G. Pallone, E. Li Bianchi, G. Gilardi, and A. Bisetti. 1996. *Strongyloides stercoralis* hyperinfection in a case of idiopathic pulmonary fibrosis. *Panminerva Med.* **38:**45–47.

62. Marlier, S., G. Menard, O. Gisserot, K. Kologo, and J. P. De Jaureguiberry. 1999. Leishmaniasis and human immunodeficiency virus: an emerging coinfection? *Med. Trop.* **59:**193–200.

63. Marra, C. M., M. R. Krone, L. A. Koutsky, and K. K. Holmes. 1998. Diagnostic accuracy of HIV-associated central nervous system toxoplasmosis. *Int. J. STD AIDS* **9:**761–764.

64. Maschke, M., U. Dietrich, M. Prumbaum, O. Kastrup, B. Turowski, U. W. Schaefer, and H. C. Diener. 1999. Opportunistic CNS infection after bone marrow transplantation. *Bone Marrow Transplant.* **23:**1167–1176.

65. Mathew, M. J., and M. J. Chandy. 1999. Central nervous system toxoplasmosis in acquired immunodeficiency syndrome: an emerging disease in India. *Neurol. India* **47:**182–187.

66. Mathis, A., R. Weber, and P. Deplazes. 2005. Zoonotic potential of the microsporidia. *Clin. Microbiol. Rev.* **18:**423–445.

67. Merzianu, M., S. M. Gorelick, V. Paje, D. P. Kottler, and C. Sian. 2005. Gastric toxoplasmosis as the presentation of acquired immunodeficiency syndrome. *Arch. Pathol. Lab. Med.* **129:**e87–e90.

68. Michiels, J. F., P. Hofman, E. Bernard, M. C. Saint Paul, C. Boissy, V. Mondain, Y. LeFichoux, and R. Loubiera. 1994. Intestinal and extraintestinal *Isospora belli* infection in an AIDS patient. A second case report. *Pathol. Res. Pract.* **190:**1089–1094.

69. Miller, R. F., M. A. Hall-Craggs, D. C. Costa, N. S. Brink, F. Scaravilli, S. B. Lucas, I. D. Wilkinson, P. J. Ell, B. F. Kendall, and M. J. Harrison. 1998. Magnetic resonance imaging, thallium-201 SPET scanning, and laboratory analyses for discrimination of cerebral lymphoma and toxoplasmosis in AIDS. *Sex. Transm. Infect.* **74:**258–264.

70. Mohandas, K., R. Sehgal, A. Sud, and N. Malla. 2002. Prevalence of intestinal parasitic pathogens in HIV-seropositive individuals in northern India. *Jpn. J. Infect. Dis.* **55:**83–84.

71. Moran, P., F. Ramos, M. Ramiro, O. Curiel, E. Gonzalez, A. Valadez, A. Gomez, G. Garcia, E. I. Melendro, and C. Ximenez. 2005. Infection by human immunodeficiency virus-1 is not a risk factor for amebiasis. *Am. J. Trop. Med. Hyg.* **73:**296–300.

72. Moreno-Camacho, A., R. Lopez-Velez, A. Munoz-Sanz, and P. Labarga-Echevarria. Intestinal parasitic infections and leishmaniasis in patients with HIV infection. *Enferm. Infecc. Microbiol. Clin.* **16:**S52–S60.

73. Morgan, U., R. Weber, L. Xiao, I. Sulaiman, R. C. Thompson, A. Ndiritu, A. Lal, A. Moore, and P. Deplazes. 2000. Molecular characterization of *Cryptosporidium* isolates obtained from human immunodeficiency virus-infected individuals living in Switzerland, Kenya, and the United States. *J. Clin. Microbiol.* **38:**1180–1183.

74. Mori, S., T. Konishi, K. Matsuoka, M. Deguchi, M. Ohta, O. Mizuno, T. Ueno, T. Okinaka, Y. Nishimura, N. Ito, and T. Nakano. 1998. Strongyloidiasis associated with nephrotic syndrome. *Intern. Med.* **37:**606–610.

75. Muller, A., G. Fatkenheuer, B. Salzberger, M. Schrappe, V. Diehl, and C. Franzen. 1998. *Strongyloides stercoralis* infection in a patient with AIDS and non-Hodgkin lymphoma. *Dtsch. Med. Wochenschr.* **123:**381–385.

76. Murakawa, G. J., T. McCalmont, J. Altman, G. H. Telang, M. D. Hoffman, G. R. Kantor, and T. G. Berger. 1995. Disseminated acanthamebiasis in patients with AIDS. A report of five cases and a review of the literature. *Arch. Dermatol.* **131:**1291–1296.

77. Navin, T. R., R. Weber, D. J. Vugia, D. Rimland, J. M. Roberts, D. G. Addiss, G. S. Visvesvara, S. P. Walhquist, S. E. Hogan, L. E. Gallagher, D. D. Juranek, D. A. Schwartz, C. M. Wilcox, J. M. Stewart, S. E. Thompson III, and R. T. Bryan. 1999. Declining CD4[+] T-lymphocyte counts are associated with increased risk of enteric parasitosis and chronic diarrhea: results of a 3-year longitudinal study. *J. Acquir. Immune Defic. Syndr. Hum. Retrovirol.* **20:**154–159.

78. Ok, U. Z., M. Cirit, A. Uner, E. Ok, F. Akcicek, A. Basci, and M. A. Ozcel. 1997. Cryptosporidiosis and blastocystosis in renal transplant recipients. *Nephron* **75:**171–174.

79. Orenstein, J. M., P. Russo, E. S. Didier, C. Bowers, N. Bunin, and D. T. Teachey. 2005. Fatal pulmonary microsporidiosis due to *Encephalitozoon cuniculi* following allogeneic bone marrow transplantation for acute myelogenous leukemia. *Ultrastruct. Pathol.* **29:**269–276.

80. Pacheco, R. S., M. C. Marzochi, M. Q. Pires, C. M. Brito, M. de F. Madeira, and E. G. Barbosa-Santos. 1998. Parasite genotypically related to a monoxenous trypanosomatid of dog's flea causing opportunistic infection in an HIV positive patient. *Mem. Inst. Oswaldo Cruz* **93:**531–537.

81. Palau, L. A., and G. A. Pankey. 1997. Strongyloides hyperinfection in a renal transplant recipient receiving cyclosporine: possible *Strongyloides stercoralis* transmission by kidney transplant. *Am. J. Trop. Med. Hyg.* **57:**413–415.

82. Parana, R., M. Portugal, L. Vitvitski, H. Cotrim, L. Lyra, and C. Trepo. 2000. Severe strongyloidiasis during interferon plus ribavirin therapy for chronic HCV infection. *Eur. J. Gastroenterol. Hepatol.* **12:**245–246.

83. Perz, J. F., F. K. Ennever, and S. M. Blancq. 1998. *Cryptosporidium* in tap water: comparison of predicted risks with observed levels of disease. *Am. J. Epidemiol.* **147:**289–301.

84. Piccolo, I., R. Causarano, R. Sterzi, M. Sberna. P. L. Oreste, C. Moioli, L. Caggese, and F. Girotti. 1999. Chorea in patients with AIDS. *Acta Neurol. Scand.* **100:**332–336.

85. Pineda, J. A., J. A. Gallardo, J. Macias, J. Delgado, C. Regordan, F. Morillas, F. Relimpio, J. Martin-Sanchez, A. Sanchez-Quijano, M. Leal, and E. Lissen. 1998. Prevalence of and factors associated with visceral leishmaniasis in human immunodeficiency virus type 1-infected patients in southern Spain. *J. Clin. Microbiol.* **36:**2419–2422.

86. Plumelle, Y., C. Gonin, A. Edouard, B. J. Bucher, L. Thomas, A. Brebion, and G. Panelatti. 1997. Effect of *Strongyloides stercoralis* infection and eosinophilia on age at onset and prognosis of adult T-cell leukemia. *Am. J. Clin. Pathol.* **107:**81–87.

87. Posada-Vergara, M. P., J. A. Lindoso, J. E. Tolezano, V. L. Pereira-Chioccola, M. V. Silva, and H. Goto. 2005. Tegumentary leishmaniasis as a manifestation of immune reconstitution inflammatory syndrome in 2 patients with AIDS. *J. Infect. Dis.* **192:**1819–1822.

88. Pozio, E., G. Rezza, A. Boschini, P. Pezzotti, A. Tamburrini, P. Rossi, M. Di Fine, C. Smacchia, A. Schiasari, E. Gattei, R. Zucconi, and P. Ballarini. 1997. Clinical cryptosporidiosis and human immunodeficiency virus (HIV)-induced immunosuppression: findings from a longitudinal study of HIV-positive and HIV-negative former injection drug users. *J. Infect. Dis.* **176:**969–975.

89. Raffi, F., J. Franck, H. Pelloux, F. Derouin, V. Reliquet, P. Ambroise-Thomas, J. P. Aboulker, C. Leport, and H. Dumon. 1999. Specific anti-toxoplasmic IgG antibody immunoblot profiles in patients with AIDS-associated *Toxoplasma* encephalitis. *Diagn. Microbiol. Infect. Dis.* **34:**51–56.

90. Resiere, D., J. M. Vantelon, P. Bouree, E. Chachaty, G. Nitenberg, and F. Blot. 2003. *Isospora belli* infection in a patient with non-Hodgkin's lymphoma. *Clin. Microbiol. Infect.* **9:**1065–1067.

91. Rezk, H., A. M. el-Shazly, M. Soliman, H. I. el-Nemr, I. M. Nagaty, and M. A. Fouad. 2001. Coccidiosis among immuno-competent and -compromised adults. *J. Egypt. Soc. Parasitol.* **31:**823–834.

92. Ribera, E., A. Fernandez-Sola, C. Juste, A. Rovira, F. J. Romero, L. Armadans-Gil, I. Ruiz, I. Ocana, and A. Pahissa. 1999. Comparison of high and low doses of trimethoprim-sulfamethoxazole for primary prevention of toxoplasmic encephalitis in human immunodeficiency virus-infected patients. *Clin. Infect. Dis.* **29:**1461–1466.

93. Rossi, P., F. Rivasi, M. Codeluppi, A. Catania, A. Tamburrini, E. Righi, and E. Pozio. 1998. Gastric involvement in AIDS associated cryptosporidiosis. *Gut* **43:**476–477.

94. Sandland, J. T., W. Kauffman, and P. M. Flynn. 1997. *Strongyloides stercoralis* infection mimicking relapse in a child with small noncleaved cell lymphoma. *Am. J. Clin. Oncol.* **20:**215–216.

95. Schmidt, W., T. Schneider, W. Heise, J. D. Schulzke, T. Weinke, R. Ignatius, R. L. Owen, M. Zeitz, E. O. Riecken, and R. Ullrich. 1997. Mucosal abnormalities in microsporidiosis. *AIDS* **11:**1589–1594.

96. Schuster, F. L., and G. S. Visvesvara. 2004. Amebae and ciliated protozoa as causal agents of waterborne zoonotic disease. *Vet. Parasitol.* **126:**91–120.

97. Schuster, F. L., and G. S. Visvesvara. 2004. Free-living amoebae as opportunistic and non-opportunistic pathogens of humans and animals. *Int. J. Parasitol.* **34:**1001–1027.

98. Selby, D. M., R. S. Chandra, T. A. Rakusan, B. Loechelt, B. M. Markle, and G. S. Visvesvara. 1998. Amebic osteomyelitis in a child with acquired immunodeficiency syndrome: a case report. *Pediatr. Pathol. Lab. Med.* **18:**89–95.

99. Sell, S. 2001. *Immunology, Immunopathology and Immunity*, 6th ed. ASM Press, Washington, D.C.

100. Setoyama, M., S. Fukumaru, T. Takasaki, H. Yoshida, and T. Kanzaki. 1997. SLE with death from acute massive pulmonary hemorrhage caused by disseminated strongyloidiasis. *Scand. J. Rheumatol.* **26:**389–391.

101. Shekhar, K. C., R. Krishnan, R. Pathmanathan, and C. S. Fook. 1997. Gastric strongyloidiasis in a Malaysian patient. *Southeast Asian J. Trop. Med. Public Health* **28:**158–160.

102. Silva, C. V., M. S. Ferreira, A. S. Borges, and J. M. Costa-Cruz. 2005. Intestinal parasitic infections in HIV/AIDS patients: experience at a teaching hospital in central Brazil. *Scand. J. Infect. Dis.* **37:**211–215.

103. Sobottka, I., D. A. Schwartz, J. Schottelius, G. S. Visvesvara, N. J. Pieniazek, C. Schmetz, N. P. Kock, R. Laufs, and H. Albrecht. 1998. Prevalence and clinical significance of intestinal microsporidiosis in human immunodeficiency virus-infected patients with and without diarrhea in Germany: a prospective coprodiagnostic study. *Clin. Infect. Dis.* **26:**475–480.

104. Sreenivas, D. V., A. Kimar, Y. R. Kumar, C. Bharavi, C. Sundaram, and K. Gayathri. 1997. Intestinal strongyloidiasis—a rare opportunistic infection. *Indian J. Gastroenterol.* **16:**105–106.

105. Steinberg, J. P., R. L. Galindo, E. S. Kraus, and K. G. Ghanem. 2002. Disseminated acanthamoebiasis in a renal transplant recipient with osteomyelitis and cutaneous lesions: case report and literature review. *Clin. Infect. Dis.* **35:**e43–e49.

106. Suvajdzic, N., I. Kranjcic-Zee, V. Jovanovic, D. Popovic, and M. Colovic. 1999. Fatal strongyloidosis following corticosteroid therapy in a patient with chronic idiopathic thrombycytopenia. *Haematologia* **29:**323–326.

107. Tanyuksel, M., and W. A. Petri, Jr. 2003. Laboratory diagnosis of amebiasis. *Clin. Microbiol. Rev.* **16:**713–729.

108. Thomas, M. C., and S. A. Costello. 1998. Disseminated strongyloidiasis arising from a single dose of dexamethasone before stereotactic radiosurgery. *Int. J. Clin. Pract.* **52:**520–521.

109. Torno, M. S., Jr., R. Babapour, A. Gurevitch, and M. D. Witt. 2000. Cutaneous acanthamoebiasis in AIDS. *J. Am. Acad. Dermatol.* **42:**351–354.

110. Tran, M. Q., R. Y. Gohh, P. E. Morrissey, L. D. Dworkin, A. Gautam, A. P. Monaco, and A. F. Yango, Jr. 2005. *Cryptosporidium* infection in renal transplant patients. *Clin. Nephrol.* **63:**305–309.

111. Tumwine, J. K., A. Kekitiinwa, S. Bakeera-Kitaka, G. Ndeezi, R. Downing, X. Feng, D. E. Akiyoshi, and S. Tzipori. 2005. Cryptosporidiosis and microsporidiosis in Ugandan children with persistent diarrhea with and without concurrent infection with the human immunodeficiency virus. *Am. J. Trop. Med. Hyg.* **73:**921–925.

112. Walker, M. D., and J. R. Zunt. 2005. Neuroparasitic infections: cestodes, trematodes, and proozoans. *Semin. Neurol.* **25:**262–277.

113. Waywa, D., S. Kongkriengdaj, S. Chaidatch, S. Tiengrim,

B. Kowadisaiburana, S. Chaikachonpat, S. Suwanagool, A. Chaiprasert, A. Curry, W. Bailey, Y. Suputtamongkol, and N. J. Beeching. 2001. Protozoan enteric infection in AIDS related diarrhea in Thailand. *Southeast Asian J. Trop. Med. Public Health* **32**(Suppl.):151–155,

114. Widmer, G., S. Tzipori, C. J. Fichtenbaum, and J. K. Griffiths. 1998. Genotypic and phenotypic characterization of *Cryptosporidium parvum* isolates from people with AIDS. *J. Infect. Dis.* **178**:834–840.

115. Wilairatana, P., P. Radomyos, B. Radomyos, R. Phraevanich, W. Plooksawasdi, P. Chanthavanich, C. Viravan, and S. Looareeuwan. 1996. Intestinal sarcocystosis in Thai laborers. *Southeast Asian J. Trop. Med. Public Health* **27**:43–46.

116. Wong, T. Y., C. C. Szeto, F. F. Lai, C. K. Mak, and P. K. Li. 1998. Nephrotic syndrome in strongyloidiasis: remission after eradication with anthelmintic agents. *Nephron* **79**:333–336.

117. Yagi, S., G. C. Booton, G. S. Visvesvara, and F. L. Schuster. 2005. Detection of *Balamuthia* mitochondrial 16S rRNA gene DNA in clinical specimens by PCR. *J. Clin. Microbiol.* **43**:3192–3197.

118. Yee, Y. K., C. S. Lam, C. Y. Yung, T. L. Que, T. H. Kwan, T. C. Au, and M. L. Szeto. 1999. Strongyloidiasis as a possible cause of nephrotic syndrome. *Am. J. Kidney Dis.* **33**:e4.

119. Yoshimura, K., T. Hara, H. Tsurumi, H. Goto, M. Tajika, Y. Fukutomi, N. Murakami, and H. Moriwaki. 1999. Non-Hodgkin's lymphoma with *Toxoplasma* encephalitis. *Rinsho Ketsueki* **40**:563–567.

120. Zamir, D., M. Amar, G. Groisman, and P. Weiner. 1999. *Toxoplasma* infection in systemic lupus erythematosus mimicking lupus cerebritis. *Mayo Clin. Proc.* **74**:575–578.

121. Zver, S., P. Cernelc, U. Mlakar, and J. Pretnar. 1999. Cerebral toxoplasmosis—a late complication of allogeneic haematopoietic stem cell transplantation. *Bone Marrow Transplant.* **24**:1363–1365.

Nosocomial and Laboratory-Acquired Infections

Nosocomial Infections

Nosocomial infections are those that are hospital acquired or hospital associated. According to some publications, nosocomial infections are estimated to complicate the course of 5 to 10% of all hospitalized patients in acute-care facilities (4, 79). Nosocomial infections are a significant cause of morbidity, mortality, and increased hospital costs, largely as a result of the increased length of stay. Over the past few years, escalating medical costs and the introduction of outpatient surgery have resulted in shorter hospital stays and much higher volumes of outpatient surgical procedures. Unfortunately, these changes have tended to mask true surgical infection rates. Also, postoperative surgical-site infections are often not apparent until after the patient's discharge (22). Since 1976, hospital accreditation by the Joint Commission on Accreditation of Healthcare Organizations requires the use of a nurse epidemiologist or infection control nurse whose primary responsibility is to identify, track, and control infections associated with hospitalization (17).

Individuals in the institution who fall within the surveillance parameters of the nurse epidemiologist include both patients and employees. Although we usually consider patients to be more prone to infections, some microorganisms can also infect healthy adults (9).

Definition: A nosocomial infection is defined as one diagnosed during or after hospitalization that was neither present nor incubating at the time of admission (4, 79).

One of the problems with this definition is the difficulty involved in accurately differentiating community-acquired from hospital-acquired infections, particularly when employees are involved. Depending on the individual's immune status, incubation periods may also vary.

Although any organism can infect a susceptible host, the organisms most often considered to cause possible nosocomial infections, particularly in the immunologically compromised, include the following:

Fungi: *Aspergillus* spp., *Candida* spp., *Mucor* spp., *Cryptococcus neoformans*, *Nocardia asteroides*

Algae: *Prototheca* spp.

Bacteria: *Escherichia coli, Klebsiella* spp., *Proteus* spp., *Pseudomonas* spp., *Staphylococcus* spp., *Legionella* spp.

Viruses: Cytomegalovirus, varicella-zoster virus, human immunodeficiency virus, rubella virus, rotavirus, enterovirus

Parasites: *Giardia lamblia, Entamoeba histolytica,* free-living amebae, *Toxoplasma gondii, Cryptosporidium* spp., the microsporidia, *Strongyloides stercoralis, Trypanosoma* spp., *Plasmodium* spp., *Babesia* spp., *Hymenolepis nana, Taenia solium*

Ectoparasites: *Pediculus* spp., *Sarcoptes scabiei,* dipterous fly larvae (myiasis)

Although most nosocomial infections are bacterial, fungal, or viral in origin, a few can be caused by parasites. The patient may actually have acquired the infection (*T. gondii, S. stercoralis*) many years before hospitalization. These latent infections can often be reactivated when the patient becomes debilitated for some reason, particularly when the immune system is involved.

Other areas within an institution must be considered, including interaction between food service personnel and patients and employees. Public health requirements for checking food handlers have varied over the years, but as a result of cost containment measures, the general trend is to refrain from special laboratory work or routine screening procedures for employees. Parasite transmission possibilities are presented in Table 20.1. Not every organism is listed; however, those that represent actual possibilities for transmission are emphasized.

Sometimes patients are placed in isolation settings (enteric precautions), which may not be appropriate or necessary for the organism or infection involved. A good example is a patient with amebic dysentery. If the patient has diarrhea or dysentery, the organism stage being passed in the stool is the trophozoite. The trophozoite stage is noninfectious for others. The cyst form, which is passed in more normal, formed stools, is the infective stage for humans. The trophozoites do not form cysts after the stool is passed outside of the body. Therefore, this patient with amebic dysentery does not pose an infection threat to others.

An example in which the need for enteric precautions would be recommended is a patient with diarrhea caused by *Cryptosporidium*. The oocysts are infective, and often large numbers of organisms are passed in the watery, diarrheic stool. There are several reports of suspected transmission of *Cryptosporidium* from patients to health care workers (5, 21, 46, 48, 50).

Nosocomial infection with blood and tissue parasites can occur by accidental needlestick, contact with blood (open wound, mucous membranes, eye), transfusion, or organ transplantation (51). Nosocomial transmission of babesiosis has also been documented (76).

Nosocomial Gastrointestinal Infections

Nosocomial gastroenteritis can be defined as an acute infectious gastrointestinal illness acquired by a hospitalized patient, specifically unexplained diarrhea for 2 days, or the onset of infectious diarrhea somewhere in the hospital or medical center setting. In general, requests for ova and parasite identification are not indicated for patients with nosocomial diarrhea (80). Patients at highest risk for gastrointestinal infections include neonates, the elderly, patients with achlorhydria, and those who are immunosuppressed (79). Risk factors can be categorized as those within the host and those outside the host. Risk factors within the host include impaired immunity, reduced gastric acidity, and alterations in the normal intestinal flora or intestinal motility. Risk factors outside the patient include those that alter host resistance or increase the possibility of colonization. The transmission of organisms is generally via the fecal-oral route; some of the organisms can survive for long periods outside the body. Parasites that have been implicated in nosocomial infections include *Cryptosporidium* spp., *E. histolytica,* and *G. lamblia*. Although little is known about nosocomial transmission of microsporidia, potentially infectious spores can be found in the hospital environment, particularly in stools from severely immunocompromised patients (e.g., those with AIDS). The presence of infective spores in human clinical specimens suggests that precautions when handling body fluids and personal hygiene measures such as hand washing may be important in preventing primary infections in the health care setting. However, comprehensive guidelines for disease prevention will require more definitive information regarding sources of infection and modes of transmission (8).

Cryptosporidium spp.

Although cryptosporidiosis is a well-recognized occupational hazard for persons exposed to naturally infected calves and other animals, cases of cryptosporidiosis have also been reported among persons exposed to experimentally infected animals (37). Five veterinary students who had direct (four) or indirect (one) contact with experimentally infected calves became ill 6 to 7 days later and had diarrhea for 1 to 13 days; one student was hospitalized. A researcher developed gastrointestinal symptoms 5 days after a rabbit, which was infected with oocysts through a gastric tube, coughed

Table 20.1 Possible parasite transmission in a health care setting (nonlaboratory)

Organism[a]	Mode of transmission	Comments
Protozoa (intestinal)		
Entamoeba histolytica[b] (C)	Accidental ingestion of infective cysts, trophozo-	These organisms could be important in a food handler
Entamoeba dispar (C)	ites, oocysts, or spores in food or water contami-	situation. Treatment of carriers and excellent hygiene
Entamoeba coli (C)	nated with fecal material (fecal-oral transmission)	measures would prevent or eliminate any problems.
Entamoeba hartmanni (C)		*Cryptosporidium* would be an example where "enteric
Endolimax nana (C)		precautions" would be appropriate (severe diarrhea with
Iodamoeba bütschlii (C)		stool containing large numbers of infective oocysts).
Blastocystis hominis[b] (C)		
Free-living amebae		
Giardia lamblia[b] (C)		
Dientamoeba fragilis[b] (T)		
Cryptosporidium spp.[b] (O)		
Cyclospora cayetanensis[b] (O)		
Isospora belli[b] (O)		
Microsporidia[b] (S)		
Helminths (intestinal)		
Enterobius vermicularis (E)	Inhalation or ingestion of infective eggs	Very common in children, asymptomatic
Strongyloides stercoralis (IL)	Skin penetration of infective larvae from stool material	Possible, particularly in a compromised patient
Hymenolepis nana (E)	Ingestion of infective eggs (fecal-oral)	Could lead to infection with the adult worm in humans
Taenia solium (E)	Ingestion of infective eggs (fecal-oral)	Could cause cysticercosis
Blood and tissue protozoa		
Leishmania spp. (cutaneous)	Direct contact or inoculation of infectious mate- rial from patient's lesion could lead to infection.	A rare possibility, but it has been documented (accidental person-to-person transmission)
Trypanosoma spp.	Blood transfusion	Significant problem in areas of endemicity
Ectoparasites		
Pediculus spp.	Transmission of lice and mites is usually seen in	Confinement in close quarters often lends itself to a
Sarcoptes scabiei	groups with very close contact (institutionalized settings).	transmission problem with lice and mites.
Dipterous fly larvae	Myiasis could occur anywhere.	Protection of wounds from flies would solve the potential problem.

[a] C, cyst; T, trophozoite (may be transmitted inside helminth eggs); O, oocyst; E, egg; IL, infective larvae; S, spore.
[b] Potentially pathogenic intestinal protozoa.

droplets of inoculum onto his face as he was removing the tube. The researcher's stool, which was first obtained for testing the day after he became ill, was positive for oocysts. A veterinary scientist developed flu-like symptoms 7 days after smelling for gastric odor to check the position of a gastric tube in an infected calf; she was unaware of other exposures to *Cryptosporidium*. She developed gastrointestinal symptoms 10 days after this exposure, and oocysts were found in a stool specimen about a week later.

Cryptosporidium can also be transmitted within the hospital setting. At least nine cases of occupational transmission of *Cryptosporidium* from human patients to health care workers have been reported (37). The infected staff members were symptomatic and had oocysts identified from stool examinations. These individuals included a nurse caring for an infected bone marrow transplant recipient, a nurse doing night duty on a ward where an infected 13-month-old boy was a patient, a nurse caring for infected patients before and after renal transplantation, and five nurses caring for an infected patient with AIDS. Patient-to-patient transmission of *Cryptosporidium* in hospitals has also been reported.

Cryptosporidiosis has been implicated as one of the more important opportunistic agents seen in patients with AIDS (18, 27, 28). Unfortunately, at the time of this writing, no totally effective therapy for cryptosporidiosis has been identified, despite testing of over 100 drugs. Consequently, finding this parasite in an immunocompromised host, especially one with AIDS, usually carries a poor prognosis. Also, reports of respiratory tract and biliary tree infections confirm that the developmental stages of this organism are not always confined to the gastrointestinal tract. The more than 4,000 published papers related to cryptosporidiosis that are currently in the literature include information on nosocomial transmission, day care center outbreaks, and a number of waterborne outbreaks.

The presence of a thin-walled autoinfective oocyst may explain why a small inoculum can lead to an overwhelming infection in a susceptible host and why immunosuppressed patients may have persistent, life-threatening infections in the absence of documentation of repeated exposure to oocysts. Oocysts undergo sporogony while they are in the host cells and are immediately infective when passed in the stool. In contrast, the oocysts of *Cyclospora cayetanensis* and *Isospora belli* do not sporulate until they are passed from the host and are exposed to oxygen and temperatures below 37°C.

Oocysts recovered in clinical specimens are difficult to see without special stains, such as the modified acid-fast, Kinyoun's, or Giemsa preparations, or the direct fluorescent-antibody, enzyme-linked immunosorbent assay, or immunochromatographic immunoassay methods (26, 30–32). The four sporozoites can be seen within the oocyst wall in some of the organisms, although they are not always visible in freshly passed specimens.

Clinical symptoms include nausea, low-grade fever, abdominal cramps, anorexia, and 5 to 10 watery, frothy bowel movements a day, which may be followed by constipation. Some patients present with diarrhea as described above, and others have relatively few symptoms, particularly later in the course of the infection. In patients with the typical watery diarrhea, the stool specimen contains very little fecal material and is mainly water and mucus flecks. Often, the organisms are entrapped in the mucus, and diagnostic procedures are performed accordingly. Generally a patient with a normal immune system will have a self-limited infection; however, patients who are compromised may have a chronic infection with a wide range of symptoms (asymptomatic to severe).

Occasionally, these patients require fluid replacement, and the diarrhea may persist for more than 2 weeks. This is particularly true in infants, in whom excessive fluid loss may last for more than 3 weeks. Failure to thrive has also been attributed to chronic cryptosporidiosis in infants. Since diarrheal illness is a major cause of morbidity and mortality in young children living in developing countries, it is likely that cryptosporidiosis plays a major role in the overall health status of these children. It has also been suggested that *Cryptosporidium* may be implicated in the respiratory disease that often accompanies diarrheal illness in malnourished children.

The duration and severity of diarrheal illness depends on the immune status of the patient. The most severely immunocompromised patients cannot self-cure, the illness becomes progressively worse with time, and the sequelae may be a major factor leading to death. The length and severity of illness may also depend on the ability to reverse the immunosuppression. In these patients, *Cryptosporidium* infections are not always confined to the gastrointestinal tract; additional symptoms have been associated with extraintestinal infections (respiratory problems, cholecystitis, hepatitis, and pancreatitis). Data for a small series of patients indicated that those with CD4 counts of 180 cells/mm^3 or greater cleared the *Cryptosporidium* infection over a period of 7 days to 1 month. These data are important in predicting the natural progression of the infection and in designing therapeutic trials.

Previously, most human cases have been diagnosed after examination of small- or large-bowel biopsy material, often by both light and electron microscopy. Although examination of flotation material by phase-contrast microscopy has proven to be an excellent procedure for the recovery and identification of *Cryptosporidium* oocysts, many laboratories have neither access to such equipment nor experience with phase-contrast microscopy. Also, since the organism is considered infectious for laboratory personnel, fixed specimens can be processed by using several acid-fast stains, many of which are very satisfactory in demonstrating the organisms in stool material (29).

Respiratory cryptosporidiosis has been found in AIDS patients. Sputum specimens from immunodeficient patients with undiagnosed respiratory illness should be submitted in 10% formalin and examined for *Cryptosporidium* oocysts by the same techniques as those used for stool samples. Infection of the gallbladder and biliary tree should result in oocysts being passed in the stool.

The use of fecal immunoassays has proven to be very helpful in providing more sensitive methods of detecting organisms in stool specimens. Fluorescent-antibody procedures with good specificity and sensitivity have been developed, as have enzyme immunoassay procedures and immunochromatographic cartridges, also with excellent specificity and sensitivity. Some of these reagents, particularly the combination direct fluorescent-antibody product used to identify both *Giardia* cysts and *Cryptosporidium* oocysts, are being widely used in water testing and outbreak situations (24, 25, 29–32).

The infections tend to be self-limiting in patients who have an intact immune system (24). One method of treating patients who are receiving immunosuppressive drugs is to discontinue such a regimen. Other approaches with specific therapeutic drugs such as spiramycin have been tried, but the results to date are somewhat controversial.

Data indicate that paromomycin may be a potential therapeutic agent for extended clinical trials. Ongoing studies of immunologic intervention hold promise, but this technique currently is not widely available.

Information on the epidemiology of cryptosporidiosis is well defined, and its potential pathogenicity in humans has been widely recognized. The organism is transmitted by oocysts that are fully sporulated and infective at the time they are passed in stool. Oocysts are resistant to some of the disinfectants routinely used in medical care facilities. They are susceptible to ammonia, 10% formalin in saline, freeze-drying, and exposure to temperatures below freezing or above 65°C for 30 min. Commercial bleach in a 50% solution is effective.

Direct person-to-person transmission is likely and may occur through direct or indirect contact with stool material. Direct transmission may occur during sexual practices involving oral-anal contact. Indirect transmission may occur through exposure to positive specimens in a laboratory setting or through exposure to contaminated surfaces, food, or water.

The epidemiologic considerations for cryptosporidiosis emphasize transmission by environmentally resistant oocysts, numerous potential reservoir hosts for zoonotic transmission, documentation of person-to-person transmission in day care centers, nosocomial transmission within the health care setting, occurrence of asymptomatic infections (infective, carrier state), widespread environmental distribution resulting in the probability of waterborne transmission, and the link between cryptosporidiosis and severe, life-threatening disease in individuals with impaired immune function.

Giardia lamblia

A "debilitating bout" of giardiasis thought to have represented patient-to-staff transmission has been reported. The case involved an orthopedic surgeon caring for two preschool-age patients with giardiasis. In changing the leg cast, it was noticeably stained with moist and dry feces. The physician became ill in early May and later had giardiasis confirmed with routine stool examinations. Although he typically washed his hands before and after changing casts, he rarely wore a mask; infection may have been caused by inhalation of infectious cysts in the plaster dust.

G. lamblia is worldwide in distribution and is apparently more prevalent in children than in adults and more common in warm climates than in cool ones. It is the most commonly diagnosed flagellate in the intestinal tract, and it may be the most commonly diagnosed intestinal protozoan in some areas of the world. Despite disagreement concerning the species names intestinalis and duodenalis,

both continue to be used to describe this organism. Since the majority of health care workers within the United States are used to "Giardia lamblia" and continue to report the presence of the organism by using this name, it will continue to be used throughout this text to refer to organisms found in humans as well as other mammals.

The most common location of the organisms is in the crypts within the duodenum. Again, for reasons that are not totally known, cyst formation takes place as the organisms move down through the colon. The trophozoite is usually described as being teardrop shaped from the front, with the posterior end being pointed. When examined from the side, the trophozoite resembles the curved portion of a spoon. The concave portion is the area of the sucking disk. The cysts may be either round or oval, and they contain four nuclei, axonemes, and median bodies. Often, some cysts appear to be shrunk or distorted, and one may see two halos, one around the cyst wall itself and one inside the cyst wall around the shrunken organism. The halo effect around the outside of the cyst is particularly visible on the permanent stained smear.

Although the organisms in the crypts of the duodenal mucosa may reach very high densities, they may not cause any pathologic changes. The organisms feed on the mucus secretions and do not penetrate the mucosa. Although organisms have been seen in biopsy material inside the intestinal mucosa, others have been seen only attached to the epithelium. In symptomatic cases, there may be irritation of the mucosal lining, increased mucus secretion, and dehydration. Other symptoms include epigastric pain, flatulence, and diarrhea with increased fat and mucus in the stool but no blood. Weight loss often accompanies these symptoms. Although there is speculation that the organisms coating the mucosal lining may act to prevent fat absorption, this does not completely explain why the uptake of other substances normally absorbed at other intestinal levels is prevented. Occasionally, the gallbladder is also involved, causing gallbladder colic and jaundice. G. lamblia has also been detected in bronchoalveolar lavage fluid.

Although patients with symptomatic giardiasis usually have no underlying abnormality of serum immunoglobulins, patients with immunodeficiency syndromes, particularly common variable hypogammaglobulinemia, have a high incidence of giardiasis. Giardiasis was the most common cause of diarrhea in these patients and was associated with mild to severe villous atrophy. Successful treatment of giardiasis led to symptomatic cure and improvement in mucosal abnormalities, with the exception of nodular lymphoid hyperplasia.

With the advent of AIDS, there was speculation that G. lamblia might be an important pathogen in this group of individuals. However, to date, clinical findings do not seem to confirm this possibility (58, 75).

Routine stool examinations are normally recommended for the recovery and identification of intestinal protozoa. However, because *G. lamblia* organisms are attached so securely to the mucosa by means of the sucking disk, a series of even five or six stools may be examined without recovery of the organisms. The organisms also tend to be passed in the stool on a cyclical basis. The Entero-Test capsule can be helpful in recovering the organisms, as can the duodenal aspirate. Although cysts can often be identified on the wet stool preparation, many infections may be missed without the examination of a permanent stained smear (14). Material from the string test (Entero-Test) or mucus from a duodenal aspirate should be examined as a wet preparation for motility; however, motility may be represented by nothing more than a slight flutter of the flagella because the organism will be caught up in the mucus. After diagnosis, the rest of the positive material can be preserved as a permanent stain.

Immunoassays to detect *Giardia* antigen in feces have also been developed (2, 30–32). The enzyme-linked immunosorbent assay is at least as sensitive as microscopic wet examinations (61). Fluorescence methods involving monoclonal antibodies have also proven to be extremely sensitive and specific in detecting *G. lamblia* in fecal specimens (77).

Unfortunately, serodiagnostic procedures for giardiasis do not yet fulfill the criteria necessary for wide clinical use, particularly since they may indicate either past or present infection. In contrast, the detection of antigen in stool or visual identification of organisms with monoclonal antibody reagents indicates current infection. With the increase in the prevalence of *Giardia* infections and awareness of particular situations such as nursery school settings, perhaps additional detection assays will emerge as rapid and reliable immunodiagnostic procedures (1). Because giardiasis may not produce any symptoms at all, demonstration of the organism in symptomatic patients may not rule out other possibilities such as peptic ulcer, celiac disease of some other etiology, strongyloidiasis, and possibly carcinoma.

If giardiasis is diagnosed, the patient should be treated. Giardiasis can be eliminated from most patients by metronidazole (40). Tinidazole has also been used and has proven to be more effective than metronidazole as a single dose (40).

In the absence of a parasitologic diagnosis, the treatment of suspected giardiasis is a common question with no clear-cut answer. The approach depends on the alternatives and the degree of suspicion of giardiasis, both of which will vary among patients and physicians. However, it is not recommended that treatment be given without a good parasitologic workup, particularly since relief of symptoms does not allow a retrospective diagnosis of giardiasis; the most commonly used drug, metronidazole, also targets organisms other than *G. lamblia*.

The third question involves treatment of a patient who is asymptomatic. Generally, it is recommended that all cases of proven giardiasis be treated because the infection may cause subclinical malabsorption, symptoms are often periodic and may appear later, and a carrier is a potential source of infection for others (40).

G. lamblia is transmitted by ingestion of viable cysts. Although contaminated food or drink may be the source, intimate contact with an infected individual may also provide the infection mechanism. This organism tends to be found more frequently in children or in groups that live in close quarters (16, 17, 19, 34, 43, 53, 56, 59, 72). Often, there are outbreaks due to poor sanitation facilities or breakdowns in sanitation as evidenced by infections in travelers and campers (42, 45, 53, 60). There is also an increase in the prevalence of giardiasis in the male homosexual population, probably as a result of anal or oral sexual practices (67, 71).

During the past few years, this infection has received much publicity. With increased travel, there has been a definite increase in symptomatic giardiasis within the United States (1). Various surveys show infection rates of 2 to 15% in various parts of the world.

Entamoeba histolytica

Although *E. histolytica* is not generally associated with nosocomial infections, reports indicate that transmission is certainly possible (11). Istre et al. reported on 36 cases of amebiasis during a 30-month period in patients who had received colonic irrigation therapy (38). Epidemiologic studies implicated the colonic irrigation machine; testing indicated that after routine cleaning, the machine was contaminated with *E. histolytica*. Patients at greatest risk were those who received colonic irrigation immediately after a patient with amebiasis and bloody diarrhea. Evidence supports the possibility of transmission if endoscopic equipment is not thoroughly cleaned and disinfected after use in a patient infected with *E. histolytica*.

Since *E. histolytica* can easily be confused with macrophages (trophozoites) and polymorphonuclear leukocytes (cysts) found in stool, it is important to recognize the possibility of false-positive results (in which other protozoa or human cells are confused with this organism). Recognizing the possibility of error is particularly important in a potential "outbreak" situation. These types of diagnostic errors, whereby large numbers of patients were diagnosed with amebiasis, have been documented; in one report, on review, 34 of 36 patients who were diagnosed with amebiasis were found to be negative for *E. histolytica* (10).

Over the years, some workers came to believe that what was called *E. histolytica* was really two separate species of *Entamoeba*, one being pathogenic and causing invasive disease (*E. histolytica*) and the other being nonpathogenic and causing mild or asymptomatic infections (*E. dispar*). Although this was controversial, current knowledge indicates that pathogenic *E. histolytica* is the etiologic agent of amebic colitis and extraintestinal abscesses while nonpathogenic *E. dispar* causes no intestinal symptoms and is not invasive in humans (65). Unfortunately, with the exception of some newly introduced immunoassay procedures, most laboratories will continue to use routine methods for the identification of organisms in the *Entamoeba* genus; unless the trophozoites contain ingested erythrocytes, identification to the species level is not possible. The organisms should be reported as *E. histolytica/E. dispar* with appropriate reporting to public health agencies. In these cases, the approach to the patient will be based on clinical findings and physician decisions regarding therapy. Diagnostic reagents for confirmation of true pathogenic *E. histolytica* and for the *E. histolytica/E. dispar* group are commercially available.

Microsporidia

Although the epidemiology of the microsporidia has not yet been thoroughly defined, it is highly likely that the spores shed in human clinical specimens (stool, urine, sputum, etc.) are immediately infectious. These infectious spores can be found in hospital and laboratory environments and are associated primarily with clinical specimens from immunocompromised patients, such as those with AIDS or transplant recipients. Precautions should be used when handling body fluids from these patients. Microsporidial infections in all body sites have been documented; this fact needs to be considered when determining the housing and care of the patient. Precautions would be very similar to those used for patients diagnosed with disseminated cryptosporidiosis, in which infectious organisms could be found in multiple clinical specimens.

Specific infection control guidelines for inpatients with microsporidiosis have yet to be developed, but available information suggests that all body fluids that may contain infectious spores should be handled accordingly. Standard precautions used within the institution and laboratory environments should be sufficient to prevent accidental infection with microsporidial spores.

Isospora (*Cystoisospora*) *belli*

I. belli is thought to be the only species of *Isospora* that infects humans, and no other reservoir hosts are recognized for this infection. Transmission is through ingestion of water or food contaminated with mature, sporulated oocysts. Sexual transmission by direct oral contact with the anus or perineum has also been postulated, although this mode of transmission is probably much less common. The oocysts are very resistant to environmental conditions and may remain viable for months if kept cool and moist; oocysts usually mature within 48 h following stool evacuation and are then infectious. Although it is not considered a "common" parasite, occasional positive specimens are seen in many laboratories. Since transmission is through the ingestion of infective oocysts, there is certainly the possibility that this parasite could be transmitted within the laboratory or in the patient care setting.

Examination of a fecal specimen for the oocysts is recommended. However, wet-preparation examination of fresh material either as the direct smear or as concentrated material is recommended rather than the permanent stained smear. The oocysts are very pale and transparent and can easily be overlooked. They can also be very difficult to see if the concentration sediment is from polyvinyl alcohol-preserved stool. The light level should be reduced, and additional contrast should be obtained with the microscope for optimal examination conditions. It is also quite possible to have a positive biopsy specimen but not recover the oocysts in the stool because of the small numbers of organisms present. These organisms are acid fast and can also be demonstrated by using auramine-rhodamine stains.

Hymenolepis nana

Although *H. nana* can be transmitted by the eggs and does not require an intermediate host to complete its life cycle, the nosocomial significance of this infection is unknown. However, it is important to remember that the eggs passed in the stool are mature and infective for patients and health care workers.

Taenia solium

Nosocomial transmission of the infective eggs of *T. solium*, causing cysticercosis, has not been documented. However, an individual infected with adult *T. solium* and passing eggs could certainly serve as a potential reservoir for serious transmission potential. It is important to remember that the eggs passed in the stool are mature and infective for patients and health care workers. This is true even when proglottids are submitted to the laboratory in preservative; the eggs within the uterine branches will remain viable and infectious.

Nosocomial Blood and Tissue Infections

Plasmodium spp.

Worldwide, transfusion malaria is the most common and significant nosocomial parasitic infection, although it is rare within the United States (7, 51). These infections have occurred after transfusion not only of whole blood but also of white blood cells, fresh plasma, and platelets (49). Nosocomial malaria infections have also originated through renal transplants (51). It is interesting that malaria infections have been transmitted from recipient to donor in a person-to-person transfusion situation in which backflow of blood has been a problem (7). Most cases of posttransfusion malaria may go unsuspected, with an incubation period of up to 4 weeks. Donors implicated in transfusion-transmitted malaria tend to be "semi-immune" with a very low parasite load; therefore, detection using antigens or PCR may not guarantee identification of the infected donor. Often the infectious dose from these donors may be 1 to 10 parasites in a unit of blood (73).

Malaria induced via contaminated needles shared among drug addicts has been well known for many years. Also, nosocomial patient-to-staff malaria transmission via needlestick injury has been documented, with most cases involving nurses or physicians who were exposed during venipuncture or preparation of blood smears (51).

A nosocomial outbreak of chloroquine-resistant *Plasmodium falciparum*, in which a multidose heparin vial was implicated as the source of infection, was reported from Venezuela. The vial was apparently contaminated by the blood of a parenteral drug abuser who had asymptomatic malaria and who had received 2 days of intravenous fluid therapy (62).

In the cases involving suspected malaria transmission by renal transplantation, the donors were strongly sero-positive or had a positive blood smear. However, all three patients were from areas where malaria was endemic, and the stress of surgery and immune system suppression could have also stimulated recrudescence of the infection (51). Transmission of malaria has also been documented through liver transplantation; unfortunately, the recipient died. Based on therapeutic toxicity for *Plasmodium* spp., the use of organs from donors infected with malaria carries risks that may limit transplantation from such donors (3).

Babesia spp.

Nosocomial transmission of babesiosis through routine blood transfusion, frozen-thawed blood, and platelets has also been documented (51). Although needlestick or open-wound transmission of *Babesia* spp. has not been reported, it is certainly possible, particularly in the health care setting. The risk of infection through contact with ticks is probably low.

Trypanosoma brucei gambiense and *T. brucei rhodesiense*

Although patients with African trypanosomiasis are usually symptomatic, those in the early stages of the infection may be asymptomatic and appear to be healthy potential blood donors. These types of patients are probably responsible for the rare transfusion cases reported. However, contaminated needlestick injuries and open-wound contact with contaminated blood are also potential means of transmission in the health care setting.

Trypanosoma cruzi

Although *T. cruzi* is normally transmitted by the feces of an infected triatomid bug, transmission of the organism via blood transfusion is the second most common means of acquiring the infection. In areas where infection is highly endemic, as many as 62% of blood donors are seropositive for *T. cruzi* antibody (54). One study from Los Angeles County of 1,027 consecutive blood donations found 2.4% to be seropositive for *T. cruzi* antibody by an indirect fluorescent-antibody assay (P. Kerndt, H. Waskin, F. Steurer, L. V. Kirchhoff, J. Nelson, I. Shulman, G. Geliert, and S. Watermen, *Program Abstr. 28th Intersci. Conf. Antimicrob. Agents Chemother.*, abstr. 669, 1988). Considering this information, concerns regarding the safety of the blood supply are reasonable and understandable (44). Transmission via kidney transplants has also been well documented (23).

The entry of large numbers of immigrants from Central America has contributed to growing concerns, particularly with respect to transfusion-related transmission of *T. cruzi* in areas of the world where the infection is not endemic or, if endemic, has not been a problem. Since there are triatomid bugs capable of becoming infected with *T. cruzi* within the southwestern United States, the presence of more people with latent infections also presents some concerns.

Transmission of *T. cruzi* through donor kidneys has been documented in Latin America, where serologic screening of organ donors and recipients is not standard practice. Three cases of recipient infection have been reported in the United States; the cases involved kidney and pancreas, kidney alone, and liver transplant from the same donor from Central America. One of the recipients died of myocarditis (3).

Leishmania donovani

Although transfusion-induced visceral leishmaniasis has been described only rarely, like other infections mentioned above, it may be underreported in areas of endemicity where confirmation of the route of infection may be difficult. Certainly, exposure to contaminated blood or animals would represent a potential source of infection.

Toxoplasma gondii

Nosocomial infection with *T. gondii*, primarily in cardiac or renal transplant patients, has been documented (51). However, the efficacy of trimethoprim-sulfamethoxazole for the prevention of toxoplasmosis transmission dramatically reduces the risk to organ recipients from infected donors. Most serious infections in immunocompromised patients result from reactivation of a latent infection. Transmission from a patient to a health care worker via needlestick injury or exposure to blood (open wound) has not been documented.

Nosocomial Infections with Ectoparasites

Ectoparasites of nosocomial significance include lice (head, body, pubic), scabies and other mites, and myiasis. Head and pubic lice and scabies are fairly common in many parts of the world, including the United States. These infestations are generally not serious problems within the health care setting, with the exception of scabies, which has the potential to cause large outbreaks.

Pediculus spp. and Phthirus pubis

Infection with lice belonging to the genus *Pediculus* are common and are usually found on one part of the body such as head, body, or pubic hair. The head louse, *Pediculus humanus capitis*, is found most often in school-age children, and nosocomial transmission is generally not a problem, with possible exceptions characterized by close patient-to-patient contact in pediatric wards or institutional settings.

Pediculus humanus corporis (body louse) is well known for its association with poverty and poor personal hygiene. Since transmission is through contact with or exchange of contaminated clothing or bedding, this does not seem to represent a large potential problem. However, the potential for transmission varies throughout the world.

Phthirus pubis (pubic louse) infestation occurs most commonly in sexually active adults, with a direct skin-to-skin transfer. It is unlikely that transmission within the health care setting would cause any problems.

Sarcoptes scabiei

Scabies is a highly contagious infestation of the itch mite, *Sarcoptes scabiei*. Symptoms are related to pruritic dermatitis, and apparently, transmission can occur through person-to-person direct skin contact or via fomites. Outbreaks of scabies in hospitals or institutions, particularly for the elderly or retarded, have been documented worldwide (52). These outbreaks are particularly severe when associated with Norwegian or crusted scabies; such patients may be infected with thousands of mites on the skin. This infection has been linked to cellular immune deficiencies, including AIDS (74). Other potential problems include a decreased awareness of itching and the use of topical corticosteroids (52). Because of the number of mites present, these patients are extremely contagious via sloughing skin, direct contact, and environmental contamination. Another potential problem is the fact that the symptoms mimic those of many other dermatologic problems. Burrows may not be evident, and hyperkeratotic, crusted, scaling, fissured plaques are present over the scalp, face, and back, with associated gross nail thickening (69). The number of documented reports of scabies, particularly in AIDS patients, suggests that nosocomial scabies may become more common in the future.

Although outbreaks of other mites, including the pigeon mite *Dermanyssus gallinae*, have been described, it is not clear whether these infestations are a problem. It has also been suggested that these cases may be misdiagnosed as scabies (52).

Myiasis

Myiasis refers to infestation of the body by larvae of a number of fly species. These cases occur when flies are attracted to patient blood or body tissue fluids and deposit eggs into an open wound or patient body opening such as the nose or ear. Often, these patients are not totally conscious, or they may be debilitated or immobile. It is generally thought that these cases are underreported for a number of reasons, some of which are related to institutional risk management decisions. A study from Brisbane, Australia, reported 14 infestations, 6 of which were nosocomial, during a period of 17 months (55).

Ophthalmomyiasis, caused by infestation of the eye by dipterous fly larvae, can result in symptoms ranging from minor irritation to blindness, disfigurement, or death. Infestations with the sheep botfly, *Oestrus ovis*, are normally associated with normal hosts, while infestations with others, including *Cochliomyia hominivorax*, seem to occur more commonly in compromised hosts (12).

Nosocomial Infections in the Pediatric Patient

Specific infections that can be acquired in the pediatric patient include those caused by *Cryptosporidium* spp., *G. lamblia*, *P. humanus capitis*, and *S. scabiei*. Although less is known about microsporidia, the infective spores could be transmitted in a hospital setting, particularly if spores are present in stool specimens, urine, or other body fluids from immunocompromised patients.

Cryptosporidium spp.

Cryptosporidium infection is discussed above in the section on nosocomial gastrointestinal infections. It is also important to remember than this organism has been isolated from other body sites, particularly in severely immunocompromised patients. Cryptosporidiosis has also been linked to infections in nursery schools and should certainly be considered in this setting.

Giardia lamblia

Infection has been well documented in the pediatric population. Giardiasis is most frequently associated with nursery school outbreaks and is thoroughly discussed above in the section on nosocomial gastrointestinal infections.

Pediculus humanus capitis

Although pediculosis is a well-known problem among young children, lice are not as likely to be spread to other patients except in chronic ambulatory-care situations such as psychiatric wards or pediatric ward playrooms. Under these circumstances, patients should be screened prior to admission.

Sarcoptes scabiei

Prevalence of the itch mite infection, or scabies, varies among young children; sexually active adolescents and adults are also susceptible to infection. The small size of the mite contributes to the problem and to possible transmission to medical staff and other patients. Although intimate contact is considered the most likely mode of transmission, fomite transmission is an alternative method, particularly with Norwegian scabies. It is generally recommended that if two or more patients and one or more staff members exhibit the symptoms of scabies (pruritus with or without crusting or scaling) or have positive skin scrapings, total treatment should be given. It is also important that posttreatment surveillance and testing be continued for at least 3 months.

Nosocomial Infections in the Compromised Patient

Compromised patients have poor host defenses, usually for one of three reasons: underlying disease (immunosuppression or immunodeficiencies in either the cellular or humoral mechanisms of the immune system or both), therapy for disease (chemotherapy, radiation), or hospital-based procedures and contact with infected persons (catheterization, broad-spectrum antibiotics, nebulizers, inhalation procedures, person-to-person contact, food sources, contaminated equipment, etc.).

In addition to *Cryptosporidium*, *E. histolytica*, *G. lamblia*, and microsporidia as mentioned above, *S. stercoralis*, free-living amebae, and *Leishmania* spp. cause severe disease in the compromised patient and can be transmitted in the hospital setting.

S. stercoralis has become extremely important, particularly with respect to compromised patients (41) and military personnel and travelers who may have been in an area of endemic infection many years before (64). More than 30 to 40 years after acquisition of the original infection, persistent undiagnosed disease can be found in these individuals. If, for any reason, they become immunocompromised, the result can be disseminated disease leading to the hyperinfection syndrome and death (33, 78).

The life cycle of *S. stercoralis* is one of the more complex parasite cycles. Human infection is acquired by skin penetration of the filariform larvae (infective larvae) from the soil. After penetration of the skin, the larvae are carried via the cutaneous blood vessels to the lungs, where they break out of the pulmonary capillaries into the alveoli. They then migrate via the respiratory tree to the trachea and pharynx, are swallowed, and enter the mucosa in the duodenum and upper jejunum. Development usually takes about 2 weeks, after which the females begin egg production. The eggs are oval and thin shelled and measure 50 to 58 μm by 30 to 34 μm (generally a bit smaller than hookworm eggs). The eggs usually hatch, and the rhabditiform larvae (noninfective larvae) pass out of the intestinal tract in the feces. These larvae are passed out onto the soil in the feces and develop into free-living male and female worms, eventually producing infective filariform larvae (egg, noninfective larvae, and infective larvae). In temperate climates, the free-living male and female worms do not develop; however, the rhabditiform larvae that pass out in the stool develop into the filariform (infective) larvae, which are then ready to infect the next host through skin penetration.

In situations in which autoinfection occurs, some of the rhabditiform larvae that are within the intestine develop into the filariform larvae while passing through the bowel. These larvae can then reinfect the host by (i) invading the intestinal mucosa, traveling via the portal system to the lungs, and returning to the intestine or (ii) being passed out in the feces and penetrating the host on reaching the perianal or perineal skin.

It is generally accepted that the life cycle presented above is correct; the skin-penetrating larvae travel from the skin to the intestine via an obligatory route that includes the blood, lungs, trachea, and upper gastrointestinal tract (the standard pulmonary route). It is also assumed that following autoinfection, the larvae follow the same route to return to the duodenum, where they mature. However, review information and data from Schad et al. (70) indicate that the pulmonary route is just one of several possible pathways to the duodenum, regardless of whether the larvae penetrated the skin or the intestine. In spite of some caveats, these studies definitely challenge the current accepted understanding of the life cycle. The pulmonary migration route for all skin-penetrating nematodes may not be as universally applicable as was once thought (66).

As judged from the life cycle, the three areas of primary involvement are the skin (cutaneous), the lungs (pulmonary), and the intestinal mucosa (intestinal).

Initial skin penetration usually causes very little reaction, although there may be some pruritus and erythema if the number of penetrating larvae is large. With repeated infections, the patient may mount an allergic response that will prevent the parasite from completing the life cycle. The larvae may be limited to skin migration or larva migrans.

Larval migration through the lungs may stimulate symptoms, depending on how many larvae are present and the intensity of the host immune response. Some patients are asymptomatic, whereas others present with pneumonia. With a heavy infective dose or in the hyperinfection syndrome, individuals often develop cough, shortness of breath, wheezing, fever, and transient pulmonary infiltrates (Loeffler's syndrome). Cases in which the larvae could even be found in the sputum have been reported (13, 36).

In heavy infections, the intestinal mucosa may be severely damaged with sloughing of tissue, although this type of damage is unusual. The symptoms may mimic peptic ulcer with abdominal pain, which may be localized in the right upper quadrant. Radiographic findings may mimic Crohn's disease of the proximal small intestine (6). In an immunocompetent patient, there is a leukocytosis with a peripheral eosinophilia of 50 to 75%, while in patients with chronic infections, the eosinophilia may be much lower. Some of these chronic infections have lasted over 30 years as a result of the autoinfective capability of the larvae (67). One case of chronic strongyloidiasis persisted for approximately 65 years (49).

Autoinfection is probably the mechanism responsible for long-term infections that persist for years after the person has left the area of endemic infection. The parasite

and host reach a status quo so that neither suffers any serious damage. If for any reason this equilibrium is disturbed and the individual becomes immunosuppressed, the infection proliferates, with large numbers of larvae being produced and found in every tissue of the body. Several conditions, including the increased use of immunosuppressive therapy, predispose an individual to the hyperinfection syndrome (16, 63, 74). In addition to suffering tissue damage from the migrating larvae, the patient may die from sepsis, primarily as a result of the intestinal flora. Other causes of death include peritonitis, brain damage, and respiratory failure. Few cases of reactive arthritis have been reported, and the number of cases may be underestimated. A case of reactive arthritis combined with uveitis associated with a long-standing, heavy *Strongyloides* infection was detected in a 32-year-old patient positive for human T-cell leukemia virus type 1. Treatment with thiabendazole and ivermectin resulted in rapid improvement (63).

Confirmation of the infection depends on recovery and identification of the adult worms, larvae, or eggs from the stool, duodenal material, or sputum. Fecal specimens may not be positive, even after performance of the routine ova and parasite examination, including the concentration procedure. The number of larvae found in the stool also varies from day to day.

Some reports indicate that sampling of duodenal contents may recover larvae if stools are negative (33), while others indicate that duodenal sampling may not be effective (15). Special techniques that have been recommended for the recovery of duodenal contents include the Entero-Test capsule (see chapter 29). There are also special concentration techniques (Baermann) and methods for larva culture (Harada-Mori, petri dish), which are discussed in detail in chapter 28. These techniques provide a means of larval recovery from large amounts of stool material, and often the filariform larvae are recovered. To differentiate the *Strongyloides* filariform larvae from those of hookworm, one should examine the tail under the microscope. *Strongyloides* filariform larvae have a slit in the tail, while hookworm filariform larvae have a pointed tail.

Agar plate cultures are also possible; stool is placed onto agar plates, and the plates are sealed to prevent accidental infections and then held for 2 days at room temperature. As the larvae crawl over the agar, they carry bacteria with them, thus creating visible tracks over the agar. The plates are examined under the microscope for confirmation of larvae, the surface of the agar is then washed with 10% formalin, and larval identification is confirmed via wet examination of the sediment from the formalin washings. Details can be found in chapter 28.

Debilitated or compromised patients should always be suspected of having strongyloidiasis, particularly if there are unexplained bouts of diarrhea and abdominal pain, repeated episodes of sepsis and/or meningitis with intestinal bacteria, or unexplained eosinophilia. However, a recent comparative study of the occurrence of *Strongyloides* infection in 554 AIDS and 142 non-AIDS patients demonstrated similar prevalences of infection in the two groups, thus indicating no significant statistical differences (20).

At present, the recommended drug of choice is thiabendazole; specific dosages are provided in chapter 25. In some cases repeated therapy is necessary, and the cure rates range from 55 to 100% (35). Patients with the hyperinfection syndrome should be hospitalized during therapy for proper monitoring.

Contact with contaminated infective soil, feces, or surface water should be avoided. Individuals found to have the infection should be treated. *All patients who are going to receive immunosuppressive drugs should be screened for strongyloidiasis before therapy.*

Laboratory Infections

Safety precautions tend to be emphasized in clinical laboratories, particularly microbiology laboratories (standard precautions; see chapter 35). However, general safety regulations pertaining to laboratory coats, hand washing, and use of disinfectants, as well as to not eating, drinking, and smoking, fail to address the specifics of possible parasite transmission.

In laboratories where extensive diagnostic testing is performed, exposure to possible infectious parasites may result from the handling of specimens, drawing of blood, various types of concentration procedures, organism cultures, and animal inoculation studies. Relevant parasites and their possible routes of infection are presented in Tables 20.2 and 20.3. Information on resistance to antiseptics and disinfectants can be found in Table 20.4 (57).

Intestinal Protozoa

Intestinal protozoa such as *G. lamblia*, *E. histolytica*, microsporidia, and the coccidia, particularly *Cryptosporidium* spp., cause human disease, and the protozoan cysts, oocysts, and spores can be accidentally transmitted from one person to another. The cysts, oocysts, and spores are resistant to many disinfectants. Of those currently available, ozone is the most effective protozoan cysticide, followed by chlorine dioxide, iodine, and free chlorine, all of which appear to be more effective than the chloramines (39, 47). Cyst forms, including oocysts, are the most resistant to chemical disinfection. Although the exact mechanisms of resistance are not thoroughly understood, these resistant

stages probably take up fewer disinfectant molecules from solution than do stages like the trophozoite.

Relatively few cases of laboratory-acquired infections with intestinal protozoa have been reported; however, a few cases have been detected. A worker who "checked in several hundred stool specimens, stamping numbers and dates on report cards, many of which had been contaminated from leaky containers," became infected with *G. lamblia* (37). The infection was confirmed by finding the flagellates in the stool. A laboratory technician who examined numerous stool specimens from a patient infected with *I. belli* became ill about a week later, and *I. belli* was confirmed to be present in his stool. While infecting a rabbit using a capsule containing about 400 *Isospora* oocysts, two researchers were sprayed in their faces with droplets when the rabbit regurgitated the material and vigorously shook its head; both individuals became ill about 2 weeks later (37).

Free-Living Amebae

A number of studies have investigated the responses of trophozoites and cysts of *Acanthamoeba* spp. to disinfectants in various solutions used to disinfect contact lenses; infections with these organisms have led to serious eye infections, particularly keratitis. These infections can be very difficult to treat if not recognized early, and eradication of the free-living amebae is also very difficult. These protozoa are capable of forming biofilms on contact lenses and could adversely affect the effectiveness of chemical agents used for disinfection.

Plasmodium spp.

In a series of 34 cases of laboratory-acquired malaria, 15 were due to *P. falciparum*, 10 were due to *P. cynomolgi*, and 9 were due to *P. vivax*. The majority of cases were reported from the United States and Europe, with 19 being vector borne, 10 being parenteral, and 5 occurring through a break in the skin. All patients were symptomatic (37, 68).

It has also been documented that research workers can be accidentally infected from a mosquito, possibly one that escaped from a mosquito colony. Strict containment measures should be in place. Another possible route of infection involves mosquito dissection, during which infective sporozoites could be accidentally injected.

Trypanosoma brucei gambiense and *T. brucei rhodesiense*

Cases of laboratory-acquired African trypanosomiasis can result from contact with blood or tissue from infected persons or animals. Six laboratory-acquired cases have

Table 20.2 Possible parasite transmission in a health care setting (laboratory)

Organisms[a]	Mode of transmission	Comments
Protozoa (intestinal)		
Entamoeba histolytica[b,c] (C) *Entamoeba dispar* (C) *Entamoeba coli* (C) *Entamoeba hartmanni* (C) *Endolimax nana* (C) *Iodamoeba butschlii* (C) *Blastocystis hominis*[b] (C) Free-living amebae *Giardia lamblia*[b,c] (C) *Dientamoeba fragilis*[b] (T) *Cryptosporidium* spp.[b,c] (O) *Cyclospora cayetanensis*[b] (O) *Isospora belli*[b,c] (O) Microsporidia[b] (S)	Accidental ingestion of infective cysts, trophozoites, oocysts, or spores in food or water contaminated with fecal material; also, direct transfer of stool material via fomites (fecal-oral transmission) Contamination of hands, eyes handling free-living ameba cultures; inhalation of infectious aerosols; splashes to eyes or mucous membranes	Transmission becomes more likely when fresh stool specimens are being processed and examined; submission of fecal specimens in preservatives or fixatives would decrease risks. Wearing gloves, using capped centrifuge tubes, and working in biological safety cabinet would decrease the risk of acquiring *Cryptosporidium* infections. *Not recommended*: use of potassium dichromate as collection fluid and use of sugar flotation on fresh stool. Immunocompromised employees must be educated regarding risk factors; they may elect not to work with certain organisms or perform certain procedures.
Helminths (intestinal)		
Enterobius vermicularis (E)[c]	Inhalation or ingestion of infective eggs	Very common in children, asymptomatic
Strongyloides stercoralis (IL)[c]	Skin penetration of infective larvae from stool material	Exposure would be very possible or even likely when working with stool cultures or concentrates for larva recovery.
Hymenolepis nana (E)	Ingestion of infective eggs (fecal-oral)	Ingestion of infective eggs can lead to the adult worm in humans.
Taenia solium (E)	Inhalation or ingestion of infective eggs (could lead to cysticercosis)	Exposure very likely when working with gravid proglottids (ink injection for identification of worm to species level).
Blood and tissue protozoa		
Leishmania spp.[c]	Direct contact or inoculation of infectious material from patient lesion; accidental inoculation of material from culture or animal inoculation studies	Culture forms and organisms from hamster would be infectious.
Trypanosoma spp.[c]	Same as for *Leishmania* spp.; also handling arthropod vectors	Cultures and special concentration techniques would represent possible means of exposure.
Plasmodium spp.[c,d]	Accidental inoculation could transmit any of the four species; mosquito transmission	Blood and culture material should always be handled carefully (avoid open cuts, etc.).
Toxoplasma gondii[c]	Inhalation or ingestion of oocysts in cat feces (veterinary situation); accidental inoculation of tachyzoites from culture, tube of blood, or animal isolation (mouse peritoneal cavity); needles, spills, mucous membrane contamination; tissue culture work	Although many people already have antibodies to *T. gondii*, indicating past exposure, there have been documented laboratory accidents in which the individual became ill because of the large infecting dose.
Ectoparasites		
Pediculus spp.	Specimens submitted on hair could be easily transmitted in the laboratory.	Careful handling and fixation of the arthropods would prevent any potential problems with transmission.
Sarcoptes scabiei	Transmission via skin scraping, etc., would be possible but not that likely.	Careful handling and preparation of specimens with KOH would tend to prevent any problems.
Dipterous fly larvae (myiasis)	Transmission could occur anywhere.	Protection from flies would solve the potential problem.

[a] C, cyst; T, trophozoite; O, oocyst; IL, infective larvae; E, egg; S, spore.
[b] Potentially pathogenic intestinal protozoa.
[c] Laboratory infections documented: *Leishmania* spp., *Trypanosoma cruzi*, *Trypanosoma brucei gambiense*, *Trypanosoma brucei rhodesiense*.
[d] Mosquito transmission.

Table 20.3 Potential exposures to laboratory-acquired parasitic infections*a*

Parenteral or aerosolization

Recapping a needle
Removing a needle from the syringe
Leaving a needle on the counter, point up
Dropping a syringe
Breaking hematocrit tube while pressing the end into clay
Performing venipuncture on agitated patient
Sudden animal movement during an inoculation procedure
Creating aerosols during tapeworm proglottid injection
Creating aerosols while working with cultures (blood parasites, free-living amebae)

Animal or vector bites

Being bitten by an infected animal (mouse, hamster)
Being bitten by infected mosquito (mosquito colony)

Skin exposure

Not wearing gloves during procedure
Not wearing laboratory coat (closed sleeves, closed front over clothes)
Accidentally touching face or eyes during handling of infectious materials
Exposing of eyes, nose, and mouth to potential aerosols

Ingestion

Mouth pipetting
Being sprayed with inoculum droplets from coughing or regurgitating animal

Other reasons for potential exposures

Working in disorganized laboratory bench setting
Working too fast
Not receiving proper training
Assuming that the agent is not infectious to humans
Assuming that the agent(s) is no longer viable
Using defective equipment

*a*Adapted from a number of sources, including reference 37.

been caused by *T. brucei gambiense*, and two have been caused by *T. brucei rhodesiense*. In these cases the route of exposure was usually parenteral, and all patients were symptomatic.

Trypanosoma cruzi

Infection can occur from exposure to the feces of infected triatomine bugs, by handling cultures or blood specimens, and possibly by inhaling aerosolized organisms. The predominant stage of the parasite in culture is the epimastigote stage; however, the infectious trypomastigotes can also be found. *T. cruzi* can infect laboratory personnel through needlestick injuries, through preexisting microabrasions of the skin, or through intact mucous membranes.

Sixty-five cases of laboratory-acquired *T. cruzi* infection have been reported, although in many cases, the actual geographic site was unknown. While the route of exposure for many of the cases was unknown, parenteral infection was identified for 11 individuals. The clinical status of the infected individuals was generally unknown; however, 24 of 47 were symptomatic, with 9 severe cases and 1 death (37, 68).

Leishmania spp.

In laboratory settings, leishmaniasis could be acquired through contact with an infected sand fly; containment measures for infected flies should be strictly followed. Transmission can also occur through contact with parasite cultures or clinical specimens from infected persons or animals (e.g., needlestick injuries or microabrasions of the skin). Although fewer parasites are found in patient blood specimens than in cultures or infected tissues, blood should be handled with care.

Twelve cases of laboratory-acquired leishmaniasis caused by six different species have been reported. Although most of the infected persons developed cutaneous leishmaniasis, sometimes with associated local lymphadenopathy, one person developed visceral leishmaniasis and one developed mucosal leishmaniasis as a sequela of cutaneous leishmaniasis. All patients were symptomatic, with two having severe disease (37).

Toxoplasma gondii

Infection can occur through ingestion of sporulated oocysts from feline fecal specimens or through skin or mucosal contact with tachyzoites or bradyzoites in tissue or culture. All *Toxoplasma* isolates should be considered pathogenic for humans. Procedures for separating oocysts from feline feces and for infecting mice have been described; fecal concentrations should be performed before oocysts sporulate and become infectious. Laboratory equipment and glassware that have been in contact with oocysts should be sterilized, since the oocysts are not easily killed by chemical exposure. Immunocompromised persons and *T. gondii*-seronegative women who are pregnant or might become pregnant should be counseled about the risks associated with *T. gondii* infection (e.g., central nervous system infection and congenital infection) and given the option of not working with live *T. gondii* and of not working in a laboratory in which others do so.

Of the 47 cases of laboratory-acquired toxoplasmosis, the majority were in the United States and Europe. The most common route of exposure was parenteral; however, in 10 of the cases no specific accident or route

Table 20.4 Resistance to antiseptics and disinfectants, from most to least resistant

Type of organism	Example(s)
Prions	Creutzfeldt-Jacob disease, bovine spongiform encephalopathy
Coccidia	*Cryptosporidium* spp.
Spores	*Bacillus, Clostridium difficile*
Mycobacteria	*Mycobacterium tuberculosis, M. avium*
Cysts	*Giardia lamblia*
Small nonenveloped viruses	Poliovirus
Trophozoites	*Acanthamoeba*
Gram-negative bacteria (nonsporulating)	*Pseudomonas*
Fungi	*Candida, Aspergillus*
Large nonenveloped viruses	Enteroviruses, adenovirus
Gram-positive bacteria	*Staphylococcus aureus, Enterococcus*
Lipid-enveloped viruses	Human immunodeficiency virus, hepatitis B virus

of exposure could be identified. Although 9 patients were asymptomatic, 38 were considered to be symptomatic if they had lymphadenopathy. Four individuals had encephalitis, two of whom had myocarditis, and one patient with both conditions died (37).

Specimen Handling

Another consideration involves the handling of specimens from patients with AIDS. Although the agent, human immunodeficiency virus, is considered to be a very fragile virus that does not survive in the environment, specimen-handling precautions are advised. Some laboratories suggest that stool specimens being processed for *Cryptosporidium* (often diagnosed in AIDS patients) be handled in a biological safety cabinet and be centrifuged in capped centrifuge tubes. These types of precautions will help diminish the chances of an accidental laboratory infection.

Summary

Although there are real possibilities for nosocomial and laboratory transmission of parasites, proof of such transmission is usually very difficult to document. Depending on the area of the country and the number of endemic parasites present in the community, documentation can be particularly difficult. In areas where many organisms are endemic, it is almost impossible to differentiate between a community-acquired infection and one acquired in the work setting.

Some of the infectious agents being handled in the laboratory present a potentially more serious problem. Although many laboratories do not handle all of these agents, their personnel may still be at risk when performing certain techniques or handling certain specimens. For these reasons, it is very important that everyone become aware of possible hazards when working in the field of diagnostic medical parasitology. The information presented in Tables 20.1 through 20.3 provides specific, practical information concerning safety precautions in the laboratory.

References

1. **Addis, D. G., J. P. Davis, J. M. Roberts, and E. E. Mast.** 1992. Epidemiology of giardiasis in Wisconsin: increasing incidence of reported cases and unexplained seasonal trends. *Am. J. Trop. Med. Hyg.* **47:**13–19.
2. **Addis, D. G., H. M. Mathews, J. M. Stewart, S. P. Walquist, R. M. Williams, R. J. Finton, H. C. Spencer, and D. D. Juranek.** 1991. Evaluation of a commercially available enzyme-linked immunosorbent assay for *Giardia lamblia* antigen in stool. *J. Clin. Microbiol.* **29:**1137–1142.
3. **Angelis, M., J. T. Cooper, and R. B. Freeman.** 2003. Impact of donor infections on outcome of orthotopic liver transplantation. *Liver Transplant.* **9:**451–462.
4. **Barrett-Connor, E., S. L. Brandt, H. J. Simon, and D. C. Dechairo (ed.).** 1978. *Epidemiology for the Infection Control Nurse.* The C. V. Mosby Co., St. Louis, Mo.
5. **Baxby, D., C. A. Hart, and C. Taylor.** 1983. Human cryptosporidiosis: a possible case of hospital cross infection. *Br. Med. J. (Clin. Res.)* **287:**1760–1761.
6. **Brasitus, T. A., R. P. Gold, R. H. Kay, A. M. Magun, and W. M. Lee.** 1980. Intestinal strongyloidiasis: a case report and review of the literature. *Am. J. Gastroenterol.* **73:**65–69.

7. Bruce-Chwatt, L. J. 1974. Transfusion malaria. *Bull. W. H. O.* **50:**337–346.

8. Bryan, R. T. 1995. Microsporidiosis as an AIDS-related opportunistic infection. *Clin. Infect. Dis.* **21:**S62–S65.

9. Burke, J. F., and G. Y. Hildick-Smith (ed.). 1978. *The Infection Prone Hospital Patient.* Little, Brown, and Co., Boston, Mass.

10. Centers for Disease Control. 1985. Pseudo-outbreak of intestinal amebiasis—California. *Morb. Mortal. Wkly. Rep.* **34:**125–126.

11. Cheng, H. S., and L. C. Wang. 1999. Amoebiasis among institutionalized psychiatric patients in Taiwan. *Epidemiol. Infect.* **122:**317–322.

12. Chodosh, J., and J. Clarridge. 1992. Ophthalmomyiasis: a review with special reference to *Cochliomyia hominivorax*. *Clin. Infect. Dis.* **14:**444–449.

13. Chu, E., W. L. Whitlock, and R. A. Dietrich. 1990. Pulmonary hyperinfection syndrome with *Strongyloides stercoralis*. *Chest* **97:**1475–1477.

14. Collins, J. P., K. F. Keller, and L. Brown. 1978. "Ghost" forms of *Giardia lamblia* cysts initially misdiagnosed as *Isospora*. *Am. J. Trop. Med. Hyg.* **27:**334–335.

15. Coutinho, J. O., J. Croce, R. Campos, V. A. Neto, and L. C. Fonseca. 1953–1954. Contribuicao para o conhecimento da estrongiloidiase humana en Sao Paulo. *Folia Clin. Biol.* **20:**141–176; **21:**19–48, 94–120.

16. Craft, J. C. 1982. *Giardia* and giardiasis in children. *Pediatr. Infect. Dis. J.* **1:**196–211.

17. Cundy, K. R., and W. Ball (ed.). 1977. *Infection Control in Health Care Facilities.* University Park Press, Baltimore, Md.

18. Current, W. L., and L. S. Garcia. 1991. Cryptosporidiosis. *Clin. Microbiol. Rev.* **3:**325–358.

19. Danciger, M., and M. Lopez. 1975. Numbers of *Giardia* in the feces of infected children. *Am. J. Trop. Med. Hyg.* **24:**237–242.

20. Dias, R. M. D. S., A. C. S. Mangini, D. M. A. G. V. Torres, S. A. G. Vellosa, M. I. P. G. Dasilva, R. M. Dasilva, M. O. A. Correa, and C. Coletti. 1992. *Strongyloides stercoralis* in patients with acquired immunodeficiency syndrome (AIDS). *Rev. Inst. Med. Trop. Sao Paulo* **34:**15–17.

21. Dryjanski, J., J. W. M. Gold, M. T. Ritchie, R. C. Kurtz, S. L. Lim, and D. Armstrong. 1986. Cryptosporidiosis. Case report in a health team worker. *Am. J. Med.* **80:**751–752.

22. Emmerson, M. 1995. Surveillance strategies for nosocomial infections. *Curr. Opin. Infect. Dis.* **8:**272–274.

23. Figueiredo, J. F. C., R. Martinez, J. C. daCosta, M. Moyses-Neto, H. J. Suaid, and A. S. Ferraz. 1990. Transmission of Chagas' disease through renal transplantation: report of a case. *Trans. R. Soc. Trop. Med. Hyg.* **84:**61–62.

24. Garcia, L. S. 1989. Intestinal coccidia and microsporidia in non-AIDS patients. *Clin. Microbiol. Newsl.* **11:**169–172.

25. Garcia, L. S., T. C. Brewer, and D. A. Bruckner. 1987. Fluorescence detection of *Cryptosporidium* oocysts in human fecal specimens by using monoclonal antibodies. *J. Clin. Microbiol.* **25:**119–121.

26. Garcia, L. S., T. C. Brewer, and D. A. Bruckner. 1989. Incidence of *Cryptosporidium* in all patients submitting stool specimens for ova and parasite examination: monoclonal antibody-IFA method. *Diagn. Microbiol. Infect. Dis.* **11:**25–27.

27. Garcia, L. S., D. A. Bruckner, and T. C. Brewer. 1988. Cryptosporidiosis in patients with AIDS. *Am. Clin. Prod. Rev.* **7:**38–41.

28. Garcia, L. S., D. A. Bruckner, T. C. Brewer, and R. Y. Shimizu. 1983. Techniques for the recovery and identification of *Cryptosporidium* oocysts from stool specimens. *J. Clin. Microbiol.* **18:**185–190.

29. Garcia, L. S., and W. L. Current. 1989. Cryptosporidiosis: clinical features and diagnosis. *Clin. Rev. Clin. Lab. Sci.* **27:**439–460.

30. Garcia, L. S., and R. Y. Shimizu. 1997. Evaluation of nine immunoassay kits (enzyme immunoassay and direct fluorescence) for detection of *Giardia lamblia* and *Cryptosporidium parvum* in human fecal specimens. *J. Clin. Microbiol.* **35:**1526–1529.

31. Garcia, L. S., and R. Y. Shimizu. 2000. Dectection of *Giardia lamblia* and *Cryptosporidium parvum* antigens in human fecal specimens using the ColorPAC combination rapid solid-phase qualitative immunochromatographic assay. *J. Clin. Microbiol.* **38:**1267–1268.

32. Garcia, L. S., R. Y. Shimizu, and C. N. Bernard. 2000. Detection of *Giardia lamblia*, *Entamoeba histolytica/Entamoeba dispar*, and *Cryptosporidium parvum* antigens in human fecal specimans using the Triage Parasite Panel enzyme immunoassay. *J. Clin. Microbiol.* **38:**3337–3340.

33. Genta, R. M. 1989. Global prevalence of strongyloidiasis: critical review with epidemiologic insights into the prevention of disseminated disease. *Rev. Infect. Dis.* **11:**755–767.

34. Giacometti, A., O. Cirioni, M. Balducci, D. Drenaggi, M. Quarta, M. De Federicis, P. Ruggeri, D. Colapinto, G. Ripani, and G. Scalise. 1997. Epidemiologic features of intestinal parasitic infections in Italian mental institutions. *Eur. J. Epidemiol.* **13:**825–830.

35. Grove, D. I. 1982. Treatment of strongyloidiasis with thiabendazole: an analysis of toxicity and effectiveness. *Trans. R. Soc. Trop. Med. Hyg.* **76:**114–118.

36. Harris, R. A., D. M. Musher, V. Fainstein, E. J. Young, and J. Clarridge. 1980. Disseminated strongyloidiasis: diagnosis made by sputum examination. *JAMA* **244:**65–66.

37. Herwaldt, B. L. 2001. Laboratory-acquired parasitic infections from accidental exposures. *Clin. Microbiol. Rev.* **14:**659–688.

38. Istre, G. R., K. Kreiss, R. S. Hopkins, G. R. Healy, M. Benziger, T. M. Canfield, P. Dickinson, T. R. Englert, R. C. Compton, H. M. Mathews, and R. A. Simmons. 1982. An outbreak of amebiasis spread by colonic irrigation at a chiropractic clinic. *N. Engl. J. Med.* **307:**339–341.

39. Jarroll, E. L., A. K. Bingham, and E. A. Meyer. 1981. Effect of chlorine on *Giardia lamblia* cyst viability. *Appl. Environ. Microbiol.* **41:**483–487.

40. Jokipii, L., and A. M. M. Jokipii. 1979. Single-dose metronidazole and tinidazole as therapy for giardiasis: success rates, side effects, and drug absorption and elimination. *J. Infect. Dis.* **140:**984–988.

41. Jones, C. A. 1950. Clinical studies in human strongyloidiasis. *Gastroenterology* **16:**743–756.

42. Kettis, A. A., and L. Magnius. 1973. *Giardia lamblia* infection in a group of students after a visit to Leningrad in March 1970. *Scand. J. Infect. Dis.* **5:**289–292.

43. Keystone, J. S., S. Karjden, and M. R. Warren. 1978. Person-to-person transmission of *Giardia lamblia* in day-care nurseries. *Can. Med. Assoc. J.* **119:**242–244.

44. Kirchhoff, L. V. 1989. Is *Trypanosoma cruzi* a new threat to our blood supply? *Ann. Intern. Med.* **111:**773–775.

45. Knaus, W. A. 1974. Reassurance about Russian giardiasis. *N. Engl. J. Med.* **291:**156.

46. Koch, K. L., D. J. Phillips, R. C. Aber, and W. L. Current. 1985. Cryptosporidiosis in hospital personnel. Evidence for person-to-person transmission. *Ann. Intern. Med.* **102:**593–596.

47. Korich, D. G., J. R. Mead, M. S. Madore, N. A. Sinclair, and C. R. Sterling. 1990. Effects of ozone, chlorine dioxide, chlorine and monochloramine on *Cryptosporidium parvum* oocyst viability. *Appl. Environ. Microbiol.* **56:**1423–1428.

48. Lebeau, B., C. Pinel, R. Grillot, and P. Ambroise-Thomas. 1998. Fungal and parasitic nosocomial infections: importance and limitations of disinfection methods. *Pathol. Biol.* **46:**335–340.

49. Leighton, P. M., and H. M. MacSween. 1990. *Strongyloides stercoralis*: the cause of an urticarial-like eruption of 65 years' duration. *Arch. Intern. Med.* **150:**1747–1748.

50. Lettau, L. A. 1991. Nosocomial transmission and infection control aspects of parasitic and ectoparasitic diseases. I. Introduction/enteric parasites. *Infect. Control Hosp. Epidemiol.* **12:**59–65.

51. Lettau, L. A. 1991. Nosocomial transmission and infection control aspects of parasitic and ectoparasitic diseases. II. Blood and tissue parasites. *Infect. Control Hosp. Epidemiol.* **12:**111–112.

52. Lettau, L. A. 1991. Nosocomial transmission and infection control aspects of parasitic and ectoparasitic diseases. III. Ectoparasites/summary and conclusions. *Infect. Control Hosp. Epidemiol.* **12:**179–185.

53. Lopez, C. E., D. D. Juranek, S. P. Sinclair, and M. A. Schultz. 1978. Giardiasis in American travelers to Madeira Island, Portugal. *Am. J. Trop. Med. Hyg.* **27:**1128–1132.

54. Lorca, M., A. Atias, B. Astorga, P. Munoz, and I. Carrere. 1983. *Trypanosoma cruzi* infections in blood banks of 12 Chilean hospitals. *Bull. Pan Am. Health Org.* **17:**269–274.

55. Lukin, L. G. 1989. Human cutaneous myiasis in Brisbane: a prospective study. *Med. J. Aust.* **150:**237–240.

56. Makhlouf, S. A., M. A. Sarwat, D. M. Mahmoud, and A. A. Mohamad. 1994. Parasitic infection among children living in two orphanages in Cairo. *J. Egypt. Soc. Parasitol.* **24:**137–145.

57. McDonnell, G., and A. D. Russell. 1999. Antiseptics and disinfectants: activity, action, and resistance. *Clin. Microbiol. Rev.* **12:**147–179.

58. Meyer, E. A. (ed.). 1990. *Giardiasis, Human Parasitic Diseases.* Elsevier Science Publishing, Inc., New York, N.Y.

59. Millett, V., M. Spencer, M. Chapin, L. Garcia, J. Yatabe, and M. Stewart. 1983. Intestinal protozoa infection in a semicommunal group. *Am. J. Trop. Med. Hyg.* **32:**54–60.

60. Moore, G. T., W. W. Gross, D. McGuire, C. S. Mollohan, N. N. Gleason, G. R. Healy, and L. H. Newton. 1969. Epidemic giardiasis at a ski resort. *N. Engl. J. Med.* **281:**402–407.

61. Nash, T. E., D. A. Herrington, and M. M. Levine. 1987. Usefulness of an enzyme-linked immunosorbent assay for detection of *Giardia* antigen in feces. *J. Clin. Microbiol.* **25:**1169–1171.

62. Navarro, P., A. Betancourt, H. Paublini, I. Medina, M. J. Nunez, and M. Dominguez. 1987. Malaria causada por *Plasmodium falciparum* como infeccion nosocomial. *Bol. Ofic. Sanit. Panam.* **102:**476–482.

63. Patey, O., R. Bouhali, J. Breuil, L. Chapuis, A. Courillon-Mallet, and C. Lafaix. 1990. Arthritis associated with *Strongyloides stercoralis. Scand. J. Infect. Dis.* **22:**233–236.

64. Pelletier, L. L., Jr. 1984. Chronic strongyloidiasis in World War II Far East ex-prisoners of war. *Am. J. Trop. Med. Hyg.* **33:**55–61.

65. Petri, W. A., Jr., C. G. Clark, and L. S. Diamond. 1994. Host-parasite relationships in amebiasis: conference report. *J. Infect. Dis.* **169:**483–484.

66. Phelps, K. R., S. S. Ginsberg, A. W. Cunningham, E. Tschachler, and H. Dosik. 1991. Adult T-cell leukemia lymphoma associated with recurrent strongyloides hyperinfection. *Am. J. Med. Sci.* **302:**224–228.

67. Phillips, S. C., D. Mildvan, D. C. William, A. M. Gelb, and M. C. White. 1981. Sexual transmission of enteric protozoa and helminths in a venereal-disease-clinic population. *N. Engl. J. Med.* **305:**603–606.

68. Pike, R. M. 1976. Laboratory-associated infections: summary and analysis of 3921 cases. *Health Lab. Sci.* **13:**105–114.

69. Sandeep, C. 1995. Cutaneous manifestations of HIV. *Curr. Opin. Infect. Dis.* **8:**298–305.

70. Schad, G. A., L. M. Aikens, and G. Smith. 1989. *Strongyloides stercoralis*: is there a canonical migratory route through the host? *J. Parasitol.* **75:**740–749.

71. Schmerin, M. J., T. C. Jones, and H. Klein. 1978. Giardiasis: association with homosexuality. *Ann. Intern. Med.* **88:**801–803.

72. Sealy, D. P., and S. H. Schuman. 1983. Endemic giardiasis and day care. *Pediatrics* **72:**154–158.

73. Seed, C. R., A. Kitchen, and T. M. Davis. 2005. The current status and potential role of laboratory testing to prevent transfusion-transmitted malaria. *Transfus. Med. Rev.* **19:**229–240.

74. Sirera, G., F. Rius, J. Romeu, et al. 1990. Hospital outbreak of scabies stemming from two AIDS patients with Norwegian scabies. *Lancet* **335:**1227.

75. Smith, P. D., H. C. Lane, V. J. Gill, J. F. Manischewitz, G. V. Quinnan, A. S. Fauci, and H. Masur. 1988. Intestinal infections in patients with the acquired immunodeficiency syndrome (AIDS). *Ann. Intern. Med.* **108:**328–333.

76. Smith, R. P., A. T. Evans, M. Popovsky, L. Mills, and A. Spielman. 1986. Transfusion-acquired babesiosis and failure of antibiotic treatment. *JAMA* **256:**2726–2727.

77. Sterling, C. R., R. M. Kutob, M. J. Gizinski, M. Verastequi, and L. Stetzenbach. 1987. *Giardia* detection using monoclonal antibodies recognizing determinants of in vitro derived

cysts, p. 219–222. *In* P. Wallis and B. Hammond (ed.), *Advances in* Giardia *Research*. University of Calgary Press, Calgary, Alberta, Canada.

78. **Tabacof, J., O. Feher, A. Katz, S. D. Simon, and R. C. Gansi.** 1991. *Strongyloides* hyperinfection in two patients with lymphoma, purulent meningitis, and sepsis. *Cancer* **68:**1821–1823.

79. **Wenzel, R. P. (ed.).** 1987. *Prevention and Control of Nosocomial Infections*. The Williams & Wilkins Co., Baltimore, Md.

80. **Yanelli, B., I. Gurevich, P. E. Schoch, and B. A. Cunhu.** 1988. Yield of stool cultures, ova and parasite tests, and *Clostridium difficile* determinations in nosocomial diarrheas. *Am. J. Infect. Control* **16:**246–249.

Immunology of Parasitic Infections

For many years, there has been extensive research in immunoparasitology directed at the identification, isolation, purification, and characterization of parasite antigens. Antigens are required for immunodiagnosis, clarification of immunopathology, quantitation of various immune responses in the host, and evaluation of potential vaccines. Antibodies to defined antigens are used for typing of reagents, the study of antigenic variability, passive immunization, and in vitro inhibition studies. Without the use of defined antigens and antibodies, a thorough understanding of the immunologic aspects of the host-parasite relationship is unlikely. Table 21.1 contains information on the various categories of parasite antigens, while Table 21.2 lists various methods of antigen analysis.

It is important to understand the factors influencing the immune responses of the host, including (i) genetics; (ii) state of the host at exposure, including nutrition, age, health status, and underlying diseases; and (iii) the size, route, and frequency of the parasite loading dose. Also, there are several types of resistance, which include natural resistance (how refractory the host is to a particular parasite), acquired resistance (either specific or nonspecific), and immunoregulation (Table 21.3).

Innate Immune System. The immune system is made up of two interrelated parts, the innate and acquired systems. The innate response is the first to be activated during the response to invading organisms and is due primarily to phagocytic cells. This response is related primarily to immediate recognition and response to pathogen invasion. Phagocytes include polymorphonuclear leukocytes or neutrophils, monocytes, and macrophages. Secondary phagocytes can include dendritic cells, fibroblasts, and epithelial cells. Macrophages are found in all tissues and represent the first line of defense against pathogenic invasion of the host (Table 21.4). In order to thrive, parasites must be able to survive the innate response. Complement activation represents a first line of defense against extracellular parasites. Once the membrane attack complex is formed, parasite lysis can occur while complement components opsonize parasites for phagocytosis. One of the most important cytokines in cell-mediated innate resistance is interleukin-12 (IL-12). IL-12 activates natural killer (NK) cells, which in turn produce gamma interferon (IFN-γ), which is responsible for activation of effector mechanisms that control replication of intracellular parasites.

Table 21.1   Parasite antigens[a]

I. Immunologic criteria

  A. Importance of the induced immune response to the parasite or host (relevant or functional antigens versus incidental antigens)

    1. Host-protective antigens[b]: related to induction and expression of host-protective immunity

    2. Immunodiagnostic antigens: related to detection of antibodies, skin test sensitivity (both immediate or delayed), detection of circulating or urinary antigens

    3. Immunopathologic antigens[c]: related to induction and expression of immunopathology

    4. Parasite-protective antigens: related to induction and expression of parasite-protective immunity

  B. Immunogenicity in different animals

    1. Natural antigens: infection of natural hosts or immunization of natural hosts with or without adjuvants

    2. Novel antigens (heteroantigens): immunization of unnatural hosts with or without adjuvants, or infection of nonpermissive hosts

  C. Type of lymphocyte involved, e.g., T-cell-stimulating carrier determinant, B-cell-stimulating haptenic determinant, allergen, T_s-stimulating determinant

II. Parasitologic criteria

  A. Origin and location

    1. Soluble exoantigens: released from living parasites, parasitized cells, or cultured cell lines and called excretory/secretory (ES), excretory/secretory/tissue turnover (EST), or metabolic antigens

    2. Soluble somatic antigens: extracted from parasites or parasitized cells; may be surface or internal

    3. Dead parasites, fragments or secretory blebs

    4. Whole living parasites

    5. Body fluids of nematodes

    6. Cystic fluids of larval cestodes

  B. Parasite population and life cycle

    1. Genus-, species-, strain-, and stage-specific antigens

    2. Molting antigens

III. Biochemical criteria

  A. Composition (protein, lipid, carbohydrate, glycoprotein, glycolipid, etc.)

  B. Characteristics (size, chain structure, determinant number and type, etc.)

  C. Molecule function (enzyme, metabolite, receptor, recognition structure, etc.)

  D. Future sources (expression of cloned DNA, chemical synthesis, stimulation of anti-idiotypic antibodies)

[a] Data from reference 4.
[b] It is not the host antigens themselves but the immune responses that these antigens induce which are host protective.
[c] Immunopathologic immune responses depend on the genotype of the host and may also have beneficial effects.

Acquired/Adaptive Immune System. Although innate immunity is critical in resistance to acute parasitic infections, the acquired, or adaptive, response is necessary for long-term protective immunity. Adaptive T- and B-cell responses following infection depend on accessory cells that present antigen, provide costimulation, and produce cytokines, all of which impact the onset, duration, strength, and type of the adaptive immune response. CD4[+] helper T cells are divided into two subsets, Th1 (type 1 response) and Th2 (type 2 response), each of which produces different cytokines and has different functionalities (Table 21.5). At times, both type 1 and type 2 responses may be used to eliminate the same parasite. These two mechanisms may be functional at the same time, during different stages of the parasite life cycle, or in different locations within the body.

Immunopathologic Reactions in the Host. It is important to have a basic understanding of the types of immunopathologic reactions in the host which are responsible for disease. These are often classified into four categories or pathways by which the host responds. They have been categorized as (i) type I reaction (anaphylactic), initiated by the antigen reacting with tissue cells (basophils and mast cells) passively sensitized by antibody produced elsewhere (release of pharmacologically active substances); (ii) type II reaction (cytotoxic or cell stimulating), initiated by the antibody reacting with an antigenic component of or antigen associated with a cell or tissue element (damage occuring in the presence of complement or certain mononuclear cells); (iii) type III reaction (antigen-antibody complexes), initiated when the antigen reacts with precipitating antibody (forming microprecipitates in and around small vessels, interfering with membrane function, and forming soluble circulating complexes); and (iv) type IV reaction (delayed, tuberculin type, cell mediated), initiated by reaction of active lymphocytes

Table 21.2   Methods of antigen analysis

Gel precipitation techniques
1. Double diffusion
2. Ouchterlony
3. Immunoelectrophoresis
4. Autoradiography of precipitin lines
5. Counterimmunoelectrophoresis
6. Counterradioimmunoelectrophoresis

Immunoprecipitation and gel analysis
1. Anti-immunoglobulins
2. Sodium dodecyl sulfate-polyacrylamide gel electrophoresis
3. Isoelectric focusing
4. Nonequilibrium pH gradient electrophoresis
5. Fluorography

Gel overlay or transfer techniques
1. Direct application of antiserum to the surface of one- or two-dimensional gels
2. Direct application of antiserum to a paper transfer of the gel
3. Passive binding to nitrocellulose sheets
4. Covalent bonding to diazobenzyloxymethyl paper
5. Filter affinity transfer

Identification of surface antigens
1. Antigen labeling (intrinsic or extrinsic)
2. Antigen solubilization
3. Antigen fractionation
4. Demonstration of cell surface location

Hybridoma-derived antibodies
1. Production of monoclonal antibodies
2. Screening procedures for hybridoma culture supernatants

Recombinant DNA
1. Clarification of genomic rearrangements related to antigenic variation
2. Identification of nucleic acid sequence data
3. Production of potentially useful parasite polypeptides

Table 21.3   Immunoregulatory systems in the host

Antibody-mediated mechanisms
Immunologic blockade
Polyclonal B-cell activation
Anti-idiotypic antibodies

Cellular mechanisms
Tolerance
Specific suppressor cells
Nonspecific suppressor cells

in response to allergen, release of lymphokines, and/or cytotoxicity (infiltration of cells). These are listed in Table 21.6. The overall immune response to the presence of any parasite is no different from the responses seen with any infectious agent, all of which begin with the organism infecting the host and being recognized by the cells of the immune system. Once the parasite is recognized as foreign by the immune system, a series of steps are initiated (Figure 21.1; Table 21.7).

Parasites have three basic characteristics that make it very difficult for the host to control them: parasite size, complicated parasite life cycles, and antigenic complexity. There are also a number of ways in which parasites avoid the host immune system responses; these include (i) location within body sites that are relatively protected from the immune response, (ii) various modifications of surface antigens, and (iii) different mechanisms that can modify the host immune response. A number of these evasion mechanisms are listed in Table 21.8.

All human parasites elicit immune responses, both humoral and cellular. Specific antibodies belonging to the immunoglobulin M (IgM), IgG, IgA, and IgE classes have been found, indicating the body's recognition of parasite antigens (Table 21.9). Cell-mediated immune responses have also been clearly documented. However, in both cases, there may be little or no correlation between responses and protection of the host.

Although the immune system functions in the same way for parasites as for other organisms, there are some dramatic differences based on the complexity of the response. As parasites progress through their normal life cycle stages, many different antigens are expressed at each different stage. This leads to the development of many different antibody-dependent and -independent responses. In some cases, the immune system of the host actually becomes confused, leading to situations where the host is harmed rather than helped.

Although the host response to protozoa is quite different from the host response to helminths, some general statements apply to both situations. When a parasite and the host interact, there are several possible outcomes: (i) the parasite fails to become established in the host; (ii) the parasite becomes established and kills the host; (iii) the parasite becomes established and the host eliminates the infection; (iv) the parasite becomes established and the host, in trying to eliminate the organism, becomes damaged itself; and (v) the parasite becomes established, and the host begins to overcome the infection but is not totally successful (12). Some representative parasites are presented in this chapter. Obviously, some human parasites have been studied more than others, and these differences are reflected in the amount of available information related to parasite immunology.

Table 21.4 Macrophage functions

Macrophage functions	Mechanisms
Protective	
Detection of pathogen invasion	Opsonic receptors for microbes
	Complement and antibody receptors
	Nonopsonic receptors
	Membrane-bound pattern recognition receptors (C-type lectins, leucine-rich proteins, scavenger receptors, and integrins)
Control of microbial spread	Phagocytosis, granuloma formation, intracellular killing
Recruitment of immune cells	Release of cytokines and other inflammatory mediators
Lymphocyte activation	Antigen presentation, costimulatory molecules, cytokines
Effector cells for cell-mediated immunity	Increased phagocytosis, increased intracellular killing, clearance of apoptotic cells
Participation in humoral immunity	Receptors for antibody and complement
Nonprotective	
Tissue damage	Extracellular release of toxic oxidants, hydrolytic enzymes
	Inflammatory mediators (TNF-α, IL-1)
	Fever, wasting, septic shock
Dissemination of pathogens	Transport throughout the host
Autoimmune disease	Presentation of microbial epitopes to lymphocytes, which cross-react with self molecules

Amebiasis

With the development of axenic culture and antigen purification techniques, *Entamoeba histolytica* infections and subsequent host immune responses have been more thoroughly investigated. However, although both humoral and cellular responses occur in amebiasis, the role of immunity is still not completely understood.

Pathogenic *E. histolytica* is now well recognized as a separate species from the morphologically identical but nonpathogenic *E. dispar*. *E. histolytica* uses different virulence factors, including a surface membrane GalNAc lectin that allows adherence and induces cytolysis plus a cysteine protease. Cysteine proteases contribute to tissue invasion, causing degradation of the extracellular

Table 21.5 Differences between Th1 and Th2 CD4$^+$ T helper cells

Th1—Type 1 Responses	Th2—type 2 responses
Produce IFN-γ and IL-2; induction of cytolytic activity by CD8$^+$ T cells; production of complement-fixing antibodies	Produce cytokines (IL-4, IL-5, IL-10, and IL-13); associated with high levels of neutralizing and cell-bound antibodies, mast cell activation, eosinophilia, suppression of type 1 responses
IL-12 associated with resistance to protozoa and susceptibility to various helminths	IL-4 plays important role in development of Th2-type responses: associated with resistance to helminths (expulsion of intestinal nematodes) and susceptibility to protozoa
Produce lymphotoxin: promote macrophage activation and generation of cell-mediated immunity	Support maturation of B cells to immunoglobulin-secreting cells (activation of humoral immune mechanisms)
Resistance to intracellular pathogens requires Th1 proinflammatory response	Resistance to extracellular pathogens requires Th2 response
Prevention: IFN-γ, IL-12, IL-10 also cooperate to keep Th2 response in check	Prevention of Th1 response from overproduction: IL-10, transforming growth factor-β, IL-4
Problems: effector molecules detrimental if produced too long: nitric oxide, reactive oxygen intermediates (IL-1, IFN-γ, TNF)	Problems: strong antibody responses = antigen/antibody complexes, complement activation, eosinophils = immediate hypersensitivity; can cause hepatic fibrosis, portal hypertension, chronic morbidity

Table 21.6 Allergic reactions responsible for disease in parasitic infections[a]

Infection	Clinical disease	Reaction type
Malaria	Nephrotic syndrome	Immune complex (III)
	Anemia	Cytotoxic (II)
Chagas' disease	Myocarditis	Cytotoxic (II)
Leishmaniasis	Spectral disease	Cell mediated (IV)
Schistosomiasis	Swimmer's itch	Anaphylaxis (I)
		Cell mediated (IV)
	Katayama fever	Immune complex (III)
	Granulomatous disease	
	Schistosoma mansoni	Cell mediated (IV)
	Schistosoma haematobium	Cell mediated (IV)
Filariasis	Elephantiasis	Cell mediated (IV)
	Tropical pulmonary eosinophilia	Anaphylaxis (I)
Echinococcosis	Leaky hydatid fluid	Anaphylaxis (I)

[a] Data from references 11 and 81.

matrix, and they participate in the induction of apoptosis (programmed cell death) in cells contacted by the amebae (7). They also assist in circumventing the host immune response through cleavage of secretory IgA and IgG and activation of complement (64). Cysteine proteases are encoded by a number of genes present in *E. histolytica* but not *E. dispar*. A detailed study of cysteine proteases may eventually provide the key to the design of specific inhibitors that could serve as new chemotherapeutic agents.

Humoral Immunity

Coproantibodies can be detected in approximately 80% of patients with amebic dysentery, with a steady decline to about 55% over a period of weeks. During the same time frame, a rise in the level of antibodies in serum can be demonstrated. These data may represent a transient secretory response followed by invasion and penetration of the intestinal mucosa with production of circulating antibody. Once invasive amebiasis has been established, a serologic response can generally be demonstrated within a week after the onset of symptoms. These titers tend to be high and can persist for several years, even after invasive disease has healed or a subclinical infection has been controlled. However, this humoral response during the invasive phase of the disease apparently has little impact on potential reinfections or in the healing process.

IgM is produced, although IgG is found at a higher concentration, with the predominant subclass being IgG2 (60). Data indicate that cleavage of IgG by the extracellular cysteine proteinase of *E. histolytica* trophozoites may limit the effectiveness of the host humoral response (78). The production of anti-ameba IgE is controversial, and its role in *E. histolytica* infections is unclear. A cytotoxic effect on the trophozoites is produced, and organisms lose motility when in contact with serum containing specific antibodies. Also, introduction of antibodies to cultures inhibits growth, and antibody-treated organisms do not induce liver abscess in susceptible animal models.

It is also interesting that sera from healthy individuals with no evidence of past or present amebiasis also have the same effects on organisms as immune sera do. Even after absorption with *E. histolytica* trophozoites, the effects persist, suggesting that they are due to causes other than the presence of natural antibodies. Complement activation is involved, via classic and alternative pathways, and apparently does not require antibodies. Activation of complement pathways results in lysis of the trophozoites. It has been reported that highly virulent strains of *E. histolytica* are more resistant to complement-mediated lysis; thus, complement may also play a role in natural resistance to infection by causing a potential reduction in invasive capacity. However, it is not known if reduced complement activity plays a role in the susceptibility of certain individuals to infection or whether it may be related to host defenses.

Although the data are limited, a suppression of humoral response to sheep erythrocytes in hamsters has been documented when extracts of *E. histolytica* were used. The effect is enhanced when the erythrocytes and amebic extract

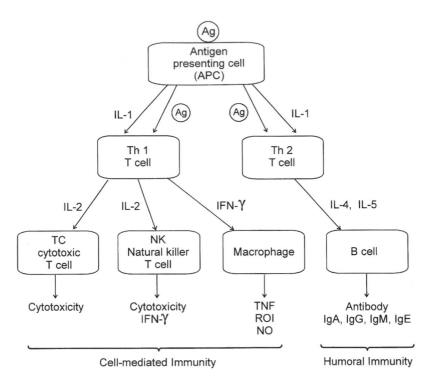

Figure 21.1 Representation of the immune response. Antigen (Ag) is recognized and processed by antigen-presenting cells (APC) and presented to Th1 or Th2 cells that carry receptors for the antigen. At the same time, the APCs release IL-1, a cytokine that activates resting Th1 and Th2 cells. Activated Th1 cells produce IL-2, which activates cytotoxic T (TC) or natural killer (NK) cells. Cytotoxic T cells kill target cells carrying the original antigen in a major histocompatibility complex-restricted way, while NK cells kill target cells nonspecifically. Another product of the Th1 cells is IFN-γ, which activates resting macrophages to become more phagocytic and to release TNF, reactive oxygen intermediates (ROI), and nitric oxide (NO). Both ROI and NO are involved in killing intracellular parasites or those in close proximity to the activated macrophage. The Th2 cells release IL-4 and IL-5, which are involved in B-cell activation and the release of antibodies specific to the antigen. IL-5 also activates eosinophils. Antibodies may cooperate with macrophages and eosinophils to serve as a bridge bringing the activated cell and the parasite together; they may also support the release of toxic molecules directly onto the surface of the parasite. *Note*: It is important to remember that some of the cytokines exert both a positive and negative influence on the immune response at the same time; this diagram is designed to show the main features of the overall immune response. (Adapted from reference 15.)

are injected on the same day; however, the primary response only is involved. Suppression of the primary response has also been documented with ovalbumin as the antigen.

Survival mechanisms allow the survival of the parasite, often in spite of both natural and immune defenses. One of these mechanisms is capping, i.e., surface redistribution of membrane antigens in amebae, which may circumvent the effect of surface-bound specific antibodies. Antibodies to the amebae bind to the organism and induce a polar redistribution of surface components. The surface becomes

an accumulation of folded membranes that extrude as a cap, containing most of the antibodies that were originally bound to the surface. The cap is then released, probably by constriction of a limiting membrane region.

In spite of the production of high levels of antibody in response to an infection with *E. histolytica*, it appears that the humoral response is not protective to the host. This is confirmed by the ease with which reinfection occurs and the fact that high antibody levels can be seen in patients who remain symptomatic with amebiasis.

Table 21.7 Immune system "alphabet soup" and definitions[a]

Term[b]	Definition	Comments
APC	Antigen-presenting cell	Processes the foreign antigen and presents it to the T helper cells
B cell	Lymphocyte; precursor of plasma cell; expresses immunoglobulins on its surface but does not release them	B lymphocyte; short life; responsible for production of immunoglobulins
T cell	Lymphocyte; helper, killer, suppressor cells	T lymphocyte; long life; responsible for cell-mediated immunity; soluble products are lymphokines
Th	T helper cell; both Th1 and Th2 carry same CD4$^+$ or CD8$^+$ surface markers	T lymphocyte; Th1 produces IL-2 and IFN-γ; Th2 produces IL-4 and IL-5; may be antagonism between products of Th1 and Th2; Th1 drives immune response toward antibody-independent immune responses; Th2 drives immune response toward antibody production
Lymphokine	Hormone-like peptide, released by activated lymphocytes	Mediates immune responses
Cytokine	Hormone-like low molecular weight protein that regulates the intensity and duration of immune responses and is involved in cell-to-cell communication	Interferon (IFN), interleukin (IL), lymphokines; released from spleen, thymus, epithelial, endothelial, and mast cells and by fibroblasts and lymphocytes
Interferon (IFN)	Small glycoproteins; classified into three groups (alpha, beta, and gamma); can regulate many cell properties and functions	IFN-α2A and -2B used as antineoplastic agents; IFN-β elaborated by fibroblasts; IFN γ elaborated by T lymphocytes in response to specific antigen or mitogenic stimulation
Interleukin (IL)	Name given to cytokines once their amino acid structure is known	IL-1: from mononuclear phagocytes; enhances proliferation of Th and growth and differentiation of B cells IL-2: from Th; causes proliferation of T cells and activated B cells (T-cell growth factor) IL-3: from monocytes, fibroblasts, and endothelial cells; increases the production of monocytes (multi-colony-stimulating factor) IL-4: from T4 lymphocytes; causes differentiation of B cells (B-cell differentiation factor) IL-5: from T cells; causes activation of B cells and differentiation of eosinophils IL-6: from fibroblasts, macrophages, and tumor cells; increases synthesis and secretions of immunoglobulins by B cells (B-cell-stimulating factor 2, IFN-β2) IL-7: from bone marrow cells; causes proliferation of B and T cells IL-8: from endothelial cells, fibroblasts, keratinocytes, macrophages, and monocytes; causes chemotaxis of neutrophils and T cells IL-9: from T cells; causes growth and proliferation of T cells IL-10: from helper T cells, B cells, and monocytes; inhibits IFN-γ secretion by T cells and mononuclear cell inflammation IL-11: from bone marrow endothelial cells, macrophages, and preadipocytes; stimulates increased concentrations of acute-phase proteins (C-reactive proteins), mannose-binding protein, serum amyloid-P component, α_1-antitrypsin, fibrinogen, ceruloplasmin, and complement components C9 and factor B in plasma IL-12: from B cells, T cells, and macrophages; induces IFN-γ gene expression in T cells and NK cells IL-13: from Th cells; inhibits mononuclear cell inflammation IL-14: from T cells; stimulates B-cell proliferation and inhibits immunoglobulin secretion IL-15: from T cells; stimulates T-cell proliferation and NK cell activation

(continued)

Table 21.7 Immune system "alphabet soup" and definitions*a* *(continued)*

Term*b*	Definition	Comments
NK cell	Natural killer cell; heterogeneous collection of lymphocytes; neither B nor T cells; activated by IL-12	Kills target cells nonspecifically; does not involve MHC*b* class I molecules; is being increasingly implicated in protection against parasitic infections
Cytotoxic T cell	T lymphocyte	Kills target cells carrying original antigen in an MHC-restricted manner; often involved in viral infections but seldom in parasitic infections
TNF	Tumor necrosis factor; a cytolytic factor that is produced by CD4 and CD8 T cells after exposure to antigen; also produced by macrophages	A polypeptide hormone; has the ability to modulate adipocyte metabolism, lyse tumor cells in vitro, and induce hemorrhagic necrosis of certain transplantable tumors in vivo
ROI, NO	Reactive oxygen intermediates; nitric oxide	Both are involved in intracellular killing of ingested parasites or destruction of parasites in close proximity to the activated macrophage

a Adapted from reference 13.
b MHC, major histocompatibility complex.

The identification and structural characterization of three amebic antigens, the serine-rich *E. histolytica* protein (SREHP), the 170-kDa subunit of the Gal/GalNAc-binding lectin, and the 29-kDa cysteine-rich protein, indicate that they all show promise as recombinant antigen-based vaccines to prevent amebiasis (20, 28, 50, 75). Recent data confirm that DNA vaccination with SREHP can provide high levels of protection against amebic liver abscess in animal models of disease (84).

Cellular Immunity

Most research on cell-mediated immunity in amebiasis has been conducted with animal models of amebic liver abscess. However, a number of different approaches to clarify the role of cellular immunity in humans have been taken. One example involved the use of skin test antigens obtained from *E. histolytica* culture. This procedure detects amebiasis long after the patient has recovered from the infection, and it has been useful in epidemiologic studies; patients with invasive amebiasis exhibit a typical delayed-type hypersensitivity reaction, although the reaction is usually negative during the acute phase of an untreated liver abscess (60).

Macrophage migration-inhibition assays have shown results similar to those obtained with skin test antigens during invasive disease. Results were negative during the early phases of liver abscesses but became positive after therapy. A number of different animal model studies also confirm that cell-mediated immunity plays an active role in the overall immune response to infection with *E. histolytica* (60).

Concanavalin A (ConA)-induced lymphocyte transformation with cells from both patients with amebic liver abscess and healthy controls is comparable; similar results have also been demonstrated with phytohemagglutinin (PHA). However, studies with infected hamsters indicated reduced reactivity to ConA and PHA; other reports

Table 21.8 Various methods used by parasites to evade the host immune response

1. Become intracellular (*Plasmodium* spp.)
2. Enter immunologically protected site soon after infection (*Plasmodium* spp.)
3. Leave site of established immune response during life cycle (*Ascaris lumbricoides*)
4. Become disguised with host antigen(s) (*Schistosoma* spp.)
5. Survive in macrophages (*Leishmania* spp., *Trypanosoma cruzi*, *Toxoplasma gondii*)
6. Live in lymphocytes (*Theileria* spp.)
7. Inactivate host lymphocytes (*Schistosoma mansoni*)
8. Cause polyclonal stimulation of lymphocytes (*Trypanosoma brucei* subspp., *Plasmodium falciparum*, *Babesia microti*, *Entamoeba histolytica*, *Trichinella spiralis*)
9. Activate macrophages (*Leishmania* spp., *Toxoplasma gondii*, *Trichinella spiralis*, *Nippostrongylus brasiliensis*)
10. Shed antigens (*Trypanosoma brucei* subspp., *Entamoeba histolytica*, *Ancylostoma caninum*)
11. Undergo antigenic variation (*Plasmodium falciparum*, *Trypanosoma brucei* subspp.)
12. Inhibit cell or antibody binding (*Schistosoma mansoni*)
13. Suppress immune system (*Trypanosoma cruzi*, *Trypanosoma brucei* subspp., *Leishmania donovani*, *Plasmodium* spp.)

Table 21.9 Immune system antibodies (neutralization, agglutination, complement activation, and facilitated opsonization)

Antibody	General comments	Functionality
IgA	Occurs as 2 subclasses (IgA1 and IgA2); can occur in monomeric, dimeric, or polymeric forms; found in secretions, where it protects the mucous surfaces of the gut, respiratory tract, and urogenital tract; occurs in breast milk	Neutralization of bacterial toxins; inhibition of invasion of the mucosa by agglutinating infectious agents or preventing attachment to the host cell
IgG	Found in serum and extravascular spaces; occurs as 4 subclasses, IgG1, IgG2, IgG3, and IgG4	Most common antibody in secondary immune response; all subclasses can agglutinate microorganisms; IgG1, IgG2, and IgG3 activate complement, with IgG3 being the most effective; IgG1 and IgG4 can cross the placenta; IgG1 and IgG3 can bind to macrophages via their Fc portions
IgM	Found in serum and extravascular spaces; pentameric immunoglobulin; single subclass	First antibody produced in response to antigenic stimulation; agglutinates microorganisms with repeated epitopes; activates complement via classic pathway
IgE	Has an affinity for mast cells, basophils, and eosinophils, to which it binds via the Fc portion	When bound to an antigen, it causes cross linking, resulting in the release of active substances from the cell to which it is bound

indicate that *E. histolytica* extracts are mitogenic for murine lymphocytes. Induced blastogenesis in lymphocytes from patients with amebic liver abscess has been shown with aqueous *E. histolytica* extracts and a subcellular antigenic fraction, thus indicating in vivo sensitization of T lymphocytes to amebic antigens (60).

According to Ortiz-Ortiz (60), all arguments in favor of protective immunity in amebiasis are based on cellular immunity, hence the emphasis on studies involving the in vitro interaction of *E. histolytica*, T lymphocytes, and macrophages. A compilation of some of these studies can be seen in Tables 21.10 and 21.11. It is apparent from these studies that cell-mediated immunity limits the clinical manifestations of amebic liver abscess and protects against recurrent infections.

Most studies reinforce the idea that cell-mediated immunity is the most important factor in acquired protective immunity. A number of factors support this conclusion: (i) cellular anergy that supports the initial *E. histolytica* invasion; (ii) increased incidence of invasive amebiasis after T-cell suppression or splenectomy; (iii) protective impact of T-cell stimulants; (iv) after recovery from amebic liver abscess, appearance and persistence of a delayed-type hypersensitivity reaction to amebic antigens; (v) transfer to immunity via sensitized T cells; and (vi) the lysis effect of cytotoxic T cells stimulated with antigen and activated macrophages.

Anergy during the course of amebic liver abscess is well documented. Animal models reveal tissue destruction with minimal inflammatory reaction surrounding these abscesses. One explanation for this anergy might be a temporary desensitization caused by circulating antigen; it is not thought to be due to decreased T-cell numbers, since

normal levels have been detected in patients with amebic liver abscesses (60).

Phagocytosis by both peripheral blood polymorphonuclear leukocytes and monocytes is reduced in various circumstances related to amebiasis. Phagocytic function in polymorphonuclear leukocytes was more depressed in patients with amebic liver abscess than in patients with other manifestations of *E. histolytica* infection or in healthy controls. In evaluating the functional capacity of monocytes, phagocytosis was significantly depressed in both liver and intestinal amebiasis and bactericidal function was depressed even more in liver amebiasis.

Studies of patients with amebic liver abscess that had been healed for a minimum of 9 weeks indicated that the CD4$^+$ (helper, inducer T-cell) lymphocyte subpopulation was smaller and the CD8$^+$ (suppressor, cytotoxic T-cell) subpopulation was increased, resulting in a lower cell ratio in patients with amebic liver abscess than in controls (71). However, when the same group of patients was checked a year later, four of five demonstrated an increased CD4$^+$ cell percentage and three of five showed an increased ratio of CD4$^+$ to CD8$^+$ cells. Since T lymphocytes from patients cured of amebic liver abscesses have been shown to destroy *E. histolytica* after activation with parasite antigens, this may have an impact on host resistance in these patients.

Giardiasis

Giardia lamblia is probably the most commonly diagnosed flagellate in the intestinal tract and may be the most commonly diagnosed intestinal protozoan in many areas of the world. Although various criteria have been used to

Table 21.10 Interaction between *Entamoeba histolytica* and T lymphocytes in vitro

Study	Comments	Reference
Virulent *E. histolytica* with lymphocytes from patients who have recovered from amebic liver abscess	Parasite lysis (also with lymphocyte supernatant fluids)	45
E. histolytica with lymphocytes, peritoneal cells, or spleen cells from infected hamsters	Parasite lysis; may have been caused by lymphokine-activated macrophages	30
Lymphocytes (treated amebic liver abscess patients) with ameba antigen or nonimmune T cells (PHA stimulated)	Parasites killed	70
E. histolytica with supernatant fluids of stimulated T cells	Inhibition of DNA and protein synthesis; not toxic	8

differentiate species of *Giardia*, there is still disagreement about classification and nomenclature. On the basis of different species names for different animal hosts, there are approximately 50 *Giardia* species. However, there are data to suggest either biological or antigenic differences among many of these species.

Studies during the last 15 years have confirmed that *Giardia* organisms elicit an immune response that probably plays a role in the variable host response generated to the presence of the parasite. Early data indicating a lower incidence of symptoms in individuals who were repeatedly exposed to the organism suggested the presence of an immune response. Subsequent studies with humans and experimental animals have confirmed the presence of both humoral and cellular immune responses to *Giardia* spp.

In giardiasis a diffuse loss of microvillous brush border, usually combined with villous atrophy, is responsible for malabsorption, which ultimately causes diarrhea. Other mucosal changes include crypt hyperplasia and increased infiltration of intraepithelial lymphocytes. The loss of brush border surface area, reduced disaccharidase activities, and increased crypt/villus ratios are mediated by

$CD8^+$ T cells, while both $CD8^+$ and $CD4^+$ small mesenteric lymph node T cells regulate the influx of intraepithelial lymphocytes (6).

Antigenic variation has been documented in giardiasis, and variation at the surface membrane of trophozoites occurs frequently. These antigens are composed of cysteine-rich proteins, and B-cell-dependent mechanisms are probably responsible for the surface antigen switch (24). Apparently, individual organisms express only one variant-specific protein at a time; the detection of multiple surface-labeled bands in some of the studies resulted from subpopulations of trophozoites expressing several different variant surface protein types (1).

Humoral Immunity

The majority of infected patients produce detectable levels of *Giardia*-specific antibodies; however, the biological role played by these antibodies in the host immune response to the infection is unclear. Diagnostic procedures have been developed for the detection of short-lived IgM anti-*Giardia* antibodies and IgG antibodies that may remain at high titers for many months after the patient

Table 21.11 Interaction between *Entamoeba histolytica* and macrophages in vitro and in vivo

Study	Comments	Reference(s)
Virulent *E. histolytica* with nonactivated macrophages	Lysis of mammalian cells; no harm to amebae	72
Virulent *E. histolytica* with activated macrophages (lymphokines induced with ameba soluble protein, ConA, or PHA)	Amebae killed; dependent on cell-cell contact, time, and oxidative and nonoxidative processes; decrease in macrophage viability, possibly owing to lysis by *E. histolytica* toxic products	72
Silica poisoning of macrophages or use of antimacrophage serum	Increased susceptibility to amebic liver abscess; larger abscesses; dissemination to other tissues (hamsters, guinea pigs)	31, 33
Bacillus Calmette-Guérin-stimulated macrophages	Decreased susceptibility to amebic liver abscess; smaller abscesses; less dissemination (to hamsters, guinea pigs)	31, 32
Congenitally athymic nude mice (*nu/nu*), macrophage dependent	Resistant to invasive amebiasis; can limit development of liver abscess	10, 76

has been treated and cured. Specific anti-*Giardia* IgM responses in serum were found in 100% of human volunteers experimentally infected with human-source *Giardia* spp. (58). Some studies indicate a correlation between the presence of antibody and symptoms in the patient. Other studies, using different antigens and antibody detection procedures, fail to confirm this correlation.

In asymptomatic patients, the presence of antibody may play a protective role or serve merely as an indicator of infection. Certainly, data indicate that people who are repeatedly exposed to the parasite will tend to be asymptomatic and will have high levels of *Giardia*-specific antibody in serum and breast milk. Patients with hypogammaglobulinemia also have a high rate of diarrhea, often due to *Giardia* spp., and children with chronic diarrhea and giardiasis have an increased incidence of hypogammaglobulinemia. Also, AIDS patients who have giardiasis but no specific antibody also have acute symptomatic infections (40). In contrast to this group of AIDS patients, those who have giardiasis with circulating *Giardia*-specific antibody may have symptoms unrelated to their *Giardia* infection.

Giardia-specific IgA may also be important in both defense against and clearance of the parasite. In one study, patients with symptomatic giardiasis had higher levels of IgA in serum than did uninfected controls and none of the patients were IgA deficient (40). The degree of protection of breast-fed infants against *Giardia* infection also depends on the level of IgA-specific antibodies in breast milk. In the intestine, IgA antibodies may influence the local immune response by inhibiting parasite adherence, tissue invasion, or penetration of the mucosa by parasite antigens (40). IgA anti-*Giardia* antibodies are the most common Ig type found in acutely ill patients with giardiasis, and they have been detected in persons exposed to a waterborne outbreak in Vermont and villagers in Thailand (40). The jejunal-plasma immune response to *Giardia* involves a decrease in the number of IgA cells and an increase in the number of IgM cells. The large population of antibody-producing plasma cells that are secreting mainly IgA is evidence of the importance of mucosa-associated lymphoid tissue in giardiasis.

Cellular Immunity

Antibodies to *Giardia* spp. work together with cellular immune mechanisms. Cellular responses to giardiasis occur at the level of the intestinal mucosa, and a number of inflammatory changes may occur during the infection. Histologic changes during giardiasis include infiltration of polymorphonuclear leukocytes into the epithelium and mononuclear leukocytes into the lamina propria, development of shortened villi, loss of the brush border, damage to epithelial cells, and an increase in epithelial cell mitosis

(40). In rare cases, giardiasis is associated with changes that mimic celiac sprue and include complete villous atrophy, dense mononuclear cell infiltration, and flattening of the epithelial cells. This accumulation of inflammatory cells in the intestinal mucosal cells of patients with giardiasis suggests that cellular responses play a role in the overall immune response to *Giardia* infection. However, while these histologic findings suggest a cellular immune response, they may also represent a nonspecific inflammatory response stimulated by the presence of an intraluminal infection.

Experimental animal studies provide evidence that T lymphocytes and Peyer's patch helper T lymphocytes play a role in the host immune response. In athymic mice, which are deficient in both T lymphocytes and Peyer's patch helper T lymphocytes, inoculation with *G. muris* results in a chronic infection with large numbers of organisms. In contrast, immunocompetent mice clear the parasite and may develop resistance to reinfection.

Reports also indicate that human mononuclear phagocytes are capable of killing *Giardia* organisms and phagocytizing *Giardia* trophozoites (40). Also, gut-associated lymphoid tissue provides the precursor cells that migrate to the mammary gland and become cells capable of secreting the Ig components of milk. Despite the number of studies on the immunology of giardiasis, the actual effect of antibody or cellular components on the organism itself is unclear. Future studies on defining specific antigens to which the host immune system responds will play a major role in our understanding of this very common infection. For reviews of current information, see references 24 and 40.

Toxoplasmosis

The gastrointestinal route appears to be the most common site of infection with *Toxoplasma gondii*, and, after disruption of the host cells, the organisms invade adjacent cells and are spread throughout the body via the lymphatics and bloodstream. Since the advent of AIDS, the impact of toxoplasmosis as a major opportunistic infection has led to extensive research in the field of immunology. Infection with *T. gondii* stimulates the development of both humoral and cellular immune responses, with the cell-mediated response being the more predominant. These immune responses of the host can dramatically affect the course and severity of the infection.

Adhesion to the host cell is facilitated by secreted proteins, and invasion uses an actin-myosin-based process. Within the cell, the parasite remains inside the parasitophorous vacuole; the organism prevents the cell from fusing with lysosomes. Eventually the tachyzoites transform into bradyzoites and encyst. Ongoing encystment depends on an ongoing type 1-mediated immune response; IFN-γ and CD4$^+$ and CD8$^+$ cells play a protective role. Although this immune response limits the tissue invasion of the parasite, it also aids the survival of *Toxoplasma* by converting it into a bradyzoite, the resistant form (26).

Humoral Immunity

In spite of the very high antibody titers to *T. gondii* often seen in human infections, humoral antibody plays a secondary role in contributing to protective immunity. The classic approach to the diagnosis of toxoplasmosis has always been based on the detection of immunoglobulins, specifically IgM and IgG. In infected humans, IgM usually appears within 2 weeks after infection whereas IgG antibodies appear later and their levels peak at approximately 2 months postinfection. The IgG titer decreases gradually to a low level and remains at this level thereafter. For this reason, the presence of IgG demonstrates proof of a prior infection with *T. gondii*. Although detection of IgM has always been considered an indicator of acute infection, the development of more sensitive diagnostic methods means that IgM can now be detected months to years after the initial acute infection. Unfortunately, these findings also complicate the interpretation of serologic test results in patients with toxoplasmosis.

IgA antibodies can be detected early in infection, and their level decreases between 3 and 9 months after infection. The detection of both IgA and IgM at the same time indicates the acute phase of toxoplasmosis and is currently used for risk evaluation of pregnant women. Apparently, IgE antibodies also appear early in infection and then their level begin to decline; the detection of IgE could also serve as an indicator for acute disease. One study indicated the presence of IgE in 86% of infected patients (62). The presence of IgE in immunosuppressed individuals appears to be correlated with the onset of complications such as adenopathies, chorioretinitis, and disease reactivation (80).

Unfortunately, in spite of what is currently known about the humoral response in toxoplasmosis, routine serologic tests are often inadequate for the diagnosis of congenital disease. Specific IgG can persist in a child for months and cannot be differentiated from maternal IgG. However, both IgM and IgA antibodies do not cross the placenta and can be detected in infected fetuses as early as the second trimester of pregnancy, assuming that the mother became infected soon after conception (19). Since children at 1 year of age have only 20% of the level of IgA in adults, early synthesis of specific IgA might also be an indicator of congenital infection (23).

A number of attempts have been made to confer resistance to *T. gondii* by passive transfer of immune homologous or heterologous antisera. The results of these studies have indicated conflicting outcomes; some investigators have reported no protection, whereas others have reported significant protection. The results appear to be related to the choice of experimental host species, specific parasite strains, and transfer of small volumes of high-titer antisera. It appears that anti-*Toxoplasma* antibody can be protective under carefully controlled conditions.

When extracellular tachyzoites are exposed to antibody in the presence of complement in vitro, their morphology is altered: the cells become vacuolated and coarse, collapse, and lyse. The changes in dye permeability in the Sabin-Feldman dye test are based on these changes seen with a combination of antibody and complement. However, the in vivo intracellular location of the parasite protects it from the circulating antibody.

Cellular Immunity

Cellular immunity is the key component of the host's immune reaction to toxoplasmosis. Macrophages, T lymphocytes, and NK cells, as well as the cytokines, are major components of the immune response (26). A number of studies have shown that there is a lower level of lymphocyte transformation in patients with acute disease than in chronically infected patients. In the presence of parasite antigens, lymphocytes from patients with evidence of infection for more than 1 year proliferate whereas lymphocytes from seronegative patients do not (26).

Animal studies have confirmed the key role of T lymphocytes in protective immunity against *T. gondii*. In infected mice, depletion of CD4$^+$ T lymphocytes reactivates toxoplasmosis in the central nervous system. These cells also play a role in the development of host resistance and in controlling brain cyst development. Apparently, in the mouse model there is a synergistic relationship between CD4$^+$ and CD8$^+$ T-cell subsets, with CD4$^+$ cells participating in the development of resistance during vaccination or acute infection and CD8$^+$ cells being involved in the maintenance of a protective immune response and the inhibition of cyst formation. CD4$^+$ and CD8$^+$ cells are the key players involved in host resistance in toxoplasmosis. The CD8$^+$ cells are activated by IL-2 secreted by the CD4$^+$ cells, and they exhibit cytotoxic activity against tachyzoites or cells infected with *T. gondii*.

Studies of patients with acute toxoplasmosis have shown a significant elevation in levels of suppressor T cells in patients with symptoms lasting more than 4 weeks. A reversed CD4$^+$/CD8$^+$ cell ratio is also seen and persists for 2 to 4 months after infection. By using flow cytometry and differential leukocyte counts, an increase in the absolute number of CD8$^+$ T cells was confirmed in patients with symptomatic toxoplasmosis, with no significant change in CD4$^+$ cell levels. Also, cloned CD4$^+$ and CD8$^+$ cells from a symptomatic patient produced less IFN-γ after stimulation with *T. gondii* antigen than did T cells from an asymptomatic patient.

Immunosuppression

Although information is still somewhat incomplete, CD8$^+$ cells can be stimulated by *T. gondii* antigen to suppress lymphocyte proliferation; both the suppression and increase in the level of the CD8$^+$ T-cell subset correlate with a more severe and prolonged clinical course (51). This study also supports the suggestion that suppressor T cells play a role in prolonging the symptoms of toxoplasmosis. Lymphadenopathy and elevated numbers of CD8$^+$ T cells have been found in patients with lymphatic toxoplasmosis; the same picture is also seen in patients with viral infections known to cause mononucleosis (Epstein-Barr virus and cytomegalovirus). Overall, the immunosuppressive effect seen in toxoplasmosis is apparently limited to the T-cell responses, while B-cell functions are unaffected.

In congenital infections, T-cell functions are also impaired; studies indicate that most congenitally infected infants demonstrate specific immunologic tolerance to *T. gondii* (55). The presence of the parasite at a critical stage of fetal development results in production of lymphocytes that are unable to distinguish immunologically between parasite and self. It is possible that increased numbers of suppressor T cells that occur during fetal life could induce tolerance by the inhibition of other lymphocytes and production of less IFN-γ and IL-2. Also, in symptomatic babies, these diminished responses to *T. gondii* antigens usually correlate with more severe disease.

In normal, unstimulated macrophages, *T. gondii* tachyzoites can continue to multiply; however, in activated macrophages, vacuoles containing living tachyzoites fuse with lysosomes and the parasites are killed. Both oxygen-dependent and oxygen-independent mechanisms are involved in parasite killing.

There is also evidence to indicate that platelets are cytotoxic for tachyzoites when direct cell-to-cell contact is established, even in the absence of antibody. These findings certainly suggest a role for platelets as one of the mechanisms in host defense against toxoplasmosis.

In two studies, a protective effect of vaccination of mice with a combination of antigen and IL-12 or IL-15 has been demonstrated. In the study using a recombinant surface antigen (rSAG1) plus IL-12, this combination reduced the parasite load in the brain by 40% (48). The second study indicated that the $\gamma\delta$ T-cell response to *T. gondii* was enhanced by the administration of exogenous IL-15 in the presence of parasite lysate antigen; the survival time after lethal challenge were lengthened (47). In this study, IgG1 and IgG2a antibody titers were increased, as was splenocyte lymphoproliferation.

African Trypanosomiasis

According to recent data, 50 million people are at risk of contracting trypanosomiasis. Also, in Africa, the extensive distribution of the vector, the tsetse fly, makes 10^7 km^2 of potential grazing land unavailable for livestock and approximately one-third of the cattle are at risk of trypanosomiasis. On the basis of isoenzyme studies, *Trypanosoma brucei gambiense* is considered a subspecies; however, it is really not clear whether *T. brucei gambiense* and *T. brucei rhodesiense* should be subspecies or variants of a single species. In this infection, the trypanosomes live in the bloodstream, where they are directly affected by the production of antibodies by the host.

Humoral Immunity

In reviewing the immune system response of the host to the presence of these parasites, it is important to understand the concept of antigenic variation. It has been long recognized that African sleeping sickness is characterized by a series of parasitemia peaks in the blood. This successive wavelike pattern is due to the appearance of trypanosomes expressing a different antigenic identity. Organisms that have changed their antigenic identity can escape the host immune response and can initiate another increase in the level of parasitemia. The success of this infection depends on the parasite's ability to select between hundreds of genes that encode the variant surface glycoprotein (VSG), or variant surface protein (VSP). Both designations are correct. As the host immune response produces high levels of antibody against one variant, parasites expressing a different VSG gene begin to proliferate. These organisms persist until they are killed by the immune response, and this pattern is repeated multiple times (22).

The factors that induce VSG switching have not been totally defined; however, the presence of host antibodies is not necessary for the changes to occur. The number of variable antigenic types that can occur in any one trypanosome population is quite large (100 or more), and more than 1,000 VSG genes have been detected. There is evidence to indicate that the VSG-specific antibodies represent a mixture of T-dependent and T-independent antibodies directed at different epitopes (66). Unfortunately, although these antibody responses play a role in the successive wavelike growth patterns of the organisms, they do not generate long-lasting immunity.

Specific changes in the humoral component of the immune response to trypanosomes include an increase in the cellularity of the spleen and lymph nodes, an increase in the proliferation of spleen cells, an increase in the number of B cells (not T cells), an increase in the IgM level, the appearance of antibodies reacting with nontrypanosome antigens, and the appearance of autoantibodies, rheumatoid factors, and immune complexes (74).

Cellular Immunity

Studies of tumor necrosis factor alpha (TNF-α) have shown that it is involved in many physiologic functions, and it has been implicated in the pathogenesis of septic shock and systemic inflammatory reactions. Increased levels have been seen in sera from patients with other parasitic infections such as malaria and leishmaniasis; in patients with malaria, increased levels of TNF-α correlate with disease severity. In a recent study, an increase in the level of TNF-α was observed in sera from patients with African trypanosomiasis compared with controls. Patients were classified as being in the early stage (blood, lymphatic) or late stage (meningoencephalitocytic) of the disease. An increase in the level of TNF-α was also noted in late-stage patients compared with those in the early stages of the disease. Data indicated that levels of TNF-α in the serum of patients with African trypanosomiasis were definitely correlated with disease severity (signs of inflammation or presence of major neurologic signs), indicating that TNF-α could be involved in at least some aspects of physiopathology in human African trypanosomiasis (*T. brucei gambiense*) (59).

Immunosuppression

It is also well documented that another key feature of African trypanosomiasis is suppression of the host immune system. Increased incidence of opportunistic infections in patients with African trypanosomiasis is well recognized and has been known since the early 1900s. Both T-cell and B-cell functions can be severely suppressed.

Animal studies have demonstrated two different pathways, one that is suppressor cell dependent and one that is suppressor cell independent. Consequently, suppression has been attributed to polyclonal activation of B cells and to the generation of suppressive cells, two different pathways that are probably coexisting mechanisms.

Prostaglandins also play a large role, not only in immunosuppression but also in other areas such as immune system modulation and pathogenesis. It has been shown that prostaglandins in cerebrospinal fluid from patients with sleeping sickness are probably responsible for sleep induction (61).

American Trypanosomiasis

American trypanosomiasis (Chagas' disease) is caused by *Trypanosoma cruzi* and is one of the major health problems in Latin American countries. It is estimated that 100 million persons are at risk of infection, of whom 16 million to 18 million are actually infected, and there are approximately 50,000 deaths per year due to Chagas' disease. In certain areas of endemicity, approximately 10% of all adult deaths are due to Chagas' disease. In addition to humans, numerous species of wild and domestic mammals can harbor the infection and serve as an enormous parasite reservoir.

Shortly after infection, both humans and animals develop an immune response to the presence of the parasites; this is clearly demonstrated by the development of parasite-specific antibodies and hypersensitivity reactions. As with many other parasite infections, these immune responses do affect the outcome of the infection; however, host protection is limited. Both antibody-mediated and cell-mediated mechanisms are involved. It has also been shown that autoantibodies directed against laminin, tubulin, and nervous tissue are present; however, whether these autoantibodies contribute to disease is somewhat unclear.

Humoral Immunity

During the initial phase of the infection, IgM is predominant, whereas later, IgG and IgA become the major antibody classes represented. Antigenic variation, which is characteristic of infections with African trypanosomes, is less common in *T. cruzi* infection. In spite of humoral and cellular immunity to *T. cruzi*, the infection is able to persist in the host.

Reviews of serologic test results show that patients can be divided into four categories: (i) uninfected, (ii) indeterminate (seropositive but asymptomatic), (iii) cardiac (symptomatic with palpitations, dyspnea, syncope, and pulmonary embolus, with findings consistent with chagasic cardiomyopathy), and (iv) mega-gastrointestinal (symptomatic with dysphagia, constipation, and abdominal pain, with documented visceromegaly of the esophagus or colon determined by barium or surgical exploration). Individuals with both cardiac and mega-gastrointestinal disease were categorized as both (9). There were no significant differences in the levels of antitrypanosomal antibodies among the various clinical groups of chronically infected individuals. Titers obtained by hemagglutination, indirect fluorescent-antibody assay, and enzyme-linked immunosorbent assay allowed a clear distinction between seropositive and seronegative individuals.

The use of recombinant *T. cruzi* antigens greatly improves the early diagnosis of congenital Chagas' disease, especially through the detection of IgM. Although detection of IgM has proven to be very helpful, additional studies related to IgA will be justified, particularly since some congenitally infected infants, who do not produce IgM, may develop specific IgA (49).

Glycoinositolphospholipids are some of the major glycolipids of the *T. cruzi* surface that activate B cells.

Apparently, *T. cruzi* glycoinositolphospholipids can directly stimulate NK cells and can induce immunoglobulin production in the absence of T cells and NK cells. These studies indicate that this *T. cruzi*-derived molecule may be one of the stimulators that lead to NK cell activation during infection (17).

Cellular Immunity

Cell-mediated immunity has also been implicated as a cause of the tissue destruction, including cardiomyopathy and megacolon, seen in Chagas' disease. In animal studies, the transfer of lymphocytes from chronically infected mice to uninfected mice can destroy cardiac cells and the T-cell lines can recreate some of the neuropathologic changes seen in chronic disease (9).

Decreased cell-mediated immune responses have been found in symptomatic patients compared with asymptomatic patients. Future work will probably be focused on T-cell suppression factors or selective cell depletion as possible mechanisms for cell-mediated immunity alterations.

CD8$^+$ T cells are critical to the control of *T. cruzi* infection and function in multiple ways, the most important being the production of IFN-γ; IL-12 is also involved in protection against *T. cruzi* (46, 54, 79). This cytokine is produced early by NK cells and later by T cells. Cytokine responses in *T. cruzi*-infected children have been partially clarified; the children were 5 to 14 years of age and were from an area of endemic infection in Paraguay. This age group tends to be the most severely affected by the early phases of infection. Acutely infected children showed a distinct Th1-type IFN-γ cytokine response to infection. The cytokine pattern seen in the seropositive, asymptomatic group was the Th0 type (expression of both IFN-γ and IL-4). It was suggested that selective induction of a Th0-type cytokine pattern is important for the development of cell-mediated and humoral responses that suppress the parasite burden, thus prolonging the onset or limiting the disease severity of chronic Chagas' disease years later (73). Also, the frequency of IFN-γ-producing T cells in patients with chronic Chagas' disease seems to be associated with a history of recent exposure and with the clinical status of the patient (43).

Reduced production of IFN-γ by CD8$^+$ T cells is associated with increased severity of Chagas' disease in humans. In chronic infection, CD8$^+$ T cells serve as effector memory cells, undergo rapid expansion, and demonstrate effector functions following reexposure to antigen. However, these responses appear to develop relatively slowly, and the effector function of the CD8$^+$ cells is compromised in some tissues, including muscle (54).

In several patients coinfected with *T. cruzi* and human immunodeficiency virus (HIV) with high parasitemia, no clinical or parasitological evidence of *T. cruzi* reactivation was seen. CD4$^+$ T-cell counts decreased over time in two patients, and the lymphocyte proliferative response to the parasite was very low in all patients. Thus, *T. cruzi* infection appears to have a long silent course in immunocompromised HIV-infected patients (15).

In group of 23 patients with AIDS and *T. cruzi* infection, 20 patients developed severe multifocal or diffuse meningoencephalitis with necrosis and hemorrhage associated with numerous tissue parasites (69). The second most severely affected site was the heart, with myocarditis in 7 of the 23 patients. Both acute myocarditis and meningoencephalitis are thought to be caused by relapses of chronic *T. cruzi* infection, but this hypothesis is not universally accepted. Since immunologic defense against *T. cruzi* is mediated mainly by T lymphocytes and since the CD4$^+$ population of T cells is depleted in AIDS patients, it appears in certain circumstances that AIDS may be responsible for reactivation of *T. cruzi* infections. In patients who do not have AIDS but have fatal, acute infections with *T. cruzi*, meningoencephalitis is uncommon and is seen most often in children younger than 4 years. Cases of Chagas' disease in immunosuppressed individuals have been described as having pseudotumoral presentations. There are expanding mass lesions in the cranial cavity, causing intracranial hypertension, and they mimic various neoplasms such as gliomas, lymphomas, or metastases (69). The United States, with the largest HIV-infected population reported by the World Health Organization up to 1987, is also known to have a significant number of individuals with Chagas' disease (35).

Studies of IL-12 indicate that this cytokine can act as a more selective amplifier of *T. cruzi* reactive cells than IL-2 does. By enhancing parasite antigen-specific immunity, IL-12 could be potentially therapeutic, with the ability to control reactivated *T. cruzi* infections associated with AIDS or other immunosuppressive conditions (18).

Immunosuppression

Suppressed delayed-type hypersensitivity has been reported to occur during acute Chagas' disease. *T. cruzi* also exerts a direct suppressive effect on human lymphocytes. Abnormalities in activated human lymphocytes may be involved in immunosuppression. These changes occur, regardless of whether the parasites and cells are in physical contact. There is also evidence that a soluble *T. cruzi* product mediates the suppressive effects. This trypanosomal immunosuppressive factor is a protein and can be purified from sera from chronic chagasic patients but not healthy individuals (42).

B cells isolated at the acute stage of infection have shown marked impairment in their response to polyclonal activators in vitro. Studies suggest that B-cell apoptosis and cell cycle arrest could be the mechanisms that control intense B-cell expansion but at the same time could be delaying the emergence of a specific immune response to *T. cruzi* (85).

Splenomegaly, lymphadenopathy, depression of immune system responses to unrelated antigens, hypergammaglobulinemia, and autoimmunity are all indications of immune system physiopathologic changes in Chagas' disease. Minoprio et al. (56) also think that specific antigenic challenges could elicit "nonspecific" immune responses greater than the antigen-specific components of the response. Underlying mechanisms might include as examples the production of cytokines, the nonspecific "stickiness" of antigens to lymphocyte surfaces, the expression of major histocompatibility complex and other cell surface components (probably IFN-γ and IL-4), and the nonspecific lymphocyte mitogens released by *T. cruzi* or superantigens which could bind to T-cell receptors and trigger accessory cell-dependent T-cell activation.

Studies have shown that there appears to be no correlation between natural killer (NK) and lymphokine-activated killer (LAK) cell functions in the same individual. Neither NK nor LAK cell functions seem to be involved in the immunosuppression associated with chronic Chagas' disease (77).

Autoimmunity

Chagas' disease cardiomyopathy (CCC) is one of the few examples of postinfectious autoimmunity, where *T. cruzi* triggers molecular mimicry-related target organ immune damage. Heart tissue destruction in CCC may be caused by autoimmune recognition of heart tissue by a mononuclear cell infiltrate, often many years after the initial infection. Indirect evidence indicates that there may be cross-reactivity between *T. cruzi* and cardiac tissue. There is also evidence for immune system recognition of cardiac myosin in CCC; the presence of an anti-cardiac myosin-cross-reactive *T. cruzi* antigen is suspected as the cause. Studies demonstrate sequence homology to a recombinant *T. cruzi* protein (B13) in several positions of cardiac myosin heavy chain. This recognition of a heart-specific *T. cruzi* cross-reactive epitope in close association with the presence of chronic heart lesions suggests the involvement of cross-reactivity between cardiac myosin and B13 in the pathogenesis of CCC (14, 38). It has been proposed that phagocytic removal of infected apoptotic cardiomyocytes, combined with signaling through innate immune receptors, may be required to initiate immune responses that damage the heart (21).

Studies demonstrate the presence of antibody to myelin basic protein in the serum of mice chronically infected with *T. cruzi*. Studies of the overlapping peptides from the myelin basic protein identified two regions responsible for the

cross-reactivity with *T. cruzi* (3). This may provide further evidence to help explain why chronic Chagas' disease is characterized by inflammatory infiltrates in myocardial and nervous tissues, with virtually no demonstrable parasites.

Another study suggests that fraction IV-*T. cruzi* (parasite antigen acidic fraction) and sciatic nerve components have some epitopes, possibly carbohydrate, in common. Thus, infection in patients with Chagas' disease could override tolerance to self components and lead to autoimmunity (29).

The scarcity of parasites in the chronic phase of the disease contrasts with the severe cardiac pathology seen in approximately 30% of chronically infected patients. Antigen-specific and non-antigen-specific mechanisms have been suggested to explain T- and B-cell activation leading to autoimmunity. However, the issue of autoimmunity versus parasite persistence as the cause of Chagas' disease pathology continues to be hotly debated.

Malaria

Human malaria is caused by infection with any of four species of *Plasmodium*, with *Plasmodium falciparum* being the most pathogenic. The other three species are *P. vivax*, *P. ovale*, and *P. malariae*. Symptoms of malaria infections include fever and malaise; however, infections with *P. falciparum* can be fatal. This is particularly true in untreated, nonimmune individuals. Despite the high fatality rate in nonimmune or partially immune individuals, people who live in areas of endemic infection for long periods generally acquire immunity and may be asymptomatic or have very mild symptoms. Why and how this slow development of immunity occurs is not completely understood, particularly when reviewing the very complex life cycle of these organisms.

The vector for malaria is the female anopheline mosquito. When the vector takes a blood meal, sporozoites contained in its salivary glands are discharged into the puncture wound. A surface protein, the circumsporozoite protein, in combination with thrombospondin-related adhesive protein allows recognition of glycosoaminoglycans in the liver, thus providing the targeting mechanism that allows the *Plasmodium* sporozoite to reach the correct liver cells. Within an hour, these infective stages are then carried via the blood to the liver, where they penetrate parenchymal cells and begin to grow, thus initiating the preerythrocytic or primary exoerythrocytic cycle. The sporozoites become round or oval and begin dividing repeatedly, resulting in large numbers of exoerythrocytic merozoites that leave the liver and invade the red blood cells (RBCs) (erythrocytic cycle). In *P. vivax* and *P. ovale*, a secondary or dormant schizogony may occur from organisms that remain quiescent in the liver until a later time (hypnozoites). Merozoite surface proteins facilitate recognition of and adhesion to the RBC.

The situation in which the RBC infection is not eliminated by the immune system or by therapy and the number of RBCs begins to increase again with subsequent clinical symptoms is called a recrudescence. All species may have a recrudescence. The situation in which the erythrocytic infection is eliminated and a relapse occurs because of a new invasion of the RBCs from liver merozoites is called a recurrence or true relapse and theoretically occurs only in *P. vivax* and *P. ovale* infections.

Once the erythrocytes and reticulocytes have been invaded, the parasites grow and feed on hemoglobin. As the nucleus begins to divide, the trophozoite is called a developing schizont. The mature schizont contains merozoites, whose number depends on the species, which are released into the bloodstream and invade RBCs, in which a new cycle of erythrocytic schizogony begins. After several erythrocytic generations, some of the merozoites begin to develop into the male and female gametocytes.

The asexual and sexual forms just described circulate in the bloodstream in infections with three species of *Plasmodium*. However, in *P. falciparum* infections, as the parasite continues to grow, the RBC membrane becomes sticky and the cells tend to adhere to the endothelial lining of the capillaries of the internal organs. Thus, only the ring forms and the gametocytes (occasionally mature schizonts) normally appear in the peripheral blood.

If gametocytes are ingested when the mosquito takes a blood meal, they mature into gametes while in the mosquito gut. The male microgametes undergo nuclear division (exflagellation), in which they break out of the RBC, become motile, and penetrate the female macrogamete; the fertilized stage is then called the zygote. The zygote becomes elongate and motile and is called the ookinete; it migrates to the mosquito midgut, secretes a thin wall, and grows into the oocyst, which extends into the insect's hemocele. Within a few days to 2 weeks, the oocyst matures and hundreds of sporozoites are formed. When the oocyst ruptures, the sporozoites are released into the hemocele and dispersed throughout the body, and some make their way into the salivary glands. When the mosquito next takes a blood meal, the sporozoites are injected with saliva into the host.

Each phase of the life cycle is associated with the expression of many stage- and species-specific proteins, many of which are found within the surface membrane of the parasite. These surface proteins are highly polymorphic and antigenically variable. Two hypotheses have been developed to explain the slow development of immunity: (i) immunity is strain specific, and so the individual would not become immune until exposed to all the major antigenic variants circulating within the area of endemicity, and (ii) acute malaria causes extreme immunosuppression, which delays the development of protective immunity. Also, antidisease immunity may be very different from antiparasite immunity. Even continuous exposure to malaria parasites rarely produces sterile immunity, and blood-stage parasites can be detected by PCR in all age groups (25). The wide range of malarial surface antigens helps explain why the development of clinical immunity requires years of repeated infections. Two mechanisms contribute to antigenic diversity, one being the genetic nucleotide replacement and recombination and the second being antigenic variation. Since different antigens are expressed during various parts of the life cycle, naturally acquired immunity tends to be stage specific. Both humoral and cellular components are required for naturally acquired immunity. CD4$^+$ and CD8$^+$ T cells are effective against intracellular liver-stage organisms, while antibodies may interfere with host cell invasion by sporozoites and merozoites. Also, antibodies to variant surface antigens expressed on the RBC surface may facilitate phagocytosis of the infected cells and prevent their adhesion to endothelial receptors.

Apparently, internal antigens are less variable and are partially conserved among different malarial species. When released during the rupture of mature schizonts, these internal antigens appear to stimulate the cytokine cascade, which leads to various symptoms and pathologic changes. (Table 21.12).

Humoral Immunity

Malarial infection leads to production of high levels of TNF and other proinflammatory cytokines; other findings include elevated immunoglobulin production, activation of complement, and redistribution of lymphocytes from the peripheral circulation to the spleen and other organs. Antibodies can confer protection against *P. falciparum* and *P. vivax*; neonates and infants are protected by maternal antibodies and, in clinical treatment trials, by immune serum or purified immunoglobulins. Animal studies involving passive transfer of whole serum or purified IgG can have a dramatic impact on the erythrocytic cycle by extending the prepatent period, reducing the parasitemia, and leading to spontaneous cure.

Levels of antimalarial antibodies in humans increase during childhood, become maximal during early adulthood, and subsequently remain fairly constant. However, there appears to be little correlation between antibody levels and protection against infection or clinical disease. Also, some of the antibodies are not directed against *Plasmodium* organisms. These antibodies may represent the degree of past exposure rather than functional immunity.

Table 21.12 Updated information on immune responses seen in malaria infections

Immune response[a]	Comments[a]	Reference
Parasite antigen-specific IL-10 and antibody responses predict accelerated parasite clearance in *P. falciparum* malaria	Parasite antigen-specific IL-10-mediated antibody responses may play a role in the control of asexual-stage parasite multiplication	53
P. falciparum-merozoite surface protein 1 antigen (MSP119): induction of small T-cell response (IL-4), significantly increased by addition of IL-2; if IFN-γ, IL-12, or neutralizing anti-IL-4 antibody present, T-cell responses abolished	Recombinant MSP119 is leading malaria vaccine candidate, can prime nonimmune human lymphocytes	27
Correlation of T-cell response and lymphokine profile with RESA peptides of *P. falciparum* containing a universal T-cell epitope and an immunopotentiator, polytuftsin	Cytokine profile is suggestive of a CD4$^+$ Th1 type of immune response; ideal for killing intracellular pathogens like plasmodia	44
TNF-α, IL-1β, IL-6 implicated in pathogenesis of severe *P. falciparum* malaria; exogenous IL-10 inhibited malarial antigen-induced cytokine production; maximal inhibition when IL-10 used in first 2 h of stimulation of PBMC	TNF-α, IL-1β, IL-6 produced within 2–4 h of stimulation; IL-10 produced after 8 h; findings suggest that IL-10 counterregulates proinflammatory response to *P. falciparum*; severe falciparum malaria may be associated with inadequate negative feedback response by IL-10	36
IFN-γ responses associated with resistance to reinfection with *P. falciparum* in young African children	Differences in Th1/Th2 cytokine balance may be linked to ability to control parasite multiplication, helping explain marked differences in both susceptibility to infection and clinical presentation	52
PfEMP-1 molecule expressed on surface of RBC infected with trophozoite or early gametocyte parasites; IgG antibodies to PfEMP-1 increased with age, mirroring decline in both prevalence and density of asexual stages and gametocytes in RBCs	Immunity to PfEMP-1 may influence malaria transmission by regulation of gametocyte production; mechanisms may be (i) controlling asexual proliferation and density and (ii) affecting gametocyte maturation	63
Cell-mediated immunity and cytokines involved in pathogenesis of malaria; MM and CM cases in children monitored	Both groups (MM, CM) had significantly higher levels of IL-6, IL-10, and TNF-α than did controls; 24 h after admission, IL-10 and IL-6 levels much higher in CM patients; findings suggest activation of monocyte-macrophage system during early stage of clinical malaria	5

[a] CM, cerebral *P. falciparum* malaria; PBMC, peripheral blood mononuclear cells; MM, mild *P. falciparum* malaria; PfEMP-1, *P. falciparum* erythrocyte membrane protein 1; RESA, ring-infected erythrocyte surface antigen.

Cellular Immunity

Animal studies indicate that T-cell-dependent immune mechanisms are required for the development of effective immunity. Although animal model results vary and may not always mimic human immune responses, it does appear that the human antimalarial response is both T-cell dependent and independent (47).

In the erythrocytic portion of the parasite life cycle, T cells are thought to serve at least two functions in immunity: (i) assisting in antibody production and (ii) producing cytokines. The role of cytokines in malaria is very complex; they affect parasite growth and clearance, and they activate macrophages to engulf infected RBCs and to kill the malarial parasites. Elevated concentrations of TNF, IL-6, and soluble TNF receptors have been found in serum in patients with severe *P. falciparum* infections.

TNF induces nitric oxide (NO) production; NO definitely plays protective roles in malaria: (i) it acts as a relaxing factor; (ii) it can downregulate cell adhesion molecules; (iii) it can reduce leukocyte-endothelial cell adherence; (iv) it can scavenge free oxygen radicals; (v) its inhibition can cause endothelial cell alterations, such as albumin leakage and platelet deposition; and (vi) it can reduce platelet aggregation and adhesion to endothelium (34). The role of NO in pathogenesis is somewhat less clear. It is possible that NO can mediate early changes in cerebral malaria, such as neurotransmission disturbances, but NO is probably not involved in the actual process of neurovascular damage.

Studies suggest that a systemic, excessive production of TNF, not compensated for by TNF receptors or a hyperproduction affecting vulnerable sites such as the brain, may exacerbate the disease and lead to vascular abnormalities and life-threatening malaria. On the other hand, TNF produced within the context of a specific immune response may be the major protective cytokine in malaria infections. This demonstrates the dual role of this particular cytokine (43).

Although many of the published research reports indicate that a number of monokines and lymphokines are released during acute malaria infections, interpretation of the relevance of cytokine levels may be somewhat misleading (68). Summary information related to the elimination of infected RBCs and mechanisms by which the parasite evades the host immune system can be seen in Table 21.13.

Immunosuppression

It has been well documented that acute malaria infections lead to a decrease in host immune system functionality. These patients have been described as being temporarily immunosuppressed. Direct evidence for this decreased immune system capability has been summarized by Riley et al. (67) and includes increased susceptibility to bacterial infections, reactivation of latent viral infections, and diminished response to some vaccines in children. Indirect evidence is that children protected by chemoprophylaxis for malaria are less susceptible to other infections and tend to respond better to routine vaccinations (67). Although this evidence indicates a lowered immunocompetency level in patients with malaria, it is not known exactly which factors influence this change. It may result directly from the level of parasitemia or from the overall impact of a febrile illness, such as malaria.

During acute malaria infections, there is a decrease in the response to parasite antigens. This may be due to immunosuppression, discussed above, or may just indicate a lack of "priming of the system" in patients with no prior exposure to the organism. Therefore, increased immune responses after cure may be due to decreased immunosuppression or increased levels of preexisting immune responses.

Several factors have been described as virulence factors with *P. falciparum*; however, formal documentation that these characteristics are always associated with virulence and significantly more likely to cause severe disease is not abundant (82). These factors include the following: (i) the multiplication capacity in which the biomass of parasites may outstrip host defense mechanisms; (ii) the RBC selectivity limiting the overall number of infected RBCs (however, this is not a big factor for *P. falciparum*); (iii) the cytoadherence and rosetting ability where all four species can cause rosetting but only *P. falciparum* can cause cytoadherence, which is thought to be linked to cerebral malaria; (iv) the potential to release cytokines; (v) antigenicity (those not recognized by the host will have an advantage); and (vi) antimalarial drug resistance, with a great deal of selective pressure against drug-sensitive parasites (82).

Numerous studies seem to emphasize that the host immune system has difficulty in controlling the infection; many unusual host responses have been described. These unusual responses may be due to host immune system overreaction, leading to inappropriate responses. The difference between protective immunity and pathology is a very delicate balance and one in which specific mechanisms are difficult to clarify (65). Specific immunopathologic complications of malaria infection are listed in Table 21.14.

Table 21.13 Mechanisms of infected red blood cell elimination and evasion of the host immune system in *P. falciparum* infections[a]

Mechanism	Occurrence in *P. falciparum* infection
Infected RBC elimination	
Opsonization	Yes
Antibody-dependent cell-mediated toxicity	Yes
NK cell activity	Yes
Cytotoxic T-cell activity	No
Evasion of host immune system	
Infected RBC knob-like membrane changes	Yes
Infected RBC surface antigenic changes	
Modified host antigens	Yes
Parasite-derived antigens	Yes
Clonal antigenic variation	Yes
Cytoadherence	
Endothelial	Yes
Autoagglutination	Yes
Rosetting of both infected and normal RBCs	Yes
Upregulation of host ligands	Yes

[a] Data from references 2 and 82.

Table 21.14   Immunopathological complications of malaria infection[a]

Pathology	Suspected mechanism[b]	Comments[b]
Hypergammaglobulinemia	Antigen-induced cytokine production (IL-6); antigenic variation; polyclonal activation	IL-6 from fibroblasts, macrophages, and tumor cells; increases synthesis and secretions of immunoglobulins by B cells; most antibodies not directed against malaria antigens; common feature of antigenic variation is presence of successive peaks of infection, each new peak being antigenically different from the previous one; partial loss of suppressor mechanisms coupled with normal CD4 helper/inducer activity may result in high levels of immunoglobulin in serum (characteristic of people living in areas of endemic malarial infection)
Immunosuppression	Macrophage dysfunction; antigenic competition; structural disruption of germinal centers; disruption of spleen function; polyclonal activation and immune system "exhaustion"	May be due to suppressive effect of malarial pigment ingested; defects in monocyte and neutrophil chemotaxis, reduced monocyte phagocytic function
Nephrotic syndrome	Immune complex deposition; autoimmunity	In nephrotic syndrome associated with *P. malariae*, malaria antigen and immune complexes can be eluted from the kidneys
Autoimmunity	Autoantibodies; antierythrocyte antibodies; dyserythropoiesis; excessive erythrophagocytosis	Defective IL-12 production in early course of infection may result in fatal anemia; uninfected RBCs destruction occurs through phagocytosis of RBCs bound to merozoites killed as a result of malaria paroxysms (thus, dyserythropoiesis may play a minimal role in anemia); hemolysis is prime cause of anemia, although destruction of parasitized RBCs is not the sole cause of hemolytic process; bone marrow suppression may be insignificant, but preexisting iron deficiency aggravates the severity of anemia; immune recognition of "self" or "modified self" RBC antigens by B or T lymphocytes may occur
Thrombocytopenia	Excessive removal of platelets; coating of platelets with malaria antigen	May also be due to platelet inactivation; increased platelet turnover; platelet-bound antibody controversial
Hyperactive malarial splenomegaly	Genetic predisposition; hypergammaglobulinemia; chronic increase of lymphocyte proliferation	Associated with increased capacity to clear RBCs by Fc receptor-mediated mechanisms and recognition of reduced deformability; spleen may also modulate cytoadherence
Burkitt's lymphoma	Coendemicity with EBV; polyclonal activation; antigen-induced cytokine production	In patients with *P. falciparum* malaria, loss of cytotoxic T-cell control of EBV in B cells, possibly due to destruction or dysfunction of a subset of CD4 cells responsible for induction of suppressor/cytotoxic CD8 cells, leads to activation and proliferation of foci of B cells containing EBV; the expanded pool and rapid turnover of these cells may increase the chances of malignant transformation, leading to the genesis of Burkitt's tumor

[a] Adapted from references 16, 37, 39, 41, 57, and 83.
[b] EBV, Epstein-Barr virus.

Summary

Although there is evidence for acquired immunity in most protozoan infections, it is often neither fully effective nor protective. As with many organisms, the fact that the host does not die of overwhelming infections supports the concept of parasite immunity.

Since the majority of helminths do not multiply within the host, the host immune system is not faced with an overwhelming infection due to parasite multiplication as seen with protozoal infections. Many helminth life cycles include both larval and adult forms, both of which stimu-late immune system reactions. However, adult and most larval forms are too large to be destroyed by antibody or by phagocytic cells. In this case, there may be some form of antibody-dependent cell-mediated cytotoxicity in which the parasites become coated with antibodies which bind eosinophils and other cells that destroy the parasites (12). Worm expulsion from the intestine is a CD4+ T-cell-mediated event associated with type 2 cytokines. There tends to be less evidence for the implication of eosinophils in host protection. While type 1 or type 2 responses may be key factors following infection with intracellular pro-

Table 21.15   Immune responses to parasites (protozoa and helminths)[a]

Parasite or infection	Comments
Protozoa	
Intestinal protozoa *E. histolytica, G. lamblia*	Most organisms in gastrointestinal tract are nonpathogenic, i.e., do not elicit immune responses *Overall immune response*: little correlation between antibody levels and protection; repeated invasion of gut wall (amebiasis) relatively common, reinfection rare in hepatic amebiasis
Leishmaniasis Self-healing dermal forms (*L. tropica, L. mexicana*) Nonhealing visceral forms (*L. donovani*)	In addition to two general categories, wide spectrum of disease, immune response slow to develop (few months to >1 yr); strong immunity to homologous strain, little antibody produced; visceral infections frequently progress to death; ~25% of patients do recover *Overall immune response*: cell mediated; parasite killing is intracellular and involves production of toxic metabolites by host cell; no apparent role for antibody or complement
African trypanosomiasis	Successive waves of parasitemia; antigens differ in each wave (antigenic variation); immune response rapid, efficient *Overall immune response*: agglutination, complement-mediated lysis; cell-mediated responses involved to a lesser degree
American trypanosomiasis	Incubation, 2–3 wk; acute phase, 4–5 wk; chronic phase, ≥20 yr; latent phase/no symptoms, >20 yr; spontaneous recovery rare (some individuals die soon after infection) *Overall immune response*: antibody involved; immunopathology is a very important aspect
Coccidiosis	Self-limiting, immune response not that important in control but very effective against reinfection with homologous organism *Overall immune response*: antibody-mediated response probably most important; dramatic differences between immunocompetent and immunosuppressed hosts (*Cryptosporidium* spp.)
Toxoplasmosis	Antibody controls infection but will not eliminate residual organisms; no protection against reinfection, but severity reduced *Overall immune response*: probably macrophage activation reinforced by antibody; dramatic differences between immunocompetent and immunosuppressed hosts
Malaria	Time span, 2–4 yr (*P. vivax, P. ovale, P. falciparum*) or lifelong (*P. malariae*) *Overall immune response*: no immune response to exoerythrocytic stages; immunity does occur to homologous strains; antibody transfer effective (mother to child); strong correlation between antibody and parasite disappearance; antibodies stage specific; T cells probably serve at least two functions (assisting in antibody production and producing cytokines)
Babesiosis	Pathogenesis not fully understood; lasting immunity after infection, antibody mediated; splenic immune properties unrelated to specific acquired immunity *Overall immune response*: humoral system-mediated immune response present; impact on host cell-mediated response more important; activated macrophages important; also evidence for some immunosuppression of the host but may not be relevant to pathogenesis
Helminths	
Schistosomiasis	No evidence of spontaneous recovery, can resist infection *Overall immune response*: primary response against skin penetration by schistosomula, effective against reinfection; six methods have been identified: (i) IgG + neutrophils, (ii) IgG + eosinophils, (iii) IgG + macrophages, (iv) IgG + eosinophils + mast cells, (v) IgE + macrophages, (vi) IgE + eosinophils + mast cells In vitro studies indicate that eosinophils bind to schistosomula, extrude hydrolytic enzymes onto surface of worm, disrupt membrane, allow eosinophils into body and detach tegument; macrophages become involved after damage by eosinophils and neutrophils; mechanisms probably similar in human disease *Most important components*: hypersensitivity responses are initiated by eggs trapped in the tissues (granulomas and local tissue damage, particularly in the liver); blockage of blood flow can cause portal hypertension and vascular abnormalities
Cestode infections	Adult forms show no evidence of acquired immunity; in larval forms, immunity is slow to develop *Overall immune response*: larval forms evade immune response; animal studies indicate eosinophils and mast cells involved in damage to cyst, with subsequent encapsulation by various cells including macrophages; early part of immune response may be similar to that seen in schistosomiasis; larval forms exert immunosuppressive effects on the host such as larval secretions inhibiting T-cell proliferative responses and IL-2 production, preventing accumulation of inflammatory cells, and interfering with normal macrophage functions

(continued)

Table 21.15 Immune responses to parasites (protozoa and helminths)[a] *(continued)*

Parasite or infection	Comments
Intestinal nematodes	Strong immune response to all intestinal nematodes (elicited by migrating larvae); self cure (spontaneous elimination of adult worms in gastrointestinal tract); worm infections elicit Th2-subset responses, cytokines generate inflammatory responses (IgE, mast cells, and eosinophils), which are powerful mediators of intestinal inflammation *Overall immune response*: some evidence of acquired immunity and resistance to reinfection; immune system mechanisms stage specific; possible attachment of antibody and perhaps complement followed by cell attachment which damages cuticle, followed by phagocytosis by macrophages; increased IgE levels seen
Toxocariasis	
Visceral	Prolonged and severe T-cell-dependent eosinophilia; rapid, vigorous, prolonged humoral response; elevated IgG anti-A and anti-B blood group antigen titers present; elevated IgM anti-IgG (rheumatoid factor) titers may occur; elevated IgE titers
Ocular	May not be able to demonstrate circulating anti-*Toxocara* antibody; eosinophilic granulomas seen in some infections
Filariasis	Both tissue forms and microfilariae can partially evade the immune system, immunity eventually develops; strong immune responses against adult worms responsible not for protection but host pathology *Overall immune response*: when parasites eliminated from blood, antibody can be found on surface of microfilariae; antibody initiates attack by various cells (IgG, IgE, eosinophils, neutrophils, like other helminth infections); IgE may be important in protection against larval stages but may also contribute to immunopathology via hypersensitivity reactions to adult worms; the T-cell anergy seen in patients heavily infected with lymphatic filariasis or with *Onchocerca* is probably due to downregulation of Th1 responsiveness by IL-10

[a] Data from reference 120.

tozoa or gut-dwelling nematodes, the mechanisms responsible for resistance to any one organism are usually quite specific for the biological characteristics of each.

General comments on a number of different parasites, both protozoan and helminth, are presented in Table 21.15. Although not all inclusive, hopefully this chapter has provided some insight into the complex interactions between parasite and host.

References

1. **Adam, D. A.** 2001. Biology of *Giardia lamblia. Clin. Microbiol. Rev.* **14:**447–475.

2. **Allred, D. R.** 1995. Immune evasion by *Babesia bovis* and *Plasmodium falciparum*: cliff-dwellers of the parasite world. *Parasitol. Today* **11:**100–105.

3. **Al-Sabbagh, A., C. A. Garcia, M. M. Diaz-Bardales, C. Zaccarias, J. K. Sakurada, and L. M. Santos.** 1998. Evidence for cross-reactivity between antigen derived from *Trypanosoma cruzi* and myelin basic protein in experimental Chagas' disease. *Exp. Parasitol.* **89:**304–311.

4. **Anders, R. F., R. J. Howard, and G. F. Mitchell.** 1982. Parasite antigens and methods of analysis, p. 28–73. *In* S. Cohen and K. S. Warren (ed.), *Immunology of Parasitic Infections.* Blackwell Scientific Publications Ltd., Oxford, United Kingdom.

5. **Batista, J. L., G. Vanham, M. Wery, and E. Van Marck.** 1997. Cytokine levels during mild and cerebral falciparum malaria in children living in a mesoendemic area. *Trop. Med. Int. Health* **2:**673–679.

6. **Buret, A. G.** 2005. Immunopathology of giardiasis: the role of lymphocytes in intestinal epithelial injury and malfunction. *Mem. Inst. Oswaldo Cruz* **100:**185–190.

7. **Campos-Rodriguez, R., and A. Jarillo-Luna.** 2005. The pathogenicity of *Entamoeba histolytica* is related to the capacity of evading innate immunity. *Parasite Immunol.* **27:**1–8.

8. **Castellanos, C., C. Ramos, and L. Ortiz-Ortiz.** 1989. Effects of gamma interferon on synthesis of DNA and proteins by *Entamoeba histolytica. Infect. Immun.* **57:**2771–2775.

9. **Cetron, M. S., F. P. Basilio, A. P. Moraes, A. Q. Sousa, J. N. Paes, S. J. Kahn, M. H. Wener, and W. C. van Voorhis.** 1993. Humoral and cellular immune response of adults from northeastern Brazil with chronic *Trypanosoma cruzi* infection: depressed cellular immune response to *T. cruzi* antigen among Chagas' disease disease with symptomatic versus indeterminate infection. *Am. J. Trop. Med. Hyg.* **49:**370–382.

10. **Chugh, A., A. Saxena, and V. K. Vinayak.** 1985. Interactions between trophozoites of *Entamoeba histolytica* and cells of the immune system. *Aust. J. Exp. Biol. Med. Sci.* **63:**1–8.

11. **Coombs, R. R. A., and P. G. H. Gell.** 1976. Classification of allergic reactions responsible for clinical hypersensitivity and disease. *In* P. G. H. Gell and R. R. A. Coombs (ed.), *Clinical Aspects of Immunology*, 3rd ed. Blackwell Scientific Publications Ltd., Oxford, United Kingdom.

12. **Cox, F. E. G.** 1982. Immunology, p. 173–204. *In* F. E. G. Cox (ed.), *Modern Parasitology.* Blackwell Scientific Publications Ltd., Oxford, United Kingdom.

13. **Cox, F. E. G., and D. Wakelin.** 1998. Immunology and immunopathology of human parasitic infections, p. 57–84. *In* F. E. G. Cox, J. P. Krier, and D. Wakelin (ed.), *Topley and Wilson's Microbiology and Microbial Infections*, 9th ed., vol. 5. Arnold, London, United Kingdom.

14. **Cunhaneto, E., M. Duranti, A. Gruber, B. Zingales, I. Demessias, N. Stolf, G. Bellotti, M. E. Pararroyo, F. Pilleggi,**

and J. Kalil. 1995. Autoimmunity in Chagas' disease cardiopathy: biological relevance of a cardiac myosin-specific epitope crossreactive to an immunodominant *Trypanosoma cruzi* antigen. *Proc. Natl. Acad. Sci. USA* **92**:3541–3545.

15. Da-Cruz, A. M., R. P. Igreja, W. Dantas, A. C. Junqueira, R. S. Pacheco, A. J. Silva-Goncalves, and C. Pirmez. 2004. Long-term follow-up of co-infected HIV and *Trypanosoma cruzi* Brazilian patients. *Trans. R. Soc. Trop. Med. Hyg.* **98**:728–733.

16. Das, B. S., N. K. Nanda, P. K. Rath, R. N. Satapathy, and D. B. Das. 1999. Anaemia in acute *Plasmodium falciparum* malaria in children from Orissa state, India. *Ann. Trop. Med. Parasitol.* **93**:109–118.

17. de Arruda Hinds, L. B., L. M. Previato, J. O. Previato, Q. Vos, J. J. Mond, and L. M. Pecanha. 1999. Modulation of B-lymphocyte and NK cell activities by glycoinositolphospholipid purified from *Trypanosoma cruzi*. *Infect. Immun.* **67**:177–180.

18. de Barrow-Mazon, S., M. E. Guariento, and I. A. Abrahamsohn. 1997. IL-12 enhances proliferation of peripheral blood mononuclear cells from Chagas' disease patients to *Trypanosoma cruzi* antigen. *Immunol. Lett.* **57**:39–45.

19. Decoster, A., F. Darcy, A. Caron, D. Vinatier, D. Houze de l'Aulnoit, G. Vittu, G. Niel, F. Heyer, B. Lécolier, M. Delcroix, J. C. Monnier, M. Duhamel, and A. Capron. 1992. Anti-P30 IgA antibodies as prenatal markers of congenital toxoplasma infection. *Clin. Exp. Immunol.* **87**:310–315.

20. Dodson, J. M., P. W. Lenkowski, Jr., A. C. Eubanks, T. F. Jackson, J. Napodano, D. M. Lyerly, L. A. Lockhart, B. J. Mann, and W. A. Petri, Jr. 1999. Infection and immunity mediated by the carbohydrate recognition domain of the *Entamoeba histolytica* Gal/GalNAc lectin. *J. Infect. Dis.* **179**:460–466.

21. DosReis, G. A., C. G. Freire de Lima, M. P. Nunes, and M. F. Lopes. 2005. The importance of aberrant T-cell responses in Chagas disease. *Trends Parasitol.* **21**:237–243.

22. Dubois, M. E., K. P. Demick, and J. M. Mansfield. 2005. Trypanosomes expressing a mosaic variant surface glycoprotein coat escape early detection by the immune system. *Infect. Immun.* **73**:2690–2697.

23. Durandy, A., and C. Griscelli. 1990. Développement des fonctions immunitaires au cours de la vie foetale. *In* G. A. Voisin, P. Edelman, N. Genetet, J. F. Bach, and C. Surear (ed.), *Immunologie de las Reproduction*. Médecine-Sciences Flammarion, Paris, France.

24. Faubert, G. 2000. Immune response to *Giardia duodenalis*. *Clin. Microbiol. Rev.* **13**:35–54.

25. Ferreira, M. U., M. da Silva Nunes, and G. Wunderlich. 2004. Antigenic diversity and immune evasion by malaria parasites. *Clin. Diagn. Lab. Immunol.* **11**:987–995.

26. Filisetti, D., and E. Candolfi. 2004. Immune response to *Toxoplasma gondii*. *Ann. Ist. Super Sanità* **40**:71–80.

27. Garraud, O., A. Diouf, I. Holm, R. Perraut, and S. Longacre. 1999. Immune responses to *Plasmodium falciparum*-merozoite surface protein 1 (MSP1) antigen. II. Induction of parasite-specific immunoglobulin G in unsensitized human B cells after in vitro T-cell priming with MSP119. *Immunology* **97**:497–505.

28. Gaucher, D., and K. Chadee. 2003. Prospect for an *Entamoeba histolytica* Gal-lectin-based vaccine. *Parasite Immunol.* **25**:55–58.

29. Gea, S., P. Ordonez, F. Cerban, D. Iosa, C. Chizzolini, and E. Vottero-Cima. 1993. Chagas' disease cardioneuropathy: association of anti-*Trypanosoma cruzi* and anti-sciatic nerve antibodies. *Am. J. Trop. Med. Hyg.* **49**:581–588.

30. Ghadirian, E., and E. Meerovitch. 1982. In vitro amoebicidal activity of immune cells. *Infect. Immun.* **36**:243–246.

31. Ghadirian, E., and E. Meerovitch. 1982. Macrophage requirement for host defense against experimental hepatic amoebiasis in hamsters. *Parasite Immunol.* **4**:219–225.

32. Ghadirian, E., E. Meerovitch, and D. F. Hartmann. 1980. Protection against amebic liver abscess in hamsters by means of immunization with amebic antigen and some of its fractions. *Am. J. Trop. Med. Hyg.* **29**:779–784.

33. Ghadirian, E., E. Meerovitch, and P. A. L. Kongshavm. 1983. Role of macrophages in host defense against hepatic amoebiasis in hamsters. *Infect. Immun.* **42**:1017–1019.

34. Grau, G. E., and S. de Kossodo. 1994. Cerebral malaria: mediators, mechanical obstruction or more? *Parasitol. Today* **10**:408–409.

35. Hagar, J. M., and S. H. Rahimtoola. 1991. Chagas' heart disease in the United States. *N. Engl. J. Med.* **325**:763–768.

36. Ho, M., T. Schollaardt, S. Snape, S. Looareesuwan, P. Suntharasamai, and N. J. White. 1998. Endogenous interleukin-10 modulates proinflammatory response in *Plasmodium falciparum* malaria. *J. Infect. Dis.* **178**:520–525.

37. Hommel, M., and H. M. Gilles. 1998. Malaria, p. 361–409. *In* F. E. G. Cox, J. P. Krier, and D. Wakelin (ed.), *Topley and Wilson's Microbiology and Microbial Infections*, 9th ed., vol. 5. Arnold, London, United Kingdom.

38. Iwai, L. K., M. A. Juliano, L. Juliano, J. Kalil, and E. Cunha-Neto. 2005. T-cell molecular mimicry in Chagas disease: identification and partial structural analysis of multiple cross-reactive epitopes between *Trypanosoma cruzi* B13 and cardiac myosin heavy chain. *J. Autoimmun.* **24**:111–117.

39. Jakeman, G. N., A. Saul, W. L. Hogarth, and W. E. Collins. 1999. Anaemia of acute malaria infections in non-immune patients primarily results from destruction of uninfected erythrocytes. *Parasitology* **119**:127–133.

40. Janoff, E. N., and P. D. Smith. 1990. The role of immunity in *Giardia* infections, p. 215–233. *In* E. A. Meyer (ed.), *Giardiasis*. Elsevier Biomedical Press, Amsterdam, The Netherlands.

41. Jarra, W. 1983. Protective immunity to malaria and anti-erythrocyte autoimmunity. *Ciba Found. Symp.* **94**:137–158.

42. Kierszenbaum, F., and M. B. Sztein. 1994. Chagas' disease, p. 53–85. *In* F. Kierszenbaum (ed.), *Parasitic Infections and the Immune System*. Academic Press, Inc., San Diego, Calif.

43. Kremsner, P. G., S. Winkler, C. Brandts, E. Wildling, L. Jenne, W. Graninger, J. Prada, U. Bienzle, P. Juillard, and G. E. Grau. 1995. Prediction of accelerated cure in *Plasmodium falciparum* malaria by the elevated capacity of tumor necrosis factor production. *Am. J. Trop. Med. Hyg.* **53**:532–538.

44. Kuman, P., and D. N. Rao. 1999. Correlation of T-cell response and lymphokine profile with RESA peptides of *Plasmodium falciparum* containing a universal T-cell epitope and an immunopotentiator, polytuftsin. *Microbiol. Immunol.* **43**:561–566.

45. Landa, L., R. Capín, and M. Guerrero. 1976. Estudios sobre immunidad celular in la amibiasis invasora, p. 654–660. *In* B. Sepúlveda and L. S. Diamond (ed.), *Amebiasis.* Instituto Mexicano del Seguro Social, Mexico City, Mexico.

46. Laucella, S. A., M. Postan, D. Martin, B. F. Hubby, M. C. Albareda, M. G. Alvarez, B. Lococo, G. Barbieri, R. J. Viotti, and R. L. Tarleton. 2004. Frequency of interferon-gamma-producing T cells specific for *Trypanosoma cruzi* inversely correlates with disease severity in chronic human Chagas disease. *J. Infect. Dis.* **189:**909–918.

47. Lee, Y. H., K. H. Ely, A. Lepage, and L. H. Kasper. 1999. Interleukin-15 enhances host protection against acute *Toxoplasma gondii* infection in T-cell receptor alpha[−/−] deficiency mice. *Parasite Immunol.* **21:**299–306.

48. Letscher-Bru, V., O. Villard, B. Risse, M. Zauke, J. P. Klein, and T. T. Kien. 1998. Protective effect of vaccination with a combination of recombinant surface antigen 1 and interleukin-12 against toxoplasmosis in mice. *Infect. Immun.* **66:**4503–4506.

49. Lorca, M., C. Veloso, P. Munoz, M. I. Bahamonde, and A. Garcia. 1995. Diagnostic value of detecting specific IgA and IgM with recombinant *Trypanosoma cruzi* antigens in congenital Chagas' disease. *Am. J. Trop. Med. Hyg.* **52:**512–515.

50. Lotter, H., T. Zhang, K. B. Seydel, S. L. Stanley, Jr., and E. Tannich. 1997. Identification of an epitope on the *Entamoeba histolytica* 170 kD lectin conferring antibody-mediated protection against invasive amebiasis. *J. Exp. Med.* **185:**1793–1801.

51. Luft, B. J., P. W. Pedrotti, E. G. Engleman, and J. S. Remington. 1987. Induction of antigen-specific suppressor T cells during acute infection with *Toxoplasma gondii*. *J. Infect. Dis.* **155:**1033–1036.

52. Luty, A. J., B. Lell, R. Schmidt-Ott, L. G. Lehman, D. Luckner, B. Greve, P. Matrousek, K. Herbich, D. Schmid, F. Migot-Nabias, P. Deloron, R. S. Nussenzweig, and P. G. Kremsner. 1999. Interferon-gamma responses are associated with resistance to reinfection with *Plasmodium falciparum* in young African children. *J. Infect. Dis.* **179:**980–988.

53. Luty, A. J., B. Lell, R. Schmidt-Ott, L. G. Lehman, D. Luckner, B. Greve, P. Matrousek, K. Herbich, D. Schmid, S. Ulbert, F. Migot-Nabias, B. Dubois, P. Deloron, and P. G. Kremsner. 1998. Parasite antigen-specific interleukin-10 and antibody responses predict accelerated parasite clearance in *Plasmodium falciparum* malaria. *Eur. Cytokine Netw.* **9:**639–646.

54. Martin, D., and R. Tarleton. 2004. Generation, specificity, and function of CD8[+] T cells in *Trypanosoma cruzi* infection. *Immunol. Rev.* **201:**304–317.

55. McLeod, R., D. Mack, and C. Brown. 1991. *Toxoplasma gondii*: new advances in cellular and molecular biology. *Exp. Parasitol.* **72:**109–121.

56. Minoprio, P., S. Itohara, C. Heusser, S. Tonegawa, and A. Coutinho. 1989. Immunobiology of *T. cruzi* infection: the predominance of parasite-nonspecific responses and the activation of TCRI T cells. *Immunol. Rev.* **112:**183–207.

57. Mohan, K., and M. M. Stevenson. 1998. Dyserythropoiesis and severe anaemia associated with malaria correlate with deficient interleukin-12 production. *Br. J. Haematol.* **103:**942–949.

58. Nash, T. E., D. A. Herrington, G. A. Losonsky, and M. M. Levine. 1987. Experimental human infections with *Giardia lamblia. J. Infect. Dis.* **156:**974–984.

59. Okomo-Assoumou, M. C., S. Daulouede, J. Lemesre, A. N'Zila-Mouanda, and P. Vincendeau. 1995. Correlation of high serum levels of tumor necrosis factor-α with disease severity in human African trypanosomiasis. *Am. J. Trop. Med. Hyg.* **53:**539–543.

60. Ortiz-Ortiz, L. 1994. Amebiasis, p. 145–162. *In* F. Kierszenbaum (ed.), *Parasitic Infections and the Immune System.* Academic Press, Inc., San Diego, Calif.

61. Pentreath, V. W. 1991. The search for primary events causing the pathology in African sleeping sickness. *Trans. R. Soc. Trop. Med. Hyg.* **85:**145–147.

62. Pinon, J. M., D. Toubas, C. Marx, G. Mougeot, A. Bonnin, A. Bonhomme, M. Villaume, F. Foudrinier, and H. Lepan. 1990. Detection of specific immunoglobulin E in patients with toxoplasmosis. *J. Clin. Microbiol.* **28:**1739–1743.

63. Piper, K. P., R. E. Hayward, M. J. Cox, and K. P. Day. 1999. Malaria transmission and naturally acquired immunity to PfEMP-1. *Infect. Immun.* **67:**6369–6374.

64. Que, X., and S. L. Reed. 2000. Cysteine proteinases and the pathogenesis of amebiasis. *Clin. Microbiol. Rev.* **13:**196–206.

65. Ramasamy, R. 1998. Molecular basis for evasion of host immunity and pathogenesis in malaria. *Biochim. Biophys. Acta* **1406:**10–27.

66. Reinitz, D. M., and T. M. Mansfield. 1990. T-cell-independent and T-cell-dependent B-cell responses to exposed variant surface glycoprotein epitopes in trypanosome-infected mice. *Infect. Immun.* **58:**2337–2342.

67. Riley, E. M., L. Hviid, and T. G. Theander. 1994. Malaria, p. 119–143. *In* F. Kierszenbaum (ed.), *Parasitic Infections and the Immune System.* Academic Press, Inc., San Diego, Calif.

68. Roberts, D. J. 1994. BSP Malaria Meeting. *Parasitol. Today* **10:**330–334.

69. Rocha, A., A. C. O. de Meneses, A. M. Da Silva, M. S. Ferreira, S. A. Nishioka, M. K. N. Burgarelli, E. Almeida, G. Turcato, Jr., K. Metze, and E. R. Lopes. 1994. Pathology of patients with Chagas' disease and acquired immunodeficiency syndrome. *Am. J. Trop. Med. Hyg.* **50:**261–268.

70. Salata, R. A., J. Cox, and J. I. Ravdin. 1987. The interaction of human T-lymphocytes and *Entamoeba histolytica*: killing of virulent amebae by lectin-dependent lymphocytes. *Parasite Immunol.* **9:**249–261.

71. Salata, R. A., A. Martínez-Palomo, H. W. Murray, L. Canales, N. Treviño, E. Segovia, C. F. Murphy, and J. I. Ravdin. 1986. Patients treated for amebic liver abscess develop cell-mediated immune responses effective in vitro against *Entamoeba histolytica. J. Immunol.* **136:**2633–2639.

72. Salata, R. A., R. D. Pearson, and J. I. Ravdin. 1985. Interaction of human leucocytes and *Entamoeba histolytica*. Killing of virulent amebae by the activated macrophage. *J. Clin. Investig.* **76:**491–499.

73. Samudio, M., S. Montenegro-James, M. Cabral, J. Martinez, A. Rojas de Arias, and M. A. James. 1998. Cytokine responses in *Trypanosoma cruzi*-infected children in Paraguay. *Am. J. Trop. Med. Hyg.* **58:**119–121.

74. **Sileghem, M., J. N. Flynn, A. Darji, P. De Baetselier, and J. Naessens.** 1994. African trypanosomiasis, p. 1–51. *In* F. Kierszenbaum (ed.), *Parasitic Infections and the Immune System.* Academic Press, Inc., San Diego, Calif.

75. **Stanley, S. L., Jr.** 1997. Progress towards development of a vaccine for amebiasis. *Clin. Microbiol. Rev.* **10:**637–649.

76. **Stern, J. J., J. R. Graybill, and D. J. Drutz.** 1984. Murine amebiasis: the role of the macrophage in host defense. *Am. J. Trop. Med. Hyg.* **33:**372–380.

77. **Stracieri, A. B. P. L., and J. C. Voltarelli.** 1995. NK and LAK functions in human chronic Chagas' disease. *Parasite Immunol.* **17:**381–383.

78. **Tran, V. Q., D. S. Herdman, B. E. Torian, and S. L. Reed.** 1998. The neutral cysteine proteinase of *Entamoeba histolytica* degrades IgG and prevents its binding. *J. Infect. Dis.* **177:**508–511.

79. **Une, C., J. Andersson, and A. Orn.** 2003. Role of IFN-alpha/beta and IL-12 in the activation of natural killer cells and interferon-gamma production during experimental infection with *Trypanosoma cruzi. Clin. Exp. Immunol.* **134:**195–201.

80. **Villena, I. D., V. Brodard, C. Queureux, B. Lerous, D. Dupouy, G. Remy, F. Foudrinier, C. Chemla, J. E.** Gomez-Marin, **and J. M. Pinon.** 1999. Detection of specific immunoglobulin E during maternal, fetal, and congenital toxoplasmosis. *J. Clin. Microbiol.* **37:**3487–3490.

81. **Warren, K. S.** 1982. Mechanisms of immunopathology, p. 116–137. *In* S. Cohen and K. S. Warren (ed.), *Immunology of Parasitic Infections.* Blackwell Scientific Publications Ltd., Oxford, United Kingdom.

82. **White, N. J.** 1998. Malaria pathophysiology, p. 371–385. *In* I. W. Sherman (ed.), *Malaria: Parasite Biology, Pathogenesis, and Protection.* ASM Press, Washington, D.C.

83. **Whittle, H. C., J. Brown, K. Marsh, M. Blackman, O. Jobe, and F. Shenton.** 1990. The effects of *Plasmodium falciparum* malaria on immune control of B lymphocytes in Gambian children. *Clin. Exp. Immunol.* **80:**213–218.

84. **Zhang, T., and S. L. Stanley, Jr.** 1999. DNA vaccination with the serine rich *Entamoeba histolytica* protein (SREHP) prevents amebic liver abscess in rodent models of disease. *Vaccine* **18:**868–874.

85. **Zuniga, E., C. Motran, C. L. Montes, F. L. Diaz, J. L. Bocco, and A. Gruppi.** 2000. *Trypanosoma cruzi*-induced immunosuppression: B cells undergo spontaneous apoptosis and lipopolysaccharide (LPS) arrests their proliferation during acute infection. *Clin. Exp. Immunol.* **119:**507–515.

22 Antibody and Antigen Detection in Parasitic Infections

Parasitic organisms infecting humans can be divided into two classes: (i) those that multiply within the host (e.g., protozoa) and (ii) those that mature within the host but never multiply (e.g., most helminths).

Infections caused by pathogenic protozoa, which multiply within the host, are similar to infections caused by bacterial, fungal, or viral pathogens. There is continuous antigenic stimulation of the host immune system as the infection progresses. In these instances, there is often a positive correlation between clinical symptoms and serologic test results.

Some helminths migrate through the body and pass through a number of developmental stages before becoming mature adults. With rare exceptions, these infections have been difficult to confirm serologically, possibly because of a limited antigenic response by the host or failure to use the appropriate antigen in the test system. Most antigens used in serologic procedures are heterogeneous mixtures that are not well defined. The result of using such antigens may be cross-reactions or inadequate sensitivity.

Although parasites and their by-products are immunogenic for the host, the host immune response is usually not protective. Any immunity that does develop is usually species specific or even strain or stage specific.

In certain parasitic infections, the standard diagnostic laboratory procedures may not be sufficient to confirm infection or specimen collection may not be practical or cost-effective. In these circumstances, alternative methods may be helpful; these include antibody, antigen, and nucleic acid detection. In some cases, serologic methods might be clinically indicated and may be very helpful, particularly if a parasitic infection is suspected and routine results are negative. However, even with the most sophisticated technology, few serologic tests for parasitic infections can be used to confirm an infection or predict the disease outcome. Interpretation of test results may also present problems, particularly when one is dealing with patients from areas of endemic infection, who may have higher baseline titers than do patients from other areas, in whom a low titer may actually be significant. Antibody detection generally indicates exposure to the parasite at some time in the past and may not necessarily reflect a current infection. This is particularly true when testing patients who have lived in an area of endemic infection for some time; their current clinical presentation may have no relationship to a positive antibody titer for a particular parasite. Although antibody levels generally decline over a period of months to years, serologic test results neither confirm nor rule out current infection or cure.

However, a positive serologic titer to a particular parasite in a patient who has had no previous exposure to the organism is clinically relevant and probably indicates recent exposure. The importance of a complete history, including both residence and travel information, is critical for accurate interpretation of serologic test results. Specific reagents available for serologic diagnostic testing are listed in Tables 22.1 and 22.2.

Although serologic test procedures have been available for many years, they are not routinely offered by most clinical laboratories for the reasons mentioned above (sensitivity, specificity, and interpretation). Standard techniques that have been used include complement fixation (CF), indirect hemagglutination (IHA), indirect fluorescent antibody (IFA), soluble-antigen fluorescent antibody, bentonite flocculation, latex agglutination (LA), double diffusion, counterimmunoelectrophoresis (CIE), immunoelectrophoresis, radioimmunoassay, and intradermal tests. Excellent reviews of these procedures, the rationale for their use, and their advantages and limitations are provided in references 49 and 91.

During the past few years, changes in the availability of certain tests have occurred, both at the Centers for Disease Control and Prevention (CDC) and at some private laboratories. The currently available antibody and antigen detection tests are listed in Table 22.1. A number of antibody and/or antigen detection tests have been developed in the research laboratory but are not yet available commercially; these include tests for *Ancylostoma caninum*, *Angiostrongylus*, *Anisakis*, *Baylisascaris*, *Gnathostoma*, microsporidia, *Opisthorchis*, and *Trypanosoma brucei* (91). Although PCR tests have been developed for most human parasites, they are generally available only through various research laboratories. Also, the only probe test currently available for parasitic infections is for the detection of *Trichomonas vaginalis*. However, as the costs of these tests decrease and automation of the various steps is developed, there will be more interest in incorporating these methods into the routine diagnostic laboratory. CDC offers a number of serologic test procedures for diagnostic purposes, some of which are not available elsewhere. Because regulations about submission of specimens may vary from state to state, each laboratory should check with its own county or state department of public health for the appropriate

Table 22.1 Serologic, antigen, and probe tests used in the diagnosis of parasitic infections

Disease	Routine antibody test(s)[a]	Antigen or probe test(s)
Protozoa		
Amebiasis	EIA	EIA, Rapid
Babesiosis	IFA	
Chagas' disease	EIA, IFA	
Cryptosporidiosis		DFA, EIA, IFA, Rapid
Giardiasis		DFA, EIA, Rapid
Leishmaniasis	IFA, EIA	
Malaria	IFA	Rapid
Microsporidiosis		IFA[b]
Toxoplasmosis	EIA, IFA, LA	
Trichomoniasis		DFA, LA, DNA probe
Helminths		
Cysticercosis	EIA, IB	
Echinococcosis	EIA, IB	
Fascioliasis	EIA, IB	
Filariasis	EIA	Rapid
Paragonimiasis	EIA, IB	
Schistosomiasis	EIA, IB	EIA
Strongyloidiasis	EIA	
Toxocariasis	EIA	
Trichinellosis	BF, EIA	

[a] BF, bentonite flocculation; DFA, direct fluorescent antibody; EIA, enzyme immunoassay; IB, immunoblot; Rapid, rapid immunochromatography (some are in cartridge formats).
[b] Reagents are not commercially available but have been developed (for microsporidia).

Table 22.2 Antibody detection kits available commercially within the United States for the diagnosis of parasitic diseases other than toxoplasmosis[a]

Disease (parasite)	Company[b]	Antibody test format[c]
Protozoa		
Amebiasis (*Entamoeba histolytica*)	Chemicon	EIA
	IVD Research	EIA
Chagas' disease (*Trypanosoma cruzi*)	Chemicon	EIA
	Hemagen Diagnostics	EIA
	IVD Research	EIA
	InBios	EIA, Rapid
Leishmaniasis, visceral (*Leishmania* spp.)	Immunetics	IB
	InBios	Rapid
	IVD Research	EIA
Helminths		
Cysticercosis (*Taenia solium*)	Chemicon	EIA
	Immunetics	IB
	IVD Research	EIA
	InBios	EIA
Echinococcosis (*Echinococcus granulosus*)	Immunetics	IB
	IVD Research	EIA
Echinococcosis (*E. multilocularis*)	Bordier Affinity	EIA
Schistosomiasis (*Schistosoma* spp.)	IVD Research	EIA
Strongyloidiasis (*Strongyloides stercoralis*)	IVD Research	EIA
Toxocariasis (*Toxocara canis*)	IVD Research	EIA
Trichinosis (*Trichinella spiralis*)	IVD Research	EIA

[a] Data from CDC.
[b] Bordier Affinity Products, Chatanerie 2, CH-1023, Crissier, Switzerland; Chemicon, 28835 Single Oak Dr., Temecula, CA 92950; Hemagen Diagnostics, 34–40 Bear Hill Rd., Waltham, MA 02154; Immunetics, 380 Green St., Cambridge, MA 02139; InBios, 562 1st Avenue South, Suite 600, Seattle, WA 98104; IVD Research, 5909 Sea Lion Place, Suite D, Carlsbad, CA 92008.
[c] IB, immunoblot; Rapid, rapid immunochromatographic.

instructions. Additional information on procedures, availability of skin test antigens, and interpretation of test results may be obtained directly from CDC by writing or calling

Serology Unit
   Parasitology Diseases Branch
Building 4, Room 1009
Mail Stop F13
Centers for Disease Control and Prevention
4770 Buford Highway
Atlanta, GA 30034
Serology: (770) 488-7760
Chagas' Disease and Leishmaniasis: (770) 488-4474
Malaria: (770) 488-7765.

General guidelines for specimen collection can be seen in Table 22.3.

Progress has been made in the development and application of molecular methods for diagnostic purposes, including the use of purified or recombinant antigens and nucleic acid probes. The detection of parasite-specific antigen is more indicative of current disease. Many of the assays were originally developed with polyclonal antibodies which were targeted to unpurified antigens

that markedly decreased the sensitivity and specificity of the tests. Immunoassays are generally simple to perform and allow a large number of tests to be performed at one time, thereby reducing overall costs. A major disadvantage of detection of antigens in stool specimens is that the method can detect only one or two pathogens at a time. One still must perform a routine ova and parasite examination to detect other parasitic pathogens. The current commercially available antigen tests have excellent sensitivity and specificity compared with routine microscopy (Tables 22.1 and 22.4).

Protozoal Infections

Amebiasis

The causative agent of amebiasis, *Entamoeba histolytica*, is found throughout the world and is a major infectious disease in developing countries. Although this parasite is considered mainly a cause of diarrhea, the trophozoite stage can readily invade tissues, resulting in the development of extraintestinal disease.

The diagnosis of infection with *E. histolytica*/*E. dispar* is usually based on finding either trophozoite or cyst stages

Table 22.3   Specimen collection and shipment for antibody and antigen detection tests

Type of test requested	Collection	Shipment
Antibody detection		All specimens can be shipped at room temperature to arrive within several days; if specimen is small (<1.0 ml), shipment with dry ice is recommended to prevent specimen evaporating during shipment; in general, acute- and convalescent-phase sera are not required; not all laboratories are equipped to perform testing on eye fluids or CSF; advisable to call to confirm availability if other than CDC
Most parasites	Serum or plasma acceptable	
Toxocariasis, toxoplasmosis	Aqueous and vitreous eye fluids accompanied by serum	
Central nervous system infections (cysticercosis, toxoplasmosis)	CSF accompanied by serum	
Antigen detection		
Fresh or formalinized stool	Routine stool collection vials	Some diagnostic kits require fresh stool; advisable to check with the laboratory prior to specimen submission

in the stool. However, unless the trophozoites contain ingested red blood cells (RBCs), morphologic differentiation between pathogenic *E. histolytica* and nonpathogenic *E. dispar* is not possible. Antigen detection may be helpful as an adjunct to routine microscopy, either for detection of the *E. histolytica*/*E. dispar* group or for detection of pathogenic *E. histolytica*. In extraintestinal amebiasis, one may or may not find organisms in the stool. The diagnosis of extraintestinal amebiasis depends largely on serologic tests.

The discovery that there are two distinct but morphologically identical species of *Entamoeba* infecting humans is important. *E. dispar* is associated with the asymptomatic carrier state and is the most prevalent species, whereas *E. histolytica* is the pathogenic form capable of invading tissue and causing symptomatic disease. Through the use of RNA and DNA probes, the genetic difference between these two species has been substantiated. At present, when a form resembling *E. histolytica* is detected in fecal material, it is not possible to differentiate it from *E. dispar* unless the trophozoite contains ingested RBCs.

Antibody Detection. If immunoglobulin G (IgG) antibodies specific to *E. histolytica* are detected, tissue invasion has occurred and may represent past or present disease. Serum antibodies can be detected in 85 to 95% or more of the patients with invasive disease. In newborns, maternal antibody disappears within 3 months, and persistence of antibody after this time may indicate active extraintestinal disease.

The most commonly used serologic tests are IHA, IFA, CF, enzyme-linked immunosorbent assay (ELISA), LA, CIE, and immunodiffusion (ID). CF tests have not been widely used because of nonspecific and variable results. IHA and LA appear to detect the same antibody. Although ELISA and IHA are two of the more sensitive tests, antibody titers persist for years, making interpretation difficult in certain patients. ELISA and IHA are useful for epidemiologic surveys. IFA and ID not only are sensitive tests but also have high positive predictive values.

After successful therapy, titers in many of the patients decrease, allowing physicians to monitor the therapeutic response. Although tests for the detection of IgM antibodies specific for *E. histolytica* have been reported, their clinical usefulness is questionable because of the lack of sensitivity. Many of the current tests use crude trophozoite extracts, whereas a few use recombinant proteins (77).

IHA has for many years been the standard for the diagnosis of extraintestinal amebiasis; titers of ≥1:256 are seen in 95% of patients with extraintestinal amebiasis, 70% of patients with active intestinal disease, and 10% of asymptomatic cyst passers. Unfortunately, titers can persist for many years after successful therapy, and so the presence of a titer does not automatically indicate active disease. However, false-positive results at titers of >1:256 rarely occur. The sensitivity of detection of specific antibodies to *E. histolytica* in serum is reported to be near 100%, which is promising for diagnosis of amebic liver abscess (ALA) (85). Serum anti-lectin IgG antibodies could be present within 1 week after the onset of symptoms in patients with amebic colitis and ALA, with a value over 95%. In most laboratories, EIA has replaced IFA; it is equally sensitive and specific.

Recombinant proteins such as the 170-kDa subunit of the galactose-inhibitable adherence lectin have been used to detect serum antibodies in more than 95% of patients with ALA. PCR and DNA hybridization assays have been used to detect specific gene sequences related to pathogenic *E. histolytica*; however, none are commercially available at this time.

Antigen Detection. Antigen-based fecal immunoassays have several significant advantages over other methods currently used for diagnosis of amebiasis: (i) some of the assays differentiate *E. histolytica* from *E. dispar*; (ii) they have excellent sensitivity and specificity; (iii) they are readily usable by most laboratory personnel; and (iv) these assays have potential use as screening tools in situations such as waterborne outbreaks. Because there are distinct genetic differences between *E. dispar* and *E. histolytica*,

Table 22.4 Antigen detection kits available commercially within the United States for the immunodetection of parasitic organisms or antigens[a]

Organism and kit name	Company[b]	Test format[c]
Cryptosporidium spp.		
Cryptosporidium	IVD Research	EIA
PARA-TECT Cryptosporidium	Medical Chemical	EIA
Cryptosporidium	Novocastra	DFA
ProSpecT	Remel	EIA
Xpect	Remel	Rapid
Cryptosporidium II	TechLab–Inverness Medical Professional Diagnostics (formerly Wampole)	EIA
Cryptosporidium spp. and *Giardia lamblia*		
ColorPAC	Becton Dickinson	Rapid
Crypto/Giardia	IVD Research	DFA, EIA
PARA-TECT Cryptosporidium/Giardia	Medical Chemical	DFA
Merifluor	Meridian Bioscience	DFA
ImmunoCardSTAT	Meridian Bioscience	Rapid
ProSpecT	Remel	EIA
Xpect	Remel	Rapid
Giardia/Cryptosporidium CHEK	TechLab–Inverness Medical Professional Diagnostics (formerly Wampole)	EIA
Cryptosporidium spp., *Giardia lamblia*, and *Entamoeba histolytica*		
Triage	BioSite	Rapid
Entamoeba histolytica		
E. histolytica/E. dispar	IVD Research	EIA
ProSpecT	Remel	EIA
Entamoeba histolytica II	TechLab–Inverness Medical Professional Diagnostics (formerly Wampole)	EIA
Giardia lamblia		
PARA-TECH Giardia	Medical Chemical	EIA
Giardia	Novocastra	DFA
ProSpecT	Remel	EIA
Xpect	Remel	Rapid
Giardia II	TechLab–Inverness Medical Professional Diagnostics (formerly Wampole)	EIA
Trichomonas vaginalis		
Affirm VPIII	Becton Dickinson	Probe
T. vaginalis	Chemicon	DFA
OSOM Trichomonas	Genzyme	Rapid
XenoStrip-Tv	Xenotope	Rapid

[a] Adapted from CDC.

[b] Becton Dickinson Diagnostic Systems, 7 Loveton Cir., Sparks, MD 21152; BioSite, 11030 Roselle St., San Diego, CA 92121; Chemicon, 28835 Single Oak Dr., Temecula, CA 92590; Genzyme Diagnostics, One Kendall Square, Cambridge, MA 02139; Inverness Medical Professional Diagnostics (formerly Wampole), P.O. Box 1001, Cranbury, NJ 08512; IVD Research, 5909 Sea Lion Place, Suite D, Carlsbad, CA 92008; Medical Chemical Corp., 19430 Van Ness Ave., Torrance, CA 90501; Meridian Bioscience, 3471 River Hills Dr., Cincinnati, OH 45244; Novocastra, 30 Ingold Rd., Burlingame, CA 94010; Remel, P.O. Box 14428, Lenexa, KS 66215; TechLab, VPI Research Park, 1861 Pratt Dr., Blacksburg, VA 24060; Xenotope Diagnostics, 3463 Magic Dr., Suite 350, San Antonio, TX 78229.

[c] DFA, direct fluorescent antibody; EIA, enzyme immunoassay; IFA, indirect fluorescent antibody; Rapid, rapid immunochromatographic.

commercial kits have been developed to detect their presence and differentiate them in clinical samples (Table 22.4). However, current antigen detection tests require the examination of fresh or frozen (not preserved) stool specimens, while many laboratories have switched to stool collection methods using various preservatives.

Other specimens in which amebic antigens have been detected include saliva, serum, and abscess fluid. Using ELISA, the adherence lectin antigen has been detected in saliva samples of ALA patients. This assay was found to be 22% sensitive and 97.4% specific. Amebic antigen in pus specimens from ALA patients was detected by ELISA (85).

A coagglutination test for the detection of circulating antigen in sera from patients with ALA has been reported (50). This is a simple and economical slide agglutination test. In patients with ALA, 45 (90%) of 50 sera were positive by the coagglutination test, while none of the 25 control sera were positive. However, two false-positive results were seen with two sera from patients with other parasitic and miscellaneous infection controls. Although this test is not yet commercially available, it could be used as a sensitive and specific rapid slide agglutination test for the presence of amebic antigen in the sera; it would also be useful in a routine clinical laboratory setting.

Babesiosis

Definitive diagnosis of babesiosis may be difficult because of the small number of infected RBCs, or the inability to discriminate between *Plasmodium falciparum* and *Babesia* spp. Babesiosis is transmitted by ticks, which can also transmit *Borrelia burgdorferi* and *Ehrlichia* spp. There are reports of coinfection with these three disease-causing agents (58). The serologic response to acute infection is marked by a rise in the titer to ≥1:1,024 during the first few weeks followed by a gradual decline to 1:16 to 1:256 over a period of 6 months. Infections caused by a strain of *Babesia* found in the western part of the United States may not yield a diagnostic titer (69) and would represent a false-negative result. When tested for babesiosis by IFA, patients infected with malaria have been noted to yield a serologic cross-reaction, generally of low titer; however, the test is 100% specific for patients with other tick-borne infections or patients with no exposure to the organism. PCR has also been used to detect infections (65). Despite the present unavailability of commercially available screening assays, some form of serologic and nucleic acid testing for *Babesia* spp. may be justified. Assuming that interactions between humans and ticks will probably increase in the future, it is likely that both new and extant tick-borne agents pose potential threats to transfusion safety (40).

Chagas' Disease

Trypanosoma cruzi infects millions of people in Central and South America and is a significant public health problem there. Screening of blood donors in areas where Chagas' disease is endemic is very important. In addition, donor centers are concerned about contamination of the blood supply when donors are migrant workers or immigrants who resided in areas where infection is endemic. Definitive diagnosis depends on demonstration of the parasite in blood or tissues. However, examination of blood may be helpful only during the initial acute disease or during chronic exacerbation. During chronic Chagas' disease, serologic testing may be the only method of detecting the infection.

Antibody Testing.   Serologic tests used for the diagnosis of Chagas' disease include CF (Guerreiro-Machado test), IFA, IHA, ELISA, and agglutination. Cross-reactions with low titers have been noted with leishmaniasis, infectious mononucleosis, lepromatous leprosy, pemphigus foliaceus, and *Trypanosoma rangeli* infections.

Although the IFA test is more sensitive than the CF test, the CF test is more specific. ELISA kits which use crude antigens have sensitivities and specificities similar to those of IFA. Use of synthetic peptides and recombinant antigens has improved the sensitivity and specificity of these diagnostic techniques (53, 64). Tri- and tetrapeptides, representing immunodominant epitopes of *T. cruzi*, have been used in the ELISA format to detect serum antibodies. These tests showed much higher rates of sensitivity and specificity and may serve as excellent alternatives to lysate for detection of anti-*T. cruzi* antibodies, as required for developing blood-screening assays (11, 43). Recombinant antigen-based assays have been used for the detection of IgM and IgA antibodies in congenital Chagas' disease and for the evaluation of therapy (56).

In one study to determine the prevalence of Chagas' disease among donors in five Mexican blood banks, two commercial serologic assays for Chagas' disease, the Abbott Chagas EIA and the Meridian Chagas' IgG ELISA, were tested. It is obvious from the data that transfusion-associated transmission of *T. cruzi* is occurring in the study areas. Serologic testing of blood donors for Chagas' disease should be performed there and in the rest of Mexico. However, the two screening assays evaluated may lack the accuracy necessary for blood donor testing when used as suggested by the manufacturers (52).

Routine blood donor screening for *T. cruzi* with a single test results in unacceptable numbers of false-negative samples in areas of highly endemic infection or in at-risk population groups. Adding a second test seems mandatory, but selection depends on local costs and feasibility (67). Because of the risk of transfusion-related infections and the lack of adequate screening tests to detect *T. cruzi*-positive donors, PCR may become the test of choice for screening (30).

Antigen Testing. Circulating-antigen detection methods have been used to detect recently acquired and congenital infections. PCR has greater sensitivity and specificity than routine methods for the diagnosis of Chagas' disease and can detect as few as one trypomastigote per 20 ml of blood (7). PCR has been very useful in diagnosing acute and chronic Chagas' disease and in monitoring therapy (48, 92).

Cryptosporidiosis

In addition to waterborne outbreaks, *Cryptosporidium* infections commonly occur in pediatric and immunocompromised individuals. Serologic tests for the diagnosis of this infection have not been useful; however, an immunoblot assay has shown specific antigen-antibody banding patterns in individuals presumed to be infected. Using the immunoblot assay as the "gold standard," new ELISAs have been shown to be more specific and, with the 27-kDa antigen ELISA, more sensitive than the crude oocyst antigen ELISA currently being used (68). Unfortunately, these reagents are not readily available. Detection of the organism in stool specimens relies on the use of special staining techniques or fecal immunoassays. Permanent stains normally used in ova and parasite examinations do not adequately stain *Cryptosporidium* spp. A number of commercially available immunoassay kits are available and are more sensitive and specific than routine microscopic examination (32–36) (Table 22.4). Stool specimens may be fresh, frozen, or fixed; however, polyvinyl alcohol-fixed specimens are usually unacceptable for use in the fecal immunoassays.

Cyclosporiasis

Currently, there are no immunodiagnostic tests available for cyclosporiasis. However, water concentrates have been subjected to community DNA extraction followed by PCR amplification, PCR sequencing, and computer database homology comparison for the detection of *Cyclospora cayetanensis* (24).

Giardiasis

Giardia lamblia can cause chronic diarrhea and malabsorption, especially in children. The organism is one of the most easily recognized parasites when detected in the stool; however, variability in the concentration of organisms in the stool makes this infection difficult to diagnose.

Antibody Detection. Serum antibodies to *G. lamblia* have been detected in infected patients; however, because of the lack of sensitivity, they cannot be recommended for diagnosis. After successful therapy, antibodies can be detected for as long as 15 months. Studies show intriguing associations between seroprevalence, outbreak-related laboratory serologic data, and patterns of parasite contamination of drinking water. However, additional studies are required to validate the serologic approach to risk assessment of waterborne parasitic infections at a community level (45).

Antigen Detection. Antigen detection in stool, duodenal fluids, and serum by DFA, IFA, ELISA, and immunoblotting methods has been reported (32–36). These tests are reliable and as sensitive as routine ova and parasite examinations. Commercial immunoassay (Table 22.4) kits are readily available. Users will have to evaluate which kit will be most useful for their own laboratories (19). Some of the methods may require fresh specimens, and stools fixed in preservatives may not be suitable. Also, some of the kits may not detect both trophozoites and cysts of *G. lamblia* but may be selective for only one life cycle stage. PCR and DNA probes have been tried on stool specimens but have had limited success because of difficulty encountered with sample preparation and lack of sensitivity.

Leishmaniasis

Leishmania spp. are obligate intracellular parasites that are transmitted to the mammalian host by the bites of infected sand flies. Leishmaniasis is mainly a zoonosis, although in certain areas of the world there is primarily human-vector-human transmission. The World Health Organization estimates that 1.5 million cases of cutaneous leishmaniasis (CL) and 500,000 cases of visceral leishmaniasis (VL) occur every year in 82 countries. Estimates indicate that approximately 350 million people are at risk for acquiring leishmaniasis, with 12 million currently infected.

Leishmaniasis refers to a diverse group of diseases, with the spectrum of disease depending on the infecting species. Disease syndromes range from self-healing cutaneous lesions to debilitating mucocutaneous infections to subclinical viscerotropic dissemination to fatal visceral involvement. Since disease presentations vary considerably, a well-defined classification system based on clinical findings is difficult and sometimes confusing. With recent outbreaks in many areas of the world, including Brazil, India, Italy, Spain, Sudan, and Kenya, leishmaniasis has become more widely recognized as an important emerging infectious disease in many developed as well as underdeveloped countries.

The taxonomy of *Leishmania* spp. is controversial and reflects the development of multiple parameters on which classification is based. Previously, organisms were classified based on the clinical disease picture and geographic locations; with the development of criteria based on biochemical and genetic data, the taxonomic groupings no longer coincide with recognized disease entities.

Antibody Detection.   A number of serologic diagnostic methods have been developed worldwide, although they are not widely available. In general, sensitivities are >90% but specificities are lower. Falset-positive results can be seen with other infectious diseases, following treatment for VL where antibodies persist for months, and in some asymptomatic patients. There are currently four tests in use, direct agglutination (DA), IFA, CIE, and ELISA. DA appears to be the best choice as a diagnostic tool, is very specific, and does not require expensive equipment or reagents. However, patients may still have positive sera for 1.5 to 5 years after treatment (61). DA has also been adapted to filter paper sampling, and the filter paper eluates compare well with their homologous sera and correlate strongly with antibody titers. However, newer studies using ELISA formats indicate excellent sensitivity and specificity and may be very helpful in early infections.

Currently, IFA is used to detect antibodies to *Leishmania* spp. but cannot differentiate among the various species. However, it can differentiate leishmaniasis from other similar diseases, although some cross-reactivity occurs with *T. cruzi*, the cause of Chagas' disease. Sensitivity is excellent for VL (>90%) but much less so for CL and mucocutaneous leishmaniasis. The humoral response in patients coinfected with human immunodeficiency virus and *Leishmania* is much lower than that in immunocompetent patients; the immunoblot method appears to be sensitive, noninvasive, and specific for the diagnosis of VL in immunocompromised patients (74).

A previously described ELISA that detects patient antibodies reactive with the recombinant *Leishmania* protein K39 (rK39) has been modified to confirm suspected kala azar and to detect asymptomatic infection in a community study in Bangladesh. The sensitivity and specificity of the modified ELISA for kala azar were 97.0% and 98.9%, respectively, for sera from the study population (54).

Antigenic properties of a chimeric recombinant antigen have been evaluated by indirect ELISA using a panel of human and canine sera previously characterized by parasitologic and/or serologic techniques. Chimeric ELISA showed 99% specificity in both human and canine control groups, while the sensitivity was higher in canine VL than in human VL. Results suggest the potential use of this new antigen for routine serodiagnosis of VL in both human and canine hosts (12).

Molecular detection methods are available in the research setting but are not yet available commercially. Detection of anti-K39 by immunochromatographic strip testing is a rapid and noninvasive method of diagnosing VL (kala azar), has excellent sensitivity (100%) and specificity (98%), and is well suited for use in field conditions (84). However, not all results using the rapid immunochromatographic strip are as promising (47). The K39 strip test is ideal for rapid and reliable field diagnosis of VL. The test has high sensitivity and specificity, but it remains positive up to 3 years after treatment.

Post-kala azar dermal leishmaniasis (PKDL) is a dermal complication of kala azar. Diagnosis of PKDL is difficult due to the low parasite burden in the lesions. The DA test was compared with those of the rK39 strip test. The sensitivities of the DA test for antileishmanial antibody detection, based on promastigote and amastigote antigens at a cutoff titer of 1:800, were 98.5% and 100%, respectively, with corresponding specificities of 96.5% and 100%. The DA test could correctly detect 100% of polymorphic cases and 95.4% of macular PKDL cases. In comparison, the rK39 strip test was able to correctly diagnose 95.6% of polymorphic and 86.0% of macular PKDL cases. A DA test based on axenic amastigote antigen provided 100% sensitivity and specificity, making it particularly useful for macular PKDL cases, which are often missed by the rK39 strip test. Thus, the DA test is simple, reliable, and inexpensive for PKDL diagnosis and has potential applicability in field conditions (79).

Antigen Detection.   Diagnosis of VL is usually confirmed by demonstration of organisms in tissue smears. However, obtaining the specimens may be difficult. Antibody-based diagnostics are limited by their inability to predict active disease. A new LA test (KAtex), which detects parasite antigen in freshly voided and boiled urine, has been evaluated for use with VL patients before and after treatment. KAtex appears to be a promising test, which in this simplified and improved format could be useful in the diagnosis of VL (83). Antigen detection in urine by KAtex is appropriate for primary diagnosis of VL, for monitoring the efficacy of treatment, and for detection of subclinical infection (72).

Circulating antigens in 35 serum samples from patients with VL were successfully identified using dot ELISA, giving 100% sensitivity. With the exception of serum samples from patients infected with *Plasmodium vivax*, which has shown cross-reactivity with *Leishmania donovani* antigens, the test appeared to be specific for the detection of circulating leishmanial antigens (8).

Malaria

Malaria is known to infect more than 250 million individuals throughout the world and kills more than 1 million people a year, most of whom are children. The vector for malaria is the female anopheline mosquito. When the vector takes a blood meal, sporozoites contained in the salivary glands of the mosquito are discharged into the puncture wound. Within an hour, these infective stages are carried via the blood to the liver, where they penetrate parenchymal cells and begin to grow, thus initiating the preerythrocytic or primary exoerythrocytic cycle. The sporozoites become round or oval and begin dividing repeatedly. This schizogony results in large numbers of exoerythrocytic merozoites. Once these merozoites leave the liver, they invade the RBCs, initiating the erythrocytic cycle. It has been reported that in *P. vivax* and *P. ovale*, a secondary or dormant schizogony may occur from organisms that remain quiescent in the liver until a later time. These resting stages are termed hypnozoites. Delayed schizogony does not occur in *P. falciparum* and probably does not occur in *P. malariae*.

Clinical symptoms of malaria include anemia, splenomegaly, and the classic paroxysm, with its cold stage, fever, and sweats. Although the febrile paroxysms strongly suggest infection, many patients who are seen in medical facilities, particularly in the early stages of the infection, do not exhibit the typical fever pattern. They may have fever or several small, random peaks each day. Since the symptoms associated with malaria are so nonspecific, the diagnosis should be considered in any symptomatic patient with a history of travel to an area where malaria is endemic. The typical paroxysm begins with the cold stage and rigors, which last 1 to 2 h. During the next few hours the patient spikes a high fever and feels very hot, and the skin is warm and dry. The last several hours are characterized by marked sweating and a subsequent drop in body temperature to normal or subnormal.

Giemsa-stained thick and thin blood films have been used to diagnose malaria for many years and have always been considered the "gold standard." However, a number of alternative methods have been developed, including different approaches to microscopy, flow cytometry, biochemical methods, immunoassay, and molecular methods. The aim of these procedures has been to reduce cost, reduce the need for expensive equipment, increase sensitivity, and provide simple, rapid methods that do not require conventional microscopy.

Antibody Detection. Serodiagnosis is not recommended for the diagnosis of malaria except in cases of transfusion reaction, where the parasitemia is so low that malaria transmission via blood transfusion cannot be documented. For a patient with repeated negative blood smears, serologic testing may be indicated, but the presence of antibody does not correlate with past or present infection.

In regions of low immunity, serologic assays offer an efficient method to identify infectious donors. The recent development of EIAs with improved sensitivity to *P. falciparum* and *P. vivax*, the predominant transfusion threats, has improved the possibility of serologic testing. Although universal serologic screening in regions where malaria is not endemic is not cost-effective, targeted screening of donors identified to be at risk based on travel-based questioning can significantly reduce wastage through reinstatement. Importantly, transfusion safety does not appear to be compromised by this approach, as evidenced by the lack of a documented transmission in France between 1983 and September 2002, where such a strategy has been used since 1976 (76).

Antigen Detection. The substitution of newer, more expensive drugs mandates rapid, accurate, and inexpensive diagnostic procedures so that directed therapy can be provided. A new *P. falciparum* antigen detection system, the ParaSight F test (Becton Dickinson), was found to be very effective in field trials. This procedure is based on an antigen capture approach and has been incorporated in a dipstick format; the entire test takes approximately 10 min. The ParaSight F+V assay improved the ParaSight F test format by incorporating a monoclonal antibody directed against a proprietary *P. vivax*-specific antigen, in addition to the antibody directed against *P. falciparum* histidine-rich protein 2 (HRP-2), which was used in the ParaSight F test. The final ParaSight F+V prototype, evaluated in 1999, had an overall sensitivity for detection of asexual *P. falciparum* parasites of 98%. The sensitivity of the device was 100% for *P. falciparum* densities of >500 parasites/µl, with a sensitivity of 83% for parasite densities of ≤500/µl. The specificity for the exclusion of *P. falciparum* was 93%. For *P. vivax*, the overall sensitivity was 87% for the final 1999 prototype. The sensitivities calculated for different levels of *P. vivax* parasitemia were 99% for parasite densities of >5,000/µl, 92% for parasite densities of 1,001 to 5,000/µl, 94% for parasite densities of 501 to 1,000/µl, and 55% for parasite densities of 1 to 500/µl. The specificity for the exclusion of *P. vivax* was 87% (29).

The second test that detects HRP-2, the ICT Malaria *Pf* (AMRAD ICT), is also in the dipstick format. This test appears to perform well and is comparable to the ParaSight F test; however, its false-positive rate appears to be lower. A rapid in vitro immunodiagnostic test, the NOW ICT Malaria Pf/Pv test kit, has been used for the detection of circulating *P. falciparum* (Pf) and *P. vivax* (Pv) antigens in whole blood. Although there was full concordance between the results of blood smear microscopy and rapid antigen testing, these techniques are most useful when there is a discrepancy with microscopy findings. Accurate and rapid identification of parasites, particularly in cases of mixed *P. falciparum* and *P. vivax* infection, is extremely important for patient management, particularly in regions of chloroquine resistance (10).

The third dipstick format procedure, OptiMAL (Flow Inc.), uses monoclonal antibodies against species-specific parasite lactate dehydrogenase (pLDH). Its sensitivity and specificity appear to be similar to those of the HRP-2 assays, with similar limitations at lower parasitemias. An advantage of this method is its use in follow-up of posttherapy patients to confirm a cure; the test detects only viable parasites. The OptiMAL-IT assay is more sensitive than the ICT Malaria Pf/Pv test for monitoring therapeutic responses after antimalarial therapy since the LDH activity ceases when the malarial parasite dies (44). In a multicenter trial, the sensitivity of the OptiMAL test was 98%; its specificity was 100%, with positive and negative predictive values of 100 and 99%, respectively. Participating hospital physicians and laboratory directors independently reported that the OptiMAL rapid malaria test was accurate, easy to use, and well accepted by the laboratory staff. The overall conclusion was that integration of the OptiMAL rapid malaria test into the U.S. health care infrastructure would provide an important and easy-to-use tool for the timely diagnosis of malaria (62).

Toxoplasmosis

Toxoplasma gondii is capable of infecting a wide range of hosts, including mammals, birds, and reptiles. The first recognized case of human toxoplasmosis was described by Janko, an ophthalmologist, in 1923. The infection can be contracted by ingestion of oocysts from fecal material (animals in the cat family, Felidae), ingestion of tissue cysts from raw or inadequately cooked meat, organ transplantation, transfusions, or transplacental transmission. Human infection occurs primarily through the ingestion of raw or undercooked infected meat.

Infection or colonization by *T. gondii* appears to be common in most parts of the world, particularly where cats are found. In the United States, approximately 30% of adults have been exposed to toxoplasmosis; however, the vast majority of these adults did not have recognizable illness.

Congenitally acquired toxoplasmosis can result in serious and debilitating sequelae, including stillbirths, defects seen at birth, and infections in infants who are frequently asymptomatic at birth. In the latter cases, sequelae of congenital infection may arise later in life, most commonly between the first and third decades (71).

The incidence of congenital toxoplasmosis is virtually unknown and may be greatly underreported. The clinical severity of congenital toxoplasmosis varies considerably with the trimester of pregnancy when the maternal infection was acquired. Infections contracted during the third trimester result in a large number of congenital infections; however, most infants show no sign of disease. Although infections acquired by the mother in the first trimester result in a smaller number of infected infants, sequelae are the most severe, including a number of stillbirths. Recent evidence suggests that most infants with subclinical infection at birth will develop signs or symptoms of disease (chorioretinitis) as they approach adulthood.

The diagnosis of acute *Toxoplasma* infection can be established by the following: isolation of tachyzoites from blood or body fluids; demonstration of tachyzoites in cytologic preparations or tissue sections; lymph node histologic testing; demonstration of cysts in such tissues as the placenta or in the fetus or neonate; antigen detection; or serologic test results. In older children or adults, the isolation of *T. gondii* from tissues may reflect only the presence of cysts. However, depending on the clinical presentation, the isolation of the organism from lymph node tissue may reflect the presence of tachyzoites.

Antibody Detection. Since isolation of the parasite is not practical for most laboratories, serologic tests are routinely used for diagnosis (Tables 22.5 and 22.6). Properly interpreted serologic test results provide excellent supportive evidence for the clinician. Classically, the detection of *Toxoplasma* infections has been based on the detection of IgG and IgM antibodies. Diagnostic reagent comparisons indicate that most are comparable in the detection of anti-*Toxoplasma* IgG but vary considerably in the detection of IgM antibodies. It has been shown that specific IgA antibodies are extremely useful in the diagnosis of congenital and acute toxoplasmosis (71). Another test for detecting acute infection involves the antigen binding avidity of IgG antibodies (55). Because of the large volume of requested serologic tests for the diagnosis of toxoplasmosis, a large variety of tests are available for the detection of antibodies. These include the Sabin-Feldman dye test, CF, IHA, IFA, agglutination, ELISA, and immunoblotting. The tests most commonly used today are IFA and ELISA for the detection of IgG and IgM antibodies and, more recently, IgA antibodies.

Serodiagnosis of acute toxoplasmosis using a recombinant form of the dense granule antigen GRA6-GST in an ELISA for IgG proved to be very useful. For discrimination between the presence and absence of acute toxoplasmosis, the assay reached a specificity of 99.6% in a group of 431 sera (70). Also, the assay showed good intra- and interassay reproducibility; however, it is not yet

Table 22.5 *Toxoplasma* kits available commercially in the United States[a]

Company[b]	IgG kits	IgM kits
Abbott Laboratories	IMx[c], AxSYM[c]	IMx, AxSYM
Bayer	Immuno 1[c], Advia[c]	Immuno1, Advia
Beckman	ACCESS[c]	ACCESS
Biokit	LA	
bioMérieux	VIDAS[c]	VIDAS
Bio-Rad	EIA	EIA
Biotecx	EIA	EIA
Biotest	EIA	EIA
Diagnostic Products Corp.	Immulite[c]	Immulite
Diamedix	EIA	EIA
DiaSorin	EIA	EIA
GenBio	EIA, IFA	EIA, IFA
Hemagen Diagnostics	EIA, IFA	EIA, IFA
Meridian Bioscience	EIA, IFA	EIA, IFA
Inverness Medical Professional Diagnostics (formerly Wampole)	EIA, IFA	EIA, IFA

[a] Information from CDC.

[b] Abbott Labs, Diagnostics Division, North Chicago, IL 60064; Bayer Diagnostics, 511 Benedict Ave., Tarrytown, NY 10591; Beckman Coulter, 4300 N. Harbor Blvd., Fullerton, CA 92834; Biokit USA, 113 Hartwell Avenue, Lexington, MA 02173; bioMérieux, 595 Anglum Dr., Hazlewood, MO 63042; Bio-Rad, 4000 Alfred Nobel Dr., Hercules, CA 94547; Biotecx Labs, 6023 S. Loop East, Houston, TX 77033; Biotest Diagnostics Corp., 66 Ford Rd., Suite 131, Denville, NJ 07834; Diagnostic Products Corp., 5700 W. 96th St., Los Angeles, CA 90045; Diamedix Corp., 2140 N. Miami Ave., Miami, FL 33127; DiaSorin, P.O. Box 285, Stillwater, MN 55082; GenBio, 15222 A. Avenue of Science, San Diego, CA 92128; Hemagen Diagnostics, 34–40 Bear Hill Rd., Waltham, MA 02154; Inverness Medical Professional Diagnostics (formerly Wampole), P.O. Box 1001, Cranbury, NJ 08512; Meridian Bioscience, 3471 River Hills Dr., Cincinnati, OH 45244.

[c] ACCESS, Advia, AxSYM, Immulite, IMx, and VIDAS are automated assays.

commercially available. In another study using recombinant antigens, the approach was effective in distinguishing *T. gondii*-infected from uninfected infants with congenital toxoplasmosis (17).

Individuals should be initially tested for the presence of *Toxoplasma*-specific antibodies to determine their immune system status. A positive IgG titer indicates infection at some point in time. Women who have a positive IgG titer before pregnancy have never been found to transmit *Toxoplasma* infection to the fetus, regardless of the antibody titer. Women who have no demonstrable antibody must be considered at risk for infection. Because the diagnosis of acute toxoplasmosis is usually based on serologic test results and is frequently considered late in the patient's clinical course, serologic titers may have already reached their peak. In these cases, examination of paired sera would be of little benefit and a single high titer may not indicate recent infection, since titers can remain high for years.

In acute and congenital toxoplasmosis, IgA and IgM antibodies are the first to be detected. Normally, IgM and IgA antibodies do not cross the placental barrier, and these antibodies have been used as the hallmark of congenital infection. The IgA and IgM titers continue to rise during the first few weeks of infection and may remain positive for a year or more (71, 91). It is possible that the IgA and IgM titers will decrease to undetectable levels by 5 or 6 weeks postinfection; however, this is not always the case, and high levels may indicate infection within the past 3 to 5 months (66). In some cases, IgM antibodies may be detected by EIA for as long as 18 months after acute acquired infection. IgM antibody methods have given both false-positive and false-negative results because of the presence of rheumatoid factor, blocking antibody, and the induction of fetal IgM antibodies against maternal immunoglobulin. The IgM test should be performed using a method such as the IgM capture ELISA to minimize nonspecific reactions (91). It is important to remember that all IgM-positive results should be verified by a reference laboratory such as the CDC or the Toxoplasmosis Laboratory, Palo Alto Medical Research Foundation, Palo Alto, Calif.

Generally, the demonstration of either IgA or IgM serum antibody is considered diagnostic; however, the use of cord serum is unreliable because of the possible presence of maternal antibody. Both maternal and infant sera should be monitored at the same time. If both sera are positive, the infant's serum should be retested within 1 to 2 weeks. An infant IgA or IgM titer which is stable or continues to rise is diagnostic of infection. Immunocompromised patients and those with ocular toxoplasmosis may have no demonstrable antibody re-

Table 22.6 Serologic tests for detection of the presence of *Toxoplasma* antibodies[a]

Antibody	Interpretation	Comments
IgG		
IgG negative	Patient has not been infected with *T. gondii*	IgG is the first test to be run in determining patient's immune status
IgG positive	Patient has been infected with *T. gondii* at some time in the past	If more information is required regarding the time of infection, IgM testing can be performed
IgM		
IgG negative, IgM positive	These results are highly suspicious	Results may represent a false-positive IgM result; redraw in 2 wk and retest both sera
IgG positive, IgM negative	Patient was infected >1 yr ago	

IgG positive, IgM low positive	May represent
	  False-positive result
	  Infection within the last 2 yr
	  New infection
		Confirm any positive IgM result at a *Toxoplasma* reference lab (CDC or Toxoplasmosis Laboratory in Palo Alto, Calif.)

IgG positive, IgM high positive	May represent
	  Infection within last 3–6 mo
	  False-positive result
		Confirm any positive IgM result as indicated above

[a] Data from reference 91.

sponse. AIDS patients with active toxoplasmosis, including central nervous system disease, generally have no IgM antibodies and a low to moderate level of (or occasionally no) IgG antibodies (57). Early treatment of the infection may considerably delay or reduce the normal host immune response.

Antigen Detection. Tests for the detection of circulating antigen have been used for the diagnosis of infection but have not been widely accepted (39, 71). Studies indicate that detection of some low-molecular-weight antigens is diagnostic of reactivated toxoplasmosis (4). These antigens can be detected even with normal dye test titers, and their detection improves the diagnosis of reactivated disease. They may be the result of the release of bradyzoites from ruptured tissue cysts. Because of the difficulties associated with reliable serologic methods for the diagnosis of toxoplasmosis, other methods such as molecular diagnostics have been used for diagnosis. PCR assays are available primarily in specialized research centers (41, 71). PCR will be a rapid and specific means of detecting the parasite in infected individuals who have failed to mount an immune response and allows for the infection to be detected much earlier in gestation.

A specific antibody and Western blot analyses have been used to demonstrate the presence of a highly reactive antigen of 36 kDa, not only in the extract of *T. gondii* tachyzoites but also in selected sera of women with confirmed laboratory and clinical signs of recent toxoplasmosis. The 36-kDa antigen was purified from *T. gondii* tachyzoites and human serum by using electro-

elution from preparative polyacrylamide gels. An ELISA format for the detection of target *Toxoplasma* antigen (TAg-ELISA) in human serum samples has recently been developed. The TAg-ELISA detected the target antigen in 88% of sera of acutely infected women, with 91% specificity among sera from noninfected women. The detection of 36-kDa *Toxoplasma* circulating antigen in human sera appears to be a promising alternative approach for the laboratory diagnosis of active *T. gondii* infection (5).

Trichomoniasis

The diagnosis of *Trichomonas vaginalis*, a sexually transmitted disease, relies primarily on direct microscopic examinations of wet mounts. Trophozoites can be detected in vaginal secretions, urethral specimens, or spun urine. The sensitivity of the direct wet mounts is low (50 to 85%) and can be improved by culture of the organism. DFA and LA reagents are now commercially available, and both can be used on vaginal swab specimens or fluids (15, 27). The automated Affirm VP Microbial Identification Test system is also available and is being used in physician office settings, clinics, and routine clinical laboratories. Using oligonucleotide probes, the system can detect *T. vaginalis* and *Gardnerella vaginalis* from a single vaginal swab. Sensitivity has been reported as 100% compared with wet preparation examination but only 80% compared with culture. Although there are still limitations for specimen storage conditions and times, one of the benefits is certainly flexibility for laboratories serving large outpatient populations.

Helminth Infections

Cysticercosis

The diagnosis of cysticercosis usually depends on surgical removal of the parasite and microscopic examination for the presence of suckers and hooks on the scolex. Multiple larvae are frequently present, and the presence of cysticerci in the subcutaneous or muscle tissues indicates that the brain is probably also involved. The calcified larvae can be readily seen on X-ray examination. Computed tomography (CT) scans or magnetic resonance imaging (MRI) techniques may reveal the presence of lesions in the brain. Although neurocysticercosis may be suspected, particularly in a patient living in an area of endemicity, the multiple neurologic syndromes and the various clinical presentations make it difficult to confirm the diagnosis, even with CT or MRI techniques. Human cysticercosis caused by *Taenia solium* is an important public health problem in developing nations.

Antibody Detection. Serologic tests used for diagnostic purposes include CF, IHA, IFA, ELISA, immunoblotting, ID, and radioimmunoassay (49, 80). IgG is the predominant Ig detected in patients with cysticercosis; IgA, IgE, and IgM antibodies are of little value in diagnosis and cannot be correlated with the patient's clinical condition. Serodiagnosis of cysticercosis has been a problem because of cross-reactions occurring in patients with coenurus, *Echinococcus*, filariasis, and *T. saginata* infections.

The CDC enzyme-linked immunoelectrotransfer blot (EITB) using purified *T. solium* antigens has been acknowledged by the World Health Organization and the Pan American Health Organization as the immunodiagnostic test of choice for the diagnosis of cysticercosis (90). It is 100% specific and has an extremely high sensitivity. The test employs purified glycoproteins, making it more sensitive than ELISA, and has eliminated the problems with nonspecificity noted above. Both serum and cerebrospinal fluid (CSF) can be used as diagnostic specimens. The test may be more sensitive with serum than CSF; therefore, there is no need to obtain CSF. For patients with multiple cysts (more than two cysticerci), the test has a sensitivity of ≥90%. Patients with multiple calcified cysts had a seropositivity of 82%, whereas patients with a single parenchymal cyst had a seropositivity of 50 to 70%. Several private laboratories now offer this test, which was previously available only at CDC. Several immunoassay kits are also commercially available (Table 22.2). The clinical tests for the diagnosis of cysticercosis cannot be used to distinguish between present and past infections, nor can they be used to monitor the progress of therapy.

Although the assay of choice for serologic detection of cysticercosis in humans and pigs is EITB, a Western blot assay that relies on the use of seven lentil-lectin-purified glycoproteins (LLGPs) derived from *T. solium* metacestodes, scarcity of native source material and the labor-intensive process of metacestode purification reduce its practicality. EITB antigens have now been reproduced in synthetic forms. Four chemically synthesized antigens were assayed individually using ELISA and Western blotting for immunoreactivity against sera from clinically defined neurocysticercosis patients. The sensitivity and specificity of all four antigens using the ELISA format did not meet the standards set by the LLGP EITB, while the results of the Western blot format were comparable to those of the LLGP EITB (75).

Specific *Taenia crassiceps* and *T. solium* antigenic peptides have been used for the serologic diagnosis of neurocysticercosis (16). Data indicate that the *T. crassiceps* ELISA and the immunoblotting test could be very useful in detecting relevant specific peptides (19 to 13 kDa) for *T. solium*.

Antigen Detection. An ELISA for the detection of antigen secreted by viable *T. solium* metacestodes has been used on pre- and posttreatment CSF samples from Peruvian patients with neurocysticercosis demonstrated by CT and EITB; the sensitivity was 86% (31). There was no difference between antigen levels in CSF taken before or 14 days after therapy. There was a significant positive correlation between antigen levels, the number of live cysts detected by CT, and the EITB results. In contrast, there was a negative correlation with the number of enhancing lesions by CT, supporting the hypothesis that enhancing lesions correspond to a late, dead stage of the parasite. This test approach using antigen specific for viable parasites has great clinical relevance, not only for providing information on the viability of the organisms but also for understanding the pathogenesis of neurocysticercosis before and after therapy.

Antigens and antibodies in sera from humans with epilepsy or taeniasis in an area of Mexico where *T. solium* cysticercosis is endemic have also been evaluated (21). Parasite-specific antigens and antibodies were more common among patients with epilepsy and taeniasis than among the controls. Antigens seemed to be associated with late-onset epilepsy, while the antibodies were associated with subcutaneous nodules. The sensitivities of both tests were low in terms of disease detection; however, the specificity and the positive predictive value of the antigen capture assay were high when used on sera from patients with late-onset epilepsy. Since late-onset epilepsy and neurocysticercosis are associated in areas of endemic infection, antigen capture assays may provide a reliable method of detecting active cases of neurocysticercosis.

Hydatid Disease

In an area of endemic infection, exposure to sheepdogs, the presence of an enlarged liver, the presence of a palpable mass in the right upper quadrant of the abdomen, expectoration of a salty-tasting fluid with hemoptysis, and a round lesion on chest radiographs all support the presumptive diagnosis of hydatid disease in either the liver or lungs. Hydatid cysts in the liver may be confused with congenital liver cysts, amebic abscess, or hepatic tumors. Pulmonary symptoms may be confused with tuberculosis. Hydatid cysts should be considered in patients with abdominal masses for which there is no clearly defined diagnosis. Many asymptomatic cysts are first discovered after radiologic studies.

Antibody Detection. Immunodiagnostic tests have been very helpful in confirming the diagnosis of echinococcosis caused by *Echinococcus granulosus* (cystic disease) or *E. multilocularis* (alveolar disease). Serum antibodies to *Echinococcus* spp. have been detected by IHA, IFA, ELISA, CF, LA, immunoelectrophoresis, immunoblotting, and ID. The immune response depends on the location and viability of the cyst. Individuals harboring calcified, senescent, or dead cysts often have a negative serologic test result. Liver cysts produce a greater serologic response in the patient than do hyaline or lung cysts, and up to 50% of patients with lung cysts are serologically negative. Cysts in the brain and spleen are also associated with a decreased immune response in the host.

A dipstick assay for the serodiagnosis of human hydatidosis using camel hydatid cyst fluid was compared to EITB and Falcon assay screening test-ELISA (FAST-ELISA). For the diagnosis of hydatidosis, the sensitivity, specificity, and diagnostic accuracy of the dipstick assay and EITB were 100, 91.4, and 95.1% while those of FAST-ELISA were 96.2, 100, and 98.4%, respectively. Since the dipstick assay is easy to perform, has a visually interpreted result within 15 min, and is both sensitive and specific, it could be an acceptable alternative for use in clinical laboratories lacking specialized equipment and the technological expertise needed for EITB and FAST-ELISA (2).

Serologic cross-reactions have been noted in low titers in patients with other parasitic diseases, liver cirrhosis, and collagen diseases (49). A test involving a selected antigen called arc-5 or the 8-kDa antigen has been developed for the specific diagnosis of *Echinococcus* infections. It is recommended that serum be screened by either ELISA or IHA and all positive reactions be confirmed by immunoblotting (91). False-positive reactions with the arc-5

antigen can occur with cysticercosis; however, use of the 8-kDa antigen in the immunoblot assay eliminates this cross-reactivity. Immunoblotting using the 8-kDa antigen provides 99% discrimination between seropositive preoperative cystic hydatid disease cases and cross-reactive noncestode parasitic infections or malignancies.

Cross-reactivity can occur between species. The use of purified *E. multilocularis* antigens with either ELISA or immunoblotting methods has been helpful in differentiating infections caused by the two species (38, 46, 88, 89). Immunoblot tests have been used to monitor the therapy of patients, whereas IHA and ELISA may continue to give positive results for years. Tests for the detection of circulating antigen have been used for diagnosis of infection, but they are not readily available.

Patients with alveolar hydatid disease are usually serology positive when heterologous *E. granulosus* or homologous *E. multilocularis* antigens are used; crude antigens tend to give nonspecific reactions in alveolar disease, as seen with cystic disease. Immunoaffinity-purified *E. multilocularis* antigens (Em2) used in an EIA method give positive results in >95% of cases of alveolar disease. Tests using Em2 are also more useful for monitoring postoperative conditions than for monitoring posttherapy effectiveness.

Antigen Detection. Efforts to standardize and evaluate the LA test as a simple test for the detection of circulating hydatid antigen in serum are under way. The latex particles were sensitized with hyperimmune hydatid antiserum raised in rabbits. The LA test showed a sensitivity of 72%, a specificity of 98%, a positive predictive value of 93%, and a negative predictive value of 91%. This is the first report of use of the LA test for the detection of hydatid antigen in serum in the diagnosis of cystic echinococcosis (23).

Fascioliasis

In *Fasciola hepatica* infections, metacercarial larvae do not cause significant pathologic damage until they begin to migrate through the liver parenchyma. The amount of damage depends on the worm burden of the host. Hyperplasia of the bile ducts occurs, possibly as a result of toxic substances produced by the larvae. Symptoms associated with this migratory phase have included fever, epigastric and right upper quadrant pain, and urticaria, although no eggs are found in the stool until the worms mature, which takes 8 weeks; some patients experience no symptoms.

Once the worms have established themselves in the bile ducts and matured, they cause considerable damage from mechanical irritation and metabolic by-products as well as obstruction. The degree of pathology depends on the number of flukes penetrating the liver. The infection produces hyperplasia of the biliary epithelium and duct fibrosis with portal or total biliary obstruction. The gallbladder undergoes similar pathologic changes and may even harbor adult worms. Adult worms may reinvade the liver parenchyma, causing abscesses.

Multiple stool examinations may be needed to detect light infections. Eggs may be detected in the stool of individuals who have eaten *F. hepatica*-infected liver, thereby yielding an erroneous laboratory result.

Antibody Detection. Serologic diagnosis using *F. hepatica* excretory-secretory antigens has been successfully used in areas of endemicity to detect patient antibodies. This approach includes both ELISA and EITB testing. Antibodies may be detected within 2 to 4 weeks after infection. Sensitivity for the FAST-ELISA was reported to be 95%, while sensitivity for the immunoblot assay using 12-, 17-, and 63-kDa antigens was 100%. However, there does appear to be some cross-reactivity with patients who have schistosomiasis. Serologic test results can be used to monitor treated patients to ensure successful therapy, after which antibody levels drop to normal within 6 to 12 months. Serologic tests are the tests of choice for the diagnosis of acute and chronic fascioliasis when few eggs can be found in the stool or when large populations are being screened.

An ELISA has been developed based on the detection of serum IgG4 antibodies reactive with *F. hepatica* cathepsin L1 (CL1). In an area of high prevalence of fascioliasis, a highly statistically significant correlation was seen when the recombinant and native proteins were used. The assays showed that 38 (59%) of the individuals tested were seropositive while only 26 were positive for *F. hepatica* eggs in the stool. All patients with egg-negative stools were also serologically negative. Also, there was no cross-reactivity with sera from patients with schistosomiasis, cysticercosis, hydatidosis, or Chagas' disease. This study indicates that recombinant CL1 shows excellent potential for use in the development of a standardized assay for the diagnosis of human fascioliasis (60). Other cysteine proteinases have also been used for the immunodiagnosis of human fascioliasis and have demonstrated excellent results (18, 20).

Immunoblotting studies showed that an 8-kDa protein reacts with sera from humans with fascioliasis but not sera from humans with other trematode infections. This result suggests that the 8-kDa protein of *F. hepatica* is one of the diagnostic antigens in human fascioliasis without cross-reaction and may be very useful in the diagnosis of fascioliasis (51).

Antigen Detection. The detection of coproantigens using an ES78 sandwich ELISA has been used; coproantigens were detected in all patients with patent infections. The data indicate that this approach is more sensitive than stool examinations for eggs, particularly in early infections (26).

Filariasis

Lymphatic Filariasis

A wide variety of serologic tests have been used for diagnostic and epidemiologic purposes in cases of human filariasis. Except for use in patients not native to the area of endemicity, immunodiagnostic tests are of limited value. Current tests lack both sensitivity and specificity, and most people from the region of endemicity have a positive serologic response. This response may be due to exposure to nonhuman filarial antigens from infected mosquitoes, and filarial antigens may cross-react with antibodies to other parasitic diseases.

Antibody Detection. The immune response of the filaria-specific IgG4 is considerably increased in patients with active filarial infections. The detection of IgG4 antibodies also reduces the cross-reactivity of nonfilarial antibodies. IgG2 production appears to be increased in patients with elephantiasis. Studies also suggest that specific antisheath antibodies are associated with an immunologic resistance mechanism that in the endemic community is expressed with highest prevalence in young individuals before the development of patent microfilaremia. However, antibody tests cannot differentiate past from present infection, nor do they show any correlation between antibodies and worm burden.

In the simple and rapid Brugia Rapid (BR) test, an immunochromatographic dipstick is used to detect IgG4 antibodies that are reactive with a recombinant *Brugia malayi* antigen. When sera from 109 individuals with *Brugia* microfilaremias (12 infected with *B. malayi* and 97 infected with *B. timori*) were investigated using the BR test, all were found to be positive. In contrast, all of the 150 sera from individuals with *Onchocerca volvulus* or *Mansonella* infections were found to be negative in BR tests. Some cross-reactions were observed with sera from individuals infected with *Wuchereria bancrofti* (3 of 12 test positive) and *Dirofilaria* (1 of 9 test positive) (28).

A rapid-format, simple, and qualitative flowthrough immunofiltration test has been developed for the identification of total IgG antibodies to recombinant filarial antigen WbSXP-1. This test system employs colloidal gold-protein A reagent as the antibody capture reagent. The sensitivity was 90.8% and 91.4% with brugian and bancroftian microfilaremic patients, respectively. The test showed minimum reactivity with *Loa loa* microfilaria-positive sera and no reactivity with *Onchocerca*-positive sera. The test was nonreactive with samples from individuals suffering from other parasitic diseases including schistosomiasis, soil-transmitted helminth infections, and protozoan infections, confirming the potential of this test as a diagnostic tool for both brugian and bancroftian lymphatic filariasis (9).

Antigen Detection. Because of the lack of sensitivity in direct-detection methods and problems with obtaining blood specimens within correct time frames, detection of circulating antigens has been used diagnostically. Antigen detection has also been used to monitor therapy. Many of these methods are not yet standardized and are not substitutes for direct-detection methods. Most are not sensitive enough to detect amicrofilaremic individuals who have clinical filariasis.

PCR assays will be a very useful diagnostic tool because they can discriminate between past and present infection and can be used to monitor therapy. In a preliminary evaluation of the diagnostic capability of PCR with diurnally collected sputum from bancroftian filariasis patients, the data were excellent; this approach appears to have excellent potential (1). New approaches for antigen detection also include (i) the ICT card test for serum specimens, (ii) the TropBio ELISA for serum specimens, and (iii) the TropBio ELISA for filter paper specimens; data from all three methods were very promising for their use as diagnostic tools in bancroftian filariasis (63, 78).

Onchocerciasis

The prepatent period of onchocerciasis in humans is 3 to 15 months. Light infections may cause no symptoms, and worms may be free in the tissue; however, in heavy infections, major disease manifestations include dermatitis, onchocercomas (subcutaneous nodules containing adult worms), lymphadenitis, and blindness. Even though there is an inflammatory reaction that causes the formation of a fibrotic capsule (onchocercoma) around the adult worm, the main pathology appears to be directed against the microfilariae.

Antibody Detection. Recombinant antigens for serologic diagnosis can help overcome the problems associated with poor sensitivity and specificity, particularly in areas where other human filarial infections overlap with those caused by *O. volvulus*. A cocktail of recombinant *O. volvulus* antigens for serologic diagnosis with the potential to predict the endemicity of onchocerciasis infection has been developed (14). When a sentinel population of 5- to 15-year-old individuals was used to compare communities, the serologic index was sufficiently sensitive to measure the changes in endemicity that would be required for control programs. However, serologic testing cannot

distinguish between current and past infection and tends to be more important in monitoring specific groups.

It has also been shown that tests based on IgE detection are more specific than those based on IgG detection. However, current tests still lack sensitivity and specificity, and the results do not correlate with past or present infection or with parasite burden.

Antigen Detection. Studies have been reported using recombinant proteins and a dipstick immunobinding assay (DSIA) for the detection of ocular microfilariae, a transblot immunobinding assay (TADA) for the detection of skin microfilariae, and a dot blot immunobinding assay (DIA) for the detection of urinary microfilariae (59). Sensitivities were 100% for DSIA, 82.5% for TADA used on samples of tears, 97% for TADA used on dermal fluids (skin snip), and 96% for DIA used on urine samples.

A recent report discussed the development and evaluation of a dot blot immunobinding assay (DIA-BA) based on the biotin-avidin binding system for the detection of *O. volvulus*-specific antigens in body fluids (87). The biotinylated probes were then used to detect *O. volvulus*-specific antigens initially blotted onto a nitrocellulose membrane. The smallest amount of blotted antigens detectable by the new test is 0.5, 1, 1, and 2 ng in urine, dermal fluid, tears, and serum samples, respectively. The test was 100% sensitive when used with urine and only 54.76% when used with serum from skin-snip-positive subjects. The specificity of the test was 100%, while the dermal fluid specificity was 97.5%. Also, there was a positive correlation between the color intensities on the blot and the skin microfilaria loads of the individuals. The DIA-BA test could be very useful for mass diagnosis of prepatent, low-level, and high-level infections due to *O. volvulus* (87).

Paragonimiasis

Humans become infected with *Paragonimus* spp. by ingesting uncooked crabs or crayfish containing metacercariae. The metacercariae excyst in the duodenum and migrate through the intestinal wall into the abdominal cavity. The larvae migrate around or through the diaphragm into the pleural cavity and the lungs. They mature to adults in the vicinity of the bronchioles, where they discharge their eggs into the bronchial secretions. Individuals with symptoms of chronic cough, vague chest pains, and hemoptysis who have resided in an area where infections are endemic and have a history of eating raw crayfish or crabs should be suspected of having paragonimiasis. *Paragonimus* eggs can be detected in sputum and in the stool, and concomitant examinations should be performed to improve the overall detection rate. In many individuals in whom the infection is eventually confirmed, small numbers of eggs are present intermittently in the sputum and feces. For patients with light infections, up to seven sputum examinations have been recommended. Frequently, pulmonary paragonimiasis is misdiagnosed as pulmonary tuberculosis. The Ziehl-Neelsen method for detecting mycobacteria destroys *Paragonimus* eggs.

Antibody Detection. In addition to routine diagnostic methods used to recover the characteristic eggs, immunodiagnostic tests have been used to diagnose pulmonary and extrapulmonary infections and to differentiate tuberculosis from paragonimiasis. CF has been used to diagnose active infections since it was noted that the test became negative soon after the death of the worms. An immunoblot test for the detection of antibodies to *P. westermani* using adult worm homogenates was found to be highly sensitive and specific; this test has been available at CDC since 1988. Only one false-positive result has been obtained with this method: a patient with *Schistosoma haematobium* infection.

EIA using adult excretory-secretory antigens was found to be very sensitive for detecting parasite-specific IgG and IgE. Pleural effusion fluid may be more suitable than serum for detection of infections. Dot ELISA has also been used to detect parasite-specific antigen in human sera. Monoclonal antibodies to detect species-specific and stage-specific antigens are highly sensitive and specific for the detection of active infections (25). The serology-based assays are available in areas of endemicity or in specialized diagnostic centers. Most of these assays use nonstandardized reagents and have not been used in clinical trials.

An intradermal test is also available for epidemiologic surveys in areas of endemicity; however, positive reactions persist for years after parasitologic cures have been effected. Purification of antigen for intradermal tests does not improve the efficacy of screening for paragonimiasis, probably due to the nature of skin sensitization evoked by many elements of the parasite but not by a few selected components. The crude worm antigen may be valuable for screening a large population, yet the diagnosis should not depend totally on this test. The final diagnosis should be made by repeated sputum examinations over time, a comprehensive patient history, serologic testing, and radiologic testing if necessary.

Schistosomiasis

Humans become infected with *Schistosoma* spp. by penetration of cercariae through intact skin rather than by the ingestion of metacercariae. Once the cercariae have successfully entered the host, the organism is termed a schistosomulum. After about 48 h, the schistosomulum migrates through the tissues and finally invades a blood vessel. On entry into the blood vessels, the schistosomulum is carried to the lungs and then the liver. Once within the liver sinusoids, the worms begin to mature into adults. Depending on the species, the adult worms are found in various veins. The worms form pairs (male and female), with the female lying in the gynecophoral canal of the male. Development of sexual maturity of female schistosomes depends on the presence of mature male worms.

The eggs are immature when first laid and take approximately 8 to 10 days to develop a mature miracidium. Egg deposition takes place intravascularly. Many of the eggs laid are swept downstream, where they become lodged in the microvasculature of the liver and other organs. About half of the eggs swept away by the bloodstream become embedded in the mesenteric venule wall. The presence of eggs in the tissues stimulates granuloma formation; the eggs die, calcify, and are absorbed by the host eventually.

Disease syndromes associated with schistosomiasis are related to the stage of infection, previous host exposure, worm burden, and host response. Syndromes include cercarial dermatitis, acute schistosomiasis (Katayama fever), and related tissue changes resulting from egg deposition.

Specific diagnosis of schistosomiasis by detection of eggs in stool or urine specimens is possible only after egg production has begun. Eggs may be found in feces as early as 5 weeks after infection. The ease of egg detection depends on the worm burden and the duration of the infection. Patients with a low worm burden or old (chronic) infections may have very few eggs in their feces or urine, and the infections may not be confirmed due to the use of insensitive diagnostic methods.

Antibody Detection. A large number of serologic tests have been used in the diagnosis of schistosomiasis; however, most have not been particularly useful because of cross-reactions with other helminth infections, continuation of elevated titers long after the completion of successful treatment, and the slow immunologic response of the host. Because of the complex life cycle, a large number of antigens have been used in serologic tests. Only a few have been useful. The most frequently used tests are the circumoval precipitin test, the cercaria-Hullen reaction, IFA, IHA, immunoblotting, EIA, and FAST-ELISA.

Currently at CDC, specimens are first tested by FAST-ELISA with *S. mansoni* adult microsomal antigen (MAMA). A positive reaction (>8 U/µl of serum) indicates infection with *Schistosoma* species. The sensitivity is 99% for *S. mansoni* infection, 95% for *S. haematobium* infection, and ≥50% for *S. japonicum* infection. The specificity is 99%. Because of the lower sensitivity for *S. haematobium* and

S. japonicum, immunoblot analyses relevant to the patient's travel history are performed. As with many other parasitic infections, the presence of antibody indicates infection at some time in the past and cannot be correlated with clinical status, parasite burden, egg production, or prognosis.

EIA methods with soluble adult worm or egg antigens for the detection of antibodies have been used as screening tests, with confirmation of infection performed by EITB tests. Individuals infected with schistosomes have an elevated IgE and IgG4 response, as is commonly seen with other helminthic infections. Few studies have investigated the persistence of antibodies over time after therapy. It is possible that high titers posttherapy reflect treatment failure or reinfection.

The dipstick dye immunoassay test (DDIA), developed in China for the detection of antibodies against *S. japonicum*, relies on soluble egg antigen (SEA) labeled with a colloidal dye. This assay is not only rapid, simple, and inexpensive but also particularly useful for screening in the field. Sensitivity and specificity results support the notion that the DDIA with *S. japonicum* SEA can safely be implemented for the diagnosis of *S. mekongi*; however, careful interpretation of results is mandatory when sample are obtained from areas where *Opisthorchis viverrini* is also endemic (93).

Antigen Detection. ELISA methods have also been used to detect circulating schistosome antigens in the sera and urine of infected patients and may be the preferred method of diagnosis. In addition to diagnosis of infection, antigen levels have been used to assess disease severity in terms of intensity of infection and to monitor therapy. Individuals treated with praziquantel experienced a rapid clearance of circulating anodic antigen (CAA) and circulating cathodic antigen (CCA); however, in individuals treated with other therapeutic drugs, clearance may take much longer. A dipstick assay based on the anti-CCA sandwich ELISA, on a nitrocellulose format, can detect CCA in the urine of patients with *S. mansoni* infection with a sensitivity of 92%. The results were at least as good as those obtained using ELISA or the Kato-Katz fecal smear examination. This rapid, simple approach may lend itself to more extensive use in future screening programs (82). The combination of urine CCA and serum CAA for the detection of circulating antigens and the combination of the *S. haematobium* adult worm microsomal antigen (HAMA) FAST-ELISA and the HAMA EITB for detecting antibodies has significantly improved the sensitivity of antibody and antigen detection (3). However, while CCA dipsticks are a good alternative, or complement, to stool microscopy for field diagnosis of intestinal schistosomiasis, they have no proven value for field diagnosis of urinary schistosomiasis (81). Another reagent strip assay

for the diagnosis of schistosomiasis based on detection of parasite antigens in the urine of infected individuals has been evaluated. The test uses the principle of lateral flow through a nitrocellulose strip of the sample mixed with a colloidal carbon conjugate of a monoclonal antibody specific for *Schistosoma* CCA. The strips and the conjugate in the dry format are stable for at least 3 months at ambient temperature in sealed packages, making the test suitable for transport and use in areas where schistosomiasis is endemic (86).

Using an SEA monoclonal antibody, circulating schistosome antigen was undetectable in serum from patients with other parasitic infections and in 94% of serum samples and 84% of urine samples from *S. mansoni*-infected patients. A significant reduction in circulating schistosome antigen levels was detected in serum and urine samples approximately 12 weeks after praziquantel therapy (73). In another study using a 63-kDa circulating antigen, the antigen was detected in 92% of urine samples from 330 *S. mansoni*-infected individuals, with a 16% false-positive result among 130 noninfected individuals (6).

Strongyloidiasis

Human infection with *Strongyloides stercoralis* is acquired by skin penetration by the filariform larvae (infective larvae) from the soil. These larvae are long and slender and may remain viable in soil or water for several days. After penetration of the skin, they are carried via the cutaneous blood vessels to the lungs, where they break out of the pulmonary capillaries into the alveoli. They then migrate via the respiratory tree to the trachea and pharynx, are swallowed, and enter the mucosa in the duodenum and upper jejunum. Development usually takes about 2 weeks, after which the females begin egg production. The eggs usually hatch, and the rhabditiform larvae (noninfective larvae) pass out of the intestinal tract in the feces. These larvae are passed out onto the soil in the feces and develop into free-living male and female worms, eventually producing infective filariform larvae (eggs, noninfective larvae, and infective larvae). In temperate climates, the free-living male and female worms do not develop; however, the rhabditiform larvae that pass out in the stool develop into the filariform (infective) larvae, which are then ready to infect the next host through skin penetration.

In situations in which autoinfection occurs, some of the rhabditiform larvae within the intestine develop into filariform larvae while passing through the bowel. These larvae can then reinfect the host by (i) invading the intestinal mucosa, traveling via the portal system to the lungs, and returning to the intestine or (ii) being passed out in the feces and penetrating the host on reaching the perianal or perineal skin.

Pathology present in strongyloidiasis can vary, both in severity and in the areas of the body involved. Some individuals remain totally asymptomatic, with the only abnormal clinical finding being a peripheral eosinophilia. Considering the life cycle of the parasite, the three areas of primary involvement would be the skin (cutaneous), the lungs (pulmonary), and the intestinal mucosa (intestinal).

Confirmation of the infection depends on recovery and identification of the adult worms, larvae, or eggs from the stool, duodenal material, or sputum. Fecal specimens may not be positive even after the routine ova and parasite examination, including the concentration procedure, is performed. The numbers of larvae found in the stool also vary from day to day.

Antibody Detection.   Although not routinely available, EIA and immunoblot methods have been investigated for use in both the diagnosis and postchemotherapy evaluation of human strongyloidiasis. These methods appeared to be both sensitive and specific, and within about 6 months after therapy they demonstrated a significant decrease in antibody levels. However, in cases where treatment was not effective, the individuals did not show a significant fall in antibody titers. More detailed studies to determine appropriate intervals for serologic evaluation and criteria for successful cure are required before these tests become more widely used.

Currently, EIA, with a sensitivity of 84 to 92%, is recommended. Some patients may be negative, particularly those who are carriers (8 to 16%), and antibody test results cannot be used to differentiate between past and current infection. It is interesting that immunocompromised patients still tend to have detectable IgG antibodies, in spite of their immune system deficiencies.

Strongyloides ratti larval extract was used for the standardization of ELISA to detect genus-specific IgE in humans with strongyloidiasis. A significant positive correlation was found between levels of *Strongyloides*-specific IgE and total IgE in sera from patients with strongyloidiasis. *S. ratti* heterologous extract could be a useful tool for detecting genus-specific IgE by ELISA, thus helping characterize the immune response profile in human strongyloidiasis (22).

Toxocariasis

Toxocara spp. are non-human-derived ascarids that are capable of undergoing limited development in the human host. Infections with the organism have been classified into two categories: visceral larva migrans (VLM), which is a systemic infection occurring in patients with a mean age of 4.6 ± 3.6 years; and ocular larva migrans (OLM),

which also occurs in children. However, concurrent infection with VLM and OLM in the same patient is rare. Eosinophilia, which is pronounced in VLM, is almost absent in OLM. The average age of patients with OLM is also older than that of patients with VLM.

Antibody Detection.   The clinical diagnosis of toxocariasis can be very difficult, and only fairly recently has a reliable immunodiagnostic test been developed. Serologic test results must be interpreted with care because of the prevalence of asymptomatic infection and the age and socioeconomic status of the individual (42). Previous serologic tests had involved adult- or larval-stage somatic extracts and whole or sectioned larvae. The EIA uses third-stage-larva secretory antigen and is superior to all other serodiagnostic tests. The test is highly specific and does not give cross-reactions with sera obtained from patients infected with other commonly occurring human parasites.

Even though EIA is an extremely sensitive test, the diagnostic predictive value for VLM is significantly different from that for OLM (91). A titer of 1:32 is considered diagnostic for VLM (sensitivity, 78%), whereas a titer of 1:8 is considered diagnostic for OLM (sensitivity, 90%). The specificity is >90% at a titer of ≥1:32. This lower predictive titer for OLM raises the possibility of a false-positive diagnosis in a patient who has an asymptomatic *Toxocara* infection and ocular disease due to other etiology. In addition to serum antibody, vitreous antibody can be measured. Titers in the vitreous fluid were equal to or greater than the serum titers. Measurement of ocular fluid antibody should increase the specificity of the EIA, yielding a better definitive diagnosis for the patient.

Trichinellosis

Human infection with *Trichinella spiralis* is initiated from the ingestion of raw or poorly cooked pork, bear, walrus, or horse meat or meat from other mammals (carnivores and omnivores) containing viable, infective larvae. The tissue is then digested in the stomach. The excysted larvae invade the intestinal mucosa, develop through four larval stages, mature, and mate by the second day. By the sixth day of infection, the female worms begin to deposit motile larvae, which are carried by the intestinal lymphatic system or mesenteric venules to the body tissues, primarily striated muscle. Deposition of larvae continues for approximately 4 weeks, with each female producing up to 1,500 larvae in the nonimmune host.

The very active muscles, which have the greatest blood supply, including the diaphragm, muscles of the larynx, tongue, jaws, neck, and ribs, the biceps, and the gastrocnemius, are invaded. The cyst wall results from the host immune response to the presence of the larvae, and the encysted larvae may remain viable for many years, although calcification can occur within less than a year. As few as 5 larvae per g of body muscle can cause death, although 1,000 larvae per g have been recovered from individuals who died from causes other than trichinosis.

Symptoms that may develop within the first 24 h include diarrhea, nausea, abdominal cramps, and general malaise, all of which may suggest food poisoning, particularly if several people are involved. Studies also indicate that the diarrhea can be prolonged, lasting up to 14 weeks (average, 5.8 weeks), with few or no muscle symptoms.

During muscle invasion, there may be fever, facial (particularly periorbital) edema, and muscle pain, swelling, and weakness. The extraocular muscles are usually the first to be involved, followed by the muscles of the jaw and neck, limb flexors, and back. Muscle damage may cause problems in chewing, swallowing, breathing, etc., depending on which muscles are involved. The most severe symptom is myocarditis, which usually develops after the third week; death may occur between the fourth and eighth weeks. Other severe symptoms, which can occur at the same time, may involve the central nervous system. Although *Trichinella* encephalitis is rare, it is life-threatening.

Antibody Detection.   Serologic tests are very helpful, with the standard two being EIA and bentonite flocculation, which are recommended for trichinosis. EIA detects antigen first and is used for routine screening, and all EIA-positive specimens are tested by bentonite flocculation for confirmation. A positive reaction in both tests indicates infection with *T. spiralis* within the last few years. Often, antibody levels are not detectable within the first month postinfection. The titers tend to peak in the second or third months postinfection and then decline over a few years. Positive reactions are detected at some point during disease progression in 80 to 100% of all patients with symptomatic disease; detectable antibody usually appears between 3 and 5 weeks after infection.

Antigen Detection.   A modified double-sandwich ELISA has been developed using polyclonal antibodies raised in rabbits and guinea pigs against larval somatic antigens of *T. spiralis*. The first detection of coproantigen was as early as the first day postinfection; the level of coproantigen gradually increased to reach its peak on the seventh day and then decreased to disappear completely during the third week postinfection. Another test, the coagglutination test, was used, and this test confirmed the previous results. Results suggest that the coproantigen

detection test could be used to confirm ongoing early *T. spiralis* infection (13).

Intradermal Tests

In the absence of reliable serologic diagnostic tests, skin tests have been used to provide indirect evidence of infection. Most skin tests have been used primarily for research and epidemiologic purposes. Some of the more widely used skin tests are the Casoni and Montenegro tests. In many cases, the antigens used are difficult to obtain and are not commercially available. The antigens are usually crude extracts that have not been standardized and are neither highly sensitive nor specific. They may provoke an immune response that complicates further serologic testing, and there is always the danger of provoking an anaphylactic reaction. In addition, there are ethical questions related to giving patients injections of nonstandardized foreign protein, particularly if the antigens were derived from in vivo materials.

Casoni Test

In the past, the Casoni test was the only means of diagnosing exposure to hydatid disease. This intradermal test, in which hydatid cyst fluid is used as the antigen, has many advantages as a result of its simplicity; its nonspecificity has been its major limitation. Diagnostic sensitivity was limited in patients who had intact or hyaline cysts. An immune response was detected more frequently in patients with liver cysts than in those with lung cysts.

False-positive reactions have occurred in patients with other parasitic and nonparasitic diseases. The Casoni antigen may also sensitize the patient, leading to antibody production, and anaphylactic reactions have also been reported. This test is rarely used for diagnostic purposes because of the ethical concerns addressed above.

Casoni's skin test and IHA are still used in Turkey. For 120 patients retrospectively studied during 1997 to 2004, results indicated that an immediate skin reaction to crude hydatid antigens was more useful than IHA (37).

Montenegro Test

The Montenegro test uses formalinized promastigotes of any species of *Leishmania* to evoke a delayed hypersensitivity reaction in infected patients. In some areas, it is the method of choice for diagnosis because of its simplicity.

Patients infected with *Leishmania donovani* (kala azar) have a negative Montenegro test during active disease. The test becomes positive after successful treatment or when there is a spontaneous cure. Problems similar to those mentioned above for the Casoni test have occurred with the Montenegro test. The skin test may induce antibody production and may also cause anaphylactic reactions.

References

1. **Abbasi, I., J. Githure, J. J. Ochola, R. Agure, D. K. Koech, R. M. Ramzy, S. A. Williams, and J. Hamburger.** 1999. Diagnosis of *Wuchereria bancrofti* infection by the polymerase chain reaction employing patients' sputum. *Parasitol. Res.* **85**:844–849.

2. **Al-Sherbiny, M. M., A. A. Farrag, M. H. Fayad, M. K. Makled, G. M. Tawfeek, and N. M. Ali.** 2004. Application and assessment of a dipstick assay in the diagnosis of hydatidosis and trichinosis. *Parasitol. Res.* **93**:87–95.

3. **Al-Sherbiny, M. M., A. M. Osman, K. Hancock, A. M. Deelder, and V. C. Tsang.** 1999. Application of immunodiagnostic assays: detection of antibodies and circulating antigens in human schistosomiasis and correlation with clinical findings. *Am. J. Trop. Med. Hyg.* **60**:960–966.

4. **Ashburn, D., M. M. Davidson, A. W. Joss, T. H. Pennington, and H. Y. Do.** 1998. Improved diagnosis of reactivated toxoplasmosis. *Mol. Pathol.* **51**:105–109.

5. **Attallah, A. M., H. Ismail, A. S. Ibrahim, L. A. Al-Zawawy, M. T. El-Ebiary, and A. M. El-Waseef.** 2006. Immunochemical identification and detection of a 36-kDa *Toxoplasma gondii* circulating antigen in sera of infected women for laboratory diagnosis of toxoplasmosis. *J. Immunoassay Immunochem.* **27**:45–60.

6. **Attallah, A. M., E. Yones, H. Ismail, S. A El Masry, A. Tabil, A. A. Elenein, and N. A. El Ghawalby.** 1999. Immunochemical characterization and diagnostic potential of a 63-kilodalton *Schistosoma* antigen. *Am. J. Trop. Med. Hyg.* **60**:493–497.

7. **Avila, H. A., D. S. Sigman, L. M. Cohen, R. C. Millikan, and L. Simpson.** 1991. Polymerase chain reaction amplification of *Trypanosoma cruzi* kinetoplast minicircle DNA isolated from whole blood lysates: diagnosis of chronic Chagas' disease. *Mol. Biochem. Parasitol.* **48**:211–222.

8. **Azazy, A. A.** 2004. Detection of circulating antigens in sera from visceral leishmaniasis patients using dot-ELISA. *J. Egypt. Soc. Parasitol.* **34**:35–43.

9. **Baskar, L. K., T. R. Srikanth, S. Suba, H. C. Mody, P. K. Desai, and P. Kilaraj.** 2004. Development and evaluation of a rapid flow-through immuno filtration test using recombinant filarial antigen for diagnosis of brugian and bancroftian filariasis. *Microbiol. Immunol.* **48**:519–525.

10. **Beg, M. A., S. S. Ali, R. Haqqee, M. A. Khan, Z. Qasim, R. Hussain, and R. A Smego, Jr.** 2005. Rapid immunochromatography-based detection of mixed-species malaria infection in Pakistan. *Southeast Asian J. Trop. Med. Public Health* **36**:562–564.

11. **Bentonico, G. N., E. O. Miranda, D. A. Silva, R. Houghton, S. G. Reed, A. Campos-Neto, and J. R. Mineo.** 1999. Evaluation of a synthetic tripeptide as antigen for detection of IgM and IgG antibodies to *Trypanosoma cruzi* in serum samples from patients with Chagas' disease or viral diseases. *Trans. R. Soc. Trop. Med. Hyg.* **93**:603–606.

12. **Boarino, A., A. Scalone, L. Gradoni, E. Ferroglio, F. Vitale, R. Zanatta, M. G. Giuffrida, and S. Rosati.** 2005. Development of recombinant chimeric antigen expressing immunodominant B epitopes of *Leishmania infantum* for serodiagnosis of visceral leishmaniasis. *Clin. Diagn. Lab. Immunol.* **12**:647–653.

13. **Boulos, L. M., I. R. Ibrahim, A. Y. Negm, and S. M. Aly.** 2001. Detection of coproantigen in early trichinellosis. *Parasite* **8**:S136–S139.

14. **Bradley, J. E., B. M. Atogho, L. Elson, G. R. Stewart, and M. Boussinesq.** 1998. A cocktail of recombinant *Onchocerca volvulus* antigens for serologic diagnosis with the potential to predict the endemicity of onchocerciasis infection. *Am. J. Trop. Med. Hyg.* **59**:877–882.

15. **Briselden, A. M., and S. L. Hillier.** 1994. Evaluation of Affirm VP microbial identification test for *Gardnerella vaginalis* and *Trichomonas vaginalis*. *J. Clin. Microbiol.* **32**:148–152.

16. **Bueno, E. C., A. J. Vaz, L. D. Machado, J. A. Livramento, and S. R. Mielle.** 2000. Specific *Taenia crassiceps* and *Taenia solium* antigenic peptides for neurocysticercosis immunodiagnosis using serum samples. *J. Clin. Microbiol.* **38**:146–151.

17. **Buffolano, W., E. Beghetto, M. Del Pezzo, A. Spadoni, M. Di Christina, E. Petersen, and N. Gargano.** 2005. Use of recombinant antigens for early postnatal diagnosis of congenital toxoplasmosis. *J. Clin. Microbiol.* **43**:5916–5924.

18. **Carnevale, S., M. I. Rodriguez, E. A. Guarnera, C. Carmona, T. Tanos, and S. O. Angel.** 2001. Immunodiagnosis of fasciolosis using recombinant procathepsin L cysteine proteinase. *Diagn. Microbiol Infect. Dis.* **41**:43–49.

19. **Church, D., K. Miller, A. Lichtenfeld, H. Semeniuk, B. Kirkham, K. Laupland, and S. Elsayed.** 2005. Screening for *Giardia/Cryptosporidium* infections using an enzyme immunoassay in a centralized regional microbiology laboratory. *Arch. Pathol. Lab. Med.* **129**:754–759.

20. **Cordova, M., L. Reategui, and J. R. Espinoza.** 1999. Immunodiagnosis of human fascioliasis with *Fasciola hepatica* cysteine proteinases. *Trans. R. Soc. Trop. Med. Hyg.* **93**:54–57.

21. **Correa, D., E. Sarti, R. Tapia-Romero, R. Rico, I. Alcantara-Anguiano, A. Salgado, L. Valdez, and A. Flisser.** 1999. Antigens and antibodies in sera from human cases of epilepsy or taeniasis from an area of Mexico where *Taenia solium* cysticercosis is endemic. *Ann. Trop. Med. Parasitol.* **93**:69–74.

22. **Costa-Cruz, J. M., J. Madalena, D. A. Silva, M. C. Sopelete, D. M. Campos, and E. A. Taketomi.** 2003. Heterologous antigen extract in ELISA for the detection of human IgE anti-*Strongyloides stercoralis*. *Rev. Inst. Med. Trop. Sao Paulo* **45**:265–268.

23. **Devi, C. S., and S. C. Parija.** 2003. A new serum hydatid antigen detection test for diagnosis of cystic echinococcosis. *Am. J. Trop. Med. Hyg.* **69**:525–528.

24. **Dowd, S. E., D. John, J. Eliopolus, C. P. Gerba, J. Naranjo, R. Klein, B. Lopez, M. de Mejia, C. E. Mendoza, and I. L. Pepper.** 2003. Confirmed detection of *Cyclospora cayetanensis*, *Encephalitozoon intestinalis* and *Cryptosporidium parvum* in water used for drinking. *J. Water Health* **1**:117–123.

25. **Eamsobhana, P., A. Yoolek, P. Punthuprapasa, and S. Suvouttho.** 2004. A dot-blot ELISA comparable to immunoblot for the specific diagnosis of human parastrongyliasis. *J. Helminthol.* **78**:287–291.

26. Espino, A. M., A. Diaz, A. Perez, and C. M. Finley. 1998. Dynamics of antigenemia and coproantigens during a human *Fasciola hepatica* outbreak. *J. Clin. Microbiol.* **36:**2723–2726.

27. Ferris, D. G., J. Hendrich, P. M. Payne, A. Getts, R. Rassekh, D. Mathis, and M. S. Litaker. 1995. Office laboratory diagnosis of vaginitis. Clinician-performed tests compared with a rapid nucleic hybridization test. *J. Fam. Pract.* **41:**575–581.

28. Fischer, P., I. Bonow, T. Supali, P. Ruckert, and N. Rahmah. 2005. Detection of filaria-specific IgG4 antibodies and filarial DNA, for the screening of blood spots for *Brugia timori*. *Ann. Trop. Med. Parasitol.* **99:**53–60.

29. Forney, J. R., C. Wongsrichanalai, A. J. Magill, L. G. Craig, J. Sirichaisinthop, C. T. Bautista, R. S. Miller, C. F. Ockenhouse, K. E. Kester, N. E. Aronson, E. M. Andersen, H. A. Quino-Ascurra, C. Vidal, K. A. Moran, C. K. Murray, C. C. DeWitt, D. G. Heppner, K. C. Kain, W. R. Ballou, and R. A. Gasser, Jr. 2003. Devices for rapid diagnosis of malaria: evaluation of prototype assays that detect *Plasmodium falciparum* histidine-rich protein 2 and a *Plasmodium vivax*-specific antigen. *J. Clin. Microbiol.* **41:**2358–2366.

30. Galel, S. A., and L. V. Kirchhoff. 1996. Risk factors for *Trypanosoma cruzi* infections in California blood donors. *Transfusion* **36:**227–231.

31. Garcia, H. H., L. J. Harrison, R. M. Parkhouse, T. Montenegro, S. M. Martinez, V. C. Tsang, R. H. Gilman, and The Cysticercosis Working Group in Peru. 1998. A specific antigen-detection ELISA for the diagnosis of human neurocysticercosis. *Trans. R. Soc. Trop. Med. Hyg.* **92:**411–414.

32. Garcia, L. S., and R. Y. Shimizu. 2000. Detection of *Giardia lamblia* and *Cryptosporidium parvum* antigens in human fecal specimens using the ColorPAC combination rapid solid-phase qualitative immunochromatographic assay. *J. Clin. Microbiol.* **38:**1267–1268.

33. Garcia, L. S., and R. Y. Shimizu. 1997. Evaluation of nine immunoassay kits (enzyme immunoassay and direct fluorescence) for detection of *Giardia lamblia* and *Cryptosporidium parvum* in human fecal specimens. *J. Clin. Microbiol.* **35:**1526–1529.

34. Garcia, L. S., R. Y. Shimizu, and C. N. Bernard. 2000. Detection of *Giardia lamblia, Entamoeba histolytica/E. dispar*, and *Cryptosporidium parvum* antigens in human fecal specimens using the EIA Triage® parasite panel. *J. Clin. Microbiol.* **38:**3337–3340.

35. Garcia, L. S., R. Y. Shimizu, S. Novak, M. Carroll, and F. Chan. 2003. Commercial assay for detection of *Giardia lamblia* and *Cryptosporidium parvum* antigens in human fecal specimens by rapid solid-phase qualitative immunochromatography. *J. Clin Microbiol.* **41:**209–212.

36. Garcia, L. S., A. C. Shum, and D. A. Bruckner. 1992. Evaluation of a new monoclonal antibody combination reagent for the direct fluorescent detection of *Giardia* cysts and *Cryptosporidium* oocysts in human fecal specimens. *J. Clin. Microbiol.* **30:**3255–3257.

37. Gonlugur, U., S. Ozcelik, T. E. Gonlugur, and A. Celiksoz. 2005. The role of Casoni's skin test and indirect haemagglutination test in the diagnosis of hydatid disease. *Parasitol. Res.* **97:**395–398.

38. Gottstein, B., P. Jacquier, S. Bresson-Hadni, and J. Eckert. 1993. Improved primary immunodiagnosis of alveolar echinococcosis in humans by enzyme-linked immunosorbent assay using Em2plus antigen. *J. Clin. Microbiol.* **31:**373–376.

39. Hafid, J., R. T. M. Sung, H. Raberin, Z. Y. Akono, B. Pozzetto, and M. Jana. 1995. Detection of circulating antigens of *Toxoplasma gondii* in human infection. *Am. J. Trop. Med. Hyg.* **52:**336–339.

40. Herwaldt, B. L., G. de Bruyn, N. J. Pieniazek, M. Homer, K. H. Lofy, S. B. Slemenda, T. R. Fritsche, D. H. Persing, and A. P. Limaye. 2004. *Babesia divergens*-like infection, Washington State. *Emerg. Infect. Dis.* **10:**622–629.

41. Hohlfeld, P., F. Daffos, J. Costa, P. Thullienz, F. Forestier, and M. Vidaud. 1994. Prenatal diagnosis of congenital toxoplasmosis with polymerase-chain-reaction test on amniotic fluid. *N. Engl. J. Med.* **331:**695–699.

42. Holland, C. V., P. O'Lorcain, M. R. Taylor, and A. Kelly. 1995. Sero-epidemiology of toxocariasis in school children. *Parasitology* **110:**535–545.

43. Houghton, R. L., D. R. Benson, L. Reynolds, P. McNeill, P. Sleath, M. Lodes, Y. A. Skeiky, R. Badaro, A. U. Krettli, and S. G. Reed. 2000. Multiepitope synthetic peptide and recombinant protein for the detection of antibodies to *Trypanosoma cruzi* in patients with treated or untreated Chagas' disease. *J. Infect. Dis.* **181:**325–330.

44. Iqbal, J., A. Siddique, M. Jameel, and P. R. Hira. 2004. Persistent histidine-rich protein 2, parasite lactate dehydrogenase, and panmalarial antigen reactivity after clearance of *Plasmodium falciparum* monoinfection. *J. Clin. Microbiol.* **42:**4237–4241.

45. Isaac-Renton, J., J. Blatherwick, W. R. Bowie, M. Fyfe, M. Khan, A. Li, A. King, M. McLean, L. Medd, W. Moorehead, C. S. Ong, and W. Robertson. 1999. Epidemic and endemic seroprevalence of antibodies to *Cryptosporidium* and *Giardia* in residents of three communities with different drinking water supplies. *Am. J. Trop. Med. Hyg.* **60:**578–583.

46. Ito, A., M. Nakao, H. Kutsumi, M. W. Lightowlers, M. Itoh, and S. Sato. 1993. Serodiagnosis of alveolar hydatid disease by Western blotting. *Trans. R. Soc. Trop. Med. Hyg.* **87:**170–172.

47. Jelinek, T., S. Eichenlaub, and T. Loscher. 1999. Sensitivity and specificity of a rapid immunochromatographic test for diagnosis of visceral leishmaniasis. *Eur. J. Clin. Microbiol. Infect. Dis.* **18:**669–670.

48. Junqueira, A. C. V., E. Chiari, and P. Wincker. 1996. Comparison of the polymerase chain reaction with two classical parasitological methods for the diagnosis of Chagas disease in an endemic region of north-eastern Brazil. *Trans. R. Soc. Trop. Med. Hyg.* **90:**129–132.

49. Kagan, I. G., and S. E. Maddison. 1992. Serodiagnosis of parasitic diseases, p. 529–543. *In* N. R. Rose, E. C. de Macario, J. L. Fahey, H. Friedman, and G. M. Penn (ed.), *Manual of Clinical Laboratory Immunology*, 4th ed. American Society for Microbiology, Washington, D.C.

50. Karki, B. M., and S. C. Parija. 1999. Co-agglutination test for the detection of circulating antigen in amebic liver abscess. *Am. J. Trop. Med. Hyg.* **60:**498–501.

51. Kim, K., H. J. Yang, and Y. B. Chung. 2003. Usefulness of 8 kDa protein of *Fasciola hepatica* in diagnosis of fascioliasis. *Korean J. Parasitol.* **41:**121–123.

52. Kirchhoff, L. V., P. Paredes, A. Lomeli-Guerrero, M. Paredes-Espinoza, C. S. Ron-Guerrero, M. Delgado-Mejia, and J. G. Pena-Munoz. 2006. Transfusion-associated Chagas' disease (American trypanosomiasis) in Mexico: implications for transfusion medicine in the United States. *Transfusion* **46**:298–304.

53. Krautz, G. M., L. M. C. Galvao, J. R. Cancado, A. Guevara-Espinoza, A. Quaissi, and A. U. Krettli. 1995. Use of a 24-kilodalton *Trypanosoma cruzi* recombinant protein to monitor cure of human Chagas' disease. *J. Clin. Microbiol.* **33**:2086–2090.

54. Kurkjian, K. M., L. E. Vaz, R. Haque, C. Cetre-Sossah, S. Akhter, S. Roy, F. Steurer, J. Amann, M. Ali, R. Chowdhury, Y. Wagatsuma, J. Williamson, S. Crawford, R. F. Breiman, J. H. Maguire, C. Bern, and W. E. Secor. 2005. Application of an improved method for the recombinant k 39 enzyme-linked immunosorbent assay to detect visceral leishmaniasis disease and infection in Bangladesh. *Clin. Diagn. Lab. Immunol.* **12**:1410–1415.

55. Lappalainen, M., P. Koskela, M. Koskiniemi, P. Ammala, V. Hiilesmaa, K. Terasmo, K. O. Raivio, J. S. Remington, and K. Hedman. 1993. Toxoplasmosis acquired during pregnancy: improved serodiagnosis based on avidity of IgG. *J. Infect. Dis.* **167**:691–697.

56. Lorca, M., C. Veloso, P. Munoz, M. I. Bahamonde, and A. Garcia. 1995. Diagnostic value of detecting specific IgA and IgM with recombinant *Trypanosoma cruzi* antigens in congenital Chagas' disease. *Am. J. Trop. Med. Hyg.* **52**:512–515.

57. Luft, B. J., and J. S. Remington. 1992. Toxoplasmic encephalitis in AIDS. *Clin. Infect. Dis.* **15**:211–222.

58. Mitchell, P. D., K. D. Reed, and J. M. Hofkes. 1996. Immunoserologic evidence of coinfection with *Borrelia burgdorferi*, *Babesia microti*, and human granulocytic *Ehrlichia* species in residents of Wisconsin and Minnesota. *J. Clin. Microbiol.* **34**:724–727.

59. Ngu, J. L., C. Nkenfou, E. Capuli, T. E. McMoli, F. Perler, J. Mbwagbor, C. Tume, O. B. Nlatte, J. Donfack, and T. Asonganyi. 1998. Novel, sensitive and low-cost diagnostic tests for 'river blindness'—detection of specific antigens in tears, urine and dermal fluid. *Trop. Med. Int. Health* **3**:339–348.

60. O'Neill, S. M., M. Parkinson, A. J. Dowd, W. Strauss, R. Angles, and J. P. Dalton. 1999. Short report. Immunodiagnosis of human fascioliasis using recombinant *Fasciola hepatica* cathepsin Li cysteine proteinase. *Am. J. Trop. Med. Hyg.* **60**:749–751.

61. Oskam, L., J. L. Nieuwenhuijs, and A. Hailu. 1999. Evaluation of the direct agglutination test (DAT) using freeze-dried antigen for the detection of anti-*Leishmania* antibodies in stored sera from various patient groups in Ethiopia. *Trans. R. Soc. Trop. Med. Hyg.* **93**:275–277.

62. Palmer, C. J., J. A. Bonilla, D. A. Bruckner, E. D. Barnett, N. S. Miller, M. A. Haseeb, J. R. Masei, and W. M. Stauffer. 2003. Multicenter study to evaluate the OptiMAL test for rapid diagnosis of malaria in U.S. hospitals. *J. Clin. Microbiol.* **41**:5178–5182.

63. Pani, S. P., S. L. Hoti, P. Vanamail, and L. K. Das. 2004. Comparison of an immunochromatographic card test with night blood smear examination for detection of *Wuchereria bancrofti* microfilaria carriers. *Natl. Med. J. India* **17**:304–306.

64. Peralta, J. M., M. G. M. Teixeira, W. G. Shreffler, J. B. Pereira, J. M. Burns, Jr., P. R. Sleath, and S. G. Reed. 1994. Serodiagnosis of Chagas' disease by enzyme-linked immunosorbent assay using two synthetic peptides as antigens. *J. Clin. Microbiol.* **32**:971–974.

65. Persing, D. H., D. Mathiesen, W. F. Marshall, S. R. Telford, A. Spielman, J. W. Thomford, and P. A. Conrad. 1992. Detection of *Babesia microti* by polymerase chain reaction. *J. Clin. Microbiol.* **20**:2097–2103.

66. Pinon, J. M., C. Chemla, I. Villena, F. Fuodrinier, D. Aubert, D. Puygauthier-Toubas, B. Leroux, D. Dupouy, C. Quereux, M. Talmud, T. Trenque, G. Potron, M. Pluot, G. Remy, and A. Bonhomme. 1996. Early neonatal diagnosis of congenital toxoplasmosis: value of comparative enzyme-linked immunofiltration assay immunological profiles and anti-*Toxoplasma gondii* immunoglobulin M (IgM) or IgA immunocapture and implications for postnatal therapeutic strategies. *J. Clin. Microbiol.* **34**:579–583.

67. Pirard, M., N. Iihoshi, M. Boelaert, P. Basanta, F. Lopez, and P. Van der Stuyft. 2005. The validity of serologic tests for *Trypanosoma cruzi* and the effectiveness of transfusional screening strategies in a hyperendemic region. *Transfusion* **45**:554–561.

68. Priest, J. W., J. P. Kwon, D. M. Moss, J. M. Roberts, M. J. Arrowood, M. S. Dworkin, D. D. Juranek, and P. J. Lammie. 1999. Detection by enzyme immunoassay of serum immunoglobulin G antibodies that recognize specific *Cryptosporidium parvum* antigens. *J. Clin. Microbiol.* **37**:1385–1392.

69. Quick, R. E., B. L. Herwaldt, J. W. Thomford, M. E. Garnett, M. L. Eberhard, M. Wilson, D. H. Spach, J. W. Dickerson, S. R. Telford, K. R. Steingart, R. Pollock, D. H. Persing, J. M. Kobayashi, D. D. Juranek, and P. A. Conrad. 1993. Babesiosis in Washington state: a new species of *Babesia*? *Ann. Intern. Med.* **119**:284–290.

70. Redlich, A., and W. A. Muller. 1998. Serodiagnosis of acute toxoplasmosis using a recombinant form of the dense granule GRA6 in an enzyme-linked immunosorbent assay. *Parasitol. Res.* **84**:700–706.

71. Remington, J. S., and J. O. Klein. 1995. *Infectious Diseases of the Fetus and Newborn Infant*, 4th ed. The W. B. Saunders Co., Philadelphia, Pa.

72. Riera, C., R. Fisa, P. Lopez, E. Ribera, J. Carrio, V. Falco, I. Molina, M. Gallego, and M. Portus. 2004. Evaluation of a latex agglutination test (KAtex) for detection of *Leishmania* antigen in urine of patients with HIV-*Leishmania* coinfection: value in diagnosis and post-treatment follow-up. *Eur. J. Clin. Microbiol. Infect. Dis.* **23**:899–904.

73. Salah, F., Z. Demerdash, Z. Shaker, A. El Bassiouny, G. El Attar, S. Ismail, N. Badir, A. S. El Din, and M. Mansour. 2000. A monoclonal antibody against *Schistosoma haematobium* soluble egg antigen: efficacy for diagnosis and monitoring of cure of *S. haematobium* infection. *Parasitol. Res.* **86**:74–80.

74. Santos Gomes, G., S. Gomes-Pereira, L. Campino, M. D. Araujo, and P. Abranches. 2000. Performance of immunoblotting in diagnosis of visceral leishmaniasis in human immunodeficiency virus-*Leishmania* sp.-coinfected patients. *J. Clin. Microbiol.* **38**:175–178.

75. Scheel, C. M., A. Khan, K. Hancock, H. H. Garcia, A. E. Gonzalez, R. H. Gilman, V. C. Tsang, and the Cysticercosis Working Group in Peru. 2005. Serodiagnosis

of neurocysticercosis using synthetic 8-kD proteins: comparison of assay formats. *Am. J. Trop. Med. Hyg.* **73:**771–776.

76. **Seed, C. R., A. Kitchen, and T. M. Davis.** 2005. The current status and potential role of laboratory testing to prevent transfusion-transmitted malaria. *Transfus. Med. Rev.* **19:**229–240.

77. **Shenai, B. R., B. L. Komalam, A. S. Arvind, P. R. Krishnaswamy, and P. V. S. Rao.** 1996. Recombinant antigen-based avidin-biotin microtiter enzyme-linked immunosorbent assay for serodiagnosis of invasive amebiasis. *J. Clin. Microbiol.* **34:**828–833.

78. **Simonsen, P. E., and S. K. Dunyo.** 1999. Comparative evaluation of three new tools for diagnosis of bancroftian filariasis based on detection of specific circulating antigens. *Trans. R. Soc. Trop. Med. Hyg.* **93:**278–282.

79. **Singh, R., B. V. S. Raju, R. K. Jain, and P. Salotra.** 2005. Potential of direct agglutination test based on promastigote and amastigote antigens for serodiagnosis of post-kala-azar dermal leishmaniasis. *Clin. Diagn. Lab. Immunol.* **12:**1191–1194.

80. **Sloan, L., S. Schneider, and J. Rosenblatt.** 1995. Evaluation of enzyme-linked immunoassay for serological diagnosis of cysticercosis. *J. Clin. Microbiol.* **33:**3124–3128.

81. **Stothard, J. R., N. B. Kabatereine, E. M. Tukahebwa, F. Kazibwe, D. Rollinson, W. Mathieson, J. P. Webster, and A. Fenwick.** 2006. Use of circulating cathodic antigen (CCA) dipsticks for detection of intestinal and urinary schistosomiasis. *Acta Trop.* **97:**219–228.

82. **Strickland, G. T., and B. L. Ramirez.** 2000. Schistosomiasis, p. 805–832. *In* G. T. Strickland (ed.), *Hunter's Tropical Medicine and Emerging Infectious Diseases*, 8th ed. The W. B. Saunders Co., Philadelphia, Pa.

83. **Sundar, S., S. Agrawal, K. Pai, M. Chance, and M. Hommel.** 2005. Detection of leishmanial antigen in the urine of patients with visceral leishmaniasis by a latex agglutination test. *Am. J. Trop. Med. Hyg.* **73:**269–271.

84. **Sundar, S., S. G. Reed, V. P. Singh, P. C. Kimar, and H. W. Murray.** 1998. Rapid accurate field diagnosis of Indian visceral leishmaniasis. *Lancet* **351:**563–565.

85. **Tanyuksel, M., and W. A. Petri, Jr.** 2003. Laboratory diagnosis of amebiasis. *Clin. Microbiol. Rev.* **16:**713–729.

86. **van Dam, G. J., J. H. Wichers, T. M. Ferreira, D. Ghati, A. van Amerongen, and A. M. Deelder.** 2004. Diagnosis of schistosomiasis by reagent strip test for detection of circulating cathodic antigen. *J. Clin. Microbiol.* **42:**5458–5461.

87. **Wembe, F. E., C. Tume, S. L. Ayong, G. Manfono, G. Lando, T. Asonganyi, and L. J. Ngu.** 2005. Development of an antigen detection dot blot assay for the diagnosis of human onchocerciasis based on the biotin-avidin binding system. *Bull. Soc. Pathol. Exot.* **98:**177–181.

88. **Wen, H., and P. S. Craig.** 1994. Immunoglobulin subclass response in human cystic and alveolar echinococcosis. *Am. J. Trop. Med. Hyg.* **51:**741–748.

89. **Wen, H., P. S. Craig, A. Ito, D. A. Vuitton, S. Bresson-Hadni, J. C. Allan, M. T. Rogan, E. Paollilo, and M. Shambesh.** 1995. Immunoblot evaluation of IgG and IgG-subclass antibody responses for immunodiagnosis of human alveolar echinococcosis. *Ann. Trop. Med. Parasitol.* **89:**485–495.

90. **Wilson, M., R. T. Bryan, J. A. Fried, D. A. Ware, P. M. Schantz, J. B. Pilcher, and V. C. Tsang.** 1991. Clinical evaluation of the cysticercosis enzyme-linked immunoelectrotransfer blot in patients with neurocysticercosis. *J. Infect. Dis.* **164:**1007–1009.

91. **Wilson, M., and P. M. Schantz.** 2000. Parasitic immunodiagnosis, p. 1117–1122. *In* G. T. Strickland (ed.), *Hunter's Tropical Medicine and Emerging Infectious Diseases*, 8th ed. The W. B. Saunders Co., Philadelphia, Pa.

92. **Wincker, P., M. Bosseno, C. Britto, N. Yaksic, M. A. Cardoso, C. M. Morel, and S. F. Breniere.** 1994. High correlation between Chagas' disease serology and PCR-based detection of *Trypansoma cruzi* kinetoplast DNA in Bolivian children living in an endemic area. *FEMS Microbiol. Lett.* **124:**419–424.

93. **Zhu, Y. C., D. Socheat, K. Bounlu, Y. S. Liang, M. Sinoun, S. Insisiengmay, W. He, M. Xu, W. Z. Shi, and R. Berquist.** 2005. Application of dipstick dye immunoassay (DDIA) kit for the diagnosis of schistosomiasis mekongi. *Acta Trop.* **96:**137–141.

Histologic Identification of Parasites

Often, histology preparations are submitted to the clinical laboratory for review, consultation, or confirmation. Sections of tissue that contain parasites may be seen infrequently in many pathology departments; thus, consultation services are needed.

This chapter is designed not as a comprehensive presentation of histologic parasitology but rather as a review of parasite morphology, potential sites in the human host, any recommended special processing or stains, and additional helpful hints. The information is arranged in table form according to body site (Table 23.1).

Specific organisms are listed within parasite groups (protozoa, helminths, etc.). The table does not include organisms that have been reported as only rarely infecting humans. Some of these more rare infections are represented by so few cases that a physician may never see one (see chapter 18). The photographs in Figures 23.1 to 23.118 illustrate representative parasitic infections in various body tissues. Although most of the specimens are from human tissues, a few are from animals to illustrate a particular parasite and representative body tissue.

Although a great deal of information is contained in this chapter, there has been no attempt to illustrate every organism and every possible tissue involved. Also, identification of parasite adults or immature forms depends on specimen submission and subsequent preparation. If the organisms and tissues are not well preserved, identification may be difficult or impossible. Distinguishing actual organisms from artifact material may also complicate the process.

Most diagnostic procedures in medical parasitology do not require extensive patient information; however, the examination of tissues and final identification of any organisms present can be simplified by reviewing as much information as possible about the patient. Specifics would include place of residence, travel history, food preferences, recent health status, health status of family members, current symptoms, outdoor activities, animal exposure, and any possible unusual environmental exposures. If specific lesions have been identified, the source of the tissue(s) must be identified, as well as the method by which the specimen was obtained. It is particularly relevant to know anything about the status of the specimen regarding adequate fixation and/or any lag time prior to fixation that might influence the overall morphology.

Some general descriptions are provided with diagrams to illustrate specific morphologic criteria used for parasite identification. Because the protozoa are unicellular organisms with one or more nuclei and various organelles, their simplicity does not require extensive descriptive material. However, the identification of various helminths and arthropods requires a more thorough understanding of parasite morphology and appropriate terminology (Table 23.2). Tables 23.3 to 23.5 contain information on staining characteristics, nuclear and cytoplasmic stains, and specific stains for carbohydrates. This information provides some basic guidelines for the understanding of how various stains are categorized, used, and interpreted. Tables 23.6 and 23.7 indicate the locations within the body where human protozoal and helminth parasites are found. Table 23.8 contains very specific information on the musculature of nematodes and can be found following Figure 23.34. One can also refer to the excellent references provided with this chapter.

Table 23.1 Histologic identification of parasites[a]

Body site	Organisms	Stain[b]	Comments[b]
Intestine	**Protozoa**		
	Entamoeba histolytica (cecum, ascending colon, sigmoid rectal region; other sites less common); trophozoites and cysts are normally passed in stool	H&E, PAS, trichrome or iron hematoxylin in stool	Differential between organism and histiocyte; in an organism, the nucleus may not be visible; in histiocytes, a portion of the nucleus is almost always visible; both trophozoites and histiocytes stain red with PAS (review nuclear details); histolysis of invaded tissue; flask-shaped ulcers; organisms seen in healthy ulcer margins, not in central, necrotic areas
	Giardia lamblia (duodenum); trophozoites and cysts are normally passed in stool; organisms can be seen in duodenal aspirates or using other sampling methods (Entero-Test)	H&E, Giemsa, trichrome or iron hematoxylin in stool	Normal to mild to severe blunting of villi; loss of the brush border; infiltration of lamina propria with lymphocytes and granulocytes; chronic inflammation; tissue invasion does not occur; trophozoites may appear as sagittal sections of small, crescentic organisms that may adhere to the epithelial surface; front views will reveal the typical teardrop shape
	Balantidium coli; trophozoites and cysts are normally passed in stool; tissue invasion may occur in the colon and appendix	H&E, trichrome or iron hematoxylin in stool (may be difficult to see on permanent stained smears; wet mount is best)	Simple hyperemia to marked ulcerations (large intestine); very large size (trophozoites); presence of cilia, large cytostome, and prominent macronucleus will distinguish these ciliates from other protozoa; due to large size, cut planes may not reveal all internal morphology; early lesion resembles that of *E. histolytica*, but the opening into the mucosa is larger
	Cryptosporidium spp. (have been recovered from mouth to anus, more common in small intestine); oocysts (4–6 μm) are passed in stool; disseminated disease can occur, with oocysts being passed in sputum	H&E, modified acid-fast stains for stool (oocysts positive)	Mild to moderate villous atrophy, increased crypt size, mild to moderate mononuclear cell infiltrates of lamina propria; organisms along brush border are small and measure 3–5 μm; other coccidia and the microsporidia develop within the cells and so are not confused with the brush border location of *Cryptosporidium* spp.
	Cyclospora cayetanensis	H&E, modified acid-fast stains for stool (oocysts positive, but staining variable)	Developmental stages have been seen in jejunal enterocytes, usually at the luminal end rather than at the brush border; tends to resemble developmental stages of *I. belli*
	Isospora belli (more common in small intestine); oocysts are passed in stool; unsporulated oocysts measure 20–33 μm long by 10–19 μm wide and may contain a single sporoblast (usually seen in cases of diarrhea)	H&E, modified acid-fast stains for stool (oocysts positive)	Same as *Cryptosporidium* spp., but development occurs within the epithelial cells; tissue eosinophilia can be seen

Organism	Stain(s)	Histologic features
Microsporidia; identification of organisms to the genus and species levels requires electron microscopy, tissue culture, immunofluorescence studies, or PCR; differentiation of spores from bacteria or small yeasts in stool is extremely difficult	H&E, tissue Gram stains, PAS, silver stains; modified trichrome and calcofluor white used for stool or urine	Developing spores seen within enterocytes; sizes of individual spores are ~1–3 μm in humans; *Enterocytozoon bieneusi* and *Encephalitozoon* (*Septata*) *intestinalis* involved, primarily seen in the compromised host; both organisms can disseminate to other organs and spores can be found in liver, lungs, kidneys, CNS, and cornea
Trypanosoma cruzi (esophagus, colon); trypomastigote may be found in peripheral blood in the early stage of infection	H&E	Disease of ganglion cells of mesenteric plexuses; can cause severe distension and thinning of visceral wall; scarring, chronic inflammation; amastigotes may shrink, nuclei and kinetoplasts may not be visible in tissue sections and may resemble *Histoplasma*; tissues most often involved include brain, cardiac, and smooth muscle cells

Nematodes

Organism	Stain(s)	Histologic features
Enterobius vermicularis (cecum, appendix, other sites; may also be found in female reproductive tract or peritoneal cavity); eggs and adult worms recovered on Scotch tape preparations or pin-worm paddles; stool not recommended	H&E	Characteristic lateral alae on worm cross sections; eggs in utero may also be visible; most often seen in appendix with little or no inflammation
Ascaris lumbricoides (small intestine, bile ducts, liver, peritoneal cavity); fertilized and unfertilized eggs passed in stool; adult worms may also be seen in stool; migrating larvae may be seen in sputum	H&E	Large size, characteristic muscular wall formed by long, irregular muscle fibers; sectioned (eggs may be seen with their thick, bumpy shells); the body wall has a thick, multilayered cuticle, fibrous-appearing hypodermis (expands to form large lateral chords, within which are seen excretory canals); short, muscular esophagus and simple intestinal tube with an irregular lumen; larval forms may be more difficult to identify
Strongyloides stercoralis (usually duodenum and upper jejunum; may be in any tissue in disseminated disease); rhabditiform (noninfectious) larvae usually seen in stool; filariform (infectious) larvae, as well as adult worms and eggs, may be seen in stool in hyperinfection cases	H&E	Sections may reveal adult females, well-segmented eggs, and numerous larvae; although the morphology of the adult female may be difficult to see in tissues, the small size of the worm and the presence of developing eggs and larvae close to the adult worm strongly suggest this diagnosis; with hyperinfection, the bowel lining becomes congested and edematous, with possible ulcerations; chronic infection may reveal bowel wall fibrosis
Hookworms (*Necator*, *Ancylostoma*) (usually upper part of small intestine); hookworm eggs normally passed in stool; eggs of both genera look the same	H&E	Relatively thick cuticle and thin hypodermis; lateral hypodermal chords are usually visible; wide-open mouth sucking the mucosa; adult *Necator* has cutting plates, while *Ancylostoma* has teeth

(continued)

Table 23.1 Histologic identification of parasites^a (*continued*)

Body site	Organisms	Stain^b	Comments^b
	Trichuris trichiura (large intestine); whipworm eggs normally passed in stool	H&E	Most sections reveal both the thicker body area ("whip handle"), which lies free in the lumen, and the slender anterior area ("whip"), which is firmly embedded into the mucosa; sectioned eggs may be visible
	Trichinella spp. (duodenum, small intestine)	H&E	Edema, chronic inflammation, hyperemia, serosa petechiae, dilatation of bowel loops; prominent Peyer's patches; swollen intestinal villi, mucin secretion; diffuse infiltration of eosinophils (lamina propria); larvae may be seen and may resemble microfilariae
	Cestodes		
	Taenia saginata (most common), *T. solium*, *Diphyllobothrium latum*; proglottids and/or eggs are normally passed in the stool	H&E	May see loose proglottids and/or eggs in appendix
	Hymenolepis nana; eggs are normally passed in the stool	H&E	Both larval and adult forms may be in intestinal sections in the human
	Echinococcus granulosus (mesentery)	H&E, acid fast	External laminated, anuclear cyst wall; hooklets are acid fast
	Trematodes		
	Schistosoma mansoni, *S. japonicum*, *S. haematobium*, less common schistosomes (*S. mekongi*, *S. intercalatum*); eggs may be seen in urine and/or stool	H&E	Adults and eggs may be seen in the mesentery or mesenteric veins; egg identification may depend on demonstration of spine in section; important to determine if egg contains miracidium larva (indicates active infection); in the wall of the intestinal tract, polyp formation may occur (more eggs found in polyps)
Liver	Protozoa		
	Leishmania donovani	H&E, PAS	Organisms are dot-like in routine section; they are larger (2–4 μm) in tissue impression smear; kinetoplast next to nucleus is very characteristic and differentiates these organisms from *Toxoplasma* (no kinetoplast); *Histoplasma* is PAS positive, while *L. donovani* is PAS negative
	Toxoplasma gondii	H&E, PAS	Resting forms (bradyzoites) are pyriform with no kinetoplast (if many organisms are seen per cyst, they may appear as dots); bradyzoites within cysts are PAS positive; tissue cysts have a very thin wall; tissue cysts of *Toxoplasma* must be differentiated from *Sarcocystis* (seen most frequently in skeletal or cardiac muscle and larger than *Toxoplasma*); when reviewing animal tissues (from dogs), *Neospora caninum* may be present and will mimic *Toxoplasma*

Organism	Stain	Comments
Microsporidia	See intestine	See information on intestine
Plasmodium falciparum, *P. vivax*, *P. malariae*, *P. ovale*	H&E	Exoerythrocytic stages (mature tissue schizonts) measure 45–80 μm and contain 15,000–40,000 merozoites per cell; malarial pigment may be seen in Kupffer cells; capillaries may be filled with parasitized RBCs (pigment is also visible here); the host does not respond to the presence of these organisms; pigment within blood vessels can be seen with routine tissue stains, but the use of polarized light may also be helpful in demonstrating the birefringent granules of malarial pigment
Nematodes		
Capillaria hepatica; eggs in human stool would be from the ingestion of parasitized livers, probably from squirrels or other animals (spurious infections—eggs disappear from the feces in a few days)	H&E, trichrome (Masson)	Liver parenchyma; eggs remain in liver and the life cycle does not continue in humans; eggs resemble those of *T. trichiura*, but the shell appears beaded or pitted; polar plugs are less pronounced than in *T. trichiura*; intense granulomatous reaction around eggs; the general egg shape, the bipolar plugs, and striations in the thick shell make this identification relatively easy
Toxocara cati or *T. canis*, *Baylisascaris* spp. (visceral larva migrans); adult worms play no role in human disease (very rare exceptions)	H&E	Liver contains largest number of larvae; any organ can be affected; eosinophilic granuloma; focal necrosis may occur; larval morphology includes single lateral ala, nonpatent intestine, and large paired excretory columns; *Baylisascaris* larvae are larger than *Toxocara* larvae and have a patent gut and a predilection for the CNS
Cestodes		
Echinococcus granulosus; adult worms and/or eggs found only in the small intestine of canids (primarily dogs)	H&E, acid fast, GMS, PAS	External laminated, anuclear cyst wall; thin nucleated inner layer; hooklets are acid fast; calcified cysts often reveal no hooklets; mature cysts contain brood capsules and protoscolices; degeneration of brood capsules results in "hydatid sand" (degenerating protoscolices and hooklets in the cyst fluid); some cysts are sterile
Echinococcus multilocularis; adult worms and /or eggs found only in the small intestine of canids (primarily dogs and foxes)	H&E, acid fast, GMS, PAS	Alveolar structure with no limiting membrane; scolices seldom present, thus no protoscolices, hooklets, or calcareous corpuscles; laminated and germinal layers are collapsed, folded, and scattered throughout the liver
Trematodes		
Schistosoma mansoni, *S. japonicum*, *S. haematobium*, less common schistosomes (*S. mekongi*, *S. intercalatum*); eggs normally passed in urine and/or stool	H&E, trichrome (Masson)	Granulomatous tissue reaction around eggs (lymphocytes, plasma cells, eosinophils); trichrome demonstrates fibrotic tissue; eggs of *S. japonicum* more numerous in liver (greater egg-laying capacity); adult worms seen most frequently in cross or transverse sections within blood vessels in the wall of the intestine or in liver; often the female is seen in copula, surrounded by the larger male body

(continued)

Table 23.1 Histologic identification of parasites[a] (*continued*)

Body site	Organisms	Stain[b]	Comments[b]
	Fasciola hepatica (bile duct); eggs normally passed in stool	H&E	Large size plus cross section of worm containing intestinal diverticula; eggs may also be seen but are usually collapsed or distorted
	Clonorchis (*Opisthorchis*) *sinesis* (bile duct); eggs normally passed in stool	H&E	May be eosinophilic infiltration and slight thickening of the duct wall; may progress to fibrosis, cirrhosis of liver, biliary obstruction, etc.; long, coiled uterus full of eggs may be seen; adult trematodes in bile ducts may suggest *C. sinensis*, *Opisthorchis* spp., *Dicrocoelium dendriticum*, and *Fasciola* spp.
Spleen	Protozoa *Leishmania donovani*	H&E, PAS	See information on liver
	Plasmodium falciparum, *P. vivax*, *P. malariae*, *P. ovale*	H&E	See information on liver
Gallbladder	This organ is seldom invaded by parasites; occasionally, *Cryptosporidium*, microsporidia, *Clonorchis* (*Opisthorchis*) *sinensis*, *Ascaris lumbricoides*, and *Fasciola hepatica* are found; schistosome eggs may also be found		Eggs and/or sections of adult trematodes may be seen; developmental stages of *Cryptosporidium* possible
Kidney	Protozoa Microsporidia	See intestine	See information on intestine
Urinary bladder	Trematode *Schistosoma haematobium*	H&E, acid fast	Seen within blood vessels, often near the bladder; early inflammation may progress to thickening, hyperplasia, fibrosis, and ulceration; carcinoma may be a complication; terminal egg spine may be visible; important to determine if egg contains miracidium; eggs may be distorted and collapsed; calcified eggs may also be seen; eggs are not acid fast, like those of *S. mansoni* and *S. japonicum*; adult worms may be seen in copula within the blood vessels
Lungs	Protozoa *Entamoeba histolytica*	H&E	Usually extension of liver abscess in the right lower lobe; may occur elsewhere; nucleus is not always visible; should be area of histolysis around organisms
	Toxoplasma gondii	H&E	Cysts look like those in the liver; tachyzoites are crescent shaped, especially when released from the cells
	Cryptosporidium spp.	H&E	Organism morphology similar to that in intestine; organisms may be seen in sputum in disseminated disease

	Microsporidia	Tissue Gram stains, PAS, silver stains, H&E	Organism morphology similar to that in intestine; spores may be found in sputum
	Nematodes *Strongyloides stercoralis, Ascaris lumbricoides,* hookworm, *Toxocara* spp., *Wuchereria bancrofti, Brugia malayi, Dirofilaria immitis*	H&E	Migrating larvae (*Strongyloides, Ascaris,* hookworm, *Toxocara* spp.) may be seen; microfilariae (*Wuchereria*) may cause an eosinophilic granuloma (pneumonitis, alveolar hemorrhage, edema)
	Trematodes *Paragonimus westermani,* other *Paragonimus* spp.; eggs can be found in sputum and/or stool	H&E	Reaction by lung tissue isolates the worm in a thick, fibrous capsule containing hemorrhagic, purulent exudate and eggs; adults typically paired, surrounded by fibrous capsule; spines on tegument and typical eggs make the diagnosis relatively easy
Heart	Protozoa *Entamoeba histolytica*	H&E, PAS	May be present; pericardial tamponade may occur
	Trypanosoma cruzi	H&E	Amastigotes are morphologically identical to those of *L. donovani* found in the macrophages of the reticuloendothelial system; inflammatory cell infiltrates, interstitial edema, and separation of myofibers
	Toxoplasma gondii	H&E	Cysts appear like those in liver (no kinetoplast as in amastigotes of *T. cruzi*)
	Microsporidia	See intestine	See information on intestine
	Plasmodium falciparum (most important), *P. vivax, P. malariae, P. ovale*	H&E	May be marked congestion of the capillaries; blocked with parasitized RBCs and malarial pigment; evidence of anoxia (cloudy swelling, fatty degeneration, myocardial infarcts) may be seen
Skeletal muscle	Protozoa *Sarcocystis* spp. (rare)	H&E, GMS	Cylindrical cysts (Miescher's tubes) formed (double outer membrane and transverse trabeculae separating the cyst into closed compartments packed with spores); silver stain will clearly reveal the external capsule and septa
	Microsporidia	PAS, AFB, GMS, H&E	Organisms in the genera *Pleistophora, Trachipleistophora,* and *Brachiola* can be found in muscle in compromised hosts (cases on record rare); atrophic and degenerating muscle fibers full of spores (*Pleistophora* in clusters of ~12 organisms); each cluster enclosed by enveloping membrane (pansporoblastic membrane); spores measure 2.8 by 3.4 µm

(continued)

Table 23.1 Histologic identification of parasites[a] *(continued)*

Body site	Organisms	Stain[b]	Comments[b]
	Nematode		
	Trichinella spp.	H&E	Diaphragm most commonly invaded; biopsy usually not performed until after day 17, when larvae present a typical, coiled appearance; old, calcified cysts may be difficult to recognize
	Cestode		
	Taenia solium (cysticercosis)	H&E	May also occur in brain, eye, muscles, heart, liver, lungs, abdominal cavity; cyst is oval or round and contains one invaginated scolex with crown of hooklets and four suckers; larva is usually surrounded by tissue reaction capsule, which is wrinkled after tissue processing (very typical appearance, often diagnostic even without cross section of scolex); cysticercus has a dark-staining spiral canal and a scolex surrounded by a thin, lightly staining bladder wall; parenchymatous portion consists of the invaginated scolex with its spiral canal, four suckers, and an armed rostellum with hooks; surface of spiral canal is deeply folded; parenchyma is usually filled with calcareous corpuscles; racemose form of cysticercosis appears as multiple sections of bladder wall tegument (scolex and spiral canal not seen)
Skin and subcutaneous tissue	Protozoa		
	Leishmania spp.	H&E, GMS	See information on liver
	Nematodes		
	Wuchereria bancrofti, Brugia malayi	H&E	Adults in lymphatic vessels (may be diagnosed in tissue sections); microfilariae in blood
	Loa loa		Adults in subcutaneous tissue; microfilariae in blood
	Onchocerca volvulus		Adults in subcutaneous nodules or in loose connective tissues from deeper areas within the body; may be diagnosed in tissue sections (fibrous nodules); microfilariae in skin; accidental infections with animal species of *Onchocerca* may occur; key features of thick cuticle, transverse ridges, and striations in the middle layer of cuticle are helpful
	Mansonella ozzardi, M. perstans, M. streptocerca, Diroflaria spp.		Adults in body cavities (abdominal, pleural) and pericardium; microfilariae in blood (*M. streptocerca* in skin)
	Hookworm larvae	H&E	Differentiate from *Strongyloides* spp. (not found in skin)

Cestodes			
	Spirometra spp. (sparganosis)	H&E	Larvae often invade subcutaneous tissues or muscles (in or about the eyes); acute inflammatory reaction, which is more severe as larva dies and is absorbed; solid-bodied sparganum lacks suckers and surrounding bladder wall seen in other cysticerci
	Sparganum proliferum	H&E	Produces a more severe reaction with metastatic branching and spread throughout the tissues; no scolex seen
Arthropods			
	Demodex folliculorum (mite)	H&E	Invades hair follicles and sebaceous glands; may be chronic inflammatory cells around mites
	Sarcoptes scabiei (itch mite)	H&E	Superficial skin layer (never below epidermis); preferred site is interdigital spaces, may be elsewhere (heavy infestations, i.e., Norwegian scabies)
	Tunga penetrans (chigoe, sand flea)	H&E	Prefers toes; causes intense itching, even pain
Lymphatic vessels	**Nematodes** *Wuchereria bancrofti, Brugia malayi*	H&E	Damage ranges from inflammatory sensitization to elephantiasis (tissue hyperplasia)
Testes	**Nematode** *Wuchereria bancrofti*	H&E	Worms may be calcified and lymphatic vessel replaced by collagen tissue
Nervous system	**Protozoa** *Naegleria fowleri, Acanthamoeba* spp., *Balamuthia mandrillaris* (leptomyxid ameba), *Sappinia diploidea*	H&E	Brain lesions demonstrate congestion or hemorrhages, marked cellular infiltration (mononuclear, polymorphonuclear), and abscess formation; trophozoites have large karyosome and no chromatin on nuclear membrane; may be difficult to differentiate *Acanthamoeba* from *Balamuthia*
	Entamoeba histolytica	H&E, PAS	Histolysis of invaded tissue; trophozoite nucleus may not always be visible
	Microsporidia, *Encephalitozoon* spp., *Vittaforma* spp., *Nosema* spp., *Brachiola* spp., *Trachipleistophora* spp., "*Microsporidium*" spp.	PAS, AFB, GMS	Spores range from 2 to 4 μm and have been found in cerebrospinal fluid; appearance in tissue similar to other body sites; eye infections reported with *Encephalitozoon, Vittaforma, Nosema,* and members of the catchall genus "*Microsporidium*"

(continued)

Table 23.1 Histologic identification of parasites[a] (*continued*)

Body site	Organisms	Stain[b]	Comments[b]
	Trypanosoma brucei gambiense, T. brucei rhodesiense	H&E	May be no reactive changes around organisms (brain); undulating membrane may be difficult to see
	Toxoplasma gondii	H&E	See information on liver; there is no cyst formation in an early acute infection (central nervous system more commonly involved)
	Plasmodium falciparum	H&E	Congested blood vessels containing parasitized RBCs and malarial pigment
Nematodes			
	Angiostrongylus cantonensis	H&E	Produces eosinophilic meningoencephalitis; cross section of worm is characterized by high muscle fibers under the cuticle with two prominent cords; cellular infiltrates composed of eosinophils and mononuclear cells
	Toxocara spp.	H&E	Granulomas or necrotic areas within periretinal inflammatory membrane
Cestodes			
	Taenia solium (cysticercosis)	H&E	May induce fibrous encapsulation; pronounced cellular reaction may occur as larva begins to die
	Echinococcus granulosus	H&E	Wall is finely laminated and has affinity for either acidic or basic dyes
Trematodes			
	Schistosoma japonicum	H&E	Hyperemia of meninges overlying granulomas in brain
	Paragonimus spp. (any area of brain)	H&E	Cysts may contain eggs, cellular debris, Charcot-Leyden crystals, eosinophils, plasma cells, and lymphocytes; granulomas form around eggs

[a] Data from references 1–6; not every body site or every parasite is included in this table.
[b] CNS, central nervous system; H&E, hematoxylin and eosin stain; GMS, Gomori methenamine-silver stain; PAS, periodic acid-Schiff; RBC, red blood cell; AFB, acid-fast bacillus.

Table 23.2 General characteristics of helminths

Characteristic	Nematodes	Cestodes	Trematodes	Acanthocephala	Pentastomes (tongue worms)
General	Elongate-cylindrical, bilaterally symmetrical with secondary triradiate symmetry at anterior end; integument nonnucleated cuticula secreted by hypodermis (may be smooth, striated, bossed, or spined); body has 3 main layers (cuticle, hypodermis, somatic muscles); four cords arise from hypodermis and project into body cavity; no circular muscles; body cavity is pseudocele (not lined with mesothelium); all viscera suspended in pseudocele; all separate sexes	Adult tapeworm consists of few to many egg-producing units (proglottids) which develop from distal end of scolex; chain called strobila; each proglottid contains male and female genital organs (hermaphroditic); food absorbed through integument; at surface is layer of microvilli (microtriches) over entire outer surface (comparable to brush border of host intestinal epithelium; primitive excretory system; tegument homogeneous and noncellular; circular, longitudinal muscles; proglottids filled with mesenchyme or parenchyma (loose meshwork of branched cells and fluid-filled spaces); calcareous corpuscles usually present in parenchyma	Flukes usually dorsoventrally flattened and hermaphroditic (other than schistosomes); have oral and ventral suckers; integument frequently covered by spines; mesenchyme below integument; circular, oblique, longitudinal muscles; alimentary canal is inverted Y shape; nutrients generally absorbed through tegument	Canals in tegument and arrangement of deeper layers of body wall distinctive; proboscis (retractile into proboscis sheath) armed with spines and body proper; separate sexes; probably related to tapeworms	Body elongate, tongue-like, cylindrical, or moniliform, with many pseudosegments but no separation into head, thorax, and abdomen; head with median subterminal mouth and two pairs of hollow, retractile claws; eyes lacking; sexes separate; striated muscles; have been considered highly modified annelids
Body cavity	Pseudocoelom—yes	No—parenchyma, containing calcareous corpuscles (unique to cestodes, prominent in larval forms)	No—filled with mesenchyme	Pseudocoelom—no; loose parenchymous matrix	Reduced—yes
Body tegument (wall)	Cuticle, hypodermis; cuticle has striations, annulations, or ridges; hypodermis has dorsal, ventral, and lateral chords; well-developed longitudinally oriented, smooth muscles	Smooth	Smooth or spinous, two suckers (oral and ventral/acetabulum); syncytial epithelium bounded externally by a plasma membrane and internally by trilaminated plasma membrane	Thick, multilayer; outer plasma membrane, thin fibrous layer; thick felt-like middle layer, very thick fibrous inner layer; structure of proboscis with hooks and spines on body surface key features; canals present in tegument	May have spines; cuticle may have prominent spines or may be smooth; subcuticular glands lie below cuticle, as do striated muscle fibers; cuticle contains chitin and has sclerotized openings that are diagnostic for pentastomids

(continued)

Table 23.2 General characteristics of helminths *(continued)*

Characteristic	Nematodes	Cestodes	Trematodes	Acanthocephala	Pentastomes (tongue worms)
Cuticle (nematodes)	3 layered (cortical, median, basal); surface may have longitudinal ridges (may be called alae); may also have inflations called bosses or plaques	Surface may be covered with hairy projections called microtriches	True cuticle and epidermis shed or fails to develop	NA[a]	NA
Hypodermis (nematodes)	Layer of cells lying between the cuticle and muscles; projects into the pseudocoelom at the dorsal, ventral, and lateral positions to form 4 chords; secretes the cuticle	NA	NA	NA	NA
Chords	Lateral chords most conspicuous; in some groups, lateral secretory/excretory columns lie within or associated with lateral chords	NA	NA	NA	NA
Muscle fibers	Longitudinally oriented, smooth muscles; no circular muscles	Longitudinal and circular smooth muscles	Longitudinal, circular, and diagonal smooth muscles	Longitudinal and circular smooth muscles	Striated muscle fibers
Body muscles	Somatic and specialized				
Somatic (nematodes)	Elongate, spindle-shaped cells; lie in rows under hypodermis, are parallel to and overlap each other; each cell has basal, contractile portion and cytoplasmic, noncontractile portion	Between tegument and subtegumental nuclei are 2 thin layers of muscle fibers; outer layer is circular fibers and inner layer is longitudinal; some differences among species	Outer layer of circular muscle, inner layer of longitudinal muscle; dorsoventral muscle fibers course throughout the parenchyma and surround internal organs	Outer circular and inner longitudinal fibers	Acidophilic glands can be seen in larval sections
Specialized (nematodes)	Associated with digestive and reproductive systems				
Body cavity contents	Contains digestive tube and tubular gonads	No	No	No digestive system	Yes, intestine large

Digestive system	Mouth, buccal cavity, esophagus, intestine, rectum	No	Oral opening, pharynx, esophagus which bifurcates into a pair of blindly ending tubular intestinal ceca	No—lacunar system; nutrients absorbed through body wall	Simple gut
Reproductive system	Separate sexes; simple tubular structures; usually 1 testis and paired ovaries	Hermaphroditic	Hermaphroditic; 2 testes and 1 ovary; exception is seen with schistosomes (separate sexes)	Separate sexes; 2 testes and 1 ovary, breaks up into "floating" ovaries	Separate sexes
Nervous system	Not relevant for identification	Not relevant for identification	Not relevant for identification	Not relevant for identification	Not relevant for identification
Excretory system	Referred to as secretory/excretory system; varies from large gland cell that opens to the exterior via a duct in ventral midline to being free in the pseudocoelom	Not relevant for identification	Bilaterally symmetrical	NA	NA

[a] NA, not applicable.

Table 23.3 Staining characteristics

Characteristic	Comments
Staining action	**Substantive:** Stain that acts immediately and directly on the tissue without the intervention of any other substance
	Adjective: Tissue is first treated with an agent, which in turn attaches the stain to the tissue (mordant staining)
	Impregnation: Involves the deposition of sensitive metallic substances over selected cells and tissue structures that are rendered visible by subsequent reduction of the metal
Staining time	**Progressive:** Using the microscope, performed by watching the degree of staining at various points during the process and stopping the process when the desired staining action has been achieved
	Regressive: Used for the best sharpness of differentiation; whole tissue is stained and then differentiated to remove excess dye from the parts that should be relatively unstained
Action on tissue	**General:** Stain that colors all parts of the tissue equally, providing no significant differentiation
	Selective: Stain that differentiates between classes of tissue or between parts of cells
Differentiation	Using regressive staining methods, the process of removing excess stain is called differentiation; this usually provides sharp staining contrasts because the hydronium and hydroxide ions in the solvents used for differentiation diffuse more rapidly than any dye ion; some of the ways in which a tissue section may be differentiated include the use of acidic or basic medium and excess mordant, buffers, or oxidizers
Staining mechanisms	**Capillarity and osmosis:** Account for penetration of the dye into the interior of the tissue
	Absorption or solution: Passage of the dye molecule from the dye bath solution to the solution in the substance being dyed
	Adsorption: Deposition of dye on the surface of the dyed material (physical surface adsorption plus salt linkages of dye to the protein chain)
Chemical factors	**Acid-base:** Based on the fact that certain cell parts are assumed to be acidic and other parts are alkaline
	Ionic strength of dye solution: Dissolved salt, either neutral or buffered, in the dye solution influences the interaction of dye and tissue; increased ionic strength decreases the staining of both acidic and basic dyes; salt ions may compete with color ion for binding site on the protein molecules
	Dye concentration: Greater amounts of dye are bound with increasing concentrations of the dye; amount of dye bound by tissue is limited by the number of available binding sites
	Fixation of tissue: When cells are fixed, the affinity for stains increases; protein molecules are reorganized so that chemical groups are more available to the dye (characteristic shared by all fixatives); greater basic dye uptake after formalin fixation; more acidic dye uptake after fixation with mercuric chloride fixatives
	Temperature: Increase in temperature increases the diffusion rate of dye molecules; increase also causes protein molecules to swell the fibers, rendering them more open to dye penetration
	Staining equilibrium: Staining is a reversible reaction; when the solution environment is changed, the equilibrium concentration in the tissue is altered

Table 23.4 Nuclear and cytoplasmic stains

Stain	Comments
Nuclear stains	
Technical classification	**Reagents used:** Dyes can be either natural or synthetic; natural dyes include cochineal, carmine, and hematoxylin; synthetic dyes include derivatives of substances found in coal tar, oxazines (safranin, basic fuchsin), thiazins (methylene blue, toluidine blue), crystal violets, and gentian violets
	How reagents are used: Depending on mordants, other solutions, etc., results will vary
Nucleic acids	**DNA (nucleus):** Feulgen reaction used to demonstrate DNA (periodic acid-Schiff stain)
	RNA (nucleolus, cytoplasm/mitochondria, ribosomes): Methyl green-pyronin used to demonstrate DNA (methyl green) and RNA (pyronin)
Cytoplasmic stains	
Definition	Usually consist of acidic dyes that contain sulfonic and carboxylic acids that combine with tissue bases, especially those proteins with an excess of the basic amino acids such as arginine, lysine, hydroxylysine, and histidine
Common cytoplasmic stains	**Picric acid:** Nitro dye used as a yellow contrast stain in collagen methods, as well as with hematoxylin
	Eosin: Used as a counterstain with hematoxylin; can be used as an aqueous or alcoholic solution
Mordant	
Definition	A substance capable of combining with a dye and the material to be dyed, thereby increasing the affinity or binding of the dye

Table 23.5 Carbohydrate stains

Carbohydrates	Stains[a]	Comments
Polysaccharides	**Iodine:** Starch and by-products (amylose and amylopectin) give blue color; glycogen and amyloid give a red-brown reaction **PAS:** glycogen strongly positive	Glycogen is the only polysaccharide that occurs in higher animals (liver cells, muscle, cartilage, certain parathyroid cells); starch occurs as carbohydrate reserves in plants; cellulose occurs in cell walls of plants
Mucopolysaccharides (acidic mucopolysaccharides, neutral mucopolysaccharides [not very relevant])	**Iodine:** Starch and by-products (amylose and amylopectin) give blue color; glycogen and amyloid give a red-brown reaction **PAS:** For carbohydrates, glycogen strongly positive, acid mucopolysaccharide does not stain, neutral mucopolysaccharide is not present in significant amounts, but chitin will stain; for lipids, glycolipids and phospholipids are PAS positive; for proteins, pale background color probably due to very small amounts of carbohydrate	Hyaluronic acid is the only example of a simple acidic mucopolysaccharide; main component of ground substance (substance that embeds collagen and elastic fibers) (eye, synovia, some pathologic conditions); for sulfated acidic mucopolysaccharides, heparin is found in a large variety of tissues (tissue mast cells are the site of production), and chondroitin sulfate is found in connective tissue and cartilage
Mucoproteins	**PAS:** Positive **Toluidine blue for mucin:** Positive	Complexes in which the protein, rather than the carbohydrate, is the principal constituent; found in blood group antigens, digestive and respiratory, etc.; the term "mucin" usually signifies mucoprotein
Glycoproteins	**PAS:** Positive	Similar to mucoproteins (hexosamine content, <4%); collagen, reticulin fibers; frequently synonymous with "mucoids"
Glycolipids	**PAS:** Positive	Contain 1 mol each of a fatty acid, sphingosine, and hexose; examples are cerebrosides and gangliosides

[a] PAS, periodic acid-Schiff.

Table 23.6 Most likely, secondary, and rare body site locations for human protozoan parasites[a]

Parasite	Life cycle stage[b]	Most likely site(s)	Secondary site(s)	Rare site(s)
Amebae				
Entamoeba histolytica	T	Colon	Liver, lungs, kidneys, skin, small intestine	Brain, heart, lymph nodes, peritoneal cavity, reproductive system, eye, spleen, bone
Entamoeba gingivalis[c]	T	Buccal cavity		
Acanthamoeba spp.	T/C	Eye, brain, skin		Lungs, bone
Balamuthia mandrillaris	T/C	Brain		
Naegleria fowleri	T	Brain		
Sappinia diploidea	T	Brain		
Flagellates				
Giardia lamblia	T	Small intestine, appendix		
Leishmania spp.	A	Skin, liver, spleen, bone marrow		Reproductive system, eye
Trypanosoma cruzi	A	Heart, bloodstream, bone marrow	Muscle, brain, colon, female reproductive system	Stomach, small intestine, lymph nodes
Ciliate				
Balantidium coli	T	Small intestine, appendix, colon		Lymph nodes, peritoneal cavity
Apicomplexans				
Cryptosporidium spp.	CO, O	Small intestine, appendix	Stomach, lungs, liver, gallbladder	
Cyclospora cayetanensis	CO, O	Small intestine		
Isospora belli	CO, O	Small intestine		Lungs, lymph nodes
Sarcocystis spp.	CO	Muscle, heart, small intestine, blood vessel walls		
Toxoplasma gondii	B, Ta	Brain, lungs, heart, lymph nodes, eye, female reproductive system, spleen	Skin, liver, gallbladder, kidneys, pancreas, small intestine, male reproductive system, bone	Bladder, ureters
Plasmodium spp.	M	Bloodstream, red blood cells	Liver	
Microsporidia				
Enterocytozoon bieneusi	S	Small intestine, colon	Lungs, liver, kidneys	
Encephalitozoon intestinalis	S	Small intestine, colon	Lungs, heart, liver, gallbladder, kidneys	
Encephalitozoon spp.	S	Small intestine, colon, eye	Lungs, heart, liver, gallbladder, kidneys, bloodstream vessels, lymph nodes	
Pleistophora	S	Muscle	Eye	
Trachipleistophora, Brachiola	S	Muscle	Eye	
Nosema	S	Eye, muscle (*Nosema*-like)		
Vittaforma corneae	S	Eye		
"Microsporidium"	S	Eye		

[a] Adapted from references 1 and 5.
[b] A, amastigotes; B, bradyzoites; C, cyst; CO, coccidian stages; M, malarial parasites; O, oocyst; S, spores and microsporidial stages; T, trophozoite; Ta, tachyzoites.
[c] If found in sputum, can be misdiagnosed as *E. histolytica* from pulmonary abscess.

Table 23.7   Most likely, secondary, and rare body site locations for human helminth parasites[a]

Parasite	Life cycle stage[b]	Most likely site(s)	Secondary site(s)	Rare site(s)	Comments
Intestinal nematodes					
Ascaris lumbricoides	A/L	Small intestine (A), lungs (L)	Liver, gallbladder, kidney, pancreas (A)	Pleural cavity	Tendency to migrate if disturbed
Enterobius vermicularis	A	Small intestine, colon	Peritoneal cavity, female reproductive system	Lungs	Extremely common; eggs highly infectious
Trichuris trichiura	A	Colon	Small intestine		"Whipworm"
Capillaria philippinensis	A	Small intestine			Eggs resemble *Trichuris*
Hookworms (*Ancylostoma duodenale, Necator americanus*)	A/L	Small intestine (A), lungs (L)			Eggs identical
Trichostrongylus spp.	A	Small intestine			Similar to hookworm disease
Strongyloides stercoralis	A/L	Small intestine (A), small intestine (L)	Lungs, colon (L)	Liver, kidneys (L)	May cause disseminated disease
Tissue nematodes					
Trichinella spp.	A/L	Small intestine (A), muscle (L)		Brain (L)	Pork, bear, walrus meat
Baylisascaris procyonis	L	Brain	Muscle, lungs, heart, liver, gallbladder		Associated with raccoons
Lagochilascaris spp.	A/L	Skin, buccal cavity		Lungs	Rare human infection
Toxocara canis	L	Eye, liver	Lungs	Brain, heart	Causes VLM and OLM
Toxocara cati	L	Eye, liver		Brain, heart	Causes VLM and OLM
Ancylostoma spp.	L	Skin	Muscle	Eye	Causes OLM
Dracunculus medinensis	A	Skin		Pericardium, eye, brain	"Fiery serpent"
Angiostrongylus cantonensis	A	Brain		Lungs, eye	Rat lungworm
Angiostrongylus costaricensis	A	Small intestine, peritoneal cavity			Abdominal disease
Gnathostoma spinigerum	L	Skin, muscle	Brain	Small intestine, eye	Larvae migrate throughout the body
Anisakis spp.	L	Stomach, small intestine		Peritoneal cavity	Associated with sushi and sashimi
Pseudoterranova decipiens	L	Stomach, small intestine		Peritoneal cavity	Associated with sushi and sashimi
Capillaria hepatica	A	Liver			No eggs found in stool
Thelazia spp.	A	Eye			Reported in California
Filarial nematodes					
Wuchereria bancrofti	A	Lymphatics		Lungs, eye	Elephantiasis
Brugia malayi	A	Lymphatics		Lungs	Elephantiasis
Brugia spp. (zoonotic)	A	Lymphatics		Lungs	
Onchocerca volvulus	A	Skin			"River blindness"
Onchocerca spp. (zoonotic)	A	Skin			
Loa loa	A	Skin, eye			African eye worm
Mansonella perstans	A	Peritoneal cavity		Eye	

(continued)

Table 23.7 Most likely, secondary, and rare body site locations for human helminth parasites[a] *(continued)*

Parasite	Life cycle stage[b]	Most likely site(s)	Secondary site(s)	Rare site(s)	Comments
Mansonella strepto-cerca	A	Skin			
Mansonella ozzardi	A	Skin			
Dirofilaria immitis	A	Lungs	Heart	Skin	"Coin" lesions
Dirofilaria repens	A	Skin, eye		Lungs	
Dirofilaria spp.	A	Skin, eye		Bloodstream	
Unusual nematodes					
Oesophagostomum spp.	A/L	Colon	Peritoneal cavity		Nodular mass lesions
Mammomonogamus spp.	A	Lungs, buccal cavity			Y-shaped worms
Dioctophyma renale	A/L	Kidney (A), skin (L)			Primarily right kidney
Tissue cestodes					
Taenia solium (cysticercus)	L	Skin, muscle, brain, eye	Heart		Pork tapeworm
Echinococcus granulosus (unilocular cystic hydatid)	L	Liver, lungs	Muscle, brain, heart, kidneys, eye, spleen, bone	Peritoneal cavity	Sheep-sheepdog cycle; hydatid disease
E. multilocularis (alveolar cystic hydatid)	L	Liver	Brain, lungs	Peritoneal cavity	Difficult to treat
Taenia/Multiceps spp. (coenurus)	L	Skin, muscle, brain, eye	Peritoneal cavity		Most often in central nervous system
Spirometra/Diphyllobothrium spp. (spargana)	L	Skin, muscle	Lungs, heart, peritoneal cavity, reproductive system, bladder	Brain, eye	Use of raw, infected tissue as poultice
Mesocestoides spp.	L	Peritoneal cavity			Approximately 20 cases
Intestinal trematodes					
Heterophyids	A/E	Small intestine	Heart (E)		Eggs almost identical
Clonorchis/Opisthorchis	A/E	Liver			Chinese liver fluke
Fasciola	A/E	Liver		Skin, peritoneal cavity	Sheep liver fluke
Paragonimus spp.	A/E	Lungs	Skin, brain (A/E); muscle (A/E)		Lung fluke
Schistosoma mansoni	A/E	Blood vessels, peritoneal cavity, mesenteries (A); colon (E)	Brain (E)	Liver, gallbladder	(E) Inferior mesenteric veins (A)
Schistosoma haematobium	A/E	Bladder, ureters, blood vessels (A); colon (E)			Vesicle, prostate, uterine plexuses (A)
Schistosoma japonicum	A/E	Peritoneal cavity, mesenteries, blood vessels	Lungs (E); brain, small intestine, colon (E)	Liver, gallbladder	(E) Superior mesenteric veins (A)
Unusual trematodes					
Alaria spp.	L	Skin	Stomach, lymph nodes, liver	Eye	Ingestion of poorly cooked frogs (legs)
Arthropods					
Demodex spp.	A/L	Skin (hair follicles)			Follicle mite
Sarcoptes scabiei	A/L	Skin			Itch mite, scabies

(continued)

Table 23.7 *(continued)*

Parasite	Life cycle stage[b]	Most likely site(s)	Secondary site(s)	Rare site(s)	Comments
Tunga penetrans	A	Skin			Chigoe, sand flea
Pentastomes	L	Lungs, liver, peritoneal cavity	Lymph nodes, lymphatics in buccal cavity	Eye	Tongue worms
Myiasis (fly larvae)					
Cochliomyia	L	Nose, ear, brain			Infects living tissue; common name is New World screwworm
Calliphora	L	Skin, intestine, GU tract			Infects dead tissue; common name is bluebottle
Cordylobia	L	Skin, usually feet			Infects subcutaneous boils; common name is Tumbu fly
Sarcophaga	L	GI tract, skin, GU tract, vagina			Infects dead or living tissue, depending on species; common name is flesh fly
Dermatobia	L	Skin, boils, brain			Infects living tissues; common name is human botfly
Cuterebra	L	Skin, eye			Infects living tissue; common name is North American botfly
Gastrophilus	L	Skin, migrates to eye			Undergoes "larva migrans" cutaneous migration; common name is horse botfly
Hypoderma	L	Skin, eye, nasopharynx			Undergoes extensive larval wandering (limbs); common name is warble fly
Oestrus	L	Nasopharynx, eye, sinuses			Infects living tissue, resembles conjunctivitis; common name is sheep nasal botfly

[a] Adapted from references 1, 4, and 5.
[b] A, adult (for filarial nematodes, includes immature adults, not microfilariae); E, eggs; GI, gastrointestinal tract; GU, genitourinary tract; L, larva(e).

Protozoa

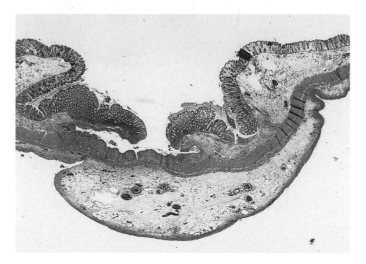

Figure 23.1 *Entamoeba histolytica.* Section of an amebic ulcer of the colon with characteristic undermining and partial destruction of the muscularis. The serosa is edematous and hyperemic. Multiple ulcers may develop and coalesce. Organisms would be found in the healthy tissue border, not in the necrotic material within the ulcer. ×7. (Armed Forces Institute of Pathology photograph.)

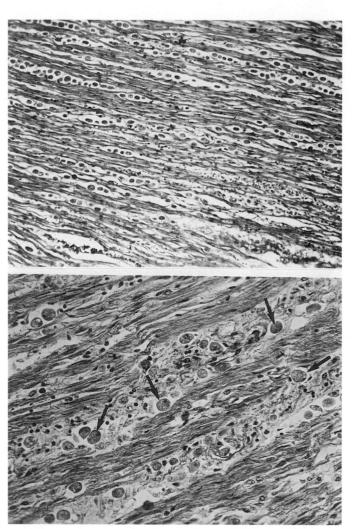

Figure 23.3 (Upper) Section of colon shows diffuse invasion of muscularis by trophozoites of *Entamoeba histolytica.* ×85. (Armed Forces Institute of Pathology photograph; contributed by Ruy Perez-Tamayo, Mexico, D.F.) (Lower) Higher magnification of the same specimen. Trophozoites have infiltrated the muscularis of the colon (arrows). ×170. (Armed Forces Institute of Pathology photograph.)

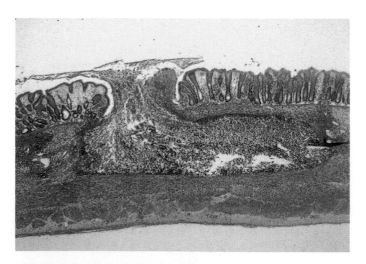

Figure 23.2 *Entamoeba histolytica.* Section of an amebic ulcer of the colon with erosion of the muscularis. The characteristic "flask-shaped" ulcer, in which the opening on the surface is much smaller than the actual ulcer below, is clearly visible. Multiple ulcers are often present and may coalesce under the surface. (From A Pictorial Presentation of Parasites: A cooperative collection prepared and/or edited by H. Zaiman.)

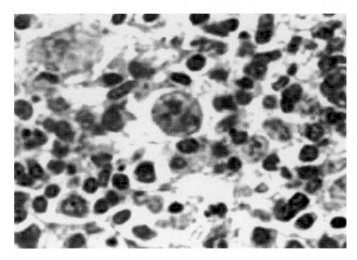

Figure 23.4 *Entamoeba histolytica*. Trophozoite in the exudate within the ulcerated dermis. The nuclear structure is visible but does not clearly show the nucleus and karyosome. The nuclear structure is more darkly staining than the cytoplasm.

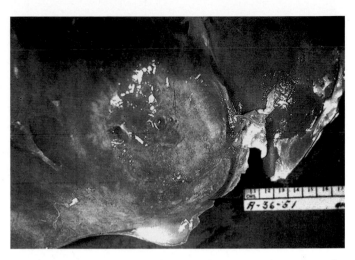

Figure 23.6 Gross image of liver abscess caused by *Entamoeba histolytica*. Note the sunken area of the abscess. (Centers for Disease Control and Prevention.)

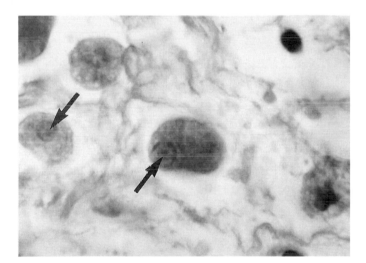

Figure 23.5 Section of colon showing the nuclei and central karyosomes (arrows) of two invading trophozoites of *Entamoeba histolytica*. ×1,224. (Armed Forces Institute of Pathology photograph.)

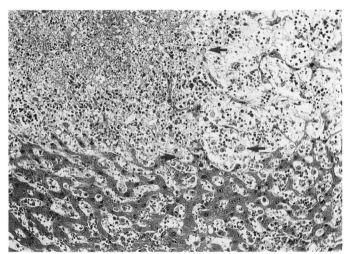

Figure 23.7 Section of liver at the margin of an amebic abscess showing several *Entamoeba histolytica* trophozoites (arrows) in addition to necrosis. ×110. (Armed Forces Institute of Pathology photograph.)

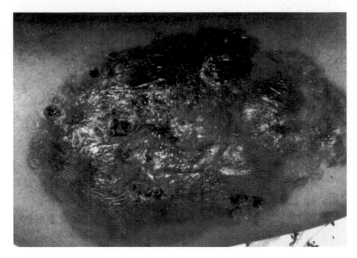

Figure 23.8 *Acanthamoeba* spp. Cutaneous abscess on arm; nonhealing ulcer, possibly as a result of a human bite wound, that was not recognized as being caused by *Acanthamoeba* spp. Therapy with routine antibiotics was ineffective. (Courtesy of George Healy, Centers for Disease Control and Prevention.)

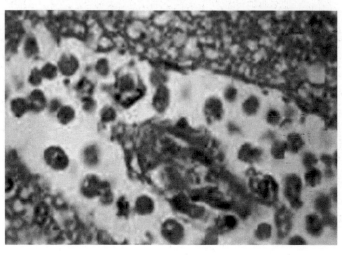

Figure 23.10 *Naegleria fowleri.* Trophozoites are seen in brain tissue. The organisms are shaped like amebae and have the large, characteristic karyosome. The staining of the karyosome is darker than that of the cytoplasm. The karyosome is not clearly visible in all amebae, and the overall image is not as clear as that in Figure 23.9. (Centers for Disease Control and Prevention.)

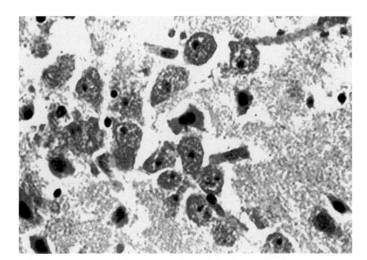

Figure 23.9 *Balamuthia mandrillaris.* Trophozoites are seen in brain tissue. The organisms are shaped like amebae and have the large, characteristic karyosome. The staining of the karyosome is much darker than that of the cytoplasm of the amebae. (Centers for Disease Control and Prevention.)

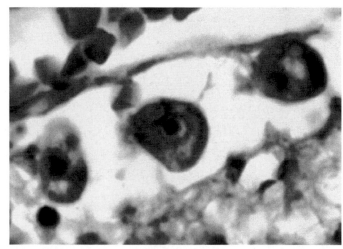

Figure 23.11 *Naegleria fowleri.* Trophozoites are seen in brain tissue (1,000× oil immersion). Note the very clearly delineated karyosome within the nucleus. Also note the vacuolated cytoplasm that is sometimes seen at higher magnifications. (Centers for Disease Control and Prevention.)

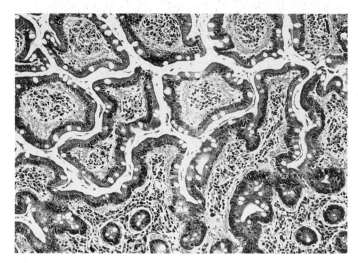

Figure 23.12 *Giardia lamblia* trophozoites seen as very small sickle-like profiles over the intestinal epithelium. ×127. At this magnification, it is very difficult to identify the organisms, but the shape and appearance of the organisms are strongly suggestive of *G. lamblia*. (Armed Forces Institute of Pathology photograph.)

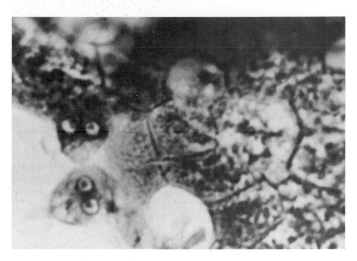

Figure 23.14 *Giardia lamblia* trophozoites seen as teardrop-shaped organisms; note the two "eyes" (nuclei) looking back at you. At this magnification (1,000× oil immersion), the nuclei, median bodies, and axonemes are visible. Duodenal smear.

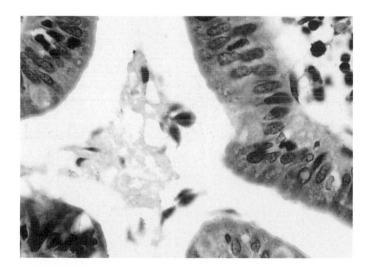

Figure 23.13 *Giardia lamblia* trophozoites seen as teardrop-shaped organisms. ×645. Although the teardrop shape is visible for some of the trophozoites, the internal structures (nuclei, median bodies, and axonemes) are not visible at this magnification. (Armed Forces Institute of Pathology photograph.)

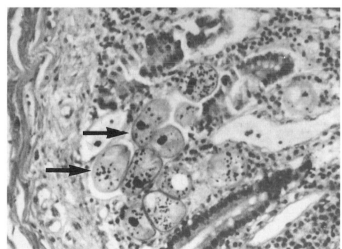

Figure 23.15 *Balantidium coli* trophozoites within the intestinal wall (arrows). Normally, the ulcers involve the mucosa and submucosa; however, invasion of the muscular layer may occur, as well as bowel wall perforation. Ulcers tend to be flask shaped (like *E. histolytica*) and shallow and have a wide opening. Extraintestinal dissemination can occur but is rare. (From A Pictorial Presentation of Parasites: A cooperative collection prepared and/or edited by H. Zaiman.)

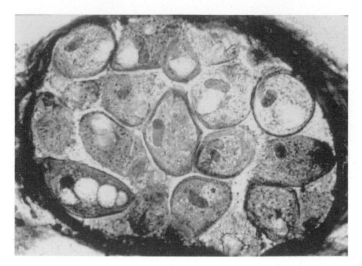

Figure 23.16 Necrotic gangrenous colon showing *Balantidium coli* trophozoites in a venule. (From A Pictorial Presentation of Parasites: A cooperative collection prepared and/or edited by H. Zaiman. Photograph courtesy of R. B. Holliman.)

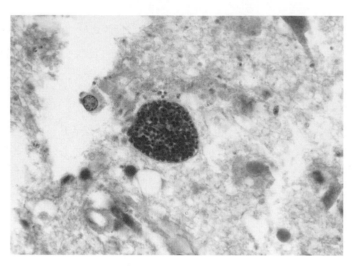

Figure 23.18 *Toxoplasma gondii* cyst in brain (note organisms within the cyst); non-AIDS patient. ×645. (Armed Forces Institute of Pathology photograph.)

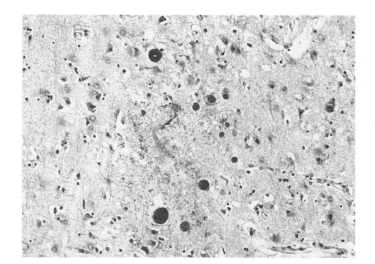

Figure 23.17 *Toxoplasma gondii* cysts in brain tissue (dark, round objects) of a non-AIDS patient. The presence of these cysts may be an accidental finding at autopsy and may not be the cause of the patient's illness. ×129. (Armed Forces Institute of Pathology photograph.)

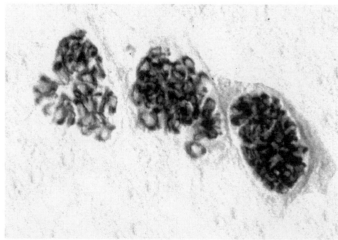

Figure 23.19 *Toxoplasma gondii* cysts in tissue. Some of the organisms appear to be more oval or crescent shaped. While these cysts occur in tissue, their presence may or may not be relevant to the patient's clinical condition.

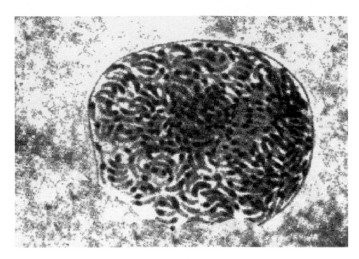

Figure 23.20 *Toxoplasma gondii* cyst in tissue. In this cyst, the individual organisms can be clearly seen. Cysts in brain tissue tend to be more round than those seen in striated muscle.

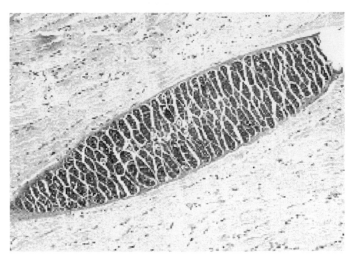

Figure 23.22 *Sarcocystis* spp. Note the septa which are visible in this low-power image. Bradyzoites of *Toxoplasma gondii* in striated muscle resemble *Sarcocystis* spp. However, the *Sarcocystis* bradyzoites are usually more rounded at both ends, and they are contained in larger cysts than those of *Toxoplasma*.

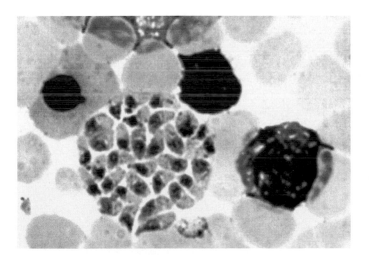

Figure 23.21 *Toxoplasma gondii* tachyzoites in bone marrow from a 5-year-old child with leukemia. Note the oval or crescent-shaped organisms. (Centers for Disease Control and Prevention.)

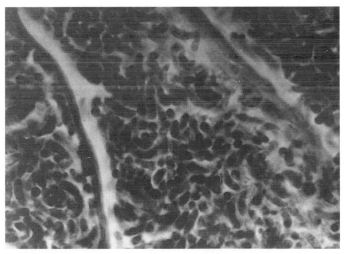

Figure 23.23 The *Sarcocystis* septa are visible, as are the crescentic spores. The sarcocyst wall varies from thin and smooth to thick and striated. Although sarcocysts may be confused with cysts of *Toxoplasma*, sarcocysts tend to be larger and contain larger bradyzoites. (From A Pictorial Presentation of Parasites: A cooperative collection prepared and/or edited by H. Zaiman.)

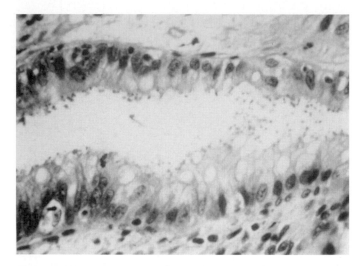

Figure 23.24 *Cryptosporidium* parasites in the rectum. Note the upper left section with organisms (dark spots) against the brush border. (From A Pictorial Presentation of Parasites: A cooperative collection prepared and/or edited by H. Zaiman. Photograph courtesy of K. Lewin.)

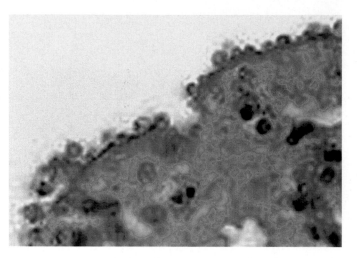

Figure 23.26 *Cryptosporidium* parasites in the colon. Note the organisms right at the edge of the brush border. The organisms are within parasitophorous vacuoles and hence are enclosed within membranes of host cell origin. The mucosal architecture tends to be abnormal, with marked shortening of the villi and hypertrophy of the crypts. (Centers for Disease Control and Prevention.)

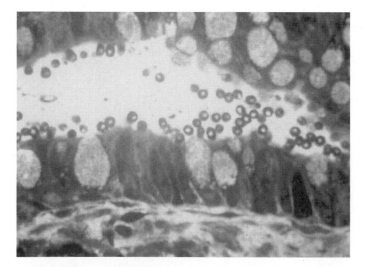

Figure 23.25 *Cryptosporidium* parasites in the rectum (higher magnification than Figure 23.24). Previously, most human cases were diagnosed after examination of small or large bowel biopsy material, often using both light and electron microscopy. However, because biopsy specimens may not originate from the infected area of the mucosa, immunoassays for the detection of antigen in the stool are being widely used. (From A Pictorial Presentation of Parasites: A cooperative collection prepared and/or edited by H. Zaiman. Photograph courtesy of K. Lewin.)

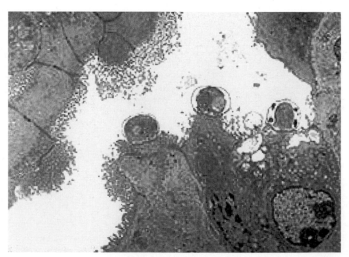

Figure 23.27 *Cryptosporidium* spp. Transmission electron microscopy of organisms within parasitophorous vacuoles on the intestinal mucosa. (U.S. Department of Agriculture.)

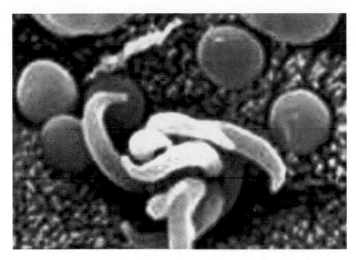

Figure 23.28 *Cryptosporidium* spp. Scanning electron microscopy of organisms at the brush border. Note the sporozoites being released from the cell. (U.S. Department of Agriculture.)

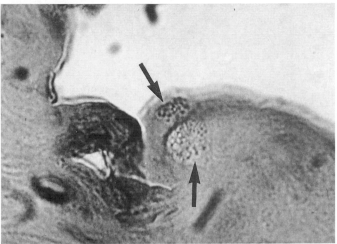

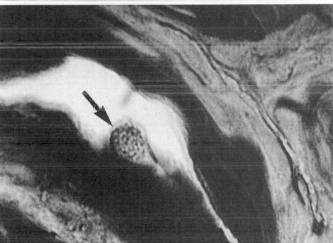

Figure 23.29 Microsporidia in a corneal lesion; note the periodic acid-Schiff-positive dot at the end of each spore (arrows). Routine histologic testing can be performed by using tissue Gram stains or silver stains. Touch preparations can be methanol fixed and stained with Giemsa stain. Plastic-embedded tissues stained with periodic acid-Schiff, silver, acid-fast, and routine hematoxylin-eosin stains generally stain better than paraffin-embedded tissues. This finding may be related to the use of formalin as a tissue fixative (1,000× oil immersion).

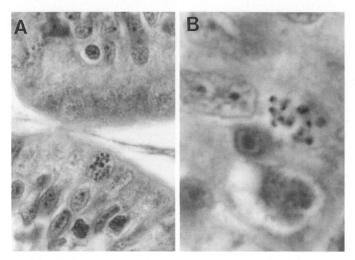

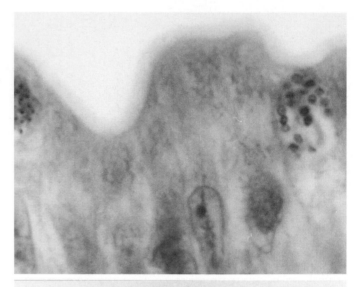

Figure 23.30 Microsporidia in the intestine; note the spores. When seen in stool, the spores measure approximately 1 to 3 µm. (A) Lower magnification; (B) higher magnification. The microsporidia multiply extensively within the host cell cytoplasm; the life cycle includes repeated divisions by binary fission (merogony) or multiple fission (schizogony) and spore production (sporogony). Both merogony and sporogony can occur in the same cell at the same time. During sporogony, a thick spore wall is formed, providing environmental protection for this infectious stage of the parasite. Microsporidia are characterized by having spores containing a polar tubule, which is an extrusion mechanism for injecting the infective spore contents into host cells. To date, seven genera have been recognized in humans: *Brachiola*, *Encephalitozoon*, *Enterocytozoon*, *Pleistophora*, *Trachipleistophora*, *Vittaforma*, and "*Microsporidium*," a catchall genus for those organisms not yet classified.

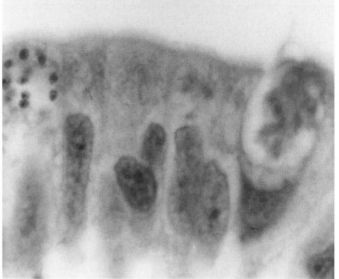

Figure 23.31 Microsporidia in various stages of development in the intestinal enterocytes (1,000× oil immersion).

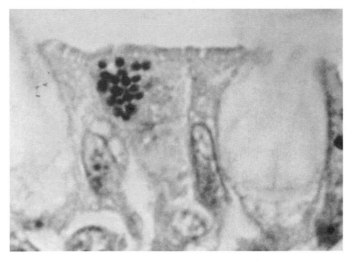

Figure 23.32 Microsporidia in various stages of development in the intestinal enterocytes. Note the cells that have evacuated their contents into the lumen of the gut (1,000× oil immersion).

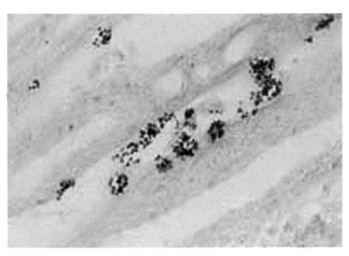

Figure 23.33 Microsporidia. Note the spores within the kidney. Dissemination to the kidneys is not uncommon, particularly with *Encephalitozoon* spp. Any patient with suspected microsporidiosis should have both stool and urine examined for spores. Both *Enterocytozoon bieneusi* and *Encephalitozoon* spp. can disseminate from the gut to other body sites, including the kidneys.

Helminths—Roundworms

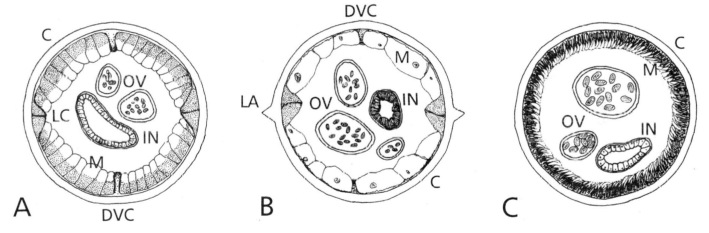

Figure 23.34 Diagram of roundworm musculature morphology in tissue. (A) Polymyarian type. (B) Meromyarian type. (C) Homomyarian type. C, cuticle; DVC, dorsal-ventral chord; IN, intestine; LA, lateral alae; LC, lateral chord; M, muscle; OV, ovary. See Table 23.8 below. (Illustration by Sharon Belkin.)

Table 23.8 Musculature of nematodes[a]

Characteristic	Polymyarian	Meromyarian	Homomyarian
No. of rows of muscle cells	More than 5	Between 2 and 5	Does not exceed 2
Size of cells	Small, uniform	Large, various shapes	Very small, uniform
Arrangement of cells	Projecting into body cavity	Irregular	Regular
No. of cells	Many in each quarter	2 or 3 in each quarter	Many in complete circle
Lateral chords	Present	Present	Absent
Parasites with this type of musculature	*Ascaris*, filaria, *Dracunculus*, *Angiostrongylus*, *Anisakis*	*Enterobius*, hookworms	*Trichuris*, *Trichinella*

[a] Adapted from references 1 to 6.

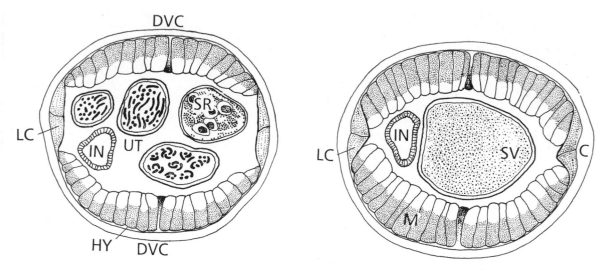

Figure 23.35   Diagram through *Loa loa* (female on the left, male on the right) (polymyarian type of musculature). Note the well-developed muscle layer divided into four bands separated by the lateral, ventral, and dorsal chords. The muscle cells project into the body cavity. C, cuticle; DVC, dorsal-ventral chord; HY, hypodermis; IN, intestine; LC, lateral chord; M, muscle; SR, seminal receptacle; SV, seminal vesicle; UT, uterus. (Illustration by Sharon Belkin.)

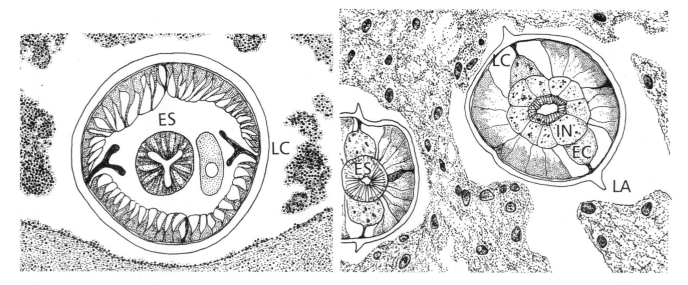

Figure 25.36   Diagram through *Anisakis* (left) and *Baylisascaris procyonis* (right) (polymyarian type of musculature). In *Anisakis*, note the well-developed muscular esophagus (ES) and the Y-shaped lateral chords (LC). In *B. procyonis* (brain tissue), one section is through the esophagus (ES) while the other is through the middle of the body, showing the excretory columns (EC), intestine (IN), and large lateral chords (LC). Note that the lateral alae (LA) are visible in each section.

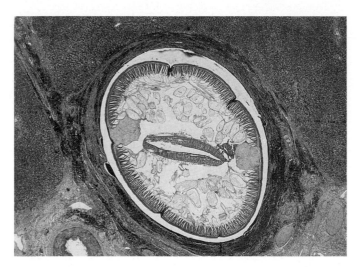

Figure 23.37 Adult *Ascaris lumbricoides* in a dilated bile duct. ×12. At this low magnification, the individual polymyarian muscle cells are somewhat more difficult to see but the large lateral chords are clearly visible. (Armed Forces Institute of Pathology photograph.)

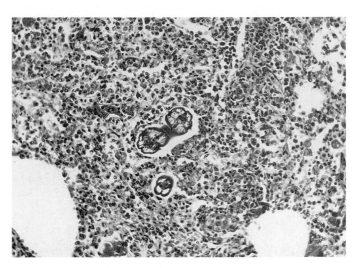

Figure 23.39 Three transverse sections of *Ascaris lumbricoides* larvae in the lung, causing bronchopneumonia. ×157. Even in this low-magnification image, the muscle structure (polymyarian) appears to be divided into quadrants by the lateral chords. (Armed Forces Institute of Pathology photograph.)

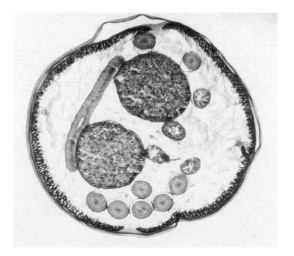

Figure 23.38 Cross section of an *Ascaris lumbricoides* adult female. The outer cuticle is partially detached from the parasite. The hypodermis and musculature are visible below the cuticle. Central to the musculature, the flattened tube is the gut and the round tubes are reproductive structures. The large round tubes are egg-containing uteri; the others are sections through ovaries and oviducts. (From A Pictorial Presentation of Parasites: A cooperative collection prepared and/or edited by H. Zaiman.)

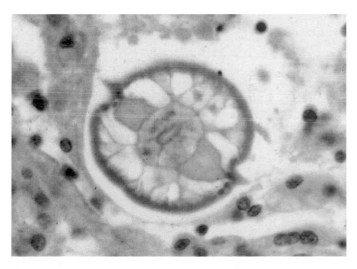

Figure 23.40 *Baylisascaris procyonis* larva (polymyarian type of musculature) and inflammation in the corpus callosum of an infected rabbit. *B. procyonis* is an ascarid normally found in raccoons, has a normal ascarid-like life cycle, and causes a very serious zoonotic disease in humans. Human infections result from ingestion of eggs that are passed in very large numbers (millions of eggs/day) in the feces of infected raccoons. Rather than developing to adult worms as occurs in the raccoon, the larvae migrate extensively throughout the body tissues, causing visceral larva migrans (VLM) and/or neural larva migrans (NLM). (From A Pictorial Presentation of Parasites: A cooperative collection prepared and/or edited by H. Zaiman. Photograph courtesy of K. R. Kozacos.)

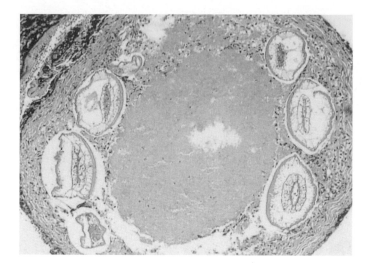

Figure 23.41 *Lagochilascaris* (polymyarian type of musculature) encapsulated larvae in the submucosa of a resected tonsil. (From A Pictorial Presentation of Parasites: A cooperative collection prepared and/or edited by H. Zaiman. Photograph courtesy of D. Botero and M. D. Little.)

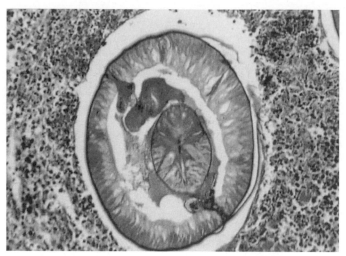

Figure 23.43 *Anisakis* in the stomach wall. Note the typical polymyarian type of musculature and the Y-shaped lateral chords. Human infection is acquired by the ingestion of raw, pickled, salted, or smoked saltwater fish. The larvae often penetrate the walls of the digestive tract (frequently the stomach), where they become embedded in eosinophilic granulomas. The throat is occasionally involved. (From A Pictorial Presentation of Parasites: A cooperative collection prepared and/or edited by H. Zaiman.)

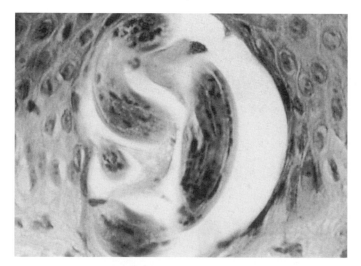

Figure 23.42 *Toxocara* larva present in the skin. Since these are not photographed from a cross section, the typical musculature arrangement is not visible. (From A Pictorial Presentation of Parasites: A cooperative collection prepared and/or edited by H. Zaiman.)

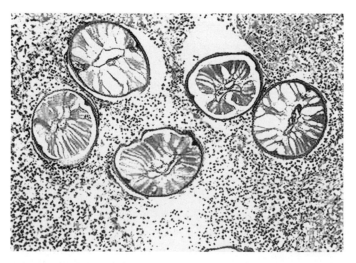

Figure 23.44 *Dirofilaria* sp. Note the polymyarian type of musculature and the clearly delineated lateral chords. Also note the longitudinal ridges in the cuticle (bumpy appearance). (Courtesy of Marjorie R. Fowler and Andrea Linscott.)

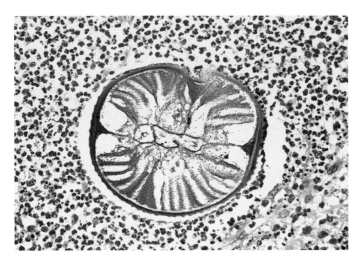

Figure 23.45 *Dirofilaria* sp. (higher magnification than Figure 23.44). Note the typical polymyarian type of musculature and the more clearly delineated longitudinal ridges on the inner surface of the cuticle. The cuticle is relatively thick and multilayered. The very prominent lateral chords are also visible; although ventral and dorsal chords are present, they are usually inconspicuous. (Courtesy of Marjorie R. Fowler and Andrea Linscott.)

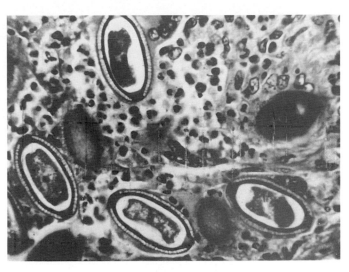

Figure 23.46 *Capillaria hepatica* eggs in liver. These eggs resemble those of *Trichuris trichiura*, but the shells in *Capillaria* eggs are striated. (From A Pictorial Presentation of Parasites: A cooperative collection prepared and/or edited by H. Zaiman.)

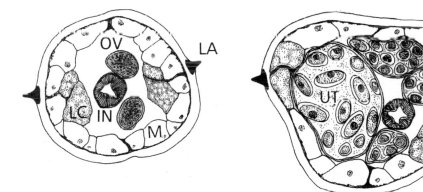

Figure 23.47 Diagram through *Enterobius vermicularis* (both images are of females) (meromyarian type of musculature). Note the very conspicuous lateral alae (LA). The muscle cells (M) are large, of various shapes, and irregular; there are fewer cells per quadrant than seen in the polymyarian type (Figure 23.35). The lateral chords (LC), intestine (IN), ovary (OV), and uterus (UT) are also visible. (Illustration by Sharon Belkin.)

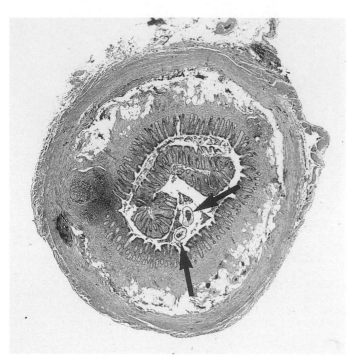

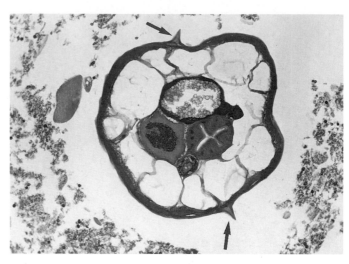

Figure 23.50 Colon containing *Enterobius vermicularis*, non-gravid female, midbody. Note the lateral alae (arrows). ×166. Also note the typical meromyarian musculature, with large, differently shaped, irregular cells (two or three cells in each quadrant). (Armed Forces Institute of Pathology photograph.)

Figure 23.48 Appendix containing two transverse sections of a female *Enterobius vermicularis* worm (arrows). ×13. Although this nematode has the meromyarian type of musculature, this magnification is too low to see any details. (Armed Forces Institute of Pathology photograph.)

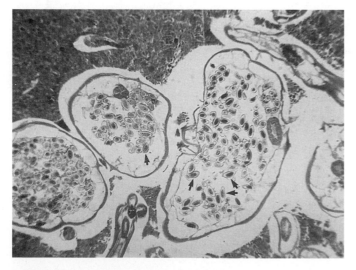

Figure 23.49 Cross section of adult *Enterobius vermicularis* worms in the appendix. Note the characteristic football-shaped eggs (arrows). The lateral alae are also visible. (From A Pictorial Presentation of Parasites: A cooperative collection prepared and/or edited by H. Zaiman.)

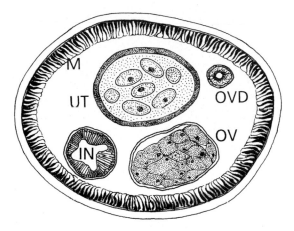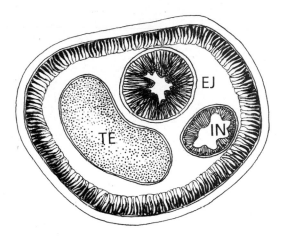

Figure 23.51 Diagram through *Trichuris trichiura* (female on the left, male on the right) (homomyarian type of musculature). Note the very small, uniform muscle cells (M) with a regular arrangement, with many cells arranged in a complete circle. Note also the absence of lateral chords. The ejaculatory duct (EJ), intestine (IN), ovary (OV), oviduct (OVD), uterus (UT), and testis (TE) can also be seen. (Illustration by Sharon Belkin.)

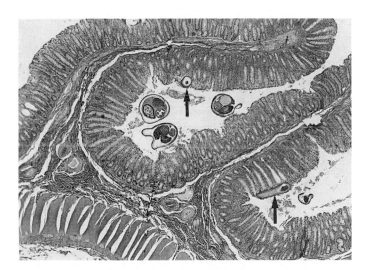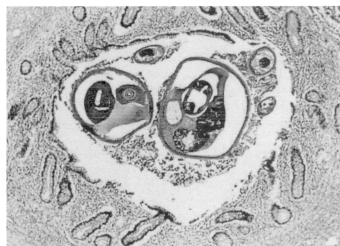

Figure 23.52 Several sections of adult male and female *Trichuris trichiura* worms in the colon, showing the narrow anterior portion of the worm within the mucosa ("whip") (arrows) and the thicker posterior portion ("handle") free in the lumen. ×15. (Armed Forces Institute of Pathology photograph.)

Figure 23.53 *Trichuris trichiura* in cross section. Note the smaller sections that represent the head end, which is embedded in the mucosa, while the larger sections represent the tail portion (whip handle).

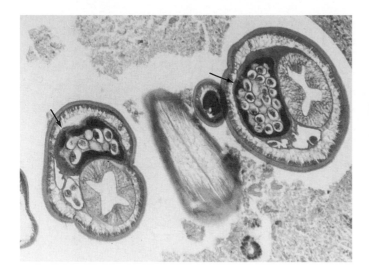

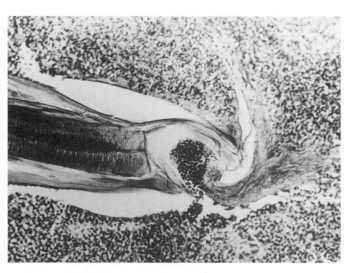

Figure 23.54 *Trichuris trichiura* in the canine cecum. Cross sections through the thin "whiplike" anterior portion and the thick posterior "handle" are seen. Note the musculature of the body wall beneath the cuticle, the gut, and the uterus filled with eggs (arrows). The musculature is of the homomyarian type, with very small, uniform, regular cells with many cells in a complete circle. No lateral chords are present. (From A Pictorial Presentation of Parasites: A cooperative collection prepared and/ or edited by H. Zaiman.)

Figure 23.56 Adult *Ancylostoma caninum* in the canine intestine. (From A Pictorial Presentation of Parasites: A cooperative collection prepared and/or edited by H. Zaiman.)

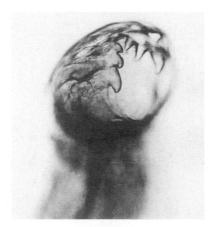

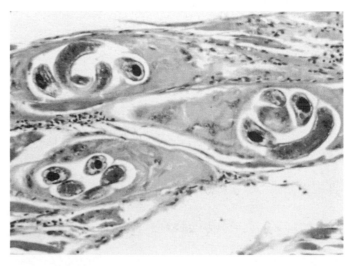

Figure 23.55 *Ancylostoma caninum*. There are three pairs of teeth in the mouth of this adult hookworm. (From A Pictorial Presentation of Parasites: A cooperative collection prepared and/ or edited by H. Zaiman.)

Figure 23.57 *Trichinella spiralis* encysted larvae in muscle. *Trichinella* forms a complex of at least five species, all of which appear to be the same morphologically but, on the basis of DNA studies and comparative features, are quite different. The five recognized *Trichinella* spp. are *T. spiralis*, *T. nativa*, *T. nelsoni*, *T. britovi*, and *T. pseudospiralis*; a sixth species, *T. papuae*, has also just been proposed. Although *Trichinella* spp. have the homomyarian type of musculature, the magnification of this image is insufficient to show the details.

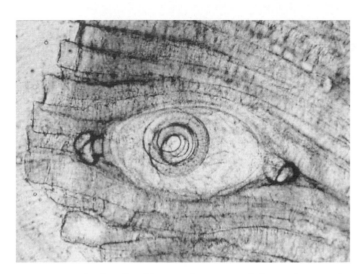

Figure 23.58 *Trichinella spiralis* encysted larva in muscle. Squash preparation of tissue biopsy specimen.

Helminths—Cestodes

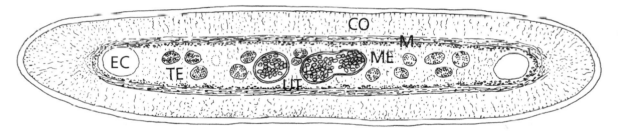

Figure 23.59 Diagram of a transverse section of a *Taenia* proglottid. Under the thick tegument, the underlying parenchyma is divided into cortical (CO) and medullary (ME) layers by a thick band of longitudinal muscles (M). The excretory columns (EC) are clearly visible. Branches of the uterus (UT) and testes (TE) are also seen. (Illustration by Sharon Belkin.)

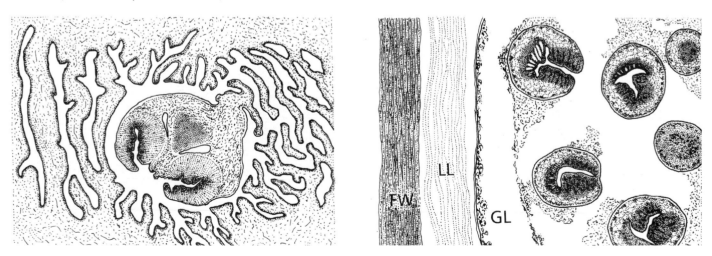

Figure 23.60 (Left) Diagram of a cysticercus of *Taenia solium*. The extensively folded spiral canal and a scolex with two large suckers are clearly visible. The denser tissue of the rostellum (between the two suckers) is also visible. In this particular section plane, the hooklets are not visible. Host tissue forms a fibrous capsule around the cysticercus. (Right) Diagram of a unilocular hydatid cyst of *Echinococcus granulosus*. The outer thick fibrous wall is visible (FW) and is produced by the host. Within the fibrous layer is the thinner laminated layer (LL), while right below that layer is the thin germinal epithelial layer (GL) from which daughter cysts and multiple protoscolices arise. Multiple protoscolices within a brood capsule are visible in this illustration. Note that the middle laminated layer is acellular and the thin germinal layer contains some calcareous corpuscles. (Illustration by Sharon Belkin.)

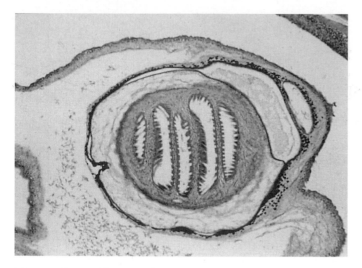

Figure 23.61 *Taenia solium* cysticercus in pig muscle. The cysticercus consists of a fluid-filled bladder containing a single protoscolex. The cysticerci (*T. solium*) have a thick bladder wall, and the rostellum has two rows of hooklets (13 each). Each cysticercus measures 5 to 15 mm long by 4 to 12 mm wide. A convoluted spiral canal leads to the rostellum. The parenchyma usually contains calcareous corpuscles. (From A Pictorial Presentation of Parasites: A cooperative collection prepared and/or edited by H. Zaiman. Photograph courtesy of P. M. Schantz.)

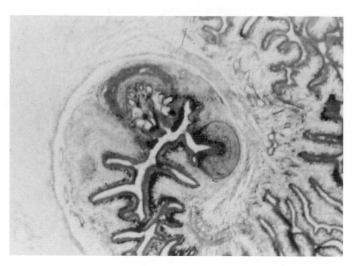

Figure 23.63 *Taenia solium* subcutaneous nodule from a patient (higher magnification than Figure 23.62). (From A Pictorial Presentation of Parasites: A cooperative collection prepared and/or edited by H. Zaiman. Photograph courtesy of K. Juniper, Jr.)

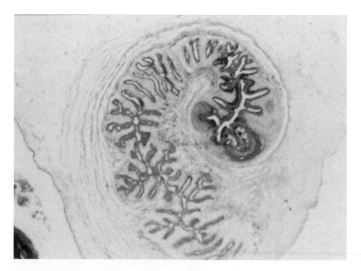

Figure 23.62 *Taenia solium* subcutaneous nodule from a patient. (From A Pictorial Presentation of Parasites: A cooperative collection prepared and/or edited by H. Zaiman. Photograph courtesy of K. Juniper, Jr.)

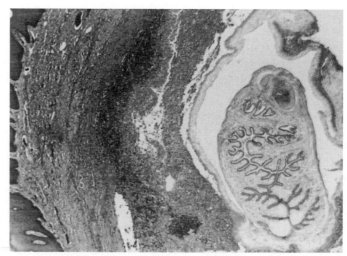

Figure 23.64 *Taenia solium* cysticercus. Low magnification.

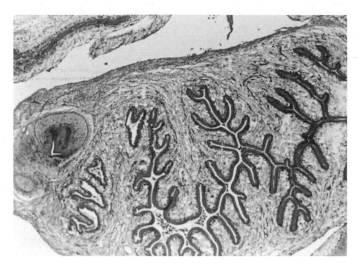

Figure 23.65 *Taenia solium* cysticercus (higher magnification than Figure 23.64; same specimen).

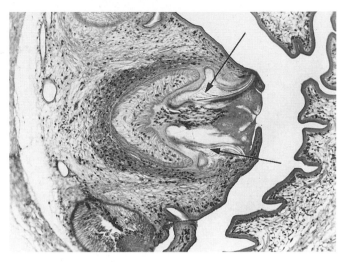

Figure 23.67 Scolex of a cysticercus showing a sucker and hooklets of *Taenia solium* (arrows). ×148. (Armed Forces Institute of Pathology photograph.)

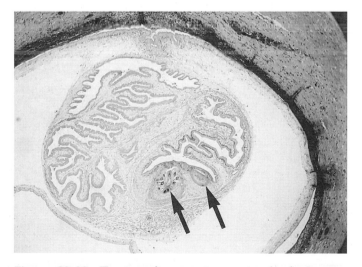

Figure 23.66 *Taenia solium* cysticercus in the brain, surrounded by fibrous tissue. The scolex is invaginated with two of the four suckers (arrows), and several hooklets are visible. The scolex is surrounded by fluid and the cyst wall. ×42. (Armed Forces Institute of Pathology photograph.)

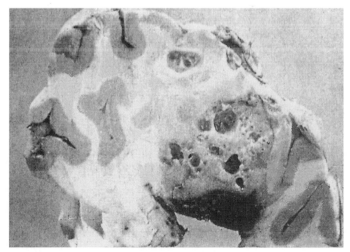

Figure 23.68 *Taenia solium* racemose cyst in the brain. Autopsy gross specimen. (From a Pictorial Presentation of Parasites: A cooperative collection prepared and/or edited by H. Zaiman.)

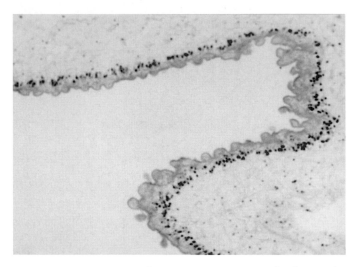

Figure 23.69 *Taenia solium* racemose cyst in the brain; the bladder wall of the racemose cysticercus has three separate layers. The tegumental surface has wartlike protuberances and is acidophilic. Beneath the tegument are small, rounded, pyknotic nuclei. The innermost layer is made up of loose connective tissue.

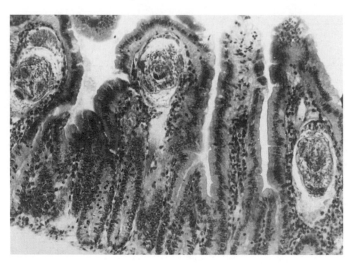

Figure 23.71 *Hymenolepis nana* cysticerci in the villi of a mouse. When mature, these cysticerci emerge, attach to the intestinal wall, and develop into the adult worm (same life cycle seen in humans). (From A Pictorial Presentation of Parasites: A cooperative collection prepared and/or edited by H. Zaiman. Photograph courtesy of B. Gueft; specimen courtesy of M. Yoeli.)

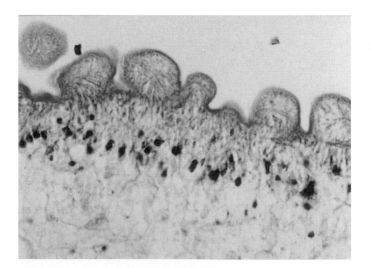

Figure 23.70 *Taenia solium* racemose cyst in the brain (higher magnification than Figure 23.69; same specimen).

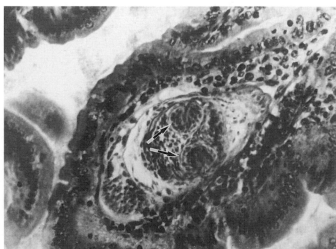

Figure 23.72 *Hymenolepis nana* cysticerci in the villi of a mouse. Note the curved suckers (arrows). (From A Pictorial Presentation of Parasites: A cooperative collection prepared and/or edited by H. Zaiman. Photograph courtesy of B. Gueft; specimen courtesy of M. Yoeli.)

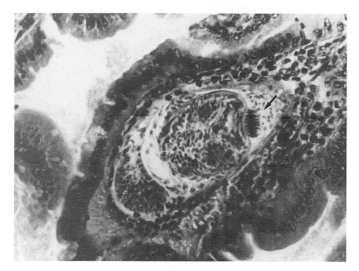

Figure 23.73 *Hymenolepis nana* cysticerci in the villi of a mouse. Note the hooklets (arrow). (From A Pictorial Presentation of Parasites: A cooperative collection prepared and/or edited by H. Zaiman. Photograph courtesy of B. Gueft; specimen courtesy of M. Yoeli.)

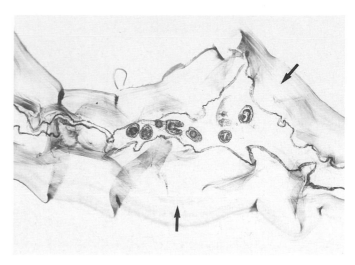

Figure 23.75 Wall of the *Echinococcus granulosus* cyst, which has a laminated membrane (arrow). The germinal layer and scolices are also visible. ×35. (Armed Forces Institute of Pathology photograph.)

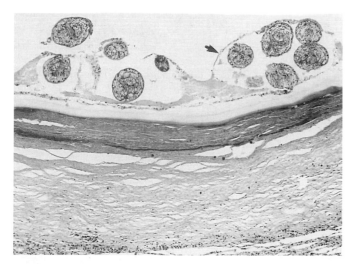

Figure 23.74 *Echinococcus granulosus* brood capsules (arrow) arising from the germinal layer. The fibrinous wall of the host and the laminated membrane are also seen. ×84. (Armed Forces Institute of Pathology photograph.)

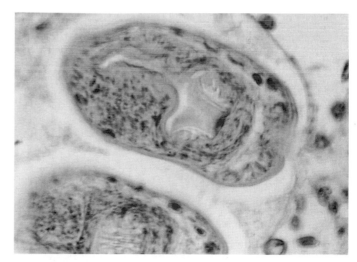

Figure 23.76 *Echinococcus granulosus*, magnification of the protoscolices.

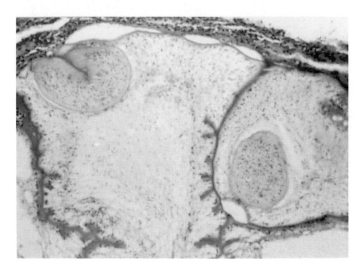

Figure 23.77 Coenurus from a human eye. *Taenia* species with multiple scolices; no daughter cysts develop in a coenurus. (From A Pictorial Presentation of Parasites: A cooperative collection prepared and/or edited by H. Zaiman. Photograph courtesy of P. J. Fripp.)

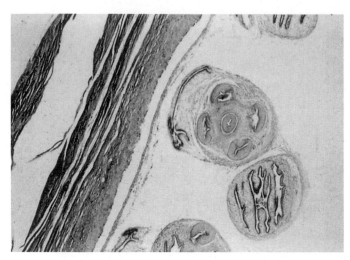

Figure 23.79 Coenurus. Note the section through all four suckers. Coenurosis is caused primarily by two species of *Taenia*, *T. multiceps* and *T. serialis*, formerly known as *Multiceps multiceps* and *M. serialis*. A coenurus consists of a viscous, fluid-filled bladder into which multiple scolices invaginate. Cysts measure from a few millimeters to a few centimeters in diameter. Protoscolices can number >100, each having four suckers and an armed rostellum. These tend to be much less common than the typical cysticerci caused by *T. solium*.

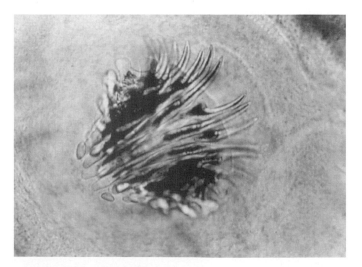

Figure 23.78 Coenurus from a human eye (higher magnification of hooklets than Figure 23.77; same specimen). (From A Pictorial Presentation of Parasites: A cooperative collection prepared and/or edited by H. Zaiman. Photograph courtesy of P. J. Fripp.)

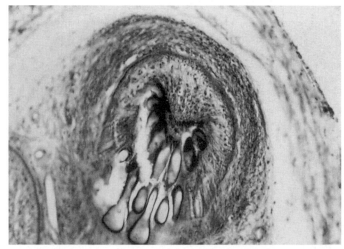

Figure 23.80 *Taenia* coenurus. Note the sectioned hooklets. (From A Pictorial Presentation of Parasites: A cooperative collection prepared and/or edited by H. Zaiman.)

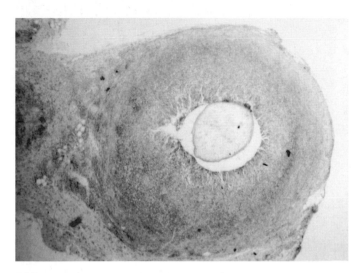

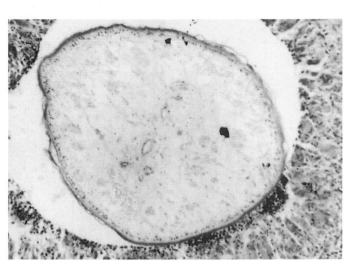

Figure 23.81 *Spirometra* sparganum larva in tissue. Spargana are larval forms of various tapeworms in the genus *Spirometra*. Spargana are white and ribbonlike in shape, range from a few millimeters to >30 cm in length, and are actively motile. The solid sparganum lacks suckers and bladder walls; there are irregular bundles of muscle fibers, typical folded tegument, and calcareous corpuscles in the parenchyma. (From A Pictorial Presentation of Parasites: A cooperative collection prepared and/or edited by H. Zaiman.)

Figure 23.82 *Spirometra* sparganum larva (higher magnification than Figure 23.81; same specimen). Some of the calcareous corpuscles are visible. They appear as small, outlined bodies within the parenchyma. (From A Pictorial Presentation of Parasites: A cooperative collection prepared and/or edited by H. Zaiman.)

Helminths—Trematodes

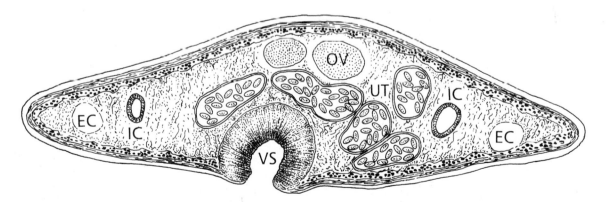

Figure 23.83 Diagram of the cross section of a trematode. Note the coiled uterus (UT) containing eggs, the lobed ovary (OV), intestinal ceca (IC), and excretory canals (EC). Also, note the large ventral sucker (VS). (Illustration by Sharon Belkin.)

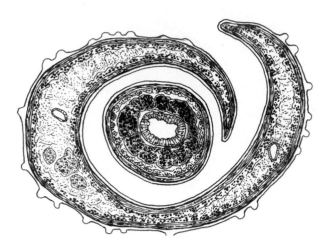

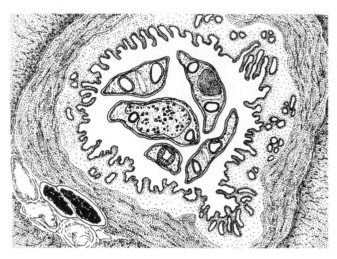

Figure 23.84 (Left) Diagram of male and female *Schistosoma* trematodes. The female worm is lying with the male worm in copulo. Note the small tuberculations on the dorsal tegument of the male worm. (Right) Sections of adult *Clonorchis sinensis* in the common bile duct. Note that in different sections, different structures are visible, including the intestinal ceca, the ovary, and parts of the testis. Also note the marked proliferation of the bile duct epithelium. (Illustration by Sharon Belkin.)

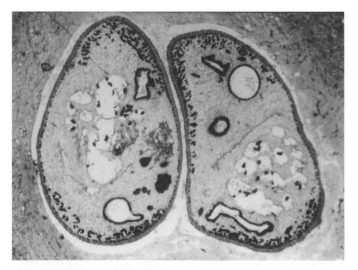

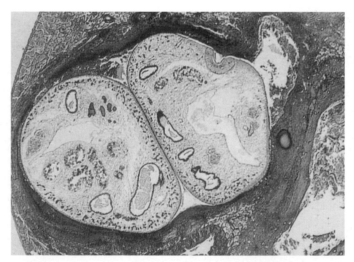

Figure 23.85 Two *Paragonimus westermani* parasites within a lung cavity. Note the dark, small spots of the vitellaria just within the tegument. The dark openings represent the ceca, with clear portions of the uterus. (From A Pictorial Presentation of Parasites: A cooperative collection prepared and/or edited by H. Zaiman.)

Figure 23.86 *Paragonimus westermani* adult worms in the lung. In this image, the vitellaria are also visible, as are portions of the uterus, the intestinal ceca, and the ventral sucker (seen in the right cross section). (From A Pictorial Presentation of Parasites: A cooperative collection prepared and/or edited by H. Zaiman.)

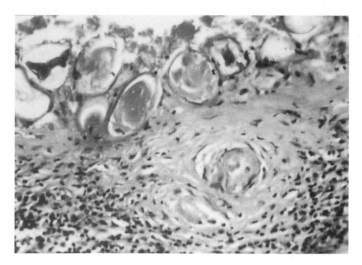

Figure 23.87 *Paragonimus westermani* eggs in the lung. Note the granuloma surrounding one of the eggs. (From A Pictorial Presentation of Parasites: A cooperative collection prepared and/or edited by H. Zaiman.)

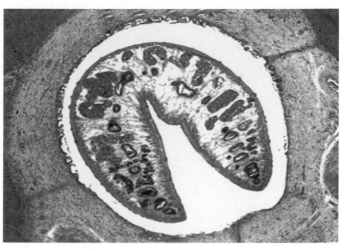

Figure 23.89 Adult *Fasciola hepatica* in the liver. Note that the intestinal ceca and portions of the testes are clearly visible. (From A Pictorial Presentation of Parasites: A cooperative collection prepared and/or edited by H. Zaiman; photograph courtesy of B.C. Walton.)

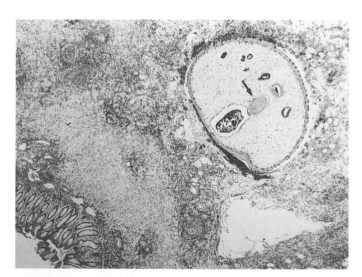

Figure 23.88 Adult *Fasciola hepatica* embedded in the intestinal wall. The parasite presents as an oval to round mass limited by a spined cuticle. The parasitic parenchyma is punctuated by multiple sections through the gut. The two centrally located tubular structures, separated by a light-staining muscular sucker (arrow), are part of the reproductive apparatus of the parasite. A marked inflammatory response is present in the adjacent host tissues, and fresh red blood cells can be seen near the parasite. (From A Pictorial Presentation of Parasites: A cooperative collection prepared and/or edited by H. Zaiman; photograph courtesy of B. C. Walton.)

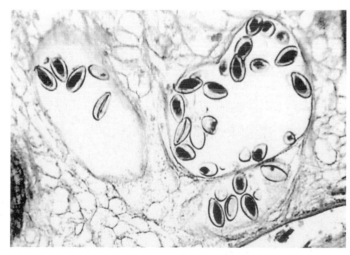

Figure 23.90 *Clonorchis sinensis* eggs in the adult worm within the bile duct. Note the operculum that is visible on the egg(s) in the left upper portion of the photograph. (From A Pictorial Presentation of Parasites: A cooperative collection prepared and/or edited by H. Zaiman.)

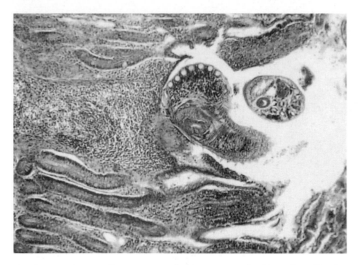

Figure 23.91 *Echinostoma* adult worm in the bowel wall. The spines around the oral sucker are seen in cross section. (From A Pictorial Presentation of Parasites: A cooperative collection prepared and/or edited by H. Zaiman.)

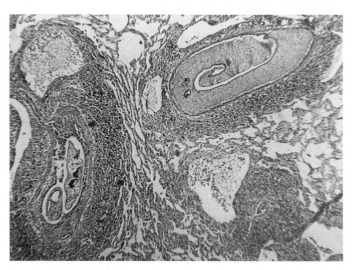

Figure 23.93 *Schistosoma* adults in the lung. In the lower left worm cross section, the smaller portion is the female worm; however, it is somewhat difficult to tell that the female worm is lying within the gynecophoral canal of the male. In the right upper section, the male worm is clearly seen. Note that the female worm is not visible in the upper left image and may not be present with the male worm (in copulo). (From A Pictorial Presentation of Parasites: A cooperative collection prepared and/or edited by H. Zaiman.)

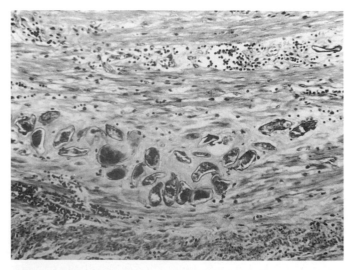

Figure 23.92 *Schistosoma japonicum* eggs in the appendix. The eggs are acid fast. (From A Pictorial Presentation of Parasites: A cooperative collection prepared and/or edited by H. Zaiman. Photograph courtesy of I. Miyazaki, Fukuoka, Japan.)

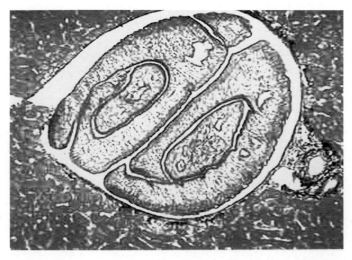

Figure 23.94 *Schistosoma* adults in tissue. Note that the smaller female worm is lying within the gynecophoral canal of the male worm. The intestinal ceca are visible within both male and female worms. (From A Pictorial Presentation of Parasites: A cooperative collection prepared and/or edited by H. Zaiman.)

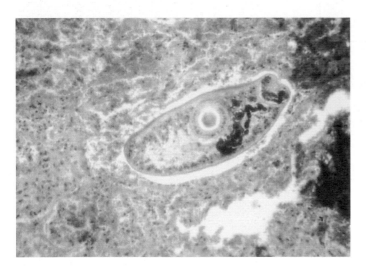

Figure 23.95 *Alaria americana* in an autopsied human lung, showing hemorrhagic pneumonia and the etiologic agent, a mesocercaria. Note the evidence of both the oral sucker and the acetabulum (ventral sucker). (From A Pictorial Presentation of Parasites: A cooperative collection prepared and/or edited by H. Zaiman. Photograph courtesy of R. S. Freeman.)

Blood Parasites

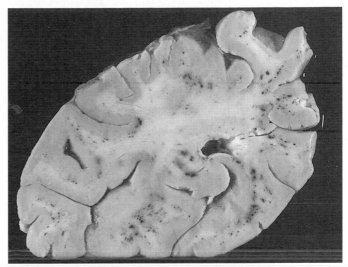

Figure 23.97 Malaria. Section through the cerebrum showing more congestion of the white matter than of the gray. (Armed Forces Institute of Pathology photograph.)

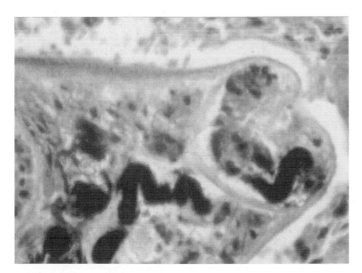

Figure 23.96 *Alaria americana* (higher magnification than Figure 23.95; same specimen). Note the oral sucker at the anterior end. (From A Pictorial Presentation of Parasites: A cooperative collection prepared and/or edited by H. Zaiman. Photograph courtesy of R. S. Freeman.)

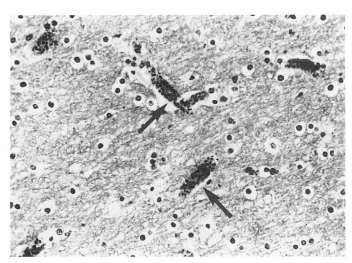

Figure 23.98 Cerebral malaria showing cerebral capillaries congested with parasitized erythrocytes (arrows). ×269. Parasitized erythrocytes stick to the endothelium and occlude the capillaries. (Armed Forces Institute of Pathology photograph.)

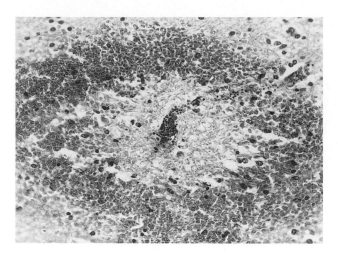

Figure 23.99 A ring hemorrhage in the brain surrounding a thrombosed vessel. The thrombus contains many parasitized erythrocytes. The erythrocytes in the hemorrhagic area do not contain parasites, a feature which contrasts the low percentage of parasitized erythrocytes in the circulating blood with the high percentage of parasitized erythrocytes in the thrombosed capillaries. ×305. (Armed Forces Institute of Pathology photograph.)

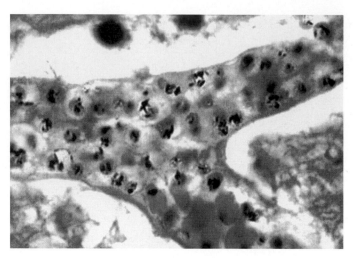

Figure 23.101 Cerebral malaria. Note the capillary containing parasitized erythrocytes. In cerebral malaria, the brain is edematous and heavy and has broadened and flattened gyri. Microscopically, the capillaries and small veins are congested and many of the erythrocytes are parasitized. Parasitized erythrocytes stick to the endothelium and occlude the capillaries. Not only are the rings visible, but pigment grains are also seen within the erythrocytes (Armed Forces Institute of Pathology photograph.)

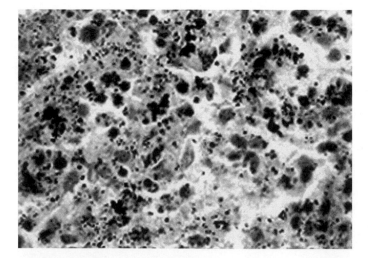

Figure 23.100 Malarial pigment present in the spleen. The spleen is enlarged to about 500 g during an acute attack and is soft, diffusely pigmented, congested and has rounded edges. The phagocytic cells lining the splenic sinuses contain malarial pigment, free parasites, and parasitized erythrocytes. During an acute attack, malarial pigment is evenly disbursed throughout the spleen.

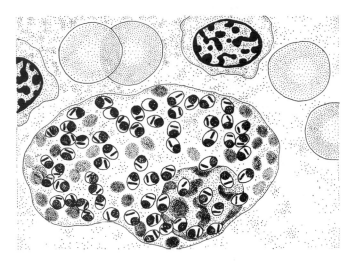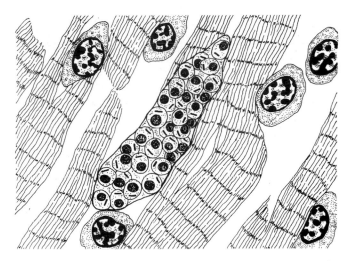

Figure 23.102   Diagram of amastigotes. (Left) *Leishmania donovani* amastigotes in a spleen smear. Each amastigote contains the dark, round nucleus and the bar-shaped kinetoplast. When cells are packed with amastigotes, often the complete morphology for each amastigote is not visible. (Right) *Trypanosoma cruzi* amastigotes in cardiac muscle. Each amastigote contains the nucleus (round, dark) and the kinetoplast (small bar). (Illustration by Sharon Belkin.)

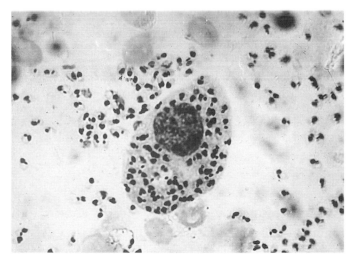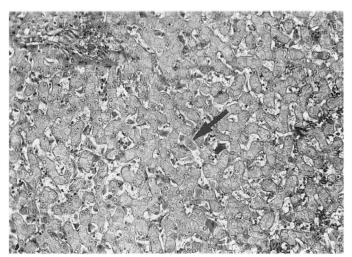

Figure 23.103   Smear of bone marrow showing many intracellular and extracellular *Leishmania* organisms. The amastigotes (Leishman-Donovan bodies) appear larger in smears than in tissue sections. The nuclei and kinetoplasts are distinct in these organisms. ×850. In some amastigotes, both the nucleus and kinetoplast (bar) are visible. (Armed Forces Institute of Pathology photograph.)

Figure 23.104   *Leishmania donovani.* Section of liver showing dilated sinusoids and greatly enlarged Kupffer cells (arrow). ×140. (Armed Forces Institute of Pathology photograph.)

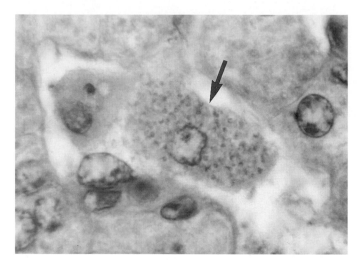

Figure 23.105   Enlargement of the Kupffer cell illustrated in Figure 23.104 (arrow). The cytoplasm contains *Leishmania* organisms. ×1,530. The individual amastigotes are packed into the cell; organism characteristics (nuclei and kinetoplasts) are not clearly visible. The amastigotes appear as very small dots. (Armed Forces Institute of Pathology photograph.)

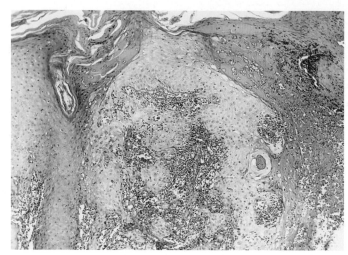

Figure 23.107   *Leishmania* spp. (cutaneous leishmaniasis). Biopsy specimen taken through the margin of an ulcer showing necrosis, hyperplastic epithelium, and inflammatory cells in the dermis. ×61. (Armed Forces Institute of Pathology photograph.)

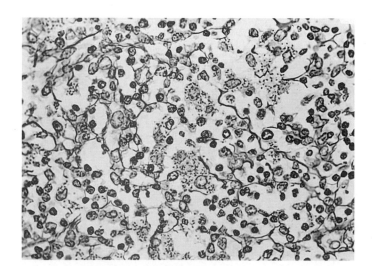

Figure 23.106   Wilder's reticulum stain of the same liver as shown in Figures 23.104 and 23.105. The small black dots within Kupffer cells are *Leishmania* organisms. ×374. At this magnification, individual organism characteristics are not visible. (Armed Forces Institute of Pathology photograph.)

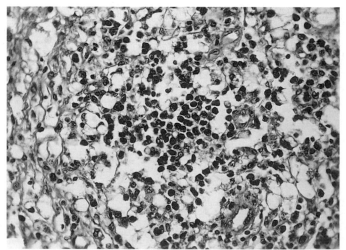

Figure 23.108   *Leishmania* spp. (cutaneous leishmaniasis). Inflammatory cell infiltrate, which is composed of histiocytes, lymphocytes, and plasma cells. Some of the histiocytes contain amastigotes. ×259. (Armed Forces Institute of Pathology photograph.)

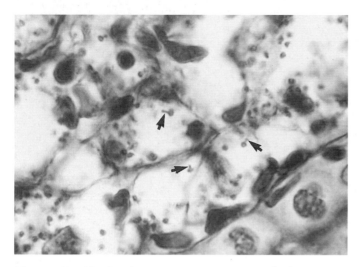

Figure 23.109 *Leishmania* spp. (cutaneous leishmaniasis). Reticulum stain showing the nuclei and kinetoplasts (arrows) of the many amastigotes within histiocytes. ×1,530. It is difficult to see the outlines of the actual amastigotes, but the nuclei and kinetoplast are visible in some of the amastigotes within the histiocytes. (Armed Forces Institute of Pathology photograph.)

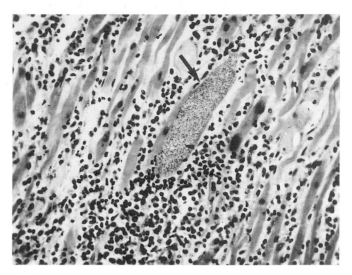

Figure 23.111 *Trypanosoma cruzi.* Acute chagasic myocarditis with a myofiber filled with amastigotes (arrow). ×400. At this magnification, the amastigotes look like very small dots. (Armed Forces Institute of Pathology. Photograph contributed by the Gorgas Memorial Laboratory.)

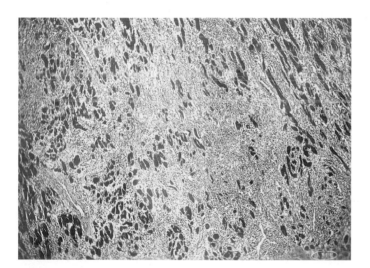

Figure 23.110 *Trypanosoma cruzi.* Chronic chagasic cardiopathy with scarring and chronic inflammation, left ventricle. ×40. (Armed Forces Institute of Pathology photograph.)

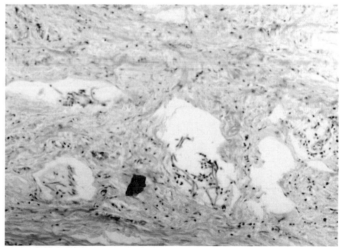

Figure 23.112 *Wuchereria bancrofti* microfilariae in the epididymis. Note the thin, threadlike structures in the lower right portion of the photograph. Various lengths are seen, depending on the actual tissue cut. Although these helminths are nematodes, they are shown with the blood parasites, since blood is the primary specimen for diagnostic purposes.

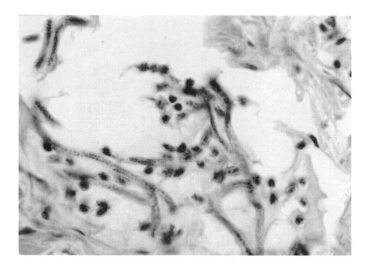

Figure 23.113 *Wuchereria bancrofti* microfilariae (higher magnification than Figure 23.112; same specimen). In this image, the "stringy" microfilariae are easily visible; there is the suggestion of cellular nuclei within the larval worm.

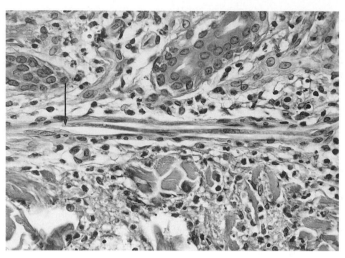

Figure 23.115 Dermis. Longitudinal section of a *Loa loa* microfilaria within a capillary shows the cephalic space (arrow). ×336. (Armed Forces Institute of Pathology photograph.)

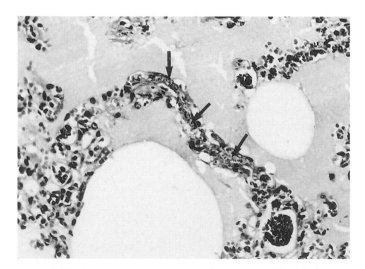

Figure 23.114 Microfilaria of *Wuchereria bancrofti* (arrows) in the lung. ×510. In this image the elongated microfilaria is seen and the cellular nuclei within the organism are also visible. (Armed Forces Institute of Pathology photograph.)

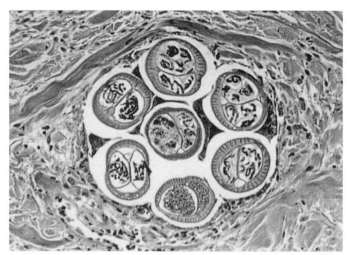

Figure 23.116 *Mansonella streptocerca* in a section of skin. Note the long, slender microfilariae in utero. (From A Pictorial Presentation of Parasites: A cooperative collection prepared and/or edited by H. Zaiman.)

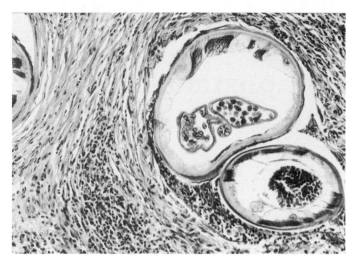

Figure 23.117 *Onchocerca volvulus* nodule containing adult worms; note the microfilariae in utero (cross sections of the slender microfilariae). (From A Pictorial Presentation of Parasites: A cooperative collection prepared and/or edited by H. Zaiman.)

References

1. **Binford, C. H., and D. H. Connor.** 1976. *Pathology of Tropical and Extraordinary Diseases.* Armed Forces Institute of Pathology, Washington, D.C.

2. **Katz, M., D. D. Despommier, and R. W. Gwadz.** 1989. *Parasitic Diseases,* 2nd ed. Springer-Verlag, New York, N.Y.

3. **Kenny, M.** 1973. *Pathoparasitology.* The Upjohn Co., Kalamazoo, Mich.

4. **Meyers, W. M., R. C. Neafie, A. M. Marty, and D. J. Wear (ed.).** 2000. *Pathology of Infectious Diseases,* vol. 1. *Helminthiases.* Armed Forces Institute of Pathology, Washington, D. C.

5. **Orihel, T. C., and L. R. Ash.** 1995. *Parasites in Human Tissues.* American Society of Clinical Pathologists, Chicago, Ill.

6. **Sun, T.** 1988. *Parasitic Disorders,* 2nd ed. The Williams & Wilkins Co., Baltimore, Md.

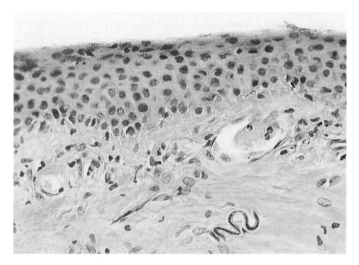

Figure 23.118 Microfilariae seen in superficial corneal stroma. ×259. (Armed Forces Institute of Pathology photograph.)

Medically Important Arthropods

Over one million species in the Animal Kingdom are found in the phylum Arthropoda. These invertebrates are characterized by the possession of an exoskeleton, an internal hemocele that is filled with hemolymph and the internal organs, and a body that is usually divided into the head, thorax, and abdomen. Arthropods develop through a process called metamorphosis, often involving three or four stages: egg, nymph, and adult (bedbugs, sucking lice, and cockroaches); egg, larva, pupa, and adult (flies, fleas, mosquitoes, and bees); or egg, larva, nymph (one or more), and adult (ticks and mites). Classification of the medically important arthropods is shown in Table 24.1 (17, 43, 47).

The importance of this group is due primarily to the morbidity and mortality of the various diseases they transmit, as well as their wide geographic distribution. Arthropods can transmit disease via mechanical transmission, in which the arthropod vector is not a part of the life cycle of the organism, and biological transmission, in which the arthropod vector is an integral part of the life cycle of the organism (multiplication or development or both in the vector). Vector-borne parasitic human infections are listed in Table 24.2. In addition to serving as vectors of various infections, certain arthropods may cause some degree of pathologic changes as a result of their bites, stings, or presence on or in the body. Examples are given in Table 24.3. Specific diagnostic tests and recommended therapy for the more common arthropods are listed in Table 24.4. Specific terms related to the identification of specific arthropods can be found in Table 24.5. The relative sizes of some of the more important arthropods can be seen in Figure 24.1.

Arthropods and Their Relationship to Disease

Biological Vectors of Microorganisms

As biological vectors, arthropods can serve as intermediate hosts for human infections or can accidentally carry microorganisms from one host to another. They can transmit various protozoa (the agents of malaria, leishmaniasis, and trypanosomiasis, and *Babesia* spp.), helminths (filarial worms, lung flukes, and tapeworms), bacteria (*Yersinia pestis* and *Francisella tularensis*), rickettsiae (the agents of typhus, trench fever, and bartonellosis), chlamydias (the agents of trachoma, lymphogranuloma venereum, and psittacosis), spirochetes (the agents of relapsing fever, yaws, and Lyme disease), and viruses (arboviruses).

Table 24.1 Classification of medically important arthropods[a]

Class	Order	Common name
Insecta	Diptera	Blackflies
		Sand flies
		Mosquitoes
		Biting midges
		Deerflies
		Botflies
		Warble flies
		Houseflies
		Tsetse flies, etc.
		Flesh flies
		Blowflies
		Kads
		Louseflies
	Hemiptera	Assassin bug
		Bedbugs
	Coleoptera	Beetles
	Siphonaptera	Fleas
	Anoplura	Sucking lice
	Mallophaga	Biting lice
	Hymenoptera	Honeybees
		Wasps
		Ants
	Blattaria	Cockroaches
Arachnida	Acari (S)	Hard ticks
		Soft ticks
		Itch mites
		Chiggers
		Follicle mites
	Araneae (S)	Spiders
	Scorpiones (S)	Scorpions
Pentastomida		Tongue worms
Diplopoda		Millipedes
Chilopoda		Centipedes
Crustacea (SP)	Copepoda (S)	Water fleas
	Decapoda	Crabs, crayfish

[a] S, subclass; SP, subphylum.

As vectors, arthropods may serve as mechanical vectors of the etiologic agent, merely transporting the organisms from one host to another or from the environment to a host. However, most agents of arthropod-borne disease use the arthropod as a biological vector, requiring a period of incubation or development in this host. Although most vectors ingest the etiologic agent of disease, they transfer the organisms to humans in several ways, including via an arthropod-derived vomit drop on food or drink or in a puncture wound. Fecal pellets may also contain the infecting organism, as can the hemolymph if the vector is crushed on the skin. Salivary secretions can also contain the infective organisms.

A number of criteria must be fulfilled before an arthropod can be considered a disease vector. The suspected vector must be known to bite humans in a geographic area where human infections are documented. Second, the geographic distribution of the suspected vector must include the distribution of the disease in humans. Third, the causative agent of disease must be isolated from wild-caught vectors and must be the same species as that causing disease in humans. Fourth, it must be shown that naturally or experimentally infected vectors can maintain the infection throughout the complete life cycle of the parasite. Finally, experimental transmission of the parasite by a bite is generally considered proof of the vector's ability to transmit the organism.

Bites and Envenomation

Most individuals have come in direct contact with arthropods which are problems because of their bites and human sensitivity to their saliva or toxins. Human contact occurs when arthropods are feeding or bite in self-defense. Generally these bites do not cause serious problems, unlike stings, for which the reaction may be more severe. The outcome depends on a number of factors, including the arthropod involved, the amount of toxic substance introduced into the body, the physical condition of the bite victim, and possible previous exposure. Individual responses can vary from mild itching and redness to life-threatening reactions.

Arthropods usually associated with the introduction of saliva during the act of biting include the Acari (ticks and mites), Anoplura (sucking lice), Araneae (spiders), Chilopoda (centipedes), Diptera (flies, mosquitoes, and biting midges), Hemiptera (true bugs: bedbugs, kissing bugs, and triatomid bugs), and Siphonaptera (fleas).

Envenomation is usually associated with stings; however, some groups can introduce venom during a bite. These include some of the spiders, particularly the widow spiders, violin spiders, and some tarantulas. Human contact with some groups of urticating caterpillars (which have body hairs that can inject venom) may also produce mild to serious skin conditions (Figures 24.2 to 24.5). Eye disease has also been linked to this type of contact (5, 20).

Tick-induced paralysis is also seen and is caused by the injection of toxin from gravid female ticks embedded in the flesh of the host. A review of 33 cases of

Table 24.2 Vector-borne human infections

Type of infection and disease	Causative agent	Vector (common name)
Protozoal infections		
Malaria	*Plasmodium* spp.	Mosquitoes
Leishmaniasis	*Leishmania* spp.	Sand flies
Chagas' disease	*Trypanosoma cruzi*	Triatomid bugs
East African trypanosomiasis	*T. brucei rhodesiense*	Tsetse flies
West African trypanosomiasis	*T. brucei gambiense*	Tsetse flies
Babesiosis	*Babesia* spp.	Ticks
Helminthic infections		
Filariasis	*Wuchereria bancrofti*	Mosquitoes
Filariasis	*Brugia malayi*	Mosquitoes
Filariasis	*Dirofilaria* spp.	Mosquitoes
Filariasis	*Mansonella perstans*	Biting midges
Filariasis	*M. streptocerca*	Biting midges
Filariasis	*M. ozzardi*	Biting midges
Onchocerciasis	*Onchocerca volvulus*	Blackflies
Loiasis	*Loa loa*	Deerflies
Dog tapeworm infection	*Dipylidium caninum*	Dog lice, dog fleas, human fleas
Rat tapeworm infection	*Hymenolepis diminuta*	Rat fleas, beetles, grain beetles
Dwarf tapeworm infection	*Hymenolepis nana*	Grain beetles (rare)

tick-induced paralysis in Washington state showed that most of the patients were female (76%) and most cases (82%) occurred in children younger than 8 years; two of the children died (14). Onset of illness occurred from 14 March to 22 June, and nearly all cases indicated probable exposure east of the Cascade Mountains. Of the 28 patients for whom information was available, 54% were hospitalized. When tick identification to species was reported, *Dermacentor andersoni* was consistently reported. Certainly tick-induced paralysis is a preventable cause of illness and death, and if diagnosed early, it requires simple, low-cost intervention (tick removal).

Stinging arthropods can be found in the following groups: Hymenoptera (ants, bees, hornets, and wasps) and the Scorpiones. Individuals vary in their reaction to these stings; certainly, anaphylaxis has been seen in cases related to bee stings. The frequency of mass bee attacks has dramatically increased in the Americas following the introduction and spread of the aggressive Africanized "killer" bee (*Apis mellifera scutellata*). Currently there is no specific therapy available. However, an antivenom for the treatment of mass bee attacks has been developed; venoms from European and Africanized bees appear to be almost, if not, identical (26). Scorpion stings are usually no more severe than bee stings; however, there are exceptions, and each year there are numerous deaths in Mexico as a result of scorpion stings. The venom is thought to be a tox-albumin that affects the nervous and pulmonary systems.

Some millipedes produce secretions that can cause contact vesicular dermatitis. These fluids can be squirted from glands located on the body segments. Blister beetles can also produce these types of fluids, leading to skin blisters within hours after contact.

Although death is rarely attributable to these types of injuries, small children and the very old may demonstrate severe reactions to both arthropod bites and stings.

The fire ant, *Solenopsis invicta*, is an example of an arthropod that has proven to be relatively dangerous, particularly in certain patients; the stings cause a fiery, burning sensation (Figure 24.6). Severe allergic reactions, particularly in children, may require immediate attention. In the southeastern United States, the continued spread of these ants has caused agricultural problems, changes in the ecosystem, and increasing numbers of people with sting sequelae, including hypersensitivity reactions, secondary infections, and rare neurologic problems. Apparently, evolutionary changes have facilitated their expansion northward into Virginia and westward into California; further expansion is also anticipated (29). Because of their resistance to natural and chemical control, fire ants can overwhelm their environment, causing destruction of land and animals. They can also cause a variety of health problems in humans, ranging from simple stings to overwhelming anaphylaxis and death. Although deaths caused by imported fire ant stings are rare, they

Table 24.3 Medically important arthropods and their potential effect on humans

Representative species	Common name	Geographic distribution	Effect(s) on humans
Insecta			
Anoplura (sucking lice)			
Pediculus humanus humanus	Body louse	Worldwide	Skin reactions to bites; vectors of rickettsiae and spirochetes
P. humanus capitis	Head louse	Worldwide	Skin reaction to bites
Phthirus pubis	Crab louse	Worldwide	Skin reaction to bites
Mallophaga (biting lice)			
Trichodectes canis	Dog louse	Worldwide	None—can transmit *Dipylidium caninum*
Hemiptera (true bugs)			
Cimex lectularius	Bedbug	Worldwide	Skin reaction to bites
C. hemipterus	Tropical bedbug	Tropical and subtropical	Skin reaction to bites
Triatoma infestans, Rhodnius prolixus, Panstrongylus megistus	Kissing bug, conenose bug	Tropical and subtropical regions of the New World	Skin reaction to bites; vectors of *Trypanosoma cruzi*, the cause of Chagas' disease
Hymenoptera (bees, wasps, ants)			
Apis mellifera	Honeybee	Worldwide	Painful sting, potential anaphylaxis
Bombus spp.	Bumblebee	Worldwide	Painful sting, potential anaphylaxis
Various genera and species of the family Vespidae	Wasp, hornets, yellow jackets	Worldwide	Painful sting, potential anaphylaxis
Solenopsis spp.	Fire ant	Tropical America, south-eastern United States	Painful bite and multiple stings, potential anaphylaxis
Diptera (flies, mosquitoes, and their relatives)			
Culicoides spp.	Biting midge	Worldwide	Serious biting pest; skin reaction to bites; vectors of several filarid nematodes
Leptoconops spp.			
Phlebotomus spp.	Sand fly	Worldwide	Skin reaction to bites; vectors of leishmanias, bacteria, and viruses
Lutzomyia spp.	Sand fly	Tropical region of New World	
Simulium spp.	Blackfly, buffalo gnat	Worldwide	Serious biting pests; skin reaction to bites; vectors of *Onchocerca* and *Mansonella* spp.
Aedes spp.	Mosquito	Worldwide	Serious biting pests; skin reaction to bites; vectors of viruses, protozoa, and filaria
Anopheles spp	Mosquito		
Culex spp.	Mosquito		
Culiseta spp.	Mosquito		
Mansonia spp.	Mosquito		
Tabanus spp.	Horsefly	Worldwide	Biting pests; painful bite followed by skin reaction
Chrysops spp.	Deerfly	Worldwide	Biting pests; vectors of tularemia and *Loa loa*
Glossina spp.	Tsetse fly	Africa	Biting pests; vectors of African trypanosomes
Siphonaptera (fleas)			
Ctenocephalides canis, C. felis	Dog flea, cat flea	Worldwide	None—can transmit *Dipylidium caninum*
Pulex irritans	Flea	Worldwide	Dermatitis, ulcerations
Xenopsylla cheopis	Rat flea	Worldwide	None—can transmit *Hymenolepis diminuta*
Tunga penetrans	Chigoe flea	Tropical and subtropical	Nodular swelling with subsequent ulceration
Orthoptera (cockroaches, grasshoppers, and their relatives)			
Blatella germanica	German cockroach	Worldwide	Mechanical disseminator of pathogens; intermediate host of certain helminths
Periplaneta americana	American cockroach	Worldwide	
Various genera	Cockroach	Worldwide	
Coleoptera (beetles)			
Various genera and species	Beetles	Worldwide	Mechanical disseminators of pathogens

(continued)

Table 24.3   Medically important arthropods and their potential effect on humans *(continued)*

Representative species	Common name	Geographic distribution	Effect(s) on humans
Various species of the family Meloidae	Blister beetles	Worldwide	Blistering of skin after contact with adult
Tenebrio spp.	Grain beetles	Worldwide	Intermediate hosts of tapeworms—*Hymenolepis* spp.
Lepidoptera (moths and butterflies)			
Various genera and species	Caterpillar	Worldwide	Urticating spines and hairs may cause reaction when handled or inhaled
Arachnida			
Scorpionida (scorpions)			
Various genera and species	Scorpion	Tropical and subtropical	Initially painful sting, often followed by systemic reactions
Araneidae (spiders)			
Lactrodectus mactans	Black widow spider	Americas	Bite usually painless, delayed systemic reaction
Lactrodectus spp.	Widow spider	Worldwide	Bite usually painless, delayed systemic reaction
Loxosceles reclusa	Brown recluse spider	Americas	Initial blister at wound site followed by sometimes extensive necrosis and slow healing
Loxosceles laeta	South American brown spider	South America	
Acarina (ticks and mites)			
Ornithodoros spp.	Soft tick	Worldwide	Skin reactions to bite; tick paralysis; vectors of relapsing fever
Amblyomma spp.	Hard tick	Worldwide	Skin reactions to bite; tick paralysis; vectors of rickettsias, viruses, bacteria, and protozoa
Dermacentor spp.			
Ixodes spp.			
Demodex folliculorum	Follicle mite	Worldwide	Found in sebaceous glands and hair follicles, occasional skin reactions
Sarcoptes scabiei	Human itch mite	Worldwide	Burrows in skin, causing severe itching
Trombicula spp.	Chigger, red bug	Worldwide	Intense itching at site of attachment; vectors of rickettsias
Crustacea			
Cyclops spp.	Copepod	Worldwide	Intermediate hosts of guinea worm and broadfish tapeworm
Diaptomus spp.			
Chilopoda			
Scolopendra spp.	Centipede	Tropical and subtropical	Painful bite
Pentastomida			
Linguatula serrata	Tongue worm	Tropical	Endoparasitic in internal organs

may become more common as the fire ant population expands (40). Two alkaloid components of imported fire ant venom cause cardiorespiratory depression and elicit seizures in rats. Such effects identify these alkaloids as toxic compounds and may explain the cardiorespiratory failure seen in some individuals who experience massive fire ant stings (23).

S. *invicta* and S. *richteri* are not the only ant species to cause serious allergic or adverse reactions. Ant species belonging to 6 different subfamilies (Formicinae, Myrmeciinae, Ponerinae, Ectatomminae, Myrmicinae, and Pseudomyrmecinae) and 10 genera (*Solenopsis, Formica, Myrmecia, Tetramorium, Pogonomyrmex, Pachycondyla, Odontomachus, Rhytidoponera, Pseudomyrmex,* and *Hypoponera*) have this capability. Awareness that species other than imported fire ants may cause severe reactions should lead to more rapid patient evaluation and treatment (31).

Table 24.4 Common arthropods: diagnostic procedures and recommended therapy

Name	Symptoms	Diagnostic procedure	Recommended therapy
Fly larvae			
Cutaneous, mucocutaneous	Painful, indolent ulcers; furuncle-like sores; dermal lesions like creeping eruption	Remove larvae, place in petri dish (blood agar plate). Identification will depend on examination of the larval stigmal plates and/or the adult fly (should often be referred to an experienced entomologist).	Local palliatives, often including antibiotics for secondary infections
Nose	Obstruction of nasal passages, facial edema, and fever		
Ear	Crawling sensations, buzzing noises, foul-smelling discharge		
Eye	Severe irritation and pain		
Intestinal	Severe nervous symptoms as well as internal irritation		Treatment with castor oil will expel these larvae
Bedbugs	Asymptomatic, edema, inflammation, indurated ring at site of feeding, asthma, anxious concern	Examination and identification of adult bug	Local palliatives and insecticides, thorough washing of all bedding
Reduviid bugs, (kissing bugs, cone-nose bugs)	Local swelling and induration, oozing of blood from puncture site, generalized urticaria (bites commonly on exposed parts of skin, face, outer area of eye, lip)	Examination and identification of adult bug	Local palliatives
Fleas			
Tunga penetrans	Adult female burrows into skin (often toes and feet), causing nodular swelling, subsequent ulceration and festering sores	Examination and identification of adult flea	Wearing shoes, flea removal aseptically with needle, bathing of wound and treating with antiseptic
Pulex irritans (human), *Xenopsylla cheopis* (rat), *Ctenocephalides canis* (dog)	Asymptomatic; roseate raised lesion, frequently edematous, indurated, or pustular (bites are more common on lower extremities)		Local palliatives, use of recommended insecticides
Lice			
Pediculus spp., *Phthirus pubis*	Cutaneous lesions, roseate elevated papule, intense pruritus, lesions may become secondarily infected by bacteria	Examination and identification of adult and/or eggs on hair shaft	Shampoo or ointment (1% lindane); dusts with 1% malathion; bedding can be fumigated with ethyl formate; remaining nits can be removed with fine-toothed comb
Ticks			
Tick paralysis	Generalized toxemia, fever (104°F), rapid ascending flaccid paralysis, difficulty in swallowing and respiration, and death; little or no change in white cell count	Careful examination of body, particularly back of the neck and along spinal column	Prompt removal of the engorging female tick will usually limit progression of paralysis (due to toxin in tick saliva)
Other ticks and bites	Inflammatory infiltration of tissues, local hyperemia, edema, hemorrhage (histolysis at the bite site as a result of salivary secretions)	Examination and identification of adult tick	Anterior portion should be grasped with forceps close to the skin and pulled *straight out*; leaving mouth parts may cause severe tissue reactions and possible secondary infections

(continued)

Table 24.4   Common arthropods: diagnostic procedures and recommended therapy *(continued)*

Name	Symptoms	Diagnostic procedure	Recommended therapy
Mites			
Sarcoptes scabiei (itch mite)	Small skin vesicles; scratching results in bleeding and scab formation followed by secondary infection	Skin scraping (avoid excoriated lesions); get material from early papule or burrow (between fingers, wrists, elbows, axillae, penis, scrotum, buttocks, backs of knees, ankles, toes, under breasts)	Topical application of 1% lindane in lotion or ointment (may require one to three applications); should be applied after scrubbing and soaking in warm water to expose lesions
Spiders			
Black widow spider	Symptoms will depend on amount of venom injected; may include abdominal cramps, hypertension, and convulsions	History and/or identification of spider	Specific antivenom is available (patients should be tested for horse serum sensitivity before use); other supportive therapy depending on symptoms
Violin spider, brown recluse spider	Bite may be painless; mild to severe pain within hours; erythema, vesicle formation, and itching followed by chills, headache, and nausea; the lesion becomes necrotic, spreads, and heals very slowly; kidney damage and death may also occur	Identification of spider	Prompt administration of steroids, early surgical excision of lesion (skin grafting may be necessary), treatment of secondary infections

Tissue Invasion

A number of arthropods are capable of actual tissue invasions, some of which may occur in the superficial tissues and some in deep body sites. These groups include Acari (ticks and mites), Diptera (flies, mosquitoes, and biting midges), Pentastomida (tongue worms) (see chapter 18), and Siphonaptera (fleas). The pentastomes are wormlike parasites that are related to the arthropods. On ingestion of eggs by humans, larvae migrate to the tissues, where they develop into nymphs and become encapsulated. Although most of these infections are asymptomatic, clinical manifestations can occur and include inflamed and enlarged lymph nodes, abdominal pain, jaundice, and ocular lesions (see chapter 18).

Various ticks, mites, and fleas tend to invade tissues at the superficial levels; these include *Sarcoptes scabiei* (itch mite), *Tunga penetrans* (chigoe flea), and members of the Argasidae (soft ticks) and the Ixodidae (hard ticks). Not only do these arthropods cause painful bites, but also they are responsible for transmission of diseases caused by bacteria, viruses, and rickettsia.

When we consider arthropod tissue invasion, we usually think of myiasis, i.e., infestation with fly larvae (maggots). Clinically, myiasis is often categorized according to the part of the body invaded: cutaneous, mucocutaneous, intestinal, and genitourinary. Some larvae develop only in living tissue, while others prefer dead or necrotic tissue. Entry sites for egg or first-stage larva deposition can include unbroken skin, open wounds or lesions, nose or ears, mouth with subsequent swallowing, and genitourinary openings. There are three general groups of myiasis-causing flies: (i) those that require a host for the development of their larval stages, (ii) those that develop in a host if entry occurs via wounds or sores but development in a host is not required, and (iii) those that accidentally invade a host but generally complete larval development without a host.

Entomophobia and Delusory Parasitosis

Fear of insects, entomophobia, is sometimes seen and may become an issue in mentally disturbed individuals (17). These patients may report insect infestations, describing in great detail the presence of tiny insects in or on the skin. Symptoms may be described as crawling or biting sensations and are often accompanied by excessive scratching by the patient.

Delusory parasitosis usually represents a more serious emotional disorder in which patients believe that they have parasites in or on the body. These patients often describe many physician visits and consults with unproductive results; laboratory findings are consistently negative. They are absolutely convinced that they have legitimate problems and continue to try and get confirmation of problems that are imaginary rather than real. They tend to bring in small bits and pieces of skin, debris, pictures, drawings, etc., none of which demonstrate true infestation. Often, these bits and pieces of debris are of nonanimal origin (string, paper, plant material, etc.). We have received complete notebooks filled with 8-by-10-in. photographs of "parasites"—none of which depict anything that can be recognized as possible arthropods

Table 24.5   Common terms used in discussing arthropods

Term	Definition
Antennae	Segmented appendages located in the lateral side of the head
Capitulum	"False-head" found in arachnids; basal portion of structure is "basis capituli"
Cell	Area of wing enclosed by veins
Cerci	Appendages (slender, two) at terminal abdominal segment
Chelicerae	Appendages (grasping or cutting) in place of jaws; found in arachnids
Claw	Appendage (sharp, curved) at end of leg
Complete metamorphosis	Cycle of development (e.g., egg, larva, pupa, adult); each stage distinct from others
Compound eye	Visual elements (large number) grouped on each side of head
Costa	Thickening of the anterior edge of the wing
Coxa	Proximal segment of the leg
Ecdysis	Shedding of larval skin during transformation from one stage to another
Emergence	Appearance of winged insect from the pupal stage
Epipharynx	Organ attached to inner part of labrum
Exoskeleton	External skeleton
Fat body	Group of cells (widely distributed, food store); contains protein, fats, glycogen
Festoons	Notches seen at posterior part of body of hard ticks
Gill	Specialized respiratory organ seen in aquatic insects
Haltere	Club-shaped structure in place of second pair of wings in Diptera
Hypopharynx	Tongue-shaped structure; lies between labrum and labium; connected to salivary duct
Hypostome	Lower lip which takes the form of an elongated structure armed with teeth or hooks in ticks
Incomplete metamorphosis	Morphology between adults and immature stages similar; stages resemble each other
Labial palpi	Pair of appendages arising from each side of the labrum
Labium	Lower lip which forms floor of mouth
Labrum	Upper lip which forms roof of the mouth
Longitudinal veins	Series of six veins following the subcosta; numbered in sequence
Maggot	Common name given to larva of some Diptera (flies)
Mandibles	Pair of upper jaws
Maxillae	Pair of lower jaws
Mesonotum	Dorsal surface of second thoracic segment
Myiasis	Infestation of human and animal tissues with fly maggots
Nymph	Immature stage in insects undergoing incomplete metamorphosis
Ocellus	Simple eye
Ovipositor	Tubular structure through which eggs are laid
Palmate hairs	Fan-shaped hairs on dorsal surface of anopheline larva
Pedipalp	Second pair of appendages which arise from cephalothorax in arachnids
Pleuron	Lateral part of a segment
Pronotal comb	Anterior segment of thorax
Pronotum	Dorsal surface of first thoracic segment
Prothorax	Anterior segment of thorax
Pseudotracheae	Small tubes found on labella (through which fluid is sucked) in houseflies and members of the Tabanidae
Pygidium	Pincushion-like structure seen on ninth segment of fleas
Scutellum	Small posterior section of the tergum
Scutum	Chitinous shield seen on the dorsum of Acarina
Sensilla	Hairs which have sensory function
Setae	Hairlike structures which are hollow internally
Spermatheca	Sperm-storing organ in lower genital tract of females
Spiracles	Respiratory openings
Subcosta	Vein lying posterior to the costa
Tergum	Dorsal plate on thoracic segment
Tissue invasion	Infestation (superficial tissues); infection (deeper tissues)
Tracheae	Ringed breathing tubes
Vector	Transmitter of infection (arthropods)

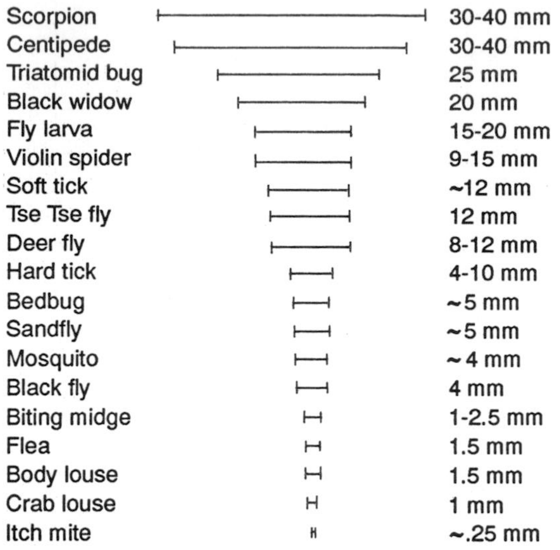

Scorpion		30-40 mm
Centipede		30-40 mm
Triatomid bug		25 mm
Black widow		20 mm
Fly larva		15-20 mm
Violin spider		9-15 mm
Soft tick		~12 mm
Tse Tse fly		12 mm
Deer fly		8-12 mm
Hard tick		4-10 mm
Bedbug		~5 mm
Sandfly		~5 mm
Mosquito		~4 mm
Black fly		4 mm
Biting midge		1-2.5 mm
Flea		1.5 mm
Body louse		1.5 mm
Crab louse		1 mm
Itch mite		~.25 mm

Figure 24.1 Relative sizes of various arthropods. (Illustration by Sharon Belkin.)

or any other type of parasite. These patients search out specific medical professionals, often by making multiple telephone calls and calling medical and other scientific societies. Unfortunately, they can benefit only from psychiatric treatment. However, patients with delusions of parasitosis

Figure 24.2 Caterpillars that sting (A and B) or cause dermatitis (C and D). (A) Io moth, *Automeris io*; (B) puss caterpillar, *Megalopyge opercularis*; (C) saddleback caterpillar, *Sibine stimulea*; (D) brown tail moth, *Euproctis chrysorrhoea*. (Illustration by Sharon Belkin.)

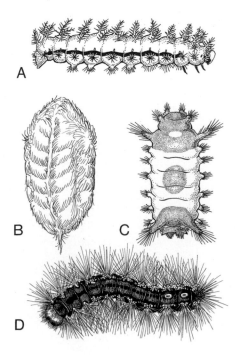

Figure 24.3 Saddleback caterpillar (*Sibine* sp.), which may cause dermatitis and necrosis. The irritant is in the barbed rigid nettling hairs. (Armed Forces Institute of Pathology photograph.)

generally reject psychiatric referral. The diagnosis of delusions of parasitosis can often be made on the basis of the history alone, but it is important to ensure that the patient does not have an organic skin disorder and that the delusion is not secondary to another mental or physical illness.

In spite of these psychiatric problems, it is still appropriate to check all possibilities: bites from unrecognized mites or other insects either at home or in the workplace or potential exposure to other allergic stimuli that may be unrelated to insect or organism contact need to be considered. Unfortunately, in many of these patients, the presentation is so convincing that it is difficult at first to recognize the situation as a mental rather than actual physical problem.

Class Insecta (Insects)

Some of the more important orders are presented in this section. Adults have a segmented body composed of the head, thorax, and abdomen; one pair of antennae; three

Figure 24.4 Caterpillars. *Monema flavescens* (left) and *Parasa* sp. (right). The small knobs on the body segments carry the stinging hairs, which are too small to be seen in the photograph. (Armed Forces Institute of Pathology photograph.)

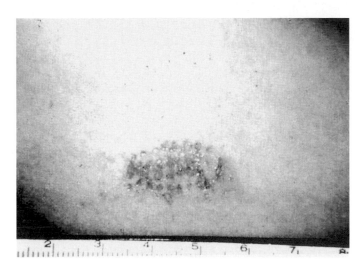

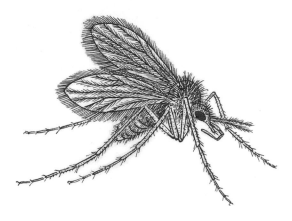

Figure 24.7 Sand fly (*Phlebotomus* and *Lutzomyia* spp., vectors of *Leishmania* spp.), ~5 mm. (Illustration by Sharon Belkin.)

Figure 24.5 *Megalopyge opercularis*, the puss caterpillar. Dermatitis is caused by venom introduced into human skin by hollow larval poisonous spines and occurs 24 h after exposure. (From A Pictorial Presentation of Parasites: A cooperative collection prepared and/or edited by H. Zaiman.)

pairs of legs; and, in some cases, wings. The exoskeleton is composed of hard plates (scleritis) separated by flexible membranes (conjunctivae or arthrodial membrane) or grooves (sutures).

Order Diptera (Flies, Mosquitoes, and Midges)

The order Diptera contains numerous kinds of flies, mosquitoes, and midges. The members of this group have two pairs of wings, one pair of which, the halteres, are vestigial and are used for balance only.

Sand Flies (Leishmaniasis)

Sand flies occur mainly in the tropics and subtropics, limited to areas where the temperatures are above 15.6°C for a minimum of 3 months of the year. They are found

Figure 24.6 Severe reaction to multiple stings from a fire ant, *Solenopsis invicta*, on the arm of a male in Mississippi. (Armed Forces Institute of Pathology photograph.)

in a number of habitats, from below sea level to 2,800 m above sea level or higher. Sand flies are small, long-legged flies that are about 5 mm long (Figure 24.7). These small flies are able to pass through ordinary mosquito netting. Two genera (*Phlebotomus* in the Old World and *Lutzomyia* in the New World) are of medical importance (Figure 24.8). Only the females feed on blood, usually that of mammals; such feeding normally occurs at night. The bite site is usually indurated and inflamed, with a wheal 1 to 2 cm in diameter, and is accompanied by pruritus. Some patients even respond with a systemic allergic reaction, including fever, nausea, and malaise. Although many sand flies can fly extended distances, they usually move in short, jerky hops along walls and ceilings. Spraying with residual insecticide is somewhat effective; however, elimination of larval habitats would be a more important control measure.

Sand flies are a geologically old group, being identified from about 120 million years ago, prior to the origin of mammals. They probably originally fed on reptiles, and their ancient status led to the evolution of different groups in the New and Old Worlds. Their main importance is as vectors of *Leishmania* spp., causing human cutaneous, mucocutaneous, and visceral disease. About 80 of 700 known sand fly species are known vectors of *Leishmania* spp. Two species of *Lutzomyia* occur in south and central Texas, as well as parts of Mexico, and are probably the vectors of cutaneous leishmaniasis in Texas. It is interesting that lectins and toxins in the plant diet of *Phlebotomus papatasi* can kill *Leishmania major* promastigotes in the sand fly and in culture; these vegetation components may decrease parasite transmission in certain areas (24). Control measures against the adult flies include the use of the bacterial insecticide, *Bacillus thuringiensis* (52).

Figure 24.8 Sand fly (*Phlebotomus* sp.). (Photograph courtesy of Duane J. Gubler, Centers for Disease Control and Prevention.)

Blackflies (Filariasis [*Onchocerca* and *Mansonella* spp.])

Blackflies (buffalo gnats, turkey gnats, Kolumbtz flies) transmit *Onchocerca volvulus* worldwide; in the Old World these vectors are limited to Africa, while in the New World the vectors are found in various parts of Central and South America. Blackflies are small flies (*Simulium* spp.), measuring 1 to 5 mm long and usually having a stout body and humped back (Figure 24.9). Only the females suck blood, and swarms may attack humans as well as domestic

Figure 24.9 Blackfly (*Simulium* sp., vector of *Onchocerca volvulus* and *Mansonella* spp.), 4 mm. (Illustration by Sharon Belkin.)

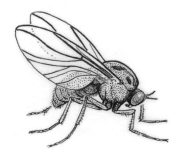

and wild mammals, usually during the daytime; the blood meal is required for egg development. Although not usually that painful, the bite is accompanied by intense itching and the development of a raised, ulcerative lesion at the site, most probably caused by the saliva in the puncture wound. Individuals who are particularly susceptible may experience severe pruritus and body swelling (blackfly fever), which is seen predominantly in children in rural areas of the northeastern United States. Similar to Hymenoptera, a Simuliidae bite occasionally results in extreme hypersensitivity and should be considered in appropriate cases.

These flies normally do not bite indoors and may fly long distances from their habitats. A massive control effort is now under way in West Africa to lower the intensity of *Onchocerca* transmission to a level of infection that does not lead to visual impairment. Because flight distances are long and the larvae develop in water, control measures are difficult at best. Development occurs in fast-running streams, often very deep; one species was found breeding at depths down to 190 ft (ca. >50 m).

Simuliids have to survive periods when the temperature is too low to sustain routine activities and when the rivers dry up during the off season. They can apparently reduce their respiratory rate and replace trehalose in the hemolymph by high polyhydric alcohols, thus remaining dormant below 4°C. Once the temperature rises, the reverse process is rapidly completed and growth resumes. These flies may not be present in sufficient numbers to be pests, but they can still be important vectors of the filarial worm causing onchocerciasis or "river blindness."

Deerflies (Filariasis [*Loa loa*])

Adult flies (*Chrysops* spp.) are medium sized to large (about the size of a housefly or larger) and stoutly built, and they have large eyes and dull-colored bodies (Figures 24.10 and 24.11). Only the female is adapted to take a blood meal. The bite site can become a painful wound that takes a long time to heal; individuals may also respond with an allergic reaction to the saliva of the fly. Day-biting flies transmit the human form of disease with *L. loa*, a disease that affects up to 13 million people. This biting time preference coincides with the peak microfilaremia in humans. The adult flies live high in the forest canopy and descend to the forest floor to feed on humans. While most blood-sucking insects are repelled by wood fire smoke, apparently in Africa the flies are attracted to wood fires. Control measures have not been particularly successful, and loiasis continues to be a significant problem in West Africa.

Tsetse Flies (African Trypanosomiasis)

An estimated 60 million people are at risk for African trypanosomiasis, with 300,000 to 500,000 new cases

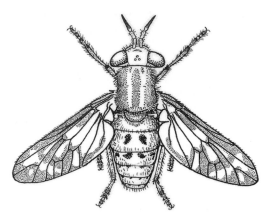

Figure 24.10 Deerfly (*Chrysops* sp., vector of *Loa loa*), 8 to 12 mm. (Illustration by Sharon Belkin.)

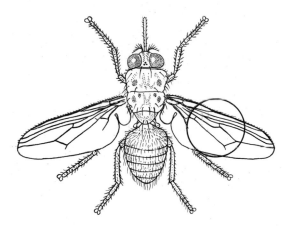

Figure 24.12 Tsetse fly (*Glossina* sp., vector of African trypanosomes), 12 mm. The characteristic "hatchet" cell is seen within the circle. (Illustration by Sharon Belkin.)

each year, primarily in the Democratic Republic of Congo, Angola, and Sudan. Tsetse flies (*Glossina* spp.) are robust, honey-brown flies measuring 6 to 15 mm long (the size of a housefly) (Figure 24.12). The genus habitat is limited to the African continent south of the Sahara. Both sexes are voracious blood feeders, and both can transmit parasites. Adult flies of the important vectors are daytime feeders and are attracted to moving animals and humans. The flies use movement and smell to find appropriate hosts. Although they can fly long distances, their flight pattern is generally limited to the closest available host and back to particular vegetation, where they spend long periods of rest. The wings have the characteristic "hatchet" cell, which is cleaver shaped (Figure 24.12).

The adults are long-lived and fly long distances, usually along a preferred flight line; they live in areas called fly belts. Some species live by rivers, and others are usually found in the savanna. They feed on many animals as well as humans. This fact, along with their long life

Figure 24.11 Deerfly (*Chrysops* sp.). (Photograph courtesy of Duane J. Gubler, Centers for Disease Control and Prevention.)

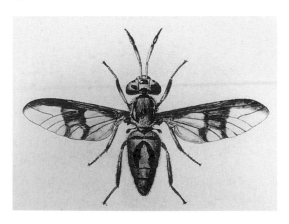

span, makes vector control difficult; it sometimes involves such drastic measures as killing off game animals in close proximity to humans or moving whole villages out of these fly belts. The use of insecticides has also proven to be somewhat successful. Wherever there are tsetse flies, it is impossible to keep cattle without regular chemotherapy. Trypanosomiasis has inhibited or totally prevented the development of a cattle-raising industry in much of tropical Africa. Currently, cattle ranches are still impractical in the savanna areas of Africa. Also, there is increasing resistance to melarsoprol, the key drug for central nervous system involvement in African trypanosomiasis.

Stableflies, Houseflies, and Horseflies (Potential Carriers of Infectious Organisms)

Flies in the genus *Stomoxys* (stableflies) (Figures 24.13 and 24.14) resemble the housefly. Both male and female flies can be tremendous pests with an impressive bite (causing a stabbing pain), which involves ripping open the skin through which the biting apparatus is inserted. Along the southeast coast of the United States, these flies breed in decaying seaweed and can be a tremendous tourist problem during fly season. Although control of the larvae has been difficult, insecticides are effective against the adult flies. Residual insecticides are initially effective, but these flies rapidly develop resistance.

Houseflies can harbor >100 different pathogens, most of which can be transmitted. They range from viruses to helminths and include protozoan cysts, oocysts, spores, and helminth eggs. There are several ways in which they can disseminate pathogens. The surface of the body can be contaminated; pathogens can be regurgitated onto food through the fly's vomit drop; or pathogens can pass through the gut of the fly and be deposited in its feces.

Figure 24.13 Stablefly (*Stomoxys* sp.), 6 to 15 mm long. (Photograph courtesy of Duane J. Gubler, Centers for Disease Control and Prevention.)

It is interesting that in areas such as northern Africa where tsetse flies are not found, domestic animal trypanosomes, *T. evansi* and *T. vivax vivax*, are transmitted mechanically by horseflies (*Tabanus*) and stableflies (*Stomoxys*). These flies can spread trypanosomes within domestic animal herds, causing epidemics with tremendous financial loss.

Biting Midges (Filariasis [*Mansonella* spp.])

Biting midges (*Culicoides* spp.) are the smallest of the biting flies, measuring between 1 and 4 mm (Figure 24.15). They are also known as "punkies," "no-see-ums," "gnats,"

Figure 24.14 Stablefly (*Stomoxys* sp.), 6 to 15 mm long. (Photograph courtesy of Duane J. Gubler, Centers for Disease Control and Prevention.)

or "flying teeth." Only the females feed on blood, and swarms may attack humans or other mammals, usually at dusk, when the air is still. Screen and netting controls are difficult because of the small size of the adult fly. The genus *Culicoides* is widely distributed throughout the world and ranges from sea level to almost 14,000 ft in Tibet. Members of the 1921 Everest expedition complained of being bitten continually from base camp almost to the summit. These biting midges can cause acute discomfort, particularly on warm, humid days; in spite of their small size, the bite often produces a severe local reaction. Females make small cuts in the skin, and the saliva contains an anticoagulant. Once the blood pools, it is sucked up. It was calculated in the early 1980s that because of the tremendous number of bites to which a human can be subjected, a person spending 1 h per day on the beach in Trinidad in the early morning would receive 38 bites infective for *Mansonella ozzardi* in the course of a year. However, with the tremendous tourist industry impact, control measures are constantly being revisited.

Mosquitoes (Malaria, Filariasis, and Arbovirus Infections)

Mosquitoes are slender, delicate insects that are distributed worldwide, number over 3,000 species, and have aquatic larval and pupal stages (Figure 24.16). They are the most important group of arthropods in terms of disease transmission and, as pests, produce an irritating dermatitis. To be effective vectors for malaria, female *Anopheles* mosquitoes must feed frequently on humans, must be relatively susceptible to the malarial gametocytes, must live long enough for the malaria parasite to complete the life cycle, and must be present in sufficient numbers to maintain transmission of the parasite (Figure 24.17). Blood feeding follows a circadian rhythm, with most species feeding at night or during the twilight hours, either right after sunset or before sunrise. The biting cycle has control implications, since mosquito nets are effective against nocturnal species, especially those that bite after midnight. Mosquitoes are not limited to ground level elevation but in the forests can be found from ground level up into the canopy, where birds and primates live.

In the tropics, breeding usually continues throughout the year. Important species within the United States include *Aedes aegypti* (yellow fever, dengue fever), *Aedes albopictus* (yellow fever, dengue fever, LaCrosse encephalitis virus), *Ochlerotatus* (*Aedes*) *sollicitans* (Eastern equine encephalitis [EEE]), *Aedes vexans* (EEE), *Anopheles quadrimaculatus* (malaria in the eastern United States in the early 20th century), *Culex nigripalpus* (St. Louis encephalitis [SLE]), *Culex quinquefasciatus* (SLE, West Nile virus [WNV]), *Culex salinarius* (SLE, WNV, EEE), and *Culex tarsalis* (Western equine encephalitis [WEE], SLE).

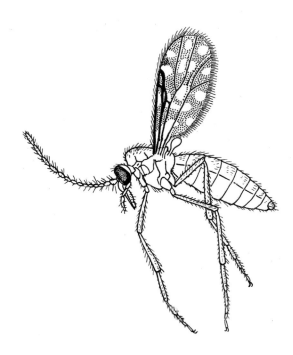

Figure 24.15 Biting midge (*Culicoides* sp., vector of *Mansonella* spp.), 1 to 2.5 mm (usually 1 to 4 mm long). (Illustration by Sharon Belkin; adapted from E. C. Faust, R. F. Russell, and R. C. June, ed., *Clinical Parasitology*, 8th ed., Lea & Febiger, Philadelphia, Pa., 1970.)

Myiasis

Myiasis is the infestation of live human and vertebrate animals with dipterous larvae which feed on the host's dead or living tissue, liquid body substances, or ingested food (Tables 24.6 and 24.7). Maggots can infest any organ or tissue accessible to fly oviposition; most cases probably occur when a female fly lands on a human host and deposits either eggs or larvae.

Human infestations have been found in the breast, urinary bladder, colon, penis, gums, brain, carcinoma of the scalp, and various skin lesions (1, 19, 25, 27, 50). It is also important to realize that nosocomial transmission can occur; myiasis cases have occurred in hospital patients,

Figure 24.16 Mosquito, ~4 mm. (Illustration by Sharon Belkin.)

Figure 24.17 *Anopheles* mosquito. (From A Pictorial Presentation of Parasites: A cooperative collection prepared and/or edited by H. Zaiman.)

often those who are debilitated (12, 13, 36, 44, 51, 58). Specimens should be killed, preserved in hot (60°C) 70% alcohol, and then submitted to the laboratory for identification. Myiasis-producing members of the Diptera of medical importance can be seen in Table 24.6 and Figure 24.18.

For business or pleasure, more people than ever before are traveling throughout the world. Since 3 to 10% of these travelers experience skin, hair, or nail disorders related to their travels, it is more and more likely that Western physicians will be expected to treat exotic infections imported from different countries. Tungiasis and furunculoid myiasis are two typical disorders of intertropical regions. These are conditions induced by the presence of arthropod larvae or eggs in the skin. It will become more important for physicians to recognize these and other conditions, many of which they may have not seen previously (54).

The identification of larvae found in human tissues is based on a number of morphologic characters, including (i) structure of the chitinized mouthparts, (ii) abdominal spiracles, and (iii) structures of the cuticle (these vary depending on the particular instar larva and the number of molts). Occasionally, if living mature larvae are submitted for examination, they can be held until they pupate and hatch; the adult flies can then be examined for identification. Doppler ultrasonography has also been used to evaluate the number of larvae present per lesion; this approach helps avoid misdiagnoses and treatment delays (41).

Pseudomyiasis refers to the accidental ingestion of dead or living fly larvae with no associated pathologic changes or symptoms. The term "sanguinivorous myiasis" has been used to refer to superficial attachment of larvae during blood sucking.

Medicolegal Forensic Entomology

Forensic entomology has become a recognized discipline that is concerned with the application of information on

Table 24.6 Diptera of medical importance in myiasis

Scientific and common names	General habits	Body site(s)[a]	Diagram
Calliphoridae			
Calliphorinae (metallic group)			A
Chrysomyia bezziana, Old World screwworm (**A**)	Living tissues	Any body site, eye, erosion of bone	
Cochliomyia hominivorax, New World screwworm (**B**)	Living tissues	Nasal, aural sites, brain penetration	B
Lucilia spp., greenbottles	Dead tissues	Skin, GI tract, GU tract	
Calliphora spp., bluebottles (**C**)	Dead tissues	Skin, GI tract, GU tract, nose	C
Phormia regina, black blowfly (**D**)	Wounds	Traumatic myiasis	D
Cynomyopsis cadaverina, blowfly	Wounds	Traumatic myiasis	
Calliphorinae (nonmetallic group)			
Auchmeromyia senegalensis, Congo floor maggot (**E**)	Sucks blood from skin	Does not penetrate, detaches after feeding	E
Cordylobia anthropophaga, Tumbu fly (**F**)	Subcutaneous boils, sloughing, gangrene	Penetrates unbroken skin, usually feet	F
Sarcophaginae			
Wohlfahrtia spp., flesh flies	Wounds, normal tissue (head)	Skin, sense organs	G
Wohlfahrtia magnifica, Old World flesh fly (**G**)	Most common in humans		
Sarcophaga haemorrhoidalis	Dead tissues	GI tract, also skin, GU tract, nose, sinuses	
Sarcophaga carnaria	Living tissues	Vagina	
Cuterebridae			
Dermatobia hominis, human botfly (**H**)	Living tissues	Skin, subcutaneous boils, brain	H
Cuterebra spp., North American botflies		Skin, eye	
Gasterophilidae			
Gastrophilus spp., horse (stomach) botflies (**I**)	"Larva migrans" cutaneous migration	Skin, migrate to eye	I
Hypodermatidae			
Hypoderma spp., warble flies (**J**)	Skin penetration, extensive larval wandering (limbs)	Skin, eye, nasopharynx, chest, back, limbs, etc.	J
Hypoderma lineatum, lesser cattle warble fly			
Hypoderma tarandi, reindeer warble fly			
Cephenemyia ulrichii, moose throat botfly		Eye	
Oestridae			K
Oestrus ovis, sheep nasal botfly (**K**)	Living tissues, resembles conjunctivitis	Nasopharynx, eye, sinuses	

[a] GI, gastrointestinal; GU, genitourinary.

Table 24.7   Species infecting the skin, eyes, nose, and ears in myiasis and examples of spiracles (stigmal plates)[a]

Representative species	Common name	Spiracle (stigmal plates)
Dermatobia hominis	Human botfly	
Gastrophilus intestinalis	Horse botfly	
Hypoderma bovis	Cattle botfly	
Oestrus ovis	Sheep botfly	
Calliphora vomitoria	Flesh fly	
Wohlfahrtia vigil	Flesh fly	
Sarcophaga fuscicauda	Flesh fly	

[a] Adapted from reference 5; illustrations by Sharon Belkin.

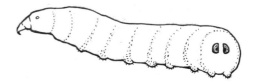

Figure 24.18 Fly larva and stigmal plates, 15 to 20 mm. (Illustration by Sharon Belkin.)

insects and other arthropods to legal issues. The most common applications deal with the determination of death, manner of death, movement of the corpse, and other aspects of the investigation. Insects occurring in carrion can be assigned to four groups. Those that feed on the carrion, including blowflies, arrive within a few hours of death. They are followed by species that feed on the body, species that feed on its inhabitants, and species that use the corpse as an extension of their natural environment. Various stages are identified in the decomposition of the corpse and are important in helping set the time of death: (i) autolytic and microbial decomposition, during which blowflies oviposit; (ii) active decomposition by insects; (iii) advanced decomposition, primarily by coleopteran beetle larvae; (iv) microbial decomposition, which takes over when coleopteran beetle larvae leave and ends in mummification; and (v) separation of the skeleton into individual bones, when the time of death can no longer be determined (9).

The entomologist must determine the time the insect attack began; many factors influence the confirmation of this time: cadaver location (whether it was exposed to sun or shade); whether the cadaver was covered, buried, or open to the elements; whether it was wet or dry; and ambient and ground temperatures. Often a pig is used to simulate the corpse conditions and to determine various rates of decomposition and insect activity; this is particularly helpful if conditions in a particular ecoclimate are not known.

Because the length of the oldest maggots recovered from a body often provides an accurate estimate of the time of death, it is important to ensure adequate fixation without subsequent shrinkage. Studies have shown that maggot shrinkage can lead to underage errors, thus leading to differences of many hours. It has been found that maggots killed in boiling water and then placed in appropriate preservatives do not shrink. The length of the larval crop, which can be useful in age estimates of post-feeding larvae, is not altered significantly when using this approach. Standardized handling and fixation of maggots at the crime scene are critical to valid and consistent interpretations (48).

It is also possible that fly pupae may contaminate forensic entomology samples at death scenes if they originate from animal carcasses or other decomposing organic material rather than from human remains. These contaminants may falsely increase postmortem estimates if no pupae or puparia are actually associated with the body (3).

To permit rapid identification of arthropods, random amplified polymorphic DNA typing has been used to support classic morphologic and forensic analysis of maggots found on a human corpse. This approach has been used to determine if maggots found on the body match those on the outside of the body bag or wrap or if they match the pupae found on the floor or the ground under the corpse. This type of methodology can discriminate between closely related flies and beetles found on corpses (7). This approach has been very helpful in setting the time of death for bodies found less than 1 month after death. At this point, larval insects can be very difficult to identify morphologically when the stages are immature and only fragments of pupae and first-instar larvae are found. An easy and objective method of insect identification is DNA typing; this new approach provides excellent information in a relatively short time (35).

Another relatively new approach involves forensic analysis of DNA recovered from a larva's gut, which can be used to identify the larval food source. This method for associating a maggot with a corpse could be useful, particularly in situations where remains were removed from a suspected crime scene, when an alternative food source is near the scene or the body, or in a chain-of-evidence dispute. However, since maggot gut content analysis represents a new approach, many of the limitations of the technique have not yet been identified. In spite of possible limitations, it appears that third-instar larvae actively feeding on a corpse could be considered an excellent source of human DNA (8, 59).

Insects can also assist in determining the cause of death and can serve as reliable alternate specimens for toxicologic analyses in the absence of tissues and fluids normally used for such purposes. The presence of drugs and/or toxins in decomposing tissues may alter the rate and patterns of development of arthropods feeding on these tissues. Insects can also accumulate heavy metals and toxic compounds, or their by-products, in their tissues (21).

Types of Myiasis Categorized According to Affected Tissues

Cutaneous and Mucocutaneous Tissues (Eye, Nose, and Ear). Larval penetration of cutaneous and mucocutaneous tissues causes various degrees of pathologic abnormalities, ranging from an irritating pruritus to invasion of the eyes, brain, nasopharynx, bones, ear canal, vagina, tongue, and open wounds. The only known therapeutic procedure, other than applying local palliatives, is surgical removal and treatment for secondary bacterial or fungal infections (if these occur). Examples of representative fly larvae and their stigmal plates are seen in Table 24.7. The stigmal plates or spiracles (respiratory openings) from mature fly larvae are characteristic for each species and can be valuable in identification. The larvae orient themselves so that these spiracles are normally at the surface of the host skin, etc., to allow oxygen exchange by the larvae.

Intestinal Tissues. Female flies often deposit eggs or larvae on food that is then eaten by humans. Occasionally, some eggs or larvae survive, become temporarily lodged in the intestinal crypts and folds, and partially or completely develop before being passed in the stool (1).

Other Sites. There are reports of larvae recovered from urine, the vagina, and the lungs (accidental inhalation of either a gravid female fly or airborne eggs) (5).

Types of Myiasis Categorized According to Tissue Viability

Specific Myiasis. The larvae attack only living tissues of humans and animals, usually cutaneous and subcutaneous areas but also the nose, mouth, sinuses, eyes, ears, anus, and vagina. There are generally six dipterous families represented in this type of infestation (Table 24.6).

Semispecific Myiasis. The larvae breed on the bodies of dead animals and attach to dermal and subdermal areas and cavities. This type of myiasis is represented by four families of dipterous flies (Table 24.6).

Accidental Myiasis. Infestations are usually accidental and occur infrequently. They are usually in the gastrointestinal tract and are acquired by ingesting food contaminated with larvae. Although the eggs are usually digested or fail to hatch, some may survive, become lodged in the crypts, and develop through part of their life cycle before leaving the body. They may also invade the ears, nose, sinuses, and urinary passages. There are approximately 15 families of flies whose members are implicated in this type of infestation.

Dipterous Flies Involved in Cases of Myiasis

Calliphorid Flies (Blowflies) (*Cochliomyia*, *Phaenicia*, and *Phormia* spp.).   The adults feed on material found in wounds or decaying flesh; females lay their eggs in batches of 200 to 300 in the flesh of dead animals. The eggs usually hatch within 24 h and feed on exudates of wounds or decaying flesh; they penetrate the tissue by using powerful oral hooks, hence the name "screwworm" (Figure 24.19) (5). Human ophthalmomyiasis has been reported, with one review specifically discussing infection with *Cochliomyia hominivorax* (11). Adult flies have been known to lay eggs on scabs, sores, scratches, pimples, or dried blood; in the healthy mucous membranes of the eyes, nose, mouth, or vagina; and in the umbilicus of neonates. Recent human infestations with *C. hominivorax* include the axilla, nose, palate, and mastoid cells. Blowflies often appear shiny or coppery green to shiny black.

Another unusual case of *Phormia regina* myiasis of the scalp was reported in 1988. *P. regina*, the black blowfly, causes facultative myiasis, usually invading wounds, benign and malignant ulcers, and other diseased tissue; rarely are normal tissues involved. However, in this case, there were no recognizable associated infected wounds or tumor. The patient was a 64-year-old homeless woman who had numerous scratches on her scalp. There is speculation that infection started with the repetitive scratching and that the open suppurative scalp wounds, which were never washed, were likely sites for egg deposition. *P. regina* is very common throughout the United States, especially during April to September.

Flesh Flies (*Sarcophaga*, *Wohlfahrtia*, and *Parasarcophaga* spp.).   The adult flies feed on feces or decaying meat or fish (5); females deposit 40 to 80 first-stage larvae. The

Figure 24.19   *Phaenicia* sp. (screwworm). (Photograph courtesy of Duane J. Gubler, Centers for Disease Control and Prevention.)

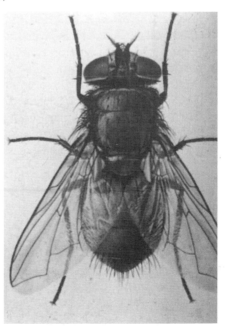

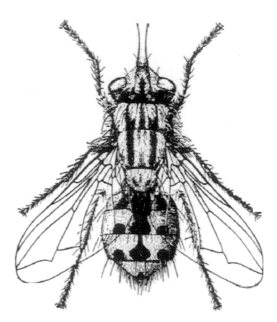

Figure 24.20 Flesh fly, which causes myiasis in humans. (Adapted from M. T. James, *U.S. Department of Agriculture Miscellaneous Publication 631*, 1947.)

Figure 24.21 Flesh fly. (Photograph courtesy of Duane J. Gubler, Centers for Disease Control and Prevention.)

larvae penetrate the tissue, thus causing problems depending on the body site (Figures 24.20 and 24.21).

Medically important genera of flesh flies include *Wohlfahrtia* and *Parasarcophaga*. *W. magnifica* never develops in carrion, but is found in wounds and natural orifices of warm-blooded animals, including humans. *W. vigil* occurs in northern North America, and the larvae are not as invasive, usually being limited to dermal tissues and causing a boil-like lesion. The first reported case of human aural myiasis caused by the flesh fly *Parasarcophaga crassipalpis* was reported in 1987. The patient was a 16-year-old, mentally retarded, nonresponsive Caucasian boy at a public institution in Adelaide, South Australia. Some blood was noticed in the left external ear canal, and 2 weeks later, dark blood was seen discharging from the ear. The physician detected movement in the ear, and that evening a single maggot emerged from the ear. No other maggots were found, and there was no tissue damage in the external ear canal.

Botflies (*Dermatobia hominis*, *Gastrophilus* spp., *Hypoderma* spp., *Cuterebra* spp., and *Oestrus ovis*) (Cutaneous, Ophthalmic, Nasal, and Aural Myiasis).

The adult female *Dermatobia hominis* fly never feeds but uses food stored up during the larval stage for egg production. Once the eggs are developed, the female fly captures a mosquito or other blood-sucking arthropod and glues the eggs to the abdomen of the arthropod. When the arthropod lands on a warm-blooded host, the young larvae emerge from the egg membranes and, within less than an hour, invade the skin. Lesions commonly occur on the unprotected skin; the tunnel produced is generally perpendicular to the surface and is occupied by a single larva. Eventually, local tissue destruction and accumulation of toxic waste produce throbbing pain and intense itching. The lesion may resemble a bacterial infection with seropurulent fluid (Figures 24.22 through 24.27). *D. hominis* is not found within the United States; however, cases are seen in travelers returning from areas of endemic infection that include parts of Mexico and Central and South America.

Although human infestation with *Gastrophilus* larvae is uncommon, the lesion is usually cutaneous. The larvae may advance in the tunnel, with the lesion resembling that of *Ancylostoma braziliense* (cutaneous larva migrans). Ophthalmomyiasis due to *Gastrophilus intestinalis* has also been reported.

Flies of the genus *Hypoderma* most commonly cause dermal myiasis in humans in the United States. The lesions are less serpiginous than those caused by *Gastrophilus* spp., particularly since the larvae bore more deeply into the subcutaneous tissues and produce a more inflamed lesion. However, since the larvae do migrate, they can appear many centimeters from the original site in a month's time. It is important to try and remove the larva before it dies and thus causes a more severe problem due to infection and pus formation.

The genus *Cuterebra* contains more than 20 species of flies native to North America, and the larvae occur

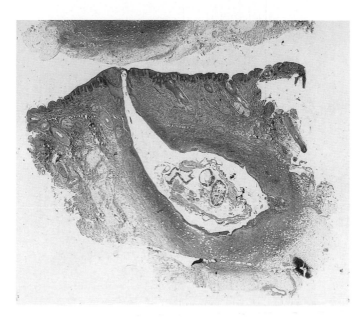

Figure 24.22 Myiasis. Nodule from the pubis of a 24-year-old woman. She was bitten while in a banana grove in Colombia 2 weeks before this nodule was removed. The fly larva is coiled in a cyst that communicates with the surface. The cyst and tract are lined in part by squamous epithelium. This lesion is characteristic of the "warble" or boil of *Dermatobia hominis*, although the species could not be determined from the section studies. (Armed Forces Institute of Pathology photograph.)

as dermal parasites of rodent and rabbit hosts. A case of cutaneous myiasis of the eyelid as a result of a *Cuterebra* larva has been reported. The patient was a 28-year-old man with a 3-week history of a swollen, tender nodule on his right upper eyelid. Apparently, 2 weeks before symptoms appeared, he was bitten on his shoulder by an insect. Redness and swelling developed and spread into the neck, face, and right upper eyelid during the following week. The patient was treated for periorbital cellulitis and a secondary *Pseudomonas* infection in the lesion, with no improvement. Two days after the lesion was finally incised, an oval larva emerged and was identified as a second-stage *Cuterebra* larva.

Gravid females of *Oestrus ovis* deposit their first-stage larvae onto the conjunctiva, the outer nares, or the lips or into the buccal cavity. These larvae have large oral hooks and spines that assist their penetration through the mucous membranes. The larvae may burrow into the eyelid, conjunctival sac, or lachrymal duct. Human ophthalmomyiasis is common in the former USSR, northern Africa, and Israel; however, it is rare in Western Europe and the United States. Invasion of the nasal sinuses may result in congestion accompanied by pruritus and frontal headache. Generally, with the exception of the eye, the larvae spontaneously leave the tissues. In cases in which the eye is involved, patients may present with ocular

Figure 24.23 Myiasis. Higher magnification of the lesion in Figure 24.22. The larva is surrounded by an intense inflammatory cell infiltrate composed of neutrophils, lymphocytes, plasma cells, and eosinophils. Magnification, ×9.7. (Armed Forces Institute of Pathology photograph.)

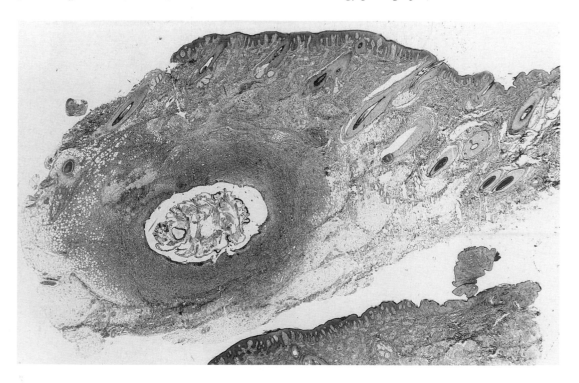

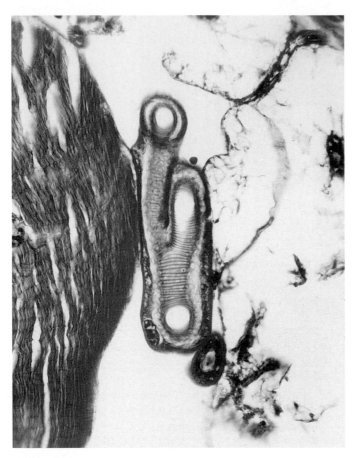

Figure 24.24   Myiasis. A portion of the spiracular system of the larva shown in Figures 24.22 and 24.23. Note the distinct tracheal rings. Magnification, ×440. (Armed Forces Institute of Pathology photograph.)

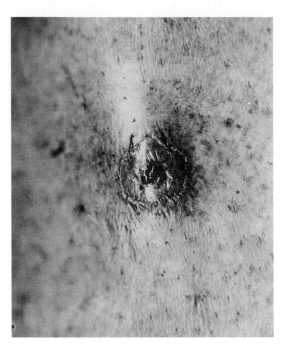

Figure 24.25   Myiasis in the skin of a leg of a patient in Panama. This is the warble of *Dermatobia hominis* of about 1 month's duration. The shiny black spot in the center is the posterior end of the larva. This lesion was over the tibia and was painful. (Armed Forces Institute of Pathology photograph.)

Figure 24.26   Myiasis. Brain of a child in Panama who died of malaria. There were several warbles on the scalp. The larva in the cavity entered through a 4-mm hole that it had bored through the anterior fontanelle. The shape and spines of the larva are consistent with *Dermatobia hominis*. (Armed Forces Institute of Pathology photograph.)

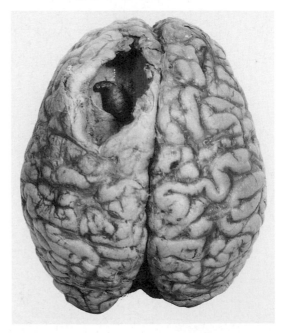

symptoms of foreign-body sensation, irritation, redness, and photophobia. Topical antibiotics and steroid eye drops can be used. Patients may also develop nasal symptoms such as sneezing, nasal discharge, and epistaxis. Larvae can be removed from the conjunctival sac under local anesthetic with no difficulty. Otolaryngology follow-up may demonstrate nasal myiasis as well. Some patients have been treated with ivermectin (34).

Order Hemiptera (True Bugs)

Bedbugs (*Cimex lectularius*)

Bedbugs measure about 5 mm long by 3 mm wide, are mahogany brown, and have only vestigial wings (Figures 24.28 and 24.29). The adult bugs may resemble unengorged ticks or very small cockroaches. Although they can mechanically transmit pathogenic microorganisms, there is no conclusive evidence that they are natural biological vectors for the transmission of organisms causing diseases.

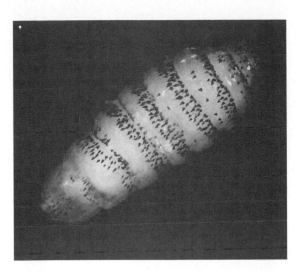

Figure 24.27 Fly larva removed from a man living in Lynchburg, Va. The patient had a pruritic "mosquito bite" on the chest, which became red and then enlarged to about 4 cm. A "head" developed, from which the patient expressed this larva. The larva was identified as *Cuterebra* sp. Magnification, ×22. (Armed Forces Institute of Pathology photograph.)

Medical problems associated with these bugs include itching and inflammation caused by their bites. Bedbugs are nocturnal and reach their peak activity before dawn; they tend to hide in cracks during the day. They respond to warmth and carbon dioxide in their search for hosts and will respond to a body that is 2°C or more above ambient temperature. They produce an aggregation pheromone that brings them together. The nocturnal biting of bedbugs can be debilitating to humans, whose sleep is interrupted every night. The presence of the bugs can be suspected by finding their specks of feces and their odor. Bedbugs are very common in Third World countries and are becoming more common even in the United States. Travelers may import bugs in their suitcases or belongings, and the bugs then "make themselves at home" in cracks and crevices, behind baseboards, and under mattresses.

Triatomid, Reduviid, Kissing, or Conenose Bugs (Chagas' Disease)

Triatomid bugs have well-developed wings, are larger than bedbugs, and have a cone-shaped head (Figures 24.30 and 24.31). These bugs are usually brown or black with orange and black markings on the abdomen where it extends beyond the folded wings. All species feed on blood, either mammalian or avian. Most species are hematophagous, feeding primarily on mammalian or avian blood. The infective forms of *Trypanosoma cruzi* (Chagas' disease) are found in the bug's feces. In areas where the infection is more prevalent, the bugs tend to defecate very soon after

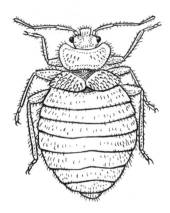

Figure 24.28 Bedbug (*Cimex* sp.), 5 mm. (Illustration by Sharon Belkin; based on an illustration from H. H. Najarian, *Textbook of Medical Parasitology*, The Williams & Wilkins Co., Baltimore, Md., 1967.)

taking a blood meal. Since the bug's saliva contains an irritant, the person tends to scratch, thus scratching in the infective forms from the bug feces. The time of the blood meal and the number of feeds during engorgement in relation to defecation are important epidemiologically and are directly related to infection potential.

The earlier a bug defecates during or after feeding, the more likely it is to be an efficient vector. Most triatomid bugs are nocturnal and feed on sleeping inhabitants of the house. They are attracted to the host by warmth, carbon dioxide, and odor. The bites are relatively painless, hence the name "kissing bug." Obviously it would not be advantageous for the person to wake up before the bugs complete their blood meal and subsequent defecation. However, some individuals have an allergic reaction to the bite approximately 24 to 48 h later; in some cases it is serious. Edema of the eyelids and conjunctivitis may occur

Figure 24.29 Bedbug (*Cimex* sp.).

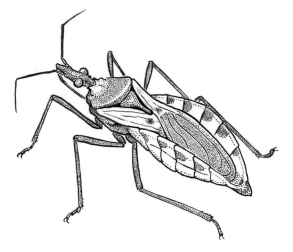

Figure 24.30 Triatomid bug (*Triatoma* sp.), 25 mm. (Illustration by Sharon Belkin.)

(Romaña's sign). The edema does not pit on pressure, and the skin is dry. Although Romaña's sign is thought to be almost pathognomonic for early Chagas' disease, in an analysis of 300 patients with early disease, unilateral periorbital edema was found in only 48.7%. Currently, it is estimated that 16 million to 18 million people are infected with *T. cruzi* and 90 million to 100 million are at risk. Although most cases occur in Mexico and Central and South America, indigenous cases have been reported from Texas, Tennessee, and California. Serologic evidence places *T. cruzi* as far north as Oklahoma. Five species of kissing bug are found within different parts of the United States: *Triatoma protracta*, *T. gerstaeckeri*, *T. sanguisuga*, *T. rubida*, and *T. rubrofasciata*.

In addition to an irritant, the bug saliva contains an anticoagulant, so that the blood remains unclotted when stored within the bug. This ensures that the blood remains fluid for easier passage into the digestive system of the

midgut. In areas of endemic infection, the number of bugs may vary tremendously in houses that harbor the triatomids. When a heavily infested house in Venezuela was torn down, nearly 8,000 triatomid bugs were recovered. It was calculated that in that house the feeding rate was 58 bugs per person per day. The loss of blood per person per month can range from 0.7 to 40 ml; however, in that particular house, it exceeded 100 ml (30).

Most reduviids kill other insects; the name "assassin bugs" is used as a general term. However, bugs associated with the transmission of *T. cruzi* are referred to as kissing bugs. While assassin bugs may bite humans, producing painful lesions, they are not associated with parasite transmission. Often the general names are used interchangeably, thus confusing the issue of actual parasite transmission.

Order Coleoptera (Beetles)

The majority of beetles, representing over 40% of all insects, are not harmful to humans. In some cases, the adult beetles contain vesicating substances such as pederin (an alkaloid within body fluids) and cantharidin (a volatile terpene found in the hemolymph). When beetles are accidentally crushed against the skin, serious blistering may occur (Figures 24.32 through 24.34). Most of these beetles are seasonal and may be found in large numbers in some fields at certain times of the year. Important beetles in the family Meloidae are in the genus *Lytta*, which contains the infamous "Spanish fly" from which the well-known aphrodisiac powder is obtained. Often, beetles in the genus *Paederus* are attracted to the light and, on contact with the skin, can secrete a highly acidic liquid that produces a painful blister. There is usually no immediate response to the fluids, but in 24 to 48 h a weal appears

Figure 24.31 Triatomid bug.

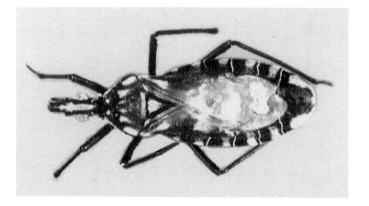

Figure 24.32 Blister beetles. (A) Margined (*Epicauta pestifera*); (B) ash gray (*Epicauta fabricii*). (Illustration by Sharon Belkin.)

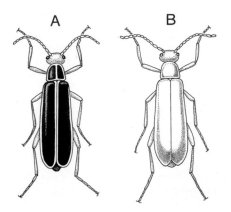

Figure 24.33 *Paederus laetus*, a beetle of Central America which elaborates a vesiculating toxin. Magnification, ×4. (Armed Forces Institute of Pathology photograph.)

Order Siphonaptera (Fleas) (*Ctenocephalides* spp., *Xenopsylla cheopis, Pulex irritans* [Human Flea], *Tunga penetrans, Nosopsyllus fasciatus, Echidnophaga gallinacea,* and "Sand Fleas")

Fleas are wingless ectoparasites of mammals and birds; whereas lice are usually host specific, fleas parasitize a number of hosts. It is this ability to move from one host to another that facilitates the transfer of disease from animals, primarily rats, to humans. Fleas are laterally flattened and have no wings. The legs are long and muscular, which allows them to jump (Figures 24.35 and 24.36). They are hosts and transmitters of several pathogenic organisms including the agents of plague and murine typhus and *Dipylidium caninum*, the dog tapeworm. Also, their bite or presence in the skin causes dermatitis or ulcerations. Medically important fleas include *Tunga penetrans* (Figures 24.37 and 24.38), *Pulex irritans, Ctenocephalides canis, C. felis,* and *Xenopsylla cheopis*.

Generally the flea bites begin with a wheal around each site within about 5 to 30 min. Itching is common, with the production of an indurated papule within about 12 to 24 h. In people who are sensitive to the bites, a delayed reaction appears in about 12 to 24 h and lasts for

and is followed by blisters. The erythema may last for several months, and the conjunctivitis can cause temporary blindness. The toxin pederin is the most complex nonproteinaceous insect secretion known and has been used in Chinese medicine for over 1,000 years.

Occasionally, reactions occur as a result of contact with the larvae of beetles in the family Dermestidae. However, contact with adult beetles in various families within the Coleoptera represents the most common situation in which humans are involved.

Figure 24.34 Contact dermatitis from staphylinid beetle, *Paederus* sp., on the body of a man in Ethiopia. (Armed Forces Institute of Pathology photograph.)

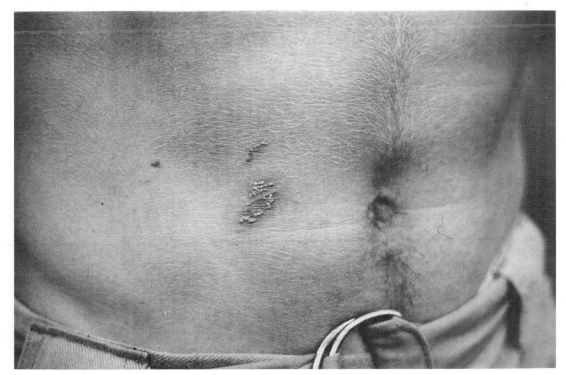

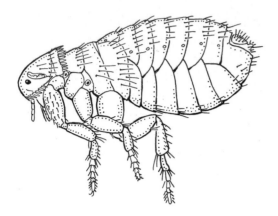

Figure 24.35 Flea, 1.5 mm. (Illustration by Sharon Belkin.)

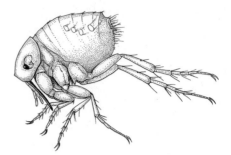

Figure 24.37 Flea (*Tunga penetrans*, chigoe flea), 1.5 mm. (Illustration by Sharon Belkin.)

a week or longer. The intense itching is often sufficient cause for the patient to consult a physician.

Cat and dog fleas, *C. felis* and *C. canis*, are major domestic pests in many parts of the world, including North America and Australia. However, the cat flea is most often found within the United States; both types of fleas can feed on either animal. Dogs tend to be more allergic to flea saliva than are cats, but both can have severe reactions to heavy infestation. Cats can remove about 50% of their fleas in a single week, but insecticide dust, spray, shampoo, or other options may be required for removal of all fleas. Flea larvae are very sensitive to drying, and outdoor environments do not generally support survival, which requires a relative humidity of >50%. In the transmission of *D. caninum*, the tapeworm eggs are ingested by the larval stages of the dog, cat, or human flea, where they develop into cysticercoid larvae. When these fleas are then ingested by the definitive host (dogs, cats, or humans), the adult worm develops within 3 to 4 weeks. Most of these infections have been found in children.

Although *Echidnophaga gallinacea* (the sticktight flea) is associated with poultry, it may also feed on humans. Chickens are commonly seen with flea "patches" around the comb, wattles, or even the eyes.

"Sand fleas" is a general term that is used for a number of arthropods, only some of which are actually fleas. In the northern United States, sand fleas are usually dog or cat fleas associated with stray animals. In the western United States, cat or human fleas associated with a number of different animals are called sand fleas. Along the coastal areas, small amphipod crustaceans found in seaweed are also called sand fleas.

Order Anoplura (Sucking Lice)

The order Anoplura contains the blood-sucking lice (*Pediculus* and *Phthirus* spp.), which have legs with claws adapted for clinging to hairs or fibers. There are three nymphal stages prior to the adult, and all stages feed on blood. In general, the biting lice respond to warmth and smell, avoid high humidity, and are negatively phototactic

Figure 24.36 Flea. (From A Pictorial Presentation of Parasites: A cooperative collection prepared and/or edited by H. Zaiman.)

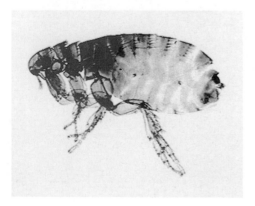

Figure 24.38 Tungiasis. Multiple and irregularly confluent tunga craters on the foot of an African. The nonpressure areas (instep and between the toes) are the most severely involved. The weight-bearing portion of the sole tends to be spared. This patient died of tetanus. (Armed Forces Institute of Pathology photograph.)

(i.e., they move toward dark objects). There are several types of human lice: the head louse, the body louse (Figures 24.39 and 24.40), and the pubic or "crab" louse (Figures 24.41 and 24.42). The head and pubic lice deposit and cement eggs onto the hair shaft (Figure 24.43), whereas the body louse deposits eggs on clothing, particularly in the seams. Although they do not specifically transmit any parasites, they do transmit epidemic typhus and louse-borne relapsing fever. Prevention involves both treatment of individuals and mass control, both of which can be difficult in situations such as jails, refugee camps, army barracks, and schools (5).

Body Lice

In a stable population of the body louse, *P. humanus*, more than two-thirds of the lice are present as eggs, about one-fourth are present as nymphs, and only 6 to 7% are present as adults. The adults measure 2 to 3 mm for the male and 2.4 to 3.6 mm for the female. Body lice can cause severe skin irritation; there may be swelling and the development of papules at the bite sites, as well as mild or severe itching. Some individuals may become sensitized to antigens introduced during louse feeding, and they can experience generalized allergic reactions. Removal of skin during itching can also lead to secondary infection. The lice tend to attach to cloth fibers along the seams inside of underwear or other places where clothes touch the body; they tend to prefer wool clothing. *P. humanus* can be removed from clothes by heating the clothes to >60°C for 15 min; this could be accomplished by placing clothes in the clothes drier for about 20 to 30 min.

Head Lice

Head lice (*P. humanus capitis*) live on the skin among the hairs on the patient's head. Their life span is about 1

Figure 24.39 Body louse (*Pediculus* sp.), 1.5 mm. (Illustration by Sharon Belkin.)

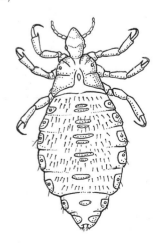

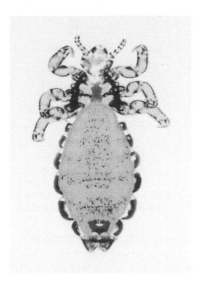

Figure 24.40 Body louse (*Pediculus* sp.).

month if they are not removed from the body, and the life cycle from egg to adult takes about 3 weeks. Washing the hair reduces the number of nymphs and adult *P. capitis*, and combing with a finely toothed comb removes most of the eggs; combing should be done when the hair is damp. During the washing process, the lice feces are also removed; they can also serve as an irritant during the infestation. It is important to remember that if one person in the family has head lice, the entire family should be treated, since others in the family may be infested but may not be aware of it. Most of the products designed for louse removal are effective, provided that directions are followed. In refugee camp situations where body lice can multiply and spread quickly and easily, DDT dust can be applied to clothing; however, resistance to DDT has arisen, and other insecticides may be required for control.

Pubic or Crab Lice

The two distinct species of *Phthiris* occur on separate hosts, the human and the gorilla. Adult pubic lice are dark gray to brown and have a typical "crab" shape. *P. pubis* occurs on the hair in the pubic and perianal areas of the human body and occasionally the axillae, eyebrows, and beard. The incidence of this infestation has been on the increase, and some think that it may be the most contagious sexually transmitted disease. However, this infection is often found in association with other venereal diseases, including gonorrhea and trichomoniasis. Transmission can occur by sleeping in the same bed as another infested individual. In patients with an infestation of *P. pubis*, there tend to be blue or slate-gray macules on the skin (maculae caeruleae); these are thought to be either altered patient blood pigments or possibly substances occurring in the salivary glands of

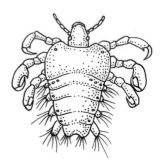

Figure 24.41 Crab louse (*Phthirus* sp.), 1 mm. (Illustration by Sharon Belkin; based on an illustration from H. H. Najarian, *Textbook of Medical Parasitology*, The Williams & Wilkins Co., Baltimore, Md., 1967.)

the lice. Once these lice have left the host, their survival time is relatively short, with death occurring in less than 48 h at 15°C. If infestation occurs on the eyelids, there may be inflammation of the eyelids. These lice live only on humans, require human blood for survival, and do not infest rooms, carpets, beds, or pets. Because the lice tend to die fairly quickly once they are off the host, it is not necessary to use insecticidal sprays or fogs in the home, work, or school environment.

Order Mallophaga (Biting and Chewing Lice)

Biting lice are morphologically similar to other lice and are not very important in human medicine, although they serve as the intermediate host for the dog tapeworm, *Dipylidium caninum*. Infestations are more common in puppies, very old dogs, or those that are sick or debilitated. None of these lice live on humans but are primarily ectoparasites of birds and mammals, feeding on feathers, hairs, and epidermal scales (5).

Figure 24.42 Crab louse. (From A Pictorial Presentation of Parasites: A cooperative collection prepared and/or edited by H. Zaiman.)

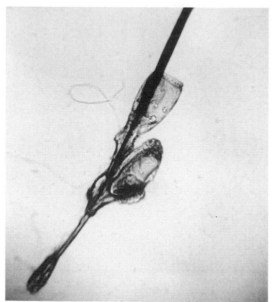

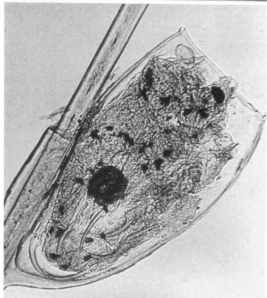

Figure 24.43 Louse. (Upper) Eggs on hair shaft; (lower) enlargement of egg on hair shaft. (Photographs courtesy of Duane J. Gubler, Centers for Disease Control and Prevention.)

Order Hymenoptera (Bees, Wasps, and Ants)

The order Hymenoptera contains insects with two pairs of membranous wings, reduced veins in the wings, and hind wings smaller than the fore pair and interlocked with the latter by means of hooklets (Figure 24.44). There are more than 100,000 described species. Many members of species within this order live together in great structured societies, and reproduction is performed either by many of the female members or, in some cases, by a single individual known as the queen. Members of the group or colony tend to attack intruders.

Bees

Although stings of members of this order normally cause only temporary pain and swelling, anaphylaxis may occur in patients who are sensitized. Within the United States, approximately 50 to 100 individuals die each year from severe anaphylactic reactions to bee and wasp stings (17). Honeybees are not necessarily aggressive, but they tend to be everywhere outdoors during the summer months. During the stinging process, pheromones are released that cause other bees in the vicinity to attack the victim. When a person has been stung, the stinger should be removed, the site should be disinfected, and ice packs should be applied to slow the spread of venom. The stinger can be removed by scraping with a knife or fingernail to prevent further venom from being injected. If the person experiences a severe local reaction, oral antihistamines and topical application of corticosteroids can be used. Systemic allergic reactions can be a life-threatening emergency requiring immediate medical help.

The spread of the Africanized honeybee to Brazil and then to other areas of South and Central America and the southern United States has been accompanied by extensive publicity. These bees can be easily irritated and provoked to protect the hive, resulting in hundreds of stings, severe reactions, and death. Although the public has received information regarding these bees, the "killer" bees are difficult to differentiate from normal honeybees, particularly visually.

Figure 24.44   (A) Bee; (B) fire ant; (C) wasp. (Illustration by Sharon Belkin.)

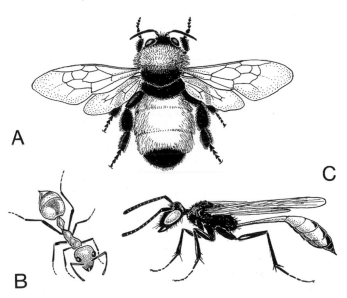

In comparison to the European honeybee, the Africanized bees tend to attack more readily, remain angry for extended periods, and display massive group stinging behavior in defense of the colony and territory. Although their stings are no more toxic than those of domestic bees, death can result from the cumulative effects of numerous stings, numbering from 400 to 1,000 stings, or from anaphylactic shock. It is very likely that we will continue to see more problems and possible deaths due to the Africanized honeybees as their geographic area continues to expand within the United States. As an example, in Venezuela in 1978, before the Africanized honeybees arrived, there were 12 deaths attributed to bee stings; however, after 1978, there were 100 deaths per year. Stings can be treated in the same way; however, in the case of multiple stings, excessive histamine release may result from venom action and not necessarily from an allergic reaction. *Physicians need to be aware of this fact and the possible need to treat for histamine overdose* (20).

Bumblebees also sting when disturbed; however, they are neither as aggressive nor as prevalent as honeybees. Treatment for stings is the same as for honeybee stings.

Fire Ants

Ants can bite as well as sting; however, most do little harm. One exception is the fire ant, *Solenopsis invicta*, which is endemic in the southeastern United States. The presence of fire ants can usually be recognized by seeing their characteristic mounds, which are earthen mounds with a diameter of 24 in. and a height of 18 to 24 in. surrounded by relatively undisturbed vegetation (Figure 24.45) (20). When even slightly disturbed, the ants tend to boil out of their mounds. The ants measure about 4 to 6 mm long and are reddish brown (red form) or black (black form). The workers attach to the skin by biting with their mandibles and then lowering their abdomen to inject the stinger; thus, the victim receives both a bite and a sting. The symptom of a sting has been described as a "burning itch." Within about 24 h, a pustule develops and may remain for about a week (Figure 24.6). Although some individuals have sustained thousands of stings with no toxic or allergic reactions, some people with hypersensitivity to fire ant venom may have severe reactions from just a few stings. At least 250 million acres within the United States are infested, now reaching as far west as New Mexico, Arizona, and California. In some heavily infested areas, 50 to 400 or more mature colonies may be found per acre. There are actually two species, *S. invicta*, the red form imported from Brazil, and *S. richteri*, the black form imported from Uruguay. The geographic range for *S. richteri* in the United States is much more limited, occurring in northeastern Mississippi, northwestern Alabama, and part of Tennessee (20).

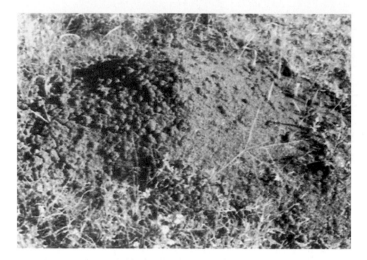

Figure 24.45   Typical mound of the fire ant; the height can reach 2 ft.

Figure 24.46   Typical flat mound of the harvester ant; the flat area with no vegetation can range from 3 to 9 ft.

Harvester Ants

Harvester ants, *Pogonomyrmex* spp., are ant species that sting both animals and humans. These ants are red to dark brown and are two or three times as large as fire ants. Harvester ant mounds are usually flat or slightly elevated, measure about 3 to 9 ft in diameter, and are surrounded by no vegetation (Figure 24.46). There is only one harvester ant species east of the Mississippi River, and it is found throughout the Southeast. In the western United States, about 20 species occur, 3 of the most serious being the red harvester ant (*P. barbatus*), the California harvester ant (*P. californicus*), and the western harvester ant (*P. occidentalis*). The reaction to harvester ant stings is not always localized but may spread along the lymphatics. This causes severe pain in the lymph nodes in the axilla or groin some time after the pain from the original sting has gone. Systemic allergic reactions tend to be uncommon.

Order Blattaria (Cockroaches)

Cockroaches are among the most ubiquitous pests throughout the world. A number of the 3,500 species have become adapted to living with humans and are referred to as "domestic" species that will breed anywhere. They are very active, nocturnal insects; anyone who has turned on the lights in the kitchen late at night remembers the surprise at seeing all the visitors! Cockroaches are flattened dorsoventrally and have long, multisegmented antennae, biting mouthparts, two pairs of wings, two cerci, and legs adapted for running (17). They feed on human food, excreta, sputa, and even bookbindings, paper, and leather.

Mechanical Transmission of Organisms.   Although certain flies and dung beetles have been implicated as mechanical vectors of *Cryptosporidium* oocysts, there is every reason to suspect that cockroaches could also do the same (22). When cockroaches feed on infected material, their mouthparts and legs become contaminated; they can then introduce these contaminants into human surroundings. Also, organisms can pass through the cockroach gut and remain fully viable. Some investigators have speculated that the cockroach may be a more significant vector of human pathogens than houseflies, transmitting viruses, bacteria, and parasites to human hosts. In one study of 234 cockroaches trapped from different sites (toilets, parlors, kitchens, and bedrooms) in houses with pit latrines and water systems, bacterial, fungal, and parasitic isolates were identical irrespective of the site and included *Escherichia coli, Klebsiella pneumoniae, Proteus vulgaris, Proteus mirabilis, Citrobacter freundii, Enterobacter cloacae, Salmonella* spp., *Pseudomonas aeruginosa, Serratia marcescens, Staphylococcus aureus, S. faecalis, S. epidermidis, Aeromonas* spp., *Candida* spp., *Rhizopus* spp., *Aspergillus* spp., *Mucor* spp., cysts of *Entamoeba histolytica*; oocysts of *Cryptosporidium* spp., *Cyclospora cayetanensis*, and *Isospora belli*; cysts of *Balantidium coli*; ova of *Ascaris lumbricoides, Ancylostoma duodenale, Enterobius vermicularis*, and *Trichuris trichiura*; and larvae of *Strongyloides stercoralis* (49). Some humans also become allergic to cockroach allergens. Control involves good sanitation, prevention of new entry, and a combination of various pesticides including bait, strips, sprays, and dusts.

Several important species include *Blatella germanica* (German cockroach), *Blatta orientalis* (Oriental cockroach), *Supella longipalpa* (brown-banded cockroach), and *Periplaneta americana* (American cockroach) (Figure

24.47). These different cockroaches prefer different habitats, with some preferring warm, moist conditions (German cockroach) and others preferring cooler locations (Oriental cockroach). The most common cockroach throughout the United States is the German cockroach, which can be found infesting most buildings. Adult German and brown-banded cockroaches are approximately 15 mm long, while the American and Oriental cockroaches measure 30 to 50 mm long.

Cockroach Allergens. In addition to the mechanical transmission of many types of bacteria, fungi, and parasites, their excrement and cast skins contain a number of antigens that cause allergic responses. Urban allergens include the cockroach, mouse, and rat. Cockroach allergy is common among inner city children with asthma, and exposure to cockroach allergen is associated with more severe disease. Cockroach allergens are one of the major etiologic risk factors for immunoglobulin E-mediated allergic respiratory illness throughout the world. A high prevalence of cockroach hypersensitivity in atopic (20 to 55%) and asthmatic (49 to 60%) populations has been documented (4, 57). In a survey of day care facilities

Figure 24.47 (Upper) American cockroach (*Periplaneta americana*) (illustration by Sharon Belkin); (lower) *Periplaneta americana*.

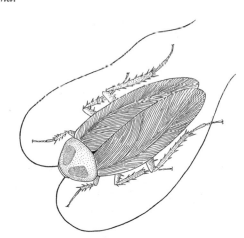

in North Carolina, detectable levels of indoor allergens were commonly found, including cockroach allergens. For many young children and day care staff, day care facilities might be a source of clinically relevant exposures to indoor allergens (2). Cockroach allergen avoidance begins with effective pest control, followed by thorough and repeated cleaning. At least 1 to 2 months is required to eliminate cockroaches, and an additional 4 to 6 months is required to remove residual allergen. Once allergen levels have been reduced, continued efforts are necessary to maintain the home free of allergen sources. The school environment can also be an important site of exposure to cat, dog, dust mite, and cockroach allergens, which have been detected in the settled dust of many schools. In industrialized nations, the school environment has a lower potential for exposure than the home environment, but schools are significant for allergic individuals whose home environment has been addressed to the extent possible. From a public health perspective, the school environment should be considered a target for primary and secondary prevention.

Class Arachnida (Ticks, Mites, Spiders, and Scorpions)

Subclass Acari (Ticks, Mites, and Chiggers)

The subclass Acari contains the soft and hard ticks, mites, and chiggers.

Soft and Hard Ticks

The soft and hard ticks are ectoparasites adapted to blood-sucking (mammals, birds, and reptiles) (18, 28, 45, 46). There is no hard dorsal plate in the soft tick, and the mouth parts are ventral to the anterior extremity (Figures 24.48 and 24.49; Table 24.8). In the hard tick, the dorsal plate or shield covers the entire surface of the male but only the anterior portion of the female (Figures 24.50 and 24.51). The dorsal surface of a tick's body often has color patterns, furrows, or sculpting that allows species determination. Ticks are responsible for the transmission of viral, bacterial, rickettsial, and protozoan (*Babesia* spp.) pathogens. The salivary secretions of some ticks may produce systemic toxemia (tick paralysis) that can cause death. Paralysis can occur from a tick attached to any part of the body, although the symptoms may be more serious when the tick injects toxin at the back of the neck or along the spinal column. Symptoms may include elevated temperature, rapid ascending flaccid paralysis, and difficulty in breathing and swallowing, and death may result. The onset may be very sudden, within 24 h, or may not appear for several days. The most likely tick involved is *Dermacentor andersoni*, the Rocky Mountain wood tick. Human cases tend to occur in the spring

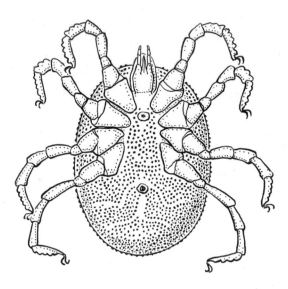

Figure 24.48 Soft tick, ~12 mm. (Illustration by Sharon Belkin; based on an illustration from H. H. Najarian, *Textbook of Medical Parasitology*, The Williams & Wilkins Co., Baltimore, Md., 1967.)

Figure 24.49 Soft tick. (Photograph courtesy of Duane J. Gubler, Centers for Disease Control and Prevention.)

and early summer, when ticks are most active and are seeking a blood meal. Tick paralysis can be particularly dangerous for children; after playing in underbrush or bushes, the child should have all clothing removed and should be carefully examined for the presence of ticks, particularly around the neck and shoulder area (under the hair). The most likely geographic area is in western Montana, Idaho, and Oregon, although cases have been reported from other states including Georgia, Tennessee, Louisiana, and the Atlantic seaboard. Ticks in the genera *Amblyomma* and *Dermacentor* have been involved in cases outside of the western United States, while in other parts of the world *Ixodes, Rhipicephalus, Hyalomma,* and *Ornithodoros* have been implicated in tick paralysis.

Argasidae (soft ticks)	Ixodidae (hard ticks)
1. Do not have dorsal shield (scutum)	1. Do have dorsal shield (scutum)
2. Capitulum is located on ventral surface and does not project anteriorly	2. Capitulum is located on anterior margin

Ticks feed by making a cut in the host epidermis, inserting the hypostome into the cut, and attaching to the host. Blood is kept flowing through the use of an anticoagulant from the salivary glands of the tick. Some hard ticks form a cement cone around the mouthparts and bite site. Female

hard ticks undergo a growth feeding phase characterized by slow continuous blood intake followed by a rapid engorgement phase occurring during the last 24 h of attachment. Hard ticks have the ability to expand considerably, often resembling a soft tick. Many hard ticks attach themselves to grasses or weeds, waiting for a host to come along; they are also attracted to carbon dioxide. Hard ticks are generally found in brushy, wooded, or weedy areas containing deer, cattle, dogs, small mammals, and other hosts. Hard ticks have to conserve moisture, and so their activity and survival are influenced by temperature as well as humidity.

Differentiation between the soft ticks and hard ticks is relatively easy once an engorged hard tick is recognized as such. However, identification to the species level among the hard ticks can be difficult. Currently there is a great deal of interest in identifying the ticks that transmit Lyme disease. In the northeastern United States, the primary vector for Lyme disease caused by *Borrelia burgdorferi* is *Ixodes scapularis*, the black-legged tick (formerly known as *I. dammini*, although it now appears that they are a single species) (38, 53). The life cycle takes 2 to 4 years. Larvae are active from August to October, when they are feeding on *Peromyscus leucopus*, the white-footed mouse. The infection rate in mice is approximately 25% and occurs from May to June. Human cases tend to peak in July. A tick needs to feed for 48 h for successful transmission of *B. burgdorferi*; although 50 to 100% of the adult *I. scapularis* ticks are infected, most are removed from the human host prior to 48 h. *I. scapularis* is found in dense brush on islands off the northeast coast and in heavily for-

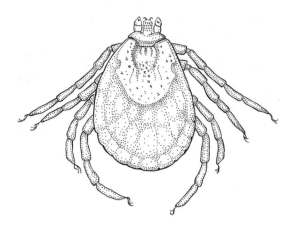

Figure 24.50 Hard tick, 4 to 10 mm. (Illustration by Sharon Belkin; based on an illustration from H. H. Najarian, *Textbook of Medical Parasitology*, The Williams & Wilkins Co., Baltimore, Md., 1967.)

ested areas inland. In Wisconsin, these ticks occur in areas between the northern coniferous-hardwood and southern deciduous forests.

In western North America, the primary vector of *B. burgdorferi* is *I. pacificus*, the western black-legged

Figure 24.51 (Upper) Hard tick (photograph courtesy of Duane J. Gubler, Centers for Disease Control and Prevention); (lower left) *Dermacentor* sp.; (lower right) *Ixodes* sp.

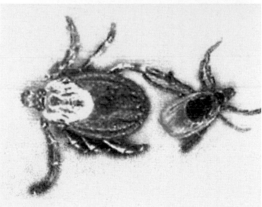

tick. These ticks range from sea level to >6,000 ft and are found from California to British Columbia and east to Nevada, Utah, and Idaho. Larvae and nymphs are active from March to June, and adults are found on humans from November to May. The infection rate of *I. pacificus* is about 1.5%. Studies looking at the prevalence of anti-arthropod saliva antibodies among residents in a high-risk community revealed a 79% positive rate compared with control groups (33). In the San Francisco Bay Area, the figure was 36%. In the study group, seropositivities for *I. pacificus* and for *B. burgdorferi* were significantly correlated. Experimental and field evaluations of two acaricides for control of *I. pacificus* in northern California appear very promising, and these substances can be a beneficial component of an overall control program (37).

The morphologic differences between *I. pacificus* and *I. scapularis* are minimal (Figures 24.53 and 24.54). Both species have no white markings on their dorsal side and no eyes or festoons. Adults are dark brown, and the mouthparts are moderately long. Occasionally, *I. scapularis* appears light brown or orangish from the dorsal view (39). The difference between the nymph and the adult can be seen in Figure 24.52. Descriptions of the more common ticks are given in Table 24.9.

Itch Mites (Mange Mites), Follicle Mites, and Dust Mites

Itch Mite. Several genera infect the skin of mammals, with *S. scabiei* being found in humans, although some mange mites of animals occasionally cause a pruritic rash

Figure 24.52 *Ixodes* spp. Diagram of the female (A) and male (B) adult ticks and the nymph (C) (bar, 3 mm). (Illustration by Sharon Belkin; based on information in reference 28.)

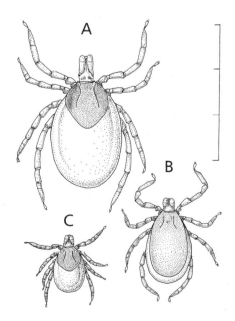

Table 24.8 Identification of the more common hard and soft ticks. Illustration by Sharon Belkin; adapted from National Communicable Disease Center *Pictorial Keys*, 1969.)

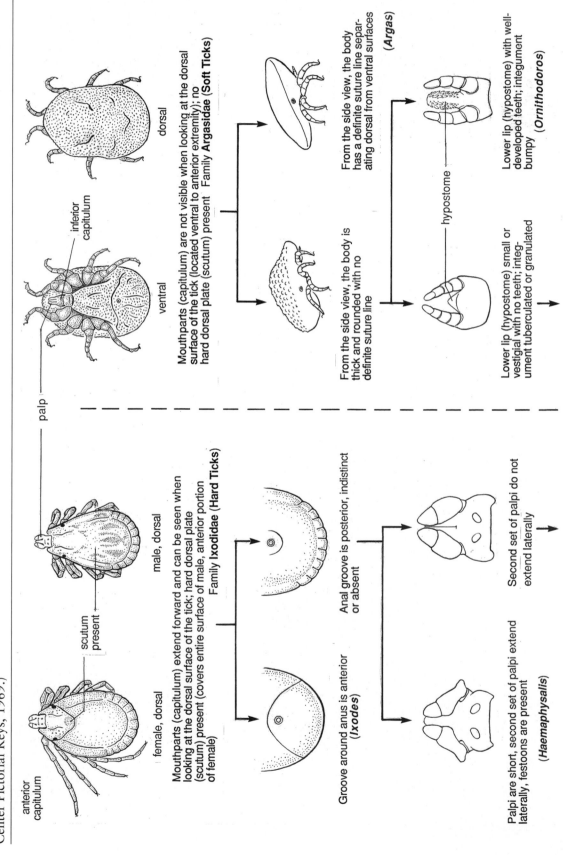

anterior capitulum

palp

inferior capitulum

female, dorsal

male, dorsal

scutum present

dorsal

ventral

Mouthparts (capitulum) extend forward and can be seen when looking at the dorsal surface of the tick; hard dorsal plate (scutum) present (covers entire surface of male, anterior portion of female) **Family Ixodidae (Hard Ticks)**

Mouthparts (capitulum) are not visible when looking at the dorsal surface of the tick (located ventral to anterior extremity); no hard dorsal plate (scutum) present **Family Argasidae (Soft Ticks)**

Groove around anus is anterior (*Ixodes*)

Anal groove is posterior, indistinct or absent

From the side view, the body is thick and rounded with no definite suture line

From the side view, the body has a definite suture line separating dorsal from ventral surfaces

(*Argas*)

hypostome

Second set of palpi do not extend laterally

Palpi are short, second set of palpi extend laterally, festoons are present

(*Haemaphysalis*)

Lower lip (hypostome) small or vestigial with no teeth; integument tuberculated or granulated

Lower lip (hypostome) with well-developed teeth; integument bumpy

(*Ornithodoros*)

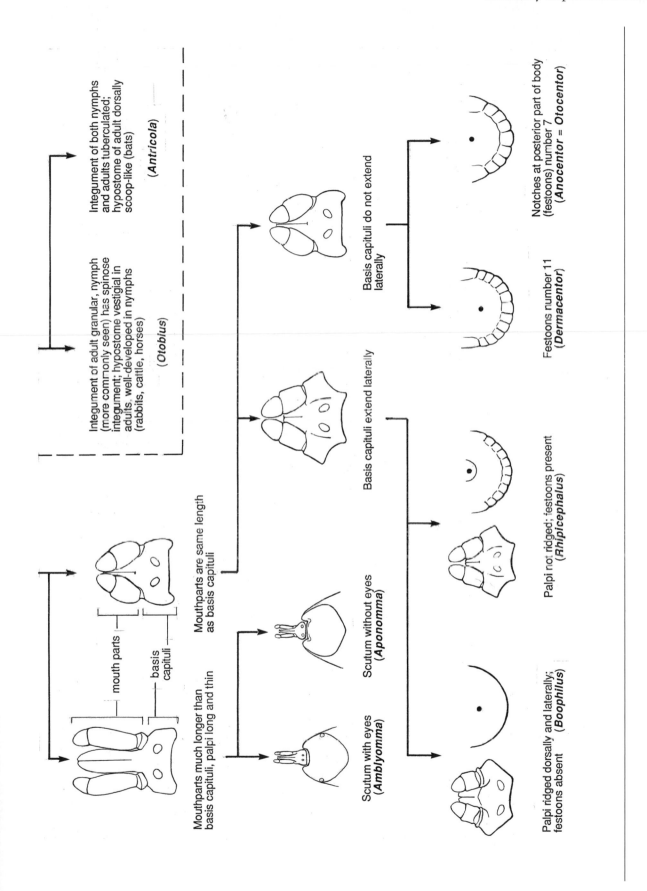

Mouthparts much longer than
basis capituli, palpi long and thin

Mouthparts are same length
as basis capituli

mouth parts

basis
capituli

Scutum with eyes
(*Amblyomma*)

Scutum without eyes
(*Aponomma*)

Basis capituli extend laterally

Basis capituli do not extend
laterally

Palpi ridged dorsally and laterally;
festoons absent   (*Boophilus*)

Palpi not ridged; festoons present
(*Rhipicephalus*)

Festoons number 11
(*Dermacentor*)

Notches at posterior part of body
(festoons) number 7
(*Anocentor = Otocentor*)

Integument of adult granular, nymph
(more commonly seen) has spinose
integument; hypostome vestigial in
adults, well-developed in nymphs
(rabbits, cattle, horses)

(*Otobius*)

Integument of both nymphs
and adults tuberculated;
hypostome of adult dorsally
scoop-like (bats)

(*Antricola*)

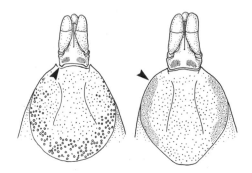

Figure 24.53 (Left) *Ixodes scapularis.* Scutum (the dorsal shield, being small in the female and almost covering the dorsal surface in the male) and the terminal capitulum, which is attached to the anterior end of the body of Ixodidae. Note that the scutum is nearly circular, with punctuations larger peripherally. The cornua (arrowhead) is small but definite for *I. scapularis.* (Right) *Ixodes pacificus.* Scutum and the terminal capitulum. Note that the scutum is oval with uniformly distributed small punctuations (arrowhead), and the cornua is absent. (Illustration by Sharon Belkin; based on information in reference 28.)

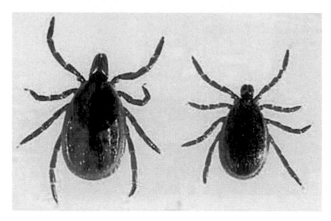

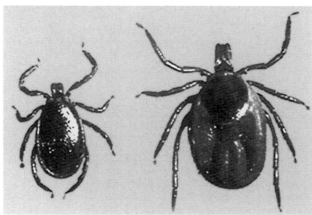

Figure 24.54 (Upper) *Ixodes scapularis.* (Left) Adult female. (Right) Adult male. Note that there are no white markings on the dorsal surface and no eyes or festoons (Table 24.8). (Lower) *Ixodes pacificus.* (Left) Adult male. (Right) Adult female.

in humans (56). *S. scabiei* is microscopic and lives in cutaneous burrows, where the fertilized female deposits eggs (Figures 24.55 and 24.56). Scabies is transmitted by close contact with infested individuals, including touching, shaking hands, sexual contact, and contact in day care centers with children and the elderly. The usual skin sites that are susceptible to infection are the interdigital spaces, backs of the hands, elbows, axillae, groin, breasts, umbilicus, penis, shoulder blades, small of the back, and buttocks (Figures 24.57 and 24.58) (5). The outstanding clinical symptom is intense itching. Scratching commonly causes weeping, bleeding, and sometimes secondary infection. A form of the infestation, called Norwegian scabies, can occur in immunosuppressed or anergic individuals; many mites are present in keratotic excrescences on the body and extremities, but pruritus is usually absent (Figures 24.59 to 24.61). This infestation is highly contagious and has been reported to be the cause of hospital epidemics. Scabies occurs worldwide, with at least 300 million cases each year. Oral ivermectin at a dose of 300 µg/kg, given as a single dose and repeated after 7 days, proved effective for the treatment and prophylaxis of scabies in an infected institutional environment (42). Specific techniques for recovery of the mites are listed below.

Skin-scraping technique. The diagnosis can be confirmed by demonstration of the mites, eggs, or scybala (fecal pellets). Because the mites are located under the surface of the skin, scrapings must be made from the infected area.

1. Place a drop of mineral oil on a sterile scalpel blade. (Mineral oil is preferred to potassium hydroxide solution or water. Mites adhere to the oil, skin scales mix with the oil, the refractility differences are greater between the mite and the oil, and the oil does not dissolve fecal pellets.)

2. Allow some of the oil to flow onto the papule.

3. Scrape vigorously six or seven times to remove the top of the papule. (There should be tiny flecks of blood in the oil.)

4. Transfer the oil and scraped material to a glass slide (an applicator stick can be used).

5. Add 1 or 2 extra drops of mineral oil to the slide and stir the mixture. Any large clumps can be crushed to expose hidden mites.

6. Place a coverslip on the slide, and examine (first on low power). The adult mites range from approximately 215 to 390 µm in length, depending on sex. The eggs are 170 µm long by 92 µm wide, and the fecal pellets are about 30 by 15 µm. The fecal pellets are yellow-brown.

Table 24.9 Descriptions of some of the more common hard ticks found in the United States

Tick	Description	Comments
Amblyomma americanum (Lone Star tick)	Reddish-brown tick species; adult stages have long mouth parts visible from above; adult females have distinct white spot on their back (scutum); males have no single spot but have inverted horseshoe-shaped markings at the posterior edge of their dorsal side	Central Texas east to the Atlantic; north to Iowa and New York; northern Mexico; probably most aggressive and common tick in southern United States; transmits tularemia, possibly ehrlichiosis
Amblyomma maculatum (Gulf Coast tick)	Somewhat similar to American dog tick, except with metallic instead of white markings, long mouth parts; adult females with metallic white or gold markings on scutum, males with numerous, mostly connected linear spots of golden white	Portions of Atlantic and Gulf Coast areas (100–200 miles inland, south into Mexico); increasingly a pest in southern United States; nuisance only
Dermacentor andersoni (Rocky Mountain wood tick)	Adults have shorter mouth parts than *Amblyomma*; usually dark brown or black with bright white markings on the scutum	Western counties of Nebraska and Black Hills of South Dakota to Cascade and Sierra Nevada Mountains; northern Arizona and northern New Mexico to British Columbia, Alberta, and Saskatchewan; transmits RMSF,[a] Colorado tick fever, tularemia; tick paralysis in United States and Canada
Dermacentor variabilis (American dog tick)	Adults dark brown or black; short, rounded mouth parts; dull or bright white markings on scutum	Medically important; primary vector of RMSF in the central and eastern United States; transmits tularemia, maybe tick paralysis; deticking dogs may lead to infection
Ixodes pacificus (western black-legged tick)	No white markings on dorsal side, no eyes or festoons, adults dark brown; moderately long mouth parts; appears almost identical to *I. scapularis*	Along Pacific coastal margins of British Columbia and the United States, possibly extending into Baja California and Mexico; most cases of Lyme disease in California are transmitted by this tick
Ixodes scapularis (black-legged tick)	Adults have no eyes, festoons, or white markings on dorsal side; dark brown but may appear light brown or orangish	Northern form in New England states and New York, south into New Jersey, Virginia, and Maryland; upper Midwest and Ontario; southern form around southern Atlantic Coast states, Texas, Oklahoma, and Mexico
Rhipicephalus sanguineus (brown dog tick)	Light to dark brown, no white markings on dorsal side; have both festoons and eyes	Most widely distributed tick; almost worldwide in Western Hemisphere; transmits RMSF; found indoors around pet bedding areas; strong tendency to crawl upward (walls of infested houses)

[a] RMSF, Rocky Mountain spotted fever.

Plastic box or petri dish method. If mineral oil preparations of skin scrapings fail to demonstrate the mites, the encrusted skin scrapings, etc., can be placed in a small plastic box or small petri dish. The container should be left undisturbed at room temperature for 12 to 24 h. Away from the living host, the mites drop to the bottom of the box or dish and can be seen with a magnifying glass or dissecting microscope.

Follicle Mites. Two species of follicle mites parasitize humans; each has its own particular habitat. *Demodex folliculorum* is found in the hair follicles above the sebaceous glands, and *D. brevis* occurs in the sebaceous glands. Both are found in the epidermis of the nose, nasolabial fold, forehead, and adjacent regions of the face; these mites have also been associated with dermatitis of the scalp. These follicle mites can be considered more of a nuisance than a true pathogen and are considered to be harmless saprophytes; symptoms may include a mild rosaceous pruritus and a fibrous tissue response around the mites. Estimates of human infection range from 25 to 100%. The two species have similar morphology, with *D. brevis* being somewhat shorter. They are elongated wormlike mites with rudimentary legs (Figures 24.62 and 24.63). Most humans acquire mites early in life, often by mother-to-child transmission.

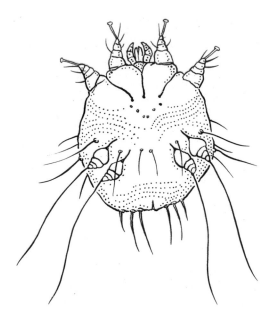

Figure 24.55   Itch mite (*Sarcoptes* sp.), ~0.25 mm. (Illustration by Sharon Belkin; based on an illustration from H. H. Najarian, *Textbook of Medical Parasitology*, The Williams & Wilkins Co., Baltimore, Md., 1967.)

House Dust Mites.   Although house dust mites in the genus *Dermatophagoides* are not commonly seen or submitted to the laboratory for confirmation, it is important to realize that these mites are directly linked to allergic reactivity that can be detected in patients with atopic eczema. Powerful allergens can be found in the mites *D. pteronyssinus* and *D. farinae*, as well as in their fecal pellets and secretions. Allergic rhinitis, asthma, and childhood eczema are attributable to these mites. The mites feed on shed human skin scales, but they also eat mold, fungal spores, pollen grains, feathers, and animal dander.

Figure 24.56   Itch mite. (From A Pictorial Presentation of Parasites: A cooperative collection prepared and/or edited by H. Zaiman.)

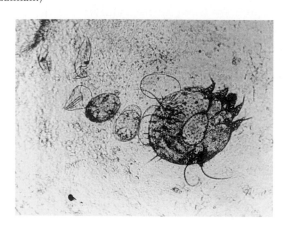

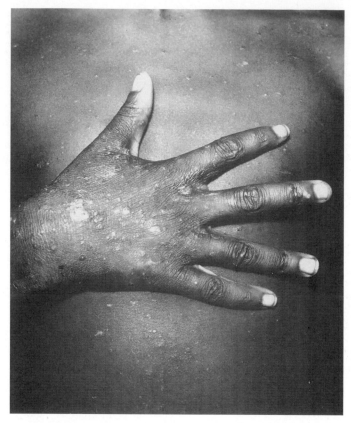

Figure 24.57   Scabies of hand, thorax, and abdomen of a Zairian child. Lesions are complicated by scratching and secondary bacterial infection. (Armed Forces Institute of Pathology photograph.)

Active treatment includes Gore-Tex bags for all bedding elements such as the mattress and box springs, a high-power vacuum cleaner, and a spray containing benzyl alcohol and tannic acid to kill mites and denature allergens. These measures are highly effective, particularly the Gore-Tex

Figure 24.58   Scabies of buttocks of a Brazilian child. There is evidence of scratching and secondary bacterial infection. (Armed Forces Institute of Pathology photograph.)

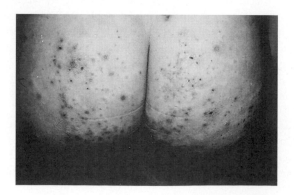

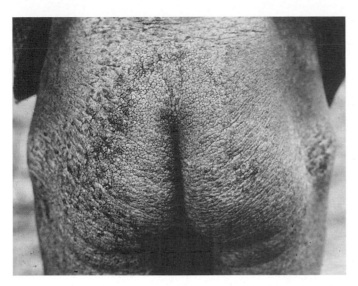

Figure 24.59 Norwegian scabies over the buttocks of a 19-year-old Zairian. Scaling and crusting were severe about the elbows and from the lower trunk to the knees. (Armed Forces Institute of Pathology photograph.)

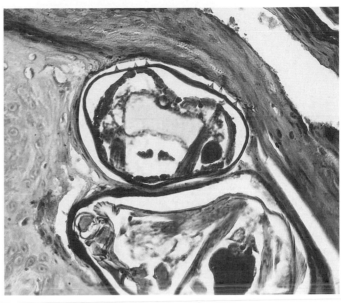

Figure 24.61 Close-up of two *Sarcoptes scabiei* mites in a section of skin shown in Figure 24.60. They are at the keratoepidermal junction. On the cuticle of the uppermost parasite, there are six dorsal spines. Magnification, ×275. (Armed Forces Institute of Pathology photograph.)

Figure 24.60 Skin from the thigh of the patient in Figure 24.59. Sections of many *Sarcoptes scabiei* mites are present in the hyperkeratotic horny layer. Magnification, ×18. (Armed Forces Institute of Pathology photograph.)

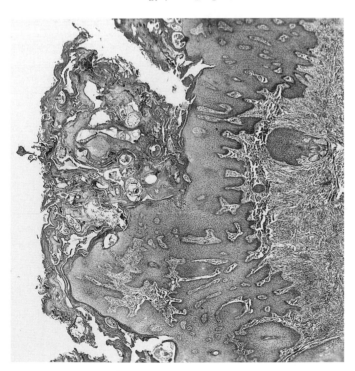

Figure 24.62 *Demodex folliculorum* (left) and *D. brevis* (right). Note the rudimentary legs.

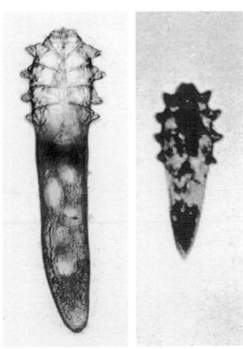

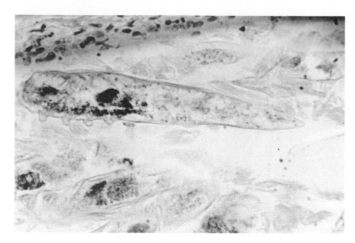

Figure 24.63 *Demodex folliculorum* in a hair follicle. (From A Pictorial Presentation of Parasites: A cooperative collection prepared and/or edited by H. Zaiman.)

bed bags (16). This approach was associated with clinical improvement in most patients with atopic eczema, particularly those who were the most severely affected. However, some of the acaricide applications are required at 2- to 3-month intervals to maintain optimal effectiveness (10). Also, treatments are different for different allergens (55).

Subclass Araneae (Spiders)

The subclass Araneae contains the spiders, only a few of which can cause pathologic changes in humans. All true spiders produce venom, although few types of venom cause more than a local irritation at the bite site. Those that can cause tissue damage are categorized as causing systemic arachnidism or primarily necrotic arachnidism.

Systemic Arachnidism (Black Widow, Other Widow Spiders)

Widow Spiders. The black widow spider (*Lactrodectus mactans* and other species) has the typical shiny black abdomen with a red hourglass shape on the ventral surface (Figures 24.64 and 24.65). However, it is important to realize that the typical hourglass-shaped marking on the underside of the abdomen may not look like the textbook description, and there are at least five species of "widow" spiders within the United States. Some black widow spiders have bright red hash marks on the dorsal surface in addition to the red hourglass on the underside. Also, the hourglass marking may appear "broken" and not complete (20). There may also be red or white marks on the dorsal side of the abdomen. The color patterns on the underside of the abdomen can vary quite a bit, with what appear to be dots, stripes, or a combination of the two. The immature spiders are tan to gray with little to no

Figure 24.64 Black widow spider (*Lactrodectus* sp.), 20 mm. (Illustration by Sharon Belkin.)

black color. The males and immature black widow spiders bite, but symptoms are mild due to the smaller amount of venom injected. The venom is neurotoxic, acting at both adrenergic and cholinergic junctions (11). Although serious systemic sequelae (abdominal cramps, hypertension, reduced heartbeat and feeble pulse, convulsions, shock, delirium, and death) can occur, uncomplicated recovery is the usual outcome (weakness, myalgia). Following recovery, the individual usually develops complete systemic immunity to subsequent bites, although there may be a local reaction. At the bite site, two small red spots may be visible, representing the skin penetration sites.

The black widow spiders tend to avoid strong light and prefer dark areas behind or underneath appliances,

Figure 24.65 Black widow spider (*Lactrodectus* sp.). (Photograph courtesy of Duane J. Gubler, Centers for Disease Control and Prevention.)

furniture, closets, etc. These spiders are very reclusive and will not bite unless provoked; generally, they attempt to escape. They are more active in the summer months. After mating, the female kills the male only if she is not well fed, thus ensuring the survival of the eggs; as many as 2,000 offspring may be produced in a single year.

The wound should be cleaned and ice packs should be applied to limit the spread of the venom. In cases of more severe reactions, the patient should seek medical attention; the use of intravenous solutions, antivenom, and tetanus vaccine are relevant considerations. Children, the elderly, and patients with underlying medical conditions may require hospitalization.

Necrotic Arachnidism (Violin or Brown Recluse Spider, Hobo Spider, Funnel Web Spider, Tarantula Spider)

Violin Spiders. The violin spider (*Loxosceles reclusa* and other species) is commonly found inside human habitation rather than in the outdoor areas where the black widow spider can be found (as well as indoors) (15). These spiders are of medium size (0.9 to 1.5 cm long), are yellowish to dark brown, and tend to have a fiddle-shaped marking on the dorsal surface (Figures 24.66 and 24.67). One of the signs of a brown recluse spider bite is the tendency of the lesion to extend downward; the direction of the lesion depends on the patient's position as vessel damage occurs. This bite presentation is apparently rare with other arthropod bites. Mild to severe pain can occur immediately after the bite or within a few hours. Extensive tissue necrosis can occur, with extended spread, very slow healing, and severe disfigurement (Figure 24.68). If a large dose of venom is injected, there may also be systemic complications, such as hemolytic anemia, hemoglobinuria, hematuria, jaundice,

Figure 24.67 Violin spider (*Loxosceles* sp.). (Armed Forces Institute of Pathology photograph.)

and fever. There is a high mortality rate associated with systemic complications, which can include hematuria, anemia, fever, rash, nausea, vomiting, coma, and cyanosis. If systemic symptoms occur, they usually appear several days after the bite; as an example, hemolysis has been documented to occur from 24 h to 3 days after a bite.

The brown recluse is nocturnal and is generally found in bathrooms, bedrooms, cellars, attics, clothing folds, boxes, and storage areas, and under rubble, etc. Like the black widow spider, it is not aggressive but will bite when bothered. These spiders are found from Minnesota to Maine south to Florida and west to Arizona and Wyoming. In South America, there is a close relative of the brown recluse, *L. laeta*, which also causes a severe necrotic lesion; these spiders have been introduced and are now well established in Los Angeles, Calif., and at

Figure 24.66 Violin spider (*Loxosceles* sp.), 9 to 15 mm. (Illustration by Sharon Belkin.)

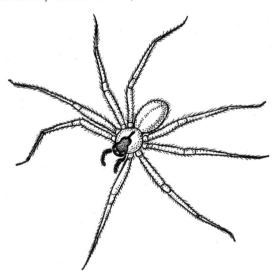

Figure 24.68 Tissue necrosis caused by *Loxosceles* sp. (Armed Forces Institute of Pathology photograph.)

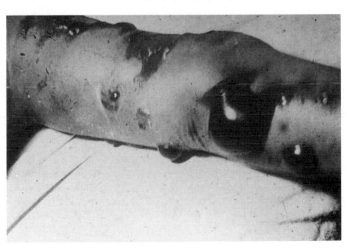

Harvard University in Cambridge, Mass. (20). Other species found mostly in the Southwest include *L. refescens*, *L. deserta*, *L. devia*, and *L. arizonica*.

As with any spider bite, if possible the spider should be collected and submitted for identification. In the case of the brown recluse, prompt medical attention is recommended. Although a specific antivenom has been used, it is not generally available. Treatment is somewhat controversial, ranging from ice packs to complete surgical excision of the bite site. Antibiotics such as cephalosporin or erythromycin have been used, as well as dapsone; however, the side effects of dapsone may also mimic symptoms resulting from the spider bite.

Hobo Spiders. Hobo spiders are large, fast spiders, about 45 mm long with distinct chevron stripes on the abdomen. They build funnel-shaped webs in or around houses. They are also now recognized as causing necrotic lesions, particularly the males, which are responsible for most bites. Reports of bites in Washington, Oregon, and Idaho continue to increase. Although the bites were originally thought to be from brown recluse spiders, this was never confirmed. It appears that the bite of these hobo spiders can cause lesions similar to those seen with bites by the brown recluse. The species involved is *Tegenaria agrestis*. Although the bite is usually painless at first, the same types of lesions and symptoms occur as seen with bites by the brown recluse. Some of these lesions can take years to heal, and systemic symptoms can also be seen.

Funnel Web Spiders. Funnel web spiders are large, dark-colored, and very aggressive; they have prominent fangs and are found in and around Sydney, Australia. They are glossy dark brown to black, while the abdomen is dark plum to black; including the legs, they measure 45 to 60 mm. They tend to build funnel webs in burrows, rotting logs, tree holes, or rock crevices; colonies can number more than 100 spiders. These are very dangerous spiders, and the bite can cause death in as little as 15 min. The bite can become very painful, and systemic symptoms such as hypotension and apnea can begin to occur within just a few minutes.

It is important for first aid to be administered immediately, including a pressure bandage and immobilization of the limb with splints. Patients are usually hospitalized and monitored for signs and symptoms indicating that venom has or has not been injected with the bite. These systemic symptoms can include mouth numbness, tongue spasms, nausea, vomiting, profuse sweating, salivation, and muscle spasms. Appropriate administration of antivenom is indicated; however, it is important to carefully monitor patients for possible allergic reactions to the antivenom.

Figure 24.69 Tarantula. (Note the hairy legs.)

Tarantula Spiders. Tarantulas occur throughout different areas of the world and become aggressive only if provoked or handled inappropriately. They are large, hairy spiders (Figure 24.69). The 30 species found in the southwestern United States have a leg span of approximately 18 cm, while the South American species found in the jungles measure 24 cm. Bites can vary from being relatively painless to causing a deep, throbbing pain that may last for hours. Individuals who are hypersensitive may present with more serious symptoms. Tarantulas sold for pets within the United States are generally species of *Abhonopelma*, which tend to be very colorful.

Subclass Scorpiones (Scorpions)

Scorpions are characterized by having large pedipalps that end in stout claws and a tail ending in a hooked stinger (Figures 24.70 through 24.72). Venom is discharged through the stinger. Humans usually come into contact with scorpions by stepping on them or touching them in the dark. Species considered to be dangerous occur throughout the world. Some species can even penetrate the soles of the feet and inject venom, causing symptoms which can be serious, particularly in children younger than 5 years (aching pain radiating from the site; lymphadenitis; generalized numbness; throbbing and twitching of fingers, toes, and earlobes; profuse sweating; glossopharyngeal involvement; vomiting; abdominal muscle spasms; convulsions; mental disturbances; partial paralysis; blindness; electrocardiographic changes; and death from respiratory paralysis).

Children younger than 5 years are most commonly exposed and have the most severe sequelae. It has been reported that in Trinidad, the death rate in exposed children can reach 25% (5). Usually, the death rate in exposed adults is very low.

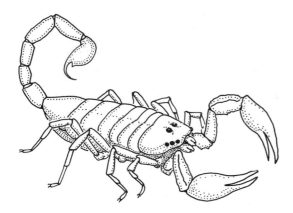

Figure 24.70 Scorpion, 30 to 40 mm. (Illustration by Sharon Belkin.)

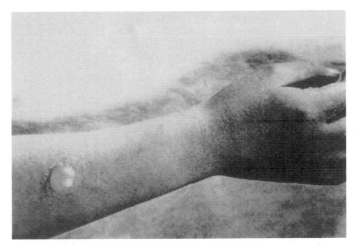

Figure 24.72 Scorpion sting. This blister has formed at the site of a scorpion sting. (Armed Forces Institute of Pathology photograph.)

Scorpions are nocturnal and generally are hidden during the day. Around homes, they are often found under houses or in attics. When actively seeking water, they may be found in kitchens or bathrooms, occasionally in the sink or bathtub. In the southwestern United States, the common striped scorpion, *Centruroides vittatus*, is found; it has a sting similar to that from a wasp or bee. *C. vittatus* is found in Texas, while *C. exilicauda* is found in southern Arizona and western New Mexico.

Most stings are not serious, and the burning sensation and wheal disappear without any intervention. However, after a sting by one of the more dangerous species, it is important to minimize the spread of the venom. Ice packs are recommended, and medical treatment may be required. Areas of the world where dangerously venomous

scorpions are found include (Old World) Algeria, Egypt, Iraq, Israel, Jordan, Morocco, South Africa, Sudan, and Turkey and (New World) Argentina, Brazil, Guyana, Mexico, Trinidad, United States (Arizona and New Mexico), and Venezuela (20).

Other Arthropods

Class Chilopoda (Centipedes)

Centipedes are usually not considered dangerous, and their importance is related to their occasional painful bites. These arthropods like to live in dark, damp environments such as debris and rubbish. They have a long, slender, segmented body with a single pair of legs per segment except for the last two segments (Figure 24.73) (6). The appendages of the first body segment are modified into poison claws. Although the poison injected is not harmful in the temperate areas of the world, the bite of *Scolopendra* spp. within the tropics and subtropics can be very painful and can cause edema, blistering, and necrotic lesions. In some cases, systemic reactions such as fever, nausea, vomiting, and headache also occur, particularly with *Scolopendra gigantea*. Rarely, some species have been recovered from the human nares and frontal sinuses or from the digestive tract; in each case, the centipede was an accidental invader of the human body (5).

Most bites are mild and self-limiting. The bite site should be thoroughly washed, with subsequent application of an ice pack and analgesics for pain. If the lesion becomes inflamed, antibiotics may be recommended.

Figure 24.71 Scorpion. Although painful, scorpion stings are rarely fatal. Exceptions are infants and children who sustain multiple stings. The venom is contained in two venom glands in the tail.

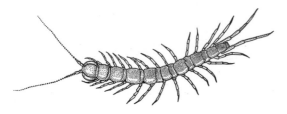

Figure 24.73 Centipede, 30 to 40 mm. (Illustration by Sharon Belkin.)

Class Diplopoda (Millipedes)

Millipedes also have a long, slender, segmented body, but unlike the centipedes, they have two pairs of legs per segment in the abdominal region (Figure 24.74). Also, they tend to be more round and "wormlike" in shape. They do not have piercing mouthparts, and they are not poisonous. However, some of the body secretions can result in a burning sensation, discoloration, and blister formation when applied to human skin. Severe conjunctivitis can also occur when the eyes come in contact with these fluids (17). Apparently, these secretions contain hydrogen cyanide as well as other aliphatic acids (5).

Class Crustacea (Copepods, Crabs, Crayfish, etc.)

The crustaceans are primarily gill-breathing arthropods of both freshwater and saltwater. Only a few are of medical importance, and these tend to be found in freshwater. Some of the members of this group serve as intermediate hosts for various human helminth parasites. Copepods, or water fleas, serve as intermediate hosts for several helminth infections. Species of *Cyclops* carry the larval forms of *Dracunculus medinensis* (Guinea worm), *Diphyllobothrium latum* (broad fish tapeworm), and *Gnathostoma spinigerum* (uncommon nematode infection). Species of *Diaptomus* are important first intermediate hosts of *D. latum*. Prevention of contamination and boiling or filtration of contaminated water will prevent these infections.

The larger crustaceans (decapods) include crabs and crayfish, which serve as the second intermediate hosts of *Paragonimus* spp. (lung fluke). Although the crabs incriminated in the transmission of *Paragonimus* spp. are freshwater forms, they can also survive in brackish water.

Figure 24.74 Millipede, 30 to 40 mm. (Illustration by Sharon Belkin.)

Thorough cooking of crabs and crayfish will help eliminate some of these infections.

In the South Pacific, prawns may be paratenic hosts of the nematode *Angiostrongylus cantonensis*, which can cause eosinophilic meningitis and cerebral damage in the human host.

Control of Arthropods of Medical Importance

Measures for the control of medically important arthropods usually involve a number of different options, including physical control, biological control, and chemical control. Although chemical means are generally the most widely used, they rarely succeed in providing a permanent solution to the problem. Often, a combination of methods provides the most effective control (32). A general dichotomous key for the identification of some of the arthropods can be found in Table 24.10, and a key to the important myiasis-producing larvae is provided in Table 24.11.

Physical Control

A number of control measures are related to the environment and may affect arthropod life cycles. An excellent example is demonstrated by the elimination of larval habitats of mosquitoes; this is basically water management. This approach could include rerouting of water, elimination of standing water, and ditch filling. With periodic monitoring, these should be long-term changes and should require little to no maintenance. Other approaches would include proper disposal of sewage, garbage, manure, and trash. Screens, bed nets, and protective clothing have become widely used and appreciated as preventive measures against flying insects. Although several other methods have been tried, these approaches are somewhat questionable, at best. One example is the use of pet products such as flea collars. Unfortunately, the use of flea collars for humans has not been approved by the Food and Drug Administration and may be dangerous due to the potential for pesticide absorption through the skin.

Biological Control

A number of programs are currently being used and include (i) use of specific viruses, bacteria, protozoa, nematodes, and fungi, all of which are pathogenic for various arthropods; (ii) genetic manipulation of arthropod populations, including release of sterile males and use of insect growth regulators; and (iii) release of actual

Table 24.10 Dichotomous key to some of the arthropods important in human disease and/or disease transmission[a]

A.	Legs:	Three or four pairs	B
	Legs:	Five or more pairs	V
B.	Legs:	Three pairs; antennae present	C
	Legs:	Four pairs; antennae absent	L
C.	Wings:	Well developed	D
	Wings:	Absent or rudimentary	L
D.	Wings:	One pair (**Diptera:** flies, mosquitoes, midges)	E
	Wings:	Two pairs	F
E.	Wings:	Scales present (**Diptera:** mosquitoes)	
	Wings:	Scales absent (**Diptera:** flies)	
F.	Mouth parts:	Sucking (long proboscis present)	G
	Mouth parts:	Chewing (long proboscis absent)	H
G.	Wings:	Scales present; proboscis coiled (**Lepidoptera:** butterflies, moths)	
	Wings:	Scales absent; proboscis not coiled (**Hemiptera:** bedbugs, kissing bugs)	
H.	Wings:	Both pairs membranous, size may vary	I
	Wings:	Front pair leathery or shell-like, covers second pair	J
I.	Wings:	Both pairs similar in size (**Isoptera:** termites)	
	Wings:	Hind wing smaller than front wing (**Hymenoptera:** bees, hornets, wasps)	
J.	Wings:	No distinct veins; front wings horny or leathery; meet as straight line in middle	K
	Wings:	Distinct veins; front wings horny or leathery; overlap in middle (**Blattaria:** cockroaches)	
K.	Abdomen:	Prominent cerci or forceps, longer than wings (**Dermaptera:** earwigs)	
	Abdomen:	No prominent cerci or forceps, covered by wings (**Coleoptera:** beetles)	
L.	Abdomen:	Three long terminal tails (**Thysanura:** silverfish and firebrats)	
	Abdomen:	No terminal tails	M
M.	Abdomen:	Narrow waist (**Hymenoptera:** ants)	
	Abdomen:	No narrow waist	N
N.	Abdomen:	Prominent pair of cerci or forceps (**Dermaptera:** earwigs) (see also K)	
	Abdomen:	No cerci or forceps	O
O.	Body:	Flattened laterally; antennae folded into head grooves (**Siphonaptera:** fleas)	
	Body:	Flattened dorsoventrally; antennae project from side of head	P
P.	Antennae:	Nine or more segments	Q
	Antennae:	Three to five segments	R
Q.	Pronotum:	Covers head (**Blattaria:** cockroaches) (see also J)	
	Pronotum:	Does not cover head (**Isoptera:** termites) (see also I)	
R.	Mouth parts:	Tubular jointed beak; three- to five-segmented tarsi (**Hemiptera:** bedbugs)	
	Mouth parts:	Retracted into head or chewing type; one- or two-segmented tarsi	S
S.	Mouth parts:	Retracted into head, adapted for sucking blood (**Anoplura:** sucking lice)	
	Mouth parts:	Chewing type (**Mallophaga:** chewing lice)	

(continued)

Table 24.10   Dichotomous key to some of the arthropods important in human disease and/or disease transmission[a] *(continued)*

T.	Body:	Oval, consisting of a single saclike region (**Acari**: ticks and mites)	
	Body:	Divided into two distinct regions, cephalothorax and abdomen	U
U.	Abdomen:	Joined to cephalothorax by slender waist (**Araneae**: spiders)	
	Abdomen:	Joined broadly to cephalothorax, stinger present (**Scorpiones**: scorpions)	
V.	Legs, swimmerets:	Five to nine pairs of legs or swimmerets; one or two pairs of antennae (**Crustacea**: copepods, crabs, crayfish)	
	Legs, swimmerets:	Ten or more pairs of legs; swimmerets absent, one pair of antennae	W
W.	Legs per body segment:	One pair (**Chilopoda**: centipedes)	
	Legs per body segment:	Two pairs (**Diplopoda**: millipedes)	

[a] Adapted from references 5, 17, 20, 28, 43, and 47.

Table 24.11   Key to the important myiasis-producing larvae[a]

1.	Body with spinous or fleshy processes laterally and dorsally or terminally	*Fannia*
	Body smooth or with short spines, but *never* having long fleshy lateral processes	2
2.	Body with a long slender tail or caudal process capable of some extension and retraction	*Eristalis*
	Body sometimes narrowed posteriorly, but *never* with a long flexible caudal process capable of some extension and retraction	3
3.	Larvae more or less grub-like; most species slightly flattened dorsoventrally	4
	Larvae maggot-like, or typical "muscoid" shape, tapering anteriorly, broadly truncate at the posterior end; cross section more or less circular at all points	5
4.	Posterior spiracular plate with three distinct slits	*Dermatobia*
	Posterior spiracular plate with many fine openings	*Hypoderma*
5.	Posterior spiracles within a well-chitinized and complete ring encircling the button area; spiracles *never* in a distinct depression	6
	Posterior spiracles with the button very slightly chitinized or absent; chitinized ring incomplete; spiracles in a distinct depression or flush with surface	8
6.	Button area with spiracular slits nearly straight	7
	Button area with spiracular slits sinuous, with at least a double curve	9
7.	Principal transverse subdivisions of spiracular slits well marked, usually not more than six; both ring and button heavily chitinized, the ring thickened into points at two places between the slits	*Calliphora*
	Transverse subdivisions of spiracular slits less distinctly marked, from 6 to 20 in number, ring and button less heavily chitinized, the ring thickened into point at only one place between the slits	*Phaenicia*
8.	Posterior spiracles in a more or less distinct pit or depression, vestigial button usually present; integument rather smooth	*Sarcophaga*
	Posterior spiracles flush with surface; integument rather spiny (Western Hemisphere)	*Cochliomyia*
9.	Posterior spiracular plates D-shaped, each slit thrown into several loops	*Musca*
	Posterior spiracular plates triangular, with rounded corners; spiracular slits S-shaped; button indistinct, centrally placed	*Stomoxys*

[a] Adapted from B. F. Prendergast (ed.), *Filth Flies, Technical Guide 30*. Armed Forces Pest Management Board, Washington, D.C., 2002 (reviewed and validated 2004).

predators such as the small fish *Gambusia affinis*, which feeds on mosquito larvae. The introduction of the irradiation technique for sterilizing males of the screwworm fly, *Cochliomyia hominivorax*, has dramatically limited the development of progeny of this myiasis-producing fly throughout the southern United States. Chemosterilants are also being used, in some cases eliminating the need for radiation techniques. Insect growth regulators generally cause reproductive dysfunction or inhibit development of the adult stage.

Chemical Control

Although chemical control measures dramatically increased after World War II, the current approach involves combination methods rather than relying on chemical means alone. Concerns have been raised about a number of chemicals, particularly in terms of their possible contamination of the food chain. Also, different chemicals and approaches vary considerably throughout the world, often depending on location and financial considerations.

Most insecticides currently in use include inorganic compounds (arsenicals and fluorine compounds), botanicals (pyrethrins from pyrethrum flowers), chlorinated hydrocarbons (DDT, lindane, chlordane, dieldrin, and methoxychlor), and organophosphorus (malathion, dicapthon, and parathion) and carbamate (Baygon, Sevin, and Bendiocarb) compounds. A great deal of emphasis has been placed on the residual action of certain insecticides, compounds that remain stable on surfaces for long periods but which are effective at very low concentrations. These residual sprays are quite different than those that are sprayed ("space sprayed") for short-term effectiveness.

Unfortunately, as with all animal populations, susceptibility and resistance to these insecticides will vary; arthropods that are resistant will survive. Crossbreeding of these survivors often tends to enhance the resistance. This has occurred with insects exposed to chlorinated hydrocarbons and organophosphorus compounds.

Insecticides can be applied as sprays (immediate action or residual), aerosols (the insecticide remains suspended in air for a long time), ultra-low-volume concentrates (essentially no solvents are used), dusts (10% insecticide in talc or pyrophyllite inert carriers), granules (inorganic carrier such as vermiculite), or bait (houseflies, cockroaches, ants). Repellents and acaricides can also be used for those who are exposed to arthropods in infested areas. True repellents do not kill the arthropods but discourage them from landing or biting.

References

1. **Aguilera, A., A. Cid, B. J. Regueiro, J. M. Prieto, and M. Noya.** 1999. Intestinal myiasis caused by *Eristalis tenax*. *J. Clin. Microbiol.* **37:**3082.

2. **Arbes, S. J., M. Sever, J. Mehta, N. Collette, B. Thomas, and D. C. Zeldin.** 2005. Exposure to indoor allergens in day-care facilities: results from 2 North Carolina counties. *J. Allergy Clin. Immunol.* **116:**133–139.

3. **Archer, M. S., M. A. Elgar, C. A. Briggs, and D. L. Ranson.** 2005. Fly pupae and puparia as potential contaminants of forensic entomology samples from sites of body discovery. *Int. J. Legal Med.* 22 Oct.:1–5. [Epub ahead of print.]

4. **Arruda, L. K.** 2005. Cockroach allergens. *Curr. Allergy Asthma Rep.* **5:**411–416.

5. **Beaver, P. C., R. C. Jung, and E. W. Cupp.** 1984. *Clinical Parasitology*, 9th ed. Lea & Febiger, Philadelphia, Pa.

6. **Beck, J. W., and J. E. Davies.** 1981. *Medical Parasitology*, 3rd ed. The C. V. Mosby Co., St. Louis, Mo.

7. **Benecke, M.** 1998. Random amplified polymorphic DNA (RAPD) typing of necrophageous insects (Diptera, Coleoptera) in criminal forensic studies: validation and use in practice. *Forensic Sci. Int.* **98:**157–168.

8. **Campobasso, C. P., J. G. Linville, J. D. Wells, and F. Introna.** 2005. Forensic genetic analysis of insect gut contents. *Am. J. Forensic Med. Pathol.* **26:**161–165.

9. **Catts, E. P., and M. L. Goff.** 1992. Forensic entomology in criminal investigations. *Annu. Rev. Entomol.* **37:**253–272.

10. **Chew, F. T., D. Y. Goh, and B. W. Lee.** 1996. Effects of an acaricide on mite allergen levels in the homes of asthmatic children. *Acta Paediatr. Jpn.* **38:**483–488.

11. **Chodosh, J., and J. Clarridge.** 1992. Ophthalmomyiasis: a review with special reference to *Cochliomyia hominivorax*. *Clin. Infect. Dis.* **14:**444–449.

12. **Ciftcioglu, N., K. Altintas, and M. Haberal.** 1997. A case of human orotracheal myiasis caused by *Wohlfahrtia magnifica*. *Parasitol. Res.* **83:**34–36.

13. **Cornet, M., M. Florent, A. Lefebvre, C. Wertheimer, C. Perez-Eid, M. J. Bangs, and A. Bouvet.** 2004. Tracheopulmonary myiasis caused by a mature third-instar *Cuterebra* larva: case report and review. *J. Clin. Microbiol.* **42:**3378.

14. **Dworkin, M. S., P. C. Shoemaker, and D. E. Anderson.** 1999. Tick paralysis: 33 cases in Washington State, 1946–1996. *Clin. Infect. Dis.* **29:**1435–1439.

15. **Fisher, R. G.** 1994. Necrotic arachnidism. *West. J. Med.* **160:**570–572.

16. **Friedmann, P. S., and B. B. Tan.** 1998. Mite elimination—clinical effect on eczema. *Allergy* **53:**97–100.

17. **Fritsche, T. R., and M. A. Pfaller.** 1995. Arthropods of medical importance, p. 1257–1273. *In* P. R. Murray, E. J. Baron, M. A. Pfaller, F. C. Tenover, and R. H. Yolken (ed.), *Manual of Clinical Microbiology*, 6th ed. American Society for Microbiology, Washington, D.C.

18. **Furman, D. P., and E. C. Loomis.** 1984. The ticks of California (Acari: Ixodida). *Bull. Calif. Insect Surv.* **25:**1–79.

19. **Gewirtzman, A., and H. Rabinovitz.** 1999. Botfly infestation (myiasis) masquerading as furunculosis. *Cutis* **63:**71–72.

20. **Goddard, J.** 2003. *Physician's Guide to Arthropods of Medical Importance*, 4th ed. CRC Press, Inc., Boca Raton, Fla.

21. **Goff, M. L., and W. D. Lord.** 1994. Entomotoxicology. A new area for forensic investigation. *Am. J. Forensic Med. Pathol.* **15**:51–57.

22. **Graczyk, T. K., M. R. Cranfield, R. Fayer, and H. Bixler.** 1999. House flies (*Musca domestica*) as transport hosts of *Cryptosporidium parvum. Am. J. Trop. Med. Hyg.* **61:** 500–504.

23. **Howell, G., J. Butler, R. D. Deshazo, J. M. Farley, H. L. Liu, N. P. Nanayakkara, A. Yates, G. B. Yi, and R. W. Rockhold.** 2005. Cardiodepressant and neurologic actions of *Solenopsis invicta* (imported fire ant) venom alkaloids. *Ann. Allergy Asthma Immunol.* **94**:380–386.

24. **Jacobson, R. L., and Y. Schlein.** 1999. Lectins and toxins in the plant diet of *Phlebotomus papatasi* (Diptera: Psychodidae) can kill *Leishmania major* promastigotes in the sandfly and in culture. *Ann. Trop. Med. Parasitol.* **93**:351–356.

25. **Johnston, M., and G. Dickinson.** 1996. An unexpected surprise in a common boil. *J. Emerg. Med.* **14**:779–781.

26. **Jones, R. G. A., R. L. Corteling, H. P. To, G. Bhogal, and J. Landon.** 1999. A novel Fab-based antivenom for the treatment of mass bee attacks. *Am. J. Trop. Med. Hyg.* **61**:361–366.

27. **Kahn, D. G.** 1999. Myiasis secondary to *Dermatobia hominis* (human botfly) presenting as long-standing breast mass. *Arch. Pathol. Lab. Med.* **123**:829–831.

28. **Keirans, J. E., and C. M. Clifford.** 1978. The genus *Ixodes* in the United States: a scanning electron microscope study and key to the adults. *J. Med. Entomol. Suppl.* **2**:1–149.

29. **Kemp, S. F., R. D. deShazo, J. E. Moffitt, D. F. Williams, and W. A. Buhner, Jr.** 2000. Expanding habitat of the imported fire ant (*Solenopsis invicta*): a public health concern. *J. Allergy Clin. Immunol.* **105**:683–691.

30. **Kettle, D. S.** 1995. *Medical and Veterinary Entomology*, 2nd ed., p. 354–356. CAB International, Wallingford, United Kingdom.

31. **Klotz, J. H., R. D. deShazo, J. L. Pinnas, A. M. Frishman, J. O. Schmidt, D. R. Suiter, G. W. Price, and S. A. Klotz.** 2005. Adverse reactions to ants other than imported fire ants. *Ann. Allergy Asthma Immunol.* **95**:418–425.

32. **Lane, R. P., and R. W. Crosskey.** 1993. *Medical Insects and Arthropods*. Chapman & Hall, London, United Kingdom.

33. **Lane, R. S., R. B. Moss, Y. P. Hsu, T. Wei, M. L. Mesirow, and M. M. Kuo.** 1999. Anti-arthropod saliva antibodies among residents of a community at high risk for Lyme disease in California. *Am. J. Trop. Med. Hyg.* **61**:850–859.

34. **MacDonald, P. J., C. Chan, J. Dickson, F. Jean-Louis, and A. Heath.** 1999. Ophthalmomyiasis and nasal myiasis in New Zealand: a case series. *N. Z. Med. J.* **112**:445–447.

35. **Malgorn, Y., and R. Coquoz.** 1999. DNA typing for identification of some species of Calliphoridae. An interest in forensic entomology. *Forensic Sci. Int.* **102**:111–119.

36. **Mielke, U.** 1997. Nosocomial myiasis. *J. Hosp. Infect.* **37**:1–5.

37. **Monsen, S. E., L. R. Bronson, J. R. Tucker, and C. R. Smith.** 1999. Experimental and field evaluations of two acaracides for control of *I. pacificus* (Acari: Ixodidae) in northern California. *J. Med. Entomol.* **36**:660–665.

38. **Oliver, J. H., Jr., M. R. Owsley, H. J. Hutcheson, A. M. James, C. Chen, W. S. Irby, E. M. Dotson, and D. K. McLain.** Conspecificity of the ticks *Ixodes scapularis* and *I. dammini. J. Med. Entomol.* **30**:54–63.

39. **Patterson, F. C., and W. C. Winn, Jr.** 2003. Practical identification of hard ticks in the parasitology laboratory. *Pathol. Case Rev.* **8**:187–198.

40. **Prahlow, J. A., and J. J. Barnard.** 1998. Fatal anaphylaxis due to fire ant stings. *Am. J. Forensic Med. Pathol.* **19**:137–142.

41. **Quintanilla-Cedillo, M. R., H. Leon-Urena, J. Contreras-Ruiz, and R. Arenas.** 2005. The value of Doppler ultrasound in diagnosis in 25 cases of furunculoid myiasis. *Int. J. Dermatol.* **44**:34–37.

42. **Ribeiro, F. A., E. Taciro, M. R. Guerra, and C. A. Eckley.** 2005. Oral ivermectin for the treatment and prophylaxis of scabies in prison. *J. Dermatol. Treat.* **16**:138–141.

43. **Richards, O. W., and R. G. Davies.** 1977. *Imm's General Textbook of Entomology*, vol. 11. *Classification and Biology*, 10th ed. Chapman & Hall, Ltd., London, United Kingdom.

44. **Sherman, R. A., G. Roselle, C. Bills, L. H. Danko, and N. Eldridge.** 2005. Healthcare-associated myiasis: prevention and intervention. *Infect. Control Hosp. Epidemiol.* **26**:828–832.

45. **Spach, D. H., W. C. Liles, G. L. Campbell, R. E. Quick, D. E. Anderson, and T. R. Fritsche.** 1993. Tick-borne diseases in the United States. *N. Engl. J. Med.* **329**:936–947.

46. **Steere, A. C.** 1989. Lyme disease. *N. Engl. J. Med.* **321:** 585–596.

47. **Strickland, G. T. (ed.).** 2000. *Hunter's Tropical Medicine*, 8th ed. The W. B. Saunders Co., Philadelphia, Pa.

48. **Tantawi, T. I., and B. Greenberg.** 1993. The effect of killing and preservative solutions on estimates of maggot age in forensic cases. *J. Forensic Sci.* **38**:702–707.

49. **Tatfeng, Y. M., M. U. Usuanlele, A. Orukpe, A. K. Digban, M. Okodua, F. Oviasogie, and A. A. Turay.** 2005. Mechanical transmission of pathogenic organisms: the role of cockroaches. *J. Vector Borne Dis.* **42**:129–134.

50. **Ugwu, B. T., and P. O. Nwadiaro.** 1999. *Cordylobia anthropophaga* mastitis mimicking breast cancer: case report. *East Afr. Med. J.* **76**:115–116.

51. **Uni, S., S. Shinonaga, Y. Nichio, A. Fukunaga, M. Iseki, T. Okamoto, N. Ueda, and T. Miki.** 1999. Ophthalmomyiasis caused by *Sarcophaga crassipalpis* (Diptera: Sarcophagidae) in a hospital setting. *J. Med. Entomol.* **36**:906–908.

52. **Wahba, M. M., I. M. Labib, and E. M. el Hamshary.** 1999. *Bacillus thuringiensis* var. *israelensis* as a microbial control agent against adult and immature stages of the sandfly, *Phlebotomus papatasi*, under laboratory conditions. *J. Egypt. Soc. Parasitol.* **29**:587–597.

53. **Wesson, D. M., D. K. McLain, J. H. Oliver, J. Piesman, and F. H. Collins.** 1993. Investigation of the validity of species status of *Ixodes dammini* (Acari: Ixodidae) using rDNA. *Proc. Natl. Acad. Sci. USA* **90**:10221–10225.

54. **Wolf, R., E. Orion, and H. Matz.** 2003. Stowaways with wings: two case reports on high-flying insects. *Dermatol. Online J.* **9**:10.

55. **Woodfolk, J. A., M. L. Hayden, N. Couture, and T. A. Platts-Mills.** 1995. Chemical treatment of carpets to reduce allergen: comparison of the effects of tannic acid and other treatments on proteins derived from dust mites and cats. *J. Allergy Clin. Immunol.* **96**:325–333.

56. **Wooley, T. A.** 1988. *Acarology: Mites and Human Welfare.* John Wiley & Sons, Inc., New York, N.Y.

57. **Wu, C. H., and M. F. Lee.** 2005. Molecular characteristics of cockroach allergens. *Cell Mol. Immunol.* **2**:177–180.

58. **Yoshitomi, A., A. Sato, T. Suda, and K. Chida.** 1997. Nasopharyngeal myiasis during mechanical ventilation. *Nippon Kyobu Shikkan Gakkai Zasshi* **35**:1352–1355.

59. **Zehner, R., J. Amendt, and R. Krettek.** 2004. STR typing of human DNA from fly larvae fed on decomposing bodies. *J. Forensic Sci.* **49**:337–340.

25 Treatment of Parasitic Infections

Chemotherapy plays a very important role not only in reducing patient morbidity and mortality but also in reducing transmission of the parasitic infection. Many of the drugs used to treat parasitic infections have serious side effects; therefore, before initiation of therapy, it is important to consider the following factors: health of the patient, parasite drug resistance, accuracy of the original dose, potential drug toxicity, and the need for follow-up examinations to monitor therapy. The mechanisms of action of most antiparasitic drugs are not well known, including those involving potential drug toxicity to the patient. Because of the limited resources of developing nations, where a majority of parasitic infections occur, there is little commercial incentive for developing effective therapeutics or vaccines. For some drugs such as triclabendazole, used extensively in veterinary practices, less is known about adverse side effects or toxicity to humans. Specific information can be obtained from the Centers for Disease Control and Prevention, CDC Drug Service, Atlanta, Ga. [day, (404) 639-3670; evenings, weekends, and holidays, (404) 639-2888]. One can also obtain drugs from CDC that are not commercially available in the United States.

Tables 25.1 and 25.2 provide information on the most commonly used drugs for treating parasitic infections (1, 4, 5, 8–10, 12–16, 19, 23, 24, 26); not all drugs are included, and some drugs listed are not available in the United States. Information on drug resistance and susceptibility testing can be seen in Table 25.3.

Albendazole (Albenza) (GlaxoSmithKline)

Preparation

The drug is available as 200-mg tablets and is used for the treatment of neurocysticercosis and hydatid disease.

Administration

The drug is given with meals. It is poorly absorbed from the gastrointestinal tract due to its low aqueous solubility. Concentrations in plasma are negligible or undetectable because it is rapidly converted to the sulfoxide metabolite prior to reaching the systemic circulation. The systemic anthelmintic activity has been attributed to the primary metabolite, albendazole sulfoxide. Oral bioavailability appears to be enhanced when albendazole is coadministered with a fatty meal (estimated fat content, 40 g) as evidenced by higher (up to

718

Table 25.1 Drugs and dosages for treating parasitic infections[a]

Organism or infection	Drug	Dosage[b]	Follow-up[b] and comments
Acanthamoeba spp. Meningoencephalitis	Amphotericin B and	1 mg/kg/day i.v., uncertain duration	Monitor clinical course; usually susceptible in vitro to pentamidine, ketoconazole (Nizoral), and flucytosine (Ancobon)
	Sulfadiazine	1 mg QID intrathecally	
	or		
	Sulfisoxazole	50 mg/kg i.v.	
	Pentamidine isethionate	4 mg/kg/day i.v.	
Keratitis	Polyhexamethylene biguanide (Baquacil) (PHMB)	0.02% applied topically hourly for 3 days, then every 4 h	Monitor clinical course; combination therapy highly recommended; PHMB available at Leiter's Park Ave. Pharmacy, San Jose, CA (800-292-6773) (www.leiterrx.com)
	plus/or		
	Propamidine isethionate (Brolene)	0.1% applied topically	
	plus		
	Neomycin-polymyxin B-gramicidin	Ophthalmic solution	
Ancylostoma caninum (eosinophilic enterocolitis)	Albendazole	400 mg once	Monitor clinical course
	or		
	Mebendazole	100 mg BID × 3 days or 500 mg once	
	or		
	Pyrantel pamoate	11 mg/kg (max 1 g) × 3 days	
Ancylostoma duodenale (hookworm)	Albendazole	400 mg once	Fecal examinations 2–4 wk posttherapy
	or		
	Mebendazole	100 mg BID × 3 days or 500 mg once	
	or		
	Pyrantel pamoate	11 mg/kg (max 1 g) × 3 days	If eggs are rare in fecal concentrations, there is no need to repeat therapy
Angiostrongylus cantonensis	Mebendazole	100 mg BID × 5 days	Monitor clinical course; may want to add corticosteroid
Angiostrongylus costaricensis	Drug of choice Mebendazole	200–400 mg TID × 10 days	Monitor clinical course; may want to add corticosteroid
	Alternative Thiabendazole	75 mg/kg/day in 3 doses × 3 days (max 3 g/day)	
Anisakis spp.	Surgical or endoscopic removal		Monitor clinical course; albendazole has also been reported to be successful

(continued)

Table 25.1 Drugs and dosages for treating parasitic infections^a (*continued*)

Organism or infection	Drug	Dosage^b	Follow-up^b and comments
Ascaris lumbricoides (roundworm)	Albendazole	400 mg once	Fecal examinations 2–4 wk posttherapy
	or Mebendazole	100 mg BID × 3 days or 500 mg once	
	or Ivermectin	150–200 µg/kg once	
Babesia spp.	Clindamycin	1.2 g BID i.v. parenterally or 600 mg TID p.o. × 7–10 days; Children: 20–40 mg/kg/day p.o. in 3 doses × 7–10 days	Monitor clinical course; repeat blood smears if symptoms persist
	plus Quinine	650 mg TID p.o. × 7–10 days; Children: 25 mg/kg/day in 3 doses × 7–10 days	
	or Atovaquone	750 mg BID p.o. × 7–10 days; Children: 20 mg/kg BID p.o. × 7–10 days	Therapeutic alternatives for persistent or relapsing infections or patients who do not tolerate clindamycin or quinine
	plus Azithromycin	600 mg daily p.o. × 7–10 days; Children: 12 mg/kg daily p.o. × 7–10 days	
Balamuthia mandrillaris (leptomyxid amebae) (amebic meningoencephalitis)	Drug of choice Pentamidine isethionate	10 µg/ml	Monitor clinical response and organisms in CSF; direct wet mounts of CSF and cultures
	Alternatives Clarithromycin (Biaxin)	500 mg TID	
	Fluconazole (Diflucan)	400 mg once daily	These 4 drugs used in combination
	Sulfadiazine	1.5 g every 6 h	
	Flucytosine (Ancobon)	1.5 g every 6 h	
Balantidium coli	Drug of choice Tetracycline	500 mg QID × 10 days; Children (≥8 yr): 40 mg/kg/day in 4 doses × 10 days (max 2 g/day)	At least 3 negative fecal examinations 1 mo after therapy; contraindicated in pregnancy and in children <8 yr
	Alternatives Iodoquinol	650 mg TID × 20 days; Children: 40 mg/kg/day in 3 doses × 20 days	
	or Metronidazole	750 mg TID × 5 days; Children: 35–50 mg/kg/day in 3 doses × 5 days	

Organism	Drug	Dosage	Comments
Diphyllobothrium latum (broad fish tapeworm)	Drug of choice Praziquantel	5–10 mg/kg once	Fecal examinations for eggs and proglottids 1 and 3 mo after therapy
	Alternative Niclosamide	2 g once Children: 50 mg/kg once	
Dipylidium caninum (dog and cat tapeworm)	Drug of choice Praziquantel	5–10 mg/kg once	Fecal examinations for eggs and proglottids 1 and 3 mo after therapy
	Alternative Niclosamide	2 g once Children: 50 mg/kg once	
Dracunculus medinensis (guinea worm)	Metronidazole	250 mg TID × 10 days Children: 25 mg/kg/day (max 750 mg/day) in 3 doses × 10 days	Monitor clinical course; slow extraction of worm; drugs not curative, but decrease inflammation
Echinococcus granulosus (hydatid cyst)	Albendazole or Surgery	400 mg BID × 1–6 mo Children: 15 mg/kg/day (max 800 mg) × 1–6 mo	Monitor clinical response to therapy; follow up with CT, ultrasonography, and radionucleotide scans
Echinococcus multilocularis (alveolar hydatid cyst)	Surgery or Albendazole	If surgery is unsuccessful or contraindicated, albendazole may be used 400 mg BID × 1–6 mo Children: 15 mg/kg/day (max 800 mg) × 1–6 mo	Monitor clinical response to therapy; follow up with CT, ultrasonography, and radionucleotide scans
Entamoeba histolytica Asymptomatic cyst passer (will be unable to morphologically differentiate *E. histolytica* from *E. dispar* on the basis of the cyst)	Iodoquinol (drug of choice) or Paromomycin Diloxanide furoate (alternative)	650 mg TID × 20 days Children: 30–40 mg/kg/day in 3 doses × 20 days (max 2 g) 25–35 mg/kg/day in 3 doses × 7 days 500 mg TID × 10 days Children: 20 mg/kg/day in 3 doses × 10 days	Fecal examinations 2 wk to 1 mo after therapy (must include permanent stained smear); EIAs for confirmation of true *E. histolytica* are available; trophozoites containing ingested RBCs can be identified as *E. histolytica* Not available commercially

(continued)

Table 25.1 Drugs and dosages for treating parasitic infections^a *(continued)*

Organism or infection	Drug	Dosage^b	Follow-up^b and comments
Mild to moderate intestinal disease	Metronidazole	500–750 mg TID × 10 days Children: 35–50 mg/kg/day in 3 doses × 7–10 days	Fecal examinations 2 wk to 1 mo after therapy (must include permanent stained smear); EIAs for confirmation of true *E. histolytica* are available; follow up with course of iodoquinol or paromomycin used to treat asymptomatic amebiasis
	or		
	Tinidazole	2 g/day × 3 days Children: 50 mg/kg (max 2 g) in 1 dose × 3 days	Tablets can be crushed and mixed with cherry syrup
Invasive amebiasis	Metronidazole	750 mg TID × 7–10 days Children: 35–50 mg/kg/day in 3 doses × 7–10 days	Monitor clinical response to therapy; follow up with CT, ultrasonography, and radionucleotide scans
	or		
	Tinidazole	2 g once daily × 5 days Children: 50 mg/kg/day (max 2 g) every day × 5 days	
Entamoeba polecki	Metronidazole	750 mg TID × 10 days Children: 35–50 mg/kg/day in 3 doses × 10 days	Fecal examinations 2 wk to 1 mo after therapy (must include permanent stained smear)
Enterobius vermicularis (pinworm)	Pyrantel pamoate	11 mg/kg once (max 1 g), repeat after 2 wk	Monitor clinical response; follow up with Scotch tape check of perianal area if symptoms persist (may take 4–6 consecutive negative tapes to rule out infection)
	or		
	Mebendazole	100 mg once, repeat after 2 wk	
	or		
	Albendazole	400 mg once; repeat after 2 wk	
Fasciola hepatica (sheep liver fluke)	Triclabendazole	10 mg/kg once or twice	Fecal examinations for eggs 1 mo after therapy; problems with availability of triclabendazole
	or		
	Bithionol	30–50 mg/kg on alternate days × 10–15 doses	
Fasciolopsis buski (giant intestinal fluke)	Praziquantel	75 mg/kg/day in 3 doses × 1 day	Fecal examinations for eggs 1 mo after therapy
Giardia lamblia (*G. intestinalis*, *G. duodenalis*)	Drug of choice Metronidazole	250 mg TID × 5 days Children: 15 mg/kg/day in 3 doses × 5 days	Fecal examinations 2 wk to 1 mo posttherapy (must include permanent stained smear or immunoassay); Entero-Test if stools are negative and patient is symptomatic; immunoassays for stool antigen available

Organism	Drug	Dose	Comments
	or Nitazoxanide	500 mg BID × 3 days; Children: 1–3 yr: 100 mg q12h × 3 days; 4–11 yr: 200 mg q12h × 3 days	
	or Tinidazole	2 g once; Children: 50 mg/kg once (max 2 g)	Tablets can be crushed and mixed with cherry syrup
	Alternatives Paromomycin	25–35 mg/kg/day in 3 doses × 7 days	
	or Furazolidone	100 mg QID × 7–10 days; Children: 6 mg/kg/day in 4 doses × 7–10 days	
	or Quinacrine	100 mg TID × 5 days; Children: 2 mg/kg TID × 5 days (max 300 mg/day)	Not available commercially
Gnathostoma spinigerum	Surgical removal plus Albendazole	400 mg BID × 21 days	Check for additional migratory swellings
	or Ivermectin	200 µg/kg/day × 2 days	
Gongylonema pulchrum	Surgical removal plus Albendazole	400 mg BID × 21 days	Check buccal cavity and esophagus
Heterophyes heterophyes (intestinal fluke)	Praziquantel	75 mg/kg/day in 3 doses × 1 day	Fecal examinations for eggs 1 mo after therapy
Hymenolepis nana (dwarf tapeworm)	Praziquantel	25 mg/kg once	Fecal examinations for eggs and proglottids 2 wk and 3 mo after therapy
	or Nitazoxanide	500 mg BID × 3 days; Children: 1–3 yr: 100 mg q12h × 3 days; 4–11 yr: 200 mg q12h × 3 days	
Isospora belli	TMP-SMX	160 mg of TMP, 800 mg of SMX (1 DS tab) BID × 10 days; Children: TMP 5 mg/kg, SMX 25 mg/kg BID × 10 days	Fecal examinations 1–2 wk after therapy
Leishmania spp.	Stibogluconate sodium (drug of choice)	20 mg of antimony/kg/day i.v. or i.m. × 28 days	For *L. donovani*, monitor clinical response to therapy; for other *Leishmania* spp, perform smear and culture of lesions 1–2 wk after therapy

(continued)

Table 25.1 Drugs and dosages for treating parasitic infections^a (*continued*)

Organism or infection	Drug	Dosage^b	Follow-up^b and comments
	or Meglumine antimonate	20 mg of antimony/kg/day × 28 days	
	or Amphotericin B	0.5–1.0 mg/kg by slow infusion daily or every 2 days for up to 8 wk	
	or Liposomal amphotericin B	3 mg/kg/day (days 1–5) and 3 mg/kg/day for days 14 and 21	
	Alternatives Pentamidine isethionate	4 mg/kg daily or every 2 days i.v. or i.m. for 15–30 doses	
	or Paromomycin	Topically twice daily × 10–20 days	
Loa loa (African eye worm)	Diethylcarbamazine (available from CDC)	6 mg/kg in 3 doses × 14 days 300 mg once/wk recommended for prevention	Blood smears, Knott concentration, and membrane filtration of blood for microfilariae 2–4 wk after therapy; for patients with no microfilariae in the blood, full doses can be given from day one; in heavy infections albendazole preferred (slower onset of action and lower risk of encephalopathy), but repeated doses may be required
Mansonella ozzardi	Ivermectin	200 µg/kg once	As above; diethylcarbamazine ineffective
Mansonella perstans	Albendazole	400 mg BID × 10 days	As above
	or Mebendazole	100 mg BID × 30 days	
Mansonella streptocerca	Diethylcarbamazine (available from CDC)	6 mg/kg/day × 14 days	See *Onchocerca volvulus*
	or Ivermectin	150 µg/kg once	
Metagonimus yokogawai (intestinal fluke)	Praziquantel	75 mg/kg/day in 3 doses × 1 day	Fecal examinations for eggs 2–4 wk after therapy
Metorchis conjunctus (North American liver fluke)	Praziquantel	75 mg/kg/day in 3 doses × 1 day	Fecal examinations for eggs 2–4 wk after therapy
Microsporidia	In some genera and species, treatment not totally effective; see also nitazoxanide under *Giardia lamblia*		
Disseminated			
Encephalitozoon hellem	Albendazole	400 mg BID	Monitor clinical response
Encephalitozoon cuniculi	Albendazole	400 mg BID	Monitor clinical response
Encephalitozoon (Septata) intestinalis	Albendazole	400 mg BID	Monitor clinical response

Organism	Drug	Dosage	Comments
Pleistophora sp.	Albendazole	400 mg BID	Monitor clinical response; may add itraconazole 400 mg p.o. once/day
Trachipleistophora hominis	Albendazole	400 mg BID	Monitor clinical response, may add itraconazole 400 mg p.o. once/day
Brachiola vesicularum	Albendazole	400 mg BID	Monitor clinical response; may add itraconazole 400 mg p.o. once/day
Intestinal			
Enterocytozoon bieneusi	Albendazole*	400 mg BID	Monitor clinical response; fecal exam for spores 2–4 wk after therapy
Encephalitozoon (Septata) intestinalis	Albendazole*	400 mg BID	Monitor clinical response; fecal exam for spores 2–4 wk after therapy
Ocular			
Encephalitozoon hellem	Albendazole plus fumagillin	400 mg BID, eye drops	Monitor clinical response
Encephalitozoon cuniculi	Albendazole plus fumagillin	400 mg BID, eye drops	Monitor clinical response
Vittaforma corneae	Albendazote plus fumagillin*	400 mg BID, eye drops	Monitor clinical response
Moniliformis moniliformis	Pyrantel pamoate	11 mg/kg once, repeat twice, 2 wk apart	Fecal exam for eggs 1 mo after therapy
Naegleria fowleri (primary amebic meningoencephalitis)	Amphotericin B	1.5 mg/kg/day i.v. in 2 doses × 3 days, then 1 mg/kg/day × 5 days	Monitor clinical response and organisms in CSF; direct wet mounts of CSF and cultures
Nanophyetus salmincola	Praziquantel	60 mg/kg/day in 3 doses × 1 day	Fecal exam for eggs 1 mo after therapy
Necator americanus (hookworm)	Albendazole or Mebendazole or Pyrantel pamoate	400 mg once / 100 mg BID × 3 days or 500 mg once / 11 mg/kg once (max 1 g) × 3 days	Fecal examinations 2–4 wk posttherapy; if eggs are rare in fecal concentrations, no need to repeat therapy
Oesophagostomum bifurcum	Albendazole	400 mg once	Monitor clinical response; pyrantel pamoate may also be effective
Onchocerca volvulus (river blindness)	Ivermectin	150 µg/kg once, repeat every 6–12 mo until asymptomatic	Skin snips for microfilariae 3–6 mo after therapy or sooner if symptoms recur
Opisthorchis viverrini (Southeast Asian liver fluke)	Praziquantel	75 mg/kg/day in 3 doses × 1 day	Fecal exam for eggs 1 mo after therapy
Paragonimus westermani (lung fluke), *Paragonimus* spp.	Drug of choice Praziquantel	75 mg/kg/day in 3 doses × 2 days	Sputum examination for eggs 2–4 wk after therapy; fecal examinations for eggs 1–2 mo after therapy
	Alternative Bithionol	30–50 mg/kg on alternate days × 10–15 doses	
Pediculus humanis (body louse)	0.5% Malathion or 1% Permethrin	Topically / Apply lotion to body, allow to remain for 10 min, then bathe	Examine clothing and bedding for nits and lice

(continued)

Table 25.1 Drugs and dosages for treating parasitic infections*a* *(continued)*

Organism or infection	Drug	Dosage*b*	Follow-up*b* and comments
	Alternatives		
	Pyrethrins with piperonyl butoxide	Topically with second application 1 wk later	
	or		
	Ivermectin	200 µg/kg × 3, days 1, 2, and 10	
Phthirus pubis (pubic louse)	0.5% Malathion	Topically	Examine clothing, bedding, and hair for nits and lice
	or		
	1% Permethrin	Apply lotion to hair, allow to remain for 10 min, then rinse off	
	Alternatives		
	Pyrethrins with piperonylbutoxide	Topically with second application 1 wk later	
	or		
	Ivermectin	200 µg/kg × 3, days 1, 2, and 10	
Plasmodium spp. Chemoprophylaxis (prevention)			
Chloroquine-sensitive areas	Chloroquine phosphate	500 mg (300 mg of base) once/wk for 1–2 wk before travel and continuing weekly for 4–6 wk after leaving area	Monitor clinical course
		Children: 5 mg of base/kg/wk once/wk, up to adult dose of 300 mg of base	
Chloroquine-resistant areas	Atovaquone/proguanil (drug of choice)	1 adult tab/day	250 mg/100 mg (1 tablet) daily, beginning 1–2 days before travel, continuing during stay and for 1 wk after leaving area
		Children:	
		11–20 kg: 1 peds tab/day	
		21–30 kg: 2 peds tabs/day	
		31–40 kg: 3 peds tabs/day	
		>40 kg: 1 adult tab/day	
	or		
	Mefloquine	250 mg once/wk	250 mg p.o. once/wk for 1–2 wk before travel and continuing weekly for 4–6 wk after leaving area
		Children:	
		5–10 kg: ⅛ tab once/wk	
		11–20 kg: ¼ tab once/wk	
		21–30 kg: ½ tab once/wk	
		31–45 kg: ¾ tab once/wk	
		>45 kg: 1 tab once/wk	
	or		
	Doxycycline	100 mg daily	
		Children: 2 mg/kg/day, up to 100 mg/day	
	Alternatives		
	Primaquine	30 mg of base daily	
		Children: 0.6 mg/kg of base daily	

	Dose	Comments
or Chloroquine phosphate	500 mg (300 mg of base) once/wk Children: 5 mg of base/kg once/wk, up to 300 mg of base	Same as for chloroquine-sensitive areas for adults and children
plus proguanil	200 mg once/day Children: <2 yr: 50 mg once/day 2–6 yr: 100 mg once/day 7–10 yr: 150 mg once/day >10 yr: 200 mg once/day	
Self-presumptive treatment		
Atovaquone/proguanil	4 adult tabs daily × 3 days Children: <5 kg: not indicated 5–8 kg: 2 peds tabs once/day × 3 days 9–10 kg: 3 peds tabs once/day × 3 days 11–20 kg: 1 adult tab once/day × 3 days 21–30 kg: 2 adult tabs once/day × 3 days 31–40 kg: 3 adult tabs once/day × 3 days >40 kg: 4 adult tabs once/day × 3 days	A traveler can be given a course of atovaquone/proguanil, mefloquine, or quinine plus doxycycline for presumptive self-treatment of febrile illness. The drug used for self-treatment should be different from that used for prophylaxis. This approach should be used in very rare circumstances when the person cannot promptly receive medical care.
or Quinine sulfate	650 mg q8h × 3–7 days Children: 30 mg/kg/day in 3 doses × 3–7 days	
plus Doxycycline	100 mg BID × 7 days Children: 4 mg/kg/day in 2 doses × 7 days	Tetracyclines contraindicated in pregnancy and in children <8 yr old
or Mefloquine	750 mg followed 12 h later by 500 mg Children: 15 mg/kg followed 12 h later by 10 mg/kg	
Treatment		
All *Plasmodium* species except chloroquine-resistant *P. falciparum* and chloroquine-resistant *P. vivax* (oral) Chloroquine phosphate	1 g (600 mg of base) p.o., then 500 mg (300 mg of base) 6 h later, then 500 mg (300 mg of base) at 24 and 48 h Children: 10 mg of base/kg (max 600 mg of base), then 5 mg of base/kg 6 h later, and then 5 mg of base at 24 and 48 h	Thick and thin blood smears

(continued)

Table 25.1 Drugs and dosages for treating parasitic infections[a] (*continued*)

Organism or infection	Drug	Dosage[b]	Follow-up[b] and comments
All *Plasmodium* spp. (parenteral)	Quinidine gluconate (parenteral drug of choice)	10-mg/kg loading dose (max 600 mg) in normal saline slowly over 1–2 h followed by continuous infusion of 0.02 mg/kg/min until start of oral therapy	Continuous EKG, blood pressure, and glucose monitoring recommended; if >48 h of parenteral treatment is required, quinine or quinidine dose should be reduced by ⅓ to ½
	or		
	Quinine dihydrochloride	20-mg/kg loading dose in 10 mg/kg 5% dextrose over 4 h followed by 10 mg/kg over 2–4 h q8h (max 1,800 mg/day) until start of oral therapy	Continuous EKG, blood pressure, and glucose monitoring recommended; if >48 h of parenteral treatment is required, quinine or quinidine dose should be reduced by ⅓ to ½
	Artemether (alternative)	3.2 mg/kg i.m., then 1.6 mg/kg/day × 5–7 days	
Chloroquine-resistant *P. falciparum*	Drugs of choice		
	Atovaquone/proguanil	2 adult tabs BID or 4 adult tabs once/day × 3 days Children: <5 kg: not indicated 5–8 kg: 2 peds tabs once/day × 3 days 9–10 kg: 3 peds tabs once/day × 3 days 11–20 kg: 1 adult tab once/day × 3 days 21–30 kg: 2 adult tabs once/day × 3 days 31–40 kg: 3 adult tabs once/day × 3 days >40 kg: 4 adult tabs once/day × 3 days	Atovaquone/proguanil is available as a fixed-dose combination tablet: (Malarone; 250 mg of atovaquone/100 mg of proguanil) and pediatric tablets (Malarone Pediatric; 62.5 mg atovaquone/25 mg proguanil). Take with food or a milky drink. Should not be given to pregnant women or patients with severe renal impairment (creatinine clearance, <30 ml/min). A few isolated reports of *P. falaciparum* resistance in Africa.
	or		
	Quinine sulfate	650 mg q8h × 3–7 days Children: 30 mg/kg/day in 3 doses × 3–7 days	Southeast Asia: resistance has increased; treat for 7 days
	plus		
	Doxycycline	100 mg BID × 7 days Children: 4 mg/kg/day in 2 doses × 7 days	Tetracyclines contraindicated in pregnancy and in children <8 yr
	or plus		
	Tetracycline	250 mg QID × 7 days Children: 6.25 mg/kg QID × 7 days	Contraindicated in pregnancy and in children <8 yr
	or plus		
	Clindamycin	20 mg/kg/day in 3 doses × 7 days	
	Alternatives		
	Mefloquine	750 mg followed 12 h later by 500 mg Children: 15 mg/kg followed 12 h later by 10 mg/kg	Allowed if pregnant

or Artesunate plus Mefloquine	4 mg/kg/day × 3 days 750 mg followed 12 h later by 500 mg Children: 15 mg/kg followed 12 h later by 10 mg/kg	Not available in U.S.
Chloroquine-resistant *P. vivax* Quinine sulfate (oral drug of choice)	650 mg q8h TID × 3–7 days Children: 25 mg/kg in 3 doses × 3–7 days	Southeast Asia: resistance increased; treat for 7 days
plus Doxycycline	100 mg BID × 7 days Children: 2 mg/kg/day × 7 days	Tetracyclines contraindicated in pregnancy and in children <8 yr
or Mefloquine	750 mg followed 12 h later by 500 mg Children: 15 mg/kg followed 12 h later by 10 mg/kg	
Alternatives Chloroquine plus Primaquine	25 mg of base/kg in 3 doses over 48 h 30 mg of base daily × 14 days Children: 0.6 mg of base/kg/day × 14 days	
Prevention of relapses (*P. vivax*, *P. ovale* only) Primaquine phosphate	30 mg of base/day × 14 days Children: 0.6 mg of base/kg/day × 14 days	Primaquine phosphate can cause hemolytic anemia, especially in patients whose red cells are deficient in G6PDH. This deficiency is more common in African, Asian, and Mediterranean people. Patients should be screened for G6PDH deficiency prior to treatment. Primaquine should not be used during pregnancy.
Sappinia diploidea (amebic meningoencephalitis) Azithromycin plus Pentamidine (i.v.) plus Itraconazole and plus Flucytosine	250 mg/day p.o., 31 wk 300 mg/day, 6 wk 200 mg BID, 25 wk 2.75 g QID, 25 wk	Combined therapy with surgical resection of CNS lesion
Sarcoptes scabiei (itch mite) Drug of choice 5% Permethrin	Topically Massage from head to soles of feet; wash off after 8–14 h	Monitor clinical course; may need mild steroid cream in sensitive individuals; may require retreatment in 10–14 days; ivermectin plus scabicide recommended for crusted scabies in immunocompromised patients

(continued)

Table 25.1 Drugs and dosages for treating parasitic infections[a] (*continued*)

Organism or infection	Drug	Dosage[b]	Follow-up[b] and comments
	Alternatives		
	Ivermectin	200 µg/kg p.o. once	Urine examinations for eggs 1 mo after therapy; bladder or rectal biopsy; treatment failure reported
	or		
	10% Crotamiton	Topically once/day × 2	
Schistosoma haematobium	Praziquantel	40 mg/kg/day in 2 doses × 1 day	Fecal examinations for eggs 1 mo after therapy; rectal biopsy
Schistosoma intercalatum	Praziquantel	40 mg/kg/day in 2 doses × 1 day	Fecal examinations for eggs 1 mo after therapy; rectal biopsy
Schistosoma japonicum	Praziquantel	60 mg/kg/day in 3 doses × 1 day	Fecal examinations for eggs 1 mo after therapy; rectal biopsy
Schistosoma mansoni	Drug of choice		
	Praziquantel	40 mg/kg/day in 2 doses × 1 day	Fecal examinations for eggs 1 mo after therapy; rectal biopsy
	Alternative		
	Oxamniquine	15 mg/kg once Children: 20 mg/kg/day in 2 doses × 1 day	Contraindicated in pregnancy; East Africa, increase dose to 30 mg/kg; South Africa, increase dose to 30 mg/kg × 2 days
Schistosoma mekongi	Praziquantel	60 mg/kg/day in 3 doses × 1 day	Fecal examinations for eggs 1 mo after therapy; rectal biopsy
Strongyloides stercoralis	Drug of choice		
	Ivermectin	200 µg/kg/day × 2 days	Fecal examinations and/or agar plate culture 1 mo after therapy
	Alternatives		
	Albendazole	400 mg BID × 7 days	
	or		
	Thiabendazole	50 mg/kg/day in 2 doses × 2 days (max 3 g/day)	This dosage may be toxic; might need to be reduced
Taenia saginata (beef tapeworm)	Drug of choice		
	Praziquantel	5–10 mg/kg p.o. once	Fecal examinations for eggs or proglottids 1 mo after therapy
	Alternative		
	Niclosamide	2 g once Children: 50 mg/kg once	Availability problems
Taenia solium (pork tapeworm)	Drug of choice		
	Praziquantel	5–10 mg/kg p.o. once	

Organism	Drug	Dosage	Comments
	Alternative Niclosamide	Same as for *T. saginata*	Same as for *T. saginata*
Toxocara canis, *T. cati* (visceral larva migrans)	Albendazole	400 mg BID × 5 days	Monitor clinical course
	or Mebendazole	100–200 mg BID × 5 days	
Toxoplasma gondii	Pyrimethamine (drug of choice)	25–100 mg/day × 3–4 wk. Children: 2 mg/kg/day × 3 days (max 25 mg/day), then 1 mg/kg/day × 4 wk	Monitor clinical course and serology; use only after first trimester
	plus Sulfadiazine	1–1.5 g QID × 3–4 wk. Children: 100–200 mg/kg/day × 3–4 wk	
	Spiramycin (alternative)	3–4 g/day × 3–4 weeks	Can be used during first-trimester infection
Trichinella spiralis, other *Trichinella* spp.	Mebendazole (drug of choice)	200–400 mg TID × 3 days, then 400–500 mg TID × 10 days	Monitor clinical course; perform muscle biopsy in severe cases; steroids for severe symptoms
	plus Corticosteroids	20–60 mg/day p.o.	Use for severe symptoms
	Albendazole (alternative)	400 mg p.o. BID × 8–14 days	
Trichomonas vaginalis	Metronidazole	2 g once or 500 mg BID × 7 days. Children: 15 mg/kg/day p.o. in 3 doses × 7 days	Pelvic examination 1 mo after therapy; sexual partners should be treated simultaneously
	or Tinidazole	2 g once. Children: 50 mg/kg once (max 2 g)	Take with food to minimize adverse side effects; can be used with metronidazole-resistant strains
Trichostrongylus spp.	Drug of choice Pyrantel pamoate	11 mg/kg once (max 1 g)	Fecal examinations 2–4 wk after therapy
	Alternatives Mebendazole	100 mg BID × 3 days	
	or Albendazole	400 mg once	
Trichuris trichiura (whipworm)	Drug of choice Mebendazole	100 mg BID × 3 days or 500 mg once	Fecal examinations 2–4 wk after therapy
	Alternative Albendazole	400 mg once	
Tropical pulmonary eosinophilia	Diethylcarbamazine (available from CDC)	6 mg/kg/day in 3 doses × 12–21 days	Monitor clinical course; if relapse occurs, retreat

(continued)

Table 25.1 Drugs and dosages for treating parasitic infections*a* (*continued*)

Organism or infection	Drug	Dosage*b*	Follow-up*b* and comments
Trypanosoma brucei gambiense (West African sleeping sickness), early stages (hemolymphatic)	**Drug of choice** Pentamidine isethionate	4 mg/kg/day i.m. × 10 days	Thick and thin blood smears 1–2 mo after therapy
	Alternative Suramin	100–200 mg (test dose) i.v., then 1 g i.v. on days 1, 3, 7, 14, and 21 Children: 20 mg/kg on days 1, 3, 7, 14, and 21	
Trypanosoma brucei rhodesiense (East African sleeping sickness), early stages (hemolymphatic)	Suramin	100–200 mg (test dose) i.v., then 1 g i.v. on days 1, 3, 7, 14, and 21 Children: 20 mg/kg on days 1, 3, 7, 14, and 21	
T. brucei gambiense and *T. brucei rhodesiense*, late stage with CNS disease	Melarsoprol	2–3.6 mg/kg/day i.v. × 3 days; after 1 wk, 3.6 mg/kg/day i.v. × 3 days; repeat again after 7 days Children: 18–25 mg/kg total over 1 mo; initial dose of 0.36 mg/kg i.v., increasing gradually to max of 3.6 mg/kg at intervals of 1–5 days for total of 9–10 doses	Cerebrospinal fluid analysis during and after therapy
	or Eflornithine	400 mg/kg/day i.v. in 4 doses × 14 days	Highly effective for *T. brucei gambiense* but variable for *T. brucei rhodesiense*
Trypanosoma cruzi (American trypanosomiasis, Chagas' disease)	Benznidazole	5–7 mg/kg/day in 2 divided doses × 30–90 days Children up to 12 yr: 10 mg/kg/day in 2 doses × 30–90 days	Thick and thin blood smears or xenodiagnosis 1–2 mo after therapy; monitor serology and EKG
	or Nifurtimox	8–10 mg/kg/day in 3–4 doses × 90–120 days Children: 1–10 yr: 15–20 mg/kg/day in 4 doses × 90 days 11–16 yr: 12.5–15 mg/kg/day in 4 doses × 90 days	
Wuchereria bancrofti	Diethylcarbamazine (available from CDC)	6 mg/kg in 3 doses × 14 days	Thick blood smears, Knott concentration, and membrane filtration of blood after 2–4 wk

a It is **very important** to thoroughly review drug information prior to use and to review published commentary concerning availability, contraindications, and monitoring requirements. Specific information can also be obtained online by checking *The Medical Letter*, August 2004: The Medical Letter, New Rochelle, N.Y. (http://www.medletter.com/freedocs/parasitic.pdf).

b QID, four times a day; BID, twice a day; TID, three times a day; max, maximum; p.o., orally; p.c., with meals; CT, computed tomography; TMP, trimethoprim; SMX, sulfamethoxazole; CSF, cerebrospinal fluid; CNS, central nervous system; i.v., intravenous; i.m., intramuscular; EIA, enzyme immunoassay; EKG, electrocardiogram. DS tab, double-strength tablet; peds, pediatric; *, see also nitazoxanide.

Table 25.2 Antiparasitic drugs[a]

Drug	Manufacturer	Trade name and comments[b]
Albendazole	GlaxoSmithKline	Albenza
Aminosidine		See Paromomycin
Amphotericin B	Apothecon, others	AmBisome
Amphotericin B	Apothecon, others	Fungizone, also available generically
Artemether	Aarenco, Belgium	Artenam; not available in the United States
Artesunate	Guilin No. 1 Factory, People's Republic of China	Not available in the United States
Atovaquone	GlaxoSmithKline	Mepron
Atovaquone-proguanil	GlaxoSmithKline	Malarone
Azithromycin	Pfizer	Zithromax
Bacitracin	Many manufacturers	
Bacitracin-zinc	Apothekernes Laboratorium A.S., Oslo, Norway	Not available in the United States
Benznidazole	Roche, Brazil	Rochagan; not available in the United States
Bithionol	Tanabe, Japan	Bitin; available through CDC Drug Service
Chloroquine HCl, chloroquine phosphate	Sanofi, others	Aralen, also available generically
Clarithromycin	Abbott	Biaxin
Clindamycin	Pfizer, others	Cleocin, also available generically
Crotamiton	Westwood-Squibb	Eurax
Dapsone	Jacobus	
Diethylcarbamazine citrate USP		Hetrazan; available through CDC Drug Service
Diloxanide furoate	Boots, United Kingdom	Furamide; not available in the United States
Doxycycline	Pfizer, others	Vibramycin; also available generically
Eflornithine (difluoromethylornithine, DFMO)	Aventis	Ornidyl; available through CDC Drug Service
Fluconazole	Roerig	Diflucan
Flucytosine	ICN	Ancobon
Fumagillin	Leiter's Park Avenue Pharmacy, San Jose, CA (800-292-6773) (www.leiterrx.com)	Fumidil-B
Furazolidone	Roberts	Furoxone; not available in the United States
Iodoquinol	Glenwood, others	Yodoxin; also available generically
Itraconazole	Janssen-Ortho	Sporanox
Ivermectin	Merck	Stromectol
Ketoconazole	Janssen, others	Nizoral; also available generically
Levamisole	Janssen	Ergamisol
Malathion	Taro	Ovide
Mebendazole	McNeil	Vermox
Mefloquine	Roche	Lariam
Meglumine antimoniate	Aventis, France	Glucantime; not available in the United States
Melarsoprol	Specia	Mel-B; available through CDC Drug Service
Metronidazole	Searle, others	Flagyl; also available generically
Miltefosine	Zentaris, Germany	Impavido; not available in the United States
Niclosamide	Bayer, Germany	Yomesan; not available in the United States
Nifurtimox	Bayer, Germany	Lampit; available through CDC Drug Service
Nitazoxanide	Romark	Alinia
Ornidazole	Roche, France	Tiberal; not available in the United States
Oxamniquine	Pfizer	Vansil; not available in the United States
Paromomycin	Monarch	Humatin
Paromomycin, topical	Teva, Israel	Leshcutan; not available in the United States
Pentamidine isethionate	Fujisawa	Pentam 300, Nebupent

(continued)

Table 25.2 Antiparasitic drugs*a* (*continued*)

Drug	Manufacturer	Trade name and comments*b*
Permethrin	GlaxoSmithKline	Nix
Permethrin	Allergan	Elimite
Polyhexamethylene biguanide	Zeneca: Leiter's Park Avenue Pharmacy, San Jose, CA (800-292-6773) (www.leiterrx.com)	Baquacil
Praziquantel	Bayer, distributed by Schering	Biltricide; not available in the United States
Primaquine phosphate USP		
Proguanil	Wyeth-Ayerst, Canada; AstraZeneca, United Kingdom	Paludrine; not available in the United States
Proguanil-atovaquone	GlaxoSmithKline	Malarone
Propamidine isethionate	Aventis, Canada	Brolene; not available in the United States
Pyrantel pamoate	Pfizer	Antiminth; not available in the United States
Pyrethrins and piperonyl butoxide	Pfizer, others	Rid; also available generically
Pyrimethamine USP	GlaxoSmithKline	Daraprim
Quinidine gluconate	Eli Lilly	Available in the United States only from the manufacturer
Quinine dihydrochloride		Not available in the United States
Quinine sulfate	Many manufacturers	
Rifampin	Aventis, others	Rifadin; also available generically
Sodium stibogluconate	GlaxoSmithKline, United Kingdom	Pentostam; available through CDC Drug Service
Spiramycin	Aventis	Rovamycine; available in the United States only from the manufacturer
Sulfadiazine		
Suramin sodium	Bayer, Germany	Germanin; available through CDC Drug Service
Thiabendazole	Merck	Mintezol
Tinidazole	Presutti: Mission Pharmaceuticals	Tindamax
TMP/Sulfa	Roche, others	Bactrim; also available generically
Triclabendazole	Novartis	Egaten; not available in the United States
Trimetrexate	U.S. Bioscience	Neutrexin

a Information from *The Medical Letter*, August 2004: The Medical Letter, New Rochelle, N.Y.
b Centers for Disease Control and Prevention (CDC), Atlanta, GA 30333; Drug Service, (404) 639-3670; evenings, weekends, holidays, (404) 639-2888; drugs not available in the United States may be available through a compounding pharmacy.

Table 25.3 Information on drug resistance and susceptibility testing for selected parasites*a*

Parasite and potential problems with susceptibility testing	Drug resistance and resistance genes	Susceptibility testing*b*
African trypanosomes (*Trypanosoma brucei gambiense, T. brucei rhodesiense*) Technically demanding tests; less relevance for field testing; potential problems with heterologous parasite populations. Lengthy time to results.	Melarsoprol, pentamidine	Cultures; assessment of viability by direct count or hydrolysis of *p*-nitrophenyl phosphate. Melarsoprol in vitro lysis assay (culture plus recording spectrophotometer for changes in absorbance of organisms over 30-min time frame). Color change assay when organisms incubated with Alamar Blue; dye added after extended incubation. Genetic markers and molecular test research ongoing.
Cryptosporidium spp.	Praziquantel, doxycycline, paromomycin	Cell line cultures (MDCK cells) exposed to drugs—studies ongoing.
Leishmania spp.	Sodium stibogluconate, meglumine antimoniate, pentamidine, amphotericin B, paromomycin, miltefosine	Promastigote cultures in standard media (Schneider's); assessment of viability by direct count, hydrolysis of *p*-nitrophenyl phosphate, conversion of MTT, reduction of Alamar Blue.

(continued)

Table 25.3   (*continued*)

Parasite and potential problems with susceptibility testing	Drug resistance and resistance genes	Susceptibility testing[b]
Technically demanding tests; less relevance for field testing; potential problems with heterologous parasite populations. Lengthy time to results.		Amastigotes adapted to in vitro drug testing in axenic and intracellular assays (now using firefly luciferase), cells lysed, substrate added, activity measured using luminometer.
		Flow cytometer also being used (isolated transfected with gene encoding GFP [fluorescent]); counting performed using flow (visceral leishmaniasis).
		Parasite-infected cells treated with drug, anti-*Leishmania* indirect fluorescent-antibody test used as detection system.
		Molecular test application research ongoing.
Plasmodium spp. A recent dose of antimalarial drug results in decreased test success; high parasite inocula result in overestimation of resistance; folate and pABA in culture medium result in antagonism of in vitro effect of antifolate drugs.	Chloroquine (*pfcrt*, *pfmdr*1), pyrimethamine, sulfadoxine, sulfadoxine-pyrimethamine and other antifolates (*dhfr*, *dhps*), atovaquone (cytochrome *b*), quinine, mefloquine, artemisinin, atovaquone-proguanil, tetracycline, primaquine	Most in vitro tests involve *Plasmodium falciparum* (asexual RBC stages—long-term cultivation possible); well cultures are plated with living organisms and various concentrations of test drug; after incubation (24–30 h), thick films are prepared from wells and Giemsa stained, and maturation of parasites is compared with that in test wells.
		A one step fluorescence assay for use in antimalarial drug screening has been developed; parasite growth is determined by using SYBR Green I, a dye with marked fluorescence enhancement upon contact with *Plasmodium* DNA.
		Parasites suspended in drug for longer periods (48–72 h); parasite growth and inhibition assessed by parasite counts, ELISA for HRP2 or pLDH; can be adapted to field use, and results very closely matched those obtained with a modified World Health Organization schizont maturation assay.
		Genetic markers for drug resistance (mutation-specific nested PCR; nested PCR followed by sequencing; RFLP; molecular beacons; real-time PCR; sequence-specific oligonucleotide probes).
		Drug impact on liver and sexual stages experimental.
Trichomonas vaginalis Potential problems with deriving axenic cultures from clinical specimens	Metronidazole, tinidazole	Testing recommended after failure of 2 standard treatment regimens; performed under both aerobic and anaerobic conditions (differences in sensitivity more pronounced when performed aerobically); well cultures are plated and examined microscopically for motile organisms (minimum lethal concentration of metronidazole: aerobic >100 µg/ml, anaerobic >3.1 µl/ml = clinical resistance).
		Search for metronidazole resistance genes and possible PCR assay ongoing.
Helminth testing	Multiple drugs	All studies more difficult; cultures in vitro limited; some drugs depend on components of immune system.

[a] Adapted from references 2, 3, 6, 7, 11, 15, 17, 18, 20–22, and 25.

[b] *dhfr*, dihydrofolate reductase; *dhps*, dihydropteroate synthetase; ELISA, enzyme-linked immunosorbent assay; GFP, green fluorescent protein; HRP2, histidine-rich protein 2; MDCK, Mardin-Darby canine kidney cells; MTT, 3-(4,5-dimethylthiazol-2-yl)-2,5-diphenyltetrazolium bromide; pABA, *p*-aminobenzoate; pLDH, parasite lactate dehydrogenase; RBC, red blood cell; RFLP, restriction fragment length polymorphism.

fivefold on average) concentrations of albendazole sulf-oxide in plasma compared to the concentration in plasma in fasted state.

Toxicity

Side effects include headache, vomiting, diarrhea, abdominal pain, increase in serum glutamic-oxaloacetic transaminase levels, and fever; migration of *Ascaris* through the mouth and nose may be seen.

Rare fatalities associated with the use of albendazole, due to granulocytopenia or pancytopenia, have been reported. Blood counts should be monitored at the beginning of each 28-day cycle of therapy and every 2 weeks during therapy with albendazole. Albendazole therapy may be continued if the total white blood cell count and absolute neutrophil count decrease appear modest and do not progress. Albendazole should not be used in pregnant women except in clinical circumstances where no alternative management is appropriate. Patients should not become pregnant for at least 1 month following cessation of albendazole therapy. If a patient becomes pregnant while taking this drug, the therapy should be discontinued immediately; the patient should be apprised of the potential hazard to the fetus.

Patients being treated for neurocysticercosis should receive appropriate steroid and anticonvulsant therapy as required. Oral or intravenous corticosteroids should be considered to prevent cerebral hypertensive episodes during the first week of anticysticercal therapy. Cysticercosis may, in rare cases, involve the retina. Before initiation of therapy for neurocysticercosis, the patient should be examined for the presence of retinal lesions. If such lesions are visualized, the need for anticysticercal therapy should be weighed against the possibility of retinal damage caused by albendazole-induced changes to the retinal lesion.

Contraindications

Due to lack of safety information, albendazole should not be given during pregnancy or lactation. Blood counts and liver function tests should be performed at the beginning of each 28-day treatment cycle and every 2 weeks.

Amphotericin B (AmBisome) (Fujisawa)

Preparation

The drug is supplied in vials containing 50 mg of amphotericin B inserted into a liposomal membrane. The product is lyophilized, and following reconstitution with sterile water, the pH is 5.0 to 6.0. It should be stored in the refrigerator and protected from light. It may be reconstituted with the addition of 12 ml of sterile water for injection USP, without a bacteriostatic agent, to make a 4-mg/ml solution. This may be further diluted by adding 1 ml of the 4-mg/ml solution to 40 ml of 5% glucose solution to make a final concentration of 0.1 mg of amphotericin per ml. The reconstituted product concentrate may be stored for up to 24 h under refrigeration. Injection of AmBisome should begin within 6 h of dilution with 5% dextrose injection. An existing intravenous line must be flushed with 5% dextrose injection prior to infusion of AmBisome. If this is not feasible, AmBisome should be administered through a separate line. AmBisome should never be reconstituted with saline, and saline must not be added to the reconstituted mixture. The mixture must not be mixed with any other drugs; this may cause precipitation of AmBisome.

Administration

AmBisome should be administered intravenously (i.v.), using a controlled infusion device, over a period of approximately 120 min. The infusion time may be reduced to approximately 60 min in patients in whom the treatment is well tolerated.

Toxicity

AmBisome-treated pediatric patients had a lower incidence of infusion-related fever, chills or rigors, and vomiting on day 1 than did amphotericin B deoxycholate-treated patients. The incidence of infusion-related cardiorespiratory events was also lower when AmBisome was used. In adult patients, AmBisome treatment resulted in a lower incidence of chills, hypertension, hypotension, tachycardia, hypoxia, hypokalemia, and various events related to decreased kidney function compared with amphotericin B deoxycholate.

Contraindications

AmBisome is contraindicated in patients who have demonstrated or have known hypersensitivity to amphotericin B deoxycholate or any other constituents of the product unless, in the opinion of the treating physician, the benefit of therapy outweighs the risk.

Amphotericin B (Fungizone) (Squibb)

Preparation

Fungizone is supplied in vials containing 50 mg of amphotericin B, 4 mg of sodium deoxycholate, and 20.2 mg of sodium phosphate as a buffer. It should be stored in the refrigerator and protected from light. It may be reconstituted by the addition of 10 ml of sterile water for injection USP, without a bacteriostatic agent, to make a 5-mg/ml solution. This may be further diluted by adding 1 ml of the 5-mg/ml solution to 50 ml of 5% glucose solution to make a final concentration of 0.1 mg of amphotericin per ml. The pH of the solution should be above 4.2. Amphotericin B solutions have been diluted with cerebrospinal fluid before intrathecal administration.

Administration

Therapy is usually initiated with a test dose of 0.25 mg/kg/day and gradually increased to a maximum dose of 1.5 mg/kg/day. It must be given slowly i.v., usually over a 6-h period. Alternate-day therapy may be better tolerated. For patients with severe life-threatening infections, therapy may have to be initiated with high doses. It may be given intrathecally in the lumbar or cisternal areas. Low doses should be given if the drug is administered intraventricularly.

Toxicity

Nausea, vomiting, fever, and phlebitis are common. Some degree of renal impairment usually occurs and is dose related. Renal function (blood urea nitrogen or serum creatine) should initially be monitored daily. Other side effects include anaphylaxis, hypokalemia, hypotension, anemia, and cardiac dysrhythmias. The side effects are much milder when amphotericin B is delivered in liposomes (AmBisome). This preparation is more expensive but makes the drug more tolerable, especially in patients with known renal failure. The main side effects of liposomal preparations are hypersensitivity reactions.

Contraindications

It may be necessary to pretreat the patient with antiemetic agents or antipyretic drugs. For patients with known hypersensitivity to the drug, it should be used only in life-threatening situations. Concomitant use of other nephrotoxic drugs should be avoided. Extreme caution should be used when the patient is taking other nephrotoxic drugs such as aminoglycosides, high-dose i.v. acyclovir, ganciclovir, foscarnet, and pentamidine. Corticosteroids and adrenocorticotropic hormone may lead to hypokalemia. The drug may increase the bone marrow toxicity of ganciclovir. Probenecid may increase the levels of amphotericin B.

Artemether (Artenam) (Aarenco, Belgium), Artesunate (Sanofi-Synthélabo/Guilin)

Preparation

Artemether is oil soluble and available in tablet, capsule, or intramuscular-injection forms. Artesunate is water soluble and can be given i.v., intramuscularly, orally, or by suppository. The activity of these artemisinin derivatives decreases after 1 to 2 h. To counter this drawback, artemisinin is given alongside lumefantrine to treat uncomplicated falciparum malaria. Lumefantrine has a half-life of about 3 to 6 days. Such a treatment is called ACT (artemisinin-based combination therapy); other examples are artemether-lumefantrine, artesunate-mefloquine, artesunate-amodiaquine, and artesunate-sulfadoxine/pyrimethamine. The artemisinin-based combinations artemether-lumefantrine and artesunate-mefloquine remain highly effective and elicit equivalent therapeutic responses in the treatment of highly drug-resistant falciparum malaria.

Administration

The drugs are given as 250-mg tablets of artemisinin and artesunate orally; they are also available as capsules containing 100 mg of artemisinin. Drugs that are available include artemether-lumefantrine (Coartem) (Novartis Pharma AG) and artesunate (Sanofi-Synthélabo/Guilin). The drugs are usually administered once a day for a minimum of 3 days.

Toxicity

The artemisinin derivatives are well tolerated, with no serious toxicity or adverse effects.

Contraindications

These drugs should be used with caution in individuals with hepatic or renal impairment. They should be avoided in pregnancy due to insufficient safety information.

Comment

Artemisia annua is a plant with a strong aroma, containing camphor and essential oils. It is a robust plant that grows in many areas of the world. However, only plants grown under special agricultural and geographic conditions contain artemisinin. The best high-yielding samples have been collected from the steep hills at altitudes over 4,500 feet around Youyang County, City of Chongqing, in Szechuan Province, China.

These drugs act mainly against the asexual erythrocytic stages of malaria. High concentrations of iron are found in red blood cells, which is also where the malaria parasites are found. When the compound enters the red blood cell, it releases free radicals, which are highly destructive to the parasites. An alternative theory suggests that artemisinins inhibit a calcium transporter in the parasite and, by specifically inhibiting this target, kill the parasites. The peroxide group in artemisinins is still crucial to activity, but the target is highly selected rather than a result of "explosion" into activity by free radicals. These drugs are also active against gametocytes by about 90% but are not effective against the intrahepatic stage of *P. vivax* and *P. ovale*.

Atovaquone (Mepron) (GlaxoSmithKline)

Preparation

Atovaquone suspension is a formulation of microfine particles of the drug. The atovaquone particles, reduced in size to facilitate absorption, are significantly smaller than those in the previously marketed tablet formulation. The suspension is for oral administration and is bright yellow with a citrus flavor, containing 150 mg of atovaquone per ml or 750 mg of atovaquone in each teaspoonful (5 ml).

Administration

The drug is given orally with meals; it is used to treat *Babesia* spp. and *Pneumocystis jiroveci* (now classified with the fungi) infections. Administering atovaquone with food enhances its absorption by approximately twofold.

Toxicity

Mepron may cause nausea, vomiting, diarrhea, and skin rashes, as well as sweating, flu syndrome, pain, sinusitis, insomnia, depression, and myalgia.

Contraindications

Mepron should not be given to anyone who has developed or has a history of life-threatening reactions to any of the components of the formulation. If it is necessary to treat patients with severe hepatic or renal impairment, caution is advised and administration should be closely monitored. This drug should be used in pregnancy only if the potential benefits outweigh the potential risks to the fetus. It should not be used during lactation.

Atovaquone-Proguanil (Malarone) (GlaxoSmithKline)

Preparation

Malarone (atovaquone and proguanil hydrochloride) is a fixed-dose combination of the antimalarial agents atovaquone and proguanil hydrochloride. Each tablet contains 250 mg of atovaquone and 100 mg of proguanil hydrochloride, and each pediatric tablet contains 62.5 mg of atovaquone and 25 mg of proguanil hydrochloride.

Administration

The drug is given orally with meals or a milky drink; atovaquone and cycloguanil (an active metabolite of proguanil) are active against the erythrocytic and exoerythrocytic stages of *Plasmodium* spp. Malarone is indicated for the prophylaxis of *P. falciparum* malaria, including in areas where chloroquine resistance has been reported. It is also indicated for the treatment of acute, uncomplicated *P. falciparum* malaria and is effective in regions where the drugs chloroquine, halofantrine, mefloquine, and amodiaquine may have unacceptable failure rates, presumably due to drug resistance.

Toxicity

Malarone may cause abdominal pain, nausea, vomiting, headache, diarrhea, asthenia, anorexia, and dizziness. About 5 to 10% of patients have transient asymptomatic elevations in transaminase and amylase levels.

Contraindications

Malarone should not be given to anyone who has developed or has a history of life-threatening reactions to any of the components of the formulation. Rare cases of anaphylaxis following treatment with atovaquone-proguanil have been reported. Due to the lack of safety data, the drug should be avoided during pregnancy or lactation or in patients with renal or hepatic impairment. Malarone has not been evaluated for the treatment of cerebral malaria or other severe manifestations of complicated malaria, including hyperparasitemia, pulmonary edema, or renal failure.

Comment

The drug interferes with two different pathways involved in the biosynthesis of pyrimidines required for nucleic acid replication. Atovaquone is a selective inhibitor of parasite mitochondrial electron transport, while the metabolite of proguanil hydrochloride, cycloguanil, is a dihydrofolate reductase inhibitor. Malarone is effective against asexual and sexual forms of *P. falciparum* and is used for prophylaxis and treatment. Unfortunately, resistance and drug failures are being reported for *P. falciparum*. Although the drug is also effective for treating *P. vivax* and *P. ovale*, primaquine is required to prevent relapses from the liver.

Benznidazole (Rochagan) (Roche, Brazil)

Preparation

Benznidazole is a 2-nitroimidazole derivative and is provided as 100-mg tablets. It is used to treat acute Chagas' disease but is not available in the United States.

Administration

The drug should be taken with meals, preferably after breakfast and after supper, to decrease the chance of stomach upset.

Toxicity

Side effects are seen in many individuals and include nausea, vomiting, abdominal pain, peripheral neuropathy, severe skin reactions, bone marrow suppression, and psychic disturbances.

Contraindications

Alcohol should be avoided during treatment. The drug has been noted to cause lymphoma in animal models and is mutagenic. Its use is not recommended during the first trimester of pregnancy. Benznidazole should not be used with anticoagulants or in patients who have renal or hepatic impairment. Although this side effect is rare, benznidazole can lower the number of white blood cells, increasing the patient's chance of getting an infection. It can also lower the number of platelets, which are necessary for proper blood clotting.

Comment

Benznidazole inhibits protein and RNA synthesis in *T. cruzi*; it causes increased phagocytosis, cytokine release, and production of reactive mitogen intermediates that lead to death of the parasites.

Bithionol (Bitin) (Tanabe, Japan) (CDC)

Preparation

Bitin is supplied as 500-mg capsules, which must be obtained from CDC. The drug is used to treat several trematode infections.

Administration

The capsules are taken orally with meals to minimize stomach irritation.

Toxicity

Patients may experience anorexia, nausea, vomiting, abdominal discomfort, diarrhea, dizziness, and headaches.

Contraindications

Bitin may cause leukopenia and toxic hepatitis, and photosensitization has occurred.

Comment

Bithionol, an investigational drug, is a phenolic substance structurally related to hexachlorophene. It is used as an alternative drug to praziquantel for treating *Paragonimus* infections. It can be used to treat extrapulmonary infections with *Paragonimus*, but multiple courses may be required for cure. Surgery may be needed to excise skin lesions or, rarely, brain cysts. Bithionol is also used to treat *Fasciola hepatica* infections. Its use is limited to treating either of these infections in patients who are unable to tolerate praziquantel due to previous reactions or in whom a previous course of praziquantel was ineffective.

Chloroquine Phosphate (Aralen) (Sanofi, Others)

Preparation

Aralen is supplied in 250- and 500-mg tablets of the diphosphate salt (500 mg of salt = 300 mg of base) and as a hydrochloride for parenteral injection in 5-ml vials containing 40 mg of base per ml. A liquid preparation (Nivaquine) is available outside the United States and is used as an antimalarial agent.

Administration

Aralen is given orally as per protocol for chemoprophylaxis or treatment of an acute attack. Pills can be ground up and placed in sweetened juice or fruit for administration to young children.

Toxicity

Side effects may be dose related and include gastrointestinal symptoms, skin rash, insomnia, headache, and nervousness. Rare reactions such as toxic psychosis and retinal damage, blood dyscrasias, and photophobia may occur.

Contraindications

Relative contraindications include liver, hematologic, and/or neurologic disease; psoriasis; and porphyria. It may be used in pregnant women when necessary for protection. Use of this drug is contraindicated in the presence of retinal or visual field changes either attributable to 4-aminoquinoline compounds or to any other etiology and in patients with known hypersensitivity to 4-aminoquinoline compounds. Parenteral chloroquine may cause hypotension, shock, and sudden death when given to young children.

Comment

Chloroquine is active against the erythrocytic forms of *Plasmodium vivax*, *P. malariae*, and susceptible strains of *Plasmodium falciparum* (but not the gametocytes of *P. falciparum*). It is not effective against exoerythrocytic forms of the parasite. Resistance of *P. falciparum* to chloroquine is widespread, and cases of *P. vivax* resistance have been reported.

Crotamiton (Eurax) (Westwood-Squibb)

Preparation

Eurax is supplied as a 10% cream or lotion and is used to treat scabies (*Sarcoptes scabiei*).

Administration

The cream or lotion is applied to the whole body surface below the chin after bathing and drying. A second application is advisable after 24 h. Reinfection from contaminated clothing and bed linens should be avoided.

Toxicity

The drug can cause irritation and sensitivity reactions, which may include rash or conjunctivitis.

Contraindications

The drug should not be used in the presence of acute exudative dermatitis and should not be applied near the eyes.

Dapsone (Jacobus)

Preparation

Dapsone is supplied as 25- and 100-mg scored tablets and used to treat *Plasmodium falciparum* (combination therapy) and *Pneumocystis jiroveci* (now classified with the fungi) infections.

Administration

Dapsone should be taken orally, and the dosage should be individually titrated.

Toxicity

Dapsone has been associated with frequent rashes, transient headaches, gastrointestinal irritation, anorexia, and infectious mononucleosis-like syndrome. The drug is carcinogenic in animal models. Complete blood counts should be performed frequently because the drug has been associated with death from agranulocytosis, aplastic anemia, and other blood dyscrasias.

Contraindications

The drug should not be given to patients with hypersensitivity to dapsone. Hemolysis and Heinz body formation may be exaggerated in individuals with a glucose-6-phosphate dehydrogenase (G6PDH) deficiency, or methemoglobin reductase deficiency, or hemoglobin M. This reaction is frequently dose related. Dapsone should be given with caution to these patients or to patients who have been exposed to other agents or conditions such as infection or diabetic ketosis capable of producing hemolysis.

Diethylcarbamazine Citrate USP (Hetrazan) (Wyeth-Ayerst) (CDC)

Preparation

Diethylcarbamazine is a piperazine derivative and is supplied in 50-mg tablets. It is used to treat filariasis and is effective against microfilariae; it is available from CDC. Hetrazan is effective against *Wuchereria bancrofti*, *Brugia malayi*, *B. timori*, *Onchocerca volvulus*, *Loa loa*, and *Mansonella streptocerca* but has little effect on *M. perstans* or *M. ozzardi*.

Administration

Tablets are scored to provide the appropriate dose. Medication should be taken after meals, adhering to a strict dosage schedule.

Toxicity

Reactions occur frequently and include headache, weakness, malaise, fever, joint pain, anorexia, nausea, and vomiting. These generally subside as therapy continues. Mild to severe reactions, including fever, rash, lymphadenopathy, conjunctivitis, leukocytosis, and eosinophilia, may occur in response to the parasite's death. Severe reaction may require steroid therapy and antihistamines.

Contraindications

Treatment of patients with heavy microfilaremia due to loiasis may cause allergic encephalitis, nephritis, or myocarditis. Patients with many microfilariae or onchocerciasis in ocular tissues may experience keratitis, chorioretinitis, and optic neuritis. Dose reductions are generally required in patients with renal impairment. The drug should probably be avoided in pregnancy but is not excreted in breast milk.

Comment

Diethylcarbamazine citrate, an investigational drug, is an anthelmintic agent that does not resemble other antiparasitic compounds. It is a synthetic organic compound which is highly specific for several parasites and does not contain any toxic metallic elements. It is used to treat patients with certain filarial infections, including lymphatic filariasis (*W. bancrofti*, *B. malayi*, or *B. timori*), tropical pulmonary eosinophilia, and loiasis (*Loa loa*). Treatment of lymphatic filariasis is problematic. Diethylcarbamazine kills microfilariae but only a variable proportion of adult worms. Another drug, ivermectin, rapidly reduces microfilaremia levels and may inhibit larval development in mosquitoes, but it does not kill adult worms and is not very effective against brugian filariasis.

Diloxanide Furoate (Furamide) (Boots, England)

Preparation

Diloxanide furoate is supplied as scored 500-mg tablets and is used to treat amebiasis; it is not available in the United States. The drug is hydrolyzed in the bowel to the active compound, diloxanide, which is >90% absorbed.

Administration

Furamide is given by mouth with repeated treatments as necessary.

Toxicity

The most common adverse effect is flatulence. Other side effects are seen, including vomiting, pruritus, and urticaria.

Contraindications

The use of Furamide is generally not recommended during pregnancy or lactation.

Comment

The drug helps clear the bowel of *Entamoeba histolytica* cysts; however, it is not effective for treating invasive disease in the bowel wall or liver. For this reason, it is often given with a 5-nitroimidazole.

Eflornithine (Difluoromethylornithine, Ornidyl) (Aventis) (CDC)

Preparation

The drug is available as a 200-mg/ml solution, and is used to treat African trypanosomiasis.

Administration

Ornidyl is given as a continuous i.v. infusion or orally; however, oral administration is usually followed by diarrhea. Eflornithine crosses the blood-brain barrier; thus it is beneficial for central nervous system disease associated with African trypanosomiasis.

Toxicity

Side effects occur in about 40% of patients; they include anemia, leukopenia, thrombocytopenia, diarrhea, abdominal discomfort, hearing loss, vomiting, dizziness, arthralgia, and numbness of the limbs.

Contraindications

Ornidyl should not be given with other i.v. drugs. It may worsen hemolytic disorders. Due to insufficient safety data, the drug should not be used during pregnancy or lactation.

Comment

Eflornithine inhibits ornithine decarboxylase, an enzyme required for the formation of polyamines, thus leading to inhibition of parasite growth. However, it is ineffective as monotherapy for *T. brucei rhodesiense*.

Fumagillin (Fumidil-B)

Preparation

The drug is supplied as a 10-mg/ml suspension in physiologic saline. Ocular lesions due to *Encephalitozoon hellem* have responded to fumagillin eye drops prepared from Fumidil-B used to control microsporidial infection in honeybees. For lesions due to *Vittaforma corneae*, topical therapy is generally not effective and keratoplasty may be required. The drug is available in limited supply from CDC and the World Health Organization.

Administration

Fumagillin should be applied topically once an hour until clinical improvement is demonstrated.

Toxicity

The soluble salt is toxic and should not be ingested; topical application is not toxic to the cornea.

Contraindications

No information is available.

Comment

Relapse may occur on discontinuation of fumagillin therapy.

Furazolidone (Furoxone) (Roberts)

Preparation

Furoxone is supplied as 100-mg tablets or as a flavored suspension of 50 mg/liter and is used to treat giardiasis. This drug is not available in the United States.

Administration

The drug is given orally. Apparently furazolidone is more effective by 50% than metronidazole in inhibiting in vitro cyst differentiation of *Giardia lamblia*.

Toxicity

Furoxone can cause polyneuritis, nausea, vomiting, dizziness, drowsiness, and general malaise. Skin rashes and agranulocytosis have occasionally been reported. Furazolidone is chemically similar to nitrofurantoin, which is well known to cause pulmonary hypersensitivity reactions; furazolidone may also induce pulmonary hypersensitivity reactions, and clinicians should be aware of this potentially serious adverse effect.

Contraindications

The drug should not be given to individuals hypersensitive to furazolidone or to infants younger than 1 month.

Hemolytic anemia can occur in patients with G6PDH deficiency. Alcohol should be avoided during treatment, because disulfiram-like reactions have been reported. Furazolidone is a monoamine oxidase inhibitor and should not be given concomitantly with other similar drugs.

Iodoquinol/Diiodohydroxyquin (Yodoxin) (Glenwood, Others)

Preparation

Iodoquinol/diiodohydroxyquin is supplied as 210- and 650-mg tablets and is used as an amebicide against *Entamoeba histolytica*. The drug is considered effective against the trophozoite and cyst forms.

Administration

The drug should be given with meals. For young children, a suspension may be made by mixing the pulverized tablet with cherry syrup to give a suspension containing 200 to 400 mg/5 ml. The drug is poorly absorbed.

Toxicity

There may be various forms of skin eruptions, urticaria, and pruritus. There may also be nausea, vomiting, abdominal cramps, diarrhea, and pruritus ani. Fever, chills, headache, vertigo, and enlargement of the thyroid have been reported. Optic neuritis, optic atrophy, and peripheral neuropathy have been reported in association with prolonged high-dosage 8-hydroxyquinoline therapy.

Contraindications

Iodoquinol should not be given to patients with known hypersensitivity to iodine or any 8-hydroxyquinoline compound, with renal or thyroid disease, or with severe liver disease not due to amebiasis. Iodoquinol should be used with caution in patients with thyroid disease. Protein-bound serum iodine levels may be increased during treatment with iodoquinol and may interfere with certain thyroid function tests. These effects may persist for as long as 6 months after discontinuation of therapy; this effect is unrelated to thyroid function. The drug should be withdrawn if hypersensitivity reactions occur. Severe neurologic damage, including optic neuritis, optic atrophy, and peripheral neuropathy, has occurred with prolonged high doses of this drug. The recommended doses and duration of therapy should not be exceeded. There may be persistent elevation of the protein-bound iodine level months after therapy. Because of the limited safety data available, this drug should probably be avoided during pregnancy or lactation.

Comment

The drug is thought to act by inactivating parasitic enzymes and inhibiting organism multiplication. Although it is effective for treating intestinal organisms, it is not effective for treating invasive disease in the bowel wall or liver; it is often combined with a 5-nitroimidazole. It has documented activity against *Balantidium coli*, *Dientamoeba fragilis*, *Giardia lamblia*, and *Blastocystis hominis*.

Ivermectin (Stromectol) (Merck)

Preparation

The drug is available in 6-mg tablets and is a semisynthetic macrolide used to treat nondisseminated strongyloidiasis and filariasis, particularly onchocerciasis. It is also now used for the treatment of ectoparasitic infestations such as scabies and lice, although it has no effect on the eggs (nits).

Administration

The drug is given orally with water 1 h before breakfast and is rapidly absorbed after ingestion.

Toxicity

The drug can cause fever, headache, rash, increased pruritus, joint and muscle pain, and tachycardia. A Mazzotti-type reaction has been seen in patients with onchocerciasis. Most adverse effects are a result of the host immune response to destruction of the parasites rather than effects of the drug itself.

Contraindications

Ivermectin should not be given to pregnant or lactating women, because it has been noted to cause birth defects in mice. Historical data have shown that microfilaricidal drugs, such as diethylcarbamazine citrate, might cause cutaneous and/or systemic reactions of varying severity (the Mazzotti reaction) and ophthalmologic reactions in patients with onchocerciasis. These reactions are probably due to allergic and inflammatory responses to the death of microfilariae. Patients treated with Stromectol for onchocerciasis may experience these reactions in addition to clinical adverse reactions possibly, probably, or definitely related to the drug itself.

Comment

Ivermectin is active against various life cycle stages of many but not all nematodes. It is active against the tissue microfilariae of *Onchocerca volvulus* but not against the adult form. The adult parasites reside in subcutaneous nodules which are infrequently palpable. Surgical excision of these nodules (nodulectomy) may be considered in the management of patients with onchocerciasis, since this procedure eliminates the microfilaria-producing adult parasites.

Ivermectin may be highly effective against *Strongyloides* and produces fewer side effects than thiabendazole. Compounds of the class bind selectively and with high affinity to glutamate-gated chloride ion channels which occur in invertebrate nerve and muscle cells. This leads to an increase in the permeability of the cell membrane to chloride ions with hyperpolarization of the nerve or muscle cell, resulting in paralysis and death of the parasite.

The drug is also effective against the microfilariae of *Wuchereria bancrofti*, *Brugia malayi*, and *Loa loa*. Rarely, patients with onchocerciasis who are also heavily infected with *L. loa* develop a serious or even fatal encephalopathy either spontaneously or following treatment with an effective microfilaricide. For these patients, the following adverse experiences have also been reported: back pain, conjunctival hemorrhage, dyspnea, urinary and/or fecal incontinence, difficulty in standing and/or walking, mental status changes, confusion, lethargy, stupor, or coma. This syndrome has been seen very rarely following the use of ivermectin; a cause-and-effect relationship has not been established. For individuals who warrant treatment with ivermectin for any reason and have had significant exposure to *L. loa*-endemic areas of West or Central Africa, pretreatment assessment for loiasis and careful posttreatment follow-up should be implemented.

Malathion (Ovide) (Taro Pharmaceuticals)

Preparation

Ovide is supplied as a 0.5% aqueous emulsion and is used to treat human louse infestations, killing both lice and nits. Malathion is an organophosphate agent which acts as a pediculicide by inhibiting cholinesterase activity in vivo.

Administration

The drug is applied as a lotion to the skin or hair. It is placed on dry hair in a sufficient amount to thoroughly wet the hair. The hair is then allowed to dry naturally and shampooed after 8 to 12 h. It is then rinsed, and a fine comb is used to remove dead lice and nits. If lice are still present 7 to 9 days later, the drug should be reapplied.

Toxicity

Malathion may cause acute renal failure. Because the potential for transdermal absorption of malathion from Ovide lotion is not known at this time, strict adherence to the dosing instructions regarding its use in children, method of application, duration of exposure, and frequency of application is required.

Contraindications

The drug should not be given to individuals exposed to organophosphorous compounds or to infants or neonates. It may cause skin irritation. Because animal reproduction studies are not always predictive of human responses, this drug should be used (or handled) during pregnancy only if clearly needed. The safety and effectiveness of Ovide lotion in children younger than 6 years has not been established in well-controlled trials.

Mebendazole (Vermox) (McNeil)

Preparation

The drug is supplied as 100-mg chewable tablets; it is a benzimidazole, broad-spectrum agent used to treat nematode infections. Mebendazole is indicated for the treatment of *Enterobius vermicularis* (pinworm), *Trichuris trichiura* (whipworm), *Ascaris lumbricoides* (common roundworm), *Ancylostoma duodenale* (common hookworm), and *Necator americanus* (American hookworm) in single or mixed infections. In humans, approximately 2% of the administered drug is excreted in urine and the remainder is excreted in the feces as unchanged drug or a primary metabolite.

Administration

Tablets are chewable, but it may be necessary to pulverize them and mix them with food for administration to young children. The drug inhibits the formation of nematode microtubules and causes glucose depletion in the worms. Periodic assessment of the hematopoietic and hepatic system is advisable during prolonged therapy.

Toxicity

Abdominal pain and diarrhea occasionally occur, particularly with the expulsion of a large number of worms in children. Rash, urticaria, and angioedema have been observed on rare occasions. Very rare reports of convulsions have been documented.

Contraindications

The drug is teratogenic in rats, but the incidence of abortion and malformations in humans does not exceed the rate in the general population. The drug, however, is not recommended in pregnancy, particularly during the first trimester, unless the benefit justifies the potential risk to the fetus. It is also not recommended during lactation. In patients with severe hepatic disease, the drug should be used with caution. There is no evidence that mebendazole, even at high doses, is effective for treating hydatid disease. There have been rare reports of neutropenia and agranulocytosis when the drug was taken for prolonged periods and at dosages substantially above those recommended.

Mefloquine Hydrochloride (Lariam) (Roche)

Preparation

The drug is supplied as 250-mg tablets equivalent to 228 g of the free base; it is a synthetic 4-quinoline methanol compound structurally related to quinine. The bioavailability of the tablet formation compared with an oral solution was over 85%. The presence of food significantly enhances the rate and extent of absorption, leading to about a 40% increase in bioavailability.

Administration

Tablets should not be ingested on an empty stomach and should be taken with at least 8 oz of water.

Toxicity

The most frequent side effects include nausea, vomiting, dizziness, diarrhea, abdominal pain, and headache. Severe seizures and delirium have been reported occasionally in roughly 1 in 200 to 1 in 1,300 patients treated for acute *Plasmodium falciparum* malaria. The drug can also potentiate dysrhythmias in patients taking beta blockers. Periodic evaluation of hepatic function should be performed during prolonged prophylaxis.

Contraindications

Use of Lariam is contraindicated in patients with a known hypersensitivity to mefloquine or related compounds (e.g., quinine and quinidine). Lariam should not be prescribed for prophylaxis in patients with active depression, a recent history of depression, generalized anxiety disorder, psychosis, schizophrenia, or other major psychiatric disorders, or in those with a history of convulsions. In patients with epilepsy, Lariam may increase the risk of convulsions. The drug should be used with caution in patients with severe hepatic or renal dysfunction. During prophylactic use, if psychiatric symptoms such as acute anxiety, depression, restlessness, or confusion occur, these may be considered prodromal to a more serious event. In these cases, the drug must be withdrawn and an alternative medication should be substituted.

In case of life-threatening, serious, or overwhelming malaria infections with P. falciparum, *patients should be treated with an i.v. antimalarial drug. Following completion of i.v. treatment, Lariam may be given to complete the course of therapy.*

Comment

Although its exact mechanism of action is not known, mefloquine is an antimalarial agent which acts as a blood schizonticide. The drug is thought to act by interfering with the digestion of hemoglobin during the erythrocytic stages of the life cycle. The drug has no activity against tissue schizonts or gametocytes. Strains of *P. falciparum* with decreased susceptibility to mefloquine can be selected in vitro or in vivo. Resistance of *P. falciparum* to mefloquine has been detected in areas of multidrug resistance in Southeast Asia. Increased incidences of resistance have also been reported in other parts of the world.

Strains of *P. falciparum* resistant to mefloquine have been detected. Lariam is indicated for the treatment of mild to moderate acute malaria caused by mefloquine-susceptible strains of *P. falciparum* (both chloroquine-susceptible and -resistant strains) or by *P. vivax*. However, patients with *P. vivax* malaria are at risk of relapse since mefloquine does not eliminate the exoerythrocytic parasites in the liver. Mefloquine is also indicated for the prophylaxis of *P. falciparum* and *P. vivax* malaria infections, including those caused by chloroquine-resistant strains of *P. falciparum*.

Mefloquine should be used during pregnancy only if the potential benefit justifies the potential risk to the fetus. Women of childbearing potential who are traveling to areas where malaria is endemic should be warned against becoming pregnant. Women of childbearing potential should also be advised to practice contraception during malaria prophylaxis with mefloquine and for up to 3 months thereafter. However, in the case of unplanned pregnancy, malaria chemoprophylaxis with mefloquine is not considered an indication for pregnancy termination. Because of the potential for serious adverse reactions in nursing infants from mefloquine, a decision should be made whether to discontinue treatment, taking into account the importance of the drug to the mother.

Melarsoprol (Mel-B, Arsobal) (Aventis) (CDC)

Preparation

Melarsoprol is supplied through the CDC Parasitic Disease Drug Service as a 3.6% solution in propylene glycol. It is used to treat African trypanosomiasis. The drug should be kept refrigerated. It is to be administered only in the hospital by or under the immediate supervision of a physician.

Administration

The drug should be given in propylene glycol as a slow i.v. infusion in the hospital. Care must be taken to avoid extravasation.

Toxicity

Abdominal pain, nausea, vomiting, and diarrhea are common. Reactive encephalopathy is not uncommon and is more likely to occur in debilitated patients and when high doses are used. The patient may also exhibit confusion, speech difficulties, restlessness, and sleep disorders. Myocardial damage, albuminuria, and hypertension have also been reported. A rare reaction is shock and subsequent death. Due to toxicity, the drug is usually reserved for the late stages of disease involving the central nervous system.

Rare side effects may include black, tarry stools; blood in urine or stools; changes in skin color of face; fainting; fast or irregular breathing; pinpoint red spots on skin; puffiness or swelling of the eyelids or around the eyes; shortness of breath; sores, ulcers, or white spots on lips or in mouth; sore throat; swollen and/or painful glands; tightness in chest; trouble in breathing; unusual bleeding or bruising; unusual tiredness or weakness; or wheezing.

Contraindications

Patients with impaired hepatic or renal function should not receive melarsoprol. Severe hemolytic reactions can occur in patients with G6PDH deficiency, and erythema nodosum may be induced in patients with leprosy. The drug should not be given during influenza epidemics. It may increase the chance of side effects in patients with fever and may increase the chance of side effects affecting the blood in patients with G6PDH deficiency.

Comment

Melarsoprol, an investigational drug, is an arsenical compound used for treating African trypanosomiasis (*Trypanosoma brucei gambiense*, *T. brucei rhodesiense*) with neurologic involvement or early African trypanosomiasis that is resistant to suramin or pentamidine. The drug apparently interferes with parasite energy generation and trophozoite multiplication.

Metronidazole (Flagyl [Searle], IVFlagyl [SCS])

Preparation

Flagyl is supplied as 375-mg tablets for oral administration and as 750-mg extended-release tablets (Flagyl ER); it is used in the treatment of amebiasis, giardiasis, trichomoniasis, and some helminth infections. It is also available in single-dose vials containing 500 mg of lyophilized powder or in 100-ml plastic containers containing a solution of 500 mg/100 ml or isotonic-buffered solution for parenteral injection. If metronidazole is being prescribed for trichomoniasis, sexual partners are often treated at the same time, even if they are asymptomatic.

Administration

Flagyl is usually given orally. When given i.v., it should be administered slowly as an intermittent or continuous infusion. Other drugs should not be added to the i.v. line during the infusion. Care must be taken to follow instructions for reconstitution of the solution, because neutralization with bicarbonate is necessary to bring the pH from 0.5–2 up to the neutral range.

Toxicity

Reactions to therapy are common, especially when the drug is given in doses appropriate for treating amebiasis. The more common side effects include anorexia, nausea, vomiting, diarrhea, epigastric distress, and abdominal cramps. A metallic taste in the mouth is often reported. Headache, dizziness, and vertigo may also occur. Ataxia, paresthesias, insomnia, and irritability may also occur. A flattening of the T waves on an electrocardiogram may be evident, and rashes and urticaria have been reported. A moderate leukopenia occasionally occurs, but levels return to normal after completion of therapy. Oral and vaginal overgrowth of yeast cells may occur. In patients receiving high doses, dark urine has occurred; this finding may be due to a metabolite of the drug. Rare reactions include blood dyscrasias, dysuria, incontinence, cystitis, dyspareunia, and fever.

Contraindications

The drug should not be given to patients with a history of blood dyscrasia, active central nervous system disease, and/or known sensitivity to metronidazole. Doses should be reduced in patients with severe liver disease. The drug should not be used in pregnancy, especially in the first trimester, unless the benefits of treatment outweigh the potential hazard to the fetus. Consumption of alcohol should be avoided during the first days of treatment and for at least 3 days after the end of the treatment because severe vomiting, abdominal cramps, and flushing may occur. Other alcohol-containing preparations (for example, elixirs, cough syrups, and tonics) may also cause problems. These problems may last for at least a day after the drug has been discontinued. Leukocyte counts with differential blood counts are recommended before and after therapy. A good alternate for use during pregnancy is paromomycin.

Miltefosine (Impavido) (Zentaris, Germany)

Preparation

The drug is given orally at 30 mg/kg twice a day; it has activity against *Leishmania* spp. It is not available within the United States. In India, it has been given at 2.5 mg/kg/day for 4 weeks, with good results for visceral leishmaniasis. Good activity has also been documented against *L. viannia panamensis*, with less activity against

L. mexicana mexicana and even lower for *L. viannia braziliensis*.

Administration

The drug is given in two divided doses or a single dose orally for 28 days. It is available as 50-mg and 10-mg capsules.

Toxicity

Most patients tolerate therapy with few or no side effects. Nausea, vomiting, and diarrhea have been reported, and there have been isolated reports of dose-related motion sickness. Some patients have reversible hepatotoxicity and nephrotoxicity. Urea and creatinine levels usually return to normal by the end of the second week of therapy. Gastrointestinal symptoms could be serious in severely ill patients who are malnourished and dehydrated.

Contraindications

The drug should not be used during pregnancy or lactation.

Comment

The mechanism is not well defined, but apparently the drug interferes with cell-signaling pathways and membrane synthesis. It also induces apoptotic cell death. It is active against the extracellular promastigote and the intracellular amastigote forms of *Leishmania* spp.

Niclosamide (Yomesan, Niclocide) (Bayer, Germany)

Preparation

Niclosamide is available in 500-mg chewable tablets (chlorosalicylamide), which is effective against infections with tapeworms (*Taenia* spp., etc). The drug is no longer available in the United States. Its use is limited, and praziquantel is now more commonly used.

Administration

The drug is given orally; however, to prevent stomach upset, it is best taken after a light meal (for example, breakfast). Niclosamide tablets should be thoroughly chewed or crushed and then swallowed with a small amount of water. If this medicine is being given to a young child, the tablets should be crushed to a fine powder and mixed with a small amount of water to form a paste.

Toxicity

Most patients tolerate therapy with few or no side effects. Nausea, vomiting, and abdominal pain occur

rarely, and there have been isolated reports of fever, dizziness, urticaria, irritability, headache, and edema of the arms.

Contraindications

Niclosamide should not be given to patients shown to be hypersensitive to the drug. Insufficient data are available for an assessment of the hazards of use of this drug during pregnancy. It is not recommended for pregnant women and children younger than 2 years.

Nifurtimox (Lampit) (Bayer, Germany) (CDC)

Preparation

Nifurtimox is a nitrofuran and is supplied as 100-mg tablets through the CDC Parasitic Disease Drug Service; it is used in the treatment of acute Chagas' disease.

Administration

The drug is given orally. Tolerance for the drug decreases with age, and children generally can tolerate higher doses over a longer period than can adults.

Toxicity

Gastrointestinal symptoms are common and may include anorexia, weight loss, abdominal pain, nausea, and vomiting. Neuropsychiatric symptoms, including convulsions, headache, drowsiness, nervousness, vertigo, paresthesia, disorientation, and memory loss, are less common but may necessitate termination of therapy.

Contraindications

The drug should not be used in pregnant women or in children younger than 1 year. Alcohol, antibiotics, and digitalis may potentiate the side effects. It should be used with caution in elderly patients with a history of neurologic or psychiatric diseases or allergic skin reactions.

Comment

Nifurtimox, an investigational drug, is a nitrofurfurylidene derivative. Its action is apparently related to its ability to cause the production of toxic reduced products of oxygen such as superoxide, hydrogen peroxide, and hydroxyl radicals. Accumulation of these products within the parasites leads to membrane damage, enzyme inactivation, and possible inhibition of protein synthesis. Nifurtimox is used to treat acute, subacute, or early chronic Chagas' disease. Although only two cases of indigenous Chagas' disease have been reported from the United States, infections may be seen in laboratory workers, immigrants, and U.S. citizens returning from Latin America, such as Peace Corps volunteers, military personnel, and missionaries. Drinking alcoholic beverages while taking this medicine may cause some unwanted effects. Other alcohol-containing preparations (for example, elixirs, cough syrups, and tonics) may also cause problems. Therefore, the patient should not drink alcoholic beverages or ingest other alcohol-containing preparations while taking nifurtimox.

Nitazoxanide (Alinia) (Romark)

Preparation

Nitazoxanide is a relatively new synthetic drug with good efficacy against a number of intestinal protozoa and helminths. Alinia tablets contain 500 mg of nitazoxanide, and Alinia for oral suspension, after reconstitution, contains 100 mg nitazoxanide per 5 ml. Nitazoxanide and its metabolite, tizoxanide, are active in vitro in inhibiting the growth of sporozoites and oocysts of *Cryptosporidium* spp. and trophozoites of *Giardia lamblia*. Nitazoxanide may also be effective in the treatment of intestinal microsporidiosis with *Enterocytozoon bieneusi* and *Encephalitozoon intestinalis*, as well as ocular infections with *Vittaforma corneae* (see dosage information under *Giardia lamblia*).

Administration

The drug should be taken orally with food. Diabetic patients and their caregivers should be aware that the oral suspension contains 1.48 g of sucrose per 5 ml. Alinia for oral suspension should be used for pediatric patients younger than 11 years.

Toxicity

The most common adverse events reported were abdominal pain, diarrhea, headache, and nausea. Other reported side effects included weakness, fever, pain, chills, flu syndrome, dizziness, somnolence, insomnia, rash, and pruritus.

Contraindications

The pharmacokinetics of nitazoxanide in patients with compromised renal or hepatic function has not been studied. Therefore, nitazoxanide must be administered with caution to patients with hepatic and/or biliary disease, renal disease, and combined renal and hepatic disease. Nitazoxanide is highly bound to plasma protein; competition for binding sites could occur with other drugs such as warfarin. Based on limited safety data, this drug should not be used during pregnancy or lactation.

Comment

The antiprotozoal activity of nitazoxanide is thought to be due to interference with the pyruvate ferredoxin oxidoreductase (PFOR) enzyme-dependent electron transfer reaction, which is essential to anaerobic energy metabolism. Studies have shown that the PFOR enzyme from *G. lamblia* directly reduces nitazoxanide by transfer of electrons in the absence of ferredoxin. The DNA-derived PFOR protein sequence of *Cryptosporidium* spp. appears to be similar to that of *G. lamblia*. Interference with the PFOR enzyme-dependent electron transfer reaction may not be the only pathway by which nitazoxanide exhibits antiprotozoal activity.

Nitazoxanide therapy has also been given for intestinal microsporidiosis at a dose of 1,000 mg twice a day for 60 consecutive days. Three posttreatment fecal examinations, including PCR, conducted over 2 months following the end of the treatment with nitazoxanide did not reveal any microsporidial spores, and the patient continued normal bowel movements. Nitazoxanide has also been reported to be effective in cell culture against *E. intestinalis* and *V. corneae*.

Paromomycin (Humatin) (Monarch)

Preparation

Humatin is supplied as 250-mg capsules and a syrup with 125 mg per 5 ml; a broad-spectrum antibiotic, it is used to treat various intestinal protozoal infections, particularly acute and chronic intestinal amebiasis. The drug is poorly absorbed after oral administration, with almost 100% of the drug being recoverable in the stool.

Administration

The drug is given orally with a full glass of water and should be taken with food.

Toxicity

Nausea, vomiting, abdominal cramps, and diarrhea have been reported, particularly in patients taking more than 3 g/day. Dizziness has also been reported.

Contraindications

The drug should not be used in patients with impaired renal function, intestinal obstruction, or ulcerative bowel lesions. Paromomycin and other aminoglycosides should be avoided in patients with myasthenia gravis. The drug should not be used during pregnancy or lactation.

Pentamidine Isethionate (Lomidine, Pentam 300, Nebupent) (Fujisawa)

Preparation

Pentamidine isethionate is supplied as 300-mg vials for i.v. or intramuscular administration. For inhalation, the vial contents are dissolved in sterile water and the inhalation solution is placed in a Respirgard II nebulizer. Saline solution must not be used to reconstitute the drug. Pentamidine isethionate is a toxic but highly effective drug used to treat the early stages of both types of African trypanosomiasis (*Trypanosoma brucei gambiense* and *T. brucei rhodesiense*). It does not cross the blood-brain barrier and is not effective in the treatment of the neurologic phases of the disease. It is also used to treat leishmaniasis and *Pneumocystis jiroveci* (now classified with the fungi) infection.

Administration

Vial contents should be dissolved in sterile water for intramuscular injection or in sterile water followed by sterile 5% glucose for i.v. administration. Pentamidine for inhalation should be administered for approximately 35 to 40 min at a rate of 5 to 7 liters/min.

Toxicity

Common side effects include hypotension, tachycardia, nausea, vomiting, and a metallic taste that occurs immediately after injection but subsides in 10 to 30 min. Hypoglycemia is occasionally seen from days 5 to 7 of therapy. Acute renal failure has occurred during the course of therapy; however, it is usually reversible in patients with normal renal function before therapy. Mild anemia, elevated liver transaminase levels, and leukopenia have been reported. Local pain and inflammation of the injection site are common, and sterile abscesses occasionally occur. Hypotension and profound shock have been reported after a single dose of the drug, and resuscitation equipment should always be readily available in the setting in which drug therapy is given. Occasional seizures and hallucinations have also been reported. Inhaled pentamidine can cause cough, bronchospasm, fatigue, bad taste in the mouth, shortness of breath, decreased appetite, dizziness, and rash. Pentamidine has also been associated with Stevens-Johnson syndrome.

Contraindications

Because pentamidine is generally used in life-threatening situations, there are no absolute contraindications to therapy. Careful observation of the patient is necessary because of the common side effects and severe drug reactions that may occur. Impaired renal function is frequent. Aerosol pentamidine should not be given with other drugs or given to patients with a history of anaphylactic reactions to inhaled or parenteral pentamidine. Some patients may develop a sudden, severe decrease in blood pressure after a dose of pentamidine. Pentamidine can decrease the white blood cell count, increasing the chance that the patient will contract certain infections. It can also lower the number of platelets, which are necessary for proper blood clotting. Both low and high blood sugar levels have also been reported. Unless the benefits outweigh the potential risks, this drug should not be used during pregnancy.

Comment

Although its mode of action remains unclear, its activity may be different for different parasites. Pentamidine (Lomidine) is used only for the early phase of the disease, since it does not cross the blood-brain barrier. It is active against *T. gambiense*, but it should not be used against *T. rhodesiense*, since primary resistance to it has been found in some areas. It should also not be used in areas where pentamidine has been widely employed for chemoprophylaxis, such as in the former Belgian Congo.

Permethrin (Nix [GlaxoSmithKline], Elimite [Allergan])

Preparation

Permethrin is supplied as a 1 or 5% cream. It is used to treat scabies (a skin infestation) and lice infestations of the head, body, and pubic area ("crabs"). Permethrin does not prevent these infestations.

Administration

The drug is available as a 1% (Nix) or 5% (Elimite) cream applied directly to hair or skin to treat lice or scabies, respectively.

If using the liquid, follow these steps: shampoo the hair using regular shampoo; thoroughly rinse and towel dry hair and scalp; allow hair to air dry for a few minutes; do not stand (or sit) in a shower or bathtub; lean over a sink to apply permethrin to the hair; shake the bottle of liquid well; thoroughly wet hair and scalp with the liquid. Cover the areas behind the ears and the back of the neck; keep permethrin on the hair for 10 minutes before rinsing it off thoroughly with water; dry hair with a clean towel; comb hair with a fine-tooth comb to remove nits (lice eggs and larvae); wash hands to remove the medication.

If using the cream, follow these steps: wash entire body or take a shower; thoroughly massage the cream into the skin over the entire body, from head to toes; leave the cream on for 8 to 14 h; wash the cream off after 8 to 14 h by taking another shower; if using product for head lice, wash combs and brushes with permethrin liquid and rinse them thoroughly with water to remove the drug.

Toxicity

Side effects are rare and include occasional burning, stinging, and numbness; increased pruritus; pain; edema; erythema; and rash.

Contraindications

The cream or lotion should not be applied near the eyes or mucous membranes. The drug may cause progressive motor neuron disease.

Polyhexamethylene Biguanide (Baquacil) (Zeneca)

Preparation

Baquacil is supplied as a 20% solution, which must be diluted with sterile saline to 0.02% for ophthalmic application. It is used to treat keratitis caused by *Acanthamoeba* spp.

Administration

Topical application is recommended at hourly intervals until clinical improvement is noted. Polyhexamethylene biguanide is available through Leiter's Park Avenue Pharmacy, San Jose, Calif. (800-292-6773; www.leiterrx.com).

Toxicity

No information is available.

Contraindications

No information is available.

Comment

While this therapy inactivates the trophozoites and cysts in *Acanthamoeba* keratitis in the majority of patients (approximately 90%), there have been notable failures, particularly when patients present late with deep stromal infection. Since no single drug can be assumed to be effective against all amebae, a potential complication is the risk of activation of dormant cysts which form in situ in *Acanthamoeba* and *Balamuthia* infections and which can lead to relapse following apparently effective treatment. This is particularly true for *Acanthamoeba* keratitis, a nonopportunistic infection of the cornea, which responds well to treatment with chlorhexidine gluconate and polyhexamethylene biguanide in combination with propamidine isethionate (Brolene), hexamidine (Desomodine), or neomycin. In a large series of *Acanthamoeba* keratitis patients with a positive microbiologic diagnosis at presentation, nearly 5% developed recurrent episodes of corneal and scleral inflammation with viable *Acanthamoeba* in the cornea despite prolonged treatment with biguanides and/or diamidines.

Praziquantel (Biltricide) (Bayer)

Preparation

Praziquantel is supplied as 150-, 500-, and 600-mg scored tablets; is a pyrazinoisoquinoline derivative; and is a synthetic heterocyclic broad-spectrum anthelmintic agent effective against all schistosome species infecting humans, as well as most other trematodes and adult cestodes. However, it is not useful for treatment of nematode infections and is not effective in treating *Fasciola hepatica*.

Administration

Praziquantel is preferably taken during a meal, particularly one with a high carbohydrate content. After oral administration, over 80% of the drug is absorbed.

Toxicity

In general, the drug is well tolerated; side effects are usually mild. Reactions include malaise, headaches, dizziness, abdominal discomfort, nausea, vomiting, fever, and rarely urticaria. When it is used in high doses, as in the treatment of central nervous system cysticercosis, reactions that have been attributed to the death of the larvae can occur. These reactions include headache, peripheral neuropathy, seizures, fever, hypotension, and shock. Anaphylaxis has also been reported. Transient hyperglycemia lasting several months after therapy has also been reported. Patients are generally hospitalized during the initiation of therapy for cysticercosis because of the potential for serious reactions such as seizures and other neurologic sequelae related to an inflammatory response.

Contraindications

The drug induces drowsiness and dizziness and so should be used with caution in patients whose vocation requires mental alertness. If the drug is given to patients with ocular cysticercosis, parasite destruction may cause irreparable eye damage. Although studies involving high doses of praziquantel in pregnant animals have failed to show any harm to the fetus, there are no controlled studies of humans. The use of praziquantel in pregnant women and in children younger than 4 years has not been well studied, and both the risks and the benefits must be considered. Praziquantel does appear in human milk, and mothers should not breast-feed during treatment and for 72 h following treatment. Concomitant administration of rifampin should be avoided.

Comment

Praziquantel causes ultrastructural changes in the helminth tegument, causing increased permeability to calcium ions. It works by causing severe spasms and paralysis of the worms' muscles. Some kinds of worms are then passed in the stool. However, the patient may not notice them since they are sometimes completely destroyed in the intestine.

Primaquine Phosphate (USP)

Preparation

The drug is supplied in tablets containing 26.3 mg of primaquine phosphate, equivalent to 15 mg of salt. It is an antimalarial agent especially effective against *Plasmodium vivax* exoerythrocytic forms, terminating relapsing vivax malaria; it is usually administered with chloroquine.

Administration

Tablets have a bitter taste and may be crushed and added to a sweet liquid or fruit to make them more palatable.

Toxicity

Patients may experience abdominal pain, cramps, or epigastric distress. High doses may result in methemoglobinemia and cyanosis. Granulocytopenia and granulocytosis, hypertension, and arrhythmia are rare. Hemolysis occurs in individuals with G6PDH deficiency.

Contraindications

The drug should not be used in patients with depressed blood counts resulting from other illnesses or concurrently with drugs, such as quinacrine, that have the potential to cause hemolysis or bone marrow depression. Because primaquine phosphate can cause hemolytic anemia, patients should be screened for G6PDH deficiency, which is most common in African, Asian, and Mediterranean people. Administration of other bone marrow depressants as hemolytic agents should be avoided. The drug should not be prescribed during pregnancy.

Propamidine Isethionate (Brolene) (Aventis, Canada)

Preparation

Brolene is provided as a 0.1% solution. It is a local anti-infective agent that is used to treat keratitis caused by *Acanthamoeba* spp., but it is not available in the United States.

Administration

Brolene is administered topically each hour for the first few days and every few hours thereafter for a minimum course of 1 month.

Toxicity

The drug may cause keratopathy; this condition is reversible but may be confused with persistent or recurrent infection.

Contraindications

No information is available.

Comment

Treatment of *Acanthamoeba* keratitis is difficult due to the organism's ability to encyst. Effective medications include topical polyhexamethylene biguanide, propamidine isethionate (Brolene), chlorhexidine digluconate 0.02%, polymixin B, neomycin, and clortrimazole 1%. In one series of 10 patients, the combination of Brolene and polyhexamethylene biguanide successfully cured all cases of *Acanthamoeba* keratitis. Cautious introduction of topical steroids along with antiamebic therapy helped resolve the inflammation and provided symptomatic relief.

Pyrantel Pamoate (Antiminth) (Pfizer)

Preparation

Antiminth is supplied in 60-ml bottles containing 50 mg of pyrantel base per ml or as tablets containing 125 mg of base and is used to treat a number of different nematode infections. However, it is associated with more side effects and is being replaced with the benzimidazoles, which are more effective.

Administration

Pyrantel pamoate may be taken with food, milk, or juice.

Toxicity

Reactions are not frequently encountered, but when they do occur, they involve primarily the gastrointestinal system. They include anorexia, nausea, vomiting, abdominal pain or cramps, diarrhea and tenesmus, and transient elevations of the transaminase enzyme levels.

Contraindications

Studies of pregnant animals have failed to show any harm to the fetus, but there is no experience with pregnant humans. There are few published data on the use of the drug in children younger than 2 years. The risks and benefits of treatment in these patients should be weighed. The drug should be used with caution in patients with liver dysfunction. Its anthelmintic effect may be antagonized by piperazine.

Comment

The drug is a neuromuscular blocking agent, resulting in spastic paralysis of the adult worms. Excellent results are obtained when treating patients with ascariasis, hookworm, and pinworm infections. While there is some activity against *Trichostrongylus*, the drug is not active against *Trichuris trichiura*.

Pyrethrin with Piperonyl Butoxide (Rid) (Pfizer, Others)

Preparation

Rid is an insecticide mixture of permethrin and piperonyl butoxide, which is used as a synergist and also to treat human louse infestations.

Administration

Rid is applied topically as a shampoo to the affected area after wetting. After application, it should be left for 10 min and then rinsed thoroughly. Hair should be combed with a fine-tooth comb to remove dead lice and nits.

Toxicity

Rid may cause allergic reactions in individuals allergic to ragweed.

Contraindications

The drug should not be used near eyes or mucous membranes.

Pyrimethamine USP (Daraprim) (GlaxoSmithKline)

Preparation

Daraprim is supplied as 25-mg tablets; it is a potent folic acid antagonist used as an antimalarial agent and in treating toxoplasmosis and several other parasitic infections. Dosages for treating toxoplasmosis approach the toxic level. Daraprim should be kept away from children and infants since they are extremely susceptible to adverse effects from an overdose and deaths have been reported.

Administration

The drug may be pulverized and mixed with sweet liquid or fruit juice. Concurrent administration of folinic acid is strongly recommended for all patients. Daraprim is indicated for the treatment of toxoplasmosis when used conjointly with a sulfonamide, providing synergy with this combination approach. Although Daraprim is used to treat acute malaria, it should not be used alone for this purpose. It is also used for chemoprophylaxis of malaria; however, there is widespread resistance throughout the world.

Toxicity

Gastrointestinal complaints (anorexia, nausea, vomiting, diarrhea, and abdominal pain and tenderness) are not uncommon. Occasional elevation of the glutamic-oxaloacetic transaminase level in serum has been reported, as have skin rashes, headache, dizziness, drowsiness, and insomnia. For patients with toxoplasmosis being treated with high doses, periodic complete blood counts, including platelet counts, are recommended. Folic acid supplementation may be necessary if the counts are depressed. Hypersensitivity reactions, occasionally severe (Stevens-Johnson syndrome, toxic epidermal necrolysis, erythema multiforme, and anaphylaxis), can occur, particularly when the drug is administered concomitantly with a sulfonamide.

Contraindications

The drug should not be used in pregnancy, and determination of human chorionic gonadotropin levels is recommended before initiation of treatment for toxoplasmosis. Caution is recommended for patients with seizure disorders and impaired hepatic or renal function. Daraprim is also contraindicated for patients with documented megaloblastic anemia due to folate deficiency.

Comment

Pyrimethamine acts against the asexual erythrocytic stage of the *Plasmodium* spp. by inhibiting the enzyme dihydrofolate reductase. The drug does not destroy gametocytes but instead arrests sporogony in the mosquito. Although the drug is active against *P. falciparum*, resistance occurs fairly rapidly and limits its use as a single agent. Therefore, the drug is often combined with a sulfonamide or sulfone, which provides sequential, synergistic inhibition of the folate biosynthesis pathway. This inhibition of folic acid biosynthesis prevents DNA replication in the parasite. Pyrimethamine is also used in combination with sulfadiazine for treating toxoplasmosis.

Quinidine Gluconate (Eli Lilly)

Preparation

Quinidine gluconate is supplied as a solution containing 80 mg/ml for injection and is used to treat severe malaria.

Administration

The drug is given i.v. in normal saline by slow infusion; 800 mg is infused in 40 ml of 5% glucose at a rate of 1 ml/min.

Toxicity

Quinidine causes gastrointestinal irritation, with nausea, vomiting, and diarrhea. Hypersensitivity may occur. Other side effects that have been reported include blurred vision, headache, belly pain, and ringing in the ears and temporary loss of hearing. Prolongation of the QT interval as indicated by an electrocardiogram, ventricular arrhythmia, hypotension, and hypoglycemia can result from the use of this drug at treatment doses.

Contraindications

The drug may induce hypotension, granulomatous hepatitis, and a lupus-like syndrome. Use should be avoided for patients with myasthenia gravis, acute infections with fever, previous hypersensitivity to quinidine, and cardiac abnormalities.

Quinine Sulfate or Quinine Dihydrochloride (Many Manufacturers)

Preparation

Quinine sulfate is supplied as 3-grain (186-mg) or 6-grain (372-mg) capsules or tablets. Quinine dihydrochloride comes in 2-ml ampoules containing 300 mg/ml (not available in the United States). It is used to treat malaria and is important in areas where i.v. quinine is not available. It is the dextrorotary optical isomer of quinine.

Administration

The oral medication is extremely bitter and may be better tolerated if mixed with sweet liquid, fruit, or cereal. For i.v. administration, medication should be diluted in normal saline at 0.5 mg/ml and administered at 1 ml/min every 8 h.

Toxicity

Mild symptoms are common and generally consist of tinnitus, headache, nausea, and disturbed vision. More severe reactions are related to inability to excrete the drug normally and to high doses. These reactions include decreased hearing, blurred vision, photophobia, diplopia, night blindness, scotomata (blind spots), mydriasis (pupil dilation), vomiting, abdominal pain, diarrhea, cutaneous flushing, sweating, rashes, facial angioedema, confusion, apprehension, excitement, delirium, and syncope.

Contraindications

Rare reactions include renal failure, acute hemolysis, hypoprothrombinemia, purpura, agranulocytosis, asthma, and ventricular tachycardia. After one dose, an idiosyncratic hypersensitivity reaction consisting of flushing, pruritus, skin rash, fever, gastric distress, dyspnea, tinnitus, and visual impairment may occur. Massive hemolysis and hemoglobinuria occasionally occur during the treatment of falciparum malaria. Quinine may aggravate the symptoms of myasthenia gravis.

Spiramycin (Rovamycine) (Aventis)

Preparation

Rovamycine is supplied as a tablet or in solution and is a macrolide antibiotic similar to leucomycin. This drug

has been tried in the treatment of cryptosporidiosis; it is available only from the manufacturer.

Administration

Spiramycin is given orally on an empty stomach or by i.v. infusion.

Toxicity

Rovamycine can cause gastrointestinal disturbances and occasional allergic reactions. More rare side effects have included bloody stools, chest pain, fever, heartburn, irregular heartbeat, nausea, recurrent fainting, stomach pain and tenderness, vomiting, and yellow eyes or skin.

Contraindications

The drug should not be given to patients with impaired liver function.

Stibogluconate Sodium (Pentostam, Solustibosan) (GlaxoSmithKline) (CDC)

Preparation

Stibogluconate sodium is available through the CDC Parasitic Disease Drug Service as a 33% solution containing 100 mg of pentavalent antimony per ml; it is used to treat leishmaniasis.

Administration

The drug is given intramuscularly or by careful, slow i.v. infusion.

Toxicity

Side effects are common and include nausea, vomiting, severe coughing, and itching. Pneumonitis, arthralgias, myalgias, bradycardia, electrocardiogram changes (T-wave depressions, increased O-T interval, and fusion of the S-T segment and T wave), abnormal liver function, headache, and mild rash also occur. Less common problems are hepatitis, hemolytic anemia, thrombocytopenia, and anaphylaxis associated with urticaria, laryngeal edema, and visceral collagens.

Contraindications

The drug should not be used in patients with myocarditis, hepatitis, liver disease, or concurrent bacterial and viral infections. Treatment should be stopped if there is progressive proteinuria, severe arthralgias, or rash. More severe reactions may include leukopenia, agranulocytosis, and electrocardiographic changes.

Comment

Pentostam (sodium stibogluconate; sodium antimony gluconate) is a pentavalent antimony compound that is used to treat all forms of leishmaniasis (cutaneous, mucocutaneous, and visceral). This drug is a well-established antileishmanial agent that has been in use for some time. It is thought to inhibit glycolysis enzymes within the organisms.

Suramin Sodium (Germanin, Antrypol, Naphuride Sodium, Bayer 205, Moranyl, Fourneau 309, Belganyl) (Bayer, Germany) (CDC)

Preparation

Suramin sodium is available through the CDC Parasitic Disease Drug Service as ampoules containing 1.0 g of powder and is a complex derivative of urea used to treat African trypanosomiasis.

Administration

A 10% solution must be prepared and used within 30 min of preparation as a slow i.v. infusion.

Toxicity

Immediate nausea, vomiting, loss of consciousness, and seizure may occur if the full i.v. dose is given to sensitive individuals. A test dose is recommended. Proteinuria is commonly seen during therapy, but it is not a parameter used to monitor therapy unless other renal function studies are abnormal. Fever is also common but is usually low grade. Rashes, localized edema, and pruritus may occur. Blepharitis, conjunctivitis, photophobia, and excessive lacrimation may occur during treatment. Kidney failure, blood dyscrasias (including pancytopenia), shock, and optic atrophy have also been reported.

Contraindications

The drug should be administered under close medical supervision and should not be used in pregnant women or in patients with hepatic or renal insufficiency. It should be used with caution in children younger than 6 years. It should not be used alone to treat African trypanosomiasis of the central nervous system.

Comment

Suramin was introduced in 1920 for the treatment of African trypanosomiasis (sleeping sickness). It is considered the drug of choice for treatment of early African trypanosomiasis with *Trypanosoma brucei rhodesiense* when the central nervous system is not involved. Pentamidine is thought to be more effective than suramin for treating early African trypanosomiasis caused by *T. brucei gambiense*. The mechanism of action of suramin is thought to be through inhibition of enzymes associated with protein synthesis and DNA metabolism.

Thiabendazole (Mintezol) (Merck)

Preparation

Mintezol is supplied as a flavored suspension containing 125 or 250 mg/5 ml, as pediatric drops containing 100 mg/ml, and as 500-mg chewable tablets. Thiabendazole is a benzimidazole that is used to treat a number of nematode infections, including strongyloidiasis, cutaneous and visceral larva migrans, trichinosis, and other intestinal roundworms.

Administration

The drug is given orally, preferably after a meal; tablets should be chewed before being swallowed.

Toxicity

Common side effects are anorexia, vomiting, and dizziness. Less frequently, diarrhea, epigastric pain, pruritus, drowsiness, lethargy, and headache occur. Side effects include Stevens-Johnson syndrome, tinnitus, visual disturbance, hypotension, perianal rash, transient rises in transaminase enzyme levels, cholestasis, parenchymal liver damage, hyperglycemia, and hematuria. Treatment may cause migration of *Ascaris* organisms to the esophagus, mouth, and nose.

Contraindications

Patients with known hypersensitivity or with renal or hepatic impairment should not use thiabendazole. The drug causes dizziness and drowsiness and should not be used by patients whose vocation requires mental alertness. It is also contraindicated as prophylaxis for pinworm infections; susceptible worm infections should be confirmed by laboratory testing prior to drug use. Due to a lack of safety data, it should be used during pregnancy and lactation only if the potential benefit justifies the potential risk to the fetus.

Comment

Thiabendazole blocks glucose uptake, depletes glucose stores, and interferes with microtubule aggregates. It also inhibits the fumarate-reductase system of helminths. However, administration of thiabendazole has been associated with side effects, and it has been replaced by other anthelmintic agents.

Tinidazole (Tindamax) (Mission Pharmaceuticals)

Preparation

Tinidazole is supplied as 150-, 200-, 300-, and 500-mg and 1-g tablets or as an oral suspension of 200 mg/ml. This drug is a synthetic nitroimidazole used to treat amebiasis and giardiasis; it was formerly at Presutti Pharmaceuticals.

Tindamax oral tablets are indicated for the treatment of trichomoniasis caused by *Trichomonas vaginalis* in both female and male patients. The organism should be identified by appropriate diagnostic procedures. Because trichomoniasis is a sexually transmitted disease with potentially serious sequelae, partners of infected patients should be treated simultaneously in order to prevent reinfection. Tindamax oral tablets are indicated for the treatment of giardiasis caused by *Giardia lamblia* in both adults and pediatric patients older than 3 years. Tindamax oral tablets are indicated for the treatment of intestinal amebiasis and amebic liver abscess caused by *Entamoeba histolytica* in both adults and pediatric patients older than 3 years. It is not indicated in the treatment of asymptomatic cyst passage.

Administration

Tinidazole is given orally with or after meals. After oral administration, it is rapidly and completely absorbed. Alcoholic beverages should be avoided while taking Tindamax and for 3 days afterward.

Toxicity

Tinidazole can cause nausea, vomiting, metallic taste, and rash. Convulsive seizures and peripheral neuropathy, the latter characterized mainly by numbness or paresthesia of an extremity, have been reported to occur in patients treated with nitroimidazole drugs including tinidazole and metronidazole. The appearance of abnormal neurologic signs demands the prompt discontinuation of Tindamax therapy. Tinidazole should be administered with caution to patients with central nervous system diseases or those with evidence of or history of blood dyscrasia.

Contraindications

As with other nitroimidazole derivatives, tinidazole may enhance the effect of warfarin and other coumarin anticoagulants, resulting in a prolongation of the prothrombin time. Alcoholic beverages and preparations containing ethanol or propylene glycol should be avoided during tinidazole therapy and for 3 days afterward because abdominal cramps, nausea, vomiting, headaches, and flushing may occur. Psychotic reactions have occurred in alcoholic patients using metronidazole and disulfiram concurrently. Because animal reproduction studies are not always predictive of the human response and because there is some evidence of mutagenic potential, the use of tinidazole during pregnancy requires that the potential benefits of the drug be weighed against the possible risks to both the mother and the fetus. Tinidazole is excreted in breast milk in concentrations similar to those seen in serum. It can be detected in breast milk for up to 72 h

following administration. Interruption of breast-feeding is recommended during tinidazole therapy and for 3 days following the last dose.

Comment

The nitro group of tinidazole is reduced by cell extracts of *Trichomonas*. The free nitro radical generated as a result of this reduction may be responsible for the antiprotozoal activity. The mechanism by which tinidazole exhibits activity against *Giardia* and *Entamoeba* species is not known. Approximately 38% of *T. vaginalis* isolates with reduced susceptibility to metronidazole also show reduced susceptibility to tinidazole in vitro. The clinical significance of such an effect is not known.

Tinidazole, like metronidazole, may interfere with certain types of determinations of serum chemistry values, such as aspartate aminotransferase, alanine aminotransferase, lactate dehydrogenase, triglyceride, and hexokinase glucose levels.

Triclabendazole (Egaten) (Novartis)

Preparation

Triclabendazole is used in veterinary practice and is also being used to treat human infections with liver and lung flukes. The drug is not available in the United States.

Administration

Egaten is given orally after a meal as a single or split dose of 10 mg/kg. A dose of 20 mg/kg is safe and effective for patients with acute fascioliasis when a single dose has failed to cure them.

Toxicity

Triclabendazole can cause abdominal pain; however, most studies reported no side effects. The tolerance of triclabendazole is considered excellent. Side effects include brief episodes of upper abdominal pain and slight fever and some mild and limited disturbances in liver function. These effects may be due to the paralysis and/or death of the flukes, resulting in the release of antigens or toxic products and the partial blockage of the bile ducts. Similar findings have been reported in various preliminary studies: triclabendazole has been reported to decrease parasite motility, but the exact mode of action of this drug is unknown. Some patients may require a second course of therapy.

Contraindications

Use of this drug may cause secondary cholangitis due to worm destruction. Triclabendazole has been reported to be nonmutagenic and nonteratogenic.

References

1. Abramowicz, M. (ed.). 2004. Drugs for parasitic infections. *Med. Lett. Drugs Ther.* **46:**1–12.

2. Alifrangis, M., S. Enosse, R. Pearce, C. Drakeley, C. Roper, I. F. Khalil, W. M. Nkya, A. M. Rønn, T. G. Theander, and I. C. Bygbjerg. 2005. A simple, high-throughput method to detect *Plasmodium falciparum* single nucleotide polymorphisms in the dihydrofolate reductase, dihydropteroate synthase, and *P. falciparum* chloroquine resistance transporter genes using polymerase chain reaction- and enzyme-linked immunosorbent assay-based technology. *Am. J. Trop. Med. Hyg.* **72:**155–162.

3. Alonso, D., J. Muñoz, J. Gascón, M. E. Valls, and M. Corachan. 2006. Short Report: failure of standard treatment with praziquantel in two returned travelers with *Schistosoma haematobium* infection. *Am. J. Trop. Med. Hyg.* **74:**342–344.

4. Anonymous. 2006. *Physicians' Desk Reference*, 60th ed. Medical Economics, Montvale, N.J.

5. Calvopina, M., R. H. Guderian, W. Paredes, M. Chico, and P. J. Cooper. 1998. Treatment of human pulmonary paragonimiasis with triclabendazole: clinical tolerance and drug efficacy. *Trans. R. Soc. Trop. Med. Hyg.* **92:**566–569.

6. Crowell, A. L., C. E. Stephens, A. Kumar, D. W. Boykin, and W. E. Secor. 2004. Evaluation of dicationic compounds for activity against *Trichomonas vaginalis*. *Antimicrob. Agents Chemother.* **48:**3602–3605.

7. Dauchy, F. A., P. Vincendeau, and F. Lifermann. 2006. Eight cases of fascioliasis: clinical and microbiological features. *Med. Mal. Infect.* **36:**42–46.

8. Del Brutto, O. H. 2005. Neurocysticercosis. *Semin. Neurol.* **25:**243–251.

9. el-Karaksy, H., B. Hassanein, S. Okasha, B. Behairy, and I. Gadallah. 1999. Human fascioliasis in Egyptian children: successful treatment with triclabendazole. *J. Trop. Pediatr.* **45:**135–138.

10. Gilbert D. N., R. C. Moellering, and M. A. Sande. 2003. *The Sanford Guide to Antimicrobial Therapy*, 33rd ed. Antimicrobial Therapy Inc., Dallas, Tex.

11. Kuhn, S., M. J. Gill, and K. C. Kain. 2005. Emergence of atovaquone-proguanil resistance during treatment of *Plasmodium falciparum* malaria acquired by a non-immune North American traveler to West Africa. *Am. J. Trop. Med. Hyg.* **72:**407–409.

12. Leang, B., L. Lynen, W. Schrooten, and J. Hines. 2005. Comparison of albendazole regimen for prophylaxis of *Strongyloides* hyperinfection in nephrotic syndrome patients on long-term steroids in Cambodia. *Trop. Doct.* **35:**212–213.

13. Moon, T. D., and R. A. Oberhelman. 2005. Antiparasitic therapy in children. *Pediatr. Clin. North Am.* **52:**917–948.

14. Moore, D. A., R. W. Girdwood, and P. L. Chiodini. 2002. Treatment of anisakiasis with albendazole. *Lancet* **360:**54.

15. Murray, H. W., J. D. Berman, C. R. Davies, and N. G. Saravia. 2005. Advances in leishmaniasis. *Lancet.* **366:**1561–1577.

16. Nduati, E. W., and E. M. Kamau. 2006. Multiple synergistic interactions between atovaquone and antifolates against *Plasmodium falciparum* in vitro: a rational basis for combination therapy. *Acta Trop.* 31 Jan. [Epub ahead of print.]

17. Nguyen-Dinh, P., and W. E. Secor. 2003. Susceptibility test methods: parasites, p. 2108–2113. *In* P. R. Murray, E. J. Baron, J. H. Jorgenson, M. A. Pfaller, and R. H. Yolken (ed.), *Manual of Clinical Microbiology*, 8th ed., ASM Press, Washington, D.C.

18. Noedl, H., C. Wongsrichanalai, and W. H. Wersndorfer. 2003. Malaria drug-sensitivity testing: new assays, new perspectives. *Trends Parasitol.* **19:**176–181.

19. Shanks, G. D., and M. D. Edstein. 2005. Modern malaria chemoprophylaxis. *Drugs* **65:**2091–2110.

20. Secor, W. E., and P. Nguyen-Dinh. 2003. Mechanisms of resistance to antiparasitic agents, p. 2098–2107. *In* P. R. Murray, E. J. Baron, J. H. Jorgenson, M. A. Pfaller, and R. H. Yolken (ed.), *Manual of Clinical Microbiology*, 8th ed. ASM Press, Washington, D.C.

21. Singh, N., and A. Dube. 2004. Fluorescent *Leishmania*: application to anti-leishmanial drug testing. *Am. J. Trop. Med. Hyg.* **71:**400–402.

22. Siripanth, C., B. Punpoowong, P. Amarapal, N. Thima, B. Eampokalap, and J. Kaewkungwal. 2004. Comparison of *Cryptosporidium parvum* development in various cell lines for screening in vitro drug testing. *Southeast Asian J. Trop. Med. Public Health* **35:**540–546.

23. Sweetman, S. (ed.). 2005. *Martindale the Complete Drug Reference*. The Pharmaceutical Press, London, United Kingdom.

24. Tripathi, R. P., R. C. Mishra, N. Dwivedi, N. Tewari, and S. S. Verma. 2005. Current status of malaria control. *Curr. Med. Chem.* **12:**2643–2659.

25. Vessière, A., A. Berry, R. Fabre, F. Benoit-Vical, and J. F. Magnaval. 2004. Detection by real-time PCR of the Pfcrt T76 mutation, a molecular marker of chloroquine-resistant *Plasmodium falciparum* strains. *Parasitol. Res.* **93:**5–7.

26. Zardi, E. M., A. Picardi, and A. Afeltra. 2005. Treatment of cryptosporidiosis in immunocompromised hosts. *Chemotherapy* **51:**193–196.

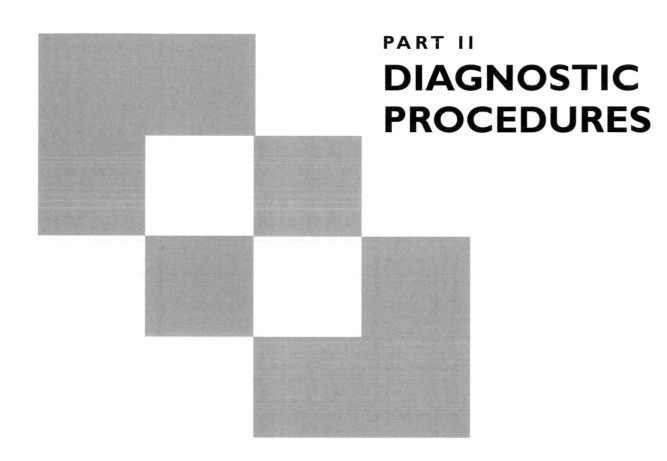

PART II

DIAGNOSTIC PROCEDURES

26

Collection, Preservation, and Shipment of Fecal Specimens

This chapter discusses various collection methods that are available for specimens suspected of containing parasites or parasitic elements. When a laboratory selects its collection methods, the decision should be based on a thorough understanding of the value and limitations of each. One of the most important aspects of specimen collection is that the final laboratory results based on parasite recovery and identification will depend on the initial fixation of the organisms (5, 9, 17, 31, 33). Unless the appropriate specimens are properly collected and processed, these infections may not be detected (7, 17, 31). Considering the current era of cost containment and review of clinical relevance of laboratory information generated, specimen rejection criteria have become more important within the context of all diagnostic microbiology procedures. Diagnostic laboratory results based on improperly collected specimens may require excessive expenditures of time and supplies and may also mislead the physician. As a part of any overall total quality management or continuous quality improvement program for the laboratory, the generation of test results must begin with stringent criteria for specimen acceptance or rejection.

Clinically relevant diagnostic parasitology testing also depends on receiving appropriate test orders from the physician. Depending on the patient's clinical condition and travel history, very specific diagnostic tests may be recommended. It is extremely important that physician clients are aware of the test order options available within the laboratory test menu and the pros and cons of each test when considered within the context of the patient's history and symptoms. Without the proper test orders, diagnostic test results may be misleading or actually incorrect. Appropriate and complete communication regarding test orders between the laboratory and physicians is mandatory for high-quality patient care.

Safety

All fresh specimens should be handled carefully, since each specimen represents a potential source of infectious material (bacteria, viruses, fungi, and parasites). Safety precautions should include awareness of the following: proper labeling of fixatives; specific areas designated for specimen handling (biological safety cabinets may be necessary under certain circumstances); proper containers for centrifugation; acceptable discard policies; appropriate policies for no eating, drinking, or smoking, etc., within the working areas; and, if applicable, correct techniques for organism culture and/or animal inoculation.

Since diagnostic parasitology work is most often performed within the microbiology division of a clinical laboratory, all general guidelines for safety would also apply. Any special precautions which apply to a particular technique are discussed in the following chapters. In general, standard precautions as outlined by the Occupational Safety and Health Act must be followed when applicable, particularly when one is handling blood and other body fluids (6).

Fresh-Specimen Collection

Procedures for the recovery of intestinal parasites should always be performed before barium is used for radiological examination. Stool specimens containing barium are unacceptable for examination, and intestinal protozoa may be undetectable for 5 to 10 days after barium is given to the patient. There are also certain substances and medications that interfere with the detection of intestinal protozoa: mineral oil, bismuth, antibiotics, antimalarial agents, and nonabsorbable antidiarrheal preparations. After administration of any of these compounds, parasitic organisms may not be recovered for a week to several weeks. The two most commonly used substances are barium and antibiotics, such as tetracycline, which modify the gastrointestinal tract flora. Specimen collection should be delayed for 5 to 10 days or at least 2 weeks after barium or antibiotics, respectively, are administered (5, 9, 23, 24, 26). The use of antibacterial therapy that affects the normal gastrointestinal tract flora will diminish the numbers of protozoa, since they feed on intestinal bacteria.

Collection of the Specimen

Fecal specimens should be collected in clean, wide-mouth containers; often a 0.5-pt (ca. 0.24-liter) waxed cardboard or plastic container with a tight-fitting lid is selected for this purpose. The fit of the lid is particularly important, both from the standpoint of accidental spillage and in order to maintain moisture within the specimen. The specimens should not be contaminated with water or urine, because water may contain free-living organisms that can be mistaken for human parasites and urine may destroy motile organisms. For safety reasons, stool specimen containers should be placed in plastic bags when transported to the laboratory for testing. Fresh specimens can also be submitted in collection vials (Figure 26.1). All fresh specimens should be carefully handled since they are potential sources of infectious organisms, including

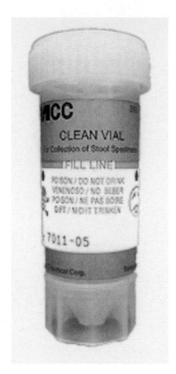

Figure 26.1 Stool collection vial; "clean vial" contains no fixatives.

bacteria, viruses, and parasites. Every specimen should be identified with the following minimal information: patient's name and identification number, physician's name, and the date and time the specimen was collected (if the laboratory is computerized, the date and time may reflect arrival in the laboratory, not the actual collection time). The specimen must also be accompanied by a request form indicating which laboratory procedures are to be performed. It would also be very helpful to have information concerning the presumptive diagnosis or relevant travel history; however, this information is rarely available, and under certain circumstances, the physician will have to be contacted for additional patient history [Example: Fever of unknown origin (FUO)—possible malaria].

Number of Specimens To Be Collected (Standard Recommendation)

It is recommended that a normal examination for stool parasites before therapy include three specimens, consisting of two specimens collected from normal movements and one collected after the use of a cathartic such as magnesium sulfate or Fleet's Phospho-Soda. A cathartic with an oil base should not be used, and a stool softener (taken either orally or as a suppository) is usually inadequate for obtaining a purged specimen. The purpose of the laxative is to stimulate some "flushing" action within the gastrointestinal tract, possibly allowing one to obtain more organisms for recovery and identification. Obviously, if the patient already has diarrhea or dysentery, the use of any laxatives would be contraindicated.

When a patient is suspected of having intestinal amebiasis, six specimens may be recommended. The examination of six specimens ensures detection of approximately 90% of amebic infections (38) (Figure 26.2). However, because of cost containment measures, the examination of six specimens is rarely requested.

Three specimens are also recommended for post-therapy examinations, and they should be collected as outlined above. However, a patient who has received treatment for a protozoan infection should be checked 3 to 4 weeks after therapy, and those treated for *Taenia* infections should be checked 5 to 6 weeks after therapy. In some cases, the physician will assume a cure for tapeworm infection unless proglottids reappear in the stool; therefore, no posttherapy specimens are submitted for examination.

Number of Specimens To Be Collected (Pros and Cons of Various Options)

During the past few years, a number of issues have surfaced regarding the collection, processing, and testing of stool specimens for diagnostic parasitology. Many of the new suggestions and options have arisen as a result of continued cost containment measures, limited reimbursement, and the elimination of mercury-based compounds for stool preservatives. The number of nonmercury

Figure 26.2 Increased detection of *Entamoeba histolytica* by using various diagnostic techniques and serial stool specimens. (Adapted by E. K. Markell, M. Voge, and D. T. John, *Medical Parasitology*, 7th ed., The W. B. Saunders Co., Philadelphia, Pa., 1992, from references 16, 22, and 29.)

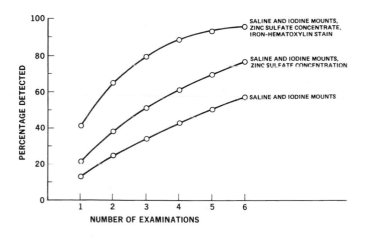

preservative choices, collection systems, concentration devices, and immunoassays has increased dramatically. Many laboratories continue to review the options, and some may be having difficulties in selecting the proper approach (2–4, 11–14, 18, 25, 27, 29, 32, 36, 39, 40, 46, 50).

It is important to realize that there are many acceptable options and that many laboratories will select different approaches. These differences should not be categorized as "right or wrong" or "acceptable or unacceptable"—they are merely different! To assume that there is only one correct approach for the examination of stool specimens is neither appropriate nor realistic. There are many parameters to consider before selecting the approach for your own laboratory. In no particular order, some of the considerations include client base, physician ordering patterns, number of specimens received per month, cost, presence or absence of appropriate equipment, current and possible methodologies (including the new immunoassays such as enzyme immunoassay [EIA], fluorescent-antibody assay [FA], and cartridge devices), availability of expert microscopists, collection options, selection of preservative-stain combinations, reimbursement issues, client education, area of the world where laboratory is located, and emphasis on the most common infections (helminth or protozoa or both).

When considering available options and laboratory test menus, it is important to make sure that the pros and cons of the approaches selected are thoroughly understood and that diagnostic tests, potential results, and reporting formats are carefully explained to all clients. As an example, if the results of a stool examination are based on a concentration sediment examination only, this information must be conveyed to the physician. Many of the intestinal protozoa are missed by this diagnostic test approach, and it is important for the physician to recognize the limitations of such testing. Most physicians receive very little, if any, exposure to medical parasitology in medical school, and many newer physicians trained as generalists or family practitioners also have limited parasitology training or experience.

In most cases, it is probably realistic to assume that patients are symptomatic if they are submitting stool specimens for diagnostic parasitology testing. In an excellent article by Hiatt et al., the premise tested was that a single stool sample from a symptomatic patient would be sufficient to diagnose infections with intestinal protozoa (19). However, with additional stool examinations for symptomatic patients, the yield of intestinal protozoa increased dramatically (*Entamoeba histolytica*, 22.7% increase; *Giardia lamblia*, 11.3% increase; and *Dientamoeba fragilis*, 31.1% increase). This publication again demonstrates the problems with performing only a single stool examination (using the ova and parasite examination). If the patient becomes asymptomatic after the first stool examination, it may be acceptable to discontinue the series of stool examinations (this should be a clinician decision).

Available options are compared to the "gold standard," which includes a series of three stool examinations (direct, concentrate, and permanent stained smear for a fresh specimen; concentrate and permanent stained smear for a preserved specimen). The single-specimen pros and cons are discussed in the previous paragraph (symptomatic versus asymptomatic patients). A suggestion has been made to pool three specimens and to perform a single concentration and a single permanent stained smear. Depending on the number of positives and negatives, this may be a viable option, but some organisms may be missed. Another testing option for three pooled specimens might be to perform a single concentration and three separate permanent stained smears. This approach would probably increase the yield of intestinal protozoa over the previous option. Another suggestion involves placing a sample of each of three stools into a single vial of preservative. This collection approach would require only a single vial, but it is very likely that the vial would be overfilled and that mixing and the ratio of fixative to stool would be inaccurate.

One could also test patient specimens by using immunoassay reagents or kits for *Giardia lamblia*, the *Entamoeba histolytica*/*E. dispar* group, *Entamoeba histolytica*, or *Cryptosporidium* spp. However, patient history information is rarely received with the clinical specimen; fecal testing based on patient risk group or situation is impossible without sufficient information. Testing of patient specimens by immunoassay procedures should be performed on request; however, client education is critical for successful implementation of this approach. Collection and testing options and the pros and cons of each are listed in Tables 26.1 and 26.2.

The Clinical Laboratory Standards Institute (formerly the National Committee for Clinical Laboratory Standards) document *Procedures for the Recovery and Identification of Parasites from the Intestinal Tract* (Approved Guideline M28-2A) was updated in 2005. Various stool collection, processing, and testing options are also included in that publication (5).

In summary, laboratories performing diagnostic parasitology testing must decide on the appropriate test methods that are relevant for their own operations based on a number of variables mentioned above. It is unrealistic to assume or state that one approach is applicable for every laboratory; however, it is important to thoroughly understand the options within your test menu and to convey this information to your clients once your approach has

been selected for implementation. Prior discussion with clients, written educational memos, meetings, and examples of revised report formats are highly recommended prior to implementation.

Collection Times

A series of three specimens as indicated above should be submitted on separate days; if possible, the specimens should be submitted every other day; otherwise, the series of three specimens should be submitted within no more than 10 days. If a series of six specimens is requested, the specimens should also be collected on separate days or within no more than 14 days. Many organisms, particularly the intestinal protozoa, do not appear in the stool in consistent numbers on a daily basis, and the series of three specimens is considered a minimum for an adequate examination (28). It is inappropriate for multiple specimens to be submitted from the same patient on the same day. One possible exception would be stool collections from a patient who has severe, watery diarrhea such that any organisms present might be missed because of the tremendous dilution factor related to fluid loss. Even under these circumstances, acceptance of more than one specimen per patient per day should not be routine but should be done only after consultation with the physician. It is also not recommended for the three specimens to be submitted one each day for three consecutive days; however, use of this collection time frame would not be sufficient cause to reject the specimens. Adequate spacing between specimens helps to provide parasite recovery within the recommended time frames.

Although the recommended number of stool specimens is three, laboratories have been more willing to accept two specimens, primarily because of cost savings and the assumption that if the patient is symptomatic, confirmation of any organisms present is just as likely to be possible from two specimens as from three specimens. However, it is important that clients understand the pros and cons of two compared with three stools. Both collection approaches are being used by diagnostic laboratories.

Specimen Type, Specimen Stability, and Need for Preservation

Fresh specimens are mandatory for the recovery of motile trophozoites (amebae, flagellates, or ciliates). The protozoan trophozoite stage is normally found in cases of diarrhea; the gastrointestinal tract contents are moving through the system too rapidly for cyst formation to occur. Once the stool specimen is passed from the body,

Table 26.1 Fecal specimens for parasites: options for collection and processing

Option	Pros	Cons
Rejection of stools from inpatients who have been in hospital for >3 days[a]	Patients may become symptomatic with diarrhea after they have been inpatients for a few days; symptoms are usually not attributed to parasitic infections but to other causes	There is always the chance that the problem is related to a health care-associated (nosocomial) parasitic infection (rare), but *Cryptosporidium* and microsporidia are possible considerations
Examination of a single stool specimen (O&P examination); data suggest that 40–50% of organisms present will be found with only a single stool exam; two O&P exams (concentration, permanent stained smear) are acceptable but are not always as good as three specimens (this may be a relatively cost-effective approach); any patient remaining symptomatic would require additional testing	Some think that most intestinal parasitic infections can be diagnosed from examination of a single stool; if the patient becomes asymptomatic after collection of the first stool, subsequent specimens may not be necessary	Diagnosis from a single stool examination depends on experience of the microscopist, proper collection, and the parasite load in the specimen; in a series of three stool specimens, it is often the case that not all three specimens are positive and/or may be positive for different organisms
Examination of a second stool specimen only after the first is negative and the patient is still symptomatic	With additional examinations, the yield of protozoa increases (*Entamoeba histolytica*, 22.7%; *Giardia lamblia*, 11.3%; and *Dientamoeba fragilis*, 31.1%) (19)	Assumes the second (or third) stool specimen is collected within the recommended 10-day time frame for a series of stools (protozoa are shed periodically); may be inconvenient for patient
Examination of a single stool and an immunoassay (EIA, FA) (*Giardia*); this approach is a mix: one immunoassay is acceptable; one O&P exam is not the best approach (review the last option below)	If the examinations are negative and the patient's symptoms subside, probably no further testing is required	Patients may exhibit symptoms (off and on), so it may be difficult to "rule out" parasitic infections with only a single stool and one fecal immunoassay. If the patient remains symptomatic, then even if two *Giardia* immunoassays are negative, other protozoa may be missed (the *Entamoeba histolytica/E. dispar* group, *Entamoeba histolytica*, *Dientamoeba fragilis*, *Cryptosporidium* spp., the microsporidia). Normally, there are specific situations where fecal immunoassays *or* O&P exams should be ordered. It is not recommended to automatically perform both the O&P and the fecal immunoassay as a stool exam for parasites.
Pool three specimens for examination; perform one concentration and one permanent stain (billing and coding issues)	Three specimens are collected over 7–10 days and may save time and expense	Organisms present in small numbers may be missed due to the dilution factor
Pool three specimens for examination; perform a single concentration and three permanent stained smears (billing and coding issues)	Three specimens are collected over 7–10 days; would maximize the recovery of protozoa in areas of the country where these organisms are most common	Might miss light helminth infection (eggs and larvae) due to the pooling of the three specimens for the concentration; however, with a permanent stain performed on each of the three specimens, this approach would probably be the next best option after the standard approach (concentration and permanent stained smear performed on every stool)
Actually collect three stools, but put a sample of stool from all three into a single vial (patient is given a single vial only)	Pooling of the specimens would require only a single vial	This would complicate patient collection and very probably result in poorly preserved specimens, especially regarding the recommended ratio of stool to preservative and the lack of proper mixing of specimen and fixative

(continued)

Table 26.1 Fecal specimens for parasites: options for collection and processing *(continued)*

Option	Pros	Cons
Perform immunoassays on selected patients[b] by FA, EIA, or rapid cartridge methods for *Giardia lamblia*, *Cryptosporidium* spp., and/or *Entamoeba histolytica*/*E. dispar* or *Entamoeba histolytica*	Would be more cost-effective than performing immunoassay procedures on all specimens; however, the information needed to group patients is often not received with specimens; client education is critical for appropriate test orders	Labs rarely receive information that would allow them to place a patient in a particular risk group, such as children <5 yr, children from day care centers (may or may not be symptomatic), patients with immunodeficiencies, and patients from outbreaks; performance of immunoassay procedures on every stool is not cost-effective, and the positive rate will be low unless an outbreak situation is involved
Perform immunoassays and O&P examinations on request[c] for *Giardia lamblia*, *Cryptosporidium* spp., and/or *Entamoeba histolytica*/*E. dispar* or *Entamoeba histolytica* (Table 26.2)	Will limit number of stools on which immunoassay procedures are performed for parasites; immunoassay results do not have to be confirmed by any other testing (such as O&P exams or modified acid-fast stains); best approach (see also ordering options in Table 26.2)	Will require education of the physician clients regarding appropriate times and patients for whom immunoassays should be ordered; educational initiatives must also include information on the test report indicating the pathogenic parasites that will not be detected by these methods; it is critical to ensure that clients know that if patients have become asymptomatic, further testing may not be required, but that if the patient remains symptomatic, further testing (O&P exams) is required; a single O&P exam may not reveal all organisms present; present plan to physicians for approval: immunoassays or O&P exams, procedure discussion, report formats, clinical relevance, limitations on each approach

[a] Two key references have addressed this issue and serve as guidelines for microbiologists in reviewing clinically relevant recommendations for specimen submission: A. J. Morris, M. L. Wilson, and L. B. Reller, Application of rejection criteria for stool ovum and parasite examinations, *J. Clin. Microbiol.* 30:3213–3216, 1992, and D. L. Siegel, P. H. Edelstein, and I. Nachamkin, Inappropriate testing for diarrheal diseases in the hospital, *JAMA* 263:979–982, 1990.

[b] It is difficult to know when an early-outbreak situation is present, where screening of all specimens for either *Giardia lamblia*, *Cryptosporidium* spp., or both may be relevant. Extensive efforts are under way to encourage communication among laboratories, water companies, pharmacies, and public health officials regarding the identification of potential or actual outbreaks. If it appears that an outbreak is in the early stages, performing the immunoassays on request can be changed to performing immunoassays on all stools.

[c] A number of variables will determine the approach to immunoassay testing and the ova and parasite examination (O&P exam) (geography, organisms recovered, positive rate, etc.).

trophozoites do not encyst but may disintegrate if not examined or preserved within a short time after passage. The time limit recommendations listed below are most relevant for the intestinal protozoa; most helminth eggs and larvae, coccidian oocysts, and microsporidian spores survive for extended periods. However, no one can predict which organisms will be present in the stool specimens; therefore, it is important to use the most conservative time frames for parasite recovery.

Liquid specimens should be examined within 30 min of passage, not 30 min from the time they reach the laboratory. If this general time recommendation of 30 min is not possible, the specimen should be placed in one of the available fixatives. Soft (semiformed) specimens may contain a mixture of protozoan trophozoites and cysts and should be examined within 1 h of passage; again, if this time frame is not possible, preservatives should be used. Immediate examination of formed specimens is not as critical; in fact, if the specimen is examined at any time within 24 h after passage, the protozoan cysts should still be intact (5, 23, 24).

In review, remember that trophozoites only are usually found in liquid specimens, both protozoan trophozoites and cysts can be recovered in soft specimens, and generally cysts only are recovered in formed specimens. The time limits mentioned above are merely guidelines; however, if fresh specimens remain unpreserved for longer times before examination, many if not all organisms may disintegrate or become distorted. Fecal specimens should never be incubated or frozen prior to examination using routine microscopy. When the acceptance criteria for specimen collection are not met, the laboratory should reject the specimen and request additional specimens.

Because there is often a time lag from the time of specimen passage until receipt in the laboratory, many clinicians, clinics, and inpatient wards use a specimen collection system that includes stool preservatives (Table 26.3). A number of commercial systems are available with many

Table 26.2 Approaches to stool parasitology: test ordering

Patient and/or situation	Test ordered[a]	Follow-up test ordered
Patient with diarrhea and AIDS (or other cause of immune deficiency) Potential waterborne outbreak (municipal water supply)	*Cryptosporidium* or *Giardia/Cryptosporidium* immunoassay	If immunoassays are negative and symptoms continue, special tests for microsporidia (modified trichrome stain) and other coccidia (modified acid-fast stain) and O&P[b] exam should be performed
Patient with diarrhea (nursery school, day care center, camper, backpacker) Patient with diarrhea and potential waterborne outbreak (resort setting)	*Giardia* or *Giardia/Cryptosporidium* immunoassay	If immunoassays are negative and symptoms continue, special tests for microsporidia and other coccidia (see above) and O&P exam should be performed
Patient with diarrhea and relevant travel history Patient with diarrhea who is a past or present resident of a developing country Patient in area of United States where parasites other than *Giardia* are found	O&P exam, *Entamoeba histolytica/E. dispar* immunoassay, confirmation for *E. histolytica*; various tests for *Strongyloides* may also be relevant (even in the absence of eosinophilia)	If exams are negative and symptoms continue, special tests for coccidia and microsporidia should be performed
Patient with unexplained eosinophilia	Although the O&P exam is recommended, the agar plate culture for *Strongyloides stercoralis* (more sensitive than the O&P exam) is also recommended	If tests are negative and symptoms continue, additional O&P exams and special tests for microsporidia and other coccidia should be performed
Patient with diarrhea (suspected food-borne outbreak)	Test for *Cyclospora cayetanensis* (modified acid-fast stain, autofluorescence)	If tests are negative and symptoms continue, special procedures for microsporidia and other coccidia and O&P exam should be performed

[a] Depending on the particular immunoassay kit used, various single or multiple organisms may be included. Selection of a particular kit depends on many variables: clinical relevance, cost, ease of performance, training, personnel availability, number of test orders, training of physician clients, sensitivity, specificity, equipment, time to result, etc. Very few laboratories will handle this type of testing exactly the same. Many options are clinically relevant and acceptable for good patient care.
[b] O&P exam, ova and parasite examination.

Table 26.3 Stool collection: pros and cons of fresh and preserved specimens[a]

Specimen	Pros	Cons
Fresh stool	1. No requirements for stool fixatives 2. Ability to see motile trophozoites 3. Lower cost (no fixative vials) 4. Can perform direct wet examination, concentration, and permanent stained smear 5. Acceptable *only* if time from stool passage to laboratory is within the time limits indicated below: in symptomatic patients, the trophozoite form of the intestinal protozoa is present and will not encyst when outside of the body but will disintegrate Time limits: liquid or watery stool, 30 min; semiformed stool, 1 h; formed stool, 24 h	1. May have excessive lag time between stool fixation or processing; trophozoites may disintegrate, thus giving a false-negative result 2. Ova and parasite examination (direct wet examination, concentration, and permanent stained smear) may be negative due to lack of organism preservation and morphology integrity
Preserved stool	1. Organism morphology is preserved when lag time between stool passage and fixation is short (this can be done by allowing patients to collect and fix stool specimens at home); once the specimen is mixed with the preservative, the delivery time to the laboratory is not critical 2. Can perform concentration and permanent stain	1. Cost of collection vials may represent a cost increase; however, overall expense may be lower because of much more accurate result (patient outcome) 2. Disposal of vials may be a problem if the laboratory is using preservatives containing mercuric chloride

[a] Most products used for specimen collection are available from any major medical supply house. FA, fluorescent antibody. Adapted from reference 9.

preservative choices; the use of such systems has become routine for many institutions, and some request a custom collection kit that may contain several types of preservatives for stool specimens, depending on the tests normally ordered by the clinicians that they service. Specific information concerning collection kit components is provided in appendix 1.

KEY POINTS

Stool Specimen Collection

1. Occupational Safety and Health Act regulations (including standard precautions) should be used for handling all specimens.
2. Interfering substances (oil-based laxatives, barium, antibiotics) should be avoided when stool specimens are collected.
3. Contamination with urine or water should be avoided.
4. Recommendation for collection: three specimens collected, one every other day or within a 10-day time frame.

   Note There are some exceptions; see Table 26.1.
5. Liquid stool: examine or preserve within 30 min of passage (trophozoites). Soft stool: examine or preserve within 1 h of passage (trophozoites and cysts). Formed stool: examine or preserve within 24 h of passage. Note that *Dientamoeba fragilis* trophozoites (no known cyst forms) can be found in formed stool specimens as well as liquid specimens.

Preservation of Specimens

Preservatives

There are a number of reasons why a lag time may occur from the time of specimen passage until examination in the laboratory (e.g., the workload in the laboratory or the transit distance or time for the specimen to reach the facility). To preserve protozoan morphology and to prevent the continued development of some helminth eggs and larvae, the stool specimens can be placed in preservative immediately after passage (by the patient using a collection kit) or once the specimen is received by the laboratory. There are several fixatives available; the more common ones, including formalin, Merthiolate (thimerosal)-iodine-formalin (MIF), sodium acetate-acetic acid-formalin (SAF), Schaudinn's fluid, polyvinyl alcohol (PVA), and the single-vial systems are discussed in detail (Table 26.4). Regardless of the fixative selected, adequate mixing of the specimen and preservative is mandatory. A flow diagram for preservation and processing is shown in Figure 26.3.

Note When selecting an appropriate fixative, keep in mind that a permanent stained smear is mandatory for a complete examination for parasites (5, 7, 9, 10, 33, 35). It is also important to remember that disposal regulations for compounds containing mercury are becoming more restrictive; each laboratory will have to check applicable state and federal regulations to help determine fixative options.

Formalin

Formalin has been used for many years as an all-purpose fixative that is appropriate for helminth eggs and larvae and for protozoan cysts, oocysts, and spores. Two concentrations are commonly used: 5%, which is recommended for preservation of protozoan cysts, and 10%, which is recommended for helminth eggs and larvae. Although 5% is often recommended for all-purpose use, most commercial manufacturers provide 10%, which is more likely to kill all helminth eggs (Figure 26.4). To help maintain organism morphology, the formalin can be buffered with sodium phosphate buffers, i.e., neutral formalin. Selection of specific formalin formulations is at the user's discretion. *Aqueous formalin will permit the examination of the specimen as a wet mount only, a technique much less accurate than a permanent stained smear for the identification of intestinal protozoa. However, the fecal immunoassays for* Giardia lamblia *and* Cryptosporidium spp. *can be performed from the aqueous formalin vial. Fecal immunoassays for the* Entamoeba histolytica/ E. dispar *group and* Entamoeba histolytica *are limited to fresh or frozen fecal specimens.* After centrifugation, special stains for the coccidia (modified acid-fast stains) and the microsporidia (modified trichrome stains) can be performed from the concentrate sediment obtained from formalin-preserved stool material. The most common formalin preparation is 10% formalin, prepared as follows:

Formaldehyde
(USP) .... 100 ml (or 50 ml for 5%)
Saline solution,
0.85% NaCl............. 900 ml (or 950 ml for 5%)

Dilute 100 ml of formaldehyde with 900 ml of 0.85% NaCl solution. (Distilled water may be used instead of saline solution.)

Note Formaldehyde is normally purchased as a 37 to 40% HCHO solution; however, for dilution, it should be considered to be 100%.

If you want to use buffered formalin, the recommended approach (5, 31) is to mix thoroughly 6.10 g of Na_2HPO_4 and 0.15 g of NaH_2PO_4 and store the dry mixture in a tightly closed bottle. Prepare 1 liter of either 10 or 5% formalin, and add 0.8 g of the buffer salt mixture.

Table 26.4 Preservatives used in diagnostic parasitology (stool specimens)

Preservative	Concentration	Permanent stained smear	Immunoassays (*Giardia lamblia*, *Cryptosporidium* spp.)	Comments
5% or 10% formalin	Yes	No	Yes	EIA, FA, cartridge
5% or 10% buffered formalin	Yes	No	Yes	EIA, FA, cartridge
MIF	Yes	Polychrome IV stain	ND	No published data
SAF	Yes	Iron hematoxylin	Yes	EIA, FA, cartridge
PVA[a]	Yes, but rarely used	Trichrome or iron hematoxylin	No	PVA interferes with immunoassays
Modified PVA[b]	Yes, but rarely used	Trichrome or iron hematoxylin	No	PVA interferes with immunoassays
Modified PVA[c]	Yes	Trichrome or iron hematoxylin	Some, but not all	PVA interferes with immunoassays
Single-vial systems[d]	Yes	Trichrome or iron hematoxylin	Some, but not all	Check with the manufacturer
Schaudinn's fluid (without PVA)[a]	Yes, but rarely used	Trichrome or iron hematoxylin	No	Mercury interferes with immunoassays

[a] These two fixatives use the mercuric chloride base in the Schaudinn's fluid; this formulation is still considered to be the "gold standard" against which all other fixatives are evaluated (organism morphology after permanent staining). Additional fixatives prepared with non-mercuric-chloride-based compounds are continuing to be developed and tested.

[b] This modification uses a copper sulfate base rather than mercuric chloride.

[c] This modification uses a zinc base rather than mercuric chloride and works well with both trichrome and iron-hematoxylin stains.

[d] These modifications use a combination of ingredients (including zinc) but are prepared from proprietary formulas. The aim is to provide a fixative that can be used for the fecal concentration, permanent stained smear, and available immunoassays (EIA, FA) for *Giardia lamblia*, *Cryptosporidium* spp., and *Entamoeba histolytica* (or the *Entamoeba histolytica/E. dispar* group).

Protozoan cysts (not trophozoites), coccidian oocysts, microsporidian spores, helminth eggs, and larvae are well preserved for long periods in 10% aqueous formalin. Hot (60°C) formalin can be used for specimens containing helminth eggs, since in cold formalin, some thick-shelled eggs (e.g., *Ascaris lumbricoides*) continue to develop, become infective, and remain viable for long periods. Several grams of fecal material should be thoroughly mixed in 5 or 10% formalin.

To collect large numbers of cysts, eggs, or larvae relatively free from other debris, the whole stool specimen is mixed in water and then strained through several layers of gauze. The suspension is allowed to sediment in a cone-shaped glass or flask for 1 h or more, and the supernatant fluid is discarded. The specimen may be washed several times in this manner before the sediment is finally fixed in hot 10% formalin, as mentioned above. When working with watery diarrhea specimens from patients with suspected cases of coccidiosis or microsporidiosis, the specimen should not be strained through gauze (oocysts and small bits of mucus may cling to the gauze); centrifugation ($500 \times g$ for 10 min) is necessary to sediment the oocysts and/or spores.

Formalin: summary

Advantages	Disadvantages
Good overall fixative for stool concentration	Does not preserve trophozoites well
Easy to prepare, long shelf life	Does not adequately preserve organism morphology for a good permanent stained smear
Formalinized stool can be used with some of the immunoassay detection kits (*Giardia lamblia*, *Cryptosporidium* spp.)	

MIF

MIF (37) is a good stain preservative for most kinds and stages of parasites found in feces; it is especially useful for field surveys. It is used with all common types of stools and aspirates; protozoa, eggs, and larvae can be diagnosed without further staining in temporary wet mounts, either made immediately after fixation or prepared several weeks later. Although some laboratories maintain that a permanent stained smear can be prepared from specimens preserved in MIF, most laboratories using such a fixative examine the material only as a wet preparation (direct smear and/or concentration sediment). For a good discussion of this technique, see reference 8.

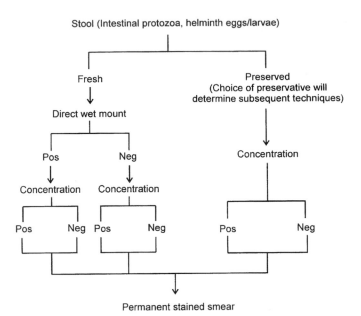

Figure 26.3 Flow diagram for preservation and processing of stool specimens. As mentioned in the text, the examination of fecal specimens using the ova and parasite examination is not considered complete unless a concentration and a permanent stained smear are examined for every specimen submitted to the laboratory. For a fresh specimen, a direct wet mount should be performed if the specimen is very soft to liquid; the complete ova and parasite examination would include the direct wet mount, the concentration, and the permanent stained smear. If the specimen is submitted in preservative, the direct wet mount should be eliminated (no motility is possible); the complete ova and parasite examination would include the concentration and the permanent stained smear (5, 9).

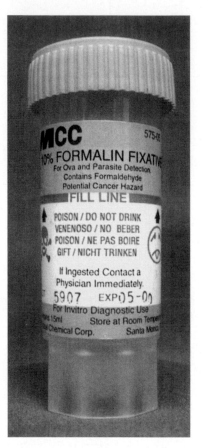

Figure 26.4 Stool collection vial containing 10% formalin; note the "FILL LINE" and the poison signs at each side of the label. When the stool specimen is added to the vial, the final ratio of stool to preservative is about 1:3.

The MIF preservative is prepared in two stock solutions, stored separately and mixed immediately before use.

Solution I (stored in a brown bottle)

Distilled water	50 ml
Formaldehyde (USP)	5 ml
Thimerosal (tincture of Merthiolate, 1:1,000)	40 ml
Glycerin	1 ml

Solution II (Lugol's Solution) (good for several weeks in a tightly stoppered brown bottle)

Distilled water	100 ml
Potassium iodide crystals (KI)	10 g
Iodine crystals (add after KI dissolves)	5 g

Combine 9.4 ml of solution I with 0.6 ml of solution II just before use.

1. Add about one-quarter teaspoon (1 g) of fresh feces to the solution, and mix with an applicator.

Fecal material should be formed or soft if egg counts are to be made later; liquid stool does not work very well for worm burden estimates.

2. Within 24 h, if undisturbed, the specimen forms three well-defined layers. The top layer, a clear orange fluid, consists mainly of formalin, Merthiolate, and water; it does not trap eggs or protozoa. The interface is a thick, pale orange or creamy yellow layer, usually 1 to 2 mm thick; this layer may trap some protozoa and helminth eggs. The bottom layer consists of deeper-staining particulate matter; eggs and protozoa are found throughout this layer.

3. With a glass pipette, MIF direct smears can be made from both the interface and bottom layers. Best results are obtained by making smears from both layers.

4. It has been suggested by some workers that a concentration technique applied to the MIF method (referred to as the MIFC or TFC method) gives

satisfactory results. This contention is debatable, and Dunn (8) suggests that it is not nearly as reliable as the MIF direct smear method.

MIF: summary

Advantages	Disadvantages
Components both fix and provide stain color	Contains mercury compounds (thimerosal)
Easy to prepare, long shelf life	Morphology of organisms on permanent stained smears generally not as good as that seen with Schaudinn's fluid or PVA (mercuric chloride base)
Very useful for field surveys	

SAF

SAF lends itself to the concentration technique, the permanent stained smear, and fecal immunoassays for *Giardia* and *Cryptosporidium* and has the advantage of not containing mercuric chloride, as is found in Schaudinn's fluid and some of the PVA fixatives (39, 49). It is a liquid fixative, much like the 10% formalin described above (Figure 26.5). The sediment is used to prepare the permanent smear, and it is frequently recommended that the stool

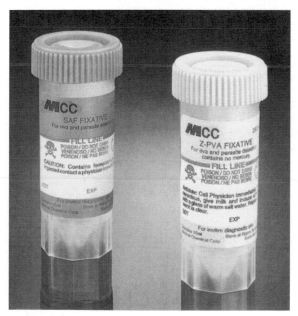

Figure 26.5 Stool collection vials, one containing SAF fixative and the other containing Z-PVA, one of the non-mercury-based fixatives. This combination of collection vials is an excellent option; concentrations and fecal immunoassays can be performed from the SAF vial, while the permanent stained smear can be performed from the Z-PVA vial. If the laboratory is going to use the iron-hematoxylin permanent stain, the permanent smear for staining could also be prepared from the SAF vial.

material be placed on an albumin-coated slide to improve adherence to the glass (Figure 26.6).

SAF is considered to be a "softer" fixative than mercuric chloride. The organism morphology is not quite as sharp after permanent staining as that of organisms originally fixed in solutions containing mercuric chloride. *The pairing of SAF-fixed material with iron hematoxylin staining provides better organism morphology than does staining SAF-fixed material with trichrome (personal observation).* Although SAF has a long shelf life and is easy to prepare, the smear preparation technique may be a bit more difficult for less experienced laboratory personnel who are not familiar with fecal specimen techniques. Laboratories that have considered using only a single preservative have selected this option (concentration, permanent stain, fecal immunoassays for *Giardia* and *Cryptosporidium*). Helminth eggs and larvae, protozoan trophozoites and cysts, and coccidian oocysts and microsporidian spores are preserved by this method. After centrifugation, special stains for the coccidia (modified acid-fast stains) and the microsporidia (modified trichrome stains) can be used with the concentrate sediment obtained from SAF-preserved stool material.

SAF fixative is prepared as follows:

Sodium acetate	1.5 g
Acetic acid, glacial	2.0 ml
Formaldehyde, 37 to 40% solution	4.0 ml
Distilled water	92.0 ml

To make Mayer's albumin, mix equal parts of egg white and glycerin. Place 1 drop on a microscope slide, and add 1 drop of SAF-preserved fecal sediment (from the

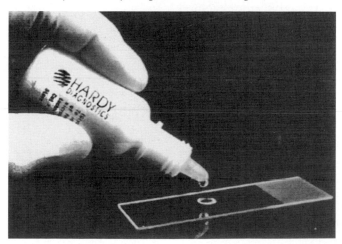

Figure 26.6 Albumin used to precoat the slide prior to the application of SAF-fixed stool concentration sediment. Once the smear is dry, it is ready for permanent staining.

concentration procedure). After mixing, allow the smear to dry at room temperature for 30 min prior to staining.

SAF: summary

Advantages	Disadvantages
Can be used for concentration and permanent stained smears	Poor adhesive properties, albumin-coated slides recommended
Contains no mercury compounds	Protozoan morphology better if iron hematoxylin stains used for permanent stained smears (trichrome fair)
Easy to prepare, long shelf life	
SAF-preserved stool can be used with the new immunoassay methods (*Giardia* and *Cryptosporidium*)	
	May be a bit more difficult to use; however, this does not seem to be a limiting factor

Schaudinn's Fluid

Schaudinn's fluid is designed to be used with fresh stool specimens or samples from the intestinal mucosal surface. Many laboratories that receive specimens from in-house patients (no problem with delivery times) often select this approach. Permanent stained smears are then prepared from fixed material. A concentration technique for Schaudinn's fluid-preserved material is also available but is not widely used.

Mercuric Chloride, Saturated Aqueous Solution
Mercuric chloride ($HgCl_2$)....................... 110 g
Distilled water....................................... 1,000 ml

Use a beaker as a water bath; boil (use a hood if available) until the mercuric chloride is dissolved; let stand for several hours until crystals form.

Schaudinn's Fixative (Stock Solution)
Mercuric chloride, saturated
    aqueous solution................................. 600 ml
Ethyl alcohol, 95% .... 300 ml

Immediately before use, add 5 ml of glacial acetic acid per 100 ml of stock solution.

Schaudinn's fluid: summary

Advantages	Disadvantages
Fixative for smears prepared from fresh fecal specimens or samples from the intestinal mucosal surfaces	Not generally recommended for use in concentration procedures
Provides excellent preservation of protozoan trophozoites and cysts	Contains mercuric chloride—disposal problem
	Poor adhesive qualities with liquid or mucoid specimens

PVA

PVA is a plastic resin that is normally incorporated into Schaudinn's fixative (1). The PVA powder is *not* a fixative but serves as an adhesive for the stool material; i.e., when the stool-PVA-fixative mixture is spread onto the glass slide, it adheres because of the PVA component. Fixation is still accomplished by the Schaudinn's fluid itself. Perhaps the greatest advantage of the use of PVA is the fact that a permanent stained smear can be prepared. Although some laboratories may perform a fecal concentration from a PVA-preserved specimen, some parasites do not concentrate well, nor do some exhibit the typical morphology that would be seen in concentration sediment from a formalin-based fixative. PVA fixative solution is highly recommended as a means of preserving cysts and trophozoites for later examination. The use of PVA fixative also permits specimens to be shipped (by regular mail service) from any location in the world to a laboratory for subsequent examination. PVA fixative is particularly useful for liquid specimens and should be used in the ratio of 3 parts PVA to 1 part fecal specimen. The formula is as follows:

PVA... 10.0 g
Ethyl alcohol, 95% .... 62.5 ml
Mercuric chloride, saturated
    aqueous .... 125.0 ml
Acetic acid, glacial.................................. 10.0 ml
Glycerin.. 3.0 ml

Mix the liquid ingredients in a 500-ml beaker. Add the PVA powder (stirring is not recommended). Cover the beaker with a large petri dish, heavy wax paper, or foil, and allow the PVA to soak overnight. Heat the solution slowly to 75°C. When this temperature is reached, remove the beaker and swirl the mixture for 30 s until a homogeneous, slightly milky solution is obtained.

PVA: summary

Advantages	Disadvantages
Can be used to prepare permanent stained smears and perform concentration techniques (see also Disadvantages)	*Trichuris trichiura* eggs and *Giardia lamblia* cysts are not concentrated as easily as from formalin-based fixatives; *Strongyloides stercoralis* larval morphology is poor (better to use formalin-based preservation); *Isospora belli* oocysts may not be visible in PVA fixative-preserved material (better to use formalin-fixed specimens)
Provides excellent preservation of protozoan trophozoites and cysts	
Long shelf life (months to years) in tightly sealed containers at room temperature	
Allows specimens to be shipped to any laboratory for subsequent examination	
Specimens preserved in PVA cannot be used with the fecal immunoassay kits	Contains mercury compounds (Schaudinn's fluid)
	May turn white and gelatinous when it begins to dehydrate or when refrigerated
	Difficult to prepare in the laboratory

Modified PVA

Although there has been a great deal of interest in developing preservatives without the use of mercury compounds, substitute compounds have not provided the quality of preservation necessary for comparable protozoan morphology on the permanent stained smear. Copper sulfate has been tried (15, 20) but does not provide results equal to those seen with mercuric chloride (15). However, zinc sulfate has proven to be a good mercury substitute and is used with trichrome stain (Figure 26.5) (16). Although zinc substitutes have become widely available, each manufacturer has a proprietary formula for the fixative.

Copper Sulfate Solution
$CuSO_4 \cdot 5H_2O$.. 20.0 g
Distilled water...................................... 1,000 ml

Add the $CuSO_4 \cdot 5H_2O$ to 1,000 ml of distilled water heated to 100°C. Mix until dissolved.

Modified PVA Fixative (Stock Solution)
Copper sulfate solution .... 600 ml
Ethyl alcohol, 95% .... 300 ml

Immediately before use, add 5 ml of glacial acetic acid per 100 ml of stock solution.

Modified PVA: summary

Advantages	Disadvantages
Can prepare permanent stained smears and perform concentration techniques	Overall protozoan morphology of trophozoites and cysts is poor when preserved in the copper sulfate-based fixative, particularly compared with organisms preserved with mercuric chloride-based fixatives
Many workers prefer the zinc substitutes over those prepared with copper sulfate	Zinc-based fixatives are a better alternative
Does not contain mercury compounds	Staining characteristics of protozoa not consistent—some good, some poor; organism identification may be more difficult, particularly with small protozoan cysts

Single-Vial Collection Systems (Other than SAF)

Several manufacturers now have available single-vial stool collection systems, similar to SAF or modified PVA methods (12). From the single vial, both the concentration and permanent stained smear can be prepared (Figures 26.7 through 26.9). It is also possible to perform fecal immunoassay procedures from some of these vials. Make sure to ask the manufacturer about all three capabilities (concentration, permanent

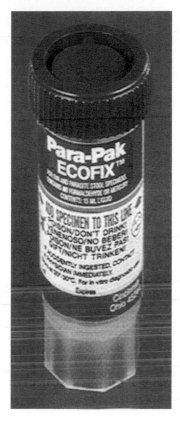

Figure 26.7   ECOFIX is an example of a fixative that represents the single-vial collection system. This fixative has been coupled with its own stain, the ECOSTAIN (12). Remember to inquire about the compatibility of all single-vial systems with the immunoassay procedures; not all single-vial preservatives are compatible with all fecal immunoassays

Single-vial systems: summary

Advantages	Disadvantages
Can be used to prepare permanent stained smears and perform concentration techniques	Overall protozoan morphology of trophozoites and cysts is not as good as that for organisms preserved with mercuric chloride-based fixatives; morphology similar to modified PVA options
Can be used in fecal immunoassay procedures (some exceptions)	
Do not contain mercury compounds	Staining characteristics of protozoa not consistent—some good, some poor; identification of *Endolimax nana* cysts may be difficult

stained smear, fecal immunoassay procedures) and for specific information indicating that there are no formula components that would interfere with any of the three methods. Like the zinc substitutes, these formulas are proprietary.

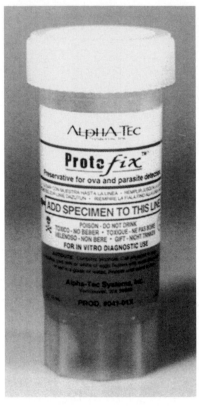

Figure 26.8 ProtoFix is a single-vial collection option. It is always important to review peer-reviewed literature regarding the results of new products compared with those previously in use.

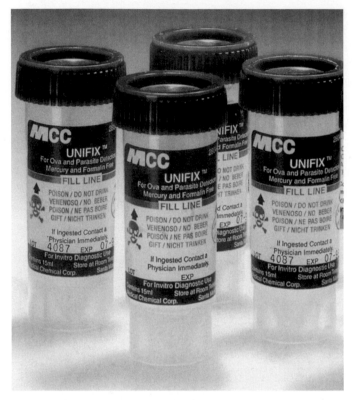

Figure 26.9 UNIFIX is a single-vial collection system; information on this product can be found at the company website (www.med-chem.com). This fixative works well with trichrome stain.

Use of Fixatives

Quality Control for Stool Fixatives

Fixatives for fecal specimens are checked for quality control by the manufacturer before sale, generally with the use of living protozoa. If you prepare your own fixatives, the following approach can be used for quality control. The specimen used for quality control presented below is designed to be used with fixatives from which permanent stained smears will be prepared (Schaudinn's fluid, PVA fixative, modified PVA fixative, SAF, or MIF). However, the same quality control specimen can also be used in a concentration; the white blood cells (WBCs) can be seen in the concentrate sediment (sedimentation concentration) or in the surface film (flotation concentration).

1. Obtain a fresh, anticoagulated blood specimen, centrifuge, and obtain a buffy coat sample (try and find a specimen with a high WBC count).
2. Mix approximately 2 g of soft, fresh fecal specimen (normal stool, containing no parasites) with several drops of the buffy coat cells.

3. Prepare several fecal smears, and fix immediately in Schaudinn's fluid to be quality controlled.
4. Mix the remaining feces-buffy coat mixture in 10 ml of PVA fixative, modified PVA fixative, SAF, or MIF preservative to be quality controlled.
5. Allow 30 min for fixation, and then prepare several fecal smears. Allow to dry thoroughly (60 min at room temperature or 30 to 60 min in an incubator [approximately 35°C]). Do not use a heat block.
6. Stain the slides by the normal staining procedure (trichrome, iron-hematoxylin).
7. After staining, if the WBCs appear well fixed and display typical morphology and color, one can assume that any intestinal protozoa placed in the same lot number of preservative would also be well fixed, provided that the fecal sample was fresh and fixed within the recommended time limits.
8. The bulk quality control specimen can be concentrated as for a normal patient specimen. If the fixative is performing correctly, the WBCs will be

visible in the concentration sediment or surface film (depending on the method used).

9. Record all quality control results. If the WBC morphology does not confirm good fixation, describe the results and indicate what corrective actions were used (repeated the test, prepared new fixative).

Procedure Notes for Use of Preservatives

1. Most of the commercially available kits have a "fill to" line on the vial label to indicate how much fecal material should be added to ensure adequate preservation of the fecal material (ratio of one part stool to three parts fixative). *However, patients often overfill the vials; remember to open the vials with the vials turned away from your face. There may be excess gas in the vials that may create aerosols once the vial lids are opened.*

2. Although the two-vial system (one vial of 5 or 10% buffered formalin [concentration] and one vial of PVA fixative [permanent stained smear]) has always been the "gold standard," laboratories are beginning to use other options, including the single-vial collection systems. Changes in the selection of fixatives are based on the following considerations:

   A. Problems with disposal of mercury-based fixatives (availability of high-temperature incineration facilities and cost) and lack of multilaboratory contracts for disposal of such products
   B. The cost of a two-vial system compared with the cost of a single collection vial
   C. Selection of specific stains (trichrome, iron hematoxylin) to use with specific fixatives
   D. Whether the newer fecal immunoassay kits can be used with stool specimens preserved in that particular fixative

Procedure Limitations for Use of Preservatives

1. Adequate fixation still depends on the following parameters:

   A. Meeting recommended time limits for lag time between passage of the specimen and fixation
   B. Use of the correct ratio of specimen to fixative (1:3)
   C. Thorough mixing of the fixative and specimen (once the specimen is received in the laboratory, any additional mixing at that time will not counteract the earlier lack of fixative-specimen mixing and contact)

2. Unless the appropriate stain is used with each fixative, the final permanent stained smear may be difficult to examine (organisms hard to see and/or identify). Examples of appropriate combinations are as follows:

A. Schaudinn's or PVA fixative with trichrome or iron-hematoxylin stain
B. SAF fixative with iron-hematoxylin stain

Shipment of Diagnostic Specimens, Biological Products, Etiologic Agents, or Infectious Substances

Biological materials shipped domestically or internationally must be packaged in compliance with hazardous-materials transport regulations. The United States has incorporated the United Nations Recommendations on the Transport of Dangerous Goods into law, making the U.S. regulations consistent with foreign shipment regulations. The U.S. regulations for packing diagnostic specimens and infectious agents for shipment were modified in 2006. The Public Health Service, Department of Transportation, and Postal Service specify requirements for packaging and shipping of biological materials. These regulations, plus those for packaging and shipping of materials via air, are similar; most carriers elect to follow the international shipping regulations within the International Air Transport Association (IATA) *Dangerous Goods Regulations* (Table 26.5).

Anyone packaging diagnostic specimens or infectious agents for shipment must receive training and be tested every 2 years. Those who complete the training and test will receive a certificate. The regulations are designed to protect all personnel who may come in contact with the package; compliance with these rules is the responsibility of the sender. Definitions of relevant terms can be found in Tables 26.5 and 26.6.

Double mailing containers should be used in shipping any parasitologic specimens other than microscope slides. The inner container is an aluminum screw-cap mailing tube that fits into an outer cardboard screw-cap mailing container. The specimen vials or tubes in the inner aluminum cylinder should be packed in cotton to absorb any moisture or material that might result from leakage or breakage. Instruction sheets, patient information sheets, etc., may be wrapped around the metal cylinder before it is placed in the outer cardboard mailer. For all packages containing infectious substances, an itemized list of contents must be enclosed between the secondary packaging and the outer packaging (21, 22, 30, 34, 41–45, 47, 48). Specific packing instructions can be found in Table 26.7, and significant changes in IATA requirements effective 2 June 2006 can be found in Table 26.8. An excellent review can be found through the *Sentinel Laboratory Guidelines for Suspected Agents of Bioterrorism, Packing and Shipping Diagnostic and Clinical Specimens, Infectious Substances, and Biological Agents*, published by the American Society for Microbiology in 2005. Training and training mate-

Table 26.5 Agencies governing transportation of dangerous goods

Governing authority	Agency[a]	Regulations
United Nations	ICAO	*Technical Instructions for the Safe Transport of Dangerous Goods by Air*
Commercial airline industry	IATA	*Dangerous Goods Regulations*
United States	DOT	*United States Hazardous Materials Uniform Safety Act*
United States	USPS	*Domestic Mail Manual. CO23 Hazardous Materials*
Canada	Transport Canada	*Transportation of Dangerous Goods Regulations*
Other nations		Individual national regulations

[a] ICAO, International Civil Aviation Organization; IATA, International Air Transport Association; DOT, Department of Transportation; USPS, U.S. Postal Service.

rial for the transportation of dangerous goods and infectious substances is available at the American Society for Microbiology (www.asm.org) and IATA (training manuals) (www.iata.com). Many of the packaging suppliers of shipping containers also have relevant training materials (SafTPak, CARGOpak).

Prepared slides, such as stained fecal smears or blood films, do not require double mailers for shipment. They may be packed in boxes, cardboard slide holders, or any other suitable container that will prevent damage or breakage. Slides should be individually wrapped in toilet tissue or facial tissue. If a number of slides are to be mailed, they can be wrapped in toilet paper as follows: place a slide on the tissue, wrap the slide several times, place the next slide on top of the first, and continue to wrap the slide several times. The series of slides will be padded and can be easily unwrapped on arrival. When you place slides in the flat cardboard containers, they need additional protection; some slides will arrive broken if this thin cardboard container is merely placed in an envelope for mailing. A plastic slide container with a snap top is an excellent option for shipping microscope slides.

Note If the slides are mounted with Permount or other mounting media, they should be completely dry before being packed.

All packages prepared for mailing should contain a complete information sheet about the specimen. An address label should be enclosed inside the package, as well as being on the outside container. All labels should be checked to ensure proper delivery. Also, remember that if you use an overnight carrier, the address must also contain a street address, not just a P.O. box or general address such as department and institution. Guidelines for packaging are provided in Table 26.7.

Documentation

The Shipper's Declaration for Dangerous Goods is a legal document that serves as a contract between the shipper and carrier; it must be accurate and legible and have no spelling errors. However, if minor discrepancies in typing or punctuation are present, they are no longer considered errors if they do not compromise safety. The document must be either handwritten by one person or typed. Two copies of the Shipper's Declaration must be completed and signed for each shipment. One copy is retained by the shipper, and the other is forwarded with the package; one copy can be a carbon copy. This document is required for shipping infectious substances and toxic substances; it is not required for shipping diagnostic specimens or biological products. If there are no dangerous goods in a shipment with dry ice, the Shipper's Declaration is not required. It is important for personnel within the laboratory to thoroughly understand these regulations and to have sample copies of appropriate documentation for referral. A checklist used by carriers to decide to accept or reject a package is widely used; a copy of this document should also be on file within the laboratory.

The current editions of the Codes of Federal Regulations can be obtained from the Superintendent of Documents, U.S. Government Printing Office, Washington, D.C. 20402. Inquiries about Postal Service publications may be directed to U.S. Postal Service, 475 L'Enfant Plaza, Washington, D.C. 20260-5365. Dangerous Goods Regulations can be obtained from Publications Assistant, International Air Transport Association, 2000 Peel Street, Montreal, Quebec, Canada, H3A 2R4.

Table 26.6   Definitions of relevant terms for packing and shipping[a]

Term	Definition
Biological product	A substance which originated from living organisms (including humans and other mammals) and has been manufactured and distributed in accordance with compliance and licensing requirements set forth by the federal government; can be classified as an infectious substance or a diagnostic (or clinical) specimen if such is appropriate. Biological products can be finished or unfinished; are intended for use in the prevention, treatment, or diagnosis of disease in humans or animals; and can be used for investigational, experimental, or development purposes. Biological products include such common items as clinical microbiology reagents and kits, serologic reagents, diagnostic reagents, and vaccines. In certain parts of the world, some licensed biological products are regarded as biohazardous and either are subject to compliance criteria specified for infectious substances or must adhere to other restrictions imposed by the government of that country.
Biological substance, category B	Any infectious substance that does not meet the criteria for a category A substance. See "Diagnostic or clinical specimen" below.
Category A substance	An infectious substance or microorganism that is transported *in a form* which, when exposure to it occurs, is capable of causing permanent disability or is life-threatening.
Category B substance	An infectious substance which does not meet category A criteria; an infectious substance *not* in a form capable of causing disability, life-threatening illness, or a fatal disease. Category B substances generally are considered to be (i) patient or clinical specimens reasonably expected to contain, or being cultured or otherwise tested for, a non-category A pathogen and (ii) cultures of microorganisms not specifically listed in category A.
Code of Federal Regulation (CFR)	U.S. laws published in the *Federal Register* and available online at http://www.gpoaccess.gov/cfr/index.html.
Culture	The result of a process by which pathogens are intentionally propagated. This definition refers to typical laboratory cultures of microorganisms grown in broth or on solid media. Typical clinical cultures may be classified as either category A or category B, depending on the organism concerned and the professional judgment of the shipper (49).
Dangerous goods	Materials which, when not properly handled and contained, can pose a risk to the health, safety, property, or environment and which are shown on the list of dangerous goods in the IATA *Dangerous Goods Regulations*.
Diagnostic or clinical specimen	A category B infectious substance; an infectious substance which does not meet the criteria for category A; generally considered to be clinical specimens such as swabs, tissue, and body fluids commonly encountered in a clinical laboratory and being cultured or otherwise tested for a pathogen.
Infectious substance	A substance which is known or reasonably expected to contain pathogens (microorganisms which can cause disease in humans and animals); material known to contain or reasonably suspected of containing a category A or B pathogen or substance; can be a class (class 6), a division (division 6.2), or a category (category A or B) of dangerous goods as defined by IATA.
Packaging	All of the materials used to contain a shipped substance and to prepare the substance for shipping; the container (receptacle) and its associated components (e.g., tubes, containers, absorbent material, boxes, and labels) used to contain and pack a substance and to ensure compliance with packing requirements.
Packing instructions	IATA-defined directions that shippers must follow to select, assemble, mark, label, and document the packing process for shipping dangerous goods, including diagnostic specimens and infectious substances; include manufacturing, testing, and performance specifications for packaging materials.
Pathogen	A microorganism (bacterium, mycobacterium, fungus, parasite, virus, plasmid, genetic element, proteinaceous infectious particle [prion], or genetically modified organism) that is known to cause or is reasonably expected to be able to cause disease in humans or animals.
Patient specimen	Material collected from humans or animals, including but not limited to excreta, secreta, blood and its components, tissue, body fluids, body organs and parts, and swabs of human material, being transported for purposes such as research, diagnosis, investigational activities, and disease treatment and prevention.
Primary specimen container	The innermost packaging containing a diagnostic specimen or infectious substance; composed of glass, metal, or plastic; must be leakproof; must be positively sealed if it contains an infectious substance.

(continued)

Table 26.6 Definitions of relevant terms for packing and shipping[a] *(continued)*

Term	Definition
Secondary specimen container	The container that contains the primary specimen container.
Shipper	Anyone who ships goods by a commercial carrier (usually an employee of a company or health care facility [e.g., laboratory staff member]); anyone who offers goods for transport to a member of IATA; anyone who completes and signs the Shipper's Declaration. The person who signs the Shipper's Declaration is the person who accepts responsibility for the accuracy of the information on the document.
Shipper's Declaration for Dangerous Goods	Shipper's Declaration: an IATA-defined and -mandated form which must accompany each shipment of dangerous goods; contains information that describes the dangerous goods; is helpful to persons who handle the shipment; must be completed by the shipper.

[a]Adapted from *Sentinel Laboratory Guidelines for Suspected Agents of Bioterrorism, Packing and Shipping Diagnostic and Clinical Specimens, Infectious Substances, and Biological Agents*, American Society for Microbiology, 2005.

Table 26.7 Comparison of packing directions for exempt human specimens and IATA packing instructions 650 and 602[a]

Requirement	Exempt human specimens[a]	650[b]	602[c]
Leakproof primary (1°) and secondary (2°) containers	Yes	Yes	Yes
Pressure-resistant 1° or 2° container	—[d]	Yes	Yes
Absorbent between 1° and 2° containers[e]	Yes	Yes	Yes
List of contents between 2° and outer package	—	Yes	Yes
Rigid outer packaging	—	Yes	Yes
Positively sealed 1° container	—	No	Yes
Name, address, and phone no. of responsible person on outer package or air waybill	—	Yes	Yes
Shipper's Declaration for Dangerous Goods	—	No	Yes
Outer packaging			
Markings and labels	—	Less	More
Strict manufacturing specifications	No	No	Yes
Quantity limits for passenger and cargo aircraft			
Maximum for each 1° container	—	1 liter (1 kg)	50 ml (50 g)
Total maximum for outer package	—	4 liters (4 kg)	50 ml (50 g)
Cost of labor and materials to pack substance	Least	More	Most

[a] Includes substances with minimal likelihood of causing disease in humans and animals and substances not likely to contain pathogens; adapted from *Sentinel Laboratory Guidelines for Suspected Agents of Bioterrorism, Packing and Shipping Diagnostic and Clinical Specimens, Infectious Substances, and Biological Agents*, American Society for Microbiology, 2005.
[b] Packing instructions for biological substances, Category B.
[c] Packing instructions for category A infectious substances.
[d] —, requirement not specified by IATA or DOT.
[e] Not required for solid substances such as tissue and solid agar media cultures or slants.

Table 26.8   Changes in IATA requirements effective 2 June 2006 (relevant for parasitology)[a]

1. *Culture* and *patient specimen* have been defined, and the definitions are user-friendly. The term *laboratory culture* is no longer used.

2. Packing instructions for patient specimens have been provided.

3. Classification of infectious substances according to risk groups has been replaced by classification of substances into either category A substance, category B substance, exempt human or animal specimen, exempt substance, and patient specimen.

4. Packing directions for exempt human or animal specimen have been provided.

5. The technical name of a substance packed according to PI 602 is no longer required after the proper shipping name marking on the outer package. For example, a package labeled "Infectious Substance, Affecting Humans (Hepatitis C Virus)" is now labeled "Infectious Substance, Affecting Humans." The technical name is still required on Shipper's Declarations.

6. Packing Instruction 650: The proper shipping name of category B substances may now be designated Diagnostic Specimen, Clinical Specimen, or Biological Substance, Category B. Beginning 1 January 2007, only Biological Substance, Category B, will be acceptable.

7. Packing Instruction 650: Quantity limits have been revised to allow up to 1 liter of liquid per primary container.

8. Packing Instruction 650: now mandates use of rigid outer containers.

9. Packing Instruction 650: Packages must be marked with a diamond symbol which contains "UN3373," and a "Diagnostic Specimen" marking adjacent to the diamond.

10. Packing Instruction 650: If an air waybill is used, the "Nature and Quantity of Goods" box must indicate the text "Diagnostic Specimen," "Clinical Specimen," or "Biological Substance, Category B," and the indication "UN 3373."

11. The "Prior arrangements as required..." statement in the Additional Handling Information section of the Shipper's Declaration is no longer required.

12. The requirement for an "air eligibility" label or marking (airplane symbol inside of a circle) has been replaced by the following certification statement on the Shipper's Declaration: "I declare that all of the applicable air transport requirements have been met."

13. Overpacks must now be labeled "Overpack" instead of "Inner Packages Comply..."

14. "Biological Substance, Category B" has been added to the IATA list of proper shipping names. Beginning 1 January 2007, this proper shipping name will replace "Clinical Specimen" and "Diagnostic Specimen."

15. Some hazardous materials (e.g. 10% formalin) used as a preservative are exempt from regulations if the quantity is ≤ 30 ml per primary container.

16. Persons who pack and ship select agents and toxins and category A agents should receive security training commensurate with their responsibilities.

[a] Adapted from *Sentinel Laboratory Guidelines for Suspected Agents of Bioterrorism, Packing and Shipping Diagnostic and Clinical Specimens, Infectious Substances, and Biological Agents*, American Society for Microbiology, 2005, and L. D. Gray and J. W. Snyder, Packing and shipping biological materials, p. 383–401, *in* D. O. Fleming and D. L. Hunt (ed.), *Biological Safety: Principles and Practices*, 4th ed., ASM Press, Washington, D.C., 2006.

References

1. **Brooke, M. M., and M. Goldman.** 1949. Polyvinyl alcohol-fixative as a preservative and adhesive for protozoa in dysenteric stools and other liquid material. *J. Lab. Clin. Med.* **34:**1554–1560.

2. **Cartwright, C. P.** 1999. Utility of multiple-stool-specimen ova and parasite examinations in a high-prevalence setting. *J. Clin. Microbiol.* **37:**2408–2411.

3. **Chan, R., J. Chen, M. K. York, N. Setijono, R. L. Kaplan, F. Graham, and H. B. Tanowitz.** 2000. Evaluation of a combination rapid immunoassay for detection of *Giardia* and *Cryptosporidium* antigens. *J. Clin. Microbiol.* **38:**393–394.

4. **Church, D., K. Miller, A. Lichtenfeld, H. Semeniuk, B. Kirkham, K. Laupland, and S. Elsayed.** 2005. Screening for *Giardia/Cryptosporidium* infections using an enzyme immunoassay in a centralized regional microbiology laboratory. *Arch. Pathol. Lab. Med.* **129:**754–759.

5. **Clinical Laboratory Standards Institute.** 2005. *Procedures for the Recovery and Identification of Parasites from the Intestinal Tract.* Approved guideline M28-2A. Clinical Laboratory Standards Institute, Villanova, Pa.

6. **Code of Federal Regulations.** 1991. Occupational exposure to bloodborne pathogens. *Fed. Regist.*, 29CFR1910.1030.

7. **Committee on Education, American Society of Parasitologists.** 1977. Procedure suggested for use in examination of clinical specimens for parasitic infection. *J. Parasitol.* **63:**959–960.

8. **Dunn, F. L.** 1968. The TIF direct smear as an epidemiological tool, with special reference to counting helminth eggs. *Bull. W. H. O.* **39:**439–449.

9. **Garcia, L. S.** 1999. *Practical Guide to Diagnostic Medical Parasitology.* ASM Press, Washington, D.C.

10. **Garcia, L. S., T. C. Brewer, and D. A. Bruckner.** 1979. A comparison of the formalin-ether concentration and trichrome-stained smear methods for the recovery and identification of intestinal protozoa. *Am. J. Med. Technol.* **45:**932–935.

11. **Garcia, L. S., and R. Y. Shimizu.** 1997. Evaluation of nine immunoassay kits (enzyme immunoassay and direct fluorescence) for detection of *Giardia lamblia* and *Cryptosporidium parvum* in human fecal specimens. *J. Clin. Microbiol.* **35:**1526–1529.

12. **Garcia, L. S., and R. Y. Shimizu.** 1998. Evaluation of intestinal protozoan morphology in human fecal specimens preserved in EcoFix: comparison of Wheatley's Trichrome stain and EcoStain. *J. Clin. Microbiol.* **36:**1974–1976.

13. **Garcia, L. S., and R. Y. Shimizu.** 1999. Detection of *Giardia lamblia* and *Cryptosporidium parvum* antigens in human fecal specimens using the ColorPAC combination rapid solid-phase qualitative immunochromatographic assay. *J. Clin. Microbiol.* **38:**1267–1268.

14. **Garcia L. S., R. Y. Shimizu, and C. N. Bernard.** 2000. Detection of *Giardia lamblia, Entamoeba histolytica/E. dispar,* and *Cryptosporidium parvum* antigens in human fecal specimens using the EIA Triage Parasite Panel. *J. Clin. Microbiol.* **38:**3337–3340.

15. **Garcia, L. S., R. Y. Shimizu, T. C. Brewer, and D. A. Bruckner.** 1983. Evaluation of intestinal parasite morphology in polyvinyl alcohol preservative: comparison of copper sulfate and mercuric chloride base for use in Schaudinn's fixative. *J. Clin. Microbiol.* **17:**1092–1095.

16. **Garcia, L. S., R. Y. Shimizu, A. Shum, and D. A. Bruckner.** 1993. Evaluation of intestinal protozoan morphology in polyvinyl alcohol preservative: comparison of zinc sulfate- and mercuric chloride-based compounds for use in Schaudinn's fixative. *J. Clin. Microbiol.* **31:**307–310.

17. **Garcia, L. S., and M. Voge.** 1980. Diagnostic clinical parasitology. I. Proper specimen collection and processing. *Am. J. Med. Technol.* **46:**459–467.

18. **Haque, R., I. K. M. Ali, S. Akther, and W. A. Petri.** 1998. Comparison of PCR, isoenzyme analysis, and antigen detection for diagnosis of *Entamoeba histolytica* infection. *J. Clin. Microbiol.* **36:**449–452.

19. **Hiatt, R. A., E. K. Markell, and E. Ng.** 1995. How many stool examinations are necessary to detect pathogenic intestinal protozoa? *Am. J. Trop. Med. Hyg.* **53:**36–39.

20. **Horen, W. P.** 1981. Modification of Schaudinn fixative. *J. Clin. Microbiol.* **13:**204–205.

21. **International Air Transport Association.** 2005. *Dangerous Goods Regulations,* 46th ed. International Air Transport Association, Montreal, Canada.

22. **International Air Transport Association.** 2005. *Dangerous Goods Regulations,* 46th ed., addendum II. 22 March 2005. International Air Transport Association, Montreal, Canada. (www.iata.org/NR/ContentConnector/CS2000/SiteInterface/sites/whatwedo/dangerousgoods/file/46rev02E.pdf.)

23. Reference deleted.

24. **Isenberg, H. D. (ed.).** 2004. *Clinical Microbiology Procedures Handbook,* 2nd ed. ASM Press, Washington, D.C.

25. **Isenberg, H. D. (ed.).** 1995. *Essential Procedures for Clinical Microbiology.* American Society for Microbiology, Washington, D.C.

26. **Jones, J. L., A. Lopez, S. P. Washquist, J. Nadle, M. Wilson, and the Emerging Infections Program FoodNet Working Group.** 2004. Survey of clinical laboratory practices for parasitic diseases. *Clin. Infect. Dis.* **38:**S198–S202.

27. **Juniper, K., Jr.** 1962. Acute amebic colitis. *Am. J. Med.* **33:**377–386.

28. **Kehl, K. S. C., H. Cicirello, and P. L. Havens.** 1995. Comparison of four different methods for detection of *Cryptosporidium* species. *J. Clin. Microbiol.* **33:**416–418.

29. **Lincicome, D. R.** 1942. Fluctuation in numbers of cysts of *Entamoeba histolytica* and *Entamoeba coli* in the stools of rhesus monkeys. *Am. J. Hyg.* **36:**321–337.

30. **Marshall, M. M., D. Naumovitz, Y. Ortega, and C. R. Sterling.** 1997. Waterborne protozoan pathogens. *Clin. Microbiol. Rev.* **10:**67–85.

31. **McVicar, J. W., and J. Suen.** 1994. Packaging and shipping biological materials, p. 239–246. *In* D. O. Fleming, J. H. Richardson, J. J. Tulis, and D. Vesley (ed.), *Laboratory Safety: Principles and Practices,* 2nd ed. ASM Press, Washington, D.C.

32. **Melvin, D. M., and M. M. Brooke.** 1982. *Laboratory Procedures for the Diagnosis of Intestinal Parasites,* 3rd ed. U.S. Department of Health, Education, and Welfare publication (CDC) 82-8282. Government Printing Office, Washington, D.C.

33. **Morris, A. J., M. L. Wilson, and L. B. Reller.** 1992. Application of rejection criteria for stool ovum and parasite examinations. *J. Clin. Microbiol.* **30:**3213–3216.

34. **Parasitology Subcommittee, Microbiology Section of Scientific Assembly, American Society of Medical Technology.** 1978. Recommended procedures for the examination of clinical specimens submitted for the diagnosis of parasitic infections. *Am. J. Med. Technol.* **44:**1101–1106.

35. **Public Health Service.** 1980. PHS interstate quarantine regulations. 42 CFR part 72, table 43.1. Interstate shipment of etiologic agents. *Fed. Regist.* **45:**28–29.

36. **Rayan, H. Z.** 2005. Microscopic overdiagnosis of intestinal amoebiasis. *J. Egypt. Soc. Parasitol.* **35:**941–951.

37. **Rosenblatt, J. E., L. M. Sloan, and S. K. Schneider.** 1993. Evaluation of an enzyme-linked immunosorbent assay for the detection of *Giardia lamblia* in stool specimens. *Diagn. Microbiol. Infect. Dis.* **16:**337–341.

38. **Sapero, J. J., and D. K. Lawless.** 1942. The MIF stain-preservation technique for the identification of intestinal protozoa. *Am. J. Trop. Med. Hyg.* **2:**613–619.

39. **Sawitz, W. G., and E. C. Faust.** 1942. The probability of detecting intestinal protozoa by successive stool examinations. *Am. J. Trop. Med.* **22:**131–136.

40. **Scholten, T. H., and J. Yang.** 1974. Evaluation of unpreserved and preserved stools for the detection and identification of intestinal parasites. *Am. J. Clin. Pathol.* **62:**563–567.

41. **Senay, H., and D. MacPherson.** 1989. Parasitology: diagnostic yield of stool examination. *Can. Med. Assoc. J.* **140:**1329–1331.

42. **Seybolt, L. M., D. Christiansen, and E. D. Barnett.** 2006. Diagnostic evaluation of newly arrived asymptomatic refugees with eosinophilia. *Clin. Infect. Dis.* **42:**363–367.

43. **U.S. Department of Transportation, Pipeline and Hazardous Materials Safety Administration.** 2005. Hazardous materials: infectious substances: harmonization with the United Nations recommendations; proposed rule (CFR 42, Parts 171, 172, 173, 175). *Fed. Regist.* **70:**29170–29187. (http://a257.g.akamaitech.net/7/257/2422/01jan20051800/edocket.access.gpo.gov/2005/pdf/05-9717.pdf.)

44. **U.S. Department of Transportation, Research and Special Programs Administration.** 2004. Harmonization with the United Nations Recommendations, International Maritime Dangerous Goods Code, and International Civil Aviation Organization's Technical Instructions; final rule (CFR 42, Parts 171, 172, et al.). *Fed. Regist.* **69:**76044–76187. (http://www.labsafety.com/refinfo/fedreg/FRPDF/122004.pdf.)

45. **U.S. Department of Transportation, Research and Special Programs Administration.** 2002. Hazardous materials: revision to standards for infectious substances and genetically modified organisms; final rule (CFR 42, Parts 171 et al.). *Fed. Regist.* **67:**53118–53144. (http://www.saftpak.com/HM226-Revised.pdf.)

46. **U.S. Postal Service.** 2006. Domestic mail manual. (http://pe.usps.com/DMMdownload.asp.)

47. **Wheatley, W.** 1951. A rapid staining procedure for intestinal amoebae and flagellates. *Am. J. Clin. Pathol.* **21:**990–991.

48. **World Health Organization.** 2004. *Transport of Infectious Substances. Background to the 17 Amendments Adopted in the 13th Revision of the United Nations Model Regulations Guiding the Transport of Infectious Substances.* World Health Organization, Geneva, Switzerland. (http://www.who.int/csr/resources/publications/WHO_CDS_CSR_LYO_2004_9/en.)

49. **World Health Organization.** 2005. *Guidance on Regulations for the Transport of Infectious Substances.* World Health Organization, Geneva, Switzerland. (http://www.who.int/csr/resources/publications/biosafety/WHO_CDS_CSR_LYO_2005_22/en.)

50. **Yang, J., and T. Scholten.** 1977. A fixative for intestinal parasites permitting the use of concentration and permanent staining procedures. *Am. J. Clin. Pathol.* **67:**300–304.

51. **Zimmerman, S. K., and C. A. Needham.** 1995. Comparison of conventional stool concentration and preserved-smear methods with Merifluor *Cryptosporidium/Giardia* direct immunofluorescence assay and ProSpecT Giardia EZ microplate assay for detection of *Giardia lamblia. J. Clin. Microbiol.* **33:**1942–1943.

Macroscopic and Microscopic Examination of Fecal Specimens

Macroscopic Examination

If the consistency of a stool specimen can be determined (formed, soft, or liquid), this information may give an indication of the organism stages that might be present. Trophozoites (potentially motile forms) of the intestinal protozoa are usually found in liquid specimens; both trophozoites and cysts might be found in a soft specimen; and the cyst forms are usually found in formed specimens. However, there are always exceptions to these general statements. Coccidian oocysts and microsporidian spores can be found in any type of fecal specimen; in the case of *Cryptosporidium* spp., the more liquid the stool, the more oocysts that are found in the specimen. Helminth eggs may be found in any type of specimen, although the chances of finding eggs in a liquid stool are reduced by the dilution factor. Tapeworm proglottids may be found on or beneath the stool on the bottom of the collection container. Adult pinworms and *Ascaris lumbricoides* are occasionally found on the surface or in the stool.

The presence of blood in or on the specimen may indicate several things and should always be reported. Dark stools may indicate bleeding high in the gastrointestinal tract, and fresh (bright red) blood most often is the result of bleeding at a lower level. In certain parasitic infections, blood and mucus may be present. Soft or liquid stool accompanied by blood is more suggestive of an amebic infection; these areas of blood and mucus should be carefully examined for the presence of trophic amebae. *Occult blood in the stool may or may not be related to a parasitic infection and could result from a number of different conditions.* Ingestion of various compounds may give a distinctive color to the stool (iron, black; barium, light tan to white).

Many laboratories prefer that stool specimens be submitted in some type of preservative. Rapid fixation of the specimen immediately after passage (by the patient) provides an advantage in terms of recovery and identification of intestinal protozoa. This advantage (preservation of organisms before distortion or disintegration) is thought to outweigh the limited motility information that might be gained by examining fresh specimens as direct wet mounts. Other laboratories still request a collection system that includes both a preserved specimen and the remainder of the fresh stool. Certainly cost is a factor, because several vials in the collection system cost more than a single vial containing preservative. Each laboratory will have to decide for itself, often basing the decision on the types of procedures ordered by the physicians who use the laboratory service, the test method selected (traditional methods, new immunoassay detection kits, or both), and the lag time between specimen collection and submission to the laboratory.

With increased emphasis on continuous quality improvement, managed-care contracts, cost containment, and the clinical relevance of diagnostic test results generated, compliance with specimen acceptance or rejection criteria has become more important and a necessary part of overall quality performance. *The generation of patient data begins with the quality of the specimen; anything that is done to compromise that quality should not be acceptable within the laboratory setting.*

If the specimen has not been preserved immediately after passage, it is important to know the age of the specimen when it reaches the laboratory. Freshly passed specimens are necessary for the detection of trophic amebae, flagellates, and ciliates. Liquid specimens must be examined within 30 min of passage (not 30 min from the time the specimen reaches the laboratory or is clocked in by the computer). Soft specimens should be examined within 1 h of passage. Immediate examination of a formed specimen is not as critical; however, if the stool cannot be examined on the day of collection, portions of the specimen should be preserved. In a routine laboratory setting, these time frames are often neither practical nor possible. Thus, the routine use of stool preservatives for diagnostic parasitology is highly recommended.

Microscopic Examination (Ova and Parasite Examination)

The microscopic examination of the stool specimen, normally called the ova and parasite examination, consists of three separate techniques: the direct wet smear, the concentration, and the permanent stained smear. Each of these methods is designed for a particular purpose and forms an integral part of the total examination (2, 3, 5, 8, 10, 14, 15, 22, 23, 25, 26, 30–32, 34, 37, 39–43, 46; J. Palmer, Letter, *Clin. Microbiol. Newsl.* 13:39–40, 1991). With increased emphasis on proper specimen collection and cost containment, the approach to the ova and parasite examination has changed somewhat during the last few years. Many laboratories are requesting that all fecal specimens be collected in preservatives prior to delivery to the laboratory to decrease the lag time between specimen passage and fixation, thus providing better organism morphology and subsequent identification.

Because preserved organisms do not exhibit motility, the direct wet smear is no longer considered a mandatory part of the routine ova and parasite examination. However, if fresh fecal specimens are delivered to the laboratory, the direct wet smear, particularly on liquid or very soft stools, should be performed.

In addition to normal specimen debris, the microscopic examination of fecal material may reveal the following:

1. Trophozoites and cysts of intestinal protozoa
2. Oocysts of coccidia and spores of microsporidia
3. Helminth eggs and larvae
4. Red blood cells (RBCs), which may indicate ulceration or other hemorrhagic problems
5. White blood cells (WBCs), specifically polymorphonuclear leukocytes (PMNs), which may indicate inflammation
6. Eosinophils, which usually indicate the presence of an immune response (which may or may not be related to a parasitic infection)
7. Macrophages, which may be present in bacterial or parasitic infections
8. Charcot-Leyden crystals, which may be found when disintegrating eosinophils are present (and may or may not be related to a parasitic infection)
9. Fungi (*Candida* spp.) and other yeasts and yeast-like fungi
10. Plant cells, pollen grains, or fungal spores, which may simulate some helminth eggs, protozoan cysts, coccidian oocysts, or microsporidial spores
11. Plant fibers or root or animal hairs, which may simulate helminth larvae

Direct Wet Smear

Normal mixing in the intestinal tract usually ensures an even distribution of organisms. However, depending on the level of infection, examination of the fecal material as a direct smear may or may not reveal organisms. The direct wet smear is prepared by mixing a small amount of stool (about 2 mg) with a drop of 0.85% NaCl; this mixture provides a uniform suspension under a 22- by 22-mm coverslip. Some workers prefer a 1.5- by 3-in. (1 in. = 2.54 cm) slide for the wet preparations rather than the standard 1- by 3-in. slide, which is routinely used for the permanent stained smear. A 2-mg sample of stool forms a low cone on the end of a wooden applicator stick. If more material is used for the direct mount, the suspension is usually too thick for an accurate examination; any sample of less than 2 mg results in the examination of too thin a suspension, thus decreasing the chances of finding organisms. If present, blood or mucus should always be examined as a direct mount. The entire 22- by 22-mm coverslip should be systematically examined with the low-power objective (10×) and low light intensity (Figure

27.1); any suspicious objects may then be examined with the high dry objective (40×). Use of an oil immersion objective (100×) on mounts of this kind is not routinely recommended unless the coverslip is sealed to the slide (a no. 1 thickness coverslip is recommended for oil immersion). For a temporary seal, a cotton-tipped applicator stick dipped in equal parts of heated paraffin and petroleum jelly should be used. Nail polish can also be used to seal the coverslip. *Many workers think that the use of the oil immersion objective on this type of preparation is impractical, especially since morphological detail is more readily seen by oil immersion examination of the permanent stained smear.* This is particularly true in a busy clinical laboratory situation.

The direct wet mount is used primarily to detect motile protozoan trophozoites. These organisms are very pale and transparent, two characteristics that require the use of low light intensity. Protozoan organisms in a saline preparation usually appear as refractile objects. If suspicious objects are seen on high dry power, at least 15 s should be allowed to detect motility of slowly moving protozoa. Application of heat by placing a hot penny on the edge of a slide may enhance the motility of trophic protozoa. Tapping on the coverslip can also stimulate the fluid to move; objects will roll over, thus providing a better view of the parasite or artifact. Helminth eggs and/or larvae, protozoan cysts, and coccidian oocysts may also be seen on the wet film, although these forms are more likely to be detected after fecal concentration procedures (Figure 27.2).

After the wet preparation has been thoroughly checked for trophic amebae, a drop of iodine can be placed at the edge of the coverslip or a new wet mount can be prepared with iodine alone (Figure 27.3). A weak iodine solution is recommended; too strong a solution may obscure the organisms. Several types of iodine are available; Lugol's and D'Antoni's are discussed here. Gram's iodine, used in bacterial work, is not recommended for staining parasitic organisms.

Figure 27.1 Method of scanning direct wet film preparation with a 10× objective. Note that the entire coverslip preparation should be examined before indicating the examination is negative. (Illustration by Nobuko Kitamura.)

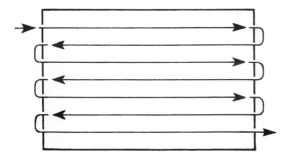

If preserved specimens are submitted to the laboratory, it is more cost-effective and clinically relevant to omit the direct smear and begin the stool examination with the concentration procedure, particularly since motile protozoa are not viable because of the prior addition of preservative. Even if parasites are seen on a direct mount of preserved stool, they would almost certainly be seen on the concentration examination as well as on the permanent stained smear (protozoa in particular). With few exceptions, intestinal protozoa should never be identified on the basis of a wet mount alone; permanent stained smears should be examined to confirm the specific identification of suspected organisms.

Saline (0.85% NaCl)

NaCl .... 0.85 g
Distilled water .... 100.0 ml

1. Dissolve the NaCl in distilled water in a flask or bottle, using a magnetic stirrer.
2. Distribute 10 ml into each of 10 screw-cap tubes.
3. Label as 0.85% NaCl with an expiration date of 1 year.
4. Sterilize by autoclaving at 121°C for 15 min.
5. When cool, store at 4°C.

D'Antoni's Iodine

Potassium iodide .... 1.0 g
Powdered iodine crystals. .... 1.5 g
Distilled water .... 100.0 ml

1. Dissolve the potassium iodide and iodine crystals in distilled water in a flask or bottle, using a magnetic stirrer.
2. The potassium iodide solution should be saturated with iodine, with some excess crystals left on the bottom of the bottle.
3. Store in a brown, glass-stoppered bottle at room temperature and in the dark.
4. This stock solution is ready for immediate use. Label as D'Antoni's iodine with an expiration date of 1 year (the stock solution remains good as long as an excess of iodine crystals remains on the bottom of the bottle).
5. Aliquot some of the iodine into a brown dropper bottle. The working solution should have a strong-tea color and should be discarded when the color lightens (usually within 10 to 14 days).

Note The stock and working solution formulas are identical, but the stock solution is held in the dark and will retain the strong-tea color while the working solution will fade and have to be periodically replaced (Figure 27.4).

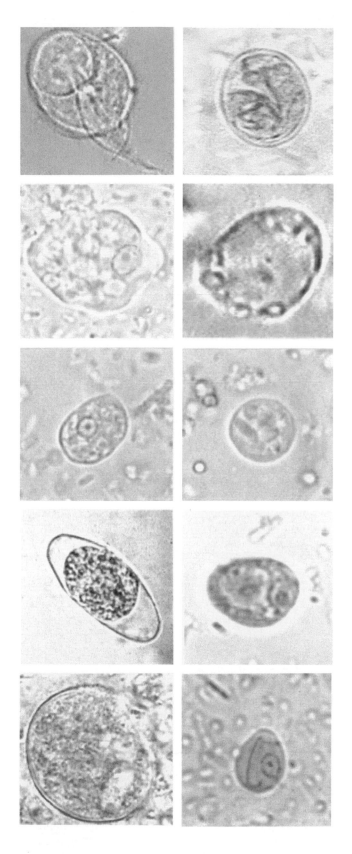

Lugol's Iodine

Potassium iodide.......................... 10.0 g
Powdered iodine crystals .... 5 g
Distilled water .... 100.0 ml

1. Follow the directions listed above for D'Antoni's iodine, including the expiration date of 1 year.
2. Dilute a portion 1:5 with distilled water for routine use (working solution).
3. Place this working solution into a brown dropper bottle. The working solution should have a strong-tea color and should be discarded when the color lightens (usually within 10 to 14 days).

Nair's Buffered Methylene Blue Stain for Trophozoites (Direct Smear)

Although not commonly used, Nair's buffered methylene blue stain is effective in showing nuclear detail in the trophozoite stages when used at a low pH; a pH range of 3.6 to 4.8 allows more active penetration of dye into the organism (35). After 5 to 10 min, the cytoplasm is stained a pale blue, with the nuclei being a darker blue; the slide should be examined within 30 min. Methylene blue (0.06% in an acetate buffer at pH 3.6) usually gives satisfactory results.

Acetate Buffer Solution Stock Solution A (0.2 M)
Acetic acid (CH_3COOH)....... 11.55 ml
Distilled water.................... 1,000.0 ml

Acetate Buffer Solution Stock Solution B (0.2 M)
Sodium acetate ($NaC_2H_3O_2$) ... 16.4 g
or
Sodium acetate
($NaC_2H_3O_2 \cdot 3H_2O$) .... 27.2 g
Distilled water.................... 1,000.0 ml

Mix the quantity of stock solutions A and B shown in the table on the following page and dilute with distilled water to a total of 100 ml.

Figure 27.2   Direct wet smear with saline. (Top row) *Giardia lamblia* trophozoite (left), *G. lamblia* cyst (right); (second row) *Entamoeba* sp. (probably *E. coli*) (left), *Blastocystis hominis* central body form (right); (third row) *Entamoeba hartmanni* trophozoite (left), *E. hartmanni* cyst (right); (fourth row) *Isospora belli* immature oocyst (left), *Iodamoeba bütschlii* cyst (right); (bottom row) *Balantidium coli* cyst (left), *Chilomastix mesnili* cyst (right).

Desired pH	Stock solution A (ml)	Stock solution B (ml)
3.6	46.3	3.7
3.8	44.0	6.0
4.0	41.0	9.0
4.2	36.8	13.2
4.4	30.5	19.5
4.6	25.5	24.5

Figure 27.3 Direct wet smear with saline and iodine. (Top) *Entamoeba coli* cyst with iodine; (middle) *E. coli* cyst with saline (left), *E. coli* cyst with iodine added (right); (bottom) *Iodamoeba bütschlii* cysts with iodine. Note that more detail can be seen once the iodine is added to the wet mount. Also, when iodine is used, the glycogen vacuole stains dark (brownish gold to brown) in the *Iodamoeba* cysts and is clearly visible.

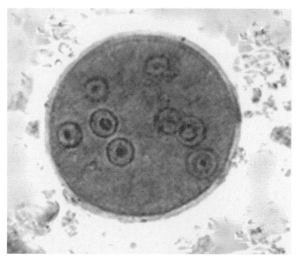

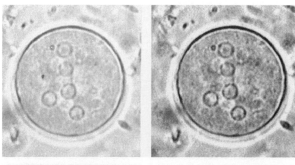

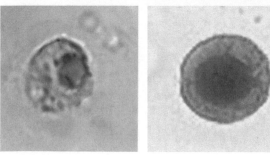

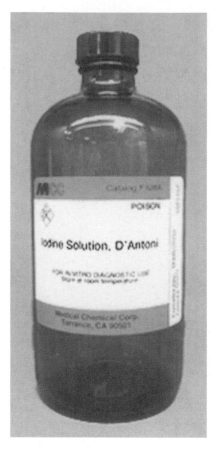

Figure 27.4 Commercially prepared D'Antoni's iodine; most commercial suppliers can provide this iodine solution.

Quality Control for Direct Smear

1. Check the working iodine solution each time it is used or periodically (once a week). The iodine and Nair's methylene blue solutions should be free of any signs of bacterial or fungal contamination.
2. The iodine should be the color of strong tea (discard if it is too light).
3. Protozoan cysts stained with iodine should contain yellow-gold cytoplasm, brown glycogen material, and paler refractile nuclei. The chromatoidal bodies may not be as clearly visible as they are in a saline mount. Human white blood cells (buffy coat cells) mixed with negative stool can be used as a quality control (QC) specimen. These human cells, when mixed with negative stool, mimic protozoan parasites. The human cells stain with the same color as that seen in the protozoa.
4. Protozoan trophozoite cytoplasm should stain pale blue and the nuclei should stain a darker blue with the methylene blue stain. Human WBCs mixed with negative stool should stain the same colors as seen with the protozoa.

5. The microscope should be calibrated (within the last 12 months), and the original optics used for the calibration should be in place on the microscope when objects are measured. Some microbiologists feel that calibration is not required on a yearly basis; however, if the microscope receives heavy use, is in a position where it can be bumped, or does not receive routine maintenance, yearly calibration is recommended. The calibration factors for all objectives should be posted on the microscope or close by for easy access.

6. All QC results should be appropriately recorded; the laboratory should also have an action plan for "out-of-control" results.

Procedure for Direct Wet Smear

1. Place 1 drop of 0.85% NaCl on the left side of the slide and 1 drop of iodine (working solution) on the right side of the slide. If preferred, two slides can be used instead of one. One drop of Nair's methylene blue can also be placed on a separate slide, although this technique is less common.

2. Take a small amount of fecal specimen (the amount picked up on the end of an applicator stick when introduced into the specimen), and thoroughly emulsify the stool in the saline and iodine preparations (use separate sticks for each).

3. Place a 22-mm coverslip (no. 1) on each suspension.

4. Systematically scan both suspensions with the 10× objective. The entire coverslip area should be examined under low power (total magnification, ×100).

5. If something suspicious is seen, the 40× objective can be used for more detailed study. At least one-third of the coverslip should be examined under high dry power (total magnification, ×400), even if nothing suspicious has been seen.

6. Another approach is to prepare and examine the saline mount and then add iodine at the side of the coverslip. The iodine will diffuse into the stool-saline mixture, providing some stain for a second examination. Remember, the iodine will kill any organisms present; thus, no motility will be seen after the iodine is added to the preparation.

Results and Patient Reports from Direct Wet Smear

Protozoan trophozoites and/or cysts and helminth eggs and larvae may be seen and identified. In a heavy infection with *Cryptosporidium* spp., *Cyclospora cayetanensis,* or *Isospora belli*, oocysts may be seen in a direct smear; however, some type of modified acid-fast stain or fecal immunoassay is normally used to detect *Cryptosporidium* spp., particularly when few oocysts are present. *Cyclospora*

oocysts are often confirmed using autofluorescence or the modified acid-fast stain. Spores of the microsporidia are too small, and the shape resembles other debris within the stool; therefore, they are not readily visible in a direct smear.

1. Motile trophozoites and protozoan cysts may or may not be identified to the species level (depending on the clarity of the morphology) and should be confirmed using the permanent stained smear.
   Examples: *Giardia lamblia* trophozoites
   Entamoeba coli cysts

2. Helminth eggs and/or larvae may be identified.
   Examples: *Ascaris lumbricoides* eggs
   Strongyloides stercoralis larvae

3. *Isospora belli* oocysts may be identified; however, *Cyclospora* and *Cryptosporidium* oocysts are generally too small to be recognized or identified without subsequent immunoassays or modified acid-fast staining.
   Example: *Isospora belli* oocysts

4. Artifacts and/or other structures may also be seen and reported as follows.

   Note These crystals and cells are quantitated; however, the quantity is usually assessed when the permanent stained smear is examined under oil immersion.
   Examples: Moderate Charcot-Leyden crystals
   Few RBCs
   Moderate PMNs

Procedure Notes for Direct Wet Smear

1. **In preserved specimens, the formalin replaces the saline and can be used as a direct smear; however, no organism motility will be visible (organisms are killed by 5 or 10% formalin). Consequently, the direct wet smear is usually not performed when the specimen (already preserved) arrives in the laboratory.** *The technical time is better spent performing the concentration and permanent stained smear. This approach is recommended for specimens submitted to the laboratory in preservative* (3).

2. As mentioned above, some workers prefer to make the saline and iodine mounts on separate slides and on 2- by 3-in. slides. Often, there is less chance of getting fluids on the microscope stage if separate slides (less total fluid on the slide and under the coverslip) or larger slides are used. Selection of slide size depends on the personal preference of laboratory personnel.

3. The microscope light should be reduced for low-power observations, since most organisms are overlooked with bright light due to limited contrast

of the internal morphology. This is particularly true when the preparation is being examined without the use of iodine. Illumination should be regulated so that some of the cellular elements in the feces show refraction. Most protozoan cysts and some coccidian oocysts are refractile under these light conditions.

Procedure Limitations for Direct Wet Smear

1. As mentioned above, because motility is lost when specimens are placed in preservatives, many laboratories are no longer performing the direct wet smear (the primary purpose is to see motility) but are proceeding directly to the concentration and permanent stained smear procedures as a better, more cost-effective use of personnel time, as well as a more clinically relevant approach.

2. Most of the time, results obtained from wet smear examinations should be confirmed by permanent stained smears. Some protozoa are very small and difficult to identify to the species level using just the direct wet smear technique. Confirmation is particularly important for *Entamoeba histolytica/ E. dispar* versus *Entamoeba coli*. Findings from the direct wet smear examination can be reported as "preliminary, based on the direct wet mount examination only," and the final report can be submitted after the concentration and permanent stain procedures are completed. However, if the labora-

tory turnaround time is less than 24 h, there is no need to send out a preliminary report; the final report can be submitted once the complete ova and parasite examination has been performed.

Concentration (Sedimentation and Flotation)

Fecal concentration has become a routine procedure as a part of the complete ova and parasite examination for parasites; it allows the detection of small numbers of organisms that may be missed by using only a direct wet smear (26, 43). There are two types of concentration procedures, sedimentation and flotation, both of which are designed to separate protozoan organisms and helminth eggs and larvae from fecal debris by centrifugation and/or differences in specific gravity (Figure 27.5) (3, 8).

Sedimentation methods (by centrifugation) lead to the recovery of all protozoa, oocysts, eggs, and larvae present; however, the concentration sediment that will be examined contains more debris. Although some workers recommend using both flotation and sedimentation procedures for every stool specimen submitted for examination, this approach is impractical for most laboratories. If one technique is selected for routine use, the sedimentation procedure is recommended as being the easiest to perform and the least subject to technical error (Figure 27.6).

A flotation procedure permits the separation of protozoan cysts, coccidian oocysts, and certain helminth eggs and larvae through the use of a liquid with a high specific

REVIEW	Direct Smear
Topic	**Comments**
Principle	To assess worm burden of patient, to provide quick diagnosis of heavily infected specimen, to check organism motility, and to diagnose organisms that might not be seen from concentration or permanent stain methods.
Specimen	Any fresh stool specimen that has not been refrigerated (liquid or soft stool).
Reagents	0.85% NaCl; Lugol's or D'Antoni's iodine; Nair's methylene blue can be used for trophozoites (not common).
Examination	Low-power examination ($\times$100) of entire 22- by 22-mm coverslip preparation (both saline and iodine); high dry power examination ($\times$400) of at least one-third of the coverslip area (both saline and iodine).
Results and Laboratory Reports	Results from the direct smear examination should often be considered presumptive; however, some organisms could be definitively identified (*Giardia lamblia* cysts and *Entamoeba coli* cysts, helminth eggs and larvae, *Isospora belli* oocysts). These reports should be categorized as "preliminary," while the final report would be available after the results of the concentration and permanent stained smear were available. However, if the testing turnaround time is <24 h, it is best to report the test results after the ova and parasite examination (fresh specimen, direct wet smear; concentration, permanent stained smear) is complete.
Procedure Notes and Limitations	Once iodine is added to the preparation, the organisms are killed and motility is lost. Specimens that arrive in the laboratory already preserved do not require a direct smear examination; the concentration and permanent stained smear should be performed instead. Direct smears are normally examined at low ($\times$100) and high dry ($\times$400) power; oil immersion examination ($\times$1,000) is not recommended (organism morphology not that clear).

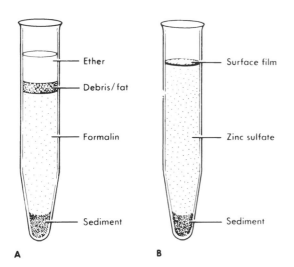

Figure 27.5 Fecal concentration procedures: various layers seen in tubes after centrifugation. (A) Formalin-ether (or ethyl acetate). The sediment should be well mixed, and a drop of sediment should be examined using the 10× low-power objective and the 40× high dry power objective. (B) Zinc sulfate (the surface film should be within 2 to 3 mm of the tube rim). Material from both the surface film and the sediment must be examined before the specimen is indicated as negative. Heavy or operculated helminth eggs do not float. (Illustration by Nobuko Kitamura.)

gravity. The parasitic elements are recovered in the surface film, and the debris remains in the bottom of the tube. This technique yields a cleaner preparation than does the sedimentation procedure; however, some helminth eggs (operculated eggs and/or very dense eggs such as unfertilized *Ascaris* eggs) do not concentrate well in the flotation method (Figure 27.7). The specific gravity may be increased, although this may produce more distortion in the eggs and protozoa. Laboratories that use only flotation procedures may fail to recover all of the parasites present; **to ensure detection of all organisms in the sample, both the surface film and the sediment should be carefully**

examined. Directions for any flotation technique must be followed exactly to produce reliable results.

Formalin-Ethyl Acetate Sedimentation Concentration

By centrifugation, the formalin-ethyl acetate sedimentation concentration procedure leads to the recovery of all protozoa, eggs, and larvae present; however, the preparation contains more debris than is found in the flotation procedure. Ethyl acetate is used as an extractor of debris and fat from the feces and leaves the parasites at the bottom of the suspension in the sediment. The formalin-ethyl acetate sedimentation concentration procedure is recommended as being the easiest to perform, allowing recovery of the broadest range of organisms, and being the least subject to technical error.

Figure 27.7 Flotation concentration. (Upper) *Fasciolopsis huski* egg (left), *Diphyllobothrium latum* egg (right). Note that both of these eggs in the top row are operculated and do not float in the zinc sulfate flotation concentration method; the opercula pop open, and the eggs fill with fluid and sink to the bottom of the tube. (Lower) Hookworm egg (left), *Trichuris trichiura* egg (right). These eggs concentrate using the flotation method and can be seen in the surface film. However, remember that both the surface film and the sediment must be examined by this method before reporting the final ova and parasite examination results.

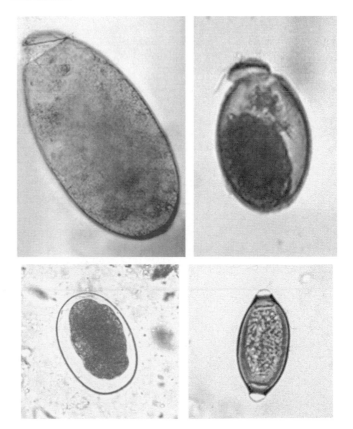

Figure 27.6 Sedimentation concentration. (Left) Unfertilized *Ascaris lumbricoides* egg. (Right) *Hymenolepis diminuta* egg.

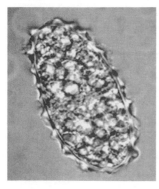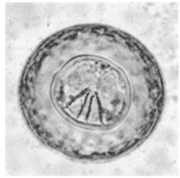

The specimen must be fresh or formalinized stool (5 or 10% buffered or nonbuffered formalin or sodium acetate-acetic acid-formalin [SAF]). Many of the single-vial preservative systems are also acceptable; however, the formulas are proprietary (e.g., UNIFIX; Medical Chemical Corp., Torrance, Calif.). Polyvinyl alcohol (PVA)-preserved specimens can also be used. However, PVA preservative formulations are rarely used for concentration methods in most laboratories but are highly recommended for the preparation of permanent stained smears.

5 or 10% Formalin
Formaldehyde (USP)
.... 100 ml (for 10%) or
.... 50 ml (for 5%)
Saline solution....... 900 ml (for 10%) or
0.85% NaCl ..... 950 ml (for 5%)

Note Formaldehyde is normally purchased as a 37 to 40% HCHO solution; however, for dilution, it should be considered to be 100%.

Dilute 100 ml of formaldehyde with 900 ml of 0.85% NaCl solution. (Distilled water may be used instead of saline solution.)

Quality Control for Sedimentation Concentration

1. Check the liquid reagents each time they are used; the formalin and saline should appear clear, without any visible contamination.
2. The microscope should be calibrated (within the last 12 months), and the objectives and oculars used for the calibration procedure should be in place on the microscope when objects are measured. The calibration factors for all objectives should be posted on the microscope or close by for easy access. Some researchers feel that a microscope does not require calibration every 12 months; however, if the microscope is moved periodically, can be easily bumped, or does not receive adequate maintenance, it should be rechecked yearly for calibration accuracy.
3. Known positive specimens should be concentrated and organism recovery should be verified at least quarterly and particularly after the centrifuge has been recalibrated. Human WBCs (buffy coat cells) mixed with negative stool can be used as a QC specimen. These human cells, when mixed with negative stool, can mimic protozoan parasites. The human cells concentrate just like human parasites such as protozoa and helminth eggs and larvae.
4. All QC results should be appropriately recorded; the laboratory should also have an action plan for "out-of-control" results.

Procedure for Sedimentation Concentration

1. Transfer 1/2 teaspoon (about 4 g) of fresh stool into 10 ml of 5 or 10% formalin in a shell vial, unwaxed paper cup, or round-bottom tube (the container may be modified to suit individual laboratory preferences). Mix the stool and formalin thoroughly, and let the mixture stand for a minimum of 30 min for fixation. If the specimen is already in 5 or 10% formalin (or SAF or other non-PVA single-vial preservatives), stir the stool-preservative mixture.
2. Depending on the amount and viscosity of the specimen, strain a sufficient quantity through wet gauze (no more than two layers of gauze or one layer if the new "pressed" gauze [e.g., Johnson & Johnson nonsterile three-ply gauze, product 7636] is used) into a conical 15-ml centrifuge tube to give the desired amount of sediment (0.5 to 1 ml) for step 3 below. Usually, 8 ml of the stool-formalin mixture prepared in step 1 is sufficient. If the specimen is received in a vial of preservative (5 or 10% formalin, SAF, or other single-vial preservatives), approximately 3 to 4 ml of the preservative-stool mixture is sufficient for testing. If the vial contains very little specimen, then the entire amount may be used in the procedure. *If the specimen contains a lot of mucus, do not strain through gauze but immediately fix in 5 or 10% formalin for 30 min and centrifuge for 10 min at 500 × g. Proceed directly to step 10.*
3. Add 0.85% NaCl or 5 or 10% formalin (some workers prefer to use formalin for all rinses) almost to the top of the tube, and centrifuge for 10 min at 500 × g. The amount of sediment obtained should be approximately 0.5 to 1 ml.
4. Decant and discard the supernatant fluid, and resuspend the sediment in saline or formalin; add saline or formalin almost to the top of the tube, and centrifuge again for 10 min at 500 × g. This second wash may be eliminated if the supernatant fluid after the first wash is light tan or clear. Some prefer to limit the washing to one step (regardless of the clarity or color of the supernatant fluid after centrifugation) to eliminate additional manipulation of the specimen prior to centrifugation. The more the specimen is manipulated and/or rinsed, the more likely it is that some organisms will be lost and accidentally discarded prior to examination.
5. Decant and discard the supernatant fluid, and resuspend the sediment on the bottom of the tube in 5 or 10% formalin. Fill the tube half full only. If the amount of sediment left in the bottom of

the tube is very small or the original specimen contained a lot of mucus, do not add ethyl acetate in step 6; merely add the formalin, spin, decant, and examine the remaining sediment.

6. Add 4 to 5 ml of ethyl acetate. Stopper the tube, and shake it vigorously for at least 30 s. Hold the tube so that the stopper is directed away from your face.

7. After a 15- to 30-s wait, carefully remove the stopper.

8. Centrifuge for 10 min at 500 × g. Four layers should result: a small amount of sediment (containing the parasites) in the bottom of the tube; a layer of formalin; a plug of fecal debris on top of the formalin layer; and a layer of ethyl acetate at the top.

9. Free the plug of debris by ringing the plug with an applicator stick; decant and discard all of the supernatant fluid. After proper decanting, a drop or two of fluid remaining on the side of the tube may run down into the sediment. Mix this fluid with the sediment.

10. If the sediment is still somewhat solid, add 1 or 2 drops of saline or formalin to the sediment, mix, add a small amount of material to a slide, add a coverslip (22 by 22 mm, no. 1), and examine.

11. Systematically scan with the 10× objective. The entire coverslip area should be examined under low power (total magnification, ×100).

12. If something suspicious is seen, the 40× objective can be used for more detailed study. At least one-third of the coverslip should be examined under high dry power (total magnification, ×400), even if nothing suspicious has been seen. As in the direct wet smear, iodine can be added to enhance morphological detail, and the coverslip can be tapped to see objects move and turn over.

Results and Patient Reports from Sedimentation Concentration

Protozoan trophozoites and/or cysts and helminth eggs and larvae may be seen and identified. Protozoan trophozoites are less likely to be seen. In a heavy infection with *Cryptosporidium* spp. or *C. cayetanensis*, oocysts may be seen in the concentrate sediment; oocysts of *I. belli* can also be seen. Spores of the microsporidia are too small, and the shape resembles that of other debris within the stool; therefore, they are not readily visible in the concentration sediment.

1. Protozoan cysts may or may not be identified to the species level (depending on the clarity of the morphology).

Examples: *Entamoeba coli* cysts
          Giardia lamblia cysts

2. Helminth eggs and/or larvae may be identified.
Examples: *Ascaris lumbricoides* eggs
          Hookworm larvae

3. *I. belli* oocysts may be identified; however, *Cyclospora* and *Cryptosporidium* oocysts are generally too small to be recognized or identified by subsequent immunoassays or modified acid-fast staining.
Example: *Isospora belli* oocysts

4. Artifacts and/or other structures may also be seen and reported as follows.

Note These crystals and cells are quantitated; however, the quantity is usually assessed when the permanent stained smear is examined under oil immersion).
Examples: Moderate Charcot-Leyden crystals
          Few RBCs
          Moderate PMNs

Procedure Notes for Sedimentation Concentration

1. The gauze should never be more than one (pressed gauze) or two (woven gauze) layers thick; more gauze may trap mucus (containing *Cryptosporidium* oocysts and/or microsporidial spores).

2. Tap water may be substituted for 0.85% NaCl throughout this procedure, although the addition of water to fresh stool causes *Blastocystis hominis* cyst (central body) forms to rupture and is not recommended. In addition to the original 5 or 10% formalin fixation, some workers prefer to use 5 or 10% formalin for all rinses throughout the procedure.

3. Ethyl acetate is widely recommended as a substitute for ether (37). It can be used in the same way in the procedure and is much safer. Hemo-De can also be used and is thought to be safer than ethyl acetate (35).

A. After the plug of debris is rimmed and excess fluid is decanted, *the sides of the tube can be swabbed with a cotton-tipped applicator stick while the tube is still upside down to remove excess ethyl acetate. This is particularly important if you are working with plastic centrifuge tubes or the plastic commercial concentrators.* If the sediment is too dry after the tube has been swabbed, add several drops of saline before preparing the wet smear for examination.

B. If there is excess ethyl acetate in the smear of the sediment prepared for examination, bubbles will be present, which will obscure the material of interest.

4. If specimens are received in SAF, begin the procedure at step 2.

5. *If specimens are received in PVA, the first two steps of the procedure should be modified as follows.*

   A. Immediately after stirring the stool-PVA mixture with applicator sticks, pour approximately half of the mixture into a tube (container optional) and add 0.85% NaCl (or 5 or 10% formalin) almost to the top of the tube.

   B. Filter the stool-PVA-saline (or formalin) mixture through wet gauze into a 15-ml centrifuge tube. Follow the standard procedure from here to completion, beginning with step 3.

6. Too much or too little sediment will result in an ineffective concentration sediment examination.

7. The centrifuge should reach the recommended speed before the centrifugation time is monitored. *However, since most laboratories have their centrifuges on automatic timers, the centrifugation time in this protocol takes into account the fact that some time will be spent coming up to speed prior to full-speed centrifugation. If the centrifugation time at the proper speed (10 min at 500 × g) is reduced, some of the organisms* (Cryptosporidium *and* Cyclospora *oocysts or microsporidian spores) may not be recovered in the sediment.*

Procedure Limitations for Sedimentation Concentration

1. Results obtained with wet smears (direct wet smears or concentration sediment wet smears) should usually be confirmed by permanent stained smears. Some protozoa are very small and difficult to identify as to species with just the direct wet smears. Also, special stains are sometimes necessary for organism identification.

2. Confirmation is particularly important for *E. histolytica*/*E. dispar* versus *E. coli*.

3. Certain organisms (*G. lamblia*, hookworm eggs, and occasionally *Trichuris* eggs) may not concentrate as well from PVA-preserved specimens as they do from those preserved in formalin. However, if enough *G. lamblia* organisms are present to concentrate from formalin, PVA should contain enough for detection on the permanent

stained smear. In clinically important infections, the number of helminth eggs present would ensure detection regardless of the type of preservative used. Also, the morphology of *Strongyloides stercoralis* larvae is not as clear from specimens in PVA as from specimens fixed in formalin.

4. For unknown reasons, *I. belli* oocysts are routinely missed in the concentrate sediment when concentrated from PVA-preserved specimens. The oocysts would be found if the same specimen were preserved in formalin rather than PVA.

5. In past publications, recommended centrifugation times have not taken into account potential problems with the recovery of *Cryptosporidium* oocysts. **There is anecdotal evidence strongly indicating that *Cryptosporidium* oocysts may be missed unless the centrifugation speed is 500 × g for a minimum of 10 min.**

6. Adequate centrifugation time and speed have become very important for recovery of microsporidial spores. In some of the earlier publications, use of uncentrifuged material was recommended. **However, we have found that centrifugation for 10 min at 500 × g definitely increases the number of microsporidial spores available for staining and subsequent examination.**

Iodine-Trichrome Stain for Sediment

A combination of Lugol's iodine solution and trichrome stain can be used to stain fecal sediment from the concentration procedure (18). Coloring the eggs and cysts yellow-brown (iodine) and the debris green (trichrome) provides contrast which facilitates the detection of parasites. The use of such an approach usually depends on personal preferences and the results of parallel trials of the current method and new methods being considered. This wet examination can be used as an adjunct procedure *but does not take the place of the unstained wet examination of the sediment.*

Quality Control for Iodine-Trichrome Stain for Sediment

1. Check the working iodine solution each time it is used or periodically (once a week). The iodine and trichrome stain solutions should be free of any signs of bacterial or fungal contamination.

2. The iodine should be the color of strong tea (discard if it is too light).

3. Protozoan cysts stained with iodine should contain yellow-gold cytoplasm, brown glycogen material, and paler refractile nuclei. The chromatoidal bodies may not be as clearly visible as in a saline mount. Human WBCs (buffy coat cells) mixed

with negative stool can be used as a QC specimen. The human cells stain with the same color as that seen in the protozoa. The background debris stains green from the components of the trichrome stain.

4. If appropriate due to extensive use and/or lack of routine maintenance, the microscope should be calibrated (within the last 12 months), and the original optics used for the calibration should be in place on the microscope when objects are being measured. The calibration factors for all objectives should be posted on the microscope or close by for easy access.

5. All QC results should be appropriately recorded; the laboratory should also have an action plan for "out-of-control" results.

Procedure for Iodine-Trichrome Stain for Sediment

1. Place 4 drops of Lugol's iodine solution into a test tube.
2. Place 4 drops of fecal concentrate into the test tube. Mix well.
3. Place 2 drops of the Lugol's iodine solution-fecal concentrate mixture from step 2 on a glass slide.
4. Add 1 drop of trichrome stain. Mix with a wooden applicator, and cover with a coverslip (22 by 22 mm, no. 1).
5. Microscopically examine the entire preparation under low power ($\times$100) and at least one-third of the area under high dry power ($\times$400).

Results and Patient Reports from Iodine-Trichrome Stain for Sediment

Protozoan trophozoites and/or cysts and some helminth eggs and larvae may be seen and identified. Lugol's iodine stains *A. lumbricoides* and *Taenia* eggs quite dark; these eggs are difficult to recognize and may be mistaken for debris, hence the need for the unstained sediment examination. In a heavy infection with *Cryptosporidium* spp., oocysts may be seen in a direct smear; however, some type of modified acid-fast stain or fecal immunoassay is normally used to detect these organisms, particularly when few oocysts are present. Oocysts of *I. belli* can also be seen in a direct smear. *Cyclospora* oocysts may not be recognized because they resemble debris. Spores of the microsporidia are too small, and the shape resembles that of other debris within the stool; therefore, they are not readily visible in a direct smear.

1. Protozoan cysts may or may not be identified to the species level (depending on the clarity of the morphology).
   Example: *Iodamoeba bütschlii* cysts

2. Helminth eggs and/or larvae may be identified.
   Example: *Hymenolepis nana* eggs

3. *I. belli* oocysts may be identified; however, *Cyclospora* and *Cryptosporidium* oocysts are generally too small to be recognized or identified by subsequent immunoassays or modified acid-fast staining.
   Example: *Isospora belli* oocysts

4. Artifacts and/or other structures may also be seen and reported as follows.

   Note These cells are quantitated; however, the quantity is usually assessed when the permanent stained smear is examined under oil immersion.

   Example: Moderate PMNs

Procedure Notes for Iodine-Trichrome Stain for Sediment

1. As mentioned above, some workers prefer to make the wet mounts on larger slides. Often, there is less chance of getting fluids on the microscope stage if larger slides are used.
2. This stain is darker than the traditional iodine stain. The microscope light should be increased over that used for the unstained wet smear examination.

Procedure Limitations for Iodine-Trichrome Stain for Sediment

1. Because this is a darker stain than routine iodine stains, it is important to also examine a saline wet smear. This is particularly important because *A. lumbricoides* and *Taenia* eggs stain too dark with the iodine and may not be recognized as helminth eggs.
2. Results obtained with wet smears should usually be confirmed by permanent stained smears. Some protozoa are very small and difficult to identify to the species level with just the direct wet smears. Confirmation is particularly important for *E. histolytica*/*E. dispar* versus *E. coli*. These findings can be reported as "preliminary report, based on direct wet smear examination only," and the final report can be submitted after the concentration and permanent stain procedures are completed. However, if the examination turnaround time is approximately 24 h or less, there is no need for a preliminary report; the final report can be submitted after completion of the concentration and permanent stained smear examinations.

Zinc Sulfate Flotation Concentration

The flotation procedure permits the separation of protozoan cysts and eggs of certain helminths from excess debris through the use of a liquid (zinc sulfate) with a high specific gravity. The parasitic elements are recovered in the surface film, and the debris and some heavy parasitic elements remain in the bottom of the tube. This technique yields a cleaner preparation than does the sedimentation procedure; however, some helminth eggs (operculated and/or very dense eggs, such as unfertilized *Ascaris* eggs) do not concentrate well in the flotation method; a sedimentation technique is recommended to detect these infections.

When the zinc sulfate solution is prepared, the specific gravity should be 1.18 for fresh stool specimens; it must be checked with a hydrometer. This procedure may be used on formalin-preserved specimens if the specific gravity of the zinc sulfate is increased to 1.20; however, this usually causes more distortion in the organisms present and is not recommended for routine clinical use. **To ensure detection of all possible organisms, both the surface film and the sediment must be examined.** For most laboratories, this is not a practical approach.

The specimen must be fresh or formalinized stool (5 or 10% buffered or nonbuffered formalin, SAF, or other non-PVA single-vial preservatives). PVA-preserved specimens can also be used; however, this approach is not commonly used or recommended.

> **Zinc Sulfate (33% Aqueous Solution)**
> Zinc sulfate .... 330 g
> Distilled water .... 670 ml

1. Dissolve the zinc sulfate in distilled water in an appropriate flask or beaker, using a magnetic stirrer.
2. Adjust the specific gravity to 1.20 by the addition of more zinc sulfate or distilled water. Use a specific gravity of 1.18 with fresh stool (nonformalinized).
3. Store in a glass-stoppered bottle with an expiration date of 24 months.

Quality Control for Flotation Concentration

1. Check the reagents each time they are used. The formalin, saline, and zinc sulfate should appear clear, without any visible contamination.
2. The microscope should be calibrated (within the last 12 months), and the objectives and oculars used for the calibration procedure should be used for all measurements on the microscope. The calibration factors for all objectives should be posted on the microscope or close by for easy access. As

mentioned above, some workers feel that recalibration of the microscope is not necessary each year; however, this would depend on the use and maintenance of that particular piece of equipment.

3. Known positive specimens should be concentrated and organism recovery should be verified at least quarterly, particularly after the centrifuge has been recalibrated. Human WBCs (buffy coat cells) mixed with negative stool can be used as a QC specimen. These human cells, when mixed with negative stool, mimic human parasites. The human cells concentrate just like human parasites such as protozoa and helminth eggs and larvae.
4. All QC results should be appropriately recorded; the laboratory should also have an action plan for "out-of-control" results.

Procedure for Flotation Concentration

1. Transfer 1/2 teaspoon (about 4 g) of fresh stool into 10 ml of 5 or 10% formalin in a shell vial, unwaxed paper cup, or round-bottom tube (the container may be modified to suit individual laboratory preferences). Mix the stool and formalin thoroughly. Let the mixture stand for a minimum of 30 min for fixation. If the specimen is already in 5 or 10% formalin (or SAF), stir the stool-formalin (or SAF) mixture.
2. Depending on the size and density of the specimen, strain a sufficient quantity through wet gauze (no more than two layers of gauze or one layer if the new "pressed" gauze [e.g., Johnson & Johnson nonsterile three-ply gauze, product 7636] is used) into a conical 15-ml centrifuge tube to give the desired amount of sediment (0.5 to 1 ml) in step 3 below. Usually, 8 ml of the stool-formalin mixture prepared in step 1 is sufficient. If the specimen is received in vials of preservative (5 or 10% formalin, SAF, or other single-vial preservatives), approximately 3 to 4 ml of the mixture is sufficient unless the specimen has very little stool in the vial. *If the specimen contains a lot of mucus, do not strain through gauze but immediately fix in 5 or 10% formalin for 30 min and centrifuge for 10 min at 500 × g. Proceed directly to step 5.*
3. Add 0.85% NaCl almost to the top of the tube, and centrifuge for 10 min at 500 × g. Approximately 0.5 to 1 ml of sediment should be obtained. Too much or too little sediment results in an ineffective concentration examination.
4. Decant and discard the supernatant fluid, resuspend the sediment in 0.85% NaCl almost to the top of the tube, and centrifuge for 10 min at 500

× *g*. This second wash may be eliminated if the supernatant fluid after the first wash is light tan or clear. Some prefer to limit the washing to one step (regardless of the color and clarity of the supernatant fluid) to eliminate additional manipulation of the specimen prior to centrifugation. The more the specimen is manipulated and/or rinsed, the more likely it is for parasitic elements to be lost.

5. Decant and discard the supernatant fluid, and resuspend the sediment on the bottom of the tube in 1 to 2 ml of zinc sulfate. Fill the tube within 2 to 3 mm of the rim with additional zinc sulfate.

6. Centrifuge for 2 min at 500 × *g*. Allow the centrifuge to come to a stop without interference or vibration. Two layers should result: a small amount of sediment in the bottom of the tube, and a layer of zinc sulfate. The protozoan cysts and some helminth eggs are found in the surface film; some operculated and/or heavy eggs are found in the sediment.

7. Without removing the tube from the centrifuge, remove 1 or 2 drops of the surface film with a Pasteur pipette or a freshly flamed (and allowed to cool) wire loop and place them on a slide. Do not use the loop as a "dipper"; simply touch the surface (bend the loop portion of the wire 90° so that the loop is parallel with the surface of the fluid). Make sure the pipette tip or wire loop is not below the surface film (Figure 27.8).

8. Add a coverslip (22 by 22 mm, no. 1) to the preparation. Iodine may be added to the preparation.

9. Systematically scan with the 10× objective. The entire coverslip area should be examined under low power (total magnification, ×100).

10. If something suspicious is seen, the 40× objective can be used for more detailed study. At least one-third of the coverslip should be examined with high dry power (total magnification, ×400), even if nothing suspicious has been seen. As in the direct wet smear, iodine can be added to enhance

morphological detail, and the coverslip can be gently tapped to observe objects moving and turning over.

Results and Patient Reports from Flotation Concentration

Protozoan trophozoites and/or cysts and some helminth eggs and larvae may be seen and identified. Heavy helminth eggs and operculated eggs do not float in zinc sulfate; they are seen in the sediment within the tube. The high specific gravity of the zinc sulfate causes the opercula to pop open; the eggs fill with fluid and sink to the bottom. Protozoan trophozoites are less likely to be seen. In a heavy infection with *Cryptosporidium* spp. or *C. cayetanensis*, oocysts may be seen in the concentrate sediment; oocysts of *I. belli* can also be seen. Spores of the microsporidia are too small, and the shape resembles other debris within the stool; therefore, they are not readily visible in the concentration sediment.

1. Protozoan cysts may or may not be identified to the species level (depending on the clarity of the morphology).
   Example: *Giardia lamblia* cysts
2. Helminth eggs and/or larvae may be identified.
   Example: Hookworm eggs
3. *I. belli* oocysts may be identified; however, *Cyclospora* and *Cryptosporidium* oocysts are generally too small to be recognized or identified by subsequent immunoassays or modified acid-fast staining.
   Example: *Isospora belli* oocysts
4. Artifacts and/or other structures may also be seen and reported as follows.
   Note These cells are quantitated; however, the quantity is usually assessed when the permanent stained smear is examined.
   Examples: Few macrophages
               Moderate PMNs

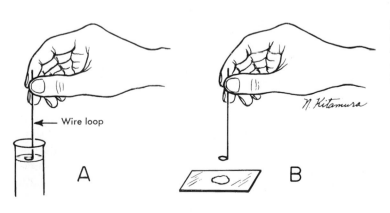

Figure 27.8   Method used to remove the surface film in the zinc sulfate flotation concentration procedure. (A) A wire loop is gently placed on (not under) the surface film. (B) The loop is then placed on a glass slide. (Illustration by Nobuko Kitamura.)

Procedure Notes for Flotation Concentration

1. The gauze should never be more than one or two layers thick; more gauze may trap mucus (containing *Cryptosporidium* oocysts, *Cyclospora* oocysts, and/or microsporidial spores). A round-bottom tube is recommended rather than a centrifuge tube.
2. Tap water may be substituted for 0.85% NaCl throughout this procedure, although the addition of water to fresh stool causes *B. hominis* cyst (central body) forms to rupture and is not recommended. In addition to the original 5 or 10% formalin fixation, some workers prefer to use 5 or 10% formalin for all rinses throughout the procedure.
3. If fresh stool is used (nonformalin preservatives), the zinc sulfate should be prepared with a specific gravity of 1.18. If formalinized specimens are to be concentrated, the zinc sulfate should have a specific gravity of 1.20.
4. If specimens are received in SAF or other single-vial preservatives, begin the procedure at step 2.
5. If fresh specimens are received, the standard procedure requires the stool to be rinsed in distilled water prior to the addition of zinc sulfate in step 4. *However, the addition of fresh stool to distilled water will destroy any* B. hominis *cysts present and is not a recommended approach.*
6. Some workers prefer to remove the tubes from the centrifuge prior to sampling the surface film. This is acceptable; however, there is more chance that the surface film will be disturbed prior to sampling.
7. Some workers prefer to add a small amount of zinc sulfate to the tube so that the fluid forms a slightly convex meniscus. A coverslip is then placed on top of the tube so that the undersurface touches the meniscus. It is left undisturbed for 5 min. The coverslip is then carefully removed and placed on a slide for examination. This approach tends to be somewhat messy, particularly if too much zinc sulfate has been added.
8. When using the hydrometer (solution at room temperature), mix the solution well. Float the hydrometer in the solution, giving it a slight twist to ensure that it is completely free from the sides of the container. Read the bottom meniscus and correct the figure for temperature, if necessary. Most hydrometers are calibrated at 20°C. A difference of 3°C between the solution temperature (room temperature) and the hydrometer calibration temperature requires a correction of 0.001, to be added if above and subtracted if below 20°C.

Procedure Limitations for Flotation Concentration

1. Results obtained with wet smears (direct wet smears or concentrated specimen wet smears) should usually be confirmed by permanent stained smears. Some protozoa are very small and difficult to identify to the species level with just the direct wet smears. Also, special stains are sometimes necessary for organism identification.
2. Confirmation is particularly important for *E. histolytica*/*E. dispar* versus *E. coli*.
3. Protozoan cysts and thin-shelled helminth eggs are subject to collapse and distortion when left for more than a few minutes in contact with the high-specific-gravity zinc sulfate. The surface film should be removed for examination within 5 min of the time the centrifuge comes to a stop. The longer the organisms are in contact with the zinc sulfate, the more distortion will be seen on microscopic examination of the surface film.
4. *Since most laboratories have their centrifuges on automatic timers, the centrifugation time in this protocol takes into account the fact that some time will be spent coming up to speed prior to full-speed centrifugation.*
5. **If zinc sulfate is the only concentration method used, both the surface film and the sediment should be examined to ensure detection of all possible organisms.**

Commercial Fecal Concentration Devices

There are a number of commercially available fecal concentration devices which may help a laboratory to standardize the concentration technique. Standardization is particularly important when personnel rotate throughout the laboratory and may not be familiar with parasitology techniques. These devices help ensure consistency, thus leading to improved parasite recovery and subsequent identification. Some of the systems are enclosed and provide a clean, odor-free approach to stool processing, features that may be important to nonmicrobiology personnel processing such specimens. Both 15- and 50-ml systems are available. It is important to remember that a maximum of 0.5 to 1.0 ml of sediment is needed in the bottom of the tube. Often, when the 50-ml systems are used, there is too much sediment in the bottom of the tube. This problem can be solved by adding less of the fecal specimen to the concentration system prior to centrifugation. Since the sediment is normally mixed thoroughly and 1 drop is taken to a coverslip for examination, good mixing may not occur if too much sediment is used. There also appears to be layering in the bottom of the tubes; again, adding less material to the concentrator at the beginning should help eliminate this problem (Figures 27.9 through 27.12).

Automated Workstation for the Microscopic Analysis of Fecal Concentrates

The FE-2 (DiaSys Corp., Waterbury, Conn.) is a countertop workstation that automates the microscopic analysis of fecal concentrates (Figure 27.13). The system automates the aspiration, resuspension, staining or diluting (based on user preference), transfer, presentation, and disposal of fecal concentrates. When the sample button is pressed, within 5 s two samples of fecal concentrate are automatically and simultaneously aspirated from the concentrate tube and transported to the glass dual-flow-cells of the Optical Slide Assembly (Figure 27.14). Based on user preference, the FE-2 also simultaneously stains or dilutes one of the two samples to be examined. After the microscopic examination of the fecal suspension within the glass viewing chambers, the flow chambers can be purged and cleaned and are ready for the next specimen. The dual-flow-cell Optical Slide Assembly is designed to fit within the stage clips of any standard upright microscope. The Optical Slide Assembly accommodates bright-field, phase-contrast, polarized-light, and other common forms of microscopy. The system can be moved from one microscope to another or can be set up as a semipermanent station for fecal concentrate microscopy. Removal to another microscope just involves removing the Optical Slide Assembly from the microscope stage.

Permanent Stained Smear

The detection and correct identification of many intestinal protozoa frequently depend on the examination of the permanent stained smear with the oil immersion lens ($100\times$ objective). These slides not only provide the microscopist with a permanent record of protozoan organisms identified but also may be used for consultations with specialists when unusual morphologic characteristics are found. Considering the morphologic variations that are possible, organisms that are very difficult to identify and do not fit the pattern for any one species may be found. A routine work flow diagram including the permanent stained smear is shown in Figure 27.15.

Although an experienced microscopist can occasionally identify certain organisms on a wet preparation, most identifications should be considered tentative until confirmed by the permanent stained slide. The smaller protozoan organisms are frequently seen on the stained smear when they are easily missed with only the direct smear and concentration methods. *For these reasons, the*

REVIEW	Concentration
Topic	**Comments**
Principle	To concentrate the parasites present, either through sedimentation or by flotation. The concentration is specifically designed to allow recovery of protozoan cysts, coccidian oocysts, microsporidian spores, and helminth eggs and larvae.
Specimen	Any stool specimen that is fresh or preserved in formalin, PVA (mercury- or non-mercury based), SAF, MIF, or the newer single-vial-system fixatives.
Reagents	5% or 10% formalin, ethyl acetate, zinc sulfate (specific gravity, 1.18 for fresh stool and 1.20 for preserved stool); 0.85% NaCl; Lugol's or D'Antoni's iodine.
Examination	Low-power examination ($\times100$) of entire 22- by 22-mm coverslip preparation (iodine recommended but optional); high dry power examination ($\times400$) of at least one-third of the coverslip area (both saline and iodine).
Results and Laboratory Reports	Often, results from the concentration examination should be considered presumptive; however, some organisms could be definitively identified (*Giardia lamblia* cysts and *Entamoeba coli* cysts, helminth eggs and larvae, *Isospora belli* oocysts). These reports should be categorized as "preliminary," while the final report would be available after the results of the concentration and permanent stained smear were available. However, if the testing turnaround time is <24 h, it is best to report the test results after the ova and parasite examination (fresh specimen, direct wet smear; concentration, permanent stained smear) is complete.
Procedure Notes and Limitations	Formalin-ethyl acetate sedimentation concentration is the most commonly used. Zinc sulfate flotation does not detect operculated or heavy eggs; both the surface film and sediment must be examined before a negative result is reported. Smears prepared from concentrated stool are normally examined at low ($\times100$) and high dry ($\times400$) power; oil immersion examination ($\times1,000$) is not recommended (organism morphology is not that clear). The addition of too much iodine may obscure helminth eggs (will mimic debris).

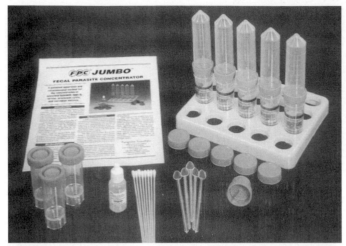

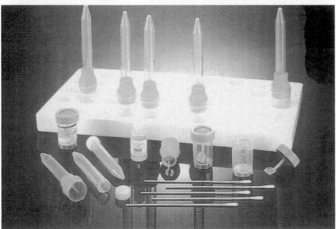

Figure 27.9 (Upper) FPC JUMBO large concentration tubes and connector system (Evergreen Scientific). (Lower) Small concentration tubes and FPC HYBRID connector system (Evergreen Scientific).

permanent stain is recommended for every stool sample submitted for a routine parasite examination. It is also important to remember that the fecal immunoassays for specific organisms have proven to be more sensitive than the routine or specialized stains for *Giardia lamblia* or *Cryptosporidium* spp. (16, 33).

There are a number of staining techniques available; selection of a particular method may depend on the degree of difficulty of the procedure and the amount of time necessary to complete the stain. The older classical method is the long Heidenhain iron hematoxylin method; however, for routine diagnostic work, most laboratories select one of the shorter procedures, such as the trichrome method or one of the modified methods involving iron hematoxylin. Other procedures are available (7, 8, 11, 13–15, 18, 20, 22, 31, 39, 42, 47; J. Palmer, Letter, *Clin. Microbiol. Newsl.* **13:**39–40, 1991), and some of them are presented here.

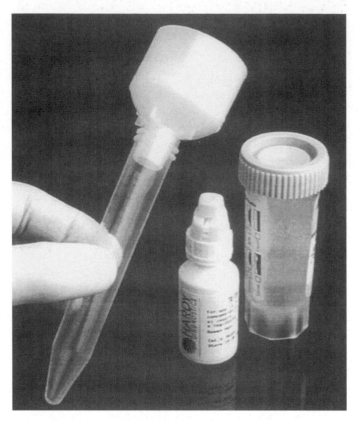

Figure 27.10 Stool collection vial and funnel used in fecal concentration (Hardy Diagnostics).

Most problems encountered in the staining of protozoan trophozoites and cysts in fecal smears occur because the specimen is too old, the smears are too dense, the smears are allowed to dry before fixation, or fixation is inadequate. There is variability in fixation in that immature cysts fix more easily than mature cysts, and *E. coli* cysts require a longer fixation time than do those of other species. It is critical that adequate mixing occur between the fecal specimen and preservative (7, 11, 13, 21, 36, 47).

Preparation of Material for Staining

Fresh Material

1. When the specimen arrives, prepare two slides with applicator sticks or brushes and *immediately* (without drying) place them in Schaudinn's fixative. Allow the slides to fix for a minimum of 30 min; overnight fixation is acceptable. The amount of fecal material smeared on the slide should be thin enough that newsprint can be read through the smear. Smears preserved in liquid Schaudinn's fixative should be placed in 70% alcohol to remove the excess fixative prior to placement in

iodine-alcohol (used for mercury-based fixatives).

2. If the fresh specimen is liquid, place 3 or 4 drops of PVA fixative (Schaudinn's fixative to which has been added a plastic powder [PVA], which serves as an adhesive to "glue" the fecal material onto the slide) on the slide, mix several drops of fecal material with the PVA fixative, spread the mixture, and allow it to dry for several hours in a 37°C incubator or overnight at room temperature.

3. Proceed with the trichrome staining procedure by placing the slides in iodine-alcohol.

PVA-Preserved Material

1. Stool specimens that are preserved in PVA fixative (mercuric chloride, copper sulfate, or zinc sulfate bases) should be allowed to fix for at least 30 min. Thoroughly mix the contents of the vial or bottle (PVA fixative/stool specimen) with two applicator sticks.

2. Pour some of the well-mixed PVA fixative-stool mixture onto a paper towel, and allow it to stand for 3 min to absorb out the excess PVA fixative. *Do not eliminate this step.*

3. With an applicator stick (or brush), apply some of the stool material from the paper towel to two slides and allow them to dry for several hours in a 37°C incubator or overnight at room temperature.

   Note The PVA fixative-stool mixture should be spread to the edges of the glass slide; this will cause the film to adhere to the slide during staining. It is also important to thoroughly dry the slides in order to prevent the material from washing off during staining.

4. The dry slides may then be placed into iodine-alcohol. There is no need to give them a 70% alcohol rinse before placing them in the iodine-alcohol, because the PVA smears are already dry (unlike the wet smears coming out of Schaudinn's fixative). The iodine-alcohol is used to remove the mercuric chloride from the slides before they are stained. When reviewing the trichrome staining procedure, you will note that this step is not required if the stool specimen is preserved in a copper sulfate- or zinc sulfate-based preservative.

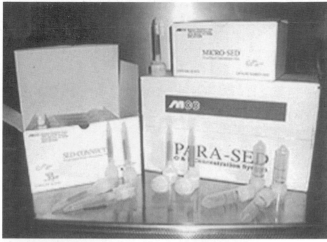

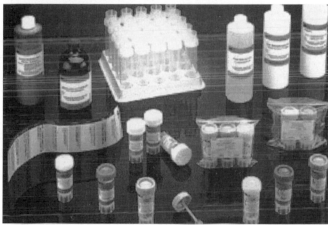

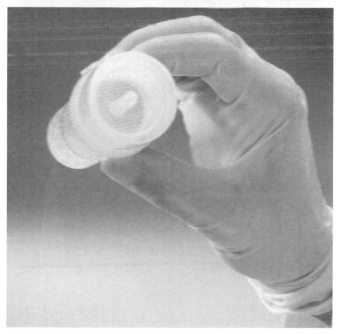

Figure 27.11 (Top) PARA-SED concentration system with both small and large tubes (Medical Chemical Corp.). (Middle) SED-CONNECT system with collection vials and various reagents (Medical Chemical Corp.). (Bottom) Filter attachment system (Medical Chemical Corp.). Note the screen, which is a substitute for the gauze in the traditional gauze/filter concentration method. See also MICRO-SED.

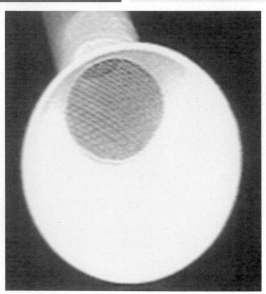

California State Department of Health Modification (2) for PVA-Preserved Material

1. Thoroughly mix the PVA-stool mixture, and strain it through damp (with tap water) gauze into 15-ml centrifuge tubes.
2. After centrifugation (2 min at 500 × g), decant and discard the PVA fixative and swab the sides of the tubes to remove any excess PVA. Although stains for *Cryptosporidium* spp. are not recommended from PVA-preserved material, **the centrifuge speed indicated here is probably not sufficient to recover the oocysts or microsporidian spores for special staining, fecal immunoassays using fluorescence, or other diagnostic procedures.**
3. Use the remaining sediment to prepare coverslip smears (not 3- by 1-in. slides), which are immediately (without drying) stained, beginning with the iodine-alcohol step of the trichrome staining procedure (for removal of mercuric chloride).
4. The staining times can be reduced to 2 to 4 min, and dehydration steps can be reduced to 1 min each.

    Note This procedure will work quite well for most stool specimens preserved in PVA fixative. However, for very liquid specimens or those containing large amounts of mucus, drying the smears before staining will be advantageous.

SAF-Preserved Material (47)

1. Thoroughly mix the SAF-stool mixture, and strain it through gauze into a 15-ml centrifuge tube.
2. After centrifugation (1 min at 500 × g), decant and discard the supernatant fluid. Although stains for coccidian oocysts and microsporidian spores are not recommended from PVA-preserved material, the centrifuge speed indicated here is probably not sufficient to recover the oocysts or spores. Centrifugation time and speed are recommended as 10 min at 500 × g. The final amount of sediment should be about 0.5 to 1.0 ml. If necessary, adjust by repeating step 1 or by resuspending the sediment in saline (0.85% NaCl) and removing part of the suspension.
3. Prepare a smear from the sediment for later staining by placing 1 drop of Mayer's albumin on the slide and adding 1 drop of SAF-preserved fecal sediment.

Figure 27.12 (Upper left) MACRO-CON concentration system (Meridian Bioscience). (Upper right) SPINCON concentration system (Meridian Bioscience). (Lower) Funnel used in fecal concentration (Meridian Bioscience).

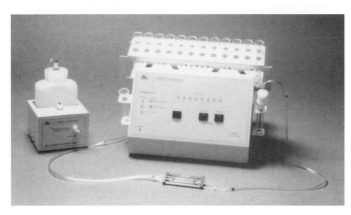

Figure 27.13 Countertop workstation that automates the microscopic analysis of fecal concentrates (DiaSys Corp.).

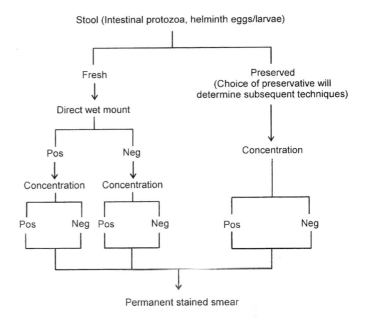

Figure 27.15 Work flow diagram for fecal specimens. The total examination includes the direct mount, concentrate, and permanent stained smear (fresh specimen) or the concentrate and permanent stained smear (preserved specimens).

Allow the smear to air dry at room temperature for 30 min prior to staining. The SAF stool smear can also be postfixed in Schaudinn's fixative prior to staining (begin the trichrome stain protocol with the 70% alcohol rinse prior to iodine-alcohol).

4. After being dried, the smear can be placed directly into 70% alcohol (step 4) of the staining procedure (the iodine-alcohol step can be eliminated).

Merthiolate-Iodine-Formalin (MIF)-Preserved Material

1. Prepare a Mayer's albumin-coated slide and allow it to air dry at room temperature for 30 min.
2. From the MIF vial (material allowed to settle in the vial for at least 1 h undisturbed), remove a portion of sediment and place this material onto

Figure 27.14 Dual-flow-cell Optical Slide Assembly (DiaSys Corp.) that fits into the stage clips of any standard upright microscope.

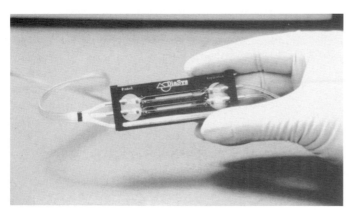

the albumin-coated slide. Allow the slide to remain flat for 5 min; if there is still fluid left on the slide after this time, stand it up and allow the excess fluid to run off onto a paper towel.

3. After all excess fluid is removed, the slides are ready for staining in polychrome IV stain.

As seen in the previous chapter, some of the single-vial fecal preservative systems are satisfactory but some do not provide morphology of sufficient quality to allow organism identification after permanent staining (21). In this particular study, when the single-vial fixatives were compared with mercuric chloride-based PVA, it was found that ECOFIX (Meridian Diagnostics, Inc., Cincinnati, Ohio) and ProtoFix (Alpha-Tec Systems, Inc., Vancouver, Wash.) were environmentally acceptable substitutes for PVA. PARASAFE (Scientific Device Laboratory, Inc., Des Plaines, Ill.) was not acceptable. Another acceptable substitute is UNIFIX (Medical Chemical Corp., Torrance, Calif.), which was equal to Z-PVA and other non-mercuric-chloride-based PVA formulations (13; "hands on" unpublished data). It is also important to remember that some manufacturers have their own stain that can be used with their particular fixative; an example would be ECOSTAIN coupled with ECOFIX (Meridian Diagnostics, Inc.) (11).

As discussed in the previous chapter, various examination options are available for very specific areas within the United States. Kellogg and Elder (22) reported that in their area (York, Pa.), examination of a single trichrome-stained smear of a concentrated stool specimen from PVA-preserved stool appeared to be a cost-effective alternative to routine analysis of both concentrated and unconcentrated specimens for parasites. It would appear from this information that in that part of the country, they tend to see mainly protozoan parasites and very few helminth infections (22).

Trichrome Stain

The trichrome technique of Wheatley (46) for fecal specimens is a modification of Gomori's original staining procedure for tissue (15). It is a rapid, simple procedure which produces uniformly well stained smears of the intestinal protozoa, human cells, yeast cells, and artifact material in about 45 min or less.

The specimen usually consists of fresh stool smeared on a microscope slide, which is immediately fixed in liquid Schaudinn's fixative, or PVA-preserved stool smeared on a slide and allowed to air dry. Although SAF- and MIF-preserved specimens can be stained with trichrome, there are other stains which are recommended for better overall results. Trichrome stains also work well with some of the single-vial preservatives such as UNIFIX (Medical Chemical Corp., Torrance, Calif.)

Trichrome Stain

    Chromotrope 2R .... 0.6 g
    Light green SF .... 0.3 g
    Phosphotungstic acid .... 0.7 g
    Acetic acid (glacial) .... 1.0 ml
    Distilled water .... 100 ml

1. Prepare the stain by adding 1.0 ml of acetic acid to the dry components. Allow the mixture to stand (ripen) for 15 to 30 min at room temperature.
2. Add 100 ml of distilled water. Properly prepared stain is purple.
3. Store in a glass or plastic bottle at room temperature. The shelf life is 24 months.

70% Ethanol plus Iodine

1. Prepare a stock solution by adding iodine crystals to 70% alcohol until a dark solution is obtained (1 to 2 g/100 ml).
2. To use, dilute the stock solution with 70% alcohol until a dark reddish brown (port wine) or strong-tea color is obtained. As long as the color

is acceptable, working solution does not have to be replaced. Replacement time depends on the number of smears stained and the size of the container (1 week to several weeks).

90% Ethanol, Acidified

    90% ethanol .... 99.5 ml
    Acetic acid (glacial) .... 0.5 ml

Prepare by combining.

70% Isopropyl or Ethyl Alcohol
100% Ethyl Alcohol (Recommended)
or 95% Ethyl Alcohol (Second Choice)
Xylene or Xylene Substitute

Quality Control for Trichrome Stain

1. Stool samples used for quality control can be fixed stool specimens known to contain protozoa or PVA-preserved negative stools to which buffy coat cells (PMNs or macrophages) have been added. A QC smear prepared from a positive PVA sample or a PVA sample containing buffy coat cells should be used when new stain is prepared or at least once each week. Cultured protozoa can also be used.
2. A QC slide should be included with a run of stained slides at least monthly; more frequent QC is recommended for those who may be unfamiliar with the method (32).
3. If the xylene becomes cloudy or there is an accumulation of water in the bottom of the staining dish containing xylene, discard the old reagents, clean the dishes, dry thoroughly, and replace with fresh 100% ethanol and xylene.
4. All staining dishes should be covered to prevent evaporation of reagents (screw-cap Coplin jars or glass lids).
5. Depending on the volume of slides stained, staining solutions will have to be changed on an as-needed basis.
6. When the smear is thoroughly fixed and the staining procedure is performed correctly, the cytoplasm of protozoan trophozoites is blue-green, sometimes with a tinge of purple. Cysts tend to be slightly more purple. Nuclei and inclusions (chromatoidal bars, RBCs, bacteria, and Charcot-Leyden crystals) are red, sometimes tinged with purple. The background material usually stains green, providing a nice color contrast with the protozoa. This contrast is more distinct than that obtained with the hematoxylin stain, which

tends to stain everything in shades of gray-blue to black.

7. If appropriate, the microscope should be calibrated (within the last 12 months), and the objectives and oculars used for the calibration procedure should be used for all measurements on the microscope. The calibration factors for all objectives should be posted on the microscope for easy access (multiplication factors can be pasted on the body of the microscope).

8. Known positive microscope slides, Kodachrome 2 × 2 projection slides, and photographs (reference books) should be available at the workstation.

9. Record all QC results; the laboratory should also have an action plan for "out-of-control" results.

Procedure for Trichrome Stain with Mercury-Based Fixatives (Figure 27.16)

Note In all staining procedures for fecal and gastrointestinal tract specimens, the term "xylene" is used in the generic sense. Xylene substitutes are recommended for the safety of all personnel performing these procedures.

1. Prepare the slide for staining as described above.
2. Remove the slide from liquid Schaudinn's fixative, and place it in 70% ethanol for 5 min.
3. Place the slide in 70% ethanol plus iodine for 1 min for fresh specimens or 5 to 10 min for PVA air-dried smears. The exposure to iodine will remove the mercuric chloride from the smear prior to staining with the actual trichrome dyes (substitution of iodine for mercury).
4. Place the slide in 70% ethanol for 5 min.* This and the next step in 70% ethanol will remove the iodine from the smear.
5. Place it in a second container of 70% ethanol for 3 min.*
6. Place it in trichrome stain for 10 min. The fecal smear no longer contains either mercuric chloride or iodine and is now ready for staining.
7. Place it in 90% ethanol plus acetic acid for 1 to 3 s. Immediately drain the rack (see Procedure Notes), and proceed to the next step. Do not allow slides to remain in this solution. This is the destaining step.
8. Dip the slide several times in 100% ethanol. Use this step as a rinse.
9. Place it in two changes of 100% ethanol for 3 min each.* This is a dehydration step.
10. Place it in xylene for 5 to 10 min.* This is a dehydration step.
11. Place it in a second container of xylene for 5 to 10

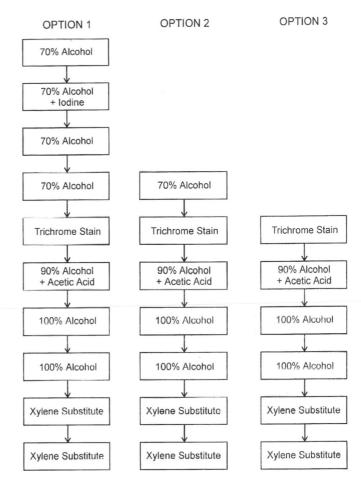

Figure 27.16 Trichrome staining. Option 1, for use with smears prepared from fixatives containing mercuric chloride. The iodine is used to remove the mercuric chloride, and the subsequent two alcohol rinse steps remove the iodine. Thus, prior to staining, both the mercuric chloride and iodine have been removed from the smear. Options 2 and 3, for use with smears prepared from fixatives containing no mercuric chloride (the user can select option 2 or 3; there is minimal to no difference).

min.* This step completes the dehydration process (removal of all water on the smear).

12. Mount the slide with a coverslip (no. 1 thickness), using mounting medium (e.g., Permount).
13. Allow the smear to dry overnight or after 1 h at 37°C.
14. Examine the smear microscopically with the 100× objective. Examine at least 200 to 300 oil immersion fields before reporting a negative result (Figure 27.17).

*Slides may be held for up to 24 h in these solutions without harming the quality of the smear or stainability of organisms.

Procedure for Trichrome Stain with Non-Mercury-Based Fixatives (Iodine-Alcohol Step and Alcohol Rinse Not Required) (Figure 27.16)

1. Prepare the slide for staining as described above.
2. Place the slide in 70% ethanol for 5 min.*
3. Place it in the trichrome stain for 10 min. Some people prefer to place the dry smear directly into the stain and eliminate step 2 from the protocol.
4. Place it in 90% ethanol plus acetic acid for 1 to 3 s. Immediately drain the rack (see Procedure Notes), and proceed to the next step. Do not allow slides to remain in this solution.
5. Dip the slide several times in 100% ethanol. Use this step as a rinse.
6. Place it in two changes of 100% ethanol for 3 min each.*
7. Place it in xylene for 5 to 10 min.*
8. Place it in a second container of xylene for 5 to 10 min.*
9. Mount with a coverslip (no. 1 thickness), using mounting medium (e.g., Permount).
10. Allow the smear to dry overnight or after 1 h at 37°C.
11. Examine the smear microscopically with the 100× objective. Examine at least 200 to 300 oil immersion fields before reporting a negative result.

*Slides may be held for up to 24 h in these solutions without harming the quality of the smear or stainability of organisms.

Note An alternative method to using mounting medium is as follows (3).

A. Remove the slide from the last container of xylene, place it on a paper towel (flat position), and allow it to air dry. Remember that some of the xylene substitutes may take a bit longer to dry than xylene itself does.
B. Approximately 5 to 10 min before you want to examine the slide, place a drop of immersion oil on the dry fecal film. Allow the oil to sink into the film for a minimum of 10 to 15 min. If the smear appears to be very refractile on examination, you have not waited long enough for the oil to sink into the film or you need to add a bit more oil to the film.
C. Once you are ready to examine the slide, place a no. 1 (22- by 22-mm) coverslip onto the oiled smear, add another drop of immersion oil onto the top of the coverslip (as you would normally do for any slide with a coverslip), and examine with the oil immersion lens (100× objective).
D. *Do not eliminate adding the coverslip*; the dry fecal material on the slide often becomes very brittle after dehydration. Without the addition of the protective coverslip, you might scratch the surface of the oil immersion lens. Coverslips are much cheaper than oil immersion objectives!

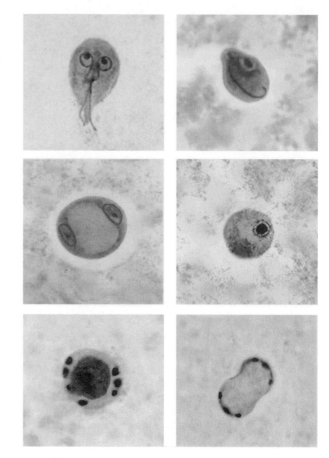

Figure 27.17 Guess the identity of these intestinal protozoa! See p. 830 for the answers.

Results and Patient Reports from the Trichrome Staining Method

Protozoan trophozoites and cysts are readily seen, although helminth eggs and larvae may not be easily identified because of excess stain retention (wet smears from the concentration procedure[s] are recommended for detection of these organisms) (Figure 27.18). Yeasts (single and budding cells and pseudohyphae) and human cells (macrophages, PMNs, and RBCs) can be identified. The following quantitation chart can be used for examination of permanent stained smears with the oil immersion lens (100× objective; total magnification, ×1,000).

Quantitation of Parasites, Cells, Yeasts, and Artifacts

Quantity	No. per 10 oil immersion fields (×1,000)
Few	≤2
Moderate	3–9
Many	≥10

1. Report the organism and stage (do not use abbreviations).
   Examples: *Entamoeba histolytica/E. dispar* trophozoites
   Giardia lamblia trophozoites
2. Quantitate the number of *B. hominis* organisms seen (rare, few, moderate, many). Do not quantitate other protozoa.
   Example: Moderate *Blastocystis hominis*
3. Note and quantitate the presence of human cells.
   Example: Moderate WBCs, many RBCs, few macrophages, rare Charcot-Leyden crystals
4. Report and quantitate yeast cells.
   Example: Moderate budding yeast cells and few pseudohyphae
5. Save positive slides for future reference. Label prior to storage (name, patient number, organisms present). Most laboratories discard their negative permanent stained smears after several weeks (rotate storage boxes and discard when full).

Procedure Notes for the Trichrome Staining Method

1. The single most important step in the preparation of a well-stained fecal smear is good fixation. If this has not been done, the protozoa may be distorted or shrunk, may not be stained, or may exhibit an overall pink or red color with poor internal morphology.
2. Slides should always be drained between solutions. Touch the end of the slide to a paper towel for 2 s

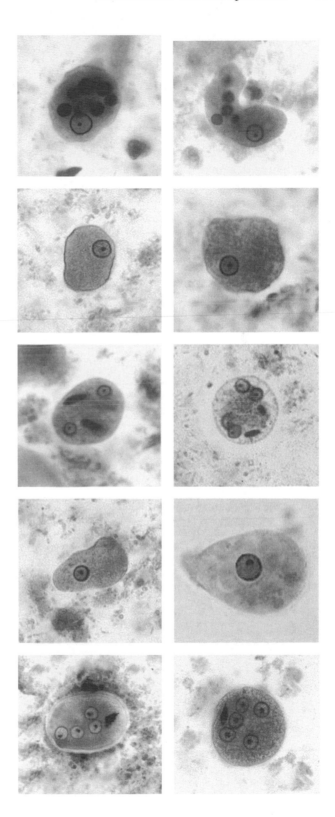

Figure 27.18 Intestinal protozoa in the genus *Entamoeba* (stained with trichrome stain). Rows, from top to bottom, show *Entamoeba histolytica* trophozoites (note ingested RBCs), *Entamoeba histolytica/E. dispar* trophozoites (the morphology does not allow differentiation between pathogenic *E. histolytica* and nonpathogenic *E. dispar*), *Entamoeba histolytica/E. dispar* cysts, *Entamoeba coli* trophozoites (note uneven nuclear chromatin), and *Entamoeba coli* cysts (note the rough ends of chromatoidal bars and nuclear characteristics).

to remove excess fluid before proceeding to the next step. This will maintain the staining solutions for a longer period. The slide can also be touched against the staining dish to drain off excess fluid before moving on to the next dish.

3. Incomplete removal of mercuric chloride (Schaudinn's fixative and PVA fixative with a mercuric chloride base) may cause the smear to contain highly refractive crystals or granules, which may prevent finding or identifying any organisms present. Since the 70% ethanol-iodine solution removes the mercury complex, it should be changed at least weekly to maintain the strong-tea color. A few minutes are usually sufficient to keep the slides in the iodine-alcohol; too long a time in this solution may also adversely affect the staining of the organisms.

4. When using non-mercury-based fixatives, the iodine-alcohol step (used for the removal of mercury) and the subsequent alcohol rinse (removal of the iodine) can be eliminated from the procedure. The smears for staining can be prerinsed with 70% alcohol and then placed in the trichrome stain, or they can be placed directly into the trichrome stain as the first step in the staining protocol (Figure 27.16).

5. Smears that are predominantly green may be due to the inadequate removal of iodine by the 70% ethanol (steps 4 and 5). Lengthening the time of these steps or more frequent changing of the 70% ethanol will help.

6. To restore weakened trichrome stain, remove the cap and allow the ethanol to evaporate (ethanol carried over on the staining rack from a previous dish). After a few hours, fresh stock stain may be added to restore lost volume. Older, more concentrated stain produces more intense colors and may require slightly longer destaining times (an extradip). PVA smears usually require a slightly longer staining time due to the presence of the plastic PVA powder.

7. Although the trichrome stain is used essentially as a "progressive" stain (that is, no destaining is necessary), best results are obtained by using the stain "regressively" (destaining the smears briefly in acidified alcohol after the initial overstaining). Good differentiation is obtained by destaining for a very short time (two dips only, approximately 2 to 3 s); prolonged destaining results in poor differentiation.

8. It is essential to rinse the smears free of acid to prevent continued destaining. Since 90% alcohol will continue to leach trichrome stain from the smears, it is recommended that after the acid-alcohol is used, the slides be quickly rinsed in 100% alcohol and then dehydrated through two additional changes of 100% alcohol.

9. In the final stages of dehydration (steps 9 to 11), the 100% ethanol and the xylenes (or xylene substitute) should be kept as free from water as possible. Coplin jars must have tight-fitting caps to prevent both evaporation of reagents and absorption of moisture. If the xylene becomes cloudy after addition of slides from the 100% ethanol, return the slides to fresh 100% ethanol and replace the xylene with fresh stock.

10. If the smears peel or flake off, the specimen might have been inadequately dried on the slide (in the case of PVA-fixed specimens), the smear may have been too thick, or the slide may have been greasy (fingerprints). However, slides generally do not have to be cleaned with alcohol prior to use.

11. On examination, if the stain appears unsatisfactory and it is not possible to obtain another slide to stain, the slide may be restained. Place the slide in xylene to remove the coverslip, and reverse the dehydration steps, adding 50% ethanol as the last step. Destain the slide in 10% acetic acid for several hours, and then wash it thoroughly first in water and then in 50 and 70% ethanol. Place the slide in the trichrome stain for 8 min, and complete the staining procedure (5).

Procedure Limitations for the Trichrome Staining Method

1. The permanent stained smear is not recommended for staining helminth eggs or larvae; they are often too dark (excess stain retention) or distorted. However, they are occasionally recognized and identified. The wet smear preparation from the concentrate is the recommended approach for identification of helminth eggs and larvae.

2. The smear should be examined with the oil immersion lens (100×) for the identification of protozoa, human cells, Charcot-Leyden crystals, yeast cells, and artifact material. Quantitation of these cells and other structures is normally done from the examination of the permanent stained smear, not the wet smear preparations (direct wet smear or concentration wet smear).

3. This high-magnification (oil immersion; total magnification, ×1,000) examination is recommended for protozoa, particularly for confirming species identification.

4. With low magnification (10× objective), one might see eggs or larvae; however, this is not recommended as a routine approach.

5. In addition to helminth eggs and larvae, *I. belli* oocysts are best seen in wet preparations

(concentration wet smears prepared from formalin-preserved, not PVA-preserved, material).

6. *Cryptosporidium* and *Cyclospora* oocysts are generally not recognized on a trichrome-stained smear (acid-fast stains or the immunoassay reagent kits are recommended). Microsporidial spores do not stain sufficiently for recognition by the regular trichrome method; modified trichrome stains are required.

Iron Hematoxylin Stain

The iron hematoxylin stain is one of a number of stains that allow one to make a permanent stained slide for detecting and quantitating parasitic organisms. Iron hematoxylin was the stain used for most of the original morphologic descriptions of intestinal protozoa found in humans (5) (Figure 27.19). On oil immersion power (×1,000), one can examine the diagnostic features used to identify the protozoan parasites. Although there are many modifications of iron hematoxylin techniques, only two methods are outlined below: the Spencer-Monroe (41) and Tompkins-Miller (42) procedures. Both methods can be used with either fresh, SAF-preserved, PVA-preserved, or single-vial system-preserved specimens.

The specimen usually consists of fresh stool smeared on a microscope slide, which is immediately fixed in Schaudinn's fixative, PVA-preserved stool smeared on a slide and allowed to air dry, SAF-preserved stool smeared on an albumin-coated slide and allowed to air dry, or single-vial-preserved stool smeared on an albumin coated slide and allowed to air dry. In some cases, the albumin is not absolutely necessary as an adhesive.

Iron Hematoxylin Stain (Spencer-Monroe Method) (41)

Solution 1

> Hematoxylin (crystal or powder) .... 10 g
> Ethanol (absolute) .... 1,000 ml

Place solution in a stoppered clear flask or bottle, and allow it to ripen in a lighted room for at least 1 week at room temperature.

Solution 2

> Ferrous ammonium sulfate
> [Fe(NH$_4$)$_2$(SO$_4$)$_2$ · 6H$_2$O] .... 10 g
> Ferric ammonium sulfate
> [FeNH$_4$(SO$_4$)$_2$ · 12H$_2$O] .... 10 g
> Hydrochloric acid (concentrated) .... 10 ml
> Distilled water .... 1,000 ml

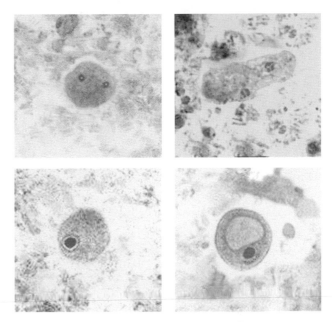

Figure 27.19 Intestinal protozoa (stained with iron hematoxylin stain). (Upper) *Dientamoeba fragilis* trophozoites (note the fragmented nuclei). (Lower left) *Iodamoeba bütschlii* trophozoite (note the large karyosome). (Lower right) *Iodamoeba bütschlii* cyst (note the large glycogen vacuole and "basket" nucleus).

Working solution

Mix equal volumes of solutions 1 and 2. The working solution should be made fresh every week.

70% ethanol plus iodine

1. Prepare a stock solution by adding iodine crystals to 70% alcohol until a dark solution is obtained (1 to 2 g/100 ml).
2. To use, dilute the stock solution with 70% alcohol until a dark reddish brown strong-tea color is obtained. As long as the color is acceptable, new working solution does not have to be made. The replacement time will depend on the number of smears stained and the size of the container (1 week to several weeks).

90% ethanol, acidified

> 90% ethanol .... 99.5 ml
> Acetic acid (glacial) .... 0.5 ml

Prepare by combining.

70% Isopropyl or Ethyl Alcohol
100% Ethyl Alcohol (Recommended)
or 95% Ethyl Alcohol (Second Choice)
Xylene or Xylene Substitute

Quality Control for Iron Hematoxylin Stain

1. Stool samples used for quality control can be fixed stool specimens known to contain protozoa or PVA-preserved negative stools to which buffy coat cells (PMNs or macrophages) have been added. A QC smear prepared from a positive PVA sample or a PVA sample containing buffy coat cells should be used when new stain is prepared or at least once each week. Cultured protozoa can also be used.

2. A QC slide should be included with a run of stained slides at least monthly; more frequent QC is recommended for those who may be unfamiliar with the method.

3. If the xylene becomes cloudy or there is an accumulation of water in the bottom of the staining dish, discard the old reagents, clean the dishes, dry them thoroughly, and replace with fresh 100% ethanol and xylene.

4. All staining dishes should be covered to prevent evaporation of reagents (screw-cap Coplin jars or glass lids).

5. Depending on the volume of slides stained, staining solutions should to be changed on an as-needed basis.

6. When the smear is thoroughly fixed and the stain is performed correctly, the cytoplasm of protozoan trophozoites will be blue-gray, sometimes with a tinge of black. Cysts tend to be slightly darker. Nuclei and inclusions (chromatoidal bars, RBCs, bacteria, and Charcot-Leyden crystals) are dark gray-blue, sometimes almost black. The background material usually stains pale gray or blue, providing some color intensity contrast with the protozoa. This contrast is less distinct than that obtained with the trichrome stain, which tends to stain everything with multiple colors (pink, red, purple, green, blue).

7. The microscope should be calibrated (within the last 12 months) (recommended but not always required, depending on the use and care of the microscope), and the objectives and oculars used for the calibration procedure should be used for all measurements on the microscope. The calibration factors for all objectives should be posted on the microscope for easy access (multiplication factors can be pasted on the body of the microscope).

8. Known positive microscope slides, Kodachrome 2 × 2 projection slides, and photographs (reference books) should be available at the workstation.

9. Record all QC results; the laboratory should also have an action plan for "out-of-control" results.

Procedure for Iron Hematoxylin Stain with Mercury-Based Fixatives

Note In all staining procedures for fecal and gastrointestinal tract specimens, the term "xylene" is used in the generic sense. Xylene substitutes are recommended for the safety of all personnel performing these procedures.

1. Prepare the slide for staining as previously described (for SAF smears or smears prepared from other non-mercury single-vial preservatives, proceed to step 4).

2. Place the slide in 70% ethanol for 5 min.

3. Place the slide in the iodine–70% ethanol (70% alcohol to which is added enough D'Antoni's iodine to obtain a strong-tea color) solution for 2 to 5 min. The iodine is designed to remove the mercury from the smear.

4. Place it in 70% ethanol for 5 min (this rinse step removes the iodine). Begin the procedure for SAF-fixed slides at this point.*

5. Wash the slide in running tap water (constant stream of water into the container) for 10 min.

6. Place the slide in iron hematoxylin working solution for 4 to 5 min.

7. Wash the slide in running tap water (constant stream of water into the container) for 10 min.

8. Place the slide in 70% ethanol for 5 min.*

9. Place the slide in 95% ethanol for 5 min.*

10. Place the slide in two changes of 100% ethanol for 5 min each.*

11. Place the slide in two changes of xylene for 5 min each.*

12. Add Permount to the stained area of the slide, and cover with a coverslip.
    Note An alternative method to using mounting medium is given on p. 804.

13. Examine the smear microscopically with the 100× objective. Examine at least 200 to 300 oil immersion fields before reporting a negative result.

*Slides may be held for up to 24 h in these solutions without harming the quality of the smear or stainability of organisms.

Procedure for Iron Hematoxylin Stain with Non-Mercury-Based Fixatives

1. Prepare the slide for staining as described above.

2. Place it in 70% ethanol for 5 min.*

3. Wash the slide in running tap water (constant stream of water into the container) for 10 min.

4. Place the slide in iron hematoxylin working solution for 4 to 5 min.

5. Wash the slide in running tap water (constant stream of water into the container) for 10 min.

6. Place the slide in 70% ethanol for 5 min.*
7. Place the slide in 95% ethanol for 5 min.*
8. Place the slide in two changes of 100% ethanol for 5 min each.*
9. Place the slide in two changes of xylene for 5 min each.*
10. Add Permount to the stained area of the slide and cover it with a coverslip.
    Note An alternative method to using mounting medium is given on p. 804.
11. Examine the smear microscopically with the 100× objective. Examine at least 200 to 300 oil immersion fields before reporting a negative result.

*Slides may be held for up to 24 h in these solutions without harming the quality of the smear or stainability of organisms.

Results and Patient Reports from the Iron Hematoxylin Staining Method

Protozoan trophozoites and cysts are readily visible, although helminth eggs and larvae may not be easily identified because of excess stain retention (wet smears from the concentration procedure[s] are recommended for detection of these organisms). Yeasts (single and budding cells and pseudohyphae) and human cells (macrophages, PMNs, and RBCs) can be identified. The following quantitation chart can be used for examination of permanent stained smears with the oil immersion lens (100× objective; total magnification, ×1,000).

1. Report the organism and stage (do not use abbreviations).
   Examples: *Entamoeba coli* trophozoites
   Dientamoeba fragilis trophozoites
2. Quantitate the number of *B. hominis* organisms seen (rare, few, moderate, many). Do not quantitate other protozoa.
   Example: Many *Blastocystis hominis*
3. Note and quantitate the presence of human cells.
   Example: Moderate WBCs, few macrophages, few RBCs, rare Charcot-Leyden crystals
4. Report and quantitate yeast cells.
   Example: Many budding yeast cells and few pseudohyphae
5. Save positive slides for future reference. Label prior to storage (name, patient number, organisms present). Most laboratories hold their negative smears for several weeks and then discard them; slide boxes that are rotated can be used to batch, store, and discard negative smears.

Procedure Notes for the Iron Hematoxylin Staining Method

1. The single most important step in the preparation of a well-stained fecal smear is good fixation. If this has not been done, the protozoa may be distorted or shrunk, may not be stained, or may exhibit an overall gray or blue-gray color with poor internal morphology.
2. Slides should always be drained between solutions. Touch the end of the slide to a paper towel for 2 s to remove excess fluid before proceeding to the next step. This will maintain the staining solutions for a longer period. The slides can also be drained against the edge of each container before being moved to the next container.
3. Incomplete removal of mercuric chloride (Schaudinn's fixative and PVA fixative prepared with a mercuric chloride base) may cause the smear to contain highly refractive crystals or granules, which may prevent finding or identifying any organisms present. Since the 70% ethanol–iodine solution removes the mercury complex, it should be changed at least weekly to maintain the strong-tea color. A few minutes are usually sufficient to keep the slides in the iodine-alcohol; too long a time in this solution may also adversely affect the staining of the organisms.
4. When using non-mercury-based fixatives, the iodine-alcohol step (used for the removal of mercury) and the subsequent alcohol rinse (used for the removal of iodine) can be eliminated from the procedure. The smears for staining can be prerinsed with 70% alcohol and then placed in the water step prior to the hematoxylin stain as the first step in the staining protocol.
5. For staining large numbers of slides, the working hematoxylin solution may be diluted and affect the quality of the stain. If dilution occurs, discard the working solution and prepare a fresh working solution.

Quantitation of Parasites, Cells, Yeasts, and Artifacts

Quantity	No. per 10 oil immersion fields (×1,000)
Few	≤2
Moderate	3–9
Many	≥10

6. The shelf life of the stock hematoxylin solutions may be extended by keeping the solutions in the refrigerator at 4°C. Because of crystal formation in the working solutions, it may be necessary to filter them before preparing a new working solution.

7. In the final stages of dehydration (steps 9 to 11), the 100% ethanol and the xylenes should be kept as free from water as possible. Coplin jars must have tight-fitting caps to prevent both evaporation of reagents and absorption of moisture. If the xylene becomes cloudy after addition of slides from the 100% ethanol, return the slides to fresh 100% ethanol and replace the xylene with fresh stock.

8. If the smears peel or flake off, the specimen might have been inadequately dried on the slide (in the case of PVA-fixed specimens), the smear may have been too thick, or the slide may have been greasy (fingerprints). However, slides generally do not have to be cleaned with alcohol prior to use.

9. On examination, if the stain appears unsatisfactory and it is not possible to obtain another slide to stain, the slide may be restained. Place the slide in xylene to remove the coverslip, and reverse the dehydration steps, adding 50% ethanol as the last step. Destain the slide in 10% acetic acid for several hours, and then wash it thoroughly first in water and then in 50 and 70% ethanol. Place the slide in the iron hematoxylin stain for 8 min, and complete the staining procedure (3, 5).

Procedure Limitations for the Iron Hematoxylin Staining Method

1. The permanent stained smear is not recommended for staining helminth eggs or larvae; these structures are often too dark (excess stain retention) or distorted. However, they are occasionally recognized and identified. The wet smear preparation from the concentrate is the recommended approach for identification of helminth eggs and larvae.

2. The smear should be examined with the oil immersion lens (100×) for the identification of protozoa, human cells, Charcot-Leyden crystals, yeast cells, and artifact material. Quantitation of these cells and other structures is normally done from the examination of the permanent stained smear, not the wet smear preparations (direct wet smear, concentration wet smear).

3. This high-magnification (oil immersion; total magnification, ×1,000) examination is recommended for protozoa, particularly for confirming species identification.

4. With low magnification (10× objective), one might see eggs or larvae; however, this is not recommended as a routine approach.

5. In addition to helminth eggs and larvae, *I. belli* oocysts are best seen in wet preparations (concentration wet smears prepared from formalin-preserved, not PVA-preserved, material).

6. *Cryptosporidium* and *Cyclospora* oocysts are generally not recognized on an iron hematoxylin-stained smear (modified acid-fast stains or the fecal immunoassay for *Cryptosporidium* spp. is recommended).

Iron Hematoxylin Stain (Tompkins-Miller Method) (42)

A longer iron hematoxylin method was described by Tompkins and Miller (42). Since differentiation of overstained slides is critical in most iron hematoxylin staining procedures, Tompkins and Miller have described a method that employs phosphotungstic acid to destain the protozoa and that gives excellent results, even in unskilled hands.

1. Prepare the slide for staining as described above (for SAF or non-mercury-based smears, proceed to step 4).

2. Place the slide in 70% ethanol for 5 min.

3. Place the slide in the iodine–70% ethanol (70% alcohol to which is added enough D'Antoni's iodine to obtain a strong-tea color) solution for 2 to 5 min.

4. Place it in 50% ethanol for 5 min. Begin the procedure for SAF- or non-mercury-fixed slides at this point.*

5. Wash the slide in running tap water (constant stream of water into the container) for 3 min.

6. Place the slide in 4% ferric ammonium sulfate mordant for 5 min.

7. Wash the slide in running tap water (constant stream of water into the container) for 1 min.

8. Place the slide in 0.5% aqueous hematoxylin for 2 min.

9. Wash the slide in tap water for 1 min.

10. Place the slide in 2% phosphotungstic acid for 2 to 5 min.

11. Wash the slide in running tap water for 10 min.

12. Place the slide in 70% ethanol (plus a few drops of saturated aqueous lithium carbonate) for 3 min.

13. Place the slide in 95% ethanol for 5 min.*

14. Place the slide in two changes of 100% ethanol for 5 min each.*

15. Place the slide in two changes of xylene for 5 min each.*

16. Add Permount to the stained area of the slide, and cover it with a coverslip.

> **Note** An alternative method to using mounting medium is as follows.
> **A.** Remove the slide from the last container of xylene, place it on a paper towel (flat position), and allow it to air dry. Remember that some of the xylene substitutes may take a bit longer to dry.
> **B.** Approximately 5 to 10 min before you want to examine the slide, place a drop of immersion oil on the dry fecal film. Allow the oil to sink into the film for a minimum of 10 to 15 min. If the smear appears to be very refractile on examination, you have not waited long enough for the oil to sink into the film or you need to add a bit more oil onto the film.
> **C.** Once you are ready to examine the slide, place a no. 1 (22- by 22-mm) coverslip onto the oiled smear, add another drop of immersion oil to the top of the coverslip (as you would normally do for any slide with a coverslip), and examine with the oil immersion lens (100× objective).
> **D.** *Do not eliminate adding the coverslip;* the dry fecal material on the slide often becomes very brittle after dehydration. Without the addition of the protective coverslip, you might scratch the surface of your oil immersion lens. Coverslips are much cheaper than oil immersion objectives!

17. Examine the smear microscopically with the 100× objective. Examine at least 200 to 300 oil immersion fields before reporting a negative result.

*Slides may be held for up to 24 h in these solutions without harming the quality of the smear or stainability of organisms.

Modified Iron Hematoxylin Stain (Incorporating the Carbol Fuchsin Step)

The following combination staining method for SAF-preserved fecal specimens was developed to allow the microscopist to screen for acid-fast organisms in addition to other intestinal parasites (Figure 27.20). For laboratories using iron hematoxylin stains in combination with SAF-fixed material and modified acid-fast stains for *Cryptosporidium*, *Cyclospora*, and *Isospora*, this modification represents an improved approach to current staining methods (J. Palmer, Letter, *Clin. Microbiol. Newsl.* **13**:39–40, 1991). This combination stain provides a saving in both time and personnel use.

Any fecal specimen submitted in SAF or other non-mercury single-vial-system preservatives can be used. Fresh fecal specimens after fixation in SAF for 30 min can also be used. This combination stain approach is not recommended for specimens preserved in Schaudinn's fixative or PVA using a mercuric chloride base.

Mayer's Albumin

Add an equal quantity of glycerin to a fresh egg white. Mix gently and thoroughly. Store at 4°C, and indicate an expiration date of 3 months. Mayer's albumin from commercial suppliers can normally be stored at 25°C for 1 year [e.g., product 756, E. M. Diagnostic Systems Inc., 480 Democrat Road, Gibbstown, NJ 08027; (800) 443-3637].

Stock Solution of Hematoxylin Stain

Hematoxylin powder .... 10 g
Ethanol (95% or 100%) .... 1,000 ml

1. Mix well until dissolved.
2. Store in a clear-glass bottle in a light area. Allow to ripen for 14 days before use.
3. Store at room temperature with an expiration date of 1 year.

Mordant

Ferrous ammonium sulfate
[Fe(NH$_4$)$_2$(SO$_4$)$_2$ · 6H$_2$O]............ 10 g
Ferric ammonium sulfate
[FeNH$_4$(SO$_4$)$_2$ · 12H$_2$O].............. 10 g
Hydrochloric acid (concentrated) .... 10 ml
Distilled water to........................ 1,000 ml

Working Solution of Hematoxylin Stain

1. Mix equal quantities of stock solution of stain and mordant.

Figure 27.20 Iron hematoxylin stain incorporating the carbol fuchsin step. Note the modified acid-fast *Cryptosporidium* oocysts and the *Giardia lamblia* cyst.

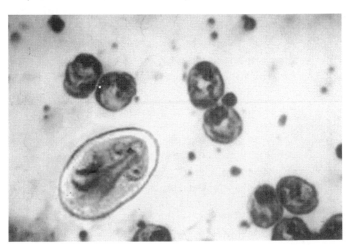

2. Allow the mixture to cool thoroughly before use (prepare at least 2 h prior to use). The working solution should be made fresh every week.

Picric Acid

Mix equal quantities of distilled water and an aqueous saturated solution of picric acid to make a 50% saturated solution.

Acid-Alcohol Decolorizer

Hydrochloric acid (concentrated).... 30 ml
Alcohol to .... 1,000 ml

70% Alcohol and Ammonia

70% alcohol.................................. 50 ml
Ammonia .... 0.5–1.0 ml

Add enough ammonia to bring the pH to approximately 8.0.

Carbol Fuchsin

1. To make basic fuchsin (solution A), dissolve 0.3 g of basic fuchsin in 10 ml of 95% ethanol.
2. To make phenol (solution B), dissolve 5 g of phenol crystals in 100 ml of distilled water. (Gentle heat may be needed.)
3. Mix solution A with solution B.
4. Store at room temperature. The solution is stable for 1 year.

Procedure for Modified Iron Hematoxylin Stain (Carbol Fuchsin Step)

1. Prepare slide.
   A. Place 1 drop of Mayer's albumin on a labeled slide.
   B. Thoroughly mix the sediment from the fecal concentration with an applicator stick.
   C. Add approximately 1 drop of the fecal concentrate to the albumin, and spread the mixture over the slide.
2. Allow the slide to air dry at room temperature (the smear appears opaque when dry).
3. Place the slide in 70% alcohol for 5 min.
4. Wash the slide in a container of tap water (not under running water) for 2 min.
5. Place the slide in Kinyoun's stain for 5 min.
6. Wash the slide in running tap water (constant stream of water into container) for 1 min.
7. Place the slide in acid-alcohol decolorizer for 4 min.*
8. Wash the slide in running tap water (constant stream of water into container) for 1 min.
9. Place the slide in iron hematoxylin working solution for 8 min.

10. Wash the slide in distilled water (in container) for 1 min.
11. Place the slide in picric acid solution for 3 to 5 min.
12. Wash the slide in running tap water (constant stream of water into container) for 10 min.
13. Place the slide in 70% alcohol plus ammonia for 3 min.
14. Place the slide in 95% alcohol for 5 min.
15. Place the slide in 100% alcohol for 5 min.
16. Place the slide in two changes of xylene for 5 min.
*This step can also be performed as follows.
   A. Place the slide in acid-alcohol decolorizer for 2 min.
   B. Wash the slide in running tap water (constant stream of water into container) for 1 min.
   C. Place the slide in acid-alcohol decolorizer for 2 min.
   D. Wash the slide in running tap water (constant stream of water into container) for 1 min.
   E. Continue the staining sequence with step 9 (iron hematoxylin working solution).

Procedure Notes for Modified Iron Hematoxylin Stain (Carbol Fuchsin Step)

1. The first 70% alcohol step acts with the Mayer's albumin to "glue" the specimen to the glass slide. The specimen may wash off if insufficient albumin is used or if the slides are too thick and are not completely dry before being stained.
2. The working hematoxylin stain should be checked each day of use by adding a drop of stain to alkaline tap water. If a blue color does not develop, prepare a fresh working stain solution.
3. The picric acid differentiates the hematoxylin stain by removing more stain from fecal debris than from the protozoa and removing more stain from the organism cytoplasm than from the nucleus. When properly stained, the background should be various shades of gray-blue and the protozoa, with medium blue cytoplasm and dark blue-black nuclei, should be easily seen.

Polychrome IV Stain

Polychrome IV stain can be used in place of trichrome for staining fecal smears by the MIF, PVA, or SAF fixative method. Both the stain and staining directions are available commercially (Devetec, Inc., P.O. Box 10275, Bradenton, FL 34282). Another source for the stain is Scientific Device Laboratory, Inc., P.O. Box 88, Glenview, IL 60025. Polychrome IV stain has been used primarily to stain permanent smears prepared from MIF-preserved fecal specimens.

REVIEW	Permanent Staimed Smears
Topic	**Comments**
Principle	To provide contrasting colors for the background debris and parasites present; designed to allow examination and recognition of detailed organism morphology under oil immersion examination (100× objective for a total magnification of ×1,000). Designed primarily to allow recovery and identification of the intestinal protozoa.
Specimen	Any stool specimen that is fresh or is preserved in PVA (mercury- or non-mercury based), SAF, MIF, or the newer single-vial-system fixatives.
Reagents	Trichrome, iron hematoxylin, modified iron hematoxylin, polychrome IV, or chlorazol black E stains and their associated solutions; dehydrating solutions (alcohols and xylenes); mounting fluid optional.
Examination	Oil immersion examination of at least 300 fields; additional fields may be required if suspect organisms have been seen in the wet preparations from the concentrated specimen.
Results and Laboratory Reports	The majority of the suspect protozoa and/or human cells could be confirmed by the permanent stained smear. These reports should be categorized as "final" and would be signed out as such (the direct wet smear and concentration examination would provide "preliminary" results). However, if the testing turnaround time is <24 h, it is best to report the test results after the ova and parasite examination (fresh specimen, direct wet smear; concentration, permanent stained smear) is complete. Therefore, the reports would not be identified as "preliminary" or "final."
Procedure Notes and Limitations	The most commonly used stains include trichrome and iron hematoxylin. Unfortunately, helminth eggs and larvae take up too much stain and usually cannot be identified from the permanent stained smear. Also, coccidian oocysts and microsporidian spores usually require other special staining methods for identification. Permanent stained smears are normally examined under oil immersion examination (×1,000), and low or high dry power is not recommended. Confirmation of the intestinal protozoa (both trophozoites and cysts) is the primary purpose of this technique.

Chlorazol Black E Stain

Chlorazol black E staining, developed by Kohn (23), is a method in which both fixation and staining occur in a single solution. This approach is used for fresh specimens, but it is not recommended for PVA-fixed material (14) because it does not include an iodine-alcohol step, which is used to remove the mercuric chloride compound found in both Schaudinn's fixative and PVA fixative prepared with mercuric chloride. The optimal staining time must be determined for each batch of fixative-stain. The length of time for which the fixative-stain can be used depends on the number of slides run through the solution within a 30-day period. If the slides appear visibly red, the solution must be changed. Although this stain is not widely used, it is another option to consider.

Specialized Stains for Coccidia (*Cryptosporidium*, *Isospora*, and *Cyclospora* Species) and the Microsporidia

Modified Kinyoun's Acid-Fast Stain (Cold Method)

Cryptosporidium spp. and *I. belli* have been recognized as causes of severe diarrhea in immunocompromised hosts but can also cause diarrhea in immunocompetent hosts (6). Oocysts in clinical specimens may be difficult to detect without special staining (29). Modified acid-fast stains are recommended to demonstrate these organisms. Unlike the Ziehl-Neelsen modified acid-fast stain, the modified Kinyoun's stain does not require the heating of reagents for staining (17). With additional reports of diarrheal outbreaks due to *Cyclospora* spp., it is also important to remember that these organisms are modified acid-fast variable and can be identified by this staining approach. Although the microsporidial spores are also acid fast, their size (1 to 2 µm) makes identification very difficult without special modified trichrome stains or the use of immunoassay reagents (1, 9, 24, 38, 44, 45).

Concentrated ($500 \times g$ for 10 min) sediment of fresh, formalin-preserved, or other single-vial fixative-preserved stool may be used. Other types of clinical specimens such as duodenal fluid, bile, and pulmonary specimens (induced sputum, bronchial washings, or biopsy specimens) may also be stained (Figure 27.21).

50% Ethanol

1. Add 50 ml of absolute ethanol to 50 ml of distilled water.
2. Store at room temperature. The solution is stable for 1 year. Note the expiration date on the label.

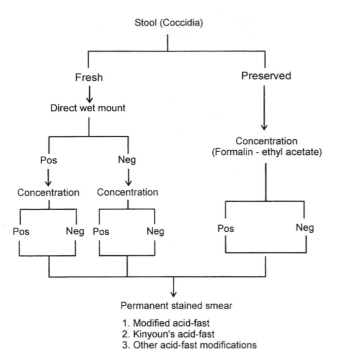

Stool (Coccidia)

Fresh

Direct wet mount

Pos

Neg

Concentration

Concentration

Pos   Neg   Pos   Neg

Preserved

Concentration
(Formalin - ethyl acetate)

Pos   Neg

Permanent stained smear

1. Modified acid-fast
2. Kinyoun's acid-fast
3. Other acid-fast modifications

Figure 27.21 Work flow diagram for fecal specimens (coccidia). The most important part of the procedure would be the concentrate (*Isospora* sp.) and the permanent stained smear, using one of the modified acid-fast techniques (*Cryptosporidium* and *Cyclospora* spp.).

Kinyoun's Carbol Fuchsin

1. Dissolve 4 g of basic fuchsin in 20 ml of 95% ethanol (solution A).
2. Dissolve 8 g of phenol crystals in 100 ml of distilled water (solution B).
3. Mix solutions A and B.
4. Store at room temperature. The solution is stable for 1 year. Note the expiration date on the label.

1% Sulfuric Acid

1. Add 1 ml of concentrated sulfuric acid to 99 ml of distilled water.
2. Store at room temperature. The solution is stable for 1 year. Note the expiration date on the label.

Loeffler Alkaline Methylene Blue

1. Dissolve 0.3 g of methylene blue in 30 ml of 95% ethanol.
2. Add 100 ml of dilute (0.01%) potassium hydroxide.
3. Store at room temperature. The solution is stable for 1 year. Note the expiration date on the label.

Quality Control for Kinyoun's Acid-Fast Stain

1. A control slide of *Cryptosporidium* from a 10% formalin-preserved specimen is included with each

staining batch run. If the *Cryptosporidium* slide stains well, any *Isospora* or *Cyclospora* oocysts present will also take up the stain, although *Cyclospora* oocysts tend to be acid-fast variable (Figure 27.22).
2. *Cryptosporidium* spp. stain pink-red. Oocysts measure 4 to 6 μm, and four sporozoites may be visible within some oocysts. The background should stain uniformly blue.
3. The specimen is also checked for adherence to the slide (macroscopically).
4. The microscope should be calibrated (within the last 12 months), and the objectives and oculars used for the calibration procedure should be used for all measurements on the microscope. The calibration factors for all objectives should be posted on the microscope for easy access (multiplication factors can be pasted on the body of the microscope). If the microscopes receive adequate maintenance and are not moved frequently, yearly recalibration may not be necessary.
5. Known positive microscope slides, Kodachrome 2 × 2 projection slides, and photographs (reference books) should be available at the workstation.
6. Record all QC results; the laboratory should also have an action plan for "out-of-control" results.

Procedure for Kinyoun's Acid-Fast Stain

1. Smear 1 to 2 drops of concentrated specimen sediment on the slide, and allow it to air dry. Do not make the smears too thick (you should be able to see through the wet material before it dries). Prepare two smears.
2. Fix with absolute methanol for 1 min.
3. Flood the slide with Kinyoun's carbol fuchsin and stain for 5 min.
4. Rinse the slide briefly (3 to 5 s) with 50% ethanol.
5. Rinse the slide thoroughly with water.
6. Decolorize with 1% sulfuric acid for 2 min or until no more color runs from the slide.
7. Rinse the slide with water. Drain.
8. Counterstain with methylene blue for 1 min.
9. Rinse the slide with water. Air dry.
10. Examine the slide with the low or high dry objective. To see the internal morphology, use an oil objective (100×).

Results and Patient Reports from Kinyoun's Acid-Fast Staining Method

The oocysts of *Cryptosporidium*, *Cyclospora*, and *Isospora* spp. stain pink to red to deep purple. Some of the four sporozoites may be visible in the *Cryptosporidium* oocysts. Some of the *Isospora* immature oocysts (entire oo-

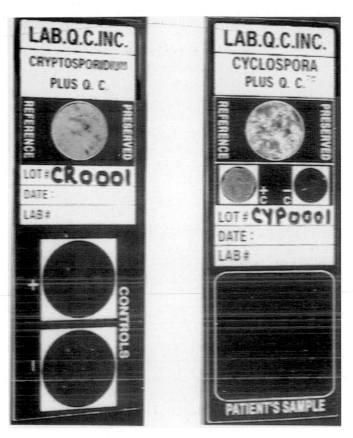

Figure 27.22 Quality control slides for performing modified acid-fast stains (*Cryptosporidium* and *Cyclospora* spp.) (Medical Chemical Corp.).

cyst) will stain, while those that are mature usually appear with the two sporocysts within the oocyst wall stained pink to purple and a clear area between the stained sporocysts and the oocyst wall. The background stains blue. If *Cyclospora* oocysts are present (uncommon unless in an outbreak situation), they tend to be approximately 8 to 10 μm, they resemble *Cryptosporidium* spp. but are larger, and they have no definite internal morphology; the acid-fast staining tends to be more variable than that seen with *Cryptosporidium* or *Isospora* spp. (Figures 27.23 and 27.24). Some of the *Cyclospora* oocysts tend to look like "wrinkled cellophane," with tremendous variation in color intensity ranging from clear to pink to red to deep purple to almost black. The stain intensity also depends on the thickness of the smear, the percentage of acid in the decolorizer (1% sulfuric acid recommended), and the length of time the smear is in contact with the decolorizer. If the patient has a heavy infection with microsporidia (immunocompromised patient), small (1- to 2-μm) spores may be seen but may not be recognized as anything other than bacteria or small yeast cells.

There is usually a range of color intensity in the organisms present; not every oocyst appears deep pink to purple. The greatest staining variation will be seen with *Cyclospora* oocysts; do not decolorize too long.

1. Report the organism and stage (oocyst). Do not use abbreviations.
   Examples: *Cryptosporidium* spp. oocysts
              Isospora belli oocysts
2. Call the physician when these organisms are identified.
3. Save positive slides for future reference. Label prior to storage (name, patient number, organisms present).

Procedure Notes for Kinyoun's Acid-Fast Staining Method

1. Routine stool examination stains (trichrome and iron hematoxylin) are not recommended; however, sedimentation concentration ($500 \times g$ for 10 min) is recommended for the recovery and identification of the coccidia after the concentration sediment has been stained with one of the modified acid-fast stains. The routine concentration (formalin-ethyl acetate) can be used to recover *Isospora* oocysts (wet sediment examination and/or modified acid-fast stains), but routine permanent stains (trichrome and iron hematoxylin) are not reliable for this purpose.
2. PVA-preserved specimens are not acceptable for staining with the modified acid-fast stain. However, specimens preserved in SAF or other non-mercury single-vial-system fixatives are perfectly acceptable.

Figure 27.23 *Cyclospora cayetanensis* stained with modified acid-fast stain. Note the range of color intensity; some oocysts resemble "wrinkled cellophane."

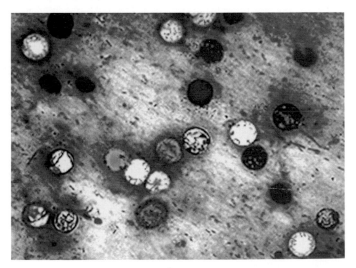

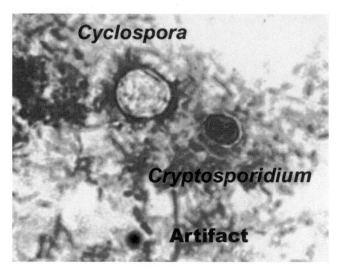

Figure 27.24 *Cyclospora* (8 to 10 μm), *Cryptosporidium* (4 to 6 μm), and artifact (~2 μm) stained with modified acid-fast stain. Note that the *Cyclospora* oocyst did not stain (modified acid-fast variable) and has the "wrinkled-cellophane" appearance, while both the *Cryptosporidium* oocyst and the artifact did stain modified acid-fast positive. These artifacts are frequently seen; it is very important to measure objects seen in the modified acid-fast smears to confirm that they are actual parasites and not artifacts.

3. Avoid the use of wet-gauze filtration (an old, standardized method of filtering stool prior to centrifugation) with too many layers of gauze that may trap organisms and not allow them to flow into the fluid to be concentrated. It is recommended that no more than two layers of gauze be used; **another option is to use the commercially available concentrators that use plastic or metal screens instead of gauze.**

4. Other organisms that stain positive include acid-fast bacteria, *Nocardia* spp., and the microsporidia (which are very difficult to find and identify even when they appear to be acid fast).

5. It is very important that smears not be too thick. Thicker smears may not adequately destain.

6. Concentration of the specimen is essential to demonstrate organisms. The number of organisms seen in the specimen may vary from many to very few; therefore, the organisms are concentrated in the sediment, which is used to prepare the smears for subsequent staining.

7. Some specimens require treatment with 10% KOH because of their mucoid consistency. Add 10 drops of 10% KOH to the sediment, and vortex until homogeneous. Rinse with 10% formalin, and centrifuge (500 × *g* for 10 min). Without decanting the supernatant, take

1 drop of the sediment and smear it thinly on a slide.

8. Commercial concentrators and reagents are available. (See Appendix 1.)

9. Concentrations of sulfuric acid of 1.0 to 3.0% are normally used. Higher concentrations remove too much stain. The use of acid-alcohol (routinely used in the Ziehl-Neelsen acid-fast staining method for the mycobacteria) decolorizes all organisms; therefore, one must use the modified decolorizer (1 to 3% H_2SO_4) for good results. In general, 1% acid is recommended; this approach will provide excellent staining results for all the coccidia.

10. There is some debate whether organisms lose their ability to take up the acid-fast stain after long-term storage in 10% formalin. Some laboratories have reported this diminished staining.

11. Specimens should be centrifuged in capped tubes, and gloves should be worn during all phases of specimen processing.

Procedure Limitations for Kinyoun's Acid-Fast Staining Method

1. Light infections with *Cryptosporidium* spp. may be missed (small number of oocysts). The fecal immunoassay methods are more sensitive.

2. Multiple specimens must be examined, since the numbers of oocysts present in the stool vary from day to day. A series of three specimens submitted on alternate days is recommended.

3. *Cyclospora* may be suspected if the organisms appear to be *Cryptosporidium* but are about twice the size (about 10 μm) (Figure 27.24). The microsporidial spores are extremely small (1 to 2 μm) and will probably not be recognized unless they are very numerous and appear to have a somewhat different morphology from the bacteria in the preparation.

Modified Ziehl-Neelsen Acid-Fast Stain (Hot Method)

Cryptosporidium and *Isospora* have been recognized as causes of severe diarrhea in immunocompromised hosts but can also cause diarrhea in immunocompetent hosts. Oocysts in clinical specimens may be difficult to detect without special staining. Modified acid-fast stains are recommended to demonstrate these organisms. Application of heat to the carbol fuchsin assists in the staining, and the use of a milder decolorizer allows the organisms to retain their pink-red color (9). With continued reports of diarrheal outbreaks due to *Cyclospora*, it is also important to remember that these organisms are acid fast and can be identified by using this staining approach (27, 28). Although the microsporidial spores are also acid fast, their size (1 to 2 μm) makes identification very difficult without special stains or the use of monoclonal antibody reagents.

Concentrated sediment of fresh, formalin-, or other non-mercury single-vial fixative-preserved stool may be used. Other types of clinical specimens such as duodenal fluid, bile, and pulmonary specimens (induced sputum, bronchial washings, or biopsy specimens) may also be stained.

Carbol Fuchsin

1. To make basic fuchsin (solution A), dissolve 0.3 g of basic fuchsin in 10 ml of 95% ethanol.
2. To make phenol (solution B), dissolve 5 g of phenol crystals in 100 ml of distilled water. (Gentle heat may be needed.)
3. Mix solution A with solution B.
4. Store at room temperature. The solution is stable for 1 year. Note the expiration date on the label.

5% Sulfuric Acid

1. Add 5 ml of concentrated sulfuric acid to 95 ml of distilled water.
2. Store at room temperature. The solution is stable for 1 year. Note the expiration date on the label.

Methylene Blue

1. Dissolve 0.3 g of methylene blue chloride in 100 ml of distilled water.
2. Store at room temperature. The solution is stable for 1 year. Note the expiration date on the label.

Quality Control for the Modified Ziehl-Neelsen Acid-Fast Staining Method

QC guidelines are the same as those for the Kinyoun's acid-fast stain and are given on p. 814.

Procedure for the Modified Ziehl-Neelsen Staining Method

1. Smear 1 to 2 drops of specimen on the slide, and allow it to air dry. Do not make the smears too thick (you should be able to see through the wet material before it dries). Prepare two smears.
2. Dry on a heating block (70°C) for 5 min.
3. Place the slide on a staining rack, and flood it with carbol fuchsin.
4. With an alcohol lamp or Bunsen burner, gently heat the slide to steaming by passing a flame under the slide. Discontinue heating once the stain begins to steam. Do not boil.

5. Allow the specimen to stain for 5 min. If the slide dries, add more stain without additional heating.
6. Rinse thoroughly with water. Drain.
7. Decolorize with 5% sulfuric acid for 30 s. (Thicker slides may require a longer destain.)
8. Rinse the slide with water. Drain.
9. Flood the slide with methylene blue for 1 min.
10. Rinse the slide with water, drain, and air dry.
11. Examine with low or high dry objective. To see internal morphology, use the oil objective (100×).

Results and Patient Reports from the Modified Ziehl-Neelsen Acid-Fast Staining Method

The oocysts of *Cryptosporidium* and *Isospora* spp. stain pink to red to deep purple. Some of the four sporozoites may be visible in the *Cryptosporidium* oocysts. Some of the *Isospora* immature oocysts (entire oocyst) stain, while those that are mature usually appear with the two sporocysts within the oocyst wall stained a pink to purple color and a clear area between the stained sporocysts and the oocyst wall (Figure 27.25). The background stains blue. If *Cyclospora* oocysts are present (uncommon), they tend to be approximately 8 to 10 μm, they resemble *Cryptosporidium* oocysts but are larger, and they have no definite internal morphology; the acid-fast staining tends to be more variable than that seen with *Cryptosporidium* or *Isospora* spp. If the patient has a heavy infection with microsporidia (immunocompromised patient), small (1- to 2-μm) spores may be seen but may not be recognized as anything other than bacteria or small yeast cells.

There is usually a range of color intensity in the organisms present; not every oocyst appears deep pink to purple. The greatest staining variation would be seen with *Cyclospora* organisms.

1. Report the organism and stage (oocyst). Do not use abbreviations.
   Examples: *Cryptosporidium* spp. oocysts
   Isospora belli oocysts
   Cyclospora cayetanensis oocysts
2. Call the physician when these organisms are identified.
3. Save positive slides for future reference. Label prior to storage (name, patient number, organisms present).

Procedure Notes for the Modified Ziehl-Neelsen Acid-Fast Staining Method

1. Routine stool examination stains (trichrome and iron hematoxylin) are not recommended; however, sedimentation concentration (500 × g for 10 min) is acceptable for the recovery and identification of

the coccidia, particularly after the concentration sediment has been stained with one of the modified acid-fast stains. The routine concentration (formalin-ethyl acetate) can be used to recover *Isospora* oocysts (wet sediment examination and/or modified acid-fast stains), but routine permanent stains (trichrome and iron hematoxylin) are not reliable for this purpose.

2. PVA-preserved specimens are not acceptable for staining with the modified acid-fast stain. However, specimens preserved in SAF or other single-vial preservatives are perfectly acceptable.

3. Avoid the use of wet-gauze filtration (an old, standardized method of filtering stool prior to centrifugation) with too many layers of gauze that may trap organisms and not allow them to flow into the fluid to be concentrated. It is recommended that no more than two layers of gauze be used. Another option is to use the commercially available concentration systems in which metal or plastic screens are used for filtration.

4. Other organisms that stain positive include acid-fast bacteria, *Nocardia* spp., and the microsporidia (which are very difficult to find and identify even when they appear to be acid fast).

5. It is very important that smears not be too thick. Thicker smears may not adequately destain.

6. Concentration of the specimen is essential to demonstrate organisms. The number of organisms seen in the specimen may vary from numerous to very few. Therefore, the sensitivity of the method is enhanced if the stains are performed on concentrated sediment.

7. Some specimens require treatment with 10% KOH because of their mucoid consistency. Add 10 drops of 10% KOH to the sediment, and vortex until homogeneous. Rinse with 10% formalin, and centrifuge ($500 \times g$ for 10 min). Without decanting the supernatant, take 1 drop of the sediment and smear it thinly on a slide.

8. Commercial concentrators and reagents are available.

9. Do not boil the stain. Gently heat until steam rises from the slide. Do not allow the stain to dry on the slide.

10. Various concentrations of sulfuric acid (0.25 to 10%) may be used; the destaining time varies according to the concentration used. Generally, a 1 or 5% solution is used. The use of acid-alcohol (routinely used in the Ziehl-Neelsen acid-fast staining method for the mycobacteria) decolorizes all organisms; therefore, the modified decolorizer must be used for good results.

11. There is some debate whether organisms lose their ability to take up the acid-fast stain after long-term storage in 10% formalin. Some laboratories have reported this diminished staining.

12. Specimens should be centrifuged in capped tubes, and gloves should be worn during all phases of specimen processing.

Procedure Limitations for the Modified Ziehl-Neelsen Acid-Fast Staining Method

1. Light infections with *Cryptosporidium* or *Cyclospora* may be missed (small number of oocysts). When available, fecal immunoassay methods for *Cryptosporidium* are more sensitive.

2. Multiple specimens must be examined, since the numbers of oocysts present in the stool vary from day to day. A series of three specimens submitted on alternate days is recommended.

3. The identification of both *Cyclospora* organisms and microsporidia may be difficult. *Cyclospora* may be suspected if the organisms appear to be *Cryptosporidium* but are about twice the size (about 8 to 10 µm). The microsporidial spores are extremely small (1 to 2 µm) and will probably not be recognized unless they are very numerous and appear to have a somewhat different morphology from the other bacteria in the preparation.

4. Often, artifact material may be seen in these stained smears (Figure 27.24). The artifacts may resemble

Figure 27.25 (Upper) Immature *Isospora belli* oocyst stained with modified acid-fast stain; (lower) *I. belli* immature oocyst (left) (note that the entire oocyst retains the stain) and *I. belli* mature oocyst containing two sporocysts (right).

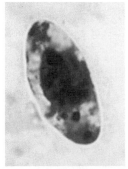
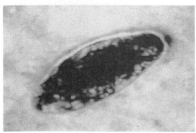

the oocysts of *Cryptosporidium* or *Cyclospora*; therefore, it is very important that any "parasites" seen in the stained smears be measured for confirmation.

There are three other stains that can be used for the coccidia, although they may not be as common as the Kinyoun or hot acid-fast method. They are the carbol fuchsin negative stain, the safranin stain, and the auramine O stain.

Carbol Fuchsin Negative Stain for *Cryptosporidium* (from W. L. Current)

1. Mix thoroughly an equal volume (3 to 10 µl) of fresh or formalin-fixed stool and Kinyoun's carbol fuchsin on a slide.
2. Spread out as a thin film, and allow to air dry at room temperature.
3. Add immersion oil directly to the stained smear, and then cover with a coverslip.
4. Observe under bright-field microscopy (×400). Everything but the oocysts stains darkly. The oocysts are bright and refractile because they contain water and everything else is oil soluble.

Rapid Safranin Method for *Cryptosporidium* (1)

1. Smear fresh or formalin-fixed feces on a slide, and allow the film to air dry at room temperature.
2. Fix briefly with one pass through the Bunsen burner flame.
3. Fix for 3 to 5 min with 3% HCl in methanol.
4. Wash with tap water (brief rinse).
5. Stain with 1% aqueous safranin for 1 min (heat until steam appears) (the authors indicate by personal communication that boiling may be beneficial).
6. Rinse in tap water.
7. Counterstain with 1% methylene blue for 30 s (0.1% aqueous crystal violet was almost as good, but malachite green was unsatisfactory).

Rapid Safranin Method for *Cyclospora*, Using a Microwave Oven (4, 44)

Another rapid safranin method uniformly stains *Cyclospora* oocysts a brilliant reddish orange. However, the fecal smears must be heated in a microwave oven before being stained. This stain is fast, reliable, and easy to perform (44).

1. Using a 10-µl aliquot of concentrated stool, prepare the smear by spreading the material thinly across the slide.
2. Allow the smear to dry on a 60°C slide warmer.
3. Cool the slide to room temperature before staining.
4. Place the slide in a Coplin jar containing acidic alcohol (3% [vol/vol] HCl in methanol), and let it stand for 5 min.
5. Wash off excess acidic alcohol with cold tap water.
6. Place the slide(s) into the Coplin jar containing the safranin solution in acidified water (pH 6.5), and microwave at full power (650 W) for 1 min. (Place the staining jar in another container to catch the overflow of stain because of boiling.)
7. Wash off excess stain with cold tap water.
8. Place the slide(s) for 1 min in a Coplin jar containing 1% methylene blue.
9. Rinse gently with cold tap water.
10. Air dry.
11. Add a coverslip to the slide by using Cytoseal 60 or other mounting medium; the immersion oil mounting method can also be used.
12. Examine the smear with low-power or high dry power objectives. To see additional morphology, use the oil immersion objective (100×).

Auramine O Stain for Coccidia (from Thomas Hänscheid)

Coccidia are acid-fast organisms and also stain well with auramine O (phenolized auramine O). The size and typical appearance of *Cryptosporidium*, *Cyclospora*, and *Isospora* oocysts enables auramine O-stained slides to be examined at low power using the 10× objective. The entire sample area can usually be examined in less than 30 s. The low cost of the reagents, the simple staining protocol, and the rapid microscopic examination also make this staining method suitable for screening unconcentrated fecal specimens.

Concentrated sediment from fresh or non-PVA-preserved stool may be used. Other stool samples may also be used, such as unconcentrated stool submitted for culture in a bacteriology transport medium. However, to increase the sensitivity of the test, small numbers of oocysts are more easily detected using concentrated stools.

Auramine O Stain

Solution 1
Auramine O .... 0.1 g
95% ethanol .... 10 ml

Solution 2

    Phenol crystals .... 3.0 g
    Distilled water................................. 87 ml

Combine solutions 1 and 2. Store in a dark bottle at room temperature for up to 3 months.

Destaining agent (0.5% acid-alcohol)

    Concentrated HCl.......................... 0.5 ml
    70% ethanol .... 100 ml

Store at room temperature for up to 3 months.

Counterstain (0.5% potassium permanganate)

    Potassium permanganate................. 0.5 g
    Distilled water.............................. 100 ml

Quality Control for the Auramine O Stain for Coccidia

QC guidelines are the same as those for the Kinyoun's acid-fast stain and are given on p. 814.

Procedure for the Auramine O Stain for Coccidia

1. Using a 10- to 20-µl aliquot of concentrated stool, prepare the smear by spreading the material across the slide.
2. Heat fix the slides either on a 65 to 75°C heat block for at least 2 h or with the flame of a Bunsen burner. However, do not overheat. Another fixation option would be to fix the slide in absolute methanol for 1 min, air dry, and then proceed with staining.
3. Cool the slide to room temperature before staining.
4. Flood the slide with the phenolized auramine O solution.
5. Allow the smear to stain for ca. 15 min. *Do not heat.*
6. Rinse the slide in water. Drain excess water from the slide.
7. Flood the slide with the destaining solution (0.5% acid-alcohol).
8. Allow to decolorize for 2 min.
9. Flood with counterstain (potassium permanganate) solution.
10. Stain for 2 min. *The timing of this step is critical.*
11. Rinse the slide in water. Drain excess water from the slide.
12. Allow the smear to air dry. Do not blot.
13. Examine the smear with a fluorescence microscope with a 10× objective and fluorescein isothiocyanate optical filters (auramine O: excitation max., ~435 nm in water; emission max.,

~510 nm). Screen the whole sample area for the presence of fluorescent oocysts. Suspicious objects can be reexamined using a 20× or 100× objective.
14. Smears can be restained by any of the carbol fuchsin (modified acid-fast) staining procedures to allow examination with light microscopy.

Results and Patient Reports from the Auramine O Staining Method

Cryptosporidium and *Cyclospora* oocysts fluoresce brightly and have a regular round appearance ("starry sky" appearance with the 10× objective). In contrast to the large majority of fluorescent artifacts, the oocysts do not stain homogeneously. Thus, the fluorescence is heterogeneously distributed in the interior of the oocyst (Figure 27.26). *Isospora* oocysts fluoresce brightly with three patterns: (i) a more or less brightly but heterogeneously stained interior of the whole oocyst, (ii) one brightly staining sporocyst, or (iii) two brightly staining sporocysts within the oocyst wall.

1. Report the organism and stage (oocyst). Do not use abbreviations.
   Examples: *Cryptosporidium* spp. oocysts
              Isospora belli oocysts
              Cyclospora cayetanensis oocysts
2. Call the physician when these organisms are identified.
3. Save positive slides for future reference. Label prior to storage (name, patient number, organisms present). These slides can be kept at room temperature in the dark, and the fluorescence remains stable for up to 3 to 4 weeks.

Procedure Notes for the Auramine O Staining Method

1. It is mandatory that positive control smears be stained and examined each time patient specimens are stained and examined.
2. For best results, examine the auramine O solution for deposits and remove them by filtration or centrifugation. This problem can also be avoided by preparing smaller volumes more frequently.
3. Slides should be observed as soon as possible after staining. However, they can be kept at room temperature in the dark, and fluorescence remains stable for up to 3 to 4 weeks.

Procedure Limitations for the Auramine O Staining Method

1. Light infections might be missed, particularly if

unconcentrated stool is used; it is always recommended that concentrated stool sediment be used for staining.

2. Use of the 40× high dry objective often causes a blurred image (fluorescent "halo" around the image, hazy contours), which appears to be the effect of interfering fluorescence from the auramine O stain located outside the plane of focus (Figure 27.26). The 100× oil immersion objective gives better-quality images. Immersion oils used for light microscopy may be autofluorescent, and special low-fluorescence immersion oil should be used.

3. If the fluorescence is not clear or definitive, a suspicious slide can be restained with a modified acid-fast stain and reexamined using light microscopy and the 100× oil immersion objective.

4. If protected from sunlight, auramine O slides can be kept on the bench at room temperature for up to 2 to 3 weeks, with only minor loss of fluorescence (photo bleaching).

Figure 27.26 (Top) Multiple microscopes with different attachments, including fluorescence (bottom right) (courtesy of Olympus America Inc.). (Middle) *Isospora belli* stained with auramine O fluorescent stain (courtesy of Thomas Hänscheid). (Bottom) *Cryptosporidium* spp. stained with auramine O fluorescent stain (left, 40× objective; right, 100× oil immersion objective) (courtesy of Thomas Hänscheid).

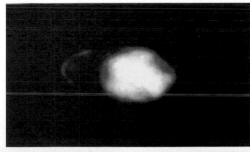

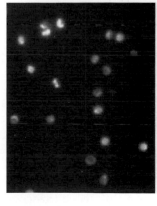 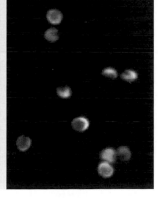

Modified Trichrome Stain for the Microsporidia (Weber—Green) (45)

The diagnosis of intestinal microsporidiosis (*Enterocytozoon bieneusi, Encephalitozoon intestinalis*) has depended on the use of invasive procedures and subsequent examination of biopsy specimens, often by electron microscopy. However, the need for a practical method for the routine clinical laboratory has stimulated some work in the development of additional methods. Slides prepared from fresh or formalin-fixed stool specimens can be stained by a chromotrope-based technique and can be examined under light microscopy. This staining method is based on the fact that stain penetration of the microsporidial spore is very difficult; therefore, the dye content in the chromotrope 2R is greater than that routinely used to prepare Wheatley's modification of Gomori's trichrome method, and the staining time is much longer (90 min) (Figures 27.27 and 27.28) (38, 45, 46). At least several of these stains are available commercially from a number of suppliers.

The specimen can be fresh stool or stool that has been preserved in 5 or 10% formalin, SAF, or the newer single-vial-system fixatives. Actually, any specimen other than tissue thought to contain microsporidia could be stained by this method.

Trichrome Stain (Modified for Microsporidia) (Weber—Green) (45)

Chromotrope 2R	6.0 g*
Fast green	0.15 g
Phosphotungstic acid	0.7 g
Acetic acid (glacial)	3.0 ml
Distilled water	100.0 ml

*(10 times the normal trichrome stain formula)

1. Prepare the stain by adding 3.0 ml of acetic acid to the dry ingredients. Allow the mixture to stand (ripen) for 30 min at room temperature.

2. Add 100 ml of distilled water. Properly prepared stain is dark purple.

REVIEW	Modified Acid–Fast Smears
Topic	**Comments**
Principle	To provide contrasting colors for both the background debris and parasites present; designed to allow examination and recognition of acid-fast characteristic of the organisms under high dry examination (40× objective for a total magnification of ×400). Designed primarily to allow recovery and identification of intestinal coccidian oocysts. Internal morphology (sporozoites) is seen in some *Cryptosporidium* oocysts under oil immersion (×1,000 magnification).
Specimen	Any stool specimen that is fresh or preserved in formalin, SAF, or the newer single-vial-system fixatives.
Reagents	Kinyoun's acid-fast stain, modified Ziehl-Neelsen stain, and their associated solutions; dehydrating solutions (alcohols and xylenes); mounting fluid optional; the decolorizing agents are less intense than the routine acid-alcohol used in routine acid-fast staining (this fact is what makes these procedures "modified" acid-fast procedures).
Examination	High dry examination of at least 300 fields; additional fields may be required if suspect organisms have been seen but are not clearly acid fast.
Results and Laboratory Reports	The identification of *Cryptosporidium* and *Isospora* oocysts should be possible; *Cyclospora* oocysts, which are twice the size of *Cryptosporidium* oocysts, should be visible but tend to be more acid-fast variable. The stain intensity depends on the thickness of the smear, the percentage of acid in the decolorizer, and the length of time the smear is in contact with the decolorizer. Although microsporidia are acid fast, their small size makes their recognition very difficult. Final laboratory results would depend heavily on the appearance of the QC slides and comparison with patient specimens.
Procedure Notes and Limitations	Both the cold and hot modified acid-fast methods are excellent for the staining of coccidial oocysts. There is some feeling that the hot method may result in better stain penetration, but the differences are probably minimal. Procedure limitations are related to specimen handling (proper centrifugation speeds and time, use of no more than two layers of wet gauze for filtration, and complete understanding of the difficulties in recognizing microsporidial spores). There is also some controversy concerning whether the organisms lose the ability to take up acid-fast stains after long-term storage In 10% formalin. The organisms are more difficult to find in specimens from patients who do not have the typical, watery diarrhea (more formed stool contains more artifact material).

3. Store in a glass or plastic bottle at room temperature. The shelf life is at least 24 months.

Acid-Alcohol

90% ethyl alcohol .... 995.5 ml
Acetic acid (glacial) .... 4.5 ml

Prepare by combining the two solutions.

Quality Control for the Modified Trichrome Staining Method (Weber—Green)

1. Unfortunately, the only way to perform acceptable QC procedures for the modified trichrome staining method is to use actual microsporidial spores as the control organisms. Obtaining these positive controls may be somewhat difficult. It is particularly important to use the actual organisms, because the spores are difficult to stain and they are very small (1 to 2.5 μm).

2. A QC slide should be included with each run of stained slides, particularly if the staining setup is used infrequently.

3. All staining dishes should be covered to prevent evaporation of reagents (screw-cap Coplin jars or glass lids).

4. Depending on the volume of slides stained, staining solutions must be changed on an as-needed basis.

5. When the smear is thoroughly fixed and the stain is performed correctly, the spores are ovoid and refractile, with the spore wall being bright pinkish red. Occasionally, the polar tube can be seen either as a stripe or as a diagonal line across the spore. The majority of the bacteria and other debris tend to stain green. However, some bacteria and debris still stain red.

6. The specimen is also checked for adherence to the slide (macroscopically).

7. The microscope should be calibrated (within the last 12 months), and the objectives and oculars used for the calibration procedure should be used for all measurements on the microscope. The calibration factors for all objectives should be posted on the microscope for easy access (multiplication factors can be pasted on the body of the microscope).

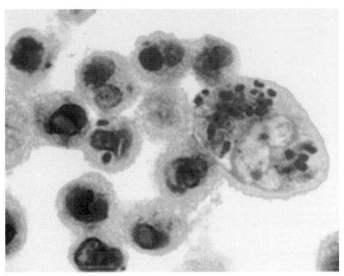

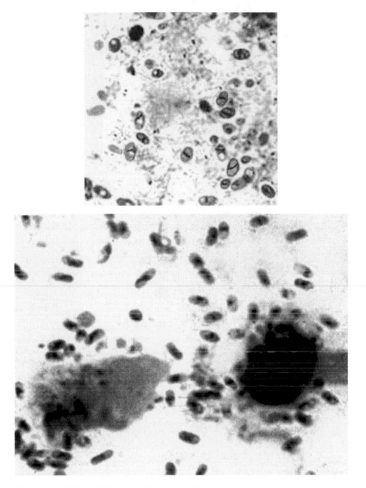

Figure 27.28 Microsporidian spores in a urine sediment, stained with a modified trichrome stain. Note that some of the spores are intracellular. A urine specimen tends to be "cleaner" than stool; therefore the spores may be easier to see and identify in urine or any specimen that contains less artifact material than stool.

Figure 27.27 (Upper) Microsporidian spores in a stool specimen concentrate sediment stained with a modified trichrome stain. The spores range from about 1.5 to 2.0 μm in diameter. Note the horizontal lines, indicating the presence of a polar tubule. (Lower) Microsporidian spores from a nasopharyngeal aspirate stained with a modified trichrome stain. Note the presence of the polar tubule within several of the spores.

Although recalibration every 12 months may not be necessary, this will vary from laboratory to laboratory, depending on equipment care and use.

8. Known positive microscope slides, Kodachrome 2 × 2 projection slides, and photographs (reference books) should be available at the workstation.

9. Record all QC results; the laboratory should also have an action plan for "out-of-control" results.

Procedure for Modified Trichrome Staining Method (Weber—Green)

1. Using a 10-μl aliquot of unconcentrated (concentrated recommended), preserved liquid stool (5 or 10% formalin or SAF or one of the non-PVA single-vial preservatives), prepare the smear by spreading the material over an area 45 by 25 mm.

2. Allow the smear to air dry.
3. Place the smear in absolute methanol for 5 min.
4. Allow the smear to air dry.
5. Place in trichrome stain for 90 min.
6. Rinse in acid-alcohol for no more than 10 s.
7. Dip slides several times in 95% alcohol. Use this step as a rinse.
8. Place in 95% alcohol for 5 min.
9. Place in 100% alcohol for 10 min.
10. Place in xylene substitute for 10 min.
11. Mount with a coverslip (no. 1 thickness), using mounting medium.
12. Examine smears under oil immersion (×1,000), and read at least 300 fields; the examination time will probably be at least 10 min per slide.

Modified Trichrome Stain for the Microsporidia (Ryan—Blue) (38)

A number of variations to the modified trichrome stain (Weber—green) have been tried in an attempt to improve the contrast between the color of the spores and the background staining. Optimal staining was achieved by modifying the composition of the trichrome solution. This stain is also available commercially from a number of suppliers.

The specimen can be fresh stool or stool that has been preserved in 5 or 10% formalin, SAF, or the newer single-vial-system fixatives. Actually, any specimen other than tissue thought to contain microsporidia could be stained by this method.

Trichrome Stain (Modified for Microsporidia) (Ryan—Blue)

Chromotrope 2R	6.0 g*
Aniline blue	0.5 g
Phosphotungstic acid	0.25 g
Acetic acid (glacial)	3.0 ml
Distilled water	100.0 ml

*(10 times the normal trichrome stain formula)

1. Prepare the stain by adding 3.0 ml of acetic acid to the dry ingredients. Allow the mixture to stand (ripen) for 30 min at room temperature.
2. Add 100 ml of distilled water, and adjust the pH to 2.5 with 1.0 M HCl. Properly prepared stain is dark purple. The staining solution should be protected from light.
3. Store in a glass or plastic bottle at room temperature. The shelf life is at least 24 months.

Acid-Alcohol

90% ethyl alcohol	995.5 ml
Acetic acid (glacial)	4.5 ml

Prepare by combining the two solutions.

Quality Control for the Modified Trichrome Staining Method (Ryan—Blue)

1. Unfortunately, the only way to perform acceptable QC procedures for this method is to use actual microsporidial spores as the control organisms. Obtaining these positive controls may be somewhat difficult. It is particularly important to use the actual organisms, because the spores are difficult to stain and are very small (1 to 2.5 µm).
2. A QC slide should be included with each run of stained slides, particularly if the staining setup is used infrequently.
3. All staining dishes should be covered to prevent evaporation of reagents (screw-cap Coplin jars or glass lids).
4. Depending on the volume of slides stained, staining solutions must be changed on an as-needed basis.
5. When the smear is thoroughly fixed and the stain is performed correctly, the spores are ovoid and refractile, with the spore wall being bright pinkish red. Occasionally, the polar tube can be seen either as a stripe or as a diagonal line across the spore. The majority of the bacteria and other debris tend to stain blue. However, some bacteria and debris still stain red.

6. The specimen is also checked for adherence to the slide (macroscopically).
7. The microscope should be calibrated (within the last 12 months), and the objectives and oculars used for the calibration procedure should be used for all measurements on the microscope. The calibration factors for all objectives should be posted on the microscope for easy access (multiplication factors can be pasted on the body of the microscope). Although recalibration every 12 months may not be necessary, this will vary from laboratory to laboratory, depending on equipment care and use.
8. Known positive microscope slides, Kodachrome 2 × 2 projection slides, and photographs (reference books) should be available at the workstation.
9. Record all QC results; the laboratory should also have an action plan for "out-of-control" results.

Procedure for the Modified Trichrome Staining Method (Ryan—Blue)

1. Using a 10-µl aliquot of concentrated (10 min at 500 × g), preserved liquid stool (5 or 10% formalin, SAF, or one of the non-PVA single-vial preservatives, zinc based), prepare the smear by spreading the material over an area 45 by 25 mm.
2. Allow the smear to air dry.
3. Place the smear in absolute methanol for 5 or 10 min.
4. Allow the smear to air dry.
5. Place in trichrome stain for 90 min.
6. Rinse in acid-alcohol for no more than 10 s.
7. Dip the slides several times in 95% alcohol. Use this step as a rinse (no more than 10 s).
8. Place in 95% alcohol for 5 min.
9. Place in 95% alcohol for 5 min.
10. Place in 100% alcohol for 10 min.
11. Place in xylene substitute for 10 min.
12. Mount with a coverslip (no. 1 thickness), using mounting medium.
13. Examine smears under oil immersion (×1,000), and read at least 300 fields; the examination time will probably be at least 10 min per slide.

Results and Patient Reports from Modified Trichrome Staining Methods (Weber or Ryan)

The microsporidial spore wall should stain pinkish to red, with the interior of the spore being clear or perhaps showing a horizontal or diagonal stripe which represents the polar tube. The background appears green or blue, depending on the method. Other bacteria, some yeast cells, and some debris stains pink to red; the shapes and sizes of the various components may be helpful in differentiating the spores from other structures. The results of this staining procedure should be reported only if the positive control smears are acceptable. The production of immunoassay reagents should provide a more specific and sensitive approach to the identification of the microsporidia in fecal specimens.

1. Report the organism.

    Examples: Microsporidial spores present. The following information can be added to the report to assist the physician in treating and following the patient:

    The organisms are most probably *Enterocytozoon bieneusi* or *Encephalitozoon intestinalis* (if from fecal specimen or urine).

    The organisms are most probably *Encephalitozoon intestinalis* (identification to species highly likely) (generally the organism involved in disseminated cases from the gastrointestinal tract to other body sites).

Procedure Notes for Modified Trichrome Staining Methods (Weber or Ryan)

1. It is mandatory that positive control smears be stained and examined each time patient specimens are stained and examined.

2. Because of the difficulty in getting the stain to penetrate the spore wall, prepare thin smears and do not reduce the staining time in trichrome. Also, make sure the slides are not left too long in the decolorizing agent (acid-alcohol). If the control organisms are too light, leave them in the trichrome longer and shorten the time to two dips in the acid-alcohol solution. Also, remember that the 95% alcohol rinse after the acid-alcohol should be performed quickly to prevent additional destaining from the acid-alcohol reagent.

3. When you purchase the chromotrope 2R, obtain the highest dye content available. Two sources are Harleco, Gibbstown, N.J., and Sigma Chemical Co., St. Louis, Mo. (dye content among the highest [85%]). Fast green and aniline blue can be obtained from Allied Chemical and Dye, New York, N.Y. See also appendix 5.

4. In the final stages of dehydration, the 100% ethanol and the xylenes (or xylene substitutes) should be kept as free from water as possible. Coplin jars must have tight-fitting caps to prevent both evaporation of reagents and absorption of moisture. If the xylene becomes cloudy after addition of slides from 100% alcohol, return the slides to fresh 100% alcohol and also replace the xylene with fresh stock.

Procedure Limitations for Modified Trichrome Staining Methods (Weber or Ryan)

1. Although this staining method stains the microsporidia, the range of stain intensity and the small size of the spores causes some difficulty in identifying these organisms. Since this procedure results in many other organisms or objects staining in stool specimens, differentiation of the microsporidia from surrounding material is still very difficult. There also tends to be some slight size variation among the spores.

2. If the patient has severe watery diarrhea, there is less artifact material in the stool to confuse with the microsporidial spores; however, if the stool is semi-formed or formed, the amount of artifact material is much greater and the spores are much harder to detect and identify. Also, the number of spores varies according to the stool consistency (generally, the more liquid the stool, the more spores will be present). However, remember that there is also a dilution factor if the stool is very liquid.

3. The workers who developed some of these procedures think that concentration procedures result in an actual loss of microsporidial spores; therefore, there is a recommendation to use unconcentrated, formalinized stool. However, there are no data indicating which centrifugation speeds, etc., were used in the study.

4. In the UCLA Clinical Microbiology Laboratory, we have generated data (unpublished) to indicate that **centrifugation at 500 × g for 10 min dramatically increases the number of microsporidial spores available for staining (from the concentrate sediment)**. This is the same method we use for centrifugation of all stool specimens, regardless of the suspected organism.

5. Avoid the use of wet-gauze filtration (an old, standardized method of filtering stool prior to centrifugation) with too many layers of gauze that may trap organisms and not allow them to flow into the fluid to be concentrated. It is recommended that no more than two layers of gauze be used. Another option is to use the commercially available concentration systems in which metal or plastic screens are used for filtration.

Modified Trichrome Stain for the Microsporidia (Kokoskin—Hot Method) (24)

Changes in temperature from room temperature to 50°C and in the staining time from 90 to 10 min have been recommended as improvements for the modified trichrome staining methods. The procedure is as follows.

1. Using a 10-μl aliquot of concentrated, preserved stool (5 or 10% formalin or SAF), prepare the smear by spreading the material over an area 45 by 25 mm.
2. Allow the smear to air dry.
3. Place the smear in absolute methanol for 5 min.
4. Allow the smear to air dry.
5. Place in trichrome stain for 10 min at 50°C.
6. Rinse in acid-alcohol for no more than 10 s.
7. Dip the slide several times in 95% alcohol. Use this step as a rinse (no more than 10 s).
8. Place in 95% alcohol for 5 min.
9. Place in 100% alcohol for 10 min.
10. Place in xylene substitute for 10 min.
11. Mount with a coverslip (no. 1 thickness), using mounting medium.
12. Examine the smear under oil immersion (×1,000) and read at least 300 fields; the examination time will probably be at least 10 min per slide.

Acid-Fast Trichrome Stain for *Cryptosporidium* and the Microsporidia

The detection of *Cryptosporidium* spp. and the microsporidia from stool specimens has depended on two separate stains. However, a method is now available that stains both organisms, an important improvement since dual infections have been demonstrated in AIDS patients (12). This acid-fast trichrome stain yields results comparable to those obtained by the Kinyoun and modified trichrome methods and considerably reduces the time necessary for microscopic examination (19, 20) (Figure 27.29). Also, it appears that modified trichrome stains and staining with fluorochromes are equally useful in the diagnosis of microsporidiosis; however, a combination of the two methods may be more sensitive in cases where the number of spores is very small (19).

The specimen can be fresh stool or stool that has been preserved in 5 or 10% formalin, SAF, or some of the newer single-vial-system fixatives. Actually, any specimen other than tissue thought to contain microsporidia could be stained by this method.

Trichrome Stain (Modified for Microsporidia) (4)
Chromotrope 2R .... 6.0 g*
Aniline blue .... 0.5 g
Phosphotungstic acid .... 0.7 g
Acetic acid (glacial) .... 3.0 ml
Distilled water .... 100.0 ml
*(10 times the normal trichrome stain formula)

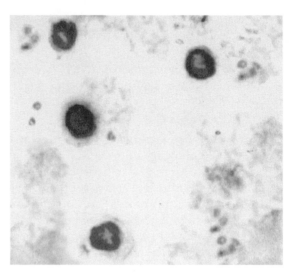

Figure 27.29 *Cryptosporidium* spp. and microsporidian spores in an acid-fast trichrome stain. Note the size differential.

1. Prepare the stain by adding 3.0 ml of acetic acid to the dry ingredients. Allow the mixture to stand (ripen) for 30 min at room temperature.
2. Add 100 ml of distilled water, and adjust the pH to 2.5 with 2.0 N HCl. Properly prepared stain is dark purple. The staining solution should be protected from light.
3. Store in a glass or plastic bottle at room temperature. The shelf life is at least 24 months.

Carbol Fuchsin Solution

Phenol solution
Phenol .... 25.0 g
Distilled water .... 500.0 ml

Saturated alcoholic fuchsin solution
Basic fuchsin .... 2.0 g
96% ethanol .... 25.0 ml

Add the mixture of phenol and water to 25.0 ml of the saturated alcoholic fuchsin solution.

Acid-Alcohol
90% ethyl alcohol .... 995.5 ml
Acetic acid (glacial) .... 4.5 ml

Prepare by combining the two solutions.

Quality Control for the Acid-Fast Trichrome Staining Method

1. Unfortunately, the only way to perform acceptable QC procedures for this method is to use actual microsporidian spores as the control organisms.

Obtaining these positive controls may be somewhat difficult. It is particularly important to use the actual organisms because the spores are difficult to stain and are very small (1 to 2.5 μm).

2. A QC slide should be included with each run of stained slides, particularly if the staining setup is used infrequently.

3. All staining dishes should be covered to prevent evaporation of reagents (screw-cap Coplin jars or glass lids).

4. Depending on the volume of slides stained, staining solutions must be changed on an as-needed basis.

5. When the smear is thoroughly fixed and the stain is performed correctly, the spores are ovoid and refractile, with the spore wall being bright pinkish red. Occasionally, the polar tube can be seen either as a stripe or as a diagonal line across the spore. The majority of the bacteria and other debris tend to stain blue; however, some bacteria and debris stain red.

6. The specimen is also checked for adherence to the slide (macroscopically).

7. The microscope should be calibrated (within the last 12 months if it has received heavy use), and the objectives and oculars used for the calibration procedure should be used for all measurements on the microscope. The calibration factors for all objectives should be posted on the microscope for easy access (multiplication factors can be pasted on the body of the microscope). Although recalibration every 12 months may not be necessary, this varies from laboratory to laboratory, depending on equipment care and use.

8. Known positive microscope slides, Kodachrome 2 × 2 projection slides, and photographs (reference books) should be available at the workstation.

9. Record all QC results; the laboratory should also have an action plan for "out-of-control" results.

Procedure for the Acid-Fast Trichrome Staining Method

1. Using a 10-μl aliquot of concentrated (10 min at 500 × *g*), preserved liquid stool (5 or 10% formalin, SAF, or one of the single-vial preservatives, zinc based), prepare the smear by spreading the material over an area 45 by 25 mm.

2. Allow the smear to air dry.

3. Place the smear in absolute methanol for 5 or 10 min.

4. Allow the smear to air dry.

5. Place in carbol fuchsin solution for 10 min (no heat required).

6. Rinse briefly with tap water.

7. Decolorize with 0.5% acid-alcohol.

8. Briefly rinse with tap water.

9. Place in trichrome stain for 30 min at 37°C.

10. Rinse in acid-alcohol for no more than 10 s.

11. Dip the slide several times in 95% alcohol. Use this step as a rinse (no more than 10 s).

12. Place in 95% alcohol for 30 s.

13. Allow the slide to air dry.

14. Examine the smear under oil immersion (×1,000) and read at least 300 fields; the examination time will probably be at least 10 min per slide.

Results and Patient Reports from the Acid-Fast Trichrome Staining Method

The microsporidial spore wall should stain pink, with the interior of the spore being clear or perhaps showing a horizontal or diagonal stripe that represents the polar tube; a vacuole may also be visible in some spores. The *Cryptosporidium* oocysts stain bright pink or violet. The results of this staining procedure should be reported only if the positive control smears are acceptable. The production of immunoassay reagents should provide a more specific and sensitive approach to the identification of the microsporidia in fecal specimens.

1. Report the organism.
   Examples: Microsporidial spores present. The following information can also be added to the report to assist the physician in treating and following the patient:
   The organisms are probably *Enterocytozoon bieneusi* or *Encephalitozoon intestinalis* (if from fecal specimens or urine).
   The organisms are probably *Encephalitozoon intestinalis* (identification to species highly likely) (generally the organism involved in disseminated cases from the gastrointestinal tract to kidneys; organisms are recovered in urine).

Procedure Notes for the Acid-Fast Trichrome Staining Method

1. It is mandatory that positive control smears be stained and examined each time patient specimens are stained and examined.

2. Because of the difficulty in achieving stain penetration through the spore wall, prepare thin smears and do not reduce the staining time in trichrome. Also, make sure the slides are not left too long in the decolorizing agent (acid-alcohol). If the control organisms are too light, leave them in the trichrome longer and shorten the time to two dips

in the acid-alcohol solution. Also, remember that the 95% alcohol rinse after the acid-alcohol step should be performed quickly to prevent additional destaining from the acid-alcohol reagent.

3. When you purchase the chromotrope 2R, obtain the highest dye content available. Two sources are Harleco, Gibbstown, N.J., and Sigma Chemical Co., St. Louis, Mo. (the dye content is among the highest [85%]). Fast green and aniline blue can be obtained from Allied Chemical and Dye, New York, N.Y. See also appendix 5.

4. In the final stages of dehydration, the 95% ethanol should be kept as free from water as possible. Coplin jars must have tight-fitting caps to prevent both evaporation of reagents and absorption of moisture.

Procedure Limitations for the Acid-Fast Trichrome Staining Method

1. Although this staining method stains the microsporidia, the range of stain intensity and the small size of the spores may still cause some difficulty in identifying these organisms. Since this proce-dure results in many other organisms or objects staining in stool specimens, differentiation of the microsporidia from surrounding material is still very difficult. There also tends to be some slight size variation among the spores.

2. If the patient has severe watery diarrhea, there is less artifact material in the stool to confuse with the microsporidial spores; however, if the stool is semiformed or formed, the amount of artifact material is much greater and so the spores are much harder to detect and identify. Also, the number of spores may vary according to the stool consistency (generally, the more liquid the stool, the more spores are present).

3. The workers who developed some of these proce-dures think that concentration procedures result in an actual loss of microsporidial spores; therefore there is a recommendation to use unconcentrated, formalinized stool. However, there are no data indicating which centrifugation speeds, etc., were used in the study.

4. In the UCLA Clinical Microbiology Laboratory, we have generated data (unpublished) to indicate that **centrifugation at 500 $\times$ g for 10 min dra-**

REVIEW	Modified Trichrome–Stained Smears
Topic	**Comments**
Principle	To provide contrasting colors for the background debris and parasites present; designed to allow examination and recognition of organism morphology under oil immersion (100$\times$ objective for a total magnification of $\times$1,000). Designed primarily to allow recovery and identification of microsporidial spores. Internal morphology (horizontal or diagonal "stripes" may be seen in some spores under oil immersion [$\times$1,000 magnification]).
Specimen	Any stool specimen that is fresh or is preserved in formalin, SAF, or one of the single-vial-system fixatives.
Reagents	Modified trichrome stain (using high-dye-content chromotrope 2R) and associated solutions; dehydrating solutions (alcohols and xylenes); mounting fluid optional.
Examination	Oil immersion examination of at least 300 fields; additional fields may be required if suspect organisms have been seen but are not clearly identified.
Results and Laboratory Reports	The identification of microsporidial spores may be possible; however, their small size will make recognition very difficult. Final laboratory results would depend heavily on the appearance of the QC slides and comparison with patient specimens. **Although extruded polar tubes have been seen in clinical samples such as sputum, cerebrospinal fluid, urine, intestinal biopsy specimens, conjunctival swab specimens, and duodenal fluid, they are rarely seen in fecal specimens.**
Procedure Notes and Limitations	Because of the difficulty in getting dye to penetrate the spore wall, this staining approach can be very helpful. Procedure limitations are related to specimen handling (proper centrifugation speeds and time, use of no more than two layers of wet gauze for filtration, and complete understanding of the difficulties in recognizing microsporidial spores owing to their small size [1 to 2.0 µm]).
Important Questions for Commercial Suppliers	Make sure to ask about specific fixatives and whether the fecal material can be stained with the modified trichrome stains and modified acid-fast stains. Also, ask if the fixatives prevent the use of any of the newer immunoassay methods now available for several of the intestinal amebae, flagel-lates, coccidia, and microsporidia (under development).

matically increases the number of microsporidial spores available for staining (from the concentrate sediment). We use the same protocol for centrifugation of all stool specimens, regardless of the suspected organism. The acid-fast trichrome procedure presented here also recommended the use of centrifuged fecal specimens.

5. Avoid the use of wet-gauze filtration (an old, standardized method of filtering stool prior to centrifugation) with too many layers of gauze that may trap organisms and not allow them to flow into the fluid to be concentrated. It is recommended that no more than two layers of gauze be used. Another option is to use the commercially available concentration systems that use metal or plastic screens for filtration.

References

1. Baxby, D., N. Blundell, and C. Hart. 1984. The development and performance of a simple, sensitive method for the detection of *Cryptosporidium* oocysts in feces. *J. Hyg.* **93**:317–323.

2. California State Department of Public Health Microbial Disease Laboratory. 1979. *PARA-1A Revised.* California State Department of Public Health, Sacramento.

3. Clinical Laboratory Standards Institute. 2005. *Procedures for the Recovery and Identification of Parasites from the Intestinal Tract.* Approved guideline M28-A2. Clinical Laboratory Standards Institute, Villanova, Pa.

4. Didier, E. S., J. M. Orenstein, A. Aldras, D. Bertucci, L. B. Rogers, and F. A. Janney. 1995. Comparison of three staining methods for detecting microsporidia in fluids. *J. Clin. Microbiol.* **33**:3138–3145.

5. Faust, E. C., J. S. D'Antoni, V. Odom, M. F. Miller, C. Peres, W. Sawitz, L. F. Thomen, J. Tobie, and J. H. Walker. 1938. A critical study of clinical laboratory technics for the diagnosis of protozoan cysts and helminth eggs in feces. *Am. J. Trop. Med.* **18**:169–183.

6. Fayer, R., and B. L. P. Ungar. 1986. *Cryptosporidium* spp. and cryptosporidiosis. *Microbiol. Rev.* **50**:458–483.

7. Fedorko, D. P., E. C. Williams, N. A. Nelson, T. D. Mazyck, K. L. Hanson, and C. P. Cartwright. 2000. Performance of Para-Pak Ultra ECOFIX compared with Para-Pak Ultra formalin/mercuric chloride-based polyvinyl alcohol for concentration and permanent stained smears of stool parasites. *Diagn. Microbiol. Infect. Dis.* **37**:37–39.

8. Garcia, L., T. Brewer, and D. Bruckner. 1979. A comparison of the formalin-ether concentration and trichrome-stained smear methods for the recovery and identification of intestinal protozoa. *Am. J. Med. Technol.* **45**:932–935.

9. Garcia, L., D. Bruckner, T. Brewer, and R. Shimizu. 1983. Techniques for the recovery and identification of *Cryptosporidium* oocysts from stool specimens. *J. Clin. Microbiol.* **18**:185–190.

10. Garcia, L. S. 1990. Laboratory methods for diagnosis of parasitic infections, p. 776–861. *In* E. J. Baron and S. M. Finegold (ed.), *Bailey & Scott's Diagnostic Microbiology*, 8th ed. The C. V. Mosby Co., St. Louis, Mo.

11. Garcia, L. S., and R. Y. Shimizu. 1998. Evaluation of intestinal protozoan morphology in human fecal specimens preserved in EcoFix: comparison of Wheatley's trichrome stain and EcoStain. *J. Clin. Microbiol.* **36**:1974–1976.

12. Garcia, L. S., R. Y. Shimizu, and D. A. Bruckner. 1994. Detection of microsporidial spores in fecal specimens from patients diagnosed with cryptosporidiosis. *J. Clin. Microbiol.* **32**:1739–1741.

13. Garcia, L. S., R. Y. Shimizu, A. C. Shum, and D. A. Bruckner. 1993. Evaluation of intestinal protozoan morphology in polyvinyl alcohol preservative: comparison of zinc sulfate- and mercuric chloride-based compounds for use in Schaudinn's fixative. *J. Clin. Microbiol.* **31**:307–310.

14. Gleason, N. N., and G. R. Healy. 1965. Modification and evaluation of Kohn's one-step staining technic for intestinal protozoa in feces or tissue. *Am. J. Clin. Pathol.* **43**:494–496.

15. Gomori, G. 1950. A rapid one-step trichrome stain. *Am. J. Clin. Pathol.* **20**:661–663.

16. Hanson, K. L., and C. P. Cartwright. 2001. Use of an enzyme immunoassay does not eliminate the need to analyze stool specimens for sensitive detection of *Giardia lamblia*. *J. Clin. Microbiol.* **39**:474–477.

17. Henriksen, S. A., and J. F. L. Pohlenz. 1981. Staining cryptosporidia by a modified Ziehl Neelsen technique. *Acta Vet. Scand.* **22**:594–596.

18. Higgins, G. V. 1988. Iodine-trichrome staining technique: a replacement of iodine solution in fecal examination. *Lab. Med.* **19**:824–825.

19. Ignatius, R., S. Henschel, O. Liesenfeld, U. Mansmann, W. Schmidt, S. Koppe, T. Schneider, W. Heise, U. Futh, E. O. Riecken, H. Hahn, and R. Ulrich. 1997. Comparative evaluation of modified trichrome and Uvitex 2B stains for detection of low numbers of microsporidial spores in stool specimens. *J. Clin. Microbiol.* **35**:2266–2269.

20. Ignatius, R., M. Lehmann, K. Miksits, T. Regnath, M. Arvand, E. Engelmann, U. Futh, H. Hahn, and J. Wagner. 1997. A new acid-fast trichrome stain for simultaneous detection of *Cryptosporidium parvum* and microsporidial species in stool specimens. *J. Clin. Microbiol.* **35**:446–449.

21. Jensen, B., W. Kepley, J. Guarner, K. Anderson, D. Anderson, J. Clairmont, W. De L'aune, E. H. Austin, and G. E. Austin. 2000. Comparison of polyvinyl alcohol fixative with three less hazardous fixatives for detection and identification of intestinal parasites. *J. Clin. Microbiol.* **38**:1592–1598.

22. Kellogg, J. A., and C. J. Elder. 1999. Justification for use of a single trichrome stain as the sole means for routine detection of intestinal parasites in concentrated stool specimens. *J. Clin. Microbiol.* **37**:835–837.

23. Kohn, J. 1960. A one stage permanent staining method for fecal protozoa. *Dapim Refuiim Med. Q. Isr.* **19**:160–161.

24. Kokoskin, E., T. W. Gyorkos, A. Camus, L. Cedilotte, T. Purtill, and B. Ward. 1994. Modified technique for efficient detection of microsporidia. *J. Clin. Microbiol.* **32**:1074–1075.

25. Lawson, L. L., J. W. Bailey, N. J. Beeching, R. G. Gurgel, and L. E. Cuevas. 2004. The stool examination reports amoeba cysts: should you treat in the face of over diagnosis and lack of specificity of light microscopy? *Trop. Doct.* **34**:28–30.

26. Levine, J. A., and E. G. Estevez. 1983. Method for concentration of parasites from small amounts of feces. *J. Clin. Microbiol.* **18:**786–788.

27. Long, E. G., A. Ebrahimzadeh, E. H. White, B. Swisher, and C. S. Callaway. 1990. Alga associated with diarrhea in patients with acquired immunodeficiency syndrome and in travelers. *J. Clin. Microbiol.* **28:**1101–1104.

28. Long, E. G., E. H. White, W. W. Carmichael, P. M. Quinlish, R. Raja, B. L. Swisher, H. Daugharty, and M. T. Cohen. 1991. Morphologic and staining characteristics of a cyanobacterium-like organism associated with diarrhea. *J. Infect. Dis.* **164:**199–202.

29. Ma, P., and R. Soave. 1983. Three-step stool examination for cryptosporidiosis in 10 homosexual men with protracted watery diarrhea. *J. Infect. Dis.* **147:**824–828.

30. Markell, E. K., and M. Voge. 1981. *Medical Parasitology*, 5th ed. The W. B. Saunders Co., Philadelphia, Pa.

31. Melvin, D. M., and M. M. Brooke. 1982. *Laboratory Procedures for the Diagnosis of Intestinal Parasites*, 3rd ed. U.S. Department of Health, Education, and Welfare publication (CDC) 82-8282. Government Printing Office, Washington, D.C.

32. Miller, J. M. 1991. Quality control of media, reagents, and stains, p. 1203–1225. *In* A. Balows, W. J. Hausler, Jr., K. L. Herrmann, H. D. Isenberg, and H. J. Shadomy (ed.), *Manual of Clinical Microbiology*, 5th ed. American Society for Microbiology, Washington, D.C.

33. Morimoto, N., C. Komatsu, M. Nishida, and T. Sugiura. 2001. Detection of *Giardia lamblia* cysts in non-fixed long-term stored human feces by direct immunofluorescence assay. *Jpn. J. Infect. Dis.* **54:**72–74.

34. Nair, C. 1953. Rapid staining of intestinal amoebae on wet mounts. *Nature* (London) **172:**1051.

35. Neimeister, R., A. Logan, B. Gerber, J. Egleton, and B. Kleger. 1987. Hemo-De as substitute for ethyl acetate in formalin-ethyl acetate concentration technique. *J. Clin. Microbiol.* **25:**425–426.

36. Pietrzak-Johnston, S. M., H. Bishop, S. Wahlquist, H. Moura, N. De Oliveira Da Silva, S. Pereira Da Silva, and P. Nguyen-Dinh. 2000. Evaluation of commercially available preservatives for laboratory detection of helminths and protozoa in human fecal specimens. *J. Clin. Microbiol.* **38:**1959–1964.

37. Ritchie, L. 1948. An ether sedimentation technique for routine stool examinations. *Bull. U.S. Army Med. Dept.* **8:**326.

38. Ryan, N. J., G. Sutherland, K. Coughlan, M. Globan, J. Doultree, J. Marshall, R. W. Baird, J. Pedersen, and B. Dwyer. 1993. A new trichrome-blue stain for detection of microsporidial species in urine, stool, and nasopharyngeal specimens. *J. Clin. Microbiol.* **31:**3264–3269.

39. Shoaib, S., A. Hafiz, and S. Tauheed. 2002. Role of trichrome staining techniques in the diagnosis of intestinal parasitic infections. *J. Pak. Med. Assoc.* **52:**152–154.

40. Smith, J. W., and M. S. Bartlett. 1985. Diagnostic parasitology: introduction and methods, p. 595–611. *In* E. H. Lennette, A. Balows, W. J. Hausler, Jr., and H. J. Shadomy (ed.), *Manual of Clinical Microbiology*, 4th ed. American Society for Microbiology, Washington, D.C.

41. Spencer, F. M., and L. S. Monroe. 1976. *The Color Atlas of Intestinal Parasites*, 2nd ed. Charles C Thomas, Publisher, Springfield, Ill.

42. Tompkins, V. N., and J. K. Miller. 1947. Staining intestinal protozoa with iron-hematoxylin-phosphotungstic acid. *Am. J. Clin. Pathol.* **17:**755–758.

43. Truant, A. L., S. H. Elliott, M. Y. Kelley, and J. H. Smith. 1981. Comparison of formalin-ethyl ether sedimentation, formalin-ethyl acetate sedimentation, and zinc sulfate flotation techniques for detection of intestinal parasites. *J. Clin. Microbiol.* **13:**882–884.

44. Visvesvara, G. S., H. Moura, E. Kovacs-Nace, S. Wallace, and M. L. Eberhard. 1997. Uniform staining of *Cyclospora* oocysts in fecal smears by a modified safranin technique with microwave heating. *J. Clin. Microbiol.* **35:**730–733.

45. Weber, R., R. T. Bryan, R. L. Owen, C. M. Wilcox, L. Gorelkin, G. S. Visvesvara, and The Enteric Opportunistic Infections Working Group. 1992. Improved light-microscopical detection of microsporidia spores in stool and duodenal aspirates. *N. Engl. J. Med.* **326:**161–166.

46. Wheatley, W. 1951. A rapid staining procedure for intestinal amoebae and flagellates. *Am. J. Clin. Pathol.* **21:**990–991.

47. Yang, J., and T. H. Scholten. 1977. A fixative for intestinal parasites permitting the use of concentration and permanent staining procedures. *Am. J. Clin. Pathol.* **67:**300–304.

The correct answers for Figure 27.17 are as follows: top left, *Giardia lamblia* trophozoite; top right, *Chilomastix mesnili* cyst; middle left, *Entamoeba coli* precyst; middle right, *Endolimax nana* trophozoite; bottom left and right, *Blastocystis hominis*.

28

Additional Techniques for Stool Examination

Among the diagnostic techniques used with stool specimens, the routine ova and parasite examination is the best known. This technique has three components: the direct wet mount, the examination of material from a stool concentrate, and the permanent stained smear. This is an excellent procedure and is recommended for most intestinal parasites. However, several other diagnostic techniques are available for the recovery and identification of parasitic organisms. Most laboratories do not routinely offer all of these techniques, but many are relatively simple and inexpensive to perform (7, 13, 14). The clinician should be aware of the possibilities and the clinical relevance of information obtained from using such techniques. Occasionally, it is necessary to examine stool specimens for the presence of scolices and proglottids of cestodes and adult nematodes and trematodes to confirm the diagnosis and/or for species identification. A method for the recovery of these stages is also described in this chapter. Although not routinely performed, tests for fecal fat and reducing substances are also included.

Culture of Larval-Stage Nematodes

Nematode infections giving rise to larval stages that hatch in soil or in tissues may be diagnosed by using certain fecal culture methods to concentrate the larvae. *Strongyloides stercoralis* larvae are generally the most common larvae found in stool specimens. However, depending on the fecal transit time through the intestine and the patient's condition, rhabditiform and, rarely, filariform larvae may be present. Also, if there is a delay in examination of the stool, embryonated ova as well as larvae of hookworm may be present. Culture of feces for larvae is useful to (i) reveal their presence when they are too scanty to be detected by concentration methods; (ii) distinguish whether the infection is due to *S. stercoralis* or hookworm on the basis of rhabditiform larval morphology by allowing hookworm egg hatching to occur, releasing first-stage larvae; and (iii) allow the development of larvae into the filariform stage for further differentiation.

The use of certain fecal culture methods (sometimes referred to as coproculture) is especially helpful for detection of light infections with hookworm, *S. stercoralis*, and *Trichostrongylus* spp. and for specific identification of parasites. The rearing of infective-stage nematode larvae also helps in the specific diagnosis of hookworm and trichostrongyle infections because the eggs of many of these species are identical and specific identifications are based on larval morphology. Additionally, such techniques are useful for obtaining a large number of infective-stage larvae for research purposes. Four culture techniques and one "enhanced-recovery" method are described in this chapter.

Harada-Mori Filter Paper Strip Culture

To detect light infections with hookworm, *S. stercoralis*, and *Trichostrongylus* spp., as well as to facilitate specific identification, the Harada-Mori filter paper strip culture technique is very useful (Figure 28.1). This filter paper test tube culture technique was initially introduced by Harada and Mori in 1955 (10) and was later modified by others (12, 23).

In a study looking at the prevalence of *S. stercoralis* in three areas of Brazil, the diagnostic efficacy of the agar plate culture method (discussed later in this chapter) was as high as 93.9% compared to only 28.5 and 26.5% by the Harada-Mori filter paper culture and fecal concentration methods, respectively, when fecal specimens were processed using all three methods (17). Among the 49 positive samples, about 60% were confirmed as positive by using the agar plate method alone. These results indicate that the agar plate approach is probably a much more sensitive diagnostic method than the other two and is recommended for the diagnosis of strongyloidiasis.

The technique requires filter paper to which fresh fecal material is added and a test tube into which the filter paper is inserted. Moisture is provided by adding water

Figure 28.1 Culture methods for the recovery of larval-stage nematodes: Harada-Mori tube method and petri dish culture method. Viable larvae are present if the specimen contains *S. stercoralis* or other nematodes. Wear gloves when handling the culture devices. (Illustration by Nobuko Kitamura.)

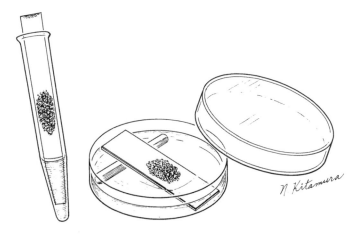

to the tube, which continuously soaks the filter paper by capillary action. Incubation under suitable conditions favors hatching of ova and/or development of larvae. Fecal specimens to be cultured should not be refrigerated, since some parasites (especially *Necator americanus*) are susceptible to cold and may fail to develop after refrigeration. *Also, caution must be exercised in handling the filter paper strip itself, since infective* Strongyloides *larvae may migrate upward as well as downward on the paper strip. Always observe standard precautions and wear gloves when performing these procedures.*

Quality Control for Harada-Mori Filter Paper Strip Culture

1. Follow routine procedures for optimal collection and handling of fresh fecal specimens for parasitologic examination.
2. Examine known positive and negative samples of stools (from laboratory animals), if available, to gain some experience in using the procedure.
3. Review larval diagrams and descriptions for confirmation of larval identification.
4. The microscope should be calibrated, and the objectives and oculars used for the calibration procedure should be used for all measurements on the microscope. If the microscope undergoes hard use or is moved around within the laboratory, it is strongly recommended that recalibration be performed every 12 months. However, if the microscope remains in the same location and receives normal use, such frequent recalibration may not be required. The calibration factors for all objectives should be posted on the microscope for easy access (multiplication factors can be pasted on the body of the microscope).
5. Record all quality control (QC) results.

Procedure for Harada-Mori Filter Paper Strip Culture

1. Smear 0.5 to 1 g of fresh feces in the center of a narrow strip of filter paper (3/8 by 5 in. [1 in. = 2.54 cm], slightly tapered at one end).
2. Add 3 to 4 ml of distilled water to a 15-ml conical centrifuge tube; identify the specimen on the tube.
3. Insert the filter paper strip into the tube so that the tapered end is near the bottom of the tube. The water level should be approximately 1/2 in. below the fecal spot. It is not necessary to cap the tube. However, a cork stopper or a cotton plug may be used.
4. Maintain the tube upright in a rack at 25 to 28°C. Add distilled water to maintain the original level

(usually evaporation takes place over the first 2 days, and then the culture becomes stabilized).

5. Keep the tube for 10 days, and check it daily by withdrawing a small amount of fluid from the bottom of the tube. Prepare a smear on a glass slide, cover the slide with a coverslip, and examine the smear with the 10× objective.

6. Examine the larvae for motility and typical morphological features to reveal whether hookworm, *Strongyloides*, or *Trichostrongylus* larvae are present.

Results and Patient Reports from Harada-Mori Filter Paper Strip Culture

Larval nematodes of hookworm, *S. stercoralis*, or *Trichostrongylus* spp. may be recovered. If *Strongyloides* organisms are present, free-living stages and larvae may be found after several days in culture.

1. Report "No larvae detected" if no larvae could be detected at the end of the incubation.
2. Report larvae detected by fecal culture.
   Example: *Strongyloides stercoralis* larvae detected by fecal culture

Procedure Notes for Harada-Mori Filter Paper Strip Culture

1. If the larvae are too active to observe under the microscope and morphologic details are difficult to see, the larvae can be heat killed within the tube or after removal to the slide; iodine can also be used to kill larvae.
2. Infective larvae may be found any time after the fourth day or even on the first day in a heavy infection. Since infective larvae may migrate upward as well as downward on the filter paper strip, *caution must be exercised in handling the fluid and the paper strip itself to prevent infection. Handle the filter paper with forceps. Wear gloves when handling the cultures.*
3. It is important to maintain the original water level to keep optimum humidity.
4. Fresh stool is required for this procedure; preserved fecal specimens or specimens obtained after a barium meal are not suitable.

Procedure Limitations for Harada-Mori Filter Paper Strip Culture

1. The Harada-Mori technique allows both parasitic and free-living forms of nematodes to develop. If specimens have been contaminated with soil or water containing these forms, it may be necessary to distinguish parasitic from free-living forms. This distinction is possible since parasitic forms are more resistant to slight acidity than are free-living forms. Proceed as follows (22, 24).

   Add 0.3 ml of concentrated hydrochloric acid per 10 ml of water containing the larvae (adjust the volume accordingly to achieve a 1:30 dilution of acid). Free-living nematodes are killed, while parasitic species live for about 24 h.

2. Specimens that have been refrigerated or preserved are not suitable for culture. Larvae of certain species are susceptible to cold environments.

Filter Paper/Slant Culture Technique (Petri Dish)

An alternative technique for culturing *Strongyloides* larvae is a filter paper/slant culture on a microscope slide placed in a glass or plastic petri dish (Figure 28.1), which was originally described by Little (19). As with previous techniques, sufficient moisture is provided by continuous soaking of the filter paper in water. Fresh stool material is placed on the filter paper, which is cut to fit the dimensions of a standard (1- by 3-in.) microscope slide. The filter paper is then placed on a slanted glass slide in a glass or plastic petri dish containing water. This technique allows direct examination of the culture system with a dissecting microscope to look for nematode larvae and free-living stages of *S. stercoralis* in the fecal mass or the surrounding water without having to sample the preparation. *Always wear gloves when performing these procedures.*

Quality Control for the Filter Paper/Slant Culture Technique (Petri Dish)

1. Follow routine procedures for optimal collection and handling of fresh fecal specimens for parasitologic examination.
2. Examine known positive and negative samples of stools (from laboratory animals), if available, to make sure that the procedure is precise.
3. Review larval diagrams and descriptions for confirmation of larval identification.
4. The microscope should be calibrated, and the objectives and oculars used for the calibration procedure should be used for all measurements on the microscope. The calibration factors for all objectives should be posted on the microscope for easy access (multiplication factors can be pasted on the body of the microscope).
5. Record all QC results.

Procedure for the Filter Paper/Slant Culture Technique (Petri Dish)

1. Cut a filter paper strip (1 by 3 in.), and smear a film of 1 to 2 g of fresh fecal material in the center of the strip.
2. Place the strip on a glass slide (1 by 3 in.). Place the slide inclined at one end of the petri dish by resting the slide on a piece of glass rod or glass tubing; identify the specimen on the dish (Figure 28.1).
3. Add water to the petri dish so that at least the bottom one-fourth of the slide is immersed in water. The stool will be kept moist by capillary action. Cover the dish, and maintain it at 25 to 28°C. As needed, add water to maintain the original level.
4. Keep the dish for 10 days. Examine daily, either with the dissecting microscope or by withdrawing a small amount of fluid and placing it on a microscope slide. Cover with a coverslip, and examine microscopically with the 10× and 40× objectives.
5. Examine any larvae recovered for typical morphologic features.

Results and Patient Reports from the Filter Paper/Slant Culture Technique (Petri Dish)

Larval nematodes of hookworm, *S. stercoralis*, or *Trichostrongylus* spp. may be recovered. If *Strongyloides* organisms are present, free-living stages and larvae may be found after several days in culture.

1. Report "No larvae detected" if no larvae could be detected at the end of incubation.
2. Report larvae detected by fecal culture.
   Example: *Strongyloides stercoralis* larvae detected
   by fecal culture

Procedure Notes for the Filter Paper/Slant Culture Technique (Petri Dish)

1. It is often difficult to observe details in rapidly moving larvae; a drop of iodine or formalin or slight heating can be used to kill the larvae.
2. Infective larvae may be found any time after the fourth day and occasionally after the first day in heavy infections. Since infective larvae may migrate anywhere on the filter paper strip, *caution must be exercised in handling the fluid and the paper strip itself to prevent infection. Wear gloves when handling the cultures.*
3. There may be infective larvae in the moisture that accumulates under the petri dish lid, so be careful not to allow the water to touch the skin when raising the lid.

4. It is important to maintain the original water level to keep optimum humidity.
5. Preserved fecal specimens or specimens obtained after a barium meal are not suitable; fresh stool specimens must be used.

Procedure Limitations for the Filter Paper/Slant Culture Technique (Petri Dish)

1. The filter paper/slant culture technique allows both parasitic and free-living forms of nematodes to develop. If specimens have been contaminated with soil or water containing these forms, it may be necessary to distinguish parasitic from free-living forms. This distinction is possible since parasitic forms are more resistant to slight acidity than are free-living forms. Proceed as follows (22, 24).

   Add 0.3 ml of concentrated hydrochloric acid per 10 ml of water containing the larvae (adjust the volume accordingly to achieve a 1:30 dilution of acid). Free-living nematodes are killed, while parasitic species live for about 24 h.
2. Specimens that have been refrigerated or preserved are not suitable for culture. Larvae of certain species are susceptible to cold environments.

Charcoal Culture

Another way to culture hookworm, *Strongyloides*, and trichostrongyle larvae is by using a granulated charcoal culture. The conditions of this culture provide an environment for larval development that mimics conditions in nature. It provides an efficient way to harvest large numbers of infective-stage larvae for use in experimental infections.

Quality Control for Charcoal Culture

1. Follow routine procedures for optimal collection and handling of fresh fecal specimens for parasitologic examination.
2. Examine known positive and negative samples of stools (from laboratory animals), if available, to make sure that the procedure is precise.
3. Review larval diagrams and descriptions for confirmation of larval identification.
4. The microscope should be calibrated, and the objectives and oculars used for the calibration procedure should be used for all measurements on the microscope. The calibration factors for all objectives should be posted on the microscope for easy access (multiplication factors can be pasted on the body of the microscope).
5. Record all QC results.

Procedure for Charcoal Culture

1. Mix 20 to 40 g of fresh fecal material in tap water until a thick suspension is obtained.
2. Add this suspension to a storage dish (4 by 3 in.) that is slightly more than half filled with no. 10 granulated hardwood charcoal. Mix thoroughly with a wooden tongue depressor until the fecal suspension is evenly distributed throughout the moistened charcoal. Water can be added to ensure that there is adequate moisture, but do not add so much water than it forms a layer on the bottom of the dish. The surface of the charcoal should glisten.
3. Cover the dish, and place it in the dark (in a drawer or cabinet).
4. Check the dish the next day to make sure that there is still sufficient moisture (i.e., that the charcoal still glistens); if water is needed, sprinkle it on the surface without further mixing.
5. Check the dish each day for moisture content. *Caution must be used because moisture will accumulate on the underface of the lid, and it may contain infective-stage larvae.*
6. Approximately 5 or 6 days after the culture has been prepared, hookworm and *Strongyloides* larvae will have reached the infective stage. This can occur earlier if the patient has a heavy infection with numerous larvae in the stool.
7. To harvest the larvae, prepare a round gauze pad of 10- to 12-layer thickness stapled at the edges and cut to fit the dish. Moisten the pad (not dripping wet), and apply it carefully with forceps so that it snugly covers the surface of the charcoal. *Do not allow your hands to touch the charcoal—the larvae will be infective!*
8. Expose the dish, with lid off, to a light source such as a gooseneck lamp. The lamp should be 6 to 8 in. from the surface of the charcoal. Make sure the lamp is not too close; the larvae can be killed by the heat.
9. After approximately 1 h, the pad can be carefully removed with forceps and inverted onto the surface of water in a pilsner glass filled with water. The gauze pad will remain at the top, and the larvae will now make their way through the pad, enter the water, and fall to the bottom of the glass, where they can be harvested with a pipette after another 30 to 60 min. With care, there will be no charcoal at the bottom of the glass, and the larvae will form a clean sediment.

Results and Patient Reports from Charcoal Culture

Larval nematodes of hookworm, *S. stercoralis*, or *Trichostrongylus* spp. may be recovered. If *Strongyloides* organisms are present, free-living stages and larvae may be found after several days in culture.

1. Report "No larvae detected" if no larvae could be detected at the end of incubation.
2. Report larvae detected by fecal culture.
   Example: *Strongyloides stercoralis* larvae detected by fecal culture

Procedure Notes for Charcoal Culture

1. It is often difficult to observe details in rapidly moving larvae; a drop of iodine or formalin or slight heating can be used to kill the larvae.
2. Infective larvae may be found any time after the fourth day and occasionally after the first day in heavy infections. Since infective larvae may be present, *use caution when handling the fluid, gauze pad, and charcoal to prevent infection. Wear gloves when handling the cultures.*
3. It is important to maintain the moisture on the charcoal to keep optimum humidity (the charcoal should glisten).
4. Preserved fecal specimens or specimens obtained after a barium meal are not suitable; fresh stool specimens must be obtained.

Procedure Limitations for Charcoal Culture

1. This technique allows both parasitic and free-living forms of nematodes to develop. If specimens have been contaminated with soil or water containing these forms, it may be necessary to distinguish parasitic from free-living forms. This distinction is possible since parasitic forms are more resistant to slight acidity than are free-living forms. Proceed as follows (22, 24).

   Add 0.3 ml of concentrated hydrochloric acid per 10 ml of water containing the larvae (adjust the volume accordingly to achieve a 1:30 dilution of acid). Free-living nematodes are killed, while parasitic species live for about 24 h.
2. Specimens that have been refrigerated or preserved are not suitable for culture. Larvae of certain species are susceptible to cold environments.

Baermann Technique

Another method of examining a stool specimen suspected of containing small numbers of *Strongyloides* larvae is the use of a modified Baermann apparatus (Figure 28.2). The Baermann technique, in which a funnel apparatus is used, relies on the principle that active larvae will migrate from a fresh fecal specimen that has been placed on a wire mesh with several layers of gauze which are in contact with tap water (2, 7, 28). Larvae migrate through the gauze into the water and settle to the bottom of the funnel, where they can be collected and examined. The main difference between this method and the Harada-Mori and petri dish methods is the greater amount of fresh stool used, possibly providing a better chance of larval recovery in a light infection. Besides being used for patient fecal specimens, this technique can be used to examine soil specimens for the presence of larvae.

Quality Control for the Baermann Technique

1. Follow routine procedures for optimal collection and handling of fresh specimens for parasitologic examination.
2. Examine known positive and negative samples of stools (from laboratory animals), if available, to make sure that the procedure is precise.
3. Review larval diagrams for confirmation of larval identification.
4. The microscope should be calibrated, and the objectives and oculars used for the calibration procedure should be used for all measurements on the microscope. The calibration factors for all objectives should be posted on the microscope for easy access (multiplication factors can be pasted on the body of the microscope).
5. Record all QC results.

Procedure for the Baermann Technique

1. If possible, use a fresh fecal specimen that has been obtained after administration of a mild saline cathartic, not a stool softener. Soft stool is recommended; however, any fresh fecal specimen is acceptable.
2. Set up a clamp supporting a 6-in. glass funnel. Attach rubber tubing and a pinch clamp to the bottom of the funnel. Place a collection beaker underneath (Figure 28.2).
3. Place a wire gauze or nylon filter over the top of the funnel, followed by a pad consisting of two layers of gauze.
4. Close the pinch clamp at the bottom of the tubing, and fill the funnel with tap water until it just soaks the gauze padding.
5. Spread a large amount of fecal material on the gauze padding so that it is covered with water. If the fecal material is very firm, first emulsify it in water.
6. Allow the apparatus to stand for 2 h or longer; then draw off 10 ml of fluid into the beaker by releasing the pinch clamp, centrifuge the fluid for

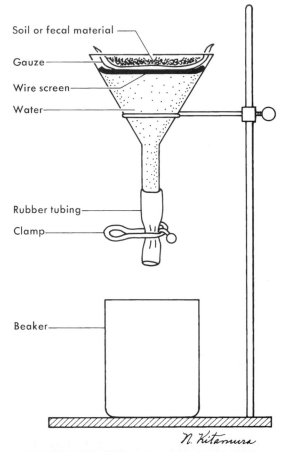

Figure 28.2 (Upper) Baermann apparatus. (Illustration by Nobuko Kitamura.) (Lower) Baermann apparatus set up in the laboratory.

2 min at 500 × g, and examine the sediment under the microscope (magnification, ×100 and ×400) for the presence of motile larvae. *Make sure that the end of the tubing is well inside the beaker before slowly releasing the pinch clamp. Infective larvae may be present; wear gloves when performing this procedure.*

Results and Patient Reports from the Baermann Technique

Larval nematodes (hookworm, *S. stercoralis*, or *Trichostrongylus* spp.) may be recovered. Both infective and noninfective *Strongyloides* larvae may be recovered, particularly in a heavy infection.

1. Report "No larvae detected" if no larvae could be detected at the end of incubation.
2. Report larvae detected by fecal culture.
   Example: *Strongyloides stercoralis* larvae detected by fecal culture

Procedure Notes for the Baermann Technique

1. It may be difficult to observe morphologic details in rapidly moving larvae; a drop of iodine or formalin or slight heating can be used to kill the larvae.
2. Infective larvae may be found any time after the fourth day and occasionally after the first day in heavy infections. *Caution must be exercised in handling the fluid, gauze pad, and beaker to prevent infection. Wear gloves when using this technique.*
3. Remember to make sure that the pinch clamp is tight until you want to release some of the water.
4. Preserved fecal specimens or specimens obtained after a barium meal are not suitable for processing by this method; fresh stool specimens must be obtained.

Procedure Limitations for the Baermann Technique

1. The Baermann technique allows both parasitic and free-living forms of nematodes to develop. If specimens have been contaminated with soil or water containing these forms, it may be necessary to distinguish parasitic from free-living forms. This distinction is possible since parasitic forms are more resistant to slight acidity than are free-living forms. Proceed as follows (22, 24).

   Add 0.3 ml of concentrated hydrochloric acid per 10 ml of water containing the larvae (adjust the volume accordingly to achieve a 1:30 dilution of acid). Free-living nematodes are killed, while parasitic species live for about 24 h.
2. Specimens that have been refrigerated or preserved are not suitable for culture. Larvae of certain species are susceptible to cold environments.
3. *Gloves should be worn when this procedure is performed.*
4. Release the pinch clamp slowly to prevent splashing; have the end of the tubing close to the bottom of the beaker for the same reason.

Modification of the Baermann Method

A simple modification of the Baermann method for diagnosis of strongyloidiasis has been developed (11). For this modification, the funnel used in the original version is replaced by a test tube with a rubber stopper, perforated to allow insertion of a plastic pipette tip (Figure 28.3). The tube containing the fecal suspension is inverted over another tube containing 6 ml of saline solution and incubated at 37°C for at least 2 h. The saline solution from the second tube is centrifuged, and the pellet is observed microscopically as a wet mount. Larvae of *S. stercoralis* can be found in the pellet. Although the method is almost identical to the original Baermann method, the amount of stool used in the modified method is smaller.

Agar Plate Culture for *Strongyloides stercoralis*

Agar plate cultures are also recommended for the recovery of *S. stercoralis* larvae and tend to be more sensitive than some of the other diagnostic methods (1, 8, 15, 16, 18, 20, 25, 29). It is important to remember that more than half of *S. stercoralis*-infected individuals tend to have low-level infections (27). The agar plate method continues to

Figure 28.3 Baermann apparatus modification. (Adapted from reference 11.)

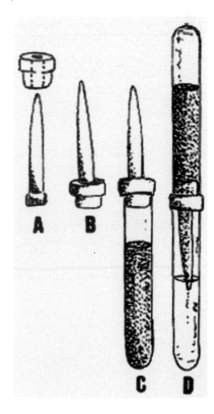

be documented as a more sensitive method than the usual direct smear or fecal concentration methods (15, 16, 20, 25). Daily search for furrows on agar plates for up to six consecutive days results in increased sensitivity for diagnosis of both *S. stercoralis* and hookworm infections. Also, a careful search for *S. stercoralis* should be performed for all patients with comparable clinical findings before a diagnosis of idiopathic eosinophilic colitis is made, because the consequent steroid treatment may have a fatal outcome by inducing widespread dissemination of the parasite (6). Human T-cell leukemia virus type 1 (HTLV-1) infection is endemic in a number of Latin American countries. HTLV-1-associated myelopathy/tropical spastic paraparesis and adult T-cell leukemia-lymphoma are emerging diseases in the region. *S. stercoralis* hyperinfection syndrome and therapeutic failure in apparently healthy patients with nondisseminated strongyloidiasis may be markers of HTLV-1 infection (9).

Stool is placed onto agar plates, and the plates are sealed to prevent accidental infections and held for 2 days at room temperature. As the larvae crawl over the agar, they carry bacteria with them, thus creating visible tracks over the agar (Figure 28.4). The plates are examined under the microscope for confirmation of the presence of larvae, the surface of the agar is then washed with 10% formalin, and final confirmation of larval identification is made via wet examination of the sediment from the formalin washings (Figure 28.5).

Occasionally, finding nematode larvae in sputum or bronchoalveolar lavage fluid specimens may be very suggestive of a potential infection with *S. stercoralis*. In Figure 28.6, larvae can be seen stained with Giemsa stain or a Gram stain. Once larvae are seen in respiratory specimens, fecal specimens can be collected for agar plate cultures to confirm strongyloidiasis.

Agar
    1.5% agar
    0.5% meat extract
    1.0% peptone
    0.5% NaCl

Note Positive tracking on agar plates has been seen on a number of different types of agar. However, the most appropriate agar formula is that listed above.

Quality Control for Agar Plate Culture for *Strongyloides stercoralis*

1. Follow routine procedures for optimal collection and handling of fresh fecal specimens for parasitologic examination.
2. Examine agar plates to ensure that there is no cracking and the agar pour is sufficient to prevent drying. Also, make sure that there is no excess water on the surface of the plates.

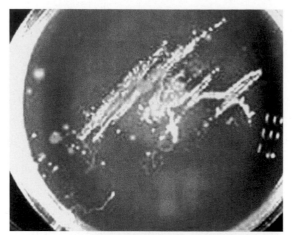

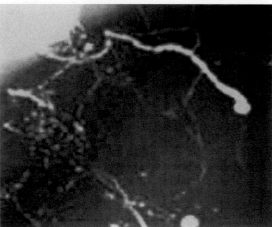

Figure 28.4 Agar plate culture method for *Strongyloides stercoralis*. (Upper) Agar plate showing bacterial growth after being distributed over the plate by the movement of the larval worms. (Lower) More random pattern of bacterial growth from the inoculation of bacteria over the agar from the movement of the larval worms.

3. Review larval diagrams and descriptions for confirmation of larval identification.
4. The microscope should be calibrated, and the objectives and oculars used for the calibration procedure should be used for all measurements on the microscope. The calibration factors for all objectives should be posted on the microscope for easy access (multiplication factors can be pasted on the body of the microscope).
5. Record all QC results (condition of agar plates).

Procedure for Agar Plate Culture for *Strongyloides stercoralis*

1. Place approximately 2 g of fresh stool in the center of the agar plate (area approximately 1 in. in diameter).

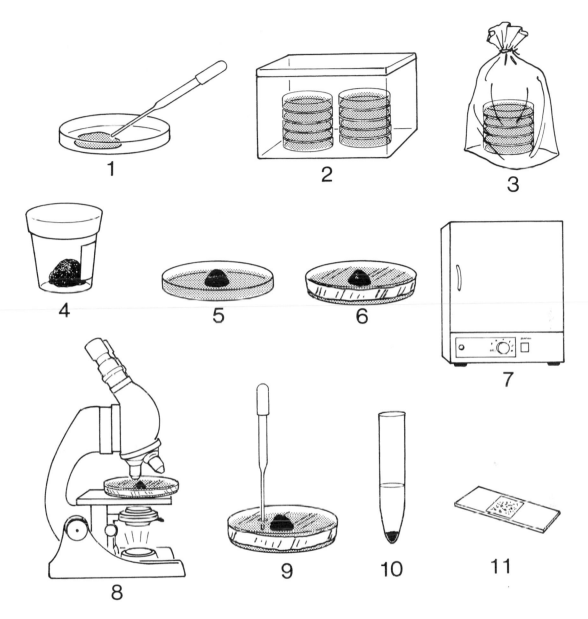

Figure 28.5 Agar culture method for *Strongyloides stercoralis*. (1) Agar plates are prepared. (2) Agar is dried for 4 to 5 days on the bench top. (3) Plates are stored in plastic bags. (4) Fresh stool is submitted to the laboratory. (5) Approximately 2 g of stool is placed onto an agar plate. (6) The plate is sealed with tape. (7) The culture plate is incubated at 26 to 33°C for 2 days. (8) The plate is examined microscopically for the presence of tracks (bacteria carried over agar by migrating larvae). (9) 10% formalin is placed onto agar through a hole made in the plastic with hot forceps. (10) Material from the agar plate is centrifuged. (11) The material is examined as a wet preparation for rhabditiform or filariform larvae (high dry power; magnification, ×400). (Illustration by Sharon Belkin.)

2. Replace the lid, and seal the plate with cellulose tape.
3. Maintain the agar plate (right side up) at room temperature for 2 days.
4. After 2 days, examine the sealed plates through the plastic lid under the microscope for microscopic colonies that develop as random tracks on the agar and evidence of larvae at the ends of the tracks away from the stool.

Note It has been documented that daily search for tracks on agar plates for up to six consecutive days results in increased sensitivity for diagnosis of both *S. stercoralis* and hookworm infections (18). When

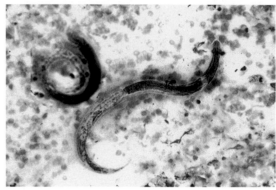

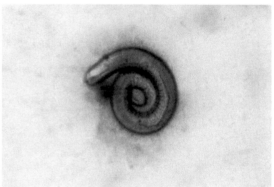

Figure 28.6 *Strongyloides stercoralis* larvae. (Upper) Giemsa-stained larvae in a specimen of bronchoalveolar lavage fluid. (Lower) Gram-stained larva in a respiratory specimen (sputum).

Results and Patient Reports from Agar Plate Culture for *Strongyloides stercoralis*

Larval nematodes of hookworm, *S. stercoralis*, or *Trichostrongylus* spp. may be recovered. If *Strongyloides* organisms are present, free-living stages and larvae may be found after several days on the agar plates.

1. Report "No larvae detected" if no larvae could be detected at the end of incubation and rinse procedure.
2. Report larvae detected by agar plate culture.
   Example: *Strongyloides stercoralis* larvae detected by agar plate culture.

Procedure Notes for Agar Plate Culture for *Strongyloides stercoralis*

1. If the larvae are too difficult to observe under the microscope and morphologic details are difficult to see, the larvae can be formalin killed within the plate and examined in the formalin-concentrated sediment.
2. Infective larvae may be found any time after the first or second day or even on the first day in a heavy infection. Since infective larvae may be present on the agar, *caution must be exercised in handling the plates once the cellulose tape is removed. Wear gloves when handling the cultures.*
3. It is important to maintain the plates upright at room temperature. Do not incubate or refrigerate them at any time; this also applies to the fresh stool specimen.
4. Fresh stool is required for this procedure; preserved fecal specimens or specimens obtained after a barium meal are not suitable.

Procedure Limitations for Agar Plate Culture for *Strongyloides stercoralis*

1. The agar plate culture technique is successful if any larvae present are viable. If the fresh stool specimen is too old, larvae may not survive and a negative result will be reported.
2. Specimens that have been refrigerated or preserved are not suitable for culture. Larvae of certain species are susceptible to cold environments.

trying to rule out strongyloidiasis in immunocompromised patients or in those who may receive immunosuppressive drugs, it is recommended that two plates be set up, one that can be examined after 2 days and one that can be examined after the full 6 days.

5. With the ends of hot forceps, make a hole in the top of the plastic petri dish.
6. Gently add 10 ml of 10% formalin through the hole onto the agar surface, and swirl to cover the surface and rinse the agar plate. Allow to stand for 30 min.
7. Remove the tape and lid of the agar plate. Pour the 10% formalin through a funnel into a centrifuge tube. Do not try and pour the formalin off directly into the centrifuge tube—the tube opening is too small, and formalin will be spilled onto the counter.
8. Centrifuge the formalin rinse fluid 5 min at $500 \times g$.
9. Prepare a wet smear preparation from the sediment and examine with a $10\times$ objective (low power) for presence of larvae. If larvae are found, confirm the identification with a $40\times$ objective (high dry power).

Egg Studies

Estimation of Worm Burdens

The only human parasites for which it is reasonably possible to correlate egg production with adult worm burdens are *Ascaris lumbricoides*, *Trichuris trichiura*, and

REVIEW Culture of Larval-Stage Nematodes	
Topic	**Comments**
Principle	Culture of feces for larvae is useful to (i) reveal their presence when they are too scanty to be detected by concentration methods; (ii) distinguish whether the infection is due to *S. stercoralis* or hookworm on the basis of rhabditiform larval morphology by allowing hookworm egg hatching to occur, releasing first-stage larvae; and (iii) allow development of larvae into the filariform stage for further differentiation.
Specimen	Any stool specimen that is fresh and has not been refrigerated can be used.
Examination	Daily checking of the fluid for the presence of larvae; hold the cultures for 10 days before submitting the final report. Agar plate cultures can be held for 6 days. When trying to rule out strongyloidiasis in immunocompromised patients or in those who may receive immunosuppressive drugs, it is recommended that two plates be set up, one that can be examined after 2 days and one that can be examined after the full 6 days.
Results and Laboratory Reports	The failure to recover larvae does not completely rule out the possibility of infection; however, the probability of infection is very low when results are negative.
Procedure Notes and Limitations	There is always the prospect of recovering infective larvae; *gloves must be worn at all times when performing these procedures and carrying out fluid examinations.* Make sure that the culture systems are kept hydrated; a certain amount of water will evaporate and be lost as a result of culture equilibration, particularly during the first couple of days.

the hookworms (*Necator americanus* and *Ancylostoma duodenale*). The specific instances in which information on approximate worm burdens is useful are when one is determining the intensity of infection, deciding on possible chemotherapy, and evaluating the efficacy of the drugs administered. With current therapy, the need for monitoring therapy through egg counts is no longer as relevant. However, several methods that can be used if necessary are discussed below. Remember that egg counts are estimates; you will obtain count variations regardless of how carefully you follow the procedure. If two or more fecal specimens are being compared, it is best to have the same individual perform the technique on both samples and to do multiple counts.

Direct-Smear Method of Beaver

The direct-smear method of Beaver is the easiest to use and is reasonably accurate when performed by an experienced technologist. In the original method, Beaver (3) used a calibrated photoelectric cell to prepare a direct smear of exactly 2 mg. For routine purposes, this is impractical, and the procedure has subsequently been modified (4) such that a direct smear of 2 mg (enough fresh fecal material to form a low cone on the end of a wooden applicator stick) of stool is prepared. Egg counts on the direct smear are reported as eggs per smear, and

the appropriate calculations can be made to determine the number of eggs per gram of stool.

Dilution Egg Count

The Stoll count (26) is probably the most widely used dilution egg-counting procedure for the purpose of estimating worm burdens. However, because of cost containment and clinical relevance (therapy is often initiated with no egg count data), most laboratories do not offer this procedure.

Stool displacement flasks for use in this procedure are available commercially. These flasks have a long neck with etched lines at 56 and 60 ml to facilitate proper filling with sodium hydroxide and fecal material. If commercial Stoll flasks are unavailable, any flask that can hold the sodium hydroxide solution, a weighed amount of 4 g of stool, and a few small glass beads can be used for a container.

1. In a calibrated Stoll flask, add 0.1 N sodium hydroxide to the 56-ml mark.
2. Add fresh fecal material to the flask so that the level of fluid rises to the 60-ml mark. This amount of feces is equivalent to 4 g of feces.
3. Add a few glass beads, and shake vigorously to make a uniform suspension. If the specimen is hard, the mixture may be placed in a refrigerator overnight before shaking to aid in mixing.

4. With a calibrated pipette, quickly remove 0.15 ml of suspension and transfer it to a slide.

5. Do not use a coverslip; place the slide on a mechanical stage, and count all of the eggs.

6. Multiply the egg count by 100 to obtain the number of eggs per gram of stool.

7. The estimate (eggs per gram) obtained will vary according to the consistency of the stool. The following correction factors should be used to convert the estimate to a formed-stool basis:

    mushy formed .... ×1.5
    mushy.. ×2
    mushy diarrheic............................ ×3

Egg counts on liquid specimens are generally unreliable; the most accurate counts are obtained with use of formed or semiformed specimens.

Modified Stoll Dilution Method

1. Fill a 15-ml centrifuge tube to the 14-ml mark with 0.1 N sodium hydroxide.
2. Add stool to bring the liquid contents up to the 15-ml mark.
3. Mix thoroughly with a wooden applicator stick.
4. If the stool is hard, allow the mixture to stand for several hours.
5. Shake to thoroughly mix, and quickly withdraw exactly 0.15 ml from the middle of the suspension.
6. Transfer the material to a slide, cover (it may be easier to count without a coverslip, since occasionally the eggs flow to the outside of the coverslip), and count the eggs in the entire preparation.
7. Multiply the egg count per preparation by 100 to give an uncorrected count of eggs per milliliter. Corrections may be made as in the Stoll dilution method for original stool consistency.

Correlation with Treatment

The following numbers have been compiled to indicate the correlation between egg counts and the need for treatment. The two helminths listed below are generally the only ones for which the egg count will determine whether the patient is treated for the initial infection. As mentioned above, with the current therapeutic agents available, this information has become less clinically relevant, at least in many parts of the world. However, it has been recognized that egg counts do not always correlate with or accurately predict the worm burden of the host.

1. For *T. trichiura*, about 30,000 eggs/g indicates the presence of several hundred worms, which may cause definite symptoms.
2. For hookworm, about 2,500 to 5,000 eggs/g usually indicates a clinically significant infection.

The effectiveness of therapy for any helminth infection may be evaluated by doing repeated egg counts after treatment. Low egg counts for *T. trichiura* and hookworm are generally reflected by a lack of clinical signs and symptoms in individuals harboring these parasites. However, the presence of even one *Ascaris* worm is potentially dangerous because of the active migrating habits of this parasite, which may result in serious clinical manifestations.

Hatching of Schistosome Eggs

When schistosome eggs (Figure 28.7) are recovered from either urine or stool, they should be carefully examined to determine viability. The presence of living miracidia within the eggs indicates an active infection that may require therapy. The viability of the miracidia can be determined in two ways: (i) the cilia of the flame cells (primitive excretory cells) may be seen on a wet smear by using high dry power and are usually actively moving, and (ii) the miracidia may be released from the eggs by a hatching procedure (Figure 28.8) (5, 7, 21, 22). The eggs usually hatch within several hours when placed in 10 volumes of dechlorinated or spring water (hatching may begin soon after contact with the water). The eggs that are recovered in the urine (24-h specimen collected with no preservatives) are easily obtained from the sediment and can be examined under the microscope to determine viability. McMullen and Beaver (21) recommended the use of a sidearm flask, but an Erlenmeyer flask is an acceptable substitute.

Figure 28.7 *Schistosoma mansoni* egg showing flame cells. (Illustration by Nobuko Kitamura; modified from E. C. Faust et al., *Craig and Faust's Clinical Parasitology*, 8th ed., Lea & Febiger, Philadelphia, Pa., 1970.)

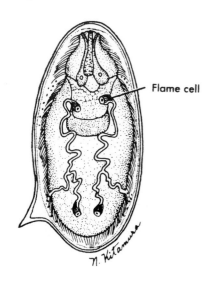

Flame cell

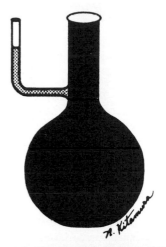

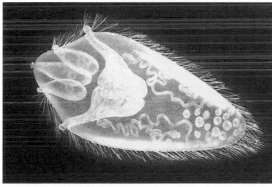

Figure 28.8 (Upper) Sidearm hatching flask used to recover miracidia from viable schistosome eggs. (Illustration by Nobuko Kitamura.) (Lower) Miracidium larva released from egg during hatching procedure. (Armed Forces Institute of Pathology photograph.)

Quality Control for Hatching of Schistosome Eggs

1. Make sure that the water used is chlorine free; chlorine kills the miracidia. You can use bottled water or leave tap water in an open pan overnight before use.
2. Check the saline solution (used to prepare the stool concentration) for the presence of any free-living organisms (flagellates or ciliates). This is normally not a problem.
3. Since it is neither realistic nor practical for the majority of laboratories to perform parallel positive-control procedures, review drawings and size measurements of schistosome eggs and/or miracidia.
4. The microscope should be calibrated, and the objectives and oculars used for the calibration procedure should be used for all measurements on the microscope. The calibration factors for all objectives should be posted on the microscope for easy access (multiplication factors can be pasted on the body of the microscope).
5. Record all QC results.

Procedure for Hatching Schistosome Eggs

1. Thoroughly homogenize a fresh stool specimen (40 to 50 g) in 50 to 100 ml of 0.85% NaCl.
2. Strain through two layers of gauze placed on a funnel. Collect material in a small beaker or 50-ml centrifuge tube.
3. Allow the suspension to settle for 1 h. Pour off and discard the supernatant fluid, and repeat this rinse process at least twice.
4. Decant the saline solution, resuspend the sediment in a small quantity of chlorine-free (spring) water (10 to 20 ml), and pour the suspension into a 500-ml sidearm flask or an Erlenmeyer flask.
5. Add chlorine-free water to the flask so that the fluid level rises to 2 to 3 cm in the side arm or to the top 2 cm of the Erlenmeyer flask. Cover the flask with aluminum foil or black paper, leaving the side arm of the flask exposed to light; if an Erlenmeyer flask is used, cover to 1 cm below the level of fluid in the neck of the flask.
6. Allow the flask to stand at room temperature for several hours or overnight in subdued light.
7. Place a bright light at the side of the flask opposite the surface of exposed water. Do not place the light against the glass, to avoid generation of excess heat. As the eggs hatch, the liberated miracidia will swim to the upper layers and collect in the side arm (or neck region of an Erlenmeyer flask). Make sure that the exposed part of the flask or side arm of the hatching flask does not get too warm.
8. Examine the illuminated area with a magnifying lens (hand lens) to look for minute white organisms swimming rapidly in a straight line (placing dark cardboard or black paper behind the flask will facilitate observation of the white miracidia against a black background).

Results and Patient Reports from Hatching Schistosome Eggs

Living miracidia may be seen; however, failure to see these larvae does not rule out schistosomiasis.

1. Report as "Miracidia of schistosomes detected, indicating the presence of viable eggs."
2. Report as "No miracidia of schistosomes detected; presence of eggs is not ruled out by this procedure."

Procedure Notes for Hatching Schistosome Eggs

1. Both urine and stool specimens must be collected without preservatives and should not be refrigerated before being processed.

REVIEW	Egg Studies
Topic	**Comments**
Principle	The specific instances in which information on approximate worm burdens is useful are when one is determining the intensity of infection, deciding on possible chemotherapy, and evaluating the efficacy of the drugs administered. With current therapy, the need for monitoring therapy through egg counts is no longer as relevant. The hatching of schistosome eggs can provide very important information for the physician; if the eggs are viable, treatment is recommended. If nothing but dead eggshells is seen, it is less critical for the patient to receive therapy.
Specimen	Any stool specimen that is fresh and has not been refrigerated can be used. For the hatching test or microscopic review for miracidium viability, it is critical that the stool specimen be processed (rinsed, centrifuged) with 0.85% NaCl rather than water, which would stimulate the eggs to hatch prematurely.
Examination	Multiple egg counts should be performed on the same specimen; it is also recommended that clinical decisions be based on a series of egg counts over time, particularly when one is monitoring therapy. All schistosome eggs recovered (but not hatched) should be examined under the microscope for moving cilia (flame cells) within the miracidium larvae.
Results and Laboratory Reports	The failure to recover nematode eggs does not completely rule out the possibility of infection; however, the probability of infection is very low when results are negative, particularly with helminths such as *T. trichiura, A. lumbricoides,* and hookworm. When schistosome eggs are reported, the viability of the eggs (miracidium larva within the eggshell) must be indicated. If dead eggshells are seen, the "nonviability" of these eggs also needs to be indicated.
Procedure Notes and Limitations	Egg counts are always subject to error; multiple counts are recommended. When performing egg counts, remember to use the stool consistency factors when calculating the total number of eggs. The failure to find schistosome eggs or to see evidence of egg viability does not rule out the possibility of schistosomiasis, particularly when a long-standing infection is suspected.

2. Hatching does not occur until the saline is removed and nonchlorinated water is added. If a stool concentration is performed, use saline throughout the procedure to prevent premature hatching.

3. Make sure that the light is not too close to the side arm or top layer of water in the Erlenmeyer flask. Excess heat kills the miracidia.

Procedure Limitations for Hatching Schistosome Eggs

1. The absence of live miracidia does not rule out the presence of schistosome eggs. Nonviable eggs or eggs that failed to hatch are not detected by this method. Microscopic examination of direct or concentrated specimens should be used to demonstrate the presence or absence of eggs.

2. Egg viability can be determined by placing some stool or urine sediment (same material as used for the hatching flask) on a microscope slide. Low-power magnification (×100) can be used to locate the eggs. Individual eggs can be examined under high dry magnification (×400); the detection of moving cilia on the flame cells (primitive excretory system) confirms egg viability.

3. Free-living ciliates may be present in soil-contaminated water. Therefore, it may be necessary to perform the following steps to differentiate those forms from the parasitic miracidia (3).

   A. Transfer a few drops of the suspension containing the organisms to a 3- by 2-in. slide, and examine under the microscope.

   B. Add a drop of weak iodine solution (weak-tea color) or dilute methylene blue (pale blue).

   C. Parasitic miracidia will stop moving, while free-living forms continue to move.

Search for Tapeworm Scolex

Since the medication used for treatment of tapeworms is usually very effective, a search for tapeworm scolices is rarely requested and no longer clinically relevant.

However, stool specimens may have to be examined for the presence of scolices and gravid proglottids of cestodes for proper species identification. This procedure requires mixing a small amount of feces with water and straining the mixture through a series of wire screens (graduated from coarse to fine mesh) to look for scolices and proglottids. *Remember to use standard precautions and wear gloves when performing this procedure.* The appearance of scolices after therapy is an indication of successful treatment. If the scolex has not been passed, it may still be attached to the mucosa; the parasite is capable of producing more segments from the neck region of the scolex, and the infection continues. If this occurs, the patient can be retreated when proglottids begin to reappear in the stool.

After treatment for tapeworm removal, the patient should be instructed to take a saline cathartic and to collect all stool material passed for the next 24 h. The stool material should be immediately placed in 10% formalin, thoroughly broken up, and mixed with the preservative (1-gal [3.8-liter] plastic jars, half full of 10% formalin, are recommended).

For additional information, see chapter 33.

Quality Control for the Tapeworm Scolex Search

1. Follow routine procedures for optimal collection and handling of fresh fecal specimens for parasitologic examination.
2. Review diagrams and sizes of proglottids and scolices of tapeworms.

Procedure for the Tapeworm Scolex Search

1. Mix a 24-h stool specimen (fresh or preserved in 10% formalin) with water, and thoroughly break up the specimen to make a watery suspension.
2. Strain the suspension (or the purged stool) through a double layer of screen wire or a sieve (a coarse-mesh screen placed over a fine-mesh screen).
3. Wash off the sediment remaining from each portion by running a slow flow of tap water over it.
4. Examine the cleansed debris with a hand lens to look for scolices and proglottids (the *Taenia* scolex is 0.5 to 1 cm long and 1 to 2 mm wide).
5. Repeat steps 3 and 4 for each portion of the suspension strained.
6. Collect the strained sediment in a glass petri dish, and place the dish over a black surface to increase the contrast of organisms against the background.
7. Observe with a magnifying hand lens, and pick pieces of worms with an applicator stick.
8. Rinse gravid proglottids and/or scolices with tap water, blot them dry on paper towels, and place them between two microscope slides separated at the edges by thin pieces of cardboard.
9. Fasten the preparation by placing rubber bands at each end of the slides so that the segments become somewhat flattened.
10. Observe under the low power of a dissecting microscope for the number of uterine branches and genital pores in the segments and the presence or absence of a rostellum of hooks on the scolex. Quite often, the scolex is broken off from the rest of the strobila (chain of proglottids) and is ~1 cm long.

Results and Patient Report from the Tapeworm Scolex Search

Tapeworm proglottids may be recovered (either singly or several attached together), and a scolex may or may not be seen.

1. Report as "A search for adult worms reveals the presence/absence of ... (finding)."
   Example: *Taenia saginata* scolex present

Procedure Notes for the Tapeworm Scolex Search

1. Remember that *Taenia solium* eggs are infective (cysticercosis), as are the eggs of *Hymenolepis nana*.
2. *Wear gloves when performing this procedure.*
3. Specimens preserved in 10% formalin are recommended.
4. If the patient has received niclosamide or praziquantel, a purged specimen is required, which should be immediately preserved in 10% formalin.
5. Wood's lamp may be used to reveal the scolices if the patient has been given quinacrine dyes; the worms will have absorbed the dye, and they fluoresce at a wavelength of 360 nm. Also, after the use of quinacrine, tapeworm proglottids appear yellow.

Procedure Limitations for the Tapeworm Scolex Search

1. Niclosamide or praziquantel therapy leads to dissolution of the tapeworm. Therefore, the scolex and other proglottids may be difficult to recover unless the patient receives a saline purge soon after taking the medication.
2. It is often difficult to identify proglottids without staining. Identification may be achieved by staining with India ink.

Qualitative Test for Fecal Fat

The microscopic examination of stool with the addition of Sudan III is a very simple, quick, and widely used technique to screen the specimen for fat. Fresh, unpreserved fecal ma-

REVIEW	Search for Tapeworm Scolex
Topic	**Comments**
Principle	The procedure requires mixing feces (from a 24-h stool specimen preserved with 10% formalin) with water and straining the mixture through a series of wire screens, graduated from coarse to fine mesh, to look for scolices and proglottids. The same procedure may be used to look for small adult nematodes and trematodes. Appearance of scolices after therapy is an indication of successful treatment.
Specimen	Any posttherapy 24-h stool specimen submitted in 10% formalin, preferably collected after the patient has taken a saline cathartic, can be used. Single stools in 10% formalin can be submitted, but the chance of finding the scolex is minimal.
Examination	A hand lens may be necessary during the search for the scolex. Also, remember to place the material in the screen over something black to provide more contrast between the dark background and the pale yellow to white scolex. The entire specimen must be strained and examined before negative results are reported.
Results and Laboratory Reports	If the scolex or proglottids are recovered, they should be identified to species level if possible, particularly if the tapeworm is *Taenia solium*.
Procedure Notes and Limitations	Remember that the eggs of *Taenia solium* (pork tapeworm) are infectious for humans (cysticercosis); *all stools should be handled carefully, and gloves should be worn at all times.* *T. solium* eggs within the uterine branches of a gravid proglottid may not be preserved, even in 10% formalin. Failure to find the tapeworm scolex does not rule out the possibility of cure, particularly when a saline cathartic has not been used.

terial is required. If there is a time delay prior to testing, the specimen should be refrigerated. Specimens collected more than 48 h earlier or specimens that are dried out should be discarded, and fresh specimens should be collected. Although this technique is quite old and the original paper may be difficult to obtain, it was published in 1961 (6b).

Sudan Black IV Stain, also known as scarlet red, was introduced by Michaelis in 1901 as a fat stain. It is a dimethyl derivative of Sudan III, which makes it a deeper and more intense stain, but it has similar physical properties and is fat soluble. This stain has been widely used as a screening method because it is easy to use and correlates well with quantitative methods.

Sudan Black IV Stain is used as a qualitative method to detect the presence of fecal fat. Normally the stool does not contain more than 20 g of fat daily. In patients with steatorrhea, fat malabsorption occurs, and greater quantities of fat are detected in the stool. This procedure, when performed carefully and consistently, is a simple method of detecting this condition in the patient (18a).

Sudan Stain
95% ethanol
Glacial acetic acid (36% [vol/vol])
Sudan III stain
  or
Sudan Black IV, certified .... 3.0 mg
  (Hardy Diagnostics, Santa Maria, Calif.)

95% ethanol .... 740 ml
Deionized water .... 260 ml

Quality Control for the Qualitative Test for Fecal Fat

1. Follow routine procedures for optimal collection and handling of fresh fecal specimens for parasitology.
2. Run a QC sample with each batch of patient tests as in procedure below.
3. As a positive control, mayonnaise is used; red-stained fat globules should be observed microscopically.
4. As a negative control, water is used; no fat globules are observed.
5. Record all QC results. If the QC results are unacceptable, the test must be repeated and documented on the corrective action sheet.

Procedure for the Qualitative Test for Fecal Fat (Sudan III)

1. Mix a small amount of stool with an equal amount of 95% ethanol in a test tube. Mix well.
2. Add 2 drops of Sudan III stain to the stool-ethanol mixture. Mix well, and let stand for a few minutes.
3. Using a pipette, remove a drop from the test tube and place it on a slide. Cover with a coverslip.

4. Using the microscope, examine the slide for globules of fat stained red (neutral fat) and needle-like crystals (fatty acid).
5. Add several drops of glacial acetic acid to the test tube. Remove a drop from the test tube and place it on a slide. Cover with a coverslip.
   A. Gently heat the slide over a flame. (Do not boil.)
   B. Observe again on the microscope for globules of fat stained red.

Results and Patient Reports for the Qualitative Test for Fecal Fat (Sudan III)

Red-stained fat globules or fatty acid crystals are seen microscopically. Report the specimen as "fat not increased" or "fat increased" depending on the number of globules seen.

1. If fewer than 100 globules/high-power field (hpf) are seen before or after heating, report as "fat not increased."
2. If more than 100 globules/hpf of fat are seen before or after heating, report as "fat increased."
3. If fatty acid crystals have been seen before heating and globules appear after heating, report as "fatty acids increased."

Procedure for the Qualitative Test for Fecal Fat (Sudan Black IV)

1. Place a small aliquot of stool suspension on a clean glass slide.
2. Mix 2 drops of 95% ethanol with the suspension on the slide.
3. Add 2 drops of Sudan Black IV Stain to the suspension on the slide, and mix well.
4. Cover the suspension with a coverslip, and examine microscopically for the presence of large orange or red droplets.

Results and Patient Reports for the Qualitative Test for Fecal Fat (Sudan Black IV)

1. Fatty acids are present as lightly staining flakes or needle-like crystals that do not stain.
2. Soaps appear as nonstaining formless flakes, coarse crystals, or rounded masses.
3. Neutral fats appear as large orange or red droplets. If 60 or more stained droplets (neutral fats) are seen per 400× (high-power) field, it is a presumptive finding that the patient has steatorrhea.

Procedure Limitations for the Qualitative Test for Fecal Fat

1. The formation of large needle-like crystals as the preparation cools after heating does not necessarily mean that the original globules were fatty acid.

Sudan III forms very short needle-like crystals in bunches as it dries.
2. Very few, if any, neutral fat globules are seen in a normal stool specimen. The presence of large amounts of neutral fat should raise suspicion that the patient has ingested mineral oil or castor oil, thus causing a false-positive result.
3. Do not count the fat that is present in vegetable cells.

Quantitation of Reducing Substances (Clinitest)

Clinitest is a reagent tablet test based on the classic Benedict's copper sulfate reduction reaction, combining ingredients with an integral heat-generating system. Clinitest provides clinically useful information about carbohydrate metabolism by determining the amount of reducing substance in urine or stool. Reducing substances convert the cupric (Cu^{2+}) to cuprous (Cu^{+}) oxide and cause a change in solution color ranging from green to orange. Unpreserved stool is required; the specimen should be placed in the refrigerator if there is a delay in testing. Specimens collected more than 48 h earlier or specimens that are dried out should be discarded, as fresh specimens should be collected. Although this is a relatively old method, there are references in the literature (6a).

Clinitest Tablets
   Clinitest reagent tablets (store tablets at room temperature in a plastic bag)
   Deionized water
   Chek-Stix positive control

Quality Control for Quantitation of Reducing Substances (Clinitest)

1. Follow routine procedures for optimal collection and handling of fresh fecal specimens for parasitology.
2. Run a QC sample with each batch of patient tests as in procedure below.
3. As a positive control, Chek-Stix is used; the development of a green, yellow, or orange color with a yellow or red precipitate is considered a positive result.
4. As a negative control, 0.5 ml of deionized water is used; blue color is considered a negative result.
5. Record all QC results. If the QC results are unacceptable, the test must be repeated and documented on the corrective action sheet.

Procedure for Quantitation of Reducing Substances (Clinitest)

1. All testing on clinical specimens should be performed in a biological safety cabinet by personnel wearing gloves and a lab coat.

2. Add 1 volume of stool to 2 volumes of deionized water, and mix thoroughly.

3. Using a disposable transfer pipette, transfer 15 drops of this suspension into a clean test tube.

4. Drop one Clinitest tablet reagent into the test tube.

5. Observe the reaction. Do not shake the tube while the chemical reaction is occurring.

6. Wait 15 s after the reaction stops, and gently shake contents to mix.

7. Compare the color of the liquid to the color chart in the package insert of the Clinitest tablet reagent.

8. Discard supplies in appropriate biohazard containers.

Results and Patient Reports for Quantitation of Reducing Substances (Clinitest)

1. Negative: clear to cloudy blue color

2. Positive: compare the liquid color to the color chart, and grade the degree of color development to the color chart (trace, 1+, 2+, 3+, or 4+). These results equate to the grams of the reducing substance present per deciliter of sample.

3. Positive: report as trace (0.25 g/dl), 1+ (0.5 g/dl), 2+ (0.75 g/dl), 3+ (1.0 g/dl), or 4+ (≥2 g/dl) Example: 1+ (0.5 g/dl)

4. Negative
Example: Negative

Procedure Limitations for Quantitation of Reducing Substances (Clinitest)

1. Clinitest is not specific for glucose and reacts with any reducing substance in stool, including lactose, fructose, galactose, and pentoses.

2. Interfering substances may affect the results. These include salicylates, penicillin, large quantities of ascorbic acid, nalidixic acid, and cephalosporins.

3. Failure to observe the reaction at all times may lead to erroneously low results if reducing substances are present in large amounts. If more than 2% sugar is present, a rapid color change may occur during boiling, causing the color to pass rapidly through bright orange to a dark brown or greenish brown.

References

1. Arakaki, T., M. Iwanaga, F. Kinjo, A. Saito, R. Asato, and T. Ikeshiro. 1990. Efficacy of agar-plate culture in detection of *Strongyloides stercoralis* infection. *J. Parasitol.* 76:425–428.

2. Baermann, G. 1917. Eine einfache Methode zur Auffindung vor Ankylostomum (Nematoden), p. 41. *In Larven in Erdproben.* Meded. Geneesk Laborat. Weltever Feestbundel.

3. Beaver, P. C. 1949. A nephelometric method of calibrating the photoelectric meter for making egg-counts by direct fecal smear. *J. Parasitol.* 35:13.

4. Beaver, P. C. 1950. The standardization of fecal smears for estimating egg production and worm burden. *J. Parasitol.* 36:451–456.

5. Chernin, E., and C. A. Dunavan. 1962. The influence of host-parasite dispersion upon the capacity of *Schistosoma mansoni* miracidia to infect *Australorbis glabratus*. *Am. J. Trop. Med. Hyg.* 11:455–471.

6. Corsetti, M., G. Basilisco, R. Pometta, M. Allocca, and D. Conte. 1999. Mistaken diagnosis of eosinophilic colitis. *Ital. J. Gastroenterol. Hepatol.* 31:607–609.

6a. Davidson, G., and M. Mullinger. 1970. Reducing substances in neonatal stool detected by Clinitest. *J. Pediatr.* 46:632–635.

6b. Drummey, B. S., J. A. Benson, Jr., and C. M. Jones. 1961. Microscopic examination of the stool for steatorrhea. *N. Engl. J. Med.* 264:85–87.

7. Garcia, L. S. 1999. *Practical Guide to Diagnostic Medical Parasitology.* ASM Press, Washington, D.C.

8. Gill, G. V., E. Welch, J. W. Bailey, D. R. Bell, and N. J. Beeching. 2004. Chronic *Strongyloides stercoralis* infection in former British Far East prisoners of war. *Q. J. Med.* 97:789–795.

9. Gotuzzo, E., C. Arango, A. de Queiroz-Campos, and R. E. Isturiz. 2000. Human T-cell lymphotropic virus-I in Latin America. *Infect. Dis. Clin. North Am.* 14:211–239.

10. Harada, U., and O. Mori. 1955. A new method for culturing hookworm. *Yonago Acta Med.* 1:177–179.

11. Hernandez-Chavarria, F., and L. Avendano. 2001. A simple modification of the Baermann method for diagnosis of strongyloidiasis. *Mem. Inst. Oswaldo Cruz* 96:805–807.

12. Hsieh, H. C. 1962. A test-tube filter-paper method for the diagnosis of *Ancylostoma duodenale, Necator americanus*, and *Strongyloides stercoralis*. *W. H. O. Tech. Rep. Ser.* 255:27–30.

13. Isenberg, H. D. (ed.). 2004. *Clinical Microbiology Procedures Handbook*, 2nd ed. ASM Press, Washington, D.C.

14. Isenberg, H. D. (ed.). 1995. *Essential Procedures for Clinical Microbiology.* American Society for Microbiology, Washington, D.C.

15. Iwamoto, T., M. Kitoh, K. Kayashima, and T. Ono. 1998. Larva currens: the usefulness of the agar plate method. *Dermatology* 196:343–345.

16. Jongwutiwes, S., M. Charoenkorn, P. Sitthichareonchai, P. Akaraborvorn, and C. Putaporntip. 1999. Increased sensitivity of routine laboratory detection of *Strongyloides stercoralis* and hookworm by agar-plate culture. *Trans. R. Soc. Trop. Med. Hyg.* 93:398–400.

17. Kobayashi, J., H. Hasegawa, E. C. Soares, H. Toma, A. R. Dacal, M. C. Brito, A. Yamanaka, A. A. Foli, and Y. Sato. 1996. Studies on prevalence of *Strongyloides* infection in Holambra and Maceio, Brazil, by the agar plate faecal culture method. *Rev. Inst. Med. Trop. Sao Paulo* 38:279–284.

18. Koga, K. S., C. Kasuya, K. Khamboonruang, M. Sukhavat, M. Ieda, N. Takatsuka, K. Kita, and H. Ohtomo. 1991. A modified agar plate method for detection of *Strongyloides stercoralis*. *Am. J. Trop. Med. Hyg.* 45:518–521.

18a. Lillie, R. D. 1997. *H. J. Conn's Biological Stains*, 9th ed. The Williams & Wilkins Co., Baltimore, Md. Reprint by Sigma Chemical Co., 1991.

19. Little, M. D. 1966. Comparative morphology of six species of *Strongyloides* (Nematoda) and redefinition of the genus. *J. Parasitol.* **52:**69–84.

20. Marchi-Blatt, J., and G. A. Cantos. 2003. Evaluation of techniques for the diagnosis of *Strongyloides stercoralis* in human immunodeficiency virus (HIV) positive and HIV negative individuals in the city of Itajai, Brazil. *Braz. J. Infect. Dis.* **7:**402–408.

21. McMullen, D. B., and P. C. Beaver. 1945. Studies on schistosome dermatitis. IX. The life cycles of three dermatitis producing schistosomes from birds and a discussion of the subfamily Bilharziellinae (Trematoda Schistosomatidae). *Am. J. Hyg.* **42:**125–154.

22. Melvin, D. M., and M. M. Brooke. 1985. *Laboratory Procedures for the Diagnosis of Intestinal Parasites*, p. 163–189. U.S. Department of Health, Education, and Welfare publication (CDC) 85-8282. U.S. Government Printing Office, Washington, D.C.

23. Sasa, M., S. Hayashi, H. Tanaka, and R. Shirasaka. 1958. Application of test-tube cultivation method on the survey of hookworm and related human nematode infection. *Jpn. J. Exp. Med.* **28:**129–137.

24. Shorb, D. A. 1937. A method of separating infective larvae of *Haemonchus contortus* (Trichostrongylidae) from free living nematodes. *Proc. Helminthol. Soc. Wash.* **4:**52.

25. Sithithaworn, P., T. Srisawangwong, S. Tesana, W. Daenseekaew, J. Sithithaworn, Y. Futimaki, and K. Ando. 2003. Epidemiology of *Strongyloides stercoralis* in north-east Thailand: application of the agar plate culture technique compared with the enzyme-linked immunosorbent assay. *Trans. R. Soc. Trop. Med. Hyg.* **97:**398–402.

26. Stoll, N. R., and W. C. Hausheer. 1926. Concerning two options in dilution egg counting: small drop and displacement. *Am. J. Hyg.* **6:**134–145.

27. Uparanukraw, P., S. Phongsri, and N. Morakote. 1999. Fluctuations of larval excretion in *Strongyloides stercoralis* infection. *Am. J. Trop. Med. Hyg.* **60:**967–973.

28. Watson, J. M., and R. Al-Hafidh. 1957. A modification of the Baermann funnel technique and its use in establishing the infection potential of human hookworm carriers. *Ann. Trop. Med. Parasitol.* **41:**15–16.

29. Zahar, J. R., J. Tankovic, E. Catherinot, P. Meshaka, and G. Nitenberg. 2006. Meningitis caused by *Enterococcus faecalis* during disseminated anguilluliasis. *Presse Med.* **35:**64–66.

Examination of Other Specimens from the Intestinal Tract and the Urogenital System

Examination for Pinworm

A roundworm parasite that has worldwide distribution and is commonly found in children is *Enterobius vermicularis*, known as the pinworm or seatworm. The adult female worm migrates out of the anus, usually at night, and deposits her eggs on the perianal area. The adult female (8 to 13 mm long) is occasionally found on the surface of a stool specimen or on the perianal skin. Since the eggs are usually deposited around the anus, they are not commonly found in feces and must be detected by other diagnostic techniques. Diagnosis of pinworm infection is usually based on the recovery of typical eggs, which are described as thick-shelled, football-shaped eggs with one slightly flattened side. Each egg often contains a fully developed embryo and will be infective within a few hours after being deposited.

The most striking symptom of this infection is pruritus, which is caused by the migration of the female worms from the anus onto the perianal skin before egg deposition. The sometimes intense itching results in scratching and occasional scarification. In most infected people, this may be the only symptom, and many individuals remain asymptomatic. Eosinophilia may or may not be present.

Infections tend to be more common in children and occur more often in females than in males. In heavily infected females, there may be a mucoid vaginal discharge, with subsequent migration of the worms into the vagina, uterus, fallopian tubes, or other body sites including the urinary tract, where they become encapsulated (1–3, 12, 22). Although tissue invasion has been attributed to the pinworm, these cases are not numerous. Symptoms that have been attributed to the pinworm infection, particularly in children, include nervousness, insomnia, nightmares, and even convulsions. In some cases, perianal granulomas may result (3).

In one report, a homosexual man presented with severe abdominal pain and hemorrhagic colitis, eosinophilic inflammation of the ileum and colon, and numerous unidentifiable larval nematodes in the stool. Using morphologic characteristics and molecular cloning of nematode rRNA genes, the parasites were identified as larvae of *E. vermicularis*; these larvae are rarely seen and are not

850

thought to cause disease. The authors stated that occult enterobiasis is widely prevalent and may be a cause of unexplained eosinophilic enterocolitis (18).

Cellulose Tape Preparations

The most widely used diagnostic procedure for pinworm infection is the cellulose tape (adhesive cellophane tape) method (10, 14, 15) (Figures 29.1 to 29.3). Several commercial collection procedures are also available. Specimens should be obtained in the morning before the patient bathes or goes to the bathroom. At least four to six consecutive negative slides should be observed before the patient is considered free of infection. Occasionally adult female pinworms are seen on the tapes or swabs.

Collection of the Specimen

1. Place a strip of cellulose tape on a microscope slide, starting 1/2 in. (1 in. = 2.54 cm) from one end and running toward the same end, continuing around this end lengthwise; tear off the strip flush with the other end of the slide. Place a strip of paper, 1/2 by 1 in., between the slide and the tape at the end where the tape is torn flush.
2. To obtain the sample from the perianal area, peel back the tape by gripping the label, and with the tape looped (adhesive side outward) over a wooden tongue depressor held against the slide and extended about 1 in. beyond it, press the tape firmly against the right and left perianal folds.
3. Spread the tape back on the slide, adhesive side down.
4. Write the name and date on the label.

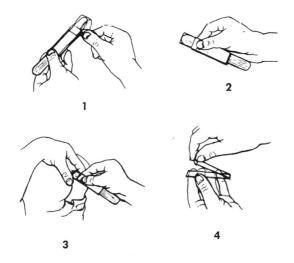

Figure 29.1 Collection of *Enterobius vermicularis* eggs by the cellulose tape method. (Illustration by Nobuko Kitamura.)

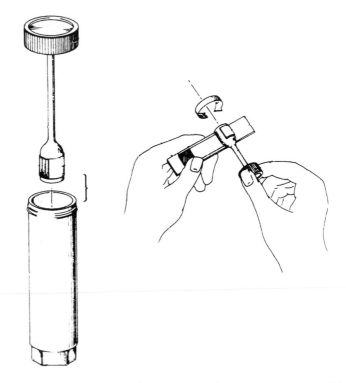

Figure 29.2 Diagram of a commercial kit (Evergreen Scientific) for use in sampling the perianal area for the presence of pinworm (*E. vermicularis*) eggs. On the left is the vial containing the sampler, which has sticky tape around the end. Once this is applied to the perianal area and eggs are picked up on the tape, the label area is placed at one end of the slide. The sticky tape is rolled down the slide and attaches to the glass. This device is easy to use and provides an area sufficient for adequate sampling. A minimum of four to six consecutive negative tapes are required to rule out a pinworm infection; most laboratories are accepting four rather than requesting the full six.

Note Do *not* use Magic transparent tape; use regular clear cellulose tape. If Magic tape is submitted, a drop of immersion oil can be placed on top of the tape to facilitate clearing.

Examination

Lift one side of the tape, apply 1 small drop of toluene or xylene, and press the tape down on the glass slide. The preparation will then be cleared, and the eggs will be visible. Examine the slide with low power and low illumination.

Anal Swabs

The anal swab technique (21) is also available for the detection of pinworm infections; however, most laboratories use the cellulose tape method because it eliminates the necessity for preparing and storing swabs. At least four

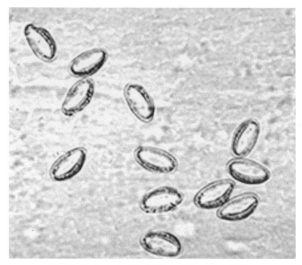

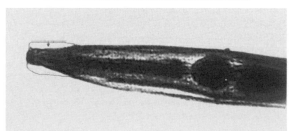

Figure 29.3 (Top) *Enterobius vermicularis* (pinworm) eggs seen in a Scotch tape preparation; note the football-shaped eggs with one side a bit more flat than the other. In some preparations, eggs are seen that contain fully developed larvae; such eggs are infective. (Middle) Adult female pinworm found on a collection device. Note the large, round esophageal bulb. (Bottom) Adult female pinworm, enlarged anterior end. Note the cephalic expansions around the end of the head, as well as the large, round esophageal bulb.

consecutive negative swabs should be obtained before the patient is considered free of infection.

Collection of the Specimen

Specimens should be obtained in the morning before the patient bathes or goes to the bathroom.

1. Prepare a mixture of 4 parts petrolatum to 1 part paraffin, and heat until liquid (just melted).
2. Dip the end of the cotton swab into the mixture,

remove the swab, and allow it to cool. If the cotton is not thoroughly coated, dip it again.
3. Store the coated swab in a 100- by 13-mm tube, and plug the end of the tube with cotton. These tubes may be stored for long periods, preferably under refrigeration.
4. Rub the swab gently over the perianal surface and into the folds. Insert the swab into the anal opening about 1/4 in., and then replace it in the tube.

Examination

1. Fill the tube containing the swab half full of xylene or xylene substitute, and let it stand for 3 to 5 min.
2. Remove the swab, and centrifuge the tube at 500 × *g* for 1 min.
3. Remove the supernatant fluid with a pipette (do not pour it off).
4. Place the sediment on a slide, and examine the material for eggs. The fluid can be examined under a coverslip, in a depression slide, or in a wax pencil circle drawn on the slide (to prevent the fluid from spreading).

Sigmoidoscopy Material

Material obtained from sigmoidoscopy can be helpful in the diagnosis of amebiasis that has not been detected by routine fecal examinations; however, a series of at least three routine stool examinations for parasites should be performed on each patient before sigmoidoscopy examination is done. Another option would be to use the immunoassay kits that are designed to detect either the *Entamoeba histolytica*/*E. dispar* group or specifically pathogenic *E. histolytica*; fresh or frozen stools are required for these kits.

Material from the mucosal surface should be aspirated or scraped and should not be obtained with cotton-tipped swabs. If swabs are the only method available, a small amount of cotton should be used on the end of the stick and should be wound tightly to prevent absorption of the sigmoidoscopy material into the cotton. At least six representative areas of the mucosa should be sampled and examined (six samples, six slides).

The specimen should be processed immediately. Usually, the amount of material is limited and should be handled properly to ensure the best examination possible. Various methods of examination are available (direct mount, several options for permanent stains). All are recommended; however, depending on the availability of trained personnel, proper fixation fluids, or the amount of specimen obtained, one or two procedures may be used. If the amount of material limits the examination to one procedure, the use of polyvinyl alcohol (PVA) fixative is highly recommended.

If the material is to be examined by any of the new immunoassay detection kits (for *Cryptosporidium* spp. or *Giardia lamblia*), 5 or 10% formalin or sodium acetate-acetic acid-formalin (SAF) fixative is recommended. However, as mentioned above, if immunoassay kits are to be used for detection of the *E. histolytica/E. dispar* group or specifically pathogenic *E. histolytica*, fresh or frozen stools are required. Many physicians performing sigmoidoscopy procedures do not realize the importance of selecting the proper fixative for material to be examined for parasites. For this reason, it is recommended that a parasitology specimen tray (containing Schaudinn's liquid fixative, PVA fixative, and 5 or 10% formalin) be provided or that a trained technologist be available at the time of sigmoidoscopy to prepare the slides. Even the most thorough examination will be meaningless if the specimen has been improperly prepared.

Direct Saline Mount

If there is no lag time after collection and a microscope is available in the immediate vicinity, some of the material should be examined as a direct saline mount for the presence of motile trophozoites. A drop of material is mixed with a drop of 0.85% sodium chloride and examined under low light intensity for the characteristic movement of amebae. It may take time for the organisms to become acclimated to this type of preparation; therefore, motility may not be obvious for several minutes. There will be epithelial cells, macrophages, and possibly polymorphonuclear leukocytes and red blood cells (RBCs), which will require a careful examination to reveal amebae (Figure 29.4).

Note Since specific identification of protozoan organisms can be difficult when only the direct saline mount is used, this technique should be used only when sufficient material is left to prepare permanent stained smears.

Morphologic differentiation between the *E. histolytica/E. dispar* group and *Entamoeba coli* can be difficult, in addition to the problem of differentiating human cells from protozoa. Also, unless trophozoites containing ingested RBCs are seen, it is impossible to differentiate organisms in the *E. histolytica/E. dispar* group from the actual pathogen, *E. histolytica*.

Permanent Stained Slide

Schaudinn's Fixative

Most of the material obtained at sigmoidoscopy can be smeared (gently) onto a slide and immediately immersed in Schaudinn's fixative. These slides can then be stained with trichrome stain and examined for specific cell morphology, either protozoan or otherwise. The procedure and staining times are identical to those for routine fecal smears.

PVA Fixative

If the material is bloody, contains a lot of mucus, or is a "wet" specimen, a few (no more than 2 or 3) drops of PVA fixative can be mixed with 1 or 2 drops of specimen directly on the slide, which is allowed to air dry (a 37°C incubator can be used) for at least 2 h before being stained. If time permits, the PVA smears should be allowed to dry overnight; they can be routinely stained with trichrome stain and examined as a permanent mount.

SAF Fixative

Material obtained at sigmoidoscopy can be placed in small amounts of SAF. After fixation for 30 min, the specimen can be centrifuged for 10 min at 500 × *g*, and smears from the small amount of sediment can be prepared for permanent staining with iron hematoxylin (trichrome stain would be the second choice). One of the organisms most strongly suspected when sigmoidoscopy is performed is *E. histolytica*, whose morphology is normally seen from the permanent stained smear; however, this identification assumes that RBCs are seen within the cytoplasm of the trophozoites (Figure 29.4). If RBCs are not seen in the trophozoite cytoplasm, the report should indicate that organisms in the *E. histolytica/E. dispar* group have been seen and identified. However, if SAF is used, both the permanent stained smear and a fecal immunoassay kit can be used. If enough material for only a single procedure is available, the permanent stained smear is recommended, particularly if the iron hematoxylin stain (incorporating the carbol fuchsin step) is used (J. Palmer, Letter, *Clin. Microbiol. Newsl.* **13:**39–40, 1991).

Figure 29.4 *Entamoeba histolytica* trophozoites. (Left) Trophozoite in wet-mount preparation. (Right) Trophozoite in permanent stained smear. Note that the nuclear and cytoplasmic characteristics are more easily seen after staining (trichrome or iron hematoxylin).

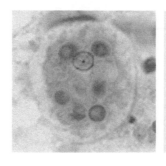 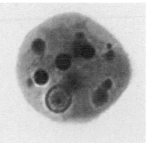

Duodenal Contents

Duodenal Drainage

In infections with *G. lamblia*, *Strongyloides stercoralis*, or *Cryptosporidium* spp., routine stool examinations may not reveal the organisms. Duodenal drainage material can be submitted for examination as direct or concentration wet mounts or permanent stained smears, techniques that may reveal the parasites (Figure 29.5). *Use standard precautions and wear gloves when handling the specimen.*

The specimen should be submitted to the laboratory in a tube containing no preservative; the amount may vary from <0.5 ml to several milliliters of fluid. The specimen may be centrifuged (10 min at 500 × *g*) and should be examined immediately as a wet mount for motile organisms (iodine may be added later to facilitate the identification of any organisms present). If the specimen cannot be completely examined within 2 h after it is taken, any remaining material should be preserved in 5 to 10% formalin. However, it is not possible to prepare a permanent stain from the formalin-preserved material. Schaudinn's fluid, PVA, modified PVA, SAF, or one of the single-vial-system preservatives is recommended for that purpose. The "falling-leaf" motility often described for *Giardia* trophozoites is rarely seen in fresh, unpreserved preparations. The organisms may be caught in mucus strands, and the movement of the flagella on the *Giardia* trophozoites may be the only subtle motility seen for these flagellates. *Strongyloides* larvae are usually very motile. Remember to keep the light intensity low.

The duodenal fluid may contain mucus; this is where the organisms, particularly *Giardia*, tend to be found. Therefore, centrifugation of the specimen is important, and the sedimented mucus should be examined. Immunoassay detection kits (*Cryptosporidium* or *Giardia*) can also be used with fresh or formalinized material.

If a presumptive diagnosis of giardiasis is made on the basis of the wet-preparation examination of the fresh specimen, the coverslip can be removed and the specimen can be fixed with either Schaudinn's fluid or PVA for subsequent staining with either trichrome or iron hematoxylin. If the amount of duodenal material submitted is very small, permanent stains can be prepared rather than using any part of the specimen for a wet-smear examination. Some investigators think that this approach provides a more permanent record, and the potential visual problems with unstained organisms, very minimal motility, and a lower-power examination can be avoided by using oil immersion examination of the stained specimen at ×1,000 magnification.

Duodenal Capsule Technique (Entero-Test)

A simple and convenient method of sampling duodenal contents that eliminates the need for intestinal intubation has been devised (5). The device consists of a length of

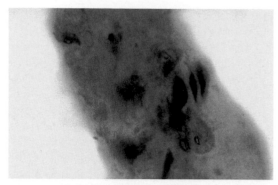

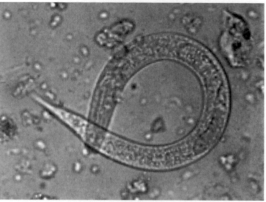

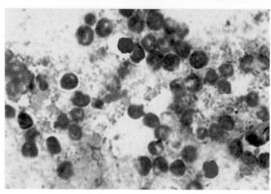

Figure 29.5 Duodenal aspirate specimens. (Top) *Giardia lamblia* in mucus from a duodenal aspirate stained with trichrome stain. Note the teardrop trophozoites (from the front view) and the darker-staining trophozoites that resemble the curved part of a spoon (side view). Also notice the number of organisms contained within the mucus. (Middle) A *Strongyloides stercoralis* rhabditiform larva in a wet mount. The genital primordial packet of cells is visible about halfway down the larva on the right side. (Bottom) *Cryptosporidium* spp. oocysts in duodenal aspirate material stained with a modified acid-fast stain.

nylon yarn coiled inside a gelatin capsule (Figure 29.6). The yarn protrudes through one end of the capsule; this end of the line is taped to the side of the patient's face. The capsule is then swallowed, the gelatin dissolves in the stomach, and the weighted string is carried by peristalsis into the duodenum. The yarn is attached to the weight by a slipping mechanism; the weight is released and passes

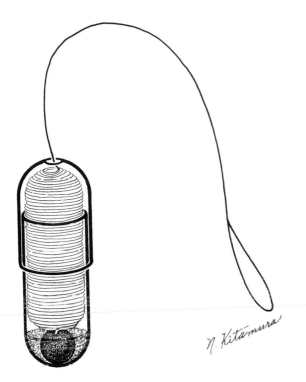

Figure 29.6 Entero-Test capsule for sampling duodenal contents. The device consists of a length of nylon yarn coiled inside a gelatin capsule. The yarn protrudes through one end of the capsule; this end of the line is taped to the side of the patient's face. The capsule is then swallowed, the gelatin dissolves in the stomach, and the weighted string is carried by peristalsis into the duodenum. (Illustration by Nobuko Kitamura.)

out in the stool when the line is retrieved after a period of 4 h. Bile-stained mucus clinging to the yarn is then scraped off (mucus can also be removed by pulling the yarn between thumb and finger) and collected in a small petri dish; *disposable gloves should be worn.* Usually 4 or 5 drops of material are obtained.

The specimen should be examined immediately as a wet mount for motile organisms (iodine may be added later to facilitate the identification of any organisms present). If the specimen cannot be completely examined within an hour after the yarn has been removed, the material should be preserved in 5 to 10% formalin or PVA-preserved mucus smears should be prepared. Organism motility is like that described above for duodenal drainage.

Note The pH of the terminal end of the yarn should be checked to ensure adequate passage into the duodenum (a very low pH means that it never left the stomach). The terminal end of the yarn should be a yellow-green color, indicating that it was in the duodenum (the bile duct drains into the intestine at this point).

Urogenital Specimens

Trichomoniasis

The identification of *Trichomonas vaginalis* (Figure 29.7) is usually based on the examination of a wet preparation of vaginal and urethral discharges and prostatic secretions or urine sediment. Multiple specimens may have to be examined before the organisms are detected. These specimens are diluted with a drop of saline and examined under low power and reduced illumination for the presence of actively motile organisms; as the jerky motility begins to diminish, it may be possible to observe the undulating membrane, particularly under high dry power. Recent data indicate that microscopic examination of a spun urine specimen performed in conjunction with microscopic examination of a vaginal fluid specimen improves the detection rate of *T. vaginalis*. While 73% of infections were detected by examination of vaginal fluid specimens and 64% were detected by examination of spun urine, the combined percentage using both methods was 85% (7). It is also important to remember that

Figure 29.7 (Upper) *Trichomonas vaginalis* trophozoite. (Illustration by Sharon Belkin.) (Lower) *T. vaginalis* trophozoites seen in a wet mount preparation.

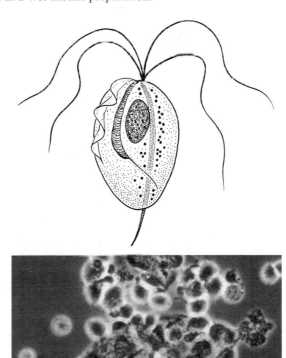

in older men with nongonococcal urethritis, diagnostic evaluation, empirical treatment, and partner management should include the possibility of infection with *T. vaginalis* (17).

Stained smears are usually not necessary for the identification of this organism. The large number of false-positive and false-negative results reported on the basis of stained smears strongly suggests the value of confirmation by observation of motile organisms from the direct mount, from appropriate culture media (4, 8, 9, 24), or from direct detection with monoclonal antibodies (11). The ability of Amies gel agar transport medium to maintain the viability of *T. vaginalis* was examined by comparison with specimens immediately inoculated into culture medium. The immediate-inoculation method detected infections in 64 (94.1%) of 68 patients, while the transport method detected infections in 62 (91.2%) of 68 patients (6). However, the data depend on the type of medium used, the time taken for transport, and the temperature at which transport was carried out. Additional information on culture options can be found in chapter 32.

T. vaginalis infection is the most prevalent sexually transmitted disease in the world. To improve diagnostic results, PCR is now being used for the detection of this infection. In a comparison of the results of PCR with those of the InPouch TV culture system in a study of 23 women who were positive, PCR detected 22 (96%) of the 23 while the wet-preparation examination detected only 12 (52%) (19). Of the 17 discrepant results, 10 were PCR positive and 7 were determined to be false positives. The sensitivity of this PCR examination was 97%, with a specificity of 98%. Other studies also report excellent results using PCR (16, 23, 26).

Filariasis

Examination of urinary sediment is indicated in certain filarial infections. The occurrence of microfilariae in urine has been reported with increasing frequency in *Onchocerca volvulus* infections in Africa. The triple-concentration technique is recommended for the recovery of microfilariae (14) (see chapter 31).

Urine is collected into a bottle, the volume is recorded, and thimerosal (1 ml/100 ml of urine) is added. The specimen is placed in a funnel fitted with tubing and a clamp; this preparation is allowed to settle overnight. The following day, 10 to 20 ml of urine is withdrawn and centrifuged. The supernatant fluid is discarded, and the sediment is resuspended in 0.85% NaCl. This preparation is again centrifuged, and 0.5 to 1.0 ml of the sediment is examined under the microscope for the presence of nonmotile microfilariae. The membrane filtration technique

used with blood (see chapter 31) can also be used with urine for the recovery of microfilariae. Administration of the drug diethylcarbamazine (Hetrazan) has been reported to enhance the recovery of microfilariae from the urine (20).

Schistosomiasis

A membrane filter technique for the recovery of *Schistosoma haematobium* eggs is also useful (25) (Figures 29.8 and 29.9).

Membrane Filter Technique

1. Collect a urine sample in a container (urine should be well mixed before step 2).
2. With a syringe, draw up 1 ml of urine into the syringe barrel.
3. Fill the rest of the barrel volume with air.
4. Attach the filter holder containing an 8-μm-pore-size Nuclepore membrane filter to the syringe.
5. Express the urine through the filter.

Figure 29.8 Membrane filtration system. (Upper) Membrane holder, which can be attached to a syringe for filtration of various types of specimens, particularly urine. (Lower) Package of membranes; different sizes with various mesh sizes are available, depending on the clinical specimen and suspected organism size.

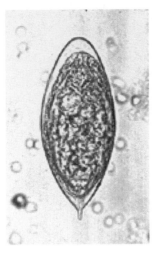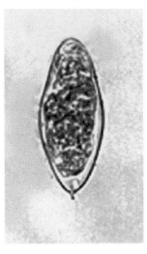

Figure 29.9 *Schistosoma haematobium* eggs from a urine filtration. Note the terminal spines on the eggs. Since the specimens were not preserved prior to filtration, determination of the viability of the eggs is possible. The viability information should be conveyed to the physician as a part of the report.

6. Remove the filter, and place it on a microscope slide face down.

7. Moisten the filter with 0.85% NaCl.

8. Examine the filter under 100× power for the presence of eggs.

The efficiency of the polycarbonate membrane filtration technique for detecting *S. haematobium* eggs in urine is increased by using a pore size of 14 µm and the suction of a water jet pump. Egg concentrations of 1 egg in >1,000 ml of urine can be detected. Viability can be assessed after filtration by staining with trypan blue. This technique is highly recommended in light infections with small numbers of eggs in which previous diagnostic methods have not confirmed the infection (13).

References

1. **al-Rufaie, H. K., G. H. Rix, M. P. Perez Clemente, and T. al-Shawaf.** 1998. Pinworms and postmenopausal bleeding. *J. Clin. Pathol.* **51:**401–402.

2. **Arora, V. K., N. Singh, S. Chaturvedi, and A. Bhatia.** 1997. Fine needle aspiration diagnosis of a subcutaneous abscess from *Enterobius vermicularis* infestation. A case report. *Acta Cytol.* **41:**1845–1847.

3. **Avolio, L., V. Avoltini, F. Ceffa, and R. Bragheri.** 1998. Perianal granuloma caused by *Enterobius vermicularis*: report of a new observation and review of the literature. *J. Pediatr.* **132:**1055–1056.

4. **Beal, C., R. Goldsmith, M. Kotby, M. Sherif, A. El-Tagi, A. Farid, S. Zakaria, and J. Eapen.** 1992. The plastic envelope method, a simplified technique for culture diagnosis of trichomoniasis. *J. Clin. Microbiol.* **30:**2265–2268.

5. **Beal, C. B., P. Viens, R. G. L. Grant, and J. M. Hughes.** 1970. A new technique for sampling duodenal contents: demonstration of upper small-bowel pathogens. *Am. J. Trop. Med. Hyg.* **19:**349–352.

6. **Beverly, A. L., M. Venglarik, B. Cotton, and J. R. Schwebke.** 1999. Viability of *Trichomonas vaginalis* in transport medium. *J. Clin. Microbiol.* **37:**3749–3750.

7. **Blake, D. R., A. Duggan, and A. Joffe.** 1999. Use of spun urine to enhance detection of *Trichomonas vaginalis* in adolescent women. *Arch. Pediatr. Adolesc. Med.* **153:**1222–1225.

8. **Borchardt, K. A.** 1994. Trichomoniasis: its clinical significance and diagnostic challenges. *Am. Clin. Lab.* **13:**20–21.

9. **Borchardt, K. A., and R. F. Smith.** 1991. An evaluation of an InPouch™ TV culture method for diagnosing *Trichomonas vaginalis* infection. *Genitourin. Med.* **67:**149–152.

10. **Brooke, M. M., and D. Melvin.** 1969. *Morphology of Diagnostic Stages of Intestinal Parasites of Man*. U.S. Department of Health, Education, and Welfare publication (HSM) 72-8116. U.S. Government Printing Office, Washington, D.C.

11. **Chang, T. H., S. Y. Tsing, and S. Tzang.** 1986. Monoclonal antibodies against *Trichomonas vaginalis*. *Hybridoma* **5:**43–51.

12. **Erhan, Y., O. Zekioglu, N. Ozdemir, and S. Sen.** 2000. Unilateral salpingitis due to *Enterobius vermicularis*. *Int. J. Gynecol. Pathol.* **19:**188–189.

13. **Feldmeier, H., U. Bienzle, M. Dietrich, and H. J. Sievertsen.** 1979. Combination of a viability test and a quantification method for *Schistosoma haematobium* eggs (filtration—trypan blue staining-technique. *Tropenmed. Parasitol.* **30:**417–422.

14. **Garcia, L. S.** 1999. *Practical Guide to Diagnostic Medical Parasitology*. ASM Press, Washington, D.C.

15. **Graham, C. F.** 1941. A device for the diagnosis of *Enterobius* infection. *Am. J. Trop. Med.* **21:**159–161.

16. **Heine, R. P., H. C. Wiesenfeld, R. L. Sweet, and S. S. Witkin.** 1997. Polymerase chain reaction analysis of distal vaginal specimens: a less invasive strategy for detection of *Trichomonas vaginalis*. *Clin. Infect. Dis.* **24:**985–987.

17. **Joyner, J. L., J. M. Douglas, Jr., S. Ragsdale, M. Foster, and F. N. Judson.** 2000. Comparative prevalence of infection with *Trichomonas vaginalis* among men attending a sexually transmitted diseases clinic. *Sex. Transm. Dis.* **27:**236–240.

18. **Liu, L. X., J. Chi, M. P. Upton, and L. R. Ash.** 1995. Eosinophilic colitis associated with larvae of the pinworm *Enterobius vermicularis*. *Lancet* **346:**410–412.

19. **Madico, G., T. C. Quinn, A. Rompalo, K. T. McKee, Jr., and C. A. Grados.** 1998. Diagnosis of *Trichomonas vaginalis* infection by PCR using vaginal swab samples. *J. Clin. Microbiol.* **36:**3205–3210.

20. **Markell, E. K., and M. Voge.** 1992. *Medical Parasitology*, 7th ed. The W. B. Saunders Co., Philadelphia, Pa.

21. **Melvin, D. M., and M. M. Brooke.** 1974. *Laboratory Procedures for the Diagnosis of Intestinal Parasites*. U.S. Government Printing Office, Washington, D.C.

22. **Ok, Y. Z., P. Ertan, E. Limoncu, A. Ece, and B. Ozbakkaloglu.** 1999. Relationship between pinworm and urinary tract infections in young girls. *APMIS* **107:**474–476.

23. **Paterson, B. A., S. N. Tabrizi, S. M. Garland, C. K. Fairley, and F. J. Bowden.** 1998. The tampon test for trichomoniasis: a comparison between conventional methods and a polymerase chain reaction for *Trichomonas vaginalis* in women. *Sex. Transm. Infect.* **74:**136–139.

24. **Perl, G.** 1972. Errors in the diagnosis of *Trichomonas vaginalis* infection. *Obstet. Gynecol.* **39:**7–9.

25. **Peters, P. A., A. A. F. Mahmoud, K. S. Warren, J. H. Ouma, and T. K. Arap Siongok.** 1976. Field studies of a rapid accurate means of quantifying *Schistosoma haematobium* eggs in urine samples. *Bull. W. H. O.* **54:**159–162.

26. **van der Schee, C., A. van Belkum, L. Zwijgers, E. van Der Brugge, E. L. O'Neill, A. Luijendijk, T. van Rijsoort-Vos, W. I. van der Meijden, H. Verbruch, and H. J. Sluiters.** 1999. Improved diagnosis of *Trichomonas vaginalis* infection by PCR using vaginal swabs and urine specimens compared to diagnosis by wet mount microscopy, culture and fluorescent staining. *J. Clin. Microbiol.* **37:**4127–4130.

Sputum, Aspirates, and Biopsy Material

Since *Pneumocystis jiroveci* (previously *P. carinii*) has been reclassified with the fungi, this organism is no longer included in the book or discussed in terms of organism, disease, pathogenesis, and diagnosis. However, other organisms such as the microsporidia can be stained using silver stains, so these methods and references have been kept in this chapter and will periodically refer to *Pneumocystis jiroveci*.

Expectorated Sputum

Although it is not one of the more common specimens, expectorated sputum may be submitted for examination for parasites. Organisms in sputum that may be detected and may cause pneumonia, pneumonitis, or Loeffler's syndrome include the migrating larval stages of *Ascaris lumbricoides*, *Strongyloides stercoralis*, and hookworm; the eggs of *Paragonimus* spp.; *Echinococcus granulosus* hooklets; and *P. jiroveci* (now classified with the fungi), *Entamoeba histolytica*, *Entamoeba gingivalis*, *Trichomonas tenax*, *Cryptosporidium* spp., and possibly the microsporidia (41). In a *Paragonimus* infection, the sputum may be viscous and tinged with brownish flecks, which are clusters of eggs ("iron filings"), and may be streaked with blood (Figure 30.1).

Collection and Examination of the Specimen

Sputum is usually examined as a wet mount (saline or iodine), using low and high dry power (×100 and ×400). The specimen is not concentrated before preparation of the wet mount. If the sputum is thick, an equal amount of 3% sodium hydroxide (or undiluted chlorine bleach) can be added; the specimen is thoroughly mixed and then centrifuged. NaOH should not be used if *Entamoeba* spp. or *T. tenax* is being sought. After centrifugation, the supernatant fluid is discarded and the sediment can be examined as a wet mount with saline or iodine. If examination has to be delayed for any reason, the sputum should be fixed in 5 or 10% formalin to preserve helminth eggs or larvae or in polyvinyl alcohol fixative to be stained later for protozoa. Usually, sputum is not a recommended specimen for the diagnosis of *P. jiroveci*; however, if such a specimen is accepted by the laboratory, other stains such as silver methenamine, Giemsa, or immunoassay reagents are used. If *Cryptosporidium* is suspected (rare), acid-fast or immunoassay techniques normally used for stool specimens can be used

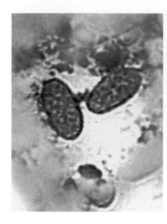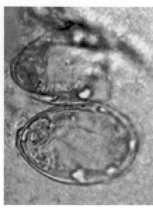

Figure 30.1 *Paragonimus* spp. eggs. (Left) Operculated eggs in a sputum specimen. (Right) Eggs photographed at a higher magnification.

(see chapters 4 and 27) (9, 15, 16, 43). Trichrome stains of material may aid in differentiating *E. histolytica* from *E. gingivalis*, and Giemsa stain may better define larvae and juvenile worms.

A sputum specimen should be collected properly so the laboratory receives a "deep sputum" sample for examination rather than a specimen that is primarily saliva from the mouth rather than from the lower respiratory passages. If the sputum is not induced, the patient can be instructed as follows.

1. Expectorated sputum specimens are collected after the patient is instructed in the appropriate measures to ensure high-quality specimens, including prior mouth washing with hydrogen peroxide and exclusion of saliva from specimens. Try to obtain the specimen early in the morning, when the chances of obtaining a deep sputum specimen are increased.
2. Transport the specimen to the laboratory in clean, closed containers as quickly as possible. Select any blood-tinged, viscous areas for sampling.
3. If the specimen is uniformly mucoid:
   A. *Wear gloves when handling specimens.*
   B. Remove a 1.0-ml portion to a 15-ml conical tube.
   C. Add 1.0 ml of mucolytic agent such as Sputalysin (Sputalysin Stat-Pack dithiothreitol solution [Behring Diagnostics, Inc.]) which has been prepared as specified by the manufacturer. This reagent can be stored unopened at room temperature until the expiration date. Date and store working solution at 4°C (include the expiration date). Discard working solution after 48 h. Prepare working solution by removing 1.0 ml from the 10-ml bottle and diluting with 9.0 ml of sterile water.

D. Incubate at room temperature for 15 min.
E. Add 2.0 ml of phosphate buffer (pH 6.8; M/15 [0.067 M]).
F. Centrifuge the material at 500 × *g* for 5 min.
G. Decant supernatant fluid, and use sediment to prepare wet mounts and smears.
H. Quality control measures should include the following.
   a. Ensure that the saline and mucolytic agent are free of contamination (particulate matter) as determined by a clear appearance. If cloudy, discard and make new working solution.
   b. Control the trichrome stain for each new set of reagents with a specimen containing blood to ensure that white cells stain with purple nuclei and blue-green cytoplasm. If cells do not stain appropriately, change reagents.
   c. With each new lot of Giemsa stain or new buffer, check the stain with a specimen containing blood to ensure that red cells stain grayish, white cell nuclei stain red-purple, and cytoplasm stains bluish. If cells do not stain appropriately, check the stock stain and buffer to find cause.
   d. Yeast-containing material may be used as a positive control for silver staining.
   e. Both *P. jiroveci*-positive and *P. jiroveci*-negative and yeast-positive material should be used for immune-specific staining (check with suppliers).
   f. Stains cannot be evaluated if controls do not stain appropriately. A stain is not within acceptable results when:
      (i) *P. jiroveci* does not stain or stains black without delineation of "parentheses."
      (ii) Other fungi and actinomycetes do not stain.
   g. Store stock stain solutions in area away from light; discard any reagent (prior to the expiration date) that is cloudy or appears contaminated.
   h. Evaluate the fluorescence microscope for correct light wavelength, using commercially available quality control slides.
   i. Record all quality control results.

Procedure for Examination of Sputum

1. *Wear gloves when performing these procedures.*
2. For expectorated sputum (no addition of mucolytic agent):
   A. Using a Pasteur pipette, place 1 or 2 drops (50 μl) on one side of a 2- by 3-in. (1 in. = 2.54

cm) glass slide, and cover with a 22- by 22-mm (no. 1) coverslip.

    B. Place a second drop on the slide, add 1 drop of saline, and cover with a coverslip.

3. Expectorated sputum (treated with a mucolytic agent) can be resuspended in 100 µl of saline; place 1 drop on a 2- by 3-in. slide, and cover with a coverslip.

4. Save the untreated specimen and remaining treated specimen for permanent smear preparation if stains are required.

5. Examine the entire 22- by 22-mm coverslip wet preparation under low light with the 10× objective to detect eggs, larvae, oocysts, or amebic trophozoites.

6. If results are inconclusive, prepare smears for permanent staining.

    A. Place 1 drop of sediment in the center of each of three 1- by 3-in. glass slides, and spread the material with the tip of the pipette.

    B. Place one slide in Schaudinn's fixative while wet; air dry the other two thoroughly.

    C. Trichrome stain the slide fixed in Schaudinn's fixative.

    D. Fix the air-dried smears in methanol; stain one with Giemsa and the other with a modified acid-fast stain. The modified trichrome stain may have to be used if infection with microsporidia is suspected (52).

    E. Put immersion oil on stained smears; examine the Giemsa-stained smear with the 10× objective and the trichrome-stained smear with the 50× oil objective if available; otherwise, use the 100× oil objective. Read at least 300 oil immersion fields (total magnification, ×1,000) before reporting the specimen as negative.

    F. The methenamine silver stain can be used; however, expectorated sputum is generally not recommended as an acceptable specimen for the recovery and identification of *P. jiroveci*.

7. Report any organisms found as follows.

    A. Give genus and species, if necessary, after confirmation with permanent stain.

    B. "No parasites found" in expectorated sputum is considered a normal/negative report; therefore, call the physician if any organisms are found.

Note Care should be taken not to confuse *E. gingivalis*, which may be found in the mouth and saliva, with *E. histolytica*, which could result in an incorrect suspicion of pulmonary abscess. *E. gingivalis* usually contains ingested polymorphonuclear leukocytes, while *E. histolytica* may contain ingested red blood cells but

not polymorphonuclear leukocytes (Figure 30.2). *T. tenax* would also be found in saliva from the mouth and thus would be an incidental finding and normally not an indication of pulmonary problems (Figure 30.3).

Induced Sputum

Concentrated stained preparations of induced sputum specimens are commonly used to detect *P. jiroveci* and differentiate trophozoite and cyst forms from other possible causes of pneumonia, particularly in AIDS patients (23, 29, 39, 47, 55). Organisms must be differentiated from other fungi such as *Candida* spp. and *Histoplasma capsulatum*. Although induced sputum specimens have been used successfully in the diagnosis of *P. jiroveci* in some institutions, they have not been useful in others (40). This difference may be due in part to careful adherence to specimen rejection criteria. If the clinical evaluation of a patient suggests *P. jiroveci* pneumonia and the induced sputum specimen is negative, a bronchoalveolar lavage fluid specimen should be evaluated with the stains presented in this protocol (2, 19, 36, 55).

Collection and Examination of the Specimen

Induced sputum specimens are collected by pulmonary or respiratory therapy staff after patients have used appropriate cleansing procedures to reduce oral contamination. Nebulizing procedures are generally determined by the respiratory therapy staff collecting the specimens. The induction protocol is critical for the success of the procedure, and it is mandatory for well-trained individuals to be involved in the recovery of organisms. Patients with *Pneumocystis* pneumonia usually have dry, nonproductive coughs. Organisms are rarely detected in expectorated sputum, which is not accepted by many laboratories as a clinically relevant specimen. In cooperation with the

Figure 30.2 (Left) *Entamoeba gingivalis* containing ingested polymorphonucluear leukocytes within large vacuoles. (Right) *Entamoeba histolytica* containing ingested red blood cells.

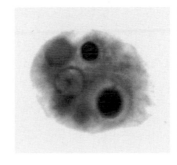
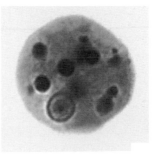

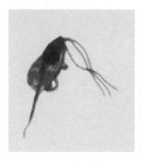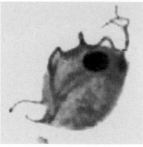

Figure 30.3 (Left) *Trichomonas tenax* from the mouth (stained with Giemsa stain). (Right) *Trichomonas vaginalis* from a genital specimen (stained with Giemsa stain). Note the large nucleus in *T. tenax* and the fact that this flagellate is somewhat smaller than *T. vaginalis*.

pulmonary staff, the laboratory processing the specimens must establish a protocol for this diagnostic procedure. Induced sputum specimens are most useful for detection of *P. jiroveci* in human immunodeficiency virus-infected individuals, because others have fewer, less readily detected organisms. Stains that can be used for organism detection include the rapid methenamine silver stain, the rapid Giemsa stain (Diff-Quik or Giemsa Plus), the Giemsa stain, and immune-specific staining (2, 5, 6, 18, 25, 27, 28, 32, 34, 35, 37, 42, 44, 51, 55).

Preparation of the Specimen Prior to Staining

1. *Wear gloves when handling specimens.*
2. Specimens which contain mucus should be treated with a mucolytic agent by adding the agent in a 1:1 ratio with the specimen, usually 2 or 3 ml, and incubating the specimen at room temperature for 15 min. Large-volume specimens usually are watery. These specimens (up to 20 to 25 ml) should be concentrated by centrifugation prior to the addition of a mucolytic agent.
3. Centrifuge the specimen at $500 \times g$ for 5 min in capped centrifuge tubes and closed carriers.
4. Decant supernatant fluids into a disinfectant solution (1:10 dilution of bleach).
5. If the sediment contains a significant amount of blood, treat a portion (one-half to one-third) with a red-cell lytic agent such as saponin or Lyse in one-half to one-third of the volume of the sediment, leave at room temperature for 5 min, and recentrifuge.
6. Decant supernatant fluid from treated specimens.
7. Use Pasteur pipettes to resuspend remaining sediment.

8. Using 1- by 3-in. glass slides, place drops of sediment in the center of each slide. For specimens treated to lyse red cells, prepare two smears, one from material before lytic treatment and one from the treated specimen.
9. Spread the drops with the pipette so that they are thin and even.
10. Air dry slides, and dip them in methanol prior to Giemsa or silver staining. Fix slides for immune-specific staining as specified in the package insert directions.

Reagents Used in Staining Procedures

1. Mucolytic agent: Sputalysin Stat-Pack dithiothreitol solution (Behring Diagnostics, Inc.). Redcell lytic agent: saponin (Aldrich, Milwaukee, Wis.; ICN Biochemicals, Costa Mesa, Calif.), Lyse (Curtin Matheson Scientific, Inc.), or Hematall LA-Hgb reagent (Fisher Scientific).
2. Rapid Giemsa stain: Diff-Quik (Baxter Scientific Products) or Wright's Dip Stat (Medical Chemical). Follow the manufacturer's instructions or use Giemsa stain.
3. Giemsa stain: azure B alcoholic stock (Harleco, Philadelphia, Pa.) diluted 1:20 with phosphate buffer containing 0.01% Triton X-100.
4. Buffers prepared from stock buffers for the Giemsa stain:
   A. Alkaline buffer NaHPO$_4$ M/15 solution is 9.5 g of Na$_2$HPO$_4$ dissolved in 1 liter of distilled water.
   B. Acid buffer NaHPO$_4$ M/15 solution is 9.2 g of NaH$_2$PO$_4$·H$_2$O dissolved in 1 liter of distilled water.

These buffers can be kept for 12 months. To prepare buffered water for the stain, add 39 ml of acid buffer and 61 ml of alkaline buffer to 900 ml of water.

Reagents for Rapid Methenamine Silver Stain (Microwave) (6)

Chromic Acid (10% Solution)
Chromic acid (CrO$_3$) .... 100.0 g
Distilled deionized water .... 1,000.0 ml

Solution is stable for up to 1 year.

Methenamine (3% Solution)
Hexamethylenetetramine, USP
  [(CH$_2$)$_6$N$_4$] .... 12.0 g
Distilled deionized water .... 400.0 ml

Solution is stable for up to 6 months.

Silver Nitrate (5% Solution)

Silver nitrate ($AgNO_3$)........................5.0 g
Distilled deionized water...............100.0 ml

Solution is stable for 1 to 6 months at 4°C.

Sodium Borate (Borax) (5% Solution)

Sodium borate
($Na_2B_4O_7 \cdot 10H_2O$) ....5.0 g
Distilled deionized water...............100.0 ml

Solution is stable for up to 1 year.

Sodium Bisulfite (1% Solution)

Sodium bisulfite ($NaHSO_3$) ....1.0 g
Distilled deionized water...............100.0 ml

Solution is stable for up to 1 year.

Gold Chloride (1% Solution)

Gold chloride ....5.0 g
Distilled deionized water...............500.0 ml

Solution is stable for up to 1 year.

Sodium Thiosulfate (5% Solution)

Sodium thiosulfate
($Na_2S_2O_3 \cdot 5H_2O$).........................50.0 g
Distilled deionized water.............1,000.0 ml

Solution is stable for up to 1 year.

Stock Light Green

Light green, S.F. (yellow)
(C.I. no. 42095)..............................0.2 g
Distilled deionized water...............100.0 ml
Glacial acetic acid
(CH_3COOH) ....0.2 ml

Solution is stable for up to 1 year.

Working Light Green

Stock light green...............................10.0 ml
Distilled deionized water..................40.0 ml

Solution is stable for up to 1 month.

Ethanol, 100% (Absolute) and 95%
Methanol, 100% (Absolute)
Xylene or Xylene Substitute

All reagents can be stored at 25°C except for the silver nitrate, which must be refrigerated. Although stable for up to 6 months, silver nitrate may have to be discarded and replaced after storage for 1 month.

Procedure for Rapid Methenamine Silver Stain (Microwave) (6)

1. Place specimen smear slides and control slides on stain rack.
2. Add 10% chromic acid to cover smears, and let stand for 10 min.
3. During this time, prepare working methenamine-silver nitrate by placing the following in order in a plastic Coplin jar: 20 ml of 3% methenamine, 1 ml of 5% silver nitrate, 1.5 ml of 5% sodium borate, and 17 ml of distilled deionized water.
4. Wash slides with distilled deionized water.
5. Cover slides with 1% sodium bisulfite for 1 min.
6. Wash slides with water, place them in a plastic Coplin jar containing methenamine-silver nitrate, and cover with cap.
7. Place in microwave oven at 50% power for approximately 35 s, rotate 90°, and heat for another 35 s; leave slides in hot liquid for 2 min. Solution should turn brown (fresh reagent may appear almost clear if the stock methenamine is freshly prepared; with time and use, the working solution will turn more brown).
8. If a water bath is used, heat it to 80°C. Place plastic Coplin jar containing stain reagents in bath for 6 min prior to placing slides in jar, add slides, and leave jar in bath for an additional 5 min. Solution should turn brown (fresh reagent may appear almost clear if the stock methenamine is freshly prepared; with time and use, the working solution will turn more brown).
9. Remove slides; wash in distilled deionized water.
10. Dip slides up and down in 1% gold chloride, wash in distilled deionized water, and place on rack again.
11. Cover slides with 5% sodium thiosulfate for 1 min.
12. Wash in distilled deionized water.
13. Cover with 0.2% fast green in acetic acid, counterstain for 1 min; use working solution.
14. Wash with distilled deionized water, and stand slides on end to drain and dry.
15. Examine with oil immersion or mount slides.
16. Fill Coplin jar used for staining with bleach (full strength), and let stand for at least 1 h.

Reagents for Rapid Methenamine Silver Stain (No Microwave) (37)

In another rapid methenamine silver-staining procedure, a 5% (rather than 10%) chromic acid solution is used, the gold chloride is present at 0.2% (rather than 1%), the sodium thiosulfate is present at 2% (rather than 5%), and the microwave oven is not used (37). Another difference from the microwave method involves fixation of the smears prior to staining. Heat fix all smears at 70°C for 10 min on a heating block for specimens that could contain *Mycobacterium tuberculosis*. Place the air-dried smears in absolute methyl alcohol for 5 min. Fix a control smear (microsporidia, fungus, or *P. jiroveci*) at the same time. It is mandatory that control smears be used every time patient specimens are stained.

Chromic Acid (5% Solution)
Chromic acid (CrO_3) .... 5 g
Distilled water .... 100 ml

Solution is stable for up to 1 year.

Methenamine (3% Solution)
Hexamethylenetetramine, USP
[$(CH_2)_6N_4$] .... 3 g
Distilled water .... 100 ml

Solution is stable for up to 1 year.

Sodium Bisulfite (1% Solution)
Sodium bisulfite ($NaHSO_3$) .... 1 g
Distilled water .... 100 ml

Solution is stable for up to 1 year.

Sodium Thiosulfate (Hypo) (2% Solution)
Sodium thiosulfate
($Na_2S_2O_3 \cdot 5H_2O$) .... 2 g
Distilled water .... 100 ml

Solution is stable for up to 1 year.

Stock Light Green
Light green, S.F. (yellow)
(C.I. no. 42095) .... 0.2 g
Distilled water .... 100 ml
Glacial acetic acid
(CH_3COOH) .... 0.2 ml

Solution is stable for up to 1 year.

Silver Nitrate (5% Solution)
Silver nitrate ($AgNO_3$) .... 5 g
Distilled water .... 100 ml

Solution is stable for 1 to 6 months at 4°C.

Sodium Borate (Borax) (5% Solution)
Borax, photographic or USP

($Na_2B_4O_7 \cdot 10H_2O$) .... 5 g
Distilled water .... 100 ml

Solution is stable for up to 1 year.

Gold Chloride (0.2% Solution)
Gold chloride, 1% solution
($AuCl \cdot HCl \cdot 3H_2O$) .... 10 ml
Distilled water .... 40 ml

This solution is stable for 1 year; however, when toning begins to fail and the organisms come out too black, it should be changed.

Note The 1% gold chloride solution is made from ampoules and is diluted as specified in the accompanying directions (dilute with tap water).

Stock Methenamine-Silver Nitrate
Silver nitrate, 5% solution .... 5 ml
Methenamine, 3% solution .... 100 ml

A white precipitate forms but immediately dissolves on shaking. Clear solutions remain stable for months at refrigerator temperature (4°C).

Working Light Green
Light green, stock solution .... 10 ml
Distilled water .... 40 ml

Solution is stable for up to 1 month.

Working Methenamine-Silver Nitrate
Borax, 5% solution .... 2 ml
Distilled water .... 25 ml

Mix, and add:

Methenamine-silver nitrate,
stock solution .... 25 ml

Caution The working methenamine-silver nitrate solution must be prepared fresh each time the stain is run. Do not try to reuse, even for a stain run immediately following one previously performed.

Procedure for Rapid Methenamine Silver Stain (No Microwave) (37)

1. While slides are fixing, fill one Coplin jar (with a lid) with 5% chromic acid. In a screw-cap Coplin jar, prepare the working methenamine-silver nitrate solution.
2. Place fixed slides into chromic acid, and put both Coplin jars into the 80°C water bath. Warm the jars and their contents in a stream of hot water for approximately 30 s before placing them in the water

bath, to prevent the jars from cracking. Place smears into the heated 5% chromic acid, and incubate them in the water bath for 2 min (oxidation step).

3. Wash slides briefly in running tap water.
4. Place slides into 1% sodium bisulfite for 30 s.
5. Rinse slides in three changes of distilled water.
6. Place slides into the jar of working methenamine-silver nitrate solution in the 80°C water bath for 4.5 min (reduction step).
7. Rinse slides with three changes of distilled water. Wipe backs of slides with a paper towel to remove excess methenamine-silver nitrate.
8. Tone in 0.2% gold chloride for 30 s.
9. Rinse in three changes of distilled water.
10. Place slides into 2% sodium thiosulfate for 30 s (removes reduced silver).
11. Rinse in three changes of distilled water.
12. Counterstain in working light green solution for 30 s.
13. Rinse in three changes of distilled water.
14. Dehydrate and clear for 30-s intervals in two changes each of 95% ethyl alcohol, 100% ethyl alcohol, and xylene or xylene substitute, respectively.
15. Mount slides in Permount.

The fungi generally stain gray to black; *P. jiroveci* exhibits a delicately stained wall, usually brownish or grayish, which is somewhat transparent. Structures described as "parentheses" are usually seen and are stained dark gray or black (Figure 30.4, upper). If the organisms are too dark, it is probably time to change the solutions (Figure 30. 4, lower). It is also important to prepare the smears so that the material is thin and evenly spread on the glass slide. Glycerin, red blood cells, and mucin stain rose taupe to dark gray. The inner parts of mycelia and hyphae stain "old rose." The background usually appears pale green. Microsporidial spores stain dark gray to black; the horizontal or diagonal "stripe" or polar filament is not visible in every spore (Figure 30.4).

Note The examination of specimens for the diagnosis of microsporidia is not often considered to be a STAT procedure and may be handled in the anatomic pathology division. The stained smears can be difficult to read and interpret, hence the need for available experienced personnel.

Rapid Giemsa Stain (Diff-Quik or Giemsa Plus)

1. Keep stain solutions in dropper bottles. Place 1 or 2 drops of red stain (solution 1) on specimen smear and control slide (normal blood film), hold for 10 s, and drain.

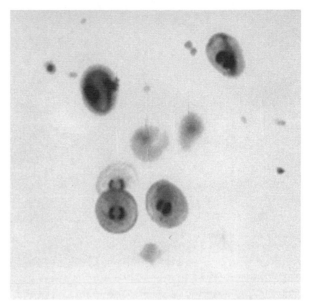

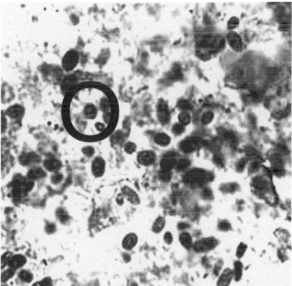

Figure 30.4 (Upper) *Pneumocystis jiroveci*. Note that the "parentheses" are visible; this is a well-stained preparation (methenamine silver stain). (Lower) Microsporidian spores. Note polar tubule within circle (methenamine silver stain).

2. Add drops of blue (solution 2), hold for 10 s, drain, and rinse very briefly with deionized water.
3. Stand slides on end to drain and air dry.
4. Slides must be examined with oil or mounted with mounting medium.

Giemsa Stain

Make Giemsa working solution fresh each day. Discard and make new solution after 10 slides have been stained in a Coplin jar. Additional directions can be found in chapter 31.

1. Place 2 ml of Giemsa (azure B) stain in a Coplin jar. Remove stain from stock stain in bottle with a clean, dry pipette.
2. Add 40 ml of phosphate buffer (pH 7.0 to 7.2) containing 0.01% Triton X-100.
3. Place fixed specimen smears and control smears in Giemsa stain for 30 min.
4. Remove slides, dip in phosphate buffer, stand slides on end, and allow to drain and air dry.
5. Examine with oil or mount slides.

Giemsa Staining for Microsporidia

1. Internal morphology may be difficult to interpret (spores with polar tubules).
2. Giemsa stain may be more relevant for specimens other than stool (urine, sputum, eye).
3. Perform modified trichrome stains on stool.

Potential Problems with Stain Interpretation

In the procedures in which the spores are stained, spores and fungi may appear very much alike. If rare organisms are seen, it may be almost impossible to differentiate microsporidia from various fungi, particularly if no budding organisms are seen.

Although a second type of stain can be used (Giemsa, which stains the spores), the organisms may be very difficult to differentiate from cellular debris. Consequently, few laboratories rely on the Giemsa stain alone (Figure 30.5).

Other methods such as fluorescent-antibody and immunoperoxidase techniques have been used for *P. jiroveci* (1, 25, 27, 32, 34, 35, 44, 55). The development of tissue culture procedures for *P. jiroveci* may lead to additional diagnostic procedures for this infection. Immunoassay reagents are also available (19, 27, 32, 34, 44, 55) and are being widely used in many diagnostic laboratories.

When noninduced sputum and the direct fluorescent-antibody technique were used, the sensitivity was 55%; this is within the range reported in the literature for the diagnosis of *Pneumocystis* pneumonia from induced sputum (39). Although detection of *P. jiroveci* has increased with the use of PCR, particularly in sputum specimens, some workers recommend that sputum samples of human immunodeficiency virus-infected patients be tested by both PCR and immunofluorescence. The results of PCR in this patient group could be misleading without careful clinical evaluation (47). However, the level of sensitivity seen using PCR indicates that these procedures should be

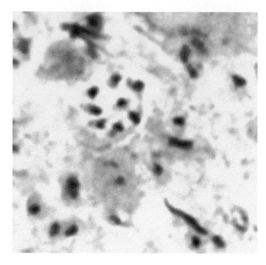

Figure 30.5 *Pneumocystis jiroveci* trophozoites within the cyst wall (stained with Giemsa stain). Note that the cyst wall is not visible when Giemsa stain is used.

considered for patients in whom *Pneumocystis* pneumonia is suggestive clinically and the specimen is negative by the immunofluorescence test (23).

Quality Control Measures

Quality control measures for all stains should include the following, and control slides must be incorporated into all stain procedures.

1. Examine a control slide for each stain prior to examination of specimen stains. The stain intensity of controls will be a guide to the stain appearance of organisms in specimens.
2. Examine stained specimen smears by systematically moving from field to field until the majority of the smear has been covered (total area for silver stain).
3. In Giemsa stains, trophozoite clumps of various sizes may be detected. In large clumps, it may be difficult to differentiate individual organisms. *Look at organisms at the edges of clumps, and look for small, more dispersed clumps.*
4. In silver- or other cyst wall-stained smears, look for the various cyst forms, including those that show dark centers, cup-shaped cysts, and cysts with foldlike lines (they look like "punched-in" ping-pong balls). If dark-staining organisms appear more oval, look carefully for budding forms, which indicate that the organisms are yeasts. *It is important to review thinner areas of the smear for microsporidial spores.*
   A. *P. jiroveci* cysts: 70% should have delicately stained walls, usually brown or gray. They

appear somewhat transparent, with structures described as "parentheses" staining black; these curved structures are usually thick (much thicker than the cyst wall) rather than thin like a line drawing.

B. Other fungi and actinomycetes: gray to black. Microsporidial spores stain dark gray to black.

C. Glycogen, mucin, and red blood cells: rose taupe to gray.

D. Background: pale green (from the light green counterstain).

5. Detection of *P. jiroveci* in specimen smears from one type of stain always suggests a careful examination of the other stain, hopefully to confirm the identification.

6. Retain all positive stained specimen slides and control slides for reference.

Reporting Smear Results

1. Report *P. jiroveci* as follows (high dry power, total magnification, ×400).

   A. "No *Pneumocystis jiroveci* seen."

   B. "*Pneumocystis jiroveci* seen" (no quantitation should be included).

2. Report other fungi that may be present as follows (low power; total magnification, ×100).

   A. "No mycotic elements seen."

   B. "Budding yeasts" or "Budding yeasts and pseudohyphae resembling *Candida* species" are quantitated as few, moderate, or many.

   C. Hyphae are reported with no quantitation:

   a. "Septate hyphae seen."

   b. "Nonseptate hyphae seen."

3. Report actinomycetes (oil immersion; total magnification, ×1,000) as follows.

   A. "No filamentous branching bacteria seen."

   B. "Filamentous branching bacteria seen" (no quantitation).

4. Report microsporidia (oil immersion; total magnification, ×1,000) as follows.

   A. "No microsporidial spores seen."

   B. "Microsporidial spores seen; identification to genus not possible."

Notes on Staining Procedures

1. Use at least two stains for detection and identification of *P. jiroveci*. With traditional histochemical stains, a trophozoite stain such as Giemsa and a cyst wall stain such as methenamine-silver nitrate are recommended.

2. There are many cyst wall stains in addition to the one described (5), and there are other modifications of the silver stain (28, 33, 37).

3. Stain effectiveness varies (36). Other counterstains may be used; a counterstain with Giemsa is useful if referral slides will be submitted for examination (51).

4. When selecting a cyst wall stain, consider stain quality, reagent stability, and potential testing frequency (including STAT requests). Toluidine blue O stains vary in dye lots and in the stability of sulfation reagents (42). In the modified procedure, the sulfation reagent made of glacial acetic acid and concentrated sulfuric acid presents disposal problems.

5. The rapid silver stain described is a modification of the procedure described by Brinn (6). Heating the chromic acid may cause nonspecific staining, making the background too dark. Also, buildup of silver on the stain container may interfere with staining. Bleach should be added to the stain jar after staining, and jars should be scrubbed periodically with a brush. Additional tips for getting good stain results include the following.

   A. Before use, inspect glassware to ensure that it does not have residual silver deposits.

   B. Use fresh reagents for each run (5% chromic acid, 1% sodium bisulfite, methenamine-silver nitrate working solution, and 2% sodium thiosulfate).

   C. Distilled deionized water must be used throughout the procedure, including slide rinses (microwave procedure).

   D. Methenamine-silver nitrate working solution must be clear; if it becomes opaque at any time, the reduction step may take longer or may not occur at all. Check the distilled deionized water source, and prepare fresh methenamine-silver nitrate working solution.

6. If mounted slides appear opaque or cloudy, the dehydration and clearing with xylene (or xylene substitute) were not adequate. Soak slides in xylene to remove the coverslips. Using fresh ethanol and xylene, repeat the dehydration steps.

7. Degenerating polymorphonuclear leukocytes may resemble *P. jiroveci*.

8. Monoclonal antibodies specific for human strain *P. jiroveci* are now available. The commercial systems vary; some are indirect stains, and some are direct stains. Reports with all systems have been variable (25, 35). Many laboratories are now using these immunospecific stains.

9. Select an organism stain and a cyst wall stain or immunospecific stain; use of a pair of stains will help avoid both false-negative and false-positive reporting.

10. In addition to organism detection, cytocentrifuge preparations of sputum (40) can be used to determine cell populations for further patient evaluation.

Aspirates

The examination of aspirated material for the diagnosis of parasitic infections may be extremely valuable, particularly when routine testing methods have failed to demonstrate the organisms. These types of specimens should be transported to the laboratory immediately after collection. Aspirates include liquid specimens collected from a variety of sites where organisms might be found. Aspirates most commonly processed in the parasitology laboratory include fine-needle aspirates and duodenal aspirates. Fluid specimens collected by bronchoscopy include bronchoalveolar lavage fluid and bronchial washings.

Procedural details for sigmoidoscopic aspirates and scrapings for the recovery of *E. histolytica* are presented in chapter 29. Techniques for preparation of duodenal aspirate material are also presented in that chapter.

Specimens Obtained from Aspiration Procedures

Fine-needle aspirates are often collected by cytopathology staff, who process the specimens, or they may be collected and sent to the laboratory directly for slide preparation and/or culture. Suggested stains are Giemsa and methenamine silver for *P. jiroveci*, Giemsa for *Toxoplasma gondii* (4), trichrome for amebae, modified acid-fast stains for *Cryptosporidium* (7, 9), and modified trichrome stains for the microsporidia (see chapter 27).

Aspirates of cysts and abscesses to be evaluated for amebae may require concentration by centrifugation, digestion, microscopic examination for motile organisms in direct preparations, and cultures and microscopic evaluation of stained preparations.

Duodenal aspirates to be evaluated for *S. stercoralis*, *Giardia lamblia*, *Cryptosporidium* spp., *Cyclospora cayetanensis*, or the microsporidia may require concentration by centrifugation prior to microscopic examination for motile organisms and permanent stains (3, 17, 48, 49).

Bone marrow aspirates to be evaluated for *Leishmania* amastigotes, *Trypanosoma cruzi* amastigotes, or *Plasmodium* spp. require Giemsa staining (21, 46).

Fluid specimens collected by bronchoscopy may be lavages or washings, with bronchoalveolar lavages preferred. Specimens usually are concentrated by centrifugation prior to microscopic examination of stained preparations. Organisms included here which may be detected are *P. jiroveci*, *Toxoplasma gondii*, and *Cryptosporidium*.

Lungs and Liver

Pneumocystosis

Although formerly classified with the sporozoa, *P. jiroveci* has been reclassified with the fungi. It is an important cause of pulmonary infections, particularly in patients who are immunosuppressed as a result of therapy or from congenital or acquired immunologic deficiencies (14, 38, 45, 50). Clinically, the infection may involve both lungs diffusely, may be localized, or may be disseminated (22, 31). For immunosuppressed patients, in whom the disease may progress very quickly, correct specimen collection and rapid diagnostic techniques are very important. The organisms can be demonstrated in stained impression smears of lung material obtained by open or brush biopsy. *P. jiroveci* can also be seen in stained smears of tracheobronchial aspirates, although examination of lung tissue is more likely to reveal the organisms. Sputum specimens are generally considered unacceptable for the recovery of *P. jiroveci*; however, in patients with severe, progressive disease, such specimens may be acceptable. To avoid the possibility of false-negative results, acceptance of sputum specimens should be carefully monitored and reviewed. Also, even for patients with fulminant disease, multiple specimens may have to be submitted to recover and identify the organisms. With a number of the stains that are available, the cyst walls stain but the cyst contents do not (Figure 30.1) (24); however, only the methenamine silver stain is discussed here (5, 32).

Amebiasis

Examination of aspirates from lung or liver abscesses may reveal trophozoites of *E. histolytica* (Figure 30.6); demonstration of the organisms is often very difficult. In many cases, serologic confirmation is recommended (54). Liver aspirate material should be taken from the margin of the abscess rather than the necrotic center (Figure 30.7). The organisms are often trapped in the viscous pus or debris and may not exhibit typical motility. The Amoebiasis Research Unit, Durban, South Africa, has recommended using proteolytic enzymes to free the organisms from the aspirate material.

The digestion technique is performed as follows.

1. A minimum of two separate portions of exudate should be removed (more than two are recommended). The first portion of the aspirate, usually yellowish white, rarely contains organisms. The last portion of the aspirated abscess material is reddish and is more likely to contain amebae. The

best material to examine is that obtained from the actual wall of the abscess.

2. Add 10 U of streptodornase per ml of thick pus; incubate this mixture for 30 min at 37°C, and shake repeatedly.

3. Centrifuge the mixture at 500 × g for 5 min. The sediment may be examined microscopically as wet mounts or used to inoculate culture media. Some of the aspirate can be mixed directly with polyvinyl alcohol on a slide and examined as a permanent stained smear.

Hydatid Disease

Aspiration of cyst material for the diagnosis of hydatid disease is a dangerous procedure and is normally performed only when open surgical techniques are used for cyst removal. Aspirated fluid usually contains hydatid sand (intact and degenerating scolices, hooklets, and calcareous corpuscles) (Figure 30.8). Some older cysts contain material that resembles curded cottage cheese, and the hooklets may be very difficult to see. Some of this material can be diluted with saline or 10% KOH; usually, scolices or daughter cysts will have disintegrated. However, the diagnosis can be made by seeing the hooklets.

Examination of hydatid cyst material is carried out as follows.

1. If the cyst material is fluid, centrifuge at 500 × g for 3 min.

2. Carefully remove some of the sediment and prepare a wet mount.

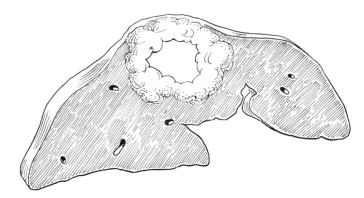

Figure 30.7 Liver abscess caused by *Entamoeba histolytica*. Amebae would be found at the advancing margin of the lesion; the last portion of the aspirated material might reveal the organisms. (Illustration by Sharon Belkin.)

3. Examine the material under low (100×) and high dry (430×) power (remember to use low light intensity, since some of the material may be very transparent).

4. If the cyst material is more viscous or solid, the material can be mixed with saline or 10% KOH and then centrifuged at 500 × g for 3 min.

5. Some of the very viscous material can be placed on a glass slide (undiluted). Another glass slide can be placed on top of the first (material will now lie between the two slides). Rub the glass slides back and forth over each other and listen for a grating sound (like grains of sand being scratched). If this occurs, you may be hearing evidence of the presence of calcified hooklets. Place the material from the smears (may be diluted with saline) under the microscope to try to confirm the presence of hooklets (which may be extremely difficult to see with only low power).

Note Remember, the absence of scolices or hooklets does not rule out the possibility of hydatid disease, since some cysts are sterile and contain no scolices and/or daughter cysts. Review of the cyst wall from pathology (tissue sections) should be able to confirm the diagnosis.

Lymph Nodes, Spleen, Liver, Bone Marrow, Spinal Fluid, Eyes, and Nasopharynx

African Trypanosomiasis, Leishmaniasis, Chagas' Disease, Primary Amebic Meningoencephalitis, Granulomatous Amebic Encephalitis, Amebic Keratitis, and Microsporidial Keratitis

Material from lymph nodes, spleen, liver, bone marrow, spinal fluid, eye specimens, or nasopharynx may be examined for the presence of parasites and should be processed as follows.

Figure 30.6 *Entamoeba histolytica* containing ingested red blood cells within the cytoplasm; morphologically this organism is *E. histolytica*, and this would be the species designation if organisms were isolated from liver abscess material.

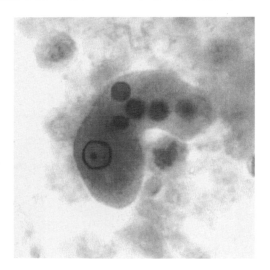

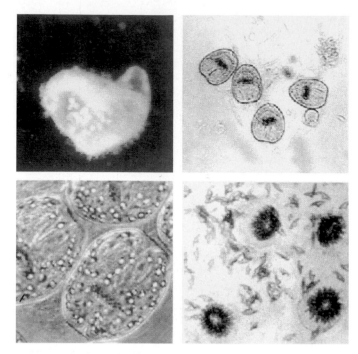

Figure 30.8 Hydatid disease (*Echinococcus granulosus*). (Upper left) Hydatid cyst with scolices budding off from the germinal layer. (Upper right) Immature scolices, with the dark area being the hooklets. (Lower left) Higher magnification of the scolices taken from the hydatid cyst fluid. (Lower right) Hooklets from disintegrating scolices. Reprinted from reference 15.

1. A portion of the fluid material can be examined under low (×100) and high dry (×400) power as a wet mount (diluted with saline) for the presence of motile organisms. Spinal fluid should not be diluted before examination.

2. Impression smears from tissues should be prepared and stained with Giemsa stain. The material is pressed between two slides, and a smear is formed when the slides are pulled apart (one across the other).

3. The smears are allowed to air dry and then processed like a thin blood film (fixed in absolute methanol and stained with Giemsa stain).

4. Appropriate culture media should be inoculated with any remaining material (see chapter 32).

5. If microsporidia are suspected, modified trichrome stains can be used; calcofluor white and immunoassay methods (currently under development) are also excellent options (8, 10–13, 20, 26, 30, 56).

Primary amebic meningoencephalitis is rare, but the examination of spinal fluid may reveal the amebae, usually *Naegleria fowleri* (Figure 30.9). Granulomatous amebic encephalitis is caused by *Acanthamoeba* spp., *Balamuthia mandrillaris*, or *Sappinia diploidea*. Uncentrifuged sedimented spinal fluid should be placed under a coverslip on a slide and observed for motile amebae; smears can also be stained with Wright's or Giemsa stain. Spinal fluid, exudate, or tissue fragments can be examined by light microscopy or phase-contrast microscopy. Care must be taken not to confuse leukocytes with actual organisms and vice versa. The appearance of the spinal fluid may vary from cloudy to purulent (with or without red blood cells), with a cell count from a few hundred to over 20,000 white blood cells (primarily neutrophils) per ml. Failure to find bacteria in this type of spinal fluid should alert one to the possibility of primary amebic meningoencephalitis. These organisms can be isolated from tissues or soil with special media (see chapter 32).

Note When spinal fluid is placed in a counting chamber, any organisms that settle to the bottom of the chamber will tend to round up and look very much like a white blood cell. For this reason, it is better to examine the spinal fluid on a slide directly under a coverslip, not in a counting chamber.

A rapid method for the diagnosis of *Acanthamoeba* or microsporidial keratitis involves the use of calcofluor white, which is a chemofluorescent dye with an affinity for the polysaccharide polymers of amebic cysts and microsporidial spores (Figures 30.10 and 30.11). The following method has proven to be very successful for examination of corneal scrapings or biopsy material (24, 53).

1. Place corneal scrapings on slides, and air dry them.
2. Fix the slides in absolute methyl alcohol for 3 to 5 min.

Figure 30.9 *Naegleria fowleri*. Note the large karyosome and lobate pseudopodia. Spinal fluid should be examined on a slide, not in a counting chamber, where protozoan trophozoites could mimic white blood cells. (Armed Forces Institute of Pathology photograph.)

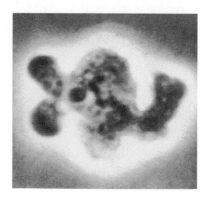

3. Add several drops of solution (0.1% calcofluor white and 0.1% Evans blue dissolved in distilled water); leave on for 5 min.

4. Turn the slide on its side, and allow excess stain to run off onto paper towels.

5. Add a coverslip, and examine under the fluorescence microscope for pale blue chemofluorescence of the amebic cysts (will not stain trophozoite). The microsporidial spores usually stain brighter; however, fluorescence can vary between 1+ and 3+.

Note Select UV irradiation with an exciter filter which transmits the 365-nm group of intense mercury spectral emission lines (Zeiss UGI or G365). View through a barrier filter which removes UV while emitting visible blue light and longer wavelengths (Zeiss no. 41 or LP420).

Cutaneous Ulcer

Leishmaniasis

Material containing intracellular *Leishmania* organisms must be aspirated from below the ulcer bed through the uninvolved skin, not from the surface of the ulcer (Figure 30.12). It is very important that the surface of the ulcer be thoroughly cleaned before specimens are taken; any contamination of the material with bacteria or fungi may prevent recovery of organisms from culture. An actual cutaneous ulcer before and after therapy can be seen in Figure 30.13.

Biopsy Material

Biopsy specimens are recommended for the diagnosis of tissue parasites. The following procedures may be used for this purpose in addition to standard histologic preparations:

Figure 30.10 *Acanthamoeba* cyst. Note the fluorescence of the cyst wall. Trophozoites do not fluoresce when calcofluor white is used.

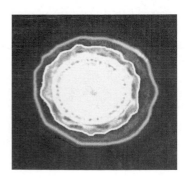

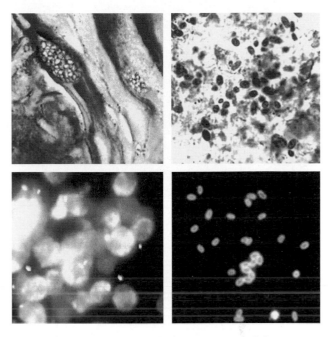

Figure 30.11 Microsporidial spores. (Upper left) Spores in a corneal scraping specimen; (upper right) spores in a fecal specimen (both images stained with silver stain). (Lower left) Microsporidial spores in a urine sediment; (lower right) spores from nasopharyngeal aspirate (both images stained using calcofluor white).

impression smears and teased and squash preparations of biopsy tissue from the skin, muscle, corneas, intestines, liver, lungs, and brain. Tissue to be examined by permanent sections or electron microscopy should be fixed as specified by the laboratories that will process the tissue. In certain cases, examination of a biopsy specimen may be the only means of confirming a suspected parasitic infection. Specimens that are going to be examined as fresh material rather than as tissue sections should be kept moist in saline and submitted to the laboratory immediately.

Figure 30.12 Proper way to aspirate material from below the ulcer bed (*Leishmania* spp.); sterile saline (0.5 to 1.0 ml) can be injected under the ulcer prior to aspiration. (Illustration by Sharon Belkin.)

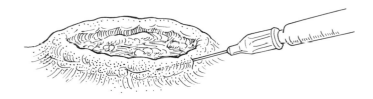

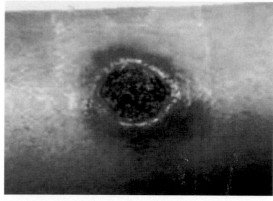

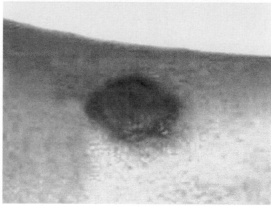

Figure 30.13 *Leishmania* cutaneous lesion. (Upper) Lesion prior to therapy; (lower) lesion after therapy.

Importance of Biopsy Specimens

Success in detection of parasites in tissue depends in part on specimen collection and on the presence of sufficient material to perform the recommended diagnostic procedures. Biopsy specimens are usually quite small and may not be representative of the diseased tissue. Examination of multiple tissue samples often improves diagnostic results. To optimize the yield from any tissue specimen, examine all areas and use as many procedures as possible. Tissues are obtained by invasive procedures, many of which are very expensive and lengthy; consequently, these specimens deserve the most comprehensive procedures possible. It is also important to remember that if the tissue specimen is not sufficient for all diagnostic procedures requested, the procedures should be prioritized after consultation with the physician.

Submission of Specimens

Tissue submitted on a sterile sponge dampened with saline in a sterile container may be used for cultures of protozoa after mounts for direct examination or impression smears for staining have been prepared. If cultures for parasites will be included, sterile slides should be used for smear and mount preparation.

1. Use sterile slides for impression smears; slides can be either autoclaved or prepared by an alternative method in which they are soaked in 95% ethyl alcohol and flamed prior to use.
2. Use sterile (autoclaved or flamed) forceps for handling tissue.
3. Place tissue in sterile petri dish to examine macroscopically and select a sample for microscopic evaluation. Provided that it is kept sterile, minced tissue can be used.
   A. If biopsy tissue is several millimeters to a centimeter in size, select the specimen from an abnormal area (granulomatous lung or ulcerated area of intestinal tissue). Generally, the tissue fragments are too small to permit such a judgment to be made.
   B. If several small pieces of tissue that look alike are submitted, use one piece. However, if they look different, use one of each type for microscopic examination. Again, sometimes this assessment is difficult to make.
4. Prepare impression smears.
   A. Blot the tissue sample on sterile toweling. If the sample size is sufficient, cut the tissue with a scalpel (very sharp cut) and use the cut surface to touch the slide.
   B. Press tissue against the sterile slide, lift, and press again. Turn the sample over, and press the other end against the slide to make two more impressions. Keep impressions close together to speed the screening process with the microscope.
   C. Air dry and fix smears in absolute methanol for 1 min for subsequent Giemsa, methenamine-silver nitrate, modified acid-fast, and modified trichrome staining. If the amount of tissue is sufficient, multiple smears should be prepared for each stain (Figure 30.14).
   D. Place wet slide in Schaudinn's fixative for subsequent trichrome staining.
   E. Fix the slide as specified by the manufacturer for immunospecific staining.
5. Teased preparations are made as follows.
   A. Place sample in the bottom of a plastic petri dish, and cover with 2 to 4 drops of saline.
   B. Gently tease tissue apart with needles, or hold the tissue with forceps while pulling apart with a scalpel or needles.
   C. Put a cover on dish, and leave at room temperature for 30 min.
6. Squash preparations are made as follows.
   A. Using a scalpel, cut selected tissue portions into very fine pieces.

B. Place a piece of tissue on a 1- by 3-in. slide, add 1 drop of saline, cover with a second 1- by 3-in. slide, and hold together with membrane clips (from a surgical supply company). If these are not available, paper clips can be used but are not as efficient. *You can also press the slides together with your fingers, but always wear gloves and be careful not to contaminate the microscope stage with tissue fluids.*

7. Skin scrapings should be submitted in a small vial.

8. Prepare cultures to demonstrate the following organisms (see chapter 32): *E. histolytica*, *Acanthamoeba* and *Naegleria* spp., and *Leishmania* spp.

9. Mouse passage for *Toxoplasma gondii*:
   A. Grind tissue in 0.85% NaCl into a fine suspension.
   B. Inject 0.2 to 0.4 ml of suspension intraperito-

neally into three to five mice of any laboratory strain weighing ~20 g.
C. Maintain mice in isolation.
D. Every day, check for signs of central nervous system dysfunction; if symptoms are detected, perform an autopsy on the animal.

Examination of Specimens

1. Examination of impression smears is summarized in Table 30.1.

2. Examine skin snips in teased preparations for detection of microfilariae of *Onchocerca volvulus* and *Mansonella streptocerca* as follows.
   A. Tease the small bit of tissue apart in a few drops of saline to release the microfilariae.
   B. Remove drops of saline to a 1- by 3-in. glass slide, cover with a no. 1 coverslip, and examine under low light for microfilariae.
   C. For a permanent preparation, place 100% methanol under the coverslip to fix filariae, partially dry, remove the coverslip, and stain with Giemsa.

3. Examine squash preparations for detection of *Trichinella* spp. in muscle microscopically at low power (×100) and under low light.

4. Examine scrapings of skin for *Sarcoptes scabiei* (scabies) microscopically at low power (×100) and under low light. Confirmation at high dry power (×400) may be necessary (Figure 30.15).

5. Inoculate cultures with ground tissue suspensions (to release organisms from the cells).
   A. Place a small tissue sample in a sterile tissue grinder (Ten Broeck or Dounce) in 0.5 ml of sterile saline, and grind until tissue is broken up but not totally liquefied.
   B. Add several drops of ground tissue to culture medium as follows.
      a. NNN medium (see chapter 32): add drops of tissue to the bottom of the slant, where they will "pool" with condensed moisture. Incubate at room temperature (isolation of *Leishmania* spp.).
      b. TYSGM-9 medium (see chapter 32): add drops of tissue to the liquid medium, add 3 drops of the starch suspension, and incubate at a 45 to 50° angle at 35°C for 48 h (isolation of *E. histolytica*).
      c. Nonnutrient agar plate seeded with bacteria (see chapter 32): add drops to center of seeded agar plate, and incubate at 35 to 37°C (isolation of *Acanthamoeba* or *Naegleria* spp.).

Figure 30.14 (Upper) *Leishmania donovani* amastigotes in a cell, stained with Giemsa stain. (Lower left) *Cryptosporidium* spp. oocysts stained with modified acid-fast stain. Note that the sporozoites are visible within some of the oocysts. (Lower right) Microsporidial spores stained with methenamine silver stain. Note the similarity to small yeast cells.

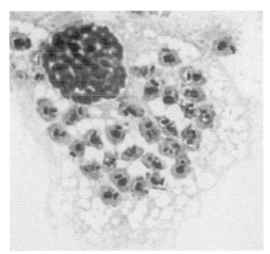

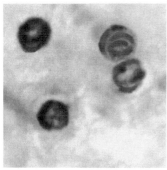
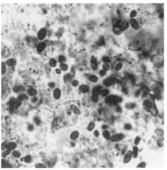

Table 30.1 Examination of impression smears

Tissue	Possible parasite	Stain
Lung	*Pneumocystis jiroveci* (now classified with the fungi)	Giemsa, methenamine silver or other cyst wall stain, optical brightening agent (calcofluor), immunospecific reagent
	Microsporidia	Modified trichrome, acid-fast stain, Giemsa, optical brightening agent (calcofluor), methenamine silver, EM[a]
	Toxoplasma gondii	Giemsa, immunospecific reagent
	Cryptosporidium spp.	Modified acid-fast stain, immunospecific reagent
	Entamoeba histolytica	Giemsa, trichrome
Liver	*Toxoplasma gondii*	Giemsa
	Leishmania donovani	Giemsa
	Cryptosporidium spp.	Modified acid-fast stain, immunospecific reagent
	Pneumocystis jiroveci (now classified with the fungi)	Giemsa, methenamine silver or other cyst wall stain, optical brightening agent (calcofluor), immunospecific reagent
	Entamoeba histolytica	Giemsa, trichrome
Brain	*Naegleria* sp.	Giemsa, trichrome
	Acanthamoeba sp.	Giemsa, trichrome
	Balamuthia mandrillaris (Leptomyxid ameba)	Giemsa, trichrome
	Sappinia diploidea	Giemsa, trichrome
	Entamoeba histolytica	Giemsa, trichrome
	Toxoplasma gondii	Giemsa, immunospecific reagent
	Microsporidia (*Encephalitozoon* spp.)	Modified trichrome, acid-fast stain, Giemsa, optical brightening agent (calcofluor), methenamine silver, EM
Skin	*Leishmania* spp.	Giemsa
	Onchocerca volvulus	Giemsa
	Mansonella streptocerca	Giemsa
	Acanthamoeba sp.	Giemsa, trichrome
Nasopharynx, sinus cavities	Microsporidia	Modified trichrome, acid-fast stain, Giemsa, optical brightening agent (calcofluor), methenamine silver, EM
	Acanthamoeba sp.	Giemsa, trichrome
	Naegleria sp.	Giemsa, trichrome
Intestine		
Small intestine	*Cryptosporidium* spp. (both small and large intestine)	Modified acid-fast stain, immunospecific reagent
	Cyclospora cayetanensis	Modified acid-fast stain
Jejunum	Microsporidia (*Enterocytozoon bieneusi*, *Encephalitozoon intestinalis*)	Modified trichrome, acid-fast stain, Giemsa, optical brightening agent (calcofluor), methenamine silver, EM
Duodenum	*Giardia lamblia*	Giemsa, trichrome
Colon	*Entamoeba histolytica*	Giemsa, trichrome
Cornea, conjunctiva	Various genera of microsporidia	Acid-fast stain, Giemsa, modified trichrome, methenamine silver, optical brightening agent (calcofluor), EM
	Acanthamoeba sp.	Giemsa, trichrome, calcofluor for cysts
Muscle	*Trichinella* spp.	Wet examination, squash preparation
	Microsporidia (*Pleistophora* sp., *Trachipleistophora* sp., *Brachiola* sp.)	Modified trichrome, acid-fast stain, Giemsa, optical brightening agent (calcofluor), methenamine silver, EM

[a] EM, electron microscopy.

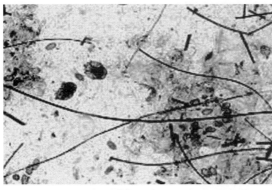

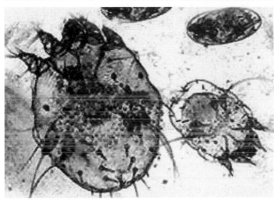

Figure 30.15 *Sarcoptes scabiei* mites from skin scrapings. (Upper) Note the two mites to the left of center. (Lower) Note the two eggs, the small nymph, and the adult mite (photographed at a higher magnification). If the light is too strong when examining the scrapings in saline, the mites will probably not be seen.

6. Examine cultures (*wear gloves at all times*).
   A. NNN agar for promastigotes of *Leishmania* spp. (see chapter 32): using a Pasteur pipette, remove a drop of fluid from interface of agar slant and culture tube, place on glass slide, cover with coverslip, and examine microscopically (×400) under low light for motile promastigotes (Figure 30.16).
   B. TYSGM-9 medium for amebae (see chapter 32): using a sterile Pasteur pipette, remove 1 or 2 drops of material from interphase of agar and overlay, place on glass slide, cover with coverslip, and examine microscopically (×100) under low light for trophozoite motility.
   C. Nonnutrient agar plate with lawn of *Escherichia coli* or *Enterobacter* sp. for detection of free-living amebae (see chapter 32): examine microscopically at low power (×100) for changes in bacterial lawn, particularly patches and tracks, indicating that the protozoan trophozoites have ingested bacteria as they move over the agar.

   D. Mouse passage for detection of *Toxoplasma gondii*:
      a. Wear canvas gloves to handle mice; sacrifice the animal(s).
      b. Pin mouse to board, spray with 70% ethyl alcohol, and open peritoneal cavity with sterile scissors.
      c. Using a sterile Pasteur pipette, remove fluid from the peritoneal cavity, and place in small tube or prepare smears.
      d. Prepare Giemsa-stained smears of peritoneal exudate, and examine microscopically at a magnification of ×1,000 for *T. gondii* tachyzoites.
      e. Place mice and all contaminated disposable materials in bag to be autoclaved and destroyed. Place nondisposable materials (scissors) in bag to be autoclaved prior to washing.
7. Correlate all examination results (wet mount, stains, and cultures) to determine presence of organisms.

Results

The majority of the protozoa are found on the permanent stained smears (impression smears, touch or squash preparations, or teased preparations). When culture is used, permanent stained smears of the culture medium or sediment may also reveal some of the protozoa. Although infrequently used, material from animals (at autopsy) can be examined as both wet and permanent stained preparations for confirmation of protozoa. Filarial infections may be confirmed by the recovery and identification of microfilariae in skin scrapings and/or biopsy specimens.

If identifications are certain, report the organisms detected; if a presumptive identification is made, confirmation by another laboratory is suggested.

Figure 30.16 *Leishmania* culture sediment. Note the promastigotes from the culture sediment. The typical "rosette" formation in the left image is frequently seen.

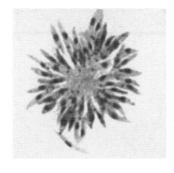

Skin

Onchocerca volvulus and *Mansonella streptocerca*

The use of skin snips is the method of choice for the diagnosis of human filarial infections with *O. volvulus* and *M. streptocerca*. Microfilariae of both species occur chiefly in the skin, although *O. volvulus* microfilariae are on rare occasions found in the blood and occasionally in the urine. For best results, the skin snip specimens should be thick enough to include the outer part of the dermal papillae. Snips may be taken in various ways. A small slice may be cut (using a razor blade) from a skin fold held between thumb and forefinger, or a slice may be taken from a small "cone" of skin pulled up by a needle. The skin snip should be so thin that significant bleeding does not occur, just a slight oozing of fluid. Corneal-scleral punches (either Holth or Walser type) have been found to be successful in taking skin snips of uniform size and depth and an average weight of 0.8 mg (range, 0.4 to 1.2 mg); this procedure is easy to perform and is painless. It has been demonstrated that in African onchocerciasis, it is preferable to take skin snips from the buttock region (above the iliac crest); in Central American onchocerciasis, the preferred skin snip sites are from the shoulders (over the scapula).

Skin snips are placed immediately in a drop of normal saline or distilled water and covered so that they will not dry; teasing the specimen with dissecting needles is not necessary but may facilitate release of the microfilariae. Microfilariae tend to emerge more rapidly in saline; however, in either fluid, the microfilariae usually emerge within 30 min to 1 h and can be examined under low-intensity light with the 10× objective of the microscope. To see definitive morphologic details of the microfilariae, allow the snip preparation to dry, fix it in absolute methyl alcohol, and stain it with Giemsa stain (Figure 30.17). For the differential diagnosis of *O. volvulus* and *M. streptocerca*, see chapter 12.

Cutaneous Amebiasis and Cutaneous Leishmaniasis

Skin biopsy specimens for the diagnosis of cutaneous amebiasis and cutaneous leishmaniasis should be processed for tissue sectioning and subsequently stained by the hematoxylin-eosin technique.

Lymph Nodes

Trypanosomiasis, Leishmaniasis, Chagas' Disease, and Toxoplasmosis

Material obtained from lymph nodes should be processed for tissue sectioning and as impression smears that should be processed as thin blood films and stained with Giemsa or other blood stains. Appropriate culture media can also be inoculated, again making sure that the specimen has been collected under sterile conditions.

Muscle

Trichinosis

The presumptive diagnosis of trichinosis is often based on a combination of facts: history of ingestion of raw or rare pork, walrus meat, or bear meat; diarrhea followed by edema and muscle pain; and the presence of eosinophilia. Usually, by the time the patient is symptomatic, the suspected food is no longer available for examination. The diagnosis may be confirmed by finding larval *Trichinella* spp. in a muscle biopsy specimen. The encapsulated larvae (which may be few) are easily seen in fresh muscle if small pieces are pressed between two slides and examined under the microscope (Figure 30.18). Larvae are usually most abundant in the diaphragm, masseter muscle, or tongue and may be recovered from these muscles at necropsy. Routine histologic sections can also be prepared.

A portion of the specimen can be digested in artificial digestive fluid at 37°C, with the larvae recovered by centrifugation; however, very young larvae might also be digested.

> **Artificial Digestive Fluid**
> Pepsin... 5 g
> Hydrochloric acid
>   (concentrated)............................. 7 ml
> Distilled water........................ 1,000 ml

1. Prepare the tissue by grinding (with a tissue grinder or commercial meat grinder or blender). The tissue should be medium grind; do not overgrind, particularly in a blender.
2. Add the tissue to the digestive fluid in the ratio of 1 part tissue to 20 parts fluid in an Erlenmeyer flask.
3. Place the flask in an incubator at 37°C, on a shaker table or magnetic stirrer if available; leave for 12 to 24 h.

Figure 30.17 Microfilariae (Giemsa stain). (Left) In this image, the microfilariae from a skin snip saline preparation are visible after staining. (Right) Microfilariae from blood; note that the characteristic nuclei are more easily seen in this preparation.

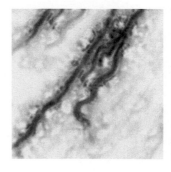 

4. Add warm water (37°C) to the digestive fluid mixture. (Add enough water to triple the original volume.)
5. Pour the diluted mixture into a Baermann funnel, and add water up to the screen.
6. Allow the mixture to stand for 1 to 2 h; the larvae will settle out in the lower part of the funnel.
7. Remove a few drops of fluid, and examine the material under the microscope (10× objective) for the presence of larvae.
8. If no larvae are present, centrifuge 50 ml and examine the sediment.

Cestode Larval Stages

Human infection with any of the larval cestodes may present diagnostic problems, and frequently the larvae are referred for identification after surgical removal. In addition to *Echinococcus granulosus* (hydatid disease) and the larval stage of *Taenia solium* (cysticercosis), other larval cestodes occasionally cause human disease. The larval stage of tapeworms of the genus *Multiceps*, a parasite of dogs and wild canids, is called a coenurus and may cause human coenurosis. The coenurus resembles a cysticercus but is larger and has multiple scolices developing from the germinal membrane surrounding the fluid-filled bladder. These larvae occur in extraintestinal locations, including the eye, central nervous system, and muscle (see chapter 14).

Human sparganosis is caused by the larval stages of tapeworms of the genus *Spirometra*, which are parasites of various canine and feline hosts; these tapeworms are closely related to the genus *Diphyllobothrium*. Sparganum

Figure 30.18 *Trichinella* spp. larvae in a squash preparation of muscle biopsy tissue; note the encysted larvae. This specimen is unstained.

larvae are elongated, ribbonlike larvae without a bladder and with a slightly expanded anterior end lacking suckers. They are usually found in superficial tissues or nodules, although they may cause ocular sparganosis, a more serious disease (see chapter 14).

The diagnosis of larval cestodes is frequently facilitated by the recognition of prominent calcareous corpuscles occurring in the tapeworm tissue; specific identification usually depends on referral to specialists.

Rectum and Bladder

Schistosomiasis

Often when a patient has an old, chronic infection or a light infection with *Schistosoma mansoni* or *S. japonicum*, the eggs may not be found in the stool and an examination of the rectal mucosa may reveal the presence of eggs. The fresh tissue should be compressed between two microscope slides and examined under the low power of the microscope (low-intensity light). Critical examination of these eggs should be made to determine whether living miracidia are still found within the egg. Treatment may depend on the viability of the eggs; for this reason, the condition of the eggs should be reported to the physician.

Mucosa from the bladder wall may reveal eggs of *Schistosoma haematobium* when they are not being recovered in the urine. As with rectal biopsy specimens, the eggs in the bladder wall should be checked for viability (Figure 30.19).

Viability Testing of Schistosome Eggs

1. With careful observation of the egg, using the 40× objective (low-intensity light), the cilia of the flame cells of the miracidium within the shell can be seen to move in a rapid flickering motion.
2. The eggs within the tissue can be removed (carefully tease the tissue apart in saline solution) and subjected to a hatching procedure. If the miracidia are released from the egg and swim to the top of the hatching flask, this movement is also proof of viability. (The procedure can be found in chapter 28.)

Digestion Procedure in Diagnosis of Schistosomiasis

Small pieces of tissue may be digested in 4% potassium hydroxide (4% sodium hydroxide may be used) for 2 to 3 h at 60 to 80°C. The material may then be concentrated by sedimentation or centrifugation and examined under the microscope for eggs (low power with low-intensity light).

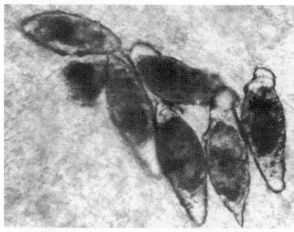

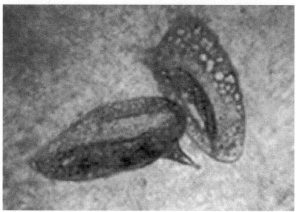

Figure 30.19 (Upper) *Schistosoma haematobium* eggs seen in a squash preparation of a bladder biopsy specimen. This specimen is unstained. (Armed Forces Institute of Pathology photograph.) (Lower) *Schistosoma mansoni* eggs from a rectal biopsy specimen.

References

1. **Arasteh, K. N., V. Simon, R. Musch, R. O. Weiss, K. Przytarski, U. M. Futh, F. Pleuger, D. Huhn, and M. P. L'age.** 1998. Sensitivity and specificity of indirect immunofluorescence and Grocott-technique in comparison with immunocytology (alkaline phosphatase anti alkaline phosphatase = APAAP) for the diagnosis of *Pneumocystis carinii* in broncho-alveolar lavage (BAL). *Eur. J. Med. Res.* **3:**559–563.

2. **Baselski, V. S., M. K. Robinson, L. W. Pifer, and D. R. Woods.** 1990. Rapid detection of *Pneumocystis carinii* in bronchoalveolar lavage samples by using cellufluor staining. *J. Clin. Microbiol.* **28:**393–394.

3. **Blanshard, C., W. S. Hollister, C. S. Peacock, D. G. Tovey, D. S. Ellis, E. U. Canning, and B. G. Gazzard.** 1992. Simultaneous infection with two types of intestinal microsporidia in a patient with AIDS. *Gut* **33:**418–420.

4. **Bottone, E. J.** 1991. Diagnosis of acute pulmonary toxoplasmosis by visualization of invasive and intracellular tachyzoites in Giemsa-stained smears of bronchoalveolar lavage fluid. *J. Clin. Microbiol.* **29:**2626–2627.

5. **Bowling, M. D., I. M. Smith, and S. L. Wescott.** 1973. A rapid staining procedure of *Pneumocystis carinii. Am. J. Med. Technol.* **39:**267–268.

6. **Brinn, N. T.** 1983. Rapid metallic histological staining using the microwave oven. *J. Histotechnol.* **6:**125–129.

7. **Clark, D. P.** 1999. New insights into human cryptosporidiosis. *Clin. Microbiol. Rev.* **12:**554–563.

8. **Croppo, G. P., G. S. Visvesvara, G. J. Leitch, S. Wallace, and D. A. Schwartz.** 1998. Identification of the microsporidian *Encephalitozoon hellem* using immunoglobulin G monoclonal antibodies. *Arch. Pathol. Lab. Med.* **122:**182–186.

9. **Current, W. L., and L. S. Garcia.** 1991. Cryptosporidiosis. *Clin. Microbiol. Rev.* **4:**325–358.

10. **Del Aguila, C., G. P. Croppo, H. Moura, A. J. Da Silva, G. J. Leitch, D. M. Moss, S. Wallace, S. B. Slemenda, N. J. Peiniazek, and G. S. Visvesvara.** 1998. Ultrastructure, immunofluorescence, Western blot, and PCR analysis of eight isolates of *Encephalitozoon* (Septata) *intestinalis* established in culture from sputum and urine samples and duodenal aspirates of five patients with AIDS. *J. Clin. Microbiol.* **36:**1201–1208.

11. **Del Aguila, C., R. Lopez-Velez, S. Fenoy, C. Turrientes, J. Cobo, R. Navajas, G. S. Visvesvara, G. P. Croppo, A. J. Da Silva, and N. J. Pieniazek.** 1997. Identification of *Enterocytozoon bieneusi* spores in respiratory samples from an AIDS patient with a 2-year history of intestinal microsporidiosis. *J. Clin. Microbiol.* **35:**1862–1866.

12. **Deplazes, P., A. Mathis, M. van Saanen, A. Iten, R. Keller, I. Tanner, M. P. Glauser, R. Weber, and E. U. Canning.** 1999. Dual microsporidial infection due to *Vittaforma corneae* and *Encephalitozoon hellem* in a patient with AIDS. *Clin. Infect. Dis.* **27:**1521–1524.

13. **Enriquez, F. J., O. Ditrich, J. D. Palting, and K. Smith.** 1997. Simple diagnosis of *Encephalitozoon* sp. microsporidial infections by using a panspecific antiexospore monoclonal antibody. *J. Clin. Microbiol.* **35:**724–729.

14. **Frenkel, J. K., M. S. Bartlett, and J. W. Smith.** 1990. RNA homology and the reclassification of *Pneumocystis. Diagn. Microbiol. Infect. Dis.* **13:**1–2.

15. **Garcia, L. S.** 1999. *Practical Guide to Diagnostic Parasitology.* ASM Press, Washington, D.C.

16. **Garcia, L. S., and R. Y. Shimizu.** 1997. Evaluation of nine immunoassay kits (enzyme immunoassay and direct fluorescence) for detection of *Giardia lamblia* and *Cryptosporidium parvum* in human fecal specimens. *J. Clin. Microbiol.* **35:**1526–1529.

17. **Garcia, L. S., R. Y. Shimizu, and D. A. Bruckner.** 1994. Detection of microsporidial spores in fecal specimens from patients diagnosed with cryptosporidiosis. *J. Clin. Microbiol.* **32:**1739–1741.

18. **Genaw, C.** 1989. Use of cresyl echt violet for the staining of *Pneumocystis carinii* as compared to Grocott's (GMS) and Giemsa methods. *J. Histotechnol.* **12:**39–40.

19. **Homer, K. S., E. L. Wiley, A. L. Smith, L. McCollough, D. Clark, S. D. Nightingale, and F. Vuitch.** 1992. Monoclonal antibody to *Pneumocystis carinii.* Comparison with silver stain in bronchial lavage specimens. *Am. J. Clin. Pathol.* **97:**619–624.

20. **Ignatius, R., S. Henschel, O. Liesenfeld, U. Mansmann, W. Schmidt, S. Koppe, T. Schneider, W. Heise, U. Futh,**

E. O. Riecken, H. Hahn, and R. Ulrich. 1997. Comparative evaluation of modified trichrome and Uvitex 2B stains for detection of low numbers of microsporidial spores in stool specimens. *J. Clin. Microbiol.* **35:**2266–2269.

21. Jelinek, T., S. Eichenlaub, and T. Loscher. 1999. Sensitivity and specificity of a rapid immunochromatographic test for diagnosis of visceral leishmaniasis. *Eur. J. Clin. Microbiol. Infect. Dis.* **18:**669–670.

22. Jules-Elysee, K. M., D. E. Stover, M. B. Zaman, E. M. Bernard, and D. A. White. 1990. Aerosolized pentamidine: effect on diagnosis and presentation of *Pneumocystis carinii* pneumonia. *Ann. Intern. Med.* **112:**750–757.

23. Khan, M. A., N. Farrag, and P. Butcher. 1999. Diagnosis of *Pneumocystis carinii* pneumonia: immunofluorescence staining, simple PCR or nPCR. *J. Infect.* **39:**77–80.

24. Kim, Y. K., S. Parulekar, P. K. W. Yu, R. J. Pisani, T. F. Smith, and J. P. Anhalt. 1990. Evaluation of calcofluor white stain for detection of *Pneumocystis carinii*. *Diagn. Microbiol. Infect. Dis.* **13:**307–310.

25. Koch, M., and W. Heizmann. 1990. Problems in the detection of *Pneumocystis carinii* by indirect immunofluorescence. *Eur. J. Clin. Microbiol. Infect. Dis.* **9:**58–59.

26. Lacey, C. J. N., A. M. T. Clarke, P. Fraser, T. Metcalfe, G. Bonsor, and A. Curry. 1992. Chronic microsporidian infection of the nasal mucosae, sinuses and conjunctivae in HIV disease. *Genitourin. Med.* **68:**179–181.

27. Magee, J. G., K. J. McDade, J. Cunningham, and V. Harrison. 1991. *Pneumocystis carinii* pneumonia—detection of parasites by immunofluorescence based on a monoclonal antibody. *Med. Lab. Sci.* **48:**235–237.

28. Mahan, C. T., and G. E. Sale. 1978. Rapid methenamine silver stain for *Pneumocystis* and fungi. *Arch. Pathol. Lab. Med.* **102:**351–352.

29. Masur, H., V. J. Gill, I. Feuerstein, A. F. Suffredini, D. Brown, H. C. Lane, R. Yarchoan, J. H. Shelhamer, and F. P. Ognibene. 1991. Effect of aerosolized pentamidine prophylaxis on the diagnosis of *Pneumocystis carinii* pneumonia by induced sputum examination in patients infected with the human immunodeficiency virus. *Am. Rev. Respir. Dis.* **144:**760–764.

30. Metcalfe, T. W., R. M. L. Doran, P. L. Rowlands, A. Curry, and C. J. N. Lacey. 1992. Microsporidial keratoconjunctivitis in a patient with AIDS. *Br. J. Ophthalmol.* **76:**177–178.

31. Metersky, M. L., and A. Catanzaro. 1991. Diagnostic approach to *Pneumocystis carinii* pneumonia in the setting of prophylactic aerosolized pentamidine. *Chest* **100:**1345–1349.

32. Midgley, J., P. A. Parsons, D. C. Shanson, O. A. N. Husain, and N. Francis. 1991. Monoclonal immunofluorescence compared with silver stain for investigating *Pneumocystis carinii* pneumonia. *J. Clin. Pathol.* **44:**75–76.

33. Musto, L., M. Flanigan, and A. Elbadawi. 1982. Ten-minute silver stain for *Pneumocystis carinii* and fungi in tissue sections. *Arch. Pathol. Lab. Med.* **106:**292–294.

34. Ng, V. L., M. A. Virani, R. E. Chaisson, D. M. Yajko, H. T. Sphar, K. Cabrian, N. Rollins, P. Charache, M. Krieger, W. K. Hadley, and P. C. Hopewell. 1990. Rapid detection of *Pneumocystis carinii* using a direct fluorescent monoclonal antibody stain. *J. Clin. Microbiol.* **28:**2228–2233.

35. Ng, V. L., D. M. Yajko, L. W. McPhaul, I. Gartner, B. Byford, C. D. Goodman, P. S. Nassos, C. A. Sanders, E. L. Howes, G. Leough, P. C. Hopewell, and W. K. Hadley.

1990. Evaluation of an indirect fluorescent-antibody stain for detection of *Pneumocystis carinii* in respiratory specimens. *J. Clin. Microbiol.* **28:**975–979.

36. Paradis, I. L., C. Ross, A. Dekker, and J. Dauber. 1990. A comparison of modified methenamine silver and toluidine blue stains for detection of *Pneumocystis carinii* in bronchoalveolar lavage specimens from immunosuppressed patients. *Acta Cytol.* **34:**513–516.

37. Pintozzi, R. L. 1978. Technical methods: modified Grocott's methenamine silver nitrate method for quick staining of *Pneumocystis carinii*. *J. Clin. Pathol.* **31:**803–805.

38. Pixley, F. J., A. E. Wakefield, S. Banerji, and J. M. Hopkin. 1991. Mitochondrial gene sequences show fungal homology for *Pneumocystis carinii*. *Mol. Microbiol.* **5:** 1347–1351.

39. Rafanan, A. L., P. Klevjer-Anderson, and M. L. Metersky. 1998. *Pneumocystis carinii* pneumonia diagnosed by non-induced sputum stained with a direct fluorescent antibody. *Ann. Clin. Lab. Sci.* **28:**99–103.

40. Rolston, K. V. I., S. Rodriguez, L. McRory, G. Uribe-Botero, R. Morice, and P. W. A. Mansell. 1988. Diagnostic value of induced sputum in patients with the acquired immunodeficiency syndrome. *Am. J. Med.* **85:**269.

41. Schwartz, D. A., R. T. Bryan, K. O. Hewanlowe, G. S. Visvesvara, R. Weber, A. Cali, and P. Angritt. 1992. Disseminated microsporidiosis (*Encephalitozoon hellem*) and acquired immunodeficiency syndrome—autopsy evidence for respiratory acquisition. *Arch. Pathol. Lab. Med.* **116:**660–668.

42. Settnes, O. S., and P. Larsen. 1979. Inhibition of toluidine blue O stain for *Pneumocystis carinii* by additives in the diethyl ether. *Am. J. Clin. Pathol.* **72:**493–494.

43. Siddons, C. A., P. A. Chapman, and B. A. Rush. 1992. Evaluation of an enzyme immunoassay kit for detecting *Cryptosporidium* in faeces and environmental samples. *J. Clin. Pathol.* **45:**479–482.

44. Stratton, N., J. Hryniewicki, S. L. Aarnaes, G. Tan, L. M. de la Maza, and E. M. Peterson. 1991. Comparison of monoclonal antibody and calcofluor white stains for the detection of *Pneumocystis carinii* from respiratory specimens. *J. Clin. Microbiol.* **29:**645–647.

45. Stringer, S. L., K. Hudson, M. A. Blase, P. D. Walzer, M. T. Cushion, and J. R. Stringer. 1989. *Pneumocystis carinii*: sequence from ribosomal RNA implies a close relationship with fungi. *Exp. Parasitol.* **68:**450–461.

46. Sundar, S., S. G. Reed, V. P. Singh, P. C. Kimar, and H. W. Murray. 1998. Rapid accurate field diagnosis of Indian visceral leishmaniasis. *Lancet* **351:**563–565.

47. Tuncer, S., S. Erguven, S. Kocagoz, and S. Unal. 1998. Comparison of cytochemical staining, immunofluorescence and PCR for diagnosis of *Pneumocystis carinii* on sputum samples. *Scand. J. Infect. Dis.* **30:**125–128.

48. Vesey, G., N. Ashbolt, E. J. Fricker, D. Deere, K. L. Williams, D. A. Veal, and M. Dorsch. 1998. The use of a ribosomal RNA targeted oligonucleotide probe for fluorescent labeling of viable *Cryptosporidium parvum* oocysts. *J. Appl. Microbiol.* **85:**429–440.

49. Visvesvara, G. S., H. Moura, E. Kovacs Nace, S. Wallace, and M. L. Eberhard. 1997. Uniform staining of *Cyclospora* oocysts in fecal smears by a modified safranin technique with microwave heating. *J. Clin. Microbiol.* **35:**730–733.

50. Wakefield, A. E., S. E. Peters, S. Banerji, P. D. Bridge,

G. S. Hall, D. L. Hawksworth, L. A. Guiver, A. G. Allen, and J. M. Hopkin. 1992. *Pneumocystis carinii* shows DNA homology with the ustomycetous red yeast fungi. *Mol. Microbiol.* **6:**1903–1911.

51. Walker, J., G. Conner, J. Ho, C. Hunt, and L. Peckering. 1989. Giemsa staining for cysts and trophozoites of *Pneumocystis carinii*. *J. Clin. Pathol.* **42:**432–434.

52. Weber, R., R. T. Bryan, R. L. Owen, C. M. Wilcox, L. Gorelkin, G. S. Visvesvara, and The Enteric Opportunistic Infections Working Group. 1992. Improved light-microscopical detection of microsporidia spores in stool and duodenal aspirates. *N. Engl. J. Med.* **326:**161–166.

53. Wilhelmus, K. R., M. S. Osato, R. L. Font, N. M. Robinson, and D. B. Jones. 1986. Rapid diagnosis of *Acanthamoeba* keratitis using calcofluor white. *Arch. Ophthalmol.* **104:**1309–1312.

54. Wilson, M., and P. M. Schantz. 2000. Parasitic immunodiagnosis, p. 1117–1122. *In* G. T. Strickland (ed.), *Hunter's Tropical Medicine and Emerging Infectious Diseases*, 8th ed. The W. B. Saunders Co., Philadelphia, Pa.

55. Wolfson, J. S., M. A. Waldron, and L. S. Sierra. 1990. Blinded comparison of a direct immunofluorescent monoclonal antibody staining method and a Giemsa staining method for identification of *Pneumocystis carinii* in induced sputum and bronchoalveolar lavage specimens of patients infected with human immunodeficiency virus. *J. Clin. Microbiol.* **28:**2136–2138.

56. Zierdt, C. H., V. J. Gill, and W. S. Zierdt. 1993. Detection of microsporidian spores in clinical samples by indirect fluorescent-antibody assay using whole-cell antisera to *Encephalitozoon cuniculi* and *Encephalitozoon hellem*. *J. Clin. Microbiol.* **31:**3071–3074.

31

Procedures for Detecting Blood Parasites

Depending on the life cycle, a number of parasites may be recovered in a blood specimen, either whole blood, buffy coat preparations, or various types of concentrations. These parasites include *Plasmodium*, *Babesia*, and *Trypanosoma* species, *Leishmania donovani*, and microfilariae. Although some organisms are motile in fresh, whole blood, species identification is normally accomplished from the examination of both thick and thin permanent stained blood films. Blood films can be prepared from fresh, whole blood collected with no anticoagulants, anticoagulated blood, or sediment from the various concentration procedures. The recommended stain of choice is Giemsa stain; however, the parasites can also be seen on blood films stained with Wright's stain, a Wright-Giemsa combination stain, or one of the more rapid stains such as Diff-Quik. Delafield's hematoxylin stain is often used to stain the microfilarial sheath; in some cases, Giemsa stain does not provide sufficient stain quality to allow differentiation of the microfilariae. A complete discussion of the proper way to examine a blood film is presented later in this chapter. This information is important and particularly relevant when one is examining proficiency testing blood films in the absence of clinical information about patient history and possible etiologic agents.

It is important to remember that standard precautions should be used at all times when blood or body fluids are handled (12). Remember that all requests for malaria diagnosis are considered STAT requests, and specimens should be collected, processed, examined, and reported accordingly.

Preparation of Thick and Thin Blood Films

Some parasites (microfilariae and trypanosomes) can be detected in fresh blood by their characteristic shape and motility, but specific identification of the organisms requires a permanent stain. Two types of blood films are recommended. Thick films allow a larger amount of blood to be examined, which increases the possibility of detecting light infections (24). However, species identification by thick film, particularly for malaria parasites, can usually be made only by experienced workers. The morphologic characteristics of blood parasites are best seen in thin films, in which the red blood cell (RBC) morphology is preserved and the size relationship between infected and uninfected red cells can be determined after staining. This characteristic is often valuable in determining the species of *Plasmodium* present from the thin blood film.

The accurate examination of thick and thin blood films and identification of parasites depend on the use of absolutely clean, grease-free slides for preparation of all blood films. Old (unscratched) slides should be cleaned first with detergent and then 70% ethyl alcohol; new slides should also be cleaned with alcohol before use. When a new box of slides is opened, the slides are coated with a substance that allows them to be pulled apart; these slides should be cleaned before use for preparation of blood films. The advantages and disadvantages of the thin and thick blood films can be seen in Table 31.1.

Blood films are usually prepared when the patient is admitted. When malaria is a possible diagnosis, after the first set of negative smears, samples should be taken at intervals of 6 to 8 h for at least three successive days. After a finger stick, the blood should flow freely; blood that has to be "milked" from the finger will be diluted with tissue fluids, decreasing the number of parasites per field. If the specimen is sent directly to the laboratory, thus eliminating laboratory-patient contact, the following approach can be used. Unless you are positive that you will receive well-prepared slides, request a tube of fresh blood (EDTA anticoagulant is preferred) and prepare the smears. For detection of stippling, the smears should be prepared within 1 h after the specimen is drawn. After that time, stippling may not be visible on stained films; however, the overall organism morphology will still be acceptable. Potential problems with anticoagulants can be seen in Table 31.2. Although blood films can be prepared from the small amount of blood left in the needle after the venipuncture collection using anticoagulant, it is not recommended for several reasons. The blood tends to clot fairly quickly in the needle, and there are safety recommendations that limit the handling of needles. Therefore, the finger stick and EDTA venipuncture are recommended for collection of specimens for blood film preparation.

The time when the specimen was drawn should be clearly indicated on the tube of blood and also on the result report. The physician will then be able to correlate the results with any fever pattern or other symptoms that the patient may have. There should also be some indication on the report that is sent back to the physician that one negative specimen does not rule out the possibility of a parasitic infection.

Note Although most laboratories use commercially available blood collection tubes, the following approach can be used when necessary. EDTA (Sequestrene) can be prepared and tubed as follows. Dissolve 5 g of EDTA in 100 ml of distilled water. Aliquot 0.4 ml into tubes, and evaporate the water. This amount of anticoagulant is sufficient for 10 ml of blood. One can also use 20 mg of EDTA (dry) per tube (20 mg/10 ml of blood). The tube should be filled with blood to provide the proper blood/anticoagulant ratio.

Thick Blood Films

Fresh Blood

To prepare the thick film, place 2 or 3 small drops of fresh blood (no anticoagulant) on an alcohol-cleaned slide. With the corner of another slide and using a circular motion, mix the drops and spread them over an area ~2 cm in diameter. Continue stirring for 30 s to prevent the formation of fibrin strands that may obscure the parasites after staining.

Anticoagulant

If blood containing an anticoagulant is used, 2 or 3 drops may be spread over an area about 2 cm in diameter; it is not necessary to continue stirring for 30 s, since fibrin strands do not form. If the blood is too thick or any grease remains on the slide, the blood may flake off during staining.

Table 31.1 Advantages of the thin and thick blood films

Thin blood film advantages	Thick blood film advantages
RBC morphology (size, shape, stippling) can be seen after fixation with methanol prior to staining (Giemsa) or as a part of the staining process (Wright). RBCs are laked in the thick film.	Larger number of parasites per field compared with the thin blood film. RBCs are laked, and so WBCs, platelets, and parasites are visible after staining.
Identification to *Plasmodium* species easier, since the parasite can be seen within the RBC. The size of the parasites within the RBCs can provide information necessary for identification to species.	Phagocytized malaria pigment may be seen within the WBCs, even with a low parasitemia.
Parasitemia can be calculated from the thin film; determination of parasitemia is mandatory for all species of *Plasmodium* and is particularly important for monitoring therapy for *P. falciparum*. Parasitemia should be reported with every set of positive blood films for malaria.	Stippling may be seen in a well-stained thick film. However, this depends on how long the blood has been in contact with anticoagulant if it is not collected as a fresh specimen (finger stick).

Table 31.2 Potential problems of the use of EDTA anticoagulant for the preparation of thin and thick blood films

Potential problem	Comments
Adhesion to the slide; blood falls off slide during staining	Incorrect ratio of anticoagulant to blood; fill tube completely with blood (7 ml or pediatric draw tube).
Distortion of parasites; same type of distortion can also be seen after blood is refrigerated (not recommended)	Prolonged storage of blood in EDTA may lead to distortion: trophozoites (*P. vivax*) and gametocytes (*P. falciparum*) tend to round up, thus mimicking *P. malariae*.
Change in ring form size	Ring forms of *P. falciparum* continue to grow and enlarge, thus resembling rings of the other species. Typical "small" rings appear larger than usual.
Use of EDTA anticoagulant (primarily used by hematology laboratories because the cellular components and morphology of the blood cells are preserved) 1. Blood smears for differentials from acceptable specimens should be prepared within 2 h of collection 2. Blood counts from acceptable venipuncture specimens should be performed within 6 h of collection Underfilling the EDTA blood collection tube can lead to erroneously low blood cell counts and hematocrits, morphologic changes to RBCs, and staining alteration; excess EDTA can shrink red cells; conversely, overfilling the blood collection tube will not allow the tube to be properly mixed and may lead to platelet clumping and clotting	EDTA prevents coagulation of blood by chelating calcium. Calcium is necessary in the coagulation cascade, and its removal inhibits and stops a series of events, both intrinsic and extrinsic, which cause clotting. In some individuals, EDTA may cause inaccurate platelet results. These anomalies, platelet clumping and platelet satellitism, may be the result of changes in the membrane structure occurring when the calcium ion is removed by the chelating agent, allowing the binding of preformed antibodies. Proper mixing of the whole-blood specimen ensures that EDTA is dispersed throughout the sample. Evacuated blood collection tubes with EDTA should be mixed by 8–10 end-to-end inversions immediately following venipuncture collection. Microcollection tubes with EDTA should be mixed by 10 complete end-to-end inversions immediately following collection. They should then be inverted an additional 20 times prior to analysis.
Loss of Schüffner's dots (stippling) in *P. vivax* and *P. ovale*	Schüffner's dots (true stippling) occur in both *P. vivax* and *P. ovale*; in the absence of stippling, identification to the species level may be more difficult or impossible.
Prolonged storage of EDTA blood (room temperature with stopper removed)	The pH, CO_2, and temperature changes may reflect conditions within the mosquito. Thus, exflagellation of the male gametocyte may occur while still in the tube of blood prior to thin and thick blood film preparation. Microgametes may be confused with *Borrelia* or may be ignored as debris.
Release of merozoites from the schizonts into the blood; normally, merozoites are not found outside of the RBCs, in contrast to *Babesia* spp., where rings may be seen outside of the RBCs	Small rings may be seen outside of the RBCs or appear to be appliqué forms, thus suggesting *P. falciparum*. It is important to differentiate these true rings (both cytoplasmic and nuclear colors) from platelets (uniform color).
Incorrect submission of blood in heparin	EDTA has less impact on parasite morphology than does heparin.

Allow the thick film to air dry (room temperature) in a dust-free area. Never apply heat to a thick film, since heat will fix the blood, causing the RBCs to remain intact during staining; the result is stain retention and inability to identify the parasites. After the thick films are thoroughly dry, they can be laked to remove the hemoglobin. Rupture of the RBCs during laking removes the RBCs from the final stained blood film; the only structures remaining on the thick film are the white blood cells (WBC), the platelets, and any parasites present. To lake the films, place them in buffer solution before staining or directly into Giemsa stain, which is an aqueous stain. If thick films are to be stained later, they should be laked before storage. Potential problems with

Table 31.3 Potential problems with thick blood film preparation and staining

Potential problem	Comments
Blood films flake off during staining	The blood film was not dry or did not dry evenly; the film was too thick; incorrect blood/anticoagulant ratio (excess EDTA); dirty or greasy slides; blood kept in EDTA too long before film preparation
Inadequate staining	Blood film too thick; staining time incorrect; stain concentration incorrect
Potential for loss of organisms	Incorrect staining technique; blood film too old prior to staining; poor environmental conditions and/or incomplete drying

the preparation and staining of thick blood films can be seen in Table 31.3.

Thin Blood Films

The thin blood film is routinely used for specific parasite identification, although the number of organisms per field is much reduced compared with the thick film. The thin film is prepared exactly as one used for a differential count (Figure 31.1). A well-prepared film is thick at one end and thin at the other (one layer of evenly distributed RBCs with no cell overlap). The thin, feathered end should be at least 2 cm long, and the film should occupy the central area of the slide, with free margins on both sides. The presence of long streamers of blood indicates that the slide used as a spreader was dirty or chipped. Streaks in the film are usually caused by dirt, and holes in the film indicate the presence of grease on the slide. After the film has air dried (do not apply heat), it may be stained. The necessity for fixation before staining will depend on the stain selected. Since Giemsa is an aqueous stain, laking of the RBCs occurs during the staining process. However, when Wright's stain is used, a fixing agent is incorporated into the stain, so that laking of the blood films must occur prior to staining.

The instrument-prepared monolayer method or coverslip methods generally do not provide the best morphology for malarial parasites within the RBCs. However, the selection of a slide preparation method can be dictated by personal preference, since a malarial infection can be diagnosed from either type of slide. Potential problems with the preparation and staining of thin blood films can be seen in Table 31.4.

Figure 31.1 Method for preparation of thin blood film. (A) Position of spreader slide; (B) well-prepared thin film. Arrows indicate the area of the slide (feather edge) used to observe accurate cell morphology. (Illustration by Nobuko Kitamura.)

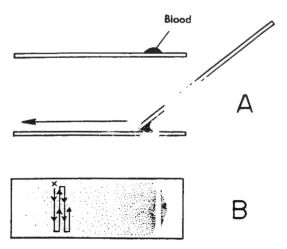

Table 31.4 Potential problems with thin blood film preparation and staining

Potential problem	Comments
Blood films are too thick	Too much blood was used; no feathered edge; blood was pulled too quickly, spreader slide was held at the wrong angle (too high)
Blood films are too thin	Blood with a low hematocrit tends to spread too thin; let the blood settle or remove some plasma; angle of the spreader slide was too low; insufficient amount of blood for film preparation
Blood films have rough edges; lack of or poor feather edge	Blood was in front of the spreader slide; uneven contact with spreader slide; spreader slide chipped; possible fibrin formation; slides greasy

Combination Thick and Thin Blood Films (on the Same Slide)

In some instances (field surveys), it is helpful to prepare slides with both a thick and a thin film on the same slide. With this type of preparation, remember the following.

1. Sufficient time must be allowed for the thick portion of the smear to dry before staining.
2. The thin film only must be fixed in absolute methanol before staining.

Combination Thick and Thin Blood Films (Can Be Stained as Either)

To prepare a slide containing blood that can be stained for either a thick or thin blood film, the following method was developed (Figure 31.2). The specimen usually consists of fresh whole blood collected by finger stick or whole blood containing EDTA collected by venipuncture less than 1 h earlier.

Visually, the smear should consist of alternating thick and thin portions throughout the length of the glass slide. One should be able to barely read newsprint through the wet or dry film. Also, the film itself should not have any clear areas or smudges, indicating that grease or fingerprints were on the glass.

Detailed Procedure

1. *Wear gloves when performing this procedure or preparing any blood films.*
2. The procedure depends on the source of the specimen.

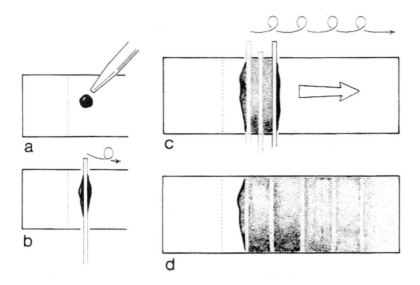

Figure 31.2 Method of thick-thin combination blood film preparation. (a) Position of the drop of EDTA-containing blood. (b) Position of the applicator stick in contact with blood and glass slide. (c) Rotation of the applicator stick. (d) Completed thick-thin combination blood film prior to staining. (Illustration by Sharon Belkin. Reprinted from L. S. Garcia, *Practical Guide to Diagnostic Parasitology*, ASM Press, Washington, D.C., 1999.)

A. Blood from a finger puncture is not recommended, since the procedure does not lend itself to "stirring" to prevent fibrin strands.

B. For blood from venipuncture, place a clean 1- by 3-in. glass microscope slide on a horizontal surface. Place a drop (30 to 40 µl) of blood onto one end of the slide about 0.5 in. from the end. Using an applicator stick lying across the glass slide and keeping the applicator in contact with the blood and glass, rotate (do not "roll") the stick in a circular motion while moving the stick down the glass slide to the opposite end. The appearance of the blood smear should be alternate thick and thin areas of blood that cover the entire slide. Immediately place the film over some small print and be sure that the print is just barely readable.

3. Allow the film to air dry horizontally and protected from dust for at least 30 min to 1 h. Do not attempt to speed the drying process by applying any type of heat, because the heat fixes the RBCs and they subsequently will not lyse (lake) in the staining process.

4. This slide can be stained as either a thick or thin blood film.

5. Label the slide appropriately.

6. If staining with Giemsa is delayed for more than 3 days or if the film is to be stained with Wright's stain, lyse the RBCs in the thick film by placing the slide in buffered water (pH 7.0 to 7.2) for 10 min, remove it from the water, and place it in a vertical position to air dry. Rapid stains are also acceptable.

Procedure Limitations

1. If the smears are prepared from anticoagulated blood that is more than 1 h old, the morphology of the parasites may not be typical and the film may wash off the slide during the staining procedure.

   If a tube of blood containing EDTA cools to room temperature and the cap has been removed, several parasite changes can occur. The parasites within the RBCs will respond as if they were now in the mosquito after being taken in with a blood meal. The morphology of these changes in the life cycle and within the RBCs can cause confusion when examining blood films prepared from this blood (11, 35, 46, 69).

   A. Stippling (Schüffner's dots) may not be visible.

   B. The male gametocyte (if present) may exflagellate and may resemble spirochetes (Figure 31.3).

   C. The ookinetes of *Plasmodium* species other than *P. falciparum* may develop as if they were in the mosquito and may mimic the crescent-shaped gametocytes of *P. falciparum.*

2. Identification to species, particularly between *P. ovale* and *P. vivax* and between the ring forms of *P. falciparum* and *Babesia* spp., may be impossible without examining one of the slides stained as a thin blood film. Also, *Trypanosoma cruzi* trypomastigotes are frequently distorted in thick films.

3. Excess stain deposition on the film may be confusing and make the detection of organisms difficult.

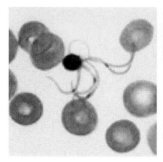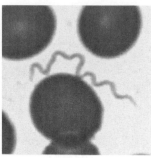

Figure 31.3 (Left) Photograph of exflagellation of the malarial microgametocyte. This may occur when anticoagulated blood is left standing at room temperature for some time prior to smear preparation. The life cycle of the parasite continues in the tube of blood as it would if the parasite had been ingested by the mosquito during a blood meal. (Right) *Borrelia* in blood. Note that the *Plasmodium* microgametes can resemble these organisms, particularly if they appear free in the background of the smear.

Fixation of Thin Blood Films: Acetone Dip for Thick Blood Films

Thin-Film Fixation. Thin blood films must be completely dry before being fixed in absolute methanol (Giemsa staining). Drying times will be extended if the humidity is high; drying films as quickly as possible produces the best results. However, they should never be heated to decrease drying times; heat tends to produce distorted parasite and blood cell morphology. Absolute methanol tends to absorb moisture from the air, so that absolute methanol for fixation should not be reused from day to day but needs to be fresh daily. Do not store this fixative in a Coplin jar. Thin blood films should be fixed and stained within 24 h; some deterioration may occur if the slides are held too long before being processed. However, the slides can be frozen for long-term storage. All thin blood films must be fixed prior to shipment to a reference laboratory; both these fixed slides and the tube of blood should be sent to the reference laboratory.

If the thin blood films are fixed for too long, some of the morphologic details may be reduced or lost. If they are fixed for too short a time, the RBCs may be distorted or partially lysed and the thin film will not be uniformly stained. If the thin films are fixed in methanol that contains water, the RBCs will be distorted and there will appear to be "holes" in the thicker areas of the films. *As a reminder, do not reuse absolute methanol for fixation; discard after each use.*

Thick-Film Acetone "Quick Dip." Although thick blood films are not fixed with absolute methanol, after the thick films are thoroughly dry, they can be dipped twice in acetone and allowed to dry before being stained.

This extra step does not interfere with RBC lysis that occurs either prior to or during staining. The acetone "quick dip" makes the thick film less likely to fall off during staining and provides a cleaner background for microscopic examination.

Buffy Coat Blood Films

For patients with suspected malaria (negative thick and thin blood films), trypanosomiasis, filariasis, or leishmaniasis, concentration procedures increase the number of organisms recovered from blood specimens. The buffy coat containing WBCs and platelets, and the layer of RBCs just below the buffy coat layer, can be used to prepare thick and thin blood films. The sensitivity of this approach is greatly enhanced over that of the thick film.

L. donovani, trypanosomes, and *Histoplasma capsulatum* (a fungus with intracellular elements resembling those of *L. donovani*) are occasionally detected in the peripheral blood. The parasite or fungus is found in the large mononuclear cells in the buffy coat (a layer of white cells resulting from centrifugation of whole citrated blood). The nuclear material stains dark red-purple, and the cytoplasm is light blue (*L. donovani*). *H. capsulatum* appears as a large dot of nuclear material (dark red-purple) surrounded by a clear halo area. Trypanosomes in the peripheral blood also concentrate with the buffy coat cells.

After centrifugation and aliquoting of the appropriate layers, some of the material can be examined as a wet mount; trypomastigotes and microfilariae may be seen as motile objects in the wet mount. After staining, *L. donovani* amastigotes may be found in the monocytes and *Plasmodium* parasites may be seen in the thick and thin films. Advantages and disadvantages of buffy coat films can be seen in Table 31.5.

Detailed Procedure

1. *Wear gloves when performing this procedure or preparing any blood films.*
2. Whole blood should be collected using EDTA anticoagulant. Although heparin can be used, if malaria films are to be prepared, EDTA is recommended.
3. Although capillary hematocrit tubes have been used in the past, the cutting and breaking of these tubes to remove the cells for film preparation is not considered a safe procedure and is *not* recommended. However, if you use a microhematocrit tube, the tube should be carefully scored and snapped at the buffy coat interface, and the white cells are prepared as a thin film. The tube can also

Table 31.5 Advantages and disadvantages of the buffy coat films

Buffy coat film advantages	Buffy coat film disadvantages
The volume of blood is considerably larger than with both thick and thin films	It takes some practice to remove the correct blood layers from the centrifuged blood in order to prepare thick and thin films
More sensitive than thick films for diagnosis of malaria	Same potential problems may occur as with traditional thick or thin blood films
Malaria pigment, which is phagocytized by the WBCs, may be seen more easily in the concentrated WBCs	Some organisms and/or stages of parasites might be damaged due to high centrifugation; however, centrifugation at 500 × g for 15 to 20 min should not cause damage
Detection of parasitemia easier for larger stages such as schizonts and gametocytes	
	Increased number of platelets may be confusing in terms of parasite differentiation

be examined before removal of the buffy coat, at the low and high dry powers of the microscope. If trypanosomes are present, the motility may be observed in the buffy coat. Microfilarial motility would also be visible.

4. Using a capillary pipette, fill a Wintrobe tube with blood containing anticoagulant (EDTA is preferred), cap the tube, and centrifuge it for 30 min at 100 × g. Another option is to centrifuge the tube of anticoagulated blood at 100 × g for 15 min, transfer that buffy coat to another tube, and centrifuge the tube at 300 × g for 30 min. After centrifugation, the tube contains three layers: plasma on top, a layer of white cells (buffy coat), and the packed RBCs on the bottom.

5. Remove and discard most of the plasma above the buffy coat, leaving a small amount on top of the buffy coat layer. Then remove the remaining plasma, buffy coat, and the RBCs *right below the buffy coat*. Transfer this aliquot to a separate tube.

6. Examine the buffy coat directly for motile trypomastigotes and microfilariae by mixing 0.5 drop of saline with 1 drop of buffy coat sediment on a microscope slide. Add a coverslip, and examine at low power (10× objective).

7. Mix the aliquot gently (avoid bubbles), and prepare thick and thin blood films on alcohol-cleaned slides.

8. Allow the films to air dry horizontally and protected from dust for at least 30 min to 1 h. Do not attempt to speed the drying process by applying

any type of heat, because the heat fixes the RBCs and they subsequently will not lyse (lake) in the staining process for the thin films.

9. Label the slide appropriately.

10. If staining with Giemsa is delayed for more than 3 days or if the film is to be stained with Wright's stain, lyse the RBCs in the thick film by placing the slide in buffered water (pH 7.0 to 7.2) for 10 min, remove it from the water, and place it in a vertical position to air dry.

Staining Blood Films

For accurate identification of blood parasites, a laboratory should develop proficiency in the use of at least one good staining method (24, 26, 65, 66). It is better to select one method that will provide reproducible results than to use several on a hit-or-miss basis. Blood films should be stained as soon as possible, since prolonged storage may result in stain retention. Failure to stain positive malarial smears within a month may result in failure to demonstrate typical staining characteristics for individual species.

The most common stains are of two types. Wright's stain has the fixative in combination with the staining solution, so that both fixation and staining occur at the same time; therefore, the thick film must be laked before staining. Giemsa stain has the fixative and stain separate; therefore, the thin film must be fixed with absolute methanol before staining.

When slides are removed from either type of staining solution, they should be dried in a vertical position. After being air dried, they may be examined under oil immersion by placing the oil directly on the uncovered blood film. If films are to be kept for a permanent record, they should be protected with a coverglass after being mounted in a medium such as Permount.

Note Blood films stained with any of the Romanowsky stains that have been mounted with Permount or other resinous mounting media are susceptible to fading of the basophilic elements and generalized loss of stain intensity. Hollander (31) has recommended the addition of 1% (by volume) 2,6-di-t-butyl-p-cresol (butylated hydroxytoluene [Sigma Chemical Co. catalog no. B1253]) to the mounting medium. Without the addition of this antioxidant, mounted stained smears eventually become pink; stains protected with this compound remain unchanged in color for many years.

Note Any slide that is protected by a coverglass and is going to be examined with an oil immersion lens must be covered by a no. 1 coverglass. If a no. 2 coverglass is used, the extra thickness may prevent the oil immersion lens from focusing properly.

Giemsa Stain

Giemsa stain is sold as a concentrated stock solution. Each new lot should be tested for optimal staining times before being used on patient specimens. If the blood cells appear to be adequately stained, the timing and stain dilution should be appropriate to demonstrate the presence of malarial and other parasites. Giemsa stain is also available as a powder for those who wish to make up their own stain. The use of prepared liquid stain or stain prepared from the powder depends on personal preference; there is apparently little difference between the two preparations. Directions for preparing the stain follow (26, 35, 36).

Note Quality control for blood film stains can be any negative or positive blood film; it is not necessary for the control slides to contain actual parasites. If the RBCs and WBCs stain correctly, any parasites present would also stain correctly with the same nuclear and cytoplasmic colors as the blood cells.

Reagents

Stock Giemsa Stain

Giemsa stain powder, certified (azure B type)	0.6 g
Methanol, absolute and certified neutral, acetone free	50 ml
Glycerin, neutral, certified	50 ml

1. Grind together small portions of stain and glycerin in a mortar, and collect mixtures in a 500- or 1,000-ml flask until all measured material is mixed.
2. Stopper the flask with a cotton plug, cover the plug with heavy paper, and place the flask in a 55 to 60°C water bath for 2 h. Make sure the water in the water bath is above the level of the stain. Shake gently at 30-min intervals.
3. After grinding the powder and glycerin in the mortar, use the 50 ml of methyl alcohol to wash the last bit of stain from the mortar; then pour the alcohol into a small, airtight bottle.
4. Remove the glycerin-stain powder mixture from the water bath, and allow it to cool to room temperature. Add alcohol washing from the mortar, and shake well.
5. Before use, filter through Whatman no. 1 paper into a brown bottle. Although the stain can be used immediately, it is better to let it stand for 2 to 3 weeks with intermittent shaking.
6. Label and store protected from light; the shelf life is 36 months, provided that results are within quality control guidelines.

Note The stock stain is stable for many years; however, it must be protected from moisture. The staining reaction is oxidative; any oxygen in water will initiate the staining reaction and destroy the stock stain. This is why the aqueous working solution of stock stain is good only for 1 day.

10% Stock Solution of Triton X-100

Triton X-100	10 ml
Distilled water	90 ml

Mix thoroughly, and store at room temperature. This solution will keep indefinitely if kept tightly stoppered.

Stock Buffers

Disodium Phosphate (Dibasic)

Na_2HPO_4, anhydrous	9.5 g
Distilled water	1,000 ml

Monosodium Phosphate (Monobasic)

$NaH_2PO_4 \cdot H_2O$	9.2 g
Distilled water	1,000 ml

Phosphate-Buffered Water (Table 31.6)
Triton-Phosphate Buffer
0.01% Triton-Buffered Water

Stock 10% aqueous Triton X-100	1 ml
Phosphate buffer	1,000 ml

0.1% Triton-Buffered Water

Stock 10% aqueous Triton X-100	10 ml
Phosphate buffer	1,000 ml

For thin blood films or a combination of thin and thick blood films, use 0.01% Triton-buffered water; for thick blood films, use 0.1% Triton-buffered water.

Liquid

The commercial liquid stain or the stock solution prepared from powder should be diluted approximately the same amount to prepare the working stain solution. Stock Giemsa liquid stain is diluted 1:10 with buffer for thin blood films; for thick films, a dilution of up to 1:50 may be used. Some people prefer to stain both thick and thin smears for a longer period in a more dilute solution. Phosphate buffer used in dilution of the stock stain should be neutral or slightly alkaline. Phosphate buffer solution may be used to obtain the right pH (Table 31.6). In some laboratories, tap water has a satisfactory pH and may be used for the entire staining procedure and the final rinse. Some workers recommend using pH 6.8 to emphasize Schüffner's dots.

Table 31.6 Phosphate buffer solutions

pH	Amt (ml) of:		
	Na$_2$HPO$_4$ (9.3 g/liter, anhydrous)	NaH$_2$PO$_4$ · H$_2$O, (9.2 g/liter)	Distilled water
6.6	37.5	62.5	900
6.8	49.6	50.4	900
7.0	61.1	38.9	900
7.2	72.0	28.0	900

Procedure for Staining Thin Films

1. Fix blood films in absolute methyl alcohol (acetone free) for 1 min.
2. Allow the slides to air dry.
3. Immerse the slides in a solution of 1 part Giemsa stock (commercial liquid stain or stock prepared from powder) to 10 to 50 parts phosphate buffer (pH 7.0 to 7.2). Stain for 10 to 60 min. *Fresh working stain should be prepared from stock solution each day.*

   Note A good general rule for stain dilution versus staining time is as follows. If the dilution is 1:20, stain for 20 min; if the dilution is 1:30, stain for 30 min, etc. However, a series of stain dilutions and staining times should be tried to determine the best dilution and time for each batch of stock stain.
4. Dip the slides briefly in phosphate-buffered water, or rinse under gently running tap water.

   Note Excessive washing will decolorize the film.
5. Drain thoroughly in a vertical position, and allow to air dry. The bottom of the slide should be wiped to remove excess stain before drying.

Procedure for Staining Thick Films

The procedure to be followed for thick films is the same as for thin films, except that the first two steps are omitted. If the slide has a thick film at one end and a thin film at the other, fix only the thin portion and then stain both parts of the film simultaneously. Normally, a stain/buffer dilution of 1:50 (vol/vol) for a staining time of 50 min is recommended for thick films. The longer staining time seems to give better results than do the shorter times that can be used for the thin film.

Results

Giemsa stain colors the components of blood as follows: RBCs, pale red; nuclei of WBCs, purple with pale purple cytoplasm; eosinophilic granules, bright purple-red; and neutrophilic granules, deep pink-purple. If malaria para-

sites are present, the cytoplasm stains blue and the nuclear material stains red to purple-red. Schüffner's dots and other inclusions in the RBCs will stain red (Figures 31.4 and 31.5). Nuclear and cytoplasmic staining characteristics of the other blood parasites such as *Babesia* spp., trypanosomes, and leishmaniae are like those of the malaria parasites. While the sheath of microfilariae may not always stain with Giemsa, the nuclei within the microfilaria itself stain blue to purple.

Wright's Stain

Wright's stain is available commercially in liquid form, ready to use, and also as a powder which must be dissolved in anhydrous, acetone-free methyl alcohol before use. Directions for preparing the stain follow.

Reagent

Wright's Stain

Wright's stain..0.9 g

Methanol, absolute and certified neutral, acetone free..........................500.0 ml

1. Grind 0.9 g of Wright's stain powder with 10 to 15 ml of methanol (anhydrous, acetone free) in a clean mortar. Gradually add methanol while grinding. As the dye is dissolved in the methanol, pour that solution off and add more methanol to the mortar. Repeat this process until the entire 500 ml of methanol has been used.
2. Store the stain in a tightly stoppered glass bottle at room temperature. Shake the bottle several times daily for at least 5 days.
3. Allow the precipitate to settle, and pour off some of the supernatant fluid into dropping bottles for use. If the stock solution has been disturbed, the supernatant fluid can be filtered through Whatman no. 1 paper into a brown bottle.
4. Label and store protected from light; the shelf life is 36 months, provided that results are within quality control guidelines.

The staining procedure requires phosphate buffer solutions (see directions in the preceding discussion of Giemsa stain); the pH required for Wright's stain is 6.6 to 6.8.

Procedure for Staining Thin Films

Since Wright's stain contains alcohol, the slides do not require fixation before staining.

1. Place a slide on a rack in a horizontal, level position, and cover the surface with stain.
2. Count the number of drops of stain needed to cover the surface. Let stand for 1 to 3 min (the optimal staining time varies with each batch of stain).

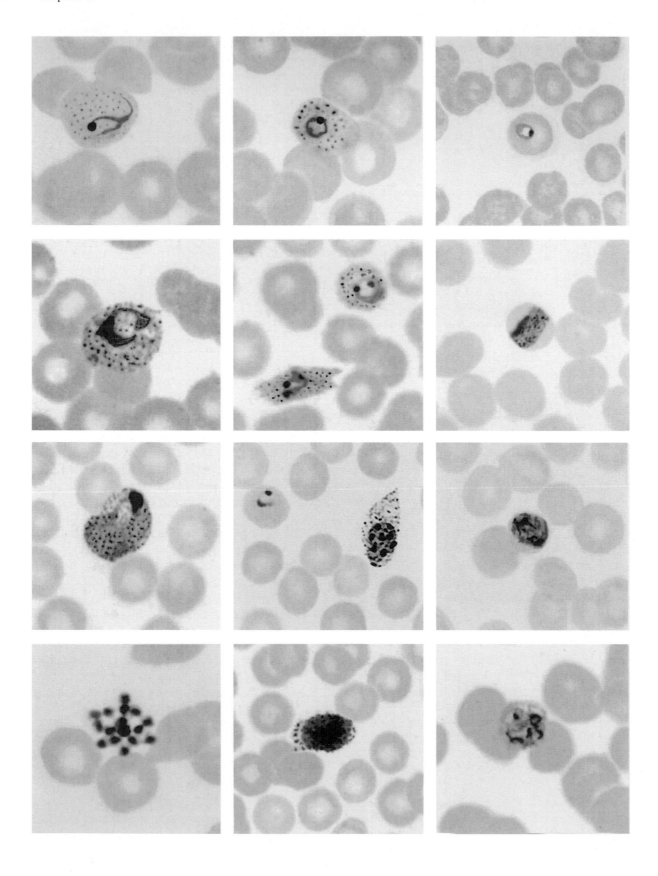

3. Add an equal number of drops of phosphate-buffered water to the slide; mix the stain and buffer by blowing on the surface of the fluid.
4. After 4 to 8 min, flood the stain from the slide with phosphate buffer. Do not pour the stain off before washing. If you do, a precipitate will be deposited on the slide.
5. Wipe the bottom of the slide to remove excess stain.
6. Allow the slide to drain and air dry.

Procedure for Staining Thick Films

Thick films stained with Wright's stain are usually inferior to those stained with Giemsa solution. Great care should also be taken to avoid excess stain precipitate on the slide during the final rinse. Before being stained, thick films must be laked in distilled water (to rupture and remove RBCs) and air dried. The staining procedure is the same as for thin films, but the staining time is usually somewhat longer and must be determined for each batch of stain.

Results

Wright's stain colors blood components as follows: RBCs, light tan, reddish, or buff; nuclei of WBCs, bright blue with contrasting light cytoplasm; eosinophilic granules, bright red; and neutrophilic granules, pink or light purple.

If malaria parasites are present, the cytoplasm stains pale blue and the nuclear material stains red. Schüffner's dots and other inclusions in the RBCs usually do not stain or stain very pale with Wright's stain. Nuclear and cytoplasmic staining characteristics of the other blood parasites such as *Babesia* spp., trypanosomes, and leishmaniae are like those of the malaria parasites (Figures 31.6 and 31.7. While the sheath of microfilariae may not always stain with Wright's stain, the nuclei within the microfilaria itself stain pale to dark blue.

General Notes on Staining Procedures

1. When large numbers of slides are being processed, remember that dry films should be stored in dust-free containers before staining, to protect fresh smears from insects.

2. If the slides cannot be stained within 48 h, thin films should be fixed in methyl alcohol and thick films should be laked in distilled water before storage. Rapid stains can also be used. Parasites will stain like the WBCs.
3. Slides should be stored at reasonably low temperatures (below 80°F [ca. 27°C]) before being stained. If the slides are exposed to high temperatures, thick films that have not been laked will become heat fixed; hemoglobin will remain fixed in the RBCs, and heavy stain retention will then prevent parasite identification. Thin films also stain poorly after exposure to high temperatures.
4. Fresh working stain should be prepared just before use. If a large number of thick films are being laked during staining, the stain should be changed after 50 slides because of the accumulation of excess hemoglobin.

Proper Examination of Thin and Thick Blood Films

Thin Blood Films

In any examination of thin blood films for parasitic organisms, the initial screen should be carried out with the low-power objective of a microscope. Microfilariae may be missed if the entire thin film is not examined. Microfilariae are rarely present in large numbers, and frequently only a few organisms occur in each thin-film preparation. Microfilariae are commonly found at the edges of the thin film or at the feathered end of the film, because they are carried to these sites during the process of spreading the blood. The feathered end of the film where the RBCs are drawn out into one single, distinctive layer of cells should be examined for the presence of malaria parasites and trypanosomes. In these areas, the morphology and size of the infected RBCs are most clearly seen.

Depending on the training and experience of the microscopist, examination of the thin film usually takes 15 to 20 min (200 to 300 oil immersion fields) at a magnification of ×1,000. Although some people use a 50× or 60× oil immersion objective to screen stained

Figure 31.4 *Plasmodium* spp. seen in stained thin blood films. (Left column from top to bottom) *Plasmodium vivax*. (1) Developing ring (note the enlarged RBC, Schüffner's dots, and ameboid ring); (2) developing trophozoite (note the enlarged RBC and Schüffner's dots); (3) large trophozoite prior to the development of schizonts (note the enlarged RBC and Schüffner's dots); (4) mature schizont with approximately 16 merozoites. (Center column from top to bottom) *Plasmodium ovale*. (1) Young ring form (note the nonameboid ring and Schüffner's dots that appear earlier than in rings of *P. vivax*); (2) developing trophozoites (note Schüffner's dots and fimbriated edges of the infected RBCs); (3) maturing schizont containing merozoites and Schüffner's dots; (4) gametocyte. (Right column from top to bottom) *Plasmodium malariae*. (1) Young ring (note the normal-sized RBC and no stippling); (2) developing trophozoite (note the normal-sized RBC and "band form" configuration of the trophozoite); (3) developing schizont (note the size of the RBC); (4) more mature schizont containing developing merozoites.

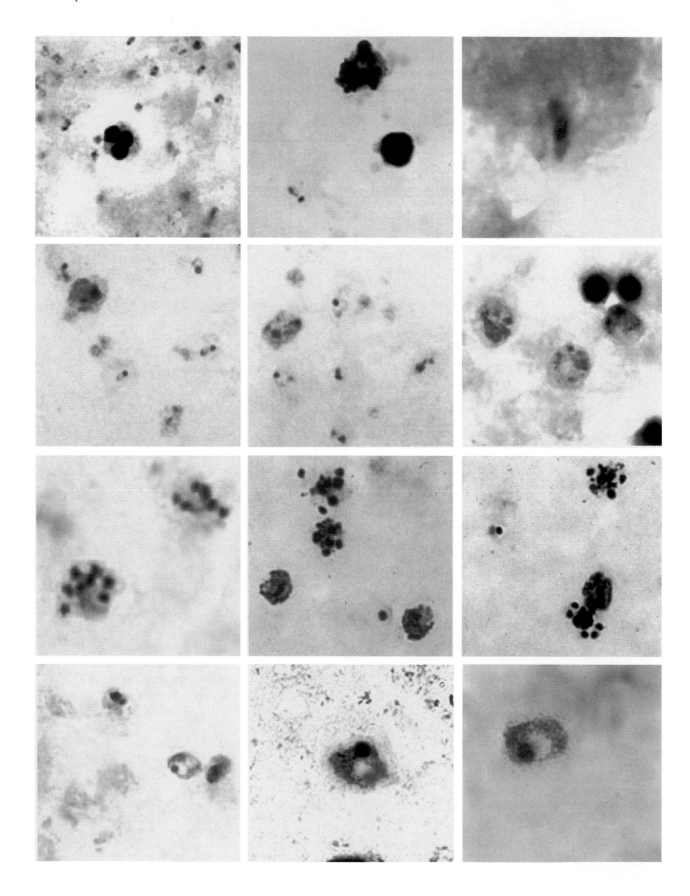

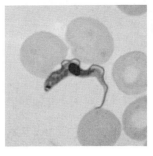
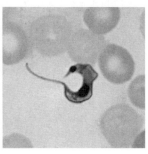

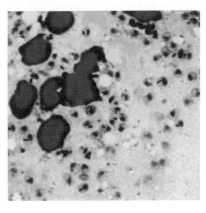

Figure 31.6 Trypomastigotes. (Left) *Trypanosoma brucei gambiense*. (Right) *Trypanosoma cruzi* (note the undulating membrane on both trypomastigotes and the larger kinetoplast in *T. cruzi*).

Figure 31.7 *Leishmania donovani* in an impression smear. Note the numerous small amastigotes containing a nucleus and bar-shaped primitive flagellum.

blood films, there is some concern that small parasites such as plasmodia, *Babesia* spp., or *L. donovani* may be missed at this smaller total magnification (×500 or ×600) compared with the ×1,000 total magnification obtained with the more traditional 100× oil immersion objective. Because people tend to scan blood films at different rates, it is important to examine a minimum number of fields, regardless of the time it takes to perform this procedure. If something suspicious has been seen in the thick film, the number of fields examined on the thin film is often considerably more than 200 to 300. The request for blood film examination should always be considered a STAT procedure, with all reports (negative as well as positive) being reported by telephone to the physician as soon as possible. Appropriate governmental agencies (local, state, and federal) should be notified within a reasonable time frame in accordance with guidelines and laws.

Diagnostic problems with the use of automated differential instruments have been reported (4, 27). Both malaria and *Babesia* infections were missed with these instruments, and therapy was delayed (Figures 31.8 and 31.9). Although these instruments are not designed to detect intracellular blood parasites, the inability of the automated systems to discriminate between uninfected RBCs and those infected with parasites may pose serious diagnostic problems.

Thick Blood Films

In the preparation of a thick blood film, the greatest concentration of blood cells will be in the center of the film. A search for parasitic organisms should be carried out initially at low magnification to detect microfilariae more readily. Examination of a thick film usually requires 5 to 10 min (approximately 100 oil immersion fields). Search for malarial organisms and trypanosomes is best done under oil immersion (total magnification, ×1,000). Intact RBCs are frequently seen at the very periphery of the thick film; such cells, if infected, may prove useful in malaria diagnosis, since they may demonstrate the characteristic morphology necessary to identify the organisms to the species level.

Determination of Parasitemia

It is important to report the level of parasitemia when blood films are reviewed and found to be positive for malaria parasites. Because of the potential for drug resistance in some of the *Plasmodium* species, particularly *P. falciparum* and *P. vivax*, it is important that every positive smear be assessed and the parasitemia reported exactly the same way on follow-up specimens as on the initial specimen. This allows the parasitemia to be monitored after therapy has been initiated. In cases where the patient is hospitalized, monitoring should be performed at 24, 48, and 72 h after initiating therapy. Generally, the parasitemia drops very quickly within the first 2 h; however, in cases of drug resistance, the level may not decrease but actually may increase over time (Figure 31.7).

Figure 31.5 *Plasmodium* spp. seen in stained thick films. (Top row from left to right) *Plasmodium falciparum*. (1) Numerous ring forms (note differences: nuclei and cytoplasm); (2) low parasitemia with a single ring form in the field; (3) crescent-shaped gametocyte (note that the RBC around the gametocyte is not visible). (Row two from left to right) *Plasmodium vivax*. (1) Ring forms and developing trophozoites; (2) some larger rings with developing trophozoite; (3) developing schizonts. (Row three from left to right) *Plasmodium malariae*. (1) Mature schizonts containing about eight merozoites (photographed at higher magnification); (2) developing schizonts and two mature schizonts; (3) single ring form and several mature schizonts. (Bottom row from left to right) *Plasmodium ovale*. (1) Developing trophozoites (note that the organisms do not appear to be ameboid like those normally seen in *P. vivax*; (2) developing trophozoite; (3) developing trophozoite (note that panels 2 and 3 are photographed at a higher magnification).

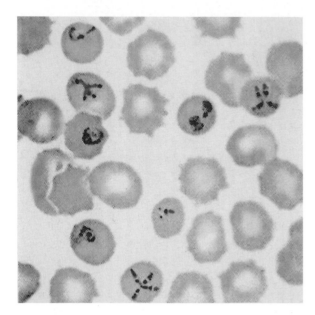

Figure 31.8 *Babesia* sp. in a thin blood film. Note the "Maltese cross" formation that is diagnostic but not always seen in blood films. The ring forms are very pleomorphic, much more so than *P. falciparum* ring forms; there tend to be numerous rings per RBC.

Malarial infections should be reported as the percentage of infected RBCs per 100 RBCs counted (0.5%, 1.0%, etc.) (11). Considering the low parasitemia frequently seen in patients within the United States, several hundred RBCs may have to be counted to arrive at an accurate count and determination of the percentage. The thin blood film must be used for this approach.

Another approach is to count the number of parasites per 100 WBCs on the smear. Either the thick or thin film can be used for this purpose. This figure can be converted to the number of parasites per microliter of blood; divide the number of parasites per 100 WBCs by 100, and multiply that figure by the number of WBCs per microliter of blood. Depending on the parasitemia, 200 or more WBCs may have to be counted, so the denominator may vary (it may be 200 or even more). In this case, blood for both the peripheral smears and cell counts must be collected at the same time.

It is critical that the same reporting method be used consistently for every subsequent set of blood films so that the parasitemia can be tracked for decrease or possible increase, indicating resistance. Also, remember that drug resistance may not become evident for a few days. The parasitemia may initially appear to decline but may then begin to increase after several days. Therefore, it is very important that patient parasitemia be monitored, particularly if an infection with *P. falciparum* has been diagnosed. Drug resistance has also been reported in *P. vivax* cases; mixed infections are also much more common than suspected.

Diagnosis of Malaria: Review of Alternatives to Conventional Microscopy

It is well known that malaria causes significant morbidity and mortality worldwide, including in countries where imported cases are seen (7, 40, 67). In many developing countries where malaria is highly endemic, diagnostic testing is often inadequate or unavailable due to a lack of trained personnel or funds or both. Although microscopic examination of Giemsa-stained thick and thin blood films remains the standard of practice, this approach is time-consuming, is based on the need for a great deal of expertise in microscopic morphology, and requires the purchase and maintenance of expensive equipment. Rapid diagnostic tests (RDTs) offer great potential to improve the diagnosis of malaria, particularly in remote areas. There are a number of new approaches to the diagnosis of malaria, including the use of fluorescent stains (QBC), dipstick antigen detection of histidine-rich protein 2 (HRP2) and parasite lactate dehydrogenase (pLDH) (ParaSight F, NOW Malaria Pf and NOW Malaria Pf/Pv, and Flow anti-pLDH *Plasmodium* monoclonal antibodies) (71), PCR, and automated blood cell analyzers. Parasitemia and its clinical correlates are given in Table 31.7, while some of the newer testing options for malaria and other blood parasites can be found in Table 31.8. Field trial data from various test methods can be seen in Table 31.9.

Histidine-rich protein 2 of *P. falciparum* (PfHRP2) is a water-soluble protein that is produced by the asexual stages and gametocytes of *P. falciparum*, expressed on the RBC membrane surface, and shown to remain in the blood for at least 28 days after the initiation of antimalarial therapy. Many RDTs are based on the detection PfHRP2, but reports from field tests have questioned their sensitivity and reliability. However, the variability in the results of PfHRP2-based RDTs may be related to the variability in the target antigen (1). This hypothesis was tested by examining the genetic diversity of PfHRP2, which includes numerous amino acid repeats, in 75 *P. falciparum* lines and isolates originating from 19 countries and testing a subset of parasites by use of two PfHRP2-based RDTs. There is extensive diversity in PfHRP2 sequences, both within and between countries. Logistic regression analysis indicated that two types of repeats were predictive of RDT detection sensitivity (87.5% accuracy), with predictions suggesting that only 84% of *P. falciparum* parasites in the Asia-Pacific region are likely to be detected at densities of ≤250 parasites/μl. Data also indicate that PfHRP3 may play a role in the performance of PfHRP2-based RDTs. These findings provide an alternative explanation for the variable sensitivity in field tests of malaria RDTs that is not due to the quality of the RDTs (1).

The persistence of parasite HRP2 in the circulation after parasite clearance has been considered a drawback

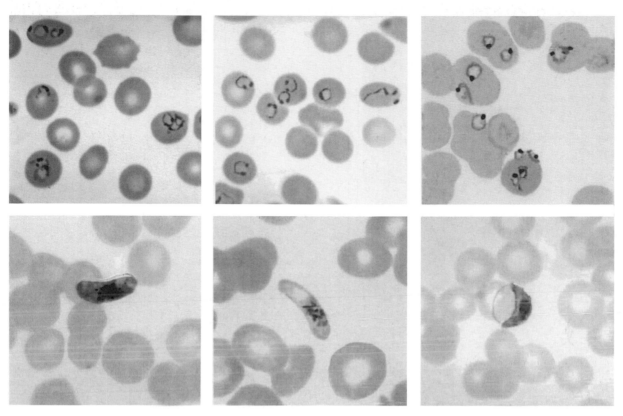

Figure 31.9 *Plasmodium falciparum.* (Upper row) Note the ring forms (multiple rings per RBC and presence of the "headphone" ring configuration). The photograph at the right is a good mimic of *Babesia* organisms, but the rings are not quite as pleomorphic as in *Babesia* spp. (Lower row) Three examples of *P. falciparum* gametocytes, two of which appear to be outside the RBC and one of which is inside.

for RDTs targeting HRP2 and a major cause of false-positive results. In one study when PCR was used as the gold standard rather than microscopy, the high rate of RDT false-positive parasitemia results in comparison with microscopy was shown to predominantly represent cases that had a parasite density below the threshold for detection by microscopy (3). Despite the generally low disease-endemic prevalence of malaria in the area, there was a high prevalence of chronic infections with low, fluctuating parasite densities that were better detected by RDT. In areas known to have low-density parasitemias, RDTs targeting HRP2 may increase the diagnostic sensitivity in comparison with microscopy. While microscopy remains the standard for comparison of the diagnostic accuracy for malaria, the limitations of microscopy, and the possibility that RDTs may have superior accuracy in some circumstances, should be taken into account when interpreting the results of diagnostic trials.

To determine the accuracy of RDTs for ruling out malaria in nonimmune travelers returning from areas where malaria is endemic, 21 studies and 5,747 individuals were surveyed from the published literature (53). Diagnostic accuracy studies of nonimmune individuals with sus-

pected malaria were included if they compared rapid tests with expert microscopic examination or PCR tests. The authors concluded that RDTs for malaria may be a useful diagnostic adjunct to microscopy in centers without major expertise in tropical medicine. Initial decisions on treatment initiation and choice of antimalarial drugs can be based on travel history and posttest probabilities after rapid testing. However, expert microscopy is still required for species identification and confirmation.

QBC Microhematocrit Centrifugation Method

Microhematocrit centrifugation with use of the QBC malaria tube (glass capillary tube and closely fitting plastic insert; CBC malaria blood tubes [Becton Dickinson, Tropical Disease Diagnostics, Sparks, Md.]) has been used for the detection of blood parasites (9, 10, 50). At the end of centrifugation of 50 to 60 μl of capillary or venous blood (5 min in a QBC centrifuge; 14,387 × g), parasites or erythrocytes containing parasites are concentrated into a small, 1- to 2-mm region near the top of the RBC column and are held close to the wall of the tube by the plastic float, thereby making them readily visible by mi-

Table 31.7 Parasitemia determined from conventional light microscopy: clinical correlation[a]

Parasitemia (%)	No. of parasites/µl	Clinical correlation
0.0001–0.0004	5–20	Number of organisms that are required for a positive thick film (sensitivity)
		Examination of 100 thick-blood-film fields (0.25 µl) may miss up to 20% of infections (sensitivity, 80–90%); at least 300 fields should be examined before reporting a negative result
		Examination of 100 thin-blood-film fields (0.005 µl); at least 300 should be examined before reporting a negative result; **both** thick and thin blood films should be examined for every specimen submitted for a suspect malaria case
		One set (thick plus thin blood films) of negative blood films does not rule out a malaria infection
0.002	100	Patients may be symptomatic below this level
0.2	10,000	Level above which immune patients will exhibit symptoms
2	100,000	Maximum parasitemia of *P. vivax* and *P. ovale* (which infect young RBCs only)
2–5	100,000–250,000	Hyperparasitemia, severe malaria[b]; increased mortality
10	500,000	Exchange transfusion may be considered; high mortality

[a] Adapted from reference 30.

[b] World Health Organization criteria for severe malaria are parasitemia of >10,000/µl and severe anemia (hemoglobin, <5 g/liter). Prognosis is poor if >20% of parasites are pigment-containing trophozoites and schizonts and/or if >5% of neutrophils contain visible pigment.

croscopy. Tubes precoated with acridine orange provide a stain which induces fluorescence in the parasites. This method automatically prepares a concentrated smear, which represents the distance between the float and the walls of the tube. Once the tube is placed into the plastic holder (Paraviewer) and immersion oil is applied onto the top of the hematocrit tube (no coverslip is necessary), the tube is examined with a 40× to 60× oil immersion objective (there must be a working distance of 0.3 mm or greater) (Figure 31.10).

Note Although a malaria infection could be detected by this method (which is much more sensitive than the thick or thin blood smear), *appropriate thick and thin blood films must be examined to accurately identify the species of the organism causing the infection.*

The range of sensitivities for the QBC test has been as low as 55% to >90% compared with microscopy. While most workers think the method is rapid, some think there are some disadvantages such as the high cost of capillary tubes and equipment, problems in species identification and quantification, technical problems, broken capillaries, and the fact that the capillaries cannot be stored for later reference. This may not be the most appropriate approach for a laboratory that receives a small number of specimens for malaria testing, but it could be helpful in other settings. In a recent study of 745 specimens from an area of low endemicity along the Colombian Pacific coast, the agreement between the QBC method and thick blood smear was reported to be 99.5% (6). An excellent review of this and other methods is provided in reference 30.

This method has been reported to have a high degree of sensitivity in the detection of cases of human filariasis (76). In one study evaluating the technique in the detection of canine *Dirofilaria immitis* as a model for human filariasis, the QBC analysis was more sensitive (55%) than the thick blood smear (39%) and was more efficient (79% versus 72%). However, accurate identification to the species level was impossible using the QBC method.

Diagnosis of tick-borne relapsing fever has also been reported using the QBC system with a high level of sensitivity, more so than the thick blood smear. This method seems to be useful for this purpose and might be considered in the management of fever in travelers returning from tropical regions (8, 74).

ParaSight F Test

A *P. falciparum* antigen detection system, the ParaSight F test (Becton Dickinson), is very effective in field trials (20, 64). This procedure is based on an antigen capture approach and has been incorporated in a dipstick format; the entire test takes approximately 10 min (Figures 31.11 and 31.12) (64). The overall performance of the ParaSight F test was reviewed in 1995 at a World Health Organization meeting looking at 15 studies with a total of 7,926 assays for the detection of *P. falciparum*. Overall, the sensitivity ranged from 84 to 94%, with specificities ranging from 81 to 99% (78). However, the test detects only *P. falciparum*, and at low parasitemias (<100/µl) the sensitivity drops to 11 to 40%. Also, since HRP2 is not present in mature gametocytes, cases where only

Table 31.8 Summary of commercially available kits for immunodetection of blood parasites, antigens, or antibodies

Organism and kit name	Manufacturer and/or distributor[a]	Type of test[b]	Sensitivity, specificity (%)	Comments[b]
Protozoa				
Plasmodium spp.				
NOW malaria test	Binax, Inc.	Rapid	*P. falciparum*, 100, 96	Antigen to *P. falciparum*
			P. vivax, 89, 98	Antigen to *P. vivax*
ParaSight F	Becton Dickinson	Rapid, HRP2	87–97, 81–99	Antigen to *P. falciparum*
Malaria antigen CELISA	Cellabs	EIA		Antigen to *P. falciparum*
Panmalaria antibody CELISA	Cellabs	EIA	83, 85	Antibody (IgG to all 4 species)
Rapimal dipstick	Cellabs	Rapid, HRP2		Antigen to *P. falciparum*
Rapimal cassette	Cellabs	Rapid, HRP2		Antigen to *P. falciparum*
Anti-pLDH *Plasmodium* monoclonal antibodies	Flow	Rapid, LDH	*P. falciparum*, 95, 100	Antigen to *P. falciparum*
			P. vivax, 93–95, 100	Antigen to *P. vivax*
Malaria Pf rapid test device	Acon	Rapid		Antigen to *P. falciparum*
Paracheck Pf dipstick, cartridge	Orchid Biomedical Systems	Rapid, HRP2	99, 100	Antigen to *P. falciparum*
Parahit-f	Span Diagnostics, Ltd.	Rapid	96–98, 99–100	Antigen to *P. falciparum*
Leishmania (visceral)				
Leishmania RapiDip InstaTest	Cortez Diagnostics	Rapid	100, NA[c]	Antibody—visceral
	Immunetics	IB		Antibody
Kalazar Detect	InBios	EIA, rapid		Antibody
Leishmania ELISA	IVD Research	EIA		Antibody
Trypanosoma cruzi (Chagas' disease)				
T. cruzi IgG CELISA	Cellabs	EIA	100, 100	Antibody
Chagas' EIA	Chemicon Intl.	EIA		Antibody
Chagas'	Hemagen	EIA	100, 100	Antibody
Trypanosoma Detect	InBios	EIA, rapid	100, 100	Antibody
Chagas' ELISA	IVD Research	EIA	100, 93	Antibody
Helminths				
Wuchereria bancrofti				
NOW Filariasis	Binax	Rapid	100, 96	Antigen
Filariasis CELISA	Cellabs	EIA		Antigen
Filariasis	TropBio	EIA		Antigen

[a] Acon Laboratories, 115 Research Dr., Bethlehem, PA 18015; Becton Dickinson Diagnostic Systems, 7 Loveton Circle, Sparks, MD 21152; Binax, 217 Read St., Portland, ME 04103; Cellabs, P.O. Box 421, Brookvale, NSW 2100, Australia; Chemicon, 28835 Single Oak Dr., Temecula, CA 92590; Cortez Diagnostics, 23961 Craftsman Road, Calabasas, CA 91302; Hemagen, 9033 Red Branch Road, Columbia, MD 21045; Flow, Inc., 6127 SW Corbett, Portland, OR 97201 (http://www.malariatest.com); Immunetics, 380 Green St., Cambridge, MA 02139; InBios, 562 1st Avenue South, Suite 600, Seattle, WA 98104; IVD Research, 5909 Sea Lion Place, Suite D, Carlsbad, CA 92008; Orchid Biomedical Systems, Plot No. 88/89, Phase II C, Verna Industrial Estate, Verna, Goa 403 722, India; Span Diagnostics Ltd., 173-B, New Industrial Estate, Udhna, Surat 394 210, India; TropBio Pty Ltd., James Cook University, Townsville, Queensland 4811, Australia.

[b] EIA, enzyme immunoassay; IgG, immunoglobulin G; IB, immunoblot.

[c] NA, not available.

gametocytes were present could be missed. There is also a possibility that some false-negative results, even at high parasitemias, are due to the lack of the HRP2 gene (72). After successful therapy, many patients continue to have circulating HRP2 antigen for 7 to 14 days after microscopic and clinical cure. Also, as many as 60% of patients positive for rheumatoid factor have a false-positive test

(34, 47). The ParaSight F test may be used in situations where no trained microscopists are available or where malaria is strongly suspected and the microscopy results are negative (5, 14, 68). This kit has proven to be very useful in many areas of the world.

In one systematic review and meta-analysis of controlled studies evaluating the diagnostic accuracy of the

Table 31.9 Field trials of antigen-detection tests for malaria

Test and country	Standard[a]	No. of samples	Sensitivity (%)	Specificity (%)	Reference	Comments, PPV, NPV (%)[a]
ParaSight F (HRP2)						
Eastern Africa	TBF	213	90–92	99	49	Lower sensitivities at ≤320/µl; high cost
Uganda	TBF	1,326	99.6 (>500/µl), 98.6 (>50/µl), 22.2 (<10/µl)	86.2	43	PPV ≥ 80; NPV ≥ 90
Germany	TBF + PCR	53	92.5	98.3	37	For nonspecialized laboratories
Canada	TBF + PCR	148	96	95	61	Similar results to ICT Malaria
Zimbabwe	TBF	123	93.9	81.2	58	Reduces unnecessary treatment
Belgium	TBF		95 (*P. falciparum*)	90 (*P. falciparum*)	75	TBF more sensitive
India	TBF	93	100	100	41	
United States	TBF + PCR	151	88 (PCR) 88 (microscopy)	97 (PCR) 83 (microscopy)	32	Ag present in 68% of samples after 7 days and in 27% on day 28
ICT Malaria Pf (HRP2)						
Honduras	TBF + PCR		100 (590/mm³)	95	63	Species identification not possible due to panspecific band
India	TBF		97	100	57	Ag present 5–15 days after therapy
Cameroon	TBF		89.1		48	1/5 pregnant women are smear negative; 94% detection using microscopy and ICT Pf
Canada	TBF + PCR	148	94	97	61	False-negative results at <100/µl
Zimbabwe	TBF	123	100	75	58	Circulating Ag after 2 wk of therapy
Belgium	TBF		95	89	75	TBF more sensitive
India	TBF	173	98.6	97.1	73	No cross-reactivity with *P. vivax*
ICT Malaria Pf/Pv (HRP2)						
Indonesia	TBF	560	95.5 (*P. falciparum*) 75 (*P. vivax*)	89.8 (*P. falciparum*) 94.8 (*P. vivax*)	71	PPV = 88.1, NPV = 96.2 (*P. falciparum*); PPV = 50, NPV = 98.2 (*P. vivax*); 96% sensitivity at >500/µl
NOW malaria test						
France	QBC, ThinBF	399	96.4 (*P. falciparum*)	97 (*P. falciparum*)	21	NPV = 99.4, perform along with blood films
Canada	TBF + PCR	256	94 (*P. falciparum*) 84 (non-*P. falciparum*) 87 (*P. vivax*) 62 (*P. ovale* and *P. malariae*)	99 (overall)	22	Useful adjunct in febrile returning travelers
Thailand	TBF, ThinBF	246	100 (*P. falciparum*) 87.3 (*P. vivax*)	96.2 (*P. falciparum*) 97.7 (*P. vivax*) 100 (non-*P. falciparum*)	77	Improved performance over NOW ICT predecessors
Anti-pLDH monoclonal antibodies (pLDH)						
Areas of endemicity	TBF	125	97 (>100/µl) 59 (<100/µl) 39 (<50/µl)		33	Comparable to microscopy at higher levels, poor at lower parasitemias; high cost a problem

(continued)

Table 31.9 *(continued)*

Test and country	Standard[a]	No. of samples	Sensitivity (%)	Specificity (%)	Reference	Comments, PPV, NPV (%)[a]
Honduras	TBF	96	94 (*P. vivax*) 88 (*P. falciparum*)	100 (*P. vivax*) 99 (*P. falciparum*)	59	Organisms missed at <100/µl; performed better than ParaSight F and ICT Malaria
Honduras	TBF + PCR		100	95	63	Species identification still a problem
The Gambia	TBF		91.3	92	13	PPV = 87.2; NPV = 94.7; sensitivity drop at <0.01%
Germany	TBF + PCR	53	88.5	99.4	37	For nonspecialized laboratories
Mexico	TBF	893	93.3 (*P. vivax*)	99.5 (*P. vivax*)	28	PPV = 96.5; NPV = 98.9
Pakistan	TBF, ThinBF	215	94.5 (*P. falciparum*) 95 (*P. vivax*)	100 (*P. falciparum*) 100 (*P. vivax*)	42	Excellent correlation with microscopy

a TBF, thick blood film; ThinBF, thin blood film; PPV, positive predictive value; NPV, negative predictive value; Ag, antigen.

ParaSight F test in comparison with light microscopy, data sources included 15,359 subjects (4,119 with *P. falciparum* malaria) in 32 studies reported in 29 publications. Overall, the ParaSight F test demonstrated 90% sensitiv- ity and 94% specificity. Both the sensitivity and specificity were significantly higher in the nonresident population than in the resident population. The posttest probability indicates that in settings of low malaria prevalence a

Figure 31.10 The QBC Malaria Test components include the ParaLens UV microscope adapter (blue-violet light module providing fiber-optic illumination with AC outlet or bulb illumination) with rechargeable battery for mobile use (attachable to any conventional microscope), the ParaFuge battery-powered centrifuge, and the ParaViewer microscope tube holder, a specially designed QBC tube viewing block that accepts standard microscope oil.

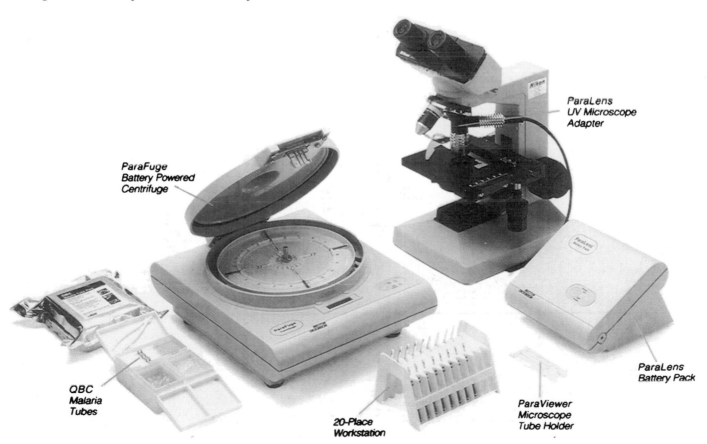

ParaSightF

- **TEST-PRINCIPLE:**
  Detection of HRP2
- **SPECIES IDENTIFIED:**
  P. falciparum
- **PRICE PER TEST:**
  depending on local
  distributers (≅US$3–5)

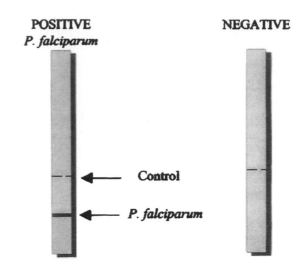

Figure 31.11   Diagram of the ParaSight F test format. (Adapted from reference 30 with permission.)

negative test almost absolutely excludes infection, while in settings of high prevalence the same result still gives a substantial chance of infection being present. The authors conclude that the ParaSight F test is a simple and accurate test for the diagnosis of *P. falciparum* infection. The test could be of particular value in the diagnosis of malaria in travelers returning from areas of endemic infection (16).

NOW Malaria Test

The ICT Malaria P.f. test (ICT Diagnostics, Brookvale, New South Wales, Australia) was the first malaria rapid diagnostic device designed for convenience of use in a booklet format with the test strip mounted on cardboard. Like other nonmicroscopic malaria rapid tests, ICT

Figure 31.12   ParaSight F test showing (from left to right) the positive test strip with a reagent control mark above the positive test result and a negative test strip with the reagent control mark.

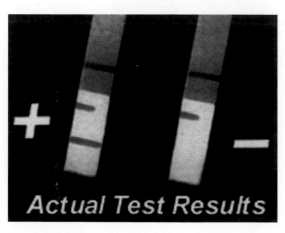

Malaria P.f. was an immunochromatographic assay. The original ICT assay detected only *P. falciparum* HRP2. By 1999, AMRAD (French's Forest, New South Wales, Australia) acquired ICT Diagnostics and continued manufacturing the product. At about the same time, the product was enhanced by adding the capability to detect non-*falciparum* malaria. The refined assay was renamed ICT Malaria P.f./P.v. This improvement was achieved by using monoclonal antibodies to capture *Plasmodium* aldolase, in addition to the HRP2 test. *Plasmodium* aldolase is an enzyme of the parasite glycolytic pathway expressed by the blood stages of *P. falciparum* as well as the non-*falciparum* malaria parasites. Monoclonal antibodies against *Plasmodium* aldolase are panspecific in their reaction and have been used in a combined *P. falciparum*-*P. vivax* immunochromatographic test that targets the panmalarial antigen along with PfHRP2.

In July 2000, AMRAD ceased the production of ICT and sold its ICT division to Binax, Inc. (Portland, Maine), where further developmental work was done to refine the test. The new ICT test was released under the name NOW ICT Malaria P.f./P.v. for Whole Blood and is presently called the NOW malaria test for whole blood (Binax, Inc.). The NOW malaria test uses an immunochromatographic format (Figure 31.13).

Prior to the development of the improved product (NOW malaria test), in a number of studies testing over 1,300 assays, the overall sensitivity ranged from 80 to 100% (25). Although some of the same issues apply to this test as to the ParaSight F test, the false-positive rate from the presence of rheumatoid factor appears to be less of an issue (29, 61).

In many areas of the world where both *P. falciparum* and *P. vivax* occur, rapid diagnostic tests for malaria

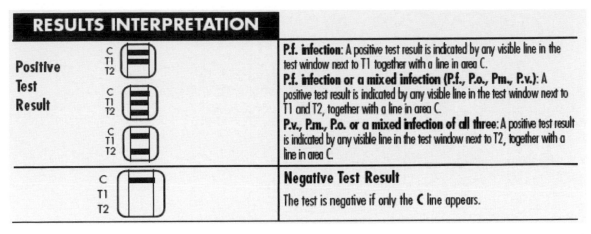

Figure 31.13 Diagram of the NOW malaria test format. (Adapted from package insert, Binax, Portland, Maine.)

must be able to differentiate the two species. It is also interesting that when a mixed-species infection with both *P. falciparum* and *P. vivax* is misdiagnosed as a single-species *P. vivax* infection, treatment for *P. vivax* can lead to a surge in *P. falciparum* parasitemia (54). The combined *P. falciparum-P. vivax* immunochromatographic test (NOW ICT Malaria Pf/Pv) was used in Indonesia, where both species occur (71). Blinded microscopy was used as the gold standard, with all discordant and 20% of concordant results cross-checked blindly. Of those with a presumptive clinical diagnosis of malaria, only 50% were parasitemic. The NOW ICT Malaria Pf/Pv test was sensitive (95.5%) and specific (89.8%) for the diagnosis of *P. falciparum* malaria, with a positive predictive value of 88.1% and a negative predictive value of 96.2%. Although the specificity and negative predictive value for the diagnosis of *P. vivax* malaria were 94.8 and 98.2%, respectively, the overall sensitivity of 75% and positive predictive value of 50% were lower than desired. With parasitemias of >500/µl, the sensitivity for the diagnosis of *P. vivax* malaria was 96%; with parasitemias of <500/µl, it was only 29%. However, by using this test, under-treatment rates would be reduced from 14.7 to 3.6% with a modest increase in the rate of overtreatment of microscopy-negative patients from 7.1 to 15.4%. Unfortunately, cost remains an obstacle to widespread implementation.

In a study using the improved product (NOW malaria test), the sensitivity for *P. falciparum* was 100% and the specificity was 96%; the sensitivity for *P. vivax* was 89%, and the specificity was 98%. Testing was performed using *P. falciparum*- and *P. vivax*-positive specimens. These results suggest improved performance over NOW ICT predecessors (Figure 31.13) (77). In another study, the NOW ICT test showed a sensitivity of 94% for the detection of *P. falciparum* malaria (96% for pure

P. falciparum infection) and 84% for non-*P. falciparum* infections (87% for pure *P. vivax* infections and 62% for pure *P. ovale* and *P. malariae* infections) compared with PCR, with an overall specificity of 99% (22). The Binax NOW ICT may represent a useful adjunct for the diagnosis of *P. falciparum* and *P. vivax* malaria in febrile returned travelers. However, the rapid test should be performed in association with more traditional methods such as examination of thick and thin blood films (2, 21).

Flow Anti-pLDH *Plasmodium* Monoclonal Antibodies

pLDH is a soluble glycolytic enzyme produced by the asexual and sexual stages of the live parasites and is present in and released from the parasite-infected erythrocytes. It has been found in all four human malaria species, and different isomers of pLDH for each of the four species exist. With pLDH as the target, a quantitative immunocapture assay, a qualitative immunochromatographic dipstick assay using monoclonal antibodies, an immunodot assay, and a dipstick assay using polyclonal antibodies have been developed.

Flow anti-pLDH *Plasmodium* monoclonal antibodies detect a malaria pLDH antigen by using an antibody incorporated into the dipstick format (62) (Figure 31.14). The test is positive only when viable parasites are present. Although the test has both a *P. falciparum*-specific and a panspecific antibody against all four species, the panspecific antibody detects only *P. vivax* with any degree of consistency (51, 59). The sensitivity and specificity of the Flow monoclonal antibodies are comparable to those of microscopy in detecting malaria infections at a parasitemia of >100/µl; however, the test failed to identify more than half of the patients with a parasitemia of <50/µl. In one study, the sensitivity of the Flow monoclonal

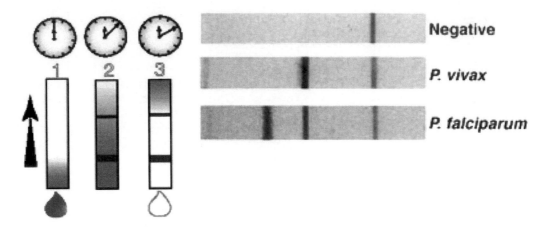

Figure 31.14 Flow anti-pLDH *Plasmodium* monoclonal antibodies rapid malaria test. (Left) Diagram of blood flow on the membrane through the timed test; note the positive control line and positive test line. (Right) Diagram of negative control, a positive *Plasmodium vivax* result, and a positive *P. falciparum* result. Note that the line for *P. vivax* is a panspecific antibody against all four species; however, the panspecific antibody has been shown to detect only *P. vivax* with any degree of consistency. (Photographs adapted from Flow's website, with permission.)

antibodies was 97% at a high parasitemia (>100/µl) but fell to 39% at <50/µl (33). Therefore, the test is similar to the HRP2 assays and should not replace conventional microscopy in the diagnosis of malaria infection (38). However, a definite advantage is the ability to confirm a cure, since the test detects only viable organisms. Also, the number of false-positive results due to rheumatoid factor is much smaller than for some of the other tests. In a comparison of results with the ICT Malaria Pf and ParaSight F tests, Flow monoclonal antibodies correctly identified *P. falciparum* malaria in patient blood samples more often than the other two procedures did. In this study, the Flow monoclonal antibodies exhibited 94% sensitivity and 100% specificity for *P. vivax* and 88% sensitivity and 99% specificity for *P. falciparum* (59). In another study, compared with PCR, the sensitivity was 93% and the specificity was 99.5%. The authors felt that the test had sufficient sensitivity and specificity to detect *P. vivax* under laboratory conditions and could also be useful for malaria diagnosis in the field in Mexico (28). As with the other rapid malaria tests, cost is always an issue; however, Flow monoclonal antibodies provide a simple, rapid, and effective test in the diagnosis of malaria, especially where well-trained microscopists are not available or the work load is too high (42).

PCR

Although the first PCR methods were developed to identify *P. falciparum*, many can now detect several or all four species. It has been well demonstrated that PCR can detect lower parasitemias than any of the traditional blood film or nontraditional dipstick methods, and it may be preferable to microscopy as the reference standard when evaluating new diagnostic tests (6, 30, 32, 33, 39, 61). One advantage of PCR is the ability to confirm false-negative results by microscopy as true positives. Another advantage is the enhanced ability to identify organisms to the species level, particularly in mixed infections, compared with microscopy. However, even PCR may miss some cases as a result of PCR inhibitors, DNA degradation, or genotypic variants. Unfortunately, the time required for PCR presents a problem for the clinical laboratory; this approach does not lend itself to routine use. With the development of automation for PCR, these tests may become much more user friendly for the routine laboratory; however, currently PCR is reserved for reference centers and special circumstances. As PCR becomes more widely used, it will be important to remember that the mode of collection and storage of blood samples may influence the sensitivity of *Plasmodium* detection. This may be critical in studies of individuals with low parasitemia or mixed infections and in comparison of data from different settings, including field settings (23).

A more recent development is the post-PCR/ligase detection reaction-fluorescent microsphere assay (LDR-FMA) (56). This assay, which uses Luminex FlexMAP microspheres, provides simultaneous, semiquantitative detection of infection by all four human malaria parasite species at a sensitivity and specificity equal to those of other PCR-based assays. In blinded studies using *P. falciparum*-infected blood from in vitro cultures, the authors identified infected and uninfected samples with 100% concordance. Also, in analyses of *P. falciparum* in vitro cultures and *P. vivax*-infected monkeys, comparisons between parasitemia and LDR-FMA signal intensity showed very strong positive correlations ($r >$ 0.95). Application of this multiplex *Plasmodium* LDR-FMA diagnostic assay will increase the speed, accuracy, and reliability of diagnosing human *Plasmodium* infections.

Automated Blood Cell Analyzers

Although the use of automated blood cell analyzers is not yet clinically relevant for the diagnosis of blood parasites, improvements in the systems may offer future information that will supplement the routine microscopy procedures currently in use (44, 70). Unfortunately, when the parasitemia is light, automation tends to have some of the same problems seen with other alternative procedures. In many cases, the changes seen using analyzers are not specific to malaria infection and could occur in many other diseases (increases in the number of large, unstained cells and thrombocytopenia).

The Cell-Dyn 4000 automated hematology analyzer (Abbott, Chicago, Ill.) has the ability to detect 91.2% of malaria patients. In one study on day 3 of follow-up posttreatment, the sensitivity was 96.7% that of microscopy. The atypical polarizing events, which indicate the presence of malarial parasites in the analyzer, were highly correlated with the levels of parasitemia in serially diluted samples of the leukocyte-depleted blood; parasites were detected down to the level of 288 $\pm$ 17.7/µl). These data suggest that the atypical light depolarization could be influenced by parasitemia and could be used as a screening method for *P. vivax* malaria patients, as well as for therapeutic monitoring (70). Microscopy is still required for species determination and parasite quantitation.

The potential of automated depolarization analysis in detecting malaria infection as part of the routine full blood count performed by the Cell-Dyn 4000 analyzer has been described previously (70). In these cases, abnormal depolarizing patterns are due to the presence of leukocyte-associated malaria hemozoin, a pigment which depolarizes the laser light. Abnormal polarizing events have also been described for samples from three individual patients infected by the nematode *Mansonella perstans*. The observed depolarizing pattern consisted of a normal depolarizing eosinophil population plus an abnormal depolarizing population that showed a close "linear" relationship between "granularity" (90° depolarization) and "lobularity" (90° polarization). This atypical population was smaller than that of normal leukocytes and thus clearly different from the patterns associated with malaria infection. Abnormal depolarization patterns of *M. perstans* clearly do not reflect leukocyte-associated malaria hemozoin. It is possible, however, that the erythrocyte-lysing agent used to facilitate leukocyte analysis by the instrument may have caused microfilaria fragmentation and thus the distinctive straight-line features of the abnormal scatter plots (15).

Diagnosis of Leishmaniasis: Review of Alternatives to Conventional Microscopy

ICT for Detection of Anti-rK-39 Antibodies

In a recent study, the diagnostic utility of an immunochromatographic test for detection of anti-rK-39 antibodies for the diagnosis of kala azar and post-kala azar dermal leishmaniasis (PKDL) was evaluated (55). Of the 120 samples tested, 57 were positive by ICT; 51 of these were diagnosed as kala azar, and 6 were diagnosed as PKDL. The controls included individuals from areas of endemic (50) and nonendemic (19) infection with malignancies, hemolytic disorders, chronic liver disease, hypersplenism, portal hypertension, metabolic disorders, or sarcoidosis. In addition, 47 sera from patients with confirmed cases of tuberculosis, malaria, typhoid, filariasis, leptospirosis, histoplasmosis, toxoplasmosis, invasive aspergillosis, amebic liver abscess, AIDS, leprosy, cryptococcosis, strongyloidiasis, and cyclosporosis, as well as from patients with collagen vascular diseases and patients with hypergammaglobulinemia, were tested to check the specificity of the test. Of the 51 patients with kala azar, all had fever lasting from <1 month to 1.5 years (median, 4.5 months). All six PKDL patients gave a history of having kala azar in the past, and their slit skin test smears were microscopically positive for Leishman-Donovan bodies. The strip test was positive in all the cases of kala azar and PKDL (estimated sensitivity, 100%), and all control sera were negative by the ICT (specificity, 100%). The rK-39 ICT is a highly sensitive and specific test and may be suitable for a rapid, cost-effective, and reliable field diagnosis of kala azar and PKDL (55).

Concentration Procedures

Cytocentrifugation Technique

Cytocentrifugation (cytospin), which uses an apparatus for concentrating cells in suspension on a microscope slide, is commonly used in most histopathology laboratories. The use of a hemolyzing and isotonic saponin solution to lyse RBCs and platelets, which contains formalin as a fixative, has led to an improved technique for the detection of *Plasmodium* spp., *L. donovani*, microfilariae, and WBCs containing malaria pigment. The concentration of the parasites present in the sediment from 100 µl of blood spread on a 6-mm-diameter circle results in good morphology that is well stained using Giemsa, Wright's, or rapid stains. This new method costs very little to perform and offers the possibility of isolating and identifying the main blood-stage parasites in the same sediment. The possible exception would be young trophozoites of *P. falciparum*, which do not concentrate well due to their small size (60).

Knott Concentration Procedure

The Knott concentration procedure is used primarily to detect the presence of microfilariae in the blood, especially when a light infection is suspected (35, 45) (Figure 31.15). The disadvantage of the procedure is that the microfilariae are killed by the formalin and are therefore not seen as motile organisms.

1. Place 1 ml of whole or citrated blood into a centrifuge tube containing 10 ml of 2% formalin. Thoroughly mix the tube contents.
2. Centrifuge for 2 min at 500 × *g* or 5 min at 300 × *g*.
3. Decant the supernatant fluid, examine some sediment as a wet mount at low (×100) and high dry (×400) power, prepare thick films from the remaining sediment, and allow the films to air dry.
4. Stain films with Giemsa, Wright's, or rapid stains.

Note Use alcohol-cleaned slides for preparation of the films made from the sediment.

Figure 31.15 *Loa loa* microfilaria at low power from a Knott concentration stained with Giemsa stain. Note that the sheath is not visible.

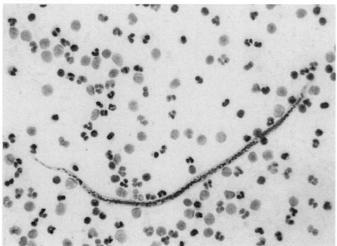

Membrane Filtration Technique

The membrane filtration technique as modified by Desowitz and others has proved highly efficient in demonstrating filarial infections when microfilaremias are of low density. It has also been successfully used in field surveys (11, 17–19, 35).

1. Draw 1 ml of fresh whole blood or anticoagulated blood into a 15-ml syringe containing 10 ml of distilled water.
2. Gently shake the mixture for 2 to 3 min to ensure that all blood cells are lysed.
3. Place a 25-mm Nuclepore filter (5-μm porosity) over a moist 25-mm filter paper pad. (This method is unsatisfactory for the isolation of *M. perstans* microfilariae because of their small size. A 3-μm-pore-size filter could be used for recovery of this organism. Other filters of similar pore size are not as satisfactory as the Nuclepore filter.) Place the filter in a Swinney filter adapter.
4. Attach the Swinney filter adapter to the syringe containing the lysed blood.
5. With gentle but steady pressure on the piston, push the lysed blood through the filter.
6. Without disturbing the filter, remove the Swinney adapter from the syringe and draw approximately 10 ml of distilled water into the syringe. Replace the adapter, and gently push the water through the filter to wash the debris from the filter.
7. Remove the adapter again, draw the piston of the syringe to about half the length of the barrel, replace the adapter, and push the air in the barrel through the filter to expel excess water.
8. To prepare the filter for staining, remove the adapter, draw the piston about half the length of the barrel, and then draw 3 ml of absolute methanol into the syringe. Holding the syringe vertically, replace the adapter and push the methanol followed by the air through the filter to fix the microfilariae and expel the excess methanol.
9. To stain, remove the filter from the adapter, place it on a slide, and allow it to air dry thoroughly. Stain with Giemsa stain as for a thick film (with 0.1% Triton X-100) or with Delafield's hematoxylin.
10. To cover the stained filter, dip the slide in toluene before mounting the filter with neutral mounting medium and a coverslip. This will lessen the formation of bubbles in or under the filter.

Gradient Centrifugation Technique

The gradient centrifugation technique is another technique for the concentration of microfilariae (52).

1. Mix 30 ml of 50% Hypaque with 14 ml of distilled water; add 1 part of this mixture to 2.4 parts of 9% Ficoll.
2. Place 4 ml of the Ficoll-Hypaque mixture in a 15-ml plastic centrifuge tube; overlay this mixture with 4 ml of heparinized venous blood.
3. Centrifuge the tube at 400 × g for 40 min.
4. Microfilariae will be found in the middle Ficoll-Hypaque layer, which separates the overlying plasma and WBC layers from the underlying RBCs.

Triple-Centrifugation Method for Trypanosomes

The triple-centrifugation procedure may be valuable in demonstrating the presence of trypanosomes in the peripheral blood when the parasitemia is light (26, 35).

1. Centrifuge anticoagulated blood for 10 min at 300 × g.
2. Remove the supernatant fluid, and transfer it to another centrifuge tube.
3. Centrifuge this fluid for 10 min at 500 × g.
4. Again, remove the supernatant fluid to another centrifuge tube.
5. Centrifuge this fluid for 10 min at 900 × g.
6. Decant the supernatant fluid, and examine the sediment as a wet preparation.
7. The sediment may be used to prepare thin films that can then be stained with any of the blood stains.

Note Remember, you are saving the supernatant fluid in step 2 for subsequent centrifugation; do not accidentally pour this fluid off for disposal. The final sediment after the third centrifugation step is used to prepare stained films for examination.

Special Stain for Microfilarial Sheath

Delafield's Hematoxylin

Some of the material that is obtained from the concentration procedures can be allowed to dry as thick and thin films and then stained with Delafield's hematoxylin, which demonstrates greater nuclear detail as well as the microfilarial sheath, if present. In addition, fresh thick films of blood containing microfilariae can be stained by this hematoxylin technique (Figures 31.16 and 31.17) (26, 35).

Preparation of Stain

Dissolve 180 g of aluminum ammonium sulfate in 1 liter of distilled water (saturated solution). Heat until dissolved. The cooled supernatant fluid is saturated aluminum alum.

Hematoxylin Solution

Hematoxylin crystals .... 4 g
95% ethyl alcohol .... 25 ml
Ammonium alum
   (saturated solution) .... 400 ml

Dissolve the hematoxylin in alcohol, and add it to the alum solution. Expose the solution to sunlight and air for ripening in a clear, cotton-plugged bottle for approximately 1 week; then filter and add the following:

Glycerin .... 100 ml
95% ethyl alcohol .... 100 ml

Age for 1 month or longer in sunlight, and then run test smears to determine whether the solution is properly aged. Nuclei should stain blue, and cytoplasm should stain different shades of red.

Procedure

1. Prepare thick films from the concentration material or from fresh blood.
2. Allow the films to air dry.

Figure 31.16 *Brugia malayi* microfilaria on a thick film stained with Delafield's hematoxylin. Note the presence of the sheath.

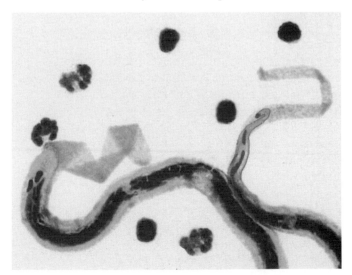

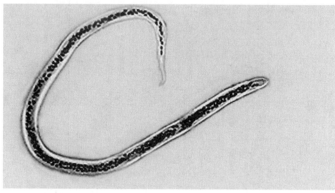

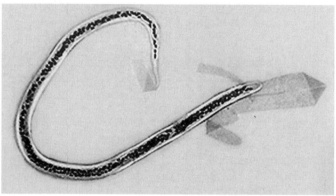

Figure 31.17 *Wuchereria bancrofti* microfilaria. (Upper) Microfilaria stained with Giemsa stain. Note that the sheath is not visible. (Lower) Microfilaria stained with Delafield's hematoxylin stain. Note that the sheath is visible.

3. Lake the films in 0.85% sodium chloride or distilled water for 15 min (not necessary for films prepared from the Knott sediment). Allow to air dry.
4. Fix all films in absolute methanol for 5 min. Allow to air dry.
5. Stain in undiluted Delafield's hematoxylin for 10 to 15 min.
6. Wash off excess stain in tap water.
7. Intensify the blue color by placing the films into tap water containing several drops of ammonia (NH$_4$OH) for several minutes.
8. Rinse again in running tap water for 5 min.
9. Air dry.
10. Mount in Permount or some other mounting medium; use a no. 1 coverglass.

References

1. Baker, J., J. McCarthy, M. Gatton, D. E. Kyle, V. Belizario, J. Luchavez, D. Bell, and Q. Cheng. 2005. Genetic diversity of *Plasmodium falciparum* histidine-rich protein 2 (PfHRP2) and its effect on the performance of Pf HRP2-based rapid diagnostic tests. *J. Infect. Dis.* **192:**870–877.
2. Beg, M. A., S. S. Ali, R. Haqqee, M. A. Khan, Z. Qasim, R. Hussain, and R. A. Smego, Jr. 2005. Rapid immunochromatography-based detection of mixed-species malaria infection in Pakistan. *Southeast Asian J. Trop. Med. Public Health* **36:**562–564.
3. Bell, D. R., D. W. Wilson, and L. B. Martin. 2005. False-positive results of a *Plasmodium falciparum* histidine-rich protein 2-detecting malaria rapid diagnostic test due to high sensitivity in a community with fluctuating low parasite density. *Am. J. Trop. Med. Hyg.* **73:**199–203.
4. Bruckner, D. A., L. S. Garcia, R. Y. Shimizu, E. J. C. Goldstein, P. M. Murray, and G. L. Lazar. 1985. Babesiosis: problems in diagnosis using autoanalyzers. *Am. J. Clin. Pathol.* **83:**520–521.
5. Carballo, A., and A. Ache. 1996. The evaluation of a dipstick test for *Plasmodium falciparum* in mining areas of Venezuela. *Am. J. Trop. Med. Hyg.* **55:**482–484.
6. Carrasquilla, G., M. Banguero, P. Sanchez, F. Carvajal, R. H. Barker, Jr., G. W. Gervais, E. Algarin, and A. E. Serrano. 2000. Epidemiologic tools for malaria surveillance in an urban setting of low endemicity along the Colombian Pacific coast. *Am. J. Trop. Med. Hyg.* **62:**132–137.
7. Causer, L. M., H. S. Bishop, D. J. Sharp, E. W. Flagg, J. F. Calderon, V. Keane, J. J. Shah, J. R. Macarthur, Jr., S. A. Maloney, M. S. Cetron, and P. B. Bloland. 2005. Rapid malaria screening and targeted treatment of United States-bound Montagnard refugees from Cambodia in 2002. *Am. J. Trop. Med. Hyg.* **72:**688–693.
8. Chatel, G., M. Gulletta, A. Matteelli, A. Marangoni, L. Signorini, O. Oladeji, and S. Caligaris. 1999. Short report: diagnosis of tick-borne relapsing fever by the quantitative buffy coat fluorescence method. *Am. J. Trop. Med. Hyg.* **60:**738–739.
9. Chiodini, P. L. 1998. Non-microscopic methods for diagnosis of malaria. *Lancet* **351:**80–81.
10. Clendennen, T. E., G. W. Long, and J. K. Baird. 1995. QBC™ and Giemsa-stained thick blood films: diagnostic performance of laboratory technologists. *Trans. R. Soc. Trop. Med. Hyg.* **89:**183–184.
11. Clinical Laboratory Standards Institute. 2000. *Laboratory Diagnosis of Blood-Borne Parasitic Diseases.* Approved guideline M15-A. Clinical Laboratory Standards Institute, Villanova, Pa.
12. Code of Federal Regulations. 1991. Occupational exposure to bloodborne pathogens. *Fed. Regist.* 29CFR1910.1030.
13. Cooke, A. H., P. L. Chiodini, T. Doherty, A. H. Moody, J. Ries, and M. Pinder. 1999. Comparison of a parasite lactate dehydrogenase-based immunochromatographic antigen detection assay (OptiMAL) with microscopy for the detection of malaria parasites in human blood samples. *Am. J. Trop. Med. Hyg.* **60:**173–176.
14. Craig, M. H., and B. L. Sharp. 1997. Comparative evaluation of four techniques for the diagnosis of *Plasmodium falciparum* infections. *Trans. R. Soc. Trop. Med. Hyg.* **91:**279–282.
15. Crespo, S., G. Palacios, S. Scott, M. Lago, S. Puente, M. Martinez, M. Baquero, and M. Subirats. 2004. Abnormal depolarizing patterns in three patients with filarial infection. *Ann. Hematol.* **83:**313–315.
16. Cruciani, M., S. Nardi, M. Malena, O. Bosco, G. Serpelloni, and C. Mengoli. 2004. Systematic review of the accuracy of the ParaSight-F test in the diagnosis of *Plasmodium falciparum* malaria. *Med. Sci. Monit.* **10:**MT81–MT88.

17. **Dennis, D. T., and B. H. Kean.** 1971. Isolation of microfilariae: report of a new method. *J. Parasitol.* **57:**1146–1147.

18. **Desowitz, R. S., and J. C. Hitchcock.** 1974. Hyperendemic bancroftian filariasis in the Kingdom of Tonga: the application of the membrane filter concentration technique to an age-stratified blood survey. *Am. J. Trop. Med. Hyg.* **23:**877–879.

19. **Desowitz, R. S., B. A. Southgate, and J. U. Mataika.** 1973. Studies on filariasis in the Pacific. 3. Comparative efficacy of the stained blood-film, counting-chamber, and membrane filtration techniques for the diagnosis of *Wuchereria bancrofti* microfilaraemia in untreated patients in areas of low endemicity. *Southeast Asian J. Trop. Med. Public Health* **4:**329–335.

20. **Dietze, R., M. Perkins, M. Boulos, F. Luz, B. Reller, and G. R. Corey.** 1995. The diagnosis of *Plasmodium falciparum* infection using a new antigen detection system. *Am. J. Trop. Med. Hyg.* **52:**45–49.

21. **Durand, F., B. Crassous, H. Fricker-Hidalgo, F. Carpentier, J. P. Brion, R. Grillot, and H. Pelloux.** 2005. Performance of the Now Malaria rapid diagnostic test with returned travelers: a 2-year retrospective study in a French teaching hospital. *Clin. Microbiol. Infect.* **11:**903–907.

22. **Farcas, G. A., K. J. Zhong, F. E. Lovegrove, C. M. Graham, and K. C. Kain.** 2003. Evaluation of the Binax NOW ICT test versus polymerase chain reaction and microscopy for the detection of malaria in returned travelers. *Am. J. Trop. Med. Hyg.* **69:**589–592.

23. **Farnert, A., A. P. Arez, A. T. Correia, A. Bjorkman, G. Snounou, and V. do Rosario.** 1999. Sampling and storage of blood and the detection of malaria parasites by polymerase chain reaction. *Trans. R. Soc. Trop. Med. Hyg.* **93:**50–53.

24. **Field, J. W., A. A. Sandosham, and Y. L. Fong.** 1963. *The Microscopical Diagnosis of Human Malaria.* 1. *A Morphological Study of the Erythrocytic Parasites in Thick Blood Films.* Institute for Medical Research, Kuala Lumpur, Malaya.

25. **Funk, M., P. Schlagenhauf, A. Tschopp, and R. Steffen.** 1999. MalaQuick versus ParaSight-F as a diagnostic aid in travellers' malaria. *Trans. R. Soc. Trop. Med. Hyg.* **93:**268–272.

26. **Garcia, L. S.** 1999. *Practical Guide to Diagnostic Parasitology.* ASM Press, Washington, D.C.

27. **Garcia, L. S., R. Y. Shimizu, and D. A. Bruckner.** 1986. Blood parasites: problems in diagnosis using automated differential instrumentation. *Diagn. Microbiol. Infect. Dis.* **4:**173–176.

28. **Gonzalez-Ceron, L., M. H. Rodriguez, A. F. Betanzos, and A. Abadia.** 2005. Efficacy of a rapid test to diagnose *Plasmodium vivax* in symptomatic patients of Chiapas, Mexico. 2005. *Salud Publica Mex.* **47:**282–287.

29. **Grobusch, M. P., U. Alpermann, S. Schwenke, T. Jeninek, and D. C. Warhurst.** 1999. False positive rapid tests for malaria in patients with rheumatoid factor. *Lancet* **353:**297.

30. **Hänscheid, T.** 1999. Diagnosis of malaria: a review of alternatives to conventional microscopy. *Clin. Lab. Haematol.* **21:**235–245.

31. **Hollander, D. H.** 1963. An oil-soluble antioxidant in resinous mounting media to inhibit fading of Romanowsky stains. *Stain Technol.* **38:**288–289.

32. **Humar, A., C. Ohrt, M. A. Harrington, D. Pillai, and K. C. Kain.** 1997. ParaSight F test compared with the polymerase chain reaction and microscopy for the diagnosis of *Plasmodium falciparum* malaria in travelers. *Am. J. Trop. Med. Hyg.* **56:**44–48.

33. **Iqbal, J., A. Sher, P. R. Hira, and R. Al-Owaish.** 1999. Comparison of the OptiMAL test with PCR for diagnosis of malaria in immigrants. *J. Clin. Microbiol.* **37:**3644–3646.

34. **Iqbal, J., A. Sher, and A. Rab.** 2000. *Plasmodium falciparum* histidine-rich protein 2-based immunocapture diagnostic assay for malaria: cross-reactivity with rheumatoid factors. *J. Clin. Microbiol.* **38:**1184–1186.

35. **Isenberg, H. D. (ed.).** 2004. *Clinical Microbiology Procedures Handbook*, 2nd ed., vol. 1, 2, and 3. ASM Press, Washington, D.C.

36. **Isenberg, H. D. (ed).** 1995. *Essential Procedures for Clinical Microbiology.* American Society for Microbiology, Washington, D.C.

37. **Jelinek, T., M. P. Grobusch, S. Schwenke, S. Steidl, F. von Sonnenburg, H. D. Nothdurft, E. Klein, and T. Loscher.** 1999. Sensitivity and specificity of dipstick tests for rapid diagnosis of malaria in nonimmune travelers. *J. Clin. Microbiol.* **37:**721–723.

38. **John, S. M., A. Sudarsanam, U. Sitaram, and A. H. Moody.** 1998. Evaluation of OptiMAL, a dipstick test for the diagnosis of malaria. *Ann. Trop. Med. Parasitol.* **92:**621–622.

39. **Johnston, S. P., N. J. Pieniazek, M. V. Xayavong, S. B. Slemenda, P. P. Wilkins, and A. J. Da Silva.** 2006. PCR as a confirmatory technique for laboratory diagnosis of malaria. *J. Clin. Microbiol.* **44:**1087–1089.

40. **Kain, K. C., M. A. Harrington, S. Tennyson, and J. S. Keystone.** 1998. Imported malaria: prospective analysis of problems in diagnosis and management. *Clin. Infect. Dis.* **27:**142–149.

41. **Kar, I., A. Eapen, T. Adak, and V. P. Sharma.** 1998. Trial with ParaSight-F in the detection of *Plasmodium falciparum* infection in Chennai (Tamil Nadu), India. *Indian J. Malariol.* **35:**160–162.

42. **Khan, S. A., M. Anwar, S. Hussain, A. H. Qureshi, M. Ahmad, and S. Afzal.** 2004. Comparison of OptiMAL malarial test with light microscopy for the diagnosis of malaria. *J. Pak. Med. Assoc.* **54:**404–407.

43. **Kilian, A. H., G. Kabagambe, W. Byamukama, P. Langi, P. Weis, and F. von Sonnenburg.** 1999. Application of the ParaSight-F dipstick test for malaria diagnosis in a district control program. *Acta Trop.* **72:**281–293.

44. **Kim, Y. R., M. Yee, S. Metha, V. Chupp, R. Kendall, and C. S. Scott.** 1998. Simultaneous differentiation and quantitation of erythroblasts and white blood cells on a high throughput clinical haematology analyser. *Clin. Lab. Haematol.* **20:**21–29.

45. **Knott, J.** 1939. A method for making microfilarial surveys on day blood. *Trans. R. Soc. Trop. Med. Hyg.* **33:**191–196.

46. **Kokoskin, E.** 2001. *The Malaria Manual.* McGill University for Tropical Diseases, Montreal, Canada.

47. **Laferi, H., K. Kandel, and H. Pichler.** 1997. False positive dipstick test for malaria. *N. Engl. J. Med.* **337:**1635–1636.

48. **Leke, R. F., R. R. Djokam, R. Mbu, R. J. Leke, J. Fogako, R. Megnekou, S. Metenou, G. Sama, Y. Zhou, T. Cadigan, M. Parra, and D. W. Taylor.** 1999. Detection of the *Plasmodium falciparum* antigen histidine-rich protein 2

in blood of pregnant women: implications for diagnosing placental malaria. *J. Clin. Microbiol.* **37:**2992–2996.

49. **Lema, O. E., J. Y. Carter, N. Nagelkerke, M. W. Wangai, P. Kitenge, S. M. Gikunda, P. A. Arube, C. G. Munafu, S. F. Materu, C. A. Adhiambo, and H. K. Mukunza.** 1999. Comparison of five methods of malaria detection in the outpatient setting. *Am. J. Trop. Med. Hyg.* **60:**177–182.

50. **Long, G. W., L. S. Rickman, and J. H. Cross.** 1990. Rapid diagnosis of *Brugia malayi* and *Wuchereria bancrofti* filariasis by an acridine orange/microhematocrit tube technique. *J. Parasitol.* **76:**278–280.

51. **Makler, M. T., R. C. Piper, and W. K. Milhous.** 1998. Lactate dehydrogenase and the diagnosis of malaria. *Parasitol. Today* **14:**376–377.

52. **Markell, E. K., M. Voge, and D. T. John.** 1992. *Medical Parasitology*, 7th ed. The W. B. Saunders Co., Philadelphia, Pa.

53. **Marx, A., D. Pewsner, M. Egger, R. Nuesch, H. C. Bucher, B. Genton, C. Hatz, and P. Juni.** 2005. Meta-analysis: accuracy of rapid tests for malaria in travelers returning from endemic areas. *Ann. Intern. Med.* **142:**836–846.

54. **Mason, D. P., and F. E. McKenzie.** 1999. Blood-stage dynamics and clinical implications of mixed *Plasmodium vivax-Plasmodium falciparum* infections. *Am. J. Trop. Med. Hyg.* **61:**367–374.

55. **Mathur, P., J. Samantaray, and N. K. Chauhan.** 2005. Evaluation of a rapid immunochromatographic test for diagnosis of kala-azar and post kala-azar dermal leishmaniasis at a tertiary care centre of north India. *Indian J. Med. Res.* **122:**485–490.

56. **McNamara, D. T., L. J. Kasehagen, B. T. Grimberg, J. Cole-Tobian, W. E. Collins, and P. A. Zimmerman.** 2006. Diagnosing infection levels of four human malaria parasite species by a polymerase chain reaction/ligase detection reaction fluorescent microsphere-based assay. *Am. J. Trop. Med. Hyg.* **74:**413–421.

57. **Mishra, B., J. C. Samantaray, and B. R. Birdha.** 1999. Evaluation of a rapid antigen capture assay for the diagnosis of falciparum malaria. *Indian J. Med. Res.* **109:**16–19.

58. **Murahwa, F. C., S. Mharakurwa, S. L. Mutambu, R. Rangarira, and B. J. Musana.** 1999. Diagnostic performance of two antigen capture tests for the diagnosis of *Plasmodium falciparum* malaria in Zimbabwe. *Cent. Afr. J. Med.* **45:**97–100.

59. **Palmer, C. J., J. F. Lindo, W. I. Klaskala, J. A. Quesada, R. Kaminsky, M. K. Baum, and A. L. Ager.** 1998. Evaluation of the OptiMAL test for rapid diagnosis of *Plasmodium vivax* and *Plasmodium falciparum* malaria. *J. Clin. Microbiol.* **36:**203–206.

60. **Petithory, J. C., F. Ardoin, L. R. Ash, E. Vandemeulebrouche, G. Galeazzi, M. Dufour, and A. Paugam.** 1997. Microscopic diagnosis of blood parasites following a cytoconcentration technique. *Am. J. Trop. Med. Hyg.* **57:**637–642.

61. **Pieroni, P., C. D. Mills, C. Ohrt, M. A. Harrington, and K. C. Kain.** 1998. Comparison of the ParaSight-F test and the ICT Malaria Pf test with the polymerase chain reaction for the diagnosis of *Plasmodium falciparum* malaria in travelers. *Trans. R. Soc. Trop. Med. Hyg.* **92:**166–169.

62. **Piper, R., J. Lebras, L. Wentworth, A. Hunt-Cooke, S. Houze, P. Chiodini, and M. Makler.** 1999. Immunocapture diagnostic assays for malaria using *Plasmodium* lactate dehydrogenase (pLDH). *Am. J. Trop. Med. Hyg.* **60:**109–118.

63. **Quintana, M., R. Piper, H. L. Boling, M. Makler, C. Sherman, E. Gill, E. Fernandez, and S. Martin.** 1998. Malaria diagnosis by dipstick assay in a Honduran population with coendemic *Plasmodium falciparum* and *Plasmodium vivax*. *Am. J. Trop. Med. Hyg.* **59:**868–871.

64. **Shiff, C. J., J. Minjas, and Z. Premji.** 1994. The ParaSight™ F test: a simple, rapid manual dipstick test to detect *Plasmodium falciparum* infection. *Parasitol. Today* **10:**494–495.

65. **Shute, P. G.** 1966. The staining of malaria parasites. *Trans. R. Soc. Trop. Med. Hyg.* **60:**412–416.

66. **Shute, P. G., and M. E. Maryon.** 1966. *Laboratory Technique for the Study of Malaria*, 2nd ed. J. & A. Churchill, Ltd., London, United Kingdom.

67. **Singh, N., A. Saxena, S. B. Awadhia, R. Shrivastava, and M. P. Singh.** 2005. Evaluation of a rapid diagnostic test for assessing the burden of malaria at delivery in India. *Am. J. Trop. Med. Hyg.* **73:**855–858.

68. **Singh, N., M. P. Singh, and V. P. Sharma.** 1997. The use of a dipstick antigen-capture assay for the diagnosis of *Plasmodium falciparum* infection in a remote forested area of central India. *Am. J. Trop. Med. Hyg.* **56:**188–191.

69. **Spudick, J. M., L. S. Garcia, D. M. Graham, and D. A. Haake.** 2005. Diagnostic and therapeutic pitfalls associated with primaquine-tolerant *Plasmodium vivax*. *J. Clin. Microbiol.* **43:**978–981.

70. **Suh, I. B., H. J. Kim, J. Y. Kim, S. W. Lee, S. S. An, W. J. Kim, and C. S. Lim.** 2003. Evaluation of the Abbott Cell-Dyn 4000 hematology analyzer for detection and therapeutic monitoring of *Plasmodium vivax* in the Republic of Korea. *Trop. Med. Int. Health.* **8:**1074–1081.

71. **Tjitra, E., S. Suprianto, M. Dyer, B. J. Currie, and N. M. Anstey.** 1999. Field evaluation of the ICT malaria P.f/P.v immunochromatographic test for detection of *Plasmodium falciparum* and *Plasmodium vivax* in patients with a presumptive clinical diagnosis of malaria in eastern Indonesia. *J. Clin. Microbiol.* **37:**2412–2417.

72. **Trarore, I., O. Koita, and O. Doumbo.** 1997. Field studies of the Parasight-F test in a malaria-endemic area: cost, feasibility, sensitivity, specificity, predictive value and the deletion of the HRP-2 gene among wild type *Plasmodium falciparum* in Mali. *Am. J. Trop. Med. Hyg.* **57:**272 (Poster SO2).

73. **Valecha, N., V. P. Sharma, and C. U. Devi.** 1998. A rapid immunochromatographic test (ICT) for diagnosis of *Plasmodium falciparum*. *Diagn. Microbiol. Infect. Dis.* **30:**257–260.

74. **van Dam, A. P., T. van Gool, J. C. Wetsteyn, and J. Dankert.** 1999. Tick-borne relapsing fever imported from West Africa: diagnosis by quantitative buffy coat analysis and in vitro culture of *Borrelia crocidurae*. *J. Clin. Microbiol.* **37:**2027–2030.

75. **Van den Ende, J., T. Vervoort, A. Van Gompel, and L. Lynen.** 1998. Evaluation of two tests based on the detection of histidine rich protein 2 for the diagnosis of imported *Plasmodium falciparum* malaria. *Trans. R. Soc. Trop. Med. Hyg.* **92:**285–288.

76. **Wang, L. C.** 1998. Evaluation of quantitative buffy coat analysis in the detection of canine *Dirofilaria immitis* infection: a model to determine its effectiveness in the diagnosis of human filariasis. *Parasitol. Res.* **84:**246–248.

77. **Wongsrichanalai, C., I. Arevalo, A. Laboonchai, K. Yingyuen, R. S. Miller, A. J. Magill, J. R. Forney, and R. A. Gasser, Jr.** 2003. Rapid diagnostic devices for malaria: field evaluation of a new prototype immunochromatographic assay for the detection of *Plasmodium falciparum*
and non-*falciparum Plasmodium. Am. J. Trop. Med. Hyg.* **69:**26–30.

78. **World Health Organization.** 1996. A rapid dipstick antigen capture assay for the diagnosis of falciparum malaria. *Bull. W. H. O.* **74:**47–54.

Parasite Recovery: Culture Methods, Animal Inoculation, and Xenodiagnosis

Culture Methods

Very few clinical laboratories offer specific culture techniques for parasites. The methods for in vitro culture are often complex, while quality control is difficult and not really feasible for the routine diagnostic laboratory. In certain institutions, some techniques may be available, particularly when consultative services are provided (reference laboratory situation) and for research purposes. However, most laboratories do not offer these techniques.

Few parasites can be routinely cultured, and the only procedures that are in general use are for *Entamoeba histolytica*, *Naegleria fowleri*, *Acanthamoeba* spp., *Trichomonas vaginalis*, *Toxoplasma gondii*, *Trypanosoma cruzi*, and the leishmanias. Often, when specimens are cultured for potential pathogens, nonpathogenic protozoa could also be recovered. These procedures are usually available only after consultation with the laboratory and on special request. For those who may be interested in trying these techniques, the several different media presented below are representative of those available. More extensive options can be found in the literature (1, 7, 24, 31, 52–55, 58, 62, 64, 65).

Cultures of parasites grown in association with an unknown microbiota are referred to as xenic cultures. A good example of this type of culture would be stool specimens cultured for *E. histolytica*. If the parasites are grown with a single known bacterium, the culture is referred to as monoxenic. An example of this type of culture would be clinical specimens (corneal biopsy specimens) cultured with *Escherichia coli* as a means of recovering species of *Acanthamoeba* and *Naegleria*. If parasites are grown as pure culture without any bacterial associate, the culture is referred to as axenic. An example of this type of culture would be the use of media for the isolation of *Leishmania* spp. or *Trypanosoma cruzi*. All three types of cultures are discussed in this chapter.

Pancreatic digests of casein are major ingredients of media used in the axenic cultivation of lumen-dwelling parasitic protozoa. Unfortunately, the digest used almost exclusively in the development of these media has not been available since the early 1980s. Many digest products have been tried in the interim with marginal results in supporting the growth of *E. histolytica*. Diamond et al. (13) have developed a casein-free medium, YI-S, consisting of a nutrient broth, vitamin mixture, and serum. This may serve as a replacement for TYI-S-33, widely used for the axenic culture of *E. histolytica* and other intestinal protozoa.

Intestinal Protozoa

Specimens include stool material, mucus, or a combination of the two. The clinical specimen(s) should be no more than 24 h old; a maximum of 2 to 3 h is recommended.

Balamuth's Aqueous Egg Yolk Infusion Medium for Amebae

Balamuth's aqueous egg yolk infusion medium is used to detect the presence of amebae. The specific solutions required are phosphate buffer and whole-liver concentrate solution (36).

Balamuth's Aqueous Egg Yolk Infusion Medium

Phosphate buffer
    (A) Potassium phosphate,
        tribasic (K_3PO_4) .... 212.27 g
        Distilled water .... 1,000 ml
    (B) Potassium phosphate,
        monobasic (KH_2PO_4) .... 136.092 g
        Distilled water .... 1,000 ml

Mix the solution in the ratio of 3 parts tribasic (A) to 2 parts monobasic (B) to make 1 M phosphate buffer stock. Dilute the stock buffer to 0.067 M before use (add 492 ml of distilled H_2O to 1 liter of 1 M phosphate buffer).

Whole-liver concentrate solution
    Liver concentrate powder
        (Wilson or Lilly) .... 5 g
    Distilled water... 100 ml

Suspend the powder in cold water, and autoclave. Filter through a Büchner funnel to remove sediment, dispense in 10-ml quantities, and reautoclave.

Preparation of Complete Medium (20)

1. Using a blender, blend 12 fresh hard-boiled egg yolks with 375 ml of 0.8% sodium chloride.
2. Autoclave at 7 lb/in^2 pressure for 10 min, repressurize slowly, and stir.
3. Autoclave again for 45 min at 7 lb/in^2 pressure.
4. Allow to cool slightly, and add distilled water to replace evaporation loss. Transfer the material to a muslin bag, and express the liquid portion, saving all the fluid. Return the volume to 375 ml with 0.8% sodium chloride.
5. Autoclave for 20 min at 121°C; cool to 5°C. Do not agitate the fluid at this point or during filtration.
6. Decant the fluid carefully through gauze into a Büchner funnel with Whatman no. 3 filter paper. Filter papers can be replaced as necessary.
7. Measure the filtrate, add an equal volume of 0.067 M phosphate buffer, and autoclave for 20 min at 121°C. After cooling, add stock liver concentrate (1 part stock liver to 9 parts medium).
8. Material can then be decanted into sterile flasks, which are stored until the medium is dispensed.

Procedure

Tubes should contain 6 to 8 ml of fluid and should be incubated for 4 days at 37°C as a sterility check. Before inoculation, a loopful of sterile rice powder or starch (can be autoclaved in a screw-cap tube) is added to each tube. To each tube, add stool material, mucus, or a combination of the two (about the size of a small pea), break it up thoroughly in the medium, and incubate at 37°C. The cultures should be checked at 2, 3, and 4 days by examining 0.1 ml of sediment under the microscope (low-intensity light) for characteristic motility. Although the initial culture may appear to be negative, subcultures may reveal organisms (Figure 32.1).

Figure 32.1 Protozoa from culture systems. (Upper left) *Entamoeba histolytica*/*E. dispar* trophozoite from liquid medium containing rice starch (note that there are no definitive erythrocytes within the cytoplasm, so that it is not possible to differentiate the true pathogen, *E. histolytica*, from the nonpathogen, *E. dispar*). (Upper right) *Naegleria fowleri* trophozoite from nonnutrient agar culture with bacterial overlay (note that this trophozoite has been stained). (Lower left) *Acanthamoeba* spp. trophozoite from nonnutrient agar culture with bacterial overlay (note the spiky acanthapodia). (Lower right) *Acanthamoeba* spp. cysts from nonnutrient agar culture with bacterial overlay (note the double hexagonal wall appearance.

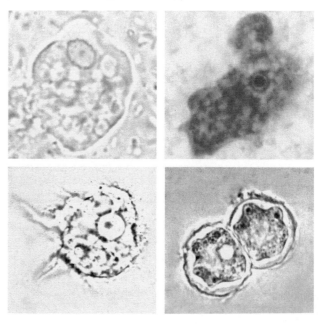

According to Dolkart and Halpern (14), the addition of gastric mucin to the egg component is reported to improve the performance of Balamuth's medium.

The addition of rice flour (prerinsed with distilled water) to Balamuth's medium is reported to support abundant growth of *Balantidium coli*.

Boeck and Drbohlav's Locke-Egg-Serum (LES) Medium for Amebae

Boeck and Drbohlav's LES medium is another culture medium used to diagnose the presence of amebae (36).

Boecke and Drbohlav's LES Medium

Locke's solution

NaCl	9.0 g
$CaCl_2$	0.2 g
KCl	0.4 g
$NaHCO_3$	0.2 g
Glucose	2.5 g
Distilled water	1,000 ml

Autoclave before storage.

Preparation of Complete Medium

1. Wash four eggs, wipe the shells with 70% alcohol, and break the eggs into a sterile flask containing glass beads.
2. Add 50 ml of Locke's solution, and shake until homogenous.
3. Dispense the medium so that a slant of 1 to 1.5 in. (1 in. = 2.54 cm) is produced in the bottom of the tube. (Tube size is not critical.)
4. Plug the tubes, and place them in a slant position in an inspissator at 70°C until the slant solidifies. Inspissator conditions may be achieved in the autoclave by leaving the door ajar (nonpressurized system).
5. Autoclave the tubes at 121°C for 20 min. Discard any damaged slants.
6. Prepare a mixture of 8 parts sterile Locke's solution to 1 part sterile inactivated human serum. Sterilize the mixture by filtration, and incubate at 37°C for 24 to 48 h as a sterility check before use. Cover the slants to a depth of <1 cm with the sterile solution, and inoculate in the same manner as for Balamuth's medium. LES medium should have a loopful of sterile rice powder added before inoculation.

Diphasic LE Medium (*Blastocystis hominis*)

LE medium is the NIH modification of Boeck and Drbohlav's medium.

Locke's Solution

NaCl	8.0 g
$CaCl_2$	0.2 g
KCl	0.2 g
$MgCl_2$	0.01 g
Na_2HPO_4	2.0 g
$NaHCO_3$	0.4 g
KH_2PO_4	0.3 g
Distilled water	1,000 ml

Autoclave for 15 min at 121°C under a pressure of 15 lb/in^2. Cool to room temperature, and remove any precipitate by filtration (Whatman no. 1 paper). Reautoclave to sterilize.

Preparation of Complete Medium

1. To prepare the egg slant, surface sterilize fresh hen eggs by flaming in 70% ethanol and break into a graduated cylinder.
2. Add 12.5 ml of Locke's solution per 45 ml of egg.
3. Emulsify in a Waring-type blender, and filter through gauze into a flask.
4. Place under vacuum to draw out all air bubbles.
5. Add 5-ml amounts of the emulsified egg to standard culture tubes (16 by 125 mm), and autoclave at 100°C for 10 min with the tubes at an angle that produces a 12- to 15-mm (ca. 0.5-in.) butt. The resulting egg slants should be free of bubbles.
6. Cool to room temperature, overlay slants with 6 ml of Locke's solution, and autoclave for 15 min at 121°C under a pressure of 15 lb/in^2.
7. Let the slants cool to room temperature, tighten the caps, and refrigerate the slants for up to 6 months.

TYI-S-33 Medium for *Entamoeba histolytica* (8)

TYI Broth

Biosate (BBL) Peptone	30.0 g
Dextrose (do not use glucose)	10.0 g
NaCl	2.0 g
L-Cysteine HCl (Sigma)	1.0 g
Ascorbic acid (J. T. Baker)	0.2 g
K_2HPO_4	1.0 g
KH_2PO_4	0.6 g
Ferric ammonium citrate (Mallinckrodt)	22.8 mg
Distilled water up to	870 ml

Note Biosate may be replaced with 20 g of Trypticase (BBL) and 10 g of yeast extract (BBL). Some lots of Biosate, Trypticase, yeast extract, or serum may inhibit growth.

1. Adjust pH to 6.8 with 1 N NaOH, and filter through Whatman no. 1 paper. Autoclave at 121°C for 15 min.
2. Cool, and add:

Heat-inactivated bovine serum
(shelf life, 6 months at 4°C) .... 100 ml
Special 107 Vitamin Mix
(below) .... 30 ml

3. Aseptically dispense 13 ml of complete medium into 16- by 125-mm screw-cap tubes. To prepare Special 107 Vitamin Mix, aseptically combine the following:

Solution 1 (see below) .... 0.4 ml
Solution 2 (see below) .... 1.2 ml
Solution 3 (see below) .... 0.4 ml
Sterile distilled water .... 18.0 ml
Diamond's TPS-1, vitamin solution,
40× (NABI) diluted to 1× .... 100.0 ml

Total volume of complete Special 107 Vitamin Mix is 120 ml.

Solution 1 (Filter Sterilized)
D,L-6,8-Thioctic acid,
oxidized (Sigma) .... 100 mg
Absolute ethanol .... 100 ml

Solution 2 (Filter Sterilized)
Vitamin B_{12} (Sigma)................................. 40 mg
Distilled water... 100 ml

Solution 3 (Filter Sterilized)
Tween 80 (Sigma)..................................... 50 g
Absolute ethanol .... 100 ml

Note The shelf life of solutions 1, 2, and 3 is 22 months.

TYI-S-33 (Keister's Modification) for *Giardia lamblia*

For TYI-S-33 (Keister's modification [7]), prepare TYI broth exactly as for *E. histolytica* with the following changes. (i) Increase the amount of L-cysteine hydrochloride to 2.0 g/liter. (ii) Add 500 mg of dehydrated bovine bile per liter. (iii) Adjust the pH to 7.0 to 7.1. (iv) Sterilize by filtration through a 0.22-μm-pore-size filter. Do not autoclave! The complete medium is made by addition of bovine serum to 10%.

TYSGM-9 Medium for *Entamoeba histolytica*

Nutrient Broth
Potassium phosphate,
dibasic (K_2HPO_4)................................. 2.8 g
Potassium phosphate,
monobasic (KH_2PO_4)........................... 0.4 g

NaCl .... 7.5 g
Casein digest peptone
(BBL, product no. 97023).................... 2.0 g
Yeast extract (BBL) .... 1.0 g
Glass-distilled water .... 970.0 ml

The nutrient broth may be stored for several months at −20°C (11, 12).

5% Tween 80 Solution

1. Vigorously stir, with a magnetic stirrer, 95 ml of glass-distilled water in a bottle.
2. Add 5 g of Tween 80 (very thick solution; must be weighed), and keep stirring for a few minutes.
3. Filter sterilize through a 0.22-μm-pore-size membrane.
4. Aseptically dispense into a number of sterile screw-cap test tubes, 10 ml per tube.
5. Label as 5% Tween 80 solution, with the preparation date and an expiration date of no longer than 1 month. Store at 4°C.

Phosphate-Buffered Saline (PBS no. 8), pH 7.2
NaCl .... 9.5 g
K_2HPO_4.. 3.7 g
KH_2PO_4.. 1.1 g
Glass-distilled water to..................... 1,000.0 ml

1. Dissolve the salts in the distilled water with a magnetic stirrer.
2. Autoclave for 15 min at 121°C.
3. When cool, label as PBS no. 8 with the preparation date and an expiration date of 3 months.

Rice Starch
For best results use rice starch obtained from British Drug Houses Ltd. or Gailard Schlesinger, Inc.

1. Dispense 500 mg of rice starch into each of several 16- by 125-mm screw-cap tubes; do not tighten the caps.
2. Place the tubes horizontally in a dry-heat sterilizer or an oven. Make sure that the rice starch is uniformly distributed loosely over the undersurface of the tubes.
3. Heat the tubes for 2.5 h at 150°C.
4. When cool, tighten caps and label as rice starch with the date of preparation and an expiration date of 3 months.

Rice Starch Suspension

1. Add 9.5 ml of sterile PBS no. 8 to each tube of rice starch.

2. Shake vigorously or use a vortex machine to uniformly suspend the rice starch at the time of use.

Stock Antibiotic Solution

1. Using a 6-ml syringe and 20-gauge needle, add 5 ml of sterile distilled water to a vial of penicillin G sodium (10^6 U).
2. Using a 6-ml syringe and 20-gauge needle, add 5 ml of sterile distilled water to a vial of streptomycin sulfate (10^6 µg/ml).
3. Shake gently, and let stand for 30 min to dissolve the antibiotics completely in the distilled water.
4. Mix the two antibiotics in a graduated flask or cylinder, and bring the volume to 125 ml with distilled water. The stock concentration of antibiotics is 8,000 U of penicillin/ml and 8,000 µg of streptomycin/ml.
5. Filter sterilize the antibiotic solution through a 0.22-µm membrane filter, dispense the filtrate into a number of sterile screw-cap vials or sterile cryovials (1 ml per vial), and label as stock antibiotic solution with the preparation date and an expiration date of 6 months. Store at 120°C in a cryovial box.

Buffered Methylene Blue Solution

Solution A, 0.2 M acetic acid
Glacial acetic acid .... 11.55 ml
Distilled water .... 988.50 ml

Add the acetic acid to the water, mix, and store in a glass-stoppered bottle. Label with the date of preparation and an expiration date of 1 year.

Solution B, 0.2 M sodium acetate
Sodium acetate
($CH_3COONa \cdot 3H_2O$) .... 16.4 g
Distilled water to .... 1,000.0 ml

Dissolve the sodium acetate in 400 ml of distilled water in a volumetric flask, bring the volume to the 1,000-ml mark, mix well, and store in a glass-stoppered bottle. Label with the date of preparation and an expiration date of 1 year.

Acetate buffer, pH 3.6
Solution A .... 46.3 ml
Solution B .... 3.7 ml
Distilled water to .... 100.0 ml

Mix solutions A and B in a volumetric flask, and bring the volume to 100.0 ml with distilled water. The pH should be 3.6. Store in a glass-stoppered bottle. Label with the date of preparation and an expiration date of 1 year.

Methylene blue stain
Methylene blue dye .... 60.0 mg
Acetate buffer
(from previous step) .... 100.0 ml

Dissolve the dye in the buffer, and store in a glass-stoppered bottle. Label with the date of preparation and an expiration date of 1 year.

Complete Medium (TYSGM-9 Medium)

1. Place 200 mg of gastric mucin (U.S. Biochemical Corp., catalog no. 16025) in a 125-ml screw-cap bottle or Erlenmeyer flask.
2. Add 97 ml of nutrient broth; using a magnetic stirrer, stir vigorously for at least 1 h or until the medium becomes clear.
3. Autoclave for 15 min at 121°C; cool to room temperature.
4. Add aseptically, in a biological safety cabinet, 5.0 ml of heat-inactivated bovine serum.
5. Add 0.1 ml of the 5% Tween 80 solution.
6. Dispense aseptically, in a biological safety cabinet, into a number of sterile 16- by 125-mm screw-cap tubes, 8 ml per tube.
7. Add 0.25 ml of rice starch solution after vigorously shaking the tube.
8. Store the tubes at 4°C for not more than 1 month.
9. The final pH of the medium should be 7.2.

Robinson's Culture
Robinson's medium is a complex medium that has nevertheless found widespread use for the isolation of enteric amebae. To prepare Robinson's medium, prepare the six following stock solutions (7).

Solution 1 (0.5% Erythromycin) (Filter Sterilized)
Prepare 0.5% erythromycin in distilled water and filter sterilize. Refrigerate.

Solution 2 (20% Bacto Peptone) (Autoclave)
Prepare 20% Bacto Peptone in distilled water. Autoclave and refrigerate.

Solution 3 (10× Phthalate Solution, Stock) (Autoclave)
Potassium hydrogen phthalate .... 102 g
40% sodium hydroxide .... 50 ml

Bring to 1 liter at pH 6.3. Autoclave for 15 min at 121°C under a pressure of 15 lb/in². Store at room temperature. Dilute 1:10 with sterile water before use.

A stock solution of phthalate-Bacto Peptone can be made by adding 1.25 ml of 20% Bacto Peptone per 100 ml of 1× phthalate solution. Store refrigerated.

Solution 4 (10× R Medium Stock) (Autoclave)

NaCl .... 25.0 g
Citric acid .... 10.0 g
Potassium phosphate, monobasic
  (KH_2PO_4) .... 25.0 g
Ammonium sulfate .... 5.0 g
Magnesium sulfate · $7H_2O$ .... 0.25 g
85% lactic acid solution........................... 20 ml

Bring to 500 ml. Dilute stock 1:10, adjusting pH to 7.0. Autoclave for 15 min at 121°C under a pressure of 15 lb/in² in 20-ml amounts.

Solution 5 (BR Medium)

To prepare BR medium, inoculate 1× R medium with a standard *Escherichia coli* strain such as O111. Incubate at 37°C for 48 h, and store at room temperature (good for several months).

Solution 6 (BRS Medium)

To prepare BRS medium, add an equal volume of heat-inactivated bovine serum to BR medium and incubate at 37°C for 24 h. Store at room temperature (good for several months).

1. To prepare agar slants, many people use screw-cap glass bijou bottles (total volume, 7 ml), but standard culture tubes also work well.
2. Autoclave a solution of 1.5% Noble agar in 0.7% NaCl–distilled water for 15 min at 121°C under a pressure of 15 lb/in².
3. Dispense in 5-ml (tube) or 3-ml (bottle) amounts, reautoclave, and slant until cool and set. For slants in tubes, use an angle that produces a 12- to 15-mm (ca. 0.5-in.) butt.
4. When cool, tighten lids and store at room temperature or refrigerated.
5. To one tube or bottle, add the following: 3 ml of 1× phthalate-Bacto Peptone, 1 ml of BRS medium, and 50 µl of erythromycin. This must be done on the same day as the inoculation. Note that although erythromycin is added to Robinson's medium at every subculture, this does not lead to a monoxenic culture as occasionally stated. Additional antibiotic treatment would be needed for this to be the case.

Xenic Culture

When initiating xenic cultures, the stool samples should be inoculated into at least two tubes, one with and the other without antibiotics. In some cases, some component of the natural bacterial flora may be helpful or even necessary for the amebae to become established (27).

1. Warm several tubes of TYSGM-9 medium in the incubator (35°C for 1 to 2 h).
2. Add 0.1 ml of stock antibiotics to each tube of medium. The final concentration of antibiotics is 100 U of penicillin per ml and 100 µg of streptomycin per ml.
3. After vortexing or vigorously shaking the tube, use a Pasteur pipette to add 3 drops of the starch suspension to each tube of the medium.
4. Place a pea-sized portion of stool sample into the bottom of the tube, and break up the stool gently with the pipette.
5. Tightly cap the tubes, and incubate at a 45 to 50° angle at 35°C for 48 h.
6. Examine the tubes with an inverted microscope and the 10× objective for the presence of amebae. Amebae, if present, are usually seen attached to the underside of the tubes, interspersed with the fecal material and rice starch. Sometimes it is necessary to gently invert the tubes in order to disperse the stool material and rice starch to uncover the amebae. If you do not have an inverted microscope, proceed to step 13.
7. If amebae are not seen, stand the tubes upright for about 30 min at 35°C.
8. With a Pasteur pipette, remove from the bottom of each tube the entire sediment and inoculate the sediment into fresh tubes containing rice starch and antibiotics.
9. Incubate as above for another 48 h.
10. Examine the tubes as before and discard the tubes if amebae are still not seen. Report patient results as negative.
    Note These patient results should not be reported unless the quality control organisms and cultures are growing, thus indicating the culture system is performing according to expected results.
11. If amebae are present in small numbers, chill the tube in ice-water for 5 min and centrifuge the tube for 5 min at 250 × g. Aspirate and discard the supernatant, and inoculate the sediment into a fresh tube as before.
12. If amebae are present in large numbers, let the tube stand upright for 30 min and remove about 0.2 ml of sediment from the bottom. Inoculate the sediment into fresh tubes as before.
13. If you do not have an inverted microscope, stand the tubes upright for about 30 min at 35°C. With a sterile pipette, remove about 0.5 ml of sediment from the bottom of the tube and place a couple

of drops onto each of two slides. Add 2 drops of methylene blue solution to one of the slides. Cover both with coverslips, and examine the slides under the microscope for amebae. Amebae may appear rounded or with pseudopodial extrusions. The nuclei may be clearly seen in the methylene blue preparation. Proceed to steps 7 through 12.

Axenic Culture

Axenic culture is used for research when strains of organisms are necessary for work requiring a culture system free from bacterial contaminants. If the axenic culture tubes become contaminated, 1,000 U of penicillin per ml and 1,000 µg of streptomycin or 50 µg of gentamicin per ml can be added to each tube. If, however, the contaminant happens to be *Pseudomonas* spp., it is probably better to discard the tube and use an uncontaminated tube for subculture purposes (20, 24).

1. Remove tubes containing TYI-S-33 medium from 4°C, and incubate at 35°C for 1 to 2 h.
2. With an inverted microscope, examine stock culture tubes of *E. histolytica* for any signs of bacterial contamination (no longer acceptable for use). Select one or several tubes showing good growth of amebae. Since the tubes are incubated in a slanted position, usually at an angle of 5 to 10°, a thick button of amebae will be seen at the bottom of the tube. Gently invert the tubes once or twice to disperse the amebae uniformly, and examine the tubes again. A majority of the amebae should be attached to the tube walls and show pseudopodial motility. If you do not have an inverted microscope, examine organisms from the bottom of the tube (as a wet smear). If you can see pseudopodial motility, proceed to step 3.
3. Immerse the tubes in a bucket of ice-cold water for about 5 to 10 min to dislodge the amebae from the tube walls. Invert the tubes several times to distribute the amebae.
4. Remove, with a Pasteur pipette, about 1.0 to 1.5 ml of culture medium; inoculate 0.5 to 1 ml into a fresh tube. Inoculate the rest of the fluid into nutrient agar, brain heart infusion, and thioglycolate broth for routine monitoring of bacterial contamination. Inoculate several tubes this way, and incubate the cultures slanted at 5 to 10° at 35°C as before.
5. If amebic growth is not good but some amebae are attached to the tube walls, remove about 10 ml of medium from the bottom with a serologic pipette and add 10 ml of fresh medium.

6. If amebic growth is not good and only few amebae are present along with a lot of debris, centrifuge the tube at 250 × *g* for 10 min, aspirate the supernatant fluid, and transfer the sediment to a fresh tube and incubate as before.

Quality Control for Intestinal Protozoan Cultures

The following control strains should be available when using these cultures for clinical specimens (24): ATCC 30925 (*Entamoeba histolytica* HU-1:CDC), ATCC 30015 (*Entamoeba histolytica* HK-9), and ATCC 30042 (*Entamoeba histolytica*-like, Laredo strain, culture at 25°C) (8).

1. Check all reagents and media (Balamuth's aqueous egg yolk infusion medium, Boeck and Drbohlav's LES medium, PBS solution no. 8, rice starch suspension, Tween 80 solution, and TYSGM-9 and TYI-S-33 media) each time they are used or periodically (once a week). The media and all solutions should be free of any signs of precipitation and bacterial and/or fungal contamination.
2. Maintain stock cultures of *E. histolytica* at 35°C (ATCC 30925 [strain HU-1:CDC] and ATCC 30015 [strain HK-9]). Maintain *E. histolytica*-like Laredo strain (ATCC 30042) at 25°C.
   A. Transfer stock culture (ATCC 30925) every other day with TYSGM-9 medium.
   B. Transfer stock culture (ATCC 30015) once every 3 days with TYI-S-33 medium.
   C. Transfer stock culture (ATCC 30042) once a month with TYSGM-9 medium.
   D. *E. histolytica* trophozoites measure 10 to 60 µm and demonstrate directional motility by extruding hyaline, finger-like pseudopodia from the cytoplasm. Cysts are not usually found in cultures.
   E. Trophozoites are uninucleate and characterized by finely granular, uniform, evenly distributed peripheral chromatin. The nucleolus is small and usually centrally located but may be eccentric.
3. Depending on its use, the microscope(s) may need to be calibrated (within the last 12 months), and the original optics used for the calibration should be in place on the microscope(s). The calibration factors for all objectives should be posted on the microscope for easy access.

Note If the tubes containing fecal material are positive for amebae after 48 h of incubation, confirm the identification with the permanent stained smear. If the tubes do not show any amebae, subculture the contents of the tubes as described above and incubate for an additional

48 h. If the tubes are still negative for amebae, report the specimen as negative and discard the tubes. Even when the culture system is within quality control guidelines, a negative culture is still not definitive in ruling out the presence of *E. histolytica*. Xenic cultures of *E. histolytica* serve only as a supplemental procedure and never replace the primary diagnosis by microscopic examination of concentration sediments and permanent stained smears. Axenic culture is used to maintain quality control strains and for research purposes.

Pathogenic Free-Living Amebae

Specimens would include cerebrospinal fluid (CSF), biopsy tissue, and autopsy tissue of the brain; for *Acanthamoeba* spp., corneal scrapings or biopsy material, contact lenses and contact lens paraphernalia such as lens cases and solutions, skin abscess material, ear discharge, or feces can also be used. All clinical specimens should be processed within 24 h; 2 to 3 h is recommended. However, eye-related specimens can be shipped by mail with apparently few problems (27, 33, 37, 48, 51, 63). The procedure for growing *Naegleria* and *Acanthamoeba* from clinical specimens involves the use of a nonnutrient agar spread with *E. coli* or some other nonmucoid bacteria. Amebas begin feeding on bacteria and soon grow to cover the agar surface in 1 to 2 days at 37°C. The presence of the protozoa can be confirmed by examining the agar surface using an inverted microscope or with a conventional microscope by inverting the plate on the stage and focusing through the agar with a 10× objective (Figure 32.1).

Diagnostic methods include direct microscopy of wet mounts of CSF or stained smears of CSF sediment, light or electron microscopy of tissues, in vitro cultivation of *Acanthamoeba*, and histologic assessment of frozen or paraffin-embedded sections of brain or cutaneous lesion biopsy material. Immunocytochemistry, chemifluorescent-dye staining, PCR, and analysis of DNA sequence variation also have been used for laboratory diagnosis. Several approaches to specimen handling and diagnostic methods can be seen in Table 32.1.

Environmental Issues

There continue to be ongoing discussions and publications regarding the association of free-living amebae and their intracellular bacterial flora, particularly within the context of environmental transmission of infection, contamination of equipment, and overall environmental concerns. The recognition that *Acanthamoeba* spp. can sequester a variety of bacteria with known potential for causing human disease suggests that these amebae serve as reservoirs for bacterial pathogens. The increase in the reported incidence of *Acanthamoeba* infections may be

due to greater recognition of the disease potential of these amebae. Also, a number of factors may account for an increased incidence of infection, such as a large number of human immunodeficiency virus-infected individuals and more patients undergoing chemotherapy or immunosuppressive therapy for organ transplantation (35). Some of these recent issues can be seen in Table 32.2.

Acanthamoeba Medium

For the isolation of *Naegleria* or *Acanthamoeba* spp. from tissues or soil samples, the following procedure is recommended (31).

Acanthamoeba Medium

Page's saline (10×)

NaCl	60 mg
$MgSO_4 \cdot 7H_2O$	2 mg
$CaCl_2 \cdot 2H_2O$	2 mg
Na_2HPO_4	71 mg
KH_2PO_4	68 mg
Distilled water	500 ml

1. Autoclave at 121°C for 15 min.
2. Store refrigerated in a glass bottle for up to 6 months.

Nonnutrient agar

Page's saline (10×)	100 ml
Difco agar	15 g
Double-distilled water	900.0 ml

1. Dissolve agar in Page's saline and distilled water with gentle heating; stir or swirl.
2. Aliquot 20 ml into screw-cap tubes (20 by 150 mm).
3. Autoclave at 15 lb/in² for 15 min; label deeps with 12-month expiration date, and store in the refrigerator.
4. Melt agar deeps, and pour into petri dishes as needed. Plates may be stored in the refrigerator for up to 3 months.

Monoxenic culture

1. Remove the nonnutrient agar plates from the refrigerator, and place them in a 37°C incubator for 30 min.
2. Add 0.5 ml of ameba saline to a slant bacterial culture of *E. coli* or *Enterobacter aerogenes*. Gently scrape the surface of the slant (do not break the agar surface). Suspend the bacteria uniformly by gently pipetting with a Pasteur pipette, and add 2 or 3 drops of this suspension to the middle of

Table 32.1   Approaches to isolation of free-living amebae[a]

Approach 1	Approach 2	Approach 3	Approach 4
Tissue sample (central nervous system tissue, corneal scrapings, skin scrapings or biopsy specimen)	Tissue sample (central nervous system tissue, corneal scrapings, skin scrapings or biopsy specimen)	CSF	Environmental sample (soil, water)
Antibiotics (Pen-Strep, gentamicin, amphotericin B)	Nonnutrient agar (plus *Escherichia coli*)	Wet mounts, stains	Nonnutrient agar (plus *Escherichia coli*)
Tissue culture (monkey kidney, rat glioma, human lung fibroblasts)	Wet mounts, stains; enflagellation (identification of *Naegleria*)	Nonnutrient agar; wet mounts, stains; enflagellation (identification of *Naegleria*)	Wet mounts, stains; enflagellation (identification of *Naegleria*)

[a]The decision to examine a specimen as a wet mount and/or permanent stained smears will depend on the amount of specimen received. If the amount is quite small, direct inoculation of nonnutrient agar seeded with bacteria is recommended. Once organisms have been isolated, they can be frozen. (Adapted from reference 53.)

Table 32.2   Issues related to free-living amebae and symbiotic bacteria

Issue	Comments	Reference
Eyewash stations and flushing regimens; amebic and bacterial concentrations temporarily decreased with various flushing regimens	The lower amebic concentrations were not sustained	5
Bath basins (whirlpool baths) in which *Hartmanella* and *Vannella* were isolated tended to harbor large number of *Legionella*	Management practices such as frequent washing of filter elements and/or frequent addition of tap water to bath basins are highly recommended to reduce microbial contaminants	29
Survival and growth of *Burkholderia cepacia* was documented in *Acanthamoeba polyphaga*	These findings should be taken into account when reviewing environmental reservoirs of *B. cepacia*	30
Legionella and amebae from the same hospital water system were cocultures; in 14 of 16 cocultivations, growth of legionellae was found	This is important for the disinfection process since bacteria within the amebae are protected and are more resistant to environmental manipulation	47
Amebae, in addition to serving as protective hosts, may play a role in the thermotolerance, invasiveness, and antibiotic resistance of bacteria	Considering the reduced immune status of many patients, this "symbiosis" of free-living amebae and bacteria might still be an underestimated but important hospital hygiene issue	66
A *Legionella* organism was isolated from a soil ameba; this bacterium is distinct from previously described species within the genus *Legionella*	Each time another different organism is isolated from free-living amebae, one wonders how many are capable of this symbiotic relationship and how many could ultimately be or become human pathogens	43
Free-living amebae are well adapted to their hostile environment and are resistant to desiccation, elevated temperatures, and various disinfectants	There is increasing interest in the potential role of free-living amebae as reservoirs and vectors of pathogenic bacteria	68
In a longitudinal study of 20 dental units, 96% of all water samples contained one or several ameba species; they were all within the *Naegleria gruberi* complex (no pathogenic strains are found within this group)	The occasional protective intracellular inclusion of legionellae in trophozoites of various species and cysts of acanthamoebae might provide an explanation for the resistance of legionellae to disinfection measures	41
The presence of adhered *Pseudomonas aeruginosa* on hydrogel contact lenses facilitated the binding of *Acanthamoeba* spp.	Cocontamination of lens systems with bacteria may be a prime factor in the development of amebic keratitis	21
Intracellular multiplication of *P. aeruginosa* has been found in free-living amebae	With their well-known resistance to chlorine, the amebae and their cysts are considered vectors for these intracellular bacteria	42
Studies indicate that the viability of *Acanthamoeba* cysts may be 25 years, during which they can maintain their invasive properties	This has tremendous environmental ramifications in terms of viability and virulence of the acanthamoebae	39

the warmed agar plate. Spread the bacteria on the surface of the agar with a bacteriological loop.

3. Inoculate the specimen on the center of the agar plate as described below.

A. For CSF samples, centrifuge the CSF at 250 × g for 10 min. With a sterile serologic pipette, carefully transfer all but 0.5 ml of the supernatant to a sterile tube and store at 4°C (for possible future use). Mix the sediment in the rest of the fluid, and, with a Pasteur pipette, place 2 or 3 drops in the center of the nonnutrient agar plate that has been precoated with bacteria. After the fluid has been absorbed, seal the plates with a 5- to 6-in. length of 1-in.-wide Parafilm strip. Incubate the plate inverted at 37°C in room air. Using a wax pencil or laboratory marker, you may want to make a circle on the underside of the plate to indicate exactly where the specimen was inoculated onto the agar.

B. For tissue samples, triturate a small piece of the tissue (brain, lung, skin abscess, corneal biopsy, or similar specimens) in a small quantity (ca. 0.5 ml) of ameba saline. Process as above. Corneal smear, ear discharge material, etc., may be placed directly on the agar surface. Incubate central nervous system tissues at 37°C (room air) and tissues from other sites at 30°C.

C. Water samples of 10 to 100 ml may be processed to isolate these amebae. First, filter the water sample through three layers of sterile gauze or cheesecloth to remove leaves, dirt, etc. Next, either (i) filter the sample through a sterile 5.0-µm cellulose acetate membrane (47 mm in diameter), invert the membrane over a nonnutrient agar plate precoated with bacteria, seal, and incubate the plates as above; or (ii) centrifuge the water sample for 10 min at 250 × g. Aspirate the supernatant, suspend the sediment in about 0.5 ml of ameba saline, and deposit this suspension in the center of the nonnutrient agar plate precoated with bacteria. Seal and incubate the plate at 37°C as before.

D. For soil samples, mix about 1 g of the soil sample with enough ameba saline (ca. 0.5 to 1 ml) to make a thick slurry. Inoculate this slurry in the center of the nonnutrient agar plate precoated with bacteria, and incubate as above.

E. For contact lens solutions, small volumes (ca. 1 to 2 ml) may be inoculated directly onto the nonnutrient agar plates precoated with bacteria. Larger volumes (2 to 50 ml) should be centrifuged as in step 3, and the sediment should be inoculated onto the center of the nonnutrient agar plate and incubated as above.

4. Using the low-power (10×) objective, examine the plates microscopically for amebae (cysts or trophozoites) every day for 10 days. Thin linear tracks (areas where amebae have ingested bacteria) might also be seen. If amebae are seen, circle that area with a wax pencil, carefully remove the Parafilm seal in a biological safety cabinet, open the lid of the petri dish, and carefully cut out the marked area from the agar with a spatula that has been heated to red hot and cooled before use to prevent contamination. Transfer the piece face down onto the surface of a fresh agar plate coated with bacteria, seal the plate with Parafilm, and incubate as before. *Naegleria* and *Acanthamoeba* spp. can be cultured by this method and, with periodic transfers, can be maintained in the laboratory indefinitely. In lieu of subcultures, the organisms can also be frozen for long-term storage. Under the microscope, the amebae resemble small uneven spots; observation for several seconds may reveal organism motility. After 4 to 5 days of incubation, the amebae begin to encyst and both trophozoites and cysts are visible. Unfortunately, *Balamuthia* spp. cannot be grown by using this system; they can be grown by using tissue culture methods on monkey kidney or lung fibroblast cell lines.

5. The enflagellation experiment is carried out as follows.

A. Examine the plates every day for signs of amebae. If present, amebae will feed on bacteria, multiply, and cover the entire surface of the plate within a few days. Once the food supply is exhausted, the amebae will differentiate into cysts.

B. Use a wax pencil to mark the area containing a large number of amebic trophozoites.

C. Using a bacteriological loop, scrape the surface of the agar at the marked area and transfer several loopfuls of the scraping to a sterile tube containing about 2 ml of sterile distilled water. Alternately, flood the surface of the agar plate with about 10 ml of sterile distilled water, gently scrape the agar surface with a loop, transfer the liquid to a sterile tube, and incubate at 37°C.

D. Periodically examine the tube with an inverted microscope for the presence of flagellates.

a. *N. fowleri*, the causal agent of primary amebic meningoencephalitis, undergoes transformation to a pear-shaped flagellate, usually with two flagella but occasionally with three or four flagella; the flagellate stage is a temporary nonfeeding stage and usually reverts to the trophozoite stage

(Figure 32.2). *N. fowleri* trophozoites are typically ameba-like and move in a sinuous way. They are characterized by a nucleus with a centrally located, large nucleolus. The trophozoites are also characterized by the presence of a contractile vacuole that appears once every 45 to 50 s and discharges its contents. The contractile vacuole looks like a hole or a dark depression inside the trophozoite and can be easily seen when examining the plate under the 10× or 40× objective. When the food supply is exhausted, *N. fowleri* trophozoites differentiate into spherical, smooth-walled cysts.

b. In contrast, *Acanthamoeba* spp., which cause keratitis and granulomatous amebic encephalitis, do not transform into the flagellate stage. *Acanthamoeba* trophozoites are characterized by the presence of fine, thorn-like processes that are constantly extended and retracted. The trophozoites produce double-walled cysts characterized by a wrinkled outer wall (ectocyst) and a polygonal, stellate, oval, or even round inner wall (endocyst). The trophozoites are also characterized by the presence of the contractile vacuole, which disappears and reappears at regular intervals (45 to 50 s).

c. The cysts of both *Acanthamoeba* and *Naegleria* spp. are uninucleate.

Figure 32.2 *Naegleria fowleri* flagellate stage. When placed in distilled water (enflagellation test), *N. fowleri*, the causal agent of primary amebic meningoencephalitis, undergoes transformation to a pear-shaped flagellate, usually with two flagella but occasionally with three or four flagella; the flagellate stage is a temporary nonfeeding stage and usually reverts to the trophozoite stage.

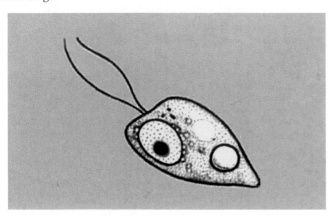

d. Immunofluorescence and immunoperoxidase tests with monoclonal or polyclonal antibodies (available at the Centers for Disease Control and Prevention) are helpful in differentiating *Acanthamoeba* and *Balamuthia* spp. in fixed tissue.

Quality Control for Pathogenic Free-Living Amebae

The following control strains should be available when using these cultures for clinical specimens: ATCC 30010 (*A. castellanii*) and ATCC 30215 (*N. fowleri*). ATCC strains of *E. coli* or *E. aerogenes* are not necessary. Any routine clinical isolate or stock organism is acceptable.

1. Check all reagents and media (ameba saline, distilled water, and nonnutrient agar plates) each time they are used or periodically (once a week).
   A. The media and all solutions should be free of any visible signs of precipitation and bacterial and/or fungal contamination.
   B. Examine the nonnutrient agar plates under the 4× objective of an inverted or binocular microscope, and make sure that no fungal contamination has occurred.

2. Maintain stock cultures of *A. castellanii* and *N. fowleri* at 25°C.
   A. Transfer stock cultures monthly, using nonnutrient agar plates prepared with Page's ameba saline and the bacterial overlay of *E. coli*.
   B. *N. fowleri* measures 10 to 35 µm and demonstrates an eruptive locomotion by producing smooth hemispherical bulges. The cyst produces smooth (7- to 15-µm) walls. The flagellate stage does not have a cytostome.
   C. *A. castellanii* is 15 to 45 µm and produces fine, tapering, hyaline projections called acanthapodia. It has no flagellate stage but produces a double-walled cyst with a wrinkled outer wall (10 to 25 µm).
   D. Trophozoites of *Naegleria* and *Acanthamoeba* spp. are uninucleate, characterized by a large, dense central nucleolus.
   E. For staining, a slide prepared from a stock strain of amebae is run in parallel with the patient slide. Staining results are acceptable when the control amebae stain well.
   F. Plate both stock cultures onto fresh media, and incubate at 37°C parallel with the patient culture. Culture results are acceptable when growth appears by day 7.
   G. Run *N. fowleri* in parallel with the patient culture being observed for enflagellation. The test results are acceptable when free-swimming,

pear-shaped flagellates with two flagella are observed in 2 to 24 h on the control slide.

3. Depending on use, the microscope(s) may have to be calibrated (within the last 12 months), and the original optics used for the calibration should be in place on the microscope(s). The calibration factors for all objectives should be posted on the microscope for easy access.

Note For patient specimens, if a plate is positive for amebae and the amebae transform into flagellates, the specimen should be reported as positive for *N. fowleri*. If the amebae do not transform into flagellates even after overnight incubation, if the trophozoites possess the characteristic acanthapodia, if they show a large centrally placed nucleolus in the nucleus on trichrome stain, and if the trophozoites differentiate into the characteristic double-walled cysts, the specimen should be reported as positive for *Acanthamoeba* sp. For contact lens solution, if the plates are positive for amebae and the amebae do not transform into flagellates but differentiate into cysts with a wrinkled outer ectocyst and an inner stellate, polygonal, oval, or round endocyst, the specimen should be reported as positive for *Acanthamoeba* sp. *Naegleria* spp. have not been isolated from contact lens solutions; however, small amebae (e.g., hartmannellid or vahlkampfiid amebae, which produce smooth-walled cysts), probably contaminants, have occasionally been isolated from these solutions. Plates inoculated with water samples are usually positive for many genera and species of small free-living amebae (freshwater is their normal habitat). Therefore, the sample should be reported as positive for small free-living amebae. The physician should be notified immediately if patient specimens are positive for *Acanthamoeba* or *Naegleria* spp.

Most patient specimens, especially CSF, should be examined microscopically as soon as they arrive in the laboratory. However, if the specimen is very clear with no visible sediment, it should be cultured without being examined microscopically.

1. After centrifugation, remove a small drop of the CSF sediment, place it on a microscope slide, cover it with a no. 1 coverslip, seal the edges of the coverslip with Vaspar (optional), and examine it immediately with the 10× or 40× objective (phase-contrast/differential interference contrast optics are preferred). If bright-field microscopy is used, reduce the illumination by adjusting the iris diaphragm.
   A. The *N. fowleri* trophozoite is highly motile and can be identified by its sinuous movement. A warmed penny may be applied to the bottom surface of the slide to activate the movement

of the trophozoite. Occasionally, a flagellate is seen traversing the field.
   B. *Acanthamoeba* trophozoites are rarely seen in the CSF. If present, they may be recognized by their characteristic acanthapodia, which are constantly extended and retracted. Both amebae, especially *Acanthamoeba* spp., may be recognized by the contractile vacuole.

2. Contact lens care solutions (opened) can be processed and examined like CSF.
3. If very small amounts of tissue are received, they should be reserved for culture. If the specimen received is visibly contaminated with bacteria and/ or fungi, the tube containing the specimen should be placed in an ice-water bath for 3 min prior to centrifugation. This causes any trophozoites attached to the walls of the tube to come off and drop into the fluid prior to centrifugation.
4. Material from the surface of a positive agar plate can be removed, fixed, and stained with trichrome for microscopic examination at a higher magnification (×1,000). Results obtained with wet mounts should always be confirmed by performing permanent stained smears (trichrome, iron hematoxylin) for nuclear characteristics to differentiate the amebae from host cells. Organisms may not be recovered if appropriate centrifugation speeds and times are not used.

Liquid Culture Media for Pathogenic Free-Living Amebae

Although *Acanthamoeba* spp. can be grown in serum-free liquid medium, *N. fowleri* requires the addition of fetal calf serum or brain extract to grow in a liquid medium. Peptone-yeast extract-glucose medium for *Acanthamoeba* spp. and Nelson's medium for *N. fowleri* are described below.

Peptone-Yeast Extract-Glucose (PYG) Medium for *Acanthamoeba* spp.

Proteose peptone (Difco)	20.0 g
Yeast extract (Difco)	2.0 g
Magnesium sulfate ($MgSO_4$)	0.980 g
Calcium chloride ($CaCl_2$)	0.059 g
Sodium citrate ($Na_3C_6H_5O_7 \cdot 2H_2O$)	1.0 g
Ferric ammonium sulfate [$Fe(NH_4)_2(SO_4)_2 \cdot 6H_2O$)]	0.02 g
Potassium phosphate, monobasic (KH_2PO_4)	0.34 g
Sodium phosphate, dibasic ($Na_2HPO_4 \cdot 7H_2O$)	0.355 g
Glucose	18.0 g
Distilled water to	1,000.0 ml

Final pH is 6.5 ± 0.2.

1. Dissolve all ingredients except CaCl$_2$ in about 900 ml of distilled water in a bottle or flask with a magnetic stirrer.
2. Add CaCl$_2$ while stirring.
3. Bring the volume to 1,000 ml with distilled water.
4. Dispense into 16- by 125-mm screw-cap tubes, 5 ml per tube.
5. Autoclave at 121°C for 15 min.
6. When tubes have cooled, label the tubes as *Acanthamoeba* medium with the preparation date and an expiration date of 3 months.
7. Store at 4°C.

Nelson's Medium for *N. fowleri*

Panmede (ox liver digest) (Difco) .... 1.0 g
Glucose .... 1.0 g
10× ameba saline (p. 917) .... 100.0 ml
Double-distilled water to .... 900.0 ml

1. Add ameba saline to distilled water to make 1× ameba saline.
2. Dissolve the ingredients in ameba saline with a magnetic stirrer.
3. Dispense into 16- by 125-mm screw-cap tubes, 10 ml per tube.
4. Autoclave for 15 min at 15 lb/in^2.
5. Cool, and label as Nelson's medium with the preparation date and an expiration date of 3 months.
6. Store at 4°C.
7. Add 0.2 ml of heat-inactivated fetal calf serum to each tube before inoculating with the amebae.

Cell Culture

Acanthamoeba, *Balamuthia*, and *Naegleria* spp. can be inoculated onto a number of different mammalian cell lines. Culture options for *Sappinia diploidea* are not yet well defined. The organisms grow very well and demonstrate cytopathic effects, similar to those seen in routine viral cultures. However, most laboratories do not offer these methods on a routine basis.

Standard Commercial Media for Growth and Isolation of *Acanthamoeba* spp.

Several studies have reported on the use of standard commercial media used for growth and isolation of *Acanthamoeba* spp. Buffered charcoal-yeast extract agar (BCYE) is an excellent commercially available culture medium for the recovery of *Acanthamoeba* spp. (49, 50). Good trophozoite recovery was obtained using BBL Trypticase soy agar (TSA) with 5% rabbit blood, TSA with 5% horse blood, and

Remel TSA with 5% sheep blood. BBL TSA with 5% horse blood or 5% rabbit blood yielded good recovery of cysts. Nonnutrient agar with either live or dead bacteria yielded good recovery of trophozoites; however, live bacteria were required for good cyst recovery (49).

Pathogenic Flagellates

Specimens from women may consist of vaginal exudate collected from the posterior fornix on cotton-tipped applicator sticks or genital secretions collected on polyester sponges. Specimens from men can include semen, urethral samples collected with polyester sponges, or urine. Urine samples collected from the patient should be the first voided specimen in the morning. It is critical that clinical specimens be inoculated into culture medium as soon as possible after collection (11, 12, 32, 40, 57). Although collection swabs can be used, there are often problems with specimens drying prior to culture; immediate processing is mandatory for maximum organism recovery. The culture method is considered to be the most sensitive for the diagnosis of trichomoniasis; however, because of the time and effort involved, some laboratories have decided to use some of the new monoclonal antibody detection kits (6). Another approach would be to use the plastic envelope method for *Trichomonas vaginalis* (InPouch TV [BIOMED Diagnostics, San Jose, Calif.]), a simplified technique for the transport and culture that is illustrated in Figures 32.3 and 32.4 (6). Because of its long shelf life, relatively low expense, and high sensitivity, studies confirm that the pouch system provides a good diagnostic culture method for *T. vaginalis* (4, 46). It has been found to be more sensitive than either Diamond's or Trichosel medium (4).

Lash's Casein Hydrolysate-Serum Medium

Lash's casein hydrolysate-serum medium is a culture medium used to diagnose *T. vaginalis* (36).

Lash's Casein Hydrolysate-Serum Medium

Casamino Acids (Difco) .... 14.0 g
Maltose .... 1.5 g
Dextrose .... 2.0 g
Sodium lactate (60% solution) .... 0.5 g

Dissolve in 500 ml of distilled water. To this solution, add:

NaCl .... 6.0 g
KCl .... 0.1 g
CaCl$_2$ .... 0.1 g

1. Adjust the pH to 6.0 with concentrated phosphoric acid. Dispense in 5-ml aliquots, plug the tubes, and autoclave at 121°C for 15 min.

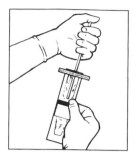

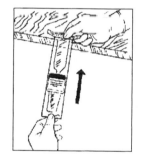

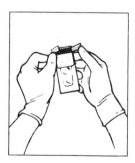

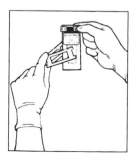

2. Prepare serum solution:
   Whole beef blood serum .... 200 ml
   Distilled water... 300 ml
   NaHCO$_3$ .... 0.1 g
3. Sterilize the serum solution by filtration.
4. The complete medium contains 5 ml of serum solution and 5 ml of basic solution for each tube.
5. Incubate the cultures at 37.5°C, and examine 24 and 48 h after inoculation.

T. vaginalis has been studied for a number of years with agar plate cultures for both cloning and diagnosis (2, 3, 28). Hollander (23) also suggests that colony morphology may be related to pathogenicity as well as being diagnostic. Many media are available for the isolation of *T. vaginalis*; some of these can be purchased commercially and have a relatively long shelf life.

CPLM (Cysteine-Peptone-Liver-Maltose) Medium

Ringer's solution
   Sodium chloride (NaCl) .... 0.6 g
   Sodium bicarbonate (NaHCO$_3$) .... 0.01 g
   Potassium chloride (KCl)......................... 0.01 g
   Calcium chloride (CaCl$_2$) ....0.01 g
   Double-distilled water to...................... 100.0 ml

Dissolve the ingredients in the order listed, and bring the volume up to 100.0 ml with distilled water.

Liver infusion
   Double-distilled water .... 330.0 ml
   Bacto Liver Infusion powder (Difco) ...... 20.0 g

1. Place the distilled water in a large beaker.
2. Add the liver infusion powder.
3. Infuse for 1 h at 50°C.
4. Raise the temperature to 80°C for 5 min to coagulate the protein.
5. Filter through a Whatman no. 1 paper with a Büchner funnel.

Figure 32.3 Illustration of the InPouch TV culture system for *Trichomonas vaginalis*. From top to bottom: (1) introduction of the specimen into the upper chamber containing a small amount of medium; (2) application of a plastic holder for microscope viewing prior to expressing medium into the lower chamber (optional); (3) transfer of a small amount of medium in the upper chamber to the lower chamber; (4) rolling down the upper chamber and sealing it with the tape; (5) plastic viewing frame used to immobilize the medium in the pouch for examination under the microscope. Specific photographs can also be seen in Figure 5.22. (Courtesy of BIOMED Diagnostics.)

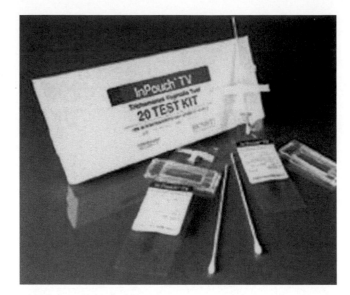

Figure 32.4 InPouch TV culture system for *Trichomonas vaginalis*. Note the pouch, swabs, and plastic pouch holder for microscopic examination of the pouch contents. (Courtesy of BIOMED Diagnostics.)

Methylene blue solution

Methylene blue	0.5 g
Glass-distilled water	100.0 ml

Mix well until dissolved.

CPLM (Cysteine-Peptone-Liver-Maltose) Complete Medium

Bacto Peptone (Difco)	32.0 g
Bacto Agar (Difco)	1.6 g
Cysteine HCl	2.4 g
Maltose	1.6 g
Bacto Liver Infusion (Difco)	320.0 ml
Ringer's solution	960.0 ml

1. With a magnetic stirrer, mix Ringer's solution and the liver infusion in a large beaker.
2. Add peptone, maltose, cysteine HCl, and agar in that order, and heat the mixture until dissolved.
3. Add 0.7 ml of aqueous methylene blue.
4. Adjust the pH to 5.8 to 6.0 with 1 N NaOH or 1 N HCl.
5. Dispense 8-ml volumes into culture tubes.
6. Autoclave at 121°C for 15 min.
7. Aseptically add 2 ml of human serum, heat inactivated at 56°C for 30 min and cooled, per tube. Horse serum is recommended as a replacement for human serum, particularly when considering safety issues such as handling human blood and blood products.
8. Label as CPLM medium with the preparation date.

9. Store at room temperature. Use as long as the amber zone indicating an anaerobic condition persists.

Diamond's TYM (Trypticase-Yeast Extract-Maltose) Complete Medium

Trypticase (BBL)	20.0 g
Yeast extract (BBL)	10.0 g
Maltose	5.0 g
L-Cysteine HCl	1.0 g
L-Ascorbic acid	0.2 g
Potassium phosphate, dibasic (K_2HPO_4)	0.8 g
Potassium phosphate, monobasic (KH_2PO_4)	0.8 g
Bacto Agar (Difco)	0.5 g
Double-distilled water	900.0 ml

1. Dissolve the buffer salts in the distilled water with a magnetic stirrer.
2. Add the remaining ingredients except the agar, in the order given, one at a time until dissolved.
3. Adjust the pH to 6.0 with 1 N HCl.
4. Add agar, and heat to dissolve.
5. Autoclave at 121°C for 15 min.
6. Cool to 45°C, and add 100 ml of bovine, sheep, or horse serum that has been inactivated for 30 min at 56°C.
7. Aseptically dispense 10-ml volumes to 16- by 125-mm screw-cap tubes.
8. Label as TYM medium with the preparation date and an expiration date of 10 days.
9. Store at 4°C.

Hollander's Modification of TYM Complete Medium

Hollander's modification of TYM complete medium differs from TYM medium by the replacement of cysteine with additional ascorbic acid and the addition of potassium chloride, potassium carbonate, and ferrous sulfate (7).

Casein digest peptone (Trypticase, BBC)	20.0 g
Yeast extract	10.0 g
Maltose	5.0 g
Ascorbic acid	1.0 g
Potassium chloride (KCl)	1.0 g
Potassium bicarbonate	1.0 g
Potassium phosphate, monobasic (KH_2PO_4)	1.0 g
Potassium phosphate, dibasic (K_2HPO_4)	0.5 g
Ferous sulfate	0.1 g
Distilled water	600 ml

Proceed as described above for Diamond's TYM complete medium.

Note Hollander also includes agar to 0.05%.

Diamond's Complete Medium (Modified by Klass)

Trypticase (BBL) .... 24.0 g
Yeast extract (BBL).................................... 12.0 g
Maltose .... 6.0 g
Cysteine HCl .... 1.2 g
Ascorbic acid.. 0.24 g
Double-distilled water .... 900.0 ml

1. Dissolve the ingredients one at a time in the order given.
2. Adjust the pH to 6.0 with 1 N HCl or 1 N NaOH.
3. Dispense in 12.5-ml aliquots to 16- by 125-mm screw-cap tubes.
4. Autoclave at 121°C for 15 min.
5. When cool (50°C), add 1 ml of sterile inactivated horse serum and 0.5 ml of antibiotic mixture to each tube.
6. Label as modified TYM medium with the date of preparation and an expiration date of 3 weeks.

Antibiotic Mixture

Sodium penicillin G .... 1,000,000 U
Streptomycin sulfate .... 1,000,000 µg
Fungizone (amphotericin B)................. 2,000 µg
Sterile double-distilled water .... 50.0 ml

1. Mix thoroughly.
   Concentration of stock solution is:
   Penicillin .... 20,000 U/ml
   Streptomycin................................. 20,000 µg/ml
   Fungizone .... 40 µg/ml
2. Dispense 1 ml of the antibiotic mixture into sterile screw-cap vials or sterile cryovials.
3. Label as antibiotic solution with the date of preparation and an expiration date of 1 year.
4. Store at −20°C in cryoboxes.

Serum Substitutions

Bovine, sheep, or horse serum can be substituted in the step involving CPLM complete medium.

Axenic Culture

1. Inoculation of culture medium.
   A. Remove tubes containing culture medium from 4°C, and incubate at 37°C for 1 to 2 h.
   B. Vigorously shake the cotton-tipped portion of the applicator stick containing the patient specimen in the medium, and then break off the tip with sterile forceps and drop it into the medium.
   C. If the material is collected on polyester sponges, drop the sponges into the medium and shake the tube.
   D. Centrifuge urine samples for 10 min at $250 \times g$, aspirate the supernatant, and inoculate the sediment into the medium.
   E. Examine the tubes daily for several days, and subculture if necessary. To subculture, first shake the tube to disperse the organisms uniformly, remove about 1 to 2 ml, and inoculate into a warmed, fresh tube.
   F. Incubate the tubes in a slanted position (45° angle).
   G. Incubate the control tubes and those containing patient material for at least 72 to 96 h.
   H. Examine the entire length of the tube. If the specimen is positive, *T. vaginalis* will be found freely swimming or attached to the tube walls.
   I. Do not report negative results until 96 h.
2. To maintain stock cultures,
   A. With an inverted microscope, examine stock culture tubes of *T. vaginalis* for any signs of bacterial contamination. Select one or more tubes showing good growth. Since the tubes are incubated in a slanted position, usually at an angle of 45°, a thick button of organisms is seen at the bottom of the tube. Gently invert the tube once or twice to disperse the trichomonads uniformly, and examine the tubes again. A large number of the organisms should be freely swimming, and a few will be attached to the tube walls.
   B. Immerse the tubes in a bucket of ice-cold water for about 5 to 10 min to dislodge the trichomonads from the tube walls. Invert the tubes several times to distribute the organisms.
   C. With a sterile Pasteur pipette, remove about 1.0 to 1.5 ml; inoculate 0.5 to 1.0 ml into a fresh tube. Inoculate the rest of the fluid into nutrient agar, brain heart infusion, and thioglycolate broth for routine monitoring of bacterial contamination. Inoculate several additional tubes this way, and incubate the cultures slanted at a 45° angle at 37°C as before.
   D. If growth is poor and only few organisms are present along with a lot of debris, centrifuge the tube at $250 \times g$ for 10 min, aspirate the supernatant, transfer the sediment to a fresh tube, and incubate as before.

Quality Control for Pathogenic Flagellates

ATCC 30001 (*T. vaginalis*) should be available as a control strain when these cultures are used for clinical specimens.

1. Check all reagents and media (at least once a week). All media, including Ringer's solution, should be free of any signs of precipitation and bacterial and/or fungal contamination.
2. Depending on use, the microscope(s) may have had to be calibrated within the last 12 months, and the original optics used for the calibration should be in place on the microscope(s). The calibration factors for all objectives should be posted on the microscope for easy access.
3. Maintain stock cultures of ATCC 30001 (*T. vaginalis*).
   A. Transfer stock cultures weekly.
      a. Stock organisms should always be cultured at the same time a patient specimen is inoculated into culture medium.
      b. If the stock organisms multiply and remain viable during the 96 h, patient results can be reported.
   B. Stain
      a. A slide prepared from a stock strain of *T. vaginalis* is run in parallel with the patient slide.
      b. Staining results are acceptable when the control organisms stain well.
4. Control organisms must be cultured each time a patient specimen is inoculated into the culture medium.

Note If no trophozoites are seen after 4 days of incubation, discard the tubes and report the culture as negative. Results of patient specimens should not be reported as positive unless control cultures are positive. Cultivation is the most sensitive method for the diagnosis of trichomoniasis. Every effort, therefore, must be made to inoculate patient materials into culture medium. However, since this method may take as long as 3 to 4 days and the patient materials occasionally contain nonviable organisms, it is imperative that microscopic examination of wet smears and/or stained smears (Giemsa) also be performed.

Axenic Cultivation of *T. vaginalis* in a Serum-Free Medium

Mammalian serum or bovine serum albumin is required for *T. vaginalis* grown under axenic conditions. However, these components inhibit several biological properties of these parasites. PACSR is a serum replacement, free of bovine serum albumin, which is used for axenic cultivation of *E. histolytica*. Studies indicate that PACSR added to several

of the standard media used for the cultivation of *T. vaginalis* can support growth in the absence of serum (38).

Flagellates of Blood and Tissue

The first types of media used to culture the blood and tissue flagellates, which are still useful for establishment of cultures, were undefined and contained a complex mixture of ingredients. Improvements led to semidefined formulations that included tissue culture media as a base and, as a next step, addition of tissue culture cells as a feeder layer to promote parasite growth. Newer developed media are completely defined, having replaced the feeder cells with various supplements. Serum, a variable component of the media, can now be replaced by various serum substitutes. Fully defined formulations are available for the cultivation of many of these organisms (55).

For the hemoflagellates found in the bloodstream, probably the most direct means of diagnosis is by examination of a Giemsa-stained blood smear under the microscope. However, if isolation of the parasite is required for confirmation, culture can be used. The easiest approach involves the use of NNN medium. Because of its relatively short shelf life (2 to 4 weeks with refrigeration) and the infrequent need for culture, most clinical laboratories would not routinely stock NNN medium. However, the CDC (Division of Parasitic Diseases, Atlanta, Ga.) can provide NNN medium for inoculation with blood or skin biopsy specimens, which can be returned to CDC for workup. In other parts of the world, major reference laboratories may provide a similar service. The procedure for isolating *Leishmania* spp. from patients with suspected cases of cutaneous, mucocutaneous, and visceral leishmaniases is basically similar to that for isolating *Trypanosome* spp. The amastigote found within host cells, or the promastigote that develops from the amastigote in culture, would be confirmatory. Punch biopsy specimens or needle aspirates serve as the inocula for NNN medium or Schneider's *Drosophila* medium with 30% fetal bovine serum. Other procedures such as splenic puncture and liver biopsy can also yield material but are invasive and less likely to be performed (55).

Specimens for culturing *Leishmania* spp. may consist of aspirates, scrapings, or biopsy material from skin lesions of patients with cutaneous leishmaniasis; bone marrow aspirates or, more rarely, splenic aspirates from visceral leishmaniasis patients; or normal skin biopsy specimens, lymph node aspirates, or pieces of liver and spleen from suspected or potential wild- or domestic-animal reservoirs. Ensure that the skin surrounding the ulcer is thoroughly cleaned and swabbed with 70% alcohol (sterile saline is not acceptable as a cleansing agent) and allowed to dry before the sample is removed. Also ensure that alcohol does not get into the ulcerated area or broken skin. A punch biopsy specimen taken from the advancing margin of the lesion is often recommended. If skin biopsy fragments are placed in transport medium (RPMI medium supplemented with 10% fetal calf serum) maintained at ambient temperature, delays in transit through the mail will still not prevent recovery of the organisms. In one study, after being received in the laboratory (transit time as long as 3 to 17 days), the biopsy specimens were ground in sterile saline and inoculated into NNN culture tubes. The tubes were incubated at 25°C and subcultured every week until the fifth week. Cultures were positive in 9 of 16 cases in all seasons and for three different *Leishmania* species. Thus, delayed culture can still yield valuable results from biopsy specimens obtained taken under field conditions (9).

It is imperative that only a few drops of bone marrow juice or spleen aspirate be inoculated into tubes. Inoculate several tubes with a few drops each rather than inoculating a single tube with a large volume (1 to 2 ml), since the serum in the specimen may contain leishmanicidal or inhibitory factors that prevent the growth of organisms. Alternatively, bone marrow juice may be centrifuged for 10 min at 250 × g and the sediment may be washed in 0.85% saline by centrifugation and then inoculated into culture tubes. Buffy coat from the blood sample rather than whole blood should be inoculated. Because the leishmanias are fastidious organisms and all isolates may not grow in any one medium, it is imperative that at least two media be used; for example, use NNN or modified Tobie's medium and Schneider's *Drosophila* medium.

Specimens for culturing *Trypanosoma cruzi* may consist of blood or the gut contents of the triatomid bug. It is advisable to use two different media such as liver infusion-tryptose (LIT) and NNN for the initial isolation of *T. cruzi*. Once growth is established, use the medium in which best growth is obtained for subculture. According to James Sullivan, LIT medium, when used as an overlay on Tobie's slants, is excellent for isolation and diagnosis (55). The major culture form is the epimastigote; occasionally, however, trypomastigotes and amastigotes are also seen.

NNN Medium (Leishmaniasis or Chagas' Disease) (22)

Bacto Agar (Difco) .... 1.4 g
Sodium chloride (NaCl)................................. 0.6 g
Double-distilled water .... 90.0 ml

1. Mix the NaCl and agar in the distilled water in a 500-ml flask.
2. Heat the mixture until the agar melts.
3. Autoclave at 121°C for 15 min.
4. Cool to about 50°C.
5. Add 10 ml of aseptically collected, defibrinated rabbit blood.
   Note Although blood collected with EDTA as the anticoagulant can be used for routine stock culture subcultures, it may not be quite as effective as defibrinated blood in isolating organisms from patient specimens. Certainly, if defibrinated blood is not available, blood collected with EDTA as the anticoagulant can be used. Make sure that the rabbit blood is fresh (no older than 10 days). It should be aseptically collected and stored at 4°C until used.
6. Dispense 4 ml into 16- by 125-mm sterile screw-cap culture tubes.
7. Place the tubes at a 10° angle (shallow slant position) until the agar sets.
8. Immediately transfer the tubes into test tube stands, and let stand upright at 4°C so that the bottom portion of the slants is covered with the water of condensation. Rapid cooling increases the water of condensation (Figure 32.5).
9. Label as NNN medium with the preparation date and an expiration date 3 weeks from the date of preparation.
10. Store at 4°C.

NNN Medium, Offutt's Modification (Leishmaniasis)

Blood Agar Base (Difco) .... 8.0 g
Double-distilled water .... 200.0 ml

1. Heat until the agar is dissolved in the distilled water in a 500-ml flask.
2. Autoclave at 121°C for 15 min.
3. Cool to about 50°C.
4. Add 15 ml of aseptically collected, defibrinated rabbit blood.
5. Dispense 4 ml into 16- by 125-mm sterile screw-cap culture tubes.
6. Place the tubes at a 10° angle (shallow slant position) until the agar sets.
7. Immediately transfer the tubes to test tube stands, and let stand upright at 4°C so that the bottom portion of the slants is covered with the water of condensation. Rapid cooling increases the water of condensation.
8. Label as Offutt's medium with the preparation date and an expiration date 3 weeks from the date of preparation.
9. Store at 4°C.

Overlay Solution (To Be Used with NNN or NNN Modified)

Sodium chloride (NaCl)................................. 4.5 g
Double-distilled water .... 500.0 ml

1. Autoclave at 121°C.
2. Dispense 4 ml aseptically into 16- by 125-mm sterile culture tubes.
3. Label as 0.9% saline with the preparation date and an expiration date 3 weeks from the date of preparation.
4. Store at 4°C.

Evan's Modified Tobie's Medium (Leishmaniasis or Chagas' Disease) (22)

Beef extract (Oxoid Lab-Lemco L29) .... 0.3 g
Bacteriological peptone (Oxoid L37)............ 0.5 g
Sodium chloride (NaCl).............................. 0.8 g
Agar (Oxoid purified)................................. 2.0 g
Double-distilled water .... 100.0 ml

1. Mix all the ingredients in the distilled water in a large beaker with a magnetic stirrer.
2. Heat the mixture until the agar melts.
3. Dispense 5 ml into 16- by 125-mm screw-cap culture tubes.
4. Autoclave at 121°C for 15 min.
5. Cool to about 50°C.
6. Add 1.2 ml of aseptically collected, defibrinated horse blood.
7. Hold the tubes upright in the palm of your hand, and roll the tubes gently to mix the blood and the agar well.
8. Place the tubes at a 10° angle (shallow slant position) until the agar sets.
9. Immediately transfer the tubes into test tube stands, and let stand upright at 4°C so that the bottom portion of the slants will be covered with the water of condensation. Rapid cooling increases the water of condensation.
10. Label as Evan's modified Tobie's medium with the preparation date and an expiration date 3 weeks from the date of preparation.
11. Store at 4°C.

Overlay Solution (To Be Used with Tobie's Medium)

Potassium chloride (KCl).............................. 0.4 g
Sodium phosphate, dibasic
   (Na$_2$HPO$_4$ · 12H$_2$O) .... 0.06 g
Potassium phosphate,
   monobasic (KH$_2$PO$_4$) .... 0.06 g
Calcium chloride (CaCl$_2$ · 2H$_2$O) .... 0.185 g
Magnesium sulfate (MgSO$_4$ · 7H$_2$O) .... 0.1 g
Magnesium chloride (MgCl$_2$ · 6H$_2$O) .... 0.1 g
NaCl.. 8.0 g
L-Proline .... 1.0 g

Phenol red .... 0.5 ml
Double-distilled water .... 1,000.0 ml

1. Place 750 ml of double-distilled water in a 1-liter beaker, and add the above ingredients one at a time in the order listed until dissolved. Use a magnetic stirrer.
2. Adjust the pH to 7.2 by adding slowly, while stirring, solid Tris.
3. Bring up the volume to 1,000 ml with distilled water.
4. Dispense 100 ml into each of several screw-cap flasks or bottles.
5. Autoclave at 121°C for 15 min.
6. Label as overlay solution with the preparation date and an expiration date of 1 month.
7. Store at 4°C.

NIH Method for Trypanosomes and Leishmaniae

Solution 1

Lean beef, desiccated............................... 25.0 g
Neopeptone.. 10.0 g
Agar .... 10.0 g
NaCl .... 2.5 g
Distilled water.. 500 ml

Solution 2 (Locke's Solution)

NaCl .... 8.0 g
KCl... 0.2 g
CaCl$_2$ .... 0.2 g
KH$_2$PO$_4$.. 0.3 g
Dextrose... 2.5 g
Distilled water... 1,000 ml

1. Infuse the beef and distilled water in a water bath at 37°C for 1 h. Heat for 5 min at 80°C.
2. Filter through filter paper (Whatman no. 2V).
3. Add the rest of the above ingredients (solution 1), and adjust the pH to 7.0 to 7.4 with NaOH.
4. Autoclave at 121°C for 20 min.
5. Cool to 45°C, and aseptically add 10% defibrinated rabbit blood.
6. Dispense 5-ml quantities in sterile tubes. Slant and cool.
7. Just before inoculation, overlay with 2 ml of sterile Locke's solution.

4 N Medium for Trypanosomes and Leishmaniae

The 4 N medium was adapted from the original formula described by Novy and McNeal (44, 45, 58).

4 N Medium for Trypanosomes and Leishmaniae

Agar base

Oxoid blood sugar base no. 2	40 g
Distilled water	1,000 ml

1. Mix the agar and water, and dissolve by autoclaving or steaming.
2. Dispense liquid agar in 5-ml aliquots into 30-ml screw-cap glass bottles.
3. Autoclave again if necessary for sterility.
4. When the medium has cooled to 45°C, add aseptically to each bottle approximately 1 ml of fresh rabbit blood and allow the agar to solidify in a slant.

Overlay for 4 N medium

1. Add 1 ml of Locke's (see NIH method for trypanosomes and leishmaniae) solution to each bottle containing 5 ml of agar.
2. Incubate the bottles at 37°C for 12 to 24 h to check for sterility.
3. Store the bottles at 4°C for 24 h or more before use.

Yaeger's LIT Medium (for Chagas' Disease) (22)

Liver infusion (Difco)	35.0 g
Tryptose (Difco)	5.0 g
NaCl	4.0 g
KCl	0.4 g
Sodium phosphate, dibasic ($Na_2HPO_4 \cdot 12H_2O$)	8.0 g
Glucose	2.0 g
Hemin (stock solution)	4.0 ml
Double-distilled water	1,000.0 ml

1. Add all the ingredients to the distilled water, and mix well with a magnetic stirrer until dissolved. Heat, if necessary, to dissolve all the ingredients.
2. Using a Whatman no. 42 filter paper in a Büchner funnel, filter with suction. Repeat this filtration one more time.
3. Adjust pH to 7.2 with 1 N NaOH or 1 N HCl.
4. Sterilize by filtration through a 0.22-µm-pore-size membrane filter.
5. Dispense 4.5 ml into each tube.
6. Label as LIT medium, with the date of preparation and an expiration date of 1 month.

Hemin Stock Solution

Hemin	100.0 mg
Triethanolamine	10.0 ml
Sterile double-distilled water	10.0 ml

1. Mix triethanolamine with water, add the mixture to a tube containing hemin, shake well, and let dissolve.

2. To make the complete medium, just before inoculation, add 0.5 ml of inactivated fetal bovine serum and 0.25 ml of antibiotic solution. The final concentrations of the antibiotics are 100 U of penicillin per ml, 100 µg of streptomycin per ml, and 0.2 µg of Fungizone per ml.

USAMRU Blood Agar Medium for Leishmaniae (67)

Blood agar base (Difco, B45)	40.0 g
Defibrinated rabbit blood	150.0 ml
Double-distilled water	1,000.0 ml

1. Dissolve 40 g of agar in 1,000 ml of distilled water by heating.
2. Cool the solution, and add 150 ml of defibrinated rabbit blood.
3. Dispense 5-ml portions into 125- by 16-mm screw-cap tubes. Position the tubes at a slant, and allow to cool, preferably on ice, to produce moisture of condensation in the tubes.
4. Incubate the tubes at 35°C for 24 h to ensure sterility. Antibiotics (penicillin and streptomycin) can be added if necessary (see the antibiotic formula in this chapter). The final concentrations of antibiotics are 100 U of penicillin per ml and 100 µg of streptomycin per ml.
5. Inoculate the medium with aspirate material or triturated tissue from biopsy specimen.

Note This medium has been especially useful for primary isolation of the *Leishmania braziliensis* complex in Latin America.

Liquid Media for Cultivation of Hemoflagellates

Hendricks et al. have reported on the highly successful use of liquid culture media for the rapid cultivation of various species of *Leishmania* and *Trypanosoma* (22). Schneider's *Drosophila* medium (GIBCO, Grand Island, N.Y.), supplemented with 30% (vol/vol) fetal calf serum, has been used in making primary isolates from human and animal infections as well as for routine maintenance of a wide variety of leishmanial and trypanosomal species in the laboratory. Both Schneider's medium and Grace's insect tissue culture medium (GIBCO) promote better growth of organisms and are less costly than the widely used blood-agar-based media. In addition, Schneider's medium may be freeze-dried for at least 2 years and reconstituted with distilled water in the field, and it will provide excellent culture results.

Stock Antibiotic Solution (To Be Used with All Media)

Sodium penicillin G	1,000,000 U
Streptomycin sulfate	1,000,000 µg

Fungizone .... 2,000 µg
Sterile double-distilled water .... 50.0 ml

1. Mix the components thoroughly. The concentration of the stock solution is:
   Penicillin.................................... 20,000 U/ml
   Streptomycin .... 20,000 µg/ml
   Fungizone................................... 40 µg/ml

2. Dispense 1 ml of the antibiotic mixture into sterile screw-cap vials or sterile cryovials.

3. Label as antibiotic solution with the date of preparation and an expiration date of 1 year.

4. Store at −20°C in cryoboxes.

Axenic Culture

1. Remove tubes containing culture medium (one of NNN, modified NNN, or Tobie's medium and one of Schneider's for *Leishmania*; one of NNN or Tobie's medium and one of LIT for *T. cruzi*) from 4°C, add fetal bovine serum and antibiotics if required, and incubate at 20 to 23°C for 1 to 2 h.

2. Inoculate the specimen (aspirate, scraping, or biopsy material from skin lesions from cutaneous leishmaniasis patients; bone marrow aspirates or splenic aspirates from visceral leishmaniasis patients; or normal skin biopsy specimens, lymph node aspirates, or pieces of liver and spleen from suspected or potential wild- or domestic-animal reservoirs) into the culture tubes. For Chagas' disease, inoculate a few drops of buffy coat into the culture tubes.

3. Add 0.5 ml of overlay (either saline or other overlay, depending on the medium). The organisms will develop in the fluid condensate and overlay at the bottom of the slant (Figure 32.5).

4. Incubate the tubes at 20 to 24°C.

5. Once every 2 to 3 days, remove a drop of medium and examine it under the low power (100×) of a microscope, preferably one equipped with phase-contrast optics.

6. If promastigotes are seen, inoculate a couple of drops of the medium into fresh culture tubes. Add a couple of drops of 0.85% saline or the overlay solution (depending on the culture medium used) to the old tube.

7. If visible contamination occurs, add antibiotics to the overlay (to contain 200 U of penicillin/ml and 200 µg of streptomycin/ml). The parasites will not proliferate if bacterial contamination is present.

8. Incubate the tubes containing the patient specimen for at least 2 weeks at 28°C or room temperature.

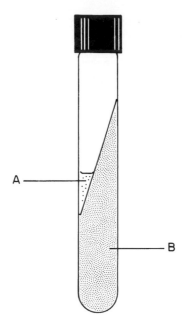

Figure 32.5 Illustration of a tube of NNN medium, used for the culture and recovery of *Leishmania* spp. (A) Fluid condensate and tissue culture medium overlay containing the developing organisms. (B) Blood agar medium. (Illustration by Sharon Belkin.)

9. If no organisms are seen even after 2 weeks of incubation, examine several drops of fluid under the microscope for promastigotes. The culture should be observed for 1 month before being signed out as negative. Transfers to fresh medium should be made once or twice a week after the culture is established.

Quality Control for Flagellates of Blood and Tissue

The following control strains should be available when using these cultures for clinical specimens: ATCC 30883 (*Leishmania mexicana*) and ATCC 30160 (*T. cruzi*).

1. Check all reagents and media at least once a week. The media should be free of any signs of precipitation and bacterial and/or fungal contamination.

2. The microscope(s) should have been calibrated within the last 12 months, and the original optics used for the calibration should be in place on the microscope(s). The calibration factors for all objectives should be posted on the microscope for easy access.

3. Maintain stock cultures of *Leishmania* spp.
   A. Transfer stock cultures weekly.
      a. Stock organisms should always be cultured at the same time a patient specimen is inoculated into culture medium.

b. If the stock organisms multiply and remain viable during the 96 h, patient results can be reported.

B. Stain
   a. Stain a slide prepared from stock culture in parallel with the patient slide.
   b. Staining results are acceptable when the control organisms stain well.

Note Cultivating the organism from suspected materials provides a definitive diagnosis, but it may take as many as 3 to 7 days. Every effort, therefore, must be made to microscopically examine wet smears and/or permanent stained smears so that appropriate therapy can be instituted without delay if findings are positive (Figure 32.6) (16, 17).

Toxoplasma gondii

The diagnosis of toxoplasmosis may be difficult, because the clinical symptoms mimic a number of various infectious and noninfectious diseases. Serologic tests that are often used for diagnosis may be insensitive in patients lacking normal immune responses. Sometimes even examination of histologic material does not reveal the organisms. With the increase in the number of laboratories

Figure 32.6 *Leishmania* spp. from culture systems. (Upper left) Wet mount of culture fluid sediment showing promastigotes of *Leishmania* spp. (Upper right) Stained smear of culture fluid sediment showing promastigotes of *Leishmania* spp. (Lower left) Stained *Leishmania* spp. promastigotes. (Lower right) Stained *Leishmania* promastigotes (note the round nucleus, bar-shaped kinetoplast, and flagella).

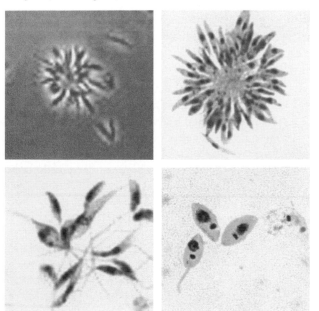

using tissue culture techniques for viral pathogens, these techniques have been used for the isolation and identification of *T. gondii*. The following procedure has been recommended for biopsy specimens, brain, liver, spleen tissue, CSF, amniotic fluid, and buffy coat preparations, and they may be particularly helpful in making the diagnosis in immunosuppressed patients (56). The procedure for buffy coat cells is as follows.

1. Collect 10 ml of blood anticoagulated with preservative-free heparin.
2. Allow the blood to sediment via gravity.
3. Remove the buffy coat (by an aseptic technique), and separate the cells from the plasma by centrifugation at 800 × g for 10 min.
4. Wash the buffy coat cells three times with Eagle's minimal essential medium (GIBCO).
5. Inoculate the washed buffy coat material onto complete human foreskin fibroblast (HFF) monolayers (in tubes and shell vials). One HFF tube and two HFF vials should be inoculated for each patient specimen.
6. Observe the cultures weekly for cytopathic effect.
7. The shell vial coverslips can be fixed and stained at 7 and 14 days postinoculation for an indirect fluorescent-antibody (IFA) assay and observed for the tachyzoites.

Note CSF, placental tissue, or other tissues can also be used to inoculate tissue culture monolayers. Uncentrifuged CSF (0.1 to 1 ml) can be used. If more than 1 ml is submitted, the specimen should be centrifuged for 10 min at 500 × g. Positive and negative controls must be tested with each set of patient specimen vials.

Three continuous cell lines (HeLa, LLC, and Vero) and three cell culture methods (culture in conventional flasks, culture in membrane-based flasks, and an automated culture system) were investigated. Overall, HeLa was the cell line of choice. Continuous passage in flasks was successful, and HeLa-derived tachyzoites can be used for the dye test, if applicable in your laboratory setting (18).

Plasmodium and Babesia spp.

Techniques for the culture of *Plasmodium falciparum* were described in 1976 (59, 61) and have been improved and modified since that time (10, 15, 19, 25, 26, 60; S. Waki, T. Miyagami, S. Nakazawa, I. Igarashi, and M. Suzuki, Letter, *Trans. R. Soc. Trop. Med. Hyg.* 78:418, 1984). Life cycle stages of the four *Plasmodium* spp. that infect humans have been established in vitro. Of these four, *P. falciparum* is the only species for which all stages have been cultured in vitro. The life cycle includes the exoerythrocytic stage (within liver cells), the erythrocytic stage (within erythrocytes or precursor reticulocytes), and the sporogonic stage (within the vector). Culture media generally consist of a basic tissue culture medium to which serum and erythrocytes are added. Most of the culture methods have been directed toward the stage found in the erythrocyte. This stage has been cultivated in petri dishes or other containers in a candle jar to generate elevated CO_2 levels or in a more controlled CO_2 atmosphere. Later developments employed continuous-flow systems to reduce the labor-intensive requirement for replenishing the system with fresh media. The exoerythrocytic and sporogonic life cycle stages have also been cultivated in vitro. Although cultivation is of great help in understanding the biology of *Plasmodium*, it does not lend itself to use for routine diagnostic purposes (54).

The availability of the microaerophilous stationary phase (MASP) culture technique, in which the parasites proliferate in a settled layer of blood cells, has provided an opportunity to study *Babesia*, a formerly obscure disease agent regarded as within the purview of veterinary parasitology, in the laboratory. A number of *Babesia* spp. have been established in continuous culture using the MASP technique. It is possible to study the basic biology of the organism—as well as host-microbe interactions, immune factors triggered by the parasite, factors involved in innate resistance of young animals to infection, and antimicrobial susceptibility—to a degree not possible before the availability of cultures. These culture systems can produce quantities of parasite nucleic acid needed for defining phylogenetic relationships, developing diagnostic methods for parasite detection in asymptomatic individuals, and producing parasite antigens and attenuated strains of *Babesia* that could be used for immunization (52).

Cryptosporidium spp.

The in vitro cultivation of *Cryptosporidium* has improved significantly in recent years. These obligate intracellular parasites colonize the epithelium of the digestive and respiratory tracts, are often difficult to obtain in significant numbers, produce durable oocysts that defy conventional chemical disinfection methods, and are persistently infectious when stored at refrigerator temperatures (4 to 8°C). While continuous culture and oocyst production have not yet been achieved in vitro, routine methods for parasite preparation and cell culture infection and assays for parasite life cycle development have been established. Parasite yields tend to be limited, but in vitro growth is sufficient to support a variety of research studies, including assessing potential drug therapies, evaluating oocyst disinfection methods, and characterizing life cycle stage development and differentiation (1).

Microsporidia

Although various microsporidia that infect humans can be identified from clinical specimens by serologic and/or molecular methods, none of these methods are commercially available. Unfortunately, in some cases microscopic examination of biopsy specimens does not yield conclusive results. It is also possible that microsporidial organisms may be present in very small numbers, which can be easily missed during routine histologic examinations. Some microsporidia such as *Encephalitozoon* and *Brachiola* spp., even when they are present in small numbers, can become established in cell cultures, thus facilitating their identification at a later time. Therefore, attempts at culturing these organisms should be made whenever possible, since many clinical laboratory personnel are familiar with cell culture methodology (64).

Animal Inoculation

Most routine clinical laboratories do not have the animal care facilities necessary to provide animal inoculation capabilities for the diagnosis of parasitic infections. Host specificity for many animal parasite species is a well-known fact which limits the types of animals available for these procedures. In certain suspect infections, animal inoculation may be requested and can be very helpful in making the diagnosis, although animal inoculation certainly does not take the place of other, more routine procedures.

Leishmania spp.

The hamster is the laboratory animal of choice for the isolation of any form of *Leishmania* spp. A generalized infection results after intraperitoneal inoculation; spleen impression smears should be examined for the presence of organisms.

1. Aspirates or biopsy material obtained under sterile conditions from cutaneous ulcers, lymph nodes, spleen, liver, bone marrow, buffy coat cells, or CSF may be used for inoculation.
2. Material (0.25 to 1 ml) obtained from these sources should be inoculated under sterile conditions by the intraperitoneal route. Material from patients with mucocutaneous leishmaniasis should be inoculated by the intranasal route or into the feet.
3. Young (either sex) hamsters (2 to 4 months old) should be used for this procedure.
4. If the animal dies several days after inoculation, splenic aspirates should be examined for the presence of organisms. The material should be

prepared as thick blood films and stained with Giemsa or other blood stains.

Note The infection develops slowly in hamsters; several months may be required to produce a detectable infection. For this reason, culture procedures are usually selected as more rapid means of parasite recovery. Intranasal lesions or those in the feet may develop very slowly; experimental animals should be kept for 9 to 12 months before a negative report is sent out.

Trypanosoma spp. (20)

Several laboratory animals can be used for the recovery of trypanosomes. Organisms are usually found in the blood of the animals within the first week after inoculation; however, if no organisms are found, the animals should be checked at the end of 2 and 4 weeks before results are reported as negative. *Trypanosomes are infectious, so extreme care should be used when examining any blood or tissue suspected to contain these organisms.*

1. Blood, lymph node aspirates, tissue, or CSF obtained under sterile conditions may be used for inoculation.
2. Material (up to 2 ml) should be inoculated intraperitoneally into guinea pigs or white rats for *Trypanosoma gambiense* and *T. rhodesiense* and into white mice for *T. cruzi*.
3. The number of organisms in the blood may vary; therefore, smears should be prepared frequently (every few days) over a 4-week period after inoculation.

Note Rats should be checked for the presence of *T. lewisi* (common parasite in rats) before inoculation to prevent a possible false-positive result.

Toxoplasma gondii

All common laboratory animals can be infected by *T. gondii*. White rats and mice are generally used. White rats develop a chronic infection that can be a useful means of maintaining a strain of organisms. Mice that have been inoculated by the intraperitoneal route develop a fulminating infection that leads to death within a few days. Tremendous numbers of organisms can be recovered from the ascitic fluid. These specimens should be handled carefully to avoid an accidental laboratory infection.

General Procedure

1. The *Toxoplasma* organisms are found throughout the body after dissemination via the bloodstream. Any body tissue or fluid can be used for animal inoculation; the most common specimens are blood, lymph node fluid, and CSF.
2. The material used for inoculation (the amount may be very small [less than 0.25 ml]) should be obtained under sterile conditions and should be inoculated via the intraperitoneal route.
3. White mice of either sex and any age can be used. The animals should be checked daily for symptoms of illness.
4. After several days, the organisms can be recovered at necropsy from the peritoneal fluid of the mouse. The material should be prepared as thick blood films and stained with Giemsa or other blood stains.
5. Blind passage of peritoneal fluid to additional mice is recommended if the original mice appear to be negative.
6. If mice survive for up to 6 months, serum can be tested for the presence of antibody.

Procedure for Tissue

1. Grind the tissue in sterile 0.85% NaCl.
2. Prepare a 10% suspension (10% tissue, 90% NaCl).
3. Inoculate six mice via the intraperitoneal route. Each mouse should receive 1 ml.
4. The mice can be reinoculated the following day with an additional 1-ml dose.
5. Check the mice daily for symptoms.
6. If the animals are not sick at the end of 6 days, examine 1 drop of peritoneal fluid from each mouse directly and then stain it with Giemsa stain.

Xenodiagnosis

Xenodiagnosis is a technique that uses the arthropod host as an indicator of infection. Uninfected reduviid bugs are allowed to feed on the blood of a patient who is suspected of having Chagas' disease (*T. cruzi* infection) (34) (Figure 32.7). After 30 to 60 days, feces from the bugs are examined over a 3-month time frame for the presence of developmental stages of the parasite, which are found in the hindgut of the vector. This type of procedure is used primarily in South America for field work, and the appropriate bugs are raised in various laboratories specifically for this purpose.

The term "xenodiagnosis" has also been applied to the diagnosis of trichinosis (*Trichinella spiralis*). Muscle tissue from a patient suspected of having the disease is fed to uninfected rats; the rats are then checked after the appropriate time for the presence of *T. spiralis* larvae, particularly in the diaphragm. This procedure is rarely requested and is not available in most clinical laboratories.

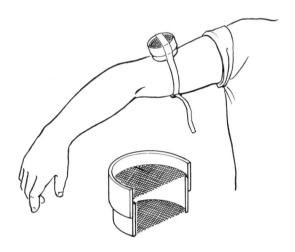

Figure 32.7 Illustration of the process of xenodiagnosis used for the diagnosis of Chagas' disease. (Illustration by Sharon Belkin.)

References

1. Arrowood, M. J. 2002. In vitro cultivation of *Cryptosporidium* species. *Clin. Microbiol. Rev.* **15**:390–400.

2. Asami, K., Y. Nodake, and T. Ueno. 1955. Cultivation of *Trichomonas vaginalis* on solid medium. *Exp. Parasitol.* **4**:34–39.

3. Beal, C., R. Goldsmith, M. Kotby, M. Sherif, A. El-Tagi, A. Farid, S. Zakaria, and J. Eapen. 1992. The plastic envelope method, a simplified technique for culture diagnosis of trichomoniasis. *J. Clin. Microbiol.* **30**:2265–2268.

4. Borchardt, K. A., M. Z. Zhang, H. Shing, and K. Flink. 1997. A comparison of the sensitivity of the InPouch TV, Diamond's and Trichosel media for detection of *Trichomonas vaginalis*. *Genitourin. Med.* **73**:297–298.

5. Bowman, E. K., A. A. Vass, R. Mackowski, B. A. Owen, and R. L. Tyndall. 1996. Quantitation of free-living amoebae and bacterial populations in eyewash stations relative to flushing frequency. *Am. Ind. Hyg. Assoc. J.* **57**:626–633.

6. Chang, T. H., S. Y. Tsing, and S. Tzang. 1986. Monoclonal antibodies against *Trichomonas vaginalis*. *Hybridoma* **5**:43–51.

7. Clark, C. G., and L. S. Diamond. 2002. Methods for cultivation of luminal parasitic protists of clinical importance. *Clin. Microbiol. Rev.* **15**:329–341.

8. Cote, R. (ed.) 1984. *American Type Culture Collection.* American Type Culture Collection, Rockville, Md.

9. Dedet, J. P., F. Pratlong, R. Pradinaud, and B. Moreau. 1999. Delayed culture of *Leishmania* in skin biopsies. *Trans. R. Soc. Trop. Med. Hyg.* **93**:673–674.

10. Desjardins, R., and J. Bowdre. 1981. In vitro cultivation of malaria parasites. *Clin. Microbiol. Newsl.* **3**:52–53.

11. Diamond, L. S. 1983. Lumen dwelling protozoa: *Entamoeba*, *Trichomonads*, and *Giardia*, p. 65–109. *In* J. B. Jensen (ed.), *In Vitro Cultivation of Protozoan Parasites*. CRC Press, Inc., Boca Raton, Fla.

12. Diamond, L. S. 1987. *Entamoeba*, *Giardia* and *Trichomonas*, p. 1–28. *In* A. E. R. Taylor and J. R. Baker (ed.), *In Vitro*

Methods for Parasite Cultivation. Academic Press, Inc., Orlando, Fla.

13. Diamond, L. S., C. G. Clark, and C. C. Cunnick. 1995. YI-S, a casein-free medium for axenic cultivation of *Entamoeba histolytica*, related *Entamoeba*, *Giardia intestinalis*, and *Trichomonas vaginalis*. *J. Eukaryot. Microbiol.* **42**:277–278.

14. Dolkart, R., and B. Halpern. 1958. A new monophasic medium for the cultivation of *Entamoeba histolytica*. *Am. J. Trop. Med. Hyg.* **7**:595–596.

15. Druilhe, P., D. Mazier, O. Brandicourt, and M. Gentilini. 1983. One-step *Plasmodium falciparum* cultivation—application to in-vitro drug testing. *Trop. Med. Parasitol.* **34**:233–234.

16. Evans, D. A. 1978. Kinetoplastida, p. 55–58. *In* A. E. R. Taylor and J. R. Baker (ed.), *Methods of Culturing Parasites In Vitro*. Academic Press, Inc., New York, N.Y.

17. Evans, D. A. 1987. *Leishmania*, p. 52–75. *In* A. E. R. Taylor and J. R. Baker (ed.), *In Vitro Methods for Parasite Cultivation*. Academic Press, Inc., Orlando, Fla.

18. Evans, R., J. M. Chatterton, D. Ashburn, A. W. Joss, and D. O. Ho-Yen. 1999. Cell-culture system for continuous production of *Toxoplasma gondii* tachyzoites. *Eur. J. Clin. Microbiol. Infect. Dis.* **18**:879–884.

19. Fairlamb, A. H., D. C. Warhurst, and W. Peters. 1985. An improved technique for the cultivation of *Plasmodium falciparum* in vitro without daily medium change. *Ann. Trop. Med. Parasitol.* **75**:7–17.

20. Garcia, L. S. 1999. *Practical Guide to Diagnostic Parasitology.* ASM Press, Washington, D.C.

21. Gorlin, A. I., M. M. Gabriel, L. A. Wilson, and D. G. Ahearn. 1996. Effect of adhered bacteria on the binding of *Acanthamoeba* to hydrogel lenses. *Arch. Ophthalmol.* **114**:576–580.

22. Hendricks, L. D., D. E. Wood, and M. E. Hajduk. 1978. Hemoflagellates: commercially available liquid media for rapid cultivation. *Parasitology* **76**:309–316.

23. Hollander, D. H. 1976. Colonial morphology of *Trichomonas vaginalis* in agar. *J. Parasitol.* **62**:826–828.

24. Isenberg, H. D. (ed.). 2004. *Clinical Microbiology Procedures Handbook*, 2nd ed., p. 7.0.1–7.10.8.2. ASM Press, Washington, D.C.

25. Jensen, J. B. 2005. Reflections on the continuous cultivation of *Plasmodium falciparum*. *J. Parasitol.* **91**:487–491.

26. Jensen, J. B., W. Trager, and J. Doherty. 1979. *Plasmodium falciparum*: continuous cultivation in a semiautomated apparatus. *Exp. Parasitol.* **48**:36–41.

27. Krogstad, D. A., G. S. Visvesvara, K. W. Walls, and J. W. Smith. 1985. Blood and tissue protozoa, p. 612–630. *In* E. H. Lennette, A. Balows, W. J. Hausler, Jr., and H. J. Shadomy (ed.), *Manual of Clinical Microbiology*, 4th ed. American Society for Microbiology, Washington, D.C.

28. Kulda, J., S. M. Honigberg, J. K. Frost, and H. D. Hollander. 1970. Pathogenicity of *Trichomonas vaginalis*. A clinical and biologic study. *Am. J. Obstet. Gynecol.* **108**:908–918.

29. Kuroki, T., S. Sata, S. Yamai, K. Yagita, Y. Katsube, and T. Endo. 1998. Occurrence of free-living amoebae and *Legionella* in whirlpool bathes. *Kansenshogaku Zasshi* **72**:1056–1063.

30. Landers, P., K. G. Kerr, T. J. Rowbotham, J. L. Tipper, P. M. Keig, E. Ingham, and M. Denton. 2000. Survival

and growth of *Burkholderia cepacia* within the free-living amoeba *Acanthamoeba polyphaga*. *Eur. J. Clin. Microbiol. Infect. Dis.* **19**:121–123.

31. **Lennette, E. H., A. Balows, W. J. Hausler, Jr., and H. J. Shadomy (ed.).** 1985. *Manual of Clinical Microbiology*, 4th ed. American Society for Microbiology, Washington, D.C.

32. **Linstead, D.** 1990. Cultivation, p. 91–111. *In* B. M. Honigberg (ed.), *Trichomonads Parasitic in Man*. Springer-Verlag, New York, N.Y.

33. **Ma, P., G. S. Visvesvara, A. J. Martinez, F. H. Theodore, P.-M. Daggett, and T. K. Sawyer.** 1990. *Naegleria* and *Acanthamoeba* infections: review. *Rev. Infect. Dis.* **12**:490–513.

34. **Maekelt, G.** 1964. A modified procedure of xenodiagnosis of Chagas' disease. *Am. J. Trop. Med. Hyg.* **13**:11–15.

35. **Marciano-Cabral, F., and G. Cabral.** 2003. *Acanthamoeba* spp. as agents of disease in humans. *Clin. Microbiol. Rev.* **16**:273–307.

36. **Markell, E., and M. Voge.** 1976. *Medical Parasitology*. The W. B. Saunders Co., Philadelphia, Pa.

37. **Martinez, A. J.** 1985. *Free-Living Amebas: Natural History, Prevention, Diagnosis, Pathology, and Treatment of the Disease*. CRC Press, Inc., Boca Raton, Fla.

38. **Mata-Cardenas, B. D., J. Vargas-Villarreal, L. Navarro-Marmolejo, and S. Said-Fernandez.** 1998. Axenic cultivation of *Trichomonas vaginalis* in a serum-free medium. *J. Parasitol.* **84**:638–639.

39. **Mazur, T., E. Hadas, and I. Iwanicka.** 1995. The duration of the cyst stage and the viability and virulence of *Acanthamoeba* isolates. *Trop. Med. Parasitol.* **46**:106–108.

40. **McMillan, A.** 1990. Laboratory diagnostic methods and cryopreservation of trichomonads, p. 297–310. *In* B. M. Honigberg (ed.), *Trichomonads Parasitic in Man*. Springer-Verlag, New York, N.Y.

41. **Michel, R., and M. Borneff.** 1989. The significance of amoebae and other protozoa in water conduit systems in dental units. *Zentbl. Bakteriol. Mickobiol. Hyg. Ser. B* **187**:312–323.

42. **Michel, R., H. Burghardt, and H. Bergmann.** 1995. *Acanthamoeba*, naturally intracellularly infected with *Pseudomonas aeruginosa*, after their isolation from a microbiologically contaminated drinking water system in a hospital. *Zentbl. Hyg. Umweltmed.* **196**:532–544.

43. **Newsome, A. L., T. M. Scott, R. F. Benson, and B. S. Fields.** 1998. Isolation of an amoeba naturally harboring a distinctive *Legionella* species. *Appl. Environ. Microbiol.* **64**:1688–1693.

44. **Novy, F. G., and W. J. McNeal.** 1903. The cultivation of *Trypanosoma brucei*. A preliminary note. *JAMA* **41**:1266–1268.

45. **Novy, F. G., and W. J. McNeal.** 1904. On the cultivation of *Trypanosoma brucei*. *J. Infect. Dis.* **1**:1–30.

46. **Ohlemeyer, C. L., L. L. Hornberger, D. A. Lynch, and E. M. Swierkosz.** 1998. Diagnosis of *Trichomonas vaginalis* in adolescent females: InPouch TV culture versus wet-mount microscopy. *J. Adolesc. Health* **22**:205–208.

47. **Pabst, U., J. Demuth, T. Gebel, and H. Dunkelberg.** 1997. Establishment of a method for determining the association between *Legionella* sp. and *Amoeba* sp. using polymerase chain reaction. *Zentbl. Hyg. Umweltmed.* **199**:568–577.

48. **Page, F. C.** 1988. *A New Key to Fresh Water and Soil Gymnamoebae*. Fresh Water Biological Association, Ambleside, England.

49. **Penland, R. L., and K. R. Wilhelmus.** 1997. Comparison of axenic and monoxenic media for isolation of *Acanthamoeba*. *J. Clin. Microbiol.* **35**:915–922.

50. **Penland, R. L., and K. R. Wilhelmus.** 1998. Laboratory diagnosis of *Acanthamoeba* keratitis using buffered charcoal-yeast extract agar. *Am. J. Ophthalmol.* **126**:590–592.

51. **Rondanelli, E. G.** 1987. *Amphizoic Amoebae: Human Pathology*. Piccin Nuova Libraria, Padua, Italy.

52. **Schuster, F. L.** 2002. Cultivation of *Babesia* and *Babesia*-like blood parasites: agents of an emerging zoonotic disease. *Clin. Microbiol. Rev.* **15**:365–373.

53. **Schuster, F. L.** 2002. Cultivation of pathogenic and opportunistic free-living amebas. *Clin. Microbiol. Rev.* **15**:342–354.

54. **Schuster, F. L.** 2002. Cultivation of *Plasmodium* spp. *Clin. Microbiol. Rev.* **15**:355–364.

55. **Schuster, F. L., and J. J. Sullivan.** 2002. Cultivation of clinically significant hemoflagellates. *Clin. Microbiol. Rev.* **15**:374–389.

56. **Shepp, D., R. Hackman, F. Conley, J. Anderson, and J. Meyers.** 1985. *Toxoplasma gondii* reactivation identified by detection of parasitemia in tissue culture. *J. Intern. Med.* **103**:218–221.

57. **Smith, R. F.** 1986. Detection of *Trichomonas vaginalis* in vaginal specimens by direct immunofluorescence assay. *J. Clin. Microbiol.* **24**:1107–1108.

58. **Taylor, A. E. R., and J. R. Baker.** 1968. *The Cultivation of Parasites In Vitro*. Blackwell Scientific Publications Ltd., Oxford, United Kingdom.

59. **Trager, W.** 1976. Prolonged cultivation of malaria parasites (*Plasmodium coatneyi* and *P. falciparum*), p. 427–434. *In* H. Van den Bossche (ed.), *Biochemistry of Parasites and Host-Parasite Relationships*. Elsevier-North Holland Biomedical Press, Amsterdam, The Netherlands.

60. **Trager, W.** 1979. *Plasmodium falciparum* in culture: improved continuous flow method. *J. Protozool.* **26**:125–129.

61. **Trager, W., and J. B. Jensen.** 1976. Human malaria parasites in continuous culture. *Science* **193**:673–675.

62. **Trager, W., and J. B. Jensen.** 2005. Human malaria parasites in continuous culture. 1976. *J. Parasitol.* **91**:484–486.

63. **Visvesvara, G. S.** 1995. Pathogenic and opportunistic free-living amebae, p. 1196–1203. *In* P. R. Murray, E. J. Baron, M. A. Pfaller, F. C. Tenover, and R. H. Yolken (ed.), *Manual of Clinical Microbiology*, 6th ed. American Society for Microbiology, Washington, D.C.

64. **Visvesvara, G. S.** 2002. In vitro cultivation of microsporidia of clinical importance. *Clin. Microbiol. Rev.* **15**:401–413.

65. **Visvesvara, G. S., and L. S. Garcia.** 2002. Culture of protozoan parasites. *Clin. Microbiol. Rev.* **15**:327–328.

66. **Walochnik, J., O. Picher, C. Aspock, M. Ullmann, R. Sommer, and H. Aspock.** 1998. Interactions of "Limax amoebae" and gram-negative bacteria: experimental studies and review of current problems. *Tokai J. Exp. Clin. Med.* **23**:273–278.

67. **Walton, B. C., J. J. Shaw, and R. Lainson.** 1977. Observations on the in vitro cultivation of *Leishmania braziliensis*. *J. Parasitol.* **63**:1118–1119.

68. **Winiecka-Krusnell, J., and E. Linder.** 1999. Free-living amoebae protecting *Legionella* in water: the tip of an iceberg? *Scand. J. Infect. Dis.* **31**:383–385.

33

Fixation and Special Preparation of Fecal Parasite Specimens and Arthropods

Fixation of Parasite Specimens and Arthropods

Adequate fixation of parasites is important not only for diagnostic procedures but also as a means of preserving positive material for personnel and student training. There are many fixatives and preservatives available; however, only the more common ones are presented here.

Although this chapter does not include commentary related to the histopathology laboratory, some information about formalin fixatives may be helpful. Concerns about the toxicity of formalin, particularly in the quantities used in a routine histology laboratory, have led to trials of alternative methods of tissue fixation that do not require formalin. Alcohol-based fixatives have been proposed as optimal for immunohistochemical and nucleic acid methods and may be useful for diagnostic light microscopy. Some laboratories have converted to use of an alcoholic fixative (containing 56% ethanol and 20% polyethylene glycol). Although comparative scores between the two were slightly in favor of the formalin, there were no significant differences in terms of tissue architecture, cell borders, cytoplasm, nuclear contours, chromatin texture, red blood cell membranes, and uniformity of staining. Alcohol-polyethylene glycol appears to be a satisfactory alternative to formalin in routine diagnostic surgical pathology (3). In another study, Histochoice produced a staining intensity that was comparable, and in many cases superior, to that of formalin (26).

Formalin-fixed and paraffin-embedded tissues present particular challenges for proteomic analysis. However, most archived tissues in pathology departments and tissue banks worldwide are available in this form. Different approaches to removal of the embedding medium and protein digestion have been developed, thus releasing tryptic peptides, which are suitable for analysis by liquid chromatography-mass spectrometry. Peptide identifications made using this approach are comparable to those from matched fresh frozen tissue. Apparently, a high level of sequence coverage can be seen for proteins under study (21).

The effect of fixation on the degradation of nuclear and mitochondrial DNA in different tissues has been examined (19). Samples of different tissues were preserved in seven fixatives for periods extending from 1 to 336 days to determine which fixatives reduce the time-dependent degradation of DNA and preserve the histologic structure. For long-term storage in combination with amplification of nuclear and mitochondrial DNA, consistent results were obtained with Carnoy's solution and glutaraldehyde. Variable results were observed for buffered formalin. In regard to comparison of the different tissues, the quantities recovered from skeletal muscles and kidneys were larger than from other tissues. These fixative studies will become even more important as molecular methods applied to fixed tissues become more common.

Molecular characterization of morphologic change requires precise tissue morphology and RNA preservation; however, traditional fixatives usually result in fragmented RNA. To optimize molecular analyses of fixed tissues, morphologic and RNA integrity in rat liver was assessed when sections were fixed in 70% neutral-buffered formalin, modified Davidson's II medium, 70% ethanol, UMFIX, modified Carnoy's solution, modified methacarn, Bouin's medium, phosphate-buffered saline, or 30% sucrose. Each sample was treated with standard or microwave fixation and standard or microwave processing, and sections were evaluated microscopically. RNA was extracted and assessed for preservation of quality and quantity (4). Modified methacarn, 70% ethanol, and modified Carnoy's solution resulted in tissue morphology providing a reasonable alternative to formalin. Modified methacarn and UMFIX best preserved RNA quality. Neither microwave fixation nor processing affected RNA integrity relative to standard methods, although morphology was somewhat improved. Modified methacarn, 70% ethanol, and modified Carnoy's solution provided acceptable tissue morphology and RNA quality with both standard and microwave fixation and processing methods. Of these three fixatives, modified methacarn provided the best results and can be considered a fixative of choice where tissue morphology and RNA integrity are being assessed in the same specimens.

PCR has also been used to identify tissue-embedded ascarid nematode larvae. Two sequences of the internal transcribed spacer regions of the rRNA gene of the ascarid parasites (ITS1 and ITS2) were compared with those registered in GenBank. PCR amplification of the ITS regions was sensitive enough to detect a single larva of *Ascaris suum* mixed with porcine liver tissue. These results suggest that even a single larva embedded in tissues from patients with larva migrans could be identified by sequencing the ITS regions (13).

Although formalin fixation is the most common storage, transportation, and preservation method for stool samples, it dramatically reduces the ability to extract DNA from stool samples for PCR-based diagnostic tests. Apparently, the deleterious effects of formalin are both time and concentration dependent and may result from fragmentation of fixed DNA during its purification. This has been seen in studies of the effect of formalin fixation on PCR of *Entamoeba histolytica* (22).

Parasites found in feces can be preserved in one of two ways (9). One procedure involves thorough mixing of the fecal specimen directly in the fixative. This mixture can then be stored at room temperature without any additional processing. There are several disadvantages of this method: (i) if mixing is not complete, some of the organisms will not be well preserved and will disintegrate during storage; (ii) if the number of organisms is small, a random aliquot may not reveal the parasites; and (iii) the proper ratio of fixative to specimen may not be correct when one is working with larger volumes of material.

The second approach is to clean and separate the organisms from excess debris and concentrate them before fixation. The specimens can be mixed with a large volume of water, passed through a series of screens (large to small pore size), and finally allowed to sediment in a conical container (pilsner glass). After approximately 1 h, the sediment can be washed several times, resedimented, and finally fixed with hot (60 to 63°C) fixative to quickly preserve and stop any further development of certain helminth (*Ascaris* and *Trichuris*) eggs. The main disadvantage of this method is extensive washing and manipulation, which may lead to some organism distortion or destruction before fixation.

The adult forms of most helminths must be relaxed prior to fixation. If this preliminary step is not performed, the worms often contract and curl up when they come in contact with the fixative, thus preventing visual examination of some of the key morphologic characteristics. Several methods, including those involving tap water, physiologic saline, and dilute menthol, are available for this purpose.

Protozoa

Definitive morphology needed for identification of protozoan trophozoites and cysts is best seen on the permanent stained smear, which can be prepared from fresh fecal specimens or from stool that has been submitted to the laboratory in fixative. No washing techniques should be used before bulk fixation, because the trophozoites will be destroyed. Immediately after collection or submission to the laboratory, the specimens should be thoroughly mixed with the fixative of choice. The ratio of fixative to stool should be at least 3 parts fixative to 1 part stool, and the fixation time should be at least 4 h (or less, depending on the amount of specimen used). The normal collection vial systems usually require a fixation time of at least 30 min. If an entire fecal specimen is used, 4 h is recommended. Individual slides can then be prepared from bulk-fixed material by a technique described by Scholten and Yang (24). This method involves centrifugation and smear preparation from the sediment. Although formalin-preserved specimens, except for sodium acetate-acetic acid-formalin-preserved fecal material, are not recommended for the preparation of permanent stained smears, formalin fixation for cyst preservation is recommended for the preservation of teaching specimens. The morphology of cysts can often be seen in formalinized wet mounts sufficiently well to identify organisms to the species level or certainly to be highly suggestive.

Streck tissue fixative has also been tested as a substitute for formalin and polyvinyl alcohol in fecal preservation. Stool samples were examined microscopically as follows: (i) in wet mounts (by bright-field and epifluorescence microscopy), (ii) in modified acid-fast-, trichrome-, and safranin-stained smears, and (iii) with two commercial test kits. Specific results showed that *Cyclospora* oocysts retained full fluorescence, modified acid-fast- and safranin-stained smears of *Cryptosporidium* and *Cyclospora* oocysts were equal in staining quality, and results were comparable in the immunofluorescence assay and enzyme immunoassay commercial kits. However, stool fixed in Streck tissue fixative and stained with trichrome showed unacceptable staining quality compared with stool fixed in polyvinyl alcohol. Thus, Streck fixative is an excellent substitute for formalin; however, modifications to the trichrome procedure will be required to improve the staining characteristics of protozoan parasites (20).

Formalin-Saline Solution

Although formalin is generally used at room temperature for routine diagnostic work, hot (60 to 63°C) 5% formalin in a ratio of 3 parts fixative to 1 part stool is recommended for bulk specimens containing intestinal protozoa. Long-term storage of protozoa is enhanced with buffered formalin; *the solution should be replaced every 6 months.* Acceptable morphology for teaching purposes can be maintained for 6 months to several years, depending on the organism and the lag time between specimen collection and fixation. The organism cytoplasm will tend to become glassy or granular with very poor nuclear definition. Cysts of *Entamoeba coli* and *Giardia lamblia* tend to maintain their morphologic characteristics for years, whereas cysts of *E. histolytica* and *E. hartmanni* do not preserve well (Figure 33.1). All of these cysts retain their ability to take up iodine in a wet, direct smear.

The standard 10% formalin will fix protozoan cysts; however, it does not preserve morphology as well as does the buffered 5% formalin.

Formalin-Saline Solution
Formaldehyde.................................5 ml
0.85% saline solution....................95 ml

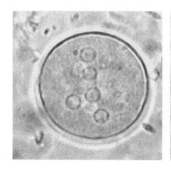 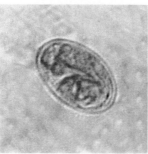

Figure 33.1 (Left) *Entamoeba coli* cyst (unstained). (Right) *Giardia lamblia* cyst (unstained).

Formaldehyde is normally purchased as a 37% HCHO solution; however, for dilution, it should be considered to be 100%.

Buffered Formalin-Saline Solution
Formaldehyde...............................400 ml
0.85% saline solution................7,600 ml
Na_2HPO_4 ....6.10 g
NaH_2PO_4 ....0.15 g

10% Formalin Solution
Formaldehyde...............................10 ml
Distilled water ....90 ml

Solutions To Induce Relaxation in Adult Helminths (1)

Tap Water or Physiologic Saline

1. Place living worms into a dish containing tap water or 0.85% NaCl solution.
2. Refrigerate the dish for 2 to 4 h.

Note Tap water generally works better than saline; also, the trematodes expel their eggs, thus allowing the internal morphology to be seen more clearly.

Dilute Menthol (17)

1. Dissolve 24 g of menthol crystals in 10 ml of 95% ethyl alcohol, and mix well.
2. Store the solution until needed.
3. Place living worms into a dish containing 100 ml of tap water or 0.85% NaCl.
4. Add 1 drop of menthol solution to the dish containing the worms.
5. Refrigerate the dish for 2 to 3 h.

Note The use of menthol accelerates relaxation of the worms. When this approach is used, the rostellum on the scolex of adult cestodes remains extruded during fixation.

Nematodes

Roundworms can be preserved with several different fixatives (Figure 33.2). Formalin is usually not recommended, since it tends to harden the tissues. Nematodes can be killed with hot water (60 to 63°C) and then transferred to a preservative, such as alcohol-glycerin or alcohol-formalin-acetic acid (AFA). It is recommended that nematode larval stages be fixed in hot water. Direct fixation in preservatives may cause the cuticle to become "sticky," and the larvae will be damaged when they adhere to the glass container (9).

Alcohol-Glycerin

Alcohol (70%) containing 5% glycerin is an excellent fixative for most nematodes and should be used hot (60 to 63°C). Specimens can be left in this original fixing solution indefinitely. The glycerin will protect the specimen if the alcohol evaporates. Fixative evaporation is another reason why fixed specimens should be routinely checked every 6 months for possible fixative replacement.

> Alcohol-Glycerin
> 95% ethyl alcohol .... 70 ml
> Distilled water .... 25 ml
> Glycerin .... 5 ml

AFA Solution

A solution of alcohol, formalin, and acetic acid can be routinely used for nematodes, trematodes, and cestodes.

Figure 33.2 Adult *Ascaris lumbricoides* nematode (male worm).

The fixative should be used hot (60 to 63°C); after fixation for 24 h, parasites can be stored in the alcohol-glycerin mixture.

> AFA Solution
> Formaldehyde .... 10 ml
> 95% ethyl alcohol .... 50 ml
> Acetic acid, glacial .... 5 ml
> Distilled water .... 45 ml

Glacial Acetic Acid

Undiluted glacial acetic acid is recommended for fixation of the smaller nematodes (Figure 33.3). They are killed instantly in an extended position. This fixative is not recommended for the larger worms, such as *Ascaris lumbricoides*. The acetic acid will clear the worm tissue so that the internal structure becomes visible. Morphologic characteristics can be seen under the microscope if the worm is placed in water on a slide. Worms fixed with acetic acid can be placed in AFA for 24 h and then into alcohol-glycerin for long-term storage.

Dilute Formalin

Although formalin is not recommended as a general fixative for nematodes, a dilute solution (1 to 2%) can be used to kill *A. lumbricoides*. Higher concentrations should not be used since the differences in osmotic pressure may cause the worms to rupture. After storage in dilute formalin for at least 24 h, the worms can be transferred to 10% formalin for long-term storage.

Note *Ascaris* eggs continue to develop in formalin (10% or less) to the infective stage and remain viable and infective for a number of months (Figure 33.4). Worms should be handled with caution, particularly if they are to be dissected.

Figure 33.3 Adult *Trichuris trichiura* nematode (note the "whiplike" appearance; this is a much smaller adult worm than *Ascaris lumbricoides*).

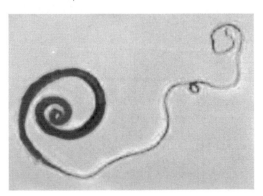

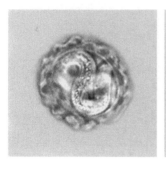
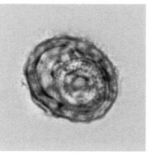

Figure 33.4 *Ascaris lumbricoides* fertilized eggs. Note the fully developed larva within each egg; these eggs were viable when photographed. In some cases, the moving larva can be seen, even in formalin-preserved specimens.

Trematodes

Most trematodes are very muscular and may contract when placed in fixatives. Living specimens should be placed in cold 0.85% saline for 30 min to several hours (depending on the size of the worm) before fixation. Specimens can be placed on a slide in a petri dish and then covered with another slide or coverglass to flatten the worm. Do not apply pressure, since doing so may distort internal organs (Figure 33.5).

Note Formalin is never recommended for the fixation of trematodes (9).

Figure 33.5 Adult *Fasciola hepatica* trematodes. (Upper) Preserved but unstained fluke; (lower) stained fluke (note the morphologic details that can be seen after staining).

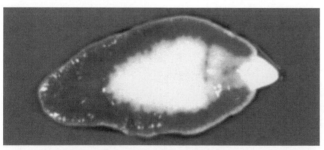

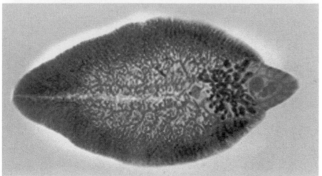

AFA fixative (see above) should be used hot (60 to 63°C); the worms can be stored in AFA or transferred to 70% alcohol for long-term storage.

Cestodes

Acid fixatives dissolve the characteristic calcareous corpuscles found in tapeworm tissue and should not be used. Since these corpuscles may be used to diagnosis tapeworm tissue in histologic sections, buffered formalin or non-acid-containing fixatives are recommended (9).

Formalin

Hot (60 to 63°C) buffered formalin is recommended for the fixation of cestodes. If the whole worms are immediately swirled around in the container of fixative, they will be rapidly fixed with minimal contraction of the proglottids.

AFA Solution

AFA fixes tapeworm tissue well; however, the acid dissolves the calcareous corpuscles. The worms can be transferred to 70% alcohol for long-term storage (Figure 33.6).

Note All tapeworms and proglottids should be handled with care, especially if the species has not been determined. The eggs of *Taenia solium* (cysticercosis) and *Hymenolepis nana* are infectious for humans (Figure 33.7).

Helminth Eggs and Larvae

Whole fecal specimens can be mixed directly with an appropriate fixative, although it is recommended that the organisms be concentrated before fixation. The use of formalin-saline as described for the fixation of protozoa is recommended for helminth eggs and larvae. If well preserved, most helminth eggs maintain their morphologic characteristics indefinitely.

Arthropods

Clinical laboratory personnel may receive arthropods for identification from various patient sources (surface of the body, stool, sputum, etc.) (6, 12). Specimens may be

Figure 33.6 *Taenia* unstained proglottid. Note that morphologic details cannot be seen in this preserved but unstained preparation.

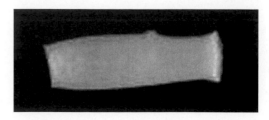

 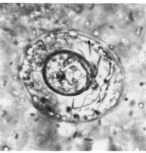

Figure 33.7 (Left) *Taenia* spp. egg (note the striated shell and hooklets seen on the oncosphere). (Right) *Hymenolepis nana* egg (note the six-hooked embryo and the polar filaments that are found between the oncosphere and the shell).

submitted when the actual source is unknown (implicated in a bite, found in the house or yard, etc.). Small, wingless insects and other arthropods should be placed in alcohol (70 to 95%), where they can remain indefinitely. Formalin fixative is not recommended for such specimens. Most flying insects should be killed in a chloroform tube or cyanide bottle and then preserved as a dry mount. Large maggots and other larvae can be killed in hot water first to prevent body shrinkage and contraction when placed in alcohol (8). A number of books also contain detailed information on the collection, preservation, and mounting of arthropods (10, 16, 25).

DNA sequencing has become much more common in surveys of archival research collections, particularly in reviewing rare taxa of arthropods. As an example, marine invertebrates have historically been maintained in ethanol following initial fixation in formalin. These collections often represent rare or extinct species or populations, provide detailed time series samples, or come from presently inaccessible or difficult-to-sample localities. Results obtained from preserved crustaceans in archival research collections indicate that in the absence of fresh or frozen tissues, archived formalin-fixed, ethanol-preserved specimens will prove a useful source of material for gene sequence data analysis by PCR and direct sequencing (7). It has also been determined that the now widespread use of critical-point drying of wasps and other insects from alcohol is advocated as a potential source of DNA from rare taxa (5).

Another problem that has been documented involves morphology changes seen using various fixatives. Thus, fixation and mounting can significantly influence the morphometric analysis of mites and other arthropods. It is recommended that morphometric studies be conducted using consistent methods to reduce experimental bias and that the methods used be reported in publications dealing with morphometric analyses (23).

Modified Berelese's medium is an all-purpose medium that kills, fixes, and preserves many arthropod specimens. No dehydration in alcohol is necessary.

Modified Berelese's Medium

Gum acacia	8 g
Distilled water	8 ml
Glycerin	5 ml
Chloral hydrate	70 g
Acetic acid, glacial	3 ml

Mounting and Staining of Parasite Specimens for Examination

Following preservation, many staining and mounting techniques for the preparation of permanent or semipermanent mounts of helminth eggs and larvae and of arthropods may be used.

Nematodes

Because the cuticle of nematodes prevents the uptake of stain, these worms are usually rendered transparent with glycerin or lactophenol. In this way, the internal morphology can be observed for identification purposes. Specific morphologic features that can be seen include the cuticle, alimentary tract, and reproductive structures. Small nematodes can be mounted in glycerin jelly. Standard mounting media containing resins and dehydrating agents are generally not satisfactory because the worms tend to collapse and become distorted during dehydration.

Glycerin Jelly Preparation

Specimens are transferred from pure glycerin directly into the glycerin jelly, a medium that will gel at room temperature, thus providing a semipermanent mount.

Glycerin Jelly (Refractive Index, 1.47)

Gelatin, granulated	10 g
Distilled water	60 ml
Glycerin, pure	70 ml
Phenol crystals, melted	0.5 ml

Dissolve the gelatin in water in a beaker in a water bath. Add the glycerin and melted phenol. Store in small, wide-mouth bottles, and refrigerate. Remelt in a hot water bath before use. Dispense with a dropper onto the slide.

1. Nematodes should first be killed in glacial acetic acid, AFA, or alcohol-glycerin.
2. Place nematodes in a small dish containing 70% alcohol–5% glycerin solution (several millimeters deep).
3. Partially cover the dish to allow gradual evaporation of the alcohol and water for approximately 24 to 36 h. The larger the nematodes, the longer the time needed to complete the evaporation.

Note The evaporation procedure should be done slowly to prevent collapse of the worm; the larger the specimen, the longer the evaporation time.

4. Transfer the worms to another dish containing a few milliliters of glycerin. The specimen should be placed just below the surface; it will eventually (within a few hours) sink to the bottom of the dish. It is then ready to be mounted.

5. Place a drop of liquefied glycerin jelly on the slide, and transfer the specimen into the drop. When the coverglass is added, the glycerin jelly should flow out to the edges of the coverglass. Allow the glycerin jelly to begin to solidify, and apply the coverglass. If the specimen is large, place the coverglass onto small pieces of broken glass to provide more depth under the coverglass.

6. Allow the preparation to gel overnight (horizontal position). The following day, the coverglass can be sealed with Vaspar or enamel paint. These preparations usually last for several years, particularly if no unnecessary pressure is applied to the coverglass.

Note If personnel are not experienced in preparation of these permanent mounts, remember that the worms can be stored indefinitely in vials of pure glycerin. They can be studied in this medium as temporary mounts and then carefully returned to the vial of glycerin.

Glycerin
Nematode specimens can be examined as temporary mounts and subsequently stored in pure glycerin. This approach is particularly helpful for larger specimens (Figure 33.8).

Lactophenol
Lactophenol is recommended when larger specimens are to be examined as temporary mounts. The worms can be placed into lactophenol from alcohol or formalin; clearing

will occur, with the length of time depending on the size of the worm. The worms can then be washed in alcohol.

Lactophenol (Refractive Index, 1.44)
Glycerin, pure .... 20 ml
Lactic acid .... 10 ml
Phenol crystals, melted .... 10 ml
Distilled water .... 10 ml

Trematodes

Trematodes are usually studied as stained whole mounts (Figure 33.9). Two stains that are recommended are carmine and hematoxylin. Many modifications are found in the literature, and most give excellent results. The stains are usually available commercially, and they can also be easily prepared in the laboratory.

Semichon's Acid Carmine
Carmine powder .... 5 g
Distilled water .... 50 ml
Acetic acid, glacial .... 50 ml

Add glacial acetic acid slowly to the water. Add the carmine powder, and heat to 95 to 100°C for 15 min. Cool and filter (stock solution). Add few drops of stock solution to 70% alcohol, making the working stain pink.

Van Cleave's Combination Hematoxylin Stain
Van Cleave's combination hematoxylin stain is a combination of Delafield's hematoxylin and Ehrlich's hematoxylin that rarely overstains trematode specimens.

Figure 33.9 Stained trematodes. (Left) *Eurytrema* sp. (Right) *Heterophyes* sp. Note that the morphologic details are clearly visible after staining.

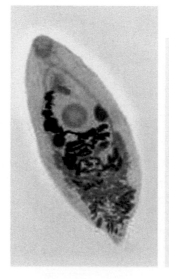

Figure 33.8 Semipermanent mounts of small nematodes. (Left) Small nematode from soil sample. (Right) Female *Enterobius vermicularis* (pinworm) nematode.

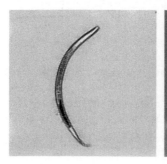
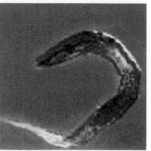

Delafield's Hematoxylin Stain

Hematoxylin powder .... 4 g
Ethyl alcohol, 95% .... 25 ml
Aluminum ammonium
sulfate [AlNH$_4$(SO$_4$)$_2$],
saturated aqueous solution
(aluminum alum solution).......... 400 ml
Glycerin.. 100 ml
Methyl alcohol .... 100 ml

Dissolve the hematoxylin in ethyl alcohol, add aluminum alum solution, and let stand for 1 week exposed to air and light (in a paper-capped container). Filter; add glycerin and methyl alcohol. Age for 6 to 8 weeks in a tightly capped bottle in the refrigerator. This is the working stain solution (no dilution is needed).

Ehrlich's Hematoxylin Stain

Hematoxylin powder .... 2 g
Ethyl alcohol, 95% .... 100 ml
Glycerin.. 100 ml
Distilled water .... 100 ml
Acetic acid, glacial........................ 10 ml
Potassium aluminum sulfate
[K$_2$SO$_4$Al$_2$(SO$_4$)$_3$ · 12H$_2$O]
(potassium alum) .... 3 g

Dissolve hematoxylin powder in alcohol; add the other ingredients. Expose to air and light for at least 2 weeks (in a paper-capped container). The solution can be ripened immediately by the addition of 0.4 g of sodium iodate (NaIO$_3$). Store refrigerated in a tightly capped bottle. This is the working solution (no dilution is needed).

Van Cleave's Hematoxylin Staining Solution

Delafield's hematoxylin stain............ 1 ml
Ehrlich's hematoxylin stain .... 1 ml
Distilled water .... 100 ml
Potassium aluminum sulfate............. 6 g

Dissolve the potassium alum in water and add the two hematoxylin stains. This is the working solution.

1. Staining should be done in dilute solutions of carmine or hematoxylin (at least overnight). Specimens stained in carmine are placed into the dilute stain from 70% ethyl alcohol. Those stained in hematoxylin are placed in the stain from water. It is best to overstain and then destain.
2. Rinse in 70% ethyl alcohol.
3. Destain in weak acid-alcohol (2 to 4 drops of concentrated HCl in 100 ml of 70% ethyl alcohol). Leach the color from tissues until they are clear but

the internal organs remain well stained. Destaining may take several minutes to hours; however, the specimens must be periodically checked to avoid overdestaining.
4. Rinse in 70% alcohol.
5. Place the specimens for 30 min to 1 h in a solution of 70% ethyl alcohol containing 1 or 2 drops of saturated aqueous Na$_2$CO$_3$, NaHCO$_3$, or LiCO$_3$. This step neutralizes the acid step and prevents continued destaining.
6. Rinse in 70% ethyl alcohol.
7. Dehydrate through 80, 95, and 100% ethyl alcohol, with 11 to 15 min for each alcohol change.
8. Clear the specimens in xylene for at least 15 min.
9. Mount in Permount or other permanent mounting medium.

Cestodes

Cestodes can also be examined as stained whole mounts, although the *Taenia* tapeworms can be examined more rapidly with India ink in a temporary mount. The same carmine and hematoxylin stains as used for trematodes can be used for cestodes.

India Ink Procedure for Tapeworm Proglottids

Identification of adult worms usually involves examination of a tapeworm proglottid. A *Taenia* proglottid must be gravid, containing the fully developed uterine branches. Using a syringe (1 ml or less) and a 25-gauge needle, India ink is injected into the central uterine stem of the proglottid, filling the uterine branches with ink, or into the uterine pore. The proglottid can then be rinsed in water or saline, blotted dry on paper towels, pressed between two slides, and examined. After the identification has been made, the proglottid can be left between the two slides (place a rubber band around the slides), dehydrated through several changes of ethyl alcohol (50, 70, 90, and 100%), cleared in two changes of xylene, and mounted with Permount for a permanent record. After the xylene step, the proglottid will be stuck to one of the two slides; do not try to remove it (it is very brittle and will crack) but merely add the Permount to the proglottid and then add the coverglass (Figure 33.10).

Euparal Mounts of Tapeworm Proglottids

Most helminth eggs mounted in Euparal (Flatters and Garnett, Ltd., Manchester, England) exhibit excellent optical and drying properties, and this mounting medium can also be used to mount tapeworm proglottids (2). Proglottids should be placed in 100% ethyl alcohol for 5 to 10 min and then pressed between two slides as described above. They should then be placed in Euparal; transparency

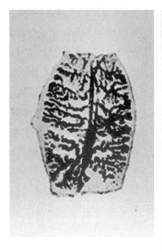

Figure 33.10 *Taenia* gravid proglottids after India ink injection of the uterine branches. (Left) *Taenia solium*. (Right) *Taenia saginata*.

will occur in 2 to 3 h. The best results have been obtained by keeping the slides in a 50°C incubator overnight.

Double-Coverglass Method for Microscopic Mounts of Cysts, Eggs, and Larvae

Slides prepared by the double-coverglass technique will last for several weeks to several years, depending on the organisms and the care taken in preparation. The procedure is as follows (9).

1. Place a small drop of 10% formalin or formalin-saline suspension (containing cysts, eggs, and/or larvae) in the center of a 22-mm round or square coverglass.
2. Using forceps, very carefully apply a smaller coverglass (12, 15, or 18 mm round or square) to the suspension so that the fluid flows to the edges of the small coverglass (with no bubbles). If excess fluid flows beyond the small coverglass, it should be blotted dry or allowed to evaporate. Do not allow it to dry too long; otherwise, bubbles will accumulate under the coverglass.
3. Place a large drop of Permount or other permanent mounting medium in the center of a 1- by 3-in. (1 in. = 2.54 cm) microscope slide. Using forceps, place the double-coverglass preparation (small coverglass side down) onto the mounting medium so that the medium flows to the edges of the large coverglass.
4. Allow the preparation to thoroughly dry in a horizontal position (this may take several days).

Mounting of Arthropods for Examination (11, 14, 15, 18)

Before being mounted, specimens in which xylene is used as a solvent require dehydration through 50, 70, and 95% ethyl alcohol. Clearing can be done in clove oil, carbol-xylene (3 parts xylene to 1 part phenol crystals), or absolute ethyl alcohol followed by xylene. The specimens should remain in each solution for at least 15 to 20 min. Specimens mounted in balsam or Permount will take several days to harden; however, those mounted in isobutyl methacrylate will dry very rapidly (within a few hours). With this type of permanent mount, ringing or sealing of the coverglass is not necessary. Some specific recommendations are presented below.

Mites

Temporary mounts can be made with a drop of 50% ethyl alcohol, which is gently heated. Clearing and extension of the specimen occur, revealing the typical morphology (Figure 33.11). Permanent mounts of living specimens can be made by placing the specimen in a drop of chloral-gum medium, adding a coverglass, and then gently heating until bubbling begins. Specimens originally preserved in alcohol must first be washed in distilled water to remove the alcohol.

Chloral-Gum Medium

Distilled water	35 ml
Chloral hydrate	30 g
Gum arabic	20 g
Glycerin	12 ml
Glucose syrup (e.g., Karo syrup)	3 ml

Fleas and Lice

Specimens may be preserved in 70% alcohol or mounted on slides for identification (Figure 33.12).

Figure 33.11 *Sarcoptes scabiei* itch mites in wet mounts. These specimens could be seen using the high dry objective (magnification, ×400).

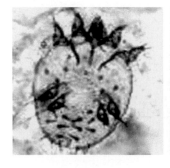
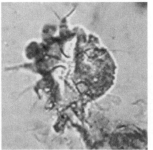

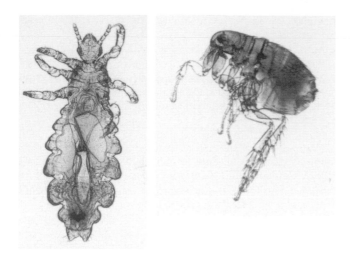

Figure 33.12 (Left) Body louse, *Pediculus humanus*. (Right) Flea, *Pulex irritans*.

1. Drop living fleas or preserved specimens into 10% KOH for a few days until sufficiently cleared.
2. Transfer for 30 min to a small volume of water containing a few drops of concentrated HCl.
3. Dehydrate in 50% ethyl alcohol for 30 min.
4. Dehydrate in 95% ethyl alcohol for 30 min.
5. Clear in beechwood creosote for 1 h, or place in several changes of absolute ethyl alcohol and clear in clove oil or xylene.
6. Mount on slides in balsam, Permount, or isobutyl methacrylate.

Ticks

Specimens should be preserved in 70% alcohol or cleared and mounted on slides. They can be fixed in an extended position by gently being pressed between two slides while being immersed in hot water. Clearing in KOH, dehydration in alcohols, and mounting with balsam, Permount,

Figure 33.13 Ticks. (Left) Example of a hard tick, *Ixodes scapularis*. (Right) Example of a soft tick, *Argas* sp.

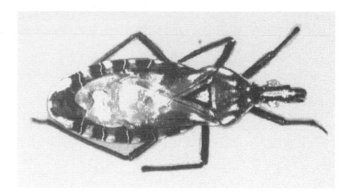

Figure 33.14 Example of a large hard-body insect, which is a triatomid bug. These large insects can be pinned, labeled, and stored in boxes containing naphthalene flakes or paradichlorobenzene.

or isobutyl methacrylate can be done as recommended for fleas (Figure 33.13).

Miscellaneous Arthropods

Spiders, scorpions, centipedes, lice, bedbugs, maggots and other larvae, nymphs, and soft-bodied insects can be preserved in 70% ethyl alcohol containing a small amount of glycerin to prevent drying and shrinkage. Containers should remain tightly sealed and should be checked periodically. Larger, hard-bodied insects can be pinned, labeled, and stored in boxes containing naphthalene flakes or paradichlorobenzene to prevent damage from mold and living insects (Figure 33.14).

Note Identification of the many species of arthropods can best be handled by an entomologist or specialist working with a particular arthropod group. Additional help with identification may be obtained at a local university entomology department, a military base (over 100 entomologists are employed by the Army and Navy), or the entomology department of natural history museums or the Smithsonian Institution.

References

1. Ash, L. R., and T. C. Orihel. 1987. *Parasites: a Guide to Laboratory Procedures and Identification*. ASCP Press, Chicago, Ill.
2. Berlin, O. G., and M. J. Miller. 1980. Euparal as a permanent mounting medium for helminth eggs and proglottids. *J. Clin. Microbiol.* **12**:700–703.
3. Bostwick, D. G., N. al Annouf, and C. Choi. 1994. Establishment of the formalin-free surgical pathology laboratory. Utility of an alcohol-based fixative. *Arch. Pathol. Lab. Med.* **118**:298–302.
4. Cox, M. L., C. L. Schray, C. N. Luster, Z. S. Stewart, P. J. Korytko, K. N. Khan, J. D. Paulauskis, and R. W. Dunstan. 2006. Assessment of fixatives, fixation, and tissue processing on morphology and RNA integrity. *Exp. Mol. Pathol.* **80**:183–191.

5. Dillon, N., A. D. Austin, and E. Bartowsky. 1996. Comparison of preservation techniques for DNA extraction from hymenopterous insects. *Insect Mol. Biol.* 5:21–24.

6. Faust, E. C., P. C. Beaver, and R. C. Jung. 1975. *Animal Agents and Vectors of Human Disease*, 4th ed. Lea & Febiger, Philadelphia, Pa.

7. France, S. C., and T. D. Kocher. 1996. DNA sequencing of formalin-fixed crustaceans from archival research collections. *Mol. Mar. Biol. Biotechnol.* 5:304–313.

8. Fritsche, T. R. 1999. Arthropods of medical importance, p. 1449–1466. *In* P. R. Murray, E. J. Baron, M. A. Pfaller, F. C. Tenover, and R. H. Yolken (ed.), *Manual of Clinical Microbiology*, 7th ed. ASM Press, Washington, D.C.

9. Garcia, L. S., and L. R. Ash. 1979. *Diagnostic Parasitology: Clinical Laboratory Manual*, 2nd ed. The C. V. Mosby Co., St. Louis, Mo.

10. Goddard, J. 2003. *Physician's Guide to Arthropods of Medical Importance*, 4th ed. CRC Press, Inc., Boca Raton, Fla.

11. Horsefall, W. R. 1962. *Medical Entomology*. The Ronald Press Co., New York, N.Y.

12. Hunter, G. W., J. C. Swartzwelder, and D. F. Clyde. 1976. *Tropical Medicine*, 5th ed. The W. B. Saunders Co., Philadelphia, Pa.

13. Ishiwata, K., A. Shinohara, K. Yagi, Y. Horii, K. Tsuchiya, and Y. Nawa. 2004. Identification of tissue-embedded ascarid larvae by ribosomal DNA sequencing. *Parasitol. Res.* 92:50–52.

14. James, M. T., and R. F. Harwood. 1969. *Herms' Medical Entomology*, 6th ed. The Macmillan Co., New York, N.Y.

15. Kettle, D. S. 1995. *Medical and Veterinary Entomology*, 2nd ed. CAB International, Wallingford, United Kingdom.

16. Lane, R. P., and R. W. Crosskey. 1993. *Medical Insects and Arthropods*. Chapman & Hall, Ltd., London, United Kingdom.

17. Malek, E. T. A. 1951. Menthol relaxation of helminths before fixation. *J. Parasitol.* 37:321.

18. Matheson, R. 1950. *Medical Entomology*, 2nd ed. Comstock Publishing Co., Ithaca, N.Y.

19. Miething, F., S. Hering, B. Hanschke, and J. Dressler. 2006. Effect of fixation to the degradation of nuclear and mitochondrial DNA in different tissues. *J. Histochem. Cytochem.* 54:371–374.

20. Nace, E. K., F. J. Steurer, and M. L. Eberhard. 1999. Evaluation of Streck tissue fixative, a nonformalin fixative for preservation of stool samples and subsequent parasitologic examination. *J. Clin. Microbiol.* 37:4113–4119.

21. Palmer-Toy, D. E., B. Krastins, D. A. Sarracino, J. B. Nadol, Jr., and S. N. Merchant. 2005. Efficient method for the proteomic analysis of fixed and embedded tissues. *J. Proteome Res.* 4:2404–2411.

22. Ramos, F., R. Zurabian, P. Moran, M. Ramiro, A. Gomez, C. G. Clark, E. I. Melendro, G. Garcia, and C. Ximenez. 1999. The effect of formalin fixation on the polymerase chain reaction characterization of *Entamoeba histolytica*. *Trans. R. Soc. Trop. Med. Hyg.* 93:335–336.

23. Reese, N. E., W. M. Boyce, I. A. Gardner, and D. M. Nelson. 1996. Fixation affects morphometric characters of *Psoroptes cuniculi* mites (Acari: Psoroptidae). *J. Med. Entomol.* 33:835–838.

24. Scholten, T. H., and J. Yang. 1974. Evaluation of unpreserved and preserved stools for the detection and identification of intestinal parasites. *Am. J. Clin. Pathol.* 62:563–567.

25. Steyskal, G. C., W. L. Murphy, and E. M. Hoover. 1987. *Insects and Mites: Techniques for Collection and Preservation*. USDA miscellaneous publication no. 1443. U.S. Department of Agriculture, Washington, D.C.

26. Vince, D. G., A. Tbakhi, A. Gaddipati, R. M. Cothren, J. F. Cornhill and R. R. Tubbs. 1997. Quantitative comparison of immunohistochemical staining intensity in tissues fixed in formalin and Histochoice. *Anal. Cell Pathol.* 15:119–129.

Artifacts That Can Be Confused with Parasitic Organisms

Although many body sites and specimens can be examined for the presence of parasites, the most difficult specimen in which to differentiate parasites from artifacts is usually fecal material. Feces consist of a number of components, including (i) undigested food residue; (ii) digestive by-products; (iii) epithelial cells, mucus, and other secretions from the digestive tract; and (iv) many types of microorganisms such as bacteria and yeasts. Considering the ratio between fecal debris and parasites, it is not surprising that many artifacts are responsible for incorrect identifications of protozoan trophozoites and cysts and of helminth eggs and larvae. Often, many yeast cells and other artifacts are confused with coccidian oocysts or microsporidial spores. Appropriate training, adherence to protocols, use of quality control measures, and availability of reference materials and consultants should help minimize identification errors.

Protozoa

A number of cells and other organisms can easily be confused with intestinal protozoa. These are listed in Tables 34.1 and 34.2 and illustrated in Figure 34.1.

Amebae

Occasionally, free-living amebae are found in feces or as contaminants in water. Morphologically, they differ from parasitic amebae in having one or more large contractile vacuoles in the trophozoite form and having very thick cyst walls. They can also be differentiated on the basis of cultivation; i.e., they are much easier to culture than the pathogenic protozoa. Amebae which have been recovered from stool material include *Entamoeba moshkovski*, *Naegleria gruberi*, *Hartmanella hyalina*, *Sappinia diploidea*, *Vahlkampfia punctata*, and *V. lobospinosa*. Contamination of specimens can be avoided by using dry collection containers, saline, or formalin for the concentration rinses and dilution of the specimen and by rapid fixation or examination of the specimen immediately after passage (4).

Table 34.1 Artifacts and other confusing cells and organisms[a]

Artifact	Resemblance	Differential characteristics of artifacts (permanent stain)		
		Saline mount	Cytoplasm	Nucleus
PMNs[b] (seen in dysentery and other inflammatory bowel diseases)	*E. histolytica/ E. dispar* cysts	Usually not a problem if cells are from fresh blood. Granules in cytoplasm. Cell border irregular	Less dense, often frothy. Border less clearly demarcated than that of ameba. May look very similar to that in protozoa	More coarse. Larger, relative to size of organism. Irregular shape and size. Chromatin unevenly distributed. Chromatin strands may link nuclei or may appear to be 4 separate nuclei
Macrophages (seen in dysentery and other inflammatory bowel diseases; may be present in purged specimens)	Amebic trophozoites, especially *E. histolytica/ E. dispar*	Nuclei larger and of irregular shape, with irregular chromatin distribution. Cytoplasm granular; may contain ingested debris. Cell border irregular and indistinct. Movement irregular; pseudopodia indistinct but may mimic protozoa	Coarse. May contain inclusions. May include RBCs; if RBCs present, will mimic *E. histolytica*	Large and often irregular in shape. Chromatin irregularly distributed. May appear to have karyosome. Will have more nuclear material per cell than protozoan trophozoite (nucleus may be absent)
Squamous epithelial cells (from anal mucosa)	Amebic trophozoites	Nucleus refractile and large. Cytoplasm smooth. Cell border distinct	Stains poorly	Large and single. Large chromatin mass may resemble karyosome
Columnar epithelial cells (from intestinal mucosa)	Amebic trophozoites	Nucleus refractile and large. Cytoplasm smooth. Cell border distinct	Stains poorly	Large with heavy chromatin on nuclear membrane. Often large central chromatin mass resembling karyosome
Blastocystis hominis (protozoan)	Protozoan cysts of other species or yeasts	Spherical to oval, 0.6–15 µm long. Central clear area. Peripheral nuclei may be confusing	Central mass may stain light or dark. Prominent wall	Peripheral nuclei may look like large karyosomes (no peripheral nuclear chromatin)
Yeasts (normal constituent of feces)	Protozoan cysts, microsporidia	Oval. Thick wall. No internal structure. Budding forms may be seen. Some are round and overlap size of protozoan cysts	Oval to round. Little internal structure. Refractile cell wall. Budding forms may be seen; will resemble microsporidia	None
Bacteria (normal constituent of feces)	Microsporidia	Small, oval; no internal structure. May have terminal spores that resemble vacuoles in microsporidian spore	Various colors on modified trichrome stains; generally pink and will resemble large, gram-negative rods that mimic microsporidian spores	None
Starch granules	Protozoan cysts	Rounded or angular. Very refractile. No internal structure. Stain pink to purple in iodine mounts	Not a problem in permanently stained slides	

a Other artifacts, such as contaminating plant cells and pollen grains, are occasionally seen. These should not be difficult to differentiate.
b Stools containing many human cells are usually soft or diarrheic; consequently, they would be unlikely to contain amebic cysts, and cells containing what appear to be four nuclei are almost certain to be PMNs.

Table 34.2 Artifacts that resemble parasites[a]

Clinical specimen and artifact	Resemblance
Stool	
Free-living amebae, flagellates, ciliates	Parasitic amebae, flagellates, ciliates
Free-living helminths, helminth eggs, or mite eggs	Helminth eggs, larvae, or adult worms
Yeast cells	*Cryptosporidium* spp., *Cyclospora* spp., microsporidia, helminth eggs, or protozoan cysts
Bacteria	Microsporidian spores
Fungi	Helminth eggs
Plant material	
Cells	Protozoan cysts, helminth eggs
Root hairs	Nematode larvae
Pollen grains	Helminth eggs (*Ascaris* or *Taenia* spp.)
Pineapple juice crystals	Charcot-Leyden crystals
Human cells	
PMNs	
Fresh bleeding (fresh cells)	Regular PMNs as in peripheral blood smear
Old blood (disintegrating cells)	*Entamoeba histolytica/E. dispar* cyst
Macrophages	*Entamoeba histolytica/E. dispar* trophozoite
RBCs	*Cryptosporidium* spp., *Cyclospora* spp.
Epithelial cells	Amebic trophozoites
Blood	
Platelets	Malaria parasites, *Babesia* spp.
Abnormal RBC inclusions (Howell-Jolly bodies, Cabot's rings)	Malaria parasites, *Babesia* spp.
Contaminants	
Yeast	Fungemia, parasitemia
Bacteria	Bacteremia
Plant or dust fibers	Microfilariae
Stain precipitate	Malaria, *Babesia* spp.
Body fluids	
Detached ciliary tufts	Flagellate or ciliate protozoa
Respiratory specimens	
Yeast	*Pneumocystis jiroveci*, *Cryptosporidium* spp.
Urine	
Pentatrichomonas hominis (from stool contamination)	*Trichomonas vaginalis*
Bacteria	Microsporidian spores

[a] Adapted from reference 7.

Some protozoa, such as *Entamoeba coli*, may contain fungi. *Sphaerita* spp. can be found in the cytoplasm, and *Nucleophaga* spp. can be found in the nucleus. *Sphaerita* spp. (sometimes called *Polyphaga* spp.) measure approximately 0.5 to 1.0 μm and are found in tightly packed clusters (Figure 34.2).

Flagellates

Flagellates can be difficult to differentiate, and free-living organisms are occasionally seen in a stool specimen that has been contaminated with water or saline containing *Bodo caudatus* and *Cercomonas longicauda*. These two organisms are classified in the same family as *Retortamonas intestinalis*.

Ciliates

Apparently, free-living ciliates are found in stagnant water, sewage, and soil and may be seen in fecal specimens contaminated with water or saline. Organisms that have been reported are *Uronema nigricans*, *Lembus pusillus*, and *Balantiophorus minutis* (mistakenly called *Balantidium minutum*) (4, 11).

Coccidia and Microsporidia

Cryptosporidium spp. and *Cyclospora* cayetanensis

Cryptosporidium measures approximately 4 to 6 μm and overlaps in size with a number of objects, including yeasts and debris that are found in stool specimens.

Figure 34.1 Various structures that may be seen in stool preparations. (Top row) Macrophage (left) and epithelial cells (right) that can be confused with *Entamoeba histolytica/E. dispar* trophozoites. (Second row) Polymorphonuclear leukocyte with a fragmented nucleus (left) and artifact (right) that can be confused with *Entamoeba* spp. cysts. (Third row) Two artifacts that can resemble protozoan cysts. (Fourth row) Yeast cells (left) and an artifact (right) that can be confused with *Cryptosporidium* spp. and *Cyclospora cayetanensis*, respectively, on positive acid-fast stains; it is important to measure the structures/organisms carefully before confirming organism identification. (Bottom row) Yeast cells (left) that can be confused with microsporidial spores (however, notice the budding cell within the circle), and artifacts (right) that can also be confused with microsporidial spores; these were thought to be bacteria.

Cyclospora cayetanensis measures approximately 8 to 10 µm and can be easily confused with other coccidia or artifacts, especially if careful measurements are not taken. Without the use of modified acid-fast stains or immunoassay detection methods, a light infection with coccidia will probably be missed; the more normal the stool consistency, the fewer oocysts and more artifacts that will be present. The oocysts can also be confused with artifact material when the staining results are of poor quality; they appear nonuniform and more like debris (Figure 34.1).

Isospora belli

When *Giardia lamblia* shrinks within the cyst wall, it can resemble *Isospora belli* in the immature form (a single developing sporoblast; the two mature sporocysts within the oocyst wall have not yet formed). Although this error is occasionally made, the easiest way to differentiate the two is by size. *Giardia* cysts measure approximately 11 to 14 µm long by 7 to 10 µm wide, and *I. belli* measures 20 to 33 µm long by 10 to 19 µm wide. This represents another situation in which measurement of the organisms can help prevent diagnostic errors.

Figure 34.2 Fungi parasitizing *Entamoeba coli*. (A) *Sphaerita* (or *Polyphaga*) sp. within the cytoplasm; (B) *Nucleophaga* sp. within the nucleus.

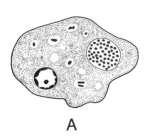

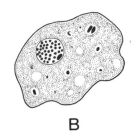

A   B

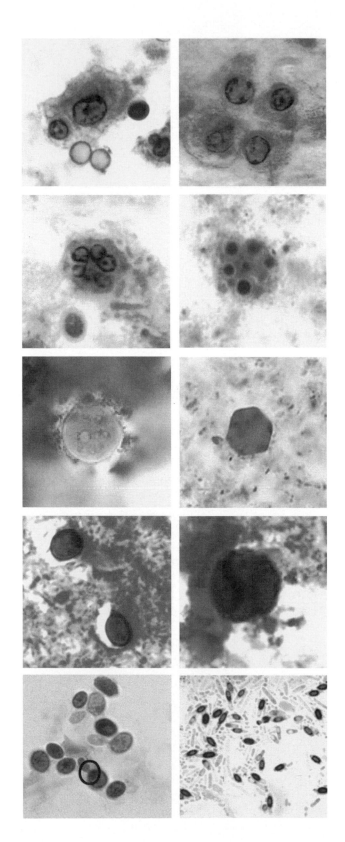

Microsporidia

Microsporidian spores in humans measure approximately 1 to 4 μm, with the majority being 1 to 2 μm. These spores are round to oval and can mimic yeast cells and bacteria. Although they do stain in a modified trichrome procedure (see chapter 27), the color is usually pale and both the size and color overlap with those of many of the yeasts or bacteria present in the specimen (Figure 34.1). The spores usually stain light to dark pink; they can be confused with bacteria looking like large gram-negative rods. Also, some of the bacilli contain terminal spores that mimic the large vacuole often seen at one end of the microsporidian spore. Size and staining color are rarely that helpful. It is important to prepare very thin smears prior to staining; the spores will be more visible, and the stain will penetrate more successfully. Without confirmation of the presence of the polar tubule in at least some of the spores (diagonal or horizontal line across the spore), it is almost impossible to confirm a microsporidial infection by using routine modified trichrome stains for fecal material.

Blood and Body Fluids

Malaria Parasites and *Babesia* spp.

One of the most common errors in examining blood smears is the incorrect identification of platelets as parasites. When mature schizonts rupture, the merozoites almost immediately penetrate another red blood cell (RBC) and are not seen outside of the RBCs. Extracellular "organisms" are almost always platelets. It is important to focus constantly when examining blood smears; the platelets are often on top of the RBCs. Another tip relates to color; parasites always have two colors, blue nuclei and red cytoplasm, and they are separate. Platelets tend to be uniform in color with no real internal structure; there are red and blue components, but the colors almost blend to form purple. Other internal structures within the RBCs such as Howell-Jolly bodies and Cabot's rings may be confusing (Figures 34.3 to 34.5). If the blood is held too long in EDTA anticoagulant or the ratio of blood to anticoagulant is incorrect, *Plasmodium* organisms can become distorted and may resemble other malaria stages or different species. Potential problems with using EDTA anticoagulant for the preparation of thin and thick blood films are disussed in chapter 31 (Table 31.2).

If a tube of blood containing EDTA cools to room temperature and the cap has been removed, the parasites can undergo several changes. The parasites within the RBCs will respond as if they were now in the mosquito after being taken in with a blood meal. The morphology of these changes in the life cycle and within the RBCs can cause confusion when examining blood films prepared from this blood (Figure 34.5).

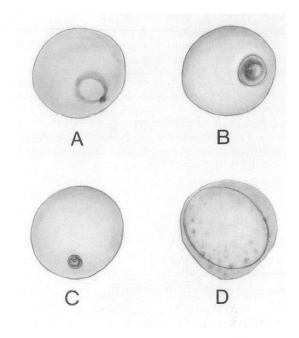

Figure 34.3   Various structures within the RBC. (A) Malarial "ring" form (early trophozoite); (B) platelet on the RBC surface; (C) Howell-Jolly body; (D) Cabot's ring. (Illustration by Sharon Belkin.)

Leishmaniae and Trypanosomes

Although they are rarely seen, intracellular amastigotes called Leishman-Donovan bodies may be found in the monocytes in peripheral blood smears from patients with visceral leishmaniasis. When they are found in a bone marrow or splenic aspirate preparation, they are easier to find, probably because at that point there may be a high index of suspicion regarding the etiologic agent. In any suspected case of visceral leishmaniasis, peripheral blood buffy coat preparations are usually examined before more extensive procedures are undertaken. The amastigotes range from 3 to 5 μm, and a defined nucleus and kinetoplast may be difficult to see, particularly if a number of organisms are packed in the cell (see chapter 8). The amastigotes may resemble *Histoplasma capsulatum* if the kinetoplast bar structure is not easily seen (Figure 34.6).

Microfilariae

Any laboratory using staining reagents must use good quality control measures to ensure that the solutions do not become contaminated with artifacts or free-living organisms. Bits of cotton fiber, lint, and other components of dust can mimic microfilariae in wet mounts or when stained. However, the artifacts do not contain any internal nuclei. Rarely, nonhuman parasites can also be

Figure 34.4 (Top row and second row) Stain deposition on the surface of uninfected RBCs that could easily be confused with developing *Plasmodium* spp. stages. (Third row) *Plasmodium falciparum* gametocytes that have rounded up and no longer appear as the typical crescent-shaped gametocytes that are normally seen (could be due to low temperatures and/or storage for too many hours in EDTA blood). (Fourth row) Developing *Plasmodium vivax* trophozoites that appear to resemble *P. falciparum* gametocytes (found on blood smears prepared from EDTA blood that had been collected more than 8 h previously). (Bottom row) RBCs containing Howell-Jolly bodies that could be confused with very small, young ring forms of *Plasmodium* spp.

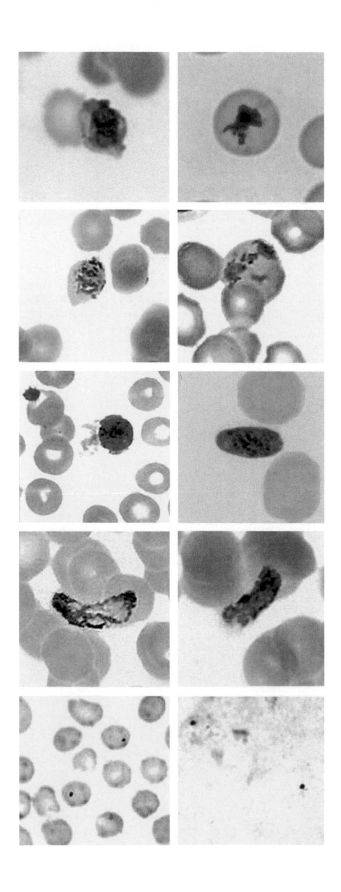

confused if they appear on the stained blood smears. One laboratory used to dry the blood films upright, leaning against a fish tank (many years ago, before more stringent safety measures were instituted). Examination of one of the stained blood films suggested that the patient had a filarial infection. Since the patient history did not support this diagnosis, further studies were performed. When samples of the fish tank water were centrifuged and examined, the "microfilariae" were found! This is just one example of the many unusual sources of artifact contamination (Figure 34.7).

Figure 34.5 Exflagellation of *Plasmodium vivax* microgametocyte; these microgametes could easily be confused with some type of spirochete. These forms were seen in blood films prepared from blood stored for longer than 12 h in EDTA prior to additional smear preparation.

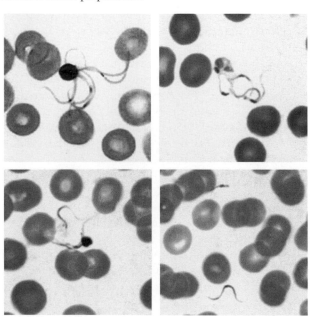

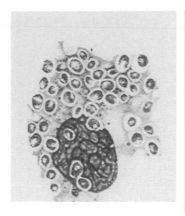
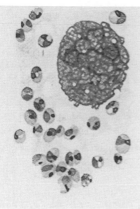

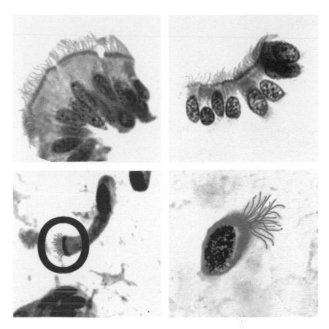

Figure 34.6 (Left) *Histoplasma capsulatum.* (Right) *Leishmania donovani.* Note that the *Leishmania* amastigotes have the bar while the *Histoplasma* amastigotes do not; *Histoplasma* also has the "halo" around the organisms.

Figure 34.8 Bronchial epithelium cells. When these cells disintegrate, the ciliary tufts may be visible and may be confused with protozoan flagellates or ciliates (detached ciliary tufts = ciliocytophthoria).

Body Fluids: Ciliated Epithelial Cells

Detached ciliary tufts (ciliocytophthoria) have been seen in a variety of body fluids (especially peritoneal and amniotic fluids; also respiratory specimens). These tufts are the remnants of ciliated epithelium that are found as a part of normal cellular turnover in a number of sites: respiratory tract and sinuses, ventricles of the brain, central canal of the spinal cord, and epithelia of the male and female reproductive tracts. The tufts are motile, measure 10 to 15 μm in diameter, and can be confused with ciliated or flagellated protozoa. However, when they are carefully examined on a stained smear, there is no internal structure like that seen in protozoa (1, 6, 9). Bronchial epithelium cells can be seen in Figure 34.8; ciliocytophthoria are anucleate remnants of these ciliated cells. The ciliated tufts can be seen on the cells in this figure.

Helminths

Adult Worms and Larvae

Plant or root hairs, such as the fuzz on peaches, may resemble nematode larvae. The root hairs tend to be clear and refractile, while the larvae pick up stain (iodine), which reveals internal structures (Figure 34.9). It is important to recognize this potential error when examining formalin-fixed specimens submitted as proficiency-testing specimens. In fecal specimens from patients with diarrhea, partially digested plant material, such as bean sprouts or other vegetable material, can mimic adult nematodes or tapeworm proglottids. Also, all stages of free-living nematodes can occur in feces or as contaminants of the water used in making fecal suspensions.

Hairworms (often called horsehair worms) belonging to the phylum Nematomorpha can be confused with human parasites (3). These adult worms are slender, measuring 10 to 50 cm long, and have a blunt, rounded anterior end. "The caudal end of the male is bifurcate or has a dorsoventral groove; that of the female is entire or trilobate" (3) (Figure 34.10). The adult worms are free living in water, while the larvae are parasites of insects. Human infection is quite rare and accidental, although in the past literature, serious health problems were attributed to "hair snakes" in the human body. Generally, human infection occurs through ingestion of free-living adult worms or adolescent worms within their insect hosts in drinking water or food.

Figure 34.7 Artifacts that can resemble microfilariae in wet mounts; these structures are not parasites but instead are some type of threads. Note that there is no internal structure visible.

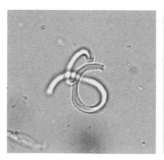
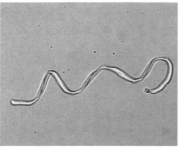

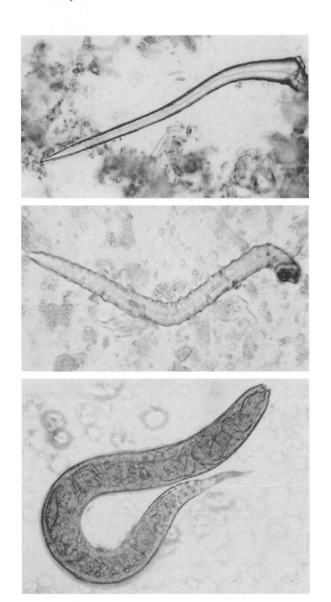

Figure 34.9 (Top) Root hair. (Middle) Root hair. Note that there is no internal structure visible within the root hairs. (Bottom) *Strongyloides stercoralis* rhabditiform larva. Note the short buccal cavity at the head end of the larva and the genital primordial packet of cells within the curved portion of the body.

Worms have been reported as passing from the urethra in several cases. They have been identified as being in the genera *Gordius*, *Chordodes*, *Parachordodes*, *Paragordius*, *Pseudogordius*, and *Neochordodes*. Worms have been recovered in vomitus, urine, and feces; often, the stated origin of the worms in the body was not well documented. In spite of the reported cases, no evidence of pathogenicity has been demonstrated (3). Often symptoms were attributed to other causes or were psychological.

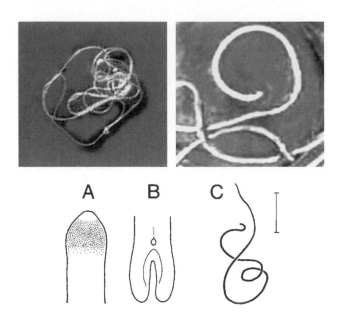

Figure 34.10 *Gordius* worms. (Upper) Adult worms (called hairworms or horsehair worms), which measure 20 to 50 cm long and are very slender. (Lower) Characteristic structure of *Gordius* worms. (A) Diagram of anterior end; (B) diagram of posterior end of a male worm in the genus *Gordius*; (C) whole worm. Bar, 1 cm. (Illustration by Sharon Belkin.)

Eggs

Plant cells tend to have thick, smooth walls and are not as symmetrical as helminth eggs (Figures 34.11 and 34.12). Some of these plant cells range from 15 to 150 μm in diameter and may be confused with *Ascaris* eggs. Pollen grains are also thick-walled, symmetrical structures that stain very darkly with iodine, may be round or trilobed, and are 15 to 20 μm in diameter (Figure 34.13). They may resemble *Taenia* eggs. Another example is the "Beaver body." This structure is *Psorospermium haeckelii*, a stage of an alga that occurs in the tissues of crayfish. It is sometimes confused with helminth eggs when found in fecal specimens from individuals who have ingested crayfish (5) (Figure 34.14). Accidental ingestion or contamination of food or water containing the eggs of plant nematodes, such as *Heterodera* species or mites or mite eggs, can also lead to confusing situations.

Human Cells

The human cells most likely to cause problems with identification are the polymorphonuclear leukocytes (PMNs) and the macrophages (Table 34.1; Figure 34.15). These cells are frequently present in patients with nonspecific inflammatory bowel disease (ulcerative colitis). Therapy for this condition often includes

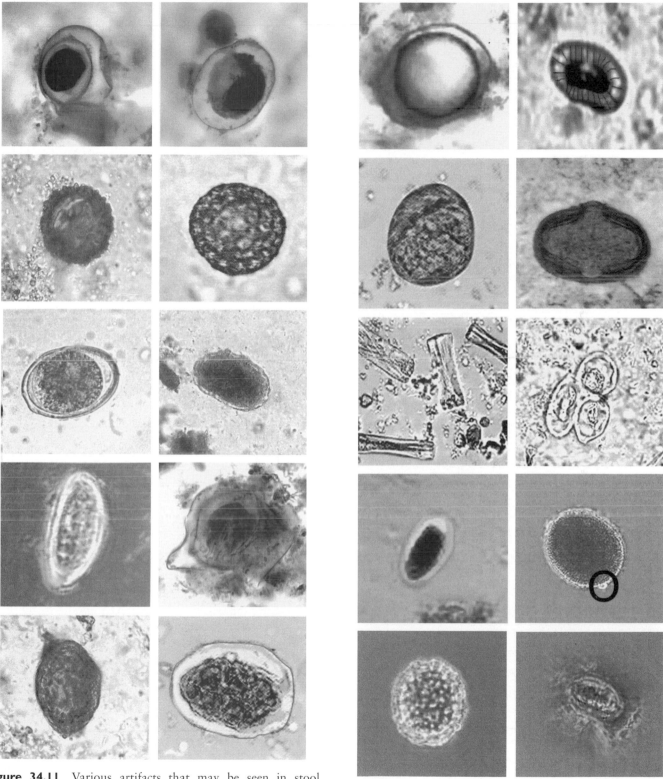

Figure 34.11 Various artifacts that may be seen in stool preparations (wet mounts or permanent stained smears). Many of these structures are pollen grains or egglike objects. Visually, they can be confused with some of the following helminth eggs: *Hymenolepis nana*, *Ascaris lumbricoides*, hookworm, and *Enterobius vermicularis*.

Figure 34.12 Various artifacts that may be seen in stool preparations (wet mounts). Note the egg-like structure in the fourth row (right). There is a small bubble (within the circle) that mimics the small knob found at the abopercular end of a *Diphyllobothrium latum* egg.

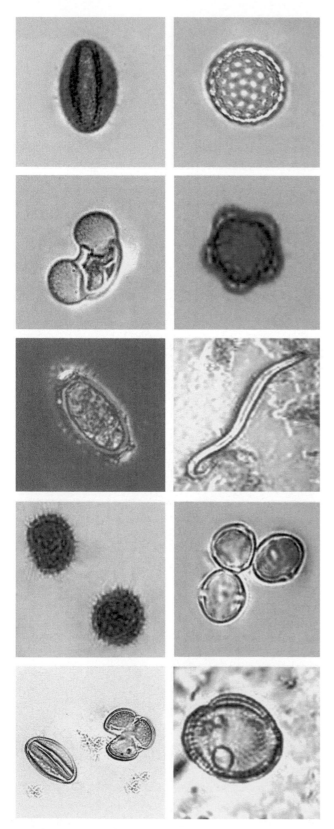

Figure 34.13 Various types of pollen grains and a root hair. These structures can mimic various helminth eggs (*Ascaris lumbricoides*, *Trichuris trichiura*), as well as nematode larvae.

immunosuppressive agents; this therapy would definitely be contraindicated in patients with amebiasis (4). These cells should be reported and quantitated (rare, few, moderate, many).

Polymorphonuclear Leukocytes

Large numbers of PMNs are often found in patients with bacterial dysentery, and they may also be present in patients with intestinal amebiasis or ulcerative colitis (Figure 34.15). These cells may be distinguished from *Entamoeba histolytica* as follows.

PMNs	*E. histolytica*/*E. dispar* (cysts)
1. Average size, 14 µm (10 to 12 µm on permanent stained smear)	1. Average size, 20 µm (less on permanent stained smear)
2. Ratio of nuclear material to cytoplasm, 1:1	2. Ratio of nuclear material to cytoplasm, 1:10–1:12 (trophozoite), 1:2–1:3 (cyst)
3. Nucleus: 2–4 segments connected by narrow, short chromatin bands. Segments may appear as separate nuclei like those of *E. histolytica*/*E. dispar* cyst—focus carefully to reveal connecting chromatin strands	3. Nucleus: round with central karyosome and peripheral chromatin
4. Granular cytoplasm	4. Uniform, agranular cytoplasm—may contain RBCs
5. Trichrome staining characteristics similar to *E. histolytica*/*E. dispar*	5. Trichrome: green cytoplasm, dark red nuclear material

Eosinophils

The identification of eosinophils in a fecal specimen usually indicates the presence of an immune response in the host. This allergic response may be caused by a parasitic infection or by other antigens such as pollen or food. Eosinophils are essentially the same size as PMNs and are characterized by the presence of large, purple-staining granules (trichrome) and usually a bilobed nucleus, which may be obscured by the granules.

Macrophages

Macrophages (monocytes) are large, mononuclear, phagocytic cells that may resemble *E. histolytica*/*E. dispar* trophozoites (Figure 34.15). These cells may be found in patients with intestinal amebiasis and ulcerative colitis and can be differentiated from amebae as follows.

Macrophages	E. histolytica/E. dispar (trophozoites)
1. Size: 30–60 µm, may be 5–10 µm less on permanent stained smear	1. Size: 12–60 µm; average, 20 µm (less on permanent stained smear)
2. Ratio of nuclear material to cytoplasm, 1:4–1:6	2. Ratio of nuclear material to cytoplasm, 1:10–1:12
3. One large nucleus that may be irregular in shape (like monocyte nucleus)	3. One nucleus, round, with central karyosome and peripheral chromatin
4. Usually contains ingested debris, PMNs, and RBCs	4. May contain RBCs and some debris; no PMNs
5. May contain red-staining round bodies and *nucleus may be absent*	5. *Nucleus always present*
6. Trichrome staining characteristics similar to *E. histolytica/E. dispar*	6. Trichrome: green cytoplasm, dark red nuclear material

Lymphocytes

Lymphocytes have a large, dense, dark-staining nucleus surrounded by very little cytoplasm. They are approximately two-thirds the size of PMNs.

Red Blood Cells

In a buffy coat preparation, some RBCs may be present. These cells measure ~7.5 µm in a wet preparation and may be present in the stool as an indication of ulceration (parasitic or nonspecific) or other vascular or hemorrhagic problems. In the trichrome stained slide, they appear as round or elongate (distortion may occur during smear preparation) red-purple bodies with no granules or inclusions and may be somewhat smaller than 7.5 µm.

Charcot-Leyden Crystals

Charcot-Leyden (CL) crystals are formed from the breakdown products of eosinophils and basophils and may be present in the stool or sputum with eosinophils or alone. They are slender crystals with pointed ends, and they stain red-purple with trichrome stain. Many different crystal sizes can be seen in the same specimen (Figure 34.16). They indicate that an immune response has taken place, but the cause may or may not be parasitic. The presence of eosinophils and/or CL crystals in the stool may not correlate with an increased eosinophilia on the peripheral blood smear.

Identification of CL crystals in body fluids and secretions is considered an indicator of eosinophil-associated allergic inflammation. The overall structural fold of CL crystal protein is similar to that of galectins, and this is the first structure of an eosinophil protein to be determined (8).

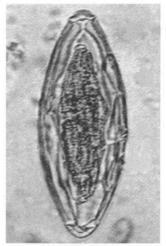

Figure 34.14 (Upper) "Beaver bodies," which are algae occasionally found in stool; (lower) leaf structure which resembles a trematode. (Courtesy of Joseph Dipersio.)

The protein exhibits weak carbohydrate binding activity for simple saccharides. There may be a potential intracellular and/or extracellular role(s) for the galectin-associated activities of CL crystal protein in eosinophil and basophil function in allergic diseases and inflammation (8).

It has also been reported that pineapple crystals sometimes mimic CL crystals in stool specimens. Apparently, these crystals can be found in fresh and canned pineapple and pineapple juice; also, they are not digested in the alimentary canal of humans. They range from 30 to 130 µm in length and 1 to 2 µm in width. They appear to have parallel edges and are pointed at both ends. In wet preparations, they can resemble CL crystals. Those who routinely examine stool specimens for parasites should probably examine some pineapple juice under the microscope to become familiar with the appearance of these crystals.

Nonhuman Elements Seen in Feces (Yeast Cells)

There are many yeast cells that may be round to oval and measure ~4 to 8 µm which can be seen in fecal material. On a wet mount, they may resemble small protozoan

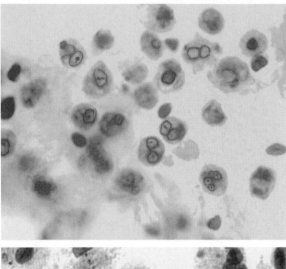

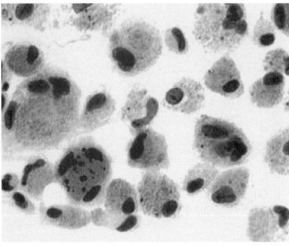

Figure 34.15 (Upper) Polymorphonuclear leukocytes. Note the lobed nuclei; if these cells have been in the stool for some time (unpreserved), the nuclei may fragment into four or five pieces, thus resembling multiple nuclei seen in amebic cysts. (Lower) Macrophages. Although these cells often resemble amebic trophozoites, the ratio of nuclear material to cytoplasm is quite different than that seen in actual protozoa.

cysts (*Endolimax nana* or *Entamoeba hartmanni*). After staining, they appear fairly uniform in color (red to green with trichrome stain) without many inclusions; if granules are seen, they are usually small but may resemble small protozoan karyosomes. Depending on the stain used, small yeast cells can be confused with coccidian oocysts or microsporidial spores. It is important to note the presence of budding yeast cells and/or pseudohyphae (clinically relevant only in freshly preserved specimens). The presence of branching pseudohyphae may be an indication of pathogenicity of the particular yeast present (usually *Candida* spp.) and should be reported. Large numbers of budding yeast cells in a fresh or freshly preserved specimen, indicating a potential source for a systemic infection,

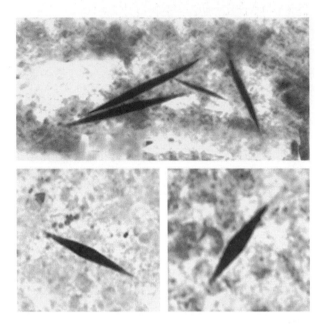

Figure 34.16 Charcot-Leyden crystals. These crystals are formed from the breakdown products of eosinophils and basophils and may be present in the stool or sputum with or without eosinophils. They tend to stain quite dark on the permanent stained fecal smears, often darker than nuclear material; and although the shape is consistent, there is a large size range in a single fecal smear or sputum mount.

particularly in immunosuppressed patients, should also be reported (Figure 34.17).

Insect Larvae

Finding insect larvae in stool is not common but may occur as a result of ingestion of whole larvae or adult insects with food. The presence of live larvae may suggest myiasis or, probably more common, contamination of the stool specimen. In these situations, it is always important to find out how and when the specimen was collected prior to submission, particularly if it was submitted as a fresh stool. Proper fixation of the suspected object is important for further identification (see chapter 33).

Spurious Infections

Spurious infections occur when individuals ingest liver from various animals. The eggs are digested free when the liver is eaten and will be passed in the stool for several days. Repeat ova and parasite examinations are recommended for several days to rule out a true infection. Examples are eggs of *Fasciola hepatica*, *Dicrocoelium dendriticum*, or *Capillaria hepatica*, which are present in the livers of cattle, sheep, and rodents, respectively. Occasionally, rarer eggs are found and may represent spurious infections acquired by eating the flesh of fish, birds, or other animals,

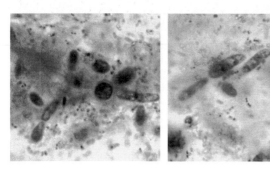

Figure 34.17 Yeast cells in permanent stained fecal smears. Depending on the size and permanent stain used (trichrome, modified acid-fast, modified trichrome), single yeast cells can often be confused with the coccidia or microsporidia.

both vertebrates and invertebrates (Figure 34.18). In a true human infection with *Capillaria hepatica*, no eggs are found in the stool; diagnosis requires histologic examination. Eggs in liver biopsy specimens can be identified on the basis of their characteristic morphology.

Delusory Parasitosis

Occasionally, clinical specimens in which the patient has placed various objects or organisms to feign parasitism are submitted for examination. These patients are usually

Figure 34.18 (Upper row) *Fasciola hepatica* egg (130 to 150 μm by 63 to 90 μm) (left, image is lower magnification than *Dicrocoelium* egg) and *Dicrocoelium dendriticum* egg (38 to 45 μm by 22 to 30 μm) (right). (Lower row) *Capillaria hepatica* eggs (51 to 68 μm by 30 to 35 μm). (Left) Eggs in liver; egg passed in the stool (resembles egg of *Trichuris trichiura*). However, note the striated shell of *C. hepatica* compared with the nonstriated shell seen in a *Trichuris* egg.

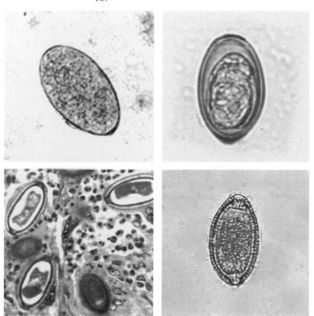

mentally disturbed and have often seen numerous physicians and submitted clinical specimens to many laboratories. The objects placed in the specimens range from pieces of thread to plant material to earthworms. Often, these patients call the laboratory with extensive histories of "parasitic infections" and seek referrals to other experts or consultants. They may bring photographs or samples of the "parasites" that they have found, all of which will be submitted for identification. These "infections" are not limited to specimens such as stool but may include skin, urine, and other samples for analysis. Often, these patients are women older than 50 years, with possible other medical problems, who present with a wide variety of symptoms. It is likely that dopaminergic and serotonergic dysfunction may play a role in delusional parasitosis; dopamine and serotonin antagonists may be relevant for the treatment of this disorder (10, 12).

This disease can become a tremendous burden both for the patient and for the family. Patients can appear to be very rational about everything else, with the exception of their beliefs concerning infection or infestation with parasites (2, 10, 12). *However, actual infections have been misdiagnosed as delusory parasitosis when appropriate diagnostic procedures have not been used and the true infections have been missed.* A thorough investigation for parasites on the patient, on pets, and in the work and home environment should be completed before assigning a diagnosis of delusory parasitosis.

References

1. **Ashfag-Drewett, R., C. Allen, and R. L. Harrison.** 1990. Detached ciliary tufts: comparison with intestinal protozoa and a review of the literature. *Am. J. Clin. Pathol.* **93:**541–545.
2. **Aw, D. C., J. Y. Thong, and H. L. Chan.** 2004. Delusional parasitosis: case series of 8 patients and review of the literature. *Ann. Acad. Med. Singapore* **33:**89–94.
3. **Beaver, P. C., R. C. Jung, and E. W. Cupp.** 1984. *Clinical Parasitology*, 9th ed. Lea & Febiger, Philadelphia, Pa.
4. **Belding, D. L.** 1965. *Textbook of Parasitology*, 3rd ed. Appleton-Century-Crofts, New York, N.Y.
5. **Garcia, L. S.** 1999. *Practical Guide to Diagnostic Parasitology*, ASM Press, Washington, D.C.
6. **Hadziyannis, E., B. Yen-Lieberman, G. Hall, and G. W. Procop.** 2000. Ciliocytophthoria in clinical virology. *Arch. Pathol. Lab. Med.* **124:**1220–1223.
7. **Isenberg, H. D.** 2004. *Clinical Microbiology Procedures Handbook*. 2nd ed., vol. 1, 2, and 3. ASM Press, Washington, D.C.
8. **Leonidas, D. D., B. L. Elbert, Z. Zhou, H. Leffler, S. J. Ackerman, and K. R. Acharya.** 1995. Crystal structure of human Charcot-Leyden crystal protein, an eosinophil lysophospholipase, identifies it as a new member of the carbohydrate-binding family of galectins. *Structure* **3:**1379–1393.
9. **Mahoney, C. A., N. Sherwood, E. H. Yap, T. P. Singleton, D. J. Whitney, and P. J. Cornbleet.** 1993. Ciliated cell

remnants in peritoneal dialysis fluid. *Arch. Pathol. Lab. Med.* **117:**211–213.

10. **Narumoto, J., H. Ueda, H. Tsuchida, T. Yamashita, Y. Kitabayashi, and K. Fukui.** 2006. Regional cerebral blood flow changes in a patient with delusional parasitosis before and after successful treatment with risperidone: a case report. *Prog. Neuropsychopharmacol. Biol. Psychiatry* 20 Jan. [Epub ahead of print.]

11. **Smith, J. W., R. M. McQuay, L. R. Ash, D. M. Melvin, T. C. Orihel, and J. H. Thompson.** 1976. *Diagnostic Medical Parasitology: Intestinal Protozoa.* American Society of Clinical Pathologists, Chicago, Ill.

12. **Wenning, M. T., L. E. Davy, G. Catalano, and M. C. Catalano.** 2003. Atypical antipsychotics in the treatment of delusional parasitosis. *Ann. Clin. Psychiatry* **15:**233–239.

35

Equipment, Supplies, Safety, and Quality System Recommendations for a Diagnostic Parasitology Laboratory: Factors Influencing Future Laboratory Practice

Some of the following information has been adapted from *Clinical Microbiology Procedures Handbook*, published by the American Society for Microbiology (34). Additional information on the following topics can also be found in that publication.

Equipment

Microscope

Good, clean microscopes and light sources are mandatory for the examination of specimens for parasites (Figure 35.1). Organism identification depends on morphologic differences, most of which must be seen under stereoscopic microscopes (magnification, $\leq \times 50$) or regular microscopes at low ($\times 100$), high dry ($\times 400$), and oil immersion ($\times 1,000$) magnifications. The use of a $50\times$ or $60\times$ oil immersion objective for scanning can be very helpful, particularly if the $50\times$ oil and $100\times$ oil immersion objectives are placed side by side. This arrangement on the microscope can help avoid accidentally getting oil on the $40\times$ high dry objective. Calibration of the microscope is discussed later in this chapter.

Types

Stereoscopic Microscope. A stereoscopic microscope is recommended for larger specimens (arthropods, tapeworm proglottids, and various artifacts). The total magnification usually varies from approximately $\times 10$ to $\times 45$, either with a zoom capacity or with fixed objectives ($0.66\times$, $1.3\times$, and $3\times$) that can be used with $5\times$ or $10\times$ oculars. Depending on the density of the specimen or object being examined, you must be able to direct the light source either from under the stage or onto the top of the stage.

Regular Light Microscope. The light microscope should be equipped with the following:

1. *Head.* A binocular head is recommended and should be equipped with a diopter adjustment to compensate for focus variation in the eyes.
2. *Oculars.* $10\times$ oculars are required; $5\times$ oculars can be helpful but are optional.

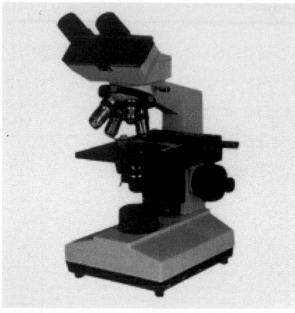

Figure 35.1 Microscopes routinely used in clinical laboratories.

3. *Objectives.* 10× (low power), 40× (high power), 100× (oil immersion). Some laboratories are currently using 50× or 60× oil immersion lenses for screening permanent stained smears. Examination with a combination of the 50× oil or 60× oil and the 100× oil immersion lenses allows screening to proceed more quickly and eliminates the problem of accidentally getting oil on the high dry objective lens when switching back and forth between the 40× (high dry) and 100× (oil immersion) objectives.

4. *Stage.* A mechanical stage for both vertical and horizontal movement is required. Graduated stages are mandatory for recording the exact location of an organism on a permanent stained smear and recommended for any facility performing diagnostic parasitology procedures. This capability is essential for consultation and teaching responsibilities. *However, remember that calibration numbers for the exact location of an organism may not be the same for different microscopes and different stages.*

5. *Condenser.* A bright-field condenser equipped with an iris diaphragm is required; however, an adjustable condenser is not required with the newer microscopes. The condenser numerical aperture should be equal to or greater than the highest objective numerical aperture.

6. *Filters.* Both clear blue glass and white ground-glass filters are recommended.

7. *Light source.* The light source, along with an adjustable voltage regulator, is usually contained in the microscope base. This light source should be aligned as specified by the manufacturer. If the light source is external, the microscope must be equipped with an adjustable mirror and an adjustable condenser containing an iris diaphragm. The light source should be a 75- to 100-W bulb.

Maintenance (54)

1. Use a camel hair brush to remove dust from all optical surfaces; remove oil and finger marks immediately from the lenses with *several thicknesses* of lens tissue. *Single-thickness lens tissue may permit corrosive acids from the fingers to damage the lens. Do not use any type of tissue other than lens tissue, otherwise you may scratch the lens. Use very little pressure, to prevent removal of the coatings on external surfaces of the lenses.*

2. Use water-based cleaning solutions for normal cleaning. If you have to use organic solvents, use them in very small amounts and only if absolutely necessary to remove oil from the lens. *Since microscope manufacturers do not agree on solvents to be used, each company's recommendations should be consulted.* One recommended solvent is 1,1,1-trichloroethane; it is good for removing immersion oil and mounting media and does not soften the lens sealers and cements. *Xylene, any alcohols, acetone, or any other ketones should never be used as cleaning fluids.*

3. After the lamp has been installed into the lamp holder, clean it with lens tissue moistened in 70% isopropyl or ethyl alcohol (to remove oil from

fingers). *Make sure that the lamp is cool and the switch is in the off position when replacing or removing the lamp.*

4. Clean the stage with a small amount of disinfectant (70% isopropyl or ethyl alcohol) when it becomes contaminated.

5. Using petroleum jelly or light grease, clean and lubricate the substage condenser slide as needed.

6. Cover the microscope when not in use. In extremely humid climates (a relative humidity of more than 50%), good ventilation is necessary to prevent fungal growth on the optical elements.

7. At least annually, schedule a complete general cleaning and readjustment to be performed by a factory-trained and authorized individual. If microscopes are in continual use, maintenance should be performed twice a year. Record all preventive maintenance and repair data (date, microscope identification number, names of company and representative, maintenance and/or repairs, part replacement, recommendations for next evaluation, estimated cost if you have such information). This information should be cumulative so that a review for each piece of equipment can be scanned quickly for continuing problems, justification for replacement requests, etc. Depending on the physical site and use of the microscope, laboratories may use different maintenance schedules.

Calibration

The identification of protozoa and other parasites depends on several factors, one of which is size. Any laboratory doing diagnostic work in parasitology should have a calibrated microscope available for precise measurements (25). Measurements are made with a micrometer disk that is placed in the ocular of the microscope; the disk is usually calibrated as a line divided into 50 units. Depending on the objective magnification used, the divisions in the disk represent different measurements. The ocular disk division must be compared with a known calibrated scale, usually a stage micrometer with a scale of 0.1- and 0.01-mm divisions. Although there is not universal agreement, it is probably appropriate to recalibrate the microscope once each year. This recommendation should be followed if the microscope has received heavy use or has been bumped or moved multiple times. Often, the measurement of red blood cells (approximately 7.5 µm) is used to check the calibrations of the three magnifications ($\times 100$, $\times 400$, and $\times 1,000$). Latex or polystyrene beads of a standardized diameter (Sigma, J. T. Baker, etc.) can be used to check the calculations and measurements. Beads of 10 and 90 µm in diameter are recommended.

Supplies

1. Ocular micrometer disk (line divided into 50 units) (any laboratory supply distributor [Fisher, Baxter, Scientific Products, VWR, etc.])

2. Stage micrometer with a scale of 0.1- and 0.01-mm divisions (Fisher, Baxter, Scientific Products, VWR, etc.)

3. Immersion oil

4. Lens paper

5. Binocular microscope with $10\times$, $40\times$, and $100\times$ objectives. Other objective magnifications may also be used ($50\times$ oil or $60\times$ oil immersion lenses).

6. Oculars of $10\times$. Some may prefer $5\times$; however, smaller magnification may make final identifications more difficult.

7. Single $10\times$ ocular to be used to calibrate all laboratory microscopes (to be used when any organism is being measured).

Note All measurements should be documented in quality control records.

Procedure

1. Unscrew the eye lens of a $10\times$ ocular, and place the micrometer disk (engraved side down) within the ocular. Use lens paper (several thicknesses) to handle the disk; keep all surfaces free of dust and lint.

2. Place the calibrated micrometer on the stage, and focus on the scale. You should be able to distinguish the difference between the 0.1- and 0.01-mm divisions. Make sure that you understand the divisions on the scale before proceeding.

3. Adjust the stage micrometer so that the "0" line on the ocular micrometer is exactly lined up on top of the 0 line on the stage micrometer.

4. After these two 0 lines are lined up, do not move the stage micrometer any farther. Look to the right of the 0 lines for another set of lines that is superimposed. The second set of lines should be as far to the right of the 0 lines as possible; however, the distance varies with the objectives being used (Figure 35.2).

5. Count the number of ocular divisions between the 0 lines and the point where the second set of lines is superimposed. Then, on the stage micrometer, count the number of 0.1-mm divisions between the 0 lines and the second set of superimposed lines.

6. Calculate the portion of a millimeter that is measured by a single small ocular unit.

7. When the high dry and oil immersion objectives are used, the 0 line of the stage micrometer will

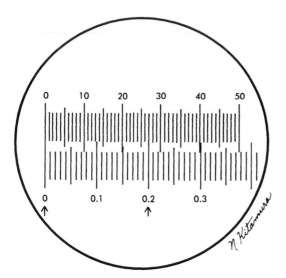

Figure 35.2 Ocular micrometer, top scale; stage micrometer, bottom scale. (Illustration by Nobuko Kitamura; modified from reference 42.)

increase in size whereas the ocular 0 line will remain the same size. The thin ocular 0 line should be lined up in the center or at one edge of the broad stage micrometer 0 line. Thus, when the second set of superimposed lines is found, the thin ocular line should be lined up in the center or at the corresponding edge of the broad stage micrometer line.

Example:

A. $\dfrac{\text{Stage reading}}{\text{ocular reading}} \times$

$\dfrac{1{,}000 \ \mu\text{m}}{1 \ \text{mm}} = $ ocular units (μm)

B. Low power (10×)

$\dfrac{0.8 \ \text{mm}}{100 \ \text{units}} \times \dfrac{1{,}000 \ \mu\text{m}}{1 \ \text{mm}} = 8.0 \ \mu\text{m}$ (factor)

C. High dry power (40×)

$\dfrac{0.1 \ \text{mm}}{50 \ \text{units}} \times \dfrac{1{,}000 \ \mu\text{m}}{1 \ \text{mm}} = 2.0 \ \mu\text{m}$ (factor)

D. Oil immersion (40×)

$\dfrac{0.05 \ \text{mm}}{62 \ \text{units}} \times \dfrac{1{,}000 \ \mu\text{m}}{1 \ \text{mm}} = 0.8 \ \mu\text{m}$ (factor)

Example: If a helminth egg measures 15 ocular units by 7 ocular units (high dry objective), using the factor of 2.0 μm for the 40× objective (example C

above), the egg measures 30 by 14 μm and is probably *Clonorchis sinensis*.

Example: If a protozoan cyst measures 23 ocular units (oil immersion objective), using the factor of 0.8 μm for the 100× objective (example D above), the cyst measures 18.4 μm.

Results. For each objective magnification, a factor will be generated (1 ocular unit = certain number of micrometers). If standardized latex or polystyrene beads or a red blood cell is measured with various objectives, the size of the object measured should be the same (or very close), regardless of the objective magnification. The multiplication factor for each objective should be posted (either on the base of the microscope or on a nearby wall or bulletin board) for easy reference. Once the number of ocular lines per width and length of the organism is measured, then, depending on the objective magnification, the factor (1 ocular unit = certain number of micrometers) can be applied to the number of lines to obtain the width and length of the organism. Comparison of these measurements with reference measurements in various books and manuals should confirm the organism identification.

Procedure Notes for Microscope Calibration

1. The final multiplication factors will only be as good as your visual comparison of the ocular 0 and stage micrometer 0 lines.
2. As a rule of thumb, the high dry objective (40×) factor should be approximately 2.5 times more than the factor obtained from the oil immersion objective (100×). The low-power objective (10×) factor should be approximately 10 times that of the oil immersion objective (100×).

Limitations of Microscope Calibration

1. After each objective has been calibrated, the oculars containing the disk and/or the objectives cannot be interchanged with corresponding objectives or oculars on another microscope.
2. Each microscope used to measure organisms must be calibrated as a unit. The original oculars and objectives that were used to calibrate the microscope must also be used when an organism is measured.
3. The objective containing the ocular micrometer can be stored until needed. This single ocular can be inserted when measurements are taken. However, this particular ocular containing the

ocular micrometer disk must have also been used as the ocular during microscope calibration.

Centrifuge

A table or floor model centrifuge to accommodate 15-ml centrifuge tubes is recommended. It is also helpful to have a centrifuge that can hold 50-ml centrifuge tubes, particularly when some of the commercial concentration systems are used. Regardless of the model, a free-swinging or horizontal head is recommended. With this type of centrifuge, the sediment is deposited evenly on the bottom of the tube and the flat surface of the sediment allows removal of the supernatant fluid from the sediment, particularly when you cannot turn the tube upside down to pour out the supernatant fluid. Most laboratories are using carrier cups that have screw-cap closures; this feature, in addition to capped centrifuge tubes, will minimize any aerosol formation and/or distribution.

Maintenance (54)

1. Before each run, visually check the carrier cups, trunnions, and rotor for corrosion and cracks. If anything is found to be defective, replace it immediately or remove the equipment from service. Check for the presence and insertion of the proper cup cushions before each run.

2. At least quarterly, check the speed at all regularly used speeds with a stroboscopic light to verify the accuracy of a built-in tachometer or speed settings. Remember to record results. Some laboratories perform this function every 6 months or yearly.

3. Following a breakage or spill and at least monthly, disinfect the centrifuge bowl, buckets, trunnions, and rotor with 10% household bleach or phenolic solution. Following disinfection, rinse the parts with warm water and perform a final rinse with distilled water. Thoroughly dry the parts with a clean absorbent towel to prevent corrosion. At least quarterly, brush the inside of the cups with mild warm soapy water and use fine steel wool to remove deposits; the cups should then be rinsed in distilled water and thoroughly dried.

4. Follow manufacturer's recommendations for preventive maintenance (lubrication).

5. Semiannually, check brushes and replace if worn to 1/4 in. (1 in. = 2.54 cm) of the spring. Also semiannually, check the autotransformer brush and replace if worn to 1/4 in. of the spring.

6. Record all information relating to preventive maintenance and repair (date, centrifuge identification number, names of company and representative, maintenance and/or repairs, part replacement, recommendations for next evaluation, estimated cost if you have such information). This information should be cumulative so that a review for each piece of equipment can be scanned quickly for continuing problems, justification for replacement requests, etc.

Fume Hood

Chemical fume hoods should be used when there is risk of exposure to hazardous fumes or splashes while preparing or dispensing chemical solutions. Airflow is generally controlled by a movable sash and should be in the range of 80 to 120 ft/min (1ft = 30.48 cm). Chemical fume hoods are certified annually. Although a fume hood is not required for diagnostic parasitology work, many facilities keep the staining setup and formalin (see below for a discussion of regulations regarding the use of formaldehyde) in a fume hood. Fume hoods may also be preferred for the elimination of odors. The placement of reagents, supplies, and equipment within the hood should not interfere with the proper airflow.

Maintenance

1. At least yearly, with the sash fully open and the cabinet empty, check the air velocity with a thermoanemometer (minimum acceptable face velocity, 100 ft/min) (7). Also, a smoke containment test should be performed with the cabinet empty to verify proper directional face velocity.

2. Lubricate the sash guides as needed.

Biological Safety Cabinet

Biological hazards are best contained within a class IIA or class IIB biological safety cabinet (BSC). BSCs operate at a negative air pressure with air passing through a HEPA filter, and the vertical airflow serves as a barrier between the cabinet and the user. Although a BSC is not required for processing routine specimens in a diagnostic parasitology laboratory, some laboratories use class I (open-face) or class II (laminar-flow) BSCs for processing all unpreserved specimens. Use of a biological safety cabinet is recommended if the laboratory is performing cultures for parasite isolation. However, remember that BSCs should not be used as fume hoods. Toxic, radioactive, or flammable vapors or gases are not removed by HEPA filters (63).

Maintenance (54)

1. After each use, disinfect the work area. Since UV radiation has very limited penetrating power, do not depend on UV irradiation to decontaminate the work surface (68). At least weekly, clean UV lamps (in the off position) with 70% isopropyl or ethyl alcohol.

2. At least annually, have class I BSCs certified. They should also be certified after installation but before use and after they have been relocated or moved. Certification should include the following and will be documented by the trained company representative (contracted to handle the BSC inspection).

    A. Measurements of the air velocity are taken at the midpoint height approximately 1 in. behind the front opening. Measurements should be made approximately every 6 in.

    B. The average face velocity should be at least 75 linear ft/min. A thermoanemometer with a sensitivity of ±2 linear ft/min should be used (53).

    C. With the cabinet containing the routine work items, such as a Bunsen burner, test tube rack, bacteriological loop and holder, etc., a smoke containment test should be performed to determine the proper directional velocity.

    D. Record the date of recertification, the names of the individual and company recertifying the cabinet, and any recommendations for future service. Any maintenance performed should also be documented in writing.

3. Replace the filters as needed.

4. On installation, have a class II BSC certified to meet Standard 49 of the National Sanitation Foundation, Ann Arbor, Mich. (53). The cabinet must also be recertified at least annually and/or when it is moved, after filters are replaced, when the exhaust motor is repaired or replaced, and when any gaskets are removed or replaced. Record the date of recertification, the names of the individual and company performing the service, and any recommendations for future service.

Refrigerator-Freezer

Any general-purpose laboratory (non-explosion-proof) or household-type refrigerator-freezer (4 to 6°C) can be used in the parasitology laboratory. Solvents with flash points below refrigeration temperature should not be stored, even in modified (explosion-proof) refrigerators.

Maintenance

1. On a daily basis, monitor and record the temperature of the refrigerator. The thermometer should be placed into a liquid to permit stable temperature recording, or thermocouples may be used.

2. On a daily basis, monitor and record the temperature of the freezer. The thermometer should be placed in antifreeze (any brand with freezing point below that of the freezer, e.g., ethylene glycol-water solutions, glycerol-water solutions, or Prestone) to permit stable temperature recording. Thermocouples may be used instead.

3. Periodically when the door is opened, check to see if the fan is operational.

4. Monthly, check the door gasket for deterioration, cracks, and proper seal. Seal problems are often seen when ice begins to build up in a freezer or the temperature is not holding. Periodically, petroleum jelly can be rubbed onto the door gasket to lubricate the material and to help maintain flexibility for a tight seal when the door is shut.

5. Semiannually, clean the condenser tubing and air grill with a vacuum cleaner.

6. Semiannually, check to ensure that the drain tubes are kept open.

7. Annually, wash the interior with a warm solution of baking soda and water (approximately 1 tablespoon/qt [ca. 13 to 14 g/0.946 liter]). Rinse with clean water, and dry. Also, wash the door gasket and water collection tray with a mild soap and water. If the gasket accumulates a black mold, scrub with 50% household bleach solution and a small brush. Rinse with clean water, and dry.

Supplies

Supplies for the diagnostic parasitology laboratory are often identical to those needed for routine work in other areas of microbiology. Although not every size of glassware used is specified, the list below should be helpful for anyone setting up a laboratory for this type of work.

Glassware

1. Disposable glass or plastic pipettes and bulbs (some sterile for culture work)

2. Pipettes, 1, 5, and 10 ml (some sterile for culture work)

3. Glass slides (1 by 3 in., or larger if preferred). Slides with rounded edges are now available (safety-"sharps")

4. Coverslips (22 by 22 mm; no. 1 or larger if preferred)

5. Beakers, 250, 500, 1,000, and 2,000 ml

6. Covered Coplin jars or staining dishes (with slide rack)

7. Graduated cylinders, 50, 100, 500, and 1,000 ml

8. Mortar and pestle (range of sizes)

9. Flasks, Erlenmeyer, 500 and 1,000 ml

10. Flasks, volumetric, 500 and 1,000 ml

11. Bottles, brown, 150 to 200 ml

12. Bottles, clear, 100, 500, and 1,000 ml

13. Bottles, airtight, 50 ml

14. Funnel (glass) to hold filter paper

15. Büchner funnel

16. Centrifuge tubes, 15 and 50 ml (some with screw caps)

17. Petri dish, plastic, sterile

18. Tubes, screw-cap, 13 by 100 mm or 16 by 125 mm (some sterile for culture work)

19. Plastic syringe, 15 ml

20. Sterile syringes (glass or plastic), 1, 10, 20, and 50 ml

Miscellaneous Supplies

1. Culture tube racks

2. Gauze (woven or pressed)

3. Applicator sticks (wood)

4. Sterile cryovials or screw-cap vials (to hold 1 ml)

5. Box for vial storage in freezer

6. Filter paper, Whatman no. 1 and Whatman no. 42

7. Sterile filtration system

8. Membrane filters (pore size, 0.22 μm) (to be used with sterile filter)

9. Nuclepore membrane filter, 25-mm, 5-μm, and 3-μm porosity

10. Swinney filter adapter (attaches to syringe, holds filter)

11. Filter paper pad, 25 mm (used to support the membrane filter in the Swinney adapter)

12. Bacteriological loop

13. Sterile syringe needles, 20 and 27 gauge

14. Vaspar

15. Parafilm (American Can Co.) or equivalent

16. Slide boxes for positive-slide storage

17. Forceps and scissors

18. Stage micrometer with scale of 0.1- and 0.01-mm divisions

19. Disk micrometer divided into 50 units

20. Biohazard container with disinfectant for proper disposal of slides, tubes, and pipettes

21. Biohazard container for proper disposal of patient specimens

ATCC Quality Control Organisms

1. An 18- to 24-h-old culture of *Escherichia coli* or *Enterobacter aerogenes*

2. ATCC 30010 (*Acanthamoeba castellanii*)

3. ATCC 30133 (*Naegleria gruberi*)

4. ATCC 30925 (*Entamoeba histolytica* HU-1:CDC)

5. ATCC 30015 (*Entamoeba histolytica* HK-9)

6. ATCC 30001 (*Trichomonas vaginalis*)

7. ATCC 30883 (*Leishmania mexicana*)

8. ATCC 30160 (*Trypanosoma cruzi*)

Note None, some, or all of these QC organisms will be required, depending on the culture procedures performed in the laboratory.

Safety: Personnel and Physical Facilities

General Precautions (5, 7, 9, 10, 20, 23, 25, 28, 45, 48–51, 53–55, 59)

1. *Be careful!* All material to be received by or discarded from the laboratory must be considered potentially pathogenic.

2. Smoking, eating, or drinking in the laboratory is not permitted.

3. Do not work with uncovered open cuts or broken skin. Cover them with a Band-Aid, finger cot, or other suitable means, such as rubber or plastic gloves.

4. Mouth pipetting of specimens is not permitted.

5. Do not create aerosols. Remember, infectious diseases, such as infectious hepatitis, may be transmitted by aerosols produced by centrifuges, stirrers, pipettors, etc. Exercise extreme care when using such equipment. Cool inoculating loops or needles before touching colonies on plates or inserting into liquid material.

6. Develop the habit of keeping your hands away from your mouth, nose, and eyes. Wash hands well with soap before leaving the laboratory.

7. Do not lay personal articles, such as eyeglasses, on the bench in your work area.

8. Laboratory coats or gowns are not to be worn outside the laboratory, particularly not to the employee lounge or cafeteria or out of the building.

9. Wipe off benches in your working area with disinfectant *before and after* each day's work. Keep your area clean at all times.

10. In case of injury or unusual incident, however slight, the supervisor in charge must be immediately notified, and a report to occupational health facility is required. Also fill out an accident report form. Major accidents must be documented and reported in detail to the supervisor and chief of microbiology. The report should indicate:

    A. Cause of the accident

    B. Type of contamination or hazard

    C. List of personnel possibly exposed and amount of exposure to possible pathogenic material

    D. Decontamination procedures taken if pathogenic material was involved

    E. Actions taken to prevent recurrence

    F. Actions taken to safeguard or monitor employees

11. Infections may be spread by a number of routes in the laboratory. The actual occurrence of an infection depends on the virulence of the infecting agent, susceptibility of the host, route of entry, inoculating dose, etc. (Table 35.1).

    A. *Airborne.* Droplets and aerosols may be formed by simply removing caps or cotton plugs or swabs from tubes. Heating liquids on needles too rapidly may also create an aerosol. Breakages in centrifuges are serious accidents (Table 35.2).

    B. *Ingestion.* Ingestion can occur through mouth pipetting, failure to wash hands after handling specimens or cultures, or smoking, eating, and drinking in the laboratory.

    C. *Direct inoculation.* Scratches, needlesticks, cuts from broken glass, or animal bites may permit direct inoculation.

    D. *Skin contact.* Some very virulent organisms, and others not so virulent, can enter through small cuts or scratches or through the conjunctiva of the eye.

    E. *Vectors.* Flies, mosquitoes, ticks, fleas, and other ectoparasites can be potential sources of infection in the laboratory, especially if animal work is performed.

12. Personnel who display risk-prone behavior or are pregnant, immunocompromised, or immunosuppressed should be restricted from performing work with highly infectious microorganisms and, in some situations, be restricted to a low-risk laboratory (62).

Handwashing

Handwashing is the most important procedure to reduce the duration of exposure to an infectious agent or chemical, to prevent dissemination of an infectious agent, and to reduce overall infection rates in a health care facility.

Table 35.1   Routes of exposure associated with diagnostic laboratory work

Route	Practice or potential cause
Ingestion	Mouth pipetting
	Splashes of infectious material into the mouth
	Contaminated articles or fingers placed in the mouth
	Eating or drinking within the working laboratory space
Inoculation	Needle sticks
	Cuts
	Animal and insect bites or scratches
Contamination of skin and mucous membranes	Spills or splashes into eyes, mouth, or nose or onto intact or nonintact skin
	Contaminated surfaces, equipment, and articles
Inhalation	Numerous procedures that can produce aerosols

Table 35.2 Laboratory functions that may produce aerosols

Laboratory function	Practice or potential cause (type of relevant culture)
Inoculating-loop manipulations	Subculturing or streaking culture plates "Cooling" a loop in culture medium Flaming a loop (Free-living ameba culture systems)
Pipette	Mixing microbiological suspensions or fresh specimens Pipette spills on hard surfaces and benches Blowing contents from pipettes (Any function involving pipettes and liquids)
Needle and syringe manipulation	Expelling air Withdrawing needle from stopper Injecting animals Spray created when needle separates from syringe (Any function involving syringe-needle combination)
Other laboratory functions	Centrifugation with open containers (centrifuge tubes and/or carriers) Blenders, shakers, sonicators, and mixing instruments (specimens, cultures) Pouring or decanting fluids (specimen processing) Opening culture containers (any parasitic culture) Spillage of infectious material Lyophilization and filtration under vacuum

Hand contamination occurs during manipulation of specimens and during contact with work surfaces, telephones, and equipment. Laboratory personnel should wash their hands immediately after removing gloves, after obvious contamination, after completion of work, before leaving the laboratory, and before hand contact with nonintact skin, eyes, or mucous membranes (63).

Handwashing sinks should be located at each entry/exit door; if possible, the faucet should be operated by a knee/foot control. If these controls are not available, the faucet should be turned on and off using a paper towel. Handwashing should be performed using soap or an antiseptic compound, starting at the wrist area and extending down between the fingers and around and under the fingernails, and rinsed from the wrists downward. Recently, the Centers for Disease Control and Prevention recommended that in addition to traditional soap and water handwashing, health care personnel can use alcohol-based gels (63).

Personal Protective Equipment (OSHA 2001 Blood Borne)

Occupational Safety and Health Administration (OSHA) standards for exposure to blood-borne pathogens require the laboratory to provide personal protective equipment (PPE) for its employees. OSHA standards require that PPE be provided, used, and maintained for all hazards found in the workplace, including biological, environmental, and chemical hazards, radioactive compounds, and mechanical irritants capable of causing injury or illness through absorption, inhalation, or physical contact. Employees must be trained in the appropriate use of PPE for a specific task, the limitations of PPE, and procedures for maintaining, storing, and disposing of PPE.

Gloves

Gloves protect the wearer from exposure to potentially infectious material and other hazardous material and are available in material designed for specific tasks. Gloves must be provided by the employer and must be the proper size and appropriate material for the task. Due to latex hypersensitivity in some workers, only powder-free latex gloves should be used or gloves should be manufactured from nitrile, polyethylene, or other material.

Protective Clothing

Laboratory workers should wear long-sleeved coats or gowns that extend below the level of the workbench, and they should be worn fully closed. The material must be fluid resistant if there is any potential for splashing or spraying. Fluid-proof clothing (plastic or plastic lined) must be worn when there is the potential for soaking by infectious material. Laboratory workers should not wear laboratory clothing outside the laboratory. All protective clothing should be changed immediately when contaminated to prevent the potentially infectious material or chemical from contacting the skin. Coats and gowns should not be taken home for cleaning but should be laundered by the institution.

Face and Eye Protection

Face and eye protection should be used when splashes or sprays of infectious material or chemicals may occur. Equipment includes goggles, face shields, and splashguards. Face shields provide the best protection for the entire face and neck, although splashguards provide an alternative method. If only goggles are worn, the user should also wear an appropriate mask to prevent contamination of mucous membranes.

Figure 35.3   Example of a commercially available disinfectant that can be used for countertops, small spills, and any hard surface that requires a disinfectant (www.med-chem.com).

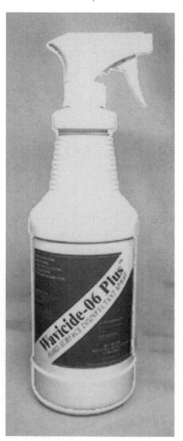

Handling Specimens (45)

1. Specimens with *gross internal contamination* must not be accepted. Place the specimen in a plastic bag to protect subsequent handlers. The test request slips should be bagged separately from the specimen (keep paperwork clean and uncontaminated). Wear gloves when handling the specimen, and wash your hands thoroughly when you finish handling the specimen. Notify the supervisor immediately about a contaminated specimen so that further corrective action may be taken.

2. Specimens to be centrifuged *must* be placed in a sealed container to prevent aerosols.

3. All specimens other than bacterial or fungal blood culture bottles to be set up for isolation of organisms must be processed in a BSC. *All* "open-system" work is to be done on an absorbent surface (i.e., a towel) in a BSC, using appropriate protective techniques. Towels should be changed daily. We recommend that the immediate working area of the towel be dampened with 2.5% amphyl or any recommended disinfectant to handle small spills (Figure 35.3).

Processing Specimens (45)

1. *All specimens are potentially pathogenic—always use careful techniques* (Tables 35.3 and 35.4).

2. All discard material used in processing of specimens is considered contaminated.

   A. Fluids used in the processing of specimens (buffers, etc.), as well as excess liquid specimens, should be poured into plastic screw-top autoclavable bottles and sterilized prior to disposal. A tongue depressor placed in the bottle can help to prevent splashing if liquid is poured down the slanting stick. These sticks can also be placed in disinfectant before being discarded. Disinfectants and incineration can also be used for decontamination of infectious materials.

   B. Reusable items (tissue grinders, bottles, etc.) are placed in an autoclave pan. Once the pan is full, place it (without a cover) in a large brown autoclave bag and staple shut. After being autoclaved, the reusable items can be cleaned.

   C. Specimen containers and centrifuge tubes are disposable. These items should be placed in plastic autoclave bags and secured with masking tape for sterilization. Slides can also be placed in containers of liquid disinfectant prior

Table 35.3 Potential routes of laboratory infection with human parasites

Organisms[a]	Mode of transmission	Comments
Intestinal protozoa		
Entamoeba histolytica[b] (C)	Accidental ingestion of infective cysts, trophozoites, oocysts, or spores in	Transmission becomes more likely when fresh stool
Entamoeba dispar (C)	food or water contaminated with fecal material. Also direct transfer of	specimens are being processed and examined; submission of
Entamoeba coli (C)	stool material via fomites (fecal/oral transmission). Cysts (10–100 are	fecal specimens in preservatives or fixatives would decrease
Entamoeba hartmanni (C)	required to initiate an infection with intestinal protozoa. (These modes	the risks. (This applies to all intestinal protozoa.)
Endolimax nana (C)	apply to all intestinal protozoa.)	
Iodamoeba bütschlii (C)		
Blastocystis hominis[b] (C)		
Giardia lamblia[b] (C)		
Dientamoeba fragilis[b] (T)		
Cryptosporidium[b] (O)	Oocysts (as few as 30) are required to initiate an infection with *Cryptosporidium* spp.	Wearing gloves, using capped centrifuge tubes, and working in BSC would decrease the risk of acquiring *Cryptosporidium* infections; the use of potassium dichromate as collection fluid and use of sugar flotation on fresh stool is *not recommended* for *Cryptosporidium*.
Cyclospora cayetanensis[b] (O)		
Isospora belli[b] (O)		
Microsporidia[b] (S)	The infective dose for microsporidial spores has not been documented; however, one assumes the number would be small.	It is well documented that microsporidial spores can be inhaled, thus initiating an infection.
Intestinal helminths		
Enterobius vermicularis (E)	Inhalation or ingestion of infective eggs	Very common in children, asymptomatic
Strongyloides stercoralis (L)	Skin penetration of infective larvae from stool material	Exposure would be very possible or likely when working with stool cultures or concentrates for recovery of helminth larvae.
Hymenolepis nana (E)	Ingestion of infective eggs (fecal/oral)	Ingestion of infective eggs can lead to the adult worm in humans.
Taenia solium (E)	Inhalation or ingestion of infective eggs could lead to cysticercosis	Exposure is very likely when working with gravid proglottids (ink injection for worm identification).
Blood and tissue protozoa		
Leishmania spp.	Direct contact or inoculation of infectious material from patient lesion; accidental inoculation of material from culture or animal inoculation studies	Culture forms or organisms from hamster would be infectious.
Trypanosoma spp.	Same as for *Leishmania* spp.	Cultures and special concentration techniques would represent possible means of exposure.
Plasmodium spp.	Accidental inoculation could transmit any of the 4 species; infectious dose would be approximately 10 infected RBCs	Blood should always be handled carefully (avoid open cuts, etc.).
Toxoplasma gondii	Inhalation or ingestion of oocysts in cat feces (veterinary situation; accidental inoculation of tachyzoites from culture, tube of blood, animal isolation (mouse peritoneal cavity)	Although many people already have antibodies to *Toxoplasma*, indicating past exposure, there have been documented laboratory accidents where the individual became ill due to the large infecting dose.

(continued)

Table 35.3 Potential routes of laboratory infection with human parasites *(continued)*

Organisms[a]	Mode of transmission	Comments
Ectoparasites		
Pediculus spp.	Specimens submitted on hair could be easily transmitted in the laboratory	Careful handling and fixation of the arthropods would prevent any potential problems with transmission
Sarcoptes scabiei	Transmission via skin scraping, etc., is possible but not very likely	Careful handling and preparation of specimens with KOH would tend to prevent any problems
Dipterous fly larvae (myiasis)	Transmission could occur anywhere	Protection from flies would solve the potential problem

[a] C, cyst; T, trophozoite; O, oocyst; IL, infective larvae; E, egg; S, spore.
[b] Potentially pathogenic intestinal protozoa.

to disposal. Remember, slides should be treated as sharps.

D. Pipettes can be placed in a covered discard pan containing 5% amphyl or placed in autoclavable containers for sterilization before discarding.

Spills

1. Disinfectants in use at present
   A. *Amphyl.* Amphyl (made up of soap, *o*-phenyl-phenol, and alcohol) is a lipophilic disinfectant with a phenol coefficient of 10.
   B. *Chlorine.* Sodium hypochlorite was reported by Klein and Deforest (36) to be virucidal against 25 viruses tested (included enteroviruses). At 0.02% available free Cl_2, all viruses tested were inactivated in 10 min (6, 9, 60, 66).
   C. *Alcohols.* Alcohols generally act by coagulating proteins and as organic solvents. Rapid evaporation may lead to inadequate "killing" time. Although there are some conflicting reports, 70% solutions generally have the best microbicidal activity. Caution is advised because of flammability.
2. Immediate actions
   A. Clear the area at once.
   B. Notify the supervisor and chief of microbiology.
   C. Assess the type of spill and degree of hazard involved.
   D. Determine the most effective and least hazardous approach to clean up and decontaminate.
3. "Dry spills" with no significant aerosol formation
   A. Flood the area with disinfectant solution.
   B. Soak up disinfectant and contaminated material with absorbent material (sand or paper towels), and dispose of it in a plastic biohazard bag or sealed container. Wear gloves for cleanup.
   C. If routine cleanup is not possible, the unit may have to be decontaminated with a sterilizing gas such as paraformaldehyde.
   D. Thoroughly wash the unit (if possible) after decontamination.
4. Liquid spills on the bench or floor
   A. If significant aerosols were formed, the area should be evacuated and not reentered for at least 1 h.
   B. Flood the area with disinfectant, and cover the spill with absorbent material (sand or paper towels). Wear gloves during cleanup.

Table 35.4   Risk assessment and exposure control plan for the parasitology laboratory[a]

Parasitology task	Exposure risk		Personal protective equipment[b]				Engineering controls[d]	
	Blood and body fluid contact	Cultured biological agent exposure	Gloves	Lab coat/ gown[c]	Face shield	Splash shield	BSC	Sharps containers readily accessible[e]
Concentrate fecal specimens, prepare smears and wet mounts	Low	BSL 2[f]	R	Coat				Pipettes/sticks
Read fecal wet mounts	Low	BSL 2	R	Coat				Slides/coverslips
Prepare thick and thin blood films	High	BSL 2	R	Coat	A	A		Slides/coverslips
Perform blood concentrations	High	BSL 2	R	Coat	A			Pipettes
Read blood concentration wet mounts	Low	BSL 2	R	Coat				Slides/coverslips
Stain and read fecal or blood slides	Low	BSL 2		Coat				Slides/coverslips
Culture fecal protozoa	High	BSL 2	R	Coat				Pipettes
Culture blood parasites	High	BSL 2	R	Coat				Pipettes
Culture free-living amebae	High	BSL 2	R	Coat	A	A		Pipettes, slides/ coverslips
Examine arthropod specimens	Low	BSL 2	R	Coat				Slides/coverslips
Perform antigen detection/PCR/ DNA probes	High	BSL 2	R	Coat	A[g]	A[g]		Pipettes

[a] Adapted from reference 34, protocol 15.3.2 (Risk Assessment).
[b] Remove PPE when leaving the laboratory. R, required; A, one of the required alternatives.
[c] Gowns with solid front and impervious to liquid. Many employers provide and launder gowns, thus eliminating the need for laboratory coats.
[d] Recapping of needles is prohibited. Tubes should be carried in racks; alternatively, plastic tubes should be used. Wash hands when leaving the laboratory.
[e] Sharps include scalpel blades, pipettes, plastic loops, sticks, needles, syringes, slides, and coverslips.
[f] BSL 2, biosafety level 2.
[g] Vortexing or other splatter-generating steps require the use of a BSC or safety shield.

C. Dispose of absorbent and contaminated material in plastic bags or sealed containers, and autoclave.

D. Thoroughly wash the area after cleanup.

5. Centrifuge spills

A. Shut off the instrument, and evacuate the area at once.

B. Do not reenter the area for at least 1 h until aerosols have settled.

C. The person entering the area to clean up should wear protective clothing, gloves, and a mask.

D. If liquids are present, soak up in absorbent material and handle as above. If liquids are not present, clean the instrument and room thoroughly before resuming work.

E. Wipe all surfaces with 2.5% amphyl or other disinfectant.

6. Spills in incubators, autoclaves, or other closed areas

A. Flood with disinfectants, and soak up liquids with absorbent material.

B. Dispose of the material as specified above, if possible.

C. If routine cleanup is not possible, the unit may have to be decontaminated with a sterilizing gas such as paraformaldehyde.

D. Thoroughly wash the unit (if possible) after decontamination.

Disposal of Contaminated Materials

1. Autoclave screw-cap tubes before cleaning or discarding.

2. Discard specimens and cultures into containers with double plastic lining. Liners should be changed when the containers are about half full.

3. Place culture plates into "Biohazard" receptacles lined with autoclavable bags.

4. Materials or containers to be reused should be autoclaved before being cleaned. Place them in sealed and clearly labeled containers to minimize hazard to others before sterilization.

5. Any breakage of glass or leakage of contaminated materials should be reported to the supervisor or chief of microbiology at once for instructions on procedures for safe cleanup.

Standard Precautions

All laboratories should adhere to the concept of standard precautions, which state that all patients and all laboratory specimens are potentially infectious and should be handled accordingly (63). This concept arose from the observation that infections are often unrecognized in patients. "Standard precautions" replaces earlier terms such as Blood and Body Fluid Precautions, Universal Precautions, and Body Substance Precautions found in OSHA documents (63). The OSHA documents place the emphasis on blood-borne pathogens such as human immunodeficiency virus (HIV) and hepatitis B and C viruses, whereas the concept of standard precautions recognizes that all infectious agents and all other potentially infectious material, except sweat, pose a risk to the health care worker (63).

OSHA identifies a number of practices that should be implemented to protect the worker from exposure to blood-borne pathogens, including an exposure control and risk assessment plan (63). Methods that should be implemented to minimize exposure to infectious agents, to shield the laboratory worker from infectious material through a set of engineering and work practice controls, and to use personal protective equipment. In addition, the OSHA regulations require that employers provide hepatitis B vaccination and postexposure evaluation and follow-up; communicate the hazards to employees; and maintain appropriate records (63). Employees who decline immunization against hepatitis B virus are required to sign a hepatitis B vaccine declination form.

Note Under circumstances in which differentiation between body fluid types is difficult or impossible, all body fluids shall be considered potentially infectious.

1. All job positions with any exposure to blood and body fluids must be listed on an exposure form, which must match up with continuing education training records from the facility.
   A. Training must be provided at no cost to the employee and during working hours.
   B. Training shall be provided as follows:
      a. At the time of initial assignment to job functions in which occupational exposure may occur
      b. Within 90 days after the effective date of the standard
      c. At least annually thereafter
2. A complete task assessment must be performed for employees who have any exposure; protective measures, including equipment, must be defined.
   A. Engineering controls
   B. Immunization programs

C. Work practices (hand washing and procedures for handling sharps)
D. Disposal and handling of infectious waste
E. Use of personal protective gloves, gowns, and goggles
F. Use of mouthpieces, resuscitation bags, or other ventilation devices
G. Use of disinfectants
H. Labeling and signs
I. Training and education programs
J. Postexposure follow-up

Some of the specifics are as follows:

A. All procedures shall be performed to minimize splashing, spraying, spattering, and generation of droplets of these substances.
B. When there is occupational exposure, the employer shall provide, at no cost to the employee, PPE such as (but not limited to) gloves (including hypoallergenic gloves, glove liners, powderless gloves, or other alternatives), gowns, laboratory coats, face shields or masks, and eye protection. Other options might include mouthpieces, resuscitation bags, pocket masks, or other ventilation devices.
C. PPE is considered "appropriate" only if it does not permit blood or other potentially infectious materials to pass through or to reach the employee's work clothes, street clothes, undergarments, skin, eyes, mouth, or other mucous membranes. The type and characteristics of outer garments depends on the tasks and degree of exposure anticipated.
D. All working surfaces shall be cleaned and decontaminated after contact with blood or other potentially infectious materials. All spills shall be immediately contained and cleaned up by appropriate professional staff or others trained and equipped to work with potentially concentrated infectious materials.
E. Specimens of blood or other potentially infectious materials shall be placed in a container which prevents leakage during collection, handling, processing, storage, transport, or shipping (46, 50, 64).
F. Mouth pipetting or suctioning of blood or other potentially infectious material is prohibited.
G. Certified BSCs or other combinations of personal protection or physical containment

devices shall be used for all activities that may pose a threat of exposure. BSCs shall be certified when installed, whenever they are moved, and at least annually.

3. Key questions that people ask concerning these rules are listed below (4).

A. How does OSHA define "occupational exposure"?

"Reasonably anticipated skin, eye, mucous membrane, or parenteral contact with blood or other potentially infectious materials that may result from the performance of an employee's duties"

B. What does "reasonably anticipated" mean?

"Actual contact would be expected during an autopsy or surgery. In these cases, blood or other potentially infectious materials come in direct contact with the employee's gloves or other protective clothing. In other cases, contact may not occur each time the task or procedure is performed, but when blood or other potentially infectious materials are an integral part of the activity, it is reasonable to anticipate that contact may result."

C. What should the exposure control plan contain?

"The Exposure Control Plan is a key provision because it requires the employer to identify the individuals who will receive the training, protective equipment, vaccination, and other provisions of the standard."

D. What is considered "appropriate" personal protective equipment?

"Gloves, gowns, lab coats, face shields or masks and eye protection, mouthpieces, resuscitation bags, pocket masks, or other ventilation devices. Such protection will be considered 'appropriate' only if it does not permit blood or other potentially infectious materials to pass through to or to reach the employee's work clothes, street clothes, undergarments, skin, eyes, mouth, or other mucous membranes under normal conditions of use and for the duration of time which the protective equipment will be used. The type and characteristics of outer garments will depend upon the tasks and degree of exposure anticipated."

E. Does this mean that protective clothing must be "impervious" or "fluid-proof"?

"Selection of personal protective equipment is performance-oriented. Selection of the type and characteristics of necessary personal protective equipment is based upon the exposure anticipated to be associated with the task. In those instances where such equipment was incapable of halting penetration of blood or other potentially infectious materials normally encountered during a procedure, then additional protection such as a plastic apron should be worn or used."

Note "Fluid-resistant" and "fluid-proof" no longer appear in the final rule.

F. How can the employer ensure that the employee uses appropriate PPE?

"It is not the intent that the employer 'watch over everyone's shoulder' for compliance, but there are reasonable policies and regulations that employees follow."

G. What does the standard say about accessibility, cleaning, repair, removal, and storage of PPE?

"The employer shall ensure that appropriate personal protective equipment in appropriate sizes is available and is accessible at the worksite; the employer shall clean, launder, and dispose of personal equipment; the employer shall repair or replace when necessary at no cost to the employee. If a garment is penetrated by potentially infectious materials, it shall be removed ASAP and replaced by clean garments. When these garments are removed, they must be placed in a designated area or container for storage, washing, decontamination or disposal."

H. What is the employer's responsibility if PPE or the hepatitis B vaccine is refused?

"The employer shall ensure that the employee uses appropriate personal protective equipment unless the employer shows that the employee temporarily and briefly declined to use personal protective equipment when, under rare and extraordinary circumstances, it was the employee's professional judgment that its use would have prevented the delivery of health care services or would have created an increased hazard to the safety of the worker or co-worker. When an employee makes this decision, the circumstance shall be investigated and documented in order to determine whether changes need to be made to prevent such reoccurrences."

If the employee declines to accept hepatitis B vaccination, a form indicating that fact must be signed by the employee and held in the file.

I. What training is required for employees?

"Training must be available at the worksite, at no cost to the employee, and during working hours. It shall be provided at the time of initial assignment to tasks where exposure may occur, within 90 days after the effective date of the Standard, and at least annually thereafter."

J. What type of training is required for personal protective equipment?

"An explanation of the use and limitations of methods that will prevent exposure is required, including engineering controls (biosafety cabinets), work practices, and personal protective equipment. Information on types, proper use, location, removal, handling, decontamination, and disposal of personal protective equipment must be presented, in addition to an explanation of the basis for selection of such controls."

Hepatitis Exposure Protocol

Laboratory personnel have a high risk of contracting hepatitis for the following reasons:

1. Frequent close personal contact with patients with hepatitis
2. Direct and frequent contact with biological specimens containing the virus
3. Frequent opportunity for accidental puncture wound with contaminated needles and sharp objects
4. Carelessness in handling specimens
5. Inadequate or unsafe disposal of contaminated needles, specimens, or other objects
6. Carelessness in proper and frequent hand washing technique

The prevention of hepatitis is the primary reason for the emphasis on infection control precautionary measures found in any safety manual (e.g., no smoking, eating, drinking, or mouth pipetting and procedures for safe specimen handling). These same procedures also apply to HIV.

All known hepatitis specimens must be so labeled; however, to minimize exposure to the virus, *every specimen should be handled as though it could transmit hepatitis*. Studies have shown that hepatitis may be transmitted to laboratory personnel in several ways:

1. Puncture or other wounds (HIV)
2. Abrasions of the skin (HIV)
3. Aerosols (inhaled into the respiratory tract)
4. By mouth (oral route)

5. Direct contact with the patient
6. Splashing material into the eyes

The following protocol should be followed in the event of exposure:

1. Immediately wash the exposed area with soap and water.
2. Report the incident to the supervisor and laboratory personnel office.
3. Consult with a physician in Employee Health Facility concerning possible recommendations (e.g., administration of immune serum globulin).

Dangerous Properties of Industrial Materials

OSHA requires each laboratory to develop a comprehensive, written chemical hygiene plan (CHP). Every hazardous chemical in the laboratory, regardless of the type of risk, volume, or concentration, must be included in the CHP. The plan should include storage requirements, handling procedures, location of OSHA-approved material safety data sheets, and the medical procedures that are to be followed if exposure occurs. The CHP must specify the clinical signs and symptoms of the environmental conditions (such as a spill) that would give the employer reason to believe that exposure had occurred. When such conditions exist, the CHP should indicate the appropriate medical attention required.

Ethyl ether (also known as ether, anesthesia ether, diethyl ether, ethyl oxide, sulfuric ether, and ethoxyethane) is so volatile that dangerous concentrations are readily built up in the laboratory atmosphere. It is highly flammable and has a tendency to form explosive peroxides. Its low flash point, low ignition temperature, wide explosive range, high volatility, and very heavy vapor (which tends to "pocket") combine to make ethyl ether an extremely serious fire hazard. The spontaneous formation of explosive peroxides presents severe risks to the user who does not apply precautionary practices. Therefore, it is mandatory to follow these safety procedures (15, 16).

1. *Date* cans of ether immediately upon receipt.
2. *Do not attempt to open* a can that is 1 year old (after time of receipt); peroxides may form in sealed cans and explode when the cans are opened. Order appropriate quantities to avoid expensive waste.
3. *Date* the can of ether when opened; do not use after 1 month. Order cans of appropriate size to avoid waste.
4. *Store opened (corked) cans* of ether close to the floor, in a cool place, on a shelf, or in a cabinet

that is *not* airtight. The heavy vapors can dissipate along the floor, with less opportunity for contact with sources of ignitions such as an open flame and electric sparks.

5. *Never store* opened cans of ether in a refrigerator, since ether still vaporizes at refrigerator temperatures. If peroxides have formed from ether, merely opening the refrigerator door may trigger an explosion. The heavy vapors that collect in an airtight refrigerator may ignite on contact with the electrical refrigerator motor. This recommendation also applies to "shielded" refrigerators (those in which the electrical systems are protected).

6. Unopened cans may be stored with opened cans (near the floor on an open shelf or nonairtight cabinet), in a storage cabinet for flammable materials, or in a storage room for volatile substances.

7. Outdate and discard cans after storage for 1 month (if opened) or 1 year (if unopened).

8. Flush empty, or almost empty, ether cans with copious amounts of water before discarding. "Empty" cans have been known to explode! Do not flush more than the contents of a 0.25-lb (ca. 0.1-kg) can down the drain; consult the Research and Occupational Safety Office or the local fire department for proper disposal of large quantities. Empty, thoroughly rinsed cans may be discarded in the regular trash.

9. Do not attempt to open any containers of uncertain age or condition or those whose cap or stopper is tightly stuck. Peroxides have been known to form under the threads of caps! These containers should be discarded.

10. Unopened outdated cans must be delivered to Research and Occupational Safety for disposal (each institution may have a comparable office or safety office).

11. *Certain chemicals can be replaced by others that are safer.* Ethyl acetate (72), with a higher flash point (4.0°C), has replaced ether (flash point of −45°C) in the formalin-ether concentration technique. Hemo-De (Medical Industries, Los Angeles, Calif., or Fisher Scientific, Los Angeles, Calif.) can also replace ethyl acetate in the concentration procedure (56). Hemo-De has a flash point of 57.8°C and is generally regarded as safe by the U.S. Food and Drug Administration. Xylene, as used in the trichrome or iron hematoxylin staining of polyvinyl alcohol (PVA)-fixed fecal smears, poses potential toxic and fire hazards. Again, Hemo-De has successfully replaced xylene in both the carbol-xylene and xylene steps of the trichrome procedure (55). Other substitutes are available as well (e.g., Hemo-Sol [Fisher Scientific]). Check with your local pathology departments or reagent suppliers for other alternatives.

12. Mercuric chloride, used in PVA and in Schaudinn's fixative, presents both a toxic hazard and a disposal problem. Copper sulfate has been suggested as a substitute for mercuric chloride (26, 33); however, protozoan morphology is not as clear and precise when this formula is used. The use of zinc rather than copper gives fairly good morphology when the trichrome stain is used (27).

Note Information on other substances is provided in Table 35.5.

Current OSHA Regulations for the Use of Formaldehyde

Formaldehyde has been in use for over a century as a disinfectant and preservative; it is also found in a number of industrial products. There is disagreement about the carcinogenic potential of lower levels of exposure, and epidemiologic studies of the effects of formaldehyde exposure among humans have given inconsistent results. Studies of industry workers with known exposure to formaldehyde report little evidence of increased cancer risk (41). It also appears that persons with asthma respond no differently from healthy individuals following exposure to concentrations of formaldehyde up to 3.0 ppm (58).

OSHA requires all workers to be protected from dangerous levels of vapors and dust. Formaldehyde vapor is the most likely air contaminant to exceed the regulatory threshold in the clinical laboratory (13, 14). Current OSHA regulations require vapor levels not to exceed 0.75 ppm (measured as a time-weighted average [TWA]) and 2.0 ppm (measured as a 15-min short-term exposure). *OSHA requires monitoring for formaldehyde vapor wherever formaldehyde is used in the workplace. The laboratory must have evidence at the time of inspection that formaldehyde vapor levels have been measured, and both 8-h and 15-min exposure must have been determined.*

If each measurement is below the permissible exposure limit and the 8-h measurement is below 0.5 ppm, no further monitoring is required as long as laboratory procedures remain constant. If the 0.5-ppm 8-h TWA or the 2.0-ppm 15-min level is exceeded, monitoring must be repeated semiannually. If either the 0.75-ppm 8-h TWA or the 2.0-ppm 15-min level is exceeded (unlikely in a

Table 35.5 Potential hazards of industrial substances[a]

Name of compound	Synonym	Hazard
Acids and bases		Extremely irritating and corrosive to the skin and mucous membranes. Can quickly cause burns and deep ulceration with ultimate scarring. Prolonged contact with dilute solutions has a destructive effect on tissue.
Ethyl acetate	Acetic ether	Absorbed through intact skin. Irritating to mucous surfaces, particularly the gums, eyes, and respiratory passages; is also mildly narcotic. Repeated or prolonged exposures cause conjunctival irritation and corneal clouding.
Formalin	Formaldehyde solutions	Severe local irritant. If swallowed, causes violent vomiting and diarrhea.
Mercury compounds		Local irritants and systemic poisons. Absorbed through intact skin. Strong allergens.
Methyl alcohol	Methanol	Absorbed through intact skin. Acutely, there is little or no danger, but with chronic use a toxic effect may develop because, once absorbed, it is eliminated very slowly, so that a cumulative effect can result with repeated low dosage.
Phenol	Carbolic acid	After skin contact, the area becomes white, wrinkled, and softened, and there is usually no immediate complaint of pain; later, intense burning is felt. Absorption is through intact skin and can be quite rapid. Upon spillage, death can occur within 30 min. In acute phenol poisoning, the main effect is on the central nervous system.
Potassium dichromate	Potassium bichromate	*Carcinogen.* Has a very corrosive action on the skin and mucous membranes. The ulcers produced are frequently painless and are slow to heal. Associated with cancer of the lungs.
o-Tolidine	3,3-Dimethylbenzidine	*Carcinogen. One of the many aromatic amines that are recognized as carcinogens to the human bladder, ureter, and renal pelvis and suspected carcinogen to the intestines, lungs, liver, and prostate.*
o-Toluidine	o-Methylaniline	*Carcinogen.* While it can be absorbed through intact skin, the main portal of entry is by inhalation. Can produce severe systemic disturbances.
Xylene	1-Dimethyl benzene	Acute toxicity may be greater than that of benzene or toluene. Absorption takes place chiefly through the lungs. Skin irritation may be serious. Extended exposure can cause gastrointestinal neurologic and tissue damage.

[a] Data from I. N. Sax, *Dangerous Properties of Industrial Materials*, 4th ed., Litton Education Publications, 1975.

clinical laboratory setting), employees must be required to wear respirators. Accidental skin contact with aqueous formaldehyde must be prevented by the use of proper clothing and equipment (gloves and laboratory coats).

The amendments of 1992 add medical removal protection provisions to supplement the existing medical surveillance requirements for employees suffering significant eye, nose, or throat irritation and for those experiencing dermal irritation or sensitization from occupational exposure to formaldehyde. In addition, these amendments establish specific hazard-labeling requirements for all forms of formaldehyde, including mixtures and solutions composed of at least 0.1% formaldehyde in excess of 0.1 ppm. Additional hazard labeling, including a warning label that formaldehyde presents a potential cancer hazard, is required where formaldehyde levels, under reasonably foreseeable conditions of use, may potentially exceed 0.5 ppm. The final amendments also provide for annual training of all employees exposed to formaldehyde at levels of 0.1 ppm or higher (12).

Note The use of monitoring badges may not be a sensitive enough method to correctly measure the 15-min exposure level. Contact the Occupational Health and Safety Office within your institution for monitoring options. Usually, the accepted method involves monitoring airflow in the specific area(s) within the laboratory where formaldehyde vapors are found.

Latex Allergy

The incidence and evidence of clinical latex sensitivity appear to be increasing since its first description in 1979. Since the introduction of standard precautions, the use of latex gloves in health care workers has increased dramatically (7). Those at risk appear to be health care workers and employees working in latex industries, patients with atopic diathesis, and patients who underwent repeated surgical procedures during childhood (69). The allergic response occurs to one or more natural rubber latex proteins, resulting in contact urticaria, angioedema, allergic rhinitis, asthma, or anaphylaxis. Early recognition of sensitization to latex is crucial to prevent the occurrence of life-threatening reactions in sensitized health care providers or patients (65). Latex allergy is now an important medical, occupational, medicolegal, and financial problem, and it is essential that policies be developed to reduce the problem (11, 40, 70). There also seems to be a relationship between allergies to natural latex rubber and allergies to avocados, chestnuts, and bananas. A high incidence of latex sensitivity has also been found in people with spina bifida (8, 21).

The use of cornstarch powder in gloves can sensitize healthy people and exacerbate the symptoms of allergic patients. The powder spreads the latex allergens into the environment (69). Accompanying this increase in latex allergy is evidence of positive circulating specific immunoglobulin E antibodies to latex (31). Unfortunately, the primary treatment for latex allergy is avoidance.

Quality Systems

The area of quality systems forms an integral part of any clinical laboratory operation. In many cases of clinical laboratory testing, various programs and guidelines for quality assurance procedures (including QC) have been thoroughly outlined. However, specific quality assurance recommendations within the parasitology section of microbiology have not been well defined in the past, and the information in this chapter should help clarify these recommendations. Although the diagnostic procedures in this area seldom yield numerical data, there are still a number of QC measures that can help to ensure accurate results.

Extent of Services

The range of services provided will vary from one laboratory to another. To comply with the Health Care Financing Administration (HCFA) guidelines, only two extent classifications (rather than the previous four) are recognized in parasitology by the College of American Pathologists (17).

1. *Extent 2.* "Limited parasitology performed. The laboratory is able to recognize the presence of parasites, including protozoa in clinical specimens, but may need to refer them for definitive identification. When permanently stained smears for intestinal protozoa are indicated, polyvinyl alcohol fixative (PVA) or other appropriate fixative is available, and such material is placed in a fixative and then referred to a reference laboratory for staining identification."

2. *Extent 3.* "Definitive identification of parasites present to the extent required to establish a correct clinical diagnosis and to aid in selection of safe and effective therapy. Laboratories are expected to

have a high degree of expertise and should be able to differentiate *Plasmodium falciparum* from other species of *Plasmodium*."

Note There is no longer a College of American Pathologists listing for extent 1 (no parasitologic procedures are performed and all specimens are submitted to a reference laboratory). However, in this situation, the laboratory would not be required to subscribe to outside proficiency testing specimens in this discipline.

Proficiency Testing

Based on the Extent of Services mentioned above and the requirements in the Clinical Laboratory Improvement Amendments of 1988 (CLIA '88), all (with rare exceptions) clinical laboratories (including physician office laboratories performing other than waived tests) must participate in an approved program of interlaboratory comparison testing that is consistent with the level and complexity of the work performed. The comprehensive microbiology and/or parasitology survey programs are recommended. The state Department of Health must be consulted to confirm which proficiency testing services fulfill the requirements related to a laboratory's particular accreditation, particularly if the laboratory accepts interstate specimens in parasitology.

There must be evidence of active review of the survey results by the laboratory director or the designated supervisor. There also must be written evidence of evaluation of the results and, if indicated, corrective action when unacceptable results have been reported. Corrective-action statements should review the correct answer and specific steps that will be taken to ensure that the same, or similar, error will not occur in the future. On the basis of the regulations within CLIA '88, HCFA or one of the HCFA-approved agencies (may include some state agencies) will be involved in the process of monitoring laboratory performance and coordinating any possible sanctions that might be involved as a result of unacceptable performance in proficiency testing programs.

Copies of the 28 February 1992 *Federal Register* containing the final CLIA regulations can be obtained by writing to Government Printing Office, Attn: New Order, P.O. Box 371954, Pittsburgh, PA 15250-7954. Specify the date of the issue requested (28 February 1992) and stock number (069-001-00042-4). A $3.50 check or money order payable to "Superintendent of Documents" or a Visa or MasterCard number and expiration date should be enclosed. Orders may be made by telephone [(202) 783-3238] or fax [(202) 512-2250].

It is recommended that all of the proficiency testing specimens be saved for future review, particularly when students are being trained or technologists, pathologists,

Table 35.6   Recommendations for the handling and examination of proficiency-testing specimens

Specimen	Method	Comments
Formalin	Direct examination	Although some directions recommend sampling the vial after mixing, if organism numbers are small, they may be missed. Another option is to allow the vial to stand on the counter for a minimum of 30 min, then take one small drop from the bottom of the vial, and prepare the coverslip (usually 22 by 22 mm). If iodine is used, make sure the overall color is not too dark; helminth eggs may appear to be debris. The color should resemble strong tea. There is normally not sufficient material to perform a concentration procedure; this is not recommended. The entire coverslip preparation should be examined with the 10× low-power objective. At least 1/4 to 1/3 of the area should be examined with the 40× high dry objective. The use of immersion oil is not recommended for these preparations.
PVA fecal smear	Permanent stain (trichrome or iron hematoxylin)	Although many laboratories have switched to nonmercury fixatives, it is important to remember that some proficiency-testing fecal specimens for permanent staining have been preserved in mercury-based fixatives. This means that *you must include the iodine and subsequent alcohol rinse steps in your staining method to remove the mercury from the smear.* If this is not done, the overall result may be very poor. Both trichrome and iron hematoxylin staining methods can be used for these smears. At least 300 fields should be examined with the 100× oil immersion objective. These stained smears can be examined with a 50× or 60× oil immersion lens, but final organism identification often requires the 100× oil immersion objective. Consult your PT agency for mercury or nonmercury use.
Formalin or fecal smear (non-PVA)	Modified acid-fast stain	Special staining is required if the patient history is suggestive of possible cryptosporidiosis. Unless there is some indication (history of immunosuppression), it is unlikely that *Cryptosporidium* is a possibility. If you perform a modified acid-fast stain, either the hot or cold method is acceptable. Make sure that the destaining step is not too harsh; do not use acid-alcohol, commonly used in the routine Ziehl-Neelsen acid-fast stain. Generally, a mild destain is recommended (1 to 3% sulfuric acid). Although the organisms can be seen with the low-power (10×) objective, it is recommended that the high dry (40×) objective be used. The 50× oil immersion objective is also an accepted approach to screening the smear.
Thin blood film	These films arrive already stained	The *entire blood film* (including the area where the drop was placed on the slide) should be carefully examined with the 10× low-power objective. Often, when these films are prepared, any microfilariae present are at the edges of the blood or in the thick portion of the film. Without this type of examination, these microfilariae may be missed. With the 100× oil immersion lens, a minimum of 300 oil immersion fields should be examined. These stained blood films can be screened with a 50× or 60× oil immersion lens, but final organism identification often requires the 100× oil immersion objective. If young malarial rings are present, they could be missed if only the lower-power oil immersion lens is used.
Thick blood film	These films arrive already stained	The entire thick blood film should be examined with the 10× low-power objective. When these films are prepared, any microfilariae present may be at the edges of the blood or in any portion of the thick film. Without this type of examination, these microfilariae may be missed. If microfilariae are seen, morphologic details can be seen using the high dry 40× objective. With the 100× oil immersion lens, a minimum of 300 oil immersion fields should be examined. These stained blood films can be screened with a 50× or 60× oil immersion lens, but final organism identification often requires the 100× oil immersion objective. If young malarial rings are present, they could be missed if only the lower-power oil immersion lens is used.
Antigen detection using fecal immunoassays	Fluorescent-antibody assays, enzyme immunoassays, or the cartridge format tests are available	Enzyme immunoassays (antigen detection, no centrifugation recommended) 1. Remember to thoroughly rinse the wells according to the instructions; do not eliminate any of the rinse steps. Make sure each well receives the total number of rinses required. 2. Make sure the stream of buffer goes directly into the wells. Use a wash bottle with a small opening, so you have to squeeze the bottle to get the fluid to squirt directly into the wells. 3. When the directions tell you to "slap" the tray down onto some paper towels to remove the last rinse fluid, make sure you slap it several times. Don't be too gentle; the cups will not fall out of the holder. 4. Before adding the last reagents, the wells should be empty of rinse buffer (not dry, but empty of excess fluid).

(continued)

Table 35.6 *(continued)*

Specimen	Method	Comments
		Fluorescence (visual identification of the organisms) 1. Since you will be looking for the actual organisms (cysts of *Giardia* and/or oocysts of *Cryptosporidium*), this test can be performed on centrifuged stool (500 × g for 10 min) to increase the sensitivity. 2. Remember to thin out the smear; it is important to make sure that the slides are thoroughly dry before adding reagents. The slides can be placed in a 35°C incubator for about 30 min to 1 h to make sure that they are dry before being processed. If the material on the wells is too thick, it may not dry thoroughly and may fall off of the glass. It is better to let them dry longer rather than for too short a time. A heat block is *not* recommended for this purpose. 3. *Gently* rinse the reagents from the wells; do not squirt directly into the wells, but allow the rinse fluid to flow over the wells. 4. Remember that not all clinical specimens provide the 3+ to 4+ fluorescence that we often see in the positive control. Also, from time to time, you may see fluorescing bacteria and/or some yeast in certain patient specimens. This is not that common, but the shapes can be distinguished from *Giardia* cysts and/or *Cryptosporidium* oocysts. The intensity of the fluorescence may vary, depending on the filters. If your filters demonstrate both the yellow-green fluorescence and the red-orange counterstain, the organisms may not appear quite as bright as when the yellow-green filter only is used. Both approaches are acceptable and may reflect personal laboratory preferences. 5. Make sure to examine the edges of the wells. Sometimes in a light infection, the edges contain organisms while the organisms in the middle of the well may be a bit more difficult to detect (thick area). Lateral-flow cartridges (antigen detection, no centrifugation recommended) 1. If the stool is too thick, the addition of reagents will not thin it out enough. If the specimen poured into the well remains too thick, the fluid will not flow up the membrane. If your specimens arrive in fixative and there is no fluid at the top of the vial overlaying the stool, this means that the vial may have been overfilled with stool. These specimens will have to be diluted with the appropriate diluent before being tested. 2. It is always important to see the control line indicated as positive all the way across the membrane, not just at the edges. 3. A positive test result is generally much lighter than the control line; this is normal. 4. At the cutoff time to read the result, any color at all visible in the test area should be interpreted as a positive. 5. Do not read/interpret the results after the time indicated in the directions; you may get a false-positive result.
Photomicrograph	Project to view	These 2 × 2 photomicrographs are generally accompanied by some type of history or measurements of the object. It is important to actually project the image. Observing the slide through a small slide holder may not reveal sufficient detail for the accurate identification of the specific organism shown in the slide.
All specimens	Data entry	It is important to remember to very carefully check your responses and entry of choices onto the appropriate forms. When being graded, your responses will be evaluated on data submitted. If the incorrect answer is marked, you may be penalized, depending on consensus and a number of other parameters. Prior to submitting data, familiarize yourself with the paperwork and information requested. It is important for the laboratory to establish some type of tracking system for these specimens. You should have a log book or something similar where the arrival date and due date for each proficiency-testing module can be clearly entered. One person should be responsible for communicating with the agency in the event of breakage, failure of the shipment to arrive, etc. Each agency will give you at the beginning of the year an anticipated schedule of dates for shipment of each set of samples for testing.

and residents are receiving review training. It is just as important to maintain negative specimens as to maintain positive ones, since many errors are made in which artifacts are identified as parasites. Specimens should be periodically checked to make sure that the volume of fixative is sufficient to prevent the specimens from drying out. These specimens should be cataloged for easy reference.

Specific recommendations for the handling and examination of proficiency testing specimens can be found in Table 35.6. These guidelines should assist the participant in the proper approach to these specimens.

CLIA '88 mandated universal regulation for all clinical laboratory testing sites in the United States, including previously unregulated sites in physician offices. Quality testing was to be achieved through a combination of total quality management and mandated minimum quality practices. Through both internal and external proficiency testing, performance standards were also mandated. After the implementation of CLIA '88, the percentage of laboratories passing proficiency testing requirements has improved and almost all laboratories are currently using quality practices for a more comprehensive quality system approach (22).

Survey results from the Laboratory Proficiency Testing Program, Toronto, Ontario, Canada, over the course of their program, which began in 1977, indicate a marked improvement in laboratory performance. They send four samples in each of four surveys per year consisting of specimens to be examined for gastrointestinal parasites. Improvement is thought to result from a combination of voluntary withdrawal by laboratories with poor performance and improved performance by the other laboratories. Extensive educational initiatives have also played a significant role in improved performance (71).

In-House Quality Control

Supervision

The QC program should be under surveillance by the chief technologist or section supervisor and should be reviewed at least once each month by the laboratory director, QC supervisor, or other supervisor designate. There should be written evidence of active review of all records (controls for routine procedures, instrument function tests, and equipment checks [temperature, humidity, systems, maintenance]). There should also be written evidence of corrective action when controls do not fall within acceptable limits. A written checking system should be in operation to help detect clerical errors, analytical errors, and/or unusual laboratory results. This system should also provide for error correction within acceptable time frames. With the emphasis on outcome-oriented measures, it is very important to have the capability to track testing through the preanalytical, analytical, and postanalytical phases (specimen integrity, processing, testing, reporting, and consultation), with primary emphasis being placed on accurate, reliable, and timely diagnostic testing for good patient outcome. This approach to laboratory inspections will be emphasized by HCFA or other HCFA-designated agencies for outside inspections and should also be emphasized in all in-house quality assurance programs.

Procedure Manuals

Although card files and wall charts are acceptable for quick review, a well-written and complete procedure manual is very important for the routine operation of the laboratory and mandatory for accreditation requirements (Table 35.7). Instructions contained in manufacturers' package inserts are helpful but do not fulfill the requirements for a procedure manual. It is highly recommended that the National Committee for Clinical Laboratory Standards (NCCLS; now the Clinical Laboratory Standards Institute [CLSI]) format be followed for protocol preparation (47). This publication covers the design, preparation, maintenance, and use of technical procedure manuals for the clinical laboratory. The procedure manual should contain for each test (i) the principle of the test; (ii) acceptable specimens (including rejection criteria) and instructions for proper specimen collection and processing; (iii) preparation of reagents and solutions; (iv) all supplies and equipment necessary to perform the test; (v) QC procedures, results, and interpretation (including corrective action if QC is out of control); (vi) the test method; (vii) possible results; (viii) the correct way to report results; (ix) procedure notes (tips); (x) limitations of the procedure; (xi) additional tables and charts; and (xii) references. Complete parasitology protocols (written according to the NCCLS guidelines) can also be found in the *Clinical Microbiology Procedures Handbook* (34).

Specimen Handling and Record Storage

All specimens must be recorded in an accession book, log book, worksheet, computer, or other comparable record. If specimens are not examined as fresh material, they should be preserved in appropriate fixatives for later examination. On the basis of current guidelines, it is recommended that specimen and test result records be kept for at least 2 years (longer for pathology records and those from the blood bank). Also, remember that when your laboratory is inspected, you must be able to retrieve those records within no more than a few hours; this is particularly important if the inspector is actually checking this parameter.

Table 35.7 Requirements for technical procedure manuals[a]

Requirement	National regulatory agency or law[b]				
	CLIA/ Medicare 1967	CAP 1978	AABB 1981[c]	JCAHO 1980[c]	CLIA 1988
General					
Manual must be written; textbook not acceptable	X	X	X 1	X	X
Complete and current; contains only those tests done in the section	X	X	X	X	X
Reviewed and approved annually	X	X	X	X	X
Changes dated and approved	X	X	X	X	X
Available at the bench	X	X	X	X	X
Specimen					
Patient preparation	X	X	X	X	X
Specimen collection	X	X	X	X	X
Specimen handling	X	X	X	X	X
Specimen storage	X	X	X	X	X
Criteria for unacceptable specimen	X	X	X	X	X
Quality control					
Preparation of controls	X	X	X-2	X	X
Mean value and upper/lower limits	X	X	X-2	X	X
Lot number and date of preparation	X	X	X-2	X	X
Remedial action to be taken with unacceptable results	X	X	X-2	X	X
Procedure					
Preparation and storage of reagents and standards	X	X	X-2	X	X
Methodology	X	X	X-2	X	X
Calibration and linearity	X	X	X-2	X	X
Derivation of results	X	X	X-2	O	X
Alternate procedure for automated tests	X	X	X-2	X	X
Normal ranges and results	X	X	X-2	X	X
Criteria for handling abnormal and alert values	X	X	X-2	X	X
Flow sheets, keys, tables	X	X	X-2	O	X
Notes	X	X	O	O	X
References	X	X	X-2	O	X

[a] Consult the regulations in question as to manual requirements for specific sections (e.g., microbiology, parasitology, hematology). For further information, see reference 47.
[b] Consult state and local regulatory agencies for additional requirements.
[c] X-1, AABB does not accept textbooks; X-2, required in procedure manuals of AABB-accredited blood banks and transfusion services by the National Committee on Inspection and Accreditation; X, listed as required; O, not listed specifically.

Test Request Requisitions

All requisitions should be designed to allow sufficient information for patient and physician identification and any clinically relevant information. Specific items should include (i) patient name, (ii) patient identification number, (iii) name of ordering physician, (iv) source of specimen, (v) specific procedure(s) ordered, (vi) time of specimen collection, and (vii) time of receipt or initial processing by the laboratory. Other helpful information, particularly in the parasitology division, is (i) suspected diagnosis, (ii) travel history, and (iii) recent medication. Since this infor-mation is often not on the original requisition, the physician may have to be contacted for clarification.

Procedure QC and Documentation

QC measures for all procedures are included in the individual protocols within this section of the book. Not only must the QC checks be performed as indicated, but also complete documentation should be generated, including specific plans for out-of-control results and problem resolution. Result expectations and normal results or values must be specifically defined and posted on any QC check sheets used in the laboratory. It is not sufficient to merely check a column that QC is within acceptable limits; you must also have on these sheets a definition of what normal or "acceptable limits" means. Dates and space for the initials or name of the person performing the QC checks must also be included. Specific procedure recommendations include the following (19, 25, 34, 42, 57).

1. The examination of formed stools should include a concentration procedure.

2. A permanent stained smear is recommended for *every stool specimen* submitted for testing (liquid, soft, or formed).

3. A direct wet mount should be performed for fresh stool specimens, in addition to the concentration and permanent stained smear. Since the direct wet mount is designed to allow the detection of organism motility, it is not necessary to perform the direct wet mount on preserved specimens or formed stools. Consequently, if the stool specimens are received in preservative, proceed directly to the concentration and permanent stained smear.

4. Both thick and thin blood films are prepared for examination for malarial parasites.

5. Stained smears are washed with buffered distilled water (pH 6.8 to 7.2).

6. Adequate numbers of oil immersion fields should be examined (at least 300) on both stained blood films and permanent stained fecal smears.

Test Result Reports

With few exceptions, the following methods of reporting are acceptable.

1. All parasites should be reported, regardless of whether they are considered nonpathogenic or pathogenic. The presence of any parasites within the intestine generally confirms that the patient has acquired the organism through fecal-oral contamination.

2. Generally, protozoa and helminths are not quantitated on the laboratory slip. However, the specific stage (i.e., trophozoites, cysts, oocysts, spores, eggs, or larvae) is indicated.

3. Exceptions to quantitation would be *Blastocystis hominis* (there may be some association between numbers and symptoms) and *Trichuris trichiura* (light infections may not be treated).

4. The complete genus and species name of the organism identified should be reported.

5. Charcot-Leyden crystals should be reported and quantitated.

6. If the specimen is fresh or freshly preserved, budding yeast cells should be mentioned and quantitated.

7. All quantitation should be consistent (Table 35.8).

8. Helminth eggs and larvae should be reported.

9. Special circumstances (additional specimens required to rule out this infection, etc.) require additional comments on the laboratory report.

10. Remember, the name of the laboratory actually performing the procedures must be indicated on the final report or elsewhere.

Reagents

All reagents should be properly labeled with (i) content, (ii) concentration, (iii) date prepared or received, (iv) lot number, (v) date placed in service, and (vi) expiration date. If reagents are prepared in-house, you should also have recorded (i) the name of person preparing the reagent and (ii) the date and results of the QC check.

The specific gravity of certain solutions (e.g., zinc sulfate) should be checked before use (this is particularly important if the solution is used infrequently). The hydrometer should have a scale large enough to differentiate between specific gravities of 1.18 for use with fresh specimens and 1.20 for use with formalinized specimens. This type of solution should be kept in a tightly stoppered bottle.

Stains should be routinely checked with control specimens for correct staining properties before use. Certainly, new lot numbers or new batches of stain should be checked. Also, if reagent use is infrequent, the stains should be checked before use. The same rule generally applies to fixatives.

With these basic recommendations in mind, the following specific QC methods can be used. While not every reagent used in diagnostic parasitology work is reviewed, these methods can be adapted to other reagents.

The examination of fecal material for ova and parasites may reveal, in addition to parasitic organisms, cells

Table 35.8   Quantitation of organisms[a]

Category	Quantitation	
	Protozoa	**Helminths**
Rare	2–5/cslp[b]	2–5/cslp[c]
Few	1/5–10 hpf	1/5–10 lpf
Moderate	1–2/hpf to 1/2–3 hpf	1–2/lpf
Many	Several/hpf	Several/lpf

[a] This quantitation (based on a 22- by 22-mm coverslip) is used by the Centers for Disease Control and Prevention, Proficiency Testing Program, Microbiology–Parasitology, 1985. Abbreviations: cslp, 22- by 22-mm coverslip; hpf, high-power field (magnification, ×400); lpf, low-power field (magnification, ×100).
[b] Total coverslip with high dry power (magnification, ×400).
[c] Total coverslip with low power (magnification, ×100).

of human origin (blood cells). Most of these cells are not usually found in the stool; however, they may be found in the presence of parasites and/or other disease-producing organisms. These cells may occur in the stool without the presence of a specific etiologic agent (allergic conditions, ulcerative colitis, etc.). The buffy coat cells (white blood cells [WBCs]) can also be used as a QC check for various fixatives. The WBCs normally found in the buffy coat portion of whole blood include polymorphonuclear neutrophils, eosinophils, lymphocytes, and monocytes (macrophages). When blood is visible in the stool (indicating the presence of fresh blood), the WBC morphology is very similar to that seen on the stained thin blood film used for a differential. However, if the blood has been in the gastrointestinal tract for some time, the cellular morphology changes slightly.

QC for Schaudinn's Fixative

1. Collect a fresh tube of blood (lavender top, EDTA anticoagulant), or get a tube from the hematology section (high WBC count, if possible).

2. Centrifuge the blood, and remove the buffy coat (layer containing WBCs).

3. Add the buffy coat to approximately 2 to 4 g of fresh, soft stool, and mix thoroughly *but gently*.

4. This stool-WBC mixture can be smeared onto several slides and immediately immersed in Schaudinn's fixative.

5. After staining, if the WBCs appear well fixed and typical morphology is visible, one can assume that any intestinal protozoa placed in the same lot number of Schaudinn's fixative would also be well fixed, provided that the fecal sample was fresh and fixed within recommended time limits.

QC for PVA Fixative

1. Mix approximately 2 g of fresh, soft stool with 10 ml of PVA (in solution, ready for use).

2. To this PVA-stool mixture, add several drops of buffy coat cells (collected as above) and mix gently.

3. After a 30-min fixation, pour a small amount of the stool-PVA-cell mixture onto a paper towel to absorb the excess PVA.

4. The material can then be smeared onto several slides, allowed to dry thoroughly (60 min at room temperature or 30 min at 35°C), and stained. After staining, if the WBCs appear well fixed and typical morphology is visible, one can assume that any intestinal protozoa placed in the same lot number

of PVA fixative would also be well fixed provided that the fecal sample was fresh and fixed within recommended time limits.

Note *Entamoeba coli* mature cysts are very difficult to fix properly and may be difficult to identify on the stained smear. For this reason, it is possible to have fixatives that are adequately QC checked but do not always yield good morphology for this particular organism. Use of a longer fixation time (60 min) sometimes produces better morphology after staining.

Note The same approach can be used for other stool preservatives (sodium acetate-acetic acid-formalin [SAF], zinc- and copper-based fixatives, and other single-vial collection systems).

QC for Stains (Trichrome, Iron Hematoxylin, etc.)

1. Slides may be prepared from a known positive fecal specimen preserved in PVA or SAF. These slides can then be stained and checked for typical organism morphology. Positive PVA-preserved samples can be purchased commercially; many slides can be prepared from one sample. QC slides (already prepared and ready for staining) are also available commercially.

2. Buffy coat cells (after appropriate fixation) may also be used to check the staining properties of each new lot number of stain.

Note Do not prepare too many slides in advance. If dry PVA fecal smears are held too long before staining, the organism or buffy coat cell morphology may be poor.

Instruments and Equipment

General requirements include the following: (i) all instruments and equipment should be on a routine preventive maintenance schedule, and records should be maintained in writing; (ii) instrument manuals and maintenance/service records should be available for technical personnel to review; (iii) all thermometers should be checked and calibrated against certified standard thermometers, and the information should be recorded in writing; and (iv) checked and calibrated thermometers should be in every refrigerator, freezer, incubator, water bath, heat block, etc., and temperatures should be recorded on each day of use. As specific requirements, (i) centrifuges should be calibrated for correct speed and the calibration should be recorded on the instrument, and (ii) all microscopes should be calibrated by using a stage and ocular micrometer, the calibrations should be posted on the instrument, and recalibrations should be performed if any ocular or objective is switched on that particular microscope. It is recommended that microscopes be recalibrated once each year, even if the oculars and objectives have not been changed. This is particularly true if the microscopes receive heavy use.

Reference Materials

Reference materials should be available for comparison with unknown organisms, refresher training, and the training of additional personnel. Ideal reference materials include formalin-preserved specimens of helminth eggs, larvae, and protozoan cysts; stained fecal smears of protozoan oocysts, cysts, and trophozoites; and positive blood smears. Color slides and atlases are recommended, although the level of microscopic focus cannot be changed. Reference books and manuals from a number of publishers are available, and selected ones should be part of the parasitology library. It is also recommended that you maintain a list of consultants who can be called in case questions arise.

Patient Outcome Measures

Patient outcome measures often refer to aspects of quality assurance that are monitored in additional to more narrow parameters such as QC. Among the issues that are appropriate to monitor on an ongoing basis are the following:

1. Submission of the appropriate specimen within the correct time frame for collection
2. Condition of the specimen when received by the laboratory (too old, volume not sufficient, contains interfering substances, etc.)
3. Turnaround time for test results
4. Appropriateness of STAT requests
5. Improvements in compliance with established guidelines for the above, particularly after continuing-education efforts
6. Clinical relevance of and documented use of test results

All of these quality assurance parameters can be monitored by using the Joint Commission on Accreditation of Healthcare Organizations (JCAHO) 10-step criteria for developing and implementing monitoring and evaluation activities (35). These 10 steps are listed below.

1. *Assign responsibility.* The designated individual identifies and assigns responsibilities to others within the department and ensures that these responsibilities are fulfilled.
2. *Delineate scope of care.* Delineating the scope of care involves identifying the diagnostic and therapeutic modalities used, times and locations where

services are provided, types of personnel providing care, and all clinical services within the laboratory.
3. *Identify important aspects of care.* To effectively use the institution's resources in quality assurance, activities selected should be those with the greatest impact on patient care (the function occurs frequently or affects large numbers of patients, serious consequences could occur if services are not provided correctly or within set time frames, and the issue has tended to cause problems for staff and/or patients in the past).
4. *Identify indicators.* Indicators are well-defined, objective variables that are used to monitor the quality and appropriateness of any selected aspect of patient care. Within the laboratory, indicators could include clinical situations in which it may be inappropriate to use a particular test, situations in which inappropriate tests are often used, correct sequencing of tests, and clinical outcomes that may indicate inappropriate test use for a given situation.
5. *Establish thresholds for evaluation.* The thresholds for evaluation must relate specifically to the indicator and establish a value below which one does not continue to monitor and above which one continues to monitor and work on improvement and reaching the threshold or moving below.

   Note Some monitoring of an indicator usually occurs before a reasonable threshold can be identified. Once the threshold has been determined, monitoring and educational efforts continue, with the goal being to reach or go below the threshold. Once the threshold has been reached, consideration may be given to lowering it even more, thus leading to additional improvements in the indicator being monitored.
6. *Collect and organize data.* The following must be determined for each indicator: data sources, data collection method(s), sampling system, time frame for data collection, and process for comparing level of performance with set thresholds.
7. *Evaluate care.* If the data indicate that the acceptable threshold of performance has not been reached, extensive review occurs to identify problems or opportunities for improvement. An analysis of patterns or trends (specific shifts, units, personnel, skills) can help identify specific areas for improvement and changes. If the threshold of acceptance has been reached, it should be reevaluated in terms of keeping it where it is or possibly lowering it further to strive for additional improvements. If the decision is made to leave the threshold where it

is, monitoring may be performed on a less frequent basis (every 6 months and then every year), just to verify that the data do not indicate that performance has declined.

8. *Take actions to solve identified problems.* The laboratory develops a plan of action that specifies who or what is expected to change, what action is appropriate, and when the changes should be complete. Three of the most common causes of problems are insufficient knowledge or poor communication, defects in the process, and poor compliance with process expectations. Frequently, educational presentations by various means (one on one, verbal presentations, written information, newsletters) can be used to help resolve the problems.

9. *Assess the actions and document improvement.* After sufficient time is allowed for improvements to occur, follow-up assessments are very important in documenting what progress has been made. This process must focus on the problem or opportunity for improvement, not on the action taken. If improvement is found, less frequent monitoring may be necessary; if the problem remains, new action should be taken with subsequent assessment for evidence of improvements.

10. *Communicate relevant information to the organizationwide quality assurance program.* It is critical that the findings not only be reviewed within the area being examined but also be conveyed and discussed at all levels of management. In addition to these discussions, documentation of such discussions is mandatory; one method often used is meeting minutes.

Continuous Quality Improvement, Total Quality Management, or 10-Step and FOCUS-PDCA for Performance Improvement Activities

Continuous quality improvement (CQI) (or total quality management) is a continuous quality improvement process that evaluates processes from a customer satisfaction point of view. The aim is continuous process improvement. This approach is a natural extension and expansion of the quality assurance programs that laboratories have been using for the past few years. CQI uses familiar tools such as check sheets, run charts, and flowcharts in new ways. Issues for review in a quality improvement program include some of the issues involved in a quality assurance program, such as timeliness of response to requests, turnaround time, and effective communication. *The problem-solving process in a CQI laboratory is done by broad-based teams with members from all affected groups inside and outside the laboratory. Team members are trained in and use group process methods for identifying problems and generating solutions. Rather than management remaining "top down and autocratic," it becomes "bottom up and independent."*

CQI laboratories focus on improving customer satisfaction; however, "customers" may be defined differently from the traditional use of the word. A CQI customer can be anyone who uses the products of the production process—a ward clerk responsible for charting results, a patient needing blood drawn for preadmission tests, or a physician waiting for a STAT result. Anyone who is an "end user" of the laboratory process is a customer and should be satisfied with the laboratory's product.

Table 35.9 Ten-step and FOCUS-PDCA grid for organizing performance improvement activities

Ten-step	FOCUS-PDCA
1. Assign responsibility — F	Find a process to improve
2. Delineate scope of care — O	Organize a team that knows the process
3. Identify important aspects of care — C	Clarify current knowledge of the process
4. Identify indicators — U	Understand sources of process variation
5. Establish thresholds for evaluation — S	Select the process improvement
6. Collect and organize data — P	Plan (plan the improvement action)
7. Evaluate care — D	Do (test the action)
8. Take action — C	Check (determine the effect of the action)
9. Assess the effectiveness of actions — A	Act (implement the action broadly)
10. Communicate findings	

JCAHO uses the term "continuous quality improvement." The concept of "quality assurance" has been changed to "quality assessment and improvement" and includes the major trends leading to quality improvement. The formal surveillance process is deemphasized in lieu of promoting activities that better reflect the principles of CQI. *However, this does not suggest that JCAHO will abandon the 10-step process.* The use of CQI requires a shift in the organization's definition of quality from "good enough if it meets standards" to "we must work continuously to improve quality." This whole process is similar to any recommendation for problem solving; however, there are specific differences to consider.

1. Empowerment of employees to become involved in analyzing, operating, recommending change, and implementing solutions is necessary for CQI to be adopted.

2. CQI focuses on continuous monitoring of processes, not just limited issues.

3. The problems identified almost always cross over multiple departments or units.

4. The key issue always focuses on customer satisfaction and how we define it, measure it, monitor it for success, fix it when things go wrong, etc.

5. Groups who work on these identified problems include members from all areas relevant to the final service product (from initial ordering or product to final delivery).

6. The identification of areas for improvement originates from all levels within the institution.

Table 35.10 Information regarding the HCFA self-survey process (CLIA): alternate quality assessment survey

Issue	Comments
Survey process	Respond to 45 "Yes/No" questions; some may require a written response, review dates, or submission of documentation. A checklist is provided to help you identify which questions require documentation. *Do not send original laboratory documents.*
"Review" questions	Questions that concern review of laboratory's policies, procedures, records, etc., refer to a periodic review that is sufficient to demonstrate effective policies and to correct/modify ineffective ones. Reviews must be documented.
"Representative sample"	This means a review of a variety of records, requisitions, etc., that is sufficient to demonstrate whether policies and procedures are being accurately and completely followed.
Time line	The form must be completed and returned within 15 days of receipt to the State Agency return address.
General laboratory information	Includes name, address, type of CLIA certificate, levels of test complexity performed (waived, provider-performed microscopy, moderate, high), and annual test volumes (all areas of the laboratory). Microbiology includes bacteriology, mycobacteriology, mycology, parasitology, virology, general immunology, and syphilis serology. Parasitology includes direct preps, ova and parasite preps, and wet preps. Key information includes any changes regarding testing, testing sites, and laboratory personnel. *Documentation must include* demonstration of qualifications of newly hired personnel or changing positions since last CLIA on-site survey (licenses, diplomas, degrees, certificates, and education and experience).
Laboratory assessment	Specific areas within this category include patient test management, QC, proficiency testing, comparison of test results (different methodologies), relationship of patient information to test results, personnel assessment, communications and complaint investigations, and quality assurance review.
Attestation statement	The director of the laboratory is required to sign in ink to attest to the accuracy and honesty of the data submitted in the questionnaire.
Checklist	This checklist is designed to help you prepare and include written documentation to accompany the completed questionnaire.
Appendix A	This list contains a list of analytes or tests; if your laboratory performs any of these procedures, you must be enrolled in a proficiency testing (PT) program for those analytes or tests. *You must be enrolled in PT for the full extent of testing being performed (e.g., Gram stain, acid-fast stain, direct antigen testing, isolation, identification, and susceptibility).*
Appendix B	This section contains guidelines for counting tests (per questions regarding annual test volumes).

7. A simplified version of the above is as follows.

   A. Identify and choose a process to fix.

   B. Analyze all aspects of the issue.

   C. Select a possible solution.

   D. Correct the process.

   E. Monitor the results.

   F. If results are satisfactory, solidify the change and move on to other issues.

   G. Who does it? A group composed of representatives of all aspects of the process under review.

Another way to look at organizing performance improvement activities can be seen in Table 35.9. This approach merges the 10-step and FOCUS-PDCA approaches; similarities are certainly more numerous than differences.

CLIA '88 Inspection Process

HCFA began implementing the performance-based survey process in February 1996. This is a self-survey process, and the form used is titled the Alternate Quality Assessment Survey (AQAS). The form is designed to be used in certain laboratories for recertification purposes under the CLIA program in lieu of an on-site survey. Laboratories that were surveyed under the CLIA program via the State Agencies and found to have exceptional performance, i.e., no or few minor deficiencies and satisfactory proficiency testing, are potential candidates for the AQAS. The survey form contains questions that reflect an outcome-oriented, quality improvement type of assessment. Samples of some of the areas included in the questionnaire can be seen in Table 35.10.

Following receipt of the completed form, the State Agency reviews the laboratory's responses and supplemental documentation. On the basis of this information, the laboratory may receive a telephone call requesting more information or may receive an on-site visit. If no problems are identified, the laboratory receives recertification for a 2-year period, provided that all relevant fees are paid. The certificate is forwarded to the laboratory. No laboratory will go longer than 4 years without an on-site survey. Approximately 5% of the laboratories are visited on-site to verify the effectiveness of this self-survey process.

The performance-based survey is a reward for exceptional performance and an incentive for improvement for those laboratories that have not yet been granted permission to use this self-survey approach. General information regarding the survey process can be obtained by calling Judith Yost (Director, Center for Laboratories, Health Standards and Quality Bureau) at HCFA [(410) 786-3531] (44).

New Quality Guidelines

Currently, there are ongoing efforts at both the national and international levels to provide guidelines for clinical laboratories in order to upgrade their QC and quality assurance programs to a broader definition—that of quality systems. Where QC and quality assurance programs monitor performance aspects of very specific operations, a quality system applies universal quality elements throughout the entire operation of the laboratory.

ISO Guidelines

ISO (International Organization for Standardization) is the world's largest developer of standards. Since 2000, the worldwide standard for a quality system in business and industry has been the ISO 9001 guidelines (the third edition is published by the International Organization for Standardization). Countries trading products across international boundaries have found the standardized requirements very beneficial in improving quality; this approach has also reduced the time and cost of multiple inspections due to the international nature of their business. In the future, there will be a universal standard for quality management in medical laboratories: ISO 15189 (29, 39).

When the large majority of products or services in a particular business or industry sector conform to International Standards, a state of industry-wide standardization can be said to exist. This standardization is achieved through consensus agreements between national delegations representing all the economic stakeholders concerned: suppliers, users, government regulators, and other interest groups, such as consumers. They agree on specifications and criteria to be applied consistently in the classification of materials, in the manufacture and supply of products, in testing and analysis, in terminology, and in the provision of services.

According to their objectives, laboratories will choose a recognition of their quality management system with an ISO 9001 certification or a recognition extended to the technical skills with an ISO 17025 or ISO 15189 accreditation. The contents of these last two documents are very similar, and both integrate requirements of the standard ISO 9001. The standard ISO 17025 is somewhat distant from clinical laboratory needs, requiring many efforts of adaptation, just like the ISO 9001 standard. The standard ISO 15189 seems to be well adapted but more constraining, considering the detailed requirements required. It necessitates a perfect control of the preanalytical phase, which is difficult to acquire in a clinical framework where the clinical specimens are not taken by the laboratory staff.

A number of advantages are associated with the use of ISO. The focus on patients has been reestablished, all processes are identified and subject to continuous improvement, and performance measurements provide an integrated picture of results. Measurements subsequently lead to improvement of quality of care and to quality system improvements. The documentation system can serve the organization's needs without leading to bureaucracy. Given the need for adequate quality management tools in health care and the need to demonstrate quality, the positive effects suggest that ISO will become more prevalent in health care organizations (67).

ISO provides technical and quality management, which results in benefits observed in daily laboratory practices. Technical requirements can include the addition of formal personnel training plans and detailed records, method development and validation procedures, measurement of method uncertainty, and a defined equipment calibration and maintenance program. Also, an expanded definition of the sample preparation process can be implemented and documented to maintain consistency in sampling. Management quality improvements often emphasize document control to maintain consistent analytical processes, improved monitoring of supplier performance, a periodic contract review process for documenting customer requirements, and a system for handling customer comments and complaints. Continuous improvement is monitored and improved through corrective and preventive action procedures and audits. Quarterly management review of corrective actions, nonconforming testing, and proficiency testing results also identifies more long-term trends. The practical benefits of these technical and management quality improvements can be seen on a daily basis in the laboratory. Faster identification and resolution of issues regarding methods, personnel or equipment, improved customer satisfaction, and overall increased laboratory business are all the result of implementing an effective quality system. Certainly, the ISO system provides one option among various quality improvement approaches (32).

CLSI (NCCLS) Model

In the 1990s, blood centers and transfusion services within the United States began adopting the concepts of quality systems into their routine operations, primarily based on direction provided by a 1993 FDAQA guideline (24). After a series of revisions, the American Association of Blood Banks (AABB) published their 10 Quality System Essentials (QSEs) (1). These have been incorporated into the most recent edition of the AABB *Standards for Blood Banks and Transfusion Services* (2).

It became very obvious that the QSEs proposed for the blood bank were applicable to the entire laboratory operation. These essentials also incorporated many of the quality requirements developed by CLIA '88, JCAHO, and the College of American Pathologists (CAP).

Recognizing that standardization would be not only valuable, but also essential for high-quality laboratory operations as a whole, the Clinical Laboratory Standards Institute (CLSI) (formerly known as the National Committee for Clinical Laboratory Standards [NCCLS]) developed a subcommittee to prepare a guideline encompassing a complete quality system for the clinical laboratory. This document was recently published as GP26-A, *A Quality System Model for Health Care*, a proposed guideline (45). A comparison of the CLSI and ISO/DIS 15189 quality system models is given in Table 35.11. It is also important to realize that the 10 rules (QSEs) apply to all stages of the work within the laboratory (Table 35.12).

The QSEs form a composite of requirements that have come from various sources, including CLIA '88, the JCAHO standards, and the CAP checklist. By placing all the similar requirements under the headings in Table 35.12, it becomes clear that they are applicable to any and all health care organizations.

Factors Influencing Future Laboratory Practice

As consolidation among payers continues through mergers, acquisitions, and other arrangements, providers of care find themselves with fewer buyers of care with whom to negotiate. This requires the development of strategies through which provider leverage can be maintained. Consolidation on the provider side has been one response. Price reductions and exclusive arrangements have been others. Response to these changes also includes affiliation with community providers to increase bargaining strength and aggressive cost reduction to allow more competitive bidding. However, many institutions are now facing the acceptance of marked reductions in reimbursement rates or terminating contracts. As additional changes occur and negotiations continue on these issues, each institution much achieve a balance between the market share realized from each contract, the revenue per service unit, and the total revenue.

Marketplace trends include continued growth in managed care, particularly in the Health Maintenance Organization sector; integration of payers, providers, suppliers, and buyers; mergers and alliances; integration between hospitals and physicians; continued pressure for provider consolidation to address excess bed capacity; continued shift from inpatient to outpatient procedures; and provider alliances based on complementing services (linkage of primary-care facility with the tertiary or quaternary expertise of another institution).

Table 35.11 NCCLS and ISO/DIS 15189 quality system models[a]

NCCLS QSEs[b]	ISO/DIS 15189[b]		Comments
1. Organization	4.1	Organization and management	Executive management, directors, managers, and supervisors must support and participate in the quality system; develop policies for QSEs, provide resources for processes required for implementation; conduct periodic reviews of the quality system, review findings, take corrective action, and plan for improvements
	4.2	Quality management system	
	4.4	Referral of examinations to other laboratories	
	4.12	Management review	
2. Personnel	5.1	Personnel	Personnel need clearly defined job descriptions and qualifications; uniform processes for orientation, training, competency assessment, performance review; training must be adequate and appropriate for the situation
3. Equipment	5.3	Laboratory equipment	All equipment must be installed, calibrated, maintained, used, and stored as specified by manufacturer guidelines; all accreditation/regulatory requirements must be followed and documented; consistent processes for installing, calibrating, using, maintaining, troubleshooting, servicing, etc., as required
4. Purchasing and inventory	4.5	External services and supplies	Processes should be in place for supplies and services used, vendor criteria, ordering, receiving, logging, inspecting, testing, and managing inventory
5. Process control	5.2	Accommodation and environmental conditions	Identification of all processes involved in workflow, need to be validated to ensure they work as expected; clearly written protocols for all laboratory levels of personnel; QC and proficiency-testing programs must be in place according to regulatory and accreditation requirements
	5.4	Preexamination procedures	
	5.5	Examination procedures	
	5.6	Assuring the quality of examination procedures	
	5.7	Postexamination procedures	
	5.8	Reporting results	
	5.9	Alterations and amendments of reports	
6. Documents and records	4.3	Document control	Written policies, protocols, etc., must be uniform in appearance and content; records include worksheets, forms, computer printouts, labels, tags, etc.; documents must be prepared, reviewed, managed, and archived
	4.10	Quality records and technical records	
7. Occurrence management	4.6	Control of nonconformities	Variance recognition, documentation, analysis, identification of underlying problems, corrective action, review; large part of risk management program
8. Internal assessment	4.11	Internal audits	Quality assurance indicators must be identified and measured for all major processes in laboratory operations; action is required when performance falls outside acceptable limits; periodic audits of operations and systems can document differences between process documentation/protocols and actual practice
9. Process improvement	4.8	Preventive actions	Identification of processes for improvement as a result of complaints, internal and external inspections, internal audits, benchmarking, customer feedback, and personnel recommendations for change; use of problem resolution process that includes problem identification, prioritization, analysis, root cause determination, development, implementation, and review of solution
	4.9	Corrective actions	
10. Service and satisfaction	4.7	Consultative services and resolution of complaints	Feedback sought from both internal and external customers; when problems are identified, resolution and review stems from process improvement and problem-solving initiatives

[a] Data from NCCLS, A *Quality System Model for Health Care*, Approved Guideline GP26-A, NCCLS, Wayne, Pa., 1959, and International Organization for Standardization/TC 212/WG 1, ISO/DIS 15189: *Quality Management in the Medical Laboratory*, International Organization for Standardization, Geneva, Switzerland, 1998.
[b] ISO/DIS, International Organization for Standardization/Draft International Standard: QSE, quality system essentials.

Table 35.12 Laboratory workflow categories

Preanalytical	Analytical	Postanalytical	Information management
Patient assessment	Testing	Result reporting	Laboratory information system (LIS)
Test request	Test results review	Post-testing specimen management	Information management
Specimen collection	Interpretation		Clinical application/interpretation and consultation
Specimen transport			
Specimen receipt			

Strategies for survival will include (i) increased ability to capture market share, preferably related to covered lives rather than patient days; (ii) improved information systems to accurately identify costs (fixed, variable by service, procedure) and track the profitability of contracts; and (iii) recognition of the importance of these issues and implementation of proactive measures to structure the organization for survival in the coming years.

Managed Care

With the increase in health care contracts and overall health care reform, a number of potential issues will affect the laboratory of the future. Each institution is actively seeking new markets, with the overall objective being to increase the patient base served by that particular institution. As these contracts increase in number, the reimbursement approach becomes one of capitated payment; the provider takes on the risk previously assumed by the third-party payer. The provider receives a certain amount for each patient per year, the assumption being that the care can be delivered at less than the reimbursement capitated amount; thus, a profit is realized. As these changes occur, there continue to be mergers and the development of large care networks. However, as more large networks are developed, there are fewer contracts on which to bid. Each institution may be forced to consider contracts that pay less than the current costs of service. Obviously, these changes serve as a tremendous incentive to deliver care at continued reduced costs. Once contracts are lost to other bidders, it may be more difficult to recapture that lost market.

Managed care is changing from a cost containment environment to a value purchasing environment. Considerations include careful examination of what is being received for the money spent, increases in requests for outcome data, and case management options. As competitiveness in the managed-care arena increases, long-standing partnerships between institutions and Health Maintenance Organizations will have no inherent merit in guaranteeing continued partnerships. More focused medical management, lower costs, and higher efficiency will determine the continuation or elimination of present agreements. The new environment will demand increased cohesiveness, cooperation, and creativity on the part of all employees within an institution.

Table 35.13 Evolution of the health care delivery system and the impact on clinical laboratory testing[a]

Cause or development	Changes
Failure to slow exponential growth of total health care expenditures	Rapid and progressive penetration of MCOs in early 1990s; diagnostic testing viewed as a "commodity" rather than medical service
Growth of managed care	Hospital-based laboratories in a financial downward spiral
Shorter LOS; reduction in admissions	Reduction in in-house testing
Managed-care-associated businesses, consolidation and networking; outreach test market, point-of-care and satellite testing	Major strategies developed for restoring tests eliminated by massive changes
Regulatory restrictions imposed by CLIA '88; expanded penetration of MCOs (reimbursement limited to a few in-house procedures)	POLs closing
Reference labs gaining more tests through contracting with MCOs	Continued reduction of test volumes in POLs and hospital laboratories

[a] CLIA '88, Clinical Laboratory Improvement Amendments of 1988; LOS, length of stay; MCO, managed-care organization; POL, physician office laboratory.

An insurance, payment, and delivery system for health care, loosely described as "managed care," is increasingly viewed as dysfunctional by consumers (i.e., patients), providers, and employers, the last of which constitute the primary source of funding for health insurance for the employed sector of the population. In the current environment of managed-care dominance within the United States, clinical laboratories continue to change their focus of operations and mission in response to the continually changing landscape. Traditional laboratories that are unwilling to change and adapt to this environment will probably not survive. A synopsis of changes since the early 1990s is provided in Table 35.13.

There appears to be a climate of blame in which the various players in the health care arena reproach each other and advance conflicting solutions. Although the "demise" or significant modification of the managed-care system is widely predicted, there is a distinct absence of any consensus about the parameters of politically or economically viable alternatives. Nonetheless, it appears that "something else" will emerge—a return of the "managed-competition" plan proposed in the early years of the Clinton presidency, a national health model resembling the Canadian system, implicit and explicit rationing of resources, or other combinations and approaches.

Financial Considerations

Capitated reimbursement has already been mentioned as one of the driving forces behind the necessity to deliver health care at continuing reduced rates. Each year, reimbursement rates decline; currently, leverage lies with the payer, not the provider. Competitive bidding has become an absolute necessity. Also, institutions continue to review their costs in much greater detail than at any time in the past. To survive, each institution must know exact costs so that it can bid appropriately and remain financially viable. Complicating this approach are the various regulatory requirements and the fact that each contract is somewhat different in terms of services and costs. Laboratories are continually reviewing their test menus, methodologies, etc., for the most clinically relevant, cost-effective approach to diagnostic testing.

Current problems related to health care costs include a growing federal deficit and concerns about the future viability of Medicare, the increasing inability of state budgets to fund rising Medicaid expenditures, an aging population that will generate increasing demands on health care resources, the proliferation of costly technological and pharmacological advances in the ability to diagnose and treat disease, and a health care insurance and delivery system unable to provide insurance and

predictable access to care for approximately 40 million Americans (3).

Decentralized Testing

In the last few years, many inpatient procedures have been shifted to the outpatient setting in response to payer per diem cost containment initiatives. Most institutions are actively involved in developing or expanding outreach programs, including laboratory services, teaching, and consultations. Another area that will continue to expand is point-of-care (POC) testing, or testing at the site of the patient (bedside, emergency room, or clinic), or alternative-site testing (home or shopping mall). With this expansion comes new technology that is advertised as "fail-safe and foolproof," some of which may be true and some of which may be exaggerated. Obviously, this approach is directly linked to patient and overall customer satisfaction related to a short turnaround time for test results. Since more and more patient care is being delivered on an outpatient basis as decentralized care, reductions in length of hospital stay and in cost continue to be issues for review.

In some institutions that service very specific patient groups, POC testing may offer distinct advantages; however, in institutions with highly efficient specimen transport systems and rapid response capabilities within the main clinical laboratory, advantages of POC testing may be minimal to none. Continued review of regulatory requirements, QC issues, proper and consistent documentation, proficiency testing, performance enhancement, and cost-effectiveness is mandatory for success. As technology moves toward microcomputerization, microchemistry, chip technology, and enhanced test menus, the use of POC will require ongoing scrutiny (18, 30, 37, 38, 43, 61).

Although the push for decentralized testing may be coming from a variety of sources, it is generally thought that the central laboratory has become more involved in managing the new testing technologies and trying to circumvent anticipated problems with testing quality and patient care.

There are real, continued concerns about test accuracy, precision, training, and QC issues; personnel training and experience; and proficiency testing by those actually performing the testing. To date, there is little scientific evidence for cost reduction and/or reduction in length of stay. There appears to be mounting evidence that in some settings, the cost is higher than that of central-laboratory testing.

Laboratory Services

Large, integrated testing laboratories may become "the laboratory of the future," particularly considering the issue of economy of scale; i.e., the more we perform large numbers of the same test, theoretically the lower the cost. Although this approach is more applicable to some areas of the laboratory (chemistry), it is being carefully evaluated even in labor-intensive areas with less automation (microbiology). As technology continues to change, the departmental limits within any laboratory (chemistry, hematology, microbiology, or blood bank) continue to become less well defined. Many laboratory operations are now structured around automated and manual methods rather than specific clinical disciplines. This approach will continue and will also affect personnel training and utilization. Review of in-house versus send-out tests will continue, particularly regarding cost containment parameters.

With continued development and implementation of computerized history algorithms, clinical pathways, and case management, more structured support for clinical decision making will also dramatically affect the laboratory. Inappropriate test-ordering patterns and overutilization will become less and less significant. The use of algorithms may also become much more common within the laboratory itself. All of these changes will be developed through the use of multidisciplinary teams, including all members of the health care delivery system (ancillary departments, nursing staff, and physicians).

New technology will continue to be evaluated for a number of parameters, including type of data, technical characteristics, diagnostic characteristics, clinical utility, and cost-benefit analysis. Laboratory design will be another key issue and will be influenced by a number of factors, including consolidation of health care facilities; continued shift to outpatient care and POC and alternative-site testing; development and use of clinical paths; and development and evaluation of new technologies. Physical plant design should consider a number of factors, such as analysis of institution mission, scope of the project, and specific objectives; written statements of needs; mechanisms for keeping people informed; careful analysis of operational needs (work flow, traffic patterns, etc.); detailed review of current and future space requirements (consolidation of instrumentation, workstations, storage capacity, and equipment location); flexibility of design (modular furniture and utilities); and safety issues.

Technological Trends

Currently, hospitals and clinics are not able to meet their expenses without continued decreases in labor costs. During the next few years, the greatest savings in laboratory costs will come from technology that leads to labor reductions. Laboratory automation has the potential to be a highly successful, cost-saving measure and has been implemented in Japan, where many laboratories employ 5 to 10 times fewer full-time equivalents than European and North American laboratories while still achieving similar

productivity results. Future expectations include the performance of all but a few esoteric tests by automated systems, the use of mass spectrometry, and performance of most routine testing on a single automated analyzer. It is very likely that specimens will arrive in the laboratory ready for testing, with all processing having been done prior to receipt by the laboratory. The specimen may contain information chips carrying patient medical history and physiologic data at the time of sampling. These chips may also allow wireless encoding and reading of specimen information. Thus, the old specimen-processing area can concentrate on customer service, provision of medical information, and client education.

As an adjunct to the highly automated core laboratory, miniature analyzers and POC testing will play a major role in testing. Multianalyte, spectroscopy-based, noninvasive sensors will provide a wide range of tests at the bedside; there is even the possibility of a wireless invasive analyzer that would reside in a device that could be injected or implanted under the skin or elsewhere in the body.

New molecular tools have been developed rapidly, especially with the excitement over the completion of the human genome project. Synthetic antibody-like molecules with high affinity and specificity have been developed from both DNA and RNA. These molecules may replace antibodies in molecular diagnostics, achieving tremendous levels of sensitivity.

Biometrics is the use of sensors to measure natural body features for positive identification. In the future, biometrics will be extremely helpful in confirming that patients and their body fluids or other specimens are correctly identified. A patient's complete medical history may be available on the Web when the patient simply places a finger into the sensor mechanism.

Nonlaboratory systems, such as robots, will be used much more commonly for delivery functions, saving a significant amount of money as well as improving service. Surgery robots that can perform complex procedures through very small incisions are currently under development, and some have been implemented.

Clinical Decision Support

A number of approaches are being developed and used to serve as "practice guidelines"—more structured approaches for both information gathering and clinical decision making related to the patient. A clinical pathway is defined as a set of interventions or actions that is used to help a patient move progressively through a clinical experience to an expected, positive outcome. One of the main reasons for developing clinical pathways is to provide a tool to evaluate clinical patient management and outcomes. Once defined sets of expectations are developed, data can be used to monitor variances. The cost of care and utilization of resources can then be monitored more closely.

By using clinical pathways for predictable clinical outcomes (conditions, diseases, and procedures selected are generally very well established and patient outcomes are consistent), many also believe that patient care is more consistent. These clinical pathways can also be used as a teaching tool for all health care personnel. The development of clinical paths must be a multidisciplinary effort, with input from all relevant members of the health care team. Many institutions are also using the process of clinical path development to foster team building and continuous quality improvement initiatives.

Regardless of names or titles, the following are becoming more widely used and accepted. They are listed from the most well-defined and predictable medical situations (clinical pathway) to very complex cases with many underlying factors and complications (case management).

1. Clinical pathways (care maps) (used for conditions, diseases, and procedures for which the patient outcome is very predictable)
2. Clinical guidelines (used for care situations in which the outcome may have a few more variables)
3. Clinical algorithms (used for patient outcome situations in which more options may be required)
4. Case management (generally used for very complex cases, for which it is very difficult to establish a predetermined pathway; the patient may have multiple medical problems and/or complications)

Personnel Issues

Many personnel issues will continue to be relevant and important. Particular emphasis will probably include (i) the use of unlicensed individuals in relationship to CLIA '88 and various states developing, implementing, and revising personnel regulations; (ii) cross-training and the use of multiskilled laboratory personnel; (iii) tremendous flexibility in covering fluctuations in workload; (iv) expansion of the use of multidisciplinary teams; (v) continued emphasis on horizontal rather than vertical management structures; and (vi) development and utilization of personnel competency assessment programs (now mandated by regulations).

Changing Demographics

The aging of the population will have a profound effect on many health care issues. Laboratory testing and monitoring will be no exception. Patient access, alternative-site testing, and clinical decision support options will play a large role in helping manage this expanding population.

Emerging Diseases

Not only do we have to deal with expanding diseases and conditions in the elderly, but also we now are faced with some serious problems related to infectious-disease threats. There is explosive population growth worldwide with expanding poverty and urban migration; international travel is increasing; and technology is rapidly changing. All of these factors affect our risk of exposure to the infectious agents in our environment. In recent years, our antimicrobial drugs have become less effective against many infectious agents, and experts in infectious diseases are concerned about the possibility of a "post-antibiotic era." As a result, our ability to detect, contain, and prevent emerging infectious diseases is in jeopardy.

References

1. **American Association of Blood Banks.** 1997. *Association Bulletin 97-4: Quality System Essentials.* American Association of Blood Banks, Bethesda, Md.
2. **American Association of Blood Banks.** 2006. *Standards for Blood Banks and Transfusion Services,* 24th ed. American Association of Blood Banks, Bethesda, Md.
3. **Bachner, P.** 2004. Future political, social, economic, and regulatory impacts on pathology and laboratory medicine, p. 763–771. *In* L. S. Garcia (ed. in chief), *Clinical Laboratory Management.* ASM Press, Washington, D.C.
4. **Baxter Scientific Products.** 1992. Special laboratory safety issue. *Stat* **Spring:**1–28.
5. **Beltrami, E. M., I. T. Williams, C. N. Shapiro, and M. E. Chamberland.** 2000. Risk and mangagement of blood-borne infections in health care workers. *Clin. Microbiol. Rev.* **13:**385–407.
6. **Berte, L. M.** 2000. New quality guidelines for laboratories. *Med. Lab. Obs.* **32:**46–50.
7. **Bloom, H. M.** 1986. Designs to simplify laboratory construction and maintenance, improve safety, and conserve energy, p. 138–143. *In* B. M. Miller, D. H. M. Groschel, J. H. Richardson, D. Vesley, J. R. Songer, R. D. Housewright, and W. E. Barkley (ed.), *Laboratory Safety: Principles and Practices.* American Society for Microbiology, Washington, D.C.
8. **Catani, A.** 1999. Latex allergy in children. *J. Investig. Allergol. Clin. Immunol.* **9:**14–20.
9. **Centers for Disease Control and Prevention.** 1985. Recommendations for preventing transmission of infection with human T-lymphotropic virus type III/lymphadenopathy-associated virus in the workplace. *Morb. Mortal. Wkly. Rep.* **45:**681–686, 691–695.
10. **Centers for Disease Control and Prevention.** 2001. Updated U.S. Public Health Service guidelines for the management of occupational exposures to HBV, HCV, and HIV and recommendations for postexposure prophylaxis. *Morb. Mortal. Wkly. Rep.* **50(RR11):**1–42.
11. **Cheng, L., and D. Lee.** 1999. Review of latex allergy. *J. Am. Board Fam. Pract.* **12:**285–292.
12. **Code of Federal Regulations.** 1992. Update 27 May 1992. Title 29, CFR 1910.1048. U.S. Government Printing Office, Washington, D.C.

13. **Code of Federal Regulations.** 1987. Update 27 May 1992. Title 29, CFR 1910.1200 and 29 CFR 1910.1296. U.S. Government Printing Office, Washington, D.C.

14. **Code of Federal Regulations.** 1989. Title 29, CFR 1910.106. U.S. Government Printing Office, Washington, D.C.

15. **Code of Federal Regulations.** 1989. Title 29, CFR 1910.1200. U.S. Government Printing Office, Washington, D.C.

16. **Code of Federal Regulations.** 1989. Title 29, CFR 1910.1450. U.S. Government Printing Office, Washington, D.C.

17. **College of American Pathologists.** 1992. *Commission on Laboratory Accreditation.* College of American Pathologists, Chicago, Ill.

18. **Collinson, P. O.** 1999. The need for a point of care testing: an evidence-based appraisal. *Scand. J. Clin. Lab. Investig. Suppl.* **230:**67–73.

19. **Committee on Education, American Society of Parasitologists.** 1977. Procedures suggested for use in examination of clinical specimens for parasitic infection. *J. Parasitol.* **63:**959–960.

20. **Denys, G. A.** 2004. Biohazards and safety, p. 15.0.1–15.7.7. *In* H. D. Isenberg (ed.), *Clinical Microbiology Procedures Handbook*, 2nd ed, vol. 3. ASM Press, Washington, D.C.

21. **Dyke, M.** Alison Bell Memorial Award. Latex sensitivity and allergy: fact and fiction. *Br. J. Theatre Nurs.* **9:**165–168.

22. **Ehrmeyer, S. S., and R. H. Laessig.** 1999. Effect of legislation (CLIA '88) on setting quality specifications for U.S. laboratories. *Scand. J. Clin. Lab. Investig.* **59:**563–567.

23. **Fleming, D. O., J. H. Richardson, J. J. Tulis, and D. Vesley** (ed.). 1995. *Laboratory Safety: Principles and Practices*, 2nd ed. ASM Press, Washington, D.C.

24. **Food and Drug Administration, Center for Biologics Evaluation and Research.** 1995. *Guideline for Quality Assurance in Blood Establishments.* Docket 91N-0405. Food and Drug Administration, Rockville, Md.

25. **Garcia, L. S.** 1999. *Practical Guide to Diagnostic Parasitology.* ASM Press, Washington, D.C.

26. **Garcia, L. S., R. Y. Shimizu, T. C. Brewer, and D. A. Bruckner.** 1983. Evaluation of intestinal parasite morphology in polyvinyl alcohol preservative: comparison of copper sulfate and mercuric chloride base for use in Schaudinn's fixative. *J. Clin. Microbiol.* **17:**1092–1095.

27. **Garcia, L. S., R. Y. Shimizu, A. C. Shum, and D. A. Bruckner.** 1992. Evaluation of intestinal protozoan morphology in polyvinyl alcohol preservative: comparison of zinc-based and mercuric chloride-based compounds for use in Schaudinn's fixative. *J. Clin. Microbiol.* **31:**307–310.

28. **Gröschel, D. H. M.** 1986. Safety in clinical microbiology laboratories, p. 32–35. *In* B. M. Miller, D. H. M. Gröschel, J. H. Richardson, D. Vesley, J. R. Songer, R. D. Housewright, and W. E. Barkley (ed.), *Laboratory Safety: Principles and Practices.* American Society for Microbiology, Washington, D.C.

29. **Haeckel, R., and M. Kindler.** 1999. Effect of current and forthcoming European legislation and standardization on the setting of quality specifications by laboratories. *Scand. J. Clin. Lab. Investig.* **59:**569–573.

30. **Halpern, N. A., and T. Brentjens.** 1999. Point of care testing informatics. The critical care-hospital interface. *Crit. Care Clin.* **15:**577–591, vi–vii.

31. **Holme, S. A., and R. S. Lever.** 1999. Latex allergy in atopic children. *Br. J. Dermatol.* **140:**919–921.

32. **Honsa, J. D., and D. A. McIntyre.** 2003. ISO 17025: practical benefits of implementing a quality system. *J. A. O. A. C. Int.* **86:**1038–1044.

33. **Horen, W. P.** 1981. Modification of Schaudinn's fixative. *J. Clin. Microbiol.* **13:**204–205.

34. **Isenberg, H. D.** 2004. *Clinical Microbiology Procedures Handbook*, 2nd ed., vol. 1, 2, and 3. ASM Press, Washington, D.C.

35. **Joint Commission for the Accreditation of Healthcare Organizations.** 1987. *Monitoring and Evaluation of Pathology and Medical Laboratory Services.* Joint Commission for the Accreditation of Healthcare Organizations, Chicago, Ill.

36. **Klein, M., and A. Deforest.** 1965. The chemical inactivation of viruses. *Fed. Proc.* **24:**319.

37. **Kost, G. J.** 1999 Knowledge optimization theory and application to point-of-care testing. *Stud. Health Technol. Inform.* **62:**189–190.

38. **Kost, G. J., S. S. Ehrmeyer, B. Chernow, J. W. Winkelman, G. P. Zaloga, R. P. Dellinger, and T. Shirey.** 1999. The laboratory-clinical interface: point-of-care testing. *Chest* **115:**1140–1154.

39. **Libber, J. C.** 1999. Effect of accreditation schemes on the setting of quality specifications by laboratories. *Scand. J. Clin. Lab. Investig.* **59:**575–578.

40. **McCracken, S.** 1999. Latex glove hypersensitivity and irritation: a literature review. *Probe* **33:**13–15.

41. **McLaughlin, J. K.** 1994 Formaldehyde and cancer: a critical review. *Int. Arch. Occup. Environ. Health* **66:**295–301.

42. **Melvin, D. M., and M. M. Brooke.** 1982. *Laboratory Procedures for the Diagnosis of Intestinal Parasites*, 3rd ed. U.S. Department of Health, Education, and Welfare publication (CDC) 82–8282. Government Printing Office, Washington, D.C.

43. **Murray, R. P., M. Leroux, E. Sabga, W. Palatnick, and L. Ludwig.** 1999. Effect of point of care testing on length of stay in an adult emergency department. *J. Emerg. Med.* **17:**811–814.

44. **Naeve, R. A.** 1994. *Managing Laboratory Personnel: the CLIA and OSHA Manual.* Thompson Publishing Group, New York, N.Y.

45. **National Committee for Clinical Laboratory Standards.** 1998. *A Quality System Model for Health Care.* Proposed guideline GP26-P. National Committee for Clinical Laboratory Standards, Wayne, Pa.

46. **National Committee for Clinical Laboratory Standards.** 1993. *Clinical Laboratory Waste Management.* Approved guideline GP5-A. National Committee for Clinical Laboratory Standards, Wayne, Pa.

47. **National Committee for Clinical Laboratory Standards.** 1996. *Clinical Laboratory Procedure Manuals*, 3rd ed. Approved guideline 3P2-3A. National Committee for Clinical Laboratory Standards, Wayne, Pa.

48. **National Committee for Clinical Laboratory Standards.** 1999. *Laboratory Diagnosis of Blood-Borne Parasitic Diseases.* Approved guideline M15-A. National Committee for Clinical Laboratory Standards, Wayne, Pa.

49. **National Committee for Clinical Laboratory Standards.** 2001. *Protection of Laboratory Workers from Occupationally Acquired Infections.* Approved standard M29-A2. National Committee for Clinical Laboratory Standards, Wayne, Pa.

50. **National Committee for Clinical Laboratory Standards.** 2002. *Clinical Laboratory Waste Management.* Approved standard GP5–2A. National Committee for Clinical Laboratory Standards, Wayne, Pa.

51. **National Committee for Clinical Laboratory Standards.** 2002. *Implementing a Needlestick and Sharps Injury Prevention Program in the Clinical Laboratory: a Report.* Standard X3-R. NCCLS, Wayne, Pa.

52. **National Institutes of Health.** 1979. *NIH Guidelines for Recombinant DNA Research Supplement: Laboratory Safety Monograph.* National Institutes of Health, Bethesda, Md.

53. **National Sanitation Foundation.** 1987. *NSF Standard no. 49 for Class II (Laminar Flow) Biohazard Cabinetry.* National Sanitation Foundation, Ann Arbor, Mich.

54. **Neimeister, R.** 1992. Introduction, p. 7.1.1–7.1.11. *In* H. D. Isenberg (ed.), *Clinical Microbiology Procedures Handbook.* American Society for Microbiology, Washington, D.C.

55. **Neimeister, R., A. L. Logan, and J. H. Egleton.** 1985. Modified trichrome staining technique with a xylene substitute. *J. Clin. Microbiol.* **22:**306–307.

56. **Neimeister, R., A. L. Logan, B. Gerber, J. H. Egleton, and B. Kleger.** 1987. Hemo-De as a substitute for ethyl acetate in formalin-ethyl acetate concentration technique. *J. Clin. Microbiol.* **25:**425–426.

57. **Parasitology Subcommittee, Microbiology Section of Scientific Assembly, American Society of Medical Technology.** 1978. Recommended procedures for the examination of clinical specimens submitted for the diagnosis of parasitic infections. *Am. J. Med. Technol.* **44:**1101–1106.

58. **Paustenbach, D., Y. Alarie, T. Kulle, N. Schachter, R. Smith, J. Swenberg, H. Witschi, and S. B. Horowitz.** 1997. A recommended occupational exposure limit for formaldehyde based on irritation. *J. Toxicol. Environ. Health* **50:**217–263.

59. **Richmond, J., and R. W. McKinney (ed.).** 1999. *Biosafety in Microbiological and Biomedical Laboratories,* 4th ed. CDC/NIH, U.S. Government Printing Office, Washington, D.C.

60. **Rutala, W. A., and D. J. Weber.** 1997. Uses of inorganic hypochlorite (bleach) in health-care facilities. *Clin. Microbiol. Rev.* **10:**597–610.

61. **Schallom, L.** 1999. Point of care testing in critical care. *Crit. Care Nurs. Clin. N. Am.* **11:**99–106.

62. **Sewell, D. L.** 1995. Laboratory-associated infections and biosafety. *Clin. Microbiol. Rev.* **8:**389–405.

63. **Sewell, D. L.** 2004. Laboratory safety, p. 446–472. *In* L. S. Garcia (ed. in chief), *Clinical Laboratory Management.* ASM Press, Washington, D.C.

64. **Snyder, J. W.** 2002. Packaging and shipping of infectious substances. *Clin. Microbiol. Newsl.* **24:**89–93.

65. **Tarlo, S. M.** 1998. Latex allergy: a problem for both healthcare professionals and patients. *Ostomy Wound Manage.* **44:**80–88.

66. **Valenti, W. M.** 1986. AIDS and the lab: infection control guidelines. *Med. Lab. Obs.* Feb:53–56.

67. **van den Heuvel, J., L. Koning, A. J. Bogers, M. Berg, and M. E. van Dijen.** 2005. An ISO 9001 quality management system in a hospital: bureaucracy or just benefits? *Int. J. Health Care Qual. Assur. Inc. Leadersh. Health Serv.* **18:**361–369.

68. **Vesley, D., and J. Lauer.** 1986. Decontamination, sterilization, disinfection, and antisepsis in the microbiology laboratory, p. 182–198. *In* B. M. Miller, D. H. M. Gröschel, J. H. Richardson, D. Vesley, J. R. Songer, R. D. Housewright, and W. E. Barkley (ed.), *Laboratory Safety: Principles and Practices.* American Society for Microbiology, Washington, D.C.

69. **Virey-Griffaton, E., M. P. Luhucher-Michel, and D. Verloet.** 2000. Natural latex allergy. Primary and secondary prevention in work environment. *Presse Med.* **29:**257–262.

70. **Wakelin, S. H., and I. R. White.** 1999. Natural rubber latex allergy. *Clin. Exp. Dermatol.* **24:**245–248.

71. **Wood, D. E., J. Palmer, P. Missett, and J. L. Whitby.** 1994. Proficiency testing in parasitology. An educational tool to improve laboratory performance. *Am. J. Clin. Pathol.* **102:**490–494.

72. **Young, K. H., S. L. Bullock, D. M. Melvin, and C. L. Spruill.** 1979. Ethyl acetate as a substitute for diethyl ether in the Formalin-ether sedimentation technique. *J. Clin. Microbiol.* **10:**852–853.

Medical Parasitology: Case Histories

Protozoal infections

Helminth infections

Blood parasite infections

Diagnostic methods

The case histories in this chapter are taken from actual practice and provide an excellent method of review for the reader. Each case includes the history, laboratory findings, and clinical course. Self-assessment questions and a comprehensive discussion supplement each history. Also, parasite images photographed from actual bench work microscopy accompany each case. The cases illustrate many kinds of potential diagnostic problems encountered within the parasitology section of the laboratory and emphasize clinical relevance and recommended approaches for specimen ordering, collection, processing, testing, and reporting. Although not every parasite known to be a human pathogen is included, a wide spectrum of organisms and diseases is presented to illustrate particular points.

Protozoal Infections

Case 1

A 30-year-old patient complained that he had had diarrhea for about a week, and specimens were submitted to the laboratory. He had previously been in excellent health with no complaints, although he did report an unexplained weight loss of several pounds. On examination of the permanent stained smears, the following images were seen (Figure 36.1). These protozoan trophozoites measured approximately 17 µm long. The permanent stained smears have been stained using Wheatley's trichrome stain modification. The test result is reported as follows: "*Entamoeba histolytica* trophozoites present."

Questions

1. Based on the diagnosis of diarrhea and the laboratory findings, how would you identify the objects seen in the permanent stained smears?
2. Was the report correct as submitted to the physician? Why or why not?
3. Is diarrhea a normal finding in an infection caused by this organism?
4. Why is it important to report the presence of this organism correctly?

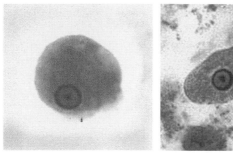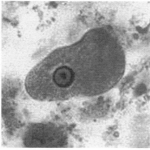

Figure 36.1 Case 1. Microscopic images (oil immersion 100× objective); permanent stained smear.

Correct Identification

The images presented in this case history are trophozoites of *Entamoeba histolytica/E. dispar*; therefore, the physician's report ("*Entamoeba histolytica* trophozoites present") was incorrect. You will notice that the amebic trophozoites in Figure 36.1 did not contain any ingested red blood cells (RBCs) within the cytoplasm. If no RBCs are seen in the cytoplasm and the trophozoites measure greater than or equal to 12 µm, the true pathogen, *E. histolytica*, cannot be differentiated from the nonpathogenic *E. dispar*. Currently, there are immunoassays available that can confirm the presence of organisms in the *E. histolytica/E. dispar* group, as well as one immunoassay that can confirm the presence of the pathogen *E. histolytica* (3, 4, 7, 8).

The correct laboratory report for the organisms seen in the case history would have been "*Entamoeba histolytica/E. dispar* trophozoites present." A computer comment could be added as follows: "Based on organism morphology, the presence of the true pathogen, *Entamoeba histolytica* (cause of amebiasis), could not be confirmed." Trophozoites of *E. histolytica*, the true pathogen, can be seen in Figure 36.2; the correct identification is confirmed by the presence of ingested RBCs within the organism cytoplasm.

If your laboratory offers the immunoassay specific for *E. histolytica*, the computer comment could be: "If

you want confirmation of the presence of *Entamoeba histolytica*, submit a fresh stool sample." The immunoassays for *E. histolytica* or the *E. histolytica/E. dispar* group require fresh or frozen stools; formalinized specimens are not acceptable for testing.

Prior to the separation of *E. histolytica* into two distinct species (*E. histolytica* and *E. dispar*), there were suggestions that some *E. histolytica* strains were pathogenic and some were not. Now that *E. dispar* has been classified as a nonpathogenic organism, totally separated from the pathogenic *E. histolytica*, treatment of *E. dispar* infection is usually not recommended. However, this recommendation is based on the separation of the two species by molecular means, not by morphology alone. Since few laboratories will be routinely identifying the two *Entamoeba* species to the level of *E. histolytica* or *E. dispar*, the clinician will have to decide on the basis of clinical findings whether the patient has true pathogenic *E. histolytica* or nonpathogenic *E. dispar* infection. The laboratory report will merely indicate the presence of the "*Entamoeba histolytica/E. dispar*" group.

Infection with *E. dispar* cannot be diagnosed on the basis of morphologic criteria alone, even from the permanent stained smear. When the organism is seen in the routine ova and parasite examination, it cannot be differentiated from *E. histolytica* unless trophozoites are seen to contain RBCs in the cytoplasm, a finding that is diagnostic for *E. histolytica*. When routine diagnostic methods and not immunoassay procedures specific for *E. histolytica* are used, morphologic criteria would provide characteristics of both; the laboratory report would indicate "*Entamoeba histolytica/E. dispar*."

Key Points for Laboratory Identification

1. When routine ova and parasite (O&P) examinations are ordered, three specimens (stool) should be submitted for the diagnosis of intestinal amebiasis.

2. Any examination for parasites in stool specimens must include the use of a permanent stained smear (even on formed stool).

3. Presumptive identification on a wet preparation must be confirmed by using the permanent stained smear. Without using *E. histolytica*-specific antigen detection tests, the presence of the organism(s) should be reported as follows. Trophozoites containing ingested RBCs within the cytoplasm should be reported as the true pathogen ("*Entamoeba histolytica* trophozoites present"). However, if cysts or trophozoites containing no ingested RBCs are seen, the report should indicate that the *Entamoeba* group is present ("*Entamoeba histolytica/E. dispar* group trophozoites and cysts present").

Figure 36.2 Case 1. *Entamoeba histolytica* trophozoites containing ingested RBCs.

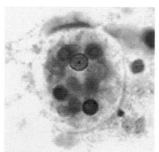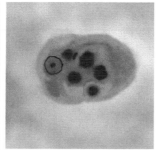

4. The six smears should be prepared at sigmoidoscopy but should not take the place of the O&P examination.

5. Immunoassay procedures can be performed to confirm the presence of the *E. histolytica*/*E. dispar* group or can be used to differentiate pathogenic *E. histolytica* from nonpathogenic *E. dispar*. Remember that some of these reagent kits may require fresh or frozen stools only; formalin-preserved stool or stool containing polyvinyl alcohol are not acceptable.

6. The serologic test for antibody may *or may not* be positive in intestinal disease and is much more likely to be positive in extraintestinal disease.

Case 2

A 28-year-old man presented with rhinosinusitis and was diagnosed as having a squamous cell carcinoma of the nasal septum. He was presumed to be a healthy and immunocompetent individual; however, he was later diagnosed with AIDS. As the diagnostic workup was expanded and other possible etiologic agents were considered, the following images were seen from agar culture plates (Figure 36.3).

These organisms have spine-like pseudopods, with the average diameter of the trophozoites being 30 μm. The nucleus has the typical large karyosome, and morphology can be seen on a wet preparation. The cysts are round with a single nucleus, which also has the large karyosome seen in the trophozoite nucleus. The double wall is usually visible, with the slightly wrinkled outer cyst wall and what has been described as a polyhedral inner cyst wall (4, 7, 12).

Questions

1. Can you describe the type of culture system that was used to support the growth of these organisms?

2. What do you think these organisms are (genus)? Why?

Figure 36.3 Case 2. Protozoan. (Left) Trophozoite with "spiky" acanthapodia; (right) cyst with hexagonal double wall.

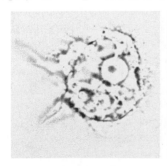 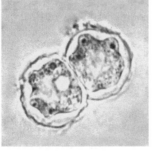

3. What morphologic characteristics support your identification?

4. What impact would AIDS have on this opportunistic infection?

Correct Identification

Assuming the possibility of infection with free-living amebae, diagnostic procedures could include agar plate cultures for the recovery of trophozoites and cysts of *Acanthamoeba*. Nonnutrient agar plates are routinely used with a bacterial overlay. The organisms are *Acanthamoeba* spp. Note the spiky acanthapodia in the trophozoite stage and the presence of the characteristic double-walled cyst form.

The protracted course in this case is characteristic of severe illness caused by *Acanthamoeba* and distinguishes this disease from primary amebic meningoencephalitis (PAM), caused by *Naegleria*, which has a much more acute presentation. Additional features that helped to distinguish the patient's *Acanthamoeba* infection from that caused by *Naegleria* include the typical trophozoite and cyst morphology obtained from the culture system.

Although this patient was diagnosed as having a squamous cell carcinoma of the nasal septum, this was an incorrect diagnosis. Once the diagnosis of AIDS was made, the workup was expanded to include possible parasitic pathogens, particularly organisms in the free-living amebae group. The final diagnosis was AIDS along with disseminated acanthamoebiasis. This case points up the importance of considering all possibilities, particularly in a patient who appears to be immunocompetent but who may actually be immunocompromised for one or more reasons.

An early diagnosis can be made if amebic infection is considered in patients with various presentations. *Acanthamoeba* spp. are now well accepted as opportunistic pathogens in immunocompromised patients with AIDS, generally those with a low CD4$^+$ cell count. Unfortunately, diagnosis of this infection requires a high index of suspicion, since both clinical and histologic findings may mimic those of disseminated fungal or algal disease or noninfectious causes such as carcinoma. Patients with AIDS and *Acanthamoeba* infection may experience general symptoms such as fever and chills, nasal congestion or lesions, neurologic symptoms, and musculoskeletal and cutaneous lesions. Some patients, especially those with AIDS, develop erythematous nodules, chronic ulcerative skin lesions, or abscesses.

The primary site of infection is thought to be the sinuses and lungs. The skin is also thought to be a possible portal of entry. In patients with disseminated disease, the duration of infection from onset to death can range from 7 days to 5 months. Since early therapy can alter the clinical outcome, recognition of this disease is critical. It is important to initiate appropriate diagnostic testing for confirmation of the causative agent as soon as possible.

Acanthamoeba spp. have been found in the upper respiratory tract in humans with and without symptoms. In cases of disseminated disease, the portal of entry remains unknown but is suspected to be the respiratory tract. However, in patients with skin lesions, the skin may be the site of primary inoculation. Although many of the reported cases have been seen in compromised patients, cases have also occurred in immunocompetent hosts.

Although *Naegleria* dissemination outside the central nervous system (CNS) may occur, it is apparently rare and is unlike the protracted situation seen in this patient. The morphology of *Naegleria* trophozoites is quite different from that of *Acanthamoeba* trophozoites, and an image can be seen below (Figure 36.4). Note the lobular pseudopodia as compared with the "spiky" pseudopodia seen in *Acanthamoeba* trophozoites (Figure 36.3).

The nonnutrient agar plates are made using Page's saline. Before use, the plates should be removed from the refrigerator and placed at 37°C for 30 min. Add about 0.5 ml of saline to a slant culture of *Escherichia coli*. Gently scrape the surface, and suspend bacteria uniformly. Add 2 or 3 drops of this suspension to the middle of the warmed agar plate. Using a loop, spread the bacteria on the surface of the agar. Depending on the specimen, cerebrospinal fluid (CSF) or other nervous system material should be incubated at 37°C; other tissues should be incubated at 30°C. After the fluid/specimen has been absorbed, seal the plates with a 5- to 6-in. strip of Parafilm; incubate the plate inverted in room air.

Using the low-power objective (10×), examine the plates for amebic cysts or trophozoites every day for 10 days. Thin, linear tracks, which represent areas where amebae have ingested the bacteria, might also be seen.

Current therapy for *Acanthamoeba* infection includes amphotericin B and sulfadiazine or sulfisoxazole or pentamidine isethionate. Treatment is somewhat problematic, and amphotericin B may be ineffective. Skin ulcers have been treated by combination therapy involving systemic intravenous pentamidine isethionate for 1 month and oral itraconazole for 8 months. The skin ulcers can also be cleaned twice daily with chlorhexidine gluconate solution and treated by topical application of 2% ketoconazole cream.

Key Points for Laboratory Identification

1. As with *N. fowleri*, beware of a false-positive Gram stain (the leukocyte count in CSF may be elevated).

2. Examine clinical material on a slide with a coverslip; do not use a counting chamber (organisms in CSF look like leukocytes).

3. Nonnutrient agar with Page's saline and an *E. coli* overlay can be tried for culture recovery.

4. Various other media containing different agar bases, some of which contain horse, sheep, or rabbit blood, are also available.

5. Tissue stains are also effective, and the use of calcofluor white to visualize the double-walled cyst is recommended.

6. Refer to chapter 32 for specific methods and recommendations.

Case 3

A 6-year-old girl was admitted to the hospital in a coma. The patient was normally a healthy child with no significant medical history. Several days prior to admission, the family had been vacationing at a lake in the southern United States. Everyone had been swimming and boating during the vacation. Several days before admission, the patient had complained of some vague upper respiratory distress and a sore throat. On the day prior to admission, her mother indicated that she had complained of a stiff neck and headache; she also had a fever. On the day of admission, she became very confused and lapsed into a coma.

On physical examination, her neck was stiff and her pupils reacted sluggishly; her temperature was 104°F. The CSF was hemorrhagic and contained many neutrophils.

Figure 36.4 Case 3. Protozoan. (Left) Trophozoites growing on a nonnutrient agar plate seeded with *Escherichia coli* bacteria; (right) stained trophozoite (note the more globular pseudopods compared with the organism in case 2).

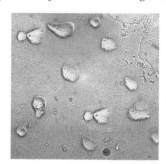
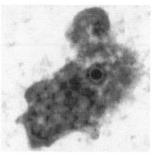

Although there was some debris on the smear, the Gram stain was read as negative. The following images were seen from culture and stained preparations (Figure 36.4) (4, 7).

Questions

1. What laboratory procedure should be performed immediately and why?
2. If the procedure was positive, what would you expect to see?
3. Treatment should be started immediately with what and why?
4. What is there about the history that should alert everyone to this disease possibility?

Correct Identification

The CSF should be examined as a wet mount (not in a counting chamber), looking for motile amebic trophozoites. The specimen can be centrifuged or examined as unspun fluid; it should not be refrigerated prior to examination. Amebic trophozoites can mimic neutrophils in a counting chamber, but characteristic motility can be seen on a slide with a coverslip. In a positive wet mount, motile trophozoites of free-living amebae, *Naegleria fowleri*, would be seen. The pseudopodia are rounded and nonspiky, and motility is visible unless the organisms are cold or dead. The only drug that appears to be effective is amphotericin B. Very few patients have survived this infection; two patients responded to amphotericin B and did survive. Rare cases have also responded to miconazole and, in one case, rifampin.

Central nervous system (CNS) infections caused by free-living amebae have been recognized only since the mid-1960s, and our understanding of this disease process is still incomplete. One type of meningoencephalitis (PAM) is a fulminant and rapidly fatal disease that mainly affects children and young adults. The disease closely resembles bacterial meningitis but is caused by *N. fowleri*, an organism found in moist soil and freshwater habitats.

There are trophozoite and cyst stages in the life cycle; the stage depends primarily on environmental conditions. Trophozoites can be found in water or moist soil and can be maintained in tissue culture or other artificial media. The trophozoites can occur in two forms, ameboid and flagellate. Motility can be observed in hanging-drop preparations from cultures of CSF; the ameboid form (the only form recognized in humans) is elongate with a broad anterior end and tapered posterior end. The size ranges from 7 to 20 μm. The diameter of the rounded forms is usually 15 μm. There is a large, central karyosome and no peripheral nuclear chromatin. The cytoplasm is somewhat granular and contains vacuoles. The ameboid-form organisms change to the flagellate form when they are transferred from culture or teased from tissue into water and maintained at a temperature of 27 to 37°C. The change may occur within a few hours or may take as long as 20 h. The flagellate form is pear shaped, with two flagella at the broad end. Motility is typical, with either spinning or jerky movements. These flagellate forms do not divide, but when the flagella are lost, the ameboid forms resume reproduction. Cysts from nature and from agar cultures look the same and have a single nucleus almost identical to that seen in the trophozoite. The shape is generally round, measuring from 7 to 10 μm, and there is a thick double wall.

PAM caused by *N. fowleri* is an acute, suppurative infection of the brain and meninges. In humans, with extremely rare exceptions, the disease is rapidly fatal. The period between contact with the organism and onset of clinical symptoms such as fever, headache, and rhinitis may vary from 2 to 3 days to as long as 7 to 15 days. The amebae may enter the nasal cavity by inhalation or aspiration of water, dust, or aerosols containing the trophozoites or cysts. The organisms then penetrate the nasal mucosa, probably through phagocytosis of the olfactory epithelium cells, and migrate via the olfactory nerves to the brain. Early symptoms include vague upper respiratory distress, headache, lethargy, and occasionally olfactory problems. The acute phase includes sore throat, stuffy blocked or discharging nose, and severe headache. Progressive symptoms include pyrexia, vomiting, and stiffness of the neck. Mental confusion and coma usually occur approximately 3 to 5 days prior to death. The cause of death is usually cardiorespiratory arrest and pulmonary edema. PAM can resemble acute purulent bacterial meningitis, and the conditions may be difficult to differentiate, particularly in the early stages. Unfortunately, if the CSF Gram stain is interpreted incorrectly, with the identification of bacteria representing a false-positive result, the antibacterial therapy has no impact on the amebae and the patient will usually die within several days. Extensive tissue damage occurs along the path of amebic invasion; the nasopharyngeal mucosa shows ulceration, and the olfactory nerves are inflamed and necrotic. Hemorrhagic necrosis is concentrated in the region of the olfactory bulbs and the base of the brain. Organisms can be found in the meninges, perivascular spaces, and sanguinopurulent exudates.

Clinical and laboratory data usually cannot be used to differentiate pyogenic meningitis from PAM, so the diagnosis may have to be reached by a process of elimination. A high index of suspicion is often mandatory for early diagnosis. Although most cases are associated with exposure to contaminated water through swimming or bathing, this is not always the case. The rapidly fatal course of 4 to 6 days after the beginning of symptoms, with an incubation period of 1 day to 2 weeks, requires early diagnosis and immediate chemotherapy if the patient is to survive. Analysis of the CSF shows decreased glucose and increased protein concentrations. Leukocytes may range from several hundred to >20,000 cells/mm^3. Gram stains and bacterial cultures of CSF are negative.

A definite diagnosis could be made by demonstration of the amebae in the CSF or in biopsy specimens. Either CSF or sedimented CSF should be placed on a slide, under a coverslip, and observed for motile trophozoites; smears can also be stained with Wright's or Giemsa stain. CSF, exudate, or tissue fragments can be examined by light microscopy or phase-contrast microscopy. Care must be taken not to mistake leukocytes for actual organisms or vice versa. It is very easy to confuse leukocytes and amebae, particularly when one is examining CSF by using a counting chamber, hence the recommendation to use just a regular slide and coverslip. Motility may vary, so the main differential characteristic is the spherical nucleus with a large karyosome.

Specimens should never be refrigerated prior to examination. When centrifuging the CSF, low speeds (250 × g) should be used so that the trophozoites are not damaged; however, centrifugation at 500 × g is not detrimental to the morphology. Although bright-field microscopy with reduced light is acceptable, phase microscopy, if available, is recommended. Use of smears stained with Giemsa or Wright's stain or a Giemsa-Wright's stain combination can also be helpful. If *Naegleria* is the causative agent, trophozoites only are normally seen. If the infecting organism is *Acanthamoeba*, cysts may also be seen in specimens from the CNS. Unfortunately, most cases of PAM are diagnosed at autopsy; confirmation of these tissue findings must include culture and/or special staining using monoclonal reagents in indirect fluorescent-antibody procedures. Organisms can also be cultured on nonnutrient agar plated with *Escherichia coli*.

Over 175 presumptive or proven cases of PAM have been reported in the literature, including cases from the United States (86 cases as of January 1998), Ireland, England, Belgium, Czechoslovakia, Australia, New Zealand, Brazil, and Zambia. Clinical patient histories indicate exposure to the organism via freshwater lakes or swimming pools shortly before onset; patients had been previously healthy with no specific underlying problems. Pathogenic *Naegleria* organisms have also been isolated from nasal passages of individuals with no history of water exposure, suggesting the possibility of airborne exposure.

The first isolations of environmental strains of pathogenic *N. fowleri* were reported from water and soil in Australia and from sewage sludge samples in India. Detection of *N. fowleri* in heated discharge water has been reported in Belgium and Poland. Since then, there have been additional reports describing isolations of virulent or avirulent strains of *N. fowleri* from the environment. Studies in Belgium have clearly indicated that *N. fowleri* strains are present in artificially heated waters (power plant warm discharge), and studies in the United States have indicated that virulent strains are also found in lakes at geographic latitudes with water temperatures of 14 to 35°C that are totally isolated from any source of thermal discharge. These data suggest that the presence of pathogenic or potentially pathogenic amebic strains may depend on both climate and modification of the natural environment. The ability of the cysts to survive under various environmental conditions has been investigated by several workers. These findings suggest that *N. fowleri* cysts produced in the warm summer months may survive the winter and are capable of growth during the following summer.

General preventive measures include public awareness of potential hazards of contaminated water. It has been recommended that warm discharge water not be used for sports and recreational purposes, particularly since DNA restriction fragment profiles of environmental strains and human isolates were homogeneous.

Only 3 patients of more than 175 with presumptive or parasitologically proven cases of PAM had survived this infection outside of the United States by the 1970s. One case within California was documented in which the patient was successfully treated with amphotericin B, miconazole, and rifampin. There was a synergistic effect with miconazole and amphotericin B in vitro; however, rifampin was ineffective. The patient received intrathecal and intravenous injections of the drug. Dexamethasone and phenytoin were also given to this patient to decrease intracranial pressure and seizure, respectively. *Naegleria* infections have also been treated successfully with amphotericin B, rifampin, and chloramphenicol; amphotericin B, oral rifampin, and oral ketoconazole; and amphotericin B alone.

Key Points for Laboratory Identification

1. Never refrigerate the specimen(s) prior to examination.

2. Beware of the false-positive Gram stain, especially since PAM commonly mimics pyogenic meningitis with the CSF containing increased numbers of leukocytes.

3. *Naegleria* trophozoites mimic leukocytes if put in a counting chamber. Motility is more likely to be

seen if a drop of CSF is placed directly on a slide and a coverglass is added.

4. Organisms can be cultured on nonnutrient agar plated with *E. coli*.

Case 4

An immunocompetent 5-year-old girl was diagnosed with diarrhea off and on for about 10 days, and two stool specimens were submitted to the laboratory for examination. After examination of the concentration sediment from both specimens, nothing was seen and the results were reported as "No parasites seen" (Figure 36.5, left). No permanent stained smears were prepared or examined. The child continued to be symptomatic with vague gastrointestinal problems and diarrhea. Another stool specimen was submitted, and both concentration sediment and permanent stained smears were examined. The parasite identified is seen in Figure 36.5 (right). This protozoan trophozoite measures approximately 13 to 16 μm long (3, 4, 7, 8, 12).

Questions

1. Would you agree or disagree with the original diagnostic approach used for the first two stool samples?
2. If one considers all the intestinal protozoa, which organisms might have been missed without the permanent stained smear examination?
3. Identification of which organism requires the permanent stained smear? Why is this true?
4. What references might be helpful in reviewing guidelines for a complete O&P examination?

Figure 36.5 Case 4. (Left) Stool fecal concentration sediment examined as a wet mount (appears to be negative); (right) protozoan trophozoite containing two nuclei, which are fragmented into several nuclear granules.

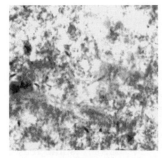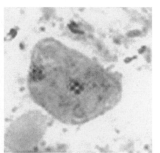

Correct Identification

The organism image presented is a trophozoite of the flagellate *Dientamoeba fragilis*; thus, the physician's report ("No parasites seen") was incorrect. The first report was based on examination of the concentration sediment wet preparation (Fig. 36.5, left), in which no organisms were seen.

D. fragilis was first seen in 1909 but not described until 1918. On the basis of electron microscopy studies, it has been reclassified as an ameba-flagellate rather than an ameba and is closely related to *Histomonas* and *Trichomonas* spp. It has a cosmopolitan distribution, and past surveys showed incidence rates of 1.4 to 19%. Much higher incidence rates have been reported for mental-institution inmates, missionaries, and Native Americans in Arizona. *D. fragilis* tends to be common in some pediatric populations, and in some studies the incidence figures are higher for patients younger than 20 years. Trophozoites can contain one or two nuclei, but no cyst form has been identified for this organism.

Although its pathogenic status is still not well defined, *D. fragilis* has been associated with a wide range of symptoms. Case reports of children infected with *D. fragilis* reveal a number of symptoms including intermittent diarrhea, abdominal pain, nausea, anorexia, malaise, fatigue, poor weight gain, and unexplained eosinophilia. The most common symptoms in patients infected with this parasite appear to be intermittent diarrhea and fatigue. In some patients, both the organism and the symptoms persist or reappear until appropriate treatment is initiated.

Diagnosis of *D. fragilis* infections depends on proper collection and processing techniques (a minimum of three fecal specimens). Although survival time for this parasite has been reported as 24 to 48 h, the survival time in terms of morphology is limited, and stool specimens must be examined immediately or preserved in a suitable fixative soon after defecation. It is particularly important that permanent stained smears of stool material be examined with an oil immersion lens (1,000×). These organisms have been recovered in formed stool; therefore, a permanent stained smear must be prepared for every stool submitted for a parasite examination.

If preserved specimens are submitted to the laboratory, it may be more cost-effective to delete the direct smear and begin the stool examination with the concentration procedure, particularly since motile protozoa are not viable because of the prior addition of preservative. Even if parasites are seen on a direct mount of preserved stool, they would almost certainly be seen on the concentration examination as well as on the permanent stained smear (protozoa in particular). With few exceptions, intestinal protozoa should never be identified on the basis of a wet mount alone; permanent stained smears should be examined to confirm the specific identification of suspected organisms.

Every stool submitted for an O&P examination must be examined using the concentration and permanent stained smear procedures. This guideline is consistent with the College of American Pathologists laboratory inspection checklist.

Key Points for Laboratory Diagnosis

1. O&P examination (fresh stool specimen, liquid and/or very soft): direct wet smear, concentration, permanent stained smear.
2. O&P examination (fresh stool specimen, formed stool): concentration, permanent stained smear.
3. O&P examination (preserved stool specimen): concentration, permanent stained smear.

All intestinal protozoan infections can be missed if the concentration is the only test to be performed. The permanent stained smear is much more sensitive than the concentration alone. To date, there are no commercial immunoassay products available for the confirmation of infection with *D. fragilis*, although they are under development.

Case 5

A 3-year-old boy was first seen in the clinic because of failure to thrive and vague symptoms such as poor appetite and failure to gain weight. There was one other child in the family, a 6-year-old brother who was in good health. The patient was apparently fine for the first couple of years. He began to have diarrhea with light-colored stools; however, this condition was not always present. Although stool examinations were performed, it was unclear exactly what diagnostic methods were used and the test results were reported as negative. The child was placed on a high-protein, high-calorie diet with vitamins and supplements. However, he showed very little improvement over a 4-month period. Based on these and other findings (barium exam showed "large dilated loops of hypotonic bowel"), the child was admitted to the hospital with a diagnosis of celiac disease.

Complete stool examinations were performed (O&P examination: direct wet mount, concentration, and permanent stained slide). The following image is a wet preparation of stool (Figure 36.6); the organism was seen using the high dry objective of the microscope. Motility was seen, but it was very limited and the organism identification could not be confirmed without completion of the O&P examination (3, 4, 7, 8). The images in Figure 36.6 were photographed at a high magnification.

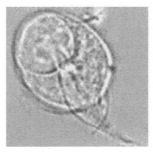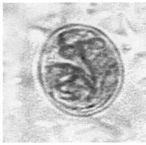

Figure 36.6   Case 5. (Left) Protozoan trophozoite seen in a wet mount; (right) protozoan cyst seen in a wet mount (note how round, distorted, and three-dimensional the cyst appears.

Questions

1. What parasites should be considered based on the patient's history?
2. What factors might be responsible for the "negative" laboratory reports?
3. After seeing the organisms in the wet mounts, could these parasites be responsible for the patient's symptoms?
4. What other diagnostic tests might be available? Would they be more or less sensitive than the O&P examination? Please explain.

Correct Identification

Figure 36.6 shows a *Giardia lamblia* trophozoite (left) and a *G. lamblia* cyst (right). The trophozoite measured approximately 8 by 12 µm, and there appeared to be minimal motility related to the flagella; the cyst measured approximately 10 µm and appeared to be somewhat round. Although this organism was seen in the wet mount, additional morphology could be seen from the concentration and permanent stained slide examinations. While the presumptive identification could be made from the concentration wet mount, confirmation of the diagnosis usually depends on examination of the permanent stained smear.

The incubation time for giardiasis is approximately 12 to 20 days. Because the acute stage usually lasts only a few days, giardiasis may not be recognized as the infection but may mimic acute viral enteritis, bacillary dysentery, bacterial or other food poisonings, acute intestinal amebiasis, or "traveler's diarrhea" (caused by toxigenic *E. coli*). However, the type of diarrhea plus the lack of blood, mucus, and cellular exudate is consistent with giardiasis. Although the organisms in the crypts of the duodenal mucosa may reach very high densities, they may not cause any pathology. The organisms feed on the mucous secretions and do not penetrate the mucosa. For some reason, in symptomatic cases there may be irritation of the mucosal lining, increased mucus secretion, and dehydration.

The acute phase is often followed by a subacute or chronic phase. Symptoms in these patients include recurrent, brief episodes of loose, foul-smelling stools; there may be increased distention and foul flatus. Between the episodes of passing mushy stools, the patient may have normal stools or may be constipated. Abdominal discomfort continues to include marked distention and belching with a rotten-egg taste. Chronic disease must be differentiated from amebiasis and other intestinal parasites such as *Dientamoeba fragilis*, *Cryptosporidium* spp., *Cyclospora cayetanensis*, *Isospora belli*, and *Strongyloides stercoralis* and from inflammatory bowel disease and irritable colon. Based on symptoms such as upper intestinal discomfort, heartburn, and belching, giardiasis must also be differentiated from duodenal ulcer, hiatal hernia, and gallbladder and pancreatic disease. Because giardiasis may not cause any symptoms at all, demonstration of the organism in symptomatic patients may not rule out other possibilities such as peptic ulcer, celiac disease, strongyloidiasis, and possibly carcinoma.

Giardiasis is one of the more common causes of traveler's diarrhea and has been recorded from all parts of the world. It has also been speculated that visitors to areas where *Giardia* is endemic are more likely to present with symptoms than are individuals who live in the area; this difference is most probably due to the development of immunity from prior, and possible continued, exposure of residents to the organism.

Both the trophozoite and the cyst are included in the life cycle of *G. lamblia*. Trophozoites divide by means of longitudinal binary fission, thus producing two daughter trophozoites. The most common location of the organisms is in the crypts within the duodenum. The trophozoites are the intestinal dwelling stage and attach to the epithelium of the host villi by means of the ventral disk. The attachment is substantial and results in disk "impression prints" when the organism detaches from the surface of the epithelium. Trophozoites may remain attached or may detach from the mucosal surface. Since the epithelial surface sloughs off the tip of the villus every 72 h, apparently the trophozoites detach at that time. For reasons which are not totally known, cyst formation takes place as the organisms move down through the colon. The trophozoites retract the flagella into the axonemes, the cytoplasm becomes condensed, and the cyst wall is secreted. As the cyst matures, the internal structures are doubled, so that when excystation occurs, the cytoplasm divides, thus producing two trophozoites. Excystation would normally occur in the duodenum.

Routine stool examinations are normally recommended for the recovery and identification of intestinal protozoa. However, in the case of *G. lamblia*, because the organisms are attached so securely to the mucosa by means of the sucking disk, a series of even five or six stool specimens may be examined without recovering the organisms. The organisms also tend to be passed in the stool on a cyclical basis. The Entero-Test capsule can be helpful in recovering the organisms, as can the duodenal aspirate. Although cysts can often be identified on the wet stool preparation, many infections may be missed without the examination of a permanent stained smear. If material from the string test (Entero-Test) or mucus from a duodenal aspirate is submitted, it should be examined as a wet preparation for motility; however, motility may be represented by nothing more than a slight flutter of the flagella because the organism is caught up in the mucus. After diagnosis, the rest of the positive material can be preserved as a permanent stain.

Procedures using immunoassays (enzyme immunoassay, fluorescent-antibody assay, cartridge formats) have also been developed to detect *Giardia* antigen in feces. The immunoassays are more sensitive than the O&P examination. Many of these newer methods are being used to diagnose patients suspected of having giardiasis or those who may be involved in an outbreak situation, particularly if the patient history is suggestive.

Transmission is by ingestion of viable cysts. Although contaminated food or drink may be the source, transmission may occur via intimate contact with an infected individual. This organism tends to be found more frequently in children or in groups that live in close quarters. Often there are outbreaks due to poor sanitation facilities or breakdowns as evidenced by infections in travelers and campers. There may also be an increase in the prevalence of giardiasis in the male homosexual population, probably due to anal and/or oral sexual practices. Because of the potential for a wild-animal reservoir and possibly other domestic-animal reservoir hosts, measures in addition to personal hygiene and improved sanitary measures have to be considered. Iodine has been recommended as an effective disinfectant for drinking water, but it must be used according to directions. Filtration systems have also been recommended, although they have certain drawbacks such as clogging.

Most experts agree that the single most effective practice that prevents the spread of infection in the child care setting is good handwashing by the children, staff, and visitors. Rubbing hands together under running water is the most important part of washing away infectious organisms. Premoistened towelettes or wipes and waterless hand cleaners should not be used as a substitute for washing hands with soap and running water. These guidelines are not limited to giardiasis but include all potentially infectious organisms.

Key Points for Laboratory Diagnosis

1. Even if a series of three stool specimens is submitted and examined correctly, the organisms may not be recovered and identified. The organisms are shed on a sporadic basis.

2. Motility on wet preparations may be difficult to see because the organisms may be caught up in mucus.

3. Any examination for parasites in stool specimens must include the use of a permanent stained smear (even on formed stool) (Figure 36.7).

4. Duodenal drainage and/or the use of the Entero-Test capsule may be very helpful in organism recovery. However, this technique does not take the place of the ova and parasite examination. Fecal immunoassays for antigen detection are now widely used and include the enzyme immunoassay, fluorescent-antibody assay, and cartridge formats. Based on the patient's history, a fecal immunoassay performed on two separate stool specimens is more appropriate and sensitive than the O&P. However, it is important to remember that patients who are carriers and do not present with acute symptoms may have insufficient numbers of organisms for a positive fecal immunoassay result, even though the immunoassays are more sensitive than the O&P.

If giardiasis is diagnosed, the patient should be treated. In the majority of cases, metronidazole is the drug of choice, but it is still considered investigational. Metronidazole is not recommended for pregnant women; paromomycin may be used to treat giardiasis in pregnancy, although it is not absorbed and not highly effective. Tinidazole has also been used and has proven to be more effective than metronidazole as a single dose. Furazolidone is another option but has been reported to be mutagenic and carcinogenic.

Figure 36.7 Case 5. (Left) *Giardia lamblia* trophozoite seen in a permanent stained smear; (right) *G. lamblia* cyst seen in a permanent stained smear (note how much more detail is seen compared to the wet mounts in Figure 36.6).

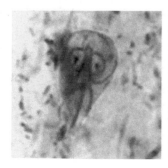 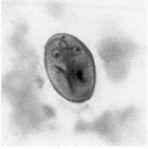

In the absence of a parasitologic diagnosis, the treatment of suspected giardiasis is a common question with no clear-cut answer. The approach depends on the alternatives and the degree of suspicion of giardiasis, both of which vary among patients and physicians. However, it is not recommended that treatment be given without a good parasitologic workup, particularly since relief of symptoms does not allow a retrospective diagnosis of giardiasis; the most commonly used drug, metronidazole, targets other organisms besides *G. lamblia*.

Should asymptomatic patients be treated? Generally, it is recommended that all cases of proven giardiasis be treated because the infection may cause subclinical malabsorption, symptoms are often periodic and may appear later, and a carrier is a potential source of infection for others. Certainly in areas of the world where infection rates, as well as the prospect of reinfection, are extremely high, the benefit-per-cost ratio would also have to be examined.

Holminth Infections

Case 6

A 28-year-old postdoctoral student presented to his physician with complaints of recurrent abdominal cramps. He had grown up in Italy, come to the United States for graduate studies, and returned home each year for the holidays. Over a period of several years he had complained of recurrent abdominal cramps; however, the causes were thought to be changes in diet, as well as the usual stress associated with graduate school. During the past few months, the symptoms had become worse and he occasionally had diarrhea as well. There was some speculation that he might have an ulcer.

On presentation, the patient appeared in no particular distress, and the examination was unremarkable. The one exception was some slight discomfort on palpation of the abdomen. A gastrointestinal series was normal; however, stool examinations revealed the following objects (Figure 36.8). These objects measured about 30 by 47 μm and were easily seen in the saline wet preparation examination. Note the suggestion of filaments that lie between the egg shell and the embryo (3, 4, 7, 8).

Questions

1. Which group of parasites includes these objects?
2. Is there anything about the object morphology that suggests the identification?
3. If you know what is causing the infection, do patients usually present with symptoms?
4. What key characteristic of the life cycle plays a role in transmission?

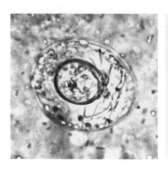

Figure 36.8 Case 6. Helminth eggs (wet mounts of concentration sediment). Note the polar filaments that lie between the oncosphere and the egg shell.

Correct Identification

These images are *Hymenolepis nana* eggs; these eggs generally measure 30 to 47 μm in diameter. They are round to oval, contain a six-hooked oncosphere (embryo), and have polar filaments that lie between the egg shell and the developing embryo.

Since the adult worms or proglottids are rarely seen in the stool, the diagnosis is based on recovery and identification of the characteristic eggs. They are most easily seen and identified from fresh specimens or those preserved in formalin-based fixatives. The thin-shelled eggs also tend to collapse on the permanent stained smear and may be difficult to identify in this type of preparation. Eggs with the characteristic thin shell, six-hooked oncosphere, and polar filaments are diagnostic. Remember, the eggs are infectious, and unpreserved stool specimens should be handled with caution.

An infection with *H. nana* may cause no symptoms, even with a heavy worm burden. Some patients complain of headache, dizziness, anorexia, abdominal pain, diarrhea, or possibly irritability. Some patients have a low-grade eosinophilia of 5% or more. When young children have heavy infections, they may have loose stools or even diarrhea containing mucus. Persistent, diffuse abdominal pain seems to be the most common symptom.

Infection is usually acquired by the ingestion of *H. nana* eggs, primarily from human stool. The eggs hatch in the stomach or small intestine, and the liberated larvae (oncospheres) penetrate the villi in the upper small intestine. The larvae develop into the cysticercoid stage in the tissue and migrate back into the lumen of the small intestine, where they attach to the mucosa. The adult worms mature within several weeks; they measure up to 40 mm long. Adult worms are rarely seen in the stool. This life cycle involves transition from the egg to the adult worm within the human; no intermediate host is required. Although accidental ingestion of the insect intermediate host can result in development of the adult worms, this mode of infection is probably rare. There is evidence that auto- or hyperinfection occurs when eggs spontaneously hatch in the small intestine and initiate a new life cycle. Although adult worms live for only about a year, some patients carry a large worm burden for many years, probably due to the autoinfection capability within the life cycle.

H. nana is the only human tapeworm in which an intermediate host is not required and transmission is from person to person. Children are usually infected more often than adults. Since infection is from person to person via the eggs, good personal hygiene is an important preventive measure.

Niclosamide has been widely used for years. However, the use of praziquantel in a single oral dose is apparently more effective and kills the cysticercoid as well as the adult worm. The recommended dose for adults and children is 25 mg/kg in a single dose (see chapter 25).

Key Points for Laboratory Diagnosis

1. Adult worms or proglottids are rarely seen in the stool.
2. Eggs with the characteristic thin shell, six-hooked oncosphere, and polar filaments are diagnostic.
3. Egg morphology is more easily seen in fresh specimens or those preserved in formalin-based fixatives (Figure 36.9).
4. The eggs are infectious, and unpreserved stool specimens should be handled with caution.

Case 7

A 47-year-old man was admitted to the hospital with complaints of severe mid-epigastric pain that had become worse over a period of about a week. This patient had

Figure 36.9 Case 6. *Hymenolepis nana* eggs or artifacts (permanent stained smear). The image on the left is probably just an artifact that resembles a helminth egg. The image on the right may or may not be an actual *H. nana* egg. On permanent stained smears, helminth egg morphology often appears distorted with very dark staining.

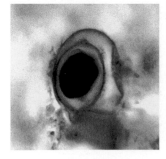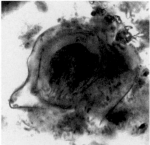

received prednisone over the course of several years, and the dose had been increased 2 months before. Previous diagnostic testing included an O&P examination, whose results were reported as "No parasites seen."

He was treated with supportive care for epigastric pain and developing pneumonia, but his condition failed to improve. He became comatose and died 3 days later. Autopsy findings included the following image (of his colon) (Figure 36.10).

Questions

1. What infection most probably matches this image?
2. Based on the history, why might there be no evidence of eosinophilia?
3. What was there about his history that could lead to complications with this parasite?
4. Does the negative O&P examination in this case make sense? Why or why not? Should a repeat O&P examination have been ordered?
5. Which other test could have been ordered? Is this test more or less sensitive that the routine O&P examination?

Correct Identification

The autopsy image shows a *Strongyloides stercoralis* rhabditiform larva in a crypt of the colon. Remember, high eosinophilia may or may not be present. In this case, no eosinophils were noted (eosinopenia can occur in the hyperinfection syndrome and is a poor prognostic sign). Additional images can be seen in Figure 36.11.

This case represents a generalized infection with *S. stercoralis*, involving the lungs, liver, peritoneum, small intestine, colon, respiratory diaphragm, heart, lymph nodes, skeletal muscle, and periadrenal and peripancreatic

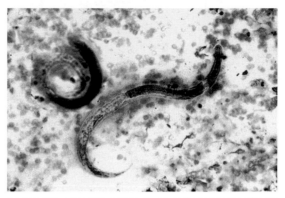
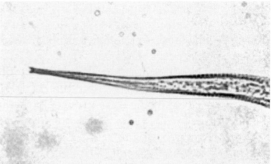
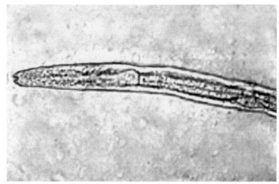
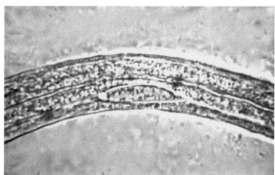

Figure 36.10 Case 7. Image from autopsy (colon). (Armed Forces Institute of Pathology photograph.)

Figure 36.11 Case 7. *Strongyloides stercoralis*. (Top) Larvae in sputum; (second from top) slit in the tail of filariform (infective) larva obtained using the Baermann concentration apparatus; (third from top) short buccal capsule and genital primordium packet of cells in a rhabditiform (noninfective) larva; (bottom) clear image of the genital primordium packet of cells in a rhabditiform larva.

fat. There was generalized peritonitis. Intact rhabditiform and filariform larvae of *S. stercoralis* were seen in the tissues and in stool specimens. It is unfortunate that additional O&P examinations or the agar plate culture were not performed during this hospital admission.

Strongyloidiasis is most often found in warm climates but can also occur in temperate and cold climates. It overlaps the same geographic range as hookworm infections. The disease is contracted by penetration of the filariform larvae into the skin by contact with infected soil, by autoinfection, or by fecal-oral contamination. Generalized disease may present in many ways: paralytic ileus, an acute surgical abdomen, asthma, and as a protein-losing enteropathy with malabsorption. It should be considered in the differential diagnosis of patients with bronchopneumonia and accompanying abdominal symptoms.

Findings of bilateral pulmonary disease were due to extensive intra-alveolar hemorrhage and interstitial pneumonia; numerous larval forms were present in the lungs. When young worms break out of the pulmonary capillary into the alveoli, hemorrhage and cellular infiltration into the air sacs and bronchioles result.

Malnutrition, lymphoma, treatment with immunosuppressive agents, and corticosteroid therapy are predisposing factors for strongyloidiasis. It is also very important to consider this infection in military personnel and travelers who may have been in an area of endemic infection many years earlier. More than 30 to 40 years after acquisition of the original infection, persistent, undiagnosed disease can be found in these individuals. If, for any reason, they become immunocompromised, the result can be disseminated disease leading to the hyperinfection syndrome and death.

In situations in which autoinfection occurs, some of the rhabditiform larvae that are within the intestine develop into the filariform larvae while passing through the bowel. These larvae can then reinfect the host by (i) invading the intestinal mucosa, traveling via the portal system to the lungs, and returning to the intestine, or (ii) being passed out in the feces and penetrating the host on reaching the perianal or perineal skin.

Pathology present in strongyloidiasis can vary both in severity and in the areas of the body involved. Some individuals may remain totally asymptomatic, with the only abnormal clinical finding being a peripheral eosinophilia.

Cutaneous Infection. Initial skin penetration usually causes very little reaction, although there may be some pruritus and erythema if there are large numbers of penetrating larvae. With repeated infections, the patient may mount an allergic response that prevents the parasite from completing the life cycle. The larvae may be limited to skin migration or larva migrans. The term larva currens ("racing larvae") was proposed in 1958 and is now generally accepted for cases of strongyloidiasis in which one or more rapidly progressing linear urticarial tracks starting near the anus are found. There is speculation that some of these cases may involve larvae of other species of *Strongyloides*. These tracks may progress as fast as 10 cm/h, with intermittent movement, usually on the thighs. Onset is sudden, and the lesions may disappear within 12 to 18 h.

Pulmonary Infection. Larval migration through the lungs may stimulate symptoms, depending on how many larvae are present and the intensity of the host immune response. Some patients are asymptomatic, while others present with pneumonia. With a heavy infective dose or in the hyperinfection syndrome, individuals often develop cough, shortness of breath, wheezing, fever, and transient pulmonary infiltrates (Loeffler's syndrome). Cases have also been reported where the larvae can be found in the sputum.

Intestinal Infection. In heavy infections, the intestinal mucosa may be severely damaged with sloughing of tissue, although this type of damage is unusual. Symptoms may mimic peptic ulcer with abdominal pain, which may be localized in the right upper quadrant. Radiographic findings may mimic Crohn's disease of the proximal small intestine. In immunocompetent patients there is a leukocytosis with a peripheral eosinophilia of 50 to 75%, while in chronic cases the eosinophilia may be much lower. Some of these chronic infections have lasted for over 30 years as a result of the autoinfective capability of the larvae. One case of chronic strongyloidiasis persisted for approximately 65 years.

Hyperinfection Syndrome. Autoinfection is probably the mechanism responsible for long-term infections that persist for years after the person has left the area of endemic infection. The parasite and host reach a status quo so that neither suffers any serious damage. If for any reason this equilibrium is disturbed and the individual becomes immunosuppressed, then the infection proliferates, with large numbers of larvae being produced and found in every tissue of the body. Several conditions, including the use of immunosuppressive therapy, predispose an individual to the hyperinfection syndrome. In addition to the actual tissue damage from the migrating larvae, the patient may die from sepsis, primarily as a result of the intestinal flora. Other causes of death include peritonitis, brain damage, and respiratory failure.

Reactive Arthritis. Few cases of reactive arthritis have been reported, and the number of cases may be underestimated. A case of reactive arthritis combined with uveitis associated with a long-standing, heavy *Strongyloides* infection has been found in a 32-year-old patient infected with human T-cell leukemia virus. Treatment with thiabendazole and ivermectin resulted in rapid improvement.

Debilitated or compromised patients should always be suspected of having strongyloidiasis, particularly if there are unexplained bouts of diarrhea and abdominal pain, repeated episodes of sepsis and/or meningitis with intestinal bacteria, or unexplained eosinophilia. However, a recent comparative study of the occurrence of *Strongyloides* in 554 AIDS and 142 non-AIDS patients demonstrated a similar prevalence of infection in the two groups, indicating that there are no significant statistical differences.

Key Points for Laboratory Identification

1. The rhabditiform (noninfective) larvae are normally recovered from the stool concentrate. However, it is important to remember that filariform (infective) larvae can also be recovered in the stool.

2. If the stool specimens are negative, examination of duodenal contents is recommended (duodenal aspirates, Entero-Test capsule).

3. Various concentrates (Baermann) and cultures (Harada-Mori, petri dish, or agar plate) can also be used for larval recovery.

4. Eggs are rarely seen in the stool but may be recovered from duodenal contents.

5. In very heavy infections, eggs (less common), larvae (both types), and adult worms may be recovered in the stool.

6. Remember that if agar plates are used, they must be dried at room temperature for 4 to 5 days to eliminate free water on the agar prior to use. The agar plate culture method is thought to be the most sensitive diagnostic procedure available (4, 7).

Thiabendazole has been used in the past; in some cases, repeated or extended (1 week) therapy may be necessary, and the cure rates range from 55 to 100%. Patients with the hyperinfection syndrome should be hospitalized during therapy for proper monitoring. Ivermectin has become more widely used in the last few years.

Contact with contaminated infective soil, feces, or surface water should be avoided. Individuals found to have the infection should be treated. All patients who are going to receive immunosuppressive drugs should be screened for strongyloidiasis before therapy.

Strongyloidiasis is endemic at least as far north as New York City. It is very likely that endemic foci also exist in other metropolitan areas within the United States, and cases similar to the one presented here could occur anywhere within the country. This was a case of disseminated strongyloidiasis presenting as acute abdominal distress in an adult who had been receiving immunosuppressive therapy.

Case 8

A 34-year-old man from Mexico went to the clinic because he found some white, rectangular objects in his stool. These objects were about an inch long and appeared to be "moving" very slowly. He had always been healthy and had come to the United States 2 years earlier. He denied experiencing gastrointestinal or other symptoms. He had lived in a rural area of Mexico, and he reported no unusual dietary habits or significant contact with animals. During the last few days, he complained of anorexia, some nausea, and abdominal cramps; however, he had no diarrhea. Laboratory findings were normal; however, the routine O&P examination revealed the following images (Figure 36.12) (3, 4, 7, 8).

Questions

1. What do you think the structures are as seen in Figure 36.12?

2. Are there any key morphologic characteristics that could help you identify this parasite?

3. What laboratory tests might be appropriate at this point?

4. Are there any standard precautions that should be

Figure 36.12 Case 8. (Upper) Structure obtained from stool specimen; (lower) objects seen in the concentration sediment wet mount (using high dry 40× objective).

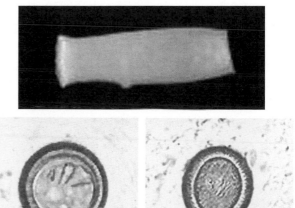

taken? If so, what are they and why might they be important? If not, why not?

Correct Identification

The helminth eggs are those of *Taenia* spp. From the egg morphology, it is impossible to identify the cestode to the species level as *T. saginata* (beef tapeworm) or *T. solium* (pork tapeworm). This structure recovered from the patient's stool specimen is a *Taenia* sp. proglottid. To more fully visualize the internal structure of the proglottid, an India ink injection procedure could be performed. Results of this procedure can be seen in Figure 36.13.

The life cycle of *T. saginata* is very similar to that of *T. solium*. Infection with the adult worm is initiated by the ingestion of raw or poorly cooked beef containing encysted *T. saginata* larvae. As with *T. solium*, the larva is digested out of the meat in the stomach and the tapeworm evaginates in the upper small intestine and attaches to the intestinal mucosa, where the adult worm matures within 5 to 12 weeks. The adult worm can reach a length of 25 m but often measures only about half this length. Although a single worm is usually found, multiple worms can be present (personal observation).

The scolex is "unarmed" and has four suckers with no hooks. The proglottids usually number 1,000 to 2,000, with the mature proglottids being broader than long and the gravid proglottids being narrower and longer. Identification to the species level is usually based on the number of main lateral uterine branches, which are counted on one side of the gravid proglottid. There are between 15 and 20 branches, with an average of 18.

Gravid proglottids often crawl from the anus during the day, when the host is most active. These proglottids may crawl under the stool when a specimen is submitted unpreserved in the stool collection carton. The eggs cannot be distinguished from those of *T. solium*. They are round to slightly oval, measure 31 to 43 μm, have a thick, striated shell, and contain the six-hooked embryo (oncosphere).

Figure 36.13 Case 8. *Taenia saginata* gravid proglottid; note the large number of lateral uterine branches when counted on one side.

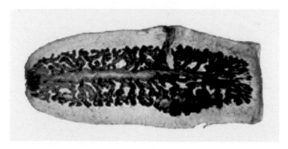

The eggs can remain viable in the soil for days to weeks. On ingestion by cattle, the oncospheres hatch in the duodenum, penetrate the intestinal wall, and are carried via the lymphatics or bloodstream, where they are filtered out in the striated muscle. They then develop into the bladder worm or cysticercus within approximately 70 days. The mature cysticercus measures 7.5 to 10 mm wide by 4 to 6 mm long and contains the immature scolex, which has no hooks (unarmed). Other animals found to harbor cysticerci include buffalo, giraffe, llama, and possibly reindeer. Actual cases of human cysticercosis with *T. saginata* are rare in the literature, and there is speculation that some reported cases have been inaccurately diagnosed.

There are usually few symptoms associated with the presence of the adult worm in the intestine. Although rare symptoms (obstruction, diarrhea, hunger pains, weight loss, and appendicitis) have been reported, the most common complaint is the discomfort and embarrassment caused by the proglottids crawling from the anus. This occurrence may be the first clue that the patient has a tapeworm infection. Occasionally, the proglottids are also seen on the surface of the stool after it is passed.

Key Points for Laboratory Identification

1. Preliminary examination of the gravid proglottid may not allow identification without clearing or injection of the uterine branches with India ink. Since there is always the possible danger that the proglottid is *T. solium*, with the inherent problem of egg ingestion and cysticercosis, all specimens should be handled with extreme caution.

2. The eggs look identical to those of *T. solium*, so that identification to the species level cannot be performed with eggs alone.

3. Although gravid proglottids of *T. saginata* are often somewhat larger than those of *T. solium*, this difference may be very minimal or impossible to detect.

4. Any patient with *Taenia* eggs recovered in the stool should be cautioned to use good hygiene. These patients should be treated as soon as possible to avoid the potential danger of accidental infection with the eggs, which may lead to cysticercosis (*T. solium*).

For therapy, the use of praziquantel or niclosamide has been recommended. The same general approach as used for *T. solium* is used to eradicate the adult worm. Therapy is usually very effective; however, if the proglottids begin to reappear in the stool or crawl from the anus, retreatment is necessary.

Case 9

A 45-year-old engineer presented to his physician with jaundice. He reported that a few days earlier he had developed a fever and also had experienced diarrhea, nausea, and vomiting. He also indicated that he had had a cough about a week earlier. A few days later, he experienced right upper quadrant pain. On the day he presented to the physician, he had noticed that his eyes were somewhat yellow. The patient works for an American-based engineering firm that has multiple contracts throughout South America; their main responsibilities include drilling for water and construction of irrigation canals in recently cleared land areas of several South American countries, including Bolivia, Argentina, Chile, and Peru. Periodically he spends time in these areas, usually eating whatever is available. Often the engineering job is adjacent to sheep- and/or cattle-raising regions.

On presentation, the patient appeared in no particular distress; however, he was jaundiced. The lungs and heart were normal. There was guarding and tenderness in the right upper quadrant. The liver was enlarged and tended down to 3 cm below the costal margin, with a span of 15 cm. There was no peripheral edema. The white blood cell count was 18,500, with 35 polymorphonuclear leukocytes, 10 lymphocytes, 3 monocytes, and 52 eosinophils. Stool examinations for parasites revealed the following image (Figure 36.14). These objects measured about 140 by 70 μm and were easily seen in the saline wet preparation examination.

Questions

1. Based on the size of the objects in Figure 36.14, what are the possibilities regarding identification?
2. What key morphologic features are helpful in making the diagnosis?
3. Are these key characteristics visible? If yes, what are they? If not, why not?

Figure 36.14   Case 9. Objects seen in the wet mount microscopic examination of concentration sediment. These objects measured about 140 by 70 μm and were easily seen in the saline wet preparation examination.

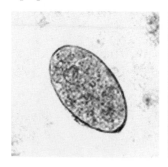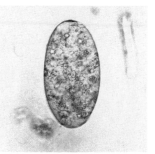

4. Are there any additional laboratory tests that might be helpful?

Correct Identification

The object in Figure 35.14 is a *Fasciola hepatica* egg; these eggs generally measure 130–150 by 63–90 μm. They are unembryonated, operculated, large, ovoid, and usually brownish yellow (3, 4, 8). However, based on morphology alone, the egg could also be *Fasciolopsis buski* or *Fasciola gigantica*.

The patient's symptoms reflect the phase of the infection, as well as the number of parasites present in the host. In the acute phase, symptoms may be present over a period of weeks to months. Metacercarial larvae do not produce significant pathologic damage until they begin to migrate through the liver parenchyma. The amount of damage depends on the worm burden of the host. Linear lesions of 1 cm or greater can be found. Hyperplasia of the bile ducts occurs, possibly as a result of toxins produced by the larvae. Symptoms associated with this migratory phase can include fever, epigastric and right upper quadrant pain, and urticaria. Asymptomatic infections appear to be more common in Peru.

In the more chronic phases of the disease, the patient generally has few to no symptoms once the flukes have lodged in the biliary passages. However, there may be some epigastric and right upper quadrant pain, diarrhea, nausea, vomiting, hepatomegaly, and jaundice. If the flukes are found in the extrahepatic biliary ducts, symptoms may mimic those seen in cholelithiasis. In the chronic phase, there tend to be some liver function abnormalities, as well as eosinophilia.

Once the worms have established themselves in the bile ducts and matured, they cause considerable damage from mechanical obstruction and metabolic by-products. The degree of pathologic change depends on the number of flukes penetrating the liver. The infection produces hyperplasia of the biliary epithelium and fibrosis of the ducts, with portal or total biliary obstruction. The gallbladder undergoes similar pathologic changes and may even harbor adult worms.

Symptoms and signs of infection during the late stages after egg production has begun are those associated with biliary obstruction and cholangitis. Acute epigastric pain, fever, pruritus, jaundice, hepatomegaly, and eosinophilia are common.

Adult worms, which may live for 9 years in the bile ducts, produce eggs that are carried by the bile fluid into the intestinal lumen and passed into the environment with the feces. The eggs are unembryonated, operculated, large, ovoid, and brownish yellow. The miracidium develops within 1 to 2 weeks and escapes from the egg to infect the snail intermediate host, *Lymnaea* sp. Cercariae are liberated from the snail after the production of a sporocyst generation and two or three rediae generations. Cercariae encyst on water vegetation, e.g., watercress. Humans are infected by ingestion of uncooked aquatic vegetation on which metacercariae are encysted. Metacercariae excyst in the duodenum and migrate through the intestinal wall into the peritoneal cavity. The larvae enter the liver by penetrating the capsule (Glisson's capsule) and wander through the liver parenchyma for up to 9 weeks. They finally enter the bile ducts, where they mature and produce eggs, which are passed out in feces. The adult worms can attain a length of >1 in. (1 in. = 2.54 cm) and a width of about 0.5 in.

Fascioliasis is a cosmopolitan disease found where there is close association of livestock, humans, and snails. The largest numbers of infections have been found in Bolivia, Ecuador, Egypt, France, Iran, Peru, and Portugal. Although animal fascioliasis is endemic throughout the Americas, human infections are rare in most countries. Reservoir hosts include herbivores such as cattle, goats, and sheep. Animal fascioliasis is a major veterinary problem in Europe. Watercress is the main source of human infection, and in many countries wild watercress is touted as a healthy natural food. In many areas of the world, animal manure is used as the primary fertilizer for cultivation of watercress.

Prevention may be accomplished by public health education in areas where infections are endemic, stressing the dangers of eating watercress grown in the wild where animals and snails are abundant. Other measures have included mollusciciding, treatment of infected animals, and draining of pasture lands.

Key Points for Laboratory Identification

1. Multiple stool specimens may be required to find the trematode eggs. In a very light infection, eggs may not be recovered; however, the patient might be asymptomatic.

2. The sedimentation concentration method should be used; because the eggs are operculated, they do not float in the zinc sulfate flotation concentration method.

3. When adding iodine to the wet preparation, do not add too much or the eggs will stain very darkly and resemble debris.

4. The wet preparation can be examined using the 10× (low-power) objective; these eggs are large enough that they can usually be seen using this magnification.

5. Eggs of *F. hepatica* can resemble those *Fasciola gigantica* (liver fluke) and *Fasciolopsis buski* (intestinal fluke).

6. Do not make the wet preparation too thick; if it is too thick, the eggs can be obscured by normal stool debris.

7. If available, antigen or antibody detection may be helpful in confirming the infection.

Although treatment with praziquantel at a dose of 25 mg/kg taken after each meal for 2 days is sometimes effective, it does not appear to be effective in treating cases of infection in Egypt. Treatment with 30 to 50 mg of biothionol/kg on alternate days for 10 to 15 doses is recommended. Triclabendazole at 10 mg/kg as a single dose is also recommended but may not be as easily obtained. The drug is given orally in single or multiple doses and has few side effects. The drug acts by inhibiting protein synthesis in *F. hepatica* and will probably become the drug of choice.

Case 10

A 14-month-old boy was admitted to the hospital with complaints of irritability, ataxia, and weakness that had developed over a period of several weeks. Just prior to admission, he was unable to sit up or walk. A complete blood count showed 35% eosinophilia; lumbar puncture revealed an elevated eosinophilia. His neurologic status did not improve. Several weeks after admission, his magnetic resonance imaging scan revealed severe cortical atrophic changes and severe diffuse white-matter degeneration. He remained in a vegetative state and died several months after his initial symptoms. He was not seropositive for *Toxocara canis*, and he had not lived in or traveled to areas where other causes of eosinophilic meningoencephalitis would be suspected. At autopsy, the following image was seen from brain tissue (Figure 36.15) (6).

Figure 36.15 Case 10. Image from autopsy (brain tissue).

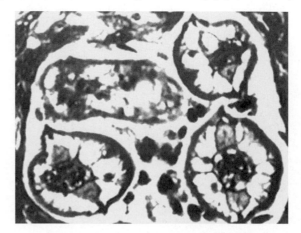

Questions

1. What are some of the other causes of eosinophilic meningoencephalitis?
2. What questions would you ask regarding possible patient history details (animal contact, play practices, etc.)?
3. Does the patient's age suggest any possible diagnostic options?
4. What preventive measures could be taken to avoid this particular infection?

Correct Identification

The image seen in Figure 36.15 reveals a histologic section of larvae consistent with the raccoon roundworm, *Baylisascaris procyonis*. On further questioning, it became clear that the child had been exposed to feral raccoons and soil contaminated with raccoon feces in his backyard. Raccoons were seen in the surrounding area, and typical *B. procyonis* eggs were found in the soil. Adult worms were also found in the raccoons (Figure 36.16); the adult worms look very much like *Ascaris lumbricoides*.

The diagnosis of *B. procyonis* encephalitis was confirmed. This patient presented with the typical clinical syndrome, including peripheral and CSF eosinophilia. On further testing, the patient demonstrated positive antibody results in indirect-immunofluorescence testing of both serum and CSF to *B. procyonis*. Both the intensity and titers increased with time. Other infections causing eosinophilic meningoencephalitis include *Toxocara* spp. larval migrans.

Key Points for Laboratory Diagnosis

1. Definitive diagnosis requires identification of the larvae in tissues; however, this can be difficult depending on the body site. Cross sections of larvae tend to measure 60 to 70 µm, and the larvae have prominent, single lateral alae and paired, conical excretory columns (smaller than central intestine).
2. Ocular examinations may reveal retinal lesions, larval tracks, or migrating larvae; these findings may provide the tentative diagnosis involving a helminth infection.
3. Using indirect fluorescent-antibody testing, enzyme-linked immunosorbent assay, and Western blotting, anti-*Baylisascaris* antibodies can be detected in CSF. Currently, the only source for serologic testing is the Department of Veterinary Pathobiology at Purdue University, West Lafayette, Ind. [K. Kazacos; phone (765) 494-7558]. Acute- and convalescent-phase titers demonstrate several-fold increases in both serum and CSF antibody levels; there is no cross-reactivity with *Toxocara*. With the availability of a reliable serologic test, there is less need to perform a brain biopsy.
4. *In most cases, the clinical history provides the main clues.*

Feral raccoons were found in the area around the house; both dirt and plant debris were contaminated with viable *B. procyonis* eggs. Ingestion of infective *B. procyonis* eggs probably occurred through ingestion of contaminated soil or other hand-to-mouth transfer from areas or articles contaminated with raccoon feces. This roundworm is a common parasite of raccoons, especially in the Midwest and the Northeast and on the west coast of the United States. Infection rates can range as high as 82% in raccoons. In many metropolitan areas, the raccoon population is quite high, and people tend to lure the animals close to homes through intentional feeding or through dog food left in the surrounding areas. Thus, contamination of the surrounding environment with raccoon feces is common. The infected animals shed an average of >25,000 eggs/g of feces; millions of eggs can be shed each day. These infective eggs are very environmentally resistant and can survive for years in the soil.

Raccoons tend to defecate in the same areas, which are called latrines (often in barns, decks, patios, etc.). These sites remain sources of eggs that are infective for both humans and other animals for long periods. Young children with a history of ingesting soil and/or feces are at greatest risk for infection; most confirmed cases have been in this age group. Individuals with developmental impairment tend to be at special risk due to a greater tendency to exhibit behaviors that lead to egg ingestion.

Unfortunately, neural larval migrans caused by *B. procyonis* carries a poor prognosis. However, high-dose albendazole given early in the infection may prevent or halt the progression of CNS disease. Treatment before larvae

Figure 36.16 Case 10. *Baylisascaris procyonis*. (Left) Typical eggs isolated from raccoons; (right) adult worms isolated from raccoons.

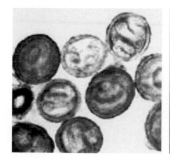

reach the CNS is beneficial. Because they diminish inflammatory reactions, steroids may be beneficial unless the CNS damage is too advanced. Although ivermectin has been used, this drug does not cross the blood-brain barrier.

In addition to direct damage caused by the migrating larvae, it is likely that eosinophil-derived neurotoxin contributes to the manifestations of encephalitis. Two features distinguish this infection from other nematode infections that cause larval migrans: the somatic migration and invasion of the CNS and the continued growth of the larvae to a large size within the CNS. Despite therapy, to date there are no documented cases of neurologically intact survivors.

Although reports of cases of eosinophilic meningoencephalitis caused by *B. procyonis* infections are rare, large populations of raccoons with high rates of *B. procyonis* infections live in close proximity to humans in many areas. Widespread contamination of the domestic environment suggests that the risk of exposure and human infection may be substantial. Prevention of infection is the most important public health measure; educating the public is essential for disease prevention.

There is no effective cure for *B. procyonis* infection; treatment is symptomatic and involves systemic corticosteroids and anthelmintic agents. Unfortunately, neural larva migrans is usually not responsive to anthelmintic therapy; by the time the diagnosis is made, extensive damage has already taken place. Drugs that can be tried include albendazole, mebendazole, thiabendazole, levamisole, diethylcarbamazine, and ivermectin. Experimental data from animal studies indicate that albendazole and diethylcarbamazine may have the best CSF penetration and larvicidal activity. Unfortunately, since the diagnosis does not take place until symptoms have begun, larval invasion of the CNS has already taken place. Therefore, treatment is usually started too late in the course of the infection.

Considering the potential outcomes associated with delay of treatment until symptoms begin, prophylactic treatment for asymptomatic children exposed to raccoons or contaminated environments is now being considered. Since the anthelmintics do not appear to be problematic, prophylaxis seems appropriate for specific individuals. Albendazole has been used in these cases; treatment has been discontinued when environmental testing has proven to be negative.

Blood Parasite Infections

Case 11

The patient was a 52-year-old man from California who had just returned from a trip to Africa. About a week after he returned, he began complaining of fever, headache, and general malaise. The first blood specimens examined using automated instrumentation revealed nothing significant. He continued to have high fevers and severe headaches, and eventually several additional blood specimens were submitted to the microbiology laboratory for examination as thick and thin blood films (4, 5, 7–9, 14). The following images were seen (Figure 36.17).

Questions

1. What parasitic infections should be considered in the diagnosis?
2. Do the images in Figure 36.17 provide a diagnosis?
3. Why do you think automation failed to indicate anything significant?
4. Does your presumptive diagnosis match the patient's history? Why or why not?

Correct Identification

At first glance, the fact that the man had just returned from a trip to Africa might suggest a parasitic infection as a possibility. The patient's symptoms were caused by a mixed infection with *Plasmodium falciparum* and *Plasmodium vivax* that he had acquired in Africa. Although he had

Figure 36.17 Case 11. Blood films stained with Giemsa stain. (Upper left) Thick blood film; (upper right) thin blood film; (lower left) thick blood film; (lower right) thin blood film.

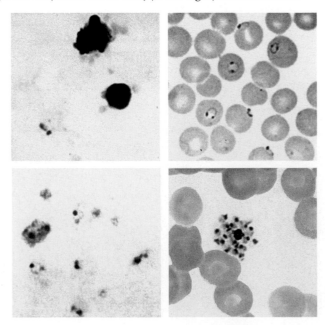

taken prophylaxis for malaria, he had not taken the medication on a regular basis and did not continue to take the medication after returning to the United States. Because he had never had any contact with malaria before (i.e., he was immunologically naive), he began to have symptoms with a very low parasitemia (>0.01%) that was undetectable by automated instrumentation. By the time his symptoms became more severe (about 2 weeks), he had numerous parasites in the peripheral blood. By the time the infection was diagnosed, the man was severely ill. Organisms seen on the blood films were

1. *Plasmodium falciparum* ring form (thick blood film)
2. *Plasmodium falciparum* ring forms (thin blood film) (note the typical "headphone" appearance of some of the ring forms)
3. *Plasmodium vivax* (thick blood film) (note the developing rings and one developing schizont)
4. *Plasmodium vivax* (thin blood film) (note the mature schizont with about 18 merozoites)

Comments on P. falciparum. *P. falciparum* tends to invade all ages of RBCs, and the proportion of infected cells may exceed 50%. Schizogony occurs in the internal organs (spleen, liver, bone marrow, etc.) rather than in the circulating blood. Ischemia caused by the plugging of vessels within these organs by masses of parasitized RBCs produces various symptoms, depending on the organ involved. It has been suggested that a decrease in the ability of the RBCs to change shape when passing through capillaries or the splenic filter may lead to the plugging of the vessels.

Onset of a *P. falciparum* malaria attack occurs 8 to 12 days after infection and is preceded by 3 to 4 days of vague symptoms such as aches, pains, headache, fatigue, anorexia, or nausea. The onset is characterized by fever, a more severe headache, and nausea and vomiting, with occasional severe epigastric pain. There may be only a feeling of chilliness at the onset of fever. The periodicity of the cycle is not established during the early stages, and the presumptive diagnosis may be totally unrelated to a possible malaria infection. If the fever does develop a synchronous cycle, it is usually a cycle of somewhat less than 48 h. True relapses from the liver do not occur, and recrudescences are rare after a year.

Severe or fatal complications of *P. falciparum* malaria can occur at any time during the infection and are related to the plugging of vessels in the internal organs; the symptoms depend on the organ(s) involved. The severity of the complications in a malaria infection may not correlate with the parasitemia seen in the peripheral blood, particularly in *P. falciparum* infections.

Disseminated intravascular coagulation is a rare complication of malaria; it is associated with high parasite burden, pulmonary edema, rapidly developing anemia, and cerebral and renal complications. Vascular endothelial damage from endotoxins and bound parasitized blood cells may lead to clot formation in small vessels. Cerebral malaria is most often seen in *P. falciparum* malaria, although it can occur in the other types as well. If the onset is gradual, the patient may become disoriented or violent or may develop severe headaches and pass into a coma. Some patients, even those who exhibit no prior symptoms, may suddenly become comatose. Physical signs of CNS involvement are quite variable, and there is no real correlation between the severity of the symptoms and the peripheral-blood parasitemia.

Although malaria is no longer endemic within the United States, this infection is considered to be life-threatening, and laboratory requests for blood smear examination and organism identification should be treated as STAT requests. Malaria is usually associated with patients having a history of travel within an area where malaria is endemic, although other routes of infection are well documented. During 2002 (the most recent data available), >45% of the malaria cases reported in the United States were caused by *P. falciparum*, the most pathogenic of the four species infecting humans.

Comments on P. vivax. The primary clinical attack usually occurs 7 to 10 days after infection, although there are strain differences, and a much longer incubation period is possible. In some patients, symptoms such as headache, photophobia, muscle aches, anorexia, nausea, and sometimes vomiting occur before organisms can be detected in the bloodstream. In other patients, the parasites can be found in the bloodstream several days before symptoms appear.

During the first few days, the patient may not exhibit a typical paroxysm pattern but, rather, may have a steady low-grade fever or an irregular remittent fever pattern. Once the typical paroxysms begin, after an irregular periodicity, a regular 48-h cycle is established. An untreated primary attack may last from 3 weeks to 2 months or longer. The paroxysms become less severe and more irregular in frequency and then stop altogether. In 50% of patients, relapses occur after weeks, months, or up to 5 years (or more).

Severe complications are rare in *P. vivax* infections, although coma and sudden death or other symptoms of cerebral involvement have been reported. These patients can exhibit cerebral malaria, renal failure, circulatory collapse, severe anemia, hemoglobinuria, abnormal bleeding, acute respiratory distress syndrome, and jaundice. Studies have confirmed that these were not mixed infections with

P. falciparum but single-species infections with *P. vivax*. However, of the two infections, *P. falciparum* causes the most severe complications.

Since *P. vivax* infects only the reticulocytes, the parasitemia is usually limited to around 2 to 5% of the available RBCs. Splenomegaly occurs during the first few weeks of infection, and the spleen progresses from being soft and palpable to hard, with continued enlargement during a chronic infection. If the infection is treated during the early phases, the spleen will return to its normal size.

Leukopenia is usually present; however, leukocytosis may be present during the febrile episodes. Concentrations of total proteins in plasma are unchanged, although the albumin level may be low and the globulin fraction may be elevated. The increase in the concentration of gamma globulins is caused by the development of antibodies. The level of potassium in serum may also be increased as a result of RBC lysis.

Diagnosis. Frequently, for a number of reasons, organism recovery and subsequent identification are more difficult than the textbooks imply. It is very important that this fact be recognized, particularly when one is dealing with a possibly fatal infection with *P. falciparum*. When requests for malarial smears are received in the laboratory, some patient history information should be made available to the laboratorian. This information should include the following:

1. Where has the patient been, and what was the date of return to the United States? ("Where do you live?")
2. Has malaria ever been diagnosed in the patient before? If so, which species was identified?
3. What medication (prophylaxis or otherwise) has the patient received, and how often? When was the last dose taken?
4. Has the patient ever received a blood transfusion? Is there a possibility of other needle transmission (drug user)?
5. When was the blood specimen drawn, and was the patient symptomatic at the time? Is there any evidence of a fever periodicity? Answers to such questions may help eliminate the possibility of infection with *P. falciparum*, usually the only species that can rapidly lead to death.

Often the diagnosis of malaria is considered, and a single blood specimen is submitted to the laboratory for examination; however, single films or specimens cannot be relied on to exclude the diagnosis, especially when incomplete prophylactic medication or therapy is used.

Incomplete use of antimalarial agents may be responsible for reducing the numbers of organisms in the peripheral blood, thus leading to a blood smear that contains few organisms, which then reflects a low parasitemia when in fact serious disease is present. Patients with a relapse case or an early primary case may also have few organisms in the blood smear.

It is recommended that both thick and thin blood films be prepared on admission of the patient, and at least 200 to 300 oil immersion fields should be examined on the thin film before a negative report is issued. Since one set of negative films does not rule out malaria, additional blood specimens should be examined over a 36-h time frame. Although Giemsa stain is recommended for all parasitic blood work, the organisms can also be seen with other blood stains such as Wright's stain or the rapid stains. Blood collected with EDTA anticoagulant is acceptable; however, if the blood remains in the tube for any length of time, true stippling may not be visible within the infected RBCs (*P. vivax*, as an example). Also, when using anticoagulants, it is important to remember that use of the proper ratio between blood and anticoagulant is necessary for good organism morphology. If the blood stands for >2 h prior to blood film preparation, organism distortion is very likely, with the morphology beginning to mimic that seen with *P. malariae*. If the blood stands for >6 h prior to blood film preparation, the organisms will disintegrate.

Malaria is one of the few parasitic infections considered to be immediately life-threatening, and a diagnosis of *P. falciparum* malaria should be considered a medical emergency because the disease can be rapidly fatal. Any laboratory providing the expertise to identify malarial parasites should do so on a 24-h basis, 7 days/week.

Characteristics of *P. falciparum* (malignant tertian malaria) include the following:

1. 36- to 48-h cycle
2. Tends to infect any cell regardless of age; thus, very heavy infection may result
3. All sizes of RBCs
4. No Schüffner's dots (Maurer's dots: may be larger, single dots, bluish)
5. Multiple rings per cell (only young rings, gametocytes, and occasional mature schizonts are seen in peripheral blood)
6. Delicate rings; may have two dots of chromatin per ring, appliqué or accolé forms
7. Crescent-shaped gametocytes

Note Without the appliqué form, Schüffner's dots, multiple rings per cell, and other developing stages, differentiation among the species can be very difficult. It is obvious that the early rings of all four species can mimic one another very easily. Remember that one set of negative blood films cannot rule out a malaria infection, and this information must be conveyed to the physician.

Characteristics of *P. vivax* (benign tertian malaria) include the following:

1. 48-h cycle
2. Tends to infect young cells
3. Enlarged RBCs
4. Schüffner's dots after 8 to 10 h
5. Delicate ring
6. Very ameboid trophozoite
7. Mature schizont contains 12 to 24 merozoites

Reporting Parasitemia.   It is important to report the level of parasitemia when blood films are reviewed and found to be positive for malaria parasites. Because of the potential for drug resistance in some of the *Plasmodium* species, particularly *P. falciparum*, it is important that every positive smear be assessed and the parasitemia reported using the exact same method on follow-up specimens as on the initial specimen. This allows the parasitemia to be monitored after therapy has been initiated. In cases where the patient is hospitalized, monitoring should be performed at 24, 48, and 72 h after initiation of therapy. Generally, the parasitemia will drop very quickly within the first 24 h; however, in cases of drug resistance, the level may appear to drop and then begin to rise again or the level may not decrease but actually increase over time.

Case 12

A 37-year-old private contractor returning from Iraq presented to his physician with a lesion on his arm. About 6 weeks earlier, he had noticed a small, red area on his arm. There was some itching, and over the next few weeks, the area ulcerated and measured about an inch in diameter. There was no drainage. He indicated that he was feeling well and had not had any fever, chills, anorexia, or weight loss. His medical history was unremarkable. The ulcer had a moist base and raised borders as seen in Figure 36.18 (1, 2, 4, 7, 9).

Questions

1. Based on the patient's history, what presumptive diagnosis might be appropriate?
2. What other diagnostic tests might be suggested at this point?
3. Based on additional diagnostic testing, what would you expect to find in order to support the presumptive diagnosis?
4. Should any special precautions be taken when

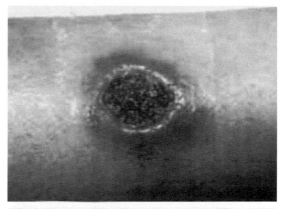

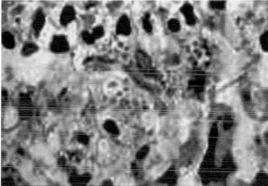

Figure 36.18   Case 12. (Upper) Lesion on patient's arm; (lower) skin biopsy specimen from margin of lesion.

processing and examining the clinical material? Why or why not?

Correct Identification

The lesion in Figure 36.18 could have been caused by a number of agents; however, some options became much more likely after the patient's travel and work history were reviewed. The patient had traveled and worked in Iraq. He did not remember being bitten at any specific

Figure 36.19   Case 12. Cutaneous leishmaniasis. (Left) Skin macrophage containing Leishman-Donovan bodies (note the nucleus and bar-shaped kinetoplast); (right) sand fly vector of leishmaniasis.

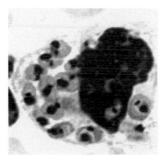

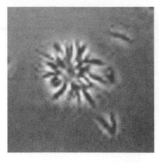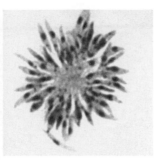

Figure 36.20 Case 12. (Left) Wet mount of organisms from culture; (right) Giemsa-stained smear of culture sediment.

time, but he did indicate that there were lots of flies (sand flies) present. The following images were obtained from a skin biopsy specimen that was examined in pathology, and the organisms were isolated in culture (Figures 36.19 and 36.20). This patient had cutaneous leishmaniasis.

It is important to remember that with cutaneous and/or mucocutaneous lesions, the only area where the parasites are generally found is the lesion itself. Unless the infection is the visceral type (*Leishmania donovani*), no organisms are found in the spleen, bone marrow, and/or liver. There may be exceptions in cutaneous or mucocutaneous types of disease in which the amastigotes are found in other parts of the reticuloendothelial system; however, this tends to be unusual.

Leishmania major, *L. tropica*, *L. aethiopica*, and, rarely, *L. infantum* cause cutaneous disease in the Old World; disease manifestations include nodular and ulcerative skin lesions. Local geographic names for cutaneous leishmaniasis include Oriental sore, Baghdad boil, Delhi boil, Biskra button, and Aleppo evil.

The lesions of Old World cutaneous leishmaniasis are very similar to those seen with New World cutaneous disease. Weeks to months after infection, an erythematous, often pruritic papule develops at the bite site. This papule may become scaly and enlarged, developing a central ulcer surrounded by a raised margin. Disease progression at this point will vary, depending on the species involved. Lesions may be single or multiple and usually occur on exposed areas of the body such as the face, hands, feet, arms, and legs. In general, all lesions on a patient have a similar appearance and progress at the same speed.

The original lesion may remain as a flattened plaque or may progress, with the surface becoming covered with fine, papery scales. These scales are dry at first but later become moist and adherent, covering a shallow ulcer. As the ulcer enlarges, it produces an exudate and may develop a crust. The edge of the lesion is usually raised.

Depending on the species, satellite lesions are common and merge with the original lesion.

In areas of the world where physicians are very familiar with leishmaniasis, the diagnosis may be made on clinical grounds. However, in other areas of the world, where the disease is rare, the condition may not be recognized as leishmaniasis. Definitive diagnosis depends on demonstrating the amastigotes in tissue specimens or the promastigotes in culture. Newer results suggest that PCR is a valuable tool for the diagnosis of leishmaniasis on a routine basis and can provide valuable epidemiologic information in areas of endemic infection. Cutaneous leishmaniasis may have to be differentiated from a number of other lesions and diseases, including basal cell carcinoma, tuberculosis, various mycoses, cheloid, and lepromatous leprosy.

The ability to detect parasites in aspirates, scrapings, or biopsy specimens depends on the number of amastigotes present, the level of the host immune response, the absence or presence of bacterial and/or fungal contamination within the lesion, and whether the specimen is collected from an active or healing lesion. If the patient has multiple lesions, specimens should be collected from the more recent or active lesions. These lesions should be thoroughly cleaned with 70% alcohol, and necrotic debris should be removed to prevent the risk of bacterial and/or fungal contamination of the specimen. Also, the specimen should be taken from the advancing margin of the lesion; the central portion of the ulcer contains nothing but necrotic debris.

Appropriate specimens would include scrapings of the lesions, as well as a collection of several punch biopsy specimens, taken from the most active lesion areas. However, fine-needle aspiration cytology has also been recommended. Biopsy specimens can be divided and used for cultures and touch preparations; some material should always be saved and submitted for routine histologic examination. The biopsy specimen can be used to make impression smears (touch preparations); these smears should be prepared after portions of the specimen have been placed in culture media by using sterile techniques. After cleaning off any excess blood, a horizontal cut is made through the core of the specimen and the cut surface is gently touched to glass slides; multiple slides should be prepared. Once material has been set up for culture and touch preparations have been made, the remainder of the tissue can be sent to the Pathology Department for routine processing.

Culture media that have been used for the recovery and growth of the leishmaniae include Novy-MacNeal-Nicolle medium and Schneider's Drosophila medium supplemented with 30% fetal bovine serum. Patient cultures should not be set unless the laboratory maintains specific organism strains for quality control checks. Both the control and patient cultures should be examined twice weekly for the first 2 weeks and once a week thereafter for up to 4 weeks before they are called negative. Promastigote stages can be detected microscopically in wet mounts taken from centrifuged culture fluid. This material can also be stained with Giemsa stain to facilitate observation at a higher magnification.

A sensitive microcapillary culture method (MCM) was developed for the rapid diagnosis of cutaneous leishmaniasis (1). The MCM is superior to the traditional culture method (TCM) as determined by the smaller inoculum size, the higher sensitivity for detection of promastigotes, and the shorter time to the emergence of promastigotes. With lesion amastigote loads from grade III to 0, the positive rates and the periods for promastigote emergence were three- to fourfold higher and faster with the MCM than with the TCM, e.g., 83 to 97% positive in 4 to 7 days versus 20 to 40% positive in 15 to 30 days ($p = 0.0001$). This MCM has the advantage of simplicity and may be suitable for diagnostic use and for parasite retrieval in many other areas of endemic infection where parasites are known to be difficult to grow.

Key Points for Laboratory Identification

1. For a complete laboratory diagnosis, microscopic examination of Giemsa-stained touch preparations and cultures is recommended.
2. Aspirates or biopsy specimens should be taken from the margin or base of the most active lesion (papule or ulcer). The lesion should be cleaned before the sample is collected, to reduce the chances of contamination with fungi or bacteria.
3. Multiple slides should be prepared for examination.
4. Tissue imprints or smears should be stained with Giemsa stain (or one of the stains commonly used for blood smears) (see chapter 31).
5. Amastigote stages should be found within macrophages or close to disrupted cells.
6. If cultures are to be performed, specimens must be taken aseptically and control organism cultures should be set up at the same time (see chapter 32).
7. Cultures should be checked weekly for 4 weeks before they are declared negative.

Most Old World cutaneous leishmaniasis lesions are self-healing over a period of a few months to several years. This healing process confers immunity to individuals living within the immediate area of endemic infection; therefore, treatment is not recommended unless disfiguring scarring is a possibility. Local care of the lesion and treatment of secondary bacterial infection are essential for healing. Antileishmanial therapy is indicated in immunocompromised hosts and patients with progressive, multiple, or critically located lesions. Pentavalent antimony compounds remain the main therapeutic option for all species. They are given intravenously, intramuscularly, or intralesionally. Cryotherapy and some systemic antifungal agents have been used successfully. Oral azoles are promising new treatments for lesions caused by *L. major*.

Case 13

A 61-year-old man presented to his physician with a history of symptoms, beginning about 3 weeks after he returned from a trip. Symptoms included general malaise followed by fever, chills, headaches, and back pain. The man lived in southern Texas and had returned from a photographic safari to Africa, including Kenya, Tanzania, Uganda, and South Africa. He had seen a physician before returning to the United States and had received some antimalarial medication. He was an active wildlife photographer and was often exposed to a number of life zones and resident plants, animals, insects, etc., both at home and on trips outside of the United States.

Although both the patient and several others in the group remembered being bitten by insects, he was the only one who was symptomatic. On examination, he was found to have several swollen areas, including a lymph node on his neck and a draining lesion on his upper arm.

Blood was drawn for routine hematology procedures, and a specimen was also drawn to be sent to the microbiology laboratory for examination as blood films stained with Giemsa stain (Figure 36.21) (4, 7, 9, 12). Based on his travel history, additional laboratory work revealed a slightly elevated white blood cell count with a relative lymphocytosis. His gamma globulin level was elevated, with a marked elevation of the immunoglobulin M (IgM) level.

Figure 36.21 Case 13. Objects seen in Giemsa-stained thin blood films. (Left) Patient smear.

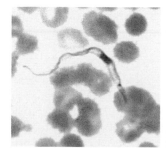

Questions

1. In Figure 36.21, what are the major differences between the two parasites seen?
2. Is there anything in the patient's history that would help you with the diagnosis? What and why?
3. How do you explain the marked elevation in the IgM level?
4. Is there anything in the life cycle that would explain the laboratory findings?

Correct Identification

The patient was suffering from African trypanosomiasis, probably caused by *Trypanosoma brucei rhodesiense* (East African trypanosomiasis) based on travel history (Kenya and Uganda), the severity of symptoms, and the overall description of the case. Although the man lived in southern Texas, where *Trypanosoma cruzi* (the cause of Chagas' disease, or American trypanosomiasis) is found, the African trypomastigotes and the American trypomastigotes look quite different.

The African trypomastigotes (both East and West African trypomastigotes) (Figure 36.21, left) have a very small kinetoplast (the dot at one end of the organism), while *T. cruzi* tends to have a very large kinetoplast that appears to go beyond the body of the organism (Figure 36.21, right). Although both organisms can actually be found in a C shape, this finding alone does not indicate infection with *T. cruzi*.

The course of infection in African trypanosomiasis occurs in three stages. Soon after inoculation of the metacyclic trypanosomes by an infected tsetse fly, an inflammatory lesion, called the trypanosomal chancre, develops at the site of inoculation. Trypomastigotes then invade the local lymphatics and later the bloodstream. After a time, the organisms invade the choroid plexus and enter the brain and CSF. There is quite a bit of variation between East African and West African trypanosomiasis. In the East African disease (caused by *T. brucei rhodesiense*), chancres are common and the hemolymphatic stage is severe, rapidly progressing to a fatal meningoencephalitis, often within months of infection. With the West African form of the disease (caused by *T. brucei gambiense*), chancres are uncommon, the hemolymphatic stage may even be inapparent, and meningoencephalitis progresses very slowly, often over several years: hence the "sleeping sickness" syndrome.

During the course of the infection, the number of trypomastigotes in the blood fluctuates. This relapsing parasitemia is due to the host immune response to the parasites. Each decline in parasite number results from the antibody-mediated destruction of trypomastigotes bearing a particular variant surface glycoprotein (VSG). Each new wave of parasitemia represents the growth of a trypomastigote population expressing an antigenically different VSG. This process of antigenic variation is a feature of all African trypomastigotes. Since each trypomastigote expresses only one VSG at a time and has as many as 1,000 different VSG genes, the number of different variable antigen types that can be expressed during an infection is quite large. The continued response of the host to the new/different variable antigen types leads to elevated gamma globulin levels, particularly the IgM levels, including in the CSF. Generally, if an individual does not have an elevated IgM level, he almost certainly does not have African trypanosomiasis.

Key Points for Laboratory Identification

1. Trypomastigotes are highly infectious, and health care workers must use blood-borne pathogen precautions.
2. Trypomastigotes may be detected in aspirates of the chancre and enlarged lymph nodes in addition to blood and CSF.
3. Concentration techniques, such as centrifugation of CSF and blood, should be used in addition to thin and thick smears (Giemsa, Wright's, or rapid stains).
4. Trypomastigotes are present in largest numbers in the blood during febrile periods.
5. Multiple daily blood examinations may be necessary to detect the parasite.
6. Blood and CSF specimens should be examined during therapy and 1 to 2 months posttherapy.

All drugs currently used in the therapy of African trypanosomiasis are toxic and require prolonged administration. Treatment should be started as soon as possible and is based on the patient's symptoms and laboratory findings. The choice of antiparasitic drug depends on whether the CNS is infected. Specific details can be found in chapters 9 and 25.

Case 14

The patient was a 9-month-old male infant living in the United States, with multiple medical problems, who was admitted to the hospital with the complaint of irregular recurrent fever of 3 months' duration. He had received several blood transfusions over a period of several months, primarily for low hemoglobin levels. He became feverish with a temperature of 102 to 104°F, which did not respond to antimicrobials. His fever appeared to peak about every 4 days. Just prior to admission, the fever

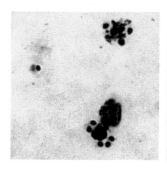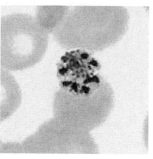

Figure 36.22 Case 14. Blood films stained with Giemsa stain. (Left) Thick blood film; (right) thin blood film.

began to occur daily. On examination of both thick and thin blood films prepared from blood collected in EDTA tube (purple top), the following images were seen (Figure 36.22). Both the thick and thin blood films were stained using Giemsa stain (4, 7, 9, 14).

Questions

1. Based on the patient's history, what parasitic infections might be suspected? Why?
2. Considering the organism morphology, is there enough detail to allow identification to the genus and species levels?
3. What differences might be seen if the blood films had been stained with Wright-Giemsa or rapid stains? Would this approach have been acceptable? Why or why not?
4. Would the request for blood films be considered a STAT request? Why or why not?

Figure 36.23 Case 14. Thin blood films stained with Giemsa stain. *Plasmodium malariae* "band" forms are evident where the developing trophozoites spread across the RBC. Note that the infected RBCs are normal to smaller than normal size compared with the uninfected RBCs.

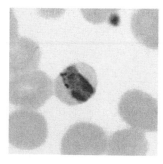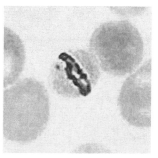

Correct Identification

The images (Figure 36.22) obtained after the first blood specimen was drawn represent *Plasmodium malariae* developing and mature schizonts in a thick blood film (left) and a mature schizont (containing merozoites) in the thin blood film (right). Note the small size of the infected RBC (typical of *P. malariae*). The number of merozoites in the mature schizont is always a clue to the species identification; in this case there appear to be about eight merozoites.

The images obtained after the second blood draw (Figure 36.23) represent developing trophozoites, which are spread across the infected RBC as "band" forms (again *P. malariae*). This configuration is not always seen but is very suggestive of *P. malariae*.

Key Characteristics of *P. malariae*

1. There is a 72-h cycle.
2. The parasite tends to infect old cells.
3. It infects small to normal-sized RBCs.
4. No Schüffner's dots are present; small dots (Ziemann's dots) may be present, but they are not considered true stippling.
5. There is a thick ring and a heavy chromatin dot.
6. "Band" form trophozoites are often present.
7. The mature schizont contains about eight merozoites, often arranged around the remaining clump of malarial pigment. This configuration of the merozoites is often called the "daisy" form.

Since this patient had received several blood transfusions, it is logical to assume that one of the transfusions was the source of the infection. Individuals can harbor *P. malariae* in the blood for many years at very low levels of parasitemia; these individuals are asymptomatic. Often, even with a complete blood donor history, the possibility of malaria infection with this organism is not suspected. Individuals have been known to harbor this parasite for up to 42 years. In spite of the patient's severe splenomegaly and splenectomy, he was considered cured after therapy with chloroquine and remained asymptomatic.

It is very important to realize that a single set of thick and thin blood films can be negative, even though the patient may be infected. In this case, both thick and thin blood films were positive. A second draw was taken to examine the thick and thin blood films for additional stages and/or evidence of a mixed infection. Venipunctures were performed for both blood draws, with the recommended EDTA anticoagulant in the lavender (purple)-top tube. It is important that the slides be prepared as quickly as possible after the blood draw to prevent the organism distortion and possible loss that can occur if the blood is allowed to stand for several hours prior to slide preparation. Remember, every request for malaria blood films should always be considered a STAT request and the laboratory coverage should be 24 h/day, 7 days/week.

Examination of the thin blood film is relatively simple when the parasitemia is high, as in this slide. However, a returning traveler with his or her first malaria infection may experience the typical clinical symptoms of high fever, chills, myalgia, and headache with a much lower parasitemia. Also, these patients may present to the emergency room with vague symptoms that do not represent the typical textbook description; they may have malaise, a steady low-grade fever, and even diarrhea. These low levels of parasitemia are often impossible to detect using thin blood film examination only. For this reason, the key to successful detection of malaria parasites in the peripheral blood is the examination of both thick and thin blood films from every patient suspected of having malaria (or any patient from whom blood is submitted to the laboratory for blood film examination).

Thick films allow a larger amount of blood to be examined, which increases the possibility of detecting light infections. Species identification from the thick-film examination, particularly in the case of malaria, may be difficult for workers with little experience examining thick blood films. The morphologic characteristics of blood parasites are best seen in thin films, particularly the relationship between the sizes of infected and uninfected RBCs. However, in cases with a low parasitemia, the identification to the species level may have to be accomplished by thick-film examination.

The accurate examination of thick and thin blood films and identification of parasites depends on the use of absolutely clean, grease-free slides for preparation of all blood films. Old (unscratched) slides should be cleaned first in detergent and then with 70% ethyl alcohol; new slides should also be cleaned with alcohol before use.

Blood films are usually prepared when the patient is admitted; in instances in which malaria is a possible diagnosis, samples should be taken at intervals of 6 to 8 h for at least three successive days after the first set of negative smears, particularly if *P. falciparum* has not been excluded as a diagnosis. Extensive therapy information is available in chapter 25.

Diagnostic Methods

Case 15

The patient was a 54-year-old man from the United States, who had received a liver transplant 14 months earlier. He had traveled throughout the world as a

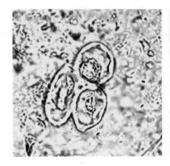

Figure 36.24 Case 15. Objects seen in a wet mount (left) and a permanent stained smear (right). The structure on the right measures approximately 16 µm, and the structures on the left measure approximately 65 µm.

professional consultant (Europe, Asia, Central and South America, Australia, New Zealand, etc.). He was diagnosed as having diarrhea, cough, and general malaise. He had a history of vague health problems since the transplant but did not seek medical attention for several weeks. The first stool specimen was submitted to the laboratory. After examination of the concentration sediment and permanent stained smears from all three specimens, the following objects were seen (Figure 36.24) (3, 4, 7, 10, 11, 13).

The structure on the right measures approximately 16 µm, and the structures on the left measure approximately 65 µm. The structure on the right is stained with the standard Wheatley's modification of the Gomori trichrome, while the structure on the left is a wet preparation of the sedimentation concentration.

After submission of the other two stool specimens, it was also decided to perform special stains for the coccidia and microsporidia. The following things were seen on the stained smears. Structures on the left measure ca. 4 to 6 µm, while those on the right measure ca. 1.5 µm (Figure 36.25).

Figure 36.25 Case 15. (Left) Objects in a concentrated fecal sediment stained with modified acid-fast stain. (Right) Objects in a concentrated fecal sediment stained with modified trichrome stain (note the diagonal or horizontal lines in some of the objects).

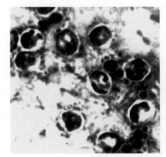
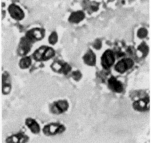

Questions

1. Based on Figure 36.24, what do you think the objects are?
2. Considering the patient's history, what types of parasites would you suspect? Why?
3. What additional laboratory tests might have been performed prior to the routine O&P examination? Do you think other laboratory tests would have been more appropriate? Why or why not?
4. Why should special stains or other testing options be considered for a patient with this type of history? What would you be looking for?

Correct Identification

The images presented after the first stool specimen represent artifact material (Figure 36.24). The structure on the right mimics an *Entamoeba coli* cyst; however, this structure is not an organism but an artifact. The structure on the left tends to resemble helminth eggs, but they are vegetable cells and not parasites. Unfortunately, even when structures are within the correct size range, the morphology may not be correct or accurate enough to identify the structure as a true parasite.

The two images obtained when special stains were performed provide excellent examples of *Cryptosporidium* spp. oocysts (Figure 36.25, left) and microsporidial spores (Figure 36.25, right), probably *Enterocytozoon bieneusi* or *Encephalitozoon intestinalis*.

Because this patient is a posttransplant (liver) patient, who has been and continues to be immunosuppressed, continuous diarrhea with negative O&P examinations should suggest the use of special stains for the coccidia (modified acid-fast stains) or microsporidia (modified trichrome stains). These diagnostic options are important, particularly if the patient is immunosuppressed and continues to be symptomatic with diarrhea.

Although there are immunoassays available for the diagnosis of cryptosporidiosis (enzyme immunoassay, fluorescent-antibody assay, cartridge formats), many laboratories continue to use the modified acid-fast stain. It is important to remember that the destaining (acid) step should be performed quickly and the acid content should not be too high. If the percentage is above 3% acid, too much color may be removed from the oocysts. Color loss can be a definite problem for the identification of *Cyclospora cayetanensis* (modified acid-fast variable) but is less of a problem with *Cryptosporidium* spp. Many laboratories are using this stain with a 1% acid destain step. Both *Cryptosporidium* and *Cyclospora* look fine when this approach is used. However, regardless of the destain step, some of the *C. cayetanensis* oocysts may appear pale

pink or even colorless. It is also recommended that these special stains (for both coccidia and microsporidia) be performed on concentration sediment to enhance sensitivity (centrifuge for 10 min at $500 \times g$).

Currently, there are no commercially available immunoassays for detection of the microsporidian spores in clinical specimens. There are a number of different modified trichrome stain methods, some hot and some cold; they all work well, so selection of a particular method often depends on personal preferences among laboratory personnel. Again, it is important to perform these stains on concentrated fecal sediment. Remember that other clinical specimens can also be stained (sputum, urine, etc.) and may contain less debris than that seen in a fecal specimen.

References

1. **Allahverdiyev, A. M., S. Uzun, M. Bagirova, M. Durdu, and H. R. Memisoglu.** 2004. A sensitive new microculture method for diagnosis of cutaneous leishmaniasis. *Am. J. Trop. Med. Hyg.* **70:**294–297.
2. **Blum, J., P. Desjeux, E. Schwartz, B. Beck, and C. Hatz.** 2004. Treatment of cutaneous leishmaniasis among travellers. *J. Antimicrob. Chemother.* **53:**158–66.
3. **Clinical Laboratory Standards Institute.** 2005. *Procedures for the Recovery and Identification of Parasites from the Intestinal Tract.* Approved guideline M28-A2. Clinical Laboratory Standards Institute, Villanova, Pa.
4. **Garcia, L. S.** 1999. *Practical Guide to Diagnostic Medical Parasitology.* ASM Press, Washington, D.C.
5. **Garcia, L. S., R. Y. Shimizu, and D. A. Bruckner.** 1986. Blood parasites: problems in diagnosis using automated differential instrumentation. *Diagn. Microbiol. Infect. Dis.* **4:**173–176.
6. **Gavin, P. J., K. R. Kazacos, and S. T. Shulman.** 2005. Baylisascariasis. *Clin. Microbiol. Rev.* **18:**703–718.
7. **Isenberg, H. D. (ed.).** 2004. *Clinical Microbiology Procedures Handbook,* 2nd ed., vol. 1, 2, and 3. ASM Press, Washington, D.C.
8. **Isenberg, H. D. (ed.).** 1995. *Essential Procedures for Clinical Microbiology.* ASM Press, Washington, D.C.
9. **National Committee for Clinical Laboratory Standards.** 2000. *Laboratory Diagnosis of Blood-Borne Parasitic Diseases.* Approved guideline M15-A. National Committee for Clinical Laboratory Standards, Wayne, Pa.
10. **Ryan, N. J., G. Sutherland, K. Coughlan, M. Globan, J. Doultree, J. Marshall, R. W. Baird, J. Pedersen, and B. Dwyer.** 1993. A new trichrome-blue stain for detection of microsporidial species in urine, stool, and nasopharyngeal specimens. *J. Clin. Microbiol.* **31:**3264–3269.
11. **Visvesvara, G. S., H. Moura, E. Kovacs-Nace, S. Wallace, and M. L. Eberhard.** 1997. Uniform staining of *Cyclospora* oocysts in fecal smears by a modified safranin technique with microwave heating. *J. Clin. Microbiol.* **35:**730–733.
12. **Warren, K. S., and A. A. F. Mahmoud.** 1990. *Tropical and Geographical Medicine,* 2nd ed. McGraw-Hill Inc., New York, N.Y.
13. **Weber, R., R. T. Bryan, R. L. Owen, C. M. Wilcox, L. Gorelkin, G. S. Visvesvara, and The Enteric Opportunistic**

Infections Working Group. 1992. Improved light-microscopical detection of microsporidia spores in stool and duodenal aspirates. *N. Engl. J. Med.* **326:**161–166.

14. **Wilcox, A.** 1960. *Manual for the Microscopical Diagnosis of Malaria in Man.* U.S. Department of Health, Education, and Welfare, Washington, D.C.

Appendixes

Although common names are frequently used to describe parasitic organisms, these names may represent different parasites in different parts of the world. To eliminate these problems, a binomial system of nomenclature is used in which the scientific name consists of the genus and species. On the basis of life histories and morphology, systems of classification have been developed to indicate the relationship among the various parasite species.

Parasites of humans are classified in five major subdivisions: Protozoa (amebae, flagellates, ciliates, sporozoans, and coccidia); Platyhelminthes, or flatworms (cestodes and trematodes); Acanthocephala, or thorny-headed worms; Nematoda, or roundworms; and Arthropoda (insects, spiders, mites, ticks, etc.). Information on Protozoa, Nematoda, Platyhelminthes, and Arthropoda is presented in appendix 1, Table A1.1. This classification scheme is designed to provide some order and meaning to a widely divergent group of organisms. No attempt has been made to include every possible organism, only those that are considered to be clinically relevant in the context of human parasitology. This information provides some insight into the parasite groupings, thus leading to a better understanding of organism morphology, parasitic infections, and the appropriate clinical diagnostic approach.

Data on the distribution of various parasites are presented in Tables A1.2 through A1.7; these tables are from E. K. Markell, M. Voge, and D. T. John, *Medical Parasitology*, 6th ed., The W. B. Saunders Co., Philadelphia, 1986. Diagnostic clinical review information and parasite prevalence estimations are presented in Tables A1.8 through A1.21.

Flowcharts for various diagnostic kits and staining tables are given in appendix 2. Common problems in parasite identification are discussed in appendix 3. Appendix 4 contains examples of quality control recording sheets. Information on sources of diagnostic products and teaching materials is provided in appendix 5. General references, a list of relevant journals and periodicals, and abstracts and bibliographic sources are provided in appendix 6. Appendix 7 contains a number of color plates, including relevant human parasites and arthropods.

Appendix 8 is one of the most important sections in this edition. This section contains information that has been published just months prior to final publication of the book. This "late-breaking" synopsis of very recent publications will assist readers in obtaining the latest information available. I encourage all readers to review this section as they read the various chapters throughout the book.

APPENDIX 1
Information Tables

Table A1.1 Classification of human parasites

I. Protozoa
  1. **Intestinal amebae**
     Entamoeba histolytica
     Entamoeba dispar[a]
     Entamoeba hartmanni
     Entamoeba coli
     Entamoeba polecki
     Endolimax nana
     Iodamoeba bütschlii
     Blastocystis hominis
  2. **Intestinal flagellates**
     Giardia lamblia[b]
     Chilomastix mesnili
     Dientamoeba fragilis
     Pentatrichomonas hominis
     Enteromonas hominis
     Retortamonas intestinalis
  3. **Intestinal ciliates**
     Balantidium coli
  4. **Intestinal coccidia and microsporidia**
     Cryptosporidium parvum
     Cryptosporidium hominis
     Cryptosporidium spp.
     Cyclospora cayetanensis
     Isospora belli
     Sarcocystis hominis
     Sarcocystis suihominis
     Sarcocystis "lindemanni"[c]
     Microsporidia
       Enterocytozoon bieneusi
       Encephalitozoon (Septata) intestinalis

  5. **Sporozoa and flagellates from blood and tissue**
     Sporozoa (causing malaria and babesiosis)
       Plasmodium vivax
       Plasmodium ovale
       Plasmodium malariae
       Plasmodium falciparum
       Babesia spp.
     Flagellates (leishmaniae and trypanosomes)
     Old World *Leishmania* species
       Leishmania (Leishmania) tropica
       Leishmania (Leishmania) major
       Leishmania (Leishmania) aethiopica
       Leishmania (Leishmania) donovani
       Leishmania (Leishmania) archibaldi
       Leishmania (Leishmania) infantum
     New World *Leishmania* species
       Leishmania (Leishmania) mexicana
       Leishmania (Leishmania) amazonensis
       Leishmania (Leishmania) pifanoi
       Leishmania (Leishmania) garnhami
       Leishmania (Leishmania) venezuelensis
       Leishmania (Leishmania) chagasi
       Leishmania (Viannia) braziliensis
       Leishmania (Viannia) colombiensis
       Leishmania (Viannia) guyanensis
       Leishmania (Viannia) lainsoni
       Leishmania (Viannia) naiffi
       Leishmania (Viannia) panamensis
       Leishmania (Viannia) peruviana
       Leishmania (Viannia) shawi
     Old World *Trypanosoma* species
       Trypanosoma brucei gambiense

(continued)

Table A1.1 *(continued)*

Trypanosoma brucei rhodesiense
New World *Trypanosoma* species
Trypanosoma cruzi
Trypanosoma rangeli

6. **Amebae and flagellates from other body sites**
   Amebae
   Naegleria fowleri
   Sappinia diploidea
   Acanthamoeba spp.
   Hartmanella spp.
   Balamuthia mandrillaris
   Entamoeba gingivalis
   Flagellates
   Trichomonas vaginalis
   Trichomonas tenax

7. **Coccidia, sporozoa, and microsporidia from other body sites**
   Coccidia
   Toxoplasma gondii
   Sporozoa
   Pneumocystis jiroveci[d]
   Microsporidia
   Nosema ocularum
   Pleistophora spp.
   Pleistophora ronneafiei
   Trachipleistophora hominis
   Trachipleistophora anthropophthera
   Brachiola vesicularum
   Brachiola (Nosema) algerae
   Brachiola (Nosema) connori
   Encephalitozoon cuniculi
   Encephalitozoon (Septata) intestinalis
   Encephalitozoon hellem
   Vittaforma corneae (Nosema corneum)
   "*Microsporidium*"[e]

II. **Nematodes (roundworms)**
   1. **Intestinal**
      Ascaris lumbricoides
      Enterobius vermicularis
      Ancylostoma duodenale
      Necator americanus
      Strongyloides stercoralis
      Strongyloides fuelleborni
      Trichostrongylus colubriformis
      Trichostrongylus orientalis
      Trichostrongylus spp.
      Trichuris trichiura
      Capillaria philippinensis
   2. **Tissue**
      Trichinella spiralis

Trichinella spp. (*T. britovi, T. murrelli, T. nativa, T. nelsoni, T. papuae, T. pseudospiralis, T. zimbabwensis*)
Toxocara canis and *Toxocara cati* (visceral and ocular larva migrans)
Ancylostoma braziliense and *Ancylostoma caninum* (cutaneous larva migrans)
Dracunculus medinensis
Angiostrongylus cantonensis
Angiostrongylus costaricensis
Gnathostoma spinigerum
Anisakis spp. (larvae from saltwater fish)
Phocanema spp. (larvae from saltwater fish)
Contracaecum spp. (larvae from saltwater fish)
Hysterothylacium
Porrocaecum spp.
Capillaria hepatica
Thelazia spp.
Ternidens diminutus

3. **Blood and tissues (filarial worms)**
   Wuchereria bancrofti
   Brugia malayi
   Brugia timori
   Loa loa
   Onchocerca volvulus
   Mansonella ozzardi
   Mansonella streptocerca
   Mansonella perstans
   Dirofilaria immitis (usually found in lung lesions; in dogs, heartworm)
   Dirofilaria spp. (may be found in subcutaneous nodules)

III. **Cestodes (tapeworms)**
   1. **Intestinal**
      Diphyllobothrium latum
      Diplogonoporus spp.
      Dipylidium caninum
      Hymenolepis nana
      Hymenolepis diminuta
      Taenia solium
      Taenia saginata
   2. **Tissue (larval forms)**
      Taenia solium
      Echinococcus granulosus
      Echinococcus multilocularis
      Echinococcus vogeli
      Echinococcus oligarthrus
      Multiceps multiceps
      Spirometra mansonoides
      Diphyllobothrium spp.

(continued)

Table A1.1   Classification of human parasites *(continued)*

IV. Trematodes (flukes)

1. **Intestinal**

   Fasciolopsis buski

   Echinostoma ilocanum

   Echinochasmus perfoliatus

   Heterophyes heterophyes

   Metagonimus yokogawai

   Gastrodiscoides hominis

   Phaneropsolus bonnei

   Prosthodendrium molenkempi

   Spelotrema brevicaeca

   Plagiorchis spp.

   Neodiplostomum seoulense

2. **Liver and lung**

   Clonorchis (Opisthorchis) sinensis

   Opisthorchis viverrini

   Opisthorchis felineus

   Dicrocoelium dendriticum

   Fasciola hepatica

   Fasciola gigantica

   Paragonimus westermani

   Paragonimus kellicotti

   Paragonimus africanus

   Paragonimus uterobilateralis

   Paragonimus miyazakii

   Paragonimus mexicanus

   Paragonimus caliensis

3. **Blood**

   Schistosoma mansoni

   Schistosoma haematobium

   Schistosoma japonicum

   Schistosoma intercalatum

   Schistosoma mekongi

   Schistosoma malayi

   Schistosoma mattheei

V. **Arthropods**

1. **Arachnida**

   Scorpions

   Spiders (black widow, brown recluse)

   Ticks (*Dermacentor, Ixodes, Argas,* and *Ornithodoros* spp.)

   Mites (*Sarcoptes* spp.)

2. **Crustacea**

   Copepods (*Cyclops* spp.)

   Crayfish, lobsters, and crabs

3. **Pentastomida**

   Tongue worms

4. **Diplopoda**

   Millipedes

5. **Chilopoda**

   Centipedes

6. **Insecta**

   Anoplura (sucking lice [*Pediculus* and *Phthirus* spp.])

   Blattaria (cockroaches)

   Hemiptera (true bugs [*Triatoma* spp.])

   Coleoptera (beetles)

   Hymenoptera (bees, wasps, etc.)

   Lepidoptera (butterflies, caterpillars, moths, etc.)

   Diptera (flies, mosquitoes, gnats, and midges [*Phlebotomus, Aedes, Anopheles, Glossina, Simulium* spp., etc.])

   Siphonaptera (fleas [*Pulex* and *Xenopsylla* spp., etc.])

---

[a] The name *Entamoeba histolytica* is used to designate the true pathogen, while the name *E. dispar* is now being used to designate the nonpathogen. However, unless trophozoites containing ingested red blood cells (*E. histolytica*) are seen, the two organisms cannot be differentiated on the basis of morphology seen in permanent stained smears of fecal specimens. Fecal immunoassays are available for identification of the *Entamoeba histolytica/E. dispar* group and for differentiating between the two species.

[b] Although some individuals have changed the species designation for the genus *Giardia* to *G. intestinalis* or *G. duodenalis*, there is no general agreement. Therefore, for this listing, the name *Giardia lamblia* is retained.

[c] When humans are intermediate hosts, preventive measures involve careful disposal of animal feces that may contain the infective sporocysts. This may be impossible in the wilderness areas, where wild animals may serve as reservoir hosts for many of the different types of organisms that have been grouped under the term *Sarcocystis* "*lindemanni.*" However, this name is no longer used.

[d] *Pneumocystis jiroveci* has now been reclassified with the fungi.

[e] This designation is not a true genus, but a "catch-all" for organisms that have not been (or may never be) identified to the genus and/or species levels (*Microsporidium ceylonensis, Microsporidium africanum*). In the future, *Microsporidium* may be accepted as an official genus within the microsporidia.

Table A1.2   Distribution of selected parasitic infections in the Americas

Category	Distribution[a]				
	North America	Mexico and Central America	Caribbean	Tropical South America	Temperate South America
Protozoa					
Leishmaniasis					
Cutaneous	X•	X	O	X	O
Mucocutaneous	O	X	O	X	X
Visceral	O	X	O	X	X
Malaria	X•	X	Hispaniola	X	X
Trypanosomiasis (*Trypanosoma cruzi*)	X•	X	O	X	X
Trematodes					
Schistosomiasis (*Schistosoma mansoni*)	O	O	X	X	X
Paragonimiasis	X•	X	O	X	O
Cestodes					
Diphyllobothriasis	X	O	O	O	X
Taeniasis (*Taenia solium*)	X•	X	O	X	O
Hydatid disease	X	X	O	X	X
Nematodes					
Filariasis					
Wuchereria bancrofti	O	X	X	X	O
Onchocerca volvulus	O	X	O	X	O

[a] X, present; O, absent; X•, sporadic or rare occurrence.

Table A1.3   Distribution of selected parasitic infections in Europe

Category	Distribution[a]				
	Northwest	Southwest	Central	Eastern	Mediterranean littoral
Protozoa					
Leishmaniasis					
Cutaneous	O	O	O	X	X
Visceral	O	O	O	O	X
Trematodes					
Schistosomiasis (*Schistosoma haematobium*)	O	Portugal	O	O	O
Clonorchiasis, opisthorchiasis	O	O	X	X	O
Cestodes					
Diphyllobothriasis	X	O	X	O	O
Hydatid disease	X	X	X	X	X

[a] X, present; O, absent.

Table A1.4   Distribution of selected parasitic infections in Africa

Category	Distribution[a]				
	North	**West**	**Central**	**East**	**South**
Protozoa					
Leishmaniasis					
Cutaneous	X	X	X	X	X
Visceral	X	X	X	X	X
Malaria	X	X	X	X	X
Trypanosomiasis					
Gambian (west)	O	X	X	O	O
Rhodesian (east)	O	O	O	X	X
Trematodes					
Schistosomiasis					
Schistosoma haematobium	X	X	X	X	X
Schistosoma mansoni	X	X	X	X	X
Paragonimiasis	X	X	X	O	O
Cestodes					
Hydatid disease	X	X	O	X	O
Nematodes					
Filariasis					
Wuchereria bancrofti	X	X	X	X	O
Loa loa	O	X	X	X	O
Onchocerca volvulus	O	X	X	X	O

[a] X, present; O, absent.

Table A1.5   Distribution of selected parasitic infections in Asia

Category	Distribution[a]			
	Southwest	Central South	Southeast	East
Protozoa				
Leishmaniasis				
Cutaneous	X	X	O	O
Visceral	O	X	O	O
Malaria	X	X	X	X
Trematodes				
Fasciolopsiasis	O	X	X	X
Clonorchiasis, opisthorchiasis	O	X	X	X
Paragonimiasis	O	X	X	X
Schistosomiasis				
Schistosoma haematobium	X	India	O	O
S. japonicum	O	O	X	X
S. mansoni	O	O	O	O
Cestodes				
Diphyllobothriasis	O	O	O	X
Hydatid disease	X	O	X	O
Nematodes				
Filariasis				
Wuchereria bancrofti	O	X	X	X
Brugia malayi	O	India	X	X

[a] X, present; O, absent.

Table A1.6   Distribution of selected parasitic infections in Oceania

Category	Distribution[a]					
	Australia	New Guinea	South Pacific	Polynesia	West Pacific	Hawaii
Protozoa						
Malaria	O	X	X	O	X	O
Trematodes						
Paragonimiasis	O	O	X	X	O	O
Nematodes						
Filariasis						
(*Wuchereria bancrofti*)	X	X	X	X	X	O
Eosinophilic meningitis	X	O	X	X	X	X

[a] X, present; O, absent.

Table A1.7 Cosmopolitan distribution of common parasitic infections (North America, Mexico, Central America, South America, Europe, Africa, Asia, and Oceania)

Category	Species
Protozoa	
Intestinal	*Blastocystis hominis*
	Cryptosporidium spp.
	Cyclospora cayetanensis
	Dientamoeba fragilis
	Entamoeba histolytica
	Entamoeba dispar[a]
	Giardia lamblia
	Isospora belli
	Microsporidia
Tissue	*Pneumocystis jiroveci*[b]
	Toxoplasma gondii
	Microsporidia
Other	*Balamuthia mandrillaris*
	Microsporidia
	Naegleria and *Acanthamoeba* spp.
	Trichomonas vaginalis
Cestodes	*Hymenolepis nana*
	Taenia saginata
Nematodes	
Intestinal	*Ascaris lumbricoides*
	Enterobius vermicularis
	Hookworm
	Strongyloides stercoralis
	Trichuris trichiura
Tissue	*Trichinella* spp.

[a] The name *Entamoeba histolytica* is used to designate the true pathogen, while the name *E. dispar* is now being used to designate the nonpathogen. However, unless trophozoites containing ingested red blood cells (*E. histolytica*) are seen, the two organisms cannot be differentiated on the basis of morphology.
[b] *Pneumocystis jiroveci* has now been reclassified with the fungi.

Table A1.8 Body sites and specimen collection[a]

Site	Specimen options	Collection method[b]
Blood	Smears of whole blood	Thick and thin films, fresh blood
	Anticoagulated blood	Anticoagulant
		EDTA (1st choice)
		Heparin (2nd choice)
Bone marrow	Aspirate	Sterile
Central nervous system	Spinal fluid	Sterile
Cutaneous ulcers	Aspirates from below surface	Sterile plus air-dried smears
	Biopsy specimen	Sterile, nonsterile to histopathology (formalin acceptable)
Eyes	Biopsy specimen	Sterile (in saline), nonsterile to histopathology
	Scrapings	Sterile (in saline)
	Contact lens	Sterile (in saline)
	Lens solution	Sterile (unopened commercial solutions not acceptable)
Intestinal tract	Fresh stool	0.5-pt (ca. 0.237-liter) waxed container
	Preserved stool[c]	5 or 10% formalin, MIF, SAF, Schaudinn's, PVA, single-vial collection systems
	Sigmoidoscopy material	Fresh, PVA, or Schaudinn's smears
	Duodenal contents	Entero-Test or aspirates
	Anal impression smear	Cellulose tape (pinworm examination)
	Adult worm or worm segments	Saline, 70% alcohol
Liver, spleen	Aspirates	Sterile, collected in 4 separate aliquots (liver)
	Biopsy specimen	Sterile, nonsterile to histopathology
Lungs	Sputum	True sputum (not saliva)
	Induced sputum	No preservative (10% formalin if time delay)
	Bronchoalveolar lavage fluid	Sterile
	Transbronchial aspirate	Air-dried smears
	Tracheobronchial aspirate	Air-dried smears
	Brush biopsy specimen	Air-dried smears
	Open-lung biopsy specimen	Air-dried smears
	Aspirate	Sterile
Muscle	Biopsy specimen	Fresh, squash preparation, nonsterile to histopathology
Skin	Scrapings	Aseptic, smear or vial
	Skin snip	No preservative
	Biopsy specimen	Sterile (in saline), nonsterile to histopathology
Urogenital system	Vaginal discharge	Saline swab, transport swab (no charcoal), culture medium, plastic-envelope culture
		Air-dried smear for FA
	Urethral discharge	Same as above
	Prostatic secretions	Same as above
	Urine	Single unpreserved specimen, 24-h unpreserved specimen, early morning

[a] Modified from H. D. Isenberg, ed., *Clinical Microbiology Procedures Handbook*, 2nd ed., ASM Press, Washington, D.C., 2004.

[b] MIF, Merthiolate-iodine-formalin; SAF, sodium acetate-acetic acid-formalin; PVA, polyvinyl alcohol; FA, fluorescent antibody.

[c] A number of new stool fixatives are now available; some use a zinc sulfate base rather than mercuric chloride. Some collection vials can be used as a single-vial system; both the concentration and permanent stained smear can be performed from the preserved stool. However, not all single-vial systems provide material that can be used for fecal immunoassay procedures.

Table A1.9 Body sites and possible parasites recovered (trophozoites, cysts, oocysts, spores, adults, larvae, eggs, amastigotes, and trypomastigotes)[a]

Site	Parasites	Site	Parasites
Blood		Intestinal tract *(cont.)*	*Enterocytozoon bieneusi*
Red cells	*Plasmodium* spp.		*Encephalitozoon* spp.
	Babesia spp.		Microsporidia
White cells	*Leishmania* spp.		*Ascaris lumbricoides*
	Toxoplasma gondii		*Enterobius vermicularis*
Whole blood or plasma	*Trypanosoma* spp.		Hookworm
	Microfilariae		*Strongyloides stercoralis*
Bone marrow	*Leishmania* spp.		*Trichuris trichiura*
	Trypanosoma cruzi		*Hymenolepis nana*
	Plasmodium spp.		*Hymenolepis diminuta*
			Taenia saginata
Central nervous system	*Taenia solium* (cysticerci)		*Taenia solium*
	Echinococcus spp.		*Diphyllobothrium latum*
	Naegleria fowleri		*Clonorchis* (*Opisthorchis*) *sinensis*
	Acanthamoeba and *Hartmanella* spp.		*Paragonimus* spp.
	Balamuthia mandrillaris		*Schistosoma* spp.
	Sappinia diploidea		*Fasciolopsis buski*
	Toxoplasma gondii		*Fasciola hepatica*
	Microsporidia		*Metagonimus yokogawai*
	Trypanosoma spp.		*Heterophyes heterophyes*
Cutaneous ulcers	*Leishmania* spp.	Liver and spleen	*Echinococcus* spp.
	Acanthamoeba spp.		*Entamoeba histolytica*
	Entamoeba histolytica		*Leishmania* spp.
Eyes	*Acanthamoeba* spp.		Microsporidia
	Toxoplasma gondii	Lungs	*Pneumocystis jiroveci*[b]
	Loa loa		*Cryptosporidium* spp.[c]
	Microsporidia		*Echinococcus* spp.
Intestinal tract	*Entamoeba histolytica*		*Paragonimus* spp.
	Entamoeba dispar	Muscles	*Taenia solium* (cysticerci)
	Entamoeba coli		*Trichinella* spp.
	Entamoeba hartmanni		*Onchocerca volvulus* (nodules)
	Endolimax nana		*Trypanosoma cruzi*
	Iodamoeba bütschlii		Microsporidia
	Blastocystis hominis	Skin	*Leishmania* spp.
	Giardia lamblia		*Onchocerca volvulus*
	Chilomastix mesnili		Microfilariae
	Dientamoeba fragilis	Urogenital system	*Trichomonas vaginalis*
	Pentatrichomonas hominis		*Schistosoma* spp.
	Balantidium coli		Microsporidia
	Cryptosporidium spp.		Microfilariae
	Cyclospora cayetanensis		
	Isospora belli		

[a] Modified from H. D. Isenberg, ed., *Clinical Microbiology Procedures Handbook*, 2nd ed., ASM Press, Washington, D.C., 2004. This table does not include every possible parasite that could be found in a particular body site. However, the most likely organisms have been listed.
[b] Now classified with the fungi.
[c] Disseminated in severely immunosuppressed individuals.

Table A1.10 Body site, specimen and procedures, recommended methods, relevant parasites, and comments

Body site	Procedures[a] and specimens	Recommended methods and relevant parasites[a]	Comments
Amniotic fluid	PCR (and/or culture): fresh material Animal inoculation (toxoplasmosis)	PCR based on the detection of highly repetitive gene sequences is the method of choice	Only applicable to confirm suspected prenatal *Toxoplasma* infections.
Blood	Microscopy[b]: thin and thick blood films; fresh blood (preferred) or EDTA-blood (fill EDTA tube completely with blood, then mix)	Giemsa stain (all blood parasites); hematoxylin-based stain (sheathed microfilariae); for malaria, thick and thin blood films are definitely recommended and should be prepared within 30–60 min of blood collection via venipuncture (other tests may be used as well); Wright-Giemsa stain or Diff-Quik (rapid stains) can also be used	Most drawings and descriptions of blood parasites are based on Giemsa-stained blood films. Although Wright's stain (or Wright-Giemsa combination stain) will work, stippling in malaria may not be visible and the organisms' colors will not match the descriptions. However, with other stains (those listed above, in addition to some of the "rapid" blood stains), the organisms should be detectable on the blood films. The use of blood collected with anticoagulant (rather than fresh) has direct relevance to the morphology of malaria organisms seen in peripheral blood films. If the blood smears are prepared after more than 1 h, stippling may not be visible, even if the correct pH buffers are used. Also, if blood is kept at room temperature (with the stopper removed), the male microgametocyte may exflagellate and fertilize the female macrogametocyte, and development continues within the tube of blood (as it would in the mosquito host). The ookinete may actually resemble a *Plasmodium falciparum* gametocyte. The microgamete may resemble spirochetes.
	Concentration methods: EDTA-blood	Buffy coat, fresh blood films for detection of moving microfilariae or trypanosomes; QBC, a screening method for blood parasites (hematocrit tube contains acridine orange), has been used for malaria, *Babesia*, trypanosomes, and microfilariae; it is usually impossible to identify malaria organisms to the species level; this requires high levels of training	
	Antigen detection: EDTA-blood for malaria, serum or plasma for circulating antigens (hemolyzed blood can interact in some tests)	Commercial test kits for malaria and some microfilariae	
	PCR: EDTA-blood, ethanol-fixed or -unfixed thin and thick blood films, coagulated blood, possible with hemolyzed or frozen blood samples	Sensitivity not higher than thick films for *Plasmodium* spp., much more sensitive for *Leishmania* (peripheral blood is used from immunodeficient patients only). Sequencing of PCR product is often used for species or genotype identification	So far no commercial tests available; high laboratory standards needed (may work with frozen, coagulated, or hemolyzed blood samples).
	Specific antibody detection: serum or plasma, anticoagulated or coagulated blood (hemolyzed blood can cause problems in some tests)	Most commonly used are EIA (many test kits commercially available), EITB (commercially available for some parasites), and IFA	Many labs are using in-house tests; only a few fully defined antigens are available; sensitivities and specificities of the tests should be documented by the lab.
Bone marrow	Biopsy specimens or aspirates		*Leishmania* amastigotes are recovered in cells of the reticuloendothelial system; if films are not prepared directly after sample collection, infected cells may disintegrate. Sensitivity of microscopy low, so it should be used only in combination with other methods.
	Microscopy: thin and thick films with aspirate collected in EDTA	Giemsa stain (all blood parasites)	
	Cultures: sterile material in EDTA or culture medium	Culture for *Leishmania* (or *Trypanosoma cruzi*)	
	PCR: aspirate in EDTA	PCR for blood parasites including *Leishmania* and *Toxoplasma* and rare other parasites	

(continued)

Table A1.10 Body site, specimen and procedures, recommended methods, relevant parasites, and comments (*continued*)

Body site	Procedures[a] and specimens	Recommended methods and relevant parasites[a]	Comments
Central nervous system	Microscopy: spinal fluid and CSF (wet examination, stained smears), brain biopsy specimen (touch or squash preparations, stained)	Stains: Giemsa (trypanosomes, *Toxoplasma*); Giemsa, trichrome, or calcofluor (amebae [*Naegleria* or *Sappinia*—PAM, *Acanthamoeba* or *Balamuthia*—GAE]); Giemsa, acid-fast, PAS, modified trichrome, silver methenamine (microsporidia) (tissue Gram stains also recommended for microsporidia in routine histologic preparations); H&E, routine histology (larval cestodes, *Taenia solium* cysticerci, *Echinococcus* spp.)	If CSF is received (with no suspect organism suggested), Giemsa would be the best choice; however, modified trichrome or calcofluor is also recommended as a second stain (amebic cysts, microsporidia). If brain biopsy material is received (particularly from an immunocompromised patient), cultivation is recommended for microsporidial isolation and PCR for identification to the species or genotype level. A small amount of the sample should always be stored frozen for PCR analyses in case the results of the other methods are inconclusive.
	Culture: sterile aspirate or biopsy material (in physiologic NaCl)	Free-living amebae (exception: *Balamuthia* does not grow in the routine agar/bacterial overlay method), microsporidia, and *Toxoplasma*	
	PCR: aspirate or biopsy material fresh, frozen, or fixed in ethanol	Protozoa and helminths, species and genotype characterization	
Cutaneous ulcers	Microscopy: aspirate, biopsy (smears, touch or squash preparations, histologic sections)	Giemsa (*Leishmania*); H&E, routine histology (*Acanthamoeba* spp., *Entamoeba histolytica*)	The most likely causative parasites would be *Leishmania*, which would stain with Giemsa. PAS could be used to differentiate *Histoplasma capsulatum* from *Leishmania* in tissue. In immunocompromised patients, skin ulcers have been documented with amebae as causative agents.
	Cultures (less common)	*Leishmania*, free-living amebae (often bacterial contaminations)	
	PCR: aspirate, biopsy material fresh, frozen, or fixed in ethanol	*Leishmania* (species identification), free-living amebae	
Eyes	Microscopy: smears, touch or squash preparations; biopsy specimens, scrapings, contact lens, sediment of lens solution	Calcofluor for cyst only (amebae [*Acanthamoeba*]); Giemsa for trophozoites and cysts (amebae); modified trichrome (preferred) or silver methenamine stain, PAS, acid-fast, (microsporidial spores), H&E, routine histology (cysticerci, *Loa loa*, *Toxoplasma*)	Some free-living amebae (most commonly *Acanthamoeba*) have been implicated as a cause of keratitis. Although calcofluor stains the cyst walls, it does not stain the trophozoites. Therefore, in suspect cases of amebic keratitis, both stains should be used. H&E (routine histology) can be used to detect and confirm cysticercosis. The adult worm of *Loa loa*, when removed from the eye, can be stained with a hematoxylin-based stain (Delafield's) or can be stained and examined by routine histology.
	Culture: fresh material (see above) in PBS supplemented with antibiotics if possible to avoid bacterial growth	Cultures: free-living amebae, *Toxoplasma*, microsporidia	
	PCR: fresh material in physiologic NaCl or PBS, ethanol or frozen	Free-living amebae, *Toxoplasma*, microsporidian species and genotype identification	Microsporidia: confirmation to the species or genotype levels can be done by PCR and sequence analyses; however, the spores could be found by routine light microscopy with modified trichrome, calcofluor, and/or tissue Gram stains.
Intestine	(i) Stool and other intestinal material		Stool fixation with formalin and formalin-containing fixatives preserves parasite morphology, allows prolonged storage (room temperature) and long transportation, and prevents hatching of *Schistosoma* eggs, but makes *Strongyloides* larval concentration difficult and impedes further PCR analyses.

	Specimen	Method	Comments
	Microscopy: stool, sigmoidoscopy material, duodenal contents (all fresh or preserved), direct wet smear, concentration methods	Concentration methods: ethyl acetate sedimentation of SAF-fixed stool samples (most protozoa); flotation or combined sedimentation flotation methods (helminth ova); agar or Baermann concentration (larvae of *Strongyloides* spp., fresh stool required)	
		Direct wet smear (direct examination of unpreserved fresh material is also used (motile protozoan trophozoites; helminth eggs, protozoan cysts may also be detected)	Taeniid eggs cannot be identified to the species level.
		Stains: trichrome or iron hematoxylin (intestinal protozoa); modified trichrome (microsporidia); modified acid-fast (*Cryptosporidium, Cyclospora, Isospora*)	Microsporidia: confirmation to the species or genotype levels requires PCR; however, modified trichrome and/or calcofluor stains can be used to confirm the presence of spores.
	Anal impression smear	Adhesive cellulose tape, no stain (*Enterobius vermicularis*)	Four to six consecutive negative tapes are required to rule out infection with pinworm (*Enterobius vermicularis*).
	Adult worms or tapeworm segments (proglottids)	Carmine stains (rarely used for adult worms or cestode segments). Proglottids can usually be identified to the genus level (*Taenia, Diphyllobothrium, Hymenolepis*) without using tissue stains	Worm segments can be stained with special stains. However, after dehydration through alcohols and xylenes (or xylene substitutes) without prior staining, the sexual organs and the branched uterine structure will be visible, allowing identification of the proglottid to the species level.
	Antigen detection (fresh or frozen material, suitability of fixation is test dependent)	Commercial immunoassays, e.g., EIA, FA, cartridge formats (*Entamoeba histolytica*, the *Entamoeba histolytica/E. dispar* group, *Giardia lamblia*, *Cryptosporidium* spp.); in-house tests for *T. solium* and *T. saginata*	Coproantigens can be detected in the prepatent period and independently from egg excretion.
	PCR: fresh, frozen, or ethanol-fixed material	No commercial tests available; primers for genus or species identification of most helminths and protozoa are published	Due to potential inhibition after DNA extraction from stool samples, concentration or isolation methods may be required prior to DNA extraction. However, new DNA isolation kits facilitate isolation of high-quality DNA from stool. Sequence analyses may be required for species or genotype identification.
	(ii) Biopsy specimens Microscopy: fixed for histology or touch or squash preparations for staining	H&E, routine histology (*Entamoeba histolytica, Cryptosporidium, Cyclospora, Isospora belli, Giardia*, microsporidia); less common findings would include *Schistosoma* spp., hookworm, or *Trichuris*	Special stains may be helpful for the identification of microsporidia: tissue Gram stains, silver stains, PAS, and Giemsa or modified acid-fast stains for the coccidia
	PCR: see above		
Liver and spleen	Biopsy specimens or aspirates		
	Microscopy: unfixed material in physiologic NaCl; fixed for histology	Examination of wet smears for *Entamoeba histolytica* (trophozoites), protoscolices of *Echinococcus* spp. or eggs of *Capillaria hepatica*; Giemsa (*Leishmania*, other protozoa and microsporidia); H&E (routine histology)	There are definite risks associated with punctures (aspirates and/or biopsy) of spleen or liver lesions (*Echinococcus*). Always keep a small amount of material frozen for PCR.
	Culture: sterile preparation of fresh material	For *Leishmania* (not common)	

(continued)

Table A1.10 Body site, specimen and procedures, recommended methods, relevant parasites, and comments *(continued)*

Body site	Procedures[a] and specimens	Recommended methods and relevant parasites[a]	Comments
	Animal inoculation: sterile preparation of fresh material	Intraperitoneal inoculation of *E. multilocularis* cyst material for viability test after long-term chemotherapy	
	PCR: fresh, frozen, or ethanol fixed	Species or genotype identification (e.g., *Echinococcus* spp.)	
Lungs	Sputum, induced sputum, nasal and sinus discharge, bronchoalveolar lavage fluid, transbronchial aspirate, tracheobronchial aspirate, brush biopsy specimen, open-lung biopsy specimen	Helminth larvae (*Ascaris, Strongyloides*), eggs (*Paragonimus, Capillaria*) or hooklets (*Echinococcus*) can be recovered in unstained respiratory specimens; stains include Giemsa for many protozoa including *Toxoplasma* tachyzoites, modified acid-fast stains (*Cryptosporidium*), and modified trichrome (microsporidia): routine histology was H&E; silver methenamine stain, PAS, acid-fast, and tissue Gram stains for helminths, protozoa, and microsporidia	Immunoassay reagents (FA) are available for the diagnosis of pulmonary cryptosporidiosis. Routine histologic procedures allow the identification of any of the helminths or helminth eggs present in the lungs. Disseminated toxoplasmosis and microsporidiosis are well documented, with organisms being found in many different respiratory specimens.
	Microscopy: unfixed material, treated for smear preparation		
	PCR: fresh, frozen, or fixed in ethanol		
Muscles	Biopsy material Microscopy: unfixed, touch and squash preparations or fixed for histology and EM	Larvae of *Trichinella* spp. can be identified unstained (species identification with single larvae by PCR); H&E, routine histology (*Trichinella* spp., cysticerci); silver methenamine stain, PAS, acid-fast, tissue Gram stains, EM (rare microsporidia)	If *Trypanosoma cruzi* is present in the striated muscle, the organisms could be identified in routine histology preparations. Modified trichrome and/or calcofluor stains can be used to confirm the presence of microsporidial spores.
	PCR: fresh, frozen, or ethanol fixed	Microsporidian identification to the species level requires subsequent sequencing	
Skin	Aspirates, skin snips, scrapings, biopsy specimens	See cutaneous ulcer (above)	Any of the potential parasites present can be identified by routine histology procedures.
	Microscopy: wet examination, stained smear (or fixed for histology or EM)	Wet preparations (microfilariae), Giemsa-stained smears or H&E, routine histology (*Onchocerca volvulus, Dipetalonema streptocerca, Dirofilaria repens*, other larvae causing cutaneous larva migrans [zoonotic *Strongyloides* spp.], hookworms], *Leishmania, Acanthamoeba* spp., *Entamoeba histolytica*, microsporidia, and arthropods (*Sarcoptes* and other mites)	
	PCR: fresh, frozen, or fixed in ethanol	Primers for most parasite species available	

Urogenital system	Vaginal discharge, saline swab, transport swab (no charcoal), air-dried smear for FA, urethral discharge, prostatic secretions, urine (single unpreserved, 24-h unpreserved, or early-morning specimens)	Giemsa, immunoassay reagents (FA) (*Trichomonas vaginalis*); Delafield's hematoxylin (microfilariae); modified trichrome (microsporidia); H&E, routine histology PAS, acid-fast, tissue Gram stains (microsporidia); direct examination of urine sediment for *Schistosoma haematobium* eggs or microfilariae	Although *T. vaginalis* is probably the most common parasite identified, there are others to consider; the most recently implicated organisms being in the microsporidian group. Microfilariae could also be recovered and stained. Fixation of urine with formalin prevents hatching of *Schistosoma* eggs. Material must be put into culture medium immediately after collection; do not cool or freeze.
	Microscopy: wet smears, smears of urine sediment, stained smears	Identification and propagation of *T. vaginalis* (commercialized plastic-envelope culture systems available); moving trophozoites can be detected using microscopy (or in Giemsa-stained smears)	
	Cultivation: vaginal or urethral discharge or swab preparations		
	Antigen detection (vaginal swab)	Commercial immunoassay (rapid format), *T. vaginalis*	
	Probe	Commercial instrumentation, *T. vaginalis*	
	PCR: fresh, frozen, or fixed in ethanol		

[a] CSF, cerebrospinal fluid; EIA, enzyme immunoassay; EITB, enzyme-linked immunoelectrotransfer blot (Western blot); EM, electron microscopy; FA, fluorescent antibody; GAE, granulomatous amebic encephalitis; GI, gastrointestinal; H&E, hematoxylin and eosin; IFA, indirect fluorescent-antibody assay; PAM, primary amebic encephalitis; PAS, periodic acid-Schiff stain; PBS, phosphate-buffered saline; QBC, quantitative buffy coat; SAF, sodium acetate-acetic acid-formalin.
[b] Many parasites or parasite stages may be detected in standard histologic sections of tissue material. However, species identification is difficult and additional examinations may be required. Usually, these techniques are not considered first-line methods. Additional methods like EM (electron microscopy) are carried out only by specialized laboratories and are not available for standard diagnostic purposes. EM examination for species identification has largely been replaced by PCR assays.

Table A1.11 Examination of tissue and body fluids[a]

Suspected causative agent	Disease(s)	Appropriate test(s)[b]	Positive result
Protozoa			
Naegleria fowleri	Primary amebic meningoencephalitis (PAM)	1. Wet examination of CSF (not in counting chamber) 2. Stained preparation of CSF sediment	Trophozoites present and identified
Acanthamoeba spp.	Amebic keratitis, chronic meningo-encephalitis (granulomatous amebic encephalitis [GAE])	1. Culture or stained smears 2. Calcofluor (cysts only) 3. Biopsy for routine histology	Trophozoites and/or cysts present and identified
Balamuthia mandrillaris	Chronic meningoencephalitis (GAE)	1. Calcofluor (cysts only) 2. Biopsy for routine histology	Trophozoites and/or cysts present and identified
Entamoeba histolytica	Amebiasis	1. Biopsy for routine histology 2. Immunoassays	Trophozoites present and identified
Giardia lamblia	Giardiasis	1. Duodenal aspirate 2. Duodenal biopsy or routine histology 3. Entero-Test capsule 4. Immunoassays	Trophozoites and/or cysts present and identified
Leishmania spp.			

Cutaneous lesions	Cutaneous leishmaniasis	1. Material from under bed of ulcer
		  A. Smear
		  B. Culture
		  C. Animal inoculation
		2. Punch biopsy
		  A. Routine histology
		  B. Squash preparation
		  C. Culture
		  D. Animal inoculation
			Amastigotes recovered in macrophages of skin or from animal inoculation; other stages recovered in culture

Mucocutaneous lesions	Mucocutaneous leishmaniasis	As for cutaneous leishmaniasis	Amastigotes recovered in macrophages of skin and mucous membranes or from inoculation of animals; other stages recovered in culture

Visceral	Visceral leishmaniasis (kala azar)	1. Buffy coat
		  A. Stain
		  B. Culture
		  C. Animal inoculation
		2. Bone marrow
		  A. Stain
		  B. Culture
		  C. Animal inoculation
		3. Liver or spleen biopsy
		  A. Routine histology
		  B. Stain
		  C. Culture
		  D. Animal inoculation
			Amastigotes recovered in cells of reticuloendothelial system; other stages recovered in culture

Organism	Disease	Specimen and procedures	Comments
Pneumocystis jiroveci[c]	Pneumocystosis	1. Open lung biopsy for histology 2. Lung needle aspirate 3. Bronchial brush 4. Transtracheal aspirate 5. Bronchoalveolar lavage 6. Induced sputum (AIDS patients) 7. Calcofluor 8. Immunoassays	Trophozoites or cysts present and identified; trophozoite- and cyst-specific stains available; immunoassay reagents available. Cysts present and identified
Toxoplasma gondii	Toxoplasmosis	1. Lymph node biopsy A. Routine histology B. Tissue culture isolation C. Animal inoculation 2. Serology	Identification of organisms plus appropriate serologic test results
Cryptosporidium spp.	Cryptosporidiosis	1. Duodenal scraping 2. Duodenal biopsy A. Stain B. Routine histology 3. Other tissue biopsy A. Routine histology B. Squash preparation 4. Sputum (modified AFB stains) 5. Immunoassays	Identification of organisms in microvillus border or in other tissues (lung and gallbladder have also been involved)
Microsporidia *Nosema* spp. *Encephalitozoon* spp. *Enterocytozoon* spp. *Pleistophora* spp. *Trachipleistophora* spp. *Brachiola* sp. *"Microsporidium"* spp. *Vittaforma corneae*	Microsporidiosis	Routine histology (fair); acid-fast, PAS, modified trichrome, Giemsa, and tissue Gram stains recommended (spores); animal inoculation not recommended (latent infections); EM or PCR may be necessary to confirm genus	These organisms (spores) have been found as insect or other animal parasites; the most common route of infection is probably ingestion. Human cases involve gastrointestinal tract, muscle, and CSF (AIDS); other body sites have also been documented; not limited to compromised patients
Helminths			
Larvae (*Ascaris lumbricoides, Strongyloides stercoralis*)	"Pneumonia"	Sputum, wet preparation	This is an incidental finding but has been reported in severe infections
Eggs (*Paragonimus* spp.)	Paragonimiasis	Sputum, wet preparation	Eggs will be coughed up and will appear as "iron filings"; eggs can also be found in stool
Hooklets (*Echinococcus* spp.)	Hydatid disease	Sputum, wet preparation	Rare finding, but hooklets can be found when the hydatid cyst is in the lungs
Onchocerca volvulus Mansonella streptocerca	Onchocerciasis	Skin	Skin snips examined in saline; microfilariae may be present
Schistosoma spp.	Schistosomiasis	1. Rectal valve biopsy 2. Bladder biopsy	Eggs present and identified

[a] Modified from H. D. Isenberg, ed., *Clinical Microbiology Procedures Handbook*, 2nd ed. ASM Press, Washington, D.C., 2004. CSF, cerebrospinal fluid; AFB, acid-fast bacillus; PAS, periodic acid-Schiff; EM, electron microscopy.
[b] PCR is now being used for the identification of a number of human parasites; however, these tests are generally limited to academic or reference centers. Routine laboratory use is not common.
[c] Now classified with the fungi.

Table A1.12 Key characteristics of protozoa of the intestinal tract and urogenital system[a]

Organism	Trophozoite or tissue stage	Cyst or other stage in specimen	Comments
Amebae			
Entamoeba histolytica (pathogenic)	Cytoplasm is clean; presence of RBCs is diagnostic, but cytoplasm may also contain some ingested bacteria; peripheral nuclear chromatin is evenly distributed with central, compact karyosome.	Mature cyst contains 4 nuclei; chromatoidal bars have smooth, rounded ends; cannot be differentiated from *E. dispar*.	Considered pathogenic; should be reported to Public Health; trophozoites can be confused with macrophages and cysts can be confused with WBCs in the stool.
Entamoeba dispar (nonpathogenic)	Morphology identical to that of *E. histolytica* (confirmed by presence of RBCs in cytoplasm). If no RBCs, molecular testing or fecal immunoassays are necessary to confirm species designation.	Mature cyst has morphology identical to that of *E. histolytica*.	Nonpathogenic; morphology resembles *E. histolytica*; these organisms should be reported as *Entamoeba histolytica/E. dispar* and reported to Public Health; immunoassay reagents are now available to identify the *Entamoeba histolytica/E. dispar* group and to differentiate pathogenic *E. histolytica* and nonpathogenic *E. dispar*; some laboratories may decide to use these reagents on a routine basis, depending on positivity rate and cost.
Entamoeba histolytica/E. dispar			Correct way to report, unless immunoassay is used to identify *E. histolytica* or trophozoites are seen with ingested RBCs (*E. histolytica*).
Entamoeba hartmanni (nonpathogenic)	Looks identical to *E. histolytica/E. dispar* but is smaller (<12 μm); RBCs are not ingested.	Mature cyst contains 4 nuclei but often has only 2; chromatoidal bars are often present and look like those of *E. histolytica/E. dispar* (size, <10 μm).	Shrinkage occurs on the permanent stain (especially in the cyst form). *E. histolytica/E. dispar* may actually be below the 12- and 10-μm cutoff limits and can be as much as 1.5 μm below the limits quoted for wet preparation measurements.
Entamoeba coli (nonpathogenic)	Cytoplasm is dirty and may contain ingested bacteria or debris; peripheral nuclear chromatin is unevenly distributed, with a large, eccentric karyosome.	Mature cyst contains 8 nuclei; more may be seen; chromatoidal bars (if present) tend to have sharp, pointed ends.	If a smear is too thick or thin and if stain is too dark or light, *E. histolytica/E. dispar* and *E. coli* can often be confused, since there is much overlap in morphology.
Endolimax nana (nonpathogenic)	Cytoplasm is clean, not diagnostic, with a great deal of nuclear variation; there may even be some peripheral nuclear chromatin. Normally only karyosomes are visible.	Cyst is round to oval, with the 4 nuclear karyosomes being visible.	There is more nuclear variation in this ameba than any others. The organisms can be confused with *Dientamoeba fragilis* and/or *E. hartmanni*.
Iodamoeba bütschlii (nonpathogenic)	Cytoplasm contains much debris; organisms are usually larger than *E. nana* but may look similar; large karyosome.	Cyst contains single nucleus (may be "basket nucleus") with bits of nuclear chromatin arranged on the nuclear membrane (the karyosome is the basket, bits of chromatin are the handle); large glycogen vacuole.	Glycogen vacuole stains brown with the addition of iodine in the wet preparation; "basket nucleus" is more common in cyst but can be seen in trophozoite; vacuole may be so large that the cyst collapses on itself.

	Trophozoites	Cyst	Comments
Flagellates			
Giardia lamblia (pathogenic)	Trophozoites are teardrop shaped from the front and like a curved spoon from the side; contain nuclei, linear axonemes, and curved median bodies.	Cysts are round to oval, containing multiple nuclei, axonemes, and median bodies.	Organisms live in the duodenum, and multiple stool specimens may be negative; additional sampling techniques (aspiration, Entero-Test) may be needed.
Chilomastix mesnili (nonpathogenic)	Trophozoites are teardrop shaped; cytostome must be visible for identification.	Cyst is lemon shaped with 1 nucleus and a curved fibril called a shepherd's crook.	The cyst can be identified much more easily than the trophozoite form. The trophozoite looks like some of the other small flagellates.
Dientamoeba fragilis (pathogenic)	Cytoplasm contains debris; may contain 1 or 2 nuclei (chromatin often fragmented into 4 dots).	No known cyst form.	Tremendous size and shape range on a single smear; trophozoites with 1 nucleus can resemble *E. nana*.
Trichomonas vaginalis (pathogenic)	Supporting rod (axostyle) is present; undulating membrane comes halfway down the organism; small dots may be seen in the cytoplasm along the axostyle.	No known cyst form.	Recovered from genitourinary system; often diagnosed at bedside with wet preparation (motility).
Pentatrichomonas hominis (nonpathogenic)	Supporting rod (axostyle) is present; undulating membrane comes all the way down the organism; small dots may be seen in the cytoplasm along the axostyle.	No known cyst form.	Recovered in stool; trophozoites may resemble other small flagellate trophozoites.
Ciliates			
Balantidium coli (pathogenic)	Very large trophozoites (50–100 μm long) covered with cilia; large bean-shaped nucleus present.	Morphology not significant with exception of large, bean-shaped nucleus.	Rarely seen in the United States; causes severe diarrhea with large fluid loss; will be seen in proficiency-testing specimens.
Coccidia			
Cryptosporidium spp. (pathogenic)	Seen in intestinal mucosa (edge of brush border), gallbladder, and lungs; present in biopsy specimens.	Oocysts seen in stool and/or sputum; organisms acid fast, measure 4–6 μm; hard to find if only a few are present.	Chronic infection in the compromised host (internal autoinfective cycle), self-cure in the immunocompetent host; numbers of oocysts correlate with stool consistency; can cause severe, watery diarrhea; oocysts are immediately infective when passed.
Cyclospora cayetanensis (pathogenic)	Experience with this organism is not extensive; may be difficult to identify in tissue; since patients are immunocompetent, biopsy specimens will probably rarely be required or requested.	Oocysts seen in stool; approximately 8–10 μm in size; are unsporulated and thus difficult to recognize as coccidia; mimic *Cryptosporidium* on modified acid-fast-stained smears.	Most infections are associated with the immunocompetent individual but may also be seen in the immunosuppressed patient; may be associated with traveler's diarrhea; oocysts are not immediately infective when passed. Within the United States, infections have been associated with contaminated food, including raspberries, basil, snow peas, and mesclun (baby lettuce leaves). PCR can detect 40 or fewer oocysts per 100 g of raspberries or basil but has a detection limit of around 1,000 per 100 g in mesclun lettuce.

(continued)

Table A1.12 Key characteristics of protozoa of the intestinal tract and urogenital system[a] (*continued*)

Organism	Trophozoite or tissue stage	Cyst or other stage in specimen	Comments
Isospora belli (pathogenic)	Seen in intestinal mucosal cells; seen in biopsy specimens; does not seem to be as common as *Cryptosporidium*.	Oocysts seen in stool; organisms are acid fast; best technique is concentration, not permanent stained smear.	Thought to be the only *Isospora* sp. that infects humans; oocysts are not immediately infective when passed.
Microsporidia *Nosema* spp. *Encephalitozoon* spp. *Pleistophora* spp. *Trachipleistophora* spp. *Brachiola* sp. *Enterocytozoon* spp. "*Microsporidium*" spp. *Vittaforma corneae*	Developing stages sometimes difficult to identify; spores can be identified by size, shape, and presence of polar tubules.	Depending on the genus involved, spores could be identified in stool or urine using the modified trichrome stain, calcofluor, or immunoassay reagents.	Spores are generally quite small (1–2.0 µm for *Enterocytozoon* spp.) and can easily be confused with other organisms or artifacts (particularly in stool). Infections tend to be present in immunosuppressed patients; however, they are not limited to this patient group.

[a] Modified from H. D. Isenberg, ed., *Clinical Microbiology Procedures Handbook*, 2nd ed. ASM Press, Washington, D.C., 2004. RBC, red blood cells; WBC, white blood cells.

Table A1.13   Key characteristics of tissue protozoa[a]

Species	Shape and size	Other features[b]
Toxoplasma gondii		
Trophozoites (tachyzoites)	Crescent shaped; 4–6 μm long by 2–3 μm wide	Found in peritoneal fluid of experimentally infected mice; intracellular forms somewhat smaller and not usually seen in humans. May be isolated in tissue culture, particularly from CSF. Diagnosis is most frequently based on clinical history and serologic evidence (acute- and convalescent-phase sera).
Cysts (bradyzoites)	Generally spherical; 200 μm to 1 mm in diameter	Occur in many tissues (approximately 25% of the U.S. population have these organisms in the tissues, indicating past infection). Many infections are asymptomatic. Infections in the compromised host are very serious and involve the CNS. In these patients, particularly those with AIDS, diagnostic serologic titers may be very difficult to demonstrate and/or interpret.
		Note Organisms identified in histologic preparations or isolated in animals or tissue culture systems may or may not be the causative agent of the symptoms.
Pneumocystis jiroveci[c]		
Trophozoites	Ameboid in shape; measure about 5 μm; nucleus is visible with Giemsa or hematoxylin stain	Patients with AIDS may have a longer incubation period (mean, around 40 days, but can be up to 1 yr). As many as 28% of these patients show normal chest X rays, and physical signs are absent or ill defined. Rales may or may not be detected. Serologic studies indicate that by age 4, ca. 80% of those tested are positive. Diagnosis is based on actual demonstration of the organism.
Cysts	Usually round; when mature contain 8 trophozoites; often measure 5 μm and contain very small trophozoites (1 μm)	Before the AIDS epidemic, the recommended procedure of choice was the open-lung biopsy. Currently, BAL, transbronchial biopsy, and collection of induced sputum specimens have been more widely used. No commercial reagents are available for serologic diagnosis. Immunoassay reagents for direct organism detection are commercially available.
Cryptosporidium spp.	Oocyst usually round, 4–6 μm; each mature oocyst contains sporozoites (infective on passage)	Oocyst is usually the diagnostic stage in stool, sputum, and possibly other body specimens. Various other stages in the life cycle can be seen in biopsy specimens taken from the gastrointestinal tract (brush border of epithelial cells and intestinal tract) and other tissues (lung and gallbladder). A number of modified acid-fast stains have been used successfully. Detection methods using immunoassay reagents are also available.
		Note Infection in immunocompetent hosts is self-limiting; however, in immunodeficient patients (AIDS), the infection is chronic owing to an autoinfective capability in the life cycle. The number of oocysts usually correlates with the symptoms (watery diarrhea indicates many oocysts in the specimen). The more normal the stool, the more difficult it is to find the oocysts. Clinicians should be aware that risk groups include animal handlers, travelers, immunocompromised individuals, children in day care centers, and anyone who comes in contact with these individuals. Since the oocysts are immediately infective, nosocomial transmission has been documented.
Cyclospora cayetanensis	Oocyst usually round, 8–10 μm; each oocyst is immature on passage; no internal morphology visible; oocysts appear as "wrinkled" cellophane	The oocyst is the diagnostic stage in stool. Various other stages in the life cycle can be seen in biopsy specimens taken from the gastrointestinal tract (within epithelial cells and intestinal tract); oocysts are not immediately infectious when passed. The morphology is similar to that of *Isospora belli*. A number of modified acid-fast stains have been used successfully to demonstrate the oocysts (quite acid-fast variable). Detection methods involving immunoassay reagents are under development.

(continued)

Table A1.13 Key characteristics of tissue protozoa*a* *(continued)*

Species	Shape and size	Other features*b*
Isospora belli	Ellipsoidal oocyst; usual range, 20–30 µm long by 10–19 µm wide; sporocysts rarely seen broken out of oocysts but measure 9–11 µm	Mature oocyst contains 2 sporocysts with 4 sporozoites each; the usual diagnostic stage in feces is the immature oocyst, containing a spherical mass of protoplasm (diarrhetic stool). Developing stages can be recovered from intestinal biopsy specimens. Oocysts are also acid fast and could be detected during acid-fast staining of stool for *Cryptosporidium*. Oocysts are often detected in the concentration sediment (wet preparation); they are not immediately infectious when passed.

Microsporidia
  Nosema spp.
  Encephalitozoon spp.
  Enterocytozoon spp.
  Pleistophora spp.
  Trachipleistophora spp.
  Brachiola sp.
  "*Microsporidium*" spp.
  Vittaforma corneae
	Spores are extremely small and have been recovered in all body organs, including the eye	These organisms have been found as insect or other animal parasites; the route of infection may be ingestion, inhalation, or direct inoculation (eyes). Histology results vary (spores are acid fast); PAS, silver, tissue Gram, and Giemsa stains are recommended for spores. Animal inoculation is not recommended; laboratory animals may carry occult infection. Electron microscopy or molecular methods may be necessary for confirmation and identification to the genus and species levels. Although difficult to diagnose, infections have been found in a large number of AIDS patients (*Enterocytozoon bieneusi*, *Encephalitozoon* [*Septata*] *intestinalis* in the intestinal tract, *Pleistophora* spp. in muscle, and various other microsporidia in other tissues). To date, it is still somewhat difficult to diagnose this infection by examining stool specimens prepared with optical brightening agents (calcofluor) or routine stains (modified trichrome stains). Diagnostic immunoassay reagents are under development.

Sarcocystis hominis, *S. suihominis*, *S. bovihominis*	Oocyst is thin walled and contains 2 mature sporocysts, each containing 4 sporozoites; the thin oocyst wall frequently ruptures; ovoidal sporocysts measure 9–16 µm long by 7.5–12 µm wide	Thin-walled oocyst or ovoidal sporocysts are found in stool. There have been reports of fever, severe diarrhea, abdominal pain, and weight loss in the compromised host, although the number of patients has been small. Infections occur from the ingestion of uncooked pork or beef. The life cycle occurs within the intestinal cells, with the eventual production of sporocysts in stool.
Sarcocystis "*lindemanni*"	Shapes and sizes of skeletal and cardiac muscle sarcocysts vary considerably	When humans accidentally ingest oocysts from other animal stool sources, sarcocysts that develop in human muscle apparently do little if any harm. These could be identified by routine histologic methods. When humans are intermediate hosts, preventive measures involve careful disposal of animal feces that may contain the infective sporocysts. This may be impossible in the wilderness areas, where wild animals may serve as reservoir hosts for many of the different types of organisms that have been grouped under the term *Sarcocystis* "*lindemanni*." However, this name is no longer used.

a Modified from H. D. Isenberg, ed., *Clinical Microbiology Procedures Handbook*, 2nd ed. ASM Press, Washington, D.C., 2004.
b CNS, central nervous system; CSF, cerebrospinal fluid; BAL, bronchoalveolar lavage; PAS, periodic acid-Schiff.
c Now classified with the fungi.

Table A1.14 Key characteristics of helminths[a]

Helminths	Diagnostic stage[b]	Comments
Nematodes (roundworms)		
Ascaris lumbricoides (pathogenic)	Egg: both fertilized (oval to round with thick, mammilated/tuberculated shell) and unfertilized (tend to be more oval or elongate, with bumpy shell exaggerated) eggs can be found in stool; adult worms: 10–12 in. found in stool; rarely (in severe infections), migrating larvae can be found in sputum	Unfertilized eggs do not float in flotation concentration method; adult worms tend to migrate when irritated (by anesthesia or high fever); hence, patients from areas of endemic infection should be checked for infection prior to elective surgery.
Trichuris trichiura (whipworm) (pathogenic)	Egg: barrel shaped with 2 clear, polar plugs; adult worm: rarely seen; eggs should be quantitated (rare, few, etc.), since light infections may not be treated	Dual infections with *A. lumbricoides* may be seen (both infections are acquired from ingestion of eggs from contaminated soil); in severe infections, rectal prolapse may occur in children or bloody diarrhea can be mistaken for amebiasis (these clinical manifestations are usually not seen in the United States).
Enterobius vermicularis (pinworm)	Egg: football shaped with one flattened side; adult worm: about 3/8 in. long, white with pointed tail; female migrates from the anus and deposits eggs on the perianal skin	Causes itching in some patients. Test of choice is Scotch tape preparation; 4–6 consecutive tapes necessary to rule out infection; symptomatic patients often treated without actual confirmation of infection; eggs become infective within a few hours.
Ancylostoma duodenale (Old World hookworm), *Necator americanus* (New World hookworm) (pathogenic)	Egg: eggs of both species are identical, oval with broadly rounded ends, thin shell, clear space between shell and developing embryo (8–16-cell stage); adult worms: rarely seen in clinical specimens	Causes blood-loss anemia on the differential smear in heavy infections in some patients. If stool remains unpreserved for several hours or days, the eggs may continue to develop and hatch; rhabditiform larvae may resemble those of *Strongyloides stercoralis*.
Strongyloides stercoralis (pathogenic)	Rhabditiform larvae (noninfective) usually found in the stool; short buccal cavity or capsule with large genital primordial packet of cells ("short and sexy"); in very heavy infections, larvae occasionally found in sputum and/or filariform (infective) larvae can be found in stool (slit in the tail)	May cause unexplained eosinophilia, abdominal pain, unexplained episodes of sepsis and/or meningitis, and pneumonia (migrating larvae) in the compromised patient. Potential for internal autoinfection can maintain low-level infections for many years (patient is asymptomatic or has eosinophilia); hyperinfection can occur in the compromised patient (leading to disseminated strongyloidiasis and death); agar plate culture is the most sensitive diagnostic method; many infections are low-level, and larvae are difficult to recover.
Ancylostoma braziliensis (dog or cat hookworm) (pathogenic)	Humans are accidental hosts; larvae wander through the outer layer of skin, creating tracks (causing severe itching and eosinophilia); no practical microbiological diagnostic tests	Cause of cutaneous larva migrans; typical setup for infection is when dogs and cats defecate in sand boxes and hookworm eggs hatch and penetrate human skin when in contact with infected sand or soil (children playing in sand box).
Toxocara cati and *T. canis* (dog and cat ascarid) (pathogenic)	Humans are accidental hosts; infection is by ingestion of dog or cat ascarid eggs in contaminated soil; larvae wander through deep tissues (including the eye); can be mistaken for cancer of the eye; serologic tests helpful for confirmation; infection causes eosinophilia	Cause of visceral larva migrans and ocular larva migrans; requests for laboratory services often originate in the ophthalmology clinic.

(continued)

Table A1.14 Key characteristics of helminths[a] *(continued)*

Helminths	Diagnostic stage[b]	Comments
Cestodes (tapeworms)		
Taenia saginata (beef tapeworm)	Scolex (4 suckers, no hooklets) and gravid proglottid (>12 branches on a single side) are diagnostic; eggs indicate *Taenia* spp. only (thick, striated shell containing a 6-hooked embryo or oncosphere); worm usually about 12–15 ft long	Adult worm causes symptoms in some individuals. Infection occurs via ingestion of raw or poorly cooked beef; usually only a single worm per patient; individual proglottids may crawl from the anus; India ink can be injected into proglottids to visualize the uterine branches for identification.
Taenia solium (pork tapeworm)	Scolex (4 suckers with hooklets) and gravid proglottid (<12 branches on a single side) are diagnostic; eggs indicate *Taenia* spp. only (thick, striated shell containing a 6-hooked embryo or oncosphere); worm is usually about 5–20 ft long	Adult worm causes gastrointestinal complaints in some individuals; cysticercosis (accidental ingestion of eggs) can cause severe central nervous system symptoms. Infection is via ingestion of raw or poorly cooked pork; usually only a single worm per patient; occasionally 2 or 3 proglottids (hooked together) are passed. India ink can be injected into proglottids to visualize the uterine branches for identification. Cysticerci are normally small and contained within an enclosing membrane; they occasionally develop as the "racemose" type in which the worm tissue grows in the body like a metastatic cancer.
Diphyllobothrium latum (broad fish tapeworm)	Scolex (lateral sucking grooves), gravid proglottid (wider than long, reproductive structures in the center "rosette"); eggs are operculated	Causes gastrointestinal complaints in some individuals. Infection is via ingestion of raw or poorly cooked freshwater fish; life cycle has 2 intermediate hosts (copepod and fish); worm may reach 30 ft long. Associated with vitamin B$_{12}$ deficiency in genetically susceptible groups (e.g., Scandinavians).
Hymenolepis nana (dwarf tapeworm)	Adult worm not normally seen; egg round to oval, thin shell, containing a 6-hooked embryo or oncosphere with polar filaments lying between the embryo and egg shell	Causes gastrointestinal complaints in some individuals. Infection is via ingestion of eggs (the only life cycle in which the intermediate host [grain beetle] can be bypassed); life cycle of egg to larval form to adult can be completed in the human. May be the most common tapeworm in the world.
Hymenolepis diminuta (rat tapeworm)	Adult worm not normally seen; egg round to oval, thin shell, containing a 6-hooked embryo or oncosphere without polar filaments lying between the embryo and egg shell	Uncommon; egg can be confused with *H. nana*. Eggs will be submitted in proficiency-testing specimens and must be differentiated from those of *H. nana*.
Echinococcus granulosus (pathogenic)	Adult worm found only in the carnivore (dog); hydatid cysts develop (primarily in the liver) when humans accidentally ingest eggs of the dog tapeworms; cyst contains daughter cysts and many scolices; laboratory should examine fluid aspirated from cyst at surgery	Humans are accidental intermediate hosts; the normal life cycle is dog to sheep, with the hydatid cysts developing in the liver, lung, etc., of the sheep. Human hosts may be unaware of their infection unless fluid leaks from the cyst (can trigger an anaphylactic reaction) or pain is felt at the cyst location.
Echinococcus multilocularis (pathogenic)	Adult worm found only in the carnivore (fox or wolf); hydatid cysts develop (primarily in the liver) when humans accidentally ingest eggs of the carnivore tapeworms; cyst grows like a metastatic cancer with no limiting membrane	Humans are accidental intermediate hosts. Prognosis is poor; surgical removal of the tapeworm tissue is very difficult. Found in Canada, Alaska, and less frequently in the northern United States but is becoming more common in the United States, where the geographic range is moving further south.

Trematodes (flukes)

Organism	Diagnosis	Comments
Fasciolopsis buski (giant intestinal fluke)	Eggs found in stool; very large and operculated (morphology like that of *F. hepatica* eggs)	Symptoms depend on worm burden; acquired from ingestion of plant material on which metacercariae have encysted (e.g., water chestnuts); worms are hermaphroditic.
Fasciola hepatica (sheep liver fluke)	Eggs found in stool; cannot be differentiated from those of *F. buski*	Symptoms depend on worm burden; acquired from ingestion of plant material on which metacercariae have encysted (e.g., water-cress); worms are hermaphroditic.
Clonorchis (*Opisthorchis*) *sinensis* (Chinese liver fluke)	Eggs found in stool; very small (<35 μm); are operculated with shoulders into which the operculum fits	Symptoms depend on worm burden; acquired from ingestion of raw fish. Eggs can be missed unless ×400 power is used for examination; eggs can resemble those of *Metagonimus yokogawai* and *Heterophyes heterophyes* (small intestinal flukes). Worms are hermaphroditic.
Paragonimus spp. (lung fluke)	Eggs coughed up in sputum (brownish "iron filings" are egg packets); can be recovered in sputum or stool (if swallowed); are operculated with shoulders into which operculum fits	Symptoms depend on worm burden and egg deposition. Infection acquired from ingestion of raw crabs. Eggs can be confused with those of *D. latum*. Infections seen in the Orient (infections with *P. mexicanus* found in Central and South America). *P. kellicotti* infections (rare) are seen in the United States. Worms are hermaphroditic but often cross-fertilize with another worm if present.
Schistosoma mansoni (blood fluke)	Eggs recovered in stool (large lateral spine); specimens should be collected with no preservatives (to allow demonstration of egg viability); worms in veins of large intestine	Acquired from skin penetration by a single cercaria from the freshwater snail. Pathologic findings are caused by host immune response to the presence of eggs in tissues; adult worms in veins cause no problems. Adult worms are separate sexes.
Schistosoma haematobium (blood fluke)	Eggs recovered in urine (large terminal spine); specimens should be collected with no preservatives (to allow demonstration of egg viability); worms in veins of bladder	Acquired from skin penetration by single cercaria from the freshwater snail; pathologic findings as with *S. mansoni*; 24-h and spot urine samples should be collected. Chronic infection has association with bladder cancer. Adult worms are separate sexes.
Schistosoma japonicum (blood fluke)	Eggs recovered in stool (very small lateral spine); specimens should be collected with no preservatives (to allow demonstration of egg viability); worms in veins of small intestine	Acquired from skin penetration by multiple cercariae from the freshwater snail; pathologic findings as with *S. mansoni*; infection usually the most severe of the 3 species because of original loading dose of infective cercariae from the freshwater snail (multiple cercariae stick together). Symptoms are associated with egg production, which is greatest in *S. japonicum* infections.

a Modified from H. D. Isenberg, ed., *Clinical Microbiology Procedures Handbook*, 2nd ed. ASM Press, Washington, D.C., 2004.
b 1 in. = 2.54 cm; 1 ft = 30.48 cm.

Table A1.15 Key characteristics of parasites found in blood[a]

Organism	Diagnostic stage	Comments
Protozoa		
Malaria parasites		
Plasmodium vivax (benign tertian malaria)	Ameboid rings; presence of Schüffner's dots, beginning in older rings (appear later than those in *P. ovale*); all stages seen in peripheral blood; mature schizont contains 16–18 merozoites	Infects young cells; 48-h cycle; large geographic range; tends to have true relapse from the residual liver stages (hypnozoites); enlarged red blood cells
Plasmodium ovale (ovale malaria)	Nonameboid rings; presence of Schüffner's dots, beginning in young rings (appear earlier than those in *P. vivax*); all stages seen in peripheral blood; mature schizont contains 8–10 merozoites; red blood cells may be oval and have fimbriated edges	Infects young cells; 48-h cycle; narrow geographic range; tends to have true relapse from residual liver stages (hypnozoites); enlarged red blood cells
Plasmodium malariae (quartan malaria)	Thick rings; no stippling; all stages seen in peripheral blood; presence of band forms and rosette-shaped mature schizont; abundant malarial pigment	Infects old cells; 72-h cycle; narrow geographic range; associated with recrudescence and nephrotic syndrome; no true relapse; normal or small red blood cells
Plasmodium falciparum (malignant tertian malaria)	Multiple rings; appliqué/accolé forms; no stippling (rare Maurer's clefts); rings and crescent-shaped gametocytes seen in peripheral blood (no other developing stages, with rare exception of mature schizont)	Infects all cells; 36–48-h cycle; large geographic range; no true relapse; most pathogenic of the 4 species; plugged capillaries can cause severe symptoms and sequelae (cerebral malaria, lysis of red blood cells, etc.)
Babesia spp.	Ring forms only (resemble *P. falciparum* rings); seen in splenectomized patients; endemic in the United States (no travel history necessary); if present, "Maltese cross" configuration is diagnostic but not always seen	Tick-borne infection; associated with Nantucket Island; infection mimics malaria; ring forms more pleomorphic than malaria; usually more rings/cell than in malaria; endemic in several areas in the United States (East and West Coasts); organisms occasionally seen outside red blood cells (unlike malaria merozoites)
Trypanosomes		
Trypanosoma brucei gambiense (West African sleeping sickness)	Trypomastigotes long, slender, with typical undulating membrane; lymph nodes and blood can be sampled; microhematocrit tube concentration helpful; examine spinal fluid in later stages of the infection	Tsetse fly vector; tends to be chronic infection, causing the real symptoms of sleeping sickness
Trypanosoma brucei rhodesiense (East African sleeping sickness)	Trypomastigotes long, slender, with typical undulating membrane; small kinetoplast; lymph nodes and blood can be sampled; microhematocrit tube concentration helpful; examine spinal fluid in later stages of the infection	Tsetse fly vector; tends to be more severe, short-lived infection (particularly in children); patient may die before progressive symptoms of sleeping sickness appear
Trypanosoma cruzi (Chagas' disease) (American trypanosomiasis)	Trypomastigotes short, stumpy, often curved in C shape; large kinetoplast; blood sampled early in infection; trypomastigotes enter striated muscle (heart, gastrointestinal tract) and transform into the amastigote form	Reduviid bug vector (kissing bug); chronic in adults but severe in young children; great morbidity associated with cardiac failure and loss of muscle contractility in heart and gastrointestinal tract

Leishmaniae

Organism	Morphology/Diagnostic features	Clinical/Vector information
Leishmania spp. (cutaneous); not actually a blood parasite but presented for comparison with *L. donovani*	Amastigotes found in macrophages of skin; presence of intracellular forms containing nucleus and kinetoplast diagnostic	Sand fly vector; organisms recovered from site of lesion only; specimens can be stained or cultured in NNN[b] and/or Schneider's medium; animal (hamster) inoculation rarely used
Leishmania braziliensis and other *Leishmania* spp. (mucocutaneous; not actually a blood parasite but presented for comparison with *L. donovani*)	Amastigotes found in macrophages of skin and mucous membranes; presence of intracellular forms containing nucleus and kinetoplast diagnostic	Sand fly vector; organisms recovered from site of lesion only; specimens can be stained or cultured in NNN and/or Schneider's medium; animal (hamster) inoculation rarely used
Leishmania donovani and other *Leishmania* spp. (visceral)	Amastigotes found throughout the reticuloendothelial system (spleen, liver, bone marrow, etc.); presence of intracellular forms containing nucleus and kinetoplast diagnostic	Sand fly vector; organisms recovered from buffy coat (rarely found), bone marrow aspirate, spleen or liver puncture (rarely performed); specimens can be stained or cultured in NNN and/or Schneider's medium; animal (hamster) inoculation rarely used; cause of kala azar

Helminths

Organism	Morphology/Diagnostic features	Clinical/Vector information
Wuchereria bancrofti	Microfilariae sheathed, clear space at end of tail; nocturnal periodicity seen; elephantiasis seen in chronic infections	Pathogenicity due to presence of adult worms; mosquito vector; microfilariae recovered in blood (membrane filtration, Knott concentrate, thick films); hematoxylin stains the sheath; Giemsa stain does not stain the sheath
Brugia malayi	Microfilariae sheathed, subterminal and terminal nuclei at end of tail; nocturnal periodicity seen; elephantiasis seen in chronic infections	Pathogenicity due to presence of adult worms; mosquito vector; microfilariae recovered in blood (membrane filtration, Knott concentrate, thick films); hematoxylin stains the sheath; Giemsa stains the sheath pink; sheath of *B. timori* does not stain with Giemsa
Loa loa (African eye worm)	Microfilariae sheathed, nuclei continuous to tip of tail; diurnal periodicity; adult worm may cross the conjunctiva of the eye	Pathogenicity due to presence of adult worms; mango fly vector; history of calabar swellings, worms in the eye; microfilariae difficult to recover from blood; hematoxylin stains the sheath; Giemsa stain does not stain the sheath
Mansonella spp.	Microfilariae unsheathed, nuclei may or may not extend to tip of tail (depending on species); nonperiodic; symptoms usually absent or mild	Pathogenicity mild and due to presence of adult worms; midge or blackfly vector; microfilariae recovered in blood (membrane filtration, Knott concentrate, thick films); hematoxylin and Giemsa stain characteristics identical
Mansonella streptocerca	Microfilariae unsheathed, nuclei extend to tip of tail; when immobile, curved like shepherd's crook; adults in dermal tissues	Pathogenicity mild and due to presence of adult worms and/or microfilariae; midge vector; microfilariae found in skin snips; microfilarial tails are split rather than blunt; hematoxylin and Giemsa stain characteristics identical
Onchocerca volvulus	Microfilariae unsheathed, nuclei do not extend to tip of tail; adults in nodules	Pathogenicity due to presence of microfilariae; blackfly vector; microfilariae found in skin snips; microfilariae migrate to optic nerve; cause of river blindness

[a] Modified from H. D. Isenberg, ed., *Clinical Microbiology Procedures Handbook*, 2nd ed. ASM Press, Washington, D.C., 2004.
[b] NNN, Novy-MacNeal-Nicolle.

Table A1.16 Diagnostic laboratory report information that should be relayed to the physician

Organisms and other findings[a]	Genus and species	Life cycle stages[a]	Quantitation[b]	Pathogenic/nonpathogenic	Viable/nonviable
Protozoa					
Intestinal	Yes	Yes (C, T)	No	Both	NA[c]
Blastocystis hominis		NA	Yes	NA	NA
Tissue	Yes	NA	No	NA	NA
Blood	Yes	Yes (G)	Yes	NA	NA
Genitourinary tract	Yes	NA	No	NA	NA
Trematodes					
Intestinal	Yes	NA	Yes	NA	NA
Liver	Yes	NA	Yes	NA	NA
Lungs	Yes	NA	Yes	NA	NA
Blood	Yes	NA	Yes	NA	Egg viability
Cestodes					
Intestinal	Yes	Yes (E, P)	NA	NA	NA
Tissue	Yes	NA	NA	NA	NA
Nematodes					
Intestinal	Yes	NA	Rarely[d]	NA	NA
Filarial	Yes	NA	Yes	NA	NA
Nonparasitic					
Yeast	NA	Yes[e]	Yes	NA	NA
Human cells (WBCs, RBCs)	NA	NA	Yes	NA	NA
Crystals (C-L, barium)	NA	NA	Yes	NA	NA

[a] C, cyst; T, trophozoite; G, gametocyte; E, egg; P, proglottid; RBC, red blood cell; WBC, white blood cell; C-L, Charcot-Leyden.
[b] May require numbers of infected cells, eggs, etc. (either as specific count or as rare, few, moderate, or many).
[c] NA, not applicable.
[d] If *Trichuris trichiura* eggs are rare or few, the infection may not be treated.
[e] The specimen must be fresh and should be examined immediately; alternatively, it should be freshly preserved in a stool collection kit *by the patient* (no lag time between stool passage and preservation); if these requirements are not met, the report (presence of budding yeasts and/or pseudohyphae or hyphae) may be misleading and should not be reported to the physician. Report budding yeast cells and/or the presence of pseudohyphae or hyphae.

Table A1.17 Pros and cons of stool specimen collection and testing options[a]

Option[b]	Pros	Cons
Rejection of stools from inpatients who have been in-house for >3 days	Data suggest that patients who begin to have diarrhea after they have been inpatients for a few days are not symptomatic from parasitic infections, but generally from other causes.	There is always the chance that the problem is related to a health care-associated (nosocomial) parasitic infection (rare), but *Cryptosporidium* and microsporidia are possible considerations.
Examination of 1 stool (O&P examination)	Data suggest that 40–50% of organisms present will be found with only 1 stool exam. Some feel that most intestinal parasitic infections can be diagnosed from examination of a single stool. If the patient becomes asymptomatic after collection of the first stool, subsequent specimens may not be necessary.	Diagnosis from examination of a single specimen depends on the experience of the microscopist, proper collection, and the parasite load in the specimen. In a series of 3 stool specimens, frequently not all 3 specimens are positive and/or may be positive for different organisms.
Use of 2 O&P exams (concentration, permanent stained smear)	The examination of 2 specimens is more sensitive than 1 examination; different organisms may be found in the 2 specimens; 1 might be negative while 1 is positive.	May not be as cost-effective as examination of 1 specimen. Not always as good as examination of 3 specimens (may be a relatively cost-effective approach); any patient remaining symptomatic would require additional testing.
Examination of a second stool specimen only after the first is negative and the patient is still symptomatic	With additional examinations, the yield of protozoa increases (*Entamoeba histolytica*, 22.7%; *Giardia lamblia*, 11.3%; *Dientamoeba fragilis*, 31.1%)	Assumes that the second (or third) stool is collected within the recommended 10-day time frame for a series of stools; protozoa are shed periodically. May be inconvenient for patient.
Examination of 1 stool specimen and an immunoassay (EIA, FA, lateral- or vertical-flow cartridge)	If the examinations are negative and the patient's symptoms subside, probably no further testing is required.	This approach is a mix: 1 immunoassay may be acceptable; however, immunoassay testing of 2 separate specimens may be required to confirm the presence of *Giardia* antigen. One O&P exam is not the best approach (review the last option below). Patients may exhibit symptoms (off and on), so it may be difficult to rule out parasitic infections. If the patient remains symptomatic, even if 2 *Giardia* immunoassays are negative, other protozoa may be missed (the *Entamoeba histolytica/E. dispar* group, *E. histolytica*, *Dientamoeba fragilis*, *Cryptosporidium* spp., the microsporidia). Normally, there are specific situations where fecal immunoassays or O&P exams should be ordered. It is not recommended to automatically perform both the O&P and fecal immunoassay as a stool exam for parasites (see Table A1.18). **Depending on the patient's history and clinical symptoms, EITHER the O&P exam OR a fecal immunoassay may be recommended, BUT GENERALLY NOT BOTH.**
Pool 3 specimens for examination; perform 1 concentration and 1 permanent stain	Three specimens are collected by the patient (3 separate collection vials) over 7–10 days; pooling by the laboratory may save time and expense.	Organisms present in small numbers may be missed due to the dilution factor once the specimens are pooled.
Pool 3 specimens for examination; perform 1 concentration and 3 permanent stained smears	Three specimens are collected by the patient (3 separate collection vials) over 7–10 days; pooling by the laboratory for the concentration would probably be sufficient for the identification of helminth eggs. Examination of the 3 separate permanent stained smears (1 from each vial) would maximize recovery of intestinal protozoa in areas of the country where these organisms are most common.	Might miss light helminth infection (eggs, larvae) due to the pooling of the 3 specimens for the concentration; however, with a permanent stain performed on each of the 3 specimens, this approach would probably be the next best option to the standard approach (concentration and permanent stained smear performed on every stool). Coding and billing would have to match the work performed; this may present some problems where work performed does not match existing codes.

(continued)

Table A1.17 Pros and cons of stool specimen collection and testing options[a] *(continued)*

Option[b]	Pros	Cons
Collection of 3 stools by the patient, with all 3 samples put into a single vial (patient given a single vial only)	Pooling of the specimens would require only a single vial.	This would complicate patient collection and very likely result in poorly preserved specimens, especially regarding the recommended ratio of stool to preservative and the lack of proper mixing of specimen and fixative.
Immunoassays performed for selected patients,[c] using methods for *G. lamblia*, *Cryptosporidium* spp., and/or the *E. histolytica/E. dispar* group or *E. histolytica*	Would be more cost-effective than performing immunoassay procedures on all specimens; however, information required to group patients is often not received with specimens. *This approach assumes that the physicians have guidance in terms of correct ordering options (see Table A1.18).*	Labs rarely receive information that would allow them to place a patient in a particular risk group: children <5 yr old, children from day care centers (may or may not be symptomatic), patients with immunodeficiencies, and patients from outbreaks. Performance of immunoassay procedures on every stool sample is not cost-effective, and the positive rate will be low unless an outbreak situation is involved
Immunoassays and O&P exams performed on request using methods for *G. lamblia*, *Cryptosporidium* spp., and/or *E. histolytica/E. dispar* group or *E. histolytica* (a number of variables will determine the approach to immunoassay testing and the O&P exam [geography, parasites recovered, positive rate, physician requests]; immunoassays and/or O&P exams should be separately ordered, reported, and billed)	This approach limits the number of stools on which immunoassay procedures are performed for parasites. Immunoassay results do not have to be confirmed by any other tests (such as O&P exams or modified acid-fast stains). If specific kit performance problems have been identified, individual laboratories may prefer to do additional testing. *However, the fecal immunoassays are more sensitive than the O&P exam and special stains (modified acid-fast stains). Also, this may be considered duplicate testing and may not be approved for reimbursement unless specifically ordered by the physician.*	Requires education of the physician clients regarding appropriate times and patients for whom fecal immunoassays should be ordered. Educational initiatives must also include information on the test report indicating the pathogenic parasites that will not be detected by these methods. It is critical to ensure that clients know that if patients have become asymptomatic, further testing may not be required. *However, if the patient remains symptomatic, further testing (O&P exams) is required.* Remember, a single O&P exam may not reveal all organisms present. Present plan to physicians for approval: immunoassays or O&P exams, procedure discussion, report formats, clinical relevance, and limitations on each approach.

[a] For further information, see the following papers. R. A. Hiatt, E. K. Markell, and E. Ng, How many stool examinations are necessary to detect pathogenic intestinal protozoa? *Am. J. Trop. Med. Hyg.* 53:36–39, 1995; K. S. C. Kehl, Screening stools for *Giardia* and *Cryptosporidium*: are antigen tests enough? *Clin. Microbiol. Newsl.* 18:133–135; 1996; A. J. Morris, M. L. Wilson, and L. B. Reller, Application of rejection criteria for stool ovum and parasite examinations, *J. Clin. Microbiol.* 30:3213–3216, 1992; D. L. Siegel, P. H. Edelstein, and I. Nachamkin, Inappropriate testing for diarrheal diseases in the hospital, *JAMA* 263:979–982, 1990.

[b] O&P exam, ova and parasite examination; EIA, enzyme-linked immunoassay; FA, fluorescent-antibody immunoassay.

[c] It is difficult to know when you may be in an early-outbreak situation where testing of all specimens for either *Giardia lamblia* or *Cryptosporidium* spp., or both, may be relevant. Extensive efforts are under way to encourage communication among laboratories, water companies, pharmacies, and public health officials regarding the identification of potential or actual outbreaks. If it appears that an outbreak is in the early stages, then performing the immunoassays on request can be changed to screening all stools.

Table A1.18 Approaches to stool parasitology: test ordering

Patient and/or situation	Test ordered[a]	Follow-up test ordered
Patient with diarrhea and AIDS or other cause of immune deficiency or Patient with diarrhea involved in a potential waterborne outbreak (municipal water supply)	*Cryptosporidium* or *Giardia/Cryptosporidium* immunoassay	If immunoassays are negative and symptoms continue, special tests for microsporidia (modified trichrome stain) and other coccidia (modified acid-fast stain) and O&P exam should be performed
Patient with diarrhea (nursery school, day care center, camper, backpacker) or Patient with diarrhea involved in a potential waterborne outbreak (resort setting)	*Giardia* or *Giardia/Cryptosporidium* immunoassay	If immunoassays are negative and symptoms continue, special tests for microsporidia and other coccidia (see above) and O&P exam should be performed
Patient with diarrhea and relevant travel history outside of the United States or Patient with diarrhea who is a past or present resident of a developing country or Patient in an area of the United States where parasites other than *Giardia* are found (large metropolitan centers such as New York, Los Angeles, Washington D.C., Miami, etc.)	O&P exam,[b] *Entamoeba histolytica/E. dispar* immunoassay; immunoassay for confirmation of *E. histolytica*; various tests for *Strongyloides* may be relevant (even in the absence of eosinophilia), particularly if there is any history of pneumonia (migrating larvae in lungs), sepsis, or meningitis (fecal bacteria carried by migrating larvae); agar culture plate is the most sensitive diagnostic approach for *Strongyloides*	The O&P exam is designed to detect and identify a broad range of parasites (amebae, flagellates, ciliates, *Isospora belli*, helminths); if exams are negative and symptoms continue, special tests for coccidia (fecal immunoassays, modified acid-fast stains, autofluorescence) and microsporidia (modified trichrome stains, calcofluor white stains) should be performed; fluorescent stains are also options
Patient with unexplained eosinophilia	Although the O&P exam is recommended, the agar plate culture for *Strongyloides stercoralis* (more sensitive than the O&P exam) is also recommended, particularly if there is any history of pneumonia (migrating larvae in lungs), sepsis, or meningitis (fecal bacteria carried by migrating larvae)	If tests are negative and symptoms continue, additional O&P exams and special tests for microsporidia (modified trichrome stains, calcofluor white stains, fluorescent stains) and other coccidia (modified acid-fast stains, autofluorescence, fluorescent stains) should be performed
Patient with diarrhea (suspected food-borne outbreak)	Test for *Cyclospora cayetanensis* (modified acid-fast stain, autofluorescence, fluorescent stains)	If tests are negative and symptoms continue, special procedures for microsporidia and other coccidia and O&P exam should be performed

[a] Depending on the particular immunoassay kit used, various single or multiple organisms may be included. Selection of a particular kit depends on many variables: clinical relevance, cost, ease of performance, training, personnel availability, number of test orders, training of physician clients, sensitivity, specificity, equipment, time to result, etc. Very few laboratories will handle this type of testing exactly the same. Many options are clinically relevant and acceptable for good patient care. It is critical that the laboratory report indicate specifically which organisms could be identified using the kit; a negative report should list the organisms relevant to that particular kit. It is important to remember that sensitivity and specificity data for all of these fecal immunoassay kits (fluorescent-antibody assay, enzyme immunoassay, cartridge formats) are comparable.

[b] O&P exam, ova and parasite examination.

Table A1.19 Pros and cons of ova and parasite examination options[a]

Testing option[b]	Pros	Cons
Stools from patients who have been inpatients for at least 3 days should not be submitted for routine examinations without consulting the laboratory	Normally, only very few parasitic infections can be acquired by an inpatient. Although rare, one of these infections is cryptosporidiosis, which has been documented in nosocomial transmission.	Etiologic agents other than parasites, as well as other causes such as therapy, are more likely to cause diarrhea in patients who have been hospitalized for 3 days or more. There are always exceptions, but this is a reasonable guideline to follow (see Table A1.17).
3 separate stools collected within 10 days; concentration and permanent stained smear on all specimens; if specimen is fresh (unpreserved), direct wet smear can be performed (soft, liquid specimens) for the detection of organism motility	Maximize recovery of all organisms, including those in small numbers; concentration for recovery of protozoan cysts, helminth eggs, and larvae; permanent stain for confirmation of protozoa and finding protozoa missed on concentration wet examination. Special stains or immunoassay reagents may be used for the recovery of coccidia and microsporidia.	Although this approach is the most labor intensive, the results are the most complete.[c]
3 separate stools collected within 10 days; concentration on pooled specimen, permanent stained smear on all specimens (performed as 3 separate examinations)	By pooling 3 specimens and performing one concentration and examination, time is saved. Since intestinal protozoa tend to be the most common parasites found in many areas of the United States, it is important to perform permanent stained smears on the 3 individual specimens.	If helminth eggs or larvae are present in small numbers, they may be missed. Protozoa that may be missed in small numbers will probably be identified on the individual permanent stained smears.
1 stool; concentration and permanent stained smear; additional specimens may be examined after results of the first specimen are reported	Saves time.	Organisms present in small numbers may be missed; the collection of specimens 2 or 3 may be outside the 10-day recommended time frame. The series of 3 specimens now becomes 3 series of a single specimen (depending on collection time frames).[c]
1 stool; *Giardia* and *Cryptosporidium* immunoassay[d]	This can be ordered after complete discussion with physicians so that they recognize what the results (negative) mean in terms of limitations; if negative with a patient who continues to be symptomatic, ova and parasite examinations can be ordered. In areas where *Giardia lamblia* is commonly found or in a suspect waterborne outbreak (*Cryptosporidium* or other organisms), this is an acceptable option. However, it is now clear that 2 specimens may have to be examined to confirm the presence of giardiasis.	If complete information on the relevance of a negative report is not given to the physician, this approach can be used incorrectly. It is very important that immunoassay procedures be thoroughly discussed in terms of the clinical relevance of results and the specificity and sensitivity of the methods (more sensitive than the routine ova and parasite examination).
1 stool; concentration performed; permanent stained smear performed *only* if something suspicious is seen on the concentration examination	This approach is thought to save time; however, it is unacceptable both technically and in terms of quality patient care.	Numerous studies have confirmed the importance of the permanent stain, not only in confirming the identification of intestinal protozoa *but also in finding organisms that were missed in the concentration examination*. Use of this technical approach leads to very misleading reported results and should be discouraged. If a laboratory is using this approach, the specimens for parasitology testing should be sent to another laboratory.

Examination approach

Direct wet smear: unpreserved soft or liquid stools; total 22- by 22-mm coverslip with 10× objective (low power); 1/4–1/3 of area with 40× objective (high dry power).

Concentration: total 22- by 22-mm coverslip with 10× objective (low power); 1/4–1/3 of area with 40× objective (high dry power).

Permanent stained smear: minimum of 300 fields with 100× objective (oil immersion); a 50× or 60× oil immersion lens can also be used in the examination process, but some small organisms will be missed without the use of the 100× oil immersion lens

This represents the most complete approach; however, the important point is to understand the pros and cons of any modifications to this approach. The most common change is the transition from fresh, unpreserved specimens being submitted to the laboratory to fecal specimens that are transported in fixatives. Once received by the laboratory, the direct wet smear is eliminated from the standard ova and parasite procedure (no motility is visible in preserved specimens) and only the concentration and permanent stained smear is performed.

Any modification(s) from the recommended procedures *must* be discussed thoroughly with your physician clients. It is critical that they understand the advantages and disadvantages of each approach. Information can be transmitted through various communication channels: telephone, newsletters, information updates, in-service presentations, verbal and written reminders, information via computer system, etc.

[a] It is mandatory that the physicians know exactly what procedures are performed for the diagnosis of parasitic infections in your specific laboratory and how the results should be interpreted in terms of clinical relevance. It is mandatory that they be given information regarding the pros and cons (limitations) of any diagnostic approach. The complete ova and parasite examination includes direct wet smear, concentration, and permanent stained smear (fresh specimen) or concentration and permanent stained smear (preserved specimen).

[b] A direct wet smear of fresh specimens (soft or liquid) should be examined for motile trophozoites. If the specimen is formed or is received in preservative, eliminate the direct wet smear examination and proceed directly to the concentration and permanent stained smear.

[c] An excellent reference is R. A. Hiatt, E. K. Markell, and E. Ng, 1995, How many stool examinations are necessary to detect pathogenic intestinal protozoa? *Am. J. Trop. Med. Hyg.* 53:36–39, 1995. The sensitivity of one stool examination was compared with that of three examinations. The additional examinations increased the percent positive results as follows: 22.7% increase for *Entamoeba histolytica*, 11.3% increase for *Giardia lamblia*, and 31.1% increase for *Dientamoeba fragilis*. Even in symptomatic patients, these data indicate that examination of a single stool specimen will miss a large number of infections by pathogenic protozoa.

[d] The ova and parasite examination and specific immunoassays should be separate, orderable, billable tests.

Table A1.20 Laboratory test reports: optional comments[a,b]

1. *Entamoeba histolytica/E. dispar group*

   Unless you see trophozoites containing ingested RBCs (true pathogen: *E. histolytica*), you cannot tell from the organism morphology whether you have actual pathogenic *E. histolytica* organisms or nonpathogenic *E. dispar* present. Report as seen above.

   Additional computer comments:

   A. Unable to determine pathogenicity from organism morphology.

   B. Unable to rule out true pathogen, *Entamoeba histolytica*.

   C. Depending on patient's clinical condition, treatment may be appropriate.

   If you have the kit reagents to differentiate the two organisms,[c] comments could also be added:

   A. If you wish to determine which of the two organisms is present, please submit a fresh stool specimen.

   B. To determine the presence or absence of pathogenic *E. histolytica*, submit a fresh stool specimen.

2. **Identification of nonpathogens**

   Comments that can be used for reporting nonpathogens include the following. However, *these statements assume that a complete stool exam was performed on multiple stools*; you may detect nonpathogens in the first examination or an incomplete examination but miss a pathogen (example: *Dientamoeba fragilis* requires the permanent stained smear for identification)

   Additional computer comments:

   A. Considered nonpathogenic; treatment not recommended.

   B. Nonpathogen; however, indication that patient has ingested something contaminated with fecal material.

3. **Reporting *Blastocystis hominis***

   Several comments are optional for reporting *Blastocystis hominis*.

   A. Clinical significance is controversial.

   B. Status as a pathogen is controversial.

   You may want to add a second comment:

   A. Other organisms capable of causing diarrhea should first be ruled out.

4. **Reporting *Plasmodium* spp.**

   If *Plasmodium* spp. are detected but you are unable to identify to the species level, report as follows

   A. *Plasmodium* spp. seen.

   B. Unable to "rule out" *Plasmodium falciparum*.

   C. Send additional blood samples approximately every 6 h for 3 days (unless malaria is no longer a consideration).

---

[a] It is important to remember that educational information for your clients is critical to the success of your test-reporting formats. The information in the table should be shared with your clients prior to changing your actual reporting formats. Your physician group may have a preference regarding additional comments. Information updates or newsletters are appropriate for this purpose.

[b] All of the comments in the table are optional, and wording can be changed to fit your circumstances. However, it is recommended that you select specific comments and try not to use "free text," so that everyone reports test results in the same way each time.

[c] It is important to remember that current fecal immunoassay kits for the detection of the *Entamoeba histolytica/E. dispar* group or differentiation between the true pathogen (*E. histolytica*) and the nonpathogen (*E. dispar*) require fresh or frozen fecal specimens; although preserved specimens (generally formalin base or some of the single-vial fixatives) can be used for the fecal immunoassays for *Giardia lamblia* or *Cryptosporidium* spp., they cannot be used for *Entamoeba* spp. testing.

Table A1.21   Estimated prevalence of parasitic diseases worldwide

Disease	Estimated prevalence figures
Protozoa	
Amebiasis	1% of population; annual deaths, 40,000–110,000
Giardiasis	200 million infected
Helminths (nematodes)	
Ascariasis	1.3 billion infected; annual deaths, 1,550; 320 million (35%) school-age children
Hookworm	1.3 billion infected; 233 million (25%) school-age children
Trichuriasis	900 million infected; 239 million (26%) school-age children
Strongyloidiasis	35 million infected
Trichostrongyliasis	5.5 million infected
Dracunculiasis	<100,000 infected
Helminths (cestodes)	65 million infected
Helminths (trematodes)	
Schistosomiasis	200 million infected; 500 million–600 million at risk
S. haematobium	100 million; 56 million (Africa only)
S. mansoni	60 million; 25 million (Africa only)
S. japonicum	1.5 million; annual deaths for all species, 500,000–1 million
Clonorchiasis and opisthorchiasis	13.5 million infected
Paragonimiasis	2.1 million infected
Fasciolopsiasis	10 million infected
Helminths (filarial infections)	
Lymphatic filariasis	128 million infected; >40 million seriously incapacitated or disfigured; one-third of the people infected with the disease live in India, one-third are in Africa, and most of the remainder are in South Asia, the Pacific, and the Americas
Onchocerciasis	17.7 million infected
Blood parasites	
Malaria	400 million–490 million; annual deaths, 2.2 million–2.5 million; 40% of world's population at risk (3.2 billion); 853,000 deaths/yr in children <5 yr old (fifth leading cause of death in this age group)
Leishmaniasis	12 million infected; 1.5 million–2 million new cases each year; 90% are in 5 countries (Bangladesh, Brazil, India, Nepal, and Sudan); 90% of all cases of mucocutaneous leishmaniasis occur in Bolivia, Brazil, and Peru; 90% of all cases of cutaneous leishmaniasis occur in Afghanistan, Brazil, Iran, Peru, Saudi Arabia, and Syria, with 1 million–1.5 million new cases reported annually worldwide
African trypanosomiasis	300,000–500,000; 100,000 new cases each year; annual deaths, 5,000
American trypanosomiasis	24 million infected; 60,000 annual deaths

Flowcharts and Staining Tables
for Diagnostic Procedures

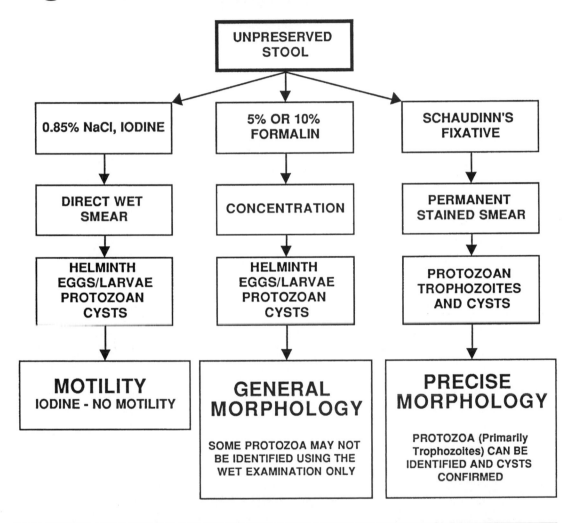

Special stains will be necessary for *Cryptosporidium* and *Cyclospora* (modified acid-fast) and the microsporidia (modified trichrome, Calcofluor). Immunoassay kits are now available for some of these organisms. If the permanent staining method (iron hematoxylin) contains a carbol fuchsin step, the coccidia will stain pink.

Flowchart A2.1 Procedure for processing fresh stool for the ova and parasite examination. Schaudinn's fixative can be prepared using the mercuric chloride base or the zinc sulfate base; the copper sulfate base does not produce organism morphologic quality equal to that seen with the mercuric chloride or zinc sulfate bases. The permanent stained smear can be stained using iron hematoxylin or trichrome stains. This approach includes the direct wet smear, concentration, and the permanent stained smear (complete ova and parasite examination for fresh specimen that is liquid or very soft).

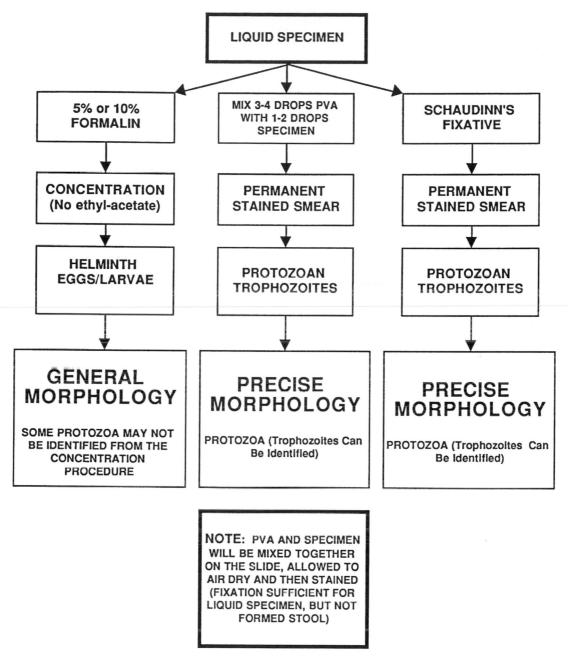

Flowchart A2.2   Procedure for processing liquid specimens for the ova and parasite examination. If the volume of the liquid stool is quite small, mixing some of the specimen with polyvinyl alcohol (PVA) fixative (Schaudinn's fixative or other fixative containing PVA plastic powder that serves as the adhesive to "glue" the fecal material onto the glass slide) directly on the slide is recommended. Make sure that the slide is thoroughly dry before staining. NOTE: PVA FIXATIVE REFERS TO ANY LIQUID FIXATIVE INTO WHICH PVA PLASTIC POWDER HAS BEEN ADDED AS AN ADHESIVE TO "GLUE" THE FECAL MATERIAL ONTO THE SLIDE.

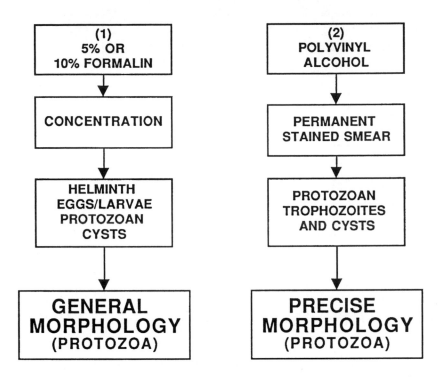

Flowchart A2.3 Procedure for processing preserved stool for the ova and parasite examination by using the traditional two-vial collection kit. This collection kit has been in use for many years. If the fecal specimen arrives in the laboratory in preservative, the direct wet mount for motility should not be performed. The complete ova and parasite examination consists of the concentration and permanent stained smear. PVA, polyvinyl alcohol; SAF, sodium acetate-acetic acid-formalin.

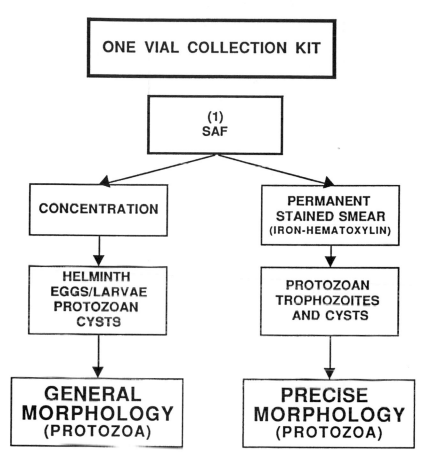

Flowchart A2.4 Procedure for processing sodium acetate-acetic acid-formalin (SAF)-preserved stool for the ova and parasite examination. SAF-preserved stool can be used with the fecal immunoassay kits (fluorescent-antibody assay, enzyme immunoassay, and cartridge formats for *Giardia lamblia* and *Cryptosporidium* spp.; kits for the true pathogen *Entamoeba histolytica* or the *Entamoeba histolytica/E. dispar* group require fresh or frozen stools); modified acid-fast stains for coccidia and modified trichrome stains can also be performed from SAF-preserved stool.

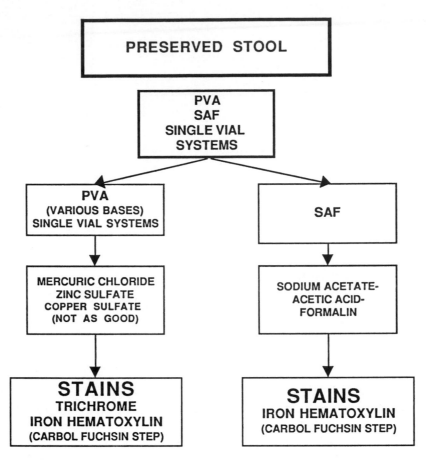

Flowchart A2.5 Use of various fixatives and their recommended stains. Fecal specimens preserved using polyvinyl alcohol (PVA) fixative (liquid fixative into which PVA plastic powder has been added) can be stained using either the iron hematoxylin or trichrome stains. However, fecal specimens preserved using sodium acetate-acetic acid-formalin appear to stain better using the iron hematoxylin approach. The single-vial preservatives (proprietary formulas) can be stained with iron hematoxylin, trichrome, or a manufacturer's stain prepared to accompany a particular fixative. *It is important to remember that parasite morphology (particularly intestinal protozoa) may not be as precise or clear as that seen when stool specimens are preserved in Schaudinn's fixative prepared with the mercuric chloride base (gold standard).*

Table A2.1   Steps in the trichrome staining procedure (mercuric chloride-based PVA-preserved stool specimens)

Steps in the procedure	Time	Comments
Prepare slide for staining		Can be prepared from Schaudinn's liquid fixative or from mercuric chloride-based polyvinyl alcohol (PVA)
70% ethanol	5 min	Rinses off excess Schaudinn's fixative
70% ethanol + iodine	5 min	Color should look like strong tea; iodine will remove mercury from fecal smears; 10 min may be better if staining a full rack of slides[a]
70% ethanol	5 min	Rinses off excess iodine
70% ethanol	5 min	Completes removal of iodine
Trichrome stain[b]	10 min	Both mercury and iodine have been removed prior to placing the smears in stain
90% ethanol + acetic acid	Several dips	This begins the destaining process; it is important to move the smears rapidly from this dish to the next so that the destaining does not continue too long
95 or 100% ethanol	Several dips	This step removes the acidified alcohol, preventing the decolorizing step from continuing
100% ethanol	3–5 min	This step begins the dehydration process, in which all moisture is removed from the smear
100% ethanol	3–5 min	Dehydration process continues
Xylene or xylene substitute	5–10 min	Dehydration process continues
Xylene or xylene substitute	5–10 min	Dehydration process completed
Permount slides or allow them to air dry with no coverslip (see Table A2.6)		Permounted slides should be allowed to dry; they can be placed on trays and put in the incubator for 30 min or overnight; room temperature overnight is also sufficient

[a] The iodine-alcohol does not require an exact concentration. The color should not be too dark or the iodine will stain the protozoa, interfering with the subsequent trichrome staining; however, the color should not be too light or the mercuric chloride in the fixative will not be removed and mercuric chloride residue in the stained fecal smear will interfere with examination. Again, a strong-tea color is preferred (do not use "port wine" as a potential color intensity guide; this color is too dark).

[b] The trichrome staining procedure is a "regressive staining method" in which the fecal smears are overstained and then destained to provide better color and contrast.

Table A2.2 Steps in the trichrome staining procedure (non-mercuric chloride-based PVA-preserved stool specimens)

Steps in the procedure	Time	Comments
Prepare slide for staining		Can be prepared from non-mercuric chloride-based Schaudinn's fixative, with polyvinyl alcohol (PVA); examples would be Schaudinn's fixative prepared with the zinc sulfate or copper sulfate bases
70% ethanol	5 min	This dish is optional; a rinse can be used for dry smears (PVA) or from liquid Schaudinn's liquid fixative
Trichrome stain[a]	10 min	Since no mercury was present on the smear, there was no need to use the iodine dish; however, when staining proficiency-testing specimens that have been preserved in mercuric chloride-based fixatives, the iodine dish and subsequent rinse steps must be added back into your staining protocol; some proficiency-testing agencies are now using fecal specimens preserved in fixatives containing no mercury; it is important to know what type of fixative was used prior to staining
90% ethanol + acetic acid	Several dips	This begins the destaining process; it is important to move the smears rapidly from this dish to the next so that the destaining does not continue too long
95 or 100% ethanol	Several dips	This step removes the acidified alcohol, preventing the decolorizing step from continuing
100% ethanol	3–5 min	This step begins the dehydration process, in which all moisture is removed from the smear
100% ethanol	3–5 min	Dehydration process continues
Xylene or xylene substitute	5–10 min	Dehydration process continues
Xylene or xylene substitute	5–10 min	Dehydration process completed
Permount slides or allow them to air dry with no coverslip (see Table A2.6)		Permounted slides should be allowed to dry; they can be placed on trays and put in the incubator for 30 min or overnight; room temperature overnight is also sufficient

[a] The trichrome staining procedure is a "regressive staining method" in which the fecal smears are overstained and then destained to provide better color and contrast.

Table A2.3   Steps in the iron hematoxylin staining procedure (mercuric chloride-based PVA-preserved stool specimens) (Spencer-Monroe method)

Steps in the procedure	Time	Comments
Prepare slide for staining		Can be prepared from Schaudinn's liquid fixative or mercuric chloride-based polyvinyl alcohol (PVA); sodium acetate-acetic acid-formalin (SAF)-preserved specimens are also recommended
70% ethanol	5 min	Rinses off excess Schaudinn's fixative
70% ethanol + iodine	5 min	Color should look like strong tea; iodine will remove mercury from fecal smears; 10 min may be better if staining a full rack of slides[a]
70% ethanol (begin the process for SAF-preserved smears or non-mercuric chloride-preserved specimens at this point)	5 min	Rinses off excess iodine
Rinse in running tap water	10 min	Use a constant stream of water running into the container
Place in iron hematoxylin working solution	4–5 min	This step is the staining process
Rinse in running tap water	10 min	Use a constant stream of water running into the container
70% ethanol	5 min	This step begins the dehydration process in which all moisture is removed from the smear
95% ethanol	5 min	Dehydration process continues
100% ethanol	3–5 min	Dehydration process continues
100% ethanol	3–5 min	Dehydration process continues
Xylene or xylene substitute	5–10 min	Dehydration process continues
Xylene or xylene substitute	5–10 min	Dehydration process completed
Permount slides or allow them to air dry with no coverslip (see Table A2.6)		Permounted slides should be allowed to dry; they can be placed on trays and put in the incubator for 30 min or overnight; room temperature overnight is also sufficient

[a] The iodine-alcohol does not require an exact concentration. The color should not be too dark, or the iodine will stain the protozoa, interfering with the subsequent trichrome staining; however, the color should not be too light or the mercuric chloride in the fixative will not be removed and mercuric chloride residue in the stained fecal smear will interfere with examination. Again, a strong-tea color is preferred (do not use "port wine" as a potential color intensity guide; this color is too dark).

Table A2.4 Steps in the iron hematoxylin staining procedure (mercuric chloride-based PVA-preserved stool specimens) (Tompkins-Miller method)

Steps in the procedure	Time	Comments
Prepare slide for staining		Can be prepared from Schaudinn's liquid fixative or mercuric chloride-based polyvinyl alcohol (PVA); sodium acetate-acetic acid-formalin (SAF)-preserved specimens are also recommended
70% ethanol	5 min	Rinses off excess Schaudinn's fixative
70% ethanol + iodine	5 min	Color should look like strong tea; iodine will remove mercury from fecal smears; 10 min may be better if staining a full rack of slides[a]
50% ethanol (begin the process for SAF-preserved smears or non-mercuric chloride-preserved specimens at this point)	5 min	Rinses off excess iodine
Rinse in running tap water	3 min	Use a constant stream of water running into the container
Place in 4% ferric ammonium sulfate mordant	5 min	This prepares the smear to take up the stain
Rinse in running tap water	1 min	Use a constant stream of water running into the container
Place in 0.5% aqueous hematoxylin	2 min	This begins the staining process
Rinse in running tap water	1 min	Use a constant stream of water running into the container
Place in 2% phosphotungstic acid	2–5 min	This provides the destaining process[b]
Rinse in running tap water	10 min	Use a constant stream of water running into the container
70% ethanol + a few drops of saturated aqueous lithium carbonate	3 min	The purpose of this step is to "blue up" the color
95% ethanol	5 min	This step begins the dehydration process in which all moisture is removed from the smear
100% ethanol	3–5 min	Dehydration process continues
100% ethanol	3–5 min	Dehydration process continues
Xylene or xylene substitute	5–10 min	Dehydration process continues
Xylene or xylene substitute	5–10 min	Dehydration process completed
Permount slides or allow them to air dry with no coverslip (see Table A2.6)		Permounted slides should be allowed to dry; they can be placed on trays and put in the incubator for 30 min or overnight; room temperature overnight is also sufficient

[a] The iodine-alcohol does not require an exact concentration. The color should not be too dark or the iodine will stain the protozoa, interfering with the subsequent trichrome staining; however, the color should not be too light or the mercuric chloride in the fixative will not be removed and mercuric chloride residue in the stained fecal smear will interfere with examination. Again, a strong-tea color is preferred (do not use "port wine" as a potential color intensity guide; this color is too dark).

[b] Like the trichrome stain, this iron hematoxylin staining procedure is a "regressive staining method" in which the fecal smears are overstained and then destained to provide better color and contrast.

Table A2.5   Steps in the iron hematoxylin staining procedure (incorporating the carbol fuchsin step)[a]

Steps in the procedure	Time	Comments
Prepare slide for staining		Can be prepared from SAF-preserved specimens (SAF sediment placed on albumin-coated slides); **specimens preserved in Schaudinn's fixative and/or PVA are not recommended (any preservative containing mercuric chloride or PVA powder is not recommended)**
70% ethanol	5 min	Helps "glue" the material onto the slide
Rinse in container of tap water	2 min	Do not use running water
Place in Kinyoun's stain	5 min	Stains for modified acid-fast characteristics
Rinse in running tap water	1 min	Use a constant stream of water running into the container
Acid-alcohol decolorizer (0.5% acid-alcohol)	4 min	This must not be too strong or *Cyclospora cayetanensis* will decolorize too much[b]
Rinse in running tap water	1 min	Use a constant stream of water running into the container
Place in iron hematoxylin working solution	8 min	This begins the staining process
Rinse in container of tap water	1 min	Do not use running water
Place in picric acid solution	3–5 min	This step differentiates the hematoxylin stain
Rinse in running tap water	10 min	Use a constant stream of water running into the container
70% ethanol + some ammonia to bring pH to 8.0	3 min	The purpose of this step is to "blue up" the color
95% ethanol	5 min	This step begins the dehydration process
100% ethanol	3–5 min	Dehydration process continues
100% ethanol	3–5 min	Dehydration process continues
Xylene or xylene substitute	5–10 min	Dehydration process continues
Xylene or xylene substitute	5–10 min	Dehydration process completed
Permount slides or allow them to air dry with no coverslip (see Table A2.6)		Permounted slides should be allowed to dry; they can be placed on trays and put in the incubator for 30 min or overnight; room temperature overnight is also sufficient

[a] This staining approach not only stains the intestinal protozoa (iron hematoxylin) but also stains the coccidia (*Cryptosporidium* spp., *Cyclospora cayetanensis*, and *Isospora belli*), which are modified acid-fast positive.

[b] This iron hematoxylin staining procedure is a "regressive staining method" in which the fecal smears are overstained and then destained to provide better color and contrast.

Table A2.6 Oil-mounted permanent stained smears (no Permount is used)

Steps in the procedure	Time	Comments
Remove stained smears from last dehydration step (xylene or xylene substitute) and place them flat on paper towels to dry	Allow to dry for 10 min	If using a xylene substitute, the smears will take longer to dry; leave on paper towels for 15–20 min
Once the fecal smears are dry, place 1 to 2 drops of immersion oil onto the dry fecal smear	3–5 min	Allow the oil to "sink into" the dry fecal smear; if the smears are somewhat thick, wait 10–15 min before microscopic examination
Place a no. 1 coverslip onto the fecal smear/oil combination	No waiting time required	A 22- by 22-mm coverslip is recommended; the addition of the coverslip will prevent damage to the oil lens when microscopic examination begins; when fecal material on the smear has been thoroughly dehydrated, the material may be hard/brittle and could scratch the oil lens if not protected
Add 1 drop of immersion oil to the top of the coverslip[a]	No waiting time required	The fecal smear is now ready to examine using the oil immersion objective (100× lens for a total magnification of ×1,000)

[a] The layers are as follows: glass slide, fecal smear, immersion oil, coverslip, immersion oil; this approach works quite well and is a substitute for the use of Permount or other mounting fluids. If a positive smear is found, the oil can be removed in xylene or xylene substitute and mounted with Permount and a coverslip in the normal way. The negative slides can be stored in a slide box that is recycled every month or so, and the old slides can then be discarded.

Table A2.7 Tips on stool processing and staining

Processing or staining step	Commentary and tips
Direct wet mount, water or saline	In general, organisms do not like distilled water; even in saline, trophozoites may not maintain adequate morphology very long; even with viable organisms, motility may be difficult to see (lower microscope light intensity); direct wet mounts should be prepared on liquid or very soft stools only (more formed stools usually do not contain any trophozoites)
Direct wet mount, iodine	If the iodine is too dark, it may interfere with helminth egg detection; eggs resemble debris and/or artifacts; internal protozoan morphology may be too dark; iodine should be the color of strong tea; iodine kills the organisms, so that no motility will be seen
Concentration-sedimentation	Rinse solutions can be water, saline, or formalin (regular 5% or 10% solution); do not rinse too many times (the supernatant fluid does *not* have to be clear); although a cleaner sediment may be easier to read, too much rinsing results in loss of organisms; a dirtier sediment is preferred to maximize organism detection; do not make coverslip preparations too thick (thinner is preferable); every centrifugation step should be at 500 × g for 10 min
Concentration–flotation	Rinse solutions can be water, saline, or formalin (regular 5% or 10% solution); do not rinse too many times (the supernatant fluid does *not* have to be clear); the more you rinse, the more likely you are to lose organisms each time; many labs use a single rinse or two at the most; it is important to use zinc sulfate of the correct specific gravity (1.18 for non-formalin-fixed material, 1.20 for specimens received in formalin); *both* the surface film and sediment must be examined (not all parasites float in the zinc sulfate method)
Permanent stained smear–smear preparation	The smear should be neither too thick nor too thin (you should be able to read newsprint through the fecal smear); if using polyvinyl alcohol (PVA)-preserved specimens, make sure that the excess PVA is removed from the stool prior to smear preparation (absorb excess PVA from specimen by placing a bit of specimen onto paper towels, wait a few minutes, and then prepare smears); smears should be thoroughly dry prior to staining (with the exception of using liquid Schaudinn's fixative); **formalin-fixed stool is not appropriate for permanent staining**
Permanent stained smear–staining	If specimen is preserved in a fixative containing mercuric chloride, it is mandatory that all mercury be removed prior to staining (iodine and subsequent alcohol steps); dehydration (moisture removal) must be complete or protozoan morphology will be difficult to see; the best dehydration is obtained with absolute ethanol; if the "commercial" absolute alcohol is used (ethanol plus other alcohols), the dehydration is not as complete, so dishes containing this solution must be changed more often (at least weekly and every few days if staining more than a few slides at a time); xylene substitutes do not dehydrate as well as xylene (dishes must be changed more often); xylene substitutes take longer to dry; however, slides must be thoroughly dry prior to mounting with immersion oil (see Table A2.6); **formalin-fixed stool is not appropriate for permanent staining**

APPENDIX 3
Common Problems in Parasite Identification

The drawings (except Figure A3.26) are reprinted from H. D. Isenberg, ed., *Clinical Microbiology Procedures* *Handbook*, 2nd ed., ASM Press, Washington, D.C., 2004. Other confusing artifacts can be seen in chapter 34.

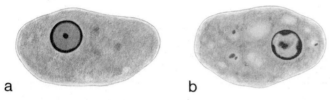

a   b

Figure A3.1 (a) *Entamoeba histolytica/E. dispar* trophozoite. Note the evenly arranged nuclear chromatin, central compact karyosome, and relatively clean cytoplasm. (b) *Entamoeba coli* trophozoite. Note the unevenly arranged nuclear chromatin, eccentric karyosome, and messy cytoplasm. These characteristics are very representative of the two organisms. (Illustration by Sharon Belkin.)

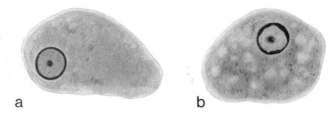

a   b

Figure A3.2 (a) *Entamoeba histolytica/E. dispar* trophozoite. Note the evenly arranged nuclear chromatin, central compact karyosome, and clean cytoplasm. (b) *Entamoeba coli* trophozoite. Note that the nuclear chromatin appears to be evenly arranged, the karyosome is central (but more diffuse), and the cytoplasm is messy, with numerous vacuoles and ingested debris. The nuclei of these two organisms tend to resemble one another (a very common finding in routine clinical specimens). (Illustration by Sharon Belkin.)

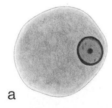
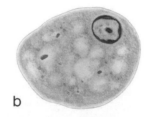

a   b

Figure A3.3 (a) *Entamoeba histolytica/E. dispar* trophozoite. Again, note the typical morphology (evenly arranged nuclear chromatin, central compact karyosome, and relatively clean cytoplasm). (b) *Entamoeba coli* trophozoite. Although the nuclear chromatin is eccentric, note that the karyosome seems to be compact and central. However, note the various vacuoles containing ingested debris. These organisms show some characteristics that are very similar (very common in clinical specimens). (Illustration by Sharon Belkin.)

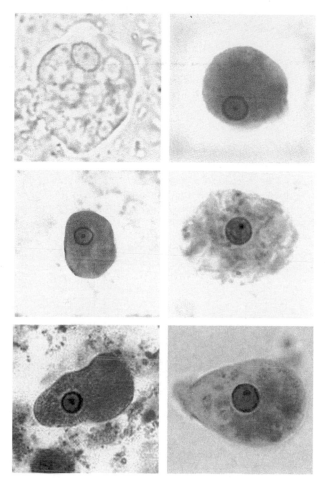

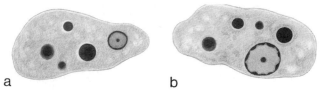

Figure A3.5 (a) *Entamoeba histolytica* trophozoite. Note the evenly arranged nuclear chromatin, central compact karyosome, and red blood cells (RBCs) in the cytoplasm. (b) Human macrophage. The key difference between the macrophage nucleus and that of *E. histolytica* is the size. Usually the ratio of nucleus to cytoplasm in a macrophage is approximately 1:6 or 1:8, while the true organism has a nucleus-to-cytoplasm ratio of approximately 1:10 or 1:12. The macrophage also contains ingested RBCs. In patients with diarrhea or dysentery, trophozoites of *E. histolytica* and macrophages are often confused, occasionally leading to a false-positive diagnosis of amebiasis when no parasites are present. Both the actual trophozoite and the macrophage may also be seen without ingested RBCs and can mimic one another. (Illustration by Sharon Belkin.)

Figure A3.4 (Top row) *Entamoeba histolytica/E. dispar* trophozoites shown in a wet preparation (left) and a permanent stained smear (right). (Middle row) *Entamoeba histolytica/E. dispar* trophozoite in permanent stained smear (left); *Entamoeba coli* trophozoite in permanent stained smear (right). (Bottom row) *Entamoeba histolytica/E. dispar* trophozoite in permanent stained smear (left); *Entamoeba coli* trophozoite in permanent stained smear (right). Note the differences in the nucleus and cytoplasm: *Entamoeba histolytica/E. dispar* has even nuclear chromatin, central karyosome, and clean cytoplasm, whereas *Entamoeba coli* has uneven nuclear chromatin, eccentric karyosome, and messy/dirty cytoplasm. However, often there are nuclear and cytoplasmic characteristics that are not specific for one species or another; too many characteristics are seen that resemble either or both species.

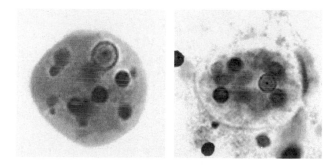

Figure A3.6 *Entamoeba histolytica* trophozoites on permanent stained smears; note the evenly arranged nuclear chromatin, central karyosome, and ingested RBCs. The presence of the ingested RBCs indicates the presence of the true pathogen, *E. histolytica*. Without the presence of ingested RBCs, the organism would have been identified as *Entamoeba histolytica/E. dispar* (it is impossible to determine pathogenicity from the morphology on the permanent stained smear).

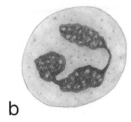

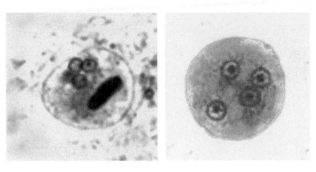

Figure A3.7 (a) *Entamoeba histolytica/E. dispar* precyst. Note the enlarged nucleus (prior to division) with evenly arranged nuclear chromatin and central compact karyosome. Chromatoidal bars (rounded ends with smooth edges) are also present in the cytoplasm. (b) Polymorphonuclear leukocyte (PMN). The nucleus is somewhat lobed (normal morphology) and represents a PMN that has not been in the gut very long. Occasionally, the positioning of the chromatoidal bars and the lobed nucleus of the PMN will mimic one another. The chromatoidal bars stain more intensely, but the shapes can overlap, as seen here. (Illustration by Sharon Belkin.)

Figure A3.9 *Entamoeba histolytica/E. dispar* cysts in permanent stained smears. Note the four nuclei and chromatoidal bar with smooth, rounded ends. From the cyst morphology, it is impossible to differentiate between the true pathogen *E. histolytica* and the nonpathogen *E. dispar.*

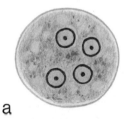
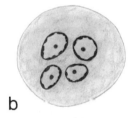

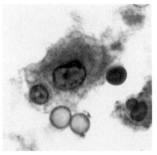
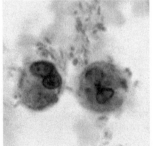

Figure A3.8 (a) *Entamoeba histolytica/E. dispar* mature cyst. Note that the four nuclei are very consistent in size and shape. (b) PMN. Note that the normal lobed nucleus has now broken into four fragments, which mimic four nuclei with peripheral chromatin and central karyosomes. When PMNs have been in the gut for some time and have begun to disintegrate, the nuclear morphology can mimic that seen in an *E. histolytica/E. dispar* cyst. However, human cells are often seen in the stool in patients with diarrhea; with rapid passage of the gastrointestinal tract contents, there will not be time for amebic cysts to form. Therefore, for patients with diarrhea and/or dysentery, if "organisms" that resemble the cell in panel b are seen, think first of PMNs, not *E. histolytica/E. dispar* cysts. (Illustration by Sharon Belkin.)

Figure A3.10 Human cells in permanent stained smears. (Left) Macrophage. Note the large size of the nucleus (the ratio of nucleus to cytoplasm is much larger than in the amebae in the genus *Entamoeba*). (Right) PMN. Note the lobed nucleus; when the lobes break apart, the "individual nuclei" can mimic those seen in amebae.

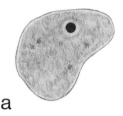
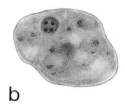

Figure A3.11 (a) *Endolimax nana* trophozoite. This organism is characterized by a large karyosome with no peripheral chromatin, although many nuclear variations are normally seen in any positive specimen. (b) *Dientamoeba fragilis* trophozoite. Normally, the nuclear chromatin is fragmented into several dots (often a "tetrad" arrangement). The cytoplasm is normally more "junky" than that seen in *E. nana*. If the morphology is typical, as in these two illustrations, differentiating between these two organisms is not very difficult. However, the morphologies of the two are often very similar. (Illustration by Sharon Belkin.)

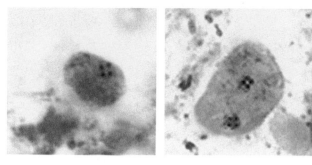

Figure A3.14 *Dientamoeba fragilis* trophozoites. (Left) Trophozoite with single nucleus fragmented into four chromatin dots. (Right) Trophozoite with two nuclei, each of which is fragmented into chromatin dots.

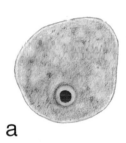
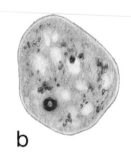

Figure A3.12 (a) *Endolimax nana* trophozoite. Note that the karyosome is large and surrounded by a "halo," with very little chromatin on the nuclear membrane. (b) *Dientamoeba fragilis* trophozoite. In this organism, the karyosome is beginning to fragment and there is a slight clearing in the center of the nuclear chromatin. If the nuclear chromatin has not become fragmented, *D. fragilis* trophozoites can very easily mimic *E. nana* trophozoites. This could lead to a report indicating that no pathogens were present when, in fact, *D. fragilis* is now considered a definite cause of symptoms. (Illustration by Sharon Belkin.)

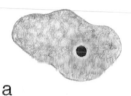
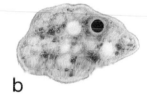

Figure A3.15 (a) *Endolimax nana* trophozoite. Note the large karyosome surrounded by a clear space. The cytoplasm is relatively clean. (b) *Iodamoeba bütschlii* trophozoite. Although the karyosome is similar to that of *E. nana*, note that the cytoplasm in *I. bütschlii* is much more heavily vacuolated and contains ingested debris. Often, these two trophozoites cannot be differentiated. However, the differences in the cytoplasm are often helpful. There is a definite size overlap between the two genera. (Illustration by Sharon Belkin.)

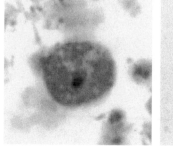
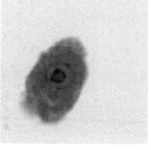

Figure A3.13 (Left) *Endolimax nana* trophozoite (note the large karyosome and no peripheral chromatin). (Right) *Dientamoeba fragilis* with a single nucleus (note the clearing within the karyosome, indicating that the chromatin is beginning to fragment into the chromatin dots).

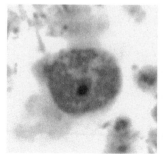
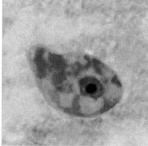

Figure A3.16 (Left) *Endolimax nana* trophozoite (as seen in Figure A3.13) (note the large single karyosome without any peripheral chromatin). (Right) *Iodamoeba bütschlii* trophozoite (note the large karyosome, some light peripheral nuclear chromatin, and the messy/dirty cytoplasm containing many vacuoles).

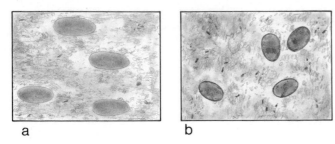

Figure A3.17 (a) RBCs on a stained fecal smear. Note that the cells are very pleomorphic but tend to be positioned in the direction in which the stool was spread onto the slide. (b) Yeast cells on a stained fecal smear. These cells tend to remain oval and are not aligned in any particular way on the smear. These differences are important when the differential identification is between *Entamoeba histolytica* containing RBCs and *Entamoeba coli* containing ingested yeast cells. If RBCs or yeast cells are identified in the cytoplasm of an organism, they must also be visible in the background of the stained fecal smear. (Illustration by Sharon Belkin.)

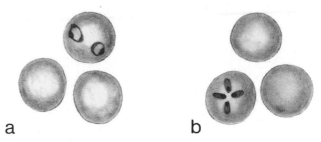

Figure A3.19 (a) *Plasmodium falciparum* rings. Note the two rings in the RBC. Multiple rings per cell are more typical of *P. falciparum* than of the other species causing human malaria. (b) *Babesia* rings. One of the RBCs contains four small *Babesia* rings. This particular arrangement is called the Maltese cross and is diagnostic for *Babesia* spp. (although this configuration is not always seen). *Babesia* infections can be confused with cases of *P. falciparum* malaria, primarily because multiple rings can be seen in the RBCs. Another difference involves ring morphology. *Babesia* rings are often of various sizes and tend to be very pleomorphic, while those of *P. falciparum* tend to be more consistent in size and shape. (Illustration by Sharon Belkin.)

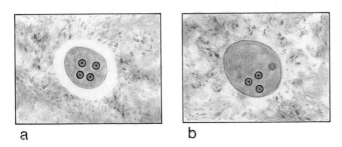

Figure A3.18 (a) *Entamoeba histolytica/E. dispar* cyst. Note the shrinkage due to dehydrating agents in the staining process. (b) *E. histolytica/E. dispar* cyst. In this case, the cyst exhibits no shrinkage. Only three of the four nuclei are in focus. Normally, this type of shrinkage is seen with protozoan cysts and is particularly important when a species is measured and identified as either *E. histolytica/E. dispar* or *Entamoeba hartmanni*. The whole area, including the halo, must be measured prior to species identification. If just the cyst is measured, the organism would be identified as *E. hartmanni* (nonpathogenic) rather than *E. histolytica/E. dispar* (possibly pathogenic). (Illustration by Sharon Belkin.)

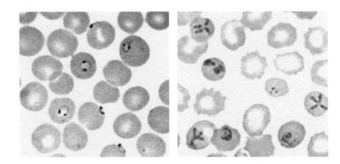

Figure A3.20 (Left) *Plasmodium falciparum* rings. Note the "clean" morphology and headphone appearance of some of the rings; also note the appliqué forms. (Right) *Babesia* spp. Note the "messy/pleomorphic" rings and the presence of the Maltese cross configuration of four rings within a single RBC; this does not occur in every species of *Babesia*.

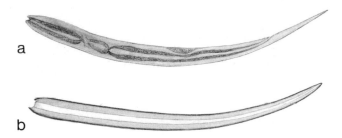

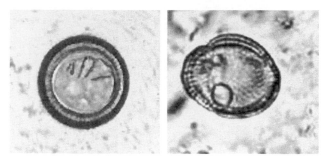

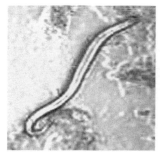
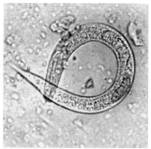

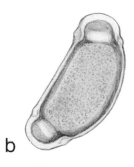

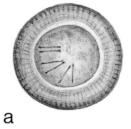
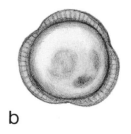

Figure A3.21 (a) *Strongyloides stercoralis* rhabditiform larva. Note the short buccal capsule (mouth opening) and the internal structure, including the genital primordial packet of cells. (b) Root hair (plant material). Note that there is no specific internal structure and the end is ragged (where it was broken off from the main plant. Plant material often mimics some of the human parasites. This comparison is one of the best examples. These artifacts are occasionally submitted as proficiency-testing specimens. (Illustration by Sharon Belkin.)

Figure A3.24 (Left) *Taenia* egg, showing the striated shell and six-hooked embryo (oncosphere) within the egg shell. (Right) Pollen grain that can resemble an actual *Taenia* egg. Note that the shape may vary, especially depending on the way the pollen grain is lying in the wet preparation.

Figure A3.22 (Left) Root hair artifact that resembles actual nematode larva. (Right) *Strongyloides stercoralis* rhabditiform larva. Note the internal structures, especially the large, genital primordium packet of cells.

Figure A3.25 (a) *Trichuris trichiura* egg. This egg is typical and is characterized by the barrel shape with thick shell and two polar plugs. (b) Bee pollen. This artifact certainly mimics the *T. trichiura* egg. However, note that the shape is somewhat distorted. This is an excellent example of a parasite look-alike that could be confusing. (Illustration by Sharon Belkin.)

Figure A3.23 (a) *Taenia* egg. This egg has been described as having a thick, radially striated shell containing a six-hooked embryo (oncosphere). (b) Pollen grain. Note that this trilobed pollen grain has a similar type of "shell" and, if turned the right way, could resemble a *Taenia* egg. This similarity represents another source of confusion between a helminth egg and a plant material artifact. When examining fecal specimens in a wet preparation, you can tap on the coverslip to get objects to move around. As they move, you can see more morphologic detail. (Illustration by Sharon Belkin.)

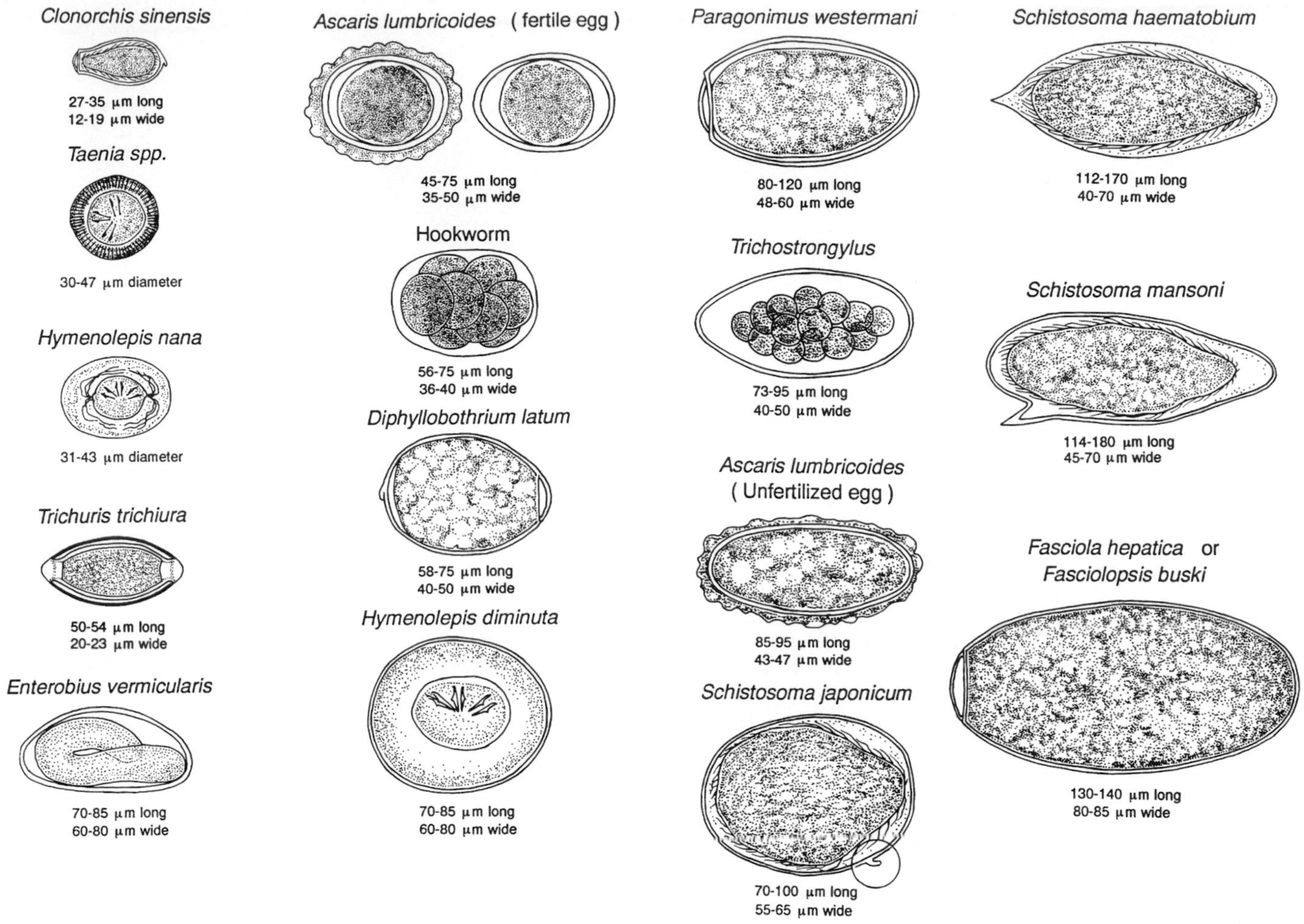

Clonorchis sinensis

27-35 μm long
12-19 μm wide

Taenia spp.

30-47 μm diameter

Hymenolepis nana

31-43 μm diameter

Trichuris trichiura

50-54 μm long
20-23 μm wide

Enterobius vermicularis

70-85 μm long
60-80 μm wide

Ascaris lumbricoides (fertile egg)

45-75 μm long
35-50 μm wide

Hookworm

56-75 μm long
36-40 μm wide

Diphyllobothrium latum

58-75 μm long
40-50 μm wide

Hymenolepis diminuta

70-85 μm long
60-80 μm wide

Paragonimus westermani

80-120 μm long
48-60 μm wide

Trichostrongylus

73-95 μm long
40-50 μm wide

Ascaris lumbricoides
(Unfertilized egg)

85-95 μm long
43-47 μm wide

Schistosoma japonicum

70-100 μm long
55-65 μm wide

Schistosoma haematobium

112-170 μm long
40-70 μm wide

Schistosoma mansoni

114-180 μm long
45-70 μm wide

Fasciola hepatica or
Fasciolopsis buski

130-140 μm long
80-85 μm wide

Figure A3.26 Relative sizes of helminth eggs.

Table A3.1 Adult nematodes and/or larvae found in stool specimens: size comparisons

Nematodes	Size
Ascaris lumbricoides	
Adult female	200–350 by 4–6 mm
Adult male	150–200 by 2–4 mm
Trichuris trichiura	
Adult female	35–50 mm
Adult male	30–45 mm
Enterobius vermicularis	
Adult female	8–13 mm
Adult male	2–5 mm
Hookworms (*Ancylostoma duodenale* and *Necator americanus*)	
Adult female	8 11 by 0.45 mm
Adult male	10–13 by 0.6 mm
Filariform larva	700 by 20 μm
Rhabditiform larva	250 by 20 μm
Strongyloides stercoralis	
Filariform larva	600 by 20 μm
Rhabditiform larva	250 by 20 μm
Trichinella spiralis	
Adult female	3–4 by 0.6 mm
Adult male	1.5 by 0.045 mm
Larva	90–100 by 6 μm
Microfilariae	
Wuchereria bancrofti	230–320 by 10 μm
Brugia malayi	170–260 by 5–6 μm
Loa loa	250–300 by 6–8.5 μm
Adult female	70 by 0.5 mm
Adult male	30–36 by 0.6 mm
Onchocerca volvulus	150–368 by 5–9 μm
Adult female	34–50 cm by 270–400 μm
Adult male	19–42 cm by 130–210 μm
Mansonella perstans	200 by 4.5 μm
Mansonella streptocerca	180–240 by 3 μm
Mansonella ozzardi	175–240 by 4.5 μm
Dracunculus medinensis	
Adult female	500–1,200 by 0.9–1.7 mm
Adult male	12–29 by 0.4 mm
Larva	500–750 by 15–25 μm
Gnathostoma spinigerum	
Adult female	25–54 mm
Adult male	11–25 mm

APPENDIX 4
Quality Control Recording Sheets

Sheets A4.1 through A4.4 are adapted from L. S. Garcia (section ed.), Parasitology section, *in* H. D. Isenberg, ed., *Clinical Microbiology Procedures Handbook*, 2nd ed., ASM Press, Washington, D.C., 2004.

1. Reagent Name:_____

2. QC Requirements (frequency):_____

3. Acceptable Criteria:_____

    Negative Control:_____

    Positive Control:_____

Date	Lot No.	Exp. Date	QC Organism	Tech	Results (A/NA)[1]	Comments Corrective Action

Comments:_____

_____

Date	Lot No.	Exp. Date	QC Organism	Tech	Results (A/NA)	Comments Corrective Action

Comments:_____

_____

Date	Lot No.	Exp. Date	QC Organism	Tech	Results (A/NA)	Comments Corrective Action

Comments:_____

_____

Date	Lot No.	Exp. Date	QC Organism	Tech	Results (A/NA)	Comments Corrective Action

Comments:_____

_____

_____

[1]A = Acceptable, NA = Not Acceptable

Sheet A4.1   Diagnostic parasitology quality control (QC) (reagents).

Manufacturer/Product: _____

QC Frequency: _____

Acceptable (A): _____

Not Acceptable (NA): _____

REAGENT	LOT #	EXP. DATE

QC RESULTS	DATE	TECH	QC RESULTS	DATE	TECH

Sheet A4.2   Diagnostic parasitology quality control (QC) (reagents)—example for multiple reagents.

1. Patient's Name/Organism trying to isolate: _____

2. Medium Number 1: _____

    Medium Number 2: _____

3. ATCC Control Strain (Indicate Organism Name and ATCC Number): _____

    _____

4. Frequency: QC strain must be set up every time patient specimen is cultured (there are no exceptions to this requirement):

MEDIUM 1: _____

Date	Medium Lot No.	Exp. Date	Tech	A[1]	NA[2]	Date Read	Comments[3] Corrective Action

Comments: _____

_____

_____

MEDIUM 2: _____

Date	Medium Lot No.	Exp. Date	Tech	A[1]	NA[2]	Date Read	Comments[3] Corrective Action

Comments: _____

_____

_____

_____

[1] Acceptable: QC Organism growth or motile trophozoites can be detected microscopically in sample of QC medium (inoculated with ATCC strain)

[2] Not Acceptable: QC Organism growth or motile trophozoites cannot be detected microscopically in sample of QC medium (inoculated with ATCC strain)

[3] Follow-up: In the event the QC culture is negative, the control strain should be resubbed to fresh media.

Sheet A4.3   Diagnostic parasitology quality control (QC) (culture)—example of a worksheet.

EQUIPMENT DESCRIPTION: _____
(Include Inventory Number)

Serial Number: _____ **Model Number:** _____ **Purchase Date:** _____

EQUIPMENT LOCATION: _____

MAINTENANCE REQUIREMENTS AND TIME FRAMES:
(List specific requirements, time frames below - use table for recording actual maintenance.)

1. _____

2. _____

3. _____

4. _____

DATE AND INITIALS	MAINTENANCE (List # of Maintenance Requirements)	COMMENTS/CORRECTIVE ACTION

REMEMBER TO ATTACH ALL RELEVANT PAPERWORK TO THIS SHEET (REPAIR INVOICES, REPLACEMENT PART INVOICES, ETC.)

Sheet A4.4   Equipment maintenance. *(continued on next page)*

DATE AND INITIALS	MAINTENANCE (List # of Maintenance Requirements)	COMMENTS/CORRECTIVE ACTION

Sheet A4.4 *(continued)*

APPENDIX 5
Commercial Supplies
and Suppliers

(See Tables A5.1 to A5.8 on the following pages.)

Table A5.1 Sources of commercial reagents and supplies[a]

Reagent or supply[b]	AJP Scientific	ALPHA-TEC	J. T. Baker	Evergreen Scientific	Fisher Scientific	Hardy Diagnostics
Specimen collection kit(s)[c]						
Formalin/PVA		X		X		X
PVA, modified (copper base)		X			X	X
PVA, modified (zinc base)		X		X	X	X
MIF		X				X
SAF		X		X	X	X
Pinworm paddles				X		X
Preservatives (bulk)						
Schaudinn's fixative solution	X	X + −				X +
PVA fixative solution	X	X				X
PVA powder[d]			X		X	
MIF solution	X	X				X
SAF solution		X				X
Concentration systems						
Formalin-ethyl acetate		X		X		X
Zinc sulfate (sp gr, 1.18/1.20)		X	X			X
Stains						
Trichrome, solution	X(W)	X(W)				X(W)
Trichrome, modified[e]	X	X(We, R)				X
Trichrome, dye powders						
Chromotrope 2R			X			
Fast green FCF			X		X	
Light green SF yellowish			X		X	
Hematoxylin solution	X					X
Hematoxylin powder			X		X	
Chlorazol black E, powder						
Giemsa, solution	X	X	X		X	X
Giemsa, powder			X		X	
Carbol fuchsin, Kinyoun's	X	X			X	X
Modified acid-fast with DMSO		X				
Auramine-rhodamine					X	
Acridine orange						
Toluidine blue O	X					
Miscellaneous						
Lugol's iodine, dilute 1:5	X	X				X
Dobell & O'Connor's or D'Antoni's iodine		X			X	X
Calcofluor white	X					
Triton X-100		X	X			X
Eosin-saline solution (1%)					X	
Mayer's albumin		X				X
Control slides or suspensions		X				X (M)

[a] Adapted from H. D. Isenberg, ed., *Clinical Microbiology Procedures Handbook*, American Society for Microbiology, Washington, D.C., 1992.

[b] PVA, polyvinyl alcohol; MIF, merthiolate-iodine-formalin; SAF, sodium acetate-acetic acid-formalin; W, Wheatley; G, Gomori; DMSO, dimethyl sulfoxide; We, Weber's modified trichrome green; R, Ryan's modified trichrome blue; M, microsporidia; X+, contains acetic acid; X−, no acetic acid; X+−, available with and without acetic acid.

[c] Some companies also have single-vial stool collection systems other than the SAF vials (non-mercuric chloride base, non-formalin). Contact those companies who are listed in the table as having routine fecal collection vials available (formalin, PVA, PVA with modified bases, MIF, SAF).

[d] Use grade with high hydrolysis and low viscosity for parasite studies.

[e] Used for the identification of microsporidian spores in stool or other specimens.

Table A5.1   *(continued)*

Harleco	Medical Chemical Corp.	Meridian Bioscience, Inc.	PML Microbiologicals	Remel	Rowley Biochemical, Inc.	Scientific Device Lab	Volu-Sol
	X	X	X	X			
	X	X	X	X			
	X	X	X	X			
	X	X	X	X			
	X	X	X	X		X	
						X	
	X +	X +		X −	X −		X −
	X	X	X	X		X	
	X	X					
	X	X	X				
	X	X				X	
	X					X	
	X(W)	X(W)	X	X(W)	X(W, G)	X(W)	X(W, G)
	X	X	X	X	X	X	X
X					X		
X					X		
X				X	X		
	X		X		X		X
X					X		
					X		
X	X				X		X
X	X				X		
X	X		X	X	X		X
			X				X
	X		X	X	X	X	X
					X		X
	X				X		
X	X	X	X	X	X		X
	X			X		X	
				X			
	X	X		X		X	
	X						
				X			
	X	X	X	X		X	

Table A5.2 Addresses of suppliers listed in Table A5.1

AJP Scientific, Inc. (Eng Scientific)
P.O. Box 1589
Clifton, NJ 07015
(800) 922-0223
(973) 472-7200
E-mail: info@engscientific.com
Products also available through various distributors

ALPHA-TEC Systems, Inc.
P.O. Box 5435
Vancouver, WA 98668
(800) 221-6058
(360) 260-2779
Fax: (360) 260-3277
Website: www.alphatecsystems.com
Products also available through various distributors

J. T. Baker
222 Red School Lane
Phillipsburg, NJ 08865
(908) 859-2151
Fax: (908) 859-9385
Website: www.jtbaker.com
Products also available through American Scientific Products or VWR

City Chemical Corp.
139 Allings Crossing Road
West Haven, CT 06516
(800) 248–2436
(203) 932–2489
E-mail: sales@citychemical.com
Chlorazol black E stain

Evergreen Scientific
2300 East 49th Street
P.O. Box 58248
Los Angeles, CA 90058-0248
(800) 421-6261
(800) 372-7300 in California
(213) 583-1331
Fax: (213) 581-2503
Website: www.evergreensci.com
Products also available through various distributors

Fisher Scientific
2761 Walnut Avenue
Tustin, CA 92681
(800) 766-7000 (national no.)
Fax: (800) 926-1166 (national no.)
Website: www.fishersci.com
Products available through regional sales offices

Hardy Diagnostics
1430 West McCoy Lane
Santa Maria, CA 93455
(800) 266-2222
Fax: (805) 346-2760
Website: www.hardydiagnostics.com
E-mail: sales@hardydiagnostics.com
Products also available through various distributors

Harleco
Voigt Global Distribution LLC
P.O. Box 412762
Kansas City, MO 64141–2762
(877) 484–3552
Website: www.vgdusa.com
Products available through American Scientific Products; product numbers remain unchanged

Medical Chemical Corp.
19430 Van Ness Avenue
Torrance, CA 90501
(800) 424-9394
Fax: (310) 787-4464
Website: www.med-chem.com
Products also available through various distributors

Meridian Bioscience, Inc.
3741 River Hills Drive
Cincinnati, OH 45244
(800) 543-1980
(513) 271-3700
Fax: (513) 271-0124
Website: www.meridianbioscience.com
Parasite Panels (teaching specimens)

MML Diagnostic Packaging
P.O. Box 458
Troutdale, OR 97060
(503) 666–8398
(800) 826–7186
Website: www.mmldiagnostics.com
Products available through various distributors

PML Microbiologicals
27120 SW 95th Ave.
Wilsonville, OR 97070
(503) 666-8398
(800) 828-7014 or (800) 547–0659, ext. 180 or 115
Fax: (800) 765-4415
Website: www.pmlmicro.com
Products also available through other distributors

Poly Science, Inc.
400 Valley Road
Warrington, PA 18976
(800) 523-2575
(215) 343-6484
Chlorozol Black E stain

Remel
12076 Santa Fe Drive
Lenexa, KS 66215
(800) 255-6730
Fax: (800) 477-5781
Website: www.remelinc.com
Products available through other distributors

(continued)

Table A5.2 *(continued)*

Rowley Biochemical, Inc.
10 Electronics Avenue
Danvers, MA 01923
(978) 739–4883
Fax: (978) 739-5640
Website: www.rowleybio.com
Products available only through company

Scientific Device Laboratory, Inc.
411 East Jarvis Avenue
Des Plaines, IL 60018
(847) 803-9495
Fax: (847) 803-8251
Website: www.scientificdevice.com
Products also available through various distributors
Cyclospora cayetanensis teaching slides available

Volu-Sol, Inc.
5095 West 2100 South
Salt Lake City, UT 84120
(800) 821-2495
(801) 974-9474
Fax: (800) 860-4317
Website: www.volusol.com
Products also available through various distributors
(Laboratory Specialists International, www.labspec.com)

Table A5.3 Sources of available reagents for immunodetection of parasitic organisms or antigens[a]

Company	Cryptosporidium (EIA)	Cryptosporidium-Giardia combination (DFA or EIA)	Entamoeba histolytica (EIA)	E. histolytical E. cispar group (EIA)	Giardia (EIA)	Crypto-Giardia (cartridge and/or separate cartridges)	Crypto-Giardia-E. histolytical E. dispar group (cartridge)	Pneumocystis jiroveci	Trichomonas vaginalis (various methods)
Antibodies, Inc.									
Biosite Diagnostics, Inc.					X		X (Triage Parasite Panel)		
Chemicon International, Inc.									X (DFA)
Dako								X (IFA), control slides also available	
Genzyme Diagnostics (Becton-Dickinson)									OSOM Trichomonas rapid test
Integrated Diagnostics (PanBio InDx)									X (latex agglutination)
Medical Chemical Corp.	X	X (DFA)			X	X			
Meridian Bioscience, Inc.	X	X (DFA)			X	X		X (DFA)	
MicroProbe Corp. (Becton-Dickinson)									X (DNA probe) (Affirm VPIII) (Trichomonas, Gardnerella, Candida)
Remel	X	X (EIA)	X		X	X			
TechLab, Inc.	X	X (DFA)	X	X	X				
Wampole (Inverness Medical)	X		X		X				

[a] Tests for the *Entamoeba histolytica/E. dispar* group or *E. histolytica* require the use of fresh or fresh, frozen stools. All other tests (with the exception of DFA for the *Cryptosporidium-Giardia* combination) can be used with fresh or preserved stools (formalinized base, *not* polyvinyl alcohol; some single-vial systems may be acceptable, so consult the manufacturer). Since the combination *Cryptosporidium-Giardia* tests (DFA) are based on visual recognition of the fluorescing oocysts and/or cysts, the specimens must not be frozen if fresh stools are used for testing. DFA, direct fluorescent-antibody assay; EIA, enzyme immunoassay.

Table A5.4 Addresses of suppliers listed in Table A5.3

Antibodies, Inc.
P.O. Box 1560
Davis, CA 95617-1560
(800) 824-8540
Fax: (530) 758-6307
Website: www.antibodiesinc.com
E-mail: antiinc@aol.com

Biosite Diagnostics, Inc.
9975 Summers Ridge Road
San Diego, CA 92121
(858) 455-4808
(888) Biosite
Fax: (858) 455-4815
Website: www.biosite.com

Chemicon International, Inc.
28835 Single Oak Drive
Temecula, CA 92590
(800) 437-7500
(951) 676–8080
Fax: (800) 437-7502
Website: www.chemicon.com

Dako
6392 Via Real
Carpinteria, CA 93013
(800) 235-5763
Fax: (800) 566-3256
Website: www.dakousa.com
E-mail: general@dakousa.com

Genzyme Corporate Offices
500 Kendall Street
Cambridge, MA 02142
(617) 252 7500
Fax (617) 252–7600
Website: www.genzyme.com

Integrated Diagnostics (PanBio InDx, Inc.)
1756 Sulphur Springs Rd.
Baltimore, MD 21227
(410) 737-8500
Fax: (410) 536-1212
E-mail: indx2@perols.com

Medical Chemical Corp.
19430 Van Ness Avenue
Torrance, CA 90501
(800) 424-9394
Fax: (310) 787-4464
Website: www.med-chem.com

Meridian Bioscience, Inc.
3471 River Hills Dr.
Cincinnati, OH 45244
(513) 271-3700
Fax: (513) 271- 0124
Website: www.meridianbioscience.com

MicroProbe Corp.
(Becton Dickinson Affirm VPIII)
1725 220th St. NE
Bothell, WA 98021
(*Trichomonas, Gardnerella, Candida*)
(201) 847-6800
Website: www.bd.com

Remel
12076 Santa Fe Drive
Lenexa, KS 66215
(800) 255-6730
Fax: (800) 477-5781
Website: www.remelinc.com

TechLab, Inc.
2001 Kraft Dr.
Blacksburg, VA 24060-6358
(800) 832-4522
(540) 953–1664
Fax: (540) 953–1665
Website: www.techlab.com

Wampole Lab (Inverness Medical)
P.O. Box 1001
Cranbury, NJ 08512
(800) 257-9525
Fax: (800) 532-0295
Website: www.invernessmedicalpd.com

Table A5.5 Commercial suppliers of diagnostic parasitology products[a]

Abbott Laboratories
100 Abbott Park Rd.
Abbott Park, IL 60064–3500
(847) 937–6100
Website: www.abbott.com
Toxoplasma (EIA for IgG, IgM)

Alexon-Trend (see Remel)

ALPHA-TEC Systems, Inc.
P.O. Box 5435
Vancouver, WA 98668
(800) 221-6058
(360) 260-2779
Fax: (360) 260-3277
Website: www.alphatecsystems.com
Control slides, reagents, stains (including polychrome IV), collection systems, concentration systems, QC suspensions

Amrad ICT (see Zenyth Therapeutics Limited)
13 Rodborough Rd.
French's Forest, NSW 2086
Australia
(612) 9453-4411
Fax: (612) 9453-4411
Website: www.amrad.com.au
Antigen detection (rapid or dipstick) for filariasis, malaria, and *P. falciparum* malaria

Bayer Diagnostics
511 Benedict Ave.
Tarrytown, NY 10591
(800) 242–2787
Website: www.bayer.com
Toxoplasma (EIA for IgG and IgM)

Beckman Coulter
4300 N. Harbor Blvd.
Fullerton, CA 92834
(800) 233-4685
Fax: (800) 643-4366
Website: www.beckman.com
Toxoplasma (EIA for IgG and IgM)

Becton Dickinson Advanced Diagnostics
2350 Qume Dr.
San Jose, CA 95131-1087
(877) 232–8995, Operator 5
Fax: (800) 954-2347
Website: www.bdfacs.com
Malaria P.f. ParaSight F dipstick, QBC acridine orange tube

Biokit USA, Inc.
113 Hartwell Ave.
Lexington, MA 02173
(800) 926-2253
Fax: (617) 861-4065
Website: www.biokitusa.com
Toxoplasma (LA for IgG)

BIOMED Diagnostics
1430 Koll Circle, Suite 101
San Jose, CA 95112
(800) 964-6466
InPouch TV (visual identification and culture system for *Trichomonas vaginalis*)

bioMérieux Vitek
595 Anglum Dr.
Hazlewood, MO 63042
(800) 638-4835
Website: www.biomerieux.com
Toxoplasma (EIA, VIDAS instrument)

Biosite Diagnostics, Inc.
9975 Summers Ridge Road
San Diego, CA 92121
(858) 455-4808
Fax: (858) 455-4815
Website: www.biosite.com
Triage Parasite Panel-Rapid (*Cryptosporidium, Giardia, Entamoeba histolytica/E. dispar*)

Biotecx Labs
6023 S. Loop East
Houston, TX 77033
(800) 535-6286
Fax: (713) 643-3743
Toxoplasma (EIA for IgG and IgM)

Biotools
Av. General Peron
2 E-28020 Madrid, Spain
(341) 571-1660
Fax: (341) 571-1232
Website: www.btools.com
Malaria (PCR, DNA)

Chemicon International, Inc.
28835 Single Oak Drive
Temecula, CA 92590
(800) 437-7500
Fax: (800) 437-7502
Website: www.chemicon.com
Trichomonas vaginalis (DFA) antigen detection

Diagnostic Products Corp.
5700 W. 96th St.
Los Angeles, CA 90045
(800) 444-5757
Fax: (800) 234-4872
Website: www.dpcweb.com
Toxoplasma (EIA, Immulite, automated)

DiaSorin
1990 Industrial Blvd.
P.O. Box 285
Stillwater, MN 55082
(800) 328-1482
Fax: (612) 779-7847
Website: www.diasorin.com
Toxoplasma (EIA for IgG and IgM)

(continued)

Table A5.5 *(continued)*

DiaSys Corp.
49 Leavenworth St.
Waterbury, CT 06702
Website: www.diasyscorp.com
Semiautomated stool concentrate examination system

Eastman Kodak Co.
Rochester, NY 14650
(800) 225-5352
(716) 458-4014
Products available only through American Scientific
Products, Fisher Scientific, VWR, and company PVA
powder, Giemsa powder

Evergreen Scientific
2300 East 49th St.
Los Angeles, CA 90058
(800) 421-6261
(213) 583-1331
Fax: (213) 581-2503
Website: www.evergreensci.com
Stool collection system, concentration system, pinworm
collection

Flow Inc.
6127 SW Corbett
Portland, OR 97201
(503) 246-2710
Fax: (503) 245-7666
Website: www.malariatest.com
E-mail: MIKEatFLOW@aol.com
Anti-pLDH *Plasmodium* monoclonal antibodies[b]

GenBio
15222-A Ave. of Science, Suite A
San Diego, CA 92128
(858) 592-9300
Website: www.genbio.com
Toxoplasma (EIA and IFA for IgG and IgM)
TORCH PANEL

Genzyme Corporate Offices
500 Kendall St.
Cambridge, MA 02142
(617) 252–7500
Fax: (617) 252–7600
ColorPac, immunochromatographic rapid immunoassay
for *Cryptosporidium-Giardia*

Hardy Diagnostics
1430 West McCoy Ln.
Santa Maria, CA 93455
(800) 266-2222
Fax: (805) 346-2760
Website: http://hardydiagnostics.com
Stains, reagents, collection system, concentration
systems, control slides

HDC Corp.
2109 O'Toole Ave., Suite M
San Jose, CA 95131
(408) 954-1909
Fax: (408) 954-0340
Website: www.hdccorp.com
Entero-Test capsules (adult and pediatric) (method of sampling upper
gastrointestinal tract)

Hemagen Diagnostics
34–40 Bear Hill Rd.
Waltham, MA 02154
(800) 436-2436
Fax: (781) 890-3748
Website: www.hemagen.com
Chagas' disease (EIA), *Toxoplasma* (EIA, IFA)

Immunetics
63 Rogers St.
Cambridge, MA 02142
(617) 492-5416
Fax: (617) 868-7879
Website: www.immunetics.com
Babesia (IB),[b] Chagas' disease (IB),[b] cysticercosis, (IB),[b] echinococcosis
(IB),[b] *Leishmania* (IB)[b]

INOVA
10180 Scripps Ranch Rd.
San Diego, CA 92131
(800) 545-9495
Fax: (619) 586-9911
Website: www.inovadx.com
Toxoplasma (EIA)

Integrated Diagnostics Inc.
(PanBio InDx, Inc.)
1756 Sulphur Springs Rd.
Baltimore, MD 21227
(410) 737-8500
Fax: (410) 536-1212
Website: www.indxdi.com
E-mail: indx2@erols.com
Trichomonas (LA) antigen detection

Interfacial Dynamics Corp.
17300 SW Upper Boones Ferry Rd., Suite 120
Portland, OR 97224
(503) 256-0076
(800) 323-4810
Fax: (503) 255-0989
E-mail: idclatex@teleport.com
Uniform-sized polystyrene microspheres (can be used to check
microscope calibrations)

Intracel
2005 NW Sammamish Rd., Suite 107
Issaquah, WA 98027
(800) 227-8357
Website: www.intracel.com
E-mail: info@intracel.com
Toxoplasma (EIA)

(continued)

Table A5.5 Commercial suppliers of diagnostic parasitology products[a] *(continued)*

IVD Research Inc.
5909 Sea Lion Place, Suite D
Carlsbad, CA 92008
(760) 929-7744
Fax: (760) 431-7759
Website: www.ivdresearch.com, www.safepath.com
(SafePath offers numerous assays for veterinary, environmental, and food safety applications)
Amebiasis (EIA), cysticercosis (EIA), echinococcosis (EIA), *Toxocara* (EIA), trichinosis (EIA), *Toxoplasma* (EIA)

KMI Diagnostics Inc.
818 51st Ave., NE, Suite 101
Minneapolis, MN 55421
(612) 572-9354
Fax: (612) 586-0748
Website: www.kmidiagnostics.com
Toxoplasma (IFA)

Medical Chemical Corp.
19430 Van Ness Avenue
Torrance, CA 90501
(800) 424-9394
Fax: (310) 787-4464
Website: www.med-chem.com
Reagents, modified trichrome stain for microsporidia, other stains, collection system, concentration system, parasitology website (current information, photographs, commonly asked questions, etc.)

Meridian Bioscience, Inc.
3741 River Hills Dr.
Cincinnati, OH 45244
(800) 543-1980
(513) 271-3700
Fax: (513) 271-0124
Stains, reagents, collection system, concentration system, *Cryptosporidium* spp. (EIA, FA), *Giardia* spp. (EIA, FA)

MicroProbe Corp.
(Becton Dickinson Affirm VPIII)
1725 220th St. NE
Bothell, WA 98021
BD (201) 847-6800
Website: www.bertec.com.tw
(*Trichomonas, Gardnerella, Candida*)
ColorPAC Rapid (*Cryptosporidium-Giardia*)

MML Diagnostic Packaging
P.O. Box 458
Troutdale, OR 97060
(503) 666-8398
(800) 826-7186
Collection system

PML Microbiologicals
27120 SW 95th Ave.
Wilsonville, OR 97070
(800) 547-0659
(503) 639-1500 in Oregon
Fax: (800) 765-4415
Website: www.pmlmicro.com
Stains, reagents, collection system, concentration system, control slides

Polysciences Inc.
400 Valley Rd.
Warrington, PA 18976
(800) 523-2575
Fax: (215) 343-0214
Website: www.polysciences.com
Pneumocystis (DFA) antigen detection

Regional Media Laboratories (Remel)
12076 Santa Fe Dr.
Lenexa, KS 66215
(800) 255-6730
Fax: (800) 477-5781
Website: www.remelinc.com
Stains, reagents, collection system, concentration system, control slides

Sanofi Diagnostics Pasteur
1000 Lake Hazeltine Dr.
Chaska, MN 55318
(800) 666-5111
Fax: (612) 368-1110
Toxoplasma (EIA, IgG, IgM), *Pneumocystis* (IFA)

Scientific Device Laboratory, Inc.
411 E. Jarvis Ave.
Des Plaines, IL 60018
(847) 803-9495
Fax: (847) 803-8251
Website: www.scientificdevice.com
Stains, stain kit, modified trichrome stain for microsporidia, fixative (formalin free), formalin-fixed protozoa and helminth eggs and larvae, control slides (*Cryptosporidium, Isospora,* microsporidia, and *Pneumocystis*), stained slides

TechLab
2001 Kraft Drive
Blacksburg, VA 24060-6358
(540) 953-1664
Fax: (540) 953-1665
Website: www.techlab.com
E-mail: techlab@techlab.com
EIA antigen detection (*Giardia, Cryptosporidium, Entamoeba histolytica, Giardia-Cryptosporidium*)

Volu-Sol, Inc. (a division of Biomune, Inc.)
5095 West 2100 South
Salt Lake City, UT 84120
(800) 821-2495
(801) 974-9474
Fax: (800) 860-4317
Website: www.volusol.com
Parasitology starter kit, stains, reagents, control slides

VWR Scientific, Inc.
P.O. Box 7900
San Francisco, CA 94120
(415) 468-7150 in Northern California
(213) 921-0821 in Southern California
Rest of United States, use either number
Stains, reagents, collection system

(continued)

Table A5.5   *(continued)*

Wampole Lab (Inverness Medical) P.O. Box 1001 Cranbury, NJ 08512 (800) 257-9525 Fax: (800) 532-0295 Website: www.invernessmedicalpd.com EIA antigen detection (*Entamoeba histolytica*, *Giardia*, *Cryptosporidium*)	Xenotope Diagnostics (contact www.genzymediagnostics.com) 3463 Magic Dr., Suite 360 San Antonio, TX 78229 (210) 582–5838 Fax: (210) 582–5839 OSOM *Trichomonas* rapid test for antigen

a Modified from H. D. Isenberg, ed., *Clinical Microbiology Procedures Handbook*, 2nd ed. ASM Press, Washington, D.C., 2004. Much of the updated immunology testing information was provided by Marianna Wilson (CDC). Abbreviations: DFA, direct fluorescent-antibody assay; EIA, enzyme immunoassay; IB, immunoblot; IFA, indirect fluorescent-antibody assay; IgG, immunoglobulin G; LA, latex agglutination; PVA, polyvinyl alcohol; QC, quality control; Rapid, rapid immunochromatographic.
b Not cleared by the Food and Drug Administration for in vitro diagnostic use.

Table A5.6   Sources of parasitologic specimens (catalogs of available materials and price lists available from the companies listed)*a*

ALPHA-TEC Systems, Inc. P.O. Box 5435 Vancouver, WA 98668 (800) 221-6058 (360) 260-2779 Fax: (360) 260-3277 Website: www.alphatecsystems.com Control slides and suspensions	Meridian Bioscience, Inc. Proficiency & Controls Department 3741 River Hills Drive Cincinnati, OH 15211 (800) 543-1980, ext. 335 (513) 271-3700, ext. 335 Fax: (513) 271-0124 Website: www.meridianbioscience.com E-mail: jross@meridianbioscience.com Parasite Panels (teaching specimens) formalin, slides (protozoa, helminths, blood parasites)
Carolina Biological Supply Co. 2700 York Road Burlington, NC 27215 (800) 334-5551 (919) 584-0381 Fax: (800) 222-7112 Website: www.carolina.com Individual slides, slide sets, CDM disk and slides	Scientific Device Laboratory, Inc. 411 E. Jarvis Avenue Des Plaines, IL 60018 (800) 448-4855 (708) 803-9495 Fax: (708) 803-8251 Website: www.scientificdevice.com Control slides and/or suspensions, *Cyclospora cayetanensis* teaching slides
McGill University Centre for Tropical Diseases National Reference Centre for Parasitology Montreal General Hospital Research Institute room R3-137 1650 Cedar Avenue Montreal, Quebec H3G 1A4 Canada Contact: evelyne.kokoskin@mcgill.ca Website: www.medicine.mcgill.ca/tropmed/txt/sales.htm (514) 934–8049 Fax: (514) 934–8347 Teaching specimens: formalin, blood films, teaching sets	Ward's Natural Science Establishment, Inc. P.O. Box 92912 Rochester, NY 14692 (800) 962-2660 Website: www.wardsci.com Slide sets, but availability may be a problem
Medical & Science Media P.O. Box 3224 Mt. Druitt Village, NSW Australia 61 2 9675 7750 Fax: 61 2 9675 7702 Website: www.msmedia.com.au Extensive lists of slides, photomicrographs, CD-ROM disks	

a Adapted from H. D. Isenberg, ed., *Clinical Microbiology Procedures Handbook*, 2nd ed. ASM Press, Washington, D.C., 2004.

Table A5.7   Sources of Kodachrome study slides (35 mm, 2 × 2) for rental[a]

Armed Forces Institute of Pathology Department of ID/Parasitic Disease Pathology 6825 16th Street, NW Bldg. 54, Room 4015 Washington, DC 20306-6000 Website: www.afip.org Study slide sets for rent, extensive parasite and tropical-diseases listings	Clinical Diagnostic Parasitology (Lynne S. Garcia) 512 12th St. Santa Monica, CA 90402 Visual Teaching Aids (set of 100 2 × 2 slides) Diagnostic Protozoa, Helminths, and Blood Parasites
American Society of Clinical Pathologists ASCP Press 2100 West Harrison Street Chicago, IL 60612-3798 (312) 738-4890 Human Parasitology, teaching slide set (Ash and Orihel, 1990), Supplement to Human Parasitology, teaching slide set (in preparation), Parasites in Human Tissues, teaching slide set (Orihel and Ash, 1996)	Medical & Science Media P.O. Box 3224 Mt. Druitt Village, NSW Australia 61 2 9675 7750 Fax: 61 2 9675 7702 Website: www.msmedia.com.au Extensive lists of slides, photomicrographs, CD-ROM disks
Carolina Biological Supply Co. 2700 York Road Burlington, NC 27215 (800) 334-5551 (919) 584-0381 Fax: (800) 222-7112 Website: www.carolina.com Individual slides, slide sets, CD-ROM disk and slides	Dr. Herman Zaiman (A Pictorial Presentation of Parasites) P.O. Box 543 Valley City, ND 58072

[a] Adapted from H. D. Isenberg, ed., *Clinical Microbiology Procedures Handbook*, 2nd ed. ASM Press, Washington, D.C., 2004.

Table A5.8   Sources of additional teaching materials, including case histories (see also chapter 36)

Centers for Disease Control and Prevention MS-F36 4770 Buford Highway NE Atlanta, GA 30341-3724 (404) 639-3331 (800) 311-3435 Website: www.dpd.cdc.gov/DPDx/HTML/Contactus.htm Extensive training materials, including the DPDx CD-ROM disk Bench Aids for teaching	Medical Chemical Corp. 19430 Van Ness Avenue Torrance, CA 90501 (800) 424-9394 Fax: (310) 787-4464 Website: www.med-chem.com Parasitology website (current information, photographs, commonly asked questions, case histories) (developed by Lynne Garcia)

APPENDIX 6
Reference Sources

GENERAL REFERENCES

1. Ash, L. R., and T. C. Orihel. 1997. *Atlas of Human Parasitology*, 4th ed. ASCP Press, Chicago, Ill.

2. Ash, J. R., and T. C. Orihel. 1991. *Parasites: a Guide to Laboratory Procedures and Identification.* ASCP Press, Chicago, Ill.

3. Ash, L. R., T. C. Orihel, and L. Savioli. 1994. *Bench Aids for the Diagnosis of Intestinal Parasites.* World Health Organization, Geneva, Switzerland.

4. Beaty, B. J., and W. C. Marquardt (ed.). 1996. *The Biology of Disease Vectors.* University Press of Colorado, Niwot.

5. Beaver, P. C., R. C. Jung, and E. W. Cupp. 1984. *Clinical Parasitology*, 9th ed. Lea & Febiger, Philadelphia, Pa.

6. Binford, C. H., and D. H. Connor. 1976. *Pathology of Tropical and Extraordinary Diseases.* Armed Forces Institute of Pathology, Washington, D.C.

7. Borchardt, K. A., and M. A. Noble (ed.). 1997. *Sexually Transmitted Diseases.* CRC Press, Inc., Boca Raton, Fla.

8. Chernin, E. 1977. A bicentennial sampler: milestones in the history of tropical medicine and hygiene. *Am. J. Trop. Med. Hyg.* **26:**1053–1104.

9. Cook, G. C., and A. Zumla (ed.). 2003. *Manson's Tropical Diseases*, 21st ed. The W. B. Saunders Co., Philadelphia, Pa.

10. Cox, F. E. G., J. P. Kreier, and D. Wakelin (ed.). 2005. Parasitology, vol. 5. *In Topley & Wilson's Microbiology and Microbial Infections*, 10th ed. Edward Arnold, London, United Kingdom.

11. Despommier, D. D., R. W. Gwadz, and P. J. Hotez (ed.). 1995. *Parasitic Diseases*, 3rd ed. Springer-Verlag, New York, N.Y.

12. Eldridge, B. F., and J. D. Edman (ed.). 2000. *Medical Entomology*, Kluwer Academic Publishers, Norwell, Mass.

13. Faust, E. C. 1949. *Human Helminthology*, 3rd ed. Lea & Febiger, Philadelphia, Pa.

14. Fayer, R. (ed.). 1997. Cryptosporidium *and Cryptosporidiosis.* CRC Press, Inc., Boca Raton, Fla.

15. Foster, W. D. 1965. *A History of Parasitology.* E. & S. Livingstone, London, United Kingdom.

16. Garcia, L. S. 1999. *Practical Guide to Diagnostic Parasitology.* ASM Press, Washington, D.C.

17. Goddard, J. 2003. *Physician's Guide to Arthropods of Medical Importance*, 4th ed. CRC Press, Inc., Boca Raton, Fla.

18. Harwood, R. F., and M. T. James. 1979. *Entomology in Human and Animal Health.* Macmillan, New York, N.Y.

19. Hoeppli, R. 1962. *Parasites and Parasitic Infections in Early Medicine and Science.* University of Malaya Press, Singapore.

20. Horsburgh, C. R., Jr., and A. M. Nelson (ed.). 1997. *Pathology of Emerging Infections.* ASM Press, Washington, D.C.

21. Horsfall, W. R. 1962. *Medical Entomology: Arthropods and Human Disease.* The Ronald Press Co., New York, N.Y.

22. Jong, E. C., and R. McMullen. 1995. *The Travel and Tropical Medicine Manual*, 2nd ed. The W. B. Saunders Co., Philadelphia, Pa.

23. Kean, B. H., K. E. Mott, and A. J. Russell. 1978. *Tropical Medicine and Parasitology Classic Investigations.* Cornell University Press, Ithaca, N.Y.

24. Kettle, D. S. 1995. *Medical and Veterinary Entomology*, 2nd ed. CAB International, Wallingford, United Kingdom.

25. Kokoskin, E. 2001. *The Malaria Manual.* McGill University Centre for Tropical Diseases, Montreal, Canada.

26. Lambert, H. P., and W. E. Farrar. 1982. *Infectious Diseases Illustrated.* The W. B. Saunders Co., Philadelphia, Pa.

27. Manson-Bahr, P. E. C., and F. I. C. Apted. 1982. *Manson's Tropical Diseases*, 18th ed. Bailliere Tindall, London, United Kingdom.

28. Markell, E., D. T. John, and W. A. Krotoski. 1999. *Medical Parasitology*, 8th ed. The W. B. Saunders Co., Philadelphia, Pa.

29. Meyers, W. M., R. C. Neafie, A. M. Marty, and D. J. Wear (ed.). 2000. *Pathology of Infectious Diseases*, vol. 1. *Helminthiases.* Armed Forces Institute of Pathology, Washington, D.C.

30. Murray, P. R., E. J. Baron, J. H. Jorgensen, M. A. Pfaller, and R. H. Yolken (ed.). 2003. *Manual of Clinical Microbiology*, 8th ed. ASM Press, Washington, D.C.

31. Nelson, A. M., and C. R. Horsburgh, Jr. 1998. *Pathology of Emerging Infections*, 2nd ed. ASM Press, Washington, D.C.

32. **Orihel, T. C., and L. R. Ash.** 1995. *Parasites in Human Tissues.* ASCP Press, Chicago, Ill.

33. **Orihel, T. C., L. R. Ash, and C. P. Ramachandran.** 1996. *Bench Aids for the Diagnosis of Filarial Infections.* World Health Organization, Geneva, Switzerland.

34. **Peters, W., and H. M. Gilles.** 1992. *Color Atlas of Tropical Medicine and Parasitology.* Year Book Medical Publishers, Inc., Chicago, Ill.

35. **Reeder, M. M., and P. E. S. Palmer.** 1981. *The Radiology of Tropical Diseases with Epidemiological, Pathological and Clinical Correlation.* The Williams & Wilkins Co., Baltimore, Md.

36. **Rondanelli, E. G., and M. Scaglia.** 1993. *Atlas of Human Protozoa.* Masson, Milan, Italy.

37. **Scarpignato, C., and P. Rampal (ed.).** 1995. *Travelers' Diarrhea: Recent Advances.* Karger Publishers, Farmington, Conn.

38. **Schmidt, G. D., and L. S. Roberts.** 1985. *Foundations of Parasitology,* 3rd ed. The C. V. Mosby Co., St. Louis, Mo.

39. **Scott, H. H.** 1939. *A History of Tropical Medicine.* Edward Arnold & Co., London, United Kingdom.

40. **Sherman, I. W. (ed.).** 1998. *Malaria.* ASM Press, Washington, D.C.

41. **Smith, K. C. V.** 1973. *Insects and Other Arthropods of Medical Importance.* John Wiley & Sons, Inc., New York, N.Y.

42. **Spencer, F. M., and L. S. Monroe.** 1976. *The Color Atlas of Intestinal Parasites,* 2nd ed. Charles C Thomas, Publisher, Springfield, Ill.

43. **Steele, J. H.** 1982. *CRC Handbook Series in Zoonoses.* CRC Press, Inc., Boca Raton, Fla.

44. **Strickland, G. T.** 2000. *Hunter's Tropical Medicine,* 8th ed. The W. B. Saunders Co., Philadelphia, Pa.

45. **Sun, T.** 1999. *Parasitic Disorders,* 2nd ed. The Williams & Wilkins Co., Baltimore, Md.

46. **Taylor, A. E. R., and J. R. Baker.** 1978. *Methods of Cultivating Parasites In Vitro.* Academic Press Inc., New York, N.Y.

47. **Von Brand, T.** 1979. *Biochemistry and Physiology of Endoparasites.* Elsevier/North Holland Publishing Co., New York, N.Y.

48. **Warren, K. S., and A. A. F. Mahmoud.** 1990. *Tropical and Geographical Medicine,* 2nd ed. McGraw Hill Book Co., New York, N.Y.

49. **Wittner, M. (ed.), and L. M. Weiss (contributing ed.).** 1999. *The Microsporidia and Microsporidiosis.* ASM Press, Washington, D.C.

50. **Yamaguchi, T.** 1981. *Color Atlas of Clinical Parasites.* Lea & Febiger, Philadelphia, Pa.

51. **Zaman, V.** 1979. *Atlas of Medical Parasitology.* Lea & Febiger, Philadelphia, Pa.

JOURNALS AND PERIODICALS

Acarologia, 1959 to date.

Acta Tropica, 1944 to date.

Advances in Parasitology, 1963 to date.

American Journal of Tropical Medicine, 1921 to 1951.

American Journal of Tropical Medicine and Hygiene, 1952 to date.

Annals of the Entomological Society of America, 1908 to date.

Annals of Tropical Medicine and Parasitology, 1907 to date.

Annual Review of Entomology, 1956 to date.

Bulletin of Entomological Research, 1910 to date.

Clinical Microbiology Reviews, 1988 to date.

Emerging Infectious Diseases, 1995 to date.

Experimental Parasitology, 1951 to date.

International Journal for Parasitology, 1971 to date.

Journal of Clinical Microbiology, 1963 to date.

Journal of Helminthology, 1923 to date.

Journal of Medical Entomology, 1964 to date.

Journal of Tropical Medicine and Hygiene, 1898 to date.

Memórias do Instituto Oswaldo Cruz, 1909 to date.

Molecular and Biochemical Parasitology, 1980 to date.

Mosquito News, 1941 to date.

Mosquito Systematics, 1969 to date.

Parasite Immunology, 1979 to date.

Parasitology, 1908 to date.

Parasitology Research, 1987 to date.

Parasitology Today, 1985 to date.

Proceedings of the Helminthological Society of Washington, 1934 to date.

Puerto Rico Journal of Public Health and Tropical Medicine, 1925 to 1950.

Systematic Parasitology, 1979 to date.

The Journal of Eukaryotic Microbiology, 1993 to date.

Transactions of the American Microscopical Society, 1892 to date.

Transactions of the Royal Society of Tropical Medicine and Hygiene, 1907 to date.

Trends in Parasitology, 2001 to date.

Tropical and Geographical Medicine, 1949 to date (continuation of *Documenta de Medicina Geographica et Tropica*).

Veterinary Parasitology, 1975 to date.

World Health Organization Monographs, Technical Reports, Reports of Expert Committees, Bulletin, and *Chronicle,* 1948 to date.

ABSTRACTS AND BIBLIOGRAPHIC SOURCES

Biological Abstracts, 1929 to date.

Excerpta Medica, section 4, *Microbiology,* 1948 to date.

Helminthological Abstracts, 1932 to date.

Index Catalogue of Medical and Veterinary Zoology (author catalog), 1932 to 1952 with current supplements.

Index Medicus, 1878 to 1889, 1903 to 1927, 1960 to date.

Protozoological Abstracts, 1977 to date.

Quarterly Cumulative Index, 1916–1926, 1927–1956.

Quarterly Bibliography of Major Tropical Diseases, 1978 to date.

Review of Applied Entomology, section B (*Medical and Veterinary*), 1913 to date.

Tropical Diseases Bulletin, 1913 to date.

WEBSITES

Parasite Images

Atlas of Medical Parasitology, Carlo Denegri Foundation: www.cdfound.to.it/default.htm

Department of Parasitology, Chiang Mai University, Thailand:
www.medicine.cmu.ac.th/dept/parasite/image.htm

Graphic Images of Parasites, Parasites and Parasitological Resources:
www.ryoko.biosci.ohio-state.edu/~parasite/home.html

Kansas State University, Division of Biology:
www.k-state.edu/parasitology/

Oklahoma State University College of Veterinary Medicine:
www.cvm.okstate.edu/~users/jcfox/htdocs/clinpara/Index.htm

Parasite Image Library:
www.dpd.cdc.gov/DPDx/HTML/Image_Library.htm

Pictorial Presentation of Parasites:
www.parasite.biology.uiowa.edu/image

University of Delaware:
www.udel.edu/medtech/diehman/medt372images.html

World Wide Web Virtual Library, Parasitology:
www.aan18.dial.pipex.com/images.htm

Parasitology Information

Centers for Disease Control and Prevention:
www.dpd.cdc.gov/dpx/

Medical Chemical Corporation:
www.med-chem.com

Literature

NCBI National Library of Medicine (PubMed):
www.ncbi.nlm.nih.gov/entrez/query.fcgi

Government Websites Related to Health Care Regulations

Centers for Medicare and Medicaid Services (CMS):
www.cms.gov

CMS Fee Schedules:
www.cms.gov/stats

CMS Program Integrity Issues:
www.the-medicare.com

Compliance Program, Fraud Alerts, Advisory Opinions, Red Book, Work Plan:
www.oig.cms.gov

Medicare Learning Network:
www.cms.gov/medlearn

National Center for Health Statistics:
www.cdc.gov/nchswww/data

National Technical Information Service:
www.ntis.gov

Office of the Inspector General (OIG) Compliance Documents for Clinical Laboratories, Hospitals, and Third Party Billing:
www.gpoaccess.gov/databases.html

Color Plates of Diagnostic Stages of Human Parasites

Plate 1 (see p. 1108) Unless so noted, the protozoa in this plate were photographed using the oil immersion lens from trichrome-stained smears. (**A**) *Entamoeba histolytica* trophozoite. Note the ingested red blood cells (RBCs) within the cytoplasm and in the background. (**B**) *Entamoeba histolytica/E. dispar* trophozoite Merthiolate-iodine-formalin [MIF], polychrome stain). Note the evenly distributed nuclear chromatin and central karyosome. Trophozoites containing no RBCs are very typical; the presence of RBCs in the cytoplasm is the exception. (From A Pictorial Presentation of Parasites: A cooperative collection prepared and/or edited by H. Zaiman.) (**C**) *Entamoeba histolytica/E. dispar* precyst. Note the enlarged nucleus (prior to division) and presence of chromatoidal bars. (**D**) *Entamoeba histolytica/E. dispar* cyst. Note the multiple nuclei and the chromatoidal bars with smooth, rounded edges. Also note the "halo" due to shrinkage. (**E**) *Entamoeba hartmanni* trophozoites. Note the absence of RBCs. The nucleus mimics that seen with *E. histolytica* (evenly distributed nuclear chromatin and central karyosome). (**F**) *Entamoeba hartmanni* cyst. Although the mature cyst contains four nuclei, cysts are often seen with one (precyst) or two. Note the numerous chromatoidal bars; these are similar to those seen with *E. histolytica/E. dispar* (smooth, rounded edges). (**G**) *Entamoeba coli* trophozoite. Note the uneven nuclear chromatin and eccentric karyosome. (Courtesy of Parasitology Training Branch, Centers for Disease Control and Prevention.) (**H**) *Entamoeba coli* cyst (iodine stain, wet mount). The presence of five or more nuclei allows the identification to be made. These cysts are usually seen more clearly in wet preparations; they may be shrunk or distorted in the permanent stained smear (see panel I). (**I**) *Entamoeba coli* cyst. Note the shrinkage and pink color; this is typical on the permanent stained smear. (From L. S. Garcia, *in* S. M. Finegold and E. J. Baron, *Diagnostic Microbiology*, 7th ed., The C. V. Mosby Co., St. Louis, Mo., 1986.) (**J**) *Entamoeba coli* cyst. This is an excellent example of an *E. coli* cyst on the permanent stained smear; the presence of five or more nuclei allows this identification. (**K**) *Endolimax nana* trophozoites. Note the large karyosome and lack of peripheral chromatin. The organism on the right is an excellent example of "nuclear variation," where the nuclear chromatin displays unusual shapes. Nuclear variation is more common in *E. nana* than any of the other protozoa. (**L**) *Endolimax nana* cyst. Note the typical oval shape (they can also be round) and the presence of four karyosomes. (**M**) *Iodamoeba bütschlii* trophozoites. Note the large karyosome,

vacuolated cytoplasm, and size (compared with *E. nana*). (**N**) *Iodamoeba bütschlii* cysts. Note the large glycogen vacuole and "basket nucleus" (chromatin granules at one edge of the nuclear membrane; karyosome appears to be the basket, with the chromatin granules being the handle [use your imagination!]). (**O**) *Blastocystis hominis*. The "central body form" of this protozoan appears in an iodine preparation, and the frame on the right contains organisms on the permanent stained smear (photographed at a lower magnification; organisms overlap sizes of other protozoa). (**P**) *Dientamoeba fragilis* trophozoites. These organisms are characterized by having either one or two nuclei (nuclear chromatin frequently fragmented). There is no known cyst form. (**Q**) *Giardia lamblia* trophozoite. Note the two nuclei, curved median bodies, and linear axonemes. (**R**) *Giardia lamblia* cysts. Although these cysts are oval, some *G. lamblia* cysts appear more round (iron hematoxylin stain). (**S**) *Giardia lamblia* cyst. The nuclei, median bodies, and axonemes are clearly visible in this cyst. Frequently, the internal structures are not this clearly seen (iron hematoxylin stain). (**T**) *Chilomastix mesnili* trophozoites and cyst. The three trophozoites are characterized by the clear "oral groove" (feeding groove) seen coming down from the nucleus at the left side of the trophozoites. Without being able to see this oral groove, the identification of this trophozoite is very difficult. In the cyst (right frame) the "shepherd's crook," the curved fibril to the right of the nucleus, can be seen. (**U**) *Trichomonas vaginalis* trophozoites (Giemsa stain). Note the undulating membrane, flagella, and linear axostyle (penetrates the end of the organism). (**V**) *Trichomonas vaginalis* trophozoite (Giemsa stain). Note the undulating membrane, flagella, and axostyle. (**W**) *Balantidium coli* trophozoite. Note the large macronucleus and cilia; this organism is from an iodine preparation and is photographed at a lower magnification. (From L. S. Garcia, *in* E. J. Baron and S. M. Finegold, *Diagnostic Microbiology*, 8th ed., The C. V. Mosby Co., St. Louis, Mo., 1990.) (**X**) *Balantidium coli* in tissue (routine histologic preparation). Note the macronucleus that is partially visible in one of the trophozoites.

Plate 2 (see p. 1109) (**A**) Charcot-Leyden crystals (trichrome stain, oil immersion). Note the characteristic shape; various sizes are present in a single specimen. The presence of these crystals indicates an immune response, not necessarily directed against parasites. (**B**) Budding yeast cells and pseudohyphae (trichrome stain, oil immersion). (C) White blood cells (polymorphonuclear leukocytes [PMNs], eosinophils; trichrome stain, oil immersion). These cells can easily be confused with actual intestinal protozoa (PMN = *E. histolytica/E. dispar* cyst; macrophage = *E. histolytica/E. dispar* trophozoite). (D) White blood cells (PMNs; trichrome stain, oil immersion). Notice how the normally lobed nucleus begins to disintegrate and break into separate pieces. When this occurs, these human cells can be confused with *E. histolytica/E. dispar* cysts (the PMN may even contain four nuclear fragments, each having a "protozoan-like" nucleus). Human white blood cells have more nuclear material per cell than do the true protozoa. (E) Macrophage (trichrome stain, oil immersion). Note the large nucleus with a karyosome. These cells can often mimic *E. histolytica/E. dispar* trophozoites. However, the ratio of nuclear material to cytoplasm in these cells is large (1/6) compared with an *E. histolytica/E. dispar* trophozoite (1/10 to 1/12). (F) *Pneumocystis jiroveci* trophozoites (Giemsa stain). With this stain, the cyst wall does not stain; however, the trophozoites within the cyst wall can be seen. (From A Pictorial Presentation of Parasites: A cooperative collection prepared and/or edited by H. Zaiman.) (G) *Pneumocystis jiroveci* cysts (Gomori methenamine silver stain). Note the "parentheses" within the cyst (two commas facing one another). (H) *Pneumocystis jiroveci* cysts (calcofluor white fluorescent stain). (Courtesy of Susan Novak.) (I) *Pneumocystis jiroveci* cysts and trophozoites (monoclonal antibody fluorescent stain). (Courtesy of Meridian Diagnostics, Inc., Cincinnati, Ohio.) (J) *Toxoplasma gondii* bradyzoites in tissue. (From A Pictorial Presentation of Parasites: A cooperative collection prepared and/or edited by H. Zaiman.) (K) *Toxoplasma gondii* tachyzoites from mouse peritoneal cavity (Giemsa stain). (From A Pictorial Presentation of Parasites: A cooperative collection prepared and/or edited by H. Zaiman.) (L) *Cryptosporidium* oocysts (modified acid-fast stain). Occasionally, the four sporozoites within the oocyst are visible. These oocysts are immediately infective when they are passed in the stool. (M) *Giardia lamblia* cyst and *Cryptosporidium* oocysts (monoclonal antibody fluorescent stain-combination product). (Courtesy of Meridian Diagnostics, Inc.) (N) *Isospora belli* oocyst (iodine stain, wet mount). These organisms can be seen in the concentration sediment from formalin preservative; morphology of this particular organism will be very poor when concentrated from polyvinyl alcohol fixative. (O) *Isospora belli* oocyst (modified acid-fast stain). If the modified acid-fast stain for *Cryptosporidium* spp. is used, *I. belli* is also visible in the stained smear. (P) *Cyclospora* oocysts (saline, wet preparation). These organisms were formerly called CLBs (cyanobacterium-like or coccidian-like organisms). (Courtesy of Earl G. Long, Centers for Disease Control and Prevention.) (Q) *Cyclospora* oocysts (autofluorescence). Note the fluorescence around the oocysts (lower magnification than that in panel P). (Courtesy of Earl G. Long.) (R) *Cyclospora* oocysts (trichrome stain). Note that the oocysts do not stain with trichrome; they appear as "ghost cells" and are approximately twice the size of *Cryptosporidium* oocysts. (Courtesy of Earl G. Long.) (S) *Cyclospora* oocysts (modified acid-fast stain). Note the variability in staining; this is common and demonstrates the difference between these organisms and *Cryptosporidium* oocysts, which stain more consistently with this stain. (Courtesy of Earl G. Long.) (T) Concentrated duodenal aspirate with pinkish-red microsporidian spores (Weber's chromotrope stain, ×1,150). (Courtesy of Rainer Weber and Ralph T. Bryan, Division of Parasitic Diseases, National Center for Infectious Diseases, Centers for Disease Control and Prevention.) (U) Microsporidian spores (polyclonal fluorescent stain [reagent courtesy of Charles

Zierdt]). In some spores, horizontal or diagonal lines (polar tubule) can be seen. In rare cases the extruded polar tubule is visible (not in this particular picture). (V) Microsporidian spores (acid-fast stain). Note the variability in staining. (Armed Forces Institute of Pathology photograph.) (W) Microsporidian spores in corneal tissue (Gomori methenamine silver stain). Note the spore outline within the tissue. (X) Microsporidian spores in corneal tissue (periodic acid-Schiff [PAS] stain). Note the spore outline within the tissue; also note the PAS-positive granule at the end of each spore.

Plate 3 (see p. 1110) (A) *Ascaris lumbricoides* unfertilized egg. Note the thick, bumpy shell. (B) *Ascaris lumbricoides* fertilized egg (iodine stain, wet mount). Note the thick, bumpy shell (more rounded off than those seen in the unfertilized egg) and more rounded shape. (C) *Enterobius vermicularis* (pinworm) eggs on tape preparation. Note the shape (football shaped with one flattened side). (D) *Trichuris trichiura* (whipworm) egg (iodine stain, wet mount). Note the barrel-shaped thick shell and polar plugs at each end of the egg. (E) Hookworm egg (iodine stain, wet mount). Note the developing embryo (usually 4- to 16-ball stage of development) and the clear space between the egg shell and developing embryo. (From L. S. Garcia, *in* E. J. Baron and S. M. Finegold, *Diagnostic Microbiology*, 8th ed., The C. V. Mosby Co., St. Louis, Mo., 1990.) (F) *Strongyloides stercoralis* rhabditiform larva (iodine stain, wet mount). Note the short mouth opening and packet of genital primordial cells. (From L. S. Garcia, *in* E. J. Baron and J. M. Finegold, *Diagnostic Microbiology*, 8th ed., The C. V. Mosby Co., St. Louis, Mo., 1990.) The right frame shows *Strongyloides stercoralis* rhabditiform larva (trichrome stain). The internal morphology is rarely seen on the permanent stained smear; the wet mount examination is recommended. (G) *Trichinella spiralis* larvae in muscle (routine histologic preparation). (From L. S. Garcia, *in* E. J. Baron and S. M. Finegold, *Diagnostic Microbiology*, 8th ed., The C. V. Mosby Co., St. Louis, Mo., 1990.) (H) Hydatid sand from hydatid cyst (immature scolices). Note the row of hooklets (dark area). (I) *Taenia solium* (pork tapeworm) scolex. Note the suckers and hooklets ("armed rostellum"). (J) *Taenia solium* (pork tapeworm) gravid proglottid (India ink injection). The number of side branches on one side (counted where they come off the main uterine stem) is normally fewer than 12 (ca. 8). The proglottids are passed singly or in small numbers attached to each other. (K) *Taenia saginata* (beef tapeworm) scolex. Note the four suckers and no hooklets. (L) *Taenia saginata* (beef tapeworm) gravid proglottid (India ink injection). The number of side branches on one side (counted where they come off the main uterine stem) is normally more than 12 (ca. 16). The proglottids are usually passed singly. (M) *Taenia* egg. Note the thick, striated shell with a six-hooked embryo within the egg shell. (N) *Diphyllobothrium latum* (broad fish tapeworm or freshwater tapeworm) scolex. Note the lateral "sucking groove." (O) *Diphyllobothrium latum* (broad fish tapeworm or freshwater tapeworm) gravid proglottid. Note that the proglottids are wider than long (opposite of *Taenia* proglottids [longer than wide]). Also note that the reproductive structures are in the middle of the proglottid ("rosette formation"). A number of attached proglottids are usually passed together. (P) *Diphyllobothrium latum* (broad fish tapeworm or freshwater tapeworm) eggs. Note that they are operculated (this is the only operculated tapeworm egg). (Q) *Hymenolepis nana* egg. These eggs are characterized by their thin shell, the presence of a six-hooked embryo within the egg shell, and the long, thin threads (polar filaments) that lie between the embryo and shell (found at the ends of the egg).

(R) *Hymenolepis diminuta* egg. These eggs are characterized by their thin shell, the presence of a six-hooked embryo within the egg shell, and no polar filaments. Unlike the magnification in the photograph, these eggs are somewhat larger than *H. nana* eggs. (S) *Clonorchis (Opisthorchis) sinensis* egg. This is one of the smallest helminth eggs (~30 μm) and is operculated. They have a small knob at the end opposite the operculum, but it may be difficult to see. (From A Pictorial Presentation of Parasites: A cooperative collection prepared and/or edited by H. Zaiman.) (T) *Paragonimus westermani* (lung fluke) egg. These eggs have an operculum which fits into the shell where the opercular "shoulders" are located (mimics a teapot rim and lid). Remember that these eggs can be found in both sputum and stool. (U) *Fasciola hepatica*/*Fasciolopsis buski* egg. These two eggs are difficult to differentiate. Both are operculated and are not embryonated when passed. This egg is photographed at a lower magnification (it is the largest of the helminth eggs). (V) *Schistosoma mansoni* egg (iodine stain, wet mount). Note the large, lateral spine. (W) *Schistosoma haematobium* egg (iodine stain, wet mount). Note the terminal spine. (X) *Schistosoma japonicum* egg (wet mount, blue filter). Note the small, lateral spine.

Plate 4   (see p. 1111) These slides are stained with Giemsa stain and are photographed using an oil immersion lens, unless stated otherwise. (A) *Plasmodium falciparum* ring forms. Note the cell with two ring forms (young trophozoites) and the appliqué or accolé form at the edge of the RBC. (B) *Plasmodium falciparum* ring forms. Note the "headphone" appearance of the ring forms (young trophozoites). (C) *Plasmodium falciparum* mature schizont and ring forms. The mature schizont is infrequently seen in the peripheral blood, but it does occur. (D) *Plasmodium falciparum* gametocytes. Note the typical crescent shape. Also, note that although the gametocytes are intracellular, the RBC membrane may not always be visible. (E) *Plasmodium vivax* early trophozoite. Note the ameboid trophozoite and the enlarged RBC. (From A Pictorial Presentation of Parasites: A cooperative collection prepared and/or edited by H. Zaiman.) (F) *Plasmodium vivax* developing trophozoites. Note the ameboid trophozoites and enlarged RBCs. (From A Pictorial Presentation of Parasites: A cooperative collection prepared and/or edited by H. Zaiman.) (G) *Plasmodium vivax* trophozoite. Note the Schüffner's dots (true stippling). (From A Pictorial Presentation of Parasites: A cooperative collection prepared and/or edited by H. Zaiman.) (H) *Plasmodium vivax* mature schizont. Note the merozoites within the mature schizont. The number of merozoites helps determine the correct malaria species. (From A Pictorial Presentation of Parasites: A cooperative collection prepared and/or edited by H. Zaiman.) (I) *Plasmodium malariae* "band" forms (developing trophozoites). The band form is typical of *P. malariae*; also note that the infected RBCs are normal size. (From A Pictorial Presentation of Parasites: A cooperative collection prepared and/or edited by H. Zaiman.) (J) *Plasmodium malariae* mature schizonts. Note the merozoites clustered around the excess malarial pigment (brownish color) (described as a "daisy" or "rosette"). (From A Pictorial Presentation of Parasites: A cooperative collection prepared and/or edited by H. Zaiman.) (K) *Plasmodium ovale* early trophozoites. Note the Schüffner's dots (true stippling) and the nonameboid trophozoite. (Parasite in left frame from A Pictorial Presentation of Parasites: A cooperative collection prepared and/or edited by H. Zaiman.) (L) *Plasmodium ovale* early trophozoites. Note the Schüffner's dots and fimbriated edges of the infected RBCs. (Parasite in left frame from A Pictorial Presentation of Parasites: A cooperative collection prepared and/or edited by H.

Zaiman.) (M) *Babesia* spp. ring forms. Note the small, multiple rings in the RBC at the top of the frame. These rings tend to be smaller and have more per cell than the ring forms in *P. falciparum* infections. (From A Pictorial Presentation of Parasites: A cooperative collection prepared and/or edited by H. Zaiman.) (N) *Babesia* spp., "Maltese cross" form. Here four elongated rings can be seen. This appearance is diagnostic for *Babesia* spp. but is not always seen. (Photograph by Zane Price.) (O) *Leishmania donovani* amastigotes (Leishman-Donovan bodies [L-D bodies]). Note the nucleus and bar (primitive flagella) in each one of the small organisms. This is a bone marrow specimen. (From A Pictorial Presentation of Parasites: A cooperative collection prepared and/or edited by H. Zaiman.) (P) *Leishmania* spp. in culture. As the organisms grow in culture, they form these "rosette" formations. (From A Pictorial Presentation of Parasites: A cooperative collection prepared and/or edited by H. Zaiman.) (Q) *Trypanosoma cruzi* trypomastigote. Note the C shape, undulating membrane, and large nucleus. (Photograph by Zane Price.) (R) *Trypanosoma cruzi* amastigotes in cardiac muscle (routine histologic preparation). Note the amastigotes clustered together. The individual amastigote looks identical to those found in leishmaniasis; however, the body site is different (reticuloendothelial system for leishmaniasis, striated muscle for Chagas' disease). (From A Pictorial Presentation of Parasites: A cooperative collection prepared and/or edited by H. Zaiman.) (S) *Trypanosoma brucei* trypomastigote. Note the slender body, no particular arrangement (unlike the typical C or U shape seen with *T. cruzi*), and small nucleus. (From A Pictorial Presentation of Parasites: A cooperative collection prepared and/or edited by H. Zaiman.) (T) *Trypanosoma brucei* trypomastigote. Note the slender body, no particular arrangement (unlike the typical C or U shape seen with *T. cruzi*), and small nucleus. (From A Pictorial Presentation of Parasites: A cooperative collection prepared and/or edited by H. Zaiman.) (U) *Wuchereria bancrofti* microfilaria (Delafield's hematoxylin stain). Note the presence of the sheath (normally not visible using Giemsa stain) and the empty tail space beyond the last nucleus. (V) *Wuchereria bancrofti* microfilaria (anterior end) (Delafield's hematoxylin stain). Note the sheath and the space between the anterior end and beginning of nuclei. (W) Microfilaria (nonsheathed). This is an example of one of the smaller microfilariae seen in a blood film; note that there is no sheath and that the nuclei are not seen as distinct dots like those seen in *W. bancrofti*. (X) *Onchocerca volvulus* nodule (routine histologic preparation, special trichrome stain). Note the cross section through the uterus of the adult worm (microfilariae visible as small curves within the uterus).

Plate 5   (see p. 1112) Trichrome stain was used for panels A through S; iron hematoxylin stain was used for panel T. (A) *Entamoeba histolytica* trophozoites (note ingested RBCs undergoing digestion). (B) *Entamoeba histolytica*/*E. dispar* trophozoite (note debris [the red object]). (C and D) *Entamoeba histolytica*/*E. dispar* cysts with chromatoidal bars (rounded ends). (E and F) Macrophages that can mimic *Entamoeba histolytica*/*E. dispar* trophozoites. (G and H) PMNs that can mimic *Entamoeba histolytica*/*E. dispar* cysts. (I) *Entamoeba coli* trophozoite with large karyosome and highly vacuolated cytoplasm. (J) *Entamoeba coli* trophozoite with uneven nuclear chromatin and eccentric karyosome. (K) *Entamoeba coli* cyst (iodine wet preparation). (L) *Entamoeba coli* cyst. (M and N) *Endolimax nana* trophozoites (large karyosome with little to no peripheral nuclear chromatin). (O) *Endolimax nana* cyst (vacuole is sometimes present, can be confused with cyst of *Iodamoeba bütschlii*). (P) *Endolimax nana* cyst with four karyosomes (the typical number) visible (also note the round rather than oval shape). (**Q**) *Iodamoeba bütschlii* trophozoite and cyst (cyst somewhat out of focus). (**R** and **S**) *Iodamoeba bütschlii* cysts with large glycogen vacuoles. (**T**) *Iodamoeba bütschlii* cyst with large glycogen vacuole and "basket" nucleus.

Plate 6   (see p. 1113) (**A**) *Giardia lamblia* trophozoites in mucus (note the teardrop shape). (**B**) *Giardia lamblia* trophozoite. (**C** and **D**) *Giardia lamblia* cysts. (**E**) *Chilomastix mesnili* trophozoite (note the cytostomal groove). (**F**) *Chilomastix mesnili* cyst (lemon shape with "shepherd's crook"). (**G**) *Pentatrichomonas hominis* trophozoite in stool (note the axostyle and typical large granules). (**H**) *Trichomonas vaginalis* trophozoite from the urogenital system (Giemsa stain). (**I** through **L**) *Dientamoeba fragilis* trophozoites (one- and two-nucleated trophozoites; note the nuclear chromatin fragmented into "dots"). (**M**) Charcot-Leyden crystals. (**N**) Macrophages, PMNs, and RBCs (no parasites are present, high dry power). (**O** and **P**) Artifacts (pollen and other plant material that can mimic helminth eggs). (**Q** through **T**) *Prototheca wickerhamii* (from culture).

Plate 7   (see p. 1114) (**A**) *Cyclospora cayetanensis* (modified acid-fast stain; courtesy of Earl Long, Centers for Disease Control and Prevention). (**B** and **C**) *Isospora belli* (modified acid-fast stain; the organism in panel B is less mature than that in panel C). (**D**) Microsporidian spores in muscle (*Pleistophora* spp.; high dry power). (**E** and **F**) Microsporidian spores in stool (modified trichrome stain; probably *Enterocytozoon bieneusi*; note the horizontal line through the spore [polar tubule] [arrow in panel E]). (**G**) Microsporidian spores in urine sediment (modified trichrome stain; probably *Encephalitozoon* [*Septata*] *intestinalis*). (**H**) Microsporidian spores in urine sediment (experimental indirect fluorescent-antibody method; probably *Encephalitozoon* [*Septata*] *intestinalis*). (**I**) Microsporidian spores in urine sediment (calcofluor white; probably *Encephalitozoon* [*Septata*] *intestinalis*). (**J** and **K**) Microsporidian spores in enterocytes (Giemsa stain). (**L**) Microsporidian spores in enterocytes (silver stain).

Plate 8   (see p. 1115) (**A** through **K**) *Plasmodium vivax* (note the presence of two rings per cell [not always limited to *Plasmodium falciparum*], enlarged RBCs, ameboid trophozoites and rings, and presence of stippling in some parasitized RBCs [Schüffner's dots]). (**L** through **S**) *Plasmodium vivax*, developing and mature schizonts. (**T**) *Plasmodium vivax*, thick blood film containing mature schizont.

Plate 9   (see p. 1116) (**A** through **J**) *Plasmodium ovale* developing trophozoites with rings (note the nonameboid growing rings, the oval shape of some RBCs, RBCs with Schüffner's dots, and RBCs with fimbriated edges). (**K** through **M**) *Plasmodium malariae* rings (note the normal-size RBCs). (**N** through **P**) *Plasmodium malariae* "band" forms. (**Q** and **R**) *Plasmodium malariae* developing schizonts. (**S** and **T**) *Plasmodium malariae* mature schizonts (note the rosette appearance of the merozoites clustered around the excess pigment).

Plate 10   (see p. 1117) (**A**) *Babesia* spp., "Maltese cross" configuration of small rings seen in some species of *Babesia*. (**B**) *Babesia* spp. (note the "band" form, which can be confused with *Plasmodium malariae*). (**C** and **D**) *Babesia* spp. with multiple rings per cell. (**E** through **I**) *Plasmodium falciparum* rings. (**J**) *Plasmodium falciparum* appliqué or accolé form. (**K**) *Plasmodium falciparum* (note the two chromatin dots and the "headphone" appearance of the ring). (**L**) *Plasmodium falciparum*, two rings per RBC. (**M** through **P**) *Plasmodium falciparum* thick blood films showing ring forms. (**Q** through **S**) *Plasmodium falciparum* micro- and macrogametocytes. (**T**) *Plasmodium falciparum*, section of capillary containing parasitized RBCs.

Plate 11   (see p. 1118) (**A**) *Dermacentor variabilis*, adult male on skin. (**B**) *Dermacentor andersoni*, male on left, female on right. (**C**) *Dermacentor variabilis*, adult females on skin. (**D**) *Argas* sp., soft tick; dorsal and ventral views. (**E**) Nit of louse on fiber. (**F**) *Phthirus pubis* (crab louse). (**G**) *Sarcoptes scabiei* (itch mite, the cause of scabies). (**H**) Human hand showing evidence of mite infestation. (Photographs from A Pictorial Presentation of Parasites: A cooperative collection prepared and/or edited by H. Zaiman.)

Plate 12   (see p. 1119) (**A**) *Cimex lectularius* (bedbug). (**B**) *Tunga penetrans* (chigoe flea), lesion on the medial aspect of a medical missionary's foot. (**C**) *Pulex irritans* (human flea). (**D**) Myiasis, ocular (species of fly unknown). (**E**) Ulcer from which a maggot was removed in a patient in Liberia. (**F**) *Dermatobia hominis* fly larva, a common cause of myiasis. (**G**) Fly larvae in a sacral decubitus ulcer in a Mexican patient confined to bed with a fractured femur. (**H**) Caterpillar rash induced by migration of caterpillar over the skin of a missionary physician in Liberia. The caterpillars usually fall from trees. (Photographs from A Pictorial Presentation of Parasites: A cooperative collection prepared and/or edited by H. Zaiman.)

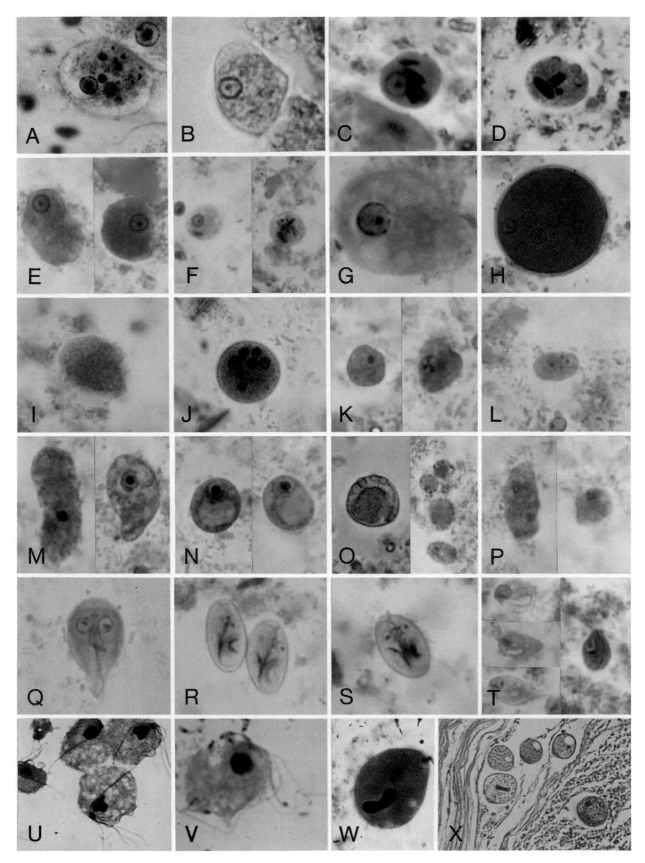

Plate 1

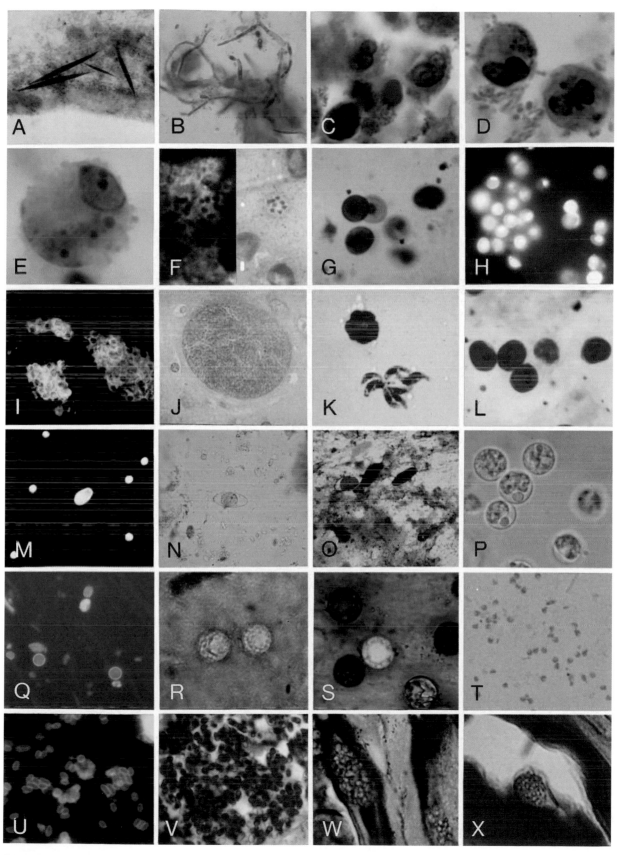

Plate 2

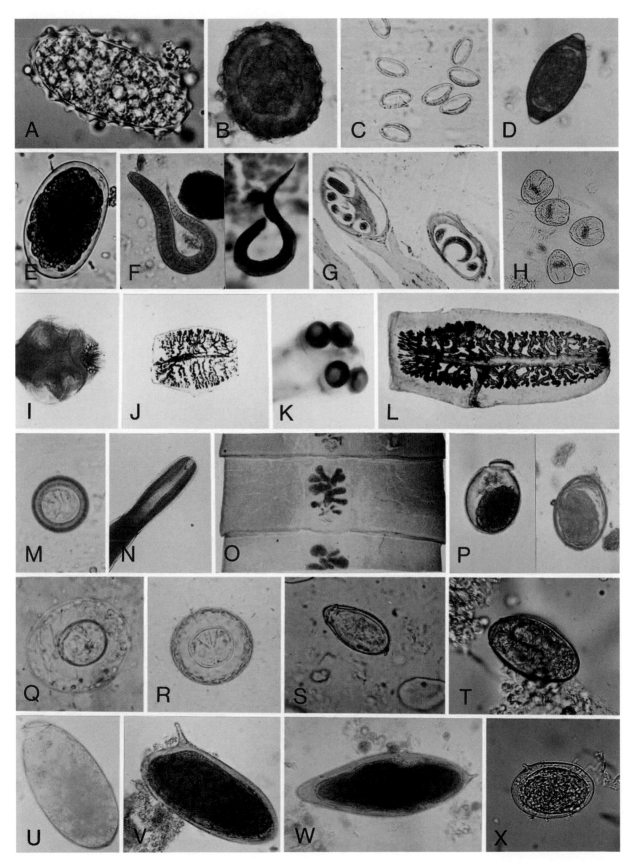

Plate 3

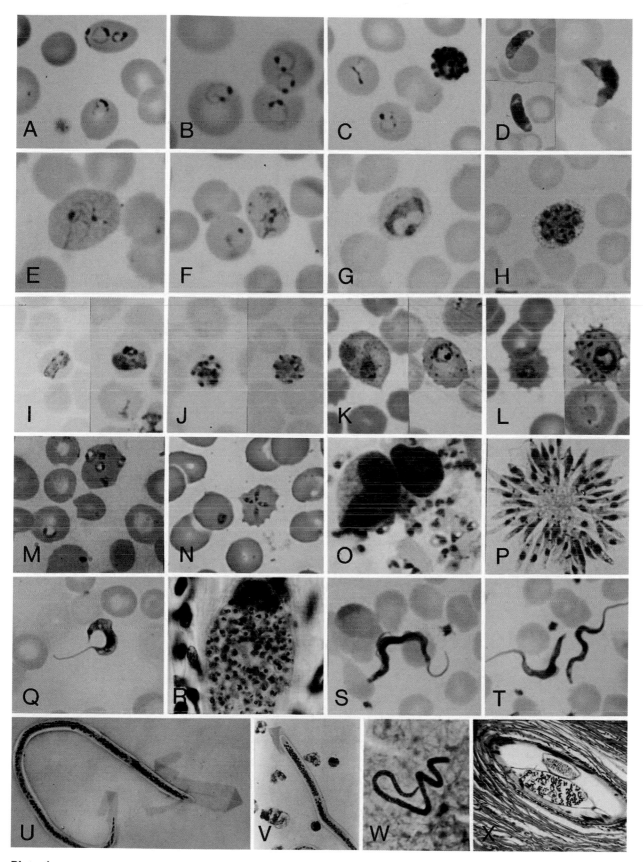

Plate 4

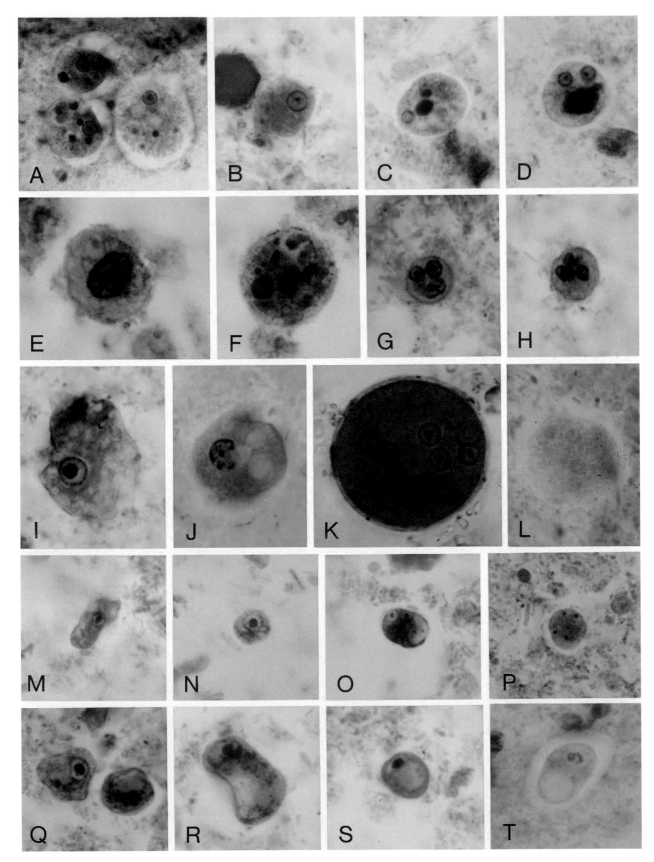

Plate 5

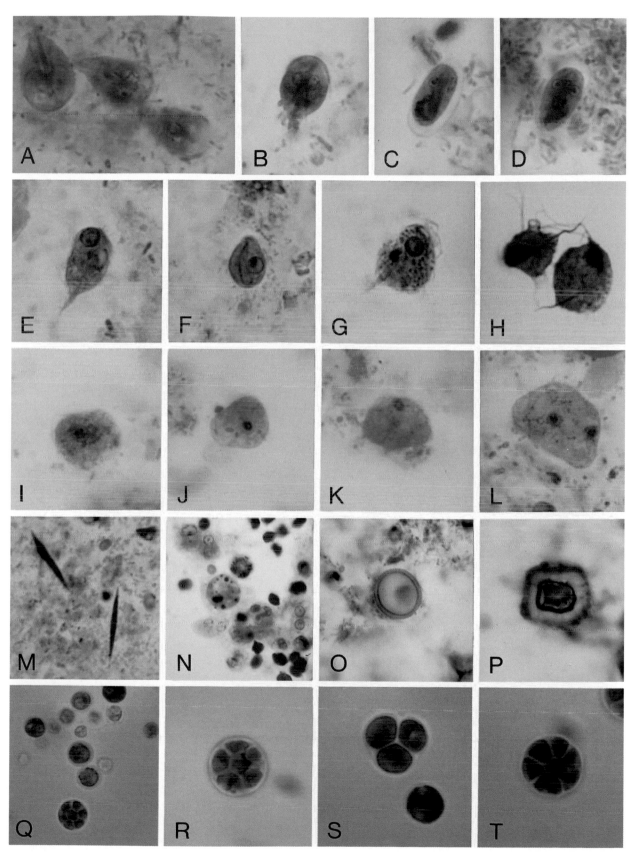

Plate 6

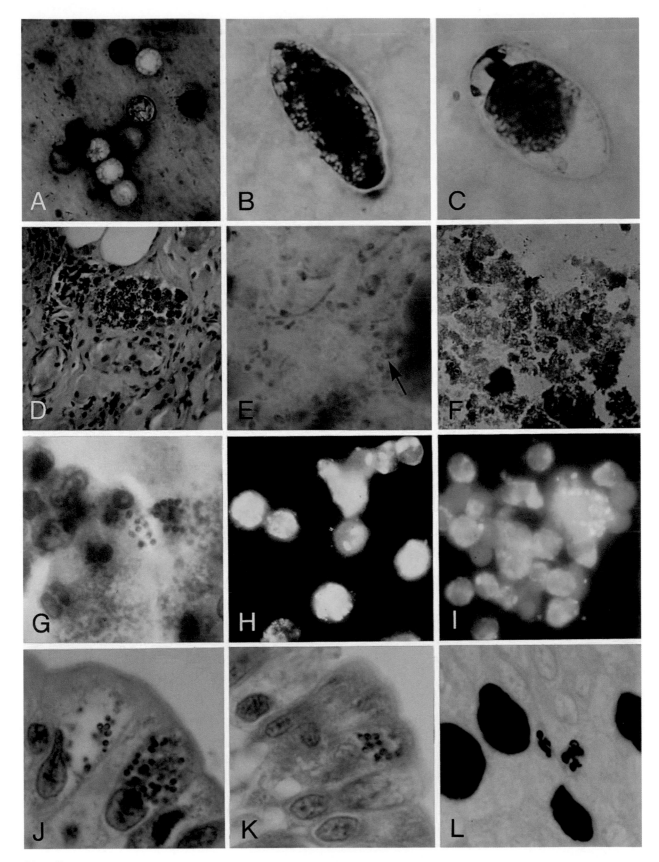

Plate 7

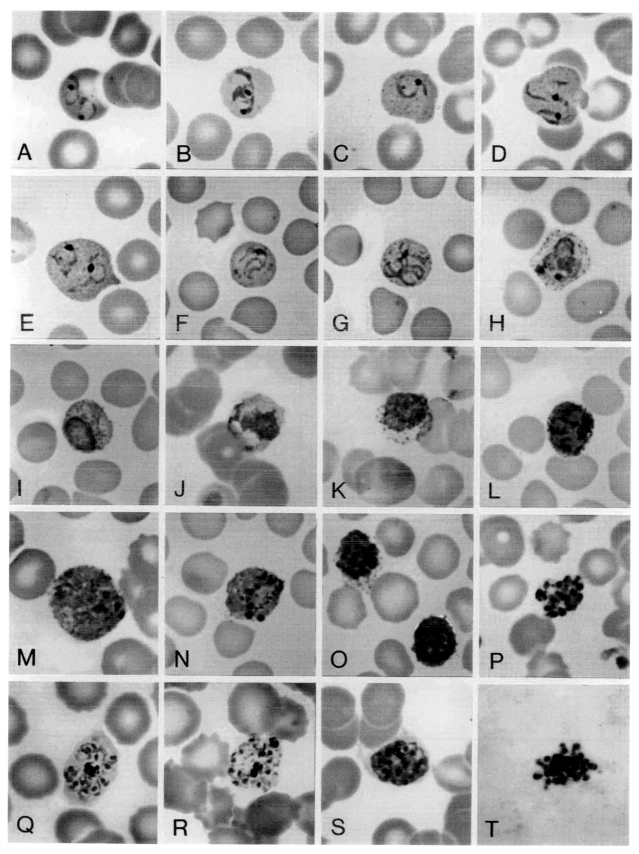

Plate 8

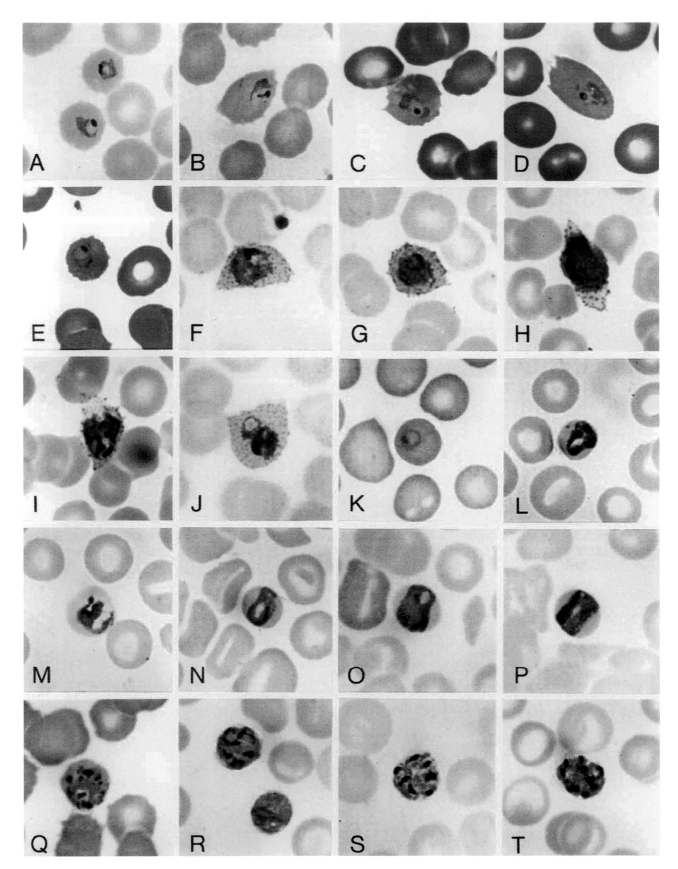

Plate 9

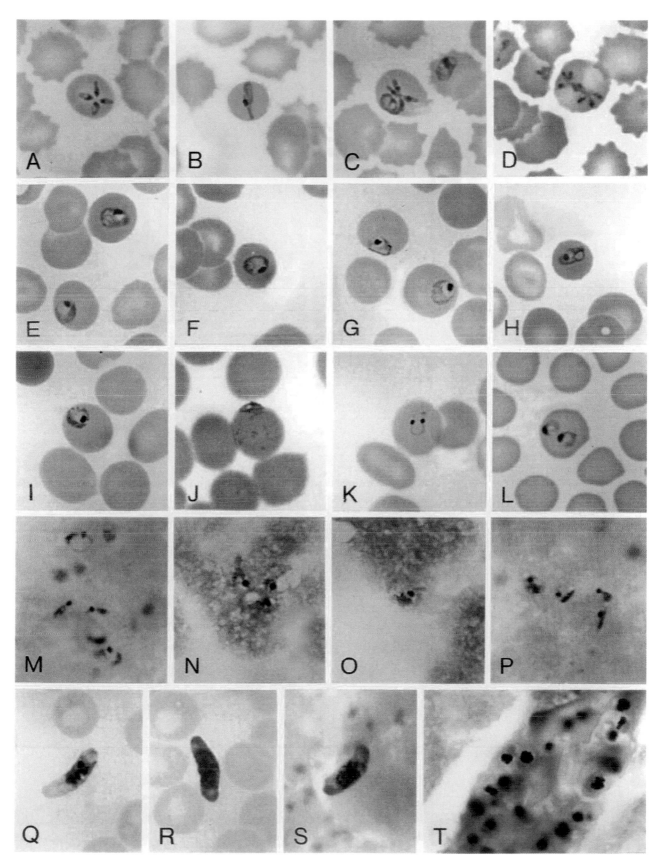

Plate 10

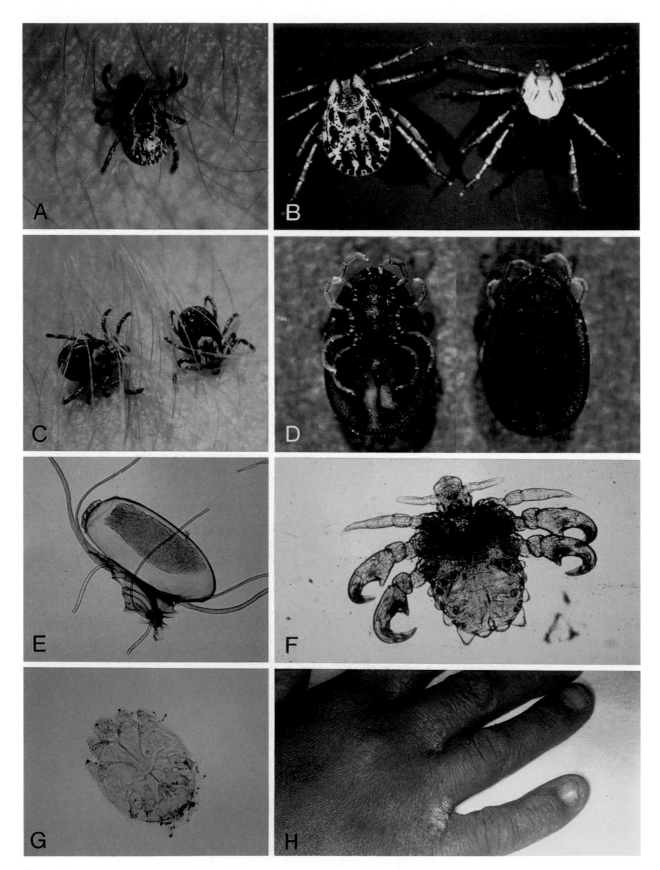

Plate 11

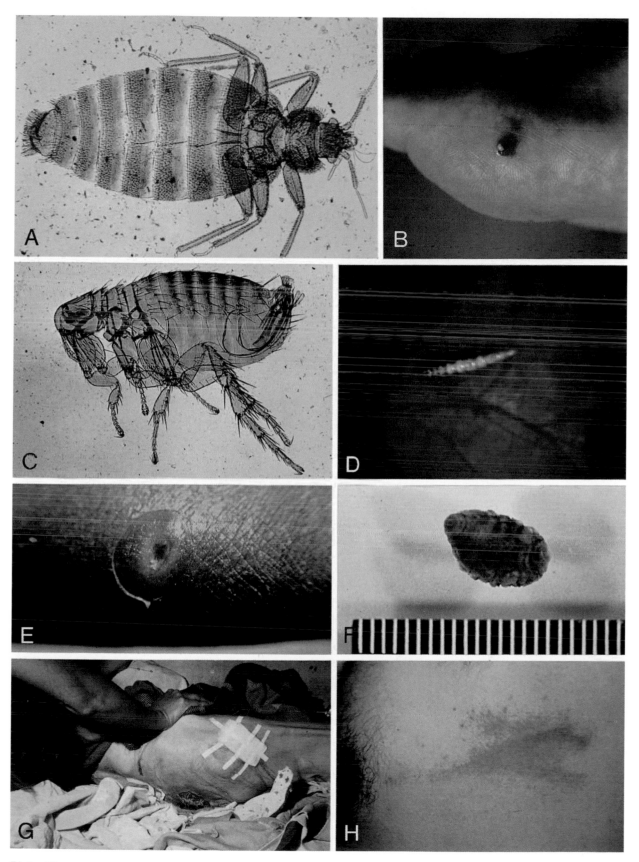

Plate 12

"Late-Breaking" Published Information

The following information was published in the last couple of months of 2005 and the first six months of 2006 and contains relevant information for the field of diagnostic medical parasitology. This is "late-breaking" published information that may be helpful for the reader.

Intestinal Protozoa (Amebae)

Entamoeba spp.

MacFarlane, R. C., and U. Singh. 2006. Identification of differentially expressed genes in virulent and nonvirulent *Entamoeba* species: potential implications for amebic pathogenesis. *Infect. Immun.* 74:340–351.

Several *Entamoeba* species and strains with differing levels of virulence have been identified. *E. histolytica* HM-1:IMSS is a virulent strain, *E. histolytica* Rahman is a nonvirulent strain, and *Entamoeba dispar* is a nonvirulent species. The authors used an *E. histolytica* DNA microarray consisting of 2,110 genes to assess the transcriptional differences between these species and strains with the goal of identifying genes whose expression correlated with a virulence phenotype. They found 415 genes expressed at lower levels in *E. dispar* and 32 genes with lower expression in *E. histolytica* Rahman than in *E. histolytica* HM-1:IMSS. Overall, 29 genes had decreased expression in both the nonvirulent species and strains than in the virulent *E. histolytica* HM-1:IMSS. Also, a number of genes with potential roles in stress response and virulence had decreased expression in either one or both nonvirulent *Entamoeba* species and strains. A number of the non-long-terminal-repeat retrotransposons (EhLINEs and EhSINEs), which modulate gene expression and genomic evolution, had lower expression in the nonvirulent species and strains than in *E. histolytica* HM-1:IMSS. These results, identifying expression profiles and patterns indicative of a virulence phenotype, may be useful in characterizing the transcriptional framework of virulence.

Moncada, D., K. Keller, S. Ankri, D. Mirelman, and K. Chadee. 2006. Antisense inhibition of *Entamoeba histolytica* cysteine proteases inhibits colonic mucus degradation. *Gastroenterology* 130:721–730.

The exact role of *Entamoeba histolytica* cysteine proteases in overcoming the colonic mucus barrier, as a prerequisite to epithelial cell disruption, is not known. The authors determined whether *E. histolytica* trophozoites expressing the antisense transcript to cysteine protease 5 (EhCP5) could degrade colonic mucin and destroy epithelial cells. Cysteine protease-deficient amebae were generated by antisense inhibition of EhCP5 and assayed for proteolytic activity against [^{35}S]cysteine labeled mucin from LS 1741 and HT-29F Cl.16E cells. Recombinant EhCP5 mucinase activity was also assessed. Disruption of an intact mucus barrier and epithelial cell invasion by amebae were measured using high-mucin-producing LS 174T and HT-29F Cl.16E monolayers or Chinese hamster ovary (CHO) cells devoid of a mucus barrier. Trophozoites with reduced cysteine protease activity were >60% more ineffective at degrading [^{35}S]cysteine-labeled colonic mucin than were wild-type amebae. However, bioactive recombinant EhCP5 degraded >45% of purified native mucin, which was specifically inhibited by the cysteine proteinase (CP) inhibitor E-64. Cysteine protease-deficient trophozoites could not overcome a protective intact mucus barrier and disrupt LS 174T or HT-29F Cl.16 cell monolayers; however, they readily adhered to and disrupted CHO monolayers devoid of a mucus barrier. These results unravel a central role for *E. histolytica* CPs as key virulence factors in disrupting an intact mucus barrier in the pathogenesis of intestinal amebiasis.

Pelosof, L. C., P. H. Davis, Z. Zhang, X. Zhang, and S. L. Stanley, Jr. 2006. Co-ordinate but disproportionate activation of apoptotic, regenerative and inflammatory pathways characterizes the liver response to acute amebic infection. *Cell. Microbiol.* 8:508–522.

It is well known that the liver has the remarkable ability to respond to injury with repair and regeneration. The protozoan parasite *Entamoeba histolytica* is the most common cause of liver abscess worldwide. The authors report a transcriptional analysis of the response of mouse liver to *E. histolytica* infection, the first study looking at acute liver infection by a nonviral pathogen. Focusing on early time points, they identified 764 genes with altered transcriptional levels in amebic liver abscess. The response to infection is rapid and complex, with concurrent increased expression of genes linked to host defense through interleukin-1 (IL-1), Toll-like receptor 2, or interferon (IFN)-induced pathways, liver regeneration via activation of IL-6 pathways, and genes associated with programmed cell death possibly through tumor necrosis factor alpha or Fas pathways. A comparison of amebic liver infection with the liver response to partial hepatectomy or toxins reveals striking similarities between amebic liver abscess and noninfectious injury in key components of the liver regeneration pathways. The response to amebic liver abscess is biased toward apoptosis compared with the response to acute liver injury from hepatectomy, toxins, or other forms of liver infection. *E. histolytica* **infection of the liver simultaneously activates inflammatory, regenerative, and apoptotic pathways, but the sum of these early responses is biased toward programmed cell death.**

Welter, B. H., R. R. Powell, R. C. Laughlin, G. C. McGugan, M. Bonner, A. King, and L. A. Temesvari. 2006. *Entamoeba histolytica*: comparison of the role of receptors and filamentous actin among various endocytic processes. *Exp. Parasitol.* 1 Feb. [Epub ahead of print.]

Uptake of iron is critical for *Entamoeba histolytica* growth, and iron-bound human transferrin (holo-transferrin) has been shown to serve as an iron source in vitro. Although a transferrin binding protein has been identified in *E. histolytica*, the mechanism by which this iron source is taken up by this pathogen is not well understood. The uptake of fluorescent dextran, holo-transferrin, and human red blood cells (hRBCs) was compared. **Both dextran and transferrin were taken up in an apparent receptor-independent fashion, in contrast to hRBCs, which were taken up in a receptor-mediated fashion.** The uptake of fluorescein isothiocyanate (FITC)-dextran and FITC-holo-transferrin differentially relied on an intact actin cytoskeleton, suggesting that their internalization routes may be regulated independently.

Blastocystis hominis

Sio, S. W., M. K. Puthia, A. S. Lee, J. Lu, and K. S. Tan. 2006. Protease activity of *Blastocystis hominis*. *Parasitol. Res.* 4 Mar. [Epub ahead of print.]

Proteases of the intestinal protozoan parasite *Blastocystis hominis* were studied for the first time by azocasein assays and gelatin sodium dodecyl sulfate-polyacrylamide gel electrophoresis (SDS-PAGE) analysis. Parasitic lysates were found to have high protease activity, and nine protease bands of low (20 to 33 kDa) and high (44 to 75 kDa) molecular masses were reported. Proteases were found to be pH dependent, and the highest proteolytic activity was observed at neutral pH. **Inhibition studies showed that *B. hominis* isolate B, like many other protozoan parasites, contains mainly cysteine proteases.**

Tan, T. C., and K. G. Suresh. 2006. Predominance of ameboid forms of *Blastocystis hominis* in isolates from symptomatic patients. *Parasitol. Res.* 98:189–193.

Blastocystis hominis is one of the most common human parasites that inhabit the intestinal tract. Conflicting reports continue to exist regarding the existence and the functional role of the ameboid forms in the life cycle of the parasite. This study investigated the presence of these forms in 20 isolates obtained from 10 symptomatic and 10 asymptomatic patients. A total of 10,000 parasite cells/ml from each isolate were inoculated into three culture tubes each containing 3 ml of Jones' medium supplemented with 10% horse serum, incubated at 37°C. The contents were examined daily for 10 days. Irregular and polymorphic ameboid forms with multiple extended pseudopodia were observed in all isolates from symptomatic patients, while none of the isolates from asymptomatic patients showed the presence of the ameboid forms. The ameboid forms were first seen on day 2, and the percentages increased from 2% to 28%, with peak percentages from day 3 to day 6. Transmission electron microscopy revealed two types of ameboid forms, one containing a large central vacuole completely filled with tiny electron-dense granules and the other containing multiple small vacuoles within the central body. The cytoplasm contained strands of electron-dense granules resembling rough endoplasmic reticulum, which is suggestive of active protein synthesis. The surface coat of the ameboid form surrounding the parasite showed uneven thickness. Acridine orange stained the central body yellow and the periphery orange, indicating activity at the level of nucleic acids. **The ameboid form could be either an indicator of pathogenicity of *B. hominis* or the form most likely to contribute to pathogenicity and to be responsible for the symptoms seen in patients.**

Intestinal Protozoa (Flagellates and Ciliates)

Dientamoeba fragilis

Stark, D., N. Beebe, D. Marriott, J. Ellis, and J. Harkness. 2006. Evaluation of three diagnostic methods, including real-time PCR, for detection of *Dientamoeba fragilis* in stool specimens. *J. Clin. Microbiol.* 44:232–235.

The authors developed a 5' nuclease (TaqMan)-based real-time PCR assay, targeting the small-subunit rRNA gene, for the detection of *Dientamoeba fragilis* in human stool specimens and compared its sensitivity and specificity to those of conventional PCR and microscopic examination by a traditional modified iron hematoxylin-staining procedure. Real-time PCR exhibited 100% sensitivity and specificity.

Giardia lamblia

Hausen, M. A., J. C. Freitas, Jr., and L. H. Monteiro-Leal. 2006. The effects of metronidazole and furazolidone during *Giardia* differentiation into cysts. *Exp. Parasitol.* 10 Feb. [Epub ahead of print.]

The search for an effective anti-*Giardia* treatment has been intense, but recurrent infections, virulence factors, and drug resistance have created obstacles to the achievement of an efficient medication. Most published reports about drug effects in *Giardia* are related to the trophozoite form, although infective cysts are continuously eliminated in the stools during the treatment. The authors analyzed the inhibitory effects of metronidazole and furazolidone on the differentiation of *Giardia* into cysts and their viability. The presence of cavities, lamellar bodies, and thread-like structures were the most frequent morphologic alterations. **The results also showed that furazolidone was more effective by 50% than metronidazole in inhibiting in vitro cyst differentiation.**

Intestinal Protozoa (Coccidia and Microsporidia)

Cryptosporidium spp.

Li, X., K. Guyot, E. Dei-Cas, J. P. Mallard, J. J. Ballet, and P. Brasseur. 2006. *Cryptosporidium* oocysts in mussels (*Mytilus edulis*) from Normandy (France). *Int. J. Food Microbiol.* 108:321–325.

Cultured mussels (*Mytilus edulis*) were collected seasonally during 1 year from three sites on the Northwestern coastal area of Normandy (France). Flesh, gills, and innerwater were examined for *Cryptosporidium* oocysts using immunomagnetic separation and the immunofluorescence assay. Oocysts were present in all samples for all sites and seasons, and flesh was the most contaminated part. Oocyst rates appeared to be related to variations in seasonal precipitation. Molecular analysis revealed that oocysts belonged to the species *Cryptosporidium parvum* (formerly genotype 2 or "bovine genotype"). Oocyst infectivity was assessed by oral administration to suckling NMRI mice, and developmental stages were observed in only one mouse infected with oocysts from one location. The detection of potentially infectious *C. parvum* oocysts of likely cattle-breeding origin in cultured edible mussels confirms their resistance to marine environments and underlines the potential risk of food-borne infection. **This work reports for the first time the presence of infectious *Cryptosporidium* oocysts in shellfish from France.**

Rossignol, J. F., S. M. Kabil, Y. el-Gohary, and A. M. Younis. 2006. Effect of nitazoxanide in diarrhea and enteritis caused by *Cryptosporidium* species. *Clin. Gastroenterol. Hepatol.* 4:320–324.

A multicenter, randomized, double-blind, placebo-controlled study was conducted with 90 outpatients aged 12 years and older from the Nile Delta region of Egypt. Patients were randomized to receive either one 500-mg nitazoxanide tablet or one matching placebo tablet, or 25 ml of nitazoxanide oral suspension (500 mg of nitazoxanide), each given twice daily for 3 days. Clinical and microbiological response rates were evaluated 4 days after completion of treatment. Of the 28 patients receiving nitazoxanide tablets, 27 (96%) responded clinically compared with 11 (41%) of 27 patients who received placebo ($P < 0.0001$). Of the 28 patients who received nitazoxanide, 26 (93%) were free of *Cryptosporidium* oocysts in each of two posttreatment stool samples compared with

only 10 (37%) of 27 patients who received placebo ($P < 0.0001$). Response rates in patients receiving the tablets and the suspension were comparable (clinical response rate for suspension, 27 [87%] of 31; microbiological response rate for suspension, 28 [90%] of 31). **These results indicate that a 3-day course of nitazoxanide is effective in treating diarrhea and enteritis caused by *Cryptosporidium* in nonimmunodeficient patients 12 years of age and older.**

Tanriverdi, S., A. Markovics, M. O. Arslan, A. Itik, V. Shkap, and G. Widner. 2006. Emergence of distinct genotypes of *Cryptosporidium parvum* in structured host populations. *Appl. Environ. Microbiol.* 72:2507–2513.

Cryptosporidium parvum isolated from cattle in northeastern Turkey and in Israel was genotyped using multiple polymorphic genetic markers, and the two populations were compared to assess the effect of cattle husbandry on the parasite's population structure. Dairy herds in Israel are permanently confined with essentially no opportunity for direct herd-to-herd transmission, whereas in Turkey there are more opportunities for transmission as animals range over wider areas and are often traded. A total of 76 *C. parvum* isolates from 16 locations in Israel and 7 farms in the Kars region in northeastern Turkey were genotyped using 16 mini- and microsatellite markers. In both countries, distinct multilocus genotypes confined to individual farms were detected; the number of genotypes per farm was larger and mixed isolates were more frequent in Turkey than in Israel. Based on the presence of distinct multilocus genotypes in individual herds, linkage disequilibrium among loci was detected in Israel. **These observations show that genetically distinct populations of *C. parvum* can emerge within a group of hosts in a relatively short time. This may explain the frequent detection of host-specific genotypes with unknown taxonomic status in surface water and the existence of geographically restricted *C. hominis* genotypes in humans.**

Microsporidia

Jordan, C. N., A. M. Zajac, K. S. Snowden, and D. S. Lindsay. 2006. Direct agglutination test for *Encephalitozoon cuniculi*. *Vet. Parasitol.* 135:235–240.

Currently, serologic diagnosis of infection with *Encephalitozoon cuniculi* is made using the indirect immunofluorescent-antibody assay (IFA) or enzyme-linked immunosorbent assay (ELISA). Although these methods are sensitive and reliable, cross-reactivity between other *Encephalitozoon* species is common, and specialized equipment is required to conduct these tests. The authors report the development of a direct agglutination test for detecting immunoglobulin G (IgG) antibodies to *E. cuniculi*. The utility of the agglutination test was examined by using CD-1 and C3H/He mice infected with *E. cuniculi* or one of two other *Encephalitozoon* species. Test sera were incubated overnight with eosin-stained microsporidial spores in round-bottom microtiter plates. In positive samples, agglutination of spores with antibodies in test sera resulted in an opaque mat spread across the well. The results indicate that the agglutination test is 86% sensitive and 98% specific for *E. cuniculi*, with limited cross-reactivity with *Encephalitozoon intestinalis*. No cross-reactivity with *Encephalitozoon hellem* was observed. The test is fast and easy to conduct, and species-specific antibodies are not required.

Jordan, C. N., J. A. Dicristina, and D. S. Lindsay. 2006. Activity of bleach, ethanol and two commercial disinfectants against spores of *Encephalitozoon cuniculi*. *Vet. Parasitol.* 136:343–346.

Hosts are infected with *Encephalitozoon cuniculi* by ingestion or inhalation of spores passed in the urine or feces, and infection is usually asymptomatic, except in young or immunocompromised hosts. Spores of *E. cuniculi* were exposed to several dilutions of commercial bleach, 70% ethanol, and dilutions of commercial disinfectants HiTor and Roccal for 10 min and then loaded onto human fibroblast cells (Hs68 cells). A 10-min exposure to these disinfectants was lethal to *E. cuniculi* spores. Additional exposure time studies were done using 0.1, 1, and 10% dilutions of bleach and 70% ethanol. Exposure of *E. cuniculi* spores to 1 or 10% bleach for 30 s rendered them noninfectious for Hs68 cells. Growth of *E. cuniculi* was observed in Hs68 cells inoculated with spores treated with 0.1% bleach for 30 s or 1, 3, or 5 min but not with spores treated for 7 min or longer. Exposure of *E. cuniculi* spores to 70% ethanol for 30 s rendered them noninfectious for Hs68 cells. **Spores of *E. cuniculi* are more sensitive to disinfectants than are coccidial oocysts and other parasite cysts. The relatively short contact time needed to kill spores indicates that disinfection of animal housing may be a viable means of reducing exposure of animals to *E. cuniculi* spores.**

Joseph, J., S. Murthy, P. Garg, and S. Sharma. 2006. Use of different stains for microscopic evaluation of corneal scrapings for diagnosis of microsporidial keratitis. *J. Clin. Microbiol.* 44:283–285.

Retrospective evaluation of potassium hydroxide plus calcofluor white (KOH+CFW), Gram, Giemsa, and modified Ziehl-Neelsen (1% H_2SO_4, cold) stains for the detection of microsporidia in corneal scrapings from 30 patients showed KOH+CFW and acid-fast stains to be the most efficient (29 of 30 [96.7%] and 28 of 30 [93.3%], respectively) in the diagnosis of microsporidial keratitis.

Singh, I., A. S. Sheoran, Q. Zhang, A. Carville, and S. Tzipori. 2005. Sensitivity and specificity of a monoclonal antibody-based fluorescence assay for detecting *Enterocytozoon bieneusi* spores in feces of simian immunodefiency virus-infected macaques. *Clin. Diagn. Lab. Immunol.* 12:1141–1144.

Enterocytozoon bieneusi is clinically the most significant among the microsporidia causing chronic diarrhea, wasting, and cholangitis in individuals with human immunodeficiency virus/AIDS. Microscopy with either calcofluor or modified trichrome stains is the standard diagnostic test for microsporidiosis but does not allow species identification. Detection of *E. bieneusi* infection based on PCR is limited to a few reference laboratories, so it is not the standard diagnostic assay. The authors have recently reported the development and characterization of a panel of monoclonal antibodies (MAbs) against *E. bieneusi*, and in this study they evaluated the specificity and sensitivity of an IFA, compared with PCR, in detecting *E. bieneusi* infection in simian immunodeficiency virus-infected macaques. **The IFA correlated with the primary PCR method, with a detection limit of 1.5×10^5 spores/g of feces, and will simplify the detection of *E. bieneusi* spores in clinical and environmental specimens and in laboratory and epidemiologic investigations.**

Protozoa from Other Body Sites

Free-Living Amebae

Clarke, D. W., and J. Y. Niederkorn. 2006. The pathophysiology of *Acanthamoeba* keratitis. *Trends Parasitol.* 22:175–180.

Acanthamoeba keratitis is a sight-threatening infection of the ocular surface that is produced by several free-living amebae of the genus *Acanthamoeba*. Infection is usually initiated by *Acanthamoeba*-contaminated contact lenses and causes exquisite pain and ulceration of the ocular surface. The pathophysiology of this infection involves an intricate series of sequential events that includes the production of several pathogenic proteases that degrade basement membranes and induce cytolysis and apoptosis of the cellular elements of the cornea, culminating in dissolution of the collagenous corneal stroma. **Targeting such proteases could lead to the development of vaccines that target the disease process rather than the pathogen itself.**

Trichomonas vaginalis

Mundodi, V., A. S. Kucknoor, T. H. Chang, and J. F. Alderete. 2006. A novel surface protein of *Trichomonas vaginalis* is regulated independently by low iron and contact with vaginal epithelial cells. *BMC Microbiol.* 6:6.

Trichomoniasis caused by *Trichomonas vaginalis* is the most common nonviral sexually transmitted disease and affects more than 250 million people worldwide. IgA has been implicated in resistance to mucosal infections by pathogens. No reports are available of IgA-reactive proteins and the role, if any, of this class of antibody in the control of this sexually transmitted disease. The availability of an IgA MAb immunoreactive to trichomonads by whole-cell ELISA was used to characterize the IgA-reactive protein of *T. vaginalis*. An IgA MAb called 6B8 was isolated from a library of MAbs reactive with surface proteins of *T. vaginalis*, which recognized a 44-kDa protein (TV44) by immunoblot analysis, and a full-length cDNA clone encoded a protein of 438 amino acids. Southern analysis revealed the gene (*tv44*) of *T. vaginalis* to be a single-copy gene. The *tv44* gene was down regulated at both the transcriptional and translational levels in iron-depleted trichomonads as well as in parasites after contact with immortalized MS-74 vaginal epithelial cells (VECs). Immunofluorescence on nonpermeabilized organisms confirmed the surface localization of TV44, and the intensity of fluorescence was reduced after parasite adherence to VECs. Also, an identical protein and gene were present in

Tritrichomonas foetus and *Trichomonas tenax.* This is the first report of a *T. vaginalis* gene (*tv44*) encoding a surface protein (TV44) reactive with an IgA MAb, and both gene and protein were conserved in human and bovine trichomonads. Further, TV44 is independently down regulated in expression and surface placement by iron and contact with VECs. TV44 is another *T. vaginalis* gene that is regulated by at least two independent signaling mechanisms involving iron and contact with VECs.

Van Der Pol, B., C. S. Kraft, and J. A. Williams. 2006. Use of an adaptation of a commercially available PCR assay aimed at diagnosis of chlamydia and gonorrhea to detect *Trichomonas vaginalis* in urogenital specimens. *J. Clin. Microbiol.* **44**:366–373.

Using reagents from a commercially available assay for *Chlamydia trachomatis* and *Neisseria gonorrhoeae*, *Trichomonas vaginalis* PCR was evaluated for detection of infection in women and men attending a sexually transmitted disease clinic. Evaluations included three primer sets, endocervical swabs, vaginal swabs and urine, and various storage conditions. The TVK3-TVK7 primer set was optimal, with sensitivities ranging from 69.5 to 96.8%. In all comparisons, *T. vaginalis* PCR performed better than routine diagnostics involving microscopy for women and culture for men (*P* > 0.05). The assay performed well for all sample types tested, and vaginal swabs were stable for up to 7 days at ambient temperature. **The use of samples prepared for, and reagents from, the *C. trachomatis*-*N. gonorrhoeae* PCR assay allowed the incorporation of *T. vaginalis* PCR diagnosis into routine clinical testing.**

Tissue Protozoa

Toxoplasma gondii

Campbell, A. L., C. L. Goldberg, M. S. Magid, G. Gondolesi, C. Rumbo, and B. C. Herold. 2006. First case of toxoplasmosis following small bowel transplantation and systematic review of the tissue-invasive toxoplasmosis following noncardiac solid organ transplantation. *Transplantation* **81**:408–417.

Toxoplasmosis prophylaxis is standard following heart and heart-lung transplantation, when an increased risk of allograft-transmitted *Toxoplasma* is well recognized. In contrast, prophylaxis and routine serologic evaluation of donors and recipients for *Toxoplasma* in noncardiac solid-organ transplantation (SOT) is not recommended. **The authors report the first case of disseminated toxoplasmosis following small bowel transplantation, presumably transmitted** via the transplanted intestine, and systematically review reported cases of toxoplasmosis in noncardiac SOT recipients to determine if current guidelines should be reconsidered. Systematic MEDLINE review was performed for tissue-invasive toxoplasmosis in noncardiac SOT recipients and analysis of clinical features, serologic status, and treatment regimens with respect to mortality. **Fifty-two cases of toxoplasmosis in noncardiac SOT recipients were identified. Eighty-six percent developed disease within 90 days of transplantation.** Presentation was nonspecific and consisted of fever (77%), respiratory distress (29%), neurologic manifestations (29%),

and bone marrow suppression (26%). Multivariate analyses demonstrated that localized disease, treatment received, and donor and recipient serostatus were predictors of survival. High-risk recipients (seropositive donor, seronegative recipient) developed disease earlier (16 days versus 31 days) and were less likely to survive than standard-risk recipients. **Toxoplasmosis following noncardiac SOT is recognized. Reduction of morbidity and mortality necessitates knowledge of donor and recipient *Toxoplasma* serostatus, prophylaxis, early diagnosis, and treatment. The findings support a reconsideration of pretransplantation evaluation and prophylaxis strategies in SOT recipients.**

De Moura, L., L. M. Bahia-Oliveira, M. Y. Wada, J. L. Jones, S. H. Tuboi, E. H. Carmo, W. M. Ramalho, N. J. Camargo, R. Trevisan, R. M. Graca, A. J. da Silva, I. Moura, J. P. Dubey, and D. O. Garrett. 2006. Waterborne toxoplasmosis, Brazil, from field to gene. *Emerg. Infect. Dis.* **12**:326–329.

Water was the suspected vehicle of *Toxoplasma gondii* dissemination in a toxoplasmosis outbreak in Brazil. A case-control study and geographic mapping of cases were performed. *T. gondii* was isolated directly from the implicated water and genotyped as SAG 2 type I.

Hu, K., J. Johnson, L. Florens, M. Fraunholz, S. Suravajjala, C. Dilullo, J. Yates, D. S. Roos, and J. M. Murray. 2006. Cytoskeletal components of an invasion machine—the apical complex of *Toxoplasma gondii*. *PLoS Pathog.* **2**:e13.

The apical complex of *Toxoplasma gondii* is thought to serve essential functions both in invasion of its host cells (including human cells) and in replication of the parasite. The understanding of apical complex function, the basis for its structure, and the mechanism for its motility are limited due to lack of knowledge of its molecular composition. The authors have partially purified the conoid-apical complex, identified ~200 proteins that represent 70% of its cytoskeletal protein components, characterized 7 novel proteins, and determined the sequence of recruitment of 5 of these proteins into the cytoskeleton during cell division. These results provide new markers for the different subcompartments within the apical complex and reveal previously unknown cellular compartments, which facilitate understanding the invasion process. **The extreme apical and extreme basal structures of this highly polarized cell originate in the same location and at the same time very early during parasite replication.**

Laibe, S., S. Ranque, C. Curtillet, F. Faraut, H. Dumon, and J. Franch. 2006. Timely diagnosis of disseminated toxoplasmosis by sputum examination. *J. Clin. Microbiol.* **44**:646–648.

The diagnosis of disseminated toxoplasmosis in a 14-year-old allogeneic bone marrow recipient with graft-versus-host disease was determined by the detection of *Toxoplasma gondii* tachyzoites in sputum smears. Sputum analysis is a valuable alternative in the clinical assessment of pulmonary toxoplasmosis, especially when conventional invasive techniques may not be practical.

Shrestha, S. P., T. Tomita, L. M. Weiss, and A. Orlofsky. 2006. Proliferation of *Toxoplasma gondii* in inflammatory macrophages in vivo is associated with diminished oxygen radical production in the host cell. *Int. J. Parasitol.* 17 Feb. [Epub ahead of print.]

While reactive oxygen species (ROS) can kill *Toxoplasma gondii* in vitro, the role played by these molecules in vivo is not known. The authors used a flow cytometry-based assay to investigate the relationship between intracellular infection and ROS production during acute peritoneal toxoplasmosis in mice. A distinct population of ROS$^+$ inflammatory macrophages was observed to increase progressively in frequency during the course of infection and to be inversely correlated with the degree of cell parasitization. These data imply that either intracellular parasites inhibit ROS synthesis or ROS-producing cells contain anti-*Toxoplasma* activity. The latter interpretation was supported by the finding that uninfected ROS-producing inflammatory macrophages were resistant to infection in vivo. However, in the same animals, ROS-producing macrophages that had previously been parasitized could readily be infected with additional parasites, suggesting that the difference in ROS production between highly infected and less highly infected cells was not due to ROS-associated killing of parasites within these cells. Also, macrophages infected with *T. gondii* in vitro and then briefly transferred to acutely infected mice up regulated ROS production that was again inversely correlated with the degree of intracellular parasitization. **These findings suggest that both ROS-associated anti-*Toxoplasma* activity and parasite-driven inhibition of ROS production underlie the observed pattern of ROS production. ROS function and parasite evasion of this function may contribute to the balance between host defense and disease progression during acute infection.**

Plasmodium and *Babesia* spp.

Plasmodium spp.

Combes, V., N. Coltel, D. Faille, S. C. Wassmer, and G. E. Grau. 2006. Cerebral malaria: role of microparticles and platelets in alterations of the blood-brain barrier. *Int. J. Parasitol.* 10 Mar. [Epub ahead of print.]

Brain lesions of cerebral malaria (CM) are characterized by a sequestration of *Plasmodium falciparum*-parasitized RBCs (PRBCs), leukocytes, and platelets within brain microvessels, by an excessive release of proinflammatory cytokines, and by disruption of the blood-brain barrier. The authors evaluated the possibility that PRBCs and platelets interact and induce functional alterations in brain endothelium. Using an in vitro model of endothelial lesion, they demonstrated that platelets can act as bridges between PRBCs and endothelial cells allowing the binding of PRBCs to endothelium devoid of cytoadherence receptors. Furthermore, platelets enhanced the cytotoxicity of PRBCs for brain endothelial cells by inducing an alteration of their monolayer integrity and increasing their apoptosis. Another aspect of inflammatory and infectious diseases is that platelets often lead to activation of vascular and blood cells, which results in an enhanced release of circulating microparticles (MP). The authors then investigated the levels of endothelial MP in plasma of Malawian children with malaria. Plasma MP numbers were markedly increased on admission only in patients with severe malaria complicated by coma. Using the experimental-mouse model of CM, the authors evaluated the pathogenic implications of MP by using genetically deficient mice in which the capacity to release circulating MP was impaired. Such mice, lacking the ABCA-1 gene, showed complete resistance to CM after infection by *Plasmodium berghei* ANKA. When purified from infected susceptible animals, MP were able to reduce the normal plasma-clotting time and to significantly enhance tumor necrosis factor release from naive macrophages. **The finding that ABCA-1 gene deletion confers complete protection against cerebral pathology, linked to impaired MP production, provides new potential targets for therapeutic amelioration of severe malaria.**

Gogtay, N. J., K. D. Kamtekar, S. S. Dalvi, S. S. Mehta, A. R. Chogle, U. Aigal, and N. A. Kshirsagar. 2006. A randomized, parallel study of the safety and efficacy of 45 mg primaquine versus 75 mg bulaquine as gametocytocidal agents in adults with blood schizonticide-responsive uncomplicated falciparum malaria [ISCRTN50134587]. *BMC Infect. Dis.* 6:16.

The World Health Organization recommends that adults with uncomplicated *Plasmodium falciparum* infection successfully treated with a blood schizonticide receive a single dose of 45 mg of primaquine (PQ) as a gametocytocidal agent. An earlier pilot study suggested that 75 mg of bulaquine (BQ), of which PQ is a major metabolite, may be a useful alternate to PQ. In a randomized, partial-blind study, 90 hospitalized adults with *P. falciparum* malaria that was blood schizonticide responsive and a gametocytemia of >5.5/µl within 3 days of diagnosis were randomized to receive single doses of either 45 mg of PQ or 75 mg of BQ on day 4. Gametocytemia was assessed on days 8, 15, 22, and 29, and gametocyte viability was determined by exflagellation on day 8. On day 8, 20 (65%) of 31 PQ recipients versus 19 (32%) of 59 BQ recipients showed persistence of gametocytes (*P* = 0.002). At day 15 and beyond, all patients were gametocyte free. On day 8, 16 of 31 PQ and 7 of 59 BQ volunteers showed gametocyte viability (*P* = 0.000065). **Based on these results, BQ is a safe, useful alternate to PQ as a *Plasmodium falciparum* gametocytocidal agent and may clear gametocytemia faster than PQ.**

Hill, A. V. 2006. Pre-erythrocytic malaria vaccines: towards greater efficacy. *Nat. Rev. Immunol.* 6:21–32.

Several new types of *Plasmodium falciparum* vaccine are now being evaluated in clinical trials. Recently, two vaccine candidates that target the pre-erythrocytic stages of the malaria life cycle—a protein particle vaccine with a powerful adjuvant and a prime-boost viral-vector vaccine—entered phase II clinical trials in the field, and the first has shown partial efficacy in preventing malarial disease in African children. The authors discuss the potential immunological basis for the encouraging partial protection induced by these vaccines and consider ways to develop more effective malaria vaccines.

Johnston, S. P., N. J. Pieniazek, M. V. Xayavong, S. B. Slemenda, P. P. Wilkins, and A. J. da Silva. 2006. PCR as a confirmatory technique for laboratory diagnosis of malaria. *J. Clin. Microbiol.* 44:1087–1089.

The authors compared a nested PCR assay and microscopic examination of Giemsa-stained blood films for detection and identification of *Plasmodium* spp. in blood specimens. PCR was more sensitive than microscopy and was capable of identifying malaria parasites at the species level when the microscopy results were equivocal.

Magill, A. J. 2006. Malaria: diagnosis and treatment of falciparum malaria in travelers during and after travel. *Curr. Infect. Dis. Rep.* 8:35–42.

Plasmodium falciparum is responsible for most of the mortality in travelers with imported malaria. Problems that occur while abroad include the inaccuracy of the microscopic diagnosis of malaria, both false positives and false negatives, when ill travelers seek care. A false-positive diagnosis can result in unnecessary parenteral injections that carry a risk of transmission of blood-borne pathogens, receipt of potentially dangerous drugs such as halofantrine, or receipt of fake, counterfeit drugs. Increased morbidity and mortality are associated with delays in diagnosis and initiation of prompt treatment for falciparum malaria. The availability of expert microscopy to confirm the diagnosis of malaria is limited. **The presence of splenomegaly and thrombocytopenia is strongly associated with malaria and would justify empiric treatment. The availability of atovaquone-proguanil, a safe and well-tolerated oral drug, should prompt a reconsideration of current treatment recommendations that discourage empiric treatment on clinical suspicion alone.**

Lee, S. A., A. Yeka, S. L. Nsobya, C. Dokomajilar, P. J. Rosenthal, A. Talisuna, and G. Dorsey. 2006. Complexity of *Plasmodium falciparum* infections and antimalarial drug efficacy at 7 sites in Uganda. *J. Infect. Dis.* 193:1160–1163.

Malaria infections in Africa frequently include multiple parasite strains. The authors examined the relationship between the number of infecting *Plasmodium falciparum* strains and the responses to three different combination therapies in 3,072 patients with uncomplicated malaria at seven sites in Uganda. Patients infected with at least three strains had almost three times the odds of treatment failure (odds ratio, 2.93 [95% confidence interval, 2.51 to 3.43]; $P < 0.001$) compared with those infected with one or two strains. **Data suggest that efforts to reduce the complexity of infection in areas of high endemicity through the use of intermittent presumptive therapy, improved case management, and reduction in transmission intensity may improve the efficacy of antimalarial therapies.**

Mita, T., A. Kaneko, I. Hwaihwanje, T. Tsukahara, N. Takahashi, H. Osawa, K. Tanabe, T. Kobayakawa, and A. Bjorkman. 2006.
Rapid selection of *dhfr* mutant allele in *Plasmodium falciparum* isolates after the introduction of sulfadoxine/pyrimethamine in combination with 4-aminoquinolines in Papua New Guinea. *Infect. Genet. Evol.* 4 Apr. [Epub ahead of print.]

To overcome the declining efficacy of the 4-aminoquinolines in Papua New Guinea, sulfadoxine-pyrimethamine (SP) has been combined with the 4-aminoquinolines as a first-line treatment for falciparum malaria since 2000. To assess how this change had affected SP-resistant gene polymorphisms, the authors determined allele frequencies of *dhfr* and *dhps* in 113 *Plasmodium falciparum* isolates from Wewak, East Sepik of Papua New Guinea, in 2002 and 2003. In *dhfr*, the double mutant (ACNRNVI) was the predominant allele, with a prevalence of 91%. The authors found a significant decrease of wild *dhfr* allele prevalence (7%) compared with that reported in the adjacent area of East Sepik called the Wosera region (57%), before the drug policy changed in 1990 to 1993. Between 2002 and 2003, the prevalence of this allele decreased from 15% to 3% ($P = 0.02$). Two distinct microsatellite haplotypes flanking *dhfr* were found in isolates with the *dhfr* double mutant, suggesting the selection of preexisting SP-resistant parasites rather than a frequent occurrence of *dhfr* mutations. The *dhfr/dhps* quartet mutations (ACNRNVI in *dhfr* and SGEAA in *dhps*) were identified in six (8%) of the isolates from 2003. This genotype, which is associated with in vivo resistance to SP, has not been reported before in Papua New Guinea. **The data suggest that isolates resistant to SP were rapidly selected despite the use of the SP combination therapy, probably because of their preexisting high level of resistance to the 4-aminoquinoline partner drug.**

Murthi, P., B. Kalionis, H. Ghabrial, M. E. Dunlop, R. A. Smallwood, and R. B. Sewell. 2006. Kupffer cell function during the erythrocytic stage of malaria. *J. Gastroenterol. Hepatol.* 21:313–318.

Previous studies using isolated perfused rat liver in vivo have suggested that during the erythrocytic phase of malaria infection, overall phagocytosis by Kupffer cells is enhanced. The aim of the present study was to further investigate the individual phagocytic capacity and prostaglandin E$_2$ secretion of isolated Kupffer cells in vitro and the immunohistochemical characteristics of Kupffer cells in vivo. Malaria was induced in male Sprague-Dawley rats ($n = 12$) by inoculation with parasitized RBCs containing *Plasmodium berghei*. Kupffer cells were isolated by centrifugal elutriation. **A significantly increased yield of Kupffer cells was obtained from malaria-infected livers compared to controls.** There was an increased internalization by phagocytosis of ^{3}H-labeled bovine serum albumin-containing latex microspheres after 60 min in malaria-infected Kupffer cells compared to controls. Prostaglandin E$_2$ secretion into the cell culture medium was significantly suppressed in malaria-infected Kupffer cells compared to controls. Staining of ED1, ED2, and proliferating-cell nuclear antigen was greater in malaria-infected livers compared to controls. **These results indicate that the number of Kupffer cells is significantly increased and their phagocytic activity on a cell-by-cell basis is enhanced during the erythrocytic stage of malaria.**

Otieno, R. O., C. Ouma, J. M. Ong'echa, C. C. Keller, T. Were, E. N. Waindi, M. G. Michaels, R. D. Day, J. M. Vulule, and D. J. Perkins. 2006. Increased severe anemia in HIV-1-exposed and HIV-1-positive infants and children during acute malaria. *AIDS* 20:275–280.

Since the primary hematologic complication in both pediatric human immunodeficiency virus type 1 (HIV-1) infection and malaria is anemia, coinfection with these pathogens may promote life-threatening severe malarial anemia (SMA). The primary objective of the study was to determine if HIV-1 exposure and/or HIV-1 infection increased the prevalence of SMA in children with acute malaria. **The effect of HIV-1 exposure and HIV-1 infection on the prevalence of SMA (hemoglobin level, <6.0 g/dl), parasitemia (parasites per microliter), and high-density parasitemia (≥10,000 parasites/µl) in children aged 2 years or younger presenting at hospital with acute *Plasmodium falciparum* malaria in a rural holoendemic malaria transmission area of western Kenya was investigated.** On enrollment, a complete hematologic and clinical evaluation was performed on all children. Malaria parasitemia was determined, and children with acute *P. falciparum* malaria were evaluated for HIV-1 exposure and infection by two rapid serologic antibody tests and HIV-1 DNA PCR, respectively. Relative to an HIV-1-uninfected group ($n = 194$), the HIV-1-exposed ($n = 100$) and HIV-1-positive ($n = 23$) groups had lower hemoglobin concentrations ($P < 0.001$ and $P < 0.001$, respectively), while parasitemia and high-density parasitemia (HDP) were equivalent among the three groups. Multivariate analyses demonstrated that the risk of SMA was elevated in HIV-1-exposed and HIV-1-positive children. The multivariate model further revealed that HIV-1 exposure or infection was not significantly associated with HDP. **Results demonstrate that both HIV-1 exposure and HIV-1 infection are associated with increased prevalence of SMA during acute *P. falciparum* infection, independent of parasite density.**

Perez-Leal, O., A. Mongui, J. Cortex, G. Yepes, J. Leiton, and M. A. Patarroyo. 2006. The *Plasmodium vivax* rhoptry-associated protein 1. *Biochem. Biophys. Res. Commun.* 341:1053–1058.

Rhoptries are cellular organelles localized at the apical pole of apicomplexan parasites. Their content is rich in lipids and proteins that are released during target cell invasion. *Plasmodium falciparum* rhoptry-associated protein 1 (RAP1) has been the most widely studied among the rhoptry proteins of this species and is considered to be a good antimalarial vaccine candidate since it displays little polymorphism and induces antibodies in infected humans. MAbs directed against RAP1 are also able to inhibit target cell invasion in vitro, and protection against *P. falciparum* experimental challenge is induced when nonhuman primates are immunized with this protein expressed in its recombinant form. This study identifies and characterizes RAP1 in *Plasmodium vivax*, the most widespread parasite species causing malaria in humans, which produces more than 80 million infections yearly, mainly in Asia and Latin America. **This new protein is encoded by a two-exon gene and is proteolytically processed** in a similar manner to its falciparum homologue, and the immunofluorescence pattern displayed is suggestive of its rhoptry localization. **Studies evaluating the protective efficacy of *P. vivax* RAP1 in nonhuman primates are recommended, taking into account the relevance of its *P. falciparum* homologue as an antimalarial vaccine candidate.**

Riley, E. M., S. Wahl, D. J. Perkins, and L. Schofield. 2006. Regulating immunity to malaria. *Parasite Immunol.* 28:35–49.

Much of the pathology of malaria infection can be immune mediated, which implies that immune responses have to be carefully regulated. The mechanisms by which antimalarial immune responses are thought to be regulated were discussed at the Malaria Immunology Workshop (Bloomberg School of Public Health, Johns Hopkins University, Baltimore, Md., February 2005). Potential regulatory mechanisms include regulatory T cells, which significantly modify cellular immune responses to various protozoan infections, including leishmaniasis and malaria; neutralizing antibodies to proinflammatory malarial toxins such as glycosylphosphatidylinositol and hemozoin; and self-regulating networks of effector molecules. Innate and adaptive immune responses are further moderated by the broader immunologic environment, which is influenced by both the genetic background of the host and coinfection with other pathogens. **A detailed understanding of the interplay between these different immunoregulatory processes may facilitate the rational design of vaccines and novel therapeutics.**

Sacci, J. B., Jr., U. Alam, D. Douglas, J. Lewis, D. L. Tyrrell, A. F. Azad, and N. M. Kneteman. 2006. *Plasmodium falciparum* infection and exoerythrocytic development in mice with chimeric human livers. *Int. J. Parasitol.* 36:353–360.

The exoerythrocytic stage of *Plasmodium falciparum* remains a difficult phase of the parasite life cycle to study. The host and tissue specificity of the parasite requires experimental infection of humans or nonhuman primates and subsequent surgical recovery of parasite-infected liver tissue to analyze this stage of the parasite's development. This type of study is impossible with humans due to obvious ethical considerations, and the cost and complexity of working with primate models have precluded their use for extensive studies of the exoerythrocytic stage. In this study the authors assessed, for the first time, the use of transgenic, chimeric mice containing functioning human hepatocytes as an alternative for modeling the in vivo interaction of *P. falciparum* parasites and human hepatocytes. Infection of these mice with *P. falciparum* sporozoites produced morphologically and antigenically mature liver stage schizonts containing merozoites capable of invading human RBCs. Additionally, when microdissection was used, highly enriched *P. falciparum* liver stage parasites essentially free of hepatocyte contamination were recovered for molecular studies. **These results obtained using mice with chimeric human livers establish a stable murine model for *P. falciparum* that will have a wide utility for assessing the biology of the parasite, potential antimalarial chemotherapeutic agents, and vaccine design.**

Singh, S. K., R. Hora, H. Belrhali, C. E. Chitnis, and A. Sharma. 2006. Structural basis for Duffy recognition by the malaria parasite Duffy-binding-like domain. *Nature* **439**:741–744.

Molecular processes that govern pathogenic features of erythrocyte invasion and cytoadherence in malaria are reliant on *Plasmodium*-specific Duffy-binding-like domains (DBLs). These cysteine-rich modules recognize diverse host cell surface receptors during pathogenesis. DBLs from parasite RBC binding proteins mediate invasion, and those from the antigenically variant *P. falciparum* erythrocyte membrane protein 1 (PfEMP1) have been implicated in cytoadherence. The simian (*P. knowlesi*) and human (*P. vivax*) malarial parasites invade human RBCs exclusively through the host DARC receptor (Duffy antigen receptor for chemokines). The authors present the crystal structure of the *P. knowlesi* DBL domain (Pkα-DBL), which binds to DARC during invasion of human erythrocytes. Pkα-DBL retains the overall fold observed in DBLs from *P. falciparum* erythrocyte binding antigen 175 (EBA-175). Mapping the residues that have previously been implicated in binding highlights a fairly flat but exposed site for DARC recognition in subdomain 2 of Pkα-DBL; this is in sharp contrast to receptor recognition by EBA-175. In Pkα-DBL, the residues that contact DARC and the clusters of residues under immune pressure map to opposite surfaces of the DBL, suggesting a possible mechanism for immune evasion by *P. vivax*. **The comparative structural analysis of Pkα-DBL and *P. falciparum* EBA-175 provides a framework for the understanding of malaria parasite DBLs and may affect the development of new prophylactic and therapeutic strategies.**

Babesia spp.

Carcy, B., E. Precigout, T. Schetters, and A. Gorenflot. 2006. Genetic basis for GPI-anchor merozoite surface antigen polymorphism of *Babesia* and resulting antigenic diversity. *Vet. Parasitol.* 18 Mar. [Epub ahead of print.]

Glycosylphosphatidylinositol anchor merozoite surface antigens (GPI-anchor MSA) are proposed to act in the invasion process of infective merozoites of *Babesia* into host RBCs. Because of their essential function in the survival of *Babesia* parasites, they constitute good candidates for the development of vaccines against babesiosis, and they have been extensively analyzed. These antigens include *Babesia bovis* variable MSA and *Babesia bigemina* gp45/gp55 proteins of the agents of bovine babesiosis from tropical and subtropical countries and the *Babesia divergens* Bd37 and *Babesia canis* Bc28 proteins of the main agents of bovine and canine babesiosis in Europe, respectively. However, they are very polymorphic antigens, and *Babesia* parasites have evolved molecular mechanisms that enable these antigens to evade the host immune system as a survival strategy. This review focuses on the genetic basis of GPI-anchor MSA polymorphism and the antigenic diversity of B-cell epitopes that might be generated in each of these *Babesia* species. No *Babesia* genome sequence is available. However, the available sequences suggest that two distinct, non-cross-reactive GPI-anchor MSA (i.e., with unique B-cell epitopes) may be required by all *Babesia* species for invasion and that these two distinct GPI-anchor MSA would be encoded by a multigene family. **The data are consistent with the ability of biological clones from *Babesia* spp. to use these multigene families for the expression of GPI-anchor MSA, either conserved (*B. canis* and *B. bovis*) or polymorphic (*B. divergens* and *B. bigemina*) in their amino acid sequence. Moreover, as a consequence for successful parasitism, the data suggest that both conserved and polymorphic GPI-anchor MSA would present unique B-cell epitopes.**

Fox, L. M., S. Wingerter, A. Ahmed, A. Arnold, J. Chou, L. Rhein, and O. Levy. 2006. Neonatal babesiosis: case report and review of the literature. *Pediatr. Infect. Dis. J.* **25**:169–173.

Jaundice, hepatosplenomegaly, anemia, and conjugated hyperbilirubinemia developed in a preterm infant with transfusion-associated neonatal babesiosis. The diagnosis was eventually made by blood smear, serology, and PCR amplification. The patient was treated with clindamycin and quinine and made a favorable recovery. The authors review nine other cases of neonatal babesiosis reported in the literature, including six that were transfusion associated, two that were congenital, and two that were tick transmitted.

Kjemtrup, A. M., K. Wainwright, M. Miller, B. L. Penzhorn, and R. A. Carreno. 2006. *Babesia conradae*, sp. nov., a small *Babesia* identified in California. *Vet. Parasitol.* 6 Mar. [Epub ahead of print.]

Small piroplasms as a cause of canine babesiosis have usually been identified as *Babesia gibsoni*. Recent genetic studies suggested that small piroplasms are more likely to be composed of at least three genotypically distinct species. In southern California, canine babesiosis caused by a small piroplasm has been documented since 1990. Morphologic characteristics of this parasite include a small (0.3 to 3.0 μm) intraerythrocytic merozoite stage with predominantly ring, pyriform, tetrad, ameboid, or anaplasmoid forms. Transmission electron microscopic images of merozoites demonstrate the presence of an apical complex consisting of an inner subplasmalemmal membrane and rhoptries. **Based on phylogenetic analyses of the 18S rRNA and the ITS-2 genes, the Californian small piroplasm isolate is more closely related to piroplasm isolates from wildlife and humans in the western United States than it is to *B. gibsoni*. Molecular and morphologic evidence supports naming the small piroplasm from southern California as a distinct species, *Babesia conradae*.**

Leiby, D. A. 2006. Babesiosis and blood transfusion: flying under the radar. *Vox Sang.* **90**:157–165.

Infectious agents impact transfusion medicine as an increasing number of pathogens are described that pose a potential blood safety risk. While the recent focus has been on newly emerging pathogens, several other pathogens remind us that other agents continue to pose threats, but fail to elicit adequate measures to prevent their transmission by blood transfusion. Perhaps foremost among this group of agents are the *Babesia* spp., which have been known to cause human disease in the United States for close to 40 years. *B. microti*, *B. divergens*, and several *Babesia*-like agents are responsible for a growing number of human babesiosis infections. At the same time, in the United States there has been a sharp rise in the number of transfusion-transmitted infections by *Babesia* spp., attributable almost exclusively to *B. microti*. **Despite the obvious public health issues posed by *Babesia* spp., options for preventing blood transfusion-related transmission remain limited. However, recognition that the *Babesia* spp. are indeed an ongoing and expanding blood safety threat will probably prove instrumental in the development of viable interventions to limit the transmission of these agents.**

Vial, H. J., and A. Gorenflot. 2006. Chemotherapy against babesiosis. *Vet. Parasitol.* 24 Feb. [Epub ahead of print.]

Some *Babesia* spp., particularly *Babesia microti* and *Babesia divergens*, can infect humans, and human babesiosis is a significant emerging tick-borne zoonotic disease. Clinical manifestations differ markedly between European and North American diseases. In clinical cases, a combination of clindamycin and quinine is administered as the standard treatment, but administration of atovaquone-azithromycin is also effective. Supportive therapy such as intravenous fluids and blood transfusions are employed when necessary. More specific fast-acting new treatments for babesiosis remain to be developed. This should be facilitated by the sequencing of the *Babesia* spp. genome and increased interest for this malaria-like parasite.

Leishmania spp.

Alborzi, A., M. Rasouli, and A. Shamsizadeh. 2006. *Leishmania tropica*-isolated patient with visceral leishmaniasis in southern Iran. *Am. J. Trop. Med. Hyg.* 74:306–307.

Visceral leishmaniasis (VL) is caused by various strains of *Leishmania donovani*, *Leishmania infantum*, and *Leishmania chagasi* with different geographic distributions. Leishmanial DNA was extracted from the slides of bone marrow aspirates and spleen punctures, which were positive for Leishman bodies from patients who were referred to the hospitals affiliated with Shiraz University of Medical Sciences in southern Iran. Differences in strains were indicated by size difference determined by PCR amplification as visualized on agarose gels. PCR results and smears had 100% correlation. **The dominant strain of *Leishmania* was *L. infantum* (63 of the 64 cases), but 1 case of *L. tropica* was also detected. VL mostly involves children younger than 2 years in Iran; therefore, infection with *L. infantum* was expected. However, this study is the first to report VL caused by *L. tropica* in Iran.**

Bensoussan, E., A. Nasereddin, F. Jonas, L. F. Schnur, and C. L. Jaffe. 2006. Comparison of PCR assays for diagnosis of cutaneous leishmaniasis. *J. Clin. Microbiol.* 44:1435–1439.

Three PCR assays were compared and validated against parasite cultures and microscopic evaluation of stained tissue smears, using 92 specimens from patients with imported and locally acquired suspected cases of cutaneous leishmaniasis (CL) in Israel and the West Bank. The kinetoplast DNA (kDNA) PCR showed the highest sensitivity (98.7%), correctly diagnosing 77 of 78 confirmed positive samples, followed by the rRNA gene internal transcribed spacer 1 (ITS1) PCR (71 of 78 positive; 91.0% sensitivity) and the spliced leader minicxon PCR (42 of 78 positive; 53.8% sensitivity). Parasite culture or microscopy alone detected 62.8% (49 of 78) or 74.4% (58 of 78) of the positive specimens, respectively, while culture and microscopy together improved the overall sensitivity to 83.3% (65 of 78). Except for the kDNA PCR that had six false positives, all other assays were 100% specific. Furthermore, restriction enzyme analysis of the ITS1 PCR product enabled identification of 74.6% of the positive samples, which included strains of *Leishmania major* (50.9%), *Leishmania tropica* (47.2%), and the *Leishmania braziliensis* complex (1.9%). **The data suggest that PCR with kDNA should be used for the diagnosis of CL and that an ITS1 PCR can be reliably used for the diagnosis of CL when rapid species identification is needed.**

Campanelli, A. P., A. M. Roselino, K. A. Cavassani, M. S. Pereira, R. A. Mortara, C. I. Brodskyn, H. S. Goncalves, Y. Belkaid, M. Barral-Netto, A. Barral, and J. S. Silva. 2006. CD4$^+$CD25$^+$ T cells in skin lesions of patients with cutaneous leishmaniasis exhibit phenotypic and functional characteristics of natural regulatory T cells. *J. Infect. Dis.* 193:1313–1322.

Endogenous regulatory T (T$_{reg}$) cells are involved in the control of infections, including *Leishmania* infection in mice. *Leishmania viannia braziliensis* is the main etiologic agent of CL in Brazil, and it is also responsible for the more severe mucocutaneous form. The authors investigated the possible involvement of T$_{reg}$ cells in the control of the immune response in human skin lesions caused by *L. viannia braziliensis* infection. Data showed that functional T$_{reg}$ cells can be found in skin lesions of patients with CL. These cells express phenotypic markers of T$_{reg}$ cells (CD25, cytotoxic T-lymphocyte-associated antigen 4, Foxp3, and glucocorticoid-induced tumor necrosis factor receptor) and are able to produce large amounts of IL-10 and transforming growth factor β. Furthermore, CD4$^+$CD25$^+$ T cells derived from the skin lesions of four of six patients with CL significantly suppressed in vitro the phytohemagglutinin-induced proliferative T-cell responses of allogeneic peripheral-blood mononuclear cells from healthy control subjects at a ratio of 1 T$_{reg}$ cell to 10 allogeneic peripheral-blood mononuclear cells. **The data suggest that functional T$_{reg}$ cells accumulate at sites of *Leishmania* infection in humans and possibly contribute to the local control of effector T-cell functions.**

Croft, S. L., S. Sundar, and A. H. Fairlamb. 2006. Drug resistance in leishmaniasis. *Clin. Microbiol. Rev.* 19:111–126.

Leishmaniasis is a complex disease, with visceral and cutaneous manifestations, and is caused by over 15 different species of the protozoan parasite genus *Leishmania*. There are significant differences in the sensitivity of these species both to the standard drugs, including pentavalent antimonials and miltefosine, and to those undergoing clinical trial, for example, paromomycin. **Over 60% of patients with VL in Bihar State, India, do not respond to treatment with pentavalent antimonials.** This is now considered to be due to acquired resistance. Although this class of drugs has been used for over 60 years for leishmaniasis treatment, it is only in the past 2 years that the mechanisms of action and resistance have been identified and related to drug metabolism, thiol metabolism, and drug efflux. **With the introduction of new therapies, including miltefosine in 2002 and paromomycin in 2005 to 2006, the authors stress the importance of a strategy to prevent the emergence of resistance to new drugs; combination therapy, monitoring of therapy, and improved diagnostics could play an essential role in this strategy.**

Genestra, M., W. J. Souza, D. Guedes-Silva, G. M. Machado, L. Cysne-Finkelstein, R. J. Bezerra, F. Monteiro, and L. L. Leon. 2006. Nitric oxide biosynthesis by *Leishmania amazonensis* promastigotes containing a high percentage of metacyclic forms. *Arch. Microbiol.* 31 Mar. [Epub ahead of print.]

Due to the diversity of its physiologic and pathophysiologic functions and general ubiquity, the study of nitric oxide (NO) has increased. The authors demonstrated that *Leishmania amazonensis* promastigotes produce NO, a free radical synthesized from L-arginine by nitric oxide synthase (NOS). A soluble NOS was purified from *L. amazonensis* promastigotes by affinity chromatography (2′, 5′-ADP-agarose), and on SDS-PAGE the enzyme migrated as a single protein band of 116.2 ± 6 kDa. Furthermore, a constitutive NOS was detected through indirect immunofluorescence using anti-cNOS and in NADPH consumption assays. **The data show that NO production, detected as nitrite in the culture supernatant, is prominent in promastigote preparations with large numbers of metacyclic forms, suggesting an association with the differentiation and infectivity of the parasite.**

Rotureau, B., C. Ravel, C. Aznar, B. Carme, and J. P. Dedet. 2006. First report of *Leishmania infantum* in French Guiana: canine visceral leishmaniasis imported from the old world. *J. Clin. Microbiol.* 44:1120–1122.

The first two cases of canine VL in French Guiana are described. One infected dog was most probably imported from France. A second dog was then infected with *Leishmania infantum* in French Guiana. **These observations support the intercontinental transportation theory for *L. infantum*.**

Rotureau, B., C. Ravel, P. Couppie, F. Pratlong, M. Nacher, J. P. Dedet, and B. Carme. 2006. Use of PCR-restriction fragment length polymorphism analysis to identify the main New World *Leishmania* species and analyze their taxonomic properties and polymorphism by application of the assay to clinical samples. *J. Clin. Microbiol.* 44:459–467

Approximately 13 characterized *Leishmania* species are known to infect humans in South America; 5 of these are transmitted in the sylvatic ecotopes of the whole French Guianan territory and are responsible for CL. Restriction

fragment length polymorphism analyses have given promising results in the diagnosis of CL. The end of the small subunit and ITS1 of the rRNA genes were sequenced and targeted by PCR/restriction fragment length polymorphism analysis in the 10 main New World *Leishmania* species from the two subgenera. The procedure was tested on 40 samples from patients with CL, and the results were compared with those of conventional methods. The results of this simple genus-specific method were in agreement with those of previous isoenzyme analyses. This method distinguished the most medically relevant *Leishmania* species with only one enzyme (RsaI) and could be **performed directly on human biopsy specimens (sensitivity, 85.7%). Performing New World *Leishmania* species typing rapidly and easily in the field constitutes a very valuable improvement for detection of *Leishmania* spp.** Revealing great diversity with several enzymes, this method could also be useful for taxonomic, ecological, and epidemiologic studies in space and time.

Rotureau, B., C. Ravel, M. Nacher, P. Couppie, I. Curtet, J. P. Dedet, and B. Carme. 2006. Molecular epidemiology of *Leishmania* (*Viannia*) *guyanensis* in French Guiana. *J. Clin. Microbiol.* 44:468–473.

Genetic polymorphism of a nuclear sequence encompassing the end of the ribosomal small subunit and ITS1 of 265 isolates from patients with CL was examined by restriction fragment length polymorphism analysis. Genotypes based on the fingerprinting phenetic integration were compared to epidemiologic, clinical, and geographic data. In agreement with previous reports, five different *Leishmania* species were identified, but *Leishmania* (*Viannia*) *guyanensis* represented 95.8% of the samples. Two distinct *L.* (*V.*) *guyanensis* populations were found to originate in two ecologically characterized regions. Higher lesional parasite densities and the need for additional treatments were significantly linked to genotype group I. Parasites of genotype group II were more likely to cause chronic and disseminated cutaneous forms in patients. **Although *L.* (*V.*) *guyanensis* was previously said not to be very polymorphic, the data indicate a significant degree of discrimination among *L.* (*V.*) *guyanensis* isolates from diverse ecological areas and with different clinical implications.**

Stark, D., S. Pett, D. Marriott, and J. Harkness. 2006. Post-kala-azar dermal leishmaniasis due to *Leishmania infantum* in a human immunodeficiency virus type 1-infected patient. *J. Clin. Microbiol.* 44:1178–1180.

The authors report the first case of post-kala azar dermal leishmaniasis due to *Leishmania infantum* in an HIV-1-infected patient in Australia. Molecular characterization of the isolate was performed using PCR/restriction fragment length polymorphism targeting both repetitive sequences from *Leishmania* nuclear DNA and repetitive kinetoplast DNA minicircles for species differentiation.

Zhang, W. W., C. Miranda-Verastegui, J. Arevalo, M. Ndao, B. Ward, A. Llanos-Cuentas, and G. Matlashewski. 2006.

Development of a genetic assay to distinguish between *Leishmania viannia* species on the basis of isoenzyme differences. *Clin. Infect. Dis.* 42:801–809.

Tegumentary leishmaniasis in Latin America is caused by *Leishmania viannia braziliensis* complex parasites. *L. braziliensis* and *Leishmania viannia peruviana* are the two predominant *Leishmania* species in Peru. *L. braziliensis* is more virulent, because it can cause mucocutaneous leishmaniasis, known as espundia, which results in severe facial destruction. Early identification of the species that causes the initial cutaneous infection would greatly help to prevent mucocutaneous leishmaniasis because it would allow more aggressive treatment and follow-up. However, because of the close genetic similarity of *L. braziliensis* and *L. peruviana*, there is no simple assay to distinguish between these species. The authors cloned the mannose phosphate isomerase gene from both *L. braziliensis* and *L. peruviana*. It is the only known isoenzyme capable of differentiating between *L. braziliensis* and *L. peruviana* in multilocus enzyme electrophoresis. Only a single nucleotide polymorphism was found between the mannose phosphate isomerase genes from *L. braziliensis* and *L. peruviana*, resulting in an amino acid change from threonine to arginine at amino acid 361. A PCR assay was developed to distinguish the single-nucleotide polymorphism of the mannose phosphate isomerase gene to allow for the specific identification of *L. braziliensis* or *L. peruviana*. This assay was validated with 31 reference strains that were previously typed by multilocus enzyme electrophoresis, successfully applied to patient biopsy samples, and adapted to a real-time PCR assay. This innovative approach combines new genetic knowledge with traditional biochemical fundamentals of multilocus enzyme electrophoresis to better manage leishmaniasis in Latin America.

Trypanosoma spp.

African Trypanosomiasis

Bisser, S., O. N. Ouwe-Missi-Oukem-Boyer, F. S. Toure, Z. Taoufig, B. Bouteille, A. Buguet, and D. Mazier. 2006. Harbouring in the brain: a focus on immune evasion mechanisms and their deleterious effects in malaria and human African trypanosomiasis. *Int. J. Parasitol.* 9 Mar. [Epub ahead of print.]

Malaria and human African trypanosomiasis represent the two major tropical vector-transmitted protozoan infections, displaying different prevalence and epidemiologic patterns. Death occurs mainly due to neurologic complications which are initiated at the blood-brain barrier level. Adapted host immune responses present differences but also similarities in blood-brain barrier/parasite interactions for these diseases and are the focus of this review. The authors describe and compare parasite evasion mechanisms, the initiating mechanisms of central nervous system (CNS) pathology, and the major clinical and neuropathologic features. They also highlight the common immune-mediated mechanisms leading to brain involvement. **In both diseases, neurologic damage is caused mainly by cytokines (gamma interferon, tumor necrosis factor-α, and IL-10), NO, and endothelial cell apoptosis.**

Blum, J., C. Schmid, and C. Burri. 2006. Clinical aspects of 2541 patients with second stage human African trypanosomiasis. *Acta Trop.* 97:55–64.

The clinical symptoms and signs of patients with second-stage human African trypanosomiasis are described for a large cohort of patients treated in a prospective multicenter, multinational study. Special emphasis is given to the influence of the disease stage (duration, number of white blood cells [WBC] in cerebrospinal fluid [CSF]) and patient age on the clinical picture. **Even though the frequencies of the symptoms and signs are highly variable among centers, the clinical picture of the disease is similar for all countries.** Headache (78.7%), sleeping disorder (74.4%), and lymphadenopathy (56.1%) are the most frequent symptoms and signs, and they are similar for all stages of the disease. Lymphadenopathy tends to be highest in the advanced second stage (59.0%). The neurologic and psychiatric symptoms increase significantly with the number of WBC in the CSF, indicating the stage of progression of the disease. Pruritus is observed in all stages and increases with the number of WBC in CSF from 30 to 55%. In children younger than 7 years, lymphadenopathy is less frequently reported (11.8 to 37.3%) than in older children or adults (56.4 to 61.2%). Fever is most frequently reported in children between 2 and 14 years of age (26.1 to 28.7%), and malnutrition is significantly more frequently observed in children of all ages (43 to 56%) than in adults (23.5%).

Joshi, P. P., V. R. Shegokar, R. M. Powar, S. Herder, R. Katti, H. R. Salkar, V. S. Dani, A. Bhargava, J. Jannin, and P. Truc. 2005. Human trypanosomiasis caused by *Trypanosoma evansi* in India: the first case report. *Am. J. Trop. Med. Hyg.* 73:491–495.

The authors report an Indian farmer who had fluctuating trypanosome parasitemia associated with febrile episodes for 5 months. Morphologic examination of the parasites indicated the presence of large numbers of trypanosomes belonging to the species *Trypanosoma evansi*, which is normally a causative agent of animal trypanosomiasis known as surra. Basic clinical and biological examinations are described, using several assays including parasitological, serologic, and molecular biological tests, all of which confirmed the infecting species as *T. evansi*. Analysis of CSF indicated no invasion of the CNS by trypanosomes. Suramin, a drug used exclusively for treatment of early-stage human African trypanosomiasis with no CNS involvement, led to an apparent cure in the patient. **This is the first reported case of human infection due to *Trypanosoma evansi*, which was probably caused by transmission of blood from an infected animal.**

Kennedy, P. G. 2006. Diagnostic and neuropathogenesis issues in human African trypanosomiasis. *Int. J. Parasitol.* 3 Mar. [Epub ahead of print.]

Human African trypanosomiasis, also known as sleeping sickness, is caused by protozoan parasites of the genus *Trypanosoma* and is a major cause of human mortality and morbidity. The East African and West African variants, caused by *Trypanosoma brucei rhodesiense* and *Trypanosoma brucei gambiense*, respectively, differ in their presentation, but the disease is fatal if untreated. **Accurate staging of the disease into the early hemolymphatic stage and the late encephalitic stage is critical because the treatment for the two stages is different. The only effective drug for late-stage disease, melarsoprol, which crosses the blood-brain barrier, is followed by a severe posttreatment reactive encephalopathy in 10% of patients, half of whom die.** There is no current consensus on the diagnostic criteria for CNS involvement, and the specific indications for melarsoprol therapy also differ. There is a pressing need for a quick, simple, cheap, and reliable diagnostic test to diagnose human African trypanosomiasis in the field and also to determine CNS invasion. **CSF and plasma analyses for patients with human African trypanosomiasis have indicated a role for both proinflammatory and counterinflammatory cytokines in determining the severity of the meningoencephalitis of late-stage disease; at least in *T. brucei rhodesiense* infection, the balance of these opposing cytokines may be critical.** Animal models have allowed a greater understanding of the more direct effect of trypanosome infection on CNS function, including the disruption of circadian rhythms, as well as the immunologic determinants of the passage of trypanosomes across the blood-brain barrier.

Kennedy, P. G. 2006. Human African trypanosomiasis— neurological aspects. *J. Neurol.* 20 Mar. [Epub ahead of print.]

Human African trypanosomiasis, also known as sleeping sickness, is a major cause of death and disability in 36 countries in sub-Saharan Africa. The disease is caused by the protozoan parasite of the *Trypanosoma* genus, which is transmitted by the bite of the tsetse fly. The two types of human African trypanosomiasis, the East African form due to *Trypanosoma brucei rhodesiense* and the West African form due to *T. brucei gambiense*, differ in their tempo of infection, but in both cases the disease is always fatal if untreated. As well as multiple systemic features seen in the early (hemolymphatic) stage of disease, the late (encephalitic) stage is associated with a wide range of neurologic features including neuropsychiatric, motor, and sensory abnormalities. **Accurate staging of the disease is absolutely essential because of the potentially fatal complications of melarsoprol treatment of late-stage disease, the most important of which is a severe posttreatment reactive encephalopathy,** whose pathogenesis is not fully understood. However, there is no universal consensus as to how late-stage disease should be diagnosed using CSF criteria, and this has been very problematic in human African trypanosomiasis. **A more recent alternative drug for late-stage gambiense disease is eflornithine. There is a need for a nontoxic oral drug for both early- and late-stage disease that would eliminate many of the staging problems.**

Kibona, S. N., L. Matemba, J. S. Kaboya, and G. W. Lubega. 2006. Drug-resistance of *Trypanosoma b. rhodesiense* isolates from Tanzania. *Trop. Med. Int. Health* **11**:144–155.

The authors screened 35 *Trypanosoma brucei rhodesiense* strains in the mouse model for sensitivity to melarsoprol (1.8,

3.6, and 7.2 mg/kg), diminazene aceturate (3.5, 7, and 14 mg/kg), suramin (5, 10, and 20 mg/kg) and isometamidium (0.1, 1.0, and 2 mg/kg). Thirteen isolates suspected to be resistant were selected for further testing in vitro and in vivo. From the in vitro testing, 50% inhibitory concentrations (IC_{50}s) were determined by a short-term viability assay, and MICs were calculated by a long-term viability assay. For in vivo testing, doses higher than those in the initial screening test were used. Two *T. brucei rhodesiense* stocks expressed resistance in vivo to melarsoprol at 5 mg/kg and at 10 mg/kg. Melarsoprol had high IC_{50}s and MICs for these strains, consistent with those for the melarsoprol-resistant reference strain. Infection with another isolate relapsed after treatment with 5 mg of melarsoprol/kg although the isolate did not appear resistant in vitro. One isolate was resistant to diminazene at 14 mg/kg, and another was resistant at both 14 and 28 mg/kg. The IC_{50}s for these two isolates were consistent with the diminazene-resistant reference strain. Infections with two isolates relapsed at a suramin dose of 5 mg/kg, although no isolate appeared resistant in the in vitro tests. Two isolates were resistant to isometamidium at 1.0 mg/kg, and the IC_{50}s were higher. Two isolates were cross resistant to melarsoprol and diminazene, and one isolate was cross resistant to suramin and isometamidium. **The reduced susceptibility of *T. brucei rhodesiense* isolates to these drugs strongly indicates that drug resistance may be emerging in northwestern Tanzania.**

MacLean, L., M. Odiit, and J. M. Sternberg. 2006. Intrathecal cytokine responses in *Trypanosoma brucei rhodesiense* sleeping sickness patients. *Trans. R. Soc. Trop. Med. Hyg.* **100**:270–275.

Intrathecal cytokine levels and blood-CSF barrier function in 91 *Trypanosoma brucei rhodesiense*-infected patients were studied. The concentrations of the cellular immune activation marker neopterin and the cytokines IL-6 and IL-10 in CSF were increased over control and posttreatment levels in all patients, with maximal levels observed in late-stage (meningoencephalitic) individuals. Analysis of concentration quotients in CSF and serum indicated that IL-10 and neopterin were derived from CNS in at least 25% of the patients. Blood-CSF barrier dysfunction occurred in 64% of late-stage patients but not in early-stage patients. **While the high level of neopterin observed in the CSF of late-stage patients is indicative of widespread cellular activation, the increased levels of IL-6 and IL-10 suggest that counterinflammatory cellular responses may be important in the regulation of neuropathogenesis in late-stage human African trypanosomiasis.**

American Trypanosomiasis

Benchimol Barbosa, P. R. 2006. The oral transmission of Chagas' disease: an acute form of infection responsible for regional outbreaks. *Int. J. Cardiol.* 4 Apr. [Epub ahead of print.]

Orally transmitted Chagas' disease is an ordinarily rare form of *Trypanosoma cruzi* transmission and is responsible for regional outbreaks. Ingestion of contaminated material is generally associated with massive parasitic transmission, ultimately leading to acute myocarditis, with more severe clinical presentation at younger ages and high death rates. Close monitoring of regional outbreaks by health agencies is mandatory to prevent recrudescence of the disease.

de Castro, A. M., A. O. Luquetti, A. Rassi, E. Chiari, and L. M. da Cunha Galvao. 2006. Detection of parasitemia profiles by blood culture after treatment of human *Trypanosoma cruzi* infection. *Parasitol. Res.* 29 Mar. [Epub ahead of print.]

The change in parasitemia profile, measured by sequential blood cultures of 27 benznidazole-treated patients compared with 13 untreated patients, during the chronic phase of Chagas' disease is described. All patients were adults (age limits, 23 to 88 years) with positive serology (three tests); 23 of them were females. All patients submitted six blood cultures, three before and three after the benznidazole treatment. The parasitemia was classified as nondetected (with three negative blood cultures), medium (one positive culture in three), or high (two or three positive cultures). Of the 8 patients with nondetected parasitemia before treatment, 7 still had the same profile and only 1 switched to medium; of the 8 with medium parasitemia, 7 shifted to nondetected and 1 shifted to high parasitemia; of the 11 patients with high parasitemia before treatment, 10 converted to nondetected and only 1 was positive. Of the 27 patients, 19 changed the parasitemia profile (70.4%), and **the rate of therapeutic failure was 11.1% (3 of 27) during the first 24 months of follow-up after treatment.** The shift to the nondetected parasitemia profile was from 8 of 27 to 24 of 27 patients for the first 2 years after benznidazole treatment. Only 46.2% (6 of 13) of the nontreated individuals changed their parasitemia profile. **The authors concluded that benznidazole has a strong trypanocidal effect (88.8%) and that the rate of therapeutic failure during the first 2 years after trypanocidal treatment is 11.1%.**

Mora, M. C., O. Sanchez Negrette, D. Marco, A. Barrio, M. Ciaccio, M. A. Segura, and M. A. Basombrio. 2005. Early diagnosis of congenital *Trypanosoma cruzi* infection using PCR, hemoculture, and capillary concentration, as compared with delayed serology. *J. Parasitol.* 91:1468–1473.

Congenital *Trypanosoma cruzi* infection is a highly pathogenic and underreported condition. Early recognition is essential for effective treatment. Umbilical cord blood from newborns ($n = 302$) born of infected mothers was analyzed by microhematocrit, hemoculture, and PCR methods. Each subject was then monitored serologically. In calibrated suspensions of *T. cruzi* in blood, the sensitivity of PCR was 27-fold higher than that of hemoculture. However, this advantage was not reflected during routine testing of samples from maternity wards, partly because of the uneven distribution of few parasites in small samples. Levels of detection of congenital infection were 2.9% (8 of 272) for microhematocrit, 6.3% (18 of 287) for hemoculture, 6.4% (15 of 235) for PCR, and 8.9% (27 of 302) for cumulated results. Evaluation against the standard of delayed serology indicates that the regular use of PCR, hemoculture, and microhematocrit with blood

samples allows rapid detection of about 90% of congenitally infected newborns from samples that can be obtained before the mother and child leave the maternity ward.

Intestinal Nematodes

Capillaria philippinensis

Intapan, P. M., W. Maleewong, W. Sukeepaisarnjaroen, and N. Morakote. 2006. Potential use of *Trichinella spiralis* antigen for serodiagnosis of human capillariasis philippinensis by immunoblot analysis. *Parasitol. Res.* 98:227–231.

Intestinal capillariasis is an emerging helminthic zoonosis caused by *Capillaria philippinensis* and is frequently fatal if not diagnosed correctly. The present study demonstrates cross-reactivity between *Trichinella spiralis* larval antigens and *C. philippinensis*-infected human sera by immunoblotting. Sera from 16 patients with proven intestinal capillariasis and 16 patients with proven trichinosis were tested. The antigenic patterns recognized by sera from patients with intestinal capillariasis varied with the molecular masses, ranging from less than 20.1 to more than 94 kDa. The immunoblotting profiles of the sera from patients with trichinosis were similar to those of the sera from patients with intestinal capillariasis. The antigenic bands with 100% reactivity were located at 36.5, 40.5, and 54 kDa, respectively. Sera from patients with trichuriasis, strongyloidiasis, opisthorchiasis, and healthy controls differed clearly from the previous two and produced very faint patterns of reactivity and attenuated bands. **This assay is potentially useful for large-scale screenings of persons at risk for *C. philippinensis* infection. However, parasitologic stool examinations from the positive patients are necessary as second-tier laboratory tests for confirming the diagnosis.**

Hookworm

Fleming, F. M., S. Brooker, S. M. Geiger, I. R. Caldas, R. Correa-Oliveira, P. J. Hotez, and J. M. Bethony. 2006. Synergistic associations between hookworm and other helminth species in a rural community in Brazil. *Trop. Med. Int. Health* 11:56–64.

The authors determined the prevalence and intensity of infections by single and multiple helminth species in an age-stratified sample of 1,332 individuals from 335 households. Hookworm was the most prevalent helminth (68.2%), followed by *Ascaris lumbricoides* (48.8%) and *Schistosoma mansoni* (45.3%). Overall, 60.6% of individuals had multiple-helminth infections. Multivariate analysis indicated significant positive associations for coinfection with hookworm and *S. mansoni* and for coinfection with hookworm and *A. lumbricoides*. Coinfections with hookworm and *A. lumbricoides* resulted in higher egg counts for both, suggesting a synergistic relationship between these species, although the authors found important age differences in this relationship. However, the intensity of *S. mansoni* or *A. lumbricoides* coinfection did not differ from that of monoinfection. These results have implications for the epidemiology, immunology, and control of multiple-helminth infections. More research is needed to examine the rates of reinfection and immune responses after chemotherapy and the extent to which the effects of polyparasitism are altered by chemotherapy.

Helde, M. R., R. D. Bungiro, L. M. Harrison, I. Hamza, and M. Capello. 2006. Dietary iron content mediates hookworm pathogenesis in vivo. *Infect. Immun.* 74:289–295.

Hookworm infection is associated with growth delay and iron deficiency anemia in developing countries. A series of experiments were designed to test the hypothesis that host dietary iron restriction mediates susceptibility to hookworm infection by using the hamster model of *Ancylostoma ceylanicum*. Animals were maintained on diets containing either 10 ppm iron (iron restricted) or 200 ppm iron (standard/high iron) and were then infected with *A. ceylanicum* third-stage larvae. Infected animals fed the standard diet exhibited statistically significant growth delay and reduced blood hemoglobin levels compared to uninfected controls on day 20 postinfection. In contrast, no statistically significant differences in weight or hemoglobin concentration were observed between infected and uninfected animals fed the iron-restricted diet. Moreover, iron-restricted animals were observed to have reduced intestinal worm burdens on days 10 and 20 postinfection compared to those of animals maintained on the standard/high-iron diet. In a subsequent study, animals equilibrated on diets containing a range of iron levels (10, 40, 100, or 200 ppm) were infected with *A. ceylanicum* and monitored for evidence of hookworm disease. Infected animals from the intermediate-dietary-iron (40- and 100-ppm) groups exhibited greater weight loss and anemia than did those in the low (10-ppm)- or high (200-ppm)-iron groups. Mortality was also significantly higher in the intermediate-dietary-iron groups. **These data suggest that severe dietary iron restriction impairs hookworm development in vivo but that moderate iron restriction enhances host susceptibility to severe disease.**

Ranjit, N., M. K. Jones, D. J. Stenzel, R. B. Gasser, and A. Loukas. 2006. A survey of the intestinal transcriptomes of the hookworms, *Necator americanus* and *Ancylostoma caninum*, using tissues isolated by laser microdissection microscopy. *Int. J. Parasitol.* 6 Mar. [Epub ahead of print.]

The gastrointestinal tracts of multicellular blood-feeding parasites are targets for vaccines and drugs. Recombinant vaccines that interrupt the digestion of blood in the hookworm gut have shown efficacy, and so the authors explored the intestinal transcriptomes of the human and canine hookworms, *Necator americanus* and *Ancylostoma caninum*, respectively. They used laser microdissection microscopy to dissect gut tissue from the parasites, extracted the RNA, and generated cDNA libraries. A total of 480 expressed sequence tags were sequenced from each library and assembled into contigs, accounting for 268 *N. americanus* genes and 276 *A. caninum* genes. Only 17% of *N. americanus* and 36% of *A. caninum* contigs were assigned gene ontology classifications. Twenty-six *N. americanus* (9.8%) and 18 *A. caninum* (6.5%) contigs did not have homologues in any databases, including dbEST. Of these novel clones, seven *N. americanus* and three *A. caninum* contigs had open reading frames with predicted secretory signal peptides. The most abundant transcripts corresponded to mRNAs encoding cholesterol and fatty acid binding proteins, C-type lectins, activation-associated secretory proteins, and proteases of different mechanistic classes, particularly astacin-like metallopeptidases. Expressed sequence tags corresponding to known and potential recombinant vaccines were identified, and these included homologues of proteases, anti-clotting factors, defensins, and integral membrane proteins involved in cell adhesion.

Strongyloides stercoralis

Arsic-Arsenijevic, V., A. Dzamic, Z. Dzamic, D. Milobratovic, and D. Tomic. 2005. Fatal *Strongyloides stercoralis* infection in a young woman with lupus glomerulonephritis. *J. Nephrol.* 18:787–790.

The authors report a case of fatal *Strongyloides stercoralis* infection in a 35-year-old woman with lupus glomerulonephritis after prolonged steroid therapy. An epidemiologic evaluation revealed that the patient originated from a rural area in Bosnia and Herzegovina, which was an area of *S. stercoralis* endemicity in the former Yugoslavia. She had severe gastrointestinal and pulmonary symptoms and a history of a 13-kg weight loss in 3 months. Histopathologic examination of large-bowel mucosa showed nematode larvae. Microscopy of stool, sputum, and urine samples confirmed *S. stercoralis*. The diagnosis was delayed because of the low suspicion index, the absence of eosinophilia, and nonspecific signs of infection, which could be a result of the underlying disease or the effects of corticosteroids. **This case highlights the importance of screening for *S. stercoralis* in patients starting immunosuppressive therapy, especially if they are from areas of endemic infection.**

Ternidens deminutus

Hemsrichart, V. 2005. *Ternidens deminutus* infection: first pathological report of a human case in Asia. *J. Med. Assoc. Thai.* 88:1140–1143.

The authors report a case of *Ternidens deminutus* infection in a 33-year-old Thai woman who was admitted to the hospital because of abdominal pain and a right lower quadrant mass. Exploratory laparotomy revealed an omental mass with attached terminal ileum. Resection of the mass together with the terminal ileum and the right-sided colon was performed. Pathologically, the omental mass was an abscess containing an immature male *Ternidens deminutus*. **The parasite usually is found in the intestine of primates in Africa and Asia. Human infection occurs when food contaminated with infective filariform larvae is ingested. The larvae molt in the intestinal wall and become adults. They pass eggs in feces. Eggs in contaminated soil hatch and become rhabditiform larvae and then infective filariform larvae. This is the first reported human case in Asia.**

Tissue Nematodes

Angiostrongylus cantonensis

Kanpittaya, J., S. Jitpimolmard, S. Tiamkao, and E. Mairiang. 2000. MR findings of eosinophilic meningoencephalitis attributed to *Angiostrongylus cantonensis*. *Am. J. Neuroradiol.* 21:1090–1094.

Eosinophilic meningoencephalitis is prevalent and widely distributed in Thailand, especially in the northeastern and central parts of the country. *Angiostrongylus cantonensis* **is one of the causative agents of fatal eosinophilic meningoencephalitis.** The nematodes cause extensive tissue damage by moving through the brain and inducing an inflammatory reaction. In six patients with eosinophilic meningoencephalitis, the clinical presentation included severe headache, clouded consciousness, and meningeal irritation. Abnormal magnetic resonance imaging (MRI) findings included prominence of the Virchow-Robin spaces, subcortical enhancing lesions, and abnormal high-T_2 signal lesions in the periventricular regions. Proton brain MR spectroscopy was performed for three patients; the results were abnormal in one patient with a severe case, showing decreased choline levels in a lesion. Small hemorrhagic tracts were found in one patient. Lesions thought to be due to microcavities and migratory tracts were found in only one patient. **MRI and MR spectroscopy findings are helpful in understanding the pathogenic mechanisms of the disease.**

Anisakiasis

De Nicola, P., L. Napolitano, N. Di Bartolomeo, M. Waku, and P. Innocenti. 2005. Anisakiasis presenting as perforated ulcer of the cecum. *G. Chir.* 26:375–377.

Symptoms of cecal anisakiasis in this patient were compatible with appendicitis. Surgery was performed, and a perforated ulcer of the cecum was found. Ileocolic resection was performed. The histologic result showed the presence of *Anisakis simplex* larva in the muscle of the cecum. The patient was discharged on day 5 without complications and at present remains asymptomatic. He had eaten uncooked anchovies some days before the onset of the disease.

Chai, J. Y., K. Darwin Murrell, and A. J. Lymbery. 2005. Fish-borne parasitic zoonoses: status and issues. *Int. J. Parasitol.* 35:1233–1254.

The fish-borne parasitic zoonoses have been limited for the most part to populations living in low- and middle-income countries, but the geographic limits and populations at risk are expanding because of growing international markets, improved transportation systems, and demographic changes such as population movements. **While many people in developed countries recognize meat-borne zoonoses such as trichinellosis and cysticercosis, far fewer are acquainted with the fish-borne parasitic zoonoses, which are mostly helminthic diseases caused by trematodes, cestodes, and nematodes. Yet these zoonoses are responsible for large numbers of human infections around the world. The list of potential fish-borne parasitic zoonoses is quite long.** However, in this review, emphasis has been placed on liver fluke diseases such as clonorchiasis, opisthorchiasis, and metorchiasis, as well as on intestinal trematodiasis (the heterophyids and echinostomes), anisakiasis (due to *Anisakis simplex* larvae), and diphyllobothriasis. The life cycles, distributions, epidemiology, clinical aspects, and, importantly, the research needed for improved risk assessments, clinical management, and prevention and control of these important parasitic diseases are reviewed.

Angiostrongylus cantonensis

Liu, I. H., Y. M. Chung, S. J. Chen, and W. L. Cho. 2006. Necrotizing retinitis induced by *Angiostrongylus cantonensis*. *Am. J. Ophthalmol.* 141:577–579.

The authors report a case of bilateral necrotizing retinitis induced by *Angiostrongylus cantonensis*. A 52-year-old Asian woman developed eosinophilic meningitis after eating several undercooked snails. One week later, sudden onset of vision loss in both eyes was noted. Widespread yellow retinal exudates were accompanied by bullous retinal detachment in both eyes. *A. cantonensis* infection was confirmed by positive ELISA of the serum and CSF and a positive Western blot test of the subretinal fluid. **After treatment with mebendazole, levamisole, and corticosteroid, these necrotizing patches regressed gradually. However, the final visual acuity was no light perception. Considering the possible outcome, *A. cantonensis* infection should be considered one of the causes of necrotizing retinitis.**

Baylisascaris procyonis

Gavin, P. J., K. R. Kazacos, and S. T. Shulman. 2005. Baylisascariasis. *Clin. Microbiol. Rev.* 18:703–718.

The raccoon roundworm, *Baylisascaris procyonis*, is the most common and widespread cause of clinical larva migrans in animals. **In addition, it is increasingly recognized as a cause of devastating or fatal neural larva migrans in infants and young children and ocular larva migrans in adults.** Humans become infected by accidentally ingesting infective *B. procyonis* eggs from raccoon latrines or articles contaminated with their feces. **Two features distinguish *B. procyonis* from other helminthes that cause larva migrans: (i) its aggressive somatic migration and invasion of the CNS and (ii) the continued growth of larvae to a large size within the CNS.** Typically, *B. procyonis* neural larva migrans presents as acute fulminant eosinophilic meningoencephalitis. Once invasion of the CNS has occurred, the prognosis is grave with or without treatment. **To date, despite anthelmintic treatment of cases of *B. procyonis* neural larva migrans, there are no documented neurologically intact survivors.** Epidemiologic study of human cases of neural larva migrans demonstrate that contact with raccoon feces or an environment contaminated by infective eggs and geophagia or pica are the most important risk factors for infection. In many regions of the United States, increasingly large populations of raccoons, with high rates of *B. procyonis* infection, live in close proximity to humans. **Although documented cases of human baylisascariasis remain relatively uncommon, widespread contamination of the domestic environment by infected raccoons suggests that the risk of exposure and human infection is probably substantial. In the absence of early diagnosis or effective treatment, prevention of infection is the most important public health measure.**

Capillaria hepatica

Klenzak, J., A. Mattia, A. Valenti, and J. Goldberg. 2005. Hepatic capillariasis in Maine presenting as a hepatic mass. *Am. J. Trop. Med. Hyg.* 72:651–653.

The authors report the first case of hepatic capillariasis in Maine. The patient was a 54-year-old male carpenter who presented with a subacute history of severe abdominal pain, fevers, and weight loss. Initial diagnostic studies suggested a hepatic mass associated with para-aortic lymphadenopathy. The patient underwent open laparotomy for resection of the mass. He was found to have an eosinophilic granuloma in the liver; further evaluation revealed degenerating *Capillaria hepatica*. The exact route of infection in this case is unknown but most probably involves accidental ingestion of soil contaminated with mature *Capillaria* eggs. This patient had a low parasite burden and did not exhibit significant peripheral eosinophilia. After treatment with thiabendazole, he recovered uneventfully.

Gnathostoma spp.

Sangchan, A., K. Sawanyawisuth, P. M. Intapan, and A. Mahakkanukrauh. 2006. Outward migration of *Gnathostoma spinigerum* in interferon alpha treated hepatitis C patient. *Parasitol. Int.* 55:31–32.

After a hepatitis C virus (HCV)-infected Thai woman received the first injected dose of pegylated alpha-2b interferon (Peg-IFN-α-2b), she developed cyclic painful swelling nodules on the right upper quadrant of her abdomen and the right anterior lower chest wall. The nodules subsided spontaneously within 1 to 2 days but were recurrent after every Peg-IFN-α-2b injection. She also experienced acute urticaria. After 9 months of therapy, an immature *Gnathostoma spinigerum* male migrated out from the skin nodule shortly after a Peg-IFN-α-2b injection as scheduled. The worm showed a head bulb bearing eight transverse rows of spines, which indicated an immature stage. It had four well-defined pairs of caudal papillae on the posterior body, which were used to identify the male gender. The painful migratory swelling and urticaria disappeared after the parasite was removed. She was continually treated and had both virologic and biochemical responses to the HCV treatment. **This case demonstrates that outward migration of *G. spinigerum* may be stimulated by the injection of Peg-IFN-α-2b.**

Toxocara spp.

Cianferoni, A., L. Schneider, P. M. Schantz, D. Brown, and L. M. Fox. 2006. Visceral larva migrans associated with earthworm ingestion: clinical evolution in an adolescent patient. *Pediatrics* 117:e336–e339.

A 16-year-old girl developed a cough, hypereosinophilia (absolute eosinophil count, 32,000/mm^3), hypergammaglobulinemia, and multiple noncavitary pulmonary nodules 1 month after having ingested an earthworm on a dare. Spirometry revealed moderate restriction and reduced gas diffusion. Parabronchial biopsy demonstrated eosinophilic organizing pneumonitis with multiple eosinophilic microabscesses, and *Toxocara* titers were elevated (>1:4,096). Ophthalmologic examination ruled out ocular larva migrans. The patient received a 10-day course of albendazole (400 mg orally twice daily) and demonstrated significant clinical improvement with resolution of cough and pulmonary function abnormalities. Her WBC count and hypergammaglobulinemia normalized within 20 days, yet eosinophils (absolute eosinophil count, 1,780/mm^3) and *Toxocara* titers

(>1:4,096) remained elevated 3 months after completion of antihelminthic therapy. In this instance, **the ingested earthworm served as the paratenic carrier of *Toxocara* larvae from the soil to the patient. This case highlights the clinical evolution of pulmonary visceral larva migrans infection caused by *Toxocara* spp. associated with a discrete ingestion in an adolescent patient. In addition, it provides a rare opportunity to define the incubation period of visceral larva migrans and emphasizes the importance of education regarding sources of *Toxocara* infection.**

Trichinella spp.

Dupouy-Camet, J. 2006. Trichinellosis: still a concern for Europe. *Eur. Surveill.* 20 Jan. [Epub ahead of print.]

Trichinellosis is a zoonotic disease caused by the ingestion of raw meat containing larvae of the nematode *Trichinella*. **Four species of *Trichinella* are found in Europe: *Trichinella spiralis* (cosmopolitan), *T. britovi* (in wildlife from mountainous areas), *T. nativa* (in wildlife from colder and northern regions), and *T. pseudospiralis* (a cosmopolitan nonencapsulating species).** Human trichinellosis causes high fever, facial edema, myositis, and eosinophilia. It can be a serious disease, particularly in elderly patients, in whom neurologic or cardiovascular complications can lead to death.

Turk, M., F. Kaptan, N. Turker, M. Korkmaz, S. El, D. Ozkaya, S. Ural, I. Vardar, M. Z. Alkan, N. A. Coskun, M. Turker, and E. Pozio. 2006. Clinical and laboratory aspects of a trichinellosis outbreak in Izmir, Turkey. *Parasite* 13:65–70.

Epidemiologic, clinical, and laboratory data were collected during an outbreak of trichinellosis, which occurred in Izmir, Turkey, between January and March 2004. The source of the infection was raw meatballs made with a mixture of uncooked beef and pork. Of 474 persons who were admitted to the Ataturk Training and Research Hospital during this period with a history of raw-meatball consumption, 154 (32.5%; 87 males and 67 females; mean age, 31 years; range, 6 to 67 years) were confirmed to have trichinellosis. **Among persons with a confirmed diagnosis, 79% had myalgia, 77% had weakness and malaise, 63% had arthralgia, 40% had jaw pain, 68% had fever, 63% had periorbital and/or facial edema, 49% had edema at the trunk and limb, 42% had abdominal pain, 40% had nausea and vomiting, 28% had diarrhea, 23% had subconjunctival hemorrhage, 25% had macular or petechial rash, 4% had subungual hemorrhage, 15% had cardiac complaints, and 0.2% had neurologic complaints.** Nine patients (5.8%) were hospitalized due to severe myalgia (*n* = 2), high fever (*n* = 3), neurologic manifestations (*n* = 1), thrombophlebitis (*n* = 2), or palmar erythema (*n* = 1). Eosinophilia was present in 88% of the patients with confirmed cases at admission. Elevated levels of serum creatine phosphokinase, lactate dehydrogenase, and aspartate aminotransferase were detected in 72%, 70%, and 16% of the patients with confirmed cases, respectively. Seroconversion occurred in most cases between weeks 4 and 6 after the infection. All of the patients with confirmed cases were treated with mebendazole. People with severe symptoms were also treated with prednisolone (60 mg/day for 3 days), and those with a moderately severe clinical pattern received a nonsteroidal anti-inflammatory drug (naproxen sodium, 550 mg/day). **All the patients recovered without any clinical sequelae.**

Zhang, G. P., J. Q. Guo, X. N. Wang, J. X. Yang, Y. Y. Yang, Q. M. Li, X. W. Li, R. G. Deng, Z. J. Xiao, J. F. Yang, G. X. Xing, and D. Zhao. 2006. Development and evaluation of an immuno-chromatographic strip for trichinellosis detection. *Vet. Parasitol.* 137:286–293.

An immunochromatographic strip was developed for the serologic detection of trichinellosis in swine. In the strip, the excretory-secretory (ES) antigen of *Trichinella* labeled with colloidal gold was used as the detector, and the staphylococcal protein A and goat anti-ES antibody were blotted on the nitrocellulose membrane for the test and control lines, respectively. Evaluation of the strip was performed by comparing 60 clinical positive blood samples detected by the artificial-digestion method with 46 serum samples from pigs infected with parasites other than *Trichinella* and 30 serum samples from parasite-free healthy pigs. The strip exhibited **high specificity and sensitivity that were closely correlated with the ELISA results. Furthermore, the dipstick assay based on the strip is rapid (10 min) and easy to perform. This suggests that the immunochromatographic strip is an acceptable alternative for clinical laboratories lacking specialized equipment as well as for field diagnosis.**

Filarial Nematodes

Babu, S., C. P. Blauvelt, V. Kumaraswami, and T. B. Nutman. 2006. Cutting edge: diminished T cell TLR expression and function modulates the immune response in human filarial infection. *J. Immunol.* 176:3885–3889.

Patent lymphatic filariasis is characterized by profound antigen-specific T-cell hyporesponsiveness with impaired IFN-γ and IL-2 production. Because T cells express a number of Toll-like receptors (TLRs) and respond to TLR ligands, the authors hypothesized that diminished T-cell TLR function could partially account for the T-cell hyporesponsiveness in filariasis. T cells expressed TLR1, TLR2, TLR4, and TLR9, and the baseline expression of TLR1, TLR2, and TLR4, but not TLR9, was significantly lower in T cells from the filaria-infected individuals than in those from the uninfected individuals (both endemic and nonendemic). TLR function was significantly diminished in the T cells from filaria-infected individuals based on decreased T-cell activation and cytokine production in response to TLR ligands. **Thus, diminished expression and function of T-cell TLR is a novel mechanism underlying T-cell immune tolerance in lymphatic filariasis.**

Gyapong, J. O., and N. A. Twum-Danso. 2006. Editorial. Global elimination of lymphatic filariasis: fact or fantasy? *Trop. Med. Int. Health* 11:125–128.

In 1997, the World Health Assembly resolved to eliminate lymphatic filariasis as a public health problem by the year 2020. By the end of 2004, almost half of the 83 countries where infection is endemic had initiated national programs, providing mass drug administration to an at-risk population of approximately 435 million individuals. This remarkable achievement is the result of an enormous amount of technical, financial, and political support from public and private sectors at the community, national, regional, and global levels. As the global program to eliminate lymphatic filariasis enters its second quarter of operations, there are substantial opportunities to be taken and critical challenges to be addressed.

Ramirez, B. L., O. M. Howard, H. F. Dong, T. Edamatsu, P. Gao, M. Hartlein, and M. Kron. 2006. *Brugia malayi* asparaginyl-transfer RNA synthetase induces chemotaxis of human leukocytes and activates G-protein-coupled receptors CXCR1 and CXCR2. *J. Infect. Dis.* 193:1164–1171.

Lymphatic filariasis is a chronic human parasitic disease in which the parasites repeatedly provoke acute and chronic inflammatory reactions in the host bloodstream and lymphatics. Excretory-secretory products derived from filariae are thought to play an important role in the development of associated immunologic conditions; however, the specific mechanisms involved in these changes are not well understood. Recently, human cytoplasmic aminoacyl-RNA synthetases, which are autoantigens in idiopathic inflammatory myopathies, were shown to activate chemokine receptors on T lymphocytes, monocytes, and immature dendritic cells by recruiting immune cells that could induce innate and adaptive immune responses. Filarial (*Brugia malayi*) asparaginyl-tRNA synthetase (AsnRS) is known to be an immunodominant antigen that induces strong human IgG3 responses. Recombinant *B. malayi* AsnRS was used to perform cellular function assays (chemotaxis and kinase activation assays). Unlike human AsnRS, parasite AsnRS is chemotactic for neutrophils and eosinophils. Recombinant *B. malayi* AsnRS but not recombinant human AsnRS induced chemotaxis of CXCR1 and CXCR2 single-receptor-transfected HEK-293 cell lines, blocked CXCL1-induced calcium flux, and induced mitogen-activated protein kinase. **The data suggest that a filarial parasite chemoattractant protein may contribute to the development of chronic inflammatory disease and that chemokine receptors may be therapeutic targets to ameliorate parasite-induced pathology.**

Intestinal Cestodes

Diphyllobothrium spp.

Year, H., C. Estran, P. Delaunay, M. Gari-Toussaint, J. Dupouy-Camet, and P. Marty. 2006. Putative *Diphyllobothrium nihonkaiense* acquired from a Pacific salmon (*Oncorhynchus keta*) eaten in France: genomic identification and case report. *Parasitol. Int.* 55:45–49.

The authors report a likely case of *Diphyllobothrium nihonkaiense* contracted in France through the consumption of a Pacific salmon imported from Canada. The species diagnosis was made by molecular analysis of two mitochondrial genes (COI and ND3). **This case is rather unusual in that**

D. nihonkaiense has never been detected along the Pacific coast of North America.

Taenia asiatica

Jeon, H. K., and K. S. Eom. 2006. *Taenia asiatica* and *Taenia saginata*: genetic divergence estimated from their mitochondrial genomes. *Exp. Parasitol.* 17 Mar. [Epub ahead of print.]

The authors conducted a differential identification of *Taenia asiatica* and *Taenia saginata* through the mapping of mitochondrial genomes and the sequencing of the *cox1* and *cob* genes. The entire mitochondrial genomes of *T. asiatica* and *T. saginata* were amplified by long-extension PCR and cloned; each was approximately 14 kb in size. Restriction maps of *T. asiatica* and *T. saginata* mitochondrial genomes were then constructed using 13 restriction enzymes. The resulting restriction patterns provided an estimation of their genetic divergence at 4.8%. **The actual sequence divergence was computed as 4.5% from the *cox1* gene and 4.1% from the *cob* gene. These results support the designation of *T. asiatica* as a separate species from *T. saginata*.**

Taenia saginata

Abuseir, S., C. Epe, T. Schnieder, G. Klein, and M. Kuhne. 2006. Visual diagnosis of *Taenia saginata* cysticercosis during meat inspection: is it unequivocal? *Parasitol. Res.* 1 Apr. [Epub ahead of print.]

A total of 267 cysts were collected from March to December 2004 from two main abattoirs in northern Germany. The cysts were classified by the usual organoleptic methods during meat inspection as *Cysticercus bovis*. The reported prevalence of cysticercosis in the two abattoirs was 0.48 and 1.08%, respectively. The cysts were examined macroscopically for their morphology and constituents and classified as viable or degenerating (dead). **The DNA was extracted from these cysts and subjected to PCR amplification to evaluate the detection methods used and to make certain that the cysts did indeed belong to *C. bovis*, as indicated at the slaughterhouses.** Two sets of primers were used with different sensitivity levels. The first, HDP1, was able to detect 200 fg of *Taenia saginata* DNA and 100 pg of *C. bovis* DNA. The other primer set, HDP2, was able to detect 1 pg of *T. saginata* DNA and 1 ng of *C. bovis* DNA. **No more than 52.4% of the samples tested positive for *C. bovis* in the PCR using both primers, while 20% of the viable cysts and 49.2% of the degenerating cysts tested negative with both primers.**

Taenia solium

Bruschi, F., M. Masetti, M. T. Locci, R. Ciranni, and G. Fornaciari. 2006. Cysticercosis in an Egyptian mummy of the late Ptolemaic period. *Am. J. Trop. Med. Hyg.* 74:598–599.

The authors describe an ancient case of cysticercosis that was discovered in an Egyptian mummy of a young woman aged about 20 years who lived in the late Ptolemaic period (second to first centuries BC). On removal and rehydration of the stomach, a cystic lesion in the stomach wall was observed by naked eye. Microscopic examination of sections of this lesion revealed a cystic structure, with a wall, with numerous projecting eversions, a characteristic feature of the larval stage (cysticercus) of the human tapeworm *Taenia solium*. Immunohistochemical testing with serum from a *T. solium*-infected human confirmed the identity of the cyst. **This finding is the oldest on record of this zoonotic parasite. This observation also confirms that in Hellenistic Egypt, the farming of swine, along with humans as an intermediate host of this parasite, occurred, and supports other archeological evidence.**

Hancock, K., S. Pattabhi, F. W. Whitfield, M. L. Yushak, W. S. Lane, H. H. Garcia, A. E. Gonzalez, R. H. Gilman, and V. C. Tsang. 2006. Characterization and cloning of T24, a *Taenia solium* antigen diagnostic for cysticercosis. *Mol. Biochem. Parasitol.* 147:109–117.

The third and final diagnostic antigen of the lentil lectin purified glycoproteins extracted from the larval stage of *Taenia solium* has been characterized, cloned, and expressed. T24 is an integral membrane protein that belongs to the tetraspanin superfamily. It migrates at a position corresponding to 24 kDa and as a homodimer at 42 kDa. Antibodies from cysticercosis patients recognize secondary-structure epitopes that are dependent on correctly formed disulfide bonds. A portion of T24, the large, extracellular loop domain, was expressed in an immunologically reactive form in insect cells. When tested in a Western blot assay with a large battery of serum samples, **this protein, T24H, has a sensitivity of 94% (101 of 107) for detecting cases of cysticercosis with two or more viable cysts, and a specificity of 98% (284 of 290). The identification and expression of T24H sets the stage for the development of an ELISA suitable for testing single samples and for large-scale serosurveys that is not dependent on the isolation and purification of antigens from parasite materials.**

Tissue Cestodes

Echinococcus spp.

Budke, C. M. 2006. Global socioeconomic impact of cystic echinococcosis. *Emerg. Infect. Dis.* 12:296–303.

Cystic echinococcosis (CE) is an emerging zoonotic parasitic disease throughout the world. Human incidence and livestock prevalence data for CE were gathered from published literature and the Office International des Epizooties databases. Disability-adjusted life years (DALYs) and monetary losses resulting from human and livestock CE were calculated from recorded human and livestock cases. Alternative values obtained by assuming substantial underreporting are also provided. **When no underreporting is assumed, the estimated human burden of disease is 285,407 DALYs or an annual loss of $193,529,740. When underreporting is accounted for, this amount rises to 1,009,662 DALYs or $763,980,979. An annual livestock production loss of at least $141,605,195 and possibly up to $2,190,132,464 is also estimated. This initial valuation demonstrates the necessity for increased monitoring and global control of CE.**

Carmena, D., A. Benito, and E. Eraso. 2006. Antigens for the immunodiagnosis of *Echinococcus granulosus* infection: an update. *Acta Trop.* **98**:774–786.

The taeniid tapeworm *Echinococcus granulosus* is the causative agent of echinococcal disease, an important zoonosis with worldwide distribution. Accurate immunodiagnosis of the infection requires highly specific and sensitive antigens to be used in immunodiagnostic assays. The choice of an appropriate source of antigenic material is a crucial point in the improvement of the diagnostic features of tests and must be based on the developmental stage of the parasite and the host. The most common antigenic sources used for the immunodiagnosis of echinococcal disease are hydatid cyst fluid, somatic extracts, and excretory-secretory products from protoscolices or adults of *E. granulosus*. Hydatid cyst fluid is the antigenic source of reference for immunodiagnosis of human hydatidosis, which is based mainly on the detection of antigens B and 5. Somatic extracts have been widely used in the serodiagnosis of *E. granulosus* infection in dogs and ruminant intermediate hosts, although in the last few years the detection of excretory-secretory products of the worm in feces (coproantigens) has become the most reliable method of detecting of the parasite in the definitive host. This review emphasizes recent advances in the identification and characterization of novel antigens with potential for the immunodiagnosis of echinococcal disease. Progress in recombinant technologies and synthetic peptides is also discussed. The paper highlights the need to search for new antigenic components with high diagnostic sensitivity and specificity, a fact that remains a crucial task in the improvement of the immunodiagnosis of the disease.

Jenkins, D. J., A. McKinlay, H. E. Duolong, H. Bradshaw, and P. S. Craig. 2006. Detection of *Echinococcus granulosus* coproantigens in faeces from naturally infected rural domestic dogs in southeastern Australia. *Aust. Vet. J.* **84**:12–16.

This study was designed to investigate the occurrence of *Echinococcus granulosus* in rural domestic dogs in farming areas around Yass, New South Wales, and Mansfield and Whitfield, Victoria. Feces were collected rectally from farm dogs voluntarily presented by their owners in four farming districts in New South Wales and two in Victoria. Feces were collected in the field, an extract was prepared from each sample, and *E. granulosus* coproantigens were detected using ELISA. Farmers were also questioned about their dog feeding and worming practices. **E. granulosus coproantigens were detected in 99 (29%) of 344 dogs from 95 farms in southeastern New South Wales and 38 (17.5%) of 217 dogs from 43 farms in Victoria.** Cross-reactions between *E. granulosus* coproantigen-trapping antibody and coproantigens in feces from dogs monospecifically infected with other species of intestinal helminthes (*Taenia ovis, T. hydatigena, T. pisiformis, Spirometra ericacei, Dipylidium caninum,* hookworm, *Toxocara canis, Trichuris vulpis*) were not evident. **Dietary and worming data revealed that many owners fed raw meat and occasionally offal from domestic livestock and wildlife to their dogs and that few owners wormed their dogs frequently**

enough to preclude the chance of patent *E. granulosus* being present in their dogs. *E. granulosus* occurs commonly in rural dogs in southeastern Australia, and an education program promoting the public health importance of responsible management of rural dogs is urgently needed.

Karabulut, K., I. Ozden, A. Poyanli, O. Bilge, Y. Tekant, K. Acarli, A. Alper, A. Emre, and O. Ariogul. 2006. Hepatic atrophy-hypertrophy complex due to *Echinococcus granulosus*. *J. Gastrointest. Surg.* **10**:407–412.

Obstruction of a major hepatic vein, major portal vein, or biliary tree branch causes atrophy of the related hepatic region and frequently hypertrophy in the rest of the liver—the atrophy-hypertrophy complex (AHC). Whether hydatid cysts can cause AHC is controversial. The records of 370 patients who underwent surgery for hepatic hydatid disease between August 1993 and July 2002 were evaluated retrospectively. Excluding six patients with previous interventions on the liver, AHC had been recorded in the operative notes of 16 patients (4.4%); for all patients, a cyst located in the right hemiliver had caused atrophy of the right hemiliver and compensatory hypertrophy of the left hemiliver. The computed tomograms of seven patients were suitable for volumetric analysis. The median (range) right and left hemiliver volumes were 334 (0 to 686) ml and 1,084 (663 to 1,339) ml, respectively. The median (range) cyst volume was 392 (70 to 1,363) ml. **AHC due to *Echinococcus granulosus* was confirmed by objective volumetric analysis. The presence of AHC should alert the surgeon to two implications. First, pericystectomy may be hazardous due to association with major vascular and biliary structures. Second, in patients with AHC, the hepatoduodenal ligament rotates around its axis; this should be considered to avoid vascular injury if a common bile duct exploration is to be performed.**

Sapkas, G. S., T. G. Machinis, G. D. Chloros, K. N. Fountas, G. S. Themistocleous, and G. Vrettakos. 2006. Spinal hydatid disease, a rare but existent pathological entity: case report and review of the literature. *South. Med. J.* **99**:178–183.

Spinal hydatid disease is a fairly common cause of spinal cord compression in countries where the infection is endemic; however, involvement of the epidural space with sparing of the vertebral column is rare. Early diagnosis and surgical decompression with total removal of the hydatid lesion, when possible, is generally considered the standard of care for this disease. The authors describe a case of massive epidural hydatid disease without involvement of the vertebral column in a 62-year-old male patient, treated with a two-stage surgical operation and administration of systemic albendazole. **The literature on the clinical features, diagnosis, treatment, and prognosis of spinal echinococcosis is reviewed.**

Yildirim, M., N. Erkan, and E. Vardar. 2006. Hydatid cysts with unusual localizations: diagnostic and treatment dilemmas for surgeons. *Ann. Trop. Med. Parasitol.* **100**:137–142.

Although the liver and lungs are by far the most common localizations for the larval cysts of *Echinococcus granulosus* in humans, the cysts may develop at other sites and cause signs and symptoms that may be easily confused with those of other illnesses. In a retrospective study in Turkey, 6 male and 11 female patients with cystic echinococcosis, each with at least one cyst in an unusual site, were investigated. **The patients, who had a mean age of 41.6 years, had cysts in the pancreas, intra-abdominal cavity, kidney, spleen, ovary, breast, mediastinum, chest wall, muscle, and/or subcutaneous tissue.** In terms of Gharbi's classification, 15 (75%) of the 20 cysts in these patients were type I and 5 (25%) were type II. Fourteen of the patients each had single cysts, two had multiple cysts, and one had an unknown number of cysts. All but one of the patients (who had a pancreatic cyst) were treated by total cystectomy. **In areas where cystic echinococcosis is endemic, any patient presenting with a cystic mass, in any tissue or organ, should be considered to have a potential case of the disease and should be carefully investigated by radioimaging and/or ultrasonography.**

Spirometra spp.

Nobayashi, M., H. Hirabayashi, T. Sakaki, F. Nichimura, H. Fukui, S. Ishizaka, and M. Yoshikawa. 2006. Surgical removal of a live worm by stereotactic targeting in cerebral sparganosis. *Neurol. Med. Chir.* **46:**164–167.

A 64-year-old man presented with generalized tonic-clonic convulsion followed by weakness of the right lower extremity. He had a medical history of hypertension, hyperlipidemia, and right cerebellar infarction. Computed tomography showed a small, high-density nodule with an enhanced perifocal low-density area in the left occipital lobe. T_1-weighted MRI showed a ring-shaped and partial string-like nodule with enhancement by gadolinium. T_2-weighted MRI showed the white matter of the left occipital lobe as high intensity. Computed tomography and MRI seemed to indicate metastatic brain tumors, although cortical atrophy and ventricular dilation were recognized. Left parietal craniotomy was performed under stereotactic targeting to obtain a definitive diagnosis. During manipulation at the center of the targeted lesion, a white, tape-like body was found and recognized to be a live worm. Serologic testing revealed strong immunopositivity against *Spirometra mansoni*. The infection route in the present case was probably through eating raw chicken meat. Cerebral sparganosis is extremely rare but should be considered in the differential diagnosis of metastatic brain tumors, especially in areas of endemic infection.

Intestinal Trematodes

Gymnophalloides seoi

Seo, M., H. Chun, J. G. Ahn, K. T. Jang, S. M. Guk, and J. Y. Chai. 2006. A case of colonic lymphoid tissue invasion by *Gymnophalloides seoi* in a Korean man. *Korean J. Parasitol.* **44:**87–89.

A 65-year-old Korean man, living in Mokpo city, Jeollanam-do, Republic of Korea, visited a local clinic complaining of right upper quadrant pain and indigestion. At colonoscopy, he was diagnosed as having a carcinoma of the ascending colon, and a palliative right hemicolectomy was performed. Subsequently, an adult fluke of *Gymnophalloides seoi* was incidentally found in a surgical pathology specimen of the lymph node around the colon. The worm was found to have invaded gut lymphoid tissue, with characteristic morphologies of a large oral sucker, a small ventral sucker, and a ventral pit surrounded by strong muscle fibers. **This is the first reported case of mucosal tissue invasion by *G. seoi* in the human intestinal tract.**

Liver and Lung Trematodes

Fasciola hepatica

Molloy, J. B., and G. R. Anderson. 2006. The distribution of *Fasciola hepatica* in Queensland, Australia, and the potential impact of introduced snail intermediate hosts. *Vet. Parasitol.* **137:**62–66.

A survey was conducted to establish the distribution of the liver fluke, *Fasciola hepatica*, in the state of Queensland, Australia, and to evaluate the impact of the introduced snail intermediate hosts, *Pseudosuccinia columella* and *Austropeplea viridis*. Serum samples from a total of 5,103 homebred cattle in 142 beef herds distributed throughout the state and 523 pooled milk samples from dairy herds from the state's major dairy regions were tested for antibodies to *F. hepatica* by ELISA. Snails were collected on infected properties around the limits of the *F. hepatica* distribution. *F. hepatica* infection was detected in 44 dairy herds and 2 beef herds. The distribution of infected herds indicates that *F. hepatica* is established only in southeastern Queensland. The distribution there was patchy, but the parasite was more widespread than suggested by an earlier survey. The predominant intermediate host species found along the northern limit of the distribution was *P. columella*. **We conclude that the introduction of *P. columella* and *A. viridis* has not yet had a major impact on the distribution of *F. hepatica* in Queensland. However, the presence of *P. columella*, which is much more adaptable to tropical habitats than the native intermediate host, *Austropeplea tomentosa*, at the northern limit of the *F. hepatica* distribution suggests that there is potential for the parasite to expand its range.**

Opisthorchis viverrini

Stensvold, C. R., W. Saijuntha, P. Sithithaworm, S. Wongratanacheewin, H. Strandgaard, N. Ornbjerg, and M. V. Johansen. 2006. Evaluation of PCR based coprodiagnosis of human opisthorchiasis. *Acta Trop.* **97:**26–30.

In this study, a recently developed PCR test for the detection of *Opisthorchis viverrini* in human fecal samples was evaluated using two parasitological methods as references. During a survey of food-borne trematodes (FBT) in the Vientiane Province, Lao PDR, 85 samples were collected and evaluated for FBT eggs by the Kato-Katz technique, the formalin ethyl acetate concentration technique (FECT), and a PCR analysis for the distinction between *O. viverrini* and other FBT. The two parasitological methods did not differ in the ability of detecting FBT eggs, and a single Kato-Katz reading was characterized by a sensitivity of 85% compared to two FECT readings. The PCR analysis gave positive results only in cases where eggs had been demonstrated by parasitological examination. However, the PCR analysis gave negative results in some samples with very high egg counts. **With a PCR sensitivity of approximately 50% in samples with fecal egg counts of >1,000, the previously reported PCR sensitivity based on in vitro studies was not supported. It is believed that technical problems rather than diagnostic reference-related issues were responsible for the relatively poor PCR performance.** Further studies should aim at optimizing DNA extraction and amplification, and future PCR evaluation should include a specificity control such as scanning electron microscopy of eggs in test samples or expulsion of adult trematodes from PCR-tested individuals.

Paragonimus spp.

Le, T. H., N. V. De, D. Blair, D. P. McManus, H. Kino, and T. Agatsuma. 2006. *Paragonimus heterotremus* Chen and Hsia (1964), in Vietnam: a molecular identification and relationships of isolates from different hosts and geographical origins. *Acta Trop.* 98:25–33.

Paragonimiasis caused by *Paragonimus heterotremus* Chen and Hsia (1964) is a newly detected disease in Vietnam. Twelve samples of *Paragonimus* (Platyhelminthes: Trematoda: Digenea: Paragonimidae) from different life stages (eggs, miracidia, metacercariae, and adults from natural and experimental hosts) and host species (crab, dog, cat, and human) were collected in different geographic locations in Vietnam. DNA sequences were obtained from each sample for partial mitochondrial cytochrome *c* oxidase subunit 1 (*cox1*) (387 bp) and the entire second ribosomal internal transcribed spacer (ITS2) (361 bp). The ITS2 sequences were identical among all specimens, including those previously reported in GenBank. For *cox1*, there were sequence differences between specimens from Vietnam (four provinces, different locations) and those from Guangxi (China) and Saraburi (Thailand). **Phylogenetic trees inferred from *cox1* and ITS2 sequences using sequence data for 15 *P. heterotremus* specimens and for other *Paragonimus* spp. revealed that all *P. heterotremus* strains originating from Vietnam, Thailand, and China form a distinct group. This information also confirms the identity of the Vietnamese specimens as *P. heterotremus*.**

Lee, E. G., B. K. Na, Y. A. Bae, S. H. Kim, E. Y. Je, J. W. Ju, S. H. Cho, T. S. Kim, S. Y. Kang, S. Y. Cho, and Y. Kong. 2006. Identification of immunodominant excretory-secretory cysteine proteases of adult *Paragonimus westermani* by proteome analysis. *Proteomics* 6:1290–1300.

Paragonimus westermani causes inflammatory lung disease in humans. The parasite excretes a host of biologically active molecules, which are thought to be involved in pathophysiologic and immunologic events during infection. Analyses of the protein profiles of the excretory-secretory products (ESP) of adult *P. westermani* revealed approximately 147 protein

spots, at least 15 of which were identified as cysteine proteases (CPs), at pHs between 4.5 and 8.5 and molecular masses between 27 and 35 kDa. An additional three CPs (designated PwCP-3, PwCp-8, and PwCp-11) were newly recognized by TOF/TOF mass spectrometry. Their molecular biological information, which had a high-level sequence homology, was elucidated. The majority of the CPs reacted strongly with sera from paragonimiasis patients. When the authors observed the chronological changes in the antibody responses of the respective CPs to canine sera collected serially at 1, 3, 5, 7, 11, and 14 weeks after experimental infection, these molecules exhibited a multiplicity of distinct immune recognition patterns. **Results clearly showed that *P. westermani* adult ESP were composed principally of excretory-secretory CPs and that these CPs may exert effects not only on host tissue degradation and nutrient uptake but also on the immune-regulating cells via synergistic and independent interactions.**

Lee, J. C., G. S. Cho, J. H. Kwon, M. H. Shin, J. H. Lim, and W. K. Kim. 2006. Macrophageal/microglial cell activation and cerebral injury induced by excretory-secretory products secreted by *Paragonimus westermani*. *Neurosci. Res.* 54:113–139.

Cerebral paragonimiasis causes various neurologic disorders including seizures, visual impairment, and hemiplegia. The excretory-secretory product (ESP) released by *Paragonimus westermani* has a CP activity and plays important roles in its migration in the host tissue and modulation of host immune responses. To gain more insight into the pathogenesis of ESP in the brain, we investigated the inflammatory reaction and cerebral injury following microinjection of ESP into rat striatum. The size of injury was maximal 3 days after microinjection of ESP and then declined to control levels as astrocytes repopulated the injury. ED1-positive monocytes and microglia were confluently found inside the injury. The mRNA expression of inducible NOS (iNOS) occurred as early as 9 h after ESP injection and then declined to control levels within 1 day. The iNOS inhibitor aminoguanidine largely decreased the expression of iNOS but did not reduce the size of lesion caused by ESP. Interestingly, however, heat inactivation of ESP caused a decrease of injury formation with no altered expression of iNOS. **The data indicate that ESP causes brain tissue injury by recruiting activated monocytes and microglia via heat-labile protease activity.**

Wongkham, C., P. M. Intapan, W. Maleewong, and M. Miwa. 2005. Evaluation of human IgG subclass antibodies in the serodiagnosis of *Paragonimiasis heterotremus*. *Asian Pac. J. Allergy Immunol.* 23:205–211.

IgG1, IgG2, IgG3, and IgG4 responses to the excretory-secretory antigens of the lung fluke *Paragonimus heterotremus* were analyzed using the immunoblotting technique in an attempt to further improve the sensitivity and specificity for serodiagnosis of human paragonimiasis. Serum samples from patients with proven paragonimiasis, from patients with other parasitic infections, from those with pulmonary tuberculosis, and from healthy counterparts were analyzed. **The results indicate that immunoblotting for the detection** of IgG4 antibodies to an excretory-secretory product of *P. heterotremus* with an approximate molecular mass of 31.5 kDa is the most reliable test. It has accuracy, sensitivity, specificity, and positive and negative predictive values of 97.6%, 100%, 96.9%, 90%, and 100%, respectively.

Blood Trematodes (Schistosomes)

Alonso, D., J. Munoz, J. Gascon, M. E. Valls, and M. Corachan. 2006. Failure of standard treatment with praziquantel in two returned travelers with *Schistosoma haematobium* infection. *Am. J. Trop. Med. Hyg.* 74:342–344.

A single 40-mg/kg dose of praziquantel continues to be the standard treatment for schistosomiasis caused by *Schistosoma mansoni* and *S. haematobium* in all clinical settings. Experimental development of drug resistance and the recent isolation of *S. mansoni* strains with a natural tolerance to high doses of praziquantel have raised concerns over the adequacy of such a dose. The authors describe two Spanish travelers with genitourinary schistosomiasis caused by *S. haematobium* in whom repeated standard treatment failed to clear the infection.

Doenhoff, M. J., and L. Pica-Mattoccia. 2006. Praziquantel for the treatment of schistosomiasis: its use for control in areas with endemic disease and prospects for drug resistance. *Expert Rev. Anti Infect. Ther.* 5:199–210.

Praziquantel became available for the treatment of schistosomiasis and other trematode-inflicted diseases in the 1970s. It was revolutionary because it could be administered orally and had very few unwanted side effects. As a result of marked reductions in the price of praziquantel, the rate at which it is used has accelerated greatly in recent years. For the foreseeable future, it will be the mainstay of programs designed to control schistosome-induced morbidity, particularly in sub-Saharan Africa, where schistosomiasis is heavily endemic. There is currently no evidence to suggest that any schistosomes have developed resistance to praziquantel as a result of its widespread use. Nevertheless, while resistance may not pose an obvious or immediate threat to the usefulness of praziquantel, complacency and a failure to monitor developments may have serious consequences in the longer term, since it will be the only drug that is readily available for the large-scale treatment of schistosomiasis. (See the published paper right above this one for comparison.)

El-Masry, S., M. Lotfy, M. El-Shahat, and G. Badra. 2006. Serum laminin assayed by Slot-Blot-ELISA in patients with combined viral hepatitis C and schistosomiasis. *Clin. Biochem.* 16 Feb. [Epub ahead of print.]

Hepatic schistosomiasis and chronic HCV infection are the most prevalent causes of hepatic fibrosis in humans. Laminin (LA) has been related to liver fibrosis and subsequent development of portal hypertension in chronic liver disease. There are no available data describing the pattern of LA in combined HCV- and schistosoma-infected patients; **the study was designed to assess the serum LA as an index of liver fibrosis in** patients with schistosomiasis and/or chronic viral hepatitis C and to evaluate a developed slot-blot ELISA as a method of estimation. This study examined four groups: group I included 34 patients with schistosomiasis, group II included 58 patients infected with HCV, group III included 68 patients with combined chronic viral hepatitis C and schistosomiasis, and group IV included 50 healthy individuals who served as a control group. The serum LA level was measured in the different groups quantitatively by ELISA and semiquantitatively by slot blot ELISA. **Significantly higher serum LA concentrations measured by ELISA were found in patients with combined chronic viral hepatitis C and schistosomiasis than in patients with either chronic viral hepatitis C or schistosomiasis alone. The serum LA concentration was significantly higher in the patient groups than the control group; it was positively correlated with fibrosis grading scores.** Semiquantitative results of serum LA concentrations using the developed slot blot ELISA were found to have approximately the same power of ELISA results in different groups. The overall sensitivity, specificity, positive predictive value, negative predictive value, and efficiency of ELISA for estimation of serum LA levels were 85.6%, 84.0%, 94.5%, 64.6%, and 90%, respectively, and the figures for slot blot ELISA were 87.5%, 82.0%, 94%, 67.2%, and 88%, respectively. The serum LA concentration was significantly increased in patients coinfected with HCV and *Schistosoma mansoni*. **The newly developed slot blot ELISA is a simple, rapid, and highly sensitive assay for detection of LA in patients with hepatic fibrosis. Moreover, all steps were performed at room temperature without the need to use expensive equipment; this may enhance the application of this assay in screening programs.**

Hoare, M., W. T. Gelson, S. E. Davies, M. Curran, and G. J. Alexander. 2005. Hepatic and intestinal schistosomiasis after orthotopic liver transplant. *Liver Transplant.* 11:1603–1607.

Schistosomiasis affects 200 million to 250 million people worldwide. Hepatic schistosomiasis is a well-recognized cause of chronic liver disease and portal hypertension. There are no previous reports of schistosomiasis after liver transplantation. **The authors report on two cases of schistosomiasis in liver transplant recipients—a case of gastric schistosomiasis and a case of hepatic schistosomiasis. A discussion of the pathology of schistosomal infection and a rationale for screening potential liver transplant recipients from areas of endemic infection is included.**

Kjetland, E. F., P. D. Ndhlovu, E. Gomo, T. Mduluza, N. Midzi, L. Gwanzura, P. R. Mason, L. Sanvik, H. Friis, and S. G. Gundersen. 2006. Association between genital schistosomiasis and HIV in rural Zimbabwean women. *AIDS* 28:593–600.

To determine the association between female genital *Schistosoma haematobium* infection and HIV, a cross-sectional study with a 1-year follow-up was undertaken. Gynecologic and laboratory investigations were performed for *S. haematobium* and HIV. Sexually transmitted infections and demographic and urogenital history were analyzed as confounders. The participants were 527 sexually active, nonpregnant, nonmenopausal women between the ages of 20 and 49 years. The setting was a rural Zimbabwean community where *S. haematobium*-related lesions were found in 46% of the women, HIV was found in 29%, and herpes simplex virus type 2 (HSV-2) was found in 65%. **In permanent residents (>3 years' residency), HIV was found in 41% of women (29 of 70) with laboratory-proven genital schistosomiasis as opposed to 26% (96 of 375) in the schistosomal ova-negative group. In multivariate analysis, *S. haematobium* infection of the genital mucosa was significantly associated with HIV seropositivity.** All seven women who became HIV positive during the study period (seroincidence, 3.1%) had signs of *S. haematobium* at baseline. **In agreement with other studies, HIV was significantly associated with HSV-2, syphilis, and human papillomavirus.** The highest HIV prevalence (45%) was found in the 25- to 29-year age group. **Women with genital schistosomiasis had an almost threefold risk of having HIV infection in this rural Zimbabwean community. Prospective studies are needed to confirm the association.**

Lamyman, M. J., D. J. Noble, S. Narang, and N. Dehalvi. 2006. Small bowel obstruction secondary to intestinal schistosomiasis. *Trans. R. Soc. Trop. Med. Hyg.* 18 Jan. [Epub ahead of print.]

Intestinal obstruction caused by chronic schistosomiasis infection is rare, with only 12 previously recorded cases in the literature. The authors report the first recorded case in a European hospital. A 36-year-old Caucasian man who was born and lived in the United Kingdom presented with small bowel obstruction. He had visited China and Indonesia 8 years previously. **At laparotomy, there was an obstructing inflammatory mass close to the ileocecal junction and several small bowel strictures. Initially he was thought to have Crohn's disease. However, subsequent histologic testing diagnosed intestinal schistosomiasis.**

Liang, S., C. Yang, B. Zhong, and D. Qiu. 2006. Re-emerging schistosomiasis in hilly and mountainous areas of Sichuan, China. *Bull. W. H. O.* 84:139–144.

Despite great strides in schistosomiasis control over the past several decades in Sichuan Province, China, the disease has reemerged in areas where it was previously controlled. The authors reviewed historical records and found that schistosomiasis had reemerged in 8 counties by the end of 2004; 7 of 21 counties with transmission control and 1 of 25 counties with transmission interruption as reported in 2001 were confirmed to have local disease transmission. **The average "return time" (from control to reemergence) was about 8 years. The onset of reemergence was commonly indicated by the occurrence of acute infections.** Survey results suggest that environmental and sociopolitical factors play an important role in reemergence. **The main challenge would be to consolidate and maintain effective control in the longer term until "real" eradication is achieved.**

Makhlouf, L. M., A.-H. Servah, A. D. Abd El-Hamid, E. M. Hussein, and R. M. Saad. 2006. INF-gamma, IL-5 and IgE profiles in chronic schistosomiasis mansoni Egyptian

patients with or without hepatitis C infection. *J. Egypt. Soc. Parasitol.* 36:177–196.

The immune response to clinical forms of chronic schistosomiasis mansoni in patients with or without HCV infection was evaluated by assays of the levels of IFN-γ and IL-5 in serum to estimate cell-mediated immunity and assays of the IgE level to estimate humoral immunity. This study examined three patient groups. G.I included 25 patients with intestinal schistosomiasis, G.II included 15 patients with hepatosplenic schistosomiasis, and G.III included 40 patients with hepatosplenic schistosomiasis coinfected with HCV. Control G.IV included 15 healthy persons matched by age and sex with the patients. G.I had high IFN-γ levels (92%) and normal IL-5 and IgE levels. The immune response was a 100% Th-1 response. G.II had high IFN-γ (26.7%), IL-5 (86.7%), and IgE (73.3%) levels. The immune response was 73.4% Th-0, 13.3% Th-1, and 13.3% Th-2. G.III had high IFN-γ (62.7%), IL-5 (100%), and IgE (92.5%) levels. The immune response was 62.5% Th-0 and 37.5% Th-2. **The shift to Th-0 and Th-2 immunity as well as the associated depression of Th-1 in the group of patients with mixed infection may be playing a role in the persistence and severity of both diseases. Such immunity defects add to the decreased response to HCV clearance.**

Silva, I. M., R. Thiengo, M. J. Conceicao, L. Rey, F. E. Pereira, and P. C. Ribeiro. 2006. Cystoscopy in the diagnosis and follow-up of urinary schistosomiasis in Brazilian soldiers returning from Mozambique, Africa. *Rev. Inst. Med. Trop. Sao Paulo* 48:39–42.

The assessment of urinary schistosomiasis in individuals returning from areas of endemic infection often requires diagnostic resources not used in areas of exposure in order to determine complications or to establish more precise criteria of cure. Cystoscopy and 24-h urine examination were performed after treatment with a single dose of praziquantel (40 mg/kg of body weight) given to 25 Brazilian military men who were part of a United Nations peace mission to Mozambique in 1994. The median age of the individuals was 29 years, and they all had a positive urine parasitological examination. The alterations detected by cystoscopy were hyperemia and granulomas in the vesical submucosa in 59.1% of the individuals and granulomas alone in 40.9%. **A vesical biopsy revealed granulomas in all patients and viable eggs in 77.3% even after a period during which the patients no longer excreted eggs in their urine. Cystoscopy after treatment, followed by biopsy and histopathologic evaluation, performed in areas where the evolution of the disease can be better monitored, was found to be a safe criterion of parasitological cure.**

Stothard, J. R., N. B. Kabatereine, E. M. Tukahebwa, F. Kazibwe, D. Rollinson, W. Mathieson, J. P. Webster, and A. Fenwick. 2006. Use of circulating cathodic antigen (CCA) dipsticks for detection of intestinal and urinary schistosomiasis. *Acta Trop.* 97:219–228.

An evaluation of a commercially available antigen capture dipstick that detects schistosome circulating cathodic antigen (CCA) in urine was conducted in representative areas in Uganda and Zanzibar where intestinal and urinary schistosomiasis, respectively, is endemic. Under field-based conditions, the sensitivity and specificity of the dipstick was 83% and 81% for detection of *Schistosoma mansoni* infections while the positive and negative predictive values were 84%. Light egg-positive infections were sometimes CCA negative, while CCA-positive infections included egg-negative children. A positive association between fecal egg output and intensity of the CCA test band was observed in Uganda. Estimating the prevalence of intestinal schistosomiasis by school with dipsticks was highly correlated ($r = 0.95$) with Kato-Katz stool examinations, typically within ±8.5%. In Zanzibar, however, dipsticks totally failed to detect *S. haematobium* despite examining children with egg-patent schistosomiasis. This was also later corroborated by surveys in Niger and Burkina Faso. Laboratory testing of dipsticks with aqueous adult worm lysates from several reference species showed correct functioning; however, dipsticks failed to detect CCA in urine from *S. haematobium*-infected hamsters. While CCA dipsticks are a good alternative, or complement, to stool microscopy for field diagnosis of intestinal schistosomiasis, they have no proven value for field diagnosis of urinary schistosomiasis. At approximately $2.6 per dipstick, they are presently too expensive to be cost-effective for wide-scale use.

Summer, A. P., W. Stauffer, S. R. Maroushek, and T. E. Nevins. 2006. Hematuria in children due to schistosomiasis in a nonendemic setting. *Clin. Pediatr.* **45**:177–181.

Infection with *Schistosoma haematobium* is common in immigrants from tropical Africa and commonly presents with painless hematuria. Since chronic, heavy infection can lead to significant morbidity, it is imperative for clinicians who serve the immigrant and refugee population to become familiar with this traditionally exotic disease. Increased awareness will allow earlier diagnosis and treatment of infection, avoiding complications and minimizing expensive and invasive diagnostic procedures.

Unusual Parasitic Infections

Nematodes

Eamsobhana, P. T. Mongkolporn, P. Punthuprapasa, and A. Yoolek. 2006. *Mammomonogamus* roundworm (Nematoda: Syngamidae) recovered from the duodenum of a Thai patient: a first and unusual case originating in Thailand. *Trans. R. Soc. Trop Med. Hyg.* **100**:387–391.

A pair of *Mammomonogamus laryngeus* roundworms in copula was recovered from the duodenum of a 72-year-old male Thai patient from Kanchanaburi Province. Eggs were also found in the stool of the patient. This is the first case of *Mammomonogamus* infection originating in Thailand, since the previous two reports from Thailand attributed the infection as originating in Malaysia. The occurrence of adult worms in the duodenum is unusual

and differs from previous findings in the larynx, posterior pharynx, tracheal wall, and bronchi.

Pampiglione, S., T. C. Orihel, A. Gustinella, W. Gatzemeier, and L. Villani. 2005. An unusual parasitological finding in a subcutaneous mammary nodule. *Pathol. Res. Pract.* **201**:475–478.

While examining some histologic sections of a clinically suspected neoplastic nodule in a woman's breast, sections of *Dirofilaria repens* were noted in the same nodule along with sections of a different nematode. The latter appeared to be a specimen possibly belonging to the genus *Anatrichosoma* (family Trichosomoididae), a parasitic group of helminths rarely found in humans.

Watthanakulpanich, D., M. T. Anantaphruti, and W. Maipanich. 2005. *Diploscapter coronata* infection in Thailand: report of the first case. *Southeast Asian J. Trop. Med. Public Health* **36** (Suppl. 4):99–101.

A 73-year-old Thai woman living in Mueang District, Saraburi Province, central Thailand, presented with numerous hookworm-like nematodes, finally revealed as *Diploscapter coronata* by fecal culture. The patient exhibited no significant clinical signs of the gastrointestinal or genitourinary systems and was generally not ill as a result of this unusual infection. Less commonly, patients have presented with symptoms and signs of *D. coronata* infection. However, potentially serious consequences can occur where people are exposed to an environment that has been contaminated with infected feces or, more specifically, infective eggs; such conditions could lead to human infection with *D. coronata* worms. This was the first reported occurrence of human *D. coronata* infection in Thailand.

Cestodes

Padgett, K. A., S. A. Nadler, L. Munson, B. Sacks, and W. M. Boyce. 2005. Systematics of *Mesocestoides* (Cestoda: Mesocestoididae): evaluation of molecular and morphological variation among isolates. *J. Parasitol.* **91**:1435–1443.

A hypothesis-based framework was used to test if three genetic strains of *Mesocestoides* (clades A, B, and C) are distinct evolutionary lineages, thereby supporting their delimitation as species. For comparative purposes, three established cestode species, *Taenia pisiformis*, *Taenia serialis*, and *Taenia crassiceps*, were assessed by the same methods. Sequence data from mitochondrial rDNA (12S) and the second internal transcribed spacer of nuclear rDNA (ITS2) revealed derived (autapomorphic) characters for lineages representing clade A ($n = 6$ autapomorphies), clade B ($n = 4$), and clade C ($n = 9$), as well as *T. pisiformis* ($n = 15$) and *T. serialis* ($n = 12$). Furthermore, multivariate analysis of morphologic data revealed significant differences among the three genetic strains of *Mesocestoides* and between *T. pisiformis* and *T. serialis*. The level of phenotypic variation within evolutionary lineages of *Mesocestoides* and *Taenia* spp. tapeworms was similar. Results from this study support recognizing *Mesocestoides* clades A, B, and C as separate species and provide evidence that clade B and *Mesocestoides vogae* are conspecific.

Trematodes

Guk, S. M., J. H. Park, E. H. Shin, J. L. Kim, A. Lin, and J. Y. Chai. 2006. Prevalence of *Gymnophalloides seoi* infection in coastal villages of Haenam-gun and Yeongam-gun, Republic of Korea. *Korean J. Parasitol.* 44:1–5.

One coastal village in Haenam-gun and two in Yeongam-gun, Jeollanam-do were surveyed for intestinal parasite infections by fecal examination. The egg-positive rates of *Gymnophalloides seoi* were high, 24.1% (14 of 58 patients) in Haenam-gun and 9.3% (11 of 118 patients) in Yeongam-gun. The egg-positive rates of heterophyids, including *Heterophyes nocens*, and of *Clonorchis sinensis* were 10.3% and 6.9% in Haenam-gun and 14.4% and 8.5% in Yeongam-gun, respectively. After praziquantel treatment and purgation, a total of 37,761 fluke specimens were recovered from 17 patients, 11 in Haenam-gun and 6 in Yeongam-gun. *Gymnophalloides seoi* was the most commonly recovered species, with 37,489 specimens in total (2,205 per person). Other recovered flukes included *Heterophyes nocens*, *Stictodora fuscata*, *Heterophyopsis continua*, *Pygidiopsis summa*, and undetermined species. These results indicate that the areas surveyed are new endemic foci of *G. seoi*.

Karadag, B., A. Bilici, A. Doventas, F. Kantarci, D. Selcuk, N. Dincer, Y. A. Oner, and D. S. Erdincler. 2005. An unusual case of biliary obstruction caused by *Dicrocoelium dendriticum*. *Scand. J. Infect. Dis.* 37:385–388.

Dicrocoelium dendriticum is a liver fluke that induces biliary obstruction. Infection usually occurs in herbivores such as sheep, goats, and deer; human infection is very rarely encountered in clinical practice. The authors report on a 65-year-old female presenting with biliary obstruction caused by *D. dendriticum*. Following treatment with triclobendazole, her symptoms disappeared, and laboratory values returned to their normal range within 6 months. Parasitosis is an important cause of biliary obstruction. For patients presenting with biliary obstruction, *D. dendriticum* should be included in the differential diagnosis.

Antibody and Antigen Detection in Parasitic Infections (General Immunology) (See Also Individual Relevant Chapters)

Protozoa

Hernandez-Marin, M., I. Hernandez-Spengler, G. Ramos-Martinez, and L. Pozo-Pena. 2006. Chimeric synthetic peptides as antigens for detection of antibodies to *Trypanosoma cruzi*. *Biochem. Biophys. Res. Commun.* 339:89–92.

Six chimeric synthetic peptides (QCha-1, QCha-2, QCha-3, QCha-4, QCha-5, and QCha-6) incorporating antigenic sequences of two immunodominant repeat B-cell epitopes of *Trypanosoma cruzi* were synthesized by conventional solid-phase peptide synthesis. The antigenic activity of these peptides was evaluated by ultramicro-ELISA by using panels of positive Chagasic sera (*n* = 82), while the specificity was evaluated by using samples from healthy blood donors (*n* = 44) and patients with other infectious diseases (*n* = 86). The antigenicity of the chimeric peptides in solid-phase immunoassays was compared with that of the monomeric peptides. Data demonstrated that the chimeric peptide QCha-5 was the most reactive. The results indicate that chimeric peptide as a coating antigen is very useful for the immunodiagnosis of Chagas' disease.

Helminths

Allan, J. C., and P. S. Craig. 2006. Coproantigens in taeniasis and echinococcosis. *Parasitol. Int.* 55(Suppl.):S75–S80.

The application of modern immunodiagnostic or molecular diagnostic techniques has improved the diagnosis of the taeniid cestode infections echinococcosis and taeniasis. One particularly promising approach is the detection of parasite-specific antigens in feces (coproantigens). This approach has been applied to both *Echinoccocus* and *Taenia* species and has gained increasingly widespread use. Taeniid coproantigen tests are based on either monoclonal or polyclonal antibodies raised against adult tapeworm antigens. These tests have the following common characteristics: they are largely genus specific, their specificity is high (>95%), parasite antigen can be detected in feces weeks prior to patency, levels of coproantigen are independent of egg output, coproantigen is stable for days at a range of temperatures (–80 to 35°C) and for several months in formalin-fixed fecal samples, and coproantigen levels drop rapidly (1 to 5 days) following successful treatment. In the genus *Taenia*, most work has been done on *Taenia solium* and coproantigen tests have reliably detected many more tapeworm carriers than has microscopy. For *Echinococcus* species, there is a broad positive correlation between test sensitivity and worm burden, with a reliable threshold level for the test of >50 worms. Characterization of taeniid coproantigens in an attempt to further improve the tests is under way. Studies indicate taeniid coproantigens include high-molecular-mass (>150 kDa), heavily glycosylated molecules with carbohydrate moieties which contribute substantially to the levels of antigen detected in feces.

Arruda, G. C., E. M. Quagliato, and C. L. Rossi. 2006. Intrathecal synthesis of specific immunoglobulin G antibodies in neurocysticercosis: evaluation of antibody concentrations by enzyme-linked immunosorbent assay using a whole cysticercal extract and cyst vesicular fluid as antigens. *Diagn. Microbiol. Infect. Dis.* 54:45–49.

The demonstration of intrathecal antibody production has proven useful for showing the involvement of the CNS in several diseases. In the present study, the intrathecal synthesis of cysticercus-specific IgG antibodies was investigated in 30 patients with neurocysticercosis based on calculation of the specific IgG antibody index (AI_{IgG}). An AI_{IgG} of ≥1.5 was considered to be indicative of intrathecal antibody production. Antibody concentrations in serum and CSF samples were evaluated by using an ELISA with two antigen preparations from *Taenia solium* cysticerci, namely, a whole cysticercal extract (TsoW) and the vesicular fluid of the parasite (TsoVF). Intrathecal, cysticercus-specific IgG antibody synthesis was observed in 21 and 23 patients (70% and 76.6%) when using the TsoW and TsoVF antigens, respectively. Detection of the intrathecal synthesis of specific antibodies may be a potentially useful tool in establishing involvement of the CNS in cysticercosis.

Arthropods

Nisbet, A. J., and J. F. Huntley. 2006. Progress and opportunities in the development of vaccines against mites, fleas, and myiasis-causing flies of veterinary importance. *Parasite Immunol.* 28:165–172.

Despite the potential benefits offered by vaccination against ectoparasites, there have been few commercial successes with this strategy in spite of sustained efforts involving increasingly sophisticated techniques. This review outlines the progress and challenges offered by recent research into vaccination against some of the major ectoparasites of veterinary importance. It also provides insight into the opportunities arising from our increased understanding of the immunology of host-parasite relationships and the potential for exploitation of this knowledge as well as knowledge arising from new genomic data provided by expressed sequence tag projects.

Nuttall, P. A., A. R. Trimnell, M. Kazimirova, and M. Labuda. 2006. Exposed and concealed antigens as vaccine targets for controlling ticks and tick-borne diseases. *Parasite Immunol.* 28:155–163.

Tick vaccines derived from Bm86, a midgut membrane-bound protein of the cattle tick, *Boophilus microplus*, are currently the only commercially available ectoparasite vaccines. Despite the introduction of these vaccines into the market in 1994 and the recognized need for alternatives to chemical pesticides, progress in developing effective antitick vaccines (and ectoparasite vaccines in general) is slow. The primary rate-limiting step is the identification of suitable antigenic targets for vaccine development. Two sources of candidate vaccine antigens have been identified: "exposed" antigens that are secreted in tick saliva during attachment and feeding on a host and "concealed" antigens that are normally hidden from the host. Recently, a third group of antigens has been distinguished that combines the properties of both exposed and concealed antigens. This latter group offers the prospect of a broad-spectrum vaccine effective against both adults and immature stages of a wide variety of tick species. It also shows transmission-blocking and protective activity against a tick-borne pathogen.

Blood Parasites

Diez, H., M. C. Lopez, T. M. Del Carmen, F. Guzman, F. Rosas, V. Velazco, J. M. Gonzalez, and C. Puerta. 2006. Evaluation of IFN-gamma production by CD8 T lymphocytes in response to the K1 peptide from KMP-11 protein in patients infected with *Trypanosoma cruzi. Parasite Immunol.* 13:55–66.

The cellular response mediated by major histocompatibility protein class I-restricted CD8[+] T cells is crucial in the control of Chagas' disease. The K1 peptide derived from *Trypanosoma cruzi* KMP-11 protein has a high binding affinity to the HLA-A*0201 molecule. Nevertheless, it is not known whether this peptide is processed and displayed as a major histocompatibility protein class I epitope during natural infection by *T. cruzi*. The aim of this study was to evaluate, by enzyme-linked immunospot assay, the ability of K1 peptide to activate CD8[+] T lymphocytes to produce IFN-γ. Therefore, CD8[+] T lymphocytes from 22 HLA-A*0201[+] individuals, 12 chronic Chagas' disease patients, and 10 uninfected controls were analyzed. **The results revealed that two of the Chagas' disease patients had IFN-γ-secreting CD8[+] T cells that were able to respond to K1 peptide with a relative frequency of 110 and 230/10[6] CD8[+] T cells. In contrast, none of the HLA-A*0201[+] uninfected controls responded to K1 peptide.** Responses to HLA-A*0201-restricted peptide from the influenza virus matrix protein were found in six Chagas' disease patients and four uninfected controls with an average frequency of 175 and 111/10[6] CD8[+] T cells, respectively. Moreover, a flow cytometric assay for degranulation showed that Chagas' disease patients who responded had K1-specific cytotoxic CD8[+] T cells. **It is shown here for the first time that the K1 peptide is efficiently processed, presented, and recognized by CD8[+] T lymphocytes during the natural course of Chagas' disease.**

Hafalla, J. C., I. A. Cockburn, and F. Zavala. 2006. Protective and pathogenic roles of CD8[+] T cells during malaria infection. *Parasite Immunol.* 28:15–24.

CD8[+] T cells play a key role in protection against preerythrocytic stages of malaria infection. Many vaccine strategies are based on the idea of inducing a strong infection-blocking CD8[+] T-cell response. **The authors summarize what is known about the development, specificity, and protective effect of malaria-specific CD8[+] T cells and report on recent developments in the field.** Although work with mouse models continues to advance our understanding of the basic biology of these cells, many questions remain to be answered, particularly regarding the roles of these cells in human infections. **Increasing evidence is also emerging for a harmful role for CD8[+] T cells in the pathology of cerebral malaria in rodent systems. Once again, the relevance of these results to human disease is one of the primary questions remaining.**

Riley, E. M., S. Wahl, D. J. Perkins, and L. Schofield. 2006. Regulating immunity to malaria. *Parasite Immunol.* 28:35–49.

The optimal outcome of a malaria infection is that parasitized cells are killed and degraded without inducing significant pathology. Since much of the pathology of malaria infection can be immune mediated, this implies that immune responses have to be carefully regulated. Potential regulatory mechanisms include regulatory T cells, which significantly modify cellular immune responses to various protozoan infections including leishmaniasis and malaria; neutralizing antibodies to proinflammatory malarial toxins such as glycosylphosphatidylinositol and hemozoin; and self-regulating networks of effector molecules. Innate and adaptive immune responses are further moderated by the broader immunological environment, which is influenced by both the genetic background of the host and coinfection with other pathogens.

Stephens, R., and J. Langhorne. 2006. Priming of CD4[+] T cells and development of CD4[+] T cell memory: lessons for malaria. *Parasite Immunol.* **28**:25–30.

CD4 T cells play a central role in the immune response to malaria. They are required to help B cells produce the antibody that is essential for parasite clearance. They also produce cytokines that amplify the phagocytic and parasitocidal response of the innate immune system, as well as dampening this response later to limit immunopathology. Therefore, understanding the mechanisms by which T helper cells are activated and the requirements for development of specific, and effective, T-cell memory and immunity is essential in the quest for a malaria vaccine. The authors summarize discussions of CD4 cell priming and memory in malaria and in vaccination and outline critical future lines of investigation. Critical parameters in T-cell activation include the cell types involved, the route of infection, and the timing, location, and cell types involved in antigen presentation. A new generation of vaccines that induce CD4 T-cell activation and memory are being developed with new adjuvants. Studies of T-cell memory focus on differentiation and factors involved in the maintenance of antigen-specific T cells and control of the size of that population. To improve the detection of T-cell memory in the field, efforts will have to be made to distinguish antigen-specific responses from cytokine-driven responses.

Stevenson, M. M., and B. C. Urban. 2006. Antigen presentation and dendritic cell biology in malaria. *Parasite Immunol.* **28**:5–14.

Dendritic cells (DCs) are important both in amplifying the innate immune response and in initiating adaptive immunity and shaping the type of T helper response. Although the role of DCs in immune responses to many intracellular pathogens has been delineated and research is under way to identify the mechanisms involved, relatively little is known about the role of DCs in immunity to malaria. The authors provide an overview and summary of previous and current studies aimed at investigating the role of DCs as antigen-presenting cells. The role of DCs in inducing innate and adaptive immunity to blood-stage malaria is discussed, and the mechanisms involved are presented. Data from studies of humans infected with *Plasmodium falciparum*, the major human parasite responsible for the high morbidity and mortality associated with malaria throughout many regions of the developing world, as well as data from experimental mouse models are presented. Overall, the data from these studies are conflicting. The possible reasons for these differences, including the use of different parasite species and parasite strains in the mouse studies, are discussed. Nevertheless, together the data have important implications for the development of an effective malaria vaccine since the selection of appropriate *Plasmodium* antigens and/or adjuvants, targeting innate immune responses involving DCs, may provide optimal protection against malaria.

Treatment (See Also Individual Relevant Chapters)

Ellekvist P., and H. Colding. 2006. Transport proteins as drug targets in *Plasmodium falciparum*. New perspectives in the treatment of malaria. *Ugeskr. Laeger.* **168**:1314–1317.

The malaria parasite *Plasmodium falciparum* infects and replicates in human RBCs. Through the use of substrate-specific transport proteins, *P. falciparum* takes up nutrients from the RBC cytoplasm. The sequencing and publication of the *P. falciparum* genome have made it possible to identify, clone, and characterize a number of these transport proteins from the parasite. Since the *P. falciparum* transport proteins differ from their human homologues, they may provide potential drug targets in the treatment of malaria. An example of a *P. falciparum* transport protein which seems promising as a drug target is the parasite's hexose transporter. Furthermore, the antimalarial drug artemisinin has been shown to interact specifically with the parasite's Ca^{2+} pump. A number of other transport proteins are also discussed as possible drug targets.

Kennedy, P. G. 2006. Diagnostic and neuropathogenesis issues in human African trypanosomiasis. *Int. J. Parasitol.* 3 Mar. [Epub ahead of print.]

Human African trypanosomiasis, also known as sleeping sickness, is caused by protozoan parasites of the genus *Trypanosoma* and is a major cause of human mortality and morbidity. The East African and West African variants, caused by *Trypanosoma brucei rhodesiense* and *Trypanosoma brucei gambiense*, respectively, differ in their presentation, but the disease is fatal if untreated. Accurate staging of the disease into the early hemolymphatic stage and the late encephalitic stage is critical as the treatment for the two stages is different. The only effective drug for late-stage disease, melarsoprol, which crosses the blood-brain barrier, is followed by a severe posttreatment reactive encephalopathy in 10% of patients, half of whom die. There is no current consensus on the diagnostic criteria for CNS involvement, and the specific indications for melarsoprol therapy also differ. There is a pressing need for a quick, simple, cheap, and reliable diagnostic test to diagnose human African trypanosomiasis in the field and also to determine CNS invasion. Analyses of CSF and plasma from patients with human African trypanosomiasis have indicated a role for both proinflammatory and counterinflammatory cytokines in determining the severity of the meningoencephalitis of late-stage disease, and, at least in *T. brucei rhodesiense* infection, the balance of these opposing cytokines may be critical. Rodent models of human African trypanosomiasis have proved very useful in modeling the posttreatment reactive encephalopathy of humans and have demonstrated the central role of astrocyte activation and cytokine balance in determining CNS disease. The use of such animal models has also allowed a greater understanding of the more direct mechanisms of the effects of trypanosome infection on CNS function including the disruption of circadian rhythms, as well as the immunologic determinants of the passage of trypanosomes across the blood-brain barrier.

Meehan, W. J., S. Badreshia, and C. L. Mackley. 2006. Successful treatment of delusions of parasitosis with olanzapine. *Arch. Dermatol.* 142:352–355.

Delusional parasitosis is a rare disorder in which patients have a fixed, false belief of being infested with parasites. It is often accompanied by a refusal to seek psychiatric care. **Delusional parasitosis is classically treated with typical antipsychotic agents, the traditional dermatologic choice being pimozide.** However, pimozide's adverse-effect profile and the need for frequent electrocardiographic monitoring make such treatment less practical. The authors describe three patients who were diagnosed as having delusional parasitosis that was successfully treated with an atypical antipsychotic agent, olanzapine (5 mg/day), which has recently been approved by the Food and Drug Administration. **Olanzapine has a more benign adverse-effect profile than typical antipsychotic agents, and its use eliminates the need for electrocardiographic monitoring.** Olanzapine therapy has been associated with such adverse effects as sedation, hyperlipidemia, weight gain, and insulin resistance, all of which were infrequent in the patients in this study. **Olanzapine is an atypical antipsychotic agent that can be used as a first-line agent in delusional parasitosis as a safer therapeutic option without a specialized monitoring regimen.**

Ramzy, R. M., M. El Setouhy, H. Helmy, E. S. Ahmed, K. M. Abd Elaziz, H. A. Farid, W. D. Shannon, and G. J. Weil. 2006. Effect of yearly mass drug administration with diethylcarbamazine and albendazole on bancroftian filariasis in Egypt: a comprehensive assessment. *Lancet* 367:992–999.

Egypt was one of the first countries to implement a national program to eliminate lymphatic filariasis based on the World Health Organization strategy of repeated rounds of mass drug administration (MDA) with diethylcarbamazine and albendazole (target population, 2.5 million in 181 localities). The authors assessed the effect of five yearly rounds of MDA on filariasis in four sentinel villages in Egypt. They studied two areas with different infection rates before MDA. The Qalubyia study area had a low infection rate because of previous treatment with diethylcarbamazine; this was typical of most filariasis-endemic villages in Egypt before the advent of MDA. The Giza study area had a high baseline infection rate. The authors undertook repeated surveys in villages for treatment compliance and tests for microfilaraemia and circulating filarial antigenemia, antibodies to filarial antigen Bm14 in schoolchildren, and infections in indoor-resting mosquitoes (assessed by PCR). MDA compliance rates were excellent (>80%). In Giza after MDA, the prevalence of microfilaraemia and circulating filarial antigenemia fell from 11.5% to 1.2% and from 19.0% to 4.8%, respectively. Corresponding rates in Qalubyia fell from 3.1% to 0% and 13.6% to 3.1%, respectively. Rates of antifilarial antibody and circulating filarial antigenemia in schoolchildren (aged

about 7 to 8 years), fell from 18.3% to 0.2% and from 10.0% to 0.4% in Giza, respectively, and from 1.7% to 0% and 1.7% to 0% in Qalubyia, respectively. Mosquito infection rates fell from 3.07% to 0.19% in Giza and from 4.37% to 0% in Qalubyia. **MDA greatly affects variables related to infection (microfilaraemia and circulating filarial antigenemia prevalence rates) and transmission (antifilarial antibodies in young children and mosquito infection rates). Our results suggest that after five rounds of MDA, filariasis is likely to have been eliminated in most localities in Egypt where the infection is endemic.**

Valecha, N., H. Joshi, A. Eapen, J. Ravinderan, A. Kumar, S. K. Prajapati, and P. Ringwald. 2006. Therapeutic efficacy of chloroquine in *Plasmodium vivax* from areas with different epidemiological patterns in India and their *Pvdhfr* gene mutation pattern. *Trans. R. Soc. Trop. Med. Hyg.* 27 Feb. [Epub ahead of print.]

Among the four human malaria parasites, drug resistance occurs mainly in *Plasmodium falciparum*. However, there are some reports of chloroquine (CQ) resistance in *P. vivax* from different geographic regions. In India, approximately 50% of a total of 2 million cases of malaria reported annually are due to *P. vivax*. CQ is the drug of choice for treatment. Since few cases of treatment failure have been reported from India, this study was undertaken to generate data systematically on the efficacy of CQ in 287 patients from different epidemiologic regions. **Cure rates for 28 days were 100%, and there was rapid parasite clearance in all age groups from all study sites. Although *P. vivax* has been reported to be inherently resistant to sulfonamide and pyrimethamine, Indian isolates exhibited only double mutations in *dhfr* in vitro.**

Medically Important Arthropods

Faulde, M., and W. Uedelhoven. 2006. A new clothing impregnation method for personal protection against ticks and biting insects. *Int. J. Med. Microbiol.* 6 Mar. [Epub ahead of print.]

The efficacy and residual activity of a factory-based, permethrin-impregnated military battle dress uniform (BDU) produced by using a new polymer-coating technique were evaluated by laboratory and field testing during deployment to Afghanistan and compared with those of two commercially available, widely used dipping methods. Residual permethrin concentrations and remaining contact toxicities on treated fabrics before laundering, after up to 100 launderings as well as after being worn out during deployment, were tested against *Aedes aegypti* and *Ixodes ricinus*. The residual amount of permethrin was considerably larger for the polymer-coating technique, with 280 mg/m remaining after 100 launderings. Polymer-coated BDUs collected for disposal after being worn out during military deployment showed equivalent or better residual knockdown efficacy against test arthropods then did the U.S. Army Illinois Department of Agriculture kit after 50 launderings, which represent the recommended baseline for reimpregnation or disposal of the impregnated fabric. **BDUs impregnated by the polymer-coating method were found to be effective throughout the lifetime of** the uniform, ensuring the protection of soldiers in the field from arthropod vectors while simultaneously decreasing logistical constraints and occupational health threats.

Pasay, C., S. Walton, K. Fischer, D. Holt, and J. McCarthy. 2006. PCR-based assay to survey for knockdown resistance to pyrethroid acaricides in human scabies mites (*Sarcoptes scabiei* var. *hominis*). *Am. J. Trop. Med. Hyg.* **74:**649–657.

Permethrin, in the form of a topical cream, is being increasingly used for community-based programs to control endemic scabies. The development of resistance has reduced the use of pyrethroids for the control of many arthropods of economic and health importance. The best-recognized form of pyrethroid resistance, known as knockdown resistance or kdr, has been linked to specific mutations in the target of these agents, the para-homologous voltage-sensitive sodium channel gene (*Vssc*). To develop tools to study resistance to pyrethroid acaricides, the authors cloned 3,711 and 6,151 bp, respectively, of cDNA and genomic fragments of the *Vssc* gene from the scabies mite, *Sarcoptes scabiei*. The sequence encompasses the major polymorphic amino acid residues associated with pyrethroid resistance. **A PCR-based strategy has been developed that enables the genotyping of individual scabies mites. This will facilitate early detection and monitoring of pyrethroid resistance in scabies mite populations under drug selection pressure.**

Rupp, M. R., and R. D. deShazo. 2006. Indoor fire ant sting attacks: a risk for frail elders. *Am. J. Med. Sci.* **331:**134–138.

The authors have previously reported 10 indoor sting attacks by imported fire ants, most of which involved frail elderly people in the southeastern United States. Since the range of these insects is expanding and attacks often attract media attention, additional attacks may have occurred and were reported in local newspapers. The authors searched the archives from 1989 until 2004 of 182 U.S. newspapers in fire ant-endemic areas in 10 states. Ten additional cases of indoor fire ant sting attacks were reported in local newspapers between 1991 and 2004. This brings the total to 16 attacks on adults and 4 on infants. Most adult attacks occurred in long-term care facilities, but three involved hospitalized patients. Morbidity ranged from nightmares to death in seven adults. One of the infants died, and two suffered long-term morbidity. Of the 20 sting victims, 6 died within 1 week of the attack. Of the 10 attacks reported in newspapers, 7 did not result in significant medical consequences, compared with only 2 of the 10 attacks in previously published reports. **Increasing numbers of indoor fire ant sting attacks are occurring in the United States, and frail elderly people and infants are at risk. They should be removed from indoor areas where ants are present until the ants are eradicated.**

Titus, R. G., J. V. Bishop, and J. S. Mejia. 2006. The immunomodulatory factors of arthropod saliva and the potential for these factors to serve as vaccine targets to prevent pathogen transmission. *Parasite Immunol.* **28:**131–141.

In general, attempts to develop vaccines for pathogens transmitted by arthropods have met with little or no success. The saliva of arthropods that transmit disease enhances the infectivity of the pathogens transmitted to the vertebrate host, and vaccinating against components of the saliva of arthropods or against antigens expressed in the gut of arthropods can protect the host from infection and decrease the viability of the arthropod. **The results of this study suggest that multisubunit vaccines that target the pathogen itself as well as arthropod salivary gland components and arthropod gut antigens may be the most effective at controlling arthropod-borne pathogens, since these vaccines would target several facets of the life cycle of the pathogen. This review covers known immunomodulators in arthropod salivary glands, instances when arthropod saliva has been shown to enhance infection, and a limited number of examples of antiarthropod vaccines, with emphasis on three arthropods: sand flies, mosquitoes, and hard ticks.**

Testing of Stool Specimens

Nunes, C. M., L. G. Lima, C. S. Manoel, R. N. Pereira, M. M. Nakano, and J. F. Garcia. 2006. Fecal specimen preparation methods for PCR diagnosis of human taeniosis. *Rev. Inst. Med. Trop. Sao Paulo* **48:**45–47.

Sample preparation and DNA extraction protocols for DNA amplification by PCR, which can be applied in human fecal samples for taeniasis diagnosis, are described. DNA extracted from fecal specimens with phenol-chloroform-isoamilic alcohol and DNAzol reagent had to be purified to generate fragments of 170 and 600 bp by HDP2-PCR. This purification step was not necessary with the QIAmp DNA stool mini kit. **The best DNA extraction results were achieved after egg disruption with glass beads, with either phenol-chloroform-isoamilic alcohol, DNAzol reagent, or the QIAmp DNA stool mini kit.**

Management Issues in Diagnostic Medical Parasitology

Conraths, F. J., and G. Schares. 2006. Validation of molecular-diagnostic techniques in the parasitological laboratory. *Vet. Parasitol.* **136:**91–98.

Diagnostic laboratories today often operate according to standard quality management procedures such as ISO/IEC 17025. This requires that only validated methods be used. Validation procedures help to document that a particular protocol used by the accredited laboratory has a guaranteed performance in that particular laboratory. Several study designs exist for validation procedures. Computer programs are available to help with the statistical analysis of validation results. The results obtained in the protocol that is to be validated can be compared to those obtained in an already established test (agreement). For a method that is used under routine conditions or for epidemiologic studies, it is necessary to assess the diagnostic sensitivity and specificity of the technique. These parameters can be estimated by comparing the method to be validated with an existing reliable method ("gold standard"). This is done by testing a standard set of well-documented samples using both techniques in parallel. Approaches using Bayes' theorem are used to perform gold standard-free validations. **Many PCR-based methods are characterized by an excellent analytical sensitivity and are thus good candidates for diagnostic tools of the required diagnostic sensitivity. However, the high level of analytical** sensitivity can also make molecular techniques susceptible to cross-contamination and carryover problems, leading to false-positive results. Moreover, the presence of inhibitors can cause false-negative results. After an initial validation, test performance must be continuously monitored, e.g., by using combined Shewhart-CUSUM control routines, and test results must be compared to those obtained by other laboratories (proficiency testing).

Glossary

Certain terms listed here do not appear in the text, but it is hoped that their inclusion will be useful. Diseases and parasites in parentheses represent some, but not all, examples that pertain to the terms listed. One of my favorite definitions is "Parasitologist: A quaint person who seeks truth in strange places" (G. D. Schmidt and L. S. Roberts, *Foundations of Parasitology*, The C. V. Mosby Co., St. Louis, Mo., 1977).

Abdominal pain Crampy abdominal pain that may be seen in amebic colitis; right upper quadrant pain in amebic abscess; severe duodenitis or jejunitis with *Strongyloides* organisms penetrating the mucosa; pain suggestive of gastric ulcer or appendicitis with anisakid larvae (penetration of gut wall).

Aberrant parasite One which is never transmitted from person to person and which develops abnormally in humans (*Echinococcus multilocularis, Angiostrongylus*).

Abscess A localized collection of pus caused by liquefaction necrosis of tissue (amebiasis, filariasis).

Acanthamoebiasis Infection by free-living soil and water amebae of the genus *Acanthamoeba* that may result in a necrotizing dermal or tissue invasion, a fulmination and usually fatal primary amebic meningoencephalitis (PAM), or a subacute or chronic granulomatous amebic encephalitis (GAE).

Acaricide A mite-destroying agent.

Accidental parasite A parasite found in other than its normal host; also called an incidental parasite.

Accolé Early ring form of *Plasmodium falciparum* found at the margin of a red cell: appliqué.

Acid-fast stain, modified Developed for staining some of the coccidia (*Cryptosporidium, Cyclospora*); must be used with gentle decolorizer (1% sulfuric acid recommended).

Acquired immunity Immunity arising from a specific immune response, humoral or cell mediated, stimulated by antigen in the body of the host (active) or in the body of another individual with the antibodies or lymphocytes transferred to the host (passive).

Acute abdomen An abdominal condition of abrupt onset usually associated with abdominal pain resulting from inflammation, perforation, obstruction, infarction, or rupture of intra-abdominal organs; surgical intervention is usually necessary (ascariasis, anisakiasis).

Aerosol A system of respirable particles dispersed in a gas, smoke, or fog that can be retained in the lungs.

AIDS Acquired immune deficiency syndrome.

Alae Pronounced, longitudinal cuticular ridges in nematodes, usually found in larval stages (*Ascaris lumbricoides*), although occasionally present in adult worms (*Enterobius vermicularis*).

Allergy A hypersensitive condition acquired by exposure to a particular allergen (helminth infections, house dust mites, ectoparasite bites or stings).

Amastigote Small, round, intracellular stage of *Leishmania* spp. and *Trypanosoma cruzi* in which the base of the flagellum is anterior to the nucleus but there is no external flagellum; also called Leishman-Donovan body, L-D body, or leishmanial stage.

Anamnestic response Immune response to a challenge or secondary antigen inoculation, marked by a more rapid and stronger manifestation of the immune reaction than after the primary immunizing dose.

Anaphylaxis Hypersensitivity produced by exposure to further doses of the same protein, usually when exposure is within less than 2 weeks (bee stings, echinococcosis).

Anemia A reduced number of erythrocytes per cubic millimeter, reduction in the amount of hemoglobin, or reduction in the volume of packed red blood cells per 100 ml of blood (malaria, hookworm, and *Diphyllobothrium latum* infections).

Anergic Absence of sensitivity to an antigen or the condition resulting from desensitization (cutaneous leishmaniasis).

Aneurysm   Circumscribed dilation of an artery connecting directly with the lumen of an artery or a cardiac chamber connecting directly with the lumen of an artery, usually due to an acquired or congenital weakness of the wall of the artery or chamber.

Anorexia   Absence of appetite.

Anthropophilic   Human-seeing or human-preferring, especially with reference to blood-sucking arthropods.

Anthropozoonosis   A zoonosis maintained in nature by animals and transmissible to humans.

Antigenic determinant   The areas on an antigen molecule that bind with antibody or specific receptor sites on the sensitized lymphocyte; these areas determine the specificity of the antibody or lymphocyte.

Anuria   Absence of urine secretion (scorpion sting).

Aphasia   Partial or complete inability to speak and/or understand spoken words.

Apoptosis   Single deletion of scattered cells by fragmentation into membrane-bound particles, which are phagocytosed by other cells; believed to be due to programmed cell death.

Appendicitis   Inflammation of the vermiform appendix (amebiasis, *Ascaris*, *Trichuris*, and *Enterobius* infections).

Appendix   Thick-walled fingerlike projection at the end of the cecum.

Appliqué   Early ring form of *Plasmodium falciparum* found at the margin of a red cell; accolé.

Ascites   Effusion of serous fluid into the abdominal cavity (schistosomiasis).

Aspirate   Fluid removed from a cavity or lesion (leishmaniasis, hydatid disease, amebiasis).

Asthma, bronchial   Disease characterized by wheezing, coughing, and difficulty in breathing caused by spasmodic contraction of the bronchi; may be the result of allergic reaction (ascariasis, strongyloidiasis, visceral larva migrans, house dust mites). Also see *Eosinophilia*.

Autochthonous   Indigenous or normally found in a particular area.

Autoinfection   Reinfection by an organism already present in the body with an increase in the number of parasites without their undergoing a cycle outside the body; self-infection (*Strongyloides stercoralis*, *Hymenolepis nana*, *Cryptosporidium* spp.).

Autopsy   Gross and microscopic postmortem examination of the organs of the body to determine the cause of death or pathologic changes.

Axostyle   Rodlike supporting structure in some flagellates that gives rigidity to the body (*Trichomonas* spp.).

Babesiosis   A disease of humans that is caused by a blood protozoan of various animals; tick borne; organism morphology similar to young rings of *Plasmodium falciparum*; more severe in compromised patients.

Band form   Older trophozoites of *Plasmodium malariae* that may stretch across the red cell in the form of a band.

Barbiero   Brazilian term for the blood-sucking hemipteran triatomid bug, *Panstrongylus megistus*, an important vector of Chagas' disease (caused by *Trypanosoma cruzi*).

Basket nucleus   Nuclear structure that may be seen in *Iodamoeba bütschlii* cysts (occasionally trophozoites); in well-stained organisms, fibrils may be seen running between the karyosome and the chromatin granules. The "basket of flowers" has been described as follows: the karyosome forms the basket, the fibrils form the stems, and the granules form the flowers.

Baylisascaris   A genus of ascarid nematodes found in the intestines of mammals, including humans.

Baylisascaris procyonis   A common roundworm of raccoons (ingestion of infective eggs from the environment) that can cause larva migrans (central nervous system). Human cases present with eosinophilic meningoencephalitis (there have been several fatal cases in the United States).

Behçet's syndrome   More common in men; involves dermatitis, erythema nodosum, thrombophlebitis, cerebral involvement; may include arthritis, uveitis, or iridocyclitis with hypopyon.

Benign tertian malaria   Malaria caused by *Plasmodium vivax*.

Bilharziasis   Another name for schistosomiasis; often used to indicate infection with either *Schistosoma mansoni* or *S. haematobium*.

Binary division   Dividing of a cell into two nearly equal daughter individuals.

Biopsy   Removal of tissue from a living person; examination, usually microscopic, used to make diagnosis (Chagas' disease, leishmaniasis, schistosomiasis, visceral larva migrans, amebiasis).

Blackwater fever   A condition in which the diagnostic symptom is the passage of reddish to black urine (containing hemoglobin), indicating massive intravascular hemolysis (*Plasmodium falciparum*).

Blepharoplast   A small, dark-staining body near the base of the flagellum that is associated with fibrillar portions of the flagellum (blood flagellates).

Botfly   Robust, hairy fly of the order Diptera, often strikingly marked in black and yellow or gray, whose larvae produce a variety of myiasis conditions in humans and various domestic animals.

Bothrium   One of the two shallow grooves on the scolex of *Diphyllobothrium latum*.

Bots   The larvae of several species of botflies.

Bradyzoite   A slowly multiplying trophozoite contained in the cyst; typical of a chronic infection with *Toxoplasma gondii*.

Buccal cavity/capsule The space between the oral opening and the beginning of the esophagus in nematodes; useful diagnostically in rhabditiform larvae (*Strongyloides stercoralis*, hookworm).

Buccal teeth/plates Toothlike structure in the mouth cavity of the adult hookworms of the genus *Ancylostoma* (in contrast to buccal plates found in the mouth cavity of the genus *Necator*).

Budding Production of daughter cell as outward growth from the original cell or germinal layer (yeastlike fungi, hydatid cyst).

Buffy coat Grayish white layer of fibrin and leukocytes that forms the upper part of the sedimented cells in whole blood after the addition of anticoagulant (examined in visceral leishmaniasis and trypanosomiasis).

Bursa Umbrella-like expansion of the cuticle at the posterior end of male nematodes; it is supported by muscular rays and aids in copulation (hookworms, trichostrongyles).

Cachexia Severe debilitated state (heavy worm infestation, visceral leishmaniasis).

Calabar swellings Transient swellings of subcutaneous tissues; associated with *Loa loa* infection.

Calcareous corpuscles Rounded masses composed of concentric layers of calcium carbonate that are found in the tissue of tapeworms, characteristic of tapeworm tissue; certain fixatives must be avoided to preserve corpuscles for identification purposes (tapeworms).

Calcifications Portions of host or parasite tissue that become hardened by the deposition of calcium salts; may be visible on radiographic studies (toxoplasmosis, cysticercosis, trichinosis, schistosomiasis).

Capitonnage Closure of a cyst cavity by using sutures (hydatid disease).

Capitulum The blood-sucking, probing, sensing, and holdfast mouthparts of a tick, including the basal supporting structure; the relative size and shape of mouthparts forming the capitulum are characteristic for the genera of hard ticks.

Carcinogen Substance capable of causing a malignant tumor in humans or animals.

Carditis Inflammation of the heart tissues (Chagas' disease).

Carrier A host that harbors a particular pathogen without showing manifestations of disease.

Casoni antigen Skin test antigen composed of sterile hydatid fluid (hydatid disease).

Cathartic A substance causing evacuation of bowels by increasing bulk, stimulating peristalsis.

Cecum A blindly ending appendage of the intestine.

Cell-mediated immunity Type of immunity that is independent of antibody and is initiated by cytotoxic T cells or by the secretion of lymphokines by T cells. The lymphokines cause other cells, usually lymphocytes and macrophages, to aggregate in the area of T-cell–antigen interaction, leading to a local reaction such as a granuloma.

Cercaria Free-living, tailed larval stage of trematodes; may infect humans by direct penetration (schistosomes); may attach to vegetation and encyst to the metacercarial stage (*Fasciola* and *Fasciolopsis* spp.), or may penetrate tissues of vertebrates or invertebrates and encyst to the metacercarial stage (*Clonorchis* spp.).

Cestode Tapeworm.

Chagas' disease Infection caused by *Trypanosoma cruzi* (American trypanosomiasis).

Chagoma Small granuloma in the skin, caused by early multiplication of *Trypanosoma cruzi* (Chagas' disease).

Charcot-Leyden crystals Slender crystals that are formed from the breakdown products of eosinophils; shaped like double, elongated pyramids with pointed ends; can be found in feces, sputum, and tissues; indicates an immune response that may or may not be related to a parasitic infection.

Chiclero ulcer Single, self-limiting cutaneous papule, nodule, or ulcer (usually on the face or ears) (leishmaniasis).

Chigger The six-legged larva of *Trombicula* species; a blood-sucking stage of mites that includes the vectors of scrub typhus.

Chorea Irregular, spasmodic, involuntary movements of the limbs or facial muscles.

Chorioretinitis Inflammation of the posterior coat and retina of the eye (toxoplasmosis).

Chromatin Deep staining DNA-containing portion of the nucleus (protozoa).

Chromatoidal bar/body Deep-staining, bar-shaped, round, or splinter-shaped inclusions found in the cytoplasm of certain amebae (*Entamoeba* spp.).

Chyluria The presence of lymph and emulsified fat in the urine (filariasis).

Cilia Small, hairlike cytoplasmic projections from a cell or organism; used for motility (*Balantidium coli*).

Cirrhosis Disruption of the normal structure of the liver; destruction of liver cells and increase in connective tissue (schistosomiasis).

CLBs These organisms were thought to be a new pathogen, possibly an oocyst, a flagellate, an unsporulated coccidian, a large *Cryptosporidium* sp., a blue-green alga (cyanobacterium-like body), or a coccidian-like body. They are now known to be coccidia in the genus *Cyclospora*.

Coenurus Tapeworm larva of the genus *Multiceps*, characterized by multiple scolices invaginated into the fluid filled bladder; no daughter cysts produced.

Colitis Inflammation of the colon.

Colon Portion of the large intestine from the cecum to the rectum.

Coma Absolute unconsciousness.

Commensalism Association in which one individual receives benefits and the other is neither helped nor harmed.

Complement A series of proteins that bind in a complex series of reactions to antibody (immunoglobulin G or M) when the antibody is itself bound to an antigen; produces cell lysis if the antibody is bound to antigens on the cell surface.

Concentration techniques Procedures, usually in fecal examinations, allowing the examination of large amounts of feces (flotation or sedimentation procedures; some available for blood specimens and urine specimens).

Concomitant immunity Type of immunity where the host harbors a population of adult worms but is immune to reinfection.

Congenital Present at, and usually before, birth (regardless of cause).

Conjunctivitis Inflammation of the membrane that lines the eyelids and covers the surface of the eyeball.

Convulsion A violent, involuntary contraction of an extensive group of muscles; disturbance of cerebral function (spider and scorpion envenomation).

Coracidium A ciliated larval tapeworm stage that occurs in the eggs of cestodes such as *Diphyllobothrium* and *Spirometra* spp. (pseudophyllidian tapeworms); this is a free-swimming stage, containing six hooks like those found in the oncospheres of other tapeworms.

Cor pulmonale Cardiopulmonary problems which may terminate in congestive heart failure; obstructive vascular disease (*Schistosoma mansoni* infection).

Corrosive Any substance that causes visible destruction of human tissue at the site of contact. The U.S. Environmental Protection Agency defines corrosivity as a substance that is highly acidic (pH < 2.1) or highly alkaline (pH > 12.4).

Creeping eruption Penetration and migration through subcutaneous tissues by skin-penetrating nematodes, resulting in intense pruritus and sometimes secondary bacterial or fungal infection (*Strongyloides* and hookworm infections, cutaneous larva migrans).

Crithidia Old term for epimastigote.

Cryptosporidiosis Intestinal infection caused by coccidia (*Cryptosporidium* spp.).

Cutaneous Pertaining to the skin.

Cuticle Outermost, three-layered portion of the body wall of nematodes (*Ascaris lumbricoides*, etc.).

Cysticercoid The solid larval stage of tapeworms containing an invaginated scolex that occurs in arthropod intermediate hosts (*Hymenolepis*, *Dipylidium*).

Cysticercosis Tissue infection with larval tapeworms in which the scolex is inverted into a fluid-filled bladder (*Taenia solium*).

Cysticercus Tapeworm larva of the family Taeniidae (includes *Taenia solium* and *T. saginata*) in which a single scolex is invaginated into a fluid-filled bladder.

Cystoenterostomy Internal drainage of pancreatic cysts into some portion of the intestinal tract (hydatid disease).

Cytoplasm The protoplasm of a cell exclusive of the nuclear material.

Daisy Colloquial term descriptive of the segmented forms (merozoites) of the mature schizont of *Plasmodium malariae*.

Decontamination A procedure that eliminates or reduces microbial or toxic agents to a safe level with respect to the transmission of infection or other adverse affects.

Definitive host Host in which the sexual reproduction of a parasite occurs.

Delayed hypersensitivity A manifestation of cell-mediated immunity, different from immediate hypersensitivity in that the maximal response is reached about 24 h or more after intradermal injection of the antigen; infection site is infiltrated by monocytes and macrophages.

Dermatitis Inflammation of the skin (filariasis, schistosomiasis, infections with *Strongyloides* and hookworm larvae, leishmaniasis, *Sarcoptes* infections).

Diarrhea Frequent passage of soft or liquid stool (no blood); may be caused by any parasite or infection normally found in any part of the intestine (*Giardia lamblia*, *Isospora belli*, *Dientamoeba fragilis*, *Balantidium coli*, *Cryptosporidium* spp., *Enterocytozoon bieneusi*, *Encephalitozoon intestinalis*, possibly other microsporidia, *Cyclospora* spp., *Leishmania* spp. causing visceral leishmaniasis, *Plasmodium falciparum*, tapeworms, *Trichinella* spp., *Schistosoma* spp., hookworm).

Didelphis A genus of marsupials, commonly called opossums, that serve as reservoir hosts of *Trypanosoma cruzi*.

Dioecious Having the male and female sexes of a species as separate individuals.

Direct smear (stool) Approximately 2-mg suspension of feces in water or saline for the purpose of examination for parasites; the primary aim is to see motility.

Disinfectant An agent intended to destroy or irreversibly inactivate all microorganisms, but not necessarily their spores, on inanimate surfaces, e.g., work surfaces or medical devices.

Diurnal Pertaining to the daylight portion of the 24-h day.

Duodenum The proximal portion of the small intestine (*Strongyloides stercoralis*, *Giardia lamblia*).

Dysentery Frequent watery stools, usually containing blood and/or mucus; associated with inflammation of the intestine, usually the colon (*Entamoeba histolytica*, *Trichuris trichiura*).

Dyspnea Difficulty in breathing.

Dysuria Painful or difficult urination.

Ecosystem The fundamental unit in ecology, comprising the living organisms and the nonliving elements that interact in a defined region.

Ectoparasite Organism that lives on or within the skin of its host (lice, mites, ticks).

Ectopic site Outside the normal location; i.e., the position of a parasite which lodges in an atypical part of the body.

Ectoplasm Outer layer of the cytoplasm of a cell.

Edema Presence of large amounts of fluid, usually in subcutaneous tissues (filariasis, trypanosomiasis, hookworm).

Elephantiasis Inflammation and obstruction of the lymphatic system, resulting in hypertrophy and thickening of the surrounding tissues, usually involving the extremities and external genitalia (filariasis).

Encapsulation Active process of walling off a parasite by the host through the formation of a connective tissue capsule (trichinosis).

Encephalitis Inflammation of the brain (Chagas' disease, trichinosis).

Encystment Formation of a resistant external wall by protozoa to enable them to survive drying and adverse environmental conditions; encysted forms are infective to humans.

Endemic Disease present in a localized community or area at all times.

Endoparasite Parasite that lives within the body of the host.

Endoplasm Inner portion of the cytoplasm of a cell.

Endospore The chitinous inner spore coat (microsporidia).

Engineering controls Controls (e.g., sharps disposal containers, self-sheathing needles, safer medical devices) that isolate or remove the hazard from the workplace.

Eosinophilia Formation of large numbers of eosinophilic leukocytes as a result of some type of immune response; usually found in helminth infections, particularly with tissue invasion (visceral larva migrans, trichinosis, filariasis, schistosomiasis, ascariasis, strongyloidiasis).

Eosinophilic cerebrospinal fluid pleocytosis Increased number of eosinophils in cerebrospinal fluid (*Angiostrongylus* and *Baylisascaris* infections; coenurosis, cysticercosis, echinococcosis, *Fasciola*, *Gnathostoma*, *Paragonimus*, *Schistosoma*, *Toxocara*, *Toxoplasma*, and *Trichinella* infections; possibly *Dirofilaria*, *Onchocerca*, and *Ascaris* infections).

Eosinophilic meningitis Inflammation of membranes of the brain or spinal cord accompanied by an increased number of eosinophils, usually associated with particular helminth infections (angiostrongyliasis, gnathostomiasis, cysticercosis, schistosomiasis).

Epidemic Disease that spreads rapidly and infects many people in a community or area (usually within a short time frame).

Epidermal Pertaining to the outer layer of the skin.

Epididymitis Inflammation of the epididymis (first part of the excretory duct of each testis) (Bancroftian filariasis).

Epimastigote Developmental stage of the family Trypanosomatidae; the base of the flagellum is in front of the nucleus, and as the flagellum passes through the body to emerge as the free flagellar portion, it is attached to the body by the undulating membrane. (Old term: crithidia.)

Erythema Diffuse or patchy redness of the skin; blanching on pressure as a result of congestion of cutaneous capillaries.

Erythrocytic cycle Developmental cycle of malarial parasites within the red blood cells.

Eukaryote A cell containing a membrane-bound nucleus with chromosomes of DNA, RNA, and proteins, with cell division involving a form of mitosis in which mitotic spindles (or some microtubule arrangement) are involved; mitochondria are present.

Evagination Protrusion of some part or organ from its normal position (tapeworm scolex in larval worms).

Exflagellation The extrusion of rapidly waving flagellum-like microgametes from microgametocytes; in human malaria parasites, this occurs in the blood meal taken by the proper anopheline vector within a few minutes after ingestion of the infected blood by the mosquito.

Exoerythrocytic cycle A part of the malarial life cycle in which the mosquito introduces sporozoites into the vertebrate host; sporozoites penetrate the parenchymal liver cells and undergo schizogony, resulting in the production of liver merozoites, which then initiate the erythrocytic cycle.

Exospore The proteinaceous outer spore coat (microsporidia).

Feces The unabsorbed residue of the digestive process, along with sloughed epithelium, mucus, and bacteria.

Festoon A distinguishing characteristic of certain hard tick species, consisting of small rectangular areas separated by grooves along the posterior margin of the dorsum of both males and females.

Fever A complex physiologic response to disease mediated by pyrogenic cytokines and characterized by a rise in core temperature, generation of acute-phase reactants, and activation of immunologic systems.

Filariform larvae Slender, infective larvae of *Strongyloides stercoralis* and hookworm.

Flame cell Primitive, ciliated excretory cell in trematodes; the movement of the cilia on this cell within the miracidium larva (within a schistosome egg) indicates egg viability.

Flesh flies Members of the order Diptera, whose larvae (maggots) develop in putrefying or living tissues.

Fluke   Trematode (*Clonorchis*, *Paragonimus*, *Fasciola*, and *Schistosoma* spp.).

Fundus, ocular   The portion of the interior of the eyeball around the posterior pole, visible through the ophthalmoscope.

Funiculitis   Inflammation of the spermatic cord (filariasis).

Furcocercous   Fork-tailed (cercaria of schistosomes).

Gametocyte   In malaria, the sexual cell (male microgametocyte or female macrogametocyte). It is present in peripheral blood; fertilization occurs in the mosquito stomach with formation of the zygote.

Gastroenteritis   Inflammation of the stomach and small and large intestines.

Genital primordium   Ovoid clump of cells that becomes the reproductive system (seen in the rhabditiform larvae of *Strongyloides stercoralis* and hookworm).

Geohelminth   Any helminth that is transmitted to humans from the soil (*Ascaris*, *Strongyloides*, *Trichuris*, hookworm).

Germicide   A general term for an agent that kills pathogenic microorganisms on inanimate surfaces.

Granuloma   Tumorlike nodule of firm tissue formed as a reaction to chronic inflammation, usually of lymphoid and epithelioid cells.

Gynecophoral groove or canal   The ventral incurved fold of the body extending from the ventral sucker to the caudal extremity, for carrying the female worm (schistosomes).

Halzoun   Presence of worms on pharyngeal mucosa; congestion of tissues accompanied by difficulty in breathing and possible asphyxiation (*Fasciola hepatica* and pentastome infection).

Harada-Mori culture   Method of incubating fecal material on a filter paper strip in a test tube containing water (cover one-third of the length of paper strip) for the purpose of culturing and recovering nematode larvae (*Strongyloides stercoralis*, hookworm).

Helminth   May refer to a nematode (roundworm), cestode (tapeworm), or trematode (fluke).

Hematuria   Passage of free hemoglobin in the urine (*Plasmodium falciparum* infection, blackwater fever).

Hemoflagellate   Any flagellated protozoan blood parasite (*Trypanosoma* spp.).

Hepatitis   Inflammation of the liver (amebiasis, schistosomiasis, infections with liver flukes).

Hepatomegaly   Enlargement of the liver.

Hermaphroditism   Presence of both male and female reproductive systems in the same individual; most trematodes and cestodes are hermaphroditic.

Heterakis   A genus of important nematode parasites. *H. gallinarum* is the cecal worm of chickens and turkeys and is the vector of *Histomonas meleagridis*, a protozoan that causes histomoniasis (a model of the possible transmission of *Dientamoeba fragilis* within helminth eggs such as *Enterobius vermicularis* and *Ascaris lumbricoides*).

Heterophyids   Small intestinal flukes parasitic in humans and animals (*Heterophyes* and *Metagonimus* spp.).

Hexacanth embryo   Six-hooked tapeworm embryo (oncosphere).

HIV   Human immunodeficiency virus.

Hives   See *Urticaria*.

HMO   Health Maintenance Organization.

Host   An organism in or on which a parasite lives.

Hydatid, alveolar   Type of hydatid cyst formed by *Echinococcus multilocularis*; budding is external, as well as internal, and there is no thick outer capsule.

Hydatid, polycystic   Type of hydatid cyst formed by *Echinococcus oligarthrus* and *E. vogeli*; the structure is between those of unilocular and alveolar types of hydatid cyst.

Hydatid, unilocular   Type of hydatid cyst formed by *Echinococcus granulosus*; single limiting membrane.

Hydatid cyst   Tapeworm larval stage of the genus *Echinococcus*; consists of a large bladder with an inner germinal layer from which daughter cysts and scolices develop (some of which break off and drop into the fluid-filled bladder).

Hydatid sand   Consists of scolices, daughter cysts, hooks, and calcareous corpuscles found in the fluid within the hydatid cyst (*Echinococcus granulosus*).

Hydatid thrill   A delicate vibration felt by the hand after quick palpation or percussion over the area of the body where the hydatid cyst is located (*Echinococcus granulosus*).

Hydrocele   Accumulation of serous fluid in a saclike cavity, particularly the scrotal sac (filariasis).

Hydrocephalus   An abnormal accumulation of fluid in the cerebral ventricles or in the subarachnoid space of the brain (cysticercosis).

Hyperparasitemia   Infection with *Plasmodium falciparum* at a density of more than 250,000 parasites/μl.

Hyperpigmentation   Intensification of pigment; most obvious in dark-skinned races (leishmaniasis, onchocerciasis).

Hypersensitivity   Enhanced state of responsiveness following sensitization to a particular antigen. There are four types, type I (anaphylaxis), type II (antibody-dependent cytotoxicity), type III (immune complex hypersensitivity), and type IV (delayed hypersensitivity).

Hypnozoite   Exoerythrocytic schizozoite of *Plasmodium vivax* or *P. ovale* in the human liver, characterized by delayed primary development; responsible for malarial true relapse.

Hypopyon   The presence of pus in the anterior chamber of the eye.

Hypostome The central unpaired holdfast organ of the tick capitulum; covered with recurved spines that serve as an anchoring device while the tick feeds.

IATA The International Air Transport Association, a body of the commercial airline industry that governs international aviation. It publishes the *Technical Instructions for the Safe Transport of Dangerous Goods by Air* and regulates dangerous goods for member airlines and anyone who tenders dangerous goods to those airlines.

Immunity The ability of an individual to resist and/or control the effects of antigens (antigen sources could be animal, plant, or mineral).

Incubation period Time span from introduction of disease-causing organisms until symptoms of the disease occur.

Induced malaria Malaria infection acquired by blood transfusions or possible sharing of needles by drug addicts (*Plasmodium* spp.).

Infectious waste Waste containing or assumed to contain pathogens of sufficient virulence and quantity that exposure to the waste by a susceptible host may result in a communicable disease.

Infestation Presence of arthropods on the surface of the body, does not refer to endoparasites (*Pediculus* spp.).

Inflammation Result of tissue reaction to injury; symptoms would be redness, pain, swelling, and fever.

Inspissate To thicken or dry a substance by removing liquids by evaporation (preparation of certain types of media used to grow some of the amebae).

Intermediate host A required host in the life cycle in which larval development takes place; this must occur before the stage is infective for the definitive host or secondary intermediate hosts.

Invagination The ensheathing, enfolding, or insertion of a structure within itself or another (tapeworm larval worms).

Jaundice A condition caused by excess bilirubin and bile pigment deposition in the skin, which may give the patient a yellow appearance, especially the eyes (*Plasmodium falciparum*).

Kala azar Another name for visceral leishmaniasis (*Leishmania donovani*).

Karyosome Concentrated clumps of chromatin material within the nucleus; position and morphology are often used to differentiate intestinal protozoa.

Kerandel's sign Delayed sensation to pain (African trypanosomiasis).

Keratitis Inflammation of the cornea (Chagas' disease, onchocerciasis, *Acanthamoeba* infection, microsporidiosis).

Keratoconjunctivitis Inflammation of the conjunctiva and of the cornea.

Kinetoplast Intensely staining rod- or disk-shaped or spherical extranuclear DNA structure found in parasitic flagellates near the base of the flagellum.

Knott technique Concentration procedure using blood and dilute formalin; designed to detect microfilaria (*Wuchereria* and *Brugia* spp.).

Kupffer cells Phagocytic epithelial cells lining the sinusoids of the liver.

Lagochilascaris minor An unusual nematode parasite of humans (ingestion of infective eggs from the environment or ingestion of infective larvae in the tissues of an intermediate host). In most cases, the worms are located in the soft tissues of the neck and throat, tonsils, mastoids, and paranasal sinuses (chronic or recurrent abscesses). Fatal encephalopathy (one case) has also been reported. All cases were in the tropical regions of the Western Hemisphere.

Larva currens Cutaneous larva migrans caused by rapidly moving larvae of *Strongyloides stercoralis*, typically extending from the anal area down the upper thighs; may also be caused by zoonotic species of *Strongyloides*.

Larva migrans, cutaneous/visceral/ocular Disease characterized by thin, red, convoluted papular or vesicular lines of eruption that extend at one end while fading at the other (dog or cat hookworm). Visceral larva migrans involves migration of larvae through the deep tissues, including the eye (dog or cat ascarids, *Toxocara* spp.).

Latex allergy Allergic reaction associated with latex glove use. The two types of allergic reactions are contact dermatitis (Type IV delayed hypersensitivity) due to chemicals used in processing latex and the more serious IgE/histamine-mediated allergy (immediate or Type I hypersensitivity) due to latex proteins.

Leishman-Donovan body Small, round, intracellular form of *Leishmania* spp. (reticuloendothelial system) and *Trypanosoma cruzi* (striated muscle); also called amastigotes or L-D bodies.

Leptomonad Old term for the promastigote stage, which is long and slender and found in the insect vector for *Leishmania* spp.; it is also recovered in artificial culture media (Novy-MacNeal-Nicolle [NNN] medium).

Leukocytosis Increase in the number of white blood cells, usually to more than 10,000/mm^3.

Leukopenia Decrease in the number of white blood cells, usually to less than 4,000/mm^3.

Lichen simplex A discrete flat papule or an aggregate of papules giving a patterned configuration resembling lichens growing on rocks.

Loeffler's syndrome Transient pulmonary infiltration; day-to-day clearing in 3 to 14 days; associated with marked peripheral eosinophilia (*Ascaris lumbricoides*).

Lumen Cavity of hollow, tubular organs, such as the intestine or blood vessels.

Lymph Plasma and white blood cells that bathe the tissue cells.

Lymphadenitis Inflammation of the lymph nodes (filariasis).

Lymphangitis Inflammation of the lymphatic vessels (filariasis).

Lymphocytosis Absolute or relative increase of lymphocytes in the blood.

Lymphokine One of several kinds of effector molecules released by T lymphocytes when an antigen to which the lymphocytes are sensitized binds to the cell surface.

Lymph varices Dilated lymph vessels secondary to lymphatic blockage (filariasis).

Macrogametocyte The mother cell producing the macrogametes, or female elements of sexual reproduction in the sporozoan and microsporidian protozoa; macrogamont.

Macronucleus Large, kidney bean-shaped nucleus in *Balantidium coli* (the shape is not always consistent).

Macrophage Motile, phagocytic, mononuclear cell that originates in the tissues and may be confused morphologically with protozoan trophozoites (particularly *Entamoeba* spp.).

Malabsorption Poor fat absorption in the upper small bowel (giardiasis).

Malaise Vague feeling of discomfort throughout the whole body; tiredness.

Malaria Benign tertian malaria, *Plasmodium vivax*; malignant tertian malaria, *Plasmodium falciparum*; ovale malaria, *Plasmodium ovale*; quartan malaria, *Plasmodium malariae*; relapsing malaria, renewal of clinical activity at some interval after the primary attack; remittent malaria, a malarial fever (usually due to *P. falciparum*) in which the temperature falls but not to the normal level during the interval between two pronounced paroxysms.

Malarial pigment Composed of hematin and excess protein left over from the metabolism of hemoglobin; appears as brownish pigment after Giemsa staining (*Plasmodium* spp.).

Malignant tertian malaria Malaria caused by *Plasmodium falciparum*.

Mast cell A connective tissue cell that contains coarse, basophilic, metachromatic granules; the cell is thought to contain heparin and histamine.

Mastocytosis Abnormal proliferation of mast cells in a variety of tissues; may be systemic, involving a variety of organs, or cutaneous (urticaria pigmentosa).

Maurer's dots or clefts Irregular dots that occur infrequently in red blood cells infected with *Plasmodium falciparum* (dots tend to be more blue after Giemsa staining than do Schüffner's dots).

Measly Pertaining to pork or beef infected with the cysticerci of the tapeworms *Taenia solium* or *T. saginata*.

Megacolon Dilation of the colon (Chagas' disease).

Megaesophagus Dilation of the esophagus (Chagas' disease).

Meningitis Inflammation of the membranes of the brain or spinal cord.

Meningoencephalitis Inflammation of the brain and its surrounding membranes (trypanosomiasis, malaria, *Naegleria* and *Angiostrongylus cantonensis* infections).

Merogony Synonym for schizogony, leading to the production of daughter cells (merozoites).

Meront Diplokaryotic cells that grow and divide into daughter cells (merozoites).

Merozoite Product of schizogonic cycle in malaria; produced in the liver (preerythrocytic cycle) and in the red blood cells (erythrocytic cycle).

Mesentery Tissue that supports the intestinal tract.

Metacercaria The infective, encysted larval form of a trematode; found within the tissues of an intermediate host or on plant material (*Paragonimus*, *Clonorchis*, and *Fasciolopsis* spp.).

Methenamine silver stain Both hot and cold methods; involves deposition of silver onto cyst wall (*Pneumocystis jiroveci* [now classified with the fungi]).

Microfilaria Embryos produced by filarial worms (nematodes); usually found in the blood or tissues of patients with filariasis (*Wuchereria*, *Brugia*, and *Onchocerca* spp.).

Microgametocyte The mother cell producing the microgametes, or male elements of sexual reproduction in the sporozoan and microsporidian protozoa; microgamont.

Micrometer (micron) Unit of measure equal to 0.001 mm; abbreviated μm.

Micron See *Micrometer*.

Micronucleus Small, dotlike nucleus found in *Balantidium coli*; often very difficult to see, even in stained preparations.

Miracidium Free-living, ciliated larva released from a trematode egg and infective for the snail intermediate host.

Monocytosis Increase in the number of monocytes in the peripheral blood; may be found in both helminth and protozoan infections.

Monoecious Both male and female reproductive organs occur in the same individual (hermaphroditic) (trematodes, cestodes).

Montenegro test Delayed-hypersensitivity skin test; injection of leishmanial antigen; read after 72 h; positive reaction in cured individuals, negative in early cases (visceral and mucocutaneous leishmaniasis).

Mordant A substance capable of combining with a dye and the material to be dyed, thereby increasing the affinity or binding of the dye.

MSDS Material Safety Data Sheet. Provides detailed information about hazards and protective measures relative to hazardous chemical substance.

Multilocular cyst Cyst containing many cavities (*Echinococcus multilocularis*).

Mutualism Association in which both parties benefit and cannot survive without each other.

Myiasis Infestation with maggots (fly larvae).

Myocarditis Inflammation of the heart muscles (trichinosis, trypanosomiasis, toxoplasmosis).

Myositis Inflammation of voluntary muscles (trichinosis, *Sarcocystis* infections).

Nanophyetus salmincola Digenetic trematode that causes infections in humans (ingestion of raw, undercooked, or smoked salmon); infections have been reported from North America. Symptoms include gastrointestinal complaints, unexplained eosinophilia, and recovery of eggs from stool specimens.

Necropsy See *Autopsy*.

Negative predictive value (NPV) (Example) The probability that the individual is not infected by *Plasmodium* when the rapid test is negative.

Nit The ovum or hatched egg of a body, head, or crab louse; it is attached to human hair or clothing by a layer of chitin.

Nocturnal Pertaining to the dark portion of a 24 h day; active at night (filariasis).

Nodule A small, hard node that can be detected by touch (onchocerciasis, coenurosis, cysticercosis, myiasis).

Nucleophaga A microsporan parasite of amebae that destroys the nucleus of its host.

Nucleus A cellular inclusion composed of chromatin; the morphology is often used to help identify intestinal protozoa (*Entamoeba* and *Dientamoeba* spp.).

Obligate parasite Parasite that must always live in contact with the host.

Occult blood Blood present in very small amounts; usually detectable only by chemical means. The specimen is usually a stool sample. The finding may or may not be related to parasitic infection.

Occupational exposure Reasonably anticipated skin, eye, mucous membrane, or parenteral contact with a hazard that may result from the performance of an employee's duties.

Omentoplasty Use of the greater omentum to cover or fill a defect (hydatid disease).

Onchocercoma Nodule containing adult worms (onchocerciasis).

Oncosphere Spherical, six-hooked tapeworm larva within the eggshell (*Taenia* and *Hymenolepis* spp.).

Ookinete The motile zygotes of the malarial parasite that penetrate the mosquito stomach to form an oocyst under the outer gut lining. In an opened tube of human blood, this stage can develop over time as the blood cools and is oxygenated (the malarial parasites "think" they are in the mosquito, and exflagellation and development of the ookinete can occur within the tube of blood; this is not common but is well documented, and the morphology can mimic the *Plasmodium falciparum* gametocyte).

Operculate egg Egg with a lid (trapdoor) at one end through which the larva escapes (*Diphyllobothrium*, *Clonorchis*, and *Paragonimus* spp.).

Operculum A lidlike structure at one end of the egg shell through which the larval form escapes (*Diphyllobothrium*, *Clonorchis*, and *Paragonimus* spp.).

OPIM See *Other potentially infectious material*

Opportunistic Denoting an organism capable of causing disease only in a host whose resistance is lowered (by other diseases or drugs, etc.); also denotes the disease.

Opsonization Modification of the surface characteristics of an invading particle or organism by binding with antibody or a nonspecific molecule to facilitate phagocytosis by host cells.

Orchitis Inflammation of a testis; may be accompanied by swelling, pain, and fever (filariasis).

OSHA Occupational Safety and Health Administration.

Other potentially infectious material (OPIM) Human body fluids including semen, vaginal secretions, urine, cerebrospinal fluid, synovial fluid, pleural fluid, pericardial fluid, peritoneal fluid, amniotic fluid, saliva, body fluids which may be contaminated with blood, unfixed tissue, HIV- or hepatitis virus-containing cell or organ cultures, blood and tissue from an infected animal, reagents, infectious waste, and cultures.

Oviparous Egg-laying.

Pancytopenia Pronounced reduction in the number of erythrocytes, all types of white blood cells, and blood platelets in the circulating blood.

Pan-malarial antigen An antigen expressed by all four of the *Plasmodium* species which are pathogenic in humans, i.e., *P. falciparum*, *P. vivax*, *P. malariae*, *P. ovale*.

Papule Small, solid elevation of the skin.

Parasite An organism living on or in, and at the expense of, another organism.

Parasitemia The presence of parasites in the circulating blood (malaria, blood parasites, microfilariae).

Parasitophorous vacuole Extracellular but intracytoplasmic vacuole, surrounded by host cell membrane (*Cryptosporidium*).

Paratenic host A host in which a parasite survives without undergoing any additional development (a transport host only) (e.g., *Lagochilascaris minor*).

Parenteral Piercing mucous membranes or the skin through such events as needlesticks, human bites, and abrasions.

Paresis Slight paralysis (cysticercosis).

Paroxysm Rapid onset or return of symptoms or increased intensity of symptoms; usually applies to the periodicity of malaria symptoms (chills, fever, sweats).

Parthenogenesis Form of sexual reproduction in which the organism develops without fertilization by the male gamete (*Strongyloides stercoralis*, parasitic generation).

Patency The development stage at which helminths begin egg production; time to maturity (intestinal helminths).

Pathogen Any organism or substance that produces a disease state.

Pemphigus Autoimmune bullous diseases; nonspecific term for blistering skin diseases.

Pentatrichomonas A genus of parasitic protozoan flagellates, formerly part of the genus *Trichomonas*. New generic name for *Trichomonas hominis*, a nonpathogenic flagellate living in the human colon.

Pericardial tamponade Condition resulting from accumulation of excess fluid in the pericardium.

Periodic acid-Schiff stain (PAS) Involves an oxidation process by periodic acid followed by Schiff reagent (detection of newly formed aldehyde groups) and sulfurous rinses; glycogen, chitin, and mucoproteins are strongly PAS positive; lipids are PAS positive (amebae, worm cuticle, part of microsporidial spore, fungi).

Periodicity Recurrence of an event at regular intervals (nocturnal periodicity of microfilariae of *Wuchereria bancrofti*; malarial symptoms).

Peristalsis Movement of food through the intestine by contractions of the intestinal musculature.

Peritonitis Inflammation of the membrane that lines the abdominal cavity and covers the viscera.

Permissible exposure limit (PEL) Maximum allowed exposure during a time-weighted average (8-h workday or 40-h workweek).

Personal protective equipment (PPE) Specialized clothing or equipment worn by an employee for protection against a hazard.

Petechiae Minute, hemorrhagic spots in the skin or mucosa.

Pica A craving for substances not fit for food, such as starch, clay, ashes, or plaster; ingestion of clay or soil can be associated with exposure to *Ascaris*, *Trichuris*, hookworm, *Strongyloides*, *Toxocara*, and *Toxoplasma* organisms.

Pinworm Human roundworm parasite of the colon and rectum, extremely common in children; diagnostic approach is with a Scotch tape preparation (*Enterobius vermicularis*).

Pipe-stem fibrosis Granulomatous reaction around schistosome eggs in periportal tissues; fibrous tissue is white and hard (*Schistosoma japonicum*, *S. mansoni*).

Plasma Liquid portion of the blood or lymph.

Plerocercoid Tapeworm larva of the genera *Diphyllobothrium* and *Spirometra*; also known as a sparganum larva.

Pneumonitis Inflammation of the pulmonary tissue (*Ascaris*, *Strongyloides*, *Pneumocystis* spp.).

Polar filament (microsporidia) A thin tubule within the spore, attached anteriorly, extending back and lying coiled within the spore; part of the extrusion apparatus that everts and inoculates the sporoplasm into a host cell (microsporidia); same as polar tubule.

Polar filaments (*H. nana*) Filaments arising from opposite poles of the oncosphere membrane of *Hymenolepis nana*; they lie between the egg shell and the oncosphere.

Polar plugs Mucoid plugs present at both ends of the egg (*Trichuris trichiura*).

Positive predictive value (PPV) (Example) The probability that an individual actually is infected by *Plasmodium* when the rapid test is positive.

Prepatent period Time between infection of the host and the beginning of egg production (intestinal helminths).

Primary amebic meningoencephalitis (PAM) An invasive, rapidly fatal cerebral infection by soil and water amebae, chiefly *Naegleria fowleri*, found in humans and other primates. The disease is characterized by high fever, neck rigidity, and symptoms associated with upper respiratory infection (cough and nausea). The brain is the primary focus, especially the olfactory lobes and cerebral cortex; organisms enter through the nasal mucosa and through the cribriform plate. Death usually occurs 2 to 3 days after onset of symptoms.

Primary container A vessel, including its closure, which contains the specimen.

Prions Infectious, abnormal host proteins that cause transmissible spongiform encephalopathies and are resistant to a number of standard disinfection and sterilization procedures.

Proglottid Tapeworm segments containing male and female reproductive systems; may be immature, mature, or gravid (*Taenia* and *Hymenolepis* spp.).

Promastigote Development stage of the family Trypanosomatidae; the base of the flagellum is anterior to the nucleus, the free flagellar portion is short, and there is no undulating membrane.

Proteinuria Protein in the urine (*Plasmodium falciparum*).

Prurigo nodularis An eruption of hard nodules in the skin, caused by rubbing and accompanied by intense itching.

Pruritus Severe itching (cutaneous larva migrans, infection with scabies, itch mite, or pinworm).

Pseudocyst Usually refers to a large number of *Toxoplasma gondii* trophozoites enclosed in a macrophage or some other host cell; parasites bound by host cell tissue not of parasite origin.

Pyrogenic Producing fever.

Pyuria Presence of pus (bacteria, white blood cells) in the urine.

Quartan malaria Malaria caused by *Plasmodium malariae.*

Quotidian malaria Type of malaria in which fevers occur every 24 h; may indicate early infection or mixed infection.

Rash A superficial skin eruption; may be macular, maculo-papular, blotchy, etc.

Rectal prolapse Extrusion of the rectal mucosa through the anus; may occur in children with heavy *Trichuris trichiura* infections. Usually no sequelae after therapy.

Rediae An elongated sac with a mouth, muscular pharynx, primitive gut, birth pore, and germinal cells, which may develop into cercariae or daughter rediae (stage within snail host: trematodes).

Regulated waste Liquid or semiliquid blood or OPIMs; contaminated items that would release blood or OPIMs in a liquid or semiliquid state if compressed; items that are caked with dried blood or OPIMs and are capable of releasing these materials during handling; contaminated sharps; and pathological and microbiological wastes containing blood or OPIMs.

Reservoir host The host of an infection in which the infectious agent multiplies and/or develops and on which the agent is dependent for survival in nature; essential host for maintenance of the infection when active transmission is not occurring.

Retinochoroiditis Inflammation of both the retina and the choroid/vascular coat of the eye (toxoplasmosis, onchocerciasis, visceral larva migrans).

Rhabditoid (rhabditiform) larvae Noninfective, thick, rod-shaped larvae of hookworm and *Strongyloides stercoralis*; rhabditoid refers to the shape of the larval esophagus.

Romaña's sign Marked edema of one or both eyes; usually dry and unilateral (Chagas' disease).

Rosette Arrangement of merozoites in mature schizont of *Plasmodium malariae* in a circle around a clump of excess pigment; also the uterine arrangement in the *Diphyllobothrium latum* proglottid.

Rostellum The anterior fixed, or invertible portion of the scolex of a tapeworm, frequently provided with a row (or several rows) of hooklets ("armed rostellum") (*Taenia solium*).

Sappinia diploidea A newly recognized human pathogen, causing amebic encephalitis.

Scabies An eruption due to the mite *Sarcoptes scabiei*; the female burrows into the skin, producing a vesicular eruption with intense pruritus between the fingers, on the male genitalia, buttocks, and elsewhere on the trunk and extremities. Norwegian scabies (Norway itch) is a severe form of scabies with innumerable mites in thickened stratus corneum. This infection has been linked with cellular immune deficiencies, including AIDS.

Schizogony A stage in the asexual cycle of the malarial parasite that occurs within the red blood cells; results in the formation of merozoites (*Plasmodium* spp.).

Schüffner's dots Tiny, red-staining granules in the cytoplasm of red blood cells infected with either *Plasmodium vivax* or *P. ovale*; true stippling (with Giemsa stain).

Scolex The head or attachment portion of a tapeworm; attachment may be by suckers or hooklets (*Taenia, Diphyllobothrium,* and *Hymenolepis* spp.).

Scutum In hard ticks, a plate that largely or entirely covers the dorsum of the male and forms an anterior shield behind the capitulum of the female or immature ticks.

Seatworm Another name for pinworm (*Enterobius vermicularis*).

Secondary container A vessel into which the primary container is placed for transport within an institution; it will contain a specimen if the primary container breaks or leaks in transit.

Segmentectomy Excision of a segment of an organ or gland (hydatid disease).

Sensitivity The proportion of true positives that are correctly identified by a test.

Sharps container A container approved for the transport of contaminated sharps.

Sheath Cuticle of a larval nematode that is retained around the body; may serve as protection. Used to help identify microfilariae (*Wuchereria* and *Brugia* spp., *Loa loa*).

Shock Marked decrease in blood pressure; rapid pulse, decreased kidney function (*Plasmodium falciparum* malaria, rupture of hydatid cyst, bee stings).

Short-term exposure limit (STEL) Maximum permitted continuous exposure to a hazardous substance (normally measured in a single 15-min period).

Sigmoidoscopy Visual examination of the rectum and sigmoid flexure of the colon with a lighted tube; often performed for suspected amebiasis cases (*Entamoeba histolytica*).

Sparganum A migrating tapeworm larva that invades the subcutaneous tissues, causing inflammation and fibrosis (*Diphyllobothrium* and *Spirometra* spp.).

Specificity The proportion of true negatives that are correctly identified by a test.

Spicule Accessory reproductive structure in male nematodes; useful in identification of species.

Spiracle A breathing aperture in arthropods.

Splenomegaly Enlargement of the spleen (leishmaniasis, schistosomiasis, malaria).

Splinter hemorrhages Effects of vasculitis in trichinosis; tiny linear hemorrhages in nail beds (larval migration).

Spore One of the stages in the life cycle of the microsporidia; contains a polar tubule apparatus used for infecting cells; the number of polar tubule coils in the spore is used in organism classification.

Sporocyst An elongated sac without a mouth or other distinct internal or external structure, formed after a miracidium infects a snail. May give rise to cercariae (schistosomes) or rediae (*Paragonimus* spp.).

Sporogony (microsporidia) Production of spores.

Sporogony (*Plasmodium*) A stage in the sexual cycle in the malarial parasite that takes place in the mosquito, with eventual production of the infective sporozoites.

Sporont The zygote stage within the oocyst wall in coccidia; gives rise to sporoblasts, which form sporocysts, within which the infective sporozoites are produced.

Sporozoite Slender, spindle-shaped organism that is the infective stage of malaria parasites; inoculated into humans by the bite of an infected female mosquito; the result of the sexual cycle in the mosquito.

Spurious parasite Organisms that parasitize other hosts passing through the human intestine and are detected in the stool after ingestion; not a true infection (ingestion of animal liver containing *Capillaria* eggs).

Standard precautions Set of precautions applied to all patients designed to reduce the risk of transmission of microorganisms in the health care setting.

Steatorrhea Malabsorption characterized by the presence of fat in the stool (giardiasis, strongyloidiasis, cryptosporidiosis, isosporiasis).

Sterilant An agent intended to destroy all microorganisms (viruses, vegetative bacteria, fungi, and large numbers of highly resistant bacterial endospores) on inanimate surfaces.

Sterilization A procedure that effectively kills a micoorganism, including bacterial spores on inanimate surfaces.

Stigmal plates Characteristic breathing apparatus in fly larvae in cases of myiasis; used to identify the specific species involved; seen at the skin opening (through which the larva receives oxygen).

Strobila The entire chain of tapeworm proglottids, excluding the scolex and neck.

Superinfection Reinfection by a parasite that is already present in the body (*Strongyloides* and *Hymenolepis* spp.).

Swimmer's itch Dermatitis caused by skin penetration of humans by cercariae of schistosomes (normally infect birds and semiaquatic animals).

Symbiosis Close association between two different organisms; living together.

Syncope Fainting or loss of consciousness due to a temporary deficiency of blood supply to the brain (*Dracunculus* infection).

Systemic Involving the entire body.

Tachycardia Rapid heartbeat, usually more than 100 beats/min (trypanosomiasis).

Tachyzoite A rapidly multiplying stage in the development of the tissue phase of certain organisms such as *Toxoplasma gondii*.

Tenesmus Straining to defecate; may be painful and unproductive.

Tetrathyridium Second larval stage in which the scolex with its four suckers is invaginated into the anterior end of a plerocercoid type of body (tapeworm *Mesocestoides*).

Thrill See *Hydatid thrill*.

Tinnitus Ringing or buzzing in the ear not resulting from an external source.

Toluidine blue stain Detects the cyst walls of *Pneumocystis jiroveci* as a purple color.

Transmission Transovarial transmission: passage of parasites or infective agents from the maternal body to eggs within ovaries (transmission of viral or rickettsial pathogens to larval mites or ticks).

Traveler's diarrhea Often related to giardiasis or cryptosporidiosis.

Trichinoscope A magnifying glass used in the examination of meat suspected of containing larvae of *Trichinella* spp.

Trichrome stains Gomori's trichrome—developed for histology; Wheatley's modification—developed for staining stool specimens; modified trichrome stains (10× chromotrope dye)—used for staining microsporidian spores.

Trophozoite The feeding, motile stage of protozoa.

Trypanosome Slender, flagellate protozoan found in the blood of humans (*Trypanosoma* spp.); also referred to as trypomastigote (new term).

Trypomastigote Newer terminology for the trypanosome stage in human blood (*Trypanosoma* spp.).

Ulcer An open sore on the skin or mucous membrane, characterized by disintegration of the tissue and often by discharge of pus (leishmaniasis, amebiasis, *Dracunculus* infection).

Unilocular Cyst containing only one cavity.

Universal precautions Set of precautions designed to reduce risk of transmission of HIV, hepatitis B virus, and other bloodborne pathogens in the health care setting. Now referred to as standard precautions.

Urethritis Inflammation of the urethra (trichomoniasis).

Urticaria An allergic skin reaction characterized by the development of wheals and intense itching and burning sensations; onset and disappearance are often sudden (filarial infections).

Uta A mild form of New World or American cutaneous leishmaniasis caused by *Leishmania peruviana*, occurring in the high

Andean valleys of Peru and Bolivia; small dermal lesions on exposed skin surfaces; the dog is an important reservoir.

Uveitis Inflammation of the uvea (the pigmented, vascular layer of the eye) (*Acanthamoeba* spp.).

Vacuole Cavity in the cytoplasm of a cell that may contain ingested bacteria, yeast cells, or debris; vacuole contents and/or morphology may be helpful in identification of some of the intestinal protozoa (*Entamoeba* spp., *Iodamoeba bütschlii*).

Vaginitis Inflammation of the vagina; prolific, irritating, green or yellowish thin discharge; there may be punctate, hemorrhagic spots (trichomoniasis).

Vector A biological vector is a vector such as the *Anopheles* mosquito for malarial agents or the tsetse fly for agents of African sleeping sickness, in which the agent multiplies before being transmitted to another host; a mechanical vector is a vector that conveys pathogens to a susceptible host without essential biological development of the pathogens in the vector (transfer of organisms on the feet or mouth parts of the housefly).

Vermin Parasitic insects, such as lice and bedbugs.

Vinegar eel Free living nematode, *Turbatrix aceti*, occasionally occurring as a contaminant in laboratory solutions.

Visceral Pertaining to the internal organs of the body, especially those within the abdominal cavity.

Visceral larva migrans Tissue migration of dog and cat ascarid larvae in humans; the life cycle cannot be completed in the human host; often characterized by high peripheral eosinophilia (*Toxocara cati* and *T. canis*).

Viviparous Discharging living young (*Trichinella*, filariae).

Whipworm Common name for roundworm (*Trichuris trichiura*).

WHO World Health Organization.

Winterbottom's sign Enlargement of the posterior cervical lymph nodes (African trypanosomiasis).

Xenodiagnosis Procedure allowing the feeding of laboratory-raised triatomid bugs (known to be infection free) on the blood of patients suspected of having Chagas' disease; after several weeks, the bug feces are checked for the intermediate stages of *Trypanosoma cruzi*; a type of concentration procedure.

Zooanthroponosis A zoonosis that is normally maintained by humans but can be transmitted to other vertebrates.

Zoonosis A disease of animals that is transmissible to humans (*Cryptosporidium* spp.).

Zygote The cell resulting from the fusion of male and female gametes; formed in the mosquito stomach (*Plasmodium* spp.).

Index